SELECTED PAPERS OF K. C. CHOU

World Scientific Series in 20th Century Physics

Published

Vol. 22 A Quest for Symmetry — Selected Works of Bunji Sakita
edited by K. Kikkawa, M. Virasoro and S. R. Wadia

Vol. 23 Selected Papers of Kun Huang (with Commentary)
edited by B.-F. Zhu

Vol. 24 Subnuclear Physics — The First 50 Years: Highlights from Erice to ELN
by A. Zichichi
edited by O. Barnabei, P. Pupillo and F. Roversi Monaco

Vol. 25 The Creation of Quantum Chromodynamics and the Effective Energy
by V. N. Gribov, G. 't Hooft, G. Veneziano and V. F. Weisskopf
edited by L. N. Lipatov

Vol. 26 A Quantum Legacy — Seminal Papers of Julian Schwinger
edited by K. A. Milton

Vol. 27 Selected Papers of Richard Feynman (with Commentary)
edited by L. M. Brown

Vol. 28 The Legacy of Léon Van Hove
edited by A. Giovannini

Vol. 29 Selected Works of Emil Wolf (with Commentary)
edited by E. Wolf

Vol. 30 Selected Papers of J. Robert Schrieffer — In Celebration of His 70th Birthday
edited by N. E. Bonesteel and L. P. Gor'kov

Vol. 31 From the Preshower to the New Technologies for Supercolliders — In Honour of Antonino Zichichi
edited by B. H. Wiik, A. Wagner and H. Wenninger

Vol. 32 In Conclusion — A Collection of Summary Talks in High Energy Physics
edited by J. D. Bjorken

Vol. 33 Formation and Evolution of Black Holes in the Galaxy
— Selected Papers with Commentary
edited by H. A. Bethe, G. E. Brown and C.-H. Lee

Vol. 35 A Career in Theoretical Physics, 2nd Edition
by P. W. Anderson

Vol. 36 Selected Papers (1945–1980) with Commentary
by Chen Ning Yang

Vol. 37 Adventures in Theoretical Physics
— Selected Papers with Commentaries
by Stephen L. Adler

Vol. 38 Matter Particled — Patterns, Structure and Dynamics
— Selected Research Papers of Yuval Ne'eman
edited by R. Ruffini and Y. Verbin

Vol. 39 Searching for the Superworld — A Volume in Honour of Antonino Zichichi on the Occasion of the Sixth Centenary Celebrations of the University of Turin, Italy
edited by S. Ferrara and R. M. Mössbauer

Vol. 40 Murray Gell-Mann — Selected papers
edited by H. Fritzsch

Vol. 41 Many-Body Theory of Molecules, Clusters, and Condensed Phases
edited by N. H. March and G. G. N. Angilella

For information on Vols. 1–21, please visit http://www.worldscibooks.com/series/wsscp_series.shtml

World Scientific Series in 20th Century Physics Vol. 42

SELECTED PAPERS OF K. C. CHOU

editor

Yue-Liang Wu

Chinese Academy of Sciences, China

World Scientific

NEW JERSEY • LONDON • SINGAPORE • BEIJING • SHANGHAI • HONG KONG • TAIPEI • CHENNAI

Published by

World Scientific Publishing Co. Pte. Ltd.

5 Toh Tuck Link, Singapore 596224

USA office: 27 Warren Street, Suite 401-402, Hackensack, NJ 07601

UK office: 57 Shelton Street, Covent Garden, London WC2H 9HE

British Library Cataloguing-in-Publication Data
A catalogue record for this book is available from the British Library.

The editors and the publisher would like to thank the following publishers for their permission to reproduce the articles found in this volume

American Institute of Physics
American Physical Society
Birhäuser Verlag
Chinese Physical Society
Editorial office of "*Progress of Theoretical Physics*"
Editorial office of "*Journal of Experimental and Theoretical Physics*"
Elsevier
Science in China Press
Springer-Verlag

SELECTED PAPERS OF K C CHOU
World Scientific Series in 20th Century Physics — Vol. 42

ISBN-13 978-981-4280-37-2
ISBN-10 981-4280-37-2

Printed in Singapore by World Scientific Printers

Foreword

Kuang-Chao Chou (Guang-Zhao Zhou) is a world-renowned theoretical physicist. Before I met him in the early nineteen seventies, I have already known his many significant research contributions for more than a decade from the late nineteen fifties. His published papers have won uniformly high praises by the international scientific community and his articles are always written with depth and elegance.

Kuang-Chao was the originator of the helicity amplitude analysis. That concept was introduced by him in 1957, followed by a full mathematical analysis of this very effective tool to his work in high energy physics research. In 1959, he perfected an important theorem in T and P violations, which states that under CPT invariance, while T reflection in not conserved, the decay branching ratios of particle and anti-particle to different states can be different even though their total decay widths are the same. In addition to his work on resonances, Kuang-Chao also pioneered the use of nuclear absorption to detect the weak magnetism in weak interactions, as well as a large body of theoretical analysis on the important photo-nuclear reactions through the use of dispersion theory. In 1960, Kuang-Chao analytically demonstrated the partial conservation of axial current (PCAC), which has been regarded as one of the cornerstones in particle physics. For this fundamental contribution, he was recognized internationally as one of the founding fathers of PCAC.

In the nineteen eighties after the Cultural Revolution, under the organization and guidance by Kuang-Chao, the younger generation of Chinese scientists were able to make many significant research works on grand unification theory, CP violation, non-linear sigma model effective Lagrangian theory, spontaneous symmetry breaking in super-symmetry, topological aspects of quantum field theory and its relation to anomaly. Many of these results (for example in topological aspects of quantum field theory and its relation to anomaly) have already attracted recognition from the international as well as Chinese physics communities. In statistical physics and condensed matter physics, the research group directed by Kuang-Chao systemized the Green's function formalism in non-equilibrium statistical mechanics. Their method was applied to the studies of laser, plasma, critical dynamics, random quenching, etc.

In addition to the remarkable achievement made by Kuang-Chao in many fields of theoretical physics, he has been one of the most important scientists in the development of science and technology in China and to the promotion of the international science exchange and cooperation. Kuang-Chao Chou served as the director of Institute of Theoretical Physics at the Chinese Academy of Sciences, the Dean of the Science School of Tsinghua University, the President of Chinese Academy of Sciences, the Chairman of the China Association for Science and Technology, the Executive Vice-President of China Commission for Promoting International Science and Technology and the vice Chairman of China Association for Peace and Disarmament. He is an Academician of the Chinese Academy of Sciences and has also been elected as Foreign Associate of the US National Academy of Sciences, Fellow of the Third World Academy of Science, and Foreign Member of USSR Academy of Sciences, and the European Academy of Arts, Sciences and Humanities.

T. D. Lee

Introduction

Zhou Guang-Zhao's scientific career is divided into several periods:

I. Early period (around 1955-1956). He was working in China.
II. DUBNA period (1957-1960). Zhou was prolific in this period, publishing many papers in the JETP. I was in the USA at the time and had studied several of his papers, especially his work on PCAC. In the USA he was famous as the most brilliant young theorist in Dubna.
III. National defence period (1961-1979). Zhou returned to China and ceased publication in scientific journals. But I had seen in a museum display in Qin-Hai Province the cover page of the design report of the first Chinese atomic bomb. It bears the names of Zhou and Deng Jia-Xian. Rumors say Zhou had made crucial contributions to that design, using analytic methods without the use of computers.
IV. Later period (1979-around 1987). Zhou was at CERN, later at the VPI in the USA for several years, then back to China. In this period Zhou resumed his interest in academic research and was active in several frontier area of theoretical physics, often with young collaborators.
V. Recent period (1987-). Zhou is busy with Academy of Science matters after the late 1980s. But he still keeps up his interest in physics research.

I had first met Zhou at a dinner in Beijing hosted by Premier Zhou En-Lai in the early 1970s, almost forty years ago. During these forty years we have become close friends. I am deeply impressed by how he had transformed himself in the 1980s from a research physicist into an influential and highly respected policy maker and administrator. His success is partly rooted in his genes, of course, but I believe, also very much in his principled Chinese cultural background.

Zhou is a first rate physicist: broad, powerful and very quick in grasping new ideas. His style of doing physics reminds me of that of Landau, Salam, and of Teller. But in personal relationship Zhou is a perfect Confucian gentleman, without the aggressive edges that characterize so many famous US, European and Russian theoretical physicists.

楊振宁

C. N. Yang

Introduction

Professor Kuang-Chao Chou (Guang-Zhao Zhou) was born in Changsha, Hunan Province in 1929. Upon graduated from the Department of Physics at Tsinghua University in 1951, he enrolled as a postgraduate student in theoretical physics at Peking University where he became as a faculty member in 1954. He then immediately carried out several research works on nuclear physics and particle physics published in Chinese journals [1c,2c,3c].

In the period 1957-1960, Prof. Kuang-chao Chou worked as a researcher at the Joint Institute for Nuclear Research (JINR), Dubna in the Soviet Union, where he published more than 30 papers and made significant contributions in several important subjects concerning symmetries. His first paper at JINR was the issue on a symmetry property of the new Gell-Mann theory [1]. He then paid special attention to the spins and parities of particles [2,3,6], which enabled him to establish the relativistic theory of reactions for the polarized particles [7,8] and especially for the massless polarized particles [13], where he introduced, for the first time, the concept of helicity amplitude and its corresponding mathematical description, which has been shown to be very powerful in analyzing high energy scattering amplitude. He studied at the first time the particle-antiparticle asymmetry (CP violation) in the hyperon decays [10,14,23], where he proved an important theorem in CP violation, i.e., when CPT is invariant but T reflection is not conserved, the decay branching ratios of particle and anti-particle to different states may be different although total decay widths are the same. He also noticed in the early time the possible symmetry properties for the π-K system [22]. Its group symmetry was realized late on as the SU(3) symmetry by Gell-Mann in 1961. The renowned important paper written by Kuang-Chao Chou was on the pseudovector current and lepton decays of baryons and mesons [26], which has been an initial paper for proving the theorem of partial conservation of axial current (PCAC) in a simple and concise way, and it has been reputed to be an outstanding contribution in the research of hadron physics. He also conducted some interesting works concerning charge symmetry properties [11,18], and a series research works on dispersion relation, photo-nuclear reaction and scattering processes of mesons and hyperons [16,20,23–25,27,28,30–34,9c,10c,11c]. He also calculated at the first time the mass difference between neutron and proton [4c]. With collaborators, he also studied the capture of muon in nucleus [5c,6c,7c,8c]. It was a very productive period for Prof. Kuang-chao Chao, his works immediately attracted special attentions worldwide, for instance, his most significant and fundamental work on PCAC [26] motivated the renowned paper on "Dynamical model of elementary particles based on analogy with superconductivity" by Y. Nambu and G. Jona-Lasinio (Phys. Rev. Vol. 122, No.1 and Vol. 124, No. 1, 1961), for which Y. Nambu won the Nobel prize in 2008. Due to those remarkable achievements, Prof. Kuang-chao Chou was highly praised by the international scientific community and became a world-renowned theoretical physicist.

The period 1960's was a golden time for particle physics. It was also the most creative period for Prof. Kuang-chao Chou carrying out the important initial works in the frontiers of particle physics. Nevertheless, it was in this important time, Prof. Kuang-chao Chou made an alternative and unusual choice that he decided to give up his research interest and the advanced working and living conditions at JINR when he learnt that the project of atomic Bomb in China met some hard problems in its first stage of design and all the relevant experts from Soviet Union had to leave China. In 1961, Prof. Kuang-chao Chou returned back to China and took part in the theoretical study and design of the atomic Bomb, which was known to be the most mysterious project. He then analyzed carefully all formalisms used in the computations and checked in detail all the numerical calculations. Eventually, he solved the hard problems caused in the design of atomic Bomb and made a decisive and right judgment based on the first physical principle. It was this breakthrough, made by Prof. Kuang-chao Chou, which played a crucial rule for the success of the first atomic Bomb (1964) in China. During the period 1961-1978, Prof. Kuang-chao Chou made important contributions in the development of atomic energy, applied physics and computational mathematics, which included the high temperature and high density physics, explosion mechanics, mechanics of radiative fluid, neutron physics, plasma physics and computational mechanics.

In 1978, Institute of Theoretical Physics (ITP) was approved by the vice Premier Deng Xiao-ping to be established at the Chinese Academy of Sciences (CAS), Prof. Kuang-chao Chou joined the Institute of Theoretical Physics. In the period 1979-1987, he was able to work again on the frontiers of theoretical physics. In the early 1980s, he worked as a guest researcher and professor in the Virginia Polytechnic Institute, USA and the European Organization for Nuclear Research.

Prof. Kuang-chao Chou made a number of important contributions in multitude research areas of theoretical physics. In the field of high energy particle physics, under his leadership, some significant achievements have been yielded at several interesting subjects, which included the gauge field theories [35-41,43,45,49,52,12c,13c,14c,15c], nonlinear sigma models in symmetric coset and curved spaces [46,47,50,51,56,64,68], U(1) anomaly and chiral dynamical theories [42,44,48,53,57,73]. The most important observations by Prof. Kuang-chao Chou and his collaborators were the gauge invariance and anomaly-free condition [58,16c,17c] of the Wess-Zumino-Witten effective action and the topological origin of gauge anomalies [60]. In the paper [59], it was firstly pointed out that Wittens effective Lagrangian for chiral field missed some important terms. Thus the topological properties of gauge fields were extensively explored in several papers [62,63,67,18c]. Late on, a simplified derivation of Chern-Simons cochain was realized [71] and it has been adopted to obtain the general Chern-Simons characteristic classes [72,19c,20c]. In particular, several interesting physical applications have been investigated, which included the derivation of the anomalous term in Virasoro algebra [69], the effective action of sigma model anomalous with external gauge fields [70], the possible origin of θ-vacuum etc. At the same time, Prof. Kuang-chao Chou also paid special attention to the phenomenology of weak interactions. In particular, an interesting observation was made on the correlation between the top quark mass and direct CP violation in kaon decays [65]. As a consequence, a heavy top quark with mass much larger than W-boson mass was predicted and emphasized, and a fourth generation quark was also motivated [66] at that time.

In addition to the gauge field theories and particle physics, Prof. Kuang-chao Chou (Guang-Zhao Zhou) has also made important contributions in the fields of statistical physics and condensed matter physics. With collaborators, the closed time path Green's functions has systematically been developed [79,80,21c,22c] and applied to the critical dynamics [81,23c,24c] and statistical physics [82,83,84,85,27c,28c]. A nice review article was published in Phys. Rep. 118, 1 (1985) which has widely been cited internationally. It has been shown that the method of closed time path Green's functions allows one to treat, in a unified way, both equilibrium and nonequilibrium systems. The method has also been applied to the random system [86,87,90] and the disorder electron system [88] as well as the Bohr-Sommerfield quantization for the fractional quantum Hall effect system [93]. The other interesting works conducted the spontaneous symmetry breaking [92] and time reversal invariance [95,25c,26c] in the nonequilibrium system as well as the influence functional [94].

In 1980s, Prof. Kuang-chao Chou has led, as the director of ITP-CAS, several key research directions at ITP, which concern: quantum field theory and particle physics, nuclear physics and plasma physics, statistical physics and condensed matter physics, gravity and astrophysics. For his outstanding contributions to theoretical physics, applied physics, atomic energy and basic research sciences in various aspects, and also for his great capability of leadership, in 1987, Prof. Kuang-chao Chou was pointed by the government to serve as the president of Chinese Academy of Sciences (CAS). Once again, he has to give up his own research interest and to make alternative contributions for leading the development of sciences and technology at the CAS.

In 1996, Prof. Kuang-chao Chou was considering to return back to the scientific research. He presented in 1995 the opening address for the 17th International Lepton-Photon Conference which was regarded as the most influenced International Conference on particle physics held in Beijing. After that conference, he was actually thinking about the most hot topics in particle physics, which involved CP violation, origin of mass, neutrino physics, supersymmetric theories, low energy phenomena of grand unification theories [74–77] and a general discussion on CP and CPT violating observable in meson decays[78]. It has been shown that the 18 unknown quantities in the standard model of particle physics can well be predicted by a minimal set of 5 parameters in a class of supersymmetric grand unification model [74], about ten relations among CP violating-phase, masses and mixing angles of quarks and leptons have been obtained, which remain consistent with the experimental data.

In 1997, Prof. Kuang-chao Chou would wish to realize his desire of doing theoretical research. Nevertheless, it was unexpected that he was elected to be the vice Chairman of Standing Committee of the National People's Congress of China and pointed again by the government to serve as the Chairman of the China Association for Sciences and Technology. Until 2006, he was honored as the honorary Chairman of the Chinese Association of Sciences and Technology. Since 2006, he has been invited to serve as the member of International Advisory Committee and the honorary Chairman of the Scientific Council for the Institute of Theoretical Physics at the Chinese Academy of Sciences.

In this volume, the most papers written by Prof. Kuang-chao Chou (Guang-Zhao Zhou) are printed. All the collected papers have been arranged into four parts according to the subjects of research areas and the languages of publishing journals. Part I (in English) and Part III (in Chinese) are the papers on Field Theories, Particle Physics and Nuclear Physics, Part II (in English) and Part IV (in Chinese) are the papers on Statistical Physics and Condensed Matter Physics. From the published papers, it is seen how Prof. Kuang-chao Chou caught up the frontiers of theoretical physics in various periods and carried out the creative research works with initial ideas and motivations, as well as how he has worked in different key research directions of theoretical physics and made significant contributions to various interesting research subjects and interdisciplinary areas.

Due to the outstanding research works and significant contributions, Prof. Kuang-chao Chou has earned numerous national and international awards which include: National Prize first order in Natural Science, Prize first order from Qiushi Foundation of HongKong, Gian Carlo Wick Commemorative gold medal from the World Federation of Scientists, the Chinese Meritorious Service Medals for Nuclear Scientists and Satellite Pioneers. He was elected in 1980 to be the Academician of

Chinese Academy of Sciences. He has also been elected as Foreign Associate of the US National Academy of Sciences, Fellow of the Third World Academy of Science, and Foreign Member of USSR Academy of Sciences, Czechoslovak Academy of Sciences, Bulgarian Academy of Sciences, Romania Academy of Sciences, Mongolian Academy of Sciences, the European Academy of Arts, Sciences and Humanities, Membre fondateur Academie Francophone d'Ingenieurs. He has been honored to be the honorary PhD of several universities which include the City College of New York in USA, McGill University in Canada, Chinese University of Hong Kong. He also served as the Executive Vice-President of China Commission for Promoting International Science and Technology, the vice Chairman and honorary Chairman of China Association for Peace and Disarmament. He was conferred Commendatore dellOrdine Al Merito Della Republica Italiana in 1993.

The editor would like to thank Dr. Shan-Gui Zhou and Dr. Ling-Mei Cheng for their invaluable assistance in preparing this collection. The assistance of Ms. Lian-Zi Wei and Ms. Ji Zhang was also very helpful for completing and publishing this volume. I would like to thank Prof. K.K. Phua for providing many useful suggestions in publishing this volume.

Yue-Liang Wu

Kavli Institute for Theoretical Physics China
Key Laboratory of Frontiers in Theoretical Physics
Institute of Theoretical Physics, Chinese Academy of Sciences

Contents

Foreword (T. D. Lee) v

Introduction (C. N. Yang) vii

Introduction (Y. L. Wu) ix

Part I Field theory, particle physics and nuclear physics 1

1. Concerning a symmetry property of the new Gell-Mann theory, 3
Zh. Eksp. Teor. Fiz. **33** (1957) 1058.
Soviet Physics JETP **6** (1958) 815.
2. Spins and parities of the hyperfragment H^4_Λ and of the K meson, 5
(with M. I. Shirokov),
Zh. Eksp. Teor. Fiz. **33** (1957) 1072.
Soviet Physics JETP **6** (1958) 828.
3. Spins and parities of the $_\Lambda H^4$ hyperfragment and K-meson, 7
(with M. I. Shirokov),
Nucl. Phys. **6** (1958) 10.
4. On non-uniqueness of nucleon-nucleon scattering phase shifts, 17
(with L. G. Zastavenko and R. M. Ryndin),
Nucl. Phys. **6** (1958) 669.
5. Phase indeterminacies in nucleon-nucleon scattering, 20
(with L. G. Zastavenko and R. M. Ryndin),
Zh. Eksp. Teor. Fiz. **34** (1958) 526.
Soviet Physics JETP **7** (1958) 363.
6. On the determination of the relative parities of elementary particles, 22
Zh. Eksp. Teor. Fiz. **34** (1958) 1027.
Soviet Physics JETP **34** (1958) 710.
7. The relativistic theory of reactions involving polarized particles, 24
(with M. I. Shirokov),
Zh. Eksp. Teor. Fiz. **34** (1958) 1230.
Soviet Physics JETP **34** (1958) 851.
8. Selection rules in reactions involving polarized particles, 31
Zh. Eksp. Teor. Fiz. **35** (1958) 783.
Soviet Physics JETP **35** (1959) 543.
9. The Shapiro integral transformation, (with L. G. Zastavenko), 33
Zh. Eksp. Teor. Fiz. **35** (1958) 1417.
Soviet Physics JETP **35** (1959) 990.

10. A note on the decay of the Σ-hyperon and its antiparticle, 39
Nucl. Phys. **9** (1958/59) 652.

11. The universal Fermi interaction and the capture of muons in hydrogen, 42
(with V. Maevskii),
Zh. Eksp. Teor. Fiz. **35** (1958) 1581.
Soviet Physics JETP **35** (1959) 1106.

12. Charge symmetry properties and representations of the extended Lorentz 44
group in the theory of elementary particles, (with V. I. Ogievetskii),
Zh. Eksp. Teor. Fiz. **36** (1959) 264.
Soviet Physics JETP **36** (1959) 179.

13. Reactions involving polarized particles of zero rest mass, 49
Zh. Eksp. Teor. Fiz. **36** (1959) 909.
Soviet Physics JETP **36** (1959) 642.

14. Some symmetry properties in processes of antihyperon production with 56
annihilation of antinucleons,
Zh. Eksp. Teor. Fiz. **36** (1959) 938.
Soviet Physics JETP **36** (1959) 663.

15. On the problem of investigating the interaction between π mesons and 57
hyperons (with L. I. Lapidus),
Zh. Eksp. Teor. Fiz. **37** (1959) 283.
Soviet Physics JETP **37** (1960) 199.

16. Electromagnetic mass of the K meson, (with V. I. Ogievetskii), 61
Zh. Eksp. Teor. Fiz. **37** (1959) 866.
Soviet Physics JETP **37** (1960) 616.

17. Dispersion relations for the scattering of γ quanta by nucleons, 63
(with L. I. Lapidus),
Zh. Eksp. Teor. Fiz. **37** (1959) 1714.
Soviet Physics JETP **37** (1960) 1213.

18. Charge symmetry properties and representations of the extended Lorentz 68
group in the theory of elementary particles, (with V. I. Ogievetsky),
Nucl. Phys. **10** (1959) 235.

19. Integral transformations of the I. S. Shapiro type for particles of zero mass, 77
(with L. G. Zastavenko),
Zh. Eksp. Teor. Fiz. **38** (1960) 134.
Soviet Physics JETP **11** (1960) 97.

20. Scattering of gamma-ray quanta by nucleons near the threshold for meson 81
production, (with L. I. Lapidus),
Zh. Eksp. Teor. Fiz. **38** (1960) 201.
Soviet Physics JETP **11** (1960) 147.

21. On the production of an electron-positron pair by a neutrino in the field of a nucleus, (with A. M. Badalyan), 88
Zh. Eksp. Teor. Fiz. **38** (1960) 664.
Soviet Physics JETP **11** (1960) 477.

22. Possible symmetry properties for the π-K system, 90
Zh. Eksp. Teor. Fiz. **38** (1960) 1015.
Soviet Physics JETP **11** (1960) 730.

23. On the decay of Σ hyperons, 92
Zh. Eksp. Teor. Fiz. **38** (1960) 1342.
Soviet Physics JETP **11** (1960) 966.

24. Dispersion relations and analysis of the energy dependence of cross sections near thresholds of new reactions, (with L. I. Lapidus), 94
Zh. Eksp. Teor. Fiz. **39** (1960) 112.
Soviet Physics JETP **12** (1961) 82.

25. Inelastic final-state interactions and near-threshold singularities, (with L. I. Lapidus), 100
Zh. Eksp. Teor. Fiz. **39** (1960) 364.
Soviet Physics JETP **12** (1961) 258.

26. On the pseudovector current and lepton decays of baryons and mesons, 106
Zh. Eksp. Teor. Fiz. **39** (1960) 703.
Soviet Physics JETP **12** (1961) 492.

27. Elastic scattering of gamma quanta by nuclei, (with L. I. Lapidus) 112
Zh. Eksp. Teor. Fiz. **39** (1960) 1056.
Soviet Physics JETP **12** (1961) 735.

28. The elastic scattering of γ rays by deuterons below the pion-production threshold, (with L. I. Lapidus), 115
Zh. Eksp. Teor. Fiz. **39** (1960) 1286.
Soviet Physics JETP **12** (1961) 898.

29. On the pion-pion resonance in the p-state, (with Ho Tso-Hsiu), 121
Zh. Eksp. Teor. Fiz. **39** (1960) 1485.
Soviet Physics JETP **12** (1961) 1032.

30. On the role of the single-meson pole diagram in scattering of gamma quanta by protons, (with L. I. Lapidus), 122
Zh. Eksp. Teor. Fiz. **41** (1961) 294.
Soviet Physics JETP **14** (1962) 210.

31. Low-energy limit of the γN-scattering amplitude and crossing symmetry, (with L. I. Lapidus), 127
Zh. Eksp. Teor. Fiz. **41** (1961) 491.
Soviet Physics JETP **14** (1962) 352.

32. On the $\overline{K}$+N→Λ(Σ)+γ process, (with L. I. Lapidus), 130
Zh. Eksp. Teor. Fiz. **41** (1961) 1310.
Soviet Physics JETP **14** (1962) 932.

33. Scattering of photons by nucleons, (with L. I. Lapidus), 133
Zh. Eksp. Teor. Fiz. **41** (1961) 1546.
Soviet Physics JETP **14** (1962) 1102.

34. A suggested experiment to determine the spin of Y1* and the parities of Λ–Σ, Λ–Y1*, and Σ–Y1*, (with Su Zhao-bin and Gao Chong-shou), 140
Scientia Sinica **12** (1963) 451.

35. The pure gauge fields on a coset space, (with Tu Tung-sheng and Yean Tu-Nan), 142
Scientia Sinica **22** (1979) 37.

36. Soliton-soliton scattering and the semi-classical approximation for the problem of scattering in three-dimensional space, (with Dai Yuan-ben), 158
Scientia Sinica **22** (1979) 281.

37. The non-topological soliton with a non-Abelian internal symmetry, (with Zhu Zhong-yuan, Dai Yuan-ben and Wu Yong-shi), 172
Scientia Sinica **23** (1980) 40.

38. On the vacuum of the pure gauge fields on coset, 193
Scientia Sinica **23** (1980) 431.

39. A model of electro-weak interaction in SU(3)×U(1) gauge theory, (with Gao Chong-shou), 203
Scientia Sinica **23** (1980) 566.

40. Electro-weak theory in SU(3), (with Gao Chong-shou), 211
Chinese Science Bulletin **25** (1980) 21.

41. Disorder parameter and duality, (with Xian Ding-chang), 216
Chinese Science Bulletin **25** (1980) 635.

42. Axial U(l) anomaly and chiral symmetry-breaking in QCD, 221
AIP. Conf. Proc. **72** (1981) 621.

43. Possible $SU(4)_c \times SU(3)_f \times U(l)$ model, (with Chong-shou Gao), 227
Phys. Rev. D **23** (1981) 2690.

44. On the quantization and the renormalization of the pure-gauge fields on the coset-space, (with Ruan Tu-nan), 239
J. Univ. Sci. Tech. China **11** (1981) 15.

45. The U(1) anomalous Ward identities and chiral dynamics, 249
Chinese Science Bulletin **27** (1982) 147.

46. New non-linear σ model on symmetric spaces, (with Song Xing-chang), 255
Commun. Theor. Phys. **1** (1982) 69.

47. The H_n sigma-models and self-dual SU(n) Yang-Mills fields in four-dimensional Euclidean space, (with Song Xing-chang), 262
Commun. Theor. Phys. **1** (1982) 185.

48. On the dynamical problems and the Fermion spectra in the Rishon model, (with Dai Yuan-ben), *Commun. Theor. Phys.* **1** (1982) 725. 281

49. Wilson loop integral and string wave functional, (with Li Xiao-yuan), *Scientia Sinica, Ser. A* **25** (1982) 264. 287

50. Bäcklund transformation, local and nonlocal conservation laws for nonlinear σ models on symmetric coset spaces, (with Song Xing-chang), *Scientia Sinica, Ser. A* **25** (1982) 716. 295

51. Local conservation laws for various nonlinear σ models, (with Song Xing-chang), *Scientia Sinica, Ser. A* **25** (1982) 825. 302

52. Composite gauge bosons in a nonabelian theory, (with Swee-Ping Chia and Charles B. Chiu), *Phys. Lett.* **109 B** (1982) 457. 311

53. Massless baryons and anomalies in chiral $(QCD)_2$, (with D. Amati and S. Yankielowicz), *Phys. Lett.* **110 B** (1982) 309. 316

54. On the determination of effective potentials in supersymmetric theories, (with D. Amati), *Phys. Lett.* **114 B** (1982) 129. 320

55. Koba-Nielsen-Olesen scaling and production mechanism in high energy collisions, (with Liu Lian-sou and Men Ta-chong), *Phys. Rev. D* **28** (1983) 1080. 323

56. Nonlinear σ model on multidimensional curved space with certain cylindrical symmetry, (with Song Xing-chang), *Commun. Theor. Phys.* **2** (1983) 971. 329

57. Kac-Moody algebra for two dimensional principal chiral models, (with Song Xing-chang), *Commun. Theor. Phys.* **2** (1983) 1391. 341

58. On the gauge invariance and anomaly-free condition of the Wess-Zumino-Witten effective action, (with Guo Han-ying, Wu Ke and Song Xing-chang), *Phys. Lett.* **134B** (1984) 67. 349

59. On Witten's effective Lagrangian for chiral field, (with Guo Han-ying, Wu Ke and Song Xing-chang), *Commun. Theor. Phys.* **3** (1984) 73. 352

60. The topological origins of Gauge anomalies, (with Guo Han-ying, Wu Ke and Song Xing-chang), *Commun. Theor. Phys.* **3** (1984) 125. 360

61. On the dynamical symmetry breaking for the N = l pure supersymmetric Yang-Mills model, (with Dai Yuan-ben and Chang Chao-hsi), *Commun. Theor. Phys.* **3** (1984) 221. 365

62. The unified scheme of the effective action and chiral anomalies in any even dimensions, (with Han-ying Guo, Xiao-yuan Li, Ke Wu and Xing-chang Song), *Commun. Theor. Phys.* **3** (1984) 491. 374

63. Symmetric and asymmetric anomalies and effective Lagrangian, (Han-ying Guo, Ke Wu and Xing-chang Song), *Commun. Theor. Phys.* **3** (1984) 593. 382

64. On the two dimensional non-linear σ model with Wess-Zumino term (I) classical theory, (with Dai Yuan-ben), *Commun. Theor. Phys.* **3** (1984) 767. 393

65. CP violation from the standard model, *Proceedings of Europhysics Topical Conference on Flavor Mixing in Weak Interactions*, (Ettore Majorana International Science Series: Physical Sciences, Vol. 20), 609. 397

66. Top quark mass and the fourth generation of quark, (with Wu Yue-liang and Xie Yan-bo), *Chinese Physics Letters* **1** (1984) 47. 424

67. Anomalies of arbitrary gauge group and the its reduction group, Einstein and Lorentz anomalies, (with Han-ying Guo and Ke Wu), *Commun. Theor. Phys.* **4** (1985) 91. 426

68. On the two dimensional non-linear σ model with Wess-Zumino term (II) quantum theory, (with Dai Yuan-ben), *Commun. Theor. Phys.* **4** (1985) 123. 437

69. Derivation of the anomalous term in Virasoro algebra by topological method, (with Wu Yue-liang and Xie Yan-bo), *Commun. Theor. Phys.* **5** (1986) 359. 442

70. Effective action of sigma model anomalous with external gauge fields, (with Yue-liang Wu and Yan-bo Xie), *Modern Phys. Lett. A* **1** (1986) 23. 448

71. A simplified derivation of Chern-Simons cochain and a possible origin of θ-vacuum term, (with Wu Yue-liang and Xie Yan-bo), *Commun. Theor. Phys.* **7** (1987) 27. 453

72. The general Chern-Simons characteristic classes and their physical applications, (with Wu Yue-liang and Xie Yan-bo), *Commun. Theor. Phys.* **8** (1987) 341. 464

73. Signature for chiral-symmetry breaking at high temperatures, (with Lay-nam Chang and Ngee-pong Chang), *Phys. Rev. D* **43** (1991) 596. 491

74. CP violation, fermion masses and mixings in a predictive SUSY SO(10)×Δ(48)×U(1) model with small tanβ, (with Y. L. Wu), *Phys. Rev. D* **53** (1996) R3492. 497

75. Low energy phenomena in a model with symmetry group SUSY SO(10)×Δ(48)×U(1), (with Wu Yue-liang), *Science in China, Ser. A*, **39** (1996) 65. 501

76. A solution to the puzzles of CP violation, neutrino oscillation, fermion masses and mixings in an SUSY GUT model with small tanβ, (with Yue-liang Wu), *Nucl. Phys. B* (Proc. Suppl.) **52A** (1997) 159. 514

77. A possible unification model for all basic forces, (with Wu Yue-liang), *Science in China, Ser. A*, **41** (1998) 324. 518

78. Searching for rephase-invariant CP- and CPT-violating observables in meson decays, (with W. F. Palmer, E. A. Paschos and Y. L. Wu), *Euro. Phys. J. C* **16** (2000) 279. 524

Part II Statistical physics and condensed matter physics 533

79. Renormalization of the closed time path Green's functions in nonequilibrium statistical field theory, (with Su Zhao-bing), *Chin. Phys.* **1** (1981) 635. 535

80. Dyson equation and Ward-Takahashi identities of the closed time path Green's function, (with Su Zhao-bing), *Chin. Phys.* **1** (1981) 645. 545

81. Closed time path Green's functions and critical dynamics, (with Zhao-bin Su, Bai-lin Hao and Lu Yu), *Phys. Rev. B* **22** (1980) 3385. 559

82. On theory of the statistical generating functional for the order parameter (I) general formalism, (with Su Zhao-bin, Hao Bai-lin and Yu Lu), *Commun. Theor. Phys.* **1** (1982) 295. 582

83. On theory of the statistical generating functional for the order parameter (II) density matrix and the field-theoretical structure of the generating functional, (with Su Zhao-bin, Hao Bai-lin and Yu Lu), *Commun. Theor. Phys.* **1** (1982) 307. 594

84. On theory of the statistical generating functional for the order parameter (III) effective action formalism for the order parameter, (with Su Zhao-bin, Hao Bai-lin and Yu Lu), *Commun. Theor. Phys.* **1** (1982) 389. 606

85. On an approximate form of the coupled equations of the order parameter with the weak electromagnetic field for the ideal superconductor, (with Su Zhao-bin), *Commun. Theor. Phys.* **1** (1982) 669. 621
86. On a dynamic theory of quenched random system, (with Su Zhao-bin and Yu Lu), *Commun. Theor. Phys.* **2** (1983) 1181. 632
87. A dynamical theory of the infinite range random Ising model, (with Su Zhao-bin and Yu Lu), *Commun. Theor. Phys.* **2** (1983) 1191. 641
88. A dynamical theory of random quenched system and its application to infinite-ranged Ising model, (with Su Zhao-bin and Yu Lu), *Proceedings of the First Asia-Pacific Physics Conference*, (World Scientific Pub. Co.), 474. 652
89. Symmetry and Ward identities for disordered electron system, (with Lin Jian-cheng and Shen Yu), *Commun. Theor. Phys.* **3** (1984) 139. 667
90. Does Parisi's solution of the Sherrington-Kirkpatrick model locate on the absolute maximum of the free energy? (with Shen Yu), *Commun. Theor. Phys.* **3** (1984) 263. 677
91. Equilibrium and nonequilibrium formalisms made unified, (with Zhao-bin Su, Bai-lin Hao and Lu Yu), *Phys. Rep.* **118** (1985) 1. 682
92. Spontaneous symmetry breaking and Nambu-Goldstone mode in a non-equilibrium dissipative system, (with Zhao-bin Su), *Prog. Theo. Phys. Supp.* **86** (1986) 34. 812
93. The canonical description and Bohr-Sommerfeld quantization for the fractional quantum Hall effect system, (with Zhao-bin Su, Han-bin Pang and Yan-bo Xie), *Phys. Lett A* **123** (1987) 249. 821
94. Influence functional and closed-time-path Green's function, (with Zhao-bin Su, Liao-yuan Chen and Xiao-tong Yu), *Phys. Rev.* **B37** (1988) 9810. 826
95. Time reversal invariance and its application to nonequilibrium stationary states, (with Z. B. Su), *Thirty Years Since Parity Nonconservation: A Symposium for T. D. Lee*, (Birkhäuser Verlag, 1988) 117. 829

Part III 场论、粒子物理与核物理 **845**
Field theory, particle physics and nuclear physics

1. 双力程核子力的讨论, 847
物理学报, 1955 年第 11 卷第 4 期第 299-316 页.
A discussion on two-range nuclear force,
Acta Physica Sinica **11** (1955) 299 (in Chinese).
2. 介子场中的三体力位能, 865
北京大学学报自然科学版, 1955 年第 1 期第 53-66 页.
Three-body potentials in meson fields,
Acta Scicentiarum Naturalum Universitis Pekinesis **1** (1955) 53 (in Chinese).
3. 由微观电磁场方程组求得宏观电磁场方程组的方法, 879
北京大学学报自然科学版, 1956 年第 2 期第 211-218 页.
Derivation of Maxwell equations from the microscopic Lorentz equations,
Acta Scicentiarum Naturalum Universitis Pekinesis **2** (1956) 211 (in Chinese).
4. 中子和质子的质量差, 887
物理学报, 1959 年第 15 卷第 5 期第 269-276 页.
Mass difference between neutron and proton,
Acta Physica Sinica **15** (1959) 269 (in Chinese).
5. 普适费米弱相互作用理论及μ介子在原子核上的俘获, 895
(与 B. 马耶夫斯基),
物理学报, 1959年15卷第7期第377-388页.
The universal Fermi theory of weak interaction and the capture of muon in atomic nuclei, (with B. Maevskii),
Acta Physica Sinica **15** (1959) 377 (in Chinese).
6. μ 介子在 He^3 原子核上的俘获, (与 朱家珍, 彭宏安), 907
物理学报, 1960 年第 16 卷第 2 期第 61-69 页.
Muon capture in He^3 nucleus, (with Chu Chia-chen and Peng Hong-an),
Acta Physica Sinica **16** (1960) 61 (in Chinese).
7. 关于μ俘获中有效赝标量耦合项的符号, (与 黄念宁), 916
物理学报, 1960 年第 16 卷第 2 期第 70-75 页.
On the sign of the effective pseudoscalar term in μ capture reactions, (with Huang Nien-ning),
Acta Physica Sinica **16** (1960) 70 (in Chinese).
8. μ 介子与轻原子核散射时的反冲效应, (与 戴元本), 922
物理学报, 1960 年第 16 卷第 2 期第 76-80 页.
The effect of recoil on the scattering of μ-mesons by light nuclei, (with Dai Yuan-ben),
Acta Physica Sinica **16** (1960) 76 (in Chinese).

9. π 介子核子碰撞产生 π 介子的色散关系, (与 戴元本), 927
物理学报, 1960 年 16 卷第 5 期第 252-262 页.
Dispersion relations for pion production in pion-nucleon collisions,
(with Dai Yuan-ben),
Acta Physica Sinica **16** (1960) 252 (in Chinese).

10. 检验 π 介子散射过程是否存在 p 态共振的一个实验方法的建议, 938
(与 何祚庥),
物理学报, 1961 年第 17 卷第 3 期第 133-134 页.
A suggestion of experiment to detect the p-resonance in π-π scattering,
(with Ho Tso-Hsiu),
Acta Physica Sinica **17** (1961) 133 (in Chinese).

11. $\pi + p \to \Lambda + \pi + K$ 反应共振—近阈效应关联的研究, (与 苏肇冰, 高崇寿), 940
物理学报, 1963 年第 19 卷第 10 期第 649-672 页.
Investigation of the correlation between resonance effect and near threshold
effect in the $\pi+p\to\Lambda+\pi+K$ reaction, (with Su Zhao-bin and Gao Chong-shou),
Acta Physica Sinica **19** (1963) 649 (in Chinese).

12. 色规范群和味规范群的相互作用之间有对称性吗? (与 高崇寿), 964
高能物理与核物理, 1980 年第 4 卷第 4 期第 536-538 页.
Is there any symmetry between flavour and colour gauge interactions?
(with Gao Chong-shou),
Phys. Energ. Forties Phys. Nucl. **4** (1980) 536 (in Chinese).

13. SU(3)中的弱电统一模型, (与 高崇寿), 967
高能物理与核物理, 1980 年第 4 卷第 5 期第 609-622 页.
Unified electro-weak model in SU(3), (with Gao Chong-shou),
Phys. Energ. Forties Phys. Nucl. **4** (1980) 609 (in Chinese).

14. 路径积分量子化的一般形式, (与 尹鸿钧, 阮图南, 杜东生), 981
中国科学技术大学学报, 1980 年第 10 卷第 2 期第 26-35 页.
A general form of path integral quantization,
(with Yin Hong-jun, Ruan Tu-nan and Du Dong-sheng),
J. Univ. Sci. Tech. China **10** (1980) 26 (in Chinese).

15. 关于 SU(3)×U(1)轻子和夸克弱电统一模型的一些讨论, (与 高崇寿), 991
高能物理与核物理, 1981 年第 5 卷第 1 期第 39-46 页.
Some discussions on the SU(3)×U(l) electro-weak unified model of leptons
and quarks, (with Gao Chong-shou),
Phys. Energ. Forties Phys. Nucl. **5** (1981) 39 (in Chinese).

16. 关于 Wess-Zumino-Witten 有效作用量的规范不变性, 999
(与 郭汉英, 吴可, 宋行长),
高能物理与核物理, 1984 年第 8 卷第 2 期第 252-255 页.
On the gauge invariance of Wess-Zumino-Witten effective action,
(with Guo Han-ying, Wu Ke and Song Xing-chang),
Phys. Energ. Forties Phys. Nucl. **8** (1984) 252 (in Chinese).

17. 对称反常，非对称反常和有效拉氏量，(与 郭汉英，吴可，宋行长)， 1004
高能物理与核物理 1984 年第 8 卷第 4 期第 508-512 页.
Symmetrical anomaly, unsymmetrical anomaly and effective Lagrangian,
(with Guo Han-ying, Wu Ke and Song Xing-chang),
Phys. Energ. Forties Phys. Nucl. **8** (1984) 508 (in Chinese).
18. 任意偶维时空的有效作用和手征反常，
(与 郭汉英，李小源，吴可，宋行长)， 1009
高能物理与核物理, 1985 年第 9 卷第 2 期第 252-256 页.
Effective action and chiral anomalies in any even dimensional space,
(with Guo Han-ying, Li Xiao-yuan, Wu Ke and Song Xing-chang),
Phys. Energ. Forties Phys. Nucl. **9** (1985) 252 (in Chinese).
19. 普遍的 Chern-Simons 链和它们的应用，(与 吴岳良，谢彦波)， 1014
高能物理与核物理, 1986 年第 10 卷第 2 期第 177-185 页.
The general Chern-Simons cochain and their application,
(with Wu Yue-liang and Xie Yan-bo),
Phys. Energ. Forties Phys. Nucl. **10** (1986) 178 (in Chinese).
20. 一种新的上边缘算子及其应用，(与 吴岳良，谢彦波)， 1023
高能物理与核物理, 1986 年第 10 卷第 3 期第 283-287 页.
A new co-boundary operator and its applications,
(with Wu Yue-liang and Xie Yan-bo),
Phys. Energ. Forties Phys. Nucl. **10** (1986) 283 (in Chinese).

Part IV 统计物理与凝聚态物理 1029
Statistical physics and condensed matter physics

21. 非平衡耗散系统定常态的 Goldstone 模，(与 苏肇冰)， 1031
物理学报, 1980 年第 29 卷第 5 期第 618-634 页.
On the Goldstone mode in the stationary state of a non-equilibrium
dissipative system, (with Su Zhao-bing),
Acta Physica Sinica **29** (1980) 618 (in Chinese).
22. 三套闭路格林函数的变换关系，(与 于渌，郝柏林)， 1048
物理学报, 1980 年第 29 卷第 7 期第 878-888 页.
Transformation properties of three sets of closed time path Green’s functions,
(with Yu Lu and Hao Bai-lin),
Acta Physica Sinica **29** (1980) 878 (in Chinese).
23. 非平衡统计场论与临界动力学 (I) 广义朗之万方程， 1059
(与 苏肇冰，郝柏林，于渌)，
物理学报, 1980 年第 29 卷第 8 期第 961-968 页.
Nonequilibrium statistical field theory and critical dynamics (I) generalized
Langevin equation, (with Su Zhao-bin, Hao Bai-lin and Yu Lu),
Acta Physica Sinica **29** (1980) 961 (in Chinese).

24. 非平衡统计场论与临界动力学 (II) 拉氏场论表述, (与 郝柏林, 于渌), 1067
物理学报, 1980 年第 29 卷第 8 期第 969-977 页.
Nonequilibrium statistical field theory and critical dynamics (II) Lagrangian field theory formulation, (with Hao Bai-lin and Yu Lu),
Acta Physica Sinica **29** (1980) 969 (in Chinese).

25. 时间反演对称和非平衡统计定常态 (I), (与 苏肇冰), 1076
物理学报, 1981 年第 30 卷第 2 期第 164-171 页.
Time reversal symmetry and non-equilibrium statistical stationary states (I), (with Su Zhao-bin),
Acta Physica Sinica **30** (1981) 164 (in Chinese).

26. 时间反演对称和非平衡统计定常态(II), (与 苏肇冰), 1084
物理学报, 1981 年第 30 卷第 3 期第 401-409 页.
Time reversal symmetry and non-equilibrium statistical stationary states (II), (with Su Zhao-bin),
Acta Physica Sinica **30** (1981) 401 (in Chinese).

27. 原子核费密多体系统平均场近似的推广, (与 苏肇冰, 于渌), 1093
物理学报, 1984 年第 33 卷第 7 期第 999-1007 页.
The generalized mean field expansions for many Fermion systems, (with Su Zhao-bing and Yu Lu),
Acta Physica Sinica **33** (1984) 999 (in Chinese).

28. 序参量—统计格林函数耦合方程组, (与 苏肇冰, 于渌), 1102
物理学报, 1984 年第 33 卷第 6 期 805-813 页.
On a set of coupled equations for the order parameters—statistical Green's functions, (with Su Zhao-bin and Yu Lu),
Acta Physica Sinica **33** (1984) 805 (in Chinese).

Part I

Field theory, particle physics and nuclear physics

CONCERNING A SYMMETRY PROPERTY OF THE NEW GELL-MANN THEORY

CHOU GUAN-CHAO

Joint Institute for Nuclear Research

Submitted to JETP editor July 6, 1957

J. Exptl. Theoret. Phys. (U.S.S.R.) 33, 1058-1059 (October, 1957)

RECENTLY Gell-Mann[1] has proposed a new theory of the interaction between elementary particles. In this theory all baryons have spin ½, and the same mechanical mass and parity. They form a supermultiplet which is split up only when moderately-strong interactions occur by way of K mesons. The Hamiltonian of the interaction with π mesons is written in the following way:

$$H_\pi = ig\{(\bar{p}\gamma_5 p - \bar{n}\gamma_5 n)\pi^0 + \sqrt{2}(\bar{p}\gamma_5 n\pi^+ + \bar{n}\gamma_5 p\pi^-) + (\bar{\Xi}^0\gamma_5\Xi^0 - \bar{\Xi}^-\gamma_5\Xi^-)\pi^0 + \sqrt{2}(\bar{\Xi}^0\gamma_5\Xi^-\pi^+ + \bar{\Xi}^-\gamma_5\Xi^0\pi^-)$$
$$+ (\bar{\Sigma}^+\gamma_5\Sigma^+ - \bar{Y}^0\gamma_5 Y^0)\pi^0 + \sqrt{2}(\bar{\Sigma}^+\gamma_5 Y^0\pi^+ + \bar{Y}^0\gamma_5\Sigma^+\pi^-) + (\bar{Z}^0\gamma_5 Z^0 - \bar{\Sigma}^-\gamma_5\Sigma^-)\pi^0 + \sqrt{2}(\bar{Z}^0\gamma_5\Sigma^-\pi^+ + \bar{\Sigma}^-\gamma_5 Z^0\pi^-)\}. \quad (1)$$

(see Ref. 1 for the notation). In our note we shall show that in addition to the so-called global symmetry discussed by Gell-Mann, there is another symmetry property of his theory even in the absence of K coupling.

In the Hamiltonian (1) we make the substitutions

$$p \to \Xi_c^-,\ n \to \Xi_c^0,\ \Xi^- \to p_c,\ \Xi^0 \to n_c,\ \Sigma^+ \to \Sigma_c^-,\ Y^0 \to Z_c^0,\ \Sigma^- \to \Sigma_c^+,\ Z^0 \to Y_c^0,\ \pi^+ \to \pi^+,\ \pi^- \to \pi^-,\ \pi^0 \to -\pi^0, \quad (2)$$

where A_c denotes the charge-conjugation operator of the field of particle A. After this substitution the Hamiltonian takes the form

$$H_\pi = ig\{(\bar{\Xi}_c^0\gamma_5\Xi_c^0 - \bar{\Xi}_c^-\gamma_5\Xi_c^-)\pi^0 + \sqrt{2}(\bar{\Xi}_c^-\gamma_5\Xi_c^0\pi^+ + \bar{\Xi}_c^0\gamma_5\Xi_c^-\pi^-) + (\bar{p}_c\gamma_5 p_c - \bar{n}_c\gamma_5 n_c)\pi^0 + \sqrt{2}(\bar{n}_c\gamma_5 p_c\pi^+ + \bar{p}_c\gamma_5 n_c\pi^-)$$
$$+ (\bar{Z}_c^0\gamma_5 Z_c^0 - \bar{\Sigma}_c^-\gamma_5\Sigma_c^-)\pi^0 + \sqrt{2}(\bar{\Sigma}_c^-\gamma_5 Z_c^0\pi^+ + \bar{Z}_c^0\gamma_5\Sigma_c^-\pi^-) + (\bar{\Sigma}_c^+\gamma_5\Sigma_c^+ - \bar{Y}_c^0\gamma_5 Y_c^0)\pi^0 + \sqrt{2}(\bar{Y}_c^0\gamma_5\Sigma_c^+\pi^+ + \bar{\Sigma}_c^+\gamma_5 Y_c^0\pi^-)\}. \quad (3)$$

The following relations exist between the field operators A and B:[2]

$$\bar{A}B = \bar{B}_c A_c,\ \bar{A}\gamma_\mu B = -\bar{B}_c\gamma_\mu A_c,\quad \bar{A}\gamma_\mu\gamma_\nu B = -\bar{B}_c\gamma_\mu\gamma_\nu A_c,\quad \bar{A}\gamma_5\gamma_\mu B = \bar{B}_c\gamma_5\gamma_\mu A_c,\ \bar{A}\gamma_5 B = \bar{B}_c\gamma_5 A_c. \quad (4)$$

With the help of (4) we can satisfy ourselves that the Hamiltonian (3) agrees with (1). In the same way it can be shown that the Hamiltonian of the interaction of baryons with photons

$$H_\gamma = -ie\{\bar{p}\gamma_\mu p - \bar{\Xi}^-\gamma_\mu\Xi^- + \bar{\Sigma}^+\gamma_\mu\Sigma^+ - \bar{\Sigma}^-\gamma_\mu\Sigma^-\}A_\mu \quad (5)$$

is invariant under the substitution (2), with the addition of the substitution $A_\mu \to A_\mu$. This symmetry property of the Gell-Mann theory provides us with selection rules for certain processes. Let us consider processes in which there are only π mesons and photons in the initial and final states. For every Feynman diagram there will be another diagram in which the lines of the inner baryons are replaced by the propagation lines of the baryons appearing in the right-hand sides of the substitution (2). For example, proton lines are replaced by the lines of the baryon described by the field operator Ξ_c^- (i.e., by the lines of an anti Ξ^--particle), antiproton lines by the lines of a Ξ^--particle, etc. If the number of outer π^0 mesons is odd, then the matrix elements of these two graphs will cancel each other when they are added. Consequently in the new Gell-Mann theory the process of the decay of a π^0 meson into two γ-rays takes place only through a K interaction.

If it is assumed that β interactions of baryons also have Gell-Mann's global symmetry, then we can obtain a few other selection rules. Consider, for example, a tensor β interaction:

$$H_\beta = g_\beta\{\bar{p}\gamma_\mu\gamma_\nu n + \bar{\Xi}^0\gamma_\mu\gamma_\nu\Xi^- + \bar{\Sigma}^+\gamma_\mu\gamma_\nu Y^0 + \bar{Z}^0\gamma_\mu\gamma_\nu\Sigma^-\}\bar{e}\gamma_\mu\gamma_\nu\nu.$$

It is clear that this Hamiltonian changes sign under the substitution (2). Thus the matrix elements for the β-decay of a π^+ meson, that is

$$\pi^+ \to e^+ + \nu + n\gamma,\ n = 0,\ 1,\ \ldots$$

through a virtual baryon-pair, cancel each other in pairs. These processes also are allowed only through a K interaction.

If only the interactions of π mesons with nucleons are taken into account, then the theoretical probabilities of the processes

$$\pi^0 \to 2\gamma \text{ (Ref. 3) and } \pi^+ \to e^+ + \nu + \gamma \text{ (Ref. 4),}$$

obtained from perturbation theory, are greater than those observed. The considerations above indicate that it is possible to reduce the discrepancy between theory and experiment by allowing also for the interactions of π mesons with all baryons.* Of course, without a study of the interaction with a K meson we are still unable to say with certainty that the new Gell-Mann theory gives better agreement with the experimental data for these processes.

In conclusion I wish to thank Professor M. A. Markov, Professor Khu Nin, V. I. Ogievetskii, and M. I. Shirokov for their interest and for discussions of the work presented here.

*This same result for the decay process $\pi^0 \to 2\gamma$ was also obtained by Gell-Mann.[1] An analogous result was obtained earlier by Kinoshita.[3]

[1] M. Gell-Mann, Phys. Rev. 106, 1296 (1957).
[2] R. H. Good, Jr., Rev. Mod. Phys. 27, 187 (1955).
[3] T. Kinoshita, Phys. Rev. 94, 1384 (1954).
[4] S. B. Treiman and H. W. Wyld, Jr., Phys. Rev. 101, 1552 (1956).

Translated by W. M. Whitney
213

SPINS AND PARITIES OF THE HYPERFRAGMENT H^4_Λ AND OF THE K MESON

CHOU GUAN-CHAO and M. I. SHIROKOV

Joint Institute for Nuclear Research

Submitted to JETP editor July 26, 1957

J. Exptl. Theoret. Phys. (U.S.S.R.) 33, 1072-1073 (October, 1957)

In this note we shall give a summary of the angular correlations in the cascades:

$$\pi^- + He^4 \rightarrow H^4_\Lambda + K,\ H^4_\Lambda \rightarrow He^4 + \pi,\ K \rightarrow \pi + \pi; \quad (1)$$

$$K^- + He^4 \rightarrow H^4_\Lambda + \pi,\ H^4_\Lambda \rightarrow He^4 + \pi \quad (2)$$

for several variants of spins in parities of H^4_Λ and K.

We assume at first that the spin of the K meson is zero. If in cascade (1) the first reaction takes place near the threshold (kinetic energy of the π^- mesons in the laboratory is 620 to 640 Mev), then it is natural to assume that H^4_Λ and K are formed predominantly in the s state if π (the product of the parities of π^-, He^4, H^4_Λ, and K) is $(-1)^i$ (i is the spin of the hyperfragment), and in the p state if $\pi = (-1)^5$ (in this case the production in the s state is forbidden). The distribution about the angle γ between the direction of the incident pions and the H^4_Λ decay products (in the system where H^4_Λ is at rest), given in the table, has been obtained under these assumptions. As can be seen, it is possible in principle to determine not only i, but also the products of the parities of H^4_Λ and K (assuming that the product of the parities of the π^- and He^4 is -1).

Parity variant	Spin variant 0	1	2
$\pi = (-1)^i$	1	$3\cos^2\gamma$	$^5/_4(1-6\cos^2\gamma+9\cos^4\gamma)$
$\pi = -(-1)^i$	Forbidden reaction	$^3/_2(1-\cos^2\gamma)$	$^{15}/_2(\cos^2\gamma-\cos^4\gamma)$

Cascade (1) contains three reactions, and it therefore permits also the determination of the spin of the K meson by the Adair method[1] (true, if $k \neq 0$, the product of the parities of H^4_Λ and K is no longer determined by the angular correlations). For this purpose one first selects such cases of the reaction $\pi^- + He^4 \rightarrow H^4_\Lambda + K$, in which the hyperfragment and the K meson make (in the center of mass system) small angles $\vartheta_f \approx 0$ with the direction of the incident pions (aligned with the z axis). If the H^4_Λ decays associated with the $K \rightarrow \pi + \pi$ decays are further selected such that the pions make small angles $\vartheta \leq \Delta\vartheta$ with the same z axis, then the distributions about γ, which are chosen on the basis of this selection and which determine the spin i, will be the same as on the threshold in the variant $\pi = (-1)^i$ (see table). If one chooses instead the decays $K \rightarrow \pi + \pi$, associated with the H^4_Λ decays along the z axis, it is possible to determine k. The correlation relative to the angle between the z axis and the direction of the decay products of K (in the system where the K meson is at rest) is given as a function of k in the same first line of the table.[1] The value of the permissible intervals of small angles ϑ_f and ϑ ($\Delta\vartheta_f$ and $\Delta\vartheta$) diminishes with increasing l'_{max} — the maximum important orbital momentum of the products of the reaction $K^- + He^4$ $H^4_\Lambda + K$ — and spin k respectively. If $l'_{max} = 1$, we get $\Delta\vartheta_f \approx 20°$.

Cascade (2) was first proposed by Dalitz,[3] and Gell-Mann[4] gives a set of correlations for this cascade. If $k = 0$, these correlations have the same form as in the table, but relative to another angle θ, see Ref. 4. We emphasize here only that these formulas must be compared with the experimental distributions, obtained for the reactions $K^- + He^4 \rightarrow H^4_\Lambda + \pi^0$ without preliminary formation of a mesonic atom. If we verify somehow that the reaction took place "in flight," one can expect that at K-meson energies up to

20 Mev (but above 0.01 Mev) the He^4 nucleus captures the K meson in the s state [or in the p state, if $\pi = -(-1)^i$] by some specific (non-electromagnetic) short-lived forces.

At greater K-meson energies the correlation formulas are the same, but are valid only for reaction products making small angles with the direction of the incident K meson.

Cascade (2) does not permit determination of the spin of the K meson, and the effect of $k \neq 0$ on the discussed correlations relative to θ can be investigated only qualitatively. If $k \neq 0$, the correlation corresponding to a given i becomes less anisotropic, smoothes out, and thereby yields too low values of the hyperfragment spins.

I express my gratitude to Professor M. A. Markov for discussions.

[1] R. K. Adair, Phys. Rev. 100, 1540 (1955).
[2] M. I. Shirokov, J. Exptl. Theoret. Phys. (U.S.S.R.) 32, 1022 (1957), Soviet Phys. JETP 5, 835 (1957).
[3] R. Dalitz, Proceedings of the Rochester Conference, 1956.
[4] M. Gell-Mann, Phys. Rev. 106, 1298 (1957).

Translated by J. G. Adashko
223

Nuclear Physics **6** (1958) 10—19; © *North-Holland Publishing Co., Amsterdam*

SPINS AND PARITIES OF THE $_{\Lambda}H^4$ HYPERFRAGMENT AND K-MESON

CHOU KUANG-CHAO and M. I. SHIROKOV

Joint Institute of Nuclear Research, Laboratory of Theoretical Physics, Dubna, U.S.S.R.

Received 24 July 1957

Abstract: If K-mesons, hyperfragments or hyperons are created in the π^-+He^4 reaction near the threshold or in the K^-+He^4 reaction at low K^--particle energies (but without mesic atom formation), the spins of these unstable particles (and in some cases their parities) can be determined from the correlations between the directions of the momenta of the particles involved in these reactions and the directions of emission of decay products of the unstable particles. Qualitative changes in the correlation formulae for the cascade $K^-+He^4 \rightarrow {}_{\Lambda}H^4+\pi^0$, ${}_{\Lambda}H^4 \rightarrow He^4+\pi^-$ are indicated which arise if one rejects the assumption that the K-meson spin is zero. The helium experiments together with the previously suggested hydrogen experiments (e.g. $\pi^-+p \rightarrow \Sigma+K$ or $K^-+p \rightarrow \Sigma+\pi$) are the only ones which are of practical use for the determination of new particle spins from angular correlations.

1. Introduction

The method employed in the present paper for the determination of the spins of unstable particles consists in the following. The cascade $a+b \rightarrow c+d$, $c \rightarrow e+f$ is studied. The products of the first cascade reaction (particles c and d) are in general polarized in some definite way. The angular distribution of the decay products of a particle depends on the magnitude of its spin and on the nature of its polarization, that is, on the wave function of its spin state. Under suitable conditions this function may be known and then for each spin value of particle c a particular distribution will exist. If the experiment is carried out under prescribed experimental conditions the particle spin can be established by comparing these theoretical distributions with the experimental one.

By correlation we understand the angular distribution of the decay products of c (and/or d) which depends on its spin state; the latter in turn may depend on the angle of emission of the created particle as a parameter. If only particle c is considered and the beam a and target b are unpolarized (or possess zero spins) the correlation parameters and arguments in the general case are three vectors: the direction of the incident beam a, direction of emission of the reaction product c, and the direction of emission of its decay products. In view of the invariance of all processes under three-dimensional rotations the correlation will in fact depend only on the relative position of these three vectors which can be specified by three angles (see below).

Dalitz [1]) earlier mentioned that measurement of the correlation for $K^- + He^4 \rightarrow {}_\Lambda H^4 + \pi^0$, ${}_\Lambda H^4 \rightarrow He^4 + \pi^-$ should yield information not only on the spin of the hyperfragment but also on the parities of the new particles. The hyperfragment spin may give information on the Λ^0 spin and on the nature of the interaction between Λ^0 and nucleon. The θ correlations for spins 0 and 1 of the ${}_\Lambda H^4$ are presented in ref. [2]) for various parities on the assumption that the K-meson spin is zero.

2. Correlation for the π+He^4 Cascade

2.1. GENERAL METHOD

General formulae for the angular distribution of the decay products of an unstable particle and for the spin state of particle c and/or d produced in a reaction of the a+b → c+d type have been obtained, for example, in ref. [3]). We shall now obtain the correlation for the simple case of the cascade $\pi^- + He^4 \rightarrow {}_\Lambda H^4 + K$, ${}_\Lambda H^4 \rightarrow He^4 + \pi^-$ and thus illustrate the general arguments presented above (and also the method used in ref. [3])).

The wave function ψ^f for the products of the reaction $\pi^- + He^4 \rightarrow {}_\Lambda H^4 + K$ can be written in the form

$$\Psi^f = \hat{S}(T, -T)\Psi^0. \tag{1}$$

Of course this does not make ψ^f a better known function; so far only the dependence of ψ^f on the dynamics of the problem (i.e., on the S-matrix of the problem) has been formally singled out and in addition it has been emphasized that ψ^f depends on ψ^0 (initial condition at time $-T$).

Since the total momentum $\mathbf{P}$ of an isolated system is conserved, we get that $\mathbf{P}_f = 0$ if $\mathbf{P}_0 = 0$. The state of the system of particles ${}_\Lambda H^4$, K or π^-, He^4 can therefore be described in their relative momentum $\mathbf{p}$ and spin projection m representation (in the c.m.s.). Rewriting (1) in Dirac's notation [4]) we obtain the following expression for the probability amplitude that after the reaction ${}_\Lambda H^4$ and K will possess a momentum $\mathbf{p}_f$ and spin projections m_i and m_k

$$\begin{aligned}(\mathbf{p}_f;\, m_i,\, m_k|\mathrm{f}) &= (\mathbf{p}_f;\, m_i,\, m_k|S|\mathbf{p})(\mathbf{p}|\mathbf{p}_0) \\ &= (\mathbf{p}_f;\, m_i,\, m_k|S|\mathbf{p}_0).\end{aligned} \tag{2}$$

The symbol $(\mathbf{p}|\mathbf{p}_0)$ signifies that prior to the reaction π^- and He^4 had a definite momentum $\mathbf{p}_0$. Integration over $\mathbf{p}$ is implied.

Note that we have employed the law of conservation of total momentum. We shall now make use of the law of conservation of total angular momentum (a.m.) $\mathbf{J}$. Rewriting (1) for the case when the initial state ψ^0 is characterized by definite values of the total a.m. quantum number J_0 and its projection M_0, we see that in order for ψ^f to describe a state with the same definite J and M the S-matrix must be diagonal in J and M:

$$(\ldots J'M'|S|\ldots JM) = (\ldots|S^J|\ldots)\cdot\delta_{J'J}\cdot\delta_{M'M}. \tag{3}$$

(It can be proved that elements that are diagonal with respect to M do not depend on M.)

We now express the elements $(\mathbf{p}_f;\ m_i,\ m_k|S|\mathbf{p}_0)$ in (2) in terms of the elements of (3):

$$\begin{aligned}(\vartheta_f\varphi_f p_f;\ m_i,\ m_k|S|\mathbf{p}_0) &= C(p_0)\cdot\sum(\vartheta_f\varphi_f|l'\mu')(ikm_i m_k|iks'm')\\ &\cdot(s'l'm'\mu'|s'l'J'M')(s'l'J'M'p_f|S|JMp_0)(JM|\vartheta_0\varphi_0).\end{aligned} \tag{4}$$

The unitary transformation function for transition from the representation in variables $\mathbf{p}_f$, m_i, m_k to the representation involving J and M is obtained as the product of familiar transformation functions: 1) the wave function of a state with definite orbital a.m. l' and its projection μ' in the representation in spherical angles ϑ_f, φ_f of the vector $\mathbf{p}_f$, that is, the spherical function $Y_{l'\mu'}(\vartheta_f,\varphi_f)\equiv(\vartheta_f\varphi_f|l'\mu')$; 2) the wave function of a state with a definite total spin s' and its projection m' in the representation in variables m_i and m_k, i.e., the so-called Clebsch-Gordan coefficient $(ikm_i m_k|iks'm')$[5]; and 3) the coefficient $(s'l'm'\mu'|s'l'J'M')$.

$C(p_0)$ is a normalization factor which is of no importance for the following exposition; i and k are the spins of particles ${}_{\Lambda}H^4$, and K. $\sum$ denotes the sum over l', μ', s', m', J', M', J and M.

Employing (3), orienting the z-axis parallel to $\mathbf{p}_0$ and taking into account that in this case

$$Y_{JM}(0,\ \varphi_0) = [2J+1/4\pi]^{\frac{1}{2}}\cdot\delta_{M,0},$$

we obtain the following final expression for the wave function of the reaction products:

$$\begin{aligned}(\mathbf{p}_f;\ m_i,\ m_k|\mathrm{f}) &= C'(p_0)\cdot\sum_{l'\mu's'm'}(\vartheta_f\varphi_f|l'\mu')(ikm_i m_k|iks'm')\\ &\cdot(s'l'm'\mu'|s'l'l0)(s'l'|S^l|l)(2l+1)^{\frac{1}{2}}.\end{aligned} \tag{5}$$

The symbol J has been replaced by l which is the orbital angular momentum of particles π^- and He^4 (and is equal to the total a.m. before the reaction).

The law of conservation of total parity (a consequence of the invariance of interactions under space reflection) is expressed by the equality of the total parities before and after the reaction. Thus l and l' should be such that $\Pi_\pi\cdot\Pi_{He}(-1)^l = \Pi_H\cdot\Pi_K(-1)^{l'}$ or $\Pi(-1)^{l+l'} = +1$ where Π is the product of intrinsic parities of all four particles participating in the reaction.

In a similar manner the angular distribution of the decay products of a polarized particle of spin i at rest can be obtained if the particle decays into two zero spin particles

$$\mathscr{F}(\vartheta, \varphi) = |\sum_m Y_{im}(\vartheta, \varphi)\psi_m|^2; \tag{6}$$

ψ_m is the spin wave function of the particle.

2.2. K-MESON SPIN ZERO AND LOW π-MESON ENERGY

Let us temporarily assume that the K-meson spin is zero; then $s' = i$. Suppose the reaction $\pi^- + He^4 \to {}_\Lambda H^4 + K$ is being studied near the threshold, that is, at π-meson kinetic energies of 620—640 MeV in the laboratory system. (π-mesons of such energies can be obtained with the Birmingham proton synchrotron.) We assume that the forces between ${}_\Lambda H^4$ and K are of the short range type (that is, these particles are created in a certain finite volume and do not interact after escaping from this volume). Since we know nothing more about these forces it will be most natural to assume that slow (in c.m.s.) ${}_\Lambda H^4$ and K-particles are created predominantly in the s-state ($l' = 0$)†. Disregarding all matrix elements $(il'|S^i|l)$ with $l' = 1, 2 \ldots$ compared with $(i0|S^i|l)$ we find that l can assume only a single value $l = i$. However, this value can be forbidden by the law of conservation of parity if $\Pi(-1)^i = -1$.

In the case when $\Pi = (-1)^i$, (5) takes the form

$$(m_i|f) = \delta_{m_i,0} \cdot C'' \cdot (i0|S_i|i)(2i+1)^{\frac{1}{2}}. \tag{7}$$

Inserting (7) in (6) and denoting ϑ by γ we get

$$\mathscr{F}_i(\gamma) \propto [P_i(\cos\gamma)]^2 \tag{8}$$

where $P_i(x)$ is an unnormalized Legendre polynomial (see ref. [6]) and γ is the angle between the direction of emission of the hyperfragment decay products (in its c.m.s.) and the chosen z-axis (that is, the direction of the incident π-mesons). The correlations for spin values $i = 1$ and 2 are presented in table 1.

TABLE 1

Spin / Parity	$i = 1$	$i = 2$
$\Pi = (-1)^i$	$3\cos^2\gamma$	$\frac{5}{4}(1 - 6\cos^2\gamma + 9\cos^4\gamma)$
$\Pi = -(-1)^i$	$\frac{3}{2}(1 - \cos^2\gamma)$	$\frac{15}{2}(\cos^2\gamma - \cos^4\gamma)$

If $\Pi = -(-1)^i$ the value $l' = 0$ will be forbidden, because in this case $l = i$ and $\Pi \cdot (-1)^i = -1$. Assuming that ${}_\Lambda H^4$ and K are chiefly formed in the p-state (l can again then assume only one value $l = i$) we obtain

† Of course it is always possible to imagine an interaction for which this will not be true (even if $l' = 0$ is not forbidden by the parity conservation law; see below). However the correlations derived below are valid for arbitrary short range forces although the angular (ϑ_f) range in which the correlations are valid may be smaller (see section 3).

$$(\mathbf{p}_f\,;\; m_i|\mathrm{f}) \propto Y_{1,-m_i}(\vartheta_f, \varphi_f)(1i-m_i m_i|1ii0). \qquad (9)$$

Besides γ, the correlation depends on two other angles (ϑ_f and $\varphi_f-\varphi$). Integrating over them we can obtain the correlation

$$I_i(\gamma) \propto \sum_{m=1}^{i} [Y_{im}(\gamma, 0)(1i-mm|1ii0)]^2$$

(see table 1).

3. Correlation for Restricted Directions of Emission

3.1. K-MESON SPIN ZERO AND HIGHER π-MESON ENERGIES

Adair [7]) has indicated another approach to the problem of determining the spin from correlations. Mathematically this method is based on the formula $Y_{l,\mu}(0, \varphi) = [2l+1/4\pi]^{\frac{1}{2}} \cdot \delta_{\mu,0}$ (the quantal expression of the perpendicularity of the angular momentum to the corresponding momentum.) We shall again assume that the K-meson spin is zero but shall not impose the restriction that the π-meson energy is close to the threshold value. We shall consider the decays of only those hyperfragments which are emitted at angles $\vartheta_f \approx 0$ or $\vartheta_f \approx \pi$, i.e., along the axis $z\|\mathbf{p}_0$. From (5) it immediately follows that $(0, \varphi_f\,;\; m_i|\mathrm{f}) \propto \delta_{m_i,0}$, i.e., the spin wave function of these hyperfragments is the same as (7)†. Thus the correlation formulae are the same as those near the threshold for the parity case $\Pi = (-1)^i$.

What is the angular range of ϑ_f in which these formulae are valid? From the deduction it follows that in this interval we should be able to neglect spherical functions $Y_{l',\mu'}(\vartheta_f, \varphi_f)$ with $\mu' \neq 0$ compared with $Y_{l',0}(\vartheta_f, \varphi_f)$. A calculation shows that for $l' = 1$ one tenth compared to unity is neglected in the ϑ_f interval (0^0-20^0). This interval should decrease with increasing l'. A consequence of this is that one is obliged to assume that the forces are of short range and that the threshold should not be considerably exceeded in the experiments. For each given energy the maximum essential value of l' should be found by assuming some range of forces (usually $\approx 10^{-13}$ cm) and the permissible angular range should be estimated.

At the threshold and for $\Pi = (-1)^i$ formula (7) is valid over the whole sphere of directions of emission of the hyperfragment (for all angles ϑ_f).

† This result can easily be explained in a descriptive manner. The projection of the total a.m. on the axis $z\|\mathbf{p}_0$ is zero before the reaction as the π^- and He^4 spins are zero and the projection of the orbital a.m. on the momentum itself is zero. Thus after the reaction the sum of the projection of the $_\Lambda H^4$ spin and the orbital a.m. l' should vanish. For hyperfragments ejected along the z-axis μ' is again the projection of the a.m. on the momentum and must therefore vanish. Therefore for such hyperfragments m_i can assume only one value: $m_i = 0$. It can be seen that it is sufficient to use only conservation of the projection of the total a.m. providing that $m_i = 0$ and emission in the direction of the z-axis is not forbidden by the laws of conservation of the total a.m. and parity. (This prohibition just occurs near the threshold for $\Pi = -(-1)^i$; see below).

As the π-meson energy is increased, the region of applicability contracts towards the poles. When $\Pi = -(-1)^i$ the angular distribution as well as the spin wave-function of hyperfragments emitted along the z-axis vanish at the threshold (see (9); $(\mathbf{p}_f;\ 0|\mathrm{f}) = 0$, as $(1i00|1ii0) = 0$). This means that in this case and near the threshold no region exists in which (7) is valid. As the energy increases, hyperfragments begin to appear also at the poles and correlation (8) is valid for them.

3.2. K-MESON SPIN NOT ZERO

The assumption that the K-meson spin is zero may be abandoned if one applies Adair's method (although, admittedly, the statistics are impaired in this case). If $k \neq 0$ the most general correlation will be a function of five angles (a function of the relative position of four vectors) $\mathscr{F}(\vartheta_f;\ \vartheta,\ \varphi;\ \vartheta_K,\ \varphi_K)$ where ϑ_K, φ_K are the angles of emission of the $K \to \pi+\pi$ decay products in the K-meson rest system; these angles, as well as ϑ and φ, are measured in a coordinate system whose axis is $z \| \mathbf{p}_0$ †.

Let us construct the correlation $\mathscr{F}(0;\ \vartheta,\ \varphi;\ 0,\ \varphi_K)$, i.e., construct the angular distribution for decay products from the following ensemble of hyperfragments: first of all hyperfragments (and K-mesons) those emitted along the z-axis are chosen; after this from these events we choose those in which the decay π-mesons from $K \to \pi+\pi$ are emitted parallel to the z-axis. Formula (6) for the $K \to \pi+\pi$ decay shows that only the components $(0,\ \varphi_f,\ p_f;\ m_i,\ 0|\mathrm{f})$ of the wave-function $(\vartheta_f,\ \varphi_f,\ p_f;\ m_i,\ m_k|\mathrm{f})$ are required for this correlation. From (5) we then deduce that $(0,\ \varphi_f,\ p_f;\ m_i,\ 0|\mathrm{f}) \propto \delta_{m_i,0}$, that is, the correlation $\mathscr{F}(0;\ \vartheta,\ \varphi;\ 0,\ \varphi_K)$ is identical to (8). The correlation $\mathscr{F}(0,\ 0,\ \varphi,\ \vartheta_K,\ \varphi_K)$ has also the same form and permits one to determine the K-meson spin. It should be noted that in the cascade $\pi+\mathrm{p} \to \Sigma+\mathrm{K}$, $\Sigma \to \mathrm{N}+\pi$, $\mathrm{K} \to \pi+\pi$ this correlation can yield a spin value k or $k-1$ and hence the measured value of the K-meson spin may turn out to be smaller than the true value [7]).

In conclusion it should be stressed that the parity can be determined experimentally only in the vicinity of the threshold and only if the K-meson spin is zero. If the π^--meson and helium parities are known it should be possible to determine the product of the K-meson and hyperfragment parities. It may be mentioned that the law of conservation of strangeness (not mentioning the laws of conservation of electric charge and number of baryons) does not permit one to determine separately the parity of each new particle.

† The so-called polarization correlation of ${}_\Lambda H^4$ and K is used. In this connection it should be mentioned that formula (1.5) in ref. [3]), p. 1033, and the corresponding paragraph are wrong.

4. Correlation for the $K^- + He^4$ Cascade

The $K^- + He^4 \rightarrow {}_{\Lambda}H^4 + \pi^0$ reaction is of the same type as that discussed above (if $k = 0$ it is symbolically written as $0+0 \rightarrow i+0$) but it is accompanied by release of energy.

The most frequent events are those involving electromagnetic capture of K^- mesons on one of the helium atom orbits (formation of a mesic atom) and subsequent (non-electromagnetic) interaction with the helium nucleus. However, this subsequent capture may proceed from the s-orbit ($l=0$) as well as from the p-orbit ($l=1$); according to theoretical estimates[8]) the probabilities for each of these events are approximately the same. The presence of $l = 1$ in the initial state is explained by the fact that a system (mesic atom) with a.m. equal to 0, 1, 2 etc., is initially formed by the long range electromagnetic forces. The probability for mesic atom formation rapidly decreases with increase of K-meson energy [9]). At an energy ≈ 0.1 MeV the cross section for mesic atom formation is approximately a million times smaller than the nuclear geometric cross section, i.e. $\approx 10^{-4}$ mb. Moreover the mesic atom is predominantly produced in the s-state when the K-meson energy exceeds the ionization energy of the mesic atom (≈ 0.01 MeV). Thus if it is ascertained in some way that the reaction takes place "in flight" †, that is, without mesic atom formation (or mesic atom formation in the s-state) the effect of long range electromagnetic forces will be negligible. Employing the generally accepted concept of a finite K-meson-nucleon interaction range it seems most natural to consider the element $(s'l'|S^0|0)$ in (5) to be larger than all other elements ($l \neq 0$) providing that the energy of the incident K-meson does not exceed 20 MeV (the interaction range in this case should be assumed to be of the order of the helium nucleus, i.e., 1.6×10^{-13} cm). For $\Pi = (-1)^i$ and $k = 0$ we then obtain from (5) for the hyperfragment wave function $(m_i|\mathrm{f}) \propto Y^*_{i,m_i}(\vartheta_f, \varphi_f)$ and for the angular distribution of the hyperfragment decay products

$$\mathscr{F}(\vartheta_f, \varphi_f; \vartheta, \varphi) \propto |\sum_m Y_{im}(\vartheta, \varphi) Y^*_{im}(\vartheta_f, \varphi_f)|^2 \propto [P_i (\cos\theta)]^2 \quad (10)$$

where θ is the angle between the directions ϑ_f, φ_f and ϑ, φ. The correlation is the same as that described by (8) but refers to a different angle.

Suppose now that comparison of experiment with theory indicates that the parity is $\Pi = -(-1)^i$. The statistics can then be expanded by considering events involving mesic atom formation, since capture from the s-orbit is forbidden. The same correlation formulae as those in the second line in table 1 are obtained if capture from the d-orbit is neglected. (It should

† For example the reaction $K^- + He^4$ occurs before the K-meson track becomes perceptibly curved or the reaction products in the laboratory system are observed to emerge in directions which do not differ by exactly 180^0.

be noted that if the hyperfragment and K-meson spins are zero the reactions $\pi^- + He^4 \rightarrow {}_{\Lambda}H^4 + K$ and $K^- + He^4 \rightarrow {}_{\Lambda}H^4 + \pi^0$ will in general be forbidden for $\Pi = -(-1)^0 = -1$.)

At low K-meson energies ($l = 0$) and $\Pi = (-1)^i$ Adair's procedure for the reaction under consideration is identical with that presented above (indeed for $l = 0$ the z-axis can be chosen parallel to $\mathbf{p}_f$ and from (4) and (3) it then follows that for the projection on this axis $(m_i|f) \propto \delta_{m_i 0}$ and (10) is immediately obtained); for $\Pi = -(-1)^i$ it does not have any region of applicability.

At higher energies Adair's procedure should be slightly modified: the z-axis is chosen parallel to $\mathbf{p}_f$ (instead of $z||\mathbf{p}_0$) and the region of applicability of (10) is then restricted by the largest significant value of l which is the orbital angular momentum *before the reaction*: those events should be chosen in which the angle between the incident K-mesons and the axis $z||\mathbf{p}_f$ is small in the c.m.s., or, what is the same, in which the hyperfragments are emitted at small angles with respect to $\mathbf{p}_0$.

It should be noted that the statistics obtained for the $\pi^- + He^4 \rightarrow {}_{\Lambda}H^4 + K_0$ and $K^- + He^4 \rightarrow {}_{\Lambda}H^4 + \pi^0$ reactions can be plotted on a single graph (if $k = 0$).

The possibility $k \neq 0$ can be accounted for only qualitatively for the cascade $K^- + He^4 \rightarrow {}_{\Lambda}H^4 + \pi^0$, ${}_{\Lambda}H^4 \rightarrow He^4 + \pi^-$ (one additional assumption being made.) If $k \neq 0$ one obtains $l' = |i-k|, |i-k|+1, \ldots i+k$ (see eq. (5).)

Since the energy release in the reaction $K^- + He^4 \rightarrow {}_{\Lambda}H^4 + \pi^0$ is not very large ($\approx$ 150 MeV) we shall assume that of all these values of l' only the lowest value occurs, that is $l' = |i-k|$ (or $l' = |i-k|+1$ in the other parity

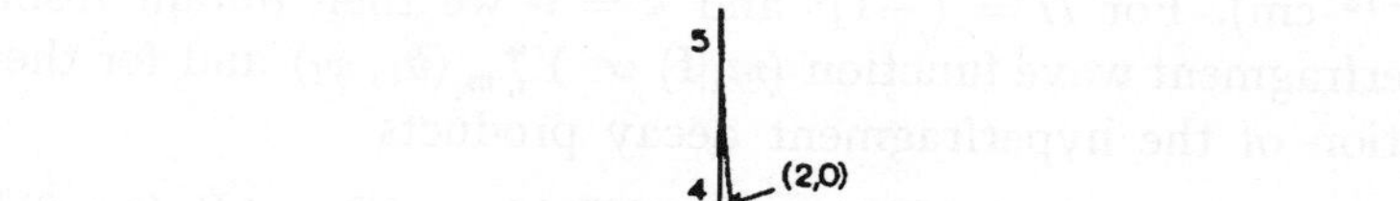

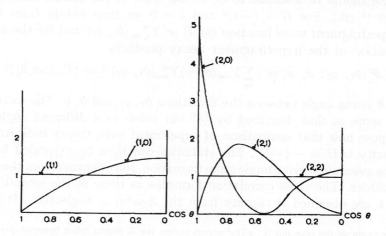

Fig. 1. Correlation in $K^- + He^4 \rightarrow {}_{\Lambda}H^4 + \pi$, ${}_{\Lambda}H^4 \rightarrow He^4 + \pi^-$ cascade for various cases (i, k) of the spin values of the hyperfragment and K-meson (the product of their parities being -1.)

case.) Clearly, the larger $|i-k|$ is, the better founded this assumption is. The calculations are now more complicated (the K-meson beam is assumed to be unpolarized.) A summary of the angular (θ) distribution for various spin values of the hyperfragment and K-meson is presented in fig. 1. (The product of their parities is -1, which corresponds to $\Pi = +1$. It should be noted that in table 1 the method of classification with respect to parity is different; thus the cases $i = 1, \Pi = -1$ and $i = 2, \Pi = +1$ are united in the same row.) Qualitatively the effect of $k \neq 0$ can be expressed as follows: $|i-k|$ or $|i-k|+1$ can be determined from the correlation. Thus $k \neq 0$ smoothes out the correlation (10) for hyperfragments (and hence lowers the measured spin value). The same conclusion can be drawn for the cascade $K^- + p \to \Sigma + K$, $\Sigma \to N + \pi$. Thus even if the θ correlation is set up for Alvarez events [10]) "in flight", isotropy, strictly speaking, would merely indicate that the hyperon and K-meson spin values are approximately equal.

5. Other Reactions

Besides the reactions discussed above, two other reactions of the $0+0 \to i+0$ type exist: the interaction of π and K-mesons with helium accompanied by formation of a ${}_{\Lambda}He^4$ hyperfragment. However at present no two-particle ${}_{\Lambda}He^4$ decays are known.

Other modes of interaction of π and K-mesons with helium are possible, such as $K^- + He^4 \to {}_{\Lambda}H^3 + n$, ${}_{\Lambda}H^3 \to He^3 + \pi^-$, and $K^- + He^4 \to \Sigma^0(\Lambda^0) + H^3$ or $K^- + He^4 \to \Sigma^- + He^3$, followed by $\Sigma \to N + \pi$. The correlation formulae for such reactions (of the $0+0 \to j+\frac{1}{2}$, $j \to \frac{1}{2}+0$ type) turn out to be the same as those for the $K^- + p \to \Sigma^- + \pi^+$ reaction [11]) so that, for example, cases of the $K^- + He^4 \to \Sigma^- + He^3$ reaction can be easily plotted on the θ correlation graph of the $K^- + p \to \Sigma^- + \pi^+$ reaction.

Besides the reactions (of the $0+0 \to i+0$, $i \to 0+0$ and $0+0 \to j+\frac{1}{2}$, $j \to \frac{1}{2}+0$ types) analyzed above and those of the $0+\frac{1}{2} \to j+0$ type † previously suggested [7, 8, 11]), no other reactions exist which are suitable for the determination of the spins of new unstable particles produced by bombarding hydrogen, deuterium, H^3, He^3 or He^4 targets with π mesons, K-mesons or nucleons. For other types of reactions (e.g. $K^- + D \to \Sigma + p$) the correlation formulae inevitably contain unknown parameters. In some cases certain assumptions regarding these parameters can be introduced which are similar to that made in section 4. However, in this case the

† Examples of such reactions are $\pi^{\mp} + p \to Y + K$ where Y is a Σ or Λ^0 particle; $K^- + p \to \overset{\circ}{\Xi}{}^- + \overset{\circ}{K}{}^+$, $\pi^- + He^3 \to {}_{\Lambda}H^3 + K^0$; and in general reactions of the type $\pi +$ nucleus (spin $\frac{1}{2}$) $\to$ hyperfragment + K, hyperfragment $\to \frac{1}{2}+0$.) Reactions involving energy release: $K^- + p \to Y + \pi$, $K^- + H^3 \to {}_{\Lambda}H^3 + \pi^-$, $\tilde{p} + He^4 \to {}_{\Lambda}H^3 + K^0$ ($\tilde{p}$ is an antiproton.)

correlation formulae cannot be expected to yield more than qualitative results and the respective reactions cannot be a reliable means of determination of the spins of new particles, although they may yield a non-trivial ($>\frac{1}{2}$) lower limit for the spin.

The authors express their appreciation to Prof. M. A. Markov for interest in this work and discussions.

Note added in the proof: **After the present paper was sent to print we learned that an identical result regarding the determination of the K-meson spin by Adair's method has been obtained by J. J. Sakurai (Phys. Rev. 107 (1957) 1119). From a point of view different from ours, Dr. Sakurai also emphasized the importance of the $\pi + He^4 \rightarrow {}_{\Lambda}H^4 + K$ reaction.**

References

1) K. Dalitz, Proceedings of the Rochester Conference (1956)
2) M. Gell-Mann, Phys. Rev. **106** (1957) 1298
3) M. I. Shirokov, JETP **32** (1957) 1022
4) P. A. M. Dirac, The Principles of Quantum Mechanics
5) A. Simon, Numerical Table of the Clebsch-Gordan Coefficients (ORNL-1718)
6) E. Jahnke and F. Emde, Tables of Functions
7) R. K. Adair, Phys. Rev. **100** (1955) 1540
8) R. Gatto, Nuovo Cimento **3** (1956) 1142
9) M. I. Podgoretsky, Ups. Fiz. Nauk **51** (1953) 253
10) L. W. Alvarez *et al.*, Nuovo Cimento **5** (1957) 1026
11) S. B. Treiman, Phys. Rev. **101** (1956) 1216

Nuclear Physics **6** (1958) 669—671; © *North-Holland Publishing Co., Amsterdam*

ON NON-UNIQUENESS OF NUCLEON-NUCLEON SCATTERING PHASE SHIFTS

L. G. ZASTAVENKO, R. M. RYNDIN and CHOU KUANG-CHAO
Joint Institute of Nuclear Research, Laboratory of Theoretical Physics, Dubna, USSR

Received 23 December 1957

The cross section for scattering of mesons by nucleons does not change upon application of the phase shift substitution proposed by Minami [1]). The two phase shift sets which are obtained from each other by means of this substitution can be differentiated either by carrying out polarization experiments [2, 3]) or by studying the energy dependence of the cross section at small energies. Below we derive similar transformations for nucleon-nucleon scattering.

Elastic nucleon-nucleon scattering can be completely described by the matrix $M(\mathbf{k}, \mathbf{k}_0; \boldsymbol{\sigma}_1, \boldsymbol{\sigma}_2)$ which defines the amplitude ψ_f of the scattered wave in terms of the initial spin state ψ_1:

$$\psi_f = M(\mathbf{k}, \mathbf{k}_0; \boldsymbol{\sigma}_1, \boldsymbol{\sigma}_2)\psi_1. \tag{1}$$

Here $\boldsymbol{\sigma}_1$ and $\boldsymbol{\sigma}_2$ are Pauli matrices of the two nucleons and $\mathbf{k}_0$ and $\mathbf{k}$ denote unit vectors directed along the direction of motion of the incident and scattered nucleons.

In order to derive the transformations under consideration we note that the scattering cross section for unpolarized nucleons $\sigma_0 = \frac{1}{4}\,\mathrm{tr}\, MM^+$ is invariant with respect to replacement of $M(\mathbf{k}, \mathbf{k}_0; \boldsymbol{\sigma}_1, \boldsymbol{\sigma}_2)$ by any of the following matrices:

$$\begin{aligned} M_1 &= \boldsymbol{\sigma}_1 \cdot \mathbf{k} M \boldsymbol{\sigma}_1 \cdot \mathbf{k}_0, \\ M_2 &= \boldsymbol{\sigma}_2 \cdot \mathbf{k} M \boldsymbol{\sigma}_2 \cdot \mathbf{k}_0, \\ M_3 &= \boldsymbol{\sigma}_1 \cdot \mathbf{k} \boldsymbol{\sigma}_2 \cdot \mathbf{k} M \boldsymbol{\sigma}_1 \cdot \mathbf{k}_0 \boldsymbol{\sigma}_2 \cdot \mathbf{k}_0. \end{aligned} \tag{2}$$

Expanding M in terms of the spherical functions $Y^{jm}_{ls}(\mathbf{k})$ which describe states with definite values of total angular momentum j, its projection m, orbital angular momentum l and spin s, we get

$$M(\mathbf{k}, \mathbf{k}_0; \boldsymbol{\sigma}_1, \boldsymbol{\sigma}_2) = \sum_{j,\, m} \frac{2\pi}{ik} \sum_{l,\, s;\, l',\, s'} Y^{jm}_{ls}(\mathbf{k}) Y^{jm+}_{l's'}(\mathbf{k}_0) R^j_{ls;\, l's'}. \tag{3}$$

The values of s and l are determined by the summation rule for angular momenta as follows:

for $s = 0$ (singlet), $l = j$,

for $s = 1$ (triplet), $l = j,\ j\pm 1$.

For a given value of j the $R^j_{ls;\,l's'}$, on account of the reversibility of motion, form a symmetrical four-row matrix R^j which satisfies the condition

$$R^{j+}R^j = -R^j - R^{j+}; \tag{4}$$

this condition follows from the unitarity of the S-matrix.

Consider, for example, the first transformation (2). Since the operator $\sigma_1\cdot\mathbf{k}$ commutes with the total angular momentum operator we have

$$\sigma_1\cdot\mathbf{k}Y^{jm}_{ls}(\mathbf{k}) = \sum_{l_1 s_1} L^{(1)j}_{l_1 s_1,\,ls}\, Y^{jm}_{l_1 s_1}(\mathbf{k}), \tag{5}$$

where $L^{(1)j}$ is a unitary matrix. Thus

$$M_1(\mathbf{k},\mathbf{k}_0;\sigma_1,\sigma_2) = \sum_{j,m}\frac{2\pi}{ik}\sum_{l,s;\,l',s'} Y^{jm}_{ls}(\mathbf{k})\,Y^{jm+}_{l's'}(\mathbf{k}_0)\,R^{(1)j}_{ls;\,l's'} \tag{6}$$

where

$$R^{(1)j} = L^{(1)j}R^jL^{(1)j+}; \tag{7}$$

$R^{(1)j}$ therefore satisfies condition (4) and possesses the same symmetry properties as R^j. The elements of $R^{(1)j}$ can thus be considered as elements of a new scattering matrix which yields the same cross section as M. The same also applies to the matrices $R^{(2)j} = L^{(2)j}R^jL^{(2)j+}$ and $R^{(3)j} = L^{(3)j}R^jL^{(3)j+}$ corresponding to the second and third transformations in (2).

The matrices $L^{(i)j}$ have the form

$$L^{(1,2)j} = \frac{i}{\sqrt{2j+1}}\begin{pmatrix} 0 & 0 & \mp\sqrt{j+1} & \mp\sqrt{j} \\ 0 & 0 & \sqrt{j} & -\sqrt{j+1} \\ \pm\sqrt{j+1} & -\sqrt{j} & 0 & 0 \\ \pm\sqrt{j} & \sqrt{j+1} & 0 & 0 \end{pmatrix},$$

$$L^{(3)j} = \begin{pmatrix} -1 & 0 & 0 & 0 \\ 0 & 1 & 0 & 0 \\ 0 & 0 & -\dfrac{1}{2j+1} & -\dfrac{2\sqrt{j(j+1)}}{2j+1} \\ 0 & 0 & -\dfrac{2\sqrt{j(j+1)}}{2j+1} & \dfrac{1}{2j+1} \end{pmatrix}. \tag{8}$$

The first column (row) refers to a singlet and the remaining ones to a triplet with the following order of l : $j,\ j+1,\ j-1$. The upper sign refers to index 1, the lower to index 2.

Since, in contrast to $\sigma_1\cdot\mathbf{k}\,\sigma_2\cdot\mathbf{k}$, the operators $\sigma_{1,2}\cdot\mathbf{k}$ do not commute with the operator of the square of total spin, $\frac{1}{4}(\sigma_1+\sigma_2)^2$, the matrices M_1 and M_2 lead to singlet-triplet transitions. The first two transformations therefore cannot be valid for collisions between identical nucleons for which singlet-triplet transitions are forbidden by the Pauli principle. This is also true for n—p scattering, providing that the hypothesis of isobaric invariance is correct.

The operators $\boldsymbol{\sigma}_i \cdot \mathbf{q}$ are operators of rotation of the ith nucleon spin through an angle π about the direction $\mathbf{q}$. This permits one to determine the transformation properties of various spin characteristics on replacement of R^j by $R^{(3)j}$. A result of this substitution, for example, is a change in the sign of the polarization P_0 arising in collisions involving unpolarized nucleons.

In conclusion, it should be noted that the quantities measured in a complete set of experiments, sufficient for the construction of the nucleon-nucleon scattering matrix [4]), do not change if along with replacement of R^j by $R^{(3)j}$ the signs of all phase shifts are reversed (complex conjugation of R^j). The non-uniqueness of the phase shifts which arises in this case can be removed by investigating the energy dependence of the cross section at low energies or by replacing one of the experiments of the complete set (e.g. K_{nn}) by a triple scattering experiment in mutually perpendicular planes.

References

1) S. Minami, Progr. Theor. Phys. **11** (1954) 213
2) S. Hayakawa, M. Kawaguchi and S. Minami, Progr. Theor. Phys. **12** (1954) 355
3) R. Ryndin and Ya. Smorodinsky, Doklady AN SSSR **103** (1955) 69
4) L. Puzikov, R. Ryndin and Ya. Smorodinsky, Nuclear Physics **3** (1957) 436

PHASE INDETERMINACIES IN NUCLEON-NUCLEON SCATTERING

L. G. ZASTAVENKO, R. M. RYNDIN, and CHOU GUAN-CHAO

Joint Institute for Nuclear Research

Submitted to JETP editor November 26, 1957

J. Exptl. Theoret. Phys. (U.S.S.R.) 34, 526-527 (February, 1958)

THE cross section for scattering of mesons by nucleons remains invariant under the phase substitution indicated by Minami.[1] The two sets of phase shifts, obtained from one another through this substitution, can only be distinguished either by means of polarization experiments,[2-3] or by analyzing the energy dependence of the cross section at low energies. We obtain below a similar transformation for the case of nucleon-nucleon scattering.

The elastic scattering of nucleons against nucleons is completely described by the scattering matrix $\mathbf{M}(\mathbf{k}, \mathbf{k}_0; \sigma_1, \sigma_2)$, which determines the amplitude χ_f of the scattered wave in terms of the initial spin state χ_i:

$$\chi_f = M(\mathbf{k}, \mathbf{k}_0; \sigma_1, \sigma_2)\,\chi_i. \tag{1}$$

Here σ_1 and σ_2 are the Pauli matrices for the two nucleons, and $\mathbf{k}_0$ and $\mathbf{k}$ denote unit vectors along the incident and scattered nucleon directions.

In order to obtain the transformation of interest, note that the scattering cross section for unpolarized nucleons $\sigma_0 = \frac{1}{4}\,\mathrm{Sp}\, MM^+$, is invariant to an exchange of $M(\mathbf{k}, \mathbf{k}_0; \sigma_1, \sigma_2)$ with one of the following matrices

$$M_1 = (\sigma_1\mathbf{k})\,M\,(\sigma_1\mathbf{k}_0), \qquad M_2 = (\sigma_2\mathbf{k})\,M\,(\sigma_2\mathbf{k}_0),$$
$$M_3 = (\sigma_1\mathbf{k})(\sigma_2\mathbf{k})\,M\,(\sigma_1\mathbf{k}_0)(\sigma_2\mathbf{k}_0). \tag{2}$$

We expand now the matrix M in terms of the spherical function $Y^{jm}_{ls}(\mathbf{k})$ describing a state of given total angular momentum j, its z-component m, orbital angular momentum l and spin s. Then

$$M(\mathbf{k}, \mathbf{k}_0; \sigma_1, \sigma_2) = \sum_{j,\,m}\frac{2\pi}{ik}\sum_{l,\,s;\,l',\,s'} Y^{jm}_{ls}(\mathbf{k})\,Y^{jm+}_{l's'}(\mathbf{k}_0)\,R^j_{ls;\,l's'}. \tag{3}$$

The values of s and l are determined from the rule for adding angular momenta, and are as follows: for $s = 0$ (singlet) $l = j$; for $s = 1$ (triplet) $l = j,\ j \pm 1$. For a given value of j, the quantities $R^j_{ls,l's'}$ form a symmetric (reversibility of the motion) four-rowed matrix R^j satisfying the condition

$$R^{j+}R^j = -R^j - R^{j+}, \tag{4}$$

which arises from the unitarity of the S-matrix.

Consider, for example, the first of the transformations (2). Since the operator $(\sigma_1\mathbf{k})$ commutes with the total angular momentum operator,

$$(\sigma_1\mathbf{k})\,Y^{jm}_{ls}(\mathbf{k}) = L^{(1)j}_{l_1s_1,\,ls}\,Y^{jm}_{l_1s_1}(\mathbf{k}). \tag{5}$$

Thus

$$M_1(\mathbf{k}, \mathbf{k}_0; \sigma_1, \sigma_2) = \sum_{j,\,m}\frac{2\pi}{ik}\sum_{l,\,s;\,l',\,s'} Y^{jm}_{ls}(\mathbf{k})\,Y^{jm+}_{l's'}(\mathbf{k}_0)\,R^{(1)j}_{ls,\,l's'}, \tag{6}$$

where

$$R^{(1)j} = L^{(1)j}R^jL^{(1)j+} \tag{7}$$

The matrix $L^{(1)j}$ is a unitary, antisymmetric, Hermitian matrix. Therefore $R^{(1)j}$ satisfies Eq. (4) and has the same symmetry properties as R^j. Therefore the elements of $R^{(1)j}$ may be considered as the elements of a new scattering matrix which leads to the same cross section as M. All this applies as well to the matrices $R^{(2)j} = L^{(2)j} \times R^jL^{(2)j+}$ and $R^{(3)j} = L^{(3)j}R^jL^{(3)j+}$, corresponding to the second and third transformation of (2).

The matrix $L^{(1)j}$ has the form

$$L^{(1,2)j} = \frac{i}{\sqrt{2j+1}}\begin{pmatrix} 0 & 0 & \mp\sqrt{j+1} & \mp\sqrt{j} \\ 0 & 0 & \sqrt{j} & -\sqrt{j+1} \\ \pm\sqrt{j+1} & -\sqrt{j} & 0 & 0 \\ \pm\sqrt{j} & \sqrt{j+1} & 0 & 0 \end{pmatrix},$$

$$L^{(3)j} = \begin{pmatrix} -1 & 0 & 0 & 0 \\ 0 & 1 & 0 & 0 \\ 0 & 0 & -\frac{1}{2j+1} & -\frac{2\sqrt{j(j+1)}}{2j+1} \\ 0 & 0 & -\frac{2\sqrt{j(j+1)}}{2j+1} & \frac{1}{2j+1} \end{pmatrix}. \tag{8}$$

The first columns (row) correspond to singlet states, the remaining to triplet states in the following ordering of l: $j,\ j+1,\ j-1$. The upper sign corresponds to superscript 1, and the lower sign to superscript 2.

Inasmuch as the operators $(\sigma_{1,2}\mathbf{k})$ in contradistinction to $(\sigma_1\mathbf{k})(\sigma_2\mathbf{k})$ do not commute with the square of the total spin operator $\frac{1}{4}(\sigma_1 + \sigma_2)^2$, the matrices M_1 and M_2 lead to singlet-triplet transitions. Therefore the first two transformations cannot take place in the case of identical nucleon collisions where singlet-triplet transitions are forbidden by the Pauli principle. This also is true for n–p scattering if isotopic invariance holds.

The operators $(\sigma_1\mathbf{q})$ represent operators which rotate the spin of the i-th nucleon by an angle π about the direction **q**. This allows to determine the transformation properties of various spin characteristics when R^j is replaced by $R^{(3)j}$. For example, this exchange leads to a change in the sign of the polarization P which takes place in the collision of unpolarized nucleons.

We finally remark that changing the sign of all the phase shifts (taking the complex conjugate of R^j) leaves the cross section unchanged, and changes the sign of $\mathbf{P}_0$. Thus a simultaneous application of this transformation with the transformation R^j into $R^{(3)j}$ leaves unchanged the cross section as well as the polarization. Therefore the two sets of elements of R obtained from one another by means of the indicated transformation, cannot be distinguished through the simplest polarization experiments (double scattering).

[1] S. Minami, Progr. Theor. Phys. **11**, 213 (1954).

[2] Hayakwa, Kawaguchi, and Minami, Progr. Theor. Phys. **12**, 355 (1954).

[3] R. Ryndin and Ia. Smorodinskii, Dokl. Akad. Nauk SSSR **103**, 69 (1955).

Translated by M. A. Melkanoff
101

ON THE DETERMINATION OF THE RELATIVE PARITIES OF ELEMENTARY PARTICLES

CHOU HUAN-CHAO (CHZHOU GUAN-CHZHAO)

Joint Institute of Nuclear Studies

Submitted to JETP editor January 16, 1958

J. Exptl. Theoret. Phys. (U.S.S.R.) **34**, 1027-1028 (April, 1958)

SINCE parity is not conserved in the weak interactions, it is of great interest to determine the parities of elementary particles by means of the strong interactions. We consider below several reactions which can be used for the determination of the relative parities of strange particles. We have in mind the following type of reaction:

$$p(n) + He^4 \rightarrow {}_{\Lambda}He^5 + K^+(K^0) \qquad (1)$$

with the assumption that the spins of ${}_{\Lambda}He^5$ and K are $\frac{1}{2}$ and 0, respectively.

This process is completely described by a spin-space matrix $M(\mathbf{n}, \mathbf{n}')$ which gives the amplitude of the diverging wave.[1] The most general form of the matrix $M(\mathbf{n}, \mathbf{n}')$ is

$$M(\mathbf{n}, \mathbf{n}') = a + b\sigma[\mathbf{n} \times \mathbf{n}'], \qquad (2)$$

when the product I of the parities of all four particles is equal to +1, and

$$M(\mathbf{n}, \mathbf{n}') = a\sigma\mathbf{n} + b\sigma\mathbf{n}', \qquad (3)$$

when I = −1. Here $\mathbf{n}$ and $\mathbf{n}'$ are unit vectors parallel to the momenta of the incident and emerging particles, respectively; a and b are certain functions of the energy and of the angle between $\mathbf{n}$ and $\mathbf{n}'$. The density matrix ρ_i of the initial state has the form

$$\rho_i = A(1 + \sigma\mathbf{P}), \qquad (4)$$

where $\mathbf{P}$ is the polarization vector of the incident particles. If the reaction takes place at threshold, or if we select only the ${}_{\Lambda}He^5$ particles emitted forward (i.e., $\mathbf{n} \parallel \mathbf{n}'$), then we can neglect the second term in Eq. (2). At the threshold, Eq. (3) takes the form $M = a\sigma\mathbf{n}$, and for $\mathbf{n} \parallel \mathbf{n}'$ we have $M = (a + b)\sigma\mathbf{n}$. The polarization vector $\mathbf{P}'$ of the ${}_{\Lambda}He^5$ particle in the final state is calculated by the formula[1]

$$\mathbf{P}' = \mathrm{Sp}(M\rho_i M^+\sigma) / \mathrm{Sp}(M\rho_i M^+). \qquad (5)$$

Substituting Eqs. (2), (3), and (4) into (5), we get

$$\mathbf{P}' = \mathbf{P}, \qquad \text{when } I = +1, \qquad (6)$$

$$\mathbf{P}' = (2(\mathbf{Pn})\,\mathbf{n} - \mathbf{P}), \qquad \text{when } I = -1. \qquad (7)$$

If the parity is not conserved in the decay of the ${}_{\Lambda}He^5$, then from the angular asymmetry of the decay one can measure the direction of polarization of the ${}_{\Lambda}He^5$ and distinguish between the possibilities (6) and (7). We emphasize that the incident beam must be polarized, and in such a way that the polarization vector is neither parallel nor perpendicular to the direction $\mathbf{n}$. The other reactions of this general type are as follows:

$\Sigma^{\pm} + He^4 \rightarrow {}_{\Lambda}He^5 + \pi^{\pm}$ at threshold, or when the ${}_{\Lambda}He^5$ emerges forward; (8)

$\Sigma^{\pm} + He^4 \rightarrow \Lambda + He^4 + \pi^{\pm}$ at threshold; (8′)

$p(n) + He^4 \rightarrow \Lambda + He^4 + K^+(K^0)$ at threshold. (1′)

In the last two reactions He^4 can be replaced by any other nucleus with spin 0 (for example, C^{12}).

We note that, in the reactions (1) and (1′), the polarization vectors of the $_\Lambda He^5$ and Λ depend only on the vectors **P** and **n** and the relative parities of K and Λ. Therefore, a case of the reaction (1′) can be simply added in with the cases of reaction (1).

The writer expresses his gratitude to M. I. Shirokov and L. G. Zastavenko for valuable advice and a discussion of the results.

[1] L. Wolfenstein and J. Ashkin, Phys. Rev. 85, 947 (1952); R. Oehme, Phys. Rev. 98, 147 (1955).

Translated by W. H. Furry
213

SOVIET PHYSICS JETP VOLUME 34(7), NUMBER 5 NOVEMBER, 1958

THE RELATIVISTIC THEORY OF REACTIONS INVOLVING POLARIZED PARTICLES

CHOU KUANG-CHAO and M. I. SHIROKOV

Joint Institute for Nuclear Research

Submitted to JETP editor December 6, 1957

J. Exptl. Theoret. Phys. (U.S.S.R.) **34**, 1230-1239 (May, 1958)

It is shown that, in the rest system, the relativistic formulas for the angular distribution and the polarization vectors and tensors for a reaction of the type $a + b \to c + d$ are essentially the same as the nonrelativistic formulas, if the spin of a particle is defined as its internal angular momentum around its center of mass. The square of this internal angular momentum is Lorentz invariant. The spins of the particles are arbitrary, and their rest masses are nonvanishing.

The main difference from the nonrelativistic case is that the description of the spin state is not the same in different Lorentz reference systems. Therefore for cascades of reactions (for example, for experiments on double scattering) corrections must be applied to the nonrelativistic formal theory. The relativistic changes in the angular correlations are indicated for successive reactions of the type $\pi + p \to Y + K$, $Y \to N + \pi$.

INTRODUCTION

FOR reactions of the type $a + b \to c + d$, formal theories are known which express the angular distribution and the state of polarization of the products of a reaction in terms of the states of polarization of the incident beam and the target and unknown parameters which are the elements of the S matrix for the process $a + b \to c + d$. The simplest example is the well-known formula for the function $f(\theta)$ which appears in the expression

$$\psi(\mathbf{r}) = e^{i\mathbf{kr}} + f(\theta)\, e^{ikr}/r$$

for the wave function of a stationary scattering process of particles of spin zero. In this case the unknown parameters are called the scattering phase shifts.

These theories are based on the use of the conservation laws (in particular the law of the conservation of the total angular momentum). Coester and Jauch[1] were the first to obtain formulas for the angular distribution and polarization in the case of particles a, b, c, d of arbitrary spins; their starting point was the explicit formulation of the conservation laws in terms of the diagonal property of the S matrix with respect to the conserved quantities. These same formulas have been obtained by Simon and Welton, but by a different method (cf., e.g., Ref. 2).

These formulas are nonrelativistic, but only because the spin state of the particles is described in the Pauli approximation (so that it has the same appearance in all Lorentz reference systems). The theory of the scattering of spinless particles is essentially relativistic. In order to obtain the angular distribution in any Lorentz system, one has only to transform $\sigma(\theta) = |f(\theta)|^2$ from the center-of-mass system into the desired system by using known formulas. What is therefore required for the relativistic generalization is the definition of the relativistic spin operator. The spin operator which we introduce satisfies all the requirements which can be demanded in terms of the concept of the spin as the intrinsic angular momentum of a particle.*

To obtain the relativistic formulas we use the method of Coester and Jauch in a form presented in a paper by one of us.[3] We emphasize that in this method one needs only the ability to describe the state of a free particle possessing spin; we do not need relativistic equations (for free particles) like the Dirac equation (which plays an essential part in Stapp's relativistic theory[4] of the scattering of particles with spin $\frac{1}{2}$).

1. THE CONSERVED PHYSICAL QUANTITIES IN A RELATIVISTIC THEORY

The conservation laws are simply an expression of the fact that the physical processes in an isolated system must not depend on the means used to de-

*Iu. M. Shirokov has informed us that he has employed this same description of the spin state (which was obtained by him from the theory of the irreducible representations of the inhomogeneous Lorentz group) to the formulation of a similar relativistic theory of polarization and correlation effects.

scribe it, in particular on the choice of the reference system. Here it is of course assumed that space-time is homogeneous and isotropic (we can suppose, however, that this assumption is contained in the concept of an isolated system). In quantum mechanics this fact is expressed by the requirement that the S matrix of a physical process must commute with ten operators, the infinitesimal displacements of the origin of the space and time coordinates providing operators P_μ, and infinitesimal space-time rotations giving operators $M_{\mu\nu}$. The fact that an operator commutes with the S matrix means that the S matrix is diagonal with respect to the eigenvalues of this operator,* and consequently that the corresponding physical quantity is conserved, i.e., remains unchanged in all internal processes.

Four conservation laws have the clear physical meaning of the conservation of the total momentum and energy $P_\mu\{P_x, P_y, P_z, iP_0\}$. Of the six other operators $M_{\mu\nu}$, three operators M_{k4} (k = 1, 2, 3) do not have an immediate physical meaning, and instead of $M_{\mu\nu}$ we shall introduce six other operators, which have the physical meaning of the coordinates of the center of mass of the physical system and its total angular momentum around its center of mass.

The properties of the center of mass follow directly from its conceptual meaning: the motion of the system as a whole can be characterized primarily (in the very first approximation) as the motion of a material point with mass equal to the rest mass (or energy) of the system and with momentum equal to the total momentum **P** of the system. The center of mass of an isolated system must therefore move uniformly in a straight line. Furthermore, in quantum mechanics we must require that the center of mass **R** actually be the coordinate operator of a certain particle, i.e., in particular, that the well known commutation relations hold between $R_x, R_y, R_z, P_x, P_y, P_z$.

We can obtain such an operator **R** in the following way. It is well known that the following are the commutation relations which must be satisfied by the operators P_i and $M_{\mu\nu}$ (cf. Ref. 6 and Sec. 3 of Ref. 5):

$$[P_i, P_j] = 0;\ [P_i, E] = 0;\ [M_i, P_j] = i\varepsilon_{ijk}P_k; \quad (1.1)$$

$$[M_i, E] = 0;\ [M_i, M_j] = i\varepsilon_{ijk}M_k; \quad (1.2)$$

$$[N_i, P_j] = i\delta_{ij}E;\ [N_i, E] = iP_i;\ [M_i, N_j] = i\varepsilon_{ijk}N_k; \quad (1.3)$$

$$[N_i, N_j] = -i\varepsilon_{ijk}M_k. \quad (1.4)$$

Notations: [A, B] = AB − BA; i, j, k take the values 1, 2, 3; and

$$\{M_1, M_2, M_3\} = \{M_{23}, M_{31}, M_{12}\};\ iN_j = M_{j4};$$

$$E = (\mathbf{P}^2 + m_0^2)^{1/2};\ \hbar = c = 1;$$

ϵ_{ijk} is a tensor antisymmetric in all its indices, with $\epsilon_{123} = 1$. It is understood that we confine ourselves to those state vectors ψ_0 which describe states with a definite rest mass m_0, i.e., for which $P_\mu P_\mu \psi_0 = -m_0^2\psi_0$, or $(P_0 - E)\psi_0 = 0$ (the time displacement operator P_0 is equivalent to the factor E). We note that since $[N_i, E] = iP_i$ the average value of N_i is a linear function of the time. Therefore N_i is "conserved" in the sense that internal processes have no effect on this time dependence.

We introduce three new operators R_x, R_y, R_z, for which

$$[R_i, R_j] = 0,\quad [R_i, P_j] = i\delta_{ij}$$

(from which if follows that $[R_i, E] = iP_i/E$). We represent **M** in the form

$$M_k = \sum_{i,j} \varepsilon_{kij}R_iP_j + J_k,$$

and find from Eq. (1.1) that $[J_i, P_j] = 0$. If we require that R_x, R_y, R_z be the components of a spatial vector (as must indeed be the case), i.e., that $[M_i, R_j] = i\epsilon_{ijk}R_k$, then $[J_i, R_j] = 0$, and J_x, J_y, J_z also form a three-dimensional vector. It then follows from Eq. (1.2) that $[J_i, J_j] = i\epsilon_{ijk}J_k$. In like fashion, representing **N** in the form

$$N_i = {}^1/_2(R_iE + ER_i) + K_i \equiv R_iE - iP_i/2E + K_i,$$

we find that $[K_i, P_j] = 0$. Therefore also $[K_i, E] = 0$, and the average value of K_i is constant in time.

We now pose the problem of expressing **M** and **N** in terms of the operators **R** and **J** which we have introduced. For **M** this has already been done. It remains only to express the spatial polar vector **K** in terms of **P** and **J**. It can be shown that if R_x, R_y, R_z are the first three components of any four-vector [i.e., for example, [N , R] = 0 if i ≠ j), then **K** cannot be expressed in terms of **J** and **P** only so as to satisfy all the commutation relations for **N**. This means that if **K** is constructed from **P** and **J** alone, then $[\mathbf{R}, \mathbf{P}]_k$ and J_k are not the spatial components of a four-dimensional tensor of the second rank.

*Let us write out AS − SA = 0 as a matrix product. In doing so, we choose a representation in which the operator A is diagonal (i.e., we label matrix elements with its eigenvalues). Then

$$A_{ik}S_{kl} - S_{im}A_{ml} = (A_i - A_l)S_{il} = 0,$$

i.e., S_{il} must be equal to zero if $i \neq l$.

The simplest **K** (namely one linear in J_x, J_y, J_z) satisfying Eqs. (1.3) and (1.4) has the form

$$\mathbf{K} = [\mathbf{PJ}](E \pm m)^{-1}$$

(cf. Refs. 7 and 6).*

Since the problem has a solution, it follows that: (1) **R** is "conserved" (in the same sence as **N**), since it can be expressed in terms of the conserved operators $M_{\mu\nu}$ (see Appendix). **R** can be called the center-of-mass operator. It is the same as definition (e) of the center-of-mass in the papers of Pryce.[8] (2) **J** is also conserved and, what is particularly important for our purpose, $J^2 = J_x^2 + J_y^2 + J_z^2$ is Lorentz invariant, since $[\mathbf{N}, J^2] = 0$. We emphasize that this is true for arbitrary $\mathbf{K} = \mathbf{K}(\mathbf{P}, \mathbf{J})$.

2. USE OF THE CONSERVATION LAWS. RELATIVISTIC DEFINITION OF THE SPIN OF A PARTICLE

The four conservation laws for the total momentum and energy and the three conservation laws for the center of mass can be expressed very simply. The argument is usually carried through in the Lorentz system of reference K_s in which the (conserved) total momentum is zero (the so-called center-of-mass system). The origin of the coordinate system can be taken at the center of mass (more precisely, at the point given by the average value of the operator **R**† for the particles a and b (or c and d). Then **J** is the total angular momentum. Since the commutation relations between J_x, J_y, J_z are the same as for the total angular momentum or for the Pauli spin matrices, the eigenvalues of $\hat{J}^2$ and $\hat{J}_z$ are respectively equal to $h^2 J(J+1)$ and $M = J, J-1, \ldots, -J$.

The conservation law for **J** is expressed by the fact that the S matrix is diagonal with respect to the eigenvalues of $\hat{J}^2$ and $\hat{J}_z$:

$$(\ldots J'M' | S | \ldots JM) = (\ldots | S^{JM} | \ldots) \delta_{J'J} \delta_{M'M} \quad (2)$$

and also the fact that $(\ldots | S^{JM} | \ldots)$ is independent of the value of M, which follows from $[J_x, S] = 0$.

To find, for example, the angular distribution of particles c and d, we need to know the explicit expression of the elements of the S matrix in the representation of the particle momenta. In order to express these elements in terms of the elements (2), we must first of all enumerate the remaining variables of a complete set (denoted in Eq. (2) by dots), which commute with each other and with J^2 and J_z.

The initial and final states of the process a + b → c + d are states of systems consisting of two free noninteracting particles. From the meaning of the S matrix, its elements are the transition amplitudes between such states. Therefore to label the elements of the S matrix we must take as our variables a complete set of quantum mechanical quantities describing the free particles a and b or c and d. The total angular momentum **J** (in the system K_s) is represented in the form $\mathbf{J} = \mathbf{j}_1 + \mathbf{j}_2$, where **j** is the total angular momentum of a single particle in K_s.

The procedure stated in Sec. 1 for obtaining the conserved angular momentum relative to the center of mass can be applied to a system of arbitrary physical nature (for example, to a system of fields). One needs only to know a concrete representation of the operators P and M. Therefore it is natural to apply this procedure to an "elementary" particle, whose physical nature is in general unknown (by the definition of "elementary"). Besides the coordinates **r** of the center of mass and the momentum **p** we then get just one conserved external characteristic of the particle, its angular momentum **s** relative to its center of mass **r**. In defining the spin s of a particle, we are only fixing precisely the concept of the spin as the intrinsic angular momentum of the particle.

Accordingly, $\mathbf{j} = [\mathbf{r} \times \mathbf{p}] + \mathbf{s}$, and in the system K_s, in which $\mathbf{p}_1 = -\mathbf{p}_2 = \mathbf{p}$, we get

$$\begin{aligned} \mathbf{J} &= [\mathbf{r}_1 \times \mathbf{p}_1] + \mathbf{s}_1 + [\mathbf{r}_2 \times \mathbf{p}_2] + \mathbf{s}_2 \\ &= [(\mathbf{r}_1 - \mathbf{r}_2) \times \mathbf{p}] + \mathbf{s}_1 + \mathbf{s}_2 \equiv \mathbf{l} + \mathbf{s}_1 + \mathbf{s}_2. \end{aligned} \quad (3)$$

In the system K_s we can now proceed in complete formal analogy with the nonrelativistic treatment to introduce the total spin operator $\mathbf{s} = \mathbf{s}_1 + \mathbf{s}_2$ (the square of which, however, is not a Lorentz-invariant quantity). The eigenfunctions of the square of this quantity, s^2, and its component s_z can be expressed in terms of the products $\psi_{i_1 m_1} \times \psi_{i_2 m_2}$ of the eigenfunctions of the squares and components of the operators $\mathbf{s}_1$ and $\mathbf{s}_2$ (the eigenvalues of s_1^2 are denoted by $h^2 i_1(i_1+1)$). Since

*We have not been able to show that no other **K**'s exist. Beginning with different considerations, L. G. Zastavenko has evidently proved the uniqueness of **K**. We are grateful to him for a discussion of this question.

†The conservation law for **R** means something more than the conservation of the average value. The requirement $[\mathbf{R}, S] = 0$ means that if the system is in a state with a definite value of **R** (we note that in the interaction representation the wave function for the external behavior of the system does not change with the time), internal processes in the system will not take it out of this state. This property does not get used explicitly, but an operator **R** of this kind is required for the definition of the conserved angular momentum **J** of the system (and of the spin of a particle, see below).

the commutation relations for $\mathbf{s}$, $\mathbf{s}_1$, $\mathbf{s}_2$ have the usual form, $[s_x, s_y] = is_z$ and so on, the coefficients in these expressions will be the well known Clebsch-Gordan coefficients $(i_1 i_2 m_1 m_2 \mid i_1 i_2 s m)$, which are also the transformation functions for the transformation from the representation in the variables i_1, i_2, m_1, m_2 into the i_1, i_2, s, m-representation (and inversely). A similar meaning attaches to the coefficient $(l s \mu m \mid l s J M)$.

We can now take as the variables denoted by dots in Eq. (2) s, l, and the absolute values of the momenta in K_s (or the total energy of the system, which in K_s is equal to an invariant, the rest mass of the system).

3. FORMULAS FOR THE CROSS-SECTION AND THE POLARIZATION VECTOR AND TENSORS. RELATIVISTIC ROTATION OF THE SPIN

We can now express the elements of the S matrix in the representation of the momenta of the particles and their spin components in terms of the elements $(p_c, s', l', J', M' \mid S \mid p_a, s, l, J, M)$. The transformation function from the representation in the variables p, s, l, J, M into the representation of the momenta and the spin components is the product of three transformation functions. We can write out the transformation as follows (cf. Ref. 9):

$$\begin{aligned}(\mathbf{p}_c;\, m_c,\, m_d \mid S \mid \mathbf{p}_a,\, m_a,\, m_b) &= (\vartheta_c \varphi_c p_c \mid l' \mu' p_c) \\ &\times (i_c i_d m_c m_d \mid i_c i_d s' m')(l' s' \mu' m' \mid l' s' J M)(s' l' \mid S^{J, E(p_a)} \mid s l) \\ &\times (l s J M \mid l s \mu m)(i_a i_b s m \mid i_a i_b m_a m_b)(l \mu p_a \mid \vartheta_a \varphi_a p_a).\end{aligned} \tag{4}$$

We have used Eq. (2) and the law of the conservation of the total energy. $\mathbf{p}_c$ and $\mathbf{p}_a$ are the momenta of particles c and d and a and b respectively in K; ϑ_c, φ_c, p_c, ϑ_a, φ_a, p_a are their spherical angles and absolute values. It is recalled that p_c is a function of p_a:

$$\sqrt{p_a^2 + \varkappa_a^2} + \sqrt{p_a^2 + \varkappa_b^2} = \sqrt{p_c^2 + \varkappa_c^2} + \sqrt{p_c^2 + \varkappa_d^2}.$$

Summation over the labels l', μ', s', m', J, M, l, μ, s, m is understood.

$$(\vartheta \varphi p \mid l \mu p_0) = 2\pi\hbar \sqrt{\frac{2R}{V}}\, \frac{i^{-l}}{p}\, Y_{l\mu}(\vartheta, \varphi)\,(p \mid p_0),$$

where $Y_{l\mu}(\vartheta, \varphi)$ is a spherical harmonic. For the other notations see Ref. 3 (in particular, Appendix II).

From the formula $\rho' = S\rho S^+$ we can now get the density matrix ρ' of the products of the reaction in the representation of their momenta and spin components (ρ is the density matrix for beam and target in the same representation). The problems of normalization and of deducing the cross-section in the system K_s are solved in just the same way as in the nonrelativistic case (cf. Ref. 3). One can also introduce in just the same way the statistical polarization tensors instead of the density matrices. All the formulas will have the same form as those of the nonrelativistic case.[3] The difference lies in the fact that the quantities m_a, m_b, m_c, m_d (or τ, ν), and also the total spin and other variables are relative to the system K_s. The same spin state has a different form in a different Lorentz system K (for example, in the laboratory system).

In order to learn how a spin state specified in K_s is described in K, we must find the transformation function from the representation in the eigenvalues of s^2 and s_z in the system K_s to the representation in the eigenvalues of $\tilde{s}^2$ and $\tilde{s}_z$, which are the square and component of the same operator, but in the system K. In view of the fact that s^2 is Lorentz invariant, the spin operator $\tilde{\mathbf{s}}$ is a vector rotated as compared with $\mathbf{s}$. Consequently, the transformation function is the same as is obtained in the solution of the problem of describing a given spin state in a rotated system of spatial axes:

$$(\tilde{m} \mid m) = D^i_{\tilde{m}m}(\Phi_2, \theta, \Phi_1) = e^{-i\tilde{m}\Phi_2} i^{m - \tilde{m}} P^i_{\tilde{m}m}(\cos\theta)\, e^{-im\Phi_1}; \tag{5}$$

The matrices $P^i_{mm}(\cos\theta)$ are defined in Ref. 10 [Eq. (22) on page 77; we note that the matrix P^1_{mn} written out explicitly on page 78 does not agree with Eq. (22) and is incorrect]. If the rotation is interpreted as a turning of a vector in a stationary coordinate system, then it consists of (a) a rotation of the vector around the z axis by the angle Φ_1, (b) a rotation around the y axis by the angle θ, and (c) a rotation around the z axis by the angle Φ_2. All these rotations are counterclockwise. In the Appendix we show how to find the axis of rotation and the rotation angle Ω of the spin vector for the transformation from K_s to K. For the transformation of the spin state of the reaction products from K_s to the laboratory system we get as the Eulerian angles of the rotation

$$\begin{gathered}\{\Phi_1,\, \theta,\, \Phi_2\} = \{+\varphi,\, \Omega,\, -\varphi\}, \\ \sin\Omega = \frac{\beta v \sin\vartheta\,(1 + \gamma + \gamma_\beta + \gamma')}{(1+\gamma)(1+\gamma_\beta)(1+\gamma')}\,\gamma\gamma_\beta,\end{gathered} \tag{6}$$

where

$$v = |\mathbf{p}|/\omega = \sqrt{\omega^2 - \varkappa^2}/\omega; \quad \gamma = \omega/\varkappa;$$

$$\gamma_\beta = (1 - \beta^2)^{-1/2}; \quad \gamma' = \omega'/\varkappa;$$

ω' is the energy of a reaction product in the laboratory system, and ϑ and φ are the spherical angles of its momentum **p** in K_s, measured in an axis system with the z axis parallel to $\boldsymbol{\beta}$ (the x and y axes are chosen arbitrarily).

In Stapp's[4] Eq. (48) for $\sin \Omega$ there is a mistake (or a missprint): it does not have the factors $\gamma\gamma_\beta$ [$\gamma^{(a)}\gamma^{(b)}$ in his notation]. If one repeats Stapp's calculations (in accordance with his arguments), the result is just the present Eq. (6). The rotation by the angle Ω must be applied counterclockwise around the vector $\boldsymbol{\beta}[\mathbf{p}]$, $\boldsymbol{\beta}$ being the velocity of the laboratory system relative to the center-of-mass system of the reaction (see Appendix).

This relativistic effect of rotation of the spin state of course does not show up at all in the transformation of the angular distribution into the laboratory system (since the angular distribution is a polarization tensor of rank zero). One needs only to carry out the ordinary (kinematic) relativistic transformation of the angles from K_s to the laboratory system. The nonrelativistic theory of the angular distribution in reactions with unpolarized beams and targets remains valid also in the relativistic domain (except for changes or increased precision in the meanings of the quantities involved in the formulas).

As for the polarization vector and the polarization tensors, they are not directly measured in the experiments. In order to measure the polarization vector for the product c of the reaction a + b → c + d, we must scatter c from a target e and measure the asymmetry in the angular distribution of the scattered particles c. Then we obtain information about the polarization vector **P′** in the center-of-mass system K'_s of the reaction c + e → c + e. The desired polarization vector is obtained from **P′** by a rotation. The angle of this rotation is found from Eq. (6). In fact, in the successive Lorentz transformations from K_s to the laboratory system (by means of the known velocity $\boldsymbol{\beta}$) and then to the system K′ (velocity $\boldsymbol{\beta}'$) a rotation occurs only in the first transformation, since the momentum $\mathbf{p}'_c$ of particle c in the laboratory system is parallel to $\boldsymbol{\beta}'$, so that $\sin \Omega_2 \sim \sin(\hat{\boldsymbol{\beta}'\mathbf{p}'_c}) = 0$. This question is analyzed in more detail in Ref. 4, and the treatment carried through there is valid for arbitrary spin.

Quite generally, the relativistic rotation of the spin is seen to be of importance only for the treatment of cascades of reactions. In the following section we shall deal with the problem of the relativistic changes of the angular correlations in successions of reactions of type a + b → c + d, c → e + f.

In conclusion we point out that in the change from K_s to the system K_0 in which a particle is at rest the description of the spin state does not undergo any changes, since here $\boldsymbol{\beta} \parallel \mathbf{p}$ and $\Omega = 0$. Therefore we can regard the quantities m_a, m_b, and so on as describing the spin states of the particles in their rest systems K_0. This interpretation is preferable to the preceding one: the spin states of particles are described by quantities whose definition does not depend on the system K_s, i.e., on what target the particle is interacting with, on what its energy is, or on the energy balance of a particular reaction.*

4. THE RELATIVISTIC ANGULAR CORRELATIONS IN CASCADES OF THE TYPE a + b → c + d, c → e + f

We shall consider first the cascade $\pi^- + p \to Y + K$, $Y \to N + \pi$, which is well known in the literature. If the first reaction occurs near threshold, the correlation in the angle γ between the directions of the incident π^- mesons and the decay nucleons provides a way of determining the spin j of the hyperon Y. This correlation can be found if one substitutes into the expression

$$F(\vartheta, \varphi) = \frac{\omega}{\sqrt{4\pi}} \sum_{q=0,2\ldots}^{2j-1} (2q+1)^{-1/2} Q(j, q) \sum_{\nu=-q}^{q} Y_{q\nu}(\vartheta, \varphi)\, \rho(q, \nu) \qquad (7)$$

for the angular distribution of the decay products of the hyperon in its rest system K_Y the concrete expression for the statistical polarization tensors $\rho(q, \nu)$ of the hyperon (cf. Refs. 12 and 13). In the center-of-mass system K_s of the reaction $\pi^- + p \to Y + K$ we have near threshold (the z axis is directed along the π^- meson beam)

$$\rho_s(q, \nu) \sim Q(j, q)\, \delta_{\nu,0}. \qquad (8)$$

The nonrelativistic correlation in the angle γ is

---

*In connection with this interpretation, however, the following misunderstanding can arise. Since there is only one system K_0 in which a particle is at rest, in any reaction the m values mean always the same thing: the spin components in the rest systems. Consequently it might seem that no transformations of the spin state are actually necessary. The point is that if we are given the velocity $\mathbf{v}_{21}$ of a system K_2 relative to K_1 and the velocity $\mathbf{v}_{32}$ of K_3 relative to K_2, then the velocity $\mathbf{v}_{31}$ (which is of course a function of $\mathbf{v}_{21}$ and $\mathbf{v}_{32}$) turns out not to be parallel to $\mathbf{v}_{13}$ if $[\mathbf{v}_{21}\mathbf{v}_{32}] \neq 0$ (cf. Ref. 11, Sec. 22). The transformation from K_1 to K_3 must have the form of a Lorentz transformation with a spatial rotation [see Eq. (58) of Ref. 11]. If the particle was at rest in K_1, then in K_3 it has the velocity $\mathbf{v}_{13}$, and by using this velocity we can go over to a system K_4 in which the particle is again at rest. Calculations show that the product of the transformations from K_1 to K_3 and then on to K_4 has the form of a pure spatial rotation: $\mathbf{s}_{(4)} = D^{-1}\mathbf{s}_{(1)}$, if $D\mathbf{v}_{13} = -\mathbf{v}_{31}$ (the space axes of the Lorentz systems K_1, K_2, K_3, K_4 are of course assumed parallel).

obtained by simply substituting Eq. (8) into Eq. (7):

$$F_{nr}(\vartheta,\varphi)\sim\sum_{q=0}^{2j-1}(2q+1)^{-1/2}Q^2(j,q)Y_{q,0}(\vartheta,0) \quad (9)$$

$$\sim\sum_{q=0}^{2j-1}Q^2(j,q)P_q(\cos\gamma).$$

In actual fact one must substitute into Eq. (7) not the $\rho_s(q,\nu)$, but the statistical tensors of the hyperon referred to the system K_Y:

$$\rho(q,\nu)=\sum_{\nu'}D^q_{\nu,\nu'}(\varphi_c,\Omega(\vartheta_c),-\varphi_c)\rho_s(q,\nu')$$

$$=\sqrt{4\pi/(2q+1)}\,Y^*_{q,\nu}(\Omega,\varphi_c)\,Q(j,q). \quad (10)$$

Here φ_c and ϑ_c are the spherical angles of the emission of the hyperon in the system K_s. The angle Ω is determined by Eq. (6), since in the transformation from K_s to the laboratory system and then on into K_Y a rotation occurs only in the passage from K_s to the laboratory system. (We note that in the experiment $F(\vartheta,\varphi)$ is obtained by translating the measured distribution from the laboratory system into K_Y, and not by going from K_s to the rest system of the hyperon).

Substituting Eq. (10) into Eq. (7), we get

$$F_r(\vartheta,\varphi)\sim\sum_{q=0}^{2j-1}(2q+1)^{-1}Q^2(j,q)\sum_{\nu=-q}^{+q}Y_{q\nu}(\vartheta,\varphi)Y^*_{q\nu}(\Omega,\varphi_c)$$

$$=(1/4\pi)\sum_{q=0}^{2j-1}Q^2(j,q)P_q(\cos\gamma_r), \quad (11)$$

where γ_r is now the angle between the direction of emergence of the decay products and the direction $\{\Omega(\vartheta_c),\varphi_c\}$. Thus the expression for the correlation has its old form, if we change the definition of the angle γ.

In the experiment we are discussing Ω does not exceed 1.5°. If we construct the distribution in γ, choosing only cases with fixed $\vartheta_c\approx 90°$ and fixed φ_c, then the difference between the nonrelativistic and the relativistic correlations can amount to 3% for $j = 3/2$ and 5% for $j = 5/2$. In the actual experimental procedure, of course, all cases of the cascade are used in the construction of the correlation $F(\gamma)$. If Eq. (11) is just integrated over φ_c, the difference between the resulting correlation $F_r(\gamma,\Omega(\vartheta_c))$ and the nonrelativistic result (9) does not exceed 0.1% over the entire range of angles γ and ϑ_c (for $j\le 5/2$).

In the case of the cascade $K^- + p \rightarrow Y + \pi$, $Y \rightarrow N + \pi$, the angular correlation again does not contain any unknown parameters and depends only on the spin j of the hyperon, if the energies of the K^- particles do not exceed 20 or 30 Mev but are large enough (> 0.1 Mev) so that mesic atoms are not formed (for more details see Ref. 9). In the center-of-mass system of the reaction $K^- + p \rightarrow Y + \pi$, with the z axis parallel to the direction of $[\mathbf{n}_K\times\mathbf{n}_Y]$, where $\mathbf{n}_K$ is the direction of the incident beam of K particles,

$$\rho_s(q,\tau)\sim Q(j,q)\,\delta_{\tau,0}.$$

Relative to this same set of axes, but in the rest system K_Y of the hyperon

$$\rho(q,\tau)=\sum_{\tau'}D^q_{\tau\tau'}(0,\Omega,0)\rho_s(q,\tau')$$

$$=\sqrt{4\pi/(2q+1)}\,Y^*_{q,\tau}(\Omega,0)\,Q(j,q). \quad (12)$$

The difference between the relativistic correlation $F_r(\vartheta,\varphi)$ and the nonrelativistic function is basically the same: substituting Eq. (12) into Eq. (7), we get the correlation $F_r(\theta)$ of the angle θ between the direction $\mathbf{n}$ of the emission of the decay products and the vector obtained by rotating $\mathbf{n}_Y$ by the angle Ω around the vector $[\mathbf{n}_K\times\mathbf{n}_Y]$ (i.e., in the plane of the reaction). The nonrelativistic correlation had the same form, but θ was the angle between $\mathbf{n}$ and $\mathbf{n}_Y$.

The correlation proposed by Adair[13] (cf. also Ref. 9) admits of energies larger than those near threshold, but does not change when treated relativistically: for this correlation one uses cases in which the hyperons are emitted at small angles with the direction of the incident beam, and then $\Omega\approx 0$.

Since the most general case of the cascade $a + b \rightarrow c + d$, $c \rightarrow e + f$, in which all the spins are arbitrary and the correlation depends on unknown parameters, is of no practical interest, we simply note without proof that the nonrelativistic form of the angular correlation can be retained. To do this one finds for each case of the cascade, using the measured angles of emission of the particle c, a particular system of coordinates belonging to this case. The angles of emission of the products from the decay of c are calculated relative to this system. The distribution in these recalculated angles has the old, nonrelativistic, form (but one has, of course, changed the rule for constructing the angular correlation from the experimental data).

APPENDIX

1. Let us express $\mathbf{s}$ in terms of $M_{\mu\nu}$ and p_μ. Let κ be the rest mass of the particle and $\omega = (p^2+\kappa^2)^{1/2}$, and

$$\mathbf{M}=[\mathbf{r}\times\mathbf{p}]+\mathbf{s},$$

$$\mathbf{N}=\mathbf{r}\omega-i\mathbf{p}/2\omega+[\mathbf{p}\times\mathbf{s}]/(\omega+\kappa). \quad (A.1)$$

The four-dimensional vector

$$\Gamma_\sigma = (1/2i)\,\varepsilon_{\mu\nu\sigma\lambda} M_{\mu\nu} p_\lambda$$

($\epsilon_{\mu\nu\sigma\lambda}$ is the completely antisymmetric tensor of the fourth rank with $\epsilon_{1234} = 1$) then has the form:

$$\Gamma = s\varkappa + (ps)\,p/(\omega + \varkappa), \quad \Gamma_4 = i\,(sp). \qquad \text{(A.2)}$$

Noting that $(\Gamma p) = \omega(sp)$, we find from Eq. (A.2):

$$s = \Gamma/\varkappa - (\Gamma p)\,p/\varkappa\omega(\omega + \varkappa). \qquad \text{(A.3)}$$

All these operator equations are to be understood as being in the momentum representation.

From the second of the relations (A.1) we now get

$$r\omega = N + ip/2\omega - [p \times \Gamma]/\varkappa(\omega + \varkappa). \qquad \text{(A.4)}$$

2. The vector $\widetilde{s}$ in the new system K, which moves relative to K_s with the velocity β (in units of the speed of light) can now be found in the following way. Substituting into the right and left members of the equations [cf. Ref. 11, Sec. 18, Eq. (25)]

$$\widetilde{\Gamma} = \Gamma + \beta\,\{(\Gamma\beta)(\gamma_\beta - 1)\,\beta^{-2} - \gamma_\beta \Gamma_{4/i}\},$$

$$\widetilde{\Gamma}_{4/i} = \gamma_\beta\,\{\Gamma_{4/i} - (\beta\Gamma)\} \qquad \text{(A.5)}$$

the expressions (A.2) for Γ and $\widetilde{\Gamma}$ in terms of s, p_μ and $\widetilde{s}$, $\widetilde{p}_\mu$, respectively, and replacing the $\widetilde{p}_\mu$ by their expressions in terms of p_μ (which have the same form (A.5)), we get the expressions for $\widetilde{s}$ in terms of s. First of all we establish the fact that $\widetilde{s}$ is a linear combination of the vectors s, β, and p. This means that the vector $\widetilde{s}$ is obtained from s by a rotation around an axis perpendicular to β and p. There remains only the determination of the magnitude and sign of the angle of rotation around this axis. For this purpose we choose a convenient set of three space axes (it is obvious that the angle of rotation cannot depend on the choice of the axes): $z \parallel \beta$, $y \parallel [\beta \times p]$. A rotation of the vector around the y axis in the counterclockwise direction through the angle Ω must have the form

$$\widetilde{s}_x = \cos\Omega s_x + \sin\Omega s_z \qquad \widetilde{s}_z = -\sin\Omega s_x + \cos\Omega s_z. \qquad \text{(A.6)}$$

Representing the expression for $\widetilde{s}$ in terms of s (in the chosen set of axes) in the form (A.6), and finding the coefficient of s_z in the expression for $\widetilde{s}_x$ (as that having the simplest form), after cumbersome calculations we get the formula (6) of Sec. 3 for $\sin\Omega$.

[1] F. Coester and J. M. Jauch, Helv. Phys. Acta **26**, 3 (1953).

[2] A. Simon, Prob. Sovr. Fiz. 1955, No. 6, p. 21 (Russian translation); Phys. Rev. **92**, 1050 (1953).

[3] M. I. Shirokov, J. Exptl. Theoret. Phys. (U.S.S.R.) 32, 1022 (1957), Soviet Phys. JETP **5**, 835 (1958).

[4] H. P. Stapp, Phys. Rev. **103**, 425 (1956).

[5] C. Møller, Dansk. Mat-Fys. Medd. **23**, No. 1 (1945).

[6] L. L. Foldy, Phys. Rev. **102**, 568 (1956).

[7] Iu. M. Shirokov, Dokl. Akad. Nauk SSSR **94**, 857 (1954).

[8] M. H. L. Pryce, Proc. Roy. Soc. **A150**, 166 (1935); **A195**, 62 (1948).

[9] Chou Kuang-Chao and M. I. Shirokov, Nuclear Phys. **6**, 10 (1958).

[10] I. M. Gel'fand and Z. Ia. Shapiro, Usp. Mat. Nauk **7**, 3 (1952).

[11] C. Møller, The Theory of Relativity. Oxford 1952.

[12] M. I. Shirokov, J. Exptl. Theoret. Phys. (U.S.S.R.) **31**, 734 (1956), Soviet Phys. JETP **4**, 620 (1957).

[13] R. K. Adair, Phys. Rev. **100**, 1540 (1956).

Translated by W. H. Furry
247

SELECTION RULES IN REACTIONS INVOLVING POLARIZED PARTICLES

CHOU KUANG-CHAO

Joint Institute for Nuclear Research

Submitted to JETP editor April 23, 1958

J. Exptl. Theoret. Phys. (U.S.S.R.) **35**, 783-785 (September, 1958)

SIMON and Welton[1] and Shirokov[2] have obtained the selection rules for a reaction of the type $a + b \rightarrow c + d$ in the form of relations between the polarization vectors and tensors. They assume that the initial state is not polarized. The present communication gives a derivation of the selection rules for any arbitrarily-polarized initial state. We shall use Shirokov's notation[2] and assume that all the particles have nonvanishing rest mass.

Consider the statistical tensors of the final state in the $a + b \rightarrow c + d$ reaction,

$$\rho'(q_c, \tau_c, q_d, \tau_d; g_c g_a^{-1}), \tag{1}$$

which depend essentially on the parameters of the rotation $g_c g_a^{-1}$ which carries the z_a, y_a, x_a coordinate system into the z_c, y_c, x_c coordinate system. The first of these systems is associated with the initial state, the z_a axis being parallel to $\mathbf{n}_a$, and the y_a axis being perpendicular to the production plane of particle a. The second of these systems has z_c parallel to $\mathbf{n}_c$, and y_c in the direction of the cross product $\mathbf{n}_a \times \mathbf{n}_c$. Here $\mathbf{n}_i$ is the unit vector along the direction of motion of particle i, and q is the rank of the statistical tensors. The spin indices τ are defined in terms of these particular coordinate systems. With this choice of coordinate systems, the Euler angles for the rotation $g_c g_a^{-1}$ are $\{-\pi, \theta_c, \pi - \varphi_c\}$, where θ_c and φ_c are the spherical angles of the unit vector $\mathbf{n}_c$ in the z_a, y_a, x_a coordinate system (see Shirokov[2]).

Let the state obtained from the initial one by space reflection be characterized by the statistical tensor ρ'_I. Under the reflection the z_a and z_c axes, chosen along the momenta of particles a and c, change direction, while the y_a and y_c axes remain invariant. The spherical angles θ_{cI} and φ_{cI} of the reflected $-\mathbf{n}_c$ vector in the reflected $\{z_a, y_a, x_a\}_I$ coordinate system are

$$\vartheta_{cI} = \vartheta_c, \quad \varphi_{cI} = -\varphi_c. \tag{2}$$

The spin operators remain invariant under reflection. If θ_c and φ_c are replaced by θ_{cI} and φ_{cI} in Eq. (1), we obtain the ρ'_I statistical tensors from ρ'; the spin indices τ of the new ρ'_I tensors must be quantized with respect to the old nonreflected z_c, y_c, x_c system. Since the reflected $\{z_c, y_c, x_c\}_I$ coordinate system differs from the initial one only by rotation through an angle π about the y_c axis, the transformation properties of the statistical tensors[2] lead to the equations

$$\rho'_I(q_c, \tau_c, q_d, \tau_d; \vartheta_c, -\varphi_c) = \sum_{\tau'_c \tau'_d} D^{q_c}_{\tau_c \tau'_c}(0, \pi, 0) D^{q_d}_{\tau_d \tau'_d}(0, \pi, 0) \times \rho'(q_c, \tau'_c, q_d, \tau'_d; \vartheta_c, \varphi_c). \tag{3}$$

Here the spin indices τ are quantized with respect to their own proper coordinate systems. Since $D^q_{\tau\tau'}(0, \pi, 0) = (-1)^{q+\tau} \delta_{\tau,-\tau'}$ (see Shirokov[2] and Gel'fand and Shapiro[3]), Eq. (3) leads to

$$\rho'_I(q_c, \tau_c, q_d, \tau_d; \vartheta_c, -\varphi_c) = (-1)^{q_c+\tau_c+q_d+\tau_d} \rho'(q_c, -\tau_c, q_d, -\tau_d; \vartheta_c, \varphi_c). \tag{4}$$

The law of parity conservation may be stated in the following way: if the initial statistical tensors ρ are replaced by the reflected tensors ρ_I, the tensors ρ'' of the products of the reaction are the tensors ρ'_I which are obtained from ρ' by Eq. (4). In other words, if the statistical tensors ρ' are written $\rho' = F(\rho)$, then

$$\rho'_I = F(\rho_I). \tag{5}$$

Equations (4) and (5) together give the most general selection rules in the form of a relation between the statistical tensors.

Let us now consider some simple examples. If the initial state of the a + b → c + d reaction is unpolarized, then $\rho = \rho_I$. Then according to (4) and (5) $\rho' = \rho'_I$, or

$$\rho'(q_c, \tau_c, q_d, \tau_d; \vartheta_c) = (-1)^{q_c+\tau_c+q_d+\tau_d}\rho'(q_c, -\tau_c, q_d, -\tau_d; \vartheta_c). \tag{6}$$

In our case the ρ' tensors do not depend on φ_c Equations (6) are the same selection rules as Simon and Welton obtained for q = 1 and the same as those obtained by Shirokov.*

Let us now consider a cascade of the form a + b → c + d followed by c + e → f + g (the incident beam a, the target b, and e are unpolarized). According to (6), $\rho = \rho_I$ in the initial state of the second reaction, and we obtain

$$\rho'(q_f, \tau_f, q_g, \tau_g; \vartheta_f, -\varphi_f) = (-1)^{q_f+\tau_f+q_g+\tau_g}\rho'(q_f, -\tau_f, q_g, -\tau_g; \vartheta_f, \varphi_f). \tag{7}$$

For the special case in which $q_f = q_g = 0$, Eq. (7) becomes

$$\sigma(\vartheta_f, -\varphi_f) = \sigma(\vartheta_f, \varphi_f). \tag{8}$$

Since φ_f is the azimuth angle of n_f in the coordinate system in which the y_c axis is directed along $n_a \times n_c$, Eq. (8) states the well known fact that the angular distribution is symmetric about the production plane of the incident particle in the second reaction of the cascade. Equations (7) may be regarded as a generalization of this assertion.

In conclusion, we remark that our selection rules can also be obtained by Shirokov's method, but the present approach is simpler.

The author expresses his gratitude to Professors M. A. Markov, M. I. Shirokov, and L. G. Zastavenko for interest and discussion of the results.

*We remark that the first and second selection rules given by Shirokov are actually two different ways of stating the same rule.

[1] A. Simon and T. A. Welton, Phys. Rev. 90, 1036 (1953).

[2] M. I. Shirokov, J. Exptl. Theoret. Phys. (U.S.S.R.) 32, 1022 (1957), Soviet Phys. JETP 5, 835 (1952).

[3] M. I. Gel'fand and Z. Ia. Shapiro, Uspekhi Mat. Nauk 7, 3 (1952).

Translated by E. J. Saletan
154

SOVIET PHYSICS JETP VOLUME 35(8), NUMBER 6 JUNE, 1959

THE SHAPIRO INTEGRAL TRANSFORMATION

CHOU KUANG-CHAO and L. G. ZASTAVENKO

Joint Institute for Nuclear Research

Submitted to JETP editor May 5, 1958

J. Exptl. Theoret. Phys. (U.S.S.R.) **35**, 1417-1425 (December, 1958)

We give an integral transformation which is related to the decomposition, into irreducible representations of the proper Lorentz group, of the representation according to which the wave function of a particle with mass M and spin s transforms.

THE problem mentioned in the abstract was stated and solved in 1955 by I. S. Shapiro.[1] Unfortunately, he used a wave-function transformation law which is incorrect for $s > 0$ (the transformation law for the spin under a pure Lorentz transformation[2]), and thus the equations he obtained can be used only for spin zero.

In this paper we solve Shapiro's problem for arbitrary spin using a wave-function transformation law previously obtained;[2] the principles by which the integral transformation is obtained are the same as those used by Shapiro.

In the first two sections we state some known facts which we shall need later.[2,3] In the third and fourth sections we derive the integral transformation.

1. IRREDUCIBLE REPRESENTATIONS OF THE PROPER LORENTZ GROUP

Let

$$k = (k_1 k_2 k_3 k) \tag{1.0}$$

be the 4-momentum of a particle with mass 0, and let

$$k_0 = (00kk). \tag{1.0'}$$

Obviously $k = R(\mathbf{n}) k_0$, where $R(\mathbf{n})$ is the rotation* which carries the third axis into the direction given by $\mathbf{n} = \mathbf{k}/k$; if θ and φ are the spherical angle coordinates of $\mathbf{n}$, we may write

$$R(\mathbf{n}) = R_3(\varphi + \pi/2) R_1(\theta), \tag{1.1}$$

where $R_3(\alpha)$ is a counterclockwise rotation about the 3-axis through an angle α. Let $k' = Sk$, where S is a transformation of the Lorentz group. We shall denote the relation between $\mathbf{n} = \mathbf{k}/k$ and $\mathbf{n}' = \mathbf{k}'/k'$ by

$$\mathbf{n}' = S\mathbf{n}; \tag{1.2}$$

then $\mathbf{n}' = S_1\mathbf{n}$ and $\mathbf{n}'' = S_2\mathbf{n}'$ implies that $\mathbf{n}'' = S_2 S_1 \mathbf{n}$. Arbitrary transformation t of the Lorentz group can be written in the form

$$t = rK, \tag{1.3}$$

where $K\mathbf{n}_0 = \mathbf{n}_0$ in the sense of (1.2), and r is a pure rotation such that $r\mathbf{n}_0 = t\mathbf{n}_0$ [again in the sense of (1.2)]. Here the choice of $\mathbf{n}_0$ is arbitrary; (1.3) obviously determines r up to an arbitrary rotation about the $\mathbf{n}_0$ direction. We shall choose $\mathbf{n}_0$ as the unit vector along the third axis; we define r by the condition $r = R_3(\varphi_1) R_1(\theta)$, which means that $\varphi_2 = 0$, and then the factorization implied by (1.3) is clearly unique.

Now let $x' = Sx$. It is well known that if $H(x) = x_0 + (\sigma \cdot \mathbf{x})$ and $b(S)$ is the matrix corresponding to S in the 2×2 representation of the Lorentz group, then $H(x') = b(x) H(x) b^*(S)$. If we now choose $x = k_0$ [see Eq. (1.0')] we see that in the 2×2 representation a K transformation of the type in (1.3) is represented by a matrix of the form

$$K = \begin{pmatrix} k_{11} & k_{12} \\ 0 & k_{22} \end{pmatrix}. \tag{1.4}$$

We now define the transformation $K(S, \mathbf{n})$ by applying (1.3) to the matrix $t = S^{-1}R(\mathbf{n})$ (the definition of $R(\mathbf{n})$ is given in (1.1)):

$$S^{-1}R(\mathbf{n}) = R(S^{-1}\mathbf{n}) K(S, \mathbf{n}) \tag{1.5}$$

[the vector $S^{-1}\mathbf{n}$ is defined in (1.2)]. Applying both sides of (1.5) to $\mathbf{n}_0$, we see that in the present case $r = R(S^{-1}\mathbf{n})$.

Further, making use of the uniqueness of the factorization given by (1.3), we find that $K(S, \mathbf{n})$ has the property that

$$K(S_2, \mathbf{n}) K(S_1, S_2^{-1}\mathbf{n}) = K(S_2 S_1, \mathbf{n}), \tag{1.6}$$

which we shall need later.

*In what follows we shall always denote by R a pure rotation (or the matrix in the appropriate function space corresponding to such a rotation); L will denote a pure Lorentz transformation.

Consider the association

$$f(\mathbf{n}) \xrightarrow{S} f'(\mathbf{n}) = \alpha(S, \mathbf{n}) f(S^{-1}\mathbf{n}). \quad (1.7)$$

Let us find the properties that $\alpha(S, \mathbf{n})$ must satisfy in order that this association form a representation of the Lorentz group. We have

$$f'(\mathbf{n}) \xrightarrow{S_1} f''(\mathbf{n}) = \alpha(S_1, \mathbf{n}) f'(S_1^{-1}\mathbf{n})$$
$$= \alpha(S_1, \mathbf{n}) \alpha(S, S_1^{-1}\mathbf{n}) f(S^{-1}S_1^{-1}\mathbf{n}).$$

It is thus necessary that

$$\alpha(S_1, \mathbf{n}) \alpha(S, S_1^{-1}\mathbf{n}) = \alpha(S_1 S, \mathbf{n}). \quad (1.8)$$

It is seen from this that we may choose as $\alpha(S, \mathbf{n})$ any matrix element $[K(S, \mathbf{n})]_{22}$ to any power [see (1.4) to (1.6)]. Thus

$$\alpha = |k_{22}|^a \exp\{i \arg k_{22} \cdot b\}.$$

Now (1.4) implies that if $k' = S^{-1}k$, then $|k_{22}|^2 = |\mathbf{k}|/|\mathbf{k}'|$. Denoting the association $\mathbf{k} \to \mathbf{n} = \mathbf{k}/|\mathbf{k}|$ by

$$|\mathbf{k}| = k(\mathbf{n}), \quad (1.9)$$

we may write $|k(S, \mathbf{n})_{22}|^2 = k(\mathbf{n})/k(S^{-1}\mathbf{n})$.

Further, let us write*

$$\arg k_{22} = {}^1/_2\, \varphi(S, \mathbf{n}), \quad (1.9a)$$

so that $\alpha(S, \mathbf{n})$ becomes

*Let us study the angle $\varphi(S, \mathbf{n})$ in more detail.

(a) $S = R$. In this case the definition of $\varphi(s, \mathbf{n})$ in (1.9a) is equivalent to

$$R^{-1}R(\mathbf{n}) = R(R^{-1}\mathbf{n}) R_3(\varphi(R, \mathbf{n})). \quad (1.9b)$$

(b) $S = L_p$ [the definition of L_p is given in (2.1)]. Let the rotation $R(L, \mathbf{n})$ be defined (clearly not uniquely) by

$$L^{-1}\mathbf{n} = R(L, \mathbf{n})\,\mathbf{n}. \quad (1.9c)$$

Then

$$L^{-1}R(\mathbf{n}) = R(L, \mathbf{n}) R^{-1}(L, \mathbf{n}) L^{-1}R(\mathbf{n})$$
$$= R(L, \mathbf{n}) R(\mathbf{n}) \{R^{-1}(\mathbf{n}) R^{-1}(L, \mathbf{n}) L^{-1}R(\mathbf{n})\}.$$

The transformation in curly brackets does not alter the direction of the 3-axis. If we take advantage of its arbitrariness and choose the rotation $R(L_p, \mathbf{n})$ to be about the axis parallel to $\mathbf{p} \times \mathbf{n}$, the k_{22} matrix element of the transformation in the brackets will be real. (To see this we note that the expression in curly brackets is equal to $R^{-1}(L_{p_0}, \mathbf{n}_0) L_{p_0}^{-1}$. Here p_0 is obtained from p by the same rotation as leads to $\mathbf{n}_0$, the unit vector in the direction of the 3-axis from $\mathbf{n}$. The proof is completed by a simple calculation involving the matrices of the 2×2 representation.) The angle $\varphi(L, \mathbf{n})$ therefore satisfies the relation

$$R(L_p, \mathbf{n}) R(\mathbf{n}) = R(R(L_p, \mathbf{n})\,\mathbf{n}) R_3(\varphi(L_p, \mathbf{n})). \quad (1.9d)$$

The angle involved in the rotation $R(L_p, \mathbf{n})$ is given in the Appendix.

$$\alpha_{m\rho}(S, \mathbf{n}) = [k(\mathbf{n})/k(S^{-1}\mathbf{n})]^{1-i\rho/2} e^{im\varphi(S,\mathbf{n})}. \quad (1.10)$$

Equations (1.7) and (1.10) define the complete set of representations of the proper Lorentz group with ρ taking on all complex values, while m takes on only integral and half-integral values.[3] It can be shown that all these representations are nondecomposable and that all except those for whic $i\rho$ is an integer are irreducible; they are all inequivalent, except for pairs one of which has parameters ρ and m, and the other of which has parameters $-\rho$ and $-m$. It is easily shown that when ρ is real, (1.7) gives a unitary representation of the Lorentz group in the sense that for such representations one can define a scalar product

$$(f_1, f_2) \equiv \int f_1^*(\mathbf{n}) f_2(\mathbf{n})\, d\Omega(\mathbf{n}),$$

which remains invariant under (1.7), i.e., which is equal to

$$(f_1', f_2') = \int \left[\frac{k(\mathbf{n})}{k(S^{-1}\mathbf{n})}\right]^2 d\Omega(\mathbf{n})\, f_1^*(S^{-1}\mathbf{n}) f_2(S^{-1}\mathbf{n}).$$

Indeed, we note that

$$[k(\mathbf{n})]^2\, d\Omega(\mathbf{n}) = [k(S^{-1}\mathbf{n})]^2\, d\Omega(S^{-1}\mathbf{n}).$$

Then writing $S^{-1}\mathbf{n} = \mathbf{m}$, we obtain $(f_1'f_2') = (f_1f_2)$. These unitary representations form the so-called first class of unitary representations of the Lorent group.

The Lorentz group also has another set of unitary representations (the second class), but for our purposes only the first will be necessary.

2. THE TRANSFORMATION OF THE SPIN UNDER PROPER LORENTZ TRANSFORMATIONS

Let L_p (or L_p) be a pure Lorentz transformation such that

$$p = L_p p_0, \quad (2.1)$$

where $p = (\mathbf{p}, E)$, and $p_0 = (0, M)$.

Let S be any Lorentz transformation. We shall write the transformation SL_p as a product of a pure Lorentz transformation and a rotation, i.e., $SL_p = LR$. It is easily seen that here $L = L_{Sp}$. One can thus consider the formula

$$SL_p = L_{Sp} R(S, p) \quad (2.2)$$

to be a definition of the rotation $R(S, p)$.

It is easily shown that

$$R(S_2S_1, (S_2S_1)^{-1}p)$$
$$= R(S_2, S_2^{-1}p) R(S_1, (S_2S_1)^{-1}p). \quad (2.3)$$

From this one easily finds that the association

$$\Psi_{sM}(p\sigma) \xrightarrow{S} \Psi'_{sM}(p\sigma) = R^s(S, S^{-1}p)_{\sigma\sigma'} \Psi_{sM}(S^{-1}\mathbf{p}, \sigma') \quad (2.4)$$

(where R^S is the matrix which represents the rotation $R(S, S^{-1}p)$ in the $(2s+1)$-dimensional irreducible representation of the three-dimensional rotation group[4]) gives a representation of the Lorentz group. This is the representation according to which the wave-function $\Psi_{sM}(\mathbf{p}, \sigma)$ of a particle with mass M and spin s transforms.

3. INTEGRAL TRANSFORMATION

Let us now break up the representation sM defined by (2.4) into irreducible representations ρm [see (1.7) and (1.10)].

For the representation sM we can define the invariant scalar product

$$\sum_\sigma \int \Psi^*(\mathbf{p}, \sigma)\Psi(\mathbf{p}, \sigma)d^3\mathbf{p}/E_p,$$

so that in decomposing sM we can obtain only unitary representations.* Thus the desired decomposition leads to the integral transformation

$$\Psi_{p\sigma} = \sum_{m=-s}^{s}\int_0^\infty d\rho \int d\Omega(\mathbf{n})\, Y_{\rho mn}(\mathbf{p}, \sigma) f_{\rho mn}, \tag{3.1}$$

$$f_{\rho mn} = \sum_{\sigma=-s}^{s}\int \frac{d^3\mathbf{p}}{E_p} Y'_{\rho mn}(\mathbf{p}, \sigma)\Psi_{p\sigma}. \tag{3.1a}$$

The kernels $Y_{\rho mn}(\mathbf{p}, \sigma)$ and Y' must satisfy the following condition. If $\Psi_{p\sigma} \rightleftarrows f_{\rho mn}$ according to (3.1) and (3.1a), and if $\Psi_{p\sigma} \xrightarrow{S} \Psi'_{p\sigma}$ according to (2.4) and $f_{\rho mn} \xrightarrow{S} f'_{\rho mn}$ according to (1.7), then $\Psi'_{p\sigma} \rightleftarrows f'_{\rho mn}$ according to (3.1) and (3.1a), for all S. We may write this as the following condition on $Y_{\rho mn}(\mathbf{p}, \sigma)$:

$$R^s(S, S^{-1}p)_{\sigma\sigma'} Y_{\rho m S^{-1}\mathbf{n}}(S^{-1}\mathbf{p}, \sigma') = [k(\mathbf{n})/k(S^{-1}\mathbf{n})]^{-1-i\rho/2} e^{im\varphi(S,\mathbf{n})} Y_{\rho mn}(\mathbf{p}, \sigma). \tag{3.2}$$

The inverse kernel $Y'_{\rho mn}(\mathbf{p}, \sigma)$ satisfies a similar condition if we write

$$Y'_{\rho mn}(\mathbf{p}, \sigma) = C_{m\rho}\bar{Y}_{\rho mn}(\mathbf{p}, \sigma),$$

and if (3.2) is satisfied.

Let us first set $\mathbf{p} = 0$ and $S = R$ in (3.2). Then $R^S(S, S^{-1}p) = R^S(R)$ is simply the matrix corresponding to the rotation R. In this case (3.2) gives

$$R^s(R)_{\sigma\sigma'} Y_{\rho m R^{-1}\mathbf{n}}(0, \sigma') = e^{im\varphi(R,\mathbf{n})} Y_{\rho mn}(0, \sigma).$$

Comparing this with (1.9b) and recalling that $R_3(\varphi)_{\sigma\sigma'} = e^{-i\sigma\varphi}\delta_{\sigma\sigma'}$, we find that

$$Y_{\rho mn}(0, \sigma) = R^s(\mathbf{n})_{\sigma m}. \tag{3.3}$$

Further, let us set $S = L_p$ in (3.2). Then

*In fact, only representations of the first class.[1]

$$R(S, S^{-1}p)_{\sigma\sigma'} = \delta_{\sigma\sigma'},\quad k(\mathbf{n})/k(S^{-1}\mathbf{n}) = M/(E_p - \mathbf{p}\cdot\mathbf{n}),$$

and we obtain

$$Y_{\rho mn}(\mathbf{p}, \sigma) = \left(\frac{E_p - \mathbf{p}\cdot\mathbf{n}}{M}\right)^{-1-i\rho/2} e^{-im\varphi(L_p,\mathbf{n})} Y_{\rho m L_p^{-1}\mathbf{n}}(0, \sigma). \tag{3.4}$$

Noting that in (1.9c) $R(L_p, \mathbf{n})\mathbf{n} = L_p^{-1}\mathbf{n}$, we can rewrite (3.4), using (3.3), in the form

$$Y_{\rho mn}(\mathbf{p}\sigma) = \left(\frac{E_p - \mathbf{p}\cdot\mathbf{n}}{M}\right)^{-1-i\rho/2} R^s_{\sigma\sigma'}(L_p, \mathbf{n}) R_{\sigma' m}(\mathbf{n}). \tag{3.5}$$

Let us now prove that the Y kernel as defined by (3.4) satisfies (3.2). We do this by inserting (3.4) into (3.2). The left side then becomes

$$R(S, S^{-1}p)_{\sigma\sigma'} R_{\sigma' m}(L_{S^{-1}p}^{-1}, S^{-1}\mathbf{n}) \exp\{-im\varphi(L_{S^{-1}p}, S^{-1}\mathbf{n})\} \times \{k(S^{-1}\mathbf{n})/k(L_{S^{-1}p}^{-1}, S^{-1}\mathbf{n})\}^{1+i\rho/2},$$

while the right side becomes

$$\exp\{im\varphi(S, \mathbf{n})\}[k(\mathbf{n})/k(S^{-1}\mathbf{n})]^{-1-i\rho/2}[k(\mathbf{n})/k(L_p^{-1}\mathbf{n})]^{1+i\rho/2} \times \exp\{-im\varphi(L_p, \mathbf{n})\} R(L_p^{-1}\mathbf{n})_{\sigma m}$$

According to (2.2), $SL_{S^{-1}p} = L\, R(S, S^{-1}p)$. Therefore

$$k(L_{S^{-1}p}^{-1}S^{-1}\mathbf{n}) = k[R^{-1}(S, S^{-1}p)L_p^{-1}\mathbf{n}] = k(L_p^{-1}\mathbf{n}),$$

since a rotation does not change the magnitude k. Thus the factors k on both sides cancel. Further,

$$R(S, S^{-1}p)_{\sigma\sigma'} R(L_{S^{-1}p}^{-1}S^{-1}\mathbf{n})_{\sigma' m}$$

$$= R(S, S^{-1}p)_{\sigma\sigma'} R[R^{-1}(S, S^{-1}p)L_p^{-1}\mathbf{n}]_{\sigma' m} = R(L_p^{-1}\mathbf{n})_{\sigma m} \times \exp\{-im\varphi[R^{-1}(S, S^{-1}p), R^{-1}(S, S^{-1}p)L_p^{-1}\mathbf{n}]\}.$$

Thus the rotations on the right and left are also equal; we are left with the angles φ. We use (2.2) to write the last angle obtained in the form

$$\varphi[R^{-1}(S, S^{-1}p), L_{S^{-1}p}^{-1}S^{-1}\mathbf{n}].$$

To this we must add $\varphi(L_{S^{-1}p}, S^{-1}\mathbf{n})$; according to (1.8) and (2.2), their sum is

$$\varphi[L_{S^{-1}p}R^{-1}(S, S^{-1}p), S^{-1}\mathbf{n}] = \varphi[S^{-1}L_p, S^{-1}\mathbf{n}].$$

Taking the angle $\varphi(S, \mathbf{n})$ from the right to the left and adding it here, we obtain $\varphi(L_p, \mathbf{n})$; the same angle remains on the right. This completes the proof.

4. CALCULATION OF $C_{m\rho}$

We have thus arrived at the mutually reciprocal integral equations

$$\Psi_{sM}(\mathbf{p}, \sigma) = \sum_m \int d\rho d\Omega(\mathbf{n})[(E_p - \mathbf{p}\cdot\mathbf{n})/M]^{-1-i\rho/2} \times R_{\sigma\sigma'}(L_p, \mathbf{n}) R_{\sigma' m}(\mathbf{n}) f_{\rho m}(\mathbf{n}); \tag{4.1}$$

$$f_{\rho m}(\mathbf{n}) = C_{m\rho} \sum_{\sigma} \int \frac{d^3\mathbf{p}}{E_p} [(E_p - \mathbf{p}\mathbf{n})/M]^{-1+i\rho/2}$$

$$\times \bar{R}_{\sigma\sigma'}(L_{\mathbf{p}}, \mathbf{n}) \bar{R}_{\sigma' m}(\mathbf{n}) \Psi_{sM}(\mathbf{p}\sigma) \qquad (4.2)$$

[see Eqs. (3.1) to (3.5)].

To find the still undetermined factor $C_{m\rho}$, let us rewrite (4.1) and (4.2) in the form

$$\delta(\rho - \rho')\delta(\mathbf{n} - \mathbf{n}')\delta_{mm'} = C_{m\rho} \int \frac{d^3\mathbf{p}}{E_p} [(E_p - \mathbf{p}\cdot\mathbf{n})/M]^{-1+i\rho/2}$$

$$\times [(E_p - \mathbf{p}\cdot\mathbf{n}')/M]^{-1-i\rho'/2}$$

$$\times \bar{R}_{\sigma\sigma'}(L_p, \mathbf{n}) \bar{R}_{\sigma' m}(\mathbf{n}) R_{\sigma\sigma''}(L_p, \mathbf{n}') R_{\sigma'' m'}(\mathbf{n}'). \qquad (4.3)$$

We now multiply both sides of (4.3) by $\bar{R}_{am'}(\mathbf{n}')$, sum over m', integrate over $d\Omega(\mathbf{n}')$, and choose $\mathbf{n} = \mathbf{n}_0$ along the 3-axis. We thus obtain

$$\delta(\rho - \rho')\delta_{ma} = C_{m\rho} \int \frac{d^3\mathbf{p}}{E_p} [(E_p - \mathbf{p}\cdot\mathbf{n}_0)/M]^{-1+i\rho/2}$$

$$\times [(E_p - \mathbf{p}\cdot\mathbf{n}')/M]^{-1-i\rho'/2} \bar{R}_{\sigma m}(L_p, \mathbf{n}_0) R_{\sigma a}(L_p, \mathbf{n}')\, d\Omega(\mathbf{n}').$$

In this expression we write $R(L_{\mathbf{p}}, \mathbf{n}') = R(\mathbf{p}/p) \times R(L_{\mathbf{p}_0}, \mathbf{n}'') R^{-1}(\mathbf{p}/p)$. Here $\mathbf{p}_0$ is directed along the 3-axis, and $\mathbf{n}''$ is obtained from $\mathbf{n}'$ by the same transformation which carries $\mathbf{p}$ into $\mathbf{p}_0$. Since $\mathbf{p}\cdot\mathbf{n}' = \mathbf{p}_0\cdot\mathbf{n}''$ and $d\Omega(\mathbf{n}') = d\Omega(\mathbf{n}'')$, we have

$$\delta(\rho - \rho')\delta_{ma} = C_{m\rho} \int \frac{d^3\mathbf{p}}{E_p} [(E_p - \mathbf{p}\cdot\mathbf{n}_0)/M]^{-1+i\rho/2}$$

$$\times [(E_p - \mathbf{p}_0\cdot\mathbf{n}'')/M]^{-1-i\rho'/2}$$

$$\times \bar{R}_{\sigma m}(L_p, \mathbf{n}_0) R_{\sigma\alpha}(\mathbf{p}/p) \bar{R}_{\gamma\beta}(\mathbf{p}/p) R_{\gamma\beta}(L_{\mathbf{p}_0}\mathbf{n}'')\, d\Omega(\mathbf{n}'').$$

We shall first perform the integration over the azimuth angles, writing

$$d\Omega(\mathbf{p}) = \sin\theta\, d\theta\, d\varphi, \quad d\Omega(\mathbf{n}'') = \sin\theta''\, d\theta''\, d\varphi''.$$

According to (1.1), $R(\mathbf{p}/p) = R_3(\varphi + \pi/2) R_1(\theta)$. According to Appendix A

$$R(L_{\mathbf{p}}, \mathbf{n}_0) = R_3(\varphi - \pi/2) R_1(\alpha(p, t)) R_3^{-1}(\varphi - \pi/2),$$

$$R(L_{\mathbf{p}_0}\mathbf{n}'') = R_3(\varphi'' + \pi/2) R_1(\alpha(p, t'')) R_3^{-1}(\varphi'' + \pi/2).$$

Here $t = \mathbf{p}\cdot\mathbf{n}/p = \cos\theta$ and $t'' = \mathbf{p}\cdot\mathbf{n}''/p = \cos\theta''$, while the angle $\alpha(p, t)$ is defined [see Eqs. (A.2) to (A.4)]. Since $R_3(\varphi)_{nm} = \delta_{nm}\exp\{-in\varphi\}$,

$$\frac{1}{(2\pi)^2}\delta(\rho - \rho')\delta_{ma} = C_{m\rho}$$

$$\times \int_0^\infty \frac{p^2 dp}{E_p} \int_{-1}^{1} dt \int_{-1}^{1} dt'' [(E_p - pt)/M]^{-1+i\rho/2}$$

$$\times [(E_p - pt'')/M]^{-1-i\rho'/2} (-1)^{m-\sigma} \sum_{\alpha,\sigma} \bar{u}^s_{\sigma m}(\alpha(p, t))$$

$$\times u^s_{\alpha a}(\alpha(p, t'')) u^s_{\sigma\alpha}(\theta) \bar{u}^s_{\alpha a}(\theta)\delta_{am};$$

here the $u^s_{mn}(x)$ are functions defined by Gel'fand and Shapiro.[4] By making use of the properties of these functions,* this equation is easily transformed to the form

$$(1/2\pi)^2\delta(\rho - \rho') = C_{m\rho} \int_0^\infty \frac{p^2 dp}{E_p} \int_{-1}^{1} dt \int_{-1}^{1} dt'' [(E_p - pt)/M]^{-1+i\rho/2}$$

$$\times [(E_p - pt'')/M]^{-1-i\rho'/2} \qquad (4.4)$$

$$\times \sum_{\alpha} u^s_{m\alpha}[\alpha(p, t) + \theta] u^s_{\alpha a}(\alpha(p, t'')) \bar{u}^s_{m\alpha}(\theta),$$

or

$$(2\pi)^{-2}\delta(\rho - \rho') = C_{m\rho} \int_0^\infty \frac{p^2 dp}{E_p} \sum_{\alpha} F_{\alpha\rho' s}(p) f_{m\alpha\rho s}(p),$$

where

$$F_{\alpha\rho' s}(p) = \int_{-1}^{1} [(E_p - pt'')/M]^{-1-i\rho'/2} u^s_{\alpha a}(\alpha(p, t''))\, dt'';$$

$$f_{m\alpha\rho s}(p) = \int_{-1}^{1} [(E_p - pt)/M]^{-1+i\rho/2} u^s_{m\alpha}[\alpha(pt) + \theta] \bar{u}^s_{m\alpha}(\theta)\, dt.$$

We now calculate $F_{\alpha\rho s}(p)$. We do this by using the power series

$$u^s_{\alpha a}(\alpha) = \sum_q a^q_{sa} x^q,$$

where $x = 1 + \cos\alpha$, and

$$a^q_{sa} = \frac{(-)^{s-q}(s+q)!}{2^q (q-\alpha)!(q+\alpha)!(s-q)!}.$$

The calculation gives

$$F_{\alpha\rho s}(p) = \frac{1}{p}\Big\{\sum_q a^q_{sa} 2^{q+1} \frac{q\cos(\lambda\rho/2) + (\rho/2)\sin(\lambda\rho/2)}{q^2 + (\rho/2)^2}$$

$$+ Q_{\alpha\rho s}(p)\Big\}. \qquad (4.6)$$

Here $\cosh\lambda = E_p/M$, and $Q_{\alpha\rho s}(p)$ can be represented in the form $A(p, \rho)\cos(\lambda\rho/2) + B(p, \rho)\sin(\lambda\rho/2)$, where A and B are functions bounded for all $p > 0$ and approaching zero as $p \to \infty$ as $1/p$ uniformly in ρ.

We now evaluate $f_{m\alpha\rho s}(p)$. To do this we make use of the series

$$u^s_{m\alpha}(\theta) = (1 + x)^{|(m+\alpha)/2|}(1 - x)^{|(m-\alpha)/2|}$$

$$\times \sum_0^{n_{\max}}{}' A_n (1 + x)^n, \qquad x = \cos\theta.$$

Using (A.5) and (A.6), we find that for $|m \pm \alpha| \neq 0$

$$^* \quad u^s_{mn}(\theta) = (-1)^{s-m} i^{n-m}[2^s(s-m)!]^{-1}$$

$$\times [(s+n)!(s-m)!/(s-n)!(s+m)!]^{1/2}(1-x)^{(m-n)/2}$$

$$\times (1+x)^{-(m+n)/2}(d/dx)^{s-n}(1-x)^{s-m}(1+x)^{s+m}, \quad x = \cos\theta;$$

$$u^s_{mn}(\theta) = u^s_{nm}(\theta) = u^s_{-n-m}(\theta);$$

$$\sum_\alpha u^s_{m\alpha}(\theta_1) u^s_{\alpha n}(\theta_2) = u^s_{mn}(\theta_1 + \theta_2).$$

$$|f_{ma\rho s}(p)| < \text{const}/p^{1+a}, \tag{4.7}$$

where a is the smaller of $|m \pm \alpha|$. We rewrite the sum in (4.4) in the form

$$\sum_{\alpha} u^s_{m\alpha}[\alpha(p,t)+\theta]\,\bar{u}^s_{m\alpha}(\theta)\,u^s_{\alpha\alpha}(\alpha')$$

$$=\sum_{\alpha} u^s_{m\alpha}[\alpha(pt)+\theta]\,\bar{u}^s_{m\alpha}(\theta)\,u^s_{mm}(\alpha')$$

$$+\sum_{\alpha\mp m} u^s_{m\alpha}(\alpha+\theta)\,\bar{u}^s_{m\alpha}(\theta)\,[u^s_{\alpha\alpha}(\alpha')-u^s_{mm}(\alpha')].$$

Because $u_{mm} = u_{-m-m}$, we see that $\alpha = -m$ does not enter into the last sum. Using the summation theorem for the u_{mm} functions, we can write (4.4) in the form

$$\delta(\rho-\rho') = (2\pi)^2 C_{m\rho}\int_0^\infty \frac{p^2dp}{E_p}\Big\{F_{m\rho' s}(p)\,F_{m-\rho s}(p)$$

$$+\sum_{|\alpha|\neq|m|} f_{ma\rho s}(p)\,[F_{a\rho' s}(p)-F_{m\rho' s}(p)]\Big\}. \tag{4.8}$$

It follows from (4.6) and (4.7) that only the product of the principal parts of $F_{m\rho s}$ and $F_{m-\rho s}$ contributes to the δ-function. All the other parts of the integral in (4.8) can only give a finite contribution (which must cancel in the sum). Thus

$$\delta(\rho-\rho')\sim(2\pi)^2C_{m\rho}\int_0^\infty d\lambda\sum_q\frac{q\cos(\lambda\rho/2)+(\rho/2)\sin(\lambda\rho/2)}{q^2+(\rho/2)^2}$$

$$\times\sum_{q'}\frac{q'\cos(\lambda\rho'/2)+(\rho'/2)\sin(\lambda\rho'/2)}{(q')^2+(\rho'/2)^2}2^{q+q'+2}a^q_{sm}a^{q'}_{sm}.$$

Since

$$\frac{1}{\pi}\int_0^\infty\cos(\lambda\rho/2)\cos(\lambda\rho'/2)\,d\lambda$$

$$=\frac{1}{\pi}\int_0^\infty\sin(\lambda\rho/2)\sin(\lambda\rho'/2)\,d\lambda=\delta(\rho-\rho')$$

and

$$\int_0^\infty\cos(\lambda\rho/2)\sin(\lambda\rho'/2)\,d\lambda\approx0,$$

we arrive at

$$\frac{1}{C_{m\rho}}=16\pi^3\sum_{qq'}\frac{[qq'+(\rho/2)^2]\,a^q_{sm}a^{q'}_{sm}2^{q+q'}}{[q^2+(\rho/2)^2]\,[(q')^2+(\rho/2)^2]}.$$

It follows from this that (see Appendix B)

$$C_{m\rho}=\frac{\rho^2+(2m)^2}{(4\pi)^3}. \tag{4.9}$$

5. CONCLUSION

Thus our final result is that if a relation of the form (4.1) exists, the inverse relation is given by (4.2) with $C_{m\rho}$ given by (4.9). In these equations $E_p = (M^2+p^2)^{1/2}$, the rotation R(n) is defined by (1.1), and $R(L_p, n)$ is defined by (1.9c) and by specifying the direction of its axis perpendicular to both **p** and **n**. By R^s_{ab} we denote the matrix elements of the (2s + 1)-dimensional irreducible representation of the three-dimensional rotation group which correspond to the rotation R^4

Equation (4.1) gives the expansion of the wave function $\Psi_{sM}(p, \sigma)$, transforming like a wave function of a particle with spin s and mass M according to the s, M representation [see (2.4)] in terms of the functions $f_{\rho m}(n)$, transforming according to the irreducible representations of the proper Lorentz group. Equation (4.2) gives these irreducible components of $\Psi_{sM}(p, \sigma)$ in terms of the function itself.

The authors express their gratitude to Professor M. A. Markov for his interest in the work.

APPENDIX A

Let $\boldsymbol{\alpha}$ be the vector of the rotation $R(L_p, n)$, so that

$$R(L_p,\mathbf{n})\,\mathbf{n}=\cos\alpha\cdot\mathbf{n}+[\boldsymbol{\alpha}\times\mathbf{n}]\sin\alpha/\alpha+\boldsymbol{\alpha}(\boldsymbol{\alpha}\cdot\mathbf{n})(1-\cos\alpha)/\alpha^2.$$

From the conditions

$$L_p^{-1}\mathbf{n}=R(L_p,\mathbf{n})\,\mathbf{n}\ \text{ and }\ \boldsymbol{\alpha}\sim[\mathbf{p}\times\mathbf{n}], \tag{A.1}$$

which determine $R(L_p, n)$, and from the formula

$$L_p^{-1}\mathbf{n}=[\mathbf{n}M+\mathbf{p}(\mathbf{p}\cdot\mathbf{n})/(E_p+M)-\mathbf{p}]\,[E_p-(\mathbf{p}\cdot\mathbf{n})]^{-1},$$

which is implied by (1.2) and (2.1), we easily arrive at

$$\frac{\boldsymbol{\alpha}}{\alpha}\sin\alpha=\frac{[\mathbf{p}\times\mathbf{n}]}{E_p+M}\Big[1+\frac{M}{E_p-(\mathbf{p}\cdot\mathbf{n})}\Big]. \tag{A.2}$$

This leads to

$$1+\cos\alpha=\frac{(E_p-\mathbf{p}\cdot\mathbf{n}+M)^2}{(E_p+M)(E_p-\mathbf{p}\cdot\mathbf{n})}, \tag{A.3}$$

$$1-\cos\alpha=\frac{p^2-(\mathbf{p}\cdot\mathbf{n})^2}{(E_p+M)(E_p-\mathbf{p}\cdot\mathbf{n})}. \tag{A.4}$$

It is now easily shown that

$$1+\cos(\alpha+\theta)=(E_p-p)(1+t)/(E-pt), \tag{A.5}$$

$$1-\cos(\alpha+\theta)=(E_p+p)(1-t)/(E-pt), \tag{A.6}$$

where $t = \cos\theta = \mathbf{p}\cdot\mathbf{n}/p$.

APPENDIX B

We shall here prove (4.9). We have

$$\sum_{qq'}a^qa^{q'}2^{q+q'}\frac{qq'+(\rho/2)^2}{[q^2+(\rho/2)^2]\,[q'^2+(\rho/2)^2]}=2\sum_q\frac{q2^qa^q}{q^2+(\rho/2)^2}\sum_{q'}\frac{2^{q'}a^{q'}}{q+q'}.$$

Comparing the equation $A_q \equiv \sum_{q'} 2^{q'}a^{q'}/(q+q')$

with $u^s_{mm} = \sum_{q'} a^{q'} x^{q'}$, we see that

$$A_q = \frac{1}{2^q}\int_0^2 dx u^s_{mm} x^{q-1}$$

$$= \frac{(-)^{s-m}}{2^{s+q}(s-m)!}\int_0^2 x^{q-m-1} dx\,(d/dx)^{s-m}(2-x)^{s-m}x^{s+m}.$$

When $q > m$, integration by parts shows simply that $A_q = 0$. When $q = m$, we easily obtain

$$A_m = (-)^{s-m}\frac{(s-m)!\,(2m-1)!}{(s+m)!} = \frac{1}{2ma^m_{sm}2^m}.$$

Thus

$$\sum_{q,q'} = \frac{1}{m^2 + (\rho/2)^2}.$$

[1] I. S. Shapiro, Dissertation, Moscow State University (1955); Dokl. Akad. Nauk SSSR **106**, 647 (1956), Soviet Phys. "Doklady" **1**, 91 (1956).
[2] M. I. Shirokov and Chou Kuang-Chao, J. Exptl. Theoret. Phys. (U.S.S.R.) **34**, 1230 (1958), Soviet Phys. JETP **7**, 851 (1958); Iu. M. Shirokov, J. Exptl. Theoret. Phys. (U.S.S.R.) **34**, 717 (1958), Soviet Phys. JETP **7**, 493 (1958); E. P. Wigner, Revs. Modern Phys. **29**, 255 (1957).
[3] M. A. Naimark, Usp. Mat. Nauk **9**, 19 (1954).
[4] I. M. Gel'fand and Z. Ia. Shapiro, Usp. Mat. Nauk **7**, 3 (1952).

Translated by E. J. Saletan
302

8.B Nuclear Physics 9 (1958/59) 652—654; © North-Holland Publishing Co., Amsterdam

A NOTE ON THE DECAY OF THE Σ-HYPERON AND ITS ANTIPARTICLE

CHOU KUANG-CHAO
Joint Institute of Nuclear Research, Dubna, USSR

Received 17 July 1958

Abstract: This note examines the information which may be derived from measurements of the branching ratios and angular correlation of Σ- and $\bar{\Sigma}$-decays.

1. Branching Ratio

Recently Okubo in a very interesting paper [1]) proposed an experiment on the measurement of the branching ratio in the decay of the Σ^+ hyperon and its antiparticle $\bar{\Sigma}^+$. A difference between their branching ratios will indicate the violation of invariance both under charge conjugation and under time reversal.

To obtain his results Okubo utilized the lowest order perturbation theory for the weak interaction. In view of the great importance of this experiment as a test of the validity of invariance under time reversal we shall show in this note that Okubo's results are quite general. They can be derived from the Lüders-Pauli theorem together with the requirement of unitarity of the S-matrix.

The probability of the decay process is characterized by the following elements of the R-matrix

$$\langle i, j, l, m|R|j, m\rangle$$

where i denotes the total isobaric spin of the nucleon-meson state; j the total angular momentum of the system which is equal to the spin of the Σ-particle; m the eigenvalue of the operator j_z; l the orbital angular momentum of the nucleon-meson state; and R is connected with the scattering matrix S by the relation $S = 1-iR$.

Okubo's derivation rests on the relation (7) in his paper [1]) which can be expressed in the form

$$\langle i, j, l, m|R|j, m\rangle_c = I_0(-1)^l \exp(2i\delta_{i,j,l})\langle i, j, l, m|R|j, m\rangle^* \qquad (1)$$

where the subscript c on the left-hand side denotes the matrix elements for the antiparticle process; I_0 the product of the inner parities of the Σ, N particles and the π-meson; $\delta_{i,j,l}$ the phase shift of the π—N scattering in the corresponding state. The relation (1) is proved in ref.[1]) by considering only the lowest order perturbation in the weak interaction.

Applying the ICT-theorem to the transition matrix elements we obtain according to ref.[2])

$$\langle i, j, l, m|R|j, m\rangle_c = -I_0(-1)^l\langle i, j, l, -m|R^+|j, -m\rangle^*$$
$$= -I_0(-1)^l\langle i, j, l, m|R^+|j, m\rangle^* \qquad (2)$$

where R^+ denotes the hermitian conjugate operator of R. The unitarity condition for the S-matrix leads to the relation

$$-i(R^+ - R) = R^+R. \qquad (3)$$

Multiplying both sides of (3) on the left by the state $\langle i, j, l, m|$ and on the right by $|j, m\rangle$ we get

$$-i\langle i, j, l, m|R^+ - R|j, m\rangle = [\exp(-2i\delta_{i,j,l}) - 1]\langle i, j, l, m|R|j, m\rangle. \qquad (4)$$

In obtaining (4) we note that the only possible intermediate state which gives a contribution to the matrix element $\langle i, j, l, m|R^+R|j, m\rangle$ is the nucleon-meson state if the electromagnetic and other weak interaction effects are neglected. This neglect does not correspond to the lowest order perturbation theory, since we are now dealing with the R matrix, whose elements are connected directly with the observed probability of the corresponding process. A simple calculation shows that the contribution of the neglected intermediate states (e.g. nucleon+meson+photon) is at least of the order $\alpha = 1/137$ times smaller than the main contribution.

Similarly for the antiparticle process we have

$$-i\langle i, j, l, m|R^+ - R|j, m\rangle_c = [\exp(-2i\delta_{i,j,l}) - 1]\langle i, j, l, m|R|j, m\rangle_c. \qquad (5)$$

The phase shifts for the $\pi - \overline{\mathrm{N}}$ scattering are taken equal to that of $\pi - \mathrm{N}$ scattering. Applying (2) to the left-hand side of (4) and (5) we get

$$\langle i, j, l, m|R^+ - R|j, m\rangle_c = I_0(-1)^l\langle i, j, l, m|R^+ - R|j, m\rangle^* \qquad (6)$$

and the relation (1) follows directly from (4), (5) and (6).

2. Angular Correlation

Measurement of the angular correlation in the decay of the Σ-particle and its antiparticle will also yield valuable information on the validity of invariance under charge conjugation and under time reversal. The simplest case will be the decay of Σ^- and its antiparticle. The asymmetry parameter α defined in ref.[3]) has the form

$$\alpha = \mathrm{Re}\, A^*B/(|A|^2 + |B|^2)$$

with

$$A = \langle \tfrac{3}{2}, \tfrac{1}{2}, 0, m|R|\tfrac{1}{2}, m\rangle$$
$$B = \langle \tfrac{3}{2}, \tfrac{1}{2}, 1, m|R|\tfrac{1}{2}, m\rangle \qquad (7)$$

where, for simplicity, the spin of the Σ-particle is assumed to be $\frac{1}{2}$. The asymmetry parameter for the antiparticle is denoted by α_c. From (1) and (7) we get

$$\frac{\alpha_c}{\alpha} = -\frac{\cos(\delta_s-\delta_p-\Delta_s+\Delta_p)}{\cos(\delta_s-\delta_p+\Delta_s-\Delta_p)} \tag{8}$$

where we have denoted the phase shifts of the s-state and the p-state by δ_s and δ_p respectively; Δ_s and Δ_p are defined by the relation

$$A = |A| \exp[i(\delta_s+\Delta_s)], \quad B = |B| \exp[i(\delta_p+\Delta_p)].$$

The π—N phase shifts are given by ref.[4]):

$$\delta_s-\delta_p = -17° \qquad (\text{at } E \approx 120 \text{ MeV}).$$

If the degree of polarization of Σ^- and its antiparticle can be determined independently and the deviation from I, T and C invariance is quite large, the deviation of α_c/α from unity may be detected by experiment. A deviation of α_c/α from unity will indicate violation of both T and C invariance.

Formula (8) retains its form when the spin of the Σ particle is $\frac{3}{2}$. In this case δ_s, δ_p, Δ_s and Δ_p should be replaced by the corresponding quantities in the states $i=\frac{3}{2}$, $j=\frac{3}{2}$, $l=1$ and $i=\frac{3}{2}$, $j=\frac{3}{2}$, $l=2$, respectively. A similar but more complicated formula holds for the decay of Σ^+ and its antiparticle.

Finally, we note that the lack of asymmetry in the decay of the Σ^--particle [5]) may be due to the accidental equality $\delta_s-\delta_p+\Delta_s-\Delta_p \approx \pm 90°$. In this case we may expect quite large asymmetry in the decay of its anti-particle.

References

1) S. Okubo, Phys. Rev. **109** (1958) 984
2) S. Watanabe, Revs. Mod. Phys. **27** (1955) 40
3) T. D. Lee, J. Steinberger, G. Feinberg, P. K. Kabir and C. N. Yang, Phys. Rev. **106** (1957) 1367
4) H. A. Bethe and F. de Hoffman, Mesons and Fields, Vol. II (1955)
5) International conference on mesons and recently discovered particles, Padova-Venice (1957)

THE UNIVERSAL FERMI INTERACTION AND THE CAPTURE OF MUONS IN HYDROGEN

CHOU KUANG-CHAO and V. MAEVSKII

Joint Institute for Nuclear Research

Submitted to JETP editor September 13, 1958

J. Exptl. Theoret. Phys. (U.S.S.R.) 35, 1581-1582 (December, 1958)

THE theory of the universal Fermi interaction with (V-A) coupling proposed by Marshak and Sudarshan[1] and by Feynman and Gell-Mann[2] is supported by all existing experimental evidence on β decay. To explain the equality of the Fermi β-decay constant and the constant for the μ decay, Feynman and Gell-Mann have put forward the hypothesis of a conserved vector current in the weak interactions. This hypothesis leads to the appearance of an anomalous magnetic-moment effect in the β decay ("β magnetism"). Calculations of the effect have been made by Gell-Mann and others[3-5] and have apparently already received experimental confirmation in β decay.[6]

The latest experimental data show that the probability of the β decay or (μ, ν) decay of hyperons is apparently considerably smaller than that given by the universal (V-A) theory.[7] But even if these data are confirmed and we are forced to abandon the universal interaction including the hyperons, nevertheless the universal interaction between (p, n), (e, ν), and (μ, ν) will remain an extremely probable hypothesis.

Since it already seems that the β decay and the decay of the μ meson can be explained within the framework of the (V-A) theory, it is of particular interest to examine the process of μ capture and test the idea of the universal (V-A) interaction in this case.

In the present note we present expressions for the probability of the capture, the angular distribution, and the polarization of the emerging neutrons in the case of capture of polarized μ^- mesons by protons, based on the assumption of the universal (V-A) coupling with conserved vector current as proposed by Gell-Mann and Feynman.

The capture probability and the angular distribution are (in units $\hbar = c = 1$):

$$w = (2\pi)^{-2}(\pi a_\mu^3)^{-1}2^{-1}G^2I\,[1 + \alpha P\,(\mathbf{n}\cdot\mathbf{s})]\,p^2\,d\Omega,$$

$$I = 1 + 3\lambda^2 + \beta(1 + \lambda^2) + \beta\mu(2\lambda + \beta\mu/2),$$

$$I\alpha = 1 - \lambda^2 + \beta(1 + \lambda^2) - \beta\mu(2\lambda + \beta\mu/2).$$

The total probability for capture is

$$\tau^{-1} = 2^{-1}G^2\pi^{-2}a_\mu^{-3}p^2I.$$

The polarization $\langle\sigma_n\rangle$ of the neutrons is given by ooo formula

$$I\,[1 + \alpha P(\mathbf{n}\cdot\mathbf{s})]\langle\sigma_n\rangle = [a + b\,(\mathbf{n}\cdot\mathbf{s})]\,\mathbf{n} + c\mathbf{s},$$

$$a = -2\,[\lambda(\lambda + 1) + \beta\lambda + \beta\mu(\lambda + \beta\mu/4)],$$

$$b = -\beta P\,[\lambda(\lambda + 1) - \mu(1 + \lambda + \beta\mu/2)],$$

$$c = 2P(\lambda - 1)\,[\lambda + \beta(\lambda + \mu)/2].$$

Here we have used the notations: $p\mathbf{n}$ is the momentum of the neutron, and $P\mathbf{s}$ is the polarization of the μ meson ($\mathbf{n}$ and $\mathbf{s}$ are unit vectors); $\lambda = -C_A/C_V$, $\beta = p/M$; $G \equiv 2^{1/2}C_V$ is the universal coupling constant; M is the nucleon mass; $\mu = \mu_p - \mu_n$ is the difference of the magnetic moments of the proton and neutron; $a_\mu = (m_\mu e^2)^{-1}$ is the radius of the mesonic K orbit in hydrogen.

Terms of the order $(v/c)^2$ have been neglected in the calculation (v is the speed of the neutron). In addition it has been assumed that in the pseudovector coupling the higher moments do not give an additional renormalization. A numerical computation (taking $G = (1.01 \pm 0.01) \times 10^{-5} M^{-2}$, $\mu = 4.7$, $\lambda = 1.24$, $\beta = 0.1$) shows that the effect of the anomalous magnetic moment increases the total probability of capture by ≈ 17 percent, and the angular correlation coefficient by a factor of about 2.4; the polarization of the neutrons is changed only slightly as compared with the results of the ordinary theory. At present calculations are being carried out on the capture of mesons by nuclei according to the Feynman — Gell-Mann theory. It can be expected that in this case also the correction introduced by the anomalous magnetic moment will be large.

[1] E. C. G. Sudarshan and R. E. Marshak, Phys. Rev. 109, 1860 (1958).

[2] R. P. Feynman and M. Gell-Mann, Phys. Rev. 109, 193 (1958).

[3] M. Gell-Mann, Test of the Nature of the Vector Interaction in β Decay (preprint).

[4] J. Bernstein and R. Lewis, The Vector Interaction in β Decay (preprint).

[5] Weinberg, Marshak, Okubo, Sudarshan, and Teutsch, Phys. Rev. Letters 1, 25 (1958).

[6] Boehm, Soergel, and Stech, Phys. Rev. Letters 1, 77 (1958).

[7] F. Eisler, R. Plano, et al., Leptonic Decay Modes of the Hyperons (preprint).

Translated by W. H. Furry
329

SOVIET PHYSICS JETP VOLUME 36 (9), NUMBER 1 JULY, 1959

CHARGE SYMMETRY PROPERTIES AND REPRESENTATIONS OF THE EXTENDED LORENTZ GROUP IN THE THEORY OF ELEMENTARY PARTICLES

V. I. OGIEVETSKII and CHOU KUANG-CHAO

Joint Institute for Nuclear Research

Submitted to JETP editor July 15, 1958

J. Exptl. Theoret. Phys. (U.S.S.R.) 36, 264-270 (January, 1959)

The extended Lorentz group, including the complete Lorentz group and charge conjugation, is considered. It is shown that the use of irreducible projective representations of this extended group requires the existence of charge multiplets. Charge symmetry and associated production of strange particles follow from the invariance under reflections and charge conjugation and from the conservation laws for the electric and baryonic charges. The Pauli-Gürsey transformation holds for free nucleons. The extension of the condition of invariance under this transformation to the case of interactions leads to isobaric invariance for strong interactions of all particles.

1. INTRODUCTION

It is known that the strongly interacting particles form charge multiplets (p, n, π^+, π^0, π^-, K^+, K^0, etc.). Particles belonging to the same multiplet have almost identical masses and identical spins, but differ in their electric charges. In agreement with experiment one adopts the hypothesis of charge symmetry and the stronger hypothesis of charge independence. In the conventional theory this is expressed by the invariance under rotations in a certain formal isobaric space. The particles of a given multiplet are considered as states of the same particle with different projections of the isobaric spin. For example, the proton and the neutron form the nucleon. The description of the nucleon makes use of a reducible eight-component representation of the full Lorentz group. We have a similar situation (reducibility of the representation of the full Lorentz group) for the other strongly interacting particles.

Now the question arises: if it is required that the elementary particles be described only by irreducible representations, would it then be possible to extend the Lorentz group and to find irreducible representations of this extended group which automatically lead to the existence of charge multiplets and charge symmetries? The solution of this question is the subject of the present paper.

We extend the Lorentz group in the following fashion.

The wave functions of quantum theory are complex functions. The operation of charge conjugation C, which takes a particle into its antiparticle, is always represented by the product of a linear operator (matrix) and the antilinear operator of complex conjugation:

$$C: \qquad \psi_c = C_0\psi^*, \tag{1}$$

where C_0 is determined such that ψ_C transforms according to the same irreducible representation of the proper Lorentz group as ψ.

Besides the proper Lorentz group L, the spatial reflections I, and the time reversal T, we also include the charge conjugation C in the extended group. Together with the conventional irreducible representations of the extended group, we also consider its irreducible projective representations.*

The importance of using the projective representations of the full Lorentz group was pointed out by Gel'fand and Tsetlin[1] in connection with the theory of parity doublets of Lee and Yang. The possibility of using projective representations is connected with the indeterminacy of the phase factor of the quantum theoretical wave function. Subsequent to Gel'fand and Tsetlin, the projective representations of the full Lorentz group were discussed

*We are given a projective representation of the group G, if an operator R(g) is given for each element g of the group G such that the operator $R(g_1g_2) = \alpha(g_1, g_2)R(g_1)R(g_2)$ corresponds to the product of group elements g_1g_2. If $\alpha(g_1, g_2) = 1$, the projective representation reduces to the conventional one. In general, $\alpha(g_1, g_2)$ can also be equal to -1. Anticommuting operators of the projective representation may thus correspond to commuting elements of the group. In particular, the conventional spinor representation is a projective representation (the operations $I = \gamma_4$ and $T = \gamma_4\gamma_5$ anticommute, whereas the spatial and time reflections commute).

by Taylor and McLennan.[2] Taylor notes the connection between these representations and the isobaric invariance. As only the full Lorentz group is considered, protons and neutrons, and $\pi^{\pm}$ and π^0 mesons, differ with respect to their spatial parities. Salam and Pais, at the Seventh Rochester Conference, also discussed the need for a new definition of the operations of space-time reflections from which the charge symmetries would follow.

In this paper we do not consider all irreducible projective representations of the extended group. We restrict ourselves to those necessary for the description of the strongly interacting particles. We shall show that multiplets, charge symmetries, and associated production of strange particles are immediate consequences of the standard conservation laws for the number of baryons and electric charge, and of the invariance with respect to the full Lorentz group and charge conjugation, if the nucleons, Ξ particles, and K mesons are described by the new projective representations of the extended group, while the remaining particles are described in the usual fashion.

In our theory the Pauli-Gürsey transformation holds for free nucleons. This transformation is connected with the isobaric invariance in a natural way. If the requirement of invariance under this transformation is extended to the interaction Lagrangian for the nucleons, the isobaric invariance for strong interactions follows for all particles.

In this theory the Lagrangian for the interaction with electromagnetic fields can be easily written down with the help of the charge operator. It appears that the electromagnetic interactions are invariant only under Wigner time reversal, but not under Schwinger time reversal.

The case of weak interactions, which do not conserve spatial parity, is more complicated and will not be discussed in the present paper.

To be definite, we shall assume that the relative parities of all baryons are identical and that the reflection of the conventional spinors is performed with the help of the operator γ_4. All bosons are considered as pseudoscalars. We start with the discussion of the nucleons.

2. THE FREE NUCLEON FIELD

It is easily shown that the requirement

$$I^2 = T^2 = C^2 = 1, \qquad (2)$$

leads, for the case of four-component spinors, to the following expressions for the operators I, T, and C:

$$\text{a) } I: \quad \psi' = \gamma_4\psi, \qquad \text{b) } T: \quad \psi' = i\gamma_4\gamma_5\psi, \qquad \text{c) } C: \quad \psi_c = i\gamma_2\psi^*, \qquad (3)$$

where T is the Schwinger time reversal for spinors,[3] and the matrices γ_i are expressed in the Pauli representation.

The following commutation relations hold between the operators I, T, and C:

$$\text{a) } IT = -TI, \quad \text{b) } IC = -CI, \quad \text{c) } TC = -CT. \qquad (4)$$

We retain relations (2), but we require that, in contrast to the conventional theory, the sign in relation (4a) is changed for nucleons, i.e., we demand that

$$\text{a) } IT = TI, \quad \text{b) } IC = -CI, \quad \text{c) } TC = -CT. \qquad (4')$$

The commutation relations (2) and (4′) can only be satisfied by 8×8 matrices:

$$\text{a) } I: \quad \psi' = \tau_3 \times \gamma_4\psi, \qquad \text{b) } T: \quad \psi' = 1 \times \gamma_4\psi, \qquad \text{c) } C: \quad \psi_c = i\tau_3 \times \gamma_2\psi^*, \qquad (5)$$

where τ are the Pauli matrices. These operators, together with the operators of the proper Lorentz group (in which γ_μ should everywhere be replaced by $1\times\gamma_\mu$), form the irreducible projective representation of the extended Lorentz group. The spinors $\psi = \begin{pmatrix}\psi_1 \\ \psi_2\end{pmatrix}$ have eight components.

In this representation the Lagrangian for the free field is uniquely determined:*

$$L = \bar{\psi}(1 \times \gamma_\mu\partial_\mu + i\tau_2 \times \gamma_5 m)\psi, \qquad (6)$$

where $\bar{\psi} = \psi^{*T}(1\times\gamma_4)$. The field equations have the form

$$1 \times \gamma_\mu\partial_\mu\psi = -i\tau_2 \times \gamma_5 m\psi. \qquad (7)$$

The Lagrangian (6) as well as Eq. (7) are invariant with respect to the two one-parameter groups of transformations

$$\psi' = \exp(i\lambda)\psi, \qquad (8)$$

$$\psi' = \exp[i\tau_1 \times \gamma_5\lambda]\psi, \qquad (9)$$

and with respect to the three-parameter group

$$\psi' = a\psi + b\tau_3 \times \gamma_5\psi_c, \qquad (10)$$

where $|a|^2 + |b|^2 = 1$.

The transformations (9) and (10) are the analogs of the Pauli transformations[4] for the neutrino. They are different only in that γ_5 is replaced by $\tau_1\times\gamma_5$ and $\tau_3\times\gamma_5$, respectively.

*According to Schwinger $L \to L^T$, where the superscript T signifies transposition of the operators of the Hilbert space.[3]

We introduce the new four-component spinors

$$\psi_p = (\psi_1 + \gamma_5\psi_2)/\sqrt{2}, \quad \psi_{pc} = (\psi_{c1} + \gamma_5\psi_{c2})/\sqrt{2},$$
$$\psi_{nc} = (-\gamma_5\psi_1 + \psi_2)/\sqrt{2}, \quad \psi_n = (\gamma_5\psi_{c1} - \psi_{c2})/\sqrt{2}, \tag{11}$$

they satisfy the ordinary Dirac equation

$$\gamma_\mu \partial_\mu \psi = -m\psi, \tag{12}$$

Under the transformation (9) we obtain

$$\psi_p' = \exp(i\lambda)\,\psi_p, \quad \psi_n' = \exp(i\lambda)\,\psi_n,$$
$$\psi_{pc}' = \exp(-i\lambda)\,\psi_{pc}, \quad \psi_{nc}' = \exp(-i\lambda)\,\psi_{nc}. \tag{13}$$

Transformation (9) may thus be viewed as a gauge transformation connected with the conservation law for the number of baryons. ψp, ψn, ψpc, and ψ_{nc} refer to the proton, neutron, antiproton, and antineutron fields respectively.

The following transformation should be related to the conservation law for the electric charge:

$$E:\ \psi' = \exp[(i/2)(1 \times 1 + \tau_1 \times \gamma_5)\lambda]\,\psi. \tag{14}$$

Indeed, under this transformation:

$$E:\ \psi_p' = \exp(i\lambda)\,\psi_p, \quad \psi_{pc}' = \exp(-i\lambda)\,\psi_{pc},$$
$$\psi_n' = \psi_n, \quad \psi_{nc}' = \psi_{nc}. \tag{15}$$

The three-parameter trnnsformation (10) is isomorphic to a rotation in the isobaric space. Indeed, if we form the conventional eight-component nucleon field $\psi_N = \begin{pmatrix}\psi_p \\ \psi_n\end{pmatrix}$, we have, under the transformation (10),

$$\psi_N' = \exp[i(\boldsymbol{\tau}\cdot\boldsymbol{\lambda})]\,\psi_N, \tag{16}$$

where λ is a vector with real components λ_1, λ_2, λ_3; τ are the usual 2×2 Pauli matrices, and

$$a = \cos|\lambda| + \frac{i\lambda_3}{|\lambda|}\sin|\lambda|; \quad b = \frac{\sin|\lambda|}{|\lambda|}(\lambda_2 - i\lambda_1). \tag{17}$$

An analogous isomorphism arising from a purely formal doubling of the number of components was pointed out by Gürsey.[5]

3. INTERACTION OF NUCLEONS WITH ORDINARY BOSONS

We first consider the interaction of nucleons with a neutral pseudoscalar field φ_0 with positive "time-parity:"

$$\begin{aligned} I&: \quad \varphi_0' = -\varphi_0,\\ T&: \quad \varphi_0' = \varphi_0,\\ C&: \quad \varphi_{0c} = \varphi_0. \end{aligned} \tag{18}$$

The requirement of invariance with respect to I, T, C, transformation (9), and E lead to a unique interaction Lagrangian (in the following we only consider interaction Lagrangians without derivatives):

$$L_0 = ig_0\bar{\psi}\tau_3 \times \gamma_5\psi\varphi_0 = ig_0(\bar{p}\gamma_5 p - \bar{n}\gamma_5 n)\,\varphi_0 = ig_0\bar{\psi}_N\tau_3\gamma_5\psi_N\varphi_0, \tag{19}$$

the meson field φ_0 couples to the protons and neutrons with a different sign, and φ_0 may be identified with the neutral π^0 meson.

If the neutral meson φ_0' were a spatial pseudoscalar, but had negative time-parity, we would uniquely obtain the interaction Lagrangian

$$L_0' = g_0'\bar{\psi}\tau_2 \times 1\psi\varphi_0 = ig_0'(\bar{p}\gamma_5 p + \bar{n}\gamma_5 n)\,\varphi_0, \tag{20}$$

φ_0' may be related to the hypothetical ρ_0 meson.

For the Lagrangian describing the interaction of nucleons with a charged pseudoscalar boson field φ:

$$\begin{aligned} I&: \quad \varphi' = -\varphi,\\ T&: \quad \varphi' = \varphi,\\ C&: \quad \varphi_c = \varphi^* \end{aligned}$$

we similarly obtain the unique expression

$$L = ig(\bar{\psi}\psi_c\varphi^* - \bar{\psi}_c\psi\varphi) = 2ig(\bar{p}\gamma_5 n\varphi^* + \bar{n}\gamma_5 p\varphi). \tag{21}$$

We may thus assume that $\varphi\,(\varphi^*)$ describes $\pi^-\,(\pi^+)$ mesons.

The charge symmetry (i.e., the possibility of the simultaneous interchange p $\rightleftarrows$ n, $\pi^+ \rightleftarrows \pi^-$, $\pi^0 \rightleftarrows -\pi^0$) of the interactions (19) and (20) is obvious. The general form of the Lagrangian for interactions with π mesons is

$$L_\pi = ig_0(\bar{p}\gamma_5 p - \bar{n}\gamma_5 n)\,\pi^0 + 2ig(\bar{p}\gamma_5 n\pi^+ + \bar{n}\gamma_5 p\pi^-). \tag{22}$$

If we now require that not only the free nucleon Lagrangian, but also the interaction Lagrangian be invariant under transformations of the three-parameter group, then $g = g_0/\sqrt{2} = g_\pi/\sqrt{2}$, and

$$L_\pi = ig_\pi\bar{\psi}_N(\boldsymbol{\tau}\cdot\boldsymbol{\pi})\,\psi_N. \tag{23}$$

Under the transformation (10) the meson fields transform according to

$$(\boldsymbol{\tau}\cdot\boldsymbol{\pi})' = \exp[i(\boldsymbol{\tau}\cdot\boldsymbol{\lambda})]\,(\boldsymbol{\tau}\cdot\boldsymbol{\pi})\exp[-i(\boldsymbol{\tau}\cdot\boldsymbol{\lambda})], \tag{24}$$

where the masses of the π^+, π^-, and π^0 mesons must be identical.

We have thus arrived at the conventional isobaric-invariant theory of the interaction between π mesons and nucleons.

4. FREE K MESONS

The conventional representation of the extended Lorentz group for bosons is exhausted by the π mesons. We shall describe the K mesons by the projective representation in which

$$I^2 = 1, \quad C^2 = 1, \quad T^2 = -1, \qquad IT = TI, \quad IC = CI, \quad TC = CT. \tag{25}$$

The simplest irreducible representation consistent with these commutation relations is two-dimensional, $\varphi = \begin{pmatrix}\varphi_1 \\ \varphi_2\end{pmatrix}$. The operators I, T, and C have the form

$$I: \; \varphi' = -\varphi, \qquad T: \; \varphi' = i\tau_2\varphi, \qquad C: \; \varphi_c = \varphi^*. \tag{26}$$

We identify the K^+ meson with φ_1, the K^0 meson with φ_2, the K^- meson with φ_1^*, and the $\overline{K}^0$ meson with φ_2^*. The conservation law for the electric charge is then connected with the transformation

$$E: \; \varphi' = \exp\left[\tfrac{1}{2}(1 + \tau_3)\lambda\right]\varphi. \tag{27}$$

Indeed, with this transformation,

$$K^{+\prime} = \exp(i\lambda) K^+; \quad K^{-\prime} = \exp(-i\lambda) K^-, \qquad K^{0\prime} = K^0, \quad \overline{K}^{0\prime} = \overline{K}^0. \tag{28}$$

The conservation law for the hypercharge corresponds to the transformation $\varphi' = \exp(i\tau_3\lambda)\varphi$, so that

$$K^{+\prime} = \exp(i\lambda) K^+, \quad K^{0\prime} = \exp(i\lambda) K^0, \qquad K^{-\prime} = \exp(-i\lambda) K^-, \quad \overline{K}^{0\prime} = \exp(-i\lambda)\overline{K}^0. \tag{29}$$

5. INTERACTION OF K MESONS WITH NUCLEONS. Λ AND Σ PARTICLES

We now turn to the investigation of the interaction of K mesons with nucleons. Since both the K mesons and the nucleons transform according to the projective representation of the extended Lorentz group, and since the baryonic charge is conserved, an additional baryon must necessarily participate in the interaction. This requirement inevitably leads to the law of associated production of strange particles. We first consider the case of a neutral baryon. We already said earlier that the relative spatial parities of all baryons are, for definiteness, assumed to be identical:

$$I: \quad Y'_0 = \gamma_4 Y_0. \tag{30}$$

For the nucleon transformation T there are two possibilities for the unknown baryon:

$$T: \quad Y'_0 = -\gamma_4\gamma_5 Y_{0c}, \tag{31}$$

$$Y'_0 = \gamma_4\gamma_5 Y_{0c}. \tag{32}$$

We have to introduce the antibaryon Y_{0c} into equations (31) and (32), since the transformation T for the nucleons anticommutes with the transformation[9] related to the conservation of the baryonic charge.

If we choose (31) for T, then the only form of the Lagrangian invariant under the transformations of the extended Lorentz group and the transformations (9) and E is

$$L = ig[\overline{\psi}(1\times\gamma_5 - \tau_1\times 1)(1\times 1 + \tau_3\times 1)\varphi\times Y_0 - \overline{\psi}_c(1\times\gamma_5 + \tau_1\times 1)(1\times 1 - \tau_3\times 1)\varphi^*\times Y_0] + \text{Herm. conj.} \tag{33}$$

In going from ψ and φ to the operators of the nucleon and K meson fields, we obtain the usual form for the Lagrangian for the interaction of nucleons with Λ_0 particles:

$$L = ig_\Lambda(\overline{p}\gamma_5\Lambda_0 K^+ + \overline{n}\gamma_5\Lambda_0 K^0) + \text{Herm. conj.} \tag{34}$$

The transformation law (32) for T leads to a Lagrangian that differs from expression (33) only by a plus sign between the two terms of expression (33). It corresponds to the Σ_0 particle:

$$L = ig_{\Sigma_0}(\overline{p}\gamma_5\Sigma_0 K^+ - \overline{n}\gamma_5\Sigma_0 K^0) + \text{Herm. conj.} \tag{35}$$

We thus arrive at the conclusion that the transformation laws for Λ_0 and Σ_0 corresponding to the transformation T for the nucleons, differ by their signs. We now consider the interaction of nucleons with charged baryons. We can find a Lagrangian which is invariant under charge conjugation, space inversion, and time reversal, and which is consistent with the conservation laws for the electric and baryonic charges, only if we require that, under T,

$$T: \quad \Sigma^+ \to -\gamma_4\gamma_5\Sigma_c^-, \quad \Sigma^- \to -\gamma_4\gamma_5\Sigma_c^+, \tag{36}$$

This implies that the masses of the Σ^+ and Σ^- particles are equal. The Lagrangian has the form

$$L = -ig[\overline{\psi}(1\times 1 - \tau_1\times\gamma_5)(1\times 1 - \tau_3\times 1)\varphi^*\times\Sigma^+ + \overline{\psi}_c(1\times 1 + \tau_1\times\gamma_5)(1\times 1 + \tau_3\times 1)\varphi\times\Sigma^-] + \text{Herm. conj.} \tag{37}$$

In conventional notation this can be written in the form

$$L = ig_{\Sigma\pm}(\overline{p}\gamma_5\Sigma^+K^0 + \overline{n}\gamma_5\Sigma^-K^+) + \text{Herm. conj.} \tag{38}$$

The charge symmetry is obvious.

If we now require that the interaction Lagrangian also be invariant under the Pauli-Gürsey type transformation (10), we obtain

$$g_{\Sigma\pm} = \sqrt{2}\, g_{\Sigma_0} = \sqrt{2}\, g_\Sigma; \tag{39}$$

The masses of Σ^+, Σ^-, and Σ^0 must be equal.

We arrive at the usual isobaric-invariant Lagrangian

$$L = ig_{\Sigma}[(\bar{p}\gamma_5\Sigma_0 K^+ - \bar{n}\gamma_5\Sigma_0 K^0)$$
$$+ \sqrt{2}\,\bar{p}\gamma_5\Sigma^+ K^0 + \sqrt{2}\,\bar{n}\gamma_5\Sigma^- K^+] + \text{Herm. conj.} \quad (40)$$

6. Ξ PARTICLES

Still another possibility remains within the framework of these representations. In the projective representation (5), we can change the sign between the terms 1×1 and $\tau_1 \times \gamma_5$ in the transformation (14) connected with the conservation law for the electric charge:

$$E:\ \psi' = \exp\left[\frac{i}{2}(1\times 1 - \tau_1 \times \gamma_5)\lambda\right]\psi. \quad (41)$$

At the same time we retain the transformation (9) connected with the conservation law for the baryonic charge. In all formulas we then simply have to replace p by Ξ^0, n by Ξ^-, K^+ by $\bar{K}^0$, and K^0 by K^-. We again have charge symmetry. If we require invariance under the Pauli transformation even in the case of interaction, we obtain the usual isobaric-invariant Lagrangians.

7. CONCLUSION

We carried out the program which we set ourselves in the beginning of this paper. We introduced the new irreducible projective representation of the extended Lorentz group. We showed that the existence of charge multiplets, charge symmetry, and associated production of strange particles are consequences of the standard conservation laws. The isobaric invariance follows from the invariance under the Pauli-Gürsey type transformation for free nucleons. This transformation is applicable, since the number of components of wave functions transforming according to projective representations necessarily had to be increased.

We did not discuss the weak interactions from this point of view. This task is much more difficult and less unambiguous due to the violation of the parity conservation laws.

The authors express their sincere gratitude to Prof. Gel'fand for valuable comments.

[1] I. M. Gel'fand and M. L. Tsetlin, J. Exptl. Theoret. Phys. (U.S.S.R.) 31, 1107 (1956), Soviet Phys. JETP 4, 947 (1957).

[2] J. C. Taylor, Nucl. Phys. 3, 606 (1957). J. A. McLennan, Jr., Phys. Rev. 109, 986 (1958).

[3] J. Schwinger, Phys. Rev. 91, 713 (1953).

[4] W. Pauli, Nuovo cimento 6, 204 (1957).

[5] F. Gürsey, Nuovo cimento 7, 411 (1958).

Translated by R. Lipperheide
37

SOVIET PHYSICS JETP VOLUME 36 (9), NUMBER 3 SEPTEMBER, 1959

REACTIONS INVOLVING POLARIZED PARTICLES OF ZERO REST MASS

CHOU KUANG-CHAO

Joint Institute of Nuclear Studies

Submitted to JETP editor May 31, 1958

J. Exptl. Theoret. Phys. (U.S.S.R.) 36, 909-918 (March, 1959)

In this paper the group-theoretical point of view is used for the description of the spin states of particles of zero rest mass. Complete sets of operators and of their eigenfunctions for a system of two particles are found in the representation of the momenta and spins. The statistical tensors for the particles produced in a reaction of the type $a + b \to c + d$ or $a \to c + d$ are obtained in the case in which one of these particles has no rest mass. The most general selection rules are derived for the reaction $a + b \to c + d$ in the form of relations between the statistical tensors, under the condition that the space and time parities are conserved. The wave functions are calculated for a system of two identical particles of zero rest mass.

1. INTRODUCTION

In reference 1 a study was made of the relativistic theory of reactions involving polarized particles with non-zero rest masses. This theory is not applicable to reactions involving photons and neutrinos. In what follows we shall call a particle (a photon, neutrino, or graviton) with zero rest mass a γ-particle. The purpose of the present paper is to study the theory of reactions involving γ-particles.

The general theory of reactions involving photons has already been extensively developed in papers of Simon[2] and of Morita and others;[3] these authors describe the spin states of photons by a vector potential, using the Lorentz supplementary condition to exclude the longitudinal and scalar components. Such a procedure is complicated, and it is difficult to extend it to cases of higher spins.

The group-theoretical description of the spin states of γ-particles has been given by Wigner and by Yu. M. Shirokov.[4] Starting from the theory of the irreducible representations of the inhomogeneous Lorentz group, they showed that the spin states of free γ-particles are completely determined by the operator

$$\hat{\Sigma} = (\mathbf{j} \cdot \mathbf{n}), \qquad (1)$$

where $\mathbf{j}$ is the total angular momentum of the γ-particle; $\mathbf{n}$ is the unit vector parallel to the momentum of the γ-particle. $\hat{\Sigma}$ is a Lorentz-invariant operator with integral and half-integral values of i. Since the total angular momentum $\mathbf{j}$ is an axial vector and $\mathbf{n}$ is a polar vector, the operator $\hat{\Sigma}$ transforms as a pseudoscalar under space reflections. Thus if a γ-particle has a definite parity, then for each momentum there exist two spin states with eigenvalues $\pm i$ of the operator $\hat{\Sigma}$ If, on the other hand, it does not have a definite parity, then it can have only one spin state (of longitudinal polarization).

In the present paper we use the operator $\hat{\Sigma}$ for the description of the spin states of γ-particles. As can be seen from Eq. (1), the operator $\hat{\Sigma}$ is closely connected with the direction of the momentum. Therefore we shall work directly in the momentum representation. We emphasize that a spin vector does not exist for a γ-particle, and the total angular momentum $\mathbf{j}$ cannot be represented in the form of the sum of an orbital angular momentum and a spin. Because of this the complete sets of operators for the description of the states of a system containing γ-particles, given in Sec. 2 of the present paper, differ from the sets generally used for a system without γ-particles. In Sec. 2 we shall find the complete sets of operators and of their eigenfunctions for systems of two particles. To obtain the formulas for the statistical tensors (s-tensors for short) of reaction products, and to obtain the selection rules under the condition that space and time parities are conserved, we use a method developed by M. I. Shirokov.[5,6]

Landau and Yang have established the existence of specific selection rules for the decay of a particle into two photons.[7] Generalized selection rules of this type have been obtained by Shapiro.[8] Because of the peculiarities of a system of two identical γ-particles it is of interest to calculate the explicit form of the wave functions of this system for a definite total angular momentum, angular momentum component, and parity; these functions can be used to obtain angular distributions and polarizations for the decay of particles into pairs of identical γ-particles. It is shown that certain selection rules remain valid when parity is not conserved in a decay.

2. COMPLETE SETS OF OPERATORS AND OF THEIR EIGENFUNCTIONS

We denote state vectors of γ-particles characterized by the momentum **p** and the spin operator $\hat{\Sigma}$ by symbols $|\mathbf{p}, \mu\rangle$, where $\mu = \pm i$ is the eigenvalue of the operator $\hat{\Sigma}$. Since $\hat{\Sigma}$ is invariant under the proper Lorentz group, μ remains unchanged by a Lorentz transformation. Therefore state vectors in different coordinate systems differ only by a phase factor

$$|\mathbf{p}, \mu\rangle \xrightarrow{L} |\mathbf{p}, \mu\rangle' = \eta_\mu (L, \mathbf{p})\, |L^{-1}\mathbf{p}, \mu\rangle, \qquad (2)$$

where L is a Lorentz transformation; $\eta_\mu(L, \mathbf{p})$ satisfies the group relation

$$\eta_\mu (L_1, \mathbf{p})\, \eta_\mu (L_2, L_1^{-1}\mathbf{p}) = \eta_\mu (L_2 L_1, \mathbf{p}).$$

The explicit form of the factor η_μ depends on the choice of the coordinate system in which the total angular momentum is quantized. If we choose the axis of quantization of the total angular momentum in such a way that the z axis is parallel to **p** and the x axis to $[\mathbf{v}\times\mathbf{p}]$, where **v** is the velocity of the new coordinate system relative to the old for the Lorentz transformation L, then by means of the formalism developed in reference 9 it can easily be shown that in this case $\eta_\mu = 1$.*

From Eq. (2) we easily find the transformation function between the state vectors of a γ-particle in two different coordinate systems 1 and 2:

$$\langle \mathbf{p}_1, \mu_1 | \mathbf{p}_2, \mu_2 \rangle$$

$$= \sqrt{p_1/p_2}\,\delta^3 \left(p_{1i} - \sum_{j=1}^{4} a_{ij}\, p_{2j}\right) \delta_{\mu_1\mu_2}, \quad i = 1, 2, 3, \qquad (3)$$

where a_{ij} is the matrix of the Lorentz transformation, and the factor $(p_1/p_2)^{1/2}$ is the square root of the Jacobian of the transformation,

$$[\partial (p_{1x},\, p_{1y},\, p_{1z})/\partial (p_{2x},\, p_{2y},\, p_{2z})]^{1/2}.$$

We note that if the rest mass of the particle is not equal to zero but its speed approaches that of light, then the angle of rotation of the spin in a Lorentz transformation (cf. reference 1) goes to zero,[10] so that the corresponding transformation function approaches the form (3).

Let us consider the state of a system of two free particles 1 and 2. The rest mass of the system of two free particles is in general not zero.* As is shown in reference 1, for such a system of particles there exists a reference system in which the total momentum is zero and the total angular momentum equals the angular momentum relative to the center of mass of the system of particles. In what follows all results are given in the center-of-mass system. They can be expressed in another coordinate system, if need be, by means of the transformation function (3).

Let us denote the total angular momentum of the system by **J**, and the total angular momentum of particle α by $\mathbf{j}_\alpha$. We suppose at first that the rest mass of particle 2 is not zero. In this case $\mathbf{j}_2$ can be represented as a sum of an orbital angular momentum $\mathbf{l}_2$ and a spin $\mathbf{s}_2$.[1] We introduce a new vector **j** in the following way:

$$\mathbf{J} = \mathbf{j}_1 + \mathbf{j}_2 = \mathbf{j} + \mathbf{s}_2, \quad \mathbf{j} = \mathbf{j}_1 + \mathbf{l}_2. \qquad (4)$$

The spin operator of particle 2 commutes with the operator **j**, by definition. Since **J** and $\mathbf{s}_2$ satisfy the usual commutation relations for angular momenta, **j** also satisfies these relations.

In the center-of-mass system a complete set of operators for the two-particle system consists of the following operators:

$$\mathbf{J}^2,\ J_z,\ \mathbf{j}^2,\ \hat{\Sigma}_1,\ \mathbf{p}. \qquad (5)$$

If particles 1 and 2 have definite parities and parity is conserved in the reaction under consideration, then it is better to replace the operator $\hat{\Sigma}_1$ in the set (5) by the space-inversion operator I.

To express observable quantities in terms of diagonal elements of the R matrix ($R = S - 1$, where S is the scattering matrix), we must find the eigenfunctions of the complete set of operators (5) in the representation of the momenta and spins of the free particles. We write these eigenfunctions in the center-of-mass system in the form

$$\langle \mathbf{n}, \mu_1, \mu_2, p\,|\,J, M, j, \mu_1', p'\rangle, \qquad (6)$$

where μ_α is the eigenvalue of the operator $(\mathbf{j}_\alpha \mathbf{n})$, M is the eigenvalue of the operator J_z, and **n** is the unit vector parallel to the direction of the relative momentum **p**.

---

*If the coordinate system is chosen as in reference 9, then $\eta_\mu(L, \mathbf{p}) = \exp[i\mu\varphi(L, \mathbf{p})]$, where $\varphi(L, \mathbf{p})$ is given by Eq. (1.91) of reference 9.

*The rest mass M of the system is given by

$$M^2 = \left(\sqrt{\mathbf{p}_1^2 + M_1^2} + \sqrt{\mathbf{p}_2^2 + M_2^2}\,\right)^2 - (\mathbf{p}_1 + \mathbf{p}_2)^2,$$

and is zero only when the rest masses M_1 and M_2 of the two particles are zero and their momenta are parallel ($\mathbf{p}_1 \parallel \mathbf{p}_2$).

The functions (6) are determined by the following conditions.

1. If we choose the z axis in the direction **n**, then

$$\langle \mathbf{k}, \mu_1, \mu_2, p \mid J, M, j, \mu_1', p'\rangle = C\delta_{\mu_1\mu_1'}\delta_{\mu_1+\mu_2, M}\delta_{pp'}, \quad (7)$$

where **k** is the unit vector in the direction of the z axis, and C is a constant which depends on J, j, and μ_α. The relation (7) follows from the equation $J_z = \hat{\Sigma}_1 + \hat{\Sigma}_2$, if the z axis is parallel to **n**.

2. Under a three-dimensional rotation the functions (6) transform in the following way (like spherical functions):

$$\langle \mathbf{n}, \mu_1, \mu_2, p \mid J, M, j, \mu_1', p'\rangle = \sum_{M'} \langle \mathbf{n}', \mu_1, \mu_2, p \mid J, M', j, \mu_1', p'\rangle D^J_{M'M}(g), \quad (8)$$

if the transformation of the unit vector **n** under the rotation is written $\mathbf{n}' = g\mathbf{n}$ (cf. Note 4 in reference 5). In what follows we shall use the Euler angles $\{\varphi_1, \vartheta, \varphi_2\}$ and the functions $D^J_{M'M}$ introduced in reference 5. We emphasize that the angle ϑ is defined here as a rotation around the y axis.

3. The total angular momentum **J** is the vector sum of the operators **j** and $\mathbf{s}_2$ (Eq. (4)).

4. The functions (6) form a complete set of orthogonal and normalized eigenfunctions. The normalization is based on a volume V with the radius R.

5. The time-reversal operator T is represented by the product of a unitary operator K and the operation of complex conjugation. The phase factor of the function (6) is determined by the condition[11]

$$K\langle \mathbf{n}, \mu_1, \mu_2, p \mid J, M, j, \mu_1', p'\rangle^* = \eta_T \langle -\mathbf{n}, \mu_1, \mu_2, p \mid J, M, j, \mu_1', p'\rangle^* = (-1)^{J+M}\langle \mathbf{n}, \mu_1, \mu_2, p \mid J, -M, j, \mu_1', p'\rangle. \quad (9)$$

Taking $\mathbf{n}' = \mathbf{k}$ in Eq. (8) and substituting Eq. (7) in the right member of Eq. (8), we get

$$\langle \mathbf{n}, \mu_1, \mu_2, p \mid J, M, j, \mu_1', p'\rangle = C\delta_{\mu_1\mu_1'}\delta_{pp'}D^J_{\mu_1+\mu_2, M}(g_\mathbf{n}), \quad (10)$$

where $g_\mathbf{n}$ is the rotation of the coordinate system zyx into the system $z_1y_1x_1$ (z_1 axis parallel to **n**). If we choose the y_1 axis in the direction of $[\mathbf{k}\times\mathbf{n}]$, the Euler angles for the rotation $g_\mathbf{n}$ are $\{-\pi, \vartheta, \pi-\varphi\}$ where ϑ and φ are the spherical angles of the unit vector **n** in the zyx coordinate system.

The constant C is determined by conditions 3, 4, and 5. Choosing the phase factor of the state vector $|\mathbf{n}, \mu_1, \mu_2, \mathbf{p}\rangle$ so that $\eta_T = (-1)^{\mu_1+\mu_2}$, and using the relation

$$D^{J*}_{\mu_1+\mu_2, M}(g_{-\mathbf{n}}) = (-1)^{J+M-\mu_1-\mu_2}D^J_{\mu_1+\mu_2, -M}(g_\mathbf{n}),$$

we find the phase factor C from Eq. (9). We thus get

$$C = \frac{2\pi\hbar\sqrt{R}}{p\sqrt{V}}\left(\frac{2j+1}{4\pi}\right)^{1/2} C^{J\mu_1+\mu_2}_{j\mu_1 s_2\mu_2}. \quad (11)$$

Under the reflection I the function (10) with the constant C from Eq. (11) transforms in the following way:

$$I\langle \mathbf{n}, \mu_1, \mu_2, p \mid J, M, j, \mu_1', p'\rangle = I_0'\eta_I\langle -\mathbf{n}, -\mu_1, -\mu_2, p \mid J, M, j, \mu_1', p'\rangle = I_0'\eta_I(-1)^{-j-s_2}\langle \mathbf{n}, \mu_1, \mu_2, p \mid J, M, j, -\mu_1', p'\rangle, \quad (12)$$

where I_0' is the product of the intrinsic parities of the two particles.

We assume that $I^2 = 1$ in Eq. (12). This condition determines the factor η_I apart from its sign: $\eta_I = \pm(-1)^{s_2+i}$, or

$$I\langle \mathbf{n}, \mu_1, \mu_2, p \mid J, M, j, \mu_1', p'\rangle = \pm(-1)^{j-i_1}I_0'\langle \mathbf{n}, \mu_1, \mu_2, p \mid J, M, j, -\mu_1', p'\rangle, \quad (13)$$

where $i_1 = |\mu_1|$. The eigenfunctions with definite parity I′ have the form

$$\langle \mathbf{n}, \mu_1, \mu_2, p \mid J, M, j, I', p'\rangle = 2^{-1/2}(1 + I'I)\langle \mathbf{n}, \mu_1, \mu_2, p \mid J, M, j, i_1, p'\rangle, \quad (14)$$

where I′ is an eigenvalue of the operator I.

We introduce new variables t:

$$I' = \pm(-1)^{j-i_1+t}.$$

Here t = 0 or t = 1. In reactions with photons one value corresponds to electric radiation, and the other to magnetic radiation (cf. reference 3). The lack of uniqueness in the sign affects only the value of t, but cannot affect the final observable results for the density matrix. Therefore in what follows we can consider the expression (13) with the plus sign only.

Substituting Eqs. (10), (11), and (13) in Eq. (14), we get the eigenfunctions of the complete set of operators $\mathbf{J}^2$, J_z, $\mathbf{j}^2$, I, and **p** in the representation of the momenta and spins:

$$\langle \mathbf{n}, \mu_1, \mu_2, p \mid J, M, j, I', p'\rangle = \frac{2\pi\hbar\sqrt{R}}{p\sqrt{V}}\left(\frac{2j+1}{4\pi}\right)^{1/2} C^{J\mu_1+\mu_2}_{j\mu_1 s_2\mu_2} D^J_{\mu_1+\mu_2, M}(g_\mathbf{n})\delta_{pp'}\sigma^t, \quad (16)$$

where

$$\mu_1 = i_1\sigma, \quad \sigma = \pm 1.$$

If the rest masses of both particles are zero, we cannot introduce the operator **j**. In this case we use the following complete set of operators:

$$\mathbf{J}^2, J_z, \hat{\Sigma}_1, \hat{\Sigma}_2, p. \quad (17)$$

The eigenfunctions of the operators (17) are found in a similar way, and have the form

$$\langle \mathbf{n}, \mu_1, \mu_2, p \,|\, J, M, \mu_1', \mu_2', p' \rangle$$
$$= \frac{2\pi\hbar \sqrt{R}}{p\sqrt{V}} \left(\frac{2J+1}{4\pi}\right)^{1/2} D^J_{\mu_1+\mu_2,\, M}(g_{\mathbf{n}})\, \delta_{\mu_1\mu_1'}\delta_{\mu_2\mu_2'}\delta_{pp'}. \quad (18)$$

The complete sets (5) and (18) can also be used for a system of two particles with nonvanishing rest masses, which is usually described by the set $\mathbf{J}^2$, $\mathrm{J_Z}$, $\boldsymbol{l}^2$, $\mathbf{S}^2$, p, where $\boldsymbol{l}$ is the orbital angular momentum of the relative motion, and $\mathbf{S} = \mathbf{i}_1 + \mathbf{i}_2$ is the sum of the spins of the two particles. Between the eigenfunctions of these sets of operators there exist unitary transformations (transformation functions)

$$\langle J, M, j, \mu, p \,|\, J', M', l, S, p' \rangle$$
$$= \delta_{JJ'}\delta_{MM'}\delta_{pp'}(2l+1)^{1/2}(2S+1)^{1/2} C^{J\mu}_{l0S\mu}\, W(lSji_2, Ji_1)$$

and

$$\langle J, M, \mu_1, \mu_2, p \,|\, J', M', l, S, p' \rangle$$
$$= \delta_{JJ'}\delta_{MM'}\delta_{pp'}\{(2l+1)/(2J+1)\}^{1/2} C^{S\mu_1+\mu_2}_{i_1\mu_1 i_2\mu_2} C^{J\mu_1+\mu_2}_{l0S\mu_1+\mu_2},$$

where W(abcd, ef) is the Racah coefficient.

3. GENERAL FORMULAS FOR THE ANGULAR DISTRIBUTIONS AND THE POLARIZATION VECTORS AND TENSORS FOR THE REACTIONS a + b → c + d AND a → c + d

Let us first consider the density matrix of the γ-particles, $<\mu_\gamma|\rho|\mu_\gamma'>$. Since μ_γ takes only the two values $\pm$i, we cannot construct the s-tensors from $<\mu_\gamma|\rho|\mu_\gamma'>$ in the ordinary way. Fano[12] has shown that for photons it is convenient to use the Stokes parmeters. This idea is easily generalized and is adopted in the present work. The Stokes parameters are related to the density matrix of the γ-particles in the following way:

$$\rho(q_\gamma \chi_\gamma) = 2^{1/2} \sum_{\sigma_\gamma, \sigma_\gamma'} (-1)^{1/2 - 1/2\sigma_\gamma'} C^{q_\gamma \chi_\gamma}_{1/2,\, 1/2\sigma_\gamma,\, 1/2,\, -1/2\sigma_\gamma'} \langle \mu_\gamma | \rho | \mu_\gamma' \rangle, \quad (20)$$

where $\rho(q_\gamma, \chi_\gamma)$ are the Stokes parameters, $\mu_\gamma = i_\gamma \sigma_\gamma$, and $\mu_\gamma' = i_\gamma \sigma_\gamma'$.

We emphasize that $\rho(1, \chi_\gamma)$ defined by Eq. (20), do not transform like a vector under rotations. The physical interpretation of the Stokes parameters given in Fano's paper for photons is also correct for any γ-particles. We shall not repeat it. In this paper the Stokes parameters will for convenience also be called s-tensors.

The calculations of the s-tensors of the reaction-product particles is made by the method of M. I. Shirokov.[5] We use the notations introduced in reference 5. If the masses of particles 1 and 2 in the initial and final states of the reaction are not zero, we use the complete set of operators $\mathbf{J}^2$, $\mathrm{J_Z}$, $\boldsymbol{l}^2$, $\mathbf{S}^2$, p, where $\boldsymbol{l}$ is the orbital angular momentum and $\mathbf{S}$ is the total spin of the two particles. We get the following results:

1. γ + b → c + d.

$$\rho'(q_c\chi_c q_d\chi_d;\, \mathbf{n}_c)$$
$$= [N_1/(4\pi)^2]\,[(2i_c+1)(2i_d+1)2^{-1}(2i_b+1)^{-1}]^{1/2}$$
$$\times \sum Y^*_{q_c\chi_c q_d\chi_d}(J_1 l_1' S_1',\ Jq'\chi',\ J_2 l_2' S_2')\, \langle S_1' l_1' \alpha | R^{J_1} | j_1 t_1 \alpha_1 \rangle$$
$$\times \langle S_2' l_2' \alpha' | R^{J_2} | j_2 t_2 \alpha_2 \rangle^*\ Y_{q_\gamma\chi_\gamma q_b\chi_b}(J_1 j_1 t_1,\ Jq\chi,\ J_2 j_2 t_2)$$
$$\times D^J_{\chi'\chi}(g_c g_\gamma^{-1})\, \rho(q_\gamma\chi_\gamma q_b\chi_b,\, \mathbf{n}_\gamma), \quad (21)$$

where

$$Y_{q_c\chi_c q_d\chi_d}(J_1 l_1' S_1',\ Jq'\chi',\ J_2 l_2' S_2')$$
$$= [(2q_c+1)(2q_d+1)]^{1/2} C^{q'\chi'}_{q_c\chi_c q_d\chi_d}$$
$$\times (-1)^{q'+\chi'} i^{\,l_1-l_2}\, X(i_c q_c i_c, S_1' q' S_2',\ i_d q_d i_d)$$
$$\times G_{\chi'}(J_1 l_1' S_1',\ Jq,\ J_2 l_2' S_2'); \quad (22)$$

$$Y_{q_\gamma\chi_\gamma q_b\chi_b}(J_1 j_1 t_1,\ Jq\chi,\ J_2 j_2 t_2) = \sum [(2J_1+1)(2J_2+1)$$
$$\times (2j_1+1)(2j_2+1)(2J+1)(2q+1)]^{1/2}$$
$$\times C^{h 2i_\gamma \chi_\gamma}_{J\chi q_b -\chi_b} (-1)^{J_1 + i_\gamma \chi_\gamma + q_b + \chi_b}$$
$$\times X(J_1 J J_2,\ j_1 h j_2,\ i_b q_b i_b)\, F_h(j_1 t_1,\ j_2 t_2 q_\chi \chi_\gamma); \quad (23)$$

$$F_h(j_1 t_1,\ j_2 t_2,\ q_\gamma 0)$$
$$= (-1)^{-i_\gamma}[1 + I_{t_1}/I_{t_2}(-1)^{h+q_\gamma}]\, C^{q_\gamma 0}_{1/2\, 1/2\, 1/2\, -1/2} C^{h0}_{j_1 i_\gamma j_2 -i_\gamma}, \quad (24)$$

$$F_h(j_1 t_1, j_2 t_2, 11) = (-1)^{j_1+j_2+h+t_2+1} C^{h-2i_\gamma}_{j_1 -i_\gamma j_2 -i_\gamma},$$

$$F_h(j_1 t_1,\ j_2 t_2,\ 1-1) = (-1)^{j_1+j_2+h+t_1} C^{h 2i_\gamma}_{j_1 i_\gamma j_2 i_\gamma};$$

$$I_{t_\alpha} = (-1)^{j_\alpha - i_\gamma + t_\alpha},\ N_1 = (2\pi h)^4 R^2 [V^2 p_b^2 p_d^2]^{-1} (p_b;\, p_d)^2.$$

The sum is taken over q′, χ′, J′, l_1', S_1', J_2, l_2', S_2', j_1, t_1, j_2, t_2, J, q, χ, q_γ, χ_γ, q_b, χ_b.

2. a + b → γ + d.

$$\rho'(q_\gamma\chi_\gamma q_d\chi_d;\, \mathbf{n}_\gamma)$$
$$= [N_1/(4\pi)^2]\,[2(2i_d+1)(2i_a+1)^{-1}(2i_b+1)^{-1}]^{1/2}$$
$$\times \sum Y^*_{q_\gamma\chi_\gamma q_d\chi_d}(J_1 j_1' t_1',\ Jq'\chi',\ J_2 j_2' t_2)$$
$$\times \langle j_1' t_1' \alpha | R^{J_1} | S_1 l_1 \alpha_1 \rangle \langle j_2' t_2' \alpha' | R^{J_2} | S_2 l_2 \alpha_2 \rangle$$
$$\times Y_{q_a\chi_a q_b\chi_b}(J_1 l_1 S_1,\ Jq\chi,\ J_2 l_2 S_2)$$
$$\times D^J_{\chi'\chi}(g_\gamma g_a^{-1})\, \rho(q_a\chi_a q_b\chi_b;\, \mathbf{n}_a). \quad (25)$$

3. γ + b → γ' + d.

$$\rho'(q_{\gamma'}\chi_{\gamma'}, q_d\chi_d; \mathbf{n}_{\gamma'}) = [N_1/(4\pi)^2]\,[(2i_d+1)(2i_b+1)^{-1}]^{1/2}$$
$$\times \sum Y^*_{q_{\gamma'}\chi_{\gamma'}q_d\chi_d}(J_1 j' t'_1, Jq'\chi', J_2 j'_2 t'_2)$$
$$\times \langle j'_1 t'_1 \alpha | R^{J_1} | j_1 t_1 \alpha_1 \rangle \langle j'_2 t'_2 \alpha' | R^{J_2} | j_2 t_2 \alpha_2 \rangle$$
$$\times Y_{q_\gamma \chi_\gamma q_b \chi_b}(J_1 j_1 t_1, Jq\chi, J_2 j_2 t_2)$$
$$\times D^J_{\chi'\chi}(g_{\gamma'} g_\gamma^{-1})\,\rho(q_\gamma \chi_\gamma q_b \chi_b; \mathbf{n}_\gamma). \tag{26}$$

4. $a \rightarrow \gamma + d$.

$$\rho'(q_\gamma \chi_\gamma q_d \chi_d, \mathbf{n}_\gamma) = (N_2/4\pi)\,[2(2i_d+1)(2i_a+1)^{-1}]^{1/2}$$
$$\times \sum Y^*_{q_\gamma \chi_\gamma q_d \chi_d}(i_a j'_1 t'_1, q_a q' \chi', i_a j'_2 t'_2) \langle j'_1 t'_1 \alpha | R^{i_a} | \alpha_1 \rangle$$
$$\times \langle j'_2 t'_2 \alpha' | R^{i_a} | \alpha_2 \rangle^* D^{q_a}_{\chi'\chi_a}(g_\gamma)\,\rho(q_a, \chi_a), \tag{27}$$

where $N_2 = (2\pi\hbar)^3 R (Vp^2)^{-1} (E_\gamma | E_a)$. The sum is taken over j'_1, t'_1, j'_2, t'_2, q_a, χ_a, q', χ'.

In particular for $q_\gamma = q_b = q_c = q_d = 0$ the formula (21) gives the angular distribution of the products of the reaction $\gamma + b \rightarrow c + d$ for unpolarized incident beam and unpolarized target. In this case our formula agrees with the result obtained by Morita et al., if we eliminate certain phase factors by means of a redefinition of the elements of the R matrix. Our results confirm the conclusion that Simon's formulas are erroneous.

4. SELECTION RULES

If parity is conserved selection rules exist in the form of relations between the s-tensors. These selection rules have been obtained in references 2, 5, 6, 13 for the case in which all the particles have nonvanishing rest masses. The results are easily extended to reactions involving γ-particles. As an example we shall consider a reaction of the type $\gamma + b \rightarrow c + d$.

The conservation of spatial parity gives the following equation:

$$I_{t_1} I_{t_2} = (-1)^{l'_1 + l'_2}. \tag{28}$$

From the properties of the D-function and Eqs. (22) — (24) it follows that:

$$D^J_{\chi'\chi}(g_c g_\gamma^{-1}) = D^J_{\chi'\chi}(-\pi, \vartheta_c, \pi - \varphi_c)$$
$$= (-1)^{\chi' - \chi} D^J_{-\chi' -\chi}(-\pi, \vartheta_c, \pi + \varphi_c), \tag{29}$$

$$Y_{q_c \chi_c q_d \chi_d}(J_1 l'_1 S'_1, Jq'\chi', J_2 l'_2 S_2) = (-1)^{l'_1 + l'_2 + J + q_c + q_d}$$
$$\times Y_{q_c, -\chi_c, q_d, -\chi_d}(J_1 l'_1 S'_1, Jq' - \chi', J_2 l'_2 S'_2), \tag{30}$$

$$Y_{q_\gamma \chi_\gamma q_b \chi_b}(J_1 j_1 t_1, Jq\chi, J_2 j_2 t_2) = (-1)^{J + q_\gamma + q_b}$$
$$\times I_{t_1} I_{t_2} Y_{q_\gamma - \chi_\gamma q_b - \chi_b}(J_1 j_1 t_1, Jq - \chi, J_2 j_2 t_2). \tag{31}$$

Equation (31) is obtained by the use of the equations $(-1)^{\gamma q} = (-1)^{\chi\gamma}$ for $\chi_\gamma = \pm 1$ and $(-1)^h = I_{t_1} I_{t_2} (-1)^{q_\gamma}$ for $\chi_\gamma = 0$. Substituting Eqs. (29), (30), and (31) in Eq. (21), we get the most general selection rules holding when parity is conserved; these rules are expressed as follows: if we replace the s-tensors $\rho(q_\gamma \chi_\gamma q_b \chi_b, \mathbf{n}_\gamma)$ of the initial state by

$$\rho_I = (-1)^{q_\gamma + 2i_\gamma \chi_\gamma + q_b + \chi_b} \rho(q_\gamma, -\chi_\gamma, q_b, -\chi_b; \mathbf{n}_\gamma),$$

then we get in the final state the s-tensors

$$\rho'_I = (-1)^{q_c + \chi_c + q_d + \chi_d} \rho'(q_c - \chi_c q_d - \chi_d; \vartheta, -\varphi).$$

The results obtained here for the reaction involving γ-particles and the writer's results[13] for the reaction without γ-particles are consistent with each other. We emphasize that the spin states of the γ-particles are described not by s-tensors, but by Stokes parameters. We shall not consider here the special cases dealt with in reference 13, for which the results have just the same form.

The conservation of time parity gives a relation between the s-tensors of the direct ($\gamma + b \rightarrow c + d$) and inverse ($c + d \rightarrow \gamma + b$) reactions. Such relations have been obtained by M. I. Shirokov[6] for the reaction without γ-particles. His formulas require only a slight change for the reactions involving γ-particles. In what follows we suppose that spatial parity is also conserved.

Repeating the calculations of reference 6 step by step and using the same notations, we get the relation for the reaction with γ-particles that corresponds to Eq. (8) of reference 6:

$$\langle q_\gamma, \chi_\gamma, q_b, \chi_b | W_I(\varphi_1, \vartheta, -\varphi_2) | q_c, \chi_c, q_d, \chi_d \rangle$$
$$= (-1)^{q_\gamma + 2i_\gamma \chi_\gamma + q_b + \chi_b + q_c + \chi_c + q_d + \chi_d} \langle q_c, -\chi_c, q_d, -\chi_d,$$
$$\times | W_D(-\pi + \varphi_2, \vartheta, \pi - \varphi_1) | q_\gamma, -\chi_\gamma, q_b, -\chi_b \rangle, \tag{32}$$

where the Euler angles $\{-\pi + \varphi_2, \vartheta, \pi - \varphi_1\}$ are those for the rotation $g_c g_\gamma^{-1}$ in the direct reaction, and $\{\varphi_1, \vartheta, -\varphi_2\}$ are those for the rotation $g_\gamma g_c^{-1}$ in the inverse reaction. We emphasize that in the direct reaction the y_c axis is not parallel to the vector $[\mathbf{n}_\gamma, \mathbf{n}_c]$, but is obtained by clockwise rotation of this vector through the angle φ_2 around the z_c axis. The angles ϑ and φ_1 are equal to the spherical angles of the vector $\mathbf{n}_c$ in the coordinate system $z_\gamma y_\gamma x_\gamma$. In the inverse reaction ϑ and $\pi + \varphi_2$ are equal to the spherical angles of the vector $\mathbf{n}_\gamma$ in the system $z_c y_c x_c$, and the y_γ axis makes the angle $\pi + \varphi_1$ with the direction of $[\mathbf{n}_c \times \mathbf{n}_\gamma]$.*

Replacing the factor F in Eq. (9) of reference 6 by

*We note that Eq. (8) of reference 6 is valid only in the specially chosen systems $z_c y_c x_c$ and $z_a y_a x_a$ in which $\varphi_1 = \varphi_2$.

$$F = [(2i_c + 1)(2i_d + 1)]^{1/2} [2(2i_b + 1)]^{-1/2} \quad (33)$$

and $(-1)^{q_a + \chi_a}$ by $(-1)^{q_\gamma + 2i_\gamma \chi_\gamma}$, we get relations corresponding to Eqs. (9) and (10) in reference 6.

Let us go on to the consideration of the special cases studied in reference 6. The relation of detailed balancing between the differential cross-sections of the direct and inverse reactions for unpolarized initial state has the well known form

$$p_\gamma^2 2(2i_b + 1)\,\sigma_D(\vartheta) = p_c^2 (2i_c + 1)(2i_d + 1)\,\sigma_I(\vartheta). \quad (34)$$

The difference $\sigma'_D(\vartheta, \varphi_1) - \sigma_D(\vartheta, \varphi_1)$ between the cross-sections with a polarized beam of γ-particles and with an unpolarized beam in the direct reaction can be expressed in the form

$$\sigma'_D(\vartheta, \varphi_1) - \sigma_D(\vartheta, \varphi_1) = -F^2 p_c^2 p_\gamma^{-2} [(-1)^{2i_\gamma} \rho_1 \rho'_{-1} + \rho_0 \rho'_0 + \rho'_{-1}\rho_1 (-1)^{2i_\gamma}] = -F^2 p_c^2 p_\gamma^{-2} (\rho' (\varphi_1, \vartheta, -\varphi_2)\,\rho), \quad (35)$$

where ρ_{-1}, ρ_0, and ρ_1 are the Stokes parameters characterizing the spin state of the incident γ-particles in the direct reactions; ρ'_{-1}, ρ'_0, and ρ'_1 are the Stokes parameters for the spin state of the γ-particles produced in the inverse reaction with unpolarized initial state.

5. SYSTEM WITH TWO IDENTICAL γ-PARTICLES

Let us construct symmetric or antisymmetric wave functions with definite parity, total angular momentum, z-component of angular momentum, and energy, for a system of two identical γ-particles. The state vector $|J, M, \mu'_1, \mu'_2, p\rangle$ transforms under reflection in the following way:

$$I\,|J, M, \mu'_1, \mu'_2, p\rangle = I'_0 \eta'_I\,|J, M, -\mu'_1, -\mu'_2, p\rangle,$$

where I'_0 is the product of the intrinsic parities of the two identical particles, which is always equal to unity; η'_I is a phase factor.

Since the state vector $|J, M, \mu'_1, \mu'_2, p\rangle$ transforms under rotations according to an irreducible representation of the rotation group with the weight J, a rotation through the angle φ around the direction n of the relative momentum of the particles gives a factor $\exp[i(\mu'_1 + \mu'_2)\varphi]$. Accordingly, in the rotated coordinate system the factor η'_I is replaced by $\eta''_I = \eta'_I \exp[2i(\mu'_1 + \mu'_2)\varphi]$. Therefore the value of the phase factor η'_I depends on the direction of the x (or y) axis in the coordinate system with the z axis parallel to n. In view of the fact that the system of two γ-particles occurs only in the final state of some reaction, we choose the coordinate system with axes $z \parallel n$, $y \parallel \zeta - (\zeta \cdot n)n$, $x \parallel [\zeta \times n]$, where ζ is the polarization vector of the particles in the initial state of the reaction. After reflection the z and x axes change their directions, but the y axis remains unchanged. The "reflected" coordinate system differs from the original system only by a rotation through the angle π around the y axis. Therefore from the relation $D^J_{\mu'\mu}(0, \pi, 0) = (-1)^{J-\mu}\delta_{\mu', -\mu}$ we get

$$I\,|J, M, \mu'_1, \mu'_2, p\rangle = I'_I \sum_{\dot\mu_1 \dot\mu_2} |J, M, \dot\mu_1, \dot\mu_2; p\rangle D^J_{\dot\mu_1 + \dot\mu_2\, \mu'_1 + \mu'_2}(0, \pi, 0) = I'_I (-1)^{J - \mu'_1 - \mu'_2}\,|J, M, -\mu_1, -\mu_2, p\rangle. \quad (36)$$

Let us now consider the operator S that replaces the first particle by the second and vice versa. In the chosen coordinate system the operator S can be expressed as the product of the operator I, which changes the direction of the relative momentum, and the operator that interchanges the spin indices μ'_1 and μ'_2. Thus

$$S\,|J, M, \mu_1, \mu_2, p'\rangle = (-1)^{J - \mu'_1 - \mu'_2}\,|J, M, -\mu_2, -\mu_1, p\rangle. \quad (37)$$

From Eqs. (36) and (37) we get the symmetric or antisymmetric wave vectors with definite parity I′, total angular momentum, z-component of angular momentum, and energy

$$|J, M, \mu, I', p\rangle = A[1 + I'I + S'S + I'S'IS]\,|J, M, \mu_1, \mu_2, p\rangle\,\delta_{|\mu'_1 + \mu'_2|, \mu}, \quad (38)$$

where $\mu = 2i$ or 0 is the eigenvalue of the operator $|\hat\Sigma_1 + \hat\Sigma_2|$, A is a normalization constant, and $S' = +1$ for particles obeying Bose statistics and $S' = -1$ for Fermi statistics. Multiplying Eq. (38) from the left by $\langle n, \mu_1, \mu_2, p|$ and using Eqs. (19), (36), and (37), we get the wave functions for this system in the momentum-spins representation.

1. $\mu = 2i$, $\mu_1 = \mu_2 = \pm i$.

$$\langle n, \mu_1, \mu_2, p\,|\,J, M, 2i, I', p'\rangle = (1/2\sqrt{2})(1 + S'I')(2\pi\hbar\sqrt{R}/p\sqrt{V})\left(\frac{2J+1}{4\pi}\right)^{1/2}\delta_{pp'} \times [D^J_{2i,M}(g_n)\,\delta_{\mu_1+\mu_2, 2i} + I'(-1)^{J+2i} D^J_{-2i,M}(g_n)\,\delta_{\mu_1+\mu_2, -2i}]. \quad (39)$$

2. $\mu = 0$, $\mu_1 = -\mu_2 = \pm i$.

$$\langle n, \mu_1, \mu_2, p\,|\,J, M, 0, I', p\rangle = (1/2\sqrt{2})(1 + S'(-1)^J)(2\pi\hbar\sqrt{R}/p\sqrt{V})[(2J+1)/4\pi]^{1/2}\delta_{pp'} \times D^J_{0M}(g_n)(\delta_{\mu_1 i}\delta_{\mu_2 -i} + I'(-1)^J \delta_{\mu_1 -i}\delta_{\mu_2 i}). \quad (40)$$

As can be seen from Eqs. (39) and (40), the wave functions are nonvanishing only when $S'I' = 1$ for $\mu = 2i$, and $S'(-1)^J = 1$ for $\mu = 0$. These are just the specific selection rules obtained by Landau and by Yang for photons and by Shapiro in the general case. We note that when parity is not conserved the selection rule $S'(-1)^J = 1$ for $\mu = 0$ remains valid, but the factor $\delta_{\mu_1 i}\delta_{\mu_2 -i} + I'(-1)^J \times \delta_{\mu_1 -i}\delta_{\mu_2 i}$, which characterizes the polarization of the system, is changed. Therefore in the decay of a particle with spin $J < 2i$ into two identical γ-particles a measurement of the correlation of the polarizations of the γ-particles not only will give the parity of this particle when parity is conserved (cf. Yang, reference 7), but also will give information about the nonconservation of parity in the decay.

The wave functions (39) and (40) can be written together in a single formula:

$$\langle \mathbf{n}, \mu_1, \mu_2, p \mid J, M, \mu, I', p' \rangle = \frac{1}{2\sqrt{2}} F_\mu \frac{2\pi\hbar\sqrt{R}}{p\sqrt{V}} \times \left(\frac{2J+1}{4\pi}\right)^{1/2} D^J_{\mu_1+\mu_2, M}(g_\mathbf{n})\, \sigma^t \delta_{|\mu_1+\mu_2|, \mu}, \qquad (41)$$

where $\sigma i = \mu_1$, $I' = (-1)^{J-2i+t}$,

$$F_\mu = (1 + S'I') \text{ for } \mu = 2i,$$
$$F_\mu = (1 + S'(-1)^J) \text{ for } \mu = 0.$$

By means of Eq. (41) one can easily calculate the angular distributions and the polarizations for the decay of a particle into two identical γ-particles. We shall not do this here.

In conclusion I express my gratitude to Professor M. A. Markov, M. I. Shirokov, and L. G. Zastavenko for interest shown in this work and a discussion of the results.

[1] Chou Kuang-Chao and M. I. Shirokov, J. Exptl. Theoret. Phys. (U.S.S.R.) **34**, 1230 (1958), Soviet Phys. JETP **7**, 851 (1958).

[2] A. Simon, Phys. Rev. **92**, 1050 (1953).

[3] Morita, Sugie, and Yoshida, Prog. Theor. Phys. **12**, 713 (1954).

[4] E. P. Wigner, Ann. of Math. **40**, 1490 (1939). Yu. M. Shirokov, J. Exptl. Theoret. Phys. (U.S.S.R.) **33**, 1208 (1957), Soviet Phys. JETP **6**, 929 (1958).

[5] M. I. Shirokov, J. Exptl. Theoret. Phys. (U.S.S.R.) **32**, 1022 (1957), Soviet Phys. JETP **5**, 835 (1957).

[6] M. I. Shirokov, J. Exptl. Theoret. Phys. (U.S.S.R.) **33**, 975 (1957), Soviet Phys. JETP **6**, 748 (1957).

[7] L. D. Landau, Dokl. Akad. Nauk SSSR **60**, 207 (1948). C. N. Yang, Phys. Rev. **77**, 242 (1950).

[8] I. S. Shapiro, J. Exptl. Theoret. Phys. (U.S.S.R.) **27**, 393 (1954).

[9] Chou Kuang-Chao and L. G. Zastavenko, J. Exptl. Theoret. Phys. (U.S.S.R.) **35**, 1417 (1958), Soviet Phys. JETP **8**, 990 (1959).

[10] E. P. Wigner, Revs, Modern Phys. **29**, 255 (1957).

[11] L. C. Biedenharn and M. E. Rose, Revs. Modern Phys. **25**, 729 (1953). R. Huby, Proc. Phys. Soc. **A67**, 1103 (1954).

[12] U. Fano, Phys. Rev. **93**, 121 (1954).

[13] Chou Kuang-Chao, J. Exptl. Theoret. Phys. (U.S.S.R.) **35**, 783 (1958), Soviet Phys. JETP **8**, 543 (1959).

Translated by W. H. Furry
163

SOME SYMMETRY PROPERTIES IN PROCESSES OF ANTIHYPERON PRODUCTION WITH ANNIHILATION OF ANTINUCLEONS

CHOU KUANG-CHAO

Joint Institute for Nuclear Studies

Submitted to JETP editor November 20, 1958

J. Exptl. Theoret. Phys. (U.S.S.R.) **36**, 938-939 (March, 1959)

LET us consider the reaction

$$\widetilde{p} + p \rightarrow \widetilde{\Sigma}^- + \Sigma^-. \quad (1)$$

We denote the amplitude for it by $f(p_i, p_f, \sigma_p, \sigma_a)$, where p_i and p_f are the relative momenta in the initial and final states, and σ_p and σ_a are the Pauli matrices of the particles and antiparticles. From invariance with respect to charge conjugation it follows that

$$f(p_i, p_f, \sigma_p, \sigma_a) = f(-p_i, -p_f, \sigma_a, \sigma_p). \quad (2)$$

If the initial state is unpolarized, then it is not hard to prove by Eq. (2) that the polarization vectors of the hyperon (P_Σ) and of the antihyperon ($P_{\widetilde{\Sigma}}$) in the final states are given by

$$P_\Sigma = P_{\widetilde{\Sigma}} = A[p_i p_f]/|[p_i p_f]|, \quad (3)$$

where A is a function of the scalar $(p_i \cdot p_f)$.

Measurement of the angular asymmetries in the decay of the Σ^- and $\widetilde{\Sigma}^-$ produced in the reaction (1) gives the ratio

$$P_\Sigma \alpha_\Sigma / P_{\widetilde{\Sigma}} \alpha_{\widetilde{\Sigma}} = \alpha_\Sigma / \alpha_{\widetilde{\Sigma}}, \quad (4)$$

where α_Σ and $\alpha_{\widetilde{\Sigma}}$ are the antisymmetry coefficients of the decays. As has been shown in reference 1, measurement of the ratio $\alpha_\Sigma/\alpha_{\widetilde{\Sigma}}$ is of great significance for testing the conservation laws associated with time reversal T and charge conjugation C; this ratio differs from unity only if T and C are not conserved in the decay.

Let us go on to the consideration of the two cases

$$\widetilde{p} + p\,(\widetilde{n} + n) \rightarrow Y_1 + \widetilde{Y}_2 + m\pi^+ + n\pi^- + l\pi^0 \quad (5)$$

$$\rightarrow \widetilde{Y}_1 + Y_2 + n\pi^+ + m\pi^- + l\pi^0. \quad (6)$$

The amplitudes for the reactions (5) and (6) are expressed in the form

$$f_1(p_i, p_f, p_\alpha^+, p_\beta^-, p_\gamma^0, \sigma_p, \sigma_a), \quad \alpha = 1, 2, \ldots m,$$
$$\beta = 1, \ldots n,$$
$$f_2(p_i, p_f, p_\beta^+, p_\gamma^-, p_\gamma^0, \sigma_p, \sigma_a), \quad \gamma = 1, \ldots l, \quad (7)$$

where $p^{\pm,0}$ are the momenta of $\pi^{\pm,0}$ mesons. From invariance with respect to C it follows that

$$f_2(p_i, p_f, p_\beta^+, p_\alpha^-, p_\gamma^0, \sigma_p, \sigma_a)$$
$$= f_1(-p_i, -p_f, p_\alpha^-, p_\beta^+, p_\gamma^0, \sigma_a, \sigma_p). \quad (8)$$

Using the relation (8), we get not only equality of the total cross-sections and the angular distributions of these two processes (σ_1 and σ_2), but also equality of the polarization vectors of the hyperons and antihyperons in the final state (P_Y and $P_{\widetilde{Y}}$). When the initial state is unpolarized we have

$$\sigma_1(p_i, p_f, p_\alpha^+, p_\beta^-, p_\gamma^0) = \sigma_2(-p_i, -p_f, p_\beta^-, p_\alpha^+, p_\gamma^0),$$
$$P_{Y_1}(p_i, p_f, p_\alpha^+, p_\beta^-, p_\gamma^0) = P_{\widetilde{Y}_1}(-p_i, -p_f, p_\beta^-, p_\alpha^+, p_\gamma^0),$$
$$P_{\widetilde{Y}_2}(p_i, p_f, p_\alpha^+, p_\beta^-, p_\gamma^0) = P_{Y_2}(-p_i, -p_f, p_\beta^-, p_\alpha^+, p_\gamma^0). \quad (9)$$

Analogous relations exist also for reactions in which K mesons and nucleons are produced.

There are also a number of selection rules for reactions of the types

$$\widetilde{n} + p\,(\widetilde{p} + n) \rightarrow Y_1 + \widetilde{Y}_2 + m\pi^+ + n\pi^- + l\pi^0$$
$$\rightarrow \widetilde{Y}_1' + Y_2' + m\pi^+ + n\pi^- + l\pi^0, \quad (10)$$

where $\widetilde{Y}_1'$ and Y_2' are obtained from Y_1 and $\widetilde{Y}_2$ by means of the operator G, which is the product of the charge-conjugation operator and a rotation through the angle π around the x axis of the isobaric space.[2] For example

$$\widetilde{\Sigma}^- = G\Sigma^+, \quad \widetilde{\Sigma}^0 = G\Sigma^0, \quad \widetilde{\Sigma}^+ = G\Sigma^-,$$

$$\widetilde{\Lambda}^0 = G\Lambda^0, \quad \widetilde{n} = Gp, \quad \pi^{\pm,0} = G\pi^{\pm,0}$$

and so on. We denote the respective amplitudes of the reactions (10) by

$$f_1(p_i, p_f, p_\alpha^+, p_\beta^-, p_\gamma^0, \sigma_p, \sigma_a) \text{ and } f_2(p_i, p_f, p_\alpha^+, p_\beta^-, p_\gamma^0, \sigma_p, \sigma_a).$$

From invariance with respect to G it follows that

$$f_1(p_i, p_f, p_\alpha^+, p_\beta^-, p_\gamma^0, \sigma_p, \sigma_a)$$
$$= \eta f_2(-p_i, -p_f, p_\alpha^+, p_\beta^-, p_\gamma^0, \sigma_a, \sigma_p), \quad (11)$$

where $\eta = \pm 1$ is a phase factor.

It is easy to show from Eq. (11) that for an unpolarized initial state

$$\sigma_1(p_i, p_f, p_\alpha^+, p_\beta^-, p_\gamma^0) = \sigma_2(-p_i, -p_f, p_\alpha^+, p_\beta^-, p_\gamma^0),$$
$$P_{Y_1}(p_i, p_f, p_\alpha^+, p_\beta^-, p_\gamma^0) = P_{\widetilde{Y}_1'}(-p_i, -p_f, p_\alpha^+, p_\beta^-, p_\gamma^0),$$
$$P_{\widetilde{Y}_2}(p_i, p_f, p_\alpha^+, p_\beta^-, p_\gamma^0) = P_{Y_2'}(-p_i, -p_f, p_\alpha^+, p_\beta^-, p_\gamma^0). \quad (12)$$

[1] Chou Kuang-Chao, Nuclear Phys. (in press).

[2] T. D. Lee and C. N. Yang, Nuovo cimento **3**, 749 (1956).

Translated by W. H. Furry
176

SOVIET PHYSICS JETP VOLUME 37 (10), NUMBER 1 JANUARY, 1960

ON THE PROBLEM OF INVESTIGATING THE INTERACTION BETWEEN π MESONS AND HYPERONS

L. I. LAPIDUS and CHOU KUANG-CHAO

Joint Institute of Nuclear Studies

Submitted to JETP editor February 27, 1959

J. Exptl. Theoret. Phys. (U.S.S.R.) 37, 283-288 (1959)

It is shown that use of the unitary property of the S matrix makes it possible to obtain some information about the scattering of π mesons by Λ and Σ hyperons from an analysis of the data on the interaction of K mesons with nucleons. The possibility of studying the π-Λ and π-Σ interactions by examining peripheral collisions of hyperons with nucleons is discussed.

THE study of the interactions of π mesons with hyperons is of special interest in connection with the determination of the symmetry properties of the interactions of π mesons with various baryons.

1. Let us consider the reactions

$$\widetilde{K} + N \to \widetilde{K} + N, \tag{1a}$$

$$\widetilde{K} + N \to \Sigma(\Lambda) + \pi, \tag{1b}$$

$$\Sigma(\Lambda) + \pi \to \Sigma(\Lambda) + \pi \tag{1c}$$

in a range of K-meson energies in which one can neglect channels in which two pions are produced. Since the elements of the S matrix for the reactions (1) are connected with each other by the condition of unitarity, the question arises as to what information about the scattering amplitudes Σ(Λ) + π → Σ(Λ) + π can be obtained by studying the cross sections and polarizations in processes (1a) and (1b). The first part of the present paper contains an attempt to answer this question.

In what follows we assume that the spin of the K meson is zero and that the hyperon spin is ½. We further assume that the interactions are invariant under space inversion, time reversal, and rotations in isotopic space.

The reactions (1) are described by elements of the T matrix (iT = S − 1) diagonal in the isotopic-spin quantum number,

$$T^0 = \begin{pmatrix} a_K^0 & a_{K\Sigma}^0 \\ a_{\Sigma K}^0 & a_\Sigma^0 \end{pmatrix}, \quad T^1 = \begin{pmatrix} a_K^1 & a_{K\Sigma}^1 & a_{K\Lambda} \\ a_{\Sigma K}^1 & a_\Sigma^1 & a_{\Sigma\Lambda} \\ a_{\Lambda K} & a_{\Lambda\Sigma} & a_\Lambda \end{pmatrix}, \tag{2}$$

where $a_K^0(a_K^1)$ is the amplitude for scattering $\widetilde{K}$ + N → $\widetilde{K}$ + N in the state with the indicated value of the isotopic spin, 0 (1); $a_{K\Sigma}^0(a_{K\Sigma}^1)$ is the amplitude for the reaction $\widetilde{K}$ + N → Σ + π in the state with the isotopic spin 0 (1), and so on.

The spin structure of the scattering amplitude a_α can be represented in the form

$$a_\alpha = A_\alpha + iB_\alpha (\sigma [\mathbf{n} \times \mathbf{n}']), \tag{3}$$

where **n** (**n′**) is the unit vector parallel to the momentum of the particles in the initial (final) state, in the center-of-mass system; A_α and B_α are two complex functions of the energy and of **n**·**n′**.

The reaction amplitude $a_{\alpha\beta}$ has the form

$$a_{\alpha\beta} = A_{\alpha\beta} + iB_{\alpha\beta} (\sigma [\mathbf{n} \times \mathbf{n}']), \tag{4}$$

when the product of the intrinsic parities of all four particles in the initial (final) states is Π = +1, and the form

$$a_{\alpha\beta} = A_{\alpha\beta}(\sigma \mathbf{n}) + B_{\alpha\beta}(\sigma \mathbf{n}'), \tag{5}$$

when Π = −1. Here $A_{\alpha\beta}$ and $B_{\alpha\beta}$ are two complex functions of the energy and of **n**·**n′**.

Let us turn to the analysis of the conditions for determining the T matrix from the experimental data. It can be seen from Eqs. (2), (3), and (4) that the number of real scalar functions involved in the matrixes T^0 and T^1 is 13 × 4 = 52. The invariance of the interaction under time reversal means that the S matrix is symmetric, and this reduces the number of functions determining the T matrix from 52 to 36. It can be shown further that when the conditions for the S matrix to be unitary are taken into account, the number of independent real functions is decreased by a factor of two and becomes 18.

The same result is obtained if we use the general formulas obtained in reference 1.

Let us now consider what information can be obtained by studying only processes (1a) and (1b). The number of real functions characterizing these processes is 5 × 4 = 20. They satisfy four relations of unitarity. Therefore only 16 of them are independent.

Reaction	Amplitude
(a) $K^- + p \to K^- + p$	$\frac{1}{2}(a_K^1 + a_K^0)$
(b) $K^- + p \to K^0 + n$	$\frac{1}{2}(a_K^1 - a_K^0)$
(c) $K_2^0 + p \to K_1^0 + p$	$\frac{1}{2}(a_K^1 - \tilde{a}_K^0)$
(d) $K_2^0 + p \to K_2^0 + p$	$\frac{1}{2}(a_K^1 + \tilde{a}_K^0)$
(e) $K^- + p \to \Lambda + \pi^0$	$a_{K\Lambda}$
(f) $K^- + p \to \Sigma^- + \pi^+$	$-(a_{K\Sigma}^0/\sqrt{6} + a_{K\Sigma}^1/2)$
(g) $K^- + p \to \Sigma^0 + \pi^0$	$a_{K\Sigma}^0/\sqrt{6}$
(h) $K^- + p \to \Sigma^+ + \pi^-$	$-(a_{K\Sigma}^0/\sqrt{6} - a_{K\Sigma}^1/2)$

The table shows 8 reactions of types (1a) and (1b) and their amplitudes. The symbol K_2^0 (K_1^0) denotes the long-lived (short-lived) K^0 meson; a_K^0 is the scattering amplitude of the K^0 mesons, which is determined in the analysis of the scattering of K^+ mesons by nucleons. In what follows we assume that the amplitude a_K^0 is already known.

In reality the reactions (c) and (d) in the table are the same process. By studying the time dependences of the scattering cross section and of the polarization after scattering (i.e., the dependences on the distance to the target), one can determine the amplitudes of reactions (c) and (d) separately.

By measuring the differential cross sections and the polarization of the nucleons in reactions (a) — (d) as shown in the table, we can completely fix the scattering amplitudes a_K^0 and a_K^1. The experimental data on the cross sections and polarizations of the hyperons in reactions (e) — (h), together with the four relations of unitarity, enable us to determine the reaction amplitudes $a_{K\Sigma}^0$, $a_{K\Sigma}^1$, and $a_{K\Lambda}$, apart from a common phase factor.

Since the expressions for the cross sections and polarizations, and also the unitarity relations for reactions (1a) and (1b) are invariant under the replacement

$$a_{K\Sigma}^0 \to e^{i\delta_0(E)} a_{K\Sigma}^0, \quad a_{K\Sigma}^1 \to e^{i\delta_0(E)} a_{K\Sigma}^1,$$
$$a_{K\Lambda} \to e^{i\delta_1(E)} a_{K\Lambda}, \tag{6}$$

we cannot determine two phase factors $e^{i\delta_0}$ and $e^{i\delta_1}$, which are functions of the energy alone. This last follows from the relations that the amplitudes (4) and (5) satisfy by virtue of the unitary property of the S matrix.

Since the number of independent real functions involved in T^0 and T^1 is 18, and 16 of them are determined apart from two phase factors through the study of processes (1a) and (1b), for the complete reconstruction of the scattering amplitudes of pions by Λ and Σ hyperons in the states with isotopic spins 0 and 1, we need to determine in addition two more real functions of the energy and $\mathbf{n}\cdot\mathbf{n}'$, and two phase factors.

For each state with total angular momentum j and orbital angular momentum $l = j \pm \frac{1}{2}$, the T^0 matrix can be written in the form

$$-i\begin{pmatrix} \rho_K^0 \exp(2i\delta_K^0) - 1 & i\rho_{K\Sigma}^0 \exp(i\delta_{K\Sigma}^0) \\ i\rho_{K\Sigma}^0 \exp(i\delta_{K\Sigma}^0) & \rho_\Sigma^0 \exp(2i\delta_\Sigma^0) - 1 \end{pmatrix}, \tag{7}$$

where the ρ_α are certain positive functions of the energy, and the δ_α are the phases of the corresponding processes.

From the conditions for unitarity of the S matrix it follows that

$$\delta_{K\Sigma}^0 = \delta_K^0 + \delta_\Sigma^0, \qquad \rho_\Sigma^0 = \rho_K^0 = \{1 - (\rho_{K\Sigma}^0)^2\}^{1/2}. \tag{8}$$

The quantities ρ_K^0, δ_K^0, $\delta_{K\Sigma}^0$ can be determined apart from a common phase factor by studying processes (1a) and (1b). The quantities ρ_Σ^0 and δ_Σ^0 are then determined to the same accuracy from the relations (8).

Thus for π-Σ scattering the difference of the phases in the various states with zero isotopic spin is completely determined by the study of the reactions with K particles.*

For the states with isotopic spin (1) we have instead of the matrix (7)

$$-i\begin{pmatrix} \rho_K e^{2i\delta_K} - 1 & \rho_{K\Sigma} e^{i\delta_{K\Sigma}} & \rho_{K\Lambda} e^{i\delta_{K\Lambda}} \\ \rho_{K\Sigma} e^{i\delta_{K\Sigma}} & \rho_\Sigma e^{2i\delta_\Sigma} - 1 & \rho_{\Sigma\Lambda} e^{i\delta_{\Sigma\Lambda}} \\ \rho_{K\Lambda} e^{i\delta_{K\Lambda}} & \rho_{\Sigma\Lambda} e^{i\delta_{\Sigma\Lambda}} & \rho_\Lambda e^{2i\delta_\Lambda} - 1 \end{pmatrix}. \tag{9}$$

Here and in what follows, we shall write instead of $\rho_\alpha^1(\delta_\alpha^1)$ simply $\rho_\alpha(\delta_\alpha)$.

From the unitarity conditions we get

$$\rho_{K\Sigma}^2 + \rho_\Sigma^2 + \rho_{\Sigma\Lambda}^2 = 1, \quad \rho_{K\Lambda}^2 + \rho_{\Sigma\Lambda}^2 + \rho_\Lambda^2 = 1, \tag{10}$$

$$\cos(2\delta_\Sigma + 2\delta_K - 2\delta_{K\Sigma}) = \frac{(\rho_K\rho_{K\Sigma})^2 + (\rho_{K\Sigma}\rho_\Sigma)^2 - (\rho_{K\Lambda}\rho_{\Sigma\Lambda})^2}{2\rho_\Sigma\rho_K(\rho_{K\Sigma})^2}, \tag{11}$$

$$\cos(\delta_{\Sigma\Lambda} + 2\delta_K - \delta_{K\Lambda} - \delta_{K\Sigma}) = \frac{(\rho_{K\Lambda}\rho_{\Sigma\Lambda})^2 + (\rho_K\rho_{K\Sigma})^2 - (\rho_{K\Sigma}\rho_\Sigma)^2}{2\rho_{K\Lambda}\rho_{\Sigma\Lambda}\rho_K\rho_{K\Sigma}}, \tag{12}$$

$$\cos(2\delta_\Lambda + 2\delta_K - 2\delta_{K\Lambda}) = \frac{(\rho_K\rho_{K\Lambda})^2 + (\rho_{K\Lambda}\rho_\Lambda)^2 - (\rho_{K\Sigma}\rho_{\Sigma\Lambda})^2}{2\rho_\Lambda\rho_K(\rho_{K\Lambda})^2}. \tag{13}$$

---

*It may turn out that in carrying out an unambiguous analysis it will be helpful to take into account Coulomb effects and the energy dependence of the S matrix at low energies.

We note that the Minami ambiguity exists for the reactions in question.

Some possibilities for determining the parity of the K meson relative to the hyperons through the analysis of the reactions (1) have recently been discussed by Amati and Vitale.[2]

It is easy to convince oneself that even when ρ_K, $\rho_{K\Sigma}$, $\rho_{K\Lambda}$, δ_K, $\delta_{K\Sigma}$, and $\delta_{K\Lambda}$ are known, the unitarity relations (10) — (13) are insufficient for the reconstruction of the matrix T^1. For this we need to know one more parameter in each state (for example, ρ_Σ).

We note that the relations (10) — (13) lead to some interesting inequalities. Noting that $\rho_\alpha > 0$ and $|\cos\theta| < 1$, we get from Eqs. (10) and (11)

$$0 < \rho_\Sigma^2 = 1 - \rho_{K\Sigma}^2 - \rho_{\Sigma\Lambda}^2 < 1 - \rho_{K\Sigma}^2, \tag{14}$$

$$|(\rho_K\rho_{K\Sigma})^2 + (\rho_{K\Sigma}\rho_\Sigma)^2 - \rho_{K\Lambda}^2(1 - \rho_{K\Sigma}^2 - \rho_\Sigma^2)| < 2\rho_\Sigma\rho_K(\rho_{K\Sigma})^2. \tag{15}$$

Let us introduce the new notations

$$\rho_{K\Sigma}^2 + \rho_{K\Lambda}^2 = a, \quad \rho_K\rho_{K\Sigma}^2 = b, \quad (\rho_K\rho_{K\Sigma})^2 - \rho_{K\Lambda}^2(1 - \rho_{K\Sigma}^2) = c.$$

Then Eq. (15) can be rewritten in the form

$$|a\rho_\Sigma^2 + c| < 2b\rho_\Sigma. \tag{15'}$$

From Eq. (15′) together with Eq. (14) we get

$$\max\left\{0;\ \frac{b}{a} - \frac{1}{a}\sqrt{b^2 - ac}\right\} < \rho_\Sigma < \min\left\{\sqrt{1 - \rho_{K\Sigma}^2};\ \frac{b}{a} + \frac{1}{a}\sqrt{b^2 - ac}\right\}. \tag{16}$$

The inequality (15′) holds only for $b^2 - ac \geq 0$. Consequently, the observable quantities ρ_K, $\rho_{K\Lambda}$, and $\rho_{K\Sigma}$ must satisfy this inequality. Similarly we have

$$\max\left\{0;\ \frac{b_1}{a_1} - \frac{1}{a_1}\sqrt{b_1^2 - a_1c_1}\right\} < \rho_{\Sigma\Lambda} < \min\left\{\sqrt{1 - \rho_{K\Sigma}^2};\ \sqrt{1 - \rho_{K\Lambda}^2};\ \frac{b_1}{a_1} + \frac{1}{a_1}\sqrt{b_1^2 - a_1c_1}\right\}, \tag{17}$$

$$\max\left\{0;\ \frac{b_2}{a_2} - \frac{1}{a_2}\sqrt{b_2^2 - a_2c_2}\right\} < \rho_\Lambda < \min\left\{\sqrt{1 - \rho_{K\Lambda}^2};\ \frac{b_2}{a_2} + \frac{1}{a_2}\sqrt{b_2^2 - a_2c_2}\right\}, \tag{18}$$

where

$$a_1 \equiv \rho_{K\Lambda}^2 + \rho_{K\Sigma}^2 = a, \qquad b_1 \equiv \rho_{K\Lambda}\rho_K\rho_{K\Sigma},$$
$$a_2 \equiv \rho_{K\Lambda}^2 + \rho_{K\Sigma}^2 = a, \qquad b_2 \equiv \rho_K(\rho_{K\Lambda})^2,$$
$$c_1 \equiv (\rho_K\rho_{K\Sigma})^2 - \rho_{K\Sigma}^2(1 - \rho_{K\Sigma}^2),$$
$$c_2 \equiv (\rho_\Lambda\rho_{K\Lambda})^2 - \rho_{K\Sigma}^2(1 - \rho_{K\Lambda}^2).$$

2. Recently Chew and Low, and Okun' and Pomeranchuk,[3] have suggested that peripheral collisions be studied as a method for determining interactions between unstable particles. We shall assume that this method can be used for the determination of the scattering amplitude for $\Sigma(\Lambda) + \pi \to \Sigma(\Lambda) + \pi$ through studies of the processes $\Sigma + N \to \Sigma(\Lambda) + N + \pi$ and $\Lambda + N \to \Sigma + N + \pi$. The key point of the method is that the amplitude for the reaction $\Sigma + N \to \Sigma(\Lambda) + N + \pi$, regarded as a function of $(p_N' - p_N)^2$, where p_N (p_N') is the four-vector momentum of the nucleon in the initial (final) state, has a pole in the nonphysical region, $(p_N' - p_N)^2 = \mu^2$ (μ is the mass of the π meson). It is shown that the virtual process

$$N \to N + \pi, \quad \Sigma(\Lambda) + \pi \to \Sigma(\Lambda) + \pi \tag{19}$$

corresponds to the pole term, whose residue is proportional to the amplitude for $\pi - \Lambda(\Sigma)$ scattering. Assuming that in the physical region near the pole the reaction $\Sigma + N \to \Sigma(\Lambda) + N + \pi$ is determined by the process (19), one can extrapolate its amplitude into the nonphysical region and separate out the residue of the pole term.

To estimate the effect of other terms in the physical region near the pole, we shall formulate certain rules that must be fulfilled if the contribution of the pole term in actually predominant in this region.

a) In the region near the pole the amplitude of the reaction $\Sigma^+ + p \to \Sigma^+ + p + \pi^0$ is equal to that of the reaction $\Sigma^- + p \to \Sigma^- + p + \pi^0$. This rule follows from the invariance of the virtual process

$$\pi^0 + \Sigma^\pm \to \pi^0 + \Sigma^\pm \tag{20}$$

under rotations in the isotopic space. Similarly it can be shown that the amplitudes for the following pairs of processes are equal:

$\Sigma^+ + p \to \Sigma^0 + p + \pi^+$ and $\Sigma^- + p \to \Sigma^0 + p + \pi^-$,
$\Lambda + p \to \Sigma^+ + n + \pi^0$ and $\Lambda + p \to \Sigma^0 + n + \pi^+$,
$\pi^+ + p \to \pi^+ + \pi^0 + p$ and $\pi^- + p \to \pi^- + \pi^0 + p$ etc.

b) Near the pole the amplitudes for the reactions

$$\Sigma^\pm(\Lambda) + p \to \Sigma^\pm(\Lambda) + p + \pi^0$$
$$\text{and}\ \tilde{\Sigma}^\pm(\tilde{\Lambda}) + p \to \tilde{\Sigma}^\pm(\tilde{\Lambda}) + p + \pi^0$$

are equal to each other. This rule follows from the invariance of the virtual process (20) under charge conjugation. In our case it is not of any great practical importance, but it can be of interest in other cases. For example, it can be shown that the amplitudes for the reactions $K^\pm + N \to K^\pm + N + \pi^0$ are equal. This equality is useful for the determination of the interaction of K mesons with π mesons, and has been noted by Okun' and Pomeranchuk.

c) If the nucleons are unpolarized in the initial state, then they remain unpolarized in the final state also.

Let us consider a reaction of the type

$$\Sigma^+ + p \to \Lambda + p + \pi^+ \tag{21}$$

in the region near the pole $(p_\Sigma - p_\Lambda)^2 = \mu^2$. We assume that the dominant process in this region is

$$\Sigma^+ \to \Lambda + \pi^+, \quad \pi^+ + p \to \pi^+ + p, \tag{22}$$

the amplitude for which is proportional to

$$\frac{\bar{u}(p_\Lambda)\Gamma u(p_\Sigma)}{(p_\Lambda - p_\Sigma)^2 - \mu^2} a_{\pi p}, \tag{23}$$

where $\Gamma = 1$ if the relative parity Π of the Σ and Λ particles is -1, and $\Gamma = \gamma_5$ if $\Pi = +1$; $a_{\pi p}$ is the amplitude for the scattering $\pi^+ + p \to \pi^+ + p$. The amplitude for the process (22) does not contain any dependence on the spin operators of the hyperons for $\Pi = -1$ (more exactly, it contains a term proportional to σ_Y, but with a small coefficient); on the other hand, if $\Pi = +1$, the amplitude is proportional to $\sigma_Y \cdot \mathbf{k}$, where $\mathbf{k}$ is the unit vector parallel to the difference

$$\mathbf{P}_\Sigma / (E_\Sigma + M_\Sigma) - \mathbf{P}_\Lambda / (E_\Lambda + M_\Lambda).$$

If in the initial state the Σ^+ is polarized (polarization vector $\mathbf{P}$), then it can be shown from Eq. (23) that in the final state the polarization vector $\mathbf{P}'$ of the Λ particle is given by

$$\mathbf{P}' = \mathbf{P} \text{ for } \Pi = -1,$$
$$\mathbf{P}' = 2(\mathbf{P}\mathbf{k})\mathbf{k} - \mathbf{P} \text{ for } \Pi = +1. \tag{24}$$

Thus if in the region where the pole term predominates one could measure the polarization of the Λ particles produced in the reaction (21) with polarized Σ, then it would be possible not only to evaluate the effect of the non-pole terms, but also to get information about the relative parity of the Λ and Σ hyperons.

In a number of cases the study of the polarization of the products from peripheral collisions can be a source of information about the parities of unstable particles.*

The writers express their deep gratitude to Professor M. A. Markov for helpful discussions.

[1] Bilen'kiĭ, Lapidus, Puzikov, and Ryndin, J. Exptl. Theoret. Phys. (U.S.S.R.) **35**, 959 (1958), Soviet Phys. JETP **8**, 669 (1959); Nucl. Phys. **7**, 646 (1958).

[2] D. Amati and B. Vitale, Nuovo cimento **9**, 895 (1958).

[3] G. F. Chew, Proc. CERN Annual Conference, 1958, Geneva, page 97; preprint, 1958. L. B. Okun' and I. Ya. Pomeranchuk, J. Exptl. Theoret. Phys. (U.S.S.R.) **36**, 300 (1959), Soviet Phys. JETP **9**, 207 (1959). G. F. Chew and F. E. Low, preprint, 1958.

[4] J. G. Taylor, Nucl. Phys. **9**, 357 (1959); preprint, 1959.

Translated by W. H. Furry
41

*The possibility of determining the parities of particles by the study of peripheral collisions without considering the polarization has been discussed recently by Taylor.[4]

ELECTROMAGNETIC MASS OF THE K MESON

CHOU KUANG-CHAO and V. I. OGIEVETSKIĬ

Joint Institute for Nuclear Research

Submitted to JETP editor May 28, 1959

J. Exptl. Theoret. Phys. (U.S.S.R.) **37**, 866-867 (September, 1959)

IN the recent experiments by Rosenfeld et al.[1] and Crawford et al.[2] it was established that the mass of the neutral K meson exceeds that of the charged K^+ meson by ~ 4.8 Mev. On the face of it the sign of this mass difference appears to contradict the concept that the K^+ and K^0 mesons are spinless particles belonging to the same charge doublet. Indeed, if the K^0 meson has no electromagnetic interactions and the mass difference is of electromagnetic origin then the electromagnetic self-mass of the charged K meson should make it heavier than the neutral one (see, e.g., reference 3). On this basis the above-mentioned authors are inclined to interpret their results as an argument in favor of the Pais hypothesis,[4] according to which the K^+ and K^0 meson do not form a charge doublet and may have different intrinsic parities.

It is shown below that there is not as yet sufficient basis for this conclusion since the mass difference can be explained, within the framework of the Gell-Mann–Nishijima multiplet scheme, by the electromagnetic interactions of the K^0 meson. Indeed, as noted in an interesting paper by G. Feinberg,[5] a spinless neutral particle which is different from its antiparticle, e.g., K^0, can interact with the electromagnetic field. This interaction results from a virtual dissociation of the K^0 meson into strongly interacting particles, for example a nucleon and an antihyperon. As a consequence the K^0 meson will have an electromagnetic structure.

In the general case the gauge-invariant electromagnetic interaction Lagrangian can be written as

$$L = -j_\mu(x) A_\mu(x) \tag{1}$$

where $j_\mu(x)$ is the operator for the total current of all interacting particles. In the β-formalism of Duffin and Kemmer the matrix element of the current taken between single K-meson states will have the form*

$$\langle p' | j_\mu(x) | p \rangle_K = -ie(2\pi)^{-3} e^{-iqx} \bar{v}(p') \beta_\mu [F_{1K}(q^2) + \tau_3 F_{2K}(q^2)] v(p),$$

$$q = p' - p, \quad \bar{v}(p') = v^+(p')(2\beta_4^2 - 1), \tag{2}$$

where p′ and p are the K-meson four-momenta in the final and initial states, $v(p')$ and $v(p)$ are the corresponding wave functions in the β-formalism, and $F(q^2)$ is the form factor satisfying

$$F_{K^+}(q^2) = F_{1K}(q^2) + F_{2K}(q^2), \quad F_{K^+}(0) = 1,$$
$$F_{K^0}(q^2) = F_{1K}(q^2) - F_{2K}(q^2), \quad F_{K^0}(0) = 0, \tag{3}$$

since the charge of the particle is $eF(0)$.

Due to interaction (1) and by taking into account (2) we find for the self-mass of the K meson†

$$\Delta m = \frac{ie}{(2\pi)^4} \frac{\bar{v}(p)}{\bar{v}v} \beta_\nu \int d^4q$$
$$\times \frac{i(\hat{p} - \hat{q}) + (\hat{p} - \hat{q})^2/2m - [(p-q)^2 + m^2]/m}{[(p-q)^2 + m^2]\, q^2} \beta_\nu [F_K(q^2)]^2 v(p),$$
$$\hat{q} = \beta_\mu q_\mu \tag{4}$$

or

$$\Delta m = \frac{ie^2}{2(2\pi)^4 m} \int d^4q \frac{[F_K(q^2)]^2}{q^2} \left\{ \frac{(2p-q)^2}{(p-q)^2 + m} - 4 \right\}. \tag{5}$$

The $F_K(q^2)$ as a function of q^2 can be determined only from an as yet nonexistent exact theory or a full analysis of future experiments. For our purposes it is sufficient to take, for example,

$$F_{K^+}(q^2) = 16m^4/(q^2 + 4m^2)^2,$$
$$F_{K^0}(q^2) = -4\lambda q^2 m^2/(q^2 + 4m^2)^2 \tag{6}$$

then from (5) and (6) we obtain for the mass difference

$$m_{K^0} - m_{K^+} = (m/8\pi^2) e^2 ({}^7/_3 \lambda^2 - 1)$$
$$= (m/2\pi) \alpha ({}^7/_3 \lambda^2 - 1). \tag{7}$$

Comparing with the experimental value of 4.8 Mev we deduce that $\lambda \approx 2$.

We note that it will be difficult to observe experimentally other effects due to the interaction under consideration.‡

Consequently it is not necessary to give up the idea that K^+ and K^0 form a charge doublet in order to explain the observed[1,2] mass difference. Both the sign and the magnitude of the difference $m_{K^0} - m_{K^+}$ could be a consequence of electromagnetic interactions.

---

*As remarked by Feinberg in the case of the π^0 meson, which is a truly neutral particle, such a matrix element would vanish as a consequence of invariance under charge conjugation.

†Expression (5) for the self-mass may also be derived from the usual theory in which the K mesons are described by second order wave equations and the electromagnetic interaction is introduced in a gauge-invariant manner by the substitution: $\partial/\partial x_\mu \to \partial/\partial x_\mu - ieF(-\Box^2) A_\mu(x)$.

‡The absence of bremsstrahlung and the difficulties involved in separating the electromagnetic and nuclear scattering for K^0 were discussed by Feinberg.[5] The most characteristic experiment would involve observation of fast δ electrons from K^0 mesons. However the K^0–e scattering cross section is very small at low energies. Consequently the effect will be vanishingly small since even a 1-Bev K meson in the laboratory system will have an energy of the order of only a few Mev in the K^0–e center-of-mass system.

---

[1] Rosenfeld, Solmitz, and Tripp, Phys. Rev. Lett. **2**, 110 (1959).

[2] Crawford, Cresti, Good, Stevenson, and Ticho, Phys. Rev. Lett. **2**, 112 (1959).

[3] S. Gasiorowicz and A. Petermann, Phys. Rev. Lett. **1**, 457 (1958).

[4] A. Pais, Phys. Rev. **112**, 624 (1958).

[5] G. Feinberg, Phys. Rev. **109**, 1381 (1958).

Translated by A. M. Bincer
166

SOVIET PHYSICS JETP VOLUME 37 (10), NUMBER 6 JUNE, 1960

DISPERSION RELATIONS FOR THE SCATTERING OF γ QUANTA BY NUCLEONS

L. I. LAPIDUS and CHOU KUANG-CHAO

Joint Institute of Nuclear Research

Submitted to JETP editor July 1, 1959

J. Exptl. Theoret. Phys. (U.S.S.R.) 37, 1714-1721 (December, 1959)

Dispersion relations for the scattering of γ quanta by nucleons with one subtraction are considered. For forward scattering six relations have been obtained which do not contain unknown constants or infrared divergencies.

1. Dispersion relations for the scattering of γ quanta by nucleons have been considered by a number of authors.[1-6] The presence of infrared divergencies, however, limits the applicability of these relations for the analysis of the experimental data.

In the present paper we derive dispersion relations in a form which is convenient for practical application. In a forthcoming paper[7] we shall use these relations in the discussion of the scattering of γ quanta near the threshold for meson production.

Before discussing the dispersion relations for the scattering of γ quanta by nucleons, we consider in somewhat more detail the kinematics of the process. Let k and k′ be the momentum four-vectors of the incident and the scattered photon, and p and p′, those of the incident and scattered nucleons. These quantities are related by the conservation law

$$k + p = k' + p'. \tag{1}$$

We introduce the notation

$$P = \tfrac{1}{2}(p + p').$$

Following Prange,[8] we choose the following four orthogonal vectors as basis vectors:

$$P' = P - (PK)K/K^2,\quad K = \tfrac{1}{2}(k + k'),$$
$$Q = \tfrac{1}{2}(k - k'),\quad N_\mu = i\varepsilon_{\mu\nu\lambda\sigma} P'_\lambda K_\nu Q_\sigma. \tag{2}$$

The scattering amplitude can be written in the form

$$\mathscr{M} = \bar{u}(p')\, e'_\mu N_{\mu\nu} e_\nu\, u(p), \tag{3}$$

$N_{\mu\nu}$ can be expressed in terms of invariant functions,

$$N_{\mu\nu} = \sum_{\sigma\sigma'} \eta^\sigma_\mu C_{\sigma\sigma'} \eta^{\sigma'}_\nu, \tag{4}$$

where η^σ are the four basis vectors of (2).

Gauge invariance requires that $e'k' = ek = 0$, $k'_\mu N_{\mu\nu} = 0$, and $N_{\mu\nu}k_\nu = 0$. As a consequence, $N_{\mu\nu}$ consists of a sum of eight invariant functions, $\mathscr{M}_1, \ldots, \mathscr{M}_8$, of the two invariants $M\nu = -PK$ and Q^2:

$$\begin{aligned} e'_\mu N_{\mu\nu} e_\nu = {} & \frac{Q^2}{M^2\nu^2 - Q^2(Q^2+M^2)}(e'P')(eP')[\mathscr{M}_1 + i\hat{k}\mathscr{M}_2] \\ & + \frac{1}{Q^2[M^2\nu^2 - Q^2(Q^2+M^2)]}(e'N)(eN)[\mathscr{M}_3 + i\hat{k}\mathscr{M}_4] \\ & - \frac{i}{M^2\nu^2 - Q^2(Q^2+M^2)}[(e'P')(eN) \\ & - (e'N)(eP')]\gamma_5[\mathscr{M}_5 + i\hat{k}\mathscr{M}_7] - \frac{i}{M^2\nu^2 - Q^2(Q^2+M^2)} \\ & \times [(e'P')(eN) + (e'N)(eP')]\gamma_5[\mathscr{M}_8 + i\hat{k}\mathscr{M}_6]. \end{aligned} \tag{5}$$

The normalization factors $Q^2/[M^2\nu^2 - Q^2(Q^2 + M^2)]$ have been introduced for convenience.

It can be shown that the following relations hold in an arbitrary system (thus, in particular, in the Breit system and in the center of mass system):

$$\frac{Q^2(e'P')(eP')}{M^2\nu^2 - Q^2(M^2+Q^2)} = \frac{(\mathbf{e}'\boldsymbol{\varkappa})(\mathbf{e}\boldsymbol{\varkappa}')}{|\boldsymbol{\varkappa}||\boldsymbol{\varkappa}'|\sin^2\theta} = \frac{(\mathbf{e}'\mathbf{k})(\mathbf{e}\mathbf{k}')}{\sin^2\theta},$$

$$\frac{(e'N)(eN)}{M^2\nu^2 - Q^2(M^2+Q^2)} = \frac{(\mathbf{e}'[\mathbf{k}\times\mathbf{k}'])(\mathbf{e}[\mathbf{k}\times\mathbf{k}'])}{\sin^2\theta} = \frac{(\mathbf{e}'\boldsymbol{\rho})(\mathbf{e}\boldsymbol{\rho})}{\sin^2\theta},$$

$$\frac{-[(e'P')(eN) \mp (e'N)(eP')]}{M^2\nu^2 - Q^2(Q^2+M^2)} = \frac{(\mathbf{e}'\mathbf{k})(\mathbf{e}\boldsymbol{\rho}) \mp (\mathbf{e}'\boldsymbol{\rho})(\mathbf{e}\mathbf{k}')}{\sin^2\theta}, \tag{6}$$

where $\boldsymbol{\kappa} = |\kappa|\mathbf{k}$; θ is the angle between $\mathbf{k}$ and $\mathbf{k}'$; $\mathbf{k}$ and $\mathbf{k}'$ are unit vectors along $\boldsymbol{\kappa}$ and $\boldsymbol{\kappa}'$: $\boldsymbol{\rho} = \mathbf{k} \times \mathbf{k}'$. We prove (6) in the Breit system, where

$$\mathbf{P} = 0, \qquad \mathbf{P}' = -(PK)\mathbf{K}/K^2 = -M\nu\mathbf{K}/Q^2. \tag{7}$$

The following formulas are easily seen to be correct:

$$|\varkappa| = k_0 = M\nu/\sqrt{Q^2 + M^2}, \tag{8}$$

$$2Q^2 = k_0^2(1 - \cos\theta), \tag{9}$$

$$k_0^2 - Q^2 = \frac{M^2\nu^2 - Q^2(Q^2+M^2)}{Q^2 + M^2} = \frac{k_0^2}{2}(1 + \cos\theta). \tag{10}$$

Multiplying (9) and (10), we obtain

$$\frac{k_0^4}{4}\sin^2\theta = \frac{Q^2[M^2\nu^2 - Q^2(Q^2+M^2)]}{Q^2 + M^2}. \tag{11}$$

With the help of (8) we write (11) in the form

$$k_0^2 \sin^2\theta = 4Q^2 [M^2\nu^2 - Q^2 (Q^2 + M^2)] / M^2\nu^2. \quad (12)$$

From (5) and (7) we obtain, finally,

$$\frac{Q^2}{M^2\nu^2 - Q^2 (Q^2 + M^2)} \frac{M^2\nu^2}{Q^4} \frac{1}{4} (\mathbf{e}'\varkappa)(\mathbf{e}\varkappa') = \frac{(\mathbf{e}'\varkappa)(\mathbf{e}\varkappa')}{k_0^2 \sin^2\theta} = \frac{(\mathbf{e}'\mathbf{k})(\mathbf{e}\mathbf{k}')}{\sin^2\theta}. \quad (13)$$

Using the formula

$$N = -\sqrt{Q^2 + M^2}\, \frac{1}{2} [\mathbf{k}\times\mathbf{k}'], \quad (14)$$

the remaining equations of (6) can be proved in an analogous fashion. As is known, the requirement of invariance under time reversal (or, as will be shown below, the requirement of "crossing symmetry") reduces the number of independent invariant functions to six ($\mathcal{M}_7 = \mathcal{M}_8 = 0$).

The general expression for the amplitude for the scattering of γ quanta by particles with spin $\frac{1}{2}$ can be written in the form[9-11]

$$\mathcal{M} = R_{1c} (\mathbf{e}\mathbf{e}') + R_{2c} (\mathbf{s}_c \mathbf{s}_c') + iR_{3c} (\sigma [\mathbf{e}'\times \mathbf{e}]) + iR_{4c} (\sigma [\mathbf{s}_c' \times \mathbf{s}_c]) + iR_{5c} [(\sigma\mathbf{k}_c)(\mathbf{s}_c'\mathbf{e}) - (\sigma\mathbf{k}_c')(\mathbf{s}_c\mathbf{e}')] + iR_{6c} [(\sigma\mathbf{k}_c')(\mathbf{s}_c'\mathbf{e}) - (\sigma\mathbf{k}_c)(\mathbf{e}'\mathbf{s}_c)], \quad (15)$$

where R_1, R_3, and R_5 describe the electric, and R_2, R_4, and R_6 describe the magnetic transitions; $\mathbf{e}$ and $\mathbf{e}'$ are the polarization vectors before and after the collision; $\mathbf{s} = \mathbf{k} \times \mathbf{e}$, $\mathbf{s}' = \mathbf{k}' \times \mathbf{e}'$, where $\mathbf{k}$ and $\mathbf{k}'$ are unit vectors in the direction of the momentum of the γ quantum before and after the scattering; the label "c" refers to the center-of-mass system.

The expression for $\mathcal{M}$ containing the terms not higher than those linear in the energy of the γ quanta[12,13] can be written in the form

$$\mathcal{M} = -e^2M^{-1} (\mathbf{e}\mathbf{e}') + ieM^{-1} (2\mu - e/2M)\, \nu_c (\sigma [\mathbf{e}'\times \mathbf{e}]) + 2\mu^2\nu_c (\sigma [\mathbf{s}_c\times \mathbf{s}_c']) + ieM^{-1}\mu\nu_c [(\mathbf{e}\mathbf{k}_c')(\sigma\mathbf{s}_c') - (\mathbf{e}'\mathbf{k}_c)(\sigma\mathbf{s}_c)]. \quad (16)$$

With the help of the relation

$$(\sigma\mathbf{s}')(\mathbf{e}\mathbf{k}') - (\sigma\mathbf{s})(\mathbf{e}'\mathbf{k}) = -2(\sigma[\mathbf{e}'\times\mathbf{e}]) + (\sigma\mathbf{k}')(\mathbf{e}\mathbf{s}') - (\sigma\mathbf{k})(\mathbf{e}'\mathbf{s}), \quad (17)$$

we can bring (16) into the form (15). Here

$$R_1^0 = -e^2/M, \; R_2^0 = 0, \; R_3^0 = -2(e/2M)^2\nu_c, \; R_4^0 = -2\mu^2\nu_c, \; R_5^0 = 0, \; R_6^0 = (e/M)\,\mu\nu_c. \quad (18)$$

We write the scattering amplitude in the Breit system in the form (15), and obtain from a comparison of (15) and (5) the relations

$$R_1 \sin^2\theta = E(\mathcal{M}_1 \cos\theta + \mathcal{M}_3)/M - k_0 (\mathcal{M}_2 \cos\theta + \mathcal{M}_4),$$

$$R_2 \sin^2\theta = -E(\mathcal{M}_3 \cos\theta + \mathcal{M}_1)/M + k_0 (\mathcal{M}_2 + \mathcal{M}_4 \cos\theta),$$

$$R_3 = k_0^2\mathcal{M}_2/2M, \quad R_4 = -k_0^2\mathcal{M}_4/2M,$$

$$R_5 \sin^2\theta = \frac{k_0^2}{2M}(\mathcal{M}_2 \cos\theta + \mathcal{M}_4) - \frac{k_0}{2M}(1 + \cos\theta)\,\mathcal{M}_5 - \frac{k_0E}{2M}(1 - \cos\theta)\,\mathcal{M}_6,$$

$$R_6 \sin^2\theta = -\frac{k_0^2}{2M}(\mathcal{M}_2 + \mathcal{M}_4 \cos\theta) + \frac{k_0}{2M}(1 + \cos\theta) \times \mathcal{M}_5 - \frac{k_0E}{2M}(1 - \cos\theta)\,\mathcal{M}_6. \quad (19)$$

It is easily seen that for $\theta \to 0$ we obtain ($E = M$, $k_0 = \nu$)

$$R_1 + R_2|_{\theta=0} = \tfrac{1}{2}[\mathcal{M}_1 - \mathcal{M}_3 - \nu(\mathcal{M}_2 - \mathcal{M}_4)],$$

$$R_3|_{\theta=0} = \nu^2\mathcal{M}_2/2M,$$

$$R_4|_{\theta=0} = -\nu^2\mathcal{M}_4/2M,$$

$$R_5 + R_6|_{\theta=0} = [\nu^2(\mathcal{M}_2 - \mathcal{M}_4) + M\nu\mathcal{M}_6]/4M. \quad (20)$$

It can be shown[1-4] from the general principles of quantum field theory that the $\mathcal{M}_i$, as functions of ν, are analytic for $\theta = 0$ ($Q^2 = 0$) and for

$$Q^2 < Q_{max}^2 = \frac{(2M + m_\pi)(6M^2 + 9Mm_\pi + 4m_\pi^2)}{4M(M + m_\pi)^2} m_\pi^2 \approx 3m_\pi^2,$$

where m_π is the mass of the π meson.

Regarding the forward scattering amplitude, one usually restricts oneself to two dispersion relations for the functions $R_1 + R_2$ and $R_3 + R_4 + 2R_5 + 2R_6$. It is seen from formulas (20) that for $\theta = 0$ we actually have four dispersion relations for $R_1 + R_2$, R_3, R_4, and $R_5 + R_6$ separately.

2. The retarded causal amplitude for the scattering of photons can be written in the form

$$\bar{u}(p')\, N_{\mu\nu}^{ret}\, u(p) = -2\pi^2 i (p_0 p_0'/M^2)^{1/2} \times \int d^4z e^{-ikz} \left\langle p' \left| \theta(z_0) \left[j_\mu\left(\frac{z}{2}\right), \left[j_\nu\left(-\frac{z}{2}\right)\right]\right| p \right\rangle. \quad (21)$$

Analogously, we can write for the advanced causal amplitude

$$\bar{u}(p')\, N_{\mu\nu}^{adv} u(p) = -(2\pi^2)\, i\, (p_0 p_0'/M^2)^{1/2} \times \int d^4z e^{-ikz} \left\langle p' \left| \theta(-z_0) \left[j_\mu\left(\frac{z}{2}\right), j_\nu\left(-\frac{z}{2}\right)\right]\right| p \right\rangle. \quad (22)$$

We define the quantities $D_{\mu\nu}$ and $A_{\mu\nu}$ by

$$D_{\mu\nu} = (N_{\mu\nu}^{ret} + N_{\mu\nu}^{adv})/2, \quad A_{\mu\nu} = (N_{\mu\nu}^{ret} - N_{\mu\nu}^{adv})/2i. \quad (23)$$

Taking the complex conjugate on both sides of (21) and recalling that j_μ is a Hermitian operator, we get

$$\beta N_{\mu\nu}^{+ret}(p'k'pk)\,\beta = N_{\mu\nu}^{ret}(p - k'p' - k). \quad (24)$$

Changing the order of j_μ and j_ν in the commutator in (21) and replacing the integration variable z by $-z$, we obtain

$$N_{\mu\nu}^{ret}(p'k'pk) = N_{\mu\nu}^{adv}(p' - kp - k'). \quad (25)$$

Substituting (5) in (24) and (25), we have

$$\mathcal{M}_{1,3}^*(\nu, Q^2) = \mathcal{M}_{1,3}(-\nu, Q^2), \quad \mathcal{M}_{2,4}^*(\nu, Q^2) = -\mathcal{M}_{2,4}(-\nu, Q^2),$$

$$\mathcal{M}_{5,6}^*(\nu, Q^2) = \mathcal{M}_{5,6}(-\nu, Q^2), \quad \mathcal{M}_7 = \mathcal{M}_8 = 0. \quad (26)$$

With the help of (26) the dispersion relations can be easily written in the form

$$D_{1,3,5,6}(\nu, Q^2) = \frac{2}{\pi} \mathrm{P}\int_0^\infty \frac{\nu' A_{1,3,5,6}(\nu', Q^2)}{\nu'^2 - \nu^2} d\nu',$$

$$D_{2,4}(\nu, Q^2) = \frac{2\nu}{\pi} \mathrm{P}\int_0^\infty \frac{A_{2,4}(\nu', Q^2)}{\nu'^2 - \nu^2} d\nu'. \quad (27)$$

Let us consider the dispersion relations for $Q^2 = 0$, when the Breit system coincides with the laboratory system and ν becomes the energy of the γ quantum in the laboratory system (l.s.). We obtain two dispersion relations by simply setting $Q^2 = 0$ in (27). Four more relations follow if we first differentiate with respect to Q^2 and then set $Q^2 = 0$. Since the dispersion relations for $\mathscr{M}_{2,4}$ in the e^2 approximation contain infrared divergencies of the type $1/\nu - 1/\nu_0$, we find that for $Q^2 = 0$ the only dispersion relations of practical use are those for the combinations

$$R_1 + R_2 \equiv L_1, \quad R_5 + R_6 \equiv L_2,$$

and for the quantities

$$R_3 \equiv L_3, \quad R_4 \equiv L_4,$$

which do not contain terms which become infinite for $\nu \to 0$. For $Q^2 = 0$ we have $L = L(\nu)$, and it can be shown by comparing (20) and (26) that

$$L_1(-\nu) = L_1(\nu), \quad L_{2,3,4}(-\nu) = -L_{2,3,4}(\nu). \quad (28)$$

We can therefore write down the following dispersion relations for these quantities

$$\operatorname{Re} L_1(\nu) - \operatorname{Re} L_1(0) = \frac{2\nu^2}{\pi} \mathrm{P} \int_{\nu_0}^{\infty} \frac{\operatorname{Im} L_1(\nu')}{\nu'(\nu'^2 - \nu^2)} d\nu',$$

$$\operatorname{Re} L_{2,3,4}(\nu) - \nu \operatorname{Re} L'_{2,3,4}(0) = \frac{2\nu^3}{\pi} \mathrm{P} \int_{0}^{\infty} \frac{\operatorname{Im} L_{2,3,4}(\nu')}{\nu'^2(\nu'^2 - \nu^2)} d\nu',$$

$$\nu_0 = m_\pi (1 + m_\pi / 2M). \quad (29)$$

We shall derive two more dispersion relations in the following section.

From (18) we find

$$\operatorname{Re} L_1(0) = -e^2/M, \quad \nu \operatorname{Re} L'_2(0) = \frac{e}{M}\mu\nu = \frac{e^2}{M}\frac{1}{2}\frac{\nu_0}{M}\lambda\frac{\nu}{\nu_0},$$

$$\mu = \frac{e}{2M}\lambda, \quad \nu \operatorname{Re} L'_3(0) = -2\left(\frac{e}{2M}\right)^2 \nu = -\frac{1}{2}\frac{e^2}{M}\frac{\nu_0}{M}\frac{\nu}{\nu_0},$$

$$\nu \operatorname{Re} L'_4(0) = -2\mu^2\nu = -\frac{1}{2}\frac{e^2}{M}\lambda^2\frac{\nu_0}{M}\frac{\nu}{\nu_0}. \quad (30)$$

The first of the relations (29) is brought into the usual form by using the optical theorem:

$$\operatorname{Re}[R_1(\nu) + R_2(\nu)] = -\frac{e^2}{M} + \frac{\nu^2}{2\pi^2} \mathrm{P} \int_{\nu_0}^{\infty} \frac{\sigma(\nu')\, d\nu'}{\nu'^2 - \nu^2}; \quad (29')$$

it then coincides with the relation obtained by Gell-Mann, Goldberger, and Thirring.[1]

3. In the center-of-mass system the quantities $R_{1c}, \ldots, R_{6c}$ can be expressed in terms of the scattering amplitudes in states with definite angular momentum and parity.

Let us denote the amplitudes for the electric dipole transitions with total angular momentum $1/2$ and $3/2$ by $\mathscr{E}_1$ and $\mathscr{E}_3$, respectively; let $\mathscr{E}_2$ be the amplitude for the electric quadrupole transition with total angular momentum $3/2$, and likewise $\mathfrak{M}_1$, $\mathfrak{M}_3$, and $\mathfrak{M}_2$ the amplitudes for the magnetic dipole and quadrupole transitions. We must further introduce the amplitudes $C'(\mathfrak{M}_3, \mathscr{E}_2)$, $C'(\mathscr{E}_2, \mathfrak{M}_3)$, $C'(\mathscr{E}_3, \mathfrak{M}_2)$, and $C'(\mathfrak{M}_2, \mathscr{E}_3)$ corresponding to the transitions from the states $\mathfrak{M}_i$ into $\mathscr{E}_k$. Invariance under time reversal requires that

$$C'(\mathfrak{M}_3, \mathscr{E}_2) = C'(\mathscr{E}_2, \mathfrak{M}_3), \quad C'(\mathscr{E}_3, \mathfrak{M}_2) = C'(\mathfrak{M}_2, \mathscr{E}_3). \quad (31)$$

Finally, using the technique of projection operators, we find for states with $j \leq 3/2$

$$R_{1c} = \mathscr{E}_1 + 2\mathscr{E}_3 + 2\mathscr{E}_2 \cos\theta - \mathfrak{M}_2,$$

$$R_{2c} = \mathfrak{M}_1 + 2\mathfrak{M}_3 + 2\mathfrak{M}_2 \cos\theta - \mathscr{E}_2,$$

$$R_{3c} = \mathscr{E}_1 - \mathscr{E}_3 + 2\mathscr{E}_2 \cos\theta + \tfrac{1}{2}\mathfrak{M}_2 + \sqrt{6}\, C'(\mathscr{E}_3, \mathfrak{M}_2),$$

$$R_{4c} = \mathfrak{M}_1 - \mathfrak{M}_3 + 2\mathfrak{M}_2 \cos\theta + \tfrac{1}{2}\mathscr{E}_2 + \sqrt{6}\, C'(\mathfrak{M}_3, \mathscr{E}_2),$$

$$R_{5c} = -\mathscr{E}_2 - \sqrt{6}\, C'(\mathfrak{M}_3 \mathscr{E}_2) \equiv -\mathscr{E}_2 - C(\mathfrak{M}_3, \mathscr{E}_2),$$

$$R_{6c} = -\mathfrak{M}_2 - \sqrt{6}\, C'(\mathscr{E}_3, \mathfrak{M}_2) \equiv -\mathfrak{M}_2 - C(\mathscr{E}_3, \mathfrak{M}_2). \quad (32)$$

Using the relations (16) and (18), we can separate out the energy dependence of these quantities for $\nu_c \to 0$ in the form

$$\mathscr{E}_1^0 + 2\mathscr{E}_3^0 = -e^2/M, \quad \mathscr{E}_1^0 - \mathscr{E}_3^0 = -[2(e/2M)^2 - e\mu/M]\,\nu_c,$$

$$\mathscr{E}_2^0 = 0, \quad \mathfrak{M}_1^0 = -4\mu^2\nu_c/3, \quad \mathfrak{M}_3^0 = 2\mu^2\nu_c/3, \quad \mathfrak{M}_2^0 = 0,$$

$$C^0(\mathscr{E}_2, \mathfrak{M}_3) = 0, \quad C^0(\mathscr{E}_3, \mathfrak{M}_2) = -e\mu\nu_c/M. \quad (33)$$

In addition to (29) we can derive two more dispersion relations by differentiating

$$R_3 = k_0^2 \mathscr{M}_2 / 2M \text{ and } R_4 = k_0^2 \mathscr{M}_4 / 2M$$

with respect to Q^2 and then setting $Q^2 = 0$. The factor k_0^2 in R_3 and R_4 goes to ν^2 for $Q^2 \to 0$ and thus compensates for the possible infrared divergency in $\mathscr{M}_2$ and $\mathscr{M}_4$.

Let us now consider the quantity

$$R_3 = \frac{k_0^2}{2M}\mathscr{M}_2 = \frac{M\nu^2}{2}\frac{\mathscr{M}_2(\nu, Q^2)}{\sqrt{Q^2 + M^2}}.$$

If $M_2(\nu, Q^2)$ is an analytic function of ν for $Q^2 < Q^2_{max}$, then R_3 and $\partial R_3/\partial Q^2$ will also be analytic functions of ν. Since the contribution of the one-nucleon state to R_3 has the form

$$D\nu\nu_b^2/(\nu_b^2 - \nu^2),$$

where D is some constant and $\nu_b = Q^2/M$, it is at once clear that

$$D\frac{\partial}{\partial Q^2}\frac{\nu\nu_b^2}{\nu_b^2 - \nu^2}\bigg|_{Q^2=0} = 0,$$

i.e., the contribution of the one-nucleon state to the dispersion relation for $\partial R_3/\partial Q^2$ for $Q^2 = 0$ reduces to zero. If we restrict ourselves to the states included in formulas (32), we have

$$\partial R_3/\partial Q^2 = -2\mathscr{E}_2(\nu)/\nu^2. \quad (34)$$

For these same states $\mathscr{E}_2(\nu)/\nu^2$ will be an analytic function of ν whose crossing symmetry agrees with the symmetry of R_3. Then the dispersion relations for $\mathscr{E}_2(\nu)/\nu^2$ can be written in the form

$$\mathrm{Re}\frac{\mathscr{E}_2(\nu)}{\nu^2}=\frac{2\nu}{\pi}\int_0^\infty\frac{\mathrm{Im}\,\mathscr{E}_2(\nu')\,d\nu'}{\nu'^2(\nu'^2-\nu^2)}=\frac{2\nu}{\pi}\int_{\nu_0}^\infty\frac{\mathrm{Im}\,\mathscr{E}_2(\nu')\,d\nu'}{\nu'^2(\nu'^2-\nu^2)}$$

or, finally,

$$\mathrm{Re}\,\mathscr{E}_2(\nu)=\frac{2\nu^3}{\pi}\int_{\nu_0}^\infty\frac{\mathrm{Im}\,\mathscr{E}_2(\nu')\,d\nu'}{\nu'^2(\nu'^2-\nu^2)}. \quad (35)$$

Combining (33) with the dispersion relation for $R_5 + R_6$, we see that $\mathscr{E}_2(\nu)$ and $C(\mathscr{E}_2, \mathfrak{M}_3)$ obey separately the dispersion relation (35).

Analogously, by considering the derivative of R_4, we can show that $\mathfrak{M}_2$ also satisfies the dispersion relation (35). For the states included in formulas (32) it was possible, in the final result, to obtain six dispersion relations for the eight quantities characterizing the scattering of γ quanta. We did not succeed in removing the difficulties connected with the presence of infrared divergencies in the other relations. The repeated differentiation with respect to Q^2 gives rise to the appearance of unknown constants, which have been calculated by some authors with the help of perturbation theory and which can be determined experimentally if sufficient experimental data are available, in analogy to the procedure used in the scattering of mesons and nucleons. In the present paper we shall not employ this method.

To carry out calculations with the help of the dispersion relations for the scattering of γ quanta by nucleons, we require rather detailed data on the amplitude for photoproduction. From the available data we may conclude that $\mathrm{Im}\,\mathfrak{M}_2 = 0$. If we further use $\mathfrak{M}_2^0 = 0$, we find

$$\mathfrak{M}_2 = 0. \quad (36)$$

With the assumptions

$$C(\mathfrak{M}_2, \mathscr{E}_3) = C(\mathscr{E}_3, \mathfrak{M}_2) = C^0(\mathscr{E}_3, \mathfrak{M}_2) = -e\mu\nu/M,$$

$$\mathfrak{M}_1 = \mathfrak{M}_1^0 = -4\mu^2\nu/3,$$

which do not contradict the available experimental data on photoproduction, the number of dispersion relations will agree with the number of functions introduced.

To calculate the imaginary parts of the amplitude we make use of the unitarity of the S matrix. For γ quantum energies below the threshold of π meson production the imaginary parts of the quantities $R_1, \ldots, R_6$ are small. Above the threshold the imaginary parts of $R_1, \ldots, R_6$ are determined by the unitarity requirement on the S matrix.

Neglecting the terms which are quadratic in the electromagnetic interaction, we obtain the relation

$$i\,[\mathscr{M}^+(-\mathbf{k}', -\mathbf{e}', -\mathbf{k}, -\mathbf{e}, -\sigma) - \mathscr{M}(\mathbf{k}', \mathbf{e}', \mathbf{k}, \mathbf{e}, \sigma)]$$
$$= \frac{\nu_c}{2\pi}\int d\Omega(\mathbf{q}_+)\,[T^+_{\gamma\to\pi^+}(\mathbf{q}_+, \mathbf{k}', \mathbf{e}', \sigma)\,T_{\gamma\to\pi^+}(\mathbf{q}_+, \mathbf{k}, \mathbf{e}, \sigma)]$$
$$+ \frac{\nu_c}{2\pi}\int d\Omega(\mathbf{q}_0)\,[T^+_{\gamma\to\pi^0}(\mathbf{q}_0, \mathbf{k}', \mathbf{e}', \sigma)\,T_{\gamma\to\pi^0}(\mathbf{q}_0, \mathbf{k}, \mathbf{e}, \sigma)], \quad (38)$$

where

$$T_{\gamma\to\pi}(\mathbf{q}, \mathbf{k}, \mathbf{e}, \sigma) = iE_1(\sigma\mathbf{e}) - M_1\{([\mathbf{k}\times\mathbf{e}]\,\mathbf{q}) - i(\sigma[[\mathbf{k}\times\mathbf{e}]\,\mathbf{q}])\}$$
$$- M_3\{2([\mathbf{k}\times\mathbf{e}]\,\mathbf{q}) + i(\sigma[[\mathbf{k}\times\mathbf{e}]\,\mathbf{q}])\}$$
$$+ \frac{i}{2}E_2\{(\sigma\mathbf{k})(\mathbf{e}\mathbf{q}) + (\sigma\mathbf{e})(\mathbf{k}\mathbf{q})\} \quad (39)$$

is the amplitude for the photoproduction of pions on a proton. For the lowest states E_1 corresponds, as is known, to a transition from a state with angular momentum $J = ½$ and negative parity accompanied by the creation of a meson in the $s_{1/2}$ state; M_1 corresponds to a transition from $J = ½$ with creation of a meson in the $p_{1/2}$ state; M_3 and and E_2 correspond to a transition from $J = 3/2^+$ with creation of a meson in the $p_{3/2}$ state. It follows from (38) that above the threshold

$$\mathrm{Im}\,R_{1c} = \nu_c\left\{|E_1|^2 + \frac{1}{3}|E_2|^2\cos\theta\right\} = \nu_c A_1,$$
$$\mathrm{Im}\,R_{2c} = \nu_c\{|M_1|^2 + 2|M_3|^2 - |E_2|^2/6\} = \nu_c A_2,$$
$$\mathrm{Im}\,R_{3c} = \mathrm{Im}\,R_{1c},$$
$$\mathrm{Im}\,R_{4c} = \nu_c\{|M_1|^2 - |M_3|^2 + |E_2|^2/12 + (E_2^*M_3 + E_2M_3^*)/2\} = \nu_c A_4,$$
$$\mathrm{Im}\,R_{5c} = -\nu_c\{|E_2|^2/6 + (E_2^*M_3 + E_2M_3^*)/2\} = \nu_c A_5, \quad \mathrm{Im}\,R_{6c} = 0. \quad (40)$$

It is easily seen with the help of (40) that the total interaction cross section for the γ quanta is

$$\sigma_t = (4\pi/\nu_c)\,\mathrm{Im}\,[R_{1c}(0°) + R_{2c}(0°)]$$
$$= 4\pi\,[|E_1|^2 + |M_1|^2 + 2|M_3|^2 - |E_2|^2 \cdot 6], \quad (41)$$

this agrees, as it should, with the total cross section for the photoproduction of π mesons.[14] Using (40) we obtain

$$\mathrm{Re}\,\mathscr{E}_2(\nu) = \frac{1}{6}\frac{2\nu^3}{\pi}\int_{\nu_0}^\infty\frac{d\nu'}{\nu'}\frac{|E_2^+|^2 + |E_2^0|^2}{\nu'^2 - \nu^2},$$

$$\mathrm{Re}\,\mathfrak{M}_3(\nu) = \frac{2}{3}\mu^2\nu + \frac{2\nu^3}{\pi}\int_{\nu_0}^\infty\frac{d\nu'}{\nu'}\frac{|M_3^+|^2 + |M_3^0|^2}{\nu'^2 - \nu^2},$$

$$\mathrm{Re}\,[\mathscr{E}_1(\nu) + 2\mathscr{E}_3(\nu)] = -\frac{e^2}{M}$$
$$+ \frac{\nu^2}{2\pi^2}\int_{\nu_0}^\infty\frac{d\nu'\,(\sigma^0 + \sigma^+)}{\nu'^2 - \nu^2} - \mathrm{Re}\,\mathscr{E}_2(\nu) - 2\mathrm{Re}\,[\mathfrak{M}_3(\nu) - \mathfrak{M}_3^0],$$

$$\mathrm{Re}\,[\mathscr{E}_1(\nu) - \mathscr{E}_3(\nu)]$$
$$= -\left[2\left(\frac{c}{2M}\right)^2 - \frac{e}{M}\mu\right]\nu + \frac{2\nu^3}{\pi}\int_{\nu_0}^\infty\frac{d\nu'}{\nu'}\frac{|E_1^+|^2 + |E_1^0|^2}{\nu'^2 - \nu^2},$$

$$\mathrm{Re}\,C(\mathfrak{M}_3, \mathscr{E}_2) = \mathrm{Re}\,C(\mathscr{E}_2, \mathfrak{M}_3) = \frac{2\nu^3}{\pi}\int_{\nu_0}^\infty\frac{d\nu'}{\nu'}\frac{(\mathrm{Re}\,E_2^*M_3)^+ + (\mathrm{Re}\,E_2^*M_3)^0}{\nu'^2 - \nu^2}. \quad (42)$$

The contribution from multiple π meson production and from the production of other particles is neglected. If it is shown by further analysis that Im $\mathfrak{M}_2 \neq 0$, it is not difficult to take this into account.

[1] Gell-Mann, Goldberger, and Thirring, Phys. Rev. **95**, 1612 (1954).

[2] N. N. Bogolyubov and D. V. Shirkov, Dokl. Akad. Nauk SSSR **113**, 529 (1957).

[3] A. A. Logunov and A. R. Frenklin, Nucl. Phys. **7**, 573 (1958).

[4] A. A. Logunov and P. Ş. Isaev, Nuovo cimento **10**, 917 (1958).

[5] T. Akiba and J. Sato, Progr. Theoret. Phys. **19**, 93 (1958).

[6] R. H. Capps, Phys. Rev. **106**, 1031 (1957) and **108**, 1032 (1957).

[7] L. I. Lapidus and Chou Kuang-Chao, JETP, in press.

[8] R. E. Prange, Phys. Rev. **110**, 240 (1958).

[9] V. I. Ritus, JETP **33**, 1264 (1957), Soviet Phys. JETP **6**, 972 (1957).

[10] L. I. Lapidus, JETP **34**, 922 (1958), Soviet Phys. JETP **7**, 638 (1958).

[11] M. Kawaguchi and S. Minami, Progr. Theoret. Phys. **12**, 789 (1954). A. A. Logunov and A. N. Tavkhelidze, JETP **32**, 1393 (1957), Soviet Phys. JETP **5**, 1134 (1957).

[12] F. E. Low, Phys. Rev. **96**, 1428 (1954).

[13] M. Gell-Mann and M. L. Goldberger, Phys. Rev. **96**, 1433 (1954).

[14] M. Gell-Mann and K. M. Watson, Ann. Rev. Nucl. Science **4**, 219 (1954).

Translated by R. Lipperheide
334

8.B Nuclear Physics 10 (1959) 235—243; ©North-Holland Publishing Co., Amsterdam

CHARGE SYMMETRY PROPERTIES AND REPRESENTATIONS OF THE EXTENDED LORENTZ GROUP IN THE THEORY OF ELEMENTARY PARTICLES

V. I. OGIEVETSKY and CHOU KUANG-CHAO
Joint Institute of Nuclear Research, Dubna USSR

Received 17 July 1958

Abstract: The extended Lorentz group, which includes the complete Lorentz group and the charge conjugation operation, is considered. It is shown that use of irreducible projective representations of the extended group requires the existence of charge multiplets. Charge symmetry and pair production of strange particles follow from invariance under reflections and charge conjugation and from the laws of conservation of electric and baryon charges. The Pauli-Gürsey transformation is valid for free nucleons. The requirement of invariance under this transformation in the case of interaction also leads to isobaric invariance for all particles in strong interactions.

1. Introduction

As is well known, strongly interacting particles can be combined into charge multiplets (p, n; π^+, π^-, π^0; K^+, K^0 etc.). Particles belonging to a given multiplet have almost equal masses and the same spin, but possess different electric charges. The hypothesis of charge symmetry and the more stringent hypothesis of charge independence are then postulated in accord with experimental data. In the usual theory an expression of this fact is invariance under rotations in a certain formal isobaric space. Particles of a given multiplet are treated as states of a particle of a given isobaric spin, the isobaric spin projections of these states being different. The proton and neutron, for example, comprise the nucleon.

The nucleon can be described by the reducible 8-component representation of the complete Lorentz group. A similar situation (reducibility of the complete Lorentz group representation) also holds for other strongly interacting particles.

The following question now arises. If one requires that elementary particles be describable only by irreducible representations, will it then be possible to extend the Lorentz group in such a way, and find such irreducible representations of this extended group, as to lead automatically to the existence of charge multiplets and yield the charge symmetry properties? This problem is examined in the present paper.

The Lorentz group will be extended in the following way.

In quantum theory the wave functions are complex. The charge conjugation operation C which changes a particle into its antiparticle can always be represented as the product of a linear operator (matrix) and complex conjugation antilinear operator:

$$C: \quad \psi_C = C_0 \psi^*, \tag{1}$$

where C_0 is defined in such a way that ψ_C is transformed according to the same irreducible representation of the proper Lorentz group as ψ.

Besides the proper Lorentz group L, *space reflections* I, *and time reversal* T *we include in the extended group the charge conjugation operation* C.

Furthermore, along with the usual irreducible representations of the extended group we shall also consider its projective irreducible representations †.

The importance of utilization of projective representations of the complete Lorentz group has been pointed out by Gelfand and Tsetlin [1]) in connection with Lee and Yang's theory of parity doublets. The possibility of application of projective representations is connected with the uncertainty of the phase factor in the quantal wave function. After Gelfand and Tsetlin, projective representations of the complete Lorentz group were also considered by Taylor and McLennan [2]). A relation between these representations and isobaric invariance is indicated in Taylor's paper [2]). Only the complete Lorentz group was considered, and hence protons and neutrons and $\pi^{\pm}$ and π^0-mesons differ only with respect to space parity. The idea that new definitions of space-time reflections are required which would lead to charge symmetries was also expressed by Salam and Pais at the 7th Rochester Conference.

In the present paper we shall not attempt to study all the irreducible projective representations of the extended group and confine our attention to those which are required for description of strongly interacting particles. It will be shown that if nucleons, Ξ-particles and K-mesons are described by the unusual, projective representations of the extended group, and all other particles are described in the usual manner, then multiplicity, charge symmetry and pair production of strange particles follow from the standard laws of conservation of baryons, electric charge and invariance under the complete Lorentz group and charge conjugation.

In the present theory the Pauli-Gürsey transformation is assumed to

† If to each element g of group G there corresponds an operator $R(g)$ such that the product of the group elements is associated with the operator

$$R(g_1 g_2) = \alpha(g_1 g_2) R(g_1) R(g_2)$$

it will be said that a projective representation of group G is given. If $\alpha(g_1, g_2) \equiv 1$, the projective representation is of the usual type. In the projective representation anticommuting operators of the representation can be associated with the commuting group elements. In particular, the customary spinor representation is projective (operations γ_4 and $\gamma_4\gamma_5$ anticommute, whereas the space and time reflections commute).

hold for free nucleons and is related in a natural manner to isobaric invariance. The requirement that the interaction Lagrangian be also invariant under this transformation for nucleons leads to isobaric invariance in strong interactions between all types of particles.

It is not difficult to write down in this theory the interaction Lagrangian for electromagnetic fields with the aid of the charge operator. It is found that Schwinger time reversal is not valid for electromagnetic interactions and only Wigner time reversal holds.

Weak interactions in which also space parity is not conserved are more complicated and will not be considered in this article.

For the sake of concreteness we shall assume that the relative space parity of all baryons is the same and that reflection of the usual spinors is performed with the aid of operator γ_4. All bosons are assumed to be pseudoscalars. We start our discussion with nucleons.

2. Free Nucleon Field

If one demands for 4-component spinors that

$$I^2 = T^2 = C^2 = 1, \tag{2}$$

it can readily be shown that the operators I, T and C can be expressed as follows:

$$I\colon\quad \psi' = \gamma_4\psi, \tag{3a}$$

$$T\colon\quad \psi' = i\gamma_4\gamma_5\psi, \tag{3b}$$

$$C\colon\quad \psi_C = i\gamma_2\psi^*, \tag{3c}$$

where T is the Schwinger spinor time reversal operation [3]) and matrices γ are expressed in the Pauli representation.

The following commutation relations hold for I, T and C:

$$IT = -TI, \tag{4a}$$

$$IC = -CI, \tag{4b}$$

$$TC = -CT. \tag{4c}$$

In contrast to the usual theory, in retaining relation (2) we require that relation (4a) change sign for nucleons, i.e., that the following condition be satisfied:

$$IT = TI, \quad \text{(a)}, \quad IC = -CI, \quad \text{(b)}, \quad TC = -CT, \quad \text{(c)}. \tag{4'}$$

Only 8×8 matrices satisfy commutation rules (2) and (4'):

$$\begin{aligned} I\colon&\quad \psi' = \tau_3\times\gamma_4\psi,\\ T\colon&\quad \psi' = 1\times\gamma_4\psi,\\ C\colon&\quad \psi_C = i\tau_3\times\gamma_2\psi^*, \end{aligned} \tag{5}$$

where τ is a Pauli matrix.

These operators together with the operators of the proper Lorentz group, in which γ_μ should everywhere be replaced by $1\times\gamma_\mu$, form a projective irreducible representation of the extended Lorentz group, the spinors ψ being 8-component.

In this representation the free field ψ has a uniquely defined Lagrangian †

$$L = \bar{\psi}(1\times\gamma_\mu\partial_\mu + i\tau_2\times\gamma_5 m)\psi, \tag{6}$$

where $\bar{\psi} = \psi^{*T}1\times\gamma_4$ and the field equations have the form

$$1\times\gamma_\mu\partial_\mu\psi = -i\tau_2\times\gamma_5 m\psi. \tag{7}$$

Lagrangian (6) as well as equation (7) are invariant under two single-parameter transformation groups

$$\psi' = \exp(i\lambda)\psi, \tag{8}$$

$$\psi' = \exp(i\tau_1\times\gamma_5\lambda)\psi \tag{9}$$

and the three-parameter group

$$\psi' = a\psi + b\tau_3\times\gamma_5\psi_C, \tag{10}$$

where $|a|^2+|b|^2 = 1$.

Transformations (9) and (10) are similar to the Pauli transformations for the neutrino [4]) and differ from them only in that γ_5 is replaced by $\tau_1\times\gamma_5$ and $\tau_3\times\gamma_5$, respectively.

If we introduce the new 4-component spinors

$$\begin{aligned} \psi_p &= \frac{1}{\sqrt{2}}(\psi_1+\gamma_5\psi_2), & \psi_{p_c} &= \frac{1}{\sqrt{2}}(\psi_{C1}+\gamma_5\psi_{C2}), \\ \psi_{n_c} &= \frac{1}{\sqrt{2}}(-\gamma_5\psi_1+\psi_2), & \psi_n &= \frac{1}{\sqrt{2}}(\gamma_5\psi_{C1}-\psi_{C2}), \end{aligned} \tag{11}$$

they will be found to satisfy the familiar Dirac equation

$$\gamma_\mu\partial_\mu\psi = -m\psi \tag{12}$$

and under transformation (9)

$$\begin{aligned} \psi'_p &= \exp(i\lambda)\psi_p, & \psi'_n &= \exp(i\lambda)\psi_n, \\ \psi'_{p_c} &= \exp(-i\lambda)\psi_{p_c}, & \psi'_{n_c} &= \exp(-i\lambda)\psi_{n_c}. \end{aligned} \tag{13}$$

Thus (9) can be regarded as a gauge transformation related to the law of conservation of baryons, and ψ_p, ψ_n, ψ_{p_c} and ψ_{n_c} refer to the proton, neutron, antiproton and antineutron fields respectively.

The transformation

$$\text{E:}\quad \psi' = \exp[\tfrac{1}{2}i(1\times 1+\tau_1\times\gamma_5)\lambda]\psi \tag{14}$$

should be related to the law of conservation of electric charge.

† According to Schwinger, under time reversal one has $L \to L^T$, where the transposition sign refers to Hilbert space operators [3]).

Indeed, under transformation E

$$\text{E:}\quad \begin{aligned} \psi'_{\text{p}} &= \exp(i\lambda)\psi_{\text{p}}, & \psi'_{\text{p}_c} &= \exp(-i\lambda)\psi_{\text{p}_c}, \\ \psi'_{\text{n}} &= \psi_{\text{n}}, & \psi'_{\text{n}_c} &= \psi_{\text{n}_c}. \end{aligned} \tag{15}$$

The three-parameter transformation (10) is isomorphic with rotation in isobaric space. Indeed, if we form the usually employed 8-component nucleon field $\psi_{\text{N}} = \binom{\psi_{\text{p}}}{\psi_{\text{n}}}$ we obtain under transformation (10)

$$\psi'_{\text{N}} = \exp[i(\boldsymbol{\tau} \cdot \boldsymbol{\lambda})]\psi_{\text{N}} \tag{16}$$

where $\boldsymbol{\lambda}$ is a vector possessing the real components $\lambda_1, \lambda_2, \lambda_3$ and the τ are the usual two-dimensional Pauli matrices and

$$a = \cos|\boldsymbol{\lambda}| + \frac{i\lambda_3}{|\boldsymbol{\lambda}|}\sin|\boldsymbol{\lambda}|, \qquad b = \frac{\sin|\boldsymbol{\lambda}|}{|\boldsymbol{\lambda}|}(\lambda_2 - i\lambda_1). \tag{17}$$

Gürsey [5]) has pointed out that an analogous isomorphism appears when the number of components is formally doubled.

3. Interaction Between Nucleons and Ordinary Bosons

Consider first the interaction between nucleons and a neutral pseudoscalar field φ_0 with positive time parity

$$\text{I:}\ \ \varphi'_0 = -\varphi_0, \qquad \text{T:}\ \ \varphi'_0 = \varphi_0, \qquad \text{C:}\ \ \varphi_{0\text{C}} = \varphi_0. \tag{18}$$

The condition of invariance under I, T and C and transformations (9) and E uniquely yields the interaction Lagrangian (in the following only the interaction Lagrangian without derivatives will be considered)

$$\begin{aligned} L &= ig_0\bar{\psi}\tau_3 \times \gamma_5\psi\varphi_0 \\ &= ig_0(\bar{\psi}_{\text{p}}\gamma_5\psi_{\text{p}} - \bar{\psi}_{\text{n}}\gamma_5\psi_{\text{n}})\varphi_0 \\ &= ig_0(\bar{\psi}_{\text{N}}\gamma_5\tau_3\psi_{\text{N}}\varphi_0), \end{aligned} \tag{19}$$

i.e., the meson-proton and meson-neutron coupling constants have different signs and hence φ_0 can be assigned to a neutral π^0-meson.

If the neutral meson φ'_0 were a space pseudoscalar but would possess a negative time parity, the interaction Lagrangian

$$\begin{aligned} L &= g'_0\bar{\psi}\tau_2 \times 1\psi\varphi'_0 \\ &= ig'_0(\bar{\psi}_{\text{p}}\gamma_5\psi_{\text{p}} + \bar{\psi}_{\text{n}}\gamma_5\psi_{\text{n}})\varphi'_0, \end{aligned} \tag{20}$$

which can be ascribed to the questionable ρ_0-meson, would follow in a unique manner.

For the interaction Lagrangian between nucleons and a charged pseudoscalar boson field φ

$$\text{I:}\ \ \varphi' = -\varphi; \qquad \text{T:}\ \ \varphi' = \varphi; \qquad \text{C:}\ \ \varphi_{\text{C}} = \varphi^*$$

and we again uniquely obtain

$$L = ig(\bar{\psi}\psi_C\varphi^* - \bar{\psi}_C\psi\varphi) = 2ig(\bar{\psi}_p\gamma_5\psi_n\varphi^* + \bar{\psi}_n\gamma_5\psi_p\varphi). \tag{21}$$

Thus it may be considered that $\varphi(\varphi^*)$ refers to $\pi^-(\pi^+)$ mesons.

Charge symmetry (i.e., the possibility of simultaneously making the substitutions $\psi_p \rightleftarrows \psi_n$, $\pi^+ \rightleftarrows \pi^-$, $\pi^0 \rightleftarrows -\pi^0$) of interactions (19) and (21) is evident. The general Lagrangian for interaction with π-mesons has the form

$$L = ig_0(\bar{\psi}_p\gamma_5\psi_p - \bar{\psi}_n\gamma_5\psi_n) + 2ig(\bar{\psi}_p\gamma_5\psi_n\pi^+ + \bar{\psi}_n\gamma_5\psi_p\pi^-). \tag{22}$$

If we now require that not only the free nucleon Lagrangian but also the nucleon interaction Lagrangian be invariant under transformations of the three-parameter group we find

$$g = g_0/\sqrt{2} = g_\pi/\sqrt{2}$$

and (23)

$$L_\pi = ig_\pi\bar{\psi}_N\gamma_5(\boldsymbol{\tau}\cdot\boldsymbol{\pi})\psi_N.$$

Under transformation (10) the meson fields transform as follows:

$$(\boldsymbol{\tau}\cdot\boldsymbol{\pi})' = \exp[i(\boldsymbol{\tau}\cdot\boldsymbol{\lambda})](\boldsymbol{\tau}\cdot\boldsymbol{\pi})\exp[-i(\boldsymbol{\tau}\cdot\boldsymbol{\lambda})]. \tag{24}$$

The π^+, π^- and π^0-meson masses should be the same in this case.

We have thus arrived at the usual isobarically invariant theory of interaction between π-mesons and nucleons.

4. Free K-Mesons

For bosons the usual representation of the extended Lorentz group encompasses only π-mesons. We shall describe K-mesons by the projective representation in which

$$I^2 = 1, \qquad T^2 = -1, \qquad C^2 = 1,$$
$$IT = +TI, \qquad IC = +CI, \qquad TC = +CT. \tag{25}$$

The simplest irreducible representation in which these commutation rules are valid is a two-dimensional one: $\varphi = \binom{\varphi_1}{\varphi_2}$ and the operators I, T and C have the form

$$I\colon\ \varphi' = -\varphi, \qquad T\colon\ \varphi' = i\tau_2\varphi, \qquad C\colon\ \varphi_C = \varphi^*. \tag{26}$$

Let us identify the K^+-meson with φ_1, the $\tilde{K}^0$-meson with φ_2, K^--meson with φ_1^*, and K^0-meson with φ_2^*.

We can then relate the law of conservation of electric charge to the transformation

$$E\colon\ \varphi' = \exp[\tfrac{1}{2}i(1+\tau_3)\lambda]\varphi. \tag{27}$$

Indeed, under this transformation

$$\begin{aligned} K^{+\prime} &= \exp(i\lambda)K^{+}, & K^{-\prime} &= \exp(-i\lambda)K^{-}, \\ K^{0\prime} &= K^{0}, & \tilde{K}^{0\prime} &= \tilde{K}^{0}. \end{aligned} \tag{28}$$

The transformation $\varphi' = \exp(i\tau_3\lambda)\varphi$ for which

$$\begin{aligned} K^{+\prime} &= \exp(i\lambda)K^{+}, & K^{0\prime} &= \exp(i\lambda)K^{0}, \\ K^{-1} &= \exp(-i\lambda)K^{-}, & \tilde{K}^{0\prime} &= \exp(-i\lambda)\tilde{K}^{0}, \end{aligned} \tag{29}$$

corresponds to the law of conservation of hypercharge.

5. K-Meson-Nucleon Interaction. Λ and Σ Particles

We now pass to a study of the interaction between K-mesons and nucleons. Since K-mesons as well as nucleons transform according to the projective representation of the extended Lorentz group and the law of conservation of baryon charge is valid, one other baryon describable by the usual representation should necessarily be involved in the interaction. This condition inevitably leads to pair production of strange particles.

We shall first consider the case of a neutral baryon. As mentioned above, for the sake of definiteness the relative space parity of all baryons is assumed to be the same:

$$\text{I:} \quad Y'_0 = \gamma_4 Y_0. \tag{30}$$

Under the T transformation for nucleons two possibilities exist for the investigated baryon:

$$\text{T:} \quad Y'_0 = -\gamma_4\gamma_5 Y_{0_C}, \tag{31}$$

$$Y'_0 = \gamma_4\gamma_5 Y_{0_C}. \tag{32}$$

The antibaryon Y_{0_C} should be introduced in equations (31) and (32), as transformation T for nucleons anticommutes with the transformation of conservation of baryon charge (9).

If the law (31) is chosen for T, it will be found that the only type of Lagrangian consistent with invariance under the extended Lorentz group is

$$\begin{aligned} L = ig[&\bar{\psi}(1\times\gamma_5-\tau_1\times 1)(1\times 1+\tau_3\times 1)\varphi\times Y_0 \\ &-\bar{\psi}_C(1\times\gamma_5+\tau_1\times 1)(1\times 1-\tau_3\times 1)\varphi^*\times Y_0]+\text{Herm. conj.} \end{aligned} \tag{33}$$

Changing from ψ and φ to the nucleon field and K-meson field operators we obtain the familiar nucleon-Λ_0 particle interaction Lagrangian

$$L = ig_\Lambda(\bar{p}\gamma_5\Lambda_0 K^{+}+\bar{n}\gamma_5\Lambda_0 K^{0})+\text{Herm. conj.} \tag{34}$$

The law (32) for T yields a Lagrangian which differs from expression (33) only in sign between the two terms in expression (33) and corresponds to the Σ_0-particle:

$$L = ig_{\Sigma_0}(\bar{p}\gamma_5\Sigma_0 K^{+}-\bar{n}\gamma_5\Sigma_0 K^{0})+\text{Herm. conj.} \tag{35}$$

We thus arrive at the conclusion that under transformation T for nucleons the transformations for Λ_0 and Σ_0 differ only by sign. Considering now the interaction between nucleons and charged baryons we find it possible to set up a Lagrangian which is invariant under charge conjugation, space and time reflections and the laws of conservation of electric and baryon charges only if

$$\text{T:}\begin{array}{l}\Sigma^+ \to -\gamma_4\gamma_5\Sigma_C{}^-,\\ \Sigma^- \to -\gamma_4\gamma_5\Sigma_C{}^+,\end{array} \tag{36}$$

which implies equality of the Σ^+ and Σ^- particle masses. The Lagrangian has the form

$$\begin{aligned}L = -ig[\bar{\psi}(1\times 1-\tau_1\times\gamma_5)(1\times 1-\tau_3\times 1)\varphi^*\times\Sigma^+ \\ +\bar{\psi}_C(1\times 1+\tau_1\times\gamma_5)(1\times 1+\tau_3\times 1)\varphi\times\Sigma^-]+\text{Herm. conj.}\end{aligned} \tag{37}$$

In the usual notation it can be represented as follows:

$$L = ig_{\Sigma^\pm}(\bar{\text{p}}\gamma_5\Sigma^+\text{K}^0+\bar{\text{n}}\gamma_5\Sigma^-\text{K}^+)+\text{Herm. conj.} \tag{38}$$

Charge symmetry is evident.

If we furthermore require that a transformation of the Pauli-Gürsey type (10) also leave the interaction Lagrangian invariant, we get

$$g_{\Sigma^\pm} = \sqrt{2}\, g_{\Sigma_0} = \sqrt{2}\, g_\Sigma, \tag{39}$$

where the Σ^+, Σ^- and Σ_0 masses should be equal and the usual isobarically invariant Lagrangian

$$\begin{aligned}L = ig_\Sigma(\bar{\text{p}}\gamma_5\Sigma_0\text{K}^+-\bar{\text{n}}\gamma_5\Sigma_0\text{K}^0 \\ +\sqrt{2}\,\bar{\text{p}}\gamma_5\Sigma^+\text{K}^0+\sqrt{2}\,\bar{\text{n}}\gamma_5\Sigma^-\text{K}^+)+\text{Herm. conj.}\end{aligned} \tag{40}$$

obtains.

6. Ξ-particles

The ideas presented above admit of one other possibility. Thus, retaining the same transformation (9) for the law of conservation of baryons one can, in the projective representation (5), change, in transformation (14) which is related to the law of conservation of electric charge, the sign between 1×1 and $\tau_1\times\gamma_5$:

$$\text{E:}\quad \psi' = \exp\,[\tfrac{1}{2}i(1\times 1-\tau_1\times\gamma_5)]\psi. \tag{41}$$

In this case one should simply replace in all formulas p by Ξ^0, n by Ξ^-, K^+ by $\text{K}^-{}_0$, K^0 by K^-. We again obtain charge symmetry properties and the requirement of invariance under Pauli-Gürsey transformations in interaction again leads to the usual isobarically invariant Lagrangians.

7. Conclusion

The plan set out at the beginning of this paper has been carried out. By introducing a new irreducible projective representation of the extended Lorentz group for nucleons and proceeding from this new representation we have been able to demonstrate that the charge symmetry properties, pair production of strong particles and multiplicity follow from the standard conservation laws whereas isobaric invariance follows from a transformation of the Pauli-Gürsey type for free nucleons. This type of invariance can be incorporated in the theory as the number of components of the wave functions which transform in accord with the irreducible representations is necessarily larger.

Weak interactions have not been studied from this viewpoint in the present paper. This is a more difficult problem and less unambiguous owing to violation of space parity.

The authors are sincerely thankful to Prof. I. M. Gelfand for valuable discussions.

References

1) I. M. Gelfand and M. L. Tsetlin, JETP **31** (1956) 1107
2) I. C. Taylor, Nuclear Physics **3** (1957) 606;
J. A. McLennan Jr., Phys. Rev. **109** (1958) 986
3) J. Schwinger, Phys. Rev. **91** (1953) 713
4) W. Pauli, Nuovo Cimento **6** (1957) 204
5) F. Gürsey, Nuovo Cimento **7** (1958) 411

SOVIET PHYSICS JETP VOLUME 11, NUMBER 1 JULY, 1960

INTEGRAL TRANSFORMATIONS OF THE I. S. SHAPIRO TYPE FOR PARTICLES OF ZERO MASS

L. G. ZASTAVENKO and CHOU KUANG-CHAO

Joint Institute for Nuclear Research

Submitted to JETP editor May 28, 1959

J. Exptl. Theoret. Phys. (U.S.S.R.) **38**, 134-139 (January, 1960)

An expansion in terms of the irreducible representations of the proper Lorentz group is given for the representation which specifies the transformation of the wave function of a particle of zero mass and of arbitrary spin.

THE correspondence

$$\Psi(\mathbf{p}, \sigma) \xrightarrow{S} \Psi'(\mathbf{p}, \sigma) = \exp\{i\sigma\varphi(S, \mathbf{k})\}\Psi(S^{-1}\mathbf{p}, \sigma), \quad (1)$$

where S is a transformation belonging to the Lorentz group, **p** transforms like the momentum vector of a particle of mass 0, $\mathbf{k} = \mathbf{p}/p$, σ is an integer or a half-integer, and $\varphi(S, \mathbf{k})$ is the angle defined by formula (1.9) in reference 1 (hereafter referred to as I) has, as can be easily seen, the group property.

It defines the transformation law for the wave function of a particle of mass zero, and with spin component σ along the direction of the momentum **p**, under transformations of the proper Lorentz group.[2]

The object of the present paper is to give an expansion of this representation in terms of the irreducible (ρ, m)-representations (cf. I) of the proper Lorentz group.

1. INTEGRAL TRANSFORMATIONS FOR PARTICLES OF MASS 0

In a manner similar to the way this was done in I we obtain the following system of mutually inverse integral transformations

$$\Psi(\mathbf{p}, \sigma) = \int d\rho \int d\Omega(\mathbf{n})\, Y_{\rho m \mathbf{n}}(\mathbf{p}, \sigma)\, f_{\rho m \mathbf{n}}, \quad (2)$$

$$f_{\rho m \mathbf{n}} = \sum_{\sigma} \int \frac{d^3\mathbf{p}}{|\mathbf{p}|} Y'_{\rho m \mathbf{n}}(\mathbf{p}, \sigma)\, \Psi(\mathbf{p}, \sigma), \quad (3)$$

where $f_{\rho m \mathbf{n}}$ transforms according to the irreducible representation (ρ, m) of the proper Lorentz group, and we obtain the following conditions for determining the functions Y and Y′:

$$Y_{\rho m S^{-1}\mathbf{n}}(S^{-1}\mathbf{p}, \sigma) = \exp\{im\varphi(S, \mathbf{n}) - i\sigma\varphi(S, \mathbf{k})\}\,[K(\mathbf{n}), K(S^{-1}\mathbf{n})]^{-1-i\rho/2}\, Y_{\rho m \mathbf{n}}(\mathbf{p}, \sigma), \quad (4)$$

where $K(\mathbf{n})$ is defined by formula (1.9), I. An analogous condition for $Y'_{\rho m \mathbf{n}}(\mathbf{p}, \sigma)$ is satisfied if we take

$$Y'_{\rho m \mathbf{n}}(\mathbf{p}, \sigma) = C_{m\rho}\overline{Y}_{\rho m \mathbf{n}}(\mathbf{p}, \sigma).$$

Both here and later a bar above a letter denotes taking the complex conjugate.

We note that the function

$$Y_{\rho m \mathbf{n}}(\mathbf{p}, \sigma) = (1/2\pi)\,\delta_{m\sigma}\,\delta(1 - (\mathbf{nk}))\, p^{-1+i\rho/2} \quad (5)$$

satisfies expression (4).

Thus, from (2) – (5) we obtain

$$\Psi(\mathbf{p}, m) = \int d\rho \int d\Omega(\mathbf{n})\, \delta(\mathbf{n} - \mathbf{k})\, p^{-1+i\rho/2} f_{\rho m \mathbf{n}}, \quad (6)$$

$$f_{\rho m \mathbf{n}} = C_\rho \int \frac{d^3\mathbf{p}}{|\mathbf{p}|}\, \delta(\mathbf{n} - \mathbf{k})\, p^{-1-i\rho/2}\, \Psi(\mathbf{p}, m). \quad (7)$$

Here we have taken into account the fact that $\delta[1 - (\mathbf{nk})] = 2\pi\delta(\mathbf{n} - \mathbf{k})$.

On substituting (7) into (6) we find that $C_\rho = 1/4\pi$ while the integral over ρ in (6) should be taken from $-\infty$ to $+\infty$. Thus, the formulas

$$\Psi(\mathbf{p}, m) = \int_{-\infty}^{+\infty} d\rho \int d\Omega(\mathbf{n})\, \delta(\mathbf{n} - \mathbf{k})\, p^{-1+i\rho/2} f_{\rho m \mathbf{n}}, \quad (8)$$

$$f_{\rho m \mathbf{n}} = \frac{1}{4\pi} \int \frac{d^3\mathbf{p}}{|\mathbf{p}|}\, \delta(\mathbf{n} - \mathbf{k})\, p^{-1-i\rho/2}\, \Psi(\mathbf{p}, m) \quad (9)$$

give a solution of the proposed problem; it turns out to be simpler than in the case of non-zero rest mass.*

2. COMPARISON OF THE RESULTS OBTAINED HERE WITH THOSE OF I

We compare the results obtained here with formulas (4.1), I and (4.2), I for $M = 0$ (M is the particle mass). A direct transition to the limit $M \to 0$ in the formulas indicated above is impossible. Instead of this we shall carry out the following formal manipulation of those formulas: we

*In the case $m = 0$ the same representation $(\rho, 0)$ is in fact contained twice in the result obtained, since the representations $(\rho, 0)$ and $(0, \rho)$ are equivalent. The transition between these two representations is given by formulas (14) and (15) with $m = 0$.

carry out the transition to the limit $M \to 0$ in the factor $R(L_p, n)$, we introduce a new variable of integration p/M (retaining for it the old notation p) and we replace the factor $\sqrt{1+p^2} - \mathbf{p}\cdot\mathbf{n}$ by $p - \mathbf{p}\cdot\mathbf{n}$. If we now introduce the components of the wavefunction having a definite component of the spin along the direction of the momentum we shall obtain

$$\tilde{\Psi}(\mathbf{p}, m) = \int_{-\infty}^{0} d\rho \int d\Omega(n) \times [p - \mathbf{pn}]^{-1+i\rho/2} e^{im\theta(\mathbf{p},\mathbf{n})} (-)^{s+m} \tilde{\varphi}_{-\rho-m\mathbf{n}}, \quad (10)$$

$$\tilde{\varphi}_{-\rho-m\mathbf{n}} = \frac{\rho^2 + (2m)^2}{(4\pi)^3} \int \frac{d^3\mathbf{p}}{|\mathbf{p}|} \times [p - \mathbf{pn}]^{-1-i\rho/2} e^{-im\theta(\mathbf{p},\mathbf{n})} (-)^{s+m} \tilde{\Psi}(\mathbf{p}, m). \quad (11)$$

Here

$$\tilde{\Psi}(\mathbf{p}, m) = \sum_{\sigma} D^s_{\sigma m}(\mathbf{k}) \tilde{\Psi}'(\mathbf{p}, \sigma), \quad (12)$$

where $\tilde{\Psi}'(\mathbf{p}, \sigma)$ is the wavefunction utilized in I; the notation $D^s_{\sigma m}(\mathbf{k})$ is also explained there. In the derivation of formulas (10) and (11) the following equation was used:

$$[D^s(\mathbf{k})^{-1} D^s(L_{\mathbf{p}}\mathbf{n}) D^s(\mathbf{n})]_{mn} \xrightarrow[M\to 0]{} \delta_{m,-n} (-1)^{s+m} e^{im\theta(\mathbf{p},\mathbf{n})}, \quad (13)$$

where $\theta(\mathbf{p}, \mathbf{n})$ is defined in Appendix B.

We shall compare the expressions (8) and (9), obtained above with (10) and (11).* From the function $f_{\rho m \mathbf{n}}$ which transforms according to the (ρ, m) representation we go over to the function $\varphi_{-\rho - m\mathbf{n}}$, which transforms according to the $(-\rho, -m)$ representation. We make use of the fact that the representations (ρ, m) and $(-\rho, -m)$ are equivalent. Therefore the function $\varphi_{-\rho,-m,\mathbf{n}}$ may be obtained from the function $f_{\rho m \mathbf{n}}$ by the following unitary transformation:

$$\varphi_{-\rho,-m,\mathbf{n}} = \int U_{\rho m}(\mathbf{n}, \mathbf{k}) f_{\rho m \mathbf{k}} \, d\Omega(\mathbf{k}), \quad (14)$$

$$f_{\rho m \mathbf{k}} = \int \overline{U}_{\rho m}(\mathbf{n}, \mathbf{k}) \varphi_{-\rho,-m,\mathbf{n}} \, d\Omega(\mathbf{n}), \quad (15)$$

where

$$U_{\rho m}(\mathbf{n}, \mathbf{k}) = \sum_{l \geq |m|} \sum_{\alpha=-l}^{l} \frac{2l+1}{4\pi} \times \frac{\Gamma(l+1+i\rho/2)}{\Gamma(l+1-i\rho/2)} \overline{D}^l_{\alpha,-m}(\mathbf{n}) D^l_{\alpha m}(\mathbf{k}). \quad (16)$$

*It is proved later that (10) and (11) are not equivalent to (8) and (9) (and are therefore incorrect). We emphasize that this by no means indicates that the results of the present paper contradict those of I; it merely means that the formal manipulation which leads to (10) and (11) is not justified.

The function U is discussed in Appendix A and satisfies the following unitarity condition:*

$$\int \overline{U}_{\rho m}(\mathbf{l}, \mathbf{n}) U_{\rho m}(\mathbf{l}, \mathbf{k}) \, d\Omega(\mathbf{l}) = \delta(\mathbf{n} - \mathbf{k}). \quad (17)$$

On substituting (14) and (15) into (8) and (9) we obtain

$$\varphi_{-\rho,-m,\mathbf{n}} = \frac{1}{4\pi} \int \frac{d^3\mathbf{p}}{|\mathbf{p}|} p^{-1-i\rho/2} U_{\rho m}(\mathbf{n}, \mathbf{k}) \Psi(\mathbf{p}, m), \quad (18)$$

$$\Psi(\mathbf{p}, m) = \int_{-\infty}^{\infty} d\rho \int d\Omega(\mathbf{n}) \overline{U}_{\rho m}(\mathbf{n}, \mathbf{k}) p^{-1+i\rho/2} \varphi_{-\rho,-m,\mathbf{n}}. \quad (19)$$

We substitute into (18) and (19) the following formula derived in Appendix A:

$$U_{\rho m}(\mathbf{n}, \mathbf{k}) = A_{\rho m} [1 - (\mathbf{nk})]^{-1-i\rho/2} Q_m(\mathbf{n}, \mathbf{k}), \quad (20)$$

where

$$A_{\rho m} = 2^{1+i\rho/2} \Gamma(m+1+i\rho/2) / 4\pi\Gamma(m - i\rho/2), \quad (21)$$

$$Q_m(\mathbf{n}, \mathbf{k}) = m \sum_{l \geq |m|} \sum_{\alpha=-l}^{l} \frac{2l+1}{l(l+1)} \overline{D}^l_{\alpha,-m}(\mathbf{n}) D^l_{\alpha m}(\mathbf{k}). \quad (22)$$

We then obtain

$$\varphi_{-\rho,-m,\mathbf{n}} = \frac{1}{4\pi} \int \frac{d^3\mathbf{p}}{|\mathbf{p}|} p^{-1-i\rho/2} A_{\rho m} \times [1 - (\mathbf{nk})]^{-1-i\rho/2} Q_m(\mathbf{n}, \mathbf{k}) \Psi(\mathbf{p}, m), \quad (23)$$

$$\Psi(\mathbf{p}, m) = \int_{-\infty}^{\infty} d\rho \int d\Omega(\mathbf{n}) p^{-1+i\rho/2} \overline{A}_{\rho m} \times [1 - (\mathbf{nk})]^{-1+i\rho/2} \overline{Q}_m(\mathbf{n}, \mathbf{k}) \varphi_{-\rho,-m}. \quad (24)$$

Since

$$|A_{\rho m}|^2 = \frac{1}{(4\pi)^2} (\rho^2 + 4m^2)$$

and

$$Q_m(\mathbf{n}, \mathbf{k}) = e^{-im\theta(\mathbf{n},\mathbf{k})} (-1)^{s+m} \quad (25)$$

(cf. Appendix C), (23) and (24) differ from (10) and (11) only in that the integral over ρ in (23) and (24) is taken between the limits from $-\infty$ to ∞.

In particular, for $m = 0$ each irreducible representation $(\rho, 0)$ occurs twice in the expansion under consideration in contrast to the case $M \neq 0$.

APPENDIX A

DEFINITION OF THE FUNCTION $U_{\rho m}(\mathbf{n}, \mathbf{k})$

It may be easily shown that from the fundamental relations (14), (15) and the transformation law for $f_{\rho m \mathbf{k}}$ and $\varphi_{-\rho,-m,\mathbf{n}}$ the following functional equation for $U_{\rho m}(\mathbf{n}, \mathbf{k})$ may be obtained:

*In particular, for $m = 0$ we obtain the following simple integral representation for the δ-function:

$$\frac{\rho^2}{(4\pi)^2} \int [1 - (\mathbf{ln})]^{-1+i\rho/2} [1 - (\mathbf{lk})]^{-1-i\rho/2} d\Omega(\mathbf{l}) = \delta(\mathbf{n} - \mathbf{k})$$

[cf. also formulas (A.4) and (A.6)].

$$U_{\rho m}(S^{-1}\mathbf{n}, S^{-1}\mathbf{k})$$
$$= U_{\rho m}(\mathbf{n},\mathbf{k})[K(\mathbf{n})K(\mathbf{k})/K(S^{-1}\mathbf{n})K(S^{-1}\mathbf{k})]^{-1-i\rho/2}$$
$$\times \exp\{im[\varphi(S,\mathbf{n})+\varphi(S,\mathbf{k})]\} \quad (A.1)$$

(the notations $K(\mathbf{n})/K(S^{-1}\mathbf{n})$ and $\varphi(S, \mathbf{n})$ are defined in I). Since the functions $D^l_{\alpha m}(\mathbf{k})$ for $l = |m|, |m|+1, \ldots$ and for fixed m form a complete system, U may be represented in the following form

$$U_{\rho m}(\mathbf{n}, \mathbf{k}) = \sum_{l\alpha}\sum_{l'\beta} X_{l\alpha l'\beta}\overline{D}^l_{\alpha,-m}(\mathbf{n})D^{l'}_{\beta m}(\mathbf{k}). \quad (A.2)$$

On taking in formula (A.1) for S the pure rotation $S = R$ we obtain on taking into account formula (1.9b), I,

$$X_{l\alpha l'\beta} = X_l\delta_{ll'}\delta_{\alpha\beta}.$$

Further, we take in (A.1) for S the infinitesimal pure Lorentz transformation L:

$$L^{-1}\mathbf{n} = (\mathbf{n}+\boldsymbol{\varphi})/[1+(\mathbf{n}\boldsymbol{\varphi})].$$

It can then be easily seen that

$$K(L^{-1}\mathbf{n})/K(\mathbf{n}) = 1 + (\boldsymbol{\varphi}\mathbf{n}).$$

Since according to (1.9d), I

$$e^{im\varphi(L,\mathbf{k})}D^l_{\beta m}(L^{-1}\mathbf{k}) = \sum_\gamma D^l_{\beta\gamma}(R(L,\mathbf{k}))D^l_{\gamma m}(\mathbf{k})$$

we obtain from (A.1) and (A.2)

$$\sum X_l\overline{D}^l_{\sigma-m}(\mathbf{n})D^l_{\alpha m}(\mathbf{k}) = [K(\mathbf{n})K(\mathbf{k})/K(L^{-1}\mathbf{n})K(L^{-1}\mathbf{k})]^{1+i\rho/2}$$
$$\times\sum X_l\overline{D}^l_{\alpha\gamma}(R(L,\mathbf{n}))\overline{D}^l_{\gamma-m}(\mathbf{n})D^l_{\alpha\delta}(R(L,\mathbf{k}))D^l_{\delta m}(\mathbf{k}). \quad (A.3)$$

The parameter of the rotation R occurring in the above is defined by formula (A.4), I:

$$D^l(R(L,\mathbf{k})) = e^{-i(\mathbf{H}\boldsymbol{\alpha})} \approx 1 - i(\mathbf{H}\boldsymbol{\alpha}), \qquad \boldsymbol{\alpha} = [\mathbf{k}\times\boldsymbol{\varphi}].$$

We then obtain from formula (A.3)

$$\sum X_l\overline{D}^l_{\alpha,-m}(\mathbf{n})D^l_{\alpha m}(\mathbf{k}) = [1-(1+i\rho/2)\boldsymbol{\varphi}(\mathbf{n}+\mathbf{k})]$$
$$\times\sum X_l\overline{\{[1-i(\mathbf{H}^l[\mathbf{n}\times\boldsymbol{\varphi}])]D^l(\mathbf{n})\}}_{\alpha,-m}$$
$$\times\{[1-i(\mathbf{H}^l[\mathbf{k}\times\boldsymbol{\varphi}])]D^l(\mathbf{k})\}_{\alpha,m}.$$

Here we must express the cyclic components of the vectors $\mathbf{n}$ and $\mathbf{k}$ in terms of the generalized spherical harmonics $D^1_{\alpha 0}(\mathbf{n})$, $D^1_{\alpha 0}(\mathbf{k})$, and we must then eliminate products of the D-functions in accordance with the following rule

$$D^1_{ab}(\mathbf{n})D^l_{cd}(\mathbf{n}) = \langle 1lac|LM\rangle\langle 1l\,bd|LN\rangle D^L_{MN}(\mathbf{n}).$$

By equating to zero the coefficient of φ we obtain (in the intermediate steps of the calculation we make use of Racah's rule for combining three Clebsch-Gordan coefficients into one):

$$X_l[l(l+1)-l'(l'+1)-i\rho] + (-)^{l-l'+1}X_{l'}[l'(l'+1)$$
$$-l(l+1)-i\rho](2l+1)/(2l'+1) = 0.$$

From this it follows that

$$X_l = C(2l+1)\Gamma(l+1+i\rho/2)/\Gamma(l+1-i\rho/2).$$

On utilizing the unitarity condition (17) already mentioned in the main text we obtain, finally, formula (16).

In order to obtain formula (20) we note that the function

$$Q_{\rho m}(\mathbf{n},\mathbf{k}) \equiv [1-(\mathbf{nk})]^{1+i\rho/2}U_{\rho m}(\mathbf{n},\mathbf{k}) \quad (A.4)$$

satisfies the same functional equation (A.1) which is satisfied also by $U_{\rho m}(\mathbf{n},\mathbf{k})$, only we must set in it $1+i\rho/2 = 0$. From this it follows that

$$Q_{\rho m}(\mathbf{n},\mathbf{k}) = A_{\rho m}\sum_{l\geqslant|m|}\frac{2l+1}{l(l+1)}\overline{D}^l_{\alpha-m}(\mathbf{n})D^l_{\alpha m}(\mathbf{k}).$$

In order to find A_ρ, we set $\mathbf{k} = -\mathbf{n}$ in (A.4).
Since

$$D^l(\mathbf{n}) = R_3(\varphi+\pi/2)R_1(\theta),$$
$$D^l(-\mathbf{n}) = R_3(\varphi+3\pi/2)R_1(\pi-\theta),$$

then

$$[D^l(\mathbf{n})^{-1}D^l(-\mathbf{n})]_{-m,m} = [R_1(-\theta)R_3(\pi)R_1(\pi-\theta)]^l_{-m,m}$$
$$= [R_1(-\pi)R_3(\pi)]^l_{-m,m} = e^{-im\pi}(-1)^{2l}(-1)^{l+m}e^{-im\pi}$$

Since

$$\sum_{l\geqslant|m|}\frac{2l+1}{l(l+1)}(-1)^{l+m}$$
$$= \sum_{l\geqslant|m|}\left(\frac{1}{l+1}(-1)^{l+m}+\frac{1}{l}(-1)^{l+m}\right) = (-1)^{2m}\frac{1}{m}$$

and, moreover,

$$\sum_{l\geqslant|m|}(-1)^{l+m}(2l+1)\Gamma(l+1+i\rho/2)/\Gamma(l+1-i\rho/2)$$
$$= \sum_{l\geqslant m}(-1)^{l+m}\left[\frac{\Gamma(l+2+i\rho/2)}{\Gamma(l+1-i\rho/2)}+\frac{\Gamma(l+1+i\rho/2)}{\Gamma(l-i\rho/2)}\right]$$
$$= (-1)^{2m}\Gamma(m+1+i\rho/2)/\Gamma(m-i\rho/2),$$

the proof of formula (20) is complete.

APPENDIX B

PROOF OF FORMULA (13)

According to formula (1.9b), I,

$$D^S(L_p\mathbf{n})D^S(\mathbf{n}) = D^S(L_p^{-1}\mathbf{n})R_3(\varphi(L_p,\mathbf{n})).$$

Further, from formula (A.2) of Appendix A in I it may be easily seen that

$$L_p^{-1}\mathbf{n}\xrightarrow[M\to 0]{}-\mathbf{k}.$$

Finally, we have,

$$[R(\mathbf{k})]^{-1} R(-\mathbf{k})$$
$$= R_1(-\vartheta) R_3(-\varphi - \pi/2) R_3(\varphi - \pi + \pi/2) R_1(\pi - \vartheta)$$
$$= R_3(-\pi) R_1(\pi),$$
$$[D^S(\mathbf{k})^{-1} D^S(-\mathbf{k})]_{mn} = [R_3(-\pi) R_1(\pi)]^S_{mn} = \delta_{n-m}(-1)^{S+m}.$$

Therefore,

$$\vartheta(\mathbf{p}, \mathbf{n}) = \lim_{M \to 0} \varphi(L_{\mathbf{p}}, \mathbf{n}). \qquad \text{(B.1)}$$

APPENDIX C

PROOF OF FORMULA (25)

In formula (A.1) we set $1 + i\rho/2 = 0$, $S = L\mathbf{p}$, with $\varphi(S, \mathbf{k}) = 0$. Further, we let M tend to zero; we then have

$$S^{-1}\mathbf{k} \to \mathbf{k}, \quad S^{-1}\mathbf{n} \to -\mathbf{k},$$

$$Q_m(S^{-1}\mathbf{n}, S^{-1}\mathbf{k}) \to Q_m(-\mathbf{k}, \mathbf{k}) = (-1)^{S+M}.$$

Thus from (A.5) and (B.1) we obtain formula (25).

[1]Chou Kuang-Chao and L. G. Zastavenko, JETP 35, 1417 (1958), Soviet Phys. JETP 8, 990 (1959).

[2]Chou Kuang-Chao, JETP 36, 909 (1959), Soviet Phys. JETP 9, 642 (1960).

Translated by G. Volkoff
21

SOVIET PHYSICS JETP VOLUME 11, NUMBER 1 JULY, 1960

SCATTERING OF GAMMA-RAY QUANTA BY NUCLEONS NEAR THE THRESHOLD FOR MESON PRODUCTION

L. I. LAPIDUS and CHOU KUANG-CHAO

Joint Institute for Nuclear Research

Submitted to JETP editor July 9, 1959

J. Exptl. Theoret. Phys. (U.S.S.R.) **38**, 201-211 (January, 1960)

The elastic scattering of γ-ray quanta near the threshold for single meson production is treated by means of dispersion relations. It is shown that when one takes into account meson production in the s state there are appreciable departures from monotonic variation with energy of the scattering amplitudes, cross sections, and other observable quantities near the threshold of the reaction. On definite assumptions about the analysis of photoproduction in the range of γ-ray energies up to 220 Mev, calculations are made of the scattering amplitude and the differential and total cross sections for elastic scattering of polarized and unpolarized γ-rays by protons, and also of the polarization of the recoil protons above the photoproduction threshold.

1. The study of the scattering of γ-ray quanta by nucleons is especially interesting near the threshold for single meson production. The region near the photoproduction threshold is of interest not only for comparisons with the predictions of dispersion relations, but also in particular in connection with the studies of departures from monotonic variation with the energy of the cross-sections (and polarizations) near the threshold of the reaction.[1] From this latter point of view the scattering of γ-ray quanta by nucleons and nuclei near the threshold for meson production is of especial interest as an example of a process going with a comparatively small cross section and being strongly perturbed above threshold by the process of intense meson production. Thus marked effects can be expected in the region near the threshold. It is clear that a sufficiently accurate experimental study of the anomalies near the threshold can be useful in understanding the process of pion production near threshold.

As a more detailed examination shows, the polarization effects are especially sensitive to the parameters characterizing the photoproduction of pions. Our main purpose here is a detailed examination of the effect of meson production on the cross section, the polarization of the recoil nucleons, and the polarization of the γ rays near the photoproduction threshold.

Phenomenological analysis and dispersion relations are used to obtain formulas useful for the analysis of experimental data. The results of the numerical calculations, which are based on definite assumptions about the analysis of the photoproduction, must be regarded as preliminary. In making the numerical estimates we have completely neglected fine-structure effects associated with the mass difference of the mesons (and of the nucleons).

There are many well known papers in which the scattering of γ-ray quanta by nucleons has been treated by various methods (see literature references in our previous paper[2]). In the present paper we have tried to manage with a minimum number of assumptions, without resorting to approximate methods, whose use is hard to justify. We consider not only the scattering cross sections for unpolarized γ rays, but also the polarization effects in the scattering. In this connection we have also considered the polarization of the γ rays.

2. Let us represent the transition matrix in the form

$$\mathcal{M} = \sum_{\mu\nu} e'_\mu N_{\mu\nu} e_\nu \equiv (\mathbf{e}'N\mathbf{e}).$$

Let us choose two coordinate systems x, y, z and x′, y′, z′ in which the z and z′ axes are parallel to the initial and final momenta of the photon, and the y and y′ axes are in the same direction. In these coordinates the functions for the spin eigenstates of the photon with the eigenvalues $S_z = \pm 1$ have the following form:

$$\zeta_1 = -(\mathbf{h} - i\mathbf{j})/\sqrt{2}, \quad \zeta_{-1} = (\mathbf{h} + i\mathbf{j})/\sqrt{2},$$
$$\zeta'_1 = -(\mathbf{h}' - i\mathbf{j})/\sqrt{2}, \quad \zeta'_{-1} = (\mathbf{h}' + i\mathbf{j})/\sqrt{2}, \quad (1)$$

where h, j, and k are unit basis vectors directed along these coordinate axes. In the general case the polarization state of the photon will be a linear combination, i.e.,

$$\mathbf{e} = c_1\zeta_1 + c_{-1}\zeta_{-1}, \tag{2}$$

where c_1 and c_{-1} are the respective probabilities (sic) of the photon states with $S_z = +1$ and $S_z = -1$.

Using the spin eigenstates as the basis of the representation, we can write the transition matrix in the form

$$\mathscr{M} = \begin{pmatrix} (\zeta_1^{\prime *}\mathbf{N}\zeta_1) & 0 & (\zeta_1^{\prime *}\mathbf{N}\zeta_{-1}) \\ 0 & 0 & 0 \\ (\zeta_{-1}^{\prime *}\mathbf{N}\zeta_1) & 0 & (\zeta_{-1}^{\prime *}\mathbf{N}\zeta_{-1}) \end{pmatrix}. \tag{3}$$

Let us further introduce the density matrix of the photon in the form

$$\rho = \begin{pmatrix} c_1c_1^* & 0 & c_1c_{-1}^* \\ 0 & 0 & 0 \\ c_{-1}c_1^* & 0 & c_{-1}c_{-1}^* \end{pmatrix}. \tag{4}$$

The density matrix ρ_f of the final state is connected with the density matrix ρ_{in} of the initial state by the relation

$$\rho_f = \mathscr{M}\rho_{in}\mathscr{M}^+. \tag{5}$$

Although in Eqs. (3) and (4) the transition matrix and the density matrix are written as three-rowed matrices, they have only four independent nonzero elements. Consequently we can represent them by means of two-rowed matrices and use the well known apparatus of the Pauli matrices.[3]

$$\mathscr{M} = \begin{pmatrix} (\zeta_1^{\prime *}\mathbf{N}\zeta_1) & (\zeta_1^{\prime *}\mathbf{N}\zeta_{-1}) \\ (\zeta_{-1}^{\prime *}\mathbf{N}\zeta_1) & (\zeta_{-1}^{\prime *}\mathbf{N}\zeta_{-1}) \end{pmatrix} = A + \mathbf{B}\boldsymbol{\sigma}_\gamma, \tag{6}$$

$$\rho = \begin{pmatrix} c_1c_1^* & c_1c_{-1}^* \\ c_{-1}c_1^* & c_{-1}c_{-1}^* \end{pmatrix} = \frac{1}{2}(1 + \boldsymbol{\sigma}_\gamma\mathbf{P}), \tag{7}$$

where P_x, P_y, and P_z are the Stokes parameters. Nonvanishing P_x and P_y correspond to linear polarization of the photons along the x and y axes, while $P_z \neq 0$ corresponds to circular polarization of the photon.

From Eq. (6) it is not hard to get

$$\begin{aligned} 2A &= (\zeta_1^{\prime *}\mathbf{N}\zeta_1) + (\zeta_{-1}^{\prime *}\mathbf{N}\zeta_{-1}) = \mathrm{Sp}\,\mathscr{M}, \\ 2B_z &= (\zeta_1^{\prime *}\mathbf{N}\zeta_1) - (\zeta_{-1}^{\prime *}\mathbf{N}\zeta_{-1}) = \mathrm{Sp}(\sigma_\gamma^z\mathscr{M}), \\ 2B_x &= (\zeta_1^{\prime *}\mathbf{N}\zeta_{-1}) + (\zeta_{-1}^{\prime *}\mathbf{N}\zeta_1) = \mathrm{Sp}\,(\sigma_\gamma^x\mathscr{M}), \\ 2iB_y &= (\zeta_1^{*}\mathbf{N}\zeta_{-1}) - (\zeta_{-1}^{\prime *}\mathbf{N}\zeta_1) = i\,\mathrm{Sp}\,(\sigma_\gamma^y\mathscr{M}) \end{aligned} \tag{8}$$

where the spurs (traces) are taken over the photon variables. The quantities A and B_i can be connected with the quantities $R_1, \ldots, R_6$, which were introduced in reference 2 (hereafter referred to as I) and which determine the matrix $\mathscr{M}$ [cf. Eq. (I, 15)]:

$$\begin{aligned} 2A &= (R_1 + R_2)(1 + \cos\theta) - i(R_3 + R_4)\sin\theta(\boldsymbol{\sigma}\mathbf{n}), \\ 2B_z &= (R_3 + R_4)\boldsymbol{\sigma}(\mathbf{k} + \mathbf{k}') + (1 + \cos\theta)(R_5 + R_6)\boldsymbol{\sigma}(\mathbf{k} + \mathbf{k}'), \\ 2iB_y &= [R_3 - R_4 - (1 - \cos\theta)(R_5 - R_6)]\boldsymbol{\sigma}(\mathbf{k} - \mathbf{k}'), \\ 2B_x &= (R_1 - R_2)(1 - \cos\theta) + i(R_3 + R_4)\sin\theta(\boldsymbol{\sigma}\mathbf{n}), \end{aligned} \tag{9}$$

where $\mathbf{n}\sin\theta = \mathbf{k}\times\mathbf{k}'$, $\cos\theta = \mathbf{k}\cdot\mathbf{k}'$.

It is easy to calculate the density matrix of the final state:

$$\begin{aligned} \rho_f &= \tfrac{1}{2}(A + \boldsymbol{\sigma}_\gamma\mathbf{B})(1 + \boldsymbol{\sigma}_\gamma\mathbf{P})(A^+ + \boldsymbol{\sigma}_\gamma\mathbf{B}^+) \\ &= \tfrac{1}{2}\{AA^+ + \mathbf{B}\mathbf{B}^+ + (A\mathbf{B}^+ + \mathbf{B}A^+)\mathbf{P} - i([\mathbf{B}\mathbf{B}^+]\mathbf{P})\} \\ &+ \tfrac{1}{2}\boldsymbol{\sigma}_\gamma\{A\mathbf{B}^+ + \mathbf{B}A^+ + i[\mathbf{B}\mathbf{B}^+] + (AA^+ - \mathbf{B}\mathbf{B}^+)\mathbf{P} + [\mathbf{B}(\mathbf{P}\mathbf{B}^+) \\ &+ (\mathbf{B}\mathbf{P})\mathbf{B}^+] + iA[\mathbf{P}\mathbf{B}^+] - i[\mathbf{P}\mathbf{B}]A^+\}. \end{aligned} \tag{10}$$

By means of the expression (10) one can calculate all observable quantities. For the interaction of unpolarized γ rays and nucleons the differential cross section will have the form

$$d\sigma/do \equiv I_0(\theta) = \tfrac{1}{2}\mathrm{Sp}\,(AA^+ + \mathbf{B}\mathbf{B}^+), \tag{11}$$

where the spur is taken over the nucleon variables. Substituting Eq. (9) in Eq. (11), we get

$$\begin{aligned} 4I_0(\theta) &= |R_1 + R_2|^2(1 + \cos^2\theta) + |R_1 - R_2|^2(1 - \cos\theta)^2 \\ &+ |R_3 + R_4|^2(3 - \cos^2\theta + 2\cos\theta) \\ &+ |R_3 - R_4|^2(3 - \cos^2\theta - 2\cos\theta) \\ &+ 2|R_5 + R_6|^2(1 + \cos\theta)^3 + 2|R_5 - R_8|^2(1 - \cos\theta)^3 \\ &+ 4\mathrm{Re}\,(R_3 + R_4)^*(R_5 + R_6)(1 + \cos\theta)^2 \\ &- 4\mathrm{Re}\,(R_3 - R_4)^*(R_5 - R_6)(1 - \cos\theta)^2. \end{aligned} \tag{12}$$

The expression for the polarization of the nucleon after the interaction of an initially unpolarized photon and nucleon can be represented in the form*

$$\begin{aligned} 2I_0(\theta)\langle\boldsymbol{\sigma}\rangle_f &= \sin\theta\,\mathbf{n}\,\mathrm{Im}\,[(R_3 + R_4)(R_1 + R_2)^*(1 + \cos\theta) \\ &- (R_3 - R_4)(R_1 - R_2)^*(1 - \cos\theta)] \\ &= 2i\,[\mathbf{k}\times\mathbf{k}']\{R_1R_4^* - R_1^*R_4 + R_2R_3^* - R_2^*R_3 \\ &+ [R_1R_3^* - R_1^*R_3 + R_2R_4^* - R_2^*R_4]\cos\theta\}. \end{aligned} \tag{13}$$

The well known fact that the cross section $I_0(\theta)$ does not change when one replaces electric transitions by magnetic appears in the fact that Eq. (12) is invariant under the simultaneous interchanges:

$$R_1 \rightleftarrows R_2, \quad R_3 \rightleftarrows R_4, \quad R_5 \rightleftarrows R_6. \tag{14}$$

It can be seen from Eq. (13) that the expression for the polarization of the recoil nucleon also remains unchanged by this transformation.

3. Let us now establish the relations between the Stokes parameters and the statistical tensor moments. As is well known, the statistical tensor moments are defined by the relations

*Eqs. (19), (23), and (24) in reference 4 contain errors.

$$T_{00} = 1/\sqrt{3}, \quad T_{10} = S_z/\sqrt{2}, \quad T_{20} = \sqrt{\tfrac{2}{3}}\left(\tfrac{3}{2}S_z^2 - 1\right),$$
$$T_{22} = \tfrac{1}{2}[S_x^2 - S_y^2 + i(S_xS_y + S_yS_x)],$$
$$T_{2-2} = \tfrac{1}{2}[S_x^2 - S_y^2 - i(S_xS_y + S_yS_x)]. \quad (15)$$

They are normalized so that

$$\mathrm{Sp}\, T_{JM}T^+_{J'M'} = \delta_{JJ'}\delta_{MM'}. \quad (16)$$

By means of these tensor moments the density matrix can be represented in the form

$$\rho_f = \rho_{00}T_{00} + \rho_{10}T_{10} + \rho_{20}T_{20} + \rho_{22}T_{22} + \rho_{2-2}T_{2-2}. \quad (17)$$

Here $\rho_{00} = 2^{1/2}\rho_{20} = 3^{-1/2}$. The parameters ρ_{JM} are connected with the Stokes parameters:

$$\rho_{10} = \sqrt{2}P_z, \quad \rho_{22} = P_x - iP_y, \quad \rho_{2-2} = P_x + iP_y. \quad (18)$$

In virtue of time-reversal invariance,[5,4] the expression for the cross section $I(\theta, \varphi)$ for scattering of a polarized γ-ray beam by unpolarized protons can be put in the form

$$I(\theta, \varphi) = I_0(\theta)[1 + 2\langle T_{22}\rangle_i \langle T_{22}\rangle_f \cos 2\varphi], \quad (19)$$

where

$$2I_0(\theta)\langle T_{22}\rangle_f = \sin^2\theta(|R_1|^2 + |R_4|^2 - |R_2|^2 - |R_3|^2), \quad (20)$$

$\langle T_{22}\rangle_i$ is the initial polarization of the γ-ray beam. We note that the expression (20) changes sign under the transformation (14).

4. For practical calculations we use the results of reference 2. The deviations from monotonic variation of the cross-section for scattering of γ rays in the immediate neighborhood of the meson-production threshold are due to the production of mesons in the s state. According to the available experimental data, the cross section for production of π^+ mesons in the s state is much larger than the cross section for production of π^0 mesons in this state. Difficult experiments had to be done even to establish the fact that π^0 mesons indeed are produced in the s state.

The energy ν of a photon in the laboratory system (l.s.) and the energy ν_c in the center-of-mass system (c.m.s.) are connected by the well known relation

$$\nu_c = \nu/\sqrt{1 + 2\nu/M}.$$

Using the expression for the total energy of the meson in the c.m.s.

$$\omega_c = (\nu + m_\pi^2/2M)/\sqrt{1 + 2\nu/M}$$

we find without difficulty that the expression for the square of the momentum of the meson produced

$$q_c^2 = \omega_c^2 - m_\pi^2 = (\nu - \nu_0)(\nu + \nu_0 - m_\pi^2/2M)/(1 + 2\nu/M);$$
$$\nu_0 = m_\pi(1 + m_\pi/2M)$$

can with some accuracy (better than 7.5 percent) be replaced by

$$q_c^2 \approx (\nu^2 - \nu_0^2)/(1 + 2\nu/M). \quad (21)$$

This then gives

$$q_c/\nu_c = (\nu^2 - \nu_0^2)^{1/2}/\nu.$$

Because there is a mass difference between neutron and proton and between π^+ and π^0 mesons the effects near threshold in the scattering of γ rays by nucleons have a "fine structure." To make reliable numerical calculations one would need a much more detailed analysis of the data on photoproduction than we now have available. Wishing to get an idea of the scale of size of the effects near threshold, we shall confine ourselves mainly to a consideration of π^+-meson production. The quantities E_1, E_{33}, and M_{33} are taken from the analysis of Watson and others.[6] The production of π^0 mesons is taken into account only in the resonance state (through M_{33}). The connection of the photoproduction amplitudes with the π-N scattering phase shifts is well known (cf., e.g., references 6, 7).* In the range of energies where we can neglect the difference between the $\nu_0(\pi^0)$ and $\nu_0(\pi^+)$ thresholds, when we sum the contributions from π^+ and π^0 mesons the terms containing the π-N scattering phase shifts cancel each other. For example

$$|M_3^+|^2 + |M_3^0|^2 = 6|M_{33}|^2 + \tfrac{3}{4}|M_{13}^{(1)} - 2\delta M_{13}^{(1)}|^2 \approx 6|M_{33}|^2$$

We have calculated the dispersion integrals by using simple expressions to interpolate the energy dependence of $|E_1|^2$ and $|M_{33}|^2$ and then integrating directly. Setting $\nu_0 = 150$ Mev, and hereafter measuring energies in terms of ν_0, in the range $1 \le \nu \le \nu_1 = 2.20$ we approximate the energy dependence of $|E_1|^2$ by the following expression:

$$|E_1|^2 \approx |E_1^+|^2 = A\sqrt{\nu^2 - 1}/\nu,$$
$$A = (3.3\cdot 10^{-15}\,\mathrm{cm/sr}^{1/2})^2\nu_0 = 0.54\, e^2/M. \quad (22)$$

It is just the contribution E_1^2 in the dispersion integrals that leads to the nonmonotonic behavior in the energy dependence of the real parts of the amplitudes. As can be seen from (I, 42), the contribution of $|E_1|^2$ is characterized by two integrals

$$\frac{2\nu^2}{\pi}\int_1^{\nu_1}\frac{|E_1|^2}{\nu'^2 - \nu^2}d\nu', \quad \frac{2\nu^3}{\pi}\int_1^{\nu_1}\frac{|E_1|^2 d\nu'}{\nu'(\nu'^2 - \nu^2)}. \quad (23)$$

Substitution of Eq. (22) in the expression (23) gives

*In the more general form of the problem one requires the parametrization of a three-rowed S matrix, which describes bot the photo-production and scattering of π mesons and also the scattering of γ rays by nucleons. For the scattering of γ rays effects of deviations from isotopic invariance can give additions to the scattering phase shifts (and to the mixing coefficients) that are by no means small.

$$\frac{2\nu^2}{\pi}\int_1^{\nu_1}\frac{|E_1|^2}{\nu'^2-\nu^2}d\nu' = \frac{2}{\pi}A\times\begin{cases}\tan^{-1}(\nu_1^2-1)^{1/2} - \dfrac{(\nu^2-1)^{1/2}}{2}\ln\left|\dfrac{(\nu_1^2-1)^{1/2}+(\nu^2-1)^{1/2}}{(\nu_1^2-1)^{1/2}-(\nu^2-1)^{1/2}}\right|, & \nu>1\\ \tan^{-1}(\nu_1^2-1)^{1/2}-(1-\nu^2)^{1/2}\tan^{-1}\sqrt{\dfrac{\nu_1^2-1}{1-\nu^2}}, & \nu<1\end{cases} \quad (24)$$

and

$$\frac{2}{\pi}\nu^3\int_1^{\nu_1}\frac{|E_1|^2}{\nu'(\nu'^2-\nu^2)}d\nu' = \frac{2}{\pi}A\nu\times\begin{cases}\left(\dfrac{\nu_1^2-1}{\nu_1^2}\right)^{1/2} - \dfrac{(\nu^2-1)^{1/2}}{2\nu}\ln\left|\dfrac{\nu(\nu_1^2-1)^{1/2}+\nu_1(\nu^2-1)^{1/2}}{\nu(\nu_1^2-1)^{1/2}-\nu_1(\nu^2-1)^{1/2}}\right|, & \nu>1\\ \left(\dfrac{\nu_1^2-1}{\nu_1^2}\right)^{1/2}-\left(\dfrac{1-\nu^2}{\nu^2}\right)^{1/2}\tan^{-1}\sqrt{\dfrac{\nu^2(\nu_1^2-1)}{\nu_1^2(1-\nu^2)}}, & \nu<1.\end{cases} \quad (25)$$

From Eqs. (24), (25), (22), (I, 42), and (I, 32) it can be seen that at the meson-production threshold the derivatives of the quantities R_1 and R_3 go to infinity (approaching threshold from the side $\nu > 1$), and the derivatives of the real parts of these quantities also go to infinity (on the side $\nu < 1$), whereas on the other side of threshold the derivatives are finite. This result is very general. Thus the dispersion relations turn out to contain specific effects near the reaction threshold like those discussed and analyzed without use of the dispersion relations by Wigner, Baz', Okun', Breit, Capps, Newton, and others.*

The use of dispersion relations makes it possible to examine in more detail the effect on the elastic scattering (or on the reaction) of the inelastic processes that occur in a certain energy range. Moreover, the interesting effects that occur in the immediate neighborhood of the reaction threshold ("local effects" which could be discussed when one does not use the method of analytic continuation given by the dispersion relations) are only a part of the total effect of the inelastic processes on the energy dependences of the quantities that characterize the elastic scattering.

From the example of the scattering of γ rays by protons we can see how the presence of the inelastic process of meson photoproduction in the energy range $\nu > 1$ affects the characteristics of the elastic scattering, including also effects for $\nu < 1$ (deviation from the Powell formula, or from Eq. (I.16) for $\gamma < 1$). The deviation from monotonic variation in Eqs. (24) and (25) is characterized by a sharp drop from the value of the function at $\nu = 1$ in the region $\nu < 1$ (with an infinite derivative at $\nu = 1$) and a slow drop in the region $\nu > 1$ (with a finite derivative at $\nu = 1$).

5. In the range of energies 330 – 500 Mev ($2.2 < \nu < 3.34$) the quantity $|E_1|^2$ is represented in the form

$$|E_1|^2 = 1.27(1-0.175\nu)^2e^2/M. \quad (26)$$

The contribution from this energy range to the values of the real parts of the amplitudes is small, if for the scattering of the γ rays we consider the energy near and below the threshold.

The analysis of the photoproduction made previously, and particularly the results of Akiba and Sato, indicate that

$$|M_3|^2 = 6|M_{33}|^2 \approx |E_2|^2 \approx \mathrm{Re}(E_2^* M_3). \quad (27)$$

For our estimates we adopt Eq. (27). The polarization of the recoil nucleons is especially sensitive to this assumption. In the energy range $1 < \nu < 2$ the quantity $|M_{33}|^2$ can be approximated by the expression

$$|M_{33}|^2 = B_0\nu(\nu^2-1)^{3/2}, \qquad B_0 = 0.009e^2/M. \quad (28)$$

Consequently,

$$|M_3|^2 = 6|M_{33}|^2 = B\nu(\nu^2-1)^{3/2}, \qquad B = 0.054e^2/M,$$

and the contribution of this expression, which describes the production of mesons in the p state, to the dispersion relations is given by the integrals

$$\frac{2\nu^2}{\pi}\int_1^{\nu_1}\frac{|E_2|^2}{\nu'^2-\nu^2}d\nu' = \frac{2B}{\pi}\nu^2\left[\frac{1}{3}(\nu_1^2-1)^{3/2}+(\nu_1^2-1)^{1/2}(\nu^2-1)\right]$$
$$+\frac{2B}{\pi}\nu^2\begin{cases}-\frac{1}{2}(\nu^2-1)^{3/2}\ln\left|\dfrac{(\nu_1^2-1)^{1/2}+(\nu^2-1)^{1/2}}{(\nu_1^2-1)^{1/2}-(\nu^2-1)^{1/2}}\right|, & \nu>1\\ (1-\nu^2)^{3/2}\tan^{-1}\sqrt{(\nu_1^2-1)/(1-\nu^2)}, & \nu<1\end{cases}, \quad (29)$$

and

$$\frac{2\nu^3}{\pi}\int_1^{\nu_1}\frac{d\nu'}{\nu'}\frac{|E_2|^2}{\nu'^2-\nu^2} = \frac{B}{\pi}\nu^3\left[\nu_1(\nu_1^2-1)^{1/2}+(\nu^2-3/2)\ln\left|\frac{\nu_1+(\nu_1^2-1)^{1/2}}{\nu_1-(\nu_1^2-1)^{1/2}}\right|\right]$$
$$-\frac{B}{\pi}\nu^3(\nu^2-1)\begin{cases}\sqrt{\dfrac{\nu^2-1}{\nu^2}}\ln\left|\dfrac{\nu(\nu_1^2-1)^{1/2}+\nu_1(\nu^2-1)^{1/2}}{\nu(\nu_1^2-1)^{1/2}-\nu_1(\nu^2-1)^{1/2}}\right|, & \nu>1\\ 2\sqrt{\dfrac{1-\nu^2}{\nu^2}}\tan^{-1}\dfrac{\sqrt{(\nu_1^2-1)/\nu_1^2}}{\sqrt{(1-\nu^2)/\nu^2}}, & \nu<1\end{cases}, \quad (30)$$

which have the characteristic feature that the second derivative with respect to the energy goes to infinity (again on the side $\nu < 1$).

*The writers plan to turn to the application of dispersion relations to this problem in another paper.

In the energy range $2 < \nu < 3.34$

$$6\,|M_{33}|^2 = 2.17\,(1 - 0.244\nu)^2 e^2 / M. \tag{31}$$

The contributions of the expressions (28) and (31) are given by integrals of the forms

$$J_1(\nu) = \frac{2\nu^3}{\pi}\int_{\nu_1}^{\nu_2} \frac{\alpha + \beta\nu' + \gamma\nu'^2}{\nu'(\nu'^2 - \nu^2)}\,d\nu'$$

$$= \frac{\nu}{\pi}\ln\left\{\left(\frac{\nu_2 - \nu}{\nu_1 - \nu}\right)^{\alpha+\beta\nu+\gamma\nu^2}\left(\frac{\nu_2 + \nu}{\nu_1 + \nu}\right)^{\alpha-\beta\nu+\gamma\nu^2}\left(\frac{\nu_1}{\nu_2}\right)^{2\alpha}\right\}, \tag{32}$$

$$J_2(\nu) = \frac{2\nu^2}{\pi}\int_{\nu_1}^{\nu_2} \frac{d\nu'\,(\alpha + \beta\nu' + \gamma\nu'^2)}{\nu'^2 - \nu^2}$$

$$= \frac{\nu}{\pi}\left\{2\gamma\nu(\nu_2 - \nu_1) + (\alpha + \gamma\nu^2)\ln\left(\frac{\nu_2 - \nu}{\nu_1 - \nu}\,\frac{\nu_1 + \nu}{\nu_2 + \nu}\right)\right.$$

$$\left. + \beta\nu\ln\left(\frac{\nu_2^2 - \nu^2}{\nu_1^2 - \nu^2}\right)\right\} \tag{33}$$

6. The energy dependences of the real parts of the amplitudes $R_1, \ldots, R_6$ (in the l.s.), calculated by means of dispersion relations, are shown in Fig. 1, a, b. The half-widths of Re (R_1) and Re (R_2)

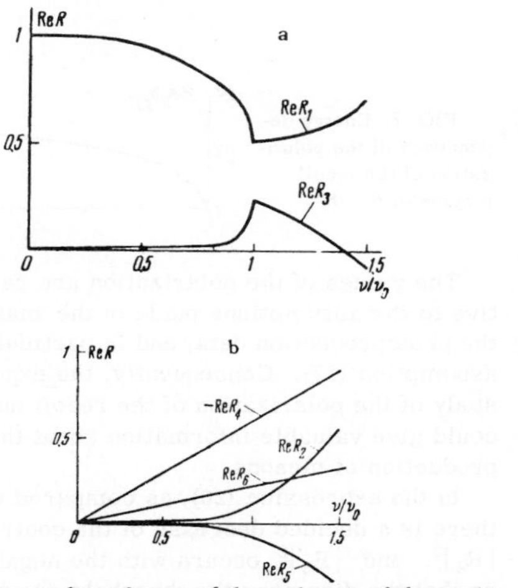

FIG. 1. Energy dependences of the real parts of the amplitudes R_1 and R_3 (a) and R_2, R_4, R_5, and R_6 (b) (the values of the functions are expressed in terms of e^2/M as a unit).

are $\nu_0/10$ and $\nu_0/20$, respectively, and are mainly due to the square of the ratio of the real part to the coefficient A in Eq. (22):

$$\varepsilon = 1 - \nu = \tfrac{1}{8}(\mathrm{Re}\,R)^2/A^2. \tag{34}$$

In a general analysis of the nonmonotonic behavior near the threshold A. I. Baz' has given for the width of the peak restrictions of the form $r_0(1-\nu^2)^{1/2} \ll 1$ (where r_0 is the interaction radius). The more detailed treatment of the present paper has automatically given the more accurate criterion (34). The effect of the inelastic processes on Re (R_3) is very strong, although the contribution of Re (R_3) to the observable quantities is small, so that the experimental study of the energy dependence of Re (R_3) is a difficult problem. The energy dependence of Re (R_4) and Re (R_6) is given with great accuracy by the general relation (I, 18). The departure from zero of Re (R_2) and Re (R_5) is entirely due to inelastic processes, but the production of mesons in the s state does not contribute to these quantities.

The differential scattering cross section (in the c.m.s.) (12) can be written in the form

$$I_0(\theta, \nu) = A_0(\nu) + A_1(\nu)\cos\theta$$

$$+ A_2(\nu)\cos^2\theta + A_3(\nu)\cos^3\theta. \tag{35}$$

The results of calculations for the scattering angles 90° and 0° are shown in Figs. 2 and 3. We at once note the marked difference between the energy dependences of the cross sections at $\theta = 0$ and at 90°.

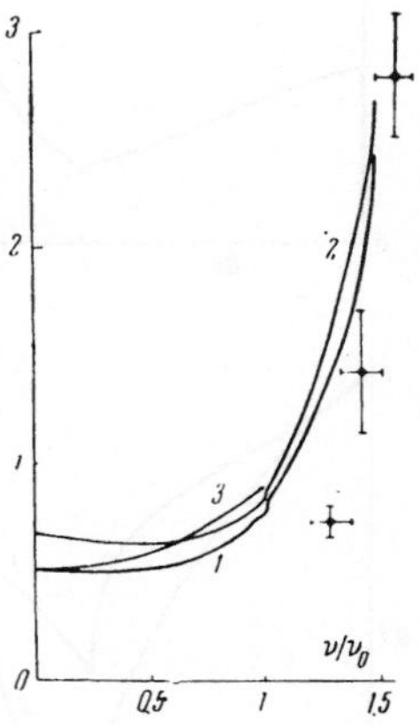

FIG. 2. Energy dependences of the differential cross section $I_0(90°)$ (curve 1), the total scattering cross section (curve 2), and the differential cross section with the dispersion part not taken into account (curve 3) (the values of the functions are expressed in terms of $(e^2/M)^2$ as a unit). Experimental data from reference 9.

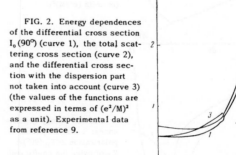

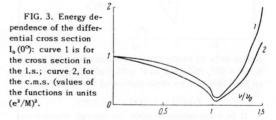

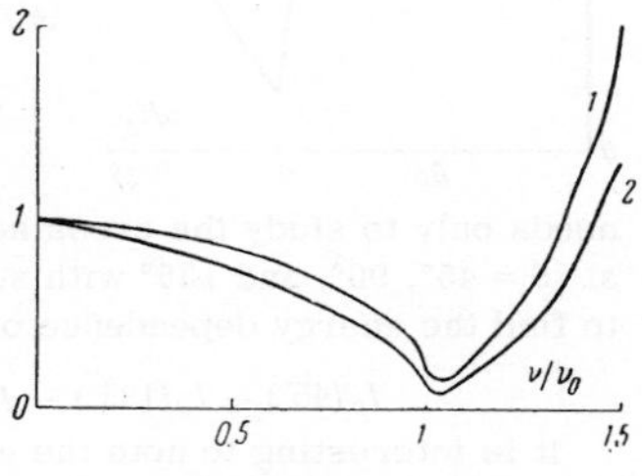

FIG. 3. Energy dependence of the differential cross section $I_0(0°)$: curve 1 is for the cross section in the l.s.; curve 2, for the c.m.s. (values of the functions in units $(e^2/M)^2$.

The function $I_0(0°, \nu)$ has been calculated earlier by Cini and Stroffolini.[8] We have improved the accuracy in the region near the threshold. Outside this region there is good agreement between the two calculations. Our results relating to $I_0(90°, \nu)$ in the energy region near 200 Mev also agree with other published calculations.[9] A new contribution is the careful treatment of the region near threshold, in which there are effects not discussed previously.

Figure 2 shows the energy dependence of the total cross section for elastic scattering, and also shows for comparison the energy dependence of the cross section calculated from Eqs. (16) and (18). The effects near threshold are practically imperceptible, but the difference between the two curves shows the general effect of inelastic processes on the elastic-scattering cross section.

The local effects are much more prominent if we calculate the difference

$$\sigma_s/4\pi - I_0(90°,\ \nu)$$

or the dependence of A_2 on the energy ν (Figs. 4 and 5). To get experimental data on A_2 one

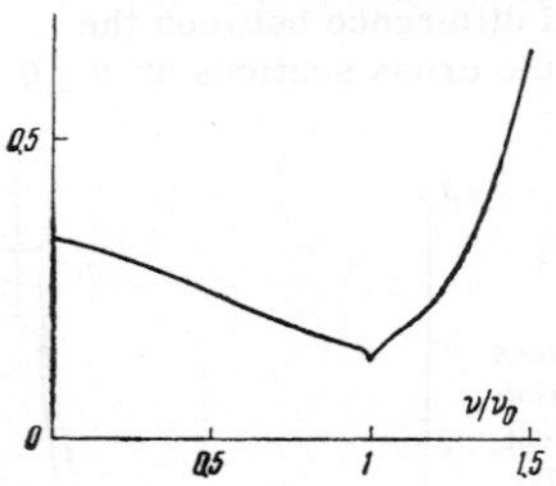

FIG. 4. Energy dependence of $2[\sigma_s/4\pi - I_0(90°)]$ (in units $(e^2/M)^2$).

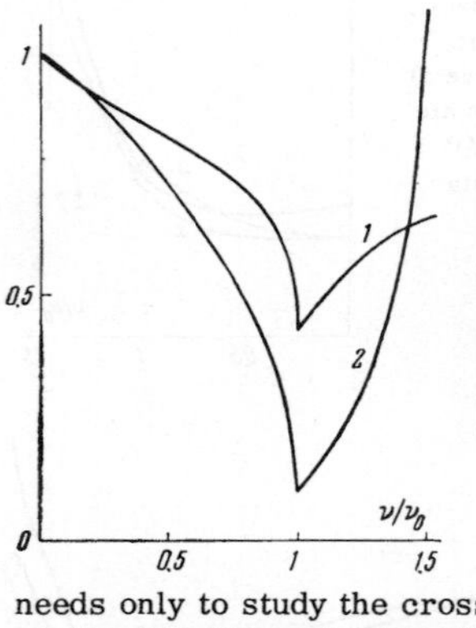

FIG. 5. Energy dependences: 1 – of the photon polarization $2\langle T_{22}(90°)\rangle$; 2 – of twice the coefficient $A_2(\theta)$ of the $\cos^2\theta$ term in the cross section (values of functions in units $(e^2/M)^2$).

needs only to study the cross sections $I_0(\theta, \nu)$ at $\theta = 45°$, 90°, and 135° with sufficient accuracy to find the energy dependence of the difference

$$I_0(45°) + I_0(135°) - I_0(90°).$$

It is interesting to note the energy dependence of the polarization of the recoil nucleon. Below the meson-production threshold the imaginary parts of the quantities $R_1, \ldots, R_6$ vanish in the e^2 approximation, the right member of Eq. (13) is zero, and there is no polarization of the recoil nucleon. Below threshold, in virtue of invariance under time reversal, the cross section for scattering by polarized protons does not differ from $I_0(\theta)$. Above the threshold for production of π mesons there is a nonvanishing polarization of the recoil nucleons. The values of the imaginary parts of the amplitudes above threshold are shown in Fig. 6. The results of calculations on the dependence of the polarization at $\theta = 90°$ (angle in c.m.s.) on the photon energy (in the l.s.) are shown in Fig. 7. It can be seen that over a rather wide range of energies, 180 – 220 Mev, the polarization reaches 20 to 25 percent.

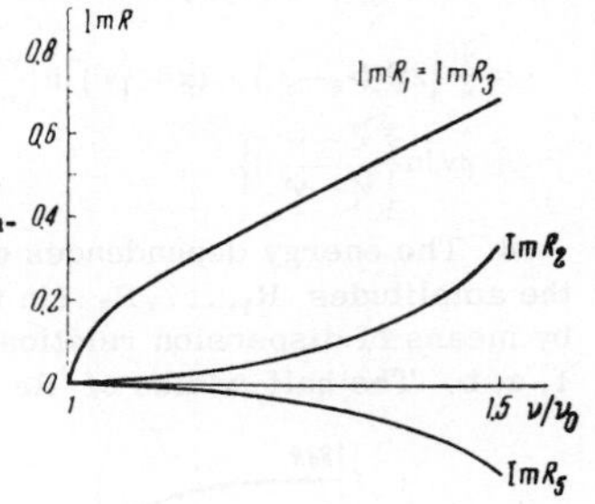

FIG. 6. Energy dependence of the imaginary parts of the amplitudes (in units e^2/M).

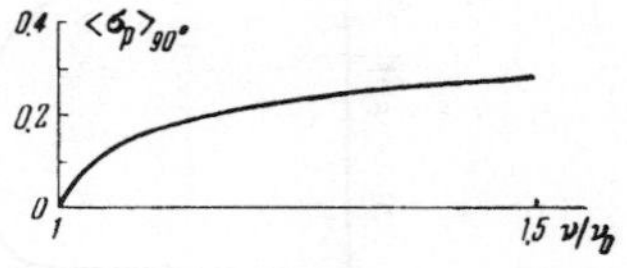

FIG. 7. Energy dependence of the polarization of the recoil protons at $\theta = 90°$.

The values of the polarization are rather sensitive to the assumptions made in the analysis of the photoproduction data, and in particular to the assumption (27). Consequently, the experimental study of the polarization of the recoil nucleons could give valuable information about the photoproduction of mesons.

In the expression (20), as compared with $I_0(\theta)$, there is a decided decrease of the contribution of $|R_4|^2$, and $|R_3|^2$ occurs with the negative sign, so that the dips near the threshold are particularly marked in the energy dependence of $\langle T_{22}(90°)\rangle$ (Fig. 5).

7. A detailed examination of the scattering of γ rays by nucleons in the region near the meson-production threshold, made by the use of dispersion relations, has made it possible to see what effect the production of mesons in the s state has on the anomalies near the threshold. The scattering of γ rays by nucleons and by nuclei is an example of the sort of process in which the energy dependence of the amplitudes is especially strongly affected by inelastic processes and the effects extend over a wide range of energies. In γ-N scattering the local effects on a number of observable

quantities are quite appreciable, but rather severe requirements are imposed on the procedures for experimental studies, expecially as regards resolution in energy, since the widths of the dips in question are of the order of 5 to 10 Mev.

The treatment given in the present paper shows that the effects near threshold are sometimes masked by the strong energy dependence of the scattering amplitudes. Therefore it seems that the most favorable conditions for the experimental study of such effects should be found at small energies, and also for the interaction of particles with small spins.

In the case of γ-N scattering, besides the contribution of the "peak" amplitudes R_1 and R_3, there are large effects from other amplitudes, particularly from R_4. The effects of these "smearing-out" factors may be smaller in the scattering of γ rays by helium nuclei (or other spinless nuclei), since in this case the transition matrix will have the form

$$M = R_1'(ee') + R_2'(ss').$$

A treatment of the scattering of γ-ray quanta by deuterons near the threshold for the photodisintegration of the deuteron, where local effects will evidently be large, will be presented in another paper.

From the point of view of the general effect of some processes on others it is interesting to analyze the photodisintegration of the deuteron in the energy range near and below the threshold for meson production. Noting the results of the calculations on the γ-N scattering, we can evidently suppose that the well known "resonance" energy dependence of the cross section for the photodisintegration of the deuteron is due to meson-production processes above threshold and can be treated by a method using dispersion relations.

It is commonly assumed that at quite high γ-ray energies the γ-N scattering cross sections will be almost entirely due to inelastic processes, i.e., to the imaginary parts of the amplitudes. In this connection it may be very interesting to study γ-N scattering, and especially the polarization of the recoil nucleons, near the thresholds of reactions of the production of new particles, such a:

$$\gamma + N \rightarrow Y + K,$$

and a number of other processes. In this case the difficulties associated with the size of the cross section and the low energy of the recoil nucleon may very probably be smaller.

The writers are deeply grateful to B. Pontecorvo and Ya. Smorodinskiĭ for helpful discussion

[1] E. Wigner, Phys. Rev. **73**, 1002 (1948). A. I. Baz', JETP **33**, 923 (1957), Soviet Phys. JETP **6**, 709 (1958). G. Breit, Phys. Rev. **107**, 1612 (1957) A. I. Baz' and L. B. Okun', JETP **35**, 757 (1958), Soviet Phys. JETP **8**, 526 (1958). R. K. Adair, Phys. Rev. **111**, 632 (1958).

[2] L. I. Lapidus and Chou Kuang-Chao, JETP **37**, 1714 (1959), Soviet Phys. JETP **10**, 1213 (1960

[3] H. A. Tolhoek, Revs. Modern Phys. **28**, 277 (1956).

[4] L. I. Lapidus, JETP **34**, 922 (1958), Soviet Phys. JETP **7**, 638 (1958).

[5] L. Wolfenstein and J. Ashkin, Phys. Rev. **85**, 947 (1952). L. Wolfenstein, Ann. Rev. Nuclear Sci. **6**, 43 (1956).

[6] Watson, Keck, Tollestrup, and Walker, Phys. Rev. **101**, 1159 (1956).

[7] E. Fermi, Suppl. Nuovo cimento **2**, 17 (1955). (Russian Transl., IIL, 1956).

[8] M. Cini and A. Stroffolini, Nuclear Phys. **5**, 684 (1958).

[9] T. Akiba and J. Sato, Progr. Theoret. Phys. **19**, 93 (1958). G. Chew, Proc. Annual Internat. Conf. on High Energy Physics at CERN, 1958, p. 93.

Translated by W. H. Furry
33

ON THE PRODUCTION OF AN ELECTRON-POSITRON PAIR BY A NEUTRINO IN THE FIELD OF A NUCLEUS

A. M. BADALYAN and CHOU KUANG-CHAO

Submitted to JETP editor November 26, 1959

J. Exptl. Theoret. Phys. (U.S.S.R.) **38**, 664-665 (February, 1960)

PRESENT experimental possibilities have allowed a rather close approach to a measurement of the cross section for scattering of a neutrino by an electron.[1] This process is a very important one for testing the theory of the universal weak interaction.

In the laboratory system, in which the electron is at rest, and for incident neutrino energy $\omega_1 \gg m$, the cross section for scattering of a neutrino by an electron is

$$\sigma_1 = (g^2/3\pi)\, m\omega_1, \tag{1}$$

i.e., a linear function of ω_1.

There is another process, $\nu + Z \rightarrow \nu + Z + e^+ + e^-$, for which the laboratory system coincides with the center-of-mass system. On one hand, it could be expected that the cross section for this process would be smaller than that for scattering, since it contains the factor $(Ze^2)^2$, and the phase volume gives an additional numerical factor $(2\pi)^{-2}$. On the other hand, the phase volume is proportional to ω_1^8, since there are three particles in the final state.

This process is described by two second-order diagrams. The calculation of the contributions of the two diagrams to the cross section leads to extremely cumbersome formulas. We shall, however, get the right order of magnitude for the total cross section if we confine ourselves to the contribution of one diagram. The differential cross section for the process then has the form

$$d\sigma_2 = \frac{16g^2 (Ze^2)^2}{\omega_1\omega_2\varepsilon_+\varepsilon_-} \frac{d\mathbf{p}_- d\mathbf{p}_+ d\mathbf{k}_2}{q^4 (2\pi)^5} \frac{(k_1k_2)}{m^2 - f^2} \times \left[2\varepsilon_+\varepsilon_- - (p_+p_-) + 2fp_+ \frac{2\varepsilon_-^2 + m^2 - (fp_-)}{m^2 - f^2}\right] \times \delta(\omega_1 - \omega_2 - \varepsilon_+ - \varepsilon_-), \tag{2}$$

where

$$f = k_1 - k_2 - p_-, \qquad q = k_1 - k_2 - p_+ - p_- .$$

Here k_1, k_2, p_+, and p_- are four-vectors that refer respectively to the neutrino in its initial and final states and to the positron and electron; ω_1, ω_2, ϵ_+, and ϵ_- are the corresponding energies.

For high energies of all the particles involved in the process the differential cross section $d\sigma_2$ has a sharp maximum near the direction of the momentum of the incident neutrino. All of the emerging particles are concentrated in a narrow cone around this direction, with angular aperture $\vartheta \approx m/\omega_1$. This follows from the fact that the denominator of the expression (2) contains the factor

$$[\omega_1\omega_2(1-\cos\vartheta_{12})+\omega_1\varepsilon_+(1-v_+\cos\vartheta_{1+}) - \omega_2\varepsilon_+(1-v_+\cos\vartheta_{2+})]$$

(ϑ_{ik} is the angle between the momenta of the i-th and k-th particles, and v_+ is the velocity of the positron), together with the fact that the effective recoil momentum of the nucleus is $q \sim m$.

The reduction of the "effective" solid angle sharply lowers the degree of the energy dependence of the total cross section. Apart from terms of second order in m/ω_1 the total cross section is

$$\sigma_2 = \frac{8g^2(Ze^2)^2\omega_1^2}{3(2\pi)^3}\alpha\left(\ln\frac{\omega_1}{m}-\beta\right), \qquad \omega_1 \gg m, \tag{3}$$

where $\alpha, \beta \sim 1$; $1 < \beta < 2$. Comparison of Eqs. (1) and (3) shows that for $Z/137 \approx 1/2$ the cross section σ_2 becomes comparable with σ_1 only for incident neutrino energy $\omega_1 \approx 10$ Mev. It is only at energies higher than this that the process of production of an electron-positron pair may become observable.

The writers express their gratitude to Ya. A. Smorodinskiĭ for his interest in this work and for a discussion of the results.

[1] C. L. Cowan, Jr. and F. Reines, Phys. Rev. 107, 528 (1957).

Translated by W. H. Furry
135

POSSIBLE SYMMETRY PROPERTIES FOR THE π-K SYSTEM

CHOU KUANG CHAO

Joint Institute for Nuclear Research

Submitted to JETP editor January 27, 1960

J. Exptl. Theoret. Phys. (U.S.S.R.) 38, 1015-1016 (March, 1960)

THE Hamiltonian describing the π-K system has the form

$$H = H_\pi + H_K + g\pi_\alpha\pi_\alpha K_\lambda^+ K_\lambda, \tag{1}$$

where H_π is the pion Hamiltonian including the $\pi\pi$ interaction, H_K is the K-meson Hamiltonian, and g is the coupling constant of the $\pi\pi$KK interaction.[1] It is assumed in (1) that the π-meson and K-meson interactions with baryons can be neglected.

The Hamiltonian (1) is invariant under rotations of the pion field operator in isospin space with the K-meson field operator held fixed. In other words, it is possible to consider the pion as an isovector in one space and the K meson as an isospinor in another space. The Hamiltonian (1) is invariant under rotations in either space.

Let us denote pion—K-meson scattering amplitudes by $f(\pi + K \rightarrow \pi + K)$. Then from the above symmetry properties we obtain the following selection rules:

1) The following scattering amplitudes are equal to each other:

$$f(\pi^\pm + K^\pm \rightarrow \pi^\pm + K^\pm) = f(\pi^0 + K^\pm \rightarrow \pi^0 + K^\pm)$$
$$= f(\pi^\pm + K^0 \rightarrow \pi^\pm + K^0) = f(\pi^\pm + \overline{K^0} \rightarrow \pi^\pm + \overline{K^0})$$
$$= f(\pi^0 + K^0 \rightarrow \pi^0 + K^0) = f(\pi^0 + \overline{K^0} \rightarrow \pi^0 + \overline{K^0}). \tag{2}$$

2) The charge-exchange amplitudes vanish:

$$f(\pi^+ + K^- \rightarrow \pi^0 + \overline{K^0}) = f(\pi^- + K^+ \rightarrow \pi^0 + K^0)$$
$$= f(\pi^+ + K^0 \rightarrow \pi^0 + K^+) = f(\pi^- + \overline{K^0} \rightarrow \pi^0 + K^-) = 0.$$

3) The $K + \overline{K} \rightarrow n\pi$ annihilation process proceeds only through the isoscalar state.

To obtain experimental verification of these selection rules, one can study the angular distribution of the products in the reaction $K + N \rightarrow K + N + \pi$, for which the one-meson term in the cross section is proportional to

$$\Delta^2(\Delta^2 + \mu^2)^{-2} |f(\pi + K \rightarrow \pi + K)|^2, \tag{3}$$

where Δ^2 is the square of the nucleon momentum transfer. Expression (3) has a maximum for $\Delta^2 = \mu^2$ in the physical region.[2] A measurement of the form of this maximum would provide information on the amplitudes $f(\pi + K \rightarrow \pi + K)$.

According to the theory of Okun' and Pomeranchuk[3] and Chew and Mandelstam[4] the scattering phase shifts in high angular momentum states are determined by diagrams with the smallest number of exchanged π mesons. If the K^+ and K^0 have the same parity then the $K + N \rightarrow K + N$ scattering phase shifts in high angular momentum states are determined by diagrams with two mesons exchanged. Consequently a phase shift analysis of the process $K + N \rightarrow K + N$ would give certain information about the amplitudes $f(\pi + K \rightarrow \pi + K)$.

A violation of these selection rules would imply that the Hamiltonian contains terms with derivatives of the form

$$g' \pi \times \frac{\partial}{\partial x_\mu} \pi \cdot K^* \tau \frac{\partial}{\partial x_\mu} K$$

or that baryon pairs play an important role in πK interactions. Since g' is not dimensionless, a new fundamental length would appear in the Hamiltonian (in the first version).

The author is grateful to Prof. M. A. Markov and V. I. Ogievetskiĭ for their interest in this work and valuable discussions.

[1] S. Barshay, Phys. Rev. **109**, 2160; **110**, 743 (1958).

[2] C. Goebel, Phys. Rev. Lett. **1**, 337 (1958).

[3] L. B. Okun' and I. Ya. Pomeranchuk, JETP **36**, 300 (1959), Soviet Phys. JETP **9**, 207 (1959).

[4] G. F. Chew, Report at the 1959 Kiev Conference (in press).

Translated by A. M. Bincer
205

ON THE DECAY OF Σ HYPERONS

CHOU KUANG-CHAO

Joint Institute of Nuclear Studies

Submitted to JETP editor January 27, 1960

J. Exptl. Theoret. Phys. (U.S.S.R.) **38**, 1342-1343 (April, 1960)

THE experimental data on the probabilities and asymmetry coefficients of the decays of Σ hyperons by various channels evidently satisfy the rule $|\Delta I| = \frac{1}{2}$. If the $|\Delta I| = \frac{1}{2}$ rule receives final experimental confirmation, it will be necessary to renounce the theory of the universal weak interaction between charged currents.[3] At present it is desirable to have more data to test this rule.

Let us denote the amplitudes for the processes $\Sigma^+ \rightarrow p + \pi^0$, $\Sigma^+ \rightarrow n + \pi^+$, and $\Sigma^- = n + \pi^-$ by A_+, A_0, and A_-, respectively, where $A = a + ib(\boldsymbol{\sigma}\mathbf{k})$; **k** is the unit vector in the direction of motion of the nucleon. The absence of asymmetry in the decays $\Sigma^\pm \rightarrow n + \pi^\pm$ means that for these processes

$$\mathrm{Re}(ab^*) = 0. \quad (1)$$

There are three ways to satisfy the condition (1): 1) $a = 0$, 2) $b = 0$, 3) the phases of a and b differ by 90°. Since the interaction of pion and nucleon in the final state is small, the third possibility violates the conservation of time parity.

Many authors[2,3] have shown that the rule $|\Delta I| = \frac{1}{2}$ holds only for $a_0 = b_- = 0$ or $a_- = b_0 = 0$. To choose from among the three cases the one that exists in nature, one could use measurements of the polarization of the nucleons from the decay of polarized $\Sigma^\pm$ particles produced in reactions $\pi^\pm + p \rightarrow \Sigma^\pm + K^+$.

Denoting the polarization vectors of nucleons and Σ hyperons by $\mathbf{P}$ and $\mathbf{P}_\Sigma$, we get

$$\mathbf{P} = \frac{2\mathrm{Re}\,(ab^*)}{|a|^2+|b|^2}\mathbf{k} + \frac{|a|^2-|b|^2}{|a|^2+|b|^2}\mathbf{P}_\Sigma + \frac{2|b|^2}{|a|^2+|b|^2}(\mathbf{P}_\Sigma\mathbf{k})\,\mathbf{k} + \frac{2\,\mathrm{Im}\,(ab^*)}{|a|^2+|b|^2}[\mathbf{k}\times\mathbf{P}_\Sigma]. \quad (2)$$

In particular $\mathbf{P} = 2(\mathbf{P}_\Sigma\mathbf{k})\mathbf{k} - \mathbf{P}_\Sigma$ for $a = 0$; $\mathbf{P}_\Sigma = \mathbf{P}$ for $b = 0$; and for the third case $\mathbf{P}$ has a component along the direction of $\mathbf{k}\times\mathbf{P}_\Sigma$.

It is obvious that a measurement of the direction of the polarization vector of the nucleons will not only give information to test the rule $|\Delta I| = \frac{1}{2}$, but can also help to choose one solution from the two that are possible ($a_0 = b_- = 0$ or $a_- = b_0 = 0$) if this rule holds.

If there is no transverse polarization of the neutrons from Σ^- decay, this means that the initial Σ^- particle is unpolarized. In this case the absence of asymmetry in the Σ^- decay does not lead to Eq. (1). The quantity $\mathrm{Re}\,(ab^*)$ can be determined from a measurement of the longitudinal polarization of the neutrons.

[1] E. C. G. Sudarshan and R. E. Marshak, Phys. Rev. 109, 1860 (1958). R. P. Feynman and M. Gell-Mann, Phys. Rev. 109, 193 (1958).

[2] F. S. Crawford, Jr. et al., Phys. Rev. 108, 1102 (1957). F. Eisler et al., Phys. Rev. 108, 1353 (1957).

[3] R. E. Sawyer, Phys. Rev. 112, 2135 (1958). G. Takeda and M. Kato, Progr. Theoret. Phys. 21, 441 (1959). B. d'Espagnat and J. Prentki, Phys. Rev. 114, 1366 (1959). B. T. Feld, Preprint. R. H. Dalitz, Revs. Modern Phys. 31, 823 (1959). M. Gell-Mann, Revs. Modern Phys. 31, 834 (1959).

Translated by W. H. Furry
256

SOVIET PHYSICS JETP VOLUME 12, NUMBER 1 JANUARY, 1961

DISPERSION RELATIONS AND ANALYSIS OF THE ENERGY DEPENDENCE OF CROSS SECTIONS NEAR THRESHOLDS OF NEW REACTIONS

L. I. LAPIDUS and CHOU KUANG-CHAO

Joint Institute for Nuclear Research

Submitted to JETP editor February 5, 1960

J. Exptl Theoret. Phys. (U.S.S.R.) **39**, 112-119 (July, 1960)

The application of dispersion relations to the analysis of the energy dependence of scattering (and reaction) amplitudes near thresholds of new reactions is discussed. General expressions are obtained which characterize the nonmonotonic behavior of forward-scattering amplitudes as functions of the energy. The energy dependence of one of the amplitudes for elastic scattering of γ-ray quanta by deuterons is examined near the threshold for photodisintegration of the deuteron.

1. An examination of the scattering of γ-ray quanta by nucleons near the threshold for pion production[1] has shown that the dispersion relations automatically lead to the appearance of discontinuities of the derivative of the real part of the amplitude, if one takes account of the energy dependence of the reaction cross section near threshold.* Within the framework of the dispersion relations the problem of the appearance of discontinuities of the derivative of the forward scattering amplitude involves the analysis of integrals of the form

$$\frac{k_0^2}{4\pi^2} \mathrm{P} \int \frac{d\omega}{k} \frac{\sigma(\omega)}{\omega \mp \omega_0},$$

where the usual notation is used, and the total cross section $\sigma(\omega)$ includes both the elastic scattering cross section $\sigma_s(\omega)$ and the inelastic interaction cross section $\sigma_c(\omega)$.

The behavior of $\sigma_c(\omega)$ near the threshold of the binary reaction

$$a + b \rightarrow c + d \tag{2}$$

involving particles with masses μ (incident), M (target), m and $\mathfrak{M}$ is given by the expression

$$\sigma_c(\omega) = Bq_c/k_c, \tag{3}$$

where q_c and k_c are the momenta before and after the collision (in the c.m.s.) and B is a constant. It is not hard to see that

$$(q_c/k_c)^2 = (\omega - \omega_t)(\omega + \omega_t - \delta)/(\omega^2 - \mu^2), \tag{4}$$

where $\omega = (k^2 + \mu^2)^{1/2}$ is the total energy of the incident particle in the laboratory system (l.s.),

$$\omega_t = \mu + [(\mathfrak{M} + m)^2 - (M + \mu)^2]/2M \tag{5}$$

is the threshold energy of the reaction (2), and

$$\delta = (\mathfrak{M}^2 + m^2 - \mu^2 - M^2)/M. \tag{6}$$

The application of dispersion relations makes it possible to examine both "local effects" near the very threshold, which in some cases lead to sharp "peaks," "dips," and "steps," and also the general influence of inelastic processes occurring in some energy range on the processes at a given energy. We recall that the unitarity relations for the S matrix make it possible to take into account the influence on a given process of other processes occurring at the same energy.

2. The study of γ-N scattering has shown that there are "local effects" in only two of the six scalar amplitudes needed to describe the transition matrix in this case. The presence of other strongly energy-dependent amplitudes hinders the analysis. For a detailed analysis of the inelastic processes it is necessary to examine the dispersion relations for nonvanishing momentum transfers Q^2, and possibly also double dispersion relations.

In the present paper we confine ourselves to the examination of dispersion relations in the total energy for $Q^2 = 0$ for a scalar function $A(\omega)$, which is the trace of the scattering matrix,

$$A(\omega) = \mathrm{Sp}\, M(\omega, Q^2 = 0), \tag{7}$$

and whose imaginary part is related to the total cross section. The contribution of inelastic processes to $D = \mathrm{Re}\, A(\omega)$ is characterized by the two

*The nonmonotonic behavior of the cross section near the threshold has been treated phenomenologically in a diploma research by G. Ustinova, and also by Capps and Holladay.[2]

integrals

$$\frac{k_0^2}{4\pi^2} \mathrm{P} \int \frac{d\omega}{k} \frac{\sigma_c^-(\omega)}{\omega - \omega_0}, \qquad (8)$$

$$\frac{k_0^2}{4\pi^2} \mathrm{P} \int \frac{d\omega}{k} \frac{\sigma_c^+(\omega)}{\omega + \omega_0}, \qquad (9)$$

where $\sigma_c^-(\omega)$ is the total cross section of reaction (2) and $\sigma_c^+(\omega)$ is the cross section of the reaction cross-symmetrical to (2).

Being interested in the energy dependence of the real part of the quantity $A(\omega)$ in Eq. (7), let us calculate the integrals

$$\frac{k_0^2}{4\pi^2} \mathrm{P} \int_{\omega_t}^{\omega_1} \frac{d\omega}{k} \frac{\sigma_c^-}{\omega - \omega_0} = \frac{k_0^2 B^-}{4\pi^2} \mathrm{P} \int_{\omega_t}^{\omega_1} \frac{d\omega}{k^2} \frac{[(\omega - \omega_t)(\omega + \omega_t - \delta)]^{1/2}}{\omega - \omega_0}$$

$$= \frac{B^- k_0^2}{4\pi^2} \Pi(\omega_0), \qquad (10)$$

$$\frac{k_0^2}{4\pi^2} \mathrm{P} \int_{\omega_t}^{\omega_1} \frac{d\omega}{k} \frac{\sigma_c^+}{\omega + \omega_0} = \frac{k_0^2 B^+}{4\pi^2} \Pi(-\omega_0); \qquad (11)$$

in these integrals the range of integration is from the threshold ω_t to ω_1, the limit of the region of the s state in σ_c in which Eq. (3) is still valid. If we introduce the notations

$$a(\omega_0) = (\omega_0 - \omega_t)(\omega_0 + \omega_t - \delta),$$

$$a(-\omega_0) = (\omega_0 + \omega_t)(\omega_0 - \omega_t + \delta),$$

$$R = (\omega_1 - \omega_t)(\omega_1 + \omega_t - \delta), \qquad (12)$$

it is not hard to show that

$$k_0^2 \Pi(\omega_0) = -\{\psi(\omega_0) - \tfrac{1}{2}(1 + \omega_0/\mu)\psi(\mu) - \tfrac{1}{2}(1 - \omega_0/\mu)\psi(-\mu)\}, \qquad (13)$$

where

$$\psi(\mu) = \sqrt{-a(\mu)}\left[\frac{\pi}{2} + \tan^{-1}\frac{(2\mu - \delta)(\omega_1 - \mu) + 2a(\mu)}{2\sqrt{-a(\mu)R}}\right],$$

$$\psi(-\mu) = \sqrt{-a(-\mu)}\left[\frac{\pi}{2} - \tan^{-1}\frac{(2\mu + \delta)(\omega_1 + \mu) - 2a(-\mu)}{2\sqrt{-a(-\mu)R}}\right]. \qquad (13')$$

For the function $\psi(\omega)$ at the point ω_0 we get the expressions

$$\psi(\omega_0) = \sqrt{-a(\omega_0)}\left[\frac{\pi}{2} + \tan^{-1}\frac{(2\omega_0 - \delta)(\omega_1 - \omega_0) + 2a(\omega_0)}{2\sqrt{-a(\omega_0)R}}\right], \quad \omega_0 \leqslant \omega_t; \qquad (14)$$

$$\psi(\omega_0) = \sqrt{a(\omega_0)} \ln\left|\frac{2a(\omega_0) + (2\omega_0 - \delta)(\omega_1 - \omega_0) + 2\sqrt{a(\omega_0)R}}{(\omega_1 - \omega_0)(2\omega_t - \delta)}\right|, \quad \omega_0 \geqslant \omega_t. \qquad (14')$$

The expression for $k_0^2 \Pi(-\omega_0)$ is obtained from (13) and (14) by the replacement $\omega_0 \to -\omega_0$.

It follows from (14) and (14′) that the first derivative has a discontinuity at the point ω_0. The resulting energy dependence has the characteristic feature that the derivative is infinite on the side $\omega_0 < \omega_t$ and has a finite value when we approach the point $\omega_0 = \omega_t$ from the region $\omega_0 < \omega_t$.

It might seem that the quantity $k_0^2 \Pi(-\omega_0)$ obtained as the result of substituting (3) and (4) in (9) would also have a discontinuity of the derivative at $\omega_0 = \omega_t - \delta$. On changing the sign of ω_0 in (14), however, we easily verify that this is not so. The values of the derivative of $k_0^2 \Pi(-\omega_0)$ calculated with approach to $\omega_0 = \omega_t$ from the two sides are identical. Thus although in a relativistic treatment cross-symmetrical inelastic processes indeed contribute to the real part of the scattering amplitude, they do not lead to nonmonotonic energy dependence of the amplitude.

The application of dispersion relations enables us to get detailed information about the magnitude and half-width of the anomaly near the threshold. The half-width ϵ of the drop in the region $\omega_0 < \omega_t$ can be estimated roughly in the following way. Near $\omega_0 = \omega_t$ ($\omega_0 < \omega_t$) the argument of the arc tangent is large; using this fact, we find

$$\psi(\omega_0) = \sqrt{-a}\,\pi. \qquad (15)$$

Defining the half-width ϵ by the condition

$$D(\omega_t - \varepsilon) = \tfrac{1}{2} D(\omega_t) \qquad (16)$$

and using Eq. (15), we then get

$$D(\omega_t) - (B/4\pi)\sqrt{(\omega_t - \omega_0)(\omega_0 + \omega_t - \delta)} = \tfrac{1}{2} D(\omega_t), \qquad (17)$$

from which we have

$$\varepsilon \approx \tfrac{1}{8}(\omega_t - \delta/2)^{-1}[4\pi D(\omega_t)/B]^2. \qquad (18)$$

In the limiting case in which the contribution to $D(\omega_t)$ not associated with the expressions (10) and (11) can be neglected,

$$D(\omega_t) = BJ(\omega_t)/4\pi^2, \qquad (19)$$

and from Eq. (13) we have the result that

$$J(\omega_t) = \tfrac{1}{2}[(1 + \omega_t/\mu)\psi(\mu) + (1 - \omega_t/\mu)\psi(-\mu)]. \qquad (20)$$

3. Let us consider the photoproduction of neutral pions

$$\gamma + p \to p + \pi^0 \qquad (21)$$

near the threshold of the reaction

$$\gamma + p \to n + \pi^+. \qquad (22)$$

In this energy range it suffices to consider the electric-dipole transition. We denote by E^0 and E^+ the transition elements for neutral and charged mesons, respectively. From the condition of unitarity and the experimental fact that Re $E^0 \approx 0$, we get the result

$$\operatorname{Im} E^0 = \sqrt{2/3}\,(\alpha_3 - \alpha_1)\operatorname{Re} E^+, \qquad (23)$$

where α_3 and α_1 are the phase shifts for π-N scattering. Substituting the experimental data for α_3, α_1, and E^+, we get

$$\operatorname{Im} E^0 = \sqrt{2/3}\,(a_3 - a_1)\, q_0^{1/2} q_+^{1/2} \sqrt{q_+/\nu} \cdot 3{,}3 \cdot 10^{-15} \text{ cm}, \qquad (24)$$

where

$$q_0^{1/2} \approx (\nu^2 - \nu_{0t}^2)^{1/4}, \quad q_+ \approx (\nu^2 - \nu_{+t}^2)^{1/2}$$

(ν is the energy of the photon, and a_3, a_1 are the scattering lengths). The anomalies at the threshold are determined by an integral of the form

$$P \int_{\nu_{+t}}^{\nu_1} d\nu\, (\nu^2 - \nu_{0t}^2)^{1/4} (\nu^2 - \nu_{+t}^2)^{1/2} / \nu^{1/2} (\nu - \nu_0), \qquad (25)$$

which undoubtedly gives "peak" singularities.

In analogy with this, in the general case we can consider the cross section of the reaction

$$a + b \to c + d \qquad (26)$$

near the threshold of the reaction

$$a + b \to e + f. \qquad (27)$$

If the threshold of reaction (27) is far from that of reaction (26), we can always find an energy range where

$$\operatorname{Im} M\,(ab \to cd) = M\,(ab \to ef)\, M^+ (ef \to cd) + \ldots = Aq + \ldots; \qquad (28)$$

here A is a weakly varying function of the energy and q is the relative momentum of the system ef. The other terms in the sum (28) are also slowly varying functions of the energy if there are no thresholds of other reactions in the neighborhood. In this case the dispersion integral has the usual form, and we can determine the magnitude and half-width of the "peak" or "dip" in the same way as for the case of scattering.

In the case of such a process as the photoproduction of pions

$$\operatorname{Im} M\,(ab \to cd) = Aqq_0^{1/2} + \ldots, \qquad (29)$$

since the threshold of the reaction ab → ef is close to that of the reaction ab → cd. In these cases the dispersion integrals are rather complicated and we have not been able to carry out the integration.

4. It thus follows from the conditions of causality and unitarity, together with Eqs. (3) and (4), that the first derivative of the real part of the scattering amplitude has a discontinuity, and that the derivative from the side $\omega_0 < \omega_t$ is infinite. That it is precisely the first derivative that is infinite is due to the form of the relations (3) and (4). The behavior of the cross section for the reaction (2), when its products are in states with nonvanishing angular momentum l, is given by the expression

$$\sigma_c^{(l)} = B^{(l)}\, [(\omega - \omega_t)(\omega + \omega_t - \delta) / (\omega^2 - \mu^2)]^{l + 1/2}. \qquad (30)$$

Substitution of (3) in (8) makes the lth derivative infinite.

It is perhaps interesting to note that, unlike the nonrelativistic treatment, this use of the dispersion relations has not required the assumption that the partial amplitudes are analytic. It has turned out that it is enough to use only the analytic character of the scattering amplitude with respect to the total energy, with a bounded value of the momentum transfer, $Q^2 < Q^2_{max}$.

As Baz' has pointed out, the unitarity of the S matrix has the consequence that as the number of channels increases the effect in each channel decreases. An analysis of γ-N scattering near the threshold for photoproduction of pions, for which there are "peak" effects in only two out of six scalar functions, has shown that there is also a smearing of the effect with increase of the spin of the particles.

An important feature of the theory of dispersion relations is the discussion of the convergence of the dispersion integrals at high energies or, what is the same thing, of the number of subtractions. The main calculations in the present paper are made for dispersion relations with one subtraction. In the case of dispersion relations without subtraction one must make the replacement

$$k_0^2 \int \frac{d\omega}{k} \frac{\sigma}{\omega - \omega_0} \to \int \frac{d\omega\, k\sigma}{\omega - \omega_0} \qquad (31)$$

With sufficiently high experimental accuracy the difference between Eq. (8) and Eq. (31) can give information about the number of subtractions.

Let us note briefly what sort of singularities can appear near the threshold of the reaction

$$a + b \to c + d + f. \qquad (32)$$

By substituting in Eq. (8) the cross section of the reaction (32) in the form

$$\sigma_c = B' k_c^{-1} P^4_{c\,max} \approx \Gamma\,(\omega - \omega_t)^2$$

(reaction products in the s state), we get

$$\int_{\omega_t}^{\omega_1} \frac{\sigma d\omega}{\omega - \omega_0} \approx \Gamma \int_{\omega_t}^{\omega_1} \frac{d\omega\,(\omega - \omega_t)^2}{\omega - \omega_0} = \Gamma \Big\{ \frac{1}{2}\,[(\omega_1 - \omega_0)^2 - (\omega_t - \omega_0)^2] + 2\,(\omega_0 - \omega_t)(\omega_1 - \omega_t) + (\omega_0 - \omega_t)^2 \ln|(\omega_1 - \omega_0)/(\omega_t - \omega_0)|\Big\}, \qquad (33)$$

which leads to a logarithmic infinity in the second derivative of $D(\omega)$ with respect to ω.

For a reaction with four particles in a final s state the quantity $(\omega_0 - \omega_t)^2 \ln|\omega_t - \omega_0|$ is replaced by $(\omega_0 - \omega_t)^5 \ln|\omega_0 - \omega_t|$. Similar behavior of the real part of the scattering amplitude appears at the thresholds of all reactions.

An example of the application of dispersion relations that is well known in the literature is the analysis of the coherent scattering of photons in the Coulomb field of a nucleus[3] (cf. also reference 4). It is not hard to convince oneself that near $\omega = 2m$ ($\gamma \equiv \omega/2m = 1$) — the threshold for production of an electron-positron pair — the real part of the scattering amplitude has an energy dependence of the type $x^k \ln x$ ($x = \gamma - 1$). To see this we have only to examine the expression for the real part of the amplitude:

$$D(\omega) = \frac{Z^2}{m}\left(\frac{e^2}{4\pi}\right)^3 \left\{\frac{1}{\gamma\pi}[2C_1(\gamma) - D_1(\gamma)] + \frac{\gamma}{27\pi}\left[\left(109 + \frac{64}{\gamma^2}\right)E_1(\gamma) - \left(67 - \frac{6}{\gamma^2}\right)\left(1 - \frac{1}{\gamma^2}\right)F_1(\gamma)\right] - \frac{1}{9\gamma^2} - \frac{9}{4}\right\}, \quad (34)$$

where

$$C_1(\gamma) = \mathrm{Re}\int_0^\gamma \frac{\arcsin x}{x}\cosh^{-1}\left(\frac{\gamma}{x}\right)dx, \quad C_1(1) = 1.62876;$$

$$D_1(\gamma) = \mathrm{Re}\int_0^\gamma \frac{\cosh^{-1}(\gamma/x)}{(1 - x^2)^{1/2}}dx, \qquad D_1(1) = 1{,}83193;$$

$$E_1(\gamma) = \begin{cases} E(\gamma^{-1}), & \gamma \geqslant 1 \\ \gamma^{-1}E(\gamma) + (\gamma - \gamma^{-1})K(\gamma), & \gamma \leqslant 1 \end{cases};$$

$$F_1(\gamma) = \begin{cases} K(\gamma^{-1}), & \gamma \geqslant 1 \\ \gamma K(\gamma), & \gamma \leqslant 1 \end{cases},$$

and $K(\gamma)$ and $E(\gamma)$ are the complete elliptic integrals of the first and second kinds. As is well known, for $1 - \gamma^2 \ll 1$ [$\Lambda = \ln(4(1 - \gamma^2)^{-1/2}$],

$$K(\gamma) = \Lambda + \tfrac{1}{4}(\Lambda - 1)(1 - \gamma^2) + \tfrac{9}{64}(\Lambda - \tfrac{7}{6})(1 - \gamma^2)^2 + \tfrac{25}{256}(\Lambda - \tfrac{37}{30})(1 - \gamma^2)^3 + \ldots,$$

$$E(\gamma) = 1 + \tfrac{1}{2}(\Lambda - \tfrac{1}{2})(1 - \gamma^2) + \tfrac{3}{64}(\Lambda - \tfrac{13}{12})(1 - \gamma^2)^2 + \ldots, \quad (35)$$

which indeed shows that the dependence is of the form $x^k \ln x$. It is not hard to check that the scattering of light by light near the threshold of the reaction $\gamma + \gamma \to e^+ + e^-$ is a process, well known in quantum electrodynamics, for which the amplitude is characterized by a "local" anomaly (cf. Figs. 2—4 in the paper by Karplus and Neuman[5])*. The amplitude for the Compton effect near the threshold of the reaction $\gamma + e \to 2e + e^+$ has the characteristic dependence $x^2 \ln x$.

5. The elastic scattering of γ rays by deuterons near the threshold for photodisintegration of the deuteron is an example of a process for which use of dispersion relations is necessary for the analysis of the anomaly near the threshold. The nonmonotonic behavior near the threshold in this case comes from the magnetic-dipole disintegration. The electric-dipole disintegration leads to appreciable changes in the energy dependence of the amplitude for elastic γ-d scattering in a certain relatively wide range of energies.

The amplitude for forward elastic γ-d scattering can be represented in the form

$$e_i' T_{ik} e_k = A\mathbf{e}'\mathbf{e} + iB\mathbf{S}[\mathbf{e}'\mathbf{e}] + \tfrac{1}{2}C[(\mathbf{Se})(\mathbf{Se}') + (\mathbf{Se}')(\mathbf{Se})] + \tfrac{1}{2}D[(\mathbf{S}[\mathbf{k}\times\mathbf{e}])(\mathbf{S}[\mathbf{k}\times\mathbf{e}']) + (\mathbf{S}[\mathbf{k}\times\mathbf{e}'])(\mathbf{S}[\mathbf{k}\times\mathbf{e}])]. \quad (36)$$

The cross section for scattering of unpolarized γ rays by unpolarized deuterons then takes the form

$$\sigma_s(0°) = |A + \tfrac{2}{3}(C + D)|^2 + \tfrac{1}{18}|C + D|^2 + \tfrac{2}{3}|B|^2 + \tfrac{1}{3}|D - C|^2, \quad (37)$$

and we have

$$k\sigma_t = 4\pi\,\mathrm{Im}(A + \tfrac{2}{3}C + \tfrac{2}{3}D).$$

By means of the dispersion relation for the quantity $L = A + \tfrac{2}{3}C + \tfrac{2}{3}D$,

$$\mathrm{Re}\,L(\omega) = -\frac{e^2}{M_d} + \frac{2\omega^2}{\pi}\mathrm{P}\int_{\omega_d}^{\infty}\frac{\mathrm{Im}\,L(\omega')}{\omega'(\omega'^2 - \omega^2)}d\omega', \quad (38)$$

where ω_d is the threshold for photodisintegration of the deuteron, let us examine the effect of inelastic processes on the energy dependence of the real part of the amplitude L. In the calculation of the dispersion integral it is convenient to use the theoretical expressions for the cross sections for photodisintegration of the deuteron (cf., e.g., reference 8).

Let us begin with the examination of the "local effects." The expression for the cross section for magnetic-dipole disintegration is

$$\sigma_c^{(m)} = \frac{2\pi}{3}\frac{e^2}{\hbar c}\left(\frac{\hbar}{Mc}\right)^2(\mu_p - \mu_n)^2\frac{(\gamma - 1)^{1/2}(1 + \sqrt{\epsilon'/|\epsilon|})^2}{\gamma[\gamma - 1 + \epsilon'/|\epsilon|]}, \quad (39)$$

where $\gamma = \omega/|\epsilon|$, ω being the energy of the photon; $|\epsilon| = 2.22$ Mev and $\epsilon' \sim 70$ kev are the binding energies of the np system in the $^3S_1$ and $^1S_0$ states; and the rest of the notation is as usual. Because of the factor $[\gamma - 1 + \epsilon'/|\epsilon|]^{-1}$ the expression (39) does not admit of the simple analytic continuation

*Regarding effects of the Coulomb interaction see papers by Baz'[6] and by Fonda and Newton.[7]

$$\varkappa = \sqrt{\gamma - 1} \rightarrow i|\varkappa|,$$

since in this case $\gamma\sigma_c^{(m)}$ goes to infinity below the threshold at $|\kappa|^2 = \epsilon'/|\epsilon|$. Substitution of Eq. (39) in the dispersion integral

$$Z_L(\gamma_0) = \varepsilon \frac{\gamma_0^2}{2\pi^2} \mathrm{P}\int_1^\infty \frac{d\gamma\sigma_c^{(m)}(\gamma)}{\gamma^2 - \gamma_0^2}$$

for $\gamma_0 \neq \delta$ gives

$$Z_L(\gamma_0) = \frac{2}{3}\frac{e^2}{2Mc^2}\frac{\varepsilon}{Mc^2}(\mu_p - \mu_n)^2(1 + \sqrt{\varepsilon'/|\varepsilon|})^2 \times \left\{\frac{\sqrt{1-\gamma_0}}{\delta - \gamma_0}\theta(1-\gamma_0) + \frac{\sqrt{1+\gamma_0}}{\delta + \gamma_0} - \frac{2}{\delta} - \frac{2\gamma_0^2\sqrt{\varepsilon'/|\varepsilon|}}{\delta(\delta^2 - \gamma_0^2)}\right\}, \quad (40)$$

where $\delta = 1 - \epsilon'/|\epsilon|$, $\theta(x) = 1$ for $x \geq 1$, and $\theta(x) = 0$ for $x < 1$; for $\gamma_0 = \delta$

$$Z_L(\delta) = \frac{2}{3}\frac{e^2}{2Mc^2}\frac{\varepsilon}{Mc^2}(\mu_p - \mu_n)^2(1 + \sqrt{\varepsilon'/|\varepsilon|})^2 \times \{\sqrt{1+\delta}/2\delta - 2/\delta + 2\sqrt{\varepsilon'/|\varepsilon|}/3\delta + \sqrt{|\varepsilon|/\varepsilon'}/2\}. \quad (41)$$

The dependence of the quantity $\Delta_L(\gamma_0) = Z_L(\gamma_0)/(e^2/2Mc^2)$ on the γ-ray energy is shown in the diagram (curve 1)

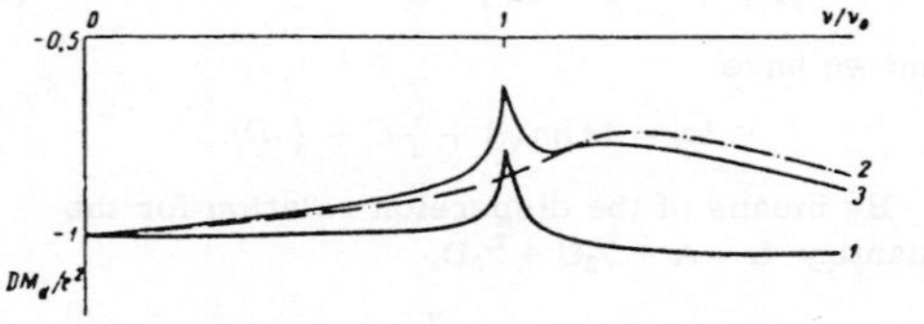

For extreme values of γ_0 we get from Eq. (40)

$$\Delta_L(\gamma_0) = \frac{2}{3}\frac{\varepsilon}{Mc^2}(\mu_p - \mu_n)^2\left(1 + \sqrt{\frac{\varepsilon'}{|\varepsilon|}}\right)^2 \times \left\{\frac{3}{8} + 2\frac{\varepsilon'}{|\varepsilon|} - \sqrt{\frac{\varepsilon'}{|\varepsilon|}}\left(1 + 2\frac{\varepsilon'}{|\varepsilon|}\right)\right\}\gamma_0^2$$

$$\approx \frac{2}{3}\frac{\varepsilon}{Mc^2}(\mu_p - \mu_n)^2\left(1 + \sqrt{\frac{\varepsilon'}{|\varepsilon|}}\right)^2\frac{1}{4}\gamma_0^2, \quad \gamma_0 \ll 1; \quad (42)$$

$$\Delta_L(\gamma_0) = -\frac{4}{3}\frac{\varepsilon}{Mc^2}(\mu_p - \mu_n)^2\left(1 + \sqrt{\frac{\varepsilon'}{|\varepsilon|}}\right)^2, \quad \gamma_0 \gg 1. \quad (43)$$

At the photodisintegration threshold with $1 - \delta = 1/30$.

$$Z_L(1) = 0.24\, e^2/2Mc^2.$$

On the side of energies smaller than the threshold energy the half-width of the peak is somewhat smaller than ϵ', i.e., about 50 or 60 kev.

The contribution of the cross section for dipole absorption

$$\sigma_c^{(d)} = 4\pi\frac{e^2}{Mc^2}\frac{\hbar c}{\varepsilon}\frac{(\gamma - 1)^{3/2}}{\gamma^3} \quad (44)$$

at $D = \mathrm{Re}\, L$ is of the form

$$\Delta_p(\gamma_0) = 2Mc^2 Z_p(\gamma_0)/e^2 = 2\{\gamma_0^{-2}[(1-\gamma_0)^{3/2}\theta(1-\gamma_0) + (1+\gamma_0)^{3/2} - 2] - 3/4\}. \quad (45)$$

The quantity $\Delta_p = \Delta_p(\gamma_0)$ is shown in the diagram by curve 2. In the limiting cases

$$\Delta_p(\gamma_0) = \tfrac{3}{32}\gamma_0^2, \quad \gamma_0 \ll 1; \quad (46)$$

$$\Delta_p(\gamma_p) = -\tfrac{3}{2}, \quad \gamma_0 \gg 1. \quad (47)$$

At the threshold for photodisintegration

$$\Delta_p(1) = 0{,}156.$$

The total effect of the dipole and magnetic-dipole disintegrations on the real part of the amplitude L is shown in the diagram by curve 3. Right at the threshold the effect of photodisintegration leads to a change of the amplitude by about 40 percent.

The contribution of photodisintegration of the deuteron to the polarizability of the deuteron can be seen from Eqs. (42) and (46). Since the value of the cross section for photodisintegration of the deuteron at high energies is larger than the sum of the expressions (39) and (44), the estimates obtained here can be regarded as lower limits on the quantities, although the contribution of high energies is small.

The treatment carried through here for one amplitude of the γ-d scattering can serve as an indication that inclusion of the effects of inelastic processes, and primarily those of the photodisintegration of the deuteron, in the analysis of elastic γ-d scattering can be important over a wide range of energies.

Similar effects must naturally occur also in the scattering of γ rays by heavier nuclei. A study of the elastic scattering of γ rays by nuclei shows[9] that for quite a number of elements the cross section for nuclear scattering of γ rays near the threshold of the reaction (γ, n) is characterized by a peak of considerable height with an energy width of about ±2 Mev, which is evidently due to nonmonotonic effects near the threshold. Further improvement of the accuracy of the experimental data on elastic scattering of γ rays and on the energy dependence of the cross sections of (γ, n) reactions near threshold is necessary for a more reliable analysis of this effect.

[1] L. I. Lapidus and Chou Kuang-Chao, JETP 37, 1714 (1959), Soviet Phys. JETP 10, 1213 (1960); JETP 38, 201 (1960), Soviet Phys. JETP 11, 147 (1960).

[2] R. H. Capps and W. G. Holladay, Phys. Rev. 99, 931 (1955), Appendix B.

[3] F. Rohrlich and R. L. Gluckstern, Phys. Rev. 86, 1 (1952).

[4] A. I. Akhiezer and V. B. Berestetskiĭ, Квантовая электродинамика (Quantum Electrodynamics), Fizmatgiz, 1959, Sec. 55.

[5] R. Karplus and M. Neuman, Phys. Rev. 83, 776 (1951).

[6] A. I. Baz', JETP 36, 1762 (1959), Soviet Phys. JETP 9, 1256 (1959).

[7] L. Fonda and R. G. Newton, Ann. Phys. 7, 133 (1959).

[8] A. I. Akhiezer and I. Ya. Pomeranchuk, Некоторые вопросы теории ядра (Some Problems of Nuclear Theory), Gostekhizdat 1958, Secs. 11 and 12.

[9] E. G. Fuller and E. Hayward, Phys. Rev. 101, 692 (1956).

Translated by W. H. Furry
23

SOVIET PHYSICS JETP VOLUME 12, NUMBER 2 FEBRUARY, 1961

INELASTIC FINAL-STATE INTERACTIONS AND NEAR-THRESHOLD SINGULARITIES

L. I. LAPIDUS and CHOU KUANG-CHAO

Joint Institute for Nuclear Research

Submitted to JETP editor, February 23, 1960

J. Exptl. Theoret. Phys. (U.S.S.R.) **39**, 364-372 (August, 1960)

It is shown that non-monotonic energy variations can occur in the energy spectrum of particle a from a reaction of the type A + B → a + C + D in the neighborhood of the threshold for the reaction C + D → E + F. As an example, we analyze the spectrum of K mesons from the reaction N + N → Λ + N + K in the region of the energy of the Λ – N pair close to the threshold for the process Λ + N → Σ + N. For the process p + p → Λ + N + K we find the energy spectrum of the K mesons when the incident nucleons are unpolarized, and the polarization of the baryons when the incident nucleons are polarized.

We discuss the non-monotonic energy variations in the spectra of particles for some other reactions. In the Appendix we analyze the production of Y – K pairs in np collisions and discuss the case of a scalar K particle.

1. INTRODUCTION

IT is known that in processes of production of particles an interaction between two of the particles formed affects the energy spectrum and angular distribution of the third particle. In certain cases, the effect of final-state interaction can be separated from the primary mechanism for production of the particles. This occurs when the effective radius for the primary interaction is much less than the radius of interaction of a pair of particles in the final state. In addition, if the interaction of the pair of particles with other emerging particles is weak, the interaction of the two particles in the final state can be characterized by a two-particle scattering length.

The theory of final-state interaction was applied by Migdal,[1] Brueckner and Watson,[2] and Paruntseva[3] to meson production in NN collisions. Recently Henley[4] and Feldman and Matthews[5] applied it to the analysis of the reaction

$$N + N \rightarrow Y + N + K. \tag{1}$$

They showed that the energy spectrum of the K mesons is strongly distorted by the effect of the YN interaction.

Karplus and Rodberg[6] generalized the theory of final-state interaction to the case where the strong interaction in the final state can lead to an inelastic process.

In the present paper we shall show that in the neighborhood of the threshold for production of the Σ hyperon certain anomalies occur in the energy spectrum of the K particles formed together with the Λ hyperons. They are a new example of near-threshold anomalies which have been extensively studied in recent years.[7]

In addition to the cross section for the new inelastic process, the shape and appearance of near-threshold anomalies depend on the spin and parity of the particles. The study of these anomalies with sufficient accuracy can enable us to determine properties of the produced particles. On the assumption that the final state of reaction (1) is described by singlet and triplet s waves of the Y-N system, we analyze in the second section of the present paper the kinematics of the reaction and obtain expressions for the energy spectrum of the K mesons and the polarization of the Λ particles and nucleons when the incident beam of nucleons is polarized.

In the third section, starting from the unitarity of the S matrix and the analyticity of the reaction amplitude, we give a general formulation of the theory of inelastic final-state interactions.

In Sec. 4 we consider local near-threshold anomalies in the energy spectrum of K mesons in the reaction N + N → Λ + N + K in the neighborhood of the threshold for formation of the Σ hyperon.

In conclusion, we mention some other similar processes and discuss the possible generalization of the method developed here to these processes.

2. KINEMATICS. PHENOMENOLOGICAL ANALYSIS.

We introduce Jacobi coordinates in the final state of the three-particle system:

$$\mathbf{R}=\frac{M_N\mathbf{r}_N+M_Y\mathbf{r}_Y+M_K\mathbf{r}_K}{M_N+M_Y+M_K},\qquad \boldsymbol{\rho}=\mathbf{r}_K-\frac{M_N\mathbf{r}_N+M_Y\mathbf{r}_Y}{M_N+M_Y},$$

$$\mathbf{r}=\mathbf{r}_N-\mathbf{r}_Y, \tag{2}$$

where M_N, M_Y and M_K are respectively the masses of the nucleon, hyperon and K meson; $\mathbf{r}_N$, $\mathbf{r}_Y$ and $\mathbf{r}_K$ are their coordinates. The momenta conjugate to $\mathbf{R}$, ρ, and $\mathbf{r}$ will be $\mathbf{p}_R$, $\mathbf{p}_Y$ and $\mathbf{q}$ respectively. The total energy E in the new variables is equal to (c.m.s.)

$$E=p_Y^2/2m_Y+q^2/2\mu+M_K+M_Y-M_N; \tag{3}$$

$$M=M_N+M_Y+M_K,\qquad m_Y=\frac{M_NM_Y}{M_N+M_Y},$$

$$\mu=\frac{M_K(M_N+M_Y)}{M_K+M_N+M_Y}. \tag{4}$$

The phase volume of the final state is expressed in terms of p_Y and q as follows:

$$dJ=m_Y\,p_Y\,d\Omega_Y\,q^2\,dq\,d\Omega_q, \tag{5}$$

where $d\Omega_y$ and $d\Omega_q$ are the solid angles for the momenta $\mathbf{p}_Y$ and $\mathbf{q}$ respectively.

To be specific, we consider the reaction

$$p+p\rightarrow\Lambda+p+K^+ \tag{6}$$

below the threshold of the reaction

$$p+p\rightarrow\Sigma^0+p+K^+. \tag{7}$$

The admissible energy for the final state of reaction (6) in the c.m.s. does not exceed 80 Mev, so that we may assume that the particles which are formed are in an s state.

Let us represent the S matrix element in the form

$$\langle\Lambda pK^+|S|pp\rangle=-2\pi i\delta(E_i-E_f)\langle\Lambda pK^+|T|pp\rangle. \tag{8}$$

If the K meson is a pseudoscalar particle, the spin structure of the T matrix has the form

$$\langle\Lambda pK|T|pp\rangle=A_\Lambda(\boldsymbol{\sigma}_1+\boldsymbol{\sigma}_2,\mathbf{k})$$
$$+B_\Lambda\{(\boldsymbol{\sigma}_1-\boldsymbol{\sigma}_2,\mathbf{k})+i([\boldsymbol{\sigma}_1\boldsymbol{\sigma}_2]\mathbf{k})\}$$
$$+C_\Lambda\{(\boldsymbol{\sigma}_1-\boldsymbol{\sigma}_2,\mathbf{k})-i([\boldsymbol{\sigma}_1\boldsymbol{\sigma}_2]\mathbf{k})\}, \tag{9}$$

where σ is the spin matrix, $\mathbf{k}$ is a unit vector along the direction of the incident proton: A_Λ, B_Λ, and C_Λ are scalar functions of the total energy E and the relative momentum P_Λ of the $\Lambda-N$ pair. Since there are two identical particles in the initial state, the elements of the T matrix must be antisymmetrized with respect to the two initial protons. It can be shown that this results in $B_\Lambda=0$.

The expression for the cross section for reaction (6) with unpolarized particles has the form

$$\frac{d\sigma}{d\Omega_\Lambda d\Omega_q dT}=(2\pi)^4\frac{E}{2(E^2-4M_N^2)^{1/2}}$$
$$\times(2m_\Lambda\mu)^{3/2}[T(T_{max}-T)]^{1/2}$$
$$\times[|A_\Lambda+C_\Lambda|^2+|A_\Lambda-C_\Lambda|^2+2|C_\Lambda|^2], \tag{10}$$

where $T=q^2/2\mu$ is the kinetic energy of the K meson with respect to the center of mass of the $\Lambda-N$ system.

If the protons in the initial state are polarized (with polarization vector $\mathbf{P}$), the polarization vector of the Λ particle in the final state, $\mathbf{P}_\Lambda$, will be

$$\mathbf{P}_\Lambda[|A_\Lambda+C_\Lambda|^2+|A_\Lambda-C_\Lambda|^2+2|C_\Lambda|^2]=2[|A_\Lambda+C_\Lambda|^2$$
$$-|C_\Lambda|^2](\mathbf{kP})\mathbf{k}+[|A_\Lambda-C_\Lambda|^2-|A_\Lambda+C_\Lambda|^2]\mathbf{P}. \tag{11}$$

The expression for the polarization of the nucleon in the final state differs from (11) by the sign in front of C_Λ.

3. ELASTIC FINAL-STATE INTERACTION

Let us look at the unitarity condition

$$\langle\Lambda pK|T-T^+|pp\rangle$$
$$=2\pi i\sum_n\langle\Lambda pK|T|n\rangle\langle n|T^+|pp\rangle\delta(E_i-E_n), \tag{12}$$

where $|n\rangle$ is a possible intermediate state lying on the same energy surface as the initial state. Let us assume that in the region of energy considered the imaginary part of the T matrix is related mainly to strong interaction in the $\Lambda-p$ system. Then we may neglect on the right side of (12) all intermediate states except for ΛpK states, and approximately replace $\langle\Lambda pK|T|\Lambda'p'K'\rangle$ by $\langle\Lambda p|T|\Lambda'p'\rangle\langle K|K'\rangle$. This means that we are neglecting the interaction between the K meson and the $\Lambda-p$ pair.

In the low energy region the matrix element $\langle\Lambda p|T|\Lambda'p'\rangle$ is equal to

$$\langle\Lambda N|T|\Lambda'N'\rangle$$
$$=(4\pi^2p_\Lambda m_\Lambda)^{-1}\left[\tfrac{1}{4}(3+\boldsymbol{\sigma}_1\boldsymbol{\sigma}_2)\alpha_3+\tfrac{1}{4}(1-\boldsymbol{\sigma}_1\boldsymbol{\sigma}_2)\alpha_1\right], \tag{13}$$

where

$$\alpha_3=e^{i\delta_3}\sin\delta_3,\qquad \alpha_1=e^{i\delta_1}\sin\delta_1, \tag{14}$$

and δ_1 and δ_3 are the scattering phases in the singlet and triplet states respectively.

Using all these assumptions and taking account of invariance under time reversal, we find from (12)

$$\operatorname{Im} A_\Lambda = \frac{\operatorname{Re} \alpha_3}{1 - \operatorname{Im} \alpha_3} \operatorname{Re} A_\Lambda - \frac{\operatorname{Im} \alpha_3}{\operatorname{Re} \alpha_3} \operatorname{Re} A_\Lambda = \tan \delta_3 \operatorname{Re} A_\Lambda ,$$

$$\operatorname{Im} C_\Lambda = \tan \delta_1 \operatorname{Re} C_\Lambda ,$$

$$A_\Lambda = (1 + i \tan \delta_3) \operatorname{Re} A_\Lambda \approx (1 + i a_3 p_\Lambda) \operatorname{Re} A_\Lambda ,$$

$$C_\Lambda = (1 + i \tan \delta_1) \operatorname{Re} C_\Lambda \approx (1 + i a_1 p_\Lambda) \operatorname{Re} C_\Lambda . \qquad (15)$$

From (15) we see that for $\delta \to 0$, i.e., in the absence of final-state interaction, the quantities A_Λ and C_Λ are real functions.

In the energy region we are considering, the matrix elements of the reaction matrix are functions of two quantities: E — the total energy, and ω — the total energy of the $\Lambda - p$ system. If all the singularities of the amplitude are associated with physical processes, then A_Λ and C_Λ as analytic functions of ω and E are representable in the form

$$(p_\Lambda a)^{-1} e^{i\delta(\omega)} \sin \delta (\omega) f(\omega) F_\Lambda (E), \qquad (16)$$

where $f(\omega)$ is an entire function which, for small values of the energy, can be replaced by a constant.

Thus we finally approximate A_Λ and C_Λ by expressions

$$A_\Lambda = \frac{1}{p_\Lambda a_3} e^{i\delta_3} \sin \delta_3 \cdot A_\Lambda^0 , \quad C_\Lambda = \frac{1}{p_\Lambda a_1} e^{i\delta_1} \sin \delta_1 \cdot C_\Lambda^0 , \qquad (16')$$

where a_3 and a_1 are the triplet and singlet Λp-scattering lengths in the s state, while A_Λ^0 and C_Λ^0 can be regarded approximately as real functions of the total energy E alone. Consequently, taking account of the unitarity of the S matrix and the analyticity of the reaction amplitude leads directly to the main result of the theory of final-state interaction (cf., for example, the paper of Gribov[8]).

By using (16) the expressions for the reaction cross section and the polarization of the Λ particles can be represented as

$$\frac{d\sigma}{dT} = (2\pi)^4 \frac{E}{2(E^2 - 4M_N^2)^{1/2}} (4\pi)^2 (2m_\Lambda \mu)^{3/2} [T(T_{max} - T)]^{1/2}$$

$$\times \left[2 \frac{\sin^2 \delta_3}{(p_\Lambda a_3)^2} |A_\Lambda^0|^2 + 4 \frac{\sin^2 \delta_1}{(p_\Lambda a_1)^2} |C_\Lambda^0|^2 \right], \qquad (17)$$

$$\mathbf{P}_\Lambda \left[\frac{\sin^2 \delta_3}{(p_\Lambda a_3)^2} |A_\Lambda^0|^2 + 2 \frac{\sin^2 \delta_1}{(p_\Lambda a_1)^2} |C_\Lambda^0|^2 \right] = \left[\frac{\sin^2 \delta_3}{(p_\Lambda a_3)^2} |A_\Lambda^0|^2 \right.$$

$$\left. + 2 A_\Lambda^0 C_\Lambda^0 \frac{\sin \delta_1 \sin \delta_3 \cos (\delta_1 - \delta_3)}{p_\Lambda^2 \cdot a_3 a_1} \right] (\mathbf{kP}) \mathbf{k}$$

$$- 2 A_\Lambda^0 C_\Lambda^0 \frac{\sin \delta_1 \sin \delta_3 \cos (\delta_1 - \delta_3)}{p_\Lambda^2 \cdot a_3 a_1} \mathbf{P} \qquad (18)$$

$p_\Lambda^2 = 2m_\Lambda (T_{max} - T)$. If we change the sign in front of C_Λ^0 on the right of equation (18), we obtain the expression for the polarization of the recoil nucleons. Expressions (17) and (10) can be considered as a generalization of the results of Henley, who neglected the dependence of the reaction matrix on spin.

From (17) and (18) we see that the investigation of the energy spectrum of K mesons and, in particular, of the polarization of Λ particles and nucleons is very important for the determination of the Λp-scattering lengths.

4. INELASTIC INTERACTION. NEAR-THRESHOLD SINGULARITIES.

As the energy is increased, the Σ channel is opened, and we may expect a change in the spectrum of K mesons and other quantities for the ΛKp channel. In this case, in the unitarity condition (8), we must consider as a possible intermediate state the state $|\Sigma NK\rangle$. We shall restrict ourselves to interaction in s-states.

As in the preceding section we assume that*

$$\langle \Lambda NK | T | \Sigma N' K' \rangle \approx \langle \Lambda N | T | \Sigma N' \rangle \langle K | K' \rangle ,$$

and use the fact that

$$\langle \Lambda N | T | \Sigma N \rangle = [4\pi^2 p_\Lambda^{1/2} p_\Sigma^{1/2} m_\Lambda^{1/2} m_\Sigma^{1/2}]^{-1}$$

$$\times \left[\tfrac{1}{4} (3 + \sigma_1 \sigma_2) \beta_3 + \tfrac{1}{4} (1 - \sigma_1 \sigma_2) \beta_1 \right], \qquad (19)$$

where the indices Λ and Σ denote quantities in the corresponding channels, while

$$p_\Sigma = [2m_\Sigma (E' - T)]^{1/2}, \quad E' = E - M_\Sigma + M_\Lambda . \qquad (20)$$

Assuming that there are no bound states of the p-Σ system, we represent the energy dependence of β_3 and β_1 in the low-energy region in the form

$$\beta_3 = b_3 p_\Sigma^{1/2} , \qquad \beta_1 = b_1 p_\Sigma^{1/2} , \qquad (21)$$

if the internal parities of Σ and Λ are the same.

The influence of the Σ channel shows itself for the Λ channel not only as an additional term in the unitarity condition (8), but also as an additional term in the matrix element of the Λp scattering matrix proportional to p_Σ:

$$\alpha_3 = \alpha_3^0 + i c_3 p_\Sigma , \qquad \alpha_1 = \alpha_1^0 + i c_1 p_\Sigma , \qquad (22)$$

where

$$c_{1,3} = (p_\Lambda / 4\pi) \sigma_{1,3}^{\Sigma,\Lambda} , \qquad (23)$$

and $\sigma_{2j}^{\Sigma,\Lambda}$ is the total cross section for the reaction $\Sigma + N \to \Lambda + N$ in the state with angular momentum j.

Using (19)-(23), we find from (8)

$$\operatorname{Im} C_\Lambda = (\operatorname{Im} C_\Lambda)_{p_\Sigma = 0} + C_\Lambda p_\Sigma ,$$

$$\operatorname{Im} A_\Lambda = (\operatorname{Im} A_\Lambda)_{p_\Sigma = 0} + A_\Lambda p_\Sigma , \qquad (24)$$

*The inclusion of terms of the type <pp|T⁺|pp> <pp|T|YN′K′>, which are small for this reaction, but are necessary in other cases, complicates the expressions but does not change the fundamental result.

where $(\delta \neq \pi/2)$

$$A'_\Lambda = A^0_\Lambda (p_\Lambda/4\pi)\,\sigma_3^{\Sigma,\Lambda}(p_\Sigma = 0)\,\tan^2\delta_3 + A^0_\Sigma b_3/\cos^2\delta_3,$$

$$C'_\Lambda = C^0_\Lambda (p_\Lambda/4\pi)\,\sigma_1^{\Sigma,\Lambda}(p_\Sigma = 0)\,\tan^2\delta_1 + C^0_\Sigma b_1/\cos^2\delta_1. \tag{25}$$

The relation (24) is valid when the kinetic energy T of the K meson is less than E′. For T > E′ the production of a real Σ particle becomes impossible, and we must replace p_Σ by ik_Σ, where $k_\Sigma = \sqrt{2m_\Sigma(T - E')}$, T > E′, so that the term which depends linearly on k_Σ appears in the real part of the reaction amplitude.

The presence of terms proportional to $p_\Sigma(T < E')$ and $k_\Sigma(T > E')$ causes the derivative with respect to the energy to become infinite both in the energy spectrum of the K mesons and in the energy dependence of the polarization of Λ particles (and nucleons). The order of magnitude of these anomalies is given by (24) and (25), and their shape depends on the relative sign of A^0_Λ, A^0_Σ, $b_{3,1}$ and δ. All four cases of anomalies which have been discussed in the literature for binary reactions can also occur in this present case.

All of the expressions in Secs. 2, 3, and 4 were given for the production of particles in pp collisions. It is not difficult to generalize them to the case of np collisions. This is done in the Appendix. We also discuss there the case of a scalar K particle.

We note that, in the general case also, the quantities which replace A'_Λ and C'_Λ have terms which are directly related to the final-state interaction, as well as terms which are not caused by it.

We emphasize that the expressions obtained in the present section refer to interaction in an s state of the final system. The relatively large mass difference of the Λ and Σ hyperons makes it difficult to apply the theory of inelastic interaction to the analysis of reaction (1), but this does not change the basic assertion that there is a non-monotonic behavior in the spectrum and the causes for its occurrence.

It was shown earlier[9] that the direct analytic continuation $p_\Sigma \to ik_\Sigma$ can not be carried out when there is a resonance in the neighborhood of the threshold. In this case, it is necessary to make use of dispersion relations. Since the analytic behavior of the reaction amplitude as a function of ω is not known, we have not carried out such an analysis. However, even if such a resonance occurs, we may expect non-monotonic variation with energy for a relative energy of the Λ–N pair equal to the threshold for the new channel.

If Σ and Λ have opposite parities, the first term of the expansion in (22) starts with p^3_Σ and only the second derivative with respect to the energy becomes infinite. Consequently, the study of threshold anomalies in the energy spectrum of K mesons with sufficiently high accuracy may prove important for determining the relative parity of the Σ and Λ particles.

5. DISCUSSION

Thus, endothermic inelastic interactions of the type C + D → E + F in the final state of the reaction A + B → a + C + D can give rise to non-monotonic variations with energy in the spectrum of the particles a, whose form can be determined from the condition of analyticity and unitarity of the S matrix. To investigate these singularities experimentally requires, of course, good accuracy and high energy resolution, but as a result of discovering them and studying them one can obtain information concerning the interaction of unstable particles, their spins and parities.

Earlier we have treated the production of hyperons and K mesons in NN collisions. We mention various other processes in which similar anomalies can occur whose study may give information concerning the interaction of unstable particles.

In the spectrum of π^+ mesons from the reaction

$$K^- + p \to \Lambda + \pi^- + \pi^+ \tag{26}$$

in the neighborhood of the threshold for

$$\pi^- + \Lambda \to \Sigma^- + \pi^0 \tag{27}$$

there will occur an anomaly whose magnitude and character will be related to K^-p scattering at low energies via the reaction amplitude (27).

In the spectrum of protons from the process for production of π mesons by K mesons

$$K^- + p \to p + \pi^0 + K^- \tag{28}$$

an anomaly may occur for an energy corresponding to the threshold for the reaction

$$\pi^0 + K^- \to \overline{K}^0 + \pi^-, \tag{29}$$

if there exist forces leading to such a reaction.

If one attempts to construct a Lagrangian for the πK interaction and does not consider interactions containing derivatives, the expression obtained

$$L_{int} = g\,(\varphi_\pi^i \cdot \varphi_\pi^i)\,(\varphi_\Theta^k \cdot \varphi_\Theta^k)$$

is invariant with respect to rotation of the isotopic spin of each of the particles, and all processes for K π scattering with charge exchange are forbidden. Under more general assumptions one does not obtain a forbiddenness for reaction (20), so that the observation of a non-monotonic variation with energy in the proton spectrum from reaction (28) would be of interest from the point of view of the study of the symmetry of the π K interaction.

Among reactions in which two π mesons participate, it is interesting to note that in the distribution of nucleons from the reactions

$$\gamma + p \to p + \pi^0 + \pi^0, \qquad \pi^- + p \to n + \pi^0 + \pi^0 \tag{30}$$

there may occur similar anomalies for a relative energy of the π^0 mesons exceeding 9 Mev, where the charge exchange reaction

$$\pi^0 + \pi^0 \to \pi^- + \pi^+ \tag{31}$$

becomes possible. Including threshold phenomena in the reaction (31) can have significant effects in the theory of $\pi\pi$ interaction at low energies.

The existence of a threshold in reaction (31) can lead to a non-monotonicity in the spectrum of charged π mesons from τ' decay.

$$\tau^{\pm} \to \pi^{\pm} + \pi^0 + \pi^0.$$

Analogously to the reactions (30), in the spectrum of nucleons from the reactions

$$\gamma + p \to p + K^- + K^+, \qquad \pi^- + p \to n + K^+ + K^- \tag{32}$$

in the neighborhood of the threshold for the reaction

$$K^+ + K^- \to \overline{K}{}^0 + K^0 \tag{33}$$

there may occur energy anomalies associated with KK interaction. Moreover, in the final state of reaction (33), there is no Coulomb interaction which might mask the non-monotonicity (cf. the paper of Newton and Fonda[10]).

In the spectrum of protons from the reaction

$$p + p \to p + p + \pi^0 \tag{34}$$

near the threshold for

$$\pi^0 + p \to n + \pi^+$$

and in the spectrum of π^+ mesons from the reaction

$$p + p \to n + p + \pi^+ \tag{35}$$

near the threshold for

$$n + p \to d + \pi^0$$

there will also be energy non-monotonicities.*

APPENDIX

A. PRODUCTION OF A PSEUDOSCALAR K MESON IN np COLLISION

In np collisions there are two possibilities for production of Λ particles

$$n + p \to \Lambda + p + K^0, \tag{A.1}$$

$$n + p \to \Lambda + n + K^+. \tag{A.2}$$

We denote the reaction amplitudes in the singlet and triplet isotopic spin states by T_0 and T_1 respectively. The reaction (A.1) is then described by the amplitude $\frac{1}{2}(T_0 + T_1)$, and the reaction (A.2) by the amplitude $\frac{1}{2}(T_1 - T_0)$. The spin dependence of the isotopic triplet amplitude is given by (9) with $B_\Lambda = 0$, while

$$\langle \Lambda NK \,|\, T_0 \,|\, NN \rangle = B_\Lambda \{(\sigma_1 - \sigma_2,\ \mathbf{k}) + i\mathbf{k}\,[\sigma_1 \times \sigma_2]\}. \tag{A.3}$$

Under the assumptions made earlier we can take account of final-state interaction by setting

$$B_\Lambda = B_\Lambda^0 (p_\Lambda a_3)^{-1} e^{i\delta_3} \sin\delta_3, \tag{A.4}$$

where B_Λ^0 is a function of the total energy E.

The expressions for the cross sections for production and polarization of the Λ particles have the following forms:

$$d\sigma = (2\pi)^4 \frac{E}{2\,[E^2 - 4M_N^2]^{1/2}} \frac{1}{4} (4\pi)^2 (2m_\Lambda\mu)^{3/2} [T\,(T_{max} - T)]^{1/2}\, dT$$
$$\times \left[\,|A_\Lambda + C_\Lambda \pm B_\Lambda|^2 + |A_\Lambda - C_\Lambda \mp B_\Lambda|^2 + 2\,|C_\Lambda \mp B_\Lambda|^2\right], \tag{A.5}$$

$$\mathbf{P}_\Lambda \left[\,|A_\Lambda + C_\Lambda \pm B_\Lambda|^2 + |A_\Lambda - C_\Lambda \mp B_\Lambda|^2 + 2\,|C_\Lambda \mp B_\Lambda|^2\right]$$
$$= 2\left[\,|A_\Lambda + C_\Lambda \pm B_\Lambda|^2 - |C_\Lambda \mp B_\Lambda|^2\right](\mathbf{kP})\,\mathbf{k}$$
$$+ \left[\,|A_\Lambda - C_\Lambda \mp B_\Lambda|^2 - |A_\Lambda + C_\Lambda \pm B_\Lambda|^2\right]\mathbf{P}. \tag{A.6}$$

The plus sign in front of B_Λ^0 holds for reaction (A.1), and the minus sign for (A.2). From (A.5) and (A.6) it is easy to obtain the "intensity rules":

$$d\sigma(np \to \Lambda pK^0) = d\sigma(np \to \Lambda nK^+), \tag{A.7}$$

$$\mathbf{P}_\Lambda(np \to \Lambda pK^0) = \mathbf{P}_\Lambda(np \to \Lambda nK^+) \quad \text{for } \mathbf{P} \parallel \mathbf{k}. \tag{A.8}$$

These relations are obtained on the assumption

---

*The scattering lengths for low energies of the π^0-p system differ from those obtained on the assumption of isotopic invariance because of the presence of non-monotonicities which violate isotopic invariance and are related to the reaction

$$\pi^0 + p \to n + \pi^+.$$

An estimate using dispersion relations gives a correction ~ 5%.

that one need only consider the s wave in the final state. They can be used for an experimental check of this assumption.

B. PRODUCTION OF A SCALAR K MESON IN NN COLLISIONS

In this case

$$\langle \Lambda NK | T_1 | NN \rangle = A_\Lambda, \quad \langle \Lambda NK | T_0 | NN \rangle = B_\Lambda (\sigma_1 \sigma_2); \tag{B.1}$$

$$A_\Lambda = A_\Lambda^0 (p_\Lambda a_1)^{-1} e^{i\delta_1} \sin \delta_1, \quad B_\Lambda = B_\Lambda^0 (p_\Lambda a_3)^{-1} e^{i\delta_3} \sin \delta_3. \tag{B.2}$$

If we introduce

$$f(NN \to \Lambda NK) = \frac{d\sigma (NN \to \Lambda NK) \cdot 8 (E^2 - 4M_N^2)^{1/2}}{dT \, [T (T_{max} - T)]^{1/2} (2\pi)^4 E (4\pi)^2 (2m_\Lambda \mu)^{3/2}},$$

the cross section and polarization of the Λ particles in all three reactions are given by

$$f(pp \to \Lambda pK^+) = |A_\Lambda^0|^2 (p_\Lambda a_1)^{-2} \sin^2 \delta_1,$$

$$\mathbf{P}_\Lambda (pp \to \Lambda pK) = \mathbf{P};$$

$$f(np \to \Lambda NK) = f(pp \to \Lambda pK^+) + 3 |B_0|^2 (p_\Lambda a_3)^{-2} \sin^2 \delta_3, \tag{B.3}$$

$$\mathbf{P}_\Lambda (np \to \Lambda NK) = (|A_\Lambda|^2 - |B_\Lambda|^2)(|A_\Lambda|^2 + 3|B_\Lambda|^2)^{-1} \mathbf{P}. \tag{B.4}$$

[1] A. B. Migdal, JETP **28**, 10 (1955), Soviet Phys. JETP **1**, 7 (1955).

[2] K. M. Watson and K. Brueckner, Phys. Rev. **83**, 1 (1951). K. M. Watson, Phys. Rev. **88**, 1163 (1952).

[3] R. P. Paruntseva, JETP **22**, 123 (1952).

[4] E. M. Henley, Phys. Rev. **106**, 1083 (1957).

[5] G. Feldman and P. T. Matthews, Phys. Rev. **109**, 546 (1958).

[6] R. Karplus and L. S. Rodberg, Phys. Rev. **115**, 1058 (1959).

[7] E. P. Wigner, Phys. Rev. **73**, 1002 (1948). A. I. Baz', JETP **33**, 923 (1957), Soviet Phys. JETP **6**, 709 (1958). G. Breit, Phys. Rev. **107**, 1612 (1957). R. G. Newton, Ann. Phys. **4**, 29 (1958). R. K. Adair, Phys. Rev. **111**, 632 (1958). L. Fonda, Nuovo cimento **13**, 956 (1959).

[8] V. N. Gribov, JETP **33**, 1431 (1957), Soviet Phys. JETP **6**, 1102 (1958); JETP **34**, 749 (1958), Soviet Phys. JETP **7**, 514 (1958); Nucl. Phys. **5**, 653 (1958).

[9] L. I. Lapidus and Chou Huang-chao, JETP **39**, 112 (1960), Soviet Phys. JETP, **12**, ((1961).

[10] L. Fonda and R. G. Newton, Ann. Phys. **7**, 133 (1959).

Translated by M. Hamermesh
74

SOVIET PHYSICS JETP VOLUME 12, NUMBER 3 MARCH, 1961

ON THE PSEUDOVECTOR CURRENT AND LEPTON DECAYS OF BARYONS AND MESONS

CHOU KUANG-CHAO

Joint Institute for Nuclear Research

Submitted to JETP editor April 7, 1960

J. Exptl. Theoret. Phys. (U.S.S.R.) 39, 703-712, (September, 1960)

By the use of the analytic properties of a certain matrix element it is shown that the result of Goldberger and Treiman regarding the decay $\pi \to \mu + \nu$ is valid for wider classes of strong interactions than those found by Feynman, Gell-Mann, and Levy, and in particular for the ordinary pseudoscalar theory with pseudoscalar coupling. A formula is obtained which can be used for an experimental test of the assumptions that are made. Lepton decays of hyperons and K mesons are also discussed.

1. INTRODUCTION

At the present time the theory of the universal V − A interaction given by Feynman and Gell-Mann and by Sudarshan and Marshak is in good agreement with all the experimental data on β decay and the decay of the μ meson.[1] The experimental ratio of the probabilities for the two types of π-meson decay, $R(\pi \to e + \nu)/R(\pi \to \mu + \nu)$, also agrees with the theoretical prediction. It may therefore be supposed that the universal V − A theory is also valid for processes of capture of μ mesons in nuclei.

One of the most important problems is the calculation of the probability of the decay $\pi \to \mu + \nu$ according to the universal V − A theory. This problem has been studied in detail in a paper by Goldberger and Treiman (G. T.)[2] by means of the technique of dispersion theory. Despite the fact that G. T. made many crude approximations, the numerical result of their work agrees almost exactly with the experimental result.

Quite recently Feynman, Gell-Mann, and Levy (F.G.L.) have reexamined this problem in a very interesting paper.[3] They have shown that the G.T. result can be obtained rigorously in certain models. Namely, let us write the Hamiltonian for β decay and μ-meson capture in the form

$$H = (g_0/\sqrt{2})(P_\alpha + V_\alpha) L_\alpha + \text{Herm. adj.} \tag{1}$$

where

$$L_\alpha = \bar{\nu}\gamma_\alpha (1 + \gamma_5) e + \bar{\nu}\gamma_\alpha (1 + \gamma_5) \mu. \tag{2}$$

P_α and V_α are the pseudovector and vector currents for the weak interactions. F.G.L. succeeded in finding three models of the strong interactions in which the following equation holds:

$$\partial_\alpha P_\alpha (x) = ia\pi (x) / \sqrt{2},$$

where a is a constant parameter and $\pi(x)$ is the pion-field operator. By using the equation (3), F.G.L. obtained the G.T. result in a simple and elegant way.

F.G.L. stated that their results would be extended later to any theory of the strong interactions. In their new theory, it is said in reference 3, there appears a form factor $\varphi(s)$, which is very complicated in the usual theory. In the opinion of F.G.L. it is only in the case of their models that it is reasonable to assume that $\varphi(s)$ is slowly varying.

After studying reference 3 we have come to the conclusion that the G.T. result is a good approximation for a wider class of strong interactions. In the present paper this conclusion is examined under the following assumptions:

1. The matrix element $\langle n | \partial_\alpha P_\alpha(0) | p \rangle$ is an analytic function of the variable $s = -(p_n - p_p)^2$.
2. If the matrix element of the commutator for equal times is equal to zero, then we can write a dispersion relation without subtraction.
3. The contribution of the nearest singularities predominates in the dispersion relation.

From our point of view the form factor $\varphi(s)$ actually is slowly varying in any theory in which there is a dispersion relation without subtraction for a certain matrix element.

In Secs. 2 and 3 a derivation of the G.T. result is presented in the most general form. It is shown that the G.T. result is also a good approximation for the ordinary pseudoscalar theory with pseudoscalar coupling. A relation is obtained between the pseudovector constant g_A for μ capture, g_A for β decay, and the pseudoscalar coupling constant f for μ capture. Since we can measure $g_{A\mu}$,

$g_{A\beta}$, and f separately, a test of this relation between the constants gives a sensitive criterion for the correctness of the assumptions made about the universality of the pseudovector coupling in the weak interaction and the analyticity of a certain matrix element.

In Sec. 4 the lepton decays of hyperons and K mesons are treated in an analogous way. From the data on the lifetime of K mesons the result is obtained that the pseudovector coupling constant g_{AY} for the β decay of hyperons is smaller than the coupling constant g_A for the β decay of neutrons.[9]

2. THE RESULT OF GOLDBERGER AND TREIMAN

Let us write

$$i\partial_\alpha P_\alpha(x) \equiv O(x). \tag{4}$$

Applying this equation to the decay $\pi \rightarrow \mu + \nu$, we get

$$\langle 0|O(0)|\pi\rangle = -q_\alpha \langle 0|P_\alpha(0)|\pi\rangle, \tag{5}$$

where q_α is the four-momentum of the pion. The matrix element $\langle 0|P_\alpha(0)|\pi\rangle$ can be expressed in the form

$$\langle 0|P_\alpha(0)|\pi\rangle = -q_\alpha F(m^2)/\sqrt{2q_0}, \tag{6}$$

where m is the mass of the pion and $F(m^2)$ is a constant parameter, which is determined by the lifetime of the pion.

Substituting Eq. (6) in Eq. (5), we get

$$\langle 0|O(0)|\pi\rangle = -m^2 F/\sqrt{2q_0}. \tag{7}$$

Let us now turn to the consideration of β decay and μ capture. In the general case the matrix element $\langle n|P_\alpha(0)|p\rangle$ is of the form

$$\langle n|P_\alpha(0)|p\rangle = \bar{u}_n\{g_A\gamma_\alpha\gamma_5 + if(p_p - p_n)_\alpha\gamma_5\}u_p, \tag{8}$$

where g_A and f are invariant functions of $s = -(p_p - p_n)^2$.

Applying the relation (4) to β decay and μ capture, we get

$$\langle n|O(0)|p\rangle = -(p_p - p_n)_\alpha \langle n|P_\alpha(0)|p\rangle. \tag{9}$$

Substituting Eq. (8) in Eq. (9), we have

$$\langle n|O(0)|p\rangle = i[2Mg_A + fs]\bar{u}_n\gamma_5 u_p. \tag{10}$$

The central problem is to find the connection between the matrix elements $\langle n|O(0)|p\rangle$ and $\langle 0|O(0)|\pi\rangle$. This can be done if we use the analyticity properties of the matrix element

$$\langle n|O(0)|p\rangle = i\bar{u}_n\gamma_5 u_p T(s), \tag{11}$$

$$T(s) = -\sqrt{2}GFm^2/(-s+m^2) + T'(s), \tag{12}$$

where G is the renormalized constant of the strong interactions of pions with nucleons, and $T'(s)$ is a function analytic in the region

$$|s| < 9m^2. \tag{13}$$

The derivation of (11) — (14) is given later, in Sec. 3. It is also shown there that $T'(s)$ is in fact a slowly varying function for small s.

In the region $|s| < m^2$ the function $T'(s)$ is approximated with good accuracy by a constant.

Let us rewrite (12) in the form

$$T(s) = -\sqrt{2}GF\varphi(s)m^2/(-s+m^2), \tag{14}$$

where

$$\varphi(s) = 1 + \alpha(s - m^2)/m^2. \tag{15}$$

Comparing (11) with (10), we get

$$2Mg_A + fs = -\sqrt{2}GF\varphi(s)m^2/(-s+m^2). \tag{16}$$

A very important fact is that the relation (16) holds for all s. Setting s = 0, we get

$$F = -2Mg_{A\beta}/\sqrt{2}G\varphi(0), \qquad g_{A\beta} = g_A(0). \tag{17}$$

This is the fundamental result of F.G.L., which was first obtained in the paper of Goldberger and Treiman.[2]

For μ capture

$$s_\mu = -Mm_\mu^3/(M+m_\mu) = -0.9m_\mu^2.$$

From (15), (16), and (17) we get*

$$2Mg_{A\mu} + f_\mu s_\mu = m^2 2Mg_{A\beta}/(-s_\mu + m^2). \tag{18}$$

Equation (18) is an exact relation between g_A, f for μ capture and g_A for β decay, which can be tested experimentally. It must be pointed out that the derivation of (18) has been carried out in the most general way, for an arbitrary value of α. As will be shown in Sec. 3, this holds for almost any theory in which the matrix element $\langle n|\partial_\alpha P_\alpha \times (0)|p\rangle$ is analytic.

Substituting the experimental values of G, $g_{A\beta}$ and F in (17), we get

$$\varphi(0) = 0.8. \tag{19}$$

From this and Eq. (15) we find

$$\alpha = 0.2. \tag{20}$$

We emphasize that the G.T. result is valid only for those theories in which the condition $\alpha \ll 1$ holds. This question is discussed in the following section.

---

*This relation is contained implicitly in a formula of Goldberger and Treiman.[4]

3. THE ANALYTICITY OF THE MATRIX ELEMENTS

Let us now turn to the calculation of the matrix element $\langle n|O(0)|p\rangle$. Using the standard method,[5] we write

$$\langle n|O(0)|p\rangle = -i\bar{u}_n \int d^4z e^{-ip_n z} \langle 0|T(\eta(z)O(0))|p\rangle$$
$$-\bar{u}_n^{\bullet}\int d^4z e^{-ip_n z}\,\delta(z_0)\,\langle 0|[\psi_n(z),\,O(0)]|p\rangle, \qquad (21)$$

where $\eta(z) = iS^{+}\delta S/\delta\bar{\psi}_n(z)$ is the current operator for the neutron field. Hereafter the equal-time commutator will be omitted; it would give an additive constant in the final expression and would not affect the analytic structure of the matrix element, for example, the locations of the poles and their residues, the branch points, and so on.

We note that

$$T(\eta(z)O(0)) = \theta(-z)[O(0),\,\eta(z)] + \eta(z)O(0),$$

where $\theta(z) = 1$ for $z_0 > 0$ and $\theta(z) = 0$ for $z_0 < 0$. The second term makes no contribution to the matrix element. Thus we have

$$\langle n|O(0)|p\rangle = -i\bar{u}_n\int d^4z e^{-ip_n z}\theta(-z)\langle 0|[O(0),\,\eta(z)]|p\rangle. \qquad (22)$$

In the coordinate system $p_p = 0$ it is easy to show by the method of Bogolyubov[6] that the function $T(s)$ in Eq. (11) has a pole at $s = m^2$ and a cut that begins at the point $s = 9m^2$. At other points it is analytic, if the following inequality holds:

$$|\operatorname{Im} p_{n0}| > |\operatorname{Im}\sqrt{p_{n0}^2 - M^2}| \qquad (p_{n0} = M - s/2M). \qquad (23)$$

Unfortunately, the inequality (23) is satisfied only in the case of imaginary nucleon mass. We assume in what follows that the analyticity of the matrix element in the variable s does not change on analytic continuation with respect to the mass variable.

The residue at the pole $s = m^2$ is easily calculated and is equal to $2^{1/2}G\langle 0|O(0)|\pi\rangle(2q_0)^{1/2}$. Thus we have

$$T(s) = \frac{\sqrt{2G}}{-s+m^2}\langle 0|O(0)|\pi\rangle\sqrt{2q_0}$$
$$+\,T'(s) = -\frac{\sqrt{2G}Fm^2}{-s+m^2} + T'(s). \qquad (24)$$

The corresponding Feynman diagram for the term with the pole is shown in the drawing.

In Eq. (24), $T'(s)$ is an analytic function with a branch point at $s = 9m^2$. The spectral resolution of the function $T'(s)$ if of the form

$$T'(s) = a_0 + \frac{s}{\pi}\int_{9m^2}^{\infty}\frac{\rho(s')}{s'(s'-s)}ds', \qquad (25)$$

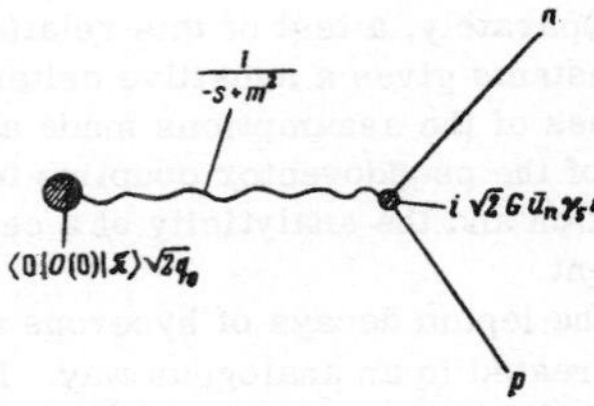

where $\rho(s')$ is the spectral function. For small s we can expand $T'(s)$ in a power series in s, which has the radius of convergence $9m^2$.

Setting

$$T'(s) = \sum_n a_n s^n, \qquad (26)$$

one can easily show that for large n

$$\lim_{n\to\infty}\left|\frac{a_{n+1}s}{a_n}\right| \leqslant \frac{|s|}{9m^2}. \qquad (27)$$

Therefore the series (26) converges rapidly in the region $|s| < m^2$.

If the spectral function does not change sign, the inequality (27) holds also for small n. In this case, for arbitrary $n > 1$ we have

$$|a_{n+1}| = \frac{1}{\pi}\left|\int_{9m^2}^{\infty}\frac{\rho(s')}{s'^{n+2}}ds'\right| \leqslant \frac{1}{9m^2}\frac{1}{\pi}\left|\int_{9m^2}^{\infty}\frac{\rho(s')}{s'^{n+1}}ds'\right| = \frac{|a_n|}{9m^2}.$$

We note that for β decay and μ capture the distance between the points $s = 0$ and $s_\mu = -0.9m_\mu^2$ is much smaller than the radius of convergence $9m^2$. Therefore with good accuracy we can replace $T'(s)$ by a single constant both for β decay and for μ capture (the error is of the order of $0.9m_\mu^2/9m^2 \approx 1/20$).

Thus we get the final result given by the formulas (14) and (15).

Let us now go on to the consideration of the quantity α. If the matrix element of the equal-time commutator is not zero, then in general one must use a dispersion relation with a subtraction. In this case the quantity α is proportional to the subtraction constant a_0, which can be very large.

If, on the other hand, the matrix element of the commutator is zero, than there is a dispersion relation without a subtraction. Then it is reasonable to suppose that the contribution of the nearest singularity predominates and the quantity α is small ($\alpha \approx 0.2 \ll 1$).

Thus it is reasonable to assume that $\varphi(s)$ is slowly varying for any theory in which there is a dispersion relation without subtraction.

Let us consider the ordinary pseudoscalar theory, with the Lagrangian

$$L = -\overline{N}(\hat{\partial} + M_0 - iG_0(\tau\pi)\gamma_5)$$
$$\times N - m_0^2\pi^2/2 - (\partial_\alpha\pi)^2/2 - \lambda_0\pi^4. \quad (28)$$

By means of the gauge transformation

$$N \rightarrow (1 + i(\tau v)\gamma_5)N, \quad \pi \rightarrow \pi + v(4M_0 + 2M)/3G_0$$

we get by the standard method, explained in reference 3,

$$P_\alpha = \overline{N}\tau\gamma_\alpha\gamma_5 N - i\partial_\alpha\pi(4M_0 + 2M)/3G_0, \quad (29)$$
$$O(x) = i\partial_\alpha P_\alpha = 2G_0\overline{N}N\pi + i\tfrac{2}{3}(M_0 - M)\overline{N}\tau\gamma_5 N$$
$$+ (m_0^2\pi + 4\lambda_0\pi^2\pi)(4M_0 + 2M)/3G_0. \quad (30)$$

We shall show in this case that the equal-time commutator for the operator O makes no contribution to the matrix element $\langle n|O(0)|p\rangle$. Let us examine the matrix element of the commutator

$$I = \langle 0|2G_0N\pi + i\tfrac{2}{3}(M_0 - M)\tau\gamma_5 N|N\rangle.$$

From symmetry properties we have

$$I = iA\tau\gamma_5 u_N. \quad (31)$$

Multiplying Eq. (31) on the left by the matrix $\tau\gamma_5$, we get

$$i3Au_N = -2i\langle 0|\eta(0)|N\rangle,$$

where

$$\eta(0) = iG_0(\tau\pi)\gamma_5 N + (M - M_0)N$$

is the current of the nucleon field. It is known that the matrix element $\langle 0|\eta(0)|N\rangle$ is equal to zero, and therefore $I = 0$.

Thus we have shown that in the ordinary pseudoscalar theory there exists the pseudovector current (29), which satisfies all the necessary requirements.

If the pseudovector current is of the ordinary form

$$P_\alpha = \overline{N}\tau\gamma_\alpha\gamma_5 N,$$

then the matrix element of the commutator is not zero, and in general there is no dispersion relation without subtraction. Even in this case there is hope that the G.T. result is valid. This question will be discussed in the Appendix.

4. LEPTON DECAYS OF HYPERONS AND K MESONS

The experimental limit for the probabilities of lepton decays of Λ and Σ hyperons is an order of magnitude smaller than the theoretical value calculated on the hypothesis that the effective coupling constants in hyperon decays are equal to those in β decays.[7] Many authors have expressed the opinion that the universality of the weak interactions evidently does not extend to strange-particle decays. Nevertheless, it is reasonable to assume the existence of a limited universality [a lepton current in the form (2)[8]].

In what follows we assume that the K meson is pseudoscalar and the V and A interactions exist for the lepton decays of strange particles. In this case the Hamiltonian for the weak decays of strange particles is of the form (1). Following the example given in Sec. 2 for the pseudoscalar theory with pseudoscalar coupling, we can construct the pseudovector current in such a form that a dispersion relation without subtraction holds for the matrix element $\langle N|\partial_\alpha P_\alpha|Y\rangle$.

Generally speaking, the matrix element for hyperon decay consists of three terms:

$$\langle N|P_\alpha(0)|Y\rangle = \bar{u}_N\{g_{AY}\gamma_\alpha\gamma_5 + i\xi_Y[(\hat{p}_N - \hat{p}_Y)$$
$$\times\gamma_\alpha - \gamma_\alpha(\hat{p}_N - \hat{p}_Y)]\gamma_5 + if_Y(p_Y - p_N)_\alpha\gamma_5\}u_Y, \quad (32)$$

from which we have

$$\langle N|O(0)|Y\rangle = i\langle N|\partial_\alpha P_\alpha|Y\rangle$$
$$= i[(M_N + M_Y)g_{AY} + f_Y s]\bar{u}_N\gamma_5 u_Y, \quad (33)$$

where $s = -(p_Y - p_N)^2$. Repeating one after another the arguments presented in Secs. 2 and 3, we easily get the following equation:

$$[(M_N + M_Y)g_{AY} + f_Y s]$$
$$= -G_{KY}F_K m_K^2/(-s + m_K^2) + T_Y(s), \quad (34)$$

where G_{KY} is the renormalized coupling constant for the KYN interaction, and F_K is a constant parameter associated with the decay of K mesons. We have further

$$\langle 0|P_\alpha(0)|K\rangle = -q_\alpha F_K/\sqrt{2q_0}. \quad (35)$$

We can determine F_K from data on the lifetime for the decay $K \rightarrow \mu + \nu$. In Eq. (34) $T_Y(s)$ is a function that is analytic in the region

$$|s| < (m_K + 2m)^2. \quad (36)$$

Let us denote by T_N the kinetic energy of the nucleon recoil in the rest system of the hyperon. Expressing s in terms of T_N, we get

$$s = (M_Y - M_N)^2 - 2M_Y T_N. \quad (37)$$

In the present case the values of s that correspond to β and μ decays are very close together, as compared with the distance between the s given by Eq. (37) and $s = (m_K + 2m)^2$. Therefore with good accuracy we can replace $T_Y(s)$ by a constant a_Y.

Thus we have

$$[(M_N + M_Y)g_{AY} + f_Y s] = -G_{KY}F_K m_K^2/(-s + m_K^2) + a_Y. \tag{38}$$

The relation (38) can be used to test the universality of the pseudovector current in lepton decays of strange particles.

Applying the dispersion theory of Goldberger and Treiman, we find for the function f_Y:

$$f_Y = -G_{KY}F_K/(-s + m_K^2) + T'_Y(s), \tag{39}$$

where $T'_Y(s)$ is a function analytic in the region (36), which with good accuracy can be replaced by a constant a'_Y.

Substituting Eq. (39) in Eq. (38), we get

$$(M_N + M_Y)g_{AY} = -G_{KY}F_K + a_Y - sa_Y. \tag{40}$$

The relation (40) is a generalization of the formula of Goldberger and Treiman for the decay of strange particles.

The experimental data on the lifetimes of K and π mesons show that $F_K^2 \ll F_\pi^2$. Therefore it can be seen from a comparison of Eqs. (40) and (16) that to accuracy $a_Y - sa'_Y$

$$g_{AY}^2 \ll g_{A\beta}^2, \tag{41}$$

even for the case in which G_{KY} and G are of the same order of magnitude. This fact was first pointed out by Sakita.[9]

We emphasize that our method can also be easily extended to the case of a scalar K meson and to other types of weak interactions (for example, S + P).

In the case in which the relative parity of K and YN is positive, we have to deal with the divergence of a vector current

$$i\partial_\alpha V_\alpha = O(x). \tag{42}$$

The matrix element $\langle N|V_\alpha(0)|Y\rangle$ is of the form

$$\langle N|V_\alpha(0)|Y\rangle = \bar{u}_N\{g_{VY}\gamma_\alpha + iC_Y[(\hat{p}_N - \hat{p}_Y)\gamma_\alpha - \gamma_\alpha(\hat{p}_N - \hat{p}_Y)] + id_Y(p_Y - p_N)_\alpha\}u_Y. \tag{43}$$

From this we have

$$\langle N|O(0)|Y\rangle = i[(M_N - M_Y)g_{VY} + d_Y s]u_N u_Y. \tag{44}$$

It is easy to repeat the remaining arguments, and the final formula will be of the form

$$(M_N - M_Y)g_{VY} + d_Y s = -G_{KY}F_K m_K^2/(-s + m_K^2) + a_Y. \tag{45}$$

Applying the dispersion theory for the function d_Y, we find

$$d_Y = -G_{KY}F_K/(-s + m_K^2) + a'_Y. \tag{46}$$

Substituting (46) in (45), we get

$$(M_N - M_Y)g_{VY} = -G_{KY}F_K + a_Y - a'_Y s. \tag{47}$$

Comparing (47) and (16), one sees that to accuracy $a_Y - sa'_Y$

$$\left(\frac{g_{VY}}{g_{A\beta}}\right)^2 \approx \left(\frac{2M_N}{M_N - M_Y}\frac{F_K G_{KY}}{F_\pi G_\pi}\right)^2 \approx 5C\left(\frac{G_{KY}}{G_\pi}\right)^2,$$

where C is of the order of unity. Therefore in the case of the scalar K meson the small probability of lepton decay of hyperons could be explained only by having the coupling constant G_{KY} for the KYN interaction be smaller than the pion-nucleon constant G_π.

We note that Λ and Σ can have different relative parities. Let us consider this case. For simplicity we call the K particle a scalar, if the relative parity of K and Λ N is positive, and a pseudoscalar if it is negative. In the case of the pseudoscalar K meson, Eq. (40) holds for the decay of Λ particles, and Eq. (47) holds for the decay of Σ particles, if we write $\langle N|P_\alpha|\Sigma\rangle$ in the form (43). In the case of the scalar K meson, conversely, Eq. (47) holds for the decay of Λ particles and Eq. (40) for Σ particles, if we write $\langle N|V_\alpha|\Sigma\rangle$ in the form (32).

We note that the relations (38) and (45) can be used for the determination of the renormalized coupling constants G_{KY}, if precise experiments are made on the decays of strange particles.

The writer expresses his hearty gratitude to Professor M. A. Markov, Ya. A. Smorodinskiĭ, and Chu Hung-Yüang, and also to Ho Tso-Hsiu and V. I Ogievetskiĭ for their interest in this work and a discussion of the results.

APPENDIX

In the usual theory the pseudovector current has the form

$$\mathbf{P}_\alpha = \bar{N}\tau\gamma_\alpha\gamma_5 N, \tag{A.1}$$

for which the divergence has been calculated in reference 3 and is given by

$$\partial_\alpha \mathbf{P}_\alpha = 2M_0\bar{N}\tau\gamma_5 N - 2iG_0\bar{N}N\pi. \tag{A.2}$$

In order to calculate the matrix element $\langle 0|\partial_\alpha P_\alpha|\pi\rangle$, we write Eq. (A.2) in the form

$$\partial_\alpha \mathbf{P}_\alpha = -i\frac{2M_0}{G_0}\mathbf{j} + i\frac{2M_0}{G_0}[(m_0^2 - m^2)\pi - 4\lambda_0\pi^2\pi] - 2iG_0\bar{N}N\pi, \tag{A.3}$$

where $\mathbf{j}$ is the meson-field current. Using the

fact that the matrix element $\langle 0|j(0)|\pi\rangle$ is equal to zero, we get

$$\langle 0|\partial_\alpha P_\alpha(0)|\pi\rangle = i2M_0G_0^{-1}\delta m^2\sqrt{Z_3}/\sqrt{2q_0}\,\cdot$$
$$-\,8M_0\lambda_0G_0^{-1}\langle 0|\pi^2\pi|\pi\rangle - 2iG_0\langle 0|\bar{N}N\pi|\pi\rangle, \quad (A.4)$$

where Z_3 is the renormalization constant for the pion wave function.

We now go on to the consideration of $\langle n|\partial_\alpha P_\alpha|p\rangle$. We rewrite Eq. (A.4) in the form

$$\partial_\alpha P_\alpha = -i\frac{4M_0+2M}{3G_0}j + O(x), \quad (A.5)$$

where the operator O is that of Eq. (30).

In the present case we can write a dispersion relation without subtraction for the matrix element $\langle n|O(0)|p\rangle$, with the term with the pole defined by Eq. (12). Thus we have

$$i\langle n|\partial_\alpha P_\alpha(0)|p\rangle = \tfrac{1}{3}(4M_0+2M)\,GG_0^{-1}d(s)F(s)\sqrt{Z_3}\,i\bar{u}_n\gamma_5u_p$$
$$+\,\langle n|O(0)|p\rangle, \quad (A.6)$$

where d(s) and F(s) are the respective form factors for the π-meson propagation function and the vertex part. If the first term in Eq. (A.6) is small in comparison with the term with the pole, then the G.T. result would hold also for the usual theory.

We assume that the first term in Eq. (A.5) predominates. Comparing Eqs. (A.5) and (A.6), we get

$$F_7 \approx \frac{2M_0}{G_0}\frac{\delta m^2}{m^2}\sqrt{Z_3}. \quad (A.7)$$

By means of Eq. (A.7) the first term in Eq. (A.6) can be expressed in the form

$$i\frac{2M_0+M}{3M_0}G\frac{Fm^2}{\delta m^2}\bar{u}_n\gamma_5u_p d(s)F(s). \quad (A.8)$$

Since in perturbation theory the quantity δm^2 diverges quadratically, it is very probable that

$$\frac{2M_0+M}{3M_0}\frac{m^2}{\delta m^2} \ll 1. \quad (A.9)$$

In this case the first term in Eq. (A.6) is actually small in comparison with the term that has the pole. Therefore it seems to us that the G.T. result is also valid for the usual theory.

It is interesting to note one more example, in which the pseudovector current has the form

$$iP_\alpha(x) = \partial_\alpha\pi(x). \quad (A.10)$$

It is easy to show directly from Eq. (A.10) that the matrix element $\langle n|P_\alpha(0)|p\rangle$ for β decay is equal to zero.

In this case the divergence of the pseudovector current is

$$i\partial_\alpha P_\alpha = m_0^2\pi - iG_0\bar{N}\tau\gamma_5N - 4\lambda_0\pi^2\pi = m^2\pi(x)$$
$$-\,j(x) = O(x) - j(x). \quad (A.11)$$

From this we get

$$\langle 0|O(0)|\pi\rangle = i\langle 0|\partial_\alpha P_\alpha(0)|\pi\rangle = m^2\sqrt{Z_3}/\sqrt{2q_0}, \quad (A.12)$$
$$i\langle n|\partial_\alpha P_\alpha|p\rangle = \langle n|O(0)|p\rangle - i\sqrt{2}G\,d(s)F(s)\sqrt{Z_3}\bar{u}_n\gamma_5u_p. \quad (A.13)$$

The term with the pole is of the form

$$\sqrt{2}G\sqrt{Z_3}m^2/(-s+m^2). \quad (A.14)$$

Comparing the expression (A.14) with the second term in Eq. (A.13), we verify that they are of the same order and cancel each other.

From this example it can be seen that for those theories in which a dispersion relation without subtraction does not hold for the matrix element $\langle n|\partial_\alpha P_\alpha(0)|p\rangle$ the G.T. result is in general not a good approximation.

[1] R. P. Feynman and M. Gell-Mann, Phys. Rev. 109, 193 (1958). E. C. G. Sudarshan and R. E. Marshak, Phys. Rev. 109, 1860 (1958).

[2] M. L. Goldberger and S. B. Treiman, Phys. Rev. 110, 1178 (1958).

[3] Feynman, Gell-Mann, and Levy, The Axial Vector Current in β Decay (preprint, 1960).

[4] L. Wolfenstein, Nuovo cimento 8, 882 (1958). M. L. Goldberger and S. B. Treiman, Phys. Rev. 111, 354 (1958).

[5] N. N. Bogolyubov and D. V. Shirkov, Введение в теорию квантованных полей (Introduction to the Theory of Quantized Fields), GITTL, 1957. [Interscience, 1959] Lehmann, Symanzik, and Zimmerman, Nuovo cimento 1, 205 (1955).

[6] Bogolyubov, Medvedev, and Polivanov, Вопросы теории дисперсионных соотношений (Problems of the Theory of Dispersion Relations) Fizmatgiz, 1958.

[7] F. S. Crawford, Jr., et al., Phys. Rev. Letters 1, 377 (1958). P. Nordin et al., Phys. Rev. Letters 1, 380 (1958).

[8] Chou Kuang-Chao and V. Maevskiĭ, JETP 35, 1581 (1958), Soviet Phys. JETP 8, 1106 (1959). R. H. Dalitz, Revs. Modern Phys. 31, 823 (1959).

[9] B. Sakita, Phys. Rev. 114, 1650 (1959).

Translated by W. H. Furry
133

SOVIET PHYSICS JETP VOLUME 12, NUMBER 4 APRIL, 1961

ELASTIC SCATTERING OF GAMMA QUANTA BY NUCLEI

L. I. LAPIDUS and CHOU KUANG-CHAO

Joint Institute of Nuclear Research

Submitted to JETP editor May 12, 1960

J. Exptl. Theoret. Phys. (U.S.S.R.) **39**, 1056-1058 (October, 1960)

The energy dependence of the cross section for elastic scattering of γ quanta near the photonuclear threshold is investigated with the help of the dispersion relation for forward scattering. The first peak in the scattering cross section is attributed to dispersion effects. Experiments required for a more detailed analysis are discussed.

1. A recent investigation of the scattering of γ quanta by deuterons below the threshold for pion production with the help of dispersion relations[1] shows that photonuclear processes have a very strong effect on the elastic scattering of γ quanta in a wide region of energies (up to ~ 100 Mev). In the present paper we discuss the scattering of low-energy γ quanta by nuclei.

Fuller and Hayward,[2] as well as Penfold and Garwin,[3] have already used dispersion relations for the analysis of the elastic scattering of γ quanta by nuclei. However, they considered only the scattering above the threshold for the (γN) reaction. Taking the formation of particles in the S state into account leads to the known non-monotonic behavior near the threshold for the (γn) reaction. The dispersion relations not only allow us to discuss the Wigner-Baz' effect, but also to take account of the general effect of inelastic processes on the energy dependence of the elastic scattering amplitude in a wide region of energies. It appears to us that, within the framework of the dispersion relations, the first peak in the scattering cross section of γ quanta[2] is related to the near-threshold effects in a natural way.

We restrict the discussion to forward scattering. In the dipole approximation, which, apparently, does not contradict the experimental data on the absorption of γ quanta by nuclei up to ~ 30 Mev,[3] our results are also valid for other scattering angles and for the total elastic scattering cross section $\sigma_s(\nu)$.

2. The scattering amplitude for γ quanta has the form

$$T = R_1(\mathbf{e}'\mathbf{e}) + R_2([\mathbf{k}\times\mathbf{e}]\times[\mathbf{k}'\times\mathbf{e}']) + (\mathbf{e}'\mathbf{e})[R_1 + (\mathbf{k}\mathbf{k}')R_2] - R_2(\mathbf{k}\mathbf{e}')(\mathbf{k}'\mathbf{e}) \tag{1}$$

for a spinless nucleus and

$$\begin{aligned} T = {} & R_1(\mathbf{e}'\mathbf{e}) + R_2([\mathbf{k}\times\mathbf{e}][\mathbf{k}'\times\mathbf{e}']) + iR_3(\boldsymbol{\sigma}[\mathbf{e}\times\mathbf{e}]) \\ & + iR_4(\boldsymbol{\sigma}[[\mathbf{k}'\times\mathbf{e}']\times[\mathbf{k}\times\mathbf{e}]]) + iR_5[(\boldsymbol{\sigma}\mathbf{k})(\mathbf{e}[\mathbf{k}'\times\mathbf{e}']) - (\boldsymbol{\sigma}\mathbf{k}')([\mathbf{k}\times\mathbf{e}]\mathbf{e}')] \\ & + iR_6[(\boldsymbol{\sigma}\mathbf{k}')([\mathbf{k}'\times\mathbf{e}']\mathbf{e}) - (\boldsymbol{\sigma}\mathbf{k})(\mathbf{e}'[\mathbf{k}\times\mathbf{e}])] \end{aligned} \tag{2}$$

for the scattering from a nucleus with spin $\frac{1}{2}$. The corresponding expressions for nuclei with larger spins are more complicated.

In the dipole approximation only R_1 is different from zero. Dispersion relations can be written down for all scalar functions R_i. The imaginary parts of the amplitudes are related to the cross sections for absorption of γ quanta in different states by the unitarity condition. The scattering amplitude for γ quanta at low energies is known, so that the real and imaginary parts of all R_i can be calculated from the data on the absorption of γ quanta in different states with the help of the dispersion relations and the unitarity condition. However, up to the present time no detailed analysis of the absorption of γ quanta by nuclei has been carried out. Therefore we restrict ourselves to the dispersion relation for $R_1 + R_2$, which has the usual form

$$\operatorname{Re}[R_1(\nu, \theta = 0^\circ) + R_2(\nu, \theta = 0^\circ)] = -\frac{Z^2e^2}{M_A} + \frac{\nu_0}{2\pi^2} \mathrm{P} \int_{\nu_t}^{\infty} \frac{\sigma_t(\nu)\, d\nu}{\nu^2 - \nu_0^2} \tag{3}$$

In the literature accessible to us (see the new review article of Wilkinson[4]) we did not find detailed information on the energy dependence of the cross section for absorption of γ quanta by nuclei. Only in the case of He_2^4 (and the deuteron) is the photodisintegration cross section known in a wide region of energies.[5] Experimental data on $\sigma_t(\nu)$ for aluminum up to ~ 30 Mev were obtained recently by Mihailović et al.[6] These authors found evidence of a fine structure in the dependence of the cross section in the region of the giant resonance. Data for the region near threshold are not available. By smoothing out these experimental data, we obtained with the help of relation (3) the energy dependence of the scattering amplitude shown schematically in Fig. 1. This behavior of the scattering amplitude is apparently characteristic for a whole group of nuclei with large $\sigma_t(\nu)$.

The photodisintegration of the deuteron leads to a decrease in the cross section for γd scattering near the threshold, as compared to the Thomson cross section. However, for other nuclei with a large absorption cross section the picture is different. Since the dispersive part of the amplitude below and slightly above the threshold is positive, the total amplitude vanishes at some energy below ν_t and then becomes positive (in the case of the deuteron[1] the amplitude does not change its sign). Above the threshold the real part of the amplitude decreases rapidly, goes through zero, and becomes negative, increasing in absolute value with increasing energy. Above the threshold, the amplitude has, of course, both a real and an imaginary part.

All this leads to an energy dependence of the scattering cross section which is shown in Fig. 2.

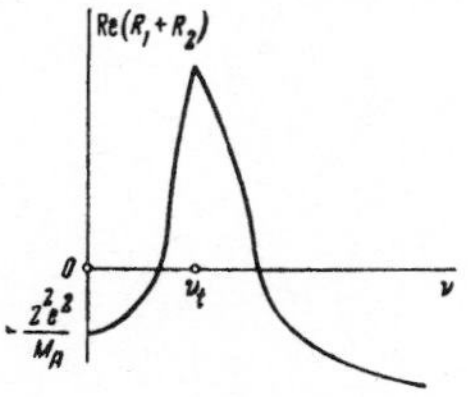

FIG. 1

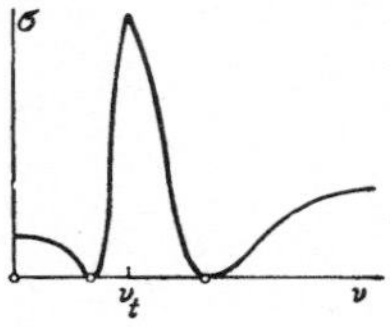

FIG. 2

In its general behavior, it agrees with the experimental data.[2] For the aluminum nucleus the scattering cross section reaches the value $\sigma_s(\nu) \approx 2 \times 10^{-28}$ cm^2 in the region of the first maximum, which is close to the experimental value. It is clear that, if the dispersion effects are not so large as to change the sign of the scattering amplitude in the threshold region, the first maximum will not appear.

The ratios between the cross sections at the maxima for different nuclei are connected with the relative role of the absorption of γ quanta in the region near threshold and in the region of the giant resonance. The fact that not only the region near threshold, but also the giant resonance, gives a contribution to the first maximum leads to an appreciable widening of this peak. The half-width for aluminum exceeds 2 Mev.

In the evaluation of the data on the absorption of γ quanta by aluminum we did not consider the effects connected with the difference in the thresholds for the (γp) and (γn) reactions, and we also neglected the resolving power of the apparatus. It seems to us that a more detailed analysis would lead to little change in the basic result.

3. The interpretation of the energy dependence of the cross section for scattering of γ quanta by nuclei proposed earlier[7] is thus apparently confirmed. Owing to the relatively poor accuracy of the experimental data on the absorption cross sections and to the absence of data on the absorption cross sections for a whole series of nuclei, it is not possible now to conduct a reliable analysis for all nuclei whose scattering properties are known.

The fruitfulness of using the dispersion relations in the analysis of the scattering of γ quanta by nuclei makes it feasible to conduct a whole series of investigations with γ quanta. First of all, it appears necessary to obtain information on the energy dependence of the absorption cross sections of γ quanta in the region near threshold as well as in the region of high energies. The consideration of the cross sections for photoproduction of pions in the calculation of $\sigma_t(\nu)$ may become necessary in the energy region of ~ 120 to 150 Mev and above. Knowing the absorption cross sections in a wide region of energies, one can obtain information on the polarizability of nuclei. For Al the polarizability

$$\alpha \equiv \left[\frac{d}{d\nu^2}\operatorname{Re}(R_1 + R_2)\right]_{\nu=0} = \frac{\hbar c}{2\pi^2} P \int_{\nu_t}^{\infty} \frac{\sigma_t(\nu)\, d\nu}{\nu^2}$$

turns out to equal $\sim 2 \times 10^{-39}$ cm^3 (the error may be as large as 50%). For He_2^4 the value is $\alpha = (0.70 \pm 0.05) \times 10^{-40}$ cm^3.

A more detailed phenomenological analysis of the absorption of γ quanta in a wide region of energies becomes inevitable if one wants to obtain information on the spin dependence of the dispersive parts of the scattering amplitudes for the γ quanta. On the other hand, the dispersion relations and the unitarity of the S matrix can be used to obtain information on the absorption cross section from the experimental data on the scattering of γ quanta by nuclei. For this purpose the inverse dispersion relations may prove to be convenient.

The authors are grateful to Ya. A. Smorodinskiĭ for useful comments.

---

[1] L. I. Lapidus and Chou Kuang-Chao, JETP (in press).

[2] E. G. Fuller and E. Hayward, Phys. Rev. 101, 692 (1956).

[3] A. S. Penfold and E. L. Garwin, Phys. Rev. 116, 120 (1959).

[4] D. H. Wilkinson, Ann. Rev. Nucl. Sci. 9, 1 (1959).

[5] A. N. Gorbunov, Tr. FIAN (Proceedings of the Physics Institute, Academy of Sciences) 13, 145 (1960).

[6] Mihailović, Prege, Kernel, and Kregar, Phys. Rev. 114, 1621 (1959).

[7] L. I. Lapidus and Chou Kuang-Chao, JETP 39, 112 (1960), Soviet Phys. JETP 12, 82 (1961).

Translated by R. Lipperheide

196

SOVIET PHYSICS JETP VOLUME 12, NUMBER 5 MAY, 1961

THE ELASTIC SCATTERING OF γ RAYS BY DEUTERONS BELOW THE PION-PRODUCTION THRESHOLD

L. I. LAPIDUS and CHOU KUANG-CHAO

Joint Institute for Nuclear Research

Submitted to JETP editor May 12, 1960

J. Exptl. Theoret. Phys. (U.S.S.R.) **39**, 1286-1295 (November, 1960)

Dispersion relations and the conditions for unitarity of the S matrix are used for the analysis of the elastic scattering of γ rays by deuterons below the threshold for pion production. The low-energy limit is examined for the scattering of γ rays by nuclei of arbitrary spin. The energy dependence of elastic γd scattering is deduced on the basis of the experimental data on the photodisintegration of the deuteron. The result differs decidely from that of the impulse approximation over a wide range of energies. It is found that it is not important to include the influence of photoproduction of pions from deuterons in the range of energy considered.

1. INTRODUCTION

THE scattering of γ rays by deuterons is an example of a process whose amplitude is decidedly affected by inelastic processes, such as the photodisintegration of the deuteron and the photoproduction of mesons. Inclusion of the influence of pion photoproduction, which is important at γ-ray energies near and above the photoproduction threshold, requires a rather detailed analysis of the processes

$$\gamma + d \to NN\pi, \qquad \gamma + d \to d + \pi^0$$

and is not dealt with here.

The influence of the photodisintegration of the deuteron on the elastic γd scattering near the threshold for photodisintegration, and the departures from monotonic variation with the energy, which lead to a sharp decrease of the cross section, have been considered previously.[1] The purpose of the present paper is to make an analysis of γd scattering on the basis of dispersion relations over a wider energy range, in which meson production still does not have much effect.

The experimental data[2] on the scattering of γ rays by deuterons in the energy range 50 — 100 Mev do not fit into the framework of the impulse approximation,[3] and this forces us to carry through an analysis that does not involve this approximation. On the other hand, the contribution to the scattering amplitude from meson-production processes falls off rapidly below the threshold for photoproduction of mesons.

We shall confine ourselves to the forward scattering. In the calculation of the dispersion integrals we take into account the cross sections for the electric-dipole and magnetic-dipole photodisintegrations. We begin with the phase-shift analysis, so as to express the imaginary parts of the scattering amplitudes in terms of the quantities that characterize the photodisintegration of the deuteron. We then consider the dispersion relations for forward scattering and the low-energy theorem. The dispersion integrals are evaluated in the range of γ-ray energies below ~ 100 Mev. The real and imaginary parts of the amplitudes are obtained, and the polarizabilities of the deuteron and of nucleons are discussed.

2. THE PHENOMENOLOGICAL ANALYSIS

As is well known, the formulas for the electric and magnetic multipole waves $\mathbf{Y}_{lm}^{(\lambda)}(\mathbf{k})$ of a photon are ($\lambda = 0, 1$)

$$\mathbf{Y}_{lm}^{(0)} = \sum_{\mu} C_{l\, m-\mu\, 1\mu}^{lm} Y_{l\, m-\mu}(\mathbf{k}) \zeta_{\mu}, \tag{1a}$$

$$\mathbf{Y}_{lm}^{(1)} = -i [\mathbf{k} \times \mathbf{Y}_{lm}^{(0)}], \tag{1b}$$

where $\mathbf{k}$ is the unit vector along the momentum of the photon in the center-of-mass system, $Y_{lm}(\mathbf{k})$ are normalized spherical functions, and

$$\zeta_{+} = -(\mathbf{i} + i\mathbf{j})/\sqrt{2},$$

$$\zeta_0 = \mathbf{k}, \qquad \zeta_{-} = (\mathbf{i} - i\mathbf{j})/\sqrt{2} \tag{2}$$

— the eigenfunctions of the photon spin — satisfy the transversality condition

$$\mathbf{k}\mathbf{Y}_{lm}^{(\lambda)}(\mathbf{k}) = 0.$$

If we write

$$\eta_{+} = -\frac{1}{\sqrt{2}}\begin{pmatrix}1\\ i\\ 0\end{pmatrix}, \quad \eta_0 = \begin{pmatrix}0\\ 0\\ 1\end{pmatrix}, \quad \eta_{-} = \frac{1}{\sqrt{2}}\begin{pmatrix}1\\ -i\\ 0\end{pmatrix} \tag{3}$$

for the spin functions of the deuteron, then for the γd system we can construct eigenfunctions of the total angular momentum J^2, the component J_z, and the parity from Eqs. (1) and (3):

$$\mathbf{Y}_{jlM}^{(\lambda)}(\mathbf{k}) = \sum_r C_{lM-r1r}^{jM}\, \mathbf{Y}_{l\,M-r}^{(\lambda)}(\mathbf{k})\,\eta_r. \tag{4}$$

In the center-of-mass system (c.m.s.) all quantities in the final state are denoted by symbols with primes; for example, $\mathbf{k}'$ denotes the direction of the photon momentum in the final state.

By means of Eq. (4) we can write the scattering matrix T in the form

$$(\mathbf{e}'T\mathbf{e}) = \sum_{\substack{JMll'\\ \lambda\lambda'}} Y_{jl'M}^{(\lambda')*}(\mathbf{k}')\, Y_{jlM}^{(\lambda)}(\mathbf{k})\, a_{jll'}^{\lambda\lambda'}, \tag{5}$$

where $\mathbf{e}$ and $\mathbf{e}'$ are the respective polarization vectors of the photon in the initial and final states.

Parity conservation requires that

$$a_{jll'}^{\lambda\lambda'} = 0 \text{ for } (-1)^{l+\lambda} \neq (-1)^{l'+\lambda'}. \tag{6}$$

Time-reversal invariance leads to the symmetry condition

$$a_{jll'}^{\lambda\lambda'} = a_{jl'l}^{\lambda'\lambda}. \tag{7}$$

The usual arguments show that for forward scattering the spin dependence of the matrix T is of the form

$$(\mathbf{e}'T\mathbf{e}) = A(\mathbf{e}'\mathbf{e}) + iB(\mathbf{S}[\mathbf{e}'\times\mathbf{e}]) + \tfrac{1}{2}C[(\mathbf{S}\mathbf{e})(\mathbf{S}\mathbf{e}') + (\mathbf{S}\mathbf{e}')(\mathbf{S}\mathbf{e})] + \tfrac{1}{2}D[(\mathbf{S}[\mathbf{k}\times\mathbf{e}])(\mathbf{S}[\mathbf{k}'\times\mathbf{e}']) + (\mathbf{S}[\mathbf{k}'\times\mathbf{e}'])(\mathbf{S}[\mathbf{k}\times\mathbf{e}])]. \tag{8}$$

Here $\mathbf{S}$ is the operator of the spin vector of the deuteron; its components S_i satisfy the Duffin-Kemmer commutation relations:

$$[S_i, S_j] = i\epsilon_{ijk}S_k, \qquad S_iS_jS_k + S_kS_jS_i = \delta_{ij}S_k + \delta_{jk}S_i. \tag{9}$$

Using the Stokes parameters to describe the polarization of the photon, as was done in our paper on γN scattering,[4] we get without difficulty from Eqs. (2) and (8):

$$(\zeta_{\pm}^{*}T\zeta_{\pm}) = A \mp B(\mathbf{S}\mathbf{k}) + \tfrac{1}{2}(D+C)[2-(\mathbf{S}\mathbf{k})^2],$$

$$(\zeta_{\pm}^{*}T\zeta_{\mp}) = \tfrac{1}{2}(D-C)\{(\mathbf{S}\mathbf{i})^2 - (\mathbf{S}\mathbf{j})^2 \mp i[(\mathbf{S}\mathbf{i})(\mathbf{S}\mathbf{j}) + (\mathbf{S}\mathbf{j})(\mathbf{S}\mathbf{i})]\}. \tag{10}$$

By means of Eq. (10) and the method developed previously[4] we can construct the density matrix of the final state and calculate all observable quantities. The unpolarized forward scattering cross section is given by

$$\sigma_0(0^\circ) = |A + \tfrac{2}{3}(C+D)|^2 + \tfrac{1}{18}|C + D|^2 + \tfrac{2}{3}|B|^2 + \tfrac{1}{3}|D-C|^2, \tag{11}$$

and we have

$$4\pi\,\mathrm{Im}\,(A + \tfrac{2}{3}C + \tfrac{2}{3}D) = q\sigma_t, \tag{11'}$$

where σ_t is the total interaction cross section, including both elastic and inelastic interactions; $\mathbf{q} = |q|\mathbf{k}$.

Let us turn to the phase-shift analysis. We include the amplitudes for electric-dipole and magnetic-dipole transitions. The magnetic-dipole transition is characterized by the matrix

$$F_j^0 = \sum Y_{j1M}^{(0)}(\mathbf{k})\, Y_{j1M}^{(0)}(\mathbf{k}) = \frac{3}{4\pi}\sum C_{1M-r\,1r}^{jM} C_{1M-r'1r'}^{jM} C_{10\,1M-r}^{1M-r} C_{10\,1M-r'}^{1M-r'}\, \eta_r\eta_{r'}^{*}\zeta_{M-r}\zeta_{M-r'}^{*}, \tag{12}$$

where we have used the fact that for forward scattering

$$Y_{lm} = \delta_{lm}\sqrt{(2l+1)/4\pi}.$$

From Eqs. (12) we easily obtain

$$(\zeta_{+}^{*}F\zeta_{+}) + (\zeta_{-}^{*}F\zeta_{-}) = \frac{3}{8\pi}[\alpha_j + \beta_j S_z^2],$$

$$(\zeta_{+}^{*}F\zeta_{+}) - (\zeta_{-}^{*}F\zeta_{-}) = \frac{3}{8\pi}\gamma_j S_z,$$

$$(\zeta_{+}^{*}F\zeta_{-}) = -\frac{3}{8\pi}\beta_j T_{2,-2} = -\frac{3}{8\pi}\beta_j\frac{1}{2}(S_x - iS_y)^2,$$

$$(\zeta_{-}^{*}F\zeta_{+}) = -\frac{3}{8\pi}\beta_j T_{2,2} = -\frac{3}{8\pi}\beta_j\frac{1}{2}(S_x + iS_y)^2. \tag{13}$$

For j = 0, 1, 2 the quantities α_j, β_j, γ_j are

$j =$	2	1	0
$\alpha_j =$	1	1	0
$\beta_j =$	$^1/_6$	$-^1/_2$	$^1/_3$
$\gamma_j =$	$^5/_6$	$-^1/_2$	$-^1/_3$

(14)

In obtaining the relations (13) we have used the relations

$$\eta_{+}\eta_{+}^{*} = \tfrac{1}{2}(S_z^2 + S_z), \quad \eta_{-}\eta_{-}^{*} = \tfrac{1}{2}(S_z^2 - S_z), \quad \eta_0\eta_0^{*} = 1 - S_z^2;$$

$$\eta_{+}\eta_{-}^{*} = T_{2,2} = \tfrac{1}{2}[S_x^2 - S_y^2 + i(S_xS_y + S_yS_x)],$$

$$\eta_{-}\eta_{+}^{*} = T_{2,-2} = \tfrac{1}{2}[S_x^2 - S_y^2 - i(S_xS_y + S_yS_x)]. \tag{15}$$

The case of the electric-dipole transition is obtained from the relations (13) by replacing ζ_{+} by $i[\zeta_{+}\times\mathbf{k}] = -\zeta_{+}$ and ζ_{-} by $i[\zeta_{-}\times\mathbf{k}] = \zeta_{-}$. Comparing Eqs. (13) and (14) with Eq. (10), we get

$$2A = \frac{3}{8\pi}\sum_j(\alpha_j + 2\beta_j)(a_j^{(m)} + a_j^{(e)}),$$

$$2B = -\frac{3}{8\pi}\sum_j\gamma_j(a_j^{(m)} + a_j^{(e)}),$$

$$C = -\frac{3}{8\pi}\sum_j\beta_j a_j^{(e)}, \qquad D = -\frac{3}{8\pi}\sum_j\beta_j a_j^{(m)}. \tag{16}$$

The condition for unitarity of the S matrix reduces to the relation

$$2\pi i\,[T^*(-\mathbf{k}', -\mathbf{k}, -\mathbf{e}', -\mathbf{e}, -\mathbf{S})_{\gamma d\to\gamma d} - T(\mathbf{k}', \mathbf{k}, \mathbf{e}', \mathbf{e}, \mathbf{S})_{\gamma d\to\gamma d}] = q\int d\Omega_{n+p}\, T^+_{\gamma d\to np} T_{\gamma d\to np}, \tag{17}$$

where q is the relative momentum of the γd system. Let us represent $T_{\gamma d\to np}$ in the form

$$T_{\gamma d\to np} = \sum Y^s_{je'M}(\mathbf{n})\, Y^{(\lambda)^*}_{jl M}(\mathbf{k})\, d^{s\lambda}_{jl'l}, \tag{18}$$

where j is the total angular momentum, l' is the orbital angular momentum in the final state, and s is the total spin of the np system.

Parity conservation requires that

$$(-1)^{l+\lambda+1} = (-1)^{l'}.$$

The quantities $d^{s\lambda}_{jl'l}$ are connected with the partial cross sections for photodisintegration:

$$24\pi\sigma^{s(m)}_{jl'} = (2j+1)\,|d^{s0}_{jl'1}|^2, \qquad 24\pi\sigma^{s(e)}_{jl'} = (2j+1)\,|d^{s1}_{jl'1}|^2. \tag{19}$$

The total cross section for photodisintegration is given by

$$24\pi\sigma_{\gamma d\to np} = \sum_{jl's}(2j+1)\,[\,|d^{s0}_{jl'1}|^2 + |d^{s1}_{jl'1}|^2]. \tag{20}$$

Substitution of Eqs. (18) and (5) in Eq. (17) gives

$$4\pi\,\mathrm{Im}\,a^{\lambda\lambda'}_{jll'}(q) = q\sum_{l''s}(d^{s\lambda}_{jl''l})^* d^{s\lambda'}_{jl''l'}, \tag{21}$$

and using Eqs. (16) and (14) we get the results

$$\mathrm{Im}\left[A + \tfrac{2}{3}(C+D)\right] = \frac{3}{16\pi}\sum_j\left(\alpha_j + \tfrac{2}{3}\beta_j\right)\left(\mathrm{Im}\,a^{(m)}_j + \mathrm{Im}\,a^{(e)}_j\right) = \frac{q}{4\pi}\sigma_{\gamma d\to np};$$

$$\mathrm{Im}\,A = (3\sigma_0 + \tfrac{6}{5}\sigma_2)\,q/4\pi,$$
$$\mathrm{Im}\,B = (\tfrac{3}{2}\sigma_0 + \tfrac{3}{4}\sigma_1 - \tfrac{3}{4}\sigma_2)\,q/4\pi,$$
$$\mathrm{Im}\,C = (-3\sigma^{(e)}_0 + \tfrac{3}{2}\sigma^{(e)}_1 - \tfrac{3}{10}\sigma^{(e)}_2)\,q/4\pi,$$
$$\mathrm{Im}\,D = (-3\sigma^{(m)}_0 - \tfrac{3}{10}\sigma^{(m)}_2)\,q/4\pi, \tag{22}$$

where σ_j denotes the partial cross section for photodisintegration in the state j, including the factor $(2j+1)$.

3. CROSSING SYMMETRY AND THE DISPERSION RELATIONS

The retarded amplitude for γd forward scattering can be written in the form

$$\langle\mu'|\mathbf{e}'N^{ret}\mathbf{e}|\mu\rangle = -2\pi^2 i\int d^4z e^{-iqz}\langle\mathbf{p}, \mu'|\Theta(z_0)\,[\mathbf{e}'\mathbf{j}(z/2), \mathbf{e}\mathbf{j}(-z/2)]\,|\mathbf{p}, \mu\rangle, \tag{23}$$

where μ and μ' are the spin indices of the deuteron.

For the advanced amplitude we have the analogous expression

$$\langle\mu'|\mathbf{e}'N^{adv}\mathbf{e}|\mu\rangle = 2\pi^2 i\int d^4z e^{-iqz}\langle\mathbf{p}', \mu'|\Theta(-z_0)\,[\mathbf{e}'\mathbf{j}(z/2), \mathbf{e}\mathbf{j}(-z/2)]\,|\mathbf{p}, \mu\rangle. \tag{24}$$

For the case of forward scattering the deuteron momentum p can be set equal to zero.

Considering the relations complex conjugate to Eqs. (23) and (24), we get

$$\langle\mu'|\mathbf{e}'N^{ret(adv)}(q)\mathbf{e}|\mu\rangle^* = \langle\mu|\mathbf{e}'N^{ret(adv)}(-q)\mathbf{e}|\mu'\rangle. \tag{25}$$

Interchanging the order of $(\mathbf{e}'\mathbf{j}(z/2))$ and $(\mathbf{e}\mathbf{j}(-z/2))$ in Eqs. (23) and (24) and changing the sign of the variable z, we arrive at the relation

$$\langle\mu'|\mathbf{e}'N^{ret(adv)}(q)\mathbf{e}|\mu\rangle = \langle\mu'|\mathbf{e}N^{adv(ret)}(-q)\mathbf{e}'|\mu\rangle. \tag{26}$$

Let us represent $N^{ret(adv)}$ in the form (8). The conditions (25) and (26) reduce to the following symmetry properties of the scalar functions A, B, C, D:

$$A^{ret(adv)}(\nu)^* = A^{ret(adv)}(-\nu), \quad B^{ret(adv)}(\nu)^* = -B^{ret(adv)}(-\nu),$$
$$C^{ret(adv)}(\nu)^* = C^{ret(adv)}(-\nu), \quad D^{ret(adv)}(\nu)^* = D^{ret(adv)}(-\nu); \tag{27}$$

$$A^{adv}(\nu) = A^{ret}(-\nu), \qquad B^{adv}(\nu) = -B^{ret}(-\nu),$$
$$C^{adv}(\nu) = C^{ret}(-\nu), \qquad D^{adv}(\nu) = D^{ret}(-\nu). \tag{28}$$

Denoting hereafter the quantities $A(\nu)$, $C(\nu)$, and $D(\nu)$ by $L_1(\nu)$, $L_2(\nu)$, and $L_3(\nu)$, respectively, and $B(\nu)$ by $L_4(\nu)$, we write the dispersion relations for the scalar functions in the form

$$\mathrm{Re}\,L_{1,2,3}(\nu_0) - \mathrm{Re}\,L_{1,2,3}(0) = \frac{2\nu_0^2}{\pi}\,\mathrm{P}\int_{\nu_d}^{\infty}\frac{d\nu\,\mathrm{Im}\,L_{1,2,3}(\nu)}{\nu(\nu^2-\nu_0^2)}, \tag{29}$$

$$\mathrm{Re}\,L_4(\nu_0) - \nu_0\,\mathrm{Re}\,L_4'(0) = \frac{2\nu_0^3}{\pi}\,\mathrm{P}\int_{\nu_d}^{\infty}\frac{d\nu\,\mathrm{Im}\,L_4(\nu)}{\nu^2(\nu^2-\nu_0^2)}, \tag{30}$$

where ν_d is the threshold for photodisintegration of the deuteron, approximately equal to the binding energy of the deuteron.

In order for it to be possible to use the relations (29) and (30) for an actual analysis, it is necessary to know $L_{1,2,3}(0)$ and $L_4'(0)$, i. e., to calculate the γd scattering amplitude in the energy region close to zero. The result of the calculations carried out in the following section is that

$$\mathrm{Re}\,L_1(0) = -e^2/M, \qquad \mathrm{Re}\,L_{2,3}(0) = 0,$$
$$\mathrm{Re}\,L_4'(0) = (\mu_0 - e/M_d)^2, \tag{31}$$

where μ_0 is the magnetic moment and M_d the mass of the deuteron.

4. THE LOW-ENERGY LIMIT FOR γd SCATTERING

Thirring, Low, Gell-Mann, and Goldberger[5] have shown that the limiting values at $\nu_0 = 0$ of the scattering amplitude and of its derivative with respect to the photon frequency are determined by the statistical properties of the system, for systems with spin $1/2$. Following a method developed by Low, we shall show that analogous results are also valid for systems with arbitrary spin.

The S matrix for the scattering of photons from the state (q, e) into the state (q′, e′) is given by the expression

$$S' = -e'_i q_{ij} e_j (4q_0 q'_0)^{-1/2}, \tag{32}$$

$$q_{ij} = \int \mathrm{P}\,[j_i(x),\ j_j(y)]\, e^{iqx - iq'x}\, dx dy. \tag{33}$$

Using the technique of Low, it is not hard to get the results*

$$g_{ij} = g_{ij}^{(0)} + A\delta_{ij} + B\varepsilon_{ijk} S_k + D(S_i S_j + S_j S_i), \tag{34}$$

$$A(\mathbf{q}'\mathbf{q}) + B(\mathbf{S}\,[\mathbf{q}'\times\mathbf{q}]) + D\,[(\mathbf{S}\mathbf{q}')(\mathbf{S}\mathbf{q}) + (\mathbf{S}\mathbf{q})(\mathbf{S}\mathbf{q}')] = q_0 q'_0 C, \tag{35}$$

$$C = \frac{(2\pi)^4}{i}\,\delta^{(4)}(p' + q' - p - q) \sum \Big[\frac{\langle \mathbf{q} - \mathbf{q}' | j_0 | \mathbf{q}\rangle \langle \mathbf{q} | j_0 | 0\rangle}{E(q) - E(0) - q_0} + \frac{\langle \mathbf{q} - \mathbf{q}' | j_0 | -\mathbf{q}'\rangle \langle -\mathbf{q}' | j_0 | 0\rangle}{E(q') - E(0) + q'_0} \Big], \tag{36}$$

$$g_{ij}^{(0)} = \frac{(2\pi)^4}{i}\,\delta^{(4)}(p' + q' - p - q) \sum \Big[\frac{\langle \mathbf{q} - \mathbf{q}' | j_i | \mathbf{q}\rangle \langle \mathbf{q} | j_j | 0\rangle}{E(q) - E(0) - q_0} + \frac{\langle \mathbf{q} - \mathbf{q}' | j_j | -\mathbf{q}'\rangle \langle -\mathbf{q}' | j_i | 0\rangle}{E(q') - E(0) + q'_0} \Big]. \tag{37}$$

The summation in Eqs. (36) and (37) is taken over the spins of the particles involved in the reaction.

Let us consider the case in which the states $|\mathbf{q}\rangle$ and so on are eigenstates of a system with spin S. For the calculation of Eqs. (36) and (37) we need the expression for the current matrix $\langle \mathbf{p}_2 | \mathbf{j} | \mathbf{p}_1\rangle$ in the low-energy region to accuracy v/c, and for $\langle \mathbf{q}' | j_0 | \mathbf{q}\rangle$ to accuracy v^2/c^2. It turns out that these matrix elements can be determined with the required accuracy on the basis of general principles.

Since $\mathbf{j}$ and j_0 are Hermitian operators and the interaction is invariant under three-dimensional rotations and time reversal, the most general form of the matrix element of the current is, in the approximation in question,

$$\langle \mathbf{p}_2 | \mathbf{j} | \mathbf{p}_1\rangle = (e/2M)(\mathbf{p}_1 + \mathbf{p}_2) + i\mu \mathbf{S}\times[\mathbf{p}_2 - \mathbf{p}_1] + c\,\{\mathbf{S}(\mathbf{S}, \mathbf{p}_1 + \mathbf{p}_2) + (\mathbf{S}, \mathbf{p}_1 + \mathbf{p}_2)\mathbf{S}\}, \tag{38}$$

$$\langle \mathbf{p}_2 | j_0 | \mathbf{p}_1\rangle = a + b(p_1^2 + p_2^2) + d(\mathbf{p}_1\mathbf{p}_2) + if(\mathbf{S}\,[\mathbf{p}_2\times\mathbf{p}_1]) + h\,[(\mathbf{S}\mathbf{p}_1)(\mathbf{S}\mathbf{p}_1) + (\mathbf{S}\mathbf{p}_2)(\mathbf{S}\mathbf{p}_2)] + g\,[(\mathbf{S}\mathbf{p}_1)(\mathbf{S}\mathbf{p}_2) + (\mathbf{S}\mathbf{p}_2)(\mathbf{S}\mathbf{p}_1)], \tag{39}$$

where e is the total charge, $\mu s = \mu_0$ is the total magnetic moment, and the quantities a, b, c, d, f, h, g are invariant constants.

Under Lorentz transformations j_1 behaves like a component of a four-vector. Being an irreducible representation of the inhomogeneous Lorentz group, the wave function $|\mathbf{p}, \mu\rangle$ transforms in the following way:

$$|\mathbf{p}, \mu\rangle \xrightarrow{L} |\mathbf{p}, \mu\rangle' = R_{\mu\mu'}(L, \mathbf{p})\, |L^{-1}\mathbf{p}, \mu'\rangle, \tag{40}$$

where $R_{\mu\mu'}(L, \mathbf{p})$ is the rotation of the spin in the Lorentz transformation, which has been treated by a number of authors.[6,7]

Let us consider two coordinate systems. In one

$$\mathbf{p}_1 = 0, \qquad \mathbf{p}_2 = \mathbf{p},$$

and in the other

$$\mathbf{p}_1 = \mathbf{q}, \qquad p_{10} = E_q = \sqrt{q^2 + M^2},$$

$$\mathbf{p}_2 = \mathbf{p} + \frac{\mathbf{q}}{E_q}\left[\frac{(\mathbf{p}\mathbf{q})}{q^2} E_q \left(1 - \frac{E_q}{M}\right) + \frac{E_q}{M} E_p\right],$$

$$p_{20} = \frac{E_q}{M}\left[E_p + \frac{(\mathbf{p}\mathbf{q})}{E_q}\right]. \tag{41}$$

The second system moves with the velocity $-\mathbf{q}/E_q$ relative to the first. For the Lorentz transformation from the first system to the second we have to accuracy v^2/c^2

$$R(L, \mathbf{p}) = 1 + i(\mathbf{S}\,[\mathbf{p}\times\mathbf{q}])/2M^2. \tag{42}$$

We also have to accuracy v/c

$$\begin{aligned} \langle \mathbf{p} + \mathbf{q} | \mathbf{j} | \mathbf{q}\rangle' &= e(\mathbf{p} + 2\mathbf{q})/2M + i\mu\,[\mathbf{S}\times\mathbf{p}] + c\,\{\mathbf{S}(\mathbf{S}, \mathbf{p} + 2\mathbf{q}) + (\mathbf{S}, \mathbf{p} + 2\mathbf{q})\mathbf{S}\} \\ &= \langle \mathbf{p} | \mathbf{j} | 0\rangle + (\mathbf{q}/M)\langle \mathbf{p} | \mathbf{j} | 0\rangle \\ &= e\mathbf{p}/2M + i\mu\,[\mathbf{S}\times\mathbf{p}] + c\,[\mathbf{S}(\mathbf{S}\mathbf{p}) + (\mathbf{S}\mathbf{p})\mathbf{S}] + a\mathbf{q}/M \end{aligned} \tag{43}$$

and to accuracy v^2/c^2

$$\begin{aligned} \langle \mathbf{p} + \mathbf{q} | j_0 | \mathbf{q}\rangle' &= a + b(p^2 + 2\mathbf{p}\mathbf{q} + 2q^2) + d(\mathbf{p}\mathbf{q} + q^2) + if(\mathbf{S}\,[\mathbf{p}\times\mathbf{q}]) + h\,\{\mathbf{S}, \mathbf{p} + \mathbf{q})(\mathbf{S}, \mathbf{p} + \mathbf{q}) + (\mathbf{S}\mathbf{q})(\mathbf{S}\mathbf{q})\} \\ &\quad + g\,\{(\mathbf{S}, \mathbf{p} + \mathbf{q})(\mathbf{S}\mathbf{q}) + (\mathbf{S}\mathbf{q})(\mathbf{S}, \mathbf{p} + \mathbf{q})\} \\ &= [1 - i(\mathbf{S}\,[\mathbf{p}\times\mathbf{q}])/2M^2]\, E_q M^{-1}\, [\langle \mathbf{p} | j_0 | 0\rangle + (qM^{-1}, \langle \mathbf{p} | \mathbf{j} | 0\rangle)] = a - ia(\mathbf{S}[\mathbf{p}\times\mathbf{q}])/2M^2 + aq^2/2M^2 \\ &\quad + bp^2 + h(\mathbf{S}\mathbf{p})(\mathbf{S}\mathbf{p}) + e(\mathbf{p}\mathbf{q})/2M^2 + i\mu(\mathbf{S}\,[\mathbf{p}\times\mathbf{q}])/M. \end{aligned} \tag{44}$$

From Eq. (43) it follows that

$$a = e, \qquad c = 0,$$

and from Eq. (44) that

$$d + 2b = e/2M^2, \qquad f = \mu/M - e/2M^2, \qquad g = h = 0.$$

*This is the most general expression for an arbitrary S, if we are not concerned with terms with energy dependence higher than linear.

Finally we have the covariant expressions

$$\langle \mathbf{p}_2 | \mathbf{j} | \mathbf{p}_1 \rangle = e(\mathbf{p}_1 + \mathbf{p}_2)/2M + i\mu \mathbf{S} \times [\mathbf{p}_2 - \mathbf{p}_1],$$

$$\langle \mathbf{p}_2 | j_0 | \mathbf{p}_1 \rangle = e + i(\mu/M - e/2M^2)(\mathbf{S}[\mathbf{p}_- \mathbf{p}_1]) + e(\mathbf{p}_1\mathbf{p}_2)/2M^2 + 2b(\mathbf{p}_1 - \mathbf{p}_2)^2, \quad (45)$$

and, as must be so, the first of these is the same as the matrix element of the current of a nonrelativistic particle interacting with a magnetic field (cf., e. g., the book of Landau and Lifshitz[8]). It turns out that the term contained in the expression (45) makes no contribution to the final result.

By means of Eq. (45) one easily gets

$$S = i(2\pi)^4\delta^{(4)}(p' + q' - p - q)(4q_0q_0')^{-1/2}\{e^2(\mathbf{e}'\mathbf{e})/M - 2ie(\mathbf{S}[\mathbf{e}'\times\mathbf{e}])M(\mu - e/2M) - (i\mu^2/q_0)(\mathbf{S}[[\mathbf{e}\times\mathbf{q}]\times[\mathbf{e}'\times\mathbf{q}']]) - (ie\mu/Mq_0)[(\mathbf{e}\mathbf{q}')(\mathbf{S}[\mathbf{q}'\times\mathbf{e}']) - (\mathbf{e}'\mathbf{q})(\mathbf{S}[\mathbf{q}\times\mathbf{e}])]\} \quad (46)$$

or for the matrix T

$$-T = e^2(\mathbf{e}'\mathbf{e})/M - 2i(e/M)\nu(\mu - e/2M)(\mathbf{S}[\mathbf{e}'\times\mathbf{e}] - i(\mu^2/\nu)(\mathbf{S}[[\mathbf{e}\times\mathbf{q}]\times[\mathbf{e}'\times\mathbf{q}']]) - (ie\mu/M\nu)[(\mathbf{e}\mathbf{q}')(\mathbf{S}[\mathbf{q}'\times\mathbf{e}']) - (\mathbf{e}'\mathbf{q})(\mathbf{S}[\mathbf{q}\times\mathbf{e}])]. \quad (47)$$

For forward scattering, in particular, we have

$$-T = e^2(\mathbf{e}'\mathbf{e})/M - i\nu(\mathbf{S}[\mathbf{e}'\times\mathbf{e}])[\mu_0/S - e/M]^2, \quad (48)$$

from which the result (31) indeed follows for S = 1.

In the energy range below the threshold for photoproduction of mesons from deuterons the terms that depend on the spin make an insignificant contribution to the cross section, since the mass of the nucleus is doubled in comparison with that of a nucleon, and the magnetic moment is much smaller.

5. RESULTS OF THE ANALYSIS. DISCUSSION

Experimental data on the photodisintegration of the deuteron are available right up to ~500 Mev.[9] The results of the calculation of Re $[L_1 + (2/3)(L_2 + L_3)]$, for which it is sufficient to know the total cross sections, are shown in Fig. 1, where the values of the real part of the amplitude are represented as fractions of e^2/M_pc^2. The photon energy is measured as a multiple of the threshold for photodisintegration of the deuteron: $\nu_0/\nu_d = \gamma_0$. The diagram also shows the energy dependence of

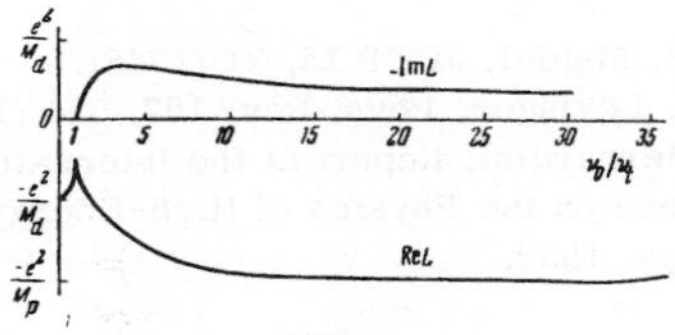

FIG. 1

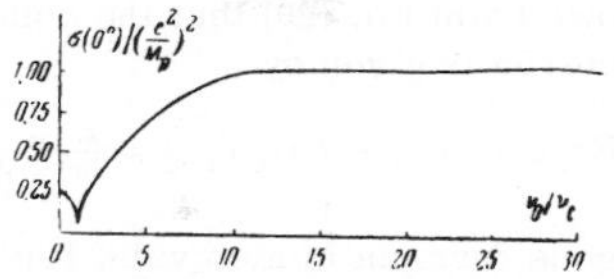

FIG. 2

the imaginary part of the quantity $L_1 + (2/3)(L_2 + L_3)$.

In the energy range considered, $\nu_0 \lesssim 100$ Mev, the dominant contribution is that of photodisintegration with $\nu \lesssim 75$ Mev. For the other amplitudes a more detailed analysis of the photodisintegration is required. If we assume that the contribution from photodisintegration with $\nu_0 \lesssim 80$ Mev is also decisive for the other amplitudes and use the analysis of de Swart and Marshak,[10] it is possible to estimate the dispersion parts of all the scalar amplitudes. But for γd scattering the spin-dependent amplitudes play a much smaller part as compared with the case of γN scattering. Figure 2 shows the energy dependence of the γd forward scattering. With increase of the γ-ray energy the γd scattering cross section at first shows a marked decrease as compared with the Thomson limit, and then ($\nu \gtrsim 4$ Mev) rapidly rises, and in the range $20 < \nu < 80$ Mev reaches values larger than $(e^2/M_dc^2)^2$ by a factor four.

Inclusion of the magnetic-dipole absorption, especially near the threshold, leads to an additional sharp "dip" of the cross section,[1] with a total half-width ~200 — 300 kev. The width of the total decrease of the cross section is considerably larger.

The large influence of inelastic processes that involve the deuteron as a whole, in addition to the processes involving the individual nucleons of the deuteron, makes it impossible to apply the impulse approximation to elastic γd scattering over a wide range of energies.* The presence of the inelastic process of photodisintegration of the deuteron has an especially strong effect on the polarizability of the deuteron. If, as A. M. Baldin has shown, the polarizability of nucleons is entirely due to the process of meson production, on the other hand the main contribution to the polarizability of the deuteron, and of nuclei in general, comes from photonuclear processes at much smaller energies.

---

*In a preprint received very recently, Schult and Capps have made a new examination of the corrections to the impulse approximation for γd scattering and have come to a similar conclusion.

It follows from Eq. (29) that the polarizability of the deuteron is given by

$$\alpha_d = \frac{d}{d\nu^2}\left[\mathrm{Re}\left(L_1 + 2/3 L_2 + 2/3 L_3\right)\right]_{\nu=0} = \frac{\hbar c}{2\pi^2} \mathrm{P} \int_{\nu_d}^{\infty} \frac{\sigma_t(\nu) d\nu}{\nu^2}. \quad (49)$$

An analogous formula is also valid for other nuclei. Dipole absorption plays the fundamental role in the total interaction cross section $\sigma_t(\nu)$. When this is taken into account Eq. (49) goes over into the well known formula of Migdal[11] (cf. also reference 12).

Substitution into Eq. (49) of the expressions (41) and (46) of reference 1 gives for the sum of the electric and magnetic polarizabilities of the deuteron

$$\alpha_e + \alpha_m = \alpha_d = \frac{e^2}{M_p c^2}\left(\frac{\hbar c}{\varepsilon}\right)^2 \left\{\frac{3}{64} + \frac{1}{12}\left(1 + \sqrt{\frac{\varepsilon'}{\varepsilon}}\right)^2 \frac{\varepsilon}{M_p c^2} \times (\mu_p - \mu_n)^2\right\} = 0.64 \times 10^{-39}\ \mathrm{cm}^3, \quad (50)$$

which agrees with the result of Levinger and Rustgi (cf. reference 12).

The presence of sizable contributions from photodisintegration in the γd elastic scattering amplitude and in the polarizability of the deuteron prevents our obtaining reliable conclusions about the polarizability of neutrons from the experimental data on the scattering of low-energy γ rays by deuterons.

To get information on the magnetic polarizability of the deuteron one must evidently have a much more detailed analysis of the photodisintegration of the deuteron and of processes of photoproduction of mesons from deuterons.*

Strictly speaking, the treatment carried out in the present paper is valid only for forward scattering. In the dipole approximation, however, the main results remain valid for other scattering angles also. But we have not made a direct comparison with the experimental data, since in the experiments[2] inelastic scattering of γ rays by deuterons,

$$\gamma + d \rightarrow n + p + \gamma,$$

was observed along with the elastic scattering. Recently A. M. Baldin (private communication) has examined the corrections to the impulse approximation in the inelastic scattering of γ rays and has arrived at the conclusion that for this process also there are appreciable corrections associated with the photodisintegration.

Thus we can evidently conclude that the results of an analysis that takes into account the photodisintegration of the deuteron (and the production of mesons), and the experimental data on the scattering of γ rays by deuterons in the energy range ~ 50 – 100 Mev are in agreement with each other. For a more reliable comparison of calculated results with experiment one first needs an analysis of the inelastic processes over a wider range of energies.

The writers are grateful to A. M. Baldin, V. I. Gol'danskii, and Ya. A. Smorodinskiĭ for numerous discussions.

*All conclusions concerning the magnetic polarizability of the proton are very sensitive to the assumptions that have to be made in the analysis of the photoproduction of pions. When one uses the analysis of Watson it follows from the results[4] that the magnetic polarizability of the proton is small. (In the case of the analysis of Watson it goes to zero.) This conclusion evidently is not in contradiction with the experimental data.[13]

[1] L. I. Lapidus and Chou Kuang-Chao, JETP **39**, 112 (1960), Soviet Phys. JETP **12**, 82 (1961).

[2] Hyman, Ely, Frisch, and Wahlig, Phys. Rev. Letters **3**, 93 (1959).

[3] R. H. Capps, Phys. Rev. **106**, 1031 (1957); **108**, 1032 (1957).

[4] L. I. Lapidus and Chou Kuang-Chao, JETP **37**, 1714 (1959), Soviet Phys. JETP **10**, 1213 (1960); JETP **38**, 201 (1960), Soviet Phys. JETP **11**, 147 (1960).

[5] F. E. Low, Phys. Rev. **96**, 1428 (1954). M. Gell-Mann and M. L. Goldberger, Phys. Rev. **96**, 1453 (1954).

[6] Chou Kuang-Chao and M. I. Shirokov, JETP **34**, 1230 (1958), Soviet Phys. JETP **7**, 851 (1958).

[7] L. Zastavenko and Chou Kuang-Chao, JETP **35**, 1417 (1958), Soviet Phys. JETP **8**, 990 (1959).

[8] L. Landau and E. Lifshitz, Quantum Mechanics, Pergamon, 1958.

[9] Barnes, Carver, Stafford, and Wilkinson, Phys. Rev. **86**, 359 (1952). J. Halpern and E. V. Weinstock, Phys. Rev. **91**, 934 (1953). Lew Allen, Jr., Phys. Rev. **98**, 705 (1955). J. C. Keck and A. V. Tollestrup, Phys. Rev. **101**, 360 (1956). Whalin, Schriever, and Hanson, Phys. Rev. **101**, 377 (1956). D. R. Dixon and K. C. Bandtel, Phys. Rev. **104**, 1730 (1956). Aleksandrov, Delone, Slovokhotov, Sokol, and Shtarkov, JETP **33**, 614 (1957), Soviet Phys. JETP **6**, 472 (1958). Tatro, Palfrey, Whaley, and Haxby, Phys. Rev. **112**, 932 (1958).

[10] J. J. de Swart, Physica **25**, 233 (1959). J. J. de Swart and R. E. Marshak, Physica **25**, 1001 (1959).

[11] A. B. Migdal, JETP **15**, 81 (1948).

[12] J. S. Levinger, Phys. Rev. **107**, 554 (1957).

[13] G. Bernardini, Report at the International Conference on the Physics of High-Energy Particles, Kiev, 1959.

Translated by W. H. Furry
240

ON THE PION-PION RESONANCE IN THE *p* STATE

HO TSO-HSIU and CHOU KUANG-CHAO

Joint Institute for Nuclear Research

Submitted to JETP editor September 30, 1960

J. Exptl. Theoret. Phys. (U.S.S.R.) **39**, 1485-1486 (November, 1960)

QUITE recently there has been great interest in the question of the existence of a p resonance (isobar) in pion-pion scattering.[1] From a study of the structure of nucleons by the method of dispersion relations, Frazer and Fulco have concluded that an isobar with mass 435 Mev and half-width 10 Mev must exist in the p state of pion-pion systems.[2] Similar results on the presence of a p resonance in pion-pion scattering have also been obtained by other authors.[3] On the other hand, the integral equations for pion-pion scattering have been derived more accurately by means of the ordinary theory of one-dimensional dispersion relations.[4] Preliminary results on the solution of these equations, obtained by means of an electronic computing machine, show that the amplitude of the p wave is very small. It is as yet unknown whether there is another solution with a large p amplitude. Therefore a direct experimental test of the presence of a p isobar in the pion-pion system is of great importance. For this purpose Chew and others[5] have suggested the reactions

$$e^+ + e^- \to \pi^+ + \pi^-, \quad e^- + e^- \to e^- + e^- + \pi^+ + \pi^-, \qquad (1)$$

studies of which could help to provide information about the interaction of pions in the p state. These processes are interesting because the theoretical interpretation of the results is simple and clear. But because of the lack of high-energy clashing electron and positron beams, it is difficult to conduct such experiments at present.

In the present note we suggest the study of the following processes:

$$\pi^\pm + He^4 \to He^4 + \pi^\pm + \pi^0, \qquad (2a)$$

$$\pi^\pm + d \to d + \pi^\pm + \pi^0, \qquad (2b)$$

$$p + p \to d + \pi^+ + \pi^0. \qquad (2c)$$

For all of these processes the initial isotopic spin is $I = 1$. Consequently, the pair of pions in the final state has the isotopic spin $I = 1$ and is in a state with odd orbital angular momentum. In the low energy region these pions are mainly in the p state.

Let us assume that there is an isobar with mass 435 Mev and half-width 10 Mev in the p state. Then in the reactions (2) the pairs of pions come from the decay of isobars that have been produced together with nuclei He^4 or d. Because of this it is to be expected that there will be a sharp maximum in the spectrum of the He^4 (or d).

Let us consider, for example, the reaction (2a). Suppose the energy of the incident pion beam is 700 Mev in the laboratory system. (l.s.). Then in the center-of-mass system (c.m.s.) one should observe a maximum in the spectrum of the He^4 at energy 11 Mev and with half-width 2 Mev. In the case of the reaction (2c) with incident beam energy 1.4 Bev in the l.s. the deuteron spectrum in the c.m.s. must have a maximum at energy 36 Mev and with half-width 3 Mev.

If the p isobar does not exist, then the shape of the spectrum of the He^4 (d) varies smoothly and is determined mainly by the statistical phase-volume factor. Therefore measurements of the spectra of the nuclei in the reactions (2) will give information about the existence of a p resonance in the pion-pion system.

We note that the process $d + d \to He^4 + \pi^0 + \pi^+ + \pi^-$ is also useful for studying the isobaric structure of pion-pion systems in the iso-scalar state.

[1] S. D. Drell, Proceedings of Annual International Conference on High Energy Physics, CERN, 1958.

[2] W. R. Frazer and J. R. Fulco, Phys. Rev. Letters **2**, 365 (1959).

[3] F. Cerulus, Nuovo cimento **14**, 827 (1959).

[4] Hsien Ting-Ch'ang, Ho Tso-Hsiu, and W. Zoellner, Preprint D-547, Joint Institute of Nuclear Studies.

[5] G. F. Chew, preprint, 1960; N. Cabibbo and R. Gatto, Phys. Rev. Letters **4**, 313 (1960); L. M. Brown and F. Calogero, Phys. Rev. Letters **4**, 315 (1960).

Translated by W. H. Furry
272

SOVIET PHYSICS JETP VOLUME 14, NUMBER 1 JANUARY, 1962

ON THE ROLE OF THE SINGLE-MESON POLE DIAGRAM IN SCATTERING OF GAMMA QUANTA BY PROTONS

L. I. LAPIDUS and CHOU KUANG-CHAO

Joint Institute for Nuclear Research

Submitted to JETP editor February 27, 1961

J. Exptl. Theoret. Phys. (U.S.S.R.) **41**, 294-302 (July, 1961)

It is shown that when the sign of the γN-scattering pole diagram connected with π^0-meson decay is correctly chosen, the contribution of the pole to the cross section for the scattering of γ quanta by protons decreases considerably. In order to obtain information on the lifetime of the π^0 meson, the precision of the experiments must be appreciably improved.

1. INTRODUCTION

A few years ago Low[1] called attention to the presence of a pole diagram connected with the decay of the neutral pion, in the amplitude of elastic scattering of γ quanta by protons. An account of this diagram, from the point of view of the double dispersion relations for γN scattering, is equivalent to an examination of the nearest singularity in Q^2. Several interesting considerations in connection with the double dispersion relations for γN scattering are contained in the paper by N. F. Nelipa and L. V. Fil'kov (preprint).* Zhizhin[2] considered a contribution of this amplitude in different states. Recently, Hyman et al[3] and in greater detail Jacob and Mathews[4] noted that the addition of the one-meson pole amplitude greatly improves the agreement between the theoretical and experimental results in the γ-quantum region from 100 to 250 Mev. This problem is considered in detail in a recently published paper by Bernadrini, Yamagata, et al.[5]

It is known that an analysis based on dispersion relations[6,7] leads to scattering cross section values greater than the experimental values in this energy region. In the present paper we wish to call attention to the sign of the pole amplitude, which is very important, since the interference terms play the principal role. From the results of Goldberger and Treiman[8] for the decay of the neutral pion, and from the dispersion relations for forward scattering, which we used previously,[7] it follows that the (relative) sign of the pole diagram differs from that used by Jacob and Mathews. Thus, the addition of the pole diagram does not improve the agreement between the theoretical and experimental results, and the discrepancy calls for a different explanation.

*The authors are grateful to Nelipa and Fil'kov, and also to Dr. Yamagata (see below), for acquainting them with their results prior to publication.

2. SCATTERING AMPLITUDE

We denote by p and p′ the nucleon momentum vectors in the initial and final states, respectively, and by q and q′ the same quantities for the γ quanta. Since they satisfy the conservation law

$$q + p = q' + p', \tag{1}$$

it is convenient to introduce the following four orthogonal vectors:

$$K = \tfrac{1}{2}(q + q'), \qquad Q = \tfrac{1}{2}(q' - q) = \tfrac{1}{2}(p - p'),$$
$$P' = P - K(PK)/K^2, \qquad N_\mu = i\varepsilon_{\mu\nu\sigma\rho} P'_\nu K_\sigma Q_\rho, \tag{2}$$

where $P = (p + p')/2$. From these four vectors we can construct two independent scalars:

$$Q^2, \qquad M\nu = -(PK). \tag{3}$$

The lengths of the vectors introduced in (2) are connected with Q^2 and $M\nu$ by the relations

$$K^2 = -Q^2, \qquad P^2 = -Q^2 - M^2,$$
$$P'^2 = P^2 - (PK)^2/K^2 = Q^{-2}[M^2\nu^2 - Q^2(Q^2 + M^2)],$$
$$N^2 = -P'^2K^2Q^2 = Q^2[M^2\nu^2 - Q^2(Q^2 + M^2)]. \tag{4}$$

The S-matrix element for γN scattering can be represented in the form

$$\langle p'q' | S | pq\rangle = \langle p'q' | pq\rangle + \frac{i}{2\pi}\delta^{(4)}(p' + q' - p - q)\frac{MN}{(p_0p_0'q_0q_0')^{1/2}}, \tag{5}$$

where

$$N = \bar{u}(p')e'_\mu N_{\mu\nu}e_\nu u(p) = 2\pi^2 i\left(\frac{p_0p_0'}{M^2}\right)^{1/2}\int d^4z e^{-i(Kz)}\left\langle p' \middle| T\left(e'\cdot j\left(\frac{z}{2}\right)\right) \times \left(e\cdot j\left(-\frac{z}{2}\right)\right)\middle| p\right\rangle. \tag{6}$$

In the center-of-mass system (c.m.s.) the differential cross section is given by the relation

$$\frac{d\sigma}{do} = \sum_{\text{spins}} \left| \frac{M}{W} N \right|^2, \tag{7}$$

where $W^2 = -(P+K)^2$ is the square of the total energy in the c.m.s.

The scattering amplitude N can be written as a sum of six invariant functions ($\hat{K} = \gamma_\mu K_\mu$):

$$e'_\mu N_{\mu\nu} e_\nu = \frac{(e'P')(eP')}{P'^2}[T_1 + i\hat{K}T_2] + \frac{(e'N)(eN)}{N^2}[T_3 + i\hat{K}T_4]$$
$$- \frac{(e'P')(eN) - (e'N)(eP')}{(P'^2N^2)^{1/2}} i\gamma_5 T_5$$
$$+ \frac{(e'P')(eN) + (e'N)(eP')}{(P'^2N^2)^{1/2}} \gamma_5 \hat{K} T_6. \tag{8}$$

In some cases it is also convenient to represent the amplitude as an operator in spin space in terms of six non-covariant functions R_i:

$$\frac{M}{W} e'_\mu N_{\mu\nu} e_\nu = R_1(ee') + R_2(s's) + iR_3(\sigma[e'e]) + iR_4(\sigma[s's])$$
$$+ iR_5[(\sigma k)(s'e) - (\sigma k')(se')]$$
$$+ iR_6[(\sigma k')(s'e) - (\sigma k)(se')], \tag{9*}$$

where $s = k \times e$, $s' = k' \times e'$; e, k and e′, k′ are the polarization of photon-momentum unit vectors before and after scattering, respectively.

3. MATRIX ELEMENT OF NEUTRAL-PION DECAY

The S matrix for the decay of the neutral pion has the form

$$\langle q'q|S|q_\pi\rangle = \frac{1}{(2\pi)^{3/2}} \frac{1}{\sqrt{2\omega_k}} (2\pi)^4 \delta^{(4)}(q_\pi - q - q')$$
$$\times \langle q'q|J(0)|0\rangle, \tag{10}$$

where q and q′ are the photon momenta; q_π is the 4-momentum of the pion; J(x) is the current of the pion field:

$$J(x) = i\frac{\delta S}{\delta\varphi(x)} S^+ = ig_0\bar{\psi}(x)\gamma_5\tau_3\psi(x) \tag{11}$$

[$\varphi(x)$ is the meson-field operator, $\psi(x)$ is the nucleon-field operator, and g_0 is the non-renormalized constant of the pion-nucleon interaction]. The Heisenberg equation for the meson field can be written in the form

$$(-\Box^2 + m_\pi^2)\varphi(x) = J(x), \tag{12}$$

and in the notation of Goldberger and Treiman[8]

$$\mathcal{M} \equiv (2\pi)^3\sqrt{4qq'}\langle q'q|J|0\rangle$$
$$= -i\varepsilon_{\mu\nu\sigma\lambda} e'_\mu e_\nu q_\sigma q'_\lambda F[(q+q')^2], \tag{13}$$

where $F(q^2)$ is the form factor. The expression for the decay S matrix contains $F(-m_\pi^2)$.

*(ee′) = e·e′ [ee′] = e × e′.

The probability of decay of the neutral pion is

$$w = \sum_{qq'ee'} (2\pi)^3 |\langle qq'|S|q_\pi\rangle|^2/VT$$
$$= \frac{1}{(2\pi)^2}\delta^{(4)}(q_\pi - q - q')\, d^3q\, d^3q' \frac{1}{8q_{\pi 0}q_0q'_0} q^2 q'^2$$
$$\times \sum_{ee'} [(es') + (e's)]^2 |F|^2. \tag{14}$$

Summing over e and e′ and integrating over the angles we obtain in the pion rest system

$$w = (m_\pi^3/64\pi)|F|^2. \tag{15}$$

The pion lifetime τ is

$$\tau = 64\pi/m_\pi^3|F|^2. \tag{16}$$

Using the dispersion technique, Goldberger and Treiman have shown that

$$F(0) = -\frac{ge^2}{4\pi^2 m_\pi}(1+\mu_p)\frac{I_0 + \rho I_1}{1 + (g^2/4\pi)I_1}, \tag{17}$$

$$\rho = [2\mu_p - (\mu_p^2 - \mu_n^2)]/(1+\mu_p), \tag{18}$$

where μ_p and μ_n are the anomalous magnetic moments of the proton and neutron, while I_0 and I_1 are positive integrals. It follows from (17) that

$$F(0)g < 0. \tag{19}$$

This sign is of importance for what is to follow.

4. SINGLE-MESON DIAGRAM FOR THE SCATTERING OF GAMMA QUANTA BY PROTONS

The S-matrix element of the pole diagram is

$$\langle p'q'|S-1|pq\rangle = ig\frac{i}{(2\pi)^3}\bar{u}(p')\gamma_5 u(p)$$
$$\times \delta^{(4)}(p' + q' - p - q)$$
$$\times (2\pi)^4 \frac{1}{(p'-p)^2 + m_\pi^2}\langle q'|J_\pi(0)|q\rangle. \tag{20}$$

It can be shown that

$$\langle q'|J_\pi(0)|q\rangle$$
$$= \frac{1}{(2\pi)^3}\frac{1}{(4q_0q'_0)^{1/2}}(-i)\varepsilon_{\mu\nu\sigma\lambda}e'_\mu e_\nu q'_\sigma q_\lambda F[(q'-q)^2]. \tag{21}$$

Since the matrix element $\langle q'|J_\pi(0)|q\rangle$ is taken for the pole at $(q'-q)^2 = -m_\pi^2$, Eq. (21) contains exactly the value of F encountered in the π^0 decay.

Substituting (21) in (20) and going to the c.m.s., we obtain

$$\langle p'q'|S-1|pq\rangle = \frac{igF}{(2\pi)^6}\frac{Mq^3}{(4q_0q'_0p_0p'_0)^{1/2}}(2\pi)^4\delta^{(4)}(p'+q'$$
$$-p-q)\frac{1}{2M}[i((\sigma k)(es') - (\sigma k')(e's))$$
$$-i((\sigma k')(es') - (\sigma k)(e's))]. \tag{22}$$

Comparing (22) with (9) we obtain for the contribution of the pole diagram

$$R_{1p} = R_{2p} = R_{3p} = R_{4p} = 0,$$

$$R_{5p} = -R_{6p} = \frac{gF}{8\pi W} \frac{q^3}{(p-p')^2 + m_\pi^2}; \tag{23}$$

from this we conclude that the contribution made to the amplitude by the pole diagram due to the exchange and decay of the pseudo-scalar neutral meson reduces to the combination

$$R_{5p} - R_{6p} = \frac{gFm_\pi}{8\pi W} \frac{q}{m_\pi} \frac{1}{1 + m_\pi^2/2q^2 - \cos\theta} \tag{24}$$

It is important to note that by virtue of (19)

$$R_{5p} - R_{6p} < 0, \tag{25}$$

if it is assumed that $F(0)$ and $F(-m_\pi^2)$ do not differ greatly.

In the expression for the cross section [formula (16) in [5]] the pole term enters in the combination

$$\tfrac{1}{2} |R_5 - R_6|^2 (1 - \cos\theta)^3$$
$$- \operatorname{Re}(R_3 - R_4)^* (R_5 - R_6) (1 - \cos\theta)^2. \tag{26}$$

The contribution of one pole diagram has the form

$$I_0^P(0) = \tfrac{1}{2} |R_5 - R_6|^2 (1 - \cos\theta)^3$$
$$= \frac{2}{m_\pi \tau} \left(\frac{q}{W}\right)^2 \frac{g^2}{4\pi} \left(\frac{1}{m_\pi}\right)^2 \frac{(1 - \cos\theta)^3}{(1 + m_\pi^2/2q^2 - \cos\theta)^3} \tag{27}$$

which agrees with the result of Jacob and Mathews.

We can expect the cross section of scattering by 90° to be reduced by addition of the pole term only when the second term in (26) is negative. Since R_4 is large and negative, owing to the large anomalous magnetic moment of the proton, $\operatorname{Re}(R_3 - R_4)$ is a positive quantity in the region of energy under consideration. Thus, the second term in (26) is positive if $R_{5p} - R_{6p} < 0$. Consequently, assuming the analysis of Goldberger and Treiman to be correct, the pole diagram does not decrease the theoretical value of the cross section, but increases it.

If we use the results of our own analysis,[7] we find that $\operatorname{Re}(R_5 - R_6)$ is determined not only by the limit theorem, but also by the amplitudes of photoproduction of E_2 and M_3. Since in this case the "isotropic" part of the contribution of the pole amplitude is automatically taken into account, it is necessary to add to the previously-obtained amplitude not all of expression (24), but only the contribution of (24) to the higher states, i.e., the difference

$$(R_5 - R_6)_p - \frac{1}{2\pi} \int (R_5 - R_6)_p \sin\theta d\theta .$$

As a result of this procedure, which is necessary in order not to violate the unitarity of the S matrix (when $\theta = 90°$), the quantity y_0^{-1} (where $y_0 = 1 + m_\pi^2/2q^2$) is replaced by

$$y_0^{-1} - \tfrac{1}{2} \ln |(y_0 + 1)/(y_0 - 1)|,$$

which leads to replacement of $2/3$ by -0.14 when $q^2 = m_\pi^2$ ($y_0 = 3/2$).

Thus, the contribution of the amplitude is decreased by a factor of 5, and the sign of the contribution changes. By virtue of this, a much higher accuracy is necessary before the connection between the amplitude of the neutral-pion decay and the amplitude of the scattering of γ quanta by protons can manifest itself. It was recently shown that the lifetime of the neutral pion is $(2.0 \pm 0.4) \times 10^{-16}$ sec,[9] which also decreases the contribution of the pole diagram.

The indeterminacies in the analysis of the photoproduction cannot influence the conclusion regarding the sign of the interference term in (24), since this sign is determined by the well known theorem for low energies. The scattering amplitude at low frequencies, first obtained by Low[10] and Gell-Mann and Goldberger,[11] is reviewed in the appendix, where it is obtained as the contribution of the single-nucleon terms (see [6]).

We note, in particular, that

$$T_5^0 = \frac{e^2}{M} (1 + \lambda) \frac{Q^2}{Q^2, M^2 - \nu^2} . \tag{28}$$

Let us give another, less rigorous but more illustrative proof of the correctness of the determination of the sign of the pole diagram.*

The matrix element $\langle q' | J_\pi(0) | q \rangle$ can be represented in the form

$$\langle q' | J_\pi(0) | q \rangle = i\varepsilon_{\mu\nu\sigma\lambda} e'_\mu e_\nu q_\sigma q'_\lambda \frac{1}{(2\pi)^3} F[(q - q')^2]$$
$$= -\frac{2Q^2}{(2\pi)^3} F \frac{(e'P')(eN) - (e'N)(eP')}{(P'^2 N^2)^{1/2}}, \tag{29}$$

so that

$$\langle p'q' | S - 1 | pq \rangle = i (2\pi)^{-2} g\delta^{(4)} (p' + q' - p - q)$$
$$\times i\bar{u}(p') \gamma_5 u(p)$$
$$\times \frac{2Q^2 F}{4Q^2 + m_\pi^2} \frac{(e'P')(eN) - (e'N)(eP')}{(P'^2 N^2)^{1/2}}; \tag{30}$$

hence

$$T_{5p} = \frac{gF}{\pi} \frac{Q^2}{4Q^2 + m_\pi^2} . \tag{31}$$

We now introduce the function

$$f(\nu, Q^2) = T_5(\nu, Q^2)/Q^2. \tag{32}$$

If we regard $f(\nu, Q^2)$ as an analytic function of Q^2 at fixed ν, we obtain from Cauchy's theorem and from (31)

*An analogous approach was used earlier[12] to obtain the Goldberger-Treiman relations.

$$f(\nu, Q^2) = \frac{gF}{\pi}\frac{1}{4Q^2 + m_\pi^2} + J_Q, \quad (33)$$

where J_Q is the dispersion interval, the lower limit of which is $4m_\pi^2$. In the region $Q^2 \ll 4m_\pi^2$, the integral in (33) is small and we can approximate $f(\nu, Q^2)$ by the expression

$$f(\nu, Q^2) \approx \frac{gF}{\pi}\frac{1}{4Q^2 + m_\pi^2}. \quad (34)$$

On the other hand, $f(\nu, Q^2)$ is also an analytic function of ν for fixed Q^2. By Cauchy's theorem with account of (28) we have

$$f(\nu, Q^2) = \frac{e^2(1+\lambda)}{M}\frac{1}{Q^4/M^2 - \nu^2} + J_\nu, \quad (35)$$

where J_ν is a second dispersion integral. In the region $2\nu \lesssim m_\pi$, the pole term will predominate and

$$f(\nu, Q^2) \approx \frac{e^2(1+\lambda)}{M}\frac{1}{Q^4/M^2 - \nu^2}. \quad (36)$$

It is obvious that (34) does not hold near $M^2\nu^2 \approx Q^2$, and (36) does not take place when $4Q^2 = -m_\pi^2$. It is still possible, however, that expressions (34) and (36) are valid simultaneously near certain values of ν and Q^2. Equating these expressions for $2\nu = m_\pi$ and $Q \approx 0$, we obtain

$$F = -4\pi e^2(1+\lambda)/gM, \quad (37)$$

which is very close to the formula of Goldberger and Treiman, obtained by an entirely different method.

Actually, from (17) we obtain for $(g^2/4\pi^2) I_1 \gg 1$

$$F = -4\pi\frac{e^2(1+\lambda)}{g}\frac{I_0 + pI_1}{I_1},$$

which coincides with (37), apart for a numerical factor.

The literature reports two different choices of the common phase for the γN scattering amplitude, one with a Thomson limit $+e^2/M$, the other with $-e^2/M$. The error in the published papers lies in the fact that the choice of the common factor in the one-meson amplitude does not correspond to the choice of the sign of the remaining amplitude.

A direct comparison of the amplitude used by Jacob and Mathews[4] with (9) shows that the functions f_i introduced in [4] are related with R_i by the equations

$$-f_1 = R_1 + R_2\cos\theta, \quad f_2 = R_2,$$
$$f_3 = R_3 + R_4\cos\theta + (R_5 + R_6)(1+\cos\theta) - (R_5 - R_6)(1-\cos\theta),$$
$$f_4 = R_4, \quad f_5 = R_4 + R_5, \quad f_6 = R_6,$$

where the difference in the common phase factor is taken into account, and from which it is clear that the sign used in [4] for the pole term differs from that proved in the present paper.

APPENDIX

SINGLE-NUCLEON TERMS IN THE DISPERSION RELATIONS

Recognizing that

$$T\left(e'\cdot j\left(\tfrac{z}{2}\right)\right)\left(e\cdot j\left(-\tfrac{z}{2}\right)\right) = \theta(z_0)\left[e\cdot j'\left(\tfrac{z}{2}\right), e\cdot j\left(-\tfrac{z}{2}\right)\right] + \left(e'\cdot j\left(\tfrac{z}{2}\right)\right)\left(e\cdot j\left(-\tfrac{z}{2}\right)\right), \quad (A.1)$$

we determine the retarded and advanced amplitudes:

$$N^{r,a} = \pm 2\pi^2 i\,(p_0p_0'/M^2)^{1/2} \times\int d^4z e^{\pm i(Kz)}\left\langle p'\left|\theta(\pm z_0)\left[e'\cdot j\left(\tfrac{z}{2}\right), e\cdot j\left(-\tfrac{z}{2}\right)\right]\right|p\right\rangle. \quad (A.2)$$

The vertex part of the current has the form

$$\langle p'|ej(0)|p\rangle = \frac{ie}{(2\pi)^3}\bar u(p')\left[\hat e + i\frac{\lambda}{4M}(\hat e(\hat p' - \hat p) - (\hat p' - \hat p)\hat e)\right]u(p) = \frac{ie}{(2)^3}u(p')\left[(1+\lambda)\hat e + \frac{i\lambda}{M}(eP)\right]u(p), \quad (A.3)$$

where ϵ is the charge of the nucleon.

The pole term has in the region of positive frequencies the form

$$A^0 = -\frac{(2\pi)^6}{4}\sum_n \delta^{(4)}(k - p + p_n)\langle p'|e\cdot j(0)|p_n\rangle \times\langle p_n|e'j(0)|p\rangle = \frac{\varepsilon^2}{4}\int d^3p_n\delta^{(4)}(k - p + p_n) \times\bar u(p')\left[(1+\lambda)\hat e + \frac{i\lambda}{M}ep'\right]u(P-K)\bar u(P-K) \times\left[(1+\lambda)\hat e' + \frac{i\lambda}{M}e'p\right]u(p). \quad (A.4)$$

Using the relations

$$\sum u(P-K)\bar u(P-K) = \frac{-i(\hat P - \hat K) + M}{2p_{n0}}, \quad (A.5)$$

$$(2p_{n0})^{-1}d^3p_n = d^4p_n\theta(p_{n0})\,\delta(p_n^2 + M^2), \quad (A.6)$$

we obtain

$$A^0 = \frac{\varepsilon^2}{4}\delta(p_n^2 + M^2)\bar u(p')\left[(1+\lambda)\hat e + \frac{i\lambda}{M}(ep')\right] \times[-i(\hat P - \hat K) + M]\left[(1+\lambda)\hat e' + \frac{i\lambda}{M}(e'p)\right]u(p). \quad (A.7)$$

We can express A^0 in terms of the fundamental invariants:

$$A^0 = \frac{(e'P')(eP')}{P'^2}A_1^0 + \frac{(e'N)(eN)}{N^2}A_2^0 + \frac{(e'P')(eN) - (e'N)(eP')}{(P'^2N^2)^{1/2}}A_3^0 + \frac{(e'P')(eN) + (e'N)(eP')}{(P'^2N^2)^{1/2}}A_4^0. \quad (A.8)$$

Comparing (A.7) and (A.8), we obtain

$$A_1^0P'^2 = \frac{\varepsilon^2}{4}\delta(p_n^2 + M^2)\bar u(p')\left[(1+\lambda)\hat P' + \frac{i\lambda}{M}(P'p')\right] \times[-i(\hat P - \hat K) + M]\left[(1+\lambda)\hat P' + \frac{i\lambda}{M}(P'p)\right]u(p). \quad (A.9)$$

It is easy to verify that

$$(P'p') = (P', P - Q) = (P'p) = P'^2. \quad (A.10)$$

$$\bar{u}(p')\,[\hat{P}'(-i(\hat{P}-\hat{K})+M)\hat{P}']\,u(p)$$
$$= \bar{u}(p')\left\{-i\hat{K}P'^2 + 2P'^2M + 2i\frac{(PK)}{K^2}\hat{K}P'^2\right\}u(p), \quad (A.11)$$

$$\bar{u}(p')\hat{P}'[-i(\hat{P}-\hat{K})+M]\,u(p) = \bar{u}(p')\Big[M\Big(iM - \frac{(PK)}{K^2}\hat{K}\Big)$$
$$- i(P^2+(PK)) + i\Big((\hat{P}\hat{K}) + \frac{(PK)}{K^2}\hat{K}\hat{P}\Big)\Big]u(p), \quad (A.12)$$

$$\bar{u}(p')\,[(-i(\hat{P}-\hat{K})+M)\hat{P}']\,u(p)$$
$$= \bar{u}(p')\Big[M\Big(iM - \frac{(PK)}{K^2}\hat{K}\Big) - i(P^2+(PK))$$
$$+ i\Big(\hat{K}\hat{P} + \frac{(PK)}{K^2}\hat{P}\hat{K}\Big)\Big]u(p), \quad (A.13)$$

$$\bar{u}(p')\,[-i(\hat{P}-\hat{K})+M]\,u(p) = \bar{u}(p')\,[i\hat{K}+2M]\,u(p). \quad (A.14)$$

Using (A.10) — (A.14) and noting that at the pole we have

$$(P-K)^2 = P^2 + K^2 - 2(PK)$$
$$= 2K^2 - 2(PK) - M^2 = -M^2$$

or that

$$K^2 = (PK), \quad (A.15)$$

we obtain

$$A_1^0 = \frac{1}{4}\varepsilon^2\delta(2K^2 - 2(PK))\,\bar{u}(p')(2M + i\hat{K})\,u(p)$$
$$= \frac{\varepsilon^2}{8M}\delta\Big(\nu - \frac{Q^2}{M}\Big)\bar{u}(p')(2M + i\hat{K})\,u(p). \quad (A.16)$$

Analogously,

$$A_2^0 = -\frac{\varepsilon^2}{8M}\delta\Big(\nu - \frac{Q^2}{M}\Big)\bar{u}(p')\,i\hat{K}\,u(p)\,(1+\lambda)^2, \quad (A.17)$$

$$(A_3^0 + A_4^0)(P'^2N^2)^{1/2}$$
$$= \frac{\varepsilon^2(1+\lambda)}{8M}\delta\Big(\nu - \frac{Q^2}{M}\Big)\bar{u}(p')\Big[\frac{P'^2}{M}\hat{N}(-iM + \hat{K})\Big]u(p), \quad (A.18)$$

$$(A_4^0 - A_3^0)(P'^2N^2)^{1/2}$$
$$= \frac{\varepsilon^2(1+\lambda)}{8M}\delta\Big(\nu - \frac{Q^2}{M}\Big)\bar{u}(p')\Big[\frac{P'^2}{M}(\hat{K} - iM)\hat{N}\Big]u(p). \quad (A.19)$$

From (A.18) and (A.19) we find

$$A_4^0(P'^2N^2)^{1/2} = \frac{\varepsilon^2(1+\lambda)}{8M}\delta\Big(\nu - \frac{Q^2}{M}\Big)(-iP'^2)\,\bar{u}(p')\,\hat{N}\,u(p),$$
$$A_3^0(P'^2N^2)^{1/2} = \frac{\varepsilon^2(1+\lambda)}{8M}\delta\Big(\nu - \frac{Q^2}{M}\Big)\frac{P'^2}{M}\bar{u}(p')\hat{N}\hat{K}u(p). \quad (A.20)$$

It can be shown that

$$\bar{u}(p')\,\hat{N}\hat{K}u(p) = (P'^2N^2)^{1/2}\,i\bar{u}(p')\,\gamma_5 u(p),$$
$$i\bar{u}(p')\,\hat{N}u(p) = K^2\bar{u}(p')\,\gamma_5\hat{K}u(p).$$

If we now take into account the fact that $(P'^2N^2)^{1/2} = P'^2Q^2$ by virtue of (4), we obtain from (A.20)

$$A_3^0 = -\frac{\varepsilon^2(1+\lambda)}{8M}\delta\Big(\nu - \frac{Q^2}{M}\Big)\,i\bar{u}(p')\,\gamma_5\,u(p),$$
$$A_4^0 = \frac{\varepsilon^2(1+\lambda)}{8M}\delta(\nu - Q^2/M)\,\bar{u}(p')\,\gamma_5\hat{K}u(p). \quad (A.21)$$

Finally from (A.16), (A.17), and (A.21) we obtain

$$T_1^0 = \frac{\varepsilon^2}{2\pi M}\frac{Q^2}{Q^4/M^2 - \nu^2}, \qquad T_2^0 = \frac{\varepsilon^2}{4\pi M}\frac{\nu}{Q^4/M^2 - \nu^2},$$
$$T_3^0 = 0, \qquad T_4^0 = -\frac{\varepsilon^2(1+\lambda)^2}{4\pi M}\frac{\nu}{Q^4/M^2 - \nu^2},$$
$$T_5^0 = MT_6^0 = \frac{\varepsilon^2(1+\lambda)}{4\pi M}\frac{Q^2}{Q^4/M^2 - \nu^2} \qquad \Big(\frac{\varepsilon^2}{4\pi} = \frac{1}{137}\Big), \quad (A.22)$$

which coincides with the previously-obtained results and has the correct signs.

In all the calculations of the single-nucleon terms it is assumed that parity is conserved in the electromagnetic interactions. The results obtained remain valid also in the presence of CP invariance.

---

[1] F. E. Low, Proc. 1958 Ann. Intern. Conf. on High Energy Physics at CERN, p. 98.

[2] E. D. Zhizhin, JETP **37**, 994 (1959), Soviet Phys. JETP **10**, 707 (1960).

[3] Hyman, Ely, Frish, and Wahlig, Phys. Rev. Lett. **3**, 93 (1959).

[4] M. Jacob and J. Mathews, Phys. Rev. **117**, 854 (1960).

[5] Bernadrini, Hanson, Odian, Yamagata, Auerbach, and Filosofo, Nuovo cimento **18**, 1203 (1960).

[6] T. Akiba and J. Sato, Progr. Theor. Phys. **19**, 93 (1958).

[7] L. I. Lapidus and Chou Kuang-chao, JETP **37**, 1714 (1959) and **38**, 201 (1960); Soviet Phys. JETP **10**, 1213 (1960) and **11**, 147 (1960).

[8] M. L. Goldberger and S. B. Treiman, Nuovo cimento **9**, 451 (1958).

[9] Glasser, Seeman, and Stiller, Bull. Am. Phys. Soc. **6**, 1 (1961).

[10] F. E. Low, Phys. Rev. **96**, 1428 (1954).

[11] M. Gell-Mann and M. L. Goldberger, Phys. Rev. **96**, 1433 (1954).

[12] Chou Kuang-chao, JETP **39**, 703 (1960), Soviet Phys. JETP **12**, 492 (1961). Bernstein, Fubini, Gell-Mann, and Thirring, Nuovo cimento **17**, 757 (1960).

Translated by J. G. Adashko
58

SOVIET PHYSICS JETP VOLUME 14, NUMBER 2 FEBRUARY, 1962

LOW-ENERGY LIMIT OF THE γN-SCATTERING AMPLITUDE AND CROSSING SYMMETRY

L. I. LAPIDUS and CHOU KUANG-CHAO

Joint Institute for Nuclear Research

Submitted to JETP editor February 27, 1961

J. Exptl. Theoret. Phys. (U.S.S.R.) 41, 491-494 (August, 1961)

The low energy limit for the γN scattering amplitude is derived with the aid of single-nucleon invariant amplitudes. Subsequent terms in ν for $Q^2 = 0$ and the expression for the limiting value of the first derivative in Q^2 as $Q^2 \to 0$ can be obtained by taking into account the conditions of crossing symmetry.

1. Low, Gell-Mann, and Goldberger showed [1] that the condition of relativistic and gauge invariance makes it possible to express the limiting value of the amplitudes for the scattering of low energy γ quanta on spin-$\frac{1}{2}$ particles and the limiting value of the derivative of the amplitude with respect to the frequency as $\nu \to 0$ in terms of the charge and magnetic moment of the particle. This result was later generalized [2] to the case of elastic scattering of γ quanta by particles with other spins and also to the case of bremsstrahlung. [3] The result for elastic scattering also holds when only CP invariance is assumed. Consideration of the single-nucleon terms in the dispersion relations for γN scattering [4-6] also leads to the limit theorem. (A similar result holds for bremsstrahlung. [7])

In the present note, we derive the limit theorem for γN scattering on the basis of the single-nucleon terms. The requirement of crossing symmetry for the invariant functions $T_i(\nu, Q^2)$ (= 1, . . . , 6) makes it possible to obtain additional terms for the limiting values of the functions $R_i(\nu, 0)$, which characterize the γN scattering matrix in the center-of-mass system, and also the limiting values of the derivatives of the amplitudes with respect to Q^2 as $\nu \to 0$. (For the definition of the quantities T_i and R_i see, e.g., [6].)

2. The invariant functions $T_i(\nu, Q^2)$ are related to the scalar functions $R_i(\nu, Q^2)$ (i = 1, . . . , 6) in the following way:

$$T_1 - T_3 = \frac{8MW^2}{(W^2 - M^2)^2}\left[\nu - \frac{W - M}{W + M}\frac{Q^2}{M}\right](R_3 + R_4) - \frac{4W}{(M + W)}\left[1 - \frac{4Q^2W^2}{(W^2 - M^2)^2}\right](R_1 + R_2),$$

$$T_2 - T_4 = \frac{8MW^2}{(W^2 - M^2)^2}\left[1 + \frac{2W}{M}\frac{Q^2}{(W + M^2)}\right](R_3 + R_4) + \frac{4W}{(W + M)^2}\left[1 - \frac{4Q^2W^2}{(W^2 - M^2)^2}\right](R_1 + R_2),$$

$$T_1 + T_3 = \frac{8MW^2}{(W^2 - M^2)^2}\left[\nu - \frac{W - M}{W + M}\frac{Q^2}{M}\right](R_3 - R_4) + \frac{16W^3Q^2}{(W + M)(W^2 - M^2)^2}(R_1 - R_2),$$

$$T_2 + T_4 = \frac{8MW^2}{(W^2 - M^2)^2}\left[1 + \frac{2W}{M}\frac{Q^2}{(W + M)^2}\right](R_3 - R_4) - \frac{16W^3Q^2}{(W + M)^2(W^2 - M^2)^2}(R_1 - R_2),$$

$$\frac{M\nu + Q^2}{W^2}T_5 = \frac{8W^2Q^2}{(W^2 - M^2)^2}(R_5 - R_6) - (R_3 - R_4),$$

$$\frac{M\nu + Q^2}{W}T_6 = \left(2 - \frac{8W^2Q^2}{(W^2 - M^2)^2}\right)(R_5 + R_6) + (R_3 + R_4), \quad (1)$$

where W is the total c.m.s. energy and ν and Q^2 are two invariants characterizing the kinematics of the process; $W^2 - M^2 = 2M\nu + 2Q^2$.

The pole terms for $T_i(\nu, Q^2)$ have the form [6]

$$T_1^0 = \frac{2e^2}{M}\frac{Q^2}{Q^4/M^2 - \nu^2}, \qquad T_2^0 = \frac{e^2}{M}\frac{\nu}{Q^4/M^2 - \nu^2}, \qquad T_3^0 = 0,$$

$$T_4^0 = -\frac{e^2(1 + \lambda)^2}{M}\frac{\nu}{Q^4/M^2 - \nu^2},$$

$$T_5^0 = MT_6^0 = \frac{e^2(1 + \lambda)}{M}\frac{Q^2}{Q^4/M^2 - \nu^2}, \quad (2)$$

where we have used the system of units in which $\hbar = c = 1$ and the magnetic moment is $\mu = e(1 + \lambda)/2M$.

For $Q^2 = 0$, it follows from (1) that

$$(T_1 - T_3)_0 = \frac{2W^2}{M\nu}(R_3 + R_4)_0 - \frac{4W}{W + M}(R_1 + R_2)_0,$$

$$(T_2 - T_4)_0 = \frac{2W^2}{M\nu^2}(R_3 + R_4)_0 + \frac{4W}{(W + M)^2}(R_1 + R_2)_0,$$

$$(T_1 + T_3)_0 = \frac{2W^2}{M\nu}(R_3 - R_4)_0, \quad (T_2 + T_4)_0 = \frac{2W^2}{M\nu^2}(R_3 - R_4)_0,$$

$$(T_5)_0 = -\frac{W^2}{M\nu}(R_3 - R_4)_0,$$

$$(T_6)_0 = \frac{W}{M\nu}[2(R_5 + R_6) + R_3 + R_4].$$

Differentiating the relations in (1) with respect to Q^2, we obtain, in the limit $Q^2 = 0$,

$$(T_1 - T_3)'_0 = \frac{2W^2}{M\nu}(R_3 + R_4)'_0 - \frac{4W}{W+M}(R_1 + R_2)'_0$$
$$- \frac{2(2W^3 + M^2W + M^3)}{M^2\nu^2(W+M)}(R_3 + R_4)_0$$
$$+ \frac{4}{W(W+M)}\left[\frac{W^4}{M^2\nu^2} - \frac{M}{W+M}\right](R_1 + R_2)_0,$$
$$(T_2 - T_4)'_0 = \frac{2W^2}{M\nu^2}(R_3 + R_4)'_0 + \frac{4W}{(W+M)^2}(R_1 + R_2)'_0$$
$$+ \frac{4}{M^2\nu^2}\left[\frac{W^3}{(W+M)^2} - M - \frac{M^2}{\nu}\right](R_3 + R_4)_0$$
$$- \frac{4}{(W+M)^2}\left[\frac{W^3}{M^2\nu^2} - \frac{1}{W} + \frac{2}{M+W}\right](R_1 + R_2)_0,$$
$$(T_1 + T_3)'_0 = \frac{2W^2}{M\nu}(R_3 - R_4)'_0$$
$$- \frac{2(2W^3 + WM^2 + M^3)}{M^2\nu^2(W+M)}(R_3 - R_4)_0$$
$$+ \frac{4W^3}{(W+M)M^2\nu^2}(R_1 - R_2)_0,$$
$$(T_2 + T_4)'_0 = \frac{2W^2}{M\nu^2}(R_3 - R_4)'_0$$
$$+ \frac{4(R_3 - R_4)_0}{M^2\nu^2}\left[\frac{W^3}{(W+M)^2} - M - \frac{M^2}{\nu}\right]$$
$$- \frac{4W^3}{(W+M)^2}\frac{(R_1 - R_2)_0}{M^2\nu^2},$$
$$(T_5)'_0 = \frac{W^2}{M\nu}\left[\frac{2W^2}{M^2\nu^2}(R_5 - R_6)_0 - (R_3 - R_4)'_0\right.$$
$$\left. + \frac{M}{W^2\nu}(R_3 - R_4)_0\right],$$
$$(T_6)'_0 = \frac{W}{M\nu}\left[-\frac{2W^2}{M^2\nu^2}(R_5 + R_6)_0\right.$$
$$+ (2R_5 + 2R_6 + R_3 + R_4)'$$
$$\left. - \frac{M+\nu}{W^2\nu}(2R_5 + 2R_6 + R_3 + R_4)\right]. \quad (4)$$

It is seen from (2) that the terms $T_1 - T_3$ and $T_2 + T_4$ do not contain poles for $Q^2 = 0$. It then follows from (1) that $(R_1 + R_2)_0$ and $(R_3 \pm R_4)_0/\nu$ are finite when $\nu \to 0$.

Since the functions $(T_2 \mp T_4)_0$ have a singularity of the form

$$\left[-\frac{e^2}{M} \mp \frac{e^2}{M}(1+\lambda)^2\right]\frac{1}{\nu},$$

it then follows from (1) that

$$\frac{(R_3 \pm R_4)_0}{\nu} = -\frac{e^2}{2M^2}[1 \pm (1+\lambda)^2] \quad (5)$$

as $\nu \to 0$, which is in accordance with the limit theorem.

Since T_5 and T_6 do not contain poles when $Q^2 = 0$, the quantity $(R_5 + R_6)/\nu$ should remain constant as $\nu \to 0$. Similarly, from the condition that $(T_1 \pm T_3)'_0$ contains a pole of the second order

$$(T_1 \pm T_3)'_{0p} = -2e^2/M\nu^2,$$

and that $\nu(T_2 - T_4)'_0$ does not contain a pole, it follows that

$$(R_1 \pm R_2)_0 = -e^2/M, \quad (6)$$

and $(R_3 \pm R_4)'_0 \to$ const and $\nu(R_1 \pm R_2)'_0 \to$ const as $\nu \to 0$.

Since

$$(T_5)'_{0p} = M(T_6)'_{0p} = -e^2(1+\lambda)/2M^2,$$

we conclude that

$$(R_5 \pm R_6)_0 = \pm e^2(1+\lambda)\nu/2M^2, \quad (7)$$

and $(2R_5 + 2R_6 + R_3 + R_4)'_0 \to$ const as $\nu \to 0$.

We see that formulas (5)—(7) obtained from consideration of the pole terms (2) contain the results of the limit theorem for $Q^2 = 0$.

3. It is of interest to note that with the aid of the conditions of crossing symmetry one can obtain additional information on the low energy limit. It follows from crossing symmetry that, for example, the quantity $T_1 - T_3$ should be an even function of ν. If in the first relation of (3) we make the substitution

$$(R_1 + R_2)_0 = -\frac{e^2}{M} + \alpha_1\nu + \ldots,$$
$$(R_3 + R_4)_0 = -\frac{e^2}{2M^2}[1 + (1+\lambda)^2]\nu + \alpha_3\nu^3 + \ldots \quad (8)$$

and take into account the fact that $W = (M^2 + 2M\nu)^{1/2} \approx M(1 + \nu/M)$ for small ν, then from the condition that there is no linear dependence on ν we obtain the relation

$$\alpha_3 M - \alpha_1 = (e^2/M)\left[\tfrac{1}{2} + (1+\lambda)^2\right]. \quad (9)$$

It follows from the requirement of crossing symmetry that the quantity $\nu(T_2 - T_4)$ should be an even function of ν.

The absence of a linear dependence of the terms on ν leads to the relation

$$\alpha_3 M = (e^2/M)\left[\tfrac{3}{2} + (1+\lambda)^2\right]. \quad (10)$$

From (8)—(10) we have

$$(R_1 + R_2)_0 = -\frac{e^2}{M}\left(1 - \frac{\nu}{M}\right) + O(\nu^3),$$
$$(R_3 + R_4)_0 = -\frac{e^2}{2M^2}\left(1 - \frac{3\nu}{M}\right)\nu$$
$$- \frac{e^2}{2M^2}(1+\lambda)^2\left(1 - \frac{2\nu}{M}\right)\nu + O(\nu^3). \quad (11)$$

The functions $T_1 + T_3$, T_5, $(T_2 + T_4)$, and T_6 should be even functions of ν. Similar considerations lead to

$$(R_3 - R_4)_0 = -\frac{e^2}{2M^2}[1 - (1+\lambda)^2]\left(1 - \frac{2\nu}{M}\right)\nu + O(\nu^3), \quad (12)$$

and

$$[2(R_5+R_6)+R_3+R_4]_0$$

$$= -\frac{e^2}{2M^2}\lambda^2\left(1-\frac{\nu}{M}\right)\nu + O(\nu^3), \quad (13)$$

$$(R_5+R_6)_0 = \frac{e^2(1+\lambda)}{2M^2}\nu$$

$$+\frac{e^2}{4M^3}[\lambda^2-3-2(1+\lambda)^2]\nu^2+O(\nu^3). \quad (13')$$

The function $(T_1 - T_3)'_0$ is an even function of ν. Inserting in (4)

$$(R_3+R_4)'_0 = \alpha'_3 + \ldots, \qquad (R_1+R_2)'_0 = \alpha'_1/\nu + \ldots,$$

we obtain from the condition that there is no term proportional to $1/\nu$

$$2M\alpha'_3 - 2\alpha'_1 = (e^2/M)[1-2(1+\lambda)^2].$$

A similar condition for the even function $\nu(T_2 - T_4)'_0$ leads to

$$2M\alpha'_3 = -(e^2/M)[3+2(1+\lambda)^2].$$

Then $\alpha'_1 = -2e^2/M^2$, and therefore

$$(R_1+R_2)'_0 = -2\frac{e^2}{M^2}\frac{1}{\nu}+O(1),$$

$$(R_3+R_4)'_0 = -\frac{e^2}{2M^2}[3+2(1+\lambda)^2]+O(\nu). \quad (14)$$

The condition that the poles of the first order in the even functions $(T_1 + T_3)'_0$ and $\nu(T_2 + T_4)$ vanish leads to

$$(R_1-R_2)_0 = -\frac{e^2}{M}\left(1-\frac{3\nu}{M}\right)+O(\nu^2),$$

$$(R_3-R_4)'_0 = \frac{e^2}{2M^3}[-3+2(1+\lambda)^2]+O(\nu). \quad (15)$$

Similar conditions for the functions $(T_5)'_0$ and $(T_6)'_0$ require that

$$(R_5-R_6)_0 = -\frac{e^2(1+\lambda)}{2M^2}\nu+\frac{e^2}{4M^3}[-2+8(1+\lambda)$$

$$+(1+\lambda)^2]\nu^2+O(\nu^3),$$

$$(2R_5+2R_6+R_3+R_4)_0 = -\frac{e^2}{2M^3}(2\lambda^2-2\lambda-1). \quad (16)$$

It should be kept in mind that the expression for the limiting energy is valid for amplitudes in the center-of-mass system. The result obtained can be useful for analysis of the scattering of γ quanta by nucleons with the aid of the dispersion relation technique.

[1] F. Low, Phys. Rev. **96**, 1428 (1954); M. Gell-Mann and M. L. Goldberger, Phys. Rev. **96**, 1433 (1954).

[2] L. I. Lapidus and Chou Kuang-Chao, JETP **39**, 1286 (1960), Soviet Phys. JETP **12**, 898 (1961).

[3] F. E. Low, Phys. Rev. **110**, 974 (1958).

[4] N. N. Bogolyubov and D. V. Shirkov, Doklady Akad Nauk SSSR **113**, 529 (1957).

[5] T. Akiba and I. Sato, Progr. Theoret. Phys. (Kyoto) **19**, 93 (1958).

[6] L. I. Lapidus and Chou Kuang-Chao, JETP **41**, 294 (1961), Soviet Phys. JETP **14**, 210 (1962).

[7] S. M. Bilen'kii and R. M. Ryndin, JETP **40**, 819 (1961), Soviet Phys. **13**, 575 (1961).

Translated by E. Marquit
90

SOVIET PHYSICS JETP VOLUME 14, NUMBER 4 APRIL, 1962

ON THE $\overline{K} + N \to \Lambda(\Sigma) + \gamma$ PROCESS

L. I. LAPIDUS and CHOU KUANG-CHAO

Joint Institute for Nuclear Research

Submitted to JETP editor May 18, 1961

J. Exptl. Theoret. Phys. (U.S.S.R.) **41**, 1310-1314 (October, 1961)

Some information on $\Lambda(\Sigma) + \pi \to \Lambda(\Sigma) + \gamma$ processes can be obtained by investigating the $\overline{K} + N \to \Lambda(\Sigma) + \gamma$ reaction. A detailed phenomenological analysis of these processes in the s state is performed. The Kroll-Ruderman theorem for photoproduction of pions on hyperons near threshold is considered.

1. One of the most important problems in elementary-particle physics is the study of interactions between unstable particles, where for lack of an unstable-particle target it becomes necessary to use indirect methods for this purpose.

We have shown earlier [1] that by using the unitarity condition for the S matrix we can establish certain relations between the matrix elements for the processes $\overline{K} + N \to \overline{K} + N$, $\overline{K} + N \to \Lambda(\Sigma) + \pi$ and $\Lambda(\Sigma) + \pi \to \Lambda(\Sigma) + \pi$ for states with arbitrary values of the angular momentum. It is therefore necessary to obtain certain information on the processes $\Lambda(\Sigma) + \pi \to \Lambda(\Sigma) + \pi$ by analyzing the cross sections and polarizations of the baryons in elastic scattering and in reactions involving K mesons and nucleons. Similar conclusions were reached later by other authors. [2,3] In [2] and [3] there is a detailed analysis of elastic scattering and interaction of K mesons with nucleons in the s state. Existing experimental data allow us to establish the phase difference of the s waves in $\pi\Sigma$ scattering with isospin I = 1 and I = 0.

In order to obtain certain information on the electromagnetic and strong interactions of hyperons, we consider in the present article the processes

$$\overline{K} + N \to \Lambda(\Sigma) + \gamma. \tag{1}$$

The S-matrix unitarity conditions cause the matrices for the processes $\pi + \Lambda(\Sigma) \to \Lambda(\Sigma) + \gamma$ to be related with the matrix elements of processes (1).

2. For simplicity we consider the reactions (1) in the s state only. We use the K-matrix method developed in [3]. For our problem it is convenient to use a symmetrical and Hermitian K matrix, expressed in terms of a T matrix with the aid of the relation

$$K = T - i\pi K\rho T = T - i\pi T\rho K, \tag{2}$$

where ρ is the density matrix of the phase volume for the intermediate states with fixed total energy. For two-particle (binary) reactions with a definite angular momentum, the matrix ρ is diagonal. In the relativistic normalization of the wave functions, the diagonal elements of the ρ matrix are

$$\rho_{nn} = M_n k/\pi E, \tag{3}$$

where k is the relative momentum of the particle in the c.m.s., M_n is the mass of the baryons in the intermediate states, and E is the total energy of the system:

$$E = (k^2 + M_n^2)^{1/2} + (k^2 + m^2)^{1/2}. \tag{4}$$

If we introduce the notation

$$K' = \pi\rho^{1/2}K\rho^{1/2}, \qquad T' = \pi\rho^{1/2}T\rho^{1/2}, \tag{5}$$

then Eq. (2) can be rewritten as

$$K' = T' - iK'T' = T' - iT'K'. \tag{6}$$

From (6) we obtain

$$T' = (1 - iK')^{-1} K' = K' (1 - iK')^{-1}. \tag{7}$$

The cross section of reaction (1) expressed in terms of the T′ matrix, in a state with definite angular momentum J and with definite parity, is

$$\sigma(i \to j) = 4\pi k_i^{-2} (J + 1/2) \, |\langle j | T' | i\rangle|^2. \tag{8}$$

Let us consider the submatrices of the introduced K and T matrices, which we denote by

$$\begin{aligned}
\alpha &= \langle \overline{K}N | K | \overline{K}N\rangle, & T_{KK} &= \langle \overline{K}N | T | \overline{K}N\rangle,\\
\beta &= \langle \overline{K}N | K | Y\pi\rangle, & T_{KY} &= \langle \overline{K}N | T | Y\pi\rangle,\\
\beta^+ &= \langle Y\pi | K | \overline{K}N\rangle, & T_{YK} &= \langle Y\pi | T | \overline{K}N\rangle,\\
\gamma &= \langle Y\pi | K | Y\pi\rangle, & T_{YY} &= \langle Y\pi | T | Y\pi\rangle,\\
\xi &= \langle \overline{K}N | K | Y\gamma\rangle, & T_{K\gamma} &= \langle \overline{K}N | T | Y\gamma\rangle,\\
\xi^+ &= \langle Y\gamma | K | \overline{K}N\rangle, & T_{\gamma K} &= \langle Y\gamma | T | \overline{K}N\rangle,\\
\eta &= \langle Y\pi | K | Y\gamma\rangle, & T_{Y\gamma} &= \langle Y\pi | T | Y\gamma\rangle,\\
\eta^+ &= \langle Y\gamma | K | Y\pi\rangle, & T_{\gamma Y} &= \langle Y\gamma | T | Y\pi\rangle,\\
\zeta &= \langle Y\gamma | K | Y\gamma\rangle, & T_{\gamma\gamma} &= \langle Y\gamma | T | Y\gamma\rangle.
\end{aligned} \tag{9}$$

We denote the submatrices of the K′ and T′ matrices by the corresponding primed letters. We neglect the matrix ζ, which is at least one order of magnitude smaller than the other matrices.

If we introduce

$$K_0 = \begin{pmatrix} \alpha & \beta \\ \beta^+ & \gamma \end{pmatrix}, \quad \delta = \begin{pmatrix} \xi \\ \eta \end{pmatrix}, \tag{10}$$

then we can write

$$K = \begin{pmatrix} K_0 & \delta \\ \delta^+ & 0 \end{pmatrix}. \tag{11}$$

From (5), (7), (10), and (11) we readily find that

$$\begin{aligned}
T'_{KK} &= (1 - iX')^{-1} X', \\
T'_{KY} &= (1 - iX')^{-1} \beta' (1 - i\gamma')^{-1} \\
&= (1 - i\alpha')^{-1} \beta' (1 - iZ')^{-1}, \\
T'_{YK} &= (1 - iZ')^{-1} \beta'^T (1 - i\alpha')^{-1} \\
&= (1 - i\gamma')^{-1} \beta'^T (1 - iX')^{-1}, \\
T'_{YY} &= (1 - iZ')^{-1} Z', \; T'_{K\gamma} \\
&= (1 - iX')^{-1} \xi' + i (1 - iX')^{-1} \beta' (1 - i\gamma')^{-1} \eta', \\
T'_{Y\gamma} &= i (1 - iZ')^{-1} \beta'^T (1 - i\alpha')^{-1} \xi' + (1 - iZ')^{-1} \eta', \\
T'_{\gamma K} &= \xi'^T (1 - iX')^{-1} + i\eta'^T (1 - i\gamma')^{-1} \beta'^T (1 - iX')^{-1}, \\
T'_{\gamma Y} &= i\xi'^T (1 - i\alpha')^{-1} \beta' (1 - iZ')^{-1} + \eta'^T (1 - iZ')^{-1},
\end{aligned} \tag{12}$$

where

$$X' = \alpha' + i\beta' (1 - i\gamma')^{-1} \beta'^T,$$
$$Z' = \gamma' + i\beta'^T (1 - i\alpha')^{-1} \beta'. \tag{13}$$

3. In our discussion it is sufficient to take into account the electromagnetic interaction in first-order perturbation theory, considering separately the contributions from the iso-scalar and iso-vector parts of the electromagnetic interaction.

We start from the iso-scalar current. In this case the total isospin is I = 0 for the Λ + γ system and I = 1 for the Σ + γ system. We denote by ξ^0_Λ, ξ^1_Σ, η^0_Λ, and η^1_Σ the matrix elements with iso-scalar current for the processes $\overline{K}$ + N → Λ(Σ) + γ and Λ(Σ) + π → Λ(Σ) + γ, respectively. In the case of the iso-vector current, the total isospin is I = 1 for the Λ + γ system and I = 0 or 1 for the Σ + γ system. The corresponding matrix elements will be denoted by ξ^1_Λ, ξ^1_Σ, ξ'^1_Σ, η^1_Λ, η^0_Σ, and η'^1_Σ.

Let us consider the channels with isospin I = 0. In this case the submatrices α, β, and γ are simply numbers. Expressions (13) are then reduced to

$$X = \alpha + i\pi\beta^2\rho_\Sigma/(1 - i\pi\rho_\Sigma\gamma) = a + ib, \tag{14}$$

where

$$a = \alpha - \pi^2\beta^2\gamma\rho_\Sigma^2/[1 + \pi^2\rho_\Sigma^2\gamma^2],$$
$$b = \pi\beta^2\rho_\Sigma/[1 + \pi^2\rho_\Sigma^2\gamma^2] > 0. \tag{15}$$

Substituting (14) in (12) we get

$$T'_{KK} = (1 - iX')^{-1} X' = \pi\rho_K (a^0 + ib^0)\,\Delta_0^{-1},$$
$$T'_{\Sigma K} = \pi^{1/2}\rho_K^{1/2} (b^0)^{1/2} e^{i\lambda_\Sigma} \Delta_0^{-1}, \tag{16}$$

where

$$\tan\lambda_\Sigma = \pi\rho_\Sigma\gamma, \qquad \Delta_0 = 1 - i\pi\rho_K (a^0 + ib^0).$$

Formulas (16) for the processes $\overline{K}$ + N → $\overline{K}$ + N and $\overline{K}$ + N → Σ + π were obtained by many authors.[3]

Let us write

$$T'_{\gamma K} = \xi^T (1 - iX')^{-1} + i\eta^T (1 - i\gamma')^{-1} \beta'^T (1 - iX')^{-1}$$
$$= \begin{pmatrix} T'_{\Lambda\gamma K} \\ T'_{\Sigma\gamma K} \end{pmatrix}, \quad \eta^T = \begin{pmatrix} \eta^0_{\Lambda\Sigma} \\ \eta^0_{\Sigma\Sigma} \end{pmatrix}, \quad \xi^T = \begin{pmatrix} \xi^0_{\Lambda K} \\ \xi^0_{\Sigma K} \end{pmatrix}. \tag{17}$$

From (14)—(17) we readily find that

$$T'_{\Lambda\gamma K} = \pi\rho_{\gamma\Lambda}^{1/2}\rho_K^{1/2} [\xi^0_{\Lambda K} + i\eta^0_{\Lambda\Sigma}\pi^{1/2}\rho_\Sigma^{1/2} (b^0)^{1/2} e^{i\lambda_\Sigma}]\,\Delta_0^{-1},$$
$$T'_{\Sigma\gamma K} = \pi\rho_{\gamma\Sigma}^{1/2}\rho_K^{1/2} [\xi^0_{\Sigma K} + i\eta^0_{\Sigma\Sigma}\pi^{1/2}\rho_\Sigma^{1/2} (b^0)^{1/2} e^{i\lambda_\Sigma}]\,\Delta_0^{-1}. \tag{18}$$

We note that $T'_{\Lambda\gamma K}$, $T'_{\Sigma\gamma K}$, and $T'_{\Sigma K}$ have almost the same energy dependence in the low-energy region, where (assuming the relative parity of the hyperons to be positive) the energy dependence of ρ_Σ, ρ_Λ, $\rho_{\gamma\Lambda}$, and $\rho_{\gamma\Sigma}$ can be neglected.

Let us proceed to examine the channels with isospin I = 1. In this case γ and β are matrices,

$$\gamma = \begin{pmatrix} \gamma_{\Lambda\Lambda} & \gamma_{\Sigma\Lambda} \\ \gamma_{\Lambda\Sigma} & \gamma_{\Sigma\Sigma} \end{pmatrix}, \quad \beta = (\beta_{\Lambda K}, \beta_{\Sigma K}). \tag{19}$$

It is easy to verify that in this case X is simply a complex number

$$X = a^1 + ib^1, \tag{20}$$

where

$$a^1 = \alpha - \pi\beta\rho_Y^{1/2} \frac{1}{1 + \gamma'^2} \gamma'\rho_Y^{1/2}\beta^T, \qquad b^1 = \pi\beta\rho_Y^{1/2} \frac{1}{1 + \gamma'^2} \rho_Y^{1/2}\beta^T \tag{21}$$

From (12), (13), and (19)—(21) it follows that

$$T'_{KK} = \pi\rho_K (a^1 + ib^1)\,\Delta_1^{-1}, \quad T'_{\Lambda K} = \pi^{1/2}\rho_K^{1/2} (b^1_{\Lambda K})^{1/2} e^{i\lambda_{\Lambda K}} \Delta_1^{-1},$$
$$T'_{\Sigma K} = \pi^{1/2}\rho_K^{1/2} (b^1_{\Sigma K})^{1/2} e^{i\lambda_{\Sigma K}} \Delta_1^{-1}, \tag{22}$$

where

$$\pi^{1/2}\rho_K^{1/2} b^{1/2}_{\Lambda K} e^{i\lambda_{\Lambda K}} \equiv \langle\Lambda\,|\,(1 - i\gamma')^{-1}\beta'^T\,|\,K\rangle,$$
$$\pi^{1/2}\rho_K^{1/2} b^{1/2}_{\Sigma K} e^{i\lambda_{\Sigma K}} \equiv \langle\Sigma\,|\,(1 - i\gamma')^{-1}\beta'^T\,|\,K\rangle,$$
$$\Delta_1 = 1 - i\pi\rho_K (a^1 + ib^1), \tag{23}$$

and the quantities $b_{\Lambda K}$ and $b_{\Sigma K}$ are related with b by the equation $b_{\Lambda K}' + b_{\Sigma K} = b$. If we represent the matrices ξ and η in the form

$$\xi = (\xi_{\Lambda K}, \xi_{\Sigma K}), \quad \eta = \begin{pmatrix} \eta_{\Lambda\Lambda} & \eta_{\Sigma\Lambda} \\ \eta_{\Lambda\Sigma} & \eta_{\Sigma\Sigma} \end{pmatrix}, \tag{24}$$

then the matrix elements $T'_{\gamma\Lambda K}$ and $T'_{\gamma\Sigma K}$ become

$$T'_{\gamma\Lambda K} = \pi\rho_{\gamma\Lambda}^{1/2}\rho_K^{1/2}\Delta_1^{-1}\,[\xi_{\Lambda K} + i\eta_{\Lambda\Lambda}\pi^{1/2}\rho_\Lambda^{1/2}b_{\Lambda K}^{1/2}e^{i\lambda_{\Lambda K}}$$
$$+ i\eta_{\Lambda\Sigma}\pi^{1/2}\rho_\Sigma^{1/2}b_{\Sigma K}^{1/2}e^{i\lambda_{\Sigma K}}], \tag{25}$$

$$T'_{\gamma\Sigma K} = \pi\rho_{\gamma\Sigma}^{1/2}\rho_K^{1/2}\Delta_1^{-1}\,[\xi_{\Sigma K} + i\eta_{\Sigma\Lambda}\pi^{1/2}\rho_\Lambda^{1/2}b_{\Lambda K}^{1/2}e^{i\lambda_{\Lambda K}}$$
$$+ i\eta_{\Sigma\Sigma}\pi^{1/2}\rho_\Sigma^{1/2}b_{\Sigma K}^{1/2}e^{i\lambda_{\Sigma K}}]. \tag{26}$$

To simplify matters we introduce new symbols

$$\alpha_\Lambda^0 = \pi^{1/2}\rho_{\gamma\Lambda}^{1/2}\,[\xi_{\Lambda K}^0 + i\eta_{\Lambda\Sigma}^0\pi^{1/2}\rho_\Sigma^{1/2}(b^0)^{1/2}e^{i\lambda_\Sigma}],$$
$$\alpha_\Sigma^0 = \pi^{1/2}\rho_{\gamma\Sigma}^{1/2}\,[\xi_{\Sigma K}^0 + i\eta_{\Sigma\Sigma}^0\pi^{1/2}\rho_\Sigma^{1/2}(b^0)^{1/2}e^{i\lambda_\Sigma}],$$
$$\alpha_\Lambda^1 = \pi^{1/2}\rho_{\gamma\Lambda}^{1/2}\,[\xi_{\Lambda K}^1 + i\eta_{\Lambda\Lambda}^1\pi^{1/2}\rho_\Lambda^{1/2}(b_{\Lambda K}^1)^{1/2}e^{i\lambda_{\Lambda K}}$$
$$+ i\eta_{\Lambda\Sigma}^1\pi^{1/2}\rho_\Sigma^{1/2}(b_{\Sigma K}^1)^{1/2}e^{i\lambda_{\Sigma K}}],$$
$$\alpha_\Sigma^1 = \pi^{1/2}\rho_{\gamma\Sigma}^{1/2}\,[\xi_{\Sigma K}^1 + i\eta_{\Sigma\Lambda}^1\pi^{1/2}\rho_\Lambda^{1/2}(b_{\Lambda K}^1)^{1/2}e^{i\lambda_{\Lambda K}}$$
$$+ i\eta_{\Sigma\Sigma}^1\pi^{1/2}\rho_\Sigma^{1/2}(b_{\Sigma K}^1)^{1/2}e^{i\lambda_{\Sigma K}}],$$
$$\alpha_\Sigma'^1 = \pi^{1/2}\rho_{\gamma\Sigma}^{1/2}\,[\xi_{\Sigma K}'^1 + i\eta_{\Sigma\Lambda}'^1\pi^{1/2}\rho_\Lambda^{1/2}(b_{\Lambda K}^1)^{1/2}e^{i\lambda_{\Lambda K}}$$
$$+ i\eta_{\Sigma\Sigma}'^1\pi^{1/2}\rho_\Sigma^{1/2}(b_{\Sigma K}^1)^{1/2}e^{i\lambda_{\Sigma K}}], \tag{27}$$

with which the cross sections of the processes (1) can be written in the following form:

Process:	Cross section:
$K^- + p \to \Lambda^0 + \gamma$; $\overline{K}^0 + n \to \Lambda^0 + \gamma$	$\dfrac{2\pi m_K}{E_K k}\left\|\dfrac{\alpha_\Lambda^0}{\Delta_0} \pm \dfrac{\alpha_\Lambda^1}{\Delta_1}\right\|^2,$
$K^- + p \to \Sigma^0 + \gamma$; $\overline{K}^0 + n \to \Sigma^0 + \gamma$	$\dfrac{2\pi m_K}{E_K k}\left\|-\dfrac{\alpha_\Sigma^0/\sqrt{3}}{\Delta_0} \pm \dfrac{\alpha_\Sigma^1}{\Delta_1}\right\|^2,$
$K^- + n \to \Sigma^- + \gamma$; $\overline{K}^0 + p \to \Sigma^+ + \gamma$	$\dfrac{2\pi m_K}{E_K k}\left\|\dfrac{\alpha_\Sigma^1 \pm \alpha_\Sigma'^1/\sqrt{2}}{\Delta_1}\right\|^2.$

Thus, the experimental investigation of the processes $\overline{K} + N \to \Lambda(\Sigma) + \gamma$ in $\overline{K}p$ and $\overline{K}d$ collisions can yield certain information on the matrix elements α_Λ and α_Σ. Naturally, this information is not sufficient to reconstitute the matrix elements ξ and η, which describe the photoproduction of mesons on hyperons. Nonetheless they may prove useful for a study of the interaction between hyperons and mesons or photons.

4. A powerful method for the analysis of strong interactions is the method of dispersion relations (d.r.), the use of which yields in many cases interesting results in the low-energy region. It can be assumed that the d.r. method is applicable to the photoproduction of mesons and hyperons. In the present paper we confine ourselves to a generalization of the Kroll-Ruderman theorem for photoproduction of pions near threshold.[4]

Let us assume that the Λ and Σ hyperons have a positive relative parity and that the K meson is pseudoscalar. If the created particles have low energies account of the electric dipole radiation is sufficient. The generalized Kroll-Ruderman theorem states that, accurate to $m_\pi/M \approx 15\%$, the matrix for the electric dipole transition is determined completely by the pion-hyperon coupling constant.

Let us write the Hamiltonian of the pion-hyperon interaction in the form

$$\mathcal{H} = ig_{\Sigma\Lambda}\overline{\psi}_\Sigma\gamma_5\psi_\Lambda\psi_\pi + ig_{\Sigma\Sigma}([\overline{\psi}_\Sigma\gamma_5\psi_\Sigma]\psi_\pi) + \text{Herm. conj.} \tag{28}$$

Following Low's method[5] we can obtain

$$\eta_{\Lambda\Sigma}^0 \sim m_\pi/M, \quad \eta_{\Sigma\Lambda}^1 \sim m_\pi/M, \quad \eta_{\Sigma\Sigma}^1 \sim m_\pi/M,$$
$$\eta_{\Sigma\Lambda}'^1 = \eta_{\Lambda\Sigma}'^1 = \sqrt{2}\,\alpha^{1/2}f_{\Sigma\Lambda}\,[1 + O(m_\pi/M)],$$
$$\eta_{\Sigma\Sigma}'^1 = \alpha^{1/2}f_{\Sigma\Sigma}\,[1 + O(m_\pi/M)], \quad \eta_{\Sigma\Sigma}^0 \approx m_\pi/M.$$

Here m_π is the pion mass, M is the hyperon mass, $\alpha = \epsilon^2/4\pi = 1/137$, and $f^2 = g^2/8\pi M$.

[1] L. I. Lapidus and Chou Kuang-chao, JETP 37, 283 (1959), Soviet Phys. JETP 10, 199 (1960).

[2] Jackson, Ravenhall, and Wyld, Nuovo cimento 9, 834 (1958). R. H. Dalitz and S. F. Tuan, Ann. of Physics 8, 100 (1959). M. Ross and G. Snow, Phys. Rev. 115, 1773 (1959).

[3] R. H. Dalitz and S. F. Tuan, Ann. of Physics 10, 307 (1960). P. T. Mathews and A. Salam, Nuovo cimento 13, 382 (1959). J. D. Jackson and H. Wyld, Nuovo cimento 13, 84 (1959).

[4] K. M. Kroll and M. A. Ruderman, Phys. Rev. 93, 233 (1954).

[5] F. E. Low, Phys. Rev. 97, 1392 (1955).

Translated by J. G. Adashko
224

SOVIET PHYSICS JETP VOLUME 14, NUMBER 5 MAY, 1962

SCATTERING OF PHOTONS BY NUCLEONS

L. I. LAPIDUS and CHOU KUANG-CHAO

Joint Institute for Nuclear Research

Submitted to JETP editor May 18, 1961

J. Exptl. Theoret. Phys. (U.S.S.R.) 41, 1546-1555 (November, 1961)

An analysis of elastic scattering of photons with energies up to 300 Mev by protons is carried out by making use of the dispersion relations method. Six dispersion relations are utilized to estimate the real parts of the amplitudes at $Q^2 = 0$. Photoproduction of pions is taken into account in a larger energy region than was done previously. Five subtraction constants are determined from the long wavelength limit and expressed in terms of the nucleon charge and magnetic moment. Differential cross sections and polarizations of the recoil nucleons are estimated. Photon-nucleon scattering at high energies is discussed.

1. Following the work of Gell-Mann, Goldberger, and Thirring[1] dispersion relations for photon-nucleon scattering, whose validity in the e^2-approximation has been rigorously proved by Logunov,[2] have been applied to the analysis of experimental data by a number of workers.[3-7] Cini and Stroffolini[3] were the first to calculate forward scattering cross sections for photons with energies up to 210 Mev. Certain qualitative peculiarities in the energy dependence of the forward scattering cross section were indicated earlier in[1], and also in[8].

Capps[4] has considered γN scattering through an arbitrary angle by taking into account a minimal number of states. In so doing he made use of some unpublished results of Gell-Mann and J. Mathews.

Akiba and Sato[5] considered scattering through nonzero angles. In order to evaluate the subtraction constants in some of the dispersion relations they made use of perturbation theory.

The authors have previously[6] considered in detail dispersion relations for all six invariant functions, that characterize the γN-scattering amplitude, and have carried out a dispersion analysis in the energy region up to 200 Mev in an approximation in which certain recoil effects were ignored. It was shown that if photoproduction of pions in S states is taken into account significant modifications are introduced in the near threshold region. These changes are such as to improve the agreement between the dispersion analysis and experiment. Near threshold the energy dependence of the amplitudes and cross sections becomes nonmonotonic.

Aside from certain differences, connected with what assumptions were made regarding the number of subtraction constants in the dispersion relations and the maximum angular momentum of the states taken into account, all the published papers turned out to have in common the inability to obtain good agreement with experimental data in the energy region near 160—200 Mev.

In a number of papers[9,7,10] an attempt was made to eliminate this discrepancy by taking into account the contribution from Low's diagram.[11] However a direct measurement of the lifetime of the π^0 meson[12] together with an analysis of the question of the sign of the pole amplitude[13] have led to the conclusion, that the inclusion of Low's amplitude cannot substantially affect the results of the analysis. In connection with these discrepancies between the analysis and the existing experimental data we carry out in this work an analysis of γN scattering based on dispersion relation, in which we take into account in addition to photoproduction of pions in S states the contribution from the high energy regions in a more careful manner; we also analyze the question of the number of subtractions in the dispersion relations and, taking nucleon recoil fully into account, estimate the previously introduced quantities $R_i(\nu)$ at $Q^2 = 0$.

2. The connection between the invariant functions $T_i(\nu, Q^2)$ and the amplitudes $R_i(\nu, Q^2)$ in the barycentric system is given by Eq. (1) of[14] (in the following[14] will be referred to as A). For the definitions of $T_i(\nu, Q^2)$ and $R_i(\nu, Q^2)$ see[13] (in the following[13] will be referred to as B). The notation in the present paper is the same as the notation in A and B. By R_i without additional marks we will understand here the amplitudes in the barycentric frame.

Since according to the optical theorem

$$\operatorname{Im}(R_1 + R_2) = \frac{\nu_c \sigma_t}{4\pi} = \frac{w^2 - M^2}{2w}\frac{\sigma_t}{4\pi} \tag{1}$$

(w is the total energy in the barycentric frame) it follows that under the assumption

$$\sigma_t(w) \to \text{const} \quad \text{as} \quad w \to \infty$$

we have asymptotically as $w \to \infty$

$$R_1 + R_2 \to w^2 \sim \nu. \tag{2}$$

Assuming further that as $w \to \infty$ all $R_i \sim \nu$, we get from A, Eq. (1) that as $w \to \infty$

$$T_1 - T_3 \to w^2, \quad T_1 + T_3 \to w^2, \quad T_5 \to w^2,$$

$$T_2 - T_4 \to w, \quad T_2 + T_4 \to \text{const}, \quad T_6 \to w. \tag{3}$$

Consequently, under the assumptions here made, the dispersion relations for T_1, T_3 and T_5 should contain one subtraction, whereas the dispersion relations for the quantities T_2, T_4 and T_6 may be written with no subtractions.

In order to estimate the amplitudes $R_1 + R_2$, R_3, R_4, and $R_5 + R_6$ it is sufficient to write dispersion relations for T_1, T_3 and T_5 at $Q^2 = 0$. At $Q^2 = 0$ the invariant

$$\nu = \nu_{lab} - Q^2/M$$

becomes, as is well known, the photon energy in the laboratory frame ν_{lab} (denoted in the following by ν).

3. As can be seen from A, Eq. (1), for forward scattering the functions T_5 and $T_2 + T_4$ reduce to $R_4 - R_3$, so that at $Q^2 = 0$ the dispersion relations for T_5 and $T_2 + T_4$ are equivalent.

Let us consider the functions

$$F_1(\nu_0) = \tfrac{1}{2}[T_1 - T_3 - \nu_0(T_2 - T_4)] = w_0(R_1 + R_2)/M,$$
$$F_2(\nu_0) = \nu_0 T_6 = w_0[R_3 + R_4 + 2R_5 + 2R_6]/M,$$
$$F_3(\nu_0) = \nu_0(T_1 + T_3)/2M = (w_0/M)^2(R_3 - R_4),$$
$$F_4(\nu_0) = \tfrac{1}{2}(T_1 - T_3)$$
$$= w_0^2(R_3 + R_4)/M\nu_0 - 2w_0(R_1 + R_2)/(M + w_0). \tag{4}$$

It is clear from the discussion above that the dispersion relations for the functions $F_1, \ldots, F_4$ should contain one subtraction. All quantities on the right side of Eq. (4) are in the barycentric frame. If one takes into account that (for $Q^2 = 0$) the amplitudes in the laboratory system are connected to the corresponding quantities in the barycentric system by

$$(R_1 + R_2)^{\pi} = w_0(R_1 + R_2)/M,$$
$$(R_4 - R_3)^{\pi} = (w_0/M)^2(R_4 - R_3), \tag{5}$$
$$[R_3 + R_4 + 2R_5 + 2R_6]^{\pi} = w_0[R_3 + R_4 + 2R_5 + 2R_6]/M,$$

then one obtains from the dispersion relations for $F_1, \ldots, F_4$

$$D_{1,4}^{lab}(\nu_0) - D_{1,4}(0) = \frac{2\nu_0^2}{\pi}\int_{\nu_t}^{\infty}\frac{d\nu}{\nu^2 - \nu_0^2}\frac{A_{1,4}(\nu)}{\nu},$$

$$D_{2,3}^{lab}(\nu_0) - \nu_0 D'_{2,3}(0) = \frac{2\nu_0^3}{\pi}\int_{\nu_t}^{\infty}\frac{A_{2,3}(\nu)\,d\nu}{\nu^2(\nu^2 - \nu_0^2)}, \tag{6}$$

where

$$D_1^{\pi} = \operatorname{Re}(R_1 + R_2)^{lab}, \quad D_2^{lab} = \operatorname{Re}[R_3 + R_4 + 2R_5 + 2R_6]^{lab},$$
$$D_3^{lab} = \operatorname{Re}(R_4 - R_3)^{lab}, \quad D_4^{lab} = \operatorname{Re} F_4(\nu);$$
$$D_1(0) = -e^2/M, \quad D'_2(0) = -2\mu_a^2,$$
$$D'_3(0) = -2[\mu^2 - (e/2M)^2], \quad D_4(0) = -(e^2/2M)\lambda(2 + \lambda),$$

and where $A_i(\nu_0)$ stands for the imaginary part of the corresponding amplitude; $\mu = e(1 + \lambda)/2M$ stands for the magnetic moment and μ_a for the anomalous magnetic moment of the nucleon.

If the elements of the amplitude for the photoproduction of pions in states with $J \le 3/2$ in the barycentric frame are denoted by E_1, E_2, E_3 (electric transitions into $1/2-$, $3/2+$ and $3/2-$ respectively), M_1, M_2, M_3 (magnetic transitions into $1/2+$, $3/2-$ and $3/2+$ respectively), then the unitarity relations lead to the equalities

$$\operatorname{Im} R_1 = \nu_c\left\{|E_1|^2 + 2|E_3|^2 + \tfrac{1}{3}|E_2|^2\cos\theta - \tfrac{1}{6}|M_2|^2\right\},$$
$$\operatorname{Im} R_3 = \nu_c\left\{|E_1|^2 + \tfrac{1}{3}|E_2|^2\cos\theta - |E_3|^2 + \tfrac{1}{12}|M_2|^2 + \operatorname{Re}(E_3^* M_2)\right\},$$
$$\operatorname{Im} R_5 = -\nu_c\left\{\tfrac{1}{6}|E_2|^2 + \operatorname{Re}(E_3^* M_3)\right\}, \tag{7}$$

which represent the generalization of the corresponding equalities in[6]. The expressions for $\operatorname{Im} R_2$ differ from those for $\operatorname{Im} R_1$ by the exchange $E_i \rightleftarrows M_i$. Analogously, the expression for $\operatorname{Im} R_4$ may be obtained from that for $\operatorname{Im} R_3$, and for $\operatorname{Im} R_6$ from $\operatorname{Im} R_5$.

In Eq. (7) we mean by the modulus of the amplitude on the right side the sum of the contributions from photoproduction of π^+ and π^0 mesons. We note that if the mass differences between mesons and between nucleons are ignored then a cancellation in the interference terms, for example in $|E_{\pi^0}|^2 + |E_{\pi^+}|^2$, occurs as a consequence of isotopic symmetry in the pion photoproduction process. At that

$$A_1(\nu) = \nu\sigma_t/4\pi = \nu\left\{|E_1|^2 + |M_1|^2 + 2|E_3|^2 + 2|M_3|^2 + \tfrac{1}{6}|M_2|^2 + \tfrac{1}{6}|E_2|^2\right\},$$

$$A_2(\nu) = \nu\left\{|E_1|^2 + |M_1|^2 + \tfrac{1}{3}|M_2|^2 + \tfrac{1}{3}|E_2|^2 - |M_3 + \tfrac{1}{2}E_2|^2 - |E_3 + \tfrac{1}{2}M_2|^2\right\},$$

$$A_3(\nu) = -\nu(w/M)\{|E_1|^2 - |M_1|^2 + |M_3 - \tfrac{1}{2}E_2|^2 - |E_3 - \tfrac{1}{2}M_2|^2\},$$

$$A_4(\nu) + (w - M)\sigma_t/4\pi = w\{|E_1|^2 + |M_1|^2 + \tfrac{2}{3}|M_2|^2 + \tfrac{2}{3}|E_2|^2 - |E_3 - \tfrac{1}{2}M_2|^2 - |M_3 - \tfrac{1}{2}E_2|^2\}. \quad (8)$$

4. As was shown by Goldberger,[1] the sum rule that follows from the nonsubtracted dispersion relations:

$$\mathrm{Re}(R_1 + R_2) \to + \frac{1}{2\pi^2}\int_{\nu_t}^{\infty} \sigma_t(\nu)\,d\nu > 0 \quad (9)$$

is in contradiction with the long wavelength limit

$$R_1 + R_2 \to -e^2/M < 0. \quad (10)$$

Consequently, nonsubtracted dispersion relations for the amplitude $R_1 + R_2$ violate the requirements of relativistic and gauge invariance on which the long wavelength limit is based.

Let us remark that possible sum rules involving the square of the magnetic moment are not in direct contradiction with the long wavelength limit when nonsubtracted dispersion relations are assumed for $F_2(\nu)$. As can be seen from Eqs. (6) and (8), of particular importance here is the contribution of the resonant state, proportional to $|M_3|^2$. The result is unchanged if one takes into account the (numerically important) contribution from photoproduction in S states, which decreases the effective contribution of $|M_3|^2$.

The sum rule for the square of the magnetic moment is very sensitive to the ratio of the photoproduction amplitudes E_2 and M_3. For certain ratios (for example for $E_2 = M_3$[5]) one can arrive at a contradiction. At the present time, however, the analysis of photoproduction is not sufficiently precise to permit the assertion that the experimental data are in contradiction with the sum rule. An increase in the accuracy of the photoproduction analysis, aimed at obtaining information about the amplitudes E_2, M_2 and E_3, would be most welcome.

The fact that unsubtracted dispersion relations give rise to definite sum rules may be of particular interest in certain processes. Thus, in the case of $\pi\pi$ scattering analogous considerations (applied to dispersion relations at $Q^2 = 0$) lead to the conclusion that the S-state scattering lengths a_0 and a_2 are positive at low energies. The same holds for πK and KK scattering.

5. If in addition to the functions introduced previously one studies properties of the functions*

$$F_5(\nu_0) = (T_2 - T_4)', \quad (11)$$

$$F_6(\nu_0) = (T_2 + T_4)', \quad (12)$$

$$F_7(\nu_0) = T_6', \quad (13)$$

one concludes that $F_{5,6}(\nu)$ are odd functions of ν and contain no poles, whereas $F_7(\nu)$ is an even function of ν with a second order pole. As $\nu \to \infty$

$$F_{5,6,7} \to \nu^{-1/2},$$

so that the dispersion relations for these functions need no subtractions.

These dispersion relations may turn out to be useful since when photoproduction in states with $J \le 3/2$ is taken into account the angular dependence of the amplitudes $R_i(\nu, Q^2)$ in the barycentric frame takes the form (cf.[6])

$$R_3 = \mathscr{E}_1 - \mathscr{E}_3 + 2\mathscr{E}_2\cos\theta + \tfrac{1}{2}m_2 + C(\mathscr{E}_3 m_2),$$

$$R_4 = m_1 - m_3 + 2m_2\cos\theta + \tfrac{1}{2}\mathscr{E}_2 + C(m_3\mathscr{E}_2),$$

$$R_5 = -\mathscr{E}_2 - C(m_3\mathscr{E}_2), \quad R_6 = -m_2 - C(\mathscr{E}_3 m_2), \quad (14)$$

and is characterized by eight functions of energy $\mathscr{E}_{1,2,3}$, $m_{1,2,3}$, $C(\mathscr{E}_3 m_2)$ $C(m_3\mathscr{E}_2)$, which can be expressed in terms of $R_i(\nu, 0)$ and $R_i'(\nu, 0)$.

It follows from Eq. (14) that if we restrict ourselves to contributions from states with $J \le 3/2$

$$R_1' = R_3' = 2\mathscr{E}_2(\partial\cos\theta/\partial Q^2)_{Q^2=0} = -4\mathscr{E}_2 w_0^2/M^2\nu_0^2,$$

$$R_2' = R_4' = -4m_2 w_0^2/M^2\nu_0^2, \quad R_5' = R_6' = 0,$$

so that

$$(R_1 + R_2)' = (R_3 + R_4)' = [R_3 + R_4 + 2(R_5 + R_6)]'. \quad (15)$$

In the long wavelength limit[14]

$$(R_1 + R_2)' = -2e^2/M^2\nu + O(1),$$

$$(R_3 + R_4)' = -e^2[3 + 2(1+\lambda)^2]/2M^3 + O(\nu),$$

$$(R_3 + R_4 + 2R_5 + 2R_6)' = -e^2(2\lambda^2 - 2\lambda - 1)/2M^3 + O(\nu). \quad (16)$$

The fact that Eq. (15) is in contradiction with the long wavelength limit (16) means that the restriction to states with $J \le 3/2$ is not a good approximation even in the low energy region. The crossing symmetry conditions introduce kinematic corrections of the order of ν/M, which corresponds to inclusion of states with higher values of J. The carrying out of the analysis with this high a precision requires the introduction of additional functions of energy and discussion of a larger number

*The prime denotes differentiation with respect to Q^2 and subsequent passage to $Q^2 = 0$.

of dispersion relations. Introduction of the Low diagram does not resolve the indicated contradiction. All estimates of the amplitudes given here were obtained with the neglect of $R_i'(\nu, 0)$.

6. The results of the calculations of the amplitudes $R_i(\nu_0)$ at $Q^2 = 0$ are shown in the figures. The energy of the photons ν_0 is given in units of the threshold energy $\nu_t = 150$ Mev, and the values of the amplitudes in units of e^2/Mc^2.

For the calculation of the forward differential scattering cross section

$$\sigma(0^\circ) = |R_1 + R_2|^2 + |R_3 + R_4 + 2R_5 + 2R_6|^2$$

the amplitudes $R_1 + R_2$ and $R_3 + R_4 + 2R_5 + 2R_6$ are sufficient.

To estimate $D_1(\nu_0)$ use was made of the data on the total cross section for the interaction of photons with protons, including the second maximum and the cross section for pion pair production. The dependence of $A_1(\nu_0)$ is shown in Fig. 1. Previously we have neglected contributions from the energy region above 500 Mev. The result of estimating the amplitude $R_1 + R_2$ is shown in Fig. 2. The main difference between this and previous results appeared in the region $1 < \nu_0 < 2$, where as a consequence of a cancellation between the long wavelength limit and dispersion terms the value of $D_1(\nu_0)$ is significantly decreased. Let us note that this is precisely the energy region that is sensitive to a change in $A_1(\nu_0)$. The second maximum in $A_1(\nu_0)$ corresponds to the second maximum in photoproduction.

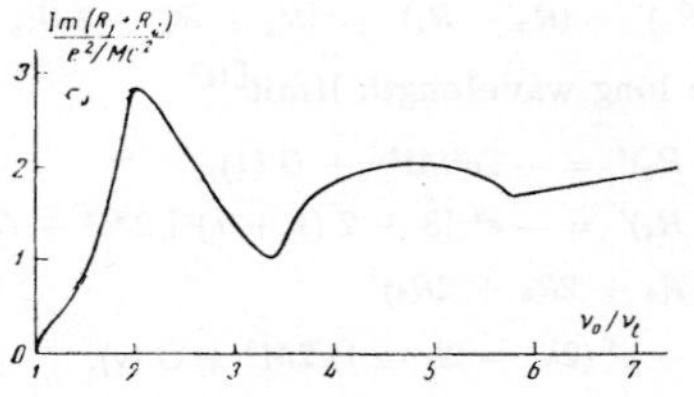

FIG. 1

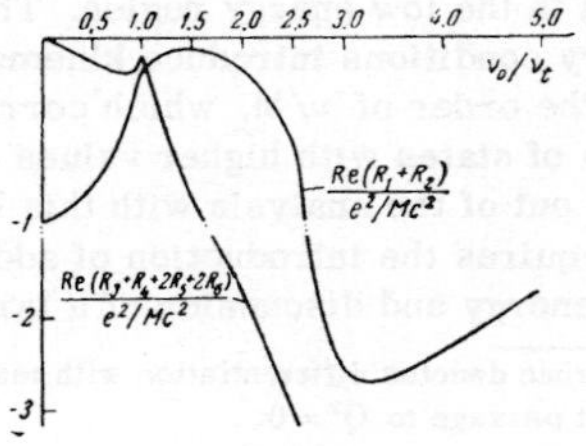

FIG. 2

For estimating real parts of the amplitudes, other than $R_1 + R_2$, which require much more detailed experimental data on photoproduction, we limit ourselves to the energy region up to 300 Mev. For the amplitude $R_1 + R_2$ it turns out to be possible to go much further, although with increasing energy the indeterminacy in the contribution from photoproduction of pairs (and larger numbers) of pions becomes appreciable.

In a number of papers[15,16] the γp scattering at 300–800 Mev has been looked upon as a diffraction process with $\mathrm{Re}\, R_i \ll \mathrm{Im}\, R_i$. The experimental study of γp scattering in the region of the second resonance is of interest as a sensitive method of investigation of the maximum itself.

If, ignoring all $\mathrm{Re}\, R_i$, we restrict ourselves to the imaginary parts of the amplitudes alone and consider only the contribution proportional to $|E_3|^2$, then we find immediately from Eq. (7) that

$$R_2 = R_4 = R_5 = R_6 = 0,$$

$$R_1 = \mathrm{Im}\, R_1 = -2\,\mathrm{Im}\, R_3 = 2\nu_t |E_3|^2,$$

whereas the differential cross section[6] is equal to

$$\sigma(0) = \tfrac{1}{8} R_1^2 (7 + 3\cos^2\theta) = \tfrac{1}{2} R_3^2 (7 + 3\cos^2\theta), \tag{17}$$

in agreement with the results of Minami.[16] The same result for the form of the angular distribution remains valid if in Eq. (7) only M_3 ($R_1 \to R_2$, $R_3 \to R_4$) is different from zero. If simultaneously E_3 and M_3 (with $\mathrm{Re}\, R_i = 0$) are different from zero then we have

$$\sigma(0) = \tfrac{1}{2}(R_3^2 + R_4^2)(7 + 3\cos^2\theta) + 10 R_3 R_4 \cos\theta. \tag{18}$$

However, as our estimates indicate, the quantities $\mathrm{Re}\,(R_1 + R_2)$ are large in the region of the second resonance and cannot be ignored. From this point of view the second resonance differs drastically from the $^3/_2$, $^3/_2$ resonance, in whose energy region

$$\mathrm{Re}\,(R_1 + R_2) \ll \mathrm{Im}\,(R_1 + R_2).$$

The results of the calculations for $R_3 \pm R_4$, $R_3 + R_4 + 2R_5 + 2R_6$ and $R_5 + R_6$ are shown in Figs. 2–4. In the evaluation of dispersion inte-

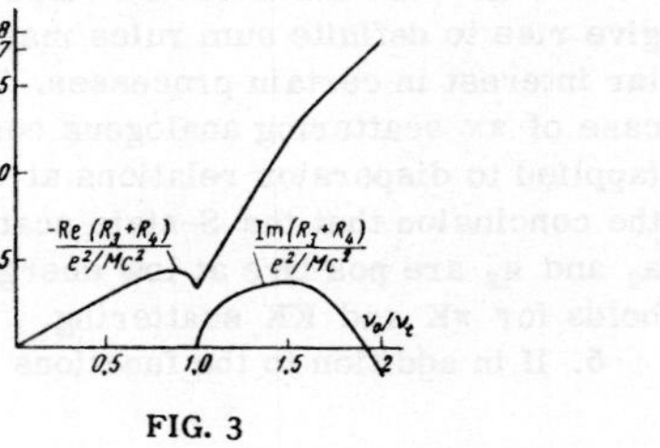

FIG. 3

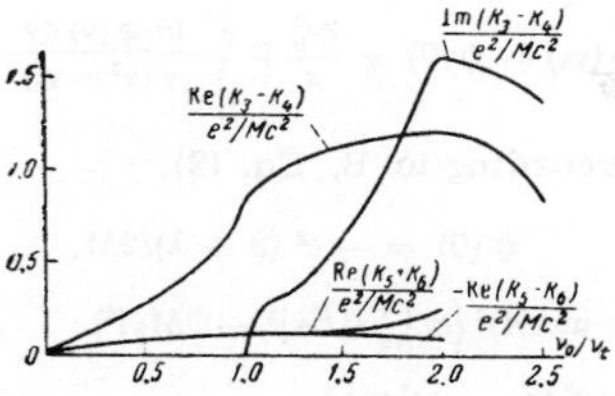

FIG. 4

grals $|E_1|^2$, $|M_3|^2$ and $|E_3|^2$ were assumed to be different from zero, and the energy dependence of $|E_1|^2$ and $|M_3|^2$ was taken from[6], whereas $|E_3|^2$ was assumed to be different from zero in the energy region $3.1 < \nu_0 < 5.8$. Let us remark that even in the absence of an imaginary part for $R_5 + R_6$ the real part of this quantity differs from its long wavelength limit, since the dispersion relations are satisfied by the invariant functions $T_i(\nu, Q^2)$.

The values of $\sigma(0°)$ are shown in Fig. 5, where we give for comparison the results of Cini and Stroffolini[3] for $\sigma_{C-S}(0°)$ in the barycentric frame. A significant difference can be seen in the near-threshold region.

7. For an estimate of $R_1 - R_2$ and $R_5 - R_6$ the

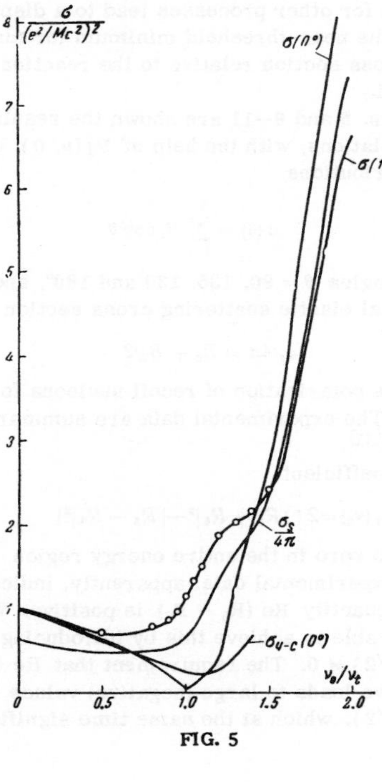

FIG. 5

dispersion relations (6) are not sufficient. Let us consider the function

$$F(\nu) = w^{-2}\varphi(\nu) = (M^2\nu^2/2w^2)\,[(T_1 + T_3)' - \nu(T_2 + T_4)']. \quad (19)$$

As can be seen from B, Eq. (4), we have

$$F(\nu) = \frac{w}{M}\left\{R_1 - R_2 - \frac{2M\nu}{w(w+M)}(R_3 - R_4)\right\}. \quad (20)$$

A study of the dispersion relation for $F(\nu)$ makes it possible to estimate $R_1 - R_2$ if $R_3 - R_4$ is known. In the energy region under consideration the coefficient of $(R_3 - R_4)$ in Eq. (20) is of the order of ν/M, however since the value of $R_3 - R_4$ is large (in comparison with $R_1 - R_2$), the second term in Eq. (20) cannot be ignored. The function $\varphi(\nu)$ introduced in Eq. (19) is an analytic function of ν with a cut along $\nu_t < \nu < \infty$, satisfying the crossing symmetry condition:

$$\varphi(\nu) = \varphi^*(-\nu). \quad (21)$$

Thus, for $\nu \ll \nu_t$ the function $\varphi(\nu)$ is a real function and

$$\varphi(\nu) \cong a + b\nu^2, \quad (22)$$

$$F(\nu) = \frac{\varphi(\nu)}{M^2 + 2M\nu} \cong \frac{a}{M^2}\left(1 - \frac{2\nu}{M}\right) + b\nu^2 + \ldots \quad (23)$$

We see that the linear term in $F(\nu)$ is fully determined by the first term in Eq. (22). It therefore follows from Eq. (20) that for small ν

$$R_1 - R_2 = -(e^2/M)(1 - 3\nu/M) + O(\nu^2)$$

and the linear term in $R_1 - R_2$ and in $F(\nu)$ are fully determined by the requirement of crossing symmetry, as is discussed in detail in B.

The function $F(\nu)$ introduced in Eq. (19) is an analytic function of ν with cuts along $\nu_t < \nu < \infty$ and $-\infty < \nu < -\nu_t$ and a (kinematic) pole at

$$w^2 = M^2 + 2M\nu = 0.$$

The requirements of crossing symmetry lead to the relation

$$F(-\nu) = \frac{M^2 + 2M\nu}{M^2 - 2M\nu} F^*(\nu),$$

and for small ν

$$F(\nu) \cong -(e^2/M)(1 - 2\nu/M) + O(\nu^2).$$

Applying the Cauchy formula to $F(\nu_0)$, for $\rho \to \infty$, along the contour shown in Fig. 6 and writing a dispersion relation with a subtraction we obtain

$$F(\nu_0) = -\frac{e^2}{M}\left(1 - \frac{2\nu_0}{M}\right) + \frac{\nu_0^2}{2\pi i}\int_C \frac{F(\nu)\,d\nu}{\nu^2(\nu - \nu_0)} = -\frac{e^2}{M}\left(1 - \frac{2\nu_0}{M}\right)$$

$$+ \frac{\nu_0^2}{\pi} P\int_{\nu_t}^{\infty} \operatorname{Im} F\left[\frac{1}{\nu - \nu_0} + \frac{M^2 + 2M\nu}{M^2 - 2M\nu}\frac{1}{\nu + \nu_0}\right]\frac{d\nu}{\nu^2}$$

$$+ \frac{\nu_0^2}{2\pi i}\int_{C_+} \frac{F(\nu + i\varepsilon)}{\nu^2(\nu - \nu_0)}d\nu + \frac{\nu_0^2}{2\pi i}\int_{C_-} \frac{F(\nu - i\varepsilon)}{\nu^2(\nu - \nu_0)}d\nu$$

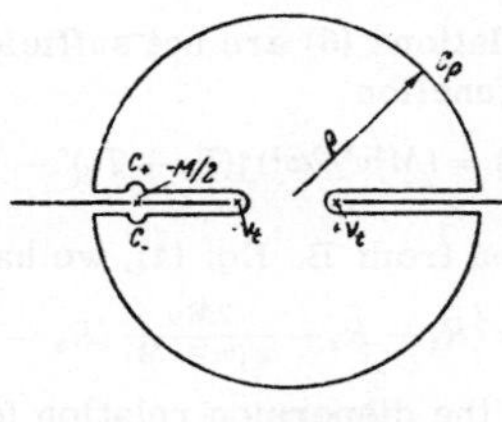

FIG. 6

and

$$\mathrm{Re}\,F(\nu_0) = -\frac{e^2}{M}\left(1 - \frac{2\nu_0}{M}\right) + K(\nu_0) + \frac{4\nu_0^3\,\mathrm{Re}\,F(M/2)}{M(\nu_0 + M/2)}; \quad (24)$$

$$K(\nu_0) = \frac{\nu_0^2}{\pi}\int_{\nu_t}^{\infty}\mathrm{Im}\,F(\nu)\left[\frac{1}{\nu - \nu_0} + \frac{M^2 + 2M\nu}{M^2 - 2M\nu}\,\frac{1}{\nu + \nu_0}\right]\frac{d\nu}{\nu^2}. \quad (25)$$

Since

$$K(M/2) = 0,$$

Re F(M/2) cannot be determined from Eq. (24), and this quantity enters as a free parameter, which must be determined starting from the experimental data. Under the restriction to photoproduction in the states with $J \leq 3/2$ only we get

$$\mathrm{Im}\,F(\nu) = \nu\Big\{\frac{M}{w}(|E_1|^2 - |M_1|^2) + 2(|E_3|^2 - |M_3|^2)\left(1 + \frac{1}{2}\frac{w - M}{w}\right) + \frac{1}{2}\frac{M}{w}(|E_2|^2 - |M_2|^2)\left(1 + \frac{1}{6}\frac{w - M}{w}\right)\Big\}. \quad (25')$$

In Fig. 7 are shown the results of estimating Re $(R_1 - R_2)$ with the help of Eq. (24) when the contribution proportional to Re F(M/2) is ignored.

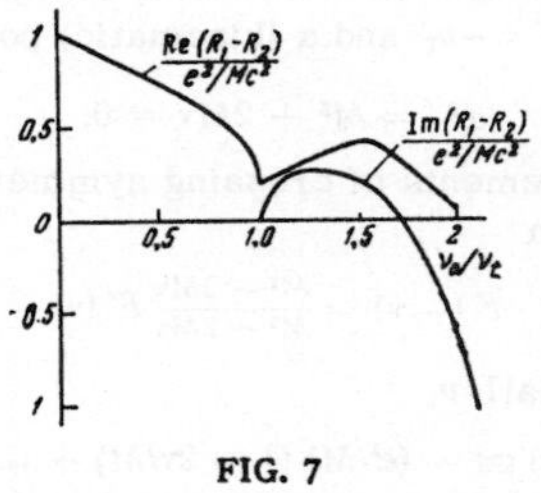

FIG. 7

For an estimate of $R_5 - R_6$ at $Q^2 = 0$, as can be seen from B, Eq. (4), it is sufficient to consider the function

$$\psi(\nu_0) = \tfrac{1}{3}\nu_0^2[T_5' + \tfrac{1}{2}(T_1 + T_3)]' = \left(\frac{w_0}{M}\right)^3 \times\left\{\frac{w_0}{\nu_0}(R_5 - R_6) + \frac{M}{w_0 + M}[R_1 - R_2 - (R_3 - R_4)]\right\}, \quad (26)$$

for which the dispersion relation has the form

$$\mathrm{Re}\,\psi(\nu_0) - \psi(0) = \frac{2\nu_0^2}{\pi}\,\mathrm{P}\int_{\nu_t}^{\infty}\frac{\mathrm{Im}\,\psi(\nu)\,d\nu}{\nu(\nu^2 - \nu_0^2)}, \quad (27)$$

where, according to B, Eq. (2),

$$\psi(0) = -e^2(2 + \lambda)/2M, \quad (28)$$

$$\mathrm{Im}\,\psi(\nu) = (w/M)^2\Big\{w\left[\tfrac{1}{6}(|E_2|^2 - |M_2|^2) + \mathrm{Re}(E_2^*M_3 - M_2^*E_3)\right] + M\nu(M + w_0)^{-1}[3(|E_3|^2 - |M_3|^2) + \tfrac{1}{4}(|E_2|^2 - |M_2|^2) + \mathrm{Re}(E_2^*M_3 - M_2^*E_3)]\Big\}. \quad (29)$$

The results of estimating Re $(R_5 - R_6)$ at $Q^2 = 0$ for Re F(M/2) = 0 are shown in Fig. 4. Estimates of the quantities $R_3 \pm R_4$ and $R_5 - R_6$, which play a dominant role in the differential cross section for $\nu_0 \gtrsim 1$, do not differ appreciably from those obtained previously.[6]

The results here obtained are of interest from the point of view of the study of the energy dependence of amplitudes near the threshold of a new reaction.[6] In that case all estimates can be carried out to the end. Let us call attention to the dependence of the amplitude Re $(R_1 + R_2)$, whose value continues to fall off also above threshold. This result indicates that a sharp energy dependence of the imaginary parts of the amplitudes above threshold may also for other processes lead to a displacement of the near-threshold minimum (or maximum) of the cross section relative to the reaction threshold.

In Figs. 5 and 8–11 are shown the results of the calculations, with the help of $R_i(\nu, 0)$, of angular distributions

$$\sigma(\theta) = \sum_{l=0}^{3} B_l \cos^l\theta$$

for the angles θ = 90, 135, 139 and 180°, and also of the total elastic scattering cross section

$$\sigma_s/4\pi = B_0 + B_2/2$$

and of the polarization of recoil nucleons for $\theta = 90°$. The experimental data are summarized in[10] and[17].

The coefficient

$$B_3(\nu_0) = 2[|R_5 + R_6|^2 - |R_5 - R_6|^2]$$

is near to zero in the entire energy region $\nu_0 \lesssim 2$.

The experimental data, apparently, indicate that the quantity Re $(R_5 - R_6)$ is positive. We were not able to achieve this by introducing Re F(M/2) ≠ 0. The requirement that Re $(R_5 - R_6)$ be positive leads to large (negative) values for Re F(M/2), which at the same time significantly

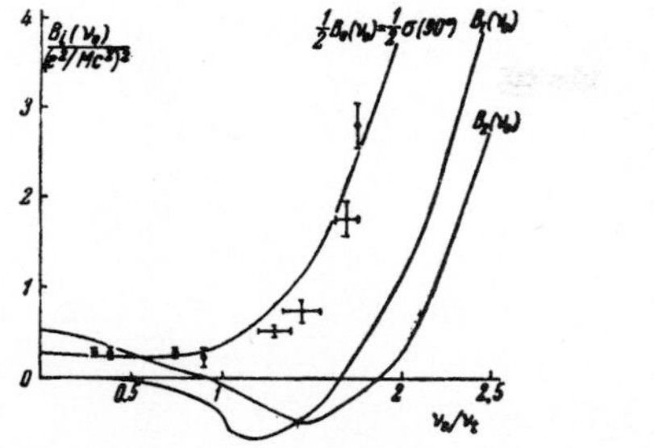

FIG. 8. Energy dependence of the coefficients in the angular distribution. The experimental points are from [9,10,17].

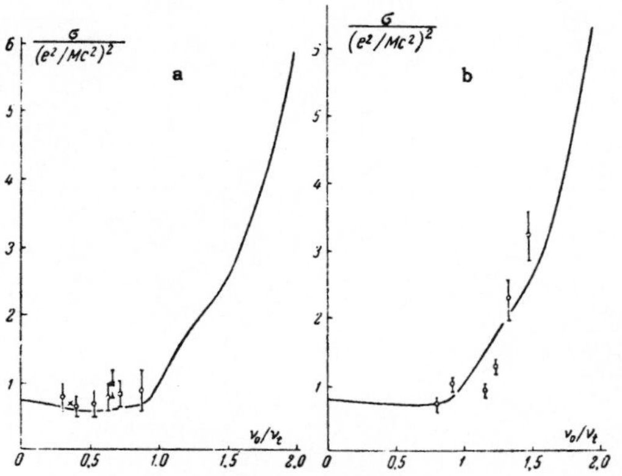

FIG. 9. Energy dependence of the scattering cross section: a – for $\theta = 135°$, b – for $\theta = 139°$. The experimental data are from [9,10,17].

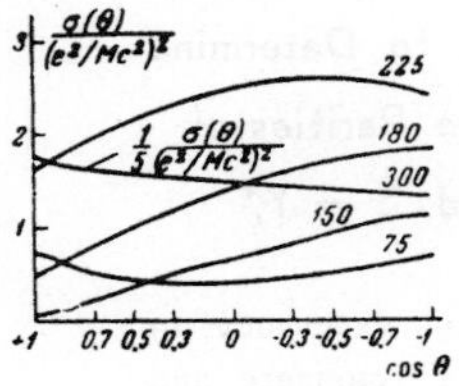

FIG. 10. Differential cross sections at different photon energies (indicated on the curves).

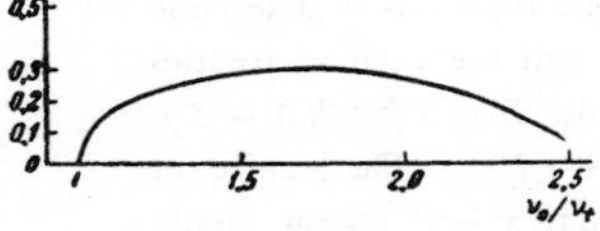

FIG. 11. Polarization of recoil protons.

increases the contribution of $|R_1 - R_2|^2$ to the cross section and does not lead to an improvement in the agreement with the experimental data.

It is necessary to remark that outside the region $1 < \nu_0 < 1.3$ a satisfactory agreement between the dispersional analysis and experimental data is obtained. In the region $1 < \nu_0 < 1.3$, which is particularly sensitive to dispersion effects, it is apparently necessary to take into account contributions from higher states, for which it is necessary to have information on pion photoproduction in a larger energy region.

[1] Gell-Mann, Goldberger, and Thirring, Phys. Rev. **95**, 1612 (1954). M. L. Goldberger, Phys. Rev. **99**, 979 (1955).

[2] A. A. Logunov, Dissertation, Joint Inst. for Nucl. Research (1959).

[3] M. Cini and R. Stroffolini, Nucl. Phys. **5**, 684 (1958).

[4] R. H. Capps, Phys. Rev. **106**, 1031 (1957); **108**, 1032 (1957).

[5] T. Akiba and I. Sato, Progr. Theor. Phys. **19**, 93 (1958).

[6] L. I. Lapidus and Chou Kuang-chao, JETP **37**, 1714 (1959) and **38**, 201 (1960), Soviet Phys. JETP **10**, 1213 (1960) and **11**, 147 (1960).

[7] M. Jacob and J. Mathews, Phys. Rev. **117**, 854 (1960).

[8] M. Gell-Mann and M. L. Goldberger, Proc. 1954 Glasgow Conf. on Nucl. and Meson Physics, Pergamon Press, London-N.Y. (1954).

[9] Hyman, Ely, Frisch, and Wahlig, Phys. Rev. Lett. **3**, 93 (1959).

[10] Bernardini, Hanson, Odian, Yamagata, Auerbach, and Filosofo, Nuovo cimento **18**, 1203 (1960).

[11] F. E. Low, Proc. 1958 Ann. Intern. Conf. on High Energy Physics at CERN, p. 98.

[12] Glasser, Seeman, and Stiller, Bull. Amer. Phys. Soc. **6**, 1 (1961).

[13] L. I. Lapidus and Chou Kuang-chao, JETP **41**, 294 (1961), Soviet Phys. JETP **14**, 210 (1962).

[14] L. I. Lapidus and Chou Kuang-chao, JETP **41**, 491 (1961), Soviet Phys. JETP **14**, 352 (1962)

[15] Y. Yamaguchi, Progr. Theor. Phys. **12**, 111 (1954). S. Minami and Y. Yamaguchi, Progr. Theor. Phys. **17**, 651 (1957).

[16] S. Minami, Photon-Proton Collision at 250-800 Mev (preprint).

[17] Govorkov, Gol'danskii, Karpukhin, Kutsenko, Pavlovskaya, DAN SSSR **111**, 988 (1956), Soviet Phys. "Doklady" **1**, 735 (1957). Gol'danskii, Karpukhin, Kutsenko, and Pavlovskaya, JETP **38**, 1695 (1960), Soviet Phys. JETP **11**, 1223 (1960). V. V. Pavlovskaya, Dissertation, Phys. Inst. Acad. Sci. (1961).

Translated by A. M. Bincer
261

PHYSICS

A Suggested Experiment to Determine the Spin of Y_1^* and the Parities of $\Lambda - \Sigma$, $\Lambda - Y_1^*$ and $\Sigma - Y_1^*$

In this note we propose an experiment to observe the final state $\Lambda - \pi$ resonance and the cusp arising from the near threshold effect of the Σ-production, and from this to determine the spin of Y_1^* ($J_{Y_1^*}$) and the relative parities of $\Lambda - \Sigma$, $\Lambda - Y_1^*$ and $\Sigma - Y_1^*$ ($P(\Lambda - \Sigma)$, $P(\Lambda - Y_1^*)$ and $P(\Sigma - Y_1^*)$). The centre of mass energy of the initial $\pi - P$ system should be around 1900 MeV (this corresponds to 1305 MeV of the incident pion in the laboratory system). Choose the events in which the energy of the kaon lies below 90 MeV and observe the correlation between resonance and the cusp. This energy region will just cover the Y_1^* resonance due to Y_1^* and the cusp due to

Σ-production. Furthermore, since the energy of the kaon is relatively low and much below the mass of K^* ($M_{K^*}\sim$880MeV,$Q_{K^*}\sim$250MeV), only the s-wave of kaon needs to be considered. Obviously, the cusp will appear in the $J = 1/2$ state with $\rho_{\Lambda\pi} = 0$ or $\rho_{\Lambda\pi} = 1$ according as $P(\Lambda - \Sigma)$ is even or odd. Since the position of the $\Sigma - \pi$ threshold lies below the $\Lambda - \pi$ resonance only by 55 MeV, and the background of the resonance is rather small (around 1 : 2 → 3 to the slope of this resonance). The cusp can only be observed when it appears in the same partial wave state as the resonance, and this will lead to a useful information about the $\Lambda - \Sigma$ relative parity and the spin of Y_1^*. Moreover, the relative angular momentum $\rho_{Y_1^*}$ of the $\Lambda - \pi$ system can be determined from the angular distribution and polarization measurements. From the above observations the spin of Y_1^* and the relative parities can be determined. We list below the different conclusion corresponding to all possible experimental situations.

(1) If the cusp is observed together with the resonance, then we have $J_{Y_1^*} = 1/2$, $P(\Sigma - Y_1^*) = -$, and if the $\rho_{Y_1^*}$ can further be determined (see (4) below), then $P(\Lambda - \Sigma) = (-)^{\rho_{Y_1^*}}$.

(2) If only the resonance is observed, and if the $J_{Y_1^*}$, $\rho_{Y_1^*}$ can further be determined (see (3) and (4) below), then

i. when $J_{Y_1^*} = 1/2$, we shall have $P(\Sigma - Y_1^*) = +$, $P(\Lambda - \Sigma) = (-)^{\rho_{Y_1^*}+1}$;

ii. when $J_{Y_1^*} = 3/2$, we may have $P(\Lambda - Y_1^*) = -$, since the $\rho_{Y_1^*} = 2$ state can be excluded in this energy.

(3) When the initial nucleon is unpolarized:

i. if $\frac{d\sigma}{d\Omega_{\Lambda\pi} dE_k}$ is isotropic and the hyperon is unpolarized, then we have $J_{Y_1^*} = 1/2$;

ii. if $\frac{d\sigma}{d\Omega_{\Lambda\pi} dE_k} \sim a\cos^2\theta + b\cos\theta + c$ and $\mathbf{P}_\Lambda \frac{d\sigma}{d\Omega_{\Lambda\pi} dE_k} \sim \sin 2\theta \frac{\mathbf{n}_{\Lambda\pi} \times \mathbf{n}_i}{|\mathbf{n}_{\Lambda\pi} \times \mathbf{n}_i|}$, then we have $J_{Y_1^*} = 3/2$, where $\cos\theta = \mathbf{n}_i \cdot \mathbf{n}_{\Lambda\pi}$, $\mathbf{n}_i$, $\mathbf{n}_{\Lambda\pi}$, and $\Omega_{\Lambda\pi}$ being respectively the incident direction of π-meson, the direction of the Λ-π relative motion, and the solid angle of the Λ-π system.

(4) When the initial nucleon is polarized and its polarization is orthogonal to $\mathbf{n}_i$, then in the case of $J_{Y_1^*} = 1/2$:

i. when $\mathbf{P}_\Lambda \frac{d\sigma}{d\Omega_{\Lambda\pi} dE_k} = \mathbf{P} \cdot \text{const.}$, we have $\rho_{Y_1^*} = 0$;

ii. when $\mathbf{P}_\Lambda \frac{d\sigma}{d\Omega_{\Lambda\pi} dE_k} \sim \mathbf{P} - 2\mathbf{n}_{\Lambda\pi}(\mathbf{n}_{\Lambda\pi} \cdot \mathbf{P})$ we have $\rho_{Y_1^*} = 1$, where $\mathbf{P}$ and $\mathbf{P}_\Lambda$ are the polarizations of the initial nucleon and the final hyperon respectively. It will be noted that (1) and (3) are two independent observations for $J_{Y_1^*}$; (2) and (4) are observations on relative parities.

Two of the authors (Su Zhao-bin and Gao Chong-shou) are indebted to Dr. Chou Kuang-chao for his kindly guidance, and also to Prof. Hu Ning for his interest and support.

Su Zhao-bin (苏肇冰)
Gao Chong-shou (高崇寿)
Chou Kuang-chao (周光召)

Peking University

Jan. 7, 1963

Vol. XXII No. 1 SCIENTIA SINICA January 1979

THE PURE GAUGE FIELDS ON A COSET SPACE

Chou Kuang-chao (周光召)
(*Institute of Theoretical Physics, Academia Sinica*)

Tu Tung-sheng (杜东生)
(*Institute of High Energy Physics, Academia Sinica*)

and Yean Tu-nan (阮图南)
(*University of Science and Technology of China*)

Received August 18, 1978.

Abstract

The concept of the pure gauge fields on a coset space is introduced. By using gauge fields on subgroup H, pure gauge fields on coset space G/H and the induced representation, a local gauge invariant Lagrangian theory on group G is constructed. The application of this theory to $SU_2 \times SU_2$ gauge theory, the σ model and the non-trivial topological property of the pure gauge field are discussed.

I. Introduction

Since the emerging of the non-Abelian gauge field theory unifying weak and electromagnetic interaction[1], the properties of the non-Abelian gauge field have been extensively investigated and the important progress has been made[2]. As is well known, even with the symmetry of the vacuum spontaneously broken and the Goldstone bosons absorbed through the Higgs mechanism, the non-Abelian gauge field theory can still be renormalized. Though in the unitary gauge, only physical particles appear, the theory is not obviously renormalizable. While in the Landau gauge the theory is manifestly renormalizable, it still contains fictitious particles, which seem to destroy unitarity. The theory is in fact both unitary and renormalizable as can be shown simply in the R_ξ gauge. The worst divergent diagrams are cancelled with each other. The undesirable effects of the fictitious particles are also cancelled.

Many low energy hadronic experiments showed that hadrons have not only SU_3 (or SU_4) symmetry but also $SU_2 \times SU_2$ (or $SU_3 \times SU_3$) chiral symmetry[3]. The pion is a pseudo-Goldstone boson resulting from the spontaneous breaking of the chiral symmetry. It was proved in [4] that when the chiral group is broken in Goldstone mode, the effective Lagrangian involving the pion field still possesses chiral symmetry in the approximation of the lowest order in breaking parameter and pion momentum. In this case, the pion field offers a non-linear realization[5] of the chiral group. If the pion field is taken as an elementary field offering a non-linear realization of the chiral group, the resultant Lagrangian has complicated non-linear terms. This Lagrangian gives some results which in the lowest order perturbation theory agree with experiments. But the theory is not renormalizable[6]. The linear σ model can be constructed, using linear representation of the chiral group. This model is renormalizable, but the pion field does not apparently have the characteristic of a non-linear representation of the chiral group. In this model, it is difficult to get results in the lowest order perturbation expansion in agreement with experiments.

If a gauge degree of freedom like that of the non-Abelian gauge field theory is introduced to make the theory gauge invariant, the theory might be manifestly renormalizable in one gauge (which is to be called the renormalizable gauge), while in another gauge (which is to be called the physical gauge) good physical results might be easily extracted by tree approximation.

For this purpose we introduce the concept of the pure gauge scalar fields on the coset space. In the case the global topological properties are trivial, the pure gauge scalar fields are the manifestation of the gauge degrees of freedom and can be eliminated by choosing a suitable gauge. By using the pure gauge scalar fields the renormalizable gauge can be connected with the physical gauge.

If the subgroup is $U(1)$, the gauge field on which is the electromagnetic field, and if monopole exists, the monopole and its electromagnetic field can be described in terms of the pure gauge scalar fields. In this way we can avoid introducing singular strings in the expression describing the vector potential of the E. M. fields of the monopole. Further the equation of motion of the monopole and the electromagnetic fields might be derived from the Lagrangian.

The plan of this paper is as follows: In section II we review briefly the induced representation of group G on its subgroup H, introduce the concept of the pure gauge field on the coset space, and discuss its transformation properties. In section III we discuss a physical system which is local gauge invariant with respect to the subgroup H. Using the pure gauge fields on the coset space we construct a local gauge invariant Lagrangian with respect to the whole group G. In section IV, for the sake of illustration, we construct an $SU_2 \times SU_2$ gauge invariant theory, which coincides with linear σ model in the renormalizable gauge and with the non-linear chiral model in the physical gauge except a few additional terms which account for the renormalizability. In section V, we discuss how to describe monopoles by means of pure gauge fields on the coset space. In section VI, we give the coset element parametrization with respect to the subgroup $U_m \times U_{n-m}$ of the group U_n. The corresponding expressions for the pure gauge fields on the coset space are also given.

II. The Induced Representation on the Subgroup and the Pure Gauge Fields on the Coset Space

Consider a transformation group G underlying a physical system. Let H denote one of the subgroups of G, and g and h designate any element in G and H respectively, namely,

$$G = \{\cdots g \cdots\}, \quad H = \{\cdots h \cdots\}.$$

ϕ denotes any representative element of the coset space with respect to subgroup H, namely,

$$G/H=\{\cdots\phi\cdots\}.$$

According to the theory of Lie group, for any element $g\in G$, there is the unique left coset decomposition,

$$g=\phi h. \tag{2.1}$$

(For right coset decomposition the discussion is similar.) Eq. (2.1) represents a decomposition of the bundle manifold $G(G/H, H)$ into its base manifold G/H and fibre H. By choosing suitable local coordinate system in these manifolds we get the corresponding parametrization of (2.1).

Similarly, for a specified $\phi_0\in G/H$, we have

$$g\phi_0=\phi(g,\phi_0)h(g,\phi_0), \tag{2.2}$$

where $\phi(g,\phi_0)$ and $h(g,\phi_0)$ belong to the coset space and the subgroup respectively. We can take $h(g,\phi_0)$ as the image of g on subgroup H through ϕ_0, that is,

$$g\xrightarrow{\phi_0}h(g,\phi_0). \tag{2.3}$$

According to (2.2), there is

$$g'g\phi_0=\phi(g'g,\phi_0)h(g'g,\phi_0). \tag{2.4}$$

Multiplying (2.2) by g' from the left-hand side, we have

$$\begin{aligned} g'g\phi_0&=g'\phi(g,\phi_0)h(g,\phi_0)\\ &=\phi(g',\phi(g,\phi_0))h(g',\phi(g,\phi_0))h(g,\phi_0). \end{aligned} \tag{2.5}$$

Comparing (2.4) with (2.5), we have

$$h(g'g,\phi_0)=h(g',\phi(g,\phi_0))h(g,\phi_0), \tag{2.6}$$

$$\phi(g'g,\phi_0)=\phi(g',\phi(g,\phi_0)). \tag{2.7}$$

We have therefore the following mappings:

g acts on ϕ_0 to give the image $g\to h(g,\phi_0)$, (2.8)

g' acts on g successively to give the image $g'\to h(g',\phi(g,\phi_0))$, (2.8)′

$g'g$ acts on ϕ_0 to give the image $g'g\to h(g'g,\phi_0)$. (2.8)″

Eq. (2.6) gives the multiplicative combination law of the images of G on H. Eqs (2.1), (2.2), (2.6), and (2.8) are the rules concerning the mapping of group G on its subgroup H through $\phi_0\in G/H$. It is defined as the induced representation[7].

Assume that $A_\mu{}^i$($i=1,2\ldots n_H$, n_H is the number of generators of subgroup H) are gauge fields on subgroup H and are transformed according to the induced representation of G on its subgroup H. Define

$$\hat{A}_\mu=-i\frac{\kappa}{2}\sum_{i=1}^{n_H}A_\mu^i\lambda_i, \tag{2.9}$$

where λ_i are generators of subgroup H and κ is the coupling constant. Introduce the pure gauge field ϕ_0 which is any specified element in G/H. Let us further define

$$\hat{B}_\mu = \phi_0(\partial_\mu + \hat{A}_\mu)\phi_0^{-1}, \tag{2.10}$$

and discuss the transformation property of $\hat{B}_\mu$. Making the local gauge transformation g of the group G as usual, we have

$$\hat{B}_\mu \to \hat{B}'_\mu = g(\partial_\mu + \hat{B}_\mu)g^{-1}. \tag{2.11}$$

Owing to (2.10) and (2.2),

$$\hat{B}'_\mu = g\{\partial_\mu + \phi_0(\partial_\mu + A_\mu)\phi_0^{-1}\}g^{-1} = \phi'(\partial_\mu + \hat{A}'_\mu)\phi'^{-1}, \tag{2.12}$$

where

$$\phi' = \phi(g, \phi_0), \quad \hat{A}'_\mu = h(g, \phi_0)(\partial_\mu + \hat{A}_\mu)h^{-1}(g, \phi_0). \tag{2.13}$$

Making local gauge transformation g' successively, from (2.5), (2.13), we get

$$\begin{aligned} \hat{B}''_\mu &= g'(\partial_\mu + \hat{B}'_\mu)g'^{-1} = g'\phi'(\partial_\mu + \hat{A}'_\mu)(g'\phi')^{-1} \\ &= \phi(g', \phi(g, \phi_0))\{\partial_\mu + h(g', \phi(g, \phi_0))h(g, \phi_0)(\partial_\mu + \hat{A}_\mu) \\ &\quad \cdot h^{-1}(g, \phi_0)h^{-1}(g', \phi(g, \phi_0))\}\phi^{-1}(g', \phi(g, \phi_0)). \end{aligned} \tag{2.14}$$

If the product $g'g$ is taken as one gauge transformation, we have

$$\begin{aligned} \hat{B}''_\mu &= g'g(\partial_\mu + \hat{B}_\mu)(g'g)^{-1} \\ &= \phi(g'g, \phi_0)\{\partial_\mu + h(g'g, \phi_0)(\partial_\mu + \hat{A}_\mu)h^{-1}(g'g, \phi_0)\}\phi^{-1}(g'g, \phi_0). \end{aligned} \tag{2.15}$$

Thus, making two gauge transformations g and g' successively is equivalent to making one gauge transformation $g'g$ according to (2.6) and (2.7). In both cases we obtain the same potential $\hat{B}''_\mu$. In the sense of induced representation, the gauge transformation is closed. That is the basis of constructing a gauge invariant Lagrangian in terms of pure gauge fields on the coset space.

III. Construction of Local Gauge Invariant Lagrangian on the Group G in Terms of Pure Gauge Fields on Coset Space

Consider a Lagrangian which is global gauge invariant on the group G and local gauge invariant on the subgroup H.

$$\mathscr{L}(\varphi(x), D_\mu^{(A)}\varphi(x), \hat{F}_{\mu\nu}^{(A)}), \tag{3.1}$$

where

$$\varphi(x) = \begin{pmatrix} \varphi_1(x) \\ \varphi_2(x) \\ \vdots \end{pmatrix}$$

are field quantities providing a basis for a representation of the group G, and

$$D_\mu^{(A)} = (\partial_\mu + \hat{A}_\mu), \tag{3.2}$$

$$\hat{F}_{\mu\nu}^{(A)} = \partial_\mu\hat{A}_\nu - \partial_\nu\hat{A}_\mu + [\hat{A}_\mu, \hat{A}_\nu] \tag{3.3}$$

are the covariant operator and the field strength of the gauge fields $\hat{A}_\mu$ on subgroup H

respectively. According to the global gauge invariant property of the Lagrangian, for any element g of G, there is

$$\mathscr{L}(g\varphi(x), gD_\mu^{(A)}\varphi(x), g\hat{F}_{\mu\nu}^{(A)}g^{-1}) = \mathscr{L}(\varphi(x), D_\mu^{(A)}\varphi(x), \hat{F}_{\mu\nu}^{(A)}). \tag{3.4}$$

Notice that, if we substitute g by $g(x)$, considering the absence of derivative of $g(x)$ in (3.4), the identity will still hold.

Put $\phi_0(x)$ as an element of the coset space G/H. Owing to (3.4), there is

$$\mathscr{L}(\phi_0(x)\varphi(x), \phi_0(x)D_\mu^{(A)}\varphi(x), \phi_0(x)\hat{F}_{\mu\nu}^{(A)}\phi_0^{-1}(x)) = \mathscr{L}(\varphi(x), D_\mu^{(A)}\varphi(x), \hat{F}_{\mu\nu}^{(A)}). \tag{3.5}$$

Introducing the new field quantities,

$$\begin{aligned} \Phi(x) &= \phi_0(x)\varphi(x), \\ \hat{B}_\mu(x) &= \phi_0(x)(\partial_\mu + \hat{A}_\mu)\phi_0^{-1}(x), \end{aligned} \tag{3.6}$$

we can easily verify

$$\begin{aligned} D_\mu^{(B)} &\equiv \partial_\mu + \hat{B}_\mu = \phi_0(x)D_\mu^{(A)}\phi_0^{-1}(x), \\ \hat{F}_{\mu\nu}^{(B)} &\equiv \partial_\mu\hat{B}_\nu - \partial_\nu\hat{B}_\mu + [\hat{B}_\mu, \hat{B}_\nu] = \phi_0(x)\hat{F}_{\mu\nu}^{(A)}\phi_0^{-1}(x). \end{aligned} \tag{3.7}$$

By using (3.6) and (3.7), Eq. (3.5) can be rewritten as

$$\mathscr{L}(\Phi(x), D_\mu^{(B)}\Phi(x), \hat{F}_{\mu\nu}^{(B)}) = \mathscr{L}(\varphi(x), D_\mu^{(A)}\varphi(x), \hat{F}_{\mu\nu}^{(A)}). \tag{3.8}$$

Now let us prove that the Lagrangian

$$\mathscr{L}(\Phi(x), D_\mu^{(B)}\Phi(x), \hat{F}_{\mu\nu}^{(B)}) \tag{3.9}$$

is a local gauge invariant on the group G. In fact, by making local gauge transformation $g(x) \in G$,

$$\begin{aligned} \Phi(x) &\to \Phi'(x) = g(x)\Phi(x), \\ \hat{B}_\mu(x) &\to \hat{B}'_\mu(x) = g(x)(\partial_\mu + \hat{B}_\mu)g^{-1}(x), \end{aligned} \tag{3.10}$$

one can easily verify that

$$\begin{aligned} D_\mu^{\prime(B)}\Phi'(x) &= g(x)D_\mu^{(B)}\Phi(x), \\ F_{\mu\nu}^{(B)\prime} &= g(x)\hat{F}_{\mu\nu}^{(B)}g(x). \end{aligned} \tag{3.11}$$

Thus, we have

$$\begin{aligned} &\mathscr{L}(\Phi'(x), D_\mu^{\prime(B)}\Phi'(x), \hat{F}_{\mu\nu}^{\prime(B)}) \\ &\quad = \mathscr{L}(g(x)\Phi(x), g(x)D_\mu^{(B)}\Phi(x), g(x)\hat{F}_{\mu\nu}^{(B)}g^{-1}(x)), \end{aligned} \tag{3.12}$$

where there is no derivative of $g(x)$. According to (3.4), (3.8) and (3.12), we have

$$\mathscr{L}(\Phi'(x), D_\mu^{(B)\prime}\Phi'(x), \hat{F}_{\mu\nu}^{(B)\prime}) = \mathscr{L}(\Phi(x), D_\mu^{(B)}\Phi(x), \hat{F}_{\mu\nu}^{(B)}). \tag{3.13}$$

That is to say, the Lagrangian (3.9) is a local gauge invariant under the group G.

From (3.5)—(3.8) it is seen that the Lagrangian (3.9) and (3.1) differ with only a gauge transformation. By making the local gauge transformation $g(x) = \phi_0^{-1}(x)$, the $\mathscr{L}$ in (3.8) will become $\mathscr{L}$ in (3.1). It is equivalent to taking the gauge condition as $\phi_0(x) = 1$ in the beginning.

Assume that the number of generators of the group $G(H)$ is $n_G(n_H)$, then we have n_G-n_H parameters describing the coset space G/H. Through the element $\phi_0(x)$ of the coset space we have introduced n_G-n_H field quantities which provide a basis for a non-linear realization of the group G, and which are scalars under Lorentz transformation. (We assume that G is an internal symmetry group).

In the usual gauge theory, n_G-n_H further vector fields on the coset space are introduced for the gauge invariance on the group G, but in the present case, only n_G-n_H scalar fields are to be introduced. It will be proved below that these scalar fields represent gauge degrees of freedom. They are not independent field quantities and can thus be eliminated through gauge transformation.

For illustration let us evaluate the conjugate momentums of $\Phi(x)$ and $\hat{B}_\mu(x)$. Choose a coordinate system in which $\phi_0(x)\partial_\mu\phi_0^{-1}(x)$ can be written as

$$\phi_0\partial_\mu\phi_0^{-1} = \sum_{j=1}^{n_G-n_H} \hat{f}_j(\alpha_i(x))\partial_\mu\alpha_j(x), \tag{3.14}$$

where $\alpha_j(x)$ $(j = 1, 2\cdots n_G - n_H)$ are new pure gauge fields introduced on the coset space. From $\hat{F}^{(B)}_{\mu\nu} = \phi_0\hat{F}^{(A)}_{\mu\nu}\phi_0^{-1}$ it is easily seen that $\hat{F}^{(B)}_{\mu\nu}$ does not involve the derivatives of ϕ_0 and $\alpha_j(x)$. Only the term $D^{(B)}_\mu\ \Phi(x)$ in $\mathscr{L}$ involves $\partial_\mu\alpha_j(x)$. Let P_{α_j} denote the conjugate momentum of $\alpha_j(x)$, then

$$P_{\alpha_j} = \frac{\delta\mathscr{L}}{i\delta D^{(B)}_4\Phi(x)}\hat{f}_j(\alpha_i(x))\Phi(x), \tag{3.15}$$

while the conjugate momentum of $\Phi(x)$ is

$$P_\Phi = \frac{\delta\mathscr{L}}{i\delta D^{(B)}_4\Phi(x)}. \tag{3.16}$$

Thus we have

$$P_{\alpha_j} = P_\Phi\hat{f}_j(\alpha_i(x))\Phi(x), \quad j = 1, 2, \cdots n_G - n_H. \tag{3.17}$$

Eq. (3.17) shows that the new Lagrangian in (3.9) gives n_G-n_H constraint conditions which are generated by gauge invariance of the Lagrangian (3.9) on the coset space G/H. These constraints make $\alpha_i(x)$ not independent.

Owing to the constraints or gauge invariance it is necessary to introduce gauge conditions in the course of the quantization of the theory described by the Lagrangian (3.9). Eq. (3.17) makes necessary the introduction of n_G-n_H gauge conditions, while for vector fields $\hat{A}_\mu$ on subgroup H, n_H gauge conditions must be introduced. Thus there are n_G gauge conditions altogether.

If we choose n_G-n_H gauge conditions as

$$\alpha_j(x) = 0, \quad j = 1, 2, \cdots n_G - n_H, \tag{3.18}$$

the Lagrangian (3.9) is restored to (3.1).

However, both (3.9) and (3.1) describe the same physical system. It is justified to ask, since the Lagrangian of (3.9) is equivalent to that of (3.1), why do we introduce the concept of the pure gauge fields on the coset space? The answer is as follows:

In studying the theory with the chiral symmetry of $SU_2 \times SU_2$ or $SU_3 \times SU_3$, we know that when elementary particles are basis of the linear representation of the chiral group, it is easy to construct a renormalizable Lagrangian. But the vacuum is not symmetrical under the chiral transformation. When the symmetry of the vacuum is spontaneously broken, π, K mesons emerge as Goldstone bosons which offer the non-linear realizations of the chiral group.

It is difficult to prove the renormalizability of the theory with non-linear Lagrangian. The simplest Lagrangian of the non-linear realization of the σ model has been proved to be non-renormalizable.

If we start from linear representation of the group G, we may write down the renormalizable Lagrangian (for instance in the form (3.1)). Assume that the vaccum has only the symmetry of the subgroup H and that the spontaneous breaking happens on the coset space G/H, then there will be n_G-n_H Goldstone scalar fields in the system[3]. These scalar fields are collective excitations of the system and should be described by the coset space element $\phi_0(x)$. In this case the pure gauge fields on the coset space $\alpha_j(x)$ represent the Goldstone fields. They form a manifold on which the non-linear realization of the group G operates. They have a clear physical meaning. We can choose in (3.9) a gauge which keeps $\alpha_j(x)$ and eliminates the surplus degrees of freedom. So a Lagrangian constructed of $\alpha_j(x)$ and other field quantities is obtained. This Lagrangian differs from the apparently renormalizable Lagrangian only by a gauge transformation. Thus the theory described by the new Lagrangian built in terms of $\alpha_j(x)$ might also be renormalizable. We call the gauge $\alpha_j(x) = 0$ the renormalizable gauge, and the gauge, in which $\alpha_j(x)$ are kept as basic field quantities with the surplus degrees of freedom eliminated, the physical gauge. Notice that $\alpha_j(x)$ have a real physical meaning only when the vacuum is broken spontaneously on coset space G/H. Only in such case the term "physical gauge" is valid. In the next section we shall discuss the Lagrangian in both kinds of gauges for the case of $SU_2 \times SU_2$ chiral group.

IV. $SU_2 \times SU_2$ Pure Gauge Fields on the Coset Space and the σ Model

As an example we discuss the connection between the $SU_2 \times SU_2$ pure gauge fields on the coset space and the σ model.

According to the general theory given in Section III, $SU_2 \times SU_2$ gauge invariant Lagrangian can be written as

$$\mathscr{L} = -\frac{1}{4} f^i_{\mu\nu} f^i_{\mu\nu} - \frac{1}{4} T_r\{(\partial_\mu \Phi^+ + \hat{B}_{\mu L}\Phi^+ - \Phi^+\hat{B}_{\mu R})(\partial_\mu \Phi + \hat{B}_{\mu R}\Phi - \Phi\hat{B}_{\mu L})\}$$
$$- V\left(\frac{1}{2} T_r \Phi^+\Phi\right) - \bar{\psi} r_\mu(\partial_\mu + \hat{B}_\mu)\psi - G\bar{\psi}\left(\frac{1+r_5}{2}\Phi + \frac{1-r_5}{2}\Phi^+\right)\psi, \quad (4.1)$$

where

$$f^i_{\mu\nu} = \partial_\mu B^i_\nu - \partial_\nu B^i_\mu + \kappa f_{ijk} B^j_\mu B^k_\nu \quad (4.2)$$

is the gauge field strength on the group G;

$$\hat{B}_\mu = \phi_0(\partial_\mu + \hat{A}_\mu)\phi_0^{-1}, \quad \phi_0 \in G/H. \quad (4.3)$$

$\hat{B}_{\mu L}$ and $\hat{B}_{\mu R}$ are related to $\hat{B}_\mu$,

$$\hat{B}_\mu = \frac{1+r_5}{2}\hat{B}_{\mu L} + \frac{1-r_5}{2}\hat{B}_{\mu R}. \quad (4.4)$$

It is easy to prove that the Lagrangian in (4.1) is gauge invariant under the following gauge transformation:

$$\begin{aligned}
&\psi \to \psi' = g\psi = e^{i\hat{a}r_5} e^{i\hat{b}}\psi,\\
&\Phi \to \Phi' = e^{-i\hat{a}} e^{i\hat{b}} \Phi e^{-i\hat{b}} e^{-i\hat{a}},\\
&\hat{B}_\mu \to \hat{B}'_\mu = e^{i\hat{a}r_5} e^{i\hat{b}}(\partial_\mu + \hat{B}_\mu)e^{-i\hat{b}} e^{-i\hat{a}r_5} = \frac{1+r_5}{2}\hat{B}'_{\mu L} + \frac{1-r_5}{2}\hat{B}'_{\mu R},\\
&\hat{B}_{\mu L} \to \hat{B}'_{\mu L} = e^{i\hat{a}} e^{i\hat{b}}(\partial_\mu + \hat{B}_{\mu L})e^{-i\hat{b}} e^{-i\hat{a}},\\
&\hat{B}_{\mu R} \to \hat{B}'_{\mu R} = e^{-i\hat{a}} e^{i\hat{b}}(\partial_\mu + \hat{B}_{\mu R})e^{-i\hat{b}} e^{i\hat{a}},\\
&\hat{a} = \boldsymbol{a}(x)\cdot\boldsymbol{\tau}/2, \quad \hat{b} = \boldsymbol{b}(x)\cdot\boldsymbol{\tau}/2, \quad \boldsymbol{\tau} \text{ are Pauli matrices.}
\end{aligned} \quad (4.5)$$

If we take the subgroup $H \equiv 1$, the identity element, the coset space $G/H \equiv G = \{\ldots g \ldots\}$, where

$$g = e^{i\hat{\beta}r_5} e^{i\hat{a}}, \quad \hat{\beta} = \boldsymbol{\beta}(x)\cdot\boldsymbol{\tau}/2, \quad \hat{a} = \boldsymbol{a}(x)\cdot\boldsymbol{\tau}/2. \quad (4.6)$$

There are then:

$$\begin{aligned}
&\hat{A}_\mu = 0,\\
&\hat{B}_\mu = e^{i\hat{\beta}r_5} e^{i\hat{a}} \partial_\mu e^{-i\hat{a}} e^{-i\hat{\beta}r_5},\\
&\hat{B}_{\mu L} = e^{i\hat{\beta}} e^{i\hat{a}} \partial_\mu e^{-i\hat{a}} e^{-i\hat{\beta}},\\
&\hat{B}_{\mu R} = e^{-i\hat{\beta}} e^{i\hat{a}} \partial_\mu e^{-i\hat{a}} e^{i\hat{\beta}}.
\end{aligned} \quad (4.7)$$

Starting from (4.1) and making the gauge transformation $g(x) = e^{-i\hat{a}}, e^{-i\hat{\beta}r_5}$, we get

$$\hat{B}'_\mu = \hat{B}'_{\mu L} = \hat{B}'_{\mu R} = 0, \quad (4.8)$$

where we have used (4.7). After gauge transformation,

$$\begin{aligned}
\mathscr{L} = &-\frac{1}{4}T_r\{(\partial_\mu\Phi'^+ + \hat{B}'_{\mu L}\Phi'^+ - \Phi'^+\hat{B}'_{\mu R})(\partial_\mu\Phi' + \hat{B}'_{\mu R}\Phi' - \Phi'\hat{B}'_{\mu L})\}\\
&- V\left(\frac{1}{2}T_r\Phi'^+\Phi'\right) - \bar{\psi}' r_\mu(\partial_\mu + \hat{B}'_\mu)\psi' - G\bar{\psi}'\left(\frac{1+r_5}{2}\Phi' + \frac{1-r_5}{2}\Phi'^+\right)\psi',
\end{aligned} \quad (4.9)$$

$$\begin{aligned}
= &-\frac{1}{4}T_r(\partial_\mu\Phi'^+\partial_\mu\Phi') - V\left(\frac{1}{2}T_r\Phi'^+\Phi'\right) - \bar{\psi}' r_\mu\partial_\mu\psi'\\
&- G\bar{\psi}'\left(\frac{1+r_5}{2}\Phi' + \frac{1-r_5}{2}\Phi'^+\right)\psi',
\end{aligned} \quad (4.10)$$

where (4.8) has been used. The Lagrangian (4.10) is just that of the linear σ model. Thus $SU_2 \times SU_2$ gauge invariant Lagrangian in the renormalizable gauge becomes the usual linear σ model.

Starting from (4.9) and making the gauge transformation

$$g'' = e^{i\hat{\theta}r_5}, \tag{4.11}$$

we have then

$$\left.\begin{aligned} &\psi' \to \psi'' = g''(x)\psi', \\ &\Phi' \to \Phi'' = e^{-i\hat{\theta}}\Phi' e^{-i\hat{\theta}} = \rho = \Phi''^{+}, \\ &\hat{B}'_\mu \to \hat{B}''_\mu = e^{i\hat{\theta}r_5}\partial_\mu e^{-i\hat{\theta}r_5}, \\ &\hat{B}'_{\mu L} \to \hat{B}''_\mu = e^{i\hat{\theta}}\partial_\mu e^{-i\hat{\theta}}, \\ &\hat{B}'_{\mu R} \to \hat{B}''_\mu = e^{-i\hat{\theta}}\partial_\mu e^{i\hat{\theta}}, \end{aligned}\right\} \tag{4.12}$$

$$\begin{aligned} \mathscr{L} \to \mathscr{L} = &-\frac{1}{4}T_r\{[\partial_\mu\rho + \rho(e^{i\hat{\theta}}\partial_\mu e^{-i\hat{\theta}} - e^{-i\hat{\theta}}\partial_\mu e^{i\hat{\theta}})] \\ &\cdot[\partial_\mu\rho - \rho(e^{i\hat{\theta}}\partial_\mu e^{-i\hat{\theta}} - e^{-i\hat{\theta}}\partial_\mu e^{i\hat{\theta}})]\} \\ &- V(\rho^2) - \bar{\psi}''r_\mu(\partial_\mu + e^{i\hat{\theta}r_5}\partial_\mu e^{-i\hat{\theta}r_5})\psi'' - G\bar{\psi}''\psi''\rho \end{aligned} \tag{4.13}$$

$$\begin{aligned} = &-\frac{1}{2}(\partial_\mu\rho)^2 - \frac{1}{4}\rho^2 T_r(\partial_\mu e^{-2i\hat{\theta}}\partial_\mu e^{2i\hat{\theta}}) - V(\rho^2) \\ &- \bar{\psi}''r_\mu(\partial_\mu + e^{i\hat{\theta}r_5}\partial_\mu e^{-i\hat{\theta}r_5})\psi'' - G\bar{\psi}''\psi''\rho. \end{aligned} \tag{4.14}$$

We call the gauge transformation (4.12) the physical gauge because in that gauge the $\hat{\theta}$ is chosen in such a way that Φ'' is diagonal. Here we assume, like the usual σ model, that

$$\Phi' = \sigma' + i\boldsymbol{\tau}\cdot\boldsymbol{\pi}', \tag{4.15}$$

where σ' is a scalar field, $\boldsymbol{\pi}'$ is a posudoscalar field and both are real; ρ is a real scalar field.

Lagrangian (4.14) can be obtained from another point of view.

Applying the canonical transformation

$$\begin{aligned} \psi' &= e^{-i\hat{\theta}r_5}\psi'', \\ \Phi' &= \rho e^{2i\hat{\theta}} \end{aligned} \tag{4.16}$$

to (4.10), we can get the same Lagrangian (4.14). Thus the Lagrangian (4.14) can be obtained from (4.1) through gauge transformation in one way and from (4.10) through canonical transformation in the other.

There is

$$\Phi' = \rho e^{2i\hat{\theta}} = \rho e^{i\boldsymbol{\tau}\cdot\boldsymbol{\theta}} = \rho\left(\cos\theta + i\boldsymbol{\tau}\cdot\frac{\boldsymbol{\theta}}{\theta}\sin\theta\right). \tag{4.17}$$

Assume that $V(\rho^2)$ has a minimum at $\rho = \rho_0$.

Put

$$\rho = \rho_0 + \rho', \tag{4.18}$$

$$\rho_0 \frac{\boldsymbol{\theta}}{\theta} \sin\theta = \boldsymbol{\pi}, \tag{4.19}$$

then there is

$$\sin^2\theta = \boldsymbol{\pi}^2/\rho_0^2, \quad \cos^2\theta = 1 - \boldsymbol{\pi}^2/\rho_0^2, \tag{4.20}$$

$$\Phi' = \rho(\sqrt{1 - \boldsymbol{\pi}^2/\rho_0^2} + i\boldsymbol{\tau}\cdot\boldsymbol{\pi}/\rho_0). \tag{4.21}$$

Comparing both sides of (4.21), we have

$$\sigma' = \rho(1 - \boldsymbol{\pi}^2/\rho_0^2)^{1/2}, \quad \boldsymbol{\pi}' = \frac{\rho}{\rho_0}\boldsymbol{\pi}. \tag{4.22}$$

Substituting (4.17)—(4.22) into (4.14), we obtain

$$\mathscr{L} = \mathscr{L}_{\rho'} + \mathscr{L}_{\pi} + \mathscr{L}_{N} + \mathscr{L}_{\rho'\pi} + \mathscr{L}_{\rho'N} + \mathscr{L}_{\pi N}, \tag{4.23}$$

with

$$\mathscr{L}_{\rho'} = -\frac{1}{2}(\partial_\mu\rho')^2 - V((\rho_0 + \rho')^2), \tag{4.24}$$

$$\mathscr{L}_{\pi} = -\frac{1}{2}(\partial_\mu\boldsymbol{\pi})^2 - \frac{1}{2}\frac{(\boldsymbol{\pi}\cdot\partial_\mu\boldsymbol{\pi})^2}{\rho_0^2 - \boldsymbol{\pi}^2}, \tag{4.25}$$

$$\mathscr{L}_{N} = -\bar{\psi}''(\gamma_\mu\partial_\mu + M)\psi'', \quad M = G\rho_0, \tag{4.26}$$

$$\mathscr{L}_{\pi N} = -\bar{\psi}''\gamma_\mu B_\mu(\rho_0\cdot\boldsymbol{\pi})\psi'', \tag{4.27}$$

$$B_\mu(\rho_0, \boldsymbol{\pi}) = -\frac{i}{2\rho_0}\gamma_5\boldsymbol{\tau}\cdot\partial_\mu\boldsymbol{\pi} - \frac{i}{2}\gamma_5\frac{(\boldsymbol{\tau}\cdot\boldsymbol{\pi})(\boldsymbol{\pi}\cdot\partial_\mu\boldsymbol{\pi})}{\rho_0\boldsymbol{\pi}^2}\left(\left(1 - \frac{\boldsymbol{\pi}^2}{\rho_0^2}\right)^{-1/2} - 1\right)$$

$$+ \frac{i}{2}\left(1 - \left(1 - \frac{\boldsymbol{\pi}^2}{\rho_0^2}\right)^{1/2}\right)\frac{\boldsymbol{\tau}\cdot(\boldsymbol{\pi}\times\partial_\mu\boldsymbol{\pi})}{\boldsymbol{\pi}^2}. \tag{4.28}$$

$$\left.\begin{aligned} \mathscr{L}_{\rho'\pi} &= -\frac{2\rho_0\rho' + \rho'^2}{2\rho_0^2}\left[(\partial_\mu\boldsymbol{\pi})^2 + \frac{(\boldsymbol{\pi}\cdot\partial_\mu\boldsymbol{\pi})^2}{\rho_0^2 - \boldsymbol{\pi}^2}\right] \\ \mathscr{L}_{\rho'N} &= -G\bar{\psi}''\psi''\rho' \end{aligned}\right\}, \tag{4.29}$$

Expanding (4.25) in terms of $\boldsymbol{\pi}^2/\rho_0^2$, we have

$$\mathscr{L}_{\pi} = -\frac{1}{2}(\partial_\mu\boldsymbol{\pi})^2 - \frac{1}{2\rho_0^2}(\boldsymbol{\pi}\cdot\partial_\mu\boldsymbol{\pi})^2 - \frac{1}{2\rho_0^2}(\boldsymbol{\pi}\cdot\partial_\mu\boldsymbol{\pi})^2\frac{\boldsymbol{\pi}^2}{\rho_0^2}\cdots$$
$$-\cdots. \tag{4.30}$$

The Lagrangian (4.30) is exactly the same as the Lagrangian of the non-linear σ model. But in (4.23) there is the additional term $\mathscr{L}_{\rho'\pi}$ which also gives a contribution to the π-π scattering. Besides, Lagrangian (4.23) is gauge invariant and can be changed into a renormalizable one through a gauge transformation. Thus the theory described by the Lagrangian (4.23) might be renormalizable.

Now, we shall derive the Goldberger-Treiman relation in the lowest order approximation from our model.

According to (4.10), the axial vector current is

$$j_{\mu 5} = -i\bar{\Phi}' r_\mu r_5 \frac{\boldsymbol{\tau}}{2} \psi' + \sigma' \partial_\mu \boldsymbol{\pi}' - \boldsymbol{\pi}' \partial_\mu \sigma'. \tag{4.31}$$

Substituting (4.16)—(4.22) into (4.31) we have

$$\begin{aligned} j_{\mu 5} = & -i\bar{\Phi}'' r_\mu r_5 \frac{1}{2} \left\{ \sqrt{1 - \frac{\boldsymbol{\pi}^2}{\rho_0^2}}\, \boldsymbol{\tau} + \left(1 - \sqrt{1 - \frac{\boldsymbol{\pi}^2}{\rho_0^2}}\right) \frac{(\boldsymbol{\tau} \cdot \boldsymbol{\pi})\boldsymbol{\pi}}{\boldsymbol{\pi}^2} \right. \\ & \left. - r_5 \frac{\boldsymbol{\tau} \times \boldsymbol{\pi}}{\rho_0} \right\} \psi'' + \rho \sqrt{1 - \frac{\boldsymbol{\pi}^2}{\rho_0^2}}\, \partial_\mu \left(\frac{\rho}{\rho_0} \boldsymbol{\pi} \right) - \frac{\rho}{\rho_0} \boldsymbol{\pi} \partial_\mu \left(\rho \sqrt{1 - \frac{\boldsymbol{\pi}^2}{\rho_0^2}} \right) \end{aligned} \tag{4.32}$$

$$\approx -i\bar{\Phi}'' r_\mu r_5 \frac{\boldsymbol{\tau}}{2} \psi'' + \rho_0 \partial_\mu \boldsymbol{\pi} + \cdots. \tag{4.33}$$

The β decay of nucleons should be determined by $G_A j_{\mu 5}$. As can be seen from (4.33), the corresponding coupling constant of π decay is

$$f_\pi = \rho_0 G_A. \tag{4.34}$$

Substituting $M_N = \rho_0 G$ into the above equation we have

$$f_\pi G = G_A M_N. \tag{4.35}$$

As will be shown in the following, G is just the π-N-N coupling constant, so that (4.35) is just the G-T relation.

In fact, in the lowest order approximation there is:

$$\mathscr{L}^{(0)}_{\pi N} = \frac{i}{2\rho_0} \bar{\Phi}'' r_\mu r_5 \boldsymbol{\tau} \cdot \partial_\mu \boldsymbol{\pi} \psi''. \tag{4.36}$$

Noticing that

$$\begin{aligned} & \partial_\mu [\bar{\Phi}'' r_\mu r_5 \boldsymbol{\tau} \psi''] \cdot \boldsymbol{\pi} - \partial_\mu [\bar{\Phi}'' r_\mu r_5 \boldsymbol{\tau} \psi'' \cdot \boldsymbol{\pi}] = -\bar{\Phi}'' r_\mu r_5 \boldsymbol{\tau} \psi'' \\ & \cdot \partial_\mu \boldsymbol{\pi} \bar{\Phi}'' r_\mu r_5 \boldsymbol{\tau} \cdot \partial_\mu \boldsymbol{\pi} \psi'' = -2\rho_0 G \bar{\Phi}'' r_5 \boldsymbol{\tau} \cdot \boldsymbol{\pi} \psi'', \end{aligned} \tag{4.37}$$

we get

$$\mathscr{L}^{(0)}_{\pi N} = -iG\bar{\Phi}'' r_5 \boldsymbol{\tau} \psi'' \cdot \boldsymbol{\pi}. \tag{4.38}$$

So, G is just the π-N-N coupling constant.

V. The Pure Gauge Fields on Coset Space and Monopole

In this section we shall discuss the example with $G = SU_N$, $H = U(1)$ and $U(1)$ gauge potential as the electromagnetic potential.

Define

$$\hat{A}_\mu = ieA_\mu \hat{I}_0, \tag{5.1}$$

where $\hat{I}_0$ is the generator of the subgroup $U(1)$. Introduce

$$\hat{B}_\mu = \phi(\partial_\mu + \hat{A}_\mu)\phi^{-1}, \tag{5.2}$$

where

$$\phi \in \frac{SU(N)}{U(1)}.$$

When the monopole exists, A_μ has a Dirac string.

It is convenient to use $\hat{B}_\mu$ and ϕ as the dynamical variables. Define

$$\mathscr{B}_\mu = \frac{1}{ieN} T_r(\hat{B}_\mu \hat{I}(x)), \tag{5.3}$$

where

$$\hat{I}(x) = \phi(x)\hat{I}_0\phi^{-1}(x). \tag{5.4}$$

For simplicity, we choose

$$\hat{I}_0 = \frac{1}{\sqrt{N-1}}\begin{pmatrix} 1 & & & & \\ & 1 & & & \\ & & \ddots & & \\ & & & 1 & \\ & & & & 1-N \end{pmatrix}. \tag{5.5}$$

then we have

$$\hat{I}_0^2 = a\hat{I}_0 + b, \quad a = \frac{2-N}{\sqrt{N-1}}, \quad b = 1. \tag{5.6}$$

For the other choice of $\hat{I}_0$, we can deal with it in the same way. Multiply both sides of (5.2) by $\hat{I}(x)$ from the right-hand side and take trace, and we obtain

$$A_\mu = \mathscr{B}_\mu - \frac{1}{ieN} T_r[(\phi\partial_\mu\phi^{-1})\hat{I}(x)]. \tag{5.7}$$

We may introduce differential forms:

$$\omega = A_\mu dx_\mu, \quad \Omega = \mathscr{B}_\mu dx_\mu. \tag{5.8}$$

According to (5.7),

$$\omega = \Omega - \frac{1}{ieN} T_r[(d\phi^{-1})\phi\hat{I}_0]. \tag{5.9}$$

$$\begin{aligned} F_A = d\omega &= \partial_\nu A_\mu dx_\nu \wedge dx_\mu \\ &= d\Omega + \frac{1}{ieN} T_r(d\phi^{-1} \wedge d\phi\hat{I}_0). \end{aligned} \tag{5.10}$$

It can be proved by using (5.6) that

$$T_r(d\hat{I}(x) \wedge d\hat{I}(x) \cdot \hat{I}(x)) = (a^2 + 4b)T_r(d\phi^{-1} \wedge d\phi\hat{I}_0). \tag{5.11}$$

Substituting (5.11) into (5.10) we have

$$F_A = \frac{1}{2} F_{\mu\nu}^{(A)} dx_\mu \wedge dx_\nu$$

$$= \frac{1}{2}(\partial_\mu \mathscr{B}_\nu - \partial_\nu \mathscr{B}_\mu) dx_\mu \wedge dx_\nu$$

$$+ \frac{1}{2ieN(a^2+4b)} T_r\{(\partial_\mu \hat{I}(x)\partial_\nu \hat{I}(x) - \partial_\nu \hat{I}(x)\partial_\mu \hat{I}(x)) \cdot \hat{I}(x)\} dx_\mu \wedge dx_\nu$$

$$= \frac{1}{2}(\partial_\mu \mathscr{B}_\nu - \partial_\nu \mathscr{B}_\mu) dx_\mu \wedge dx_\nu$$

$$+ \frac{1}{2ieN} T_r\{(\partial_\mu \phi^{-1}\partial_\nu \phi - \partial_\nu \phi^{-1}\partial_\mu \phi)\hat{I}_0\} dx_\mu \wedge dx_\nu. \tag{5.12}$$

Thus we obtain

$$F^{(A)}_{\mu\nu} = \partial_\mu A_\nu - \partial_\nu A_\mu$$

$$= \partial_\mu \mathscr{B}_\nu - \partial_\nu \mathscr{B}_\mu + \frac{1}{ieN(a^2+4b)} T_r\{(\partial_\mu \hat{I}(x)\partial_\nu \hat{I}(x) - \partial_\nu \hat{I}(x)\partial_\mu \hat{I}(x))\hat{I}(x)\}$$

$$= \partial_\mu \mathscr{B}_\nu - \partial_\nu \mathscr{B}_\mu + \frac{1}{ieN} T_r\{(\partial_\mu \phi^{-1}\partial_\nu \phi - \partial_\nu \phi^{-1}\partial_\mu \phi)\hat{I}_0\}. \tag{5.13}$$

In the above equations, $\phi = \phi(x)$ can be regarded as a mapping of a space point $\boldsymbol{x}$ to a point in the coset space $SU(N)/U(1)$.

Take a sphere S^2 with infinite radius and consider the mapping

$$\phi(x):\ S^2 \to \frac{SU(N)}{U(1)}. \tag{5.14}$$

If $\phi(x)$ is equivalent to non-identity element of the homotopy group $\pi_2(SU(N)/U(1))$, there will be monopole in the system. The contribution of the monopole to the field strength is described by the second term on the right-hand side in (5.12) and (5.13). Integrating F_A over the space surface, we obtain the magnitude of the magnetic charge, namely,

$$\oint F_A = 4\pi g. \tag{5.15}$$

We illustrate the details for the special case of $G = SU(2)$.

Assume that the monopole is at the origin of the coordinate system. Choose

$$\phi(x) = e^{i\theta \cdot \boldsymbol{n\tau}/2}, \quad \boldsymbol{n} = \sin\varphi \boldsymbol{i} - \cos\varphi \boldsymbol{j}, \tag{5.16}$$

where (θ, φ) are the polar and azimuthal angles of the space point $\boldsymbol{r}$ respectively. $\boldsymbol{i}, \boldsymbol{j}$ are the unit vectors along the positive direction of x and y axes respectively. Then we have

$$\hat{I}(x) = \phi\tau_3\phi^{-1} = \hat{\boldsymbol{r}} \cdot \boldsymbol{\tau} = \frac{\boldsymbol{r}}{r} \cdot \boldsymbol{\tau}, \quad \hat{\boldsymbol{r}} = \frac{\boldsymbol{r}}{r}. \tag{5.17}$$

It is easy to prove that

$$T_r(d\hat{I}(x) \wedge d\hat{I}(x)\hat{I}(x)) = 4i \sin\theta d\theta d\varphi. \tag{5.18}$$

Then there is

$$\oint F_A = \frac{-i}{4eN}\oint T_r(d\hat{I}(x)\wedge d\hat{I}(x)\hat{I}(x)) = \frac{2\pi}{e}. \tag{5.19}$$

Combining the above with $\oint F_A = 4\pi g$, we obtain the quantization condition,

$$eg = \frac{1}{2}. \tag{5.20}$$

If the monopole is at the point $\mathbf{y}(t)$, we should choose $\phi(x)$ in some other way.

In summary, in the presence of the monopole, we can divide A_μ into two parts: $\mathscr{B}_\mu$, the electromagnetic field potential not originating from magnetic charge, and $\phi(x)$, the pure gauge field on the coset space describing the contribution of the magnetic charge. $\mathbf{y}(t)$ here can be taken as a dynamical variable. Using (5.13) and the Lagrangian,

$$\mathscr{L} = -\frac{1}{4}F^A_{\mu\nu}F^A_{\mu\nu} = -\frac{1}{4}(F^B_{\mu\nu} + G_{\mu\nu})(F^B_{\mu\nu} + G_{\mu\nu}). \tag{5.21}$$

where

$$F^B_{\mu\nu} = \partial_\mu\mathscr{B}_\nu - \partial_\nu\mathscr{B}_\mu,$$

$$G_{\mu\nu} = \frac{1}{ieN}T_r\{(\partial_\mu\phi^{-1}\partial_\nu\phi - \partial_\nu\phi^{-1}\partial_\mu\phi)\hat{I}_0\}.$$

we can derive the equation of motion of $\mathscr{B}_\mu(x)$ and $\mathbf{y}(t)$ and thereby the interaction of the magnetic charge with the electromagnetic field. However the self-energy of the monopole is divergent. Thus the monopole is point-like in this theory.

VI. The Decomposition of the Group U_n and the Corresponding Pure Gauge Fields on the Coset Space

We represent the group $U_n = \{\cdots g\cdots\}$, with

$$g = e^{i\Theta},\ \Theta^+ = \Theta,\ \Theta = \begin{pmatrix}\alpha_1 & \beta\\ \beta^+ & \alpha_2\end{pmatrix},\ \alpha_1^+ = \alpha_1,\ \alpha_2^+ = \alpha_2, \tag{6.1}$$

where α_i are square submatrices, and β is a rectangular sub-matrix. In that case we can choose

$$h = e^{i\begin{pmatrix}\alpha_1 & 0\\ 0 & \alpha_2\end{pmatrix}} = \begin{pmatrix}e^{i\alpha_1} & 0\\ 0 & e^{i\alpha_2}\end{pmatrix}\in H,$$

$$\phi = e^{i\begin{pmatrix}0 & \beta\\ \beta^+ & 0\end{pmatrix}} = \cos\begin{pmatrix}0 & \beta\\ \beta^+ & 0\end{pmatrix} + i\sin\begin{pmatrix}0 & \beta\\ \beta^+ & 0\end{pmatrix}\in G/H, \tag{6.2}$$

where cosine and sine functions are defined as the Taylor expansions. Thus, we have

$$\phi = \begin{pmatrix}\cos\sqrt{\beta\beta^+} & 0\\ 0 & \cos\sqrt{\beta^+\beta}\end{pmatrix}\begin{pmatrix}1 & i\frac{\tan\sqrt{\beta\beta^+}}{\sqrt{\beta\beta^+}}\beta\\ i\frac{\tan\sqrt{\beta^+\beta}}{\sqrt{\beta^+\beta}}\beta_+ & 1\end{pmatrix}. \tag{6.3}$$

Define

$$\varphi = i\frac{\tan\sqrt{\beta\beta^+}}{\sqrt{\beta\beta^+}}\beta = i\beta\frac{\tan\sqrt{\beta^+\beta}}{\sqrt{\beta^+\beta}},$$

$$\varphi^+ = -i\beta^+\frac{\tan\sqrt{\beta\beta^+}}{\sqrt{\beta\beta^+}} = -i\frac{\tan\sqrt{\beta^+\beta}}{\sqrt{\beta^+\beta}}\beta^+. \tag{6.4}$$

Substituting (6.4) into (6.3) we obtain

$$\phi = \begin{pmatrix} \frac{1}{\sqrt{1+\varphi\varphi^+}} & 0 \\ 0 & \frac{1}{\sqrt{1+\varphi^+\varphi}} \end{pmatrix}\begin{pmatrix} 1 & \varphi \\ -\varphi^+ & 1 \end{pmatrix} = \begin{pmatrix} 1 & \varphi \\ -\varphi^+ & 1 \end{pmatrix}\begin{pmatrix} \frac{1}{\sqrt{1+\varphi\varphi^+}} & 0 \\ 0 & \frac{1}{\sqrt{1+\varphi^+\varphi}} \end{pmatrix},$$

$$\phi^{-1} = \begin{pmatrix} \frac{1}{\sqrt{1+\varphi\varphi^+}} & 0 \\ 0 & \frac{1}{\sqrt{1+\varphi^+\varphi}} \end{pmatrix}\begin{pmatrix} 1 & -\varphi \\ \varphi^+ & 1 \end{pmatrix} = \begin{pmatrix} 1 & -\varphi \\ \varphi^+ & 1 \end{pmatrix}\begin{pmatrix} \frac{1}{\sqrt{1+\varphi\varphi^+}} & 0 \\ 0 & \frac{1}{\sqrt{1+\varphi^+\varphi}} \end{pmatrix}. \tag{6.5}$$

Thus, the left coset decomposition of the group U_n is

$$g = \phi h = \begin{pmatrix} \frac{1}{\sqrt{1+\varphi\varphi^+}}e^{i\alpha_1} & \varphi\frac{1}{\sqrt{1+\varphi^+\varphi}}e^{i\alpha_2} \\ -\varphi^+\frac{1}{\sqrt{1+\varphi\varphi^+}}e^{i\alpha_1} & \frac{1}{\sqrt{1+\varphi^+\varphi}}e^{i\alpha_2} \end{pmatrix}. \tag{6.6}$$

Put

$$g = \begin{pmatrix} A & B \\ C & D \end{pmatrix}, \tag{6.7}$$

then there is :

$$g\phi = \begin{pmatrix} (A-B\varphi^+)\frac{1}{\sqrt{1+\varphi\varphi^+}} & (A\varphi+B)\frac{1}{\sqrt{1+\varphi^+\varphi}} \\ (C-D\varphi^+)\frac{1}{\sqrt{1+\varphi\varphi^+}} & (C\varphi+D)\frac{1}{\sqrt{1+\varphi^+\varphi}} \end{pmatrix}$$

$$= \phi' h' = \begin{pmatrix} \frac{1}{\sqrt{1+\varphi'\varphi'^+}}e^{i\alpha_1'} & \varphi'\frac{1}{\sqrt{1+\varphi'^+\varphi'}}e^{i\alpha_2'} \\ -\varphi'^+\frac{1}{\sqrt{1+\varphi'\varphi'^+}}e^{i\alpha_1'} & \frac{1}{\sqrt{1+\varphi'^+\varphi'}}e^{i\alpha_2'} \end{pmatrix}. \tag{6.8}$$

Comparing the two matrices in (6.8), we have

$$\varphi \xrightarrow{g} \varphi' = (A\varphi+B)\frac{1}{C\varphi+D},$$

$$\varphi^+ \xrightarrow{g} \varphi^{+\prime} = -(C-D\varphi^+)\frac{1}{A-B\varphi^+}. \tag{6.9}$$

Thus the action of G on the coset space is equivalent to fractional linear transformation on the field φ.

The gauge invariant Lagrangian is still of the form (3.1).

It is easy to show that

$$\hat{B}_\mu = \phi(\partial_\mu + \hat{A}_\mu)\phi^{-1} = \begin{pmatrix} \frac{1}{\sqrt{1+\varphi\varphi^+}} & 0 \\ 0 & \frac{1}{\sqrt{1+\varphi^+\varphi}} \end{pmatrix}$$

$$\cdot \begin{pmatrix} \varphi\partial_\mu\varphi^+ - \sqrt{1+\varphi\varphi^+}\,\partial_\mu\sqrt{1+\varphi\varphi^+} + \hat{A}_\mu^{(1)} + \varphi\hat{A}_\mu^{(2)}\varphi^+, & -\partial_\mu\varphi - \hat{A}_\mu^{(1)}\varphi + \varphi\hat{A}_\mu^{(2)} \\ \partial_\mu\varphi^+ - \varphi^+A_\mu^{(1)} + \hat{A}_\mu^{(2)}\varphi^+, & \varphi^+\partial_\mu\varphi - \sqrt{1+\varphi^+\varphi}\,\partial_\mu\sqrt{1+\varphi^+\varphi} + \hat{A}_\mu^{(2)} + \varphi^+\hat{A}_\mu^{(1)}\varphi \end{pmatrix}$$

$$\cdot \begin{pmatrix} \frac{1}{\sqrt{1+\varphi\varphi^+}} & 0 \\ 0 & \frac{1}{\sqrt{1+\varphi^+\varphi}} \end{pmatrix}, \tag{6.10}$$

where

$$\hat{A}_\mu = -\frac{iK}{2}\sum_{i\in H}\lambda_i A_\mu^i = \begin{pmatrix} \hat{A}_\mu^{(1)} & 0 \\ 0 & \hat{A}_\mu^{(2)} \end{pmatrix}. \tag{6.11}$$

The existence of expressions $\frac{1}{\sqrt{1+\varphi\varphi^+}}$ and $\frac{1}{\sqrt{1+\varphi^+\varphi}}$ in (6.10) will bring complicated non-linear interaction terms into the theory.

References

[1] Weinberg, S.: *Phys. Rev. Letters,* V19 (1967), 1264; Salam, A.: *Elementary Particle Theory* (ed. N. Svartholm) (1968).
[2] For instance, Abers, E. S. & Lee, B. W.: Gauge Theories, *Phys. Reports,* **V9c** (1973).
[3] Baton & Laurens.: *Nucl. Phys.,* **B3** (1967), 349; Malamud, E. & Schein, P.: *Proc. of Argonne Conference* (1969), 108; Deinet, W. et al.: *Phys. Letters,* **30B** (1969), 359.
[4] Lee, B. W.: *Chiral Dynamics,* Gordon and Breach.
[5] Weinberg, S.: *Phys. Rev. Letters,* V18 (1967), 188.
[6] Taylor, K.: *Phys. Rev.,* **D3** (1971), 1846.
[7] Coleman, S., Wess, J. & Zumino, B.: *Phys. Rev.,* **177** (1969), 2239.
[8] Goldstone, J.: *Nuovo Cimento,* **19** (1961), 15.

Vol. XXII No. 3 SCIENTIA SINICA March 1979

SOLITON-SOLITON SCATTERING AND THE SEMI-CLASSICAL APPROXIMATION FOR THE PROBLEM OF SCATTERING IN THREE-DIMENSIONAL SPACE

CHOU KUAN-CHAO (周光召), AND DAI YUAN-BEN (戴元本)

(*Institute of Theoretical Physics, Academia Sinica*)

Received March 20, 1978.

ABSTRACT

The soliton-soliton scattering amplitude in three-dimensional space is obtained in the semi-classical approximation. The results obtained are also applicable to the problem of the scattering of ordinary particles and generalize the previous results in this respect.

I. INTRODUCTION

The quantum theory of the soliton has attracted much attention in particle physics in recent years. However, so far few researches have been carried out for the problem of soliton-soliton scattering. For one-dimensional scattering, applying the W. K. B. approximation Jackiw and Woo[1] obtained a relation between the phase-shift δ and the time-delay Δ_{cl} in the classical solution. In the W. K. B. approximation one has

$$2\delta(E) = \lim_{\substack{x_f\to\infty \\ x_i\to-\infty}} \int_{x_i}^{x_f} dx' \left(\sqrt{2m(E-U(x'))} - \sqrt{2mE}\right),$$

where E and m are the energy and the mass of the particle respectively, and U is the potential acting on it. It then follows that,

$$2\frac{d}{dE}\delta(E) = \lim_{\substack{x_f\to\infty \\ x_i\to-\infty}} \left[\int_{x_i}^{x_f} \frac{dx'}{v(x',E)} - \frac{x_f - x_i}{v(E)}\right] = \Delta_{cl}(E), \tag{1}$$

where $v(x', E)$ is the velocity of the particle with energy E at the point x', $v(E)$ is the velocity of the free particle with the same energy. So far as we know, no concrete result for the three-dimensional scattering problem has been published in the literature. The meaning of the soliton-soliton scattering solution of the three-dimensional classical field equation in quantum theory remains to be clarified.

The present article is devoted mainly to the problem of the soliton-soliton scattering in three-dimensional space. In Sec. 2 the needed formulas of the canonical quantization for the problem of the scattering of solitons are described and the general form of the scattering amplitude with the neglect of the effect of the exchange of mesons is given. In Sec. 3 the three-dimensional soliton-soliton scattering amplitude is obtained in the semi-classical approximation. There are two relations between the phase of the scattering amplitude and the dynamical quantities of the corresponding classical orbit, which are generalizations of (1) in the case of three-dimensional space. Results obtained there elucidate the meaning of the classical scattering solution in quantum theory. The formulas obtained in this section are also applicable to the scattering of ordinary particles. Since the assumption of central symmetry of force is not needed in the derivation, the main result is more general than that in previous semi-classical researches. In Appendix B it is shown that for the scattering in one dimension a generalized form of the formula (1) can be derived without appealing to the W.K.B. approximation.

II. The Amplitude of the Soliton-Soliton Scattering

Suppose that there is an N component scalar field φ (φ^i, $i = 1, 2, \ldots N$) and the classical equation of motion of which,

$$\Box\varphi - V'(\varphi) = 0,$$

has solutions $\varphi_{cl}(\boldsymbol{x} - \boldsymbol{Z}(t),\ \boldsymbol{z}(t))$ corresponding to the soliton-soliton scattering, in which $\boldsymbol{Z} = \frac{1}{2}(\boldsymbol{x}_1 + \boldsymbol{x}_2)$, $\boldsymbol{z} = \boldsymbol{x}_1 - \boldsymbol{x}_2$, where $\boldsymbol{x}_1$ and $\boldsymbol{x}_2$ are the coordinates of the centers of the two solitons. Suppose that

$$\varphi_{cl}(\boldsymbol{x} - \boldsymbol{Z}(t),\ \boldsymbol{z}(t)) \to \varphi_{cl}(\boldsymbol{x} - \boldsymbol{X}_1(t)) + \varphi_{cl}(\boldsymbol{x} - \boldsymbol{X}_2(t))$$

as $t \to -\infty$, and that

$$\varphi_{cl}(\boldsymbol{x} - \boldsymbol{Z}(t),\ \boldsymbol{z}(t)) \to \varphi_{cl}(\boldsymbol{x} - \boldsymbol{X}'_1(t)) + \varphi_{cl}(\boldsymbol{x} - \boldsymbol{X}'_2(t)) \tag{2}$$

as $t \to +\infty$, in which $\varphi_{cl}(\boldsymbol{x} - \boldsymbol{x}(t))$ is a single-soliton solution and

$$\boldsymbol{X}_i(t) = \boldsymbol{X}_i(0) + \boldsymbol{u}_i t,$$
$$\boldsymbol{X}'_i(t) = \boldsymbol{X}'_i(0) + \boldsymbol{u}'_i t. \qquad (i = 1, 2) \tag{3}$$

According to the scheme of canonical quantization[2] one introduces operators $\hat{\boldsymbol{Z}}(t)$ and $\boldsymbol{Z}(t)$ corresponding to respectively with $z(t)$ and $\boldsymbol{Z}(t)$ with u_1 and u_2 considered as parameters. Quantized field operators $\hat{\varphi}(\boldsymbol{x}, t)$ can be expanded as

$$\hat{\varphi}(\boldsymbol{x}, t) = \varphi_{cl}(\boldsymbol{x} - \hat{\boldsymbol{Z}}(t), \hat{\boldsymbol{z}}(t)) + \sum_n \hat{q}_n(t)\, \psi_n(\boldsymbol{x} - \hat{\boldsymbol{Z}}(t), \hat{\boldsymbol{z}}(t)), \tag{4}$$

where $\hat{q}_n(t)$ terms describe the production and the annihilation of the "mesons". In this article we shall neglect these terms corresponding to the excitation of the meson freedom. Substituting $\hat{\varphi}(\boldsymbol{x}, t) = \varphi_{cl}(\boldsymbol{x} - \hat{\boldsymbol{Z}}(t), \hat{\boldsymbol{z}}(t))$ into the expressions

$$L = \int d^3\boldsymbol{x} \left[-\frac{1}{2}(\partial_\mu \varphi)^2 - V(\varphi) \right],$$

$$H = \int d^3\boldsymbol{x}\left[\frac{1}{2}\left(\frac{\partial\varphi}{\partial t}\right)^2 + \frac{1}{2}\sum_i\left(\frac{\partial\varphi}{\partial x_i}\right)^2 + V(\varphi)\right]$$

of the Lagrangian and the Hamiltonian of the field, one finds that

$$\hat{H} = \frac{1}{2}\hat{P}_i M_{ij}^{-1}(\hat{\boldsymbol{z}})\hat{P}_j + \frac{1}{2}\hat{p}_i m_{ij}^{-1}(\hat{\boldsymbol{z}})\hat{p}_j + U(\hat{\boldsymbol{z}}), \tag{5}$$

where $\hat{P}_i$ and $\hat{p}_i$ are canonical momenta conjugate to $\hat{Z}_i$ and $\hat{z}_i$ respectively,

$$\hat{U} = \int d^3\boldsymbol{x}[V(\varphi_{cl}(\boldsymbol{x} - \hat{\boldsymbol{Z}}(t), \hat{\boldsymbol{z}}(t))) + \sum_i(\partial_i\varphi_{cl}(\boldsymbol{x} - \hat{\boldsymbol{Z}}(t), \hat{\boldsymbol{z}}(t)))^2],$$

$$\hat{M}_{ij} = \int d^3\boldsymbol{x}\left[\frac{\partial}{\partial Z_i}\varphi_{cl}(\boldsymbol{x} - \hat{\boldsymbol{Z}}(t), \hat{\boldsymbol{z}}(t))\right]\left[\frac{\partial}{\partial Z_j}\varphi_{cl}(\boldsymbol{x} - \hat{\boldsymbol{Z}}(t), \hat{\boldsymbol{z}}(t))\right],$$

$$\hat{m}_{ij} = \int d^3\boldsymbol{x}\left[\frac{\partial}{\partial z_i}\varphi_{cl}(\boldsymbol{x} - \hat{\boldsymbol{Z}}(t), \hat{\boldsymbol{z}}(t))\right]\left[\frac{\partial}{\partial z_j}\varphi_{cl}(\boldsymbol{x} - \hat{\boldsymbol{Z}}(t), \hat{\boldsymbol{z}}(t))\right], \tag{6}$$

$$m_{ij}(\boldsymbol{z}) \to \text{constant}, \quad M_{ij}(\boldsymbol{z}) \to \text{constant}, \quad U(\boldsymbol{z}) \to U_0 \tag{7}$$

as $|\boldsymbol{z}| \to \infty$.

Formally, the formula (5) is not relativistic. Nevertheless, with $\hat{U}$, $\hat{M}_{ij}$ and $\hat{m}_{ij}$ depending on the parameters u_1 and u_2, it can be shown that, after taking matrix element the relativistic formulas of the momentum and the energy can be satisfied with appropriately chosen u_1 and u_2, similar to that discussed in [2].

Now we turn to consider the quantum-mechanical scattering process of two solitons. Let $\boldsymbol{P}_1$, $\boldsymbol{P}_2$ and $\boldsymbol{P}_1'$, $\boldsymbol{P}_2'$ be the momenta of the two solitons in the initial state and the final state respectively. Let $|\boldsymbol{Z}, \boldsymbol{z}, t\rangle$ denote the eigenstate of the operators $\hat{\boldsymbol{Z}}(t)$ and $\hat{\boldsymbol{z}}(t)$ in the Heisenberg representation and $|\boldsymbol{Z}, \boldsymbol{z}, 0\rangle \equiv |\boldsymbol{Z}, \boldsymbol{z}\rangle$. Apart from a possible phase-factor irrelevant to the final results,

$$|\boldsymbol{Z}, \boldsymbol{z}, t\rangle = e^{\frac{i}{\hbar}\hat{H}t}|\boldsymbol{Z}, \boldsymbol{z}\rangle.$$

Therefore, the propagator is

$$\langle\boldsymbol{Z}', \boldsymbol{z}', t'|\boldsymbol{Z}, \boldsymbol{z}, t\rangle = \langle\boldsymbol{Z}', \boldsymbol{z}'|e^{-\frac{i}{\hbar}\hat{H}(t'-t)}|\boldsymbol{Z}, \boldsymbol{z}\rangle. \tag{8}$$

Let $|\boldsymbol{P}_1, \boldsymbol{P}_2\ t\rangle$ denote the eigenstate of the operators $\hat{\boldsymbol{p}}_1(t)$, $\hat{\boldsymbol{p}}_2(t)$ in the Heisenberg representation corresponding to the eigenvalue $\boldsymbol{P}_1$ and $\boldsymbol{P}_2$ and $|\boldsymbol{P}_1, \boldsymbol{P}_2, 0\rangle \equiv |\boldsymbol{P}_1, \boldsymbol{P}_2\rangle$.

$$|\boldsymbol{P}_1, \boldsymbol{P}_2, t\rangle = e^{\frac{i}{\hbar}\eta(\boldsymbol{P}_1, \boldsymbol{P}_2, t)} \cdot e^{\frac{i}{\hbar}\hat{H}t}|\boldsymbol{P}_1, \boldsymbol{P}_2\rangle, \tag{9}$$

where $e^{\frac{i}{\hbar}\eta}$ is a phase-factor. In the problem of the scattering process, $|\boldsymbol{z}(t)| \to \infty$ as $t \to \pm\infty$. It then follows from (7) that

$$\frac{d}{dt}\hat{\boldsymbol{P}}_i(t) = \frac{i}{\hbar}[\hat{H}, \hat{\boldsymbol{P}}_i] \sim 0, \quad i = 1, 2.$$

so $|\boldsymbol{P}_1, \boldsymbol{P}_2, t\rangle$ can differ from $|\boldsymbol{P}_1, \boldsymbol{P}_2, t + \Delta t\rangle$ at most by a phase-factor and can, therefore, be chosen to be independent of t when $t \to \pm\infty$. It can be seen from (9) that one should take

$$e^{\frac{i}{\hbar}\eta(P_i,t)} = e^{-\frac{i}{\hbar}E(P_i)t}. \tag{10}$$

It follows from (8)—(10) that the scattering matrix element can be written as

$$\begin{aligned}{}_{in}\langle \boldsymbol{P}'_1, \boldsymbol{P}'_2|S|\boldsymbol{P}_1, \boldsymbol{P}_2\rangle_{in} &= \lim_{\substack{t'\to\infty\\ t\to-\infty}} \langle \boldsymbol{P}'_1, \boldsymbol{P}'_2, t'|\boldsymbol{P}_1, \boldsymbol{P}_2, t\rangle = \frac{1}{(2\pi\hbar)^6}\lim_{\substack{t'\to\infty\\ t\to-\infty}} \iiiint \\ &\cdot \exp\frac{i}{\hbar}(\boldsymbol{P}\cdot\boldsymbol{Z} - \boldsymbol{P}'\cdot\boldsymbol{Z}' + \boldsymbol{p}\cdot\boldsymbol{z} - \boldsymbol{p}'\cdot\boldsymbol{z}')\exp\frac{i}{\hbar}(E't' - Et) \\ &\cdot \langle \boldsymbol{Z}', \boldsymbol{z}'|\exp\frac{i}{\hbar}[\hat{H}(t-t')]|\boldsymbol{Z}, \boldsymbol{z}\rangle\, d^3\boldsymbol{Z}'\, d^3\boldsymbol{Z}\, d^3\boldsymbol{z}'\, d^3\boldsymbol{z},\end{aligned} \tag{11}$$

where $\boldsymbol{P}$, $\boldsymbol{p}$ and $\boldsymbol{P}'$, $\boldsymbol{p}'$ are the total momentum and the relative momentum of the initial state and the final state respectively, E and E' are the total energy of the initial state and the final state. Let

$$K(\boldsymbol{Z}', \boldsymbol{z}'; \boldsymbol{Z}, \boldsymbol{z}; t'-t) \equiv \langle \boldsymbol{Z}', \boldsymbol{z}'|\exp\frac{i}{\hbar}[\hat{H}(t-t')]|\boldsymbol{Z}, \boldsymbol{z}\rangle. \tag{12}$$

$$\because [\hat{\boldsymbol{P}}, \hat{H}] = 0$$

$$\begin{aligned}\therefore \frac{\hbar}{i}\frac{\partial}{\partial Z'_i}K(\boldsymbol{Z}', \boldsymbol{z}'; \boldsymbol{Z}, \boldsymbol{z}; t'-t) &= \langle \boldsymbol{Z}', \boldsymbol{z}'|\hat{P}_i \exp\frac{i}{\hbar}[\hat{H}(t-t')]|\boldsymbol{Z}, \boldsymbol{z}\rangle \\ &= \langle \boldsymbol{Z}', \boldsymbol{z}'|\exp\frac{i}{\hbar}[\hat{H}(t-t')]\hat{P}_i|\boldsymbol{Z}, \boldsymbol{z}\rangle = -\frac{\hbar}{i}\frac{\partial}{\partial Z_i}K(\boldsymbol{Z}', \boldsymbol{z}'; \boldsymbol{Z}, \boldsymbol{z}; t'-t).\end{aligned}$$

The above equality shows that $K(\boldsymbol{Z}', \boldsymbol{z}'; \boldsymbol{Z}, \boldsymbol{z}; t'-t)$ is independent of $\boldsymbol{Z}' + \boldsymbol{Z}$. Making the Fourier transformation

$$\begin{aligned}&K(\boldsymbol{Z}', \boldsymbol{z}'; \boldsymbol{Z}, \boldsymbol{z}; t'-t) \\ &= \frac{1}{(2\pi\hbar)^3}\int K(\boldsymbol{P}''; \boldsymbol{z}', \boldsymbol{z}; t'-t)\exp\left[\frac{i}{\hbar}\boldsymbol{P}''\cdot(\boldsymbol{Z}'-\boldsymbol{Z})\right] d^3\boldsymbol{P}'',\end{aligned} \tag{13}$$

and substituting it into (11), one finds that

$$\begin{aligned}&\lim \langle \boldsymbol{P}'_1, \boldsymbol{P}'_2, t'|\boldsymbol{P}_1, \boldsymbol{P}_2 t\rangle \\ &= \delta^3(\boldsymbol{P} - \boldsymbol{P}')\frac{1}{(2\pi\hbar)^3}\iint \exp\frac{i}{\hbar}(\boldsymbol{p}\cdot\boldsymbol{z} - \boldsymbol{p}'\cdot\boldsymbol{z}' + E't' - Et) \\ &\cdot K(\boldsymbol{P}; \boldsymbol{z}', \boldsymbol{z}; t'-t)\, d^3\boldsymbol{z}\, d^3\boldsymbol{z}'.\end{aligned} \tag{14}$$

It can be shown from (26) that $K(\boldsymbol{Z}', \boldsymbol{z}'; \boldsymbol{Z}, \boldsymbol{z}; t'-t)$ satisfies the Schrödinger equation

$$\begin{aligned}&\hbar i\frac{\partial}{\partial t'}K(\boldsymbol{Z}', \boldsymbol{z}'; \boldsymbol{Z}, \boldsymbol{z}; t'-t) \\ &= \left[-\frac{\hbar^2}{2}\frac{\partial}{\partial Z'_i}M_{ij}^{-1}\frac{\partial}{\partial Z'_j} - \frac{\hbar^2}{2}\frac{\partial}{\partial z'_i}m_{ij}^{-1}\frac{\partial}{\partial z'_j} + U(\boldsymbol{z}')\right]K(\boldsymbol{Z}', \boldsymbol{z}'; \boldsymbol{Z}, \boldsymbol{z}; t'-t),\end{aligned} \tag{15}$$

and the initial condition

$$K(\boldsymbol{Z}', \boldsymbol{z}'; \boldsymbol{Z}, \boldsymbol{z}; 0) = \delta^3(\boldsymbol{Z} - \boldsymbol{z}')\delta^3(\boldsymbol{z} - \boldsymbol{z}'). \tag{16}$$

Substituting (13) into (15) and (16) one finds that $K(\boldsymbol{P};\ \boldsymbol{z}',\ \boldsymbol{z};\ t' - t)$ satisfies the equation of motion

$$-\frac{\hbar}{i}\frac{\partial}{\partial t'}K(\boldsymbol{P};\ \boldsymbol{z}', \boldsymbol{z};\ t' - t) = \hat{H}_{\text{eff}}(\boldsymbol{P}, \boldsymbol{z}')\,K(\boldsymbol{P};\ \boldsymbol{z}', \boldsymbol{z};\ t' - t), \tag{17}$$

$$\hat{H}_{\text{eff}}(\boldsymbol{P},\ \boldsymbol{z}') = -\frac{\hbar^2}{2}\frac{\partial}{\partial z_i'}\,m_{ij}^{-1}(\boldsymbol{z}')\,\frac{\partial}{\partial z_j'} + \bar{U}(\boldsymbol{P}, \boldsymbol{z}'),$$

$$\bar{U}(\boldsymbol{P},\ \boldsymbol{z}') = U(\boldsymbol{z}') + \frac{1}{2}P_i M_{ij}^{-1}(\boldsymbol{z}')P_j, \tag{18}$$

and the initial condition

$$K(\boldsymbol{P};\ \boldsymbol{z}', \boldsymbol{z};\ 0) = \delta^3(\boldsymbol{z} - \boldsymbol{z}'). \tag{19}$$

It follows from (17) and (19) that

$$K(\boldsymbol{P};\ \boldsymbol{z}', \boldsymbol{z};\ t' - t) = \langle \boldsymbol{z}' |\ \exp\frac{i}{\hbar}\ \hat{H}_{\text{eff}}(t - t')\,|\boldsymbol{z}\rangle. \tag{20}$$

III. Semi-Classical Approximation

Now we proceed to solve the equation (17) to the first order term in $\hbar$ with the semi-classical approximation. Let us try the solution of the following form:

$$K(\boldsymbol{P};\ \boldsymbol{z}', \boldsymbol{z};\ t' - t) = Af(\boldsymbol{z}', \boldsymbol{z}, t' - t)\exp\left[\frac{i}{\hbar}\ S_{cl}(\boldsymbol{z}', \boldsymbol{z};\ t' - t)\right], \tag{21}$$

where f does not depend on $\hbar$ and S_{cl} is the action of the classical orbit which is expressed by

$$S_{cl} = \int_t^{t'} L_{\text{eff}}(t'')\,dt'',$$

$$L_{\text{eff}} = \boldsymbol{p}\cdot\dot{\boldsymbol{z}} - H_{\text{eff}} = \frac{1}{2}\ \dot{z}_i\,m_{ij}(\boldsymbol{z})\ \dot{z}_j - \bar{U}(\boldsymbol{P}, \boldsymbol{z}).$$

Comparing the coefficients of $\hbar^0$ one obtains the equation

$$\frac{1}{2}\ m_{ij}^{-1}(\boldsymbol{z}')\ \frac{\partial S_{cl}}{\partial z_i'}\frac{\partial S_{cl}}{\partial z_j'} + \bar{U}(\boldsymbol{P}, \boldsymbol{z}') + \frac{\partial S_{cl}}{\partial t'} = 0. \tag{22}$$

Since this is a classical Hamilton-Jacobi equation, Equation (17) is satisfied to order $\hbar^0$. Comparing the coefficient of $\hbar$ one obtains the equation

$$-\frac{\partial f}{\partial t'} = m_{ij}^{-1}(\boldsymbol{z}')\frac{\partial S_{cl}}{\partial z_j'}\frac{\partial f}{\partial z_i'} + \frac{1}{2}\left[m_{ij}^{-1}(\boldsymbol{z}')\frac{\partial^2 S_{cl}}{\partial z_i'\partial z_j'} + \frac{\partial m_{ij}^{-1}}{\partial z_i'}\frac{\partial S_{cl}}{\partial z_j'}\right]f. \tag{23}$$

The above equation can be written as

$$\frac{\partial f^2}{\partial t'} + \frac{\partial}{\partial z'_i}(v_i f^2) = 0,$$

where v is the classical velocity of the particle. Written in this form the equation is the same as that in the corresponding semi-classical formula for the scattering of ordinary particles.

For the problem of ordinary particles, m_{ij} is a constant diagonal matrix. In this case, it is already known that Eq. (23) has the solution[3]:

$$f = \left\|\frac{\partial^2 S_{cl}(\boldsymbol{z}', \boldsymbol{z}; t' - t)}{\partial z_i \partial z'_j}\right\|^{\frac{1}{2}}. \tag{24}$$

In Appendix A it is verified that (24) remains to be the solution of (23) in the general case. The constant A in the formula (21) can be determined by means of the condition (19). When $\epsilon = t' - t$ is small, $S_{cl} \simeq \frac{1}{2\epsilon}(z' - z)_i\, m_{ij}(z' - z)_j$. From (19) one obtains

$$A \int \|m_{ij}\|^{\frac{1}{2}}\, \epsilon^{-\frac{3}{2}} \exp\left[\frac{i}{\hbar}\frac{1}{\epsilon}(z' - z)_i\, m_{ij}(z' - z)_j\right] d^3\boldsymbol{z}' = 1.$$

Therefore,

$$A = \frac{1}{(2\pi\hbar_i)^{3/2}}.$$

Taking account of the fact that there may be more than one classical orbits connecting the points $\boldsymbol{z}$ and $\boldsymbol{z}'$, one obtains finally

$$K(\boldsymbol{P}; \boldsymbol{z}', \boldsymbol{z}; t' - t) = \frac{1}{(2\pi\hbar i)^{3/2}} \sum \left\|\frac{\partial^2 S_{cl}}{\partial z_i \partial z'_j}\right\|^{\frac{1}{2}} \exp\left[\frac{i}{\hbar} S_{cl}(\boldsymbol{z}', \boldsymbol{z}; t' - t)\right]. \tag{25}$$

Substituting (25) into (14) one finds that

$$\lim_{\substack{t \to -\infty \\ t' \to +\infty}} \langle \boldsymbol{P}'_1, \boldsymbol{P}'_2, t' | \boldsymbol{P}_1, \boldsymbol{P}_2, t\rangle = \frac{(-i)^{\frac{3}{2}}}{(2\pi\hbar)^{9/2}}\, \delta^3(\boldsymbol{P} - \boldsymbol{P}') \iint \exp\frac{i}{\hbar}(\boldsymbol{p}\cdot\boldsymbol{z} - \boldsymbol{p}'\cdot\boldsymbol{z}' - Et + E't') \left\|\frac{\partial^2 S_{cl}}{\partial z_i \partial z'_j}\right\|^{\frac{1}{2}} \exp\frac{i}{\hbar} S_{cl}(\boldsymbol{z}', \boldsymbol{z}; t' - t)\, d^3\boldsymbol{z}\, d^3\boldsymbol{z}'. \tag{26}$$

We now apply the saddle point approximation to carry out the integration with respect to $\boldsymbol{z}$ and $\boldsymbol{z}'$. Integrating first with respect to $\boldsymbol{z}$, one obtains a factor,

$$\int \exp\frac{i}{\hbar}[S_{cl}(\boldsymbol{z}', \boldsymbol{z}; t' - t) + \boldsymbol{p}\cdot\boldsymbol{z}] \left\|\frac{\partial^2 S_{cl}}{\partial z_i \partial z'_j}\right\|^{\frac{1}{2}} d^3\boldsymbol{z} = (2\pi\hbar i)^{\frac{3}{2}} \left\|\frac{\partial^2 S_{cl}}{\partial z_i \partial z'_j}\right\|^{\frac{1}{2}}_{\boldsymbol{z}^0} \cdot \left\|\frac{\partial^2 S_{cl}}{\partial z_i \partial z_j}\right\|^{-\frac{1}{2}}_{\boldsymbol{z}^0} \exp\frac{i}{\hbar} S_{cl}(\boldsymbol{z}, \boldsymbol{p}; t' - t), \tag{27}$$

where

$$S_{cl}(\boldsymbol{z}', \boldsymbol{p}; t' - t) = S_{cl}(\boldsymbol{z}', \boldsymbol{z}^0; t' - t) + \boldsymbol{p}\cdot\boldsymbol{z}^0, \tag{28}$$

$$-\frac{\partial S_{cl}(\boldsymbol{z}', \boldsymbol{z}^0; t' - t)}{\partial z_i^0} = p_i. \tag{29}$$

Since the relations (28) and (29) constitute a Legendre transformation, one has

$$\frac{\partial S_{cl}(\boldsymbol{z}', \boldsymbol{p}; t'-t)}{\partial p_k} = z_k^0(\boldsymbol{z}, \boldsymbol{p}, t'-t). \tag{30}$$

On integrating with respect to $\boldsymbol{z}'$, account must be taken of the condition of conservation of energy of the classical solution. It follows from (28) and (29) that

$$\frac{\partial S_{cl}(\boldsymbol{z}', \boldsymbol{p}; t'-t)}{\partial z_k'} = \frac{\partial S_{cl}(\boldsymbol{z}', \boldsymbol{z}^0; t'-t)}{\partial z_k'} \equiv p_{cl}'. \tag{31}$$

Recalling that $m_{ij} \to$ constant, $\bar{U} \to$ constant, as $t \to -\infty$ and $t' \to +\infty$, one obtains from the Hamilton-Jacobi equation and (29), (31) that

$$|\boldsymbol{p}_{cl}'| = |\boldsymbol{p}|. \tag{32}$$

Choosing the third coordinate axis of $\boldsymbol{z}'$ to be parallel to the direction of $\boldsymbol{p}'$, one may determine the saddle point of the integral with respect to z_1' and z_2' by the following conditions:

$$\frac{\partial S_{cl}(\boldsymbol{z}', \boldsymbol{p}; t'-t)}{\partial z_1'} = \frac{\partial S_{cl}(\boldsymbol{z}', \boldsymbol{p}; t'-t)}{\partial z_2'} = 0, \tag{33}$$

which fixes the direction of $\boldsymbol{p}_{cl}'$ to be parallel to the third axis. From (32)

$$\frac{\partial S_{cl}(\boldsymbol{z}', \boldsymbol{p}; t'-t)}{\partial z_3'} = p. \tag{34}$$

Taking derivatives of the above equality one obtains that

$$\frac{\partial^2 S_{cl}(\boldsymbol{z}', \boldsymbol{p}; t'-t)}{\partial z_3' \partial z_i'} = 0, \qquad i = 1, 2, 3.$$

So the saddle point method is needed only for the integration with respect to z_1' and z_2'. In the neighbourhood of the saddle point $S_{cl}(\boldsymbol{z}', \boldsymbol{p}; t'-t)$ is linear in z_3'. Therefore, after carrying out the integration with respect to z_3', one obtains, for the S matrix element, a factor $2\pi\hbar\delta(p-p')$ which guarantees the conservation of energy. So one obtains that

$$\begin{aligned}
&\int \exp\frac{i}{\hbar}[S_{cl}(\boldsymbol{z}', \boldsymbol{p}; t'-t) - p'z_3'] \left\| \frac{\partial^2 S_{cl}(\boldsymbol{z}', \boldsymbol{z}; t'-t)}{\partial z_i \partial z_j'} \right\|_{\boldsymbol{z}^0}^{\frac{1}{2}} \cdot \left\| \frac{\partial^2 S_{cl}(\boldsymbol{z}', \boldsymbol{z}; t'-t)}{\partial z_i \partial z_j} \right\|_{\boldsymbol{z}^0}^{-\frac{1}{2}} \\
&\cdot d^3\boldsymbol{z}' = (2\pi\hbar i)(2\pi\hbar)\,\delta(p-p') \left[\left\| \frac{\partial^2 S_{cl}(\boldsymbol{z}', \boldsymbol{z}; t'-t)}{\partial z_i \partial z_j'} \right\|^{\frac{1}{2}} \right. \\
&\cdot \left. \left\| \frac{\partial^2 S_{cl}(\boldsymbol{z}', \boldsymbol{z}; t'-t)}{\partial z_i \partial z_j} \right\|^{-\frac{1}{2}} \cdot \left\| \frac{\partial^2 S_{cl}(\boldsymbol{z}', \boldsymbol{p}; t'-t)}{\partial z_{\perp i}' \partial z_{\perp j}'} \right\|^{-\frac{1}{2}} \right]_{\boldsymbol{z}^0, z_1'^0, z_2'^0} \\
&\cdot \exp\frac{i}{\hbar} S_{cl}(\boldsymbol{p}', \boldsymbol{p}; t'-t),
\end{aligned} \tag{35}$$

where

$$S_{cl}(\boldsymbol{p}', \boldsymbol{p}; t'-t) = S_{cl}(\boldsymbol{z}', \boldsymbol{p}; t'-t) - \boldsymbol{p}' \cdot \boldsymbol{z}', \tag{36}$$

$$\frac{\partial S_{cl}(\boldsymbol{z}', \boldsymbol{p}, t'-t)}{\partial z'_j} = p'_j. \tag{37}$$

Substituting (27), (35) into (26), collecting all the factors and using the Legendre transformation relations (28), (29), (36) and (37), one obtains

$$_{in}\langle \boldsymbol{P}'_1, \boldsymbol{P}'_2 | S | \boldsymbol{P}_1, \boldsymbol{P}_2 \rangle_{in} = \frac{i}{2\pi\hbar} \delta^3(\boldsymbol{P}-\boldsymbol{P}')\,\delta(p-p') \sum \left[\left\| \frac{\partial^2 S_{cl}(\boldsymbol{z}', \boldsymbol{z};\, t'-t)}{\partial z'_j\, \partial z_i} \right\|^{\frac{1}{2}} \right.$$
$$\left. \cdot \left\| \frac{\partial^2 S_{cl}(\boldsymbol{z}', \boldsymbol{z};\, t'-t)}{\partial z_i\, \partial z_j} \right\|^{-\frac{1}{2}} \cdot \left\| \frac{\partial^2 S_{cl}(\boldsymbol{z}', \boldsymbol{p}, t'-t)}{\partial z'_{\perp i}\, \partial z'_{\perp j}} \right\|^{-\frac{1}{2}} \exp \frac{i}{\hbar} S_{cl}(\boldsymbol{p}', \boldsymbol{p}, E) \right], \tag{38}$$

where

$$S_{cl}(\boldsymbol{p}', \boldsymbol{p}, E) = S_{cl}(\boldsymbol{z}', \boldsymbol{z};\, t'-t) + \boldsymbol{p}\cdot\boldsymbol{z} - \boldsymbol{p}'\cdot\boldsymbol{z}' + E(t'-t),$$
$$\frac{\partial S_{cl}(\boldsymbol{z}', \boldsymbol{z};\, t'-t)}{\partial z'_j} = p'_j,$$
$$\frac{\partial S_{cl}(\boldsymbol{z}', \boldsymbol{z};\, t'-t)}{\partial z_i} = -p_i,$$
$$\frac{\partial S_{cl}(\boldsymbol{z}', \boldsymbol{z};\, t'-t)}{\partial (t'-t)} = -E. \tag{39}$$

Since the above relations constitute a Legendre transformation, one has

$$\frac{\partial S_{cl}(\boldsymbol{p}', \boldsymbol{p}, E)}{\partial p'_j} = -z'_j,$$
$$\frac{\partial S_{cl}(\boldsymbol{p}', \boldsymbol{p}, E)}{\partial p_i} = z_i,$$
$$\frac{\partial S_{cl}(\boldsymbol{p}', \boldsymbol{p}, E)}{\partial E} = t'-t. \tag{40}$$

In the formula (38) Σ denotes the summation of the contributions of all classical orbits with the initial momentum $\boldsymbol{p}$ and the final mementum $\boldsymbol{p}'$.

Now we calculate the product of the determinants in the formula (38),

$$I = \left\| \frac{\partial^2 S_{cl}(\boldsymbol{z}', \boldsymbol{z};\, t'-t)}{\partial z'_j, \partial z_i} \right\| \cdot \left\| \frac{\partial^2 S_{cl}(\boldsymbol{z}', \boldsymbol{z};\, t'-t)}{\partial z_i\, \partial z_j} \right\|^{-1} \cdot \left\| \frac{\partial^2 S_{cl}(\boldsymbol{z}', \boldsymbol{p}, t'-t)}{\partial z'_{\perp i}\, \partial z'_{\perp j}} \right\|^{-1}$$
$$= \frac{\partial(p'_j, z'_k)}{\partial(z_i, z'_k)} \cdot (-1) \frac{\partial(z_j, z'_k)}{\partial(p_i, z'_k)} \cdot \frac{\partial(p_j, z'_k)}{\partial(p_i, p'_{\perp j}, z'_3)} = -\frac{\partial(p'_j, z'_k)}{\partial(p_i, p'_{\perp j}, z'_3)}$$
$$= -\frac{\partial(p'_3, z'_1, z'_2)}{\partial(p_1, p_2, p_3)}.$$

Let $\tilde{p}_i$ denote the three components of the vector $\boldsymbol{p}$ in the direction of the coordinate axes $\boldsymbol{e}_i'(i=1,2,3)$ of $\boldsymbol{z}'$, then

$$\frac{\partial(\tilde{p}_1, \tilde{p}_2, \tilde{p}_3)}{\partial(p_1, p_2, p_3)} = 1,$$

$$\therefore I = -\frac{\partial(p'_3, z'_1, z'_2)}{\partial(\tilde{p}_1, \tilde{p}_2, \tilde{p}_3)}. \tag{41}$$

Let θ denote the scattering angle and b denote the collision parameter as $t' \to \infty$. It can be seen from Fig. 1 that

$$\tilde{p}_3 = p\cos\theta, \quad \tilde{p}_1 = -p\sin\theta\cos\varphi_p, \quad \tilde{p}_2 = -p\sin\theta\sin\varphi_p, \tag{42}$$

$$z'_1 = b\cos\varphi, \quad z'_2 = b\sin\varphi, \quad p'_3 = p. \tag{43}$$

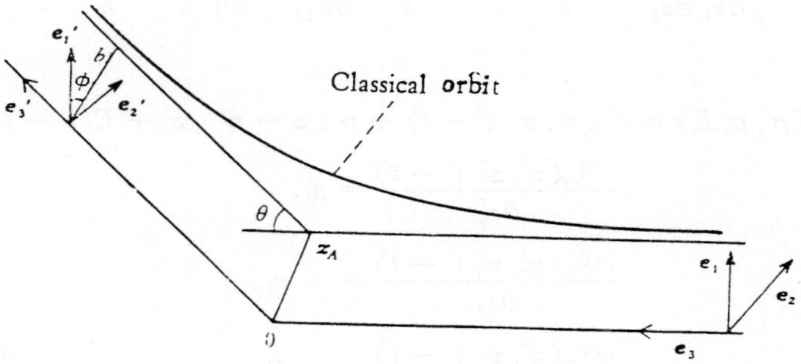

Fig. 1

Therefore,

$$I = -\frac{\partial(p'_3, z'_1, z'_2)}{\partial(\tilde{p}_1, \tilde{p}_2, \tilde{p}_3)} = -\frac{\partial(z'_1, z'_2, p'_3)}{\partial(\cos\theta, \varphi_p, p)}\frac{\partial(\cos\theta, \varphi_p, p)}{\partial(\tilde{p}_1, \tilde{p}_2, \tilde{p}_3)} = -\frac{1}{p^2}\frac{\partial(z'_1, z'_2)}{\partial(\cos\theta, \varphi_p)}. \tag{44}$$

Substituting the formula (44) into the formula (38), one obtains finally that

$$_{in}\langle \boldsymbol{P}'_1, \boldsymbol{P}'_2 | S | \boldsymbol{P}_1, \boldsymbol{P}_2 \rangle_{in} = -\frac{1}{2\pi\hbar}\delta^3(\boldsymbol{P} - \boldsymbol{P}')\delta(E - E')\frac{1}{E}\sum\left[\frac{\partial(z'_1, z'_2)}{\partial(\cos\theta, \varphi_p)}\right]^{\frac{1}{2}}$$
$$\cdot \exp\frac{i}{\hbar}S_{cl}(\boldsymbol{p}', \boldsymbol{p}, E). \tag{45}$$

Formula (45) is the general formula in the semi-classical approximation. It is also applicable to the scattering of ordinary particles. If the classical orbits satisfy the condition

$$\varphi_p = \varphi, \quad \frac{\partial b}{\partial \varphi} = 0, \tag{46}$$

(45) can be reduced to

$$_{in}\langle \boldsymbol{P}'_1, \boldsymbol{P}'_2 | S | \boldsymbol{P}_1, \boldsymbol{P}_2 \rangle_{in} = -\frac{1}{2\pi\hbar}\delta^3(\boldsymbol{P} - \boldsymbol{P}')\delta(E - E')\frac{1}{E}\left(b\frac{\partial b}{\partial\cos\theta}\right)^{\frac{1}{2}}$$
$$\cdot \exp\frac{i}{\hbar}S_{cl}(\boldsymbol{p}', \boldsymbol{p}, E). \tag{47}$$

When there is only one classical orbit having contribution, it follows from (47) that

$$\frac{d\sigma}{d\Omega} = b\left|\frac{\partial b}{\partial \cos\theta}\right|. \tag{48}$$

which is just the well-known formula of the scattering cross-section of classical particles in a centrally symmetric force field.

We now turn to discuss the phase-factor in (45). In the case in which rotational invariance holds, one can write $S_{cl}(\boldsymbol{p}', \boldsymbol{p}, E) = S_{cl}(p', p, E, \tau)$, where $\tau = (\boldsymbol{p}' - \boldsymbol{p})^2$ is the momentum transfer squared. Noting that $\boldsymbol{p}'$, $\boldsymbol{p}$ and E are considered as independent variables in (40), one finds that

$$-z_j' = \frac{\partial S_{cl}(\boldsymbol{p}', \boldsymbol{p}, E)}{\partial p_j'} = \left(\frac{p_j'}{p'}\frac{\partial}{\partial p'} + 2(p_j' - p_j)\frac{\partial}{\partial \tau}\right) S_{cl}(p', p, E, \tau),$$

$$z_i = \frac{\partial S_{cl}(\boldsymbol{p}', \boldsymbol{p}, E)}{\partial p_i} = \left(\frac{p_i}{p}\frac{\partial}{\partial p} - 2(p_i' - p_i)\frac{\partial}{\partial \tau}\right) S_{cl}(p', p, E, \tau),$$

$$t' - t = \frac{\partial S_{cl}(\boldsymbol{p}', \boldsymbol{p}, E)}{\partial E}. \tag{49}$$

In the center of mass system,

$$\boldsymbol{z}' = \boldsymbol{z}_A + \frac{2\boldsymbol{p}'}{E}\left(t' - \frac{\Delta_{cl}}{2}\right),$$

$$\boldsymbol{z} = \boldsymbol{z}_A + \frac{2\boldsymbol{p}}{E}\left(t + \frac{\Delta_{cl}}{2}\right), \tag{50}$$

where Δ_{cl} is the time-delay, namely, the difference of the time elapsed for the classical particle travelling along the orbit and the time for the free particle with the same energy moving along the asymptotics of the orbit. Taking into consideration the relation $(\boldsymbol{p}' \mp \boldsymbol{p})\cdot\boldsymbol{p} = \boldsymbol{p}'\cdot(\boldsymbol{p} \pm \boldsymbol{p}')$, it can be found from (49) and (50) that

$$(p_i' - p_i)\left(\frac{\partial S_{cl}}{\partial p_i'} - \frac{\partial S_{cl}}{\partial p_i}\right) = \frac{1}{p}(\boldsymbol{p}' - \boldsymbol{p})\cdot\boldsymbol{p}'\left(\frac{\partial S_{cl}}{\partial p'} + \frac{\partial S_{cl}}{\partial p}\right) + 4\tau\frac{\partial S_{cl}}{\partial \tau}$$

$$= -2(\boldsymbol{p}' - \boldsymbol{p})\cdot\boldsymbol{z}_A - \frac{2}{E}(\boldsymbol{p}' - \boldsymbol{p})\cdot\boldsymbol{p}'\left(\frac{\partial S_{cl}}{\partial E} - \Delta_{cl}\right),$$

$$(p_i' + p_i)\left(\frac{\partial S_{cl}}{\partial p_i'} + \frac{\partial S_{cl}}{\partial p_i}\right) = \frac{\boldsymbol{p}\cdot(\boldsymbol{p} + \boldsymbol{p}')}{p}\left(\frac{\partial S_{cl}}{\partial p} + \frac{\partial S_{cl}}{\partial p'}\right)$$

$$= \frac{2\boldsymbol{p}'\cdot(\boldsymbol{p} + \boldsymbol{p}')}{E}\left(-\frac{\partial S_{cl}}{\partial E} + \Delta_{cl}\right).$$

From these two formulas it follows that

$$\frac{\partial S_{cl}}{\partial \tau} = -\frac{(\boldsymbol{p}' - \boldsymbol{p})\cdot\boldsymbol{z}_A}{2\tau}, \tag{51}$$

$$\frac{\partial S_{cl}}{\partial E} + \frac{E}{2p}\left(\frac{\partial S_{cl}}{\partial p'} + \frac{\partial S_{cl}}{\partial p}\right) = \Delta_{cl}. \tag{52}$$

Writing $S_{cl}(p', p, E, \tau)$ as $S_{cl}(E, \tau)$ by means of the relations $p = p' = \sqrt{\frac{E^2}{4} - M^2}$

in the center of mass system, (52) is reduced to

$$\frac{\partial S_{cl}(E,\tau)}{\partial E}=\Delta_{cl}, \tag{53}$$

which is the generalization of formula (1) in the case of three dimensions.

IV. Discussion

For the scattering of ordinary particles, the formula (47) was obtained previously in the case of centrally symmetric force field[4]. It was obtained in [4] after a series of approximations with the W. K. B. approximation to the partial wave amplitude as the starting point. Since the conservation of angular momentum is used in the derivation, it is not applicable to the general case of non-centrally-symmetric force field. It should be noticed that in the derivation of the formula (45) from (11) in the present article the assumption of centrally symmetric force field is not needed. In so doing, Formula (45) is more general than the previous results, and the approximations used in the derivation are simpler than those used in [4].

The results obtained in this article is applicable only to the case in which there is at least one classical orbit with the initial momentum $\boldsymbol{p}$ and the final momentum $\boldsymbol{p}'$. If no such classical orbit exists, one must use other methods, for example, the method of complex path. The factor $\partial(z'_1,z'_2)/\partial(\cos\theta,\varphi_p)$ in Formula (45) has, in general, singular lines or singular surfaces. If the classical orbit passes these singular points, there will be an additional phase-factor $\exp i(n\pi/2)$ in the scattering amplitude as has been discussed in previous works.

In the present article the effect of the exchange of "mesons" on the scattering of solitons is neglected. The exchange of "mesons" also causes the fluctuation of the solitons around the classical orbit and can in general contribute to the scattering amplitude in the order of approximation taken. This problem needs further investigation. In the adiabatic approximation its effect can be represented by an additional term of the order $\hbar$ in the potential $\bar{U}$.

The soliton in the three-dimensional space usually has an internal quantum number, for example, the charge. Introducing the collective coordinate of the $U(1)$ group and the corresponding canonical momentum, it is not difficult to generalize the results obtained above to this case. For the soliton-soliton scattering without exchanging the charge the same formula (45) is obtained in this case.

Appendix A

Taking derivatives of the Hamilton-Jacobi equation (22) one finds that

$$m^{-1}_{ij}\frac{\partial^2 S_{cl}}{\partial z'_i\partial z'_k}\frac{\partial^2 S_{cl}}{\partial z'_j\partial z_l}+\frac{\partial m^{-1}_{ij}}{\partial z'_k}\frac{\partial S_{cl}}{\partial z'_i}\frac{\partial^2 S_{cl}}{\partial z'_j\partial z_l}+m^{-1}_{ij}\frac{\partial S_{cl}}{\partial z'_i}\frac{\partial^3 S_{cl}}{\partial z'_j\partial z'_k\partial z_l}$$

$$+\frac{\partial^3 S_{cl}}{\partial t\partial z_l\partial z'_k}=0. \tag{A.1}$$

Let

$$S_{jl} \equiv \frac{\partial^2 S_{cl}}{\partial z_j' \partial z_l}, \qquad T_{kj} \equiv m_{ij}^{-1} \frac{\partial^2 S_{cl}}{\partial z_i' \partial z_k'} + \frac{\partial m_{ij}^{-1}}{\partial z_k'} \frac{\partial S_{cl}}{\partial z_i'}, \tag{A.2}$$

then (A.1) can be written as

$$\nabla_{tz} S_{kl} \equiv \left(\frac{\partial}{\partial t} + m_{ij}^{-1} \frac{\partial S_{cl}}{\partial z_i'} \frac{\partial}{\partial z_j'} \right) S_{kl} = - T_{kj} S_{jl}. \tag{A.3}$$

From (24)

$$\nabla_{tz} f = \frac{1}{2f} \nabla_{tz} (S_{1i} S_{2j} S_{3k} \epsilon_{ijk}) = \frac{1}{2f} [(\nabla t_z S_{1i}) S_{2j} S_{3k} + S_{1i} (\nabla_{tz} S_{2j}) S_{3k} + S_{1i} S_{2j} (\nabla_{tz} S_{3k})] \epsilon_{ijk}.$$

With (A. 3) and (A. 2) the above equation can be reduced to

$$\nabla_{tz} f = -\frac{1}{2f} \left(\sum_{i=1}^{3} T_{ii} \right) \|S_{jk}\| = -\frac{f}{2} \left(m_{ij}^{-1} \frac{\partial^2 S_{cl}}{\partial z_i' \partial z_j'} + \frac{\partial m_{ij}^{-1}}{\partial z_j'} \frac{\partial S_{cl}}{\partial z_i'} \right). \tag{A.4}$$

This is just Eq. (23).

Appendix B

Let there be classical soliton-soliton scattering solutions $\varphi_{cl}(x - z(t), z(t))$ in one-dimensional space. In the system of vanishing total momentum,

$$\varphi_{cl}(x - Z(t), z(t)) \to \varphi_{cl}\left(\frac{x - X_1(t)}{\sqrt{1-u^2}} \right) + \varphi_{cl}\left(\frac{x - X_2(t)}{\sqrt{1-u^2}} \right), \quad \text{as } t \to -\infty,$$

$$\varphi_{cl}(x - Z(t), z(t)) \to \varphi_{cl}\left(\frac{x - X_1'(t)}{\sqrt{1-u^2}} \right) + \varphi_{cl}\left(\frac{x - X_2'(t)}{\sqrt{1-u^2}} \right), \quad \text{as } t \to +\infty,$$

where

$$X_i(t) = X_{iA} \pm u\left(t + \frac{\Delta_{cl}}{2} \right), \qquad X_i'(t) = X_{iA} \pm u\left(t - \frac{\Delta_{cl}}{2} \right). \tag{B.1}$$

According to the scheme of canonical quantization of solitons as described in Sec. 2, after neglecting the terms corresponding to the excitation of the meson freedom, the field operator $\hat{\varphi}(x, t)$ can be expressed as

$$\hat{\varphi}(x, t) = \varphi_{cl}(x - \hat{Z}(t), \hat{z}(t)). \tag{B.2}$$

On the analogy of Formula (5), in the case of one-dimensional space one finds

$$\hat{H} = \frac{1}{2} \hat{P} M^{-1}(\hat{z}) \hat{P} + \frac{1}{2} \hat{p} m^{-1}(\hat{z}) \hat{p} + U(\hat{z}), \tag{B.3}$$

where $\hat{P}$ and $\hat{p}$ are the canonical momenta corresponding to $\hat{Z}$ and $\hat{z}$ respectively. In the center of mass system, the term containing $\hat{P}$ can be dropped. $\hat{z}(t)$ and $\hat{p}(t)$ satisfy the quantum-mechanical Hamiltonian equations of motion,

$$\hat{\dot{z}} = i[\hat{H}, \hat{z}], \qquad \hat{\dot{p}} = i[\hat{H}, \hat{p}]. \tag{B.4}$$

When $|z| \to \infty$, $m(z) \to m_0$, $U(z) \to U_0$,

$$H \to H_0 = \frac{1}{2m_0} p^2 + U_0. \tag{B.5}$$

As is usually done in theories of scattering, one can introduce "in" field, "out" field and the corresponding S operator with the relation

$$\varphi^{\text{out}}(x, t) = S^+ \varphi^{\text{in}}(x, t)\, S. \tag{B.6}$$

$\hat{z}^{\text{in}}(t)$, $\hat{z}^{\text{out}}(t)$ and the canonical momenta $\hat{p}^{\text{in}}$, $\hat{p}^{\text{out}}$ conjugate to them satisfy equations

$$\hat{\dot{z}}^{\text{in}} = i[\hat{H}_0^{\text{in}}, \hat{z}^{\text{in}}], \qquad \hat{\dot{p}}^{\text{in}} = [\hat{H}_0^{\text{in}}, \hat{p}^{\text{in}}] = 0,$$

$$\hat{\dot{z}}^{\text{out}} = i[\hat{H}_0^{\text{out}}, \hat{z}^{\text{out}}], \qquad \hat{\dot{p}}^{\text{out}} = i[\hat{H}_0^{\text{out}}, \hat{p}^{\text{out}}] = 0. \tag{B.7}$$

From these equations and (B. 5) we get

$$\hat{z}^{\text{in}}(t) = \hat{z}^{\text{in}}(0) + \frac{\hat{p}^{\text{in}}}{m_0} t, \qquad \hat{z}^{\text{out}}(t) = \hat{z}^{\text{out}}(0) + \frac{\hat{p}^{\text{out}}}{m_0} t. \tag{B.8}$$

From the conservation of energy,

$$S^+ \hat{H}_0^{\text{in}} S = \hat{H}_0^{\text{in}} = \hat{H}_0^{\text{out}},$$

$$\therefore\ (\hat{p}^{\text{out}})^2 = (\hat{p}^{\text{in}})^2.$$

Corresponding to the classical solution (B. 1), we postulate that

$$\hat{p}^{\text{out}} = \hat{p}^{\text{in}}. \tag{B.9}$$

This equality is equivalent to neglecting the quantum-mechanical reflective wave. (In the case of identical particles, the reflective wave need not be considered. The exchange term automatically involves the reflective wave.) From this equality and the canonical commutation relation one finds that

$$[\hat{z}^{\text{out}}(t), \hat{p}^{\text{in}}] = [\hat{z}^{\text{out}}(t), \hat{p}^{\text{out}}] = i.$$

So $\hat{z}^{\text{out}}(t)$ can be written as

$$\hat{z}^{\text{out}}(t) = \hat{z}^{\text{in}}(t) - \frac{\hat{p}^{\text{in}}}{m_0} \Delta(\hat{p}^{\text{in}}). \tag{B.10}$$

Comparing this formula with (B. 8) it can be seen that $\Delta(\hat{p}^{\text{in}})$ is the time-delay operator. It follows from (B. 9) that S commutes with $\hat{p}^{\text{in}}$ and, therefore, is a function of $\hat{p}^{\text{in}}$ only. Taking into account the unitarity of the S operator, S can be written as

$$S = e^{2i\delta(\hat{p}^{\text{in}})} \tag{B.11}$$

It follows from (B. 10) and (B.11) that

$$e^{-2i\delta(\hat{p}^{\text{in}})}\hat{z}^{\text{in}}(t)\,e^{2i\delta(\hat{p}^{\text{in}})} = \hat{z}^{\text{in}}(t) - 2\frac{d}{d\hat{p}^{\text{in}}}\delta(\hat{p}^{\text{in}}) = \hat{z}^{\text{in}}(t) - m_0^{-1}\hat{p}_0^{\text{in}}\Delta(\hat{p}^{\text{in}}).$$

$$\therefore\ 2\frac{d}{d\hat{p}^{\text{in}}}\delta(\hat{p}^{\text{in}}) = m_0^{-1}\hat{p}^{\text{in}}\Delta(\hat{p}^{\text{in}}). \tag{B.12}$$

Sandwiching this equality between the eigenstates of the momentum, one obtains that

$$2\frac{d}{dE}\delta(E) = \Delta(E). \tag{B.13}$$

$\Delta(\hat{p}^{\text{in}})$ is determined by the quantum-mechanical Hamiltonian equations of motions (B. 4). If the semi-classical approximation is used, $\Delta(E) \simeq \Delta_{cl}(E)$, where $\Delta_{cl}(E)$, is determined by the classical Hamiltonian equations of motion. Eq. (B. 13) then becomes identical to (1).

References

[1] Jackiw, R. & Woo, G.: *Phys. Rev.*, **D12** (1975), 1643.
[2] Christ, N. H. & Lee, T. D.: *Phys. Rev.*, **D12** (1975), 1606.
[3] Gutzwiller, M. C.: *J. Math. Phys.*, 8 (1967), 1979.
[4] Ford, K. W. & Wheeler, J. A.: *Ann. Phys.*, 7 (1959), 287;
Berry, M. V. & Mount, K. E.: *Rep. Prog. Phys.*, 35 (1972), 315;
Knoll, J. & Shaffer, R.: *Ann. Phys.*, 97 (1976), 307.

Vol. XXIII No. 1 SCIENTIA SINICA January 1980

THE NON-TOPOLOGICAL SOLITON WITH A NON-ABELIAN INTERNAL SYMMETRY

ZHOU GUANGZHAO (CHOU KUAN-CHAO 周光召),
ZHU ZHONGYUAN (CHU CHUNG-YUAN 朱重远),
DAI YUANBEN (戴元本) AND WU YONGSHI (吴泳时)
(*Institute of Theoretical Physics, Academia Sinica*)
Received September 9, 1978.

ABSTRACT

The non-topological soliton with a non-Abelian internal symmetry in $3+1$ dimensional space-time is examined. The case of $SU(2)$ internal symmetry is discussed in detail. Existence and stability of the classical single-soliton solution are investigated with the example of a concrete model. The quantization of the single-soliton solution is carried out with the method of collective coordinates. It is pointed out that the quantized soliton may possess, in addition to ordinary isotopic-spin quantum numbers $\mathscr{I}$ and $\mathscr{I}_3$, a new quantum number $\mathscr{I}'_3$. A new Lorentz covariant method for the quantization of the moving soliton is proposed.

I. INTRODUCTION

There are soliton solutions in many nonlinear field theories. Physically, they represent extended objects. The possibility of using these soliton solutions to describe hadrons has been explored in many literatures[1]. Early investigations were concentrated on $1+1$ dimensional nonlinear scalar field theories. In several models of this type, static soliton solutions can be found out in analytical form. In physical $3+1$ dimensional space-time, according to Derrick's theorem, there is no static solution for a system of scalar fields. However, if the system of nonlinear scalar fields has an internal symmetry G, it is possible that soliton solutions of the form $\exp\{i\alpha_i(t)T_i\}\cdot\varphi_c(\boldsymbol{x})$ exist, where T_i's are generators of the group G in the representation to which the scalar field φ belong and $\alpha_i(t)$ are group parameters which are nevertheless time-dependent. The case of an Abelian group G was investigated carefully by Friedberger, Lee and Sirlin[2]. The case of a non-Abelian group G is more complicated. In this article we shall examine the soliton of the system of scalar fields with a non-Abelian internal symmetry, taking the iso-spin $SU(2)$ group as an example.

In Sec. II, equations determining the classical soliton solution and the general form of these classical solutions are discussed. The cases of $I=1/2$ and $I=1$ representations of the group $SU(2)$ are discussed in detail.

In Sec. III, a $G=SU(2)$ model analogous to the $U(1)$ model discussed in [2] is considered as an example. Existence of classical single-soliton solutions is proved and stability of these solutions is discussed for this model. It is shown that only solutions in which $\varphi_i(\boldsymbol{x})$ has merely components corresponding to $I_3=\pm I_{3\max}$, are stable.

In Sec. IV, the collective coordinate method (taking group parameters $\alpha_i(t)$ as collective coordinates) proposed by Christ and Lee[3] is used to quantize the classical single-soliton solution. Consistency of the perturbation theory starting from the semi-classical approximation is discussed. A new treatment of the problem of Lorentz covariance for the moving soliton is proposed. It is pointed out that in the semi-classical approximation the collective motion of the soliton in the iso-spin space in a theory of iso-spinor φ field is one of a spherically symmetric top, so that the corresponding quantized single-soliton state has three isotopic-spin quantum numbers $\mathscr{I}$, $\mathscr{I}_3$ and $\mathscr{I}'_3$, instead of two. Meanwhile, in the case of the φ field in $I=1$ representation the motion of the soliton in the isotopic-spin space corresponds to a plane rotator.

Finally, the physical implications of the results obtained are discussed in Sec. V.

II. General Discussion of Classical Soliton Solutions

1. *General Formulation*

Let us consider a system of scalar field $\varphi(x)$ which is a representation of the group G. Generally speaking, the representation may be reducible. Under the operation of the element g of the group G,

$$\begin{aligned} \varphi(x) &\to \varphi'(x) = U(g)\varphi(x) \equiv g\varphi(x), \\ \varphi^+(x) &\to \varphi'^+(x) = \varphi^+(x)U^{-1}(g) \equiv \varphi^+(x)g^{-1}, \end{aligned} \tag{1}$$

where $U(g)$ is the unitary matrix representation of g. In the following, $U(g)$ will be denoted simply by g. The Lagrangian of the system is

$$\mathscr{L} = -\partial_\mu\varphi^+\partial_\mu\varphi - V(\varphi, \varphi^+), \tag{2}$$

where the nonlinear potential $V(\varphi, \varphi^+)$ is invariant under the group G.

We now may seek the classical soliton solution of the form

$$\varphi(x) = g(t)\varphi_c(\boldsymbol{x}), \tag{3}$$

where $g(t) \equiv g(\alpha_i(t))$ in which α_i's depend only on t, while $\varphi_c(\boldsymbol{x})$ depends only on the space-coordinates $\boldsymbol{x}$. Substituting (3) into the Euler-Lagrange equation,

$$\partial_\mu\partial_\mu\varphi(x) = \frac{\partial V(\varphi, \varphi^+)}{\partial\varphi^+}, \tag{4}$$

corresponding to the Lagrangian (2) and multiplying the both sides of the equation by $g^{-1}(t)$, one obtains the equation,

$$[i\dot{\Omega}(t) - (\Omega(t))^2]\varphi_c(\boldsymbol{x}) = \nabla^2\varphi_c(\boldsymbol{x}) - \frac{\partial V(\varphi_c, \varphi_c^+)}{\partial\varphi_c^+}, \tag{5}$$

where

$$\dot{\Omega}(t) \equiv \frac{d\Omega(t)}{dt},$$

$$i\Omega(t) \equiv g^{-1}\frac{\partial g}{\partial\alpha_k}\dot{\alpha}_k = g^{-1}\frac{\partial g}{\partial t}. \tag{6}$$

The requirement that Eq. (5) has solutions, leads to certain conditions satisfied by $g(t)$. These conditions can also be derived from the condition that the solution (3) should satisfy the principle of a least action. Substituting (3) into (2) one finds

$$L^{(c)} = \int d^3\boldsymbol{x}\mathscr{L} = \frac{1}{2} M_{ik}^{(c)} \dot{\alpha}_i \dot{\alpha}_k - \int d^3\boldsymbol{x} \overline{V}(\varphi_c, \varphi_c^+), \tag{7}$$

where

$$M_{ik}^{(c)} = \int d^3\boldsymbol{x} \varphi_c^+(\boldsymbol{x}) \left(\frac{\partial g^{-1}}{\partial \alpha_i} \frac{\partial g}{\partial \alpha_k} + \frac{\partial g^{-1}}{\partial \alpha_k} \frac{\partial g}{\partial \alpha_i} \right) \varphi_c(\boldsymbol{x}), \tag{8}$$

$$\overline{V}(\varphi, \varphi^+) = \nabla\varphi^+ \cdot \nabla\varphi + V(\varphi, \varphi^+). \tag{9}$$

Noting that, from the invariance of the potential V under G, we have

$$V(\varphi, \varphi^+) = V(\varphi_c, \varphi_c^+).$$

$\alpha_i(t)$'s are considered as generalized coordinates whose corresponding canonical momenta are

$$P_k^{(c)} = \frac{\partial L^{(c)}}{\partial \dot{\alpha}_k} = M_{kk'}^{(c)} \dot{\alpha}_{k'}. \tag{10}$$

In order to see the physical meaning of $P_k^{(c)}$, we may write

$$dg \cdot g^{-1} = i I_k d\alpha_k, \quad \frac{\partial g}{\partial \alpha_k} = i I_k g, \tag{11}$$

where I_k is the generator corresponding to the group parameter α_k. It is then not difficult to prove that when $\varphi(x)$ takes the form of (3),

$$P_k^{(c)} = -i \int d^3\boldsymbol{x} \left\{ \varphi^+(x) I_k \frac{\partial \varphi(x)}{\partial t} - \frac{\partial \varphi^+(x)}{\partial t} I_k \varphi(x) \right\}. \tag{12}$$

Therefore, $P_k^{(c)}$ is nothing but the component of the conserved isotopic-spin corresponding to the generator I_k.

The Hamiltonian corresponding to (7) is

$$H^{(c)} = \frac{1}{2} M_{kk'}^{(c)^{-1}} P_k^{(c)} P_{k'}^{(c)} + \int d^3\boldsymbol{x} \overline{V}(\varphi_c, \varphi_c^+). \tag{13}$$

Evidently, $H^{(c)}$ is just the energy E_c when $\varphi(x)$ takes the form of (3), and then

$$H^{(c)} = E_c = \int d^3\boldsymbol{x} \left[\frac{\partial \varphi^+}{\partial t} \frac{\partial \varphi}{\partial t} + \nabla\varphi^+ \cdot \nabla\varphi + V(\varphi, \varphi^+) \right]. \tag{14}$$

Noting that $V(\varphi_c, \varphi_c^+)$ does not depend on α_k, we have the Euler-Lagrange equation of α_k in the form,

$$\frac{d}{dt} (M_{kk'}^{(c)} \dot{\alpha}_{k'}) = \frac{1}{2} \frac{\partial M_{ll'}^{(c)}}{\partial \alpha_k} \dot{\alpha}_l \dot{\alpha}_{l'}. \tag{15}$$

This is the necessary condition for $g(t)\varphi_c(\boldsymbol{x})$ to be a solution of the equation of motion.

2. *The Case of* $G = SU(2)$, $I = 1/2$

For definiteness, consider the case of $G = SU(2)$ with φ belonging to $I = 1/2$ representation. Take the Euler angles (φ, θ, ψ) as parameters of the element g of the $SU(2)$ group (the angle φ should not be confused with the field $\varphi(x)$),

$$g = e^{iI_3\varphi} e^{iI_2\theta} e^{iI_3\psi}, \tag{16}$$

where $I_a = \frac{1}{2}\sigma_a$. Using the equality

$$g^{-1}dg = i\{[\cos\theta I_3 + \sin\theta(\cos\psi I_1 + \sin\psi I_2)]d\varphi + (-\sin\psi I_1 + \cos\psi I_2)d\theta + I_3 d\psi\} \tag{17}$$

and the anticommutation relation for $I = 1/2$ representation,

$$I_a I_b + I_b I_a = \frac{1}{2}\delta_{ab}, \tag{18}$$

we can easily find from (8) that

$$M^{(c)}_{\varphi\varphi} = M^{(c)}_{\theta\theta} = M^{(c)}_{\psi\psi} = M_{I0},$$
$$M^{(c)}_{\varphi\psi} = M_{I0}\cos\theta, \quad M^{(c)}_{\varphi\theta} = M^{(c)}_{\psi\theta} = 0, \tag{19}$$

where

$$M_{I0} = \frac{1}{2}\int d^3\boldsymbol{x}\varphi_c^+(x)\varphi_c(\boldsymbol{x}). \tag{20}$$

Therefore the Lagrangian (7) can be written as

$$L^{(c)} = \frac{1}{2}M_{I0}(\dot{\theta}^2 + \dot{\varphi}^2 + \dot{\psi}^2 + 2\dot{\varphi}\dot{\psi}\cos\theta) - \int d^3\boldsymbol{x}\overline{V}(\varphi_c, \varphi_c^+), \tag{21}$$

which is formally the same as the Lagrangian of a spherically symmetric top. The generalized momenta (10) corresponding to $\varphi(t)$, $\theta(t)$ and $\psi(t)$ are

$$P^{(c)}_\varphi = M_{I0}(\dot{\varphi} + \dot{\psi}\cos\theta), \quad P^{(c)}_\theta = M_{I0}\dot{\theta}, \quad P^{(c)}_\psi = M_{I0}(\dot{\psi} + \dot{\varphi}\cos\theta), \tag{22}$$

respectively. The Hamiltonian corresponding to (21) is

$$H^{(c)} = \frac{1}{2M_{I0}}\left[P^{(c)2}_\theta + \frac{1}{\sin^2\theta}(P^{(c)2}_\varphi + P^{(c)2}_\psi - 2P^{(c)}_\varphi P^{(c)}_\psi\cos\theta)\right] + \int d^3\boldsymbol{x}\overline{V}(\varphi_c, \varphi_c^+). \tag{23}$$

Recalling that the ordinary isotopic-spin components of the field $\varphi(x)$ are

$$\mathcal{I}_a = -i\int d^3\boldsymbol{x}\left[\varphi^+(x)I_a\frac{\partial\varphi(x)}{\partial t} - \frac{\partial\varphi^+(x)}{\partial t}I_a\varphi(x)\right], \quad (a = 1, 2, 3) \tag{24}$$

we can readily verify that

$$P^{(c)}_\varphi = \mathcal{I}_3,$$
$$P^{(c)2}_\theta + \frac{1}{\sin^2\theta}[P^{(c)2}_\varphi + P^{(c)2}_\psi - 2P^{(c)}_\varphi P^{(c)}_\psi\cos\theta] = \mathcal{I}_a\mathcal{I}_a = \mathcal{I}^2. \tag{25}$$

Introducing the moving frame in the iso-spin space in connection with Euler angles (φ, θ, ψ), we find that the generator of the rotation about the third axis of the moving frame is

$$I_3' = gI_3g^{-1} = -\sin\theta(\cos\varphi I_1 - \sin\varphi I_2) + \cos\theta I_3. \tag{26}$$

It can be easily verified that the component $\mathscr{I}_3'$ of the isotopic-spin of the field $\varphi(x)$ along the third axis of the moving frame equals P_ψ:

$$\mathscr{I}_3' = -i\int d^3\boldsymbol{x}\left[\varphi^+(x)I_3'\frac{\partial\varphi(x)}{\partial t} - \frac{\partial\varphi^+(x)}{\partial t}I_3'\varphi(x)\right] = P_\psi^{(c)}. \tag{27}$$

In the present case, Eq. (12) is reduced to

$$\begin{aligned} &\ddot{\theta} + \dot{\varphi}\dot{\psi}\sin\theta = 0, \\ &\ddot{\varphi} + \ddot{\psi}\cos\theta - \dot{\theta}\dot{\psi}\sin\theta = 0, \\ &\ddot{\psi} + \ddot{\varphi}\cos\theta - \dot{\theta}\dot{\varphi}\sin\theta = 0. \end{aligned} \tag{28}$$

Substituting (17) into (6) one finds the expression of $\Omega(t)$ as

$$\Omega(t) = \sum_{i=1}^{3}\omega_i I_i, \tag{29}$$

where

$$\begin{aligned} \omega_1 &= -\dot{\theta}\sin\varphi + \dot{\varphi}\sin\theta\cos\psi, \\ \omega_2 &= \dot{\theta}\cos\psi + \dot{\varphi}\sin\theta\sin\psi, \\ \omega_3 &= \dot{\psi} + \dot{\varphi}\cos\dot{\theta}. \end{aligned} \tag{30}$$

The Euler equation, $\dot{\Omega}(t) = 0$ of the spherically symmetric top can be easily derived from Eq. (28). With this result, Eq. (5) is reduced to the form

$$\left[\nabla^2 - \frac{\partial V(\varphi_c^+\varphi_c)}{\partial(\varphi_c^+\varphi_c)} + \frac{\omega^2}{4}\right]\varphi_c(\boldsymbol{x}) = 0, \tag{31}$$

where $\omega^2 = \sum_{i=1}^{3}\omega_i^2$ is constant.

If $\varphi(x)$ is the direct-sum of $I = 1/2$ and $I = 0$ representations, similar results can be obtained for the $I = 1/2$ part of $\varphi(x)$. In the next section we shall prove that Eq. (31) has soliton solutions in a concrete model.

3. *The Case of* $G = SU(2)$, $I = 1$

Let us consider the real field belonging to the $I = 1$ representation of $G = SU(2)$, in which the components of $\varphi(x)$ in the rectangular basis in the iso-spin space are real. As an Ansatz, $\boldsymbol{\varphi}_c(\boldsymbol{x})$ in (3) is required to have the form (cf. the next paragraph),

$$\boldsymbol{\varphi}_c(\boldsymbol{x}) = f_c(\boldsymbol{x})\boldsymbol{\varphi}_0, \tag{32}$$

where $\boldsymbol{\varphi}_0$ is a constant iso-vector. Suitably choosing the basis in the iso-spin space (or making suitable $SU(2)$ transformations to $\boldsymbol{\varphi}_0$) $\boldsymbol{\varphi}_0$ can be brought into the form,

$$\begin{aligned} \varphi_c^1(\boldsymbol{x}) &= \varphi_c^2(\boldsymbol{x}) = 0, \\ \varphi_c^3(\boldsymbol{x}) &= f_c(\boldsymbol{x}). \end{aligned} \tag{33}$$

For the real field,

$$\begin{aligned} L &= \int d^3\boldsymbol{x}\mathscr{L} = \int d^3\boldsymbol{x}\left[-\frac{1}{2}\partial_\mu\varphi\cdot\partial_\mu\varphi - V(\varphi)\right] \\ &= \frac{1}{2}M_{ik}^{(c)}\dot{\alpha}_i\dot{\alpha}_k - \int d^3\boldsymbol{x}\bar{V}(\varphi_c), \end{aligned} \tag{34}$$

where

$$M_{ik}^{(c)} = \frac{1}{2}\int d^3\boldsymbol{x}\varphi_c(\boldsymbol{x})\left(\frac{\partial g^{-1}}{\partial\alpha_i}\frac{\partial g}{\partial\alpha_k} + \frac{\partial g^{-1}}{\partial\alpha_k}\frac{\partial g}{\partial\alpha_i}\right)\varphi_c(\boldsymbol{x}), \tag{35}$$

$$\bar{V}(\varphi_c) = \frac{1}{2}\nabla\varphi_c\cdot\nabla\varphi_c + V(\varphi_c). \tag{36}$$

From (33), one finds that

$$I_3\varphi_c(\boldsymbol{x}) = 0. \tag{37}$$

Using (17) we can easily obtain that

$$\begin{aligned} &M_{\varphi\varphi}^{(c)} = M_{I0}\sin^2\theta, \qquad M_{\theta\theta}^{(c)} = M_{I0}. \\ &M_{ik}^{(c)} = 0 \quad \text{otherwise}, \end{aligned} \tag{38}$$

where

$$M_{I0} = \int d^3\boldsymbol{x}(f_c(\boldsymbol{x}))^2. \tag{39}$$

Therefore,

$$L^{(c)} = \frac{1}{2}M_{I0}(\dot{\theta}^2 + \dot{\varphi}^2\sin^2\theta) - \int d^3\boldsymbol{x}\bar{V}(\varphi_c), \tag{40}$$

$$H^{(c)} = \frac{1}{2M_{I0}}\left(P_\theta^{(c)^2} + \frac{1}{\sin^2\theta}P_\varphi^{(c)^2}\right) + \int d^3\boldsymbol{x}\bar{V}(\varphi_c), \tag{41}$$

where

$$P_\varphi^{(c)} = M_{I0}\dot{\varphi}\sin^2\theta, \qquad P_\theta^{(c)} = M_{I0}\dot{\theta}. \tag{42}$$

Let $\mathscr{I}_a (a = 1, 2, 3)$ be ordinary components of the isotopic-spin of the φ field, it can be easily proved that

$$\begin{aligned} &P_\varphi^{(c)} = \mathscr{I}_3, \\ &P_\theta^{(c)^2} + \frac{1}{\sin^2\theta}P_\varphi^{(c)^2} = \mathscr{I}_a\mathscr{I}_a = \mathscr{I}^2. \end{aligned} \tag{43}$$

It follows from (37) that $P_\varphi^{(c)} \equiv \mathscr{I}_3' = 0$ in this case.

In the present case Eq. (15) is reduced to

$$\begin{aligned} &\ddot{\theta} - \dot{\varphi}^2\sin\theta\cos\theta = 0, \\ &\frac{d}{dt}(\dot{\varphi}\sin^2\theta) = 0. \end{aligned} \tag{44}$$

Starting from (44) one can prove that

$$\dot{\omega}_1(t) = -\omega_2(t)\omega_3(t),$$
$$\dot{\omega}_2(t) = \omega_1(t)\omega_3(t), \tag{45}$$

where ω_i are defined by (31) and (32). On the other hand, (44) can be derived from (45) independent of the value of $\omega_3(t)$. The reason is that, because of (37), $\psi(t)$ in (16) is not a real dynamical variable of the system, so $\omega_3(t)$ can take any value without changing $\varphi(t)$ and $\theta(t)$ in Eq. (44). Without any loss of generality we can take $\omega_3(t) = 0$. Therefore (45) is reduced to $\dot{\Omega}(t) = 0$ and Eq. (45) is reduced to

$$\left[\nabla^2 - 2\frac{\partial V(f_c^2)}{\partial f_c^2} + \omega^2\right] f_c(\boldsymbol{x}) = 0, \tag{46}$$

where ω^2 is a constant. For the model considered in the next section this equation has soliton solutions.

Note that in the present case that is different from the $I = 1/2$ case, the motion of the soliton in the iso-spin space is a plane rotator.

4. *Discussions of the Case of High-dimensional Representations*

With the case of high-dimensional representations, the general discussion is more complicated. However, the discussion can be simplified in certain special cases.

Recall that in the two cases discussed above we have $\dot{\Omega}(t) = 0$. Generally speaking, we can only ascertain that $(i\dot{\Omega}(t) - (\Omega(t))^2)\varphi_c(\boldsymbol{x})$ on the left-hand side of (5) is independent of t. If we require further that

$$i\dot{\Omega}(t) - (\Omega(t))^2 = \text{constant matrix}, \tag{47}$$

it can be derived from the Hermitian property of $\Omega(t)$ that $\dot{\Omega}(t) = 0$. From (6) one finds

$$g(t) = g(0)\exp(i\Omega t), \tag{48}$$

where $g(0)$ is independent of t. Therefore, Eq. (5) is reduced to

$$(\nabla^2 + \Omega^2)\varphi_c(\boldsymbol{x}) = \frac{\partial V(\varphi_c, \varphi_c^+)}{\partial \varphi_c^+}. \tag{49}$$

Note that the solution $\varphi(x) = g(t)\varphi_c(\boldsymbol{x})$ is an invariant under the transformation

$$\varphi_c(\boldsymbol{x}) \to g_0\varphi_c(\boldsymbol{x}), \qquad g(t) \to g(t)g_0^{-1}, \tag{50}$$

where g_0 is independent of t. But under (50) $\Omega \to g_0\Omega g_0^{-1}$. Therefore we can transform Ω to a diagonal matrix by g_0 suitably chosen. Denoting the components of $\varphi(x)$ in an irreducible representation of $SU(2)$ by φ^i and diagonal elements of Ω by $\omega_i (i = 1, 2, \cdots, N)$, Eq. (49) takes the form

$$(\nabla^2 + \omega_i^2)\varphi_c^i(\boldsymbol{x}) = \frac{\partial V(\varphi_c^+\varphi_c)}{\partial(\varphi_c^+\varphi_c)}\varphi_c^i(\boldsymbol{x}), \tag{51}$$

in which some $\varphi_c^i(\boldsymbol{x})$ may be zero. If there are M indices $i(M \leqslant N)$ for which $\varphi_c^i(\boldsymbol{x}) \neq 0$, and ω_i^2 are not all equal, the system of Eq. (51) is difficult to be satisfied. Even if it can be satisfied, the energy of the corresponding soliton solution is expected to be not the lowest. In the case of equal ω_i^2's, nonzero $\varphi_c^i(\boldsymbol{x})$'s satisfy the same equation in (51). It is expected that for the lowest energy solution, nonzero $\varphi_0^i(\boldsymbol{x})$'s are all proportional to the same function $f_c(\boldsymbol{x})$. In the model considered in the next section, the situation is indeed so. Therefore, for the lowest-energy stable soliton solution, we shall confine ourselves to the case in which $\Omega^2 \doteq \omega^2 1_M$, i. e. Ω^2 is proportional to a unit matrix in the subspace in which $\varphi_c^i(\boldsymbol{x}) \neq 0$. In this case, Eq. (51) is simplified to

$$(\nabla^2 + \omega^2)\varphi_c^i(\boldsymbol{x}) = \frac{\partial V(\varphi_c^+ \varphi_c)}{\partial(\varphi_c^+ \varphi_c)} \varphi_c^i(\boldsymbol{x}). \tag{52}$$

For reasons stated above we shall consider only solutions of this equation of form

$$\varphi_c^i(\boldsymbol{x}) = f_c(\boldsymbol{x})\varphi_0^i, \tag{53}$$

where φ_0^i is constant for $i = 1, 2, \cdots, M$ and $\varphi_0^i = 0$ otherwise.

For $SU(2)$ group, in the representation in which I_3 is diagonal (spherical basis), Ω, being diagonal, can only have the form $\Omega = \omega I_3$. Because of the condition $\Omega^2 \doteq \omega^2 1_M$, $\varphi_c^i(\boldsymbol{x})$ can only have two nonzero components corresponding to $I_3 = \pm I_{3i}$. In the following, we shall consider mainly classical soliton solutions of this form.

III. Properties of Classical Soliton Solutions

Classical properties of non-topological solitons with an internal $U(1)$ symmetry was discussed in detail in [1, 2]. In this section we shall generalize them to the case of the iso-spin $SU(2)$ group.

As an example, let us consider the following Lagrangian density:

$$\mathscr{L} = -\partial_\mu \varphi^+ \partial_\mu \varphi - \frac{1}{2}\partial_\mu \chi \partial_\mu \chi - V_1(\varphi^+\varphi, \chi) - V_2(\chi), \tag{54}$$

$$V_1 = f^2\chi^2\varphi^+\varphi, \qquad V_2 = \frac{1}{8}g^2(\chi^2 - \chi_v^2)^2,$$

where $\varphi(x)$ belongs to some representation of the $SU(2)$ group and χ is a real scalar field. The equations of motion corresponding to (54) are

$$-\Box\varphi + \frac{\partial V_1}{\partial(\varphi^+\varphi)}\varphi = 0, \tag{55}$$

$$-\Box\chi + \frac{\partial V_1}{\partial \chi} + \frac{\partial V_2}{\partial \chi} = 0. \tag{56}$$

According to the discussions in the above section, we shall concern ourselves only with soliton solutions of the form,

$$\begin{aligned} \varphi(\boldsymbol{x}, t) &= g(0)e^{+i\omega I_3 t}\varphi_c(\boldsymbol{x}), \\ \chi(\boldsymbol{x}, t) &= \chi(\boldsymbol{x}), \end{aligned} \tag{57}$$

where I_3 is diagonal. The isotopic-spin of this solution is

$$\mathscr{I}_a = \sum_{i,j} \omega(I_{3i} + I_{3j}) e^{-i\omega(I_{3i}-I_{3j})t} \int d^3\boldsymbol{x} \varphi_c^i(\boldsymbol{x}) \times [g(0)^{-1} I_a g(0)]^{ij} \varphi_c^j(\boldsymbol{x}),$$

where I_{3i} is the eigenvalue of I_3 corresponding to $\varphi_c^i(\boldsymbol{x})$. Since $\mathscr{I}_a$, being conserved, must be independent of t, the above equation can be reduced to

$$\mathscr{I}_a = \sum_i 2\omega I_{3i} \int d^3\boldsymbol{x} \varphi_c^{+i}(\boldsymbol{x}) [g(0)^{-1} I_a g(0)]^{ii} \varphi_c^i(\boldsymbol{x}).$$

From the relation

$$g^{-1}(0) I_a g(0) = \sum_{a'} R_{aa'} I_{a'},$$

where R is a matrix of rotation determined by $g(0)$, and the fact that only nonzero diagonal elements are those of I_3, the equation of $\mathscr{I}_a$ can be rewritten as

$$\mathscr{I}_a = \sum_i 2\omega I_{3i}^2 R_{a3} \int d^3\boldsymbol{x} \varphi_c^{+i}(\boldsymbol{x}) \varphi_c^i(\boldsymbol{x}).$$

Define the total isotopic-spin as

$$\mathscr{I} \equiv \sqrt{\mathscr{I}_a \mathscr{I}_a} = \sum_i 2\omega I_{3i}^2 \int d^3\boldsymbol{x} \varphi_c^{+i}(\boldsymbol{x}) \varphi_c^i(\boldsymbol{x}), \tag{58}$$

which is evidently independent of $g(0)$. The energy of the field corresponding to (57) is

$$E = \int d^3\boldsymbol{x} \left\{ \nabla\varphi_c^+(\boldsymbol{x}) \cdot \nabla\varphi_c(\boldsymbol{x}) + \frac{1}{2}(\nabla\chi)^2 + V_1 + V_2 + \omega^2 \varphi_c^+(\boldsymbol{x}) I_3^2 \varphi_c(\boldsymbol{x}) \right\}, \tag{59}$$

which is also independent of $g(0)$.

Substituting (57) into (55), equations satisfied by φ_c and χ are found to be

$$\left\{ -\nabla^2 + \frac{\partial V_1}{\partial(\varphi_c^+\varphi_c)} - \omega^2 I_{3i}^2 \right\} \varphi_c^i(\boldsymbol{x}) = 0,$$

$$\left\{ -\nabla^2\chi + \frac{\partial V_1}{\partial\chi} + \frac{\partial V_2}{\partial\chi} \right\} = 0. \tag{60}$$

It can be seen from (58)—(60) that solutions of the form (57) with different $g(0)$'s have the same energy, isotopic-spin and equations of motion. Therefore, we can consider only the solution with $g(0)=1$ as the representative of these solutions. Furthermore, the energy, isotopic-spin and equations of motion are all invariant under arbitrary rotations in the subspace spanned by $\varphi_c^i(\boldsymbol{x})$ and $\varphi_c^{-i}(\boldsymbol{x})$. Therefore, we need only to discuss the case in which $\varphi_c^i(\boldsymbol{x}) = 0$ for $I_{3i} < 0$.

It is convenient to introduce dimensionless quantities by

$$\varphi_c(\boldsymbol{x}) = \frac{\mu}{g} B(\boldsymbol{\rho}), \quad \chi(\boldsymbol{x}) = \frac{\mu}{g} A(\boldsymbol{\rho}),$$

$$K = \frac{m}{\mu} = \frac{f}{g}, \qquad \nu = \frac{\omega}{\mu},$$

$$m = f\chi_\nu, \qquad \mu = g\chi_\nu,$$

$$\boldsymbol{\rho} = \mu \boldsymbol{x}.$$

(58—60) can now be written as

$$\mathscr{I} = \frac{2\nu}{g^2}\int d^3\rho B^+ I_3^2 B, \tag{61}$$

$$E = \frac{\mu}{g^2}\int d^3\rho \left\{\nabla_\rho B^+ \nabla_\rho B + \frac{1}{2}(\nabla_\rho A)^2 + K^2 A^2 B^+ B + \frac{1}{8}(A^2-1)^2\right\} + \frac{1}{2}\omega\mathscr{I}, \tag{62}$$

$$\left[-\nabla_\rho^2 + 2K^2 B^+ B + \frac{1}{2}(A^2-1)\right]A = 0,$$

$$[-\nabla_\rho^2 + K^2 A^2 - \nu^2 I_3^2]B = 0. \tag{63}$$

It can be easily proved that (63) can be obtained from the variation of (62) with respect to A and B at fixed $\mathscr{I}$.

1. "*Free*" *Solutions—Solutions Independent of Space Coordinates*

Putting $A(\boldsymbol{\rho}) =$ constant and $B(\boldsymbol{\rho}) =$ constant inside the cubic box of the volume V in Eq. (63) the condition for non-vanishing B^i is found to be

$$A^2 = \nu^2 I_{3i}^2 / K^2.$$

It follows that other components must be zero. Denoting this solution by $B_{(i)}$, one has

$$B_{(i)}^j = \delta_i^j B,$$

and

$$B^* B = \frac{1}{4K^2}(1 - A^2).$$

Evidently, corresponding to each I_{3i}, there is a solution with the total isotopic-spin

$$\mathscr{I} = \frac{V\nu I_{3i}^2}{2K^2 g^2}\left(1 - \frac{\nu^2 I_{3i}^2}{K^2}\right),$$

and the energy

$$E = \omega\mathscr{I} + \frac{\mu K^4 g^2 \mathscr{I}^2}{2V\nu^2 I_{3i}^4}.$$

In the limit $V \to \infty$, $B^* \cdot B \to 0$, $A \to 1$ and $\omega I_{3i} \to m$. we have

$$E = \frac{m\mathscr{I}}{I_{3i}}. \tag{64}$$

This formula tells us that for fixed $\mathscr{I}$ "free" solutions corresponding to different I_{3i}'s differ in their energies. The larger the I_{3i} the lower the energy. The solution with the least energy has only one component corresponding to $I_{3\max}$ different from zero.

Are these solutions all classically stable? To answer this question, one must calculate the second variation of the energy. It is easy to prove that

$$(\delta^2 E)_{\mathscr{I}} = \frac{\mu}{g^2}\int d^3\rho \Psi^+ H \Psi + \frac{2\nu^3\mu}{\mathscr{I} g^4}\left[\int d^3\rho(\Psi^+ I_3^2 B + B^+ I_3^2 \Psi)\right]^2, \tag{65}$$

where

$$H = \begin{pmatrix} -\frac{1}{2}\nabla_\rho^2 + K^2B^+B + \frac{3A^2-1}{4}, & 2K^2AB^+ \\ 2K^2AB, & -\nabla_\rho^2 + K^2A^2 - \nu^2 I_3^2 \end{pmatrix}, \tag{66}$$

$$\Psi = \begin{pmatrix} \delta A \\ \delta B \end{pmatrix}, \qquad \Psi = (\delta A, \delta B^+). \tag{67}$$

Substituting in (65) the "free" solution of A and B and taking the limit $V \to \infty$, we find that only the diagonal terms of H do not vanish. The eigenvalues of H are continuous spectra starting from $\frac{1}{4}(3A^2-1)$ and $(K^2A^2 - \nu^2 I_{3j}^2)$ respectively. If for some solution $B_{(l)}(l \neq I_{3\max})$ we choose

$$\delta B_{(l)}^i \begin{cases} \neq 0 & \text{for } i = I_{3\max}, \\ = 0 & \text{for } i \neq I_{3\max}, \end{cases}$$

$$\delta A = 0,$$

then we have

$$(\delta^2 E)_{\mathscr{I}} = \frac{\mu}{g^2}\int d^3\rho(\delta B_{(l)}^+)(\nu^2 I_{3l}^2 - \nu^2 I_{3\max}^2)\delta B_{(l)} < 0.$$

This means the "free" solution with $l \neq I_{3\max}$ is classically unstable. For the "free" solution with $l = I_{3\max}$, H has no negative eigenvalue. Therefore this solution is classically stable.

2. *Existence of Classical Soliton Solutions*

In this section we shall prove that stable $SU(2)$ soliton solutions exist. For this purpose, take the following trial solution:

$$A = \begin{cases} 0, & \rho \leqslant \mu R, \\ 1 - \exp\left[-\frac{\rho - \mu R}{\mu a}\right], & \rho \geqslant \mu R, \end{cases}$$

$$B = \begin{cases} b_{(l)}^i/\rho \cdot \sin \omega I_{3l}/\mu, & \rho \leqslant \mu R, \\ 0, & \rho \geqslant \mu R, \end{cases}$$

$$\omega I_{3l} R = \pi, \quad b_{(l)}^i = \delta_l^i b_{(l)}. \tag{68}$$

This trial solution satisfies the equation of motion for $\rho \leqslant \mu R$ and has the correct behaviour for $\rho \gg \mu R$. The isotopic-spin and the energy of the trial solution (68) are

$$\mathscr{I} = \frac{4\pi^2}{g^2} b_{(l)}^+ I_3^2 b_{(l)},$$

$$E = \omega\mathscr{I} + \frac{\pi\mu^4}{6g^2}\left\{R^3 + \frac{11}{4}\cdot R^2 d + \frac{89}{24}Rd^2 + \frac{635}{288}d^3 + \frac{6}{\mu^2 d}\right.$$
$$\left.\cdot\left(R^2 + Rd + \frac{1}{2}d^2\right)\right\} \tag{69}$$

respectively. For solutions of the form (68), the larger I_3 is, the smaller ω will be. It can be seen from (69) that for fixed $\mathscr{I}$ the solution with $I_{3l} = I_{3\,\max}$ has the lowest energy. For this branch of trial solutions, we have

$$\omega I_{3\,\max} = \frac{\pi}{R}.$$

It can be proved by the method used in [2] that at least for ω sufficiently small ($R \gg d - 0(\mu^{-1})$) and

$$\frac{\mathscr{I}}{I_{3\,\max}} \geqslant \frac{1}{2g^2}\left(\frac{4\pi\mu}{3m}\right)^4,$$

the energy of the trial solution is lower than that of the "free" solution. This means, when the isotopic-spin of the field becomes larger than certain critical value, stable soliton solutions must exist.

3. *Soliton Solutions for ω Value Near to m/I_{3l}*

In the $U(1)$ case, the energy of the soliton solution approaches that of the "free" solution as $\omega \to m$. For the $SU(2)$ case considered in this section, there is a series of "free" solutions with $\omega \to m/I_{3l}$. There is also a series of corresponding soliton solutions. If we solve the equation in the approximation $\omega \to m/I_{3l}$ and substitute it into the expressions of the isotopic-spin and the energy, we find $\mathscr{I} \to \infty$ and

$$E_l \simeq \frac{m}{I_{3l}}\mathscr{I} + \frac{8m}{I_{3l}\mathscr{I}}\left(\frac{\pi I_{3l} M_2}{g^2 K^4}\right)^2 + \cdots \tag{70}$$

as $\omega \to m/I_{3l}$, where $M_2 > 0$. It can be seen by comparing this formula with (64) that there is indeed a series of soliton solutions whose energy approaches to that of the corresponding "free" solution from above as $\omega \to m/I_{3l}$ and $\mathscr{I} \to \infty$.

4. *Stability of Soliton Solutions*

It can be easily shown that soliton solutions with $I_{3l} \neq I_{3\max}$ are classically unstable. In fact, if we choose

$$\delta B^i_{(l)} = \begin{cases} \delta\lambda \cdot B^i_{(l)}, & \text{for } i = I_{3\,\max}, \\ 0, & \text{for } i \neq I_{3\,\max}, \end{cases}$$
$$\delta A = 0,$$

in (65), we can find

$$(\delta^2 E)_{\mathscr{I}, B_{(l)}, A} = \frac{\mu}{g^2}\int d^3\rho\, \delta B^+(-\nabla^2 + K^2 A^2 - \nu^2 I_3^2)\delta B$$
$$= \frac{\mu}{g^2}\int d^3\rho\, \delta B^+(\nu^2 I_{3l}^2 - \nu^2 I_{3\,\max}^2)\delta B < 0.$$

This proves the above conclusion.

As to the branch of the soliton solution $B_{(I=I_{3\max})}$ it has been shown that (i) there is at least one solution in this branch whose energy becomes lower than that of the lowest "free" solution when $\mathscr{I}$ becomes larger than certain critical value and ω becomes sufficiently small, (ii) energies of soliton solutions in this branch approach to that of free solutions from above as $\omega \to \mathscr{I}/I_{3\max}$ and $\mathscr{I} \to \infty$. Therefore, the E-$\mathscr{I}$ curve must have the form as shown in Fig. 1.

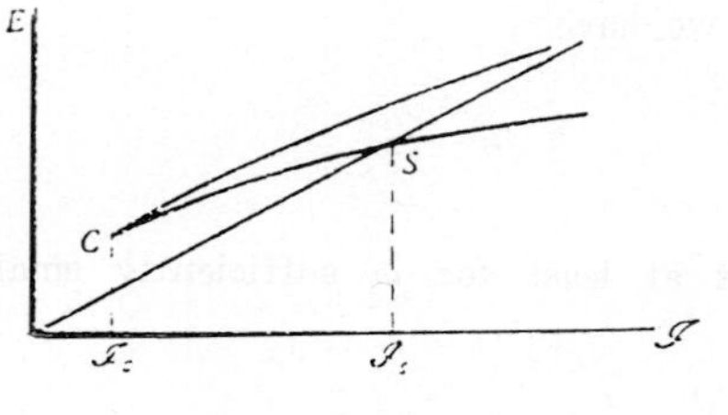

Fig. 1. E-$\mathscr{I}$ curve.

It is very similar to the E–Q curve of $U(1)$[2]. For $\mathscr{I} < \mathscr{I}_s$, the energy of the lower branch of the soliton solution in the figure is the absolute minimum. Therefore it is stable. For the segment CS, the energy of the soliton is higher than that of the "free" solution. However, just like the $U(1)$ case discussed in [2], it is the local minimum and is thus still classically stable. In fact, Theorems 1 and 2 in [2] can be directly brought here and only minor changes need to be made in the last part of Theorem 3, which do not affect the final results. The details will not be described here.

IV. Quantization of Single-soliton Solutions

1. *Canonical Quantization With Collective Coordinates*

Let us assume that classical single-soliton solutions of the form (3) exist and then examine their quantization. Generalizing the method of [2,3], we may write

$$\varphi(x) = g(t)\left[\varphi_c(\boldsymbol{x}) + \sum_n q_n(t)\varphi_n(\boldsymbol{x})\right], \tag{71}$$

where $\varphi_n(\boldsymbol{x})$'s are orthogonal solutions of the following eigenequation,

$$[\nabla^2 + \omega^2 - V''(\varphi_c)]\varphi_n(\boldsymbol{x}) = \omega_n^2\varphi_n(\boldsymbol{x}).$$

corresponding to nonzero eigenvalues ($\omega_n \neq 0$). They are required to satisfy the following conditions,

$$\int d^3\boldsymbol{x}\varphi_n^+(\boldsymbol{x})g^{-1}\frac{\partial g}{\partial \alpha_k}\varphi_c(\boldsymbol{x}) = 0. \tag{72}$$

$$\int d^3\boldsymbol{x}\varphi_n^+(\boldsymbol{x})\frac{\partial}{\partial \boldsymbol{x}}\varphi_c(\boldsymbol{x}) = 0, \tag{72'}$$

in order that the zero-modes corresponding to G invariance and translational invariance can be removed. $q_n(t)$ can be made real by suitably choosing $\varphi_n(\boldsymbol{x})$. By using these formulas the single-soliton Lagrangian is found to be

$$L = \frac{1}{2}\sum_{k,k'}\dot{\alpha}_k M_{kk'}\dot{\alpha}_{k'} + \frac{1}{2}\sum_{n,n'}\dot{q}_n M_{nn'}\dot{q}_{n'} + \sum_{k,n}\dot{\alpha}_k M_{kn}\dot{q}_n$$
$$- \int d^3\boldsymbol{x}\bar{V}\left(\varphi_c + \sum_n q_n\varphi_n\right), \tag{73}$$

where

$$M_{kk'} = \int d^3\boldsymbol{x} \left(\varphi_c^+ + \sum_n q_n \varphi_n^+\right)\left(\frac{\partial g^{-1}}{\partial \alpha_k}\frac{\partial g}{\partial \alpha_{k'}} + \frac{\partial g^{-1}}{\partial \alpha_{k'}}\frac{\partial g}{\partial \alpha_k}\right)$$

$$\cdot \left(\varphi_c + \sum_n q_n \varphi_n\right), \tag{74}$$

$$M_{nn'} = \int d^3\boldsymbol{x} (\varphi_n^+ \varphi_{n'} + \varphi_{n'}^+ \varphi_n) = 2\delta_{nn'}, \tag{75}$$

$$M_{kn} = \sum_{n'} q_{n'} \int d^3\boldsymbol{x} \left\{\varphi_n^+ g^{-1} \frac{\partial g}{\partial \alpha_k} \varphi_{n'} - (n \longleftrightarrow n')\right\}, \tag{76}$$

$$\overline{V}(\varphi) = \nabla\varphi^+ \cdot \nabla\varphi + V(\varphi). \tag{77}$$

Then, the canonical momenta conjugate to $\alpha_k(t)$ and $q_n(t)$ are

$$P_k = M_{kk'}\dot{\alpha}_{k'} + M_{kn}\dot{q}_n, \tag{78}$$

$$\pi_n = 2\dot{q}_n + M_{kn}\dot{\alpha}_k. \tag{79}$$

Then, canonical quantization amounts to replacing generalized coordinates $\alpha_k(t)$ (collective coordinates of the group parameters), $g_n(t)$ and corresponding conjugate momenta $P_k(t)$ and $\pi_n(t)$ by operators satisfying canonical commutation relations. According to the rule given in [3], the quantum Hamiltonian corresponding to (73).is

$$\hat{H} = \frac{1}{2}\hat{J}^{-1}\hat{P}_A \hat{M}_{AA'}^{-1}\hat{J}\hat{P}_{A'} + \int d^3\boldsymbol{x}\overline{V}\left(\varphi_c + \sum_n \hat{q}_n \varphi_n\right), \tag{80}$$

where

$$J = \sqrt{\det M_{AA'}},$$

$$\hat{P} = \begin{pmatrix} \hat{P}_k \\ \hat{\pi}_n \end{pmatrix}, \qquad M_{AA'} = \begin{pmatrix} M_{kk'} & M_{kn'} \\ M_{nk'} & \delta_{nn'} \end{pmatrix}.$$

2. *Perturbation Theory*

Expand $\hat{H}$ to a power series of $\hat{q}_n$ as follows:

$$\hat{H} = \hat{H}_0 + \hat{H}_1 + \hat{H}_2 + \cdots, \tag{81}$$

where $\hat{H}_l$ contains terms of the lth order in $\hat{q}_n$. For definiteness, let us take $G = SU(2)$ and consider the case of φ belonging to $I = 1/2$ representation. It is easy to prove that

$$M_{\varphi\varphi} = M_{\theta\theta} = M_{\psi\psi} = M_I, \qquad M_{\varphi\psi} = M_I \cos\theta,$$
$$M_{\varphi\theta} = M_{\psi\theta} = 0, \tag{82}$$

where

$$M_I = \frac{1}{2}\int d^3\boldsymbol{x}\left(\varphi_c^+ + \sum_n q_n \varphi_n^+\right)\left(\varphi_c + \sum_n q_n \varphi_n\right). \tag{82'}$$

Remembering that M_{kn} is linear in q_n, we can take $\hat{H}_0$ and $\hat{H}_1$ in (81) as,

$$\hat{H}_0 = \frac{1}{2M_{I0}}\hat{\mathscr{I}}^2 + \int d^3\boldsymbol{x}\overline{V}(\varphi_c), \tag{83}$$

$$\hat{H}_1 = \sum_n q_n(t) \int d^3\boldsymbol{x} \left[\frac{\delta \hat{H}_0}{\delta \varphi_c(\boldsymbol{x})} \varphi_n(\boldsymbol{x}) + \varphi_n^+(\boldsymbol{x}) \frac{\delta \hat{H}_0}{\delta \varphi_c^+(\boldsymbol{x})}\right], \tag{84}$$

where M_{I0} is given by (20) and

$$\hat{\mathscr{I}}^2 = \frac{1}{\sin\theta} \hat{P}_\theta \sin\theta \hat{P}_\theta + \frac{1}{\sin^2\theta}(\hat{P}_\varphi^2 + \hat{P}_\psi^2 - 2\cos\theta \hat{P}_\varphi \hat{P}_\psi) \tag{85}$$

is the total isotopic-spin operator. Neglecting the quantum correction of absorption and radiation of mesons in the single soliton state as the lowest order of the semi-classical approximation, we can take only the $\hat{H}_0$ term. It follows from (83) that

$$[\hat{H}_0, \hat{\mathscr{I}}^2] = [\hat{H}_0, \hat{P}_\varphi] = [\hat{H}_0, \hat{P}_\psi] = 0. \tag{86}$$

Therefore, in the lowest order approximation, $\hat{H}_0$, $\hat{\mathscr{I}}^2$, $\hat{P}_\varphi$ and $\hat{P}_\psi$ are commutative conserved quantities in the single-soliton state.

Let $|E_0; \mathscr{I}, M', M\rangle$ be the simultaneous eigenvector of the operators $\hat{H}_0$, $\hat{\mathscr{I}}^2$, $\hat{P}_\varphi$ and $\hat{P}_\psi$, corresponding to eigenvalues E_0, $\mathscr{I}(\mathscr{I}+1)\hbar^2$, $M'\hbar$ and $M\hbar$ respectively. The wave function of this state denoted by

$$\langle \varphi, \theta, \psi | E_0, \mathscr{I}, M', M\rangle = \sqrt{2\mathscr{I}+1}\, D^{\mathscr{I}}_{M'M}(\varphi, \theta, \psi), \tag{87}$$

satisfies the following eigen-equations

$$\left[\frac{1}{\sin\theta}\frac{\partial}{\partial\theta}\left(\sin\theta\frac{\partial}{\partial\theta}\right) + \frac{1}{\sin^2\theta}\left(\frac{\partial^2}{\partial\varphi^2} + \frac{\partial^2}{\partial\psi^2} - 2\cos\theta\frac{\partial^2}{\partial\varphi\partial\psi}\right)\right]$$

$$\cdot D^{\mathscr{I}}_{M'M}(\varphi, \theta, \psi) = -\mathscr{I}(\mathscr{I}+1) D^{\mathscr{I}}_{M'M}(\varphi, \theta, \psi),$$

$$-i\frac{\partial}{\partial\varphi} D^{\mathscr{I}}_{M'M}(\varphi, \theta, \psi) = M' D^{\mathscr{I}}_{M'M}(\varphi, \theta, \psi),$$

$$-i\frac{\partial}{\partial\psi} D^{\mathscr{I}}_{M'M}(\varphi, \theta, \psi) = M D^{\mathscr{I}}_{M'M}(\varphi, \theta, \psi). \tag{88}$$

From the well-known result in the theory of angular momentum, the solution of these equations is

$$D^{\mathscr{I}}_{M'M}(\varphi, \theta, \psi) = e^{iM'\varphi} d^{\mathscr{I}}_{M'M}(\theta) e^{iM\varphi}. \tag{89}$$

Thus the wave function of the single-soliton state $|E_0; \mathscr{I}, M', M\rangle$ in the representation (φ, θ, ψ) is formally the same as that of a spherically symmetric top. It follows from (83) that the energy of this state is

$$E_0 = \frac{1}{2M_{I0}}\mathscr{I}(\mathscr{I}+1)\hbar^2 + \int d^3\boldsymbol{x}\,\overline{V}(\varphi_c). \tag{90}$$

We now proceed to prove that it is always possible to choose a classical single-soliton solution such that $\mathscr{I}^2$, $\mathscr{I}_3$, $\mathscr{I}'_3$ and E_c of this solution are exactly equal to the corresponding eigenvalues of a given single-soliton state $|E_0; \mathscr{I}, M', M\rangle$.

For this purpose, note that the classical equations of motion (28) of $g(t)$ are equivalent to

$$\begin{aligned}&\dot{\theta}^2+\dot{\varphi}^2+\dot{\psi}^2+2\dot{\varphi}\dot{\psi}\cos\theta=\omega^2,\\&\dot{\varphi}+\dot{\psi}\cos\theta=\omega_3',\\&\dot{\psi}+\dot{\varphi}\cos\theta=\omega_3,\end{aligned} \tag{91}$$

where ω^2, ω_3 and ω_3' are constants. It is easy to verify that

$$\mathscr{I}^2=M_{I0}\omega^2, \qquad \mathscr{I}_3=M_{I0}\omega_3', \qquad \mathscr{I}_3'=M_{I0}\omega_3. \tag{92}$$

It follows from (23) that the energy of the classical solution is

$$E_c=\frac{1}{2M_{I0}}\omega^2+\int d^3\boldsymbol{x}\overline{V}(\varphi_c). \tag{93}$$

Therefore, if the classical solution is chosen to satisfy the relations

$$\begin{aligned}&M_{I0}^2\omega^2=\mathscr{I}(\mathscr{I}+1)\hbar^2,\\&M_{I0}\omega_3'=M'\hbar,\\&M_{I0}\omega_3=M\hbar,\end{aligned} \tag{94}$$

its energy and isotopic-spin $\mathscr{I}^2$, $\mathscr{I}_3$ and $\mathscr{I}_3'$ will equal corresponding eigenvalues of the quantum single-soliton state in the lowest approximation.

When (94) is satisfied, it can be derived from (31) that

$$\frac{\delta\hat{H}_0}{\delta\varphi_c^+(\boldsymbol{x})}|E_0;\mathscr{I},M',M\rangle=0. \tag{95}$$

It follows from this equation and (84) that

$$\hat{H}_1|E_0;\mathscr{I},M',M\rangle=0, \tag{96}$$

which effectively eliminates the meson q_n tadpole diagram in the single-soliton sector. This is just the condition for consistency of the perturbation theory discussed in [1, 2].

For the real field $\varphi(x)$ belonging to the $I=1$ representation the discussion can proceed parallel to that described above. In this case we have

$$\hat{H}_0=\frac{1}{2M_{I0}}\hat{\mathscr{I}}^2+\int d^3\boldsymbol{x}\overline{V}(f_c), \tag{97}$$

$$\hat{H}_1=\sum_n\hat{q}_n(t)\int d^3\boldsymbol{x}\varphi_n^T(\boldsymbol{x})\frac{\delta\hat{H}_0}{\delta f_c(\boldsymbol{x})}, \tag{98}$$

where M_{I0} is given by (39), while the total isotopic-spin operator

$$\hat{\mathscr{I}}^2=\frac{1}{\sin\theta}\hat{P}_\theta\sin\theta\hat{P}_\theta+\frac{1}{\sin^2\theta}\hat{P}_\varphi^2 \tag{99}$$

can be obtained from (85) by putting $\hat{P}_\psi=0$. Thus, if we put the corresponding quantum number M and ω_3 in the classical solution to zero, (85)—(96) can be taken over to this case. In particular, in the lowest (semi-classical) approximation the quantum single-soliton state can be characterized by the eigenvalues of $\hat{H}_0$, $\hat{\mathscr{I}}^2$ and $\hat{P}_\varphi$ and denoted as $|E_0;\mathscr{I},M'\rangle$. The energy and wave function of this state are

$$E_0 = \frac{\hbar^2}{2M_{I0}} \mathscr{I}(\mathscr{I}+1) + \int d^3\boldsymbol{x}\overline{V}(f_c),$$

$$\langle\varphi, \theta | \mathscr{I}, M'\rangle = Y_{\mathscr{I}M'}(\theta, \varphi) \tag{100}$$

respectively. The latter is formally the same as the wave function of a plane rotator.

3. *Lorentz Covariance*

Having discussed the single-soliton state at rest, we now proceed to examine the moving soliton state to illustrate the Lorentz covariance of the theory. A new method will be used to treat this problem. Let us take again the example of $G = SU(2)$, $I = 1/2$.

Let us assume that the soliton moves along the z-axis with velocity u. Denote the space-time coordinates in the rest system of the soliton and those in the laboratory system by $(\boldsymbol{x}', t')$ and $(\boldsymbol{x}, t)$ respectively. They are related by the Lorentz transformation:

$$\begin{aligned} x &= x', & y &= y', \\ z &= \gamma(z' + ut), & t &= \gamma(t' + uz'), \\ z' &= \gamma(z - ut), & t' &= \gamma(t - uz), \end{aligned} \tag{101}$$

where $\gamma = (1-u^2)^{-\frac{1}{2}}$. Let us introduce new space-time coordinates describing the motion of the center of mass of the soliton. They are denoted by $(\boldsymbol{X}', T')$ and $(\boldsymbol{X}, T)$ in the rest system of the soliton and in the laboratory system respectively. In the rest system of the soliton,

$$T' = t' \tag{102}$$

is the proper time of the soliton. Under the Lorentz transformation, $(\boldsymbol{X}', T')$ are transformed as

$$\begin{aligned} X &= X', & Y &= Y', \\ Z &= \gamma(Z' + uT'), & T &= \gamma(T' + uZ'), \\ Z' &= \gamma(Z - uT), & T' &= \gamma(T - uZ). \end{aligned} \tag{103}$$

Originally, T' and T are defined only on the world-line of the center of mass of the soliton. We have used (102) to extend the definition of T' to arbitrary space-time points not on the world-line of the center of mass of the soliton keeping T and T' related by the transformation (103). Thus T is also defined on arbitrary points. The relation between t and T so defined is

$$T = \gamma(T' + uZ') = \gamma(t' + uZ') = \gamma^2(t - uz) + \gamma uZ'. \tag{104}$$

Therefore

$$\left.\frac{\partial T}{\partial t}\right|_z = \gamma^2, \quad \frac{\partial T}{\partial z} = -\gamma^2 u, \quad \frac{\partial T}{\partial x} = \frac{\partial T}{\partial y} = 0. \tag{105}$$

Note that, from (104) and (101) we still have $T = t$ on the world-line of the center of mass of the soliton.

In the rest system of the soliton the classical soliton solution is $g(t')\varphi_c(\boldsymbol{x}'-\boldsymbol{X}')$. In the laboratory system we have

$$\varphi(\boldsymbol{x},t)=g(\gamma(t-uz))\varphi_c(x-X,y-Y,\gamma(z-ut)-Z'). \tag{106}$$

Substituting T for t and using (101)—(103) we find that

$$\begin{aligned}\varphi(\boldsymbol{x},T)&=g(T/\gamma)\varphi_c(x-X,\ y-Y,\ \gamma^{-1}(z-Z(T)))\\&\equiv g\varphi_c(\gamma^{-1};\boldsymbol{x}-\boldsymbol{X}(T)),\end{aligned} \tag{107}$$

where an irrelevant additional term $-uZ'$ has been omitted in the argument of $g(T/\gamma)$.

Using (x,y,z,t) as coordinates, the action of the system can be written as

$$A=\int d^3\boldsymbol{x}\,dt\,\mathscr{L}=\gamma^{-2}\int d^3\boldsymbol{x}\,dT\,\mathscr{L}(\varphi(\boldsymbol{x},T))=\int dT L(\alpha(T),\boldsymbol{X}(T)). \tag{108}$$

Using the relations

$$\frac{\partial\varphi}{\partial t}=\gamma^2\frac{\partial\varphi}{\partial T}=\gamma^2\left[\frac{\partial g}{\partial\alpha_i}\varphi_c(\gamma^{-1};\boldsymbol{x}-\boldsymbol{X}(T))\dot{\alpha}_i+g\frac{\partial\varphi_c(\gamma^{-1},\boldsymbol{x}-\boldsymbol{X}(T))}{\partial Z}\dot{Z}\right],$$

$$\frac{\partial\varphi}{\partial x_a}=g\frac{\partial\varphi_c(\gamma^{-1};\boldsymbol{x}-\boldsymbol{X}(T))}{\partial x_a},\qquad a=1,2.$$

$$\frac{\partial\varphi}{\partial z}=g\frac{\partial\varphi_c(\gamma^{-1},\boldsymbol{x}-\boldsymbol{X}(T))}{\partial z}\gamma^2-\gamma^2u\dot{\alpha}_i\frac{\partial g}{\partial\alpha_i}\varphi_c(\gamma^{-1};\boldsymbol{x}-\boldsymbol{X}(T)),$$

with

$$\dot{\alpha}_i=\frac{d\alpha_i}{dT},\qquad \dot{Z}=\frac{dZ(T)}{dT},$$

we can easily obtain

$$L=\frac{\gamma}{2}\dot{\alpha}_iM_{ik}^{(c)}\dot{\alpha}_k+\frac{\gamma}{2}M_0\dot{Z}^2-\frac{M_{I0}}{2\gamma}\omega^2-\frac{1}{2}\left(\gamma+\frac{1}{\gamma}\right)M_0. \tag{109}$$

In the derivation of (109), the fact has been used that Eq. (31) satisfied by $\varphi_c(\boldsymbol{x})$ is the extremum of the variation of the following functional,

$$J[\varphi_c]=\int d^3\boldsymbol{x}'\left[\frac{\partial\varphi_c^+}{\partial x_a'}\frac{\partial\varphi_c}{\partial x_a'}+V(\varphi_c^+\varphi_c)-\frac{\omega^2}{4}\varphi_c^+\varphi_c\right].$$

It follows from the virial theorem that

$$\int d^3\boldsymbol{x}'V(\varphi_c^+\varphi_c)=-\frac{1}{3}\int d^3\boldsymbol{x}'\frac{\partial\varphi_c^+}{\partial x_a'}\frac{\partial\varphi_c}{\partial x_a'}+\frac{\omega^2}{4}\int d^3\boldsymbol{x}'\varphi_c^+\varphi_c,$$

$$(a=1,2,3).$$

$\boldsymbol{M}_0$ in (109) is defined as

$$M_0=\frac{2}{3}\int d^3\boldsymbol{x}'\frac{\partial\varphi_c^+(\boldsymbol{x}')}{\partial x_a'}\frac{\partial\varphi_c(\boldsymbol{x}')}{\partial x_a'}.$$

Substituting $\gamma = [1-(\dot{Z}(T))^2]^{-\frac{1}{2}}$ into (59), we obtain that

$$L = -\left(M_0 + \frac{1}{2} M_{I0}\omega^2\right)\sqrt{1-\dot{Z}^2} + \frac{1}{2}\frac{\dot{\alpha}_i M_{ik}^{(c)}\dot{\alpha}_k}{\sqrt{1-\dot{Z}^2}}. \tag{110}$$

It follows that the canonical momenta conjugate to $\alpha_i(T)$ and $Z(T)$ are

$$P_i = \frac{\partial L}{\partial \dot{\alpha}_i} = \frac{1}{\sqrt{1-\dot{Z}^2}} M_{ik}^{(c)}\dot{\alpha}_k = M_{ik}^{(c)}\alpha'_k, \tag{111}$$

$$P_z = \frac{\partial L}{\partial \dot{Z}} = \left(M_0 + \frac{1}{2} M_{I0}\omega^2\right)\frac{\dot{Z}}{\sqrt{1-\dot{Z}^2}} + \frac{1}{2}\frac{\dot{\alpha}_i M_{ik}^{(c)}\dot{\alpha}_k \dot{Z}}{(1-\dot{Z}^2)^{3/2}}$$

$$= \left(M_0 + \frac{1}{2} M_{I0}\omega^2 + \frac{1}{2}\alpha'_i M_{ik}^{(c)}\alpha'_k\right)\frac{\dot{Z}}{\sqrt{1-\dot{Z}^2}}, \tag{112}$$

where $\alpha'_i = \dfrac{d\alpha_i(T')}{dT'}$. From (110) the classical Hamiltonian is found to be

$$H = \frac{M_0 + \frac{1}{2} M_{I0}\omega^2}{\sqrt{1-\dot{Z}^2}} + \frac{1}{2}\frac{\dot{\alpha}_i M_{ik}^{(c)}\dot{\alpha}_k}{(1-\dot{Z}^2)^{3/2}}$$

$$= \sqrt{\left(M_0 + \frac{1}{2} M_{I0}\omega^2 + \frac{1}{2} P_i M_{ik}^{(c)-1} P_k\right)^2 + P_z^2}\,. \tag{113}$$

Denoting the Hamiltonian of the soliton at rest by H', we have from (112) and (113)

$$P_z = \frac{\dot{Z}}{\sqrt{1-\dot{Z}^2}} H', \qquad H = \sqrt{H'^2 + P_z^2}\,. \tag{114}$$

Therefore the quantum Hamiltonian (neglecting the meson terms) is

$$\hat{H}_0 = \sqrt{\left(M_0^2 + \frac{1}{2} M_{I0}\omega^2 + \frac{1}{2M_{I0}}\hat{\mathscr{I}}^2\right)^2 + \hat{P}_z^2}, \tag{115}$$

where $\hat{\mathscr{I}}^2$ is given by (85). Evidently, $\hat{P}_z$, $\hat{\mathscr{I}}^2$, $\hat{P}_\varphi$ and $\hat{P}_\psi$ are commutative conserved operators. Therefore, the moving single-soliton state can be characterized by their eigenvalues and denoted by $|E_0, P_z; \mathscr{I}, M', M\rangle$. Evidently, the energy eigenvalue E_0 and momentum eigenvalue P_z satisfy the relativistic relation,

$$E_0 = \sqrt{E_0'^2 + P_z^2}\,, \tag{116}$$

where E'_0 is the energy (90) of the rest soliton. Note that because of (111) the eigenvalues of the isotopic-spin operators $\hat{\mathscr{I}}^2, \hat{P}_\varphi, \hat{P}_\psi$ in the rest system of the soliton are the same as those in the laboratory system.

To take account of the meson terms, one needs only to replace $\varphi_c(\gamma^{-1};\boldsymbol{x}-\boldsymbol{X}(T))$ by

$$\varphi_c(\gamma^{-1};\boldsymbol{x}-\boldsymbol{X}(T)) + \sum q_n(T/\gamma)\varphi_n(\gamma^{-1};\boldsymbol{x}-\boldsymbol{X}(T)).$$

(111)—(113) remain valid up to terms that are linear in q_n, so also (114) holds true to this order. Therefore, the terms linear in q_n are eliminated from $\hat{H}$ if they are eliminated from $\hat{H}'$. Thus the perturbation theory is again applicable.

Apparently, this method to deal with the problem of Lorentz covariance of the quantum single-soliton state is general. It is not restricted to the $I = 1/2$ representation or the $SU(2)$ group.

V. Discussion

From the results obtained above it can be concluded that the method of collective coordinates and canonical quantization can be used to formulate a consistent quantum theory of the soliton in an $SU(2)$ symmetric theory, at least, for the single-soliton sector. Because of the noncommutative character of the group and the difficulty inherent in a relativistic formulation of the two-body problem, some problems remain to be solved to establish a relativistic quantum theory of two solitons.

Among the results obtained above an interesting point is that for soliton solutions of the $I = 1/2$ scalar field, apart from the total isotopic-spin $\mathscr{I}$, the third component of the isotopic-spin $\mathscr{I}_3$, there is an additional quantum number $\mathscr{I}'_3$ so that the degeneracy of the energy level is $(2\mathscr{I} + 1)^2$. This quantum number is related to the charge of the soliton. In fact if the solution of (31) is assumed to be of the form

$$f(\boldsymbol{x})\begin{pmatrix} c_1 \\ c_2 \end{pmatrix},$$

the classical soliton solution can always be brought into the form,

$$\varphi(x) = g(t)\begin{pmatrix} f(\boldsymbol{x}) \\ 0 \end{pmatrix}$$

by a suitable $SU(2)$ transformation. Substituting the above equation into the expression of the charge of field

$$Q = -i\int(\varphi^+\dot{\varphi} - \dot{\varphi}^+\varphi)d^3\boldsymbol{x},$$

we get

$$Q = 2P_\psi = 2\mathscr{I}'_3.$$

After quantization, the charge of the soliton can take integral values $2\mathscr{I}$, $2\mathscr{I} - 2$, $\cdots -2\mathscr{I} + 2$, $-2\mathscr{I}$. So far there has been no experimental evidence for existence of multiplets of this type. Do they not exist in nature really? This is still a point worth noticing when one looks at experimental results.

In the center-of-mass system solitons with nonzero spin can be treated with the method similar to that used to treat the isotopic-spin in this article. However, the Wigner rotation of spin and other complications will appear when the Lorentz transformation is performed. Since the soliton is not a point, it may also have a third spin quantum number S'_3 in addition to S_3. Another interesting point is that in certain cases the soliton with finite mass may have a divergent moment of inertia. In these cases all spin-states are degenerate. Therefore when the soliton interacts with other particles it can absorb arbitrary angular momentum without changing its state. Just like that the momentum appears to be non-conserved in the case of an infinitely heavy particle interacting with other particles, the angular momentum would appear to be non-conserved in experimental observations. Nevertheless, this does not imply that the isotropy of space is violated.

References

[1] For a review of early works, see *Phys. Rep.*, **23C** (1976), No. 3.
[2] Friedberg, R., Lee, T. D. & Sirlin, A.: *Phys, Rev.*, **D13** (1976), 2739.
[3] Christ, N. & Lee, T. D.: *Phys. Rev.*, **D12** (1975), 1606.

Vol. XXIII No. 4 SCIENTIA SINICA April 1980

ON THE VACUUM OF THE PURE GAUGE FIELDS ON COSET

ZHOU GUANGZHAO (CHOU KUANG-CHAO 周光召)
(*Institute of Theoretical Physics, Academia Sinica*)

Received February 3, 1979.

ABSTRACT

The topological properties of the vacuum states for the pure gauge fields on coset are studied. When the homotopy group $\pi_3(G/H)$ of the coset manifold is different from zero, a winding number operator can be constructed. It is possible to introduce a θ_c-vacuum on coset. The constrained conditions restrict the value of θ_c to be zero in physical states.

I. INTRODUCTION

The study of topological properties of the vacuum in theories of non-Abelian gauge fields has attracted a good deal of attention in recent years. For pure Yang-Mills fields in the temporal gauge $A_0 = 0$, it has been shown that there exists a topological non-trivial vacuum state $A_\mu \neq 0$. The Euclidean instantons are believed to be the realization of tunneling between different vacuum states. The physical vacuum is the so-called θ-vacuum[1].

Although θ-vacuum is a gauge invariant concept, its realization is quite different for different choices of the gauge conditions. In Coulomb gauge, for example, Gribov has proved that there are topological nontrivial vacuum states with half integer winding number[2,3]. It is suggested in [8] that, in Coulomb gauge, the change of winding number for the vacuum states should be described by a transition function on the intersection of two gauge patches that cover the whole compacted Euclidean space.

In a previous paper (hereafter call I)[4], the concept of pure gauge fields on the coset G/H of a compact Lie group G with respect to its subgroup H, is introduced. With the help of this concept, we can construct a local gauge invariant Lagrangian under the group G, which contains vector gauge fields only on the subgroup H.

After the introduction of the pure gauge fields on the coset, a question naturally arises concerning the topological properties of their vacuum states. It is the aim of the present paper to study this problem. Our conclusion is: If the homotopy group $\pi_3(G/H) \neq 0$, there exist topological non-trivial vacuum states for the pure gauge fields on the coset. It is possible to introduce an analogous θ_c-vacuum with a definite value of $\theta_c = 0$.

The paper is organized as follows: In Sec. II, we study the gauge transformation for the pure gauge fields on the coset. An operator $T(g)$ on the Hilbert space which induces the gauge transformation g, is constructed. The physical states of the system are subject to the constraint conditions derived from the gauge invariance of the theory. In Sec. III, topological properties of the vacuum states are studied in the temporal gauge. The θ-vacuum is introduced and its value is determined for the physical states. Finally in Sec. IV we shall discuss the results obtained.

II. Gauge Transformation

Let G be a compact Lie group, H its subgroup, with elements denoted by g and h respectively. An element $g \in G$ can always be decomposed into a product of two elements belonging to the left coset G/H and the subgroup H respectively:

$$g = \phi h, \quad \phi \in G/H, \quad h \in H. \tag{2.1}$$

Let $\phi_0(x)$ be a given function valued in the coset, then for an arbitrary element $g \in G$ we have

$$g\phi_0 = \phi(g, \phi_0)\, h(g, \phi_0), \tag{2.2}$$

where $\phi(g, \phi_0) \in G/H$ and $h(g, \phi_0) \in H$ are nonlinear realizations of the group G[7]. They satisfy the following relations,

$$\phi(g'g, \phi_0) = \phi(g', \phi(g, \phi_0)), \tag{2.3}$$

$$h(g'g, \phi_0) = h(g', \phi(g, \phi_0))h(g, \phi_0) \tag{2.4}$$

for arbitrary g', $g \in G$.

It is more convenient to use local coordinates to parametrize the group manifolds and the coset space. For infinitesimal gauge transformation we write

$$g = 1 + i\hat{I}_j\xi_j, \quad h = 1 + i\hat{I}_j h_j, \tag{2.5}$$

where $\hat{I}_j$, $j = 1, \cdots, n_G$ are generators of the group G. We assume that the first n_H generators $\hat{I}_j$ belong to the subgroup H. ξ_j, $j = 1, \cdots, n_G$; h_j, $j = 1, \cdots, n_H (h_j = 0$, $j > n_H)$ are infinitesimal parameters that characterize the group elements g and h respectively. For finite group elements the same letters $\xi_j(h_j)$ shall be used to parametrize them. The local coordinates for the coset space will be denoted by α_j, $j = n_H + 1, \cdots, n_G$, and the local coordinates for the functions $\phi(g, \phi_0)$ and $h(g, \phi_0)$ by $R_j(g, \phi_0)$ $j = n_{H+1}, \cdots, n_G$ and $h_j(g, \phi_0)$, $j = 1, \cdots, n_H$ respectively. When g is the infinitesimal gauge transformation (2.5), we have

$$R_j(1 + i\hat{I}_j\xi_j, \phi_0) = \alpha_{0j} + R_{j,k}(\alpha_0)\, \xi_k,$$

$$h_j(1 + i\hat{I}_j\xi_j, \phi_0) = h_{j,k}(\alpha_0)\, \xi_k, \tag{2.6}$$

to the first order in ξ_k. In (2.6) α_{0j} are the coordinates of the element ϕ_0. In the following repeated group indices are summed from 1 to n_G, so we must stick to the convention that $h_j = 0 \ \forall\ j > n_H$ and $R_j = 0 \ \forall\ j \leqslant n_H$.

As a nonlinear realization of the group G, the functions $R_{j,k}$ and $h_{j,k}$ satisfy the following realization of the commutation relations:

$$R_{l,k}\frac{\partial}{\partial\alpha_{0l}}R_{i,j}-R_{l,j}\frac{\partial}{\partial\alpha_{0l}}R_{i,k}=if_{jkm}R_{i,m} \tag{2.7}$$

and

$$[\hat{H}_j,\hat{H}_k]+R_{l,j}\frac{\partial}{\partial\alpha_{0l}}\hat{H}_k-R_{l,k}\frac{\partial}{\partial\alpha_{0l}}\hat{H}_j=if_{jkm}\hat{H}_m, \tag{2.8}$$

where

$$\hat{H}_j=\hat{I}_ih_{i,j}(\alpha_0), \tag{2.9}$$

and f_{jkm} are the structure constants of the group G.

Consider now a system with fields $\Phi(x)$ which form the basis of a linear representation of the group G. In the absence of the gauge fields the Lagrangian has the form

$$\mathscr{L}_0(\Phi(x),\quad \partial_\mu\Phi(x)), \tag{2.10}$$

which is assumed to be global invariant, and we must introduce gauge fields. In I we have constructed a gauge field $\hat{B}_\mu$ on the group G which consists of a vector gauge field $\hat{A}_\mu$ on the subgroup and a pure gauge field $\phi_0(x)$ on the coset,

$$\hat{B}_\mu(x)=\phi_0(x)(\partial_\mu+\hat{A}_\mu(x))\phi_0^{-1}(x),\quad \phi_0(x)\in G/H. \tag{2.11}$$

When $\hat{A}_\mu(x)$ transform under $g(x)\in G$ in the following nonlinear way,

$$g(x):\ \hat{A}_\mu(x)\to\hat{A}'_\mu(x)=h(g,\phi_0)(\partial_\mu+\hat{A}_\mu(x))h^{-1}(g,\phi_0), \tag{2.12}$$

it is easily verified that $\hat{B}_\mu$ transform as a usual gauge fields under the group G,

$$g(x):\ \hat{B}_\mu(x)\to\hat{B}'_\mu(x)=g(x)(\partial_\mu+\hat{B}_\mu(x))g^{-1}(x). \tag{2.13}$$

With the help of these gauge fields we can construct a local invariant Lagrangian from (2.10),

$$\mathscr{L}(\Phi(x),\ D_\mu^{(B)}\Phi(x),\ F_{\mu\nu}^{(B)})=\mathscr{L}_0(\Phi(x),\ D_\mu^{(B)}\Phi(x))-\frac{1}{4a}\mathrm{tr}\{F_{\mu\nu}^{(B)}F^{(B)\mu\nu}\}, \tag{2.14}$$

where

$$\left.\begin{aligned}D_\mu^{(B)}&=\partial_\mu+\hat{B}_\mu,\\ F_{\mu\nu}^{(B)}&=\partial_\mu\hat{B}_\nu-\partial_\nu\hat{B}_\mu+[\hat{B}_\mu,\hat{B}_\nu]=\phi_0(x)\,F_{\mu\nu}^{(A)}\phi_0^{-1}(x),\\ F_{\mu\nu}^{(A)}&=\partial_\mu\hat{A}_\nu-\partial_\nu\hat{A}_\mu+[\hat{A}_\mu,\hat{A}_\nu].\end{aligned}\right\} \tag{2.15}$$

In (2.14) a is the normalization constant determined by the condition,

$$\mathrm{tr}\{\hat{I}_j\hat{I}_k\}=a\delta_{jk}. \tag{2.16}$$

In the Lagrangian (2.14) we can choose $\Phi(x)$, $\hat{A}_\mu(x)$ and $\alpha_{0l}(x)$, $l=n_H+1,\cdots,n_G$, which parametrize the coset function $\phi_0(x)$ as independent dynamical variables. The Lagrangian (2.14) is invariant under the infinitesimal gauge transformation $g(\xi)=1+i\hat{I}_j\xi_j$ with the corresponding changes of the fields,

$$\left.\begin{aligned}&\delta\Phi(x)=i\hat{I}_k\Phi(x)\xi_k,\\ &\delta A_\mu^l(x)=f_{jil}A_\mu^i(x)h_{j,k}(\alpha_0)\xi_k-\partial_\mu(h_{l,k}(\alpha_0)\xi_k),\quad l=1,\cdots,n_H\\ &\delta\alpha_{0j}(x)=iR_{j,k}(\alpha_0)\xi_k,\quad j=n_H+1,\cdots,n_G.\end{aligned}\right\} \tag{2.17}$$

In (2.17), $A^l_\mu(x)$ are the components of the gauge field $\hat{A}_\mu$,

$$\hat{A}_\mu = -iA^l_\mu \hat{I}_l, \quad A^l_\mu = 0 \ \forall\, l > n_H. \tag{2.18}$$

The operator which produces this gauge transformation can be written as

$$\hat{T}(g(x)) = \exp\{i\int d^3x \hat{I}(g(x), x)\}. \tag{2.19}$$

For infinitesimal gauge transformation $g(x)$ in (2.5), the generator $\hat{I}(g(x), x)$ is equal to

$$\begin{aligned}\hat{I}(1 + i\hat{I}_j\xi_j, x) &= i\hat{P}_\Phi(x)\hat{I}_k\hat{\Phi}(x)\xi_k(x) + i\hat{P}_{\alpha_{0l}}(x)R_{l,k}(\hat{\alpha}_0)\xi_k(x) \\ &- \hat{P}_{A^l_\nu}(x)f_{mjl}\hat{A}^j_\nu(x)h_{m,k}(\hat{\alpha}_0)\xi_k(x) + \hat{P}_{A^l_\nu}(x)\partial_\nu(h_{l,k}(\hat{\alpha}_0)\xi_k(x)),\end{aligned} \tag{2.20}$$

where $\hat{P}_\Phi(x)$, $\hat{P}_{\alpha_{0l}}(x)$ and $\hat{P}_{A^l_\nu}(x)$ are respectively the conjugate momentums of the quantized fields $\hat{\Phi}(x)$, $\hat{\alpha}_{0l}(x)$ and $\hat{A}^l_\nu(x)$. The commutation relations are

$$\left.\begin{aligned}[\hat{\Phi}(x), \hat{P}_\Phi(y)]_{x_0=y_0} &= i\delta^3(\mathbf{x}-\mathbf{y}), \\ [\hat{\alpha}_{0l}(x), \hat{P}_{\alpha_{0m}}(y)]_{x_0=y_0} &= i\delta_{lm}\delta^3(\mathbf{x}-\mathbf{y}), \\ [\hat{A}^l_\mu(x), \hat{P}_{A^m_\nu}(y)]_{x_0=y_0} &= i\delta_{\mu\nu}\delta_{lm}\delta^3(\mathbf{x}-\mathbf{y}).\end{aligned}\right\} \tag{2.21}$$

Actually these quantized fields are not independent. There are constrained conditions due to the gauge invariance of the Lagrangian. We shall regard these constraints as weak conditions which are satisfied only when a physical state is acted upon.

As we have mentioned in I that the derivative terms $\partial_\mu\alpha_{0l}(x)$ appear only in $D^{(B)}_\mu\Phi(x)$, the momentum $P_{\alpha_{0l}}(x)$ must be related to $P_\Phi(x)$ through the constraint conditions,

$$\hat{P}_{\alpha_{0l}}(x) = \hat{P}_\Phi(x)\hat{f}_l(\alpha_0)\hat{\Phi}(x), \quad l = n_H + 1, \cdots, n_G, \tag{2.22}$$

where $\hat{f}_l(\alpha_0)$ is determined by the relation,

$$\phi_0(x)\partial_\mu\phi_0^{-1}(x) = \hat{f}_l(\alpha_0)\partial_\mu\alpha_{0l}(x). \tag{2.23}$$

From the group relation,

$$\phi(g, \phi_0) = g\phi_0 h^{-1}(g, \phi_0), \tag{2.24}$$

and (2.23), we get for arbitrary $g \in G$,

$$\begin{aligned}\hat{f}_j(\phi(g, \phi_0))\partial_\mu R_j(g, \phi_0) &= \phi(g, \phi_0)\partial_\mu\phi^{-1}(g, \phi_0) \\ &= g\phi_0(h^{-1}(g, \phi_0)\partial_\mu h(g, \phi_0))\phi_0^{-1}g^{-1} + g\hat{f}_j(\phi_0)g^{-1}\partial_\mu\alpha_{0j} + g\partial_\mu g^{-1}.\end{aligned} \tag{2.25}$$

Equating terms proportional to $\partial_\mu\xi_k(x)$ on both sides of (2.25) we get

$$\hat{f}_j(\alpha_0)R_{j,k}(\alpha_0) = \phi_0\hat{I}_l\phi_0^{-1}h_{l,k}(\alpha_0) - \hat{I}_k. \tag{2.26}$$

Using this relation we can rewrite the constraint equations (2.22) in the following form,

$$\hat{P}_{\alpha_{0j}}R_{j,k} + \hat{P}_\Phi\hat{I}_k\hat{\Phi} = \hat{P}_\Phi\phi_0\hat{I}_l\phi_0^{-1}\hat{\Phi}h_{l,k}(\hat{\alpha}_0). \tag{2.27}$$

When the generator (2.20) acts upon a physical state, it can be simplified with the help of the constraints (2.27). Thus we obtain

$$\hat{I}(1+i\hat{I}_k\xi_k,\ x)|\text{phys}\rangle = I(h(1+iI_k\xi_k,\ \phi_0),\ x)|\text{phys}\rangle, \tag{2.28}$$

where

$$\begin{aligned}\hat{I}(h(1+i\hat{I}_j\xi_j,\phi_0),x) &= i\hat{P}_\psi \hat{I}_l \hat{\phi} h_{l,k}\xi_k \\ &\quad - \hat{P}_{A^l_\nu} f_{mjl} A^j_\nu h_{l,k}\xi_k - \hat{P}_{A^l_\nu}\partial_\nu(h_{l,k}\xi_k).\end{aligned} \tag{2.29}$$

In (2.29) we have redefined the matter fields as

$$\hat{\phi}(x) = \phi_0^{-1}(x)\Phi(x), \quad \hat{P}_\psi(x) = \hat{P}_\Phi(x)\phi_0(x). \tag{2.30}$$

It is clear from (2.29) that $I(h,x)$ is a generator in the subgroup H with $\hat{\phi}(x)$ and $A^l_\nu(x)$ as independent dynamical variables. For general gauge transformation $g(x)$ one can integrate (2.28) to give

$$\hat{T}(g(x))|\text{phys}\rangle = \hat{T}(h(g(x),\ \phi_0))|\text{phys}\rangle. \tag{2.31}$$

Eq. (2.31) just means that in the present theory a gauge transformation $g(x)$ on physical states is equivalent to a gauge transformation on the subgroup H with element $h(g(x),\phi_0)$ which is a nonlinear realization of the group G.

III. The Vacuum States

After the introduction of the pure gauge fields on a coset, a natural question is how to define their vacuum states. As the vacuum states are described by pure gauges in the usual non-Abelian gauge theories, we can transcribe most of the reasonings from the existing theories to the present case.

For simplicity we shall work in the temporal gauge $B_0=0$. The winding number operator is defined as,

$$\hat{N} = \frac{1}{12\pi^2 a}\int d^3x \epsilon_{ijk}\,\text{tr}\,[\hat{B}_i(x)\hat{B}_j(x)\hat{B}_k(x) + 3\hat{B}_i(x)F^{(B)}_{jk}(x)], \tag{3.1}$$

where a is determined by (2.16). Under a gauge transformation $g(x)$ the operator $\hat{N}$ transforms in the following way,

$$g(x):\ \hat{N}\to\hat{N}' = T(g)\hat{N}T^+(g) = \hat{N}+\nu(g), \tag{3.2}$$

where

$$\begin{aligned}\nu(g) &= \frac{1}{12\pi^2 a}\int d^3x\epsilon_{ijk}\,\text{tr}\,[g\partial_i g^{-1} g\partial_j g^{-1} g\partial_k g^{-1}] \\ &\quad + \frac{1}{4\pi^2 a}\oint d\sigma_i \epsilon_{ijk}\,\text{tr}\,[\partial_j g^{-1} g\hat{B}_k].\end{aligned} \tag{3.3}$$

In (3.3) the second term on the right-hand side is a surface integral at the boundary of the three-dimensional volume. $\nu(g)$ is the winding number induced by the gauge transformation $g(x)$. If $g(x)$ satisfies the boundary condition,

$$g(\mathbf{x})\to 1, \quad \text{as} \quad |\mathbf{x}|\to\infty, \tag{3.4}$$

it may be regarded as a map of the compacted three-dimensional space S^3 onto the group manifold G:

$$g(x):\ S^3 \to G. \tag{3.5}$$

In this case $\nu(g)$ is the Kronecker index of the mapping (3.5) and takes the value of an integer.

In the following we shall consider two examples that make our discussion clear. First, let us take the group G to be a compact simple Lie group and the subgroup H to be an Abelian group $U(1)$. It is well known[5] that the homotopy group

$$\pi_3(G) = Z, \quad \pi_3(U(1)) = 0. \tag{3.6}$$

The vacuum states corresponding to the vector gauge fields $\hat{A}_\mu$ on the subgroup are topologically trivial in the present case. However, we can still find a gauge transformation $g(x)$ in the group G which has the winding number $\nu(g)$ equal to one.

Let $|n\rangle$ be the eigenstate of the operator $\hat{N}$ with eigenvalue n, $\hat{T}^+(g)|n\rangle$ will be an eigenstate of $\hat{N}$ with eigenvalue $n + \nu(g)$. As $\hat{T}(g)$ is a gauge transformation which commutes with all gauge invariant observables, the physical states should be eigenstates of the unitary operator $\hat{T}(g)$. From states $|n\rangle$ with definite winding number we can construct eigenstates,

$$|\theta_c\rangle = \sum_n \exp\{in\theta_c\}|n\rangle \tag{3.7}$$

of the operator $\hat{T}(g)$ with eigenvalue $e^{i\theta_c\nu(g)}$. This is the θ_c-vacuum for the pure gauge fields on the coset.

In the present case not all values of θ_c are compatible with the constraint condition (2.30).

Since $\pi_3(H) = 0$, $\hat{T}(h)$ cannot change the winding number n, therefore we can normalize the states such that

$$\hat{T}(h)|\text{phys}\rangle = |\text{phys}\rangle. \tag{3.8}$$

From the constraint equations (2.31) and (3.8), we conclude that $\theta_c = 0$ on physical states.

As a second example we take the group G to be the chiral $SU(3)_L \otimes SU(3)_R$ and the subgroup H to be the parity-conserved $SU(3)$. We shall express the elements $g \in G$ in the following form,

$$g = \phi(\alpha)\,h(\beta), \tag{3.9}$$

where

$$\begin{aligned} \phi(\alpha) &= \exp\left\{\frac{i}{2}\,\gamma_5\lambda_j\alpha_j(x)\right\}, \\ h(\beta) &= \exp\left\{\frac{i}{2}\,\lambda_j\,\beta_j(x)\right\} \end{aligned} \tag{3.10}$$

are elements in the coset and the subgroup respectively. In (3.10) λ_j, $j = 1, \cdots, 8$

are the usual Gell-Mann matrices of $SU(3)$. We can also express these elements in terms of the right-handed and left-handed group elements,

$$\phi(\alpha) = g_L(\alpha)g_R^{-1}(\alpha), \quad h(\beta) = g_L(\beta)g_R(\beta), \tag{3.11}$$

where

$$\left.\begin{aligned} g_L(\xi) &= \exp\left\{\frac{1}{4} i(1+\gamma_5)\hat{\lambda}_j\xi_j\right\}, \\ g_R(\xi) &= \exp\left\{\frac{1}{4} i(1-\gamma_5)\hat{\lambda}_j\xi_j\right\}. \end{aligned}\right. \tag{3.12}$$

It is possible to define two winding number operators corresponding respectively to the group $SU(3)_L$ and $SU(3)_R$,

$$\left.\begin{aligned} \hat{N}_L &= \frac{1}{12\pi^2 a}\int d^3x \varepsilon_{ijk} \operatorname{tr}[\hat{B}_{iL}\hat{B}_{jL}\hat{B}_{kL} + 3\hat{B}_{iL}\hat{F}_{jk}^{(B_L)}], \\ \hat{N}_R &= \frac{1}{12\pi^2 a}\int d^3x \varepsilon_{ijk} \operatorname{tr}[\hat{B}_{iR}\hat{B}_{jR}\hat{B}_{kR} + 3\hat{B}_{iR}\hat{F}_{jk}^{(B_R)}], \end{aligned}\right\} \tag{3.13}$$

where $\hat{B}_{\mu R}$ and $\hat{B}_{\mu L}$ are respectively the corresponding gauge fields on the right-handed and left-handed subgroup of G. They are related to the gauge fields $\hat{B}_\mu$ through the relation,

$$\hat{B}_\mu = \frac{1}{2}(1+\gamma_5)\hat{B}_{\mu L} + \frac{1}{2}(1-\gamma_5)\hat{B}_{\mu R}. \tag{3.14}$$

Under a gauge transformation $g = g_L \cdot g_R$ it is easily verified that

$$g:\ \hat{N}_{L,R} \to \hat{N}'_{L,R} = \hat{N}_{L,R} + \nu_{L,R}(g) = \hat{N}_{L,R} + \nu(g_{L,R}), \tag{3.15}$$

where $\nu(g_{L,R})$ is given by the same formula (3.3).

From (3.11) we easily obtain,

$$\nu_L(h) = \nu_R(h), \quad \nu_L(\phi) = -\nu_R(\phi). \tag{3.16}$$

Let $|n_L, n_R\rangle$ be eigenstates of the operators $\hat{N}_L$ and $\hat{N}_R$ with eigenvalues n_L and n_R respectively,

$$\left.\begin{aligned} \hat{N}_L|n_L, n_R\rangle &= n_L|n_L, n_R\rangle, \\ \hat{N}_R|n_L, n_R\rangle &= n_R|n_L, n_R\rangle. \end{aligned}\right\} \tag{3.17}$$

We can always find an element $h \in H$ such that $\nu_L(h) = \nu_R(h) = 1$ and an element $\phi \in G/H$ such that $\nu_L(\phi) = -\nu_R(\phi) = 1$. Then under the gauge transformations described by these h and ϕ, we have

$$\left.\begin{aligned} \hat{T}^+(h)|n_L, n_R\rangle &= |n_L+1, n_R+1\rangle, \\ \hat{T}^+(\phi)|n_L, n_R\rangle &= |n_L+1, n_R-1\rangle. \end{aligned}\right\} \tag{3.18}$$

Let us now introduce two winding numbers corresponding to the gauge transformations h and ϕ respectively:

$$n = \frac{1}{2}(n_L + n_R), \quad n_C = \frac{1}{2}(n_L - n_R). \tag{3.19}$$

In terms of these new winding numbers we can write the state $|n_L, n_R\rangle$ as $|n, n_C\rangle$ and have

$$\left.\begin{aligned}\hat{T}^{+}(h)|n, n_C\rangle &= |n+1, n_C\rangle, \\ \hat{T}^{+}(\phi)|n, n_C\rangle &= |n, n_C+1\rangle.\end{aligned}\right\} \tag{3.20}$$

The eigenstates of the operators $\hat{T}(h)$ and $\hat{T}(\phi)$ can be written as

$$|\theta, \theta_C\rangle = \sum_{n,\, n_C} \exp\{i(n\theta + n_C\theta_C)\}|n, n_C\rangle.$$

This is the θ-vacuum states for the present example.

Like the first example, $\hat{T}(\phi)$ is equivalent to $\hat{T}(h(\phi, \phi_0))$ when a physical state is acted upon. Since $\hat{T}(h)$ cannot change the winding number n_C, we conclude that θ_C can be normalized to zero for physical states.

In general cases, if H is a semi-simple Lie group, $\pi_2(H) = 0$, one can prove with the help of the exact sequence[5]

$$\longrightarrow \pi_3(H) \xrightarrow{i_*} \pi_3(G) \xrightarrow{P_*} \pi_3(G/H) \xrightarrow{d} \pi_2(H) \xrightarrow{d} \underset{\substack{\| \\ 0}}{\pi_2(H)} \longrightarrow$$

that

$$\pi_3(G/H) = \pi_3(G)/\pi_3(H). \tag{3.21}$$

When $\pi_3(G/H) \neq 0$, the pure gauge fields on the coset will have topological nontrivial vacuum states. On account of the constraint conditions (2.30) the θ_C-vacuum introduced on the coset has the value $\theta_C = 0$. Therefore it does not cause CP non-conservation.

Before the conclusion of the present section we shall briefly discuss the problem of vacuum tunneling. As an illustration we choose $G = SU(2)$ and H to be the identity element. In this simple case the coset is the whole group.

In the temporal gauge it is well known[1] that the change of the winding number between the future and the past vacuum states is equal to the second Chern class or the Pontryagin number,

$$\begin{aligned} q &= \nu(t = +\infty) - \nu(t = -\infty) \\ &= \frac{1}{32\pi^2 a} \int d^4x \operatorname{tr}\{\hat{F}^{(B)*}_{\mu\nu} \hat{F}^{(B)\mu\nu}\}, \end{aligned} \tag{3.22}$$

where ${}^*\hat{F}_{\mu\nu} = \frac{1}{2}\varepsilon_{\mu\nu\rho\tau}\hat{F}^{\rho\tau}$ is the dual tensor of $\hat{F}_{\mu\nu}$.

In our case the field

$$\hat{B}_\mu(x) = g(x)\,\partial_\mu g^{-1}(x)$$

is a pure gauge field. On those points where $g(x)$ is regular we always have

$$F^{(B)}_{\mu\nu} = 0. \tag{3.23}$$

The contribution to the integral (3.22) comes only from those points where $g(x)$ is singular. On these singular points $\hat{B}_\mu(x)$ may represent point instantons. One example will be the Polyakov-'t Hooft instanton with vanishing size, which can be expressed in Euclidean space as[6],

$$g(x) = \frac{x_4 + i\boldsymbol{\sigma}\cdot\mathbf{x}}{\sqrt{x^2}}. \tag{3.24}$$

This $g(x)$ has point singularity at $x=0$. The field $\hat{B}_\mu$ with $g(x)$ given by (3.24) will contribute unit Pontryagin number when substituted into the integral (3.22).

In the present theory one should allow the existence of such point instantons (i.e. gauge transformations with point singularities), which cause tunneling between vacuum states with different winding numbers.

IV. Discussions

We have shown in previous sections that it is possible to define a θ_C-vacuum for the pure gauge fields on a coset G/H in the temporal gauge when the homotopy group $\pi_3(G/H)$ is different from zero. On the other hand, one can fix the vacuum uniquely by choosing, for instance, the gauge condition $\alpha_l = 0$. Is this fact contradictory with those analysis in the temporal gauge? Let us recall that similar situation happens in ordinary non-Abelian gauge field theories. In Coulomb gauge, when the vector potential $\hat{A}_\mu$ is subject to the boundary conditions $\lim_{|\boldsymbol{x}|\to\infty} |\boldsymbol{x}|^{3/2}\hat{A}_i = 0$, the classical vacuum is uniquely $\hat{A}_i = 0$[8]. Jackiw, Muzinich and Rebbi have analyzed in detail this question of how to describe vacuum in Coulomb gauge[8]. They concluded that the Coulomb gauge description of a gauge field with non-vanishing Pontryagin index cannot be single-valued. Sufficiently large potential would evolve discontinuously in time a sudden transition between gauge equivalent transverse configurations.

We have mentioned in the end of the previous section that even in temporal gauge $\hat{B}_\mu$ must have singularities corresponding to point instantons when the Pontryagin index is non-vanishing. Transformation from temporal gauge to the gauge $\phi_0(x) = 1$ is possible only by singular gauge transformation. Thus we expect that the matter fields $\Phi(x)$ and the vector gauge field $\hat{B}_\mu$ will evolve singularities in the gauge $\phi_0(x) = 1$. As point instantons have finite action, there is no compelling reason to exclude such singular configurations.

If these singular configurations were not allowed in the theory, the present formulation would be completely equivalent to the usual one physically. The vacuum states with different winding numbers on the coset represent ground states of unconnected world in this case.

We have demonstrated in the present work and in I that point instantons and point monopoles can be described by singularities of pure gauge fields on coset. The existence of such topological particles makes the present formulation different from the usual theory. It is not clear whether such theory of pure gauge fields with singularities is applicable to reality. However, it has some peculiarities. Besides the θ_C-vacuum discussed here, it may generate additional anomalies in the conservation of certain currents. We shall discuss these problems elsewhere.

References

[1] Callan, C. G., Dashen, R. F. & Gross, D. J., *Phys. Lett.*, **63B** (1976), 334; Jackiw, R. & Rebbi, C., *Phys. Rev. Lett.*, **37** (1976), 172; 't Hooft, G., *Phys. Rev. Lett.*, **37** (1976), 8; *Phys. Rev.*, **D14** (1976), 3432.

[2] Gribov, V. N., *Nucl. Phys.*, **B139** (1978), 1.

[3] Sciuto, S., *Phys. Lett.*, **71B** (1977), 129; Ademollo, M., Napolitano, E. & Sciuto, S., *Nucl. Phys.*, **B134** (1978), 477.

[4] Zhou Guangzhao, Du Dongsheng & Ruan Tunan, *Scientia Sinica*, **22** (1979), 37.

[5] Steenrod, N., *Topology of Fibre Bundles*, Prin. Univ. Press, (1951); Boya, L. J., Carinena, J. F. & Mateos, J., *Fort. d. Phys.*, **26** (1978), 175.

[6] Belavin, A. A., Polykov, A. M., Schwartz, A. S. & Tyupkin, Y. S., *Phys. Lett.*, **59B** (1975), 85. 't Hooft G., *Phys. Rev. Lett.*, **37** (1976), 8.

[7] Coleman, S., Wess, J. & Zumino, B., *Phys. Rev.*, **177** (1969), 2239; Callan, C. G. Jr., *et al.*, *Phys. Rev.*, **177** (1969), 2247; Salam, A. & Strathdee, J., *Phys. Rev.*, **184** (1969), 1750.

[8] Jackiw, R., Muzinich, I. & Rebbi, C., *Phys. Rev.*, **D17** (1978), 1576; Nicole, D. A., *Nuclear Phys.*, **B139** (1978), 151.

Vol. XXIII No. 5 SCIENTIA SINICA May 1980

A MODEL OF ELECTRO-WEAK INTERACTION IN $SU(3) \times U(1)$ GAUGE THEORY

Zhou Guangzhao (Chou Kuang-chao, 周光召)
(*Institute of Theoretical Physics, Academia Sinica*)
and Gao Chongshou (高崇寿)
(*Department of Physics, Beijing University*)

Received August 20, 1979.

Abstract

The helicity mixed representation is used to construct a model of electro-weak interaction in $SU(3)\times U(1)$ gauge theory. It reduces in the low energy region to the simple $SU(2)\times U(1)$ model in the limit of $\sin^2\theta_W = 1/4$. When $\sin^2\theta_W$ is slightly less than 1/4, there is a small correction in the neutral current sector which can be tested.

I. Introduction

Recent neutrino-induced elastic and deep inelastic neutral current data are in agreement with expectations based on the simple $SU(2) \times U(1)$ gauge model of Weinberg-Salam [1,2]. The Weinberg angle θ_W is shown to be slightly less than 30°.

In a previous note (hereafter call I) a model in $SU(3)$ has been proposed for the electro-weak interactions of leptons. The essential point is to arrange the left-handed and right-handed leptons in a single triplet without the introduction of supergroup structure[3,4]. $\sin^2\theta_W$ is shown to be 1/4 [3].

In the present paper we offer another model in the same spirit which takes quark into account. The gauge group is chosen to be $SU(3) \times U(1)$. Although this is not a simple group and another coupling constant g' is introduced, one can show that $\sin^2\theta_W \leqslant 1/4$. In the limiting case where $\sin^2\theta_W = 1/4$, the charged and neutral currents are identical with that based on the simple $SU(2) \times U(1)$ gauge model. For $\sin^2\theta_W$ slightly less than 1/4 there is a small correction in the neutral current sector which can be tested in experiments.

As in I, a global $U(1)$ symmetry is preserved after spontaneous symmetry breaking. This symmetry provides a conserved quantum number called weak strangeness. There are nine gauge bosons in the model, four of which are the usual $W^\pm, Z^\circ$ and the photon. The remaining five contain a neutral Z_2, a charged pair $V^\pm$ and a doubly charged pair $U^{\pm\pm}$ bosons. Two triplets of Higgs are used to generate masses of the gauge bosons and the fermions. After spontaneous symmetry breaking four heavy Higgs with two neutral ones χ°, ϕ° and a charged pair $\chi^\pm$ remain. On account of the conservation of weak strangeness, $V^\pm, U^{\pm\pm}$ and $\chi^\pm$ mesons can be produced only in pairs and the lightest of these mesons might be a stable particle.

The paper is organized as follows: In Sec. II the transformation properties of fields are studied. A representation of the $SU(3)$ algebra, which mixes the γ_5 and the internal group generators, is used for fermions. In Sec. III, the mass spectrum of vector gauge bosons is obtained through Higgs mechanism and the conservation of weak strangeness is established. The eletro-weak interaction of the fermions is discussed in Sec. IV. Deviation in neutral currents from the Weinberg-Salam model is obtained. Finally we shall discuss the results obtained.

II. Transformation Properties

The generators $\hat{I}_i$, $i=1,\cdots,8$ of the $SU(3)$ group can be decomposed into two sets $\hat{I}_a$ and $\hat{I}_\alpha$. There are many possibilities. One possible choice which will be used in the following is $a=1,3,8,4,6$ and $\alpha=2,5,7$. Other choices give similar results. Define

$$\hat{I}_a^{(\varepsilon)}=\varepsilon\hat{I}_a,\quad \hat{I}_\alpha^{(\varepsilon)}=\hat{I}_\alpha. \tag{2.1}$$

where ε commutes with $\hat{I}_i$ and satisfies the relation $\varepsilon^2=1$. One easily verifies that the Lie algebra for $\hat{I}_i^{(\varepsilon)}$ is the same as that for $\hat{I}_i$. Four possible choices for ε : $\varepsilon=+1$, $-1, \gamma_5, -\gamma_5$ will be used below. The generator of the $U(1)$ gauge group will be denoted by $\hat{Y}$.

Besides the conservation of various fermion numbers there is another global $U(1)$ symmetry whose generator will be denoted by $\hat{S}$. This global $U(1)$ will combine with an Abelian subgroup in $SU(3)\times U(1)$ to give a new conservation number S_W after spontaneous symmetry breaking. This new quantum number S_W is called weak strangeness in I.

Each generation (ν_e, e), (ν_μ, μ) or (ν_τ, τ) of leptons forms half of a triplet in $SU(3)$ with $Y=0$. Quarks of each color and isotopic doublet form half of a $Y=2/3$ triplet plus half of a $Y=2/3$ singlet in $SU(3)$. Transformation properties for fermions are

$$\begin{aligned}&\text{for triplet:} && \psi \longrightarrow \psi'=U^{(S)}(\xi_j(x))e^{i\hat{Y}\theta(x)}e^{i\hat{S}\eta}\psi;\\ &\text{for singlet:} && u\longrightarrow u'=e^{i\hat{Y}\theta(x)}e^{i\hat{S}\eta}u.\end{aligned} \tag{2.2}$$

where

$$U^{(\varepsilon)}(\xi_j(x))=\exp\{i\hat{I}_j^{(\varepsilon)}\xi_j(x)\}. \tag{2.3}$$

Here $\varepsilon=+,-,5,-5$ stand for $\varepsilon=+1,-1,+\gamma_5,-\gamma_5$ respectively. In Eq. (2.2) $\xi_j(x)$, $j=1,\cdots 8$, $\theta(x)$ and η are group parameters for the local $SU(3)\times U(1)$ and the global $U(1)$ respectively. The generator $\hat{S}$ will take values $\frac{1}{6}\gamma_5-\frac{1}{2}$ for lepton triplet, $\frac{1}{6}\gamma_5-\frac{5}{6}$ for quark triplet and $-\frac{1}{2}\gamma_5-\frac{5}{6}$ for quark singlet.

There are nine gauge fields $A_\mu^j(x)$, $j=1,\cdots,8$ of $SU(3)$ and $B_\mu(x)$ of $U(1)$. Define

$$\hat{A}_\mu^{(\varepsilon)}=igA_\mu^j(x)\hat{I}_j^{(\varepsilon)}, \tag{2.4}$$

Then $\hat{A}_\mu^{(\epsilon)}$ transforms under $SU(3)$ in the following way,

$$\hat{A}_\mu^{(\epsilon)} \longrightarrow \hat{A}_\mu^{(\epsilon)\prime} = U^{(\epsilon)}(\xi_j(x))(\partial_\mu + A_\mu^{(\epsilon)}(x))U^{(\epsilon)+}(\xi_j(x)). \tag{2.5}$$

Two triplets of Higgs fields Φ_1 with $Y=0$ and Φ_2 with $Y=-1$ are used. They transform as

$$\Phi \longrightarrow \Phi' = U^{(+)}(\xi_j(x))e^{i\hat{Y}\theta(x)}e^{i\hat{S}\eta}\Phi. \tag{2.6}$$

The generator $S=-\frac{1}{3}$ for Φ_1, $S=\frac{5}{3}$ for Φ_2 and $S=0$ for gauge vector bosons.

The covariant derivatives for the fermions and the Higgs are easily constructed. They are

$$\begin{aligned} &\text{for fermion triplet:} && D_\mu\psi = \left(\partial_\mu + \hat{A}_\mu^{(5)} + \frac{i}{2}g'\hat{Y}B_\mu\right)\psi; \\ &\text{for fermion singlet:} && D_\mu u = \left(\partial_\mu + \frac{i}{2}g'\hat{Y}B_\mu\right)u; \\ &\text{for Higgs triplet:} && D_\mu\Phi = \left(\partial_\mu + \hat{A}_\mu^{(+)} + \frac{i}{2}g'\hat{Y}B_\mu\right)\Phi. \end{aligned} \tag{2.7}$$

III. Symmetry Breaking, Conservation of Charge and Weak Strangeness

The vacuum expectation values of the Higgs are taken to be

$$\langle\Phi_1\rangle = \begin{pmatrix} v_1 \\ 0 \\ 0 \end{pmatrix}, \quad \langle\Phi_2\rangle = \begin{pmatrix} 0 \\ 0 \\ v_2 \end{pmatrix}. \tag{3.1}$$

One local $U(1)$ and one global $U(1)$ symmetry remain unbroken. Their generators are

$$\text{for charge:} \quad \hat{Q} = \hat{I}_3 - \sqrt{3}\,\hat{I}_8 + \hat{Y}, \tag{3.2}$$

$$\text{for weak strangeness:} \quad \hat{S}_W = \frac{2}{\sqrt{3}}\hat{I}_8 + \hat{Y} + \hat{S}, \tag{3.3}$$

which are conserved quantum numbers. After spontaneous symmetry breaking all physical particles are eigenstates of charge $\hat{Q}$ and weak strangeness $\hat{S}_W$.

For triplet fermions we must replace $\hat{I}_8$ by $\frac{1}{2}\gamma_5\hat{\lambda}_8$ and get from (3.3):

$$\hat{S}_W = \frac{1}{\sqrt{3}}\gamma_5\hat{\lambda}_8 + \frac{1}{6}\gamma_5 + \hat{Y} = \frac{1}{2}\gamma_5\begin{pmatrix} 1 & & \\ & 1 & \\ & & -1 \end{pmatrix} + \hat{Y}. \tag{3.4}$$

We use the quantum number S_W to classify all three components in a triplet. This implies that the helicity components with the same S_W be either (L, L, R) or $(R,$

R, L). The now observed leptons and quarks are then chosen to be

$$\psi_l = \begin{pmatrix} \nu_{eL} \\ e_L \\ e_R \end{pmatrix}, \quad \psi_q = \begin{pmatrix} u_L \\ d'_L \\ d'_R \end{pmatrix}, \quad u_R \quad \text{etc.} \tag{3.5}$$

The other half components are weak strange fermions, which will get heavy masses by a suitable choice of additional Higgs fields and an additional singlet lepton field. The problem of mass spectrum for fermions will be discussed elsewhere.

The charges and the weak strangeness of the particles participating in low-energy weak interactions are given in Table 1.

Table 1

	ν_{e_L}	e_L	e_R	u_L	d'_L	d'_R	u_R	γ	W^-	W^+
Q	0	−1	−1	2/3	−1/3	−1/3	2/3	0	−1	1
S_W	0	0	0	0	0	0	0	0	0	0
	V^-	V^+	U^{--}	U^{++}	Z_1	Z_2	ϕ^0	χ^0	χ^-	χ^+
Q	−1	1	−2	2	0	0	0	0	−1	1
S_W	1	−1	1	−1	0	0	0	0	1	−1

The mass terms of the vector gauge bosons are easily derived as,

$$\frac{1}{2} g^2 |v_1|^2 \left(W^+W^- + V^+V^- + \frac{2}{3} Z^{02} \right)$$

$$+ \frac{1}{2} g^2 |v_2|^2 (V^+V^- + U^{++}U^{--})$$

$$+ \frac{1}{4} g^2 |v_2|^2 \left(\frac{1}{\sqrt{3}} Z^0 + \frac{1}{\sin\varphi} Z'^0 \right)^2, \tag{3.6}$$

where

$$W^\pm = \frac{1}{\sqrt{2}} (A^1 \mp iA^2), \quad V^\pm = \frac{1}{\sqrt{2}} (A^4 \pm iA^5),$$

$$U^{\pm\pm} = \frac{1}{\sqrt{2}} (A^6 \pm iA^7), \quad Z^0 = \frac{1}{2} (A^8 + \sqrt{3} A^3),$$

$$Z'^0 = -\sin\varphi \left(\frac{1}{2} (A^3 - \sqrt{3} A^8) \right) + \cos\varphi B, \tag{3.7}$$

and

$$\sin\varphi = \frac{g}{\sqrt{g^2 + g'^2}}, \quad \cos\varphi = \frac{g'}{\sqrt{g^2 + g'^2}}. \tag{3.8}$$

From (3.6) and (3.7) we obtain:

(i) The photon field $A = \cos\varphi \left(\frac{1}{2} (A^3 - \sqrt{3} A^8) \right) + \sin\varphi B$ is massless as required;

(ii) The masses of the charged vector bosons are

$$m_W^2 = \frac{1}{2} g^2 |v_1|^2, \quad m_U^2 = \frac{1}{2} g^2 |v_2|^2,$$
$$m_V^2 = m_W^2 + m_U^2. \tag{3.9}$$

Let us introduce a new parameter

$$v \equiv \left|\frac{v_2}{v_1}\right|^2 = \frac{m_U^2}{m_W^2}, \tag{3.10}$$

which will be useful in the following.

The Z and Z' mesons are not eigenstates of the mass matrix. Let the true neutral vector bosons be Z_1 and Z_2, which are related to Z and Z' by a rotation.

$$Z = \cos\alpha Z_1 + \sin\alpha Z_2,$$
$$Z' = -\sin\alpha Z_1 + \cos\alpha Z_2. \tag{3.11}$$

Diagonizing the mass matrix we have

$$m_{Z_1}^2 = m_Z^2 \left[1 + \frac{1}{4} v \left(1 - \frac{\sqrt{3}\,\mathrm{tg}\,\alpha}{\sin\varphi}\right)^2\right].$$
$$m_{Z_2}^2 = m_Z^2 \left[\mathrm{tg}^2\alpha + \frac{1}{4} v \left(\mathrm{tg}\,\alpha + \frac{\sqrt{3}}{\sin\varphi}\right)^2\right], \tag{3.12}$$

where

$$m_Z^2 = \frac{4}{3} m_W^2 \cos^2\alpha, \tag{3.13}$$

$$\mathrm{tg}\,2\alpha = \frac{2\sqrt{3}\, v \sin\varphi}{3v - (4 + v)\sin^2\varphi}. \tag{3.14}$$

One interesting limiting case is[1)]

$$\sin\varphi \ll 1, \quad \frac{\sin^2\varphi}{v} \ll 1. \tag{3.15}$$

In this limiting case, Eq. (3.14) becomes

$$\mathrm{tg}\,\alpha = \frac{1}{\sqrt{3}} \sin\varphi \left(1 + \frac{4}{3v}\sin^2\varphi + O(\sin^4\varphi)\right). \tag{3.16}$$

Substituting Eq. (3.16) into Eqs. (3.12) and (3.13), we obtain,

$$m_{Z_1}^2 = m_Z^2 \left(1 + \frac{4}{9v}\sin^4\varphi + O(\sin^6\varphi)\right),$$
$$m_{Z_2}^2 = m_Z^2 \cdot \frac{3v}{4\sin^2\varphi}\left(1 + \frac{2}{3}\sin^2\varphi + O(\sin^4\varphi)\right) \gg m_{Z_1}^2 \text{ for small } \frac{\sin^2\varphi}{v}, \tag{3.17}$$

1) We note that $\mathrm{tg}\,\alpha = \frac{1}{\sqrt{3}} \sin\varphi$, $m_{Z_1}^2 = m_W^2/\cos\theta_W$ and m_U^2, m_V^2, $m_{Z_2}^2 \to \infty$ in the limiting case $v \to \infty$ for arbitrary value of $\sin\varphi$. The neutral current $J_{N_1\mu}$ between fermions can be written as $J_{N_1\mu} = J_\mu^3 - \sin^2\theta_W J_\mu^{\mathrm{e.m.}}$. Therefore, all observable results in this limiting case are exactly the same as those in the Weinberg-Salam model.

$$m_Z^2 = \frac{4}{3} m_W^2 \left(1 - \frac{1}{3} \sin^2 \varphi + O(\sin^4 \varphi)\right). \tag{3.18}$$

Comparing the above with the mass formula in the Weinberg-Salam model,

$$m_Z^2 = m_W^2 / \cos^2 \theta_W , \tag{3.19}$$

we obtain to the order $\sin^2\varphi$ that

$$\sin^2 \theta_W = \frac{1}{4} \cos^2 \varphi \leqslant \frac{1}{4}. \tag{3.20}$$

This is an encouraging result since recent experiments require that $\sin^2\theta_W \leqslant \frac{1}{4}$. However, this result is model-dependent, one obtains different values of $\sin^2 \theta_W$ when different representations for fermions and Higgs are chosen as is shown elsewhere[5].

After spontaneous symmetry breaking four heavy Higgs remain. They are related to the triplets Φ_1 and Φ_2 in the form,

$$\Phi_1 = \begin{pmatrix} \phi^0 \\ 0 \\ 0 \end{pmatrix}, \quad \Phi_2 = \begin{pmatrix} \chi^- \\ 0 \\ \chi^0 \end{pmatrix},$$

where ϕ^0 and χ^0 are real fields and χ^- is complex. Φ_2 has no coupling with fermions Only ϕ^0 field in Φ_1 has direct coupling with fermions and generates masses of them after symmetry breaking.

IV. Interactions Between Fermions and Gauge Vector Bosons

The gauge interaction Lagrangian for the fermions can be written as

$$\begin{aligned} \mathscr{L}_{\text{int}} &= \bar{\psi}_q \gamma^\mu \left(\partial_\mu + \hat{A}_\mu^{(5)} + \frac{i}{2} g' \hat{Y} B_\mu\right) \psi_q \\ &+ \bar{u}_R \gamma^\mu \left(\partial_\mu + \frac{i}{2} g' \hat{Y} B_\mu\right) u_R \\ &+ \bar{\psi}_l \gamma^\mu (\partial_\mu + \hat{A}_\mu^{(5)}) \psi_l , \end{aligned} \tag{4.1}$$

where

$$\psi_q = \begin{pmatrix} u_L \\ d'_L \\ d'_R \end{pmatrix}, \quad \psi_l = \begin{pmatrix} \nu_{e_L} \\ e_L \\ e_R \end{pmatrix}, \text{ etc.}$$

As $\bar{L}\gamma^\mu R = \bar{R}\gamma^\mu L = 0$, one can immediately see that $A_\mu^{4,5,6,7}$ do not couple directly to fermions. It is also easy to find that

$$\begin{aligned} \bar{\psi} \gamma^\mu \hat{A}_\mu^{(5)} \psi &= ig \bar{\psi} \frac{\gamma^\mu}{2} (\gamma_5 \hat{\lambda}_1 A_\mu^1 + \hat{\lambda}_2 A_\mu^2 + \gamma_5 \hat{\lambda}_3 A_\mu^3 + \gamma_5 \hat{\lambda}_8 A_\mu^8) \psi \\ &= ig \bar{\psi} \frac{\gamma^\mu}{2} (\hat{\lambda}_1 A_\mu^1 + \hat{\lambda}_2 A_\mu^2 + \hat{\lambda}_3 A_\mu^3 + \hat{\lambda}'_8 A_\mu^8) \psi, \end{aligned} \tag{4.2}$$

where λ_i, $i = 1, \cdots, 8$ are the usual Gell-Mann matrices for $SU(3)$ and

$$\lambda_8' = \frac{1}{\sqrt{3}} \begin{pmatrix} 1 & & \\ & 1 & \\ & & 2 \end{pmatrix}. \qquad (4.3)$$

λ_8' is just the generator used by Néeman, Fairlie and others in their graded $SU(2|1)$ gauge group formulation. It occurs here simply as a result of the difference in the action of γ_5 on the left-handed and the right-handed fermions, i.e.,

$$\gamma_5 L = L, \qquad \gamma_5 R = -R.$$

In terms of the physical gauge fields found in the previous section, it is easily verified that the interactions with $W^\pm$ and the photon have the usual form with

$$e = \frac{1}{2} g \cos\varphi, \qquad (4.4)$$

which indicates that $\sin^2\theta_W = \frac{1}{4}\cos^2\varphi$ in the charged current sector. This is in agreement with that obtained from the mass of Z_1 meson in the limiting case of small $\sin^2\varphi$.

The interaction Lagrangian with the neutral vector bosons has the form,

$$i g \bar{\psi}\gamma^\mu \Big\{ Z_{1\mu}\Big[\frac{1}{2\sqrt{3}}(2\hat{\lambda}_3 - \hat{Q} + \hat{Y})\cos\alpha + \frac{1}{2}\Big(\sin\varphi\,\hat{Q} - \frac{1}{\sin\varphi}\hat{Y}\Big)\sin\alpha\Big]$$
$$+ Z_{2\mu}\Big[\frac{1}{2\sqrt{3}}(2\hat{\lambda}_3 - \hat{Q} + \hat{Y})\sin\alpha - \frac{1}{2}\Big(\sin\varphi\,\hat{Q} - \frac{1}{\sin\varphi}\hat{Y}\Big)\cos\alpha\Big]\Big\}\psi. \qquad (4.5)$$

In the limiting case of small $\sin\varphi$, this interaction can be greatly simplified. The effective Lagrangian for the interchange of one neutral vector boson between fermions can be written as,

$$\mathscr{L}_{\text{eff}} = 4\frac{G_F}{\sqrt{2}}\Big[J_{N_1\mu}J^\mu_{N_1} + \frac{4}{9v}J_{N_2\mu}J^\mu_{N_2}\Big], \qquad (4.6)$$

where

$$J_{N_1\mu} = J^3_\mu - \sin^2\theta_W J^{\text{e.m.}}_\mu - \frac{4}{3v}(1 - 4\sin^2\theta_W)J^Y_\mu, \qquad (4.7)$$

$$J_{N_1\mu} = J^Y_\mu \cos^2\theta_W + (1 - 4\sin^2\theta_W)(J^3_\mu - J^{\text{e.m.}}_\mu) \qquad (4.8)$$

with

$$\sin^2\theta_W = \frac{1}{4}\cos^2\varphi$$

are two neutral currents coupled to the Z_1 and Z_2 mesons. In the case of $\sin^2\theta_W = 1/4$, the first neutral current has exactly the same form as in the simple gauge theory, while the second one is a pure vector current for quarks which is difficult to observe owing to the interference with strong interaction. Recent experiments tell us that $\sin^2\theta_W$ is slightly less than $1/4$. The present model provides a small correction (depending on the value v) in the neutral currents of fermions. This small effect could be measured by more accurate experiments.

V. Discussions

We have extended the previous $SU(3)$ model for leptons to a $SU(3)\times U(1)$ model for both quarks and leptons. Results like the mass spectrum and the Weinberg angle are model-dependent. However, there are some common features which make a whole class of such models different from the usual ones. Firstly, there are gauge bosons V and U which have no direct coupling with yet observed fermions. They only couple ordinary fermions with weak strange fermions which are assumed to be heavy in the present model. Thus in the low-energy region they have practically no influence on the electro-weak processes. These gauge bosons can be produced through virtual photons or Z bosons in electron positron colliding beams. The pair production cross section for these mesons in sufficiently high energy colliding beams will be of the same order as that for the $W^{\pm}$ bosons if the phase volume correction is taken into account.

Secondly, it is possible to define a new quantum number called weak strangeness in this class of models. The V,U gauge bosons, some fermions and some Higgs scalars are weak strange particles. They are produced only in pairs and the lightest one will be stable if the conservation of weak strangeness is exact.

The model formulated in the present paper could also be tested in accurate experiments on neutral current interactions.

There are two problems that remain unsolved. The first one concerns the cancellation of triangular anomalies. It is not possible to cancel anomaly associated with leptons by that with quarks in the present model since both leptons and quarks have the same representation in $SU(3)$ and $SU(3)$ is not anomaly free. Some heavy fermions, yet unobserved, with γ_5 replaced by $-\gamma_5$ in their transformation matrices must be added to cancell the anomalies. It might be possible that the original Lagrangian has left-right symmetry with equal number of these left and right multiplets. The yet unobserved fermions get their heavy masses only after spontaneous symmetry breaking.

The second problem concerns the determination of the coupling constant g', perhaps, in a grand unified $SU(6)$ theory. This problem is currently under investigation and will be reported later.

We would like to thank our colleagues at the Institute of Theoretical Physics, the Institute of High Energy Physics and Beijing University for their interest shown in this work.

References

[1] Weinberg, S., *Phys. Rev. Lett.*, 19 (1967), 1264; Salam, A., *Proc. 8th Nobel Symposium*, Stockholm, (1968); Glashow, S. L., Illiopoulos, J. & Maiani, L., *Phys. Rev.*, **D2** (1970), 1285.

[2] Tittel, K., Baltay C. & Weinberg, S., Talks given in 19th international conference on high energy physics, Tokyo, (1978).

[3] Chou Kwang-chao & Gao Chong-shou, *Kexue Tongbao*, **25** (1980), 21.

[4] Néeman, Y., *Phys. Lett.*, **81B** (1979), 190; Fairlie, D. B., *Phys. Lett.*, **82B** (1979), 97; Squires, E. J., *Phys. Lett.*, **82B** (1979), 395; Taylor, J. G., *Phys. Lett.*, **83B** (1979), 331; *ibid.*, **84B** (1979), 79; Dondi, P. H. & Jarvis, P. D., *Phys. Lett.*, **84B** (1979), 75.

[5] Lee, B. W. & Weinberg, S., *Phys. Rev. Lett.*, **38** (1977), 1237; Lee, B. W. & Schrock, R. E., *Phys. Rev.*, **D17** (1978), 2410.

Vol. 25 No. 1-2 KEXUE TONGBAO January 1980

ELECTRO-WEAK THEORY IN $SU(3)$

ZHOU GUANGZHAO (CHOU KWANGCHAO 周光召)
(*Institute of Theoretical Physics, Academia Sinica*)

AND GAO CHONGSHOU (高崇寿)
(*Department of Physics, Beijing University*)

Received August 20, 1979.

Ne'eman and Fairlie[1,2] have recently attempted to embed the $SU(2)\times U(1)$ electro-weak group[3] into a supersymmetric $SU(2|1)$ gauge theory. Fairlie and others[2,4] further extended the graded gauge fields over a space time manifold with more than 4 dimensions. Such an embedding would fix the Weinberg angle to be $\sin^2\theta_w = 1/4$ and if it was performed in an extended space time or super space time manifold. the Higgs mass could thus be determined. The number of extra dimensions is equal to twice the number of lepton triplets. Otherwise, the lepton masses will be the same. The common feature of this kind of theory is such that it contains unphysical particles with unfamiliar statistical properties.

In the present note, helicity mixed representation is used to construct a model of unified electro-weak theory for leptons in a gauge theory of simple group $SU(3)$. Each generation of two component leptons forms half a triplet in $SU(3)$.

The generators of the group $SU(3)$ $\hat{I}_i$, $i=1,\cdots,8$ can be decomposed into two sets $\hat{I}_a$ and $\hat{I}_\alpha$ where a and α are listed in Table 1.

Table 1

Case	a	α
I	1 3 4 6 8	2 5 7
II	1 3 5 7 8	2 4 6
III	2 3 5 6 8	1 4 7
IV	2 3 4 7 8	1 5 6
V	4 5 6 7	1 2 3 8
VI	1 2 4 5	3 6 7 8
VII	1 2 6 7	3 4 5 8

We now define new generators $\hat{I}_i^{(\epsilon)}$ by

$$\hat{I}_a^{(\epsilon)} = \epsilon \hat{I}_a, \quad \hat{I}_\alpha^{(\epsilon)} = \hat{I}_\alpha, \tag{1}$$

where ϵ commutes with $\hat{I}_i^{(\epsilon)}$ and satisfies the relation

$$\epsilon^2 = 1. \tag{2}$$

It can be easily verified that the Lie algebras for $\hat{I}_i^{(\epsilon)}$ are the same as those for $\hat{I}_i$. Four possible choices for $\epsilon: \epsilon = +1, -1, +\gamma_5, -\gamma_5$ will be used in the following. They are denoted by $+$, $-$, 5, -5 respectively.

Besides the local $SU(3)$ symmetry, a global $U(1)$ symmetry with generator $\hat{S}$ is introduced in the present model. The lepton triplets are transformed in the following way

$$\psi \to \psi' = U^{(5)}(\xi_j(x))\, e^{i\hat{S}\eta}\, \psi,$$
$$\bar{\psi} \to \bar{\psi}' = \bar{\psi} e^{-i\hat{S}\eta}\, U^{(-5)+}(\xi_j(x)), \tag{3}$$

where $U^{(\epsilon)}(\xi_j(x)) = \exp\{iI_j^{(\epsilon)}\xi_j(x)\}$ are the transformation matrices in $SU(3)$, $\xi_j(x)$ and η are the group parameters for local $SU(3)$ and global $U(1)$ group respectively. The generator $\hat{S} = \frac{1}{6}\gamma_5$ for lepton triplets.

Gauge invariant kinetic energy for the leptons has the form

$$\bar{\psi}\gamma^{\mu}(\partial_{\mu} + \hat{A}_{\mu}^{(5)})\,\psi, \tag{4}$$

where $\hat{A}_{\mu}^{(\epsilon)\prime} = igA_{\mu}^{j}\hat{I}_j^{(\epsilon)}$ are the gauge potentials. They will be transformed as

$$\hat{A}_{\mu}^{(\epsilon)} \to \hat{A}_{\mu}^{(\epsilon)\prime} = U^{(\epsilon)}(\partial_{\mu} + \hat{A}_{\mu}^{(\epsilon)})\, U^{(\epsilon)\dagger}. \tag{5}$$

There are many possible choices of the Higgs fields. Here as an example we take it to be a 3×3 matrix Φ which can be transformed in the following way:

$$\Phi \to \Phi' = e^{-\frac{i}{3}\eta} U^{(-)}(\xi_j(x))\, \Phi U^{(+)\dagger}(\xi_j(x)). \tag{6}$$

Their covariant derivatives are

$$D_{\mu}\Phi = \partial_{\mu}\Phi + A_{\mu}^{(-)}\Phi - \Phi A_{\mu}^{(+)}. \tag{7}$$

For concreteness we shall use a and α listed in the first case in Table 1. The first four cases give essentially the same results. In this case the Higgs field can be chosen as

$$\Phi^{(6)} = \frac{1}{\sqrt{6}}\,\phi^0 + \phi^a \hat{I}_a, \tag{8}$$

which is a six-dimensional representation of the gauge group. Another possible choice is a three-dimensional representation $\Phi^{(3)} = \phi^{\alpha}\hat{I}_{\alpha}$.

The invariant Lagrangian has the form

$$\mathscr{L} = -\frac{1}{4}F_{\mu\nu}^{i}F^{i\mu\nu} + \bar{\psi}\gamma^{\mu}(\partial_{\mu} + A_{\mu}^{(5)})\psi$$
$$+ \frac{1}{2}t_r\{(D_{\mu}\Phi)^{\dagger}(D^{\mu}\Phi)\} + \frac{1}{2}f\bar{\psi}\Phi(1+\gamma_5)\psi$$
$$+ \frac{1}{2}f^{*}\bar{\psi}\Phi^{+}(1-\gamma_5)\psi - V(\Phi, \Phi^{+}). \tag{9}$$

If there are only $SU(3)$ invariants of even power of Φ in the self interaction potential $V(\Phi, \Phi^{+})$, it is easily verified that the Lagrangian (9) is also invariant under the global $U(1)$ group defined above.

The vacuum expectation value of the Higgs scalar $\Phi^{(6)}$ is taken to be

$$\langle \Phi^{(6)} \rangle = v\hat{\lambda}_6, \tag{10}$$

where $\hat{\lambda}_i$ $i = 1, \ldots, 8$ are the Gell-Mann matrices for $SU(3)$. The electromagnetic gauge group $U_{E,M}(1)$ and a global $U(1)$ with generator $\hat{S}_W$ remain unbroken. The conserved charge $\hat{Q}$ and the weak strangeness $\hat{S}_W$ corresponding to these two unbroken $U(1)$ symmetry are respectively,

$$\hat{Q} = \hat{I}_3 - \sqrt{3}\,\hat{I}_8, \tag{11}$$

$$\hat{S}_w = \hat{S} + \frac{2}{\sqrt{3}}\hat{I}_8. \tag{12}$$

The masses of the vector bosons are found to be

$$m_Z^2 = \frac{4}{3} m_W^2, \quad m_V^2 = m_W^2, \quad m_U^2 = 4m_W^2, \tag{13}$$

where $V^{\pm}$ and $U^{\pm\pm}$ are four new vector bosons corresponding to

$$V^{\pm} = \frac{1}{\sqrt{2}}(A^4 \pm iA^5),$$

$$U^{\pm\pm} = \frac{1}{\sqrt{2}}(A^6 \pm iA^7). \tag{14}$$

The photon field A and the neutral vector boson Z^0 are related to the gauge fields A^3 and A^8 in the following way:

$$A = \frac{1}{2}(A^3 - \sqrt{3}A^8), \quad Z^0 = \frac{1}{2}(\sqrt{3}\,A^3 + A^8). \tag{15}$$

The charge and the weak strangeness for gauge bosons are listed in Table 2.

Table 2

Particles	$W^{\pm}$	A	Z^0	$V^{\pm}$	$U^{\pm\pm}$
Q	± 1	0	0	± 1	± 2
S_w	0	0	0	∓ 1	∓ 1

We require that the weak strangeness for the yet to be observed leptons be ½. This ensures that the helicity of the lepton triplet be of the form (L, L, R) after spontaneous symmetry breaking. There are $S_w = -½$ components of the form (R, R, L) in ψ. It is possible to make their masses very heavy by a suitable choice of additional Higgs fields. We assign to the components for the first generation of leptens as

$$\psi = \begin{pmatrix} \nu_L \\ e_L \\ e_R \end{pmatrix}. \tag{16}$$

The charge operator for leptons now takes the form

$$\hat{Q} = I_3^{(5)} - \sqrt{3}\, I_8^{(5)} = \frac{1}{2}\gamma_5(\lambda_3 - \sqrt{3}\,\lambda_8). \tag{17}$$

When a lepton triplet of the form (16) is acted upon, the charge operator becomes

$$\hat{Q} = \frac{1}{2}(\lambda_3 - \sqrt{3}\,\lambda_8') = \begin{pmatrix} 0 & & \\ & -1 & \\ & & -1 \end{pmatrix}, \tag{18}$$

where

$$\lambda_8' = \frac{1}{\sqrt{3}}\begin{pmatrix} 1 & & \\ & 1 & \\ & & 2 \end{pmatrix}. \tag{19}$$

We note that λ_8' is just the generator used by Ne'eman, Fairlie and others in their graded gauge group formulation. It occurs here simply as a result of the difference in the action of γ_5 on left-handed and right-handed leptons, i.e.

$$\gamma_5 L = L, \quad \gamma_5 R = -R,$$

The weak and electromagnetic interactions for leptons are now easily obtained from the Lagrangian. It is easily verified that these interactions have exactly the conventional form in the Weinberg-Salam model with $\sin^2\theta_w = 1/4$. The weak strange vector bosons do not interact with the present observed leptons directly. They only couple $S_w = 1/2$ leptons with the $S_w = -1/2$ heavy leptons. Therefore, they have practically no influences on the low-energy weak interaction.

Five heavy scalars remain in the present choice of Higgs fields. They can have arbitrary masses when suitable self-interaction potential is chosen. One of the Higgs scalars is a real neutral field, with zero weak strangeness. Its vacuum expectation value is ν given in (10) which generates masses for both vector bosons and electrons. The other four Higgs are weak strange particles which do not couple to leptons directly. Two of these Higgs are neutral and the other two are doubly charged.

Conservation of weak strangeness ensures the pair production of weak strange particles. The weak strange bosons of the lightest mass would be stable particles if this conservation law is exact. The $V^\pm$ and $U^{\pm\pm}$ vector bosons can be produced through the mediation of other heavy vector bosons and photons. The production cross section for a $V^\pm$ pair is the same as that for a $W^\pm$ pair in an electron-positron annihilation reaction with sufficiently high energy. These characteristics may be of help in the experiments looking for these weak strange particles.

When terms like $\det\Phi + \det\Phi^+$ were allowed in the Higgs self-interaction potential, the conservation of weak strangeness would be broken and all heavy bosons would become unstable. Nevertheless, even in this case, the decay amplitude of the V-particle (assumed to be the lightest of the weak strange particles) is of the order of magnitude $O(g^3)$ in contrast with $O(g)$ for the decay of W-particles.

If $\Phi^{(3)}$ is chosen to be the Higgs, a residual $SU(2)$ symmetry remains unbroken and the double charged U-particles remain massless together with the photon. A linear combination of $\Phi^{(3)}$ and $\Phi^{(6)}$ will not change the mass of V-particle but left the U-particle mass in the range $0 \leqslant m_u \leqslant 2m_w$. All other conclusions still hold in these cases.

There is no difficulty in adding further generations of leptons like (ν_μ, μ) and (ν_τ, τ) in the present theory.

The extension of the present model to include quarks is a problem yet to be solved. The difficulty lies in the different charge assignments for leptons and quarks. There are two possibilities: Either composite model in terms of prequarks should be tried, or the gauge group $SU(3)$ should be enlarged. Both of these approaches have their own problems. Some of the preliminary results on this subject will be reported in other papers.

We would like to thank our colleagues at the Institute of Theoretical Physics, and the Institute of High Energy Physics, Academia Sinica and Beijing University for their interest in this work.

References

[1] Ne'eman, Y., *Phys. Lett.*, **81B**(1979), 190.
[2] Fairlie, D. B., *ibid.*, **82B**(1979), 97.
[3] Weinberg, S., *Phys. Rev. Lett.*, **19**(1967), 1264;
Salam, A., *Proc. 8th. Nobel Symposium*, Stockholm, 1968;
Glashow, S. L., Illiopoulos J. & Maiani. L., *Phys. Rev.*, **D2**(1970), 1285.
[4] Squires, E. J., *Phys. Lett.*, **82B** (1979), 395;
Jaylor, J. G., *ibid.*, **83B**(1979), 331; **84B**(1979), 79;
Dondi, P. H. & Jarvis, P. D., *ibid.*, **84B**(1979), 75.

Vol. 25 No. 8 KEXUE TONGBAO August 1980

DISORDER PARAMETER AND DUALITY

ZHOU GUANGZHAO (CHOU KUANG-CHAO 周光召)
(*Institute of Theoretical Physics, Academia Sinica*)

AND XIAN DINGCHANG (HSIEN TING-CHANG 冼鼎昌)
(*Institute of High Energy Physics, Academia Sinica*)

Received August 17, 1979.

In studying the problem of quark confinement, t' Hooft introduced a loop dependent disorder parameter $B(C')$, which depends on loop C' and is related to the Wilson loop operator $A(C)$ of the non-Abelian gauge field $\hat{A}_\mu$ by the algebraic equation[1]

$$B(C')A(C) = A(C)B(C')z. \tag{1}$$

For the $SU(N)$ group, z is the element of the center of the group $Z(N)$. t' Hooft defined the disorder parameter $B(C')$ by singular gauge transformations. The authors of Refs. [1—3] have pointed out the eletromagnetic duality property between $B(C')$ and $A(C)$. By introducing dual potential and defining the Wilson loop operator $\tilde{A}(C')$ as composed of dual potential, Halpern[2] proved that in the Abelian case, $\tilde{A}(C')$ is a realization of $B(C')$ in the algebraic equation (1) and guessed that a corresponding $\tilde{A}(C')$ could be introduced in the non-Abelian case. However, in all these works, no direct proof has been given for the explicit electromagnetic duality property between $B(C')$ as defined by the singular gauge transformations, and $A(C)$. The purpose of this paper is to discuss directly the duality property between $B(C')$ and $A(C)$ from the definition of $B(C')$.

In order to illustrate our method, let us discuss first the Abelian case. By gauge transformation, the gauge potential A_μ will become

$$A'_\mu = A_\mu + \frac{1}{e}\,\partial_\mu\lambda, \tag{2}$$

whose value is path-dependent when λ is singular. Now by defining the field strength corresponding to this singular gauge transformation as

$$F^{st}_{\nu\mu} = \frac{1}{2e}\,[\partial_\mu, \partial_\nu]\lambda, \quad \mu, \nu = 0, 1, 2, 3, \tag{3}$$

and by integrating Eq. (3) over a surface Σ bounded by a loop C, one obtains

$$\frac{1}{e}\oint \partial_\mu\lambda dy^\mu = \iint_\Sigma F^{st}_{\mu\nu} dy^\mu \wedge dy^\nu. \tag{4}$$

Let us consider the case of two 3-dimensional loops C and C' at the same specific time, the coordinates of them are respectively

$$\begin{aligned} C:\ & \mathbf{y} = \mathbf{y}(\theta), \quad 0 \leqslant \theta \leqslant 2\pi, \quad \mathbf{y}(0) = \mathbf{y}(2\pi); \\ C':\ & \mathbf{z} = \mathbf{z}(\theta'), \quad 0 \leqslant \theta' \leqslant 2\pi, \quad \mathbf{z}(0) = \mathbf{z}(2\pi). \end{aligned} \tag{5}$$

Suppose $H_k^{st} = \frac{1}{2} \varepsilon_{kjl} F_{jl}^{st}(k, j, l = 1, 2, 3)$ to be the magnetic field strength formed by the magnetic force lines along the direction of C', then it follows from Eq. (4)

$$\lambda(2\pi) - \lambda(0) = ne\Phi, \tag{6}$$

where n is the number of times loop C encircling loop C', counterclockwise. For the Abelian case, the values of λ at the same point reached by different paths can differ by an integer factor of 2π, hence,

$$\lambda(2\pi) - \lambda(0) = 2\pi mn, \tag{7}$$

where m is also an integer. From Eqs. (6) and (7) the quantization condition of the magnetic flux

$$e\Phi = 2\pi m \tag{8}$$

is obtained.

The above-mentioned field strength along C' may be expressed as

$$F_{ij}^{st}(\mathbf{x}, C') = \oint_{C'} dz^l \varepsilon_{ijl} \delta^3(\mathbf{x} - \mathbf{z}(\theta'))\Phi. \tag{9}$$

And from Eq. (3) the following equation for λ can be derived:

$$\frac{1}{e} \partial_j \lambda(\mathbf{x}, C') = \int d^3x' G(\mathbf{x} - \mathbf{x}') \partial^l F_{jl}^{st}(\mathbf{x}', C'), \tag{10}$$

where i, j, l, assume the spatial indices 1, 2, 3, only; $G(\mathbf{x})$ satisfies the following equation

$$\nabla^2 G(\mathbf{x}) = \delta^3(\mathbf{x}).$$

In the gauge of $A_0 = 0$, the disorder operator $B(C')$ is defined as the following operator of singular gauge transformation

$$B(C') = \exp\left\{ i \int \hat{F}_{0j}(\mathbf{x})\, \partial_j \left(\frac{1}{e} \lambda(\mathbf{x}, C') \right) d^3x \right\}, \tag{11}$$

where $\hat{F}_{0j}$ represents the electric field strength in the Heisenberg representation and is the canonically conjugate operator of the gauge potential $\hat{A}_j$; $\partial_j \left(\frac{1}{e} \lambda(\mathbf{x}, C') \right)$ is given by (10). Carrying out the variation of Eq. (11) with respect to the coordinate $\mathbf{z}(\theta')$ of loop C', one obtains

$$B^{-1}(C') \delta B(C') = i \int \hat{F}_{0j} \partial_j \left(\frac{1}{e} \delta\lambda(\mathbf{x}, C') \right) d^3x, \tag{12}$$

and from (9) and (10) it follows

$$\partial_j \left(\frac{1}{e} \delta\lambda(\mathbf{x}, C') \right) = \int d^3x' G(\mathbf{x} - \mathbf{x}') \partial^l \delta F_{jl}^{st}(\mathbf{x}', C'), \tag{13}$$

and

$$\delta F_{ji}^{st}(\mathbf{x}, C') = \epsilon_{jli} \frac{\partial}{\partial z^k} \delta^3(\mathbf{x} - \mathbf{z}) \Phi dz^k \wedge dz^l. \tag{14}$$

Substituting (14) into (13), and then the result into (12); by partial integration and using the Gauss's theorem $\partial_k \hat{F}_{0k} = 0$, one finds

$$B^{-1}(C')\delta B(C') = \frac{i}{2} \Phi \epsilon_{klj} \hat{F}_{0j}(z) dz^k \wedge dz^l$$

$$= \frac{i}{2} \Phi \hat{F}_{kl}^d(z) dz^k \wedge dz^l, \tag{15}$$

where $\hat{F}_{kl}^d = \frac{1}{2} \epsilon_{kl\mu\nu} \hat{F}^{\mu\nu}$ is the conjugate field strength.

In the gauge of $A_0 = 0$, the Wilson loop operator $A(C)$ is

$$A(C) = \exp\left\{ie \oint_C \hat{A}_i(y) dy^i\right\}. \tag{16}$$

Now from the variation of $A(C)$ with respect to the coordinate $\mathbf{y}(\theta)$ of loop C, it is easy to derive

$$A^{-1}(C)\delta A(C) = \frac{i}{2} e \hat{F}_{kl}(y) dy^k \wedge dy^l \tag{17}$$

and comparing (15) with (17), one finds the following electric and magnetic duality properties:

$$e \longleftrightarrow \Phi; \quad \hat{F}_{kl} \longleftrightarrow \hat{F}_{kl}^d; \quad \mathbf{y}(\theta) \longleftrightarrow \mathbf{z}(\theta'); \quad B(C') \longleftrightarrow A(C). \tag{18}$$

It should be pointed out that this duality property holds only for a pure electromagnetic field system, since the sourceless Gauss's theorem has been used in the proof.

Now let us discuss the non-Abelian case and the gauge group is chosen to be $SU(N)$. First of all, we have to look for the singular gauge transformations which define the disorder operator. Many authors[4] have pointed out that in the non-Abelian case, corresponding to the gauge invariant magnetic string, a unit vector $\hat{I}(x)$ of Lie algebra should be introduced, which remains invariant under parallel displacement:

$$D_\mu \hat{I}(x) = \partial_\mu \hat{I}(x) + e[\hat{A}_\mu(x), \quad \hat{I}(x)] = 0, \quad \mathrm{tr}\,(\hat{I}^2(x)) = 2. \tag{19}$$

And the field strength projection in the direction along $\hat{I}(x)$

$$F_{\mu\nu} = \frac{1}{2} \mathrm{tr}\,(\hat{F}_{\mu\nu} \hat{I}) \tag{20}$$

contains the gauge invariant magnetic string.

The group parameter $\lambda_a(x, C')$ characterizing the singular gauge transformation is defined as

$$\hat{\lambda}(x, C') = \lambda_a(x, C') \hat{I}_a = \hat{I}(x) \lambda(x, C'), \tag{21}$$

where $\hat{I}_a$, $a = 1, \cdots n_G$ are the generators of the group G; $\lambda(x, C')$ is a singular function defined by Eq. (10). In the $A_0 = 0$ gauge, the disorder parameter is defined as the singular gauge transformation by the following expression:

$$B(C') = \exp\left\{ i \int \hat{F}^a_{0j} \left(D_j \left(\frac{1}{e} \hat{\lambda} \right) \right)_a d^3x \right\}. \tag{22}$$

By using the invariance property of $\hat{I}(x)$ under parallel displacement, from (22) it follows

$$B(C') = \exp\left\{ i \int \hat{F}_{0j} \partial_j \left(\frac{1}{e} \lambda(x, C') \right) d^3x \right\}, \tag{23}$$

where $\hat{F}_{0j}$ is the gauge invariant electric field strength defined by (20). It is easy to see that (23) and the corresponding operator in the Abelian gauge field theory possess the same form, hence, by variation with respect to the coordinate of loop C', one finds

$$B^{-1}(C')\delta B(C') = \frac{i}{2} \Phi F^d_{jk}(z) dz^j \wedge dz^k, \tag{24}$$

where

$$F^d_{jk} = \frac{1}{2} \operatorname{tr} \{\hat{F}^d_{jk} \hat{I}\} \tag{25}$$

represents the dual strength.

In the $A_0 = 0$ gauge, the Wilson loop operator is

$$A(C) = \operatorname{tr}\left(T \exp\left\{ ie \oint_C \hat{A}_i(y) dy^i \right\} \right), \tag{26}$$

where T represents the operator of ordering along loop C. It is easy to prove[5]

$$B(C')A(C)B^{-1}(C') = A(C)z,$$

where

$$z = \exp\{ie\Phi \hat{I}(x)\} \tag{27}$$

represents the elements in the center $Z(N)$. By carrying out the variation of $A(C)$ with respect to the coordinate of loop C, Nambu et al.[6] have obtained

$$A^{-1}(C)\delta A(C) = \frac{i}{2} e\hat{F}_{jk}(y, C) dy^j \wedge dy^k, \tag{28}$$

where

$$F_{jk}(y, C) = \frac{1}{2} \operatorname{tr} \{\hat{F}_{jk}(y) \hat{\phi}(y, C)\}, \tag{29}$$

while

$$\hat{\phi}(y, C) = A^{-1}(C) \hat{I}_a \operatorname{tr}\left\{ T \left(\hat{I}_a(y) \exp\left\{ ie \oint_C \hat{A}_j(y) dy^j \right\} \right) \right\}. \tag{30}$$

The $\hat{I}_a(y)$ in Eq. (30) represents the $\hat{I}_a$ at point y of loop C. Generally speaking, $\hat{\phi}(y, C)$ is a functional of loop C. However, it has been proved[6] that $\hat{\phi}(y, C)$ is invariant under parrallel displacement along the direction of loop C, though it does not possess this property along other directions.

By comparing (24) with (28), it can be seen that there is some sort of duality property between $B(C')$ and $A(C)$. But this duality property is not as perfect as that in the Abelian case because of the fact that $\hat{I}(x)$ and $\phi(y, C)$ are not completely similar in character. This may not be as surprising as it appears since it is not always possible to find out an appropriate dual potential[7] in the non-Abelian gauge field theory. After the submission of this work for publication, we noticed that Mandelstam[8] had considered some problems similar to that discussed in the present paper.

References

[1] 't Hooft G., *Nucl. Phys.*, **B138** (1978), 1.

[2] Halpern, M. B., *Phys. Rev.*, **D19** (1979), 517.

[3] Yoneya, T., *Nucl. Phys.*, **B144** (1978), 195; Englert, F. & Windey, P., *Phys. Rep.*, **49** (1979), 173.

[4] 't Hooft G., *Nucl. Phys.*, **B79** (1974), 276; Polyakov, A. M., *JETP Lett.*, **20** (1974), 20; Ezawa, Z. F. & Tze, H. C., *Nucl. Phys.*, **B100** (1975), 1; 段一士，葛墨林，侯伯宇，《物理学报》，**25** (1976), 514; Arafune, J., Freund, P. G. O. & Ooebel, C. J., *J. Math. Phys.*, **16** (1975), 433.

[5] Corrigan, E. & Olive, D., *Nucl. Phys.*, **B110** (1976), 237.

[6] Nambu, Y., *Phys. Lett.*, **80B** (1979), 372; Corrigan, E. & Hasslacher, B., *Physl Lett.*, **81B** (1979), 181; Gervais, J. L. & Neveu, A., *Phys. Lett.*, **80B** (1979), 255.

[7] Gu Chao-hao & C. N. Yang, *Sci. Sin.*, **18** (1975), 483; Brandt, R. A. & Neri, F., *Nucl. Phys.*, **B145** (1978), 221.

[8] Mandelstam, S., *Phys. Rev.*, **D19** (1979), 2391.

Reprinted with permission from *AIP. Conf. Proc.* **72** (1980) 621.

AXIAL U(1) ANOMALY AND CHIRAL SYMMETRY-BREAKING IN QCD

K. C. Chou*
Virginia Polytechnic Institute, Blacksburg, VA

It is well known that the absence of the ABJ anomaly is necessary for the corresponding gauge field theory to be renormalizable[1]. This condition places severe restrictions on the choice of the possible gauge group, and the representation for fermions, as we have just heard from A. Zee. I would like to report, on the other hand, some consequences of the presence of ABJ anomaly in certain global current conservation equations. This is a work done in collaboration with L. N. Chang.

One important question in QCD is the origin of chiral symmetry breaking. This problem is related to the structure of the theory for large distances, where perturbation theory cannot be used. It has long been conjectured that topologically non-trivial gauge field configurations play some significant role in explaining both confinement and chiral symmetry breaking[2].

Recently, Coleman and Witten[3], and Veneziano[4], have analyzed the question within the context of $1/N_c$ expansion. In particular, Coleman and Witten argue that the axial anomaly and confinement already imply chiral symmetry breaking. Their argument makes no essential use of the non-trivial topological configurations, but relies instead on the absence of analyticity structure in the axial vector vertex, brought about by the anomaly. Veneziano reaches the same conclusion, using the $1/N_c$ expansion and by

*On leave from Institute for Theoretical Physics, Peking.

consideration of the fluctuation of topological charge density in the pure Y-M field sector.

In this talk I want to point out that their conclusion on the necessity of chiral symmetry breaking can be obtained without recourse to the $1/N_c$ expansion, if proper attention is paid to the quantum fluctuations in the topological charge density.

We start by recalling that in the presence of non-trivial topological configurations, one could incorporate the θ-vacuum caused by the resultant phenomenon of tunneling by augmenting the conventional Lagrangian with an additional term

$$\mathcal{L} = -(\bar{q}\gamma^{\mu}D_{\mu}q) - \frac{1}{4}F^{i}_{\mu\nu}F^{i\mu\nu} - \theta\nu(x) - J_a(x)\,O_a(x)$$

$$\nu(x) = \frac{g^2}{32\pi^2}F^{i}_{\mu\nu}{*F}^{i\mu\nu}, \quad D_{\mu} = \partial_{\mu} - g\hat{A}_{\mu} \tag{1}$$

$${*F}_{\mu\nu} = \frac{1}{2}\epsilon_{\mu\nu\sigma\rho}F^{\sigma\rho} \quad .$$

In equation (1), θ is the parameter characterizing the topological structure of the vacuum, while $J_a(x)$ are external sources coupled to various combinations of quark currents $O_a(x)$. Since we are interested in chiral symmetry breaking, our attention will be focused on the densities $\bar{q}(1\pm\gamma_5)q \equiv O_{\pm}$ and $\bar{q}\gamma^{\mu}(1\pm\gamma_5)q \equiv O^{\mu}_{\pm}$. In the following θ will be considered as a function of x in some intermediate steps of calculation.

The Lagrangian of (1) has an apparent $U(N_f) \times U(N_f)$ chiral symmetry, with N_f flavours, if we set $J_a(x) = 0$. However, due to the ABJ anomaly in the axial current, θ will change to $\theta + \sqrt{2N_f}\,\xi_5$ under the abelian chiral phase transformation $q(x) \to \exp[i\gamma_5\xi_5]q(x)$.

The best way to study the chiral symmetry structure of (1)

when $J(x) \neq 0$ is through the effective generating functional, which we shall now define. The generating functional, W, for the connected Green's functions implied by (1) can be expressed as

$$W[J,\theta] = -i \ln Z$$
$$Z = \int DqD\bar{q}DA_\mu \, \exp\{i\int d^4x \, (\mathcal{L}(x) + \Delta\mathcal{L}(x)] \quad . \tag{2}$$

Here $\Delta\mathcal{L}(x)$ includes the gauge fixing and compensating terms necessary to give meaning to the A_μ integration. The classical fields U(x) can now be defined by direct differentiation

$$U\ (x) = - \frac{\delta W}{\delta J_\pm(x)} \quad . \tag{3}$$

As a result of the axial anomaly, and the formal invariance properties of (1), the generating functional W has to satisfy a Ward identity of the form

$$\partial_\mu \frac{\delta W}{\delta J_{\mu\pm}(x)} = iJ_\pm \frac{\delta W}{\delta J_\pm(x)} \pm N_f \frac{\delta W}{\delta\theta(x)} \quad . \tag{4}$$

This Ward identity can be satisfied by any functional with the following local invariance property

$$W[J\ (x),\ J_{\mu\pm}(x),\ \theta(x)]$$
$$= W[J_\pm(x)\ e^{\pm i\sqrt{\frac{2}{N_f}}\,\xi_5(x)},\ J_{\mu\pm}(x) \pm \partial_\mu\left(\sqrt{\frac{2}{N_f}}\,\xi_5(x)\right)$$
$$\cdot\ \theta(x) - \sqrt{2N_f}\,\xi_5(x)] \tag{5}$$

where $\xi_5(x)$ is an arbitrary function of x.

We define now the generating functional Γ by using a Legendre transformation on the sources of the scalar currents $O_\pm$

$$\begin{aligned}\Gamma[U_\pm(x),\ J_{\mu\pm}(x),\ \theta(x)] \\ = W[J_\pm(x),\ J_{\mu\pm}(x),\ \theta(x)] \\ + \int d^4x\ \left(J_+(x)U_+(x) + J_-(x)\ U_-(x)\right)\end{aligned} \tag{6}$$

Then the Ward identity for the axial U(1) symmetry implies that Γ is invariant under the local transformation

$$\begin{aligned} U_\pm &\to U_\pm e^{+i\sqrt{\frac{2}{N_f}}\ \xi_5(x)} \\ J_{\mu\pm} &\to J_{\mu\pm}(x) + \partial_\mu\sqrt{\frac{2}{N_f}}\ \xi_5(x) \\ \theta(x) &\to \theta(x) - \sqrt{2N_f}\ \xi_5(x)\end{aligned} \tag{7}$$

Or in other words, Γ is a functional of the form

$$\Gamma[U_\pm e^{+i\frac{1}{N_f}\theta},\ J_{\mu\pm}(x) \pm \partial_\mu(\frac{1}{N_f}\ \theta)] \tag{8}$$

Note that the classical fields $U_\pm(x)$, which are the vacuum expectation values of the corresponding quark bilinear fields, can be determined through the relation

$$\frac{\delta\Gamma}{\delta U_\pm(x)} = J_\pm(x) \tag{9}$$

Any nontrivial solution to (9) when $J_\pm = 0$, $J_{\mu\pm} = 0$ and $\theta(x) =$ constant, would signal the existence of spontaneous chiral symmetry breaking.

Since $U_+ = U_-^*$ we can write them in the form

$$U_\pm + U\, e^{\pm i\sqrt{\frac{2}{N_f}}\ \frac{\eta'}{f_\pi}} \tag{10}$$

where η' can be interpreted as the vacuum expectation value of the η'-meson field which corresponds to the axial U(1) pseudo Goldstone field in chiral dynamics.

Notice that Γ is an even function of $\theta - \frac{\sqrt{2N_f}}{f_\pi}\eta'$ as a consequence of (8), and the symmetry under space reflection. It is easily proved that the CP conserving solution of eq. (9) at $J_\pm = 0$, $J_{\pm\mu} = 0$, $\theta(x) = \theta$ is

$$\eta' = \frac{f_\pi}{\sqrt{2N_f}}\,\theta \,. \tag{11}$$

Thus all the physical quantity evaluated at this point will be θ independent and CP conserving.

Nevertheless, the nth order derivative of Γ with respect to θ when η' is kept fixed is the Green's function of $n\nu(x)$'s where diagrams with one particle lines of η and U are removed.

$$\frac{\delta^n\Gamma}{\delta\theta^n}\bigg|_{\eta',u} = (-i)^n \int d^4x_1 \ldots d^4x_n \, \langle 0|T\left(\nu(x_1)\ldots\nu(x_n)\right)|0\rangle_{\text{I.P.I.}} \tag{12}$$

It has been shown by one of us[5] that $\frac{\delta^2\Gamma}{\delta\theta^2}\Big|_{\mu'u}$ is proportional to the η' meson mass. This result is a generalization of the result given first by Witten[6] in the leading $1/N_c$ expansion approximation.

The point we wish to emphasize is that (12) can be non-vanishing only if $U \neq 0$, so that if any of the moments defined in (12) were to be nonzero, chiral symmetry would be spontaneously broken. This is the main conclusion of my talk. Now the right hand side of (12) represents a quantum correlation function of the topological charge density, and there is no general reason for (12) to vanish for n even. The case for n odd can be excluded, of course, in the chiral limit when CP is a good symmetry. We therefore argue that QCD will, in general, induce the spontaneous symmetry breaking of flavor chiral symmetry.

The picture we are presenting may therefore be summarized as follows: Owing to the presence of instantons, the QCD vacuum acquires an additional parameter θ. In the absence of any external spontaneous chiral symmetry breaking, like those induced by Higgs couplings, the chiral phases of the quarks will automatically refer themselves to θ. This is the direct consequence of the axial anomaly. However, large scale quantum fluctuations of the topological charge density requires such phases to be defined globally, which can only occur if the chiral symmetry is spontaneously broken. Hence quantum corrections to topologically non-trivial gauge configurations induce spontaneous chiral symmetry breaking.

REFERENCES

1. D. J. Gross and R. Jackiw, Phys. Rev. D6, (1972) 477. C. P. Korthals Altes and M. Perrottet, Phys. Lett. 39B, (1972) 546.
2. G. 't Hooft, Phys. Rev. D14 (1976) 3432. D. G. Caldi, Phys. Rev. Lett. 39 (1977) 121. C. G. Callan, R. F. Dashen and D. J. Gross, Phys. Rev. D17 (1978) 2717.
3. S. Coleman and E. Witten, Phys. Rev. Lett. 45 (1980) 100.
4. G. Veneziano, CERN TH-2872 (1980).
5. K. C. Chou, ASITP-80-005 (1980).
6. E. Witten, Nucl. Phys. B156 (1979) 269.

PHYSICAL REVIEW D VOLUME 23, NUMBER 11 1 JUNE 1981

Possible $SU(4)_c \times SU(3)_f \times U(1)$ model

Chong-Shou Gao
Stanford Linear Accelerator Center, Stanford University, Stanford, California 94305
*and Department of Physics, Beijing University, Beijing, People's Republic of China**

Kuang-Chao Chou
Institute of Theoretical Physics, Academia Sinica, Beijing, People's Republic of China

(Received 14 October 1980)

An anomaly-free model of strong and electroweak interactions involving leptons and quarks in the $SU(4)_c \times SU(3)_f \times U(1)$ gauge theory is constructed. After spontaneous symmetry breaking, it reduces to quantum chromodynamics for strong interactions and a broken $SU(3)\times U(1)$ model for electroweak interactions. As a limiting case it gives the same results as those of the Weinberg-Salam model in the low-energy region. The Weinberg angle is bounded by $\sin^2\theta_W < 1/4$ and becomes slightly less than 30° in the limiting case. Below the mass scale of $SU(4)_c$ breaking there exists an inequality between the Weinberg angle and the strong coupling constant, which is consistent with experiments. A correction to the neutral current of the Weinberg-Salam model is suggested. A new conserved quantum number is introduced in this model and there exist several new fermions with masses lighter than 160 GeV. The Kobayashi-Maskawa expression of Cabibbo mixing for quarks may be obtained in the model, generalized to include several generations of fermions.

I. INTRODUCTION

Recent neutrino-induced neutral-current experiments are in agreement with the expectations based on the Weinberg-Salam model.[1] The Weinberg angle θ_W is found to be $\sin^2\theta_W = 0.230 \pm 0.009$, averaged over the various experiments.[2] Beyond the Weinberg-Salam model one may ask the following.

(1) Is there any symmetry higher than $SU(2)_L \times U(1)$ for the electroweak interaction?

(2) Does $\sin^2\theta_W$ being slightly less than $\frac{1}{4}$ have special physical meaning?

(3) How can the Weinberg-Salam model be unified with the strong interaction?

If the answer to the first question is "no," the next problem to be solved is the grand unification of the Weinberg-Salam model with the strong interaction. In this way one may construct a model of grand unification, such as the SU(5) model suggested by Georgi and Glashow.[3]

If one thinks that the answer to the first question is "yes," this leads to another question: What kind of higher symmetry might this be? There are several considerations which may become the motivations to choose the higher symmetry group: (i) left-right symmetry before spontaneous symmetry breaking, (ii) quark-lepton unification, and (iii) a sensible prediction for the empirical Weinberg angle.

Among the many possibilities meeting these criteria is the $SU(3)\times U(1)$ group for the electroweak interactions. In a previous paper[4] a model with the $SU(3)\times U(1)$ gauge group was proposed. This model is left-right symmetric before spontaneous symmetry breaking and anomaly free. It gives the same results as those of the Weinberg-Salam model in the low-energy region in a limiting case. The Weinberg angle is bounded by $\sin^2\theta_W \leq \frac{1}{4}$ in this model and $\sin^2\theta_W$ becomes slightly less than $\frac{1}{4}$ in the limiting case.

In this paper we discuss a way of unifying the strong and electroweak interactions by embedding this $SU(3)\times U(1)$ model into a larger one, $SU(4)_c \times SU(3)\times U(1)$, where the main results of Ref. 4, including the interesting property of $\sin^2\theta_W \leq \frac{1}{4}$, are preserved and several further consequences are obtained.

Before discussing this model we will briefly analyze the construction of the $SU(3)\times U(1)$ group, which will be helpful in understanding the motivation of an extension to $SU(4)\times SU(3)\times U(1)$.

When one embeds the $SU(2)_L \times U(1)$ model into an $SU(3)\times U(1)$ model, a naive requirement is that ν_L and e_L correspond to the first two components of a left-handed triplet and e_R correspond to the third component of the right-handed triplet of the SU(3) group. There are two possibilities.

Case A. ν_L and e_L belong to the representation $\underline{3}_L$ and e_R belongs to the representation $\underline{3}_R$ of the SU(3) group. This case has been investigated by Lee, Weinberg, Shrock, Segre, Weyers, and many others[5] in detail.

Case B. ν_L and e_L belong to the representation $\underline{3}_L$ while e_R belongs to the representation $\underline{3}^*_R$, the conjugate representation of $\underline{3}$. This case is investigated in Ref. 4.

These two cases lead to different consequences, as summarized in Table I. In the expressions for the charge operator, $\hat{I}_3$ and $\hat{I}_8$ are the third and the eighth generators of the SU(3) group, respect-

TABLE I. Comparison of the two possible schemes in the SU(3)× U(1) group for electroweak interactions.

	Case A	Case B
Charge operator	$\hat{Q}=\hat{I}_3+(1/\sqrt{3})\hat{I}_8+\hat{Y}$	$\hat{Q}=\hat{I}_3-\sqrt{3}\hat{I}_8+\hat{Y}$
$\sin^2\theta_W$	$\frac{3}{4}\frac{1}{1+(3g^2/g'^2)}$	$\frac{1}{4}\frac{1}{1+(g^2/g'^2)}$
Boundary	$<\frac{3}{4}$	$<\frac{1}{4}$
New conserved quantum number	No	Weak strangeness
Additional heavy particles	Yes	Yes
With exotic charges	No	Some of them
To be embedded into an SU(6) model	Easily	Cannot

ively, while Y is the generator of the U(1) group. Of course the Y assignment of the fermion multiplets are different in these two cases. In case B, there exists an additional conserved quantum number called the weak strangeness S_W, coming from an unbroken U(1) symmetry after spontaneous symmetry breaking. There are several heavy particles to be discovered in this case too. But most of them have nonvanishing values of S_W while the known particles in the Weinberg-Salam model have $S_W=0$. We may call these particles with $S_W \neq 0$ the weak strange particles. Some of them have "exotic" values of charge, for example, $Q=2$ for a heavy vector boson and $Q=\frac{5}{3}$ for a heavy quark.

The most interesting characteristic of the case B is that the upper bound on the Weinberg angle is close to the measured value. In addition, the existence of the conservation of weak strangeness gives many new physical predictions for the high-energy electroweak interactions, and is therefore interesting in its own right. However, because the left-handed and right-handed fermions belong to different representations of the SU(3) group, this kind of model cannot be embedded into a simple SU(6) model of grand unification. So we have to study other ways of connecting this kind of model with the strong interaction. One attractive idea is that the color group may be an SU(4) group and the lepton number may be treated as the fourth color as suggested by Pati and Salam.[6] Adopting this idea, we will extend this kind of SU(3)×U(1) model to an $SU(4)_c \times SU(3) \times U(1)$ model.

II. FUNDAMENTAL STRUCTURE OF THE MODEL

The local gauge groups considered in this model are the SU(4) color group, the SU(3) flavor group, and the U(1) group. Their generators, the corresponding gauge fields, and the coupling constants are denoted by

$$\hat{I}'_j,\ C^j_\mu,\ j=1,\ldots,15,\ g''\text{ for SU(4)},$$

$$\hat{I}_i,\ A^i_\mu,\ i=1,\ldots,8,\ g\text{ for SU(3)},$$

$$\hat{F},\ B_\mu,\ g'\text{ for U(1)},$$

respectively. Besides the local symmetry there is another global U(1) symmetry whose generator will be denoted by $\hat{S}$. This global U(1) will combine with an Abelian subgroup in SU(4)×SU(3) ×U(1) to give a new conserved quantum number S_W after spontaneous symmetry breaking. We will use four numbers in parentheses (m, n, F, S) to denote the representations for these four groups, respectively. For example, ($\underline{4}$, $\underline{3}$, -1, 1) means the representation $\underline{4}$ for the SU(4), the representation $\underline{3}$ for the SU(3), $F=-1$ for the local U(1), and $S=1$ for the global U(1) groups.

For simplicity we discuss the model involving only one generation of fermions. It can easily be extended to the case involving several generations. The fermions form four left-handed multiplets and four right-handed multiplets:

$$\psi_L:\ (\underline{4},\ \underline{3},\ 1,\ 1),\quad \psi_R:\ (\underline{4},\ \underline{3^*},\ 1,\ -1),$$

$$S_L:\ (\underline{4},\ \underline{1},\ 1,\ -3),\quad S_R:\ (\underline{4},\ \underline{1},\ 1,\ 3),$$

$$\psi'_L:\ (\underline{4},\ \underline{3^*},\ -1,\ -1),\quad \psi'_R:\ (\underline{4},\ \underline{3},\ -1,\ 1),$$

$$S'_L:\ (\underline{4},\ \underline{1},\ -1,\ 3),\quad S'_R:\ (\underline{4},\ \underline{1},\ -1,\ -3). \quad (2.1)$$

After spontaneous symmetry breaking, one SU(3) symmetry, one local U(1) symmetry, and one global U(1) symmetry remain unbroken. The unbroken SU(3) group is a subgroup on the first three dimensions of the SU(4) group and becomes the color gauge group for quarks. The generators of the local U(1) and the global U(1) groups are the charge

$$\hat{Q}=\hat{I}_3-\sqrt{3}\,\hat{I}_8+(\tfrac{2}{3})^{1/2}\hat{I}'_{15}+\tfrac{1}{2}\hat{F} \tag{2.2}$$

and the weak strangeness

$$\hat{S}_W=\frac{2}{\sqrt{3}}\hat{I}_8-\tfrac{1}{2}\hat{F}+\tfrac{1}{6}\hat{S}\,, \tag{2.3}$$

respectively.

Since quarks are degenerate for three colors and the color index is unimportant in many discussions, we may omit it and express the fermion multiplets as

$$\psi_L=\begin{bmatrix}u_L\\ d_L\\ h_L\\ \\ \nu_L\\ e_L\\ E_L\end{bmatrix},\quad \psi_R=\begin{bmatrix}g_R\\ h_R\\ d_R\\ \\ N_R\\ E_R\\ e_R\end{bmatrix},\quad \psi'_L=\begin{bmatrix}d'_L\\ u'_L\\ w_L\\ \\ e'_L\\ \nu'_L\\ \xi_L\end{bmatrix},\quad \psi'_R=\begin{bmatrix}x_R\\ w_R\\ u'_R\\ \\ \zeta_R\\ \xi_R\\ \nu'_R\end{bmatrix}, \tag{2.4}$$

$$S_L=\begin{bmatrix}g_L\\ \\ N_L\end{bmatrix},\quad S_R=\begin{bmatrix}u_R\\ \\ \nu_R\end{bmatrix},\quad S'_L=\begin{bmatrix}x_L\\ \\ \zeta_L\end{bmatrix},\quad S'_R=\begin{bmatrix}d'_R\\ \\ e'_R\end{bmatrix}, \tag{2.5}$$

where the symbols in the upper half of each column vector denote three colors of quarks while the symbols in the lower half of each column vector denote leptons. The quantum-number assignments of various states are listed in Table II.

This model is anomaly free; the proof can be carried out in the same way as in Ref. 4.

TABLE II. Charge and weak-strangeness quantum number for fermions.

		Q	S_W
ψ, S	ν	0	0
	e	−1	0
	N	0	−1
	E	1	−1
	u	$\frac{2}{3}$	0
	d	$-\frac{1}{3}$	0
	g	$\frac{2}{3}$	−1
	h	$\frac{5}{3}$	−1
ψ', S'	ν'	0	0
	e'	−1	0
	ξ	−2	1
	ζ	−1	1
	u'	$\frac{2}{3}$	0
	d'	$-\frac{1}{3}$	0
	w	$-\frac{4}{3}$	1
	x	$-\frac{1}{3}$	1

III. SPONTANEOUS SYMMETRY BREAKING

Six multiplets of the Higgs fields are introduced to realize the spontaneous symmetry breaking. Their transformation properties are

$$\Phi_A\colon\ (\underline{1},\ \underline{3},\ 0,\ -2),\quad \Phi_D\colon\ (\underline{1},\ \underline{3},\ -2,\ -2),$$

$$\Phi_B\colon\ (\underline{1},\ \underline{6^*},\ 0,\ -2),\quad \Phi_E\colon\ (\underline{4},\ \underline{1},\ 1,\ 3), \tag{3.1}$$

$$\Phi_C\colon\ (\underline{15},\ \underline{3},\ 0,\ -2),\quad \Phi_F\colon\ (\underline{1},\ \underline{3},\ 2,\ 4),$$

respectively.

The self-interaction potential of the Higgs fields is chosen to be

$$\begin{aligned}V=\sum_{l=A}^{F}\,&[-a_l\,\mathrm{tr}\Phi_l^\dagger\Phi_l+b_l(\mathrm{tr}\Phi_l^\dagger\Phi_l)^2]+c(\mathrm{tr}\Phi_A^\dagger\Phi_D)(\mathrm{tr}\Phi_D^\dagger\Phi_A)\\ &+d[\phi_{Di}^*\phi_{Dj}\phi_{B\{ik\}}^*\phi_{Al}+\phi_{Di}\phi_{Dj}^*\phi_{B\{ik\}}\phi_{Al}^*]\epsilon_{jkl}+e(\phi_{E\mu}^*\phi_{E\nu}\phi_{C\mu\nu i}^*\phi_{Ai}+\phi_{E\mu}\phi_{E\nu}^*\phi_{C\mu\nu i}\phi_{Ai}^*)\\ &+f(\phi_{Ai}\phi_{Dj}\phi_{Fk}+\phi_{Ai}^*\phi_{Dj}^*\phi_{Fk}^*)\epsilon_{ijk}\,,\end{aligned} \tag{3.2}$$

where a's, b's, c, d, e, and $f>0$, ϕ_{Ai}, $\phi_{B\{ik\}}$, $\phi_{C\mu\nu i}$, ϕ_{Di}, $\phi_{F\mu}$, and ϕ_{Fi} are the components of Φ_A, Φ_B, Φ_C, Φ_D, Φ_E, and Φ_F, respectively. As shown in Appendix A, it leads to a stable minimum for V. The vacuum expectation values of the Higgs fields may be taken to be

$$\langle\phi_A\rangle_0 = v_A\begin{bmatrix}1&0&0&0\\0&1&0&0\\0&0&1&0\\0&0&0&1\end{bmatrix}\otimes\begin{pmatrix}1\\0\\0\end{pmatrix},\quad \langle\phi_D\rangle_0 = v_D\begin{bmatrix}1&0&0&0\\0&1&0&0\\0&0&1&0\\0&0&0&1\end{bmatrix}\otimes\begin{pmatrix}0\\0\\1\end{pmatrix},$$

$$\langle\phi_B\rangle_0 = \frac{v_B}{\sqrt{2}}\begin{bmatrix}1&0&0&0\\0&1&0&0\\0&0&1&0\\0&0&0&1\end{bmatrix}\otimes\begin{pmatrix}0&0&0\\0&0&1\\0&1&0\end{pmatrix},\quad \langle\phi_E\rangle = v_E\begin{bmatrix}0\\0\\0\\1\end{bmatrix}\otimes\begin{pmatrix}1&0&0\\0&1&0\\0&0&1\end{pmatrix}, \tag{3.3}$$

$$\langle\phi_C\rangle_0 = \frac{v_C}{\sqrt{12}}\begin{bmatrix}1&0&0&0\\0&1&0&0\\0&0&1&0\\0&0&0&-3\end{bmatrix}\otimes\begin{pmatrix}1\\0\\0\end{pmatrix},\quad \langle\phi_F\rangle_0 = v_F\begin{bmatrix}1&0&0&0\\0&1&0&0\\0&0&1&0\\0&0&0&1\end{bmatrix}\otimes\begin{pmatrix}0\\1\\0\end{pmatrix},$$

respectively, where all v's are positive and determined by the coefficients in V. Owing to the stability of the minimum, no pseudo-Goldstone particle appears after spontaneous symmetry breaking and all the remaining Higgs particles are massive. They may be rather heavy by a suitable choice of the coefficients in the self-interaction potential. One may easily verify that the color SU(3) symmetry, the electromagnetic U(1) symmetry, and the weak strange U(1) symmetry remain unbroken after spontaneous symmetry breaking.

There are 24 gauge bosons in this model. After spontaneous symmetry breaking all gauge bosons other than eight gluons and the photon get masses. The mass terms of the vector bosons have the form

$$\tfrac{1}{2}g^2(v_A{}^2+v_B{}^2+v_C{}^2+v_F{}^2)W^+W^- + \tfrac{1}{2}g^2(v_A{}^2+v_B{}^2+v_C{}^2+v_D{}^2)V^+V^- + \tfrac{1}{2}g^2(4v_B{}^2+v_D{}^2+v_F{}^2)U^{++}U^{--}$$

$$+\tfrac{1}{2}g''^2\left((\tfrac{8}{3}v_C{}^2+v_E{}^2)\sum_{i=1}^{3}C_i^{2/3}C_i^{-2/3}+\tfrac{1}{4}g^2(v_A{}^2+v_B{}^2+v_C{}^2)\right|A^3+\frac{1}{\sqrt{3}}A^8\left|^2+\tfrac{1}{4}g^2v_D{}^2\left|\frac{2}{\sqrt{3}}A^8+\frac{2g'}{g}B\right|^2\right.$$

$$+\tfrac{1}{4}g^2v_E{}^2\left|\frac{g'}{g}B-\left(\frac{3}{2}\right)^{1/2}\frac{g''}{g}C^{15}\right|^2+\tfrac{1}{4}g^2v_F{}^2\left|-A^3+\frac{1}{\sqrt{3}}A^8+\frac{2g'}{g}B\right|^2, \tag{3.4}$$

where

$$W^{\pm}=\frac{1}{\sqrt{2}}(A^1\mp iA^2),\quad V^{\pm}=\frac{1}{\sqrt{2}}(A^4\pm iA^5),$$

$$U^{\pm\pm}=\frac{1}{\sqrt{2}}(A^6\pm iA^7),\quad C_1^{\pm2/3}=\frac{1}{\sqrt{2}}(C^9\mp iC^{10}), \tag{3.5}$$

$$C_2^{\pm2/3}=\frac{1}{\sqrt{2}}(C^{11}\mp iC^{12}),\quad C_3^{\pm2/3}=\frac{1}{\sqrt{2}}(C^{13}\mp iC^{14}).$$

The masses of these particles are

$$m_W{}^2=\tfrac{1}{2}g^2(v_A{}^2+v_B{}^2+v_C{}^2+v_F{}^2),$$

$$m_V{}^2=\tfrac{1}{2}g^2(v_A{}^2+v_B{}^2+v_C{}^2+v_D{}^2),$$

$$m_U{}^2=\tfrac{1}{2}g^2(4v_B{}^2+v_D{}^2+v_F{}^2), \tag{3.6}$$

$$m_C{}^2=\tfrac{1}{2}g''^2(\tfrac{8}{3}v_C{}^2+v_E{}^2),$$

respectively. From (3.6) we get the inequalities

$$m_V{}^2<m_U{}^2+m_W{}^2,\quad m_U{}^2<m_V{}^2+3m_W{}^2. \tag{3.7}$$

Three neutral gauge bosons will get masses from the last four terms in (3.4); we denote them by Z_1, Z_2, and Z_3, respectively. Eight gluons and the photon remain massless. The electromagnetic field is

$$A=\frac{1}{(1+1/\lambda^2+1/\mu^2)^{1/2}}\left(\tfrac{1}{2}A^3-\frac{\sqrt{3}}{2}A^8+\frac{1}{\lambda}C^{15}+\frac{1}{\mu}B\right), \tag{3.8}$$

where

$$\lambda=\sqrt{6}\,\frac{g''}{g},\quad \mu=2\frac{g'}{g}. \tag{3.9}$$

The quantum numbers of these bosons are listed in Table III. The bosons indicated by an asterisk are new particles introduced in this model. The six $C_i^{\pm2/2}$ are the so-called leptoquarks. They have fractional charges and couple quarks to leptons. The V and U bosons have nonvanishing weak strangeness. We will call this kind of particles

TABLE III. Charge and weak-strangeness quantum number for gauge bosons.

	Gluons	$C_i^{2/3}$	$C_i^{-2/3}$	γ	Z_1	Z_2	Z_3
Q	0	$\frac{2}{3}$	$-\frac{2}{3}$	0	0	0	0
S_W	0	0	0	0	0	0	0
	QCD	*	*	WS	WS	*	*

	W^+	W^-	V^+	V^-	U^{++}	U^{--}
Q	1	-1	1	-1	2	-2
S_W	0	0	-1	1	-1	1
	WS	WS	*	*	*	*

the weak strange particles.

An interesting limiting case is

$$g^2 \ll g'^2,\ g''^2, \quad \text{i.e., } \lambda^2,\ \mu^2 \gg 1,$$
$$v_F^{\,2} \ll v_A^{\,2}+v_B^{\,2}+v_C^{\,2} \ll v_D^{\,2},\ v_E^{\,2}. \tag{3.10}$$

The physical meaning of the limiting case is that the coupling constant of the SU(3) group is much smaller than those of the SU(4) color group and the U(1) group; the mass scales of $SU(4)_c$ and $SU(3)_f$ breaking are much higher than that for the secondary $SU(2)_f$ breaking.

In this limiting case the masses of the three massive neutral bosons are approximately

$$m_{Z_1}^{\ 2}=\tfrac{2}{3}g^2(v_A^{\,2}+v_B^{\,2}+v_C^{\,2})\left[1-\frac{1}{3}\left(\frac{1}{\lambda^2}+\frac{1}{\mu^2}\right)\right],$$

$$m_{Z_2}^{\ 2}=\tfrac{1}{4}g^2\left\{\mu^2 v_D^{\,2}+\frac{\lambda^2+\mu^2}{4}v_E^{\,2}-\left[\left(\mu^2 v_D^{\,2}+\frac{\lambda^2+\mu^2}{4}v_E^{\,2}\right)^2-\lambda^2\mu^2 v_D^{\,2}v_E^{\,2}\right]^{1/2}\right\}, \tag{3.11}$$

$$m_{Z_3}^{\ 2}=\tfrac{1}{4}g^2\left\{\mu^2 v_D^{\,2}+\frac{\lambda^2+\mu^2}{4}v_E^{\,2}+\left[\left(\mu^2 v_D^{\,2}+\frac{\lambda^2+\mu^2}{4}v_E^{\,2}\right)^2-\lambda^2\mu^2 v_D^{\,2}v_E^{\,2}\right]^{1/2}\right\}.$$

Z_1 can be expressed as

$$Z_1\approx\tfrac{1}{2}(\sqrt{3}\,A^3+A^8)-\frac{1}{\sqrt{3}}\left(\frac{1}{\mu}B+\frac{1}{\lambda}C^{15}\right)+\frac{1}{2\sqrt{3}}\left(\frac{1}{\lambda^2}+\frac{1}{\mu^2}\right)(A^3-\sqrt{3}\,A^8). \tag{3.12}$$

Its dominant component is just $Z=\frac{1}{2}(\sqrt{3}\,A^3+A^8)$, so it may be treated as the particle corresponding to the Z boson in the Weinberg-Salam model. Comparing with the mass formula in the Weinberg-Salam model

$$M_Z^{\ 2}=M_W^{\ 2}/\cos^2\theta_W, \tag{3.13}$$

we obtain to the first order of approximation for the $1/\lambda^2$ and $1/\mu^2$ expansion that

$$\sin^2\theta_W=\tfrac{1}{4}\left[1-\left(\frac{1}{\lambda^2}+\frac{1}{\mu^2}\right)\right]. \tag{3.14}$$

The limiting condition (3.10) ensures that $\sin^2\theta_W$ is slightly less than $\frac{1}{4}$ and is consistent with experiment.

Both Z_2 and Z_3 are much heavier than Z_1. Their dominant components are B and C^{15}. For two special cases the simple expressions simplify.

Case A. When $\mu^2 v_D^{\,2}\gg\frac{1}{4}(\lambda^2+\mu^2)v_E^{\,2}$, we get

$$m_{Z_2}^{\ 2}\approx\tfrac{1}{8}g^2\lambda^2 v_E^{\,2},\quad m_{Z_3}^{\ 2}\approx\tfrac{1}{2}g^2\mu^2 v_D^{\,2}, \tag{3.15}$$

$$Z_2\approx C^{15},\quad Z_3\approx B. \tag{3.16}$$

Case B. When $\mu^2 v_D^{\,2}\ll\frac{1}{4}(\lambda^2+\mu^2)v_E^{\,2}$, we get

$$m_{Z_2}^{\ 2}\approx\tfrac{1}{2}g^2\frac{\lambda^2\mu^2}{\lambda^2+\mu^2}v_D^{\,2},\quad m_{Z_3}^{\ 2}\approx\tfrac{1}{8}g^2(\lambda^2+\mu^2)v_E^{\,2}, \tag{3.17}$$

$$Z_2\approx\frac{1}{(\lambda^2+\mu^2)^{1/2}}(\lambda B+\mu C^{15}),$$

$$Z_3\approx\frac{1}{(\lambda^2+\mu^2)^{1/2}}(-\mu B+\lambda C^{15}). \tag{3.18}$$

The masses of gauge bosons are shown in Fig. 1. It manifests the existence of three mass scales, which are related to the breaking of $SU(4)_c$, $SU(3)_f$, and $SU(2)_f$, respectively.

The interaction Lagrangian among the fermions and the Higgs fields has the form

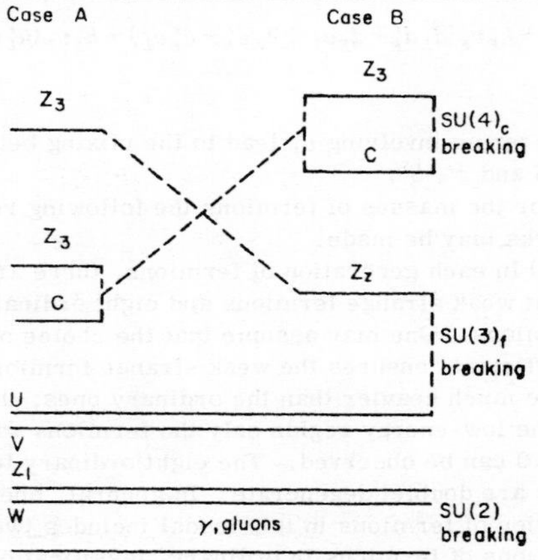

FIG. 1. Mass spectrum for the gauge bosons.

$$\mathfrak{L}_{FH} = \sum_{i=A,B,C} [f_i(\overline{\psi}_R\Phi_i\psi_L + \overline{\psi}_L\Phi_i^\dagger\psi_R) + f_i'(\overline{\psi}_L'\Phi_i\psi_R' + \overline{\psi}_R'\Phi_i^\dagger\psi_L')] + \sum_{i=A,C} [h_{i_1}(\overline{S}_R\Phi_i^\dagger\psi_L + \overline{\psi}_L\Phi_i\overline{S}_R) + h_{i_1}'(\overline{S}_L'\Phi_i^\dagger\psi_R' + \overline{\psi}_R'\Phi_i S_L')]$$
$$+ \sum_{i=A,C} [h_{i_2}(\overline{S}_L\Phi_i\psi_R + \overline{\psi}_R\Phi_i^* S_L) + h_{i_2}'(\overline{S}_R'\bar{\Phi}_i\psi_R' + \overline{\psi}_L'\Phi_i^* S_R')] + h_F(\overline{S}_R'\Phi_F^\dagger\psi_L\Phi_F S_R') + h_F'(\overline{S}_R\bar{\Phi}_F\psi_L'\Phi_F^* S_R) , \tag{3.19}$$

where Φ_A and Φ_C in the first terms are expressed in matrix form as

$$\langle\Phi_A\rangle_0 = \frac{1}{\sqrt{2}}\begin{bmatrix}1&0&0&0\\0&1&0&0\\0&0&1&0\\0&0&0&1\end{bmatrix}\begin{pmatrix}0&0&0\\0&0&1\\0&-1&0\end{pmatrix}, \quad \langle\Phi_C\rangle_0 = \frac{1}{\sqrt{24}}\begin{bmatrix}1&0&0&0\\0&1&0&0\\0&0&1&0\\0&0&0&-3\end{bmatrix}\begin{pmatrix}0&0&0\\0&0&1\\0&-1&0\end{pmatrix}. \tag{3.20}$$

After spontaneous symmetry breaking the mass terms of the fermions are derived from (3.3), (3.19), and (3.20). They have the form

$$\frac{1}{\sqrt{2}}\left(-f_Av_A + f_Bv_B - \frac{1}{\sqrt{12}}f_Cv_C\right)(\overline{d}_Ld_R + \overline{d}_Rd_L) + \frac{1}{\sqrt{2}}\left(-f_A'v_A + f_B'v_B - \frac{1}{\sqrt{12}}f_C'v_C\right)(\overline{w}_Lw_R + \overline{w}_R\overline{w}_L)$$
$$+\frac{1}{\sqrt{2}}\left(-f_Av_A + f_Bv_B + \frac{3}{\sqrt{12}}f_Cv_C\right)(\overline{e}_Le_R + \overline{e}_Re_L) + \frac{1}{\sqrt{2}}\left(-f_A'v_A + f_B'v_B + \frac{3}{\sqrt{12}}f_C'v_C\right)(\overline{\xi}_L\xi_R + \overline{\xi}_R\xi_L)$$
$$+\frac{1}{\sqrt{2}}\left(f_Av_A + f_Bv_B + \frac{1}{\sqrt{12}}f_Cv_C\right)(\overline{h}_Lh_R + \overline{h}_Rh_L) + \frac{1}{\sqrt{2}}\left(f_A'v_A + f_B'v_B + \frac{1}{\sqrt{12}}f_C'v_C\right)(\overline{u}_L'u_R' + \overline{u}_R'u_L')$$
$$+\frac{1}{\sqrt{2}}\left(f_Av_A + f_Bv_B - \frac{3}{\sqrt{12}}f_Cv_C\right)(\overline{E}_LE_R + \overline{E}_RE_L) + \frac{1}{\sqrt{2}}\left(f_A'v_A + f_B'v_B - \frac{3}{\sqrt{12}}f_C'v_C\right)(\overline{\nu}_L'\nu_R' + \overline{\nu}_R'\nu_L')$$
$$+\left(h_{A1}v_A + \frac{1}{\sqrt{12}}h_{C1}v_C\right)(\overline{u}_Lu_R + \overline{u}_Ru_L) + \left(h_{A1}'v_A + \frac{1}{\sqrt{12}}h_{C1}'v_C\right)(\overline{x}_Lx_R + \overline{x}_Rx_L) + \left(h_{A1}v_A - \frac{3}{\sqrt{12}}h_{C1}v_C\right)(\overline{\nu}_L\nu_R + \overline{\nu}_R\nu_L)$$
$$+\left(h_{A1}'v_A - \frac{3}{\sqrt{12}}h_{C1}'v_C\right)(\overline{\zeta}_L\zeta_R + \overline{\zeta}_R\zeta_L) + \left(h_{A2}v_A + \frac{1}{\sqrt{12}}h_{C2}v_C\right)(\overline{g}_Lg_R + \overline{g}_Rg_L) + \left(h_{A2}'v_A + \frac{1}{\sqrt{12}}h_{C2}'v_C\right)(\overline{d}_L'd_R' + \overline{d}_R'd_L')$$
$$+\left(h_{A2}v_A - \frac{3}{\sqrt{12}}h_{C2}v_C\right)(\overline{N}_LN_R + \overline{N}_RN_L) + \left(h_{A2}'v_A - \frac{3}{\sqrt{12}}h_{C2}'v_C\right)(\overline{e}_L'e_R' + \overline{e}_R'e_L')$$
$$+ h_Fv_F(\overline{d}_Ld_R' + \overline{d}_R'd_L + \overline{e}_Le_R' + \overline{e}_R'e_L) + h_F'v_F(\overline{u}_L'u_R + \overline{u}_Ru_L' + \overline{\nu}_L'\nu_R + \overline{\nu}_R\nu_L') . \tag{3.21}$$

The terms involving v_F lead to the mixing between ψ, S and ψ', S'.

For the masses of fermions the following remarks may be made.

(1) In each generation of fermions, there are eight weak strange fermions and eight ordinary fermions. One may assume that the choice of coefficients ensures the weak strange fermions to be much heavier than the ordinary ones; thus in the low-energy region only the fermions with $S_W = 0$ can be observed. The eight ordinary fermions are doublet degenerate. In general, one generation of fermions in this model includes two generations of fermions in ordinary classification. The existence of the τ lepton and b quark implies that there are at least two generations existing in this model. This means that one may expect the existence of a fourth generation of fermions in ordinary classification.

(2) We will discuss the generalized Cabibbo mixing of fermions. If we have n degenerate states d_i, $i = 1, \ldots, n$, the mass term of these states can be expressed as

$$\overline{d}_{iL}m_{ij}d_{jR} + \overline{d}_{jR}m_{ji}^*d_{iL} . \tag{3.22}$$

From mass matrix $M = (m_{ij})$, one may construct two Hermitian matrices MM^+ and M^+M and diagonalize them by means of certain unitary transformations U and V,

$$UMM^\dagger U^\dagger = VM^\dagger MV^\dagger = (m_1^2 \ldots m_n^2)\ . \tag{3.23}$$

This implies that the transformation

$$d_L \to U^\dagger d_L \quad d_R \to V^\dagger d_R \tag{3.24}$$

will diagonalize the mass matrix M. Since the charged weak current is left-handed, only U relates itself to the generalized Cabibbo mixing. If one uses $U_{-1/3}$ and $U_{2/3}$ to denote the U matrices for $-\frac{1}{3}$ charged and $\frac{2}{3}$ charged quarks, respectively, then the unitary matrix $U = U_{-1/3}\, U^\dagger_{2/3}$ describes the Cabibbo mixing among quarks. There are n^2 parameters appearing in the U matrix. $2n$-1 of them can be eliminated by the choice of relative phases and $n(n-1)/2$ of them can be related to the rotation angles in an n-dimensional space. So, there are $n(n-3)/2+1$ phases appearing in the general expression of mixing, which may lead to the CP violation in weak interaction for $n \geq 3$ and is just the description given by Kobayashi-Maskawa.[7]

In this model n is an even number. For $n=2$ if one denotes the mass matrix as

$$M = \begin{pmatrix} \alpha & \gamma \\ \delta & \beta \end{pmatrix}, \tag{3.25}$$

then the eigenvalues of the mass matrix are

$$m = \tfrac{1}{2}\{[(\alpha+\beta)^2 + (\gamma-\delta)^2]^{1/2} + [(\alpha-\beta)^2 + (\gamma+\delta)^2]^{1/2}\ , \tag{3.26}$$

and the mixing angles have the forms

$$\tan^2\theta_L = \frac{2(\alpha\delta+\beta\gamma)}{\alpha^2-\beta^2+\gamma^2-\delta^2}\ ,$$

$$\tan^2\theta_R = \frac{2(\alpha\gamma+\beta\delta)}{\alpha^2-\beta^2-\gamma^2+\delta^2}\ . \tag{3.27}$$

If one identifies the four ordinary quarks in this case as u, d, c, and s, the Cabbibo angle θ_C can be expressed as

$$\theta_C = \theta_{L-1/3} - \theta_{L2/3}\ . \tag{3.28}$$

In this model, $\delta=0$ holds and α, β, γ's can be obtained from (3.21). Since there exist at least two generations in this model, the mixing should be for the $n=4$ case.

(3) There exist right-handed neutrinos in this model. Since they do not couple with charged leptons via the W boson, there is no contradiction with experiment. Besides, there may be the Cabibbo-type mixing among massive neutrons, which will manifest itself in neutrino oscillations. This model predicts that the oscillation among different flavors of neutronos may happen while the oscillation between ν and $\bar{\nu}$ is forbidden. This prediction is consistent with recent experiments.[8]

IV. INTERACTIONS BETWEEN THE FERMIONS AND THE GAUGE BOSONS

The gauge interaction Lagrangian for fermions can be written as

$$\begin{aligned}\mathcal{L} = {} & \bar{\psi}_L \gamma^\mu (\partial_\mu + ig\hat{I}_i A^i_\mu + ig''\hat{I}'_j C^j_\mu + ig'\tfrac{1}{2}B_\mu)\psi_L + \bar{\psi}_R\gamma^\mu(\partial_\mu - ig\hat{I}^*_i A^i_\mu + ig''\hat{I}'_j C^j_\mu + ig'\tfrac{1}{2}B_\mu)\psi_R \\ & + \bar{S}_L\gamma^\mu(\partial_\mu + ig''\hat{I}'_j C^j_\mu + ig'\tfrac{1}{2}B_\mu)S_L + \bar{S}_R\gamma^\mu(\partial_\mu + ig''\hat{I}'_j C^j_\mu + ig'\tfrac{1}{2}B_\mu)S_R \\ & + \bar{\psi}'_L\gamma^\mu(\partial_\mu - ig\hat{I}^*_i A^i_\mu + ig''\hat{I}'_j C^j_\mu - ig'\tfrac{1}{2}B_\mu)\psi'_L + \bar{\psi}'_R\gamma^\mu(\partial_\mu + ig\hat{I}_i A^i_\mu + ig''\hat{I}'_j C^j_\mu - ig'\tfrac{1}{2}B_\mu)\psi'_R \\ & + \bar{S}'_L\gamma^\mu(\partial_\mu + ig''\hat{I}'_j C^j_\mu - ig'\tfrac{1}{2}B_\mu)S'_L + \bar{S}'_R\gamma^\mu(\partial_\mu + ig''\hat{I}'_j C^j_\mu - ig'\tfrac{1}{2}B_\mu)S'_R\ .\end{aligned} \tag{4.1}$$

We shall discuss the interactions between the $S_W=0$ fermions and the gauge vector bosons first. For this purpose, the terms involving $S_W \neq 0$ fermions are neglected and the following substitutions are made in (4.1).

$$\psi_L \to \begin{bmatrix} u_L \\ d_L \\ \cdots \\ \\ \nu_L \\ e_L \\ \cdots \end{bmatrix},\quad \psi_R \to \begin{bmatrix} \cdots \\ \cdots \\ d_R \\ \\ \cdots \\ \cdots \\ e_R \end{bmatrix},\quad \psi'_L \to \begin{bmatrix} d'_L \\ u'_L \\ \cdots \\ \\ e'_L \\ \nu'_L \\ \cdots \end{bmatrix},\quad \psi'_R \to \begin{bmatrix} \cdots \\ \cdots \\ u'_R \\ \\ \cdots \\ \cdots \\ \nu'_R \end{bmatrix}, \tag{4.2}$$

$$S_L \to \begin{bmatrix} \cdots \\ \\ \cdots \end{bmatrix},\quad S_R \to \begin{bmatrix} u_R \\ \\ \nu_R \end{bmatrix},\quad S'_L \to \begin{bmatrix} \cdots \\ \\ \cdots \end{bmatrix},\quad S'_R \to \begin{bmatrix} d'_R \\ \\ e'_R \end{bmatrix},$$

Only the W boson appears in the charged weak interaction; this is reasonable on account of the conservation of the weak strangeness. The charged weak interaction has the form

$$\begin{aligned} & i\frac{g}{\sqrt{2}}(\bar{u}_L\gamma^\mu W^+_\mu d_L + \bar{d}_L\gamma^\mu W^-_\mu u_L - \bar{u}'_L\gamma^\mu W^+_\mu d'_L \\ & \quad - \bar{d}'_L\gamma^\mu W^-_\mu u'_L) \\ & + i\frac{g}{\sqrt{2}}(\bar{\nu}_L\gamma^\mu W^+_\mu e_L + \bar{e}_L\gamma^\mu W^-_\mu \nu_L \\ & \quad - \bar{\nu}'_L\gamma^\mu W^+_\mu e'_L - \bar{e}'_L\gamma^\mu W^-_\mu \nu'_L), \end{aligned} \tag{4.3}$$

where the terms involving quarks imply the sum over three colors.

The interactions involving C_μ^j, $j=1,\ldots,8$ are just the color-SU(3) interactions of quantum chromodynamics (QCD). The interactions involving C_μ^j, $j=9,\ldots,14$ give the transition between quarks and leptons. Since the masses of $C_i^{\pm 2/3}$ are very heavy, they can appear only at rather high energies.

The neutral interactions consist of four terms involving a photon, Z_1, Z_2, and Z_3, respectively. The interaction involving the photon gives the electric charge to be

$$e=\tfrac{1}{2}g\,\frac{1}{(1+(1/\lambda^2)+1/\mu^2)^{1/2}}\,. \tag{4.4}$$

Comparing with another definition of the Weinberg angle in the Weinberg-Salam model

$$e=g\sin\theta_W \tag{4.5}$$

we obtain

$$\sin^2\theta_W=\frac{1}{4}\,\frac{1}{1+(1/\lambda^2)+1/\mu^2}<\frac{1}{4} \tag{4.6}$$

in contrast to (3.16). We can treat (3.15) and (4.5) as two definitions of the Weinberg angle. In the limiting case, when Z_1 is much lighter than Z_2, they lead to the same expression for $\sin\theta_W$, as is evident from (3.16) and (4.6).

From the expression (4.6) for θ_W, a lower bound on the strong coupling constant may be obtained:

$$\alpha_s>\frac{2}{3}\,\frac{\alpha}{(1-4\sin^2\theta_W)}\,. \tag{4.7}$$

However, α_s is a running constant. It decreases as Q^2 increases, so that inequality (4.7) should hold for any energy below the mass scale of the SU(4) breaking. Using the experimental value $\sin^2\theta_W=0.230\pm0.009$,

$$\alpha_s>0.06, \tag{4.8}$$

consistent with the experimental estimation of α_s.

We may use the inequality (4.8) to estimate the mass scale of SU(4) breaking. If we use the estimation of $\alpha_s\approx 0.23$ for $Q\sim 30$ GeV in the formula

$$\alpha_s(Q^2)=\frac{12\pi}{(33-2N_f)\ln(Q^2/\Lambda^2)}\,, \tag{4.9}$$

the upper limit of this mass scale may be estimated as $\sim 10^5$ GeV and $\sim 10^6$ GeV for $N_f=5$ and 6, respectively. Of course, the estimation value will increase as the number of flavors increases.

Now we discuss the neutral currents in this model. The interaction involving Z_1 manifests itself very much like the usual neutral current. In the limiting case of (3.10), to lowest order it can be expressed as

$$\mathcal{L}_{\rm eff}=4\,\frac{G_F}{\sqrt{2}}\,J_{1\mu}J_1^\mu\,, \tag{4.10}$$

where

$$J_{1\mu}\approx J_\mu^3-\sin^2\theta_W J_\mu^{\rm em}. \tag{4.11}$$

It is just the well-known formula in the Weinberg-Salam model. To the next order of approximation J_1 becomes

$$J_{1\mu}=J_\mu^3-\sin^2\theta_W J_\mu^{\rm em}-\frac{2}{3}\,\frac{v_A{}^2+v_B{}^2+v_C{}^2}{v_D{}^2}\,\frac{1}{\lambda^2+\mu^2}\,\frac{\mu}{\lambda}\left[\frac{\lambda}{\mu}\left(\frac{2}{3}\right)^{1/2}J_\mu'^{15}-\frac{\mu}{\lambda}\left(\frac{2}{3}\right)^{1/2}J_\mu^F\right]. \tag{4.11$'$}$$

Since $(v_A{}^2+v_B{}^2+v_C{}^2)/v_D{}^2\approx m_W{}^2/m_V{}^2\ll 1$ and $\lambda^2\gg 1$, the correction term in (4.11′) is smaller in magnitude than the main terms.

The effective Lagrangian involving Z_2 and Z_3 can be expressed as

$$\mathcal{L}_{\rm eff}=4\,\frac{G_F}{\sqrt{2}}\,(r_2J_{2\mu}J_2^\mu+r_3J_{3\mu}J_3^\mu) \tag{4.12}$$

with $r_3\ll r_2\ll 1$.

When $\mu^2v_D{}^2\gg\frac{1}{4}(\lambda^2+\mu^2)v_E{}^2$, $J_{2\mu}$, r_2, $J_{3\mu}$, and r_3 take the forms

$$J_{2\mu}=(\tfrac{2}{3})^{1/2}J_\mu^{15},\quad r_2=\tfrac{1}{4}\lambda^2\,\frac{m_W{}^2}{m_{Z_2}{}^2}\,,$$
$$J_{3\mu}=J_\mu^F,\quad r_3=\frac{\mu^2}{16}\,\frac{m_W{}^2}{m_{Z_3}{}^2}=\frac{m_W{}^2}{16m_V{}^2}\,. \tag{4.13}$$

Since $r_3\ll r_2$, the Z_2 interaction is more important than the Z_3 interaction. Since the effective Z_1 charge and Z_2 charge of the neutrino are $\frac{1}{2}$ and $-\frac{1}{2}$, respectively, the contribution of Z_2 interaction to the neutrino-induced neutral-current experiment can be described effectively as a correction for the $J_{1\mu}$ current

$$J_{1\mu}\to J_{1\mu}-r_2J_{2\mu}=J_{1\mu}-r_2(\tfrac{2}{3})^{1/2}J_\mu'^{15}. \tag{4.14}$$

One may note that the correction term in (4.11′) is much smaller than that from Z_2 current and only the Z_2 correction should be considered. Since the effective Z_2 charges of quarks and leptons are $\frac{1}{6}$ and $-\frac{1}{2}$, respectively, this correction can be observed in more accurate experiments.

When $\mu^2v_D{}^2\ll\frac{1}{4}(\lambda^2+\mu^2)v_E{}^2$ we get

$$J_{2\mu}=\tfrac{1}{2}J_\mu^F+(\tfrac{2}{3})^{1/2}J_\mu'^{15},\quad r_2=\frac{1}{4}\,\frac{\lambda^2\mu^2}{\lambda^2+\mu^2}\,\frac{m_W{}^2}{m_{Z_2}{}^2}=\frac{m_W{}^2}{4m_V{}^2}\,,$$
$$J_{3\mu}=\left(\frac{2}{3}\right)^{1/2}\frac{\lambda}{\mu}\,J_\mu'^{15}-\frac{1}{2}\,\frac{\mu}{\lambda}\,J_\mu^F,\quad r_3=\frac{1}{4}\,\frac{\lambda^2\mu^2}{\lambda^2+\mu^2}\,\frac{m_W{}^2}{m_{Z_3}{}^2}\,. \tag{4.15}$$

The correction term in (4.11′) is much smaller than that from the Z_2 current in this case. Since the effective Z_2 charge is of the form

$$Q'_{Z_2} = \frac{1}{2} F + (\tfrac{2}{3})^{1/2} I'^{15} \qquad (4.16)$$

which gives different values to two kinds of fermions, we may use them to distinguish these two kinds of fermions, ψ, S and ψ', S'. The effective Z_2 charges for different fermions are given in Table IV. Owing to the mixing between these two kinds of fermions, this effect might not be explicit. However, one may expect the existence of some difference between them.

Two interesting remarks may be made.

(1) In the tree approximation there is no restriction on the fermion masses. If one calculates the radiative corrections, the stability of the vacuum will give a bound to the fermion masses.[9,10] Using the formula given in Ref. 9 one finds that the masses of E, ξ, h, and w are bounded by

$$m < 2\left[\frac{3}{4}\left(1+\frac{1}{4\cos^4\theta_W}\right)\right]^{1/4} m_W \approx 160 \text{ GeV}. \qquad (4.17)$$

The masses of other weak strange fermions are bounded by

$$m_N, m_\zeta < \left[\frac{3}{2}\left(1+\frac{1}{4\cos^4\theta_W}\right)\right]^{1/4} m_W + \left(\frac{1}{123}\right)^{1/4} 2|\sin\theta_W| \left(\frac{\alpha_s}{\alpha}\right)^{1/2} m_W, \qquad (4.18)$$

$$m_\varepsilon, m_x < \left[\frac{3}{2}\left(1+\frac{1}{4\cos^4\theta_W}\right)\right]^{1/4} m_W + \left(\frac{27}{41}\right)^{1/4} 2|\sin\theta_W| \left(\frac{\alpha_s}{\alpha}\right)^{1/2} m_W. \qquad (4.19)$$

These estimations depend on the value of α_s. If one takes $\alpha_s \approx 0.20$ in (4.18) and (4.19), the upper limits become $m_N, m_\zeta < 210$ GeV and $m_\varepsilon, m_x < 440$ GeV, respectively. However, (4.17) means that there exist at least four weak strange fermions in one generation lighter than 160 GeV. This prediction can be verified experimentally.

(2) Though the quantum numbers F for ψ and S are different from those for ψ' and S', the radiative decay between two degenerate states is allowed in this model. For example, if e and e' mix with each other and form two leptons e_1 and e_2, $m_{e_1} < m_{e_2}$, one may calculate the partial width of the decay mode $e_2 \to e_1 + \gamma$. In the tree approximation it is forbidden. But in the one-loop approximation, it is allowed. Using the method given in Ref. 11 one may obtain the branching ratio to be

$$R \equiv \frac{\Gamma(e_2 \to e_1 \gamma)}{\Gamma(e_2 \to e_1 \bar{\nu}_1 \nu_2)} = \frac{3\alpha}{\pi^2}\left(\frac{m_W}{m_Y}\right)^4 \eta, \qquad (4.20)$$

where η is a function of the mixing angles. For the case B discussed in Sec. III, η can be expressed as

$$\eta = (\tfrac{1}{3}s_+ + c_+ s_+ + c_- s_-)^2 + (s_- + \tfrac{5}{3} c_- s_+ + \tfrac{1}{3} c_+ s_-)^2, \qquad (4.21)$$

where

$$s_+ = \tfrac{1}{2}(\sin 2\theta_L + \sin 2\theta_R), \quad s_- = \tfrac{1}{2}(\sin 2\theta_L - \sin 2\theta_R), \qquad (4.22)$$

$$c_+ = \tfrac{1}{2}(\cos 2\theta_L + \cos 2\theta_R), \quad c_- = \tfrac{1}{2}(\cos 2\theta_L - \cos 2\theta_R),$$

θ_L and θ_R are the left-handed and the right-handed mixing angles, respectively. For several special values of mixing angles, η may vanish. But in general η is of the order of unity. If one takes $\eta \sim 1$ and identifies e_2 and e_1 as the muon and electron, respectively, the lower limit of m_Y can be obtained as

$$m_Y > 28 m_W, \qquad (4.23)$$

from the experimental upper limit $R < 3.6\times10^{-9}$. This estimation is consistent with the requirement of the limiting case discussed above,

$$\frac{m_Y{}^2}{m_W{}^2} \approx \frac{v_D{}^2}{v_A{}^2 + v_B{}^2 + v_C{}^2} \gg 1.$$

If one identifies e_2 as τ instead of μ, the lower limit of m_Y will be much lighter than that obtained from μ.

V. ADDITIONAL CONSERVATION LAWS

In this model there exist several additional conservation laws. We discuss them separately.

A. The conservation of quark number

Many models of grand unification predict that the proton decays. But in this model the proton may be stable. We will give a simple proof of this. One may introduce a new global U(1) symmetry in this model, whose generator is denoted by $\hat{R}$, and the R assignments of the various multiplets being

$$R = 1 \text{ for } \psi, S, \psi', S', \Phi_E,$$

$$R = 0 \text{ for } A^i_\mu, B_\mu, C^j_\mu, \Phi_A, \Phi_B, \Phi_C, \Phi_D, \Phi_F.$$

TABLE IV. The effective Z_2 charges for different kinds of fermions.

ψ, S	lepton	0
	quark	$\frac{2}{3}$
ψ', S'	lepton	-1
	quark	$-\frac{1}{3}$

One may easily verify that the invariance of this global symmetry holds before spontaneous symmetry breaking. After spontaneous symmetry breaking it is broken too. But it will combine with $\hat{I}'_{15}$ and form a new global U(1) symmetry whose generator is the quark number

$$\hat{N} = (\tfrac{3}{2})^{1/2}\hat{I}'_{15} + \tfrac{3}{4}\hat{R}.$$

$\hat{N}$ is conserved after spontaneous symmetry breaking. Since the particles with nonvanishing N are, for $N=1$, quarks, $C_i^{2/3}$, and several particles in Φ_C and Φ_B, and for $N=-1$, antiquarks, $C_i^{-2/3}$, and several particles in Φ_C and Φ_E^+, and all these bosons with nonvanishing N are heavy, the proton may be stable.

One may note that the global U(1) symmetry R is not an intrinsic one. The possibility of introducing such a symmetry depends on the existence of the U(1) symmetries S and F, which rules out the existence of terms involving $\epsilon_{\mu\nu\lambda\sigma}$ structure in the Higgs potential. So the stability of the proton depends on the U(1) symmetries S and F. If, for example S is partially conserved, the proton may be long-lived.

B. The conservation of weak strangeness

In this model there exist four weak strange vector bosons and eight weak strange fermions in each generation. They are listed below.

	$S_W=-1$	$S_W=1$
vector boson	V^+, U^{++}	V^-, U^{--}
lepton	N^0, E^+	ξ^{-2}, ζ^{-1}
quark	$g^{2/3}, h^{5/3}$	$w^{-4/3}, x^{-1/3}$

All of them are massive and probably heavier than the known particles. The conservation of weak strangeness requires that (1) weak strange particles can be produced only in pairs; (2) they are weakly decaying and the decay chains end in a final state with the lightest weak strange particle; (3) the lightest weak strange particle is stable; and (4) the weak strange bosons do not couple directly with the ordinary fermion pair.

Since the weak strange vector bosons are much heavier than the W boson while some of the weak strange fermions are lighter than 160 GeV, the lightest weak strange particle should be a fermion. It may be produced in high-energy e^+e^- collisions and manifests itself as either a stable lepton or a stable hadron. The weak strange fermions interact with each other via the right-handed current coupling with the W bosons. The transition between weak strange quark and lepton will be suppressed by the propagator of V or U. This means that, for example, if the lightest weak strange particle is a lepton, the width of the lightest weak strange quark ought to be smaller than that estimated in the ordinary way.

It is interesting that the masses of all weak strange fermions with exotic charges (ξ^{--}, $h^{5/3}$, and $w^{-4/3}$) are bounded by $m < 160$ GeV, so that existence of these particles can be observed in the e^+e^- experiment. However, owing to the weak-strangeness conservation, the existence of weak strange particles will not have significant influence on weak-interaction processes below the threshold for production of the weak strange particles.

It has to be noted that the existence of the global U(1) symmetry is not intrinsic, and is derived from the special structure of the Higgs potential. Otherwise, there might be several terms violating the weak-strangeness conservation, which will then become partial.

VI. SUMMARY AND REMARKS

The main results of this model can be summarized as follows.

(1) The electroweak interaction can be connected to the strong interaction via the SU(4)×SU(3)× U(1) model. In this model there exist the same number of left-handed and right-handed multiplets of the fermions as before spontaneous symmetry breaking.

(2) This model is anomaly free.

(3) As a limiting case, it gives the same results as those of the Weinberg-Salam model and is in agreement with experiment. The Weinberg angle θ_W is bounded by the relation $\sin^2\theta_W \leq \frac{1}{4}$ and the neutral current for ordinary fermions reduces to that of the Weinberg-Salam model in the limiting case.

(4) There are some deviations from the Weinberg-Salam model concerning the predictions in neutral currents. The additional terms in neutral currents are colorless and flavorless, may possibly be different between the two kinds of fermions, and can be verified by more accurate experiments.

(5) A new conserved quantum number S_W, called the weak strangeness, is introduced in the present model. The model predicts many particles with nonvanishing weak strangeness. All of them are likely to be heavier than known fermions. They can be produced only in pairs and the lightest weak strange particle is stable. Some of the weak strange particles have unusual charges and can easily be identified experimentally.

(6) There exists a relation between the Weinberg angle and the strong coupling constant. A more accurate value of the Weinberg angle can be used to get the bound of the strong coupling constant and to estimate the upper limit for the mass scale

of symmetry breaking.

(7) Quark number may be conserved and the proton may be stable in this model.

(8) There are two types of fermions introduced in this model. They will mix with each other after spontaneous symmetry breaking, giving use to Cabibbo-type mixing of fermions.

(9) The neutrinos may be massive and there may exist oscillations among different kinds of neutrinos. The oscillation between neutrino and antineutrino is excluded in this model.

ACKNOWLEDGMENTS

This work was supported in part by the U. S. Department of Energy under Contract No. DE-AC03-76SF00515. We would like to thank Professor S. Drell, Dr. Y. S. Tsai, Dr. H. Quinn, Dr. W. B. Atwood, Dr. A. Ogawa, and Dr. C. L. Ong, for carefully reading the manuscript and for valuable discussions. We thank Dr. S. Wolfram for valuable discussion about fermion masses.

APPENDIX: THE FORM OF THE HIGGS POTENTIAL

There are six Higgs multiplets introduced in this model. Their transformation properties are

$$\begin{aligned}
&\Phi_A:\ (\underline{1},\underline{3},0,-2)\,,\\
&\Phi_B:\ (\underline{1},\underline{6}^*,0,-2)\,,\\
&\Phi_C:\ (\underline{15},\underline{3},0,-2)\,,\\
&\Phi_D:\ (\underline{1},\underline{3},-2,-2)\,,\\
&\Phi_E:\ (\underline{4},\underline{1},1,3)\,,\\
&\Phi_F:\ (\underline{1},\underline{3},2,4)\,.
\end{aligned} \tag{A1}$$

We will discuss the form of the Higgs potential, which can realize the spontaneous symmetry breaking we need, and the stability of the breaking.

The potential will have the standard potentials for every Higgs multiplet,

$$V = V_0 + V_I\,, \tag{A2}$$

$$V_0 = \sum_{l=A}^{F} [-a_l \operatorname{tr}\Phi_l^\dagger \Phi_l + b_l(\operatorname{tr}\Phi_l^\dagger\Phi_l)^2]\,,$$

where $a_l, b_l > 0$. In Eq. (A2) the minimum of V_0 takes place at $\operatorname{tr}\Phi_l^\dagger\Phi_l = a_l/2b_l$.

For simplicity, we will introduce the following notation. ξ is an n vector in an n-dimensional complex space with the components ξ_i, $i=1,\ldots,n$. $R(\xi)$ is used to denote

$$R(\xi) = (\xi_i^* \xi_i)^{1/2}\,. \tag{A3}$$

For two vectors ξ and ζ in the same linear space, the scalar product of them can be expressed as

$$(\xi,\zeta) = \xi_i^\dagger \zeta_i = R(\xi)R(\zeta)e^{i\phi}\cos\theta\,. \tag{A4}$$

We will use the indices $i,j,k,\ldots$ for the SU(3) group and the indices $\mu,\nu,\lambda,\ldots$ for the SU(4) group.

In order to remove the degeneracy of spontaneous symmetry breaking the additional term V_I of the Higgs potential must be introduced. We introduce a correlation term of Φ_A and Φ_D as

$$c(\operatorname{tr}\Phi_A^\dagger\Phi_D)(\operatorname{tr}\Phi_D^\dagger\Phi_A)\,. \tag{A5}$$

According to (A4) this becomes

$$cR^2(\Phi_A)R^2(\Phi_D)\cos^2\theta = cv_A^2 v_D^2\cos^2\theta\,.$$

If $c > 0$ the minimum takes place at $\cos^2\theta = 0$. This means that there are nonvanishing vacuum expectation values for Φ at different components of Φ_A and Φ_D. We may use a tranformation to ensure that the nonvanishing vacuum expectation values take place for the $i=1$ and $i=3$ components of Φ_A and Φ_D respectively. We make both v_A and v_D positive by suitable choice of the phases.

We further introduce an additional term as

$$d[\phi_{Di}^*\phi_{Dj}\phi_{B\{ik\}}^*\phi_{Al}\epsilon_{jkl} + \phi_{Di}\phi_{Dj}^*\phi_{B\{ik\}}\phi_{Al}^*\epsilon_{jkl}]\,, \tag{A6}$$

where $d > 0$. Since the minimum takes place at $\langle\phi_{Di}\rangle_0 = v_D\delta_{i3}$ and $\langle\phi_{Ai}\rangle_0 = v_A\delta_{i1}$, it becomes effectively

$$-2d\,v_D^2 v_A \operatorname{Re}\phi_{B\{32\}}\,.$$

It becomes a minimum as $\operatorname{Re}\phi_{B\{32\}} > 0$. However, the self-potential V_0 makes the restriction that the components of Φ_B agree,

$$\operatorname{tr}\Phi_B^\dagger\Phi_B = v_B^2\,.$$

This means that the minimum takes place at

$$\phi_{B\{32\}} = v_B > 0$$

which is just adopted in the model.

Now we introduce the additional term involving Φ_A, Φ_C, and Φ_E as

$$e[\phi_{E\mu}^*\phi_{E\nu}\phi_{C\mu\nu i}^*\phi_{Ai} + \phi_{E\mu}\phi_{E\nu}^*\phi_{C\mu\nu i}\phi_{Ai}^*] \tag{A7}$$

with the coefficient $e > 0$. Using a transformation in the SU(4) group to ensure that $\langle\phi_{E\mu}\rangle_0 = v_E\delta_{\mu 4}$ with $v_E > 0$ it becomes

$$2ev_E^3 v_A \operatorname{Re}\phi_{C441}\,.$$

The components of Φ_C can also be denoted by the index $\alpha = 1,\ldots,15$ instead of μ and ν. We change the notation as $\phi_{\mu\nu i} \to \phi_{(\alpha)i}$, then we have $\phi_{C441} \to -\sqrt{3}/2\phi_{C(15)1}$ and this additional term becomes

$$-\sqrt{3e}\,v_E^2 v_A \operatorname{Re}\phi_{C(15)1}\,.$$

Using the same argument discussed above we obtain that the minimum takes place at

$$\phi_{C(15)1} = v_C > 0\,.$$

The additional term involving Φ_A, Φ_D, and Φ_F is chosen to be

$$f(\phi_{Ai}\phi_{Dj}\phi_{Fk}+\phi^*_{Ai}\phi^*_{Dj}\phi^*_{Fk})\epsilon_{ijk}\,, \tag{A8}$$

where $f>0$. Since the minimum takes place at $\langle\phi_A\rangle_0=v_A\delta_{i1}$ and $\langle\phi_{Di}\rangle=v_D\delta_{i3}$, it becomes effectively

$$-2fv_Av_D\,\mathrm{Re}\phi_{F2}$$

and becomes a minimum at $\mathrm{Re}\phi_{F2}=v_F>0$.

In summary, if the self-interaction potential of the Higgs fields has the form

$$V=\sum_{l=A}^{E}[-a_l\,\mathrm{tr}\Phi_l^\dagger\Phi_l+b_l\,(\mathrm{tr}\Phi_l^\dagger\Phi_l)^2]+c(\mathrm{tr}\Phi_A^\dagger\Phi_D)(\mathrm{tr}\Phi_D^\dagger\Phi_A)+d(\phi^*_{Di}\phi_{Dj}\phi^*_{B(ik)}\phi_{Al}+\phi_{Di}\phi^*_{Dj}\phi_{B(ik)}\phi^*_{Al})\epsilon_{jkl}$$
$$+e(\phi^*_{E\mu}\phi_{E\nu}\phi^*_{C\mu\nu i}\phi_{Ai}+\phi_{E\mu}\phi^*_{E\nu}\phi_{C\mu\nu i}\phi^*_{Ai})+f(\phi_{Ai}\phi_{Dj}\phi_{Fk}+\phi^*_{Ai}\phi^*_{Dj}\phi^*_{Fk}) \tag{A9}$$

with a's, b's, c, d, e, and $f>0$, then it leads to a minimum capable of generating the spontaneous symmetry breaking adopted in this model.

Now we discuss the stability of the spontaneous symmetry breaking. There are eight-, six-, and four-dimensional degeneracies appearing in the choice of the breaking components for Φ_E, Φ_D, and Φ_A, respectively. These $7+5+3=15$ superfluous components can be removed by the choice of gauge, making 15 gauge bosons massive. So no pseudo-Goldstone particle will appear after the spontaneous symmetry breaking, and all remaining components of the Higgs fields will get mass. In other words, the spontaneous symmetry breaking is stable.

*Permanent address.

[1]S. Weinberg, Phys. Rev. Lett. 19, 1264 (1967); A. Salam, in *Elementary Particle Theory: Relativistic Groups and Analyticity (Nobel Symposium No. 8)*, edited by N. Svartholm (Almqvist and Wiksell, Stockholm, 1968); S. L. Glashow, J. Iliopoulos, and L. Maiani, Phys. Rev. D 2, 1285 (1970).

[2]K. Winter, in *Proceedings of the 1979 International Symposium on Lepton and Photon Interactions at High Energies, Fermilab*, edited by T. B. W. Kirk and H. D. I. Abarbanel, (Fermilab, Batavia, Illinois, 1980), p. 258.

[3]H. Georgi and S. L. Glashow, Phys. Rev. Lett. 32, 438 (1974).

[4]K. C. Chou and C. S. Gao, Report No. SLAC-PUB-2445, 1979 (unpublished); Sci. Sin. 23, 566 (1980).

[5]B. W. Lee and S. Weinberg, Phys. Rev. Lett. 38, 1237 (1977); B. W. Lee and R. E. Shrock, Phys. Rev. D 17, 2410 (1978); M. Yoshimura, Prog. Theor. Phys. 57, 237 (1977); T. Moriya, *ibid.* 59, 2028 (1978); H. Komatsu, *ibid.* 59, 2013 (1978); K. Ishikawa *et al.*, *ibid.* 59, 227 (1978); K. Inoue *et al.*, *ibid.* 60, 627 (1978); G. Segre and J. Weyers, Phys. Lett. 65B, 243 (1976); K. Inoue *et al.*, Prog. Theor. Phys. 58, 1914 (1977).

[6]J. C. Pati and A. Salam, Phys. Rev. D 10, 275 (1974).

[7]M. Kobayashi and K. Maskawa, Prog. Theor. Phys. 49, 652 (1973).

[8]F. Reines *et al.*, Phys. Rev. Lett. 45, 1307 (1980).

[9]S. Coleman and E. Weinberg, Phys. Rev. D 7, 1888 (1973).

[10]H. D. Politzer and S. Wolfram, Phys. Lett. 82B, 242 (1979).

[11]J. D. Bjorken, K. Lane, and S. Weinberg, Phys. Rev. D 16, 1474 (1977).

第11卷 第2期
1981年6月

中国科学技术大学学报
JOURNAL OF CHINA UNIVERSITY OF SCIENCE AND TECHNOLOGY

Vol.11 No.2
June, 1981

On the Quantization and the Renormalization of the Pure-Gauge Fields on the Coset-Space

Zhou Guang-zhao
(Chou Kuang-chao)

(*Institute of Theoretical Physics, Academia Sinica*)

Ruan Tu-nan

(*Department of Modern Physics, China University of Science and Technology*)

Abstract

By using Faddeev-Popov trick, the path-integral quantization of the pure gauge fields on the coset space is realized. The invariance under B. R. S. transformation of this theory is demonstrated, from which the Ward-Slavnov identities are deduced. Since this theory is renormalizable in the gauge $\phi_0=1$, $F_A=0$. it is also renormalizable in other gauges. This is verified by means of the gauge independence of the S-matrix.

1. Introduction

In reference [1] the concept of pure gauge fields on a coset space has been proposed. First we consider a Lagrangian which is gauge invariant under the subgroup H of a group G,

$$\mathscr{L}^{(A)} = \mathscr{L}\,(\varphi(x), D_\mu^{A}\varphi(x), F_{\mu\nu}^{A}(x)), \tag{1.1}$$

where $\varphi(x)$ forms the basis of a linear representation of group G; $\hat{A}_\mu$ are the gauge fields on the subgroup H. Let $\hat{I}_l$ $(l=1,2,\cdots,n_H)$ be the generators of the

Received Nov. 6, 1980.

subgroup H, which satisfy the following commutation relations,

$$[\hat{I}_l, \hat{I}_m] = i f_{lmn} \hat{I}_n, \quad (l, m, n = 1, 2, \ldots, n_H). \tag{1.2}$$

Then $\hat{A}_\mu$, D_μ^{A} and $F_{\mu\nu}^{A}$ introduced in Eq. (1.1) can be written as

$$\hat{A}_\mu = -ig\hat{I}_l A^l_\mu, \quad D^A_\mu = \partial_\mu + \hat{A}_\mu, \tag{1.3}$$

$$F^A_{\mu\nu} = \partial_\mu \hat{A}_\nu - \partial_\nu \hat{A}_\mu + [\hat{A}_\mu, \hat{A}_\nu].$$

It is possible to construct a Lagrangian which is gauge invariant on the group G with vector gauge fields only on the subgroup H,

$$\mathscr{L}^{(B)} = \mathscr{L}(\Phi(x), D^B_\mu \Phi(x), \hat{F}^B_{\mu\nu}(x)), \tag{1.4}$$

Where

$$\Phi(x) = \phi_0(x)\varphi(x),$$

$$\hat{B}_\mu(x) = \phi_0(x)(\partial_\mu + \hat{A}_\mu(x))\phi_0^{-1}(x),$$

$$D^B_\mu = \phi_0(x) D^A_\mu \phi_0^{-1}(x),$$

$$\hat{F}^B_{\mu\nu}(x) = \phi_0(x)\hat{F}^A_{\mu\nu}(x)\phi_0^{-1}(x). \tag{1.5}$$

In Eq. (1.5), $\phi_0(x)$ is a function valued on the coset G/H. Obviously the difference between the Lagrangian $\mathscr{L}^{(A)}$ and $\mathscr{L}^{(B)}$ is just a gauge transformation on the group G. That is, $\mathscr{L}^{(B)}$ becomes $\mathscr{L}^{(A)}$ when a gauge transformation with $g(x) = \phi_0^{-1}(x)$ is performed. In other words, $\mathscr{L}^{(B)}$ is equivalent to $\mathscr{L}^{(A)}$ under the gauge condition $\phi_0(x) = 1$. So $\hat{B}_\mu$ is not an independent field variables, and we take the independent field variables to be $\phi_0(x)$, $\Phi(x)$ and $\hat{A}_\mu(x)$.

In order to restrict the gauge freedom in quantization; one must choose a definite gauge condition. We first choose the gauge $\phi_0 = 1$, write the path integral, then carry out the quantization in an arbitrary gauge by using the Faddeev-Popov trick. The effective action is invariant under B. R. S. transformation, and from it the Ward-Takahashi identities can be derived. If it can be shown that the S-matrix is invariant under variation of the gauge condition, then the renormalization in an arbitrary gauge can be proved. This is because the theory with Lagrangian $\mathscr{L}^{(A)}$ is renormalizable, so the invariance of the S-matrix implies that the theory with Lagrangian $\mathscr{L}^{(B)}$ is also renormalizable.

2. Path-Integral Quantization in the Renormalizable Gauge

If we choose the gauge condition $\phi_0(x) = 1$, then $\varphi(x)$ and $\hat{A}_\mu(x)$ are ordinary field variables. The Lagrangian $\mathscr{L}^{(A)}$ clearly describes an ordinary field theory which can be quantized by the path-integral formalism[2] and is renormalizable. We call this gauge the renormalizable gauge. The generating functional in this gauge is

$$Z=\int[d\varphi dA_\mu]\delta(F_A)\Delta_A\exp\left\{i\int d^4x\mathscr{L}^{(A)}\right\}, \tag{2.1}$$

where the gauge conditions are

$$F_{Al}(\varphi,A_\mu)=0,\quad (l=1,2,\cdots n_H). \tag{2.2}$$

The number of gauge conditions in Eq. (2.2) is equal to the number of generators of the subgroup H, i. e. n_H. For example, $F_{Al}=\partial_\mu A_\mu^l$, $(l=1,2,\cdots,n_H)$, is the coulomb gauge. The Faddeev-Popov determinant Δ_A satisfies the following condition,

$$\Delta_A\int d\mu(h)\delta(F_A^h)=1. \tag{2.3}$$

with

$$F_A^h=F_A(h\varphi,h(\partial_\mu+\hat{A}_\mu)h^{-1}), \tag{2.4}$$

where $d\mu(h)$ is the invariant measure on the subgroup H, i. e.

$$d\mu(h)=d\mu(hh')=d\mu(h'h). \tag{2.5}$$

Choosing a local coordinate system h_l $(l=1,2,\cdots,n_H)$ to parametrize the subgroup element h,

$$h=e^{i\hat{I}_lh_l}, \tag{2.6}$$

we have

$$hdh^{-1}=-i\hat{I}_l dh_m h_{lm},\quad h_{lm}=h_{lm}(h). \tag{2.7}$$

Define the invariant length on the subgroup H manifold as,

$$dS_H^2=2\mathrm{tr}(hdh^{-1})^+(hdh^{-1})=dh_l dh_m h_{nl}h_{nm}, \tag{2.8}$$

from which the invariant measure of the subgroup H is obtained,

$$d\mu(h)=dh_1dh_2\cdots dh_{n_H}\cdot\det(h_{lm}), \tag{2.9}$$

with the help of the Eq. (2.5) we can verify that Δ_A is an invariant quantity on the subgroup H, $\Delta_A^h=\Delta_A$ and we have

$$\Delta_A=\det M^{(A)},\quad M_{lm}^A(x,y)=\left.\frac{\delta F_{Al}^h(x)}{\delta h_{lm}(y)}\right|_{h=e}. \tag{2.10}$$

Let us choose α_i, $(i=n_H+1,\cdots,n_G)$, as the local coordinates which parametrize the coset element $\phi_0(x)$,

$$\phi_0(x)=e^{i\hat{k}_i\alpha_i(x)}, \tag{2.11}$$

where $\hat{k}_i$ $(i=n_H+1,\cdots,n_G)$ are generators of the coset G/H. So $\phi_0=1$ corresponds to $\alpha_i=0$ $(i=n_H+1,\cdots,n_G)$ from which we can define

$$\delta(\phi_0-1)=\prod_{i=n_H+1}^{n_G}\delta(\alpha_i)\frac{1}{\sqrt{g(\alpha)}}. \tag{2.12}$$

The invariant measure $d\mu(\phi_0)$ should satisfy the following relations

$$d\mu(\phi_0)=d\mu(\phi(g,\phi_0)),$$

$$\int d\mu(\phi_0)\delta(\phi_0-1)=1. \tag{2.13}$$

Inserting the left hand side of (2.13) into Eq. (2.1), the generating functional becomes

$$Z=\int[d\mu(\phi_0)d\Phi dA_\mu]\delta(\phi_0-1)\delta(F_A)\Delta_A\exp\left\{i\int d^4x\mathscr{L}^{(8)}\right\}, \tag{2.14}$$

where $F_A=F_A(\Phi,A_\mu)$, $\Delta_A=\Delta_A(\Phi,A_\mu)$. Let us now consider how to choose $d\mu(\phi_0)$. If $d\mu(g)$ is an invariant measure of the group G, then for fixed g' we must have

$$d\mu(g)=d\mu(gg')=d\mu(g'g), \tag{2.15}$$

Putting $g=\phi h$, we have $g'g=\phi(g',\phi)h(g',\phi)h$. Furthermore, we define $d\mu(\phi)$ by the relation $d\mu(g)=d\mu(\phi)d\mu(h)$. From this relation we have

$$d\mu(g'g)=d\mu(\phi(g',\phi))d\mu(h(g'\phi)h), \tag{2.16}$$

where $d\mu(h)$ is the invariant measure of the subgroup H. Therefore by fixing g' and ϕ one can obtain $d\mu(h(g',\phi)h)=d\mu(h)$ and from which it is deduced

$$d\mu(\phi)=d\mu(\phi(g',\phi)). \tag{2.17}$$

Comparing Eq (2.13) with Eq. (2.17) we conclude that $d\mu(\phi)=d\mu(g)/d\mu(h)$ is the invariant measure on the coset space. Because of the fact that coset is not closed under group multiplication, it follows that

$$\phi d\phi^{-1}=-i\sum_i\hat{f}_i(\alpha)d\alpha_i, \tag{2.18}$$

where

$$\hat{f}_i(\alpha)=f_{il}(\alpha)\hat{l}_l+f_{ij}(\alpha)\hat{k}_j \tag{2.19}$$

has components valued on the algebra of the subgroup H. By means of the relation

$$gdg^{-1}=\phi(hdh^{-1})\phi^{-1}+\phi d\phi^{-1},$$

Eqs. (2.7) and (2.18) the invariant length on the group G manifold can be expressed in the form

$$\begin{aligned}dS^2&=2\mathrm{tr}(gdg^{-1})^+(gdg^{-1})\\&=dh_ldh_mh_{nl}h_{nm}+g_{ij}d\alpha_id\alpha_j+2h_{nl}g_{nj}dh_ld\alpha_j,\end{aligned} \tag{2.20}$$

where

$$g_{ij}(\alpha)=2\mathrm{tr}\hat{f}_i(\alpha)\hat{f}_j(\alpha);\quad g_{il}(\alpha)=g_{li}(\alpha)=2\mathrm{tr}\hat{f}_i(\alpha)\phi\hat{I}_l\phi^{-1}. \tag{2.21}$$

The invariant measure on group G is then given by

$$d\mu(g)=\sqrt{g(\alpha)}\det(h_{lm})dh_1\cdots dh_{n_H}d\alpha_{n_{H+1}}\cdots d\alpha_{n_G}, \tag{2.22}$$

$$d\mu(\phi)=\sqrt{g(\alpha)}d\alpha_{n_{H+1}}\cdots d\alpha_{n_G},$$

where

$$g(\alpha)=\det\begin{pmatrix} g_{ij} & g_{il} \\ g_{lj} & \delta_{ln}\end{pmatrix}, \tag{2.23}$$

and $\sqrt{g}$ is the Lee-Yang factor.

3. The Path-Integral Quantization in Arbitrary Gauge

Now let us turn to a gauge different from the renormalizable gauge $\phi_0=1$, $F_A=0$. The gauge conditions are

$$F_{Ba}(\phi_0,\Phi,A_\mu)=0,\quad (a=1,2,\cdots n_G), \tag{3.1}$$

where n_G is the number of generators of the group G. The Faddeev-Popov functional $\triangle_B(\phi_0,\Phi,A_\mu)$, which satisfies

$$\triangle_B\int d\mu(g)\delta(F_B^\xi)=1,\quad \triangle_B^\xi=\triangle_B, \tag{3.2}$$

where

$$F_B^\xi=F_B(\phi(g,\phi_0),g\Phi,h(g,\phi_0)(\partial_\mu+\hat{A}_\mu)h^{-1}(g,\phi_0)), \tag{3.3}$$

can easily be found. In the following, the group element g will be represented by a set of local coordinates ξ_a $(a=1,2,\cdots,n_G)$ in the form

$$g=e^{i\hat{X}_a\xi_a},\quad gdg^{-1}=-i\hat{X}_a d\xi_b\theta_{ab}(g), \tag{3.4}$$

where $\hat{X}_a$ $(a=1,2,\cdots,n_G)$ are generators of the group G, satisfying the commutation relations

$$[\hat{X}_a,\hat{X}_b]=if_{abc}\hat{X}_c. \tag{3.5}$$

Introducing the invariant length on the group manifold of G as

$$dS^2=2\mathrm{tr}(gdg^{-1})^+(gdg^{-1})=d\xi_a d\xi_b\theta_{ca}\theta_{cb}, \tag{3.6}$$

we define the invariant measure in the following form

$$d\mu(g)=d\xi_1 d\xi_2\cdots d\xi_{n_G}\cdot\det(\theta_{ab}). \tag{3.7}$$

In terms of the local coordinates ξ_a, the Faddeev-Popov determinant has the form

$$\triangle_B=\det M^{(B)},\quad M_{ab}^B(x,y)=\frac{\delta F_{Ba}^\xi(x)}{\delta\xi_b(y)}\bigg|_{\xi=0}. \tag{3.8}$$

Inserting the left hand side of Eq. (3.2) into Eq. (2.14), we have

$$Z=\int [d\mu(\phi_0)d\Phi dA_\mu]\delta(\phi_0-1)\delta(F_A)\Delta_A \exp\left\{i\int d^4x\mathscr{L}^{(B)}\right\}\int d\mu(g)\delta(F_B^g)\Delta_B. \tag{3.9}$$

Performing a transformation in the integrating variables:

$$\phi_0\to\phi(g,\phi_0),\quad \Phi\to g\Phi,\quad \hat{A}_\mu\to h(g,\phi_0)(\partial_\mu+\hat{A}_\mu)h^{-1}(g,\phi_0), \tag{3.10}$$

we obtain from Eq. (3.9)

$$Z=\int [d\mu(\phi_0)d\Phi dA_\mu]\delta(F_B)\Delta_B \exp\left\{i\int d^4x\mathscr{L}^{(B)}\right\}$$

$$\cdot\int d\mu(g^{-1})\delta(\phi(g^{-1},\phi_0)-1)\delta(F_A^{h(g^{-1},\phi_0)})\Delta_A. \tag{3.11}$$

Decomposing $d\mu(g^{-1})$ into $d\mu(g^{-1})=d\mu(\phi(g^{-1},\phi_0))d\mu(h(g^{-1},\phi_0))$, it is easily verified that

$$\int d\mu(g^{-1})\delta(\phi(g^{-1},\phi_0)-1)\delta(F^{h(g^{-1},\phi_0)})\Delta_A=1. \tag{3.12}$$

which is substituted into Eq. (3.11) to obtain

$$Z=\int [d\mu(\phi_0)d\Phi dA_\mu]\delta(F_B)\Delta_B \exp\left\{i\int d^4x\mathscr{L}^{(B)}\right\}. \tag{3.13}$$

This is the generating functional in an arbitrary gauge $F_B=0$.

4. Properties of the Transformation

Under the left multiplication of a group element $g\epsilon G$, the element ϕ_0 on the coset space transforms as

$$g\phi_0=\phi(g,\phi_0)h(g,\phi_0). \tag{4.1}$$

For an infinitesimal gauge transformation

$$g=1+i\hat{X}_a\xi_a, \tag{4.2}$$

by using Eqs. (2.6) and (2.11) we get

$$h_l(g,\phi_0)=h_{la}(\alpha)\xi_a,\quad h_{la}(\alpha)=\left.\frac{\partial h_l(g,\phi_0)}{\partial\xi_a}\right|_{\xi=0}, \tag{4.3}$$

$$\alpha_i(g,\phi_0)=\alpha_i+R_{ia}(\alpha)\xi_a,\quad R_{ia}(\alpha)=\left.\frac{\partial\alpha_i(g,\phi_0)}{\partial\xi_a}\right|_{\xi=0}.$$

The infinitesimal form of the group relations

$$h(g'g,\phi_0)=h(g',\phi(g,\phi_0))h(g,\phi_0), \tag{4.4}$$

and

$$\phi(g'g,\phi_0)=\phi(g',\phi(g,\phi_0)), \tag{4.5}$$

can be written respectively as

$$f_{lmn}h_{ma}(\alpha)h_{ab}(\alpha)+R_{ja}(\alpha)\frac{\partial h_{lb}(\alpha)}{\partial\alpha_j}-R_{jb}(\alpha)\frac{\partial h_{la}(\alpha)}{\partial\alpha_j}=f_{abc}h_{lc}(\alpha), \tag{4.6}$$

and

$$R_{ja}(\alpha)\frac{\partial R_{ib}(\alpha)}{\partial \alpha_j} - R_{jb}(\alpha)\frac{\partial R_{ia}(\alpha)}{\partial \alpha_j} = f_{abc}R_{ic}(\alpha), \tag{4.7}$$

from which we can deduce the transformation of the field variables under the action of $g \in G$,

$$\delta A^l_\mu(x) = f_{lmn}A^m_\mu(x)h_{na}(x)\xi_a(x) + \frac{1}{g}\partial_\mu(h_{la}(x)\xi_a(x)),$$

$$\delta\Phi(x) = i\hat{X}_a\Phi(x)\xi_a(x), \tag{4.8}$$

$$\delta\alpha_i(x) = R_{ia}(x)\xi_a(x).$$

For the sake of concise presentation let us take the notation introduced by Lee[5] and let ϕ_i stand for A^l_μ, Φ_σ or α_i. In this notation Eq. (4.8) becomes

$$\delta\phi_i = D^a_i(\phi)\xi_a, \quad D^a_i(\phi) = \frac{\delta\phi_i}{\delta\xi_a}, \tag{4.9}$$

where

$$D^a_{l\mu}(x_i,x_a) = f_{lmn}A^m_\mu(x_i)h_{na}(x_i)\delta(x_i - x_a) + \frac{1}{g}\partial^{(i)}_\mu(h_l(x_i)\delta(x_i - x_a)),$$

$$D^a_\sigma(x_i,x_a) = i(\hat{X}_a)_{\sigma\rho}\Phi_\rho(x_i)\delta(x_i - x_a), \tag{4.10}$$

$$D^a_i(x_i,x_a) = R_{ia}(x_i)\delta(x_i - x_a).$$

By using Eqs. (4.6), (4.7) and (4.10), the following relations are easily deduced

$$\frac{\delta D^a_i}{\delta\phi_j}D^b_j - \frac{\delta D^b_i}{\delta\phi_j}D^a_j = -f^{abc}D^c_i, \tag{4.11}$$

where

$$f^{abc} = f_{abc}\delta(x_a - x_b)\delta(x_b - x_c). \tag{4.12}$$

The condition for the invariance of the Lagrangian under the action of the group G can be rewritten as

$$\left.\frac{\delta\mathscr{L}}{\delta\xi_a}\right|_{\xi=0} = D^a_i\frac{\delta\mathscr{L}}{\delta\phi_i} = 0. \tag{4.13}$$

We now introduce the anticommutative scalar fields ξ_a, η_a $(a = 1, 2, \cdots, n_G)$[3] and use the technique of Lee and Zinn-Justin[5] to obtain the effective action

$$S_{eff} = S_B + \frac{1}{2\alpha}F^2_{Ba} + \xi_a M^{(B)}_{ab}\eta_b. \tag{4.14}$$

Finally the generating functional has the from

$$Z = \int [d\phi d\xi d\eta]e^{iS_{eff}[\phi,\xi,\eta]}, \tag{4.15}$$

where ξ_a and η_a are the Faddeev-Popov ghost fields.

The effective action (4.14) can be shown to be invariant under the B. R. S. transformation[4]

$$\delta\phi_i = D_i^a \eta_a \delta\lambda\,, \tag{4.16}$$

$$\delta\eta_a = \frac{1}{2} f^{abc} \eta_b \eta_c \delta\lambda\,,$$

$$\delta\xi_a = \frac{1}{\alpha} F_{Ba} \delta\lambda\,, \tag{4.17}$$

whera $\delta\lambda$ is an anticommutative C-number. By means of Eq. (4.11) we can demonstrate that

$$\delta^2\phi_i = 0\,,\quad \delta^2\eta_a = 0\,,\quad \delta^2\xi_a = 0\,, \tag{4.18}$$

from which the invariance of the effective action under B. R. S. transformation follows without difficulty, i. e.

$$\delta S_{eff}[\phi,\xi,\eta] = 0\,, \tag{4.19}$$

$$[d\phi d\xi d\eta] = [d\phi' d\xi' d\eta']\,. \tag{4.20}$$

Thus the generating functional Z is invariant under B. R. S. transformation, namely

$$\delta Z = 0\,. \tag{4.21}$$

5. Ward-Slavnov Identity

Define the generating functional as

$$Z[J] = \int [d\phi d\xi d\eta] \exp\{iS_{eff}[\phi,\xi,\eta] + iJ_i\phi_i\}, \tag{5.1}$$

By using the integral formula for Grassmann variables

$$\int dC_i C_j = \delta_{ij}\,. \tag{5.2}$$

we have

$$\int [d\phi d\xi d\eta]\, \xi_a \exp\{iS_{eff} + iJ_i\phi_i\} = 0\,. \tag{5.3}$$

Upon performing B. R. S. transformation, we get

$$\int [d\phi d\xi d\eta]\, (\xi_a + \delta\xi_a) \exp\{iS_{eff} + iJ_i\phi_i + iJ_i\delta\phi_i\} = 0\,. \tag{5.4}$$

Eqs. (5.3) and (5.4) lead to

$$\int [d\phi d\xi d\eta]\, (i\xi_a J_i \delta\phi_i + \delta\xi_a) \exp\{iS_{eff} + iJ_i\phi_i\} = 0\,, \tag{5.5}$$

Substituting B. R. S. transformation (4.16), (4.17) into Eq. (5.5) we obtain the Ward-Slavnov identity

$$\int [d\phi d\xi d\eta]\left(iJ_i D_i^b(\phi)\xi_a\eta_b+\frac{1}{\alpha}F_{Ba}[\phi]\right)\exp\{iS_{eff}+iJ_i\phi_i\}=0. \tag{5.6}$$

This was first derived by Slavnov and Taylor[5] for ordinary gauge theories. It is also true for theories with pure-gauge fields on coset. We define

$$\begin{aligned} Z_{ba}[J] &= -i\int [d\phi d\xi d\eta]\xi_a\eta_b \exp\{iS_{eff}+iJ_i\phi_i\} \\ &= \int [d\phi d\xi d\eta] M_{ba}^{B\,-1}[\phi]\exp\{iS_{eff}+iJ_i\phi_i\} \\ &= M_{ba}^{(B)-1}\left[\frac{\delta}{i\delta J}\right]Z[J] \end{aligned} \tag{5.7}$$

or

$$M_{ab}^{B}\left[\frac{\delta}{i\delta J}\right]Z_{bc}[J]=\delta_{ac}Z[J], \tag{5.8}$$

then Ward-Slavnov identity (5.6) can be rewritten in the form

$$\frac{1}{\alpha}F_{Ba}\left[\frac{\delta}{i\delta J}\right]Z[J]-J_i D_i^b\left[\frac{\delta}{i\delta J}\right]Z_{ba}[J]=0. \tag{5.9}$$

6. The Gauge Independence of the S-Matrix

In the renormalizable gauge the generating functional is[5]

$$Z_A[K]=\int [d\varphi dA_\mu]\delta(F_A)\Delta_A \exp\{iS_A+iK_i\varphi_i\}, \tag{6.1}$$

where φ_i represents the ordinary field variables $\varphi_a(x)$ and $A_\mu^i(x)$. For general gauge conditions, the generating functional is

$$Z_B[J]=\int [d\mu(\phi_0)d\Phi dA_\mu]\delta(F_B)\Delta_B \exp\{iS_B+iJ_i\phi_i\}. \tag{6.2}$$

Substituting the functional (3.12) in to (6.2), we have

$$\begin{aligned} Z_B[J]=\int [d\mu(\phi_0)d\Phi dA_\mu]\delta(F_B)\Delta_B \exp\{iS_B+iJ_i\phi_i\} \\ \cdot\int d\mu(g^{-1})\delta(\phi(g^{-1},\phi_0)-1)\delta(F_A^{h(g^{-1},\phi_0)})\Delta_A. \end{aligned} \tag{6.3}$$

which upon using a transformation on the integration variables,

$$\phi_0'=\phi(g^{-1},\phi_0),\qquad \Phi'=g^{-1}\Phi,\qquad \hat{A}_\mu'=h(g^{-1},\phi_0)(\partial_\mu+\hat{A}_\mu)h^{-1}(g^{-1},\phi_0), \tag{6.4}$$

becomes

$$Z_B[J]=\int [d\mu(\phi_0)d\Phi dA_\mu]\delta(\phi_0-1)\delta(F_A)\Delta_A e^{iS_B}\int d\mu(g)\delta(F_B^g)\Delta_B \exp[iJ_i\phi_i^g]. \tag{6.5}$$

If the gauge condition F_B^g is solved for g at $g=g_0$, the integration over g in

Eq. (6.5) can be performed and we obtain

$$Z_B[J] = \int [d\mu(\phi_0) d\Phi dA_\mu] \delta(\phi_0 - 1) \delta(F_A) \Delta_A \exp\{iS_B + iJ_i \phi_i^g\}$$

$$= \int [d\varphi dA_\mu] \delta(F_A) \Delta_A \exp\{iS_A + iJ_i \phi_i(\varphi)\}$$

$$= \exp\left\{ iJ_i \phi_i \left[\frac{\delta}{i\delta K} \right] \right\} Z_A[K] \bigg|_{K=0},$$

or

$$Z_B[J] = \exp\left\{ iJ_i \phi_i \left[\frac{\delta}{i\delta K} \right] \right\} Z_A[K] \bigg|_{K=0}, \tag{6.6}$$

where $\phi_i(\varphi) = \phi_i^{g_0}|_{\phi_0=1}$. The formula (6.6) can be used to yield relationship between the Green's functions in the general gauge and that in the renormalizable gauge. On the mass shell the two generating functionals have no other differences except for changes in the renormalization constants of the external lines. So the S-matrix is the same when the gauge condition is changed. Thus we have demonstrated that the theory of pore-gauge field on coset space can be quantized and is renormalizable in arbitrary gauge provided that it is so in the gauge $\phi_0 = 1$, $F_A = 0$.

References

[1] Chou Kuang-chao, Tu Tung-sheng, Ruan Tu-nan, *scientia Sinica*, **XXII** (1979), No. 1, 37.

[2] V. N. Popov and L. D, Faddeev, *Phys. Letts*, **25B** (1967), 29.

[3] F. A. Berezin, The method of second quantiza (Academic Press, New York, 1966), 49.

[4] C. Becchi, A. Rouet and R. Stora, *Comm. Math. Phys.*, 42 (1975), 127.

[5] B. W. Lee, Methods in Field Theory (Session 28 of Les Houchos 1975), 79, and the references quoted there.

陪集空间纯规范场的量子化和重整化

周光召 (中国科学院理论物理研究所)

阮图南 (中国科学技术大学近代物理系)

摘　　要

本文利用 Faddeev-Popov 技巧完成了陪集空間純規范場的路径积分量子化．証明了理論是 B. R. S. 变换下不变的．由此导出了 Ward-Slavnov 恆等式．由于理論在規范条件 $\phi_0=1$，$F_A=0$ 下是可重整化的，它在其他規范下也是可重整化的．这可由 S 矩陣的規范无关来証明．

Vol. 27 No. 2 KEXUE TONGBAO February 1982

THE $U(1)$ ANOMALOUS WARD IDENTITIES AND CHIRAL DYNAMICS

Zhou Guangzhao (周光召)
(*Institute of Theoretical Physics, Academia Sinica*)

Received August 20, 1980.

It is generally believed that the QCD Lagrangian has N_f^2 conserved axial currents except for quark mass terms. In the world of zero bare quark masses for the first L flavours, the corresponding conservation laws are spontaneously broken and L^2 pseudo-Goldstone bosons are thought to be generated. For three quark flavours the absence of the ninth light pseudoscalar in the real world is a well-known puzzle first pointed out by Glashow[1] and studied subsequently by Wienberg[2], Kogut and Susskind[3] and many others[4–8].

In 1976 it was observed by G't Hooft[4] that instantons might help resolve the paradox through the anomaly in the axial $U(1)$ channel. However, Crewther[5] has shown in a series of papers that instantons with integer Pontryagin number are impossible to satisfy the anomalous Ward identities in WKB approximation.

Recently, a very interesting proposal based on the analysis of anomaly in the framework of the $1/N_c$ approximation has been made by Witten[6]. According to his opinion the $U(1)$ problem can be solved at the next-to leading order of its $1/N_c$ expansion, due to quantum fluctuations in the topological charge density.

In this note we argue that the idea underlying Witten's proposal is a general one not necessarily connected with the $1/N_c$ expansion. To P^2 order in the low energy limit both the mass and the wave function renormalization constant for the η'' meson are determined by the quantum fluctuations in the topological charge density.

In the world of zero bare quark masses, the Lagrangian has the form

$$\mathscr{L} = -\frac{1}{4} F^a_{\mu\nu} F^{a\mu\nu} + \bar{q} i \not{D} q - \theta(x)\nu(x) - J(x)O(x), \tag{1}$$

where

$$\nu(x) = \frac{g^2}{32\pi^2} F^a_{\mu\nu} {}^*F^{a\mu\nu}$$

is the topological charge density. In Eq. (1) a set of hermitian composite fields $O_i(x)$ is introduced with $J_i(x)$ as their external sources. The Lagrangian (1) is invariant under $U(L) \times U(L)$ chiral group with L quark flavours, except for the source term $J(x)\ O(x)$. We shall put $\theta(x) = \theta$, $J_i(x) = 0$ only at the final stage of the calculation.

The currents for the $U(L) \times U(L)$ symmetry are

$$j^\mu = \bar{q}\gamma^\mu \frac{1}{2}\hat{\lambda}_j q, \quad j^\mu_{5j} = \bar{q}\gamma^\mu\gamma_5 \frac{1}{2}\hat{\lambda}_j q, \tag{2}$$

where $\hat{\lambda}_j$ indicates the Gell-Mann matrices with $\lambda_0 = \sqrt{2/L}$.

We shall include the currents to the composite fields and write their external sources as $J_{\mu j}$ and $J_{\mu 5j}$ respectively. Sometimes we use $O(x)$ and $J(x)$ to denote composite fields other than the currents and their sources.

The generating functional for the connected Green's functions is defined as

$$W[J(x)] = \sum_n \frac{(-i)^{n-1}}{n!}\int d^4x_1\cdots d^4x_n J_{i1}(x_1)\cdots J_{in}(x_n) \cdot \langle T(O_{i1}(x_1)\cdots O_{in}(x_n))\rangle_c, \tag{3}$$

where $\hat{O}_i$ indicates all the operators written in Heisenberg picture.

The vertex functional $\Gamma[O(x), J_{\mu j}(x), J_{\mu 5j}(x)]$ is related to $W[J(x), J_{\mu i}(x), J_{\mu 5i}(x)]$ by a Legrendre transformation

$$\Gamma[O(x), J_{\mu i}(x), J_{\mu 5i}(x)] = W[J(x), J_{\mu i}(x), J_{\mu 5i}(x)] - \int J_i(x)O_i(x)d^4x, \tag{4}$$

where $O_i(x)$ is determined by

$$O_i(x) = \frac{\delta W}{\delta J_i(x)}. \tag{5}$$

In Eq. (4) the sources relating to the currents are not Legrendre-transformed. From Eq. (4) it is easily convinced that

$$\frac{\delta\Gamma}{\delta J_{\mu i}(x)} = \frac{\delta W}{\delta J_{\mu i}(x)}, \quad \frac{\delta\Gamma}{\delta J_{\mu 5i}(x)} = \frac{\delta W}{\delta J_{\mu 5i}(x)}. \tag{6}$$

We assume that the composite fields $O_i(x)$ form a linear representation of the chiral group. The Ward identities for the generating functional have the forms

$$\partial_\mu \frac{\delta W}{\delta J_{\mu j}(x)} = iJ(x)\hat{T}_j \frac{\delta W}{\delta J(x)},$$
$$\partial_\mu \frac{\delta W}{\delta J_{\mu 5 i}(x)} = iJ(x)\hat{T}_{5j} \frac{\delta W}{\delta J_{\mu 5i}} + \sqrt{2L}\langle\hat{\nu}(x)\rangle\delta_{jo}. \tag{7}$$

In (7), $\hat{T}_j$ and $\hat{T}_{5j}$ are the representation matrices of the group generators.

The axial $U(1)$ anomaly term $\sqrt{2L}\ \langle\hat{\nu}(x)\rangle$ can also be written as $\sqrt{2L}\ \frac{\delta W}{\delta\theta(x)}$. These Ward identities have been justified by Crewther and used in literature to study the $U(1)$ problem.

It is more convenient to start from the Ward identities for the generating functional of the vertex functions. They have the form

$$\partial_\mu \frac{\delta\Gamma}{\delta J_{\mu i}(x)} = -i\frac{\delta\Gamma}{\delta O(x)}\hat{T}_i O(x) + iJ_{\mu k}(x) f_{kji}\frac{\delta\Gamma}{\delta\Gamma_{\mu i}(x)}$$

$$+ iJ_{\mu 5k}(x) f_{kji}\frac{\delta\Gamma}{\delta J_{\mu 5i}(x)}, \tag{8}$$

$$\partial_\mu \frac{\delta\Gamma}{\delta J_{\mu 5i}(x)} = -i\frac{\delta\Gamma}{\delta O(x)}\hat{T}_{5i} O(x) + J_{\mu k}(x) d_{kji}\frac{\delta\Gamma}{\delta J_{\mu 5i}(x)}$$

$$+ J_{\mu 5k}(x) d_{kji}\frac{\delta\Gamma}{\delta J_{\mu i}(x)} + \sqrt{2L}\frac{\delta\Gamma}{\delta\theta(x)}\delta_{jo}. \tag{9}$$

The Ward identities (8) and (9) are satisfied by any functionals of $O(x)$, $J_{\mu j}(x)$, $J_{\mu 5j}(x)$ and $\theta(x)$ and are invariant under the following infinitesimal local gauge transformation

$$\begin{aligned}
&O(x) \to (1 + i\alpha_j(x)\hat{T}_j + i\alpha_{5j}(x)\hat{T}_{5j})O(x),\\
&J_{\mu j}(x) \to J_{\mu j}(x) - iJ_{\mu k}(x) f_{kij}\alpha_i(x) - J_{\mu 5k}(x) d_{kij}\alpha_{5i}(x) - \partial_\mu\alpha_j(x),\\
&J_{\mu 5j}(x) \to J_{\mu 5j}(x) - iJ_{\mu 5k}(x) f_{kij}\alpha_i(x) - J_{\mu k}(x) d_{kij}\alpha_{5i}(x) - \partial_\mu\alpha_{5j}(x),\\
&\theta(x) \to \theta(x) - \sqrt{2L}\alpha_{5o}.
\end{aligned} \tag{10}$$

Therefore the vertex functional is invariant under the above local gauge transformations.

In the following we shall choose the composite field $O(x)$ to be the $2L^2$ scalar and pseudoscalar biquark densities $\bar{q}\frac{\hat{\lambda}_j}{2}q$ and $i\bar{q}\gamma_5\frac{\hat{\lambda}_j}{2}q$. They can be represented by an $L \times L$ matrix

$$\hat{U} = \hat{\lambda}_j\bar{q}\frac{1}{2}(1+\gamma_5)\hat{\lambda}_j q \tag{11}$$

and its hermitian conjugate, which transforms under $U(L)\times U(L)$ into

$$\hat{U} \to V_L\hat{U}V_R^+, \tag{12}$$

where V_L and V_R are unitary matrices belonging to the left-handed and right-handed $U(L)$ groups respectively. The vacuum average value of $\hat{U}$ will be denoted by U. For $L = 3$ one can construct only four $SU(3)\times SU(3)$ invariants from U and its hermitian conjugate U^+. They are $\mathrm{tr}(UU^+)$, $\mathrm{tr}(UU^+UU^+)$, $\det(UU^+)$ and $\det U/\det U^+$, the first three of which are automatically $U(3)\times U(3)$ invariants. Under the axial $U(1)$, In((det $U/\det U^+$) transform in the following way

$$\ln(\det U/\det U^+) \to \ln(\det U/\det U^+) + i2\sqrt{2L}\alpha_{5o} \tag{13}$$

and $\theta - \frac{i}{2}\ln(\det U/\det U^+)$ is an invariant.

Within the limit of long wavelength and low frequency, the vertex functional should be a function of these invariants only. Hence after putting the external sources to zero and $\theta(x) = \theta$, we have

$$\Gamma = F\left[\mathrm{tr}\,(UU^+),\ \mathrm{tr}\,(UU^+UU^+),\ \det(UU^+),\ \theta - \frac{i}{2}\ln(\det U/\det U^+)\right]. \tag{14}$$

The symmetry $U(L) \times U(L)$ is broken spontaneously down to the parity conserving $U(L)$. It is argued in [9] that to the second order in energy momentum P of the pseudoscalar Goldstone bosons

$$UU^+ = \frac{1}{2} f_\pi^2 \tag{15}$$

holds after a suitable normalization. In chiral symmetry limit of zero bare quark masses we can always put

$$U = \frac{1}{\sqrt{2}} f_\pi \exp\{i\hat{\lambda}_j \pi_j(x)/f_\pi\}, \tag{16}$$

where π_0 is the η' boson in this notation.

From Eq. (16) we see that

$$\theta - \frac{i}{2}\ln(\det U/\det U^+) = \theta + \frac{\sqrt{2L}}{f_\pi}\eta'. \tag{17}$$

The gauge invariance conditions (10) and (13) then imply that the vertex functional for η' and θ should be constructed from the invariant

$$\theta(x) + \frac{\sqrt{2L}}{f_\pi}\eta'(x) \tag{18}$$

and the covariant derivative

$$D_\mu\eta' = \partial_\mu\eta' + f_\pi J_{\mu 5 0}. \tag{19}$$

The effective Lagrangian $L_{\mathrm{eff}}(x)$ is related to the vertex functional by

$$\Gamma = -\int \mathscr{L}_{\mathrm{eff}}(x)d^4x \tag{20}$$

and can be constructed from the invariance properties required by the Ward identities. To the order of P^2, it has the form

$$\begin{aligned}\mathscr{L}_{\mathrm{eff}} = {} & \frac{1}{2} A\left(\theta + \frac{\sqrt{2L}}{f_\pi}\eta'\right) D_\mu\eta' D^\mu\eta' \\ & + \frac{1}{2} f_\pi A_{\eta\theta}\left(\theta + \frac{\sqrt{2L}}{f_\pi}\eta'\right)\left(\partial_\mu\theta + \frac{\sqrt{2L}}{f_\pi}\partial_\mu\eta'\right) D^\mu\eta' \\ & + \frac{1}{2} f_\pi A_\theta\left(\theta + \frac{\sqrt{2L}}{f_\pi}\eta'\right)\left(\partial_\mu\theta + \frac{\sqrt{2L}}{f_\pi}\partial_\mu\eta'\right)^2 \\ & - E\left(\theta + \frac{\sqrt{2L}}{f_\pi}\eta'\right).\end{aligned} \tag{21}$$

The wave function renormalization constant for the η' meson is equal to

$$Z_{\eta'} = (A(\theta) + \sqrt{2L}A_{\eta\theta}(\theta) + 2LA_{\theta}(\theta))^{-1} \tag{22}$$

and the mass of the η' meson is

$$m^2_{\eta'} = \frac{2L}{f^2_\pi}\frac{d^2E}{d\theta^2}Z_{\eta'}. \tag{23}$$

From Eqs. (20) and (21) one obtains

$$\int \frac{\delta^2\Gamma}{\delta\theta(x)\delta\theta(o)} e^{-iP\cdot x}d^4x\,|_{\theta(x)=\theta} = -i\int \langle T(\hat{v}(x)\hat{v}(o))\rangle_\theta e^{-iP\cdot x}d^4x$$

$$= \frac{d^2E}{d\theta^2} - P^2f^2_\pi A_\theta(\theta) + O(P^4). \tag{24}$$

Therefore, both $\frac{d^2E}{d\theta^2}$ and $A_\theta(\theta)$ are related to the quantum fluctuatuions of the topological charge density in the θ-vacuum.

We also find from Eqs. (20) and (21) that

$$\int \frac{\delta^2\Gamma}{\delta J_{\mu 5o}(x)\delta\theta(o)} e^{-iP\cdot x}d^4x\,|_{\theta(x)=\theta} = -i\int \langle T(\hat{j}^\mu_{5o}(x)\hat{v}(o))\rangle_\theta e^{-iP\cdot x}d^4x$$

$$= \frac{i}{2}P^\mu f^2_\pi A_{\eta\theta}(\theta) + O(P^3), \tag{25}$$

and

$$\int \frac{\delta^2\Gamma}{\delta J_{\mu 5o}(x)\delta\eta'(o)} e^{-iP\cdot x}d^4x\,|_{\theta(x)=\theta} \equiv iP^\mu f'_\eta$$

$$= iP^\mu f_\pi(A(\theta) + \sqrt{L/2}A_{\eta\theta}(\theta)) + O(P^3). \tag{26}$$

It is easily verified that in the $1/N_c$ expansion

$$A_\theta(\theta) \simeq O(1/N_c), \quad A_{\eta\theta}(\theta) = O(1/N_c^2), \tag{27}$$

and

$$f_{\eta'}(\theta) \simeq f_\pi(1 + O(1/N_c)).$$

Therefore to the leading order in $1/N_c$, Eq. (23) reduces to Witten's formula

$$Z_{\eta'} = 1, \quad m^2_{\eta'} = \frac{2L}{f^2_\pi}\left(\frac{d^2E}{d\theta^2}\right)_{\text{no quarks},\ \theta=0}. \tag{28}$$

To the next order in $1/N_c$ we may neglect $A_{\eta\theta}$ in Eq. (22) and find

$$Z_{\eta'} = (f_{\eta'}/f_\pi + 2LA_\theta(\theta))^{-1}, \tag{29}$$

where $A_\theta(\theta)$ is related to the quantum fluctuations of the topological charge density by Eq. (24).

It is interesting to note that in general Z_η is not equal to $f_\pi/f_{\eta'}$.

A full analysis of the effective Lagrangian for low energy pseudoscalar mesons with bare quark masses taken into account can be carried through without difficulty. Results on θ periodicity and the like can be obtained in the general case in accord with those given by Di Vecchia, Veneziano and Crewther[5,7]. These questions will be discussed elsewhere.

References

[1] Glashow, S. L., in *Hadrons and Their Interactions*, Academic Press Inc. New York, (1968), 83.
[2] Weinberg, S., *Phys. Rev.*, **D11**(1975), 3583.
[3] Kogut, J. & Susskind, L., *ibid.*, **D11**(1975), 3594.
[4] G't Hooft, *ibid.*, **37**(1976), 8.
[5] Crewther, R. J., *Riv. Nuovo Cimento*, **2**(1979), 83; *Phys. Lett.*, **70B**(1977), 349; *CERN*, TH-2791, (1979).
[6] Witten, E., *Nucl. Phys.*, **B156**(1979), 269.
[7] Veneziano, G., *ibid.*, **B159**(1979), 213; Di Vechia, P., *Phys. Lett.*, **85B**(1979), 357; Di Vecchia, P. & Veneziano, G., *CERN* TH-2814, (1980).
[8] Nath, P. & Arnowitt, R., *CERN*, TH-2818, (1980).
[9] Lee, B. W., *Chiral Dynamics*, Gordon & Breach, New York, 1972.

Comm. in Theor. Phys. Vol. 1, No. 1 (1982) 69-75

NEW NONLINEAR σ MODELS ON SYMMETRIC SPACES

CHOU Kuang-chao (周光召)
(Institute of Theoretical Physics, Academia Sinica)
SONG Xing-chang (宋行长)
(Institute of Theoretical Physics, Peking University)
Received July 9, 1981.

Abstract

Two new nonlinear σ models, defined on the symmetric coset spaces GL(n,c)⊗GL(n,c)/GL(n,c) and GL(n,c)/U(n) respectively, are formulated in this paper. The latter may be useful in discussing the four-dimensional Yang-Mills fields.

In previous papers[1,2], we have studied various classical nonlinear σ models taking values on symmetric coset spaces and have shown generally how to use the duality symmetry to deduce the infinite sets of conserved currents, both local and nonlocal. In these papers, a unifying point of view is proposed to connect varieties of integrable σ models which have attracted our attention in recent years. The same point has also been noticed by Eichenherr and Forger[3], who have proved that, for compact global symmetry groups, the σ model has the dual symmetry if and only if the field takes values in a symmetric space. The main feature of our formulation is to use the so-called "gauge-covariant current" $\mathcal{K}_\mu$ [3] as the central role to formulate the equation of motion, the duality symmetry and to deduce the conserved currents. The important advantage of this formulation is its unification and simplicity.

In this note, we shall extend our study to other models, such as the GL(n,c) principal model, and a new model H_n which takes values in the symmetric space GL(n,c)/U(n).

For the purpose of application some of the notations used previously will be changed slightly.

As mentioned previously[2,3], the U(n) principal field defined on the homogeneous space U(n)/{1} can be re-formulated as a σ model on symmetric space U(n)⊗U(n)/U(n). Similarly we consider the model on symmetric space GL(n,c)⊗GL(n,c)/GL(n,c). An element of the group G=GL(n,c)⊗GL(n,c) is expressed as

$$g = (g_L, g_R) \tag{1}$$

with both g_L and g_R belonging to GL(n,c) independently. The multiplication rule+) is

$$g_1 g_2 = (g_{1L} g_{2L},\ g_{2R} g_{1R}). \tag{2}$$

+) *The g_R defined here is just the g_R^+ in the Ref. [2].*

++) *A factor -i appearing in Ref.[2] is omitted here. Sometimes the Lorentz index μ is suppressed.*

An element of the subgroup $\mathscr{h} \varepsilon$ H = GL(n,c) is represented by ·

$$\mathscr{h} = (h\ ,\ h^{+}) \tag{3}$$

which is invariant with respect to the involution σ

$$\sigma(g_L\ ,\ g_R) = (g_R^{+}\ ,\ g_L^{+}). \tag{4}$$

Then G/H is a symmetric space with involution σ , whose element is

$$\Phi = (\phi\ ,\ \phi^{+-1}). \tag{5}$$

Now any element of G can be decomposed into the product of a subgroup element and a left coset element g= $\Phi\mathscr{h}$. This means that

$$g_L = \phi h\ , \qquad g_R = h^{+}\phi^{+-1}. \tag{6}$$

Following Ref. [1,2] for any $\phi \varepsilon$ GL(n,c) define$^{++)}$

$$\Omega = \Phi^{-1}\partial\Phi = (\phi^{-1}\partial\phi\ ,\ \partial\phi^{+-1}\phi^{+})$$

$$= \mathscr{H} + \mathscr{K} = (H + K\ ,\ H^{+} - K^{+}). \tag{7}$$

Here $\mathscr{H}$ and $\mathscr{K}$ take values in the Lie algebra of subgroup H and coset G/H respectively, i.e.,

$$\mathscr{H} = (H\ ,H^{+})\ , \qquad \mathscr{K} = (K\ ,-K^{+}) \tag{8}$$

and

$$H = \tfrac{1}{2}(\phi^{-1}\partial\phi + \phi\partial\phi^{-1})\ , \qquad K = \tfrac{1}{2}(\phi^{-1}\partial\phi - \phi\partial\phi^{-1}) \tag{8'}$$

Introducing the covariant derivative defined as

$$\mathscr{D}_\mu\Phi = \partial_\mu\Phi - \Phi\mathscr{H}_\mu = (\partial_\mu\phi - \phi H_\mu\ ,\ \partial_\mu\phi^{+-1} - H_\mu^{+}\phi^{+-1})\ ,$$

$$\mathscr{D}_\mu\Phi^{-1} = \partial_\mu\Phi^{-1} + \mathscr{H}_\mu\Phi^{-1} = (\partial_\mu\phi^{-1} + H_\mu\phi^{-1}\ ,\ \partial_\mu\phi^{+} + \phi^{+}H_\mu^{+})\ , \tag{9}$$

and

$$g = \phi^{2}\ , \tag{10}$$

we have

$$\mathscr{K}_\mu = \Phi^{-1}\mathscr{D}_\mu\Phi = -(\mathscr{D}_\mu\Phi^{-1})\Phi \tag{11}$$

and then

$$-\frac{1}{2}T_r\mathcal{K}_\mu\mathcal{K}^\mu=\frac{1}{2}T_r(\mathcal{D}_\mu\Phi^{-1})\mathcal{D}^\mu\Phi$$
$$=-\frac{1}{2}t_r(K_\mu K^\mu+K_\mu^+K^{+\mu})\,. \tag{12}$$

with

$$-t_r\,K_\mu K^\mu=\frac{1}{4}t_r\,\partial_\mu g^{-1}\partial^\mu g\,. \tag{12'}$$

Now the current can be written as+)

$$\mathcal{J}=2\Phi\mathcal{K}\Phi^{-1}$$
$$=2(\phi K\phi^{-1},-\phi^+K^+\phi^{+-1})$$
$$=(J_L\,,\;J_R)\,, \tag{13}$$

where

$$J_L=\phi(\phi^{-1}\partial\phi-\phi\partial\phi^{-1})\phi^{-1}=-g\partial g^{-1}\,,$$
$$J_R=\phi^+(\phi^+\partial\phi^{+-1}-\phi^{+-1}\partial\phi^+)\phi^{+-1}=g^+\partial g^{+-1}\,. \tag{14}$$

So setting

$$Q=\Phi^2=(\phi^2\,,\;\phi^{2+-1})=(g\,,\;g^{+-1})\,. \tag{15}$$

then

$$Q^{-1}=(g^{-1}\,,\;g^+)\,,$$

We can rewrite Eq. (13) as

$$\mathcal{J}_\mu=2\Phi\mathcal{K}_\mu\Phi^{-1}=-Q\partial_\mu Q^{-1}\,. \tag{16}$$

It can then be easily shown that

$$\partial_\mu\mathcal{J}_\mu=2\Phi(\mathcal{D}_\nu\mathcal{K}_\mu+[\mathcal{K}_\nu,\mathcal{K}_\mu])\Phi^{-1}\,, \tag{17}$$

and that

$$\partial_\mu\mathcal{J}_\nu-\partial_\nu\mathcal{J}_\mu-[\mathcal{J}_\mu\,,\;\mathcal{J}_\nu]$$
$$=2\Phi(\mathcal{D}_\mu\mathcal{K}_\nu-\mathcal{D}_\nu\mathcal{K}_\mu)\Phi^{-1}=0\,. \tag{18}$$

Here the covariant derivative of $\mathcal{K}$ is defined as

$$\mathcal{D}_\mu\mathcal{K}_\nu=\partial_\mu\mathcal{K}_\nu+[\mathcal{H}_\mu\,,\;\mathcal{K}_\nu]\,. \tag{19}$$

The Lagrangian for this model is just Eq. (12);

$$\mathcal{L}=-T_r\mathcal{K}_\mu\mathcal{K}^\mu=\frac{1}{2}T_r(\mathcal{D}_\mu\Phi^{-1})\mathcal{D}^\mu\Phi$$
$$=\frac{1}{8}T_r(\partial_\mu Q^{-1})\partial^\mu Q\,, \tag{20}$$

which is invariant under the global group GL(n,c) ⊗ Gl(n,c) and the gauge group GL(n,c). The field equation reads

+) *In contrast with that in Ref.[2], a factor 2 is added here.*

$$\mathcal{D}_\mu \mathcal{D}^\mu \Phi - (\mathcal{D}_\mu \Phi)\Phi^{-1}(\mathcal{D}^\mu \Phi) = 0 \qquad (21)$$

or equivalently,

$$\mathcal{D}^\mu \mathcal{K}_\mu = 0 . \qquad (22)$$

This implies that

$$\partial^\mu \mathcal{J}_\mu = 0 ,$$

or

$$\partial^\mu J^L_\mu = 0 , \qquad \partial^\mu J^R_\mu = 0 . \qquad (23)$$

This means that $\mathcal{J}_\mu$ is conserved and $\mathcal{K}_\mu$ is covariantly conserved.

If we treat GL(n,c) ⊗ GL(n,c)/GL(n,c) model as in Ref.[2], we obtain the Lagrangian form $L = \frac{1}{2}\mathrm{Tr}\mathcal{H}_\mu\mathcal{H}^\mu = \frac{1}{2}\mathrm{tr}\partial_\mu q \cdot \partial^\mu q^{-1}$, $q \in GL(n,c)$ which can be considered as GL(n,c) principal model. But it can easily be seen that this L is not hermitian. If we add the hermitian conjugate L^+ to L, we then obtain eq.(12). So the method given here has taken into account the hermitian property by itself.

For the case in which g_L and g_R are unitary, and so are h and ϕ, we obtain

$$q^+ = q^{-1} \qquad (24)$$

and

$$H^+ = -H, \qquad K^+ = -K. \qquad (25)$$

Then the Lagrangian turns out to be

$$\mathcal{L} = -\mathrm{tr} K_\mu K^\mu = \frac{1}{4}\mathrm{tr}\,\partial_\mu g^{-1}\partial^\mu g , \qquad (26)$$

and our model reduces to the U(n) principal chiral field as discussed before.[2]

Another important case we are interested in is a new model which takes values in the symmetric space GL(n,c)/U(n). As mentioned in Eq. (3), the element of the group G=GL(n,c) takes the form

$$g = (g_L , g_L^+), \qquad g_L \in GL(n,c). \qquad (27)$$

The elements of the subgroup $\hbar \in H=U(n)$ and the coset $\Phi \in G/H$ can be represented by

$$\hbar = (h , h^+), \qquad h \in U(n) . \qquad (28)$$

and

$$\Phi = (\phi , \phi^+) , \qquad \phi^+ = \phi, \ \phi \in GL(n,c) . \qquad (29)$$

respectively. The involution operator which makes H invariant is

$$\sigma(g_L , g_L^+) = (g_L^{+-1}, g_L^{-1}) . \qquad (30)$$

Now define

$$\Omega = \Phi^{-1}\partial\Phi = (\phi^{-1}\partial\phi\,,\, -\phi^{+}\partial\phi^{+-1})$$
$$= \mathcal{H} + \mathcal{K} = (H+K\,,\, -H+K) \tag{31}$$

for any $\phi \in GL(n,c)$, and we have

$$H = \tfrac{1}{2}(\phi^{-1}\partial\phi + \phi^{+}\partial\phi^{+-1}),$$
$$K = \tfrac{1}{2}(\phi^{-1}\partial\phi - \phi^{+}\partial\phi^{+-1}). \tag{32}$$

Introducing the covariant derivatives

$$\mathcal{D}_\mu\Phi = \partial_\mu\Phi - \Phi\mathcal{H}_\mu = (\partial_\mu\phi - \phi H_\mu\,,\, \partial_\mu\phi^{+} + H_\mu\phi^{+})$$
$$\mathcal{D}_\mu\Phi^{-1} = \partial_\mu\Phi^{-1} + \mathcal{H}_\mu\Phi^{-1} = (\partial_\mu\phi^{-1} + H_\mu\phi^{-1}\,,\, \partial_\mu\phi^{+-1} - \phi^{+-1}H_\mu) \tag{33}$$

and the hermitian matrices

$$g = \phi\phi^{+}\ , \qquad Q = \Phi\Phi^{+} = (g, g), \tag{34}$$

the Lagrangian for this model can be written as

$$\mathcal{L} = -\frac{1}{2}\mathrm{Tr}\,\mathcal{K}_\mu\mathcal{K}^\mu = \frac{1}{2}\mathrm{Tr}(\mathcal{D}_\mu\Phi^{-1})(\mathcal{D}^\mu\Phi) = \frac{1}{8}\mathrm{Tr}(\partial_\mu Q^{-1})\partial^\mu Q$$
$$= -\mathrm{tr}\,K_\mu K^\mu = \frac{1}{4}\mathrm{tr}(\partial_\mu g^{-1})\partial^\mu g. \tag{35}$$

The remaining procedure can proceed according to the standard strategy as before. All the statements and the formulae among g, $\mathcal{K}$, Φ and Q (i.e., Eqs. (16) through (18)) remain to be true. But their component form has to be modified, since in present model, the second component of Φ is ϕ^{+} instead of ϕ and q is hermitian here. We list these formulae as follows:

$$J_L = 2\phi K\phi^{-1} = \phi(\phi^{-1}\partial\phi - \phi^{+}\partial\phi^{+-1})\phi^{-1}$$
$$= -(\phi\phi^{+})\partial(\phi\phi^{+})^{-1} = -g\partial g^{-1}, \tag{16'a}$$

$$J_R = 2\phi^{+-1}K\phi^{+} = \phi^{+-1}(\phi^{-1}\partial\phi - \phi^{+}\partial\phi^{+-1})\phi^{+}$$
$$= (\phi\phi^{+})^{-1}\partial(\phi\phi^{+}) = g^{-1}\partial g; \tag{16'b}$$

$$\partial_\nu J^L_\mu = 2\phi(D_\nu K_\mu + [K_\nu, K_\mu])\phi^{-1}, \tag{17'a}$$

$$\partial_\nu J^R_\mu = 2\phi^{+-1}(D_\nu K_\mu - [K_\nu, K_\mu])\phi^{+}; \tag{17'b}$$

$$\partial_\mu J^L_\nu - \partial_\nu J^L_\mu - [J^L_\mu, J^L_\nu]$$
$$= 2\phi(D_\mu K_\nu - D_\nu K_\mu)\phi^{-1} = 0, \tag{18'a}$$

and

$$\partial_\mu J^R_\nu - \partial_\nu J^R_\mu + [J^R_\mu, J^R_\nu]$$
$$= 2\phi^{+-1}(D_\mu K_\nu - D_\nu K_\mu)\phi^{+} = 0. \tag{18'b}$$

Here the covariant derivative of $\mathcal{K}$ is again defined as in Eq. (19)

and

$$D_\mu K_\nu = \partial_\mu K_\nu + [H_\mu, K_\nu]. \tag{19'}$$

The equation of motion, Eqs. (21), (22) and (23), also stand as before. Another equation which can be derived from the integrability of Ω is

$$\partial_\mu H_\nu - \partial_\nu H_\mu + [H_\mu, H_\nu] = [K_\nu, K_\mu]. \tag{36}$$

The left-hand side of this equation is nothing but the curvature (field strength) $F_{\mu\nu}$ taking values in the subalgebra u(n). From Eqs. (19') and (22), it can be shown that for two-dimensional space-time,

$$D^\mu F_{\mu\nu} \equiv \partial^\mu F_{\mu\nu} + [H^\mu, F_{\mu\nu}] = 0. \tag{37}$$

This means that $F_{\mu\nu}$ satisfies the sourceless Yang-Mills equation.

We can see from Eq. (32) that $H^+=-H$, $K^+=K$ and then Eq.(16') implies $J_L^+=J_R$ and

$$J_\mu^L = g J_\mu^R g^{-1}. \tag{38}$$

Taking account of the curvatureless property (18') we get

$$\begin{aligned} \partial_\nu J_\mu^L &= g(\partial_\nu J_\mu^R + [J_\nu^R, J_\mu^R])g^{-1} \\ &= g(\partial_\mu J_\nu^R)g^{-1}. \end{aligned} \tag{39}$$

These equations can be combined to be written as

$$\begin{aligned} g_\mu^+ &= Q^{-1} g_\mu Q, \\ \partial_\nu g_\mu^+ &= Q^{-1}(\partial_\mu g_\nu) Q. \end{aligned} \tag{40}$$

In this model, different from that in other models, the matrix q appearing in the Lagrangian is hermitian, rather than unitary. We may call it H_n model which has its great importance in connection with the four-dimentional Yang-Mills fields.[4]

We would like to point out here that it is not necessary to write down our model in the present form. As is well-known, an arbitrary nonsingular complex matrix g can always be written as the product of the form ϕh, where ϕ is hermitian and h is unitary. The decomposition is unique. Thus {h} forms the unitary subgroup U(n), and {ϕ} forms the symmetric coset space GL(n,c)/U(n). Separating the antihermitian and hermitian parts from $\phi^{-1}\partial\phi$ and identifying them as H and K respectively, we obtain Eq.(32) once again. From them all other quantities can be defined and relations among these quantities can be derived as well. But the formula tion described above has the advantage that pairs of quantities are grouped together and so the symmetry is more transparent. This property is very useful in practical applications.

To sum up, we can see from Eqs.(20) and (35) and the earlier discussion[1,2] that the Lagrangian

$$\mathcal{L}=\frac{1}{2}t_r(\partial_\mu g^{-1})\partial^\mu g . \tag{41}$$

in which q(x) belongs to some compact Lie group in matrix representation, describes a variety of nonlinear models. Different models correspond to q(x) varying over various subsets of the group G, GL(n,c) or U(n).

For the cases in which the matrix q(x) varies over the whole group G, we obtain the principal models. When q(x) is restricted to be hermitian, it yields our H_n model. A number of the so-called "second class" models[5], such as 0(N) models, CP_n models, HP_n models and Grassmann models, can be obtained when the matrices q(x) vary over some subset F of the group G in such a way that

$$q(x) = u^{-1\,*}(x)\; q_o u(x). \tag{42}$$

Here the star operation is an involution automorphism, q_o is a given fixed element of G which satisfies $q_o\, q_o^*$ =const. and u is an arbitrary element of G. For all these models the Lagrangian can be brought into the form

$$\mathcal{L} \sim -t_r K_\mu K^\mu . \tag{43}$$

and the duality symmetry stands for two-dimensional space-time. Then following the standard procedure both local and nonlocal conservation laws can be obtained.[1,2]

References

1. *K.C. Chou and X.C. Song, ASITP-80-008, PUTP-80-003.*
2. *K.C. Chou and X.C. Song, ASITP-80-010, PUTP-80-004.*
3. *H.Eichenherr and M. Forger, Nucl. Phys., B155 (1979) 381.*
4. *K.C. Chou and X.C. Song, ASITP-81-013, to be published.*
5. *E. Brézin, S. Hikami and J. Zinn-Justin, Nucl. Phys., B165 (1980) 528.*

Commun. in Theor. Phys. *Vol. 1, No. 2 (1982) 185-203*

THE H_n SIGMA-MODELS AND SELF-DUAL SU(n) YANG-MILLS FIELDS IN FOUR-DIMENSIONAL EUCLIDEAN SPACE

CHOU Kuang-chao(周光召)
(Institute of Theoretical Physics, Academia Sinica)
SONG Xing-chang(宋行长)
(Institute of Theoretical Physics, Peking University)

Received July 5, 1981.

Abstract

The H_n σ-models, i.e., non-linear σ-models taking values on the symmetric coset space SL(n,c)/SU(n), both in two-dimensional and four-dimensional Euclidean space, are formulated. The relations with self-dual SU(n) Yang-Mills fields are also discussed.

I. Introduction

In recent years there has been considerable progress in the investigation of the two-dimensional non-linear sigma models, or chiral fields,[1-3] both on classical level and on quantum level.

The main interest in these models comes from their remarkable similarity to four-dimensional non-Abelian gauge theories. Like the Yang-Mills theories, these non-linear models are self-coupled systems which are renormalizable[4] and asymptotically free[5]. They also possess some kind of instantaneous solutions,[3,6] and exhibit dynamical mass generation[7]. All these properties are believed to be important characteristics of four-dimensional gauge theories.

More recently, the parallelism between the gauge theories and the non-linear σ-models is much more strengthened by the discovery of the Backlund transformation[8,9] for the self-dual Yang-Mills fields, of a system of linear differential equations comprising the self-dual equations as their integrability condition[10], and especially of the realization that the classical Yang-Mills equations can be formulated in the loop space as chiral equations[11]. This parallelism could be put forward furthermore and may lead to new and useful computation approaches to Yang-Mills theories by analogy with the σ-models.

One of the distinguishing features for the two-dimensional σ-models is the existence of a new class of unconventional conservation laws, both local and nonlocal, and infinite in number. Generally the nonlocal conservation laws are constructed by using either the linear eigenvalue system[12], the zero-curvature conditions for the conserved currents[13], or the dual symmetry[14]. Less is known about the local ones which can also be derived either from the parametric Backlund transformations[15] or from the dual symmetry[16]. The physical meaning of such conservation laws has not been completely clarified. However it has been shown that local as well as nonlocal charges can properly be defined as operators in the quantized σ-models[17]. Acting upon the asymptotic states these conserved charges give an infinite number of strong constraints on the S matrix and suppress the particle production in some models[18]. This fact leads to the factorization of the S matrix into two-body amplitudes[19] which can be calculated exactly by N^{-1} expansion for O(N) σ-models[20].

For high-dimensional gauge theories it has been conjectured that analogous nonlocal conservation laws will provide some relation between long- and short-distance behavior[21] and cause strong constraints on the structure of S matrix as well. Pohlmeyer[8] and Prasad et al.[22] have obtained such nonlocal conservation laws for (anti) self-dual Yang-Mills fields in four-dimensions from "dual" symmetry and curvatureless condition[13] respectively. Prasad et al. have also constructed a parametric Backlund transformation for the self-dual Yang-Mills fields, but they could not deduce the local conservation laws from these transformations as was done for the principal chiral field[15]. Meanwhile Pohlmeyer has indicated that, apart from infinite numbers of nonlocal continuity equation, there exists another set of continuity equations which are the analogs of the local ones in two dimensions. He has obtained the generating functional for the second set. But in contrast to the two-dimensional case, the results obtained involve nonlocal expression of the field. So, up to now, it still leaves open the question whether there exist corresponding local currents for the self-dual Yang-Mills theories in four-dimensions or not.

In this paper we will study the H_n σ-models (which we have proposed sometime earlier[23]) further, and discuss the relation between these models and the four-dimensional self-dual Yang-Mills fields. Our discussion shows that at least for a restricted set of the solutions, the four-dimensional self-dual Yang-Mills fields do have an infinite number of local conservation laws.

The plan of this paper is as follows. In Sec. II, by a brief review of the form of the four-dimensional Euclidean space, we fix our notations and conventions. Sec. III is devoted to the H_n σ-models, i.e., the models defined on the symmetrical coset space SL(n,c)/SU(n). For both two-dimensional case and four-dimensional case, the equation of motion, the possible dual symmetry and the conservation laws deduced from this symmetry are given. We turn to discuss the self-dual Yang-Mills fields in Sec. IV and show that it bears a close resemblance to the self-dual H_n σ-models. A short discussion about the results is given in Sec. V.

II. Self-dual conditions and SU(2)×SU(2) formulation

In a four-dimensional Euclidean space, consider the transformation of coordinates from $x^\mu (\mu = 1,2,3,4)$ to $y^\alpha = (y, z, \bar{y}, \bar{z})$ defined by[24]

$$\sqrt{2}y = x_1 + ix_2 , \qquad \sqrt{2}\bar{y} = x_1 - ix_2 ,$$
$$\sqrt{2}z = x_3 - ix_4 , \qquad \sqrt{2}\bar{z} = x_3 + ix_4 . \tag{2.1}$$

The invariant length of x^μ is

$$x^\mu x^\mu = (y\bar{y} + z\bar{z} + \bar{y}y + \bar{z}z)\ .$$

So the metric in the new variable has only the following nonvanishing components

$$g_{y\bar{y}} = g_{\bar{y}y} = g_{z\bar{z}} = g_{\bar{z}z} = 1\ . \tag{2.2}$$

For any four-vector A^μ, the covariant components in the new coordinates are

$$\sqrt{2}A_y = \sqrt{2}A^{\bar{y}} = A_1 - iA_2\ ,\quad \sqrt{2}A_{\bar{y}} = \sqrt{2}A^{y} = A_1 + iA_2\ ,$$
$$\sqrt{2}A_z = \sqrt{2}A^{\bar{z}} = A_3 + iA_4\ ,\quad \sqrt{2}A_{\bar{z}} = \sqrt{2}A^{z} = A_3 - iA_4\ . \tag{2.3}$$

The same is true of the derivative vector

$$\sqrt{2}\,\partial_y = \sqrt{2}\frac{\partial}{\partial y} = \partial_1 - i\partial_2\ ,\quad \sqrt{2}\,\partial_{\bar{y}} = \sqrt{2}\frac{\partial}{\partial \bar{y}} = \partial_1 + i\partial_2\ ,$$

$$\sqrt{2}\,\partial_z = \sqrt{2}\frac{\partial}{\partial z} = \partial_3 + i\partial_4\ ,\quad \sqrt{2}\,\partial_{\bar{z}} = \sqrt{2}\frac{\partial}{\partial \bar{z}} = \partial_3 - i\partial_4\ .$$

Any anti-symmetrical tensor $F_{\mu\nu} = -F_{\nu\mu}$ can be decomposed into a self-dual part and an anti-self-dual part

$$F_{\mu\nu} = S_{\mu\nu} + T_{\mu\nu}$$

$$S_{\mu\nu} = \frac{1}{2}(F_{\mu\nu} + \tilde{F}_{\mu\nu})\ ,\qquad T_{\mu\nu} = \frac{1}{2}(F_{\mu\nu} - \tilde{F}_{\mu\nu}) \tag{2.4}$$

where $\tilde{F}_{\mu\nu}$ is the dual tensor of $F_{\mu\nu}$

$$\tilde{F}_{\mu\nu} = \frac{1}{2}\varepsilon_{\mu\nu\lambda\rho}F_{\lambda\rho}\ .\qquad (\varepsilon_{1234} = 1) \tag{2.4'}$$

In the new coordinates

$$\begin{aligned}
&S_{yz} = 0\ , && T_{yz} = F_{yz}\ ,\\
&S_{y\bar{y}} = \tfrac{1}{2}(F_{y\bar{y}} - F_{z\bar{z}})\ , && T_{y\bar{y}} = \tfrac{1}{2}(F_{y\bar{y}} + F_{z\bar{z}})\ ,\\
&S_{y\bar{z}} = F_{y\bar{z}}\ , && T_{y\bar{z}} = 0\ ,\\
&S_{z\bar{y}} = F_{z\bar{y}}\ , && T_{z\bar{y}} = 0\ ,\\
&S_{z\bar{z}} = \tfrac{1}{2}(F_{z\bar{z}} - F_{y\bar{y}})\ , && T_{z\bar{z}} = \tfrac{1}{2}(F_{y\bar{y}} + F_{z\bar{z}})\ ,\\
&S_{\bar{y}\bar{z}} = 0\ , && T_{\bar{y}\bar{z}} = F_{\bar{y}\bar{z}}\ .
\end{aligned} \tag{2.5}$$

So the self-dual condition $T_{\mu\nu} = 0\ (F_{\mu\nu} = \tilde{F}_{\mu\nu})$ [24,8] implies that

$$F_{yz} = F_{\bar{y}\bar{z}} = 0\ , \tag{2.6a}$$
$$F_{y\bar{y}} + F_{z\bar{z}} = 0\ . \tag{2.6b}$$

Now any four-vector A^μ can be arranged into a matrix form by introducing (Y. Brihaye et al., Ref.[8])

$$A = \frac{1}{\sqrt{2}}\sigma_\mu A^\mu \tag{2.7a}$$

Then

$$A^{\mu}=\frac{1}{\sqrt{2}}S_p\bar{\sigma}^{\mu}A\,. \tag{2.7b}$$

where $\sigma_\mu=(\vec{\sigma},-iI)$, $\bar{\sigma}^\mu=(\vec{\sigma},iI)=\sigma_\mu^+$, $\vec{\sigma}$ are Pauli matrices and I is a 2×2 unit matrix. For instance, the coordinates x^μ can be represented by the matrix form

$$X=\frac{1}{\sqrt{2}}\sigma_\mu x^\mu=\begin{pmatrix} z & \bar{y} \\ y & -\bar{z}\end{pmatrix}\equiv Y\,. \tag{2.7a'}$$

This notation exhibits clearly the local isomorphism between SO(4) and SU(2)×SU(2). An SO(4) rotation

$$x_\mu\to x'_\mu=R_{\mu\nu}x_\nu$$

corresponds to the transformation

$$Y\to Y'=MYN\,, \tag{2.8}$$

where M and N are two independent SU(2) matrices whose generators are

$$I_K^{(+)}=\frac{1}{2}(I_{ij}+I_{K4})\,;\qquad I_K^{(-)}=\frac{1}{2}(I_{ij}-I_{K4}) \tag{2.8'}$$

respectively $(i,j,k=1,2,3\ \text{cycl.})$, and $I_{\mu\nu}=-I_{\nu\mu}$ $(\mu,\nu=1,2,3,4)$ are the generators of SO(4). In other words, Eqs.(2.8) imply that an SO(4) vector x^μ transforms like a$(\frac{1}{2},\frac{1}{2})$ representation under SU(2)×SU(2).

From Eqs.(2.7), the scalar product of two vectors A_μ and B_μ can be expressed as

$$(A,B)\equiv A_\mu B_\mu=S_p\bar{A}B=S_pA\bar{B}\,, \tag{2.9}$$

with $\bar{A}=\frac{1}{\sqrt{2}}\bar{\sigma}^\mu A_\mu$ ($\bar{A}=A^+$ for A_μ real. + denotes hermite conjugation). Here Sp denotes the matrix trace in ordinary space (SU(2)⊗SU(2) space). Meanwhile $A\bar{B}$ and $\bar{A}B$ transform under SU(2)×SU(2) as

$$A\bar{B}\to M(A\bar{B})M^+\,;\qquad \bar{A}B\to N^+(\bar{A}B)N\,, \tag{2.10}$$

respectively. This means that the traceless matrices $A\bar{B}-\frac{1}{2}SpA\bar{B}$ and $\bar{A}B-\frac{1}{2}Sp\bar{A}B$ transform as (1,0) and (0,1) representations respectively.

As is well-known, under SU(2)×SU(2), an anti-symmetric tensor $F_{\mu\nu}$ transforms as (1,0)⊕(0,1) representation. Here (1,0) and (0,1) are just the self-dual and anti-self-dual parts respectively. So if $F_{\mu\nu}=A_\mu B_\nu-A_\nu B_\mu$, we have

$$S=A\bar{B}-\frac{1}{2}Sp\,A\bar{B}\qquad\qquad T=\bar{A}B-\frac{1}{2}Sp\,\bar{A}B$$

$$=\begin{pmatrix}\frac{1}{2}(F_{y\bar{y}}-F_{z\bar{z}}), & F_{\bar{z}y} \\ -F_{\bar{z}\bar{y}}\,, & -\frac{1}{2}(F_{y\bar{y}}-F_{z\bar{z}})\end{pmatrix};\ =\begin{pmatrix}\frac{1}{2}(F_{y\bar{y}}+F_{z\bar{z}}), & F_{zy} \\ F_{\bar{y}\bar{z}}\,, & -\frac{1}{2}(F_{y\bar{y}}+F_{z\bar{z}})\end{pmatrix}. \tag{2.11}$$

In general the field tensorccan always be written in matrix form as F = S⊕T with S and T defined by Eqs.(2.11) and transformed according to Eqs.(2.10).

Therefore, the self-dual condition (2.6) T = 0 is a vector equation covariant under $I^{(-)}$ spin rotation.

Now setting

$$U_K(\theta)=\exp\left\{i\tfrac{1}{2}\theta\sigma_K\right\}=\cos\theta/2+i\sigma_K\sin\theta/2\ . \tag{2.12}$$

we see from Eqs.(2.8) and (2.8') that $M=U_k(\theta)$ represents the synchro-rotation in planes (i,j) and (k,4) while $N=U_k(\theta)$ the contra-rotation in planes (i,j) and (k,4). For example, a rotation of angle θ in (1,2) plane is represented by the transformation with $M=N^{-1}=U_3(\theta)$ and a rotation in (2,4) plane by one with $M=N=U_2(\theta)$. Whereas the transformation

$$Y\rightarrow Y\,U_2(2\theta) \tag{2.13}$$

implies that

$$\begin{pmatrix} x_3 \\ x_1 \end{pmatrix}\rightarrow\begin{pmatrix}\cos\theta, & -\sin\theta \\ \sin\theta, & \cos\theta\end{pmatrix}\begin{pmatrix} x_3 \\ x_1\end{pmatrix};\quad \begin{pmatrix} x_2 \\ x_4 \end{pmatrix}\rightarrow\begin{pmatrix}\cos\theta, & \sin\theta \\ -\sin\theta, & \cos\theta\end{pmatrix}\begin{pmatrix} x_2 \\ x_4\end{pmatrix}. \tag{2.14}$$

III. H_n sigma-models

In a preceding paper[22], we have discussed the H_n σ-models, i.e., models defined on the symmetric coset space SL(n,c)/SU(n), in which the induced Maurer-Cartan 1-form is $d\Omega=\Omega_\mu dx^\mu$, and

$$\Omega_\mu=(\Omega^L_\mu\ ,\ \Omega^R_\mu)$$

$$\Omega^L_\mu\equiv K_\mu+H_\mu=\phi^{-1}\partial_\mu\phi\ ,\qquad \Omega^R_\mu\equiv K_\mu-H_\mu=(\partial_\mu\phi^+)\phi^{+-1}, \tag{3.1}$$

where $\phi\in SL(n,c)$. For real space, H_μ and K_μ are anti-hermitian and hermitian respectively, so H_μ belongs to the subalgebra SU(n) and K_μ to its complement algebra on the coset SL(n,c)/SU(n). Under the gauge transformation

$$\phi\rightarrow\phi h\ ,\qquad h=h(x)\in SU(n) \tag{3.2}$$

H_μ and K_μ transform as

$$H_\mu\rightarrow h^{-1}H_\mu h+h^{-1}\partial_\mu h\ , \tag{3.2a}$$

$$K_\mu\rightarrow h^{-1}K_\mu h\ . \tag{3.2b}$$

This means H_μ is a composite gauge potential and K_μ an ordinary vector field which transforms covariantly. We may notice that the combinations $H_\mu\pm K_\mu$ transform just as H_μ does. The covariant derivatives for the fields φ etc. are defined as

$$\Delta_\mu\phi\equiv\partial_\mu\phi-\phi H_\mu\ ,\qquad \Delta_\mu\phi^+\equiv\partial_\mu\phi^++H_\mu\phi^+$$

$$\Delta_\mu\phi^{-1}\equiv\partial_\mu\phi^{-1}+H_\mu\phi^{-1};\qquad \Delta_\mu\phi^{+-1}\equiv\partial_\mu\phi^{+-1}-\phi^{+-1}H_\mu \tag{3.3}$$

and then K_μ may be rewritten as

$$K_\mu = \phi^{-1}\Delta_\mu\phi = -(\Delta_\mu\phi^{-1})\phi = (\Delta_\mu\phi^{-1})\phi^{+-1} = -\phi^{+}\Delta_\mu\phi^{+-1}, \tag{3.4}$$

whereas the covariant derivative for K_μ is

$$\Delta_\nu K_\mu \equiv \partial_\nu K_\mu + [H_\nu, K_\mu]. \tag{3.3'}$$

The Maurer-Cartan integrability conditions

$$\partial_\mu \Omega^L_\nu - \partial_\nu \Omega^L_\mu + [\Omega^L_\mu, \Omega^L_\nu] = 0; \tag{3.5a}$$

and

$$\partial_\mu \Omega^R_\nu - \partial_\nu \Omega^R_\mu - [\Omega^R_\mu, \Omega^R_\nu] = 0. \tag{3.5b}$$

imply

$$\partial_\mu H_\nu - \partial_\nu H_\mu + [H_\mu, H_\nu] + [K_\mu, K_\nu] = 0; \tag{3.6a}$$

$$\Delta_\mu K_\nu - \Delta_\nu K_\mu = 0. \tag{3.6b}$$

The currents defined as ($g = \phi\phi^+$)

$$\mathcal{J}_\mu = (J^L_\mu, J^R_\mu);$$

$$J^L_\mu = 2\phi K_\mu \phi^{-1} = -g\partial_\mu g^{-1}, \tag{3.7a}$$

$$J^R_\mu = 2\phi^{+-1}K_\mu\phi^+ = g^{-1}\partial_\mu g, \tag{3.7b}$$

have the important property of being curvatureless

$$\partial_\mu J^L_\nu - \partial_\nu J^L_\mu - [J^L_\mu, J^L_\nu] = 2\phi(\Delta_\mu K_\nu - \Delta_\nu K_\mu)\phi^{-1} = 0; \tag{3.8a}$$

$$\partial_\mu J^R_\nu - \partial_\nu J^R_\mu + [J^R_\mu, J^R_\nu] = 2\phi^{+-1}(\Delta_\mu K_\nu - \Delta_\nu K_\mu)\phi^+ = 0. \tag{3.8b}$$

For two-dimensional models, the Lagrangian density can be written as (tr is the trace sign for the matrices in the internal space)

$$\mathcal{L} = -tr\, K_\mu K^\mu = tr\,(\Delta_\mu\phi^{-1})\Delta^\mu\phi = \frac{1}{4}tr(\partial_\mu g^{-1})\partial^\mu g. \tag{3.9}$$

Then the field equations read

$$\partial^\mu J^L_\mu = \partial^\mu J^R_\mu = 0. \tag{3.10a}$$

or equivalently

$$\Delta^\mu K_\mu = 0. \tag{3.10b}$$

Introducing the light cone coordinates for the Minkowski case ($\mu = 0,1$)

$$\sqrt{2}\,\xi = x^0 + x^1 , \qquad \sqrt{2}\,\eta = x^0 - x^1$$

or the complex coordinates for the Euclidean case ($\mu = 1,2$)

$$\sqrt{2}\,\xi = x^1 - i \quad x^2 = \sqrt{2}\,\bar{y} , \qquad \sqrt{2}\,\eta = x^1 + i \quad x^2 = \sqrt{2}\, y ,$$

Eq.(3.10) becomes

$$\Delta_\xi K_\eta + \Delta_\eta K_\xi = 0 . \tag{3.10b'}$$

Since Eq.(3.6b) takes the form

$$\Delta_\xi K_\eta - \Delta_\eta K_\xi = 0 , \tag{3.6b'}$$

it follows from Eqs.(3.10b') and (3.6b') that

$$\Delta_\xi K_\eta = \Delta_\eta K_\xi = 0 , \tag{3.11}$$

and the dual symmetry

$$\begin{aligned} K_\xi &\longrightarrow \gamma K_\xi , \qquad & K_\eta &\longrightarrow \gamma^{-1} K_\eta ; \\ H_\xi &\longrightarrow H_\xi , \qquad & H_\eta &\longrightarrow H_\eta , \end{aligned} \tag{3.12}$$

holds from which the conservation laws, both local and nonlocal, can be found according to the standard strategy[25]. The results obtained for the principal chiral fields can be applied to the present H_n models. (The only difference lies in the fact that K_μ is anti-hermitian in the former case but hermitian in the latter case.) The nonlocal conserved currents were given by Ogielski et al. in Ref.[15]. Besides, we obtained the explicit form[15,16] of the local ones which can be written as

$$\partial_\xi j_\eta^{(K)} + \partial_\eta j_\xi^{(K)} = 0 , \qquad K = 0, 1, 2, \cdots$$

with

$$j_\xi^{(0)} = \mathrm{tr}(K_\xi K_\xi)^{\frac{1}{2}} , \qquad j_\eta^{(0)} = 0 ;$$

$$j_\xi^{(1)} = 0, \qquad j_\eta^{(1)} = 0 ;$$

and

$$\begin{aligned} j_\xi^{(2)} &= -\tfrac{1}{4}(\lambda_i - \lambda_j)^{-1} (\Delta_\xi B_0)_{ij} (\Delta_\xi B_0)_{ji} , \\ j_\eta^{(2)} &= \lambda_i^{-1} (K_\eta)_{ij} (K_\xi)_{ji} , \end{aligned} \tag{3.13}$$

where

$$B_0 = (K_\xi K_\xi)^{-\frac{1}{2}} K_\xi = K_\xi (K_\xi K_\xi)^{-\frac{1}{2}} ,$$

and $\lambda_i > 0$ (i=1,2...n) are eigenvalues of the positive definite hermitian matrix $(K_\xi K_\xi)^{\frac{1}{2}} \equiv \Lambda$.

For four-dimensional Euclidean space (μ=1,2,3,4), H_μ is still the SU(n) gauge field. By using Eq.(3.6b), the field equation (3.10b) takes the form

$$\Delta_y K_{\bar{y}} + \Delta_z K_{\bar{z}} = 0 \quad . \tag{3.14}$$

This equation together with Eqs.(3.6) describes the whole system of four-dimensional H_n models. Different from the two-dimensional case in which the gauge field H_μ is sourceless[15], Eq.(3.6b) shows that H_μ is not sourceless in four-dimensional case.

Now we restrict ourselves to discussing the so-called self-dual H_n models, i.e., models in which the anti-self-dual part of the field strength vanishes. Then Eq. (3.6a) splits into the following form

$$\begin{aligned}
&F^H_{yz} = -[K_y, K_z] = 0, \qquad && F^H_{\bar{z}\bar{y}} = -[K_{\bar{z}}, K_{\bar{y}}], \\
&F^H_{\bar{y}\bar{z}} = -[K_{\bar{y}}, K_{\bar{z}}] = 0, && F^H_{\bar{y}z} = -[K_{\bar{y}}, K_z], \\
&F^H_{y\bar{y}} + F^H_{z\bar{z}} = -[K_y, K_{\bar{y}}] - [K_z, K_{\bar{z}}] = 0, && F^H_{y\bar{y}} - F^H_{z\bar{z}} = -[K_y, K_{\bar{y}}] + [K_z, K_{\bar{z}}].
\end{aligned} \tag{3.15}$$

Combined with Eqs,(3.6b) and (3.14), Eqs.(3.15) describe a system of four-dimensional self-dual H_n models. From Eqs.(3.14), (3.15) and (3.7), we can show that

$$\partial_y J^L_{\bar{y}} + \partial_z J^L_{\bar{z}} = 2\phi(\Delta_y K_{\bar{y}} + \Delta_z K_{\bar{z}} + [K_y, K_{\bar{y}}] + [K_z, K_{\bar{z}}])\phi^{-1} = 0, \tag{3.16a}$$

$$\partial_y J^R_{\bar{y}} + \partial_z J^R_{\bar{z}} = 2\phi^{+-1}(\Delta_y K_{\bar{y}} + \Delta_z K_{\bar{z}} - [K_y, K_{\bar{y}}] - [K_z, K_{\bar{z}}])\phi^{+} = 0. \tag{3.16b}$$

This means that $J^L_{\bar{u}}$ and $J^R_{\bar{u}}$ $(\bar{u} = \bar{y}, \bar{z})$ are "quasi-conserved" and (3.8) imply that they are curvatureless. So according to the argument in Ref.[22], we can construct an infinite number of nonlocal continuity equations by the inductive procedure proposed by Brezin et al.[13]

Now we turn to the construction of local currents. Define a current

$$J_{\bar{u}} = \mathrm{tr}\, K_{\bar{u}} B, \qquad (\bar{u} = \bar{y}, \bar{z}) \tag{3.17}$$

which is a scalar in the internal space, and then we have

$$\begin{aligned}
\partial_u J_{\bar{u}} &= \mathrm{tr}(\partial_u K_{\bar{u}})B + \mathrm{tr}\, K_{\bar{u}} \partial_u B \\
&= -\mathrm{tr}([H_u, K_{\bar{u}}]\, B) + \mathrm{tr}\, K_{\bar{u}} \partial_u B \\
&= \mathrm{tr}\, K_{\bar{u}} \Delta_u B .
\end{aligned}$$

So $J_{\bar{u}}$ is "quasi-conserved" if and only if $\Delta_u B$ is trace orthogonal to $K_{\bar{u}}$. This can be achieved by assuming that

$$\Delta_y B = a K_{\bar{z}} + bBK_{\bar{z}}B + C_1[B, K_{\bar{y}}] + C_2\{B, K_{\bar{z}}\}, \tag{3.18a}$$

$$\Delta_z B = -aK_{\bar{y}} - bBK_{\bar{y}}B + C_1[B, K_{\bar{z}}] - C_2\{B, K_{\bar{y}}\}. \tag{3.18b}$$

The compatibility condition for B gives

$$C_1 = 0 \tag{3.19}$$

but leaves a, b and c_2 arbitrary. For simplicity we may choose $c_2 = 0$, $b=-a=1$. Notice that Eqs.(3.6b), (3.14) and (3.15) are invariant under the transformation

$$X_{\bar u} \to \gamma X_{\bar u} \qquad X_u \to X_u \quad (u = y, z) \tag{3.20}$$

X_μ stands for each vector of ∂_μ, H_μ and K_μ. If we define the new current

$$J_u(\gamma) = tr\, K_u(\gamma)\, B(\gamma)$$

corresponding to Eq.(3.17) and let $B(\gamma)$ satisfy the same set of equations as B does (with the replacement(3.20)), $J_{\bar u}(\gamma)$ will also be conserved. Now expanding $B(\gamma)$ as the power series in γ

$$B(\gamma) = \sum_{k=0}^{\infty} \gamma^{-k}\, b_k\,, \tag{3.21}$$

substituting it into Eq.(3.18a) and collecting all terms of the same order in γ we obtain the following set of equations

$$K_{\bar z} = b_0 K_z b_0\,, \tag{3.22a}$$

$$\Delta_y b_k = \sum_{j=0}^{k+1} b_j K_{\bar z} b_{k+1-j}\,, \quad (k = 0, 1, 2 \cdots) \tag{3.22b}$$

from which an infinite number of local continuity equations can be derived.

$$\partial_u j_{\bar u}^{(k)} = 0,$$

$$j_{\bar u}^{(k)} = tr\,(b_k K_{\bar u}), \quad (k = 0, 1, 2, \ldots) \tag{3.23}$$

Eqs.(3.22b) can be rewitten as

$$b_0 K_{\bar z} b_k + b_k K_{\bar z} b_0 = \Delta_y b_{k-1} - \sum_{j=1}^{k-1} b_j K_{\bar z} b_{k-j} \equiv C_{k-1}, \quad (k = 1, 2, \cdots). \tag{3.22b'}$$

Eqs.(3.22a) and Eqs.(3.22b') have already been solved in Ref.[16] under the assumption that the positive definite hermitian matrix $K_{\bar z} K_{\bar z}$ is invertible so a hermitian nonsingular matrix Λ can be defined as the square root of $K_{\bar z} K_{\bar z}$ such that its eigenvalues $\lambda_i > 0$.

The solution for b_o is

$$b_0 = (K_{\bar z} K_z)^{-1/2} K_{\bar z} = K_{\bar z} (K_z K_{\bar z})^{-1/2}$$

$$b_0 K_{\bar z} = K_z b_0 = (K_z K_{\bar z})^{1/2} \equiv \Lambda\,, \quad b_0^2 = 1, \tag{3.24a}$$

and in the diagonal representation of Λ the recursive formulas for b_k (k=1,2...) are

$$(b_k)_{ij} = (\lambda_i + \lambda_j)^{-1} (C_k)_{ij}\,, \qquad (i, j = 1, 2 \cdots n). \tag{3.24b}$$

Then the first three local currents read

$$j_{\bar{y}}^{(0)} = t_r\, k_{\bar{y}} k_{\bar{z}} \Lambda^{-1}, \qquad j_{\bar{z}}^{(0)} = t_r \Lambda ;$$

$$j_{\bar{y}}^{(1)} = (2\lambda_i)^{-1} (k_{\bar{y}})_{ij} (\Delta_y b_0)_{ji}, \quad j_{\bar{z}}^{(1)} = 0 ;$$

$$j_{\bar{y}}^{(2)} = (2\lambda_i)^{-1} (k_{\bar{y}})_{ji} (\Delta_y b_1 - b_1 k_{\bar{z}} b_1)_{ij}, \tag{3.25}$$

$$j_{\bar{z}}^{(2)} = \frac{1}{2} (b_0)_{ji} (\Delta_y b_1 - b_1 k_{\bar{z}} b_1)_{ij} .$$

Now let us come back to the general case where the field strength $F^H_{\mu\nu}$ does not satisfy the self-dual conditions. The symmetry (3.20) still stands and the current $J_{\bar{u}}$ can also be defined as in (3.17). So do Eqs.(3.18). Instead of Eq.(3.19) the compatibility condition for B now gives

$$c_1 = 0 \ ; \qquad ab - c_2^2 = 0. \tag{3.26}$$

We may choose for simplicity

$$c_1 = 0 , \qquad a = b = c_2 = 1 ,$$

and proceed with the same strategy as before. Then instead of Eq.(3.22a) we obtain

$$0 = k_{\bar{z}} + b_0 k_{\bar{z}} b_0 + b_0 k_{\bar{z}} + k_{\bar{z}} b_0 ,$$

which gives

$$b_0 = -1 . \tag{3.27}$$

Now corresponding to k=0,1 case of Eq.(3.22b) we have

$$0 = b_0 k_{\bar{z}} b_1 + b_1 k_{\bar{z}} b_0 + \{ b_1 , k_{\bar{z}} \} ,$$

$$\Delta_y b_1 = b_1 k_{\bar{z}} b_1 + b_0 k_{\bar{z}} b_2 + b_2 k_{\bar{z}} b_0 + \{ b_2 , k_{\bar{z}} \} .$$

The first one is an identity while the second one gives

$$-\Delta_y b_1^{-1} = k_{\bar{z}} , \tag{3.28}$$

so the solution of b_1 involves nonlocal expression. And then the continuation equations will be nonlocal for $k \geq 1$. This result is similar to that given by Pohlmeyer.[8]

At the end of this section, we would like to show the relation between our H_n models and Pohlmeyer's model. As mentioned before, $q = \phi\phi^+$ is an nxn hermitian matrix with unit determinant. For n=2 we may set

$$q = q_0 + \vec{q} \cdot \vec{\tau} ,$$

where $\vec{\tau}$ are Pauli matrices in internal space. Hermitian means $(q_0, \vec{q})$ are real and unit determinant implies that

$$1 = \det q = q_0^2 - \vec{q}^2 = q\bar{q} = \bar{q}q ,$$

with

$$\bar{g} = g_0 - \vec{g}\cdot\vec{\tau} = g^{-1}.$$

The Lagrangian densidy (3.9) then becomes

$$\mathcal{L} \propto tr\,(\partial_\mu g^{-1})\,\partial^\mu g = \partial_\mu g_0\,\partial^\mu g_0 - \partial_\mu \vec{g}\cdot\partial^\mu \vec{g}$$

This means that our H_2 model turns out to be Pohlmeyer's SO(1,3)/SO(3) non-linear σ-models.[8] So H_n models can be considered as the generalization of $H_2 \equiv SO(1,3)/SO(3)$ σ-model.

IV. Self-dual SU(n) Yang-Mills fields

Let us consider the SU(n) Yang-Mills theories in Euclidean four-dimensional real space which satisfy the self-dual conditions.

The first two of the self-dual equations, Eqs.(2.6a), can immediately be integrated to give[24]

$$A_y = \phi^{-1}\phi_y, \qquad A_{\bar{y}} = \phi^{+}\phi_{\bar{y}}^{+-1}$$
$$A_z = \phi^{-1}\phi_z; \qquad A_{\bar{z}} = \phi^{+}\phi_{\bar{z}}^{+-1}, \tag{4.1}$$

where the reality of the gauge potential A_μ, i.e.,

$$A_\mu^{+} = -A_\mu \quad (A_{\bar{u}} = -A_u^{+}, \quad u = y, z) \tag{4.2}$$

has been taken into account, $\phi = \phi(y,z,\bar{y},\bar{z}) \in SL(n,c)$ and $\phi_y = \partial_y\phi$, etc. A gauge transformation is the replacement

$$\phi \longrightarrow \phi h, \qquad h = h(y,z,\bar{y},\bar{z}) \in SU(n), \tag{4.3}$$

under which the gauge potential A_μ and gauge field strength $F_{\mu\nu} = \partial_\mu A_\nu - \partial_\nu A_\mu + [A_\mu, A_\nu]$ transform as

$$A_\mu \longrightarrow h^{-1}(A_\mu + \partial_\mu)h,$$
$$F_{\mu\nu} \longrightarrow h^{-1}F_{\mu\nu}h.$$

It has been pointed out that, by introducing a positive hermitian matrix[8]

$$g \equiv \phi\phi^{+} \in SL(n,c),$$

which has the very important property of being invariant under the gauge transformation (2.3), the only non-vanishing field strength can be expressed in the form

$$F_{u\bar{v}} = -\phi^{+}(g^{-1}g_u)_{\bar{v}}\,\phi^{+-1}$$
$$= -\phi^{-1}(g_{\bar{v}}\,g^{-1})_u\,\phi, \tag{4.4}$$

and the third of the self-dual equations, Eq.(2.6b), now becomes

$$(g^{-1} g_y)_{\bar{y}} + (g^{-1} g_z)_{\bar{z}} = 0, \tag{4.5a}$$

or

$$(g_{\bar{y}} g^{-1})_y + (g_{\bar{z}} g^{-1})_z = 0. \tag{4.5b}$$

or equivalently

$$g_{\bar{y}y} + g_{\bar{z}z} = g_{\bar{y}} g^{-1} g_y + g_{\bar{z}} g^{-1} g_z . \tag{4.5c}$$

Since F_{zy}, $(F_{z\bar{y}} + F_{z\bar{z}})/\sqrt{2}$ and $F_{\bar{z}\bar{y}}$ are three independent components transforming like the $I^{(-)}$ spin vector $T = (T_{-1}, T_0, T_1)$, Eqs.(2.6) and then Eqs. (4.1) and (4.5) are invariant separately under the SU(2)×U(1) subgroup of SO(4), i.e., under the transformation (2.8) with

$$M = U_k , \quad N = 1 \quad (k = 1, 2, 3), \tag{4.6a}$$

and

$$M = 1 , \quad N = U_3 . \tag{4.6b}$$

Besides, as Eq.(4.5) is homogeneous, it is also invariant under the dilation of the coordinates (λ is a real parameter different from zero)

$$x_\mu \longrightarrow \lambda x_\mu . \tag{4.7}$$

Moreover, there exists a series of internal "Lorentz" transformations[8]

$$\phi \longrightarrow w\phi ,$$

$$g \longrightarrow w g w^+ , \qquad w = w(\bar{y}, \bar{z}) \in SL(n, C) \tag{4.8}$$

under which the gauge potential A_μ, and therefore $F_{\mu\nu}$, remain unchanged. Another transformation which leaves (4.1) and (4.5) invariant is the replacement[8]

$$g \longrightarrow g^{-1T} . \tag{4.9}$$

(T means transpose here.) For SU(2) case it can be considered as the special one of (4.8) by setting $w = w^+ = \tau_2$, since

$$\tau_2 \tau_\mu \tau_2 = \bar{\tau}_\mu^{\ T} , \qquad (\mu = 0, 1, 2, 3)$$

$$g \longrightarrow \tau_2 g \tau_2 = \bar{g}^T = (\det g)(g^{-1})^T ,$$

where $\tau_\mu = (I, \vec{\tau})$ and $\bar{\tau}_\mu = (I, -\vec{\tau})$.

The above description indicates an important fact that there exists a remarkable similarity between self-dual Yang-Mills theories and the H_n models discussed in the preceding section. The fields ϕ are just those discussed in H_n models and so is the gauge transformation (4.3). Only the definitions of the gauge potentials in these two theories are somewhat different. Instead of Eqs.(3.1) we have

$$A_u = \Omega^L_u = H_u + K_u , \qquad A_{\bar{u}} = -\Omega^R_{\bar{u}} = H_{\bar{u}} - K_{\bar{u}} , \quad (u = y, z) \tag{4.10}$$

from Eqs.(4.1). Similar to Eqs.(3.3), the covariant derivatives should be defined as

$$D_{\bar{u}}\phi = \partial_{\bar{u}}\phi - \phi A_{\bar{u}} \quad . \qquad D_u\phi^+ \equiv \partial_u\phi^+ + A_u\phi^+ \ .$$

$$D_{\bar{u}}\phi^{-1} = \partial_{\bar{u}}\phi^{-1} + A_{\bar{u}}\phi^{-1} \ ; \qquad D_u\phi^{+-1} \equiv \partial_u\phi^{+-1} - \phi^{+-1}A_u \ ; \tag{4.11}$$

whereas

$$D_{\bar{u}}\phi^{+-1} = \partial_{\bar{u}}\phi^{+-1} - \phi^{+-1}A_{\bar{u}} = 0 \ , \quad D_u\phi^{-1} \equiv \partial_u\phi^{-1} + A_u\phi^{-1} = 0 \ ,$$

$$D_{\bar{u}}\phi^+ \equiv \partial_{\bar{u}}\phi^+ + A_{\bar{u}}\phi^+ = 0 \ , \quad D_u\phi \equiv \partial_u\phi - \phi A_u = 0 \ . \tag{4.12}$$

as can easily be seen from Eqs.(4.1) immediately. Corresponding to Eq.(3.4) define

$$C_{\bar{u}} = \phi^{-1}(D_{\bar{u}}\phi) = -(D_{\bar{u}}\phi^{-1})\phi \qquad C_u = (D_u\phi^+)\phi^{+-1} = -\phi^+ D_u\phi^{+-1}$$

$$= \Omega^L_{\bar{u}} - A_{\bar{u}} = 2K_{\bar{u}} \ , \qquad = A_u + \Omega^R_u = 2K_u \ ; \tag{4.13}$$

and then Maurer-Cartan integrability conditions imply that

$$F_{yz} = D_y C_z - D_z C_y - [C_y, C_z] = 0, \tag{4.14a}$$

$$F_{\bar{y}\bar{z}} = -D_{\bar{y}} C_{\bar{z}} + D_{\bar{z}} C_{\bar{y}} - [C_{\bar{y}}, C_{\bar{z}}] = 0, \tag{4.14b}$$

$$F_{u\bar{v}} = D_{\bar{u}} C_v = D_v C_{\bar{u}} \ . \quad (u, v = y, z) \tag{4.14c}$$

The currents take the form

$$J^L_{\bar{u}} = \phi C_{\bar{u}}\phi^{-1} = g_{\bar{u}} g^{-1} \qquad J^L_u = \phi C_u \phi^{-1} = g_u g^{-1} \ ; \tag{4.15a}$$

$$= (D_{\bar{u}}\phi)\phi^{-1} = -\phi D_{\bar{u}}\phi^{-1} \ ,$$

$$J^R_u = \phi^{+-1} C_u \phi^+ = g^{-1} g_u \qquad J^R_{\bar{u}} = \phi^{+-1} C_{\bar{u}}\phi^+ = g^{-1} g_{\bar{u}} \ . \tag{4.15b}$$

$$= \phi^{+-1} D_u\phi^+ = -(D_u\phi^{+-1})\phi^+ ,$$

Combining all these results, we get

$$F_{yz} = \phi^{-1}(\partial_y J^L_z - \partial_z J^L_y - [J^L_y, J^L_z])\phi$$

$$= \phi^+(\partial_y J^R_z - \partial_z J^R_y + [J^R_y, J^R_z])\phi^{+-1}$$

$$= 2(\Delta_y K_z - \Delta_z K_y)$$

$$= D_y C_z - D_z C_y - [C_y, C_z] = 0 . \tag{4.16a}$$

$$F_{\bar{y}\bar{z}} = -\phi^{-1}(\partial_{\bar{y}} J^L_{\bar{z}} - \partial_{\bar{z}} J^L_{\bar{y}} - [J^L_{\bar{y}}, J^L_{\bar{z}}])\phi$$

$$= -\phi^+(\partial_{\bar{y}} J^R_{\bar{z}} - \partial_{\bar{z}} J^R_{\bar{y}} + [J^R_{\bar{y}}, J^R_{\bar{z}}])\phi^{+-1}$$

$$= -2(\Delta_{\bar{y}} K_{\bar{z}} - \Delta_{\bar{z}} K_{\bar{y}})$$

$$= -D_{\bar{y}} C_{\bar{z}} + D_{\bar{z}} C_{\bar{y}} - [C_{\bar{y}}, C_{\bar{z}}] = 0 , \tag{4.16b}$$

and

$$F_{\bar{u}v} = \phi^{-1}(\partial_v J^L_{\bar{u}})\phi \qquad = \phi^{+}(\partial_{\bar{u}} J^R_v)\phi^{+-1}$$
$$= 2(\Delta_v K_{\bar{u}} + [K_v, K_{\bar{u}}]) \qquad = 2(\Delta_{\bar{u}} K_v - [K_{\bar{u}}, K_v])$$
$$= D_v C_{\bar{u}} \qquad = D_{\bar{u}} C_v$$
$$= \phi^{-1} D_v D_{\bar{u}} \phi \qquad = (D_{\bar{u}} D_v \phi^{+})\phi^{+-1}, \tag{4.16c}$$

Therefore, besides the form given in Eqs.(4.5) which can be rewitten as

$$\partial_y J^L_{\bar{y}} + \partial_z J^L_{\bar{z}} = 0, \tag{4.5a'}$$

or

$$\partial_{\bar{y}} J^R_y + \partial_{\bar{z}} J^R_z = 0, \tag{4.5b'}$$

the third of the self-dual conditions can also be expressed as

$$\Delta_y K_{\bar{y}} + \Delta_z K_{\bar{z}} + [K_y, K_{\bar{y}}] + [K_z, K_{\bar{z}}] = 0, \tag{4.17a}$$

or

$$D_y C_{\bar{y}} + D_z C_{\bar{z}} = 0, \tag{4.17b}$$

or

$$(D_y D_{\bar{y}} + D_z D_{\bar{z}})\phi = 0. \tag{4.17c}$$

The last form Eq.(4.17c) can be considered as the shortened form of covariant Laplacian equation first proposed by Yang.[24] It is clear that in some two-dimensional planes (e.g. 1,3 plane), Eq.(4.17a) reduces to Eq.(3.10b'). So self-dual gauge theories reduce to two-dimensional H_n models.

Now we turn back to discuss the covariance property of the self-dual conditions under the spatial rotation SO(4). We have mentioned in the last section that $T = [F_{zy}, \frac{1}{\sqrt{2}}(F_{y\bar{y}} + F_{z\bar{y}}), F_{\bar{z}\bar{y}}]$ as a whole is a $I^{(-)}$ spin vector and the self-dual conditions $T = 0$ are covariant under SU(2)×SU(2) ≅ SO(4). As indicated in Eqs.(4.6), out of the six independent rotations of SO(4), four[≅SU(2)×U(1)] do leave the equations "invariant". So it is sufficient to consider one extra rotation, say+)

$$M = 1, \quad N = U_2(2\theta) = \begin{pmatrix} \cos\theta, & \sin\theta \\ -\sin\theta, & \cos\theta \end{pmatrix} \tag{4.18}$$

under which the coordinates transform as

$$y \to y' = y\cos\theta + \bar{z}\sin\theta, \qquad z \to z' = z\cos\theta - \bar{y}\sin\theta \tag{4.19}$$

and the gauge potential A and T, the anti-self-dual part of the field strength F, should transform as

$$A \to A' = AN, \qquad T \to T' = N^{+}TN. \tag{4.20}$$

i.e.,

+) *Since $U_1(2\theta) = U_3(\frac{\pi}{2})\, U_2(2\theta)\, U_3(-\frac{\pi}{2})$ and U_3 leaves the equation "invariant", the equation will be covariant under the rotation of U_1 if it is covariant under U_2.*

$$A_y \longrightarrow A_{y'} = A_y \cos\theta + A_{\bar{z}} \sin\theta\,,$$

$$A_z \longrightarrow A_{z'} = A_z \cos\theta - A_{\bar{y}} \sin\theta\,; \tag{4.20a}$$

and

$$F_{z'y'} = F_{zy}\cos^2\theta + F_{\bar{z}\bar{y}}\sin^2\theta + \frac{1}{2}(F_{y\bar{y}} + F_{z\bar{z}})\sin 2\theta\,,$$

$$F_{y'\bar{y}'} + F_{z'\bar{z}'} = (F_{y\bar{y}} + F_{z\bar{z}})\cos 2\theta + (F_{\bar{z}\bar{y}} - F_{zy})\sin 2\theta\,, \tag{4.20b}$$

$$F_{\bar{z}'\bar{y}'} = F_{\bar{z}\bar{y}}\cos^2\theta + F_{zy}\sin^2\theta - \frac{1}{2}(F_{y\bar{y}} + F_{z\bar{z}})\sin 2\theta\,.$$

In the rotated coordinates A' should behave just in the same way as A does in the original coordinates, so it can be expressed in the form

$$A_{u'} = g^{-1} g_{u'}\,, \qquad A_{\bar{u}'} = g^{-1} g_{\bar{u}'}^{+-1}\,, \qquad g \in SL(n,c) \tag{4.21}$$

Letting

$$g = W\phi\,, \qquad W = W(y', z', \bar{y}', \bar{z}') \in SL(n,c) \tag{4.22}$$

we have

$$g^{-1} g_{u'} = \phi^{-1}\phi_{u'} + \phi^{-1}(W^{-1}W_{u'})\phi\,.$$

from which we can easily see that A will transform as in Eq.(4.20a) if and only if[+)]

$$W^{-1}W_{y'} = -\sin\theta\, J_{\bar{z}}^{L}\,, \qquad W^{-1}W_{z'} = \sin\theta\, J_{\bar{y}}^{L}. \tag{4.23}$$

The converse statement is also true: The system (4.23) is compatible if and only if Eq.(4.5a') is satisfied. Therefore Eq.(4.23) is just the system of linearized equations for the self-dual gauge theories, which is equivalent to that given by Zakharov and Belavin,[10] and to the "duality-like" symmetry equations given by Pohlmeyer.[8]

Just follow the procedure applied to the two-dimensional non-linear σ-models in a previous paper;[14] and Eq.(4.23) can be integrated to give

$$W = W_0(\bar{y}', \bar{z}')\,\tilde{P}\exp\left\{-\sin\theta\int^{(y',z')}(J_{\bar{z}}^{L}\,dy' - J_{\bar{y}}^{L}\,dz')\right\}, \tag{4.24}$$

where $\tilde{P}$ denotes the anti-order operator along the path of integral. $W_0 \varepsilon\, SL(n,c)$ is the function of $\bar{y}'$ and $\bar{z}'$ only and can be taken to unit matrix by a proper transformation of the type (4.8) in the new coordinate system. Now Eq.(4.23) can be rewritten as

---

+) *From (4.23)*

$$\phi^{-1}(W^{-1}W_{y'})\phi = -\sin\theta\,\phi^{-1}J_{\bar{z}}^{L}\phi = -\sin\theta\, C_{\bar{z}}\,; \qquad \phi^{-1}(W^{-1}W_{z'})\phi = \sin\theta\,\phi^{-1}J_{\bar{y}}^{L}\phi = \sin\theta\, C_{\bar{y}}\,.$$

Then

$$g^{-1}g_{y'} = \cos\theta\,\phi^{-1}\phi_y + \sin\theta\,\phi^{-1}\phi_{\bar{z}} - \sin\theta\, C_{\bar{z}} = \cos\theta\, A_y + \sin\theta\, A_{\bar{z}}\,;$$

$$g^{-1}g_{z'} = \cos\theta\,\phi^{-1}\phi_z - \sin\theta\,\phi^{-1}\phi_{\bar{y}} + \sin\theta\, C_{\bar{y}} = \cos\theta\, A_z - \sin\theta\, A_{\bar{y}}\,.$$

$$W_y = -\,tg\,\theta\,(W_{\bar z} + W J^L_{\bar z}),$$
$$W_z = tg\,\theta\,(W_{\bar y} + W J^L_{\bar y}).$$

Then the existence of W implies $W_{yz} - W_{zy} = 0$, i.e.,

$$(W_{\bar y} + W J^L_{\bar y})_y + (W_{\bar z} - W J^L_{\bar z})_z = 0. \tag{4.25}$$

Expanding W in powers of parameter $\xi = tg\theta$, an infinite number of nonlocal continuity equations given by Pohlmeyer[8] follows. Using an inductive procedure Ogielski et al.[22] have also constructed a set of nonlocal conservation laws.

The following important fact can also be seen from Eq.(4.24). Unlike $W^{-1}W_u$, which are local functionals of field q as in Eq.(4.23), $W^{-1}W_{\bar u}$ may involve some kind of complicated integrals. This means that C_u and $C_{\bar u}$ will undergo nonlocal transformations under the rotation (4.18). So it is impossible to obtian the local conservation laws by applying to C_u and $C_{\bar u}$ the "duality-like" symmetry linked with rotation (4.18). However, in spite of this nonlocal transformation property, the existence of W in Eq.(4.24) preserves the locality of the covariant derivatives of C_u and $C_{\bar u}$, so that the Maurer-Cartan identities (4.14) and the self-dual condition (4.17b) are still satisfied by themselves. This can be checked by direct calculation.

From the above discussion we can see that great similarity exists between the four-dimensional H_n models and the classical Yang-Mills fields. Both theories deal with the same sort of field ϕ or q which undergo the same kind of gauge transformations and are subject to the same set of constraint equations— Maurer-Cartan identities, Eqs.(3.6) or (4.14). Their dynamical equations, the field equation (3.14) in the former and the self-dual equation (4.17) in the latter, are similar in form but different in content (since $A_\mu \neq H_\mu$). The main difference lies in the fact that ϕ in the former is an exact scalar while in the latter ϕ behaves as an ordinary scalar only for the transformations of the subgroup SU(2)×U(1). Rotations like the one considered in Eq.(4.18) induce local SL(n,c) transformations in internal space. Therefore the self-dual SU(n) gauge fields contain more solutions than that of the corresponding H_n models.

From the formal resemblance between Eqs.(3.14) and (4.17), we can see that the symmetry (3.20) holds and the scalar current

$$j^B_{\bar u} = tr\, C_{\bar u}\, B$$

can also be defined in gauge theories as in Eq.(3.17). This current will be "quasi-conserved" ($\partial_u j^B_{\bar u} = 0$) if and only if $D_u B$ is trace orthogonal to $C_{\bar u}$. Then similar to Eq.(3.18) we assume that $D_u B$ has the form

$$D_y B = a C_{\bar z} + b B C_{\bar z} B + C_1 [B, C_{\bar y}] + C_2 \{B, C_{\bar z}\};$$

$$D_z B = -a C_{\bar y} - b B C_{\bar y} B + C_1 [B, C_{\bar z}] - C_2 \{B, C_{\bar y}\},$$

and the constraints imposed by the compatibility condition are again

$$C_1 = 0\,, \qquad a\,b - C_2^2 = 0\,.$$

Making the choice $a = b = c_2 = 1$, expaning $B(\gamma)$ in powers of γ

$$B(\gamma) = \sum_{k=0}^{\infty} \gamma^k b_k\,,$$

inserting it into Eq.(4.27a) and collecting the terms of the same order in γ, we obtain a set of equations to determine b_k. Just like the situation in Eq.(3.28) the equations for b_k $(k \geq 1)$ yield nonlocal solutions from which follows another set of nonlocal continuity equations. This result is similar to that given by Pohlmeyer[8]. Similar equations can also be derived from the parametric Backlund transformation.+)

For those solutions of the self-dual SU(n) gauge theories which satisfy an additional condition

$$F^H_{y\bar{z}} = 0\,,$$

(and then $F^H_{\bar{y}z} = 0$, by taking hermitian conjugate), we see that the compatibility condition gives

$$C_1 = 0$$

as in Eq.(3.19), but leaves a,b and c_2 arbitrary. Then following the same procedure as given in Eqs.(3.20) through (3.24), we obtain an infinite number of local conservation laws as in Eqs.(3.25) (with the replacement $\Delta_u \to D_u$, & $K_u \to C_u$, etc.) The self-dual H_n sigma-models discussed in the preceding section are the special cases for this kind of solutions.

V. Summary

In this paper the H_n σ-models both in two-dimensional and four-dimensional Euclidean space are formulated, and the classical self-dual Yang-Mills fields are analysed in terms of the quantities defined in the H_n models. We have shown that a close resemblance exists between these two theories. In some two-dimensional planes the self-dual gauge fields reduce to the H_n models in two-dimensional space. And in whole four-dimensional Euclidean space, the self-dual gauge fields have a great similarity with the H_n models, partially in content and partially in form. The essential difference lies in the fact that the field ϕ in the H_n models is an exact scalar under the spacial rotations of SO(4) while, in the self-dual gauge theories, ϕ is a scalar only for the transforma-

---

+) *From Eqs.(14) of Ref.[9], the following continuity equation*

$$0 = \partial_y\, \mathrm{tr}\,(q^{-1}_{\bar{y}}\, q' + q'^{-1}\, q_{\bar{y}}) + \partial_z\, \mathrm{tr}\,(q^{-1}_{\bar{z}}\, q' + q'^{-1}\, q_{\bar{z}})$$

can be derived, where q is a solution of the field equation and q' another solution which dependes on a parameter β implicitly. Expanding q' in powers of γ ($q' = \sum_{k=0}^{\infty} q^{(k)} \gamma^{(k)}$, $\gamma^2 = (1-\beta)/(1+\beta)$), we obtain an infinite set of conservation laws. However the BT gives the solution of $q^{(k)}$ $(k \geq 1)$ being nonlocal functionals of q, $\partial_\mu q$, etc.

tions of the subgroup SU(2)×U(1) and the rotations other than SU(2)×U(1) give rise to non-linear local SL(n,c) transformations on ϕ in internal space.

By using the known structures of two-dimensional H_n models as a guide, an infinite number of nonlocal conservation laws in matrix form has been obtained for both four-dimensional H_n models and the self-dual Yang-Mills fields. The "duality" symmetry often used in two-dimensional non-linear σ-models gets a new feature of being coordinate dependent in four-dimensional self-dual gauge theories. Another set of symmetry can be defined in four-dimensional theories from which another set of infinite continuity equations in trace form, the analogy of the two-dimensional local conservation laws, can be constructed. The meaning of this set of continuity equations has yet to be clarified. This set of equations turn out to be local expression for a restricted set of solutions, e.g., the four-dimensional self-dual H_n models.

It might be worthwhile to study corresponding problems for the Yang-Mills theories in loop space form and to investigate the associated quantum theories.

References

1. K. Pohlmeyer, Comm. Math. Phys., 46(1976) 207.
2. M. Luscher and K. Pohlmeyer, Nucl. Phys., B137(1978) 46;
   V.E. Zaharov and A.V. Mikhailov, (Sov. Phys.) JETP 47(1978) 1017;
   H. Eichenherr, Nucl. Phys., B146(1978) 215; B155(1979) 544;
   H. Eichenherr and M. Forger, Nucl. Phys., B155(1979) 381;
   V.L. Golo and A.M. Perelomov, Phys. Lett., B79(1978) 112;
   M. Dubois-Violette and Y. Georgelin, Phys. Lett., B82(1979) 251.
3. A. d'Adda, M. Luscher and P. di Vecchia, Nucl. Phys., B146(1978) 63;
   M. Luscher, Phys. Lett., B78(1978) 465;
   E. Witten, Nucl. Phys., B149(1979) 285.
4. A.A. Migdal, (Sov. Phys.) JETP 42(1975) 743;
   E. Brezin, J. Zinn-Justin and J.C. Le Guillou, Phys. Rev., D14(1976) 3615; 4976;
   S. Hikami, Prog. Theor. Phys., 62(1979) 226;
   A. Mckane and M. Stone, Oxford Univ. preprint (1979).
5. A.M. Polyakov, Phys. Lett., B59(1975) 79;
   E. Brezin, S. Hikami and J. Zinn-Justin, Nucl. Phys., B165(1980) 528.
6. A.A. Belavin and A.M. Polyakov, (Sov. Phys.) JETP Lett., 22(1975) 245.
7. D. Gross and A. Neveu, Phys. Rev., D10(1974) 3235;
   E. Brezin, and J. Zinn-Justin, Phys. Rev., Lett., 36(1976) 691;
   W. Bardeen, B. Lee and R. Shrock, Phys. Rev., D14(1976) 985.
8. E. Corrigan, D.B. Fairlie, R.G. Yates and P. Goddard, Comm. Math. Phys., 58(1978) 223;
   Y. Brihaye, D.B. Fairlie, J. Nuyts and R.G. Yates, Journ. Math. Phys., 19(1978) 2528;
   M.K. Prasad, A. Sinha and L.L. Chau Wang, Phys. Rev. Lett., 43(1979) 750;
   K. Pohlmeyer, Comm. Math. Phys., 72(1980) 317.
9. For a review see L.L. Chau Wang, in Proc. of the 1980 Guangzhou Conf. on Theoretical Particle Physics, 1980, p1082.
10. A.A. Belavin and V.E. Zahkarov, Phys. Lett., B73(1978) 53.

11. A.M. Polyakov, Phys. Lett., B82(1979) 249;
S.Y. Wu, Physica Energiae Fortis et Physica Nuclearis, 3(1979) 382;
I. Ya. Aref'eva Lett. Math. Phys., 3(1979) 241;
L.L. Chau Wang, talk at the XXth Int. Conf. on High Energy Physics, Wisconsin, July, 1980.
12. M. Luscher and Pohlmeyer, Ref.(2);
A.T. Ogielski, Phys. Rev., D21(1980) 406.
13. E.Brezin, C. Itzykson, J. Zinn-Justin and J.B. Zuber, Phys. Lett., B82(1979) 442;
H.J. de Vega, Phys. Lett., B87(1979) 233.
14. H. Eichenherr and M. Forger, Ref.(2);
K.C. Chou and X.C. Song, ASITP preprint 80-008, to be published in Scientia Sinica.
15. K. Pohlmeyer, Ref.(1);
A.T. Ogielski, M.K. prasad, A. Sinha and L.L. Chau Wang, Phys. Lett., B91(1980) 387;
K.C. Chou and X.C. Song, ASITP.preprint 81-009.
16. H. Eichenherr, Phys. Lett., B90(1980) 121;
K. Scheler, Phys. Lett., B93(1980) 331;
F. Gursey and X.C. Tze, Ann. of Phys., 128(1980) 29;
K.C. Chou and X.C. Song, ASITP preprint 80-010, to be published in Scientia Sinica.
17. M. Luscher, Nucl. Phys., B135(1978) 1.
18. A.M. Polyakov, Trieste preprint IC/77/122 (1977).
19. D. Iagolnitzer, Phys. Rev., D18(1978) 1275.
20. A.B. Zamolodchikov and Al. B. Zamolodchikov, Nucl. Phys., B133(1978) 525.
21. A.M. Polyakov, Ref.(11).
22. M.K. Prasad, A. Sinha and L.L. Chau Wang, Phys. Lett., B87(1979) 237.
23. K.C. Chou and X.C. Song, ASITP preprint 81-003, Commun. in Theor. Phys. 1 (1982) 69.
24. C.N. Yang, Phys. Rev. Lett., 38(1977) 1377.
25. K.C. Chou and X.C. Song, Ref.(14) and (16).

Commun. in Theor. Phys. (Beijing, China) Vol. 1, No. 6 (1982) 725-730

ON THE DYNAMICAL PROBLEMS AND THE FERMION SPECTRA IN THE RISHON MODEL

CHOU Kuang-chao (周光召)

CERN and Institute of Theoretical Physics, Academia Sinica Beijing, China

DAI Yuan-ben (戴元本)

Institute of Theoretical Physics, Academia Sinica

Received May 6, 1982

Abstract

We propose to introduce a strong coupling $U(1)_Y$ axial gauge boson into the rishon model which may explain dynamical problems concerning preservation of chiral symmetries and existence of solutions to anomaly conditions. Mechanisms through which exotic particles gain masses are discussed. It is found that exotic particles such as color octet leptons and color sextet quarks may gain masses ranging from 10^1 to 10^3. Some phenomenological aspects of these exotic particles are discussed.

Among various composite models proposed for leptons and quarks the rishon model [1] has several special interesting features including economy and fitting in well investigated gauge theories. However, there are embarrassing problems of dynamical nature for this model among which are the following:

1. Within leptons and quarks the color $SU(3)_C$ interaction is much weaker than the hyper-color $SU(3)_{HC}$ interaction. In the approximation $\alpha_C=0$ the theory is similar to QCD with 6 flavours. However, it is necessary to assume that some chiral symmetries are not broken by the $SU(3)_{HC}$ interaction in contrast to QCD where chiral symmetries are assumed broken spontaneousely.

2. There is no flavour number n independent solution to 't-Hooft anomaly conditions [2] for the $\alpha_C=0$ limit $SU(3)_{HC}$ gauge theory with $SU(n)_L \times SU(n)_R \times U(1)_V$ flavour symmetry. On the other hand, it seems unlikely that the weak $SU(3)_C$ force can play an essential role in binding light fermions.

3. It is assumed that there are quarks in the configuration $u_L=(t_L t_L)V_L$ but none in the configuration $(t_R t_R)V_L$ though space parts of these wave functions can have the same symmetry. Note also that the fundamental Lagrangian of the $SU(3)_{HC} \times SU(3)_C$ gauge theory has the symmetry $t_L \to t_R$ with V_L, V_R fixed.

4. There are exotic particles (ttt) in 8 and 10 and (ttv) in 6* and 15 representations of $SU(3)_C$ which are assumed to be heavy. How can it be achieved and how large are the masses of these particles?

5. In order to avoid contradiction with the observed long life time of the proton the scale Λ of the $SU(3)_{HC}$ interaction must be assumed to be

larger than 10^6 GeV at least[3]. What can be the origin of the scale of the weak interaction $G_F^{-\frac{1}{2}} \approx 300$GeV if $\Lambda >> G_F^{-\frac{1}{2}}$? In this note we shall suggest some ideas which may provide partial solutions to these problems.

As a possible soluton of the problems 1 and 2 we propose to introduce an additional $U(1)_Y$ gauge field interacting with the current

$$Y_\mu = i\bar{t}\gamma_\mu\gamma_5 t - i\bar{V}\gamma_\mu\gamma_5 V$$

into the theory. The local symmetry of the theory is assumed to be

$$SU(3)_{HC} \times SU(3)_C \times U(1)_{B-L} \times U(1)_Y \ .$$

We have used $U(1)_{B-L}$ to replace $U(1)_Q$ in the original rishon model. The photon is assumed to emerge in the later stage when the B-L gauge boson is mixed with composite weak interaction gauge bosons in the low energy effective theory. Since contributions to $U(1)_Y$ anomaly of t and V cancel each other the theory is renormalizable. The global symmetry of the theory is $U(1)_R \times Z_{12}$, where Z_{12} is the unbroken discrete subgroup of $U(1)_X$ with

$$X_\mu = i\bar{t}\gamma_\mu\gamma_5 t + i\bar{V}\gamma_\mu\gamma_5 V.$$

We assume that the $U(1)_Y$ gauge coupling constant g_Y is comparable to $SU(3)_{HC}$ coupling constant g_{HC}. Since the $U(1)_Y$ interaction between the pair $(\bar{t}_R t_L)$ or $(\bar{V}_R V_L)$ is repulsive, it may prevent the scalar composites condensating and thereby keep some of the chiral symmetries unbroken.

In order to avoid contradicting the low energy phenomenology the $U(1)_Y$ symmetry must be broken at a large scale Λ'. This can be realized by the condensate $(V_L V_L V_R)^2$ corresponding to a Majorana mass term of ν_R which is required also from phenomenology of the rishon model[4]. We assume that the scale Λ' lies within the range $\Lambda > \Lambda' >> G_F^{-\frac{1}{2}}$. To see that the $U(1)_Y$ interaction prefers this condensation to $(\bar{t}t)$ or $(\bar{V}V)$ note that the average $U(1)_Y$ interaction energy per rishon in a composite of n rishons is

$$g_Y^2 \frac{1}{2n}\left(y^2 - \sum_{i=1}^{n} y_i^2\right)\Lambda' \ , \qquad (1)$$

where y and y_i are Y charges of the composite and constituents respectively. This interaction energy is negative for the composites $(V_L V_L V_R)^2$ while positive for $(\bar{t}t)$ and $(\bar{V}V)$. The corresponding quantity for the $SU(3)_{HC}$ interaction is

$$g_{HC}^2 \frac{1}{2n}\left(C - \sum_{i=1}^{n} C_i\right)\Lambda' \ , \qquad (2)$$

where C and C_i are the Casimir operators of $SU(3)_{HC}$. From Eqs.(1), (2) and other considerations it seems plausible to assume that the condensate $(V_L V_L V_R)^2$ is favored over other possible Y breaking condensates. This condensation also breaks $U(1)_{B-L}$, $U(1)_R$ and Z_{12}. Being ν_R Majorana mass term it transforms as $(\underline{6}, \underline{6}^*)$ under $SU(3)_{CL} \times SU(3)_{CR}$. However, it is easy to see that effective interactions produced by this condensation in the rishon number conserving sector preserve Z_{12} and transform as representations of zero triality with respect to both $SU(3)^V_{CL}$ and $SU(3)^V_{CR}$ such as $(\underline{1},\underline{1}),(\underline{8},\underline{8}),(\underline{10},\underline{10}^*)+(\underline{10}^*,\underline{10})\ldots$, where $SU(3)^V_C$ is the $SU(3)_C$ group restricted to V rishons. Therefore rishons are still kept massless at the scale Λ'. Among the H.C. singlet composites of three rishons there are only some vvv states that gain Dirac masses from the $SU(3)_{CL}\times SU(3)_{CR}$ breaking effects of this condensation (see below).

Therefore, in the first stage of the discussion on the spectra of three rishon bound states in which $U(1)_{B-L}$ and $SU(3)_C$ interactions and the condensation at the scale Λ' are neglected the global hyperflavour symmetry of the theory can be taken as †

$$SU(3)_{CL}\times SU(3)_{CR}\times U(1)_{B-L}\times U(1)_R\times Z_n\ . \tag{3}$$

If we want to require the flavour number independence of the number of the multiplets of massless composites we are led to the anomaly conditions for the theory

$$SU(n)_L\times SU(n)_R\times U(1)_R\times U(1)_{B-L} \tag{4}$$

with fermions in the representations

$$t_L=(\underline{n}\,,\underline{1}\,,\tfrac{1}{3},\tfrac{1}{3})\,,\qquad t_R=(\underline{1}\,,\underline{n}\,,\tfrac{1}{3},\tfrac{1}{3})\,,$$

$$V_L=(\underline{n}^*,\underline{1}\,,\tfrac{1}{3},-\tfrac{1}{3})\,,\qquad V_R=(\underline{1}\,,\underline{n}^*,\tfrac{1}{3},-\tfrac{1}{3})\,.$$

This problem was first discussed by authors of [5]. But they did not give a convincing argument for using the symmetry (3). Possible representations for H.C. singlet composites consist of the following:

ttt , $B-L=1$, ttv , $B-L=\frac{1}{3}$.

$$q_{1+}=(\square\square\square\,,\underline{1})\,,\qquad q_4=(\square\,,\underline{1})\,,$$

$$q_{1-}=\left(\begin{smallmatrix}\square\\\square\\\square\end{smallmatrix}\,,\underline{1}\right),\qquad q_{5+}=(\square^*,\square\square)\,,$$

$$q_{5-}=\left(\square^*,\begin{smallmatrix}\square\\\square\end{smallmatrix}\right),$$

$$q_{2+}=(\square\,,\square\square)\,,$$

$$q_{2-}=\left(\square\,,\begin{smallmatrix}\square\\\square\end{smallmatrix}\right),\qquad q_6=\left(n-1\left\{\begin{smallmatrix}\square\square\\\square\\\vdots\\\square\end{smallmatrix}\right.\,,\square\right),$$

$$q_3=\left(\begin{smallmatrix}\square\square\\\square\end{smallmatrix}\,,\underline{1}\right),\qquad q_{7+}=\left(n-1\left\{\begin{smallmatrix}\square\square\square\\\square\\\vdots\\\square\end{smallmatrix}\right.\,,\underline{1}\right),$$

$$q_{7-}=\left(n-1\left\{\begin{smallmatrix}\square\square\\\square\square\\\vdots\\\square\end{smallmatrix}\right.\,,\underline{1}\right).$$

Other composites correspond to representations q_i' which are parity conjugate to q_i and complex conjugate representations q_i^* and $q_i'^*$ obtained from q_i and q_i' by the replacement $t\leftrightarrow v$ (with opposite sign of B-L). Let us denote the indices corresponding to q_i by ℓ_i. From left-right symmetry(accompanied by the change of sign of g_Y which is irrelevant for bound states) of the the-

† *Here the SU(3) part is the unbroken subgroup of the* $SU(3)^t_L\times SU(3)^t_R\times SU(3)^V_L\times SU(3)^V_R$ *symmetry group of the Lagrangian.*

ory, we have $\ell_i = -\ell'_i$. The anomaly conditions for 3 $SU(n)_L$ and $SU(n)_L \times SU(n)_L \times U(1)_{B-L}$ triangle diagrams are readily satisfied with $\ell_i = \ell_i^*$. The anomaly condition for $SU(n)_L \times SU(n)_L \times U(1)_R$ triangle diagrams is reduced to

$$\begin{aligned} &\ell_4 + (n^2-3)\ell_3 + (n^2+1)\ell_6 + \sum_{\pm}\tfrac{1}{2}(n\pm 2)(n\pm 3)\ell_{1\pm} \\ &-\sum_{\pm}\tfrac{1}{2}n(n\pm 3)(\ell_{2\pm}+\ell_{5\pm}) + \sum_{\pm}\tfrac{1}{2}(3n\pm 1)(n\pm 2)\ell_{7\pm} = 1 . \end{aligned} \tag{5}$$

In contrast to the situation in the original rishon model, there are a lot of n independent solutions for the anomaly conditions[5].

If we insist that the space part of the internal wave functions of light fermions should be symmetrical we should have

$$\ell_{1\pm} = 0 .$$

For the illustration of mechanisms through which the exotic particles may get masses let us consider the simple n independent solution

$$\begin{aligned} &\ell_{1\pm} = 0 , \quad \ell_{2\pm} = \ell_4 = 3 , \quad \ell_3 = \ell_{5\pm} = \ell_6 = 0 , \\ &\ell_{7\pm} = 1 . \end{aligned} \tag{6}$$

This solution has the interesting property that only composites with negative $U(1)_Y$ interaction energy exist. Besides three generations of leptons and quarks with

$$\begin{aligned} &e_L^+ = (t_R t_R) t_L , \quad &\nu_L = (V_R V_R) V_L , \\ &u_L = (t_L t_L) V_L , \quad &\bar{d}_L = (V_L V_L) t_L , \end{aligned} \tag{7}$$

the solution (6) contains particles (ttt), (vvv) in <u>8</u>, <u>10</u>, <u>10</u>* and (ttt), (vvv) in <u>6</u>, <u>6</u>*, <u>15</u>, <u>15</u>*.

These particles gain masses from QCD chiral symmetry breaking. From the investigations in strong coupling approximation of lattice gauge theories[6] and coherent state variation calculations[7] it seems that the chiral breaking condensation in QCD is a direct result of the strong interaction between the fermion pair which has nothing to do with the confinement[10]. If this is correct the scale of chiral symmetry breaking is not Λ_C but M determined by

$$\frac{\alpha_c(M)}{\pi} C \simeq 1 , \tag{8}$$

where C is the value of $SU(3)_C$ Casimir operator. From Eq.(6), components of q_{7+} in <u>15</u> and components of q_{2+} in <u>10</u> of $SU(3)_C$ representations may gain masses of the order of 10GeV if they are proportional to $\langle\bar{\psi}\psi\rangle_0 \sim M^3$.

In this case the only possible way in which the remaining exotic particles <u>6</u>* of q_{7-} and <u>8</u> of $q_{2\pm}$ can gain needed masses is the electroweak interaction.

As an illustration of different mechanisms to make <u>8</u> and <u>6</u> massive let us relax the requirement of n independence and consider the following solution to the anomaly condition for n=3

$$\ell_{1\pm}=0\,,\qquad \ell_{2-}=\ell_4=3\,,\qquad \ell_{2+}=-3\,,$$
$$\ell_3=0\,,\qquad \ell_{5+}=-1\,,\qquad \ell_{5-}=1\,,\qquad\qquad (9)$$
$$\ell_6=0\,,\qquad \ell_{7+}=-2\,,\qquad \ell_{7-}=0\,.$$

The interaction of two rishons through the exchange of one color gluon contains the $SU(3)_{CL}\times SU(3)_{CR}$ breaking term

$$\bar{\Psi}_L\lambda^a\gamma_\mu\Psi_L\,\bar{\Psi}_R\lambda^a\gamma_\mu\Psi_R \qquad (10)$$

which transforms as (8,8) in $SU(3)_{CL}\times SU(3)_{CR}$ symmetry. This term causes mixing of components of different representations of $SU(3)_{CL}\times SU(3)_{CR}$ which belong to the same representation of $SU(3)_C$ and have the same x charge[8]. As a result of this mixing, among particles in the solution (9) three left-handed 8 in q_{2-} (q^*_{2-}) are mixed with three right-handed 8 in q_{2+} (q^*_{2+}) and large masses of $O(\frac{\alpha_C}{\pi}\Lambda)$ are gained. In addition, 8 in q^*_{2+} and q^*_{2-} can be mixed by the chiral breaking effect of the condensation $(V_L V_L V_R)^2$ mentioned previously and gain masses of $O(\Lambda')$. These massive particles form parity doublets. Therefore they give no contribution to mass corrections and magnetic moments of ordinary leptons. Similarly, the 3 and 3* components of $q_{5\pm}$ and $q^*_{5\pm}$ get masses of $O(\frac{\alpha_C}{\pi}\Lambda)$.

The 6* components in q_{5-} can gain masses from the color gluon correction as shown in the following diagram. From gauge invariance the gluon vertex in this diagram corresponds in the lowest dimension operator product to the effective Lagrangian

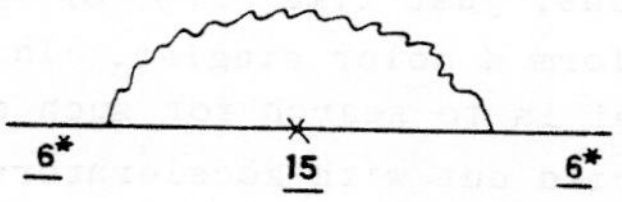

$$\frac{f}{\Lambda}g_c\,\bar{\Psi}_{15}\sigma_{\mu\nu}\Psi_6 F_{\mu\nu}\,, \qquad (11)$$

where $F_{\mu\nu}$ is the color field and f is a constant. Since the product of left-handed q_{5-} and the complex conjugate of right-handed q_{5+} belonging to (3,6*) × (3,* 3*) contains (1,8), f does not contain the suppressing factor due to $SU(3)_{CL}\times SU(3)_{CR}$ symmetry. Since q_{5+} and q_{5-} have the same x charge, (11) is not suppressed by Z_{12} either. Therefore f is not much smaller than 1. The self-energy integral formally diverges quadratically. The form-factor of composite particles provides a cut-off at the scale Λ. If the effective range of integration is Λ we should have

$$M_6\simeq C(\underline{6})\,\frac{\alpha_c}{\pi}f^2 M_{15}. \qquad (12)$$

On the other hand, the mass insertion on the internal line of the diagram is expected to decrease as the momentum becomes larger than the scale M_{15}. The estimation (12) can be invalidated if the mass insertion is switched off too fast so that the effective range of integration is not Λ by M_{15}. Anyhow, it seems possible that exotic particles in 6 gain masses of the order of M_{15}.

Note that the above discussions on masses of exotic particles do not necessarily depend on the existence of the $U(1)_Y$ gauge boson. It follows from these discussions that, in general, there are exotic particles not much heavier than 10 GeV.

Possible physical effects of colored leptons in low energy phenomenology was discussed in [9]. From SU(3) gauge invariance the effective interaction between colored leptons L and ordinary leptons ℓ in the lowest dimension is of the form

$$g_c \frac{f'}{\Lambda} \bar{L}^a \sigma_{\mu\nu} \ell F^a_{\mu\nu} , \tag{13}$$

where f' contains additional suppressing factors due to chiral invariances. Another low dimension effective interaction is of the form

$$\frac{1}{\Lambda^2} \bar{L} \Gamma L \bar{\ell} \Gamma \ell . \tag{14}$$

In low energy processes these couplings are very weak. It turns out that the existence of colored leptons with $\Lambda > 10^6$GeV and $M > 40$GeV does not contradict known experimental facts. The colored leptons in $\underline{8}$ and quarks in $\underline{6}$ discussed above can be produced in e^+-e^- or $\bar{P}$-P collisions. They decay into ordinary leptons and quarks by emitting a gluon with a narrow decay width of the order of

$$\Gamma \simeq \alpha_c (M^2) f'^2 \frac{M^3}{\Lambda^2} . \tag{15}$$

For $\Lambda > 10^6$GeV, Γ is extremely small. Before decaying to ordinary leptons and quarks they should form exotic mesons or baryons. Low lying exotic baryons should have long life-time also. Low lying exotic mesons decay mainly into gluons, just like J/ψ or Υ. The colored lepton can also combine with a gluon to form a color singlet. In our opinion, the most feasible test of the rishon model is to search for such exotic particles in experiments which can be carried out with accelerators of the next generation.

The above discussions have not provided an explanation of the problem of the scale of the weak interaction. However, in view of the complicated spectra of the theory, it seems possible that a scale of $G_F^{-\frac{1}{2}} \cong 300$GeV can emerge from the theory.

References

1. *H. Harari, Phys. Lett. 86B (1979) 83. M.A. Shupe, Phys. Lett. 86B (1979) 87.*
   H. Harari and N. Seiberg, Phys. Lett. 98B (1981) 269, the Rishon Model, WIS-81/38.
2. *G.'t-Hooft, Cargese Lectures (1979).*
3. *H. Harari, R.N. Mohapatra and N. Seiberg. Ref. TH-3123-CERN.*
4. *H. Harari and N. Seiberg, Phys. Lett. 100B (1981) 41.*
5. *S. King, Phys. Lett. 107B (1981) 201.*
6. *J-M. Blairon, R. Brout, F. Englert and J. Greensite, Nucl. Phys. B180 (1981) 439.*
   J. Greensite and J. Primack, Nucl. Phys. B180 (1981) 170.
   H. Kluberg-Stern, A. Morel, O. Napoly and B. Peterson, Nucl. Phys. B190 (1981) 504.
7. *J.R. Finger and J.E. Mandula, Quark Pair Condensation and Chiral Symmetry Breaking in QCD.*
8. *S. Weinberg, Color and Electroweak Forces as a Source of Quark and Lepton Masses. Dept. of Phys., University of Texas preprint. 1981.*
9. *C.B. Chiu and Y.B. Dai, Phys. Lett. 108B (1982) 341.*
10. *J. Kogut, M. Stone, H.W. Wyld, J. Shigemitsu, S.H. Shenker and D.K. Sinclair. The Scale of Chiral Symmetry Breaking in Quantum Chromodynamics, ILL-(TH)-82-5.*

Vol. XXV No. 3 SCIENTIA SINICA (Series A) March 1982

WILSON LOOP INTEGRAL AND STRING WAVE FUNCTIONAL

ZHOU GUANGZHAO (CHOU KUANG-CHAO 周光召) AND LI XIAOYUAN (李小源)

(*Institute of Theoretical Physics, Academia Sinica*)

Received July 3, 1980.

ABSTRACT

In this paper we show that the Wilson loop integral corresponds to an external source term in the generating functional for Green's function with a discussion of its effects. This implies that some approximations in the previous derivation of the string-like equation may not be applicable. The calculation of the field strength in the presence of this external source may be more useful in the study of confinement.

Recently, many authors[1-4] have discussed the possible connection between the Wilson loop integral[5] and the string wave functional[6], and attempted to show that under certain approximations the Wilson loop integral satisfies a string-like equation if the gauge fields meet certain constraints. Meanwhile, some authors have argued that how to formulate exactly a quantum field theory for such objects is far from clear[2], and there is still a great gap in the attempt to derive the string theory from a Yang-Mills theory. In this paper, we show that the Wilson loop integral corresponds to an external source in the vacuum and its effect is discussed. We would like to point out that some approximations in the previous derivations of the string-like equation may not be applicable in the confining vacuum. And we also suggest that the calculation of the field strength in the presence of this external source may be useful in discussing the confinement problem. For simplicity, it is appropriate to consider the example of the Abelian gauge field at first.

In the usual Euclidean space path integral formalism the Wilson loop average $A[C]$ can be defined as

$$A[C] = N\int [d\varphi(x)]e^{\int \mathscr{L}(\varphi(x))d^4x\, ig\oint_C A_\mu(x)dx^\mu}, \tag{1}$$

where $\mathscr{L}(\varphi(x))$ is the Lagrangian of the system, $\varphi(x)$ stands for all the fields including the gauge field; $A_\mu(x)$ is the vector gauge potential, g is the coupling constant, N is a normalization factor and C denotes a large close curve which is parameterized by

$$x^\mu = x^\mu(s), \quad s_0 \leqslant s \leqslant s_1, \quad x^\mu(s_0) = x^\mu(s_1). \tag{2}$$

By artificially introducing the δ function

$$A_\mu(x) = \int d^4y A_\mu(y)\delta^{(4)}(y - x(s)), \tag{3}$$

the integral around the closed curve C, $g\ \oint_C A^\mu(x)\ dx_\mu/ds\ ds$, can be converted into an integral over the whole space, then we have

$$g\oint_C A^\mu(x)\frac{dx_\mu}{ds}ds=\int d^4y A^\mu(y) j_\mu(y), \tag{4}$$

where

$$j_\mu(y, C)\equiv g\oint_C \frac{dx_\mu}{ds}\delta^{(4)}(y-x(s))ds. \tag{5}$$

Thus the Wilson loop average can be written as

$$A[C]=\int[d\varphi]e^{\int d^4y[\mathcal{L}(\varphi(y))+j_\mu(y,C)A^\mu(y)]}.$$

It is nothing else than the ordinary generating functional for the gauge field $Z[J_\mu]$ evaluated at the specific external source $J_\mu(y)=J_\mu(y)=j_\mu(y;C)$. Namely,

$$A[C]=Z[J_\mu]|_{J_\mu=j_\mu}. \tag{6}$$

It should be stressed that, in the usual generating functional approach, the external source $J_\mu(x)$ is only the auxiliary fields, because all the Green functions are obtained from the functional derivatives of the generating functional by taking all $J_\mu(x)=0$. But now, the external source $j_\mu(x)$ is properly introduced by the Wilson loop integral itself, so it may play a substantial role.

Therefore the Wilson loop average can be considered not only as a functional of the curves C, but also as a functional of the external source $J_\mu(x)$. Now it is very easy to obtain the functional derivatives of $A[C]$ with respect to the curve $x^\mu(s)$,

$$\frac{\delta A[C]}{\delta x^\mu(s)}=\int d^4y\left.\frac{\delta Z[J]}{\delta J^\nu(y)}\right|_{J^\nu(y)=j^\nu(y)}\frac{\delta j^\nu(y,C)}{\delta x^\mu(s)}, \tag{7}$$

where

$$\frac{1}{Z[J]}\frac{\delta Z[J]}{\delta J^\nu(y)}=A_\nu(y,J) \tag{8}$$

is the vacuum expectation value of the vector gauge potential $A_\mu(x)$ in the presence of the external source $J_\mu(y)$. And it is not difficult to show that

$$\frac{\delta j^\nu(y,C)}{\delta x^\mu(s)}=g\left\{\delta^\nu_\mu\frac{dx^\sigma}{ds}\frac{\delta}{\delta y^\sigma}-\frac{dx^\nu}{ds}\frac{\delta}{\delta y^\mu}\right\}\delta^{(4)}(y-x(s)). \tag{9}$$

Inserting Eqs. (8) and (9) into Eq (7), we immediately find the first-order curve functional derivatives

$$\frac{\delta A[C]}{\delta x^\mu(s)}=gF_{\mu\nu}(x(s), j(C))\frac{dx^\nu}{ds}A[C], \tag{10}$$

where

$$F_{\mu\nu}(x(s), j(C))\equiv\frac{\partial}{\partial x^\nu}A_\mu(x,j)-\frac{\partial}{\partial x^\mu}A_\nu(x,j). \tag{11}$$

At first glance, Eq. (10) seems to be a well known result for a long time[7]. But this is not the case. In fact, the $F_{\mu\nu}\ (x, j(C))$ now is the vacuum expectation value of the gauge field strength $\hat{F}_{\mu\nu}(x)$ in the presence of the specific external source $j_\mu(C)$ which is not vanishing even for pure gauge field. And $F_{\mu\nu}(x, j(C))$ obeys an equation with the external source

$$\frac{\partial}{\partial x^\mu} F^{\mu\nu}(x, j(C)) = -j^\nu(x, C), \tag{12}$$

as the average of the terms arising from gauge fixing condition is vanishing. In what follows, we will show that it is just this property that makes the result different from those of the previous derivations.

Let us differentiate Eq. (10) with respect to $x_\mu(s')$ again, we obtain

$$\frac{\delta^2 A[C]}{\delta x_\mu(s')\delta x^\mu(s)} = g^2\left[F_{\mu\nu}(x(s), j)\frac{dx^\nu}{ds}F^{\mu\sigma}(x(s'), j)\frac{dx^\sigma}{ds'}\right]A[C]$$

$$+ g\frac{\delta}{\delta x_\mu(s')}[F_{\mu\nu}(x(s), j)]\frac{dx^\nu}{ds}A[C], \tag{13}$$

and

$$\frac{\delta}{\delta x_\mu(s')}F_{\mu\nu}(x(s), j) = \delta(s-s')\partial^\mu F_{\mu\nu}(x(s), j)$$

$$+\int \frac{\delta F_{\mu\nu}}{\delta J^\sigma(y)}\Big|_{j^\sigma \cdot j^\sigma}\frac{\delta j^\sigma(y)}{\delta x_\mu(s')}d^4y$$

$$= -\delta(s-s')j_\nu(x(s)) + g\left\langle \hat{F}_{\mu\nu}\hat{F}^{\mu\sigma}\frac{dx_\sigma}{ds'}\right\rangle_{cj}, \tag{14}$$

where $\langle\quad\rangle_{cj}$ stands for the connected parts of the Green functions with the external source $j_\mu(x)$. Inserting Eq. (14) into Eq. (13), we come to the final result

$$\frac{\delta^2 A[C]}{\delta x_\mu(s')\delta x^\mu(s)} = \left\{g^2[F_{\mu\nu}(x(s), j)\frac{dx^\nu}{ds}F^{\mu\sigma}(x(s'), j)\frac{dx^\sigma}{ds'} + \left\langle\hat{F}_{\mu\nu}\frac{dx^\nu}{ds}F^{\mu\sigma}\frac{dx_\sigma}{ds'}\right\rangle_{cj}\right.$$

$$\left. - g\delta(s-s')j_\nu(x(s), C)\frac{dx^\nu}{ds}\right\}A[C]. \tag{15}$$

Eq. (15) is an exact equation for the Wilson loop average. When $s' \to s$, the second-order functional derivatives of the $A[C]$ involve singularities. In this case, it is appropriate to define

$$\frac{\delta^2 A[C]}{\delta x_\mu(s)\delta x^\mu(s)} = \frac{1}{2\varepsilon}\int_{s-\varepsilon}^{s+\varepsilon}\frac{\delta^2 A[C]}{\delta x_\mu(s')\delta x^\mu(s)}ds', \tag{16}$$

here ε is an infinitesimal quantity.

Now let us give some comments. In order to derive the string-like equation, the authors of Refs. [1—4] usually assume that

(1) The right side of Eq. (15) is completely independent of any external sources J. It implies that the gauge connection satisfies the full source-free field equations

$$D_\mu F^{\mu\nu} = 0 \quad \text{or} \quad \frac{dx_\nu(s)}{ds}D_\mu F^{\mu\nu} = 0, \tag{17}$$

and means that only the vacuum expectation value in the absence of the external sources are considered.

(2) It is possible to use rather special classical field configuration to generate automatically

$$\left\langle \hat{F}_{\mu\nu}(x(s)) \frac{dx^\nu}{ds} \hat{F}^{\mu\sigma}(x(s')) \frac{dx_\sigma}{ds'} \right\rangle_c A[C]$$

$$= \langle \hat{F}_{\mu\nu}(x(s)) \hat{F}^{\mu\sigma}(x(s')) \rangle_c \frac{dx^\nu}{ds} \frac{dx_\sigma}{ds'} A[C]. \tag{18}$$

(3)

$$\langle \hat{F}_{\mu\nu}(x(s)) \hat{F}^{\mu\sigma}(x(s)) \rangle f \delta^\sigma_\nu, \tag{19}$$

where $f = 1/4 \langle \hat{F}_{\mu\nu} \hat{F}^{\mu\nu} \rangle$ is a constant or at most a slowly varying function of x_μ.

As a result of these approximations, $A[C]$ would obey a string-like equation.

$$\frac{\delta^2 A[C]}{\delta x_\mu(s) \delta x^\mu(s)} = g^2 f \left(\frac{dx_\nu}{ds} \right)^2 A[C]. \tag{20}$$

From the above discussions, it is very clear that the assumption (1) may not be true. This is because the Wilson loop integral itself automatically induces the specific external source $j_\mu(C)$ which is not vanishing even for full source-free gauge field. Furthermore, it forces us to consider the vacuum expectation value of the field strength in the presence of this specific external source.

In order to further illustrate the above point, we may see what would happen for the solution of the string-like equation in a confining vacuum. In this case

$$A[C] = e^{-\alpha \mathscr{S}} \tag{21}$$

is in accord with the Wilson confinement criterion, and $\mathscr{S}$ in Eq. (21) is the least area enclosed by a closed curve C and is parametrized by the equation

$$x^\mu = x^\mu(\sigma, \tau); \ \alpha \text{ is a constant } \alpha = g\sqrt{f} \text{ and } f > 0.$$

By differentiating Eq. (21) with respect to curve $x_\mu(s)$, we obtain

$$\frac{\delta A[C]}{\delta x^\mu(s)} = -\alpha \frac{\delta \mathscr{S}}{\delta x^\mu(s)} A[C], \tag{22}$$

where

$$\frac{\delta \mathscr{S}}{\delta x^\mu(s)} = n_{\mu\nu} \frac{dx^\nu}{ds}, \tag{23}$$

and

$$n_{\mu\nu} = \left. \frac{\partial(x^\mu, x^\nu)}{\partial(\sigma, \tau)} \right|_{\substack{\text{at the boundary point} \\ x^\mu(s) \text{ of the area } \mathscr{S}}} \Bigg/ \left| \frac{\partial(x^\mu, x^\nu)}{\partial(\sigma, \tau)} \right|. \tag{24}$$

Checking Eq. (10) against Eq. (22), we have

$$g F_{\mu\nu}(x(s), j) \frac{dx^\nu}{ds} = -\alpha \frac{\delta \mathscr{S}}{\delta x^\mu(s)} = -\alpha n_{\mu\nu} \frac{dx^\nu}{ds} \tag{25}$$

in the confining vacuum.

Lüscher has shown that[6]

$$\left(\frac{\delta \mathscr{S}}{\delta x^\mu(s)}\right)^2 = \left(\frac{dx^\nu}{ds}\right)^2 \tag{26}$$

and

$$\frac{\delta^2 \mathscr{S}}{\delta x_\mu(s)\delta x^\mu(s)} = \lim_{\epsilon\to 0}\int_{s-\epsilon}^{s+\epsilon} \frac{\delta^2 \mathscr{S}}{\delta x_\mu(s')\delta x^\mu(s)}\, ds' = 0. \tag{27}$$

From Eqs. (22), (26) and (27), we get the following equation

$$\frac{\delta^2 A[C]}{\delta x_\mu(s)\delta x^\mu(s)} = \alpha^2 \left(\frac{dx^\nu}{ds}\right)^2 A[C]. \tag{28}$$

At first glance, Eq. (28) seems to be only a copy of Eq. (20) with $\alpha = g\sqrt{f}$, where f stands for the vacuum expectation value of the field strength squared in the absence of external sources, and only depends on the fluctuation of the vacuum itself (as the vacuum expectation value of the field strength is zero in the absence of external sources). Also f may be relevant to instantons for the self-dual gauge fields. Actually this is not so. Eq. (25) has already indicated that the field strength $F_{\mu\nu}(x(s), j)$ should not vanish, otherwise there would be no solution in a confining vacuum. And this is possible only in the case where the external sources are not neglected. In other words, in order to make the Wilson loop average $A[C]$ satisfy Eq. (28), α is inevitably dependent on the field strength induced by the external source $j_\mu(C)$ in the confining vacuum. And it is not right to consider only the fluctuation of the vacuum and to set $F_{\mu\nu}$ to zero. This indicates again that the assumption (1) may not be applicable.

Let us now consider the case of no confinement. For a large closed curve C, the Wilson confinement criterion implies that

$$A[C] = e^{-\beta L}, \tag{29}$$

where L is the perimeter of the curve C, and β is a constant. By differentiating Eq. (29) with respect to the curve $x^\mu(s)$ one has

$$\frac{\delta A[C]}{\delta x^\mu(s)} = -\beta \frac{\delta L}{\delta x^\mu(s)} A[C] = -\beta \frac{dx_\mu/ds}{[(dx^\nu/ds)^2]^{1/2}} A[C]. \tag{30}$$

Therefore, without confinement we should have

$$gF_{\mu\nu}(x(s), j)\frac{dx_\nu}{ds} = -\beta \frac{dx_\mu/ds}{[(dx^\nu/ds)^2]^{1/2}}. \tag{31}$$

Let us differentiate Eq. (27) again, we come to an equation satisfied by the Wilson loop average in the case of no confinement

$$\frac{\delta^2 A[C]}{\delta x^\mu(s)\,\delta x_\mu(s)} = \beta^2 \left[\left(\frac{dx^\nu}{ds}\right)^2\right]^{1/2} A[C]. \tag{32}$$

It is very interesting to note that Eq. (32) will be the same as Eq. (28) in form if s is taken as the length of the closed curve C (i.e. $(dx^\nu/ds)^2 = 1$).

From the above, we conclude that the calculation of the field strength $F_{\mu\nu}(x(s), j)$ in the presence of the external source $j(C)$ induced properly by the Wilson loop integral itself may be useful in the discussion of the confinement problem. In fact, we have

$$gF_{\mu\nu}(x(s), j)\,\frac{dx^\nu}{ds} = \begin{cases} -\beta\,\dfrac{ax_\mu/ds}{[(dx^\nu/ds)^2]^{1/2}}, & \text{without confinement} \\ -\alpha\,\dfrac{\delta\mathscr{S}}{\delta x^\mu(s)}, & \text{with confinement} \end{cases}$$

α is relevent to the tension of the theory.

for a large closed curve C.

The above discussions can be extended to the case of non-Abelian gauge fields. In that case let us introduce a set of auxiliary parallel field $\phi_\alpha(s)$ along the closed curve C forming the fundamental representation of the gauge group G. Their Heisenberg equation of motion and commutation relations are

$$\frac{d\Phi(s)}{ds} = g\hat{A}(s)\Phi(s), \tag{33}$$

$$\frac{d\hat{\Phi}^+(s)}{ds} = -g\hat{\Phi}^+(s)\hat{A}(s),$$

$$[\hat{\Phi}_\alpha(s), \hat{\Phi}_\beta^+(s')]_{s=s'} = \delta_{\alpha\beta}, \tag{34}$$

where $\Phi^+(s)$ is the Hermitian conjugate of $\Phi(s)$ and

$$\hat{A}(s) = A_\mu^i(x(s))\,\frac{dx^\mu}{ds}\,\hat{I}_i. \tag{35}$$

In Eq. (35) $\hat{I}_i(i=1, \ldots \eta_G)$ are the generators of the group G. The vacuum for the fields $\Phi(s)$ is chosen to be $|0\rangle$ which satisfies the condition

$$\hat{\Phi}(s)|0\rangle = 0. \tag{36}$$

Eq. (33) can be solved to give

$$\hat{\Phi}_\alpha(s) = W_{\alpha\gamma}[s, s']\hat{\Phi}_\gamma(s'), \quad \hat{\Phi}_\alpha^+(s) = \hat{\Phi}_\gamma^+(s')W_{\gamma\alpha}[s', s], \tag{37}$$

where

$$W_{\alpha\gamma}[s, s'] = \left\{P\exp\left\{g\int_{s'}^{s}\hat{A}(s'')ds''\right\}\right\}_{\alpha\gamma}. \tag{38}$$

The Wilson loop can be identified as

$$W[C] = W_{\alpha\alpha}[s_1, s_0]. \tag{39}$$

Consider now Green's function

$$\langle P(\hat{\Phi}_\alpha(s)\hat{\Phi}_\beta^+(s'))\rangle = \langle 0|P(\hat{\Phi}_\alpha(s)\hat{\Phi}_\beta^+(s'))|0\rangle = \theta(s-s')W_{\alpha\beta}[s, s']. \tag{40}$$

Hence the Wilson integral is equal to Green's function

$$W[C] = \langle tr\{\hat{\Phi}(s_1)\hat{\Phi}^+(s_0)\}\rangle. \tag{41}$$

It is easily verified that the equations of motion (33) are derived from the Lagrangian

$$L_\Phi = \Phi^+\frac{d\Phi}{ds} - \Phi^+\hat{A}(s)\Phi(s),$$

with the parameter s acting as the proper time.

Write it in a path integral form, Green's function (41) becomes

$$W[C] = \int [d\varphi][d\Phi][d\Phi^+] \exp\left\{-\int \mathscr{L}_A(x)d^4x - \int_{s_0}^{s_1} L_\Phi(s)ds\right\}$$

$$\cdot \Phi_a(s_1)\Phi_a^+(s_0). \tag{42}$$

This form is analogous to that given by Gervais and Neveu[9]. They take the Φ fields to be Fermions. But this is irrelevant and one may as well choose Φ to be bosons.

Eq. (42) can also be written in the following form,

$$W[C] = \langle P\{Z[\hat{j}^i_\mu(x, C)]\hat{\Phi}_a(s_1)\hat{\Phi}^+_a(s_0)\}\rangle, \tag{43}$$

where $Z[\hat{j}^i_\mu]$ is the generating functional for the gauge field A^i_μ with a specified external source $\hat{j}^i_\mu(x, C)$ induced by the Wilson integral,

$$\hat{j}^i_\mu(x, C) = g \oint \delta^{(4)}(x - x(s))\frac{dx_\mu}{ds}\hat{I}_i(s), \tag{44}$$

where

$$\hat{I}_i(s) = \hat{\Phi}^+(s)\hat{I}_i\hat{\Phi}(s). \tag{45}$$

In Eq. (44), the fields $\Phi(s)$ are written in the interaction picture and satisfy the free field equation $d\Phi(s)/ds = 0$.

The first variation of $W[C]$ with respect to the curve C can be easily calculated and equals

$$\frac{\delta W[C]}{\delta x^\mu(s)} = g\left\{\frac{\partial}{\partial x^\nu}\langle P(\hat{A}^i_\mu(x(s))\hat{I}_i(s)\hat{\Phi}_a(s_1)\hat{\Phi}^+_a(s_0))\rangle - (\nu \longleftrightarrow \mu)\right\}\frac{dx^\nu}{ds},$$

$$= g\left\langle P(\hat{F}^i_{\mu\nu}(x(s))\hat{I}_i(s)\Phi_a(s_1)\hat{\Phi}^+_a(s_0))\right\rangle\frac{dx^\nu}{ds}, \tag{46}$$

where $\hat{F}^i_{\mu\nu}(x(s))$ is the non-Abelian gauge field strength. The nonlinear term in $\hat{F}^i_{\mu\nu}(x(s))$ arises because of the parallel nature of the fields $\hat{I}_i(s) = \hat{\Phi}^+(s)\hat{I}_i\hat{\Phi}(s)$ under displacement along the curve C.

Eq. (46) may also be expressed as

$$\frac{\delta W[C]}{\delta x^\mu(s)} = gF_{\mu\nu}(x, C)W[C]\frac{dx^\nu}{ds}, \tag{47}$$

where

$$F_{\mu\nu}(x, C)W[C] = \langle P(\hat{F}^i_{\mu\nu}(x(s))\hat{I}_i(s)\hat{\Phi}_a(s_1)\hat{\Phi}^+_a(s_0))\rangle. \tag{48}$$

The nonvanishing of $F_{\mu\nu}(x, C)$ is intimately connected to the presence of the external sources induced by the Wilson integral just as in the Abelian case.

A second variation of $W[C]$ gives an equation similar to that of (28) and it is also illegitimate to evaluate the coefficient in front of $W[C]$ on the right side of that equation by neglecting the external sources induced by Wilson integral.

After the completion of this work, we are informed by Professor G. Parisi that Makeenko has also stressed the importance of the external source term induced by the Wilson integral in deriving the string-like equation from a Yang-Mills theory[10].

We would like to thank Prof. G. Parisi for sending us a preprint of the work of Makeenko.

References

[1] Gervais, J. L. & Neveu. A., *Phys. Lett.*, **80B** (1979), 255;
Nambu, Y., *ibid.*, **80B** (1979), 372;
Gliozzi, F., Regge, T. & Virasaro, M. A., *ibid.*, **81B** (1979), 178;
Polyakov, A. M., *ibid.*, **82B** (1979), 247.

[2] Corrigan, E. & Hasslacher, B., *ibid.*, **81B** (1979), 181.

[3] Durand, L. & Mendal, E., *ibid.*, **85B** (1979), 241.

[4] Eguchi, T., *ibid.*, **87B** (1979), 91;
Foerster, D., *ibid.*, **87B** (1971), 87;
Weingarten, D., *ibid.*, **87B** (1979), 97;
Makeenko, Yu. M. & Migdal, A. A., *ibid.*, **88B** (1979), 135.

[5] Wilson, K. G., *Phys. Rev.*, **D10** (1974), 2445.

[6] Marshall, C. & Ramond, P., *Nucl. Phys.*, **B85** (1975), 375.

[7] Mandelstam, S., *Ann. Phys.*, **19** (1962), 1.

[8] Lüscher, M., *Phys. Lett.*, **90B** (1980), 277.

[9] Gervais, J. L. & Neveu, A., *Nucl. Phys.*, **B163** (1980), 189.

[10] Makeenko, Yu. M., Institute of Theoretical and Experimental Physics (Moscow), Preprint. ИТОФ-141,

Vol. XXV No. 7 SCIENTIA SINICA (Series A) July 1982

BÄCKLUND TRANSFORMATION, LOCAL AND NONLOCAL CONSERVATION LAWS FOR NONLINEAR σ-MODELS ON SYMMETRIC COSET SPACES

ZHOU GUANGZHAO (CHOU KUANG-CHAO 周光召)
(*Institute of Theoretical Physics, Academia Sinica, Beijing*)

AND

SONG XINGCHANG (宋行长)
(*Peking University*)

Received January 9, 1981; revised June 27, 1981.

ABSTRACT

Two-dimensional nonlinear σ-models defined on symmetrical coset spaces are considered. The duality symmetry is used to construct a Bäcklund transformation which depends on a continuous parameter γ. Conserved current depending on γ is obtained. Expanding the current in powers of γ around $\gamma=1$, one gets an infinite number of nonlocal conserved currents. A modified form of conserved current depending on γ is also obtained. When this current is expanded in powers of γ or γ^{-1}, one gets two sets of infinite number of local conserved currents.

It is well known that all two-dimensional nonlinear σ-models on symmetric coset spaces have an infinite number of nonlocal conservation laws[1–4]. In contrast, much less is known about the local ones. The existence of an infinite number of local conservation laws has been established for $O(N)$ σ-models by Pohlmeyer[5], $O(N)$ principal chiral fields by Cherednik[6], and CP_n models by Eihenherr[7] and Scheler[8]. Recently Ogielski, Prasad, Sinha and Chau Wang have developed a parametric Bäcklund transformation thereby the local conservation laws can be obtained in principle[9].

In the present paper the duality symmetry[7,8,10] is used to construct a generalized Bäcklund transformation for nonlinear σ-models defined on symmetric coset spaces. The transformation contains a continuous parameter γ. Conserved current depending on γ is constructed which yields an infinite number of nonlocal conserved currents when expanded in powers of γ around $\gamma=1$. A slight modification of the γ-dependent current expanded in powers of γ around $\gamma=0$ or $\gamma^{-1}=0$ gives two sets of infinite number of local conserved currents. The main advantage of the present formulation is its simplicity and generality. We shall give a brief account in the present paper and details will be published elsewhere.

Consider a compact Lie group G with subgroup H. Any element $g_0 \in G$ can be decomposed into product of a left coset element $\phi_0 \in G/H$ and an element of the subgroup, $h_0 \in H$, i.e.

$$g_0 = \phi_0 h_0. \tag{1}$$

Under the left action of a group element g, the coset element ϕ_0 becomes

$$\left.\begin{aligned} g\phi_0 &= \phi(g, \phi_0)h(g, \phi_0), \\ \phi(g, \phi_0) &\in G/H, h(g, \phi_0) \in H. \end{aligned}\right\} \tag{2}$$

The transformation $\phi_0 \to \phi\ (g, \phi_0)$ forms a nonlinear representation of the group G[11].

The basic dynamical fields of a nonlinear σ-model are the local coordinate functions $\pi_j(x)$, $j = 1, 2, \cdots n_G - n_H$ of the coset manifold G/H: $\phi(x) = \phi(\pi(x))$. Let

$$\hat{Q}_\mu = -i\phi^{-1}(x)\partial_\mu\phi(x) = \hat{H}_\mu + \hat{K}_\mu, \tag{3}$$

where $\hat{H}_\mu$ and $\hat{K}_\mu$ are valued in the Lie algebra of the subgroup H and its complement, the subset of the Lie algebra corresponding to the coset G/H, respectively. When the coset manifold G/H is a symmetric space, $\hat{H}_\mu$ and $\hat{K}_\mu$ satisfy the following integrability conditions,

$$\left.\begin{aligned} \partial_\mu\hat{H}_\nu - \partial_\nu\hat{H}_\mu + i[\hat{H}_\mu, \hat{H}_\nu] &= -i[\hat{K}_\mu, \hat{K}_\nu], \\ \partial_\mu\hat{K}_\nu + i[\hat{H}_\mu, \hat{K}_\nu] &= \partial_\nu\hat{K}_\mu + i[\hat{H}_\nu, \hat{K}_\mu]. \end{aligned}\right\} \tag{4}$$

A Lagrangian invariant under the global transformation of the group G has the form,

$$\mathscr{L} = \frac{1}{2}\operatorname{tr}\{\hat{K}_\mu\hat{K}^\mu\}. \tag{5}$$

Under a local transformation of the group G, it is easily verified that $\hat{H}_\mu$ and $\hat{K}_\mu$ transform as follows:

$$\hat{H}_\mu \to h(g, \phi)(-i\partial_\mu + \hat{H}_\mu)h^{-1}(g, \phi) - i(h\phi^{-1}(g^{-1}\partial_\mu g)\phi h^{-1})_H, \tag{6}$$

$$\hat{K}_\mu \to h(g, \phi)\hat{K}_\mu h^{-1}(g, \phi) - i(h\phi^{-1}(g^{-1}\partial_\mu g)\phi h^{-1})_K, \tag{7}$$

where $(\quad)_H$ and $(\quad)_K$ denote projection onto the Lie algebra of the subgroup H and its complement corresponding to the coset G/H respectively.

For infinitesimal transformation we have

$$g = 1 + i\hat{I}_j\alpha_j(x). \tag{8}$$

Conserved current can be constructed and has the form

$$\left.\begin{aligned} \hat{j}_\mu &= \frac{\partial\mathscr{L}}{\partial\partial^\mu\alpha_j(x)}\hat{I}_j = \phi(x)\hat{K}_\mu(x)\phi^{-1}(x), \\ &\partial^\mu\hat{j}_\mu(x) = 0. \end{aligned}\right\} \tag{9}$$

The conserved equation can also be written as

$$\partial^\mu\hat{K}_\mu(x) + i[\hat{H}^\mu, \hat{K}_\mu] = 0. \tag{10}$$

From Eqs. (3) and (4) one easily obtains

$$\partial_\mu\hat{j}_\nu - \partial_\nu\hat{j}_\mu - 2i[\hat{j}_\mu, \hat{j}_\nu] = 0. \tag{11}$$

In the light cone coordinates $\sqrt{2}\,\xi = (t + x)$ and $\sqrt{2}\,\eta = (t - x)$ the conserved equation (10) and the integrability conditions (4) can be combined to give

$$\left.\begin{aligned}&\partial_\xi \hat{K}_\eta = i[\hat{K}_\eta, \hat{H}_\xi], \quad \partial_\eta \hat{K}_\xi = i[\hat{K}_\xi, \hat{H}_\eta],\\ &\partial_\xi \hat{H}_\eta - \partial_\eta \hat{H}_\xi + i[\hat{H}_\xi, \hat{H}_\eta] = -i[\hat{K}_\xi, \hat{K}_\eta].\end{aligned}\right\} \tag{12}$$

In terms of the current $\hat{j}_\mu$ we have

$$\left.\begin{aligned}&\partial_\xi \hat{j}_\eta + \partial_\eta \hat{j}_\xi = 0,\\ &\partial_\xi \hat{j}_\eta - \partial_\eta \hat{j}_\xi - 2i[\hat{j}_\xi, \hat{j}_\eta] = 0.\end{aligned}\right\} \tag{13}$$

Eqs. (12) and (13) will be our starting equations.

It can be easily verified that Eq. (12) is invariant under the dual transformation,

$$\left.\begin{aligned}&\hat{K}_\xi \to \hat{K}_\xi(\gamma) = \gamma \hat{K}_\xi, \quad \hat{K}_\eta \to \hat{K}_\eta(\gamma) = \gamma^{-1} \hat{K}_\eta,\\ &\hat{H}_\xi \to \hat{H}_\xi(\gamma) = \hat{H}_\xi, \quad \hat{H}_\eta \to \hat{H}_\eta(\gamma) = \hat{H}_\eta,\end{aligned}\right\} \tag{14}$$

where γ is any parameter different from zero.

The dual symmetry implies immediately that

$$\hat{Q}_\mu(\gamma) = \hat{H}_\mu(\gamma) + \hat{K}_\mu(\gamma) \tag{15}$$

is a pure gauge and may be set equal to $-ig^{-1}(\phi, \gamma)\partial_\mu g(\phi, \gamma)$ where $g(\phi, \gamma)$ can be decomposed into product,

$$g(\phi, \gamma) = \phi(\pi'(\pi, \gamma))h(\phi, \gamma) = \phi'(\phi, \gamma)h(\phi, \gamma). \tag{16}$$

Now define

$$\begin{aligned}\hat{Q}'_\mu &= -i\phi'^{-1}(\phi, \gamma)\partial_\mu \phi'(\phi, \gamma) = \hat{H}'_\mu + \hat{K}'_\mu\\ &= h(\phi, \gamma)(-i\partial_\mu + \hat{H}_\mu(\gamma))h^{-1}(\phi, \gamma) + h(\phi, \gamma)\hat{K}_\mu(\gamma)h^{-1}(\phi, \gamma).\end{aligned} \tag{17}$$

Eq. (17) tells us that $\hat{H}'_\mu$ and $\hat{K}'_\mu$ can be obtained from $\hat{H}_\mu(\gamma)$ and $\hat{K}_\mu(\gamma)$ by a gauge transformation in the subgroup H. Since Eqs. (12) are gauge invariant, we expect that $\hat{H}'_\mu = \hat{H}_\mu(\pi'(\pi, \gamma))$, $\hat{K}'_\mu = \hat{K}_\mu(\pi'(\pi, \gamma))$ satisfy the same set of Eqs. (12). Therefore if π_j is a solution of Eqs. (12), $\pi'_j(\pi, \gamma)$ constructed by Eq. (16) is also a solution of Eqs. (12). This is the generalized Bäcklund transformation we have obtained.

A conserved current depending on $\pi'(\pi, \gamma)$ can thus be constructed and has the form,

$$\left.\begin{aligned}&\hat{j}_\mu(\pi') = \phi' \hat{K}'_\mu \phi'^{-1} = g(\phi, \gamma)\hat{K}_\mu(\gamma)g^{-1}(\phi, \gamma),\\ &\partial^\mu \hat{j}_\mu(\pi') = 0.\end{aligned}\right\} \tag{18}$$

Of course, $\hat{j}_\mu(\pi') = \hat{j}_\mu(\pi, \gamma)$ satisfies the same set of Eq. (13) as $\hat{j}_\mu(\pi)$ does.

Rewriting Eq. (15) in the form,

$$-ig^{-1}(\pi, \gamma)\partial_\mu g(\pi, \gamma) = -i\phi^{-1}\partial_\mu \phi + \hat{K}_\mu(\gamma) - \hat{K}_\mu, \tag{19}$$

and integrating it, we get

$$g(\pi, \gamma) = \tilde{P}\left(\exp\left\{i\int^x (\hat{j}_\mu(\gamma) - \hat{j}_\mu)dx^\mu\right\}\right)\phi(\pi), \tag{20}$$

where $\tilde{P}$ is an anti-ordering operator along the path and

$$\hat{j}_\mu(\gamma) = \phi \hat{K}_\mu(\gamma)\phi^{-1}. \tag{21}$$

Substituting Eq. (20) into Eq. (18) we obtain a conserved current depending on a continuous parameter γ,

$$\hat{j}_\mu(x,\gamma) = \tilde{P}\left(\exp\left\{i\int^x (\hat{j}_\mu(\gamma) - \hat{j}_\mu)dx^\mu\right\}\right)\hat{j}_\mu(\gamma)P\exp\left\{-i\int^x (\hat{j}_\mu(\gamma) - \hat{j}_\mu)dx^\mu\right\}. \quad (22)$$

Eq. (22) expanded in powers of γ around $\gamma = 1$ yields an infinite number of non-local conserved currents which coincides with that obtained in Refs. [1—4].

In order to get the local conserved currents from Eq. (22) we shall study the structure of the current conservation equation first. Let us define

$$j^A_\mu(x) = \mathrm{tr}\{\hat{j}_\mu(x)\hat{A}(x)\} = \mathrm{tr}\{\hat{K}_\mu(x)\hat{B}(x)\}, \quad (23)$$

where $\hat{A}(x)$ is a function valued in the Lie algebra of G, while $\hat{B} = \phi^{-1}\hat{A}\phi$. Taking account of the equation of motion (9) or (10), one can easily obtain

$$\partial^\mu j^A_\mu(x) = \mathrm{tr}\{\hat{j}_\mu(x)\partial^\mu\hat{A}(x)\} = \mathrm{tr}\{\hat{K}_\mu(x)(\partial^\mu\hat{B}(x) + i[\hat{H}^\mu(x), \hat{B}(x)])\}. \quad (24)$$

j^A_μ will be conserved if $\partial^\mu\hat{A}(x)$ is trace orthogonal to $\hat{j}_\mu(x)$ or equivalently $\partial^\mu\hat{B}(x) + i[\hat{H}^\mu(x), \hat{B}(x)]$ is trace orthogonal to $\hat{K}_\mu(x)$.

There are many possible choices of $\hat{A}(x)$ or $\hat{B}(x)$ in which the conservation condition of j^A_μ will be satisfied. For instance we may take $\hat{B}(x)$ satisfying the equation,

$$\partial_\mu\hat{B}(x) = i[\hat{B}(x), \hat{H}_\mu(x)] + a\varepsilon_{\mu\nu}\hat{K}^\nu(x) - b\varepsilon_{\mu\nu}\hat{B}(x)\hat{K}^\nu(x)\hat{B}(x)$$
$$+ ic_1[\hat{B}(x), \hat{K}_\mu(x)] + c_2\varepsilon_{\mu\nu}\{\hat{B}(x), \hat{K}^\nu(x)\}. \quad (25)$$

It is easily verified that for arbitrary constants a, b, c_1 and c_2, the conservation condition is satisfied, and the integrability condition for $\hat{B}(x)$ imposes a constraint among these parameters:

$$ab + c_1^2 + c_2^2 = 1. \quad (26)$$

Here we consider two simplest cases.

(1) $c_1 = 1, a = b = c_2 = 0$. Then Eq. (25) turns out to be

$$\left.\begin{aligned} \partial_\xi\hat{B}(x) &= i[\hat{B}, \hat{H}_\xi] + i[\hat{B}, \hat{K}_\xi], \\ \partial_\eta\hat{B}(x) &= i[\hat{B}, \hat{H}_\eta] + i[\hat{B}, \hat{K}_\eta]. \end{aligned}\right\} \quad (27)$$

This means

$$\hat{A}(x) = \hat{A}_0 = \mathrm{const},$$

and then the following normalization condition must be satisfied:

$$\mathrm{tr}\hat{B}^2(x) = \mathrm{tr}\hat{A}^2(x) = \mathrm{const}. \quad (28)$$

Duality symmetry then implies that

$$j^A_\mu(\gamma) = \mathrm{tr}\{\hat{K}_\mu(\gamma)\hat{B}(\gamma)\} \quad (29)$$

is also a conserved current if $\hat{B}(\gamma)$ satisfies the following equations:

$$\partial_\xi\hat{B}(\gamma) = i[\hat{B}(\gamma), \hat{H}_\xi] + i\gamma[\hat{B}(\gamma), \hat{K}_\xi], \quad (30a)$$

$$\partial_\eta\hat{B}(\gamma) = i[\hat{B}(\gamma), H_\eta] + i\gamma^{-1}[\hat{B}(\gamma), \hat{K}_\eta]. \quad (30b)$$

Now expanding $\hat{B}(\gamma)$ in powers of γ^{-1}

$$\hat{B}(\gamma) = \sum_{n=0}^{\infty} \hat{B}_n \gamma^{-n}, \tag{31}$$

and substituting it into Eq. (30a), one gets

$$[\hat{B}_n, \hat{K}_\xi] = -[\hat{B}_{n-1}, \hat{H}_\xi] - i\partial_\xi \hat{B}_{n-1}. \tag{32}$$

Evidently the first coefficient

$$\hat{B}_0 = (\mathrm{tr}\hat{K}_\xi \hat{K}_\xi)^{-\frac{1}{2}} \hat{K}_\xi \tag{33}$$

satisfies both Eq. (32) with $n = 0$ and the normalization condition (28).

Explicit expressions for $\hat{B}_n (n \geqslant 1)$ can be obtained from Eq. (32) for CP_1 model or $O(3)$ chiral model. For these models Eq. (29) gives an infinite number of local conserved currents when $\hat{B}(\gamma)$ is expanded in powers of γ^{-1} as in Eq. (31).

$$\left.\begin{aligned} &\partial_\xi j_{n\eta} + \partial_\eta j_{n\xi} = 0, \\ &j_{n\xi} = \mathrm{tr}\{\hat{K}_\xi \hat{B}_n\}, \quad j_{n\eta} = \mathrm{tr}\{\hat{K}_\eta \hat{B}_{n-2}\}, \end{aligned}\right\} \tag{34}$$

with the first one equal to

$$j_{0\xi} = (\mathrm{tr}\hat{K}_\xi \hat{K}_\xi)^{\frac{1}{2}}, \quad j_{0\eta} = 0. \tag{35}$$

Another set of conserved currents can be obtained if one expands $\hat{B}(\gamma)$ in powers of γ.

In general, Eq. (32) does not necessarily yield local solution for $\hat{B}_n$ owing to the constraint equations,

$$\mathrm{tr}\{\hat{K}_\xi^m (i\partial_\xi \hat{B}_{n-1} + [\hat{B}_{n-1}, \hat{H}_\xi])\} = 0, \tag{36}$$

held for arbitrary integer m.

When the duality symmetry with $\gamma = -1$ is applied to this case one will obtain the case in which $c_1 = -1, a = b = c_2 = 0$.

(2) $c_1 = c_2 = 0$, a $b = 1$. Eq. (25) takes the form

$$\left.\begin{aligned} \partial_\xi \hat{B} &= i[\hat{B}, \hat{H}_\xi] + a\hat{K}_\xi - a^{-1}\hat{B}\hat{K}_\xi \hat{B}, \\ \partial_\eta \hat{B} &= i[\hat{B}, \hat{H}_\eta] - a\hat{K}_\eta + a^{-1}\hat{B}\hat{K}_\eta \hat{B}. \end{aligned}\right\} \tag{37}$$

The normalization condition (28) is compatible with Eq. (37) if the additional condition

$$\mathrm{tr}\{\hat{K}_\mu (\hat{B}^3 - a^2 \hat{B})\} = 0 \tag{38}$$

is held simultaneously. For CP_n models this condition coincides with Eq. (28).

The local conserved currents can be generated from Eq. (29) with $\hat{B}(\gamma)$ satisfying the following equations:

$$\partial_\xi \hat{B}(\gamma) = i[\hat{B}(\gamma), \hat{H}_\xi] + a\gamma \hat{K}_\xi - a^{-1}\gamma \hat{B}(\gamma)\hat{K}_\xi \hat{B}(\gamma), \tag{39a}$$

$$\partial_\eta \hat{B}(\gamma) = i[\hat{B}(\gamma), \hat{H}_\eta] - a\gamma^{-1}\hat{K}_\eta + a^{-1}\gamma^{-1}\hat{B}(\gamma)\hat{K}_\eta \hat{B}(\gamma). \tag{39b}$$

Expanding $\hat{B}(\gamma)$ in powers of γ^{-1} we get from Eq. (39a) that

$$a^2\hat{K}_\xi = \hat{B}_0\hat{K}_\xi\hat{B}_0. \tag{40}$$

$$a^{-1}\sum_{l=0}^{n} \hat{B}_l\hat{K}_\xi\hat{B}_{n-l} = i[\hat{B}_{n-1}, \hat{H}_\xi] - \partial_\xi\hat{B}_{n-1}. \tag{41}$$

It is evident that $\hat{B}_0 = a$ is a solution of Eq. (40). But this trivial solution is not useful since it gives $j_{0\mu}$ zero identically.

For CP_n models and their generalization, the $U(n+m)/U(n)\times U(m)$ models, or their orthogonal analogues, the Lie algebra valued function $\hat{K}$ on coset space can be decomposed into two parts:

$$\hat{K} = \hat{K}^{(+)} + \hat{K}^{(-)}, \tag{42}$$

such that

$$\hat{K}^{(-)} = (\hat{K}^{(+)})^\dagger, \quad \hat{K}^{(+)}\hat{K}^{(+)} = \hat{K}^{(-)}\hat{K}^{(-)} = 0. \tag{43}$$

with both $\hat{K}^{(+)}\hat{K}^{(-)}$ and $\hat{K}^{(-)}\hat{K}^{(+)}$ belonging to the Lie algebra of the subgroup. For these models, Eq. (39) split into two branches, a positive one and a negative one. The positive branch of Eq. (39a) reads

$$\partial_\xi\hat{B}^{(+)}(\gamma) = i[\hat{B}^{(+)}(\gamma), \hat{H}_\xi] + a\gamma\hat{K}_\xi^{(+)} - a^{-1}\gamma\hat{B}^{(+)}(\gamma)\hat{K}_\xi^{(-)}\hat{B}^{(+)}(\gamma). \tag{44}$$

From this equation and the normalization condition mentioned above, the results of Ref. [8] for CP_n models can be reproduced.

Now expanding $B(\gamma)$ in powers of γ^{-1} we get from Eq. (44) that

$$a^2\hat{K}_\xi^{(+)} = \hat{B}_0^{(+)}\hat{K}_\xi^{(-)}\hat{B}_0^{(+)}. \tag{45}$$

$$a^{-1}\sum_{j=1}^{n} \hat{B}_j^{(+)}\hat{K}_\xi^{(-)}\hat{B}_{n-j}^{(+)} = i[\hat{B}_{n-1}^{(+)}, \hat{H}_\xi] - \partial_\xi\hat{B}_{n-1}^{(+)}. \tag{46}$$

Eq. (45) can be solved to give

$$\hat{B}_0^{(+)} = a(\hat{K}_\xi^{(+)}\hat{K}_\xi^{(-)})^{-\frac{1}{2}}\hat{K}_\xi^{(+)}. \tag{47}$$

which yields the first conserved current

$$j_{0\xi} = a\,\mathrm{tr}\{(\hat{K}_\xi^{(+)}\hat{K}_\xi^{(-)})^{\frac{1}{2}}\}, \quad j_{0\eta} = 0. \tag{48}$$

in accord with Ref. [9]. From Eq. (46) with $n = 1$, we can deduce the second current

$$j_{1\xi} = j_{1\eta} = 0. \tag{49}$$

The constant a is a scaling parameter which can be put equal to 1. Higher order local currents can be obtained from Eq. (46) (or (41) more generally) by solving linear matrix equations. The solutions of these equations for concrete models will be discussed elsewhere.

References

[1] Golo, V. I. & Perelomov, A. M., *Letts. in Math. Phys.*, **2** (1978), 477; Polyakov, A. M., in "Collective Effects in Condensed Media", *Proc. of 14th Winter School of Theoretical Physical in Karpacz, Wroclav*, 1978.

[2] Zaharov, V. E. & Mikhailov, A. V., *(Sov. Phys.) JETP*, **47** (1978), 1017.

[3] Luscher, M. & Pohlmeyer, K., *Nucl. Phys.*, **B137** (1978), 46.
[4] Eichenherr, H. & Forger, M., *ibid.*, **B155** (1979), 381; Brezin, E., Itzykson, C., Zinn-Justin, J. & Zuber, J. B., *Phys. Lett.*, **82B** (1979), 442; Ogielski, A. T., *Phys. Rev.*, **D21** (1980), 406.
[5] Pohlmeyer, K., *Commun. Math. Phys.*, **46** (1976), 207.
[6] Cherednik, I. V., *Theor. Math. Phys.*, **38** (1979), 120.
[7] Eichenherr, H., *Phys. Lett.*, **90B** (1980), 121.
[8] Scheler, K., *ibid.*, **93B** (1980), 331.
[9] Ogielski, A. T., Prasad, M. K., Sinha, A. & Chau Wang, L. L., *ibid.*, **91B** (1980), 387.
[10] Flume, R. & Meyer, S., *ibid.*, **85B** (1979), 353.
[11] Coleman, S., Wess, J. & Zumino, B., *Phys. Rev.*, **177** (1969), 2239; Callan, C. G. Jr. et al., *ibid.*, **177** (1969), 2247; Salsm, A. & Strathdee, J., *ibid.*, **184** (1969), 1750.

Vol. XXV No. 8 SCIENTIA SINICA (Series A) August 1982

LOCAL CONSERVATION LAWS FOR VARIOUS NONLINEAR σ-MODELS

ZHOU GUANGZHAO (CHOU KUANG-CHAO 周光召)
(*Institute of Theoretical Physics, Academia Sinica, Beijing*)
AND SONG XINGCHANG (宋行长)
(*Peking University*)

Received January 9, 1981; revised June 27, 1981.

ABSTRACT

On the basis of the formulation given in a preceding paper, we derive an infinitive series of conserved local currents explicitly for the two-dimensional classical σ-models on the complex Grassmann manifold $U(m+n)/U(m)\otimes U(n)$ and for $U(N)$ principal chiral field.

In a previous paper[1], by means of the duality symmetry, we constructed a Bäcklund transformation depending on a continuous parameter γ, and gave a formulation to deduce both the local and nonlocal conservation laws for two-dimensional nonlinear σ-models on symmetric coset spaces. Generally the Lagrangian for the nonlinear σ-model has the form

$$\mathscr{L} = \frac{1}{2}\langle \hat{K}_\mu, \hat{K}^\mu\rangle, \tag{1}$$

which is invariant under the global transformation of a compact Lie group G while local-invariant under the gauge transformation of a closed subgroup H of G. Here $\langle , \rangle$ denotes the G invariant inner product. For the field $g(x)$ taking values in G ($g(x)$ is equivalent to $\phi(x)$ in [1] up to a gauge transformation), we define

$$\hat{\Omega}_\mu = -ig^{-1}(x)\partial_\mu g(x) = \hat{H}_\mu + \hat{K}_\mu, \tag{2}$$

where $\hat{H}_\mu$ and $\hat{K}_\mu$ are valued in the Lie algebra of the subgroup H and its orthogonal complement corresponding to the coset G/H respectively. Both $\hat{H}_\mu$ and $\hat{K}_\mu$ are hermitian when $g(x)$ is unitary. Define the covariant derivatives as

$$D_\mu g(x) = \partial_\mu g(x) - ig(x)\hat{H}_\mu(x),$$
$$(\overline{D_\mu g(x)} = \partial_\mu g^+(x) + i\hat{H}_\mu(x)g^+(x) \text{ for } g(x) \text{ unitary}). \tag{3}$$

then

$$\hat{K}_\mu = -ig^{-1}(x)D_\mu g(x)(= i\overline{D_\mu g(x)}g(x)). \tag{4}$$

So $\hat{H}_\mu$ is the composite gauge field whereas $\hat{K}_\mu$ is a vector field which transforms covariantly under the gauge transformation. The Lagrangian (1) can be written as

$$\mathscr{L} = \frac{1}{2}\langle D_\mu g(x), D^\mu g(x)\rangle, \tag{5}$$

and the field equation can be cast into the form,

$$D_\mu D^\mu g - D_\mu g g^{-1} D^\mu g = 0. \tag{6}$$

or

$$D^\mu \hat{K}_\mu \equiv \partial^\mu \hat{K}_\mu + i[\hat{H}^\mu, \hat{K}_\mu] = 0. \tag{6'}$$

And the integrability condition for $\hat{\Omega}_\mu$ gives

$$\left.\begin{aligned} &D_\mu \hat{K}_\nu - D_\nu \hat{K}_\mu = 0, \\ &\partial_\mu \hat{H}_\nu - \partial_\nu \hat{H}_\mu + i[\hat{H}_\mu, \hat{H}_\nu] = -i[\hat{K}_\mu, \hat{K}_\nu]. \end{aligned}\right\} \tag{7}$$

The conserved current $\hat{J}_\mu$ takes the form

$$\hat{J}_\mu = g(x)\hat{K}_\mu g^{-1}(x) = -i(D_\mu g)g^{-1}(= ig\overline{D_\mu g}). \tag{8}$$

and satisfies equations

$$\left.\begin{aligned} &\partial^\mu \hat{J}_\mu = 0, \\ &\partial_\mu \hat{J}_\nu - \partial_\nu \hat{J}_\mu - 2i[\hat{J}_\mu, \hat{J}_\nu] = 0. \end{aligned}\right\} \tag{9}$$

In order to get the local conserved currents a scalar current

$$j^A_\mu = \langle \hat{J}_\mu, \hat{A}(x)\rangle = \langle \hat{K}_\mu, \hat{B}(x)\rangle \tag{10}$$

is defined by introducing the G valued function $\hat{A}(x)$, and $\hat{B}(x) = g^{-1}A(x)g$. j^A_μ is conserved if $\partial^\mu \hat{A}$ is orthogonal to $\hat{J}_\mu(x)$ or $D^\mu \hat{B}$ orthogonal to $\hat{K}_\mu(x)$. Such function $\hat{A}(x)$ or $\hat{B}(x)$ exists and two simplest cases were considered in [1]. In the first case,

$$D_\mu \hat{B}(x) = i[\hat{B}(x), \hat{K}_\mu(x)]. \tag{11}$$

Generally, $\hat{B}(x)$ will take values in the whole Lie algebra of G and a normalization condition

$$\langle \hat{B}(x), \hat{B}(x)\rangle = \text{const} \tag{12}$$

must be satisfied. In the second case,

$$D_\mu \hat{B}(x) = \varepsilon_{\mu\nu}(a\hat{K}^\nu - a^{-1}\hat{B}\hat{K}^\nu\hat{B}). \tag{13}$$

The solution $\hat{B}(x)$ may be restricted to the subset of the Lie algebra corresponding to the coset G/H and the normalization condition (12) is compatible with (13) if additional condition

$$\langle \hat{K}_\mu, a^2\hat{B} - \hat{B}^3\rangle = 0 \tag{14}$$

is held simultaneously.

It is evident that Eqs. (6) and (7) are invariant under the duality transformation $\left(\text{in light cone coordinates } \xi = \dfrac{t+x}{\sqrt{2}},\ \eta = \dfrac{t-x}{\sqrt{2}}\right)$.

$$\left.\begin{aligned} &\hat{K}_\xi \to \hat{K}_\xi(\gamma) = \gamma\hat{K}_\xi, \quad \hat{K}_\eta \to \hat{K}_\eta(\gamma) = \gamma^{-1}\hat{K}_\eta. \\ &\hat{H}_\xi \to \hat{H}_\xi(\gamma) = \hat{H}_\xi, \quad \hat{H}_\eta \to \hat{H}_\eta(\gamma) = \hat{H}_\eta. \end{aligned}\right\} \tag{15}$$

Then duality transformation implies that

$$j^A_\mu(\gamma) = \langle \hat{K}_\mu(\gamma), \hat{B}(\gamma)\rangle \tag{16}$$

is also conserved if $\hat{B}(\gamma)$ satisfies Eq. (11) or (13) with $\hat{K}_\mu$ replaced by $\hat{K}_\mu(\gamma)$. These equations can be solved by expanding $\hat{B}(\gamma)$ in powers of γ^{-1}.

$$\hat{B}(\gamma) = \sum_{l=0}^{\infty} \hat{B}_l \gamma^{-l}. \tag{17}$$

For the second case, we obtain

$$a^2 \hat{K}_\xi = \hat{B}_0 \hat{K}_\xi \hat{B}_0 . \tag{18a}$$

$$\hat{B}_0 \hat{K}_\xi \hat{B}_l + \hat{B}_l \hat{K}_\xi \hat{B}_0 = -a D_\xi \hat{B}_{l-1} - \sum_{j=1}^{l-1} \hat{B}_j \hat{K}_\xi \hat{B}_{l-j} . \tag{18b}$$

Then Eq. (16) gives an infinite set of local conserved currents.

$$j_{l\xi} = \langle \hat{K}_\xi , \hat{B}_l \rangle , \quad j_{l\eta} = \langle \hat{K}_\eta , \hat{B}_{l-2} \rangle . \tag{19}$$

Now we discuss the solutions of Eq. (18) and deduce the conserved currents (19) for several concrete models.

First consider the σ-models on the complex Grassmann manifold $U(m+n)/U(m) \otimes U(n)$, which contain the $O(N)$ invariant σ-models as well as the CP_n models as special examples. Field $g(x)$ takes values in $G = U(m+n)$ and $\hat{K}_\mu$ takes values on the subset of the algebra corresponding to the coset $G/H = U(m+n)/U(m) \otimes U(n)$.

$$\hat{K}_\mu = \begin{pmatrix} & k_\mu^+ \\ k_\mu & \end{pmatrix} \tag{20}$$

where $k_\mu (k_\mu^+)$ is an $n \times m (m \times n)$ matrix. Then $\hat{K}_\xi \hat{K}_\xi$ takes values on the subalgebra of $H = U(m) \otimes U(n)$. (so is the $(\hat{K}_\xi \hat{K}_\xi)^{1/2}$ by a proper definition of square root) i. e.

$$\hat{K}_\xi \hat{K}_\xi = \begin{pmatrix} k_\xi^+ k_\xi & \\ & k_\xi k_\xi^+ \end{pmatrix}, \quad (\hat{K}_\xi \hat{K}_\xi)^{1/2} = \begin{pmatrix} (k_\xi^+ k_\xi)^{1/2} & \\ & (k_\xi k_\xi^+)^{1/2} \end{pmatrix}. \tag{21}$$

with $k_\xi^+ k_\xi$ and $k_\xi k_\xi^+$ $m \times m$ and $n \times n$ matrices respectively. When $m < n$, we may assume $k_\xi^+ k_\xi$ to be invertible (both $k_\xi^+ k_\xi$ and $k_\xi k_\xi^+$ may be assumed to be invertible when $m = n$), and define a matrix,

$$\hat{E}\begin{pmatrix} & e^+ \\ e & \end{pmatrix} = \begin{pmatrix} & (k_\xi^+ k_\xi)^{-1/2} k_\xi^+ \\ k_\xi (k_\xi^+ k_\xi)^{-1/2} & \end{pmatrix} \tag{22}$$

which has the properties,

$$\left.\begin{aligned} &\hat{E}^+ = \hat{E}, \quad \hat{E}^{k+2} = \hat{E}^k (k = 1, 2 \cdots), \\ &\hat{E} \hat{K}_\xi = \hat{K}_\xi \hat{E}, \quad \hat{E} \hat{K}_\xi \hat{E} = \hat{K}_\xi . \end{aligned}\right\} \tag{23}$$

So

$$\hat{B}_0 = \hat{E} \tag{24}$$

is a solution of Eq. (18a) (with a=1) which also satisfies the normalization condition (12) and the additional condition (14).

Since $\hat{K}_\xi \hat{K}_\xi$ is a semi-positive definite hermitian matrix it can be brought into a diagonal form:

$$\hat{K}_\xi \hat{K}_\xi = \begin{pmatrix} \lambda_1^2 & & & & & \\ & \ddots & & & & \\ & & \lambda_m^2 & & & \\ & & & \mu_1^2 & & \\ & & & & \ddots & \\ & & & & & \mu_n^2 \end{pmatrix} \tag{25}$$

by an $U(m)\otimes U(n)$ transformation, and $\lambda_\alpha > 0(\alpha = 1, 2\cdots m)$ as $k_\xi^+ k_\xi$ is invertible by assumption. Some of the eigenvalues $\mu_a(\alpha = 1, 2\cdots n)$ must be zero. In this representation we have (sometimes we ignore the index ξ in unconfused cases).

$$\left.\begin{aligned} &K_{\alpha\beta}K_{\beta c} = \lambda_\alpha^2\delta_{\alpha c}, \quad K_{ab}K_{b\gamma} = \mu_a^2\delta_{a\gamma}, \\ &E_{\alpha\beta} = \lambda_\alpha^{-1}K_{\alpha\beta}, \quad E_{ab} = K_{ab}\lambda_b^{-1}, \\ &E_{\alpha a}E_{ab} = \delta_{\alpha b}; \end{aligned}\right\} \tag{26}$$

or in matrix forms

$$\left.\begin{aligned} &k^+k = \Lambda^2, \quad kk^+ = \mu^2, \\ &e^+ = \Lambda^{-1}k^+, \quad e = k\Lambda^{-1}, \\ &e^+e = I, \end{aligned}\right\} \tag{26'}$$

where I is an $m\times m$ unit matrix, $\Lambda = \mathrm{diag}\,(\lambda_1\cdots\lambda_m)$ and $\mu = \mathrm{diag}\,(\mu_1\cdots\mu_n)$.

Now from Eq. (18b)

$$\hat{B}_0\hat{K}_\xi\hat{B}_l + \hat{B}_l\hat{K}_\xi\hat{B}_0 = -aD_\xi\hat{B}_{l-1} - \sum_{j=1}^{l-1}\hat{B}_j\hat{K}_\xi\hat{B}_{l-j} = :\hat{C}_{l-1}: \tag{18'}$$

multiplying $\hat{K}_\xi$ from right and left respectively, we get

$$(\hat{E}\hat{K})(\hat{B}_l\hat{K}) + (\hat{B}_l\hat{K})(\hat{E}\hat{K}) = (\hat{C}_{l-1}\hat{K}), \quad (\hat{K}\hat{E})(\hat{K}\hat{B}_l) + (\hat{K}\hat{B}_l)(\hat{K}\hat{E}_0) = (\hat{K}\hat{C}_{l-1});$$

$$\lambda_a(B_lK)_{ab} + (B_lK)_{ab}\lambda_b = (C_{l-1}K)_{ab}, \quad \lambda_a(KB_l)_{ab} + (KB_l)_{ab}\lambda_b = (KC_{l-1})_{ab}.$$

Therefore

$$(B_lK)_{ab} = \frac{(C_{l-1}K)_{ab}}{\lambda_a + \lambda_b}, \quad (KB_l)_{ab} = \frac{(KC_{l-1})_{ab}}{\lambda_a + \lambda_b}. \tag{27}$$

Substituting them into Eq. (18b'), we obtain the solution,

$$(B_l)_{a\beta} = \frac{1}{\lambda_a}\left[(C_{l-1})_{a\beta} - (C_{l-1}E)_{ab}\frac{\lambda_b}{\lambda_a + \lambda_b}E_{b\beta}\right], \tag{28a}$$

$$(B_l)_{ab} = \left[(C_{l-1})_{ab} - E_{aa}\frac{\lambda_a}{\lambda_a + \lambda_b}(EC_{l-1})_{ab}\right]\frac{1}{\lambda_b}. \tag{28b}$$

The local conserved currents can then be obtained from Eq. (19) and the first two are

$$j_{0\xi} = t_r\hat{K}_\xi\hat{B}_0 = K_{\alpha\beta}E_{\beta\alpha} + K_{ab}E_{ba} = 2t_r(k_\xi^+k_\xi)^{1/2}, \quad j_{0\eta} = 0; \tag{29a}$$

$$j_{1\xi} = t_r\hat{K}_\xi\hat{B}_1 = \sum_a\left(\frac{(C_0K)_{aa}}{2\lambda_a} + \frac{(KC_0)_{aa}}{2\lambda_a}\right) = -\frac{1}{2}[D(e^+e)]_{aa} = 0,$$

$$j_{1\eta} = 0. \tag{29b}$$

Solution (27) has another property, that is,

$$t_r\hat{B}_0\hat{B}_1 = \sum_a \frac{1}{2\lambda_a}[(C_0E)_{aa} + (EC_0)_{aa}]$$
$$= -\sum_a \frac{1}{2\lambda_a}[D(e^+e)]_{aa} = 0. \tag{30}$$

as required by the normalization condition (12) for $\hat{B}(\gamma)$.

$$t_r\left(\sum_{j=0}^{l} \hat{B}_j\hat{B}_{l-j}\right) = \text{const}\ \delta_{l0}. \tag{12'}$$

Taking account of the condition (30), we can deduce the second conserved current from (27) after some calculation.

$$j_{2\xi} = t_r\hat{K}_\xi\hat{B}_2 = \frac{1}{2}(C_1E + EC_1)_{aa} = -t_r\left(B_1^2K_\xi B_0 + \frac{1}{2}B_1K_\xi B_1B_0\right)$$
$$= -\sum_{ab}\frac{1}{\lambda_a}[(C_0^2)_{aa} - (C_0E)_{ab}(EC_0)_{ba}] - \sum_{ab}\frac{\lambda_a}{(\lambda_a+\lambda_b)^2}(C_0E)_{ab}(EC_0)_{ba}$$
$$= -\sum_{ab}\frac{1}{\lambda_a}[(D_\xi E)^2_{aa} - (D_\xi EE)_{ab}(ED_\xi E)_{ba}]$$
$$-\sum_{ab}\frac{\lambda_a}{(\lambda_a+\lambda_b)^2}(D_\xi EE)_{ab}(ED_\xi E)_{ba}. \tag{31a}$$

$$j_{2\eta} = t_r(\hat{K}_\eta\hat{B}_0) = \sum_a \lambda_a^{-1}(K_\xi K_\eta + K_\eta K_\xi)_{aa}. \tag{31b}$$

Alternatively the $U(m+n)/U(m)\otimes U(n)$ σ-model can also be defined as a locally $U(m)$ invariant theory of an $(m+n)\times n$ matrix field $z(x) = z_{ia}(x)$ (which is the first m columns of field $g(x)$, $i=1,2,\cdots,m+n$; $a=1,2,\cdots,m$) with $\bar{z}z=I$($\bar{z}$ is the hermitian conjugate of z), or the corresponding projector field[12].

$$P = z\bar{z}(P_{ij} = z_{ia}\bar{z}_{aj}). \tag{32}$$

The Lagrangian (1) can be written as

$$\mathscr{L} = \frac{1}{2}t_r\hat{K}_\mu\hat{K}^\mu = \frac{1}{2}t_r\overline{D_\mu g}D^\mu g$$
$$= \frac{1}{2}t_r\partial_\mu P\partial^\mu P = t_r\overline{D_\mu z}D^\mu z. \tag{33}$$

Here the covariant derivative for z field is defined as

$$D_\mu z = \partial_\mu z - izA_\mu,\quad \overline{D_\mu z} = \partial_\mu\bar{z} + iA_\mu\bar{z}, \tag{34}$$

with the $U(m)$ gauge field defined as

$$A_\mu = -i\bar{z}\partial_\mu z,\quad (A_\mu)_{ab} = (\hat{H}_\mu)_{ab}. \tag{35}$$

For a matrix X transforming as $\bar{z}z$, the covariant derivative is defined as

$$D_\mu X = \partial_\mu X + i[A_\mu, X]. \tag{34'}$$

The field equation is

$$[P, \partial_\mu\partial^\mu P] = 0,$$

or

$$D_\mu D^\mu z + z\overline{D_\mu z}D^\mu z = 0. \tag{36}$$

It can be shown after some algebraic operation, that

$$k_\xi^+ k_\xi = \overline{D_\xi z} D_\xi z = \Lambda^2. \tag{37a}$$

$$k_\xi^+ D_\xi k_\xi = \overline{D_\xi z} D_\xi D_\xi z, \tag{37b}$$

$$\overline{D_\xi k_\xi} k_\xi = \overline{D_\xi D_\xi z} D_\xi z, \tag{37c}$$

$$\overline{D_\xi k_\xi} D_\xi k_\xi = \overline{D_\xi D_\xi z} D_\xi D_\xi z - (\overline{D_\xi z}\, D_\xi z)^2. \tag{37d}$$

and

$$D_\xi \Lambda^2 = \overline{D_\xi D_\xi z}\, D_\xi z + \overline{D_\xi z} D_\xi D_\xi z. \tag{37e}$$

After a lengthy deduction, Eqs. (31) can be rewritten as

$$\begin{aligned} j_{2\xi} = & -t_r \Lambda^{-3} [\overline{D_\xi D_\xi z}\, D_\xi D_\xi z - \Lambda^4 - \overline{D_\xi D_\xi z} D_\xi z \Lambda^{-2} \overline{D_\xi z} D_\xi D_\xi z] \\ & - \sum_{ab} \lambda_a^{-1} \lambda_b^{-2} (\lambda_a + \lambda_b)^{-3} (\overline{D_\xi D_\xi z} D_\xi z)_{ab} (\overline{D_\xi z} D_\xi D_\xi z)_{ba} \\ & + \sum_{ab} \lambda_a^{-1} \lambda_b^{-1} (\lambda_a + \lambda_b)^{-3}\, (D_\xi \Lambda^2)_{ba} (D_\xi \Lambda^2)_{ba}. \end{aligned} \tag{38a}$$

$$j_{2\eta} = t_r \Lambda^{-1} (\overline{D_\xi z} D_\eta z + \overline{D_\eta z} D_\xi z). \tag{38b}$$

For CP_n model. Λ is a single component quantity depending only on ξ and can be taken equal to 1 by using a proper conformal transformation[3], so $D_\xi \Lambda^2 = 0$ and Eqs. (38) reduce to

$$j_{2\xi} = -\Lambda^{-3} (\overline{D_\xi D_\xi z} D_\xi D_\xi z - \Lambda^4) + \frac{3}{4} \Lambda^{-5} \overline{D_\xi D_\xi z} D_\xi z \overline{D_\xi z} D_\xi D_\xi z. \tag{39a}$$

$$j_{2\eta} = \Lambda^{-1} (\overline{D_\xi z} D_\eta z + \overline{D_\eta z} D_\xi z). \tag{39b}$$

Then the result coincides with results given by Eichenherr[4] and by Scheler[4].

Now we turn to discuss the $U(N)$ principal chiral field[5]. An element of group $G = U_L(N) \otimes U_R(N)$ is expressed as

$$g = (g_L, g_R), \quad g_L^+ g_L = g_R^+ g_R = 1. \tag{40}$$

with the multiplication rule

$$g_1 g_2 = (g_{1L} g_{2L}, g_{1R} g_{2R}). \tag{41}$$

Any element of G can be decomposed into a product of an element of the subgroup, $h \in H = U_{L+R}(N)$,

$$h = (h, h)$$

with a left coset element $\Phi \in G/H$,

$$\Phi = (\phi, \phi^{-1}).$$

Therefore $g = \Phi h$ means

$$g_L = \phi h, \quad g_R = \phi^{-1} h. \tag{42}$$

Following Eq. (2), we define

$$\mathscr{H} + \mathscr{K} \equiv -i\Phi^{-1}\partial\Phi = -i(\phi^{-1}\partial\phi, \phi\partial\phi^{-1}) = (H+K, H-K). \tag{43}$$

Here

$$\left.\begin{aligned}&\mathscr{H} = (H, H), \quad \mathscr{K} = (K, -K),\\ &H = -\frac{i}{2}(\phi^{-1}\partial\phi + \phi\partial\phi^{-1}), \quad K = -\frac{i}{2}(\phi^{-1}\partial\phi - \phi\partial\phi^{-1}).\end{aligned}\right\} \tag{43'}$$

Now the Cartan inner product of two elements can be written as

$$\langle g_1, g_2\rangle = T_r g_1^+ g_2 = t_r g_{1L}^+ g_{2L} + t_r g_{1R}^+ g_{2R}. \tag{44}$$

Then the Lagrangian of a principal model takes the form

$$\begin{aligned}\mathscr{L} &= \frac{1}{2} T_r \mathscr{K}_\mu \mathscr{K}^\mu = \frac{1}{2} T_r \mathscr{D}_\mu \Phi \mathscr{D}^\mu \Phi = t_r K_\mu K^\mu\\ &= \frac{1}{4} t_r \partial_\mu \phi^2 \partial^\mu \phi^{-2} = \frac{1}{4} t_r \partial_\mu q \partial^\mu q^{-1}.\end{aligned} \tag{45}$$

Here the covariant derivative for the field Φ or Φ^+ is defined as

$$\left.\begin{aligned}&\mathscr{D}_\mu \Phi \equiv \partial_\mu \Phi - i\Phi \mathscr{H} = (\partial_\mu \phi - i\phi H_\mu, \partial_\mu \phi^{-1} - i\phi^{-1} H_\mu),\\ &\mathscr{D}_\mu \Phi \equiv \partial_\mu \Phi^+ + i\mathscr{H}\Phi^+ = (\partial_\mu \phi^+ + iH_\mu \phi^+, \partial_\mu \phi + iH_\mu \phi),\end{aligned}\right\} \tag{46}$$

and

$$q = \phi^2 \tag{47}$$

is also unitary. The equation of motion from Lagrangian (45) is

$$\mathscr{D}^\mu \mathscr{K}_\mu \equiv \partial^\mu \mathscr{K}_\mu + i[\mathscr{H}^\mu, \mathscr{K}_\mu] = 0. \tag{48}$$

and the integrability condition implies

$$\left.\begin{aligned}&\mathscr{D}_\mu \mathscr{K}_\nu = \mathscr{D}_\nu \mathscr{K}_\mu,\\ &\partial_\mu \mathscr{H}_\nu - \partial_\nu \mathscr{H}_\mu + i[\mathscr{H}_\mu, \mathscr{H}_\nu] = -i[\mathscr{K}_\mu, \mathscr{K}_\nu].\end{aligned}\right\}. \tag{49}$$

So the duality symmetry (15) stands.

According to Eq. (8) the conserved current $\mathscr{J}$ can be written as

$$\mathscr{J} = \Phi \mathscr{K} \Phi^{-1} = (\phi K \phi^{-1}, -\phi^{-1} K \phi) = (J_L, J_R), \tag{50}$$

and

$$J_L = -\frac{i}{2}\phi(\phi^{-1}\partial\phi - \phi\partial\phi^{-1})\phi^{-1} = \frac{i}{2}\phi^2\partial(\phi^{-2}) = \frac{i}{2} q\partial q^{-1}, \tag{51a}$$

$$J_R = \frac{i}{2}\phi^{-1}(\phi^{-1}\partial\phi - \phi\partial\phi^{-1})\phi = \frac{i}{2}\phi^{-2}\partial\phi^2 = \frac{i}{2} q^{-1}\partial q. \tag{51b}$$

By introducing G valued function $\mathscr{A}(x)$ or equivalently $\mathscr{B}(x) = \Phi^{-1}\mathscr{A}(x)\Phi$, the quantity

$$\mathscr{J}_\mu^A = T_r \mathscr{J}_\mu \mathscr{A} = T_r \mathscr{K}_\mu \mathscr{B} = t_r K_\mu (B_L - B_R) \tag{52}$$

will be conserved if $\mathscr{D}^\mu \mathscr{B}$ is orthogonal to $\mathscr{K}_\mu$. When we take

$$\mathscr{D}_\mu \mathscr{B} = a\varepsilon_{\mu\nu}\mathscr{K}^\nu - a^{-1}\varepsilon_{\mu\nu}\mathscr{B}\mathscr{K}^\nu\mathscr{B},$$

the solution may only take values on the subset of the Lie algebra corresponding to the coset, i. e.

$$\mathscr{B} = (B_L, B_R) = (B, -B) \tag{53}$$

with B satisfying Eq. (13). Then Eq. (18) holds as before, and an infinitive set of local conserved currents is again given by Eq. (19). In present case we can make the positive definite hermitian matrix $K_\xi K_\xi$ diagonal by an $U(N)$ transformation[6],

$$K_\xi K_\xi = \begin{pmatrix} \lambda_1^2 & & & \\ & \lambda_2^2 & & \\ & & \ddots & \\ & & & \lambda_N^2 \end{pmatrix} \tag{54}$$

and assume it to be invertible so that $\lambda_i > 0$ ($i = 1, 2, \cdots, N$). Then we have the solution of Eq. (18a) for B_0 (the parameter a is put into 1),

$$B_0 = (K_\xi K_\xi)^{-1/2} K_\xi = K_\xi (K_\xi K_\xi)^{-1/2} \tag{55}$$

with the properties,

$$B_0 K_\xi = K_\xi B_0 = (K_\xi K_\xi)^{1/2}, \quad B_0^2 = 1, \tag{56}$$

and the recurrence formulas similar to Eqs. (27) and (28) are

$$(B_l K)_{ij} = \frac{(C_{l-1} K)_{ij}}{\lambda_i + \lambda_j}, \quad (K B_l)_{ij} = \frac{(K C_{l-1})_{ij}}{\lambda_i + \lambda_j}. \tag{57}$$

$$(B_l)_{ij} = \frac{(C_{l-1})_{ij}}{\lambda_i + \lambda_j}. \tag{58}$$

The zeroth conserved current follows from Eq. (55).

$$j_{0\xi} = t_r K_\xi B_0 = t_r (K_\xi K_\xi)^{1/2}, \quad j_{0\eta} = 0, \tag{59}$$

and the first current can be deduced from Eq. (57).

$$\begin{aligned} j_{1\xi} &= t_r K_\xi B_1 = \frac{1}{2} \lambda_i^{-1} (K_\xi C_0)_{ii} \\ &= -\frac{1}{2} (B_0)_{ij} (D_\xi B_0)_{ji} = -\frac{1}{4} \partial_\xi B_0^2 = 0, \\ j_{1\eta} &= 0. \end{aligned} \tag{60}$$

These results coincide with those of Ref. [6]. Similar to Eq. (30) for Grassmann model, we can also obtain an equation

$$t_r B_0 B_1 = -\frac{1}{2} \lambda_i^{-1} (B_0)_{ij} (D_\xi B_0)_{ji} = 0. \tag{61}$$

as required by the normalization condition for $B(\gamma)$. From Eq. (61) the second conserved local current can be obtained by substituting B_1 from Eq. (58) into Eq. (57) for $l = 2$.

$$\begin{aligned} j_{2\xi} &= t_r K_\xi B_2 \\ &= 1/2 t_r C_1 B_0 && \text{(57) with } l = 2 \\ &= -1/2 t_r (D_\xi B_1 + B_1 K_\xi B_1) B_0 && \text{(18b}') \text{ with } l = 2 \end{aligned}$$

$$
\begin{aligned}
&= -1/2t_r D_\xi(B_1B_0) + 1/2t_r B_1 D_\xi B_0 - 1/2t_r B_1 K_\xi B_1 B_0 \\
&= -1/2\partial_\xi t_r(B_1B_0) - 1/2t_r B_1 K_\xi B_1 B_0 \\
&\quad - 1/2t_r B_1(B_0 K_\xi B_1 + B_1 K_\xi B_0) && (18\mathrm{b}') \text{ with } l = 1 \\
&= -t_r(B_1^2 K_\xi B_0 + 1/2 B_1 K_\xi B_1 B_0) && (61) \text{ and } (56) \\
&= -1/2(\lambda_i + \lambda_j)^{-1}[(C_0)_{ij}(C_0)_{ji} + 1/2(C_0B_0)_{ij}(C_0B_0)_{ji}] && (58) \text{ and } (57) \\
&= -1/2(\lambda_i + \lambda_j)^{-1} \\
&\quad \times [(D_\xi B_0)_{ij}(D_\xi B_0)_{ji} + 1/2(D_\xi B_0 B_0)_{ij}(D_\xi B_0 B_0)_{ji}] && (18\mathrm{b}') \text{ with } l = 1 \\
&= -1/4(\lambda_i + \lambda_j)^{-1}(D_\xi B_0)_{ij}(D_\xi B_0)_{ji}.
\end{aligned}
$$

$$
\begin{aligned}
j_{2\eta} &= t_r K_\eta B_0 \\
&= \lambda_i^{-1}(K_\eta)_{ij}(K_\xi)_{ji}. && (62)
\end{aligned}
$$

References

[1] Chou, K. C. & Song, X. C., *Scientia Sinica* (Series A), **25** (1982), 716; further references are listed there.

[2] Eichenherr, H. & Forger, M., *Nucl. Phys.*, **B155** (1979), 381.

[3] Pohlmeyer, K., *Commun. Math. Phys.*, **46** (1976), 207.

[4] Eichenherr, H., *Phys. Lett.*, **90B** (1980), 121; Scheler, K., *ibid.*, **93B** (1980), 331.

[5] Zaharov, V. E. & Mikhailov, A. V., (*Sov. Phys.*) *JETP*, **47** (1978), 1017.

[6] Ogielski, A. T., Prasad, M. K., Sinha, A. & Chau Wang, L. L., *Phys. Lett.*, **91B** (1980), 387.

Volume 109B, number 6 PHYSICS LETTERS 11 March 1982

COMPOSITE GAUGE BOSONS IN A NONABELIAN THEORY ☆

Swee-Ping CHIA [1] and Charles B. CHIU
Center for Particle Theory, Department of Physics, University of Texas, Austin, TX, 78712, USA
and
Kuang-Chao CHOU [2]
Institute of Theoretical Physics, Academia Sinica, Beijing, People's Republic of China

Received 22 December 1981

We contend that the recent Weinberg–Witten no-go theorem does not necessarily rule out the possibility of gauge bosons being composite objects. A dynamical model for composite SU(3) gauge bosons is presented.

Recently Weinberg and Witten [1] gave a surprisingly simple theorem. Among other things they show that, in a theory in which there is a conserved Lorentz covariant vector current, there cannot exist massless vector particles, whether elementary or composite. Their arguments are based on the Lorentz covariance of the current [2]. This no-go theorem has cast some doubts on the line of approach of many authors [3], wherein one considers the possibility of non-abelian gauge bosons as composite objects. We contend, however, that this theorem does not necessarily rule out the possibility that gauge bosons can be bound states of some elementary fermion fields. In fact, we find that if the massless bound state is neutral with respect to the charged Noether current defined in terms of the fermion fields, then the no-go theorem is bypassed, and this bound state can exist. In this paper, we illustrate this point with an explicit example.

Consider a theory with four-fermion vector interaction. At the fundamental level there is, for definiteness say, a global SU(3)-color symmetry. The lagrangian density is given by

$$\mathcal{L} = \bar{\psi}(i\not{\partial} - m)\psi - \tfrac{1}{2}G(\bar{\psi}\gamma^{\mu}\tfrac{1}{2}\lambda^{a}\psi)(\bar{\psi}\gamma_{\mu}\tfrac{1}{2}\lambda^{a}\psi). \quad (1)$$

The corresponding SU(3) Noether current is

$$j^{a}_{\mu} = -g\bar{\psi}\gamma_{\mu}\tfrac{1}{2}\lambda^{a}\psi\,, \quad (2)$$

where $g = a\sqrt{G}$ is a dimensionless coupling constant and "a" some energy scale. The quantity j^{a}_{μ} is a conserved four-vector current. For this theory, it would appear that, by applying the argument of ref. [1], it is not possible to obtain a massless octet-color-gluon as a bound state of the fermion–antifermion system. To arrive at this conclusion, one first observes that it is not possible to construct a nonvanishing covariant matrix element of the Noether current of eq. (2) between two massless vector bound states. In the spirit of ref. [1], one would then assert that the massless vector bound state should not exist.

However, we find that the following situation may also be admissible. As we shall explicitly show, the four-fermion theory of eq. (1) without taking any specific limit could be formally equivalent to a theory with a massive vector boson mediating the current of eq. (2). Upon taking the strong-coupling limit, the massive vector boson is converted into a massless vector boson and a massless scalar boson. In this limit, the massless scalar boson, which contributes to the breaking of local gauge invariance, decouples completely from the rest of the system. In

☆ This work is supported in part by the US Department of Energy.

[1] On leave from Department of Physics, University of Malaya, Kuala Lumpur, Malaysia.

[2] Present address: Theory Division, CERN, Geneva 23, Switzerland.

turn the remaining system acquires a local gauge invariance and the massless vector boson can be identified as the gauge boson. We will further show that the matrix element of j^a_μ of eq. (2) between two massless bound states is zero, showing that the bound states are neutral with respect to the current. This illustrates our contention that the neutrality of the gauge boson to the original current could be used as a dynamical mechanism to bypass the no-go theorem, and to allow the gauge boson to be a composite object.

We now proceed to present the details of our arguments. We first write the generating functional for eq. (1) as:

$$W = \int D[\psi]\,\partial[\bar\psi]\,D[A^a]\exp\Big(\mathrm{i}\int \mathrm{d}^4x \times [\bar\psi(\mathrm{i}\partial\!\!\!/ - m - gA\!\!\!/)\psi + \tfrac{1}{2}a^2 A^a_\mu A^{a\mu} + \bar\psi\eta + \bar\eta\psi]\Big). \tag{3}$$

where η and $\bar\eta$ are four-component Grassmann color sources. Here the auxiliary octet field A^a_μ is introduced [4]. And we have used the notation $A_\mu = \frac{1}{2}A^a_\mu\lambda^a$. Integrating over the fermion fields in eq. (3), one gets:

$$W \sim \int D[A^a]\exp\Big(\mathrm{i}\int \mathrm{d}^4x \times [\mathcal{L}_\mathrm{g} + \bar\eta(\mathrm{i}\partial\!\!\!/ - m - gA\!\!\!/)^{-1}\eta + \tfrac{1}{2}a^2 A^a_\mu A^{a\mu}]\Big), \tag{4}$$

where

$$\mathcal{L}_\mathrm{g} = -\tfrac{1}{4}(\tfrac{1}{6}g^2 I_0)A^a_{\mu\nu}A^{a\mu\nu} + R(gA_{\mu\nu}), \tag{5}$$

with

$$A^a_{\mu\nu} = \partial_\mu A^a_\nu - \partial_\nu A^a_\mu - gf^{abc}A^b_\mu A^c_\nu, \tag{6}$$

and

$$I_0 = -\frac{\mathrm{i}}{4\pi^2}\int \mathrm{d}^4p\,(p^2 - m^2 + \mathrm{i}\epsilon)^{-2} = \frac{1}{2(4-n)} + \cdots. \tag{7}$$

In eq. (5), we have explicitly written down only the divergent term. In order to maintain manifest gauge invariance, we have used dimensional regularization with the cutoff prescription of keeping the $(4-n)$ factor finite, where n is the dimension. For our demonstration below we could have also used the Pauli–Villars regularization procedure, keeping the mass of the ghost fermion finite. The remaining finite terms are collected in $R(gA_{\mu\nu})$. Owing to the gauge invariant cutoff procedure adopted, the remainder is a function of $gA_{\mu\nu}$.

Eqs. (4) and (5) differ from those of earlier work [4] [‡1] in that, with our gauge invariant regularization prescription, the quadratic term of the vector field is absent in $\mathcal{L}_\mathrm{g}$ of eq. (5). In earlier work, one assumes the cancellation of the quadratic terms, which could be motivated [‡1] by regarding A_μ to be a Nambu–Goldstone boson associated with the breakdown of Lorentz symmetry. Since this cancellation procedure involves a nongauge invariant cutoff prescription, we do not adhere to this point of view here.

Without taking any specific limit, we assume that eqs. (4) and (5) together imply the presence of a massive bound state [‡2]. More specifically we denote the renormalized quantities by:

$$\tilde A^a_\mu = Z_3^{-1/2}A^a_\mu \quad\text{and}\quad g_\mathrm{R} = Z_3^{1/2}g, \tag{8}$$

where

$$Z_3 = 6/(g^2 I_0). \tag{9}$$

Here Z_3 takes into account renormalization with respect to the effects of fermions only. The theory now corresponds to an effective theory with a vector meson of mass $M = aZ_3^{1/2}$.

We shall now take the strong-coupling limit, $g \to \infty$, and in turn $Z_3 \to 0$. The theory now passes from the massive case to the massless case. At this point it is useful to perform some formal manipulation to display explicitly what happens to the degrees of freedom as the theory passes from the massive case to the massless case. Let us go back to eq. (3). We first introduce a set of new fields: $B_\mu = \frac{1}{2}B^a_\mu\lambda^a$, $\phi = \frac{1}{2}\phi^a\lambda^a$, and ψ_p and a new source η_p, by means of the SU(3) local transformation $U = \exp(-\mathrm{i}g\phi/a)$. In particular,

$$A_\mu = U[B_\mu + L^a(\phi)\,\partial_\mu\phi^a]\,U^\dagger, \quad \psi = U\psi_\mathrm{p} \text{ and } \eta = U\eta_\mathrm{p} \tag{10}$$

‡1 The original argument for the abelian case can be found for example in ref. [5].

‡2 For the abelian case, see ref. [6].

Volume 109B, number 6 PHYSICS LETTERS 11 March 1982

where

$$UL^a \partial_\mu \phi^a U^\dagger = -(\mathrm{i}/g) U \partial_\mu U^\dagger . \qquad (11)$$

Substituting these into eq. (3), the lagrangian becomes

$$\mathcal{L} = \bar{\psi}_\mathrm{p}(\mathrm{i}\not\partial - m - g\not B)\psi_\mathrm{p} + \bar{\psi}_\mathrm{p}\eta_\mathrm{p} + \bar{\eta}_\mathrm{p}\psi_\mathrm{p}$$
$$+ a^2 \,\mathrm{tr}[(B_\mu + L^a \partial_\mu \phi^a)(B^\mu + L^a \partial_\mu \phi^a)] . \qquad (12)$$

The path integral measure is also changed accordingly:

$$\mathcal{D} = D[\psi] D[\bar{\psi}] D[A^a]$$
$$= D[\psi_\mathrm{p}] D[\bar{\psi}_\mathrm{p}] D[B^a] D[\phi^a]\, \delta(\phi^a) . \qquad (13)$$

The $\delta(\phi^a)$ constraint in eq. (13) can be converted into a gauge condition on B-fields, for definiteness say the axial gauge condition: $n_\mu B^\mu = 0$, where n_μ is a space-like four-vector. After some formal minipulation, the measure (13) can be written as

$$\mathcal{D} \sim D[\psi_\mathrm{p}] D[\bar{\psi}_\mathrm{p}] D[B^a]\, \delta(n \cdot B^a) D[\phi^a]\, \Delta(\phi^a), \qquad (14)$$

where $\Delta(\phi^a)$ is a determinant involving only ϕ-fields. We record here that although the axial gauge is chosen explicitly, we could have also chosen the Lorentz gauge and obtained a gauge invariant measure which is also frame independent.

Finally after integrating over ψ_p and $\bar{\psi}_\mathrm{p}$, the generational function becomes

$$W \sim \int D[B^a] D[\phi^a]\, \Delta(\phi^a) \delta(n \cdot B^a) \exp\left(\mathrm{i} \int \mathrm{d}^4 x \mathcal{L}\right). \qquad (15)$$

where

$$\mathcal{L} = -\tfrac{1}{4}(\tfrac{1}{6} g^2 I_0) B^a_{\mu\nu} B^{a\mu\nu} + R(g B_{\mu\nu})$$
$$+ \bar{\eta}_\mathrm{p}(\mathrm{i}\not\partial - m - g\not B)^{-1} \eta_\mathrm{p}$$
$$+ a^2\, \mathrm{tr}[(B_\mu + L^a \partial_\mu \phi^a)(B^\mu + L^a \partial^\mu \phi^a)] . \qquad (16)$$

Using Z_3 and g_R defined earlier and $\widetilde{B}^a = B^a Z_3^{-1/2}$, we can rewrite the langrangian as

$$\mathcal{L} = -\tfrac{1}{4}\widetilde{B}^a_{\mu\nu}\widetilde{B}^{a\mu\nu} + R(g_\mathrm{R}\widetilde{B}_{\mu\nu}) + \bar{\eta}_\mathrm{p}(\mathrm{i}\not\partial - m - g_\mathrm{R}\not{\widetilde{B}})^{-1}\eta_\mathrm{p}$$
$$+ a^2\, \mathrm{tr}[(Z_3^{1/2}\widetilde{B}_\mu + L^a \partial_\mu \phi^a)(Z_3^{1/2}\widetilde{B}^\mu + L^a \partial^\mu \phi^a)] . \qquad (17)$$

Taking the limit $g \to \infty$, which implies $Z_3 \to 0$,

$$\mathcal{L} \to -\tfrac{1}{4}\widetilde{B}^a_{\mu\nu}\widetilde{B}^{a\mu\nu} + R(g_\mathrm{R}\widetilde{B}_{\mu\nu}) + \bar{\eta}_\mathrm{p}(\mathrm{i}\not\partial - m - g_\mathrm{R}\not{\widetilde{B}})^{-1}\eta_\mathrm{p}$$
$$+ a^2\, \mathrm{tr}(L^a L^b)\, \partial_\mu \phi^a \partial^\mu \phi^b . \qquad (18)$$

It is to be noted that so long as Z_3 is nonzero, ϕ is not an independent field. It can always be transformed away through the redefinition of the field. On the other hand, in the limit $Z_3 = 0$, ϕ can no longer be transformed away. So among the three degrees of freedom for the massive vector boson, two of them are transferred to the massless vector boson and the remaining one to the scalar boson. Furthermore, the coupling term between $\widetilde{B}^a_\mu$ and ϕ^a is proportional to $Z_3^{1/2}$. In the limit $Z_3 = 0$, ϕ^a completely decouples from $\widetilde{B}^a_\mu$. Also in this limit the $a^2\, \mathrm{tr}(L^a L^b) \partial_\mu \phi^a \partial^\mu \phi^b$ term is the only term in eq. (18) which is not local gauge invariant. Owing to the ϕ-decoupling mechanism, the remaining effective theory has now a local gauge invariance.

One can explicitly check that for the QCD theory with a local gauge symmetry, the generating functional after integrating over the fermion fields gives, in the strong-coupling limit, an expression which is identical to eq. (18) without the ϕ-term. Thus the four-fermion interaction theory defined by eq. (1) has at least in a formal sense generated a gauge boson.

We now turn to the Noether current of eq. (2). From eq. (3),

$$\langle j^a_\mu \rangle = \langle -g \bar{\psi} \gamma_\mu \tfrac{1}{2}\lambda_a \psi \rangle = -\frac{\mathrm{i}}{W} \int D[\psi] D[\bar{\psi}] D[A^b]$$
$$\times \exp\left(\mathrm{i} \int \mathrm{d}^4 x \frac{a^2}{2} A^b_\mu A^{b\mu}\right)$$
$$\times \frac{\partial}{\partial A^a_\mu} \exp\left(\mathrm{i} \int \mathrm{d}^4 x [\bar{\psi}(\mathrm{i}\not\partial - m - g\not A)]\psi + \bar{\eta}\psi + \bar{\psi}\eta\right). \qquad (19)$$

Integrating over ψ and $\bar{\psi}$, and taking the derivative with respect to A^a_μ, we obtain

$$\langle j^a_\mu \rangle_\psi = j_\mu(A_\mu) = -a^2 A^a_\mu = -a^2 \langle A^a_\mu \rangle_\psi . \qquad (20)$$

We have used the Euler–Lagrange equation to arrive at eq. (20). The same result can also be obtained formally through integration by parts. In eq. (20) A^a_μ is the field operator in the A_μ sector with a fermion field having been averaged out.

The relation of eq. (20) can be traced to stem from the original constraint associated with the auxiliary

Volume 109B, number 6 PHYSICS LETTERS 11 March 1982

field A_μ. After our integrating over fermion fields and the identification of the massive bound state, this constraint is now promoted to a current-field identity. It can be shown that j^a_μ is still a conserved current, as it should be [‡3].

Eq. (20) also enables us to calculate the matrix element of j^a_μ between two massive $\widetilde{A}$-bound states. For instance, to lowest order in perturbation theory,

$$\langle \widetilde{A}^c(p_2)|j^{a\mu}(0)|\widetilde{A}^b(p_1)\rangle \approx -a^2 Z_3^{1/2}\, g_R f^{abc} \times [(p_1+p_2)^\mu \epsilon_2^\dagger\cdot\epsilon_1 - \epsilon_1^\mu p_1\cdot\epsilon_2^\dagger - \epsilon_2^{\dagger\mu} p_2\cdot\epsilon_1] . \qquad (21)$$

This expression displays its Lorentz covariance explicitly.

Eq. (20) together with eq. (10) gives:

$$j^a_\mu = -\tfrac{1}{2}a^2 Z_3^{1/2}\langle \mathrm{tr}(U^\dagger\lambda^a U\lambda^b)\widetilde{B}^b_\mu\rangle - a^2\langle \mathrm{tr}(U^\dagger\lambda^a U L^b)\partial_\mu\phi^b\rangle . \qquad (22)$$

Now we consider the limit $Z_3 = 0$. In this limit, the first term in eq. (22) vanishes [‡4]. So j^a_μ becomes a function of pure ϕ-fields. On the other hand, from eq. (17) one sees that in this limit ϕ decouples from B. Due to this ϕ-decoupling mechanism, the second term of eq. (22) will not contribute to the matrix elements of j^a_μ between two $\widetilde{B}$-states. We have succeeded in demonstrating that in the strong-coupling limit:

$$\langle \widetilde{B}^b|j^a_\mu|\widetilde{B}^c\rangle \to 0 , \qquad (23)$$

for all b and c. In other words, the massless bound state $\widetilde{B}$ is indeed neutral with respect to the original Noether current.

It is interesting to note that in the limit $Z_3 = 0$, from eq. (22)

$$j^a_\mu = -a^2\langle \mathrm{tr}\,\lambda^a U(-\mathrm{i}/g)\partial_\mu U^\dagger\rangle . \qquad (24)$$

which is still a conserved four-vector current [7] and it is associated with the global SU(3) invariance. However, the important point is that it is now decoupled from $\widetilde{B}$ and the fermions, ψ_p.

On the other hand, in the effective theory of $\widetilde{B}$ and ψ_p of the non-ϕ sector, there is a new Noether current:

$$j^{a\mu}_{\text{non-}\phi} = -g_R\bar{\psi}_p\gamma^\mu\tfrac{1}{2}\lambda^a\psi_p + g_R f^{abc}\widetilde{B}^b_\nu\widetilde{B}^{c\mu\nu} . \qquad (25)$$

In general, the matrix elements of $j^{a\mu\nu}_{\text{non-}\phi}$ between two B-states do not vanish. But now this current is no longer a Lorentz covariant four-vector. So the no-go theorem of ref. [1] does not apply.

In closing, let us reiterate the main points here. Our starting four-fermion theory has only the global SU(3)-color symmetry. The corresponding Noether current is well defined. It is a four-vector and is conserved. After taking into account the effect of fermion loop in renormalization, this theory in its strong-coupling limit, formally contains a massless vector boson $\widetilde{B}_\mu$ together with a massless scalar boson ϕ. In this limit the original Noether current can be identified with a pure ϕ contribution as in eq. (24). But in the strong coupling limit, ϕ decouples from the rest of the system. Consequently $\widetilde{B}_\mu$ is neutral with respect to the original Noether current. On the other hand, in the non-ϕ sector, we have a local gauge invariance with the massless vector bound state acting as the gauge boson. A new Noether current can now be defined. We have thus demonstrated our contention that the vanishing of the matrix elements of the Noether current defined by the original fermion theory, taken between massless vector bound states, does not necessarily preclude the possibility of the existence of composite gauge bosons.

We wish to thank Professor E.C.G. Sudarshan, Professor S. Weinberg and Dr. Xerxes Tata for stimulating discussions and for their invaluable suggestions. Two of us (SPC and KCC) would like to thank Professor E.C.G. Sudarshan and colleagues in the Center for Particle Theory for the hospitality during their visit.

[‡3] This can be verified using fermion equations of motion together with a generalization of the current-field identity and the antisymmetric properties of f^{abc}.

[‡4] As Z_3 approaches zero, taking into account the rapid oscillation in the phase factor of U, we find $\langle \mathrm{tr}(U^\dagger\lambda^a U\lambda^b) \times B^b_\mu\rangle \to \tfrac{1}{2}\langle B^a_\mu\rangle$. So the first term on the right-hand side of eq. (22) is proportional to $Z_3^{1/2}$.

References

[1] S. Weinberg and E. Witten, Phys. Lett. 96B (1980) 59.
[2] See also: E.C.G. Sudarshan, Phys. Rev. D24 (1981) 1591.

[3] See for examples: H. Terazawa, Y. Chikashige and K. Akama, Phys. Rev. D15 (1977) 480;
T. Eguchi, Phys. Rev. D17 (1978) 611;
F. Cooper, G.S. Guralnik and N.J. Snyderman, Phys. Rev. Lett. 40 (1978) 1620;
C.C. Chiang, C.B. Chiu, E.C.G. Sudarshan and X. Tata, Is the QCD gluon a composite object?, Phys. Rev. D, to be published; and references quoted therein.
[4] See for examples: K. Kikkawa, Prog. Theor. Phys. 56 (1976) 947;
T. Eguchi, Phys. Rev. D14 (1976) 2755;
H. Terazawa, Y. Chikashige and K. Akama, Phys. Rev. D15 (1977) 480.
[5] J.D. Bjorken, Ann. Phys. (NY) 24 (1963) 174;
G.S. Guralnik, Phys. Rev. 136 (1964) B1404.
[6] I. Bialynicki-Birula, Phys. Rev. 130 (1963) 465.
[7] A.T. Ogielski, M.K. Prasad, A. Sinha and L.L. Chau Wang, Phys. Lett. 91B (1980) 387.

Volume 110B, number 3,4 PHYSICS LETTERS 1 April 1982

MASSLESS BARYONS AND ANOMALIES IN CHIRAL $(QCD)_2$

D. AMATI, Kuang-Chao CHOU [1] and S. YANKIELOWICZ
CERN, Geneva, Switzerland

Received 25 January 1982

The spectrum of chiral SU(N) gauge theories in two dimensions is shown to consist of free massless mesons and baryons that may be visualized as composites of free massless quarks and antiquarks. Baryons do not contribute to anomalies which are saturated by the mesons. The singular nature of the chiral limit is recognized.

A dynamical framework is needed in order to understand how a confining theory treats chirality and anomalies: whether by realizing chirality and saturating anomalies with massless bound fermions [1] (baryons) or by saturating them with Goldstone bosons if chiral symmetry is spontaneously broken.

In a $1/N$ approach the second scenario is realized [2], but this is no surprise because that approach is unwarranted for a dynamics with massless fermion bound states. Attention was then directed to two-dimensional models [$SU(N)_2$] where the dynamics is accessible, hoping of course to be able to distinguish peculiarities of two dimensions (particularly pathological as far as symmetry breaking is concerned) from dynamical features expected to survive in four dimensions.

't Hooft did [3] the first analysis of $SU(N)_2$ with a $1/N$ expansion in a light-cone gauge. He identified a colour singlet meson bound state that in the chiral limit ($m = 0$ where m is the quark mass) becomes massless and free. This massless meson saturates the U(1) anomaly as expected in a spontaneously broken scenario. But as previously mentioned, the $1/N$ expansion is unsuitable for detecting massless fermions.

For $m = 0$ and for any N, the conservation of both vector and axial currents implies $\Box j_\mu = 0$ and thus the existence of a corresponding massless decoupled meson that saturates trivially the appropriate anomaly. In a recent paper [4] these states have been explicitly constructed and it has been argued that no massless baryon states could exist because they would spoil the anomaly condition.

On the other hand, working in a gauge first proposed by Baluni [5] the spectrum for finite m was identified [6] and shown to contain both bound mesons and baryons with masses going to zero for $m \to 0$. This is a common feature of two-dimensional models [7]. It was also recognized [7] that these baryons would necessarily spoil the anomaly condition (as a function of N) in the chiral limit if baryon form factors would satisfy some smoothness conditions.

We therefore find a few apparent contradictions. On the one hand the baryons present for arbitrarily small m could not disappear from the Hilbert space and therefore it should be possible to find them also in the strict $m = 0$ chiral theory. On the other hand, if they are there they should not contribute to anomalies and this implies a singular behaviour of form factors such as $m \to 0$. We shall indeed show that both statements are correct and that there is no contradiction left.

The picture that arises in the chiral limit is very trivial (free massless mesons and baryons) and is totally due to the two dimensionality that allows free massless quarks to produce free massless bound states. These two-dimensional gauge theories realize chirality in such a trivial way that they shed no light on dynamical mechanisms for chirality realization or breaking in four dimensions.

In what follows we shall discuss the single flavour

[1] On leave from Institute of Theoretical Physics, Academia Sinica, Beijing, People's Republic of China.

case, the generalization to arbitrary N_f being straightforward.

Let us first identify the massless baryon for the $m = 0$ case following a method analogous to that used in ref. [4] to identify the boson. We shall therefore work in the $A_- = 0$ light-cone gauge in which the gauge field A_+ may be totally eliminated in terms of a Coulomb interaction and in which only the left-handed (actually a left mover) quark field $\psi_+ = \frac{1}{2}(1+\gamma_5)\psi$ is dynamical in the chiral $m = 0$ case under discussion. In this gauge, the role of the hamiltonian is played by the generator of displacements of $x^+ = 1/\sqrt{2}(x^0 + x^1)$ given by

$$\hat{P}_+ = -\frac{1}{4}\int dx^- \, dy^- \, dy^+ \, J^{+a}(x) \times \delta(x^+ - y^+)\,|x^- - y^-|\,J^{+a}(y)\,, \tag{1}$$

with

$$J^{+a} = \sqrt{2}\, g \psi_+^\dagger I^a \psi_+\,, \tag{2}$$

where I^a, $a = 1, ..., N^2 - 1$ are the SU(N) group generators. Energy momentum conservation ensures that $\hat{P}_+$ is independent of x^+ and one may take $x^+ = 0$ in studying the bound-state spectrum.

We expand $\psi_+(x)$ in terms of the creation and annihilation operators:

$$\psi_+(x) = \frac{1}{2^{1/4}}\int_0^\infty \frac{dP_-}{\sqrt{2\pi}}\left[e^{-iP_-x^-}\hat{b}(p_-) + e^{iP_-x^-}\hat{d}^+(p_-)\right]\,, \tag{3}$$

and write $\hat{P}_+$ in the form:

$$\begin{aligned}\hat{P}_+ = \int_0^\infty dq_- \, dp_- \int_{-\infty}^{\infty} V(k_-^2)\, dk_- \, &[\hat{b}_i^+(q_-)\hat{b}_j(q_- + k_-) \\ &+ \hat{b}_i^+(q_-)\hat{d}_j^+(-q_- - k_-) + \hat{d}_i(q_-)\hat{b}_j(-q_- + k_-) \\ &+ \hat{d}_i(q_-)\hat{d}_j^+(q_- - k_-)](I^a)_{ij}(I^a)_{mn} \\ &\times [\hat{b}_m^+(p_-)\hat{b}_n(p_- - k_-) + \hat{b}_m^+(p_-)\hat{d}_n^+(-p_- + k_-) \\ &+ \hat{d}_m(p_-)\hat{b}_n(-p_- - k_-) + \hat{d}_m(p_-)\hat{d}_n^+(p_- + k_-)]\,.\end{aligned} \tag{4}$$

The specific form of the Coulomb potential

$$V(k_-) = (g^2/4\pi)\, \mathrm{P}(k_-^{-2})\,, \tag{5}$$

where P stands for the principal value, plays no role in the following calculations. It is easily shown that the physical vacuum is the usual perturbative state $|0\rangle$ defined by $\hat{b}_i(p_-)|0\rangle = \hat{d}_i(p_-)|0\rangle = 0$ with the vacuum energy normalized to zero.

An eigenstate of $\hat{P}_+$ with eigenvalue P_+ and fixed P_- will have mass

$$M^2 = 2P_+P_-\,. \tag{6}$$

The exact zero-mass boson state found in ref. [4] has the form

$$|\varphi, p_-\rangle = \frac{P_-}{\sqrt{N}}\int_0^1 d\alpha \sum_{i=1}^{N} \hat{b}_i^+(\alpha p_-)\hat{d}_i^+((1-\alpha)p_-)|0\rangle \tag{7}$$

with a normalization condition

$$\langle\varphi, p'_-|\varphi, p_-\rangle = p_-\delta(p_- - p'_-)\,. \tag{7'}$$

Similarly, the massless baryon state we are looking for will be given by

$$|B, p_-\rangle = N_B \int_0^1 d\alpha_1 \ldots d\alpha_N \, \hat{b}_1^+(\alpha_1 p_-) \ldots \hat{b}_N^+(\alpha_N p_-) \times \delta(1 - \alpha_1 - \ldots - \alpha_N)|0\rangle \tag{8}$$

with the normalization constant

$$N_B = \left(P_-^N \int_0^1 d\alpha_1 \ldots d\alpha_N \, \delta(1 - \alpha_1 - \ldots - \alpha_N)\right)^{1/2} \tag{9}$$

Indeed, using the identities

$$(I^a)_{ij}(I^a)_{mn} = A\delta_{ij}\delta_{mn} + B\delta_{in}\delta_{jm}\,, \tag{10}$$

with

$$NA + B = 0\,, \tag{11}$$

and

$$A - B = -(N-1)^{-1}\, C_2(N)\,, \tag{12}$$

where $C_2(N)$ is the second Casimir invariant, one can easily prove that

$$\hat{P}_+|B, p_-\rangle = 0\,. \tag{13}$$

Therefore $|B, p_-\rangle$ represents an exact massless baryon state.

For quarks not in the fundamental representation of the SU(N) group, it is still possible to have massless bound baryon states provided the relations (11) and (12) are satisfied.

A direct calculation for the matrix elements of the

singlet U(1) current yields

$$\langle 0|J^+(q)|\varphi, p_-\rangle = \sqrt{N}\, q_-\, \delta(q_- - p_-)\, \delta(q_+)\,, \qquad (14)$$

and

$$\langle B, p'_-|J^+(q)|B, p_-\rangle = N(\min(p_-, p'_-)/\max(p_-, p'_-))^{N/2}$$
$$\times (p_- p'_-)^{1/2}\, \delta(q_- - p_- + p'_-)\, \delta(q_+)\,. \qquad (15)$$

The spectrum as well as the matrix elements found above are not a surprise. In order to see why, let us first recall that the light cone gauge $\hat{A}_- = 0$ implies that the right-handed quark field $\psi_- = \frac{1}{2}(1 - \gamma_5)\,\psi$ remains free. Moreover, in two dimensions a system of free massless quarks moving in the same direction has zero mass.

Had we chosen the opposite light-cone gauge condition $\hat{A}_+ = 0$ the roles would be reversed, ψ_+ would be free and the gauge invariant states (7) and (8) would appear as massless mesons and baryons constituted of free quarks. Their singlet current matrix elements are those of a free theory and being gauge-invariant quantities should coincide with those calculated before in the $\hat{A}_- = 0$ gauge. And, indeed, eqs. (14) and (15) represent free theory matrix elements. Moreover, $B\bar{B}$ will also appear as a massless bound state satisfying

$$\langle 0|J^+(q)|B, p_-; \bar{B}, p'_-\rangle = 0\,, \qquad (16)$$

as it is also easy to verify in the $\hat{A}_- = 0$ gauge. Eq. (16) implies of course that baryons do not contribute to the anomaly.

By further exploiting this separation of left and right worlds it is easy to see that the chiral theory consists wholly of free colourless mesons and baryons.

This very simple result leads to a fact that at first sight is puzzling. Usually the matrix elements of (15) and (16) are considered to be the same, each obtained by a different analytic continuation. We will show that this concept of analytic continuation is untenable in this two-dimensional chiral theory and we will identify the origin of this apparent contradiction. As already stressed, the left and right worlds are completely separated in the chiral theory and the physical states are composed of purely right movers or purely left movers, massless quarks of antiquarks. Therefore all physical states have zero square momentum and there is no invariant momentum transfer other than zero entering the theory. Thus the very concept of analytic continuation loses meaning.

This is well seen by considering the theory for $m \neq 0$ where analytic continuation of form factors has been established, and witnessing how the limit $m \to 0$ is non-analytic. Indeed, for $m/g \ll 1$ and working in the Baluni gauge [5] it has been established [6] that the spectrum is described by a sine-Gordon theory with a hamiltonian given, for the single flavour case we are discussing, by

$$H = \tfrac{1}{2}\pi^2 - \tfrac{1}{2}(\partial_x\varphi)^2 - \alpha[1 - \cos(\beta\varphi)]\,, \qquad (17)$$

where

$$\alpha = 4m^2 N_c\,, \quad \beta = (4\pi/N_c)^{1/2}\,. \qquad (18)$$

The baryon is identified with the soliton state with mass

$$M_s = 8\sqrt{\alpha}/\beta = (16/\sqrt{\pi})\, m N_c\,, \qquad (19)$$

that vanishes for $m \to 0$.

The U(1) current is given by

$$j_\mu = -(\beta/2\pi)\, \epsilon_{\mu\nu}\, \partial^\nu \varphi\,, \qquad (20)$$

and its matrix element between soliton states defines a U(1) form factor which has been exactly calculated [8]. It depends analytically on a parameter θ related to the momentum transfer t by $t = 4m^2 \cosh^2\theta/2$. For $t = 0$ $(m \neq 0)$ the form factor is equal to N (i.e., the charge). For $m \to 0$, $t \to 0$ irrespectively of θ which obviously implies a singular limit. We see therefore that for the chiral case the matrix elements (15) and (16) are not analytically related.

The limit $m = 0$ in the Baluni gauge is indeed subtle. The hamiltonian (17) loses its non-linear term which generates the topological excitations. It acquires, however, a fermionic zero mode [9] whose contribution to the current matrix elements is proportional to $\delta(q)$. It ensures that the baryon charge is N [i.e., eq. (15)] that does not contribute to the anomaly [i.e., eq. (16)] which is a $\delta(q^2)$ effect.

In everything said before we considered $N > 1$. The singular behaviour of the limit $m \to 0$ is also manifested if we consider the continuation to $N \leqslant 1$. In this region, corresponding to $4\pi \leqslant \beta^2 \leqslant 8\pi$ [cf. eq. (18)], the sine-Gordon theory of eq. (17) is equivalent to a repulsive massive Thirring model whose coupling g is related to β through [10]

$$1 + 2g/\pi \equiv \lambda = 8\pi/\beta^2 - 1\,.$$

In this region $(\lambda < 1, g < 0)$ there are no bound states and the soliton of the sine-Gordon equation corresponds

Volume 110B, number 3,4 PHYSICS LETTERS 1 April 1982

to the Thirring fermion. This is reflected in the observation that the time-like form factors of the sine-Gordon theory in this region have a cut starting at $4m^2$, unlike the meson pole which was present for $\lambda > 1$ which goes into the unphysical sheet [8]. Note that for $N = 1$ (i.e., $\lambda = 1$) we obtain a free fermion theory. Thus for $\lambda \leqslant 1$ the intermediate states contributing to the current correlation function will be fermion–antifermion pairs. Therefore, in the limit $m \to 0$ the anomaly is trivially saturated by the original quark. This apparent change of responsibility in going from $\lambda > 1$ to $\lambda \leqslant 1$ in the anomaly saturation has little meaning. As long as $m \neq 0$ a free $q\bar{q}$ system is basically different from a meson bound state, but for $m \to 0$ they become indistinguishable. This is obviously a peculiarity of two dimensions.

We see therefore that $SU(N)_2$, which has a rich light spectrum for $m \neq 0$, has a trivial chiral ($m = 0$) limit. There, the left and right moving worlds are separated and each includes free mesons and baryons which are trivial composites of massless quarks and antiquarks moving parallely. As it is easy to visualize from this trivial picture, anomalies are saturated by mesons (collinear massless $q\bar{q}$ states!) while baryon–antibaryons do not contribute because they cannot be created by the quark currents from the vacuum. This trivial way of realizing chirality is clearly a characterization of two dimensions and is therefore of little help in understanding how chirality is realized or broken in four-dimensional gauge theories.

References

[1] G. 't Hooft, in: Proc. Cargèse School (1979); see also: Y. Frishman, A. Schwimmer, T. Banks and S. Yankielowicz, Nucl. Phys. B177 (1981) 157.

[2] S. Coleman and E. Witten, Phys. Rev. Lett. 45 (1980) 100; G. Veneziano, Phys. Lett. 95B (1980) 90.

[3] G. 't Hooft, Nucl. Phys. B75 (1974) 461.

[4] W. Büchmüller, S.T. Love and R.D. Peccei, MPI-PAE/PTh, 70/81 (1981).

[5] V. Baluni, Phys. Lett. 90B (1980) 407.

[6] P.J. Steinhart, Nucl. Phys. B176 (1980) 100; D. Amati and E. Rabinovici, Phys. Lett. 101B (1981) 407.

[7] S. Elitzur, Y. Frishman and E. Rabinovici, Phys. Lett. 106B (1981) 403.

[8] M. Karowski and P. Weisz, Nucl. Phys. B139 (1978) 455.

[9] T. Banks, D. Horn and H. Neuberger, Nucl. Phys. B108 (1976) 119.

[10] S. Coleman, Erice lecture notes (1975).

Volume 114B, number 2,3 PHYSICS LETTERS 22 July 1982

ON THE DETERMINATION OF EFFECTIVE POTENTIALS IN SUPERSYMMETRIC THEORIES

D. AMATI and Kuang-chao CHOU [1]
CERN, Geneva, Switzerland

Received 8 April 1982

We propose a renormalization procedure for dynamically generated effective actions. We show that, as expected, it leads to no spontaneous supersymmetry breaking if this is unbroken at the tree level. We also understand why the usually adopted renormalization prescription has led in some models to an apparent supersymmetry breaking for an unacceptable negative-energy vacuum state.

Application of the well-established $1/N$ expansion to some supersymmetric models seemed to indicate the possibility of spontaneous dynamical breaking of supersymmetry violating general properties such as positivity of the ground-state energy or the index theorem [‡1]. This would shed negative light on $1/N$ techniques which, nevertheless, seem applicable to supersymmetric theories. This puzzling result is well illustrated by Zanon's model [2] where a non-supersymmetric minimum was found for a negative value of the effective potential.

In this note we wish to show that this apparent contradiction stems from an unappropriate renormalization procedure [3] adopted in the evaluation of effective potentials. Moreover, if the renormalization is correctly performed, the effective potential, in terms of the dynamical fields, vanishes at the origin and is otherwise positive, thus confirming that supersymmetry will be preserved by radiative corrections if it is not broken at the tree level [4].

We shall use Zanon's model to illustrate the correct renormalization procedure and find the loophole in that used in ref. [2]. The model consists of $N+1$ chiral supermultiplets ϕ and ϕ_i, $i = 1, ..., N$, described by the action [‡2]

$$S=\int d^4x\, d^4\theta\, (\bar\phi_i\phi_i + \bar\phi\phi) - \int d^2x\, d^2\theta\, \{[\tfrac{1}{2}m\phi^2 + \tfrac{1}{2}m_0\phi_i^2 + (g/\sqrt{N})\phi\phi_i^2] + \text{h.c.}\}$$

The ϕ_i fields appearing only bilinearly may be integrated over, thus leading to an effective action depending only on ϕ.

The well-known fact that these theories need only a wave function renormalization suggests a rescaling

$$\phi \to (NZ)^{1/2}\phi\,, \quad g \to Z^{-1/2}g\,, \quad m \to Z^{-1}m\,. \quad (1)$$

Fields, coupling constants and masses will now represent renormalized quantities in terms of which the integration of S over ϕ_i leads to

$$S_{\text{eff}} = N(ZS_\phi - \tfrac{1}{2}\log\det\mathcal{A})\,, \quad (2)$$

where

$$S_\phi = \int d^4x\, d^4\theta\, \bar\phi\phi - \frac{m}{2}\int d^4x\, d^2\theta\, \phi^2\,, \quad (3)$$

and

$$\det\mathcal{A} = \det\mathcal{U}\,\det{}^{-1}\mathcal{B}\,,$$

[1] On leave from Institute of Theoretical Physics, Academia Sinica, Beijing, People's Republic of China.

[‡1] The index theorem states that no spontaneous breaking of supersymmetry could happen if the numbers of the zero energy fermionic and bosonic states are unequal, see ref. [1].

[‡2] We have introduced a mass m for the fields ϕ_i. It avoids infrared problems. The final potential is infrared free so that m may be set to zero if one wishes to make a comparison with ref. [2].

Volume 114B, number 2,3 PHYSICS LETTERS 22 July 1982

$$\mathcal{U} = \begin{bmatrix} -2gF & -\Box & -2gA-m_0 & 0 \\ -\Box & -2g^*F^* & 0 & -2g^*A^*-m_0^* \\ -2gA-m_0 & 0 & 0 & -1 \\ 0 & -2g^*A^*-m_0^* & -1 & 0 \end{bmatrix},$$

$$\mathcal{B} = \begin{bmatrix} -\frac{1}{2}\mathrm{i}gA\sigma_2 & \frac{1}{4}\mathrm{i}\sigma^\mu\partial^\mu \\ \frac{1}{4}\mathrm{i}\sigma^{\mu\mathrm{T}}\partial^\mu & -\frac{1}{2}\mathrm{i}g^*A^*\sigma_2 \end{bmatrix}, \qquad (4)$$

$A(x)$ and $F(x)$ being the lowest component and the auxiliary field of the supermultiplet ϕ, respectively.

The usual way to evaluate the exact formal expression of eq. (2) is to develop it around constant fields. In so doing, one has to regularize ultraviolet divergences. We adopt a Pauli–Villars prescription which suffices for the case under analysis, and call M the regulator mass.

We now evaluate the log det term in eq. (2) through its $1/N$ expansion around constant fields $A = a, F = f$, $\chi = 0$. We shall see a singular behaviour in M for $M \to \infty$ not only in the potential (i.e., $-\mathcal{L}_{\text{eff}}$ for constant fields) but also in the coefficients of the A and χ kinetic terms. We thus write

$$\mathcal{L}_{\text{eff}} = (Z + C_A)A^*\Box A + (Z + C_\chi)\tfrac{1}{2}\mathrm{i}\bar{\chi}\sigma^\mu\partial^\mu\chi - V + R\,, \qquad (5)$$

where R is regular in M and contains derivatives in the fields A, F and χ higher than those explicitly written in eq. (5). V will have an expression of the form

$$V = -Zf^*f + maf + m^*a^*f^* + V_1\,, \qquad (6)$$

where C_A, C_χ and V_1 are explicit functions of M, a and f. V_1 is given by

$$V_1 = (1/64\pi^2)[(\alpha^2+\beta)^2\ln(\alpha^2+\beta) + (\alpha^2-\beta)^2\ln(\alpha^2-\beta) - 2\alpha^4\ln\alpha^2 - 2\beta^2\ln(M^2+\alpha^2) - 3\beta^2]\,, \qquad (7)$$

with

$$\alpha^2 = |2ga + m_0|^2\,, \quad \beta = 2|gf|\,. \qquad (8)$$

C_A has the form

$$C_A = (|g|^2/8\pi^2)\ln[(M^2+\alpha^2)/\alpha^2] + \text{terms regular in } M\,. \qquad (9)$$

The usual renormalization prescription [3] consists of determining Z by imposing a condition on V at an arbitrary specific value of f and a that defines a renormalization scale μ^2. This implies

$$Z = 1 - (|g|^2/8\pi^2)\ln M^2/\mu^2\,. \qquad (10)$$

Surely enough this is a legitimate renormalization condition in the sense that it eliminates all M^2 singularities in eq. (5) and in particular in the potential V. But if the auxiliary field f is eliminated in terms of a by

$$\partial V/\partial f = 0\,, \qquad (11)$$

then $V = V(a, f(a))$ acquires negative values, in particular for sufficiently large a. Moreover, if there is a stationary point,

$$\mathrm{d}V(a, f(a))/\mathrm{d}a = 0\,, \qquad (12)$$

away from the origin, it happens for negative values of V. This was the stationary point detected in ref. [2] and to which was assigned the responsibility of spontaneous symmetry breaking. In order to understand the reason for this apparent inconsistency with positivity of the ground-state energy implied by supersymmetry, let us analyze the kinetic terms of eq. (5). Using (9) and (10) we find, after setting $M = \infty$

$$Z + C_A = 1 + (|g|^2/8\pi^2)\ln\mu^2/\alpha^2 + \ldots\,, \qquad (13)$$

an expression which becomes negative for large values of a. The same happens for $Z + C_\chi$ and it is possible to see that the negativity of both kinetic terms is correlated with the negativity of the potential. Thus the stationary point of ref. [2] is a property of a ghost potential and cannot be identified with a stationary ground state. This rather unpleasant description may be circumvented by an alternative renormalization prescription showing clearly that supersymmetry is unbroken in this theory.

A wave function renormalization controls the normalization of the corresponding kinetic term. We are thus naturally led to normalize it at the background field values (a and f in our notation) that we wish to consider. The minimum conditions on the potential thus obtained determine those values for which linear terms in the fluctuations are absent, thus allowing the identification of a and f at the minimum with $\langle A\rangle$ and $\langle F\rangle$, respectively.

We could determine Z from $Z + C_A = 1$, but to avoid useless complications with the regular terms in eq. (9), let us choose

$$Z = 1 - (|g|^2/8\pi^2) \ln M^2/\alpha^2 \,, \qquad (14)$$

which eliminates from $Z + C_A$, and therefore from the A kinetic term, the logarithmic term which was at the origin of its negative values for the choice of Z in eq. (10).

Eq. (1) shows that if Z depends on a, the renormalized parameters g and m depend on a through Z. This dependence is the usual one, i.e.,

$$g^{-2}(\alpha) = g^{-2}(\alpha_0) + (1/8\pi^2) \ln \alpha_0^2/\alpha^2 \,. \qquad (15)$$

Eqs. (6), (7) and (14) imply

$$V = -|f|^2 + maf + m^*a^*f^* + (1/64\pi^2)[(\alpha^2+\beta)^2 \ln(\alpha^2+\beta) + (\alpha^2-\beta)^2 \ln(\alpha^2-\beta) - 2\alpha^4 \ln \alpha^2 - 2\beta^2 \ln \alpha^2 - 3\beta^2] \,. \qquad (16)$$

It is easy to see that

$$V - f\,\partial V/\partial f - f^*\,\partial V/\partial f^* = |f|^2 + (1/64\pi^2)\{(\alpha^4-\beta^2)\ln[(\alpha^4-\beta^2)/\alpha^4] + \beta^2\} \geqslant 0 \,, \qquad (17)$$

for all α and $\beta \leqslant \alpha^2$, the equality sign in (17) holding only for $\beta = 0$. It is then clear that on the line

$$\partial V/\partial f = \partial V/\partial f^* = 0 \,, \qquad (18)$$

which expresses f in terms of a, the potential $V = V(a, f(a))$ is a positive function of a with its absolute minimum at $a = 0$ where V vanishes. Therefore $\langle A\rangle = \langle F\rangle = 0$ and supersymmetry is unbroken.

We see therefore that with our renormalization prescription, we succeeded in describing the theory for all choices of the background field a with a positive potential and in terms of fluctuations which have positive kinetic terms. On the other hand, in order to describe the same theory at different values of a, we found that the coupling constant depends on a, as described in eq. (15).

If we call g_0 the coupling constant at $a = 0$, we find

$$g^2(\alpha) = g_0^2/[1 + (g_0^2/8\pi^2) \ln m_0^2/\alpha^2] \,, \qquad (19)$$

which shows a Landau-type pole at $\alpha^2 = m_0^2 \exp(8\pi^2/g_0^2)$ characteristic of a non-asymptotic free theory like the one analyzed here.

To summarize, we have shown that the renormalization prescription of refs. [2] and [3] generates negative potentials together with negative kinetic terms. Ghosts try to increase their potential energy so that a negative stationary point is energetically unfavourable as compared with a zero potential configuration. Thus, even in this language, we understand why supersymmetry is not broken in the model of ref. [2]. Moreover, we have shown how to avoid this pathological ghost interpretation through an alternative renormalization prescription that leads to bona fide fields with positive kinetic energy and with non-negative potentials as required by supersymmetry.

We may wonder why the generally adopted prescription of renormalizing the interaction at a fixed scale does not work for supersymmetric theories while it is perfectly applicable to usual theories as $\lambda\varphi^4$ [5]. In conventional theories the coupling constant is renormalized independently of the wave function and thus any prescription for the first one cannot influence the kinetic terms which are controlled by the second. In supersymmetry the only renormalization is the wave function one and therefore is determined by the kinetic terms. Or, at least if a definition of the interaction is introduced which leads to negative kinetic energy, it is impossible to appeal to another independent renormalization to correct that sign.

We wish to acknowledge fruitful discussions with L. Girardello, J. Iliopoulos, R. Cahn, G. Veneziano and S. Yankielowicz.

References

[1] E. Witten, Lecture notes at Trieste (1981); S. Cecotti and L. Girardello, Phys. Lett. 110B (1982) 39.
[2] D. Zanon, Phys. Lett. 104B (1981) 127.
[3] M. Hug, Phys. Rev. D14 (1976) 3548; D16 (1977) 1733.
[4] L. O'Raifeartaigh and G. Parravicini, Nucl. Phys. B111 (1976) 516;
W. Lang, Nucl. Phys. B114 (1976) 123.
[5] S. Coleman and E. Weinberg, Phys. Rev. D7 (1973) 1888.

PHYSICAL REVIEW D VOLUME 28, NUMBER 5 1 SEPTEMBER 1983

Koba-Nielsen-Olesen scaling and production mechanism in high-energy collisions

Chou Kuang-chao
Institute of Theoretical Physics, Academia Sinica, Beijing, China

Liu Lian-sou* and Meng Ta-chung
Institut für Theoretische Physik der Freie Universität Berlin, Berlin, Gemany
(Received 14 March 1983)

An analysis of the existing data on photoproduction and electroproduction of protons is made. Koba-Nielsen-Olesen (KNO) scaling is observed in both cases. The scaling function of the nondiffractive γp processes turns out to be the same as that for nondiffractive hadron-hadron collisions, but the scaling function for deep-inelastic e^-p collisions is very much different from that for e^-e^+ annihilation processes. Taken together with the observed difference in KNO scaling functions in e^-e^+ annihilation and nondiffractive hadron-hadron processes these empirical facts provide further evidence for the conjecture: The KNO scaling function of a given collision process reflects its reaction mechanism. Arguments for this conjecture are given in terms of a semiclassical picture. It is shown that, in the framework of the proposed picture, explicit expressions for the above-mentioned KNO scaling functions can be derived from rather general assumptions.

I. INTRODUCTION

The recent CERN $p\bar{p}$ collider experiments,[1] in which Koba-Nielsen-Olesen (KNO) scaling[2] has been observed, have initiated considerable interest[3,4] in studying the implications of this remarkable property. A physical picture has been proposed in an earlier paper[4] to understand the KNO scaling in the above-mentioned experiments,[1] and in pp[5] and e^+e^- reactions.[6] It is suggested in particular that the qualitative difference between the KNO scaling function in e^+e^- annihilation and those in nondiffractive hadron-hadron collisions is due to the difference in reaction mechanisms.

In this paper we report on the result of a systematic analysis of high-energy γp and e^-p data[7,8] as well as that of a theoretical study of the possible reaction mechanisms of these and other related processes. We show the following.

(A) KNO scaling is valid also in high-energy γp and e^-p processes. The scaling functions for nondiffractive γp and low-Q^2 (invariant momentum-transfer squared) e^-p processes are the same as for nondiffractive hadron-hadron collisions, but the scaling function for deep-inelastic e^-p collisions is very much different from that for e^-e^+ annihilation processes.

(B) The KNO scaling function for e^-e^+ annihilation,

$$\psi(z)=6z^2\exp(-\alpha z^3),\quad \alpha^{1/3}=\Gamma(\tfrac{4}{3})\,, \tag{1}$$

and that for nondiffractive hadron-hadron collisions,

$$\psi(z)=16/5(3z)^5\exp(-6z) \tag{2}$$

(here $z=n/\langle n\rangle$, n is the charged multiplicity and $\langle n\rangle$ is its average value), can be obtained from the basic assumptions of the proposed physical picture using statistical methods.

(C) The similarities and differences between observed KNO scaling functions mentioned in (A) can be understood in the framework of the proposed picture.

II. KNO SCALING IN γp AND e^-p PROCESSES

We studied photoproduction and electroproduction of protons at incident energies above the resonance region. We made a systematic analysis of the existing data[7,8] and found that: there is KNO scaling in e^-p as well as in γp processes. (See Figs. 1 and 2.) The KNO scaling function for nondiffractive γp processes and that for e^-p at low momentum transfer are the same as that for nondiffractive hadron-hadron collisions. (See Fig. 1.) The KNO scaling function for deep-inelastic e^-p collisions is very much different from that for e^+e^- annihilation processes. (See Fig. 2.)

The similarity between the KNO scaling function in nondiffractive γp (and low-momentum-transfer e^-p) and that in nondiffractive hadron-hadron processes is not very surprising. In fact, it shows nothing else but the well-known fact[9] that real (or almost real) photons at high energies behave like hadrons.

But does the difference in KNO scaling functions in e^-e^+ and deep-inelastic e^-p processes indicate that the reaction mechanisms of these two kinds of processes are qualitatively different from each other?

Before we try to answer this question, let us first examine in more detail the relationship between KNO scaling functions and reaction mechanisms in e^-e^+ annihilation and in nondiffractive hadron-hadron collisions.

III. e^-e^+ ANNIHILATION: FORMATION AND BREAKUP OF ELONGATED BAG

The KNO scaling function in e^-e^+ annihilation processes is shown in Fig. 3. It is sharply peaked at $n/\langle n\rangle\approx 1$ (n is the multiplicity of the charged hadrons

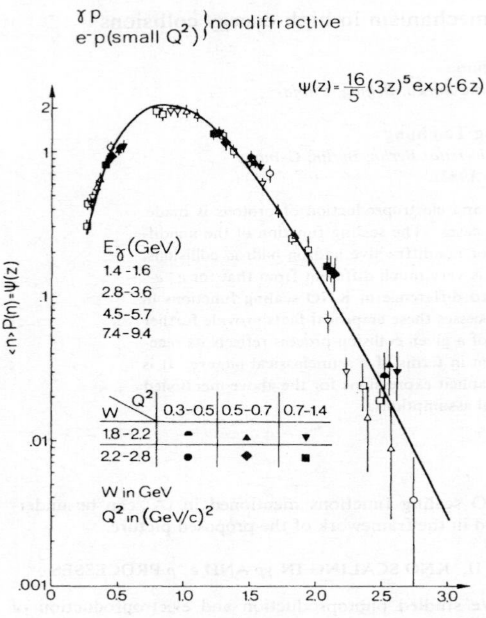

FIG. 1. The scaled multiplicity distribution for nondiffractive γp and low-momentum-transfer $e^- p$ reactions. The experimental data are taken from Refs. 7 and 8, respectively. The curve is obtained from Eq. (2).

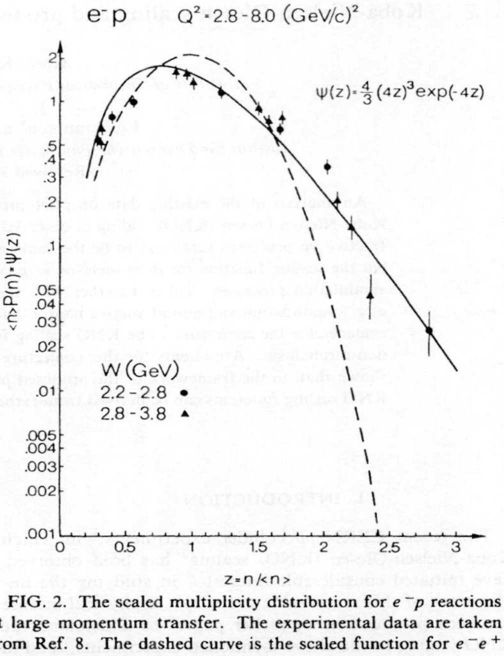

FIG. 2. The scaled multiplicity distribution for $e^- p$ reactions at large momentum transfer. The experimental data are taken from Ref. 8. The dashed curve is the scaled function for $e^- e^+$ annihilation processes, shown in Fig. 3. The solid curve is the scaled function given in Eq. (24).

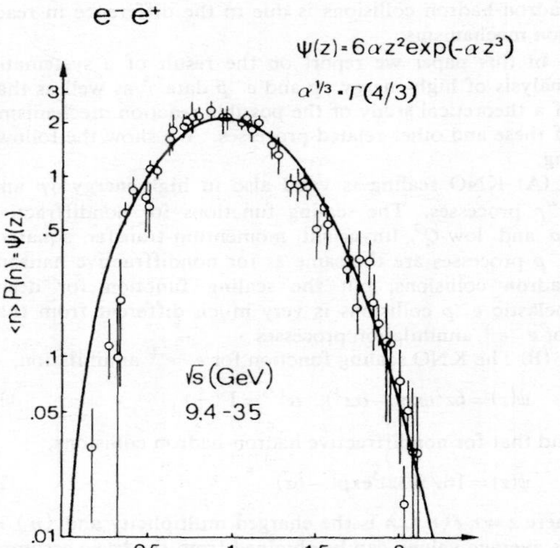

FIG. 3. The scaled multiplicity distribution for $e^- e^+$ annihilation processes. The experimental data are taken from Ref. 6. The curve is the scaling function given by Eq. (1).

and $\langle n \rangle$ is its mean value) and can be approximated by a Gaussian.[4,10] It can be qualitatively understood[4] as follows: In $e^- e^+$ annihilation processes, the electron and the positron can be considered as pointlike particles. Hadronization takes place when the colliding e^- and e^+ hit each other so *violently* that *the entire amount* of the initial energy and momentum is deposited into a single system which subsequently breaks up into pieces.

It is clear that, without further specifications, this picture would be too general and too crude to account for all the characteristic features of the $e^- e^+$ annihilation processes. In fact, the observed two-jet structure[11] and the central rapidity (along the jet axis) plateau[12] suggests that in such a collision event the compound system formed after the violent collision should have the form of a long tube which eventually breaks up. (See Fig. 4.)

A number of models[13] have been proposed in the literature to account for the above-mentioned jet and plateau structures. The most natural and most successful ones among them seem to be those based on the Schwinger mechanism.[14] We recall that Schwinger[14] observed that the quantum vacuum of a gauge field theory may be so polarizable that the charge can be completely screened. The vector meson of the gauge theory then acquire a mass. It has been pointed out by Casher *et al.*[15] that this mechanism can be used to understand why isolated quarks are

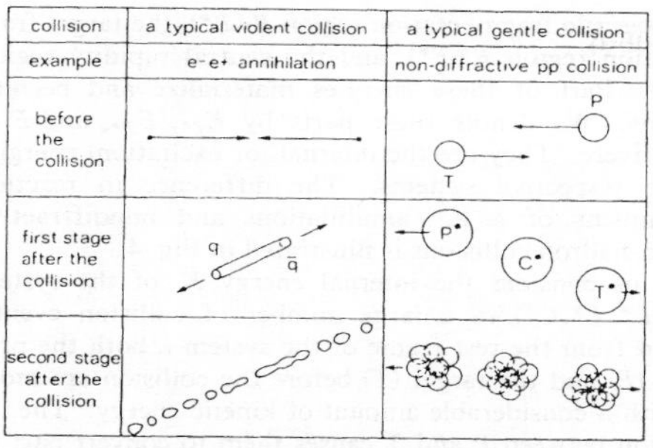

FIG. 4. Two qualitatively different types of high-energy collisions are illustrated; and one characteristic example in each case is given.

not seen in e^-e^+ annihilation processes, where the decay of the virtual photon into a quark-antiquark ($q\bar{q}$) pair is generally accepted to be true. In fact, it is envisaged that as q and $\bar{q}$ of the original $q\bar{q}$ pair move apart (at almost the velocity of light) a color-electric field is developed, and a number of polarized pairs (secondary $q\bar{q}$ pairs) are formed between them. Now, since the gluon exchanges allow q ($\bar{q}$) of arbitrarily high subenergy to interact with finite probability, an "inside-outside" cascade[16] takes place and as a consequence only color-singlet hadrons are produced. Quantitative comparisons between experiments[11,12] and the Lund model,[17] which is a semiclassical model that incorporates all the relevant features of the Schwinger model,[14] have been made.[18] The agreement seems to be very impressive.

Is it possible to understand the observed KNO scaling behavior in e^-e^+ annihilation processes in models based on the Schwinger mechanism? We now show that this question—which does not seem to have been asked before—can be answered in the affirmative.

In order to study multiplicity distributions in this framework, we need to know the relationship between the observed multiplicity of charged hadrons and the properties of the elongated bag. The following points are of particular importance in establishing this relationship:

(α) Since the number of sub-bags at the final stage of a given event is nothing else but the total multiplicity of hadrons in that event, it seems plausible to assume that the total multiplicity of charged hadrons (n) is proportional to the final length (l) of the elongated bag in every event,

$$n = (\bar{\gamma}/l_0)l \ . \tag{3}$$

Here, l_0 is the average length of the elongated hadron-bags in their "rest frames" and $\bar{\gamma}$ is the inverse of the average Lorentz contraction factors of the hadrons along the jet axis. This means, we have assumed that $\bar{\gamma}/l_0$ depends only on the total c.m. energy $\sqrt{s}$, provided that n is not too small compared to $\langle n \rangle$. Obviously, Eq. (3) is in accordance with the following empirical facts[6,19]: (i) The overwhelming part of the produced hadrons are pions of approximately the same transversed momentum with respect to the jet axis; (ii) The multiplicity of charged hadrons is distributed mainly around its average value $\langle n \rangle$ which is rather high at incident energies where KNO scaling has been observed (e.g., $\langle n \rangle \approx 7$ and 14 at $\sqrt{s} = 10$ and 30 GeV, respectively).

We note that the final length l is determined by the first breakup of the elongated bag. This is essentially a kinematical effect which can be readily demonstrated in terms of the one-dimensional Lund model[17] as shown in Fig. 5. The generalization from the one-dimensional string to a three-dimensional elongated bag does not influence the arguments used to reach this conclusion. The reason is: In the present model, the existence of a $q\bar{q}$ pair is not a sufficient, but a necessary condition for the breakup.

(β) As the original quark-antiquark pair fly apart, their kinetic energy is converted into volume and surface energies. Secondary $q\bar{q}$ pairs are produced and the elongated bag begins to split when the bag reaches a certain length such that it is energetically more favorable to do so. Note that the collective effect due to color interaction is a substantial part of the bag concept. Hence, it is expected that the probability of bag splitting should depend on the *global* rather than the *local*[20] properties of the entire system. We shall assume, for the sake of simplicity, that the elongated bag is uniform in the longitudinal direction, and that the probability df/dl_1 for a bag of length l to break somewhere (at l_1, say, where $0 < l_1 < l$) is proportional to l, that is approximately proportional to the total energy U of the elongated bag.[21] This means

$$f(l) = \int_0^l \frac{df}{dl_1} dl_1 \ , \tag{4}$$

$$\frac{df}{dl_1} = \lambda l \ , \tag{5}$$

where λ is a constant. It should be mentioned that we do

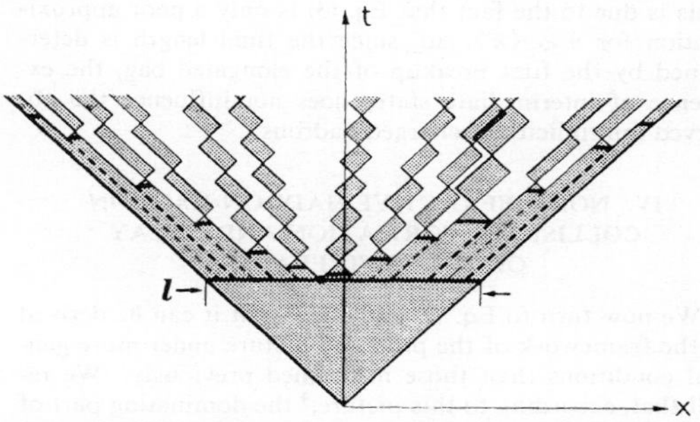

FIG. 5. The one-dimensional Lund model (see Ref. 17) is used to demonstrate that in e^-e^+ annihilation processes the final length of the elongated bag is determined by the first breakup. Here, t and x denote the time and space coordinates, respectively. Note that the generalization from the one-dimensional string to a three-dimensional elongated bag does not influence the arguments used to reach this conclusion.

not know why the above-mentioned l dependence [Eq. (5)] should be *linear*. What we know for the moment is: By assuming a power behavior l^k for df/dl_1, the experimental data require $k=1$.

(γ) Having obtained the probability $f(l)$ for an elongated bag of length l to break up, the density function for the l-distribution $P(l)$ can be calculated in the following way: Consider N events, among which in $N(l)$ of them the bag has reached the length without breaking and dN of them will break up in the interval $(l, l+dl)$, then

$$dN = -f(l)N(l)dl \ . \qquad (6)$$

It follows from Eqs. (4), (5), and (6)

$$P(l) = \frac{dN}{dl} = Cl^2\exp(-\lambda l^3/3) \ , \qquad (7)$$

where C is a normalization constant. The corresponding density function for multiplicity distribution $P(n)$ is therefore [see Eq. (3)]

$$P(n) = An^2\exp(-Bn^3) \ , \qquad (8)$$

The constants A and B are determined by the usual normalization conditions[22]:

$$\int_0^\infty P(n)dn = 2 \ , \qquad (9)$$

$$\int_0^\infty nP(n)dn = 2\langle n\rangle \ . \qquad (10)$$

From Eqs. (8), (9), and (10), we have

$$\langle n\rangle P(n) = \psi(n/\langle n\rangle) \ , \qquad (11)$$

where $\psi(z)$ is given by Eq. (1). Comparison with experiments[6] is shown in Fig. 3.

The following should be pointed out: (a) The KNO scaling behavior is obtained as a direct consequence of Eqs. (8), (9), and (10). (b) The elongated-bag model, which is obviously consistent with the physical picture discussed in Ref. 4, is more specific and gives a better description (than the Gaussian approximation) of the existing data. (c) There is a discrepancy between model and data for $z \lesssim 0.3$. This is due to the fact that Eq. (3) is only a poor approximation for $n \ll \langle n\rangle$. (d) since the final length is determined by the first breakup of the elongated bag, the existence of intermediate states does not influence the observed multiplicity of charged hadrons.

IV. NONDIFFRACTIVE HADRON-HADRON COLLISIONS: FORMATION AND DECAY OF THREE-FIREBALLS

We now turn to Eq. (2) and show that it can be derived in the framework of the proposed picture under more general conditions than those mentioned previously. We recall that, according to this picture,[4] the dominating part of the high-energy inelastic hadron-hadron collision events are nondiffractive. The reaction mechanism of such processes can be summarized as follows: Both the projectile hadron (P) and the target hadron (T) are spatially extended objects with many degrees of freedom. They go through each other during the interaction and distribute their energies in three distinct kinematical regions in phase space: the projectile fragmentation region $R(P^*)$, the target fragmentation region $R(T^*)$, and the central rapidity region $R(C^*)$. Part of these energies materialize and become hadrons. We denote these parts by E_{P^*}, E_{T^*}, and E_{C^*}, respectively. They are the internal (or excitation) energies of the respective systems. The difference in reaction mechanisms of e^-e^+ annihilations and nondiffractive hadron-hadron collisions is illustrated in Fig. 4.

Let us consider the internal energy E_i of the system i $(i=P^*,T^*,C^*)$ in a large number of collision events. Viewed from the rest frame of the system i, both the projectile (P) and the target (T) before the collision are moving with a considerable amount of kinetic energy. The interaction between P and T causes them to convert part of their kinetic energies into internal energies of the systems P^*, T^*, and C^*. Hence, each system i has two energy sources so that E_i can be expressed as

$$E_i = E_{iP} + E_{iT}, \quad i = P^*,T^*,C^* \ , \qquad (12)$$

where E_{iP} and E_{iT} are the contributions from the source P and that from the source T, respectively. Note that the two sources are independent of each other, and that among the nine variables in Eq. (12) six of them are completely random. Let $F_P(E_{iP})$ be the probability for the system i to receive the amount E_{iP} from P, and $F_T(E_{iT})$ is that for the system i to receive E_{iT} from T, then the probability for the system i to obtain E_{iP} from P and E_{iT} from T is the product $F_P(E_{iP})F_T(E_{iT})$. Physically, it is very likely that the system i completely forgets its history as soon as the system is formed. This means, the probability for the system i to obtain E_{iP} from P and E_{iT} from T depends only on the sum $E_{iP}+E_{iT}$. That is

$$F(E_i) = F_P(E_{iP})F_T(E_{iT}) \ , \qquad (13)$$

where E_i and E_{iP} and E_{iT} are related to one another by Eq. (12). Hence

$$\frac{d}{dE_{iP}}[\ln F_P(E_{iP})] = -\frac{d}{dE_{iP}}[\ln F_T(E_i - E_{iP})] \ , \qquad (14)$$

that is

$$F_P(E_{iP}) = A_P\exp(-BE_{iP}) \ , \qquad (15)$$

$$F_T(E_i - E_{iP}) = A_T\exp[-B(E_i - E_{iP})] \ , \qquad (16)$$

where A_P, A_T, and B are constants.

In order to obtain the *total* probability $P(E_i)$ for the system i to be in a state characterized by a given energy E_i, without asking the question "How much of E_i is contributed from P and how much of it from T?" we have to integrate over all the possible values of E_{iP} and E_{iT} under the condition given in Eq. (12). That is

$$P(E_i) = \int dE_{iP}dE_{iT}\delta(E_i - E_{iP} - E_{iT})F_P(E_{iP})F_T(E_{iT}) \ . \qquad (17)$$

It follows from Eqs. (15), (16), and (17)

$$P(E_i) = CE_i\exp(-BE_i) \ , \qquad (18)$$

where the constants B and C are determined by the normalization conditions

$$\int P(E_i)dE_i = 1 \ , \quad (19)$$

$$\int E_i P(E_i)dE_i = \langle E_i \rangle \ . \quad (20)$$

Hence

$$\langle E_i \rangle P(E_i) = 4E_i / \langle E_i \rangle \exp(-2E_i/\langle E_i \rangle) \ , \quad (21)$$

which is Eq. (11) of Ref. 4.

The multiplicity ($n_{\rm ND}$) distribution for nondiffractive hadron-hadron collisions given by Eq. (2) is obtained by taking into account (for details, see Ref. 4)

$$n_i/\langle n_i \rangle = E_i/\langle E_i \rangle \ , \quad i = P^*, \ T^*, \text{ and } C^* \quad (22)$$

and

$$n_{\rm ND} = n_{C^*} + n_{P^*} + n_{T^*} \ . \quad (23)$$

Note that z in Eq. (2) stands for $n_{\rm ND}/\langle n_{\rm ND} \rangle$.

It should be emphasized that the simple relationship, $E_i/n_i = \text{constant}$ ($i = C^*, P^*, T^*$), is an idealization. In reality, fluctuation in n_i for a given E_i is expected. Such effects have been taken into account by assuming that the KNO distribution for e^-e^+ annihilation (which can be approximated by a Gaussian; see Ref. 4) is due to the fluctuation of $n_{ee}/\langle n_{ee} \rangle$ about 1, and that the fluctuations of n_i about $\langle n_i \rangle$ is of the same magnitude. These fluctuations are folded into the distributions obtained from the three-fireball model for hadron-hadron processes (a detailed discussion on this point is given in the preliminary version of our paper; see Ref. 20 of Ref. 4). Comparison between data and results of that calculation shows, however, that the effect is negligible in first-order approximation.

V. A POSSIBLE REACTION MECHANISM FOR DEEP-INELASTIC e^-p PROCESSES

We now come back to the question raised at the end of Sec. II.

According to the conventional picture[23] for deep-inelastic e^-p collisions, one of the colored quarks inside the proton is hit so violently that it tends to fly away from the rest (to which it is bounded by the confining forces). As a consequence quark-diquark jet structure is expected.[23] Hence, it is natural to believe that also in this case elongated bags[15] or strings[17] are formed which hadronize. In fact, compared with the above-mentioned model for e^-e^+ annihilation, the only difference would be that the bags, tubes, or strings end with quark and diquark, instead of quark and antiquark. If this were true, the KNO scaling function for deep-inelastic e^-p processes would be the same as that for e^-e^+ annihilation.

The qualitative difference in KNO scaling functions of e^-e^+ and deep-inelastic e^-p collisions is probably because the virtual photon in e^-p processes behaves differently as that mentioned in the conventional picture.

Once we accept that (a) the virtual photon in deep-inelastic e^-p collisions cannot fragment like a hadron in hadron-hadron collisions because it has an energy deficiency compared with its momentum,[24] and (b) the proton is a spatially extended object with many internal degrees of freedom (possibly a large number of colored gluons and sea quarks in addition to the colored valence quarks) such that various colorless objects can be formed in an excited proton, it seems natural to conjecture that deep-inelastic e^-p processes take place as follows: The virtual photon in such collision processes interacts with a part of the proton gently in the sense that it "picks up" a certain amount of colorless matter in order to fragment.[25]

Note that by picking up a certain amount of colorless matter from the proton, the virtual photon becomes a real physical object. The fragmentation products of this object are nothing else but the "current fragments" observed in lepton-nucleon reactions.[26] This conjecture can be readily tested experimentally. Because, if it is correct, we should see: First, the average multiplicity $\langle n \rangle$ does not depend on Q^2 (the invariant momentum transfer). Second, $\langle n \rangle$ depends on W (the total energy of the hadronic system) in the same way as the average multiplicity in hadron-hadron collisions depends on $\sqrt{s}$ (the total c.m. energy). Third, the rapidity distribution in single-particle inclusive reactions shows a dip in the central rapidity region (near $y_{\rm c.m.} = 0$) at sufficiently high incident energies. This is because the center of the current fragments (formed by the virtual photon and the colorless matter it picked up from the photon) and that of the residue target (the rest of the target proton) move away from the central region in opposite directions. Fourth, the KNO scaling function is

$$\psi(z) = 4/3(4z)^3 \exp(-4z). \quad (24)$$

This is because, according to the proposed picture[4] the two fragmenting systems mentioned above act independently, and the KNO scaling function of each system is [See Eq. (21)]

$$\psi(z) = 4z \exp(-2z) \ . \quad (25)$$

That is, the mechanism of e^-p deep-inelastic scattering can be described as the formation and decay of two fireballs. Here we have assumed, by analogy with the nondiffractive hadron-hadron collision, that the average multiplicities of the two fireballs are equal.

The first and the second points are well-known experimental facts.[27] In connection with the third point, we see that the rapidity distribution in neutrino-proton reactions at $W > 8$ GeV clearly shows the expected dip. (See, e.g., Fig. 10 of Ref. 26.) Corresponding data for electron-proton reactions at comparable energies is expected to exhibit the same characteristic feature. Last but not least, Fig. 2 shows that Eq. (24) (the fourth point mentioned above) is indeed in agreement with the data.

The conclusion that there should be two fireballs in the intermediate stage of deep-inelastic e^-p collisions can also be reached without referring to the properties of the virtual photons, provided that such collisions take place as follows: The pointlike electron goes through the spatially extended proton, gives part of its energy and momentum to a colorless subsystem of the proton and separates this subsystem from the rest. While the incident electron is only deflected due to the interaction, the two separated subsystems of the proton become excited and subsequently decay. It should also be pointed out that, if this conjecture is correct, we expect to see only one central fireball in the $e^-e^+ \to e^-e^+X$ processes at sufficiently large momentum transfer. In that case the corresponding KNO scaling function should be the same as that given in Eq. (25). It would be very interesting to see whether this and other consequences of the proposed reaction mechanism will agree with future experiments.

ACKNOWLEDGMENT

This work was supported in part by Deutsche Forschungsgemeinschaft Grant No. Me-470-4/1.

*On leave from Hua-Zhong Teachers' College, Wuhan, and Peking University, Beijing, China.

[1]K. Alpgard *et al.*, Phys. Lett. 107B, 315 (1981); G. Arnison *et al.*, *ibid.* 107B, 320 (1981); 123B, 108 (1983); K. Alpgard *et al.*, *ibid.* 121B, 209 (1983); and the papers cited therein.

[2]Z. Koba, H. B. Nielsen, and P. Olesen, Nucl. Phys. B40, 317 (1972).

[3]See, e.g., S. Barshay, Phys. Lett. 116B, 197 (1982); T. T. Chou and C. N. Yang, *ibid.* 116B, 301 (1982); Y. K. Lim and K. K. Phua, Phys. Rev. D 26, 1785 (1982); C. S. Lam and P. S. Yeung, Phys. Lett. 119, 445 (1982); F. W. Bopp, Report No. SI-82-14, 1982 (unpublished).

[4]Liu Lian-sou and Meng Ta-chung, Phys. Rev. D 27, 2640 (1983).

[5]See P. Slattery, Phys. Rev. D 7, 2073 (1973); C. Bromberg *et al.*, Phys. Rev. Lett. 31, 1563 (1973); D. Bogert *et al.*, *ibid.* 31, 1271 (1973); S. Barish *et al.*, Phys. Rev. D 9, 268 (1974); J. Whitmore, Phys. Rep. 10C, 273 (1974); A. Firestone *et al.*, Phys. Rev. D 10, 2080 (1974); W. Thomé *et al.*, Nucl. Phys. B129, 365 (1977); W. M. Morse *et al.*, Phys. Rev. D 15, 66 (1977); R. L. Cool *et al.*, Phys. Rev. Lett. 48, 1451 (1982); and the papers cited therein.

[6]See, e.g., R. Felst, in *Proceedings of the 1981 International Symposium on Lepton and Photon Interactions at High Energies, Bonn,* edited by W. Pfeil (Physikalisches Institut, Universität Bonn, Bonn, 1981), p. 52 and the papers cited therein.

[7]R. Erbe *et al.*, Phys. Rev. 175, 1669 (1968); R. Schiffer *et al.*, Nucl. Phys. B38, 628 (1972); J. Ballam *et al.*, Phys. Rev. D 5, 545 (1972); 7, 3150 (1973); H. H. Bingham *et al.*, *ibid.* 8, 1277 (1973). See also H. Meyer, in *Proceedings of the Sixth International Symposium on Electron and Photon Interactions at High Energy, Bonn, Germany, 1973,* edited by H. Rollnik and W. Pfeil (North-Holland, Amsterdam, 1974), p. 175.

[8]V. Eckardt *et al.*, Nucl. Phys. B55, 45 (1973); J. T. Dakin *et al.*, Phys. Rev. Lett. 30, 142 (1973); Phys. Rev. D 8, 687 (1973); L. Ahrens *et al.*, Phys. Rev. Lett. 31, 131 (1973); Phys. Rev. D 9, 1894 (1974); P. H. Carbincius *et al.*, Phys. Rev. Lett. 32, 328 (1974); C. K. Chen *et al.*, Nucl. Phys. B133, 13 (1978).

[9]J. D. Bjorken, in *Proceedings of the Third International Symposium on Electron and Photon Interactions at High Energies* (SLAC, Stanford, 1967), p. 109; H. T. Nieh, Phys. Rev. D 1, 3161 (1970).

[10]K. Goulianos *et al.*, Phys. Rev. Lett. 48, 1454 (1982).

[11]See, e.g., D. Haidt, in *Proceedings of the 1981 International Symposium on Lepton and Photon Interactions at High Energies, Bonn,* (Ref. 6), p. 558 and references given therein.

[12]See, e.g., W. Hofmann, *Jets of Hadrons,* Vol. 90 of *Springer Tracts in Modern Physics* (Springer, Berlin, 1981) and papers cited therein.

[13]See, e.g., W. Hofmann (Ref. 12), pp. 25 and 47 and the papers cited therein.

[14]J. Schwinger, Phys. Rev. 125, 397 (1962); 128, 2425 (1962).

[15]A. Casher *et al.*, Phys. Rev. D 10, 732 (1974).

[16]J. D. Bjorken, in *Current Induced Reactions,* proceedings of the International Summer Institute on Theoretical Particle Physics, Hamburg, 1975, edited by J. G. Körner, G. K. Kramer, and D. Schildknecht (Springer, Berlin, 1976).

[17]B. Andersson *et al.*, Z. Phys. C 1, 105 (1979).

[18]See, e.g., D. Fournier, in *Proceedings of the 1981 International Symposium on Lepton and Photon Interactions at High Energies, Bonn* (Ref. 6) p. 91 and the papers cited therein.

[19]See, e.g., Refs. 6, 11, and 12 and the papers cited therein.

[20]See, e.g., Ref. 17.

[21]In first-order approximation the energy of the elongated bag at a given instant is proportional to its length at that moment. This energy can be considered as the total potential energy of the $q\bar{q}$ system. We recall: Hadron-spectroscopy strongly suggests that the interaction inside a hadron can be described by a linear potential (the potential energy is directly proportional to the distance) between q and $\bar{q}$, provided that they can be considered as a static source and sink of color flux. This is, e.g., the case when the quark-antiquark pairs are heavy ($c\bar{c}$, $b\bar{b}$, etc.). Now, in the case of e^-e^+ annihilation processes, since the *primary* q and $\bar{q}$ are *always* on the two ends of the elongated bag while they separate, it is *always* possible to envisage the existence of an instantaneous static source and a corresponding sink in each bag. Hence we can assume the existence of such linear potentials for *all kinds* of *primary* $q\bar{q}$ pairs.

[22]See, e.g., Refs. 1, 5, 6, and 10 and the papers cited therein.

[23]See, e.g., Hofmann (Ref. 12), p. 71 and papers cited therein.

[24]T. T. Chou and C. N. Yang, Phys. Rev. D 4, 2005 (1971).

[25]This colorless matter is not chargeless. It consists probably of a large number of sea quarks. Note also that the existence of such processes does not necessarily contradict the underlying picture of the quark-parton model which may be true to the impulse approximation.

[26]See, e.g., N. Schmitz, in *Proceedings of the 1979 International Symposium on Lepton and Photon Interactions at High Energies, Fermilab,* edited by T. B. W. Kirk, and H. D. I. Abarbanel (Fermilab, Batavia, Illinois, 1980), p. 259.

[27]See, e.g., Chen *et al.* (the last paper of Ref. 8).

Commun. in Theor. Phys. (Beijing, China) Vol. 2, No. 2 (1983) 971-982

NONLINEAR σ-MODEL ON MULTIDIMENSIONAL CURVED SPACE WITH CERTAIN CYLINDRICAL SYMMETRY

CHOU Kuang-chao (周光召)
Institute of Theoretical Physics, Academia Sinica, Beijing, China
and SONG Xing-chang (宋行长)
Institute of Theoretical Physics, Physics Department, Peking University
Beijing, China

Received September 20, 1982

Abstract

A dual transformation is found for a class of nonlinear σ-model defined on a multidimensional curved space with cylindrical symmetry. The system is invariant under a proper combination of the dual transformation and the general coordinate transformation. An infinite number of nonlocal conservation laws as well as the Kac-Moody algebra follow directly from the dual transformation. A Bäcklund transformation that generates new solutions from a given one can also be constructed.

I. Introduction

In recent years considerable progress has been made in the investigation of the two-dimensional σ-model or chiral model[1], which possesses a lot of rather interesting and mutually connected properties like the soliton solutions[2], the Bäcklund transformation[3], the infinite number of conservation laws[4,5] associated with a hidden symmetry[6-8] and the close similarity to the self-dual Yang-Mills theory in four dimensions. Besides, it has also been shown that the chiral field equation in three-dimensional cylindrical symmetric case bears a resemblance to the Ernst equation in general relativity. Moreover, in the so-called super-unification theory an important role has been played by the non-linear σ-model though the four-dimensional version of which has not been fully investigated yet. Therefore it is meaningful to extend our earlier work on two-dimensional nonlinear σ-model to the case of higher dimensions with certain cylindrical symmetry.

First let us recall some of the important results on the ordinary two-dimensional nonlinear sigma model. These results are formulated in the K-form which has the advantage of uniformity and simplicity and will be used throughout this paper.

The Lagrangian for the ordinary two-dimensional non-linear sigma model can be written in one of the following forms[5].

$$\mathcal{L} = -\frac{1}{2}\mathrm{Tr}\, K_\mu K^\mu \tag{1.1a}$$

$$= \frac{1}{2}\mathrm{Tr}\, D_\mu \varphi^{-1} D^\mu \varphi \tag{1.1b}$$

$$= \frac{1}{8}\mathrm{Tr}\, \partial_\mu q\, \partial^\mu q\ , \tag{1.1c}$$

where the physical field ϕ takes values in the symmetric coset space G/H with its covariant differentiation defined as

$$D_\mu \varphi = \partial_\mu \varphi - \varphi H_\mu \, , \qquad D_\mu \varphi^{-1} = \partial_\mu \varphi^{-1} + H_\mu \varphi^{-1} \, . \tag{1.2}$$

In Eq.(1.2) the "composite" gauge potential H_μ and the "covariant current" K_μ are defined through the following equations:

$$\Omega_\mu \equiv H_\mu + K_\mu = \varphi^{-1} \partial_\mu \varphi \, , \tag{1.3}$$

$$H_\mu = \frac{1}{2}(\Omega_\mu + \sigma \Omega_\mu \sigma) \, , \tag{1.3a}$$

$$K_\mu = \frac{1}{2}(\Omega_\mu - \sigma \Omega_\mu \sigma) \tag{1.3b}$$

$$= \varphi^{-1} D_\mu \varphi = -(D_\mu \varphi^{-1}) \varphi \, , \tag{1.3c}$$

whereas σ is the involution operator with respect to which the symmetric space G/H is defined[7] (i.e., $\sigma h \sigma = h$, $\forall h \in H$; $\sigma \phi \sigma = \phi^{-1}$, $\forall \phi \in G/H$; $\sigma^2 = 1$). The gauge invariant field q is defined by[7]

$$q = \phi \sigma \phi^{-1} \, , \qquad q^2 = 1 \, . \tag{1.4}$$

The Euler-Lagrange equation of motion reads

$$D^\mu K_\mu \equiv \partial^\mu K_\mu + [H^\mu, K_\mu] = 0 \, , \tag{1.5a}$$

or

$$D^\mu D_\mu \varphi - (D^\mu \varphi) \varphi^{-1} (D_\mu \varphi) = 0 \, , \tag{1.5b}$$

or

$$\partial^\mu \partial_\mu q - (\partial^\mu q) q (\partial_\mu q) = 0 \, . \tag{1.5c}$$

The compatibility condition for Ω_μ gives for symmetric coset space

$$\partial_\mu H_\nu - \partial_\nu H_\mu + [H_\mu, H_\nu] = -[K_\mu, K_\nu] \, , \tag{1.6a}$$

$$D_\mu K_\nu - D_\nu K_\mu = 0 \, . \tag{1.6b}$$

These two equations are known as the Gauss-Coddazi equations.

In light cone coordinates

$$\sqrt{2}\, \xi = t - x \, , \qquad \sqrt{2}\, \eta = t - x \, , \tag{1.7}$$

Eqs.(1.5a) and (1.6) can be written in the form

$$D_\xi K_\eta = D_\eta K_\xi = 0 \, , \tag{1.8a}$$

$$\partial_\xi H_\eta - \partial_\eta H_\xi + [H_\xi, H_\eta] + [K_\xi, K_\eta] = 0 \, . \tag{1.8b}$$

Explicitly, the dual transformation

$$K_\xi \longrightarrow K'_\xi = \gamma K_\xi \, , \qquad K_\eta \longrightarrow K'_\eta = \gamma^{-1} K_\eta \, ; \tag{1.9a}$$

$$H_\xi \longrightarrow H'_\xi = H_\xi \, , \qquad H_\eta \longrightarrow H'_\eta = H_\eta \, . \tag{1.9b}$$

with γ as an arbitrary constant, leaves the Lagrangian and the equations (1.8) invariant. From this transformation an infinite number of conservation laws, both local and nonlocal, can be deduced[5]. Under the transformation (1.9), the physical field ϕ and/or q transforms as

$$\varphi(x) \longrightarrow \psi(\gamma, [\varphi]) \varphi(x) \bar{h}^{-1}(\gamma, [\varphi]) = \varphi'(\gamma, x) \, , \tag{1.10a}$$

$$q(x) \longrightarrow \psi(\gamma, [\varphi]) q(x) \psi^{-1}(\gamma, [\varphi]) = q'(\gamma, x) \, , \tag{1.10b}$$

where the transformation functional (amplitude) satisfies the linearized equations.

$$\partial_\xi \psi(\gamma,[\varphi]) = \frac{1}{2}(\gamma-1)\psi(\gamma,[\varphi])J_\xi[\varphi] \quad , \tag{1.11a}$$

$$\partial_\eta \psi(\gamma,[\varphi]) = \frac{1}{2}(\gamma^{-1}-1)\psi(\gamma,[\varphi])J_\eta[\varphi] \quad , \tag{1.11b}$$

and J_μ is the Noether current for global G symmetry:

$$\begin{aligned} J_\mu[\varphi] &= 2\varphi K_\mu[\varphi]\varphi^{-1} = -2\varphi D_\mu \varphi^{-1} \\ &= -q\partial_\mu q \ . \end{aligned} \tag{1.12}$$

In a recent note[8] we have shown that, by choosing the essential parameter α and the boundary condition for the transformation amplitude $\psi(\gamma(\alpha), [\varphi(x)])$ properly, the one-parameter family of the dual transformation (1.9) forms an Abelian group A(1)

$$\gamma(\alpha)\gamma(\alpha') = \gamma(\alpha+\alpha') = \gamma(\alpha')\gamma(\alpha) \quad , \tag{1.13}$$

and $\{\psi\}$ gives a nonlinear realization of this group

$$\psi(\alpha,[\varphi'(\alpha',x)])\psi(\alpha',[\varphi(x)]) = \psi(\alpha+\alpha',[\varphi(x)]) = \psi(\alpha',[\varphi(\alpha,x)])\psi(\alpha,[\varphi(x)]) \quad . \tag{1.14}$$

It can be further shown that[8] when combining the element of A(1), $a(\alpha) \in A(1)$, with that of the internal symmetry $g \in G$ to form the product $a^{-1}(\alpha) g\, a(\alpha)$, it remains to be an element in G. This fact implies that the physical system is invariant under the semi-direct product group $G \wedge A(1)$ with G as the normal subgroup, and from this important feature follows the Kac-Moody algebra[8].

In this paper we will discuss the possibility to define a dual-like transformation for the non-linear sigma model in curved space and give a brief account of the results associated with this transformation.

II. Dual-like transformation

Consider an N+2 —dimensional curved space, the metric of which can be brought to the form

$$ds^2 = e^{2\Gamma(t,r)}(dt^2-dr^2) + \Delta^{2/N}(t,r)g_{\alpha\beta}dx^\alpha dx^\beta \quad , \tag{2.1}$$

where $\Gamma(t,r)$ and $\Delta(t,r)$ are functions of the time t and the first space coordinate r, while $g_{\alpha\beta}$ depends only on x^α, $\alpha,\beta=3,4\ldots N+2$. The action of the cylindrically symmetric non-linear sigma model can be written as

$$\begin{aligned} S &= \int g^{\mu\nu}\,\mathrm{Tr}(\partial_\mu q\,\partial_\nu q)\sqrt{-g}\,d^{N+2}x \\ &\propto \int \mathrm{Tr}(K_\xi K_\eta)\Delta(\xi,\eta)\,d\xi\, d\eta \quad , \end{aligned} \tag{2.2}$$

where $\sqrt{2}\xi=t+r$ and $\sqrt{2}\eta=t-r$. The physical field $\phi(x)$ is assumed to be independent of the last N coordinates x^α, $\alpha=3,4\ldots N+2$, and K_ξ and K_η are defined as in Eq.(1.3).

The main problem we will discuss is: does there exist a hidden symmetry for the model (2.2) in the curved space similar to the one in two-dimensional flat space? The answer is yes if the reduced metric $\Delta(\xi,\eta)$ takes a special form, namely, $\Delta(\xi,\eta)=f(\xi)-g(\eta)$. The argument goes as follows.

The equation of motion given by the action (16) reads

$$D^{\mu}(K_{\mu}\Delta)\equiv D_{\xi}(K_{\eta}\Delta)+D_{\eta}(K_{\xi}\Delta)=0 \quad . \tag{2.3}$$

Combined with the integrability condition as in Eq.(1.6) it gives

$$D_{\xi}K_{\eta}=D_{\eta}K_{\xi}=-\frac{1}{2}\left(\frac{\Delta_{\xi}}{\Delta}K_{\eta}+\frac{\Delta_{\eta}}{\Delta}K_{\xi}\right) \quad , \tag{2.4a}$$

$$H_{\xi,\eta}-H_{\eta,\xi}-[H_{\xi},H_{\eta}]=[K_{\xi},K_{\eta}] \quad , \tag{2.4b}$$

where we have used the abbreviation $\Delta_{\xi}=\partial_{\xi}\Delta$ etc..

Consider a transformation of the form

$$\begin{aligned} &H_{\xi}\to H'_{\xi}=H_{\xi} \quad , \qquad H_{\eta}\to H'_{\eta}=H_{\eta} \quad ; \\ &K_{\xi}\to K'_{\xi}=\gamma(\xi,\eta)K_{\xi} \quad , \qquad K_{\eta}\to K'_{\eta}=\gamma^{-1}(\xi,\eta)K_{\eta} \quad ; \\ &\Delta(\xi,\eta)\to \Delta'(\xi,\eta) \quad , \end{aligned} \tag{2.5}$$

with $\gamma(\xi,\eta)$ and $\Delta'(\xi,\eta)$ to be determined later. Under this transformation Eq.(2.4b) is invariant and Eq.(2.4a) becomes

$$D_{\xi}K'_{\eta}=D_{\eta}K'_{\xi}=-\frac{1}{2}\left(\frac{\Delta'_{\xi}}{\Delta'}K'_{\eta}+\frac{\Delta'_{\eta}}{\Delta'}K'_{\xi}\right) \quad . \tag{2.4a'}$$

Substituting Eq.(2.5) into Eq.(2.4a') and comparing the coefficients of K_{ξ} and K_{η} on both sides, we have

$$2\gamma\gamma_{\eta}-(\gamma^{2}-1)\frac{\Delta_{\eta}}{\Delta}=0=2\gamma^{-1}\gamma^{-1}_{\xi}-(\gamma^{-2}-1)\frac{\Delta_{\xi}}{\Delta} \quad ,$$

with the solution

$$\Delta_{+}\equiv-\frac{\Delta}{\gamma^{2}-1}=f(\xi) \quad , \qquad \Delta_{-}\equiv\frac{\Delta}{\gamma^{-2}-1}=g(\eta) \quad , \tag{2.6a, 2.6b}$$

and then

$$\Delta=\Delta_{+}-\Delta_{-}=f(\xi)-g(\eta) \quad , \tag{2.7}$$

where $f(\xi)$ and $g(\eta)$ are arbitrary functions of ξ and η respectively. From Eq.(2.7) we see that the transformation (2.5) can be well defined if and only if $\Delta(\xi,\eta)$ can be decomposed into a sum of separative functions of ξ and η, i.e., $\Delta=f(\xi)-g(\eta)$. **Moreover it is evident that such a kind of decomposition is not unique.** If, for instance, $\Delta_{+}=f(\xi)$ and $\Delta_{-}=G(\eta)$ is a reasonable decomposition, then $\Delta_{+}=f(\xi)+k$ and $\Delta_{-}=g(\eta)+k$ (k const.) is also a possible choice. Then Eqs.(2.6a,b) give γ as the functional of $f(\xi)$ and $g(\eta)$ and the function of the parameter k

$$\gamma^{2}=\frac{\Delta_{-}}{\Delta_{+}}=\frac{g(\eta)+k}{f(\xi)+k} \quad . \tag{2.8}$$

The relation between $\Delta(\xi,\eta)$ and $\Delta'(\xi,\eta)$ is also determined by Eq.(2.4a')

$$\frac{\Delta'_{\xi}}{\Delta'}=\gamma^{2}\frac{\Delta_{\xi}}{\Delta}=\frac{g(\eta)+k}{f(\xi)+k}\,\frac{f_{\xi}}{f(\xi)-g(\eta)} \quad , \tag{2.9a}$$

$$\frac{\Delta'_{\eta}}{\Delta'}=\gamma^{-2}\frac{\Delta_{\eta}}{\Delta}=\frac{f(\xi)+k}{g(\eta)+k}\,\frac{g_{\eta}}{g(\eta)-f(\xi)} \quad . \tag{2.9b}$$

Substituting the expressions for γ^2 and Δ given in Eqs.(2.7) and (2.8) into

(2.9), we obtain by integration

$$\Delta'(\xi,\eta)=C\,\frac{f(\xi)-g(\eta)}{(f(\xi)+k)(g(\eta)+k)} \quad . \tag{2.10}$$

The integration constant C can be determined by the boundary condition

$$k\longrightarrow\infty\ ,\ \gamma\longrightarrow 1\ ,\ \Delta'\longrightarrow\Delta\ ,\ C=k^2\ . \tag{2.11}$$

Therefore

$$\Delta'(\xi,\eta)=f'(\xi)-g'(\eta)\ ;$$

$$\frac{1}{f'(\xi)}=\frac{1}{f(\xi)}+\frac{1}{k}\ ,\quad \frac{1}{g'(\eta)}=\frac{1}{g(\eta)}+\frac{1}{k}\ . \tag{2.10'}$$

Therefore we see that, under the condition (2.7), the dual-like transformation for the cylindrically symmetric non-linear sigma model is specified by Eqs.(2.5), (2.8) and (2.10') with $k^{-1}=\alpha$ as the group parameter. $\alpha=0$ $(k\to\infty)$ corresponds to the identical transformation $\gamma=1$, and two transformations with parameters α and α' give another transformation with $\alpha''=\alpha+\alpha'$:

$$\begin{array}{ccccc}\Delta(\xi,\eta) & \xrightarrow{\alpha} & \Delta'(\xi,\eta) & \xrightarrow{\alpha'} & \Delta''(\xi,\eta) \\ =f(\xi)-g(\eta) & & =f'(\xi)-g'(\eta) & & =f''(\xi)-g''(\eta)\ ,\end{array} \tag{2.12}$$

$$f''(\xi)^{-1}=f'(\xi)^{-1}+\alpha'=f(\xi)^{-1}+(\alpha+\alpha')\quad .$$

This gives an Abelian group A(1) with $\alpha=k^{-1}$ as the group parameter.

Besides the group parameter α, the transformation parameter γ also depends on the reduced metric $\Delta(\xi,\eta)$, as given in Eq.(2.8). Later on we will denote it as

$$\gamma=\gamma(\alpha,\Delta)\quad . \tag{2.13}$$

For the succeeding operation of α and α', we have

$$\begin{aligned}\gamma(\alpha',\Delta')\gamma(\alpha,\Delta)&=\left[\frac{g'(\eta)+k'}{f'(\xi)+k'}\,\frac{g(\eta)+k}{f(\xi)+k}\right]^{1/2}\\ &=\left[\frac{g(\eta)+kk'/(k+k')}{f(\xi)+kk'/(k+k')}\right]^{1/2}=\gamma(\alpha'',\Delta)\quad .\end{aligned} \tag{2.14}$$

Then we see from Eq.(2.5) that

$$K''_\xi=\gamma(\alpha',\Delta')K'_\xi=\gamma(\alpha',\Delta')\gamma(\alpha,\Delta)K_\xi=\gamma(\alpha'',\Delta)K_\xi\ , \tag{2.15a}$$

$$K''_\eta=\gamma^{-1}(\alpha',\Delta')K'_\eta=\gamma(\alpha',\Delta')\gamma^{-1}(\alpha,\Delta)K_\eta=\gamma^{-1}(\alpha'',\Delta)K_\eta\ . \tag{2.15b}$$

This means that the equation of motion takes a similar form under the operation of the Abelian group A(1).

It must be pointed out that the Lagrangian (2.2) is not invariant under the dual-like transformation we have considered above. So the dual-like transformation itself is not a symmetry of our system. But this shortcoming can be cured by an appropriate general coordinate transformation which we will discuss in Sec. IV. The Lagrangian (2.2) and the equation of motion remain invariant under the combined operation of these two transformations which form a symmetry group of the physical system under consideration.

III. Linearized equation and conserved currents

From Eq.(2.5) we can see immediately that, just as the case in ordinary model, the quantity

$$\Omega_\mu = H'_\mu + K'_\mu \, , \qquad \mu = \xi, \eta$$

is a pure gauge, so that it can be expressed as

$$\Omega'_\mu = g^{-1}\partial_\mu g \tag{3.1}$$

with g being an element in G. Then by setting $g = \psi\phi$, we have

$$\psi^{-1}\partial_\mu \psi = \phi(K'_\mu - K_\mu)\phi^{-1}.$$

Therefore the field $\phi(x)$ or $q(x)$ undergoes the transformation under Eq.(2.5)

$$\varphi(x) \longrightarrow \psi(\gamma, [\varphi(x)])\, \varphi(x)\, \bar{h}^{-1}(\gamma, [\varphi(x)]) = \varphi'(\gamma, x) \, , \tag{3.2a}$$

$$q(x) \longrightarrow \psi(\gamma, [\varphi(x)])\, q(x)\, \psi^{-1}(\gamma, [\varphi(x)]) = q'(\gamma, x) \, . \tag{3.2b}$$

and the transformation amplitude $\psi(\gamma, [\phi(x)])$ satisfies the equation

$$\partial_\xi \psi(\gamma, [\varphi(x)]) = \frac{1}{2}(\gamma - 1)\, \psi(\gamma, [\varphi(x)])\, M_\xi [\varphi(x)] \, , \tag{3.3a}$$

$$\partial_\eta \psi(\gamma, [\varphi(x)]) = \frac{1}{2}(\gamma^{-1} - 1)\, \psi(\gamma, [\varphi(x)])\, M_\eta [\varphi(x)] \, . \tag{3.3b}$$

Eqs.(3.2) and (3.3) have almost the same forms as Eqs.(1.10) and (1.11) except that γ is space dependent as given in Eq.(2.13). The quantity M_μ in Eq.(3.3) is defined as

$$M_\mu[\varphi] = 2\varphi K_\mu[\varphi]\varphi^{-1} = -2\varphi D_\mu \varphi^{-1} = -q\partial_\mu q \, , \tag{3.4}$$

which satisfies the curvatureless condition

$$M_{\xi,\eta} - M_{\eta,\xi} + [M_\xi, M_\eta] = 0 \, . \tag{3.5}$$

The Noether current for the global G symmetry now becomes

$$J_\mu[\varphi] = \Delta M_\mu[\varphi] = 2\Delta\varphi K_\mu[\varphi]\varphi^{-1} \, , \tag{3.6}$$

which is conserved

$$\partial^\mu J_\mu[\varphi] \equiv \partial_\xi J_\eta[\varphi] + \partial_\eta J_\xi[\varphi] = 0 \, , \tag{3.7}$$

as can be verified directly from the equation of motion (2.3).

It can be shown by a straightforward deduction that the compatibility condition for Eqs.(3.3) is just the curvatureless condition (3.5) for $M_\mu[\phi]$ and the conservation equation (3.7) for $J_\mu[\phi]$. So Eqs.(3.3) are nothing but the linearization system of equations in the inverse scattering problem, with α being the spectral parameter.

Similar to Eq.(1.14), from Eqs.(2.14) and (3.3) we can deduce the relation

$$\psi(\gamma'(\alpha',\Delta), [\varphi'(\alpha, x)])\, \psi(\gamma(\alpha,\Delta), [\varphi(x)]) = \psi(\gamma(\alpha+\alpha',\Delta), [\varphi(x)]) \, . \tag{3.8}$$

Again, $\{\psi\}$ as a whole forms a nonlinear realization of the Abelian group A(1).

It has been noticed in Sec. II that after applying the dual-like transformation (2.5)

$$\Delta(\xi,\eta) \longrightarrow \Delta'(\xi,\eta) \, , \quad K_\xi \longrightarrow K'_\xi = \gamma K_\xi \, , \quad K_\eta \longrightarrow K'_\eta = \gamma^{-1} K_\eta \, ,$$

the transformed quantities Δ' and K'_μ satisfy an equation of motion (2.4a') just like Eq.(2.4a) is satisfied by Δ and K_μ. The transformed Noether current can be defined by

$$\begin{aligned} J'_\mu[\varphi'] &= 2\Delta'\varphi' K_\mu[\varphi']\varphi'^{-1} \\ &= 2\Delta' g K'_\mu g^{-1} , \end{aligned} \tag{3.9}$$

where Δ', K'_μ and ϕ' are given by Eqs.(2.5),(2.10') and (3.2a). $K_\mu[\phi']$ is the same functional as $K_\mu[\phi]$ in Eq.(1.3c) with ϕ replaced by ϕ' and is equivalent to K'_μ up to a gauge transformation in the subgroup H, i.e.,

$$K_\mu[\varphi'] = \bar{h}\, K'_\mu\, \bar{h}^{-1}$$

with $\bar{h}$ taking value in the subgroup H as defined in Eq.(3.2). Then it is straightforward to show that $J'_\mu[\phi']$ is conserved.

$$\partial^\mu J'_\mu[\varphi'] = \partial_\xi J'_\eta[\varphi'] + \partial_\eta J'_\xi[\varphi'] = 0 \ . \tag{3.10}$$

Now since $J'_\mu[\phi']$ contains α as an arbitrary parameter Eq.(3.10) yields an infinite number of conservation laws when it is expanded in powers of the spectral parameter α.

IV. General coordinate transformation and standard forms

Thus far we have only considered the dual-like transformation which is coordinate dependent. It is evident that the sigma model in curved space can also undergo general coordinate transformations.

It has been pointed out in Sec. II that the dual-like transformation can be defined if and only if the reduced metric $\Delta(\xi,\eta)$ takes the particular form

$$\Delta(\xi,\eta) = f(\xi) - g(\eta)$$

or equivalently satisfies the Laplacian equation

$$\Delta_{\xi\eta} = 0 \tag{4.1}$$

It is clear that by applying an appropriate general coordinate transformation, various models can be cast into four standard forms in accordance with the form of $\Delta(\xi,\eta)$:

(i) Δ=const. We obtain the ordinary two-dimensional sigma model;

(ii) $\Delta=f(\xi)\neq$const. By applying a transformation $\xi'=f(\xi)$ the equation of motion (2.4a) turns out to be

$$D_{\xi'} K_\eta = D_\eta K_{\xi'} = -\frac{1}{2}\xi'^{-1} K_\eta \ .$$

Therefore we obtain a model corresponding to $\Delta'=\xi'$;

(iii) $\Delta=-g(\eta)\neq$const. By using a similar transformation $\eta'=g(\eta)$, a model corresponding to $\Delta'=-\eta'$ is achieved;

(iv) $\Delta=f(\xi)-g(\eta)$ with neither $f(\xi)$ nor $g(\eta)$ being a constant. In this case, by making use of a transformation $\xi'=f(\xi)$, $\eta'=g(\eta)$, we obtain

$$\Delta \longrightarrow \xi' - \eta' = \sqrt{2}\, r' \ . \tag{4.2}$$

This is the cylindrically symmetric sigma model in 2+1—dimensional spacetime.

As an application of the above general coordinate transformation, let us reformulate the inverse scattering equation discussed in the last section. Hereafter we assume that the model has been brought to the standard form so that $\Delta(\xi,\eta)$ takes the form

$$\Delta(\xi,\eta)=\xi-\eta=\sqrt{2}\,r\,.$$

The corresponding Lax pair has the form

$$\partial_\xi\Psi^{-1}=\frac{1}{2}(1-\gamma)M_\xi\Psi^{-1},\quad \partial_\eta\Psi=\frac{1}{2}(1-\gamma^{-1})M_\eta\Psi^{-1}\,. \tag{4.3}$$

Introducing a function ζ as

$$\zeta(\xi,\eta;k)=\frac{1}{\sqrt{2}}(\sqrt{\xi+k}+\sqrt{\eta+k})^2\,, \tag{4.4}$$

then it follows by a simple calculation that

$$\frac{1-\gamma}{2}=\frac{\gamma}{\zeta+\gamma}\,,\quad \frac{1-\gamma^{-1}}{2}=-\frac{\gamma}{\zeta-\gamma}\,. \tag{4.5}$$

Therefore Eqs.(4.3) become

$$\partial_\xi\Psi^{-1}=\frac{\gamma}{\zeta+\gamma}M_\xi\Psi^{-1},\quad \partial_\eta\Psi^{-1}=-\frac{\gamma}{\zeta-\gamma}M_\eta\Psi^{-1}\,. \tag{4.3'}$$

This is just the pair of equations given by Mikhailov and Yarimechuk[9] which has been used to construct the soliton-like solutions.

It is well known that, besides being considered as the light cone coordinates of the variables t and r as given in Eq.(1.7), (ξ,η) can also be used as the complex combination of two Euclidean coordinates, say, z and r

$$\sqrt{2}\,\xi=iz-r\,,\quad \sqrt{2}\,\eta=iz-r=-\sqrt{2}\,\bar{\xi}\,. \tag{4.6}$$

Therefore, the same equations in case (iv), can be used equally well to describe the static axially symmetric model in 3-dimensional space. In this case our form of the linear scattering problem associated with the sigma model, Eq.(4.3), coinsides with that given by Bais and Sasaki[10]. The spectral parameter s given in Ref.[10] is related to α we used above by the formula

$$\alpha=i2\sqrt{2}\,S\,. \tag{4.7}$$

For the static axially symmetric model, by changing the variables from (ξ,η) to (r,z), and introducing $\omega=i\zeta$ instead of ζ into Eq.(4.4) we can show that

$$\omega=\omega(r,z;\lambda)=\lambda-z+\sqrt{r^2-(\lambda-z)^2}\,, \tag{4.8}$$

with

$$\lambda=i\sqrt{2}\,k\,. \tag{4.9}$$

Now it is easy to find that

$$\frac{\partial\omega}{\partial r}=\frac{2\omega r}{\omega^2+r^2}\,,\quad \frac{\partial\omega}{\partial z}=-\frac{2\omega^2}{\omega^2+r^2}\,. \tag{4.10}$$

Passing to the variables (r,z), Eq.(4.3) can be rewritten in the form

$$\partial_r\Psi^{-1}=\frac{rJ_r+\omega J_z}{r^2+\omega^2}\Psi^{-1}\,,\quad \partial_z\Psi^{-1}=\frac{rJ_z-\omega J_r}{r^2+\omega^2}\Psi^{-1}\,, \tag{4.11}$$

where ω given by Eq.(4.8) is a function of the coordinates r and z and of the spectral parameter λ. The solution of Eq.(4.11) may be considered as a functional of ω as well as a function of (r,z), i.e.,

$$\psi^{-1} = \psi^{-1}(\omega(r,z); r,z) \tag{4.12}$$

From this point of view the differentiation with respect to the (r,z) must be a compound one:

$$\frac{\partial}{\partial r} = \left(\frac{\partial}{\partial r}\right)_\omega + \frac{\partial \omega}{\partial r}\frac{\partial}{\partial \omega} = \left(\frac{\partial}{\partial r}\right)_\omega + \frac{2\omega r}{\omega^2+r^2}\frac{\partial}{\partial \omega}, \tag{4.13a}$$

$$\frac{\partial}{\partial z} = \left(\frac{\partial}{\partial z}\right)_\omega + \frac{\partial \omega}{\partial z}\frac{\partial}{\partial \omega} = \left(\frac{\partial}{\partial z}\right)_\omega - \frac{2\omega^2}{\omega^2+r^2}\frac{\partial}{\partial \omega}. \tag{4.13b}$$

These two differentiation operators have been denoted as $D_r(D_2)$ and $D_z(D_1)$ respectively by Belinsky and Zakharov in their paper on the gravity field equation[11]. From our formulation it is straightforward to show the commutativity of these two operators:

$$[D_r, D_z] \equiv [\partial_r, \partial_z] = 0, \tag{4.14}$$

and the equivalence of the pair of equations[11]

$$D_r \psi^{-1} = \frac{rJ_r + \omega J_z}{r^2+\omega^2}\psi^{-1},$$
$$D_z \psi^{-1} = \frac{rJ_z - \omega J_r}{r^2+\omega^2}\psi^{-1} \tag{4.15}$$

to the linearzation system of equations we have obtained in the last section. We emphasize here once again that in our formulation, ψ is a transformation functional which carries one field into another satisfying a similar equation of motion.

As another example of utilization of the general coordinate transformation we give here the Bäcklund transformation of our sigma model. Suppose we start with a model in the standarc form

$$\Delta = \xi - \eta = \sqrt{2}\, r$$

and $\varphi(\xi,\eta)$ is a solution of this model. By using the dual-like transformation, $\phi'(\xi,\eta)$ as defined in Eq.(3.2a) is a solution to the model with the reduced metric

$$\Delta'(\xi,\eta) = \frac{k\xi}{\xi-k} - \frac{k\eta}{\eta-k}. \tag{4.16}$$

It is possible to carry the reduced metric back into Δ by means of a coordinate transformation (c.t)

$$\Delta' \rightarrow \Delta = \xi - \eta. \tag{4.17}$$

Therefore we obtain another solution

$$\varphi'\left(\frac{-k\xi}{\xi-k}, \frac{-k\eta}{\eta-k}\right) \tag{4.18}$$

corresponding to the original $\Delta=\xi-\eta$. This is the Bäcklund transformation we have obtained. Evidently, it is just the combined transformation

$$\begin{array}{ccccc} \Delta & \xrightarrow{d.t.} & \Delta' & \xrightarrow{c.t.} & \Delta. \\ \varphi(\xi,\eta) & \longrightarrow & \varphi'(\xi,\eta) & \longrightarrow & \varphi'\left(\frac{-k\xi}{\xi-k}, \frac{-k\eta}{\eta-k}\right) \end{array}$$

which leaves the equation of motion (2.4) (and hence the Lagrangian (2.2)) invariant.

V. Kac-Moody algebra

In Sections II and III we have shown that by taking $\alpha=k^{-1}$ as the parameter, the dual-like transformations as given by Eqs.(2.5), (2.12) and (2.14) form an one parameter Abelian group A(1):

$$\gamma(\alpha',\Delta')\,\gamma(\alpha,\Delta)=\gamma(\alpha'+\alpha,\Delta)\ , \tag{5.1}$$

and that by choosing the boundary condition properly the set of the transformation functionals $\{\psi\}$ becomes a nonlinear realization for this group

$$\psi(\alpha';\varphi'(\alpha,\xi,\eta))\,\psi(\alpha;\varphi(\xi,\eta))=\psi(\alpha'+\alpha;\varphi(\xi,\eta))\ , \tag{5.2}$$

From Eq.(5.2) we have

$$\psi(-\alpha;\varphi'(\alpha,\xi,\eta))=\psi^{-1}(\alpha;\varphi(\xi,\eta))\ . \tag{5.3}$$

Under the action of the internal symmetric group, $g\in G$, the field $\phi(\xi,\eta)$ undergoes transformation

$$\begin{aligned}\phi(\xi,\eta)\xrightarrow{g}\varphi(\xi,\eta)&\equiv\varphi(g,[\varphi(\xi,\eta)])\\&=g\,\varphi(\xi,\eta)\,h^{-1}(g,[\varphi])\ .\end{aligned} \tag{5.4}$$

Here again $h(g,[\phi])$ is an element taking value in the subgroup H , and depends on both g and ϕ itself. Then following the same procedure as given in Ref.[8], we have

$$\varphi(\xi,\eta)\xrightarrow{\alpha}\varphi'(\alpha;\varphi)\xrightarrow{g}\varphi(g,[\varphi'])\xrightarrow{-\alpha}\varphi'(-\alpha,\varphi(g,[\varphi'])).$$

and we can prove that the product

$$\psi(-\alpha;\phi(g,[\varphi'(\alpha;\varphi)]))\,g\,\psi(\alpha;[\varphi])\equiv\tilde{g}(\alpha;g) \tag{5.5}$$

is another global element taking value in the group G. This implies that the system we discussed has the symmetry group $G\wedge A(1)$ with the internal group G as the normal subgroup. Then it follows that if the generators of G (ξ_j being the parameters of G)

$$I_j=-i\left.\frac{\partial g}{\partial\xi_j}\right|_{\xi=0} \tag{5.6}$$

satisfy the commutation relation

$$[I_j,I_k]=if_{jkl}\,I_l\ , \tag{5.7}$$

then the operators defined by

$$I_j(\alpha)=-i\left.\frac{\partial\tilde{g}(\alpha;g)}{\partial\xi_j}\right|_{\xi=0} \tag{5.8}$$

satisfy the similar relation

$$[I_j(\alpha),I_k(\alpha)]=if_{jkl}\,I_l(\alpha) \tag{5.9}$$

from which the compact form of a Kac-Moody algebra can be deduced [8].

VI. Conclusion

1. For chiral field defined on curved space with a special form of reduced metric Δ as given in Eq.(2.7), the dual-like transformation can be defined which gives rise to a pair of equations known as the linearization system in the inverse scattering problem and reduces to the ordinary dual transformation in the case of flat space. The physical system is invariant under the proper combination of the dual-like transformation and the general coordinate transformation.

2. From this symmetry, the infinite number of nonlocal conservation laws as well as Kac-Moody algebra can be deduced.

3. In the standard form, the model describes the cylindrically symmetric or the static axially symmetric chiral field. The equation of motion for the present model reveals great similarity to the axially symmetric Yang-Mills theory in four dimensions and to the Ernst equation in general relativity. The equation obtained can be considered as a generalization (with arbitrary group G) of the Ernst equation, which is completely integrable. The solutions can be found by the standard inverse scattering method as discussed in Ref.[9] and by other generating techniques used in the theory of gravity[12].

References

1. K. Pohlmeyer, Comm. Math. Phys., 46 (1976) 207; M. Lüscher and K. Pohlmeyer, Nucl. Phys. B137 (1978) 46; V.E. Zakharov and A.V. Mikhailov, (Sov. Phys.) JETP 47 (1978) 1071; H. Eichenherr, Nucl. Phys., B146 (1978) 215; B155 (1979) 544; H. Eichenherr and M. Forger, ibid, B155 (1979) 381; V.L. Golo, and A.M. Perelomov, Phys. Lett., B79 (1978) 112.
2. A.A. Belavin and A.M. Polyakov, (Sov. Phys.) JETP Lett., 22 (1975) 245; A. d'Adda, M. Lüscher and P. di Vecchia, Nucl. Phys., B146 (1978) 63; A. Lüscher, Phys. Lett., B78 (1978) 465; E. Witten, Nucl. Phys., B149 (1979) 285; D. Gross, Nucl. Phys., B132 (1978) 439; F. Gürsey and X.C. Tze, Ann. of Phys., 128 (1980) 29; B.Y. Hou, and P. Wang, Phys. Lett., 93B (1980) 415.
3. E. Corrigan et al., Comm. Math. Phys., 58 (1978) 223; Y. Brihaye et al., Jour, Math. Phys., 19 (1978) 2528; M.K. Prasad et al., Phys. Rev. Lett., 43 (1979) 750; K. Pohlmeyer, Comm. Math. Phys., 72 (1980) 317.
4. E. Brezin, et al., Phys. Lett., B82 (1978) 422; H.J. de Vega, ibid., B87 (1979) 233; A.T. Ogielski, Phys. Rev., D21 (1980) 406; I.V. Cherednik, Theor. Math. Phys., 38 (1979) 210; M.K. Prasad, et al., Phys. Lett., B87 (1979) 237.
5. K.C. Chou and X.C. Song, Scientia Sinica, 25A (1982) 716; 825.
6. L. Dolan and A. Roos, Phys. Rev., D22 (1980), 2018; B.Y. Hou, Yale preprint YRP 80-92 (1980); B.Y. Hou et al., Phys. Rev., D24 (1981) 2238; L. Dolan, Phys. Rev. Lett., 47 (1981) 1371; M.L. Ge and Y.S. Wu, Phys. Lett., B108 (1981) 411; C. Devchand and D.B. Fairlie, Nucl. Phys., B194 (1982), 232; L.L. Chau et al., Phys. Rev., D25 (1982) 1080; 1086; K. Ueno and Y. Nakamura RIMS-375; 376 (1981).
7. B.Y. Hou, AS-ITP-81-016.
8. K.C. Chou and X.C. Song, to be published in Commun. in Theor. Phys.
9. A.V. Mikhailov and A.I. Yarimchuk, Nucl. Phys. B202 (1982) 508.
10. F.A. Bais and R. Sasaki, Nucl. Phys., B195 (1982) 522.

11. *V.E. Zahkarov and V.A. Belinsky, (Sov. Phys.) JETP 50 (1980) 1.*

12. *W. Kinnersley, J. Math. Phys. 18 (1977), 1529; W. Kinnersley and D.H. Chitre, ibid, 18 (1977) 1538; 19 (1978) 1926; 19 (1978) 2017; I. Hauser and F.J. Ernst, Phys. Rev. D20 (1978), 362; 20 (1979) 978; J. Math. Phys., 21 (1980), 1126; 21 (1980) 1418; V.E. Zakharov and A.V. Shabat, Funct. Anal. Its Appl., 13 (1979) 13; R. Sasaki, Phys. Lett., 73A (1979) 77; Proc. Roy. Soc. Lond., A373 (1980) 373. R. Sasaki and R.K. Bullough, ibid., A376 (1981) 401.*

Note added in proof: The similarity of the stationary, axially symmetric Einstein equations to the nonlinear sigma models has been discussed by D. Maison (J. Math. Phys., 20 (1979) 871.). The results are somewhat similar to parts of our discussions given in Sec. IV. The authors would like to express their thanks to Prof. B.Y. Hou for calling their attention to this paper.

Commun. in Theor. Phys. (Beijing, China) *Vol.2, No.5 (1983) 1391-1398*

KAC-MOODY ALGEBRA FOR TWO DIMENSIONAL PRINCIPAL CHIRAL MODELS

CHOU Kuang-chao (周光召)

Institute of Theoretical Physics, Academia Sinica, Beijing, China

and SONG Xing-chang (宋行长)

Institute of Theoretical Physics, Peking University, Beijing, China

Received July 26, 1982 †

Abstract

A Darboux transformation depending on single continuous parameter t is constructed for a principal chiral field. The transformation forms a nonlinear representation of the group for any fixed value of t. Part of the kernel in the Riemann-Hilbert transform is shown to be related to the Darboux transformation with its generators forming a Kac-Moody algebra. Conserved currents associated with the Kac-Moody algebra of the linearized equations and the Nöether current for the group transformations with fixed value of t are obtained.

I. Introduction

The technology of constructing Kac-Moody algebra in a two dimensional principal chiral model[1-8] has developed rapidly over the past few years. Some beautiful mathematical results and a great deal of insight into the nature of the hidden symmetry inherent in two dimensional integrable systems have been achieved. In spite of all these developments, the conserved currents obtained in Refs. [1-8] consist of only half of the Kac-Moody algebra. The older derivation[1-6] uses two different values of the parameters l and l', which makes the whole procedure too cumbersome. In papers [7-8] the regular Riemann-Hilbert (R-H) transformations were used to demonstrate the algebraic structure of the linearized equations in a much simpler and more elegant way. However, no relation between the transformation of the linearized equations and that of the basic field was given in Refs. [7-8].

In the present note a Darboux transformation is constructed that forms a nonlinear representation of the chiral group containing a continuous parameter t. The R-H kernel H(t) is assumed to be analytic inside an annular region $a\le|t|\le b$ in the complex plane and expanded into a Laurent series. That part of the Laurent expansion analytic at $t=\pm1$ is shown to be related to the Darboux transformation whose generators form a Kac-Moody algebra. The transformation of the basic field is obtained with the same function at a particular value of t. The conserved current associated with the Kac-Moody algebra of the linearized equations and the Nöether current for the group transformations

---

† *Revised May 17, 1983.*

with a fixed value of t are obtained.

II. Principal chiral field and its linearized equations

The basic field $q(x)$ of the principal chiral model is a function taking values on a simple compact group G. The current defined as

$$j_\mu=q^{-1}\partial_\mu q \tag{2.1}$$

is both curvatureless

$$\partial_\mu j_\nu-\partial_\nu j_\mu+[j_\mu,j_\nu]=0, \tag{2.2}$$

and conserved

$$\partial^\mu j_\mu=0. \tag{2.3}$$

In the light-cone coordinates

$$\sqrt{2}\xi=t+x,\quad \sqrt{2}\eta=t-x.$$

Eqs.(2.2) and (2.3) can be written in the form

$$\partial_\xi j_\eta-\partial_\eta j_\xi+[j_\xi,j_\eta]=0,$$
$$\partial_\xi j_\eta+\partial_\eta j_\xi=0.$$

It is very convenient to work with the dual current

$$\tilde{j}^\mu=\varepsilon_{\mu\nu}j^\nu \quad (\tilde{j}_\xi=-j_\xi,\ \tilde{j}_\eta=j_\eta), \tag{2.4}$$

and the 1-form for both j_μ and $\tilde{j}_\mu$

$$j=j_\mu dx^\mu=j_\xi d\xi+j_\eta d\eta,$$
$$\tilde{j}=\tilde{j}_\mu dx^\mu=-j_\xi d\xi+j_\eta d\eta, \tag{2.5}$$

in the following where the curvaturelessness and conservation of j_μ can be rewritten as

$$dj+j\wedge j=0, \tag{2.2'}$$
$$d\tilde{j}=0. \tag{2.3'}$$

The Lax pair for the model takes the form

$$d\psi(t,j)=\psi(t,j)\Omega(t,j), \tag{2.6}$$

where

$$\Omega(t,j)=-\frac{t}{1-t}j_\xi d\xi+\frac{t}{1+t}j_\eta d\eta=\frac{t}{1-t^2}(\tilde{j}-tj), \tag{2.7}$$

with the compactibility condition

$$0=d^2\psi=\psi(d\Omega+\Omega\wedge\Omega)$$
$$=\psi\{-\frac{t^2}{1-t^2}(dj+j\wedge j)+\frac{t^2}{1-t^2}d\tilde{j}\}, \tag{2.8}$$

which implies both the curvaturelessness and the conservation of the current j as given in Eqs.(2.2') and (2.3').

III. Darboux transformations

Let $\psi(t,j)$ be a function valued in the group G such that the new amplitude

$$\psi(t,j;\chi) \equiv \psi(t,j)\chi(t,j) \tag{3.1}$$

satisfies the same equation as $\psi(t,j)$ does, i.e.,

$$d\psi(t,j;\chi) = \psi(t,j;\chi)\Omega(t,j;\chi). \tag{3.2}$$

Then the new potential $\Omega(t,j;\chi)$ must be a "gauge transformed form" of the old one $\Omega(t,j)$

$$\Omega(t,j;\chi) = \chi^{-1}(t,j)d\chi(t,j)+\chi^{-1}(t,j)\Omega(t,j)\chi(t,j) . \tag{3.3}$$

If the t dependence of $\Omega(t,j;\chi)$ is of the same form as $\Omega(t,j)$, i.e.,

$$\Omega(t,j;\chi) = -\frac{t}{1-t}j_\xi(j;\chi)d\xi + \frac{t}{1+t}j_\eta(j;\chi)d\eta, \tag{3.4}$$

the new current

$$j(j;\chi) = j_\xi(j;\chi)d\xi+j_\eta(j;\chi)d\eta \tag{3.5}$$

will be both curvatureless and conserved. The relation (3.1) is often called Darboux transformation.

IV. Riemann-Hilbert transformations

Solution $\psi(t,j)$ of the linearized equations (2.6) can in general have singularities representing solitons in the complex t-plane other than $t=\pm 1$. To find the desired transformation function $\chi(t,j)$ in Eq.(3.1), we start with a fundamental solution $\psi(t,j)$ to the linearized system (2.6) which is analytic in an annular region in the complex t-plane. In the following it is useful to construct a function

$$H(t,j,u(t)) = \psi^{-1}(t,j)u(t)\psi(t,j) \tag{4.1}$$

with $u(t)$ a group element independent of space-time and analytic in the same annular region in the complex t-plane. For an infinitesimal group transformation

$$u(t) = 1+w(t), \tag{4.2}$$

we have

$$\begin{aligned} H(t,j,u(t)) &= 1+h(t,j,w(t)), \\ h(t,j,w(t)) &= \psi^{-1}(t,j)w(t)\psi(t,j). \end{aligned} \tag{4.3}$$

By assuming the annular region in which $h(t,j,w(t))$ is analytic to be $a\le|t|\le b$ with

$$0<a\le|t|\le b<1, \tag{4.4}$$

it is possible to expand $h(t,j,w(t))$ as a Laurent series

$$\begin{aligned} h(t,j,w(t)) &= \sum_{n=0}^{\infty} h_{-n}(j,w)t^{-n}+ \sum_{n=1}^{\infty} h_n(j,w)t^n \\ &= \bar{h}_-(t,j,w(t))+h_+(t,j,w(t)), \end{aligned} \tag{4.5}$$

where

$$\bar{h}\_(t,j,w) = \sum_{n=0}^{\infty} h_{-n}(j,w)t^{-n} \tag{4.6}$$

is analytic in the region $|t|\geq a$, including the points $t=\pm 1$.

Both the kernal $H(t,j,u)$ and $h(t,j,w)$ satisfy an equation of the form

$$dH(t,j,u) = [H(t,j,u), \Omega(t,j)]. \tag{4.7}$$

If follows that $\quad d\bar{H}\_(t,j,u) = [H(t,j,u), \Omega(t,j)]\_ ,$ $\quad$ (4.7')

where $[H,\Omega]\_$ denotes that part of the Laurent expansion with zero or negative powers of t.

Since $\Omega(t,j)$ given in Eq.(2.7) is analytic in the region $|t|<1$, we can expand it near $t=0$ into a Taylor series

$$\begin{aligned}\Omega(t,j) &= -\sum_{n=0}^{\infty} t^{n+1}(j_\xi d\xi+(-1)^{n+1}j_\eta d\eta) \\ &= \sum_{n=0}^{\infty} t^{n+1}\Omega^{(n+1)}.\end{aligned} \tag{4.8}$$

In the region where both $h(t,j,w)$ and $\Omega(t,j)$ are analytic we have

$$\begin{aligned}&d\bar{h}\_(t,j,w)+[\Omega(t,j), \bar{h}\_(t,j,w)] \\ &=-[\Omega(t,j), h(t,j,w)]\_+[\Omega(t,j), \bar{h}\_(t,j,w)] \\ &=\sum_{n=0}^{\infty}\sum_{m=0}^{n} t^{n+1-m}[\Omega^{(n+1)}, h_{-m}] \\ &=-\sum_{n=0}^{\infty}\sum_{m=0}^{\infty} t^{n+1}[j_\xi d\xi+(-1)^{n+1-m}j_\eta d\eta, h_{-m}] \\ &=-\frac{t}{1-t}[j_\xi d\xi, \bar{h}\_(1,j,w(1))]+\frac{t}{1+t}[j_\eta d\eta, \bar{h}\_(-1,j,w(-1))].\end{aligned} \tag{4.9}$$

Eq.(4.9) clearly indicates that the t dependence of the quantity

$$d\bar{h}\_(t,j,w)+[\Omega(t,j), \bar{h}\_(t,j,w)]$$

is the same as that of $\Omega(t,j)$. We can therefore use this fact to build an infinitesimal transformation χ.

To first order in $w(t)$ we put

$$\chi(t,j,u(t)) = 1+\bar{h}\_(t,j,w(t)). \tag{4.10}$$

It is easy to find from Eq.(3.3) that

$$\begin{aligned}\Omega(t,j,u(t)) &= \Omega(t,j)+d\bar{h}\_(t,j,w)+[\Omega(t,j), \bar{h}\_(t,j,w)] \\ &= -\frac{t}{1-t}j_\xi(j,u(t))d\xi+\frac{t}{1+t}j_\eta(j,u(t))d\eta\end{aligned} \tag{4.11}$$

with

$$\begin{aligned}j_\xi(j,u) &= j_\xi+[j_\xi, \bar{h}\_(1,j,w(1))], \\ j_\eta(j,u) &= j_\eta+[j_\eta, \bar{h}\_(-1,j,w(-1))].\end{aligned} \tag{4.12}$$

As $\Omega(t,j,u(t))$ depends on $u(t)$, we have modified our notation adopted in Sec.II from $\Omega(t,j,\chi)$ to $\Omega(t,j,u)$ and from $j(j;\chi)$ to $j(j,u)$.

With the help of Eqs.(4.10) and (4.11) it is straitforward to verify that the transformed amplitudes given in Eq.(3.1) is a solution of the linerized equations (3.2) with $j(u)$ in Eq.(4.12) both curvatureless and conserved.

V. Group properties of the transformations

Under successive transformations by applying u_0 first and u_1 next

$$\psi(t,j) \xrightarrow{u_0} \psi(t,j,u_0) = \psi(t,j)\chi(t,j,u_0), \tag{5.1}$$

$$\begin{aligned}\psi(t,j,u_0) \xrightarrow{u_1} \psi(t,j(u_0),u_1) &= \psi(t,j,u_0)\chi(t,j(u_0),u_1) \\ &= \psi(t,j)\chi(t,j,u_0)\chi(t,j(u_0),u_1),\end{aligned} \tag{5.2}$$

a nonlinear representation of the group will be formed if the condition

$$\chi(t,j,u_1u_0) = \chi(t,j,u_0)\chi(t,j(u_0),u_1) \tag{5.3}$$

is satisfied. For infinitesimal transformations

$$u_0 = 1+w_0, \quad u_1 = 1+w_1,$$

χ's are given by Eq.(4.10) and the condition (5.3) turns out to be a commutation relation

$$\begin{aligned}&\chi(t,j,u_0)\chi(t,j(u_0),u_1)-\chi(t,j,u_1)\chi(t,j(u_1),u_0) \\ &=\bar{h}\_(t,j,[w_1,w_0]).\end{aligned} \tag{5.3'}$$

From infinitesimal transformation Eq.(4.3) it is easily found that

$$h(t,j,w_1)h(t,j,w_0) = h(t,j,w_1w_0), \tag{5.4}$$

and

$$\begin{aligned}h(t,j(u_0),w_1) &= \psi^{-1}(t,j(u_0))w_1(t)\psi(t,j(u_0)) \\ &= \chi^{-1}(t,j,u_0)\psi^{-1}(t,j)w_1(t)\psi(t,j)\chi(t,j,u_0) \\ &= \chi^{-1}(t,j,u_0)h(t,j,w_1)\chi(t,j,u_0) \\ &= h(t,j,w_1)+[h(t,j,w_1), \bar{h}\_(t,j,w_0)].\end{aligned} \tag{5.5}$$

After successive transformations of u_0 and u_1, we obtain to order w_0, w_1 and w_0w_1 the following

$$\begin{aligned}&\chi(t,j,u_0)\chi(t,j(u_0),u_1) \\ &=[1+\bar{h}\_(t,j,w_0)][1+\bar{h}\_(t,j,w_1)] \\ &=1+\bar{h}\_(t,j,w_0)+\bar{h}\_(t,j,w_1)+[h(t,j,w_1), \bar{h}\_(t,j,w_0)] \\ &\quad+\bar{h}\_(t,j,w_0)\bar{h}\_(t,j,w_1),\end{aligned} \tag{5.6a}$$

and

$$\begin{aligned}&\chi(t,j,u_1)\chi(t,j(u_1),u_0) \\ &=1+\bar{h}\_(t,j,w_1)+\bar{h}\_(t,j,w_0)+[h(t,j,w_0), \bar{h}\_(t,j,w_1)] \\ &\quad+\bar{h}\_(t,j,w_1)\bar{h}\_(t,j,w_0).\end{aligned} \tag{5.6b}$$

The commutation relation that guarantees the group properties is

$$\begin{aligned}\bar{h}\_(t,j,[w_1,w_0]) &= [\bar{h}\_(t,j,w_0), \bar{h}\_(t,j,w_1)] \\ &+[h(t,j,w_1), \bar{h}\_(t,j,w_0)]\_- [h(t,j,w_0), \bar{h}\_(t,j,w_1)]\_ ,\end{aligned} \tag{5.3''}$$

which can be easily proved as follows:

$$\begin{aligned}&\bar{h}\_(t,j,[w_1,w_0]) = [h(t,j,w_1), h(t,j,w_0)]\_ \\ &= [\bar{h}\_(t,j,w_1), \bar{h}\_(t,j,w_0)] \\ &\quad+h_+(t,j,w_1), \bar{h}\_(t,j,w_0)]\_+ [\bar{h}\_(t,j,w_1), h_+(t,j,w_0)]\_\end{aligned}$$

$$=-[\bar{h}_-(t,j,w_1),\ \bar{h}_-(t,j,w_0)] +[h(t,j,w_1),\ \bar{h}_-(t,j,w_0)]_-+[\bar{h}_-(t,j,w_1),\ h(t,j,w_0)]_-\ .$$

Therefore the infinitesimal transformations form a Lie algebra. To be more explicit let us take

$$w_0(t)=w_0^a I^a t^k,\quad w_1(t)=w_1^b I^b t^\ell, \tag{5.7}$$

with I^a the generators of the group G and form the commutator

$$[w_0(t),\ w_1(t)]=[w_0(0),\ w_1(0)]t^{k+\ell}, \tag{5.8}$$

where k and l are positive or negative integers. With this choice of $w(t)$ it is easily seen that our transformation generates an algebra isomorphic to the Kac-Moody algebra $G\times C(t^{-1},t)$.

After this work was finished, we learned that the authors of Refs. [7] and [8] have given similar proof to the same problem. We decide to publish our results because of the simplicity of our approach. In our proof neither auxiliary quantities such as

$$G(\ell',\ell)=\sum_{m,n=0}^{\infty} G^{(m,n)}\ell'^m\ell^n=\frac{1}{\ell'-\ell}\{\ell'-\ell\quad(\ell')^{-1}\quad(\ell)\}$$

nor transformations corresponding to different values of t are needed.

VI. Transformation of the basic field

Putting $t=\infty$ in Eq.(2.7) we obtain immediately

$$\Omega(\infty,j)=j. \tag{6.1}$$

Substituting Eq.(6.1) into Eq.(2.6) and using Eq.(2.1) we find

$$\psi(\infty,j)=vq, \tag{6.2}$$

with v a constant element of the group G. Therefore the corresponding transformation for the field $q(x)$ is nothing but $\chi(\infty,j,u)$, i.e.,

$$q \xrightarrow{u} q\chi(\infty,j,u)\equiv q\chi(j,u). \tag{6.3}$$

Furthermore, in the limit $t\to\infty$, the gauge transformation (3.3) for the potential $\Omega(t,j)$ reduces to the one for the current j

$$j(j,u)=\chi^{-1}(j,u)d\chi(j,u)+\chi^{-1}(j,u)j\chi(j,u). \tag{6.4}$$

According to Eq.(5.3) the function $\chi(t,j,u)$ forms a nonlinear representation of the group G for any fixed value of t, in particular, for $t=\infty$. So the same is true of the transformation on the basic field $q(x)$

$$\chi(j,u,u_0)=\chi(j,u_0)\chi(j(u_0),u_1). \tag{6.5}$$

By virtue of Eqs.(4.10) and (4.5), only the zero power term in the expansion of $h(t,j,w)$ appears in $\chi(j,u)\equiv\chi(\infty,j,u)$

$$\chi(j,u)=1+h_0(j,w). \tag{6.6}$$

For particular choice $w(t)=w(0)t^k=w^a I^a t^k$

with $w(0)=w^a I^a$ kept fixed and k integers, we change the notation from $h(t,j,w)$ to $h(t,j,k)$ etc. and denote $h(t,j,k=0)$ by $\Lambda(t,j)$

$$h(t,j,0)=\Lambda(t,j)=\psi^{-1}(t,j)w^a I^a \psi(t,j). \tag{6.7}$$

Expanding $\Lambda(t,j)$ into a Laurent series

$$\Lambda(t,j)=\sum_{n=-\infty}^{+\infty} \lambda_n(j)t^n, \tag{6.8}$$

we get

$$h(t,j,k)=\sum_{n=-\infty}^{+\infty} \lambda_n(j)t^{n+k}, \tag{6.8'}$$

from which it follows that

$$\chi(j,0)=1+h_0(j,0)=1+\lambda_0(j), \tag{6.9}$$

$$\chi(j,k)=1+h_0(j,k)=1+\lambda_{-k}(j). \tag{6.9'}$$

Therefore for any fixed value of integer k the transformation $\chi(j,k)$ on the field $q(x)$ is generated by only one term $\lambda_{-k}(j)$ in the expansion of $\Lambda(t,j)$. Multiplying t^{-k} on $\lambda_{-k}(j)$ and summing over k, we see that $\Lambda(t,j)$ itself generates a t-dependent transformation on the field $q(x)$, i.e.,

$$q \longrightarrow q+q\Lambda(t,j). \tag{6.10}$$

This is nothing but the transformation considered in Refs. [2], [5] and [7], from which the hidden symmetry is extracted. These authors have chosen a particular class of fundamental solutions such that $\psi(t,j)$ is analytic in a circle with the center at the origin $t=0$. In this case the negative powers disappear in the expansion of $\Lambda(t,j)$. Hence in their treatment the transformation corresponding to k=positive integer is trivial and the nontrivial algebra associated with its transformation consists of only half of the Kac-Moody algebra.

VII. Some comments

1. If the analytic region of $h(t,j,w(t))$ is

$$1<a\leq|t|\leq b<\infty, \tag{4.4'}$$

an inverse transformation that changes t to $1/t$ can be performed and a similar analysis leading to Kac-Moody algebra can be carried out without any difficulty.

2. On the classical level no central charges appear in the Kac-Moody algebra obtained so far. It is of great interest to look for them in the quantum version of the chiral model.

3. For given k the infinitesimal variation of the current j_μ Eq.(4.12) can be further put into the form .

$$\delta j_\xi(k)=[j_\xi,\ \bar{h}_-(1,j,k)] \\ =\partial_\xi\lambda_{-k}(j)+[j_\xi,\ \lambda_{-k}(j)]=\partial_\xi\lambda_{-k+1}(j), \quad (7.1a)$$

$$\delta j_\eta(k)=[j_\eta,\ \bar{h}_-(-1,j,k)] \\ =\partial_\eta\lambda_{-k}(j)+[j_\eta,\ \lambda_{-k}(j)]=-\partial_\eta\lambda_{-k+1}(j) \quad (7.1b)$$

from which the conservation of the transformed current $j_\mu+\delta j_\mu(k)$ is obvious.

On the other hand, the Nöether current corresponding to the transformation (6.10) has the form[9]

$$J_\xi(t)=\frac{1+t}{1-t}\,\psi(t,j)j_\xi\psi^{-1}(t,j), \quad (7.2a)$$

$$J_\eta(t)=\frac{1-t}{1+t}\,\psi(t,j)j_\eta\psi^{-1}(t,j). \quad (7.2b)$$

Multplying with the generator w^aI^a and taking trace we obtain from Eqs.(7.2a) and (7.2b) the k-th order current by expanding into power series in t

$$J_\xi^{(k)}=\mathrm{Tr}\{j_\xi(-h_0(j,-k)+2\bar{h}_-(1,j,-k))\}, \quad (7.3a)$$

$$J_\eta^{(k)}=\mathrm{Tr}\{j_\eta(-h_0(j,-k)+2\bar{h}_-(-1,j,-k))\}. \quad (7.3b)$$

The relationship between the two sets of conserved currents, further implications of the transformation acting upon the fields and the integral form of our transformations will be discussed in a seperate article[10].

References

1. L. Dolan and A. Ross, Phys. Rev., D22 (1980) 2018.
2. B.Y. HOU, Yale preprint 80-29; B.Y. HOU, M.L. GE and Y.S. WU, Phys. Rev., D24 (1981) 2238.
3. L. Dolan, Phys. Rev. Lett., 47 (1981) 371.
4. C. Devchand and D.B. Fairlie, Nucl. Phys., B194 (1982) 232.
5. L.L. CHAU, Y.S. WU, B.Y. HOU and M.L. GE, Scientia Sinica, A25 (1982) 907.
6. K. Ueno, RIMS preprint 374 (1981).
7. K. Ueno and Y. Nakamura, Phys. Lett., 117B (1982) 208.
8. Y.S. WU, CPT-82 p.1423 (1982).
9. See, for example, K.C. CHOU and X.C. SONG, Scientia Sinica, A25 (1982) 716; B.Y. HOU, AS-ITP-81-016 (1981).
10. K.C. CHOU and X.C. SONG, in preparation.

Volume 134B, number 1,2 PHYSICS LETTERS 5 January 1984

ON THE GAUGE INVARIANCE AND ANOMALY-FREE CONDITION OF THE WESS–ZUMINO–WITTEN EFFECTIVE ACTION

CHOU Kuang-chao, GUO Han-ying, WU Ke
Institute of Theoretical Physics, Academia Sinica, PO Box 2735, Beijing, China

and

SONG Xing-chang
Institute of Theoretical Physics, Peking University, Beijing, China

Received 17 August 1983
Revised manuscript received 3 October 1983

A global anomaly-free condition and a non-abelian gauge invariant Wess–Zumino–Witten effective action with less terms have been found by a systematical method rather than by trial and error. The condition requires the difference between the left- and right-handed Chern–Simon five-forms wrt the gauge group must vanish and it turns out to be the usual condition in the local sense.

A few months ago, Witten [1] has proposed a new framework to show the global aspects of the Wess–Zumino chiral effective action [2]. In addition to Witten's intriguing results, we would like to point out that in order to gauge an arbitrary subgroup $H \subseteq SU(3)_L \times SU(3)_R$, the global symmetry of the action, a global anomaly-free condition should be satisfied by the gauged group H. This condition requires that the left-handed and right-handed Chern–Simon five-forms associated with the group H must be equal to each other and it turns out in the local sense to be the usual anomaly-free condition presented by the perturbative computations at the quark level. Instead of the trial and error Noether method we would also propose a systematical way to gauging the global symmetries of the action. In the case of non-abelian symmetries, a gauge invariant effective action with less terms is found. Although the difference between Witten's action and ours concerns only with some rare processes, it would be quite interesting to find in principle a way to discriminate them.

Let us start from the ungauged Wess–Zumino–Witten effective action [‡1]

$$I(U) = -\frac{F_\pi^2}{16}\int_{S^4} d^4x \,\mathrm{tr}\,\partial_\mu U\,\partial_\mu U^{-1} + n\Gamma(U)\,. \qquad (1)$$

$$\Gamma(U) := -\frac{1}{240\pi^2}\int_Q d\Sigma^{ijklm}$$
$$\times\,\mathrm{tr}(U^{-1}\partial_i U\,U^{-1}\partial_j U\,U^{-1}\partial_k U\,U^{-1}\partial_l U\,U^{-1}\partial_m U). \qquad (2)$$

Under the transformation of the gauge group $H_L \times H_R \subseteq SU(3)_L \times SU(3)_R$, the field $U(x)$, being an element of SU(3), transforms as

$$U(x) \to U'(x) = L(x)U(x)R^{-1}(x)\,, \quad L\,(R) \in H_{L(R)}\,. \qquad (3)$$

Our purpose is to search for an action

$$\tilde{I}(A_\mu, U) = -\frac{F_\pi^2}{16}\int_{S^4} d^4x\,\mathrm{tr}\,D_\mu U\,D_\mu U^{-1} + n\tilde{\Gamma}(A_\mu, U)\,, \qquad (4)$$

such that it is gauge invariant under certain physical

[‡1] We take the notations and conventions in ref. [1], except those explained in the present letter.

0.031-9163/84/$ 03.00 © Elsevier Science Publishers B.V.
(North-Holland Physics Publishing Division)

Volume 134B, number 1,2 PHYSICS LETTERS 5 January 1984

conditions. In the expression (4), DU is the covariant derivative of U defined as

$$DU = dU + A_L U - UA_R , \quad A_{L(R)} = A^{\sigma}_{L(R)} T^{\sigma}_{L(R)} , \quad (5)$$

and the term $\widetilde{\Gamma}$ consists of $\Gamma(U)$ together with the least amount of other terms defined on the compactified spacetime $M \sim S^4$.

The results obtained are the following:

Under such necessary and sufficient condition that the left-handed Chern–Simon five-form Π_L and the right-handed one Π_R are equal to each other

$$\Pi_L = \Pi_R , \quad (6)$$

where the coefficient of the Chern–Simon five-form is defined by

$$\Pi_{ijklm} = -(24\pi^2)^{-1} \times \mathrm{tr}(F_{ij}F_{kl}A_m - \tfrac{1}{2}F_{ij}A_k A_l A_m + \tfrac{1}{10}A_i A_j A_k A_l A_m). \quad (7)$$

the expression

$$\widetilde{\Gamma}(A_\mu, U) = \Gamma(U) + \frac{1}{48\pi^2}\int d^4x\, \epsilon^{\mu\nu\alpha\beta} W_{\mu\nu\alpha\beta} \quad (8)$$

is gauge invariant under the transformation (3) and then $\widetilde{I}$ defined by (4) with expression (8) is the gauge invariant effective action, where

$$\begin{aligned} W_{\mu\nu\alpha\beta} = \mathrm{tr}\{ & [(-A_{\mu L}U_{\nu L}U_{\alpha L}U_{\beta L} + \partial_\mu A_{\nu L}A_{\alpha L}U_{\beta L} \\ & + A_{\mu L}\partial_\nu A_{\alpha L}U_{\beta L}) + (L\to R)] \\ & + \partial_\mu A_{\nu L}UA_{\alpha R}U^{-1}U_{\beta L} + U\partial_\mu A_{\nu R}U^{-1}A_{\alpha L}U_{\beta L} \\ & - \tfrac{1}{2}[A_{\mu L}U_{\nu L}A_{\alpha L}U_{\beta L} - (L\to R)] \\ & + A_{\mu L}UA_{\nu R}U^{-1}U_{\alpha L}U_{\beta L} - UA_{\mu R}U^{-1}A_{\nu L}U_{\alpha L}U_{\beta L} \\ & - A_{\mu L}\partial_\nu A_{\alpha L}UA_{\beta R}U^{-1} - \partial_\mu A_{\nu L}A_{\alpha L}UA_{\beta R}U^{-1} \\ & + A_{\mu R}\partial_\nu A_{\alpha R}U^{-1}A_{\beta L}U + \partial_\mu A_{\nu R}A_{\alpha R}U^{-1}A_{\beta L}U \\ & + A_{\mu L}UA_{\nu R}U^{-1}A_{\alpha L}U_{\beta L} + A_{\mu R}U^{-1}A_{\nu L}UA_{\alpha R}U_{\beta R} \\ & + [A_{\mu L}A_{\nu L}A_{\alpha L}U_{\beta L} + (L\to R)] - A_{\mu L}A_{\nu L}A_{\alpha L}UA_{\beta R}U^{-1} \\ & + A_{\mu R}A_{\nu R}A_{\alpha R}U^{-1}A_{\beta L}U - \tfrac{1}{2}A_{\mu L}UA_{\nu R}U^{-1}A_{\alpha L}UA_{\beta R}U^{-1}\}. \end{aligned} \quad (9)$$

If the condition (6) is not obeyed, the variation of $\widetilde{\Gamma}$ under the gauge transformation (3) becomes

$$\begin{aligned} \Delta\widetilde{\Gamma} = & -\Gamma(L) + \Gamma(R) + \frac{1}{24\pi^2}\int d^4x\, \epsilon^{\mu\nu\alpha\beta} \\ & \times \mathrm{tr}[A_{\mu L}\partial_\nu A_{\alpha L}L^{-1}\partial_\beta L + \tfrac{1}{2}A_{\mu L}A_{\nu L}A_{\alpha L}L^{-1}\partial_\beta L \\ & + \tfrac{1}{2}A_{\mu L}A_{\nu L}L^{-1}\partial_\alpha LL^{-1}\partial_\beta L + \tfrac{1}{4}A_{\mu L}L^{-1}\partial_\nu LA_{\alpha L}L^{-1}\partial_\beta L \\ & - \tfrac{1}{2}A_{\mu L}L^{-1}\partial_\nu LL^{-1}\partial_\alpha LL^{-1}\partial_\beta L - (L\to R)] . \end{aligned} \quad (10)$$

where $\Gamma(L)$ and $\Gamma(R)$ are the same as $\Gamma(U)$ defined by (2) as long as U is replaced by L and R respectively. In the case of the gauge transformation being infinitesimal

$$L = 1 + \epsilon_L , \quad R = 1 + \epsilon_R , \quad \epsilon_{L(R)} = \epsilon^{\sigma}_{L(R)} T^{\sigma}_{L(R)} , \quad (11)$$

the variation $\Delta\widetilde{\Gamma}$ becomes an infinitesimal one as well

$$\delta\widetilde{\Gamma} = \frac{-1}{24\pi^2}\int d^4x\, \epsilon^{\mu\nu\alpha\beta} \times \mathrm{tr}\{\epsilon_L[\partial_\mu A_{\nu L}\partial_\alpha A_{\beta L} + \tfrac{1}{2}\partial_\mu(A_{\nu L}A_{\alpha L}A_{\beta L})] - (L\to R)\}. \quad (12)$$

which is just Witten's result in agreement with the perturbative computations at the quark level of the anomalous variation of the effective action under an infinitesimal gauge transformation. This correspondence shows that the condition (6) is in fact a global extension of the usual perturbative anomaly-free condition.

Furthermore, in comparison with Witten's gauge invariant effective action in the non-abelian case, the action $\widetilde{I}$ presented here contains less terms than Witten's. It is easily shown that the difference between the two actions is invariant under the gauge transformation (3).

In the case of the gauge group H = U(1), the electromagnetic group, the action I is the same as Witten's, i.e. we have

$$\begin{aligned} \widetilde{\Gamma}^{em}(A_\mu, U) = & \Gamma(U) + \frac{1}{48\pi^2}\int d^4x\, \epsilon^{\mu\nu\alpha\beta} \\ & \times \mathrm{tr}\{-A_\mu Q(U_{\nu L}U_{\alpha L}U_{\beta L} + U_{\nu R}U_{\alpha R}U_{\beta R}) \\ & + 2F_{\mu\nu}A_\alpha(Q^2U_{\beta L} + Q^2U_{\beta R} + QUQU^{-1}U_{\beta L})\}. \end{aligned} \quad (13)$$

We now illustrate how to reach the results.

At first, along the standard road from global to local symmetries, one can easily write down a gauge invariant generalization of the effective action (1)

$$\hat{I} = -\frac{F_\pi^2}{16}\int d^4x \,\mathrm{tr}\, D_\mu U\, D_\mu U^{-1} + n\hat{\Gamma}(A_i, U)\,. \qquad (14)$$

$$\hat{\Gamma}(A_i, U) := -\frac{1}{240\pi^2}\int_Q d\Sigma^{ijklm}$$
$$\times\, \mathrm{tr}(U^{-1}D_iU\,U^{-1}D_jU\,U^{-1}D_kU\,U^{-1}D_lU\,U^{-1}D_mU). \qquad (15)$$

which follows straightforwardly from replacing derivatives of U by covariant derivatives of U, $\mathrm{d}U \to \mathrm{D}U$ on both the spacetime $M \sim S^4$ and the disc Q. Although $\hat{I}$ is obviously gauge invariant, it cannot be used as a physical gauge invariant effective action, because it can not be reduced to one defined on the physical spacetime $M\sim S^4 = \partial Q$. A deeper observation, however, shows that the $\hat{\Gamma}$ term can be decomposed into two parts

$$\hat{\Gamma}(A_i, U) = \widetilde{\Gamma}(A_\mu, U)$$
$$+\int_Q d\Sigma^{ijklm}(\Pi^L_{ijklm} - \Pi^R_{ijklm}) - \frac{1}{48\pi^2}\int_Q d\Sigma^{ijklm}$$
$$\times\,\mathrm{tr}(F^L_{ij}D_kUU^{-1}D_lUU^{-1}D_mUU^{-1} - 2F^L_{ij}F^L_{kl}D_mUU^{-1}$$
$$-F^L_{ij}D_kUF^R_{lm}U^{-1} + F^R_{ij}U^{-1}D_kUU^{-1}D_lUU^{-1}D_mU$$
$$-2F^R_{ij}F^R_{kl}U^{-1}D_mU - F^R_{ij}U^{-1}D_kUU^{-1}F^R_{lm}U)\,. \qquad (16)$$

A similar expression holds in the abelian case,

$$\hat{\Gamma}^{\mathrm{em}}(A_i, U) = \widetilde{\Gamma}^{\mathrm{em}}(A_\mu, U) - \frac{1}{48\pi^2}\int_Q d\Sigma^{ijklm}$$
$$\times\,\mathrm{tr}\,[F_{ij}Q(D_kUU^{-1}D_lUU^{-1}D_mUU^{-1}$$
$$+U^{-1}D_kUU^{-1}D_lUU^{-1}D_mU)$$
$$-2F_{ij}F_{kl}(Q^2D_mUU^{-1} + Q^2U^{-1}D_mU$$
$$+QUQU^{-1}D_mUU^{-1})]\,. \qquad (17)$$

The first part of $\hat{\Gamma}$ consists of the expression of $\widetilde{\Gamma}$ and the integral of the difference between the left- and the right-handed Chern–Simon five-forms (in the abelian case, the Chern–Simon five-form vanishes automatically) and the second part itself is gauge invariant on Q. Thus, one could pick up the first part of $\widetilde{\Gamma}$ to define a physical gauge invariant effective action $\widetilde{I}$ as is shown in the expressions (4) and (8) if and only if the difference between the two Chern–Simon five-forms vanishes which is just the necessary and sufficient global anomaly-free condition (6).

The form of $\widetilde{\Gamma}$ is, in general, not unique. Arbitrary addition of gauge invariant terms defined in $M \sim S^4$ will change $\widetilde{\Gamma}$. Our construction of $\widetilde{\Gamma}$ is unique in the sense that after introducing the minimum coupling on Γ we have made a maximum subtraction of gauge invariant terms on Q. The resulting $\widetilde{\Gamma}$ thus found contains less terms than that of Witten's.

We thank the referee for drawing our attention to a preprint by B. Zumino. Y.S. Wu and A. Zee. They independently discovered the relation between the Chern–Simon secondary class and the gauge anomaly.

References

[1] E. Witten, Global aspects of current algebra, Princeton preprint (1983).

[2] J. Wess and B. Zumino, Phys. Lett. 37B (1971) 95.

Commun. in Theor. Phys. (Beijing, China) Vol.3, No.1 (1984) 73-80

ON WITTEN'S EFFECTIVE LAGRANGIAN FOR CHIRAL FIELD

CHOU Kuang-chao(周光召), GUO Han-ying(郭汉英)
and WU Ke(吴 可)
Institute of Theoretical Physics, Academia Sinica
P. O. Box 2735, Beijing, China
and
SONG Xing-chang(宋行长)
Institute of Theoretical Physics, Peking University, Beijing, China

Received October 4, 1983

Abstract

Wess and Zumino's effective Lagrangian for the chiral field is derived by a new method which shows the direct relationship between the Lagrangian and the difference of the left handed and right handed Chern-Simons topological invariants. The resulting Lagrangian is explicitly left and right antisymmetric and equal to the one obtained in a previous letter by a different method.

In a previous letter[1] the effective Lagrangian for the chiral fields proposed firstly by Wess and Zumino[2] and then by Witten[3] was shown to be connected with the Chern-Simons topological invariants in higher dimensions. The effective Lagrangian constructed in Ref.[1] has less terms than that of Witten's after correcting some misprints and errors in Ref.[3].

In the present work we shall derive the same effective Lagrangian by a new method which shows direct connection with the Chern-Simons topological invariants. The result obtained coincides with the one given in Ref.[1]. The advantage of the present method consists in that the Lagrangian is written explicitly in a left-right antisymmetric form through the introduction of a matrix γ_5. This is precisely why our Lagrangian contains less terms than that of Witten's because in the latter there are left-right symmetrical terms. Since Witten's formula in its original form is not correct, we shall give a corrected one in the Appendix.

Before constructing our Lagrangian we shall briefly review some known facts about the principal chiral model and the Chern-Simons topological invariants.

The element of the principal chiral group $G \otimes G$ will be denoted by

$$g=\begin{pmatrix} g_L & 0 \\ 0 & g_R \end{pmatrix} , \qquad g_L\in G, \qquad g_R\in G \quad . \tag{1}$$

The principal chiral field U(x) is an element belonging to G, which transforms under g in the following form

$$U(x)\longrightarrow U'(x)=g_L U(x) g_R^{-1} \quad . \tag{2}$$

It is more convenient to define the chiral field as

$$Q(x)=\begin{pmatrix} 0 & U(x) \\ U^{-1}(x) & 0 \end{pmatrix} , \tag{3}$$

which forms a regular representation of the chiral group

$$Q(x)\longrightarrow Q'(x)=gQ(x)g^{-1}$$

$$=\begin{pmatrix} 0 & g_L U(x) g_R^{-1} \\ g_R U^{-1}(x) g_L^{-1} & 0 \end{pmatrix} . \tag{4}$$

Moreover, it is idempotent, i.e.,

$$Q^2(x)=1,$$

or

$$Q^{-1}(x)=Q(x). \tag{5}$$

Let

$$A_\mu=A_\mu^\alpha \hat{I}_\alpha=\begin{pmatrix} A_\mu^L & 0 \\ 0 & A_\mu^R \end{pmatrix}$$

be the gauge potential of the chiral group where $\hat{I}_\alpha$ $\alpha=1,\ldots,2N$ are generators of the group. The gauge potential A_μ transforms under a local transformation g(x) as

$$A_\mu(x)\longrightarrow A_\mu'=g(A_\mu(x)+\partial_\mu)g^{-1} \quad . \tag{6}$$

Using the chiral field Q(x) it is possible to define a new gauge field in the following way

$$A_\mu^\sigma(x)=Q(x)(A_\mu(x)+\partial_\mu)Q(x)$$

$$= \begin{pmatrix} U(A^R_\mu+\partial_\mu)U^{-1} & 0 \\ 0 & U^{-1}(A^L_\mu+\partial_\mu)U \end{pmatrix}, \tag{7}$$

which is the local involution or the local space reflection of the original gauge field A_μ, and transforms under local gauge transformation just as A_μ does in Eq.(6).

It is easily verified that the field strength of the involutionary gauge field

$$F^\sigma_{\mu\nu}=Q(x)F_{\mu\nu}Q(x). \tag{8}$$

Now we define a 5-form

$$\Pi_5(A)=\frac{1}{240\pi^2}\mathrm{Tr}\left\{\gamma_5(A\wedge A\wedge A\wedge A\wedge A-5F\wedge A\wedge A\wedge A+10F\wedge F\wedge A)\right\}, \tag{9}$$

where

$$\gamma_5=\begin{pmatrix} 1 & 0 \\ 0 & -1 \end{pmatrix}, \tag{10}$$

and

$$A=A_\mu dx^\mu, \tag{11}$$

$$F=dA+A\wedge A. \tag{12}$$

The 5-form $\Pi_5(A)$ is 2π times the difference of the left handed and right handed Chern-Simons 5-forms. The exterior differential of $\Pi_5(A)$ is 2π times the difference of the left handed and right handed third Chern class

$$d\Pi_5(A)=\frac{1}{24\pi^2}\mathrm{Tr}(\gamma_5F\wedge F\wedge F). \tag{13}$$

A similar 5-form $\Pi_5(A^\sigma)$ for the involutionary gauge field A^σ can also be defined as

$$d\Pi_5(A^\sigma)=\frac{1}{24\pi^2}\mathrm{Tr}(\gamma_5F^\sigma\wedge F^\sigma\wedge F^\sigma). \tag{14}$$

With the help of the relation

$$Q(x)\gamma_5 Q(x)=-\gamma_5 \tag{15}$$

and Eq.(8) we obtain

$$d(\Pi_5(A)+\Pi_5(A^\sigma))=0 \tag{16}$$

for arbitrary gauge field A defined on the chiral group. Therefore $\Pi_5(A)+\Pi_5(A^\sigma)$ is a closed form and in the homology trivial region it can be locally written as an exact form

$$\Gamma_{+5}=\frac{1}{2}(\Pi_5(A)+\Pi_5(A^\sigma))=d\omega_4(A,A^\sigma). \tag{17}$$

Next we study the transformation properties of the Chern-Simons topological invariants. Under a gauge transformation(6) we can write

$$A\longrightarrow A'=g(A+X)g^{-1} ,$$

$$A^\sigma\longrightarrow A^{\sigma\prime}=g(A^\sigma+X)g^{-1} , \tag{18}$$

where

$$X=dg^{-1}g . \tag{19}$$

By a straightforward calculation we get

$$\delta\Pi_5(A)=d\zeta_4(A,X)+\Gamma^\circ(X), \tag{20}$$

where

$$\Gamma^\circ(X)=\frac{1}{240\pi^2}Tr\left\{\gamma_5 X\wedge X\wedge X\wedge X\wedge X\right\} , \tag{21}$$

and

$$\zeta_4(A,X)=\frac{1}{48\pi^2}Tr\left\{\gamma_5[(F\wedge A+A\wedge F)\wedge X-A\wedge A\wedge A\wedge X -\frac{1}{2}A\wedge X\wedge A\wedge X-A\wedge X\wedge X\wedge X]\right\}. \tag{22}$$

Therefore

$$\delta(\Pi_5(A)-\Pi_5(A^\sigma))=\delta\Gamma_{-5} =d(\zeta_4(A,X)-\zeta_4(A^\sigma,X)). \tag{23}$$

It is possible to find a 4-form $\xi_4(A,A^\sigma)$ such that under gauge transformation (18)

$$\delta\xi_4(A,A^\sigma)=\zeta_4(A^\sigma,X)-\zeta_4(A,X)+\text{total divergence}. \tag{24}$$

If such a $\xi_4(A,A^\sigma)$ exists,

$$\tilde{\Gamma}_{-5}=\Pi_5(A)-\Pi_5(A^\sigma)+d\xi_4 \tag{25}$$

will be gauge invariant up to a total divergence.

The form $\xi_4(A,A^\sigma)$ can be easily found with the result

$$\xi_4(A,A^\sigma)=\frac{1}{48\pi^2}\mathrm{Tr}\Big\{\gamma_5[(F^\sigma\wedge A^\sigma+A^\sigma\wedge F^\sigma)\wedge A$$

$$-(F\wedge A+A\wedge F)\wedge A^\sigma-A^\sigma\wedge A^\sigma\wedge A^\sigma\wedge A+A\wedge A\wedge A\wedge A^\sigma+\frac{1}{2}A^\sigma\wedge A\wedge A^\sigma\wedge A]\Big\}, \tag{26}$$

where

$$\delta\xi_4=\zeta_4(A^\sigma,X)-\zeta_4(A,X)+d\xi_3\ , \tag{27}$$

$$\xi_3=\frac{1}{48\pi^2}\mathrm{Tr}\Big\{\gamma_5[(A\wedge A^\sigma-A^\sigma\wedge A)\wedge X]\Big\}\ . \tag{28}$$

Now we define an effective Lagrangian $\tilde{\Gamma}_5$ on the 5-dimensional disc Q with the 4-dimensional space-time as its boundary

$$\begin{aligned}\tilde{\Gamma}_5&=-\frac{1}{2}[\Pi_5(A)+\Pi_5(A^\sigma)]+\frac{1}{2}d\xi_4(A,A^\sigma)\\&=\frac{1}{2}d[\xi_4(A,A^\sigma)-\omega_4(A,A^\sigma)]\\&=\frac{1}{2}\tilde{\Gamma}_{-5}-\Pi_5(A).\end{aligned} \tag{29}$$

$\tilde{\Gamma}_5$ is a closed form and so the effective Lagrangian in the ordinary space-time is an integer times

$$\frac{1}{2}(\xi_4(A,A^\sigma)-\omega_4(A,A^\sigma))^*\ , \tag{30}$$

where "*" is the dual operator.

As noted independently by both the authors[1] and Zumino et al.,[4] under an infinitesimal gauge transformation

$$\delta\Pi_5(A)=\mathrm{Tr}\{\gamma_5(G\wedge X)\}\ , \tag{31}$$

where G is the standard anomaly given by Bardeen[5] and others[6]

$$G=dA\wedge dA+\frac{1}{2}d(A\wedge A\wedge A). \tag{32}$$

Using Eqs.(29) and (31) we find

$$\begin{aligned}\delta\Gamma_5&=-\delta\Pi_5(A)\\&=-\mathrm{Tr}\{\gamma_5(G\wedge X)\}+\text{total divergence}.\end{aligned} \tag{33}$$

It is now clear that $\tilde{\Gamma}_5$ multiplied by some integer can be chosen as the effective Lagrangian of the chiral field interacting with the external gauge field because it satisfies the anomalous Ward identity in the form of Eq.(33). It will be consistent only when the holonomy group is anomaly free.

Inserting the explicit form of A^σ Eq.(7) into $\Pi_5(A^\sigma)$ and $\xi_4(A,A^\sigma)$ we obtain after a lengthy but straightforward calculation the effective Lagrangian $\tilde{\Gamma}_5$.

$$
\begin{aligned}
\tilde{\Gamma}=&\frac{1}{240\pi^2}\int_Q \mathrm{Tr}(U^R_\wedge U^R_\wedge U^R_\wedge U^R_\wedge U^R)+\frac{1}{48\pi^2}\int_{S^4}\mathrm{Tr}W,\\
W=&\Big\{[A^L_\wedge U^L_\wedge U^L_\wedge U^L-A^L_\wedge dA^L_\wedge U^L-dA^L_\wedge A^L_\wedge U^L-\frac{1}{2}A^L_\wedge U^L_\wedge A^L_\wedge U^L\\
&-A^L_\wedge A^L_\wedge A^L_\wedge U^L]-[(L\longleftrightarrow R)]\Big\}\\
&-dA^L_\wedge UA^R U^{-1}_\wedge U^L+dA^R_\wedge U^{-1}A^L U_\wedge U^R\\
&+A^L_\wedge UA^R U^{-1}_\wedge U^L_\wedge U^L-A^R_\wedge U^{-1}A^L U_\wedge U^R_\wedge U^R\\
&-[A^L_\wedge dA^L+dA^L_\wedge A^L]_\wedge UA^R U^{-1}+[A^R_\wedge dA^R+dA^R_\wedge A^R]_\wedge U^{-1}A^L U\\
&-A^L_\wedge UA^R U^{-1}_\wedge A^L_\wedge U^L+A^R_\wedge U^{-1}A^L U_\wedge A^R_\wedge U^R\\
&-A^L_\wedge A^L_\wedge A^L_\wedge UA^R U^{-1}+A^R_\wedge A^R_\wedge A^R_\wedge U^{-1}A^L U\\
&-\frac{1}{2}A^L_\wedge UA^R U^{-1}_\wedge A^L_\wedge UA^R U^{-1}\ ,\\
U^R=&U^{-1}dU,\qquad U^L=UdU^{-1}
\end{aligned}
\tag{34'}
$$

in accordance with the result given in Ref.[1] by a different method.

It is seen step by step in our derivation that the effective Lagrangian is left and right antisymmetric. This is not so in Witten's trial and error method. Besides our Lagrangian contains less terms than that of Witten's (See Appendix).

Appendix

In this Appendix we will explain that Witten's effective Lagrangian[3] (24) in its original form is, however, not correct. Apart from the misprint some terms that are necessary for the gauge invariance of the effective potential are missing. We have found that the following two additional terms have to be added to the Witten's formula

$$A_{\mu L}A_{\nu L}A_{\alpha L}U_{\beta L}+A_{\mu R}A_{\nu R}A_{\alpha R}U_{\beta R}\ . \tag{A.1}$$

After adding these two terms and correcting the misprint Witten's formula becomes

$$Z_{\mu\nu\alpha\beta}=-Tr[A_{\mu L}U_{\nu L}U_{\alpha L}U_{\beta L}+(L\longleftrightarrow R)]$$

$$+Tr[(\partial_\mu A_{\nu L}A_{\alpha L}+A_{\mu L}\partial_\nu A_{\alpha L})U_{\beta L}+(L\longleftrightarrow R)]$$

$$+Tr[\partial_\mu A_{\nu R}U^{-1}A_{\alpha L}\partial_\beta U+A_{\mu R}U^{-1}\partial_\nu A_{\alpha L}\partial_\beta U]-\frac{1}{2}Tr[A_{\mu L}U_{\nu L}A_{\alpha L}U_{\beta L}$$

$$-(L\longrightarrow R)]+Tr[A_{\mu L}UA_{\nu R}U^{-1}U_{\alpha L}U_{\beta L}-A_{\mu R}U^{-1}A_{\nu L}UU_{\alpha R}U_{\beta R}]$$

$$+Tr[(\partial_\mu A_{\nu R}A_{\alpha R}+A_{\mu R}\partial_\nu A_{\alpha R})U^{-1}A_{\beta L}U$$

$$-(\partial_\mu A_{\nu L}A_{\alpha L}+A_{\mu L}\partial_\nu A_{\alpha L})UA_{\beta R}U]$$

$$+Tr[A_{\mu R}U^{-1}A_{\nu L}UA_{\alpha R}U_{\beta R}+A_{\mu L}UA_{\nu R}U^{-1}A_{\alpha L}U_{\beta L}]$$

$$+Tr[A_{\mu L}A_{\nu L}U\partial_\alpha A_{\beta R}U^{-1}+A_{\mu R}A_{\nu R}U^{-1}\partial_\alpha A_{\beta L}U]$$

$$+Tr[A_{\mu R}A_{\nu R}A_{\alpha R}U^{-1}A_{\beta L}U-A_{\mu L}A_{\nu L}A_{\alpha L}UA_{\beta R}U^{-1}$$

$$+A_{\mu L}A_{\nu L}UA_{\alpha R}A_{\beta R}U^{-1}+\frac{1}{2}A_{\mu R}U^{-1}A_{\nu L}UA_{\alpha R}U^{-1}A_{\beta L}U]$$

$$+Tr[A_{\mu L}A_{\nu L}A_{\alpha L}U_{\beta L}+(L\longleftrightarrow R)]^{+}\ . \qquad (A.2)$$

To see the necessity for the addition of these terms, the easiest way is to check the gauge invariance in the special case where A_R(or A_L) is zero. In this case only five terms (with the last one added by us) remain. They are

$$\widetilde{\Gamma}=\frac{1}{240\pi^2}\int_Q d\Sigma^{ijkm\ell}Tr(U^{-1}\partial_i UU^{-1}\partial_j UU^{-1}\partial_k UU^{-1}\partial_m UU^{-1}\partial_\ell U)$$

$$+\frac{1}{48\pi^2}\int_{S^4}d^4x\varepsilon^{\mu\nu\alpha\beta}Tr[-A_\mu\partial_\nu UU^{-1}\partial_\alpha UU^{-1}\partial_\beta UU^{-1}+A_\mu\partial_\nu A_\alpha\partial_\beta UU^{-1}$$

$$+\partial_\mu A_\nu A_\alpha\partial_\beta UU^{-1}-\frac{1}{2}A_\mu\partial_\nu UU^{-1}A_\alpha\partial_\beta UU^{-1}+A_\mu A_\nu A_\alpha\partial_\beta UU^{-1}]. \qquad (A.3)$$

Under the infinitesimal transformation

$$U\longrightarrow\widehat{U}=(1+\varepsilon)U, \qquad \varepsilon=\varepsilon_L^a T_L^a\ ,$$

$$U^{-1}dU\longrightarrow\widehat{U}d\widehat{U}=U^{-1}dU+U^{-1}d\varepsilon U,$$

$$dUU^{-1}\longrightarrow d\widehat{U}\widetilde{U}^{-1}=dUU^{-1}+d\varepsilon+\varepsilon dUU^{-1}-dUU^{-1}\varepsilon\ ,$$

$$A_\mu\longrightarrow\widetilde{A}_\mu=A_\mu+\varepsilon A_\mu-A_\mu\varepsilon-\partial_\mu\varepsilon\ , \qquad (A.4)$$

the change of the effective Lagrangian is just the anomaly

† *All the formulae in this Appendix are written in the notation of our previous paper*[1].
By changing A to iA one obtains the Witten's formulae.

$$\delta\tilde{\Gamma}=\frac{1}{24\pi^2}\int_{S^4}d^4x\varepsilon^{\mu\nu\alpha\beta}\mathrm{Tr}\varepsilon\partial_\mu(A_\nu\partial_\alpha A_\beta+\tfrac{1}{2}A_\nu A_\alpha A_\beta) \tag{A.5}$$

as claimed by Witten. Without the last term added the change of the effective Lagrangian is

$$\frac{1}{48\pi^2}\int_{S^4}d^4x\varepsilon^{\mu\nu\alpha\beta}\mathrm{Tr}[A_\mu A_\nu\partial_\alpha\varepsilon\partial_\beta UU^{-1}+\partial_\mu\varepsilon A_\nu A_\alpha\partial_\beta UU^{-1}$$
$$+A_\mu\partial_\nu\varepsilon A_\alpha\partial_\beta UU^{-1}+2\varepsilon\partial_\mu(A_\nu\partial_\alpha A_\beta)], \tag{A.6}$$

which obviously is not gauge invariant under the anomaly free condition.

Now it is clear that Witten's formula contains four more terms than ours

$$W_{\mu\nu\alpha\beta}=Z_{\mu\nu\alpha\beta,\mathit{Witten's}}-\mathrm{Tr}(A_{\mu L}A_{\nu L}U\partial_\alpha A_{\beta R}U^{-1}+A_{\mu R}A_{\nu R}U^{-1}\partial_\alpha A_{\beta L}U$$
$$+A_{\mu R}U^{-1}\partial_\nu A_{\alpha L}\partial_\beta U+A_{\mu L}A_{\nu L}UA_{\alpha R}A_{\beta R}U^{-1}$$
$$-\partial_\mu A_{\nu L}UA_{\alpha R}U^{-1}\partial_\beta UU^{-1}). \tag{A.7}$$

The difference between Witten's and ours can be combined in a gauge invariant form modula an exact differential as mentioned by us in a previous letter.[1] The result is

$$\int_{S^4}d^4x\varepsilon^{\mu\nu\alpha\beta}W_{\mu\nu\alpha\beta}=\int_{S^4}d^4x\varepsilon^{\mu\nu\alpha\beta}(Z_{\mu\nu\alpha\beta}-\mathrm{Tr}(F_{\mu\nu R}U^{-1}F_{\alpha\beta L}U)). \tag{A.8}$$

References

1. *K.C.CHOU, H.Y.GUO, K.WU and X.C.SONG "On the Gauge Invariance and Anomaly-free Condition of Wess-Zumino-Witten Effective Action", Phys.Lett., 134B(1984)67.*
2. *J.Wess and B.Zumino, Phys. Lett., 37B(1971)95.*
3. *E.Witten, Nucl. Phys., B223(1983)422.*
4. *B.Zumino. Y.S.WU and A.Zee, "Chiral Anomalies, Higher Dimensions, and Differential Geometry", preprint (Revised) 10048-18(1983)P3.*
5. *W.A.Bardeen, Phys. Rev., 184(1969)1848.*
6. *D.J.Gross and R.Jackiw, Phys.Rev., D6(1972)477.*

Commun. in Theor. Phys. (Beijing China) Vol.3; No.1 (1984) 125-129

THE TOPOLOGICAL ORIGINS OF GAUGE ANOMALIES

CHOU Kuang-chao(周光召),GUO Han-ying(郭汉英)
and WU Ke(吴 可)
Institute of Theoretical Physics, Academia Sinica,
P.O. Box 2735, Beijing, China
and
SONG Xing-chang(宋行长)
Institute of Theoretical Physics, Peking University, Beijing, China

Received October 4, 1983

Abstract

In addition to the recent discovery that the Chern-Simons secondary classes are the topological origins generating special unitary gauge group anomalies, it is shown that some special orthogonal gauge group anomalies can also be generated, from the Pontrjagin and the Euler secondary classes. The common differential geometrical and topological background are explained as well.

Recently, it has been discovered independently both by the present authors [1] and by Zumino et al.[2] that the gauge anomalies associated with the special unitary groups have profound topological origin, the Chern-Simons secondary classes. In addition to this discovery, we would like to show in the present note that gauge anomalies of the special orthogonal groups have also similar topological origins, the Pontrjagin secondary classes and the Euler secondary classes.

We will first in this note explain the differential geometrical and topological background of the relation between the anomalies and the Chern-Simons secondary classes and then extend this consideration to the cases of the gauge fields of the special orthogonal groups.

Let us start from the SU(3) gauge field defined on a 6-dimensional sphere S^6. Similar to the well known monopole and instanton analysis, we introduce two neighborhoods $H_+ \equiv S^6-\{s\}$, $H_- \equiv S^6-\{N\}$ to cover the S^6, where the $\{s\}$ and $\{N\}$ is the south and the north pole respectively, and the equator of S^6, the unit five sphere $S^5_o \subset H_+ \cap H_-$. According to the classification theorem of bundles over the sphere,[3] the characteristic map, which is defined as the transition function ϕ: $H_+ \cap H_- \longrightarrow SU(3)$ restricted on the equator, $U=\phi|_{S^5_o}:(S^5_o, y_o) \longrightarrow (SU(3),e)$, where y_o is a reference point and e the unit element of the gauge group, gives rise to the classification of the bundle based upon the homotopy group π_5(SU(3))=Z which is generated by

$$\Gamma(S_o^5,U)=\frac{1}{480\pi^3}\int_{S_o^5}\mathrm{Tr}\,\Theta\wedge\Theta\wedge\Theta\wedge\Theta\wedge\Theta\ ,\quad \Theta=dUU^{-1}. \qquad (1)$$

On the other hand, the bundle can also be classified by means of the third Chern number C_3[3], the integral of the third Chern class over the S^6

$$C_3=\frac{1}{48\pi^3}\int_{S^6}\mathrm{Tr}(\Omega\wedge\Omega\wedge\Omega),\qquad \Omega=dA+A\wedge A, \qquad (2)$$

where A is the SU(3) gauge field 1-form and Ω the field strength 2-form. And this classification is equivalent to the previous one.

This equivalence can easily be seen in the following way. By definition, the third Chern class can locally be written as an exterior differential of the Chern-Simons secondary class on H_+ and H_- respectively, so we have

$$C_3=\int_{H_+}d\Pi_5^c(A_+,\Omega_+)+\int_{H_-}d\Pi_5^c(A_-,\Omega_-), \qquad (3)$$

where Π_5^c is the Chern-Simons 5-form

$$\Pi_5^c(A,\Omega)=\frac{1}{48\pi^3}\mathrm{Tr}(\Omega\wedge\Omega\wedge A-\frac{1}{2}\Omega\wedge A\wedge A+\frac{1}{10}A\wedge A\wedge A\wedge A\wedge A). \qquad (4)$$

By taking the equator S_o^5 to be the common boundary of the H_+ and H_- and using the Stokes formula, we have

$$C_3=\int_{S_o^5}(\Pi_5^c(A_+,\Omega_-)-\Pi_5^c(A_-,\Omega_-)). \qquad (5)$$

Making a gauge transformation with respect to the U(x)

$$A_-\xrightarrow{U}A_+=U^{-1}A_-U+U^{-1}dU=:U^{-1}DU,$$

$$\Omega_-\xrightarrow{U}\Omega_+=U^{-1}\Omega U \qquad (6)$$

and substituting A_+,Ω_+ by the transformed ones, we can prove that

$$C_3=\int_{S_o^5}dG^c(A,\Omega,U)+\Gamma(S_o^5,U), \qquad (7)$$

where

$$G^c=\frac{1}{48\pi^3}\mathrm{Tr}\Big\{A\wedge dA\wedge\Theta+\frac{1}{2}A\wedge A\wedge A\wedge\Theta-\frac{1}{2}A\wedge A\wedge\Theta\wedge\Theta$$
$$-\frac{1}{4}A\wedge\Theta\wedge A\wedge\Theta-\frac{1}{2}A\wedge\Theta\wedge\Theta\wedge\Theta\Big\}\ . \qquad (8)$$

Again by the Stokes formula and $\partial S_o^5=\phi$, we obtain

$$C_3=\Gamma(S_o^5,U). \qquad (9)$$

This result shows that the two classifications are equivalent to each other.

Instead of the unit five sphere S_o^5, the equator of S^6, let us now consider a 5-dimensional disc Q^5 whose boundary is taken to be the compactified spacetime $M^4 \sim S^4$. If the disc is embedded in an S^5 whose volume is, by normalization as Witten has done,[4] 2π times the volume of the equator S_o^5, it is easy to obtain the following expression

$$2\pi \int_{Q^5} [\Pi_5^C(U^{-1}DU, U^{-1}\Omega U) - \Pi_5^C(A,\Omega)] = \Gamma(Q^5, U) + 2\pi \int_{S^4} G \quad , \tag{10}$$

where

$$\Gamma(Q^5, U) = \frac{1}{240\pi^2} \int_{Q^5} \mathrm{Tr}(\Theta\wedge \ldots \Theta\wedge) \quad , \tag{11}$$

which is the same as the one introduced by Witten[4] to show the global properties of the effective action theory. The expression (10) shows that the $\Gamma(Q^5,U)$ term is essentially linked to the Chern-Simons secondary topological invariants as long as the gauge field defined on $M^4 \sim S^4$ can be continued onto the disc Q^5 without any topological obstruction. From the point of view of gauge transformation, the expression (10) shows the gauge variation of the Chern-Simons 5-form $\Pi_5^C(A\ \Omega)$. If the gauge transformation function $U(x)$ is infinitesimal

$$U(x) = 1 + \varepsilon(x), \qquad \Theta = d\varepsilon(x), \tag{12}$$

the variation becomes an infinitesimal one as well

$$2\pi\delta \int_{Q^5} \Pi_5^C(A,\Omega) = \frac{1}{24\pi^2} \int_{S^4} \mathrm{Tr}\left\{\varepsilon[dA\wedge dA + \tfrac{1}{2} d(A\wedge A\wedge A)]\right\} \ . \tag{13}$$

Eqs.(10) and (13) are in fact the gauge anomalies with a minus sign in finite and infinitesimal forms respectively.[1,2] This fact means that the Chern-Simons secondary class is the topological origin of usual purturbative chiral anomalies with respect to the SU(3) gauge field defined on 4-dimentional spacetime $M^4 \sim S^4$.[5]

From the point of view of differential geometry and topology, the above considerations can easily be extended to the SU(k) gauge field defined on an arbitrary even dimensional compactified spacetime $M^{2n-2} \simeq S^{2n-2}$ and a 2n-1-dimensional oriented disc Q^{2n-1}, $\partial Q^{2n-1} = S^{2n-2}$. In fact, we can write

$$2\pi \int_{Q^{2n-1}} [\Pi_{2n-1}^C(U^{-1}DU, U^{-1}\Omega U) - \Pi_{2n-1}^C(A,\Omega)] = \Gamma(Q^{2n-1}, U) + 2\pi \int_{S^{2n-2}} G_{2n-2}^C \tag{14}$$

which is the general form of the finite chiral anomaly of SU(k) gauge field, where G_{2n-2}^C is a certain (2n-2)-form defined on the spacetime $M^{2n-2} \simeq S^{2n-2}$.

We now turn to the chiral anomalies associated with the SO(m) gauge fields.

As is well known, the SO(m) gauge fields on the 4-dimensional spacetime do not have anomalies except for m=6.[6] This fact can be understood with the help of differential geometry and topology.

It is well known that there are two characteristic classes related to the SO(m) gauge fields, the Pontrjagin class P and the Euler class χ.[3] The first can only be defined over 4n-dimensional compact manifolds and the second can be defined over 2n-dimensional ones as the Chern class does. By difinition, each of them can locally be written as an exterior differential of corresponding secondary characteristic classes. And under the SO(m) gauge transformations, the variations of those secondary classes are also divided into a surface term plus a term corresponding to the winding number. If we consider an odd-dimensional disc Q with boundary taken to be the even-dimensional compactified spacetime, we can obtain the expressions similar to the expression (14) for the gauge variations of both Pontrjagin secondary class Π^{P}_{4n-1} and Euler secondary class Π^{χ}_{2n-1} as follows:

$$2\pi\int_{Q^{4n-1}}(\Pi^{P}_{4n-1}(R^{-1}DR,R^{-1}\Omega R)-\Pi^{P}_{4n-1}(A,\Omega)=\Gamma(Q^{4n-1},R)+2\pi\int_{S^{4n-2}}G^{P}_{4n-2} \tag{15}$$

and

$$2\pi\int_{Q^{2n-1}}(\Pi^{\chi}_{2n-1}(R^{-1}DR,R^{-1}\Omega R)-\Pi^{\chi}_{2n-1}(A,\Omega))=\Gamma(Q^{2n-1},R)+2\pi\int_{S^{2n-2}}G^{\chi}_{2n-2}, \tag{16}$$

where R is an SO(m) gauge transformation function. Corresponding infinitesimal variations can also be obtained if the gauge function is taken to be an infinitesimal one.

Based upon similar considerations in the case of SU(k) gauge anomalies generated by the Chern-Simons secondary classes, it is reasonable to expect that Eqs.(15) and (16) give rise to the SO(m) gauge anomalies generated by the Pontrjagin secondary classes and the Euler secondary classes respectively. In order to verify this point, let us explore the SO(m) gauge anomalies on the 4-dimensional compactified spacetime $M^4 \sim S^4$. From the expression (15), it follows that there do not exist SO(m) gauge anomalies generated by the Pontrjagin secondary classes, since the Pontrjagin secondary classes cannot be defined on a 5-dimensional disc Q^5 at all. On the other hand, SO(6) anomaly can be generated by the Euler secondary invariant 5-form defined on Q^5 because this 5-form corresponds to third Euler class on a S^6 whose vector bundle takes SO(6) as the structure group. Furthermore, explicit calculation shows that in the case of infinitesimal gauge transformation the expression (16) gives rise to

$$-2\pi\delta\int_{Q^5}\pi^{\chi}_{5}(A,\Omega)=\frac{-1}{192\pi^2}\int_{S^4}\varepsilon_{\alpha_1\alpha_2\alpha_3\alpha_4\alpha_5\alpha_6}[\varepsilon^{\alpha_1\alpha_2}d\omega^{\alpha_3\alpha_4}d\omega^{\alpha_5\alpha_6} + \frac{1}{2}\varepsilon^{\alpha_1\alpha_2}d(\omega^{\alpha_3\alpha_4}\omega^{\alpha_5\ell}\omega^{\ell\alpha_6})] , \tag{17}$$

which is just the infinitesimal SO(6) gauge anomaly in agreement with the SU(4) gauge anomaly.

Thus, we conclude that Eqs.(15) and (16) can be regarded as the general expressions for the SO(m) gauge anomalies on the compactified spacetimes of certain dimensions. In the case of 4n-2 dimensional spacetimes (n=1,2,...), there exist certain special orthogonal gague group anomalies generated by the corresponding Pontrjagin secondary classes. In the case of 2n-2 dimensional spacetimes (n=1,2,3,...), the SO(2n) gauge anomaly can be generated by the Euler secondary invariant (2n-1)-forms defined on the 2n-1 dimensional disc Q^{2n-1}.

References

1. *CHOU Kuang-chao, GUO Han-ying, WU Ke and SONG Xing-chang, Phys. Lett., 134B(1984)67; Commun. in Theor. Phys. (Beijing, China), this issue.*
2. *B. Zumino, Y.-S. WU and A. Zee, Preprint, (Revised)40018-18 (1983)P3.*
3. *N. Steenrod, The Topology of Fibre Bundles, Princeton Univ. Press, 1951; S. Kobayashi and K. Nomizu, Foundations of Differential Geometry , Vol.II, Intersc. Pub., 1969.*
4. *E. Witten, Nucl. Phys., B223(1983)422.*
5. *W.A. Bardeen, Phys. Rev., 184(1969)1848; D.J. Gross and R. Jackiw, Phys. Rev., D6(1972)477.*
6. *H.Georgi and S. Glashow, Phys. Rev., D6(1972)429.*
7. *S.S. Chern, Ann. Math., 46(1945)674.*

Commun. in Theor. Phys.(Beijing) *Vol.3, No.2 (1984) 221-229*

ON THE DYNAMICAL SYMMETRY BREAKING FOR THE N=1 PURE SUPERSYMMETRIC YANG-MILLS MODEL

CHOU Kuang-chao(周光召), DAI Yuan-ben(戴元本)
and CHANG Chao-hsi(张肇西)

Institute of Theoretical Physics, Academia Sinica,
P.O.Box 2735, Beijing, China

Received November 15, 1983

Abstract

An approximate effective Lagrangian of composite operators in superfield formalism for the N=1 pure supersymmetric Yang-Mills model is obtained with the help of renormalization group equations for the generating functional. While the supersymmetry is always kept unbroken by this effective Lagrangian we find that the chiral symmetry may and may not be unbroken.

The potential application of SUSY to GUTS and composite model building makes it more and more stimulating to explore and to understand the properties of SUSY.One of the important problems is about the possible dynamical symmetry breaking e.g.whether the super-symmetry and the chiral symmetry, both or either,are broken dynamically or not,and if they are indeed broken then how and where to break.One approach to these problems is to analyse the effective Lagrangian for those composite fields which may condense. This approach has already been adopted recently to treat supersymmetric Yang-Mills(SSYM) models with or without matter fields by many authors[1,2,8].

The work[1] by Veneziano and Yankielowicz is based on the anomalous Ward identities for scale, chiral and superconformal transformation.[6] Their conclusion is that the chiral symmetry is broken dynamically but the supersymmetry is not for N=1 pure supersymmetric Yang-Mills model. In this article, the N=1 pure supersymmetric Yang-Mills model is to be examined, using renormalization group analysis i.e. an equation for the effective action is obtained, which can be interpreted as the anomalous Ward identity for certain scale transformations. And we shall see that the anomalous dimensions of the composite fields enter into this anomalous Ward identity. This is in contrast with the approach adopted in Ref.[1], where the composite fields are scaled according to their naïve dimensions. Our approach bears some resemblance to that used in Refs.[5] and [1] for QCD and SSYM. In addition, under certain approximation the equation can be solved and the solutions can be determined up to some free constant coefficients, which do not affect the final qualitative conclusion, so that some interesting results, different from those in Ref.[1], can be drawn.

Now let us look at the N=1 pure supersymmetric Yang-Mills model with the non-Abelian group $SU(N_c)$ ($N_c=3$ for SUSY QCD). We will start with the gauge covariant chiral superfield W_α, which is related to the vector superfield V by

$$W_\alpha=-\frac{1}{4}\bar{D}^2 e^{-gV} D_\alpha e^{gV} \tag{1}$$

and $V=V^a T^a$, T^a are the adjoint representation matrices of $SU(N_c)$ generators, and $Tr(T^a T^b)=\delta^{ab}$, g is the coupling constant. Let

$$S=\frac{1}{2g^2} Tr W^\alpha W_\alpha \ . \tag{2}$$

In W-Z gauge

$$V=-\Theta\sigma^m\bar{\Theta}V_m(x)+i\Theta^2\bar{\Theta}\bar{\lambda}(x)-i\bar{\Theta}^2\Theta\lambda(x)+\frac{1}{2}\Theta^2\bar{\Theta}^2 D(x) \ ,$$

then

$$S=-\frac{1}{2}\lambda^{a\alpha}\lambda^a_\alpha-i(\lambda^a\Theta)D^a(y)-\Theta\sigma^{mn}V^a_{mn}\lambda^a$$

$$+\frac{1}{2}D^a(x)D^a(x)\Theta^2-\frac{1}{4}V^{amn}V^a_{mn}\Theta^2-i\Theta^2\lambda^{a\alpha}\sigma^m_{\alpha\dot{\alpha}}D^{ab}_m\bar{\lambda}^{b\dot{\alpha}} \ , \tag{3}$$

where D^a is the auxiliary field, the Yang-Mills field strength is

$$V^{mn}=\partial_m V_n-\partial_n V_m-\frac{i}{2}g[V_m, V_n] \tag{4}$$

and $\lambda^{a\alpha}(\bar{\lambda}^{b\dot{\alpha}})$ is the Weyl spinor component. D^{ab}_m is the covariant derivative

$$D^{ab}_m=\partial_m\delta^{ab}-\frac{1}{2}gC^{abc}V^c_m \ , \tag{5}$$

C^{abc} is the structure constant of the group $SU(N_c)$.

In order to investigate the problem of condensation for the component fields of the superfield S, chiral superfields J and J^+ are introduced as external sources of S and S^+. The Lagrangian for the pure Yang-Mills theory including the gauge fixing term, ghost fields C, $\bar{C}$ and the external source terms is

$$\mathcal{L}_J=\frac{1}{2}\int d^2\Theta S+\frac{1}{2}\int d^2\bar{\Theta}S^+-\frac{1}{16\alpha}\int d^4\Theta Tr D^2 V\bar{D}^2 V$$

$$+\int d^4\Theta Tr(\bar{C}'-C')L_{\frac{gV}{2}}[(\bar{C}+C)+(\coth L_{\frac{gV}{2}})(C-\bar{C})]$$

$$+\int d^2\Theta JS+\int d^2\bar{\Theta}J^+S^+ + \text{counter terms}$$

$$\equiv\mathcal{L}_{Jc}+\text{counter terms}, \tag{6}$$

where

$$L_{\frac{gV}{2}}X\equiv\frac{1}{2}g[\,V,X]\ . \tag{7}$$

The generating functional is defined by

$$\mathcal{Z}[J,J^{+}]=\exp iW[J,J^{+}]=\int[\mathcal{D}V][\mathcal{D}C][\mathcal{D}\bar{C}]\exp\left\{i\int d^4x\mathcal{L}_J\right\}\ . \tag{8}$$

Let us assume that the theory is made finite by the following renormalization

$$\begin{aligned} &g_o=Z_g g \\ &V_o=Z_3^{\frac{1}{2}}V\ , \qquad \alpha_o=Z_3\alpha\ , \\ &C_o=\tilde{Z}^{\frac{1}{2}}C\ , \qquad \bar{C}_o=\tilde{Z}^{\frac{1}{2}}\bar{C}\ , \\ &J_o=Z(J)^{-1}J\ , \qquad J_o^{+}=Z(J^{+})^{-1}J^{+}\ . \end{aligned} \tag{9}$$

Since S is a hard operator and the chiral superfield J is dimensionless, Z(J) is in general not a constant but may depend on J due to multiple insertions of the operator S. From Eq.(9) we can derive the following renormalization group equation for the generating functional $\mathcal{Z}$:

$$\begin{aligned} -\mu\frac{\partial}{\partial\mu}\mathcal{Z}[J,J^{+},g,\alpha,\mu]=\Big\{&\beta\frac{\partial}{\partial g}-\gamma_3\alpha\frac{\partial}{\partial\alpha} \\ &+\int d^4x(\gamma J)\frac{\delta}{\delta J}+\int d^4x(\gamma^{+}J^{+})\frac{\delta}{\delta J^{+}}\Big\}\mathcal{Z}\ , \end{aligned} \tag{10}$$

where μ is the scale parameter introduced by renormalization,

$$\gamma_3=\mu\frac{\partial}{\partial\mu}\ln Z_3\ , \qquad \gamma=\mu\frac{\partial}{\partial\mu}\ln Z(J) \tag{11}$$

and $\int(\gamma J)\frac{\delta}{\delta J}\equiv\int(\gamma J)_i\frac{\delta}{\delta J_i}$ with J_i and $(\gamma J)_i$ denoting components of corresponding chiral superfields.

Since the external source J is gauge invariant, it is not difficult to see

$$\frac{\partial}{\partial\alpha}\mathcal{Z}=0. \tag{12}$$

The equation of motion of the V superfield reads

$$\int[\mathcal{D}V]\int d^4xV\frac{\delta}{\delta V}\mathcal{L}_J e^{i\int\mathcal{L}_J d^4x}=0\ . \tag{13}$$

Combining Eqs.(12), (13) with Eq.(10) we obtain

$$-\mu\frac{\partial}{\partial\mu}\mathcal{Z}=\int[\mathcal{D}V][\mathcal{D}C][\mathcal{D}\bar{C}]\frac{\beta}{g}(g\frac{\partial}{\partial g}-2\alpha\frac{\partial}{\partial\alpha}-\int d^4xV\frac{\delta}{\delta V})(i\int d^4x\ \mathcal{L}_J)e^{i\int d^4x\mathcal{L}_J}$$

$$+\int d^4x[(\gamma J)\frac{\delta}{\delta J}+h.c.]\mathcal{Z}. \qquad (14)$$

Writing $\mathcal{L}_J$ in the normal product form

$$\mathcal{L}_J=N[\int d^4x\ \mathcal{L}_{JC}], \qquad (15)$$

we obtain the equation

$$-\mu\frac{\partial}{\partial\mu}\mathcal{Z}=\int[\mathcal{D}V][\mathcal{D}C][\mathcal{D}\bar{C}](-\frac{2\beta}{g}iN\int d^4xS_F)e^{i\int d^4x\mathcal{L}_J}$$

$$+\int d^4x[(\gamma-\frac{2\beta}{g})J\frac{\delta}{\delta J}+h.c.]\mathcal{Z} \qquad (16)$$

from Eq.(14), here and later the subscript F of S_F means to take F-component of S. Let us introduce $\Gamma[S_c, S_c^+]$ through the Legendre transformation

$$\Gamma[S_c,S_c^+]=W[J,J^+]-\int d^4x(d^2\Theta JS_c+h.c.) \qquad (17)$$

with

$$S_c=\frac{\delta W[J,J^+]}{\delta J}=\langle N[S]\rangle_J. \qquad (18)$$

Since

$$\mu\frac{\partial}{\partial\mu}\Gamma[S_c,S_c^+]|_{J,J^+}=\mu\frac{\partial}{\partial\mu}\Gamma[S_c,S_c^+]|_{S_c,S_c^+}$$

$$+(\int d^4x\mu\frac{\partial S_c}{\partial\mu}|_{J,J^+}\cdot\frac{\partial}{\partial S_c}\Gamma[S_c,S_c^+]+h.c.)$$

$$=\mu\frac{\partial}{\partial\mu}\Gamma[S_c,S_c^+]|_{S_c,S_c^+}-\int d^4x(d^2\Theta J\mu\frac{\partial S_c}{\partial\mu}|_{J,J^+}+h.c.). \qquad (19)$$

From Eq.(16) we find that

$$-\mu\frac{\partial}{\partial\mu}\Gamma[S_c,S_c^+]=-\frac{2\beta}{g}\int d^4xS_{cF}$$

$$-\int d^4x[(\gamma-\frac{2\beta}{g})S_c\frac{\delta}{\delta S_c}+h.c.]\cdot\Gamma[S_c,S_c^+]. \qquad (20)$$

In the neighbourhood of the minima

$$\frac{\delta\Gamma}{\delta S_c}=\frac{\delta\Gamma}{\delta S_c^+}=0,$$

we can make the approximation

$$\gamma=\gamma(J)=\gamma\left(\frac{-\delta\Gamma}{\delta S_c}\right)\doteq\gamma(0)\equiv\gamma_0 \ . \tag{21}$$

Nothing is lost in this approximation if one uses it to find the minima of Γ only. Under this approximation, Eq.(19) is reduced to the form:

$$-\mu\frac{\partial}{\partial\mu}\Gamma[S_c,S_c^+]=-\frac{2\beta}{g}\int d^4xS_{cF}$$
$$-\int d^4x[\tilde{\gamma}S_c\frac{\delta}{\delta S_c}+h.c.]\ \Gamma[S_c,S_c^+] \ , \tag{22}$$

where $\tilde{\gamma}\equiv\gamma_0-\frac{2\beta}{g}$. At $J=J^+=0$, we have

$$\frac{\delta\Gamma}{\delta S_c}=\frac{\delta\Gamma}{\delta S_c^+}=0 \ ,$$

the first term on the right-hand side of Eq.(22) becomes the trace anomaly of the J=0 theory.[7]

$\Gamma[S_c,S_c^+]$ consists of two pieces, the F part and the D part as follows:

$$\Gamma=(\int d^4xd^2\theta\Gamma_f+h.c.)+\int d^4xd^4\theta\Gamma_d. \tag{23}$$

With the help of the dimension counting rule, equation (22) can be rewritten into the following two equations

$$3\Gamma_f-(3+\tilde{\gamma})\int d^4xS_c\frac{\delta}{\delta S_c}\Gamma_f=\frac{\beta}{g}S_c \tag{24}$$

and

$$2\Gamma_d-[(3+\tilde{\gamma})\int d^4xS_c\frac{\delta}{\delta S_c}\Gamma_d+h.c.]=0 \ . \tag{25}$$

Eqs.(24) and (25) have the structure of anomalous Ward identities for some scale transformations. If we make the following infinitesimal scale transformations in Γ

$$X\to e^{-\varepsilon}x, \qquad \Theta\to e^{-\varepsilon/2}\Theta \ , \qquad S_c\to e^{\varepsilon(3+\tilde{\gamma})}S_c \ , \tag{26}$$

Eqs.(25) and (26) imply that the change of Γ will be

$$\delta\Gamma=\varepsilon\frac{-2\beta}{g}\int d^4xS_{cF} \ . \tag{27}$$

Therefore, if we consider (unjustifiably) the $\frac{2\beta}{g}\int d^4xS_{cF}$ term as the trace anomaly of present J=0 theory as done in Ref.[1], the superfield S_c in the effective action Γ should be scaled as if it has the dimension $3+\tilde{\gamma}$.

As we are interested only in slowly varying field we can confine ourselves to the solution of Eqs.(24) and (25) local in x. A class of solution in this case is

$$\Gamma_f=-\beta(g\tilde{\gamma})^{-1}S_c+C_f S_c^{\frac{3}{3+\tilde{\gamma}}} \tag{28}$$

and

$$\Gamma_d=C_d(S_c S_c^+)^{\frac{1}{3+\tilde{\gamma}}}, \tag{29}$$

where C_f and C_d are constants with an appropriate dimension. In the theory there is only one dimensional parameter μ (the renormalization parameter), so C_f and C_d can be represented as $C_f=C_f^\circ\mu^{\frac{3\tilde{\gamma}}{3+\tilde{\gamma}}}$ and $C_d=C_d^\circ\mu^{\frac{2\tilde{\gamma}}{3+\tilde{\gamma}}}$ respectively where C_f° and C_d° are dimensionless constants. Now, in this approximation, Γ can be considered as the effective Lagrangian $\mathcal{L}_{eff}$ of this SUSY Yang-Mills model, hence we are in a position to analyse the symmetries of vacuum determined by the effective Lagrangian, what we are interested in mainly, so as to examine if the supersymmetry and/or the chiral symmetry are broken.

We'll consider two different cases separately, case I) $C_f=0$ and case II) $C_f\neq 0$.

In case I) the effective Lagrangian is

$$\mathcal{L}_{eff}=-\beta(g\tilde{\gamma})^{-1}(S_{cF}+S_{cF}^+)+C_d[(S_c S_c^+)^{\frac{1}{3+\tilde{\gamma}}}]_D, \tag{30}$$

where the foot indices F and D imply taking F-term and D-term respectively. It is easy to read out the scalar potential

$$V(\phi,L)=2\beta(g\tilde{\gamma})^{-1}L-\mathcal{C}_d\phi^{\frac{4+2\tilde{\gamma}}{3+\tilde{\gamma}}}L^2, \tag{31}$$

where in W-Z gauge $\phi=-\frac{1}{2}\lambda^{a\alpha}\lambda^a_\alpha$, $L=-\frac{1}{4}V^{amn}V^a_{mn}$ and

$$\mathcal{C}_d=(\frac{1}{3+\tilde{\gamma}})^2C_d . \tag{32}$$

In Eq.(31) we have assumed that the parity is not broken spontaneously, i.e., $\phi=\phi^*$ and $L=L^*$. From $\frac{\partial V}{\partial L}=0$, we obtain

$$L=L_{01}\equiv\beta(g\tilde{\gamma})^{-1}\mathcal{C}_d^{-1}\phi^{\frac{4+2\tilde{\gamma}}{3+\tilde{\gamma}}} \tag{33}$$

and

$$V(\phi,L)|_{L=L_{01}}=\beta^2(g\bar{\gamma})^{-2}\ C_d^{-1}\phi^{\frac{4+2\bar{\gamma}}{3+\bar{\gamma}}}\ . \quad (34)$$

From the requirement of positive definiteness of the kinetic energy terms of the effective Lagrangian we have $C_d>0$. Assuming $\frac{4+2\bar{\gamma}}{3+\bar{\gamma}}>0$ (This is certainly true for sufficiently large μ at which the renormalized coupling constant $g(\mu)$ is small) it is easy to see that the minimum of Eq.(34) is

$$V^{min}(\phi,L)=0 \quad (35)$$

at

$$L=0 \quad \text{and} \quad \phi=0. \quad (36)$$

That is in this case the supersymmetry and chiral symmetry are both unbroken.

In case II), the effective Lagrangian is

$$\mathcal{L}_{eff}=-\beta(g\bar{\gamma})^{-1}(S_{cF}+S^+_{cF})+C_f(S_c^{\frac{3}{3+\bar{\gamma}}})_F$$

$$+C_f[(S_c^+)^{\frac{3}{3+\bar{\gamma}}}]_F+C_d[(S_cS_c^+)^{\frac{1}{3+\bar{\gamma}}}]_D \quad (37)$$

and the scalar potential is

$$V(\phi,L)=2\beta(g\bar{\gamma})^{-1}L-\mathcal{C}_f^{-\frac{\bar{\gamma}}{3+\bar{\gamma}}}L-C_d\phi^{-\frac{4+2\bar{\gamma}}{3+\bar{\gamma}}}\ , \quad (38)$$

where

$$\mathcal{C}_f=\frac{6}{3+\bar{\gamma}}C_f\ . \quad (39)$$

From $\frac{\partial V}{\partial L}=0$, we obtain

$$L=L_{o2}=\beta(g\bar{\gamma})^{-1}\ C_d^{-1}\phi^{\frac{4+2\bar{\gamma}}{3+\bar{\gamma}}}(1-\frac{\bar{\gamma}g}{2\beta}\ \mathcal{C}_f\phi^{-\frac{\bar{\gamma}}{3+\bar{\gamma}}}) \quad (40)$$

and

$$V(\phi,L)|_{L=L_{o2}}=\beta^2(g\bar{\gamma})^{-2}\ C_d^{-1}\ \phi^{\frac{4+2\bar{\gamma}}{3+\bar{\gamma}}}(1-\frac{\bar{\gamma}g}{2\beta}\ \mathcal{C}_f\phi^{-\frac{\bar{\gamma}}{3+\bar{\gamma}}})^2\ . \quad (41)$$

There is one minimum for the $V(\phi,L)|_{L=L_{o2}}$ of Eq.(41) ($V^{min}(\phi,L)=0$) at

$$\phi=\phi_o=0 \quad (\text{and } L_{o2}=0). \quad (42)$$

For $\frac{2\beta}{\tilde{\gamma}g} C_f^{-1} > 0$,there is another minimum ($V^{min}(\phi,L)=0$) at

$$\phi=\phi_1=\left(\frac{2\beta}{\tilde{\gamma}g} C_f^{-1}\right)^{\frac{3+\tilde{\gamma}}{\tilde{\gamma}}} \qquad (\text{and } L_{o2}=0) . \tag{43}$$

Both the degenerate minima are reached at $L_{o2}=0$, with one of them at $\phi=0$ and the other at $\phi\neq0$. This means the supersymmetry is unbroken no matter which minimum is taken and the chiral symmetry may be broken (when $\phi=\phi_1$) and may not be broken (when $\phi=\phi_o$).

If the non-renormalization theorem[3] valid for arbitrary order in perturbation theory can be used here we should expect $C_f=0$, i.e., case I), the chiral symmetry as well as supersymmetry is unbroken as shown above.

More generally, the solutions of Eqs.(24) and (25) may have terms containing D acting on S_c or S_c^+ but we would not consider them at this moment in this paper as is done in Ref.[2]. In addition, the solution of Eq.(25) may also have terms of the form $C_{d_i}(S_c^{a_i}S_c^{+b_i}+h.c.)$ provided $a_i+b_i=\frac{2}{3+\tilde{\gamma}}$, but they can only exchange C_d (See Eq.(31)) in the scalar potential so they too will not affect the conclusion. Therefore the scalar potential Eqs.(31) and (38) are quite general.

Finally, results obtained in Ref.[9] indicate that in SSYM theory, apart from the renormalization made in Ref.[9], the superfield V may need additional non-linear renormalization. As this additional renormalization is equivalent to a gauge transformation we can expect that it does not affect the gauge invariant quantity discussed here.

In summary, we conclude that in consistency with Witten's index theorem the supersymmetry is unbroken for the N=1 pure Yang-Mills supersymmetric model but the equation obtained here admits of both, chiral symmetry preserving and chiral symmetry breaking, solutions.

References

1. G.Veneziano and S. Yankielowicz, Phys.Lett., 113B(1982)231.
2. T.R.Taylor, G.Veneziano and S.Yankielowicz, Nucl.Phys., B218(1983)493;
M.E.Peskin, Preprint SLAC-Pub.-3061;
A.Davis, M.Dine and N.Seiberg, Phys.Lett., 125B(1983)487;
H.P.Nilles, Phys.Lett., 129B(1983)103.
3. B.Zumino, Nucl.Phys., B89(1975)535;
P.West, Nucl.Phys., B106(1976)219;
W.Lang, Nucl.Phys., B114(1976)123;
M.T.Grisarn, M.Rocek and W.Siegel, Nucl.Phys., B159(1979)429.
4. J.Wess and J.Bagger, "Supersymmetry and Supergravity", Princeton Lecture Notes(1981).
5. R.Fukada and Y.Kazama, Phys.Rev.Lett., 45(1980)1142;
Phys. Rev., D21(1980)485.
6. S.Ferrara and B.Zumino, Nucl.Phys., B87(1975)207;

T.Curtright, Phys.Lett., 71B(1977)185;

L.F.Abbott, M.T.Grisaru and H.J.Schnitzer, Phys.Rev., D16(1977)2995; Phys.Lett., 71B(1977)161;

M.Grisaru, in "Recent Developments in Gravitation" (Carge se 1980) eds.M.Levy and S.Deser.

7. J.C.Collins, A.Duncan and S.D.Joglekar, Phys. Rev., D16(1977)438.

8. Y.Kazama, Preprint KUNS 683 HE (TH) 83/09.

9. O.Piguet and K. Sibold, Nucl.Phys., B197(1982)257, 272.

Commun. in Theor. Phys. (Beijing, China) Vol.3, No.4 (1984) 491-498

THE UNIFIED SCHEME OF THE EFFECTIVE ACTION AND CHIRAL ANOMALIES IN ANY EVEN DIMENSIONS

Kuang-chao CHOU(周光召), Han-ying GUO(郭汉英)
Xiao-yuan LI(李小源) and Ke WU(吴 可)
Institute of Theoretical Physics, Academia Sinica, P.O.Box 2735, Beijing, China
and
Xing-chang SONG†(宋行长)
Institute for Theoretical Physics, State University of New York at Stony Brook, Stony Brook, New York 11794, USA

Received April 29, 1984

Abstract

Based on the Weil homomorphism method, a unified scheme in which all the important topological properties of the pseudoscalar Goldstone boson and gauge fields in even dimensional space are described in one remarkably compact form is given. These properties include the effective action, the skyrmion anomalous current, Abelian anomaly, symmetric and asymmetric non-Abelian chiral anomalies, and anomaly free conditions.

A year ago E. Witten discussed the global aspects of current algebra[1] and proposed that the Wess-Zumino chiral effective action[2] can be described in a new mathematical framework. He pointed out that this action obeys a priori quantization law, analogous to Dirac's quantization of magnetic charge, and incoporates in current algebra both perturbative[3] and non-perturbative anomalies[4]. Applications to the standard weak interaction model require an arbitrary subgroup of global flavor symmetries to be gauged. However, the standard road to gauging global symmetry of the Wess-Zumino action is not available since its topological nature is unknown. In Witten's original work the author did this by resorting to the trial and error Noether method.

In their series of works[5-8] CHOU-GUO-WU-SONG suggested that one can as usual introduce the minimal gauge coupling in 5-dimensional space first, then the gauge invariant Wess-Zumino action can be obtained by a series of non-trivial but systematic mathematical manipulations.[5] In the processes a deep connection among the manifestly gauge invariant 5-forms, the Chern-Simons secondary topological invariants, anomaly free condition and the gauge invariant Wess-Zumino action has been found. Remarkably, the structure of symmetric chiral anomalies[9] can be determined by studying the gauge transformation properties of the Chern-Simons secondary topological invariants without having to evaluate any Feynman diagram[5,10]. Furthermore it has been shown[6] that the Bardeen's

† *Permanent address: Department of Physics, Peking University, Beijing, China.*

asymmetric non-Abelian anomalies[11] and the corresponding counterterms can be obtained if one does Goldstone boson expansions in the gauge invariant effective Wess-Zumino action functional. The relation between the symmetric and asymmetric non-Abelian anomalies has also been found[7,12]. Recently, it has been suggested[13] that the gauge invariant Wess-Zumino action functional in 2n-dimensional space can be obtained by properly constructing the manifestly gauge invaraint (2n+1)-forms which satisfy the Abelian anomalous Ward identity in (2n+2)-dimensional space. The problem about the uniqueness of the gauge-invariant Wess-Zumino effective action has also been discussed.[14]

In this note, we would like to suggest a unified scheme of the gauge invariant Wess-Zumino effective action functional, Abelian and non-Abelian chiral anomalies, the anomaly free condition and the manifestly gauge invariant generalized skyrmion anomalous current[15] in any even dimensional space. The description is based on the Weil homomorphism method in differential geometry[16]. It turns out that the Abelian anomaly in (2n+2)-dimensional space, the gauge invariant Wess-Zumino effective action, the symmetric and asymmetric non-Abelian anomalies and anomaly free conditions in 2n-dimensional space can all be given in a remarkably compact form. The formalism further reveals the intrinsic connection among these objects.

We start by considering the theory with a chiral $SU(N)_L \times SU(N)_R$ symmetry which is spontaneously broken down to the diagonal group SU(N). Under an $SU(N)_L \times SU(N)_R$ transformation by unitary matrices (g_L, g_R), U(x) transforms as

$$U(x) \rightarrow g_L U(x) g_R^{-1} . \tag{1}$$

In order to obtain a gauge invariant action functional under the gauge group $H \subseteq SU(N)_L \times SU(N)_R$, the gauge covariant derivative

$$DU = dU + A_L U - U A_R \tag{2}$$

should be introduced, where $A_{L(R)}$ is the gauge connection 1-form of $H_{L(R)}$, and the field strength 2-form is

$$F_{L(R)} = dA_{L(R)} + A^2_{L(R)} \tag{3}$$

respectively. Under the gauge transformation

$$\begin{aligned} &U \rightarrow h_L(x) U h_R^{-1}(x) , \qquad (h_L, h_R) \in H_L \times H_R , \\ &A_{L(R)} \rightarrow h_{L(R)}(x)(d + A_{L(R)}) h^{-1}_{L(R)}(x) . \end{aligned} \tag{4}$$

We have

$$F_{L(R)} \rightarrow h_{L(R)}(x) F_{L(R)} h^{-1}_{L(R)}(x) ,$$

$$DU \to h_L(x)DU\,h_R^{-1}(x)\ . \tag{5}$$

In particular, if we define

$$\eta \equiv U^{-1}DU = U^{-1}(d+A_L)U - A_R = {}^U\!A_L - A_R\ , \tag{6}$$

where ${}^U\!A_L(x)$ is the "gauge transformed" gauge field 1-form

$${}^U\!A_L(x) = U^{-1}(d+A_L(x))U\ , \tag{7}$$

then, under the gauge transformation (Eq.(4)) we have

$$\begin{aligned} &{}^U\!A_L(x) \to h_R(x)(d+{}^U\!A_L(x))h_R^{-1}(x)\ , \\ &{}^U\!F_L(x) \to h_R(x)\,{}^U\!F_L(x)h_R^{-1}(x)\ , \\ &\eta \to h_R(x)\eta h_R^{-1}(x)\ , \end{aligned} \tag{8}$$

where the "gauge transformed" field strength 2-form ${}^U\!F_L(x)$ is defined as

$${}^U\!F_L = d\,{}^U\!A_L + {}^U\!A_L^2\ . \tag{9}$$

Eq.(8) implies that the gauge transformation property of ${}^U\!A_L$ is exactly the same as those of A_R, and η is the gauge covariant 1-form.

Let us consider the interpolation between A_R and ${}^U\!A_L$,

$$A_t = A_R + t\eta, \qquad 0 \leq t \leq 1\ , \tag{10}$$

then the field strength 2-form

$$F_t = dA_t + A_t^2 = t\,{}^U\!F_L + (1-t)F_R + (t^2-t)\eta\ . \tag{11}$$

Let $Q_{2n+1}({}^U\!A_L,\ A_R)$ be a (2n+1)-form

$$Q_{2n+1}({}^U\!A_L, A_R) = \alpha_{n+1}(n+1)\int_0^1 \delta t\, Tr[\eta F_t^n]\ , \tag{12}$$

where α_{n+1} is the normalization constant

$$\alpha_{n+1} = \frac{i^{n+1}}{(2\pi)^{n+1}(n+1)!}\ . \tag{13}$$

Then using the Chern-Weil formula we immediately have

$$dQ_{2n+1}({}^U\!A_L, A_R) = C_{n+1}({}^U\!F_L) - C_{n+1}(F_R) \tag{14}$$

where $C_{n+1}(F)$ is the (n+1)-th Chern class of A. $Q({}^{U}A_L, A_R)$ defined by Eq.(12) is constructed out of the gauge covariant 1-form and 2-forms ${}^{U}F_L$ and F_R, and thus is manifestly gauge invariant. In the reference [10] the Chern-Simons secondary topological invariant is defined as the object which consists of only gauge fields and thus is not gauge invariant. It should cause no confusion.

Using the Chern-Weil identity Eq.(14) can be written as

$$d\left[Q_{2n+1}({}^{U}A_L,A_R)-\Pi_{2n+1}({}^{U}A_L,{}^{U}F_L)+\Pi_{2n+1}(A_R,F_R)\right]=0 \ , \tag{15}$$

where $\Pi_{2n+1}(A,F)$ is the Chern-Simons secondary topological invariant of A. This means

$$Q_{2n+1}({}^{U}A_L,A_R)-\Pi_{2n+1}({}^{U}A_L,{}^{U}F_L)+\Pi_{2n+1}(A_R,F_R) \tag{16}$$

should be a closed form. In what follows we would like to give its expression in a compact and explicit form.

Let us define a connection A_{ab} depending upon two parameters a and b with $0\leqslant a,b\leqslant 1$

$$A_{ab}=a\,{}^{U}A_L+bA_R \quad . \tag{17}$$

Denote the corresponding field strength as

$$F_{ab}=dA_{ab}+A^2_{ab} \tag{18}$$

and consider the integral

$$I\equiv\alpha_{n+1}(n+1)\int Tr\left[(\delta a\,{}^{U}A_L+\delta bA_R)F^n_{ab}\right] \tag{19}$$

over a one dimensional path which is clockwise triangle in the (a,b) plane going from the origin to the point (0,1) to (1,0) and back to the origin. The straightforward calculation gives

$$I=-\Pi_{2n+1}({}^{U}A_L,{}^{U}F_L)+\Pi_{2n+1}(A_R,F_R)+Q_{2n+1}({}^{U}A_L,A_R) \ . \tag{20}$$

On the other hand, one can apply Stokes' theorem to the integral Eq.(19) and transform it into an integral over the inside of the triangle. It follows that

$$I=d\left\{(n+1)n\alpha_{n+1}\int_0^1\delta a\int_0^{1-a}\delta b STr(A_R\,{}^{U}A_L F^{n-1}_{ab})\right\} \ . \tag{21}$$

Using the same trick, one can prove that

$$\Pi_{2n+1}({}^{U}A_L, {}^{U}F_L)=\Pi_{2n+1}(A_L,F_L)+\Pi_{2n+1}(U^{-1}dU,0)+ \\ +d\left\{(n+1)n\alpha_{n+1}\int_0^1\delta v\int_0^{1-v}\delta w\mathrm{STr}(UdU^{-1}A_L F_{vw}^{n-1})\right\}, \tag{22}$$

where the field strength 2-form

$$F_{vw}=dA_{vw}+A^2_{vw} \tag{23}$$

and

$$A_{vw}=vA_L+w(UdU^{-1}) \tag{24}$$

is the connection 1-form depending upon two parameters v and w with $0\leqslant v,w\leqslant 1$. From Eqs.(20), (21) and (22) we come to our final result

$$Q_{2n+1}({}^{U}A_L, A_R)=\Pi_{2n+1}(A_L,F_L)-\Pi_{2n+1}(A_R,F_R)+\Pi_{2n+1}(U^{-1}dU,0) \\ +d\left\{(n+1)n\alpha_{n+1}\int_0^1\delta v\int_0^{1-v}\delta w\mathrm{STr}(UdU^{-1}A_L F_{vw}^{n-1})\right\} \\ +d\left\{(n+1)n\alpha_{n+1}\int_0^1\delta a\int_0^{1-a}\delta b\mathrm{STr}(A_R{}^{U}A_L F_{ab}^{n-1})\right\}. \tag{25}$$

Eq.(25) summarizes all the important topological properties of the pseudoscalar Goldstone boson and gauge fields in (2n+2)- and 2n- dimensional spaces, as will be shown later.

1) Consider

$$*J=*(J_\mu dx^\mu)=Q_{2n+1}({}^{U}A_L, A_R)+l_{2n+1}, \tag{26}$$

where "*" is the Hodge star operator and l_{2n+1} is an arbitrary manifestly gauge invariant and exact (2n+1)-forms which consists of η and F_R, ${}^{U}F_L$. Substituting Eq.(26) into Eq.(25) and exterior differentiating it, we have

$$d*J=C_{n+1}(F_L)-C_{n+1}(F_R). \tag{27}$$

A straightforward calculation indicates in the 4-dimensional space (n=1) J defined by Eq.(26) is none other than the gauge invariant anomalous skyrmion current form[17]. Therefore

$$J=-*\left[\alpha_{n+1}(n+1)\int_0^1\delta t\mathrm{Tr}(\eta F_t^n)+l_{2n+1}\right] \tag{28}$$

can be considered as the generalized gauge invariant skyrmion anomalous current 1-form in (2n+2)-dimensional space. Obviously, this current is not conserved. It suffers from the Abelian anomaly which is simply given by Eq.(27). Note that J is constructed out of only boson fields without referring to underlying fermions. However, the subtle point is that the skyrmion, the soliton of the non-linear sigma model, can be quantized as a fermion when the number of colors of the underlying fermion theory is odd[15].

2) Define

$$\Gamma(U,Q^{2n+1})=2\pi i\int_{Q^{2n+1}}\Pi_{2n+1}(U^{-1}dU,0), \tag{29}$$

where Q^{2n+1} is a (2n+1)-dimensional manifold, $\partial Q^{2n+1}=M^{2n}$ is a 2n-dimensional compactified space. For the case of n=2, $\Gamma(U,Q^5)$ is just the Wess-Zumino term[2] proposed by Witten[1]. Thus $\Gamma(U,Q^{2n+1})$ can be considered as the generalized Wess-Zumino term in M^{2n}. Now the left-hand side of Eq.(25) is manifestly gauge invariant while the Chern-Simons secondary topological invariants are not. A necessary and sufficent condition for getting the gauge invariant Wess-Zumino term in M^{2n} is the gauge variations satisfy

$$\Delta\Pi_{2n+1}(A_L,F_L)=\Delta\Pi_{2n+1}(A_R,F_R)\ . \tag{30}$$

If this condition can be satisfied, the generalized gauge invariant Wess-Zumino term in M^{2n} can be defined as

$$\tilde{\Gamma}(U,Q^{2n+1})=\Gamma(U,Q^{2n+1})+2\pi i\int_{M^{2n}}W_{2n}\ , \tag{31}$$

where the 2n-form W_{2n} is given by Eq.(25)

$$W_{2n}=(n+1)n\alpha_{n+1}\int_0^1\delta a\int_0^{1-a}\delta b\ STr(A_R{}^U A_L F_{ab}^{n-1})$$

$$+(n+1)n\alpha_{n+1}\int_0^1\delta v\int_0^{1-v}\delta w\ STr(UdU^{-1}A_L F_{vw}^{n-1})\ . \tag{32}$$

For the case of n=2,3, the gauge invariant Wess-Zumino terms[5,12,13] in M^4 and M^6 can be given by the straightforward calculations from $\tilde{\Gamma}(U,Q^5)$ and $\tilde{\Gamma}(U,Q^7)$ respectively. Furthermore, the zero order term and first order term in Goldstone boson field give the Bardeen's counterterm R_3 and asymmetric non-Abelian anomalies[6]. This indicates that $\tilde{\Gamma}(U,Q^5)$ satisfies the non-Abelian anomalous Ward indentity and the Wess-Zumino consistency condition[2]. The same conclusions are also true for $\tilde{\Gamma}(U,Q^{2n+1})$ in M^{2n}. In particular, it can be proved that the Bardeen's counterterm $R_3(M^{2n})$ and asymmetric anomalies $G_A(M^{2n})$ in M^{2n} are[18]

$$\left.\begin{aligned}R_3(M^{2n})&=(n+1)n\alpha_{n+1}\int_0^1\delta v\int_0^{1-v}\delta w STrA_RA_LE^n\\ G_A^a(M^{2n})&=2\pi i(n+1)n\alpha_{n+1}\int_0^1\delta v(1-v)STr\lambda^a d[A_LF^{n-1}(vA_L)]\ ,\end{aligned}\right\} \tag{33}$$

$$+2\pi i(n+1)n\alpha_{n+1}\int_0^1\delta v\int_0^{1-v}\delta w$$
$$\mathrm{STr}\lambda^a\{[d(E^{n-1}A_R)+A_LE^{n-1}A_R+E^{n-1}A_RA_L]$$
$$+(n-1)[E^{n-2}A_RA_LF(vA_L)-F(vA_L)E^{n-2}A_RA_L]$$
$$+(n-1)(v-v^2)d[A_LE^{n-2}A_RA_L-E^{n-2}A_RA_L^2]$$
$$+(n-1)vw[d(E^{n-2}A_RA_LA_R)+A_LE^{n-2}A_RA_LA_R$$
$$+E^{n-2}A_RA_LA_RA_L-d(A_RE^{n-2}A_RA_L)-A_LA_RE^{n-2}A_RA_L-A_RE^{n-2}A_RA_L^2]\} \quad (34)$$

respectively, where

$$E=F(vA_L)+F(wA_R)+vw(A_LA_R+A_RA_L) .$$

3) If Eq.(30) for cancellation of anomalies is not obeyed, then $\widetilde{\Gamma}(U,Q^{2n+1})$ is not gauge invariant. From Eq.(25) the variation of $\widetilde{\Gamma}(U,Q^{2n+1})$ under a gauge transformation Eq.(4) does not vanish but is

$$\Delta\widetilde{\Gamma}(U,Q^{2n+1})+2\pi i\Delta\Big(\Pi_{2n+1}(A_L,F_L)-\Pi_{2n+1}(A_R,F_R)\Big)=0 . \quad (35)$$

Using the formula analogous to Eq.(22), it is not difficult to find that

$$\Delta\widetilde{\Gamma}(U,Q^{2n+1})=-\Gamma(h_L,Q^{2n+1})+\Gamma(h_R,Q^{2n+1})$$
$$-2\pi i d\Big\{(n+1)n\alpha_{n+1}\int_0^1\delta v\int_0^{1-v}\delta w\mathrm{STr}(h_L^{-1}dh_LA_LF_{vw}^{n-1})-(L\longleftrightarrow R)\Big\}. \quad (36)$$

The general expression of the infinitesimal variation $\delta\widetilde{\Gamma}(U,Q^{2n+1})$ can also be given. For the case of n=2, $\delta\widetilde{\Gamma}(U,Q^5)$ is just the Gross-Jackiw's symmetric anomalies[9], and $\Delta\widetilde{\Gamma}(U,Q^5)$ is likely the Witten's SU(2) non-perturbative anomaly[4]. Thus, Eq.(22) can be considered as the global version of the anomaly free condition in M^{2n}. It is not only the generalization of the anomaly free condition in the perturbative theory, but also includes the condition of free of non-perturbative anomaly. On the other hand, the anomaly-free condition Eq.(20) is equivalent to the condition

$$C_{n+1}(F_L)=C_{n+1}(F_R) , \quad (37)$$

which is nothing but the Abelian anomaly free condition in M^{2n+2}.

In summary, we have presented in this note a unified scheme in which all the important topological properties of the pseudoscalar Goldstone boson and gauge fields in (2n+2)-and 2n-dimensional spaces are described in a remarkably compact form. These properties include the gauge invariant skyrmion anomalous current, the Abelian anomaly in (2n+2) dimensional space, the gauge invariant

Wass-Zumino action, symmetric and asymmetric non-Abelian anomalies and anomaly free conditions in 2n dimensional space. Eq.(25) clearly embodies the deep connection among them.

Finally, we would like to mention that the Abelian anomaly and the non-perturbative SU(2) anomaly[4] are both related to the famous Atiyah-Singer theorem[16], as is well known. Thus there should be some direct or indirect relation between the non-Abelian anomalies and the Atiyah-Singer index theorem. We will discuss this point elsewhere.

References and Footnotes

[1] E. Witten, Nucl. Phys. B223(1983)422.

[2] J.Wess and B.Zumino, Phys. Lett. 37B(1971)95.

[3] S.Adler, Phys. Rev. 177(1969)2426;
J.Bell and R.Jackiw, Nuovo Cim. 60A(1969)47.

[4] E.Witten, Phys. Lett. 117B(1982)324.

[5] K.C.CHOU, H.Y.GUO, K.WU and X.C.SONG, Phys. Lett. 134B(1984)67.

[6] K.C.CHOU, H.Y.GUO, K.WU and X.C.SONG, to be published in Physica Energiae Fortis et Physica Nuclearis (1984) (in Chinese).

[7] K.C.CHOU, H.Y.GUO, K.WU and X.C.SONG, Commun. in Theor. Phys. 3(1984)73.

[8] K.C.CHOU, H.Y.GUO, K.WU and X.C.SONG, Commun. in Theor. Phys. 3 (1984)125;
Stony Brook preprint AS-SB-84-18.

[9] D.Gross and R.Jackiw, Phys. Rev. D6(1972)477.

[10] B.Zumino, Les Houches lectures 1983, LBL-16747 (1983);
B.Zumino. Y.S. Wu and A.Zee, to be published in Nucl. Phys. B;
L.Bonora and P.Pasti, Phys. Lett. 132B(1983)75.

[11] W.A.Bardeen, Phys. Rev. 184(1969)1848.

[12] K.Kawai and S.-H.H. Tye, Cornell preprint CLNS-84/595;
also, O. Kaymakcalan, S. Rajeev and J. Schechter, Syracuse preprint SU-4222-278(Dec. 1983).

[13] Y.P.KUANG, X.LI, K.WU and Z.Y.ZHAO, Commun. in Theor. Phys. ,this issue.

[14] C.H.CHANG, H.Y.GUO and K.WU, Institute of Theoretical Physics Academia Sinica preprint AS-ITP-84-016 (April, 1984).

[15] T.H.R. Skyrme, Proc. Roy. Soc. (London) A260(1961)127;
J.Goldstone and F.Wilczek, Phys. Rev. Lett. 47(1981)986;
E.Witten, Nucl. Phys. B233(1983)433;
A.Zee, Phys. Lett. 135B(1984)307;
F.Wilczek and A.Zee, Phys. Rev. Lett. 51(1983)2250.

[16] For example, T. Eguchi, P.B. Gilkey and A.J.Hanson, Phys. Rep. 66(1980)213.

[17] For example, A Zee, reference [15].

[18] A.Andrianov, L. Bonora and P. Pasti, Padova University preprint PD 30/83 (October, 1983).

Commun. in Theor. Phys. (Beijing, China) *Vol.3, No.5 (1984) 593-603*

SYMMETRIC AND ASYMMETRIC ANOMALIES AND EFFECTIVE LAGRANGIAN

Kuang-chao CHOU(周光召)
Han-ying GUO(郭汉英) and Ke WU(吴可)
Institute of Theoretical Physics, P.O. Box 2735
Beijing, China

and

Xing-Chang SONG†(宋行长)

Institute for Theoretical Physics
State University of New York at Stony Brook
Stony Brook, NY 11794, U.S.A.

Received June 7, 1984

Abstract

The properties of the effective Lagrangian for the low-energy Goldstone-gauge field system constructed from the Chern-Simons topological invariant are further discussed. Both the symmetric anomaly and the asymmetric anomaly are connected with this Lagrangian.

In a recent paper[1], Witten has reformulated the non-linear effective Lagrangian of QCD, which was first considered by Wess and Zumino[2], by adding to the ordinary chiral field action Γ_0 a Wess-Zumino like term Γ, a five dimensional integral over a volumn D whose surface is the image of the physical four-dimensional space M on the group manifold. This Lagrangian, when gauged by introducing flavor gauge fields, may precisely describe all effects of anomalies in low-energy processes containing Goldstone bosons. Γ_0 term is gauged as usual by minimal coupling while the gauged form of WZ term Γ is obtained by trial-and-error method[1]

$$\tilde{\Gamma}=\Gamma-\frac{i}{48\pi^2}\int Z_4 \quad . \tag{1}$$

Starting from a different point of view and making use of the properties of the Chern-Simons topological invariant, the present authors have proposed a systematical method of gauging the WZ term[3,4]. The result is similar to but different from that of Witten's

$$\hat{\Gamma}=\Gamma-\frac{i}{48\pi^2}\int_M W_4 \quad . \tag{2}$$

Under the infinitesimal gauge transformation, both $\hat{\Gamma}$ and $\tilde{\Gamma}$[5] give correct

† *On leave from the Institute of Theoretical Physics, Peking University, Beijing, China.*

anomalies G, the anomalies in the left-right (anti-) symmetric form, which appeared at an intermediate state in Bardeen's work[6] and is in agreement with the computation of the anomalies at the quark level by Gross and Jackiw[7].

As is well known, in order to obtain the normal Ward identity for the vector current, Bardeen has further introduced several counter terms[6] so that the anomaly only appears in axial-vector part. We shall call this form of the anomalies the asymmetric form G_a for convenience. The main purpose of the present note is to discuss the relation of the anomalies, both in the symmetric form G and the asymmetric form G_a to our effective Lagrangian.

To begin with, we give our notations. The canonical form of the principal chiral fields is used by introducing the double matrix[4]

$$Q=\begin{pmatrix} & U \\ U^{-1} & \end{pmatrix}, \quad U=\exp\left\{\frac{2i}{F_\pi}\sum_{\alpha=1}^{8}\lambda^\alpha\pi^\alpha\right\} . \tag{3}$$

And the gauge potentials (1-form) and strengths (2-form) are also grouped into

$$A=\begin{pmatrix} A^L & \\ & A^R \end{pmatrix}, \quad F=\begin{pmatrix} F^L & \\ & F^R \end{pmatrix}, \tag{4}$$

where

$$A^L=\frac{\lambda_L^\alpha}{2i}A^\alpha_{L\mu}dx^\mu , \qquad F^L=dA^L+A^L\wedge A^L,$$

and the same is true of right-handed quantities. The gauge transformation takes the form

$$Q \to gQg^{-1} , \quad A \to g(A+d)g^{-1} , \qquad g(x)=\begin{pmatrix} g^L(x) & \\ & g^R(x) \end{pmatrix} . \tag{5}$$

The covariant derivative of Q fields is defined as

$$DQ=dQ+[A, Q] . \tag{6}$$

And for infinitesimal gauge transformation

$$g=1+\varepsilon, \qquad \varepsilon(x)=\begin{pmatrix} \varepsilon^L(x) & \\ & \varepsilon^R(x) \end{pmatrix} ,$$

we have

$$\delta Q=[\varepsilon,Q]\ ,\qquad \delta DQ=[\varepsilon,DQ]\ ,$$
$$\delta A=-d\varepsilon+[\varepsilon,A]\ ,\qquad \delta F=[\varepsilon,F]\ . \tag{7}$$

Now the gauged form of the chiral field action Γ_0 can be written as

$$\hat{\Gamma}_0=\frac{F_\pi^2}{16}\int d^4x\ tr(D_\mu U)(D^\mu U^{-1})$$
$$=\frac{F_\pi^2}{32}\int Tr(DQ)\wedge(*DQ)\ , \tag{8}$$

where tr denotes the trace for single matrix and Tr for double matrix, and "*" is the Hodge star (dual form) in four dimensions. The WZ term[1]

$$\Gamma=\frac{-i}{240\pi^2}\int tr(U^{-1}dU)\wedge(U^{-1}dU)\wedge(U^{-1}dU)\wedge(U^{-1}dU)\wedge(U^{-1}dU)$$
$$=\frac{-i}{480\pi^2}\int Tr\gamma_5\{K\wedge K\wedge K\wedge K\wedge K\} \tag{9}$$

is indeed an integral of Chern-Simons 5-form of the pure gauge fields $U^{-1}dU$, i.e., $\pi_5(U^{-1}dU)$, over the region D, and[4]

$$\bar{W}_4=Tr\gamma_5\Big\{\frac{1}{4}(AQ\wedge AQ\wedge AQ\wedge AQ)+(dA\wedge A+A\wedge dA+A\wedge A\wedge A)\wedge QAQ$$
$$+dA\wedge QA\wedge dQ+(dA\wedge A+A\wedge dA+A\wedge A\wedge A)\wedge K+AQ\wedge AQ\wedge A\wedge K$$
$$+\frac{1}{2}A\wedge K\wedge A\wedge K-AQ\wedge AQ\wedge K\wedge K-A\wedge K\wedge K\wedge K\Big\}\ , \tag{10}$$

where

$$K=QdQ=\begin{pmatrix} UdU^{-1} & \\ & U^{-1}dU \end{pmatrix}\ ,\qquad \gamma_5=\begin{pmatrix} -1 & \\ & 1 \end{pmatrix}\ . \tag{11}$$

It can be easily seen that both Γ and $\bar{W}_4$ are odd under the left-right exchange

$$A^L \longleftrightarrow A^R\ ,\qquad U \longleftrightarrow U^{-1}\ , \tag{12a}$$

which can be expressed as the involution

$$Q\longrightarrow \sigma Q\sigma\ ,\qquad A\longrightarrow \sigma A\sigma\ ,\qquad \sigma=\begin{pmatrix} & 1 \\ 1 & \end{pmatrix}\ . \tag{12b}$$

Meanwhile, $\hat{\Gamma}_0$ in Eq.(8) is even under the same transformation. This fact indicates that, just like the ungauged action $\Gamma_{eff}=\Gamma_0+n\Gamma$ considered by Witten[1], the gauged action $\hat{\Gamma}_{eff}=\hat{\Gamma}_0+n\hat{\Gamma}$ with $\hat{\Gamma}$ given in Eq.(2) has no superfluous symmetries other than the ones given by QCD[4].

On the other hand, Witten's result $\tilde{\Gamma}$[5] contains more terms than those in our $\hat{\Gamma}$. The difference between $\tilde{\Gamma}$ and $\hat{\Gamma}$ can be seen from the relation

$$Z_4-\bar{W}_4=\frac{1}{2}\mathrm{Tr}\ FQ\wedge FQ\ , \tag{13}$$

which contains no γ_5 so that it is even under transformation (12b). Therefore, $\tilde{\Gamma}$ even has no definite parity under space reflection.

Since Eq.(13) is gauge invariant, the variation of $\hat{\Gamma}$ or $\tilde{\Gamma}$ gives the same anomalies under infinitesimal gauge transformation[8]

$$\delta\hat{\Gamma}=\delta\tilde{\Gamma}=\int_M \mathrm{Tr}\varepsilon G=\int_M \mathrm{tr}(\varepsilon^R G^R-\varepsilon^L G^L)\ ,$$

$$G=\frac{i}{48\pi^2}\gamma_5 d(2A\wedge dA+A\wedge A\wedge A)\ , \tag{14}$$

which is the left-right antisymmetric form given in Refs.[6] and [7], as is mentioned above, whereas Bardeen's asymmetric anomaly, which only couples to the divergence of the axial-vector current, is

$$G_a^\alpha=\frac{1}{4\pi^2}\mathrm{tr}\left\{\frac{\lambda^\alpha}{2}\left[F^V\wedge F^V+\frac{1}{3}F^A\wedge F^A-\frac{4}{3}(A\wedge A\wedge F^V-A\wedge F^V\wedge A+F^V\wedge A\wedge A)+\frac{8}{3}A\wedge A\wedge A\wedge A\right]\right\}, \tag{15}$$

where the 2-forms F^V and F^A are defined as

$$F^V=dV+V\wedge V+A\wedge A,\qquad F^A=dA+V\wedge A+A\wedge V, \tag{16}$$

and related to F^L and F^R through the relations

$$A^R=V+A,\qquad A^L=V-A\ . \tag{17}$$

It seems that the effective Lagrangians obtained from topological consideration[1,3,4,8] always yield the symmetric anomalies under infinitesimal gauge transformation. Then, how about the asymmetric one? Does it have anything to do with our effective Lagrangian? Indeed, as an effective Lagrangian, $\hat{\Gamma}$ must contain all anomalous interactions among gauge fields and Goldstone fields. In particular it must be the asymmetric anomaly which couples to the single Goldstone vertex. This important feature can be revealed through an appropriate chiral transformation applied to the effective Lagrangian.

As a matter of fact, in the early paper on the consistency condition for chiral anomalies[2], Wess and Zumino already obtained their effective Lagrangian by such a consideration. The result is

$$\Gamma[\xi,A]=\frac{1-\exp\xi.\bar{Y}_A}{\xi\bar{Y}_A}\xi G_a[A]\ , \tag{18}$$

where $\xi^\alpha=\frac{2}{F_\pi}\pi^\alpha$, $\bar{Y}_A$ are the local chiral operators acting upon the gauge fields and

$$\xi G_a = \int d^4x \xi^\alpha(x) G_a^\alpha(x) \ .$$

From (18), it follows immediately that

$$\Gamma[\xi,A] = \xi G_a[A] + O(\xi^2) \ , \tag{19}$$

i.e., the terms linear in π fields are just the asymmetric form of the chiral anomaly.

Now we would like to see whether the effective Lagrangian $\hat{\Gamma}$ or $\tilde{\Gamma}$ has the same property. Expanding U and integrating by parts, we have

$$\hat{\Gamma}[\xi,A] = \hat{\Gamma}[0,A] + \xi\hat{\Gamma}_1[A] + O(\xi^2) \tag{20}$$

with

$$\hat{\Gamma}[0,A] = \frac{-i}{48\pi^2}\int \mathrm{tr}\Big\{(A^R \wedge dA^R + dA^R \wedge A^R)\wedge A^L - (A^L \wedge dA^L + dA^L \wedge A^L)\wedge A^R + (A^R\wedge A^R\wedge A^R\wedge A^R) - (A^L\wedge A^L\wedge A^L\wedge A^L) - \frac{1}{2}A^L\wedge A^R\wedge A^L\wedge A^R\Big\} \tag{20a}$$

and

$$\xi\hat{\Gamma}_1 = \frac{-i}{48\pi^2}\int \mathrm{tr}\{i\xi^\alpha\lambda^\alpha[R\wedge R\wedge R\wedge L + R\wedge L\wedge L\wedge L + R\wedge L\wedge R\wedge L + (R \longleftrightarrow L)\} \ . \tag{20b}$$

Here we have introduced the abbreviation symbols

$$R \equiv d + A^R \ , \qquad L \equiv d + A^L$$

to simplify the lengthy expression. By inserting relations (16) and (17) into Eq.(15) and completing a lengthy but straightforward calculation, we can show that the integrand on the right-hand side of Eq.(20b) is nothing but the product $\xi^\alpha G_a^\alpha$, i.e.,

$$\xi\hat{\Gamma}_1 = \int \xi^\alpha G_a^\alpha \ , \quad \hat{\Gamma}_1 = *G_a^\alpha \ . \tag{21}$$

In comparison with Wess-Zumino's effective Lagrangian (18), our $\hat{\Gamma}[\xi,A]$ has the same first order piece $\xi \cdot G_a$ but is different from Eq.(19) by a zeroth order part $\hat{\Gamma}[0,A]$, which has the following interesting properties

$$\frac{\delta\hat{\Gamma}[0,A]}{\delta\varepsilon^L} + \frac{\delta\hat{\Gamma}[0,A]}{\delta\varepsilon^R} = 0 \ , \tag{22a}$$

$$d\left(\frac{\delta\hat{\Gamma}[0,A]}{\delta A^L}\right)_F + d\left(\frac{\delta\hat{\Gamma}[0,A]}{\delta A^R}\right)_F = G^L[A] - G^R[A] \ . \tag{22b}$$

Indeed, $\hat{\Gamma}[0,A]$ is nothing but the counter term which is used in Bardeen's paper[6] to remove the anomaly from Ward identity of vector current, i.e., $(2\varepsilon_A = \varepsilon^R - \varepsilon^L)$

$$\delta\hat{\Gamma}[\xi,A]-\delta\hat{\Gamma}[0,A]=\int_M \mathrm{tr}\,\varepsilon_A G_a[A] \ . \tag{23}$$

As for Witten's Lagrangian $\tilde{\Gamma}$, it is not difficult to show from relation (12) that

$$\tilde{\Gamma}[0,A]=\tilde{\Gamma}[0,A]+\xi\cdot\tilde{\Gamma}_1[A]+O(\xi^2)$$

with

$$\tilde{\Gamma}[0,A]=\hat{\Gamma}[0,A]-\frac{i}{48\pi^2}\int \mathrm{tr}(F^L{}_\wedge F^R) \ ,$$

$$\tilde{\Gamma}_1^\alpha[A]=\hat{\Gamma}_1^\alpha[A]+\frac{1}{48\pi^2}\mathrm{tr}\{\lambda^\alpha(F^R{}_\wedge F^L-F^L{}_\wedge F^R)\} \ . \tag{24}$$

Therefore, neither Eq.(21) nor Eqs.(22) and (23) are satisfied for $\tilde{\Gamma}$.

It is worth notice from the expressions in Eqs.(20) that Bardeen's anomaly G_a, although asymmetric in V and A, is symmetric in A^L and A^R, whereas the subtraction term $\hat{\Gamma}[0,A]$ is antisymmetric. This shows that the subtracted effective Lagrangian $\hat{\Gamma}[\xi,A]-\hat{\Gamma}[0,A]$ producing Bardeen's "asymmetric" anomaly, as well as the unsubtracted one $\hat{\Gamma}[\xi,A]$ producing the "symmetric" form of anomalies, is odd under the left-right exchange (12) and then even under the real parity transformation.

The above results can be further understood through the following analysis.

Let us first assume that the whole system of the Goldstone and gauge fields can be well described by a Lagrangian density in four dimensions as usual. $\mathcal{L}$ can be considered as functional of Q,DQ,A and F, as well as functional of Q,∂Q, ∂A and A. We use the notation

$$\frac{\delta\mathcal{L}}{\delta A}\Big|_{Q,\partial Q,\partial A}=\frac{\delta\mathcal{L}}{\delta A} \tag{25a}$$

to denote the functional derivative of $\mathcal{L}$ with respect to A keeping Q,∂Q and ∂A fixed, while

$$\left(\frac{\delta\mathcal{L}}{\delta A}\right)_{Q,DQ,F} \equiv \left(\frac{\delta\mathcal{L}}{\delta A}\right) \tag{25b}$$

to denote the functional derivative of keeping Q,DQ and F fixed. Then we have

$$\frac{\delta\mathcal{L}}{\delta Q}=\left(\frac{\delta\mathcal{L}}{\delta Q}\right)+\frac{\delta D_\mu Q}{\delta Q}\left(\frac{\delta\mathcal{L}}{\delta D_\mu Q}\right)=\left(\frac{\delta\mathcal{L}}{\delta Q}\right)-\left[A_\mu,\left(\frac{\delta\mathcal{L}}{\delta D_\mu Q}\right)\right] \tag{26a}$$

$$\frac{\delta\mathcal{L}}{\delta\partial_\mu Q}=\left(\frac{\delta\mathcal{L}}{\delta D_\mu Q}\right) , \tag{26b}$$

$$\frac{\delta\mathcal{L}}{\delta A_\nu}=\left(\frac{\delta\mathcal{L}}{\delta A_\nu}\right)+\frac{\delta D_\mu Q}{\delta A_\nu}\left(\frac{\delta\mathcal{L}}{\delta D_\mu Q}\right)+\frac{\delta F_{\rho\sigma}}{\delta A_\nu}\left(\frac{\delta\mathcal{L}}{\delta F_{\rho\sigma}}\right)$$

$$=\left(\frac{\delta\mathcal{L}}{\delta A_\nu}\right)+\left[Q,\left(\frac{\delta\mathcal{L}}{\delta D_\nu Q}\right)\right]-2\left[A_\mu,\ \frac{\delta\mathcal{L}}{\delta F_{\mu\nu}}\right]\ , \tag{26c}$$

$$\frac{\delta\mathcal{L}}{\delta\partial_\mu A_\nu}=2\left(\frac{\delta\mathcal{L}}{\delta F_{\mu\nu}}\right)\ . \tag{26d}$$

The assumption made here is a reasonable one since, as is pointed out in Ref.[4], although $\hat{\Gamma}=\int_D\hat{\rho}$ is defined at the beginning as an integral over a five dimensional hyperspace, $\hat{\rho}$ itself can be locally expressed as an exact form $\hat{\rho}=d*\hat{\mathcal{L}}$, where $*\hat{\mathcal{L}}$ is the dual form of an appropriate 4-dimensional Lagrangian density. The part of $\hat{\rho}$ containing gauge fields has already been expressed as an exact form as in (10). The other part $\pi_5(U^{-1}dU)$ is a closed 5-form so it can also be put into $\pi_5(U^{-1}dU)=d*\mathcal{L}_{cs}$ locally, therefore, the effective Lagrangian corresponding to $\hat{\Gamma}$

$$\hat{\mathcal{L}}=\mathcal{L}_{cs}-\frac{i}{48\pi^2}{*\bar{W}_4} \tag{27}$$

exists in principle. And the total Lagrangian is

$$\mathcal{L}=\mathcal{L}_0+n\hat{\mathcal{L}}+\mathcal{L}_F\ , \tag{28}$$

where $\hat{\mathcal{L}}_0$ is the term corresponding to $\hat{\Gamma}_0$ as in Eq.(8), $\mathcal{L}_F$ the free Lagrangian of the gauge fields, and n an integer[1] equal to the number of colors in QCD which will be set to one for simplicity in the following discussions.

Now, just like the case for odd dimensions[9], since $\mathcal{L}$ contains all kinds of interaction, including the anomalies, it should give the correct equations of motion , i.e.,

$$0=\left[Q,\ \frac{\delta\mathcal{L}}{\delta Q}-\partial_\mu\frac{\delta\mathcal{L}}{\delta\partial_\mu Q}\right]=\left[Q,\left(\frac{\delta\mathcal{L}}{\delta Q}\right)-D_\mu\left(\frac{\delta\mathcal{L}}{\delta D_\mu Q}\right)\right] \tag{29a}$$

for chiral fields (with constraints $Q^2=1$), and

$$0=\frac{\delta\mathcal{L}}{\delta A_\nu}-\partial_\mu\frac{\delta\mathcal{L}}{\delta\partial_\mu A_\nu}=\left(\frac{\delta\mathcal{L}}{\delta A_\nu}\right)+\left[Q,\left(\frac{\delta\mathcal{L}}{\delta D_\nu Q}\right)\right]-D_\mu 2\left(\frac{\delta\mathcal{L}}{\delta F_{\mu\nu}}\right) \tag{29b}$$

for gauge field.

Secondly, since the infinitesimal variation of $\hat{\Gamma}$ gives correct anomalies G, and G itself can be expressed as a total divergence of an appropriate anomalous current J_a[7], we see

$$\delta\hat{\Gamma}=\int_D d(\varepsilon G)=\int_M \varepsilon G=-\int_M \varepsilon d*J_a\ ,$$

from which we have

$$\frac{\delta\hat{\mathcal{L}}}{\delta\varepsilon}=-\partial_\mu J_a^\mu=*G\ . \tag{30}$$

Thirdly, as indicated in Ref.[4], for Lagrangian which has been put into the global invariant form instead of considering the infinitesimal gauge transformation

$$A \to g(A-d\varepsilon)g^{-1}, \qquad Q \to gQg^{-1}, \qquad F \to gFg^{-1},$$

G can also be generated by taking the variation

$$A \to A+\delta A, \; Q, \; DQ, \text{ and } F \text{ fixed},$$

i.e.,

$$G=\frac{\delta\hat{\rho}}{\delta d\varepsilon}=-\left(\frac{\delta\hat{\rho}}{\delta A}\right),$$

from which

$$J_a=\left(\frac{\delta\hat{\mathcal{L}}}{\delta A}\right). \tag{31}$$

Then, from Eqs.(30) and (31), we have

$$\frac{\delta\hat{\mathcal{L}}}{\delta\varepsilon}=-\partial_\mu J^\mu_a=-\partial_\nu\left(\frac{\delta\hat{\mathcal{L}}}{\delta A_\mu}\right). \tag{32}$$

Now, consider the variation of the total Lagrangian (28) under the infinitesimal gauge transformation (7)

$$\frac{\delta\mathcal{L}}{\delta\varepsilon}=\frac{\delta Q}{\delta\varepsilon}\frac{\delta\mathcal{L}}{\delta Q}+\frac{\delta\partial_\mu Q}{\delta\varepsilon}\frac{\delta\mathcal{L}}{\delta\partial_\mu Q}+\frac{\delta A_\nu}{\delta\varepsilon}\frac{\delta\mathcal{L}}{\delta A_\nu}+\frac{\delta\partial_\mu A_\nu}{\delta\varepsilon}\frac{\delta\mathcal{L}}{\delta\partial_\mu A_\nu}$$
$$=\left[Q,\frac{\delta\mathcal{L}}{\delta Q}-\partial_\mu\frac{\delta\mathcal{L}}{\delta\partial_\mu Q}\right]+\left[A_\nu,\frac{\delta\mathcal{L}}{\delta A_\nu}-\partial_\mu\frac{\delta\mathcal{L}}{\delta\partial_\mu A_\nu}\right]$$
$$+\partial_\mu\left[Q,\frac{\delta\mathcal{L}}{\delta\partial_\mu Q}\right]+\partial_\mu\left[A_\nu,\frac{\delta\mathcal{L}}{\delta\partial_\mu A_\nu}\right]. \tag{33a}$$

Taking account of the equations of motion (29) and using the relations (26), we see

$$\frac{\delta\mathcal{L}}{\delta\varepsilon}=\partial_\mu\left[Q,\frac{\delta\mathcal{L}}{\delta\partial_\mu Q}\right]+\partial_\mu\left[A_\nu,\frac{\delta\mathcal{L}}{\delta\partial_\mu A_\nu}\right], \tag{33b}$$

$$=D_\mu\left[Q,\left(\frac{\delta\mathcal{L}}{\delta D_\mu Q}\right)\right]+\left[A_\nu,\left(\frac{\delta\mathcal{L}}{\delta A_\nu}\right)+[F_{\mu\nu},\left(\frac{\delta\mathcal{L}}{\delta F_{\mu\nu}}\right)]\right]. \tag{33c}$$

In particular, the part concerning the chiral fields is

$$\frac{\delta \mathcal{L}}{\delta \varepsilon}\Big|_{A,\partial A} = \frac{\delta Q}{\delta \varepsilon}\frac{\delta \mathcal{L}}{\delta Q} + \frac{\delta \partial_\mu Q}{\delta \varepsilon}\frac{\delta \mathcal{L}}{\delta \partial_\mu Q}$$

$$= \left[Q,\ \frac{\delta \mathcal{L}}{\delta Q} - \partial_\mu \frac{\delta \mathcal{L}}{\delta \partial_\mu Q}\right] + \partial_\mu\left[Q,\ \frac{\delta \mathcal{L}}{\delta \partial_\mu Q}\right] \tag{34a}$$

or

$$\left(\frac{\delta \mathcal{L}}{\delta \varepsilon}\right)_{AF} = \frac{\delta Q}{\delta \varepsilon}\left(\frac{\delta \mathcal{L}}{\delta Q}\right) + \frac{\delta D_\mu Q}{\delta \varepsilon}\left(\frac{\delta \mathcal{L}}{\delta D_\mu Q}\right)$$

$$= \left[Q,\ \left(\frac{\delta \mathcal{L}}{\delta Q}\right) - D_\mu\left(\frac{\delta \mathcal{L}}{\delta D_\mu Q}\right)\right] + D_\mu\left[Q, \left(\frac{\delta \mathcal{L}}{\delta D_\mu Q}\right)\right] . \tag{34b}$$

Therefore, because of the equations of motion, the meson current

$$J^\mu[Q] = \left[Q,\ \frac{\delta \mathcal{L}}{\delta \partial_\mu D}\right] \tag{35}$$

is conserved if and only if $\delta \mathcal{L}/\delta\varepsilon|_{A,F}=0$, or is covariantly conserved if and only if $(\delta \mathcal{L}/\delta\varepsilon)_{A,F}=0$. For ungauged Wess-Zumino-Witten Lagrangian, since $\mathcal{L} = \mathcal{L}_0 + \mathcal{L}_{cs}$ is invariant under the global guage transformation, we obtain the conservation equation

$$\partial_\mu\left\{\frac{F_\pi^2}{8}(Q\partial^\mu Q) + \frac{i}{16\pi^2}\varepsilon^{\mu\nu\alpha\beta}(Q\partial_\nu Q)(Q\partial_\alpha Q)(Q\partial_\beta Q)\gamma_5\right\} = 0 . \tag{36a}$$

To the first order in π fields, this gives

$$-i\frac{F_\pi}{2}\partial_\mu(\gamma_5 \partial^\mu \pi) = 0 . \tag{36b}$$

For gauged chiral fields without WZ term, $\mathcal{L} = \hat{\mathcal{L}}_0$ being invariant under local gauge transformations, it yields the covariant conservation law

$$\frac{F_\pi^2}{16}D_\mu[Q,\ D^\mu Q] = 0 . \tag{37}$$

And for the total Lagrangian (29), the current (35) is neither conserved nor covariantly conserved, and the Euler-Lagrange equation can be written in the form

$$D_\mu\left[Q,\ \frac{\delta \hat{\mathcal{L}}_0}{\delta D_\mu Q}\right] = \left[Q,\ \frac{\delta \hat{\mathcal{L}}}{\delta Q} - \partial_\mu \frac{\delta \hat{\mathcal{L}}}{\delta \partial_\mu Q}\right] \equiv S[Q,A] , \tag{38}$$

where we have collected all the terms from the normal part of the Lagrangian, $\hat{\mathcal{L}}_0$, into current and terms from the anomalous part, $\hat{\mathcal{L}}$, into source. The current appearing in Eq.(38) is just the current in Eq.(37) and the one in Eq.(36b) to the first order. It can be shown by straightforward calculation that the leading order of the source S[Q,A] (terms containing no π's) is just Bardeen's asymmetric anomaly (apart from an overall constant coming from the coefficient in the expression of U), i.e.,

$$S[\sigma,A]=G_a[A] \ . \tag{39}$$

This is just what we found in Eq.(21).

Similarly, Eq.(33b) implies that the total current defined as

$$J_\mu=J_\mu[Q]+J_\mu[A]=\left[Q,\ \frac{\delta\mathcal{L}}{\delta\partial_\mu Q}\right]+\left[A_\nu,\ \frac{\delta\mathcal{L}}{\delta\partial_\mu A_\nu}\right] \tag{40}$$

will be conserved if and only if $\delta\mathcal{L}/\delta\varepsilon=0$. Now, from relation (32) and the fact that the normal part of the total Lagrangian, $\hat{\mathcal{L}}_0+\mathcal{L}_F$, is gauge invariant, we see J_μ is not conserved in general.

$$\partial_\mu J^\mu=\frac{\delta\mathcal{L}}{\delta\varepsilon}=\frac{\delta\hat{\mathcal{L}}}{\delta\varepsilon}=-\partial_\mu\left(\frac{\delta\mathcal{L}}{\delta A_\mu}\right) \ . \tag{41}$$

And the conserved current is

$$\tilde{J}_\mu=J_\mu+J^a_\mu \ , \qquad \partial^\mu \tilde{J}_\mu=0 \ . \tag{42}$$

Eq.(33c) gives nothing new but the covariant conservation law (34b) as discussed above.

$$D_\mu J^\mu[Q]=\left(\frac{\delta\mathcal{L}}{\delta\varepsilon}\right)_{A,F}=\frac{\delta\mathcal{L}}{\delta\varepsilon}-\left[A_\mu,\left(\frac{\delta\mathcal{L}}{\delta A_\mu}\right)\right]-\left[F_{\mu\nu},\left(\frac{\delta\mathcal{L}}{\delta F_{\mu\nu}}\right)\right] \ .$$

Notice that when a term $\mathcal{L}_s$ depending only on the gauge fields is subtracted from the Lagrangian $\mathcal{L}$, both the definition of $J_\mu[A]$ and the form of the anomalies $*G=-\partial_\mu J^\mu_a$ will be changed simultaneously. However, the covariant divergence of the meson current, $D_\mu J^\mu[Q]$, remains the same since $\mathcal{L}_s$ is irrelevant to the Goldstone fields. For Bardeen's counter-term $\hat{\Gamma}[0,A]$, this is checked by a direct calculation.

As is well known, as a consistent gauge theory, the anomalies G must be set to zero. This gives the anomaly free condition as discussed by Gross and Jackiw[7]. Alternatively, Lagrangian (29) may not be complete and there may exist some new sector (e.g. lepton sector) which also couples to the gauge fields and gives other anomalies to cancel the contribution of G. However, in this case, the role of the asymmetric anomaly G_a does not change. It still appears as part of the source coupled to the Goldstone fields, provided the Goldstone fields have nothing to do with the fields in the new sector.

In conclusion, the effective Lagrangian introduced in our previous paper[3,4], i.e., $\hat{\Gamma}$ in Eqs.(2),(9) and (10), is a correct one for describing the low-energy interaction among Goldstone mesons and gauge fields. It has the correct symmetries of QCD and contains the information of both symmetric and asymmetric anomalies. The anomalies coupled to the whole system of Goldstone and gauge fields, as shown in Eq.(41), may be the symmetric ones or the asymmetric one. It depends on the scheme of renormalization or, in other words, on the subtraction made or not, and gives rise to the anomaly-free condition. It must be the asymmetric anomaly which is coupled to the Goldstone fields just like the Ábelian chiral anomaly $\pi^0 F^*F$ in any consistent theory. This is not affected by the subtraction scheme or by the existence of the new sector.

Obviously, the results of this note can be generalized to the case of any even dimensions without difficulty.

Note Added in Proof

After completing this work, we received two papers from O. Kaymakcalan, S. Rajeev and J. Schechter[10] and H. Kawai and S.H.H. Tye[11]. in which the same effective action $\hat{\Gamma}[\xi,A]$ (first presented in Ref.[3]) are obtained by different methods and the fact that $\hat{\Gamma}[0,A]$ is just Bardeen's counter-term is also noticed. We thank Professor H.T. Nieh for calling our attention to these papers.

Acknowledgements

One of the authors (X-C.S) deeply thanks Professor C.N. YANG for his concern and inspiration. He also greatly acknowledges the financial help obtained from a Fung King-Hey fellowship through the Committee for Educational Exchange with China. Thanks are also due to Professor H.T. Nieh for his helpful discussions and careful reading of the manuscript.

This work was supported in part by NSF Grant #PHY 81-09110 A-01.

Footnotes and References

[1] *E. Witten, Nucl. Phys. B223(1983)422.*

[2] *J. Wess and B. Zumino, Phys. Lett. 37B(1971)95.*

[3] *K.C. CHOU, H.Y. GUO. K. WU and X.C. SONG, Phys. Lett. 134B(1984)67.*

[4] *K.C. CHOU, H.Y. GUO, K. WU and X.C. SONG, Commun. in Theor. Phys. 3(1984)73.*

[5] *Here we mean the result presented in the Appendix of Ref.[4], which is obtained by correcting some misprints and adding some missed terms. It produces correct anomalies and takes the form nearest to the original one in Eq.(24) of Ref.[1].*

[6] *W.A. Bardeen, Phys. Rep. 184(1969)1848; R.W. Brown, C.C. Shih, and B.L. Young, Phys. Rev. 186(1969)1491.*

[7] *D.J. Gross and R. Jackiw, Phys. Rev. D6(1972)477.*

[8] *The fact that the infinitesimal variation of the Chern-Simons 5-form gives correct anomalies has been independently noticed by B. Zumino, Y.S. Wu and A. Zee, University of Washington preprint 40048-17, to be published in Nucl. Phys. See also B. Zumino, lectures given at Les Houches, LBL-16746, Aug.1983.*

[9] *See R. Jackiw, Lecture given at Les Houches, MIT preprint CTP #1108, Aug. 1983. For original discussion about equation of motion and conservation law in anomalous case, see S. Adler in Lectures on Elementary Particles and Quantum Field Theory, edited by S. Deser, M. Grisaru and H. Pendleton (MIT Press, Cambridge, MA 1970).*

[10] *O. Kaymakcalan, S. Rajeev and J. Schechter, Syracuse preprint SU-4222-278, Dec. 1983.*

[11] *H. Kawai and S.H.H. Tye, Cornell preprint CLNS-84/595, Jan. 1984.*

Commun. in Theor. Phys. (Beijing, China) Vol.3, No.6 (1984) 767-710

ON THE TWO-DIMENSIONAL NON-LINEAR σ MODEL WITH WESS-ZUMINO TERM

I. Classical Theory

CHOU Kuang-chao(周光召) and DAI Yuan-ben(戴元本)
Institute of Theoretical Physics, Academia Sinica, P.O.Box 2735, Beijing, China

Received July 9, 1984

Abstract

The classical solutions of the two-dimensional non-linear σ model with Wess-Zumino term are shown to be equivalent to that without the Wess-Zumino term under a suitable transformation.

It is now well known that two-dimensional non-linear σ models possess a lot of rather interesting and mutually connected properties like the soliton solutions, the Bäcklund transformation and the infinite number of conservation laws. The non-linear σ model is of interest physically because it describes the low energy behavior of Goldstone bosons produced by the spontaneous symmetry breaking of an underlying fermion theory with chiral symmetry. It was pointed out by Witten[1] that the usual effective Lagrangian for the σ model has more discrete symmetries than the underlying fermion theory. Besides, it does not contain the term due to the anomaly caused by the triangular diagram (or bouble diagram in two dimensions) of the fermion loop. To cure these defects a second term called Wess-Zumino term has to be added to the effective Lagrangian which has compact form only in n+1 dimensions for a σ model in n dimensions. Then it was shown by Witten and others[2,3] that the two-dimensional SU(N) chiral model with Wess-Zumino term added is equivalent to a free massless fermion theory when the coupling constants satisfy a special condition.

In the present note we shall study only the classical behaviors of a two-dimensional σ model with Wess-Zumino term added. The quantum theory will be dealt in a separate paper. It will be shown in the following that the classical theory of two-dimensional σ model with Wess-Zumino term is almost independent of the two coupling constants. Classical solutions from models with different coupling constants are shown to be equivalent to each other by a transformation. The special condition that makes the model equivalent to a free massless fermion theory becomes a singular point of the general transformation mentioned above.

The action for the two-dimensional σ model with Wess-Zumino term added has the following form

$$I=\frac{1}{4\lambda^2}\int d^2x\ Tr(\partial_\mu g\partial^\mu g^{-1})$$
$$+\frac{n}{24\pi}\int_D d^3y\ \epsilon_{ijk}Tr(g\partial^i g^{-1}\cdot g\partial^j g^{-1}\cdot g\partial^k g^{-1}), \tag{1}$$

where the field g(x) is an element of a simple compact group G; n is an integer and λ is the coupling constant. D is a domain in three dimensions whose boundary is the two-dimensional space time under consideration. The action is invariant under the chiral transformation

$$g(x)\longrightarrow g_L g(x)g_R^{-1}\ ,$$

where g_L and g_R are two independent elements of the group G.

The equation of motion for the field g(x) can be written in terms of a pure gauge field

$$A_\mu(x)=g^{-1}(x)\partial_\mu g(x) \tag{2}$$

in the following form

$$\frac{1}{2\lambda^2}\partial^\mu A_\mu-\frac{n}{8\pi}\epsilon^{\mu\nu}\partial_\mu A_\nu=0\ . \tag{3}$$

The integrability condition for g(x) is

$$\partial_\mu A_\nu-\partial_\nu A_\mu+[A_\mu,\ A_\nu]=0\ , \tag{4}$$

when a gauge field A_μ is found to satisfy Eqs.(3) and (4), the field g(x) can be found by integrating Eq.(2). It is more convenient to work in the light cone coordinates

$$\left.\begin{aligned}\xi&=\frac{1}{\sqrt{2}}(t+x)\ ,\\ \eta&=\frac{1}{\sqrt{2}}(t-x)\ ,\end{aligned}\right\} \tag{5}$$

in which Eqs.(3) and (4) become

$$(1-a)\partial_\xi A_\eta+(1+a)\partial_\eta A_\xi=0 \tag{3'}$$

and

$$\partial_\xi A_\eta-\partial_\eta A_\xi+[A_\xi,\ A_\eta]=0\ , \tag{4'}$$

where

$$a=\frac{n\lambda^2}{4\pi} \tag{6}$$

is the only parameter of the model. Witten studied the case where

$$a=\pm1$$

and showed that the model with coupling constants λ and n satisfying this special condition is equivalent to a free massless fermion theory when the group G is SU(N)[2].

The usual non-linear σ model corresponds to the case

$$a=0,$$

where the equation of motion (3) and curvatureless condition have the form

$$\left.\begin{aligned} &\partial_\xi A_\eta+\partial_\eta A_\xi=0 \ , \\ &\partial_\xi A_\eta-\partial_\eta A_\xi+[A_\xi,\ A_\eta]=0 \ . \end{aligned}\right\} \qquad (7)$$

Now we start to prove the equivalence of models with a different parameter a. For this purpose let

$$\left.\begin{aligned} &B_\xi\equiv(1+a)A_\xi \\ \text{and}\qquad &B_\eta\equiv(1-a)A_\eta \ , \end{aligned}\right\} \qquad (8)$$

which satisfies a conservation equation from Eq.(3')

$$\partial_\eta B_\xi+\partial_\xi B_\eta=0 \ . \qquad (9)$$

One can easily show that B_μ is also curvatureless since

$$\begin{aligned} \partial_\xi B_\eta-\partial_\eta B_\xi&=(1-a)\partial_\xi A_\eta-(1+a)\partial_\eta A_\xi \\ &=\partial_\xi A_\eta-\partial_\eta A_\xi-a(\partial_\xi A_\eta+\partial_\eta A_\xi) \ . \end{aligned} \qquad (10)$$

Using the equation of motion for A_μ

$$\begin{aligned} 0&=(1-a)\partial_\xi A_\eta+(1+a)\partial_\eta A_\xi \\ &=(\partial_\xi A_\eta+\partial_\eta A_\xi)-a(\partial_\xi A_\eta-\partial_\eta A_\xi) \ , \end{aligned} \qquad (11)$$

we obtain from Eq.(10)

$$\begin{aligned} \partial_\xi B_\eta-\partial_\eta B_\xi&=(1-a^2)(\partial_\xi A_\eta-\partial_\eta A_\xi) \\ &=(1-a^2)[A_\eta,\ A_\xi]=[B_\eta,\ B_\xi] \ . \end{aligned} \qquad (12)$$

Therefore B is a pure gauge field satisfying the conservation equation just like the $A_\mu(x,a)$ for the parameter a=0. This means that all the classical solutions of the two-dimensional σ model with $a\neq0$ can be obtained from that with a=0 by the transformation Eq.(8).

It is noted that the transformation Eq.(8) is singular at a=±1 which is the case considered by Witten and others[2,3]. In that special case the finite solutions at a=±1 corresponds only to part of the solutions at a=0 that are either left-handed or right-handed moving (i.e., either $B_\xi=0$ or $B_\eta=0$).

Since we have established the equivalence of the classical solutions of two-dimensional σ model with different coupling constants except for the singular points a=±1, we conclude that all the classical behaviors are the same for these models. It is also easy to prove directly that they all contain hidden symmetries making the infinite number of conserved charges form a Kac-Moody algebra.

As pointed out above, only part of the solutions for a=0 correspond to all the finite solutions at a=±1, and it is interesting to speculate that the model with a≠±1 contains massless fermion in the tree approximation as part of its particle content. We shall return to this question in a paper on the quantum theory of these models.

References

[1] E. Witten, Nucl. Phys. B223(1983)442.

[2] E. Witten, Commun. Math. Phys. 92(1984)455.

[3] P.Di Vecchia, B.Durhuus and J.L.Petersen, "The Wess-Zumino Action in two Dimensions and Non-Abelian Bosonization", preprint NBI-HE-84-02(1984); P.Di Vecchia and P.Rossi, "On the Equivalence between the Wess-Zumino Action and the Free Fermi Theory in two Dimensions" preprint TH 3808-CERN (1984).

CP VIOLATION FROM THE STANDARD MODEL

Kuang-chao Chou

Institute of Theoretical Physics
Academia Sinica
Beijing, China

INTRODUCTION

Two decades have passed since the first observation of CP violation in kaon decay[1]. The subject is still not well understood and the progress is rather slow compared with what has been achieved in the other branches of weak interactions.

As we all know now, nature has chosen the standard $SU(2) \times U(1)$ gauge model to describe physics at an energy scale below 100 GeV. Both W and Z^0 bosons have already been seen within the error predicted by the theory[2]. It is therefore of great interest to accomodate CP violation in gauge theories which seem to be the most promising ways from a theoretical point of view.

As Kobayashi and Maskawa[3] (K-M) first pointed out, CP violation can occur in the standard model through complex phases in mass matrix with more than two generations. For three generations favored by the present experiment there is only one phase causing CP violation. Could this single phase be sufficient to explain all the CP violation effects? It is certainly welcome if it could. However, it can not be answered a priori. The origin of CP violation is closely related to that of the masses and the number of generations, which in turn are described by physics at much higher energy scales. It would not be a surprise if some new ingredients had to be added to solve the CP problem. We shall wait and see.

Since there were excellent review papers not long ago[4], it is unnecessary for me to repeat all the known results to you. What I would like to report is a recent analysis of CP violation in the K-M model after the measurement of the unexpected long lifetime of the b-quarks.

The outline of this talk is as follows:

I. Parametrization of the K-M matrix;

II. Physics of ε and ε' in kaon systems;

III. Neutral particle-antiparticle mixing and CP violation in $B^0-\bar{B}^0$ system;

IV. CP violation in partial decay rates of particles and antiparticles;

V. Concluding remarks.

I. PARAMETRIZATION OF THE K-M MATRIX

For three generations of quark the K-M matrix containing three angles and one phase is usually expressed in the following form

$$V = \begin{pmatrix} V_{ud} & V_{us} & V_{ub} \\ V_{cd} & V_{cs} & V_{cb} \\ V_{td} & V_{ts} & V_{tb} \end{pmatrix},$$

$$= \begin{pmatrix} c_1 & s_1c_3 & s_1s_3 \\ -s_1c_2 & c_1c_2c_3-s_2s_3e^{i\delta} & c_1c_2s_3+s_2c_3e^{i\delta} \\ -s_1s_2 & c_1s_2c_3+c_2s_3e^{i\delta} & c_1s_2s_3-c_2c_3e^{i\delta} \end{pmatrix} . \qquad (1.1)$$

where $c_i(s_i)$, i = 1,2,3, are the cosine (sine) of the angle θ_i. The Cabbibo angle θ_1 is determined to be[4]

$$s_1 = .227^{+.0104}_{-.0110} \quad . \tag{1.2}$$

Recent measurements on b quark lifetime and the branching ratio $\Gamma_{b\to u}/\Gamma_{b\to c}$ have put stringent bounds on the matrix elements $|V_{cb}/V_{ub}|$. Their values can be found in the talks given by Lee-Franzini and Kleinknecht in this conference.

$$|V_{cb}| = 0.0435 \pm 0.0047 \quad , \tag{1.3}$$

$$|V_{ub}/V_{cb}| \leq 0.119 \quad . \tag{1.4}$$

Both $|V_{cb}|$ and $|V_{ub}/V_{cb}|$ have been reduced from the 1983 values[5]

$$|V_{cb}| = .059^{+.016}_{-.009} \quad \text{and} \quad |V_{ub}/V_{cb}| \leq .14 \quad .$$

The fact that $|V_{cb}|$ is small and of the order of $s_1{}^2$ can be used to simplify the K-M matrix. In a first order approximation where $\mathrm{Re}V_{ij}$ are correct to order $s_1{}^3$ and $\mathrm{Im}V_{ij}$ to order $s_1{}^5$ we have

$$V = \begin{pmatrix} c_1 & s_1 & s_1 s_3 \\ -s_1 & c_1 - s_2 s_3 e^{i\delta} & s_3 + s_2 e^{i} \\ -s_1 s_2 & s_2 + s_3 e^{i\delta} & -e^{i\delta} \end{pmatrix} \quad . \tag{1.5}$$

Writing V_{ts} and V_{cb} in the following form

$$V_{ts} = s_2 + s_3 e^{i\delta} = |V_{ts}| e^{i\delta_{ts}} \quad , \tag{1.6}$$

and

$$V_{cb} = s_3 + s_2 e^{i\delta} = |V_{cb}| e^{i\delta_{cb}} \quad , \tag{1.7}$$

it is easily proved that

$$V_{ts} = e^{i\delta} V_{cb}^{*} \quad . \tag{1.8}$$

Hence we obtain

$$|V_{ts}| = |V_{cb}| = (s_2^2+s_3^2+2s_2s_3\cos\delta)^{1/2} \tag{1.9}$$

and

$$\delta = \delta_{ts} + \delta_{cb} \quad . \tag{1.10}$$

One can now redefine the phases of the b and t quarks by a transformation

$$q_b \to e^{-i\delta_{cb}} q_b \ , \tag{1.11}$$

$$q_t \to e^{i\delta_{ts}} q_t \ , \tag{1.12}$$

and get from Eqs. (1.5)-(1.12) a form first suggested by Wolfenstein[6]

$$V = \begin{pmatrix} c_1 & s_1 & s_1s_3e^{-i\delta_{cb}} \\ -s_1 & c_1-s_2s_3e^{i\delta} & |V_{cb}| \\ -s_1s_2e^{-i\delta_{ts}} & -|V_{cb}| & 1 \end{pmatrix} \quad . \tag{1.13}$$

The phases δ_{ts}, δ_{cb} are related to δ by the following relations:

$$s_2\sin\delta = |V_{cb}| \sin\delta_{cb} \ , \tag{1.14}$$

$$s_3\sin\delta = |V_{cb}| \sin\delta_{ts} \quad . \tag{1.15}$$

The advantage of the present form for the K-M matrix is that ImV_{ij} is always proportional to a common factor

$$X_{cp} = s_2 s_3 \sin\delta \quad , \tag{1.16}$$

which is the appropriate parameter measuring CP violation effects in various processes. A similar but rigorous representation of the K-M matrix was obtained recently by Chau and Keung[7].

Since both s_2 and s_3 are proportional to $|V_{cb}|$, it is more convenient in numerical calculations to scale it out. We write

$$s_3 \equiv \alpha \left|V_{cb}\right| = \frac{1}{s_1} \left|V_{ub}\right| \quad , \tag{1.17}$$

$$s_3 \sin\delta \equiv \beta \left|V_{cb}\right| \quad . \tag{1.18}$$

From the experimental bound Eq. (1.4) and the value of s_1 we find

$$\alpha \leq .524 \quad , \tag{1.19}$$

$$|\beta| = |\sin\delta_{ts}| \leq \alpha \quad . \tag{1.20}$$

With given α and β, s_2 can be solved from Eq. (1.9) to be

$$s_2 = (\sqrt{1-\beta^2} \pm \sqrt{\alpha^2-\beta^2}) \left|V_{cb}\right| \quad , \tag{1.21}$$

where we have adopted the convention of positive s_2 and s_3. The solution s_2 with positive (negative) sign in the bracket in Eq. (1.21) corresponds to $\cos\delta < 0$ ($\cos\delta > 0$) in Eq. (1.9).

II. PHYSICS OF ε AND ε' IN KAON SYSTEMS

So far CP violation has been observed only in neutral kaon systems. The ratio of the amplitudes for $K_L \to 2\pi$ and $K_S \to 2\pi$

is given in the standard notation as

$$\eta_{+-} = \varepsilon + \varepsilon'/(1 + \omega/\sqrt{2}) \ , \tag{2.1}$$

$$\eta_{00} = \varepsilon - 2\varepsilon'/(1 - \sqrt{2}\omega) \ , \tag{2.2}$$

where $\omega = ReA_2/ReA_0$ is known to be approximately .05. In terms of the amplitudes A_I for $K^0 \to 2\pi(I)$, with I the final state isospin, the mass matrix element M_{12}, ε and ε' can be expressed in the following forms

$$\varepsilon = \frac{1}{\sqrt{2}} e^{i\pi/4}\left(\frac{ImM_{12}}{\Delta M} + \xi_0\right) \ , \tag{2.3}$$

and

$$\varepsilon' = \frac{1}{\sqrt{2}} \omega(\xi_2 - \xi_0) e^{i(\delta_2-\delta_0)} \ , \tag{2.4}$$

where δ_I are final state interaction phases and

$$\xi_I = \frac{ImA_I}{ReA_I} \ . \tag{2.5}$$

The experimental value of ε is well established to be $Re\varepsilon = 0.00162 \pm .000088$ while that of ε' is still uncertain. As will be discussed later, accurate measurement of $|\varepsilon'/\varepsilon|$ is extremely important in our understanding of the origin of CP violation.

The mass matrix element for K^0-$\bar{K}^0$ transition consists of a short distance part usually identified to be the contribution from the box diagram and a long distance part determined by the low energy intermediate states

$$M_{12} = M_{12}^{sd} + M_{12}^{soft} \ . \tag{2.6}$$

The mass difference of K_L and K_S is related to ReM_{12} by the relation

$$\Delta M_K = M_S - M_L = 2\,\mathrm{Re}M_{12}$$

$$= 2(\mathrm{Re}M_{12}^{sd} + \mathrm{Re}M_{12}^{soft}) \quad . \tag{2.7}$$

The real part of M_{12}^{soft} has been estimated long ago[8]. Its value is very sensitive to the small parameter that breaks the SU(3) symmetry. Even the sign of $\mathrm{Re}M_{12}^{soft}$ can not be determined reliably. The box diagram contribution to 2 $\mathrm{Re}M_{12}^{sd}$ is dominated by charm quark exchange and consists only 1/4 ~ 3/4 of ΔM_K.[5] Since $\mathrm{Re}M_{12}$ has nothing to do with CP violation, I shall use in the following the experimental value of ΔM_K in evaluating the parameter ε. However, I would like to remark that if one finds eventually that

$$\Delta M_K \neq 2\left(\mathrm{Re}M_{12}^{box} + \mathrm{Re}M_{12}^{soft}\right),$$

it will indicate the existence of new $\Delta S = 2$ short distance interaction besides the box diagram, and thus, possible new sources of CP violation.

The imaginary part of M_{12}^{soft} can be estimated by current algebra and Penguin diagram dominance. In this approximation

$$\frac{\mathrm{Im}M_{12}^{soft}}{\mathrm{Re}M_{12}^{soft}} = -\,2\xi_0 \quad . \tag{2.8}$$

Using Eqs. (2.7) and (2.8) it is possible to eliminate the soft part of M_{12} and rewrite Eq. (2.3) in the form[9]

$$\varepsilon = \frac{1}{\sqrt{2}}\, e^{\frac{i\pi}{4}} \left(\frac{\mathrm{Im}M_{12}^{sd}}{\Delta M} + 2\xi_0\, \frac{\mathrm{Re}M_{12}^{sd}}{\Delta M}\right) \quad , \tag{2.9}$$

where M_{12}^{sd} is calculated by the box diagram[10]

$$M_{12}^{box} = \frac{G_F{}^2 B_K f_K{}^2 m_K}{12\pi^2} \{\eta_{tt}\lambda_t{}^2 m_t{}^2 f(\frac{m_t{}^2}{m_W{}^2})$$

$$+ \eta_{cc}\lambda_c{}^2 m_c{}^2 + 2\eta_{ct}\lambda_c\lambda_t m_c{}^2 \ell n \frac{m_t{}^2}{m_c{}^2}\} \quad , \tag{2.10}$$

where

$$\lambda_i = V_{is} V_{id}^* \, , \tag{2.11}$$

$$f(x) = \frac{1}{(1-x)^2}\left(1 - \frac{11}{4}x + \frac{1}{4}x^2\right) - \frac{3\,x^2 \,\ell n\, x}{2(1-x)^3} \, , \tag{2.12}$$

and $\eta_{tt} = 0.6$, $\eta_{cc} = 0.7$, $\eta_{ct} = 0.4$ are the QCD correction factors. The factor B_K accounts for the uncertainty in determining the matrix element

$$\langle K^0 | [\bar{s}\gamma_\mu(1+\gamma_5)d]^2 | \bar{K}^0 \rangle = -\frac{4}{3} B_K f_K{}^2 m_K \; . \tag{2.13}$$

Current algebraic estimation tells us that B_K is around 1/3, while some lattice calculations[11] give the value about 1, close to the vacuum-insertion value. We shall keep B_K to be a parameter in the following calculations.

Eqs. (2.9)-(2.10) have been used to predict the minimum top quark mass when the K-M angles are given, or the other way around, to set lower bound on the CP violation parameter $X_{cp} = s_2 s_3 \sin\delta$ when the top quark mass is assumed[5,12-15]. In these calculations the second term in Eq. (2.9) proportional to ξ_0 was neglected and the 1983 experimental values of $|V_{cb}|$ and $|V_{ub}/V_{cb}|$ were used.

The parameter ξ_0 could be estimated by using experimental value of $ReA_0 \approx A_0$ and Penguin diagram value of ImA_0. It is found[13-14] that

$$\xi_0 = s_2 s_3 \sin\delta \; c_6 \frac{G_F}{\sqrt{2}} s_1 \frac{\langle 2\pi(I=0)|Q_6|K^0\rangle}{A_0} \qquad (2.14)$$

where Q_6 is a (V−A)×(V+A) Penguin operator with Wilson coefficient c_6 in the effective Hamiltonian for Penguin diagram. $c_6 = \mathrm{Im}c_6/s_2 s_2 \sin\delta$ is estimated in the leading logarithmic approximation to all orders in the strong interaction to be −0.1 and is quite stable against the choice of parameters.

For the matrix element $\langle 2\pi(I=0)|Q6|K^0\rangle$ shall follow the analysis of Gilman and Hagelin[13] to use the bag model value for a conservative estimation. One finds finally that

$$\xi_0 = -\ .54\ s_2 s_3 \sin\delta \left|\frac{c_6}{0.1}\right| \left|\frac{\langle 2\pi(I=0)|Q_6|K^0\rangle}{1.4\ \mathrm{GeV}^3}\right| . \qquad (2.15)$$

A similar estimation gives $\xi_2 \approx 0$. The parameter $|\varepsilon'/\varepsilon|$ can then be determined to be

$$|\varepsilon'/\varepsilon| = 8.4\ s_2 s_3 \sin\delta \left|\frac{c_6}{0.1}\right| \left|\frac{\langle 2\pi(I=0)|Q_6|K^0\rangle}{1.4\ \mathrm{GeV}^3}\right| \qquad (2.16)$$

Now a combined analysis of $|\varepsilon|$ and $|\varepsilon'/\varepsilon|$ can be made with the help of Eqs. (2.9), (2.10), (2.15) and (2.16) when $s_3(\alpha)$, $s_2\sin\delta(\beta)$, $|V_{cb}|$ and B_K are given. The results are given in Tables 1-7 and Figs. 1-3, where both m_t and $|\varepsilon'/\varepsilon|$ are shown to be functions of α and β. The minimum top quark mass occurs at the point where $s_3(\alpha)$ saturates its upper bound and δ in the second quadrant where s_2 is larger. Its value increases as B_K and $|V_{cb}|$ decrease. For $B_K = 0.33$ and $|V_{cb}| < .059$ the minimum top quark mass is already over 60 GeV. It is also noted that $|\varepsilon'/\varepsilon|$ is large at the point where m_t is minimum. The present experimental value of $|\varepsilon'/\varepsilon| = -\ .003\pm.015$ is barely consistent with the one calculated at the point of minimum top quark mass.

TABLE 1. Values of M_t, $|\varepsilon'/\varepsilon|$, X_{B_d} and X_{B_s} of Eq. (3.10) as functions of $s_2/|V_{cb}|$ for $B_K = 0.33$, $|V_{cb}| = 0.0388$ and $|V_{ub}/V_{cb}| = .119$

$s_2/\|V_{cb}\|$	.852	1.031	1.110	1.171	1.222	1.266	1.305	1.339	1.395	1.440
M_t (GeV)	325.3	254.3	235.8	226.0	221.0	219.3	220.2	223.4	236.9	162.4
$\|\varepsilon'/\varepsilon\|$	.0057	.0065	.0066	.0066	.0065	.0063	.0061	.0058	.0051	.0043
X_{B_d}	2.10	2.10	2.17	2.26	2.38	2.52	2.70	2.90	3.45	4.28
X_{B_s}	56.1	38.3	34.2	32.0	30.96	30.6	30.8	31.5	34.4	40.07

TABLE 2. Same as Table 1 except $|V_{cb}| = .0435$

$s_2/\|V_{cb}\|$	.852	1.031	1.110	1.771	1.222	1.266	1.305	1.339	1.395	1.440
M_t (GeV)	233.6	180.9	167.6	160.7	157.3	156.3	157.3	159.9	170.3	188.9
$\|\varepsilon'/\varepsilon\|$	.0071	.0082	.0083	.0083	.0082	.0079	.0076	.0073	.0064	.0054
X_{B_d}	1.58	1.57	1.62	1.69	1.79	1.90	2.03	2.19	2.63	3.27
X_{B_s}	42.4	28.7	25.5	24.0	23.2	22.96	23.2	23.8	26.2	30.6

TABLE 3. Values of M_t, $|\varepsilon'/\varepsilon|$, X_{B_d} and X_{B_s} as functions of $s_2/|V_{cb}|$ for $B_K = 0.33$, $|V_{cb}| = 0.0482$ and $|V_{ub}/V_{cb}| = .119$

$s_2/\|V_{cb}\|$	.852	1.031	1.110	1.171	1.222	1.266	1.305	1.369	1.419	1.458
M_t (GeV)	171.3	132.1	122.4	117.6	115.3	114.8	115.7	121.4	132.5	150.9
$\|\varepsilon'/\varepsilon\|$	.0087	.0100	.0103	.0102	.0100	.0097	.0094	.0084	.0073	.0060
X_{B_d}	1.21	1.19	1.22	1.28	1.35	1.44	1.55	1.83	2.26	2.92
X_{B_s}	32.4	21.7	19.2	18.1	17.5	17.4	17.6	19.0	21.8	42'607

TABLE 4. Same as Table 3 except $|V_{cb}| = .0588$

$s_2/\|V_{cb}\|$	.852	1.031	1.110	1.171	1.222	1.266	1.339	1.395	1.440	1.474
M_t (GeV)	91.9	70.4	65.3	63.0	62.1	62.3	64.8	70.3	79.1	93.4
$\|\varepsilon'/\varepsilon\|$	.0130	.0150	.0153	.0152	.0149	.0145	.0133	.0117	.0099	.0079
X_{B_d}	.68	.64	.65	.68	.72	.78	.93	1.16	1.52	2.08
X_{B_s}	18.1	11.6	10.2	9.62	9.40	9.43	10.09	11.6	14.2	18.6

TABLE 5. Values of M_t, $|\varepsilon'/\varepsilon|$, X_{B_d} and X_{B_s} as functions of $s_2/|V_{cb}|$ for $B_K = 1$, $|V_{cb}| = 0.0388$ and $|V_{ub}/V_{cb}| = .119$

$s_2/\|V_{cb}\|$	.852	1.031	1.110	1.171	1.222	1.305	1.369	1.419	1.458	1.499
M_t (GeV)	113.4	81.9	76.1	73.6	72.9	75.7	82.0	84.2	111.8	161.3
$\|\varepsilon'/\varepsilon\|$	.0057	.0065	.0066	.0066	.0065	.0061	.0055	.0047	.0039	.0025
X_{B_d}	.41	.34	.37	.39	.42	.50	.61	.85	1.18	2.21
X_{B_s}	11.0	6.3	5.8	5.5	5.43	5.72	6.30	8.20	10.80	24.9

TABLE 6. Same as Table 5 except $|V_{cb}| = .0435$

$s_2/\|V_{cb}\|$	.852	1.031	1.110	1.171	1.222	1.305	1.395	1.440	1.474	1.499
M_t (GeV)	71.5	49.5	45.1	43.7	44.0	47.5	58.4	69.8	86.8	114.7
$\|\varepsilon'/\varepsilon\|$	.0071	.0082	.0083	.0083	.0082	.0076	.0064	.0054	.0043	.0031
X_{B_d}	.24	.190	.188	.197	.217	.28	.46	.67	1.01	1.64
X_{B_s}	6.5	3.47	2.95	2.79	2.81	3.24	4.66	6.25	9.02	14.14

TABLE 7. Values of M_t, $|\varepsilon'/\varepsilon|$, X_{B_d} and X_{B_s} as functions of $s_2/|V_{cb}|$ for $B_K = 1$, $|V_{cb}| = 0.0482$ and $|V_{ub}/V_{cb}| = .119$

$s_2/\|V_{cb}\|$	.852	1.031	1.110	1.171	1.222	1.305	1.369	1.419	1.458	1.487
M_t (GeV)	42.9	24.22	20.8	20.5	21.8	26.9	33.9	42.8	54.6	71.7
$\|\varepsilon'/\varepsilon\|$	.0087	.0100	.0103	.0102	.0100	.0094	.0084	.0073	.0060	.0046
X_{B_d}	.12	.062	.054	.058	.071	.12	.21	.34	.55	.92
X_{B_s}	3.30	1.14	0.85	.82	.93 6	1.38	2.14	3.29	5.02	8.06

TABLE 8. Same as Table 7 except $|V_{cb}| = .0435$ and Re $M_{12}^{sd}/\Delta M = 0.36$

$s_2/\|V_{cb}\|$	.852	1.031	1.110	1.171	1.222	1.266	1.305	1.338	1.395	1.440
M_t (GeV)	82.8	60.4	55.7	53.8	53.5	54.3	55.9	58.3	65.2	75.7
$\|\varepsilon'/\varepsilon\|$	.0071	.0082	.0083	.0083	.0082	.0079	.0076	.0073	.0064	.0054
X_{B_d}	.316	.267	.272	.285	.308	.340	.378	.429	.559	.767
X_{B_s}	8.44	4.88	4.29	4.04	4.00	4.11	4.31	4.65	5.58	7.18

TABLE 9. Values of M_t, $|\varepsilon'/\varepsilon|$, X_{B_d} and X_{B_s} as functions of $s_2/|V_{cb}|$ for $B_K = 1$, $|V_{cb}| = 0.0482$, $|V_{ub}/V_{cb}| = .119$ and Re $M_{12}^{sd}/\Delta M = .36$.

$s_2/\|V_{cb}\|$	.852	1.031	1.110	1.171	1.222	1.266	1.305	1.339	1.395	1.440
M_t (GeV)	54.7	37.7	34.3	33.3	33.6	34.8	36.6	38.9	45.3	54.2
$\|\varepsilon'/\varepsilon\|$	.0087	.0100	.0103	.0102	.0100	.0097	.0094	.0089	.0079	.0066
X_{B_d}	.189	.143	.139	.147	.163	.186	.217	.256	.365	.530
X_{B_s}	5.05	2.61	2.19	2.08	2.111	2.25	2.47	2.77	3.64	4.96

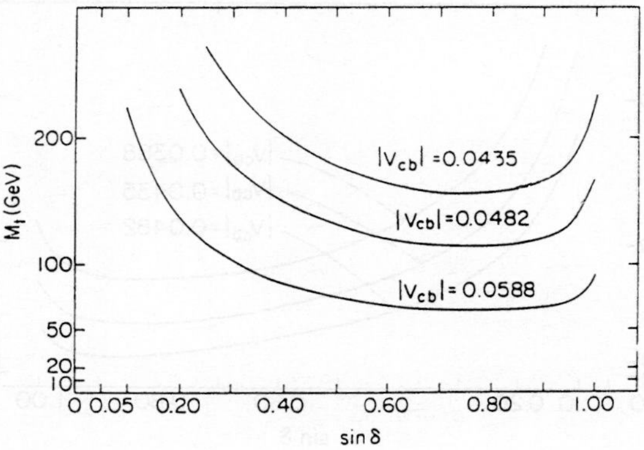

Fig. 1 Top quark mass for $B_K = .33$, $|V_{ub}/V_{cb}| = .119$.

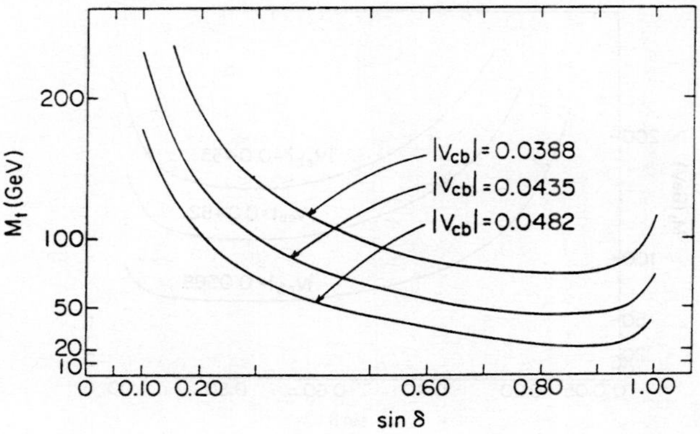

Fig. 2 Top quark mass for $B_K = 1$ and $|V_{ub}/V_{cb}| = .119$.

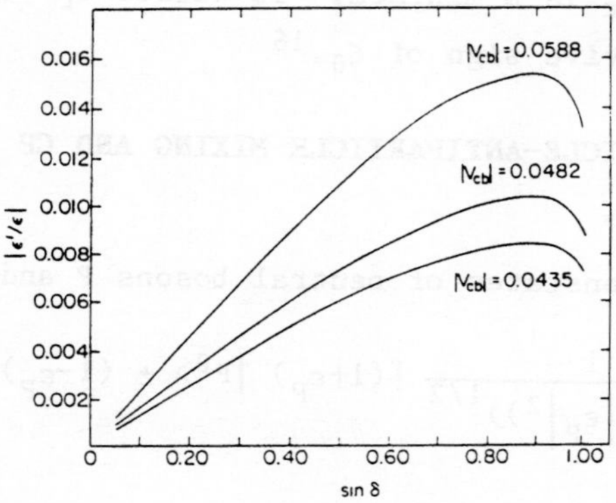

Fig. 3 $|\varepsilon'/\varepsilon|$ as function of $\sin\delta$.

To get a definite conclusion we need more accurate measurement on $|\varepsilon'/\varepsilon|$, lower bound on $|V_{ub}/V_{cb}|$ and upper bound on m_t. More reliable theoretical evaluations of B_K, $c_6\langle 2\pi(I=0)|Q_6|K^0\rangle$ and M_{12}^{soft} are also required. What we could say at the present is that new sources of CP violation besides the K-M phase δ might exist if m_t is found to be around 40 GeV, $|\varepsilon'/\varepsilon| \leq .005$ and $B_K \leq 0.6$.

The second term proportional to ξ_0 in ε is also estimated and the result is given in Tables 8-9. For $B_K = 1$, $2\,\mathrm{Re}M_{12}^{box} \simeq 3/4\ \Delta M_K$, the effect of this term is appreciable at the point where m_t is a minimum. It raises m_t roughly by 30% owing to the negative sign of ξ_0.[16]

III. NEUTRAL PARTICLE-ANTIPARTICLE MIXING AND CP VIOLATION IN B^0-$\bar{B}^0$ SYSTEM.

The mass eigenstates of neutral bosons P and $\bar{P}$ are

$$|P_{\substack{1\\2}}\rangle = \frac{1}{(2(1+|\varepsilon_P|^2))^{1/2}}\left[(1+\varepsilon_P)\,|P^0\rangle \pm (1-\varepsilon_P)\,|\bar{P}^0\rangle\right] , \qquad (3.1)$$

where ε_P is determined by the mass matrix elements relating P^0 and $\bar{P}^0$.

$$\frac{1-\varepsilon_P}{1+\varepsilon_P} = \frac{H_{21}}{H_{12}} , \qquad (3.2)$$

where $H_{ij} = M_{ij} - i\Gamma_{ij}$ are the mass matrix elements of the neutral P^0-$\bar{P}^0$ system.

For a state being P^0 at t = 0, $|\phi(t=0)\rangle = |P^0\rangle$ and later at time t

$$\left|\phi(t)\right> = f_+(t)\left|P^0\right> + \frac{1-\varepsilon_P}{1+\varepsilon_P} f_-(t)\left|\bar{P}^0\right> , \tag{3.3}$$

where

$$f_\pm(t) = \frac{1}{2}\left[e^{-im_1t-\Gamma_1t/2} \pm e^{-im_2t-\Gamma_2t/2}\right] . \tag{3.4}$$

Mixing of P^0 and $\bar{P}^0$ is necessary for the observation of CP violation effects. There are two cases of maximal mixing. In the kaon case $\delta\Gamma/\Gamma \approx 1$, either with the K^0 or $\bar{K}^0$ to begin with, it will quickly end up as K_L, which is almost an equal mixture of K^0 and $\bar{K}^0$. The second possibility occurs when $\delta m/\Gamma \simeq 1 \gg \delta\Gamma/\Gamma$ In this case, before decaying, the state oscillates quickly between P^0 and $\bar{P}^0$ and appears as an equal mixture of P^0 and $\bar{P}^0$.

Due to the simple fact that the decay width Γ for the D and T particles are K-M angle nonsuppressed, yet δm and $\delta\Gamma$ are always K-M angle suppressed; the values of $\delta m/\Gamma$ and $\delta\Gamma/\Gamma$ are both small and the observation of CP violation in neutral D and T systems is extremely difficult.

The situation is different in B systems. There one expects large mixing effects and possibly large CP violation effects[5,17-18]. One special feature of the B systems is that the complex parameter ε_B is almost imaginary and of the order 1. The imaginary nature follows from the fact that the phase of M_{12} is the same as that of Γ_{12} and the condition $M_{12} \gg \Gamma_{12}$.[1]
Therefore the observable effects, depending on the $Re\varepsilon_B$ such as the fractional difference of same-sign dilepton production and the asymmetry in semileptonic decay, are very small.

As has been emphasized by Bigi, Carter and Sanda[17] that nonleptonic on shell transition in the bottom sector might produce CP asymmetries of the order 10^{-1}-10^{-2}, whereas the effects due to CP impurities in the mixing is less than 10^{-3}. The effects fall into two categories. The asymmetry for initially pure B^0 and $\bar{B}^0$ states to decay into the same final state f is found to be

$$A = \frac{-xa\ \sin 2\phi}{1+y\ \cos 2\phi}\ , \tag{3.5}$$

where

$$x = \frac{\Delta M}{\Gamma}\ ,\quad y = \frac{\Delta\Gamma}{2\Gamma}\ , \tag{3.6}$$

$$a = \frac{1-y^2}{1+x^2}\ . \tag{3.7}$$

The angle ϕ is defined as

$$\lambda = \left(\frac{1-\varepsilon_B}{1+\varepsilon_B}\right)\frac{M}{\bar{M}} = -\ e^{-2i\phi} \tag{3.8}$$

with M and $\bar{M}$ the amplitudes

$$M = \langle f|H_W|B^0\rangle\ \ ,\quad \bar{M} = \langle f|H_W|\bar{B}^0\rangle\ . \tag{3.9}$$

The second type of experiment produces a $B^0\bar{B}^0$ pair with charge parity C and measures the decays of one B meson into a lepton and the other to a hadronic state f. The CP asymmetry was computed to be

$$A = \frac{-2xa^2\ \sin^2\phi}{1+y^2+y\cos 2\phi} \qquad \text{for } C = \text{even}\ ;$$

$$= 0 \qquad \text{for } C = \text{odd}\ . \tag{3.10}$$

For both B_s and B_d systems the parameter y is small compared to unity and can be neglected in Eqs. (3.5) and (3.10). The parameters x_s and x_d for the B_s and B_d systems are calculated using the box diagram value for ReM_{12} and the experimental value for the B meson lifetime[7,18]. Results of the calculated x are shown in Tables 1-9, which depend very much on the parameters used.

For B_d system the angle ϕ depends on the decay channel

$$tg\phi = \frac{s_3 \sin\delta}{s_2 + s_3 \cos\delta} , \qquad \text{for } b \to c \text{ decays ;}$$

$$tg\phi = tg\delta , \qquad \text{for } b \to u \text{ decays .} \qquad (3.11)$$

For B_s system the angle ϕ is very small for the b→c channel and

$$tg\phi = \frac{s_2 \sin\delta}{s_3 + s_2 \cos\delta} , \qquad \text{for } b \to u \text{ decays .} \qquad (3.12)$$

The asymmetries for B_d system can reach 20-40% in a certain range of parameters. However, if $|\varepsilon'/\varepsilon|$ is found to be small in future experiments, sinδ will become small and the CP violation effects in B^0-$\bar{B}^0$ system will also be small.

IV. CP VIOLATION IN PARTIAL DECAY RATES OF PARTICLES AND ANTIPARTICLES

To begin with I shall recall some general remarks made by Pais and Treiman[19] concerning the constraints imposed by CPT invariance. Let f_i be a set of final states connected by strong or electromagnetic interactions into which the particle p can decay and $\bar{f}_i$ the corresponding conjugate states. CPT invariance requires the partial width of p and $\bar{p}$ to be equal

$$\sum_i \Gamma(p \to f_i) = \sum_i \Gamma(\bar{p} \to \bar{f}_i) , \qquad (3.13)$$

provided this set of final states has no strong or electromagnetic interactions with the other decaying channels. If the set consists of only a single state no information about CP violation can be found. For two final states the decaying amplitudes of the particle have the form

$$a = f M_1 + g M_2 , \qquad (3.14)$$

where f and g are complex coupling constants in the effective weak interaction while M_i, i = 1,2 are different amplitudes leading to the same final state. The corresponding amplitudes for the antiparticles are

$$\bar{a} = f M_1^* + g M_2^* . \tag{3.15}$$

we can easily calculate the asymmetry to be

$$A = \frac{2\,\mathrm{Im}(f^* g)\,\mathrm{Im}(M_1^* M_2)}{|fm_1|^2 + |gM_2|^2 + 2\,\mathrm{Re}(f^* g)\,\mathrm{Re}(M_1^* M_2)} \tag{3.16}$$

In the case of strange particle decay the asymmetry is found to be very small[20]. In this case

$$\frac{\mathrm{Im}(M_1^* M_2)}{|M_1|^2 + |M_2|^2} \propto \sin(\delta_1 - \delta_2) \text{ where } \delta_i \; i = 1,2$$

are final state interaction phases and $\mathrm{Im}(f^* g) \propto s_1^2 s_2 s_3 \sin\delta$, both of which are very small. As the real parts of the corresponding coupling constants f and g in kaon decays are of the order 1, the asymmetry of partial rates for strange decays is very small of the order $10^{-5} - 10^{-6}$.

In general the number of the final hadronic channels increases rapidly as one progresses from strange to bottom decays. The final state interactions are much stronger and $2\,\mathrm{Im}(M_1^* M_2)/(|M_1|^2 + |M_2|^2)$ might reach several percent in B-decays. For those decaying channels where $|f|^2$ and $|g|^2$ are comparable with $\mathrm{Im}(f^* g) \propto s_1^2 s_2 s_3 \sin\delta$, the asymmetry can be quite big. Examples[21] are given in Table 10. Chau and Cheng proved a theorem using quark diagram techniques to show that the CP violation effects in the asymmetry of partial decay rates of particles and antiparticles are always proportional to $s_1^2 s_2 s_3 \sin\delta$ which is a very small number of the order of

10^{-5} - 10^{-6}. The apparent large asymmetry in the B-decays is due to the smallness of the denominator in Eq. (3.16). Therefore for those decay channels where the asymmetry is large the branching ratio is always small. The number of events required to observe the CP violation effects in B-decay is still large, of the order of 10^5-10^6. It is not an easier task than the observation of CP violation in kaon decays.

V. CONCLUDING REMARKS

1) Measurements of $|\varepsilon'/\varepsilon|$ and the lower limit of $|V_{ub}/V_{cb}|$ as well as the discovery of the top quark are crucial in finding a consistent picture of CP violation within the standard model. More careful calculation of B_K, $C_6\langle 2\pi/2=0|Q_6|K^0\rangle$ and M_{12}^{soft} are also necessary. A low value of m_t less than 40 GeV together with B_K around 0.33 might rule out the K-M model with three generations of quarks. In this case the K-M phases might still be the only source of CP violation if more than three generations are found. A reduction of the upper limit of $|\varepsilon'/\varepsilon|$ not only pushes the theoretical m_t up but also reduces the value of $\sin\delta$, thus making the observation of other CP violation effects more difficult.

2) CP impurities in B^0-$\bar{B}^0$ mass eigenstates are very small. Therefore observations on the fractional difference of same-sign diplepton production and the asymmetries in the semileptonic decays are very difficult.

3) The mixing for $B_s{}^0$ systems is large, but the asymmetries in the $B_s{}^0(b\to c)$ channels are small. The mixing for $B_d{}^0$ systems and the asymmetries in the decay channels $B_d{}^0(b\to c)$ may reach several percent in a certain range of parameters.

4) CP violation effects in partial decay rates of particles and antiparticles are small for strange particle decays. It is much larger for charged B decays. However the number of events

TABLE 10 **Asymmetry and number of events needed to observe CP violation in D and B decays.** Taken from Chau and Cheng[21].

Reaction	Amplitudes	Δ_{tree}	$\Delta_{Penguin}$	Br	No. of Events Needed
$B_u^- \to \pi^0 K^-$	$\frac{1}{\sqrt{2}}[V_{cb}V^*_{cs}(e)$	0	-6×10^{-2}	3.8×10^{-5}	7.3×10^{6}
	$+ V_{ub}V^*_{us}(a+b+d+e)]$	(0)	(-1.6×10^{-2})	1.4×10^{-4}	(7.3×10^{6})
$\to D^-D^{0*}$	$V_{cb}V^*_{cd}(a+e)$	-1.6×10^{-2}	1×10^{-3}	3×10^{-3}	1.3×10^{6}
	$+ V_{ub}V^*_{ud}(d+e)$	(−0.86)	(7.3×10^{-4})	(4.1×10^{-3})	(3.3×10^{2})
$\to K^-J/\psi$	$V_{cb}V^*_{cs}(b+e) + V_{ub}V^*_{us}(\tilde{d}+e)$	-3×10^{-4}	-	5×10^{-3}	2.2×10^{9}
	$+ \sum_{i=u,c,t} V_{ib}V^*_{is}(\tilde{e})$				
$B_c^- \to K^-\bar{D}^0$	$V_{cb}V^*_{cs}(d+e)$	0.39	-7×10^{-2}	2×10^{-5}	3.5×10^{6}
	$+ V_{ub}V^*_{us}(a+e)$	(6.2×10^{-2})	(7.2×10^{-5})	(1.8×10^{-2})	(1.4×10^{4})
$\to \pi^-\bar{D}^0$	$V_{cb}V_{cd}(d+e)$	-4×10^{-2}	3.7×10^{-3}	2.3×10^{-4}	2.7×10^{6}
	$+ V_{ub}V^*_{ud}(a+e)$	(−0.86)	(1×10^{-3})	(8.5×10^{-4})	(1.6×10^{3})
$\to \pi^0 D^-$	$\frac{1}{\sqrt{2}}[V_{cb}V^*_{cd}(d+e)$	−0.17	-2.2×10^{-2}	1.3×10^{-5}	2.7×10^{6}
	$+ V_{ub}V^*_{ud}(b+e)]$	(−0.86)	(-5.7×10^{-3})	(4.8×10^{-5})	2.8×10^{4}
$F^+ \to K^0\pi^+$	$V_{ud}V^*_{cd}(a+e)$	-3.6×10^{-4}	-3.3×10^{-5}	1.8×10^{-3}	4.5×10^{9}
	$+ V_{us}V^*_{cs}(d+e)$	(-2×10^{-3})	(-1.7×10^{-5})	(3.4×10^{-3})	(7.4×10^{7})

needed to observe the CP violation effects is large, of the order 10^5-10^6.

ACKNOWLEDGMENT

This talk is the result of a collaboration with Wu Yue-liang and Xie Yan-bo. Discussions with Profs. Li Xiao-yuan, Chu Chen-yuan and L.-L. Chau have helped enormously in improving my understanding of the problem. I would like to thank Mrs. Isabell for her kindness and support in typing the manuscript.

REFERENCES

1. J.H. Christenson, J.W. Cronin, V.L. Fitch and R. Turlay, Phys. Rev. Lett. 13 (1964) 138.

2. G. Arnison et al., Phys. Lett. 126B (1983) 398; ibid 129B (1983) 273;
P. Bagnaia et al., Phys. Lett. 129B (1983) 130.

3. M. Kobayashi and T. Maskawa, Prog. Theor. Phys. 49 (1973) 652.

4. For a recent review, see L.-L. Chau, "Quark Mixing in Weak Interactions", Phys. Rep. 95, No. 1 (1983).

5. L.-L. Chau and W.-Y. Keung, preprint BNL-23811 (1983).

6. L. Wolfenstein, Phys. Rev. Lett. 51 (1984) 1945.

7. L.-L. Chau and W.-Y. Keung, BNL preprint (1984).

8. C. Itzykson, M. Jacob and G. Mahoux, Nouvo Cim. Suppl. 5 (1967) 978.

9. J.S. Hagelin, Phys. Lett. 117B (1982) 441.

10. T. Inami and C.S. Lim, Prog. Theor. Phys. 65 (1981) 297.

11. N. Cabibbo and C. Martinelli, TH-3774 CERN (1983);
R.C. Brower, G. Maturana, M.B. Gavela and R. Gupta, HUTP-84/A004 NUB # 2625 (1983).

12. P.H. Ginsparg, S.L. Glashow and M.B. Wise, Phys. Rev. Lett. 50 (1983) 1415.

13. F.J. Gilman and J.S. Hagelin, preprint SLAC-PUB-3226 (1983).

14. P.H. Ginsparg and M.B. Wise, Phys. Lett. 127B (1983) 265.

15. Pham Xuan-Yen and Vu Xuan-Chi, preprint, PAR LPTHE 82/28 (1983).

16. K.C. Chou, Y.L. Wu and Y.B. Xie, preprint ASITP-84-005 (1984).

17. J.S. Hagelin, Nucl. Phys. B193 (1981) 123;
B. Carter and A.I. Sanda, Phys. Rev. Lett. 45 (1980) 952;
Phys. Rev. D23 (1981) 1567;
L.I. Bigi and A.I. Sanda, Nucl. Phys. B193 (1981) 85;
Ya. I. Azimov and A.A. Ioganssen, Yad. Fix. 33 (1981) 383,
[Sov. J. of Nucl. Phys. 33 (1981) 205].

18. L.I. Bigi and A.I. Sanda, preprint NSF-ITP 83-168 (1983);
L. Wolfenstein, preprint CMU-HEG 83-9; NSF-ITP-83-146 (1983);
E.A. Paschos and U. Türke, preprint NSF-ITP-83-168 (1983).

19. A. Pais and S.B. Treiman, Phys. Rev. D12 (1975) 2744.

20. L.L. Chau and W.Y. Keung, Phys. Rev. D29 (1984) 592.

21. J. Bernabeu and C. Jarlskog, Z. Phys. C8 (1981) 233;
L.L. Chau and H.Y. Cheng, BNL preprint (1984).

DISCUSSION

PAVLOPOULOS:

Does a t-quark mass of 60-70 GeV lead to a hopelessly small ε'/ε?

CHOU:

No, if B_K is around 0.33, ε'/ε can reach .01 for $m_t \simeq$ 60-70 GeV.

HITLIN:

If $|\varepsilon'/\varepsilon|$ is small and $\sin\delta$ is of the appropriate value, then the t quark mass could be greater than the W or Z mass. Could enough $t\bar{t}$ pairs be produced at the SPS collider to make the decay $t \to W + b$ a plausible explanation of the recently reported events at the SPS Collider?

CHOU:

We better keep in mind such possibilities.

Nov. 1984 CHINESE PHYSICS LETTERS Vol.1, No.2

TOP QUARK MASS AND THE FOURTH GENERATION OF QUARK

CHOU Kuang-chao, WU Yue-liang, XIE Yan-bo
(Institute of Theoretical Physics, Academia Sinica, Beijing)

(Received 9 August 1984)

To explain CP violation in Kaon system in the light of the recently measured b-lifetime within the framework of three generations of quarks, the top quark mass has to be greater than 50GeV if the current algebra value of the factor B_K is adopted and $|\varepsilon'/\varepsilon| < 0.01$. In this letter a fourth generation of quark is considered which can fit the present experimental data on CP violation, and $K_L \to \mu^+\mu^-$ decay rate for top quark mass is around 40GeV. The mass of the new charge 2/3 quark is predicted to be over 100 GeV.

The unexpected long lifetime of quark in the picosecond range[1] has changed the whole picture on the CP violation and the mass difference in the K_L-K_S system. In the Kobayashi-Maskawa theory[2] with three generations of quark the mixing angles θ_2 and θ_3 determined from the b decays are found to be very small[3]. Consequently, the contribution due to top quark exchange to the box diagram of the $K°$-$\bar{K}°$ mass matrix element is small unless top quark mass is large. For top quark mass less than 1 TeV there will be no appreciable effect to the K_L-K_S mass difference by top quark exchange.

All theoretical caculation of the box diagram contains matrix element relating the $K°$ and $\bar{K}°$ states to the product of quark operators. Current algebra evaluation[4] of this matrix element differs from its vaccuum insertion value by a factor of $B_K = 0.33 \pm 0.17$.

In a previous work[5] we have said the following: New sources of CP violation besides the single K-M phase might exist if m_t is found to be around 40GeV, $|\varepsilon'/\varepsilon| \leq 0.005$ and $B_K \leq 0.6$. Just a few weeks ago a new particle of mass around 40GeV, a candidate of the top quark, was announced by UAL group in CERN[6]. The measurement of ε'/ε has also reduced its upper bound[7]. Now we are facing the possibility that something new might be needed in order to explain all the existing experimental data. The simplest possibility is to add a new generation of quark.

Standard model with four generations of quark has been studied by Oakes[8]. He pointed out that a relation $M_{t'} = M_t = 40$GeV exists if Buras analysis[9] on K_L-K_S mass difference and $K_L \to \mu^+\mu^-$ decay rate is used. Buras has established an upper bound of about 40GeV on top quark mass. However, in Buras calculation the K_L-K_S mass difference is completely attributed to the top quark exchange which requires large mixing angles θ_2 and θ_3 contradicting the b-lifetime measurement. Actually, the main contribution to the K_L-K_S mass difference comes from the charm quark exchange.

Therefore the upper bound of top quark mass and the relation Mt'=Mt found in this paper are not justified. A more careful estimation can raise the upper bound of Mt up to at least 140GeV.

We have redone the calculation of CP violation parameters and in Kaon decay and the $K_L \to \mu^+\mu^-$ decay rate based on four generations of quark. There are now six mixing angles and three complex phases. By varying these Kobayashi-Maskawa parameters a satisfactory explanation of all existing data can be found with the top quark mass chosen to be 40GeV. The mass of the new charge 2/3 quark is found to be around or even greater than 100GeV, for $B_K < 0.6$, $|\varepsilon'/\varepsilon| = 0.008$, $\tau_b = 1.5$, the matrix elements V_{cd} and V_{cs} are taken to be approximately equal to that in the case of three generations, other parameters can vary freely. The K_L-K_S mass difference still can not be accounted for only by the short distance contribution from the quarks. A detail account of the calculation will be presented elsewhere.

We would like to thank Profs. L.L. Chan, G. Kane, J.T. He for informing us the experimental results on $|\varepsilon'/\varepsilon|$ and top quark mass.

REFERENCES

[1] Preprint, SLAC-PUB-3323 (1984).
[2] M. Kobayashi and T. Maskawa, Prog. Theor. Phys., 49(1973), 652.
[3] Chou Kuang-chao, Wu Yue-liang and Xie Yan-bo Preprint, AS-ITP-84-005 (1984)
[4] J. Donoghue, E. Colowich, and B. Holstein, Phys. Lett., 119B(1982), 412.
[5] K.C. Chou, Invite Talk at Europhysics conference held at Erice,Sicily, March 1984.
[6] Private Communication.
[7] Private Communication.
[8] Andrzej J. Buras, Phys. Rev. Lett., 46(1981), 1354.
[9] R.J. Oakes, Phys. Rev., D26(1982), 1128

Commun. in Theor. Phys. (Beijing, China) Vol.4, No.1 (1985) 91-101

ANOMALIES OF ARBITRARY GAUGE GROUP AND ITS REDUCTION GROUP, EINSTEIN AND LORENTZ ANOMALIES

Kuang-chao CHOU(周光召), Han-ying GUO(郭汉英)
and Ke WU(吴 可
Institute of Theoretical Physics, Academia Sinica,
P.O. Box 2735, Beijing, China

Received July 4, 1984

Abstract

Based upon the properties of the characteristic classes and their Chern-Simons secondary characteristic classes, the "Abelian" anomalies in M^{2n+2}, the Euler-Heisenberg effective actions in M^{2n+1}, as well as the non-Abelian anomalies in M^{2n} for arbitrary gauge group and its reduction subgroup have been investigated thoroughly and the application to the gravitational anomalies is made. It is shown that the "Abelian" anomalies of such groups are equal to each other, their Euler-Heisenberg actions are also closely related to each other, and their non-Abelian anomalies are also equivalent if their common generating functional can be taken as a counter-term. For the gravitational anomalies we present the common generating functional for both non-Abelian Einstein and Lorentz anomalies in M^{4n+2} and show the relationship between them.

I. Introduction

In a previous note[1] we have generalized the investigations on some aspects of gauge anomalies[2,3] and shown that the dual of Abelian or singlet anomalous current 1-form in M^{2n+2} with respect to the gauge fields of left- and right-handed special unitary groups is actually a peculiar Chern-Simons 2n+1-form[4,5] which is gauge invariant. We have also shown that from the decomposition formula of this Chern-Simons secondary characteristic class in M^{2n+1} follow the anomaly free condition, the generating functional, the general expressions for the non-Abelian anomalies, and the gauge invariant Wess-Zumino-Witten type effective Lagrangian[6,7] in M^{2n} taken to be the boundary of M^{2n+1}. In addition to these results, it is shown[8,9] that the anomalous current in M^{2n+1} is the functional derivative of the Chern-Simons invariant, i.e., the Euler-Heisenberg effective action, with respect to the gauge potential. Therefore, it is found that there exist very profound connections among these subjects in different dimensional spacetimes and they share a common topological background—the Chern-Simons secondary characteristic classes.

We will point out in the present paper that these results can be extended to gauge theories with arbitrary gauge groups, especially, the theories with special orthogonal gauge groups which not only will deepen our investigation on the topological origin of the anomaly associated with the special orthogonal gauge group[2], but also will relate to the gravitational anomalies[10-12].

It is well known that in the case of quantum spinor field in external gravitational field in M^{4n+2} one should take into account both general coordinate transformations(i.e., the Einstein transformations) and the local Lorentz transformations. As far as the tangent space at a fixed point is concerned, the Einstein transformations are actually the gauge transformations of the group GL(4n+2, R) and the Lorentz transformations are those of the group SO(4n+2, R). Because the local orthogonal frame can be regarded as an element of the coset space GL(4n+2, R)/SO(4n+2, R) and its (double) covariant derivative is equal to zero, the gauge potential of GL(4n+2, R), i.e., the Riemann-Christoffel connection for torsion free gravitational field, can always be reduced to the Lorentz gauge potential, i.e., the Ricci rotating coefficients. Corresponding to these two sorts of transformations, one should deal with two sorts of anomalies, namely the Einstein anomaly and the Lorentz anomaly. It is clear that there should be some link between them. It will be proved in the following that, in general, there exists a link between the anomaly of a gauge group G and that of its reduction subgroup H via the coset space G/H.

In the present paper, we will first extend the results given in Ref.[1] to the arbitrary gauge groups. Secondly we will explore the relation between the anomalies of a gauge group and its reduction group in detail. Then we will analyse Einstein anomaly and Lorentz anomaly and point out that not only they share a common generating functional in the case of non-Abelian anomalies and can be expressed in terms of the local frame and the gauge potential of the counterpart but they are equivalent if the common generating functional can be taken as a counterterm[f1].

II. The Anomalies in Arbitrary Dimensional Spacetimes

Let U(x) be an element of a Lie group G or its non-singular matrix representation. Under the action of the gauge group $G_L \times G_R$, U(x) transforms as

$$U(x)\to g_L U(x) g_R^{-1} , \qquad (g_L, g_R)\in G_L\times G_R \tag{2.1}$$

and its covariant derivative is defined as

$$DU=dU+A_L U-UA_R , \tag{2.2}$$

where A_H(H=L,R) is the gauge potential 1-form with respect to group G_H and the corresponding field strength 2-form is

$$F_H=dA_H+A_H^2 ,$$

(wedge symbol suppressed). Under the gauge transformations (2.1)

$$A_H \to g_H d g_H^{-1} + g_H A_H g_H^{-1} ,$$
$$F_H \to g_H F_H g_H^{-1} , \tag{2.3}$$
$$DU \to g_L DU g_R^{-1} .$$

If the anomalous current 1-form consisting of field U(x) and gauge fields in 2n+2 dimensional Euclidean compactified spacetime M^{2n+2} satisfies the Abelian or singlet anomalous divergence equation

$$d{*}J(U,A_L,A_R)=P(F_L^{n+1})-P(F_R^{n+1}) , \tag{2.4}$$

where * is the Hodge star, $P(F^{n+1})$ is a homogenous invariant polynomial[4,5] of degree 2n+2 consisting of the field strength 2-forms, then by means of our method presented in Ref.[1], one can easily obtain the general solution of the singlet anomalous current. Introducing

$$\left.\begin{aligned} &A_t=A_R+tU^{-1}DU, \qquad 0\leqslant t\leqslant 1, \\ &F_t=tU^{-1}F_LU+(1-t)F_R-t(1-t)(U^{-1}DU)^2 , \end{aligned}\right\} \tag{2.5}$$

and making use of the relation between the characteristic polynomials and Chern-Simons secondary characteristic classes[4,5]

$$\left.\begin{aligned} &dQ(A',A)=P(F'^{\,n+1})-P(F^{n+1}) , \\ &Q^{(2n+1)}(A',A)=(n+1)\int_0^1 dt\ P\{(A'-A)[d[A+t(A'-A)] \\ &\qquad +[A+t(A'-A)]^2]^n\} , \end{aligned}\right\} \tag{2.6}$$

it follows that

$$\left.\begin{aligned} &J(U,A_L,A_R)=J_{min}(U,A_L,A_R)+\mathcal{T}(U,A_L,A_R) , \\ &{*}J_{min}(U,A_L,A_R)=Q(U^{-1}A_LU+U^{-1}dU,A_R)=(n+1)\int_0^1 dt P(U^{-1}DU,F_t^n) , \\ &{*}\mathcal{T}(U,A_L,A_R)=d\eta(U,A_L,A_R) , \end{aligned}\right\} \tag{2.7}$$

where ${*}\mathcal{T}(U,A_L,A_R)$ is an arbitrary gauge invariant exact form.

Following the method of Refs.[1,3], one can prove that the Chern-Simons secondary characteristic classes satisfy

$$Q(A',A)=-Q(A,A') ,$$
$$Q(A',A)=Q(A',0)-Q(A,0)+dG(A',A) , \tag{2.8}$$

where

$$G^{(2n)}(A',A)=n(n+1)\int_0^1 dt\int_0^{1-t} ds P\{AA'F^{n-1}(tA'+sA)\} . \tag{2.9}$$

Then it follows that

$$\left.\begin{aligned}&Q(U^{-1}A_LU+U^{-1}dU,A_R)=Q(A_L,0)-Q(A_R,0)+Q(U^{-1}dU,0)+dW(U,A_L,A_R),\\&dW(U,A_L,A_R)=d\{G(U^{-1}A_LU+U^{-1}dU,A_R)+G(A_L,UdU^{-1})\} .\end{aligned}\right\} \tag{2.10}$$

In fact, $Q(U^{-1}A_LU+U^{-1}dU,A_R)$ can be regarded as the gauge invariant Euler-Heisenberg effective Lagrangian in 2n+1-dimensional Euclidean compactified spacetime M^{2n+1}, and the corresponding dynamical current defined by the functional derivative of the Lagrangian with respect to the gauge potential is the anomalous current in M^{2n+1}. It should be pointed out that the gauge theory with topological mass term in 3-dimensions proposed by Deser, Jackiw and Templeton[8] and the recent discovery on the anomaly in odd dimensions[9] can be generalized in terms of such a gauge invariant Euler-Heisenberg effective Lagrangian and corresponding anomalous current.

Let us now turn to the case in 2n-dimensional Euclidean compactified spacetime M^{2n} which can be dealt with the boundary of M^{2n+1}, i.e., $\partial M^{2n+1}=M^{2n}$. From Eq.(2.10) in such an M^{2n+1} it follows that under the necessary and sufficient anomaly free condition in M^{2n}

$$Q(A_L,0)=Q(A_R,0) , \tag{2.11}$$

the gauge invariant object

$$\tilde{\Gamma}(U,A_L,A_R)=2\pi\int_{M^{2n+1}}Q^{2n+1}(U^{-1}dU,0)+2\pi\int_{\partial M^{2n+1}=M^{2n}}W^{2n}(U,A_L,A_R) \tag{2.12}$$

gives rise to the Wess-Zumino type effective anomalous action with gauge fields in M^{2n}. Otherwise

$$\tilde{\Gamma}(U,A_L,A_R) \quad\text{or}\quad 2\pi\int_{M^{2n+1}}[Q(A_L,0)-Q(A_R,0)] \tag{2.13}$$

defines the generating functional of the non-Abelian anomalies in the left and right antisymmetric form and under the gauge transformation (2.1) they generate the finite form of such anomalies as follows:

$$A_{L-R}(g_H,A_H)=2\pi\int_{M^{2n+1}}Q(g_L^{-1}dg_L,0)+2\pi\int_{M^{2n}}G(A_L,g_L^{-1}dg_L)-(L\to R) , \tag{2.14}$$

whose infinitesimal form is

$$\left.\begin{aligned} A_{L-R}(\varepsilon_H, A_H) &= \int_{M^{2n}} \varepsilon_{aL} G^a_L(A_L, T^a_L) - (L \leftrightarrow R) \ , \\ G^a_L(A_L, T^a_L) &= 2\pi n(n+1) \int_0^1 dt(1-t) P\{T^a_L d(A_L F^{n-1}(tA_L))\} \ . \end{aligned}\right\} \quad (2.15)$$

The finite form (2.14) contains the global informations about the anomalies such as so-called non-purturbative anomalies[7].

We can also define the generating functional of the non-Abelian anomalies in the axial current form rather than the (L,R)-antisymmetric form and obtain the finite anomalies of such a kind. In fact, in the case the gauge groups are special unitary groups, $W^{(4)}(U=1, A_L, A_R)$ is the counter term R_3 introduced by Bardeen[13] to transfer the anomaly from the (L,R)-antisymmetric form into the axial current one in 4-dimensional spacetime and correspondingly $W^{(2n)}(U=1, A_L, A_R)$ should be such anomaly transfer term for the gauge fields of arbitrary group in M^{2n}.

We have so far completed the generalization of our analysis on the anomalies with respect to the special unitary gauge group in arbitrary dimensions[1] to the arbitrary gauge groups. We have seen that the Chern-Simons secondary characteristic classes play a very important role in the course of analyses.

III. Anomalies of a Gauge Group and Its Reduction Group

We now establish some relations between the anomalies of a gauge group G and its reduction group H as well as the coset space G/H.

Let $g \in G$, $h \in H$ and a coset field $\phi(x) \in G/H$. In general, we have decomposition rule and transformation property

$$g = \phi h \ , \quad (3.1)$$

$$\phi \to \phi' = g\phi h^{-1} \ . \quad (3.2)$$

If G and H are gauged, the covariant derivative of $\phi(x)$-field should be defined as

$$D\phi = d\phi + \Gamma\phi - \phi\gamma \ , \quad (3.3)$$

where Γ and γ is the gauge potential 1-form of group G and H respectively. It is well known[14] that the necessary and sufficient condition for local reduction of Γ to γ is the covariant derivative (3.3) to be vanishing, then we have

$$\left.\begin{aligned} \Gamma &= \phi\gamma\phi^{-1} + \phi d\phi^{-1} \equiv {}^{\phi}\gamma \ , \\ \gamma &= \phi^{-1}\Gamma\phi + \phi^{-1}d\phi \equiv {}^{\phi^{-1}}\Gamma \ . \end{aligned}\right\} \quad (3.4)$$

Furthermore, it can be seen that the transformation (3.2) and covariant derivative (3.3) are similar to Eqs.(2.1) and (2.2) respectively, we are able to establish a relation between the anomalies of G and H, the reduction group of G, with the help of the results obtained in the last section.

Let us start with the singlet anomalies in M^{2n+2}. If there does not exist the reduction relation between Γ and γ, the singlet anomalies themselves appear as anomalous divergence of a singlet anomalous current

$$\left.\begin{aligned} d{*}J(\phi,\Gamma,\gamma)&=P(F^{n+1}(\Gamma))-P(F^{n+1}(\gamma)) ,\\ {*}J(\phi,\Gamma,\gamma)&=Q(\phi^{-1}\Gamma,\gamma)+d\eta(\phi,\Gamma,\gamma) . \end{aligned}\right\} \tag{3.5}$$

However, if there does exist the reduction relation (3.4) between Γ and γ, it follows that

$$Q(\phi^{-1}\Gamma,\gamma)=0 , \tag{3.6}$$

$$P(F^{n+1}(\Gamma))=P(F^{n+1}(\gamma)) . \tag{3.7}$$

This means that the singlet anomalies of group G and its reduction group H are completely equivalent to each other.

Similar to Eq.(2.10), in general, we have

$$\begin{aligned} Q(\phi^{-1}\Gamma,\gamma)&=Q(\Gamma,0)-Q(\gamma,0)+Q(\phi^{-1}d\phi,0)+dW(\phi,\Gamma,\gamma) ,\\ dW(\phi,\Gamma,\gamma)&=d(G(\phi^{-1}\Gamma,\gamma)+G(\Gamma,\phi d\phi^{-1})) . \end{aligned} \tag{3.8}$$

Under the reduction relation (3.4), it follows that

$$Q(\Gamma,0)=Q(\gamma,0)-Q(\phi^{-1}d\phi,0)-dG(\Gamma,\phi d\phi^{-1}) , \tag{3.9}$$

where $G(\Gamma,\phi d\phi^{-1})$ is given by Eq.(2.9). If we take the Chern-Simons secondary characteristic classes as the Euler-Heisenberg effective Lagrangian in M^{2n+1}, the Eq.(3.9) gives rise to a relation between the two Euler-Heisenberg actions with respect to the gauge fields of G and of its reduction group H. Obviously, when the scalar field $\phi(x)$ as an element of the coset space G/H has non-trivial topology and the boundary of M^{2n+1} is not empty, i.e., $\partial M^{2n+1}\neq\emptyset$, the two Euler-Heisenberg actions are not equivalent to each other.

When the boundary of manifold M^{2n+1} is the spacetime M^{2n} under consideration, according to the analyses in the last section

$$2\pi\int_{M^{2n+1}} Q(\Gamma,0) \quad\text{and}\quad 2\pi\int_{M^{2n+1}} Q(\gamma,0) \tag{3.10}$$

are the generating functionals of the non-Abelian anomalies of G and H in $M^{2n}=\partial M^{2n+1}$ respectively. Because the gauge potential γ of the subgroup H can be kept invariant under the gauge transformation of group G and vice versa, one can also take

$$\tilde{\Gamma}(\phi,\Gamma,\gamma)=2\pi\int_{M^{2n+1}} Q(\phi^{-1}d\phi,0)+2\pi\int_{M^{2n}} G(\Gamma,\phi d\phi^{-1})$$

$$=2\pi\int_{M^{2n+1}} (Q(\gamma,0)-Q(\Gamma,0)) \tag{3.11}$$

as the common generating functional for the non-Abelian anomalies of both group G and reduction group H in M^{2n}, i.e., under the gauge transformation of G, it generates the non-Abelian anomaly of G, while under the transformation of H, it generates the anomaly of H. However, the expressions for two non-Abelian anomalies are not the same, because the transformation properties of $\tilde{\Gamma}(\phi,\Gamma,\gamma)$ under the two sorts of transformation are not the same. In fact, the finite forms of two anomalies are

$$A(g,\Gamma)=2\pi\int_{M^{2n+1}} Q(g^{-1}dg,0)+2\pi\int_{M^{2n}} G(\Gamma,g^{-1}dg)\ , \tag{3.12}$$

$$A(h,\gamma)=2\pi\int_{M^{2n+1}} Q(hdh^{-1},0)+2\pi\int_{M^{2n}} G(\gamma,hdh^{-1})\ . \tag{3.13}$$

In the case of infinitesimal transformations , i.e., $g=1+\varepsilon(g)$, $h=1+\varepsilon(h)$, the two anomalies become the infinitesimal ones

$$A(\varepsilon_{(g)},\Gamma)=2\pi\int_{M^{2n}} \varepsilon_{a(g)} G^a_{(g)}(\Gamma,T^a_{(g)}),\quad a=1,\dots,\dim G; \tag{3.14}$$

$$A(\varepsilon_{(h)},\gamma)=-2\pi\int_{M^{2n}} \varepsilon\alpha_{(h)} G^\alpha_{(h)}(\gamma,T^\alpha_{(h)}),\quad \alpha=1,\dots,\dim H; \tag{3.15}$$

where $T^a_{(g)}, T^\alpha_{(h)}$ are the generators of the Lie algebras of group G and subgroup H respectively, $G^a_{(g)}(\Gamma,T^a_{(g)})$ and $G^\alpha_{(h)}(\gamma,T^\alpha_{(h)})$ are given by the second equation of (2.15), for example,

$$G^a_{(g)}(\Gamma,T^a_{(g)})=2\pi n(n+1)\int_0^1 dt(1-t)P\{T^a_{(g)}d(\Gamma F^{n-1}(t\Gamma))\}\ . \tag{3.16}$$

It is clear that neither finite form nor infinitesimal one the non-Abelian anomalies of group G and of reduction group $H\subset G$ are not the same.

It should be pointed out that making use of the decomposition rule (3.1) of an element g of group G and the reduction relation (3.4) between the gauge potentials Γ and γ, the non-Abelian anomalies of G (the G-anomaly) can always be reexpressed with the element of coset space G/H, ϕ, the element of subgroup H,h, and the gauge potential γ; and conversely, the H-anomalies can also be rewritten in terms of ϕ, g, and Γ. Does this fact mean that the two anomalies are equivalent to each other? One may suppose whether the common generating functional of the anomalies can be redefined so as to eliminate either the G-anomaly or the H-anomaly or transfer one to its counterpart. In order to transfer, for example, the G-anomaly to the H-anomaly one needs to regard

$$2\pi\int_{M^{2n+1}} Q(\Gamma,0)=2\pi\int_{M^{2n+1}} Q({}^{\phi}\gamma,0) \tag{3.17}$$

as the G-anomaly transfer term other than the generating functional and to add this term to the common generating functional $\tilde{\Gamma}(\phi,\Gamma,\gamma)$ so that under the gauge transformation of G the new generating functional is invariant. However, from the definition (3.11) of $\tilde{\Gamma}(\phi,\Gamma,\gamma)$, such a new generating functional is nothing but the generating functional of H-anomaly. In other words, the anomaly transfer of this sort works only in the sense that their common generating functional $\tilde{\Gamma}(\phi,\Gamma,\gamma)$ can be taken as counterterm.

IV. Einstein Anomaly and Lorentz Anomaly

For the sake of simplicity, we merely analyse in the present paper the Einstein anomalies and the Lorentz anomalies associated with the tangent space at a given point in gravitational field without torsion, i.e., the anomaly of general linear group and that of its reduction group via local frame, the special orthogonal group. In this case, we can directly make use of the results obtained in last section. And the generalizations to the Einstein transformations involving a Lie derivative besides the gauge transformations and to the covariant forms of the anomalies[12] as well as to the case with torsion can also be done without difficulties and obscurities in principle.

We start with the singlet anomalies in M^{4n+4}[f3]. It is plain that the singlet Einstein anomaly and Lorentz one are determined by the n+1st-Pontrjagin classes $P(F^{2n+2}(\Gamma))$ and $P(F^{2n+2}(\gamma))$ with respect to the groups GL(4n+4, R) and SO(4n+4, R) in M^{4n+4} respectively[f4]

$$P(F^{2n+2})=\beta_{n+1}\,\mathrm{Tr}(F^{2n+2})\ ,\quad \beta_{n+1}=((2\pi)^{2n+2}(2n+2)!)^{-1} \tag{4.1}$$

and Γ is taken to be the Riemann-Christoffel and γ the Ricci connection 1-form without torsion in M^{4n+4},

$$\Gamma=\Gamma_k dx^k\ ,\qquad \Gamma_k=(\Gamma^i_{jk})_{i,j=1,\dots,\dim M};$$

$$\gamma=\gamma_k dx^k\ ,\qquad \gamma_k=(\gamma^a_{bk})_{a,b=1,\dots,\dim M}.$$

Because there exists the reduction relation between Γ and γ via the local moving frame

$$E=(e_a^j)_{a,j=1,\ldots,\dim M}; \qquad E^{-1}=(e_j^a)_{a,j=1,\ldots,\dim M};$$

we have

$$DE=dE+\Gamma E-E\gamma=0 \ , \tag{4.2}$$

so the two singlet anomalies are the same.

In M^{4n+3}, we also have the relation between the corresponding Chern-Simons secondary classes

$$Q(\Gamma,0)=Q(\gamma,0)-Q(E^{-1}dE,0)-dG(\Gamma,EdE^{-1}) \tag{4.3}$$

with Γ,γ and E taken to be the objects in M^{4n+3}, and

$$Q(\Gamma,0)=2\beta_{n+1}(n+1)\int_0^1 dt\mathrm{Tr}(\Gamma F^{2n+1}(t\Gamma)), \tag{4.4}$$

$$G(\Gamma,EdE^{-1})=\beta_{n+1}(2n+2)(2n+1)\int_0^1 dt\int_0^{1-t}ds$$
$$\mathrm{Tr}\{EdE^{-1},\Gamma,F^{2n}(t\Gamma+sEdE^{-1})\} \ . \tag{4.5}$$

The expressions, similar to Eq.(4.4), for $Q(\gamma,0)$ and $Q(E^{-1}dE,0)$ can also be written. As is analysed in the last section, the Eq.(4.3) gives rise to the relation between the Euler-Heisenberg effective actions of gauge groups GL(4n+3,R) in M^{4n+3}.

Finally, we analyse the non-Abelian Einstein anomaly and Lorentz anomaly in M^{4n+2} and assume that M^{4n+2} can be regarded as the boundary of M^{4n+3}, i.e. $M^{4n+2}=\partial M^{4n+3}$[f2]. According to the analyses in last section, we can define

$$\tilde{\Gamma}(E,\Gamma,\gamma)=2\pi\int_{M^{4n+3}} Q(E^{-1}dE,0)+2\pi\int_{M^{4n+2}} G(\Gamma,EdE^{-1}) \tag{4.6}$$

as the common generating functional of both Einstein and Lorentz anomalies with Γ,γ, and E taken to be the objects in M^{4n+2}. Under the gauge transformations of GL(4n+2,R) and SO(4n+2,R), it generates Einstein anomaly and Lorentz anomaly respectively. Their finite forms are of

$$A_E(g,\Gamma)=2\pi\int_{M^{4n+3}} Q(g^{-1}dg,0)+2\pi\int_{M^{4n+2}} G(\Gamma,g^{-1}dg), \quad g\in GL(4n+2,R); \tag{4.7}$$

$$A_L(h,\gamma)=2\pi\int_{M^{4n+3}} Q(hdh^{-1},0)+2\pi\int_{M^{4n+2}} G(\gamma,hdh^{-1}), \quad h\in SO(4n+2,R); \tag{4.8}$$

and $G(\Gamma,g^{-1}dg)$, $G(\gamma,hdh^{-1})$ have expressions similar to Eq.(4.5). The infinitesimal forms of the anomalies can also be easily written as

$$A_E(\varepsilon_{(g)}\Gamma)=2\pi\int_{M^{4n+2}} \varepsilon_{a(g)}G^a_{(g)}(\Gamma,T^a_{(g)}), \quad a=1,\ldots,\dim GL(4n+2,R), \tag{4.9}$$

$$A_L(\varepsilon_{(h)}\gamma)=-2\pi\int_{M^{4n+2}}\varepsilon_{\alpha(h)}G^{\alpha}_{(h)}(\gamma,T^{\alpha}_{(h)}),\quad \alpha=1,\ldots,\dim SO(4n+2,R), \tag{4.10}$$

where

$$G^{a}_{(g)}(\Gamma,T^{a}_{(g)})=\beta_{n+1}(2n+2)(2n+1)2\pi\int_0^1 dt(1-t)\mathrm{Tr}\{T^{a}_{(g)}d(\Gamma F^{2n}(t\Gamma))\}, \tag{4.11}$$

$$G^{\alpha}_{(h)}(\gamma,T^{\alpha}_{(h)})=\beta_{n+1}(2n+2)(2n+1)2\pi\int_0^1 dt(1-t)\mathrm{Tr}\{T^{\alpha}_{(h)}d(\gamma F^{2n}(t\gamma))\}\ , \tag{4.12}$$

which follow directly from Eq.(3.16) and the definition (4.1) of the Pontrjagin class.

Obviously, because of the reduction condition (4.2), the Einstein anomaly either in finite or in infinitesimal form can be rewritten in terms of the local frame and the objects consisting of the Lorentz anomaly and vice versa. Therefore, it is possible to redefine the generating functional so as to eliminate one of the two anomalies, such as the Lorentz anomaly and to transfer it to its counterpart. However, because of the same reason analysed in the last section, this requires that the common generating functional can be taken as a counterterm and only in this sense the gravitational and Lorentz anomalies are equivalent.

Footnotes

[f1] After we finished this work, we received an interesting preprint by Bardeen and Zumino[12]. Besides many important points about consistent and covariant anomalies as well as gravitational anomalies they explained, they also claimed that the Einstein and Lorentz anomalies in non-Abelian case are equivalent.

[f2] This means that M^{2n} has no boundary, like 2n-dimensional sphere S^{2n} etc., otherwise M^{2n} cannot be regarded as the boundary of an M^{2n+1}. We will leave the case of M^{2n} with boundary for further investigation.

[f3] The reasons why gravitational anomalies appear only in M^{4n+4} and M^{4n+2} for Abelian and non-Abelian case respectively are the same as those explained in Ref.[12].

[f4] The polynomial P, in the general case of Dirac particle in gravitational field, should be taken as $\hat{A}(M)ch(F)$, the product of A roof genus $\hat{A}(M)$ and the Chern character ch(F). In this paper, for simplicity, we take only the Pontrjagin classes as the polynomial P.

References

[1] K.C. CHOU, H.Y. GUO, X.Y. LI, K. WU and X.C. SONG, Commun. Theor. Phys., 3(1984)491.

[2] K.C. CHOU, H.Y. GUO, K. WU and X.C. SONG, Phys. Lett., 134B(1984)67; Commun. Theor. Phys., 3(1984)73; 125.

[3] B. Zumino, "Les Houches lectures 1983"; to be published by North-Holland, ed. R. Stora and B. DeWitt;
B. Zumino, Y.S. WU and A. Zee, Nucl. Phys., B239(1984)477.

[4] S.S. Chern, "Complex Manifolds without Potential Theory". Springer-Varlag, (1979).

[5] T. Eguchi, P.B. Gilkey and A.J. Hanson, Phys. Reports, 66(1980)213.

[6] J. Wess and B. Zumino, Phys. Lett., 37B(1971)95.

[7] E. Witten, Nucl. Phys., B223(1983)422.

[8] S. Deser, R. Jackiw and S. Templeton, Phys. Rev. Lett., 48(1982)975; Ann. Phys., (N.Y) 140(1982)372.

[9] A.J. Niemi and G.W. Semenoff, Phys. Rev. Lett., 51(1983)2077; A.N. Redlich, ibid., 52(1984)18.

[10] L. Alvarez-Gaume and E. Witten, Nucl. Phys., B234(1984)269.

[11] L.N. CHANG and H.T. NIEH, Stony Brook preprint ITP-SB-84-25.

[12] W.A. Bardeen and B. Zumino, Nucl. Phys. B244(1984)421.

[13] W.A. Bardeen, Phys. Rev., 184(1969)1848.

[14] S. Kobayashi and K. Nomizu, "Foundations of Differential Geometry", Vol.1, Interscience(1963).

Commun. in Theor. Phys. (Beijing, China) *Vol.4, No.1 (1985)* 123-127

ON THE TWO DIMENSIONAL NON-LINEAR σ MODEL WITH WESS-ZUMINO TERM II QUANTUM THEORY

CHOU Kuang-chao(周光召) and DAI Yuan-ben(戴元本)
Institute of Theoretical Physics, Academia Sinica, P.O.Box 2735, Beijing, China

Received August 30, 1984

Abstract

The canonical quantization is carried out for the two dimensional non-linear σ model with Wess-Zumino term. It is shown that the currents in this model satisfy a Kac-Moody algebra for arbitrary values of the coupling constant.

Last year the two dimensional non-linear σ model with Wess-Zumino term was investigated in several articles. Using the canonical quantization in the light-cone formulation Witten[1] found that currents in this model at the special value $\lambda_0^2 = \frac{4\pi}{n}$ of the coupling constant satisfy a Kac-Moody algebra which is of the same form as that for currents in the zero-mass free fermion theory. Taking into consideration the uniqueness of the representation of the Kac-Moody algebra for n=1, this result implies an isomorphism between Hilbert spaces for these two models at n=1. Witten conjectured that at these special values of constants this model is equivalent to the zero-mass free fermion model. This problem was also investigated in Refs.[2,3] with the method of functional integration. It was found that the physical equivalence between these two theories at $n = \frac{4\pi}{\lambda_0^2} = 1$ is true only in some restricted sense. In Ref. [4] we found some simple isomorphism for classical solutions of this model at different values of the constants λ_0 and n. In this article canonical quantization of this model for arbitrary values of constants λ_0 and n is carried out. We find that the currents in this model at arbitrary values of λ_0 and n≠0 suitably modified satisfy a Kac-Moody algebra which is of the same form as that for currents in the zero-mass free fermion theory. This result establishes an isomorphism between the quantum theories of these two models at n=1 and arbitrary values of λ_0.

The action of the model is of the following form

$$S=\int d^2x(\mathcal{L}_0+n\gamma) \quad , \tag{1}$$

where

$$\mathcal{L}_0 = \frac{1}{4\lambda_0^2}\mathrm{Tr}(\partial^\mu g^{-1}\cdot\partial_\mu g) \ , \tag{2}$$

$$\Gamma=\int d^2x\gamma = \frac{1}{24\pi}\int_D d^3x\varepsilon^{ijk}\mathrm{Tr}(g^{-1}\partial_i g\cdot g^{-1}\partial_j g\cdot g^{-1}\partial_k g) \ , \tag{3}$$

where D is a three dimensional disk with two dimensional space-time S^2 as its boundary and g is an element of the O(N) matrix group. The equation of motion derived from this Lagrangian is the following

$$\frac{1}{2\lambda_0^2}\partial^\mu A_\mu - \frac{n}{8\pi}\varepsilon^{\mu\nu}\partial_\mu A_\nu=0 \ , \tag{4}$$

where

$$A_\mu=g^{-1}\partial_\mu g \tag{5}$$

satisfies the curvatureless condition

$$\partial_\mu A_\nu-\partial_\nu A_\mu+[A_\mu,A_\nu]=0 \ . \tag{6}$$

For canonical quantization of the theory one needs to know the Poisson brackets of field quantities. One can derive from the commutation relations written down below that the constraint $g^T g=I$ commutes with the Hamiltonian. Therefore, the constraint is of the first kind and can be imposed on physical states without the need of introducing Dirac brackets. As a result we can write down the Poisson brackets as if the matrix elements g_{ab} are independent generalized coordinates.

Let

$$\pi_{ab} = \frac{\partial\mathcal{L}}{\partial\dot{g}_{ab}} = \pi_{oab}+n\frac{\partial\gamma}{\partial\dot{g}_{ab}} \tag{7}$$

where

$$\pi_0 = \frac{\partial\mathcal{L}_0}{\partial\dot{g}} = \frac{1}{2\lambda_0^2}(\partial_0 g^{-1})^T \ . \tag{8}$$

Taking into consideration that γ is linear in $\dot{g}$ and changes sign under the replacement $(\partial_x g_{ab},\dot{g}_{cd})\to(\dot{g}_{ab},\partial_x g_{cd})$ we can obtain the following formulas for the Hamiltonian and the momentum of the fields

$$H=\int\mathrm{Tr}(\pi^T\dot{g}-\mathcal{L})dx=\int\mathrm{Tr}(\pi_0^T\dot{g}-\mathcal{L}_0)dx$$
$$= -\frac{1}{4\lambda_0^2}\int dx\mathrm{Tr}(A_0^2+A_1^2) \ , \tag{9}$$

$$P=-\int\mathrm{Tr}(\pi^T\partial_x g)dx=-\int\mathrm{Tr}(\pi_0^T\partial_x g)dx$$
$$=\frac{1}{2\lambda_0^2}\int dx\mathrm{Tr}(A_0A_1) \ . \tag{10}$$

From the Poisson brackets between g_{ab} and the conjugate momenta in the discrete space of lattice spacing ϵ

$$P_{ab}=\frac{\partial L}{\partial g_{ab}}=\epsilon\pi_{ab}\ ,$$

we find the Poisson brackets between g_{ab} and π_{oab} as the following

$$\{g_{ab}(x_n),\ g_{cd}(x_m)\}=0\ , \tag{11}$$

$$\{g_{ab}(x_n),\ \pi_{ocd}(x_m)\}=\delta_{ac}\delta_{bd}\delta_{mn}\epsilon^{-1}\ , \tag{12}$$

$$\{P_{oab}(x_n),\ P_{ocd}(x_m)\}=\Big\{P_{ocd}(x_m),\ \epsilon n\frac{\partial}{\partial\dot{g}_{ab}(x_n)}\sum_{\ell}\gamma_\ell\Big\}$$
$$+\Big\{\epsilon n\frac{\partial}{\partial\dot{g}_{cd}(x_m)}\sum_{\ell}\gamma_\ell,\ P_{oab}(x_n)\Big\}$$
$$=n\epsilon\Big(\frac{\partial^2}{\partial g_{ab}(x_n)\partial\dot{g}_{cd}(x_m)}-\frac{\partial^2}{\partial g_{cd}(x_m)\partial\dot{g}_{ab}(x_n)}\Big)\sum_{\ell}\gamma_\ell\ , \tag{13}$$

where

$$\gamma_\ell=\gamma(g(x_\ell),\ \partial_x g(x_\ell),\ \dot{g}(x_\ell))\ ,$$

$$\partial_x g(x_\ell)=\frac{1}{\epsilon}(g(x_{\ell+1})-g(x_\ell))\ .$$

The evaluation of the right-hand side of Eq.(13) is facilitated by noting that the change of Γ under the variation of g is the following,

$$\frac{\delta\Gamma}{\delta g_{ab}(x_n)}=\Big(\frac{\partial}{\partial g_{ab}(x_n)}-\frac{d}{dt}\frac{\partial}{\partial\dot{g}_{ab}(x_n)}\Big)\epsilon\sum_{\ell}\gamma_\ell$$
$$=\Big(\frac{\partial}{\partial g_{ab}(x_n)}-\dot{g}_{cd}(x_m)\frac{\partial^2}{\partial g_{cd}(x_m)\partial\dot{g}_{ab}(x_n)}\Big)\epsilon\sum_{\ell}\gamma_\ell\ . \tag{14}$$

From Eqs.(13) and (14) we find

$$\{P_{oab}(x_n),\ P_{ocd}(x_m)\}=n\frac{\partial}{\partial\dot{g}_{cd}(x_m)}\frac{\delta\Gamma}{\delta g_{ab}(x_n)}\ . \tag{15}$$

Using the formula

$$\frac{\partial\Gamma}{\delta g_{ab}(x_n)}=-\epsilon\frac{1}{8\pi}\varepsilon^{\mu\nu}(\partial_\mu g^{-1}(x_n)\partial_\nu g(x_n)g^{-1}(x_n))_{ba}\ , \tag{16}$$

we obtain that

$$\{\pi_{oab}(x_n);\pi_{ocd}(x_m)\}=-\frac{n}{8\pi}(\partial_x g^{-1}_{ad}g^{-1}_{cb}$$

$$-g^{-1}_{ad}\partial_x g^{-1}_{cb})\delta_{mn}\epsilon^{-1}\ . \tag{17}$$

It is noted that using the formulas (9), (10) for H and P and Poisson brackets (11), (12) and (17) the effect of the Γ term in the theory appears explicitely only in the modification of the Poisson brackets between two π_0's in Eq.(17).

In the limit of continuous space we find from Eqs.(11), (12) and (20) the following Poisson brackets of A_μ

$$\{TrT^A\tilde{A}_1(x),\ TrT^B\tilde{A}_1(y)\}=0\ , \tag{18}$$

$$\{TrT^A\tilde{A}_1(x),\ TrT^B\tilde{A}_0(y)\}=Tr[T^A,T^B]\tilde{A}_1(x)\delta(x-y)$$

$$-\frac{1}{8\pi}Tr(T^AT^B)\delta'(x-y)\ , \tag{19}$$

$$\{TrT^A\tilde{A}_0(x),TrT^B\tilde{A}_0(y)\}=Tr[T^A,T^B](\tilde{A}_0(x)-n\tilde{A}_1(x))\ , \tag{20}$$

where T^A are generators of the O(N) group and

$$\tilde{A}_0=\frac{1}{2\lambda_0^2}A_0\ ,\qquad \tilde{A}_1=\frac{1}{8\pi}A_1\ . \tag{21}$$

Using Eqs.(18)-(21) it can be easily verified that the Poisson equations for A_1 and A_0 are identical with the curvatureless condition (6) and the Lagrangian equation of motion (4) respectively.

We now introduce the current

$$J_+=2i(\tilde{A}_0+n\tilde{A}_1) \tag{22}$$

and the current J_- conjugate to J_+ with respect to the space reflection

$$J_-=-2ig(\tilde{A}_0-n\tilde{A}_1)g^{-1}\ . \tag{23}$$

For $n\neq 0$ J_+'s and J_-'s for all g and x are distinct. The equal time commutation relations between J_+ and J_- obtained from the correspondence to Poisson brackets (18), (19) and (20) are the following

$$[TrT^AJ_+(x),\ TrT^BJ_+(y)]=2Tr[T^A,T^B]J_+(x)\delta(x-y)$$

$$-\frac{ni}{\pi}Tr(T^AT^B)\delta'(x-y)\ , \tag{24}$$

$$[TrT^AJ_-(x),\ TrT^BJ_-(y)]=2Tr[T^A,T^B]J_-(x)\delta(x-y)$$

$$+\frac{ni}{\pi}Tr(T^AT^B)\delta'(x-y)\ , \tag{25}$$

$$[TrT^AJ_-(x),\ TrT^BJ_+(y)]=0\ . \tag{26}$$

Commutation relations (24), (25) and (26) are identical in form with that for currents in the n-flavor zero-mass free fermion model and form the direct sum of two Kac-Moody algebras.

At the special value of the coupling constant, we have from Eqs.(9) and (10)

$$H\pm P = \frac{\lambda_0^2}{4}\int dx \mathrm{Tr} J_\mp^2 \quad . \tag{27}$$

Since operators J_+ and J_- commute it follows from Eq.(27) that they depend only on $x_+ = \frac{1}{\sqrt{2}}(t+x)$ and $x_- = \frac{1}{\sqrt{2}}(t-x)$ respectively. After the rescaling $J_\pm \to \sqrt{2} J_\mp$, the equal time commutation relations (24), (25) and (26) reduce to that obtained by Witten in Ref.[1] using light-cone formalism for the special value $\lambda_0^2 = \frac{4\pi}{n}$.

As noted by Witten[1] since there is no operator commuting with all $J_\pm(x)$'s, the Hilbert space of the model is an irreducible representation of the direct sum of two Kac-Moody algebras spanned by $J_+(x)$ and $J_-(x)$. Since the representation of Kac-Moody algebra is essentially unique for n=1 our results imply an isomorphism between the Hilbert space of the non-linear σ model at n=1 and arbitrary value of λ_0 and that of the one-flavor zero-mass free fermion model. This, however, does not imply the complete physical equivalence of these two models, because formulas for physical quantities like H and P are not identical in form for these two models.

Note that J_- is actually the charge density of the current

$$-ig\left(\frac{1}{\lambda_0^2}A_\mu - \frac{n}{4\pi}\epsilon_{\mu\nu}A^\nu\right)g^{-1} \quad , \tag{28}$$

which is conserved (modula anomalies) by the equation of motion (4). One can derive commutation relations between time and space components of currents (28) from Eqs.(18), (19) and (20). These commutation relations are similar to those in the current algebra of the fermion theory except that the anomaly appears not only in the commutators of a space component and a time component but also in the commutators of two time components of currents.

References

[1] E. Witten, Comm. Math. Phys., 92(1984)455.

[2] P.Di Vecchia, B. Durhuus and J.L. Petersen, Preprint of the Niels Bohr Institute NBI-HE-84-02; P. Di Vecchia and P. Rossi, CERN Preprint Ref. TH. 3808-CERN.

[3] A.M. Polyakov and P.B. Wiegmann, Phys. Lett., B141(1984)223.

[4] CHOU Kuang-chao and DAI Yuan-ben, Commun. Theor. Phys., 3(1984)767.

Commun. in Theor. Phys. (Beijing, China) *Vol.5, No.4(1986) 359-364*

DERIVATION OF THE ANOMALOUS TERM IN VIRASORO ALGEBRA BY TOPOLOGICAL METHOD

WU Yue-liang(吴岳良), XIE Yan-bo(谢彦波)
and
ZHOU Guang-zhao(周光召)
Institute of Theoretical Physics, Academia Sinica,
P.O. Box 2735, Beijing, China

Received October 21, 1985

Abstract

A 2-cocycle is constructed which gives the anomalous term in the Virasoro algebra.

Recently a unified mathematical scheme in terms of cocycles was developed to describe the anomalies characterizing chiral fermions interacting with a Yang-Mills field[1,2]. Specifically, the 2-cocycle descending from the Chern density with two higher numbers of dimensions is the anomalous fixed time commutator of gauge group generators[3,4].

It is well known in the string theory that the Virasoro algebra has an anomalous term[5]. In this case the generators L_n form a conformal algebra in the classical theory

$$[L_n, L_m]=(m-n)L_{n+m} , \qquad n=...-1,0,1,..., \tag{1}$$

when the string is quantized an additional anomalous term of the form

$$\delta_{n,-m} \frac{C}{12}(n^3-n) \tag{2}$$

appears in the commutator.

Physically, no gauge field has been introduced in the string theory. An interesting question is whether the above anomaly can be described by a 2-cocycle induced by an fictitious gauge field introduced purely as a tool of calculation. This note is addressed to this question.

The Virasoro algebra with anomalous term has the form

$$[L_n, L_m]=(m-n)L_{n+m}+C_{m,n} , \tag{3}$$

where $C_{m,n}$ are antisymmetric

$$C_{m,n}=-C_{n,m} \tag{4}$$

and satisfy an equation due to Jacobi indentity

$$(m-n)C_{m+n,\ell}+(n-\ell)C_{n+\ell,m}+(\ell-m)C_{\ell+m,n}=0 \ . \tag{5}$$

Solutions of Eqs.(4) and (5) can be easily found. They are

$$C_{m,n}=(m-n)b(m+n)+d\delta_{m+n,0}\,m^3 \ , \tag{6}$$

where b(m+n) is an arbitrary function of (m+n). The commutation relations (3) now become

$$[L_n,\ L_m]=(m-n)(L_{n+m}+b(m+n))+d\delta_{n+m,0}m^3 \ . \tag{7}$$

Making the change

$$L_n \longrightarrow L_n-b(n) \ , \tag{8}$$

we can rewrite Eq.(7) in the following form

$$[L_n,\ L_m]=(m-n)L_{m+n}+dm^3\delta_{m+n,0} \ . \tag{9}$$

The center in Virasoro algebra

$$\frac{C}{12}\delta_{m+n,0}(m^3-m) \tag{10}$$

can be cast into the form of Eq.(9) by a change of the generator

$$L_0 \longrightarrow L_0+\frac{C}{24} \tag{11}$$

and

$$d=\frac{C}{12} \ . \tag{12}$$

Our next problem is to find a 2-cocycle which produces this anomalous center term. For this purpose we consider a representation of L_n

$$L_n=z^n(z\frac{\partial}{\partial z}+an+f(z)) \ . \tag{13}$$

The arbitrary function $f(z)$ can be eliminated by a transformation of the wave function $\phi(z)$

$$\phi(z) \longrightarrow e^{-\int\frac{f(z)}{z}dz}\phi(z) \ . \tag{14}$$

Therefore without loss of generality we can take $f(z)=0$.

Under a finite transformation the wave function $\phi(z)$ changes into

$$\phi'(z)=e^{i\sum_n \varepsilon_n L_n}\phi(z)=U_\varepsilon(z)\phi(z') \ , \tag{15}$$

where ε_n ($n=\ldots-1,0,1,\ldots$) are group parameters and

$$L_n=z^n(z\frac{\partial}{\partial z}+an) \ , \tag{16}$$

The coordinate

$$z'=e^{\sum_n \varepsilon_n z^{n+1}\frac{\partial}{\partial z}}z \tag{17}$$

can be obtained in the following way. Let us define a function $\tau(z)$ such that

$$\frac{d\tau}{dz}=\frac{1}{\sum_n \epsilon_n z^{n+1}} \ , \tag{18}$$

then

$$\sum_n \varepsilon_n z^{n+1}\frac{\partial}{\partial z}=\frac{\partial}{\partial \tau} \ . \tag{19}$$

The operator

$$e^{\sum_n \varepsilon_n z^{n+1}\frac{\partial}{\partial z}}=e^{\frac{\partial}{\partial \tau}} \tag{20}$$

is a displacement operator of the coordinate τ. Therefore

$$z'=e^{\frac{\partial}{\partial \tau}}z(\tau)=z(\tau(z)+1) \ . \tag{21}$$

As a special case we take

$$\varepsilon_\ell=\delta_{\ell,n}\varepsilon \ . \tag{22}$$

It is easily found that

$$z'=z(1-n\varepsilon z^n)^{-\frac{\ell}{n}}=z_\varepsilon \ . \tag{23}$$

In this special case the phase factor U(z) can also be found as

$$U_\varepsilon(z)=\frac{1}{(1-n\varepsilon z^n)^a} \ , \tag{24}$$

which is seen to satisfy the group relation

$$U_{\varepsilon'}(z)U_\varepsilon(z_{\varepsilon'})=U_{\varepsilon+\varepsilon'}(z) \ . \tag{25}$$

The presence of the phase factor U(z) indicates that besides the coordinate transformation a U(1) gauge field A(z) is needed to define a covariant derivative. Under the conformal transformation described by the Virasoro algebra the gauge potential transforms as

$$A(z)\longrightarrow A'(z')=A(z)+U_\varepsilon^{-1}(z)dU_\varepsilon(z) \ . \tag{26}$$

Here we use the exterior form and differentiation in a complex manifold.

To calculate the commutator anomaly we have to find the 2-cocycle descending from the Chern density in four dimensions

$$\omega_4=-\frac{1}{8\pi^2}\mathrm{Tr}F^2 \ , \tag{27}$$

where

$$F=dA+A^2 \tag{28}$$

is the field strength.

Using the standard procedure developed in Refs.[1-4] the anomalous term is proportional to

$$\frac{1}{8\pi^2}\int vdv \ , \tag{29}$$

where

$$v=U_\varepsilon d_\xi U_\varepsilon^{-1} \approx \sum_n \left\{ d\xi^{(1)}(\varepsilon_{1n}+\xi^{(2)}\varepsilon_{2n})+\xi^{(1)}d\xi^{(2)}\varepsilon_{2n} \right\} \cdot anz^n , \quad (30)$$

where $\xi^{(i)}$(i=1,2) are two parameters varying from 0 to 1 and ε_{in} are infinitesimal group parameters.

It is easily found that

$$vdv= \sum_{n,m} m.am \cdot anz^{n+m-1} dz\, d\xi^{(1)} d\xi^{(2)} (\varepsilon_{1n}\varepsilon_{2m}-\varepsilon_{1m}\varepsilon_{2n}) . \quad (31)$$

Integrating Eq.(31) over $\xi^{(i)}$ (i=1,2) and z we obtain for the nontrivial 2-cocycle

$$\frac{1}{8\pi^2}\int vdv = \frac{a^2\ell}{8\pi} m^3 \delta_{n+m,0} 2\varepsilon_{1n}\varepsilon_{2m} ,$$

where ℓ is the number of times that the line integral wraps around z=0. The factor $\frac{1}{2}a^2\ell m^3 \delta_{n+m,0}$ is the anomalous center term for the Virasoro algebra which has the same form deduced by directly solving the Jacobi indentity. This is not surprising since Eq.(5) is an infinitesimal form of the coclosed condition for a 2-cocycle. What we have done in this paper is to show that the solution of the coclosed condition can be obtained by constructing a fictitious gauge field and using the standard procedure starting from the Chern density.

References

[1] *B. Zumino, Y.S. Wu and A. Zee, Nucl. Phys. B239(1984)477; B. Zumino, Nucl. Phys. B253(1985)477.*

[2] *H.Y. GUO, B.Y. HOU, S.K. WANG and K. WU, Commun. Theor. Phys. 4(1985)233-251 (Beijing, China); K.C. Chou, Y.L. Wu and Y.B. Xie, Anomalies and the General Chern-Simons Cochain, "Proceedings of the Tenth Hawaii Conference in High Energy Physics" (1985).*

[3] *L.D. Faddeev, Phys. Lett. 145B(1984)81; J. Mickelsson, Commun. Math. Phys. 97(1985)361; S.Y. Jo, MIT preprint CIP, 1236(1985).*

[4] *R. Jackiw, Phys. Rev. Lett. 54(1985)159; B. Grossman, Phys. Lett. 152B(1985)92; K.C. Chou, Y.L Wu and Y.B. Xie, preprint AS-ITP-85-026; Y.S. Wu and A. Zee, Phys. Lett. 152B(1985)98; Bo-yu Hou and Bo-yuan Hou, Chinese Phys. Lett. (in press).*

[5] A. Neveu and J.H. Schwarz, Nucl. Phys. B31(1971)86; P.Ramond, Phys. Rev. D3(1971)2415; P. Goddard, J. Goldstone, C. Rebb and C.B. Thorn, Nucl. Phys. B56(1973)109.

Modern Physics Letters A Vol. 1 No. 1 (1986) 23–27

EFFECTIVE ACTION OF SIGMA MODEL ANOMALIES WITH EXTERNAL GAUGE FIELDS

YIE-LIANG WU, YAN-BO XIE and GUANG-ZHAO ZHOU
Institute of Theoretical Physics, Academia Sinica
P.O. Box 2735, Beijing, People's Republic of China

Received 28 January 1986

The nonlinear sigma model describes Goldstone bosons originating from spontaneous symmetry breaking. A set of local counterterms is found to shift the anomaly of the nonlinear sigma model to that of the original model with fermions interacting with external gauge fields. The 't Hooft consistency conditions are matched automatically.

Nonlinear sigma models and their effective actions are of special interest in recent years. They are useful not only in the context of chiral dynamics but also in the geometrical interpretation and possible compactification of some superstring theories.

We consider in this note those sigma models that arise from spontaneous symmetry breaking of a theory with symmetry group G down to a subgroup H. The original theory has anomalies when its fermions interact with chiral gauge fields. After symmetry breaking the fermions will interact with reduced gauge fields and produce different anomalies. Local counterterms then have to be added to shift the anomalies back to that of the original theory. A simple derivation will be given in this note of these counterterms including the interaction of the Goldstone bosons with the external gauge fields that has been neglected in the Refs. 1 and 2.

Consider a theory with symmetry group G spontaneously broken down to a subgroup H. The effective theory then describes Goldstone boson fields in the direction of the coset space G/H. Any group element $g \in G$ can be decomposed into the product of an element φ belonging to the coset G/H and an element h of the subgroup H:

$$g = \varphi h. \tag{1}$$

The Goldstone fields are then described by a function $\varphi(x) \in G/H$. Under a group transformation the element $g\varphi(x)$ can again be decomposed in the form of Eq. (1)

$$g\varphi(x) = \varphi(g, \varphi(x))h(g, \varphi(x)), \tag{2}$$

where $\varphi(g, \varphi(x)) \in G/H$ and $h(g, \varphi(x)) \in H$ satisfy the group multiplication law

$$\varphi(g', \varphi(g, \varphi(x))) = \varphi(g'g, \varphi(x)),$$
$$h(g', \varphi(g, \varphi(x)))h(g, \varphi(x)) = h(g'g, \varphi(x)). \tag{3}$$

Therefore the Goldstone field $\varphi(x)$ transforms as

$$\varphi(x) \to \varphi'(x) = g^{-1}\varphi(x)h(g, \varphi(x)). \tag{4}$$

The 1-form $\varphi(x)\,d\varphi^{-1}(x)$ is valued in the Lie algebra of G which can be decomposed into two parts

$$\varphi(x)\,d\varphi^{-1}(x) = H + K, \tag{5}$$

where H is a 1-form valued in the Lie algebra of the subgroup and K that of the coset G/H. They transform as

$$H \to H' = h(g, \varphi(x))(H + d)h^{-1}(g, \varphi(x)) \tag{6}$$

and

$$K \to K' = h(g, \varphi(x))Kh^{-1}(g, \varphi(x)). \tag{7}$$

The kinetic energy of the Goldstone boson fields is constructed to be

$$\mathrm{tr}(K_\mu K^\mu), \tag{8}$$

which is seen to be invariant under the transformation Eq. (7). Gauge fields have to be added if the symmetry group is local. In this case Eq. (5) changes to the following form

$$\varphi^{-1}(x)(d + A)\varphi(x) = H + K, \tag{9}$$

where A is a gauge potential transforming as

$$A \to A' = g(d + A)g^{-1}. \tag{10}$$

When fermions and gauge fields are present, they are described by a Lagrangian

$$\bar{\psi}_L i\gamma^\mu(\partial_\mu + A_\mu)\psi_L \tag{11}$$

before symmetry breaking, where left-handed ψ_L and A are representations of the group G. After symmetry breaking the fermions and the gauge fields become

$$\psi_{hL}(x) = \varphi^{-1}(x)\psi_L(x),$$
$$A_h(x) = \varphi^{-1}(x)(A(x) + d)\varphi(x), \tag{12}$$

and the corresponding Lagrangian takes the form

$$\bar{\psi}_{hL} i\gamma^\mu(\partial_\mu + A_{h\mu}(x))\psi_{hL}\,. \tag{13}$$

The effective action after symmetry breaking has to produce the same anomalies as the original action. This requires adding compensating terms to the effective Lagrangian of the nonlinear sigma model. As in the case of the Wess-Zumino-Witten effective action, a closed form can be found only in the space of one higher dimension.

Consider a form ρ_{2n-1} in $(2n-1)$ dimensions. What conditions does it have to satisfy in order to describe a Lagrangian in a $(2n-2)$-dimensional manifold M which is considered to be the only boundary of the $(2n-1)$-dimensional manifold S? There are three of them:

(i) Locally ρ_{2n-1} should be an exact form so that

$$d\rho_{2n-1} = 0\,.$$

(ii) Globally it should not depend on the manifold S chosen so far as it has the same boundary M. This implies that ρ_{2n-1} integrated over any closed manifold should equal 2π times an integer:

$$\int_{S_1} \rho_{2n-1} - \int_{S_2} \rho_{2n-1} = \oint \rho_{2n-1} = 2\pi m \tag{14}$$

such that the uniqueness of the quantum amplitude $e^{i\Gamma}$ is guaranteed.

(iii) Under an infinitesimal group transformation it produces the difference of the current anomalies of the original theory with the reduced sigma model.

Let us introduce a double matrix notation and write

$$Q(x) = \begin{pmatrix} 0 & \varphi(x) \\ \varphi^{-1}(x) & 0 \end{pmatrix}, \qquad \eta = \begin{pmatrix} 1 & 0 \\ 0 & -1 \end{pmatrix}, \tag{15}$$

$$\mathscr{A}(x) = \begin{pmatrix} A(x) & 0 \\ 0 & \varphi^{-1}(d + A)\varphi \end{pmatrix}. \tag{16}$$

We have

$$Q^{-1}(x) = Q(x), \tag{17}$$

$$\mathscr{A}^Q = Q(\mathscr{A}(x) + d)Q(x) = \mathscr{A}(x), \tag{18}$$

and

$$\mathscr{F}^Q = Q\mathscr{F}Q = d\mathscr{A}^Q + \mathscr{A}^{Q^2} = \mathscr{F} = d\mathscr{A} + \mathscr{A}^2. \tag{19}$$

The Chern density

$$C_{2n}{}^{\eta}(\mathscr{F}) = \alpha_n \operatorname{tr}(\eta \mathscr{F}^n), \qquad \alpha_n = \frac{i^n}{(2\pi)^n n!}, \tag{20}$$

vanishes since

$$Q\eta Q = -\eta. \tag{21}$$

The Chern-Simons secondary class descending from the Chern density in Eq. (20) is

$$C_{2n}{}^{\eta}(\mathscr{F}) = d\Omega_{2n-1,1}{}^{\eta}(\mathscr{A}, 0), \tag{22}$$

where

$$\Omega_{2n-1,1}{}^{\eta}(A, B) = -n\alpha_n \int_0^1 dt \operatorname{tr}((A - B)F_t{}^{n-1}\eta), \tag{23}$$

with

$$F_t = dA_t + A_t{}^2 \tag{24}$$

and

$$A_t = tA + (1-t)B. \tag{25}$$

It is easily seen that

$$\Omega_{2n-1,1}{}^{\eta}(\mathscr{A}, 0) = \Omega_{2n-1,1}(A, 0) - \Omega_{2n-1,1}(A_h, 0), \tag{26}$$

whose variation under a group transformation produces the difference of the current anomalies before and after symmetry breaking. From Eq. (18) we obtain

$$\Omega_{2n-1,1}{}^{\eta}(\mathscr{A}, 0) = \Omega_{2n-1,1}{}^{\eta}(\mathscr{A}^Q, 0) = -\Omega_{2n-1,1}{}^{\eta}(\mathscr{A}, Q\,dQ^{-1}). \tag{27}$$

Using the identity derived from the generalized Chern-Simons characteristic classes[3]

$$\Omega_{2n-1,1}{}^{\eta}(A, A_1) - \Omega_{2n-1,1}{}^{\eta}(A, A_2) + \Omega_{2n-1,1}{}^{\eta}(A_1, A_2) = d\Omega_{2n-2,2}{}^{\eta}(A, A_1, A_2), \tag{28}$$

where

$$\Omega_{2n-2,2}{}^{\eta}(A, A_1, A_2) = -\frac{1}{2}n(n-1)\alpha_n \int_0^1 dt \int_0^1 dt_1 \,\mathrm{str}(M_t{}^2 F_t{}^{n-2}\eta), \tag{29}$$

$$F_t = dA_t + A_t{}^2,\ M_t = d_t A_t,$$

and

$$A_t = tA + t_1 A_1 + (1 - t - t_1)A_2;$$

we get

$$\Omega_{2n-1,1}{}^{\eta}(\mathscr{A}, 0) = \frac{1}{2}(\Omega_{2n-1,1}{}^{\eta}(\mathscr{A}, 0) - \Omega_{2n-1,1}{}^{\eta}(\mathscr{A}, Q\,dQ^{-1}))$$

$$= \frac{1}{2}(\Omega_{2n-1,1}{}^{\eta}(Q\,dQ, 0) + d\Omega_{2n-2,2}{}^{\eta}(\mathscr{A}, 0, Q\,dQ^{-1}))$$

$$= \Omega_{2n-1,1}(\varphi^{-1}\,d\varphi, 0) + \frac{1}{2}d\Omega_{2n-2,2}{}^{\eta}(\mathscr{A}, 0, Q\,dQ^{-1}). \tag{30}$$

The first term in Eq. (30) is a closed form whose integral over the $(2n-1)$-dimensional manifold is the winding number of the mapping described by the coset function $\varphi(x)$. Therefore $2\pi \int \Omega_{2n-1,1}{}^{\eta}(\mathscr{A}, 0)$ satisfies all the conditions required and can be chosen to be the compensating term for the effective action of the nonlinear sigma model. When the external gauge fields are switched off, we get the same result obtained in Refs. 1 and 2, which is described by the first term of Eq. (30). Our result is more general since it contains interaction of the Goldstone bosons with the gauge fields, described by the second term of Eq. (30). From our derivation it is also apparent that if the original theory is free from anomalies so is the nonlinear sigma model arising from spontaneous symmetry breaking. If some fermions become heavy after symmetry breaking the low energy effective theory will be consistent only when these fermions do not contribute anomalies. One could therefore omit them from the effective action. The compensating term obtained above is still valid and 't Hooft's consistency conditions are automatically incorporated.

References

1. L. Alvarez-Gaumé and P. Ginsparg, Harvard preprint HUTP 85/A015 (1985).
2. J. Bagger, D. Nemeschansky and S. Yankielowicz, SLAC PUB3698 and Harvard preprint HUTP 85/A042 (1985).
3. H.-Y. Guo, B.-Y. Hou, S.-K. Wang and K. Wu, *Commun. Theor. Phys.*, **4** (1985) 233.

Commun. in Theor. Phys. (Beijing, China) Vol.7, No.1 (1987) 27-37

A SIMPLIFIED DERIVATION OF CHERN-SIMONS COCHAIN AND A POSSIBLE ORIGIN OF θ-VACUUM TERM

CHOU Kuang-chao(周光召), WU Yue-liang(吴岳良)
and XIE Yan-bo(谢彦波)

Institute of Theoretical Physics, Academia Sinica,
P.O. Box 2735, Beijing, China

Received June 24, 1985

Abstract

In this paper, it is shown that the cohomology of generalized secondary classes, the Faddeev type cohomology and the generalized gauge transformation can be easily obtained by expanding the Chern form according to the degree of the forms in its submanifolds and using the closed property of the Chern form. It is also shown that a -vacuum term in the effective Lagrangian arises when gauge field in the group manifold is present.

I. Introduction

Recently, H.Y. Guo et al.[1] have introduced higher order characteristic classes and cocycles which generalize the Weyl homomorphyism described by the coboundary of the Chern class. About the same time Faddeev[2] has constructed another kind of higher order cocycles based on the group manifold, which have been elaborated in Ref.[3]. In this paper, a simplified derivation of Chern-Simons cochain is given. We expand the Chern form according to the degree of the forms in its submanifolds. Only by using the closed property of the Chern form, can one obtain a Chern-Simons cochain which represents a descent relation between both submanifolds. In some special cases, one can easily obtain familiar results given in Refs.[1,2,3]. In Sec.II we show the case of the cohomology of the generalized secondary classes. In Sec.III, we give the Faddeev type cohomology. In Sec.IV, the generalized gauge transformation and the Chern form gauge potential are discussed. There the θ parameter in the θ-vacuum is interpreted as a line integral of the gauge potential in the group manifold.

II. Generalized Secondary Characteristic Class

The Chern density and the Chern-Simons secondary class are now familiar to theoretical physicists. In a 2n-dimensional space with gauge potential

$$A=A_\mu(x)dx^\mu \tag{2.1}$$

and curvature

$$F=dA+A^2 \quad , \tag{2.2}$$

the Chern density is defined as

$$\Omega_{2m}(F)=\frac{i^m}{m!(2\pi)^m}\,\mathrm{Tr}(F^m) \quad , \tag{2.3}$$

which is a 2m-form. Using Bianchi identity

$$dF=[F,A] \tag{2.4}$$

it is easily verified that $\Omega(F)$ is closed

$$d\Omega_{2m}(F)=0 \tag{2.5}$$

and can be expressed as

$$\Omega_{2m}(F)=d\Omega^0_{2m-1}(A,F) \quad , \tag{2.6}$$

where

$$\left.\begin{aligned} &\Omega^0_{2m-1}(A,F)=m\int_0^1 dt\,\Omega_{2m-1}(A,F_t^{m-1}) \quad , \\ &F_t=tdA+t^2A^2 \end{aligned}\right\} \tag{2.7}$$

is the Chern-Simons secondary class.

In a recent paper[1] the authors introduced generalized secondary characteristic classes relating to the higher order cochain and cohomology. We shall show in this note that their results can be easily deduced from Eq.(2.5) by the general property that the Chern density is a closed form.

Consider a manifold consisting of two submanifolds, one of which is the ordinary space manifold of dimension N_1 with coordinates x^μ, $\mu=1,2...N_1$. The other submanifold is one of parameters $\xi^i, i=1,2,...N_2$, The exterior differential operators are

$$d_x = dx^\mu \frac{\partial}{\partial x^\mu} \tag{2.8}$$

and

$$d_\xi = d\xi^i \frac{\partial}{\partial \xi^i} .$$

The gauge potential $\mathscr{A}$ can be decomposed into two parts

$$\mathscr{A} = A(x,\xi) + B(x,\xi) \; , \tag{2.9}$$

where

$$\left.\begin{aligned} &A(x,\xi) = A_\mu(x,\xi) dx^\mu \\ &\text{and} \\ &B(x,\xi) = B_i(x,\xi) d\xi^i \; . \end{aligned}\right\} \tag{2.10}$$

The curvature $\mathscr{F}$ now becomes

$$\mathscr{F} = (d_x + d_\xi)(A+B) + (A+B)^2 = F+G+M \; , \tag{2.11}$$

where

$$\left.\begin{aligned} &F = d_x A + A^2 \; , \\ &G = d_\xi B + B^2 \; , \\ &\text{and} \\ &M = d_x B + d_\xi A + AB + BA \; . \end{aligned}\right\} \tag{2.12}$$

The Chern denisty

$$\Omega_{2n}(\mathscr{F}) = \Omega_{2n}(F+G+M) \tag{2.13}$$

satisfies Eq.(2.5), i.e.,

$$(d_x + d_\xi)\Omega_{2n}(\mathscr{F}) = 0 . \tag{2.14}$$

By expanding $\Omega_{2n}(\mathscr{F})$ according to the degree of the forms in $d\xi^i$, we have

$$\Omega_{2n}(\mathscr{F}) = \sum_{m=0}^{2n} \Omega_{2n-m,m}(A,B) \; , \tag{2.15}$$

where $\Omega_{2n-m,m}(A,B)$ is a form of degrees 2n-m in x^μ and of degrees m in ξ^i.

Substituting Eq.(2.15) into Eq.(2.14) and comparing the degrees of the form, we obtain

$$\left.\begin{aligned} &d_x\Omega_{2n,0}(F)=0\ ,\\ &d_\xi\Omega_{2n,0}=-d_x\Omega_{2n-1,1}\ ,\\ &\vdots\\ &d_\xi\Omega_{2n-m,m}=-d_x\Omega_{2n-m-1,m+1}\ ,\\ &\vdots\\ &d_\xi\Omega_{0,2n}=0\ . \end{aligned}\right\}\tag{2.16}$$

As a special case, we choose

$$\left.\begin{aligned} &B(x,\xi)=0\ ,\\ &A(x,\xi)=A^{(0)}(x)+\sum_i\xi^i(A^{(i)}(x)-A^{(0)}(x))\ , \end{aligned}\right\}\tag{2.17}$$

then

$$\left.\begin{aligned} &G=0\ ,\\ &M=d_\xi A=\sum_i d\xi^i(A^{(i)}(x)-A^{(0)}(x))\ . \end{aligned}\right\}\tag{2.18}$$

By a change of the notation

$$\left.\begin{aligned} &\xi^i\longrightarrow t_i\ ,\\ &M\longrightarrow H\ ,\\ &\mathscr{F}\longrightarrow F+M\longrightarrow F_0\ , \end{aligned}\right\}\tag{2.19}$$

Eq.(2.16) can be easily seen to be just the theorem 1 proved in Ref.[1]. The cohomology and generalized secondary characteristic classes then follow by merely integrating over simplexes in the submanifold ξ.

III. Faddeev Type Cohomology

In this section we show that the cohomology of gauge groups in Faddeev's approach[2] can be deduced as a special case of Eqs.(2.6) and (2.7)

$$\Omega_{2n}(\mathscr{F})=(d_x+d_\xi)\Omega^0_{2n-1}(A\ ,\ B).\tag{3.1}$$

Expanding the Chern-Simons secondary class according to the degrees of the form in $d\xi^i$

$$\Omega^0_{2n-1}(A\ ,B\ )=\sum_{m=1}^{2n}\Omega^0_{2n-m,m-1}(A,B). \tag{3.2}$$

Substituting Eqs.(2.15) and (3.2) into Eq.(3.1) and comparing the degrees of the forms, we have

$$\left.\begin{aligned}
&\Omega_{2n,0}=d_x\Omega^0_{2n-1,0}\ ,\\
&\vdots\\
&\Omega_{2n-m,m}=d_x\Omega^0_{2n-m-1,m}+d_\xi\Omega^0_{2n-m,m-1}\ ,\\
&\vdots\\
&\Omega_{0,2n}=d_\xi\Omega^0_{0,2n-1}\ .
\end{aligned}\right\} \tag{3.3}$$

As a special case, we choose

$$\left.\begin{aligned}
&B(x,\xi)=U^{-1}(x,\xi)d_\xi U(x,\xi)\ ,\\
&A(x,\xi)=U^{-1}(x,\xi)(A(x)+d_x)U(x,\xi)\ ,
\end{aligned}\right\} \tag{3.4}$$

where $U(x,\xi)$ is an element belonging to the gauge group G. Then

$$\left.\begin{aligned}
&M=0,\\
&G=0,\\
&F=U^{-1}(x,\xi)F(x)U(x,\xi)\ ,\\
&F(x)=d_xA(x)+A^2(x)\ .
\end{aligned}\right\} \tag{3.5}$$

In this special case the sequence Eqs.(3.3) become

$$\left.\begin{aligned}
&\Omega_{2n,0}=d_x\Omega^0_{2n-1,0}\ ,\\
&\vdots\\
&d_\xi\Omega^0_{2n-m-1,m}=-d_x\Omega^0_{2n-m-2,m+1}\ ,\\
&\vdots\\
&d_\xi\Omega^0_{0,2n-1}=0\ .
\end{aligned}\right\} \tag{3.6}$$

Eqs.(3.6) are just the results obtained in Ref.[3]. Therefore one can choose

$$U(x,\xi)=h(\xi^1,g_1(x)h(\xi^2,g_2(x)h(\xi^3\ldots)\ldots), \tag{3.7}$$

where

$$\left.\begin{aligned} &h(\xi^i,g_i(x))=e^{\xi^i u_i(x)}\ ,\\ &g_i(x)=e^{u_i(x)}\in G\ ,\\ &h(0,g_i)=1,\quad h(1,g_i)=g_i\ . \end{aligned}\right\} \tag{3.8}$$

The Faddeev type cohomology then follows by integrating over the simplexes in the submanifold ξ consisting of points

$$\left.\begin{aligned} &P_i=(\xi^1,\xi^2\ldots,\xi^{i-1},\xi^i,\xi^{i+1},\ldots)\\ &\quad=(1,1,\ldots,1,0,1,\ldots),\\ &i=1,2,\ldots,\\ &P_0=(1,1,\ldots,1,1,1,\ldots). \end{aligned}\right\} \tag{3.9}$$

IV. Gauge Fields in the Group Space and the θ-vacuum

In this section, we consider another special case

$$\left.\begin{aligned} &G=d_\xi B+B^2\neq 0\ ,\\ &M=d_\xi A+d_x B+AB+BA=0\ ,\\ &F=d_x A+A^2\neq 0\ . \end{aligned}\right\} \tag{4.1}$$

In this case the sequence Eqs.(3.3) become

$$\left.\begin{aligned} &\Omega_{2n,0}=d_x\Omega_{2n-1,0},\\ &\vdots\\ &d_\xi\Omega^0_{2n-2k-1,2k-2}=-d_x\Omega^0_{2n-2k,2k-1}\ ,\\ &d_\xi\Omega^0_{2n-2k,2k-1}=-d_x\Omega^0_{2n-2k-1,2k}+\Omega_{2n-2k,2k}\ ,\\ &\vdots\\ &d_\xi\Omega^0_{0,2n-1}=0\ , \end{aligned}\right\} \tag{4.2}$$

where $\Omega_{2n-2k,2k}$ satisfies

$$\left.\begin{aligned} d_x\Omega_{2n-2k,2k}&=0 \ , \\ d_\xi\Omega_{2n-2k,2k}&=0 \ . \end{aligned}\right\} \qquad (4.3)$$

This case can also be regarded as a generalized gauge transformation. For example, by choosing

$$\left.\begin{aligned} B(x,\xi)&=C(\xi)+U^{-1}(x,\xi)d_\xi U(x,\xi)=C(\xi)+V(x,\xi) \ , \\ A(x,\xi)&=U^{-1}(x,\xi)(A(x)+d_x)U(x,\xi) \ , \end{aligned}\right\} \qquad (4.4)$$

we have

$$\left.\begin{aligned} G&=d_\xi B+B^2=d_\xi C(\xi) \ , \\ F&=U^{-1}(x,\xi)F(x)U(x,\xi) \ , \\ M&=d_\xi A+d_x V+VA+AV=0 \ , \end{aligned}\right\} \qquad (4.5)$$

where $C(\xi)$ is Abelian gauge potential only depending on ξ, so Chern form $\Omega_{2n-2k,2k}$ can be written as

$$\left.\begin{aligned} \Omega_{2n-2k,2k} &= \frac{i^n}{n!(2\pi)^2} P\ \mathrm{Tr}(F^{n-k}G^k) \\ &= \frac{i^n}{(n-k)!\ k!(2\pi)^n}(d_\xi C(\xi))^k \mathrm{Tr}F^{n-k} \\ &= d_\xi\Omega'_{2n-2k,2k-1} \ , \\ \Omega'_{2n-2k,2k-1} &= \frac{i^n}{(n-k)!k!(2\pi)^n} C(\xi)(d_\xi C(\xi))^{k-1}\mathrm{Tr}F^{n-k} \ . \end{aligned}\right\} \qquad (4.6)$$

Introducing a new Chern density

$$\bar{\Omega}^0_{2n-2k,2k-1}=\Omega^0_{2n-2k,2k-1}+\Omega'_{2n-2k,2k-1} \ , \qquad (4.7)$$

one can see that the sequence Eqs.(4.2) become

$$\left.\begin{aligned} \Omega_{2n,0}&=d_x\Omega^0_{2n-1,0} \ , \\ &\vdots \\ d_\xi\Omega^0_{2n-2k+1,2k-2}&=-d_x\bar{\Omega}^0_{2n-2k,2k-1} \ , \end{aligned}\right. \qquad (4.8)$$

$$d_\xi \bar{\Omega}^0_{2n-2k,2k-1} = -d_x \Omega_{2n-2k-1,2k} ,$$
$$\vdots$$
$$d_\xi \Omega^0_{0,2n-1} = 0 .$$

We now extend the definition of effective action in $x \in M^{2n-2}$ which should have no boundary.

$$\bar{\Gamma}(B,A) = 2\pi \int_{x\in M^{2n-2}} \int_{\xi \in M^1} \bar{\Omega}^0_{2n-2,1}(A,B)$$
$$= \tilde{\Gamma}(A,U) + \Gamma'(A,\theta) , \qquad (4.9)$$

where

$$\tilde{\Gamma} = 2\pi \int_{x\in M^{2n-2}} \int_{\xi\in M^1} \Omega^0_{2n-2,1} . \qquad (4.10)$$

It is just the Wess-Zumino-Witten effective action in M^{2n-2} manifold.

$$\Gamma'(A,0) = 2\pi \int_{x\in M^{2n-2}} \int_{\xi\in M^1} \Omega'_{2n-2,1}$$
$$= \theta \frac{i^{n-1}}{(n-1)!(2\pi)^{n-1}} \int_{x\in M^{2n-2}} \mathrm{Tr} F^{n-1} , \qquad (4.11)$$

where

$$\theta = i \int_{\xi\in M^1} C(\xi) , \qquad (4.12)$$

when $n=3$

$$F = \frac{1}{2} F^a_{\mu\nu} T^a dx^\mu \wedge dx^\nu$$

$$\Gamma'(\theta,A) = -\frac{1}{8\pi^2}\theta \int_{x\in M^4} \mathrm{Tr}F^2 = \theta \frac{1}{64\pi^2} \int d^4x \varepsilon_{\mu\nu\rho\sigma} F^{a\mu\nu} F^{a\rho\sigma} . \qquad (4.13)$$

It is just the ordinary θ-vacuum term in a four-dimensional Euclidean space.

From Eq.(4.12), one can see that the parameter θ for θ-vacuum is the line integration of Abelian gauge potential which corresponds to the phase factor (or 1-cocycle) of the representation of the translation group presented by gauge covariant operator in the ξ-space

$$\left.\begin{aligned} &T(\xi)=\exp(\xi\cdot D)\ , \\ &D=d_\xi+B\ , \\ &\psi'(x,\xi)=\Gamma(\xi)\psi(x,0)=\exp\left(\int_0^\xi B(x,\xi)\right)\psi(x,\xi) \\ &\qquad=\exp(i\theta)U(x,\xi)\psi(x,\xi)\ , \qquad U(x,\xi)\in G\ . \end{aligned}\right\} \quad (4.14)$$

We now introduce a new gauge potential and field strength in ξ-space through the Chern 1-form $\Omega_{2n-1,1}$ or $\Omega^0_{2n-2,1}$ and 2-form $\Omega_{2n-2,2}$ or $\Omega^0_{2n-3,2}$ by integrating over x when the ordinary gauge potential $B(x,\xi)$ vanishes. One will see that they are just the θ-vacuum and topological mass term comparing with that discussed in Ref.[5]. From Eqs.(3.3) one has

$$\int_{x\in M^{2n-1}}\Omega_{2n-1,1}(A,0)=\int_{x\in M^{2n-1}}d_\xi\Omega^0_{2n-1,0}(A,0) + \int_{x\in M^{2n-1}}d_x\Omega^0_{2n-2,1}(A,0). \quad (4.15)$$

Consider a boundaryless manifold M^{2n-1} by Stokes' theorem, it can be rewritten as

$$\int_{x\in M^{2n-1}}\Omega_{2n-1,1}(A,0)=d_\xi\int_{x\in M^{2n-1}}\Omega^0_{2n-1,0}(A,0)\ , \quad (4.16)$$

the Chern form gauge potential is defined as

$$\mathscr{B}=\theta\int_{x\in M^{2n-1}}\Omega_{2n-1,1}(A,0)=\theta d_\xi\int_{x\in M^{2n-1}}\Omega^0_{2n-1,0}(A,0). \quad (4.17)$$

It is a pure U(1) gauge potential. Consider the 1-cocycle

$$\begin{aligned}\beta_1&=\int_{\xi\in M'}\mathscr{B}=\theta\int_{\xi\in M'}d_\xi\int_{x\in M^{2n-1}}\Omega^0_{2n-1,0}(A(x,\xi),0) \\ &=\theta\int_{x\in M^{2n-1}}[\Omega^0_{2n-1,0}(A(x,\xi'),0)-\Omega^0_{2n-1,0}(A(x,\xi),0)]. \end{aligned}\quad (4.18)$$

If M^{2n-1} can be dealt with as the boundary of an M^{2n} and

$$A(x,\xi')=g^{-1}A(x,\xi)g+g^{-1}d_xg\ ,$$

then $$\beta_1=\theta\int_{x\in S^{2n}}\Omega_{2n,0}(A(x,\xi)) \quad (4.19)$$

where $S^{2n}=M_+^{2n}\cup M_-^{2n}$ $\qquad \partial M_+^{2n}=-\partial M_-^{2n}=M^{2n-1}$

when n=2, β_1 is just the ordinary θ-vacuum term.

If M^{2n-1} have boundary $\partial M^{2n-1}=M^{2n-2}$ by Stokes' theorem, Eq.(4.15) becomes

$$\int_{x\in M^{2n-2}=\partial M^{2n-1}} \Omega^0_{2n-2,1}(A(x,\xi),0)$$
$$=\int_{x\in M^{2n-1}} \Omega_{2n-1,1}(A,0)-d_\xi\int_{x\in M^{2n-1}} \Omega^0_{2n-1,0} \quad . \tag{4.20}$$

The Chern form gauge potential is defined as

$$\mathscr{B}=\alpha\int_{M^{2n-2}} \Omega^0_{2n-2,1}(A(x,\xi),0)$$
$$=\alpha\left[\int_{x\in M^{2n-1}} \Omega_{2n-1,1}(A,0)-d_\xi\int_{x\in M^{2n-1}} \Omega^0_{2n-1,0}(A,0)\right], \tag{4.21}$$

the field strength is

$$\mathscr{G}=d_\xi\mathscr{B}=-\alpha\int_{x\in M^{2n-2}} \Omega_{2n-2,2}(A(x,\xi),0) \ . \tag{4.22}$$

Consider n=2,

$$\left.\begin{aligned} &M^2=\partial M^3=S^2_x \ , \\ &\mathscr{B}=\alpha\int_{S^2_x} \Omega^0_{2,1}(A,0) \ , \\ &\mathscr{G}=-\alpha\int_{S^2_x} \Omega_{2,2}(A,0) \ . \end{aligned}\right\} \tag{4.23}$$

From the definition

$$\Omega_{2,2}(A,B)=-\frac{1}{8\pi^2}\mathrm{Tr}(d_\xi A+d_x B+AB+BA)^2 \ , \tag{4.24}$$

it is easily checked that

$$\left.\begin{aligned} &\Omega^0_{2,1}(A,0)=\Omega^0_{2,1}(A_g,g^{-1}d_\xi g)-\frac{1}{8\pi^2}\mathrm{Tr}g^{-1}d_\xi g(g^{-1}d_x g)^2 \ , \\ &A_g=g^{-1}Ag+g^{-1}d_x g \ . \end{aligned}\right\} \tag{4.25}$$

Considering the integral of G over S^2_ξ in ξ-space and supposing that it is topological nontrivial, we then have

$$\int_{S^2_\xi}\mathscr{G}=\int_{D^2_\xi} d_\xi\int_{S^2_x}\alpha(\Omega^0_{2,1}(A,0)-\Omega^0_{2,1}(A_g,g^{-1}d_\xi g))$$
$$=\frac{\alpha}{8\pi^2}\int_{S^1_\xi}\int_{S^2_x}\mathrm{Tr}(g^{-1}d_\xi g(g^{-1}d_x g)^2) \ . \tag{4.26}$$

On the other hand, if there exists a monopole in ξ-space, in accordance with the monopole quantization condition one obtains

$$2\pi n=3\alpha w(g) , \tag{4.27}$$

where

$$\begin{aligned} w(g) &= \frac{1}{24\pi^2}\int_{S_\xi^1}\int_{S_x^2}\mathrm{Tr}\; g^{-1}d_\xi g(g^{-1}d_x g)^2 \\ &= \frac{1}{24\pi^2}\int_{S^3}\mathrm{Tr}\; g^{-1}d_\xi g(g^{-1}d_x g)^2 \; . \end{aligned} \tag{4.28}$$

When wind number $w(g)=1$, we have

$$\alpha = \frac{2}{3}\pi n \; , \qquad n=1,2,\ldots \; , \tag{4.29}$$

corresponds to the topological mass.

References

[1] H.Y. Guo, K. Wu, B.Y. Hou, and S.K. Wang, "Anomalies, Cohomology and Generalized Secondary Classes", preprint AS-ITP-84-044, (1984).

[2] L.D. Faddeev, Phys.Lett. 145B(1984)81.

[3] B. Zumino, "Cohomology of Gauge Groups: Cocycles and Schwinger Term", Santa Barbara NSF-ITP preprint, (1984).

[4] K.C. CHOU, Y.L. WU and Y.B. XIE, Commun. Theor. Phys. 5(1986)95.

[5] Y.S. Wu and A. Zee, "Abelian Gauge Structure inside Non-Abelian Gauge Theorems", preprint 40048-38(1984).

Commun. in Theor. Phys. (Beijing, China) Vol.8, No.3 (1987) 341-367

THE GENERAL CHERN-SIMONS CHARACTERISTIC CLASSES AND THEIR PHYSICAL APPLICATIONS

WU Yue-liang(吴岳良), XIE Yan-bo(谢彦波)
and ZHOU Guang-zhao(周光召)
Institute of Theoretical Physics, Academia Sinica,
P.O. Box 2735, Beijing, China

Received January 30, 1986

Abstract

Simple derivation of the general Chern-Simons characteristic classes is presented. Application to construct current conservation anomalies and effective action is discussed in detail. The effective action of nonlinear σ model where the fermions are interacting with external gauge fields is obtained. The central term of Virasoro algebra is also derived by topological method. Possible existence of a gauge potential in the group manifold and its relation to θ-vacuum is indicated.

I. Introduction

Anomalies due to quantum loops of fermions appear in the conservation equations and the commutator equations. They have played a strikingly important role in the development of particle physics and quantum field theory. Just mention one example, the renormalizability of gauge theories enforces the condition of cancellation of chiral anomalies, leading to the fascinating correlation between quarks and leptons and the gauge group in the superstring theory.

The fact that the expression of chiral anomalies obtained in the lowest order perturbation is exact must have a deep reason. Indeed it was realized only recently that anomalies have topological origin and represent the non-trivial elements of the cohomology of the gauge fields. Now it is possible to calculate anomalies and the effective action by differential geometric method based on the Chern-Simons characteristic classes without ever calculating a Feynman diagram.

In this article we shall review briefly some recent developments related to the work done in China[1]. It is in the direction of using cohomology of gauge groups to study the anomalies and the effective action. For alternative views and presentations please read those excellent articles by Faddeev, Jackiw, Witten, Zumino and many others[2-5].

II. Mathematical Preliminaries

Points, lines, triangles, tetrahedrons and their generalization to higher dimensions are called simplices. An n-simplex is constructed from n+1 distinct ordered points $a,a_1,\ldots,a_n$ in a manifold and will be denoted by the symbol $(a,a_1,\ldots,a_n)$. A zero simplex is a single point; a 1-simplex (a,a_1) is a line bounded by two zero simplices (a) and (a_1) with definite orientation. We define the boundary operator ∂ such that

$$\partial(a,a_1)=(a_1)-(a) \ . \tag{2.1}$$

In general, the boundary of an n-simplex consists of n+1 (n-1)-simplices. The boundary operator ∂ is defined to be

$$\partial(a,a_1,\ldots,a_n)=(a_1,a_2,\ldots,a_n)-(a,a_2,\ldots,a_n)$$
$$+\ldots+(-1)^n(a,a_1,\ldots a_{n-1}). \tag{2.2}$$

The orientation of the boundary is taken into account by the permutation factor $(-1)^\ell$. It is easily verified that the boundary of a boundary vanishes, i.e.

$$\partial^2=0 \ . \tag{2.3}$$

A real function of n-simplex $f(a,a_1,\ldots,a_n)$ is called an n-cochain. It is possible to define a coboundary operator dual to the boundary operator in the following way

$$\Delta f(a,a_1,\ldots,a_n)=f(a_1,\ldots,a_n)-f(a,a_2,\ldots,a_n)$$
$$+\ldots+(-1)^n f(a,a_1,\ldots,a_{n-1}). \tag{2.4}$$

It is also satisfied the nilpotent condition

$$\Delta^2=0 \ . \tag{2.5}$$

An n-cochain $f(a,a_1,\ldots,a_n)$ is called a cocycle if its coboundary is zero

$$\Delta f(a,a_1,\ldots,a_{n+1})=0 \ . \tag{2.6}$$

It is called coexact if it is the coboundary of an (n-1)-cochain

$h(a,a_1,\ldots,a_{n-1})$

$$f(a,a_1,\ldots,a_n)=\Delta h(a,a_1,\ldots,a_n). \quad (2.7)$$

Two cochains f and g which differ by a coexact cochain are equivalent

$$f\sim g \quad \text{if} \quad f-g=\Delta h. \quad (2.8)$$

The equivalent classes form a group called cohomology group.

We are interested in the manifold composed of gauge fields represented as points A_i. Like in the case of monopole it is not possible in general to describe a topological nontrivial object by one gauge field regular everywhere. We have to divide the space into several covers with one gauge field in each cover, and in their intersections the gauge fields are related by gauge transformations so that the field strength defined by these gauge fields are gauge equivalent. Therefore, the cochain $f(A,\ldots A_n)$ with A_i equals the gauge transformation of A

$$A_i={g_i}A=g_i(A+d)g_i^{-1} \quad (2.9)$$

is of particular interest where $g_i(x)$ are group elements. In this case the cochain can be written in another form

$$f(A,A_1,\ldots A_n)=\tilde{f}(A;g_1,\ldots g_n). \quad (2.10)$$

Either form will be used in the following, where the ~ will be omitted for simplicity. A cochain is called invariant if it satisfies the condition

$$f({}^gA,\ {}^gA_1,\ldots,{}^gA_n)=f(A,A_1,\ldots,A_n) \quad (2.11)$$

or

$$f({}^gA;gg_1,\ldots,gg_n)=f(A;g_1,\ldots,g_n). \quad (2.12)$$

To illustrate the usefulness of these mathematical concepts in physics, let us consider the effective action of an underlying theory with massless fermions interacting with external vector and axial vector gauge fields defined on a non-Abelian chiral group. The effective action cannot be a group invariant since the conservation law is broken by chiral anomalies. Let the quantum amplitude be

$$Z=e^{i\Gamma(A)}, \quad (2.13)$$

where Γ is the effective action and A the external gauge fields. Under a group transformation the quantum amplitude will be changed by a phase

$$U(g)Z(A)=e^{i\Omega_1(A;g)}Z({}^gA). \tag{2.14}$$

The infinitesimal variation of the phase Ω_1 (A;g) in the group manifold gives the anomaly for current conservations. The group multiplication condition

$$U(g_1)U(g_2)=U(g_1g_2) \tag{2.15}$$

implies that $\Omega_1(A;g)$ is a cocycle

$$\begin{aligned}&\Delta\Omega_1(A;g_1,g_2)\\&=\Omega_1({}^{g_1}A;g_2)-\Omega_1(A;g_1g_2)+\Omega_1(A;g_1)\\&=0.\end{aligned} \tag{2.16}$$

If $\Omega_1(A;g)$ is the trival element of the cocycle condition, it is coexact such that

$$\Omega_1(A;g)=\Omega_0({}^gA)-\Omega_0(A). \tag{2.17}$$

In this case, we would redefine the quantum amplitude

$$Z(A)\longrightarrow Z'(A)=Z(A)e^{i\Omega_0(A)} \tag{2.18}$$

to get an invariant effective action

$$\Gamma_{inv}=\Gamma(A)+\Omega_0(A) \tag{2.19}$$

and eliminate the anomalies by adding counter terms. This is not the case if Ω_1 is a nontrivial element of the cohomology. Therefore, the chiral anomaly appeared in the conservation equation of the current is determined by the cohomology of the group manifold. The coboundary condition Eq.(2.16) is a global form of the Wess-Zumino consitency condition for the anomalies.

A second example concerns the 2-cocycles and the Schwinger term in the commutation relation of the currents. Consider a representation of the group on a functional of the gauge fields

$$U(g)\psi(A)=V(A;g)\psi({}^gA) \quad , \tag{2.20}$$

where V(A;g) forms a projective representation

$$V(A;g_2)V({}^{g_2}A;g_1)=e^{i\Omega_2(A;g_1g_2)}\ V(A;g_1g_2). \tag{2.21}$$

The condition for the group multiplication satisfying the associative law

$$(U(g_1)U(g_2))U(g_3)=U(g_1)(U(g_2)U(g_3)) \tag{2.22}$$

requires that the phase $\Omega_2(A;g_1,g_2)$ is a 2-cocycle. In some cases $\Omega_2(A;g_1,g_2)$ is a matrix instead of a real function. In that case the cocycle condition has to be modified to incorporate the noncommutative nature of the phase Ω_2, when Ω_2 is a real function as in the case of Abelian gauge theory, the cocycle condition can be easily worked out to be

$$\begin{aligned}\Delta\Omega_2(A;g_1,g_2,g_3)&=\Omega_2(A;g_2,g_3)-\Omega_2(A;g_1g_2,g_3)\\&\quad+\Omega_2(A;g_1,g_2g_3)-\Omega_2({}^{g_3}A;g_1,g_2)\\&=0.\end{aligned} \tag{2.23}$$

If $\Omega_2(A;g_1,g_2)$ is coexact such that

$$\Omega_2(A;g_1,g_2)=W_1(A;g_1)-W_1(A;g_1g_2)+W_1({}^{g_1}A;g_2)\ , \tag{2.24}$$

the condition Eq.(2.23) satisfies automatically and Ω_2 is a trivial element of the cohomology. In this case a change of the phase

$$V(A;g)\longrightarrow V'(A;g)=V(A,g)e^{-iW_1(A;g)} \tag{2.25}$$

will eliminate the phase and V'(A;g) forms an ordinary representation.

A nontrivial phase $\Omega_2(A;g_1,g_2)$ contributes an anomalous term to the commutation relation of the generators of the group defined by the V(A;g) representation.

Higher order cocycles are studied in the literature. The 3-cocyele is related to the breakdown of the Jacobi identity. We shall not discuss them further.

In differential geometry the exterior differentiation d is a coboundary operator satisfying the nilpotent codition

$$d^2=0. \tag{2.26}$$

The gauge potential A is a one-form

$$A=A_\mu^\alpha(x)\hat{I}_\alpha dx^\mu \ , \tag{2.27}$$

where $A_\mu^\alpha(x)$ are real functions and $\hat{I}_\alpha$ the group generator. The field strength is defined to be a two-form

$$F=dA+A^2 \tag{2.28}$$

with A^2 the exterior product of two gauge potentials. F satisfies the Bianchi identity

$$dF=FA-AF. \tag{2.29}$$

An infinitiesimal gauge transformation is defined to be

$$\delta A=-AB-BA-dB, \tag{2.30}$$

where B is the exterior differential in the group manifold and B is a one-form usually defined to be a pure gauge

$$B=g^{-1}\delta g \ . \tag{2.31}$$

In this paper we shall keep B to be an arbitrary one-form in the group space and define a dual field strength

$$G=\delta B+B^2 \tag{2.32}$$

satisfying a dual Bianchi identity

$$\delta G=[G,B]. \tag{2.33}$$

The condition for the gauge potentials A and B to be single valued requires that

$$\delta F=[F,B] \ , \tag{2.34}$$

$$dG=[G,A] \ , \tag{2.35}$$

and

$$[F,G]=0. \tag{2.36}$$

III. Chern Class and the General Chern-Simons Cochain

It is well known in mathematics that the Chern class gives a representation of the cohomology classes by the curvature forms of a connection in the fibre bundle. In this section a simplified derivation of the general Chern-Simons characteristic classes will be briefly described.

The Chern density is a 2n-form

$$C_{2n}(F)=\alpha_n \mathrm{Tr}(F^n) \ , \tag{3.1}$$

where

$$\alpha_n=i^n/n!(2\pi)^n \tag{3.2}$$

is a normalization constant. Using Eq.(2.29), it is easily verified that $C_{2n}(F)$ is closed for arbitrary gauge potential

$$dC_{2n}(F)=0 \tag{3.3}$$

in so far as F is defined by Eq.(2.28). We shall show in the following that Eq.(3.3) is sufficient for us to derive all the results connected to a whole chain of higher order characteristic classes from which nontrivial elements of the cohomology can be found.

Consider a manifold consisting of two submanifolds, one of which is the ordinary space with coordinates x^μ and the other a manifold of auxiliary parameters t^i. The exterior differential operator is

$$d=d_x+d_t \tag{3.4}$$

where

$$d_x=dx^\mu \partial/\partial x^\mu \ , \tag{3.5}$$

and

$$d_t=dt^i \partial/\partial t^i \ . \tag{3.6}$$

Any gauge potential $\mathscr{A}$ can be decomposed into two parts

$$\mathscr{A}(x,t)=A(x,t)+B(x,t) \ , \tag{3.7}$$

where $A(x,t)$ is a one-form in x^μ only and $B(x,t)$ that in t^i. The field strength $\mathscr{F}$ now becomes

$$\mathscr{F}=(d_x+d_t)\mathscr{A}+\mathscr{A}^2=F+M+G \ , \tag{3.8}$$

where

$$F=d_xA+A^2 \ , \tag{3.9}$$

$$M=d_tA+d_xB+AB+BA \ , \tag{3.10}$$

and

$$G=d_tB+B^2 \ . \tag{3.11}$$

The Chern density

$$C_{2n}(\mathscr{F})=C_{2n}(F+M+G) \tag{3.12}$$

satisfies the Eq.(3.3)

$$(d_x+d_t)C_{2n}(F+M+G)=0. \tag{3.13}$$

Expanding $C_{2n}(F+M+G)$ according to the degrees of the forms in dt^i we have

$$C_{2n}(F+M+G)=\sum_{m=0}^{2n} C_{2n-m,m}(A,B), \tag{3.14}$$

where $C_{2n-m,m}(A,B)$ is a form of degrees 2n-m in dx^μ and m in dt^i. Substituting Eq.(3.14) into Eq.(3.13) and comparing the degrees of both sides we obtain

$$\left.\begin{aligned} &d_x C_{2n}(F)=0 \ , \\ &d_t C_{2n}(F)=-d_x C_{2n-1,1} \ , \\ &\vdots \\ &d_t C_{2n}(G)=0 \ . \end{aligned}\right\} \tag{3.15}$$

Now choose the gauge potential to be

$$\left.\begin{aligned} &B(x,t)=0 \ , \\ &A(x,t)=\sum_i t^i A^{(i)}(x) \ . \end{aligned}\right\} \tag{3.16}$$

The boundary of the t-submanifold is determined by the condition

$$\sum_i t^i=1 \ . \tag{3.17}$$

In this special case

$$G=0 \ , \tag{3.18}$$

$$M=\sum_i dt^i A^{(i)} \ ,$$

$$\left(\sum_i dt^i=0\right) \tag{3.19}$$

and

$$F=F(t^i,d_x A^{(i)},A^{(i)}). \tag{3.20}$$

It is easily found that

$$C_{2n-m,m}=\alpha_n\binom{n}{m}\,\mathrm{Str}(F^{n-m}M^m) \ , \tag{3.21}$$

where Str means symmetrization of the operators inside the trace operation.

The Chern-Simons secondary characteristic class can be derived by taking only one parameter $t=t^{(1)}=1-t^{(0)}$ and integrating the equation.

$$d_t C_{2n}(F)=-d_x C_{2n-1,1}$$

along a straight line (a one simplex) from t=0 to t=1. In terms of t we have

$$M=dt(A^{(1)}-A^{(0)}), \tag{3.22}$$

$$F=(1-t)F^{(0)}+tF^{(1)}+t(t-1)(A^{(1)}-A^{(0)})^2\equiv F_t \ , \tag{3.23}$$

and

$$C_{2n-1,1}=n\alpha_n tr((A^{(1)}-A^{(0)})F_t^{n-1})dt. \tag{3.24}$$

After integration we obtain

$$C_{2n}(F^{(1)})-C_{2n}(F^{(0)})=-d_x\int_0^1 n\alpha_n Tr((A^{(1)}-A^{(0)})F_t^{n-1})dt$$
$$=-d_x\Omega_{2n-1,1}(A^{(1)},A^{(0)}). \tag{3.25}$$

This is the well-known formula of Weyl homomorphism which states that the Chern density of two different gauge fields are cohomologous in space manifold.

Taking $A^{(0)}(x)=0$ and $A^{(1)}(x)=A(x)$ in Eq.(3.25) it follows that

$$C_{2n}(F)=-d\Omega_{2n-1,1}(A,0)$$
$$=d\Omega_{2n-1}(A), \tag{3.26}$$

where

$$\Omega_{2n-1}(A)=-n\alpha_n\int_0^1 Tr(AF_t^{n-1})dt \tag{3.27}$$

with

$$F_t=tF+t(t-1)A^2 \ . \tag{3.28}$$

Ω_{2n-1} is the Chern-Simons secondary characteristic class.

In the general case we choose m+1 parameters t^i and form an m-simplex Sm with m+1 vertices

$$a_0=(1,0,\ldots,0) \ ,$$

$$\left.\begin{array}{l} a_1=(0,1,\ldots,0), \\ \vdots \\ a_m=(0,0,\ldots,1), \end{array}\right\} \tag{3.29}$$

whose boundary consisting of m+1 (m-1)-simplices $S^{(i)}_{m-1}$ is specified by the equation $\sum_i t^i=1$ or

$$\begin{aligned} &\partial S_m(a_0,\ldots,a_m) \\ &=S^{(0)}_{m-1}(a_1,\ldots,a_m)-S^{(1)}_{m-1}(a_0,a_2,\ldots,a_m) \\ &+\ldots+(-1)^m S^{(m)}_{m-1}(a_0,\ldots,a_{m-1}). \end{aligned} \tag{3.30}$$

Integrating the equation

$$d_t C_{2n-m+1,m-1}=-d_x C_{2n-m,m} \tag{3.31}$$

over the m-simplex $S_m(a_0,\ldots,a_m)$ we obtain by Stoke's theorem that

$$\begin{aligned} &\Delta\Omega_{2n-m+1,m-1}(A^{(0)},\ldots,A^{(m)}) \\ &=\sum_i(-1)^i\int_{S^{(i)}_{m-1}} C_{2n-m+1,m-1} \\ &=-d_x\Omega_{2n-m,m}(A^{(0)},\ldots,A^{(m)}), \end{aligned} \tag{3.32}$$

where Δ is the coboundary operator and

$$\Omega_{2n-l,l}=\int_{S_l} C_{2n-l,l} \tag{3.33}$$

is the integral of $C_{2n-l,l}$ over the l-simplex in the parameter space. $\Omega_{2n-l,l}$ is called the lth Chern-Simons characteristic class which satisfies a sequence of descent equations starting from the Chern density.

For gauge potential regular everywhere on a compact space manifold M_{2n-m+1} of dimension 2n-m+1 the Eq.(3.32) can be integrated over M_{2n-m+1}. The right-hand side of Eq.(3.32) vanishes after integration and therefore

$$\Delta\int_M \Omega_{2n-m+1,m-1}=0\ , \tag{3.34}$$

or $\int_M \Omega$ is a (m-1)-cocycle.

In case when nontrivial topological object exists, the gauge potential cannot be regular everywhere. We have to divide the space manifold into several covers

$$M=\sum_j M_j \ , \tag{3.35}$$

In each cover M_j a regular gauge field $A^{(j)}(x)$ can be defined. In the intersection $M_i \cap M_j$ the gauge fields $A^{(i)}$ and $A^{(j)}$ are related by a gauge transformation

$$A^{(i)}=g_{ij}(A^{(j)}+d)g_{ij}^{-1} \ , \tag{3.36}$$

so that the invariants of the gauge fields are uniquely defined over the whole space manifold. We shall show that the cocycle condition Eq.(3.34) is no longer valid when the transition function $g_{ij}(x)$, which is a mapping of the boundary $\partial M_i=-\partial M_j$ to the group manifold, has nontrivial homotopy.

As an illustration, suppose two covers M_1 and M_2 with $\partial M_1=-\partial M_2$ are sufficient to define a gauge potential. From Eq.(3.26), we have

$$\int_M C_{2n}(F)=\sum_i \int_{M_i} C_{2n}(F)=\int_{\partial M_2}(\Omega_{2n-1,1}(A^{(1)},0) -\Omega_{2n-1,1}(A^{(2)},0)). \tag{3.37}$$

It is seen from the definition that $\Omega_{2n-1,1}(A,B)$ is a gauge-invariant cochain

$$\Omega_{2n-1,1}({}^gA,{}^gB)=\Omega_{2n-1,1}(A,B). \tag{3.38}$$

Therefore from Eqs.(3.36) and (3.38) we get

$$\Omega_{2n-1,1}(A^{(2)},0)=\Omega_{2n-1,1}(A^{(1)},g_{12}^{-1}dg_{12}). \tag{3.39}$$

Now choose $A^{(0)}=0$, $A^{(2)}=g_{12}^{-1}dg_{12}$ in Eq.(3.32) with m=2

$$\Omega_{2n-1,1}(A^{(1)},A^{(2)})-\Omega_{2n-1,1}(A^{(0)},A^{(2)})+\Omega_{2n-1,1}(A^{(0)},A^{(1)}) =-d_x\Omega_{2n-2,2}(A^{(0)},A^{(1)},A^{(2)}), \tag{3.40}$$

we obtain

$$\Omega_{2n-1,1}(A^{(1)},0)-\Omega_{2n-1,1}(A^{(1)},g_{12}^{-1}dg_{12}) =-\Omega_{2n-1,1}(0,g_{12}^{-1}dg_{12})+d_x\Omega_{2n-2,2}(A^{(0)},A^{(1)},A^{(2)}). \tag{3.41}$$

Integrating Eq.(3.40) over ∂M_1 it follows from Eq.(3.37) that

$$\int_M C_{2n}(F)$$

$$=-\int_{\partial M_2}\Omega_{2n-1,1}(0,g_{12}^{-1}dg_{12})$$
$$=-n\alpha_n\int_0^1 dt\cdot t^{2n-1}(t-1)^{2n-1}\int_{\partial M_2}Tr(g_{12}^{-1}dg_{12})^{2n-1}$$
$$=-n\alpha_n B(2n,2n)\int_{\partial M_2}Tr(g_{12}^{-1}dg_{12})^{2n-1}$$
$$=\text{winding number.} \tag{3.42}$$

If the transition function g_{12} is a nontrivial element of the homotopy group Π_{2n-1}, the integral of the Chern density will be equal to the winding number of the mapping of a (2n-1)-dimensional sphere to the group manifold.

IV. Group Cohomology of Faddeev

In order to get the group cohomology of Faddeev, we start from Eq.(3.26)

$$C_{2n}(F)=d_x\Omega_{2n-1,1}(A,0)\equiv d_x W_{2n-1}(A). \tag{4.1}$$

This relation is also true for arbitrary gauge field and manifolds. Let us apply this equation to a manifold consisting of the direct product of a space and a group submanifold with the coordinates x^μ and ξ^i respectively. The gauge field is

$$\mathscr{A}(x,\xi)=A(x,\xi)+B(x,\xi), \tag{4.2}$$

where $A(x,\xi)$ is a one-form in the space manifold and $B(x,\xi)$ is that in the group manifold. The field strength is

$$\mathscr{F}=(d_x+d_\xi)\mathscr{A}+\mathscr{A}^2=F+M+G\ , \tag{4.3}$$

where

$$\left.\begin{aligned}&F=d_xA+A^2\ ,\\&M=d_\xi A+d_xB+AB+BA,\\&G=d_\xi B+B^2\ .\end{aligned}\right\} \tag{4.4}$$

Eq.(4.1) then has the form

$$C_{2n}(F+M+G)=(d_x+d_\xi)W_{2n-1}(A,B). \tag{4.5}$$

Expanding both the Chern density and Chern-Simons secondary class according to the degrees of $d\xi^i$, we have

$$C_{2n}(F+M+G)=\sum_m C_{2n-m,m}(A,B) \ , \tag{4.6}$$

and

$$W_{2n-1}(A,B)=\sum_m W_{2n-m,m-1}(A,B). \tag{4.7}$$

Substituting Eqs.(4.6) and (4.7) into Eq.(4.5) and comparing the degrees of $d\xi^i$ on both sides we obtain

$$\left.\begin{aligned} &C_{2n,0}=d_x W_{2n-1,0} \\ &\vdots \\ &C_{2n-m,m}=d_x W_{2n-m-1,m}+d_\xi W_{2n-m,m-1} \ , \\ &\vdots \\ &C_{0,2n}=d_\xi W_{0,2n-1} \ . \end{aligned}\right\} \tag{4.8}$$

As a special case we choose

$$\left.\begin{aligned} &A(x,\xi)=U^{-1}(x,\xi)(A(x)+d_x)U(x,\xi) \ , \\ &B(x,\xi)=U^{-1}(x,\xi)d_\xi U(x,\xi)+C(\xi) \ , \end{aligned}\right\} \tag{4.9}$$

where $U(x,\xi)$ is a group element and $C(\xi)$ is an Abelian gauge potential in the group manifold. Then

$$\left.\begin{aligned} &F(x,\xi)=U^{-1}(x,\xi)F(x)U(x,\xi) \ , \\ &M(x,\xi)=0 \ , \\ &G(x,\xi)=d_\xi C(\xi) \ . \end{aligned}\right\} \tag{4.10}$$

We see that the conditions Eq.(2.34)-(2.36) are satisfied. From the expansion (4.6) we have

$$C_{2n,0}(A,B)=C_{2n}(F(x)) \tag{4.11}$$

$$C_{2n-m,m}(A,B)=\delta_{m,k}\binom{n}{k}\alpha_n(d_\xi(C(\xi))^k \mathrm{Tr}(F^{n-k}(x)). \tag{4.12}$$

An m-simplex in the group manifold will be defined with m+1 vertices corresponding to m+1 group elements g_0, $g_0g_1,\ldots,g_0,\ldots,g_m$. In order to do so we define a function

$$h(\xi,g(x))=e^{(1-\xi)u(x)} \tag{4.13}$$

such that

$$h(1,\ g(x))=1 \ , \tag{4.14}$$

and

$$h(0, g(x))=e^{u(x)}\equiv g(x) . \tag{4.15}$$

The function $h(\xi,g(x))$ maps a line segment $(1,0)$ in the ξ axis to a line running from the identity element to $g(x)$ in the group manifold. Now choose

$$U(x,\xi)=h(\xi^0,g_0(x)h(\xi^1,g_1(x)h(\ldots h(\xi^m,g_m(x)))\ldots). \tag{4.16}$$

The simplex in the ξ space consisting of the following m+1 vertices

$$\begin{aligned} a_i&=(\xi^0,\ldots,\xi^{i-1},\xi^i,\xi^{i+1},\ldots,\xi^m) \\ &=(0,\ldots,0,1,0,\ldots,0) \end{aligned} \tag{4.17}$$

will map into a simplex in the group manifold with vertices

$$1,\ g_0,\ g_0g_1,\ldots,g_0g_1\ldots g_{m-1} . \tag{4.18}$$

From Eq.(4.12) the term $C_{2n-2,2}$ can be written as a total differential in ξ

$$C_{2n-2}=-d_\xi W'_{2n-2,1} ,$$

where

$$W'_{2n-2,1}=-n\alpha_n C(\xi)\mathrm{Tr}(F^{n-1}) = \frac{-i}{2\pi}C(\xi)C_{2n-2}(F(x)). \tag{4.19}$$

The descent Eq.(4.8) can now be written as

$$\left.\begin{aligned} &C_{2n}(F(x))=d_x\Omega_{2n-1,0}(A(x)), \\ &d_\xi W_{2n-1,0}=-d_x W_{2n-2,1} , \\ &\quad\vdots \qquad , \end{aligned}\right\} \tag{4.20}$$

where

$$\overline{W}_{2n-2,1}=W_{2n-2,1}+W'_{2n-2,1} . \tag{4.21}$$

Integrating Eq.(4.20) over simplices in ξ manifold, we get equations similar to Eq.(3.32).

V. The Effective Action of Wess-Zumino-Witten

Consider a system of quark field ψ interacting with color gauge field C corresponding to color group $SU(N_c)$ and some external chiral flavor gauge fields A_L and A_R. The Lagrangian has the form

$$\mathscr{L}=i\bar{\psi}(\partial\!\!\!/+\mathbb{C}+\mathbb{A})\psi-\frac{1}{2g_c^2}\mathrm{Tr}C_{\mu\nu}^2\ , \tag{5.1}$$

where we have used a double matrix notation for convenience

$$\psi=\begin{pmatrix}\psi_L\\ \psi_R\end{pmatrix}\ ,\qquad A=\begin{pmatrix}A_L & 0\\ 0 & A_R\end{pmatrix}. \tag{5.2}$$

The color gauge field C_μ is diagonal and $C_{\mu\nu}$ is its field strength. The flavor currents are covariantly conserved on the classical level

$$\mathscr{D}^\mu J_\mu^i=0,\qquad J_\mu^i=\bar{\psi}\gamma_\mu\lambda^i\psi\ , \tag{5.3}$$

where λ^i i=1,...,N are the flavor group generating matrices. However, the quantum corrections break the conservation equation and produce the current anomaly

$$\mathscr{D}^\mu J_\mu^i=G^i=\frac{N_c}{24\pi^2}\varepsilon^{\mu\nu\sigma\tau}\mathrm{Tr}\gamma_5\lambda^i\partial_\mu(A_\nu\partial_\rho A_\sigma+\frac{1}{2}A_\nu A_\rho A_\sigma), \tag{5.4}$$

where N_c is the number of colors and $\gamma_5=\begin{pmatrix}1 & 0\\ 0 & -1\end{pmatrix}$.

There are different forms of current anomaly. The one presented in Eq.(5.4) is called the symmetric form which contributes to both vector and axial vector current conservations. It is possible to add local counter term to the effective Lagrangian so that the vector current is conserved and the anomalies are shifted completely to the axial vector current. This form of anomaly is called unsymmetrical one.

When color degrees of freedom have been integrated, we get an effective theory describing the low energy phenomenology. The effective action Γ under infinitesimal falvor group transformation should reproduce the anomaly

$$\delta_\xi\Gamma=\mathrm{Tr}(Gv), \tag{5.5}$$

where

$$\left.\begin{aligned}&v=g^{-1}\delta_\xi g=\hat{I}_i(\xi)d\xi^i\ ,\\ &G=G^i\hat{I}_i(\xi)\end{aligned}\right\} \tag{5.6}$$

with $\hat{I}_i$ the generators of the group. Under exterior differentiation

$$\delta_\xi v=-v^2=-\frac{1}{2}[\hat{I}_i(\xi),\hat{I}_j(\xi)]d\xi^i d\xi^j=-\frac{1}{2}f_{ijk}\hat{I}_k(\xi)d\xi^i d\xi^j, \tag{5.7}$$

where f_{ijk} is the structure constant of the flavor group. From Eq.(5.5) we have

$$\delta_\xi^2\Gamma=0=\mathrm{Tr}(\delta_\xi G v+G\delta_\xi v)$$
$$=\mathrm{Tr}(\delta_\xi G\hat{I}_i(\xi)d\xi^i+\frac{1}{2}Gf_{ijk}\hat{I}_k(\xi)d\xi^i d\xi^j), \qquad (5.8)$$

hence

$$(\delta_\xi G)^i=-\frac{1}{2}f_{ijk}d\xi^i G^k. \qquad (5.9)$$

This is the Wess-Zumino consistency condition that has to be satisfied by the current anomaly .

The descent Eq.(4.20)

$$d_\xi\overline{W}_{2n-2,1}=-d_x W_{2n-3,2} \qquad (5.10)$$

implies that $d_\xi\overline{W}_{2n-2,1}$ integrated over a 2-simplex in the group manifold is homologous to a cocycle whose space integral can be indentified with the phase Ω_1 of the effective action Γ as dicussed in Eq.(2.14). Eq.(5.10), its infinitiesimal form, integrating over a (2n-2)-compact space manifold has the form of Eq.(5.5) for an anomaly in (2n-2)-space dimensions. Therefore $\overline{W}_{2n-2,1}$, with suitable field contents, may be identified as the current anomaly which has been verified by theoretical calculations.

In order to get an explicit form of the effective action one has to perform an integration in the group manifold for $\overline{W}_{2n-2,1}$. This has been done first by Wess-Zumino[9]. They could not find a close form in 2n-2 dimensions. It has been realized later by Witten that a close form is possible only in one dimension higher[5].

Consider a form ρ_{2n-1} in 2n-1 dimensions. What conditions it has to satisfy in order to describe a Lagrangian in (2n-2)-dimensional manifold M which is considered to be the only boundary of a (2n-1)-dimensional manifold S? There are three of them:

i). Locally, ρ_{2n-1} should be an exact form so that

$$\rho_{2n-1}=d_x\mathscr{L}_{2n-2}.$$

ii). Globally, it should not depend on the manifold S choosen so far as it has the same boundary M. This implies that ρ_{2n-1} integrated over any closed manifold should be equal to 2π times integer

$$\int_{S_1}\rho_{2n-1}-\int_{S_2}\rho_{2n-1}=\oint_{S_1+S_2}\rho_{2n-1}=2\pi n \qquad (5.11)$$

such that the uniqueness of the quantum amplitude $e^{i\Gamma}$ is guaranteed.

iii). Under an infinitesimal group transformation, it produces the correct current anomaly

$$d_\xi \rho_{2n-1} = -2\pi d_x W_{2n-2,1} \,. \tag{5.12}$$

The third condition is easily seen to be satisfied by $2\pi W_{2n-1,1}$ which, however, does not satisfy the first two. To find ρ_{2n-1} satisfying all three conditions, a Goldstone field U(x) transforming as

$$U(x) \longrightarrow g_L U(x) g_R^{-1} \tag{5.13}$$

has to be used. In the matrix notation we write

$$Q(x) = \begin{pmatrix} 0 & U(x) \\ U^{-1}(x) & 0 \end{pmatrix} = Q^{-1}(x) \;; \tag{5.14}$$

which transforms as

$$Q(x) \longrightarrow g Q(x) g^{-1} \tag{5.15}$$

with g defined to be

$$g = \begin{pmatrix} g_L & 0 \\ 0 & g_R \end{pmatrix}. \tag{5.16}$$

Now define a new gauge field

$$A^Q = Q(A+d)Q^{-1} = \begin{pmatrix} U(A_R+d)U^{-1} & 0 \\ 0 & U^{-1}(A_L+d)U \end{pmatrix}, \tag{5.17}$$

where A is the one defined in Eq.(5.2). Both A and A transform in the same way

$$\left.\begin{aligned} &A \longrightarrow g(A+d)g^{-1} \;, \\ &A^Q \longrightarrow g(A^Q+d)g^{-1} \;. \end{aligned}\right\} \tag{5.18}$$

It is easily found that

$$F^Q = QFQ^{-1} \,. \tag{5.19}$$

Chern density

$$C_{2n}^{(5)}(F) = \alpha_n \mathrm{Tr}(\gamma_5 F^n) = C_{2n}(F_L) - C_{2n}(F_R) \;, \tag{5.20}$$

which is the difference of the Chern densities for the left-handed and right-handed gauge fields. From this Chern density one can derive the sequence of equations

$$C_{2n}^{(5)}(F)=d_x\Omega_{2n-1,1}^{(5)}(A,0) \; ; \tag{5.21}$$

where $\Omega_{2n-1,1}^{(5)}$ describes correctly the current anomaly indicated in Eq.(5.4).

Since

$$Q^{-1}\gamma_5 Q=-\gamma_5 \; , \tag{5.22}$$

it follows that

$$C_{2n}^{(5)}(F^Q)=-C_{2n}^{(5)}(F) \; . \tag{5.23}$$

Therefore

$$\Omega_{2n-1,1}^{(5)}(A,0)+\Omega_{2n-1,1}^{(5)}(A^Q,0) \tag{5.24}$$

is a close form. From gauge invariance and Eq.(5.22), we have

$$\Omega_{2n-1,1}^{(5)}(A^Q,0)=-\Omega_{2n-1,1}^{(5)}(A,Q^{-1}dQ). \tag{5.25}$$

Using Eq.(3.41), we get

$$\begin{aligned}
&\Omega_{2n-1,1}^{(5)}(A,0)+\Omega_{2n-1,1}^{(5)}(A^Q,0)\\
&=\Omega_{2n-1}^{(5)}(A,0)-\Omega_{2n-1,1}^{(5)}(A,Q^{-1}dQ)\\
&=\Omega_{2n-1,1}^{(5)}(Q^{-1}dQ,0)-d_x\Omega_{2n-2,2}(A,0,Q^{-1}dQ) \; ,
\end{aligned} \tag{5.26}$$

where

$$\begin{aligned}
&\Omega_{2n-1,1}^{(5)}(Q^{-1}dQ,0)\\
&=\alpha_n B(n,n)\mathrm{Tr}\{\gamma_5(Q^{-1}dQ)^{2n-1}\}\\
&=2\alpha_n B(n,n)\mathrm{Tr}\{(U^{-1}dU)^{2n-1}\}
\end{aligned} \tag{5.27}$$

is a closed form. Integration of $\Omega_{2n-1,1}^{(5)}(Q^{-1}dQ,0)$ is an even integer corresponding to the winding number of a mapping from the 2n-1 sphere to the coset manifold $G_L\times G_R/G_{L+R}$.

At the same time we have

$$\Omega_{2n-1,1}^{(5)}(A,0)-\Omega_{2n-1,1}^{(5)}(A^Q,0)$$

$$=\Omega^{(5)}_{2n-1,1}(A,A^{Q})-d_x\Omega_{2n-2,2}(A,0,A^{Q}) \ , \qquad (5.28)$$

where $\Omega^{(5)}_{2n-1,1}(A,A^{Q})$ is a gauge-invariant function. Combining Eqs.(5.26) and (5.28) we have

$$\Omega^{(5)}_{2n-1,1}(A,0)$$
$$=\frac{1}{2}\{\Omega^{(5)}_{2n-1,1}(Q^{-1}dQ,0)+\Omega^{(5)}_{2n-1,1}(A,A^{Q})$$
$$-d_x[\Omega_{2n-2,2}(A,0,Q^{-1}dQ)+\Omega_{2n-2,2}(A,0,A^{Q})]\} . \qquad (5.29)$$

We now define

$$\rho_{2n-1}=2\pi[\Omega^{(5)}_{2n-1,1}(A,0)-\frac{1}{2}\Omega^{(5)}_{2n-1,1}(A,A^{Q})]$$
$$=\pi[\Omega^{(5)}_{2n-1,1}(Q^{-1}dQ,0)=d_x[\Omega^{(5)}_{2n-2,2}(A,0,Q^{-1}dQ)$$
$$+\Omega^{(5)}_{2n-2,2}(A,0,A^{Q})]] . \qquad (5.30)$$

It is easily verified that ρ_{2n-1} thus defined satisfies all three conditions stated above. The effective action related to the anomaly can therefore be written in the form

$$\Gamma^{a}_{eff}=\int_{\substack{S\\ \partial S=M}}\rho_{2n-1}$$
$$=2\pi\int_{\substack{S\\ \partial S=M}}\Omega_{2n-1,1}(U^{-1}dU,0)$$
$$=-\pi\int_{M}[\Omega^{(5)}_{2n-2,2}(A,0,Q^{-1}dQ)+\Omega^{(5)}_{2n-2,2}(A,0,A^{Q})] . \qquad (5.31)$$

The only term in a (2n-2)-dimensional effective Lagrangian that does not have local expression in 2n-2 dimension is the one corresponding to the local expression of a winding number expressed in one dimension higher.

When the flavor symmetry $SU(N_f)_L\times SU(N_f)_R$ breaks spontaneously into $SU(N_f)_{L+R}$ in the low energy limit, pion-like Goldstone bosons are created in the direction of the coset generators. These chiral Goldstone boson fields are described by the matrix U(x) introduced above. We write

$$U(x)=\exp\{\frac{i}{f_\pi}\hat{\lambda}_i\pi_i(x)\} \ ,$$

where $\hat{\lambda}_i$, $i=1,..,N_f$, are Gell-Mann matrices and $\pi_i(x)$ are the pion-like Goldstone field. Expanding the effective action Eq.(5.31) in power series of the pion fields, we get the self-interaction between pion fields and their interactions with the gauge fields.

It is interesting to note that the term linear in the pion field $\pi_i(x)$ is proportional to the unsymmetrical current anomaly for chiral currents[5].

If an Abelian gauge field $C(\xi)$ is taken into consideration and the effective Lagrangian in constructed from $\bar{W}_{2n-2,1}$ defined in Eq.(4.21), we obtain from Eq.(4.19) that

$$\Gamma=\Gamma_{W-Z}+\Gamma_{\theta} ,$$

where

$$\Gamma_{\theta}=2\pi\int_{x\in M}\int W'_{2n-2,1}(A,B)=\theta\int C_{2n-2}(F)$$

with

$$\theta=i\int_{\xi}C(\xi).$$

This is a possible candidate for the θ-vacuum term. Since θ is a line integral of a gauge potential in group space, it is uniquely defined only when its field strength $G=d_{\xi}C(\xi)$ vanishes. Therefore $C(\xi)$ is a pure gauge $C(\xi)=d\theta(\xi)$ and $\theta=\theta(1)-\theta(0)$. We know that the axial U(1) transformation produces a θ-vacuum term in the effective Lagrangian, which causes the strong CP problem, a serious trouble in particle physics. It could be that the axial U(1) transformation is closely related to a gauge transformation of a gauge potential in the group manifold so that both contribute a θ-vacuum term while the two terms cancel each other.

VI. The Effective Action of a Nonlinear σ Model Interacting with External Gauge Field

Consider a theory with symmetry group G spontaneously broken down to a subgroup H. The effective theory then describes Goldstone boson fields in the direction of the coset space G/H. Any group element $g\in G$ can be decomposed into the product of an element ϕ belonging to the coset G/H and an element h of the subgroup H.

$$g=\phi\cdot h . \tag{6.1}$$

The Goldstone fields are then described by a function $\phi(x)\in G/H$. Under a group transformation the element $g\phi(x)$ can again decompose in the form of Eq.(6.1)

$$g\cdot\phi(x)=\phi(g,\phi(x))h(g,\phi(x)), \tag{6.2}$$

where $\phi(g,\phi(x))\in G/H$ and $h(g,\phi(x))\in H$ satisfy the group multiplication law

$$\left.\begin{aligned}&\phi(g',\phi(g,\phi(x)))=\phi(g'g,\phi(x)),\\&h(g',\phi(g,\phi(x)))h(g,\phi(x))=h(g'g,\phi(x)).\end{aligned}\right\} \tag{6.3}$$

Therefore the Goldstone field $\phi(x)$ transforms as

$$\phi(x)\longrightarrow\phi'(x)\equiv\phi(g,\phi(x))=g\phi(x)h^{-1}(g,\phi(x)). \tag{6.4}$$

The one-form $\phi^{-1}(x)d\phi(x)$ is valued in the Lie algebra of G which can be decomposed into two parts

$$\phi^{-1}(x)d\phi(x)=H+K, \tag{6.5}$$

where H is a one-form valued in the Lie algebra of the subgroup H and K is that of the coset G/H.

Under a global transformation of g, we have from Eqs.(6.4) and (6.5) that

$$H\longrightarrow H'=h(g,\phi(x))(H+d)h^{-1}(g,\phi(x)) \tag{6.6}$$

and

$$K\longrightarrow K'=h(g,\phi(x))Kh^{-1}(g,\phi(x)). \tag{6.7}$$

The kinetic energy of the Goldstone boson fields in constructed to be

$$\frac{1}{f^2}\int Tr(K_\mu K^\mu)\ , \tag{6.8}$$

which is easily seen to be an invariant with the nonlinear realization of the field $\phi(x)$ described by Eq.(6.4).

When fermions and external gauge fields are present, they are described by a Lagrangian

$$\bar{\psi}_L i\gamma^\mu(\partial_\mu+A_\mu)\psi_L\ . \tag{6.9}$$

Before symmetry breaking, left-handed ψ_L and A are representations of the group G. After the symmetry breaking the fermions and the gauge fields become

$$\psi_{hL}(x)=\phi^{-1}(x)\psi_L(x)$$

$$A_h(x)=\phi^{-1}(x)(A(x)+d)\phi(x) \tag{6.10}$$

and the corresponding Lagrangian takes the form

$$\bar{\psi}_{hL} i\gamma^{\mu}(\partial_{\mu}+A_{h\mu}(x))\psi_{hL} \tag{6.11}$$

The effective action after the symmetry breaking has to produce the same anomaly as the original action. This requires additional terms depending on $\phi(x)$ and $A(x)$ to the effective Lagrangian for the nonlinear σ model. These additional terms have to satisfy the same conditions discussed in the prevous sections.

Let us introduce a double matrix notation as before and write

$$Q(x)=\begin{pmatrix} 0 & \phi(x) \\ \phi^{-1}(x) & 0 \end{pmatrix}, \qquad \eta=\begin{pmatrix} 1 & 0 \\ 0 & -1 \end{pmatrix} \tag{6.12}$$

and

$$\mathscr{A}(x)=\begin{pmatrix} A(x) & 0 \\ 0 & \phi^{-1}(x)(A(x)+d)\phi(x) \end{pmatrix}, \tag{6.13}$$

we have

$$Q^{-1}(x)=Q(x), \quad Q\eta Q=-\eta, \tag{6.14}$$

$$\mathscr{A}^{Q}=Q(\mathscr{A}(x)+d)Q(x)=\mathscr{A}(x) \tag{6.15}$$

and

$$\mathscr{F}=Q\mathscr{F}Q=\mathscr{F}^{Q}. \tag{6.16}$$

The Chern density

$$C^{\eta}_{2n}(\mathscr{F})=\alpha_n \mathrm{Tr}(\eta\mathscr{F}^{n}) \tag{6.17}$$

vanishes since

$$\mathrm{Tr}(\eta\mathscr{F}^{n})=\mathrm{Tr}(\eta\mathscr{F}^{Qn})=\mathrm{Tr}(\eta Q\mathscr{F}^{n}Q)=-\mathrm{Tr}(\eta\mathscr{F}^{n})=0. \tag{6.18}$$

The Chern-Simons secondary class descending from the Chern density Eq.(6.17) is

$$\Omega^{\eta}_{2n-1,1}(\mathscr{A},0)=\Omega_{2n-1,1}(A,0)-\Omega_{2n-1,1}(A_h,0), \tag{6.19}$$

whose variation under group transformation produces the difference of the current anomaly before and after the symmetry breaking. From Eq.(6.15) we have

$$\begin{aligned}&\Omega^{\eta}_{2n-1,1}(\mathscr{A},0)\\ &=\Omega^{\eta}_{2n-1,1}(\mathscr{A}^{Q},0)=-\Omega^{\eta}_{2n-1,1}(\mathscr{A},QdQ).\end{aligned} \tag{6.20}$$

Hence

$$\begin{aligned}&\Omega^{\eta}_{2n-1,1}(\mathscr{A},0)\\ &=\tfrac{1}{2}(\Omega^{\eta}_{2n-1,1}(\mathscr{A},0)-\Omega^{\eta}_{2n-1,1}(\mathscr{A},QdQ))\\ &=\tfrac{1}{2}(\Omega^{\eta}_{2n-1,1}(QdQ,0)+d\Omega^{\eta}_{2n-2,2}(\mathscr{A},0,QdQ))\\ &=\Omega_{2n-1,1}(\phi^{-1}d\phi,0)+\tfrac{1}{2}d\Omega^{\eta}_{2n-2,2}(\mathscr{A},0,QdQ).\end{aligned} \tag{6.21}$$

The first term in Eq.(6.21) is a closed form whose integral over (2n-1)-manifold is a winding number of the mapping described by the coset function $\phi(x)$. Therefore, $2\pi\int_S\Omega^{\eta}_{2n-1,1}(\mathscr{A},0)$ satisfies all the conditions required and can be choosen to be the additional term of the effective action for the nonlinear σ model. When the external gauge fields are switched off, we get the same result obtained in Ref.[6]. Our result is more general since it contains interaction of the Goldstone boson with the gauge field described by the term $\int_M\Omega^{\eta}_{2n-2,2}(A,0,QdQ)$. From our derivation it is also apparent that if the original theory has no anomaly so does the effective theory. The t'Hooft consistency condition is automatically incorporated.

VII. Central Term in Virasoro Algebra

As discussed in the previous sections, a 2-cocycle can be constructed from the Chern density with two higher numbers of dimensions. This 2-cocycle is proportional to the anomalous term in the commutator of gauge group generators[2].

It is well known in the string theory that the Virasoro algebra has an anomalous term called the central term[7]. We shall show in this section how this central term can be obtained by constructing a 2-cocycle appropriate to this case.

The generators L_n of the Virasoro string form a conformal algebra in the classical theory

$$[L_n,L_m]=(m-n)L_{n+m},\qquad n,m=\ldots,-1,0,1,\ldots\ . \tag{7.1}$$

When the string is quantized, an additional term of the form

$$\delta_{n,-m}\,\frac{C}{12}(n^3-n) \tag{7.2}$$

appears in the commutator.

Physically, no gauge field has been introduced in the string theory. An interesting question is whether the above anomaly can be described by a 2-cocycle induced by an fictitious gauge field introduced purely as a tool of calculations.

The Virasoro algebra with anomalous term has the form

$$[L_n,L_m]=(m-n)L_{n+m}+C_{m,n}\ , \tag{7.3}$$

where $C_{m,n}$ are antisymmetric

$$C_{m,n}=-C_{n,m} \tag{7.4}$$

and satisfy an equation due to Jacobi identity

$$(m-n)C_{m+n,\ell}+(n-\ell)C_{n+\ell,m}+(\ell-m)C_{\ell+m,n}=0\ . \tag{7.5}$$

Solutions of Eqs.(7.4) and (7.5) can be easily found. They are

$$C_{m,n}=(m-n)b(m+n)+d\delta_{m+n,0}m^3\ , \tag{7.6}$$

where b(n+m) is an arbitrary function of (m+n). The commutation relations (7.3) now become

$$[L_n,L_m]=(m-n)(L_{n+m}+b(n+m))+d\delta_{n+m,0}m^3\ . \tag{7.7}$$

Making the change

$$L_n\longrightarrow L_n-b(n)\ , \tag{7.8}$$

we can rewrite Eq.(7.7) in the following form

$$[L_n,L_m]=(m-n)L_{n+m}+dm^3\delta_{m,-n}\ . \tag{7.9}$$

The center in Virasoro algebra

$$\frac{C}{12}\delta_{m+n,0}(m^3-m) \tag{7.10}$$

can be cast into the form of Eq.(7.9) by a change of the generator

$$L_0\longrightarrow L_0+\frac{C}{24}\ , \tag{7.11}$$

and

$$d = \frac{C}{12} \quad . \tag{7.12}$$

Our next problem is to find a 2-cocycle which produces this anomalous center term. For this purpose we consider a representation of L_n

$$L_n = z^n\left(z\frac{\partial}{\partial z} + an + f(z)\right) \quad . \tag{7.13}$$

The arbitrary function f(z) can be eliminated by a transformation of the wave function $\phi(z)$

$$\phi(z) \longrightarrow e^{-\int f(z)/z \cdot dz}\phi(z) \quad . \tag{7.14}$$

Therefore without loss of generality we can take f(z)=0.

Under a finite transformation the wave function $\phi(z)$ changes into

$$\phi'(z) = e^{i\sum\limits_n \varepsilon_n L_n}\phi(z) = U_\varepsilon(z)\phi(z') \quad , \tag{7.15}$$

where ε_n(n=...,1,0,1...) are group parameters and

$$L_n = z^n\left(z\frac{\partial}{\partial z} + an\right) \quad . \tag{7.16}$$

The coordinate

$$z' = e^{\sum\limits_n \varepsilon_n z^{n+1}\frac{\partial}{\partial z}} z \tag{7.17}$$

can be obtained in the following way. Let us define a function $\tau(z)$ such that

$$\frac{d\tau}{dz} = \frac{1}{\sum\limits_n \varepsilon_n z^{n+1}} \quad , \tag{7.18}$$

then

$$\sum_n \varepsilon_n z^{n+1}\frac{\partial}{\partial z} = \frac{\partial}{\partial \tau} \quad . \tag{7.19}$$

The operator

$$e^{\sum\limits_n \varepsilon_n z^{n+1}\frac{\partial}{\partial z}} = e^{\frac{\partial}{\partial \tau}} \tag{7.20}$$

is a displacement operator of the coordinate τ. Therefore

$$z' = e^{\frac{\partial}{\partial \tau}} z(\tau) = z(\tau(z)+1) \quad . \tag{7.21}$$

As a special case we take

$$\varepsilon_\ell=\delta_{\ell,n}\varepsilon \ . \tag{7.22}$$

It is easily found that

$$z'=z(1-n\varepsilon z^n)^{-1/n}=z_\varepsilon \ . \tag{7.23}$$

In this special case the phase factor U(z) can also be found as

$$U_\varepsilon(z)=\frac{1}{(1-n\varepsilon z^n)^a} \ , \tag{7.24}$$

which is seen to satisfy the group relation

$$U_{\varepsilon'}(z)U_\varepsilon(z_{\varepsilon'})=U_{\varepsilon+\varepsilon'}(z) \ . \tag{7.25}$$

The presence of the phase factor U(z) indicates that besides the coordinate transformation, a U(1) gauge field A(z) is needed to define a covariant derivative. Under the conformal transformation described by the Virasoro algebra, the gauge potential transforms as

$$A(z)\longrightarrow A'(z)=A(z')+U_\varepsilon^{-1}(z)dU_\varepsilon(z) \ . \tag{7.26}$$

To calculate the commutator anomaly we have to find the 2-cocycle descending from the Chern density in four dimensions. From Eq.(3.34), the anomalous term is proportional to

$$\int\Omega_{2,2}=\frac{1}{8\pi^2}\int vdv \tag{7.27}$$

where

$$v=U_\varepsilon d_\xi U_\varepsilon^{-1}$$
$$=\sum_n \{d\xi^{(1)}(\varepsilon_{1n}+\xi^{(2)}\varepsilon_{2n})+\xi^{(1)}d\xi^{(2)}\varepsilon_{2n}\}anz^n \ . \tag{7.28}$$

Here $\xi^{(i)}$, i=1,2, are two parameters varying from 0 to 1 and ε_{1n} are infinitesimal group parameters.

It is eassily found that

$$vdv=\sum_{n,m} mamanz^{n+m-1}dz\xi^{(1)}d\xi^{(1)}d\xi^{(2)}$$
$$(\varepsilon_{1n}\varepsilon_{2m}-\varepsilon_{1m}\varepsilon_{2n}). \tag{7.29}$$

Integrating Eq.(7.29) over $\xi^{(i)}(i=1,2)$ and z, we obtain for the nontrivial 2-cocycle

$$\frac{1}{8\pi^2}\int vdv = \frac{a^2\ell}{8\pi}m^3\delta_{n+m,0}2\varepsilon_{1n}\varepsilon_{2m} \quad , \tag{7.30}$$

where ℓ is the number of times which is determined by the line integral wrapping around z=0. The factor $a^2\ell m^3\delta_{n+m,0}/2$ is the anomalous center term for the Virasoro algebra which has the same form deduced by directly solving the Jacobi identity. This is not surprising since Eq.(7.5) is an infinitesimal form of the coclosed condition fc a 2-cocycle.

References

[1] K.C. CHOU, H.Y. GUO, K. WU and X.C. SONG, Commun. Theor. Phys.(Beijing) 3(1984)125,129; K.C. CHOU, H.Y. GUO, X.Y. LI, K. WU and X.C. SONG, Commun. Theor. Phys. (Beijing) 3(1984)491,498; K.C. CHOU, H.Y. GUO, K. WU and X.C. SONG, Commun. Theor. Phys.(Beijing)3(1984)593,603; K.C. CHOU, H.Y. GUO and K. WU, Commun. Theor. Phys. 4(1985)91, 101; H.Y. HUO, B.Y. GOU, S.K. WANG and K. WU, Commun. Theor. Phys. 4(1985)233,251.

[2] L.D. Faddeev, Phys. Lett. 145B(1984)81.

[3] R. Jackiw, Phys. Rev. Lett. 54(1985)159.

[4] B. Zumino, Y.S. Wu and A. Zee, Nucl. Phys. B239(1984)477; B. Zumino, Nucl. Phys. B253(1985)477.

[5] E. Witten, Nucl. Phys. B156(1983)422; L. Alvarez Gaume and E. Witten, Nucl. Phys. B234(1983)269.

[6] L. Alvarez Gaume and P. Ginsparg, HUTP preprint 85/A015(1985); J. Bagger, D. Nemeschansky and S. Yankielowicz, SLAC PUB3698 HUTP 85/A042 (1985).

[7] A. Neveu and J.H. Schwarz, Nucl. Phys. B31(1971)86; P. Ramond, Phys. Rev. D3(1971)2415; P. Goddard, J. Goldstone, C. Rebbi and C.B. Thorr, Nucl. Phys. B56(1973)109.

[8] W.A. Bardeen, Phys. Rev. 184(1969)1848; D.J. Gross and R. Jackiw, Phys. Rev. D6(1972)477.

[9] J. Wess and B. Zumino, Phys. Lett. 37B(1971)95.

PHYSICAL REVIEW D VOLUME 43, NUMBER 2 15 JANUARY 1991

Signature for chiral-symmetry breaking at high temperatures

Lay-Nam Chang
Physics Department, Virginia Polytechnic Institute and State University, Blacksburg, Virginia 24061

Ngee-Pong Chang
Department of Physics, The City College of The City University of New York, New York, New York 10031

Kuang-Chao Chou
Institute of Theoretical Physics, Academia Sinica, Beijing, China

(Received 11 July 1990)

In this paper we study the temperature dependence of chiral-symmetry breaking. Our real-time calculation shows that the thermal fermion propagator has a Lorentz-invariant massive particle pole even as $\langle \mathrm{vac}|\bar{\psi}\psi|\mathrm{vac}\rangle_\beta$ vanishes for $T > \Lambda_c e^{2/3}$. The traditional signature of chiral-symmetry breaking is found only after transforming the Dirac field to the new chiral basis at high temperatures.

I. INTRODUCTION

Ever since the pioneering study[1] by Nambu and Jona-Lasinio (NJL) of the dynamical chiral-symmetry breaking in quantum field theory, it has been recognized that $\langle \bar{\psi}\psi \rangle$ gives the signature of chiral-symmetry breaking in the theory. The question that we shall address in this paper is whether this remains the signature for chiral-symmetry breaking at higher temperatures.

Naively, we would have thought that all we need do is continue to evaluate the vacuum expectation value of $\langle \bar{\psi}\psi \rangle$ in the new thermal vacuum and if it vanishes we would conclude that chiral symmetry is restored at high temperatures. The problem comes in, however, when we begin to study the thermal fermion propagator and discover the existence of a Lorentz-invariant massive pole that persists at high temperatures even as m_r, the tree Lagrangian fermion mass, approaches zero. In the same $m_r \to 0$ limit, however, $\langle \mathrm{vac}|\bar{\psi}\psi|\mathrm{vac}\rangle_\beta$ vanishes for $T > T_c$ where $T_c = \Lambda_c e^{2/3}$. The conflict is resolved when we recognize the need for a transform of ψ at high temperatures to correctly describe the Lorentz-invariant massive particle.

II. NJL VACUUM

To appreciate the problem, we review the picture of the chiral-broken vacuum as first formulated by NJL. (We assume that we have integrated over the gluon degrees of freedom in QCD and are here discussing the effective theory involving only fermions.) In the presence of dynamical interactions, the naive vacuum $|0\rangle$ may no longer be the lowest-energy eigenstate of the Hamiltonian. The new ground state of the Hamiltonian $|\mathrm{vac}\rangle$ is the analog of the BCS ground state with quark and antiquark pairings. If $a_{p,s}$ and $b_{p,s}$ are the annihilation operators for the massless quarks and antiquarks, respectively, with helicities $s=\pm$ for the R,L states, then the new ground state is given by

$$|\mathrm{vac}\rangle = \prod_{\mathbf{p},s} (\cos\theta_p - s\,\sin\theta_p a^\dagger_{p,s} b^\dagger_{-p,s})|0\rangle \; , \tag{1}$$

where θ_p is related to the dynamical mass m acquired by the quarks

$$\tan 2\theta_p = \frac{m}{p} \tag{2}$$

and p here is the magnitude of the momentum.

We will show in this section how the NJL ground state may be understood as a rotation in chiral space and demonstrate how it affects the Dirac equation. The original massless Dirac field $\psi(\mathbf{x},0)$ is taken here to have the expansion at time $t=0$:

$$\psi(\mathbf{x},0) = \frac{1}{\sqrt{V}} \sum_{p,s} \left[\begin{pmatrix} \xi_L(\mathbf{p})a_{p,L} \\ \xi_R(\mathbf{p})a_{p,R} \end{pmatrix} + \begin{pmatrix} \xi_R(\mathbf{p})b^\dagger_{-p,R} \\ -\xi_L(\mathbf{p})b^\dagger_{-p,L} \end{pmatrix} \right] e^{i\mathbf{p}\cdot\mathbf{x}} \tag{3}$$

$$\equiv \frac{1}{\sqrt{V}} \sum_{p} \psi(\mathbf{p}) e^{i\mathbf{p}\cdot\mathbf{x}} \; , \tag{4}$$

where ξ_s are helicity eigenstates satisfying

$$\sigma\cdot\hat{\mathbf{p}}\xi_s(\mathbf{p}) = s\xi_s(\mathbf{p}) \; . \tag{5}$$

Note that the Fourier component fields $\psi(\mathbf{p})$ are in general time dependent. For the purposes of our discussion here, we focus on the time slice at $t=0$.

The NJL chiral-broken ground state can be obtained by an infinite chain of $SU(2)_p$ chiral rotations around the (parity-conserving) two-axis by angle $2\theta_p$:

$$|\mathrm{vac}\rangle = \prod_{\mathbf{p}} e^{2i\theta_p X_2(\mathbf{p})}|0\rangle \tag{6}$$

$$\equiv \prod_{\mathbf{p}} \mathcal{R}_p(\theta_p)|0\rangle \tag{7}$$

$$= \Omega|0\rangle \; . \tag{8}$$

For simplicity, we shall refer to this rotation around the two-axis as the chiral-2 rotation.

A. SU(2)$_p$ algebra

The $X_i(\mathbf{p})$ are the generators

$$X_3(\mathbf{p})=-\tfrac{1}{2}\sum_s s(a^\dagger_{p,s}a_{p,s}+b^\dagger_{-p,s}b_{-p,s}) , \tag{9}$$

$$X_2(\mathbf{p})=\frac{i}{2}\sum_s s(a^\dagger_{p,s}b^\dagger_{-p,s}-b_{-p,s}a_{p,s}) , \tag{10}$$

$$X_1(\mathbf{p})=\tfrac{1}{2}\sum_s (a^\dagger_{p,s}b^\dagger_{-p,s}+b_{-p,s}a_{p,s}) . \tag{11}$$

They satisfy the SU(2)$_p$ algebra at each momentum $\mathbf{p}$:

$$[X_i(\mathbf{p}),X_j(\mathbf{p}')]=i\epsilon_{ijk}X_k(\mathbf{p})\delta_{p,p'} . \tag{12}$$

And if we further take the sum over the momenta and form the global generators

$$X_i\equiv\sum_p X_i(\mathbf{p}) , \tag{13}$$

we see that they form the global SU(2) algebra

$$[X_i,X_j]=i\epsilon_{ijk}X_k . \tag{14}$$

Note that we are here dealing with the infinite-dimensional representation of the global SU(2) algebra.

B. Transforming the Dirac equation

Under the chiral-2 rotation in the Hilbert space, the annihilation operators and the massless Dirac field transform as

$$a_{p,s}\rightarrow A_{p,s} , \tag{15}$$

$$b_{p,s}\rightarrow B_{p,s} , \tag{16}$$

where

$$A_{p,s}\equiv\mathcal{R}_p(\theta_p)a_{p,s}\mathcal{R}_p^{-1}(\theta_p) \tag{17}$$

$$=\cos\theta_p a_{p,s}+s\sin\theta_p b^\dagger_{-p,s} , \tag{18}$$

$$B_{p,s}\equiv\mathcal{R}_p(\theta_p)b_{p,s}\mathcal{R}_p^{-1}(\theta_p) \tag{19}$$

$$=\cos\theta_p b_{p,s}-s\sin\theta_p a^\dagger_{-p,s} , \tag{20}$$

and

$$A_{p,s}|\text{vac}\rangle=B_{k,s}|\text{vac}\rangle=0 . \tag{21}$$

The new operators $A_{p,s},B_{p,s}$ describe the excitons with dynamically generated mass m that propagate in the new medium.

Under the Hilbert-space chiral-2 rotation, the massless Dirac field of Eq. (3) transforms in an obvious way into

$$\psi(\mathbf{x},0)\rightarrow\tilde{\Psi}(\mathbf{x},0) , \tag{22}$$

where

$$\tilde{\Psi}(\mathbf{x},0)=\frac{1}{\sqrt{V}}\sum_p\left[\begin{bmatrix}\xi_L(\mathbf{p})A_{p,L}\\ \xi_R(\mathbf{p})A_{p,R}\end{bmatrix}+\begin{bmatrix}\xi_R(\mathbf{p})B^\dagger_{-p,R}\\ -\xi_L(\mathbf{p})B^\dagger_{-p,L}\end{bmatrix}\right]e^{i\mathbf{p}\cdot\mathbf{x}} . \tag{23}$$

Following NJL, we may treat the massive modes as approximate eigenstates of the total Hamiltonian, and consider the time evolution of the massive modes as if they were free particles. With the explicit representation given in Eq. (23) it is straightforward to derive, using the representation

$$\gamma=\begin{bmatrix}0 & -i\sigma\\ i\sigma & 0\end{bmatrix}, \quad \gamma_4\equiv i\gamma_0=\begin{bmatrix}0 & \mathbf{I}\\ \mathbf{I} & 0\end{bmatrix}, \tag{24}$$

the free field equation for the chiral rotated $\tilde{\Psi}$. It is the spatially nonlocal equation

$$\left[\frac{\sqrt{-\nabla^2+m^2}}{\sqrt{-\nabla^2}}\gamma\cdot\nabla+\gamma_0\frac{\partial}{\partial t}\right]\tilde{\Psi}=0 . \tag{25}$$

This nonlocal Dirac equation is strange because it appears to show chiral invariance, and yet we know by construction that the chiral-rotated $\tilde{\Psi}$ describes the massive particle free field. It is also not Lorentz invariant. But in our context of temperature field theory this latter objection plays no role. It is therefore reassuring to find that there is a similarity transformation acting on the components of $\tilde{\Psi}$ that transforms away the nonlocality in the Dirac equation.

If we work with the Fourer components of $\tilde{\Psi}$ in momentum space, and define the similarity transformation

$$\tilde{\Psi}(\mathbf{p},t)\equiv e^{-i\theta_p\gamma\cdot\hat{\mathbf{p}}}\Psi(\mathbf{p},t) , \tag{26}$$

then the nonlocal equation for $\tilde{\Psi}$ implies the usual massive Dirac equation for Ψ:

$$\left[\gamma\cdot\nabla+\gamma_0\frac{\partial}{\partial t}+m\right]\Psi(\mathbf{x},t)=0 . \tag{27}$$

From Eq. (26), we can show in our representation of Dirac matrices that $\Psi(\mathbf{x},t)$ has the usual expansion ($p_0=\sqrt{p^2+m^2}$)

$$\Psi(\mathbf{x},t)=\frac{1}{\sqrt{V}}\sum_{\mathbf{p},s}(U_{p,s}A_{p,s}e^{-ip_0t}+V_{-p,s}B^\dagger_{-p,s}e^{+ip_0t})e^{i\mathbf{p}\cdot\mathbf{x}} , \tag{28}$$

where the massive spinors are given explicitly by

$$U_{p,L}=\begin{bmatrix}\cos\theta_p\xi_L(\mathbf{p})\\ \sin\theta_p\xi_L(\mathbf{p})\end{bmatrix} , \tag{29}$$

$$U_{p,R}=\begin{bmatrix}\sin\theta_p\xi_R(\mathbf{p})\\ \cos\theta_p\xi_R(\mathbf{p})\end{bmatrix} , \tag{30}$$

$$V_{-p,L}=\begin{vmatrix}\sin\theta_p\xi_L(\mathbf{p})\\ -\cos\theta_p\xi_L(\mathbf{p})\end{vmatrix}, \tag{31}$$

$$V_{-p,R}=\begin{vmatrix}\cos\theta_p\xi_R(\mathbf{p})\\ -\sin\theta_p\xi_R(\mathbf{p})\end{vmatrix}, \tag{32}$$

and satisfy the Dirac equations

$$(i\gamma\cdot\mathbf{p}-i\gamma_0\cdot p_0+m)U_{p,s}=0\ , \tag{33}$$

$$(i\gamma\cdot\mathbf{p}-i\gamma_0\cdot p_0-m)V_{p,s}=0\ . \tag{34}$$

Equation (26) is in fact an example of a transformation of the generic Foldy-Wouthuysen type (more precisely, a Cini-Touschek transformation).[2]

The result of this section may therefore be summarized in the transformation law of the massless Dirac field under the Hilbert-space chiral-2 rotation

$$\Omega\psi(\mathbf{x},0)\Omega^{-1}=e^{-\theta\gamma\cdot\nabla/\sqrt{-\nabla^2}}\Psi(\mathbf{x},0)\ , \tag{35}$$

where θ is here the differential operator in three-dimensional space implied by the momentum-space equation (2). The Cini-Touschek similarity transformation is the analog of the $\mathcal{D}_{\alpha\beta}(\Lambda)$ similarity transform of ψ that results from a Hilbert-space Lorentz transformation. Loosely speaking, a Hilbert-space chiral-2 rotation induces on the ψ a Cini-Touschek similarity transform. The nonlocality of the similarity transform reflects the infinite-dimensional nature of the SU(2) representation involved here.

The nonlocality of the $\tilde{\Psi}$ field equation is also a warning that it is the wrong basis on which to discuss the chirality of the theory. Indeed, Eq. (25) gives the false indication of chiral conservation. It is only after the nonlocal Dirac equation has been straightened out that one can test for chiral-symmetry breaking under the chiral X_3 rotations.

As we shall see, at high temperatures, the thermal radiative corrections lead to a nonlocal Dirac equation for the massive particle pole in the Green's function. Proper physical interpretation of the signature of chiral breaking requires that we do a Cini-Touschek transformation to get rid of the nonlocality in the Dirac equation of the renormalized particle pole in the Green's function. After this similarity transformation, the inherent chiral-symmetry breaking due to temperature effects becomes evident.

Before we close this section, we note the equality

$$\psi(\mathbf{x},0)=\Psi(\mathbf{x},0)\ . \tag{36}$$

This may be verified by direct substitution of the inverse transformation to Eqs. (18) and (20) into the expansion for ψ as given in Eq. (3).

At zero temperature, then, the correct signal for chiral-symmetry breaking is to use either ψ or Ψ and calculate the expectation values of $\bar{\psi}\psi$ or $\bar{\Psi}\Psi$ with respect to the full vacuum. The equality between the two Heisenberg operators at $t=0$, Eq. (36), guarantees that the vacuum expectations values obtained by the two different ψ's agree:

$$\langle \text{vac}|\bar{\psi}\psi|\text{vac}\rangle=\langle \text{vac}|\bar{\Psi}\Psi|\text{vac}\rangle\ . \tag{37}$$

III. HIGH-TEMPERATURE RESULT

In an earlier work,[3] we had reported the result of a real-time temperature-dependent field-theory calculation of dynamical chiral-symmetry breaking at high temperatures. We found that, for QCD, dynamical symmetry breaking persists at high temperatures. In this section, we present an analysis of the result and point out the close connection between the zero-temperature chiral rotation of the NJL vacuum and the Cini-Touschek transformation needed in the renormalization of the temperature-dependent Fermion two-point function.

We perform our calculation in real time.[4] Our technique is to introduce into the Lagrangian an explicit mass term for the fermion, put the system in a heat bath, use renormalization-group analysis to sum over higher loops, and study the *critical* limit as $m_r\to 0$. If in this limit, the thermal fermion propagator shows a Lorentz-invariant massive particle pole, then we say that dynamical symmetry breaking has occurred. At zero temperature,[5] we found that the chiral flip part of $S_r^{-1}(p)$, for $p_\mu p_\mu$ in some finite domain, actually survives the critical $m_r\to 0$ limit, thus signaling the bifurcation in chiral-symmetry breaking. In this section, we look for the temperature dependence of this chiral-symmetry breaking.

The result of the real-time thermal field-theory calculation may be put in the form (our results for A,B agree with the one-loop calculation of Weldon,[6] except that we have also included $\ln T/m$ terms; Weldon does not introduce an explicit mass term, and thus did not look for a perturbative root around the original m_r pole)

$$S_\beta^{-1}(\mathbf{p}^2,p_0^2T)=i\gamma\cdot\mathbf{p}(1+A)-i\gamma_0 p_0(1+B)+m_r(1+C)\ , \tag{38}$$

where A,B,C are functions of p, p_0, and T. In terms of the parameters $|\mathbf{p}|\equiv m_r\sinh\xi, p_0\equiv m_r\cosh\xi$, we have, using the Feynman gauge, and in the limit of $T^2/m_r^2\gg 1$,

$$A=\lambda_r C_f\left[-\left[\ln\frac{m_r^2}{\mu^2}-2\right]-\frac{2\pi^2T^2}{m_r^2}\frac{(\xi\coth\xi-1)}{(\sinh\xi)^2}-\ln\frac{T^2}{m_r^2}\right], \tag{39}$$

$$B=\lambda_r C_f\left[-\left[\ln\frac{m_r^2}{\mu^2}-2\right]-\frac{2\pi^2T^2}{m_r^2}\frac{\xi}{\sinh\xi\cosh\xi}-\ln\frac{T^2}{m_r^2}\right], \tag{40}$$

$$C=\lambda_r C_f\left[-4\left[\ln\frac{m_r^2}{\mu^2}-\frac{3}{2}\right]-4\ln\frac{T^2}{m_r^2}\right]. \tag{41}$$

Here we have dropped terms that are of order 1 as $T^2/m_r^2\to\infty$. Also we have defined $\lambda_r=g_r^2/(16\pi^2)$ and the relation $T_a\cdot T_a\equiv C_f\mathbf{I}$.

The Lorentz-invariant massive particle pole in the thermal fermion Green's function occurs at $p_0=\sqrt{p^2+\mathcal{M}^2}$, where perturbatively

$$\mathcal{M}^2 = \lim_{m_r \to 0} m_r^2 \left\{ 1 + \lambda_r C_f \left[-6 \left[\ln \frac{m_r^2}{\mu^2} - \frac{4}{3} \right] + \frac{4\pi^2 T^2}{m_r^2} - 6 \ln \frac{T^2}{m_r^2} + \text{const} \right] + \cdots \right\} . \tag{42}$$

The critical limit $m_r \to 0$ is taken using the fixed-point theorem of bifurcation theory. [See the discussion following Eq. (68) in Ref. 3.] As was shown in Ref. 3, this mass survives the critical limit as $m_r \to 0$, so that it is a temperature-dependent dynamical mass:

$$\mathcal{M}^2 \underset{T \to \infty}{\sim} \frac{2\pi^2}{3} \frac{T^2}{\ln \frac{T^2}{\Lambda_c^2}} . \tag{43}$$

Donoghue and Holstein[7] considered the case $m_r \neq 0$ and also found in the one-loop perturbative calculation the Lorentz-invariant massive pole at high temperatures. Our technique goes beyond one loop by using the renormalization-group analysis to sum over higher loops.

The existence of this particle mass at high temperature is already a good signal that chiral-symmetry breaking persists at high temperatures. What we want to study is whether the traditional signature of chiral-symmetry breaking is still good at high temperatures, viz.,

$$\lim_{m_r \to 0} \langle \text{vac} | \bar{\psi}\psi | \text{vac} \rangle_\beta \neq 0 . \tag{44}$$

A potential conflict with the usual notion of chiral-symmetry breaking arises when we study the chiral flip part of S_β^{-1} [Eq. (38)] and directly take the limit as $m_r \to 0$. This would be consistent with the idea that we simply evaluate the $\langle \text{vac} | \bar{\psi}\psi | \text{vac} \rangle_\beta$ in the thermal vacuum without doing any more renormalization than the minimal ones needed at $T=0$. In our case, our result in Eq. (38) is in Feynman gauge, which explains why even at $T=0$ the coefficient of $\gamma_0 p_0$ is not unity. So we renormalize S_β^{-1} by setting the coefficient of $\gamma_0 p_0$ equal to unity when $T=0$. We do this by multiplying it by the $T=0$ wave-function renormalization z_2:

$$z_2 S_\beta^{-1}|_{\text{chiral flip}} = \lim_{m_r \to 0} m_r \left\{ 1 + \lambda_r C_f \left[-3 \left[\ln \frac{T^2}{\mu^2} - \frac{4}{3} \right] \right] + \cdots \right\} \tag{45}$$

$$= \lim_{m_r \to 0} m_r (\lambda_r y)^{-6C_f/b} \tag{46}$$

where

$$z_2 = 1 - \lambda_r C_f \left[\ln \frac{m_r^2}{\mu^2} - 2 \right] + \cdots , \tag{47}$$

$$y = \frac{1}{\lambda_r} + \frac{b}{2} \left[\ln \frac{T^2}{\mu^2} - \frac{4}{3} \right] \tag{48}$$

$$= \frac{b}{2} \left[\ln \frac{T^2}{\Lambda_c^2} - \frac{4}{3} \right] . \tag{49}$$

Here y is a renormalization-group invariant. Each term in the perturbative series is valid so long as $T^2 \gg m_r^2$. Since in the end $m_r \to 0$, it would appear that the series should be valid for all T. However, the positivity requirement for y due to the representation, Eq. (46), shows that we need to impose the condition $T > \Lambda_c e^{2/3}$ for the validity of the sum.

When we now take the critical limit $m_r \to 0$, y to one-loop renormalization-group accuracy does not depend on m_r and does not approach the $y=0$ fixed point [see the discussion following Eq. (68) in Ref. 3], and so the chiral flip part does not survive the critical limit. It vanishes for $T > \Lambda_c e^{2/3}$. The only exception is when the temperature T is at the critical temperature[8,9]

$$T_c = \Lambda_c e^{2/3} . \tag{50}$$

At that point, our one-loop renormalization-group analysis fails since it diverges even before the critical limit. We need to go to two-loop renormalization-group analysis to study further the order of the phase transition at T_c.

Based only on the calculation thus far, we would conclude that chiral-symmetry breaking goes away for high temperatures, viz., for $T > T_c$. And yet the same calculation shows a Lorentz-invariant massive particle pole in the thermal fermion Green's function, which is a signal that chiral-symmetry breaking has occurred. To reconcile between the presence of a massive particle and the vanishing of the traditional signature of chiral-symmetry breaking, we proceed with the analysis of the thermal fermion Green's function.

The coefficients A, B, C in Eq. (38) are nonlinear functions of the momenta. At the particle pole $p_0 = \sqrt{p^2 + \mathcal{M}^2}$ the residue of the fermion thermal Green's function is not like the usual

$$(-i\gamma \cdot \mathbf{p} + i\gamma_0 p_0 + \mathcal{M}) , \tag{51}$$

but instead has the form

$$[-i\gamma \cdot \mathbf{p}(1+\mathcal{A}) + i\gamma_0 p_0 (1+\mathcal{B}) + m_r(1+\mathcal{C})] , \tag{52}$$

where $\mathcal{A}, \mathcal{B}, \mathcal{C}$ are nonpolynomial functions of $p^2 \equiv |\mathbf{p}|^2$, obtained by evaluating A, B, C on the particle mass shell, $p_0 = \sqrt{p^2 + \mathcal{M}^2}$. The Dirac equation for the massive particle is then the peculiar one:

$$[i\gamma\cdot\mathbf{p}(1+\mathcal{A})-i\gamma_0\sqrt{p^2+\mathcal{M}^2}(1+\mathcal{B})+m_r(1+\mathcal{C})]\tilde{u}=0\ . \tag{53}$$

If we naively form the renormalized field operator by

$$\tilde{\Psi}_R=\frac{1}{\sqrt{V}}\sum_{p,s}(\tilde{u}_{p,s}A_{p,s}e^{i\mathbf{p}\cdot\mathbf{x}-i\sqrt{p^2+\mathcal{M}^2}t}+\tilde{v}_{p,s}B^{\dagger}_{p,s}e^{-i\mathbf{p}\cdot\mathbf{x}+i\sqrt{p^2+\mathcal{M}^2}t})\ , \tag{54}$$

we would find that this renormalzed field does not satisfy the usual Dirac equation, but instead obeys the nonlocal equation

$$\left[(1+\mathcal{A})\gamma\cdot\nabla+(1+\mathcal{B})\gamma_0\frac{\partial}{\partial t}+m_r(1+\mathcal{C})\right]\tilde{\Psi}_R=0\ . \tag{55}$$

Based on our experience in the preceding section, it is clear that it would be dangerous to conclude anything about chiral-symmetry breaking based on this $\tilde{\Psi}_R$. Thermal radiative corrections have induced some chiral-2 rotation in the vacuum structure and we must find the generalized Cini-Touschek transformation that can remove the associated nonlocality. If we introduce

$$\tilde{u}=e^{i\Theta_p\gamma\cdot\hat{\mathbf{p}}}U\ , \tag{56}$$

then the Dirac equation for U is "straightened out" to the usual massive one

$$(i\gamma\cdot\mathbf{p}-i\gamma_0\sqrt{p^2+\mathcal{M}^2}+\mathcal{M})U=0\ , \tag{57}$$

provided

$$\tan 2\Theta_p=\frac{p[m_r(1+\mathcal{C})-\mathcal{M}(1+\mathcal{A})]}{p^2(1+\mathcal{A})+m_r\mathcal{M}(1+\mathcal{C})}\ . \tag{58}$$

The renormalization of the thermal field-theory propagator should thus involve an extra Cini-Touschek transformation on top of the wave-function renormalization factor

$$\psi(x)=\sqrt{Z_{2\beta}}e^{\Theta\gamma\cdot\nabla}\Psi_R(x)\ , \tag{59}$$

where Θ is the differential operator in three-dimensional space as implied by Eq. (58).

Accordingly, we have

$$S_\beta=Z_{2\beta}e^{i\Theta_p\gamma\cdot\hat{\mathbf{p}}}S_{\beta R}e^{i\Theta_p\gamma\cdot\hat{\mathbf{p}}}\ , \tag{60}$$

and as a result

$$S_{\beta R}^{-1}=i\gamma\cdot\mathbf{p}(1+A')-i\gamma_0 p_0(1+B')+m_r(1+C')\ , \tag{61}$$

where

$$1+A'=(1+\mathcal{B})^{-1}\left[(1+A)\cos 2\Theta_p+\frac{m_r}{p}\sin 2\Theta_p\right]\ , \tag{62}$$

$$1+B'=(1+\mathcal{B})^{-1}(1+B)\ , \tag{63}$$

$$1+C'=(1+\mathcal{B})^{-1}[m_r(1+C)\cos 2\Theta_p-p(1+A)\sin 2\Theta_p]\ , \tag{64}$$

so that, at the massive particle pole, the coefficient of γ_0 is properly normalized. Here

$$Z_{2\beta}^{-1}=1+\mathcal{B}(\mathbf{p}^2,T^2,m^2)\ . \tag{65}$$

We now study the chiral flip part of the renormalized inverse thermal Green's function, $S_{\beta R}^{-1}$, and take the critical limit as $m_r\to 0$. If the chiral flip part survives this limit, we then can properly claim that chiral symmetry persists at high temperatures.

Perturbatively, the chiral flip part of $S_{\beta R}^{-1}$ is given by

$$S_{\beta R}^{-1}|_{\text{chiral flip}}=m_r\left\{1+\lambda_r C_f\left[-3\left[\ln\frac{m_r^2}{\mu^2}-\frac{4}{3}\right]+\frac{2\pi^2T^2}{m_r^2}-3\ln\frac{T^2}{m_r^2}+\text{const}\right]+\cdots\right\}\ , \tag{66}$$

so that to first order in renormalization-group analysis, the chiral flip part of the renormalized inverse Green's function is simply the Lorentz-invariant physical mass of the particle pole, $\mathcal{M}$. As shown in our earlier work, this mass survives the critical limit $m_r\to 0$. Chiral-symmetry breaking persists at high temperatures. Our conclusion therefore is that at high temperatures the traditional signature of chiral-symmetry breaking reappears only when one uses the transformed Ψ_R field, and we calculate its vacuum expectation value

$$\langle \text{vac}|\bar{\Psi}_R\Psi_R|\text{vac}\rangle_\beta\neq 0\ . \tag{67}$$

Unfortunately, our results at this stage cannot be used to evaluate this thermal vacuum expectation value. We need a study of the thermal propagator in the full complex p_0 plane.

IV. CONCLUSION

In conclusion, we note once again that the traditional signature for chiral-symmetry breaking $\langle \text{vac}|\bar{\psi}\psi|\text{vac}\rangle_\beta$ is an inadequate indicator of chiral-symmetry breaking at high temperatures. Our calculations with dynamical symmetry breaking in QCD at high temperatures show a persistent Lorentz-invariant massive particle pole in the thermal fermion propagator at high temperatures—and this in spite of a vanishing $\langle \text{vac}|\bar{\psi}\psi|\text{vac}\rangle_\beta$ at high temperatures.

Thermal radiative corrections induce a further chiral-2 rotation of the vacuum structure, and the resulting renormalized Dirac field undergoes a generalized Cini-Touschek similarity transformation. In the transformed basis, the traditional signature for chiral-symmetry breaking reappears.

ACKNOWLEDGMENTS

This work was written up while one of us (N.P.C.) was visiting KEK, Japan, and he wishes to thank Dr. K. Higashijima for some stimulating conversations and the Theory Group for the warm hospitality. This work was supported in part by the NSF U.S.-China Cooperative Program, by grants from the NSF, the Department of Energy, and from PSC-BHE of the City University of New York.

[1]Y. Nambu and G. Jona-Lasinio, Phys. Rev. **122**, 345 (1961); **124**, 246 (1961).

[2]M. Cini and B. Touschek, Nuovo Cimento 7, 422 (1958). The Cini-Touschek transformation is a special case of a general class of Foldy-Wouthuysen transformations; see L. L. Foldy and S. A. Wouthuysen, Phys. Rev. **78**, 29 (1950).

[3]L. N. Chang, N. P. Chang, and K. C. Chou, in *Third Asia-Pacific Physics Conference*, proceedings, Hong Kong, 1988, edited by Y. W. Chan, A. F. Leung, C. N. Yang, and K. Young (World Scientific, Singapore, 1988).

[4]For a comprehensive review of the subject, the formalism, as well as citation of earlier work, see Zhou Guang-zhao (K. C. Chou), Su Zhao-bin, Hao Bai-lin, and Yu Lu, Phys. Rep. **118**, 1 (1985). For other equivalent approaches, see H. Umezawa, H. Matsumoto, and M. Tachiki, *Thermofield Dynamics and Condensed States* (North-Holland, Amsterdam, 1982); R. L. Kobes, G. W. Semenoff, and N. Weiss, Z. Phys. C **29**, 371 (1985); A. Niemi and G. W. Semenoff, Nucl. Phys. **B230**, 181 (1984); Ann. Phys. (N.Y.) **152**, 105 (1984).

[5]L. N. Chang and N. P. Chang, Phys. Rev. Lett. **54**, 2407 (1985); Phys. Rev. D **29**, 312 (1984); see also N. P. Chang and D. X. Li, *ibid.* **30**, 790 (1984).

[6]H. A. Weldon, Phys. Rev. D **26**, 2789 (1982).

[7]J. F. Donoghue and B. R. Holstein, Phys. Rev. D **28**, 340 (1983); **29**, 3004(E) (1984); J. F. Donoghue, B. R. Holstein, and R. W. Robinett, Ann. Phys. (N.Y.) **164**, 233 (1985). See also R. Pisarski, Nucl. Phys. **A498**, 423C (1989). G. Barton, Ann. Phys. (N.Y.) **200**, 271 (1990), has a very nice discussion of the physical origin of the Lorentz invariance of the massive particle pole.

[8]It is interesting to note that lattice gauge simulations have found a chiral transition temperature of $T_c/\Lambda_{\overline{\rm MS}}$ of around 2 (for a nucleon mass of 940 MeV) (see Ref. 9), where $\overline{\rm MS}$ denotes the modified minimal subtraction scheme.

[9]For a review, see A. Ukawa, in *Lattice '89*, proceedings of the International Symposium, Capri, Italy, 1989, edited by R. Petronzio *et al.* [Nucl. Phys. B (Proc. Suppl.) (in press)].

PHYSICAL REVIEW D VOLUME 53, NUMBER 7 1 APRIL 1996

CP violation, fermion masses and mixings in a predictive SUSY SO(10)×Δ(48)×U(1) model with small tanβ

K. C. Chou
Chinese Academy of Sciences, Beijing 100864, China

Y. L. Wu
Department of Physics, Ohio State University, Columbus, Ohio 43210

(Received 8 November 1995)

Fermion masses and mixing angles are studied in an SUSY SO(10)×Δ(48)×U(1) model with small tanβ. Thirteen parameters involving masses and mixing angles in the quark and charged lepton sector are successfully predicted by a single Yukawa coupling and three ratios of VEV's caused by necessary symmetry breaking. Ten relations among the low energy parameters have been found with four of them free from renormalization modifications. They could be tested directly by low energy experiments.

PACS number(s): 12.15.Ff, 11.30.Er, 12.10.Dm, 12.60.Jv

The standard model (SM) is a great success. Eighteen phenomenological parameters in the SM, which are introduced to describe all the low energy data, have been extracted from various experiments although they are not yet equally well known. Some of them have an accuracy of better than 1%, but some others less than 10%. To improve the accuracy for these parameters and understand them is a big challenge for particle physics. The mass spectrum and the mixing angles observed remind us that we are in a stage similar to that of atomic spectroscopy before Balmer. Much effort has been made along this direction. The well-known examples are the Fritzsch ansatz [1] and Georgi-Jarlskog texture [2]. A general analysis and review of the previous studies on the texture structure was given by Raby in [3]. Recently, Babu and Barr [4], and Mohapatra [5], and Shafi [6], Hall and Raby [7], Berezhiani [8], Kaplan and Schmaltz [9], Kusenko and Shrock [10] constructed some interesting models with texture zeros based on supersymmetric (SUSY) SO(10). Anderson, Dimopoulos, Hall, Raby, and Starkman [11] presented a general operator analysis for the quark and charged lepton Yukawa coupling matrices with two zero textures "11" and "13." The 13 observables in the quark and charged lepton sector were found to be successfully fitted by only six parameters with large tanβ. Along this direction we have shown [12] that the same 13 parameters can be successfully described, in an SUSY SO(10)×Δ(48)×U(1) model with large values of tan$\beta \sim m_t/m_b$, by only five parameters with three of them determined by the symmetry-breaking scales of U(1), SO(10), SU(5), and $SU(2)_L$. Ten parameters in the neutrino sector could also be predicted, though not unique, with one additional parameter.

In this Rapid Communication we shall present, based on the symmetry group SUSY SO(10)×Δ(48)×U(1), an alternative model with small values of tan$\beta \sim 1$ which is of phenomenological interest in testing the Higgs sector in the minimum supersymmetric standard model (MSSM) at colliders [13]. The dihedral group Δ(48), a subgroup of SU(3), is taken as the family group. U(1) is family-independent and is introduced to distinguish various fields which belong to the same representations of SO(10)×Δ(48). The irreducible representations of Δ(48) consisting of five triplets and three singlets are found to be sufficient to build an interesting texture structure for fermion mass matrices. The symmetry Δ(48)×U(1) naturally ensures the texture structure with zeros for Yukawa coupling matrices, while the coupling coefficients of the resulting interaction terms in the superpotential are unconstrained by this symmetry. To reduce the possible free parameters, the universality of coupling constants in the superpotential is assumed; i.e., all the coupling coefficients are assumed to be equal and have the same origins from perhaps a more fundamental theory. We know in general that universality of charges occurs only in gauge interactions due to charge conservation, like the electric charge of different particles. In the absence of strong interactions, family symmetry could keep the universality of weak interactions in a good approximation after breaking. In our case there are so many heavy fermions above the grand unified theory (GUT) scale and their interactions are taken to be universal in the GUT scale where family symmetries have been broken. It can only be an ansatz at the present moment where we do not know the answer governing the behavior of nature above the GUT scale. As the numerical predictions on the low energy parameters so found are very encouraging and interesting, we believe that there must be a deeper reason that has to be found in the future.

Choosing the structure of the physical vacuum carefully, the Yukawa coupling matrices which determine the masses and mixings of all quarks and leptons are given by

$$\Gamma_u^G = \frac{2}{3}\lambda_H \begin{pmatrix} 0 & \frac{3}{2} z'_u \epsilon_P^2 & 0 \\ \frac{3}{2} z_u \epsilon_P^2 & -3 y_u \epsilon_G^2 e^{i\phi} & -\frac{\sqrt{3}}{2} x_u \epsilon_G^2 \\ 0 & -\frac{\sqrt{3}}{2} x_u \epsilon_G^2 & w_u \end{pmatrix} \tag{1}$$

and

$$\Gamma_f^G=\frac{2}{3}\lambda_H\frac{(-1)^{n+1}}{3^n}\begin{pmatrix} 0 & -\frac{3}{2}z'_f\epsilon_P^2 & 0 \\ -\frac{3}{2}z_f\epsilon_P^2 & 3y_f\epsilon_G^2e^{i\phi} & -\frac{1}{2}x_f\epsilon_G^2 \\ 0 & -\frac{1}{2}x_f\epsilon_G^2 & w_f \end{pmatrix} \tag{2}$$

for $f=d,e$, and

$$\Gamma_\nu^G=\frac{2}{3}\lambda_H\frac{1}{5}\frac{(-1)^{n+1}}{15^n} \times\begin{pmatrix} 0 & -\frac{15}{2}z'_\nu\epsilon_P^2 & 0 \\ -\frac{15}{2}z_\nu\epsilon_P^2 & 15y_\nu\epsilon_G^2e^{i\phi} & -\frac{1}{2}x_\nu\epsilon_G^2 \\ 0 & -\frac{1}{2}x_\nu\epsilon_G^2 & w_\nu \end{pmatrix} \tag{3}$$

for Dirac-type neutrino coupling, where the integer n reflects the possible choice of heavy fermion fields above the GUT scale. $n=4$ is found to be the best choice in this set of models for a consistent prediction on top and charm quark masses. This is because, for $n>4$, the resulting value of $\tan\beta$ becomes too small, as a consequence, the predicted top quark mass will be below the present experimental lower limit. For $n<4$, the values of $\tan\beta$ will become larger, the resulting charm quark mass will be above the present upper bound. λ_H is a universal coupling constant expected to be of order one. $\epsilon_G\equiv v_5/v_{10}$ and $\epsilon_P\equiv v_5/M_P$ with M_P, v_{10}, and v_5 being the vacuum expectation values (VEV's) for $U(1)\times\Delta(48)$, SO(10), and SU(5) symmetry breaking, respectively. ϕ is the physical CP phase arising from the VEV's. The assumption of maximum CP violation implies that $\phi=\pi/2$. x_f, y_f, z_f, and w_f ($f=u,d,e,\nu$) are the Clebsch factors of SO(10) determined by the directions of symmetry breaking of the adjoints **45**'s. The three directions of symmetry breaking have been chosen as $\langle A_X\rangle=v_{10}\mathrm{diag}(2,2,2,2,2)\otimes\tau_2$, $\langle A_z\rangle=v_5\mathrm{diag}(-\frac{2}{3},-\frac{2}{3},-\frac{2}{3},-2,-2)\otimes\tau_2$, $\langle A_u\rangle=v_5\mathrm{diag}(\frac{2}{3},\frac{2}{3},\frac{2}{3},\frac{1}{3},\frac{1}{3})\otimes\tau_2$. The Clebsch factors associated with the symmetry-breaking directions can be easily read off from the U(1) hypercharges of the adjoints **45**'s and the related effective operators which are obtained when the symmetry $SO(10)\times\Delta(48)\times U(1)$ is broken and heavy fermion pairs are integrated out and decoupled:

$$W_{33}=(\tfrac{2}{3}\lambda_H)\tfrac{1}{2}\mathbf{16}_3\frac{\sqrt{3}}{\sqrt{1+2\left(\frac{v_{10}}{A_X}\right)^{2(n+1)}}}\left(\frac{v_{10}}{A_X}\right)^{n+1} \times\mathbf{10}_1\left(\frac{v_{10}}{A_X}\right)^{n+1}\frac{\sqrt{3}}{\sqrt{1+2\left(\frac{v_{10}}{A_X}\right)^{2(n+1)}}}\mathbf{16}_3, \tag{4}$$

$$W_{32}=(\tfrac{2}{3}\lambda_H)\frac{\sqrt{3}}{2}\epsilon_G^2\mathbf{16}_3\frac{\sqrt{3}}{\sqrt{1+2\left(\frac{v_{10}}{A_X}\right)^{2(n+1)}}}\left(\frac{v_{10}}{A_X}\right)^{n+1}\left(\frac{A_z}{v_5}\right) \times\left(\frac{v_{10}}{A_X}\right)\mathbf{10}_1\left(\frac{v_{10}}{A_X}\right)\left(\frac{A_z}{v_5}\right)\left(\frac{v_{10}}{A_X}\right)^{n+1}\mathbf{16}_2,$$

$$W_{22}=(\tfrac{2}{3}\lambda_H)\tfrac{3}{2}\epsilon_G^2\mathbf{16}_2\left(\frac{v_{10}}{A_X}\right)^n\left(\frac{A_u}{v_5}\right)\left(\frac{v_{10}}{A_X}\right)\mathbf{10}_1\left(\frac{v_{10}}{A_X}\right)\left(\frac{A_u}{v_5}\right) \times\left(\frac{v_{10}}{A_X}\right)^n\mathbf{16}_2e^{i\phi},$$

$$W_{12}=(\tfrac{2}{3}\lambda_H)\tfrac{3}{2}\epsilon_P^2\mathbf{16}_1\left[\left(\frac{v_{10}}{A_X}\right)^{n-3}\mathbf{10}_1\left(\frac{v_{10}}{A_X}\right)^{n-3}+\left(\frac{v_{10}}{A_X}\right)^n\left(\frac{A_u}{v_5}\right) \times\left(\frac{v_{10}}{A_X}\right)\mathbf{10}_1\left(\frac{v_{10}}{A_X}\right)\left(\frac{A_z}{v_5}\right)\left(\frac{v_{10}}{A_X}\right)^{n+1}\right]\mathbf{16}_2.$$

The factor $1/\sqrt{1+2(v_{10}/A_X)^{2(n+1)}}$ arising from the mixing is equal to $1/\sqrt{3}$ for the up-type quark and almost unity for other fermions due to suppression of large Clebsch factors in the second term of the square root. The relative phase (or sign) between the two terms in the operator W_{12} has been fixed. The resulting Clebsch factors are $w_u=w_d=w_e=w_\nu=1$, $x_u=5/9$, $x_d=7/27$, $x_e=-1/3$, $x_\nu=1/5$, $y_u=0$, $y_d=y_e/3=2/27$, $y_\nu=4/45$, $z_u=1$, $z_d=z_e=-27$, $z_\nu=-15^3=-3375$, $z'_u=1-5/9=4/9$, $z'_d=z_d+7/729\simeq z_d$, $z'_e=z_e-1/81\simeq z_e$, $z'_\nu=z_\nu+1/15^3\simeq z_\nu$.

An adjoint **45** A_X and a 16-dimensional (16D) representation Higgs field Φ ($\overline{\Phi}$) are needed for breaking SO(10) down to SU(5). Another two adjoint **45**'s A_z and A_u are needed to break SU(5) further down to the standard model $SU(3)_c\times SU_L(2)\times U(1)_Y$. From the Yukawa coupling matrices given above, the 13 parameters in the SM can be determined by only four parameters: a universal coupling constant λ_H and three ratios of the VEV's; $\epsilon_G=v_5/v_{10}$, $\epsilon_P=v_5/M_P$, and $\tan\beta=v_2/v_1$. In obtaining physical masses and mixings, renormalization group (RG) effects should be taken into account. As most Yukawa couplings in the present model are much smaller than the top quark Yukawa coupling $\lambda_t^G\sim 1$, in a good approximation, we will only keep top quark Yukawa coupling terms in the RG equations and neglect all other Yukawa coupling terms. The RG evolution will be described by three kinds of scaling factors. η_F ($F=U,D,E,N$) and R_t arise from running the Yukawa parameters from the GUT scale down to the SUSY-breaking scale M_S which is chosen to be close to the top quark mass, i.e., $M_S\simeq m_t\simeq 170$ GeV. They are defined by $\eta_F(M_S)=\Pi_{i=1}^3(\alpha_i(M_G)/\alpha_i(M_S))^{c_i^F/2b_i}$ ($F=U,D,E,N$) with $c_i^U=(\frac{13}{15},3,\frac{16}{3})$, $c_i^D=(\frac{7}{15},3,\frac{16}{3})$, $c_i^E=(\frac{27}{15},3,0)$, $c_i^N=(\frac{9}{25},3,0)$, $b_i=(\frac{33}{5},1,-3)$, and $R_t^{-1}=\exp[-\int_{\ln M_S}^{\ln M_G}(\lambda_t(t)/4\pi)^2dt]=[1+(\lambda_t^G)^2K_t]^{-1/12}$, where $K_t=3I(M_S)/4\pi^2$ with $I(M_S)=\int_{\ln M_S}^{\ln M_G}\eta_U^2(t)dt$. The numerical value for I taken from Ref. [15] is 113.8 for $M_S\simeq m_t=170$ GeV. Other RG scaling factors are derived by running Yukawa couplings below M_S. $m_i(m_i)=\eta_i m_i(M_S)$ for ($i=c,b$) and $m_i(1\ \mathrm{GeV})=\eta_i m_i(M_S)$ for ($i=u,d,s$). The physical top quark mass is given by $M_t=m_t(m_t)[1+\frac{4}{3}\alpha_s(m_t)/\pi]$. Using the well-measured charged lepton masses $m_e=0.511$ MeV, $m_\mu=105.66$ MeV, and $m_\tau=1.777$ GeV we obtain four important RG scaling-independent predictions:

TABLE I. Output parameters and their predicted values with $\alpha_s(M_Z)=0.113$ and input parameters: $m_e=0.511$ eV, $m_\mu=105.66$ MeV, $m_\tau=1.777$ GeV, and $m_b=4.25$ GeV.

Output parameters	Output values	Data [14]	Output parameters	Output values
M_t [GeV]	182	180 ± 15	$J_{CP}=A^2\lambda^6\eta$	2.68×10^{-5}
$m_c(m_c)$ [GeV]	1.27	1.27 ± 0.05	α	86.28°
m_u(1 GeV) [MeV]	4.31	4.75 ± 1.65	β	22.11°
m_s(1 GeV) [MeV]	156.5	165 ± 65	γ	71.61°
m_d(1 GeV) [MeV]	6.26	8.5 ± 3.0	$\tan\beta=v_2/v_1$	2.33
$\lvert V_{us}\rvert=\lambda$	0.22	0.221 ± 0.003	$\epsilon_G=v_5/v_{10}$	2.987×10^{-1}
$\frac{\lvert V_{ub}\rvert}{\lvert V_{cb}\rvert}=\lambda\sqrt{\rho^2+\eta^2}$	0.083	0.08 ± 0.03	$\epsilon_P=v_5/M_P$	1.011×10^{-2}
$\frac{\lvert V_{td}\rvert}{\lvert V_{ts}\rvert}=\lambda\sqrt{(1-\rho)^2+\eta^2}$	0.209	0.24 ± 0.11	λ_t^G	1.30
$\lvert V_{cb}\rvert=A\lambda^2$	0.0393	0.039 ± 0.005 [19]	-	-
B_K	0.90	0.82 ± 0.10 [17,16]	-	-
$f_B\sqrt{B}$ [MeV]	207	200 ± 70 [18,16]	-	-
Re$(\varepsilon'/\varepsilon)$	$(1.4\pm 1.0)\times 10^{-3}$	$(1.5\pm 0.8)\times 10^{-3}$80	-	-

$$|V_{us}|=|V_{us}|_G\simeq 3\sqrt{\frac{m_e}{m_\mu}}\left(\frac{1+\left(\frac{16}{675}\frac{m_\tau}{m_\mu}\right)^2}{1+9\frac{m_e}{m_\mu}}\right)^{1/2}=0.22, \tag{5}$$

$$\left|\frac{V_{ub}}{V_{cb}}\right|=\left|\frac{V_{ub}}{V_{cb}}\right|_G\simeq\left(\frac{4}{15}\right)^2\frac{m_\tau}{m_\mu}\sqrt{\frac{m_e}{m_\mu}}=0.083, \tag{6}$$

$$\left|\frac{V_{td}}{V_{ts}}\right|=\left|\frac{V_{td}}{V_{ts}}\right|_G\simeq 3\sqrt{\frac{m_e}{m_\mu}}=0.209, \tag{7}$$

$$\frac{m_d}{m_s}\left(1-\frac{m_d}{m_s}\right)^{-2}=9\frac{m_e}{m_\mu}\left(1-\frac{m_e}{m_\mu}\right)^{-2}=0.044, \tag{8}$$

and six RG scaling-dependent predictions:

$$|V_{cb}|=|V_{cb}|_G R_t\simeq\frac{15\sqrt{3}-7}{15\sqrt{3}}\frac{5}{4\sqrt{3}}\frac{m_\mu}{m_\tau}R_t=0.0391\left(\frac{0.80}{R_t^{-1}}\right), \tag{9}$$

$$m_s(1\ \text{GeV})=\frac{1}{3}m_\mu\frac{\eta_s}{\eta_\mu}\eta_{D/E}=159.53\left(\frac{\eta_s}{2.2}\right)\left(\frac{\eta_{D/E}}{2.1}\right)\ \text{MeV}, \tag{10}$$

$$m_b(m_b)=m_\tau\frac{\eta_b}{\eta_\tau}\eta_{D/E}R_t^{-1}=4.25\left(\frac{\eta_b}{1.49}\right)\left(\frac{\eta_{D/E}}{2.04}\right)\left(\frac{R_t^{-1}}{0.80}\right)\ \text{GeV}, \tag{11}$$

$$m_u(1\ \text{GeV})=\frac{5}{3}\left(\frac{4}{45}\right)^3\frac{m_c}{m_\mu}\eta_u R_t^3 m_t=4.23\left(\frac{\eta_u}{2.2}\right)\left(\frac{0.80}{R_t^{-1}}\right)^3\left(\frac{m_t(m_t)}{174\ \text{GeV}}\right)\ \text{MeV}, \tag{12}$$

$$m_c(m_c)=\frac{25}{48}\left(\frac{m_\mu}{m_\tau}\right)^2\eta_c R_t^3 m_t=1.25\left(\frac{\eta_c}{2.0}\right)\left(\frac{0.80}{R_t^{-1}}\right)^3\left(\frac{m_t(m_t)}{174\ \text{GeV}}\right)\ \text{GeV}, \tag{13}$$

$$m_t(m_t)=\frac{\eta_U}{\sqrt{K_t}}\sqrt{1-R_t^{-12}}\frac{v}{\sqrt{2}}\sin\beta=174.9\left(\frac{\sin\beta}{0.92}\right)\left(\frac{\eta_U}{3.33}\right)\left(\sqrt{\frac{8.65}{K_t}}\right)\left(\frac{\sqrt{1-R_t^{-12}}}{0.965}\right)\ \text{GeV}, \tag{14}$$

where the miraculus numbers in the above relations are due to the Clebsch factors. The scaling factor R_t or coupling $\lambda_t^G=(1/\sqrt{K_t})\sqrt{1-R_t^{-12}}/R_t^{-6}$ is determined by the mass ratio of the bottom quark and τ lepton. $\tan\beta$ is fixed by the τ lepton mass via $\cos\beta=m_\tau\sqrt{2}/\eta_E\eta_\tau v\lambda_\tau^G$.

The above 10 relations are our main results which contain only low energy observables. As an analogy to the Balmer series formula, these relations may be considered as empirical at the present moment. They have been tested by the existing experimental data to a good approximation and can be tested further directly by more precise experiments in the future.

In numerical predictions we take $\alpha^{-1}(M_Z)=127.9$, $s^2(M_Z)=0.2319$, $M_Z=91.187$ GeV, $\alpha_1^{-1}(m_t)=58.59$, $\alpha_2^{-1}(m_t)=30.02$, and $\alpha_1^{-1}(M_G)=\alpha_2^{-1}(M_G)=\alpha_3^{-1}(M_G)\simeq 24$ with $M_G\sim 2\times 10^{16}$ GeV. For $\alpha_s(M_Z)=0.113$, the RG scaling factors have values $(\eta_{u,d,s}, \eta_c, \eta_b, \eta_{e,\mu,\tau}, \eta_U, \eta_D/\eta_E\equiv\eta_{D/E}, \eta_E, \eta_N)$ = (2.20, 2.00, 1.49, 1.02, 3.33, 2.06, 1.58, 1.41). The corresponding predictions on fermion masses and mixings thus obtained are found to be remarkable. Our numerical predictions for $\alpha_s(M_Z)=0.113$ are given in Table I with four input parameters: three charged lepton masses and bottom quark mass $m_b(m_b)=4.25$ GeV, where B_K and $f_B\sqrt{B}$ in Table I are two important hadronic parameters and extracted from K^0-$\bar{K}^0$ and B^0-$\bar{B}^0$ mixing parameters ε_K and x_d. Re$(\varepsilon'/\varepsilon)$ is the direct CP-violating parameter in kaon decays, where large uncertanties mainly arise from the hadronic matrix elements. α, β, and γ are three angles of the unitarity triangle in the Cabibbo-Kobayashi-Maskawa (CKM) matrix. J_{CP} is the rephase-invariant CP-violating quantity.

It is amazing that nature has allowed us to make predictions on fermion masses and mixings in terms of a single Yukawa coupling constant and three ratios of the VEV's determined by the structure of the physical vacuum and understand the low energy physics from the GUT scale physics. It has also suggested that nature favors maximal spontaneous CP violation. A detailed analysis including the neutrino sector will be presented in a longer paper [20]. In comparison with the models with large $\tan\beta \sim m_t/m_b$, the present model has provided a consistent picture on the 13 parameters in the SM with better accuracy. Besides, ten relations involving fermion masses and CKM matrix elements are obtained with four of them independent of the RG scaling effects. The two types of the models corresponding to the large and low $\tan\beta$ might be distinguished by testing the MSSM Higgs sector at colliders as well as by precisely measuring the ratio $|V_{ub}/V_{cb}|$ since this ratio does not receive radiative corrections in both models. It is expected that more precise measurements from CP violation and various low energy experiments in the near future could provide crucial tests on the ten realtions obtained in the present model.

Y.L.W. would like to thank Professor S. Raby and Professor G. Steigman for useful discussions. Y.L.W. was supported in part by U.S. Department of Energy Grant No. DOE/ER/01545-662.

[1] H. Fritzsch, Phys. Lett. **70B**, 436 (1977).

[2] H. Georgi and C. Jarlskog, Phys. Lett. **86B**, 297 (1979).

[3] S. Raby, Ohio State University Report No. OHSTPY-HEP-T-95-024, 1995 (unpublished).

[4] K.S. Babu and S.B. Barr, Phys. Rev. Lett. **75**, 2088 (1995).

[5] K.S. Babu and R.N. Mohaptra, Phys. Rev. Lett. **74**, 2418 (1995).

[6] K.S. Babu and Q. Shafi, Phys. Lett. B **357**, 365 (1995).

[7] L.J. Hall and S. Raby, Phys. Rev. D **51**, 6524 (1995).

[8] Z.G. Berezhiani, Phys. Lett. B **355**, 178 (1995).

[9] D. Kaplan and M. Schmaltz, Phys. Rev. D **49**, 3741 (1994); M. Schmaltz, *ibid.* **52**, 1643 (1995).

[10] A. Kusenko and R. Shrock, Phys. Rev. D **49**, 4962 (1994).

[11] G. Anderson *et al.*, Phys. Rev. D **49**, 3660 (1994).

[12] K.C. Chou and Y.L. Wu, Science in China (Sci. Sin.) **39A**, 65 (1996).

[13] See, for example, J. Ellis, presented at the 17th International Symposium on Lepton-Photon Interactions, Beijing, China, 1995 (unpublished).

[14] CDF Collaboration, F. Abe *et al.*, Phys. Rev. Lett. **74**, 2626 (1995); D0 Collaboration, S. Abachi *et al.*, *ibid.* **74**, 2632 (1995); J. Gasser and H. Leutwyler, Phys. Rep. **87**, 77 (1982); H. Leutwyler, Nucl. Phys. **B337**, 108 (1990); Particle Data Group, L. Montanet *et al.*, Phys. Rev. D **50**, 1173 (1994).

[15] V. Barger, M.S. Berger, T. Han, and M. Zralek, Phys. Rev. Lett. **68**, 3394 (1992).

[16] J. Shigemitsu, in *Proceedings of the XXVII International Conference on High Energy Physics*, Glasgow, Scotland, 1994, edited by P.J. Bussey and I.G. Knowles (IOP, Bristol, 1995).

[17] C. Bernard and A. Soni, in *Lattice '94*, Proceedings of the International Symposium, Bielefeld, Germany, 1994 edited by F. Karsch *et al.* [Nucl. Phys. B (Proc. Suppl.) **42** (1995)]; S.R. Sharpe, in *Lattice '93*, Proceedings of the International Conference, Dallas, Texas, edited by T. Draper *et al.* [*ibid.* **34**, 403 (1994)].

[18] C. Allton, Report No. hep-lat/9509084, 1995 (unpublished), and references therein.

[19] R. Patterson, in *Proceedings of the XXVII International Conference on High Energy Physics* [16], Vol. 1, p. 149; M. Neubert, Report No. CERN-TH/95-107, hep-ph/9505238, 1995 (unpublished); A. Ali and D. London, Report No. DESY 95-148, UdeM-GPP-TH-95-32, hep-ph/9508272, 1995 (unpublished).

[20] K.C. Chou and Y.L. Wu, Chinese Academy of Sciences and Ohio State University Report No. CAS-HEP-T-96-03/004, OH STPY-HEP-T-96-008, 1996.

Vol. 39 No. 1 SCIENCE IN CHINA (Series A) January 1996

Low energy phenomena in a model with symmetry group SUSY $SO(10)\times\Delta(48)\times U(1)$

ZHOU Guangzhao (K.C. CHOU 周光召)
(Chinese Academy of Sciences, Beijing 100864, China)
and WU Yueliang (吴岳良)
(Department of Physics, Ohio State University, Columbus, Ohio 43210, USA)

Received September 4, 1995

Abstract Fermion masses and mixing angles including that of neutrinos are studied in a model with symmetry group SUSY $SO(10)\times\Delta(48)\times U(1)$. Universality of Yukawa coupling of superfields is assumed. The resulting texture of mass matrices in the low energy region depends only on a single coupling constant and VEVs caused by necessary symmetry breaking. 13 parameters involving masses and mixing angles in the quark and charged lepton sector are successfully described by only five parameters with two of them determined by the scales of $U(1)$, $SO(10)$ and $SU(5)$ symmetry breaking compatible with the requirement of grand unification and proton decay. The neutrino masses and mixing angles in the leptonic sector are also determined with the addition of a Majorana coupling term. It is found that LSND $\bar{\nu}_\mu\rightarrow\bar{\nu}_e$ events, atmospheric neutrino deficit and the mass limit put by hot dark matter can be naturally explained. Solar neutrino puzzle can be solved only by introducing sterile neutrino with one additional parameter. More precise measurements of $\alpha_s(M_Z)$, $V_{cb}, V_{ub}/V_{cb}$, m_b, m_t, as well as various CP violation and neutrino oscillation experiments will provide crucial tests of the present model.

Keywords: symmetry group, low energy.

The standard model (SM) is a great success. To understand the origin of the 18 free parameters (or 25 if neutrinos are massive) is a big challenge to high energy physics. Many efforts have been made in this direction. It was first observed by Gatto *et al.*, Cabbibo and Maiani[1] that the Cabbibo angle is close to $\sqrt{m_d/m_s}$. This observation initiated the investigation of the texture structure with zero elements[2] in the fermion Yukawa coupling matrices. A general analysis and review of the previous studies on the texture structure was given by Raby1). In refs. [3, 4] Anderson *et al.* presented an interesting model based on SUSY $SO(10)$ and $U(1)$ family symmetries with two zero textures '11' and '13' followed naturally. Though the texture '22' and '32' are not unique they could fit successfully the 13 observables in the quark and charged lepton sector with only six parameters[3].

In this paper we will follow their general considerations and make the following modifications:

* Project supported in part by Department of Energy Grant # DOE/ER/01545-655.

1) For a recent review, see Raby, S., Ohio State Univ., Preprint, OHSTPY-HEP-T-95-024.

(i) We will use a discrete dihedral group $\Delta(3n^2)$ with $n=4$, a subgroup of $SU(3)$, as our family group instead of $U(1)$ used in ref.[5]. This kind of dihedral group was first used by Kaplan and Schmaltz[6] with $n=5$. This group, having only triplet and singlet irreducible representations, suits well our purposes.

(ii) We will assume universality of Yukawa coupling before symmetry breaking so as to reduce the possible free parameters. In this kind of theories there are very rich structures above the GUT scale with many heavy fermions and scalars. All heavy fields must have some reasons to exist and interact with each other which we do not understand at this moment. So we will just take the universality of coupling constants at the GUT scale as a working assumption and not worry about the possible radiative effects. If the phenomenology is all right, one has to be more serious to find a deeper reason for it.

(iii) We will choose some symmetry breaking directions different from those in refs. [3, 4] to ensure the needed Clebsch coefficients in order to eliminate further arbitrariness of the parameters.

This paper is organized as follows: In sec. 1, we will present the results of the Yukawa coupling matrices. The resulting masses and CKM quark mixings are also presented. In sec. 2, neutrino masses and CKM-type mixings in the lepton sector are presented. All existing neutrino experiments are discussed and found to be understandable in the present model. In sec. 3, the model with superfields and superpotential is explicitly presented. Conclusions and remarks are presented in the last section.

1 Yukawa coupling matrices

With the above considerations, a model based on group SUSY $SO(10)\times\Delta(48)\times U(1)$ with a single coupling constant is constructed. Here $U(1)$ is family-independent, introduced to distinguish various fields which belong to the same representations of $SO(10)\times\Delta(48)$. Yukawa coupling matrices which determine the masses and mixings of all quarks and charged leptons are obtained by carefully choosing the structure of the physical vacuum. We find

$$\Gamma_f^G=\frac{2}{3}\lambda_H\begin{pmatrix}0 & \frac{3}{2}z_f\varepsilon_P^2 & 0\\ \frac{3}{2}z_f\varepsilon_P^2 & 3y_f\varepsilon_G^2 e^{i\varphi} & \frac{\sqrt{3}}{2}x_f\varepsilon_G^2\\ 0 & \frac{\sqrt{3}}{2}x_f\varepsilon_G^2 & w_f\end{pmatrix} \tag{1}$$

for $f=u,d,e$, and

$$\Gamma^G_\nu=\frac{2}{3}\lambda_H\begin{pmatrix} 0 & \frac{3}{2}\frac{1}{|z_\nu|} & 0 \\ \frac{3}{2}\frac{1}{|z_\nu|} & 3\frac{y_\nu}{z_\nu^2}\frac{\varepsilon_G^2}{\varepsilon_P^2}e^{i\varphi} & \frac{\sqrt{3}}{2}\frac{x_\nu}{|z_\nu|}\frac{\varepsilon_G^2}{\varepsilon_P} \\ 0 & \frac{\sqrt{3}}{2}\frac{x_\nu}{|z_\nu|}\frac{\varepsilon_G^2}{\varepsilon_P} & w_f \end{pmatrix} \tag{2}$$

for Dirac type neutrino coupling. λ_H is the universal coupling constant expected to be of order one. $\varepsilon_G=v_5/v_{10}$ and $\varepsilon_P\equiv v_5/M_P$ with M_P, v_{10} and v_5 being the VEVs for $U(1)$, $SO(10)$ and $SU(5)$ symmetry breaking, respectively. x_f, y_f, z_f and $w_f (f=u, d, e, \nu)$ are the Clebsch factors of $SO(10)$ determined by the directions of symmetry breaking of the adjoints 45s. The following three directions have been chosen for symmetry breaking, namely $\langle A_X\rangle=v_{10}\,\mathrm{diag}\,(2,2,2,2,2)\otimes\tau_2$; $\langle A_z\rangle=v_5\,\mathrm{diag}\left(\frac{2}{3},\frac{2}{3},\frac{2}{3},-2,-2\right)\otimes\tau_2$ and $\langle A_u\rangle=v_5\,\mathrm{diag}\left(\frac{2}{3},\frac{2}{3},\frac{2}{3},\frac{1}{3},\frac{1}{3}\right)\otimes\tau_2$. The resulting Clebsch factors are: $w_u=w_d=w_e=w_\nu=1$; $x_u=-7/9$, $x_d=-5/27$, $x_e=1$, $x_\nu=-1/15$; $y_u=0$, $y_d=y_e/3=2/27$, $y_\nu=4/45$; $z_u=1$, $z_d=z_e=27$, $z_\nu=-15^3=-3375$. φ is the physical CP phase[1)] arising from the VEVs. The Clebsch factors associated with the symmetry breaking directions can be easily read off from effective operators which are obtained when the heavy fermion pairs are integrated out and decoupled

$$\begin{cases} W_{33}=\left(\frac{2}{3}\lambda_H\right)\frac{1}{2}\,16_3\,10_1\,16_3, \\[2ex] W_{32}=\left(\frac{2}{3}\lambda_H\right)\frac{\sqrt{3}}{2}\varepsilon_G^2\,16_3\left(\frac{A_z}{v_5}\right)\left(\frac{v_{10}}{A_X}\right)10_1\left(\frac{v_{10}}{A_X}\right)\left(\frac{A_z}{v_5}\right)\frac{1}{\sqrt{1+\varepsilon_P^2\left(\frac{A_X}{v_{10}}\right)^6}}16_2, \\[2ex] W_{22}=\left(\frac{2}{3}\lambda_H\right)\frac{3}{2}\varepsilon_G^2\,16_2\frac{1}{\sqrt{1+\varepsilon_P^2\left(\frac{A_X}{v_{10}}\right)^6}}\left(\frac{A_u}{v_5}\right)\left(\frac{v_{10}}{A_X}\right)10_1\left(\frac{v_{10}}{A_X}\right)\left(\frac{A_u}{v_5}\right)\frac{1}{\sqrt{1+\varepsilon_P^2\left(\frac{A_X}{v_{10}}\right)^6}}16_2, \\[2ex] W_{12}=\left(\frac{2}{3}\lambda_H\right)\frac{3}{2}\varepsilon_P^2\,16_1\frac{1}{\sqrt{1+\varepsilon_P^2\left(\frac{A_X}{v_{10}}\right)^6}}\left(\frac{A_X}{v_{10}}\right)^3 10_1\left(\frac{A_X}{v_{10}}\right)^3\frac{1}{\sqrt{1+\varepsilon_P^2\left(\frac{A_X}{v_{10}}\right)^6}}16_2, \end{cases} \tag{3}$$

where the factor $1\Big/\sqrt{1+\varepsilon_P^2\left(\frac{A_X}{v_{10}}\right)^6}$ arises from mixing. The ε_P^2 term in the square root is negligible for quarks and charged leptons, but it becomes dominant for the neutrinos due to the large Clebsch factor z_ν. In obtaining the Γ_f^G matrices, some small terms arising from

1) We have rotated away other possible phases by a phase redefinition of the fermion fields.

mixings between the chiral fermion 16_i and heavy fermion pairs $\psi_j(\overline{\psi}_j)$ are neglected. They are expected to change the numerical results no more than a few percent. The factor $1/\sqrt{3}$ associated with the third family is due to the maximum mixing between the third family fermion and heavy fermions. This set of effective operators which lead to the aforementioned Yukawa coupling matrices Γ_f^G is quite unique. Uniqueness of the structure of operator W_{12} was first observed by Anderson *et al.*[4] from the mass ratios of m_e/m_μ and m_d/m_s. The effective operator W_{33} is also fixed at the GUT scale[7,8,3] in the case of large $\tan\beta$. There is only one candidate for effective operator W_{22} when the direction of breaking is chosen to be A_u, with Clebsch factors satisfying y_u:y_d:y_e=0:1:3[9] so as to obtain a correct mass ratio m_μ/m_s. The three parameters λ_H, ε_G and ε_P are determined by the three measured mass ratios m_b/m_τ, m_μ/m_τ and m_e/m_τ. Thus, the mass ratio m_c/m_t and the CKM mixing elements V_{cb} and V_{ub} put strong constraint to a unique choice of the symmetry breaking direction A_z for effective operator W_{32}. Unlike many other models in which W_{33} is assumed to be a renormalizable interaction before symmetry breaking, the Yukawa couplings of all the quarks and leptons (both heavy and light) in the present model are generated in the GUT scale after the breakdown of the family group and $SO(10)$. Therefore, the initial conditions of renormalization group (RG) evolution will be set in the GUT scale for all the quark and lepton Yukawa couplings. Consequently, one could avoid the possible Landau pole and flavor changing problems encountered in many other models due to RG running of the third family Yukawa couplings from the GUT scale to the Planck scale. The hierarchy among the three families is described by the two ratios ε_G and ε_P. Mass splittings between quarks and leptons as well as between the up and down quarks are determined by the Clebsch factors of $SO(10)$. From the GUT scale down to low energies, RG evolution has been taken into account. Top-bottom splitting in the present model is mainly attributed to the hierarchy of the VEVs v_1 and v_2 of the two light Higgs doublets in the weak scale.[1)]

An adjoint 45 A_X and a 16-D representation Higgs field $\Phi(\overline{\Phi})$ are needed for breaking $SO(10)$ down to $SU(5)$. Adjoint 45 A_z and A_u are needed to break $SU(5)$ further down to the standard model $SU(3)_c\times SU_L\times U(1)_Y$.

The numerical predictions for the quark, lepton masses and quark mixings are presented in table 1(b) while the input parameters and their values are given in table 1(a). RG effects have been considered following the standard scheme[7,4] by integrating the full two-loop RG equations from the GUT scale down to the weak scale using $M_{\rm SUSY}\approx M_{\rm WEAK}\approx M_t\approx 180\,{\rm GeV}$. From the weak scale down to the lower energy scale, three loops in QCD and two loops in QED are taken into consideration. SUSY threshold effects are not considered in detail here since the spectrum of particles is not yet determined. The bottom

1) In the case of small $\tan\beta$, i.e. $v_2\sim v_1$, the top-bottom splitting could be caused by the Yukawa couplings. We shall discuss the alternative interesting case eleswhere.

quark mass may receive corrections as large as 30%[8] due to large $\tan\beta$. However, it could be reduced by taking a suitable spectrum of superparticles. Therefore, one should not expect to have precise predictions until the spectrum of the particles is well determined. The strong coupling constant $\alpha_s(M_Z)$ is taken to be a free parameter with values given by the present experimental bounds $\alpha_s(M_z)=0.117\pm0.005$[11] in the following.

Table 1(a) Parameters and their values as a function of the strong coupling $\alpha_s(M_Z)$ determined by m_b, m_τ, m_μ, m_e and $|V_{us}|=\lambda$.

$\alpha_s(M_Z)$	φ	$\varepsilon_G=v_5/v_{10}$	$\varepsilon_P\equiv v_5/\overline{M}_P$	$\tan\beta$
0.110	73.4°	2.66×10^{-1}	0.89×10^{-2}	51
0.115	77.5°	2.51×10^{-1}	0.83×10^{-2}	55
0.120	81.5°	2.34×10^{-1}	0.77×10^{-2}	58

Table 1(b) Observables and their predicted values with the values of the parameters given in table 1(a).

Input		Output with $\alpha_s(M_Z)$	0.110	0.115	0.120
$m_b(m_b)$/GeV	4.35	m_t [GeV]	165	176	185
m_τ/GeV	1.78	$m_c(m_c)$ [GeV]	1.14	1.30	1.37
m_μ/MeV	105.6	m_s(1 GeV) [MeV]	152	172	197
m_e/MeV	0.51	m_d(1 GeV) [MeV]	6.5	7.2	8.0
$\|V_{us}\|\simeq\lambda$	0.22	m_u(1 GeV) [MeV]	3.3	4.3	6.1
		$\|V_{cb}\|\simeq A\lambda^2$	0.045	0.045	0.043
		$\left\|\frac{V_{ub}}{V_{cb}}\right\|\simeq\lambda\sqrt{\rho^2+\eta^2}$	0.053	0.056	0.063
		$\left\|\frac{V_{td}}{V_{cb}}\right\|\simeq\lambda\sqrt{(1-\rho)^2+\eta^2}$	0.201	0.199	0.198

From table 1(a), one knows that the model has large $\tan\beta$ solution with $\tan\beta\equiv v_2/v_1\sim m_t/m_b$. CP violation is near the maximum with a phase $\varphi\sim80°$. The vacuum structure between the GUT scale and Planck scale has a hierarchic structure $\varepsilon_G\equiv v_5/v_{10}\sim\lambda=0.22$ and $\varepsilon_P\equiv v_5/\overline{M}_P\sim\lambda^3$. Assuming $(\overline{M}_P/M_P)^2\approx\alpha_G\approx1/24\sim\lambda^2$ (here α_G is the unified gauge coupling, M_P is the Planck mass), we have

$$\left\{\begin{array}{l}\overline{M}_P=2.5\times10^{18}\,\mathrm{GeV},\\ v_{10}\approx(0.86\pm0.16)\times10^{17}\,\mathrm{GeV},\\ v_5\equiv M_G\approx(2.2\pm0.2)\times10^{16}\,\mathrm{GeV},\end{array}\right. \tag{4}$$

where the resulting value for the GUT scale agree well with the one obtained from the gauge coupling unification. $\overline{M}_P$ is also very close to the reduced Planck scale $\hat{M}_P=M_P/\sqrt{8\pi}=2.4\times10^{18}$ GeV and may be regarded as the scale for gravity unification.

It is evident in table 1(b) that the predictions on fermion masses and Cabbibo-Kobayashi-Maskawa (CKM) mixing angles fall in the range allowed by the experimental data[10—12]:

$$m_\tau = 1777\ \text{MeV},\quad m_\mu = 105.6\ \text{MeV},\quad m_e = 0.51\ \text{MeV},$$
$$m_b(m_b) = (4.15-4.35)\ \text{GeV},\quad m_s(1\ \text{GeV}) = (105-230)\ \text{MeV},\quad m_d(1\ \text{GeV}) = (5.5-11.5)\ \text{MeV},$$
$$m_t(m_t) = (157-191)\ \text{GeV},\quad m_c(m_c) = (1.22-1.32)\ \text{GeV},\quad m_u(1\ \text{GeV}) = (3.1-6.4)\ \text{MeV}, \tag{5}$$

and

$$V = \begin{pmatrix} V_{ud} & V_{us} & V_{ub} \\ V_{cd} & V_{cs} & V_{cb} \\ V_{td} & V_{ts} & V_{tb} \end{pmatrix} = \begin{pmatrix} 0.9747-0.9759 & 0.218-0.224 & 0.002-0.005 \\ 0.218-0.224 & 0.9738-0.9752 & 0.032-0.048 \\ 0.004-0.015 & 0.03-0.048 & 0.9988-0.9995 \end{pmatrix}. \tag{6}$$

The model also gives a consistent prediction for the B^0-$\bar{B}^0$ mixing and CP violation in kaon decays (a detailed analysis will be presented elsewhere).

It is of interest to expand the above fermion Yukawa coupling matrices Γ_f^G in terms of the parameter $\lambda = 0.22$ (the Cabbibo angle), which was found in ref. [13] to be very useful for expanding the CKM mixing matrix. With the input values given in table 1(a), we find

$$\Gamma_u^G \approx \frac{2}{3}\lambda_H \begin{pmatrix} 0 & 0.97\lambda^6 & 0 \\ 0.97\lambda^6 & 0 & -0.89\lambda^2 \\ 0 & -0.89\lambda^2 & 1 \end{pmatrix};\quad \Gamma_d^G \approx \frac{2}{3}\lambda_H \begin{pmatrix} 0 & -1.27\lambda^4 & 0 \\ -1.27\lambda^4 & 1.39\lambda^3 e^{i0.86\pi/2} & -0.97\lambda^3 \\ 0 & -0.97\lambda^3 & 1 \end{pmatrix}; \tag{7}$$

$$\Gamma_e^G \approx \frac{2}{3}\lambda_H \begin{pmatrix} 0 & -1.27\lambda^4 & 0 \\ -1.27\lambda^4 & 0.92\lambda^2 e^{i0.86\pi/2} & 1.16\lambda^2 \\ 0 & 1.16\lambda^2 & 1 \end{pmatrix};\quad \Gamma_\nu^G \approx \frac{2}{3}\lambda_H \begin{pmatrix} 0 & 0.86\lambda^5 & 0 \\ 0.86\lambda^5 & 0.85\lambda^7 e^{i0.86\pi/2} & -1.14\lambda^6 \\ 0 & -1.14\lambda^6 & 1 \end{pmatrix}$$

for $\alpha_s(M_Z) = 0.115$.

2 Neutrino masses and mixings

To find the neutrino masses and mixings will be crucial tests of the model. Many unification theories predict a see-saw type mass[14] $m_{\nu_i} \sim m_{u_i}^2 / M_N$ with $u_i = u, c, t$ being up-type quarks. For $M_N \approx (10^{-3}—10^{-4})\ M_{\text{GUT}} \approx 10^{12}—10^{13}$ GeV, one has

$$m_{\nu_e} < 10^{-7}\ \text{eV},\quad m_{\nu_\mu} \sim 10^{-3}\ \text{eV},\quad m_{\nu_\tau} \sim (3-21)\ \text{eV}. \tag{8}$$

In this case solar neutrino anomalous could be explained by $\nu_e \to \nu_\mu$ oscillation, and the mass of ν_τ is in the range relevant to hot dark matter. However, LSND events and atmospheric neutrino deficit cannot be explained in this scenario.

By choosing Majorana type Yukawa coupling matrix differently, one can construct many models of neutrino mass matrix. We shall present one here, which is found to be of interest with the following texture:

$$M_N^G = \lambda_H v_{10} \frac{\varepsilon_P^4}{\varepsilon_G^2} \begin{pmatrix} 0 & 0 & \frac{1}{2} z_N \\ 0 & y_N & 0 \\ \frac{1}{2} z_N & 0 & w_N \end{pmatrix}. \tag{9}$$

The corresponding effective operators are given by

$$\begin{cases} W_{33}^N = \lambda_H \frac{v_{10}}{2} \frac{\varepsilon_P^4}{\varepsilon_G^2} 16_3 \left(\frac{A_u}{v_5}\right)\left(\frac{\bar{\Phi}}{v_{10}/\sqrt{2}}\right)\left(\frac{\bar{\Phi}}{v_{10}/\sqrt{2}}\right)\left(\frac{A_{B-L}}{v_5}\right) 16_3, \\ W_{13}^N = \lambda_H \frac{v_{10}}{2} \frac{\varepsilon_P^4}{\varepsilon_G^2} 16_1 \left(\frac{A_u}{v_5}\right)\left(\frac{\bar{\Phi}}{v_{10}/\sqrt{2}}\right)\left(\frac{\bar{\Phi}}{v_{10}/\sqrt{2}}\right)\left(\frac{A_{3R}}{v_5}\right) 16_3, \\ W_{22}^N = \lambda_H \frac{v_{10}}{2} \frac{\varepsilon_P^4}{\varepsilon_G^2} 16_2 \left(\frac{A_u}{v_5}\right)\left(\frac{\bar{\Phi}}{v_{10}/\sqrt{2}}\right)\left(\frac{\bar{\Phi}}{v_{10}/\sqrt{2}}\right)\left(\frac{A_u}{v_5}\right) 16_2, \end{cases} \tag{10}$$

where w_N, y_N and z_N are Clebsch factors with $w_N=4/3$, $y_N=16/9$, $z_N=2/3$. They are determined by additional 45s A_{B-L} and A_{3R} with $\langle A_{B-L}\rangle = v_5 \mathrm{diag}\left(\frac{2}{3}, \frac{2}{3}, \frac{2}{3}, 0, 0\right)\otimes\tau_2$ and $\langle A_{3R}\rangle = v_5 \mathrm{diag}\left(0, 0, 0, \frac{1}{2}, \frac{1}{2}\right)\otimes\tau_2$. The 45 A_{B-L} is also necessary for doublet-triplet mass splitting[15] in the Higgs 10_1.

The light neutrino mass matrix is given via see-saw mechanism as follows:

$$M_\nu = \Gamma_\nu^G (M_N^G)^{-1} (\Gamma_\nu^G)^\dagger v_2^2$$

$$= M_0 \begin{pmatrix} \frac{3}{4}\frac{z_N}{|z_\nu|}\frac{1}{y_N} & \frac{3}{2}\frac{z_N}{z_\nu^2}\frac{y_\nu}{y_N}\frac{\varepsilon_G^2}{\varepsilon_P^2}e^{-i\varphi} & -\frac{\sqrt{3}}{4}\frac{z_N}{|z_\nu|}\frac{x_\nu}{y_N}\frac{\varepsilon_G^2}{\varepsilon_P} \\ \frac{3}{2}\frac{z_N}{z_\nu^2}\frac{y_\nu}{y_N}\frac{\varepsilon_G^2}{\varepsilon_P^2}e^{i\varphi} & -3\frac{w_N}{|z_\nu|z_N} - \sqrt{3}\frac{x_\nu}{|z_\nu|}\frac{\varepsilon_G^2}{\varepsilon_P} & 1 \\ -\frac{\sqrt{3}}{4}\frac{z_N}{|z_\nu|}\frac{x_\nu}{y_N}\frac{\varepsilon_G^2}{\varepsilon_P} & 1 & \frac{3}{2}\frac{z_N}{z_\nu^2}\frac{y_\nu}{y_N}\frac{\varepsilon_G^2}{\varepsilon_P^2} \end{pmatrix}$$

$$= 2.1\lambda_H \begin{pmatrix} 0.73\lambda^6 & 0.73\lambda^8 e^{-i0.86\pi/2} & -0.97\lambda^7 \\ 0.73\lambda^8 e^{i0.86\pi/2} & -0.86\lambda^4 & 1 \\ -0.97\lambda^7 & 1 & 0.73\lambda^8 \end{pmatrix} \tag{11}$$

with $M_0 = \frac{2}{3}\frac{\varepsilon_G^2}{\varepsilon_P^4}\frac{1}{|z_\nu|z_N}\frac{1}{\eta_\nu}\frac{v_2}{v_{10}} v_2 \lambda_H$. Here η_ν is the RG evolution factor estimated to be $\eta_\nu \approx 1.35$. Diagonalizing the above mass matrix, we obtain the masses of light Majorana neutrinos:

$$\begin{cases} \frac{m_{\nu_e}}{m_{\nu_\mu}} = \frac{3}{4}\frac{z_N}{|z_\nu|}\frac{1}{y_N} = 0.83\times 10^{-4}, \\ \frac{m_{\nu_\mu}}{m_{\nu_\tau}} = 1 - 3\frac{w_N}{|z_\nu| z_N} - \sqrt{3}\frac{x_\nu}{|z_\nu|}\frac{\varepsilon_G^2}{\varepsilon_P} \approx 0.998, \\ m_{\nu_\tau} \approx M_0 \approx 2.1\lambda_H\,\mathrm{eV}. \end{cases} \tag{12}$$

The three heavy Majorana neutrinos have masses

$$\begin{cases} \frac{m_{N_1}}{m_{N_2}} = \frac{z_N^2}{4w_N y_N} = 0.047, \qquad \frac{m_{N_2}}{m_{N_3}} = \frac{y_N}{w_N} = 1.33, \\ m_{N_3} = \frac{\varepsilon_P^4}{\varepsilon_G^2} w_N v_{10}\lambda_H \approx 0.64\times 10^{10}\lambda_H\ \mathrm{GeV}. \end{cases} \tag{13}$$

The CKM-type lepton mixing matrix is predicted to be

$$V_{LEP} = V_\nu V_e^\dagger = \begin{pmatrix} V_{\nu_e e} & V_{\nu_e \mu} & V_{\nu_e \tau} \\ V_{\nu_\mu e} & V_{\nu_\mu \mu} & V_{\nu_\mu \tau} \\ V_{\nu_\tau e} & V_{\nu_\tau \mu} & V_{\nu_\tau \tau} \end{pmatrix} = \begin{pmatrix} 0.9976 & 0.068 & 0.000 \\ -0.051 & 0.748 & 0.665 \\ 0.045 & -0.664 & 0.748 \end{pmatrix}. \tag{14}$$

CP-violating effects are found to be small in the lepton mixing matrix. As a result we find

(i) a $\nu_\mu(\bar{\nu}_\mu) \to \nu_e(\bar{\nu}_e)$ short wave-length oscillation with

$$\Delta m_{e\mu}^2 = m_{\nu_\mu}^2 - m_{\nu_e}^2 \approx (4-6)\,\mathrm{eV}^2, \qquad \sin^2 2\theta_{e\mu} \approx 1.8\times 10^{-2}, \tag{15}$$

which is consistent with the LSND experiment[6]

$$\Delta m_{e\mu}^2 = m_{\nu_\mu}^2 - m_{\nu_e}^2 \approx (4-6)\,\mathrm{eV}^2, \qquad \sin^2 2\theta_{e\mu} \approx 1.8\times 10^{-2}\text{—}3\times 10^{-3}; \tag{16}$$

(ii) a $\nu_\mu(\bar{\nu}_\mu) \to \nu_\tau(\bar{\nu}_\tau)$ long-wave length oscillation with

$$\Delta m_{\mu\tau}^2 = m_{\nu_\tau}^2 - m_{\nu_\mu}^2 \approx (1.6-2.4)\times 10^{-2}\,\mathrm{eV}^2, \qquad \sin^2 2\theta_{\mu\tau} \approx 0.987, \tag{17}$$

which could explain the atmospheric neutrino deficit[17]:

$$\Delta m_{\mu\tau}^2 = m_{\nu_\tau}^2 - m_{\nu_\mu}^2 \approx (0.5-2.4)\times 10^{-2}\,\mathrm{eV}^2, \qquad \sin^2 2\theta_{\mu\tau} \approx 0.6-1.0, \tag{18}$$

with the best fit[17]

$$\Delta m_{\mu\tau}^2 = m_{\nu_\tau}^2 - m_{\nu_\mu}^2 \approx 1.6\times 10^{-2}\,\mathrm{eV}^2, \qquad \sin^2 2\theta_{\mu\tau} \approx 1.0. \tag{19}$$

However, $(\nu_\mu - \nu_\tau)$ oscillation will be beyond the reach of CHORUS/NOMAD and E803.

(iii) Two massive neutrinos ν_μ and ν_τ with

$$m_{\nu_\mu} \approx m_{\nu_\tau} \approx (2.0-2.4)\ \mathrm{eV}, \tag{20}$$

which fall in the range required by possible hot dark matter[18].

In this case, solar neutrino deficit has to be explained by oscillation between ν_e and a sterile neutrino[19] ν_s. Since strong bounds on the number of neutrino species both from the invisible Z^0-width and from primordial nucleosynthesis[20,21] require the additional neutrino to be sterile (singlet of $SU(2)\times U(1)$, or singlet of $SO(10)$ in the GUT $SO(10)$ model). Masses and mixings of the triplet sterile neutrinos can be chosen by introducing an additional singlet scalar with VEV $v_s \approx 450$ GeV. We find

$$\begin{cases} m_{\nu_s} = \lambda_H v_s^2 / v_{10} \simeq 2.4\times 10^{-3}\,\text{eV} , \\ \sin\theta_{es} \simeq \dfrac{m_{\nu_L\nu_s}}{m_{\nu_s}} = \dfrac{v_2}{2v_s}\dfrac{\varepsilon_P}{\varepsilon_G^2} \simeq 3.8\times 10^{-2}, \end{cases} \tag{21}$$

with the mixing angle consistent with the requirement necessary for primordial nucleosynthesis[22] given by ref. [20]. The resulting parameters

$$\Delta m_{es}^2 = m_{\nu_s}^2 - m_{\nu_e}^2 \simeq 5.8\times 10^{-6}\,\text{eV}^2, \qquad \sin^2 2\theta_{es} \simeq 5.8\times 10^{-3} \tag{22}$$

are consistent with the values[19] obtained by fitting the experimental data:

$$\Delta m_{es}^2 = m_{\nu_s}^2 - m_{\nu_e}^2 \approx (4-9)\times 10^{-6}\,\text{eV}^2 , \qquad \sin^2 2\theta_{es} \approx (1.6-14)\times 10^{-3}. \tag{23}$$

This scenario can be tested by the next generation solar neutrino experiments in Sudhuray Neutrino Observatory (SNO) and Super-Kamiokanda (Super-K), both planning to start operation in 1996. By measuring neutral current events, one could identify $\nu_e \to \nu_s$ or $\nu_e \to \nu_\mu(\nu_\tau)$ since the sterile neutrinos have no weak gauge interactions. By measuring seasonal variation, one can further distinguish the small-angle MSW[22] oscillation from vacuum mixing oscillation.

3 Superpotential for fermion Yukawa interactions

Non-Abelian discrete family symmetry $\Delta(48)$ is important in the present model for constructing interesting texture structures of the Yukawa coupling matrices. It originates from the basic considerations that all three families are treated on the same footing at the GUT scale, namely the three families should belong to an irreducible triplet representation of a family symmetry group. Based on the well-known fact that the masses of the three families have a hierarchic structure, the family symmetry group must be a group with at least rank three if the group is a continuous one. However, within the known simple continuous groups, it is difficult to find a rank three group which has irreducible triplet representations. This limitation of the continuous groups is thus avoided by their finite and disconnected subgroups. A simple example is the finite and disconnected group $\Delta(48)$, a subgroup of $SU(3)$.

The generators of the $\Delta(3n^2)$ group consist of the matrices

$$E(0,0)=\begin{pmatrix} 0 & 1 & 0 \\ 0 & 0 & 1 \\ 1 & 0 & 0 \end{pmatrix} \tag{24}$$

and

$$A_n(p,q)=\begin{pmatrix} e^{i\frac{2\pi}{n}p} & 0 & 0 \\ 0 & e^{i\frac{2\pi}{n}q} & 0 \\ 0 & 0 & e^{-i\frac{2\pi}{n}(p+q)} \end{pmatrix}. \tag{25}$$

It is clear that there are n^2 different elements $A_n(p,q)$ since if p is fixed, q can take n different values. There are three different element types:

$$A_n(p,q), \qquad E_n(p,q)=A_n(p,q)E(0,0), \qquad C_n(p,q)=A_n(p,q)E^2(0,0)$$

in the $\Delta(3n^2)$ group; therefore the order of the $\Delta(3n^2)$ group is $3n^2$. The irreducible representations of the $\Delta(3n^2)$ groups consist of (i) $(n^2-1)/3$ triplets and three singlets when $n/3$ is not an integer and (ii) $(n^2-3)/3$ triplets and nine singlets when $n/3$ is an integer.

The character of the triplet representations can be expressed as[5]

$$\begin{cases} \Delta_T^{m_1m_2}(A_n(p,q))=e^{i\frac{2\pi}{n}[m_1p+m_2q]}+e^{i\frac{2\pi}{n}[m_1q-m_2(p+q)]}+e^{i\frac{2\pi}{n}[-m_1(p+q)+m_2p]} \\ \Delta_T^{m_1m_2}(E_n(p,q))=\Delta_T^{m_1m_2}(C_n(p,q))=0 \end{cases} \tag{26}$$

with $m_1, m_2=0,1,\cdots,n-1$. Note that $(-m_1+m_2,-m_1)$ and $(-m_2,m_1-m_2)$ are equivalent to (m_1,m_2).

One will see that $\Delta(48)$ (i.e. $n=4$) is the smallest of the dihedral group $\Delta(3n^2)$ with sufficient triplets for constructing interesting texture structures of the Yukawa coupling matrices.

The irreducible triplet representations of $\Delta(48)$ consist of two complex triplets $T_1(\overline{T}_1)$ and $T_3(\overline{T}_3)$ and one real triplet $T_2=\overline{T}_2$ as well as three singlet representations. Their irreducible triplet representations can be expressed in terms of the matrix representation

$$\begin{cases} T_1^{(1)}=\mathrm{diag}(i,1,-i), & T_1^{(2)}=\mathrm{diag}(1,-i,i), & T_1^{(3)}=\mathrm{diag}(-i,i,1); \\ T_2^{(1)}=\mathrm{diag}(-1,1,-1), & T_2^{(2)}=\mathrm{diag}(1,-1,-1), & T_2^{(3)}=\mathrm{diag}(-1,-1,1); \\ T_3^{(1)}=\mathrm{diag}(i,-1,i), & T_3^{(2)}=\mathrm{diag}(-1,i,i), & T_3^{(3)}=\mathrm{diag}(i,i,-1). \end{cases} \tag{27}$$

The matrix representations of $\overline{T}_1^{(i)}$ and $\overline{T}_3^{(i)}$ are the Hermician conjugates of $T_1^{(i)}$ and $T_3^{(i)}$. With this representation, we can explicitly construct the invariant tensors.

All three families with $3\times16=48$ chiral fermions are unified into a triplet 16-dimensional spinor representation of $SO(10)\times\Delta(48)$. Without losing generality, one can assign the three chiral families into the triplet representation T_1, which may be simply

Table 2 Decomposition of the product of two triplets, $T_i \otimes T_j$ and $T_i \otimes \overline{T}_j$ in Δ_{48} $(SU(3))$.

$\Delta(48)$	1	$\overline{1}$	2	3	$\overline{3}$
1	$\overline{1}\,\overline{1}2$	$A3\overline{3}$	$\overline{1}3\overline{3}$	$12\overline{3}$	$12\overline{3}$
2	$\overline{1}3\overline{3}$	$13\overline{3}$	$A23$	$1\overline{1}\,\overline{3}$	$1\overline{1}3$
3	123	$\overline{1}23$	$1\overline{1}\,\overline{3}$	$\overline{2}\,\overline{3}\,\overline{3}$	$A1\overline{1}$

Triplets T_i and $\overline{T}_i$ are simply denoted by i and $\overline{i}$, respectively. For example $T_1 \otimes \overline{T}_1 = A \otimes T_3 \otimes \overline{T}_3 \equiv A3\overline{3}$. here A represents a singlet.

denoted by $\hat{1}6 = 16_i T_1^{(i)}$. All the fermions are assumed to obtain their masses through a single 10_1 of $SO(10)$ into which the needed two Higgs doublets are unified. The model could allow a triplet sterile neutrino with small mixings with the ordinary neutrinos. A singlet scalar near the electroweak scale is necessary for generating small masses of the sterile neutrinos.

Superpotentials which lead to the above texture structures (eqs.(1), (2) and (9)) with zeros and effective operators (eqs.(3) and (10)) are found to be

$$\begin{aligned} W_Y = & \sum_{a=0}^{3} \psi_{a1} 10_1 \psi_{a2} + \overline{\psi}_{22} \chi \psi_{13} + \overline{\psi}_{21} \chi_2 \psi_{13} + \overline{\psi}_{32} \chi \psi_{23} + \overline{\psi}_{31} \chi_3 \psi_{23} \\ & + \overline{\psi}_{02} \chi \psi_{33} + \overline{\psi}_{01} \chi_0 \psi_{32} + \overline{\psi}_{33} A_X \psi_3 + \overline{\psi}_3 A_X \psi_2 + \overline{\psi}_2 A_X \psi_1 \\ & + (\overline{\psi}_{11} \chi_1 + \overline{\psi}_{12} \chi + \overline{\psi}_{13} A_z + \overline{\psi}_{23} A_u + \overline{\psi}_1 Y) \hat{1}6 \\ & + \sum_{a=0}^{3} \sum_{j=1}^{3} S_G \overline{\psi}_{aj} \psi_{aj} + \sum_{i=1}^{2} (\overline{\psi}_{i3} A_X \psi_{i3} + S_I \overline{\psi}_i \psi_i) + S_I \overline{\psi}_{33} \psi_{33} + S_P \overline{\psi}_3 \psi_3 \end{aligned} \tag{28}$$

for the fermion Yukawa coupling matrices,

$$\begin{aligned} W_R = & \sum_{i=1}^{3} (\psi'_{i1} 10_3 \psi'_{i2} + \overline{\psi}'_{i1} \chi'_i \psi + \overline{\psi}'_{i2} \chi' \psi'_{i3} + \overline{\psi}'_{i1} A_i \psi') + (\overline{\psi}' X + \overline{\psi} A_u) \hat{1}6 + \overline{\Phi} 10_3 \overline{\Phi} \\ & \sum_{i=1}^{3} \sum_{j=1}^{2} S_G \overline{\psi}'_{ij} \psi'_{ij} + S_P (\sum_{i=1}^{3} \overline{\psi}'_{i3} \psi'_{i3} + \overline{\psi} \psi + \overline{\psi}' \psi') \end{aligned}$$

for the right-handed Majorana neutrinos, and

$$W_S = \overline{\psi}_1' 10_1 \psi_2' + \overline{\psi}_1' \Phi v_s + \overline{\psi}_2' \varphi_s \hat{1}6 + (\overline{v}_s \varphi_s N_s + h.c.) + S_I \overline{N}_s N_s \tag{29}$$

for the sterile neutrino masses and their mixings with the ordinary neutrinos.

In the above superpotentials, each term is ensured by the $U(1)$ symmetry. An appropriate assignment of $U(1)$ charges for the various fields is implied. All ψ fields are triplet 16-D spinor heavy fermions, where the fields $\psi_{i3}\{\overline{\psi}_{i3}\}$, $\psi'_{i3}\{\overline{\psi}'_{i3}\}$, $\psi_i\{\overline{\psi}_i\}$ $(i=1,2,3)$, $\psi'_1\{\overline{\psi}'_1\}$, $\psi'_2\{\overline{\psi}'_2\}$, $\psi\{\overline{\psi}\}$ and $\psi'\{\overline{\psi}'\}$ belong to $(16, T_1)$ $\{(\overline{16}, \overline{T}_1)\}$ representations of $SO(10) \times \Delta_{48}$ $(SU(3))$; $\psi_{11}\{\overline{\psi}_{11}\}$ and $\psi_{12}\{\overline{\psi}_{12}\}$ belong to $(16, T_2)$ $\{(\overline{16}, T_2)\}$; $\psi_{i1}\{\overline{\psi}_{i1}\}$ and $\overline{\psi}_{i2}$ $\{\psi_{i2}\}$ $(i=2,3,0)$ belong to $(16, T_3)$ $\{(\overline{16}, \overline{T}_3)\}$; $\psi'_{i1}\{\overline{\psi}'_{i1}\}$ and $\overline{\psi}'_{i2}\{\psi'_{i2}\}$ $(i=1,2)$ belong to $(16, \overline{T}_3)$ $\{(\overline{16}, T_3)\}$; $\psi'_{31}\{\overline{\psi}'_{31}\}$ and $\psi'_{32}\{\overline{\psi}'_{32}\}$ belong to $(16, T_2)$ $\{(\overline{16}, T_2)\}$; X, Y, S_I, S_P and φ_s are singlets of $SO(10) \times \Delta_{48}$ $(SU(3))$. v_s and N_s are $SO(10)$ singlet and $\Delta(48)$ triplet fermions. 10_3 is an additional $SO(10)$ 10-representation heavy scalar. All $SO(10)$ singlet χ fields are triplets of

$\Delta(48)$, where χ_1, χ_2, χ_3, χ_0, χ belong to triplet representations $\overline{T}_3$, T_3, $\overline{T}_1$, T_2, $\overline{T}_3$, respectively; χ_1', χ_2', χ_3', χ' belong to triplet representations $\overline{T}_1$, T_2, T_3, T_3, respectively. With the above assignment for various fields, one can check that once the triplet field χ develops VEV only in the third direction, i.e. $\langle\chi^{(3)}\rangle \neq 0$, and χ' develops VEV only in the second direction, i.e. $\langle\chi^{(2)}\rangle \neq 0$, the resulting fermion Yukawa coupling matrices at the GUT scale will be automatically forced, due to the special features of $\Delta(48)$, into an interesting texture structure with four non-zero textures '33', '32', '22' and '12' which are characterized by χ_1, χ_2, χ_3, and χ_0, respectively, and the resulting right-handed Majorana neutrino mass matrix is forced into three non-zero textures '33', '13' and '22' which are characterized by χ_1', χ_2', and χ_3', respectively. It is seen that five triplets are needed. of which one triplet is necessary for unity of the three family fermions, and four triplets are required for obtaining the needed minimal non-zero textures.

The symmetry breaking scenario and the structure of the physical vacuum are considered as follows:

$$SO(10)\times\Delta(48)\times U(1) \xrightarrow{\overline{M}_P} SO(10)\times\Delta(48) \xrightarrow{v_{10}} SU(5)\times\Delta(48)$$
$$\xrightarrow{v_5} SU(3)_c\times SU(2)_L\times U(1)_Y \xrightarrow{v_1, v_2} SU(3)_c\times U(1)_{em} \tag{30}$$

and

$\langle S_P\rangle = \overline{M}$, $\langle X\rangle = v_{10} = \langle S_I\rangle$, $\langle\Phi^{(16)}\rangle = \langle\overline{\Phi}^{(16)}\rangle = v_{10}/\sqrt{2}$, $\langle Y\rangle = v_5 = \langle S_G\rangle$, $\langle\chi^{(3)}\rangle = \langle\chi_a^{(i)}\rangle = \langle\chi'^{(2)}\rangle = \langle\chi_j'^{(i)}\rangle = v_5$ with $(i=1, 2, 3;\ a=0, 1, 2, 3;\ j=1, 2, 3)$, $\langle\chi^{(1)}\rangle = \langle\chi^{(2)}\rangle = \langle\chi'^{(1)}\rangle = \langle\chi'^{(3)}\rangle = 0$, $\langle\varphi_s\rangle = v_s \approx 450$ GeV, $\langle H_2\rangle = v_2 = v\sin\beta$ with $v = \sqrt{v_1^2 + v_2^2} = 246$ GeV.

4 Conclusions

It is amazing that nature has allowed us to make predictions in terms of a single Yukawa coupling constant and a set of VEVs determined by the structure of the vacuum and to understand the low energy physics from the Planck scale physics. The present model has provided a consistent picture for the 28 parameters in SM model with massive neutrinos. The neutrino sector is of special interest for further study. Though the recent LSND experiment, atmospheric neutrino deficit, and hot dark matter could be simultaneously explained in the present model, solar neutrino puzzle can only be understood by introducing an $SO(10)$ singlet sterile neutrino. It is expected that more precise measurements from various low energy experiments in the near future could provide crucial tests on the present model.

Acknowledgement Wu Yueliang would like to thank Institute for Theoretical Physics, Chinese Academy of Sciences, for its hospitality and partial support during his visit.

References

1 Gatto, R., Sartori, G., Tonin, M., Weak self-masses, Cabbibo angle and Breoken *SU*(2), *Phys., Lett.*, 1968, B28:128.

2 Weinberg, S., Rabi, I. I., Discrete flavor symmetries and a formula for the Cabbibo angle, *Phys. Lett.*, 1977, B40: 418.

3 Raby, S., Ohio State Univ., *Preprint*, *OHSTPY-HRP-T-95-24*.

4 Dimopoulos, S. Anderson, G., Raby, S., *et al.*, Predictive ansatz for fermion mass matrices in sypersymmetric grand unified theories, *Phys. Rev.*, 1992, D45:4192.

5 Hall, L.J., Raby, S., On the generality of certain predictions for quark mixing, *Phys. Lett.*, 1993, D135:164.

6 Fairbairn, W.M., Fulto, T., Klink, W.H., Finite and disconnected subgroups of *SU*(3) and their application to the elementary-particle spectrum, *J. Math. Phys.*, 1964, 5:1038.

7 Kaplan, D., Schmaltz, M., Flavor unification and discrete non-Abelian symmetries, *Phys. Rev.*, 1994, D49:3741.

8 Ananthanarayan, B., Lazarides, G., Shafi, Q., Top-quark-mass prediction from supersymmetric grand unified theories, *Phys. Rev.* Lett., 1991, D44:1613.

9 Hall, L., Rattazzi, R., Sarid, U., The top quark mass in supersymmetric *SO*(10) unification, *Phys. Rev.*, 1994, D50: 7048.

10 Geogi, H., Jarlskog, C., A new lepton-quark mass relation in a unified theory, *Phys. Lett.*, 1979, B86.297.

11 Particle Data Group, Evidence for top quark production in bar $\{p\}p$ collisions at sqrt $\{s\}=1.8$ TeV, *Phys. Rev.*, 1994, D50:2966.

12 Abachi, S., Observation of the top quark, *Phys. Rev. Lett.*, 1995, 74:263.

13 Gasser, J., Leutwyler, H., Quark masses, *Phys. Rep.*, 1982, 87:77

14 Wolfenstein, L., Parametrization of the Kobayashi-Maskawa matrix, *Phys. Rev. Lett.*, 1983, 51:1945.

15 Gell-Mann, M., Ramond, P., Slansky R., in *Supergravity* (ed. van Nieuwenhuizen, F., Freedman, D.), Amsterdam: North Holland, 1979, 315.

16 Dimpopoulos, S., Wilczek, F., in *Proceedings of Erice Summer School* (ed. Zichichi, A.) 1981.

17 Athanassopoulos, C., Candidate events in a search for oscillations, *Phys. Rev. Lett.*, 1995.

18 Fukuda, Y., Atmospheric ratio in multi-GeV energy range, *Phys. Lett.*, 1994 335B: 237.

19 Primack, J., Holtzman, J., Klypin, A. *et al.*, Cold hot dark matter cosmology, *Phys. Rev. Lett.*, 1995, 74:2160.

20 Caldwell, D.O., Mohapatra, R.N., Cold hot dark matter cosmology in the light of solar and atmospheric neutrino oscillations, *Phys. Rev.*, 1993, D48:3259.

21 Walker, T., Steigman, G., Schramm, D.N. *et al.*, A new look at neutrino limits from big bang nucleosynthesis, *Preprint OSUTA-2/95*, hep-ph/9502400.

22 Wolfenstein, L., Neutrino oscillation in matter, *Phys. Rev.*, 1978, D17:2369.

NUCLEAR PHYSICS B PROCEEDINGS SUPPLEMENTS

ELSEVIER Nuclear Physics B (Proc. Suppl.) 52A (1997) 159–162

A solution to the puzzles of CP violation, neutrino oscillation, fermion masses and mixings in an SUSY GUT model with small $\tan\beta$

K.C. Chou [OT1] and Yue-Liang Wu [OT1*]

[a]Chinese Academy of Sciences, Beijing 100864, China

[b]Department of Physics, Ohio State University, 174 W. 18th Ave., Columbus, OH 43210, USA

CP violation, fermion masses and mixing angles including that of neutrinos are studied in an SUSY SO(10)×Δ(48)× U(1) model with small $\tan\beta$. It is amazing that the model can provide a successful prediction on twenty three observables by only using four parameters. The renormalization group (RG) effects containing those above the GUT scale are considered. Fifteen relations among the low energy parameters are found with nine of them free from RG modifications. They could be tested directly by low energy experiments.

The standard model (SM) is a great success. But it cannot be a fundamental theory. Eighteen phenomenological parameters have been introduced to describe the real world, all of unkown origin. The mass spectrum and the mixing angles observed remind us that we are in a stage similar to that of atomic spectroscopy before Balmer. In this talk, we shall present an interesting model based on the symmetry group SUSY $SO(10) \times \Delta(48) \times U(1)$ with small values of $\tan\beta \sim 1$ which is of phenomenological interest in testing the Higgs sector in the minimum supersymmetric standard model (MSSM) at Colliders[1]. For a detailed analysis see ref. [2]. The dihedral group $\Delta(48)$, a subgroup of SU(3), is taken as the family group. $U(1)$ is family-independent and is introduced to distinguish various fields which belong to the same representations of $SO(10) \times \Delta(48)$. The irreducible representations of $\Delta(48)$ consisting of five triplets and three singlets are found to be sufficient to build an interesting texture structure for fermion mass matrices. The symmetry $\Delta(48) \times U(1)$ naturally ensures the texture structure with zeros for Yukawa coupling matrices, while the coupling coefficients of the resulting interaction terms in the superpotential are unconstrainted by this symmetry. To reduce the possible free parameters, the universality of Yukawa coupling constants in the superpotential is assumed, i.e., all the coupling coefficients are assumed to be equal and have the same origins from perhaps a more fundamental theory.

Choosing the structure of the physical vacuum carefully, the Yukawa coupling matrices which determine the masses and mixings of all quarks and leptons are given at the GUT scale by

$$\Gamma_u^G = \lambda_u \begin{pmatrix} 0 & \frac{3}{2} z'_u \epsilon_P^2 & 0 \\ \frac{3}{2} z_u \epsilon_P^2 & -3 y_u \epsilon_G^2 e^{i\phi} & -\frac{\sqrt{3}}{2} x_u \epsilon_G^2 \\ 0 & -\frac{\sqrt{3}}{2} x_u \epsilon_G^2 & w_u \end{pmatrix}$$

$$\Gamma_f^G = \lambda_f \begin{pmatrix} 0 & -\frac{3}{2} z'_f \epsilon_P^2 & 0 \\ -\frac{3}{2} z_f \epsilon_P^2 & 3 y_f \epsilon_G^2 e^{i\phi} & -\frac{1}{2} x_f \epsilon_G^2 \\ 0 & -\frac{1}{2} x_f \epsilon_G^2 & w_f \end{pmatrix}$$

$$\Gamma_\nu^G = \lambda_\nu \begin{pmatrix} 0 & -\frac{15}{2} z'_\nu \epsilon_P^2 & 0 \\ -\frac{15}{2} z_\nu \epsilon_P^2 & 15 y_\nu \epsilon_G^2 e^{i\phi} & -\frac{1}{2} x_\nu \epsilon_G^2 \\ 0 & -\frac{1}{2} x_\nu \epsilon_G^2 & w_\nu \end{pmatrix}$$

with $\lambda_u = 2\lambda_H/3$, $\lambda_f = \lambda_u(-1)^{n+1}/3^n$ $(f = d, e)$ and $\lambda_\nu = \lambda_f/5^{n+1}$. Here the integer n reflects the possible choice of heavy fermion fields above the GUT scale. $n = 4$ is found to be the best choice in this set of models for a consistent prediction on top and charm quark masses. This is because for $n > 4$, the resulting value of $\tan\beta$ becomes too small, as a consequence, the predicted top quark

*Supported in part by the US Department of Energy Grant # DOE/ER/01545-675. Permanent address: Institute of Theoretical Physics, Chinese Academy of Sciences, Beijing 100080, China

PII: S0920-5632(96)00553-1

mass will be below the present experimental lower limit. For $n < 4$, the values of $\tan\beta$ will become larger, the resulting charm quark mass will be above the present upper bound. $\lambda_H = \lambda_H^0 r_3$, $\epsilon_G \equiv (\frac{v_5}{v_{10}})\sqrt{\frac{r_2}{r_3}}$ and $\epsilon_P \equiv (\frac{v_5}{\bar{M}_P})\sqrt{\frac{r_1}{r_3}}$ are three parameters. Where λ_H^0 is a universal coupling constant expected to be of order one, r_1, r_2 and r_3 denote the ratios of the coupling constants of the superpotential at the GUT scale for the textures '12', '22' ('32') and '33' respectively. They represent the possible renormalization group (RG) effects running from the scale $\bar{M}_P$ to the GUT scale. Note that the RG effects for the textures '22' and '32' are considered to be the same since they are generated from a similar superpotential structure after integrating out the heavy fermions and concern the fields which belong to the same representations of the symmetry group, this can be explicitly seen from their effective operators W_{22} and W_{32} given below. $\bar{M}_P$, v_{10}, and v_5 being the vacuum expectation values (VEVs) for $U(1) \times \Delta(48)$, SO(10) and SU(5) symmetry breaking respectively. ϕ is the physical CP phase arising from the VEVs. The assumption of maximum CP violation implies that $\phi = \pi/2$. x_f, y_f, z_f, and w_f $(f = u, d, e, \nu)$ are the Clebsch factors of $SO(10)$ determined by the directions of symmetry breaking of the adjoints **45**'s. The Clebsch factors associated with the symmetry breaking directions can be easily read off from the U(1) hypercharges of the adjoints **45**'s and the related effective operators which are obtained when the symmetry $SO(10) \times \Delta(48) \times U(1)$ is broken and heavy fermion pairs are integrated out:

$$
\begin{aligned}
W_{33} &= \lambda_3 16_3 \eta_X \eta_A 10_1 \eta_A \eta_X 16_3 \\
W_{32} &= \lambda_2 16_3 \eta_X \eta_A \left(\frac{A_z}{A_X}\right) 10_1 \left(\frac{A_z}{A_X}\right) \eta_A 16_2 \\
W_{22} &= \lambda_2 16_2 \eta_A \left(\frac{A_u}{A_X}\right) 10_1 \left(\frac{A_u}{A_X}\right) \eta_A 16_2 e^{i\phi} \\
W_{12} &= \lambda_1 16_1 [\left(\frac{v_5}{\bar{M}_P}\right)^2 \eta'_A\, 10_1\, \eta'_A \\
&+ \left(\frac{v_{10}}{\bar{M}_P}\right)^2 \eta_A \left(\frac{A_u}{A_X}\right) 10_1 \left(\frac{A_z}{A_X}\right) \eta_A] 16_2
\end{aligned}
$$

where $\lambda_i = \lambda_H^0 r_i$, $\eta_A = (v_{10}/A_X)^{n+1}$, $\eta'_A = (v_{10}/A_X)^{n-3}$. The factor $\eta_X = 1/\sqrt{1+2\eta_A^2}$ arises from mixing, and provides a factor of $1/\sqrt{3}$ for the up-type quark. It remains almost unity for the down-type quark and charged lepton as well as neutrino due to the suppression of large Clebsch factors in the second term of the square root. The relative phase (or sign) between the two terms in the operator W_{12} has been fixed. The three directions of symmetry breaking have been chosen as $< A_X >= 2v_{10}\, diag.(1,\, 1,\, 1,\, 1,\, 1) \otimes \tau_2$, $< A_z >= 2v_5\, diag.(-\frac{1}{3},\, -\frac{1}{3},\, -\frac{1}{3},\, -1,\, -1) \otimes \tau_2$, $< A_u >= v_5/\sqrt{3}\, diag.(2,\, 2,\, 2,\, 1,\, 1) \otimes \tau_2$. The resulting Clebsch factors are $w_u = w_d = w_e = w_\nu = 1$, $x_u = 5/9$, $x_d = 7/27$, $x_e = -1/3$, $x_\nu = 1/5$ $y_u = 0$, $y_d = y_e/3 = 2/27$, $y_\nu = 4/225$, $z_u = 1$, $z_d = z_e = -27$, $z_\nu = -15^3 = -3375$, $z'_u = 1 - 5/9 = 4/9$, $z'_d = z_d + 7/729 \simeq z_d$, $z'_e = z_e - 1/81 \simeq z_e$, $z'_\nu = z_\nu + 1/15^3 \simeq z_\nu$.

An adjoint **45** A_X and a 16-dimensional representation Higgs field Φ ($\bar{\Phi}$) are needed for breaking SO(10) down to SU(5). Another two adjoint 45s A_z and A_u are needed to break SU(5) further down to the standard model $SU(3)_c \times SU_L(2) \times U(1)_Y$. From the Yukawa coupling matrices given above , the 13 parameters in the SM can be determined by only four parameters: a universal coupling constant λ_H and three parameters: ϵ_G, ϵ_P and $\tan\beta = v_2/v_1$.

The neutrino masses and mixings cannot be uniquely determined as they rely on the choice of the heavy Majorana neutrino mass matrix. The following texture structure with zeros is found to be interesting for the present model

$$
M_N^G = M_R \begin{pmatrix} 0 & 0 & \frac{1}{2} z_N \epsilon_P^2 e^{i\delta_\nu} \\ 0 & y_N & 0 \\ \frac{1}{2} z_N \epsilon_P^2 e^{i\delta_\nu} & 0 & w_N \epsilon_P^4 \end{pmatrix}
$$

The corresponding effective operators are

$$
\begin{aligned}
W_{13}^N &= \lambda_1^N 16_1 (\frac{A_z}{v_5})(\frac{\bar{\Phi}}{v_{10}})(\frac{\bar{\Phi}}{v_{10}})(\frac{A_u}{v_5}) 16_3\, e^{i\delta_\nu} \\
W_{22}^N &= \lambda_2^N 16_2 (\frac{A_z}{A_X})(\frac{\bar{\Phi}}{v_{10}})(\frac{\bar{\Phi}}{v_{10}})(\frac{A_z}{A_X}) 16_2 \\
W_{33}^N &= \lambda_3^N 16_3 (\frac{A_u}{v_5})^2 (\frac{A_z}{v_5})(\frac{\bar{\Phi}}{v_{10}})(\frac{\bar{\Phi}}{v_{10}})(\frac{A_u}{v_5})^2 16_3
\end{aligned}
$$

with

$M_R = \lambda_H v_{10}^2 \epsilon_P^4 \epsilon_G^2 / \bar{M}_P$, $\lambda_2^N = \lambda_H v_{10} \epsilon_P^4 / \bar{M}_P$, $\lambda_1^N = \lambda_2^N \epsilon_P^2 \epsilon_G^2$ and $\lambda_3^N = \lambda_1^N \epsilon_P^2$. It is then not difficult to read off the Clebsch factors $y_N = 9/25$,

$z_N = 4$ and $w_N = 256/27$. The CP phase δ_ν is assumed to be maximal $\delta_\nu = \pi/2$.

In obtaining physical masses and mixings, renormalization group (RG) effects should be taken into account. The initial conditions of the RG evolution are set at the GUT scale since all the Yukawa couplings of the quarks and leptons are generated at the GUT scale. As most Yukawa couplings in the present model are much smaller than the top quark Yukawa coupling $\lambda_t^G \sim 1$, in a good approximation, we will only keep top quark Yukawa coupling terms in the RG equations and neglect all other Yukawa coupling terms. The RG evolution will be described by three kinds of scaling factors. η_F $(F = U, D, E, N)$ and R_t arise from running the Yukawa parameters from the GUT scale down to the SUSY breaking scale M_S which is chosen to be close to the top quark mass, i.e., $M_S \simeq m_t \simeq 170$ GeV. They are defined by $\eta_F(M_S) = \prod_{i=1}^{3}\left(\frac{\alpha_i(M_G)}{\alpha_i(M_S)}\right)^{c_i^F/2b_i}$ $(F = U, D, E, N)$ with $c_i^U = (\frac{13}{15}, 3, \frac{16}{3})$, $c_i^D = (\frac{7}{15}, 3, \frac{16}{3})$, $c_i^E = (\frac{27}{15}, 3, 0)$, $c_i^N = (\frac{9}{25}, 3, 0)$, $b_i = (\frac{33}{5}, 1, -3)$, and $R_t^{-1} = exp[-\int_{\ln M_S}^{\ln M_G}(\frac{\lambda_t(t)}{4\pi})^2 dt] = [1 + (\lambda_t^G)^2 K_t]^{-1/12}$, where $K_t = \frac{3I(M_S)}{4\pi^2}$ with $I(M_S) = \int_{\ln M_S}^{\ln M_G}\eta_U^2(t)dt$ with $M_S \simeq m_t = 170$GeV. Other RG scaling factors are derived by running Yukawa couplings below M_S. $m_i(m_i) = \eta_i\ m_i(M_S)$ for $(i = c, b)$ and $m_i(1GeV) = \eta_i\ m_i(M_S)$ for $(i = u, d, s)$. The physical top quark mass is given by $M_t = m_t(m_t)\left(1 + \frac{4}{3}\frac{\alpha_s(m_t)}{\pi}\right)$. The scaling factor R_t or coupling $\lambda_t^G = \frac{1}{\sqrt{K_t}}\frac{\sqrt{1-R_t^{-12}}}{R_t^{-6}}$ is determined by the mass ratio of the bottom quark and τ lepton. $\tan\beta$ is fixed by the τ lepton mass via $\cos\beta = \frac{m_\tau\sqrt{2}}{\eta_B\eta_\tau v\lambda_\tau^G}$. In numerical predictions we take $\alpha^{-1}(M_Z) = 127.9$, $s^2(M_Z) = 0.2319$, $M_Z = 91.187$ GeV, $\alpha_1^{-1}(m_t) = 58.59$, $\alpha_2^{-1}(m_t) = 30.02$ and $\alpha_1^{-1}(M_G) = \alpha_2^{-1}(M_G) = \alpha_3^{-1}(M_G) \simeq 24$ with $M_G \sim 2 \times 10^{16}$ GeV. For $\alpha_s(M_Z) = 0.113$, the RG scaling factors have values $(\eta_{u,d,s}, \eta_c, \eta_b, \eta_{e,\mu,\tau}, \eta_U, \eta_D/\eta_E \equiv \eta_{D/E}, \eta_E, \eta_N) = (2.20, 2.00, 1.49, 1.02, 3.33, 2.06, 1.58, 1.41)$. The corresponding predictions on fermion masses and mixings thus obtained are found to be remarkable. Our numerical predictions for $\alpha_s(M_Z) = 0.113$ are given in table 1 with four input parameters. Where B_K and $f_B\sqrt{B}$ in table 1 are two important hadronic parameters and extracted from $K^0 - \bar{K}^0$ and $B^0 - \bar{B}^0$ mixing parameters ε_K and x_d. $Re(\varepsilon'/\varepsilon)$ is the direct CP-violating parameter in kaon decays, where large uncertanties mainly arise from the hadronic matrix elements. α, β and γ are three angles of the unitarity triangle in the Cabibbo-Kobayashi-Maskawa (CKM) matrix. J_{CP} is the rephase-invariant CP-violating quantity. The light neutrino masses and mixings are obtained via see-saw mechanism $M_\nu = \Gamma_\nu^G(M_N^G)^{-1}(\Gamma_\nu^G)^\dagger v_2^2/(2R_t^{-6}\eta_N^2)$. The predicted values for $|V_{us}|$, $|V_{ub}|/|V_{cb}|$, $|V_{td}|/|V_{ts}|$, m_d/m_s, $|V_{\nu_\mu e}|$, $|V_{\nu_\mu\tau}|$, $|V_{\nu_e\tau}|$ as well as m_{ν_e}/m_{ν_μ} and m_{ν_μ}/m_{ν_τ} are RG-independent.

From the results in table 1, we observe the following: **1**. a $\nu_\mu(\bar{\nu}_\mu) \to \nu_\tau(\bar{\nu}_\tau)$ long-wave length oscillation with $\Delta m_{\mu\tau}^2 = m_{\nu_\tau}^2 - m_{\nu_\mu}^2 \simeq 1.5 \times 10^{-2} eV^2$ and $\sin^2 2\theta_{\mu\tau} \simeq 0.987$ could explain the atmospheric neutrino deficit[3]; **2**. Two massive neutrinos ν_μ and ν_τ with $m_{\nu_\mu} \simeq m_{\nu_\tau} \simeq 2.45$ eV fall in the range required by possible hot dark matter[4]; **3** a short wave-length oscillation with $\Delta m_{e\mu}^2 = m_{\nu_\mu}^2 - m_{\nu_e}^2 \simeq 6\ eV^2$ an $\sin^2 2\theta_{e\mu} \simeq 1.0 \times 10^{-2}$ is consistent with the LSND experiment[5]. **4**. $(\nu_\mu - \nu_\tau)$ oscillation will be beyond the reach of CHORUS/NOMAD and E803. However, $(\nu_e - \nu_\tau)$ oscillation may become interesting as a short wave-length oscillation with $\Delta m_{e\tau}^2 = m_{\nu_\tau}^2 - m_{\nu_e}^2 \simeq 6\ eV^2$ and $\sin^2 2\theta_{e\tau} \simeq 1.0 \times 10^{-2}$; **5**. Majorana neutrino allows neutrinoless double beta decay $(\beta\beta_{0\nu})$[6]. The decay rate is found to be $\Gamma_{\beta\beta} \simeq 1.0 \times 10^{-61}$ GeV which is below to the present upper limit; **6**. solar neutrino deficit has to be explained by oscillation between ν_e and a sterile neutrino ν_s [7](singlet of SU(2)× U(1), or singlet of SO(10) in the GUT SO(10) model). Masses and mixings of the triplet sterile neutrinos can be chosen by introducing an additional singlet scalar with VEV $v_s \simeq 336$ GeV. They are found to be $m_{\nu_s} = \lambda_H v_s^2/v_{10} \simeq 2.8 \times 10^{-3}$eV and $\sin\theta_{es} \simeq m_{\nu_L\nu_s}/m_{\nu_s} = v_2\epsilon_P/(2v_s\epsilon_G^2) \simeq 3.8 \times 10^{-2}$. The resulting parameters $\Delta m_{es}^2 = m_{\nu_s}^2 - m_{\nu_e}^2 \simeq 6.2 \times 10^{-6}\ eV^2$ and $\sin^2 2\theta_{es} \simeq 5.8 \times 10^{-3}$; are consistent with the values [7] obtained from fitting the

Table 1
Output observables and model parameters and their predicted values with $\alpha_s(M_Z) = 0.113$ and input parameters: $m_e = 0.511$ eV, $m_\mu = 105.66$ MeV, $m_\tau = 1.777$ GeV, and $m_b(m_b) = 4.25$ GeV.

Output	Output	Data[8]	Output	Output
M_t [GeV]	182	180 ± 15	$J_{CP}/10^{-5}$	2.68
$m_c(m_c)$ [GeV]	1.27	1.27 ± 0.05	α	86.28°
m_u(1GeV) [MeV]	4.31	4.75 ± 1.65	β	22.11°
m_s(1GeV) [MeV]	156.5	165 ± 65	γ	71.61°
m_d(1GeV) [MeV]	6.26	8.5 ± 3.0	m_{ν_τ} [eV]	2.4515
$\|V_{us}\| = \lambda$	0.22	0.221 ± 0.003	m_{ν_μ} [eV]	2.4485
$\frac{\|V_{ub}\|}{\|V_{cb}\|}$	0.083	0.08 ± 0.03	m_{ν_e} [eV]/10^{-3}	1.27
$\frac{\|V_{td}\|}{\|V_{ts}\|}$	0.209	0.24 ± 0.11	m_{ν_s} [eV]/10^{-3}	2.8
$\|V_{cb}\| = A\lambda^2$	0.0393	0.039 ± 0.005	$\|V_{\nu_\mu e}\|$	-0.049
λ_t^G	1.30	-	$\|V_{\nu_e \tau}\|$	0.000
$\tan\beta = v_2/v_1$	2.33	-	$\|V_{\nu_\tau e}\|$	-0.049
ϵ_G	0.2987	-	$\|V_{\nu_\mu \tau}\|$	-0.707
ϵ_P	0.0101	-	$\|V_{\nu_e s}\|$	0.038
B_K	0.90	0.82 ± 0.10	M_{N_1} [GeV]	~ 333
$f_B\sqrt{B}$ [MeV]	207	200 ± 70	M_{N_2} [GeV]/10^6	1.63
Re$(\varepsilon'/\varepsilon)/10^{-3}$	1.4 ± 1.0	1.5 ± 0.8	M_{N_3} [GeV]	333

experimental data.

It is amazing that nature has allowed us to make predictions on fermion masses and mixings in terms of a single Yukawa coupling constant and three ratios determined by the structure of the physical vacuum and understand the low energy physics from the GUT scale physics. It has also suggested that nature favors maximal spontaneous CP violation. It is expected that more precise measurements from CP violation, neutrino oscillation and various low energy experiments in the near future could provide a good test on the present model and guide us to a more fundamental model.

ACKNOWLEDGEMENTS: YLW would like to thank professor R. Mohapatra for a kind invitation to him to present this work at the 4th SUSY96 conference held at University of Maryland, May 29- June 1, 1996.

REFERENCES

1. G. Kane, in this proceedings; see also J. Ellis, talk given at 17th Intern. Symposium on Lepton-Photon Interactions, 10-15 August, 1995, Beijing, China.
2. K.C. Chou and Y.L. Wu, Phys. Rev. **D53** (1996) R3492; hep-ph/9511327 and hep-ph/9603282, 1996.
3. Y. Fukuda *et al.*, Phys. Lett. **335B**, 237 (1994).
4. D.O. Caldwell, in this proceedings; J. Primack *et al.*, Phys. Rev. Lett. 74 (1995) 2160.
5. C. Athanassopoulos *et al.*, Phys. Rev. Lett., (1996) nucl-ex/9504002 (1995).
6. For a recent review see, R.N. Mohapatra, Maryland Univ. Report No. UMD-PP-95-147, hep-ph/9507234.
7. D.O. Caldwell and R.N. Mohapatra, Phys. Rev. D **48**, 1993) 3259; J. Peltoniemi, D. Tommasini, and J.W.F. Valle, Phys. Lett. **298B** (1993) 383.
8. CDF Collaboration, F. Abe et al., Phys. Rev. Lett. **74**, 2626 (1995); D0 Collaboration, S. Abachi et al., Phys. Rev. Lett. **74**, 2632 (1995); J. Gasser and H. Leutwyler, Phys. Rep. **87**, 77 (1982); H. Leutwyler, Nucl. Phys. **B337**, 108 (1990); Particle Data Group, Phys. Rev. **D50**, 1173, (1994).

Vol. 41 No. 3 SCIENCE IN CHINA (Series A) March 1998

A possible unification model for all basic forces *

WU Yueliang (吴岳良)

(Institute of Theoretical Physics, Chinese Academy of Sciences, Beijing 100080, China)

and ZHOU Guangzhao (K. C. Chou, 周光召)

(Chinese Academy of Sciences, Beijing 100086, China)

Received November 10, 1997

Abstract A unification model for strong, electromagnetic, weak and gravitational forces is proposed. The tangent space of ordinary coordinate 4-dimensional spacetime is a submanifold of a 14-dimensional internal spacetime spanned by four frame fields. The unification of the standard model with gravity is governed by gauge symmetry in the internal spacetime.

Keywords: unification, internal spacetime, SO(1,13), gravity, frame fields.

One of the great theoretical endeavours in this century is to unify gravitational force characterized by the general relativity of Einstein[1,2] with all other elementary particle forces (strong, electromagnetic and weak) described by Yang-Mills gauge theory[3]. One of the difficulties arises from the no-go theorem[4] which was proved based on a local relativistic quantum field theory in 4-dimensional spacetime. Most of the attempts to unify all basic forces involve higher-dimensional spacetime, such as Kaluza-Klein Yang-Mills theories[5,6], supergravity theories[7,8] and superstring theories[9-12]. In the Kaluza-Klein Yang-Mills theories, in order to have a standard model gauge group as the isometry group of the manifold, the minimal number of total dimensions has to be 11[13]. Even so, the Kaluza-Klein approach is not rich enough to support the fermionic representations of the standard model due to the requirement of the Atiyah-Hirzebruch index theorem. The maximum supergravity has SO(8) symmetry, its action is usually also formulated as an $N = 1$ supergravity theory in 11-dimensional spacetime. Unfortunately, the SO(8) symmetry is too small to include the standard model. Consistent superstring theories have also been built based on 10-dimensional spacetime. In superstring theories, all the known particle interactions can be reproduced, but millions of vacua have been found. The outstanding problem is to find which one is the true vacuum of the theory.

In this paper we will consider an alternative scheme. Firstly, we observe that quarks and leptons in the standard model[14-16] can be unified into a single 16-dimensional representation of complex chiral spinors in SO(10)[17,18]. Each complex chiral spinor belongs to a single 4-dimensional representation of SO(1,3). In a unified theory, it is an attractive idea to treat these 64 real spinor components on the same footing, i.e. they have to be a single representation of a larger group. It is therefore natural to consider SO(1,13) as our unified group and the gauge potential of SO(1,13) as the fundamental interaction that unifies the four basic forces (strong, electromag-

* Project supported in part by the Outstanding Young Scientist Fund of China.

netic, weak and gravitational) of nature. Secondly, to avoid the restrictions given by no-go theorem and other problems mentioned above, we consider that the ordinary coordinate spacetime remains to be a 4-dimensional manifold S_4 with metric $g_{\mu\nu}(x)$, $\mu, \nu = 0, 1, 2, 3$. At each point P: x^μ, there is a d-dimensional flat space M_d with $d > 4$ and signature $(1, -1, \cdots, -1)$. We assume the tangent space T_4 of S_4 at point P to be a 4-dimensional submanifold of M_d spanned by four vectors $e_\mu^A(x)$ $\mu = 0, 1, 2, 3$; $A = 0, 1, \cdots, d-1$ such that

$$g_{\mu\nu}(x) = e_\mu^A(x)\, e_\nu^A(x)\, \eta_{AB}. \tag{1}$$

where $\eta_{AB} = \mathrm{diag}(1, -1, \cdots, -1)$ can be considered as the metric of the flat space M_d. We shall call $e_\mu^A(x)$ the generalized vierbein fields or simply the frame fields. Once the frame fields $e_\mu^A(x)$ are given, we can always supplement them with another d-4 vector fields $e_m^A(x) \equiv e_m^A(e_\mu^A(x))$, $m = 1, 2, \cdots, d-4$ such that

$$e_\mu^A(x)\, e_m^B(x)\, \eta_{AB} = 0, \quad e_m^A(x)\, e_n^B(x)\, \eta_{AB} = g_{mn}, \tag{2}$$

where $g_{mn} = \mathrm{diag}(-1, \cdots, -1)$. $e_m^A(x)$ can be uniquely determined up to an SO(d-4) rotation. In the flat manifold M_d we can use $e_\mu^A(x)$ and $e_m^A(x)$ to decompose it into two orthogonal manifolds $T_4 \otimes C_{d-4}$, where C_{d-4} will be considered to be the internal space describing SO(d-4) internal symmetry besides the spin and is spanned by the d-4 orthonormal vectors $e_m^A(x)$. In the new frame system of M_d the metric tensor is of the form

$$\begin{pmatrix} g_{\mu\nu}(x) & 0 \\ 0 & g_{mn} \end{pmatrix}. \tag{3}$$

With $e_\mu^A(x)$ and $e_m^A(x)$, we can now define the covariant vectors as $e_A^\mu(x)$ and $e_A^m(x)$ satisfying

$$\begin{aligned} e_A^\mu(x)\, e_\nu^A(x) &= g_\nu^\mu, & e_A^m(x)\, e_n^A(x) &= g_n^m, \\ e_A^\mu(x)\, e_m^A(x) &= 0, & e_A^m(x)\, e_\mu^A(x) &= 0. \end{aligned} \tag{4}$$

Under general coordinate transformations and the rotations in M_d, $e_\mu^A(x)$ transforms as a covariant vector in ordinary coordinate spacetime and a vector in the M_d rotation, $e_m^A(x)$ transforms as a covariant vector in the C_{d-4} rotation and a vector in the M_d rotation. For a theory to be invariant under both general coordinate transformations and local rotations in the flat space M_d, it is necessary to introduce affine connection $\Gamma_{\mu\nu}^\rho(x)$ for general coordinate transformations and gauge potential $\Omega_\mu^{AB}(x) = -\Omega_\mu^{BA}(x)$ for d-dimensional rotation SO(1, d-1) in M_d. These transformations are connected by the requirement that T_4 has to be the submanifold of M_d spanned by four vectors $e_\mu^A(x)$ at point P and $e_\mu^A(x)$ should be a covariantly constant frame and satisfy the condition

$$D_\mu e_\rho^A = \partial_\mu e_\rho^A - \Gamma_{\mu\rho}^\tau e_\tau^A + g_U \Omega_{\mu B}^A e_\rho^B = 0. \tag{5}$$

It is then easily verified that

$$D_\mu g_{\rho\sigma} = \partial_\mu g_{\rho\sigma} - \Gamma_{\mu\rho}^\lambda g_{\lambda\sigma} - \Gamma_{\mu\sigma}^\lambda g_{\rho\lambda} = 0, \tag{6}$$

$$D_\mu e_A^\rho = \partial_\mu e_A^\rho + \Gamma_{\mu\sigma}^\rho e_A^\sigma - g_U \Omega_{\mu A}^B e_B^\rho = 0. \tag{7}$$

With the above considerations, we can now construct an invariant action under general coordinate transformations in the ordinary coordinate spacetime and the local SO (1, d-1) group symmetry in M_d with eq. (5) as a constraint. In addition, the action is required to have no dimensional parameters and to be renormalizable in the sense of the power counting. The general form of the action which satisfies these requirements is

$$S_B = \int d^4x \sqrt{-g}\Big\{ -\frac{1}{4} F_{\mu\nu}^{AB} F_{\rho\sigma}^{CD} g^{\mu\rho} g^{\nu\sigma} \eta_{AC}\eta_{BD} - \frac{1}{2}\xi\phi^2 F_{\mu\nu}^{AB} e_A^{\mu} e_B^{\nu} + \frac{1}{2} g^{\mu\nu}\partial_\mu\phi\partial_\nu\phi + \frac{1}{4}\lambda\phi^4 \tag{8}$$
$$+ \zeta F_{\mu\nu}^{AB} F_{\rho\sigma}^{CD} g^{\mu\rho}\eta_{AC} e_B^{\nu} e_D^{\sigma} + a_1 F_{\mu\nu}^{AB} F_{\rho\sigma}^{CD} e_C^{\mu} e_D^{\nu} e_A^{\rho} e_B^{\sigma}$$
$$+ a_2 F_{\mu\nu}^{AB} F_{\rho\sigma}^{CD} e_C^{\mu} e_B^{\nu} e_A^{\rho} e_D^{\sigma} + a_3 F_{\mu\nu}^{AB} F_{\rho\sigma}^{CD} e_A^{\mu} e_B^{\nu} e_C^{\rho} e_D^{\sigma}\Big\},$$

where $\Phi(x)$ is a scalar field introduced to avoid the dimensional coupling constants. $a_i\,(i=1,2,3)$, ζ, ξ and λ are dimensionless parameters. $F_{\mu\nu}^{AB}$ is the field strength defined in a standard way:

$$F_{\mu\nu}^{AB} = \partial_\mu \Omega_\nu^{AB} - \partial_\nu \Omega_\mu^{AB} + g_U\left(\Omega_{\mu C}^{A}\Omega_\nu^{CB} - \Omega_{\nu C}^{A}\Omega_\mu^{CB}\right). \tag{9}$$

The tensor F_μ^A is defined as $F_\mu^A = F_{\mu\nu}^{AB} e_B^{\nu}$.

Using the frame fields $e_A^{\mu}(x)$ and $e_m^{A}\,(e_\mu^{A}(x))$, we can decompose $\Omega_\mu^{AB}(x)$ into three parts $e_A^{\sigma}(x)\,\Omega_\mu^{AB}(x)\,e_B^{\rho}(x)\ (\rho,\sigma=0,1,2,3)$ which describe the gravity, and $e_A^{m}(x)\,\Omega_\mu^{AB}(x)\,e_B^{n}(x)$ which characterize gauge interactions, as well as $e_A^{m}(x)\,\Omega_\mu^{AB}(x)\,e_B^{\sigma}(x)$ which connect gravity with gauge interactions. From the constraints of eq. (5), we obtain

$$g_U e_{\sigma A}(x)\,\Omega_\mu^{AB}(x)\,e_B^{\rho}(x) = \Gamma_{\mu\sigma}^{\rho} - e_{\sigma A}\partial_\mu e^{\rho A}, \tag{10}$$
$$g_U e_{mA}(x)\,\Omega_\mu^{AB}(x)\,e_B^{\sigma}(x) = - e_{mA}\partial_\mu e^{\sigma A}. \tag{11}$$

Similarly, we can reexpress $e_A^{m}(x)\,\Omega_\mu^{AB}(x)\,e_B^{n}(x)$ as

$$e_A^{n}(x)\,\Omega_\mu^{AB}(x)\,e_B^{m}(x) = A_\mu^{mn}(x) - \frac{1}{2g_U}\left(e_A^{n}\partial_\mu e^{mA} - e_A^{m}\partial_\mu e^{nA}\right), \tag{12}$$

where $A_\mu^{mn}(x) = - A_\mu^{nm}(x)\ (m, n=1, \cdots, d-4)$ is the gauge potential for $(d-4)$-dimensional rotation SO$(d-4)$ in C_{d-4}.

Note that not all the gauge fields $\Omega_\mu^{AB}(x)$ are simply new propagating fields due to the constraints $D_\mu e_\rho^{A}=0$. By counting the constraint equations $(4\times 4\times d)$, unknown $\Omega_\mu^{AB}(x)$ (with $4d(d-1)/2$ degrees of freedom) and $e_\mu^{A}(x)$ (with $4\times d$ degrees of freedom) as well as $\Gamma_{\mu\sigma}^{\rho}$(with 40 degrees of freedom for the symmetric parts $\Gamma_{(\mu\sigma)}^{\rho} = \Gamma_{(\sigma\mu)}^{\rho}$ and 24 degrees of freedom for antisymmetric parts $\Gamma_{[\mu\sigma]}^{\rho} = - \Gamma_{[\sigma\mu]}^{\rho}$), one sees that besides the antisymmetric parts $\Gamma_{[\mu\sigma]}^{\rho}$, the independent degrees of freedom are $(4d+4(d-4)(d-5)/2)$. These independent degrees of freedom coincide with the degrees of freedom of the frame fields $e_\mu^{A}(x)$ and the gauge fields $A_\mu^{mn}(x)$ of the group SO$(d-4)$. In addition, the gauge conditions in the coset SO$(1, d-1)$/SO$(d-4)$ lead to additional constraints $(4d-10)$. Thus the independent degrees of freedom are reduced to $(10+4(d-4)(d-5)/2)$ which exactly match with the degrees of freedom of the metric tensor $g_{\mu\nu}(x)$ and the gauge fields $A_\mu^{mn}(x)$ of the group SO$(d-4)$. For $d=14$, the resulting independent degrees of freedom of the fields are sufficient to describe the four basic forces, where the general relativity of the Einstein theory is described by the metric tensor. Photon, W-bosons and gluons, that mediate the electromagnetic, weak and strong interactions respectively, are different manifestations of the gauge potential $A_\mu^{mn}(x)$ of the symmetry group SO(10)[17,18]. The curvature tensor $R_{\mu\nu\sigma}^{\rho}$ and the Ricci tensor $R_{\nu\sigma} = R_{\mu\nu\sigma}^{\rho} g_\rho^{\mu}$ as well as the scalar curvature $R = R_{\nu\sigma}g^{\nu\sigma}$ are simply related to the field strength $F_{\mu\nu}^{AB}$ via $R_{\mu\nu\sigma}^{\rho} = g_U F_{\mu\nu}^{AB} e_A^{\rho} e_{\sigma B}$, $R_{\nu\sigma} = g_U F_{\mu\nu}^{AB} e_A^{\mu} e_{\sigma B}$ and $R = g_U F_{\mu\nu}^{AB} \cdot e_A^{\mu} e_B^{\nu}$. It is not difficult to check that $R_{\mu\nu}^{AB} R_{AB}^{\mu\nu} = F_{\mu\nu}^{mn} F_{mn}^{\mu\nu} + g_U^{-2} R_{\mu\rho\nu\sigma} R^{\mu\rho\nu\sigma}$, and $R_\mu^{A} R_A^{\mu} = g_U^{-2} R_{\mu\rho} R^{\mu\rho}$, where $F_{\mu\nu}^{mn}(x)$ is the field strength of the gauge potential $A_\mu^{mn}(x)$:

$$F_{\mu\nu}^{mn} = \partial_\mu A_\nu^{mn} - \partial_\nu A_\mu^{mn} + g_U\left(A_{\mu q}^{m} A_\nu^{qn} - A_{\nu q}^{m} A_\mu^{qn}\right). \tag{13}$$

With these relations, the action S_B can be simply reexpressed as

$$S_B = \int d^4x \sqrt{-g}\left\{ -\frac{1}{4} F^{mn}_{\mu\nu} F^{\mu\nu}_{mn} + \frac{1}{2}\partial_\mu\phi\partial^\mu\phi + \frac{1}{4}\lambda\phi^4 - \frac{1}{2}\zeta g_U^{-1}\phi^2 R + g_U^{-2}\left[\left(a_1 - \frac{1}{4}\right) R_{\mu\rho\nu\sigma}R^{\mu\rho\nu\sigma} + (a_2 + \xi) R_{\mu\rho}R^{\mu\rho} + a_3 R^2\right]\right\}, \tag{14}$$

which has the same form as the action of a multiplicatively renormalized unified gauge theory including the so-called R^2-gravity and a renormalizable scalar matter field as well a nonminimal gravitational-scalar coupling.

In the real world, there exist three generations of quarks and leptons. Each generation of the quarks and leptons has 64 real degrees of freedom. These degrees of freedom will be represented by the 64 components of a single Weyl fermion $\Psi_+(x)$ belonging to the fundamental spinor representation of SO(1,13). The action for fermions is given by

$$S_F = \int d^4x \sqrt{-g}\left\{ \frac{1}{2}\overline{\Psi}_+ e^\mu_A \Gamma^A \left(i\partial_\mu + g_U \Omega^{BC}_\mu \frac{1}{2}\Sigma_{BC}\right)\Psi_+ + h.\ c.\right\}, \tag{15}$$

where Σ_{AB} are the generators of the SO(1, d-1) in the spinor representations and given by $\Sigma_{AB} = \frac{i}{4}[\Gamma_A, \Gamma_B]$. Γ^A are the gamma matrices that obey $\{\Gamma^A, \Gamma^B\} = 2\eta^{AB}$. Note that the resulting total action $S = S_B + S_F$ is simple, but it is nontrival for fermionic interactions since the gauge potentials $\Omega^{AB}_\mu(x)$ are related to the independent degrees of freedom $A^{mn}_\mu(x)$ and $e^A_\mu(x)$ by some nontrival relations given in eqs. (10) —(12). In particular, the supplemented frame fields $e^A_m(x)$ will have a highly nonlinear dependence on the frame fields $e^A_\mu(x)$.

Now let us consider the conservation laws under the general coordinate transformations and local rotation SO(1, d-1). Under the local rotation $\Psi_+(x) \to e^{-\frac{1}{2}i\omega^{AB}\Sigma_{AB}}\Psi_+(x)$, it is not difficult to find the conservation law as

$$D_\mu(\sqrt{-g}S^\mu_{AB}) - \sqrt{-g}\,T_{[AB]} \equiv 0 \tag{16}$$

with

$$S^\mu_{AB} = g_U \frac{1}{4}\overline{\Psi}_+ e^\mu_C \{\Gamma^C, \Sigma_{AB}\}\Psi_+, \tag{17}$$

$$T_{[AB]} = -i\frac{1}{2}\left[\overline{\Psi}_+ e^\mu_A \Gamma_B D_\mu \Psi_+ - (D_\mu\overline{\Psi}_+) e^\mu_A \Gamma_B \Psi_+ - (A \leftrightarrow B)\right]. \tag{18}$$

The general coordinate transformations lead to the well-known energy-momentum conservation law as

$$D^\nu(\sqrt{-g}\,T_{\mu\nu}) \equiv \sqrt{-g}\,F^{AB}_{\mu\nu}S^\nu_{AB} \tag{19}$$

with

$$T_{\mu\nu} = g_{\mu\nu}\mathbf{L} - i\frac{1}{2}e^A_\mu\left[\overline{\Psi}_+\Gamma_A D_\nu\Psi_+ - (D_\nu\overline{\Psi}_+)\Gamma_A\Psi\right]. \tag{20}$$

Using the covariantly constant frame fields $e^A_\mu(x)$, we can project S^μ_{AB} and $T_{[AB]}$ into

$$S^\mu_{\rho\sigma} = S^\mu_{AB}e^A_\rho e^B_\sigma, \tag{21}$$

$$T_{[\rho\sigma]} = T_{[AB]}e^A_\rho e^B_\sigma = T_{\rho\sigma} - T_{\sigma\rho}. \tag{22}$$

The angular momentum conservation law becomes

$$D_\mu(\sqrt{-g}S^\mu_{\rho\sigma}) - \sqrt{-g}\,T_{[\rho\sigma]} \equiv 0. \tag{23}$$

It is then easy to show that these two conservation laws (eqs. (19) and (23)) are essentially the same as those occurring in special relativity by noticing the following relations:

$$-T_{[\rho\sigma]} = \partial_\mu L^\mu_{[\rho\sigma]} \equiv \partial_\mu (x_\rho T^\mu_\sigma - x_\sigma T^\mu_\rho). \quad (24)$$

Here $L^\mu_{\rho\sigma}$ is the orbital angular momentum tensor and $J^\mu_{\rho\sigma} \equiv S^\mu_{\rho\sigma} + L^\mu_{\rho\sigma}$ represents the total angular momentum tensor.

From simple ideas we have provided a unification model for strong, electromagnetic, weak and gravitational forces and constructed the action without dimensional parameters as the basis for quantum theory of all the basic forces of the elementary particles. Such a theory is conjectured to be multiplicatively renormalizable though it may remain an effective theory of a more fundamental theory. One can find a formal proof of the renormalizability of the R^2-gravity with a scalar field in ref. [19]. It was known that in the general relativity only the Einstein equations have been tested to be in good agreement with known experimental data at the classical level. Thus the general relativity of the Einstein theory may be interpreted as a classical theory in the low energy limit, so that the Einstein-Hilbert and cosmological terms may be induced as a result of the low energy limit[19,20]. For instance, these terms may result from a spontaneous symmetry breaking. Finally, we would like to comment on the so-called unitary problem due to the appearance of the higher derivative terms within the framework of perturbation theory. The higher derivative terms become important as the energy scale goes up to near the Planck scale, at that scale gravitational interaction becomes strong so that the treatment by perturbatively expanding the metric fields is no longer suitable. From gauge theories' points of view, the local Lorentz group is not a compact group; not all the components of the gauge fields are physical one; additional conditions have to be introduced to eliminate those unphysical components. This is similar to the case of gauge theories in which gauge conditions have been used to eliminate the umphysical modes (time and longitudinal modes for massless gauge fields). Therefore, to solve the so-called unitary problem in gravitational interactions, a nonperturbative treatment or an alternative approach has to be developed. We shall further study this problem in our future investigations.

We hope that the present model has provided us a new insight for unifying all the basic forces within the framework of quantum field theory. Though both the ideas and the resulting model are simple, more theoretical work and experimental efforts are needed to test whether they are the true choice of nature.

References

1 Einstein, A., *Sitz. Preuss. Akad. Wiss.*, 1915, 778, 884.
2 Einstein, A., Die Grundlagen der allgemeinen Relativitätstheorie, *Ann. Phys. Lpz.*, 1916, 49: 769.
3 Yang, C. N., Mills, R. L., Conservation of isotopic spin and isotopic gauge invariance, *Phys. Rev.*, 1954, 96: 191.
4 Coleman, S., Mandula, J., *Phys. Rev.*, 1967, 159: 1251.
5 Kaluza, Th., On the problem of unity in physics, *Sitz. Preuss. Akad. Wiss.*, 1921, Kl: 966.
6 Klein, O., Quantentheorie und fünfdimensionale Relativitätstheorie, *Z. Physik*, 1926, 37: 895.
7 Freedman, D., Ferrara, S., Van Nieuwenhuizen, P., Progress toward a theory of supergravity, *Phys. Rev.*, 1976, D13: 3214.
8 Deser, S., Zumino, B., Consistent supergravity, *Phys. Lett.*, 1976, 62B: 335.
9 Green, M., Schwartz, J. H., Covariant description of superstrings, *Phys. Lett.*, 1984, 149B: 117.
10 Candelas, P., Horowitz, G. T., Strominger, A. et al., Vacuum configurations for superstrings, *Nucl. Phys.*, 1985, B258: 46.
11 Gross, D., Harvey, J., Martinec, E. et al., Heterotic string, *Phys. Rev. Lett.*, 1985, 54: 502.
12 Green, M. B., Schwartz, J. H., Witten, E., *Superstring Theory*, Cambridge: Cambridge University Press, 1987.
13 Witten, E., Search for a realistic Kaluza-Klein theory, *Nucl. Phys.*, 1981, B186: 412.
14 Glashow, S. L., Partial-symmetries of weak interactions, *Nucl. Phys.*, 1961, 22: 579.

15 Weinberg, S., A model of leptons, *Phys. Rev. Lett.*, 1967, 19: 1264.
16 Salam, A., in *Proceedings of the Eight Nobel Symposium, on Elementary Particle Theory, Relativistic Groups, and Anerlyticit, Stochholm, Sweden*, 1968 (ed. Svartholm, N.). Stockholm: Almqvist and Wikell, 1968.
17 Georgi, H., in *Particles and Fields* 1974 (ed. Carlson, C.). New York: Amer. Inst. of Physics, 1975.
18 Fritzsch, H., Minkowski, P., *Ann. Phys.*, 1975, 93: 193.
19 Stelle, K., Renormalization of higher-derivative quantum gravity, *Phys. Rev.*, 1977, D16: 953.
20 Voronov, B.L., Tyutin, I.V., *Yad. Fiz. (J. Nucl. Phys.)*, 1984, 39: 998.

Eur. Phys. J. C 16, 279–287 (2000)
Digital Object Identifier (DOI) 10.1007/s100520000383

THE EUROPEAN PHYSICAL JOURNAL C

Searching for rephase-invariant CP- and CPT-violating observables in meson decays

K.C. Chou[1], W.F. Palmer[2,a], E.A. Paschos[3,b], Y.L. Wu[1,3,c]

[1] Institute of Theoretical Physics, Chinese Academy of Sciences, Beijing 100080, P.R. China
[2] Department of Physics, Ohio-State University, Columbus, OH 43210, USA
[3] Institut für Physik, Universität Dortmund, 44221 Dortmund, Germany

Received: 28 January 2000 / Published online: 8 June 2000 – © Springer-Verlag 2000

Abstract. We present a general model-independent and rephase-invariant formalism that cleanly relates CP and CPT noninvariant observables to the fundamental parameters. Different types of CP and CPT violations in the K^0-, B^0-, B_s^0- and D^0-systems are explicitly defined. Their importance for interpreting experimental measurements of CP and CPT violations is emphasized. In particular, we show that the time-dependent measurements allow one to extract a clean signature of CPT violation.

1 Introduction

For the discrete symmetries of nature, violations have been observed for C, P and the combined CP symmetries[1–5]. In fact two types of CP violation have now been established in the K-meson system. It remains an active problem of research to observe CP asymmetries in heavier mesons. In addition there is new interest in investigations of properties of the CPT symmetry[6]. Up to now, there are only bounds on CPT-violating parameters[7], which are sensitive to the magnitude of amplitudes, but tests of the relative phases have not yet been carried out.

In this article we present tests of CPT and CP, separately, and discuss which measurements distinguish between the various symmetry breaking terms. In addition, we derive formulae which are manifestly invariant under rephasing of the original mesonic states. The hope is to call attention to several measurements which will be accessible to experiments in the future.

Our paper is organized as follows: In Sect. 2. we present a complete set of parameters characterizing CP, T and CPT nonconservation arising from the mass matrix, i.e., the so-called indirect CP-, T- and CPT-violation. A set of direct CP-, T- and CPT-violating parameters originating from the decay amplitudes are defined in Sect. 3. In Sect. 4, we defined all possible independent observables and relate them directly to fundamental parameters which are manifestly rephasing invariant and can be applied to all meson decays. The various types of CP and CPT violation are classified, indicating how one can extract purely CPT or CP violating effects. In Sect. 5, we investigate in detail the time evolution of mesonic decays and introduce several time-dependent CP- and CPT-asymmetries which allow one to measure separately the indirect CPT- and CP-violating observables as well as direct CPT- and CP-violating observables. In particular, we show how one can extract a clean signature of CPT violation from asymmetries in neutral meson decays. In Sect. 6, we apply the general formalism to the semileptonic and nonleptonic K-meson decays and show how many rephasing invariant CP and CPT observables can be extracted separately. Our conclusions are presented in the last section.

2 CP- and CPT-violating parameters in mass matrix

Let M^0 be the neutral meson (which can be K^0 or D^0 or B^0 or B_s^0) and $\bar{M}^0$ its antiparticle. The evolution of M^0 and $\bar{M}^0$ states is dictated by

$$\frac{d}{dt}\begin{pmatrix} M^0 \\ \bar{M}^0 \end{pmatrix} = -i\begin{pmatrix} H_{11} & H_{12} \\ H_{21} & H_{22} \end{pmatrix}\begin{pmatrix} M^0 \\ \bar{M}^0 \end{pmatrix} \qquad (1)$$

with $H_{ij} = M_{ij} - i\Gamma_{ij}/2$ the matrix elements, and M_{ij}, Γ_{ij} being the dispersive and absorptive parts, respectively.

The eigenvalues of the Hamiltonian are

$$H_1 = H_{11} - \sqrt{H_{12}H_{21}}\,\frac{1-\Delta_M}{1+\Delta_M}\,,$$
$$H_2 = H_{22} + \sqrt{H_{12}H_{21}}\,\frac{1-\Delta_M}{1+\Delta_M}\,. \qquad (2)$$

with

$$\frac{1-\Delta_M}{1+\Delta_M} = \left[1 + \frac{\delta_M^2}{2} - \delta_M\sqrt{1+\frac{\delta_M^2}{4}}\right]^{1/2}, \qquad and$$

[a] palmer@mps.ohio-state.edu
[b] paschos@hal1.physik.uni-dortmund.de
[c] ylwu@itp.ac.cn

$$\delta_M = \frac{H_{22} - H_{11}}{\sqrt{H_{12}H_{21}}} \tag{3}$$

We note already that δ_M is invariant under rephasing of the states M^0 and $\bar{M}^0$. The eigenfunctions of the Hamiltonian define the physical states. Following Bell and Steinberger[8], M^0 and $\bar{M}^0$ mix with each other and form two physical mass eigenstates

$$M_1 = p_S|M^0> + q_S|\bar{M}^0>, \quad M_2 = p_L|M^0> - q_L|\bar{M}^0> \tag{4}$$

with normalization $|p_S|^2 + |q_S|^2 = |p_L|^2 + |q_L|^2 = 1$. The coeficients are given by

$$\frac{q_S}{p_S} = \frac{q}{p}\frac{1+\Delta_M}{1-\Delta_M} \equiv \frac{1-\epsilon_S}{1+\epsilon_S}, \quad \frac{q_L}{p_L} = \frac{q}{p}\frac{1-\Delta_M}{1+\Delta_M} \equiv \frac{1-\epsilon_L}{1+\epsilon_L}$$
$$\frac{q}{p} = \sqrt{\frac{H_{21}}{H_{12}}} \equiv \frac{1-\epsilon_M}{1+\epsilon_M} \tag{5}$$

We have also introduced the paramters $\epsilon_{S,L,M}$ following [9]. In the CPT conserving case they reduce to the known parameter ϵ_M. Thus we have a complete description of the physical states in terms of the mass matrix, and the time evolution is determined by the eigenvalues:

$$H_1 = M_1 - i\Gamma_1/2; \qquad H_2 = M_2 - i\Gamma_2/2 \tag{6}$$

and is given simply by

$$M_1 \to e^{-iH_1t}M_1; \qquad M_2 \to e^{-iH_2t}M_2 \tag{7}$$

We discuss next several properties related to the symmetries of the system. The parameters δ_M and $|q/p|$ are rephasing invariant and so are also other parameters defined in terms of them. CPT invariance requires $M_{11} = M_{22}$ and $\Gamma_{11} = \Gamma_{22}$, and implies that $\delta_M = 0$. Thus the difference between q_S/p_S and q_L/p_L represents a signal of CPT violation. In other words, Δ_M different from zero indicates CPT violation.

CP invariance requires the dispersive and absorptive parts of H_{12} and H_{21} to be, respectively, equal and implies $q/p = 1$. Also if T invariance holds, then independently of CPT symmetry, the dispersive and absorptive parts of H_{12} and H_{21} must be equal up to a total relative common phase, implying $|q/p| = 1$. Therefore a $Re\epsilon_M$ different from zero describes CP and T nonconservation and can be present even when CPT is conserved. Finally, two parameters, ϵ_M describing CP violation with T nonconservation and Δ_M characterizing CPT violation with CP nonconservation, are related to ϵ_S and ϵ_L via

$$\epsilon_S = \frac{\epsilon_M - \Delta_M}{1 - \epsilon_M\Delta_M}; \qquad \epsilon_L = \frac{\epsilon_M + \Delta_M}{1 + \epsilon_M\Delta_M} \tag{8}$$

and reduce to those given in [9] when neglecting the quadratic term $\epsilon_M\Delta_M$. This is a complete set of parameters describing CP, T and CPT nonconservation which originates in the mass matrix (indirect). In the next section we discuss additional parameters originating in the decay amplitudes (direct) as well as from the mixing between mass matrix and decay amplitudes (mixed-induced).

3 CP- and CPT-violating parameters in decay amplitudes

Let H_{eff} be the effective Hamiltonian which contains CPT-even $H_{eff}^{(+)}$ and CPT-odd $H_{eff}^{(-)}$ parts, i.e.,

$$H_{eff} = H_{eff}^{(+)} + H_{eff}^{(-)} \tag{9}$$

with

$$(CPT)H_{eff}^{(\pm)}(CPT)^{-1} = \pm H_{eff}^{(\pm)} \tag{10}$$

Let f denote the final state of the decay and $\bar{f}$ its charge conjugate state. The decay amplitudes of M^0 are defined as

$$g \equiv <f|H_{eff}|M^0> = \sum_i (A_i + B_i)e^{i\delta_i}$$
$$\equiv \sum_i (|A_i|e^{i\phi_i^A} + |B_i|e^{i\phi_i^B})e^{i\delta_i},$$
$$\bar{h} \equiv <\bar{f}|H_{eff}|M^0> = \sum_i (C_i + D_i)e^{i\delta_i}$$
$$\equiv \sum_i (|C_i|e^{i\phi_i^C} + |D_i|e^{i\phi_i^D})e^{i\delta_i} \tag{11}$$

with A_i and C_i being CPT-conserving amplitudes

$$<f|H_{eff}^{(+)}|M^0> \equiv \sum_i A_i e^{i\delta_i}, \quad <\bar{f}|H_{eff}^{(+)}|M^0> \equiv \sum_i C_i e^{i\delta_i} \tag{12}$$

and B_i and D_i being CPT-violating amplitudes

$$<f|H_{eff}^{(-)}|M^0> \equiv \sum_i B_i e^{i\delta_i}, \quad <\bar{f}|H_{eff}^{(-)}|M^0> \equiv \sum_i D_i e^{i\delta_i}. \tag{13}$$

Here we have used the notation of [10] for the amplitude g, and have introduced a new amplitude $\bar{h}$. The second amplitude is absent when one considers only K-meson decays and neglects possible violation of $\Delta S = \Delta Q$ rule as was the case in [10]. This is because the K-meson decays obey $\Delta S = \Delta Q$ rule via weak interactions of the standard model. The reason is simple since the strange quark can only decay to the up quark. In the case of B-, B_s- and D-meson systems both amplitudes g and $\bar{h}$ exist via the W-boson exchange of weak interactions since both b-quark and c-quark will have two different transitions due to CKM quark mixings, i.e., $b \to c, u$ and $c \to s, d$ (for explicit decay modes see the classification for the processes given in Sect. 5). ϕ_i^I ($I = A, B, C, D$) are weak phases and δ_i are strong phases from final state interactions. The subscrpts $i = 1, 2, \cdots$ denote various strong interacting final states, such as the different isospin states. For CP transformation, we adopt the phase convention

$$CP|M^0> = |\bar{M}^0>, \qquad CP|\bar{M}^0> = |M^0>, \tag{14}$$

It is then not difficult to show that the decay amplitudes of the charge conjugate meson $\bar{M}^0$ have the following form

$$\bar{g} \equiv <\bar{f}|H_{eff}|\bar{M}^0> = \sum_i (A_i^* - B_i^*)e^{i\delta_i}$$

$$\equiv \sum_i (|A_i|e^{-i\phi_i^A} - |B_i|e^{-i\phi_i^B})e^{i\delta_i},$$

$$h \equiv < f|H_{eff}|\bar{M}^0 >= \sum_i (C_i^* - D_i^*)e^{i\delta_i}$$

$$\equiv \sum_i (|C_i|e^{-i\phi_i^C} - |D_i|e^{-i\phi_i^D})e^{i\delta_i} . \quad (15)$$

In analogy to the indirect CP- and CPT-violating parameters $\epsilon_{S,L,M}$ from mass matrix, we define now parameters containing direct CP and CPT violations

$$\varepsilon'_M \equiv \frac{1-h/g}{1+h/g},\ \bar{\varepsilon}'_M \equiv \frac{1-\bar{g}/\bar{h}}{1+\bar{g}/\bar{h}};$$

$$\varepsilon''_M \equiv \frac{1-\bar{g}/g}{1+\bar{g}/g},\ \bar{\varepsilon}''_M \equiv \frac{1-h/\bar{h}}{1+h/\bar{h}} \quad (16)$$

For final states which are CP conjugate, i.e., $|\bar{f}>= CP|f>$ $= |f>$, the relations $h = \bar{g}$ and $\bar{h} = g$ hold. and thus the four parameters are reduced to two independent ones: $\varepsilon'_M = \varepsilon''_M$ and $\bar{\varepsilon}'_M = \bar{\varepsilon}''_M$.

The symmetry properties of the amplitudes are as follows. If CP is conserved, independently of CPT symmetry, one has $\bar{g}/g = 1$ and $h/\bar{h} = 1$, which implies

$$A_i = A_i^* , \qquad C_i = C_i^* , \qquad B_i = -B_i^* , \qquad D_i = -D_i^*$$

in other words:

$$\phi_i^A = \phi_i^C = 0 , \qquad \phi_i^B = \phi_i^D = \pi/2 ,$$

namely, A_i and C_i are real, while B_i and D_i are imaginary.

Similarly T invariance exchanges the initial and final states and implies, independently of CPT symmetry,

$$A_i = A_i^* , \qquad C_i = C_i^* , \qquad B_i = B_i^* , \qquad D_i = D_i^*$$

or

$$\phi_i^A = \phi_i^C = 0 , \qquad \phi_i^B = \phi_i^D = 0 ,$$

namely, all the amplitudes must be real. Finally, conservation of CPT requires $B_i = 0$ and $D_i = 0$. We summarize the results for the amplitudes in Table 1.

Reading across the first row of the table we have the conditions for CP conservation. with T conservation (first column) and without T-conservation (second column). The relations $B_i = -B_i^*$ and $D_i = -D_i^*$ imply T-violation in the presence of CP conservation. The second row of the table gives the conditions when T is conserved. with CP conservation (first column) or without CP conservation (second column). This is a complete set of amplitude with the C_i and D_i amplitudes introduced for the first time here. As a consequence, two more CP- and CPT-violating parameters $\bar{\varepsilon}_M$ and $\bar{\varepsilon}_M$ in (16) are needed.

In summary of this section, we have the following conclusions. *Values for $Re\varepsilon''_M$ and $Re\bar{\varepsilon}''_M$ different from zero describe CP nonconservation independently of T and CPT symmetries. The presence of $B_i's$ and $D_i's$ indicate simultaneous nonconservation of: CPT and either of CP or T. Zero ε''_M and $\bar{\varepsilon}''_M$ with nonzero $Im\varepsilon'_M$ and $Im\bar{\varepsilon}'_M$ implies T nonconservation. Finally, zero B_i and D_i. and complex A_i and C_i signal CPT conservation with CP and T violations.* Note that the latter case is more difficult to establish experimentally since it requires the observation of a relative phase between two amplitudes distinguished with the help of specific quantum numbers. This was the case with the ϵ'/ϵ parameter in K-meson decays.

4 Rephase invariant CP- and CPT-violating observables

The ε-type parameters defined in (5) and (16) can not be related to physical observables since they are not rephasing invariant. Let us introduce CP- and CPT-violating observables by considering the ratio.

$$\hat{\eta}_f \equiv \frac{q_S}{q_L}\frac{< f|H_{eff}|M_2 >}{< f|H_{eff}|M_1 >} = \frac{q_S}{q_L}\frac{p_L}{p_S}\frac{1 - r_f^L}{1 + r_f^S} \quad (17)$$

which enters to the time evolution of the decay amplitudes (see 27 and 28). The parameters $q_{S,L}$ and $p_{S,L}$ were defined in Sect. 2, and we also introduce the notation

$$r_f^S = (q_S/p_S)(h/g)$$

with a similar definition for r_f^L. Note that the factor q_S/q_L is necessary for the normalization and also rephase invariance, which has not been always included in the literature. In the CPT-conserving case [11] this factor is equal to unity. One can simply see from the definitions in (3)-(5) that $\hat{\eta}_f$ is rephasing invariant. The factor $q_S p_L/p_S q_L =$ $(1 + \Delta_M)^2/(1 - \Delta_M)^2$ is rephase-invariant since Δ_M has this property. The ratios $r_f^{L,S} = (q_{L,S}/p_{L,S})(h/g)$ are also rephase-invariant. To see that, let us make a phase redefinition $|M^0 >\to e^{i\phi}|M^0 >$, then $|\bar{M}^0 >\to e^{-i\phi}|\bar{M}^0 >$, $H_{12} \to e^{-2i\phi}H_{12}$ and $H_{21} \to e^{2i\phi}H_{21}$, as well as $h \to$ $e^{-i\phi}h$ and $g \to e^{i\phi}g$, thus $(q_S/p_S, q_L/p_L) \to e^{2i\phi}$ $(q_S/p_S, q_L/p_L)$ and $h/g \to e^{-2i\phi}h/g$, which makes $r_f^{L,S} =$ $(q_{L,S}/p_{L,S})(h/g)$ manifestly rephase-invariant.

It is seen that the rephase-invariant quantities $r_f^{L,S}$ and $\hat{\eta}_f$ are given by the product of complex parameters arising from the mass mixing $(q_{L,S}/p_{L,S})$ and from amplitudes (h/g). To separately define the rephase-invariant CP- and CPT-violating observables originating from the mass mixing and from the amplitudes, some algebra is neccesary[1], but it is not difficult to show that $\hat{\eta}_f$ can be rewritten as

$$\hat{\eta}_f \equiv \frac{1}{1 - \eta_\Delta}\left[\eta_\Delta + \frac{a_{\epsilon_S} + \hat{a}_{\epsilon'} + i\,\hat{a}_{\epsilon_S + \epsilon'}}{2 + a_{\epsilon_S}\hat{a}_{\epsilon'} + \hat{a}_{\epsilon_S\epsilon'}}\right] \quad (18)$$

where we have used the definitions

$$a_{\epsilon_S} = \frac{1 - |\frac{q_S}{p_S}|^2}{1 + |\frac{q_S}{p_S}|^2} = \frac{2Re\epsilon_S}{1 + |\epsilon_S|^2} = \frac{a_\epsilon - a_\Delta}{1 - a_\epsilon a_\Delta},$$

$$a_{\epsilon_L} = \frac{1 - |\frac{q_L}{p_L}|^2}{1 + |\frac{q_L}{p_L}|^2} = \frac{2Re\epsilon_L}{1 + |\epsilon_L|^2} = \frac{a_\epsilon + a_\Delta}{1 + a_\epsilon a_\Delta} \quad (19)$$

$$\eta_\Delta = \frac{2\Delta_M}{1 + \Delta_M^2} = \frac{a_\Delta + ia'_\Delta\sqrt{1 - a_\Delta^2 - a'^2_\Delta}}{1 - a'^2_\Delta}$$

[1] The algebra is described in [11]

Table 1.

	CPT-conservation		CPT-Violation		
CP-conservation	$A_i = A_i^*$	$C_i = C_i^*$	$B_i = -B_i^*$	$D_i = -D_i^*$	imply T-violation
T-conservation	$A_i = A_i^*$	$C_i = C_i^*$	$B_i = B_i^*$	$D_i = D_i^*$	imply CP-violation
	CP & T conservation				

with

$$a_\epsilon = \frac{1-|q/p|^2}{1+|q/p|^2} = \frac{2Re\epsilon_M}{1+|\epsilon_M|^2}, \quad (20)$$

$$a_\Delta = \frac{2Re\Delta_M}{1+|\Delta_M|^2}, \qquad a'_\Delta = \frac{2Im\Delta_M}{1+|\Delta_M|^2}$$

The definitions of $\hat{a}_{\epsilon'}$, $\hat{a}_{\epsilon_S+\epsilon'}$ and $\hat{a}_{\epsilon_S\epsilon'}$ are given in the appendix. The reader should note that quantities without a hat contain either only CP or only CPT nonconserving effects, and with a hat contain both CP- and CPT-nonconserving effects.

As a_ϵ, $\hat{a}_{\epsilon'}$, $\hat{a}_{\epsilon+\epsilon'}$ and $\hat{a}_{\epsilon\epsilon'}$ (for their definitions see appendix) are all rephase-invariant, so are also $\hat{a}_{\epsilon_S+\epsilon'}$ and $\hat{a}_{\epsilon_S\epsilon'}$. Note that only three of them are independent since $(1-a_\epsilon^2)(1-\hat{a}_{\epsilon'}^2) = \hat{a}_{\epsilon+\epsilon'}^2 + (1+\hat{a}_{\epsilon\epsilon'})^2$. Another rephase-invariant direct CP and CPT noninvariant observable is defined as

$$\hat{a}_{\epsilon''} = \frac{1-|\bar{g}/g|^2}{1+|\bar{g}/g|^2} = \frac{2Re\varepsilon''_M}{1+|\varepsilon''_M|^2} = \frac{a_{\epsilon''}+a_{\varepsilon\Delta}+a'_{\Delta\Delta}}{1+a'_{\varepsilon\Delta}+a_{\Delta\Delta}} \quad (21)$$

where the definitions for $a_{\epsilon''}$, $a_{\varepsilon\Delta}$, $a'_{\Delta\Delta}$, $a'_{\varepsilon\Delta}$ and $a_{\Delta\Delta}$ are presented in the appendix. Analogously, one has

$$\hat{\eta}_{\bar{f}} \equiv \frac{q_S < \bar{f}|H_{eff}|M_2>}{q_L < \bar{f}|H_{eff}|M_1>} = \frac{1}{1-\eta_\Delta} \times \left[\eta_\Delta + \frac{a_{\epsilon_S}+\hat{a}_{\bar{\varepsilon}'}+i\,\hat{a}_{\epsilon_S+\bar{\varepsilon}'}}{2+a_{\epsilon_S}\hat{a}_{\bar{\varepsilon}'}+\hat{a}_{\epsilon_S\bar{\varepsilon}'}}\right] \quad (22)$$

and

$$\hat{a}_{\bar{\varepsilon}''} = \frac{1-|\bar{h}/h|^2}{1+|\bar{h}/h|^2} = \frac{2Re\bar{\varepsilon}''_M}{1+|\bar{\varepsilon}''_M|^2} = \frac{a_{\bar{\varepsilon}''}+a_{\bar{\varepsilon}\bar{\Delta}}+a'_{\bar{\Delta}\bar{\Delta}}}{1+a'_{\bar{\varepsilon}\bar{\Delta}}+a_{\bar{\Delta}\bar{\Delta}}} \quad (23)$$

with $\bar{\Delta}_i = D_i/C_i$.

One of the interesting cases occurs when the final states are CP eigenstates, i.e., $f^{CP} = f$, and in this case $h = \bar{g}$ (or $C = A$ and $D = B$). As a consequence, we find

$$\hat{a}_{\epsilon'} = \hat{a}_{\epsilon''}, \qquad a_{\epsilon'} = a_{\epsilon''}$$

$$\hat{a}_{\epsilon+\epsilon'} = \frac{1}{1+a'_{\varepsilon\Delta}+a_{\Delta\Delta}}[a_{\epsilon+\epsilon'}+a_{\epsilon+\epsilon'_\Delta}+a_{\epsilon+\epsilon'_{\Delta\Delta}}] \quad (24)$$

where the explicit definitions for $a_{\epsilon+\epsilon'}$, $a_{\epsilon+\epsilon'_\Delta}$ and $a_{\epsilon+\epsilon'_{\Delta\Delta}}$ are again given in the appendix.

To see explicitly how many rephase invariant CPT and CP observables may be separately measured from experiments, let us consider the case for which the final states are CP eigenstates and suppose that the violations are small so that one could only keep the linear terms of the rephase invariant CPT- and CP-violating observables. With this consideration, the observable $\hat{\eta}_f$ is simplified

$$\hat{\eta}_f \simeq \frac{1}{2}[a_\epsilon + a_{\epsilon'} + a_\Delta + a_{\epsilon\Delta} + a'_{\Delta\Delta} + i(a_{\epsilon+\epsilon'} + a'_\Delta + a_{\epsilon+\epsilon'_{\Delta\Delta}} + a_{\epsilon+\epsilon'_\Delta})] \quad (25)$$

where the definitions for all the rephase invariant quantities are given in the appendix. Those with index Δ are the CPT-violating observables, the others are CP-violating ones which have been discussed in [11].

The formalism so far involves many equations which include CP and CPT violation effects either separately or mixed together. It has several advantages in comparison with other articles[12, 10]:

1. The formalism is more general than the ones reported in the literature and can be applied not only to the K-meson decays but also all other heavier meson decays.
2. All observables are manifestly rephasing invariant and well defined by directly relating to the hadronic mixing matrix elements and decay amplitudes of mesons.
3. All possible independent observables are classified, which enables one to separately measure different types of CPT- and CP-violating observables and to extract purely CPT or CP violation effects.
4. The formalism is more elegantly designed for extracting various rephase invariant CPT- and CP- violating observables from time-dependent measurements of meson decays, which will be discussed in detail in the next section.

We have thus defined all possible rephase-invariant CP and CPT noninvariant observables in terms of eight parameters related to CP and CPT breaking quantities arising either from mixing or phases of amplitude. The eight parameters are classified as follows: ϵ_M is an indirect CP-violating parameter and Δ_M the indirect CPT-violating parameters; the parameters ε''_M and $\bar{\varepsilon}''_M$ will be decomposed into four parameters, ϵ''_M, $\bar{\epsilon}''_M$, Δ_i and $\bar{\Delta}_i$, where ϵ''_M and $\bar{\epsilon}''_M$ define direct CP-violating paramters, Δ_i and $\bar{\Delta}_i$ describe direct CPT-violating parameters. ε'_M and $\bar{\varepsilon}'_M$ contain the ratio of the two decay amplitudes and can be associated with direct CP and CPT violation, as well as the interference between indirect and direct CP and CPT violations. All the CP and CPT violations can be well defined and in general classified into the following types:

1. purely indirect CP and CPT violations which are given by the rephase-invariant CP-violating observable a_ϵ and CPT-violating observables a_Δ and a'_Δ.

2. purely direct CP and CPT violations which are characterized by the rephase-invariant CP-violating observables $a_{\epsilon''}$ and $a_{\bar{\epsilon}''}$ and CPT-violating observables $a_{\varepsilon\Delta}$, $a'_{\varepsilon\Delta}$, $a_{\Delta\Delta}$, $a'_{\Delta\Delta}$, $a_{\bar{\varepsilon}\bar{\Delta}}$, $a'_{\bar{\varepsilon}\bar{\Delta}}$, $a_{\bar{\Delta}\bar{\Delta}}$ and $a'_{\bar{\Delta}\bar{\Delta}}$.
3. Mixed-induced CP and CPT violations which are described by CP-violating observables $a_{\epsilon+\epsilon'}$ and $a_{\epsilon+\bar{\epsilon}'}$ and CPT-violating observables $a_{\epsilon+\epsilon'_\Delta}$, $a_{\epsilon+\epsilon'_{\Delta\Delta}}$, $a_{\epsilon+\bar{\epsilon}'_\Delta}$ and $a_{\epsilon+\bar{\epsilon}'_{\Delta\Delta}}$.

For the case that the final states are CP eigenstates, one has $\hat{a}_{\epsilon'} = \hat{a}_{\epsilon''} = \hat{a}_{\bar{\epsilon}'} = \hat{a}_{\bar{\epsilon}''}$. Thus, in this case $\hat{a}_{\epsilon'}$ and $\hat{a}_{\bar{\epsilon}'}$ also indicate purely direct CP and CPT violations. When the final states are not CP eigenstates, $\hat{a}_{\epsilon'}$ and $\hat{a}_{\bar{\epsilon}'}$ do not, in general, provide a clear signal of direct CP violation although they contain direct CP and CPT violations. Their deviation from the values $\hat{a}_{\epsilon'} = \pm 1$, 0 and $\hat{a}_{\bar{\epsilon}'} = \mp 1$, 0 can arise from different CKM angles, final state interactions, or different hadronic form factors, but not necessarily from CP and CPT violations.

5 Extraction of CP- and CPT-violating observables

In order to measure the rephase-invariant observables defined above, we consider the proper time evolution[13, 14] of the neutral mesons

$$|M^0(t)> = \sum_{i=1}^{2} \xi_i e^{-i(m_i - i\Gamma_i/2)t}|M_i>$$

$$|\bar{M}^0(t)> = \sum_{i=1}^{2} \bar{\xi}_i e^{-i(m_i - i\Gamma_i/2)t}|M_i> \tag{26}$$

with $\xi_1 = q_L/(q_S p_L + q_L p_S)$ and $\xi_2 = q_S/(q_S p_L + q_L p_S)$ for a pure M^0 state at $t = 0$ as well as $\bar{\xi}_1 = p_L/(q_S p_L + q_L p_S)$ and $\bar{\xi}_2 = -p_S/(q_S p_L + q_L p_S)$ for a pure $\bar{M}^0$ state at $t = 0$. Thus the decay amplitudes of M^0 and $\bar{M}^0$ at the time t will be given by

$$\mathcal{A}(t) = < f|M^0(t)> = \frac{< f|M_1>}{p_S}\frac{1-\eta_\Delta}{2} \times \left(e^{-iH_1 t} + \hat{\eta}_f e^{-iH_2 t}\right); \tag{27}$$

$$\bar{\mathcal{A}}(t) = < \bar{f}|\bar{M}^0(t)> = \frac{< \bar{f}|M_1>}{q_S}\frac{1-\eta_\Delta}{2} \times \left(\frac{1+\eta_\Delta}{1-\eta_\Delta} e^{-iH_1 t} - \hat{\eta}_{\bar{f}} e^{-iH_2 t}\right) \tag{28}$$

It follows now that the time-dependent decay rates are

$$\begin{aligned}
\Gamma(M^0(t) \to f) &\propto |\mathcal{A}(t)|^2 = (|g|^2 + |h|^2)\frac{2 + a_{\epsilon_S}\hat{a}_{\epsilon'} + \hat{a}_{\epsilon_S\epsilon'}}{1 + a_{\epsilon_S}} \\
&\times e^{-\Gamma t}\left\{\left[\frac{1 + a_{\epsilon_S}\hat{a}_{\epsilon'} + (a_{\epsilon_S} + \hat{a}_{\epsilon'})Re\eta_\Delta + \hat{a}_{\epsilon_S+\epsilon'}Im\eta_\Delta}{2 + a_{\epsilon_S}\hat{a}_{\epsilon'} + \hat{a}_{\epsilon_S\epsilon'}} - Re\eta_\Delta + |\eta_\Delta|^2\right]\cosh(\Delta\Gamma t)\right. \\
&+ \left[\frac{1 + a_{\epsilon_S\epsilon'} - (a_{\epsilon_S} + \hat{a}_{\epsilon'})Re\eta_\Delta - \hat{a}_{\epsilon_S+\epsilon'}Im\eta_\Delta}{2 + a_{\epsilon_S}\hat{a}_{\epsilon'} + \hat{a}_{\epsilon_S\epsilon'}} - Re\eta_\Delta\right] \\
&\times \sinh(\Delta\Gamma t) + \left[\frac{(a_{\epsilon_S} + \hat{a}_{\epsilon'})(1 - Re\eta_\Delta) - \hat{a}_{\epsilon_S+\epsilon'}Im\eta_\Delta}{2 + a_{\epsilon_S}\hat{a}_{\epsilon'} + \hat{a}_{\epsilon_S\epsilon'}} + Re\eta_\Delta - |\eta_\Delta|^2\right]\cos(\Delta mt) + \left[\frac{\hat{a}_{\epsilon_S+\epsilon'}(1 - Re\eta_\Delta)}{2 + a_{\epsilon_S}\hat{a}_{\epsilon'} + \hat{a}_{\epsilon_S\epsilon'}}\right. \\
&\left.\left. + \frac{(a_{\epsilon_S} + \hat{a}_{\epsilon'})Im\eta_\Delta}{2 + a_{\epsilon_S}\hat{a}_{\epsilon'} + \hat{a}_{\epsilon_S\epsilon'}} + Im\eta_\Delta\right]\sin(\Delta mt)\right\}
\end{aligned} \tag{29}$$

and

$$\begin{aligned}
\Gamma(\bar{M}^0(t) \to \bar{f}) &\propto |\bar{\mathcal{A}}(t)|^2 = (|\bar{g}|^2 + |\bar{h}|^2)\frac{2 + a_{\epsilon_S}\hat{a}_{\bar{\epsilon}'} + \hat{a}_{\epsilon_S\bar{\epsilon}'}}{1 + a_{\epsilon_S}} \\
&\times e^{-\Gamma t}\left\{\left[\frac{1 + a_{\epsilon_S}\hat{a}_{\bar{\epsilon}'} + (a_{\epsilon_S} + \hat{a}_{\bar{\epsilon}'})Re\eta_\Delta + \hat{a}_{\epsilon_S+\bar{\epsilon}'}Im\eta_\Delta}{2 + a_{\epsilon_S}\hat{a}_{\bar{\epsilon}'} + \hat{a}_{\epsilon_S\bar{\epsilon}'}} - Re\eta_\Delta - |\eta_\Delta|^2\right]\cosh(\Delta\Gamma t)\right. \\
&+ \left[\frac{1 + a_{\epsilon_S\bar{\epsilon}'} - (a_{\epsilon_S} + \hat{a}_{\bar{\epsilon}'})Re\eta_\Delta - \hat{a}_{\epsilon_S+\bar{\epsilon}'}Im\eta_\Delta}{2 + a_{\epsilon_S}\hat{a}_{\bar{\epsilon}'} + \hat{a}_{\epsilon_S\bar{\epsilon}'}} - Re\eta_\Delta\right] \\
&\times \sinh(\Delta\Gamma t) - \left[\frac{(a_{\epsilon_S} + \hat{a}_{\bar{\epsilon}'})(1 + Re\eta_\Delta) + \hat{a}_{\epsilon_S+\bar{\epsilon}'}Im\eta_\Delta}{2 + a_{\epsilon_S}\hat{a}_{\bar{\epsilon}'} + \hat{a}_{\epsilon_S\bar{\epsilon}'}} + Re\eta_\Delta + |\eta_\Delta|^2\right]\cos(\Delta mt) - \left[\frac{\hat{a}_{\epsilon_S+\bar{\epsilon}'}(1 + Re\eta_\Delta)}{2 + a_{\epsilon_S}\hat{a}_{\bar{\epsilon}'} + \hat{a}_{\epsilon_S\bar{\epsilon}'}}\right. \\
&\left.\left. - \frac{(a_{\epsilon_S} + \hat{a}_{\bar{\epsilon}'})Im\eta_\Delta}{2 + a_{\epsilon_S}\hat{a}_{\bar{\epsilon}'} + \hat{a}_{\epsilon_S\bar{\epsilon}'}} - Im\eta_\Delta\right]\sin(\Delta mt)\right\}
\end{aligned} \tag{30}$$

where $\Delta\Gamma = \Gamma_2 - \Gamma_1$ and $\Delta m = m_2 - m_1$. Here we have omitted the integrals from the phase space. Similarly, one can easily write down the decay rates $\Gamma(M^0(t) \to \bar{f})$ and $\Gamma(\overline{M}^0(t) \to f)$, and then the time-dependent CP and CPT asymmetries are defined by the difference between two decay rates. In addition, in studies of the time dependence one can isolate each of four-terms. One can introduce several asymmetries from the decay rates $\Gamma(M^0(t) \to f)$, $\Gamma(\overline{M}^0(t) \to \bar{f})$, $\Gamma(M^0(t) \to \bar{f})$ and $\Gamma(\overline{M}^0(t) \to f)$. Obviously, the time dependences contains a lot of information. Therefore studies of time evolution can eliminate the various components (hamonics) in $\cos(\Delta mt)$, $\sin(\Delta mt)$, $\cosh(\Delta\Gamma t)$ and $\sinh(\Delta\Gamma t)$. We now proceed to apply the above general analysis to specific processes. As in the [11], we may classify the processes into four scenarios:

i) $M^0 \to f$ ($M^0 \not\to \bar{f}$), $\overline{M}^0 \to \bar{f}$ ($\overline{M}^0 \not\to f$), this is the case when f and $\bar{f}$ are not a common final state of M^0 and $\overline{M}^0$. Examples are: $M^0 \to M'^- \bar{l}\nu$, $\bar{M}^0 \to M'^+ l\bar{\nu}$; $B^0 \to D^- D_s^+$, $D^- K^+$, $\pi^- D_s^+$, $\pi^- K^+$. $\bar{B}^0 \to D^+ D_s^-$, $D^+ K^-$, $\pi^+ D_s^-$, $\pi^+ K^-$; $B_s^0 \to D_s^- \pi^+$, $D_s^- D^+$, $K^- \pi^+$, $K^- D^+$; $\overline{B}_s^0 \to D_s^+ \pi^-$, $D_s^+ D^-$, $K^+ \pi^-$, $K^+ D^-$. This scenario also applies to charged meson decays.

ii) $M^0 \to (f = \bar{f},\ f^{CP} = f) \leftarrow \overline{M}^0$, this is the decay to a common final state which is CP eigenstate. Such as $B^0(\bar{B}^0)$, $D^0(\bar{D}^0)$, $K^0(\bar{K}^0) \to \pi^+\pi^-$, $\pi^0\pi^0$, $\cdots$. For the final states such as $\pi^-\rho^+$ and $\pi^+\rho^-$, although each of them is not a CP eigenstate of $B^0(\bar{B}^0)$ or $D^0(\bar{D}^0)$, one can always decompose them into CP eigenstates as $(\pi\rho)_\pm = (\pi^-\rho^+ \pm \pi^+\rho^-)$ with $CP(\pi\rho)_\pm = \pm(\pi\rho)_\pm$. This reconstruction is meaningful since $\pi^-\rho^+$ and $\pi^+\rho^-$ have the same weak phase as they contain the same quark content.

iii) $M^0 \to (f,\ f \not\to f^{CP}) \leftarrow \overline{M}^0$. i.e., the final states are common final states but are not charge conjugate states. For example. $B^0(\bar{B}^0) \to K_S J/\psi$, $B_s^0(\bar{B}_s^0) \to K_S\phi$ and $D^0(\bar{D}^0) \to K_S\pi^0$, $K_S\rho^0$.

iv) $M^0 \to (f\ \&\ \bar{f},\ f^{CP} \neq f) \leftarrow \overline{M}^0$, i.e.. both f and $\bar{f}$ are the common final states of M^0 and $\overline{M}^0$, but they are not CP eigenstates. This is the most general case. For example. $B^0(\bar{B}^0) \to D^-\pi^+, \pi^-D^+$; $D^-\rho^+, \rho^-D^+$; $B_s^0(\bar{B}_s^0) \to D_s^-K^+, K^-D_s^+$; $D^0(\bar{D}^0) \to K^-\pi^+, K^+\pi^-$.

In this paper, we will only elaborate on the first two scenarios. In scenario i), the amplitudes h and $\bar{h}$ are zero, thus $\hat{a}_{\epsilon'} = -\hat{a}_{\bar{\epsilon}'} = 1$, $\hat{a}_{\epsilon+\epsilon'} = 0 = \hat{a}_{\epsilon+\bar{\epsilon}'}$ and $\hat{a}_{\epsilon\epsilon'} = -1 = \hat{a}_{\epsilon\bar{\epsilon}'}$. For this case, the time-dependent rates of (29) and (30) will become very simple.

$$\begin{aligned}\Gamma(M^0(t) \to f) &\propto |A(t)|^2 = |g|^2 e^{-\Gamma t} \cdot \{(1 + |\eta_\Delta|^2) \\ &\times \cosh \Delta\Gamma t - 2Re\eta_\Delta \sinh \Delta\Gamma t + (1 - |\eta_\Delta|^2) \\ &\times \cos \Delta m t + Im\eta_\Delta \sin \Delta m t\} \\ \Gamma(\bar{M}^0(t) \to \bar{f}) &\propto |\bar{A}(t)|^2 = |\bar{g}|^2 e^{-\Gamma t} \cdot \{(1 + |\eta_\Delta|^2) \\ &\times \cosh \Delta\Gamma t + 2Re\eta_\Delta \sinh \Delta\Gamma t + (1 - |\eta_\Delta|^2) \\ &\times \cos \Delta m t - Im\eta_\Delta \sin \Delta m t\} \end{aligned} \tag{31}$$

It is not difficult to show that the other two time-dependent decay rates which are not allowed at $t = 0$, can happen at a later t, because the M^0 develops an $\bar{M}^0$ component through mixing. They can be simply expressed as

$$\begin{aligned}\Gamma(M^0(t) \to \bar{f}) &\propto \frac{g^2 + |\bar{g}|^2}{2}(1 - \hat{a}_{\epsilon''})\left(\frac{1 - a_{\epsilon_S}}{1 + a_{\epsilon_S}}\right)\frac{(1 - a_\Delta)^2}{1 - a'^2_\Delta} \\ &\cdot e^{-\Gamma t}(cosh\Delta\Gamma t - \cos \Delta m t) \\ \Gamma(\bar{M}^0(t) \to f) &\propto \frac{g^2 + |\bar{g}|^2}{2}(1 + \hat{a}_{\epsilon''})\left(\frac{1 + a_{\epsilon_L}}{1 - a_{\epsilon_L}}\right)\frac{(1 - a_\Delta)^2}{1 - a'^2_\Delta} \\ &\cdot e^{-\Gamma t}(cosh\Delta\Gamma t - \cos \Delta m t) \end{aligned} \tag{32}$$

With these four decay rates, we can define three asymmetries which have the following simple forms when neglecting the quadratic and high order terms of the CP and CPT violating parameters (i.e., a^2_Δ, a'^2_Δ, $a_\epsilon a^2_\Delta$)

$$A_{CP+CPT}(t) = \frac{\Gamma(M^0(t) \to f) - \Gamma(\overline{M}^0(t) \to \bar{f})}{\Gamma(M^0(t) \to f) + \Gamma(\overline{M}^0(t) \to \bar{f})} \tag{33}$$

$$\simeq a_{\epsilon''} + a_{\varepsilon\Delta} + \frac{-a_\Delta \sinh \Delta\Gamma t + a'_\Delta \sin \Delta m t}{cosh\Delta\Gamma t + cos\Delta m t}$$

$$A'_{CP+CPT}(t) = \frac{\Gamma(\overline{M}^0(t) \to f) - \Gamma(M^0(t) \to \bar{f})}{\Gamma(\overline{M}^0(t) \to f) + \Gamma(M^0(t) \to \bar{f})} \tag{34}$$

$$\simeq a_{\epsilon''} + a_{\varepsilon\Delta} + 2a_\epsilon$$

$$A''_{CP+CPT}(t) = \frac{\Gamma(M^0(t) \to f) - \Gamma(\overline{M}^0(t) \to f)}{\Gamma(M^0(t) \to f) + \Gamma(\overline{M}^0(t) \to f)} \tag{35}$$

$$\simeq \frac{\cos \Delta m t - a_\epsilon \cosh \Delta\Gamma t - a_\Delta \sinh \Delta\Gamma t + a'_\Delta \sin \Delta m t}{\cosh \Delta\Gamma t - a_\epsilon \cos \Delta m t - a_\Delta \sinh \Delta\Gamma t + a'_\Delta \sin \Delta m t}$$

Their exact expressions can be found in the appendix. From the time-dependent measurements of the above asymmetrics, one shall be able to extract all observables: Δm, $\Delta\Gamma$, a_ϵ, a_Δ, a'_Δ and $\hat{a}_{\epsilon''}$.

From the above asymmetries. we easily arrive at the following important observations:

1. *As long as the experimental measurements show that the asymmetry $A_{CP+CPT}(t)$ is not a constant and depends on time, it provides a clean signature of indirect CPT violation from mixings.*
2. For the semileptonic decays $M^0 \to M'^- l\nu$ and also for the decay modes in which the final state interactions are absent, one has $a_{\epsilon''} = 0$, $a'_{\Delta\Delta} = 0$, $a'_{\varepsilon\Delta} = 0$ and $\hat{a}_{\epsilon''} = a_{\varepsilon\Delta}/(1 + a_{\Delta\Delta})$, thus nonzero $\hat{a}_{\epsilon''}$ will represent direct CPT violation from amplitudes. For this case, we come to a strong conclusion that *once the asymmetry $A_{CP+CPT}(t)$ is not zero, then CPT must be violated.*
3. By combining measurement of the above asymmetries from semileptonic and nonleptonic decays, it allows one, in principle, to separately measure the indirect CP-violating observable a_ϵ and the direct CP-violating observable $a_{\epsilon''}$ as well as the indirect CPT-violating observables a_Δ and a'_Δ, and the direct CPT-violating observable $a_{\epsilon\Delta}$.

We now discuss scenario ii) in which $\bar{h} = g$ and $h = \bar{g}$, thus $a_{\epsilon'} = a_{\epsilon''} = a_{\bar{\epsilon}'} = a_{\bar{\epsilon}''}$ and $a_{\epsilon+\epsilon'} = a_{\epsilon+\bar{\epsilon}'}$. When neglecting the quadratic and high order terms and using the relations and definitions for the rephase-invariant observables, the time-dependent asymmetry is simply given by

$$\begin{aligned}A_{CP+CPT}(t) \simeq &-(a_\epsilon + a_\Delta) + e^{-\Delta\Gamma t}[(a_\epsilon + a_\Delta + \hat{a}_{\epsilon'}) \\ &\times \cos(\Delta m t) + (a'_\Delta + \hat{a}_{\epsilon+\epsilon'})\sin(\Delta m t)]\end{aligned} \tag{36}$$

(The exact expression is given in the appendix.)

From the above time-dependent evolution $A_{CP+CPT}(t)$ one is able to extract three physical quantities: one of them is the direct CP and CPT noninvariant observable $\hat{a}_{\epsilon'}$ and the other two are the combinations of CP and CPT noninvariant observables $(a_\epsilon + a_\Delta)$ and $(a'_\Delta + \hat{a}_{\epsilon+\epsilon'})$. Combining these measurements with scenario (i), in which the indirect CP and CPT noninvariant observables a_ϵ, a_Δ and a'_Δ are expected to be determined, one will be able to extract the mixed-induced CP and CPT noninvariant observable $\hat{a}_{\epsilon+\epsilon'}$. Thus, studies of scenarios (i) and (ii) allow us to separate the three types of CP and CPT violations.

6 CP and CPT violation in K-meson system

The formalism and analyses presented above are general and can be used for all neutral meson systems. As a specific application, we are going to consider the K-meson system. From semileptonic decays of $K^0 \to \pi^- + l^+ + \nu_l$ and $\bar{K}^0 \to \pi^+ + l^- + \bar{\nu}_l$, from (33) and (34), the time-dependent measurements of the asymmetries lead to

$$A^{K_{l3}}_{CP+CPT}(t) = \frac{\Gamma(K^0(t) \to \pi^- l^+ \nu_l) - \Gamma(\overline{K}^0(t) \to \pi^+ l^- \bar{\nu}_l)}{\Gamma(K^0(t) \to \pi^- l^+ \nu_l) + \Gamma(\overline{K}^0(t) \to \pi^+ l^- \bar{\nu}_l)}$$

$$\simeq a_{\varepsilon\Delta} + \frac{-a_\Delta \sinh \Delta\Gamma t + a'_\Delta \sin \Delta m_K t}{cosh\Delta\Gamma t + cos\Delta m_K t} , \quad (37)$$

$$A'^{K_{l3}}_{CP+CPT}(t) = \frac{\Gamma(\overline{K}^0(t)\to\pi^- l^+\nu_l) - \Gamma(K^0(t)\to\pi^+ l^- \bar{\nu}_l)}{\Gamma(\overline{K}^0(t)\to\pi^- l^+\nu_l) + \Gamma(K^0(t)\to\pi^+ l^- \bar{\nu}_l)}$$
$$\simeq a_{\varepsilon\Delta} + 2a_\epsilon \quad (38)$$

where the direct CP-violating parameter $a_{\epsilon''}$ is expected to be small as the final state interactions are electromagnetic. It is then clear that *non-zero asymmetry* $A^{K_{l3}}_{CP+CPT}(t)$ *is a clean signature of CPT violation*. Its time evolution allows us to extract direct CPT-violating observable $a_{\varepsilon\Delta}$ and indirect CPT-violating observables a_Δ and a'_Δ. The combination of the two asymmetries $A'^{K_{l3}}_{CP+CPT}(t)$ and $A^{K_{l3}}_{CP+CPT}(t)$ further helps us to extract indirect CP-violating observable a_ϵ.

In the nonleptonic decays with final states being CP eigenstates. the asymmetry $A_{CP+CPT}(t)$ is given in terms of the observables $\hat{a}_{\epsilon'}$ and $\hat{a}_{\epsilon+\epsilon'}$ which concern both CP and CPT violations. In general, it is hard to clearly separate CP violation from CPT violation in the decay amplitudes, but it would be of interest to look for possibilities of establishing CPT violation arising from the decay amplitudes. For the K-meson system, there are two unique decay modes $K^0(\bar{K}^0) \to \pi^+\pi^-$ and $\pi^0\pi^0$ which are related via isospin symmetry. Their time-dependent asymmetries are given by

$$A^{(\pi^+\pi^-)}_{CP+CPT}(t) \simeq -(a_\epsilon + a_\Delta) + e^{-\Delta\Gamma t}[(a_\epsilon + a_\Delta + \hat{a}^{(+-)}_{\epsilon'})$$
$$\times \cos(\Delta m_K t) + (a'_\Delta + \hat{a}^{(+-)}_{\epsilon+\epsilon'}) \sin(\Delta m_K t)] \ . \quad (39)$$

$$A^{(\pi^0\pi^0)}_{CP+CPT}(t) \simeq -(a_\epsilon + a_\Delta) + e^{-\Delta\Gamma t}[(a_\epsilon + a_\Delta + \hat{a}^{(00)}_{\epsilon'})$$
$$\times \cos(\Delta m_K t) + (a'_\Delta + \hat{a}^{(00)}_{\epsilon+\epsilon'}) \sin(\Delta m_K t)] \ . \quad (40)$$

It is seen that since the indirect CP-violating observable a_ϵ and indirect CPT-violating observables a_Δ and a'_Δ can be extracted from asymmetries in the semileptonic decays, we then can extract the direct CP- and CPT-violating observables $\hat{a}^{(+-)}_{\epsilon'}$ and $\hat{a}^{(00)}_{\epsilon'}$ as well as mixed-induced CP- and CPT-violating observables $\hat{a}^{(+-)}_{\epsilon+\epsilon'}$ and $\hat{a}^{(00)}_{\epsilon+\epsilon'}$. We now discuss how to extract pure CPT or CP violation effects by using isospin symmetry.

When neglecting high order terms, we have

$$\hat{a}_{\epsilon'} \simeq a_{\epsilon'} + a'_{\Delta\Delta} + a_{\varepsilon\Delta}, \quad \hat{a}_{\epsilon+\epsilon'} \simeq a_{\epsilon+\epsilon'} + a_{\epsilon+\epsilon'_{\Delta\Delta}} + a_{\epsilon+\epsilon'_\Delta} \quad (41)$$

Note that their dependence on the final states are understood. Using the isospin symmetry, we find

$$A^{(+-)} = \sqrt{\frac{2}{3}}a_0 + \sqrt{\frac{1}{3}}a_2$$
$$A^{(00)} = \sqrt{\frac{1}{3}}a_0 - \sqrt{\frac{2}{3}}a_2 \quad (42)$$

with $A^{(+-)}$ and $A^{(00)}$ the amplitudes for the decay modes $K^0(\bar{K}^0) \to \pi^+\pi^-$ and $K^0(\bar{K}^0) \to \pi^0\pi^0$ respectively, where a_0 and a_2 correspond to the isospin $I = 0$ and $I = 2$ amplitudes. The same decomposition holds for $B^{(+-)}$ and $B^{(00)}$ amplitudes[2]. Considering the fact that $\omega = |A_2|/|A_0| \simeq 1/22 << 1$ due to the $\Delta I = 1/2$ rule, we obtain

$$\hat{a}^{(+-)}_{\epsilon'} \simeq a_{\epsilon'} + a'_{\Delta\Delta} + \tilde{a}_{\varepsilon\Delta} + a^0_{\varepsilon\Delta},$$
$$\hat{a}^{(00)}_{\epsilon'} \simeq -2a_{\epsilon'} - 2a'_{\Delta\Delta} - 2\tilde{a}_{\varepsilon\Delta} + a^0_{\varepsilon\Delta}. \quad (43)$$

and

$$\hat{a}^{(+-)}_{\epsilon+\epsilon'} \simeq a^0_{\epsilon+\epsilon'} + a^0_{\epsilon+\epsilon'_{\Delta\Delta}} + \tilde{a}_{\epsilon+\epsilon'} + \tilde{a}_{\epsilon+\epsilon'_{\Delta\Delta}} + a_{\epsilon+\epsilon'_\Delta}$$
$$\hat{a}^{(00)}_{\epsilon+\epsilon'} \simeq a^0_{\epsilon+\epsilon'} + a^0_{\epsilon+\epsilon'_{\Delta\Delta}} - 2\tilde{a}_{\epsilon+\epsilon'} - 2\tilde{a}_{\epsilon+\epsilon'_{\Delta\Delta}} - 2a_{\epsilon+\epsilon'_\Delta} \quad (44)$$

with

$$a^0_{\varepsilon\Delta} = 2Re\Delta_0 = 2Re\left(\frac{B_0}{A_0}\right),$$

$$\tilde{a}_{\varepsilon\Delta} = 2Re[\frac{A_2}{A_0}(\Delta_2 - \Delta_0)]\cos(\delta_0 - \delta_2)$$

$$a^0_{\epsilon+\epsilon'} = 2\frac{Im\epsilon_K}{1+|\epsilon_K|^2}Re\left(\frac{A_0^*}{A_0}\right) + 2\frac{1-|\epsilon_K|^2}{1+|\epsilon_K|^2}Im\left(\frac{A_0^*}{A_0}\right)$$

$$\tilde{a}_{\epsilon+\epsilon'} \simeq 4[\frac{Im\epsilon_K}{1+|\epsilon_K|^2}Re\left(\frac{A_2^*}{A_0}\right) + \frac{1-|\epsilon_K|^2}{1+|\epsilon_K|^2}$$
$$\times Im\left(\frac{A_2^*}{A_0}\right)]\cos(\delta_0 - \delta_2) \quad (45)$$

$$a^0_{\epsilon+\epsilon'_{\Delta\Delta}} = -2\frac{Im\epsilon_K}{1+|\epsilon_K|^2}Re\left(\frac{A_0^*}{A_0}\Delta_0^{*2}\right)$$
$$-2\frac{1-|\epsilon_K|^2}{1+|\epsilon_K|^2}Im\left(\frac{A_0^*}{A_0}\Delta_0^{*2}\right)$$

$$\tilde{a}_{\epsilon+\epsilon'_{\Delta\Delta}} \simeq -4[\frac{Im\epsilon_K}{1+|\epsilon_K|^2}Re\left(\frac{A_2^*}{A_0}\Delta_0^*\Delta_2^*\right)$$
$$+\frac{1-|\epsilon_K|^2}{1+|\epsilon_K|^2}Im\left(\frac{A_2^*}{A_0}\Delta_0^*\Delta_2^*\right)]\cos(\delta_0 - \delta_2)$$

where we have neglected quadratic terms of $\omega = |A_2/A_0|$. Note that the above results hold for any choice of phase conventions. It is then obvious that

$$a^0_{\varepsilon\Delta} = \frac{2}{3}\hat{a}^{(+-)}_{\epsilon'} + \frac{1}{3}\hat{a}^{(00)}_{\epsilon'} \quad (46)$$

which shows that once the asymmetries $\hat{a}^{(+-)}_{\epsilon'}$ and $\hat{a}^{(00)}_{\epsilon'}$ are measured, their combination given above will allow one to extract a clean signature of CPT violation arising from the decay amplitudes. Where the values of $\hat{a}^{(+-)}_{\epsilon'}$ and $\hat{a}^{(00)}_{\epsilon'}$ can be simply extracted from the asymmetry $A_{CP+CPT}(t)$ at $t = 0$ in (39). It is noticed that when $|\Delta_0| << 1$, i.e., $|a^0_{\epsilon+\epsilon'_{\Delta\Delta}}| << |a^0_{\epsilon+\epsilon'}|$ (while Δ_2 could remain at the order of one). one has

$$a^0_{\epsilon+\epsilon'} \simeq \frac{2}{3}\hat{a}^{(+-)}_{\epsilon+\epsilon'} + \frac{1}{3}\hat{a}^{(00)}_{\epsilon+\epsilon'} \quad (47)$$

which indicates that by measuring $\hat{a}^{(+-)}_{\epsilon+\epsilon'}$ and $\hat{a}^{(00)}_{\epsilon+\epsilon'}$ one may extract the direct-indirect mixed-induced CP violation.

[2] Note that normalization of $A^{(00)}$ is smaller by a factor $\sqrt{2}$ than the usual one occuring in literature.

7 Conclusions

In summary, we have developed the general model-independent and rephase-invariant formalism for testing CP- and CPT-noninvariant observables in meson decays. The formalism presented in previous articles for CPT is based on the density matrix approach[15]. In our article, we present a complete time-dependent and rephase-invariant formulation in terms of amplitudes. The rephase invariance of all CP and CPT noninvariant observables is maintained throughout the calculation. All possible independent observables have been classified systematically, which is more general and complete than the published results and can be used for all meson decays. This enables one to separately measure different types of CPT- and CP-violating observables and to neatly distinguish effects of CPT from CP violation. The formalism which involves many and elaborate definitions is directly related to fundamental parameters and can prove advantageous in establishing CPT-violating parameters from time-dependent measurements of meson decays. Several time-dependent CPT- and CP- asymmetries have been introduced, which led to some interesting observations:

i). *As long as measurements of the asymmetry $A_{CP+CPT}(t)$ in the neutral meson decays (classified in the scenario i) in Sect. 5) is not a constant but depends on time, one can conclude that CPT invariance is broken due to mixing*.

ii). For the semileptonic decays $M^0 \to M'^- l\nu$, one may come to a strong statement that *once the asymmetry $A_{CP+CPT}(t)$ is not zero, then CPT must be violated.* Among the decays the semileptonic decays are the more representative and perhaps the easiest to measure.

iii). A combined measurement of several time-dependent CPT- and CP- asymmetries from semileptonic and nonleptonic decays is necessary in order to isolate separately the indirect and direct CPT- and CP-violating effects.

Extraction of a clean signature on CPT, CP and T violation will play an important role in testing the standard model and local quantum field theory and in addition provides an interesting window for probing new physics. For all these reasons, this topic attracts a lot of attention[16]. We hope that the general rephase-invariant formalism presented in this paper will be useful for further studies of CPT, CP and T in the neutral meson systems produced at B-factories, the Φ-factory[17] and colliders.

Acknowledgements. W.P was supported in part by the US department of Energy, Division of High Energy Physics, under Grant DOE/ER/0 1545-778. Two of us (E.A.P and Y.L.W) thanks Bundesministerium für Bildung, Wissenschaft, Forschung und Technologie (BMBF), 057D093P(7), Bonn, FRG, and DFG Antrag PA-10-1 for the financial support. Y.L.W acknowledges the support by the NSF of China under Grant 19625514.

Appendix

Here we collect some useful formuli. The definitions for the rephase-invariant observables:

$$\hat{a}_{\epsilon'} = \frac{1-|h/g|^2}{1+|h/g|^2} = \frac{2Re\varepsilon'_M}{1+|\varepsilon'_M|^2} \, .$$

$$\hat{a}_{\epsilon_S-\epsilon'} = \frac{-4Im(qsh/psg)}{(1+|qs/ps|^2)(1+|h/g|^2)}$$

$$= \frac{1}{1-a_\epsilon a_\Delta}\left[\hat{a}_{\epsilon+\epsilon'}\sqrt{1-a_\Delta^2-a'^2_\Delta} - a'_\Delta(1+\hat{a}_{\epsilon\epsilon'})\right]$$

$$\hat{a}_{\epsilon_S\epsilon'} = \frac{4Re(qsh/psg)}{(1+|qs/ps|^2)(1+|h/g|^2)} - 1$$

$$= \frac{1}{1-a_\epsilon a_\Delta}\left[\hat{a}_{\epsilon\epsilon'}\sqrt{1-a_\Delta^2-a'^2_\Delta} + a'_\Delta\hat{a}_{\epsilon+\epsilon'} + (\sqrt{1-a_\Delta^2-a'^2_\Delta}-1) + a_\epsilon a_\Delta\right] . \qquad (A.1)$$

with

$$\hat{a}_{\epsilon+\epsilon'} = \frac{-4Im(qh/pg)}{(1+|q/p|^2)(1+|h/g|^2)}$$

$$= \frac{2Im\epsilon_M(1-|\varepsilon'_M|^2)+2Im\varepsilon'_M(1-|\epsilon_M|^2)}{(1+|\epsilon_M|^2)(1+|\varepsilon'_M|^2)} \, ,$$

$$\hat{a}_{\epsilon\epsilon'} = \frac{4Re(qh/pg)}{(1+|q/p|^2)(1+|h/g|^2)} - 1$$

$$= \frac{4Im\epsilon_M\, Im\varepsilon'_M - 2(|\epsilon_M|^2+|\varepsilon'_M|^2)}{(1+|\epsilon_M|^2)(1+|\varepsilon'_M|^2)} . \qquad (A.2)$$

Rephase invariant observables for purely CP and CPT violation

$$a_{\epsilon''} = \frac{|\sum_i A_i e^{i\delta_i}|^2 - |\sum_i A_i^* e^{i\delta_i}|^2}{|\sum_i A_i e^{i\delta_i}|^2 + |\sum_i A_i^* e^{i\delta_i}|^2}$$

$$= -\frac{2\sum_{ij} A_i A_j^* \sin(\delta_i-\delta_j)}{|\sum_i A_i e^{i\delta_i}|^2 + |\sum_i A_i^* e^{i\delta_i}|^2} \, ,$$

$$a_{\varepsilon\Delta} = \frac{2\sum_{i,j} A_i A_j^*(\Delta_i+\Delta_j^*)\cos(\delta_i-\delta_j)}{|\sum_i A_i e^{i\delta_i}|^2 + |\sum_i A_i^* e^{i\delta_i}|^2} \, ,$$

$$a'_{\varepsilon\Delta} = \frac{2i\sum_{i,j} A_i A_j^*(\Delta_i+\Delta_j^*)\sin(\delta_i-\delta_j)}{|\sum_i A_i e^{i\delta_i}|^2 + |\sum_i A_i^* e^{i\delta_i}|^2} \, , \qquad (A.3)$$

$$a_{\Delta\Delta} = \frac{2\sum_{i,j} A_i A_j^* \Delta_i \Delta_j^* \cos(\delta_i-\delta_j}{|\sum_i A_i e^{i\delta_i}|^2 + |\sum_i A_i^* e^{i\delta_i}|^2} \, ,$$

$$a'_{\Delta\Delta} = \frac{2i\sum_{i,j} A_i A_j^* \Delta_i \Delta_j^* \sin(\delta_i-\delta_j)}{|\sum_i A_i e^{i\delta_i}|^2 + |\sum_i A_i^* e^{i\delta_i}|^2}$$

with $\Delta_i = B_i/A_i$. Here Δ_i are rephase-invariant quantities and characterize direct CPT violation in the decay amplitudes.

$$a_{\epsilon-\epsilon'} = \frac{2Im\epsilon_M(1-|\epsilon'_M|^2)+2Im\epsilon'_M(1-|\epsilon_M|^2)}{(1+|\epsilon_M|^2)(1+|\epsilon'_M|^2)}$$

$$a_{\epsilon+\epsilon'_\Delta} = \frac{2Im\epsilon_M(1-|\epsilon'_\Delta|^2)+2Im\epsilon'_\Delta(1-|\epsilon_M|^2)}{(1+|\epsilon_M|^2)(1+|\epsilon'_\Delta|^2)} \, , \qquad (A.4)$$

$$a_{\epsilon+\epsilon'_{\Delta\Delta}} = \frac{2Im\epsilon_M(1-|\epsilon'_{\Delta\Delta}|^2)+2Im\epsilon'_{\Delta\Delta}(1-|\epsilon_M|^2)}{(1+|\epsilon_M|^2)(1+|\epsilon'_{\Delta\Delta}|^2)} \, .$$

with

$$\frac{1-|\epsilon'_M|^2}{1+|\epsilon'_M|^2} = \frac{2\sum_{i,j} Re(A_i A_j)\cos(\delta_i-\delta_j)}{|\sum_i A_i e^{i\delta_i}|^2 + |\sum_i A_i^* e^{i\delta_i}|^2} ;$$

$$\frac{1-|\epsilon'_\Delta|^2}{1+|\epsilon'_\Delta|^2} = -\frac{2\sum_{i,j} Im[A_i A_j(\Delta_i-\Delta_j)]\sin(\delta_i-\delta_j)}{|\sum_i A_i e^{i\delta_i}|^2 + |\sum_i A_i^* e^{i\delta_i}|^2} .$$

$$\frac{1-|\epsilon'_{\Delta\Delta}|^2}{1+|\epsilon'_{\Delta\Delta}|^2} = -\frac{2\sum_{i,j} Re[A_i A_j(\Delta_i\Delta_j)]\cos(\delta_i-\delta_j)}{|\sum_i A_i e^{i\delta_i}|^2 + |\sum_i A_i^* e^{i\delta_i}|^2} .$$

$$\frac{2Im\epsilon'_M}{1+|\epsilon'_M|^2} = -\frac{2\sum_{i,j} Im(A_i A_j)\cos(\delta_i-\delta_j)}{|\sum_i A_i e^{i\delta_i}|^2 + |\sum_i A_i^* e^{i\delta_i}|^2} ; \qquad (A.5)$$

$$\frac{2Im\epsilon'_\Delta}{1+|\epsilon'_\Delta|^2} = -\frac{2\sum_{i,j} Re[A_i A_j(\Delta_i-\Delta_j)]\sin(\delta_i-\delta_j)}{|\sum_i A_i e^{i\delta_i}|^2 + |\sum_i A_i^* e^{i\delta_i}|^2} ;$$

$$\frac{2Im\epsilon'_{\Delta\Delta}}{1+|\epsilon'_{\Delta\Delta}|^2} = \frac{2\sum_{i,j} Im[A_i A_j(\Delta_i\Delta_j)]\cos(\delta_i-\delta_j)}{|\sum_i A_i e^{i\delta_i}|^2 + |\sum_i A_i^* e^{i\delta_i}|^2} .$$

The exact expressions for the time-dependent CP and CPT asymmetries in the scenario i):

$$A_{CP+CPT}(t) = \frac{\Gamma(M^0(t)\to f) - \Gamma(\overline{M}^0(t)\to \bar{f})}{\Gamma(M^0(t)\to f) + \Gamma(\overline{M}^0(t)\to \bar{f})} ; \qquad (A.6)$$

$$= \frac{\hat{a}_{\epsilon''} + 2\mathcal{A}_{CPT}(t)/[(1+|\eta_\Delta|^2)\cosh\Delta\Gamma t + (1-|\eta_\Delta|^2)\cos\Delta m t]}{1 + 2\hat{a}_{\epsilon''}\mathcal{A}_{CPT}(t)/[(1+|\eta_\Delta|^2)\cosh\Delta\Gamma t + (1-|\eta_\Delta|^2)\cos\Delta m t]} .$$

$$A'_{CP+CPT}(t) = \frac{\Gamma(\overline{M}^0(t)\to f) - \Gamma(M^0(t)\to \bar{f})}{\Gamma(\overline{M}^0(t)\to f) + \Gamma(M^0(t)\to \bar{f})} . \qquad (A.7)$$

$$(\hat{a}_{\epsilon''} + \frac{2a_\epsilon}{1+a_\epsilon^2})/(1 + \frac{2a_\epsilon}{1+a_\epsilon^2}\hat{a}_{\epsilon''}) ,$$

$$A''_{CP+CPT}(t) = \frac{\Gamma(M^0(t)\to f) - \Gamma(\overline{M}^0(t)\to f)}{\Gamma(M^0(t)\to f) + \Gamma(\overline{M}^0(t)\to f)} ; \qquad (A.8)$$

$$= \frac{\frac{1-a_\Delta^2-a_\Delta'^2+a_\epsilon a_\Delta'^2}{(1-a_\epsilon)(1-a_\Delta'^2)}\cos\Delta m t - \frac{a_\epsilon - a_\Delta^2}{(1-a_\epsilon)(1-a_\Delta'^2)}\cosh\Delta\Gamma t + \mathcal{A}_{CPT}(t)}{\frac{1-a_\epsilon a_\Delta^2}{(1-a_\epsilon)(1-a_\Delta'^2)}\cosh\Delta\Gamma t - \frac{a_\epsilon(1-a_\Delta^2-a_\Delta'^2)+a_\Delta'^2}{(1-a_\epsilon)(1-a_\Delta'^2)}\cos\Delta m t + \mathcal{A}_{CPT}(t)} ,$$

with

$$\mathcal{A}_{CPT} = -\frac{a_\Delta}{1-a_\Delta'^2}\sinh\Delta\Gamma t + \frac{a'_\Delta\sqrt{1-a_\Delta^2-a_\Delta'^2}}{1-a_\Delta'^2}\sin\Delta m t . \qquad (A.9)$$

and in the scenario ii):

$$A_{CP+CPT}(t) = \frac{\hat{\Delta}_m(t) + \Delta_{CPT}(t) - a_{\epsilon_S}(\hat{\Delta}_\gamma(t) + \Delta'_{CPT}(t))}{\hat{\Delta}_\gamma(t) + \Delta'_{CPT}(t) - a_{\epsilon_S}(\hat{\Delta}_m(t) + \Delta_{CPT}(t))} , \qquad (A.10)$$

with

$$\hat{\Delta}_m(t) = (a_{\epsilon_S} + \hat{a}_{\epsilon'})\cos(\Delta m t) + \hat{a}_{\epsilon_S+\epsilon'}\sin(\Delta m t) \qquad (A.11)$$

$$\hat{\Delta}_\gamma(t) = (1 + a_{\epsilon_S}\hat{a}_{\epsilon'})\cosh(\Delta\Gamma t) + (1 + \hat{a}_{\epsilon_S\epsilon'})\sinh(\Delta\Gamma t) ,$$

and

$$\Delta_{CPT}(t) = (2 + a_{\epsilon_S}\hat{a}_{\epsilon'} + \hat{a}_{\epsilon_S\epsilon'})[\frac{a_\Delta}{1-a_\Delta'^2} \times (\cos\Delta m t - e^{\Delta\Gamma t}) + \frac{a'_\Delta}{1-a_\Delta'^2}\sin\Delta m t]$$

$$\Delta'_{CPT}(t) = -[\frac{a_\Delta}{1-a_\Delta'^2}(a_{\epsilon_S} + \hat{a}_{\epsilon'}) + \frac{a'_\Delta}{1-a_\Delta'^2}\hat{a}_{\epsilon_S-\epsilon'}] \times (\cos\Delta m t - e^{-\Delta\Gamma t}) + [\frac{a'_\Delta}{1-a_\Delta'^2}(a_{\epsilon_S} + \hat{a}_{\epsilon'}) - \frac{a_\Delta}{1-a_\Delta'^2}\hat{a}_{\epsilon_S+\epsilon'}]\sin\Delta m t , \qquad (A.12)$$

$$-(2 + a_{\epsilon_S}\hat{a}_{\epsilon'} + \hat{a}_{\epsilon_S\epsilon'})\frac{a_\Delta^2 + a_\Delta'^2}{(1-a_\Delta'^2)^2}(\cos\Delta m t - \cosh\Delta\Gamma t)$$

Note that when CPT is conserved, $\Delta_{CPT}(t) = \Delta'_{CPT} = 0$.

References

1. T.D. Lee, C.N. Yang, Phys. Rev. **104**, 254 (1956)
2. C.S. Wu, E. Ambler, R.W. Hayward, D. Hoppes, R.P. Hudson, Phys. Rev. **105**, 1413 (1957)
3. R.L. Garwin, L.M. Lederman, M. Weinrich, Phys. Rev. **105**, 1415 (1957)
4. J.I. Friedman, V.L. Telegdi, Phys. Rev. **105**, 1681 (1957)
5. R. Christenson, J. Cronin, V.L. Fitch, R. Turlay, Phys. Rev. Lett. **13**, 138 (1964)
6. J. Schwinger, Phys. Rev. **82**, 914 (1951); G. Lueders, Dansk. Mat. Fys. Medd. **28**, 17 (1954); W. Pauli, Niels Bohr and the Development of Physics (Pergamon, New York, 1955)
7. For recent analyses see for example: C.O. Dib, R.D. Peccei, Phys. Rev. D **46**, 2265 (1992); C.D. Buchanan, et al., Phys. Rev. D **45**, 4088 (1992)
8. J.S. Bell, J. Steinberger, Proc. Oxford Int. Conf. on elementary particles, 1965, p. 195
9. J.W. Cronin, Rev. Mod. Phys. **53**, 373 (1981)
10. V.V. Barmin et al., Nucl. Phys. B **247**, 293 (1984)
11. W.F. Palmer, Y.L. Wu, Phys. Lett. B **350**, 245 (1995)
12. see for example, L. Lavoura, Ann. Phys. **207** 428 (1991), and references therein
13. T.D. Lee, C.S. Wu, Annu. Rev. Nucl. Sci. **16**, 471 (1966)
14. E.A. Paschos, R. Zacher, Z. Phys. C **28** 521 (1985); For a review see for example, E.A. Paschos, U. Türke, Phys. Rep. **178** 147 (1989)
15. J. Ellis, J.S. Hagelin, D.V. Nanopoulos, M. Srednichi, Nucl. Phys. B **241**, 381 (1984); P. Huet, M. E. Peskin, Nucl. Phys. B **434**, 3 (1995) J. Ellis, J.L. Lopez, N.E. Mavromatos, D.V. Nanopoulos, Phys. Rev. D **53**, 3846 (1996)
16. L. Wolfenstein, Phys.Rev.Lett. **83** 911 (1999); L. Lavoura, hep-ph/9911209; A. I. Sanda, hep-ph/9902353; L. Lavoura, J. P. Silva, hep-ph/9902348; P. Huet, hep-ph/9607435 ; J. Ellis, N.E. Mavromatos, D.V. Nanopoulos, hep-ph/9607434; R. Adler, et al (CPLEAR Collaboration), J. Ellis, J. Lopez, N. Mavromatos, D. Nanopoulos, hep-ex/9511001; N. Mavromatos, T. Ruf (for the collaboration: J. Ellis, J. Lopez, N. Mavromatos, D. Nanopoulos, and the CPLEAR Collaboration), hep-ph/9506395
17. See for example, L. Maiani, in 'The Second DAΦNE Physics Handbook', edited by L. Maiani, G. Pancheri, N. Paver, p.3-26; G. D'Ambrosio, G. Isidoni, A. Pugliese, ibid., p.63-95; and references therein

Part II

Statistical physics and condensed matter physics

Renormalization of the closed–time-path Green's functions in nonequilibrium statistical field theory

ZHOU Guang-zhao and SU Zhao-bing
Institute of Theoretical Physics, Academia Sinica, Beijing

(Received January 24 1979)
Phys. Energ. Fortis Phys. Nucl. 3 (3), 304–313 (May-June 1979)

The problem of renormalization of the closed–time-path Green's function in nonequilibrium statistical field theory is studied. Under some reasonable assumptions on the high-energy behavior of the initial correlation functions, it is found that the same counterterms which eliminate the ultraviolet divergences in the usual field theory can also make the closed–time-path Green's functions free of ultraviolet divergences. The renormalization-group equation satisfied by the closed–time-path vertex functions is obtained and the Callan-Symanzik coefficient functions are shown to be the same as in the usual field theory.

PACS numbers: 11.10.Gh, 11.10.Ef

I. INTRODUCTION

In recent years, with the development of high-energy particle physics and high-energy astrophysics, the general interest in finite-temperature field theory has been aroused.[1] Many articles have already discussed the renormalization of finite-temperature field theory, and the renormalization-group equation satisfied by the thermal Green's function has already been deduced.[2]

In this paper, we shall discuss the problems of ultraviolet divergence and renormalization of the closed–time-path Green's function, which is a method introduced by Schwinger[3] and Keldysh[3] in studying nonequilibrium statistical processes of a system. With this method, we can discuss not only the properties of physical variables in the ground state (vacuum state) and thermal equilibrium state, but also the properties of stationary states far from thermal equilibrium and transport processes while approaching equilibrium. We expect this method to be further developed in high-energy particle physics and high-energy astrophysics.

In Sec. II, we shall review briefly the perturbation theory[4] of the closed–time-path Green's function and introduce equations and symbols necessary in later proofs. In Sec. III, we shall prove that, under suitable assumptions on the ultraviolet behavior of the initial correlation function, the closed–time-path Green's function is renormalizable. As the usual ultraviolet divergences of the vacuum expectation of the Green's function are canceled, so are the ultraviolet divergences of the closed–time-path Green's function. In Sec. IV, the renormalization-group equation satisfied by the closed–time-path Green's function is discussed.

In this paper, we adopt the system of units with $\hbar = c = 1$ and take the space-time metric as

$$g_{00} = - g_{11} = - g_{22} = - g_{33} = 1.$$

II. PERTURBATION THEORY

Let $\varphi_B(x)$ represent the bare field of the system. $\varphi_B(x)$ can have many components: scalar fields, spinors, gauge fields, the ghost fields introduced by gauge-fixing conditions, etc. In general, the various components of $\varphi_B(x)$ are not explicitly indicated; therefore one must take heed that many equations are cast in abbreviated forms: For example, $\varphi_B(x)J_B(x)$ represents in fact the sum of the products of the various components. The Lagrangian of the system is

$$\mathscr{L} = \mathscr{L}(\varphi_B(x), m_B, \lambda_B) \tag{2.1}$$

where m_B is the bare mass of the particle and λ_B the bare interaction constant. We assume that

 0273-429X/81/010635-10$05.00

the Lagrangian is renormalizable in the usual sense for ordinary field theory. Introducing the renormalized field $\varphi(x)$, mass m, and interaction constant λ in the usual sense, we let

$$\varphi_B(x) = Z_\varphi^{1/2}\varphi(x), \quad m_B = Z_m m, \quad \lambda_B = Z_\lambda \lambda, \tag{2.2}$$

where Z_φ, Z_m, and Z_λ are the renormalization factors which contain all the ultraviolet divergences of the field theory. The Lagrangian may be rewritten as

$$\mathscr{L}(\varphi_B, m_B, \lambda_B) = \mathscr{L}_0(\varphi(x), m) - V(\varphi(x), \lambda, \delta m, \delta\lambda) - \varphi(x)J(x) \tag{2.3}$$

where $\mathscr{L}_0$ is the free Lagrangian with physical mass m, $V(\varphi(x), \lambda, \delta m, \delta\lambda)$ is the interaction term which contains the divergent counterterms (represented by δm and $\delta\lambda$). If gauge field are present in the field theory, V contains also the gauge-fixing term and the unitarity-compensating ghost terms. $J(x)$ is the external source.

The closed–time-path Green's function of $\varphi(x)$ is defined as

$$G_p(x_1, \ldots, x_l) = -i^{l-1} t_r\{T_p(\hat{\varphi}(x_1)\cdots\hat{\varphi}(x_l))\hat{\rho}\} \tag{2.4}$$

where $\hat{\varphi}(x)$ is the field operator in the Heisenberg picture and $\hat{\rho}$ the density matrix of the system. The subscript p denotes the closed path along the time axis, running from $-\infty$ to $+\infty$ (positive branch) and then from $+\infty$ to $-\infty$ (negative branch). The time t_i in the coordinates x_i ($i = 1,2,\ldots,l$) traces out the closed path, and T_p represents the operator along the closed path p. In Eq. (2.4), if we replace $\hat{\varphi}(x)$ with $\hat{\varphi}_B(x)$, we obtain the closed–time-path Green's function of the bare field $\hat{\varphi}_B(x)$.

The generating functional of the closed–time-path Green's functions is introduced as

$$Z[J(x)] = t_r\{T_p(\exp\{-i\int_p \hat{\varphi}(x)J(x)d^4x\})\hat{\rho}\}, \tag{2.5}$$

the integration being carried out along the closed path p. In order to avoid complete cancellation, the external sources $J(x_+)$ and $J(x_-)$ on the positive and negative branches, respectively, are taken to be different. Taking the functional derivative with respect to $J(x)$, we obtain, from Eq. (2.5),

$$G_p(x_1, \ldots, x_l) = i\left.\frac{\delta^l Z[J(x)]}{\delta J(x_l)\cdots\delta J(x_1)}\right|_{J(x)=0}. \tag{2.6}$$

In taking the functional derivative, we adopt the convention that, when $J(x)$ is an anticommuting c-number, the derivative operation acts from the right.

Transforming into the interaction picture, the generating functional can be written as

$$Z[J(x)] = t_r\{T_p(\exp\{-i\int_p [V(\hat{\varphi}_I(x)) + \hat{\varphi}_I(x)J(x)]\})\hat{\rho}\} \tag{2.7}$$

where $\hat{\varphi}_I(x)$ is the field operator in the interaction picture, satisfying the equation of motion for a free field,

$$\partial_\mu \frac{\delta\mathscr{L}_0}{\delta\partial_\mu\varphi(x)} - \left.\frac{\delta\mathscr{L}_0}{\delta\varphi(x)}\right|_{\varphi(x)=\hat{\varphi}_I(x)} = 0. \tag{2.8}$$

Equation (2.8) is a linear homogeneous equation in $\hat{\varphi}_I(x)$ and can be written as

$$\Delta^{-1}(\partial_\mu)\hat{\varphi}_I(x) = 0 \tag{2.9}$$

where $\Delta^{-1}(\partial_\mu)$ is a derivative operator, $\partial_\mu \equiv \partial/\partial x^\mu$; for examples, when $\hat{\varphi}_I(x)$ is a scalar field,

$$\Delta^{-1}(\partial_\mu) = -\partial^2 + m^2, \tag{2.10}$$

where $\partial^2 = \partial_\mu \partial^\mu$ is the d'Alembertian operator.

With the interaction term taken out of the trace, Eq. (2.7) becomes

$$Z[J(x)] = \exp\left\{-i\int_p V\left(i\frac{\delta}{\delta J(x)}\right)d^4x\right\} t_r\left\{T_p\left(\exp\left\{-i\int_p \hat{\varphi}_I(x)J(x)d^4x\right\}\right)\hat{\rho}\right\}. \quad (2.11)$$

If Wick's Theorem in field theory is generalized slightly, it can be easily proved that

$$T_p(\exp\{-i\int_p \hat{\varphi}_I(x)J(x)\,d^4x\}) = Z_0[J(x)]:\exp\{-i\int_p \hat{\varphi}_I(x)J(x)d^4x\}: \quad (2.12)$$

where Wick's normal-order notation (: :) implies that all operators enclosed within the double dots are so arranged that all the creation operators are to the left of the annihilation operators. The generating functional $Z_0[J(x)]$ of the free field is

$$Z_0[J(x)] = \exp\left\{-\frac{i}{2}\iint_p J(x)\Delta_p(x-y)J(y)d^4xd^4y\right\} \quad (2.13)$$

where $\Delta_p(x-y)$ is the vacuum expectation of the closed–time-path Green's function of the free field,

$$\Delta_p(x-y) = -i\langle 0|T_p(\hat{\varphi}_I(y)\hat{\varphi}_I(x))|0\rangle, \quad (2.14)$$

satisfying the equation

$$\Delta^{-1}(\partial_{x\mu})\Delta_p(x-y) = \delta_p^4(x-y). \quad (2.15)$$

Here $\delta_p^4(x-y)$ is the δ function in the closed path integral, which satisfies the relation

$$\int_p f(y)\delta_p^4(x-y)d^4y = f(x) \quad (2.16)$$

for any arbitrary function $f(x)$ defined on the closed path, regardless of whether x is on the positive or negative branch.

Substituting Eq. (2.12) into Eq. (2.11), we get

$$Z[J(x)] = \exp\left\{-i\int_p V\left(i\frac{\delta}{\delta J(x)}\right)d^4x\right\} Z_0[J(x)]N[J(x)], \quad (2.17)$$

where

$$N[J(x)] = t_r\{:\exp\{-i\int_p \hat{\varphi}_I(x)J(x)d^4x\}:\hat{\rho}\} \quad (2.18)$$

is known as the initial correlation functional, which is related to the density matrix describing the initial state of the system. When the system is at the vacuum state,

$$N[J(x)] = 1.$$

For any initial state, we expand $N[J(x)]$ by iteration, getting

$$N[J(x)] = \exp\{-iW_N[J(x)]\}, \quad (2.19)$$

$$W_N[J(x)] = \sum_{l=1}^{\infty}\frac{1}{l!}\int\cdots\int_p N(x_1,\ldots,x_l)J(x_1)\cdots J(x_l)d^4x_1\cdots d^4x_l, \quad (2.20)$$

where

$$N(x_1,\ldots,x_l) = (-i)^{l-1}t_r\{:\hat{\varphi}_I(x_1)\cdots\hat{\varphi}_I(x_l):\hat{\rho}\}_c; \quad (2.21)$$

the subscript c on the right-hand side of Eq. (2.21) indicates that this is an lth-order iterated term with all the lower-order correlation terms deleted. If the system is in thermal equilibrium, then its statistical distribution is Gaussian and all iterated terms for $l>2$ vanish.

When Eqs. (2.19) and (2.20) are substituted into Eq. (2.17) and the interaction term V is expanded, the perturbation-theory expression is obtained. There are only two differences between the perturbation expansion of the closed–time–path Green's function and that of ordinary field theory. Firstly, the time integration of the closed-time-path Green's function is along a closed path. Thus when resolution into positive and negative branches is made, each Feynman diagram resolves into 2^n diagrams according to whether each of the n vertices are on the positive or negative branch. At the same time, we multiply each vertex on the negative branch by a factor -1, so that all the time integration is from $-\infty$ to $+\infty$. The other difference is that the closed–time-path Green's function can be used to describe a system in any initial state and is not restricted only to the vacuum state. The effect of the initial state is expressed by the correlation function $N(x_1,\ldots,x_l)$.

One can readily prove that the correlation function obeys

$$\Delta^{-1}(\partial x_i^{\mu})N(x_1,\ldots,x_l) = 0 \quad i = 1,\ldots,l. \tag{2.22}$$

Eq. (2.22) implies that the correlation function contributes only when it is on the mass shell. Moreover, since Eq. (2.18) does not contain the time-ordered–product operator T_p, the values assumed by the correlation function on different branches of the closed path are the same.

III. CANCELLATION OF ULTRAVIOLET DIVERGENCES OF THE CLOSED–TIME-PATH GREEN'S FUNCTION AND RENORMALIZATION

Besides the assumption made in Sec. II that the Lagrangian is renormalizable in the usual sense, we introduce the following restriction to the correlation function describing the initial state of the system:

In Hilbert space, under the Fock representation of free particles, the matrix elements of the density matrix can be represented by

$$\langle \boldsymbol{p}_1',\ldots,\boldsymbol{p}_{N'}'|\hat{\rho}|\boldsymbol{p}_1,\ldots,\boldsymbol{p}_N\rangle, \tag{3.1}$$

where $\mathbf{p}_i$ is the momentum of the ith particle. For simplicity, the internal degrees of freedom of the particles have been suppressed. Because of the restriction on the physical states, there will naturally not be any particle with infinite momentum. As a result, we assume that, when the momentum $|\mathbf{p}_i|$ (or $|\mathbf{p}_i'|$) approaches infinity, the matrix element (3.1) should approach zero extremely rapidly; for example, when the system is in thermal equilibrium, the density-matrix element approaches zero according to the Gaussian distribution. To be more accurate, we assume that, for any positive integer l,

$$\lim_{|\boldsymbol{p}_i|\to\infty} |\boldsymbol{p}_i|^l\langle \boldsymbol{p}_1',\ldots,\boldsymbol{p}_{N'}'|\hat{\rho}|\boldsymbol{p}_1,\ldots,\boldsymbol{p}_N\rangle = 0. \tag{3.2}$$

Equation (3.2) indicates that the Fourier component $N(p_1,\ldots,p_l)$ of the correlation function $N(x_1,\ldots,x_l)$ with x_i on either of the branches approaches zero faster than $|\mathbf{p}_i|^{-1}$, as $|\mathbf{p}_i|\to\infty$. We have pointed out in Sec. II that $N(p_1,\ldots,p_l)$ is not vanishing only on the mass shell; thus the product of $N(p_1,\ldots,p_l)$ and any finite polynomial of p_i integrated over d^4p_i converges.

From the above discussion, we know that, in the perturbation expansion, for those Feynman diagrams containing $N(x_1,\ldots,x_l)$, if the integrand is convergent before integration over $x_1,\ldots,x_l$, no new divergences can emerge from the integration over x_i just because of the presence of $N(x_1,\ldots,x_l)$. But this does not mean that a single Feynman diagram containing $N(x_1,\ldots,x_l)$ is not divergent, because it is quite possible that, before integration over $x_1,\ldots,x_l$, the integrand is already divergent because of other closed loops which it contains. What we want to prove for the renormalizability of the closed–time-path Green's function is that, when all the diagrams containing $N(x_1,\ldots,x_l)$ are summed, their divergences cancel one another. In order to prove this, let us first analyze the structure of the Feynman diagrams representing the vacuum state.

With use of the commutation relation

$$\frac{\delta}{\delta J(z)}\exp\left\{-\frac{i}{2}\iint_p J(x)\Delta_p(x-y)J(y)d^4xd^4y\right\}$$

$$=\exp\left\{-\frac{i}{2}\iint_p J(x)\Delta_p(x-y)J(y)d^4xd^4y\right\}\left(\frac{\delta}{\delta J(z)}-i\int_p J(u)\Delta_p(u-z)d^4u\right), \tag{3.3}$$

the generating functional (2.17) can be written as

$$Z[J(x)]=Z_0[J(x)]\exp\left\{-i\int_p V\left(i\frac{\delta}{\delta J(x)}+\int_p J(y)\Delta_p(y-x)d^4y\right)d^4x\right\}N[J(x)]. \tag{3.4}$$

In order to clarify Eq. (3.4), we first discuss the vacuum state. There

$$N[J(x)]=1,$$

and

$$Z[J(x)]=Z_V[J(x)],$$

$$Z_V[J]=Z_0[J(x)]\exp\left\{-i\int_p V\left(i\frac{\delta}{\delta J(x)}+\int_p J(y)\Delta_p(y-x)\,d^4y\right)d^4x\right\}. \tag{3.5}$$

Expanding the last term in Eq. (3.5) in terms of $J(x)$, and letting

$$Z_V[J]=Z_0[J]\sum_{l=0}^{\infty}\int\cdots\int_p d^4x_1\cdots d^4x_l d^4y_1\cdots d^4y_l$$

$$\times\frac{1}{l!}J(y_1)\cdots J(y_l)\Delta_p(y_1-x_1)\cdots\Delta_p(y_l-x_l)f_V(x_1,\ldots,x_l), \tag{3.6}$$

we can then readily prove that

$$\int\cdots\int_p d^4x_1\cdots d^4x_l\Delta_p(y_1-x_1)\cdots\Delta_p(y_l-x_l)f_V(x_1,\ldots,x_l)$$

$$=\langle 0|T_p(\hat\varphi(y_l)\cdots\hat\varphi(y_1))|0\rangle_V. \tag{3.7}$$

The subscript V denotes that the l vertices at $y_1,\ldots,y_l$ are interaction connected; i.e., the diagrams representing the contribution by the free-field generating functional $Z_0[J(x)]$ are to be deleted from the vacuum expectation of the Green's function. It is necessary to point out that Eq. (3.7) includes all V-connected Feynman diagrams with l external lines.

Using Eq. (2.15) satisfied by $\Delta_p(y-x)$, we get, from Eq. (3.7),

$$f_V(x_1,\ldots,x_l)=\Delta^{-1}(\partial_{x_1})\Delta^{-1}(\partial_{x_2})\cdots\Delta^{-1}(\partial_{x_l})\langle 0|T_p(\hat\varphi(x_l)\cdots\hat\varphi(x_1))|0\rangle_V. \tag{3.8}$$

Before proceeding further, we need to differentiate the situation on the positive branch from that on the negative. We start our discussion by first putting all the x_i on the positive branch; then

$$f_V(x_{1+},\ldots,x_{l+})=\Delta^{-1}(\partial x_1)\cdots\Delta^{-1}(\partial x_l)\langle 0|T(\hat\varphi(x_l)\cdots\hat\varphi(x_1))|0\rangle_V. \tag{3.9}$$

On the right-hand side of Eq. (3.9), we have no need to identify x_i and T is just the usual time-ordering operator. Transforming into the interaction picture, we have

$$\langle 0|T(\hat\varphi(x_l)\cdots\hat\varphi(x_1))|0\rangle_V=\langle 0|\hat S^+T(\hat\varphi_I(x_l)\cdots\hat\varphi_I(x_1)\hat S)|0\rangle_V \tag{3.10}$$

where $\hat S$ is the S matrix of the field theory. Since the vacuum state is stable, acting on the vacuum, $\hat S$ gives only a phase factor,

$$\hat S|0\rangle=e^{iL}|0\rangle \tag{3.11}$$

where L comes from the contribution of closed-loop diagrams. Although it is possibly a divergent number, it can be removed through renormalization, giving no physical effect at all. Thus

$$f_V(x_{1+},\cdots,x_{l+}) = \Delta^{-1}(\partial_{x_1})\cdots\Delta^{-1}(\partial_{x_l})e^{-iL}\langle 0|\,T(\hat{\varphi}_I(x_l)\cdots\hat{\varphi}_I(x_1)\hat{S})\,|0\rangle_V; \tag{3.12}$$

the last term on the right is the Green's function of the ordinary field theory, which we denote by $G(x_1,\ldots,x_l)$. The closed-loop diagrams of pure vacuum in G contribute a factor e^{iL} canceling the factor e^{-iL} above. This demonstrates that all contribution of closed-loop diagrams in f_V have been deleted. In momentum space, Eq. (3.12) can be written as

$$\begin{aligned} f_{V+}(p_1,\cdots,p_l) &= \Delta^{-1}(-ip_1)\Delta^{-1}(-ip_2)\cdots\Delta^{-1}(-ip_l)e^{-iL}G(p_1,\cdots,p_l) \\ &= \int\cdots\int e^{i\sum_{j=1}^{l}p_j\cdot x_j} f_{V+}(x_1,\cdots,x_l)d^4x_1\cdots d^4x_l \end{aligned} \tag{3.13}$$

where the positive sign on the subscript of f denotes that all x_i are on the positive branch. We know that $G(p_1,\ldots,p_l)$ represents the Green's function with l external lines carrying momenta $p_1,\ldots,p_l$, with every external line connected with the interaction term. It has been proved in field theory that, if the system contains scalar fields, spinors, and gauge fields, and the dimension of each term in the Lagrangian is less than or equal to 4, then when the Green's function $G(p_1,\ldots,p_l)$ is expanded in terms of closed-loop diagrams, the ultraviolet divergences in each order of the closed loop diagrams are canceled by counterterms, and also, when $|\mathbf{p}_i|$ increases, the contribution from each order of the closed-loop diagrams increases only according to some finite power of $|\mathbf{p}_i|$. However, when all orders of the diagrams are summed, is it possible that the series does not converge and instead leads to new divergence? Since the answer to this question is still unclear at the present, we limit ourselves merely to the proof that when the closed-loop Green's function is expanded in terms of the number of loops, each term is convergent.

When all the x_i are on the negative branch, similarly we have

$$f_V(x_{1-},\cdots,x_{l-}) = \Delta^{-1}(\partial_{x_1})\cdots\Delta^{-1}(\partial_{x_l})e^{+iL}\tilde{G}(x_1,\cdots,x_l) \tag{3.14}$$

where

$$\tilde{G}(x_1,\cdots,x_l) = \langle 0|\,\tilde{T}(\hat{\varphi}_I(x_l)\cdots\hat{\varphi}_I(x_1)\hat{S}^+)\,|0\rangle,$$

and $\tilde{T}$ is the anti-time-ordering operator (operator of an earlier time stands to the left). It can be easily proved that

$$\tilde{G}(x_1,\cdots,x_l) = \langle 0|\,T(\hat{\varphi}_I^+(x_1)\cdots\hat{\varphi}_I^+(x_l)\hat{S})\,|0\rangle^*, \tag{3.15}$$

where $\tilde{G}(x_1,\ldots,x_l)$ is the complex conjugate of the Green's function of the ordinary field theory. In momentum space, for each order of the closed-loop diagrams, $\tilde{G}$ has no divergences, and when the momenta increase, $\tilde{G}$ increases at the most according to some finite power of the momenta. Thus

$$f_{V-}(p_1,\cdots,p_l) = \Delta^{-1}(-ip_1)\cdots\Delta^{-1}(-ip_l)e^{iL}\tilde{G}(p_1,\cdots,p_l) \tag{3.16}$$

is not divergent.

Finally we discuss the situation when some x_i are on the positive branch and some on the negative branch. It can be easily proved that

$$\begin{aligned} &f_V(x_{1+},\ldots,x_{j+},x_{j+1-},\ldots,x_{l-}) = \Delta^{-1}(\partial_{x_1})\cdots\Delta^{-1}(\partial x_l) \\ &\times\langle 0|\,\tilde{T}(\hat{\varphi}_I(x_l)\cdots\hat{\varphi}_I(x_{j+1})S^+)T(\hat{\varphi}_I(x_j)\cdots\hat{\varphi}_I(x_1)S)\,|0\rangle_V. \end{aligned} \tag{3.17}$$

With use of the method of expanding Feynman diagrams in the discussion of the unitarity condition, it can easily be proved that

$$\begin{aligned} f_V(x_{1+},\cdots,x_{j+};\,x_{j+1-},\cdots,x_{l-}) &= \sum_{k=0}^{\infty}\frac{(i)^k}{k!}\int\cdots\int d^4z_1\cdots d^4z_k d^4u_1\cdots d^4u_k \\ \times f_{V-}(z_k,\cdots,z_1;x_{j+1},\cdots,x_l)\Delta_-(z_k-u_k)&\cdots\Delta_-(z_1-u_1)f_{V+}(x_1,\cdots,x_j,\,u_1,\cdots,u_k) \end{aligned} \tag{3.18}$$

where

$$\Delta_-(z-u) = -i\langle 0|\hat{\varphi}_I(z)\hat{\varphi}_I(u)|0\rangle, \tag{3.19}$$

satisfying the equation

$$\Delta^{-1}(\partial_z)\Delta_-(z-u) = 0. \tag{3.20}$$

Going to the momentum representation, from Eq. (3.18), we get

$$f_{V+-}(p_1,\dots,p_j;\ p_{j+1},\dots,p_l)$$
$$= \sum_{k=0}^{\infty}\frac{(i)^k}{k!}\int\cdots\int\frac{d^4q_1}{(2\pi)^4}\cdots\frac{d^4q_k}{(2\pi)^4}f_{V-}(-q_k,\dots,-q_1,p_{j+1},\dots,p_l)$$
$$\times\Delta_-(q_1)\cdots\Delta_-(q_k)f_{V+}(p_1,\dots,p_j,q_1,\dots,q_k), \tag{3.21}$$

where $\Delta_-(q)$ contributes only on the mass shell as a result of Eq. (3.20), and it can be readily verified that $\Delta_-(q)$ is nonvanishing only when $q_0>0$. From the conservation of energy-momentum (the vacuum expectation of the Green's function is translationally invariant), the integrand of Eq. (3.21) is nonvanishing only when

$$q_1+\cdots+q_k = p_{j+1}+\cdots+p_l = -(p_1+\cdots+p_j). \tag{3.22}$$

Thus when integration over d^4q_i is performed, there is the restriction

$$|\mathbf{q}_i| \leqslant q_{i0} < P_0 \tag{3.23}$$

where $P_0 = p_{j+1,0}+\cdots+p_{l0}$ is a constant which should be greater than zero as a result of the positivity of q_{i0}; otherwise Eq. (3.22) vanishes. In the above, we have proved that both f_{V-} and f_{V+} are free of divergences and, because of Eq. (3.23), the integral over q_i is a finite one. Thus each term in the expansion of Eq. (3.21) converges. Although we are not sure whether the sum of the series is convergent, yet up to a finite number of loops the series in Eq. (3.21) consists of a finite number of terms and is therefore finite.

We have shown in the above discussion of the vacuum expectation of closed–time-loop Green's function that no new ultraviolet divergences emerge. In the following, we want to consider the general case.

For any initial condition, the generating functional (3.4) can be written as

$$Z[J(x)] = Z_0[J(x)]\sum_{l=0}^{\infty}\int\cdots\int_p d^4x_1\cdots d^4x_l\frac{1}{l!}f_V(x_1,\dots,x_l)$$
$$\times:\left(i\frac{\delta}{\delta J(x_1)}+\int_p J(y_1)\Delta_p(y_1-x_1)d^4y_1\right)\cdots\left(i\frac{\delta}{\delta J(x_l)}+\int_p J(y_l)\Delta_p(y_l-x_l)d^4y_l\right):$$
$$\times N[J(x)] \tag{3.24}$$

where, when inside the symbol : :, the operator $\delta/\delta J(x_i)$ is on the right of $J(y_i)$. In other words, within the symbol : :, $\delta/\delta J(x_i)$ operates only on $N[J(x)]$ which follows it. At the same time, it can be easily shown that, the function $f_V(x_1,\dots,x_l)$ in Eq. (3.24) is the same function given by Eq. (3.8), which depends only on the vacuum expectation of the Green's function.

In Eq. (3.24), when $i\delta/\delta J(x_j)$ operates on $N[J(x)]$, some correlation functions $N(x_1,\dots,x_k)$ appear. Thus when the generating functional $Z[J(x)]/Z_0[J(x)]$ is expanded in terms of $J(x)$, its coefficients are the product of $f_V(x_1,\dots,x_l)$ and many correlation functions $N(x_1,\dots,x_k)$ integrated over some coordinates. We have already proved that, for some fixed number of closed loops, no matter whether x_i are on the positive branch or the negative branch, the Fourier components $f_V(p_1,\dots,p_l)$ of f_V are free of divergences and increase as some power of $|\mathbf{p}_i|$ as $|\mathbf{p}_i|$ increases. At the same time, the Fourier components $N(p_1,\dots,p_k)$ of the initial correlation functions approach zero faster than any power of $|\mathbf{p}_i|^{-1}$ as $|\mathbf{p}_i|$ increases. Thus, when f_V if multiplied by many correlation functions and then integrated over some coordinates, no new ultraviolet divergences will be produced. Therefore, we can make the conclusion that, as long as the number of closed loops is fixed, the number of Feynman diagrams is finite, and the sum of these diagrams will not lead to ultraviolet divergence. Whether the sum of the whole series is convergent or not is a problem that should be attacked separately.

From the above discussion, we see that, if the Green's function of an ordinary field theory is free of divergences, the closed–time-path Green's function in nonequilibrium statistical mechanics is also free of divergences, and also the counterterms in the Lagrangian and the renormalization factors are the same as those of an ordinary field theory. We can make use of these properties to deduce the renormalization-group equation[5] satisfied by the closed–time-path Green's function.

IV. THE RENORMALIZATION-GROUP EQUATION SATISFIED BY THE CLOSED–TIME-PATH GREEN'S FUNCTION

In the same way as in ordinary field theory, we can introduce connected closed–time-path Green's function and closed–time–path vertex function. Their generating functionals are denoted by $W[J(x)]$ and $\Gamma[\varphi_c(x)]$ respectively, which obey

$$W[J(x)] = i \ln Z[J(x)],$$
$$\Gamma[\varphi_c(x)] = W[J(x)] - \int_p \varphi_c(x) J(x) d^4x, \tag{4.1}$$

where $\varphi_c(x)$ is the vacuum expectation of the field $\hat{\varphi}(x)$,

$$\varphi_c(x) = \frac{\delta W[J(x)]}{\delta J(x)}. \tag{4.2}$$

The closed–time-path vertex function can be written as

$$\Gamma_p(x_1, \ldots, x_l; \mu, m, \lambda, \zeta_i) = \frac{\delta^l \Gamma[\varphi_c(x)]}{\delta\varphi_c(x_1)\cdots\delta\varphi_c(x_l)}. \tag{4.3}$$

In the vertex function (4.3), we have included its dependent physical variables explicitly. Besides the mass m and the interaction constant λ, it depends on the choice of the renormalization point μ and the physical variable ζ_i denoting the initial correlation function. If we let Γ_{PB} denotes the bare vertex function, then there exists the relation

$$\Gamma_p(x_1, \ldots, x_l; \mu, m, \lambda, \zeta_i) = Z_\varphi^{+l/2} \Gamma_{PB}(x_1, \ldots, x_l, m_B, \lambda_B, \zeta_i), \tag{4.4}$$

where Z_φ is the wave-function renormalization factor. When $\varphi(x)$ has many components, with the jth component corresponding to x_j, we have

$$Z_\varphi^{+l/2} = \left(\prod_j Z_{\varphi j}\right)^{+1/2}. \tag{4.5}$$

From Eq. (2.2), we know that

$$m_B = Z_m m, \quad \lambda_B = Z_\lambda \lambda$$

where Z_φ, Z_m, and Z_λ are functions of μ and the interaction constant λ. (Here we have adopted the method of dimensional regularization developed by 't Hooft to cancel ultraviolet divergences. This method leads to a mass-independent renormalization and therefore the Z's do not depend[6] on m.)

Differentiating Eq. (4.4) with respect to μ (with m_B and λ_B fixed), we arrive at the renormalization-group equation

$$\left[\mu\frac{\partial}{\partial\mu}+\beta(\lambda)\frac{\partial}{\partial\lambda}+\gamma_m(\lambda)m\frac{\partial}{\partial m}-\gamma_\Gamma(\lambda)\right]\Gamma_p^{(l)}=0 \tag{4.6}$$

where we have used $\Gamma_p^{(l)}$ to denote $\Gamma_p(x_1,\ldots,x_l,\mu,\lambda,m,\zeta_i)$, and

$$\beta(\lambda)=\mu\frac{\partial}{\partial\mu}\lambda\Big|_{\lambda_B,\,m_B}=-\lambda\mu\frac{\partial}{\partial\mu}\ln Z_\lambda\Big|_{\lambda_B,\,m_B},$$
$$\gamma_m(\lambda)=\mu\frac{\partial}{\partial\mu}\ln m\Big|_{\lambda_B,\,m_B}=-\mu\frac{\partial}{\partial\mu}\ln Z_m\Big|_{\lambda_B,\,m_B}, \tag{4.7}$$
$$\gamma_\Gamma(\lambda)=\frac{-l}{2}\mu\frac{\partial}{\partial\mu}\ln Z_\varphi\Big|_{\lambda_B,\,m_B}$$

are the usual Callan-Symanzik coefficients, which are the same as those given by the ordinary field theory.

When we fix x_j on the positive or the negative branch and transform to momentum space, we get altogether 2^l different vertex functions, everyone of which obeys Eq. (4.6). Thus if we let

$$\Gamma_s^{(l)}(p_1,\ldots,p_l,\mu,\lambda,m,\zeta_i),\quad s=1,2,\ldots,2^l,$$

be these vertex functions, they obey

$$\left[\mu\frac{\partial}{\partial\mu}+\beta(\lambda)\frac{\partial}{\partial\lambda}+\gamma_m(\lambda)m\frac{\partial}{\partial m}-\gamma_\Gamma(\lambda)\right]\Gamma_s^{(l)}=0,\quad s=1,2,\ldots,2^l. \tag{4.8}$$

We separate the physical variable ζ_i denoting the initial correlation function into two parts: η_i, which is dimensionless; and ξ_i, which carries the dimension of a mass. The latter can be the temperature, chemical potential, etc. For those physical variables with still higher dimensions, we can raise them to a certain power and thus reduce them to physical variables of the type ξ_i. According to dimensional analysis, we have

$$\Gamma_s^{(l)}[Kp_1,\ldots,Kp_l,K\mu,\lambda,Km,\eta_i,K\xi_i]=K^{D_\Gamma}\Gamma_s^{(l)}(p_1,\ldots,p_l,\mu,\lambda,m,\eta_i,\xi_i), \tag{4.9}$$

where D_Γ is the canonical dimension of the vertex function $\Gamma_s^{(l)}$. From Eq. (4.9), it can be easily shown that

$$\left[K\frac{\partial}{\partial K}+\mu\frac{\partial}{\partial\mu}+m\frac{\partial}{\partial m}+\xi_i\frac{\partial}{\partial\xi_i}-D_\Gamma\right]\Gamma_s^{(l)}(Kp_1,\ldots,Kp_l,\mu,\lambda,m,\eta_i,\xi_i)=0. \tag{4.10}$$

Subtracting Eq. (4.8) from Eq. (4.10) and removing $\mu\partial/\partial\mu$, we arrive at

$$\left[K\frac{\partial}{\partial K}+\xi_i\frac{\partial}{\partial\xi_i}-\beta(\lambda)\frac{\partial}{\partial\lambda}+(1-\gamma_m(\lambda))m\frac{\partial}{\partial m}+\gamma_\Gamma-D_\Gamma\right]$$
$$\times\Gamma_s^{(l)}(Kp_1,\ldots,Kp_l,\mu,\lambda,m,\eta_i,\xi_i)=0. \tag{4.11}$$

Eq. (4.11) is the Callan-Symanzik equation satisfied by the closed–time-path Green's function.

When Eq. (4.11) is solved, it leads to the same results[4] obtained by Kislinger and Morley using a finite-temperature Green's function. For example, a non-Abelian gauge field possesses the property of asymptotic freedom not only when the momenta are large but also when ξ_i is large (high temperature or high chemical potential).

[1]D. A. Kirzhnits and A. D. Linde, Phys. Lett. **42 B**, 471 (1971); D. A. Kirzhnits, JETP Lett. **15**, 529 (1972); S. Weinberg, Phys. Rev. **D 9**, 3357 (1974); L. Dolan and R. Jackiw, Phys. Rev. **D 9**, 3320 (1974); M. B. Kislinger and P. D. Morley, Phys. Rev. **D 13**, 2765 (1976).

[2]M. B. Kislinger and P. D. Morley, Phys. Rev. **D 13**, 2771 (1976); S. Weinberg, Phys. Rev. **D 9**, 3357 (1974); L. Dolan and R. Jackiw, Phys. Rev. **D 9**, 3320 (1974).

[3]J. Schwinger, J. Math. Phys. **2**, 407 (1961); L. V. Keldysh JETP **20**, 1018 (1965); R. A. Craig, J. Math. Phys. **9**, 605 (1968); R. Mills, *Propagators for many particle systems* (Gordon and Breach, New York,1969); V. Korenman, Ann. Phys. (N.Y.) **39**, 72 (1966); V. L. Berezinskii, JETP **26**, 137 (1968); G. Niklasson and A. Sjölander, Ann. Phys. (N. Y.) **49**, 249 (1968); C. P. Enz, *The many body problem* (Plenum, New York, 1969); R. Sandström, Phys. Status Solidi **38**, 683 (1970); C. Caroli, R. Combescot, P. Nozières, and D. Saint-James, J. Phys. C **4**, 916 (1971).

[4]A. G. Hall, Mol. Phys. **28**, 1 (1974); J. Phys. **A 8**, 214 (1974).

[5]K. Symanzik, Commun. Math. Phys. **23**, 49 (1971); C. G. Callan, Phys. Rev. **D 5**, 3202 (1972).

[6]G. 't Hooft, Nucl. Phys. **B61**, 455 (1973); J. C. Collins and A. J. Macfarlane, Phys. Rev. **D 10**, 1210 (1974); S. Weinberg, Phys. Rev. **D 8** 3497 (1973).

Translated by King Yuen Ng
Edited by Stanley Wu-Wei Liu

Dyson equation and Ward-Takahashi identities of the closed–time-path Green's function

ZHOU Guang-zhao and SU Zhao-bing
Institute of Theoretical Physics, Academia Sinica, Beijing

(Received 24 January 1979)
Phys. Energ. Fortis Phys. Nucl. 3 (3), 314–326 (May–June 1979)

The Dyson equation satisfied by the closed–time-path Green's function of the order parameters is considered. The transport equation for the number density of the quasiparticles is written down in a general but simple form. With use of the path-integral formulation for the generating functional of these Green's function, the Ward-Takahashi identites are deduced.

PACS numbers: 11.10.Lm, 11.10.Ef, 11.10.Np

I. INTRODUCTION

In recent years, in the study of high-energy particle physics, many problems concerning collective cooperative phenomena have been raised, e.g., the problem of the phase change of the vacuum state, the problem about the confinement of quarks, the problem of solitons, etc. All of these are not single-particle phenomena, but rather phenomena of collective motion due to the interactions of an infinite number of degrees of freedom. It is expected that, under the interactions of high-energy particles, many degrees of freedom can be excited. When one deals with the motion of these degrees of freedom, the method of nonequilibrium statistical field theory should be used.

In our view, the closed–time-path Green's-function formulism developed by Schwinger[1] and Keldysh[1] provides an effective method to study nonequilibrium statistical field theory. This method is very similar to the method of Green's function in field theory and, with only small alteration, nearly all methods in field theory can be applied. At the same time, the closed–time-path Green's function contains the statistical correlation for any initial condition; therefore, it can also be used to study the properties of the ground state, the thermal equilibrium state, transport processes, and the properties of stationary states far from equilibrium.

In our opinion, the study of nonequilibrium statistical field theory has come under higher and higher demand in the development of high-energy particle physics and astrophysics, and this is a direction that deserves emphasis. Taking solitons as an example, we note that in field theory all solitons are classical solutions of the Euler equation of the primitive Lagrangian. But the solitons observed in solid states physics, like vortex lines in superconductivity, are not classical solutions of the primitive Lagrangian. They are the classical solutions of the equation of an order parameter statistically averaged over all other degrees of freedom. In future particle physics, solitons described by order parameters and not by fundamental fields may also be possible. The main purpose of writing this paper is to arouse the interest of our colleagues in high-energy particle physics in this direction. In Sec. II, the Dyson equation satisfied by the closed–time-path Green's function is studied. This probelm has already been studied in Ref. 1 and therefore not all of our results are new. However, it is worthwhile writing them out here for our readers because our equations are more general and simpler than those in other references; our transport equation which is in a rather simple and general form deserves special mention.

The second-order Green's function in field theory is the Feynman propagator G_F; all the rest are unimportant. Of the second-order closed–time-path Green's functions, there are three independent ones that are very important: They are the second-order retarded Green's function G_r, the second-order advanced Green's function G_a, and the number density, n, of quasiparticles. In field theory, there is no attenuation when a particle propagates in vacuum; however, there is in general attenuation in the propagation of quasiparticles. This is why G_r and G_a are considered as two independent quantities, which contain not only the spectrum of propagation (dispersion) but also attenuation. The Dyson equation of the second-order closed–time-path Green's function can determine not only the propagation and attenuation of quasiparticles, but also the transportation of the number density, n, of quasiparticles.

In Sec. III, the Ward-Takahashi (W-T) identities of the closed–time-path Green's function are discussed. We use the Feynman path-integral method, when the Lagrangian possesses the global symmetry of a Lie group G, to derive the W-T identities satisfied by the closed–time-path Green's function. The situation when the symmetry is broken spontaneously[2] is also discussed. We have proved that, in general, the equation

$$\delta \Gamma / \delta Q_c(x) = 0 .$$

satisfied by the vacuum expectation $Q_c(x)$ of the order parameter $\hat{Q}(x)$ determined by the generating functional of the vertex functions, does not possess stable solutions of the form

$$Q_c(x) = Q_0(\mathbf{x}) e^{-i\omega t}, \quad \omega \neq 0 .$$

But this does not mean that soliton solution or laser-type solutions do not occur in the system; it only says that, because of quantum effects, the wave packet formed by solitons cannot be stable and it must spread and attenuate. Thus in order to search for soliton solutions, we must at first determine the classical equation satisfied by the order parameter $Q(x)$.

In the last section, we discuss the W-T identities satisfied by the closed–time-path Green's function when the Lagrangian possesses local gauge invariance.

II. CLOSED–TIME-PATH GREEN'S FUNCTION

Let $\hat{Q}(x)$ and $\hat{\rho}$ represent, respectively, in the Heisenberg picture, the physical variable and the density matrix designating the initial conditions of the system. Here $\hat{Q}(x)$ can be a fundamental field; it can also be a composite operator made up of fundamental fields. If $\hat{Q}(x)$ contains more than one component, we interpret x as $\{x_\mu, i\}$ for $\mu = 0,1,2,3$ and $i = 1,2,\ldots,n$, with x_μ denoting space-time coordinates and i the various components. The closed–time-path Green's function of the physical variable $Q(x)$ is defined as

$$G_p(x_1, \cdots, x_l) = (-i)^{l-1} t_r\{T_p(\hat{Q}(x_1)\cdots\hat{Q}(x_l))\hat{\rho}\}. \tag{2.1}$$

where the subscript p denotes the closed path varying on the time axis from $t = -\infty$ to $t = +\infty$ (called the t_+ branch) and then from $t = +\infty$ back to $-\infty$ (called the t_- branch); T_p is the order operator on the closed path p; the time coordinates of $x_1,\ldots,x_l$ can be any point on the closed path. When all $x_1,\ldots,x_l$ are on the t_+ branch, T_p is the same as the usual time-ordering operator in field theory, and

$$G_p(x_{1+}, \cdots, x_{l+}) = (-i)^{l-1} t_r\{T(\hat{Q}(x_1)\cdots\hat{Q}(x_l))\hat{\rho}\}. \tag{2.2}$$

The Green's function $G_p(x_{1+},\ldots,x_{l+})$ is the vacuum expectation of the operator $T(\hat{Q}(x_1)\cdots\hat{Q}(x_l))$. It differs from the Green's function of ordinary field theory in that the former is an average over any initial condition (described by the density matrix $\hat{\rho}$) while the latter is an average over the vacuum state. When $x_1,\ldots,x_j$ lie on the t_- branch and $x_{j+1},\ldots,x_l$ on the t_+ branch, we have

$$G_p(x_{1-}, \ldots, x_{j-}, x_{j+1+}, \ldots, x_{l+}) = (-i)^{l-1} t_r\{\tilde{T}(\hat{Q}(x_1)\cdots\hat{Q}(x_j))T(\hat{Q}(x_{j+1})\cdots\hat{Q}(x_l))\hat{\rho}\} \tag{2.3}$$

where T is the anti-time-ordering operator (operator of a later time stands to the right).

Let us introduce the generating functional of the closed–time-path Green's functions

$$Z[h(x)] = t_r\{T_p(\exp\{-i\int_p h(x)\hat{Q}(x)d^4x\})\hat{\rho}\}. \tag{2.4}$$

where $\hat{Q}(x)$ is the operator in the Heisenberg picture without taking into account of the external source $h(x)$. The integration in Eq. (2.4) is made along the closed path p, the external source $h(x)$ being different when situated on the t_+ branch and the t_- branch. Differentiating $Z[h(x)]$ with respect to $h(x)$, we get

$$i\frac{\delta Z[h(x)]}{\delta h(x)} = t_r\{T_p(\hat{Q}(x)\exp\{-i\int_p h(y)\hat{Q}(y)d^4y\})\hat{\rho}\}. \tag{2.5}$$

When $h(x_+) = h(x_-) = h(x)$,

$$T_p(\hat{Q}(x)\exp\{-i\int_p h(y)\hat{Q}(y)d^4y\}) = U^+(t)\hat{Q}(x)U(t) \equiv \hat{Q}_h(x) \tag{2.6}$$

where

$$U(t) = \exp\{-i\int_{-\infty}^{t} h(y)\hat{Q}(y)d^4y\},$$

and $\hat{Q}_h(x)$ is the operator in the Heisenberg picture when the external source term $\int h(y)\hat{Q}(y)d^3y$ is included in the Hamiltonian. We thus have

$$\frac{\delta Z[h(x)]}{\delta h(x)} = -it_r\{\hat{Q}_h(x)\hat{\rho}\},$$
$$\frac{\delta^l Z[h(x)]}{\delta h(x_1)\cdots\delta h(x_l)} = (-i)^l t_r\{T_p(\hat{Q}_h(x_1)\cdots\hat{Q}_h(x_l))\hat{\rho}\}. \tag{2.7}$$

Those readers who are not familiar with the closed–time-path Green's functions can learn from Eq. (2.6) why we have to introduce a closed path. Only in this way can we guarantee the operator to be always in the Heisenberg picture; otherwise, on the right-hand side of Eq. (2.7), after application of the trace operator t_r there will be an additional factor $U(t_i)$, t_i being the latest time for $x_1,\ldots,x_l$. This factor makes the ordinary Green's functions in any initial state unrelated to the averages of physical variables.

Let us introduce the generating functionals of the closed–time-path connected Green's functions and vertex functions,

$$W[h(x)] = i\ln Z[h(x)],$$
$$\Gamma[Q(x)] = W[h(x)] - \int_p h(x)Q(x)d^4x, \tag{2.8}$$

where the average of $\hat{Q}_h(x)$ is denoted by

$$Q(x) = \frac{\delta W[h(x)]}{\delta h(x)}. \tag{2.9}$$

Here, we do not require $h(x_+) = h(x_-)$, but directly define

$$\hat{Q}_h(x) = T_p(\hat{Q}(x)\exp\{-i\int_p h(x)\hat{Q}(x)d^4x\}). \tag{2.10}$$

When we solve for the observables, we need to take $h(x_+) = h(x_-)$; then $Q(x_+) = Q(x_-)$ will be satisfied automatically.

Similar to field theory, we have

$$\frac{\delta\Gamma[Q]}{\delta Q(x)} = \mp h(x). \tag{2.11}$$

In the following, whenever an equation contains upper and lower signs, the upper one applies when $\hat{Q}(x)$ is a boson operator while the lower one applies when $\hat{Q}(x)$ is a fermion operator. When

$\hat{Q}(x)$ is a fermion operator, both $h(x)$ and $Q(x)$ are anticommuting c-numbers and all derivatives operate from the left.

Differentiating Eq. (2.9) with respect to $Q(y)$ and Eq. (2.11) with respect to $h(y)$, we arrive at two equations,

$$\begin{aligned}\int_p G_{pc}(x, y)d^4y\Gamma_p(y, z) &= -\delta_p^4(x-z),\\ \int_p \Gamma_p(x, y)d^4yG_{pc}(y, z) &= -\delta_p^4(x-z)\end{aligned} \tag{2.12}$$

where the second-order connected Green's function is

$$\begin{aligned}G_{pc}(x, y) &= \frac{\delta^2 W}{\delta h(x)\delta h(y)} \equiv -i\langle T_p(\hat{Q}_h(x)\hat{Q}_h(y))\rangle,\\ \langle T_p(\hat{Q}_h(x)\hat{Q}_h(y))\rangle &= \frac{1}{Z[h]}t_r\{T(\hat{Q}_h(x)\hat{Q}_h(y))\hat{\rho}\} - Q(x)Q(y)\end{aligned} \tag{2.13}$$

and the second-order vertex function is

$$\Gamma_p(x, y) = \frac{\delta^2\Gamma}{\delta Q(x)\delta Q(y)}. \tag{2.14}$$

If $Q(x)$ consists of more than one component, then the integral over d^4y implies integration over space-time points and sum over the indices of the components. Also, $\delta_p^4(x-y)$ will be a δ function involving the other components. It is defined on the closed path p with the property that

$$\int_p f(y)\delta_p^4(y-x)d^4y = f(x) \tag{2.15}$$

where $f(x)$ is any arbitrary continuous function defined on the positive and negative branches.

Because x and y can be on the positive branch or the negative branch, when $G_{pc}(x,y)$ is expressed as a function of a single time, it becomes four functions which can be written as a matrix

$$\hat{G} = \begin{pmatrix} G_F & G_+ \\ G_- & G_{\tilde{F}} \end{pmatrix} \tag{2.16}$$

where G_α, with $\alpha = F, +, -$, and $\widetilde{F}$, are themselves also matrices,

$$\begin{aligned}(G_F)_{x,y} &= G_F(x, y) = G_p(x_+, y_+) = -i\langle T(\hat{Q}_h(x)\hat{Q}_h(y))\rangle,\\ (G_+)_{x,y} &= G_+(x, y) = G_p(x_+, y_-) = \mp i\langle \hat{Q}_h(y)\hat{Q}_h(x)\rangle,\\ (G_-)_{x,y} &= G_-(x, y) = G_p(x_-, y_+) = -i\langle \hat{Q}_h(x)\hat{Q}_h(y)\rangle,\\ (G_{\tilde{F}})_{x,y} &= G_{\tilde{F}}(x, y) = G_p(x_-, y_-) = -i\langle \tilde{T}(\hat{Q}_h(x)\hat{Q}_h(y))\rangle.\end{aligned} \tag{2.17}$$

If $\hat{Q}(x)$ is a self-conjugate operator, it can be easily proved that

$$\hat{G}^+ = -\hat{\eta}_1\hat{G}\hat{\eta}_1 \tag{2.18}$$

where $\hat{\eta}_i$ are the Pauli matrices,

$$\hat{\eta}_1 = \begin{pmatrix} 0 & 1 \\ 1 & 0 \end{pmatrix}, \quad \hat{\eta}_2 = \begin{pmatrix} 0 & -i \\ i & 0 \end{pmatrix}, \quad \hat{\eta}_3 = \begin{pmatrix} 1 & 0 \\ 0 & -1 \end{pmatrix}.$$

Equation (2.18) can also be written as

$$\begin{aligned}G_F^+ &= -G_{\tilde{F}},\\ G_+^+ &= -G_-.\end{aligned} \tag{2.18'}$$

Using the definition of G_α, we can also readily prove

$$G_F + G_{\tilde{F}} = G_+ + G_-. \tag{2.19}$$

From Eqs. (2.18) and (2.19), we know that, among the G_α's, only three Hermitian matrices are linearly independent. From G_α, we can also introduce the retarded Green's function G_r and the advanced Green's function G_a:

$$\begin{aligned} G_r &= G_F - G_+ = G_- - G_{\tilde{F}}, \\ G_a &= G_F - G_- = G_+ - G_{\tilde{F}}. \end{aligned} \tag{2.20}$$

Similarly, six vertex functions F_α, with $\alpha = F, +, -, \tilde{F}, r$, and a, can be defined. In matrix representation, let

$$\hat{\Gamma} = \begin{pmatrix} \Gamma_F & \Gamma_+ \\ \Gamma_- & \Gamma_{\tilde{F}} \end{pmatrix}. \tag{2.21}$$

Equation (2.12) can be written as

$$\hat{\Gamma}\hat{\eta}_3\hat{G} = -\hat{\eta}_3, \quad \hat{G}\hat{\eta}_3\hat{\Gamma} = -\hat{\eta}_3. \tag{2.22}$$

From Eq. (2.22), it can be easily proved that, when G_α satisfies relations (2.18) and (2.19), Γ_α will also satisfy similar relations,

$$\hat{\Gamma}^+ = -\hat{\eta}_1\hat{\Gamma}\hat{\eta}_1, \quad \Gamma_F + \Gamma_{\tilde{F}} = \Gamma_+ + \Gamma_-, \tag{2.23}$$

and

$$G_r = -\Gamma_r^{-1}, \quad G_a = -\Gamma_a^{-1}. \tag{2.24}$$

With use of Eq. (2.23), $\hat{\Gamma}$ can be represented by the three linearly independent Hermitian matrices so that

$$\hat{\Gamma} = -iB\hat{\eta} - A\hat{\eta}_2 - D\hat{\eta}_3, \tag{2.25}$$

where

$$\begin{aligned} \hat{\eta} &= \begin{pmatrix} 1 & 1 \\ 1 & 1 \end{pmatrix} = \hat{1} + \hat{\eta}_1, \\ B &= \frac{i}{2}(\Gamma_F + \Gamma_{\tilde{F}}) = \frac{i}{2}(\Gamma_+ + \Gamma_-), \\ D &= \frac{1}{2}(\Gamma_{\tilde{F}} - \Gamma_F) = -\frac{1}{2}(\Gamma_r + \Gamma_a), \\ A &= \frac{i}{2}(\Gamma_- - \Gamma_+) = -\frac{i}{2}(\Gamma_r - \Gamma_a), \end{aligned} \tag{2.26}$$

and the matrices B, D, and A are all Hermitian. From Eqs. (2.26) and (2.24), we get

$$\begin{aligned} G_r &= -\Gamma_r^{-1} = \frac{1}{D + iA}, \\ G_a &= -\Gamma_a^{-1} = \frac{1}{D - iA} \end{aligned} \tag{2.27}$$

designating D as the dispersive part and A the attenuation part.

If the self-conjugate operator $\hat{Q}(x)$ consists of only one component, in a uniform system, its Green's function is only a function of the relative coordinate $x - y$ and, in momentum representation, we have

$$G_r(k) = \frac{1}{D(k) + iA(k)}. \tag{2.28}$$

When attenuation $A(k)$ is small, the poles of the retarded Green's function are determined by

$$D(k) = 0. \tag{2.29}$$

If the energy is determined as $k_0 = \omega(\mathbf{k}) > 0$, then a quasiparticle with energy $\omega(\mathbf{k})$ propagates in the system. The amplitude of the quasiparticle is attenuated exponentially in propagation, with an attenuation constant

$$\gamma = \left.\frac{A(k)}{\frac{\partial D}{\partial k_0}}\right|_{k_0=\omega(\mathbf{k})}. \tag{2.29'}$$

From Eq. (2.22), we can also solve for $\hat{G}$ and obtain

$$\hat{G} = -\frac{1}{2}(\Gamma_r^{-1}\hat{N}_r - \hat{N}_a\Gamma_a^{-1}) = \frac{1}{2}(G_r\hat{N}_r - \hat{N}_aG_a) \tag{2.30}$$

where

$$\begin{aligned} \hat{N}_r &= N\hat{\eta} - i\eta_2 + \hat{\eta}_3 = \hat{\eta}(N + \hat{\eta}_3), \\ \hat{N}_a &= N\hat{\eta} - i\eta_2 - \hat{\eta}_3 = (N - \hat{\eta}_3)\hat{\eta}, \end{aligned} \tag{2.31}$$

and N is a matrix determined by the following equation

$$N\Gamma_a - \Gamma_r N = 2iB, \tag{2.32}$$

or

$$ND - DN = i(NA + AN) - 2iB. \tag{2.32'}$$

At the poles of G_r and G_a, there exists in the system a quasiparticle carrying energy $\omega(\mathbf{k})$. Its Green's function G_F can be expressed in terms of the quasiparticle number density matrix n,

$$G_F = G_r(1 \pm n) \mp nG_a. \tag{2.33}$$

Comparison of Eqs. (2.33) and (2.30) shows the relation between N and particle density n,

$$N|_{\text{at the poles of } G_r} = 1 \pm 2n, \tag{2.34}$$

In terms of n, Eq. (2.32') can be written as

$$nD - Dn = i(nA + An) \pm i(A - B) = i(nA + An) \pm \Gamma_+, \tag{2.35}$$

which is in fact the transport equation satisfied by the particle number density n. After neglecting the noncommutative parts of n and A, the right-hand side of Eq. (2.35) can be written as

$$\pm(1 \pm n)\Gamma_+ - n\Gamma_-$$

which is just the collision term on the right-hand side of the transport equation, Γ_+ (Γ_-) being directly proportional to the rate of emission (absorption) of quasiparticles.

In the following, we consider the case when $\hat{Q}$ is of single component. In an approximately uniform system, the matrices $D(x,y)$, etc., can be written as

$$D\left(X + \frac{1}{2}z,\ X - \frac{1}{2}z\right), \quad X = \frac{1}{2}(x + y), \quad z = (x - y),$$

which are slowly varying functions of X. In the momentum representation of z, expanding in terms of X_μ up to the lowest order of $\partial/\partial X_\mu$, we have, at the poles of G_r,

$$(nD - Dn)(k, X) = -i\left(\frac{\partial D(k, X)}{\partial k_\mu}\frac{\partial n(k, X)}{\partial X^\mu} - \frac{\partial D(k, X)}{\partial X_\mu}\frac{\partial n(k, X)}{\partial k^\mu}\right)$$
$$= -i\frac{\partial D}{\partial k_0}\left(\frac{\partial n}{\partial X_0} + \mathbf{v}\cdot\nabla_X n + \frac{\partial \omega}{\partial X_\mu}\frac{\partial n}{\partial k^\mu}\right). \quad (2.36)$$

In writing down Eq. (2.36), we have used the pole condition [Eq. (2.29)] of G_r to obtain the velocity of the quasiparticle,

$$\mathbf{v} = \nabla_k \omega = -\frac{\nabla_k D}{\frac{\partial D}{\partial k_0}}, \quad (2.37)$$

and

$$\frac{\partial \omega}{\partial X_\mu} = -\frac{\frac{\partial D}{\partial X_\mu}}{\frac{\partial D}{\partial k_0}}.$$

From Eqs. (2.36) and (2.35), we can obtain the transport equation satisfied by the quasiparticle number density in an approximately uniform system:

$$\frac{\partial n}{\partial X_0} + \mathbf{v}\cdot\nabla_X n + \frac{\partial \omega}{\partial X_\mu}\frac{\partial n}{\partial k^\mu} = \frac{1}{\frac{\partial D}{\partial k_0}}\{\pm i\Gamma_+(1 \pm n) - i\Gamma_- n\}. \quad (2.38)$$

In a uniform system, both n and ω do not change with X_μ and the right-hand side of Eq. (2.38) becomes zero so that

$$\frac{1 \pm n}{n} = \frac{\Gamma_-}{\pm \Gamma_+}, \quad (2.39)$$

which is just the Einstein relation of detailed balance.

In order to understand the physical meaning of Γ_+ and Γ_-, let us take a scalar field as an example. Let $\hat{Q}(x)$ be the scalar field $\hat{\varphi}(x)$, which satisfies the equation of motion,

$$(\partial^2 + m^2)\hat{\varphi}(x) = \hat{j}(x). \quad (2.40)$$

It can be easily proved that the second-order vertex function is

$$\Gamma_p(x, y) = (\partial_x^2 + m^2)\delta_p^4(x - y) + \Sigma_p(x, y), \quad (2.41)$$

where $\Sigma_p(x,y)$ is the self-energy part, which can be represented as

$$\Sigma_p(x, y) = -it_r\{T_p(\hat{j}(x)\hat{j}(y))\hat{\rho}\}_{1.P.I.}, \quad (2.42)$$

with $\{\ \}_{1.P.I}$ denoting the one-particle–irreducible part.

Let $|n\rangle$ be a complete set of orthogonal eigenstates of the energy-momentum operators and other operators commuting with them. In a uniform system, because of translational invariance, $\Sigma_p(x - y)$ is a function of $x - y$ only, and, in momentum space, it becomes

$$i\Sigma_-(k) = \int e^{ik\cdot(x-y)}t_r\{\hat{j}(x)\hat{j}(y)\hat{\rho}\}_{1.P.I.}d^4(x - y)$$
$$= \sum_{n,m}|\langle n|\hat{j}(0)|m\rangle|^2_{1.P.I.}\rho_{nn}(2\pi)^4\delta^4(k - p_l + n). \quad (2.43)$$

When $k_0 > 0$, the right-hand side of Eq. (2.43) is directly proportional to the quasiparticles absorption cross section at energy-momentum k. To be more precise,

$$i\Sigma_-(k) = 2|\mathbf{k}|\sigma_{\text{absorption}}(\mathbf{k}) = \frac{2k_0}{(2\pi)^3} W_{\text{absorption}}(k), \tag{2.44}$$

where $W_{\text{absorption}}(k)$ is the probability of quasiparticles absorption per unit time. In the same way it can be shown that, when $k_0 > 0$,

$$i\Sigma_+(k) = \int e^{ik\cdot(x-y)} t_r\{\hat{j}(y)\hat{j}(x)\hat{\rho}\}_{1.P.I.} = \frac{2k_0}{(2\pi)^3} W_{\text{emission}}(k) \tag{2.44'}$$

where $W_{\text{emission}}(k)$ is the probability of quasiparticles emission per unit time.

From Eq. (2.41), one readily verifies that

$$\Gamma_\pm(x, y) = \Sigma_\pm(x, y).$$

Thus $i\Gamma_\pm(k)$ has the same physical meaning as $i\Sigma_\pm(k)$, and Eq. (2.39) can be written in the form of the ordinary relation of detailed balance:

$$(1+n)W_{\text{emission}} = nW_{\text{absorption}}. \tag{2.39'}$$

In the above, we have made rather detailed discussion of the situation of a single-component operator. But we believe that, in the multicomponent case, n still possesses the physical meaning of quasiparticle number density and Eq. (2.35) can be considered as some generalization of the transport equation.

Before closing this section, we want to point out that, because of causality, the retarded Green's function $G_r(x,y)$, in the momentum representation corresponding to the relative coordinate $x-y$, should be analytic in the upper half of the k_0 complex plane. If $\gamma > 0$ in Eq. (2.29'), the amplitude of the quasiparticle attenuates when moving in a dissipative system, and the demand of analyticity is fulfilled automatically. If $\gamma < 0$, the quasiparticle propagates in a proliferous system where its amplitude increase, and we have to take a path of integration to pass over the poles of G_r in the upper half plane from above in order to guarantee the preservation of causality.

III. THE W-T IDENTITIES AND SPONTANEOUS SYMMETRY BREAKING

Suppose that the Lagrangian of the system is globally invariant under a Lie group G; the symmetry leads to a set of W-T identities satisfied by the closed–time-path Green's function. The group G may include the space-time symmetry as a subgroup. Let $\varphi(x)$ be a fundamental field and $Q(x)$, a function of $\varphi(x)$, be the order parameter of interest. There are many components on $\varphi(x)$ and $Q(x)$, which form a unitary representation of the basis of G. Under an infinitesimal transformation of G, $\varphi(x)$ and $Q(x)$ transform as

$$\begin{aligned} \varphi(x) &\to \varphi'(x) = \varphi(x) + \delta\varphi(x), \\ \delta\varphi(x) &= \zeta_\alpha(i\hat{I}^{(0)}_\alpha - x^\mu_\alpha(x)\partial_\mu)\varphi(x) = i\hat{I}_\alpha\varphi(x)\zeta_\alpha, \end{aligned} \tag{3.1}$$

and

$$\begin{aligned} Q(x) &\to Q'(x) = Q(x) + \delta Q(x), \\ \delta Q(x) &= \zeta_\alpha(i\hat{L}^{(0)}_\alpha - x^\mu_\alpha\partial_\mu)Q(x) = i\hat{L}_\alpha Q(x)\zeta_\alpha \end{aligned} \tag{3.2}$$

where ζ_α are the n_G infinitesimal parameters of the group G; $\hat{I}^{(0)}_\alpha$ and $\hat{L}^{(0)}_\alpha$ are the matrix representation of the generators acting on $\varphi(x)$ and $Q(x)$ and are Hermitian. Under the above transformation of G, $x^\mu_\alpha(x)$ and the space-time point x_μ are related by

$$x^\mu \to x^{\mu\prime} = x^\mu + x^\mu_\alpha(x)\zeta_\alpha. \tag{3.3}$$

In the following, we let ζ_α be infinitesimal functions of x. It can then be easily proved that the Lagrangian transforms as

$$\mathscr{L}(\varphi'(x))\frac{d^4x}{d^4x'}=\mathscr{L}(\varphi(x'))+\left(\frac{\delta\mathscr{L}}{\delta\varphi(x)}-\partial_\mu\frac{\delta\mathscr{L}}{\delta\partial_\mu\varphi(x)}\right)\delta\varphi(x)$$
$$+\partial_\mu(j^\mu_\alpha(x)\zeta_\alpha(x)) \tag{3.4}$$

where

$$j^\mu_\alpha(x)=i\frac{\delta\mathscr{L}}{\delta\partial_\mu\varphi(x)}\hat{I}_\alpha\varphi(x)-\mathscr{L}x^\mu_\alpha(x) \tag{3.5}$$

is the current in direction α. If the Lagrangian is globally invariant under the group G, then

$$\partial_\mu j^\mu_\alpha(x)=i\left(\partial_\mu\frac{\delta\mathscr{L}}{\delta\partial_\mu\varphi(x)}-\frac{\delta\mathscr{L}}{\delta\varphi(x)}\right)\hat{I}_\alpha\varphi(x). \tag{3.6}$$

Equation (3.6) indicates that the current $j^\mu_\alpha(x)$ is conserved when $\varphi(x)$ is a solution of the Euler equation. With use of Eq. (3.6), Eq. (3.4) can be written as

$$\mathscr{L}(\varphi'(x))\frac{d^4x}{d^4x'}=\mathscr{L}(\varphi(x'))+j^\mu_\alpha(x)\partial_\mu\zeta_\alpha(x). \tag{3.7}$$

Equation (3.7) gives the transformation of the Lagrangian under a local transformation when the Lagrangian is only globally invariant under G.

We now proceed to derive the W-T identities satisfied by the closed–time-path Green's function, making use of Eq. (3.7).

Employing the method used in field theory, one can readily prove that the generating functional of the closed–time-path Green's functions can be written in the form of a Feynman path integral. Introducing the external sources $J(x)$ and $h(x)$ of $\varphi(x)$ and $Q(x)$, respectively, the generating functional can be represented by

$$Z[J(x),h(x)]=N\int[d\varphi(x)]\exp\{i\int_p[\mathscr{L}(\varphi(x))-J(x)\varphi(x)$$
$$-h(x)Q(x)]d^4x\}\langle\varphi(\mathbf{x},t_+=-\infty)|\hat{\rho}|\varphi(\mathbf{x},t_-=-\infty)\rangle, \tag{3.8}$$

where N is the normalization constant. The only difference from ordinary field theory is that the Feynman path integral has to be integrated along the closed path p, with the boundary conditions at two ends determined by matrix elements of the density matrix $\hat{\rho}$. The variable of integration of Eq. (3.8) is then changed from $\varphi(x)$ to $\varphi'(x)$ which is obtained through a local transformation of the group G, the parameters $\zeta_\alpha(x)$ being infinitesimal functions satisfying the boundary conditions

$$\zeta_\alpha(\mathbf{x},t_\pm=-\infty)=0,$$
$$\lim_{|\mathbf{x}|\to\infty}\zeta_\alpha(\mathbf{x},t)=0. \tag{3.9}$$

Under such a unitary transformation, the measure $[d\varphi(x)]$ does not change, and, through Eq. (3.9), neither do the matrix elements of the density. As a result, we get

$$\partial_\mu j^\mu_\alpha\left(\varphi(x)=i\frac{\delta}{\delta J(x)}\right)Z[J(x),h(x)]=\left[J(x)\hat{I}_\alpha\frac{\delta}{\delta J(x)}\right.$$
$$\left.+h(x)\hat{L}_\alpha\frac{\delta}{\delta h(x)}\right]Z[J(x),h(x)]. \tag{3.10}$$

Let us introduce the generating functionals of connected Green's functions and vertex functions:

$$W[J(x), h(x)] = i \ln Z[J(x), h(x)], \tag{3.11}$$

and

$$\Gamma[\varphi_c(x), Q_c(x)] = W[J(x), h(x)] - \int_p (J(x)\varphi_c(x) + h(x)Q_c(x)d^4x) \tag{3.12}$$

where

$$\varphi_c(x) = \frac{\delta W}{\delta J(x)}, \quad Q_c(x) = \frac{\delta W}{\delta h(x)}. \tag{3.13}$$

With use of the commutation relation

$$i\frac{\delta}{\delta J(x)} Z = Z\left[\varphi_c(x) + i\frac{\delta}{\delta J(x)}\right], \tag{3.14}$$

Eq. (3.10) can be written as

$$\partial_\mu j_a^\mu \left(\varphi_c(x) + i\frac{\delta}{\delta J(x)}\right) = -i[J(x)\hat{I}_a\varphi_c(x) + h(x)\hat{L}_a Q_c(x)]. \tag{3.15}$$

Equation (3.15) is then the desired set of W-T identities. It has the same form as that in ordinary field theory, but here, x can be any point on the closed path p.

With use of the representation of the generating functional of connected Green's functions, Eq. (3.15) can be written as

$$\partial_\mu j_a^\mu \left(\frac{\delta W}{\delta J(x)} + i\frac{\delta}{\delta J(x)}\right) = -i\left[J(x)\hat{I}_a\frac{\delta W}{\delta J(x)} + h(x)\hat{L}_a\frac{\delta W}{\delta h(x)}\right]. \tag{3.16}$$

If we differentiate Eq. (3.16) with respect to $J(x)$ and $h(x)$ and then let $J(x)$ and $h(x)$ approach zero, the W-T identities corresponding to the closed–time-path Green's functions of all orders can be obtained.

With use of the representation of the generating functional of vertex functions, Eq. (3.15) can be written as

$$\partial_\mu j_a^\mu \left(\varphi_c(x) + i\int_p \left(\frac{\delta J(x)}{\delta\varphi_c(y)}\right)^{-1}\frac{\delta}{\delta\varphi_c(y)}\right) = \pm i\left[\frac{\delta\Gamma}{\delta\varphi_c(x)}\hat{I}_a\varphi_c(x)\right] + i\frac{\delta\Gamma}{\delta Q_c(x)}\hat{L}_a Q_c(x). \tag{3.17}$$

In Eq. (3.17), we have assumed that the order parameter $Q(x)$ is a boson operator, although the fundamental fields may also include fermion operators. By differentiating Eq. (3.17) with respect to $\varphi_c(x)$ and $Q_c(x)$, the W-T identities corresponding to vertex functions of all orders can be derived.

Making use of Eq. (3.17), we now discuss the spontaneous breaking of the symmetry, to see the difference between the W-T identities of the closed–time-path Green's functions and those identities of the Green's functions in ordinary field theory. Let us assume that, when the external sources $J(x)$ and $h(x)$ are zero, the equation

$$\frac{\delta\Gamma}{\delta\varphi_c(x)} = 0, \quad \frac{\delta\Gamma}{\delta Q_c(x)} = 0 \tag{3.18}$$

has the constant solution $\varphi_c(x) = 0$ but with $Q_c(x_+) = Q_c(x_-) = Q_{co} \neq 0$. This corresponds to a state with symmetry spontaneously broken. Differentiating Eq. (3.17) with respect to $Q(y)$, letting $\varphi_c = 0$ and $Q_c(x_+) = Q_c(x_-) = Q_{co}(x)$, and then integrating over x along the closed path p, we arrive at

$$0 = \int_p \frac{\delta^2 \Gamma}{\delta Q_c(y) \delta Q_c(x)} \hat{I}_\alpha Q_{c0}(x) d^4x. \tag{3.19}$$

In the single-time representation, it can be written as

$$\int \Gamma_r(y, x) \hat{I}_\alpha Q_{c0}(x) d^4x = 0, \tag{3.20}$$

where

$$\Gamma_r(y, x) = \Gamma_p(y_+, x_+) - \Gamma_p(y_+, x_-) = \Gamma_F(y, x) - \Gamma_+(y, x)$$

is the second-order retarded vertex function. Writing Eq. (3.20) as matrices, we get

$$\Gamma_r \cdot \hat{I}_\alpha Q_{c0} = 0, \tag{3.21}$$

with $\hat{I}_\alpha Q_{c0}$ considered as a vector. Equation (3.21) shows that $\hat{I}_\alpha Q_{c0}$ is the eigenvector of the matrix Γ_r corresponding to eigenvalue zero.

If H is the subgroup which leaves Q_{c0} invariant under the operations of the group G and if H has n_H generators $\hat{I}_\alpha$, with $\alpha = 1,...,n_H$, then we have

$$\hat{I}_\alpha Q_{c0} = 0, \quad \alpha = 1, \ldots, n_H. \tag{3.22}$$

In the coset G/H generated by the subgroup H, we have $\hat{I}_\alpha Q_{c0} \neq 0$, for $\alpha = n_{H+1},...,n_G$. Thus, in Eq. (3.21), there are only $n_G - n_H$ nonzero eigenvectors.

Taking the complex conjugate of Eq. (3.21), we have

$$Q_{c0}^* \hat{I}_\alpha \cdot \Gamma_a = 0, \tag{3.23}$$

where Γ_a is the advanced Green's function. In term of the dispersive part D and the attenuation part A, we get

$$Q_{c0}^* \hat{I}_\beta \cdot D \cdot \hat{I}_\alpha Q_{c0} = Q_{c0}^* \hat{I}_\beta \cdot A \cdot \hat{I}_\alpha Q_{c0} = 0. \tag{3.24}$$

From the discussion in Sec. II, we know that $G_r = \Gamma_r^{-1}$; the zeros of Γ_r are therefore the poles of G_r. Thus, the second-order retarded Green's function G_r has $n_G - n_H$ lossless collective excitations (poles). When $Q_{c0}(x) = Q_{c0}$ is a constant solution, in the momentum representation corresponding to the relative coordinate $x - y$, these collective excitations occur at energy-momentum $p_\mu = 0$ and are therefore the ordinary Goldstone excitations.

Another important question is whether Eq. (3.18) possesses a stable solution for the order parameter,

$$Q_c(x) = Q_0(\mathbf{x}) e^{-i\omega t}. \tag{3.25}$$

When $Q_0(x)$ is nonvanishing only in a given region of space, this solution is called a soliton-type solution. If

$$Q_0(\mathbf{x}) = \exp\{i\mathbf{k} \cdot \mathbf{x}\},$$

then it is called a laser-type solution. Since these solutions lead to the spontaneous breaking of translational invariance (soliton) and phase-shift invariance (laser), G_r possesses poles in the direction of $\nabla_x Q_c(x)$ as a result of the W-T identities. Physically, these poles indicate that, because of the specification of the coordinates of the soliton, the fluctuations of its momentum will diverge; and, in the same way, because of the specification of the phase of the laser, the fluctuations of the particle number will diverge. But we know that momentum and particle number should be conserved and even if a wave packet is formed at a certain time, it cannot remain stable. A stable wave packet leads to poles in G_r, which will in turn result in the divergence of many physical variables (e.g., energy). Since this is not allowed, Eq. (3.18) cannot possess any stable solution. From the above discussion, we learn that it is the quantum effect that leads to the spreading of the wave packet. But this does not rule out the existence of soliton-type and laser-type solutions of the classical equation of the order parameter. Thus we should first solve for these solutions from the classical equation, introduce the corresponding collective coordinates, quantize these coordinates, and then study the related quantum effect. The effect of the W-T identities in laser-type solutions will be discussed in another paper.

IV. THE W-T IDENTITIES IN A GAUGE THEORY[3]

Let G be a compact Lie group, with n_G generators $\hat{I}_i$ which obey the commutation relation

$$[\hat{I}_j, \hat{I}_k] = i f_{jkl} \hat{I}_l \tag{4.1}$$

where f_{jkl} are the structure constants of G, which can be made real and totally antisymmetric by a suitable choice of $\hat{I}_j$. We assume that G is an internal symmetry group which does not affect the space-time coordinates.

If the system is gauge invariant under group G, its fields can be divided into gauge fields $A^l_\mu(x)$ (with $l = 1,\ldots,n_G$ and $\mu = 0,1,2,3$) and matter field $\varphi_a(x)$ (with $a = 1,\ldots,n$). Here, $A^l_\mu(x)$ is a vector field which forms a canonical representation of G while $\varphi_a(x)$ consists of scalar fields and spinors which form unitary representations of G. Let ω_l be the infinitesimal parameters of the group G. Under an infinitesimal transformation

$$g(x) = 1 + i\hat{I}_l \omega_l(x), \tag{4.2}$$

$\varphi_a(x)$ and $A^i_\mu(x)$ transform as

$$\begin{aligned} &\varphi_a(x) \to \varphi'_a(x) = \varphi_a(x) + \delta\varphi_a(x), \\ &\delta\varphi_a(x) = i t^l_{ab} \omega_l(x) \varphi_b(x), \end{aligned} \tag{4.3}$$

$$\begin{aligned} &A^i_\mu(x) \to A^{i\prime}_\mu(x) = A^i_\mu(x) + \delta A^i_\mu(x), \\ &\delta A^i_\mu(x) = f_{ikl} A^k_\mu(x) \omega_l(x) + \partial_\mu \omega_i(x), \end{aligned} \tag{4.4}$$

where t^l_{ab} is the representation matrix of $\hat{I}_l$ on φ_a.

If the system is gauge invariant under the group G, its Lagrangian $L_{\text{inv}}(\varphi_a(x), A^i_\mu(x))$ transforms according to Eqs. (4.3) and (4.4) as

$$L_{\text{inv}}(\varphi'_a(x), A^{i\prime}_\mu(x)) = L_{\text{inv}}(\varphi_a(x), A^i_\mu(x)). \tag{4.5}$$

where $\omega_l(x)$ is any infinitesimal function.

In field theory, we have to fix the gauge condition and quantize[4] the system in that definite gauge. We write the gauge condition as

$$F_l(\varphi_a(x), A^i_\mu(x)) = 0, \quad l = 1, \ldots, n_G, \tag{4.6}$$

and perform the quantization under this condition. This is equivalent to starting off from the following effective Lagrangian:

$$\begin{aligned} L_{\text{eff}}(\varphi_a(x), A^i_\mu(x), c_l(x), \bar{c}_l(x)) &= L_{\text{inv}}(\varphi_a, A^i_\mu) \\ &- \frac{1}{2} F_l(\varphi_a, A^i_\mu) F_l(\varphi_a, A^i_\mu) + \int \bar{c}_l(x) M_{ll'}(x, y) c_{l'}(y) d^4y \end{aligned} \tag{4.7}$$

where

$$\begin{aligned} M_{ll'}(x, y) &= \frac{\delta F_l(\varphi'_a(x), A^{i\prime}_\mu(x))}{\delta \omega_{l'}(y)} = i \frac{\partial F_l}{\partial \varphi_a(x)} t^{l'}_{ab} \varphi_b(x) \delta^4(x - y) \\ &+ \frac{\partial F_l}{\partial A^i_\mu(x)} [f_{ikl'} A^k_\mu(x) \delta^4(x - y) + \partial_\mu \delta^4(x - y) \delta_{il'}], \end{aligned} \tag{4.8}$$

and $c_l(x)$ and $\bar{c}_l(x)$ are the Faddeev-Popov ghost fields which are anticommuting scalars.

As in field theory, in gauge theory we represent the generating functional of the closed–time-path Green's functions in the form of a path integral

$$Z[J_{\mu i}(x), J_a(x), \eta_l(x), \bar{\eta}_l(x)] = N\int [d\varphi_a(x)][dA^i_\mu(x)][dc_l(x)][d\bar{c}_l(x)]$$
$$\times \exp\{i\int_p [L_{\rm eff}(\varphi_a, A^i_\mu, c_l, \bar{c}_l) - J_{\mu i}A^{\mu i} - J_a\varphi_a - \bar{\eta}_l c_l - \eta_l\bar{c}_l]d^4x\}$$
$$\times \langle\psi(\mathbf{x}, t_+ = -\infty)|\hat{\rho}|\psi(\mathbf{x}, t_- = -\infty)\rangle \quad (4.9)$$

where $J_{\mu i}(x)$, $J_a(x)$, $\eta_l(x)$, and $\bar{\eta}_l(x)$ are external sources, η_l and $\bar{\eta}_l$ being anticommuting c-numbers, $\psi(x)$ represents the rest of the fields, and N is the normalization constant. For the fields of all nonphysical particles, $\hat{\rho}$ corresponds to their vacuum state.

Becci, Ronet, and Stora[5] have proved that the effective Lagrangian is invariant under the following global supertransformation (BRS transformation):

$$\begin{aligned}
\delta\varphi_a(x) &= it^l_{ab}\varphi_b(x)c_l(x)\delta\lambda \equiv D^l_a(\varphi_b)c_l(x)\delta\lambda,\\
\delta A^i_\mu(x) &= (f_{ikl}A^k_\mu(x) + \delta_{il}\partial_\mu)c_l(x)\delta\lambda \equiv D^l_{\mu i}(A^k_\nu)c_l\delta\lambda,\\
\delta c_l(x) &= -\frac{1}{2}f_{lik}c_i(x)c_k(x)\delta\lambda,\\
\delta\bar{c}_l(x) &= -F_l[\varphi_a, A^i_\mu]\delta\lambda,
\end{aligned} \quad (4.10)$$

where $\delta\lambda$ is an anticommuting c-number.

Equation (4.10) is the global supertransformation which leaves the effective Lagrangian $L_{\rm eff}$ invariant. We can employ the method in Sec. III to deduce the W-T identities for the corresponding closed–time-path Green's function. The W-T identities are of the same form as those in ordinary field theory except that the space-time coordinates can be any point on the closed path. In the following, we shall only present the results without making any further detailed discussion.

Let us introduce the following new external source term

$$-K_a(x)D^l_a(\varphi_b(x))c_l(x) - K^\mu_j(x)D^l_{\mu j}(A^k_\nu(x))c_l(x) - \frac{1}{2}L_l(x)f_{lik}c_i(x)c_k(x) \quad (4.11)$$

into the Lagrangian, denoting the generating functional of the closed–time-path Green's functions by

$$W[J_a, J^\mu_j, \eta_l, \bar{\eta}_l, K_a, K^\mu_j, L_l] = -i\ln Z[J_a, J^\mu_j, \eta_l, \bar{\eta}_l, K_a, K^\mu_j, L_l] \quad (4.12)$$

and introduce the generating functional of the vertex functions,

$$\Gamma[\varphi_a, A^i_\mu, c_l, \bar{c}_l, K_a, K^\mu_j, L_l] = W[J_a, J^\mu_j, \eta_l, \bar{\eta}_l, K_a, K^\mu_j, L_l]$$
$$-\int_p (J_a\varphi_a + J^\mu_j A^j_\mu + \bar{\eta}_l c_l + \eta_l\bar{c}_l)d^4x, \quad (4.13)$$

where

$$\begin{aligned}
\varphi_a(x) &= \frac{\delta W}{\delta J_a(x)}, \quad A^j_\mu(x) = \frac{\delta W}{\delta J^\mu_j(x)},\\
c_l(x) &= \frac{\delta W}{\delta\bar{\eta}_l(x)}, \quad \bar{c}_l(x) = \frac{\delta W}{\delta\eta_l(x)}.
\end{aligned} \quad (4.14)$$

In the above differentiation, K_a, K^μ_j, and L_l remain invariant.

Under the BRS transformation, the W-T identites can be written as

$$\frac{\delta\Gamma}{\delta K_a(x)}\frac{\delta\Gamma}{\delta\varphi_a(x)} + \frac{\delta\Gamma}{\delta K^\mu_j(x)}\frac{\delta\Gamma}{\delta A^j_\mu(x)} + \frac{\delta\Gamma}{\delta L_l(x)}\frac{\delta\Gamma}{\delta c_l(x)} + F_l\frac{\delta\Gamma}{\delta\bar{c}_l(x)} = 0. \quad (4.15)$$

From Eq. (4.15), the W-T identities corresponding to closed–time-path vertex functions of any order can be easily derived. Their applications will not be discussed here.

[1]J. Schwinger, J. Math. Phys. **2**, 407 (1961); L. V. Keldysh, JETP **20**, 1018 (1965); R. A. Craig, J. Math. Phys. **9**, 605 (1968); R. Mills, *Propagators for many particle systems* (Gordon and Breach, New York, 1969); V. Korenman, Ann. Phys. (N.Y.) **39**, 72 (1966); V. L. Berezinskii, JETP, **26**, 137 (1968); G. Niklasson and A. Sjölander, Ann. Phys. (N.Y.) **49**, 249 (1968); C. P. Enz, *The many body problem* (Plenum, New York, 1969); R. Sandström, Phys. Status Solidi **38**, 683 (1970); C. Caroli, R. Combescot, P. Nozieres, and D. Saint-James, J. Phys. C **4**, 916 (1971); A. G. Ha, J. Phys. A **8**, 214 (1974); Mol. Phys. **28**, 1 (1974).
[2]J. C. Ward, Phys. Rev. **78**, 1824 (1950); Y. Takahashi, Nuovo Cimento **6**, 370 (1957).
[3]A. A. Slavnov, Teor. Math. Fiz. **10**, 153 (1972); J. C. Taylor, Nucl. Phys. **B 33**, 436 (1971).
[4]L. D. Faddeev and V. N. Popov, Phys. Lett. **25B**, 30 (1970); G. 't Hooft, Nucl. Phys. **B33**, 173 (1971); B. W. Lee and J. Zinn-Justin, Phys. Rev. **D 5**, 3121 (1972); **D 5**, 3137 (1972).
[5]C. Becchi, A. Rouet, and R. Stora, Phys. Lett. **52B**, 344 (1974); Ann. Phys. (N.Y.) **98**, 287 (1976).

Translated by King Yuen Ng
Edited by Stanley Wu-Wei Liu

PHYSICAL REVIEW B VOLUME 22, NUMBER 7 1 OCTOBER 1980

Closed time path Green's functions and critical dynamics

Guang-zhao Zhou, Zhao-bin Su, and Bai-lin Hao
Institute of Theoretical Physics, Academia Sinica, Beijing, China

Lu Yu*
Lyman Laboratory, Physics Department, Harvard University, Cambridge, Massachusetts 02138

(Received 30 October 1979)

The closed time path Green's function (CTPGF) formalism is applied to the critical dynamics. The related results for the CTPGF approach are briefly reviewed. Three different forms of CTPGF's are defined, transformations from one to another form and other useful computation rules are given. The path integral presentation of the generating functional for CTPGF's is used to derive the Ward-Takahashi identities under both linear and nonlinear transformations of field variables. The generalized Langevin equations for the order parameters and conserved variables are derived from the vertex functional on the closed time path. The proper form of the equations for the conserved variables, including automatically the mode coupling terms, is determined according to the Ward-Takahashi identities and the linear response theory. All existing dynamic models are reobtained by assuming the corresponding symmetry properties for the system. The effective action for the order parameters is deduced by averaging over the random external field. The Lagrangian formulation of the statistical field theory is obtained if the random field one-loop approximation and the second-order approximation of order-parameter fluctuations on different time branches are both taken. The various possibilities of improving the current theory of critical dynamics within the framework of CTPGF's are discussed. The problem of renormalization for the finite-temperature field theory is considered. The whole theoretical framework is also applicable to systems near the stationary states far from equilibrium, whenever there exists an analog of the potential function ("free energy").

I. INTRODUCTION

The closed time path Green's function (CTPGF) formalism, developed by Schwinger[1] and Keldysh,[2] has been applied to a number of problems.[3] As pointed out by Zhou and Su,[4–6] this technique is quite effective in investigating the nonequilibrium statistical field theory. They used this method to analyze the Goldstone mode in the steady state for nonequilibrium dissipative systems such as unimode lasers in the saturation region.[7] In this article we apply the CTPGF formalism to study systems near equilibrium phase transition point. The complete system of equations for critical dynamics, including automatically the mode coupling terms and the Lagrangian formulation of the field theory are derived in a unified way. This provides a microscopic justification for the semiphenomenological models in critical dynamics and indicates various possibilities for improving the existing theory.

In the vicinity of the phase transition point the long-wave fluctuations dominate. Since the corresponding correlation length is much greater than the thermal wavelength, the quantum effect is irrelevant. However, in the quasiparticle description, such "purely" classical field theory does not correspond to the Boltzmann limit, but approaches the "super-Bose" case, i.e., the quasiparticle distribution $n \propto T/\epsilon$ where T is the absolute temperature (with $\hbar = C = k_B = 1$) and ϵ is the energy for the elementary excitation. Such a statistical field theory (or fluctuation field theory) has very close analogy with the usual quantum field theory.

In our viewpoint, the CTPGF formalism is a natural theoretical framework for studying such statistical field theory. Assuming the equilibrium density matrix for CTPGF we obtain automatically the ordinary quantum field theory for the low-temperature limit ($T << \epsilon$) and the existing static critical phenomena theory for the high-temperature case ($T >> \epsilon$) (see Appendix A). If the high-temperature limit near equilibrium state at the critical point is taken, the complete system of equations to describe the critical dynamics follows naturally, as will be shown later in this paper. By introducing the "response fields," noncommutative with the basic fields, Martin, Siggia, and Rose[8] constructed a classical statistical field theory (the MSR field theory) in close analogy with quantum field theory. As will be shown below the structure of the MSR field theory becomes clearer in the framework of CTPGF's.

In Sec. II we briefly summarize the related results for the CTPGF's, some of which are believed to be new, while others are known or have been published elsewhere.[4–6] A more or less complete list of formulas is given for reference convenience and to make up for the deficiency that papers[4–6] were published only in Chinese. The perturbation theory and the generating functional formalism for CTPGF's are outlined. Three different forms of general multipoint CTPGF's are defined and the transformations from one to another are described. Some computation rules which greatly simplify the usually complicated calculations involved when using the CTPGF are derived. These seem to be highly desirable especially in view of the fact that the technical complexity is one of the causes why the CTPGF approach has not found applications as wide as it deserves. These algebraic identities are shown to be the consequences of some basic properties of the CTPGF, opening new perspectives not inherent in the ordinary Green's-function formalism. The related properties of the two-point functions are outlined. The Feynman path integral presentation for the generating functional of CTPGF's is used to deduce the Ward-Takahashi identities under both linear and nonlinear transformations of the fields.

In Sec. III a short account of the existing theory of critical dynamics is given. The generalized Langevin equation, the mode coupling, and the Lagrangian formulation of the classical field theory are briefly reviewed to facilitate the comparison with the results in subsequent sections.

In Sec. IV the generalized Langevin equation for macrovariables is derived from the equation satisfied by the generating functional for vertex functions in the CTPGF formalism by differentiating the micro- and macro-time scales of variation and averaging over the micro-time scale. In general form this is true for both order parameters and conserved variables. The essential point is to determine the transport coefficient matrix $\gamma^{-1}(t)$ connecting these quantities. The proper form of the equation for conserved variables can be deduced from the Ward-Takahashi identities and the linear-response theory. Comparing this form of equation with the general one yields two blocks of the $\gamma^{-1}(t)$ matrix, one of which couples the conserved variables together and the other couples the conserved variables with the order parameters. The other two blocks of $\gamma^{-1}(t)$, one of which connects the order parameters and the other one connects the order parameters with the conserved variables are determined through symmetry considerations. It is important to emphasize that mode coupling terms appear naturally in these equations. They are not "introduced from the outside," as in the existing theory. Applications of the general theory to particular dynamic models are outlined.

In Sec. V the path integral formulation for the CTPGF's is used to derive the effective action for order parameters. Through Fourier transformation of the path integral the generating functional in the random external fields is introduced. Averaging over random fields yields the effective action, the general properties of which are also discussed. The most plausible trajectories are described by the time dependent Ginzburg-Landau equations, i.e., generalized Langevin equations without random forces. Fluctuations around the most plausible trajectories are considered. There is a possible new way of describing fluctuations in the CTPGF approach arising from the fact that field operators may take different values on positive and negative time branches. It turns out that with the one loop approximation of random fields which is equivalent to the Gaussian averaging, and with second-order fluctuations on different time branches the existing Lagrangian formulation of the classical statistical field theory, i.e, the MSR theory reappears.

In Sec. VI we summarize the main results obtained and discuss the potential possibilities of the CTPGF approach with regard to improving the existing theory of critical dynamics.

In Appendix A the problem of renormalization in the finite-temperature field theory is discussed. It is emphasized that near the phase transition point the leading infrared divergence has to be separated before the ultraviolet renormalization may be carried out. The necessity of using noncommutative operators to describe the time evolution of classical field theory is also discussed.

In Appendix B a proof is given for two theorems of Sec. II dealing with transformations among different forms of CTPGF's. Further useful examples and some technical details are described.

Throughout this paper we deal mainly with the applications of the CTPGF approach to dynamic critical phenomena, but it is clear from the presentation that the whole theoretical framework is also applicable to systems near stationary states far from equilibrium, provided the long-wave fluctuations are dominant.

II. SUMMARY OF THE CTPGF FORMALISM

A. Definitions and generating functionals

For simplicity we shall consider only multicomponent Hermitian Bose fields $\varphi(x)$. Extension to more general cases is obvious. The Lagrangian density can be written as

$$\mathcal{L}=\mathcal{L}_0(\varphi) - V(\varphi) - \varphi(x)J(x) \quad , \qquad (2.1)$$

where $J(x)$ is the external field.

CTPGF for $\varphi(x)$ is defined as

$$G_p(1 \cdots n) = (-i)^{n-1} \mathrm{tr}[T_p(\hat{\varphi}(1) \cdots \hat{\varphi}(n)\hat{\rho}] \ , \tag{2.2}$$

where $\hat{\varphi}(i)$ and $\hat{\rho}$ are the field operators and density matrix in the Heisenberg representation, index p indicates a closed time path consisting of positive $(-\infty, +\infty)$ and negative $(+\infty, -\infty)$ branches. The time variable t can take values on either branch. T_p is the time-ordering operator along the closed time path.

The generating functional for the CTPGF's is defined as

$$Z(J(x)) = \mathrm{tr}\left\{ T_p\left[\exp\left(-i \int_p \hat{\varphi}(x) J(x) \right) \right] \hat{\rho} \right\} , \tag{2.3}$$

where the integral is taken over the closed time path and the integration variable d^4x is omitted. In Eq. (2.3) the external fields on the positive and negative branches $J(x+)$ and $J(x-)$ are assumed to be different.

Taking functional derivatives with respect to $J(x)$ we obtain from Eq. (2.3)

$$G_p(1 \cdots n) = i \left. \frac{\delta^n Z(J(x))}{\delta J(1) \cdots \delta J(n)} \right|_{J=0} . \tag{2.4}$$

In the interaction representation the generating functional (2.3) can be rewritten as

$$Z(J(x)) = \mathrm{tr}\left\{ T_p\left[\exp\left(-i \int_p [V(\varphi_I(x)) + \varphi_I(x) J(x)] \right) \right] \hat{\rho} \right\} , \tag{2.5}$$

where $\varphi_I(x)$ satisfies the Euler equation for the free fields. The interaction term can be taken from behind the trace operator to obtain

$$Z(J(x)) = \exp\left[-i \int_p V\left(i \frac{\delta}{\delta J(x)} \right) \right] \times \mathrm{tr}\left\{ T_p\left[\exp\left(-i \int_p \varphi_I(x) J(x) \right) \right] \hat{\rho} \right\} . \tag{2.6}$$

It is easy to show by generalization of the Wick theorem that

$$T_p\left[\exp\left(-i \int_p \varphi_I(x) J(x) \right) \right] = Z_0(J(x)) : \exp\left[-i \int_p \varphi_I(x) J(x) \right] : , \tag{2.7}$$

where : : means the normal product and $Z_0(J(x))$ is the generating functional for the free field

$$Z_0(J) = \exp\left[-\frac{i}{2} \int \int_p J(x) \Delta_p(x-y) J(y) \right] , \tag{2.8}$$

Δ_p being a free propagator.

Substituting Eq. (2.7) into Eq. (2.6), we obtain

$$Z(J(x)) = \exp\left[-i \int_p V\left(i \frac{\delta}{\delta J(x)} \right) \right] Z_0(J(x)) N(J(x)) \tag{2.9}$$

with

$$N(J(x)) = \mathrm{tr}\left[: \exp\left(-i \int_p \varphi_I(x) J(x) \right) : \hat{\rho} \right] \tag{2.10}$$

as the correlation functional for the initial state. $N(J(x))$ can be expanded into a series of successive cumulants

$$N(J(x)) = \exp[-iW_N(J(x))] \ , \tag{2.11}$$

$$W_N(J(x)) = \sum_{n=1}^{\infty} \frac{1}{n!} \int_p \cdots \int_p N(1 \cdots n) J(1) \cdots J(n) \ , \tag{2.12}$$

where

$$N(1 \cdots n) = (-i)^{n-1} \mathrm{tr}[:\varphi_I(1) \cdots \varphi_I(n):\hat{\rho}]_c \ . \tag{2.13}$$

It is worthwhile to mention that correlation functions give contributions only on the mass shell and that they have the same value on different time branches because there is no time ordering operator T_p in the definition (2.10).

The perturbation theory in the CTPGF approach has the same structure as in the ordinary quantum field theory with the exception that the time integral is taken over the closed path, so every Feynman diagram is decomposed into 2^n diagrams, where n is the number of vertices. The presence of the initial correlations $N(12 \cdots)$ which vanish for the vacuum state constitutes another difference from the ordinary theory. In principle we can take into account all orders of initial correlations, but actually we shall limit ourselves to the second cumulants.

It can be shown quite generally that the counter terms for the usual quantum field theory alone are enough to remove all ultraviolet divergences for the CTPGF's under the reasonable assumptions concerning the initial correlations.[6] We shall not touch this question here, but it should be mentioned that near the phase transition point the infrared singularities have to be separated first, so that the ultraviolet renormalization for the CTPGF's in this case is different from that for the ordinary field theory (see Appendix A). The generating functional for the connected CTPGF's is defined as

$$W(J(x)) = i \ln Z(J(x)) \ , \tag{2.14}$$

$$G_p^c(1 \cdots n) = \left. \frac{\delta^n W}{\delta J(1) \cdots \delta J(n)} \right|_{J=0} = (-i)^{n-1} \langle T_p[\varphi(1) \cdots \varphi(n)] \rangle_c \ . \tag{2.15}$$

where $\langle\ \rangle_c$ stands for $\mathrm{tr}(\cdots\hat{\rho})$, with the connected parts to be taken only.

The normalization condition for the generating functional is

$$Z(J(x))|_{J_+(x)=J_-(x)=J(x)}=1\ , \tag{2.16}$$

$$W(J(x))|_{J_+(x)=J_-(x)=J(x)}=0\ . \tag{2.17}$$

It is essential to point out that this condition does not require $J(x)$ itself to vanish in contrast to the ordinary Green's function formalism. We shall frequently make use of this basic property below.

As in the usual field theory, we perform the Legendre transformation for the generating functional

$$\Gamma(\varphi_c(x))=W(J(x))-\int_p J(x)\varphi_c(x)\ , \tag{2.18}$$

where

$$\varphi_c(x)=\delta W(J(x))/\delta J(x)\ . \tag{2.19}$$

As a consequence of Eq. (2.17), it follows from $J_+(x)=J_-(x)$ that

$$\varphi_{c+}(x)=\varphi_{c-}(x)\ . \tag{2.20}$$

From the definition (2.18) we have

$$\delta\Gamma(\varphi_c)/\delta\varphi_c(x)=-J(x)\ . \tag{2.21}$$

This is the basic equation for the vertex functional, from which the generalized Langevin equation will be derived.

Taking the functional derivative of Eq. (2.19) with respect to $\varphi_c(x)$ and that of Eq. (2.21) with respect to $J(x)$, we obtain

$$\int_p G_p^c(x,y)\Gamma_p(y,z)=-\delta_p^4(x-z)\ ,$$
$$\int_p \Gamma_p(x,y)G_p^c(y,z)=-\delta_p^4(x-z)\ , \tag{2.22}$$

where the two-point vertex function

$$\Gamma_p(12)\equiv\frac{\delta^2\Gamma(\varphi_c(x))}{\delta\varphi_c(1)\delta\varphi_c(2)}\ . \tag{2.23}$$

Actually Eq. (2.22) is the Dyson equation for the CTPGF's, from which the kinetic equation for the distribution N and the energy spectrum and dissipation for quasiparticles can be deduced.[3-5] Here $\delta_p(x-y)$ is the δ function on the closed time path. It is defined for arbitrary functions along the closed time path that

$$\int_p f(y)\delta_p(y-x)=f(x)\ , \tag{2.24}$$

where x can take values on either branch of time.

Up to now we have considered CTPGF's for the basic fields $\varphi(x)$, but all the above statements about φ can be repeated word for word for all the composite operators $Q(\varphi(x))$.

B. Transformations of three sets of CTPGF

In the CTPGF approach we have to deal with three different forms of functions:

(a) Functions on the closed time path $G_p(12\cdots n)$ with subscript p, which appear under the integrals over the closed time path and are used for a concise writing of formulas. (b) The tensor functions $\hat{G}(12\cdots n)$ with time arguments on positive or negative time branches which appear under the integrals over the single time axis $(-\infty,+\infty)$ and are used for constructing the perturbation theory [in what follows we shall specify them by the Greek subscripts $G_{\alpha\beta\cdots\rho}(12\cdots n)$ with $\alpha,\beta\cdots=\pm$, etc.]. (c) The retarded, advanced, and correlation functions, representing the physical quantities $\tilde{G}(12\cdots n)$ which will be denoted by the Latin subscripts $\tilde{G}_{ij\cdots}(12\cdots)$ with $_{ij\cdots}=1,2$. Either of tensors $\hat{G}$ and $\tilde{G}$ has 2^n components.

Some of the relationships among these functions were given before,[2-4] but our main point is to generalize them to the multipoint function case.

To start with the transformations we specify first our notation. The Pauli matrices are written as

$$\sigma_1=\begin{pmatrix}0&1\\1&0\end{pmatrix},\quad \sigma_2=\begin{pmatrix}0&-i\\i&0\end{pmatrix},\quad \sigma_3=\begin{pmatrix}1&0\\0&-1\end{pmatrix}\ .$$

σ_3 will appear frequently together with $\hat{G}$ and σ_1 with $\tilde{G}$. The real orthogonal matrix

$$Q=\frac{1}{\sqrt{2}}(1-i\sigma_2)=\frac{1}{\sqrt{2}}\begin{pmatrix}1&-1\\1&1\end{pmatrix}\ ,$$
$$Q^T=Q^{-1}=\frac{1}{\sqrt{2}}\begin{pmatrix}1&1\\-1&1\end{pmatrix} \tag{2.25}$$

is used for the transformations between $\hat{G}$ and $\tilde{G}$.

The multipoint step function Θ is defined as

$$\Theta(1,2,\cdots n)=\begin{cases}1, & \text{if } t_1>t_2\cdots>t_n\ ,\\ 0, & \text{otherwise}\ .\end{cases} \tag{2.26}$$

It is the product of the two-point step functions

$$\Theta(1,2,\cdots n)=\Theta(1,2)\Theta(2,3)\cdots\Theta(n-1,n) \tag{2.27}$$

and can be used to define the T product

$$T(\varphi(1)\varphi(2)\cdots\varphi(n))=\sum_{\mathscr{P}_n}\Theta(p_1,p_2\cdots p_n)\varphi(p_1)\varphi(p_2)\cdots\varphi(p_n)\ , \tag{2.28}$$

where summation goes over all possible permutations.

The step functions satisfy some relations such as

the normalization condition

$$\sum_{\mathcal{P}_n} \Theta(p_1, p_2 \cdots p_n) = 1 \tag{2.29}$$

and the sum rule ($n > m$)

$$\Theta(1,2,\cdots m) = \sum_{P_n(1,2\cdots m)} \Theta(p_1,p_2\cdots p_n) , \tag{2.30}$$

where $P_n(1,2 \cdots m)$ means p_n with 1 preceding 2, 2 preceding 3, etc.

Going from G_p to $\hat{G}$ we have only to assign separately the "+" or "−" subscript in accordance to the value of the time argument. Take two-point function for example we have

$$\hat{G}(12) = \begin{bmatrix} G_{++}(12) & G_{+-}(12) \\ G_{-+}(12) & G_{--}(12) \end{bmatrix} = \begin{bmatrix} G_F & G_+ \\ G_- & G_{\tilde{F}} \end{bmatrix} \tag{2.31}$$

with

$$G_F(12) = -i\langle T(\varphi(1)\varphi(2))\rangle = i\frac{\delta^2 Z(J)}{\delta J_+(1)\delta J_+(2)}\bigg|_{J=0} ,$$

$$G_+(12) = -i\langle \varphi(2)\varphi(1)\rangle = i\frac{\delta^2 Z(J)}{\delta J_+(1)\delta J_-(2)}\bigg|_{J=0} ,$$

$$G_-(12) = -i\langle \varphi(1)\varphi(2)\rangle = i\frac{\delta^2 Z(J)}{\delta J_-(1)\delta J_+(2)}\bigg|_{J=0} , \tag{2.32}$$

$$G_{\tilde{F}}(12) = -i\langle \tilde{T}(\varphi(1)\varphi(2))\rangle = i\frac{\delta^2 Z(J)}{\delta J_-(1)\delta J_-(2)}\bigg|_{J=0} ,$$

where $\tilde{T}$ is the inverse time ordering operator.

The transformation proposed by Keldysh[2] for two-point functions

$$\tilde{G}(12) = Q\hat{G}(12)Q^{-1} \tag{2.33}$$

if written in components

$$G_{ij}(12) = Q_{i\alpha}Q_{j\beta}G_{\alpha\beta}(12) \tag{2.34}$$

can be generalized directly to the multipoint case

$$G_{i_1 i_2 \cdots i_n}(12\cdots n) = 2^{n/2-1}Q_{i_1\alpha_1}Q_{i_2\alpha_2}\cdots Q_{i_n\alpha_n}G_{\alpha_1\cdots\alpha_n}(12\cdots n) . \tag{2.35}$$

The inverse transformation is given by

$$G_{\alpha_1\alpha_2\cdots\alpha_n}(12\cdots n) = 2^{1-n/2}Q^T_{\alpha_1 i_1}Q^T_{\alpha_2 i_2}\cdots Q^T_{\alpha_n i_n}G_{i_1 i_2\cdots i_n}(12\cdots n) . \tag{2.36}$$

We shall see below that Eq. (2.35) contains all possible retarded, advanced, and correlation functions, associated directly with the physical quantities. The expediency of such choice of numerical coefficient becomes clearer somewhat later.

The specific features of the CTPGF's in the form of tensors can be characterized by two theorems, proof of which and more involved examples will be postponed to the Appendix B.

a. Theorem 1. The component of $\tilde{G}$ with all subscripts equal to 1 vanishes, i.e.,

$$\tilde{G}_{11\cdots 1}(12\cdots n) = 0 . \tag{2.37}$$

As consequences of this theorem for one-point and two-point functions we have

$$G_1(x) = 0,\quad G_+(x) = G_-(x) ,\quad G_{11}(xy) = 0,\quad G_{++}(xy) + G_{--}(xy) = G_{+-}(xy) + G_{-+}(xy) \tag{2.38}$$

or

$$G_F + G_{\tilde{F}} = G_+ + G_- . \tag{2.39}$$

b. Theorem 2. The other components of $\tilde{G}$ can be expressed as

$$\tilde{G}_{2\cdots 2(k),1\cdots 1(n-k)}(12\cdots n) = (-i)^{n-1}\sum_{P\begin{vmatrix}12\cdots n\\ p_1p_2\cdots p_n\end{vmatrix}}{}' \Theta(p_1p_2\cdots p_n)\langle((\cdots(\phi(p_1),\phi(p_2)),\cdots),\phi(p_n))\rangle ,$$

where

$$(\ldots,\varphi(p_i)) = \begin{cases} [\ ,\varphi(p_i)], & \text{if } k+1 \le p_i \le n \\ \{\ ,\varphi(p_i)\}, & \text{if } 1 \le p_i \le k , \end{cases} \tag{2.40}$$

the prime over summation indicating that the cases $k+1 \le p_i \le n$ are excluded from the possible permutations. The component $\tilde{G}_{2\cdots 2}(1\cdots n)$ corresponds to the case $n = k$. All the other components of $\tilde{G}$ are defined as the results of the symmetry properties of the CTPGF's

$$G_{\cdots 1\cdots 2\cdots}(\cdots i\cdots j\cdots) = G_{\cdots 2\cdots 1\cdots}(\cdots j\cdots i\cdots) . \tag{2.41}$$

For the two-point functions we obtain

$$G_{21}(xy) \equiv G_r(xy) = -i\Theta(t_x,t_y)\langle[\varphi(x),\varphi(y)]\rangle ,$$

$$G_{12}(xy) \equiv G_a(xy) = -i\Theta(t_y,t_x)\langle[\varphi(y),\varphi(x)]\rangle ,$$

$$G_{22}(xy) \equiv G_c(xy) = -i\langle\{\varphi(x),\varphi(y)\}\rangle , \quad (2.42)$$

or in the matrix form

$$\tilde{G}(xy) = \begin{bmatrix} 0 & G_a \\ G_r & G_c \end{bmatrix} . \quad (2.43)$$

Making use of Eq. (2.39), we obtain

$$G_r = G_F - G_+ = G_- - G_{\tilde{F}} ,$$

$$G_a = G_F - G_- = G_+ - G_{\tilde{F}} , \quad (2.44)$$

$$G_c = G_F + G_{\tilde{F}} = G_+ + G_- .$$

The inverse transformation is given by Eq. (2.36) and can be written as

$$\hat{G}(xy) = \tfrac{1}{2}G_c(xy)\begin{bmatrix} 1 & 1 \\ 1 & 1 \end{bmatrix} + \tfrac{1}{2}G_r(xy)\begin{bmatrix} 1 & -1 \\ 1 & -1 \end{bmatrix} + \tfrac{1}{2}G_a(xy)\begin{bmatrix} 1 & 1 \\ -1 & -1 \end{bmatrix} . \quad (2.45)$$

The first theorem is valid for both connected and disconnected CTPGF's while the second one in general form is applicable only to the disconnected functions. Further details will be given in Appendix B.

These relationships among different forms of the CTPGF's will be found quite useful in the applications. Here we shall illustrate them by several simple examples and derive some additional computational rules.

The simplest case for the connection of the CTPGF's "in sequence" is the integral of two single point functions

$$\int_p J_p(x)\varphi_p(x) = \int_{-\infty}^{\infty} \hat{J}\sigma_3\hat{\varphi} = 2\int_{-\infty}^{\infty} \tilde{J}\sigma_1\tilde{\varphi} , \quad (2.46)$$

where $\hat{J} = (J_+, J_-)$, $\hat{\varphi}^T = (\varphi_1, \varphi_2)$, etc. For short, in what follows we shall omit the symbol of integration. The integral for G_p can be understood only in the coordinate presentation, while that for $\hat{G}$ and $\tilde{G}$ can be written in both coordinate and momentum space.

The linear response to the external source can be written as

$$R_p(1) = G_p(12)J_p(2)$$

or

$$\hat{R}(1) = \hat{G}(12)\sigma_3\hat{J}(2) \quad (2.47)$$

or

$$\tilde{R}(1) = \tilde{G}(12)\sigma_1\tilde{J}(2) .$$

If we take $J_+(x) = J_-(x) = J(x)$, then

$$\begin{bmatrix} R_1 \\ R_2 \end{bmatrix} = \begin{bmatrix} 0 \\ G_r J \end{bmatrix} , \quad (2.48)$$

where the retarded function does appear naturally.

For the product of two-point functions we obtain

$$D_p(12) = A_p(13)B_p(32) ,$$

$$\hat{D}(12) = \hat{A}(13)\sigma_3\hat{B}(32) , \quad (2.49)$$

$$\tilde{D}(12) = \tilde{A}(13)\sigma_1\tilde{B}(32) ,$$

or in components

$$\begin{bmatrix} 0 & D_a \\ D_r & D_c \end{bmatrix} = \begin{bmatrix} 0 & A_aB_a \\ A_rB_r & A_rB_c + A_cB_a \end{bmatrix} \quad (2.50)$$

This rule can be generalized to the multiple product with

$$Z_p = A_p^{(1)}A_p^{(2)} \cdots A_p^{(n)} ,$$

$$\hat{Z} = \hat{A}^{(1)}\sigma_3\hat{A}^{(2)} \cdots \sigma_3\hat{A}^{(n)} , \quad (2.51)$$

$$\tilde{Z} = \tilde{A}^{(1)}\sigma_1\tilde{A}^{(2)} \cdots \sigma_1\tilde{A}^{(n)} .$$

The last equation can be written in components as

$$Z_r = A_r^{(1)}A_r^{(2)} \cdots A_r^{(n)} ,$$

$$Z_a = A_a^{(1)}A_a^{(2)} \cdots A_a^{(n)} , \quad (2.52)$$

$$Z_c = \sum_{k=1}^{n} A_r^{(1)} \cdots A_r^{(k-1)}A_c^{(k)}A_a^{(k+1)} \cdots A_a^{(n)} .$$

Similarly, by use of the inverse transformation (2.36) and (2.45) we obtain

$$Z_\mu = \sum_{k=1}^{n} A_r^{(1)} \cdots A_r^{(k-1)}A_\mu^{(k)}A_a^{(k+1)} \cdots A_a^{(n)} , \quad (2.53)$$

where $\mu = +-$ or $-+$.

If the multipoint functions stand under the integration, attention has to be paid to the order of the variables. For example the three-point vertex function

$$\Gamma_p(123) = i\Gamma_p(14)\Gamma_p(25)\Gamma_p(36)G_p(456)$$

becomes

$$\hat{\Gamma}(123) = i(\hat{\Gamma}\sigma_3)(14)(\hat{\Gamma}\sigma_3)(25)(\hat{\Gamma}\sigma_3)(36)\hat{G}(456) \quad (2.54)$$

and then

$$\tilde{\Gamma}(123) = i(\tilde{\Gamma}\sigma_1)(14)(\tilde{\Gamma}\sigma_1)(25)(\tilde{\Gamma}\sigma_1)(36)\tilde{G}(456)$$

or in components

$$\Gamma_{111} = 0 ,$$

$$\Gamma_{211} = i\Gamma_r\Gamma_a\Gamma_a G_{211} ,$$

$$\Gamma_{221} = i(\Gamma_c\Gamma_r\Gamma_a G_{121} + \Gamma_r\Gamma_c\Gamma_a G_{211} + \Gamma_r\Gamma_r\Gamma_a G_{221})$$

No additional numerical coefficient appears after transformation in any inherent to the theory relations among the multipoint CTPGF's. For example, the four-point vertex function $\Gamma_p^{(4)}$ is related to the amputated Green's functions W_p as[9]

$$\Gamma_p^{(4)} = -W_p^{(4)} + 3W_p^{(3)} G_p^{(2)} W_p^{(3)} .$$

It can be transformed into

$$\hat{\Gamma}^{(4)} = -\hat{W}^{(4)} + 3\hat{W}^{(3)} \sigma_3 \hat{G}^{(2)} \sigma_3 \hat{W}^{(3)}$$

and

$$\tilde{\Gamma}^{(4)} = -\tilde{W}^{(4)} + 3\tilde{W}^{(3)} \sigma_1 \tilde{G}^{(2)} \sigma_1 \tilde{W}^{(3)} .$$

Contrary to this, a numerical factor may appear in some relations obtained by an artificial contraction. For example, the relation

$$A_p(12) = B_p(134) C_p(342)$$

becomes

$$\tilde{A}(12) = \tfrac{1}{2} \tilde{B}(134)(\sigma_1)_{33'}(\sigma_1)_{44'} \tilde{C}(3'4'2)$$

with the coefficient $\frac{1}{2}$. These examples justify the choice of the numerical constant $2^{n/2-1}$ in the transformation formula from $\hat{G}$ to $\tilde{G}$ [Eq. (2.35)].

The δ function on the closed time path $\delta_p(x-y)$ can be written in the matrix form as

$$\hat{\delta}(x-y) = \delta(x-y)\sigma_3 \tag{2.55}$$

after transformation it becomes

$$\tilde{\delta}(x-y) = Q\hat{\delta}(x-y)Q^{-1} = \delta(x-y)\sigma_1 . \tag{2.56}$$

They are the time derivatives for the step functions on the closed time path, which in the matrix form are

$$\hat{\Theta}(12) = \begin{bmatrix} \Theta(12) & 0 \\ 1 & \Theta(21) \end{bmatrix} \tag{2.57}$$

and

$$\tilde{\Theta}(12) = \begin{bmatrix} 0 & -\Theta(21) \\ \Theta(12) & 1 \end{bmatrix} . \tag{2.58}$$

The Dyson equation for the CTPGF's Eq. (2.22) can be rewritten as

$$\hat{G}\sigma_3\hat{\Gamma} = \hat{\Gamma}\sigma_3\hat{G} = -\hat{\delta} , \quad \tilde{G}\sigma_1\tilde{\Gamma} = \tilde{\Gamma}\sigma_1\tilde{G} = -\tilde{\delta} . \tag{2.59}$$

It is interesting to note that all characteristic features of $\hat{G}$ and $\tilde{G}$ are "transmitted" automatically to $\hat{\Gamma}$ and $\tilde{\Gamma}$. In particular, we have

$$\Gamma_{11} = 0, \quad \Gamma_{++} + \Gamma_{--} = \Gamma_{+-} + \Gamma_{-+} , \tag{2.60}$$

$$\tilde{\Gamma} = \begin{bmatrix} 0 & \Gamma_a \\ \Gamma_r & \Gamma_c \end{bmatrix} , \tag{2.61}$$

$$\Gamma_a = -G_a^{-1}, \quad \Gamma_r = -G_r^{-1}, \quad \Gamma_c = G_r^{-1} G_c G_a^{-1} . \tag{2.62}$$

It is easy to show from the symmetry properties that for the Hermitian Bose field

$$\hat{G} = (\hat{G})^T = -\sigma_1 \hat{G}^* \sigma_1 = -\sigma_1 \hat{G}^\dagger \sigma_1 , \tag{2.63a}$$

which after the Fourier transformation becomes

$$\hat{G}(k) = [\hat{G}(-k)]^T = -\sigma_1 \hat{G}^*(-k) \sigma_1 = -\sigma_1 \hat{G}^\dagger(k) \sigma_1 \tag{2.63b}$$

or

$$\tilde{G}(k) = \tilde{G}^T(-k) = -\sigma_3 \tilde{G}^*(-k) \sigma_3 = -\sigma_3 \tilde{G}^\dagger(k) \sigma_3 , \tag{2.63c}$$

where T means transposition, $*$ complex conjugation, and $\dagger$ Hermitian conjugation. All these properties are transmitted to $\hat{\Gamma}$ and $\tilde{\Gamma}$ through Eq. (2.59). Similarly, the specific features of multipoint Green's functions are conveyed to the corresponding vertex functions through relations like Eq. (2.54).

The transitivity of the CTPGF's also holds for some connection "in parallel," i.e., the product of several CTPGF's connecting two points, which itself is the constituent of a Green's function. Consider for example

$$S_p(12) \equiv G_p^3(12) . \tag{2.64}$$

It may be a self-energy part of G_p. In fact by use of Eq. (2.39) and $G_r(12)G_a(12) = 0$ we obtain

$$G_{++}^3(12) + G_{--}^3(12) = G_{+-}^3(12) + G_{-+}^3(12) . \tag{2.65}$$

Moreover, the matrix

$$\tilde{S}(12) = \frac{1}{4} \begin{bmatrix} 0 & G_a(G_a^2 + 3G_c^2) \\ G_r(G_r^2 + 3G_c^2) & G_c[G_c^2 + 3(G_r + G_a)^2] \end{bmatrix} \tag{2.66}$$

behaves much like a simple G.

C. Further properties of two-point functions

By use of Eqs. (2.59)–(2.61) and (2.36) the two point vertex function $\hat{\Gamma}$ can be presented as

$$\hat{\Gamma} = -iB(I + \sigma_i) - A\sigma_2 - D\sigma_3 \tag{2.67}$$

with

$$B = \tfrac{1}{2} i(\Gamma_F + \Gamma_{\tilde{F}}) = \tfrac{1}{2} i(\Gamma_+ + \Gamma_-) , \tag{2.68a}$$

$$D = \tfrac{1}{2}(\Gamma_{\tilde{F}} - \Gamma_F) = -\tfrac{1}{2}(\Gamma_r + \Gamma_a) , \tag{2.68b}$$

$$A = \tfrac{1}{2} i(\Gamma_- - \Gamma_+) = \tfrac{1}{2} i(\Gamma_r - \Gamma_a) , \tag{2.68c}$$

where B, D, and A, are Hermitian matrices in the multicomponent fields or in the coordinate presentation. Equations (2.68b) and (2.68c) can be rewritten as

$$\Gamma_r = -G_r^{-1} = -D - iA \ , \qquad (2.69a)$$

$$\Gamma_a = -G_a^{-1} = -D + iA \ . \qquad (2.69b)$$

We shall call D the dispersion part, which determines the energy spectrum of the quasiparticle, and call A the dissipative part, which describes the decay rate of the elementary excitation.

The solution of the Dyson equation (2.59) can be presented as

$$\hat{G} = -\tfrac{1}{2}(\Gamma_r^{-1}\hat{N}_r - \hat{N}_a\Gamma_a^{-1}) = \tfrac{1}{2}(G_r\hat{N}_r - \hat{N}_aG_a) \ , \qquad (2.70)$$

with

$$\hat{N}_r = (I+\sigma_1)(N+\sigma_3), \ \hat{N}_a = (N-\sigma_3)(I+\sigma_1) \ , \qquad (2.71)$$

where N is a matrix satisfying the equation

$$N\Gamma_a - \Gamma_r N = 2iB \ , \qquad (2.72a)$$

or

$$ND - DN = i(NA + AN) - 2iB \ . \qquad (2.72b)$$

The causal propagator of the quasiparticle with energy $\epsilon(\vec{k})$, corresponding to the pole of G_r and G_a, can be presented through the density operator n as

$$G_F = G_r(1+n) - nG_a \ . \qquad (2.73)$$

By comparison with Eq. (2.70) we obtain

$$N|_{\text{pole}} = 1 + 2n \ . \qquad (2.74)$$

In terms of n Eq. (2.72b) becomes

$$nD - Dn = i(nA + An) + i(A - B) = i(nA + An) + \Gamma_+ \qquad (2.75)$$

which is the kinetic equation for the quasiparticle density n.[5] The right-hand side of Eq. (2.75) can be presented as

$$(1+n)\Gamma_+ - n\Gamma_- \ , \qquad (2.76)$$

if the noncommutativity of n and A is ignored. It corresponds to the collision term in the kinetic equation which vanishes in the thermal equilibrium; i.e.,

$$\frac{\Gamma_-}{\Gamma_+} = \frac{1+n}{n} = e^{\beta\epsilon(\vec{k})} \ . \qquad (2.77)$$

It can be shown[5] that $i\Gamma_\pm = i\Sigma_\pm$ where $\Sigma_\pm$ is the proper self-energy part, which itself is proportional to the probability of emission (+) or absorption (−) of the quasiparticles per unit time so

$$i\Gamma_\pm > 0 \ . \qquad (2.78)$$

Equation (2.77) shows that in thermal equilibrium the probability of absorption is greater than that of emission as expected.

D. Path integral presentation and Ward-Takahashi identities

Suppose the Lagrangian of the system is globally invariant under the Lie group G which may contain the space-time symmetry group as its subgroup. Let $\varphi(x)$ be the basic fields, $Q(x)$ order parameters, which are functions of $\varphi(x)$. Both $\varphi(x)$ and $Q(x)$ have several components forming the bases of the unitary representation for G.

Under the infinitesimal transformations of G

$$\varphi(x) \to \varphi'(x) = \varphi(x) + \delta\varphi(x) \ ,$$
$$\delta\varphi(x) = \zeta_\alpha[i\hat{I}_\alpha^0 - X_\alpha^\mu(x)\partial_\mu]\varphi(x) = i\hat{I}_\alpha\varphi(x)\zeta_\alpha \ , \qquad (2.79)$$

and

$$Q(x) \to Q'(x) = Q(x) + \delta Q(x) \ ,$$
$$\delta Q(x) = \zeta_\alpha[i\hat{L}_\alpha^0 - X_\alpha^\mu(x)\partial_\mu]Q(x) = i\hat{L}_\alpha Q(x)\zeta_\alpha \ , \qquad (2.80)$$

where ζ_α are a total of n_G infinitesimal parameters for group G and $\hat{I}_\alpha^0$, $\hat{L}_\alpha^0$ are Hermitian representation matrices for the generators of G. $X_\alpha^\mu(x)$ are associated with the transformations of coordinates

$$X^\mu \to X^{\mu\prime} = X^\mu + X_\alpha^\mu(x)\zeta_\alpha \ . \qquad (2.81)$$

It can be shown easily that the Lagrangian function transforms as

$$\mathcal{L}(\varphi'(x))\frac{d^4x}{d^4x'} = \mathcal{L}(\varphi(x')) + \left[\frac{\delta\mathcal{L}}{\delta\varphi(x)} - \partial_\mu\frac{\delta\mathcal{L}}{\delta\partial_\mu\varphi(x)}\right]\delta\varphi(x) + \partial_\mu[j_\alpha^\mu(x)\zeta_\alpha(x)] \ , \qquad (2.82)$$

where

$$j_\alpha^\mu(x) = i\frac{\delta\mathcal{L}}{\delta\partial_\mu\varphi(x)}\hat{I}\varphi(x) + \mathcal{L}X_\alpha^\mu(x) \qquad (2.83)$$

is the current in direction α. If the Lagrangian is invariant under the global transformation of G it follows that

$$\partial_\mu j_\alpha^\mu(x) = i\left[\partial_\mu\frac{\delta\mathcal{L}}{\delta\partial_\mu\varphi(x)} - \frac{\delta\mathcal{L}}{\delta\varphi(x)}\right]\hat{I}_\alpha\varphi(x) \ . \qquad (2.84)$$

The Eq. (2.84) shows that the currents $j_\alpha^\mu(x)$ are conserved if $\varphi(x)$ is the solution of the Euler-Lagrangian equation. By use of Eq. (2.84), Eq. (2.82) can be rewritten as

$$\mathcal{L}(\varphi'(x))\frac{d^4x}{d^4x'} = \mathcal{L}(\varphi(x')) + j_\alpha^\mu(x)\partial_\mu\zeta_\alpha(x) \qquad (2.85)$$

This is the transformation of the Lagrangian under

the local action of G, if it is invariant under the global action of the same group.

The generating functional for the CTPGF's can be presented in the form of a Feynman path integral by the well-known procedure in the field theory

$$Z(h(x),J(x)) = N\int [d\varphi(x)]\exp\left\{i\int_p [\mathcal{L}(\varphi(x)) - h(x)\varphi(x) - J(x)Q(x)]\right\}\langle\varphi(\vec{x},t_+=-\infty|\hat{\rho}|\varphi(\vec{x},t_-=-\infty)\rangle \tag{2.86}$$

N being the normalization constant. What is different from the path integral in the ordinary field theory is that the integration is carried out over the closed time path and that the boundary conditions are determined by the density matrix $\hat{\rho}$.

Transforming integration variable in Eq. (2.86) from $\varphi(x)$ to $\varphi'(x)$ under the local action of group G with infinitesimal parameters $\zeta_\alpha(x)$, satisfying boundary conditions

$$\zeta_\alpha(\vec{x},t_\pm=-\infty)=0\ ,\quad \lim_{|\vec{x}|\to\infty}\zeta_\alpha(\vec{x},t)=0\ , \tag{2.87}$$

taking into account that the measure $[d\varphi(x)]$ does not change under the unitary transformation and that the matrix element of $\hat{\rho}$ remains the same as a result of Eq. (2.87), we obtain

$$\partial_\mu j_\alpha^\mu\left\{\varphi(x)=i\frac{\delta}{\delta h(x)}\right\}Z(h(x),J(x))=\left\{h(x)\hat{I}_\alpha\frac{\delta}{\delta h(x)}+J(x)\hat{L}_\alpha\frac{\delta}{\delta J(x)}\right\}Z(h(x),J(x))\ . \tag{2.88}$$

By use of the commutation relation

$$i\frac{\delta}{\delta h(x)}Z=Z\left\{\varphi_c(x)=i\frac{\delta}{\delta h(x)}\right\} \tag{2.89}$$

Eq. (2.88) can be rewritten as

$$\begin{aligned}\partial_\mu\langle j_\alpha^\mu(x)\rangle &= \partial_\mu j_\alpha^\mu[\varphi_c(x)+i\delta/\delta h(x)]\\ &= -i[h(x)\hat{I}_\alpha\varphi_c(x)+J(x)\hat{L}_\alpha Q_c(x)]\ .\end{aligned} \tag{2.90}$$

This is the required Ward-Takahashi identity which has the same form as in the usual field theory, but here x can take arbitrary value on the closed time path.

Introducing the generating functional for the connected CTPGF's

$$W(h(x),J(x))=i\ln Z(h(x),J(x))\ , \tag{2.91}$$

and the vertex function

$$\Gamma(\varphi_c(x),Q_c(x))=W(h(x),J(x)) - \int_p [h(x)\varphi_c(x)+J(x)Q_c(x)]\ , \tag{2.92}$$

where

$$\varphi_c(x)=\delta W/\delta h(x),\quad Q_c(x)=\delta W/\delta J(x)\ , \tag{2.93}$$

we obtain from Eq. (2.90)

$$\partial_\mu j_\alpha^\mu\left\{\frac{\delta W}{\delta h(x)}+i\frac{\delta}{\delta h(x)}\right\} = -i\left\{h(x)\hat{I}_\alpha\frac{\delta W}{\delta h(x)}+J(x)\hat{L}_\alpha\frac{\delta W}{\delta J(x)}\right\} \tag{2.94}$$

and

$$\partial_\mu j_\alpha^\mu\left[\varphi_c(x)+i\int_p\left[\frac{\delta h(x)}{\delta\varphi_c(y)}\right]^{-1}\frac{\delta}{\delta\varphi_c(y)}\right] = i\left\{\frac{\delta\Gamma}{\delta\varphi_c(x)}\hat{I}_\alpha\varphi_c(x)+\frac{\delta\Gamma}{\delta Q_c(x)}\hat{L}_\alpha Q_c(x)\right\}\ . \tag{2.95}$$

Taking derivatives with respect to $h(x)$, $J(x)$ in Eq. (2.94) and then putting them to zero, we obtain successive WT identities for all orders of CTPGF's. The similar procedure in Eq. (2.95) with respect to $\varphi_c(x)$, $Q_c(x)$ will yield WT identities for the vertex functions.

The equations for the vertex functional Γ in the vanishing external field

$$\delta\Gamma/\delta\varphi_c(x)=0,\quad \delta\Gamma/\delta Q_c(x)=0 \tag{2.96}$$

can be used to discuss the spontaneous symmetry breaking and the Goldstone mode.[5] It is worthwhile to note that with the fluctuation effects being taken into account, the Eq. (2.96) does not have stable solitonlike solution

$$Q_c(x)=Q_0(\vec{x})e^{-i\omega t}\ , \tag{2.97}$$

where $Q_0(\vec{x})$ is different from zero in a limited domain of space, or the laser type solution with

$$Q_0(\vec{x})=e^{i\vec{k}\cdot\vec{x}}\ . \tag{2.98}$$

Up to now we have considered only the linear transformations of fields under the action of symmetry group. In critical dynamics the nonlinear transformations are also needed.

Suppose $\varphi_i(x)$ are basic fields, transforming under

the action of an internal symmetry group G (i.e., the space-time coordinates are not involved) like

$$\varphi_i(x) \to \varphi_i' = \varphi_i + A_{i\alpha}(\varphi)\zeta_\alpha , \quad (2.99)$$

where ζ_α remain infinitesimal group parameters, but contrary to the previous case, here $A_{i\alpha}(\varphi)$ may be arbitrary function of φ.

If the Lagrangian is invariant under the global transformation of G, we have

$$\mathcal{L} \to \mathcal{L}' = \mathcal{L} + j_\alpha^\mu \partial_\mu \zeta_\alpha(x) , \quad (2.100)$$

where

$$j_\alpha^\mu = \frac{\delta \mathcal{L}}{\delta \partial_\mu \varphi_i} A_{\alpha i}(\varphi) . \quad (2.101)$$

The Ward-Takahashi identity in this case can be derived also from the path integral presentation of the generating functional, but an additional term comes from the Jacobian of transformation. We have

$$\int [d\varphi_i] \exp\left[i \int_p (\mathcal{L} - J\varphi) \right] \langle |\rho| \rangle = \int [d\varphi_i'] \exp\left\{ i \int_p [\mathcal{L} + j_\alpha^\mu \partial_\mu \zeta_\alpha - J_i A_{\alpha i}(\varphi)\zeta_\alpha - J\varphi] \right\} \langle |\rho| \rangle$$

$$= \int [d\varphi_i] \left[1 + \frac{\partial A_{\alpha i}(\varphi)}{\partial \varphi_i} \zeta_\alpha \right] \left[1 + \int_p [j_\alpha^\mu \partial_\mu \zeta_\alpha - J_i A_{\alpha i}(\varphi)\zeta_\alpha] \right]$$

$$\times \exp\left[i \int (\mathcal{L} - J\varphi) \right] \langle |\rho| \rangle , \quad (2.102)$$

where

$$\langle |\rho| \rangle \equiv \langle \varphi(\vec{x}, t_+ = -\infty) | \rho | \varphi(\vec{x}, t_- = -\infty) \rangle .$$

It follows from Eq. (2.102) that

$$\partial_\mu j_\alpha^\mu \left\{ \varphi_{ic}(x) + i \frac{\delta}{\delta J_i(x)} \right\} = \frac{\partial A_{\alpha i}}{\partial \varphi_i} \left\{ \varphi_{jc} + i \frac{\delta}{\delta J_j(x)} \right\} - J_i A_{\alpha i} \left\{ \varphi_{jc} + i \frac{\delta}{\delta J_j(x)} \right\} . \quad (2.103)$$

If the loop correction terms are neglected, Eq. (2.103) turns out to be

$$\partial_\mu \langle j_\alpha^\mu(x) \rangle = \partial A_{\alpha i} / \partial \varphi_i - J_i A_{\alpha i} . \quad (2.104)$$

This equation will be used to obtain the nonlinear mode coupling term in the generalized Langevin equation.

III. SUMMARY OF CRITICAL DYNAMICS

There was a recent comprehensive review on the critical dynamics.[10] We give here a brief summary of the basic results to specify the notations and to facilitate the comparison with our results.

The properties of the system near critical point are described in terms of order parameters and conserved variables forming a set of macrovariables $Q = \{Q_i, i = 1, 2 \cdots n\}$. The time evolution of these stochastic variables obeys the generalized Langevin equation

$$\partial Q_i(t)/\partial t = K_i(Q) + \xi_i(t) , \quad (3.1)$$

where the random force $\xi_i(t)$, reflecting the effects of all degrees of freedom, not included in $\{Q_i\}$, is assumed to be Gaussian distributed, i.e.,

$$\langle \xi_i(t) \rangle = 0 , \quad \langle \xi_i(t) \xi_j(t') \rangle = 2\sigma_{ij} \delta(t - t') . \quad (3.2)$$

The right-hand side function $K_i(Q)$ of Eq. (3.1) consists of two parts

$$K_i(Q) = -\sigma_{ij} \delta F / \delta Q_j + V_i(Q) , \quad (3.3)$$

where the free energy $F \equiv F(Q)$ as a functional of Q is dependent on concrete models. The static equilibrium condition $\delta F/\delta Q_i = 0$ appears to be the Ginzburg-Landau equation. Therefore Eq. (3.1) without random force ξ_i is called sometimes the time-dependent Ginzburg-Landau equation (TDGL for short). In principle, the coefficient matrix σ_{ij} may have both symmetric and antisymmetric parts. The symmetric part describes the relaxation, while the antisymmetric one describes the canonical motion. If only relaxation effects are considered, σ_{ij} may be taken symmetric. According to the fluctuation dissipation theorem, the same matrix σ_{ij} does appear both in Eqs. (3.2) and (3.3). In diagonalized form σ_i = constant (dissipative relaxation) for the nonconserved Q_i, and $\sigma_i = -D_i \nabla^2$ (diffusion relaxation, D_i being the diffusion constant) for the conserved Q_i.

The dissipative coupling of different modes can be described by means of the interaction terms in the free-energy functional, but the reversible mode coupling appears as stream term $V_i(Q)$ in Eq. (3.3). Usually it takes the form[11]

$$V_i(Q) = \lambda \sum_j \left\{ \frac{\partial}{\partial Q_j} A_{ij}(Q) - A_{ij}(Q) \frac{\delta F}{\delta Q_j} \right\} , \quad (3.4)$$

where the antisymmetric tensor A_{ij} is formed from the commutators or the Poisson brackets. As a rule, the linear approximation is accepted, i.e.,

$$A_{ij}=f_{ijk}Q_k \ , \tag{3.5}$$

where f_{ijk} are structure constants for the underlying symmetry group. The form of the expression (3.4) itself makes the conservation equation for probability to be satisfied, i.e.,

$$(\partial/\partial Q_i)[V_i(Q)e^{-F}]=0 \ , \tag{3.6}$$

which means that the $\{Q_i\}$ space is divergence free, ensuring e^{-F} to be the stationary distribution with detailed balance.

It can be seen that the Langevin equation (3.1) is flexible enough to embody all possible factors. To our knowledge Eq. (3.1) is "assembled" by different reasonable arguments, so that it remains a kind of phenomenological model.

The widely accepted approach in the theory of dynamical critical phenomena is to construct the perturbation theory by iterating Eq. (3.1).[12] Since there are two different kinds of "constituent parts"—response and correlation functions, the structure of the perturbation theory becomes quite complicated. The compact presentation of such perturbation procedure is given by the MSR field theory[8] mentioned before. In analogy with the static theory of original K. G. Wilson's formulation, Eq. (3.1) can be used to carry out the renormalization transformation to derive the recurrent formulas and to calculate the critical exponents.

For the last several years the critical dynamics has been reformulated using the field-theoretical approach.[13,14]

The Gaussian stochastic process $\xi_i(t)$ can be presented by a stochastic functional.[15] Equation (3.1) can be considered as a mapping of the Gaussian process $\xi_i(t)$ onto a more complicated process $Q_i(t)$. Performing such nonlinear transformation of the Gaussian stochastic functional yields the functional description for process $Q_i(t)$.[16] A more direct way is to start from the normalization condition for δ functions under the path integral

$$\int [dQ]\delta\left[\frac{\partial Q}{\partial t}-K(Q)-\xi\right]\Delta(Q)=1 \ . \tag{3.7}$$

Since the argument of δ function is not Q, but the whole expression (3.1) it is necessary to insert the Jacobian $\Delta(Q)$ for the nonlinear transformation from ξ_i to Q_i. Neglecting multiplicative constant $\Delta(Q)$ turns out to be[16]

$$\Delta(Q)=\exp\left[-\tfrac{1}{2}\int\frac{\delta K(Q)}{\delta Q}dx\right] \ , \tag{3.8}$$

where $dx=d\vec{x}dt$ is the four-dimensional integration element. In what follows we shall omit dx for short.

Presenting the δ functions in Eq. (3.7) in terms of the continuous integral, we obtain

$$\int [dQ]\left[\frac{d\hat{Q}}{2\pi}\right]\exp\left\{\int\left[i\hat{Q}\left[\frac{\partial Q}{\partial t}-K(Q)-\xi\right]-\frac{1}{2}\frac{\delta K}{\delta Q}\right]\right\}=1 \ . \tag{3.9}$$

The insertion of factor

$$\exp\left\{-i\int[J(x)Q(x)+\hat{J}(x)\hat{Q}(x)]\right\}$$

into the integral (3.9) yields the generating functional for the averages of all possible products of the field operators (in theory of probability it is called characteristic or moment-generating functional):

$$Z_\xi(J,\hat{J})=\int[dQ]\left[\frac{d\hat{Q}}{2\pi}\right]\exp\left\{\int\left[i\hat{Q}\left[\frac{\partial Q}{\partial t}-K(Q)-\xi\right]-\frac{1}{2}\frac{\delta K}{\delta Q}-i(JQ+\hat{J}\hat{Q})\right]\right\} \ , \tag{3.10}$$

with the obvious normalization condition

$$Z_\xi[0,0]=1 \ . \tag{3.11}$$

The random force $\xi(t)$ obeys the Gaussian distribution

$$W(\xi)\propto\exp(-\tfrac{1}{2}\xi\sigma^{-1}\xi) \ , \tag{3.12}$$

where σ^{-1} is the inverse of the correlation matrix σ. Taking average in Eq. (3.10) over ξ, we obtain the Lagrangian formulation of the generating functional for the classical statistical field theory

$$Z(J,\hat{J})=\int[dQ]\left[\frac{d\hat{Q}}{2\pi}\right]\exp\left\{\int\left[-\frac{1}{2}\hat{Q}\sigma\hat{Q}+i\hat{Q}\left[\frac{\partial Q}{\partial t}-K(Q)\right]-\frac{1}{2}\frac{\delta K}{\delta Q}-iJQ-i\hat{J}\hat{Q}\right]\right\} \ . \tag{3.13}$$

In their original paper MSR introduced the "response fields," $\hat{Q}$ in our notation, noncommutative with the basic fields, to simplify the structure of the perturbation theory and the renormalization procedure. As in the ordinary field theory, the noncommutativity of the variables is not evident under the path integration. The introduction of the $\hat{Q}$ fields doubles the number of operators. In the CTPGF approach the time path is divided into positive and negative branches, so the number of operators is also doubled. As will be shown in Sec. V, the Lagrangian formulation of the MSR field theory follows naturally from the CTPGF formalism. In can be seen also that the noncommutativity of operators is not an artificial formal trick, but a necessity to describe properly the statistical fluctuations.

The Gaussian integration over $\hat{Q}$ in Eq. (3.13) can be carried out to yield

$$Z[J]=\int [dQ]\exp\left\{\int\left[-\frac{1}{2}\left(\frac{\partial Q}{\partial t}-K(Q)-\hat{J}\right)\sigma^{-1}\left(\frac{\partial Q}{\partial t}-K(Q)-\hat{J}\right)-\frac{1}{2}\frac{\delta K}{\delta Q}-iJQ\right]\right\} . \tag{3.14}$$

This expression was first obtained as a stochastic functional in Ref. 16, but it turns out that Eq. (3.13) is the more convenient starting point for the critical dynamics.[13,14]

IV. FUNDAMENTAL SYSTEM OF EQUATIONS IN CRITICAL DYNAMICS

Generally speaking, both order parameter and conserved variables when regarded as macrovariables are composite operators of the basic fields. We shall specify them by somewhat different notation. The nonconserved order parameters may be written as

$$Q_{ci}(x),\quad i=1,2,\ \ldots,n ,$$

whereas the conserved variables

$$Q_{c,n+\alpha}(x)\equiv q_\alpha(x),\quad \alpha=1,2,\ \ldots,m ,$$

where q_α corresponds to the average of the zeroth component for the conserved current

$$q_\alpha=\langle j_\alpha^0\rangle . \tag{4.1}$$

Without sacrifice of generality both Q_{ci} and q_α can be taken to be Hermitian.

Introducing the generating functional of the CTPGF vertex functions for the composite operators $\Gamma(Q_c)$, we obtain the equations which require to be satisfied by Q_c [Eq. (2.21)]; i.e.,

$$\delta\Gamma/\delta Q_{ci}(x)=-J_i(x),\quad i=1,2,\ \ldots,n+m . \tag{4.2}$$

After taking the variational derivatives one puts $J_i(x+)=J_i(x-)=J_i(\bar{x},t)$, from which it follows [from Eq. (2.19)] that $Q_{ci}(x+)=Q_{ci}(x-)=Q_i(\bar{x},t)$, where $J_i(\bar{x},t)$, $Q_i(\bar{x},t)$ are functions defined on the usual time axis $(-\infty,+\infty)$. We next show that Eqs. (4.2) lead to the generalized TDGL equations under the assumption that Q_i are smoothly varying functions of time.

Suppose the macrovariables $Q_i(\bar{x},\tau)$ to be known at the moment τ. At the moment t following closely after τ the left-hand side of Eqs. (4.2) can be expanded. If x sits on the positive time branch, we have

$$-J_i(\bar{x},t)=\left.\frac{\delta\Gamma}{\delta Q_{ci}(x+)}\right|_{Q_{ci}(x+)=Q_{ci}(x-)=Q_i(\bar{x},\tau)}+\int\Gamma_{rij}(x,y)[Q_j(\bar{y},t_y)-Q_j(\bar{y},\tau)]dy , \tag{4.3}$$

where $\Gamma_{rij}(x,y)$ are two-point retarded vertex functions after taking $Q_{c+}=Q_{c-}=Q$. If x sits on the negative branch of time, the same is true due to Eq. (2.60) $\Gamma_r=\Gamma_{++}-\Gamma_{+-}=\Gamma_{-+}-\Gamma_{--}$. Since $Q_j(\bar{y},t)$ in Eq. (4.3) varies smoothly with time, to the first order of $(t_y-\tau)$ we have

$$Q_j(\bar{y},t_y)-Q_j(\bar{y},\tau)=(t_y-\tau)\partial Q_j(\bar{y},\tau)/\partial\tau . \tag{4.4}$$

Substituting Eq. (4.4) back into Eq. (4.3) and taking into account that in the limit $t\equiv t_x\rightarrow\tau$

$$\gamma_{ij}(\bar{x},\bar{y},\tau)\equiv-\lim_{t_x\rightarrow\tau}\int dt_y(t_y-t_x)\Gamma_{rij}(\bar{x},t_x,\bar{y},t_y)=i\frac{\partial}{\partial k_0}\Gamma_{rij}(\bar{x},\bar{y},k_0,\tau)|_{k_0=0} , \tag{4.5}$$

where $\Gamma_{rij}(\vec{x},\vec{y},k_0,\tau)$ are Fourier transforms with respect to (t_x-t_y) taken at the average time $T=\frac{1}{2}(t_x+t_y)\approx\tau$, we obtain

$$\gamma(\tau)\frac{\partial Q(\tau)}{\partial\tau}=\frac{\delta\Gamma}{\delta Q_{c+}}\bigg|_{Q_{c+}=Q_{c-}=Q}+J(t) \quad . \tag{4.6}$$

Here the matrix notation is used and $\gamma_{ij}(\vec{x},\vec{y},\tau)$ are considered to be matrix elements with subscripts $i\,\vec{x}$ and $j\,\vec{y}$.

For the moment let

$$I_i(\vec{x},\tau)\equiv\frac{\delta\Gamma}{\delta Q_{c+}}\bigg|_{Q_{c+}=Q_{c-}=Q(\tau)} \tag{4.7}$$

and we calculate the functional derivative of I_i, considering it as a functional of functions $Q(\vec{x},\tau)$ with three-dimensional argument $\vec{x}$

$$\frac{\delta I_i(\vec{x},\tau)}{\delta Q_j(\vec{y},\tau)}=\int d\vec{z}\,d\tau_z\left[\frac{\delta^2\Gamma}{\delta Q_{ci}(x+)\delta Q_{ck}(z+)}\frac{\delta Q_{ck}(z+)}{\delta Q_j(\vec{y},\tau)}-\frac{\delta^2\Gamma}{\delta Q_{ci}(x+)\delta Q_{ck}(z-)}\frac{\delta Q_{ck}(z-)}{\delta Q_j(\vec{y},\tau)}\right]\bigg|_{Q_{c+}=Q_{c-}=Q(\tau)} ,$$

where

$$\frac{\delta Q_{ck}(z)}{\delta Q_j(\vec{y},\tau)}\bigg|_{Q_{c+}=Q_{c-}=Q(\tau)}=\delta_{jk}\delta^{(3)}(\vec{y}-\vec{z}) \quad .$$

Thus we obtain

$$\delta I_i(\vec{x},\tau)/\delta Q_j(\vec{y},\tau)=\Gamma_{++ij}(\vec{x},\vec{y},k_0=0,\tau)-\Gamma_{+-ij}(\vec{x},\vec{y},k_0=0,\tau) \quad ,$$

where the $k_0=0$ components of Fourier transforms appear as in Eq. (4.5). It can be shown in the same way that

$$\delta I_j(\vec{y},\tau)/\delta Q_i(\vec{x},\tau)=\Gamma_{++ji}(\vec{y},\vec{x},k_0=0,\tau)-\Gamma_{+-ji}(\vec{y},\vec{x},k_0=0,\tau)=\Gamma_{++ij}(\vec{x},\vec{y},-k_0=0,\tau)-\Gamma_{-+ij}(\vec{x},\vec{y},-k_0=0,\tau) \quad ,$$

where the symmetry properties of Γ following from Eqs. (2.59) and (2.63) are used. The difference

$$\frac{\delta I_i(\vec{x},\tau)}{\delta Q_j(\vec{y},\tau)}-\frac{\delta I_j(\vec{y},\tau)}{\delta Q_i(\vec{x},\tau)}=\lim_{k_0\to 0}[\Gamma_{-+ij}(\vec{x},\vec{y},-k_0,\tau)-\Gamma_{+-ij}(\vec{x},\vec{y},k_0,\tau)] \tag{4.8}$$

vanishes due to Eq. (2.77), i.e.,

$$\Gamma_{+-}=\exp(-\beta k_0)\Gamma_{-+}$$

near thermal equilibrium, so that there exists a functional $F(Q_i(\vec{x},\tau))$ with

$$I_i(\vec{x},\tau)=-\delta F/\delta Q_i(\vec{x},\tau) \quad . \tag{4.9}$$

Equation (4.6) can be rewritten as

$$\gamma(\tau)\partial Q(\tau)/\partial\tau=-\delta F/\delta Q(\tau)+J(\tau) \quad . \tag{4.10}$$

If the macrovariables $Q(\tau)$ do not change with time in the external field J, i.e, in the stationary state, then

$$\delta F/\delta Q=J \quad . \tag{4.11}$$

Hence F is the effective free energy of the system and Eq. (4.11) is the Ginzburg-Landau equation, determining the stationary distribution of macrovariables.

For systems in stationary states far from equilibrium expression (4.8) is equal to zero only if the dissipative part of the vertex function $A=\frac{1}{2}i(\Gamma_{-+}-\Gamma_{+-})$ Eq. (2.68c) satisfies the condition

$$\lim_{k_0\to 0}A_{ij}(\vec{x},\vec{y},k_0,t)=0 \quad . \tag{4.12}$$

In this case I_i can also be written as a variational derivative of the free energy or effective potential. Some of the stationary states satisfying the so-called "potential conditions," provided by the detailed balance, as discussed by Graham and Haken,[17] must belong to this category.

In the vicinity of all stationary states with the potential functions F Eq. (4.10) constitutes the system of time-dependent GL equations, but they are much more general than the TDGL equations in the usual sense since the mode coupling terms are also included.

It is usually customary to multiply Eq. (4.10) by the inverse matrix $\gamma^{-1}(\tau)$ to obtain

$$\frac{\partial Q(\tau)}{\partial\tau}=\gamma^{-1}(\tau)\left[-\frac{\delta F}{\delta Q(\tau)}+J(\tau)\right] \quad . \tag{4.13}$$

Using the symmetry properties of the vertex functions, following from Eqs. (2.59) and (2.63)

$$\tilde{\Gamma}(k)=\tilde{\Gamma}^T(-k)=-\sigma_3\tilde{\Gamma}^*(-k)\sigma_3=-\sigma_3\tilde{\Gamma}^\dagger(k)\sigma_3 \ , \tag{4.14}$$

it can be shown that the real part of Γ, is an even function of k_0, while the imaginary part is an odd one, so $\gamma(t)$ is a real matrix according to the definition (4.5).

In accordance with the numeration of the subscripts given at the beginning of this section, the matrix $\gamma(\tau)$ can be divided into four blocks. Two of them, corresponding to the conserved variables $\gamma_{\alpha\bar{x},\beta\bar{y}}$ and $\gamma_{\alpha\bar{x},j\bar{y}}$, can be fixed completely by comparison with the WT identities. For the general case, the proper form of the two blocks associated with the order parameters can be determined only by the symmetry considerations. This will be discussed below.

It is worthwhile to point out that Eq. (4.4) is equivalent to the Markovian approximation. In principle, the original Eq. (4.2) contains in itself the possibility of considering the memory effects.

Under the action of the symmetry group G of the system the conserved variables transform as the generators I^α of the group, i.e.,

$$q_\alpha \to q'_\alpha = q_\alpha + i f_{\alpha\beta\gamma}\zeta_\beta q_\gamma \ , \tag{4.15}$$

where $f_{\alpha\beta\gamma}$ are the structure constants of the group and ζ_β are the infinitesimal parameters of transformation. The order parameters Q_i transform as some representation $\hat{L}$ of group G

$$Q_i \to Q'_i = Q_i + iL^\alpha_{ij}\zeta_\alpha Q_j \ . \tag{4.16}$$

As shown in Sec. II D, if the Lagrangian of the system is invariant under the global symmetry transformations, the WT identities (2.95) are valid on the closed time path. In the present case, Eq. (2.95) can be written as

$$\langle \partial_\mu j^\mu_\alpha(\varphi)\rangle = i\left[\frac{\delta\Gamma}{\delta Q_{ci}(x)}L^\alpha_{ij}Q_{cj}(x)+f_{\alpha\beta\gamma}\frac{\delta\Gamma}{\delta q_\beta(x)}q_\gamma(x)\right] \ , \tag{4.17}$$

where, as before, $\Gamma\equiv\Gamma(Q_{ci},q_\alpha)$ is the generating functional of the vertex CTPGF's for the composite operators. Putting $Q_{c+}=Q_{c-}=Q(\tau)$ and let $j^\mu_\alpha\equiv\langle j^\mu_\alpha(\varphi)\rangle$ in Eq. (4.17), we obtain

$$\partial q_\alpha/\partial\tau = \nabla\vec{j}_\alpha - i[J_i(\vec{x},\tau)L^\alpha_{ij}Q_j(\vec{x},\tau) + f_{\alpha\beta\gamma}J_\beta(\vec{x},\tau)q_\gamma(\vec{x},\tau)] \ , \tag{4.18}$$

where J, Q, q, etc., are functions defined on the usual time axis $(-\infty,+\infty)$.

To determine the conserved currents $\vec{j}_\alpha$ we perform the following manipulations in analogy with the procedure used to deal with the thermal perturbations in the linear-response theory. By introducing an additional artificial external source ΔJ, superposed on the original source J, the system is forced to come into the stationary state $\partial q_\alpha/\partial t=0$. In this case, $\vec{j}_\alpha$ in Eq. (4.18) changes to $\vec{j}'_\alpha$, so that

$$\nabla\vec{j}'_\alpha - i[(J_i+\Delta J_i)L^\alpha_{ij}Q_j + f_{\alpha\beta\gamma}(J_\beta+\Delta J_\beta)q_\gamma]=0 \ . \tag{4.19}$$

Since the system is in the stationary state, we can use Eq. (4.11), i.e.,

$$\delta F/\delta Q_i = J_i+\Delta J_i \ , \quad \delta F/\delta q_\alpha = J_\alpha + \Delta J_\alpha \ , \tag{4.20}$$

to replace the source terms J_i, J_α by the functional derivatives of the free energy F. According to the linear-response theory the difference between $\vec{j}'_\alpha$ and the conserved current $\vec{j}_\alpha$ without an artificial source ΔJ can be written as

$$\vec{j}'_\alpha = \vec{j}_\alpha - l_{\alpha\beta}\nabla(\delta F/\delta q_\beta - J_\beta) \ , \tag{4.21}$$

where $l_{\alpha\beta}$ are the linear transport coefficients. Substituting Eq. (4.21) into Eq. (4.19) yields

$$\nabla\vec{j}_\alpha = l_{\alpha\beta}\nabla^2\left[\frac{\delta F}{\delta q_\beta}-J_\beta\right]+i\left[\frac{\delta F}{\delta Q_i}L^\alpha_{ij}Q_j+f_{\alpha\beta\gamma}\frac{\delta F}{\delta q_\beta}q_\gamma\right] \ . \tag{4.22}$$

Inserting the expression (4.22) for $\nabla\vec{j}_\alpha$ back into Eq. (4.18) leads to the equation of motion of the conserved variables

$$\frac{\partial q_\alpha}{\partial\tau} = l_{\alpha\beta}\nabla^2\left[\frac{\delta F}{\delta q_\beta}-J_\beta\right] + i\left[\frac{\delta F}{\delta Q_i}-J_i\right]L^\alpha_{ij}Q_j + f_{\alpha\beta\gamma}\left[\frac{\delta F}{\delta q_\beta}-J_\beta\right]q_\gamma \ . \tag{4.23}$$

Comparing Eq. (4.23) with the general equations for macrovariables (4.13) in the case of conserved variables, one determines two blocks of γ^{-1} matrix

$$[\gamma^{-1}(\tau)]_{\alpha\bar{x},\beta\bar{y}} = -[l_{\alpha\beta}\nabla^2_x + i f_{\alpha\beta\gamma}q_\gamma(\vec{x},\tau)]\delta^{(3)}(\vec{x}-\vec{y}) \ , \tag{4.24}$$

$$[\gamma^{-1}(\tau)]_{\alpha\bar{x},i\bar{y}} = -iL^\alpha_{ij}Q_j(\vec{x},\tau)\delta^{(3)}(\vec{x}-\vec{y}) \ . \tag{4.25}$$

Now we turn to the equations for the order parameters. If the order parameters Q_i form the irreducible representation of the group G and take small values near the critical point, γ^{-1} can be expanded into a power series of Q_i. It follows from the symmetry property that

$$[\gamma^{-1}(\tau)]_{i\bar{x},j\bar{y}} = \delta_{ij}\sigma_{\bar{x},\bar{y}} + \cdots \ , \tag{4.26}$$

where $\sigma_{\bar{x},\bar{y}}$, not depending on Q_i, are determined by the kinetic and dissipative characteristics of the system. Similarly, by symmetry consideration another expansion can be written

$$[\gamma^{-1}(\tau)]_{i\bar{x},\alpha\bar{y}} = i f L^\alpha_{ij}Q_j(\vec{x},\tau)\delta^{(3)}(\vec{x}-\vec{y}) + \cdots \ . \tag{4.27}$$

where the ellipsis represents higher terms of Q_i and f is invariant under group transformations and can only be a numerical constant in the lowest order of Q_i. To determine the value of f we consider the limit of vanishing dissipation. In this case the antisymmetric part of the matrix $\gamma(\tau)$ in Eq. (4.10) is dominant. The same is true for the inverse matrix in Eq. (4.13). Comparison of Eqs. (4.27) and (4.25) gives

$$f=1 \ . \tag{4.28}$$

Since the first term of Eq. (4.25) is independent of dissipation, Eq. (4.28) remains valid in its presence. Generally speaking the expansions (4.26) and (4.27) may contain other terms, including the crossover interaction of the dissipative and canonical motion. We shall not touch this problem here. In the approximation discussed above we obtain the following equations for the order parameters:

$$\frac{\partial Q_i}{\partial \tau}=\sigma\left[\frac{\delta F}{\delta Q_i}-J_i\right]-iL_{ij}^{\alpha}Q_j\left[\frac{\delta F}{\delta q_\alpha}-J_\alpha\right] \ . \tag{4.29}$$

Equations (4.29) and (4.23) form the fundamental system of equations for the critical dynamics. In the CTPGF approach J, coming from $J_+=J_-$, is the real physical external field. It may contain the additional random fields, representing the effects of degrees of freedom, not included in the macrovariables Q_i. Therefore we call this system of equations the generalized Langevin equations.

The essential point of the above given derivation is that the mode coupling terms naturally appear in the generalized Langevin equations. Moreover, they actually have the form of Eqs. (3.4) and (3.5). To be more exact, the representation matrix L_{ij}^{α} appears in the coupling terms between the order parameter and the conserved variable, while the structure constants $f_{\alpha\beta\gamma}$ appears only in coupling terms among conserved variables. In the linear approximation of Q_i, the first term of Eq. (3.4) gives no contribution. To be concrete, we divide the matrix A into four blocks. The reversible coupling among order parameters may be ignored, so that $A_{ij}=0$. The coupling terms with conserved variables in the equations for the order parameters $A_{i\alpha}=iL_{ij}^{\alpha}Q_j$ are independent of q_α, so that $\partial A_{i\alpha}/\partial q_\alpha=0$. The two other blocks, $A_{\alpha i}=-iL_{ij}^{\alpha}Q_j$ and $A_{\alpha\beta}=-if_{\alpha\beta\gamma}q_\gamma$ appearing in the equations for the conserved variables also give zero contribution due to the antisymmetric character of the representation matrix $\hat{L}$ and the structure constants.

To get the term with derivatives in Eq. (3.4) we have to start from the nonlinear WT identities derived also in Sec. II D. In fact, by use of Eq. (2.104) we can repeat the derivation for the linear case and obtain the first term of Eq. (3.4). What we want to emphasize is that Kawasaki's formula (3.4) corresponds to the tree approximation in Eq. (2.103), so in principle we can go further. Another point is that the derivative terms, coming from the Jacobian, appear only for the basic fields, but not for the composite operators which also transform the derivative terms for the composite operators. Firstly, they may come from the loop corrections, secondly, and what is more likely in our opinion, they appear as a result of changes of measure in the path integral of the effective action (see the next section).

As a concrete example consider the simplest model of the isotropic antiferromagnet, i.e., model G in Ref. 10. This system consists of two densities, a nonconserved order parameter $\vec{Q}$ which is a three-component vector representing the staggered magnetization and a conserved density $\vec{q}$, also a three-component vector representing the total magnetization of the system.

From the commutation relations

$$[q_\alpha,q_\beta]=ig_0\epsilon_{\alpha\beta\gamma}q_\gamma \ , \ \ [Q_i,q_\alpha]=ig_0\epsilon_{i\alpha j}Q_j \ , \tag{4.30}$$

the structure constants and the representation matrix can be determined immediately

$$f_{\alpha\beta\gamma}=ig_0\epsilon_{\alpha\beta\gamma} \ , \ \ L_{ij}^{\alpha}=-ig_0\epsilon_{i\alpha j} \ , \tag{4.31}$$

where $\epsilon_{\alpha\beta\gamma}$ and $\epsilon_{i\alpha j}$ are fully antisymmetric unit tensors. Substituting Eq. (4.31) into Eqs. (4.23) and (4.29), and bringing together the external field term and the derivative of the free energy, i.e., changing $F\to F-J_iQ_i-J_\alpha Q_\alpha$, we obtain

$$\frac{\partial Q_i}{\partial\tau}=-\sigma\frac{\delta F}{\delta Q_i}+g_0\epsilon_{\alpha ij}Q_j\frac{\delta F}{\delta q_\alpha} \ ,$$

$$\frac{\partial q_\alpha}{\partial\tau}=l_{\alpha\beta}\nabla^2\frac{\delta F}{\delta q_\beta}-g_0\epsilon_{\alpha ij}\frac{\delta F}{\delta Q_i}Q_j-g_0\epsilon_{\alpha\beta\gamma}\frac{\delta F}{\delta q_\beta}q_\gamma \ . \tag{4.32}$$

By taking $\sigma=\Gamma_0$, $l_{\alpha\beta}=\lambda_0\delta_{\alpha\beta}$ and changing to the vector notation, we retrieve the system of equations for model G.[10]

$$\frac{\partial\vec{Q}}{\partial\tau}=-\Gamma_0\frac{\delta F}{\delta\vec{Q}}+g_0\vec{Q}\times\frac{\delta F}{\delta\vec{q}} \ ,$$

$$\frac{\partial\vec{q}}{\partial\tau}=\lambda_0\nabla^2\frac{\delta F}{\delta\vec{q}}+g_0\vec{Q}\times\frac{\delta F}{\delta\vec{Q}}+g_0\vec{q}\times\frac{\delta F}{\delta\vec{q}} \ . \tag{4.33}$$

The models A, B, and C are much simpler due to the absence of the reversible mode coupling. The other models such as E, F, H, and J models[10] and the SSS model[18] can be treated in the same way. We shall not repeat these simple calculations here.

V. LAGRANGIAN FORMULATION OF STATISTICAL FIELD THEORY

Suppose $\hat{\varphi}_i$, $i=1,2,\cdots l$, are the basic fields of the system, $\hat{Q}_i(\hat{\varphi}), i=1,2,\cdots n+m$, are composite operators representing the order parameters and conserved variables. Some of the basic fields may be order parameters also (as in the case of lasers). For simplicity we take all of them to be Hermitian Bose operators. In what follows operators will not be distinguished by special notations since their meaning is clear from the context.

Assuming the randomness of the initial phase, the density matrix is diagonal at the moment $\tau=\tau_0$:

$$\langle\varphi'(\vec{x},\tau_0)|\rho|\varphi''(\vec{x},\tau_0)\rangle = P(\varphi'(\vec{x}),\tau_0)\delta(\varphi'(\vec{x},\tau_0)-\varphi''(\vec{x},\tau_0)) \ . \tag{5.1}$$

The initial distribution of the macrovariables $Q_i(x)$ is given by

$$P(Q_i(\vec{x}),\tau_0) = \mathrm{tr}[\delta(Q_i(x)-Q_i(\varphi(x)))\rho] = \int [d\varphi(x)]\delta(Q_i(x)-Q_i(\varphi(x)))P(\varphi(\vec{x}),\tau_0) \ . \tag{5.2}$$

The generating functional for $Q_i(\phi(x))$ [Eq. (2.86)] under the assumption Eq. (5.1) can be written as

$$\begin{aligned} Z(J(x)) &= \exp[-iW(J(x))] \\ &= \mathrm{tr}\left\{T_p\left[\exp\left[-i\int J(x)Q(\varphi(x))\right]\right]\rho\right\} = N\int [d\varphi(x)]\exp\left[-i\int [\mathcal{L}(\varphi(x))-JQ(\varphi(x))]\right]\delta(\varphi_+-\varphi_-) \ , \end{aligned} \tag{5.3}$$

where

$$\delta(\varphi_+-\varphi_-) \equiv \int d\varphi'(x)\delta(\varphi(\vec{x},\tau_+=\tau_0)-\varphi'(\vec{x}))\delta(\varphi(\vec{x},\tau_-=\tau_0)-\varphi'(\vec{x}))P(\varphi'(\vec{x}),\tau_0) \ . \tag{5.4}$$

Multiplying the right-hand side of Eq. (5.3) by the normalization factor of the δ function on the closed time path

$$\int [dQ]\delta(Q_+-Q_-)\delta(Q(x)-Q(\varphi(x)))=1 \ , \tag{5.5}$$

changing the order of integration to replace $Q(\varphi(x))$ by $Q(x)$ and using the formula

$$\delta(Q(x)-Q(\varphi(x))) = \int\left[\frac{dI}{2\pi}\right]\exp\left[i\int_p [Q(x)-Q(\varphi(x))]I(x)\right] \ , \tag{5.6}$$

we can rewrite Eq. (5.3) as

$$Z(J) = N\int [dQ]\exp\left[iS_{\mathrm{eff}}(Q)-i\int_p JQ\right]\delta(Q_+-Q_-) \ , \tag{5.7}$$

where

$$e^{iS_{\mathrm{eff}}(Q)} \equiv \int\left[\frac{dI}{2\pi}\right]\exp\left[i\int_p QI-iW(I)\right] \ . \tag{5.8}$$

Here we are performing the direct and inverse Fourier transformations of the path integral. Since the continuous integration is taken over $I(x)$, $W(I)$ can be considered as the generating functional in the random external fields. Calculating the integral by Wentzel-Kramers-Brillouin (WKB) procedure in the one-loop approximation which is equivalent to the Gaussian averaging over the random fields, we obtain the effective action $S_{\mathrm{eff}}(Q)$ for macrovariables.

This is for the case when macrovariables are composite operators. The same is true, if all or part of macrovariables are basic fields themselves. A new field can also be introduced by using the δ function. Even if the initial distribution is multiplicative for different components

$$P(\varphi',\tau_0) = \prod_{i=1}^{n} P_i(\varphi_i',\tau_0) \ ,$$

the Fourier transformations for the path integral have to be carried out simultaneously for all fields, since for the general case the Lagrangian of the system cannot be presented as a superposition of contributions from different components.

We now turn to discuss the general properties of the effective action $S_{\mathrm{eff}}(Q)$.

The generating functional for CTPGF's in the case of Hermitian Bose fields satisfies the relations

$$W(J_+(x),J_-(x))|_{J_+(x)=J_-(x)}=0 \ , \tag{5.9}$$

$$W^*(J_+(x),J_-(x)) = -W(J_-(x),J_+(x)) \ . \tag{5.10}$$

Taking successive functional derivatives of Eq. (5.9) and putting $J_+(x)=J_-(x)$ we get a number of relations between CTPGF's. It is easy to show by use of Eqs. (5.8) and (5.10) that

$$S^*_{\mathrm{eff}}(Q_+(x),Q_-(x)) = -S_{\mathrm{eff}}(Q_-(x),Q_+(x)) \ , \tag{5.11}$$

so S_{eff} is purely imaginary for $Q_+(x)=Q_-(x)$. Putting $Q_\pm(x)=Q+\Delta Q_\pm$ and taking functional expansion of Eq. (5.11) around Q, we obtain relations among functional derivatives of different order at the point Q:

$$\frac{\delta S}{\delta Q_+(x)}=\left[\frac{\delta S}{\delta Q_-(x)}\right]^* , \tag{5.12}$$

$$S_{Fij}(x,y)=S_{Fji}(y,x)=-S^*_{\tilde{F}ji}(y,x) ,$$
$$S_{\pm ij}(x,y)=S_{\mp ji}(y,x)=-S^*_{\mp ji}(y,x) , \tag{5.13}$$

where

$$S_{Fij}(x,y)\equiv\frac{\delta^2 S}{\delta Q_{i+}(x)\delta Q_{j+}(y)} , \tag{5.14}$$

etc.

If the system is invariant under the symmetry group G, i.e., the Lagrangian and the initial distribution do not change under

$$\varphi_i(x)\to\varphi_i^g(x)=U_{ij}(g)\varphi_j(x) ,$$
$$Q_i(\varphi)\to Q_i^g(\varphi)=V_{ij}(g)Q_j(\varphi) ,$$

then

$$W(J^g(x))=W(J(x)),\quad J_i^g=J_i(x)V_{ji}^\dagger(g) ,$$
$$S_{\rm eff}(Q^g(x))=S_{\rm eff}(Q(x)),\quad Q_i^g(\varphi)=V_{ij}(g)Q_j(\varphi) .$$

This is true if $S_{\rm eff}$ is calculated exactly. In fact, the symmetry properties of $S_{\rm eff}$, although related to that of the original Lagrangian, may be different from the latter due to the averaging procedure.

If the lowest order of WKB, i.e., the tree approximation is taken in Eq. (5.8), it follows that

$$Q=\delta W/\delta I , \tag{5.15}$$
$$S_{\rm eff}(Q)=-\Gamma(Q) . \tag{5.16}$$

In this case $S_{\rm eff}$ inherits all the properties of the generating functional $\Gamma[Q]$ for the vertex CTPGF's, i.e.,

$$S_{\rm eff}(Q,Q)=0 , \tag{5.17}$$

$$\delta S_{\rm eff}/\delta Q_+|_{Q_+=Q_-=Q}=\delta S_{\rm eff}/\delta Q_-|_{Q_+=Q_-=Q} , \tag{5.18}$$

$$S_F+S_{\tilde{F}}=S_++S_- , \tag{5.19}$$

$$\frac{\delta^l S_{\rm eff}}{\delta Q_{i1}(1)\cdots\delta Q_{il}(l)}=i^{l-1}\langle T_p[Q_{i1}(1)\cdots Q_{il}(l)]_{\rm 1PI} , \tag{5.20}$$

where 1PI means one particle irreducible. According to Eqs. (5.16) and (2.78),

$$-iS_\pm(k)>0 \tag{5.21}$$

after the Fourier transformation.

Near thermal equilibrium we have from Eq. (2.77)

$$S_{-ij}(k)-S_{+ij}(k)\underset{k_0\to 0}{\to}-\beta k_0 S_{-ij}(k) . \tag{5.22}$$

Up to now we have discussed only the general properties of the effective action $S_{\rm eff}(Q)$. In principle this can be derived from the microscopic generating functional W by averaging over the random external fields; it can also be constructed phenomenologically in accordance with the required symmetry properties. We shall now show that in the one-loop approximation in the path integral over dI and to the second order macrovariable fluctuations on positive and negative time branches, the current formulation of MSR field theory[13,14] is retrieved.

To calculate the integral (5.8) we expand the exponential factor around the saddle point, given by Eq. (5.15)

$$E\equiv\int_p QI-W=-\Gamma-\frac{1}{2}\int_p \Delta I W^{(2)}\Delta I+\cdots . \tag{5.23}$$

According to the computation rule described in Sec. II B, E can be rewritten as

$$E=-\Gamma-\frac{1}{2}\int\Delta\hat{I}^T\sigma_3\hat{W}^{(2)}\sigma_3\Delta\hat{I} , \tag{5.24}$$

where

$$\hat{W}^{(2)}=\begin{bmatrix}W_{++} & W_{+-}\\ W_{-+} & W_{--}\end{bmatrix},\quad \Delta\hat{I}=\begin{bmatrix}\Delta I_+\\ \Delta I_-\end{bmatrix} . \tag{5.25}$$

The result of the Gaussian integration, accurate to a constant multiplier is

$$e^{iS_{\rm eff}(Q)}=e^{-i\Gamma(Q)}|\det(\sigma_3\hat{W}^{(2)}\sigma_3)|^{-1/2} . \tag{5.26}$$

From the Dyson equation (2.59) we have

$$iS_{\rm eff}[Q]=-i\Gamma(Q)+\frac{1}{2}\operatorname{tr}\ln\hat{\Gamma}^{(2)} , \tag{5.27}$$

where

$$\hat{\Gamma}^{(2)}=\begin{bmatrix}\Gamma_{++} & \Gamma_{+-}\\ \Gamma_{-+} & \Gamma_{--}\end{bmatrix}$$

is the two-point vertex function. By use of the transformation formula (2.33) we have

$$|\det\hat{\Gamma}^{(2)}|=|\det\tilde{\Gamma}^{(2)}|=|\det\Gamma_r||\det\Gamma_a|=|\det\Gamma_r|^2 , \tag{5.28}$$

where

$$\Gamma_r(x,y)=\delta^2\Gamma/\delta Q(x)\delta\Delta(y) , \tag{5.29}$$

$$Q(x)=\frac{1}{2}[Q_+(x)+Q_-(x)] ,$$
$$\Delta(x)=Q_+(x)-Q_-(x) . \tag{5.30}$$

As shown in Sec. IV

$$\left.\frac{\delta\Gamma}{\delta\Delta(y)}\right|_{\Delta=0}=\frac{1}{2}\left.\left[\frac{\delta\Gamma}{\delta Q_+(y)}+\frac{\delta\Gamma}{\delta Q_-(y)}\right]\right|_{\Delta=0}$$
$$=-\gamma\frac{\partial Q}{\partial t}-\frac{\delta F}{\delta Q(y)} . \tag{5.31}$$

It can be seen from comparing Eq. (5.31) with Eqs. (3.1) and (3.7) that $\delta^2\Gamma/\delta Q\,\delta\Delta$ is just the transformation matrix, accurate to the coefficient matrix γ, from ξ_i to Q_i, so the Jacobian can be calculated in the same way. Taking into account that the square power in Eq. (5.28) exactly cancels out the coefficient $\frac{1}{2}$ in Eq. (5.27) we have finally

$$iS_{\text{eff}}(Q) = -i\Gamma(Q) - \frac{1}{2}\int \frac{dK}{\delta Q}dx \ , \tag{5.32}$$

where

$$K = -\gamma^{-1}\delta F/\delta Q \tag{5.33}$$

In the path integral (5.7) the most plausible path is determined by the equations

$$\delta S_{\text{eff}}(Q)/\delta Q_{\pm} = J_{\pm}(x) \ , \tag{5.34}$$

$$Q(\bar{x}, \tau_+ = \tau_0) = Q(\bar{x}, \tau_- = \tau_0) \ . \tag{5.35}$$

Taking $J_+ = J_- = J$, in the tree approximation of the path integral over $I(x)$, we obtain

$$\frac{\delta S_{\text{eff}}(Q)}{\delta Q} = J(x) = \gamma\frac{\partial Q}{\partial t} + \frac{\delta F}{\delta Q} \ , \tag{5.36}$$

which follows from Eqs. (4.6) and (5.16). This is just the TDGL equation.

We now consider the fluctuations around the most plausible trajectories. In the CTPGF approach, in addition to the fluctuations in the usual sense, field variables are permitted to take different values in positive and negative time branches. Changing variables in the path integral (5.7) to the usual time axis by introducing $Q(x)$ and $\Delta(x)$ according to Eq. (5.30) the effective action S_{eff} can be expanded as

$$S_{\text{eff}}(Q_+(x), Q_-(x)) = S_{\text{eff}}(Q(x), Q(x)) + \frac{1}{2}\int \left[\frac{\delta S_{\text{eff}}}{\delta Q_+} + \frac{\delta S_{\text{eff}}}{\delta Q_-}\right]\Delta(x) + \frac{1}{8}\int \Delta(x)(S_{++} + S_{+-} + S_{-+} + S_{--})(x,y)\Delta(y) + \cdots \ .$$

Denoting

$$\frac{1}{4}i(S_{++} + S_{+-} + S_{-+} + S_{--})(x,y) \equiv -\gamma(x)\sigma(x,y)\gamma(y) \tag{5.37}$$

and using Eqs. (5.17), (5.32), and (5.36) we obtain

$$e^{-iW(J(x))} = \int [dQ(x)][d\Delta(x)]\exp\left[-\frac{1}{2}\int \Delta(x)\gamma(x)\sigma(x,y)\gamma(y)\Delta(y) + i\int \gamma(x)\left[\frac{\partial Q}{\partial \tau} + \frac{\delta F}{\delta Q}\right]\Delta(x) - \frac{1}{2}\int \frac{\delta K}{\delta Q} - i\int (J_\Delta Q + J_0\Delta)\delta(\Delta(\bar{x}, \tau_0))\right] \ , \tag{5.38}$$

where

$$J_\Delta = J_+(x) - J_-(x), \quad J_0 = \frac{1}{2}[J_+(x) + J_-(x)] \ .$$

If we take $J_\Delta = J$ and change variables $\gamma(x)\Delta(x) \to \hat{Q}(x)$, $J_0\gamma^{-1} \to \hat{J}$ the generating functional for the MSR field theory [Eq. (3.13)] is retrieved. The Gaussian integration over $\Delta(x)$ gives

$$e^{-iW(J(x))} = N\int [dQ(x)]\exp\left\{-\frac{1}{2}\int \left[\frac{\partial Q(x)}{\partial \tau} + \frac{1}{\gamma}\left[\frac{\delta F}{\delta Q(x)} - \hat{J}(x)\right]\right]\sigma^{-1}(x,y) \times \left[\frac{\partial Q(y)}{\partial \tau} + \frac{1}{\gamma}\left[\frac{\delta F}{\delta Q(y)} - \hat{J}(y)\right]\right] + \frac{1}{2}\int \frac{\delta}{\delta Q}\left[\frac{1}{\gamma}\frac{\delta F}{\delta Q}\right] - i\int JQ\right\} \ , \tag{5.39}$$

which is the generating functional (3.14). It is interesting to point out that $\hat{J} = \frac{1}{2}(J_+ + J_-)$ corresponds to the physical external field, while $J = J_+ - J_-$ corresponds to the formal source field used for construction of generating functional.

It can be seen by comparison of Eqs. (5.39) and (3.14) that $\sigma(x,y)$ presents the correlation matrix for random forces. If Q is a smooth function of x, σ can be taken as constant

$$\sigma = -(i/4\gamma^2)(S_F + S_{\bar{F}} + S_+ + S_-)(k = 0) \ . \tag{5.40}$$

By use of Eq. (5.19), valid in the tree approximation, Eq. (5.40) can be rewritten as

$$\sigma = -(i/2\gamma^2)(S_+ + S_-) \quad . \tag{5.41}$$

According to the definition of γ [Eq. (4.5)], i.e.,

$$\gamma = \lim_{k_0 \to 0} i(\partial/\partial k_0)\Gamma_r \ ,$$

taking into account that only the dissipative part gives any contribution, and using Eqs. (5.16) and (5.22) which are valid near the thermal equilibrium state, we obtain

$$\gamma = \lim_{k_0 \to 0} \frac{\partial}{\partial k_0} A = \frac{i}{2} \lim_{k_0 \to 0} \frac{\partial}{\partial k_0}(\Gamma_- - \Gamma_+)$$
$$= \tfrac{1}{2} i\beta\Gamma_- = -\tfrac{1}{4} i\beta(S_+ + S_-) \quad . \tag{5.42}$$

Comparing Eq. (5.42) with Eq. (5.41) yields the fluctuation dissipation theorem

$$\sigma = 2/\beta\gamma \ ,$$

which in the ordinary notation is given by

$$\langle \xi(\tau)\xi(\tau') \rangle = 2\Gamma_0 kT\delta(\tau - \tau'), \quad \Gamma_0 = 1/\gamma \ . \tag{5.43}$$

For simplicity we derive here the generating functional for the single component Q. Extension to the multicomponent case is obvious.

VI. DISCUSSIONS

Summarizing the main results of this paper, we come to the following conclusions:

(i) The CTPGF approach is a natural theoretical framework for statistical field theory to describe systems with dominating long-wavelength fluctuations such as dynamical critical phenomena. By use of the CTPGF's the generalized Langevin equations for order parameters and conserved variables with mode coupling terms included in a natural way and the Lagrangian formulation of the classical field theory are deduced from a unified point of view. The perturbation theory of CTPGF in terms of $\hat{G}$ functions has the same structure as that for the ordinary field theory so it is simpler to deal with. In the current theory of critical dynamics and MSR field theory the perturbation expansion is constructed in terms of $\tilde{G}$ functions with two different types of propagators, i.e., the retarded and correlation functions. Therefore, the structure of such perturbation theory is more complicated. Another advantage of the CTPGF formalism is that the causality is guaranteed automatically. It does not need to be verified order by order, as in the existing theory.[19]

(ii) The noncommutativity of field operators, not obvious in the path integral formulation, is not a mathematical trick, but a necessity to describe the time evolution of the statistical field theory. Even if the infrared divergence of the terms, coming from the noncommutativity of operators, is lower than that for other functions, these terms are still needed when considering the time-dependent phenomena, since the infrared divergence of the response function is weaker than that for the correlation function (see Appendix A).

(iii) It can be seen from the calculations in this paper what kind of approximations are assumed in the existing theory of critical dynamics and what possible ways may be used to improve the current theory.

(a) In the existing theory the transport coefficient matrix for the coupling terms with conserved variables is assumed to be antisymmetric; i.e., only canonical motion is considered. It is possible to analyze the crossover effects of dissipative and canonical motions which may occur, in principle, in the framework of CTPGF's.

(b) The one-loop approximation in the path integral over the random fields corresponds to the Gaussian averaging. It is possible to go beyond the Gaussian approximation by calculating higher-loop corrections in the framework of CTPGF's.

(c) The current theory of critical dynamics corresponds to the second-order approximation of $\Delta(x)$, the fluctuations on positive and negative time branches. In principle higher-order corrections can be calculated. It may be more convenient to calculate directly the path integral for Q_+ and Q_-, not introducing $\Delta(x)$ explicitly.

(iv) The renormalization of the existing Lagrangian field theory is quite complicated.[13,14] One of the causes of such complexity lies in the fact that the number of vertices and primitive divergences is much greater than the number of coupling constants and also that Q and $\hat{Q}$ have different dimensions. It seems that the renormalization procedure will be simpler in terms of $\hat{G}$ functions, since different components of Green's function matrix have the same infrared divergence.

ACKNOWLEDGMENTS

One of the authors (L.Y.) would like to express his sincere gratitude to Professor H. Ehrenreich, Professor B. I. Halperin, Professor P. C. Martin, Professor D. Nelson, and Professor T. T. Wu for the kind hospitality they have extended to him at Harvard and for helpful discussions with them. He is also indebted for Professor A. Aharony, Dr. R. Bruinsma, Dr. B. L. Hu, Professor A. Jaffe, Dr. R. Morf, Dr. H. Sampolinsky, Professor M. Stephen, and Dr. A. Zippelius for interesting discussions. He is supported in part by the NSF through Grant No. DMR-77-10210.

APPENDIX A: RENORMALIZATION OF THE FINITE-TEMPERATURE FIELD THEORY

As shown by Zhou and Su[6] for the general case, the counter terms introduced in quantum field theory for $T=0$ K are enough to remove all ultraviolet divergences for the CTPGF's at any temperature. Other authors (see references cited in Ref. 6) come to the same conclusion for finite temperature field theory without resorting to the CTPGF formalism. This result is reasonable from the physical point of view since the statistical average does not change the properties of systems at very short distances and therefore does not contribute new ultraviolet divergences. What we should like to point out is that in considering the phase transitionlike phenomena it is necessary to separate first the leading infrared divergent term and then to carry out the ultraviolet renormalization which is different from that for the usual quantum field theory.

To be concrete, consider the relativistic scalar Bose field, the CTPGF propagators for which can be written as[4]

$$G_{++}(k) \equiv \Delta_F(k) = \frac{1}{k^2-m^2+i\epsilon} - 2\pi i n(\vec{k})\delta(k^2-m^2) \ ,$$

$$G_{-+}(k) \equiv \Delta_-(k) = -2\pi i \delta(k^2-m^2)[\Theta(k_0)+n(\vec{k})] \ ,$$

$$G_{+-}(k) \equiv \Delta_+(k) = -2\pi i \delta(k^2-m^2)[\Theta(-k_0)+n(\vec{k})] \ ,$$

$$G_{--}(k) \equiv \Delta_{\bar{F}}(k) = \frac{-1}{k^2-m^2-i\epsilon} - 2\pi i n(\vec{k})\delta(k^2-m^2) \ , \quad \text{(A1)}$$

where

$$n(\vec{k}) = \{\exp[\epsilon(\vec{k})/T]-1\}^{-1}, \quad \epsilon(\vec{k}) = (\vec{k}^2+m^2)^{1/2} \ . \quad \text{(A2)}$$

Near the phase transition point $m \approx 0$.

$$\epsilon(\vec{k})/T << 1, \quad n(\vec{k}) \approx T/\epsilon(\vec{k}) >> 1 \quad \text{(A3)}$$

for the long-wavelength excitations. Since the $n(\vec{k})$ terms appear together with the δ function, i.e., on the mass shell, the integration over frequencies can be carried out automatically, so the infrared divergence of these terms is higher than that for other terms by one order of magnitude. Therefore the marginal space dimension for renormalizability for finite temperature φ^4 theory is $d_c=4$, not $d_c=4-1$ as in the case of the ordinary field theory. This is what is meant by saying "quantum system in d dimensions corresponds to the classical system in $d+1$ dimensions."

What has been said above can be verified explicitly by calculating the primitive divergent diagrams for mass, vertex, and wave function renormalization, carrying out the frequency integration, and taking the high-temperature limit $T >> \epsilon(\vec{k})$ to retrieve the results which are identical with that of the current theory of critical phenomena.[20] It is much easier to verify this by use of Matsubara Green's functions, retaining only terms $\omega_n=0$ in the frequency summation.

Some investigators of finite-temperature field theory improperly use the renormalization constants for the $T=0$ K case to study phase transition related phenomena. Since the high-temperature limit has been taken for the case of phase transition both relativistic and quantum effects are unimportant. The only possible exception is phase transition near $T=0$ K, where both statistical and quantum fluctuations play their parts. As far as phase transition is concerned the ordinary field models cannot give anything new beyond the current theory of critical phenomena. (The situation for the nonabelian gauge models is not quite clear.)

The noncommutativity of operators is not essential for the static phenomena, that implies the four propagators in Eq. (A1) may be replaced by the correlation function $-2\pi i n(\vec{k})\delta^2(K^2-m^2)$. This is not the case for dynamic phenomena. The first term of G_{++} and G_{--} comes from the inhomogeneous term of Green's function equation, i.e., the commutator. If only the leading infrared divergent terms are retained the four propagators become equal to one another, so that the retarded function

$$G_r = G_{++} - G_{+-} = 0 \ .$$

Therefore the retarded Green's functions are less infrared divergent than the correlation functions. To treat them properly, the noncommutativity of operators has to be taken into account, even though this is a "purely" classical field theory. It is easy to show that all these properties illustrated with the free propagators remain true for the renormalized propagators.

As mentioned in the Introduction, the high-temperature limit of statistical field theory corresponds to the "super Bose" limit, but not the Boltzmann limit. Usually we consider the classical fields to be commutative, since unity can be ignored in comparison with n which is large. If unity cannot be neglected in the phenomena under study we have to start with the noncommutative operators. This is one of the reasons why statistical field theory and quantum field theory have so close an analogy. The physical implications of this analogy are discussed elsewhere.[21]

APPENDIX B: FURTHER RESULTS ON TRANSFORMATIONS OF CTPGF

The main results concerning the transformations of different forms for CTPGF's are described in Sec. II B. Here we shall prove the theorems (2.37) and (2.40), illustrate them by more complicated examples, and discuss the transformations for connected Green's functions.

It is more convenient for some cases to introduce the spinor notation. Let

$$Q \equiv \begin{bmatrix} \eta^T \\ \xi^T \end{bmatrix}, \quad Q^T \equiv (\eta, \xi) \; ; \tag{B1}$$

i.e.,

$$Q_1 = \eta^T = (1/\sqrt{2})(1,-1) \; , \quad Q_2 = \xi^T = (1/\sqrt{2})(1,1) \; ,$$

$$Q_1^T = \eta = \frac{1}{\sqrt{2}}\begin{bmatrix} 1 \\ -1 \end{bmatrix} \; , \quad Q_2^T = \xi = \frac{1}{\sqrt{2}}\begin{bmatrix} 1 \\ 1 \end{bmatrix} \; . \tag{B2}$$

The normalization condition

$$QQ^T = I$$

is expressed as

$$\eta^\alpha \eta_\alpha = \xi^\alpha \xi_\alpha = 1 \; , \quad \eta^\alpha \xi_\alpha = \xi^\alpha \eta_\alpha = 0 \; , \tag{B3}$$

where η^α, ξ^α are components of η^T and ξ^T, etc.

In spinor notation the transformation from $\hat{G}$ to $\tilde{G}$ [Eq. (2.35)] is

$$G_{2\cdots 2(k),1\cdots 1(n-k)}(12\cdots n) = 2^{n/2-1}\xi^{\alpha_1}\cdots\xi^{\alpha_k}\eta^{\alpha_{k+1}}\cdots\eta^{\alpha_n}G_{\alpha_1\cdots\alpha_n}(12\cdots n) \; , \tag{B4}$$

while the inverse transformation from $\tilde{G}$ to $\hat{G}$ [Eq. (2.36)] is

$$\begin{aligned} G_{\alpha_1\alpha_2\cdots\alpha_n}(12\cdots n) = 2^{1-n/2}(&\xi_{\alpha_1}\cdots\xi_{\alpha_n}G_{22\cdots 2} + \eta_{\alpha_1}\xi_{\alpha_2}\cdots\xi_{\alpha_n}G_{12\cdots 2} + \xi_{\alpha_1}\eta_{\alpha_2}\cdots\xi_{\alpha_n}G_{212\cdots 2} + \cdots \\ &+\xi_{\alpha_1}\cdots\xi_{\alpha_{n-1}}\eta_{\alpha_n}G_{2\cdots 21} + \cdots + \eta_{\alpha_1}\cdots\eta_{\alpha_{n-1}}\xi_{\alpha_n}G_{1\cdots 12} \\ &+\eta_{\alpha_1}\cdots\xi_{\alpha_{n-1}}\eta_{\alpha_n}G_{1\cdots 21} + \cdots + \xi_{\alpha_1}\eta_{\alpha_2}\cdots\eta_{\alpha_n}G_{21\cdots 1}) \; . \end{aligned} \tag{B5}$$

The generating functional $Z(J(x))$ can be expanded as

$$\begin{aligned} Z(J(x)) &= \sum_{n=0}^{\infty}\frac{1}{n!}\int_p\cdots\int_p \frac{\delta^n Z}{\delta J(1)\cdots\delta J(n)}\bigg|_{J=0} J_p(1)\cdots J_p(n) \\ &= 1 - i\sum_{n=1}^{\infty}\frac{1}{n!}\int_p\cdots\int_p G_p(1\cdots n)J_p(1)\cdots J_p(n) \\ &= 1 - i\sum_{n=1}^{\infty}\frac{1}{n!}\int_{-\infty}^{\infty}\cdots\int_{-\infty}^{\infty} G_{\alpha_1\cdots\alpha_n}(1\cdots n)(\sigma_3\hat{J})_{\alpha_1}\cdots(\sigma_3\hat{J})_{\alpha_n} \; . \end{aligned} \tag{B6}$$

If we take $J_+(x) = J_-(x) = J(x)$, Eq. (B6) becomes

$$Z(J(x)) = 1 - i\sum_{n=1}^{\infty}\frac{1}{n!}\int_{-\infty}^{\infty}\cdots\int_{-\infty}^{\infty} G_{\alpha_1\cdots\alpha_n}(1\cdots n)\eta^{\alpha_1}\cdots\eta^{\alpha_n}J(1)\cdots J(n) \; . \tag{B7}$$

According to the normalization condition (2.16)

$$Z(J(x))\big|_{J_+(x)=J_-(x)=J(x)} = 1$$

and considering the arbitrariness of $J(x)$ we obtain

$$\tilde{G}_{11\cdots 1}(12\cdots n) = 2^{n/2-1}\eta^{\alpha_1}\eta^{\alpha_2}\cdots\eta^{\alpha_n}G_{\alpha_1\alpha_2\cdots\alpha_n}(12\cdots n) = 0 \tag{B8}$$

thus the first theorem (2.37) has been proved.

It follows from Eq. (B8) for three-point functions that

$$G_{+++} + G_{+--} + G_{-+-} + G_{--+} = G_{---} + G_{++-} + G_{+-+} + G_{-++} \; . \tag{B9}$$

Similarly, for four-point functions we have

$$G^{(+)} \equiv \frac{1}{2}\sum_{\alpha\beta\gamma\delta}(1+\alpha\beta\gamma\delta)G_{\alpha\beta\gamma\delta} = G^{(-)} \equiv \frac{1}{2}\sum_{\alpha\beta\gamma\delta}(1-\alpha\beta\gamma\delta)G_{\alpha\beta\gamma\delta} \; ; \tag{B10}$$

i.e, the sums of terms with the same signature are equal.

To prove the second theorem we first multiply Eq. (B4) by the normalization condition of step functions (2.29) to obtain

$$G_{2\cdots 2(k),1\cdots 1(n-k)}(1\cdots n)=(-i)^{n-1}2^{n/2-1}\sum_{P\begin{pmatrix}1\cdots n\\ p_1\cdots p_n\end{pmatrix}}\Theta(p_1\cdots p_n)\xi^{\alpha_1}\cdots\xi^{\alpha_k}\eta^{\alpha_{k+1}}\cdots\eta^{\alpha_n}$$
$$\times\langle T_p(\varphi_{\alpha_1}(1)\cdots\varphi_{\alpha_n}(n))\rangle\ . \tag{B11}$$

Let

$$\zeta^{p_i}=\begin{cases}\xi^{p_i}, & \text{if } 1\le p_i\le k,\\ \eta^{p_i}, & \text{if } k+1\le p_i\le n,\end{cases} \tag{B12}$$

we have

$$\Theta(p_1\cdots p_n)\xi^{\alpha_1}\cdots\xi^{\alpha_k}\eta^{\alpha_{k+1}}\cdots\eta^{\alpha_n}\langle T_p(\varphi_{\alpha_1}(1)\cdots\varphi_{\alpha_n}(n))\rangle$$
$$=\Theta(p_1\cdots p_n)\zeta^{p_1}\cdots\zeta^{p_n}\langle T_p(\varphi_{p_1}(p_1)\cdots\varphi_{p_n}(p_n))\rangle \tag{B13}$$

because of the definition of the Θ function (2.26) and the symmetry of CTPGF.

Since the time ordering in the usual sense has already been fixed by the Θ function the action of the T_p operator is reduced to

$$\zeta^{p_n}\langle T_p(\varphi_{p_1}(p_1)\cdots\varphi_{p_n}(p_n))\rangle=\frac{1}{\sqrt{2}}\langle(T_p(\varphi_{p_1}(p_1)\cdots\varphi_{p_{n-1}}(p_{n-1})),\ \varphi(p_n))\rangle\ ,$$

where

$$(\ ,\ )=\begin{cases}[\ ,\ ], & \text{if } \zeta^{p_n}=\eta^{p_n},\\ \{\ ,\ \}, & \text{if } \zeta^{p_n}=\xi^{p_n}.\end{cases} \tag{B14}$$

Such a process of getting rid of T_p can be continued up to the last step to get zero, if $\zeta^{p_1}=\eta^{p_1}$ or

$$2^{-n/2+1}((\cdots(\varphi(p_i),\varphi(p_2))\cdots),\varphi(p_n))\ ,$$

if $\zeta^{p_1}=\xi^{p_1}$. We see that the factor $2^{-n/2+1}$ exactly cancels out the numerical coefficient $2^{n/2-1}$, so proving the theorem (2.40). In Sec. II B we have considered the two-point functions. As a further example we have for $n=3$

$$G_{211}(123)=(-i)^2\sum_{p\in\begin{bmatrix}23\\ ij\end{bmatrix}}\Theta(1ij)\langle[[1,i],j]\rangle\ ,$$
$$G_{221}(123)=(-i)^2\sum_{p\in\begin{bmatrix}12\\ ij\end{bmatrix}}(\Theta(ij3)\langle[\{i,j\},3]\rangle+\Theta(i3j)\langle\{[i,3],j\}\rangle)\ , \tag{B15}$$
$$G_{222}(123)=(-i)^2\sum_{p\in P_3}\Theta(ijk)\langle\{\{i,j\},k\}\rangle=(-i)^2\langle\{\{1,2\},3\}\rangle\ ,$$

and for $n=4$

$$G_{2111}(1234)=(-i)^3\sum_{p\in\begin{bmatrix}234\\ ijk\end{bmatrix}}\Theta(1ijk)\langle[[[1,i],j],k]\rangle\ ,$$
$$G_{2211}(1234)=(-i)^3\sum_{p\in\begin{bmatrix}12\\ ij\end{bmatrix}\begin{bmatrix}34\\ kl\end{bmatrix}}(\Theta(ijkl)\langle[[\{i,j\},k],l]\rangle+\Theta(ikjl)\langle[\{[i,k],j\},l]\rangle+\Theta(iklj)\langle\{[[i,k],l],j\}\rangle)\ ,$$
$$G_{2221}(1234)=(-i)^3\sum_{p\in\begin{bmatrix}123\\ ijk\end{bmatrix}}(\Theta(ijk4)\langle[\{\{i,j\},k\},4]\rangle+\Theta(ij4k)\langle\{[\{i,j\},4],k\}\rangle+\Theta(i4jk)\langle\{\{[i,4],j\},k\}\rangle)\ , \tag{B16}$$
$$G_{2222}(1234)=(-i)^3\langle\{\{\{1,2\},3\},4\}\rangle\ ,$$

where for short we write $i \equiv \varphi(i)$, etc.

It is interesting to note that although all the possible combinations of retarded, advanced, and correlation functions are realized in real time, they are defined by the CTPGF approach in a quite natural way.

The first theorem (2.37) is valid also for the connected CTPGF, since we can repeat word for word the proof starting from Eq. (2.17). This is not the case for the second theorem, where some complications appear. It follows from the definitions of disconnected [Eq. (2.4)] and connected [Eq. (2.15)] CTPGF's that

$$G_p^c(1) = G_p(1) \ , \tag{B17a}$$

$$G_p^c(1,2) = G_p(1,2) + iG_p(1)G_p(2) \ , \tag{B17b}$$

$$G_p^c(1,2,3) = G_p(1,2,3) + i[G_p(1)G_p(2,3) + G_p(2)G_p(1,3) + G_p(3)G_p(1,2)] - 2G_p(1)G_p(2)G_p(3) \ldots \ . \tag{B17c}$$

It can be shown by use of Eq. (B17) that the formulas for all "purely" retarded functions remain true, as for example

$$G_{21}^c(1,2) = G_{21}(1,2) = -i\Theta(1,2)\langle[1,2]\rangle \ , \quad G_{211}^c(1,2,3) = G_{211}(1,2,3) \ , \quad G_{2111}^c(1,2,3,4) = G_{2111}(1,2,3,4) \ldots \tag{B18}$$

These functions are similar to r functions used to construct the Lehmann-Symanzik-Zimmermann (LSZ)[22] axiomatic field theory, which are the same for both connected and disconnected functions. All other functions are modified, for example,

$$G_c^c(12) = -i(\langle\{1,2\}\rangle - \{\langle 1\rangle, \langle 2\rangle\}) \ , \tag{B19}$$

$$\begin{aligned} G_{122}^c(123) = (-i)^2 \sum_{p_2} \{ & \Theta(p_2,p_3,1)(\langle[\{p_2,p_3\},1]\rangle - \langle[\langle p_2\rangle p_3,1]\rangle - \langle[p_2\langle p_3\rangle,1]\rangle) \\ & + \Theta(p_2,1,p_3)(\langle\{[p_2,1],p_3\}\rangle - \langle[p_2,1]\rangle\langle p_3\rangle)\} \ , \end{aligned} \tag{B20}$$

$$G_{222}^c(123) = (-i)^2 \sum_{p_3} \Theta(p_1p_2p_3)(\langle\{\{p_1,p_2\},p_3\}\rangle - \langle\{p_1,p_2\}\rangle\langle p_3\rangle + 2\{\{\langle p_1\rangle, \langle p_2\rangle\}, \langle p_3\rangle\}) \ , \tag{B21}$$

$$G_{2211}^c(1234) = G_{2211} + 2i[G(1)G_{211}(234) + G(2)G_{211}(134)] + 2i[G_r(13)G_r(24) + G_r(14)G_r(23)] \ , \tag{B22}$$

$$\begin{aligned} G_{2221}^c(1234) = {} & G_{2221}(1234) + 2i[G(1)G_{221}(234) + G(2)G_{221}(134) + G(3)G_{221}(124)] \\ & + 2i[G_c(12)G_r(34) + G_c(13)G_r(24) + G_c(23)G_r(14)] \\ & - 8[G(1)G(2)G_r(34) + G(1)G(3)G_r(24) + G(2)G(3)G_r(14)] \ . \end{aligned} \tag{B23}$$

---

*On leave from the Institute of Theoretical Physics, Academia Sinica, Beijing, China.

[1] J. Schwinger, J. Math. Phys. (N.Y.) 2, 407 (1961).

[2] L. V. Keldysh, Zh. Eksp. Teor. Fiz. 47, 1515 (1964) [Sov. Phys. JETP 20, 1018 (1965)].

[3] See, e.g., D. Dubois, in *Lectures in Theoretical Physics*, edited by W. E. Brittin (Gordon and Breach, New York, 1967), Vol. IX C; V. Korenman, Ann. Phys. (N.Y.) 39, 72 (1966); D. Langreth, in *Linear and Nonlinear Electronic Transport in Solids*, edited by J. Devreese and V. Van Doren (Plenum, New York, 1976); A.-M. Tremblay, B. Patton, P. C. Martin, and P. Maldaque, Phys. Rev. A 19, 1721 (1979).

[4] Zhou Guang-zhao and Su Zhao-bin, Progress in Statistical Physics (Kexue, Beijing, to be published in Chinese), Chap. 5.

[5] Zhou Guang-zaho and Su Zhao-bin, Physica Energiae Fortis et Physica Nuclearis (Beijing) 3, 314 (1979).

[6] Zhou Guang-zhao and Su Zhao-bin, Physica Energiae Fortis et Physica Nuclearis (Beijing) 3, 304 (1979).

[7] Zhou Guang-zhao and Su Zhao-bin, Acta Phys. Sin. (in press).

[8] P. C. Martin, E. D. Siggia, and H. A. Rose, Phys. Rev. A 8, 423 (1973).

[9] Hao Bai-lin, in Progress in Statistical Physics (Kexue, Beijing, to be published in Chinese), Chap. 1.

[10] P. C. Hohenberg and B. I. Halperin, Rev. Mod. Phys. 49, 435 (1977).

[11] K. Kawasaki, in *Critical Phenomena*, edited by M. S. Green (Academic, New York, 1971).

[12] See for example, B. I. Halperin, P. C. Hohenberg, and S. K. Ma, Phys. Rev. B 10, 139 (1974); S. K. Ma and G. F. Mazenko, *ibid.* 11, 4077 (1975).

[13] H. K. Janssen, Z. Phys. B 23, 377 (1976); R. Bausch, H. K. Janssen, and H. Wagner, *ibid.* 24, 113 (1976).

[14] C. De Dominicis and L. Peliti, Phys. Rev. B 18, 353 (1978).

[15] L. Onsager and S. Machlup, Phys. Rev. 91, 1505, 1512 (1953).

[16] R. Graham, Springer Tracts Mod. Phys. 66, 1 (1973).

[17] R. Graham and H. Haken, Z. Phys. 243, 289 (1971).

[18] L. Sasvari, F. Schwabl, and P. Szepfalusy, Physica (Utrecht) 81, 108 (1975).

[19] J. Deker and F. Haake, Phys. Rev. A 11, 2043 (1975).

[20] E. Brezin, J. C. Le Guillou, and J. Zinn-Justin, in *Phase Transitions and Critical Phenomena*, edited by C. Domb and M. S. Green (Academic, New York, 1976), Vol. VI.

[21] Yu Lu and Hao Bai-lin, Wuli (in press).

[22] H. Lehmann, K. Symanzik, and W. Zimmermann, Nuovo Cimento 6, 319 (1957).

Commun. in Theor. Phys. (Beijing), China) Vol. 1, No. 3 (1982) 295-306

ON THEORY OF THE STATISTICAL GENERATING FUNCTIONAL FOR THE ORDER PARAMETER(I) — GENERAL FORMALISM

ZHOU Guang-zhao (周光召) SU Zhao-bin (苏肇冰)
HAO Bai-lin (郝柏林) YU Lu (于 渌)
(Institute of Theoretical Physics,
Academia Sinica, Beijing, China)

Received December 28, 1981.

Abstract

A theoretical scheme using closed time-path Green's functions is proposed to describe the quantum statistical properties of the order parameter in terms of a generating functional. The dynamic evolution is generated by a driving source, while the statistical correlation by a fluctuation source. The statistical causality is shown to hold explicitly and to give rise to a number of important consequences. The problem of determining the quantum statistical properties for the order parameter is reduced to finding a solution of the functional equation for it.

I. Introduction

It is well known that special tricks have to be used in the standard Green's function techniques[1] to describe systems with broken symmetry such as superconductivity or Bose-Einstein condensation. On the other hand, a systematic theoretical scheme has been developed in the quantum theory of gauge fields to treat the order parameter.[2] The generating functional formalism developed there has been applied successfully to the classical theory of static critical phenomena.[3] To treat the time-dependent phenomena Martin, Siggia and Rose(MSR) have constructed a noncommutative classical field theory closely analogous to the quantum field theory. The MSR theory has been reformulated into a Lagrangian formalism[5,6] and has been used extensively in studying dynamic critical phenomena.[7] But, the physical meaning of the "response field" introduced in that theory is not clear. There are also some attempts of quantum generalization of Langevin and Fokker-Planck equations for describing order parameter.[8] It seems difficult to use those kinds of semi-phenomenological theories to study nonuniform systems and composite order parameter. The necessity of constructing a unified quantum theory to describe nonuniform, nonequilibrium systems with broken symmetry is obvious from the current research on condensed matter and laser phenomena. In some cases (e.g. nonequilibrium superconductivity) the dynamic coupling of the order parameter with the elementary excitations is essential. It is almost impossible to incorporate this kind of coupling into the theoretical schemes mentioned above.

We have shown previously[9] that a unified microscopic description of both equilibrium and nonequilibrium systems is possible by combining the closed time-path Green's functions (CTPGF) with the techniques of generating functional. The transformation properties of CTPGF are studied in Refs.[10,11] and the MSR theory appears to be a "physical" representation of CTPGF in the classical (Rayleigh-Jeans) limit. We prove[12] the existence of a generalized potential functional for the order parameter in a nonequilibrium stationary state for systems with time reversal symmetry. The time-dependent Ginsburg-Landau (TDGL) equations for order parameters and conserved densities are derived[11-13] with both time reversible and irreversible parts, also the Lagrangian formulation in critical dynamics is obtained in a low order approximation of CTPGF. It seems to us that CTPGF formalism is a good candidate for the general quantum theory of nonequilibrium systems discussed above.

In the present series of papers we propose to construct a quantum theory of the statistical generating functional for order parameter using CTPGF which is capable of describing both aspects of the Liouville problem: dynamic evolution and statistical correlation. The theoretical scheme is general enough to incorporate nonuniform, nonequilibrium systems with either simple or composite order parameter. We will give a practical prescription for constructing the generating functional being applied to concrete systems. The dynamic coupling of the order parameter with quasi-particles will be treated in future publications.

The rest of the first paper in the present series is organized as follows. In Sec. II the generating functional for the order parameter is constructed explicitly. The dynamic evolution is generated by a driving source which might be the actual external field, while the statistical correlation is generated by a fluctuation source which vanishes on the completion of calculation. The statistical causality and its consequences are studied in Sec. III. In Sec. IV a statistical functional equation for the order parameter on a generalized manifold which is equivalent to the Langevin equation to certain extent is derived and the problem of determining the statistical properties of the order parameter is reduced therefore to finding a solution of this functional equation. The final section contains a few concluding remarks. Throughout this series of papers we will use the system of units with $\hbar = C = k_B = 1$, but we might come back to the ordinary units if necessary.

II. The statistical generating functional for order parameter

Suppose $\hat{Q}_\alpha(X)$, $\alpha = 1,2...N$, are a set of Hermitian order parameters in Heisenberg picture which might be basic field variables or composite operators and $h_\alpha(X)$ are corresponding C-number real external sources. Both sets are defined on a closed time-path consisting of a positive branch $(-\infty, +\infty)$ and a negative branch $(+\infty, -\infty)$. The CTPGF generating functional for order parameters can be written as[9,10]:

$$Z_p[h(x)] = \mathrm{Tr}\left\{\hat{\rho}\, T_p \exp\left(i\int_p d^4x\, h(x)\hat{Q}(x)\right)\right\}, \tag{2.1}$$

where $\hat{\rho}$ is the density matrix in Heisenberg picture, T_p is time-ordering operator along the closed time-path. Tr has the usual meaning of trace operation in Hilbert space. $\int_p d^4X\cdots$ means 4-dimensional integration along the closed time-path. We note in passing that the letter "*p*" will often indicate quantities defined on closed path or operations along it. We have also omitted the index α for $\hat{Q}_\alpha(X)$ and $h_\alpha(X)$ and a summation over α is implicitly understood.

Introduce index σ for the time branch as:

$$h^\sigma(x) \equiv h(x_\sigma), \tag{2.2}$$

$$\hat{Q}^\sigma(x) \equiv \hat{Q}(x_\sigma), \tag{2.3}$$

with $\sigma = +$ or $-$, indicating whether the space-time point is at the positive or at the negative branch. Eq.(2.2) defines two independent C-number functions from a single function defined on the closed path. However, Eq.(2.3) defines two operators $\hat{Q}^\sigma(X)$ in the ordinary space-time with different time-ordering properties, but representing the same quantity. Under the action of T_p the operator $\hat{Q}^-(X)$ should always precede the operator $\hat{Q}^+(X)$, the $\hat{Q}^+$ operators are time-ordered in the usual way among themselves, while the $\hat{Q}^-$ operators are anti-time-ordered (denoted by $\tilde{T}$). After time-ordering operations $\hat{Q}^+$ and $\hat{Q}^-$ operators are set to be identical. Introducing, furthermore,

$$\eta_\sigma \equiv \begin{cases} 1, & \text{if } \sigma = + \\ -1, & \text{if } \sigma = - \end{cases}$$

$$\xi_\sigma \equiv 1 \quad \text{for } \sigma = + \text{ and } -, \tag{2.4}$$

and defining

$$\hat{Q}_c(x) = \frac{1}{2}\xi_\sigma \hat{Q}^\sigma(x), \qquad \hat{Q}_\Delta(x) = \eta_\sigma \hat{Q}^\sigma(x), \tag{2.5a}$$

$$h_c(x) = \frac{1}{2}\xi_\sigma h^\sigma(x), \qquad h_\Delta(x) = \eta_\sigma h^\sigma(x), \tag{2.5b}$$

we have

$$\hat{Q}^\sigma(x) = \xi_\sigma \hat{Q}_c(x) + \frac{1}{2}\eta_\sigma \hat{Q}_\Delta(x), \tag{2.6a}$$

$$h^\sigma(x) = \xi_\sigma h_c(x) + \frac{1}{2}\eta_\sigma h_\Delta(x). \tag{2.6b}$$

The 4-dimensional integration over the closed path is transformed into the ordinary integration as

$$\int_p d^4x \cdots = \int d^4x_\sigma \eta_\sigma \cdots \tag{2.7}$$

According to Eqs. (2.4)-(2.7) and taking into account that

$$\frac{\delta h(1)}{\delta h(2)} = \delta_p^4(1-2), \tag{2.8}$$

$$\frac{\delta h^\sigma(1)}{\delta h^\rho(2)} = \delta_{\sigma\rho}\delta^4(1-2), \tag{2.9}$$

where 1,2 are short notations for X_1 and X_2, we find that

$$\frac{\delta}{\delta h^\sigma(x)} = \eta_\sigma \frac{\delta}{\delta h(x_\sigma)} = \frac{1}{2}\xi_\sigma \frac{\delta}{\delta h_c(x)} + \eta_\sigma \frac{\delta}{\delta h_\Delta(x)}, \tag{2.10}$$

$$\frac{\delta}{\delta h_c(x)} = \xi_\sigma \frac{\delta}{\delta h^\sigma(x)} = \eta_\sigma \frac{\delta}{\delta h(x_\sigma)}, \tag{2.11}$$

$$\frac{\delta}{\delta h_\Delta(x)} = \frac{1}{2}\eta_\sigma \frac{\delta}{\delta h^\sigma(x)} = \frac{1}{2}\xi_\sigma \frac{\delta}{\delta h(x_\sigma)}. \tag{2.12}$$

The closed time-path δ-function in Eq. (2.8) is defined as

$$\int_p d^4x\, \delta_p^4(x-y) = 1 . \tag{2.13}$$

Hereafter a summation over repeated index is always understood except for special reservation like in Eq. (2.10) where no summation over σ is assumed.

Using Eqs. (2.2), (2.3), (2.6), and (2.7) the generating functional defined by Eq. (2.1) can be rewritten as

$$Z[h_\Delta(x), h_c(x)] = Z_p[h(x)]$$
$$= \mathrm{Tr}\{\hat{\rho}\, T_p \exp[i\int d^4x (h_\Delta(x)\hat{Q}_c(x) + h_c(x)\hat{Q}_\Delta(x))]\}. \tag{2.14}$$

Expanding Z over h_c and h_Δ in the usual space-time we find

$$Z[h_\Delta(x), h_c(x)] = \sum_{m,\ell=0}^{\infty} \frac{1}{m!\,\ell!}\int d_1 \cdots d m\, d\bar{1} \cdots d\bar{\ell}$$
$$\times \left[\frac{\delta^{m+\ell} Z[h_\Delta(x), h_c(x)]}{\delta h_\Delta(1)\cdots\delta h_\Delta(m)\,\delta h_c(\bar{1})\cdots\delta h_c(\bar{\ell})}\right]\Bigg|_{h_\Delta=h_c=0} \cdot\ h_\Delta(1)\cdots h_\Delta(m)\, h_c(\bar{1})\cdots h_c(\bar{\ell}), \tag{2.15}$$

where

$$\frac{(-i)^n \delta^n Z[h_\Delta(x), h_c(x)]}{\delta h_\Delta(1)\cdots\delta h_\Delta(m)\,\delta h_c(m+1)\cdots\delta h_c(n)}\Bigg|_{h_\Delta=h_c=0}$$
$$= \mathrm{Tr}\{\hat{\rho}\, T_p[\hat{Q}_c(1)\cdots\hat{Q}_c(m)\,\hat{Q}_\Delta(m+1)\cdots\hat{Q}_\Delta(n)]\}. \tag{2.16}$$

It can be shown[10] using Eqs. (2.2), (2.3) and (2.5) that Eq. (2.16) can be rewritten as

$$\mathrm{Tr}\{\hat{\rho}\, T_p[\hat{Q}_c(1)\cdots\hat{Q}_\Delta(n)]\}$$
$$= \sum_{P\left(\frac{1}{\bar{1}}\cdots\frac{n}{\bar{n}}\right)}' 2^{-m+1}\,\theta(\bar{1},\cdots,\bar{n})\,\mathrm{Tr}\{\hat{\rho}(\cdots((\hat{Q}(\bar{1}),\hat{Q}(\bar{2})),\hat{Q}(\bar{3})),\cdots\hat{Q}(\bar{n}))\}, \tag{2.17}$$

where

$$\theta(\bar{1},\bar{2},\cdots\bar{n}) = \begin{cases} 1, & \text{if } t_{\bar{1}} > t_{\bar{2}} \cdots > t_{\bar{n}}, \\ 0, & \text{otherwise}, \end{cases} \tag{2.18}$$

and satisfies the normalization condition

$$\sum_{P\left(\begin{smallmatrix}1,\,2,\,\cdots\,n\\ \bar{1},\,\bar{2},\,\cdots\,\bar{n}\end{smallmatrix}\right)} \theta\,(\,\bar{1},\,\bar{2},\,\cdots\,\bar{n})=1 \tag{2.19}$$

Σ' means summation over all permutations $P(\begin{smallmatrix}1,2,\cdots n\\ \bar{1},\bar{2},\cdots \bar{n}\end{smallmatrix})$ except for cases $m+1\le \bar{1}\le n$. We use here also a simplified notation given by

$$(\cdots,\hat{Q}(\bar{i}))=\begin{cases}\{\cdots,\hat{Q}(\bar{i})\}_+, & \text{if } 1\le\bar{i}\le m,\\ [\cdots,\hat{Q}(\bar{i})]_-, & \text{if } m+1\le\bar{i}\le n.\end{cases} \tag{2.20}$$

Now we have obtained a new expression for the generating functional given by Eqs. (2.15)-(2.17) with explanation following them. Although this definition is fully equivalent to the original one given by Eq. (2.1), it allows of an entirely new interpretation. The generating functional is now defined in the ordinary space-time, while the closed time-path as an auxiliary tool disappears altogether.

To associate with our previous papers we indicate that different representations of CTPGF and transformations from one to the other have been studied in Refs. [10] and [11]. The "natural" representation of CTPGF is defined as

$$\begin{aligned}\hat{G}_{\alpha\beta\cdots\gamma}(1\,2\cdots n)&=(-i)^{n-1}\,T_r\{\hat{\rho}\,T_p(\hat{Q}^{\alpha}(1)\,\hat{Q}^{\beta}(2)\cdots\hat{Q}^{\gamma}(n))\}\\ &=i\,(-1)^n\left.\frac{\delta^n Z[h^+(x),h^-(x)]}{\delta h^{\alpha}(1)\,\delta h^{\beta}(2)\cdots\delta h^{\gamma}(n)}\right|_{h(x)=0},\end{aligned} \tag{2.21}$$

which is very convenient for constructing the perturbation theory. By means of an orthogonal transformation[10,11] we obtain the "physical" representation $\tilde{G}_{ij\cdots s}(1,2\ldots n)$ with $i,j\cdots=1,2$. In the present notation these $\tilde{G}$ functions can be written as

$$\tilde{G}_{\underbrace{22\cdots}_{K}\underbrace{11\cdots}_{n-K}}(1\,2\cdots n)=(-i)^{n-1}\,T_r\{\hat{\rho}\,T_p(\hat{Q}_c(1)\cdots\hat{Q}_c(K)\hat{Q}_\Delta(K+1)\cdots\hat{Q}_\Delta(n))\}, \tag{2.22}$$

which ought to be compared with Eq.(2.16). The components of $\tilde{G}$—retarded, advanced, correlation functions and so on—are closely related to physically observable quantities. As is seen above, these functions are functional derivatives of the same generating functional $Z[h(X)]$ with respect to the driving source $h_c(X)$ and the fluctuation source $h_\Delta(X)$. These two independent functions will generate, correspondingly, the statistical average of the commutator—dynamic response, and the average of anticommutator—statistical correlation. We will further discuss the physical consequences of this important fact and compare our formalism with the previous theories in the next two sections.

Now we define the generating functional for the connected part as

$$W[h_\Delta(x),\,h_c(x)]\equiv -i\,\ln Z[h_\Delta(x),\,h_c(x)], \tag{2.23}$$

and introduce the vertex functional by a Legendre transformation

$$\Gamma[Q_\Delta(x),Q_c(x)]\equiv W[h_\Delta(x),\,h_c(x)]-\int d^4x\,(h_\Delta(x)Q_c(x)+h_c(x)Q_\Delta(x)), \tag{2.24}$$

where

$$Q_c(x) \equiv \frac{\delta W[h_\Delta(x), h_c(x)]}{\delta h_\Delta(x)}, \tag{2.25}$$

$$Q_\Delta(x) \equiv \frac{\delta W[h_\Delta(x), h_c(x)]}{\delta h_c(x)}. \tag{2.26}$$

It follows from Eqs. (2.24)-(2.26) that

$$\frac{\delta \Gamma[Q_\Delta(x), Q_c(x)]}{\delta Q_c(x)} = -h_\Delta(x), \tag{2.27}$$

$$\frac{\delta \Gamma[Q_\Delta(x), Q_c(x)]}{\delta Q_\Delta(x)} = -h_c(x). \tag{2.28}$$

It is easy to check from Eqs. (2.1)-(2.7) and Eqs. (2.10)-(2.12) that

$$W[h_\Delta(x), h_c(x)] = W_p[h(x)], \tag{2.29}$$

$$\Gamma[Q_\Delta(x), Q_c(x)] = \Gamma_p[Q(x)], \tag{2.30}$$

where

$$W_p[h(x)] \equiv -i \ln Z_p[h(x)], \tag{2.31}$$

$$\Gamma_p[Q(x)] \equiv W_p[h(x)] - \int_p d^4x\, h(x) Q(x) \tag{2.32}$$

with

$$Q(x) = \frac{\delta W_p[h(x)]}{\delta h(x)}, \tag{2.33}$$

$$\frac{\delta \Gamma_p[Q(x)]}{\delta Q(x)} = -h(x). \tag{2.34}$$

It is worthwhile to emphasize that Eqs. (2.14), (2.29) and (2.30) tell us that the generating functionals originally defined on the closed time-path are generating functionals for Green's functions closely related to the observable quantities, if they are expressed in appropriate functional arguments. This is the starting point of the present formalism.

III. Basic properties of the generating functional

In this section we will discuss some fundamental properties of the generating functional for the order parameters and their physical implications.

First of all we indicate some basic relations implied by Eq. (2.17) and the convention following it as given by

$$Z[h_\Delta(x), h_c(x)]\Big|_{h_\Delta(x)=0} = 1, \tag{3.1}$$

or

$$\frac{\delta^n Z[h_\Delta(x), h_c(x)]}{\delta h_c(1) \cdots \delta h_c(n)}\Bigg|_{h_\Delta(x)=0} = 0, \quad for \quad n \geq 1, \tag{3.2}$$

$$W[h_\Delta(x), h_c(x)]\Big|_{h_\Delta(x)=0} = 0, \tag{3.3}$$

or

$$\frac{\delta^n W[h_\Delta(x), h_c(x)]}{\delta h_c(1) \cdots \delta h_c(n)}\Bigg|_{h_\Delta(x)=0} = 0 \tag{3.4}$$

and
$$\Gamma\left[Q_\Delta(x),\ Q_c(x)\right]\Big|_{Q_\Delta(x)=0} = 0, \tag{3.5}$$

or
$$\left.\frac{\delta^n \Gamma\left[Q_\Delta(x),\ Q_c(x)\right]}{\delta Q_c(1)\cdots\delta Q_c(n)}\right|_{Q_\Delta(x)=0} = 0 . \tag{3.6}$$

In deriving Eqs. (3.1), (3.3) and (3.5) we have used the normalization condition
$$T_r\{\hat{\rho}\} = 1 . \tag{3.7}$$

Also, we have used a special case of Eqs. (2.26) and (3.4), implying
$$Q_\Delta(x)\Big|_{h_\Delta(x)=0} = \left.\frac{\delta W\left[h_\Delta(x),\ h_c(x)\right]}{\delta h_c(x)}\right|_{h_\Delta(x)=0} = 0 . \tag{3.8}$$

In our previous papers[10,11] Eqs. (3.2), (3.3) have been named normalization conditions for the generating functionals and have been shown to give rise to a number of important consequences, in particular, Eqs. (3.2), (3.4) and (3.6 which in the previous notations look like
$$\tilde{G}_{11\cdots1}(1\,2\cdots n) = 0, \tag{3.9}$$

$$\Gamma_{11\cdots1}(1\,2\cdots n) = 0. \tag{3.10}$$

It is worthwhile to note that Eqs. (3.1)-(3.10) are valid even in the presence of an external field $h_c(X) \neq 0$.

Furthermore, three causality relations equivalent to each other follow from Eqs. (2.17), (2.20), and (2.23)-(2.28) as given by
$$\left.\frac{\delta^n Z\left[h_\Delta(x),\ h_c(x)\right]}{\delta h_\Delta(1)\cdots\delta h_\Delta(m)\,\delta h_c(m+1)\cdots\delta h_c(n)}\right|_{\substack{h_\Delta(x)=0\\ h_c(x)=0}} = 0 \tag{3.11}$$

$$\left.\frac{\delta^n W\left[h_\Delta(x),\ h_c(x)\right]}{\delta h_\Delta(1)\cdots\delta h_\Delta(m)\,\delta h_c(m+1)\cdots\delta h_c(n)}\right|_{\substack{h_\Delta(x)=0\\ h_c(x)=0}} = 0 , \tag{3.12}$$

and
$$\left.\frac{\delta^n \Gamma\left[Q_\Delta(x),\ Q_c(x)\right]}{\delta Q_\Delta(1)\cdots\delta Q_\Delta(m)\,\delta Q_c(m+1)\cdots\delta Q_c(n)}\right|_{\substack{Q_\Delta(x)=0\\ Q_c(x)=Q(x)}} = 0 \tag{3.13}$$

provided any of t_j, $m+1 \leq j \leq n$ is greater than all of t_i, $1 \leq i \leq m$. The function $Q(X)$ in Eq. (3.13) is defined as
$$Q(x) \equiv Q_c(x)\Big|_{\substack{h_\Delta(x)=0\\ h_c(x)=0}} = T_r\{\hat{\rho}\,\hat{Q}(x)\} . \tag{3.14}$$

Examining relation (3.11) we notice that in summation over (2.17) the cases of the leading time ($t_{\bar{l}}$ in Eq. (2.18)) being t_j, $m+1 \leq j \leq n$ are excluded. Eq. (3.12) is a direct consequence of Eqs. (2.23) and (3.11), while Eq. (3.13) follows from the relation between vertex functions and the amputated Green's functions.[3,14]

It is important to emphasize that a causality sequence of the statistical Green's functions is fixed by Eqs. (3.11)-(3.13), namely, the space-time point associated with $h_c(y)$, $Q_c(y)$ cannot precede the space-time point associated with $h_\Delta(X)$ and $Q_\Delta(X)$, since the former is the cause while the latter is the consequence. The usual retarded Green's function or retarded product is a special case of Eqs. (3.11) and (3.12).

We indicate here also some useful product relations implied by Eq. (2.17) and the definition of θ-function as given by Eq. (2.18). For example, we have

$$\tilde{G}_{12}(1\,2)\,\tilde{G}_{21}(1\,2) = 0, \tag{3.15}$$

$$\begin{aligned}\tilde{G}_{211}\tilde{G}_{112} &= \tilde{G}_{211}\tilde{G}_{121} = \tilde{G}_{211}\tilde{G}_{122} = 0,\\ \tilde{G}_{112}\tilde{G}_{221} &= \tilde{G}_{112}\tilde{G}_{121} = \tilde{G}_{121}\tilde{G}_{212} = 0.\end{aligned} \tag{3.16}$$

It is easy to see that the general rule is as follows:

$$\tilde{G}_{i_1\cdots i_{m_1}\cdots i_{m_2}\cdots i_n}(1\cdots n)\,\tilde{G}_{j_1\cdots j_{m_1}\cdots j_{m_2}\cdots j_n}(1\cdots n) = 0, \tag{3.17}$$

provided $i_{m_1} = i_{m_2} = \cdots = i_{m_r} = 2$ and the rest of i are equal to 1 while $j_{m_1} = j_{m_2} = \cdots = j_{m_r} = 1$ and the rest of j are equal to either 1 or 2.

Now we discuss the physical meaning of $h_\Delta(X)$, $h_c(X)$, $Q_\Delta(X)$ and $Q_c(X)$. Apart from Eq. (3.8) we obtain

$$h_\Delta(x)\Big|_{Q_\Delta(x)=0} = 0 \tag{3.18}$$

from Eqs. (2.27) and (3.6), and also

$$\begin{aligned}Q_c(x)\Big|_{h_\Delta(x)=0} &= \mathrm{Tr}\{\hat{\rho}\,T_p[\hat{Q}_c(x)\exp(i\textstyle\int d^4x\,h_c(x)Q_\Delta(x))]\}\\ &= \mathrm{Tr}\{\hat{\rho}(\tilde{T}\exp(-i\int_{-\infty}^{t} d^4y\,h_c(y)\hat{Q}(y)))\hat{Q}(x)(T\exp(i\int_{-\infty}^{t} d^4y\,h_c(y)\hat{Q}(y)))\}\end{aligned} \tag{3.19}$$

from Eqs. (2.14) and (2.25). Eqs. (3.8) and (3.18) tell us that the relations $h_\Delta(X) = 0$ and $Q_\Delta(X) = 0$ are equivalent to each other, while Eqs. (3.14) and (3.19) tell us that $Q_c(X)$ under the condition $h_\Delta(X) = 0$ is the statistical average of order parameter in the driving field $h_c(X)$ and that $Q(X)$ is the expectation value without an external field. As mentioned in Sec. II the functional derivative $\delta/\delta h_c(X)$ generates the commutator of the order parameter describing the dynamic evolution in the quantum mechanical sense, while the functional derivative $\delta/\delta h_\Delta(X)$ generates the anticommutator describing the correlation in the statistical mechanical sense. Although the physically observable quantities are defined at the functional manifold $h_\Delta(X) = 0$ or $Q_\Delta(X) = 0$, these functional arguments are needed in addition to $h_c(X)$ and $Q_c(X)$ for a complete description of the statistical systems. These two complementary aspects of the Liouville problem—dynamic evolution and statistical correlation—forming the essential elements of the semi-phenomenological Langevin-Fokker-Planck theory have been grasped in the CTPGF formalism in a natural way as specified above.

As an illustrative example we give here the response of statistical correlations to external field up to an arbitrary order as given by

$$Q_c(x)\Big|_{h_\Delta(x)=0} = \sum_{n=0}^{\infty} i^n \int_{-\infty}^{t} d1 \int_{-\infty}^{t_1} d2 \cdots \int_{-\infty}^{t_{n-1}} dn\, h_c(1) h_c(2) \cdots h_c(n)$$

$$\times \mathrm{Tr}\left\{\hat{\rho}\left[\cdots\left[\left[\hat{Q}(x), \hat{Q}(1)\right], \hat{Q}(2)\right]\cdots, \hat{Q}(n)\right]\right\}, \tag{3.20}$$

$$\frac{\delta^2 W[h_\Delta, h_c]}{\delta h_\Delta(1)\, \delta h_\Delta(2)}\Bigg|_{h_\Delta(x)=0} = \mathrm{Tr}\left\{\hat{\rho}\, T_p\left[\hat{Q}_c(1)\, \hat{Q}_c(2) \exp\left(i\int d^4x\, h_c(x) \hat{Q}_\Delta(x)\right)\right]\right\}.$$

$$- Q_c(1)\Big|_{h_\Delta(x)=0} \quad Q_c(2)\Big|_{h_\Delta(x)=0}$$

$$= \sum_{n=0}^{\infty} \frac{i^n}{2}\left[\sum_{i=0}^{n} \int_{t_2}^{t_1} d\bar{1} \int_{t_2}^{\tau_1} d\bar{2} \cdots \int_{t_2}^{\tau_{i-1}} d\bar{i} \int_{-\infty}^{t_2} d\overline{i+1} \cdots \int_{-\infty}^{\tau_{n-1}} d\bar{n}\, h_c(\bar{1}) \cdots h_c(\bar{n})\right.$$

$$\times \mathrm{Tr}\left\{\hat{\rho}\left[\cdots\left\{\left[\cdots\left[\hat{Q}(1), \hat{Q}(\bar{1})\right], \cdots \hat{Q}(\bar{i})\right], \hat{Q}(2)\right\}, \hat{Q}(\overline{i+1})\right]\cdots \hat{Q}(\bar{n})\right]\right\}$$

$$\left. + \sum_{i=0}^{n} \quad 1 \longrightarrow 2 \right] - Q_c(1)\Big|_{h_\Delta(x)=0} \cdot Q_c(2)\Big|_{h_\Delta(x)=0} \tag{3.21}$$

The details of the nonlinear response theory and a possible generalization of the fluctuation - dissipation theorem are discussed elsewhere[15].

Some of the relations discussed in this section have been obtained previously by other authors in the classical limit[5,6,16], but we believe that the present derivation is much more straightforward and complete.

IV. The statistical functional equation

It is interesting to compare the generating functional for the order parameter constructed from CTPGF as expressed by Eqs. (2.14) and (2.16) with the classical stochastic functional, the so-called Onsager-Machlup functional,[17,4-6] quite extensively used in critical dynamics. As common features and interrelations between these two formalisms we find the following:

(i) Two external sources are needed for describing the statistical properties of each order parameter in both formalisms—one for dynamic evolution, the other for statistical fluctuation, although the retarded response and the statistical correlation are defined differently in the quantum and the classical cases.

(ii) The retarded response and the statistical correlation are related by the fluctuation - dissipation theory (FDT) in both formalisms and we have shown[12] the FDT in the classical stochastic system [6,16,18] to be the Rayleigh - Jeans limit of its quantum counterpart.

(iii) The expansion coefficients of the generating functionals in these two formalisms, i.e., the expectation values of different combinations for order parameter products, satisfy the same causality relations as given by Eqs.(3.11)-(3.13). [5,16]

(iv) We have shown[10,13] that, for order parameters slowly varying with time, the Lagrangian formulation of the classical stochastic functional[5,6] is reobtained from its quantum counterpart in the one loop approximation with the second order correlation retained if the potential condition for CTPGF is satisfied.[12]

In spite of the overall similarity of these two formalisms there is one important difference: In the CTPGF formalism the physical field $h_c(X)$ is coupled with $Q_\Delta(X)$ and the fluctuation source $h_\Delta(X)$ is coupled with $Q_c(X)$ while both sources in the classical stochastic formalism ($\ell(t)$ and $\hat{\ell}(t)$ in the notations of Ref.[6] are artificial and an additional driving term has to be introduced in the Fokker-Planck equation to incorporate the coupling with the external physical field. As is seen previously the physical interpretation of $h_c(X)$ and $h_\Delta(X)$ as driving and fluctuation sources correspondingly is possible just because of this apparently "twisted" coupling $h_c(X)\hat{Q}_\Delta(X) + h_\Delta(X)\hat{Q}_c(X)$ which is derived straightforwardly in CTPGF formalism and is not conceivable in the original stochastic formalism.

In support of the correctness of this coupling form we notice that, if the projection operator onto the ground state is chosen as the density matrix, then the functional derivative

$$\left.\frac{\delta Z[h_\Delta(x),\, h_c(x)]}{\delta h_\Delta(x)}\right|_{h_\Delta(x)=0}$$

is just the generating functional for the vacuum expectations of LSZ[19] retarded products in quantum field theory.

In summary we can say that the generating functional for the order parameter constructed from CTPGF is general and effective enough to contain physically meaningful information. Both the purely dynamic evolution described by retarded products in LSZ field theory and the classical stochastic process appear to be special cases of this more comprehensive formalism. Therefore, the problem of determining the quantum statistical properties of the order parameter is reduced to finding the CTPGF generating functional for it. In principle, if the generating functional for the vertex functions $\Gamma[Q_\Delta(X), Q_c(X)]$ is known, one can find the formal solutions from Eqs. (2.27) and (2.28) as

$$Q_\Delta(x) = Q_\Delta[x;\, h_\Delta(y),\, h_c(y)], \tag{4.1}$$

$$Q_c(x) = Q_c[x;\, h_\Delta(y),\, h_c(y)], \tag{4.2}$$

and then take the functional derivatives according to Eqs. (2.25) and (2.26) to obtain all Green's functions in the limit $h_\Delta(X) = h_c(X) = 0$. But practically, the formal solutions as given by Eqs. (4.1) and (4.2) are difficult to find. One prefers instead to take consecutive functional derivatives of Eqs.(2.27) and (2.28) to obtain an infinite hierarchy of coupled equations

$$\left.\frac{\delta\Gamma[Q_\Delta, Q_c]}{\delta Q_\Delta(x)}\right|_{\substack{Q_\Delta(x)=0\\ Q_c(x)=Q(x)}} = 0, \tag{4.3}$$

$$\int d^4y\left(\frac{\delta^2\Gamma[Q_\Delta,Q_c]}{\delta Q_\Delta(1)\delta Q_c(y)}\;\frac{\delta^2 W[h_\Delta,h_c]}{\delta h_\Delta(y)\,\delta h_c(2)}\right)_{\substack{Q_\Delta(x)=0\\ Q_c(x)=Q(x)}}=-\delta^4(1-2), \tag{4.4}$$

$$\int d^4y\left(\frac{\delta^2\Gamma[Q_\Delta,Q_c]}{\delta Q_\Delta(1)\,\delta Q_c(y)}\;\frac{\delta^2 W[h_\Delta,h_c]}{\delta h_\Delta(y)\,\delta h_\Delta(2)}+\frac{\delta^2\Gamma[Q_\Delta,Q_c]}{\delta Q_\Delta(1)\,\delta Q_\Delta(y)}\;\frac{\delta^2 W[h_\Delta,h_c]}{\delta h_c(y)\,\delta h_\Delta(2)}\right)_{\substack{Q_\Delta(x)=0\\ Q_c(x)=Q(x)}}=0 \tag{4.5}$$

and so on for different Green's functions of order parameter

$$\frac{\delta^n W[h_\Delta,h_c]}{\delta h_\Delta(1)\cdots\delta h_\Delta(m)\,\delta h_c(m+1)\cdots\delta h_c(n)}.$$

By solving these equations consecutively one can find all Green's functions needed for the order parameter.

Analogous to the Langevin equation in the semi-phenomenological theory of stochastic processes, the functional equations (2.27) and (2.28) contain all statistical information. Usually in phenomenological approach, a random force term is introduced into the dynamic equation for the order parameter to make it a random variable, while in CTPGF formalism a new degree of freedom $Q_\Delta(X)$ is introduced for each order parameter to describe its statistical behavior. Although the complete statistical information is contained in Eqs. (2.27) and (2.28) defined on the manifold with both $Q_\Delta(X)$ and $Q_c(X)$ different from zero, the observable quantities are defined on the submanifold with $Q_\Delta(X)=0$ like the physically observable quantities which are given by statistical averages of the random variable in the Langevin approach. Therefore, in a certain sense, the functional equations (2.27) and (2.28) are equivalent to the Langevin equation. However, there is one important difference. The random force term in the Langevin equation is postulated independently while the high order statistical correlations in CTPGF formalism are determined self-consistently from Eqs. (2.27) and (2.28), which are complete in principle.

V. Conclusion

To sum up, we have established in this paper a theoretical framework for describing all physically meaningful statistical information for the order parameter by combining CTPGF formalism with the generating functional techniques. This framework is general enough to include both equilibrium and non-equilibrium, both uniform and non-uniform systems. It is a quantum theory in nature but incorporates the classical limit as a special case, if the low frequency, long wave-length excitations dominate as in phase-transition-like phenomena.

The practical value of such a theoretical scheme depends to a great extent upon whether a concrete and effective way can be found to construct the generating functional of the vertex functions for the order parameter $\Gamma[Q_\Delta(X),Q_c(X)]$. This is the main point we are concerned with in this series of papers. Since a generalized Wick theorem has been established[9,20] for CTPGF, the generating functionals defined at the closed time-path as given by Eqs. (2.1), (2.31)-(2.34) are very similar to those defined in the ordinary quantum field theory[2], the main difference being the presence of the density matrix $\hat{\rho}$. In the second paper of this series we will analyze in some detail the role of density matrix to adapt the field-theoretical techniques to the statistical systems.

References

1. *See, for example, A.L. Fetter, J.D. Walecka,"Quantum Theory of Many Particle Systems", McGraw-Hill, New York, (1971).*
2. *E.S. Abers, B.W. Lee, Phys. Rep. 9C (1973), 1.*
3. *E. Brezin, J.C. Le Guillou and J. Zinn-Justin, in"Phase Transitions and Critical Phenomena", Vol VI, eds, C. Domb, M.S. Green, Academic (1976), p. 127.*
4. *P.C. Martin, E.D. Siggia, and H.A. Rose, Phys. Rev. A8 (1973), 423.*
5. *R. Bausch, H.K. Janssen, and H. Wagner, Zeit. Phys. B24((1976), 116.*
6. *C. De Dominicis and L. Peliti, Phys. Rev. B18 (1978), 353.*
7. *P.C. Hohenberg and B.I. Halperin, Rev. Mod. Phys. 49 (1977), 435.*
8. *See, for example: R. Zwanzig, in"Lectures in Theoretical Physics", Vol. III, eds. W. Britten et al., Wiley, N.Y. (1961);*
   H. Mori, Progr. Theor. Phys. 33 (1965), 423;
   W. Louisell,"Quantum Statistical Properties of Radiation", Wiley, N.Y. (1973);
   H. Haken, Rev. Mod. Phys. 47 (1975), 67.
9. *ZHOU Guang-zhao, SU Zhao-bin, Ch. 5 in"Progress in Statistical Physics", eds. HAO Bai-lin et al., KEXUE (Science Press), Beijing, 1981.*
10. *ZHOU Guang-zhao, SU Zhao-bin, HAO Bai-lin and YU Lu, Phys. Rev. B22 (1980), 3385.*
11. *ZHOU Guang-zhao, YU Lu, HAO Bai-lin, Acta Physica Sinica, 29 (1980), 878.*
12. *ZHOU Guang-zhao, SU Zhao-bin, Acta Physica Sinica, 30 (1981), 164, 401*
13. *ZHOU Guang-zhao, SU Zhao-bin, HAO Bai-lin and YU Lu, Acta Physica Sinica, 29 (1980) 961;*
    ZHOU Guang-zhao, HAO Bai-lin, and YU Lu, ibid, 29 (1980), 969.
14. *HAO Bai-lin, Ch. 1."in Progress in Statistical Physics", See Ref.[9]*
15. *HAO Bai-lin, Physica, A(in press);*
    HAO Bai-lin et al., to be published.
16. *J. Deker and G. Haake, Phys. Rev. A11 (1975), 2043.*
17. *R. Graham,"Springer Tracts Mod. Phys". 66 (1973), 1.*
18. *G. Agarwal, Zeit. Phys. 252 (1972), 25;*
    S.K. Ma, G. Mazenko, Phys. Rev. B11 (1975), 4077.
19. *H. Lehmann, K. Symanzik, W. Zimmermann, Nuovo Cimento, 6 (1957), 319.*
20. *A. Hall, J. Phys. A8 (1975), 214.*

Commun. in Theor. Phys.(Beijing, China) Vol. 1, No. 3 (1982) 307-318

ON THEORY OF THE STATISTICAL GENERATING FUNCTIONAL FOR THE ORDER PARAMETER (II) —DENSITY MATRIX AND THE FIELD-THEORETICAL STRUCTURE OF THE GENERATING FUNCTIONAL

ZHOU Guang-zhao (周光召) SU Zhao-bin (苏肇冰)
HAO Bai-lin (郝柏林) YU Lu (于渌)
(Institute of Theoretical Physics, Academia Sinica)

Received December 28, 1981.

Abstract

Several equivalent expressions for the generating functional of the order parameter are derived in this paper to elucidate how the statistical properties are incorporated in its field-theoretical structure. It is shown that in the process of determining the generating functional the dynamic evolution and the statistical information can be separated to a certain extent which facilitates solving the problem. The whole procedure is greatly simplified if the statistical correlation is Gaussian or a generalized fluctuation-dissipation theorem (FDT) holds. As an example, the Gaussian character of the thermal equilibrium distribution is justified and the Matsubara technique along the imaginary time axis is extended to the real axis to provide a complete perturbation scheme for the closed time-path Green's functions (CTPGF).

I. Introduction

In the first paper of the present series[1] (as (I) hereafter) we have proposed a general formalism of the generating functional for the order parameter using CTPGF. In this second paper we will concentrate on the field-theoretical structure of the generating functional and the role of the density matrix to explore the possibility of using field-theoretical techniques in determining the generating functional for the order parameter.

In Sec. II we derive two general expressions for the generating functional which serve as a starting point for further discussion. It is clear from the structure of these expressions that the effects of the density matrix or the statistical properties of the system can be reduced to adding a new term to the effective action.

In Sec. III the thermal equilibrium situation as an important special example is considered and an explicit expression for the CTPGF generating functional with Gaussian distribution is derived. Furthermore, the Feynman rules of the Matsubara technique developed originally for Green's functions at the imaginary time axis are extended to the real time axis to give perturbation expansion rules for the real time Green's functions.

In the final Sec. IV we show that in principle the process of determining the generating functional for the order parameter can be divided into two steps: To find first a formal generating functional defined on the closed time-path without statistical information and then to include the correct statistical properties in the functional obtained. Such separation of dynamics from statistics to a certain extent would facilitate the transplantation of the field-theoretical techniques and provide a basis for a unified approach to the equilibrium and non-equilibrium phenomena. If the statistical correlation is Gaussian or a generalized FDT holds (as in a nonequilibrium stationary state), the proper generating functional for the order parameter is obtained from the functional without statistical information if the CTPGF vacuum propagators are replaced by the CTPGF propagators complying with the density matrix.

Throughout this paper we will use notations adopted in (I) and send our reader there for detailed explanation.

II. Two equivalent expressions for the CTPGF generating functional

In this section we will derive two equivalent expressions for the CTPGF generating functional to analyze the role of the density matrix and to prepare a basis for the actual determination of the generating functional.

For clarity of presentation we will consider a multi-component nonrelativistic complex field, either Bose or Fermi. As a convention the lower sign will always correspond to the Fermi case. The n-component field operator is represented by $\hat{\psi}_b^+$, $\hat{\psi}_b$, $b=1,2,\cdots,n$ while the action of the system is given by

$$I[\hat{\psi}^+,\hat{\psi}]=I_0[\hat{\psi}^+,\hat{\psi}]+I_{int}[\hat{\psi}^+,\hat{\psi}], \tag{2.1}$$

where I_o is the free part, I_{int} -nonlinear interaction. All these quantities are defined on the closed time-path. Suppose

$$I_0[\hat{\psi}^+,\hat{\psi}]=\int_p d_1 d_2\,\hat{\psi}^+(1)\,S_0^{-1}(1,2)\,\hat{\psi}(2)\equiv\hat{\psi}^+S_0^{-1}\hat{\psi}\,, \tag{2.2}$$

where[2]

$$[S_0^{-1}]_{\sigma_1\sigma_2}=\frac{1}{2}\xi_{\sigma_1}\eta_{\sigma_2}S_{0r}^{-1}+\frac{1}{2}\eta_{\sigma_1}\xi_{\sigma_2}S_{0a}^{-1}+\frac{1}{2}\xi_{\sigma_1}\xi_{\sigma_2}S_{0c}^{-1}\,, \tag{2.3}$$

$$S_{0r}^{-1}(1,2)=S_{0a}^{-1}(1,2)=\left(i\frac{\partial}{\partial t}+\frac{\nabla^2}{2m}\right)\delta^4(1-2), \tag{2.4}$$

$$S_{0c}^{-1}(1,2)=0, \tag{2.5}$$

σ_1, σ_2 being indices of the time branch for arguments 1 and 2 correspondingly. Furthermore, assume the order parameters $\hat{Q}_\alpha(x)$ to be Hermitian composite operators

$$\hat{Q}_\alpha(x)\equiv Q_\alpha[\hat{\psi}^+(x),\hat{\psi}(x)],\quad \alpha=1,2\cdots N\,. \tag{2.6}$$

Introducing external sources $J_b(x)$, $J_b^+(x)$ for $\hat{\psi}_b^+(x)$, $\hat{\psi}_b(x)$ and $h_\alpha(x)$ for

$\hat{Q}_\alpha(x)$, the CTPGF generating functional can be written as[1]

$$Z_p[h;\ J^+J] = \mathrm{Tr}\{\hat{\rho}\, T_p \exp[i(J^+\hat{\varphi} + \hat{\varphi}^+ J + h\hat{Q})]\} \tag{2.7}$$

$$= \exp\{ihQ[\mp i\frac{\delta}{\delta J},\ -i\frac{\delta}{\delta J^+}]\} Z_o[J^+, J] \tag{2.8}$$

with

$$Z_p[J^+J] = \mathrm{Tr}\{\hat{\rho}\, T_p[i(J^+\hat{\varphi} + \hat{\varphi}^+ J)]\} \tag{2.9}$$

as the generating functional for the basic field. We denote, as before, quantities defined on the closed time-path or operations along it by index "p". Letters with "^" will mean operators in Heisenberg picture except for special reservation. Usually we drop the subscripts b, α and imply a summation over them. The short notations

$$J^+\hat{\varphi} \equiv \int_p d^4x\, J_b^+(x)\hat{\varphi}_b(x),$$

$$\hat{\varphi}^+ J \equiv \int_p d^4x\, \hat{\varphi}_b^+(x) J_b(x),$$

$$h\hat{Q} \equiv \int_p d^4x\, h_\alpha(x)\hat{Q}_\alpha[\hat{\varphi}^+(x), \hat{\varphi}(x)]$$

and so on are always understood.

Take $J(x) = J^+(x) = 0$ in Eqs.(2.7) and (2.8), the CTPGF generating functional $Z_p[h; J^+, J]$ will degenerate into the statistical generating functional for the order parameter $\hat{Q}_\alpha(x)$ as

$$Z_p[h(x)] = Z_p[h(x),\ J^+(x), J(x)]\Big|_{J^+(x)=J(x)=0} \tag{2.10}$$

Eqs.(2.7)-(2.10) reduce the problem of determining the role of the density matrix in the generating functional for the order parameter to the problem of investigating its role in the CTPGF generating functional for the constituent field.

Rewriting Eq.(2.9) in the incoming interaction picture and using the Wick theorem generalized to the case of CTPGF[3,4], one can readily show[2,3] that

$$Z_p[J^+, J] = \exp\{iI_{int}[\mp i\frac{\delta}{\delta J}, -i\frac{\delta}{\delta J^+}]\}\exp\{i[-J^+S_oJ + W_p^N(J^+,J)]\}\ , \tag{2.11}$$

where

$$J^+S_oJ \equiv \int_p d1\,d2\, J^+(1)S_o(1,2)J(2), \tag{2.12}$$

$S_o(1,2)$ being the bare CTPGF satisfying

$$\int_p d^4x\, S_o(1,x)S_o^{-1}(x,2) = \int_p d^4x\, S_o^{-1}(1,x)S_o(x,2) = \delta_p(1-2) \tag{2.13}$$

and

$$\exp\{iW_p^N[J^+,J]\} = \mathrm{Tr}\{\hat{\rho} : \exp[i(J^+\hat{\varphi}_I + \hat{\varphi}_I^+ J)]:\}\ , \tag{2.14}$$

i.e.,

$$W_p^N[J^+,J] = \sum_{m,n=1}^{\infty} \frac{1}{m!\,n!}\int_p d1 \ldots dm\, d\bar{1} \ldots d\bar{n}$$
$$\times J^+(1)\cdots J^+(m) W_p^{N(m,n)}(1\ldots m,\ \bar{n}\cdots\bar{1}) J(\bar{n})\cdots J(\bar{1}) \tag{2.15}$$

with

$$W_p^{N(m,n)}(1\cdots m,\bar{n}\cdots\bar{1})=i^{m+n-1}\,Tr\{\hat{\rho}:\hat{\psi}_I(m)\cdots\hat{\psi}_I^+(\bar{1})\cdots\hat{\psi}_I^+(\bar{n}):\}_c . \tag{2.16}$$

In Eqs.(2.14) and (2.16), $\hat{\psi}_I^+(x)$ and $\hat{\psi}_I(x)$ are operators in the incoming interaction picture, $:\cdots:$ means normal product.

Since the distinction of time ordering along the positive and negative time branches does not make sense under the normal product, one can easily show that

$$Tr\{\hat{\rho}:\exp[i(J^+\hat{\psi}_I+\hat{\psi}_I^+J)]:\}=Tr\{\hat{\rho}:\exp[i(J_\Delta^+\hat{\psi}_I+\hat{\psi}_I^+J_\Delta)]:\} \tag{2.17}$$

in notations of (I), i.e.,

$$W_p^N[J^+,J]=\overline{W}^N[J_\Delta^+,J_\Delta] \tag{2.18}$$

with

$$\overline{W}^N[J_\Delta^+,J_\Delta]=\sum_{m,n=1}^{\infty}\frac{1}{m!n!}\int d1\cdots dm\,d\bar{1}\cdots d\bar{n}$$
$$\times J_\Delta^+(1)\cdots J_\Delta^+(m)\,\overline{W}^{N(m,n)}(1\cdots m,\bar{n}\cdots\bar{1})\,J_\Delta(\bar{n})\cdots J_\Delta(\bar{1}), \tag{2.19}$$

where

$$\overline{W}^{N(m,n)}(1\cdots m,\bar{n}\cdots\bar{1})=i^{m+n-1}\,Tr\{\hat{\rho}:\hat{\psi}_I(m)\cdots\hat{\psi}_I(1)\hat{\psi}_I^+(\bar{1})\cdots\hat{\psi}_I^+(\bar{n}):\}_c . \tag{2.20}$$

It is important to note that the functional at the left-hand side of Eqs.(2.17) and (2.18) is defined at the closed time-path, while that at the right-hand side is defined in the ordinary space-time. Similarly, Eq.(2.16) is an expression at the closed time-path and, therefore, each argument can take value of either positive or negative time branch, while Eq.(2.20) is an expression defined in the usual space-time.

Furthermore, taking into account that in the incoming picture the field operators satisfy the free field equation, one can easily find that

$$\int_p di'\,\overrightarrow{S_0^{-1}}(i,i')\,W_p^{N(n,m)}(1\cdots i'\cdots m,\bar{n}\cdots\bar{1})=0, \tag{2.21}$$

$$\int_p di'\,W_p^{N(n,m)}(1\cdots m,\bar{n}\cdots i'\cdots\bar{1})\,\overleftarrow{S_0^{-1}}(i',\bar{i})=0, \tag{2.22}$$

or, equivalently,

$$\int di'\,\overrightarrow{S_{0r}^{-1}}(i,i')\,\overline{W}^{N(n,m)}(1\cdots i'\cdots m,\bar{n}\cdots\bar{1})=0, \tag{2.23}$$

$$\int di'\,\overline{W}^{N(n,m)}(1\cdots m,\bar{n}\cdots i'\cdots\bar{1})\,\overleftarrow{S_{0a}^{-1}}(i',\bar{i})=0. \tag{2.24}$$

Since the moment $t=-\infty$ is chosen as the starting point of the closed time-path, the initial condition for the statistical system is fixed at this moment. Therefore, we are not allowed to integrate by parts in respect to the operator $\partial/\partial t$ arbitrarily. The correct direction of action is indicated by an arrow in Eqs.(2.21)-(2.24) to incorporate appropriately the initial condition.

Substituting Eq.(2.18) into Eq.(2.11) and taking into account Eq.(2.8), we obtain

$$Z_p[h;J^+J]=\exp\{i(h\hat{Q}[\mp i\frac{\delta}{\delta J},-i\frac{\delta}{\delta J^+}]+I_{int}[\mp i\frac{\delta}{\delta J},-i\frac{\delta}{\delta J^+}])\}$$
$$\times\exp\{i(-J^+S_0J+\overline{W}^N[J_\Delta^+,J_\Delta])\}, \tag{2.25}$$

which is the first expression for CTPGF generating functional we derive in this section. Eq.(2.25) specifies the generalized Feynman rules for CTPGF and shows clearly how the density matrix contributes to CTPGF in terms of $\overline{W}^N[J_\Delta^+, J_\Delta]$ in view of the global structure of the perturbation theory. It tells us that the density matrix affects directly only the correlation functions of the constituent field variables describing the statistical fluctuations (corresponding to $\frac{\delta}{\delta J_\Delta^+}$, $\frac{\delta}{\delta J_\Delta}$). So far as $\overline{W}^{N(m,n)}$ satisfy Eqs.(2.23) and (2.24), the contribution of the density matrix can be expressed in terms of the initial conditions(sometimes called boundary conditions) for the statistical Green's functions.

Now we derive another expression for the CTPGF generating functional. Using the following equality (up to an unimportant constant factor)

$$\exp(-iJ^+S_0J) = \int_p [d\psi^+][d\psi] \exp[i(\psi^+ S_0^{-1}\psi + J^+\psi + \psi^+ J)], \tag{2.26}$$

it is easy to show that

$$\begin{aligned} &\exp\{i(-J^+S_0J + W_p^N[J^+,J])\} \\ &= \int_p [d\psi^+][d\psi] \exp\{i(J^+\psi + \psi^+ J)\} \exp\{iW_p^N[\pm i\tfrac{\delta}{\delta\psi}, i\tfrac{\delta}{\delta\psi^+}]\} \exp(i\psi^+ S_0^{-1}\psi) \end{aligned} \tag{2.27}$$

if the path integration is taken by parts. Taking into account that

$$\frac{\delta}{\delta\psi} \exp(i\psi^+ S_0^{-1}\psi) = \exp(i\psi^+ S_0^{-1}\psi)\left(\frac{\delta}{\delta\psi} \pm i\psi^+ \overleftarrow{S}_0^{-1}\right) \tag{2.28}$$

$$\frac{\delta}{\delta\psi^+} \exp(i\psi^+ S_0^{-1}\psi) = \exp(i\psi^+ S_0^{-1}\psi)\left(\frac{\delta}{\delta\psi^+} + i\overrightarrow{S}_0^{-1}\psi\right), \tag{2.29}$$

Eq.(2.27) can be transformed into

$$\begin{aligned} &\exp\{i(-J^+S_0J + W_p^N[J^+,J])\} \\ &= \int_p [d\psi^+][d\psi] \exp\{i(J^+\psi + \psi^+J + \psi^+ S_0^{-1}\psi)\} \\ &\times \exp\{iW_p^N[\pm i\tfrac{\delta}{\delta\psi} - \psi^+\overleftarrow{S}_0^{-1}, i\tfrac{\delta}{\delta\psi^+} - \overrightarrow{S}_0^{-1}\psi]\}. \end{aligned} \tag{2.30}$$

Using Eq.(2.3) and the convention of (I) we find that

$$\eta_\sigma \frac{\delta}{\delta\psi(x_\sigma)} = \frac{\delta}{\delta\psi_c(x)}, \tag{2.31}$$

$$\eta_\sigma \frac{\delta}{\delta\psi^+(x_\sigma)} = \frac{\delta}{\delta\psi_c^+(x)}, \tag{2.32}$$

$$\eta_\sigma \int d^4y\, \psi_\rho^+(y)\eta_\rho \overleftarrow{S}_{0\rho\sigma}^{-1}(y,x) = \int d^4y\, \psi_\Delta^+(y)\overleftarrow{S}_{0r}^{-1}(y,x), \tag{2.33}$$

$$\eta_\sigma \int d^4y\, \overrightarrow{S}_{0\sigma\rho}^{-1}(x,y)\eta_\rho\psi_\rho(y) = \int d^4y\, \overrightarrow{S}_{0a}^{-1}(x,y)\psi_\Delta(y), \tag{2.34}$$

and obtain from Eq.(2.18)

$$\exp\{i W_p^N[\pm i\frac{\delta}{\delta\psi} - \psi^+\overleftarrow{S}_o^{-1}, \cdot i\frac{\delta}{\delta\psi^+} - \overrightarrow{S}_o^{-1}\psi]\}$$
$$= \exp\{i\overline{W}^N[-\psi_\Delta^+\overleftarrow{S}_{or}^{-1}, -\overrightarrow{S}_{oa}^{-1}\psi_\Delta]\} \tag{2.35}$$

and

$$\exp\{i W_p^N[-\psi^+\overleftarrow{S}_o^{-1}, -\overrightarrow{S}_o^{-1}\psi]\}$$
$$= \exp\{i\overline{W}^N[-\psi_\Delta^+\overleftarrow{S}_{or}^{-1}, -\overrightarrow{S}_{oa}^{-1}\psi_\Delta]\} \tag{2.36}$$

considering ψ_c^+, ψ_Δ^+ as well as ψ_c, ψ_Δ as independent functional arguments.

Substituting Eqs.(2.35), (2.36) into Eq.(2.30) and putting the result obtained into Eq.(2.11) we get finally

$$Z_p[h; J^+J]$$
$$= \int_p[d\psi^+][d\psi]\exp\{i(I_o[\psi^+,\psi] + I_{int}[\psi^+,\psi] + J^+\psi + \psi^+J + hQ[\psi^+,\psi])\}$$
$$\times\exp\{i W_p^N[-\psi^+\overleftarrow{S}_o^{-1}, -\overrightarrow{S}_o^{-1}\psi]\} \tag{2.37}$$

as the second expression for the CTPGF generating functional for the order parameter—a path integral presentation. It is easy to rederive Eq.(2.25) from Eq.(2.37), so these two expressions are equivalent to each other. This path integral representation is different from what we obtained previously for the CTPGF generating functional[3], in so far as the contribution of the density matrix is expressed here as an additional term in the action given by $W_p^N[-\psi^+\overleftarrow{S}_o^{-1}, -\overrightarrow{S}_o^{-1}\psi]$. According to Eq.(2.36) this term does depend only upon the field variables ψ_Δ, ψ_Δ^+ describing the statistical fluctuation, but not upon the field variables describing the dynamic evolution. On the other hand, it is clear that $W_p^N[-\psi^+\overleftarrow{S}_{o=}^{-1}, -\overrightarrow{S}_o^{-1}\psi]$ has nonvanishing contribution to the generating functional only at the initial moment $t=-\infty$. Expanding W_p^N in accord with Eq.(2.15) and integrating $\overleftarrow{S}_o^{-1}$, $\overrightarrow{S}_o^{-1}$ by parts we find that only terms corresponding to the complete differential contribute, because the expansion coefficients satisfy Eqs.(2.21) and (2.22). Since the functional integral is taken over a closed time-path, starting and ending at $t=-\infty$, $W_p^N[-\psi^+\overleftarrow{S}_o^{-1}, -\overrightarrow{S}_o^{-1}\psi]$ has nonvanishing contribution only at these end points.

Before further analyzing Eqs.(2.25) and (2.37), the derivation of which is the main subject in this section, we apply first these equations to an important special example: the contribution of the density matrix to the CTPGF generating functional in thermal equilibrium.

III. CTPGF generating functional in thermal equilibrium

As an important special case we will derive an explicit expression for the CTPGF generating functional in thermal equilibrium, i.e., for the density matrix given by

$$\hat{\rho} = \hat{\rho}_{th} = \exp\{-\Omega - \beta(\hat{H} - \mu\hat{N})\}, \tag{3.1}$$

$$T_r\left\{\hat{\rho}_{th}\right\} = 1, \tag{3.2}$$

where $\hat{H}$ is the total Hamiltonian of the system, $\hat{N}$-operator of particle number, μ-chemical potential, $\exp(-\Omega)$-normalization constant, or the inverse of the partition function.

Substituting Eqs.(2.17) and (2.18) into Eq.(2.14) we find that

$$\exp(i\,\overline{W}_{th}^{N}[J_\Delta^+, J_\Delta]) = T_r\left\{\hat{\rho}_{th} : \exp\left[i(J_\Delta^+\hat{\psi}_I + \hat{\psi}_I^+ J_\Delta)\right]:\right\}. \tag{3.3}$$

where $\hat{\psi}_I$, $\hat{\psi}_I^+$ at the right-hand side are operators in the incoming picture. It is known for the operator $\hat{A}_I(t)$ in the incoming picture that[5]

$$\exp(i\hat{H}\tau)\hat{A}_I(t)\exp(-i\hat{H}\tau) = \hat{A}_I(t+\tau). \tag{3.4}$$

It is essential to note that $\hat{H}$ in Eq.(3.4) is the total Hamiltonian. If an analytic continuation

$$\tau \longrightarrow -i\beta \tag{3.5}$$

is carried out we find that

$$\exp(\beta\hat{H})\hat{A}_I(t)\exp(-\beta\hat{H}) = \hat{A}_I(t-i\beta). \tag{3.6}$$

Taking into account that for complex fields the operator of particle number

$$\hat{N} \sim \int d^3x\,\hat{\psi}^+(x)\hat{\psi}(x) = \int d^3x\,\hat{\psi}_I^+(x)\hat{\psi}_I(x) \tag{3.7}$$

is a conserved quantity, it is easy to prove that

$$\hat{\rho}_{th}^{-1}\,\hat{A}_I(t)\,\hat{\rho}_{th} = \exp\left\{-i\beta\frac{\partial}{\partial t} - \lambda\beta\mu\right\}\hat{A}_I(t), \tag{3.8}$$

where

$$\lambda = +1, \qquad \text{if } \hat{A}_I(t) = \hat{\psi}_I^+(x),$$

$$\lambda = -1, \qquad \text{if } \hat{A}_I(t) = \hat{\psi}_I(x), \tag{3.9}$$

$$\lambda = 0, \qquad \text{if } \hat{A}_I(t) \text{ is Hermitian.}$$

With Eq.(3.8) we can apply, as done by Gaudin[6], the following identity

$$\begin{aligned} &T_r\left\{(\hat{\rho}\hat{A}(1) \mp (\pm)^n\hat{A}(1)\hat{\rho})\hat{A}(2)\cdots\hat{A}(n)\right\} \\ &= T_r\left\{\hat{\rho}[\hat{A}(1),\hat{A}(2)]_\mp\hat{A}(3)\cdots\hat{A}(n)\right\} \pm T_r\left\{\hat{\rho}\hat{A}(2)[\hat{A}(1),\hat{A}(3)]_\mp\hat{A}(4)\cdots\hat{A}(n)\right\} \\ &+ \cdots (\pm)^{n-2}T_r\left\{\hat{\rho}\hat{A}(2)\cdots\hat{A}(n-1)[\hat{A}(1),\hat{A}(n)]_\mp\right\} \end{aligned} \tag{3.10}$$

to the right-hand side of Eq.(3.3) to obtain

$$\mathrm{Tr}\{\hat{\rho}_{th} : \exp[i(J_\Delta^+\psi_I + \psi_I^+ J_\Delta)]:\}$$
$$= \mathrm{Tr}\{\hat{\rho}_{th}\} <0|: \exp(\hat{F}_I)\exp[i(J_\Delta^+\hat{\psi}_I + \hat{\psi}_I^+ J_\Delta)]:|0>, \quad (3.11)$$

where

$$\hat{F}_I = \pm\int d^4x d^4y \frac{\delta}{\delta\hat{\psi}_I^+(x)}\left(1 \mp \exp(-i\beta\frac{\partial}{\partial x_0} - \beta\mu)\right)^{-1}[\hat{\psi}_I^+(x), \psi_I(y)]_\mp \frac{\delta}{\delta\hat{\psi}_I(y)}, \quad (3.12)$$

$$[\hat{A}, \hat{B}]_\mp = \hat{A}\hat{B} \mp \hat{B}\hat{A}. \quad (3.13)$$

In deriving Eq.(3.11) the properties of the normal product and the requirement of particle number conservation are taken into account properly. Note that for nonrelativistic complex fields the operator $\hat{\psi}_I(x)$ contains only the positive frequency part, while $\hat{\psi}_I^+(x)$ the negative frequency part, so we find

$$[\hat{\psi}_I(1), \hat{\psi}_I^+(2)]_\mp = iS_0^{-+}(1,2), \quad (3.14)$$

where S_0^{-+} is defined by Eq.(2.3). Substituting Eq.(3.14) into Eq.(3.12), we find that

$$\hat{F}_I = \pm i\int d^4x d^4y \frac{\delta}{\delta\hat{\psi}_I(x)} S_0^N(x,y) \frac{\delta}{\delta\hat{\psi}_I^+(y)}, \quad (3.15)$$

with

$$S_0^N(x,y) = -\left(1 \mp \exp(i\beta\frac{\partial}{\partial x_0} - \beta\mu)\right)^{-1} S_0^{-+}(x,y). \quad (3.16)$$

Now operating $\exp(\hat{F}_I)$ upon $\exp\{i(J_\Delta\hat{\psi}_I + \hat{\psi}_I^+ J_\Delta)\}$ in Eq.(3.11) and substituting the result obtained into Eq.(3.3), we find that

$$\overline{W}_{th}^N[J_\Delta^+, J_\Delta] = -\int d^4x d^4y J_\Delta^+(x) S_0^N(x,y) J_\Delta(y). \quad (3.17)$$

According to Eq.(2.18) and the convention of (I) we have

$$W_{th,p}^N[J^+, J] = -\int_p d^4x d^4y J^+(x) S_{o,p}^N(x,y) J(y), \quad (3.18)$$

where the closed time-path matrix

$$S_{o,p}^N(x,y) = S_0^N(x,y)\begin{pmatrix}1 & 1\\ 1 & 1\end{pmatrix}. \quad (3.19)$$

Eqs.(3.17) and (3.18) constitute the contribution of the density matrix to the CTPGF generating functional in thermal equilibrium we consider in this section. As shown by these explicit expressions for the functionals $W_{th,p}^N[J^+,J]$ and $\overline{W}_{th}^N[J_\Delta^+, J_\Delta]$, the equilibrium state is a Gaussian process, the statistical properties of which are described completely by two-point correlation function $S_{o,p}^N(x,y)$.

Substituting Eq.(3.18) into Eqs.(2.11) and (2.37) we obtain the following expressions for the CTPGF generating functional in thermal equilibrium:

$$Z_p[h;\, J^+J] = \exp\left\{ i\left(hQ\left[\mp i\frac{\delta}{\delta J},\, -i\frac{\delta}{\delta J^+}\right] + I_{int}\left[\mp i\frac{\delta}{\delta J},\, -i\frac{\delta}{\delta J^+}\right]\right)\right\}$$

$$\times \exp(-iJ^+G_0J) \tag{3.20}$$

and

$$Z_p[h;\, J^+J]$$

$$= \int_p [d\psi^+][d\psi] \exp\left\{ i\left(\psi^+ G_0^{-1}\psi + I_{int}[\psi^+,\psi] + J^+\psi + \psi^+J + hQ[\psi^+,\psi]\right)\right\}, \tag{3.21}$$

where

$$G_0(1,2) = S_0(1,2) + S_{0,p}^N(1,2), \tag{3.22}$$

$$G_0^{-1}(1,2) = S_0^{-1}(1,2) - \int_p d\bar{1}\, d\bar{2}\, \overleftarrow{S}_0^{-1}(1,\bar{1})\, S_{0,p}^N(\bar{1},\bar{2})\, \overrightarrow{S}_0^{-1}(\bar{2},2). \tag{3.23}$$

It is easy to check that $G_0(1.2)$ and $G_0^{-1}(1.2)$ defined by Eqs.(3.22) and (3.23) are reciprocal to each other. It is also easy to show from Eqs.(3.16),(3.19) and (3.22) that in Fourier representation

$$G_0^{-+}(p) = \pm \exp[\beta(p_0 - \mu)]\, G_0^{+-}(p) \tag{3.24}$$

and

$$G_{0c}(p) = \frac{\mathrm{cth}\,[\beta(p_0-\mu)/2]}{\mathrm{th}\,[\beta(p_0-\mu)/2]}\left(G_{0r}(p) - G_{0a}(p)\right), \tag{3.25}$$

which are the well-known[7,8] equilibrium FDT satisfied by statistical Green's functions.

Since the equilibrium state is a stationary state, the final statistical Green's functions obtained after taking functional derivatives should be translationally invariant in time. In this procedure the only role of the second term in Eq.(3.23) is to produce $S_{0,p}^N(1.2)$ term in $G_0(1,2)$ and cannot appear independently in the final result. Therefore, we might not distinguish $\overleftarrow{S}_0^{-1}$ and $\overrightarrow{S}_0^{-1}$ from the beginning and take

$$G_0^{-1}(1,2) = S_0^{-1}(1,2) \tag{3.26}$$

instead of Eq.(3.23). In that case the first term of Eq.(3.22) can be considered as an inhomogeneous solution of the equation

$$S_0^{-1}\, G_0 = 1 \tag{3.27}$$

to describe the dynamics, while the second term of Eq.(3.22) can be considered as a homogeneous solution of Eq.(3.27) satisfying the equilibrium FDT as given by Eqs.(3.24) and (3.25). We can thus rewrite Eq.(3.21) as

$$Z_p[h;\, J^+J]$$

$$= \int_p [d\psi^+][d\psi] \exp\left\{ i\left(I_0[\psi^+,\psi] + I_{int}[\psi^+,\psi] + J^+\psi + \psi^+J + hQ[\psi^+,\psi]\right)\right\}. \tag{3.28}$$

In fact, not differentiating $\vec{S}_0^{-1}$ and $\overleftarrow{S}_0^{-1}$, Eq.(3.26) can be obtained directly from Eq.(3.20) by Fourier transformation of the path integral as given by Eq.(2.26).

An interesting result follows from Eqs.(3.20) and (3.28) as expressions for the CTPGF generating functional in thermal equilibrium: If all space-time arguments of functions and 4-dimensional integrations are extended to the closed time-path, then the CTPGF generating functional has the same structure as that for the ordinary quantum field theory. Formally, the density matrix disappears, the statistical information being contained entirely in the propagator $G_0(1,2)$, defined at the closed time-path and satisfying FDT, as given by Eq.(3.24).

Eqs.(3.20) and (3.26) extend directly the Feynman rules for Matsubara Green's functions at the imaginary time axis to the real time axis. Several authors have discussed the possible generalization of the Feynman-Wick expansion for the Matsubara functions. Some of them[9] have analytically continued the integration path from $(0, -i\beta)$ to complex time plane to derive the expansion rules for the real time retarded Green's functions which are too involved to be of practical value. The others[10] have tried to derive perturbation expansion for the closed time-path similar to ours, but the end-points of the generalized contour they have used differ by $i\beta$ which gives rise to some difficulty in justification. This difficulty is avoided by using the incoming picture in our derivation. Eqs.(3.20) and (3.26) show the Feynman rules we established for CTPGF are formally identical to those in the ordinary quantum field theory. Using the transformations derived by us previously[2] which connect the CTPGF with the real time statistical Green's functions, the perturbation expansion for the real time functions can be easily derived.

IV. The field-theoretical structure of the statistical generating functional for the order parameter

To "transplant" the techniques of effective action for determining the generating functional in quantum field theory to solving the statistical generating functional for the order parameter, we need to clarify the difference of the field-theoretical structure for these two cases. In Sec. II we have obtained two equivalent expressions for the statistical generating functional showing that the difference consists in an additional term expressing the contribution of the density matrix — $W^{(N)}$-functional for the statistical case. In Sec. III we have shown further that these two generating functionals have identical structure, the contribution of the density matrix being incorporated in the CTPGF bare propagator satisfying FDT. In this section we will analyze the results obtained in Sec. II to discuss the structure of CTPGF generating functional in the general case.

For a general nonequilibrium process, the CTPGF generating functional given by Eq.(2.37) can be rewritten as

$$Z_p[h;J^+J]=\exp\left\{iW_p^N\left[\pm i\frac{\delta}{\delta J}\overleftarrow{S}_0^{-1}, i\vec{S}_0^{-1}\frac{\delta}{\delta J^+}\right]\right\}Z_p^{(0)}[h;J^+J], \tag{4.1}$$

where

$$Z_p^{[0]}[h;J^+,J]$$

$$= \int_p [d\psi^+][d\psi] \exp\{i(I_0[\psi^+,\psi] + I_{int}[\psi^+,\psi] + J^+\psi + \psi^+ J + hQ[\psi^+,\psi])\} \tag{4.2}$$

is the CTPGF generating functional for the ground state. Since $Z_p^{[0]}[h;J^+,J]$ has exactly the same structure at the closed time-path as the ordinary field theory, we can first determine $Z_p^{[0]}[h;J^+,J]$ by a standard field-theoretical technique and then find $Z_p[h;J^+,J]$ from $Z_p^{[0]}[h;J^+,J]$ in accord with Eq.(4.1). Such a procedure is similar to what is used for solving Liouville problem in classical statistical mechanics. The dynamics of the system is described by $Z_p^{[0]}[h;J^+,J]$, while the initial statistical distribution is given by $\exp\{iW_p^N[\pm i\frac{\delta}{\delta J}\overleftarrow{S}_0^{-1}, i\overrightarrow{S}_0^{-1}\frac{\delta}{\delta J^+}]\}$.

Many interesting nonequilibrium phenomena can be described by a Gaussian process, i.e.,

$$W_p^{N(m,n)}(1,2\cdots m;\bar{n}\cdots\bar{1}) = 0 \tag{4.3}$$

except for

$$W_p^{N(1,1)}(x,y) \neq 0. \tag{4.4}$$

In such a case, we obtain an explicit formula of the CTPGF generating functional for the order parameter, if the equilibrium expression for $S_{o,p}^N$ in Eqs.(3.20)-(3.23), given by Eq.(3.19), is replaced by $W_p^{N(1,1)}(x,y)$ i.e., to let

$$S_{o,p}^N(x,y) = -W_p^{N(1,1)}(x,y). \tag{4.5}$$

It is obvious that the CTPGF generating functional in such a case has the same structure at the closed time-path as for the ordinary field theory.

On the other hand, it is clear from Eq.(2.8) that the CTPGF generating functional for the order parameter can be determined as some combination of the generating functional for constituent field in a way independent of the statistical properties. This means that if we write down the generating functional for the order parameter

$$Z_p[h] \equiv \exp\{iW_p[h]\} = Z_p[h;J^+,J]\Big|_{J^+=J=0} \tag{4.6}$$

with

$$\frac{\delta W_p[h]}{\delta h(x)} = Q(x) \tag{4.7}$$

and

$$\Gamma_p[Q] = W_p[h] - hQ, \tag{4.8}$$

and also the analogous expressions for the case without the statistical information as given by $Z_p^{[0]}$, $W_p^{[0]}$, $\Gamma_p^{[0]}$ and $Z_p^{[0]}[h;J^+,J]$ we find according to Eq.(2.8) that the structure of the generating functional for the order parameter is identical to each other in these two cases. Furthermore, Eq.(2.25) tells us how the density matrix contributes to the Green's functions of the constituent field in view of the homogeneous equations for free field (See Eqs.(2.21) and (2.22)). These $W_p^N(m,n)$ functions can be specified as initial conditions for statistical correlations. It is difficult to formulate these initial conditions in the general case, but many systems we are interested in are existent in stationary or quasi-stationary state (i.e., stationary at a micro-long, macro-short time scale), in which case the effect of the density matrix is realized through the generalized FDT.[8] In the next paper we will illustrate this point by an example of open systems.

To summarize we come to the conclusion that the determination of the CTPGF generating functional for the order parameter can be divided into two steps: to "forget" first about the density matrix and to find the generating functional without statistical information and then to include the latter in the second step. In general cases, this can be done by using Eq.(4.1). The whole procedure is greatly simplified if the statistical correlation is Gaussian or a generalized FDT can be formulated. This division into two steps facilitates the practical calculations and provides a unified approach to both equilibrium and nonequilibrium problems. In the next paper we will discuss a practical prescription for determining the generating functional and illustrate it by simple examples.

References

1. ZHOU Guang-zhao, SU Zhao-bin, HAO Bai-lin, YU Lu, the preceding paper.
2. ZHOU Guang-zhao, SU Zhao-bin, HAO Bai-lin, YU Lu, Phys. Rev. B22 (1980), 3385.
3. ZHOU Guang-zhao, SU Zhao-bin, Ch. 5 in "Progress in Statistical Physics", eds. HAO Bai-lin et al., KEXUE (Science Press), Beijing, 1981.
4. A. Hall, J. Phys. A8 (1975), 214.
5. See, for example, P. Roman, "Advanced Quantum Theory", Addison Wesley, 1965.
6. M. Guadin, Nucl. Phys. 15 (1960), 89.
7. ZHOU Guang-zhao, SU Zhao-bin, Physica Energiae Fortis et Nuclearis (Beijing), 3 (1979), 314.
8. ZHOU Guang-zhao, SU Zhao-bin, Acta Physica Sinica, 30 (1981), 164,401.
9. See, for example, I.E. Dzyloshinski, Zh. Eks. Teo. Fis. (JETP) 42 (1962), 1126; G. Baym, A. Sessler, Phys. Rev. 131 (1963), 2345.
10. See, for example, R. Mills, "Propagators for Many-Particle Systems", Gordon and Breach, N.Y. 1968.

Commun. in Theor. Phys. (Beijing, China) *Vol. 1, No. 4 (1982)* *389-403*

ON THEORY OF THE STATISTICAL GENERATING FUNCTIONAL FOR THE ORDER PARAMETER (III) —EFFECTIVE ACTION FORMALISM FOR THE ORDER PARAMETER

ZHOU Guang-zhao (周光召) SU Zhao-bin (苏肇冰)
HAO Bai-lin (郝柏林) YU Lu (于 渌)
Institute of Theoretical Physics, Academia Sinica , Beijing, China

Received December 28, 1981

Abstract

A practical scheme is proposed in this paper to determine the quantum statistical properties of the order parameter in the framework of the closed time-path Green's functions (CTPGF). As a microscopic quantum theory, this formalism is applicable in principle to both equilibrium and nonequilibrium phenomena and is capable of dealing with statistical systems both above and below their phase transition point. This formalism can be used to discuss the statistical properties of simple as well as composite order parameters in either uniform or nonuniform systems. As simple illustrations and check for the theory, the proposed prescription is applied to the contact-interaction model for superconductivity and a unimode laser system coupled with two-energy-level bound electrons.

I. Introduction

This is the third paper in the present series on the theory of the statistical generating functional for the order parameter in the framework of CTPGF. In the first paper [1] (as [I] hereafter) a general formalism is proposed to construct the generating functional for the order parameter which embodies the statistical properties of the system. In the second paper [2] (as [II] hereafter) it is shown that the determination of the generating functional for the order parameter can be divided into two steps: first to find the CTPGF generating functional formally without statistical information and then to incorporate the appropriate statistical information at the second stage. In this paper we propose a practical prescription to determine the quantum statistical properties of the order parameter.

In Sec. II we describe the general procedure of the effective action method for the order parameter, the key steps of which being the construction of the effective action for the composite operators and a loop expansion for the vertex functions. As seen from the presentation in the next section, this technique is applicable to a large class of systems with either simple or composite order parameter expressed as a quadratic form of the constituent field. Within this framework, the equilibrium and the nonequilibrium phenomena are treated on an equal footing, the difference being embodied in some cases in the fluctuation-dissipation theorem (FDT). Moreover, this formalism can be used to discuss the space and time variations of the order parameter and related quantities.

To illustrate the specific features of the present formalism we apply it to some known examples. As an application to equilibrium system with composite order parameter we discuss the four-Fermion contact interaction model for superconductivity and reobtain the BCS gap equation. As an example of nonuniform case we consider superconducting system close to the transition point with weak nonuniformity and rederive the Ginzburg-Landau equation without external magnetic field. Both these two cases are considered in Sec. III. Finally, as an example of far from equilibrium phenomena we discuss in Sec. IV the coupled two-level electron-unimode laser system and reobtain the result of semi-classical Lamb theory.

As illustrated by these examples, the statistical information is incorporated into our formalism as initial values for correlations in some analogy with the ordinary Liouville problem in quantum statistics. However, as a general scheme, our formalism is applicable to systems near thermal equilibrium (See Sec. III of [II] [2]) as well as open systems far from equilibrium.

It is worthwhile to mention that the treatment of equilibrium superconductivity and nonequilibrium laser system in the present framework is analogous to each other to a great extent. Although some authors have emphasized the similarity between these two systems [3], as far as we know, there is no other microscopic theoretical scheme to consider these two systems from a unified point of view.

Throughout this paper we will use notations adopted in (I) and (II) and ask our reader to refer to those papers for details.

II. Effective action method

In this section we describe the effective action method for practical determination of the statistical properties for the order parameter. As in (II), we consider an n-component nonrelativistic complex field ψ_b, ψ_b^+, $b=1,2...n$, for clarity of presentation, but the result obtained is completely general.

According to (II)-(2.37), the CTPGF generating functional of the system can be written as

$$Z_p[h;J^+_, J]=\int_p [d\psi^+][d\psi]\exp\{i(I_0[\psi^+,\psi]+I_{int}[\psi^+,\psi])\}$$
$$\times\exp\{i(J^+\psi+\psi^+J+hQ[\psi^+,\psi]+W_p^N[-\psi^+\tilde{S}_0^{-1},-\tilde{S}_0^{-1}\psi])\} \tag{2.1}$$

where $I_0[\psi,\psi]$ is the free part of the ψ-field action, $I_{int}[\psi^+,\psi]$ the nonlinear interaction. The order parameters

$$Q_\alpha(x)\equiv Q_\alpha[\psi^+(x),\psi(x)],\quad \alpha=1,2\cdots N \tag{2.2}$$

are composite operators of the constituent field $\psi_b(x)$. $J_b(x)$, $J_b^+(x)$ and $h_\alpha(x)$ are external sources for $\psi_b^+(x)$, $\psi_b(x)$ and $Q_\alpha(x)$ correspondingly,

$W_p^N[-\psi^+\bar{S}_0^{-1}, -\bar{S}_0^{-1}\psi]$ being the contribution from the density matrix. The generating functional for the order parameter is given by

$$Z_p[h] \equiv \exp(iW_p[h]) = Z[h; J^+, J]\Big|_{J=J^+=0} \tag{2.3}$$

with

$$\frac{\delta W_p[h]}{\delta h(x)} = Q(x) \tag{2.4}$$

and

$$\Gamma_p[Q] = W_p[h] - hQ \tag{2.5}$$

with

$$\frac{\delta \Gamma_p[Q]}{\delta Q(x)} = -h(x). \tag{2.6}$$

Similarly, we can define CTPGF generating functionals without statistical information as

$$Z_p^{(o)}[h; J^+, J] = Z_p[h; J^+, J]\Big|_{W_p^N=0}, \tag{2.7}$$

$$Z_p^{(o)}[h] = Z_p^{(o)}[h; J^+, J]\Big|_{J^+=J=0} = Z_p[h]\Big|_{W_p^N=0}, \tag{2.8}$$

$$W_p^{(o)}[h] = -i\ln Z_p^{(o)}[h] = W_p[h]\Big|_{W_p^N=0}, \tag{2.9}$$

$$\Gamma_p^{(o)}[Q] = W_p^{(o)}[h] - hQ = \Gamma_p[Q]\Big|_{W_p^N=0}, \tag{2.10}$$

where $h(x)$ and $Q(x)$ satisfy equations

$$\frac{\delta W_p^{(o)}[h]}{\delta h(x)} = Q(x), \tag{2.11}$$

$$\frac{\delta \Gamma_p^{(o)}[Q]}{\delta Q(x)} = -h(x). \tag{2.12}$$

Using the results obtained in (I), (II) and generalizing the steepest descent technique of evaluating the generating functional for the irreducible vertex functions in the usual quantum field theory [1] to the closed time-path, we can formulate the following practical rules for determining all Green's functions for order parameter.

(i) To determine the effective action for the order parameter at the closed time-path, which is formally defined as

$$Z_p^{(o)}[h(x)] = \int_p [dQ] \exp\{i(I^{eff}[Q] + hQ)\} \tag{2.13}$$

with

$$\exp(iI^{eff}[Q]) = \int_p [dh'] \exp\{i(W_p^{(o)}[h'] - h'Q)\}. \tag{2.14}$$

In practice, the effective action is found by changing variables in the path integration and comparing the result obtained with Eq.(2.13). In the next section we will give a concrete example. Alternatively, the effective action can be defined as

$$\exp(iI^{eff}[Q]) = \int_p [d\psi^+][d\psi]\,\delta(Q(x) - Q[\psi^+,\psi])$$
$$\times\exp\{i(I_0[\psi^+,\psi] + I_{int}[\psi^+,\psi] + J^+\psi + \psi^+J)\}\Big|_{J^+=J=0} \,. \tag{2.15}$$

(ii) To generalize the steepest descent method of evaluating the functional integral[4] to the closed time-path to obtain the following expansion for the vertex functional (without statistical information)

$$\Gamma_p^{(0)}[Q] = I^{eff}[Q] + \frac{i}{2}\mathrm{Tr}\,\ln\frac{\delta^2 I^{eff}[Q]}{\delta Q\,\delta Q} + \Gamma_{2p}^{(0)}[Q], \tag{2.16}$$

where $\Gamma_{2p}^{[0]}[Q]$ is the sum of one-particle irreducible (1PI) vacuum diagrams for an equivalent system with action given by

$$I_{eqv}[q] = I^{eff}[Q+q] - I^{eff}[Q] - q\frac{\delta I^{eff}(Q)}{\delta Q}, \tag{2.17}$$

where $q(x)$ is field variable for the equivalent system with free propagator defined as

$$\Delta_{eqv}^{(0)\,-1}(x,y) \equiv \frac{\delta^2 I^{eff}[Q]}{\delta Q(x)\,\delta Q(y)}\,. \tag{2.18}$$

We will call the first term of Eq.(2.16) the tree approximation, the second term one loop contribution and the rest high loop correction. It is worthwhile to note that unlike the quantum field theory, where the field variable itself is the order parameter in most cases, the effective action $I^{eff}[Q]$ and, therefore, $I_{eqv}[q]$, might have an infinite number of interaction vertices. but, only a finite number of vertices contributes to $\Gamma_P^{[0]}$ up to finite loops of 1PI diagrams.

(iii) To include the correct statistical information into the vertex functional given by Eq.(2.16). As discussed in (II), this can be done in most of practically interesting cases by using generalized FDT.

(iv) To determine the statistical Green's functions for the order parameter from the vertex functional $\Gamma_P[Q]$ according to the statistical functional equations (I)-(2.27), (2.28), or the equivalent to the hierarchy of equations (I)-(4.3)-(4.5).

(v) For some systems the functional integration (2.15) is difficult to carry out explicitly, but the effective action $I^{eff}[Q]$ can be found by transformation of path integral for $Z_P^{[0]}[h; J^+, J]$ as

$$Z_p^{(0)}[h;J^+,J] = \exp\left(\frac{i}{2}h\,M_0 h\right)Z_p^{\prime(0)}[h;J^+,J], \tag{2.19}$$

$$Z_p^{\prime(0)}[h;J^+,J]\Big|_{J^+=J=0} = \int_p [dQ]\exp\{i(I^{\prime eff}[Q] + hQ)\}\,. \tag{2.20}$$

It is important to note that M_0 in Eq.(2.19) is a quantity independent of h, J^+ and J (in the example of Sec.III, $M_0 = -2g^{-1}$). Let us define

$$Z_p^{\prime [0]}[h] \equiv \exp\{iW_p^{\prime [0]}[h]\} = Z_p^{\prime [0]}[h; J^+, J]\Big|_{J^+=J=0}, \tag{2.21}$$

It is obvious from Eq.(2.19) that

$$W_p^{[0]}[h] = W_p^{\prime [0]}[h] + \frac{1}{2} h M_0 h, \tag{2.22}$$

i.e. $W_p^{[0]}$ and $W_p^{\prime [0]}$ will generate exactly the same connected Green's functions except for the second order CTPGF which differ from each other by M_0. It is easy to show that a relation similar to Eq.(2.22),

$$W_p[h] = W_p^{\prime}[h] + \frac{1}{2} h M_0 h, \tag{2.23}$$

is also true for the connected functional with statistical information, if Eq.(II)-(4.1) is taken into account. Since W_p and $W_p^{\prime}$ have identical statistical properties, we will not distinguish quantities with and without "$\prime$" in subsequent discussions.

The above-said 5 points summarize what we call the effective action method for the order parameter. As we see, this is a practical prescription, if we know the explicit expressions for the original action $I[\psi^+, \psi]$ and the order parameter $Q[\psi^+, \psi]$. Usually, the elementary interaction vertices in the original action have 3 or 4 external lines, while the order parameters are polynomials of the constituent field variables not higher than the second power. It is easy to see that no difficulty arises in applying the described scheme to such systems. Some of the technical details will be explained in the illustrative examples given in the next two sections, where for short we will no longer distinguish by notation the CTPGF generating functionals with and without statistical information.

III. Application one – Contact interaction model for super-conductivity

In this section we apply the effective action method to the four-Fermion contact interaction model of superconductivity as an example of equilibrium system with complex composite order parameter. We also consider the superconductor close to the critical temperature with weak spatial nonuniformity to illustrate how proposed formalism is applied to inhomogeneous systems.

The basic field variable is a complex Fermi operator with spin index $\psi_b(x)$, $b = \uparrow$ or $\downarrow$. The original action is given by

$$I[\psi^+, \psi] = I_0[\psi^+, \psi] + I_{int}[\psi^+, \psi], \tag{3.1}$$

$$I_0[\psi^+, \psi] = \int_p d1 d2\, \psi^+(1) S_0^{-1}(1,2) \psi(2), \tag{3.2}$$

$$I_{int}[\psi^+, \psi] = \frac{g}{2}\int_p dx\, \psi_a^+(x) \psi_b^+(x) \psi_b(x) \psi_a(x), \tag{3.3}$$

where

$$S_0^{-1}(1,2)=\left(i\frac{\partial}{\partial\tau}+\frac{1}{2m}\nabla^2+\mu\right)\delta_p(1-2)\begin{pmatrix}1&0\\0&1\end{pmatrix} \tag{3.4}$$

is diagonal in the spin space, m-electron mass. μ -chemical potential, g-coupling constant. The order parameter of the system is given by Gorkov composite operators $\chi(x)$ and $\chi^*(x)$ defined as:

$$\chi(x)=\psi_\downarrow(x)\psi_\uparrow(x), \tag{3.5}$$

$$\chi^*(x)=\psi_\uparrow^+(x)\psi_\downarrow^+(x). \tag{3.6}$$

Introducing the external sources $\eta_b(x)$, $\eta_b^+(x)$, $h(x)$ and $h^*(x)$, the CTPGF generating functional for $\psi(x)$ and $\chi(x)$ can be written as[2]

$$\begin{aligned}Z_p[h^*,h;\eta^+,\eta]=&\int_p[d\psi^+][d\psi]\exp\{i(I_0[\psi^+,\psi]+I_{int}[\psi^+,\psi])\}\\&\times\exp\{i(\eta^+\psi+\psi^+\eta+h^*\chi+\chi^*h+W_p^N[\psi^+\overleftarrow{S}_0^{-1},\overrightarrow{S}_0^{-1}\psi])\}.\end{aligned} \tag{3.7}$$

According to rules given in Sec.II, we "forget" first W_p^N in Eq.(3.7) and carry out the integration over $\psi(x)$ to obtain the effective action. Note that $\psi(x)$ is anti-commuting Grassmann number under the path integration. Generalizing the Gaussian integration formula to the closed time-path, we obtain

$$\begin{aligned}&\exp\{\frac{i}{2}g\int_p d^4x\,\psi_a^+(x)\psi_b^+(x)\psi_b(x)\psi_a(x)\}\\&=Const\int_p[d\Phi^*][d\Phi]\exp\{i(-\frac{1}{g}\Phi^*\Phi+\Phi^*\psi_\downarrow\psi_\uparrow+\psi_\uparrow^+\psi_\downarrow^+\Phi)\}.\end{aligned} \tag{3.8}$$

Substituting Eq.(3.8) into Eq.(3.7), we find up to an unimportant constant factor that

$$\begin{aligned}Z_p[h^*,h;\eta^+,\eta]=&\int_p[d\Phi^*][d\Phi]\exp(-\frac{i}{g}\Phi^*\Phi)\\&\times\int_p[d\psi^*][d\psi]\exp\{i[\psi^+S_0^{-1}\psi+(\Phi^*+h^*)\psi_\downarrow\psi_\uparrow+\psi_\uparrow^+\psi_\downarrow^+(\Phi+h)+\eta^+\psi+\psi^+\eta]\}.\end{aligned} \tag{3.9}$$

To carry out the integration over Fermion field in Eq.(3.9) we use the Nambu spinor presentation as

$$\Psi(x)=\begin{pmatrix}\psi_\uparrow(x)\\\psi_\downarrow^+(x)\end{pmatrix}\qquad\Psi^+(x)=(\psi_\uparrow^+(x),\ \psi_\downarrow(x)) \tag{3.10}$$

and the corresponding source

$$\zeta(x)=\begin{pmatrix}\eta_\uparrow(x)\\-\eta_\downarrow^+(x)\end{pmatrix}\qquad\zeta^+(x)=(\eta_\uparrow^+(x),\ -\eta_\downarrow(x)) \tag{3.11}$$

to transform Eq.(3.9) as

$$Z_p[h^*, h;\ \zeta^+, \zeta] = \int_p [d\chi^*][d\chi] \exp\{i(-g\chi^*\chi + h^*\chi + \chi^* h - \frac{1}{g} h^* h)\}$$
$$\times \int_p [d\Psi^+][d\Psi] \exp\{i(\Psi^+ G^{-1}\Psi + \Psi^+\zeta + \zeta^+\Psi)\}\,, \tag{3.12}$$

where

$$G^{-1}(1,2) = G_0^{-1}(1,2) + gK(1,2), \tag{3.13}$$

$$G_0^{-1}(1,2) = \left(i\frac{\partial}{\partial t} + \sigma_3\left(\frac{1}{2m}\nabla^2 + \mu\right)\right)\delta_p(1-2), \tag{3.14}$$

$$K(1,2) = (\sigma_+\chi(1) + \sigma_-\chi^*(1))\,\delta_p(1-2), \tag{3.15}$$

σ_3, σ_+, σ_- being Pauli matrices in spin space. Carrying out the path integration over $d\Psi^+$ and $d\Psi$ in Eq.(3.12) yields

$$Z_p[h^*, h;\ \zeta^+, \zeta] = \exp(-\frac{i}{g}h^* h)\int_p [d\chi^*][d\chi]$$
$$\times \exp\{i(I_0[\chi^*,\chi] + I_{det}[\chi^*,\chi] + h^*\chi + \chi^* h - \zeta^+ G\zeta)\}\,, \tag{3.16}$$

where

$$I_0[\chi^*,\chi] = -g\int_p d^4x\, \chi^*(x)\chi(x)\,, \tag{3.17}$$

$$I_{det}[\chi^*,\chi] = -i\,\mathrm{Tr}\,\ln G^{-1}, \tag{3.18}$$

$$\zeta^+ G\zeta = \int_p d1 d2\ \zeta^+(1) G(1,2)\zeta(2)\,, \tag{3.19}$$

Tr means trace operation in both spin and coordinate space. The electron propagator G satisfies condition

$$\int_p d^4x\, G(1,x) G^{-1}(x,2) = \int_p d^4x\, G^{-1}(1,x)\, G(x,2) = \delta_p(1-2). \tag{3.20}$$

Comparing Eq.(3.16) with Eqs.(2.19) and (2.20) we find that Z_p has the expected form. According to discussion on point (v) in the preceding section, the effective action for $\chi(x)$ at the closed time-path is given by

$$I^{eff}[\chi^*,\chi] = I_0[\chi^*,\chi] + I_{det}[\chi^*,\chi] \tag{3.21}$$

up to a correction to second order CTPGF given by Eq.(2.23).

To illustrate the basic idea of the effective action method we will confine ourselves to the mean field approximation for the generating functional. According to Eqs.(2.16), (3.21), (3.17), (3.18) (I)-(2.30) and the convention about the notation, we find the vertex generating functional for the order parameter $\chi(x)$ in the mean field approximation as

$$\Gamma[\chi_\Delta^*, \chi_\Delta;\ \chi_c^*, \chi_c] \cong I^{eff}[\chi_\Delta^*, \chi_\Delta;\ \chi_c^*, \chi_c]$$
$$= I_0[\chi_\Delta^*, \chi_\Delta;\ \chi_c^*, \chi_c] + I_{det}[\chi_\Delta^*, \chi_\Delta;\ \chi_c^*, \chi_c]\ , \tag{3.22}$$

where

$$I_0[\chi_\Delta^*, \chi_\Delta ; \chi_c^*, \chi_c] = -g\int d^4x \left(\chi_\Delta^*(x)\chi_c(x) + \chi_c^*(x)\chi_\Delta(x)\right) , \tag{3.23}$$

$$I_{det}[\chi_\Delta^*, \chi_\Delta ; \chi_c^*, \chi_c] = -i \, Tr \, \ln G^{-1}, \tag{3.24}$$

$$\begin{aligned}[G^{-1}(1,2)]_{\alpha\beta} &= [G_0^{-1}(1,2)]_{\alpha\beta} + \\ &+ g[\sigma_+(\xi_\alpha \chi_c(1) + \tfrac{1}{2}\eta_\alpha \chi_\Delta(1)) + \sigma_-(\xi_\alpha \chi_c^*(1) + \tfrac{1}{2}\eta_\alpha \chi_\Delta^*(1))]\eta_\beta \delta_{\alpha\beta} \delta^4(1-2),\end{aligned} \tag{3.25}$$

α, β being closed time-path indices.

As seen from Eq.(3.16) the $G(1,2)$ function appearing in the mean field approximation of CTPGF generating functional for the order parameter as given by Eq.(3.22) is exactly the second order CTPGF for electron fields $\Psi(x)$, $\Psi^+(x)$ in the mean field approximation. According to discussion of Sec.II, the contribution of the equilibrium density matrix can be embodied in FDT satisfied by propagator G. In the Fourier presentation of relative coordinates, FDT is given as[6,7]

$$G_c[X,p] = (G_r[X,p] - G_a[X,p])\, th(\beta p_0/2) , \tag{3.26}$$

where $X = \frac{1}{2}(x_1 + x_2)$ while G_c, G_a, G_r are defined as

$$(G[X,p])_{\alpha\beta} = \tfrac{1}{2}\xi_\alpha \eta_\beta G_r[X,p] + \tfrac{1}{2}\eta_\alpha \xi_\beta G_a[X,p] + \tfrac{1}{2}\xi_\alpha \xi_\beta G_c[X,p]. \tag{3.27}$$

In Eq.(3.26) we assume G as well as the order parameter to be slowly varying functions of X.

The statistical vertex generating functional for the order parameter defined by Eqs.(3.22)-(3.27) contains in principle all statistical information for the order parameter in the mean field approximation. Here we will only discuss its statistical expectation value. In accord with Eqs. (I)-(4.3) and (3.22), the general equation for the expectation value of the order parameter is given by

$$\left.\frac{\delta I^{eff}[\chi_\Delta^*, \chi_\Delta ; \chi_c^*, \chi_c]}{\delta \chi_\Delta^*(x)}\right|_{\substack{\chi_\Delta^*(x) = \chi_\Delta(x) = 0 \\ \chi_c^*(x) = \chi^*(x) \\ \chi_c(x) = \chi(x)}} = 0 \tag{3.28}$$

and its complex conjugation.

We first consider stationary, homogeneous system where $\chi(x)$, $\chi^*(x)$ are independent of x and Green's functions are translationally invariant. Substituting Eqs.(3.22)-(3.25) into Eq.(3.28) we obtain the order parameter equation for uniform superconductor in the mean field approximation as

$$-g\chi - i\frac{g}{2}\mathrm{Tr}\{\sigma_- G_c(x,x)\} = 0, \tag{3.29}$$

where Tr means trace operation only in the spin space. Introducing notation

$$\Delta(x) \equiv g\chi(x), \tag{3.30}$$

$$\varepsilon(\vec{p}) = \frac{p^2}{2m} - \mu, \tag{3.31}$$

and solving G_a, G_r from Eq.(3.20) in momentum representation, we obtian G_c from FDT given by Eq.(3.26). Substituting G_c thus obtained into Eq.(3.29) yields the famous BCS gap equation as

$$\frac{g}{2}\int\frac{d^3p}{(2\pi\hbar)^3}\frac{\mathrm{th}\frac{\beta}{2}\sqrt{\varepsilon^2(\vec{p})+|\Delta|^2}}{\sqrt{\varepsilon^2(\vec{p})+|\Delta|^2}} - 1 = 0. \tag{3.32}$$

Hereafter in this section we switch onto the usual units.

Now we discuss the stationary system with weak inhomogeneity. Assume that the scale of space variation obeys the inequality

$$|\Delta\vec{x}| \gg \hbar/p_F, \tag{3.33}$$

where P_F is Fermi momentum of electron. Moreover, the following conditions are satisfied by the system

$$(T_c - T)/T_c \ll 1, \tag{3.34}$$

$$\Delta \ll T_c \sim T, \tag{3.35}$$

where $T = \beta^{-1}$ is the temperature of the system, T_c — critical temperature defined by[8]

$$gN(0)\,\ln\left(\frac{2\hbar\omega_D e^{\gamma}}{\pi T_c}\right) = 1 \tag{3.36}$$

with

$$N(0) = \frac{mp_F}{2\pi^2\hbar^3}. \tag{3.37}$$

as the density of states at the Fermi surface, ω_D being Debye frequency of phonon, γ — Euler constant. Since G_0^{-1} is translationally invariant according to Eq.(3.13), the nonuniformity of the system comes from a slow space variation of the order parameter χ as follows from Eqs.(3.22)-(3.25). For such a system we can expand the functional arguments $\chi(\vec{y})$, $\chi^*(\vec{y})$ in Eq.(3.28) around $\chi(\vec{x})$, $\chi^*(\vec{x})$ using $O(h/|\Delta\vec{x}|\,|\Delta\vec{P}|)$ as small parameter as we did before[7,9]. Taking into account Eqs.(3.33) and (3.35) to neglect quadratic term in $\partial\Delta(\vec{x})/\partial\vec{x}$, we find that

$$\left.\frac{\delta I^{eff}}{\delta \chi_\Delta^*(\vec{x})}\right|_{\substack{\chi_\Delta^*=\chi_\Delta=0,\\ \chi_c^*=\chi^*,\ \chi_c=\chi_c}}$$

$$-i\hbar\left(\frac{\partial\Gamma_{11}^r[\vec{x},\vec{q}]}{\partial\vec{q}}\right)_{\substack{q=0\\ \Delta=0}}\cdot\frac{\partial\chi(\vec{x})}{\partial\vec{x}}-\frac{\hbar^2}{2}\left(\frac{\partial^2\Gamma_{11}^r[\vec{x},\vec{q}]}{\partial\vec{q}\,\partial\vec{q}}\right)_{\substack{q=0\\ \Delta=0}}:\frac{\partial^2\chi(\vec{x})}{\partial\vec{x}\,\partial\vec{x}} \tag{3.38}$$

$$-i\hbar\left(\frac{\partial\Gamma_{12}^r[\vec{x},\vec{q}]}{\partial\vec{q}}\right)_{\substack{q=0\\ \Delta=0}}\cdot\frac{\partial\chi^*(\vec{x})}{\partial\vec{x}}-\frac{\hbar^2}{2}\left(\frac{\partial^2\Gamma_{12}^r[\vec{x},\vec{q}]}{\partial\vec{q}\,\partial\vec{q}}\right)_{\substack{q=0\\ \Delta=0}}:\frac{\partial^2\chi^*(\vec{x})}{\partial\vec{x}\,\partial\vec{x}}=0\ ,$$

where

$$\Gamma_{11}^r[x,q]=\int d^4y\,\exp\left(\frac{i}{\hbar}qy\right)\frac{\delta^2 I^{eff}}{\delta\chi_\Delta^*(x+\frac{y}{2})\,\delta\chi_c(x-y/2)} \tag{3.39}$$

$$=-g+\frac{i\hbar g^2}{2}\xi_\alpha\eta_\beta\int\frac{d^4p}{(2\pi\hbar)^4}\,\mathrm{Tr}\left\{\sigma_+G^{\alpha\beta}[x,q+p]\sigma_-G^{\beta\alpha}[x,p]\right.$$

$$\Gamma_{12}^r[x,q]=\int d^4y\,\exp\left(\frac{i}{\hbar}qy\right)\frac{\delta^2 I^{eff}}{\delta\chi_\Delta^*(x+y/2)\,\delta\chi_c^*(x-y/2)} \tag{3.40}$$

$$=i\hbar\frac{g^2}{2}\xi_\alpha\eta_\beta\int\frac{d^4p}{(2\pi\hbar)^4}\,\mathrm{Tr}\left\{\sigma_+G^{\alpha\beta}[x,q+p]\sigma_+G^{\beta\alpha}[x,p]\right.$$

Obviously, the first term in Eq.(3.38) is proportional to the left-hand side of Eq.(3.32) with Δ replaced by $\Delta(\vec{x})$. Considering $\Delta(\vec{x})$ as a small quantity and keeping its third power we obtain by the well-known technique[8]

$$\left.\frac{\delta I^{eff}}{\delta \chi_\Delta^*(\vec{x})}\right|_{\substack{\chi_\Delta^*=\chi_\Delta=0\\ \chi_c^*=\chi^*,\ \chi_c=\chi,}}$$

$$\cong gN(0)\,\Delta(\vec{x})\,\ln\left(T_c/T\right)-\frac{7\beta^2}{8\pi^2}\zeta(3)gN(0)\,\Delta(\vec{x})|\Delta(\vec{x})|^2\ . \tag{3.41}$$

Substituting Eqs.(3.39) and (3.40) into the four coefficents of Eq.(3.38) with $\Delta=0$ in Eq.(3.13) yields

$$\left.\frac{\partial\Gamma_{11}^r[\vec{x},\vec{q}]}{\partial\vec{q}}\right|_{\substack{q=0\\ \Delta=0}}=0\,, \tag{3.42}$$

$$\left.\frac{\partial^2\Gamma_{11}^r[\vec{x},\vec{q}]}{\partial\vec{q}\,\partial\vec{q}}\right|_{\substack{q=0\\ \Delta=0}}=-\frac{7}{12}\frac{g^2\beta^2u}{\pi^2 m}\zeta(3)N(0)\overleftrightarrow{I}\ , \tag{3.43}$$

$$\left.\frac{\partial\Gamma_{12}^r[\vec{x},\vec{q}]}{\partial\vec{q}}\right|_{\substack{q=0\\ \Delta=0}}=0\,, \tag{3.44}$$

$$\left.\frac{\partial\Gamma_{12}^r[\vec{x},\vec{q}]}{\partial\vec{q}\,\partial\vec{q}}\right|_{\substack{q=0\\ \Delta=0}}=0\,. \tag{3.45}$$

Finally, we obtain by putting Eqs.(3.41)-(3.45) into Eq.(3.38)

$$\frac{\hbar^2}{4m}\frac{7\beta^2\mu}{6\pi^2}\zeta(3)\nabla^2\Delta(\vec{x})+\Delta(\vec{x})\ln\frac{T_c}{T}-\frac{7\beta^2}{8\pi^2}\zeta(3)\Delta(\vec{x})\left|\Delta(\vec{x})\right|^2=0 \tag{3.46}$$

which is the microscopic form of the Ginzburg-Landau equation in the zero external magnetic field.

IV. Application two: Two-energy-level electron — unimode laser system

In this section we apply the effective action method for the order parameter to the two-energy-level electron — unimode laser system as an example of far from equilibrium systems.

We describe the positive and negative frequency parts of the vector potential of the unimode laser field by $a(x)$ and $a^*(x)$ correspondingly, while $\psi_1(x)$ and $\psi_2(x)$ represent bound electrons at upper energy level E_1^0 and lower energy level E_2^0. The action of the system is given as[10]

$$I[a^*,a;\psi^+,\psi]=I_0[a^*,a]+I_0[\psi^+,\psi]+I_{int}[a^*,a;\psi^+,\psi], \tag{4.1}$$

$$I_0[a^*,a]=\int_p d1d2\,a^*(1)\Delta_0^{-1}(1,2)a(2), \tag{4.2}$$

$$I_0[\psi^+,\psi]=\int_p d1d2\,\psi^+(1)S_0^{-1}(1,2)\psi(2) \tag{4.3}$$

$$I_{int}[a^*,a;\psi^+,\psi]=-ig\int_p d^4x\,\psi^+(x)(\sigma_+a(x)-\sigma_-a^*(x))\psi(x), \tag{4.4}$$

where

$$\Delta_0^{-1}(1,2)=\left(i\frac{\partial}{\partial t}-K_0\right)\delta_p(1-2). \tag{4.5}$$

K_0 is the photon energy, g—coupling constant of unimode electro-magnetic field with bound electron. The two-level electron system is represented by a spinor

$$\psi(x)=\begin{pmatrix}\psi_1(x)\\ \psi_2(x)\end{pmatrix} \tag{4.6}$$

with free propagator given by

$$S_0^{-1}(1,2)=\left(i\frac{\partial}{\partial t_1}+\frac{1}{2M}\nabla_1^2-\frac{E_1^0+E_2^0}{2}-\sigma_3\frac{E_1^0-E_2^0}{2}\right)\delta_p(1-2), \tag{4.7}$$

where σ_+, σ_- and σ_3 are Pauli matrices in the space of energy levels, M—atomic mass.

In this example, the basic fields themselves $a(x)$ and $a^*(x)$ constitute the order parameters we are interested in. Introducing external sources $j(x)$, $j^*(x)$, $\eta(x)$ and $\eta^+(x)$ corresponding to $a^*(x)$, $a(x)$, $\psi^+(x)$ and $\psi(x)$ the CTPGF generating functional is written as[1]

$$\begin{aligned}Z_p[j^*,j;\eta^+,\eta]=&\int_p[da^*][da][d\psi^+(x)][d\psi(x)]\\ &\times\exp\{i(I_0[a^*,a]+I_0[\psi^+,\psi]+I_{int}[a^*,a;\psi^+,\psi]\}\\ &\times\exp\{(j^*a+a^*j+\eta^+\psi+\psi^+\eta+W_p^N[-a^*\Delta_0^{-1}-\Delta_0^{-1}a;-\psi^+\overleftarrow{S_0^{-1}},-\overrightarrow{S_0^{-1}}\psi]\}.\end{aligned} \tag{4.8}$$

According to the discussion in Sec.II, we can neglect for a moment the functional W_P^N in Eq.(4.8) to carry out first the integration over $[d\psi^+][d\psi]$ which yields

$$Z_p[j^*, j; \eta^+, \eta] = \int_p [da^*][da] \times \exp\{i(I_o[a^*, a] + I_{det}[a^*, a] + j^* a + a^* j - \eta^+ S \eta)\}, \tag{4.9}$$

where

$$I_{det}[a^*, a] = -i \, Tr \, S^{-1}, \tag{4.10}$$

$$S^{-1}(1,2) = S_o^{-1}(1,2) - ig(\sigma_+ a(1) - \sigma_- a^*(1))\, \delta_p(1-2), \tag{4.11}$$

$$\int_p d^4x \, S^{-1}(1,x) S(x,2) = \int_p d^4x \, S(1,x) S^{-1}(x,2) = \delta_p(1-2), \tag{4.12}$$

Tr means here trace operation in both closed path coordinate space and the space of energy levels.

Comparing Eq.(4.9) with Eqs.(2.8) and (2.13) we find the effective action for the order parameters a(x) and a*(x) to be

$$I^{eff}[a^*, a] = I_o[a^*, a] + I_{det}[a^*, a]. \tag{4.13}$$

Like the preceding example we will discuss only the mean field approximation for the statistical generating functional to illustrate the basic idea of the formalism. According to Eqs.(2.16), (4.13), (4.2), (4.10), (I)-(2.30) and the notation adopted, the CTPGF generating functional for a(x) and a*(x) in mean field approximation is given by

$$\Gamma[a_\Delta^*, a_\Delta; a_c^*, a_c] \cong I^{eff}[a_\Delta^*, a_\Delta; a_c^*, a_c] = I_o[a_\Delta^*, a_\Delta; a_c^*, a_c] + I_{det}[a_\Delta^*, a_\Delta; a_c^*, a_c], \tag{4.14}$$

where

$$I_o[a_\Delta^*, a_\Delta; a_c^*, a_c] = \int d1 d2 \{a_\Delta^*(1) \Delta_{or}^{-1}(1,2) a_c(2) + a_c^*(1) \Delta_{oa}^{-1}(1,2) a_\Delta(2)\}, \tag{4.15}$$

$$I_{det}[a_\Delta^*, a_\Delta; a_c^*, a_c] = -i \, Tr \ln S^{-1}, \tag{4.16}$$

$$[S^{-1}(1,2)]_{\alpha\beta} = [S_o^{-1}(\alpha,\beta)]_{\alpha\beta} - ig[\sigma_+(\xi_\alpha a_c(1) + \tfrac{1}{2}\eta_\alpha a_\Delta(1)) - \sigma_-(\xi_\alpha a_c^*(1) + \tfrac{1}{2}\eta_\alpha a_\Delta^*(1))]\eta_\beta \delta_{\alpha\beta} \delta^4(1-2) \tag{4.17}$$

Δ_{or}^{-1}, Δ_{oa}^{-1} being defined by

$$[\Delta_o^{-1}(1,2)]_{\alpha\beta} = \tfrac{1}{2}\xi_\alpha \eta_\beta \Delta_{or}^{-1}(1,2) + \tfrac{1}{2}\eta_\alpha \xi_\beta \Delta_{oa}^{-1}(1,2), \tag{4.18}$$

with α, β as closed path indices.

Unlike Eq.(4.7) we assume in Eq.(4.17) that

$$S_o^{-1}(1,2)$$

$$=\left(i\frac{\partial}{\partial t_1}+\frac{1}{2M}\nabla_1^2-\frac{E_1^o+E_2^o}{2}-\sigma_3\frac{E_1^o\ E_2^o}{2}\right)\delta_p(1-2)-\Sigma(1,2), \tag{4.19}$$

where

$$\Sigma(1,2)=\begin{pmatrix}\Sigma_1(1,2) & 0\\ 0 & \Sigma_2(1,2)\end{pmatrix}, \tag{4.20}$$

i.e. the CTPGF renormalization correction $\Sigma(1,2)$ is included which is diagonal in energy representation in the lowest approximation.

As seen from Eq.(4.9) the function S(1,2) appearing in the CTPGF generating functional for the order parameter as given by Eq.(4.14) is the renormalized CTPGF propagator for bound electron. A macroscopically varying in time laser system is a quasi-stationary state compared with the laser frequency. In the present case people usually neglect the feedback interaction of the laser field upon the distribution of electrons over energy levels. Therefore, in accord with the argument presented in Sec.II, the statistical information is contained in the generalized FDT [5,6] satisfied by $S_0(1,2)$, the CTPGF electron propagator without laser order parameter being incorporated.
In momentum presentation the required FDT can be written as [5,6]

$$S_{oc}[P]=S_{or}[P](1-2N[P])-(1-2N[P])S_{oa}[P], \tag{4.21}$$

where S_{or}, S_{oa} and S_{oc} as retarded, advanced and correlation Green's function correspondingly are defined by

$$[S_o(1,2)]_{\alpha\beta}=\frac{1}{2}\xi_\alpha\eta_\beta S_{or}(1,2)+\frac{1}{2}\eta_\alpha\xi_\beta S_{oa}(1,2)+\frac{1}{2}\xi_\alpha\xi_\beta S_{oc}(1,2) \tag{4.22}$$

with

$$N[P]=\begin{pmatrix}N_1[P] & 0\\ 0 & N_2[P]\end{pmatrix}, \tag{4.23}$$

$N_1[P]$, $N_2[P]$ are translational momentum distribution functions of bound electrons. Since they are usually fixed by external parameters in the laser system, we do not indicate explicitly their coordinate dependence in Eqs.(4.21)-(4.23). Expressing the renormalized self-energy part of electron in momentum representation in terms of the energy level shift

$$\Delta E_i[P]=\frac{1}{2}\left(\Sigma_i^r[P]+\Sigma_i^a[P]\right) \tag{4.24}$$

and the line-width

$$\gamma_i[P]=\frac{i}{2}\left(\Sigma_i^r[P]-\Sigma_i^a[P]\right), \tag{4.25}$$

we rewrite Eq.(4.21), the FDT for nonequilibrium stationary state, as

$$\frac{1}{2}\Sigma_i^c[P]=-i\,\gamma_i[P](1-2N_i[P]), \tag{4.26}$$

where Σ_i^r, Σ_i^a and Σ_i^c are defined by

$$(\Sigma_i[P])_{\alpha\beta} = \frac{1}{2}\xi_\alpha \eta_\beta \Sigma_i^r[P] + \frac{1}{2}\eta_\alpha \xi_\beta \Sigma_i^a[P] + \frac{1}{2}\xi_\alpha \xi_\beta \Sigma_i^c[P], \tag{4.27}$$

$i = 1,2$ being index for energy level. We define moreover for convenience

$$E_i[P] \equiv \frac{p^2}{2M} + E_i^o + \Delta E_i . \tag{4.28}$$

The statistical vertex generating functional for the order parameter defined by Eqs.(4.14)-(4.16), (4.21) and (4.27) contains in principle complete statistical information of order parameter in mean field approximation. Take, for example, the expectation values of the order parameters as given by

$$a(x) = Tr\{\hat{\rho}\ \hat{a}(x)\}, \tag{4.29}$$

$$a^*(x) = Tr\{\hat{\rho}\ \hat{a}^*(x)\}. \tag{4.30}$$

According to Eqs.(I)-(4.30) and (4.14), they satisfy the following general equation

$$\left.\frac{\delta I^{eff}[a_\Delta^*, a_\Delta; a_c^*, a_c]}{\delta a_\Delta^*(x)}\right|_{\substack{a_\Delta^* = a_\Delta = 0 \\ a_c^*(x) = a^*(x) \\ a_c(x) = a(x)}} = 0 \tag{4.31}$$

and its complex conjugation. Substitution of Eqs.(4.14)-(4.17) into (4.31) yields

$$\left(i\frac{\partial}{\partial t} - K_o\right)a(x) + \frac{g}{2}Tr\{\sigma_- S_c(x,x)\} = 0, \tag{4.32}$$

where

$$S_c(x,y) = \frac{1}{2}\xi_\alpha \xi_\beta [S(x,y)]_{\alpha\beta} \tag{4.33}$$

and trace is taken here only in the space of energy levels. Expressing $(\Sigma[P])_{\alpha\beta}$ in Eq.(4.19) in terms of $E_i[P]$, $\gamma_i[P]$ and $N_i[P]$ by use of Eqs.(4.24)-(4.26) and (4.28), we can find $S_c(x,x)$ from Eq.(4.12) with Eqs.(4.17),(4.19) being considered. Finally, we can transform Eq.(4.32) into

$$\left(i\hbar\frac{\partial}{\partial t} - K_o\right)a(x) + g^2 a(x)\int\frac{d^3p}{(2\pi\hbar)^3}\frac{N_1(\vec{P}) - N_2(\vec{P})}{K_o - E_1(\vec{P}) + E_2(\vec{P}-\vec{K}) + i(\gamma_1+\gamma_2)}$$

$$\times \frac{1}{1 + \frac{g^2|a(K)|^2}{\gamma_1\gamma_2}\mathcal{L}(K_o - E_1(\vec{P}) + E_2(\vec{P}-\vec{K}))}, \tag{4.34}$$

where $\vec{k}$ is the wave vector for the unimode laser field and

$$\mathcal{L}(E)=\frac{(\gamma_1+\gamma_2)^2}{E^2+(\gamma_1+\gamma_2)^2}. \tag{4.35}$$

Eq.(4.34) is just the result of semi-classical Lamb theory of laser[11], if the nondiagonal dissipation is neglected.

The unimode laser system coupled with two-level bound electrons discussed in this section is a typical example of open systems. As mentioned in the Introduction, since our formalism is set up in some analogy with the usual Liouville problem in quantum statistics, the statistical information appears in it as initial conditions. This example shows, however, that this formalism is applicable to nonequilibrium, open systems as well. The field variables of a statistical system can be divided into two groups, one of which, the internal parameters, is associated with the subsystem of interest itself (such as the vector potential field in our example), the other one, the external parameters, being connected with field variables outside the subsystem (e.g. the electron variables in our case). The procedure of eliminating the external parameters by means of path integration to obtain the effective action $I^{eff}[a_\Delta^*, a_\Delta; a_c^*, a_c]$ embodies itself the characteristic feature of open system — the statistical information has to be "injected" constantly into the subsystem considered by external parameters. In the example discussed here this injection is realized through the contribution of electron propagator $S_0(1,2)$ to the effective action of the laser field. Therefore, the effective action $I^{eff}[a_\Delta^*, a_\Delta; a_c^*, a_c]$ contains not only the statistical information of the laser field itself but also, and more importantly, the statistical information about the inverted distribution of electrons.

References

1. ZHOU Guang-zhao, SU Zhao-bin, HAO Bai-lin, YU Lu, Commun. in Theor. Phys. (Beijing), 1(1982), 295.
2. ZHOU Guang-zhao, SU Zhao-bin, HAO Bai-lin, YU Lu, Commun. in Theor. Phys. (Beijing). 1(1982), 307.
3. See, for example, H. Haken, in Laser Handbook, Vol. 1, eds. F. Arrechi et al., North-Holland Amsterdam, (1972).
4. R. Jackiw, Phys. Rev. D9 (1974), 1686.
5. ZHOU Guang-zhao, SU Zhao-bin, Physica Energiae Fortis et Physica Nuclearis (Beijing) 3 (1979), 314.
6. ZHOU Guang-zhao, SU Zhao-bin, Acta Physica Sinica, 30(1981), 164, 401.
7. ZHOU Guang-zhao, SU Zhao-bin, Ch. 5 in Progress in Statistical Physics, eds. HAO Bai-lin et al., KEXUE (Science Press), Beijing, 1981.
8. See, for example, A.L. Fetter, J.D. Walecka, Quantum Theory of Many-Particle Systems, McGraw-Hill, N.Y. 1971.
9. Guang-zhao ZHOU, Zhao-bin SU, Bai-lin HAO, Lu YU, Phys. Rev., B22(1980), 3385.
10. ZHOU Guang-zhao, SU Zhao-bin, Acta Physica Sinica, 29(1980), 618.
11. M. Sargent III, M. Scully and W. Lamb, Laser Physics, Addison-Wesley, Reading, MASS.(1974).

Commun. in Theor. Phys. (Beijing, China) Vol. 1, No. 6 (1982) 669-679

ON AN APPROXIMATE FORM OF THE COUPLED EQUATIONS OF THE ORDER PARAMETER WITH THE WEAK ELECTROMAGNETIC FIELD FOR THE IDEAL SUPERCONDUCTOR

SU Zhao-bin (苏肇冰) CHOU Kuang-chao (周光召)
Institute of Theoretical Physics, Academia Sinica, Beijing, China

Received May 10, 1982

Abstract

A simplified derivation of the macroscopic electrodynamic equations of Umezawa, Mancini et al. for superconductors is given in the framework of the closed time path Green's functions (CTPGF) using generalized Ward-Takahashi identities. It is shown that the forms of the equations obtained are the same for both thermoequilibrium and nonequilibrium stationary states provided the electromagnetic field is weak and its effect on the modulus of the order parameter can be neglected. The statistical behavior of the states is completely specified in the equations by parameters which can be calculated by the method of CTPGF.

I. Introduction

The interaction between the phase of the order parameter and the macroscopic electromagnetic field (hereafter E.M.F.) always plays an important role in superconductor physics. Both the Meissner effect and the Josephson effect are brilliant examples of macroscopic quantum phenomena. With the rapid development of experimental techniques, the research relating to these topics is entering the realm beyond thermal equilibrium, for which the Ginzburg-Landau theory [1] or the Bogoliubov-de Gennes theory [2] of equilibrium superconductor becomes insufficient in some cases. It is reasonable to restart from a microscopic point of view to explore a new formalism for the interaction between order parameter and E.M.F., which is suitable for both equilibrium and nonequilibrium phenomena.

In the seventies, a set of macroscopic electrodynamic equations for ground state superconductor, different from the Ginzburg-Landau equation which is effective only near the critical point, is given by Umezawa, Mancini et al. [3], using the mapping and transformation technique on the Hilbert space within a field-theoretical formalism of ground state superconductors. These equations relate the phase of the order parameter, especially its singularities, to the macroscopic E.M.F. explicitly and are capable of treating a number of macroscopic quantum phenomena for ideal gound state superconductors. Because of the complexity of their formalism, it seems hard to clarify the assumptions and approximations involved, and therefore, difficult to generalize their formalism to a unified theory valid both for finite temperatures and stationary states in nonequilibrium.

As a primary check of the "theory of statistical generating functional for the order parameter"[4], we have rederived the Ginzburg-Landau equation near the critical point[4]. In this paper, inspired by Umezawa, Mancini et al., a set of interacting weak E.M.F.-order parameter equations for ideal superconductors is derived in the framework of Ref. [4], with its E.M.F.-phase of order parameter part formally identical to that of Ref. [3]. We show that this set of equations is an approximate form of the statistical functional equations for the order parameter[4]. Moreover, the equation for the modulus of the order parameter is incorporated into our formalism naturally and statistical information is specified by the parameters of the equations implicitly in accord with the CTPGF theory. If the intensity of the E.M.F. is not too strong and the superconductor is far from the critical point, these equations are shown to be valid not only for the case of T=0 ground state, but also for the case of T≠0 thermal equilibrium or some nonequilibrium stationary states. All parameters of the equations corresponding to different statistical situations can be calculated by the method given in Ref. [4]. In our derivation, a crucial role is played by the gauge symmetry induced Ward-Takahashi (W-T) identities for the vertex generating functional. The general form of the equations follows almost immediately with a transparent physical interpretation of our procedure which is in some sense the nonrelativistic version of Higgs mechanism for U(1) gauge symmetry when the Goldstone field has singularities.

In Sec.II the formulation of the problem is given with the approximations stated explicitly and the general form of the equations for the E.M.F. and the order parameter are derived. In Sec.III a set of coupled E.M.F.-order parameter equations for the ideal superconductor is derived by virtue of the W-T identities generalized to the closed time-path[5]. Finally in Sec.IV, as an example for comparing our results with the known ones and for illustrating the method of evaluating the parameters developed in Ref. [4], we calculate in the long wave-length limit the temperature dependence of the two basic parameters: density of the superconducting electron pairs and phase velocity of the phase excitation of the order parameter.

II. Formulation of the problem and the approximate form of the statistical functional equations for the macroscopic variables

For superconductor interacting with the E.M.F., the interesting macroscopic variables are the vector potential of the E.M.F.

$$A^{\mu}(x) = \mathrm{Tr}\{\hat{\rho}\hat{A}^{\mu}(x)\}, \quad \mu = 0, 1, 2, 3, \tag{2.1}$$

and the order parameter of the superconductor

$$\begin{aligned}\chi(x) &= \frac{1}{\sqrt{2}}(\chi_1(x) - i\chi_2(x)) \equiv \Xi(x)\exp[-i\,\Theta(x)] = \mathrm{Tr}\{\hat{\rho}\hat{\psi}_\downarrow(x)\hat{\psi}_\uparrow(x)\},\\ \chi^*(x) &= \frac{1}{\sqrt{2}}(\chi_1(x) + i\chi_2(x)) \equiv \Xi(x)\exp[i\,\Theta(x)] = \mathrm{Tr}\{\hat{\rho}\hat{\psi}_\uparrow^+(x)\hat{\psi}_\downarrow^+(x)\},\end{aligned} \tag{2.2}$$

where $\hat{\rho}$ is the normalized density matrix in Heisenberg picture, $\hat{A}^{\mu}(x)$ being the Heisenberg operators of the vector potential and $\hat{\psi}_{\uparrow}(x)$, $\hat{\psi}_{\downarrow}(x)$ - Heisenberg electron field operators with up, down spin indices. A comparison with the conventional symbols yields

$$\chi(x) = \frac{1}{g}\Delta(x), \tag{2.3}$$

where $\Delta(x)$ is the energy gap parameter, g the coupling constant of contact interaction for weak coupling superconductor. We choose gauge condition

$$\mathcal{D}_\mu(\partial) A^\mu(x) = 0 \tag{2.4}$$

for the vector potential. In Eq.(2.4), $D_\mu(\partial)$ is a linear differential operator, and "∂" is an abbreviation of $\partial_\mu = \frac{\partial}{\partial x^\mu}$. The concrete form of $D_\mu(\partial)$ will be determined later. In our paper, we take the metric tensor as

$$g_{\mu\nu} = \begin{pmatrix} 1 & & & \\ & -1 & & \\ & & -1 & \\ & & & -1 \end{pmatrix}, \qquad \mu, \nu = 0, 1, 2, 3, \tag{2.5}$$

and the transformation properties of related variables are defined according to Ref.[6]. There is another variable $B(x)$, the so-called ghost field, needed in the generating functional formalism, which corresponds to a Lagrange multiplier for the gauge condition. In accordance with Ref.[4], introducing the irreducible vertex functional of the macroscopic variables defined along the closed time-path $\Gamma_p \equiv \Gamma_p[A^\mu(x), \Xi(x), \Theta(x), B(x)]$, we will have the functional equations satisfied by the macroscopic variables $A^\mu(x)$, $\chi(x)$, $\chi^*(x)$, $B(x)$

$$\left.\frac{\delta\Gamma_p[A^\nu, \Xi, \Theta, B]}{\delta A^\mu(x)}\right|_{t^+=t^-} = 0 \quad , \tag{2.6}$$

$$\left.\frac{\delta\Gamma_p[A^\nu, \Xi, \Theta, B]}{\delta \Theta(x)}\right|_{t^+=t^-} = 0 \quad , \tag{2.7}$$

$$\left.\frac{\delta\Gamma_p[A^\nu, \Xi, \Theta, B]}{\delta B(x)}\right|_{t^+=t^-} = 0 \quad , \tag{2.8}$$

$$\left.\frac{\delta\Gamma_p[A^\nu, \Xi, \Theta, B]}{\delta \Xi(x)}\right|_{t^+=t^-} = 0 \quad . \tag{2.9}$$

Suppose the intensity of the E.M.F. is not too strong and the superconductor is far from the critical point. We can neglect the reaction of the E.M.F. to the modulus of the order parameter owing to the finite nonvanishing gap, and then, assume that

$$\Xi(x) = \Xi = \frac{1}{g}|\Delta| \quad \textit{independent of } \vec{x}, t \;, \tag{2.10}$$

and the equation Eq.(2.9) degenerates into an equation for $|\Delta|$ without the E.M.F. and with $\Theta(x)$ taken to be constant

$$\left.\frac{\delta\Gamma_p[\Xi]}{\delta\Xi(x)}\right|_{t^+=t^-} = 0 \quad , \tag{2.11}$$

where

$$\Gamma_p[\Xi] = \Gamma_p[A^\nu, \Xi, \Theta, B]\Big|_{\substack{A^\nu = B = 0 \\ \Theta = \text{Const.}}} \tag{2.12}$$

Now, the equations for the order parameter interacting with E.M.F. decouple into two parts: the one Eq.(2.11) being the equation for the modulus of the order parameter and the other, Eqs.(2.6)-(2.8), the equations of the E.M.F. vector potential coupled with the phase of the order parameter. We solve the former equation first and then put its solution into the latter one as an input parameter. As the solution of Eq.(2.11) is a problem for the superconductor itself, we will focus on Eqs.(2.6)-(2.8) in the remaining part of this paper.

Since we have assumed that the intensity of the E.M.F. is not too strong, we may linearize the functional arguments $A^\mu(x)$, $\Theta(x)$ and $B(x)$ linearly in Eqs.(2.6)-(2.8) according to the CTPGF technique given in Ref.[5] as

$$\int d^4y \left\{\Gamma^r_{\mu\nu}(x,y)A^\nu(y) + \Gamma^r_{\mu\theta}(x,y)\Theta(y) + \Gamma^r_{\mu b}(x,y)B(y)\right\} = 0 , \tag{2.13}$$

$$\int d^4y \left\{\Gamma^r_{\theta\nu}(x,y)A^\nu(y) + \Gamma^r_{\theta\theta}(x,y)\Theta(y) + \Gamma^r_{\theta b}(x,y)B(y)\right\} = 0 , \tag{2.14}$$

$$\int d^4y \left\{\Gamma^r_{b\nu}(x,y)A^\nu(y) + \Gamma^r_{b\theta}(x,y)\Theta(y) + \Gamma^r_{bb}(x,y)B(y)\right\} = 0 , \tag{2.15}$$

where

$$\Gamma^r_{ij}(x,y) = \frac{1}{2}\xi_{\sigma_1}\eta_{\sigma_2}\frac{\delta^2\Gamma_p[A^\nu, \Xi, \Theta, B]}{\delta\phi_i(x_{\sigma_1})\,\delta\phi_j(y_{\sigma_2})}\Bigg|_{\substack{\bar{A}^\mu = B = \Theta = 0 \\ \Xi = \text{Solution of Eq. (2.11)} \\ t^+ = t^-}} \tag{2.16}$$

are the 2-point retarded vertex functions of the corresponding variables $A^\mu(x)$: $\mu=0,1,2,3$, $\Theta(x)$, $B(x)$ denoted by ϕ_i, ϕ_j. In Eq.(2.16), σ_1 and σ_2, taking value "+" or "-", are the closed time-path indices, and

$$\xi_\pm = 1 , \qquad \eta_\pm = \pm 1 . \tag{2.17}$$

Obviously, by taking Eq.(2.10) into consideration, all the 2-point retarded vertex functions appearing in Eqs.(2.13)-(2,15) are translationally invariant in the linear approximation, i.e.,

$$\Gamma^r_{ij}(x,y) = \Gamma^r_{ij}(x-y) \tag{2.18}$$

In the following, as in the BCS theory, we restrict ourselves to dealing with only ideal superconductors, i.e., we will neglect the dissipative part of the 2-point retarded vertex functions. On account of $A^\mu(x)$, $\Theta(x)$ and $B(x)$ being real, all these functions appearing in Eqs.(2.13)-(2.15) are symmetric[5], i.e.,

$$\Gamma^r_{ij}(x,y) = \Gamma^r_{ji}(y,x) , \tag{2.19}$$

where i, j run over $\mu=0,1,2,3,\vartheta$, b independently of each other. Along such a line, the time variation scale of the system should be slow compared with that for exciting the ignored relaxation processes. Consequently, expanding $\Gamma^r_{\vartheta\vartheta}(x-y)$ to the second power of $\frac{\partial}{\partial t}$, taking account of Eqs.(2.18), (2.19) and the local invariance under 3-dimensional rotation, it is easy to derive the general form of $\Gamma^r_{\vartheta\vartheta}(x-y)$ as

$$\Gamma^r_{\vartheta\vartheta}(x-y) = -\frac{1}{\lambda^2}\left(\frac{\hbar c}{2e}\right)^2 Z_\vartheta^{-1}[-\nabla^2]\left(\partial_0^2 - \frac{v^2[-\nabla^2]}{c^2}\nabla^2 + \frac{m_\vartheta^2 c^2}{\hbar^2}\right)\delta^4(x-y)$$

$$= -\frac{1}{\lambda^2}\left(\frac{\hbar c}{2e}\right)^2 Z_\vartheta^{-1}[-\nabla^2]\left(\mathcal{D}(\partial) + \frac{m_\vartheta^2 c^2}{\hbar^2}\right)\delta^4(x-y), \tag{2.20}$$

where λ and m_ϑ are the constants to be determined with length and mass scales respectively, $Z_\vartheta[-\nabla^2]$ - the unknown dimensionless renormalization operator, $v^2[-\nabla^2]$ - the unknown operator with dimension of velocity square,

$$\nabla^2 \equiv -\partial_i\partial^i = \left(\frac{\partial}{\partial x^1}\right)^2 + \left(\frac{\partial}{\partial x^2}\right)^2 + \left(\frac{\partial}{\partial x^3}\right)^2 \tag{2.21}$$

being the 3-d Laplacian, c, $\hbar$ and e being the velocity of light, planck constant and the algebraic value of electron charge. In Eq.(2.20) we have also introduced a symbol

$$\mathcal{D}(\partial) \equiv \partial_0^2 - \frac{v^2[-\nabla^2]}{c^2}\nabla^2 \tag{2.22}$$

Now, we will fix the gauge condition (2.4) according to Eq.(2.20) by setting

$$\mathcal{D}^\mu(\partial) = \left(\partial^0, \frac{v^2[-\nabla^2]}{c^2}\partial^i\right) \tag{2.23}$$

in Eq.(2.4) with

$$\partial_\mu \mathcal{D}^\mu(\partial) = \mathcal{D}(\partial). \tag{2.24}$$

Along the same line, we can expand $\Gamma^r_{\mu\nu}(x-y)$ to the second power of $\frac{\partial}{\partial t}$. Since the transverse part of vector potential may behave differently from the scalar and the longitudinal part, we can introduce the transverse projection operator $L^{(T)}_{\mu\nu}$ and longitudinal-scalar projection operator $L_{\mu\nu}$ as

$$g_{\mu\nu} = L^{(T)}_{\mu\nu} + L_{\mu\nu}, \tag{2.25}$$

$$L^{(T)}_{\mu\nu} = \left(\begin{array}{c|c} 0 & \\ \hline & g_{ij} + \partial_i\frac{1}{\nabla^2}\partial_j \end{array}\right), \quad L_{\mu\nu} = \left(\begin{array}{c|c} 1 & \\ \hline & -\partial_i\frac{1}{\nabla^2}\partial_j \end{array}\right), \tag{2.26}$$

where μ, ν run over time-space coordinate indices 0,1,2,3, while i,j run over space coordinate indices 1,2,3 only. Moreover, by taking into account the gauge invariance, local invariance of 3D rotation and Eqs.(2.18), (2.19), it is also easy to derive the general form for $\Gamma^r_{\mu\nu}(x-y)$ as

$$\Gamma^{r}_{\mu\nu}(x-y) = \Big\{ Z_L^{-1}[-\nabla^2]\frac{1}{4\pi}(\Box g_{\mu\rho} - \partial_\mu\partial_\rho)L^{\rho}_{\ \nu} + Z_T^{-1}[-\nabla^2]\frac{1}{4\pi}(\Box g_{\mu\rho} - \partial_\mu\partial_\rho)L^{(T)\rho}_{\ \ \nu} + \tilde{\Pi}_{\mu\nu}[\vec{\partial}]\Big\}\delta^4(x-y)$$
$$= \Big\{ Z_L^{-1}[-\nabla^2]\left(\frac{v^2[-\nabla^2]}{c^2}\right)^{-1}\frac{1}{4\pi}(\mathcal{D}(\partial)\eta_{\mu\rho} - \mathcal{D}_\mu(\partial)\mathcal{D}_\rho(\partial))L^{\rho}_{\ \nu} + Z_T^{-1}[-\nabla^2]\frac{1}{4\pi}(\Box g_{\mu\rho} - \partial_\mu\partial_\rho)L^{(T)\rho}_{\ \ \nu} + \tilde{\Pi}_{\mu\nu}[\vec{\partial}]\Big\}\delta^4(x-y), \tag{2.27}$$

where

$$\tilde{\Pi}_{\mu\nu}[\vec{\partial}] = \tilde{\Pi}_T[-\nabla^2]L^{(T)}_{\mu\nu} + \left(\begin{array}{c|c} \tilde{\Pi}_{00}[-\nabla^2] & \\ \hline & -\tilde{\Pi}_L[-\nabla^2]\partial_i\frac{1}{\nabla^2}\partial_j \end{array}\right), \tag{2.28}$$

$Z_T[-\nabla^2]$ ($Z_L[-\nabla^2]$) being the dimensionless renormalization operators for the transverse(longitudinal-scalar)part of the vector potential, $\tilde{\Pi}_T[-\nabla^2]$, $\tilde{\Pi}_L[-\nabla^2]$ and $\tilde{\Pi}_{00}[-\nabla^2]$ being the corresponding polarization corrections. Furthermore, for convenience of comparison with Umezawa, Mancini et al.[3], a "curved" metric tensor is introduced into Eq.(2.27) as

$$\eta_{\mu\nu} \equiv \eta_{\mu\nu}[-\nabla^2] \equiv \left(\begin{array}{c|c} g_{00} & \\ \hline & g_{ij}\frac{v^2[-\nabla^2]}{c^2} \end{array}\right) \tag{2.29}$$

with

$$(\Box g_{\mu\rho} - \partial_\mu\partial_\rho)L^{\rho}_{\ \nu} = \left(\frac{v^2[-\nabla^2]}{c^2}\right)^{-1}(\mathcal{D}(\partial)\eta_{\mu\rho} - \mathcal{D}_\mu(\partial)\mathcal{D}_\rho(\partial))L^{\rho}_{\ \nu} \tag{2.30}$$

as an easily verifiable identity.

In this section, Eqs.(2.11), (2.13)-(2.15) and expressions (2.20), (2.27) are derived under two main assumptions, i.e., smallness of E.M.F. and idealness of superconductor. Before continuing our discussion, we would like to emphasize that we have made no assumption on the statistical properties of the system, i.e., on the density matrix involved in Eqs.(2.1) and (2.2); or equivalently, no explicit statistical assumption on the Eq. (2.11) and the coefficients in the Eqs.(2.13)-(2.15). Therefore, there are no limitations for the superconductor to be in the ground state.

III. W-T identities, Goldstone-Higgs mechanism and coupled equations for the E.M.F.-phase of order parameter

Two of the generalized W-T identities defined on the closed time-path for the irreducible vertex generating functional can be written as[5]

$$\partial^\mu\frac{\delta\Gamma_p[A^\nu,\Xi,\Theta,B]}{\delta A^\mu(x)} - \frac{2e}{\hbar c}\frac{\delta\Gamma_p[A^\nu,\Xi,\Theta,B]}{\delta\Theta(x)} + \mathcal{D}^+(\partial)B(x) = 0, \tag{3.1}$$

$$\frac{\delta\Gamma_p[A^\nu,\Xi,\Theta,B]}{\delta B(x)} - \mathcal{D}_\mu(\partial)A^\mu(x) = 0, \tag{3.2}$$

where, according to Eqs.(2.22) and (2.24),

$$\mathcal{D}^+(\partial) \equiv -\partial_\mu\mathcal{D}^\mu(-\partial) = \mathcal{D}(\partial). \tag{3.3}$$

The factor 2e of the second term in Eq.(3.1) is due to the fact that $\chi(x)$ is the product of two electron field variables. It is important to note that Eqs.(3.1) and (3.2) follow straightforwardly from the gauge invariance principle without any approximation such as the requirement for $\Xi(x)$ being constant. Taking functional derivatives along the closed time-path $\frac{\delta}{\delta A^{\mu}(y)}$, $\frac{\delta}{\delta \Theta(y)}$ and $\frac{\delta}{\delta B(y)}$ of Eqs.(3.1) and (3.2) successively, and then, returning to the ordinary space-time, we can readily verify that

$$\partial_x^{\mu} \Gamma_{\mu\nu}^{r}(x,y) = \frac{2e}{\hbar c}\Gamma_{\vartheta\nu}^{r}(x,y) , \tag{3.4}$$

$$\partial_x^{\mu} \Gamma_{\mu\vartheta}^{r}(x,y) = \frac{2e}{\hbar c}\Gamma_{\vartheta\vartheta}^{r}(x,y) , \tag{3.5}$$

$$\partial_x^{\mu} \partial_y^{\nu} \Gamma_{\mu\nu}^{r}(x,y) = \left(\frac{2e}{\hbar c}\right)^2 \Gamma_{\vartheta\vartheta}^{r}(x,y) , \tag{3.6}$$

$$\partial_x^{\mu} \Gamma_{\mu b}^{r}(x,y) - \frac{2e}{\hbar c}\Gamma_{\vartheta b}^{r}(x,y) + \mathcal{D}(\partial_x)\delta^4(x-y) = 0 , \tag{3.7}$$

and

$$\Gamma_{b\mu}^{r}(x,y) = \mathcal{D}_{\mu}(\partial_x)\delta^4(x-y) , \tag{3.8}$$

$$\Gamma_{b\vartheta}^{r}(x,y) = 0 , \tag{3.9}$$

$$\Gamma_{bb}^{r}(x,y) = 0 . \tag{3.10}$$

Combining Eq.(2.8) with (3.2), we reobtain the gauge condition

$$\mathcal{D}_{\mu}(\partial) A^{\mu}(x) = 0 \tag{3.11}$$

as required by consistency.

By applying Eqs.(3.4)-(3.11) to expressions (2.20), (2.27) and Eqs.(2.13)-(2.15) and taking account of Eqs.(2.18), (2.19), it is easy to prove the following theorems consecutively by simple algebraic calculations.

Theorem I. If $\Gamma_{\vartheta\vartheta}^{r}(x-y)$ is expressed by Eq.(2.20), then

$$m_{\vartheta}^2 = 0 , \tag{3.12}$$

or equivalently

$$\Gamma_{\vartheta\vartheta}^{r}(x-y) = -\frac{1}{\lambda^2}\left(\frac{\hbar c}{2e}\right)^2 Z_{\vartheta}^{-1}[-\nabla^2]\, \mathcal{D}(\partial_x)\delta^4(x-y) . \tag{3.13}$$

Theorem II. If $\Gamma_{\mu\nu}^{r}(x-y)$ is expressed by Eq.(2.27), then

$$\Gamma_{\mu\nu}^{r}(x-y) = \frac{1}{4\pi}\Big\{ Z_L^{-1}[-\nabla^2]\left(\frac{v^2[-\nabla^2]}{c^2}\right)^{-1}\Big[\Big(\mathcal{D}(\partial) + \frac{M_L^2[-\nabla^2]c^2}{\hbar^2}\Big)\eta_{\mu\rho} - \mathcal{D}_{\mu}(\partial)\mathcal{D}_{\rho}(\partial)\Big] L_{\nu}^{\rho} + Z_T^{-1}[-\nabla^2]\Big(\Box + \frac{M_T^2[-\nabla^2]c^2}{\hbar^2}\Big) L_{\mu\nu}^{(T)} \Big\}\, \delta^4(x-y) , \tag{3.14}$$

$$\Gamma_{\mu\vartheta}^{r}(x-y) = -\left(\frac{\hbar c}{2e}\right)\frac{1}{\lambda^2} Z_{\vartheta}^{-1}[-\nabla^2]\,\mathcal{D}_{\mu}(\partial_x)\delta^4(x-y) , \tag{3.15}$$

where

$$\frac{M_L^2(-\nabla^2)c^2}{\hbar^2} = \frac{4\pi}{\lambda^2}\frac{v^2[-\nabla^2]}{c^2} Z_L[-\nabla^2]\, Z_{\vartheta}^{-1}[-\nabla^2] , \tag{3.16}$$

$$\frac{M_L^2[-\nabla^2]c^2}{\hbar^2} = 4\pi\, Z_T[-\nabla^2]\, \widehat{\pi}_T[-\nabla^2] \sim \textit{undetermined} . \tag{3.17}$$

Theorem III. If the coefficients of Eqs.(2.13)-(2.15) satisfy Eqs.(2.18), (2.19), (2.20) and (2.27), then these equations can be transformed into

$$Z_L^{-1}[-\nabla^2]\left(\mathcal{D}(\partial)+\frac{M_L^2[-\nabla^2]c^2}{\hbar^2}\right)L_{\mu\nu}A^{\nu}(x)+Z_T^{-1}[-\nabla^2]\left(\Box+\frac{M_T^2[-\nabla^2]c^2}{\hbar^2}\right)L^{(T)}_{\mu\nu}A^{\nu}(x)$$
$$=4\pi\left(\frac{\hbar c}{2e}\right)\frac{1}{\lambda^2}\frac{v^2[-\nabla^2]}{c^2}Z_\vartheta^{-1}[-\nabla^2]\,\partial_\mu\left(\textcircled{H}(x)+\frac{2e}{\hbar c}\lambda^2 Z_\vartheta[-\nabla^2]B(x)\right), \tag{3.18}$$

$$\mathcal{D}(\partial)\,\textcircled{H}(x)=0\,, \tag{3.19}$$

$$\mathcal{D}(\partial)\,B(x)=0\,, \tag{3.20}$$

where Eq.(3.18) can also be broken up into two equations:

$$\left(\mathcal{D}(\partial)+\frac{M_L^2[-\nabla^2]c^2}{\hbar^2}\right)L_{\mu\nu}A^{\nu}(x)=$$
$$=\left(\frac{\hbar c}{2e}\right)\frac{4\pi}{\lambda^2}\frac{v^2[-\nabla^2]}{c^2}Z_L[-\nabla^2]Z_\vartheta^{-1}[-\nabla^2]L_\mu^{\ \nu}\,\partial_\nu\left(\textcircled{H}(x)+\frac{2e}{\hbar c}\lambda^2 Z_\vartheta[-\nabla^2]B(x)\right), \tag{3.21}$$

$$\left(\Box+\frac{M_T^2[-\nabla^2]c^2}{\hbar^2}\right)L^{(T)}_{\mu\nu}A^{\nu}(x)$$
$$=\left(\frac{\hbar c}{2e}\right)\frac{4\pi}{\lambda^2}\frac{v^2[-\nabla^2]}{c^2}Z_T[-\nabla^2]Z_\vartheta^{-1}[-\nabla^2]L^{(T)\nu}_{\mu}\,\partial_\nu\left(\textcircled{H}(x)+\frac{2e}{\hbar c}\lambda^2 Z_\vartheta[-\nabla^2]B(x)\right), \tag{3.22}$$

by using the projection operator technique. Obviously, the difference between the transverse and longitudinal waves will vanish when the wave vector tends to zero. Consequently we have

$$Z_L[0]=Z_T[0]\equiv Z[0]\,, \tag{3.23}$$

$$\frac{M_L^2[0]c^2}{\hbar^2}=\frac{M_T^2[0]c^2}{\hbar^2}=\frac{4\pi}{\lambda^2}\frac{v^2[0]}{c^2}Z[0]\,Z_\vartheta^{-1}[0]. \tag{3.24}$$

Up to an unimportant term $B(x)$ on the right-hand side of Eqs.(3.18)(or (3.21) and (3.22)), Eqs.(3.18)-(3.20), or equivalently, Eqs.(3.19)-(3.22), with the gauge condition Eq.(3.11), have just the same form as those given by Umezawa, Mancini et al. for the ground state superconductor[3]. Here, these equations are derived straightforwardly by applying the W-T identities to the linear approximation of the statistical functional equations for the macroscopic variables under certain explicitly stated assumptions. Moreover, the statistical informations are implied in Eq.(2.11) and in the coefficients of Eqs.(3.18)-(3.20) which are defined according to the corresponding 2-point retarded vertex functions as expressed by Eqs.(2.20), (2.27) and (2.13)-(2.15). Consequently, our formalism is naturally suitable not only for the ground state superconductor, but also for the thermal equilibrium cases at finite temperatures, even for certain nonequilibrium stationary states if the approximations stated in Sec. II are satisfied. Furthermore, all parameters corresponding to different statistical situations can be systematically calculated according to Ref.[4] in principle.

Following the physical interpretation of the 2-point retarded vertex functions, the dispersion relation for the phase excitation of the order parameter, i.e., the so-called Goldstone excitation, is prescribed by the coefficient of the $\delta^4(x-y)$ term in Eq.(2.20), with $v[-\nabla^2]$ being the phase velocities of such excitations. Then, theorem I is just the conventional Goldstone theorem: the rest mass of the Goldstone mode is zero. Theorem II with the related Eqs. (3.14), (3.16) and (3.24) manifests transparently how the E.M.F., primarily the longitudinal part, is acquiring a mass from the phase field of the order parameter. As pointed out in Ref.[3], the right-hand side of Eq.(3.18) (or (3.19) and (3.20)) will give trivial contributions unless the phase of the order parameter behaves singularly. In other words, the singular phase field will give rise to observable effects other than the mass of the E.M.F. Therefore, in a sense, theorems I-III may be interpreted as a generalized form of nonrelativistic Higgs mechanism allowing singular Goldstone field.

IV. Calculation of the long wave-length limits of the density of superconducting electron pairs and the phase velocity of the Goldstone excitation for superconductors in thermal equilibrium

The long wave-length limit of the transverse mass of the macroscopic E.M.F. is an important observable related to the static penetrating depth which is often expressed in terms of the density of superconducting electron pairs. For the thermal equilibrium system, one has

$$\frac{M_T^2[0]C^2}{\hbar^2} \equiv \frac{4\pi(2e)^2 n_S[T]}{(2m)\,C^2} = 4\pi\,\frac{2e^2 n_S[T]}{mc^2}\,, \tag{4.1}$$

where m is the electron mass, $T=\beta^{-1}$ the temperature and $n_S[T]$ the density of superconducting electron pairs as a function of temperature. Besides, the phase velocity of the Goldstone mode is also an interesting quantity discussed by many workers [7]. These two quantities are the main parameters of Eqs.(3.19)-(3.22) in the long wave-length limit. In order to compare our formalism with the known results of the other theories or experiments on the one hand, and as a simple illustration for deriving Eq.(2.11) and calculating the parameters in Eqs.(3.19)-(3.22) according to Ref. [4] on the other, in this section we will calculate the above-mentioned two parameters for the thermal equilibrium superconductor-E.M.F. system in weak coupling approximation.

According to Eqs.(2.11), (2.12) and the relevant discussion, the derivation of Eqs.(2.11) is a problem for superconductor itself not related to E.M.F.. This problem has already been solved for the thermal equilibrium superconductor with simplified weak contact interaction. Then, the remaining task is to calculate the two parameters mentioned.

Transforming $\Gamma^r_{\vartheta\vartheta}(x-y)$ into its Fourier representation as

$$\Gamma^r_{\vartheta\vartheta}(x-y) = \int \frac{d^4 q}{(2\pi\hbar)^4}\, \exp\left[-\frac{i}{\hbar}q(x-y)\right]\Gamma^r_{\vartheta\vartheta}[q]\,, \tag{4.2}$$

on account of Eqs.(3.13) and (2.22), we have immediately

$$\Gamma^r_{\vartheta\vartheta}[q] = \frac{1}{4e^2\lambda^2}\, Z_\vartheta^{-1}\left[\frac{\vec{q}^{\,2}}{\hbar^2}\right]\left(q_0^2 - v^2\left[\frac{\vec{q}^{\,2}}{\hbar^2}\right]\vec{q}^{\,2}\right). \tag{4.3}$$

Then, with Eqs.(3.24), (4.1) and (4.3), it is easy to verify that

$$n_S[T] = -\frac{1}{3} m Z[0]\, S_p \left[\frac{\partial^2 \Gamma^r_{\mathscr{A}\mathscr{A}}[\xi]}{\partial \vec{\xi}\, \partial \vec{\xi}}\right]_{\xi=0}, \tag{4.4}$$

$$v^2[0] = -\frac{1}{3} S_p \left[\frac{\partial^2 \Gamma^r_{\mathscr{A}\mathscr{A}}[\xi]}{\partial \vec{\xi}\, \partial \vec{\xi}}\right]_{\xi=0} \Bigg/ \left[\frac{\partial^2 \Gamma^r_{\mathscr{A}\mathscr{A}}[\xi]}{\partial \xi_0\, \partial \xi_0}\right]_{\xi=0}, \tag{4.5}$$

where S_p means tracing over the second rank tensor in the 3-dimensional space. Since we have made the assumption that the intensity of the E.M.F. is not too strong, we can ignore its nonlinearity and feedback to the modulus of the order parameter. Consistently, the renormalization effect of the E.M.F. in the long wave-length limit should not be important, and we may set

$$Z[0] \equiv Z_L[0] \equiv Z_T[0] \cong 1. \tag{4.6}$$

Besides, in accordance with (2.16), $\Gamma^r_{\mathscr{A}\mathscr{A}}(x-y)$ is independent of $A_\mu(x)$, $B(x)$ and $\Theta(x)$. Consequently, the calculations of $n_s[T]$ and $v^2[0]$ are also problems for the superconductor itself. Therefore, by making use of the standard technique of CTPGF[5] and relevant results of literature [4], after some typical calculations for equilibrium contact coupling superconductor, it is easy to derive

$$\left[\frac{\partial^2 \Gamma^r_{\mathscr{A}\mathscr{A}}[\xi]}{\partial \vec{\xi}\, \partial \vec{\xi}}\right]_{\xi=0} = \overleftrightarrow{I}\, \frac{\mu}{3m} |\Delta|^2 N(0) \int d\mathcal{E}(\vec{p})\, \frac{1}{E(\vec{p})} \frac{d}{d\mathcal{E}(\vec{p})} \left(\frac{th \frac{\beta E(\vec{p})}{2}}{E(\vec{p})}\right), \tag{4.7}$$

$$\left[\frac{\partial^2 \Gamma^r_{\mathscr{A}\mathscr{A}}[\xi]}{\partial \xi_0\, \partial \xi_0}\right]_{\xi=0} = \frac{1}{2} |\Delta|^2 N(0) \int d\mathcal{E}(\vec{p})\, \frac{1}{E^3(\vec{p})}\, th \frac{\beta E(\vec{p})}{2}, \tag{4.8}$$

where

$$\mu = \frac{p_F^2}{2m}, \tag{4.9}$$

$$|\Delta| = g\Xi, \tag{4.10}$$

$$\mathcal{E}(\vec{p}) = \frac{\vec{p}^{\,2}}{2m} - \mu, \tag{4.11}$$

$$E(\vec{p}) = \sqrt{\mathcal{E}^2(\vec{p}) + |\Delta|^2}, \tag{4.12}$$

$$N(0) = \frac{m p_F}{2\pi^2 \hbar^3} \tag{4.13}$$

where P_F is the Fermi momentum of electron, $|\Delta|$ the energy gap, $N(0)$ the density of states at the Fermi surface. Substituting Eqs.(4.6), (4.7) and (4.8) into Eqs.(4.4) and (4.5), we have

$$n_S[T] = n_{S0} - \frac{\mu}{3} |\Delta|^2 N(0) \int d\mathcal{E}(\vec{p})\, \frac{1}{E(\vec{p})} \frac{d}{d\mathcal{E}(\vec{p})} \left(\frac{th \frac{\beta E(\vec{p})}{2} - 1}{E(\vec{p})}\right) \tag{4.14}$$

$$= n_{S0} - \frac{\mu}{3} |\Delta|^2 \int \frac{d^3 p}{(2\pi\hbar)^3} \frac{1}{E(\vec{p})} \frac{d}{d\mathcal{E}(\vec{p})} \left(\frac{th \frac{\beta E(\vec{p})}{2} - 1}{E(\vec{p})}\right), \tag{4.15}$$

$$v^2[0] = n_S[T] \Bigg/ \frac{m}{2} |\Delta|^2 N(0) \int d\mathcal{E}(\vec{p})\, \frac{1}{E^3(\vec{p})}\, th \frac{\beta E(\vec{p})}{2}. \tag{4.16}$$

For T=0 as a special case, we have

$$n_s[T=0] = n_{s0} = \frac{p_F^3}{6\pi^2\hbar^3} = \int_0^{p_F} \frac{d^3p}{(2\pi\hbar)^3} \,, \tag{4.17}$$

$$v^2[0]\Big|_{T=0} = \frac{p_F^2}{3m^2} = \frac{1}{3}v_F^2 \,, \tag{4.18}$$

where v_F is the electron velocity at Fermi surface. Eqs.(4.17) and (4.18) are identical with the known results in literature [7,8]. But, we have not found other calculation for the temperature dependence of the phase velocity for the Goldstone mode to compare with. It is interesting to note that Eq. (4.15) coincides with the corresponding result derived from Werthamer's extension [9] of Ginzburg-Landau theory. Making the following approximations

$$p_F^2 \int p^2 dp \cong \int p^4 dp \,, \tag{4.19}$$

$$|\Delta|^2 \frac{1}{E(\vec{p})} \frac{d}{dE(\vec{p})} \left(\frac{th\frac{\beta E(\vec{p})}{2} - 1}{E(\vec{p})} \right) \cong \frac{d}{dE(\vec{p})} \left(th\frac{\beta E(\vec{p})}{2} - 1 \right) \tag{4.20}$$

in the sense of dominant contribution to the integration $\int d\varepsilon(\vec{p})$ we immediately reobtain the corresponding results from BCS theory [8]

$$n_s[T] = n_{s0} - \frac{1}{6\pi^2 m\hbar^3} \int_0^\infty p^4 dp \left(-\frac{df(E(\vec{p}))}{dE(\vec{p})} \right) , \tag{4.21}$$

where

$$f(E(\vec{p})) = \frac{1}{e^{\beta E(\vec{p})} + 1} \,. \tag{4.22}$$

Acknowledgement

The authors wish to thank Prof. CAI Jian-hua and Prof. WU Hang-sheng for calling this problem to their attention and are also grateful to Prof. YU Lu and Prof. CHEN Shi-gang for helpful discussions and a careful reading of the manuscript.

References

1. *V. Ginzburg, L. Landau, Zh. Eksp. Teor. Fiz. 20(1950), 1064.*
2. *P. de Gennes, "Superconductivity of Metals and Alloys", Benjamin, N.Y. 1966.*
3. *See, for example, L. Leplae, F. mancini, H. Umezawa, Phys. Rep. 10C(1974), 151.*
   H. Matsumoto, H. Umezawa, Fortschr. Phys. 24(1976), 357.
   M. Fusco-Girard, F. Mancini, M. Marinaro, Fortschr. Phys. 28(1980), 355.
4. *ZHOU Guang-zhao, SU Zhao-bin,* HAO Bai-lin *and YU Lu, Commun. in Theor. Phys., (Beijing), 1(1982), 295,307,389.*
5. *ZHOU Guang-zhao, SU Zhao-bin, Ch. 5. in "Progr. in Statistical Physics", eds. Hao Bai-lin et al., KEXUE (Science)Press , Beijing, 1981.*
6. *J. Bjorken, S. Drell, "Relativistic Quantum Fields", McGraw-Hill, N.Y. 1981.*
7. *J. Schriefer, "Theory of Superconductivity", Benjamin, Reading, Mass. 1964, and references therein.*
8. *See, for example, A. Fetter, J. Walecka, "Quantum Theory of Many-Particle Systems", Mc Graw-Hill, N.Y. 1971.*
9. ***N. Werthamer, Phys. Rev. 132(1963), 663. N. Werthamer, in "Superconductivity" Vol.1 edi. R. Parks, Marcel Dekker, INC., N.Y. 1969.***

Commun. in Theor. Phys. (Beijing, China) Vol. 2, No. 4 (1983) 1181-1189

ON A DYNAMIC THEORY OF QUENCHED RANDOM SYSTEM

SU Zhao-bin (苏肇冰), YU Lu (于 渌)
and ZHOU Guang-zhao (周光召)
Institute of Theoretical Physics, Academia Sinica, Beijing, China

Received March 10, 1983

Abstract

A dynamical theory for quenched random system is developed in the framework of CTPGF. In steady states the results obtained coincide with those following from the quenched average of the free energy. The order parameter q, a matrix in general, becomes an integral part of the second order connected CTPGF. An equation to determine q is derived from the Dyson-Schwinger equation in this formalism. Some general properties of the CTPGF in a quenched random system are discussed.

I. Introduction

In quenched random systems, part of the degrees of freedom describing impurities are frozen into a nonequilibrium but random configuration. This could be accomplished by sudden cooling of a sample in thermal equilibrium to a state with much lower temperature. The impurities are then frozen into a configuration separated by high potential barriers from an equilibrium one. Diffusion through the potential barriers will cause the nonequilibrium state to vary very slowly in time.

As pointed out by Brout [1], the space average of an observable A in a quenched random system can be replaced by the ensemble average over the impurity degrees of freedom J.

$$\overline{A}=\int A(J)\,P(J)\,dJ \quad , \tag{1}$$

where P(J) is the distribution function. Most of the previous workers [2] considered quenched random systems as if they were static. In this approach one has to evaluate quenched average of the free energy which is proportional to the logarithm of the partition function. It is a formidable task and an enormous machinery of n-replica method is introduced. This method has been applied extensively to systems like spin glass [3-10].

Recently, several authors [11-16] have proposed dynamic theories of quenched random systems in the study of spin glass based on the MSR statistical field theory [17]. The advantage of the dynamic theory is that it provides means for averaging out the quenched randomness without using the unphysical replica trick. The results obtained so far can be reproduced by the replica method with special pattern of replica symmetry breaking, which is itself a static theory [18]. Therefore the full content of the dynamic theory is still waiting to be uncovered.

The aim of the present paper is to establish a dynamic theory for the quenched random system using the closed time path Green's function method (CTPGF)[19]. CTPGF is a very general method especially suited to study slowly varying nonequilibrium processes. In it are incorporated automatically causality and fluctuation dissipation theorem (FDT). The order parameter q introduced by Edwards and Anderson[3] appears naturally in the second order Green's function. The new result obtianed in the present paper is a Dyson-Schwinger equation for the order parameter q. For slowly varying processes it is sufficient to use semiclassical approximation, the one employed in the transport equation. In this way a differential equation that describes the time evolution of the order parameter is obtained. In this paper only the general properties of CTPGF and the Dyson-Schwinger equation are studied. Application to long range quenched Ising model will be presented in a subsequent paper.

The paper is organized as follows: In Sec.II we introduce CTPGF for a quenched random system. It is proved that as the system approaches equilibrium, there exists a free energy which is the quenched average of the free energy with fixed random degrees of freedom. In Sec.III a Dyson-Schwinger equation for the order parameter q is deduced and simplified in the semiclassical approximation, Sec.IV contains concluding remarks.

II. CTPGF for a quenched random system

We shall use in the following those symbols and the language adopted in the theory of CTPGF without further explanation. The unfamiliar readers are referred to Ref.[19]. Suppose the dynamical field variable of our system is $\sigma(x)$. The action on a closed time path P has the form

$$I=\int_p d^dx\, d^dy\, \sigma(x)\, \Gamma_p^{(0,0)}(x,y)\sigma(y)-\int_p d^dx\, V(\sigma(x),J_i)\,,$$
$$+\int_p d^dx(\sigma(x)h(x)+\sigma(x)j(x))+I_{heat\ reservoir}\,, \tag{2.1}$$

where $h(x)$ is the external field; J_i are random variables with given distribution functions. The $\sigma(x)j(x)$ term represents the interaction of the dynamical field with the reservoir consisting of a set of harmonic oscillators for instance. If there are more than one dynamical fields, $\sigma(x)$ should be considered as a vector with many components.

We shall use path integral to evaluate the generating functional of CTPGF. After integrating over the field variables describing heat reservoir, we get the averaged generating functional

$$\Xi[h(x)]=\int P(J)\,\Xi[h(x),J]\,dJ\equiv\langle\,\Xi[h,J]\,\rangle_J \tag{2.2}$$

with

$$\Xi[h(x),J]=\int[d\sigma]\,e^{iI_{eff}}\langle t_+=t_0|\hat{\rho}|\,t_-=t_0\rangle\,, \tag{2.3}$$

where

$$I_{eff}=\int_p \sigma(x)\Gamma_p^{(0)}(x,y)\sigma(y)d^dx\,d^dy-\int_p V(\sigma(x),J_i)d^dx+\int_p\sigma(x)h(x)d^dx\,. \tag{2.4}$$

The system is supposed to be prepared at time $t=t_0$ by suddenly cooling to the temperature of the heat reservoir. In Eq.(2.4) the closed time path starts from $t=t_0$ to $t=+\infty$ (positive branch) and runs back from $t=+\infty$ to $t=t_0$ (negative branch). $\Gamma_p^{(0)}(x-y)$ is the second order vertex function obtained after integrating over the reservoir degrees of freedom, i.e.,

$$\Gamma_p^{(0)}(x,y)=\Gamma_p^{(0,0)}(x,y)+\Sigma_p^{(0)}(x,y)$$

with the self-energy part $\Sigma_p^{(0)}(x,y)$ determined by the interaction $\sigma(x)j(x)$ with the reservoir. It is easily proved that $\Gamma_p^{(0)}$ satisfies the FDT

$$\Gamma_c^{(0)}(k)=i\coth\frac{\beta k_0}{2}\,\mathrm{Im}\,\Gamma_r^{(0)}(k)\,. \tag{2.5}$$

Here $\Gamma_c^{(0)}$ and $\Gamma_r^{(0)}$ are the correlation and the retarded vertex function, respectively.

Introducing the generating functional for the connected CTPGF

$$\overline{Z}[h(x)]=\exp\{i\overline{W}[h(x)]\} \tag{2.6}$$

and

$$Z[h(x),J]=\exp\{iW[h(x),J]\}\,, \tag{2.7}$$

it is possible to obtain the connected CTPGF by direct differentiation. We have the averaged field

$$\overline{\sigma}(x)=\frac{\delta\overline{W}}{\delta h(x)}\,, \tag{2.8}$$

the connected CTPGF

$$\overline{G}_p(x_1,\ldots\ldots x_n)=\frac{\delta^n\overline{W}}{\delta h(x_1)\cdots\cdots\delta h(x_n)} \tag{2.9}$$

and the corresponding ones from $W[h(x),J]$. Eq.(2.2) implies that

$$\overline{Z}\,\overline{\sigma}(x)=\langle Z\sigma(x,J)\rangle_J\,. \tag{2.10}$$

It is a very important property of $Z[h,J]$ and $\overline{Z}[h(x)]$ that they are equal to unity in the physical limit when the external field $h(x)$ on the positive branch is identified with that on the negative branch. Therefore, the observed field

$$\overline{\sigma}(x)=\langle\sigma(x,J)\rangle_J \tag{2.11}$$

satisfying the requirement for a quenched average.

Differentiating Eq.(2.10) with respect to $h(y)$ and setting $h(x_+)=h(x_-)$ in the physical limit, we obtain

$$\overline{G}_p(x,y)+i\overline{\sigma}(x)\overline{\sigma}(y)=\langle G_p(x,y;J)+i\sigma(x,J)\sigma(y,J)\rangle_J \tag{2.12}$$

or

$$\overline{G}_p(x,y)=\langle G_p(x,y;J)\rangle_J+iq(x,y) \tag{2.13}$$

with the matrix

$$q(x,y)=\langle\sigma(x,J)\sigma(y,J)\rangle_J-\overline{\sigma}(x)\overline{\sigma}(y)\,. \tag{2.14}$$

Edwards and Anderson have defined an order parameter in spin glass

$$q_{EA}=\lim_{t\to\infty}\langle\sigma(0,J)\sigma(t,J)\rangle \tag{2.15}$$

which is closely related to the matrix $q(x,y)$ deduced in Eq.(2.14). For hermitian field operator $\hat{\sigma}(x)$ its average $\sigma(x,J)$ is a real function identical on the two time branches in the physical limit. Hence the matrix $q(x,y)$ is real, symmetri and equal on the two branches

$$q(x,y)=q(y,x)=q^*(x,y), \tag{2.16}$$

$$q(x_+,y_+)=q(x_+,y_-)=q(x_-,y_+)=q(x_-,y_-). \tag{2.17}$$

From Eqs.(2.13) and (2.17) we obtain for the retarded, the advanced and the correlation Green's functions the following relations:

$$\overline{G}_r(x,y)=\langle G_r(x,y;J)\rangle_J\ , \tag{2.18}$$

$$\overline{G}_a(x,y)=\langle G_a(x,y;J)\rangle_J \tag{2.19}$$

and

$$\overline{G}_c(x,y)=\langle G_c(x,y;J)\rangle_J+iq(x,y). \tag{2.20}$$

The appearence of the matrix $q(x,y)$ is a consequence of the quenched average over the random variables J_i. It characterizes the behavior of a quenched random system. By successive differentiation one can easily deduce higher order Green's functions. There one can find new matrices describing quenched random systems. We shall not discuss them in the present paper. However, it is easily proved that for a higher order ratarded (or advanced) Green's function we always have

$$\overline{G}_{r\cdots\cdots r}(x_1,\cdots\cdots x_n)=\langle G_{r\cdots\cdots r}(x_1,\cdots\cdots x_n;J)\rangle_J\ . \tag{2.21}$$

After a sufficient long time the system is expected to reach a steady but not necessarily equilibrium state, where the field $\sigma(x,J)$ is no longer time-dependent. In a previous work[19] we have shown that if

$$\mathrm{Im}\int G_r(\tfrac{t}{2},-\tfrac{t}{2};J)dt=0 \tag{2.22}$$

there exists a free energy $F(h,J)$ such that

$$\sigma(h,J)=-\frac{\delta F}{\delta h}. \tag{2.23}$$

For a system in equilibrium with a heat reservoir condition (2.22) is always satisfied. Hence the free energy $F(h,J)$ exists and is equal to the one derived from the partition function. Eqs.(2.18) and (2.22) imply that

$$\mathrm{Im}\int \overline{G}_r(\tfrac{t}{2},-\tfrac{t}{2})dt=0\ , \tag{2.24}$$

which guarantees the existence of a free energy $\overline{F}(h)$ such that

$$\overline{\sigma}=-\frac{\delta\overline{F}}{\delta h}. \tag{2.25}$$

The equality (2.11) now becomes

$$\frac{\partial\overline{F}}{\partial h}=\langle\frac{\delta F(h,J)}{\delta h}\rangle_J\ . \tag{2.26}$$

For a smooth distribution function $P(J)$ with finite moments the order of differentiation and averaging can be changed. Integrating Eq.(2.26) we get

$$\overline{F}(h) = \langle F(h,J)\rangle_J + \text{terms independent of } h. \tag{2.27}$$

As our formalism is quite general it is possible to introduce compound field variables and their corresponding external fields. The temperature can also be treated as an external field coupled to the energy as a compound field variable. In this way we can exhaust all variables in the free energy and obtain finally

$$\overline{F}(h) = \langle F(h,J)\rangle_J \tag{2.28}$$

except for an unimportant constant. Thus the quenched average of the free energy follows from this formalism. This means that results obtained in the dynamical theory will approach that of the static theory when a steady state is reached.

Before discussing the time evolution equation for the quenched matrix q(x,y) let us review some general properties of the connected CTPGF. When the random variables J_i are fixed, the Green's functions of a system in equilibrium with a heat reservoir satisfy the FDT. Let

$$G_\alpha(X+\tfrac{1}{2}x, X-\tfrac{1}{2}x) = \int \frac{d^dk}{(2\pi)^d} e^{-ik_\mu x^\mu} \tilde{G}_\alpha(k, X), \tag{2.29}$$

where $\alpha=r$, a or c. The metric used is $k_\mu x^\mu = k_0x_0 - k_i x_i$, where x_0 is the time component. X_μ $\mu=0,1,\ldots d-1$ are space time variables of macroscopic scale. Similarly, we define

$$\tilde{q}(k,X) = \int d^dx\, e^{ik\cdot x} q(X+\tfrac{1}{2}x, X-\tfrac{1}{2}x), \tag{2.30}$$

which could be considered as the classical counterpart of the quantum matrix q(x,y). The FDT now reads as follows:

$$\tilde{G}_c(k,X;J) = i\,\mathrm{cth}\frac{\beta k_0}{2}\,\mathrm{Im}\,\tilde{G}_r(k,X;J), \tag{2.31}$$

where β^{-1} is the temperature of the heat reservoir. Substituting Eq.(2.31) into Eq.(2.20), we have

$$\overline{\tilde{G}}_c(k,X) = i\,\mathrm{cth}\frac{\beta k_0}{2}\,\mathrm{Im}\,\overline{\tilde{G}}_r(k,X) + i\tilde{q}(k,X). \tag{2.32}$$

This is the FDT satisfied by the quench averaged Green's function. In high temperature approximation Eq.(2.32) can be rewritten in the form

$$\overline{\tilde{G}}_c(k,X) = \frac{2}{\beta k_0}\, i\,\mathrm{Im}\,\overline{\tilde{G}}_r(k,X) + i\tilde{q}(k,X). \tag{2.33}$$

The retarded Green's function $G_r(k,X)$ is analytic in the upper k_0-plane. Its real and imaginary parts obey the dispersion relations. If $G_r(k,X)$ tends to zero as $k_0\to\infty$ it is possible to use the unsubtracted dispersion relation

$$\mathrm{Re}\,\overline{\tilde{G}}_r(k_0,\vec{k},X) = \frac{1}{\pi}\int \frac{\mathrm{Im}\,\overline{\tilde{G}}_r(k_0',\vec{k},X)}{k_0'-k_0}\,dk_0'. \tag{2.34}$$

Using Eq.(2.33) in the high temperature limit we get

$$\mathrm{Re}\,\overline{\tilde{G}}_r(k_0=0,\vec{k},X) = \overline{\tilde{G}}_r(k_0=0,\vec{k},X).$$

$$=\beta\int\frac{dk_0}{2\pi}\left(-i\bar{G}_c(k_0,\vec{k},X)-q(k_0,\vec{k},X)\right). \tag{2.35}$$

For a long ranged Ising model when the space dependence of the Green's function can be neglected, Eq.(2.33) becomes Fischer's relation [20]

$$\tilde{\bar{G}}_r(k_0=0,t)=\beta(1-q(t,t)-\bar{\sigma}^2(t)) . \tag{2.36}$$

We see that the validity of Fischer's relation depends crucially on the high frequency behavior of the retarded Green's function such that the unsubtracted dispersion relation holds.

III. Dyson-Schwinger equation for the quenched matrix q.

We begin with the generating functional for the vertex CTPGF.

$$\bar{\Gamma}[\bar{\sigma}(x)]=\bar{W}[h(x)]-\int_p h(x)\bar{\sigma}(x)d^dx \tag{3.1}$$

with $\bar{\sigma}(x)$ determined by Eq.(2.8). From Eq.(3.1) it is easily found that

$$\frac{\delta\bar{\Gamma}}{\delta\bar{\sigma}(x)}=-h(x) . \tag{3.2}$$

Higher order vertex function can be defined by successive differentiation

$$\bar{\Gamma}_p(x_1,\cdots\cdots x_n)=\frac{\delta^n\bar{\Gamma}}{\delta\bar{\sigma}(x_1)\cdots\cdots\delta\bar{\sigma}(x_n)} . \tag{3.3}$$

The Dyson-Schwinger equation for the second order connected CTPGF follows from Eq.(3.2) by differentiation

$$\int_p d^dz\,\bar{\Gamma}_p(x,z)\,\bar{G}_p(z,y)=\int_p d^dz\,\bar{G}_p(x,z)\,\bar{\Gamma}_p(z,y)=-\delta_p(x-y), \tag{3.4}$$

where $\delta_p(x-y)$ is a δ-function defined on the closed time path. The equation satisfied by the physical Green's function can be obtained from Eq.(3.4) by setting the external field $h(x)$ to be equal on the two branches of time. Eq.(3.4) then reduces to three equations satisfied by the retarded, the advanced and the correlated Green's functions in the following matrix form

$$\bar{\Gamma}_r\bar{G}_r=\bar{G}_r\bar{\Gamma}_r=-1 , \tag{3.5}$$

$$\bar{\Gamma}_a\bar{G}_a=\bar{G}_a\bar{\Gamma}_a=-1 , \tag{3.6}$$

and

$$\bar{\Gamma}_r\bar{G}_c=-\bar{\Gamma}_c\bar{G}_a . \tag{3.7}$$

In an ordinary system near thermal equilibrium Eqs.(3.5) and (3.6) determine the energy spectrum and the life time (dissipation) of quasi-particle excitation while Eq.(3.7) becomes the transport equation for the quasi-particle distribution in the semiclassical limit. We shall show in a forthcomining paper that in a quenched random system the time evolution of the matrix q follows from Eq.(3.7).

From Eq.(3.5) one finds that

$$i\,\mathrm{Im}\,\overline{G}_r = \frac{1}{2}(\overline{G}_r - \overline{G}_a) = i\,\overline{G}_r\,\mathrm{Im}\,\overline{\Gamma}_r\,\overline{G}_a \quad . \tag{3.8}$$

Now define a new matrix

$$Q \equiv -i\left[\overline{\Gamma}_c - i\,\mathrm{cth}\frac{\beta k_0}{2}\,\mathrm{Im}\,\overline{\Gamma}_r\right] , \tag{3.9}$$

which in general is a functional of the matrix q. Then it is easily found from Eqs.(3.7), (3.8), (3.9) and (2.33) the Dyson-Schwinger equation for q

$$\overline{\Gamma}_r q = -Q\overline{G}_a \quad . \tag{3.10}$$

This is a matrix equation which can be simplified in the semiclassical approximation. The Hermitian conjugate of Eq.(3.10) reads

$$q\overline{\Gamma}_a = -\overline{G}_r Q \quad . \tag{3.11}$$

Separating the Hermitian and anti-Hermitian parts of Eq.(3.10) we get two equations

$$\overline{\Gamma}_r q + q\overline{\Gamma}_a = -Q\overline{G}_a - \overline{G}_r Q \quad , \tag{3.12}$$

$$\overline{\Gamma}_r q - q\overline{\Gamma}_a = -Q\overline{G}_a + \overline{G}_r Q \quad . \tag{3.13}$$

To proceed further let us write

$$\widetilde{\overline{G}}_r = \mathrm{Re}\,\widetilde{\overline{G}}_r + i\,\mathrm{Im}\,\widetilde{\overline{G}}_r \quad , \tag{3.14}$$

then from Eqs.(3.5) and (3.6)

$$\overline{\Gamma}_r = \frac{-1}{|\widetilde{\overline{G}}_r|^2}(\mathrm{Re}\,\widetilde{\overline{G}}_r - i\,\mathrm{Im}\,\widetilde{\overline{G}}_r) \quad , \tag{3.15}$$

$$\overline{\Gamma}_a = \frac{-1}{|\widetilde{\overline{G}}_r|^2}(\mathrm{Re}\,\widetilde{\overline{G}}_r + i\,\mathrm{Im}\,\widetilde{\overline{G}}_r) \quad . \tag{3.16}$$

In the semiclassical approximation we shall replace the product of two matrices A and B

$$AB = 1/2(AB+BA)+1/2(AB-BA)$$

by the classical expression

$$\widetilde{A}\widetilde{B} - \frac{i}{2}\left(\frac{\partial\widetilde{A}}{\partial k_\mu}\frac{\partial\widetilde{B}}{\partial X^\mu} - \frac{\partial\widetilde{A}}{\partial X^\mu}\frac{\partial\widetilde{B}}{\partial k_\mu}\right) = \widetilde{A}\widetilde{B} - \frac{i}{2}\left\{\widetilde{A},\widetilde{B}\right\}_{P.B.} \quad . \tag{3.17}$$

The validity of the semiclassical approximation is controlled by the condition

$$\left|\frac{1}{\widetilde{O}}\frac{\partial^2\widetilde{O}}{\partial k_\mu \partial X^\mu}\right| << 1 \quad , \tag{3.18}$$

where $\widetilde{O}$ may be either of the functions $\widetilde{A}$ and $\widetilde{B}$.

In this approximation Eqs.(3.12) and (3.13) become

$$\begin{aligned}\widetilde{q} - \widetilde{Q}|\widetilde{\overline{G}}_r|^2 = \frac{1}{2|\widetilde{\overline{G}}_r|^2}\Big\{ &\left(\frac{\partial\widetilde{q}}{\partial X_\mu} - |\widetilde{\overline{G}}_r|^2\frac{\partial\widetilde{Q}}{\partial X_\mu}\right)\left(\mathrm{Im}\,\widetilde{\overline{G}}_r\frac{\partial\,\mathrm{Re}\,\widetilde{\overline{G}}_r}{\partial k^\mu} - \mathrm{Re}\,\widetilde{\overline{G}}_r\frac{\partial\,\mathrm{Im}\,\widetilde{\overline{G}}_r}{\partial k^\mu}\right) \\ &- \left(\frac{\partial\widetilde{q}}{\partial k_\mu} - |\widetilde{\overline{G}}_r|^2\frac{\partial\widetilde{Q}}{\partial k_\mu}\right)\left(\mathrm{Im}\,\widetilde{\overline{G}}_r\frac{\partial\,\mathrm{Re}\,\widetilde{\overline{G}}_r}{\partial X^\mu} - \mathrm{Re}\,\widetilde{\overline{G}}_r\frac{\partial\,\mathrm{Im}\,\widetilde{\overline{G}}_r}{\partial X^\mu}\right)\Big\}\end{aligned} \tag{3.19}$$

and

$$\frac{\partial|\widetilde{\overline{G}}_r|^2}{\partial k_\mu}\left(\frac{\partial\widetilde{q}}{\partial X^\mu} - |\widetilde{\overline{G}}_r|^2\frac{\partial\widetilde{Q}}{\partial X^\mu}\right) = \frac{\partial|\widetilde{\overline{G}}_r|^2}{\partial X^\mu}\left(\frac{\partial\widetilde{q}}{\partial k_\mu} - |\widetilde{\overline{G}}_r|^2\frac{\partial\widetilde{Q}}{\partial k_\mu}\right) \quad . \tag{3.20}$$

Eqs.(3.19) and (3.20) are written for systems with only one dynamical field. The generalization to multicomponent system is obvious though tedious.

For a homogeneous system in steady state all the functions are independent of the macro space-time variables X_μ. In this case Eqs.(3.19) and (3.20) reduce to a single equation

$$\tilde{q} = \tilde{Q}\,|\tilde{G}_r|^2 \qquad (3.21)$$

The functions $\tilde{Q}$ and $\tilde{\tilde{G}}_r$ can be calculated by the field theoretical method when the model Lagrangian is given. They are functionals of q. Hence Eq. (3.21) could be used to determine the equilibrium value of the matrix $\tilde{q}(k)$.

In perturbation theory to first order Q is proportional to q in some cases, say spin glass without external magnetic field

$$\tilde{Q} = \lambda \tilde{q} \qquad (3.22)$$

$q=0$ is now a solution of Eq.(3.21). A nontrivial solution with $q \neq 0$ exhibiting a new quenched phase might exist if the condition

$$\lambda\,|\tilde{G}_r|^2 = 1 \qquad (3.23)$$

could be satisfied.

In a subsequent paper we shall apply the general results obtained to the long range quenched Ising model. We shall show there the condition (3.23) can never be satisfied owing to the stability condition that must be obeyed by the retarded Green's function G_r. Therefore the spin glass is either in a nonsteady state or not characterized by the parameter q.

IV. Conclusions

In the present paper a dynamic theory of quenched random system using the method of CTPGF has been developed. In the steady limit the results predicted by the dynamical theory approach those by the quenched average of the free energy. Edwards-Anderson order parameter which is in general a matrix, appears naturally in the second order connected CTPGF. An equation that determines the time evolution of the order parameter q has been derived from the Dyson-Schwinger equation for the second order connected CTPGF. This equation is in general a matrix equation and is simplified in the semiclassical approximation to be two partial differential equations of the first order.

It is certainly true that $\tilde{q}(k;X)$ has a sharp peak at $k=0$ and is a very slowly varying function of X. In a homogeneous system with no space dependence q will be only a function of the frequency ω and the macroscopic time t. In the literature $\tilde{q}(\omega,t)$ is often approximated by

$$\tilde{q}(\omega, t) = q_0(t)\,\delta(\omega)\ .$$

In the dynamic theories published so far we have not found any work on how $q(t)$ evolves with time t. We shall try to fill this gap by using the equations obtained in this paper. In a subsequent paper we shall try to solve these equations in the long range Ising model and obtain an explicit dependence of q on ω and t.

References

1. R. Brout, Phys. Rev., 115 (1959) 824.
2. See e.g. P.W. Anderson and T.C. Lubensky, in Lectures at Ecole de Physics on "Ill Condensed Matter", Les Houches 1978, edited by R. Balian, R. Maynard and G. Toulouse, (North Holland, (1979)).
3. S.F. Edwards and P.W. Anderson, J. Phys., F5 (1975) 965.
4. B. Sherrington and S. Kirkpatrick, Phys. Rev. Lett., 35 (1975) 1792; Phys. Rev., B17 (1978) 4384.
5. J.R.L. de Almeida and D.J. Thouless, J. Phys., All (1978) 983.
6. A. Blandin, M. Gabay and T. Garel, J. Phys., A13 (1980) 403.
7. E. Pytte and J.' Rudnick., Phys. Rev., B19 (1979) 3603.
8. A.J. Bray and M.A. Moore., Phys. Rev. Lett., 41 (1978) 1068.
9. D.J. Thouless, P.W. Anderson and R.G. Palmer., Phil. Mag., 35 (1977) 593.
10. G. Parisi, Phys. Rev. Lett., 43 (1979) 1754; J. Phys., A13 (1980) 1101, L1887, L115.
11. S.K. Ma and J. Rudnick, Phys. Rev. Lett., 40 (1978) 589.
12. C. De Dominicis, Phys. Rev., B18 (1978) 4913; Lecture notes in Physics, Vo.104, P.253, edited by C.P. Enz (Springer Verlag, Berlin, I979).
13. J.A. Hertz and R.A. Klemm, Phys. Rev. Lett., 41 (1978) 1397; 46 (1981) 496.
14. H. Sompolinsky and A. Zippelius, Phys. Rev. Lett., 47 (1981) 359.
15. H. Sompolinsky, Phys. Rev. Lett., 47 (1981) 935.
16. W. Kinzel and K.H.Fischer, Solid State Comm., 23 (1977) 687.
17. P.C. Martin, E.D. Siggia and H.A. Rose, Phys. Rev., A8 (1973) 423.
18. C. De Dominicis, M. Gabay and H. Orland, J. Phys. Lett., 42 (1981) L523.
19. ZHOU Guang-zhao, SU Zhao-bin, HAO Bai-lin and YU Lu, Phys. Rev., B22 (1980) 3385, ZHOU Guang-zhao and SU Zhao-bin, Lecture notes on "Progress in Statistical Physics", Ch. V. (1981, in Chinese).
20. K.H.Fischer, Phys. Rev. Lett., 34 (1975) 1438.

Commun. in Theor. Phys. (Beijing, China) *Vol. 2, No. 4 (1983) 1191-1201*

A DYNAMICAL THEORY OF THE INFINITE RANGE RANDOM ISING MODEL

SU Zhao-bin (苏肇冰), YU Lu (于 渌)
and ZHOU Guang-zhao (周光召)
Institute of Theoretical Physics, Academia Sinica, Beijing, China

Received March 10, 1983

Abstract

The dynamics of spin glass is studied in the framework of CTPGF. A marginal stability line is found on the q-χ plane. Below T_c with $h<h_c$, the time evolution of the order parameter follows Fischer's line exponentially to the stability boundary and then decreases in power law along the boundary to its fixed point. The Langevin equation for the spin $\sigma(t)$ is no longer valid along the stability boundary. The susceptibility is calculated in perturbation theory and found to be in good agreement with those predicted by the projection hypothesis. The general validity of the projection hypothesis is justified in the present formalism.

I. Introduction

The simplest model describing spin glass is the long range random Ising model with Hamiltonian [1,2]

$$H=-\frac{1}{2}\sum_{i\neq j} J_{ij}\sigma_i\sigma_j-\sum_i h_i\sigma_i \quad , \tag{1.1}$$

where the exchange J_{ij} are random variables with a Gaussian distribution

$$P(J_{ij})=(2\pi N/J^2)^{-\frac{1}{2}}\exp\{-N(J_{ij}-J_0/N)^2/2J^2\} \quad . \tag{1.2}$$

The spin variables σ_i take the values ±1, and N is the number of nearest neighbors. The quenched average over J_{ij} is carried out on physical variables like free energy, which is the logarithm of the partition function. The method extensively used consists of calculating the average partition function of n replicated systems and taking the limit $n\to0$

$$\langle \ln Z\rangle_J=\lim_{n\to0}\frac{1}{n}(\langle Z^n\rangle_J-1) \quad . \tag{1.3}$$

The model was studied in the mean field limit by Sherrington and Kirkpartick[2]. They found a nontrivial solution for the Edwards-Anderson order parameter q [1]. This solution is however unstable below T_c [3,4] and also yields unphysical negative entropy near $T=0$ [2]. Later using a particular scheme of replica symmetry breaking Parasi[5] found a more satisfactory solution, in which the $n\times n$ matrix of the order parameter was represented by a continuous function $q(x)$, $0\le x\le1$, in the $n\to0$ limit. However, in replica theory the physical meaning of the various order parameters remains unclear.

An alternative approach based on dynamics has been proposed by Ma and Rudnick [6] and developed subsequently by others. The dynamical theory provides not only a means for calculating the quenched average without the unphysical replica trick but also better understanding of the spin glass state, especially so since many of the low temperature properties of real spin glass are dynamic in nature.

Earlier dynamic theories [6-8] still failed in describing the spin glass below T_c. The solution exhibits the same instability encountered in the static replica theory. Recently Sompolinsky and Zippelius [9] have used a soft spin version of the random Ising model and defined the dynamics of the random system by a Langevin equation. The formalism adopted was developed by De Dominicis [7] using the functional integral method of Martin, Siggia and Rose [11]. In Ref. [9] they introduced a la Sommers [12] two order parameters q and Δ. Using this approach together with some physically plausible assumptions about the time dependence of q and Δ, a static solution has been constructed [10] which agrees in general features with Parisi's replica results [5]. This solution has subsequently been derived from the replica theory by a special scheme of replica symmetry breaking [13]. In a certain sense Sompolinsky's solution is an ingenious guess rather than a derivation from a dynamic theory. The physical assumptions involved are very difficult to justify.

In a series of papers Parisi and Toulouse [14] proposed a simple projection hypothesis for the mean field theory of the spin glass phase. They showed that a drastically simple extrapolation procedure, projecting physical properties from the instability line onto the spin glass phase, reproduces many of the expected features coming from the Monte-Carlo simulation and replica symmetry breaking scheme of Parisi. Though the projection hypothesis is very simple and elegant, its theoretical explanation is still lacking.

In a previous paper [15] we have studied the general properties of the Green's function for a random quenched system in the framework of the closed time path Green's function (CTPGF). An equation for the order matrix $q(x,y)$ has been obtained and simplified in the semiclassical approximation. The aim of the present paper is to apply the general results obtained therein to the infinite range random Ising model. The main results obtained are:

1. A marginal stability line in the q-χ plane is obtained where χ is the magnetic susceptibility of the system.

2. Above T_c Fischer's relation, a line in the q-χ plane, is inside the stable region. The order parameter q tends exponentially in time to the fixed point $q=q_0$.

3. Below T_c the Fischer's line intercepts the stability boundary at a point q_1. The order parameter will decay along Fischer's line quickly (exponentially in time) to q_1 and then decay further along the boundary slowly (with power law) to q_0.

4. The static fixed point corresponding to the Sherrington and Kirkpatrick solution lies on the Fischer's line in the unstable region. The only stable fixed point is $q=0$.

5. The projection hypothesis can be justified in the present formalism.

The paper is organized as follows: In Sec.II the second order connected Green's function and vertex function are studied in a soft spin version of the infinite range random Ising model. The stability condition and the low frequency behavior of the Green's function are discussed. The time evolution of the order parameter $\tilde{q}(\omega,t)$ is studied in Sec.III.. Sec.IV gives the susceptibility in the external magnetic field. Sec.V contains concluding remarks and a brief discussion of the projection hypothesis.

II. The model and Green's functions

For the sake of simplicity we consider here a soft spin version of the random Ising model defined by

$$H=-\frac{1}{2}\sum_{i\neq j} J_{ij}\sigma_i\sigma_j+\sum_i\left(\frac{1}{2}\gamma_0\sigma_i^2+u\sigma_i^4-h_i\sigma_i\right). \tag{2.1}$$

The length of the soft spin σ_i is allowed to vary continuously from $-\infty$ to ∞. The exchange J_{ij} are random variables with a Gaussian distribution

$$P(J_{ij})=(2\pi N/J^2)^{-\frac{1}{2}}\exp\{-NJ_{ij}^2/2J^2\}, \tag{2.2}$$

where N is the number of the nearest neighbors.

Taking into account the interaction with heat reservoir and averaging over the random variables J_{ij} we can write the generating functional of the CTPGF in the following form

$$Z[h_i(t)]=\int[d\sigma_i]\,e^{iS_{eff}}\,\langle t_+=t_0|\hat{\rho}|t_-=t_0\rangle, \tag{2.3}$$

where

$$S_{eff}=\sum_j\left\{\int_p\frac{1}{2}\sigma_j(t)\Gamma_{p0}(t,t')\sigma_j(t')dt\,dt'-\int_p(u\sigma_i^4(t)+h_i(t)\sigma_i(t))dt\right\}+i\frac{J^2}{4N}\sum_{i\neq j}\int_p\sigma_i(t)\sigma_j(t)dt\int_p\sigma_i(t')\sigma_j(t')dt'. \tag{2.4}$$

The notations used here are essentially those in Ref.[15]. However, the bar over the quenched average physical variables has been dropped for simplicity. Any matrix $A(t,t')$ can be represented by its Fourier transform $\tilde{A}(\omega,\frac{t+t'}{2})$ in the relative time $t-t'$ where

$$\tilde{A}(\omega,t)=\int d\tau\,e^{i\omega\tau}A\left(t+\frac{\tau}{2},t-\frac{\tau}{2}\right). \tag{2.5}$$

In this notation system the low frequency approximation for Γ_0 has the forms

$$\tilde{\Gamma}_{\gamma_0}(\omega,t)=(-\gamma_0+i\omega/\Gamma_0) \tag{2.6}$$

and

$$\tilde{\Gamma}_{c_0}(\omega,t)=i\frac{2}{\beta\Gamma_0}, \tag{2.7}$$

where β^{-1} is the temperature of the reservoir.

In the infinite range limit with $N\to\infty$ the matrix $G_{ij}(t,t')$ could be approximated by $\delta_{ij}G(t,t')$. In this case the second order vertex function can be calculated with the diagram expansion. It is found that

$$\tilde{\Gamma}_r(\omega, t) = -r + i\omega/\Gamma + J_r^2(q)\tilde{G}_r(\omega, t) - \tilde{\Sigma}_r(\omega, t), \tag{2.8}$$

where r and Γ are renormalized quantities that could be functions of the time t. To lowest order perturbation in u we get

$$J_r^2(q) = J^2 + 288u^2(q^2 + 2q\sigma^2), \tag{2.9}$$

where J is also renormalized and $q(t,t)$ is the order parameter. In obtaining the second term in Eq.(2.9) we make the approximation that $\tilde{q}(\omega,t)$ has a sharp peak at $\omega=0$. In Eq.(2.8) $\tilde{\Sigma}_r(\omega,t)$ is the self-energy part with the first two terms in the expansion of ω and the term proportional to $\tilde{G}_r$ is excluded. They are included in the first three terms in Eq.(2.8). Therefore, we have

$$\tilde{\Sigma}_r(\omega=0, t) = 0 . \tag{2.10}$$

To the same order of approximation we have calculated the correlated vertex function

$$\begin{aligned} \tilde{\Gamma}_c(\omega, t) = {} & i\frac{2}{\beta\omega}\,\mathrm{Im}\,\tilde{\Gamma}_r(\omega,t) + iJ_c^2(q)\tilde{q}(\omega,t) \\ & + iJ^2\tilde{\Delta}(\omega,t) - \tilde{\Sigma}_c(\omega,t) . \end{aligned} \tag{2.11}$$

Here $\tilde{\Delta}(\omega,t)$ is defined to be

$$\tilde{\Delta}(\omega, t) = \int d\tau\, e^{i\omega\tau}\, \sigma(t+\tfrac{1}{2}\tau)\,\sigma(t-\tfrac{1}{2}\tau) , \tag{2.12}$$

where $\sigma(t)$ is the average value of the spin which could be different from zero if an external magnetic field is applied. It is easily proved that $\tilde{\Delta}(\omega,t)$ has a sharp peak at $\omega=0$. $\tilde{\Sigma}_c(\omega,t)$ is the remaining self-energy part that does not have a sharp peak at $\omega=0$. To lowest order of perturbation we find

$$J_c^2(q) = J^2 + 96u^2(q^2 + 2q\sigma^2) . \tag{2.13}$$

It is to be noted here that

$$q(t, t) = \int \frac{d\omega}{2\pi}\,\tilde{q}(\omega, t) . \tag{2.14}$$

In the low frequency limit the matrix Q defined in Ref.[15] can be approximated by

$$\begin{aligned} \tilde{Q}(\omega, t) &= -i\left\{\tilde{\Gamma}_c(\omega,t) - i\frac{2}{\beta\omega}\,\mathrm{Im}\,\tilde{\Gamma}_r(\omega,t)\right\} \\ &= J_c^2(q)\,\tilde{q}(\omega,t) + J^2\tilde{\Delta}(\omega,t) . \end{aligned} \tag{2.15}$$

Only the terms that have a sharp peak at $\omega=0$ are retained.

Next we shall study the stability of this model. In perturbation theory stability should hold order by order. To insure stability in the low frequency limit $\mathrm{Im}\tilde{\Gamma}_r$, $-\mathrm{Im}\tilde{\Sigma}_r$ and $\tilde{\Gamma}_r(\omega=0,t)$ must be positive. From Eq.(2.8) it is easily deduced that

$$\mathrm{Im}\,\tilde{\Gamma}_r = \frac{\frac{\omega}{\Gamma} - \mathrm{Im}\,\tilde{\Sigma}_r}{1 - J_r^2(q)\,|\tilde{G}_r|^2} . \tag{2.16}$$

Hence we get the stability condition

$$1 \geqslant J_r^2(q)\left|\tilde{G}_r(\omega,t)\right|^2 . \tag{2.17}$$

In the zero frequency limit Eq. (2.17) becomes

$$1 \geqslant J_r^2(q)\chi^2 , \tag{2.18}$$

where $\chi=\tilde{G}\ (\omega=0,t)$ is the susceptibility of the system. The marginal stability line

$$1 = J_r^2(q)\chi^2 \tag{2.19}$$

in the q-χ plane constitutes the boundary of the stability region.*

The general analysis given in Ref. [15] tells us that the nontrivial static fixed point in the absence of a magnetic field is determined by the equation

$$q=(J_c^2(q)\chi^2)\tilde{q} . \tag{2.20}$$

From the explicit calculation in perturbation theory Eqs.(2.9) and (2.13) we find that the line representing static fixed point Eq.(2.20) apart from $q=0$ lies outside the stability region in the q-χ plane since $J_r^2(q)>J_c^2(q)$ for all values of q. We conclude therefore that this fixed point cannot be reached and ultimately the order parameter $q(t,t)$ or $\tilde{q}(\omega,t)$ will tend to zero as the time $t\to\infty$.

Before we go to the next section to study the time evolution of the order parameter, let us study the low frequency behavior of the retarded Green's function $\tilde{G}_r(\omega,t)$. For this purpose we write

$$\tilde{G}_r(\omega,t)=\chi-\Delta\tilde{G}_r , \tag{2.21}$$

where $\Delta\tilde{G}_r(\omega,t)\to 0$ as $\omega\to 0$; Eq.(2.8) regarded as an equation for $\Delta\tilde{G}_r$ can be solved in the form

$$\Delta\tilde{G}_r=\frac{1}{2J_r^2(q)\chi}\Big\{1-J_r^2(q)\chi^2-(\tilde{\Sigma}_r+i\frac{\omega}{\Gamma})\chi- \Big[(1-J_r^2(q)\chi^2-(\tilde{\Sigma}_r+i\frac{\omega}{\Gamma})\chi)^2-4J_r^2(q)(\tilde{\Sigma}_r+\frac{i\omega}{\Gamma})\chi^3\Big]^{1/2}\Big\} , \tag{2.22}$$

When ω is small $\Delta\tilde{G}_r$ can be put into the following form

$$\Delta\tilde{G}_r=-\gamma(t)\,|\omega|^\nu\left(\mathrm{ctg}\frac{\pi\nu}{2}-i\,\mathrm{sgn}\,\omega\right) , \tag{2.23}$$

where $\gamma(t)$ and ν are to be determined. Following an analysis given by Sompolinsky and Zippelis[9] it is easily found that the lowest power of ω in the self-energy part $\tilde{\Sigma}_r$ is $\omega^{2\nu}$ for $\nu\leq\frac{1}{2}$. From Eqs.(2.22) and (2.23) we obtain similar results as in Ref. [9] that $\nu=1$ if the system is inside the stability region while $\nu\leq\frac{1}{2}$ if it is on the boundary of the stability region. We shall use these results to derive the time evolution of $\tilde{q}(\omega,t)$ in the next section.

* *In the unstable region* χ *becomes complex, but* $|\chi|^{-2}=J_r^2(q)$.

III. Time evolution of the order parameter $\tilde{q}(\omega,t)$ near critical temperature

In Ref. [15] we derived an equation for $\tilde{q}(\omega,t)$ in the semi-classical approximation. We shall apply it to the model studied in previous sections. Neglecting the external magnetic field the equation for q in the linear approximation reads

$$\frac{\partial\tilde{q}}{\partial t}=\frac{1-J^2|\tilde{G}_r|^2}{1+J^2|\tilde{G}_r|^2}\,\frac{\dfrac{\partial|\tilde{G}_r|^2}{\partial t}}{\dfrac{\partial \mathrm{Re}\tilde{G}_r}{\partial\omega}\dfrac{\partial \mathrm{Im}\tilde{G}_r}{\partial t}-\dfrac{\partial \mathrm{Im}\tilde{G}_r}{\partial\omega}\dfrac{\partial \mathrm{Re}\tilde{G}_r}{\partial t}}\,\tilde{q}\;. \tag{3.1}$$

In the low frequency limit we take the approximation

$$\frac{\partial \mathrm{Re}\tilde{G}_r}{\partial\omega}\frac{\partial \mathrm{Im}\tilde{G}_r}{\partial t}-\frac{\partial \mathrm{Re}\tilde{G}_r}{\partial t}\frac{\partial \mathrm{Im}\tilde{G}_r}{\partial\omega}\simeq-\frac{\partial\chi}{\partial t}\frac{\partial \mathrm{Im}\tilde{G}_r}{\partial\omega} \tag{3.2}$$

and

$$\frac{\partial|\tilde{G}_r|^2}{\partial t}\simeq 2\chi\frac{\partial\chi}{\partial t}\;. \tag{3.3}$$

Substituting Eqs.(3.2) and (3.3) into Eq.(3.1) it follows that

$$\frac{\partial\tilde{q}}{\partial t}=-2\,\frac{1-J^2\chi^2}{1+J^2\chi^2}\,\frac{\chi}{\dfrac{\partial \mathrm{Im}\tilde{G}_r}{\partial\omega}}\,\tilde{q}\;. \tag{3.4}$$

In the stability region we have

$$1-J^2\chi^2>0 \tag{3.5}$$

and

$$\frac{\partial \mathrm{Im}\tilde{G}_r}{\partial\omega}=\frac{1}{\Gamma}\chi^2>0\;. \tag{3.6}$$

For Ising model the susceptibility χ is expressed in terms of $q(t,t)$ in the stability region by the Fischer's relation

$$\chi=\beta(1-q(t,t))\;. \tag{3.7}$$

Integrating Eq.(3.4) over ω we get an equation for $q(t,t)$

$$\frac{\partial q}{\partial t}=-2\,\frac{1-J^2\beta^2(1-q)^2}{1+J^2\beta^2(1-q)^2}\,\frac{\Gamma}{\beta(1-q)}\,q\;. \tag{3.8}$$

Above the critical temperature $\beta_c=1/J$, the only fixed point of Eq.(3.8) is $q=0$. Near $q=0$ the order parameter q will tend exponentially in time to the fixed point

$$q\simeq q_{in}\,\exp\left\{-\frac{t}{\tau_0}\right\}, \tag{3.9}$$

where

$$\tau_0=\frac{1+J^2\beta^2}{2(1-J^2\beta^2)}\,\frac{\beta}{\Gamma}=\frac{\beta_c^2+\beta^2}{2(\beta_c^2-\beta^2)}\,\frac{\beta}{\Gamma}\;. \tag{3.10}$$

Below T_c in the linear approximation Eq.(3.8) has another fixed point on the marginal stability line

$$q_1=1-\frac{\beta_c}{\beta}\;. \tag{3.11}$$

The order parameter tends to this point with the time dependence

$$q(t,t)=1-\frac{\beta_c}{\beta}+a\,\exp\left\{-\frac{t}{\tau_1}\right\}, \tag{3.12}$$

where
$$\tau_1 = \frac{\beta_c^2}{2(\beta-\beta_c)\Gamma} \; . \tag{3.13}$$

It is noted that both τ_0 and τ_1 have a simple pole at the critical temperature.

In the linear approximation below T_c the order parameter q will reach a finite value q_1 indicating the existence of the spin glass phase. However, this is not true when the higher order effect is taken into consideration. The true fixed point lies on the Fischer's line in the unstable region. After hitting the marginal stability line at $q=q_1$ the order parameter $q(t,t)$ will vary slowly further along the marginal stability line down to $q=0$.

To lowest order of nonlinear correction the two equations for $\tilde{q}(\omega,t)$ in the low frequency limit have the following forms
$$\frac{\partial\tilde{q}}{\partial t} = -2\frac{1-J_c^2(q)\chi^2}{1-J_r^2(q)\chi^2}\frac{\chi}{\frac{\partial \mathrm{Im}\tilde{G}_r}{\partial\omega}}\tilde{q} \tag{3.14}$$
and
$$\frac{\partial\tilde{q}}{\partial t}\frac{\partial|\tilde{G}_r|^2}{\partial\omega} = \frac{\partial\tilde{q}}{\partial\omega}\frac{\partial|\tilde{G}_r|^2}{\partial t} \; . \tag{3.15}$$

Along the marginal stability line χ and q are related by
$$\chi^2 = \frac{1}{J_r^2(q)} \; . \tag{3.16}$$

Using the explicit form of $J_r(q)$ and $J_c(q)$ given in Eqs.(2.9) and (2.13) it is easily found from Eqs.(3.14) and (3.16) that $\partial\tilde{q}/\partial t \sim q^2(t,t)\tilde{q}$ which is small for small values of q. In the low frequency limit we have from Eq.(2.23)
$$\tilde{G}_r = \chi - \gamma(t)|\omega|^\nu\left(\mathrm{ctg}\frac{\pi\nu}{2} - i\,\mathrm{sgn}\,\omega\right) \tag{3.17}$$
with $\gamma(t)>0$. Since
$$1-J_c^2(q)\chi = \frac{192u^2q^2}{J^2+288u^2q^2} > 0 \tag{3.18}$$
along the line (3.16), we see from Eq.(3.14) that q will continuously drop to $q=0$, for however small but finite ω.

Neglecting terms like $\partial\chi/\partial t \sim q\frac{\partial q}{\partial t} \sim q^4$ we can integrate Eq.(3.15) to get
$$\tilde{q}(\omega,t) = \tilde{q}(|\omega|\,\gamma^{\frac{1}{\nu}}(t)) \equiv \tilde{q}(x) \tag{3.19}$$

and
$$q(t,t) = \int\frac{d\omega}{2\pi}\tilde{q}(\omega,t) = \gamma^{-\frac{1}{\nu}}(t)\int q(x)\frac{dx}{2\pi} \equiv \alpha\gamma^{-\frac{1}{\nu}}(t) \; . \tag{3.20}$$

Using Eqs.(3.19) and (3.20) we obtain from Eq.(3.14)
$$\frac{d\tilde{q}}{dx}\gamma^{\frac{3}{\nu}-1}\frac{d\gamma}{dt} = -\frac{192u^2}{J^3x^\nu}\alpha^2\tilde{q} \; . \tag{3.21}$$

The right-hand side of Eq.(3.21) is a function of x only. Hence we have
$$\gamma^{\frac{3}{\nu}-1}\frac{d\gamma}{dt} = \frac{\nu}{3}a = \mathrm{const.} \; . \tag{3.22}$$

Integrating the above equation we get

$$\gamma(t)=(at+b)^{\frac{\nu}{3}}, \tag{3.23}$$

where a and b are two constants. Substituting Eq.(3.23) into Eq.(3.20) it follows that

$$q(t,t)=\alpha(at+b)^{-\frac{1}{3}}. \tag{3.24}$$

Therefore, q(t,t) decreases according to a power law in time and a>0.

Substituting Eq.(3.22) into Eq.(3.21) and integrating it, we obtain

$$\tilde{q}(\omega,t)=\tilde{q}_0\exp\left\{-\frac{3}{\nu(1-\nu)a}\cdot\frac{192u^2}{J^3}\alpha^2\left[|\omega|(at+b)^{\frac{1}{3}}\right]^{1-\nu}\right\}. \tag{3.25}$$

We note that $\tilde{q}(\omega,t)$ has a peak at $\omega=0$ but it is not of the form $\tilde{q}_0\delta(\omega)$.

IV. Susceptibility and the order parameter with small external magnetic field

In the following we shall use the unit $J=T_c=1$ and take the value of u to be 1/12 below T_c. The stability condition Eq.(2.19) now takes the form

$$1\geqslant(1+2q^2+4q\sigma^2)\chi^2. \tag{4.1}$$

Inside the stability region the Fischer's relation holds

$$\chi=\beta(1-q(t,t)-\sigma^2(t)), \tag{4.2}$$

where $\sigma(t)$ is the average value of the spin depending on the external magnetic field.

Above T_c the whole line (4.2) lies inside the stability region. The static order parameter q_0 and the average spin σ are related by the equation

$$q_0=\left(q_0+\sigma^2+\frac{2}{3}q_0^3+2q_0^2\sigma^2\right)\chi^2 \tag{4.3}$$

with χ given in Eq.(4.2). For small σ we can solve Eq.(4.3) for q_0

$$q_0=\frac{\beta^2}{1-\beta^2}\sigma^2+\cdots\cdots. \tag{4.4}$$

The susceptibility is therefore

$$\begin{aligned}\chi=\frac{d\sigma}{dh}&=\beta-\frac{\beta}{1-\beta^2}\sigma^2+\cdots\cdots\\&=\beta-\frac{\beta^3}{1-\beta^2}h^2+\cdots\cdots.\end{aligned} \tag{4.5}$$

At the critical point $\beta=1$ Eq.(4.3) is still valid. In this case we have

$$q_0=\frac{\sigma}{\sqrt{2}}-\frac{19}{24}\sigma^2+\cdots\cdots \tag{4.6}$$

and

$$\begin{aligned}\chi&=1-\frac{\sigma}{\sqrt{2}}-\frac{5}{24}\sigma^2+\cdots\cdots\\&=1-\frac{h}{\sqrt{2}}+\frac{1}{24}h^2+\cdots\cdots.\end{aligned} \tag{4.7}$$

Below T_c there exists a critical σ_c corresponding to a critical external field h_c. Above h_c the static fixed point is still lying in the stable region. The critical σ_c can be calculated by Eq.(4.3) and the equation

$$\chi = \beta(1-q_0-\sigma_c^2) = \frac{1}{(1+2q_0^2+4q_0\sigma_c^2)^{1/2}} \quad . \tag{4.8}$$

Near critical temperature it is found that

$$\sigma_c^2 = \frac{4}{3}\tau^3[1+3\tau] \tag{4.9}$$

or

$$h_c^2 = \frac{4}{3}\tau^3[1+3\tau] \quad , \tag{4.10}$$

where $\tau = 1 - \frac{1}{\beta}$.

For $h<h_c$, the static fixed point is no longer on the Fischer's line. It is on the boundary of the stability region. We find therefore

$$q_0 = (\frac{3}{4}\sigma^2)^{1/3} + \cdots\cdots \tag{4.11}$$

and

$$\begin{aligned} \chi &= 1-q_0^2 - \cdots\cdots \\ &= 1-(\frac{3}{4}\sigma^2)^{2/3} - \cdots\cdots \\ &= 1-(\frac{3}{4}h^2)^{2/3} - \cdots\cdots \end{aligned} \tag{4.12}$$

Formulas (4.5), (4.7), (4.10) and (4.12) are in qualitative agreement with the predictions from the projection hypothesis.

We have shown in Sec.III that below T_c the order parameter q will fast decay to the intersection point of the Fischer's line with the boundary of the stability region and then decrease slowly along the boundary to $q=q_0$. This point of interesection depends on the external magnetic field and forms a line on the q-h plane, which is just the stability line found by Almeida and Thouless in the static approach[3]. The corresponding order parameter q_1 is determined by the equation

$$\beta(1-q_1-\sigma^2) = \frac{1}{(1+2q_1^2+4q_1\sigma^2)^{1/2}} \quad . \tag{4.13}$$

Near critical temperature Eq.(4.13) can be solved for q to get

$$q_1 = \tau + \tau^2 + \tau^3 - \sigma^2 + \cdots\cdots \quad , \tag{4.14}$$

which agrees with Parisi's $q(x)$ at $x=1$.

The results obtained so far, though valid only for small q because of the perturbation calculation, indicate strongly that the projection hypothesis is very close to truth. We shall justify this hypothesis theoretically in the following section.

V. Discussions and conclusions

Starting from the general formalism of CTPGF we have studied the infinite range random Ising model in some detail. We found in the q-χ plane a marginal stability line. Above T_c the Fischer's line lies entirely inside the stable region. The order parameter $q(t,t)$ tends exponentially in time to its stationary value along the Fischer's line. The low frequency behavior of the vertex function is normal with $\mathrm{Im}\tilde{\Gamma}_r=\omega/\Gamma$. In this case a Langevin equation for the spin $\sigma(t)$ could be written down from the general formalism.

Below T_c the whole Fischer's line is inside the stable region only when the external magnetic field $h>h_c$. In this case the physical picture remains the same as in the case $T>T_c$. In case $h<h_c$, the Fischer's line will intersect the stability boundary at a point q_1. The fixed points on the Fischer's line are all situated in the unstable region. Hence the time evolution of q is divided into two stages. Firstly q decays exponentially along Fischer's line to point q_1. Then it decreases further with a $-1/3$ power law to its stationary point q_0 along the stability boundary. Since the low frequency behavior of the vertex function is abnormal on the boundary with $\mathrm{Im}\tilde{\Gamma}_r \sim \chi^{2}\gamma|\omega|^{\nu}\,\mathrm{sgn}\,\omega$, $\nu\le 1/2$ it is not possible to write a Langevin equation for the spin $\sigma(t)$ at the second stage.

In our general formalism the stability boundary in the q-χ plane is temperature independent. Therefore, the stationary point (q_0,χ_0) is temperature independent. As a consequence, the magnetization is also temperature independent and the entropy does not vary with the magnetic field. This is just the content of the projection hypothesis stated in Ref. [14]. It follows naturally from our theory.

Parisi and Toulouse have also identified the Edwards-Anderson order parameter with the intersection point q_1 which is determined in the whole termperature range by the equation

$$1=\langle\chi^2\rangle_z=\frac{\beta^2}{\sqrt{2\pi}}\int_{-\infty}^{\infty}dz\,e^{-\frac{z^2}{2}}\,\mathrm{sech}^4\beta(z\sqrt{q_1}+h)\,. \tag{5.1}$$

This equation can also be derived from our formalism. Therefore, the thermodynamic properties predicted by our theory will agree with those by projection hypothesis.

Comparing our results with those of Sompolinsky [10] it is natural to conjecture that the function $q(x)$, $0\le x\le 1$ corresponds to our $q(t,t)$ varying along the stability boundary from q_1 to q_0. Not only does the present theory give a clear physical meaning to the function $q(t,t)$ but also an equation that can be solved explicitly for the time evolution of the order parameter $q(t,t)$.

The present formalism can be applied to other quenched random systems, which we hope to present in subsequent publications.

References

1. S.F. Edwards and P.W. Anderson, J. Phys., F5 (1975) 965.
2. D. Sherrington and S. Kirkpatrick, Phys. Rev. Lett., 35 (1975) 1972.
3. T.R.L De Almeida and D.J. Thouless, J. Phys., A11 (1978) 983.
4. E. Pytte and J. Rudnick, Phys. Rev., B19 (1979) 3603.
5. G. Parisi, Phys. Lett., A73 (1979) 203; Phys. Rev. Lett., 43 (1979) 1574; J. Phys., A13 (1980) 1887.
6. S.K. Ma and J. Rudnick, Phys. Rev. Lett., 40 (1978) 589.
7. C. De Dominicis, Phys. Rev., B18 (1978) 4913; Lecture notes in Physics, V.104 p.253, edited by C.P. Enz (Springer Verlag, Berlin, 1979).
8. J.A. Hertz and R.A. Klemm, Phys. Rev. Lett., 21 (1978) 1397; 46 (1981) 496.
9. H. Sompolinsky and A. Zippelius, Phys. Rev. Lett., 47 (1981) 359.
10. H. Sompolinsky, Phys. Rev. Lett., 47 (1981) 935.
11. P.C. Martin, E.D. Siggia and H.A. Rose, Phys. Rev., A8 (1973) 423.
12. H.J. Sommers, Z. Physik, B31 (1978) 301; ibid B33 (1979) 173.
13. C. de Dominicis, M. Gabay and H. Orland, J. Phys. Lett., 42 (1981) L-523.
14. G. Parisi and G. Toulouse, J. Phys. Lett., 41 (1980) 361; J. Vannimenns, G. Toulouse and G. Parisi, J. Physique, 42 (1981) 565.
15. SU Zhao-bin, YU Lu and ZHOU Guang-zhao, "On a dynamic theory of quenched random system", Commun, in Theor. Phys., this issue.

474

A Dynamical Theory of Random Quenched System and Its Application to Infinite-Ranged Ising Model*

Su Zhao-bin   Yu Lu   Zhou Guang-zhao

(Institute of Theoretical Physics, Beijing, China)

Abstract

A dynamical thoery for quenched random systems is developed in the framework of the closed time-path Green's functions (CTPGF). The order parameter q, a matrix in general, appears naturally as an integral part of the second order connected CTPGF. An equation to determine q is derived from the Dyson-Schwinger equation. The formalism developed is applied to the study of the long-ranged random Ising model. A boundary line is found on the q-$|\chi|$ plane. It is argued that the spin-glass phase is characterized by the fixed point lying on the stability boundary. The magnetization is calculated in perturbation and is found to be in good agreement with those predicted by the projection hypothesis. The general validity of the projection hypothesis is discussed.

I. Introduction

Much progress has been made in recent years on the understanding of the spin-glass (SG) phase in magnetic systems with infinite-ranged random exchange. A mean field theory (MFT) with order parameter being a continuous function $q(x)$, $0 \leq x \leq 1$, has been derived by use of a particular scheme of replica symmetry breaking or a dynamical approach[1-10]. The MFT is free of instabilities although the physical meaning of the order parameter $q(x)$ and the origin of the apparent violation of the fluctuation-dissipation theorem (FDT) are still under intensive investigation[11-13]. On the other hand, a drastically simple extrapolation procedure projecting physical properties from the marginal stability line onto the SG phase has been shown by Parisi and Toulouse[14] to reproduce all the nice features coming from the Monte-Carlo simulation and the MFT. However, a theoretical explanation of this very simple and elegant projection hypothesis is still lacking.

In this note we investigate properties of the infinite-ranged random Ising model in the framework of the closed time path Green's functions (CTPGF)[15]. The order parameter, a matrix $q(t,t')$ in general, appears naturally as a part of the second order connected CTPGF which satisfies the Dyson-Schwinger equation. The FDT is assumed to be satisfied by the CTPGF before averaging over the random exchange. The Fischer's relation is found to be violated in the SG phase[16]. The validity of Fischer's relation depends on the existence of the FDT and the use of a unsubtracted dispersion relation for the retarded CTPGF. It is more likely that a subtracted dispersion relation is necessary on the stability boundary and in the unstable region. We believe, this substraction rather than the violation of the FDT is the real cause for the breakdown of the Fischer's relation in the SG phase.

In the present formalism the dynamical behavior and the static properties of the order parameter $q(t,t')$ are determined completely by the Dyson-Schwinger equation which can be solved approximately for small q in the low frequency limit. The physical picture obtained can be displayed more clearly on the q-$|\chi|$ plane where χ is the magnetic susceptibility of the system. A physical boundary is found on this plane.

Above T_c, the Fischer relation, a line on the q- $|\chi|$ plane, lies completely inside the stable region and the order parameter q tends exponentially in time to its fixed point $q=q_o$. Below T_c, the Fischer line intersects the boundary at the point $q=q_1$, which is a fixed point only when the external magnetic field $h=h_c$. In case $h<h_c$, after reaching $q=q_1$ along the Fischer line, the order parameter q will decay further along the boundary down to its fixed point $q=q_o$. In a rough apprriximation the decay along the boundary is found to obey a power law. In this formalism the physical boundary on the q- $|\chi|$ plane is shown to be temperture independent. Therefore, the fixed point on the boundary characterizing the SG phase is temperature independent. As a consequence, the magnetization M(h) is also temperature independent and the entropy is independent of the external magnetic field. This is just the assumption of the projection hypothesis which follows naturally from our theory.

The rest of the paper is organized as follows: In Sec. II the general properties of CTPGF for a random quenched system are presented. The stability condition and the physical boundary are discussed in Sec. III for the infinite-ranged random Ising model. In Sec. IV the susceptibility is calculated for small external magnetic field and is compared with what follows from the projection hypothesis. The dynamical evolution of the order parameter is also briefly discussed. The final Sec. V contains some concluding remarks.

II. CTPGF for a quenched random system

For the sake of simpicity we study the soft-spin version of the Edwards-Anderson SG model defined by the Hamiltonian

$$H=-\frac{1}{2}\sum_{i\neq j} J_{ij}\,\sigma_i\sigma_j-\sum_i h_i\sigma_i+V(\sigma_i), \quad (2.1)$$

where the interaction J_{ij} are random Gaussian variables with zero average and mean square fluctuation J^2/N, N being the number of the neighbors.

The generating functional of CTPGF with the interaction kept fixed can be represented by a path integral in the following form

$$Z[h,J_{ij}]=\int[d\sigma]\,e^{iI_{eff}}\langle t_+=t_o|\hat{\rho}|t_-=t_o\rangle \quad (2.2)$$

where

$$I_{eff} = \int_p \sigma_i(x)\,\Gamma_p^{(0)}(x-y)\,\sigma_i(y)\,d^dx\,d^dy - \int \left(V(\sigma_i, J_{ij}) - h_i(x)\sigma_i(x)\right)d^dx \quad (2.3)$$

and $\hat{\rho}$ is the density matrix. In Eq.(2.3) the closed time-path starts from $t=t_0$ to $t=\infty$ (positive branch) and runs back from $t=\infty$ to $t=t_0$ (negative branch). $\Gamma_p^0(x-y)$ is the second order vertex function obtained after integrating over the reservoir degrees of freedom. It satisfies the FDT with the temperture of the reservoir

$$\tilde{\Gamma}_c^{(0)}(k) = i\,cth\frac{\beta k_0}{2}\,I_m\,\tilde{\Gamma}_r^{(0)}(k), \quad (2.4)$$

where $\Gamma_c^{(0)}$ and $\Gamma_r^{(0)}$ are the correlation and the retarded vertex functions respectively.

The advantage of using CTPGF for random systems is that the quenched average can be performed directly on the generating functional instead of its logarithm. This is possible owing to an important property of the generating functional $Z[h,J_{ij}]$, namely, it equals unity in the physical limit when the external magnetic field h(x) on the positive branch is identified with that on the negative branch.

Introducing the averaged generating functional

$$\bar{Z}[h] = \int dJ_{ij}\,P(J_{ij})\,Z[h,J_{ij}] \equiv \langle Z[h,J_{ij}]\rangle_J \quad (2.5)$$

it is possible to calculate the connected CTPGF by a direct differentiation. Eq.(2.5) then implies that

$$\bar{Z}\,\bar{\sigma}_i(h) = \langle Z\,\sigma_i(h,J_{ij})\rangle_J. \quad (2.6)$$

In the physical limit both Z and $\bar{Z}$ equal unity and the observed magnetization satisfies the requirement of a quenched average

$$M[h] = \frac{1}{N}\sum_i \bar{\sigma}_i[h] = \frac{1}{N}\sum_i \langle \sigma_i[h,J_{ij}]\rangle_J. \quad (2.7)$$

Differentiating Eq.(2.6) w.r.t. h(x) and setting $h(x_+)=h(x_-)$ we obtain

$$\overline{G}_{pij}(t,t') = \langle G_{pij}(t,t';J_{kl})\rangle + iq_{ij}(t,t'), \tag{2.8}$$

where the order parameter matrix

$$q_{ij}(t,t') = \langle \sigma_i(t,J_{ij})\sigma_j(t,J_{ij})\rangle_J - \overline{\sigma}_i(t)\,\overline{\sigma}_j(t) \tag{2.9}$$

is real, symmetric and equal to each other on the two branches,i.e.,

$$q_{ij}(t,t') = q_{ji}(t',t) = q^*_{ij}(t,t'), \tag{2.10}$$

$$q_{ij}(t_+,t'_+) = q_{ij}(t_+,t'_-) = q_{ij}(t_-,t'_+) = q_{ij}(t_-,t'_-). \tag{2.11}$$

From Eqs.(2.8) and (2.11) follow the retarded, the advanced and the correlated Green's functions

$$G_{rij}(t,t') = \langle G_{rij}(t,t';J_{ij})\rangle_J, \tag{2.12}$$

$$G_{aij}(t,t') = \langle G_{aij}(t,t';J_{ij})\rangle_J \tag{2.13}$$

and

$$G_{cij}(t,t') = \langle G_{cij}(t,t';J_{ij})\rangle_J + iq_{ij}(t,t'). \tag{2.14}$$

The order parameter matrix q(x,y) which characterizes the behavior of a quenched random system, appears in this formalism as an integral part of the second order CTPGF. By sucessive differentiation one can easily deduce higher order CTPGFs with additional order parameter matrices which we shall not discuss in this note.

Before discussing FDT let us take the Fourier transform w.r.t. the relative coordinates and write the two-point matrices in the form of Wigner's distribution in phase space

$$\widetilde{G}_{ij}(\omega,t) = \int d\tau\, e^{i\omega\tau} G_{ij}(t+\tfrac{\tau}{2}, t-\tfrac{\tau}{2}). \tag{2.15}$$

After a very short time of microscopic scale the Green's functions of a system with the interaction J_{ij} kept fixed must satisfy the FDT

$$\tilde{G}_c(\omega, t, J_{ij}) = i\, cth \frac{\beta\omega}{2} I_m \tilde{G}_r(\omega, t, J_{ij}), \tag{2.16}$$

where β^{-1} is the temperature of the reservoir. Substituting Eq.(2.16) into Eq.(2.14) it follows that

$$\tilde{\bar{G}}_c(\omega, t) = i\, cth \frac{\beta\omega}{2} I_m \tilde{\bar{G}}_r(\omega, t) + i\tilde{q}(\omega, t) \tag{2.17}$$

which can be regarded as the FDT for the quenched random system.

The retarded Green's function $G_r(k,X)$ is analytic in the upper k_o-plane. Its real and imaginary parts satisfy a dispersion relation. In the stable region it is usually assumed that a unsubtracted dispersion relation holds, i.e.,

$$Re\, \tilde{\bar{G}}_r(\omega, t) = \frac{1}{\pi}\int d\omega' \frac{I_m \tilde{\bar{G}}_r(\omega', t)}{\omega' - \omega}. \tag{2.18}$$

However, this can not be true in the unstable region where either the real part or the imaginary part changes sign. Using FDT in the high temperature approximation we get from Eqs.(2.17) and (2.18) the relation

$$Re\,\tilde{\bar{G}}_r(0,t) = \tilde{\bar{G}}_r(0,t) = \beta\,[-i\bar{G}_c(t,t) - q(t,t)]. \tag{2.19}$$

For a long-ranged random Ising model where the space dependence can be neglected, the Fischer relation [Eq.(2.19)] becomes

$$\tilde{\bar{G}}_r(0,t) = \beta\,[1 - q(t,t) - \bar{\sigma}^2(t)]. \tag{2.20}$$

It should be stressed that the Fischer relation is valid only in the stable region.

Now we are going to study the equation governing the dynamic behavior of the order parameter matrix q(x,y). We begin with the vertex functional

$$\Gamma[\bar{\sigma}(x)] = \bar{W}[h(x)] - \int_p h(x)\,\bar{\sigma}(x)\, d^d x \tag{2.21}$$

where $\bar{\sigma}(x)$ is the averaged magnetization. The vertex CTPGF can be defined by successive differentiation of the vertex functional w.r.t. $\bar{\sigma}(x)$, i.e.,

$$\bar{\Gamma}[\bar{\sigma}(x)] = \bar{W}[h(x)] - \int_p h(x)\bar{\sigma}(x) d^d x. \tag{2.22}$$

The Dyson-Schwinger equation for the second order CTPGF has the following form

$$\begin{aligned}\int_p d^d z\, \bar{\Gamma}_p(x,z)\bar{G}_p(z,y) &= \int_p \bar{G}_p(x,z)\bar{\Gamma}_p(z,y) d^d z \\ &= -\delta_p(x-y).\end{aligned} \tag{2.23}$$

where $\delta_p(x-y)$ is a δ-function defined on the closed time-path. In the physical limit Eq.(2.23) reduces to three equations for the retarded, the advanced and the correlated Green's functions in the following matrix form

$$\bar{\Gamma}_r \bar{G}_r = \bar{G}_r \bar{\Gamma}_r = -1. \tag{2.24}$$

$$\bar{\Gamma}_a \bar{G}_a = \bar{G}_a \bar{\Gamma}_a = -1 \tag{2.25}$$

and

$$\bar{\Gamma}_r \bar{G}_c = -\bar{\Gamma}_c \bar{G}_a \tag{2.26}$$

Now define a new matrix

$$\tilde{Q} \equiv -i[\tilde{\bar{\Gamma}}_c - i\,cth\frac{\beta\omega}{2} Im\tilde{\bar{\Gamma}}_r] \tag{2.27}$$

which is a functional of the order parameter matrix q and vanishes together with q. Then the Dyson-Schwinger equation for q is easily found from Eqs.(2.24), (2.26) and (2.27) to be

$$\bar{\Gamma}_r q = -Q\bar{G}_a. \tag{2.28}$$

This is a matrix equation which can be simplified in the semiclassical approximation where the product of two matrices AB is replaced by the classical expression

$$\tilde{A}\tilde{B} + \frac{i}{2}\left(\frac{\partial\tilde{A}}{\partial k_\mu}\frac{\partial\tilde{B}}{\partial x^\mu} - \frac{\partial\tilde{A}}{\partial x^\mu}\frac{\partial\tilde{B}}{\partial k_\mu}\right). \tag{2.29}$$

Here $\tilde{A}$ and $\tilde{B}$ are the corresponding Wigner's distribution functions.

III. The stability condition and the physical boundary of the model given by the Hamiltonian Eq.(2.1)

After averaging over the random variables J_{ij} the generating functional takes the following form

$$Z[h(t)] = \int [d\sigma_i]\, e^{iS_{eff}} \langle t_+ = t_0 | \hat{\rho} | t_- = t_0 \rangle, \tag{3.1}$$

where

$$S_{eff} = \sum_j \Big\{ \int_p \Big[\tfrac{1}{2} \sigma_j(t) \Gamma_p^{(0)}(t-t') \sigma_j(t') dt dt' - u\, \sigma_j^4(t) dt + h_j(t) \sigma_j(t) dt \Big] \Big\} + i \frac{J^2}{4N} \sum_{i \neq j} \int_p \sigma_i(t) \sigma_j(t) dt \int_p \sigma_i(t') \sigma_j(t') dt' \tag{3.2}$$

The bar over the quench-averaged quantities has been dropped for simplicity. In the infinite-ranged limit with $N \to \infty$ the matrix $G_{ij}(t,t')$ can be approximated by $\delta_{ij} G(t,t')$. In this case the second order vertex function can be calculated without difficulty in a diagrammatic expansion. It is found that

$$\tilde{\Gamma}_r(\omega,t) = -r + i\omega/\gamma + J_r^2(q) \tilde{G}_r(\omega,t) - \tilde{\Sigma}_r(\omega,t), \tag{3.3}$$

where r and γ are renormalized quantities that could depend on q, σ and the time t. In the limit where J tends to zero r is the inverse of the magnetic susceptibility which is proportional to the temperature and increases as the magnetization σ increases. We shall assume that r will keep this qualitative behavior even in a random spin system. In Eq.(3.3) $\tilde{\Sigma}(\omega,t)$ is the self-energy part with the first two terms in the expansion of ω and the term proportional to G_r being excluded. Therefore, we have

$$\tilde{\Sigma}_r(\omega=0,t) = 0. \tag{3.4}$$

To the lowest order perturbation in u we find

$$J_r^2(q) = J^2 + 288\,u^2(q^2 + 2q\sigma), \tag{3.5}$$

where J is also remormalized and q=q(t,t) is the order parameter. In obtaining the second term in Eq.(3.5) we have made the approximation that $\tilde{q}(\omega,t)$ has a sharp peak at ω =o.

To the same order of approximation we find the correlated vertex function

$$\tilde{\Gamma}_c(\omega,t) = i\,\frac{2}{\beta\omega}\,\mathrm{Im}\,\tilde{\Gamma}_r(\omega,t) + i\,J_c^2(q)\,\tilde{q}(\omega,t) + i\,J^2\tilde{\Delta}(\omega,t) - \tilde{\Sigma}_c(\omega,t), \tag{3,6}$$

where $\tilde{\Delta}(\omega,t)$ is defined to be

$$\tilde{\Delta}(\omega,t) = \int d\tau\, e^{i\omega\tau}\,\sigma(t+\tfrac{\tau}{2})\,\sigma(t-\tfrac{\tau}{2}) \tag{3.7}$$

which is also sharply peaked at ω =o. In Eq.(3.6) $\tilde{\Sigma}_c$ is the remaining self-energy part that does not have a sharp peak at ω =o. $J_c^2(q)$ can be easily calculated in perturbation to be

$$J_c^2(q) = J^2 + 96\,u^2(q^2 + 2q\sigma). \tag{3.8}$$

In the low frequency limit the matrix Q defined in Eq.(2.27) has therefore the form

$$\tilde{Q}(\omega,t) = J_c^2(q)\,\tilde{q}(\omega,t) + J^2\tilde{\Delta}(\omega,t). \tag{3.9}$$

Here only terms that have a sharp peak at ω =o are retained. It is noted that $J_r^2(q)$ is greater than $J_c^2(q)$ for all values of q. This fact is very important in the following.

In the zero frequency limit the Dyson-Schwinger equation for the retarded Green's function [Eq.(2.24)] has the form

$$\chi^{-1} = r - J_r^2(q)\,\chi, \tag{3.10}$$

where $\chi = G_r(o,t)$ is the susceptibility of the system. This equation can be solved to give

$$\chi = \frac{1}{2J_r^2(q)}\left[r - \sqrt{r^2 - 4J_r^2(q)}\right]. \tag{3.11}$$

The susceptibility increases as r decreases and reaches its maximum at $r=2J_r(q)$ where $\chi^{-1}=J_r(q)$. Further decrease of r will make χ complex and the system unstable. Therefore, the stability region is bounded by the inequality

$$\chi J_r(q) < 1 \tag{3.12}$$

It is easily seen from Eq.(3.11) that in the unstable region

$$|\chi|^2 J_r^2(q) = 1 \tag{3.13}$$

which is a curve on the $q-|\chi|$ plane. On this plane all stable points are situated in a region bounded from above by the curve [Eq.(3.13)] consisting of marginally stable and unstable points. In the stable region q and χ are related by the Fischer relation. Hence the physical state of the random system can only evolve either along the Fischer's line when it is stable, or along the boundary Eq.(3,13) when it is marginally stable or unstable.

Before turning to the next section let us briefly mention the low frequency behavior of the retarded Green's function. For this purpose write

$$\tilde{G}_r(\omega,t) = \chi - \alpha(t)|\omega|^\nu\left(cth\frac{\pi\nu}{2} - i\,sgn\,\omega\right), \tag{3.14}$$

where $\alpha(t)$ and ν are to be determined. An analysis similar to that given in Ref.9 shows that $\nu \leq 1/2$ if the state is marginally stable and $\nu=1$ otherwise.

IV. Susceptibility and the order parameter q in small external magnetic field

In the following we shall take the value of u to be 1/12 in the units $J=T_c=1$. From Eqs.(2.28) and (3.7)-(3.9) the static fixed point for q is determined by the equation

$$q_0 = |\chi|^2 \left(q_0 + \sigma^2 + \frac{2}{3} q_0^3 + 2 q_0^2 \sigma^2 \right), \tag{4.1}$$

where $q_0 = \text{Lim}\ q(t,t)$.

Above T_c, the whole Fischer line lies inside the stable region and we have

$$\chi = \beta (1 - q_0 - \sigma^2). \tag{4.2}$$

For small external magnetic field we can solve Eqs.(4.1) and (4.2) for q_0

$$q_0 = \frac{\beta^2}{1-\beta^2} \sigma^2 + \cdots \tag{4.3}$$

The susceptibility is therefore

$$\chi = \frac{d\sigma}{dh} = \beta - \frac{\beta}{1-\beta^2} \sigma^2 + \cdots \tag{4.4}$$

$$= \beta - \frac{\beta^3}{1-\beta^2} h^2 + \cdots$$

At the critical temperature Eq.(4.2) is still valid. In this case we have

$$q_0 = \frac{\bar{\sigma}}{\sqrt{2}} - \frac{19}{24} \sigma^2 + \cdots \tag{4.5}$$

and

$$\chi = 1 - \frac{h}{\sqrt{2}} + \frac{1}{24} h^2 + \cdots \tag{4.6}$$

Below T_c, there exists a critical external field h_c, above which the static fixed point in still lying in the stable region. The critical field h_c can be calculated and is found to be

$$h_c^2 = \frac{4}{3}\tau^3(1+3\tau+\cdots) \tag{4.7}$$

near T_c where $\tau = 1-1/\beta$.

For $T<T_c$ and $h<hc$ the intersection of the Fischer line with the fixed point equation (4.1) is above the physical boundary which can never be reached. This false fixed point is just the one found before[2] which yields negative entropy at low temperatures. The only possible fixed point in this case is on the boundary where the solution is marginally stable. In these new units the boundary is described by the equation

$$\chi = (1+2q_0^2+4q_0\sigma^2)^{-1/2} \tag{4.8}$$

Solving Eqs.(4.1) and (4.8) we find

$$q_0 = (\tfrac{3}{4}\sigma^2)^{1/3} + \cdots \tag{4.9}$$

and

$$\begin{aligned} \chi &= 1 - q_0^2 + \cdots \\ &= 1-(\tfrac{3}{4}\sigma^2)^{2/3} + \cdots \\ &= 1-(\tfrac{3}{4}h^2)^{2/3} + \cdots \end{aligned} \tag{4.10}$$

All the results obtained in Eqs.(4.3)-(4.6) and (4.9)-(4.10) agree with those predicted by the projection hypothesis.

Now we shall give a brief account of the dynamical behavior of the order matrix $\tilde{q}(\omega,t)$. Details will be presented elsewhere. In the semi-classical approximation stated in the end of Sec. II the Dyson-Schwinger equation for $\tilde{q}(\omega,t)$ has the form

$$\frac{\partial \tilde{q}}{\partial t} = \frac{1-J_c^2(q)|\tilde{G}_r|^2}{1+J_c^2(q)|\tilde{G}_r|^2}\,\frac{\dfrac{\partial|\tilde{G}_r|^2}{\partial t}}{\dfrac{\partial \mathrm{Re}\tilde{G}_r}{\partial\omega}\dfrac{\partial \mathrm{Im}\tilde{G}_r}{\partial t}-\dfrac{\partial \mathrm{Re}\tilde{G}_r}{\partial t}\dfrac{\partial \mathrm{Im}\tilde{G}_r}{\partial\omega}}\,\tilde{q} \tag{4.11}$$

where the external magnetic field is neglected for simplicity.

Above T_c, the only fixed point is q=o. We can linearize the Eq.(4. near q=o and find out the relaxation time

$$\tau_o = \frac{\beta_c^2+\beta^2}{2(\beta_c^2-\beta^2)}\,\frac{\beta}{\gamma} \tag{4.12}$$

Below T_c, the Fischer line intersects the boundary at the point $q=q_1$. In the linearized approximation q_1 is also a fixed point of the equation (4.11). The order parameter q(t,t) will tend to q_1 with a relaxation tim

$$\tau_1 = \frac{\beta_c^2}{2(\beta-\beta_c)\gamma}\,. \tag{4.13}$$

Note that both τ_o and τ_1 have simple poles at the critical temperature.

Though in the linear approximation below T_c the order parameter q tends to a finite value q_1 indicating the existence of a spin-glass phase it is not true when higher order effects are taken into consideration. q is not a true fixed point and the order parameter q is still time depende The true fixed point is again q=o on the physical boundary. After hittin the boundary at $q=q_1$ the order parameter q(t,t) must vary further along t boundary to its fixed point q=o. The time dependence of q(t,t) along the boundary is not clear as it might pass through some unstable regions.

We have shown in the end of Sec.III that the low frequency dependence of the retarded Green's function is drastically different on the marginal stability line. The time evolution of $\tilde{q}(\omega,t)$ on the marginal stability line can be obtained by substituting Eqs.(3.14) and (4.8) into Eq.(4.11). For small q we have

$$\frac{\partial\tilde{q}(\omega,t)}{\partial t} = -\frac{4}{3}\,\frac{g^2}{\alpha\nu}\,|\omega|^{1-\nu}\,\tilde{q}(\omega,t). \tag{4.14}$$

where α is the coefficient in the low frequency expansion of the retarded Green's function which is positive. From Eq.(4.14) we find immediately that $\tilde{q}(o,t)$ does not change with time and $\tilde{q}(\omega,t)$ for $\omega \neq o$ tends to zero as time t goes to infinity.

If the ω =o part of the order parameter has nonzero measure and it constitutes q_{EA} part of q_1, then the order parameter q will reach q_{EA} finally along the marginal stability line. An important question to be solved is to find the value of q_{EA} which certainly depends on the dynamic processes with infinitely long relaxation time in the $N \rightarrow \infty$ limit.

V. Discussions

Starting from the general formalism of CTPGF we have studied the infinite-ranged random Ising model in some detail. Both the unstable and the marginally stable states are found to be lying on a boundary line in the q- $|\chi|$ plane. Above T_c, the fixed point is on the Fischer line inside the stable region. In this case the low frequency behavior of the vertex function is normal and the order parameter q(t,t) tends exponentially in time to its fixed point along the Fischer line.

Below T_c and h_c there are no fixed point on the Fischer line inside the stable region. In this case the fixed point is lying on the boundary in a marginally stable state. In the presence of the persistent external magnetic field h the order parameter q will decay exponentially along the Fischer line to the intersection point q_1 and then decreases further along the boundary down to the fixed point q_o. The magnetization calculated at the state q_o agrees with that follows from the prejection hypothesis. In the absence of external magnetic field q_o=0. If the ω =0 part of the order parameter q(t,t) constitutes a finite part of q_1, say q_{EA}, the system will finally reach a steady state with $q=q_{EA}$.

It is noted that the boundary line on the q-$|\chi|$ plane is temperature independent. Therefore, the fixed point below T_c is temperature independent. As a consequence, the magnetization is also temperature independent and the entropy does not vary with the magnetic field. This is just the assumption of the projection hypothesis stated in[14]. It follows naturally from the present formalism.

Comparing our results with those of Parisi and Sompolinsky it is natural to conjecture that the function q(x), $0 \leq x \leq 1$, corresponds to our $q(t,t)$ varying along the boundary from q_1 to q_0. Not only does the present theory give a clear physical meaning to the order parameter $q(t,t)$ but it also derives an equation for q that can be solved in principle to get the time evolution of the system.

Commun. in Theor. Phys. (Beijing, China) Vol.3, No.2 (1984) 139-148

SYMMETRY AND WARD IDENTITIES FOR DISORDERED ELECTRON SYSTEMS

LIN Jian-cheng(林建成) SHEN Yu(沈 宇)
and ZHOU Guang-zhao(周光召)
Institute of Theoretical Physics, Academia Sinica,
P. O. Box 2735, Beijing, China

Received November 20, 1983

Abstract

We study the field theory approach to Anderson localization in the framework of closed time path Green's function (CTPGF). The theory is found to be invariant under an Sp(2) group. Ward identities related to this symmetry are derived. A non-linear σ model arises as a consequence of the dynamical symmetry breaking caused by the imaginary part of the retarded Green's function.

I. Introduction

Recent experiment on quantum Hall effect clearly indicates the existence of extended states for two dimensional disordered electronic system subject to a magnetic field.[1-3] This exciting fact challenges the existing theory of localization which predicts a mobility edge in 2+ε dimensions in the absence of a magnetic field.[4-5] Several years ago Wegner[6] suggested a field theory model of non-interacting disordered electron gas dealing with the scaling properties near the mobility edge where the conductance plays the role of the coupling constant of a non-linear σ model. This model was further developed by many others[7-9] and derived formally from field theory by Mckane and Stone.[5] Extension to the case with magnetic field in two dimensions is also made recently by Pruisken.[10]

To achieve quenched average over the random potential all previous papers on the field theory approach adopt the unphysical replica trick. With n replicated systems an $O(n_+, n_-)$ or $U(2n)$ symmetry is found. The critical behavior near the mobility edge from the side consisting of extended states is governed by a Goldstone mode due to the spontaneous breakdown of the replica symmetry. The Goldstone mode is described by an $O(n_+, n_-)$ or $U(2n)$ non-linear σ model where the integer n has to be analytically continued to zero in order to get the physical results from the calculation.

Although the replica trick is very successful in many disordered systems, its physical meaning is obscure and mysterious. Sometimes it may lead to ambiguous results possibly due to the non-uniqueness of the analytic continuation of a function f(n) to f(n=0) from values defined only on a discrete set of integers n. Even the simplest amorphous system of the Ising spin glass is not well understood in the replica language.

The aim of the present series of papers is to reformulate the field theory for the disordered system avoiding the use of the replica trick. This is possible in the framework of CTPGF where a rather successful theory of infinite range Ising spin glass has been developed recently.[11] The present paper is the first one of this series applying CTPGF to the study of localization and discussing mainly the general symmetry properties of the Green's functions. An Sp(2) group symmetry is found and the corresponding Ward identities are derived. Order parameter and symmetry breaking pattern are briefly discussed. Detailed analysis will be given in subsequent papers.

In the following notations adopted in CTPGF will be widely used. Readers not familiar with CTPGF can consult papers[12] for further explanations.

The paper is organized as follows: In Sec.II the model is formulated in CTPGF. The symmetry and the Ward identities will be given in Sec.III. Finally we discuss the symmetry breaking and the resulting non-linear σ model.

II. Green's functions and their generating functionals

We are concerned with the effect of disorder on the Green's functions of a non-interacting electron gas moving in external fields. The Lagrangian of the system in the second quantized language can be written in the following form:

$$\mathcal{L}=\psi^+(x)\left\{i\frac{\partial}{\partial t}+e\phi(x)-\frac{1}{2m}(-i\nabla+e\vec{A}(x))^2-V\right\}\psi(x) \ , \qquad (2.1)$$

where V(x) is a random potential with Gaussian distribution. $\vec{A}(x)$ and $\phi(x)$ are the electromagnetic potentials which will be considered as external fields in the following.

The energy spectrum specifying the nature of the electronic state either to be extended or local is determined by the imaginary part of the retarded Green's function defined on the vacuum state. In CTPGF the retarded, the advanced and the correlated Green's functions are closely related and form a single Green's function defined on a closed time path p. They can be derived from the generating functional for CTPGF. With fixed potential V(x) the generating functional has the form

$$Z[J,V]=\int[d\psi][d\psi^+]\exp\left\{i\int_p[\mathcal{L}(x)+I(x)+\psi^+J(x)+J^+\psi(x)]d^dx\right\} , \qquad (2.2)$$

where the time integration is along a closed path P running from $t_+=-\infty$ to $+\infty$ (positive branch) and turning back from $t_-=+\infty$ to $-\infty$ (negative branch); $J(x)$ and $J^+(x)$ are external sources and the integration variables $\psi(x)$ and $\psi^+(x)$ are anticommuting Grassmann variables. At the boundary of integration where $t_\pm = -\infty$ is the vacuum state whose influence on the path integral is to provide correct analytic behavior (imaginary parts) for the various Green's functions. The effect is summarized in the term $I(x)$ which will be specified later.

To proceed further it is more convenient working in the single time formalism and expressing the functions $\psi(x)$, $J(x)$ etc. as column vectors

$$\hat{\psi}(x)=\begin{pmatrix}\psi_+(x)\\ \psi_-(x)\end{pmatrix}, \qquad \hat{J}(x)=\begin{pmatrix}J_+(x)\\ J_-(x)\end{pmatrix} \qquad \text{etc.} \tag{2.3}$$

where $\psi_+(x)$ ($\psi_-(x)$) are the function $\psi(x)$ on the positive (negative) branch of time integration.

In single time formalism the term $I(x)$ has the form

$$I(x)=i\eta\hat{\psi}^+(1-\hat{\sigma}_1+i\hat{\sigma}_2)\hat{\psi} , \tag{2.4}$$

where η is a positive infinitesimal number and $\hat{\sigma}_i$=1,2,3 are the Pauli matrices. This term provides correct analytic behavior for the retarded, the advanced and the Feynman Green's functions. Then the generating functional Eq.(2.2) becomes

$$Z[\hat{J},V]=\int[d\hat{\psi}][d\hat{\psi}^+]\exp\Big\{i\int_P\{\hat{\psi}^+[(i\frac{\partial}{\partial t}+e\phi -\frac{1}{2m}(-i\nabla+e\vec{A})^2-V)\hat{\sigma}_3+i\eta(1-\hat{\sigma}_1+i\hat{\sigma}_2)]\hat{\psi} +\hat{\psi}^+\hat{\sigma}_3\hat{J}+\hat{J}^+\hat{\sigma}_3\hat{\psi}\}d^dx\Big\} . \tag{2.5}$$

Now the time integration is performed from $t=-\infty$ to $t=+\infty$.

It has been shown in Ref.[11] that the quenched average over random potential V can be performed directly on the generating functional Z. Therefore we obtain

$$\bar{Z}[\hat{J}]=\int[dV]P[V]Z[\hat{J},V] , \tag{2.6}$$

where P(V) is the distribution functional for the random potential which is in general taken to be Gaussian.

The generating functional for the connected Green's functions $\bar{W}[\hat{J}]$ is defined to be

$$\bar{W}[\hat{J}]=-i\ln\bar{Z}[\hat{J}] . \tag{2.7}$$

The vacuum average of the functions $\hat{\psi}$ and $\hat{\psi}^+$ are then

$$\psi_c(x)=\hat{\sigma}_3\frac{\delta\overline{W}}{\delta\hat{J}^+(x)} \quad , \qquad \hat{\psi}^+(x)=\frac{\delta\overline{W}}{\delta\hat{J}(x)}\hat{\sigma}_3 \ , \tag{2.8}$$

where we have adopted the convention that differentiation of the Grassmann variables $\delta/\delta\hat{J}(x)$ and $\delta/\delta\hat{\psi}(x)$ is acting from the right while that of $\delta/\delta\hat{J}^+(x)$ and $\delta/\delta\hat{\psi}^+(x)$ is from the left in order to avoid the possible confusion due to the anticommutability of the Grassmann variables. The second connected Green's functions can also be obtained as

$$\hat{G}(x,y)=-\hat{\sigma}_3\frac{\delta^2\overline{W}}{\delta\hat{J}^+(x)\delta\hat{J}(y)}\hat{\sigma}_3\Big|_{\hat{J}(x)=\hat{J}^+(x)=0}$$

$$=\begin{pmatrix} G_{++}(x,y) & G_{+-}(x,y) \\ G_{-+}(x,y) & G_{--}(x,y) \end{pmatrix} . \tag{2.9}$$

The retarded, the advanced and the correlation Green's function are three independent combinations of the four components defined on the positive and negative branches of the time axis. They are

$$G_r(x,y)=G_{++}(x,y)-G_{+-}(x,y)$$

$$=G_{-+}(x,y)-G_{--}(x,y) \quad ,$$

$$G_a(x,y)=G_{++}(x,y)-G_{-+}(x,y)$$

$$=G_{+-}(x,y)-G_{--}(x,y) \quad , \tag{2.10}$$

$$G_c(x,y)=G_{++}(x,y)+G_{--}(x,y)$$

$$=G_{+-}(x,y)+G_{-+}(x,y) \quad .$$

Moreover we can construct the generating functional for the vertex (one particle irreducible) function.

$$\overline{\Gamma}(\hat{\psi}_c)=\overline{W}[\hat{J}]-\int(\hat{\psi}_c^+\hat{\sigma}_3\hat{J}+\hat{J}^+\hat{\sigma}_3\hat{\psi}_c)d^dx \ . \tag{2.11}$$

It is easily found that

$$\frac{\delta\overline{\Gamma}}{\delta\hat{\psi}_c}=-\hat{J}^+\hat{\sigma}_3 \ , \qquad \frac{\delta\overline{\Gamma}}{\delta\hat{\psi}_c^+}=-\hat{\sigma}_3\hat{J} \ , \tag{2.12}$$

and the second order vertex function

$$\hat{\Gamma}(x,y)=\hat{\sigma}_3\frac{\delta^2\bar{\Gamma}}{\delta\hat{\psi}_c^+(x)\delta\hat{\psi}_c(y)}\hat{\sigma}_3\Big|_{\hat{\psi}_c=\hat{\psi}_c^+=0}\ , \tag{2.13}$$

which satisfies the Dyson equation with the connected Green's function $\hat{G}(x,y)$ in the form

$$\int\hat{G}(x,z)\hat{\sigma}_3\hat{\Gamma}(z,y)d^dz=\hat{\sigma}_3(x-y),$$

$$\int\hat{\Gamma}(x,z)\hat{\sigma}_3G(z,y)d^dz=\hat{\sigma}_3\hat{\delta}(x-y)\ . \tag{2.14}$$

In the tree level the second order vertex function equals

$$\hat{\Gamma}(x,y)=\hat{\sigma}_3[i\frac{\partial}{\partial t}+e\phi-\frac{1}{2m}(-i\nabla+e\vec{A})^2]\delta(x-y)$$

$$+i\eta(1-\hat{\sigma}_1+i\hat{\sigma}_2)\delta(x-y)\ . \tag{2.15}$$

Correspondingly

$$\Gamma_r(x,y)=[i\frac{\partial}{\partial t}+e\phi-\frac{1}{2m}(-i\nabla+e\vec{A})^2+i\eta]\delta(x-y)\ ,$$

$$\Gamma_a(x,y)=-[i\frac{\partial}{\partial t}+e\phi-\frac{1}{2m}(-i\nabla+e\vec{A})^2-i\eta]\delta(x-y)\ ,$$

and

$$\Gamma_c(x,y)=2i\eta\delta(x-y)\ . \tag{2.16}$$

It is seen from Eq.(2.16) that terms proportional to η give the correct analytic behavior of the vertex functions.

III. Symmetry and Ward identities

We shall restrict our discussion to Wegner model where the correlation of the random potential between different energy shells vanishes, i.e. we require

$$<V(\vec{x},E)>=0\ ,$$

$$<V(\vec{x},E)V(\vec{y},E')>=\gamma\delta(\vec{x}-\vec{y})\delta(E-E')\ . \tag{3.1}$$

Furthermore we assume the external fields to be time independent. Then it is possible to make Fourier transform in the action integral such that

$$S=\int d^{d-1}x dE\Big\{\hat{\psi}^{+}(\vec{x},E)\hat{\sigma}_3[E+e\phi-\frac{1}{2m}(-i\nabla+e\vec{A})^2-V]\hat{\psi}(\vec{x},E)$$

$$+i\eta\hat{\psi}^{+}(\vec{x},E)(1-\hat{\sigma}_1+i\hat{\sigma}_2)\hat{\psi}(\vec{x},E)$$

$$+\hat{\psi}^{+}\hat{\sigma}_3\hat{J}(\vec{x},E)+\hat{J}^{+}\hat{\sigma}_3\hat{\psi}(\vec{x},E)\Big\}\ . \qquad (3.2)$$

Apart from the source terms and the small terms proportional to η this action has a global Sp(2) symmetry keeping

$$\hat{\psi}^{+}\hat{\sigma}_3\hat{\psi}$$

invariant. The function $\hat{\psi}(x)$ forms a two dimensional representation of the Sp(2) group transforming as

$$\hat{\psi}(x)\longrightarrow\hat{\psi}'(x)=U\hat{\psi}(x)\ ,$$

$$\hat{\psi}^{+}(x)\longrightarrow\hat{\psi}'^{+}(x)=\hat{\psi}^{+}(x)U^{+}\ , \qquad (3.3)$$

where

$$U=\exp\Big\{\hat{\sigma}_1\lambda_1+\hat{\sigma}_2\lambda_2+i\hat{\sigma}_3\lambda_3\Big\} \qquad (3.4)$$

satisfies the condition

$$U\hat{\sigma}_3U^{+}=U^{+}\hat{\sigma}_3U=\hat{\sigma}_3\ . \qquad (3.5)$$

In Eq.(3.4) λ_i, i=1,2,3 are group parameters.

The term proportional to η does not respect this symmetry. Like the small external magnetic field that helps to choose the direction of symmetry breaking in ferromagnets the term (2.4) can be considered as a small external field inducing the breakdown of the Sp(2) symmetry. Actually the Sp(2) symmetry is spontaneously broken by the dynamical generation of the imaginary part of the retarded (advanced)Green's functions.

If we make an infinitesimal transformation with group parameters $\lambda_i(E)$ of the integration variables in the path integral of the generating functional $\bar{Z}[\hat{J}]$ we obtain three Ward identities corresponding to the three generators of the Sp(2) group. They are

$$2i\eta\int d^{d-1}x\mathrm{Tr}\Big\{(1+\hat{\sigma}_1)[-i\frac{\delta^2\bar{W}}{\delta\hat{J}^{+}(\vec{x},E)\delta\hat{J}(\vec{x},E)}+\frac{\delta\bar{W}}{\hat{J}^{+}(\vec{x},E)}\frac{\delta\bar{W}}{\delta\hat{J}(\vec{x},E)}]\Big\}$$

$$=\int d^{d-1}x[\hat{J}^{+}(\vec{x},E)\hat{\sigma}_1\frac{\delta\bar{W}}{\delta\hat{J}^{+}(\vec{x},E)}+\frac{\delta\bar{W}}{\delta\hat{J}(\vec{x},E)}\hat{\sigma}_1\hat{J}(\vec{x},E)], \qquad (3.6)$$

$$2i\eta\int d^{d-1}x\mathrm{Tr}\Big\{(1-i\sigma_2)[-i\frac{\delta^2\bar{W}}{\delta\hat{J}^{+}(\vec{x},E)\delta\hat{J}(\vec{x},E)}+\frac{\delta\bar{W}}{\delta\hat{J}^{+}(\vec{x},E)}\frac{\delta\bar{W}}{\delta\hat{J}(\vec{x},E)}]\Big\}$$

$$=\int d^{d-1}x[\hat{J}^{+}(\vec{x},E)\hat{\sigma}_2\frac{\delta\bar{W}}{\delta\hat{J}^{+}(\vec{x},E)}+\frac{\delta\bar{W}}{\delta\hat{J}(\vec{x},E)}\hat{\sigma}_2\hat{J}(\vec{x},E)], \tag{3.7}$$

and

$$2i\eta\int d^{d-1}x\ \mathrm{Tr}\left\{(\hat{\sigma}_1-i\hat{\sigma}_2)[-i\frac{\delta^2\bar{W}}{\delta\hat{J}^{+}(\vec{x},E)\delta\hat{J}(\vec{x},E)}+\frac{\delta\bar{W}}{\delta\hat{J}^{+}(\vec{x},E)}\ \frac{\delta\bar{W}}{\delta\hat{J}(\vec{x},E)}]\right\}$$

$$=i\int d^{d-1}x[\frac{\delta\bar{W}}{\delta\hat{J}(\vec{x},E)}\hat{\sigma}_3\hat{J}(\vec{x},E)-\hat{J}^{+}(\vec{x},E)\hat{\sigma}_3\frac{\delta\bar{W}}{\delta\hat{J}^{+}(\vec{x},E)}]\ . \tag{3.8}$$

Ward indentities for various Green's functions can be derived from Eqs. (3.6)-(3.8) by differentiating with respect to the external sources $\hat{J}$ and $\hat{J}^{+}$ and then putting them to zero. As a special example we shall show how the dynamical generation of an imaginary part of the retarded Green's function will break the symmetry.

It is convenient to introduce two vectors

$$\hat{\xi}=\begin{pmatrix}1\\1\end{pmatrix} \qquad \text{and} \qquad \hat{\eta}=\begin{pmatrix}1\\-1\end{pmatrix}\ . \tag{3.9}$$

then

$$\hat{\xi}\hat{\xi}^{+}=1+\hat{\sigma}_1\ , \qquad \hat{\xi}\hat{\eta}^{+}=\hat{\sigma}_3-i\hat{\sigma}_2\ ,$$

$$\eta\hat{\xi}^{+}=\hat{\sigma}_3+i\hat{\sigma}_2 \tag{3.10}$$

and

$$\hat{\eta}\hat{\eta}^{+}=1-\hat{\sigma}_1\ .$$

The Green's function can be put into the form

$$\hat{G}(\vec{x},\vec{y},E)=-\hat{\sigma}_3\frac{\delta^2\bar{W}}{\delta\hat{J}^{+}(\vec{x},E)\delta\hat{J}(\vec{x},E)}\hat{\sigma}_3$$

$$=\frac{1}{2}\left\{G_c\hat{\xi}\hat{\xi}^{+}+G_f\hat{\eta}\hat{\eta}^{+}+G_r\hat{\xi}\hat{\eta}^{+}+G_a\hat{\eta}\hat{\xi}^{+}\right\}\ , \tag{3.11}$$

where

$$G_r=\frac{1}{2}(G_{++}+G_{-+}-G_{+-}-G_{--}),$$

$$G_a=\frac{1}{2}(G_{++}+G_{+-}-G_{-+}-G_{--}), \tag{3.12}$$

and

$$G_c=\frac{1}{2}(G_{++}+G_{--}+G_{+-}+G_{--})$$

are the retarded, the advanced and the correlated Green's function respectively while

$$G_f=\frac{1}{2}(G_{++}+G_{--}-G_{-+}-G_{+-}) \tag{3.13}$$

is a fictitious Green's function which always vanishes in the physical limit when the external sources are put equal to zero. In that case Eq.(3.12) reduces to Eq.(2.10). From Eq.(3.11) we deduce

$$G_r=-\mathrm{Tr}[(\hat{\sigma}_3-i\hat{\sigma}_2)\frac{\delta^2\bar{W}}{\delta\hat{J}^+(\vec{x},E)\delta\hat{J}(\vec{y},E)}]$$

$$=-\eta_\alpha^+\frac{\delta^2\bar{W}}{\delta\hat{J}_\alpha^+(\vec{x},E)\delta\hat{J}_\beta(\vec{y},E)}\xi_\beta \tag{3.14}$$

and

$$G_a=-\xi_\alpha^+\frac{\delta^2\bar{W}}{\delta\hat{J}_\alpha^+(\vec{x},E)\delta\hat{J}_\beta(\vec{y},E)}\eta_\beta \tag{3.15}$$

Now we are ready to derive a Ward identity for the imaginary part of the retarded Green's function.

Taking the derivative $\delta^2/\delta\hat{J}_\alpha^+(\vec{y},E)\delta\hat{J}_\alpha(\vec{y},E)$ on both sides of Eq.(3.7) we obtain

$$\mathrm{Im}G_r(\vec{y},\vec{y},E)=\mathrm{Tr}\left\{\hat{\sigma}_2\frac{\delta^2\bar{W}}{\delta\hat{J}^+(\vec{y},E)\delta\hat{J}(\vec{y},E)}\right\}\Big|_{\hat{J}=\hat{J}^+=0}$$

$$=i\eta\int d^{d-1}x\,\mathrm{Tr}\Big\{(1-i\hat{\sigma}_2)[-i\frac{\delta^4\bar{W}}{\delta\hat{J}_\alpha^+(\vec{y}E)\delta\hat{J}_\alpha(\vec{y}E)\delta\hat{J}^+(\vec{x},E)\delta\hat{J}(\vec{x},E)}$$

$$+\frac{\delta^2\bar{W}}{\delta\hat{J}_\alpha(\vec{y},E)\delta\hat{J}^+(\vec{x}.E)}\frac{\delta^2\bar{W}}{\delta\hat{J}_\alpha^+(\vec{y},E)\delta\hat{J}(\vec{x},E)}]\Big\}\Big|_{\hat{J}=\hat{J}^+=0} \tag{3.16}$$

As is well known, $\mathrm{Im}G_r(\vec{y},\vec{y},E)$ is proportional to the density of the state $\rho(E)$ at energy E. It is different from zero certainly for extended states and possibly for localized states as $\eta\to0$. Therefore the Sp(2) symmetry is spontaneously broken in these cases. Mckane and Stone pointed out in Ref.[5] that there are two ways to satisfy the Ward identity. For extended states dynamic generation of the imaginary part of the retarded Green's function caused by the breakdown of the Sp(2) symmetry leads to the existence of Goldstone modes, which have long range correlation and govern the critical behavior from the side of extended states. For localized states there should be no Goldstone mode described by a pole at the origin in the momentum plane. To satisfy the Ward identity the integrand of the integral on the right hand side of Eq.(3.16) has to be divergent as $\eta\to0$ before the integration. Though the interpretation of Mckane and Stone is given in an entirely different theory based on replica trick we expect it to be true also in our formalism.

IV. Discussions

The order parameter to break the Sp(2) symmetry is the imaginary part of the retarded or advanced Green's funciton. A Goldstone mode on the extended state will be generated by the spontaneous breaking of the Sp(2) symmetry. To describe this Goldstone mode it is convenient to introduce a composite matrix field

$$\hat{q}(\vec{x},E)=\hat{\psi}(\vec{x},E)\hat{\psi}^{+}(\vec{x},E) \quad , \tag{4.1}$$

whose vacuum expectation value is the second order Green's function. Under the Sp(2) group the field $\hat{q}$ transforms as

$$\hat{q} \longrightarrow U\hat{q}U^{+} \quad . \tag{4.2}$$

The vacuum expectation value of $\hat{q}$ is

$$\langle\hat{q}\rangle=a(1-\hat{\sigma}_1+i\hat{\sigma}_2)+b\hat{\sigma}_3 , \tag{4.3}$$

where the diagonal part a describes the imaginary part of the G_{++} and G_{--} Green's functions and the part b their real part. Since U satisfies condition Eq.(3.5) we see that the real part $b\hat{\sigma}_3$ does not break the symmetry while the imaginary part a breaks the Sp(2) symmetry. Therefore Goldstone modes will be dynamically generated by the condensation of the $\hat{q}$ field. We shall show in a subsequent paper that these Goldstone fields satisfy a non-linear σ model equation whose coupling constant is proportional to the conductance of the disordered electron system. We expect that the critical behavior near the mobility edge will be governed by the long range renormalization effect of these Goldstone modes. This problem is now under study and will be reported later.

References

1. *K. Kon Kiltzing, G. Dorda and M. Pepper, Phys. Rev. Lett.,* *45(1980)494.*

2. *D. J. Thouless, J. Phys.,* *C14(1981)3475.*

3. *Aoki H and Kamimura, Solid State Commun.,* *21(1977)45.*

4. *F. J. Wegner, Z. Phys.,* *B25(1976)327.*

5. *A. J. Mckane and M. Stone, Ann. of Phys.,* *131(1981)36.*

6. J. Wegner, Z. Phys., B35(1979)207.
7. E. Brezin, S. Hikami and J. Zinn-Justin, Nucl. Phys., B165(1980)528.
8. S. Hikami, Prog. Theor. Phys., 64(1980)1466.
9. S. Hikami, Phys. Lett., B98(1981)208.
10. A. Pruisken (Preprint)
11. SU zhao-bin, YU Lu, ZHOU Guang-zhao, Commun. in Theor. Phys. (Beijing, China) 2(1983)1181, 1191.
12. ZHOU Guang-zhao, SU Zhao-bin, HAO Bai-lin and YU Lu, Phys. Rev., B22(1980)3385.

Commun. in Theor. Phys. (Beijing, China) Vol.3, No.2 (1984) 263-267

DOES PARISI'S SOLUTION OF THE SHERRINGTON-KIRKPATRICK MODEL LOCATE ON THE ABSOLUTE MAXIMUM OF THE FREE ENERGY?

SHEN Yu(沈宇) and ZHOU Guang-zhao(周光召)

Institute of Theoretical Physics, Academia Sinica,
P.O. Box 2735, Beijing, China

Received November 20, 1983

Abstract

In the spin glass phase near the critical temperature we find in a particular replica symmetry breaking pattern an order parameter with free energy greater than that of the Parisi's.

Earlier solutions of the infinite ranged spin glass model of Sherrington and Kirkpatrick (SK)[1,2] lead to unphysical results such as negative entropy and are indeed unstable. The first satisfactory solution in the replica approach was given by Parisi, who has introduced a replica symmetry breaking scheme involving an order parameter function q(x) with $0<x<1$[3]. This solution was found to be locally marginal stable[4] and has remarkable physical properties.

In the replica approach the quenched average of the free energy is carried out by first calculating the partition function of n replicated systems, taking average and then letting n analytically continue to zero. The order parameter is taken to be the correlation of spins in two different replicated systems

$$Q_{\alpha\beta}=\sum_{i=1}^{N}[<S_i^{(\alpha)}S_i^{(\beta)}>-\delta_{\alpha\beta}]/N\ , \tag{1}$$

where $\alpha,\beta=1,\ldots n$ are replica indices. The limit, for n going to zero, of this matrix $Q_{\alpha\beta}$ is the physical order parameter, which in the Parisi's solution is approximated by a function q(x) with x a parameter defined on the interval (0,1).

Although the free energy of n replicated systems with $n>2$, considered as a function of the order parameter $Q_{\alpha\beta}$, will tend to its minimum as the stable point of equilibrium, this is not so when n goes to zero. Strikingly enough, it is the maximum of the free energy considered as a function of $Q_{\alpha\beta}$ in the $n\to 0$ limit that will be the stable point of equilibrium. It is verified that Parisi's q(x) is situated at a local maximum of the free energy.

Is Parisi's solution an absolute maximum of the free energy? This question will be studied in the present note. The answer is no. We shall show in the following that there exists particular choice of $Q_{\alpha\beta}$ in the $n\to 0$ limit whose free energy is greater than that of Parisi's.

We shall limit our discussion to the case near the critical temperature T_c where analytic analysis is possible. Near T_c the matrix $Q_{\alpha\beta}$ is small (proportional to $\tau=T_c-T$) so that we can use the power expansion given in Ref.[3]. One finds the free energy[2,5]

$$F(Q)=\lim_{n\to 0}(-\tau \mathrm{Tr}(Q^2)+\frac{1}{3}\mathrm{Tr}(Q^3)+y\Sigma_{\alpha,\beta}Q^4_{\alpha,\beta})/n, \qquad (2)$$

where $Y=-\frac{1}{4}<0$ and the replica symmetry is broken. We will follow Parisi and look for a maximum of F(Q) with $Q_{\alpha\beta}=O(\tau)$

Parisi defines a self-similar iterative procedure to build $Q_{\alpha\beta}$:

$$Q_{\alpha\alpha}=0 , \qquad (3)$$

$$Q_{\alpha\beta}=q_i \quad \text{if} \quad I(\alpha/m_i)=I(\beta/m_i) \quad \text{and}$$

$$I(\alpha/m_{i+1})=I(\beta/m_{i+1}) , \qquad (4)$$

where I(x) denotes the integer part of the number x. In Eq.(4) $q_i(i=0,...K)$ are real numbers and $m_i(i=1,...K)$ are integer numbers such that m_{i-1}/m_i is an integer ($i\geq 1$) with $m_0=1, m_{k+1}=n$. Parisi assumed that as $n\to 0$ the order of m_i remains unchanged so that we have

$$1>m_1>m_2>\ldots>m_K>0 .$$

It is noted that m_i-m_{i+1} changes sign in the process of analytic continuation of n to zero. By defining the function

$$q(x)=q_i \quad \text{for} \quad m_i>x>m_{i-1} ,$$

it is easily found that

$$\lim_{n\to 0}\frac{1}{n}\mathrm{Tr}(Q^2)=\Sigma_0^K(m_i-m_{i+1})q_i^2$$
$$=-\int_0^1 dxq^2(x) .$$

It is this negative sign on the right-hand side that changes the free energy from minimum to maximum. In this way Parisi found that

$$F(q,\tau)=\int_0^1 dx[\tau q^2-yq^4-\frac{1}{3}xq^3(x)-q^2(x)\int_x^1 q(y)dy]. \qquad (5)$$

By variation of q(x) Parisi found a maximum of $F(q,\tau)$ near T_c in the spin glass phase where

$$q(x)=\begin{cases}\frac{x}{3} & \text{for} \quad x>3\tau \ , \\ \tau & \text{for} \quad x<3\tau \ ,\end{cases} \qquad (6)$$

and

$$F(\tau)=\frac{1}{3}\tau^3+O(\tau^4) \ . \qquad (7)$$

In the following we shall calculate the free energy in a different replica breaking scheme and compare it with that of Parisi's to the first non-trivial order in τ. Since $Q_{\alpha\beta}$ is of the order near T_c and the free energy starts with τ^3, it is sufficient to calculate the free energy up to term Tr Q^3. Therefore we shall look for the maximum of

$$F(Q,\tau)=\lim_{n\to 0}(-\tau \mathrm{Tr}Q^2 +\frac{1}{3}\mathrm{Tr}(Q^3))/n \ . \qquad (8)$$

Let us take $n=2n_p$ where n_p is considered to be an integer and write the n×n matrix Q as

$$Q=Q_p \times \begin{pmatrix}\alpha & 1 \\ 1 & \alpha\end{pmatrix} , \qquad (9)$$

where Q_p is an n×n matrix which is parametrized according to Parisi and is approximated by a function Q(x) 0<x<1, in the $n_p\to 0$ limit. In Eq.(9) α is an arbitrary real constant. From Eq.(9) one easily deduces

$$\mathrm{Tr}(Q^2)=\mathrm{Tr}(Q_p^2)\cdot 2(1+\alpha^2) \ ,$$
$$\mathrm{Tr}(Q^3)=\mathrm{Tr}(Q_p^3)\cdot 2(\alpha^3+3\alpha) \ . \qquad (10)$$

Substituting Eq.(10) into Eq.(8) we get the free energy

$$F(q,\tau,\alpha)=(\alpha^3+3\alpha)\int_0^1[\frac{1+\alpha^2}{\alpha^3+3\alpha}\tau q^2(x)$$
$$-\frac{1}{3}xq^3(x)-q^2(x)\int_x^1 q(y)dy]dx$$
$$=(\alpha^3+3\alpha)F_P(q, \frac{1+\alpha^2}{\alpha^3+3\alpha}\tau) \ , \qquad (11)$$

where $F_p(q,\tau)$ is the free energy of Parisi's to the same approximation. The maximum of the free energy $F(q,\tau,\alpha)$ is therefore

$$F(q,\tau,\alpha)=\frac{1}{3}\frac{(1+\alpha^2)^3}{(\alpha^3+3\alpha)^2}\tau^3+O(\tau^4) \ . \tag{12}$$

It is now easily seen that the free energy obtained in Eq.(12) is greater than that of Eq.(7) when α is chosen to be sufficiently small, say α=1/3.

One notices from Eq.(11) that the expression (12) will be correct only when the expansion parameter

$$\frac{1+\alpha^2}{\alpha^3+3\alpha} \tag{13}$$

is small. Therefore one should not let $\alpha\to0$ in Eq.(12) to get the absurd that that $F(q,\alpha,\tau)\to\infty$ as $\alpha\to0$. In a reasonable range $\alpha\sim1/3$ the expansion is valid for sufficiently small τ and we expect the resulting $F(q,\tau,\alpha)$ Eq.(12) to be correct.

In the Appendix the free energy is analyzed when the number of replica n is a large even integer (n>4). There one has to find the minimum of the free energy. It is shown that Parisi's solution is not an absolute minimum for n large and $n\tau$ small.

Appendix

From Eq.(2) the free energy for n replicated systems near critical temperature has the form

$$nF(Q,n,\tau)=(-\tau\mathrm{Tr}(Q^2)+\tfrac{1}{3}\mathrm{Tr}(Q^3)+y\sum_{\alpha\beta}Q^4_{\alpha\beta}). \tag{A.1}$$

We shall look for the minimum of this function for n an even integer greater than 4 and τ small. In this case the term $y\Sigma Q^4_{\alpha\beta}$ can be neglected and the Parisi's solution reduces to the symmetric solution of SK in the first non-trivial approximation so that

$$Q_{SK}: \qquad Q_{\alpha\beta}=\begin{cases}0 & \text{for} \quad \alpha=\beta \ , \\ q_n & \text{for} \quad \alpha\neq\beta \ .\end{cases} \tag{A.2}$$

Therefore we get

$$\mathrm{Tr}(Q^2)=n(n-1)q^2 \ ,$$
$$\mathrm{Tr}(Q^3)=n(n-1)(n-2)q^3 \ . \tag{A.3}$$

The minimum of $nF(Q,n,\tau)$ appears at

$$q=\frac{2\tau}{n-2} \tag{A.4}$$

with

$$nF(Q,n,\tau)=-\frac{4}{3}\tau^3\frac{n(n-1)}{(n-2)^2} \quad . \tag{A.5}$$

This is Parisi's solution. In the particular case of $n=2n_p$ Eq.(A.5) becomes

$$2n_pF(Q,2n_p,\tau)=-\frac{2}{3}\tau^3\frac{n_p(2n_p-1)}{(n_p-1)^2} \quad . \tag{A.6}$$

Now consider a particular scheme of replica symmetry breaking by taking for $n=2n_p$

$$Q=Q_{SK}\times\begin{pmatrix} \alpha & 1 \\ 1 & \alpha \end{pmatrix}, \tag{A.7}$$

we find the free energy to be

$$2n_pF'(Q,2n_p,\tau)=-\frac{8}{3}\frac{(1+\alpha^2)^3}{(\alpha^3+3\alpha)^2}\tau^3\frac{n_p(n_p-1)}{(n_p-2)^2} \quad . \tag{A.8}$$

It is easily verified that

$$2\frac{n_p(n_p-1)}{(n_p-2)^2}>\frac{n_p(2n_p-1)}{(n_p-1)^2} \qquad \text{for} \quad n_p>2$$

and

$$2\frac{(1+\alpha^2)^3}{(\alpha^3+3\alpha)^2}>1 \qquad \text{for all } \alpha \; .$$

Therefore the free energy F' is smaller than that of Parisi's. Hence we conclude that Parisi's solution is not situated at the absolute minimum of the free energy for n replicated systems with n an even integer greater than 4.

References

1. *D.Sherrington and S.Kirkpatrick, Phys. Rev. Lett.,* 35*(1975)1972;*
D.Sherrington and S.Kirkpatrick, Phys. Rev., B17*(1978)4385.*
2. *A.J.Bray and M.A.Moore, Phys. Rev. Lett,* 41*(1978)1068.*
3. *G.Parisi, J. Phys.,* A13*(1980)L115; J. Phys.,* C13*(1980)403;* A13*(1980) 1887; Philos. Mag.,*41*(1980)677; Phys. ReP.,* 67*(1980) 25.*
4. *C.De Dominicis and I.Kondor, preprint.*
5. *F.Pytte and T. Rudnik, Phys. Rev.,* B19*(1979)3603.*

PHYSICS REPORTS (Review Section of Physics Letters) 118, nos. 1 & 2 (1985) 1–131. North-Holland, Amsterdam

EQUILIBRIUM AND NONEQUILIBRIUM FORMALISMS MADE UNIFIED

Kuang-chao CHOU, Zhao-bin SU,* Bai-lin HAO and Lu YU

Institute of Theoretical Physics, Academia Sinica, P.O. Box 2735, Beijing, China

Received 5 June 1984

Contents:

1. Introduction 3
   1.1. Why closed time-path? 3
   1.2. Few historical remarks 4
   1.3. Outline of the paper 5
   1.4. Notations 6
2. Basic properties of CTPGF 7
   2.1. Two-point functions 7
   2.2. Generating functionals 12
   2.3. Single time and physical representations 18
   2.4. Normalization and causality 24
   2.5. Lehmann spectral representation 28
3. Quasiuniform systems 31
   3.1. The Dyson equation 31
   3.2. Systems near thermoequilibrium 34
   3.3. Transport equation 39
   3.4. Multi-time-scale perturbation 44
   3.5. Time dependent Ginzburg–Landau equation 46
4. Time reversal symmetry and nonequilibrium stationary state (NESS) 48
   4.1. Time inversion and stationarity 49
   4.2. Potential condition and generalized FDT 52
   4.3. Generalized Onsager reciprocity relations 53
   4.4. Symmetry decomposition of the inverse relaxation matrix 55
5. Theory of nonlinear response 58
   5.1. General expressions for nonlinear response 58
   5.2. General considerations concerning multi-point functions 62
   5.3. Plausible generalization of FDT 67
6. Path integral representation and symmetry breaking 70
   6.1. Initial correlations 71
   6.2. Order parameter and stability of state 76
   6.3. Ward–Takahashi identity and Goldstone theorem 79
   6.4. Functional description of fluctuation 82
7. Practical calculation scheme using CTPGF 89
   7.1. Coupled equations of order parameter and elementary excitations 90
   7.2. Loop expansion for vertex functional 92
   7.3. Generalization of Bogoliubov–de Gennes equation 96
   7.4. Calculation of free energy 99
8. Quenched random systems 103
   8.1. Dynamic formulation 104
   8.2. Infinite-ranged Ising spin glass 109
   8.3. Disordered electron system 114
9. Connection with other formalisms 119
   9.1. Imaginary versus real time technique 119
   9.2. Quantum versus fluctuation field theory 123
   9.3. A plausible microscopic derivation of MSR field theory 125
10. Concluding remarks 127
Note added in proof 128
References 128

* Current address: Department of Physics, City College of New York, New York, NY 10031, U.S.A.

Abstract:
In this paper we summarize the work done by our group in developing and applying the closed time-path Green function (CTPGF) formalism, first suggested by J. Schwinger and further elaborated by Keldysh and others. The generating functional technique and path integral representation are used to discuss the various properties of the CTPGF and to work out a practical calculation scheme. The formalism developed provides a unified framework for describing both equilibrium and nonequilibrium phenomena. It includes the ordinary quantum field theory and the classical fluctuation field theory as its limiting cases. It is well adapted to consider the symmetry breaking with either constituent or composite order parameters. The basic properties of the CTPGF are described, the two-point functions are discussed in some detail with the transport equation and the time dependent Ginzburg–Landau equation derived as illustrations. The implications of the time-reversal symmetry for stationary states are explored to derive the potential condition and to generalize the fluctuation–dissipation theorem. A system of coupled equations is derived to determine self-consistently the order parameter as well as the energy spectrum, the dissipation and the particle distribution for elementary excitations. The general formalism and the useful techniques are illustrated by applications to critical dynamics, quenched random systems, theory of nonlinear response, plasma, nuclear many-body problem and so on.

1. Introduction

1.1. Why closed time-path?

The field-theoretical technique, introduced into the many-body theory since the late fifties, has proved to be highly successful in studying the ground state, the thermoequilibrium properties and the linear response of the system to the external disturbance [1–3]. However, only limited progress has been made in investigating the nonequilibrium properties beyond the linear response by using the field-theoretical methods. To appreciate the difficulties encountered here, let us recall some basic ingredients of the field-theoretical approach.

The Green function is defined as an average of the time ordered product of Heisenberg field operators over some state which we do not specify for the moment, i.e.,

$$G(t_1, t_2) = -\mathrm{i}\langle T(\hat{A}(t_1)\hat{B}(t_2))\rangle \,. \tag{1.1}$$

By introducing the interaction picture, (1.1) can be rewritten as

$$G(t_1, t_2) = -\mathrm{i}\langle S^\dagger T(\hat{A}_\mathrm{I}(t_1)\hat{B}_\mathrm{I}(t_2)S)\rangle \,, \tag{1.2}$$

where the S matrix is defined as

$$S \equiv U(\infty, -\infty) = T\exp\Big(-\mathrm{i}\int_{-\infty}^{\infty} \mathscr{H}^\mathrm{I}_\mathrm{int}(t)\,\mathrm{d}t\Big)\,, \tag{1.3}$$

with the interacting part of the Hamiltonian $\mathscr{H}^\mathrm{I}_\mathrm{int}(t)$ in the interaction picture.

If we are interested in the ground-state properties, then

$$S|0\rangle = \mathrm{e}^{\mathrm{i}L}|0\rangle, \qquad \langle 0|S^\dagger = \langle 0|\mathrm{e}^{-\mathrm{i}L}\,, \tag{1.4}$$

where L is a phase factor contributed by the vacuum fluctuations and can be set equal to zero, if the renormalized ground state is considered. Therefore, we can easily get rid of $S^\dagger$ in (1.2) so that the powerful arsenal of the quantum field theory can be used without major changes in the many-body theory at zero temperature.

For systems in thermoequilibrium at different from zero temperature, we cannot relate observable quantities directly to the elements of the S matrix, but the density matrix in this case take the following form:

$$\hat{\rho} = \exp[\beta(\mathscr{F} - \hat{\mathscr{H}})] , \tag{1.5}$$

where β is the inverse temperature, $\mathscr{F}$ the free energy, $\hat{\mathscr{H}}$ the Hamiltonian. If we consider β as an imaginary time it, $\hat{\rho}$ behaves like an evolution operator $\exp(-i\hat{\mathscr{H}}t)$. The well-known Matsubara technique [4–7] has been successfully developed by making use of this property.

However, it is not easy to handle the $S^{\dagger}$ term in (1.2), if a general nonequilibrium state is considered. An intelligent way out was suggested by J. Schwinger in 1961 [8]. Let us imagine a time-path p which goes from $-\infty$ to $+\infty$ and then returns back from $+\infty$ to $-\infty$. We can then define a generalized S_p matrix along this closed time-path (as we call it)

$$S_p \equiv T_p \exp\left\{-i \int_p \mathscr{H}_{\rm int}(t)\, dt\right\} , \tag{1.6}$$

where T_p is the time-ordering operator along this path p. It is identical to the standard T operator on the positive branch $(-\infty, +\infty)$ and represents $\tilde{T}$ – an anti-time-ordering operator on the negative branch $(+\infty, -\infty)$. Also, any point at the negative branch is considered as a later instant than any time at the positive branch. Equipped with such generalized S_p matrix we can define the Green function along the closed time-path p as

$$\begin{aligned} G_p(t_1, t_2) &= -i\langle T_p(\hat{A}(t_1)\hat{B}(t_2))\rangle \\ &= -i\langle T_p(\hat{A}_I(t_1)\hat{B}_I(t_2)S_p)\rangle . \end{aligned} \tag{1.7}$$

Although for physical observables the time values t_1, t_2 are on the positive branch, both positive and negative branches will come into play at intermediate steps of calculation if a self-consistent formalism is intended.

The introduction of the closed time-path appears at the first glance as a purely formal trick to restore the mathematical analogy with the quantum field theory. Actually, it has deeper motivation. In particle physics, people are mostly interested in scattering processes for which the S matrix providing the probability of transition from the in-states to the out-states, is the most suitable framework. In statistical physics, however, we are mainly concerned with the expectation value of physical quantities at finite time t. It is thus natural to introduce the S_p matrix along the closed time-path p going from the state at $-\infty$ along t-axis to the $+\infty$ state and returning back to the $-\infty$ state (see $S^{\dagger}$ in (1.22)). This way we can establish a direct connection of S_p with observable quantities. As we will see later, the great merits of the closed time-path Green function (CTPGF) formalism more than justify the technical complications occurring due to the introduction of the additional negative time branch.

1.2. Few historical remarks

After Schwinger's initiative in 1961 [8], the closed time-path formalism has been elaborated and developed further by Keldysh and many others [9–19]. Some people used to call it Keldysh formalism. For the recent 20 years, this technique has been used to attack a number of interesting problems in statistical physics and condensed matter theory such as spin system [20], superconductivity [21–24], laser [25], tunneling and secondary emission [26–32], plasma [33, 34], other transport processes [35–38] and so on. For some of these systems like laser, the application of the CTPGF formalism is essential because the standard technique cannot be used directly for far from equilibrium situations, whereas for some of the others the CTPGF approach is used mainly due to its technical convenience. It is our impression, however, that the potential advantages of this formalism have not yet been fully exploited, partly because of its apparent technical complexity.

For the last few years we have combined the generating functional technique and the path integral representation, widely used in the quantum field theory [39], with the CTPGF approach and have developed a unified framework to describe both equilibrium and nonequilibrium systems with symmetry breaking and dynamical coupling between the order parameter and the elementary excitations [40–49]. To check the formalism developed and to explore its potentiality we have applied it to a number of problems including critical dynamics, quenched random systems, nonlinear response theory, superconductivity, laser, plasma, nuclear matter, quasi-one-dimensional conductor, and so on [40–57]. Although most of these problems in principle can be also discussed using other techniques, the logical simplicity and the flexibility, the unified approach to equilibrium and nonequilibrium processes as well as the deep insight one can get make the CTPGF formalism promising and encouraging.

1.3. Outline of the paper

In this paper we would like to summarize some of the results obtained by our group in developing and applying the CTPGF formalism. Because of the limitation of space we will only outline the main features along with some useful techniques of the CTPGF approach and illustrate them by few examples. Since the major part of our papers was published either in Chinese or in not easily accessible English journals, we will attempt to make this article self-contained as much as possible. Nevertheless, we should warn the reader that some part of this review is still descriptive and sketchy. A brief summary of the CTPGF formalism was given by us earlier [58], but this paper is much more extended and complete. Since we are mainly summarizing our own results, the contributions of other authors in developing and applying the CTPGF approach may not be emphasized as they should be. We apologize to them for any possible omissions or underestimates. To keep the integrity of presentation we will not distinguish carefully what was known before and what is new.

The topics to be covered in this review can be seen from the table of contents. We will not repeat them here. Few remarks, however, are in order. Section 2 is mainly tutorial, but the subsection on normalization and causality is important for further discussion. Section 3 is devoted to a detailed discussion of the two-point functions. The differentiation of the micro- and macro-time scales described there is very useful. In section 4 the potential condition and the fluctuation-dissipation theorem (FDT) are discussed from a microscopic point of view. The theory of nonlinear response which may be important for future applications is outlined in section 5. Section 6 is devoted to the consideration of the symmetry breaking and the Ward–Takahashi (WT) identities. We believe that the CTPGF formalism is advantageous in studying systems with broken symmetry. We also mention there the additional way of describing fluctuations available in the CTPGF approach. The unified framework of treating the dynamical coupling between the order parameter and the elementary excitations mentioned before, is given in section 7. We could start from this formalism at the very beginning, but the present more inductive exposition is probably more convenient for the reader. In section 8 we show that the quenched average may be carried out directly on the generating functional in the CTPGF formalism and the replica trick can be thus avoided. The connections with other formalisms are described in section 9. Readers, familiar with them might have a look at this section before the others. An experienced and busy reader could get a rough idea about the CTPGF approach by a quick scanning of sections 2 and 6–9.

1.4. Notations

Throughout this paper we will use the units $\hbar = k_B = c = 1$ except for few paragraphs where the Planck constant $\hbar$ is written out explicitly to emphasize the quasiclassical nature of expansion.

The metric tensor we use is given by

$$g^{00} = -g^{11} = -g^{22} = -g^{33} = 1 \tag{1.8}$$

with the scalar product and the d'Alembertian

$$\Box \equiv \partial_\mu \partial^\mu = \partial_t^2 - \nabla^2 \tag{1.9}$$

defined correspondingly.

The Fourier transformation with respect to the relative coordinates $x-y$ is defined as

$$G(x, y) = G\left[\frac{x+y}{2}, x-y\right] = \int \frac{d^{d+1}p}{(2\pi)^{d+1}} \exp[-ip\cdot(x-y)]\tilde{G}\left(\frac{x+y}{2}, p\right), \tag{1.10}$$

where d is the space dimension and $p\cdot x = p_0 t - \boldsymbol{p}\cdot\boldsymbol{r}$. The tilde "$\tilde{}$" will be omitted wherever no confusion occurs.

The formalism presented in this paper can be applied to a broad class of fields including non-Abelian gauge fields, but in most cases we will illustrate it by a real boson field, either relativistic or nonrelativistic (e.g. phonons), and a nonrelativistic complex boson or fermion field. The former will be denoted by $\hat{\varphi}(x)$, whereas the latter by $\hat{\psi}(x)$ and $\hat{\psi}^\dagger(x)$. Wherever a double sign $\pm$ or $\mp$ appears, the upper case will always correspond to the boson field, while the lower one corresponds to the fermion field.

As a rule, the field operator is not distinguished by the caret "$\hat{}$" which itself is used in some cases to denote a two-component vector or a 2×2 matrix.

Also, for simplicity we introduce an abbreviated notation for integration

$$J\varphi \equiv \int J(x)\varphi(x) \equiv \int d^dx\, dt\, J(x)\varphi(x). \tag{1.11}$$

The form at the right is used only in exceptional cases, while the middle one most frequently.

The Pauli matrices are defined as

$$\sigma_1 = \begin{pmatrix} & 1 \\ 1 & \end{pmatrix}, \quad \sigma_2 = \begin{pmatrix} & -i \\ i & \end{pmatrix}, \quad \sigma_3 = \begin{pmatrix} 1 & \\ & -1 \end{pmatrix}. \tag{1.12}$$

2. Basic properties of CTPGF

As mentioned in the Introduction, this section is mainly tutorial. To get familiar with the concepts and notations used in the CTPGF formalism we start from two-point functions (section 2.1) in close contact with the ordinary Green functions. We then define the generating functional and discuss the perturbation theory (section 2.2). The single time representation and the physical representation as well as the transformation from one to another are discussed in section 2.3. Furthermore, the consequences of the normalization and the causality are outlined in section 2.4. Finally, the Lehmann spectral representation is described in section 2.5.

2.1. Two-point functions

The two-point Green functions are most useful in practical applications and hence their properties have been most thoroughly investigated. In this section we first define the two-point CTPGF and then discuss their connection with the ordinary retarded, advanced and correlation functions along with the causality relations. As we will see later, they are special cases of much more general relations following from the normalization condition of the generating functional and the causality. The explicit expressions will be given for free propagators in thermoequilibrium systems.

2.1.1. Definition

The two-point CTPGF for a complex field $\psi(x)$ is defined as

$$G(x, y) \equiv -\mathrm{i}\,\mathrm{Tr}\{T_p(\psi(x)\psi^\dagger(y))\hat{\rho}\}$$
$$\equiv -\mathrm{i}\langle T_p(\psi(x)\psi^\dagger(y))\rangle\,, \tag{2.1}$$

where $\psi(x)$, $\psi^\dagger(y)$ are Heisenberg operators, $\hat{\rho}$ the density matrix, T_p the time ordering operator as discussed in the Introduction. Inasmuch as x, y can assume values on either positive or negative time branches, $G(x, y)$ can be presented as a 2×2 matrix

$$\hat{G}(x, y) \equiv \begin{pmatrix} G_{++} & G_{+-} \\ G_{-+} & G_{--} \end{pmatrix} \equiv \begin{pmatrix} G_{\mathrm{F}} & G_{+} \\ G_{-} & G_{\tilde{\mathrm{F}}} \end{pmatrix}, \tag{2.2}$$

with

$$G_{\mathrm{F}}(x, y) \equiv -\mathrm{i}\langle T(\psi(x)\psi^\dagger(y))\rangle\,. \tag{2.3a}$$

$$G_{+}(x, y) \equiv \mp\mathrm{i}\langle\psi^\dagger(y)\psi(x)\rangle\,, \tag{2.3b}$$

$$G_{-}(x, y) \equiv -\mathrm{i}\langle\psi(x)\psi^\dagger(y)\rangle\,, \tag{2.3c}$$

$$G_{\tilde{\mathrm{F}}}(x, y) \equiv -\mathrm{i}\langle\tilde{T}(\psi(x)\psi^\dagger(y))\rangle\,. \tag{2.3d}$$

Here G_{F} is the usual Feynman causal propagator, whereas the other three are new in the CTPGF formalism. Sometimes $G_{\tilde{\mathrm{F}}}$ defined as expectation value of anti-time-ordering product, is called anti-causal propagator. Using the step function

$$\theta(x, y) = \begin{cases} 1, & \text{if } t_x > t_y \\ 0, & \text{otherwise} \end{cases}, \tag{2.4}$$

eqs. (2.3a) and (2.3d) can be rewritten as

$$G_F(x, y) = -i\theta(x, y)\langle\psi(x)\psi^\dagger(y)\rangle \mp i\theta(y, x)\langle\psi^\dagger(y)\psi(x)\rangle\,,$$
$$G_{\tilde F}(x, y) = -i\theta(y, x)\langle\psi(x)\psi^\dagger(y)\rangle \mp i\theta(x, y)\langle\psi^\dagger(y)\psi(x)\rangle\,. \tag{2.5}$$

These four functions are not independent of each other. There is an algebraic identity

$$G_{++} + G_{--} = G_{+-} + G_{-+}\,,$$
or
$$G_F(x, y) + G_{\tilde F}(x, y) = G_+(x, y) + G_-(x, y)\,, \tag{2.6}$$

following from the normalization of the step function

$$\theta(x, y) + \theta(y, x) = 1\,. \tag{2.7}$$

In what follows we will call CTPGFs defined by relations like (2.3) as "single time" representation and denote them by a tensor $G_{\alpha\beta\cdots\rho}(12\cdots n)$ with Greek subscripts $\alpha, \beta, \ldots, \rho = \pm$. As a whole, the tensor itself is written as $\hat{G}$.

2.1.2. Physical representation

The CTPGFs defined above are most convenient for calculations, but more direct contact with measurable quantities is established via the "physical" representation defined as

$$G_r(x, y) \equiv -i\theta(x, y)\langle[\psi(x), \psi^\dagger(y)]_\mp\rangle\,, \tag{2.8a}$$
$$G_a(x, y) \equiv i\theta(y, x)\langle[\psi(x), \psi^\dagger(y)]_\mp\rangle\,, \tag{2.8b}$$
$$G_c(x, y) \equiv -i\langle\{\psi(x), \psi^\dagger(y)\}\rangle\,, \tag{2.8c}$$

where G_r, G_a and G_c are retarded, advanced and correlation functions, correspondingly. In this definition,

$$[\psi(x), \psi^\dagger(y)]_+ = \{\psi(x), \psi^\dagger(y)\} = \psi(x)\psi^\dagger(y) + \psi^\dagger(y)\psi(x)\,.$$

It is straightforward to check that these functions are related to the CTPGF in single time representation as follows:

$$G_r = G_F - G_+ = G_- - G_{\tilde F}\,, \tag{2.9a}$$
$$G_a = G_F - G_- = G_+ - G_{\tilde F}\,, \tag{2.9b}$$
$$G_c = G_F + G_{\tilde F} = G_+ + G_-\,. \tag{2.9c}$$

The inverse relations are given by

$$\hat{G} = \tfrac{1}{2}G_r\begin{pmatrix}1 & -1\\ 1 & -1\end{pmatrix} + \tfrac{1}{2}G_a\begin{pmatrix}1 & 1\\ -1 & -1\end{pmatrix} + \tfrac{1}{2}G_c\begin{pmatrix}1 & 1\\ 1 & 1\end{pmatrix}. \tag{2.10}$$

If we introduce two-component vectors

$$\xi \equiv \begin{pmatrix}1\\ 1\end{pmatrix}, \qquad \xi^\dagger \equiv (1, 1), \qquad \eta \equiv \begin{pmatrix}1\\ -1\end{pmatrix}, \qquad \eta^\dagger \equiv (1, -1), \tag{2.11}$$

(2.10) can be rewritten as

$$\hat{G} = \tfrac{1}{2}G_r\xi\eta^\dagger + \tfrac{1}{2}G_a\eta\xi^\dagger + \tfrac{1}{2}G_c\xi\xi^\dagger, \tag{2.12a}$$

or in components

$$G_{\alpha\beta} = \tfrac{1}{2}G_r\xi_\alpha\eta_\beta + \tfrac{1}{2}G_a\eta_\alpha\xi_\beta + \tfrac{1}{2}G_c\xi_\alpha\xi_\beta. \tag{2.12b}$$

Sometimes it is convenient to introduce a matrix form for the physical functions

$$\tilde{G} = \begin{pmatrix}0 & G_a\\ G_r & G_c\end{pmatrix}. \tag{2.13}$$

The transformations (2.10) and (2.12) can then be presented as

$$\hat{G} = Q^{-1}\tilde{G}Q, \qquad \tilde{G} = Q\hat{G}Q^{-1}, \tag{2.14}$$

using the orthogonal matrix

$$Q = \frac{1}{\sqrt{2}}\begin{pmatrix}1 & -1\\ 1 & 1\end{pmatrix} = \frac{1}{\sqrt{2}}(1 - i\sigma_2), \qquad Q^\dagger = Q^{-1} = \frac{1}{\sqrt{2}}\begin{pmatrix}1 & 1\\ -1 & 1\end{pmatrix}, \tag{2.15}$$

which was first introduced by Keldysh [9]. In what follows we will call $\tilde{G}$ the CTPGF in physical representation and denote their components by $G_{ij\cdots l}(12\cdots n)$ with Latin subscripts $i, j, \ldots, n = 1, 2$. In the case of two-point functions,

$$G_{11} = \tfrac{1}{2}(G_F + G_{\tilde{F}} - G_+ - G_-) = 0, \tag{2.16a}$$
$$G_{12} = G_a = \tfrac{1}{2}(G_F - G_- + G_+ - G_{\tilde{F}}), \tag{2.16b}$$
$$G_{21} = G_r = \tfrac{1}{2}(G_F - G_+ + G_- - G_{\tilde{F}}), \tag{2.16c}$$
$$G_{22} = G_c = \tfrac{1}{2}(G_F + G_{\tilde{F}} + G_+ + G_-). \tag{2.16d}$$

We see thus G_{11} is always zero and the other equations of (2.16) are identical to those of (2.9) by virtue of the identity (2.6).

It is obvious from definition (2.8) that

$$G_{12}(x, y) \equiv G_a(x, y) = 0, \quad \text{if} \quad t_x > t_y; \qquad G_{21}(x, y) \equiv G_r(x, y) = 0, \quad \text{if} \quad t_y > t_x, \tag{2.17}$$

and also that

$$G_{12}(x, y) \cdot G_{21}(x, y) = 0, \tag{2.18}$$

because

$$\theta(x, y)\theta(y, x) = 0. \tag{2.19}$$

As will be shown later, almost all that has been said here, e.g., $G_{11} = 0$, the transformation of single and physical representations, the causality relations (2.17), (2.18), etc., can be easily generalized to the multi-point functions using the generating functional technique. However, before going on to describe this technique itself we give here the explicit expressions for the free propagators.

2.1.3. Free propagators

The Lagrangian of the free fermion field is given by

$$\mathscr{L}_0 = \int \psi^\dagger(x)\left(\mathrm{i}\frac{\partial}{\partial t} + \frac{\nabla^2}{2m}\right)\psi(x), \tag{2.20}$$

where m is the particle mass. The single time CTPGFs are defined by (2.3). To distinguish it from the Bose case we will use the letter S instead of G. If the system is in thermoequilibrium, the free propagator can be evaluated immediately from the definition. In Fourier space the CTPGFs turn out to be

$$S_{0p}(p) = \begin{pmatrix} S_F & S_+ \\ S_- & S_{\tilde{F}} \end{pmatrix},$$

$$S_F(p) = \frac{1 - n(p)}{p_0 - p^2/2m + \mathrm{i}\varepsilon} + \frac{n(p)}{p_0 - p^2/2m - \mathrm{i}\varepsilon}$$

$$= \frac{1}{p_0 - p^2/2m + \mathrm{i}\varepsilon} + 2\pi\mathrm{i}n(p)\delta(p_0 - p^2/2m), \tag{2.21a}$$

$$S_+(p) = 2\pi\mathrm{i}n(p)\delta(p_0 - p^2/2m), \tag{2.21b}$$

$$S_-(p) = -2\pi\mathrm{i}(1 - n(p))\delta(p_0 - p^2/2m), \tag{2.21c}$$

$$S_{\tilde{F}}(p) = -\frac{n(p)}{p_0 - p^2/2m + \mathrm{i}\varepsilon} - \frac{1 - n(p)}{p_0 - p^2/2m - \mathrm{i}\varepsilon}$$

$$= \frac{-1}{p_0 - p^2/2m - \mathrm{i}\varepsilon} + 2\pi\mathrm{i}n(p)\delta\,(p_0 - p^2/2m). \tag{2.21d}$$

where

$$n(p) = \frac{1}{\exp[(p_0 - \mu)/T] + 1} \tag{2.22}$$

is the Fermi distribution with μ as the chemical potential. If $n(p)$ is set equal to zero, we recover the propagator for the "pure" vacuum. It is interesting to note that the additional term $2\pi i n(p)\delta(p_0 - p^2/2m)$ is the same for all components of $\hat{G}$. The reason for this will be clear from the next section.

In accord with (2.9) we find the physical functions to be

$$S_r(p) = \frac{1}{p_0 - p^2/2m + i\varepsilon}, \tag{2.23a}$$

$$S_a(p) = \frac{1}{p_0 - p^2/2m - i\varepsilon}, \tag{2.23b}$$

$$S_c(p) = -2\pi i(1 - 2n(p))\delta(p_0 - p^2/2m). \tag{2.23c}$$

We note in passing that the retarded and the advanced Green functions S_r, S_a do not depend on the particle distribution $n(p)$.

Similarly, for the Hermitian boson field described by the Lagrangian

$$\mathscr{L}_0 = \tfrac{1}{2}\int (\partial_\mu\varphi(x)\partial^\mu\varphi(x) - m^2\varphi^2(x)) \tag{2.24}$$

we have

$$\Delta_p(x, y) \equiv -i\langle T_p(\varphi(x)\varphi(y))\rangle \equiv \begin{pmatrix} \Delta_F(x, y) & \Delta_+(x, y) \\ \Delta_-(x, y) & \Delta_{\tilde{F}}(x, y) \end{pmatrix}. \tag{2.25}$$

In the Fourier space the free boson CTPGFs are

$$\Delta_F(p) = \frac{1}{p_0^2 - \omega^2(p) + i\varepsilon} - 2\pi i f(p)\delta(p_0^2 - \omega^2(p)), \tag{2.26a}$$

$$\Delta_+(p) = -2\pi i(\theta(p_0) + f(p))\delta(p_0^2 - \omega^2(p)), \tag{2.26b}$$

$$\Delta_-(p) = -2\pi i(\theta(-p_0) + f(p))\delta(p_0^2 - \omega^2(p)), \tag{2.26c}$$

$$\Delta_{\tilde{F}}(p) = -\frac{1}{p_0^2 - \omega^2(p) - i\varepsilon} - 2\pi i f(p)\delta(p_0^2 - \omega^2(p)), \tag{2.26d}$$

where

$$f(p) = \frac{1}{\exp[\omega(p)/T] - 1} \tag{2.27}$$

is the Bose distribution and

$$\omega(p) = \sqrt{p^2 + m^2} \tag{2.28}$$

is the particle energy.

If $f(p) = 0$, we recover the standard boson propagator of the quantum field theory [39]. Also, the additional term proportional to $f(p)$ is the same for all components of $\hat{\Delta}$.

The corresponding retarded, advanced and correlation functions are given by

$$\Delta_r(p) = \frac{1}{p_0^2 - \omega^2(p) + 2i\varepsilon p_0}, \tag{2.29a}$$

$$\Delta_a(p) = \frac{1}{p_0^2 - \omega^2(p) - 2i\varepsilon p_0}, \tag{2.29b}$$

$$\Delta_c(p) = -2\pi i(1 + 2f(p))\delta(p_0^2 - \omega^2(p)). \tag{2.29c}$$

It can be shown [40] that the expressions for fermion and boson propagators (2.21), (2.26) remain the same for inhomogeneous, nonequilibrium systems provided $n(p)$ and $f(p)$ are replaced by their nonequilibrium counterparts – Wigner distributions $n(X, p)$, $f(X, p)$ in the external field, where

$$X = (x + y)/2 \tag{2.30}$$

is the center of mass coordinates.

2.2. Generating functionals

For interacting fields we can construct the perturbation expansion in full analogy with the quantum field theory. The Wick theorem can be generalized to the CTPGF case, most conveniently by using the generating functional. For simplicity, we consider real bosons. The extension to other systems is obvious.

2.2.1. Definition of $Z[J]$

The Lagrangian of the system is given by

$$\mathscr{L} = \mathscr{L}_0(\varphi) + \int (J(x)\varphi(x) - V(\varphi)), \tag{2.31}$$

where $\mathscr{L}_0(\varphi)$ is given by (2.24), $V(\varphi)$ the self-interaction and $J(x)$ the external source. The generating functional for CTPGF is defined as

$$Z[J(x)] \equiv \mathrm{Tr}\Big\{T_p\Big[\exp\Big(\mathrm{i}\int_p J(x)\varphi(x)\Big)\Big]\hat{\rho}\Big\}, \tag{2.32}$$

where the integration path p and the time ordering product along it T_p have been already defined in the Introduction. In general, the external source on the positive branch $J_+(x)$ and the negative branch $J_-(x)$ are assumed to be different. They will be set equal to each other or both to zero at the end of calculation.

The n-point CTPGF is defined as

$$G_p(1\cdots n) \equiv (-\mathrm{i})^{n-1}\,\mathrm{Tr}[T_p(\varphi(1)\cdots\varphi(n))\hat{\rho}] = \mathrm{i}(-1)^n \left.\frac{\delta^n Z[J(x)]}{\delta J(1)\cdots\delta J(n)}\right|_{J=0}. \tag{2.33}$$

2.2.2. Generalized Wick theorem

In the incoming interaction picture (2.32) can be rewritten as

$$Z[J(x)] = \mathrm{Tr}\Big\{T_p\Big[\exp\Big(-\mathrm{i}\int_p (V(\varphi_\mathrm{I}(x)) - J(x)\varphi_\mathrm{I}(x))\Big)\Big]\hat{\rho}\Big\}, \tag{2.34}$$

where the in-field $\varphi_\mathrm{I}(x)$ satisfies the free equation of motion. The interaction term can be then taken from behind the trace operator to obtain

$$Z[J(x)] = \exp\Big[-\mathrm{i}\int_p V\Big(\frac{1}{\mathrm{i}}\frac{\delta}{\delta J(x)}\Big)\Big]\,\mathrm{Tr}\Big\{T_p\Big[\exp\Big(\mathrm{i}\int_p J(x)\varphi_\mathrm{I}(x)\Big)\Big]\bar{\rho}\Big\}. \tag{2.35}$$

It is easy to show by generalizing the Wick theorem that [40]

$$T_p\Big[\exp\Big(\mathrm{i}\int_p J(x)\varphi_\mathrm{I}(x)\Big)\Big] = Z_0[J(x)]\!:\exp\Big[\mathrm{i}\int_p J(x)\varphi_\mathrm{I}(x)\Big]\!:, \tag{2.36}$$

where : : means normal product and $Z_0[J(x)]$ is the generating functional for the free field

$$Z_0[J(x)] = \exp\Big\{-\frac{\mathrm{i}}{2}\int_p\int J(x)G_{0p}(x-y)J(y)\Big\}, \tag{2.37}$$

G_{0p} being the free propagator given by (2.26) with $f(p)=0$. Substituting (2.36) into (2.35) we obtain

$$Z[J(x)] = \exp\Big[-\mathrm{i}\int_p V\Big(\frac{1}{\mathrm{i}}\frac{\delta}{\delta J(x)}\Big)\Big]Z_0[J(x)]N[J(x)], \tag{2.38}$$

with

$$N[J(x)] = \mathrm{Tr}\Big[: \exp\Big(\mathrm{i}\int J(x)\varphi_{\mathrm{I}}(x)\Big) : \hat{\rho}\Big] \tag{2.39}$$

as the correlation functional for the initial state. $N[J(x)]$ can be expanded into a series of successive cumulants

$$N[J(x)] = \exp(\mathrm{i}\, W_p^N[J(x)])\,, \tag{2.40}$$

$$W_p^N[J(x)] = \sum_{n=1}^{\infty} \frac{1}{n!} \int_p \cdots \int_p W_p^n(1\cdots n) J(1)\cdots J(n)\,, \tag{2.41}$$

where

$$W_p^n(1\cdots n) = \mathrm{i}^{n-1}\,\mathrm{Tr}[:\varphi_{\mathrm{I}}(1)\cdots\varphi_{\mathrm{I}}(n):\hat{\rho}]_{\mathrm{c}}\,. \tag{2.42}$$

It is worthwhile to note that correlation functions contribute to the propagator only on the mass shell because φ_{I} satisfies the homogeneous equation

$$(p_0^2 - \omega^2(p))\varphi_{\mathrm{I}}(p) = 0\,, \tag{2.43}$$

where $\omega(p)$ is the boson energy (2.28). Also, the definition (2.39) is independent of the time branch, i.e., each CTPGF component will get the same additional term as we have seen in the last section on the example of free propagator.

Therefore, the perturbation expansion in the CTPGF approach has a structure identical to that of the quantum field theory except that the time integration is carried out over the closed path consisting of positive and negative branches. Rewritten in single time representation, each n-point function (also the corresponding Feynman diagram) is decomposed into 2^n functions (diagrams). The presence of initial correlations $W_p^n(1\cdots n)$ which vanish for the vacuum state, constitutes another difference from the ordinary field theory. In principle, all orders of correlations can be taken into account, but in most cases we will limit ourselves to the second cumulant.

It can be shown quite generally that the counter terms of the quantum field theory at zero temperature are enough to remove all ultraviolet divergences for the CTPGFs under reasonable assumption concerning the initial correlations [41]. We will not elaborate further on this point here, but it should be mentioned that near the phase transition point the infrared singularities have to be separated first so that the ultraviolet renormalization for CTPGFs in this case is different from that of the ordinary field theory (see section 9.2).

2.2.3. Connected and vertex generating functionals

The generating functional for the connected CTPGF is defined as

$$W[J(x)] = -\mathrm{i}\ln Z[J(x)]\,, \tag{2.44}$$

$$G_p^c(1\cdots n)=(-1)^{n-1}\left.\frac{\delta^n W[J(x)]}{\delta J(1)\cdots\delta J(n)}\right|_{J=0}$$
$$=(-\mathrm{i})^{n-1}\langle T_p(\varphi(1)\cdots\varphi(n))\rangle_c \tag{2.45}$$

where $\langle\ \rangle_c$ stands for $\mathrm{Tr}(\cdots\hat{\rho})$ with the connected parts taken only. The $n=1$ case corresponds to the expectation value

$$\varphi_c(x)=\delta W[J(x)]/\delta J(x)=\langle\varphi(x)\rangle_J \tag{2.46}$$

of the field operator in the presence of the external source. Therefore, $\varphi_c(x)$ is also a functional of $J(x)$.

Performing the Legendre transformation upon $W[J]$, we obtain the vertex functional

$$\Gamma[\varphi_c(x)]=W[J(x)]-\int_p J(x)\varphi_c(x)\,, \tag{2.47}$$

which depends on $\varphi_c(x)$ explicitly as well as implicitly via $J(x)$ by eq. (2.46). It follows then from (2.46) and (2.47) that

$$\delta\Gamma[\varphi_c(x)]/\delta\varphi_c(x)=-J(x)\,. \tag{2.48}$$

This is the basic equation of the CTPGF formalism from which we will derive a number of important consequences.

The general n-point vertex function, or one particle irreducible (1PI) function, is defined as

$$\Gamma_p(1\cdots n)=\delta^n\Gamma[\varphi_c(x)]/\delta\varphi_c(1)\cdots\delta\varphi_c(n)\,. \tag{2.49}$$

2.2.4. Dyson equation

As an immediate consequence of the definition for the generating functionals $W[J(x)]$ and $\Gamma[\varphi_c(x)]$ we derive here the Dyson equation.

Taking functional derivative of (2.48) with respect to $J(y)$ and using (2.45) and (2.46)

$$\frac{\delta}{\delta J(y)}\left(\frac{\delta\Gamma[\varphi_c(x)]}{\delta\varphi_c(x)}\right)=\int_p\frac{\delta\varphi_c(z)}{\delta J(y)}\frac{\delta^2\Gamma[\varphi_c(x)]}{\delta\varphi_c(z)\delta\varphi_c(x)}$$
$$=\int_p\frac{\delta^2 W[J]}{\delta J(y)\delta J(z)}\frac{\delta^2\Gamma[\varphi_c]}{\delta\varphi_c(z)\delta\varphi_c(x)}=-\frac{\delta J(x)}{\delta J(y)},$$

we obtain the Dyson equation

$$\int_p G_p(y,z)\Gamma_p(z,x)=\delta_p(x-y)\,. \tag{2.50a}$$

Similarly, by varying (2.46) with respect to $\varphi_c(y)$ we find

$$\int_p \Gamma_p(x, z) G_p(z, y) = \delta_p(x - y). \tag{2.50b}$$

Here

$$G_p(x, y) \equiv -\delta^2 W[J(x)]/\delta J(x) \delta J(y) \tag{2.51}$$

is the two-point connected Green function (the subscript c is suppressed), whereas

$$\Gamma_p(x, y) \equiv \delta^2 \Gamma[\varphi_c]/\delta\varphi_c(x)\delta\varphi_c(y) \tag{2.52}$$

is the two-point vertex function containing only 1PI part. The δ-function on the closed path δ_p is defined as

$$\int_p \delta_p(x - y) f(y) = f(x). \tag{2.53}$$

In the single time representation

$$\int_p \mathrm{d}t = \int_{-\infty}^{\infty} \mathrm{d}t_+ - \int_{-\infty}^{\infty} \mathrm{d}t_-, \tag{2.54}$$

where the minus sign in the second term comes from the definition of the closed time-path. The negative branch goes from $+\infty$ to $-\infty$. To satisfy eq. (2.53), it should be that

$$\delta_p(x - y) = \begin{cases} \delta(x - y), & \text{if both } x, y \text{ on positive branch,} \\ -\delta(x - y), & \text{if both } x, y \text{ on negative branch,} \\ 0, & \text{otherwise.} \end{cases} \tag{2.55}$$

In matrix notation δ_p can be written as

$$\hat{\delta}(x - y) = \delta(x - y)\sigma_3. \tag{2.56}$$

The Dyson equation (2.50) in single time representation is thus

$$\int \hat{\Gamma}(x, z)\sigma_3 \hat{G}(z, y) = \sigma_3 \delta(x - y), \qquad \int \hat{G}(x, z)\sigma_3 \hat{\Gamma}(z, y) = \sigma_3 \delta(x - y). \tag{2.57}$$

The transformation properties for the two-point connected Green function are the same as those discussed in section 2.1.2, except that a disconnected part proportional to $-i\varphi_c(x)\varphi_c(y)$ should be subtracted from all $\hat{G}$ functions, whereas a term $-2i\varphi_c(x)\varphi_c(y)$ must be subtracted only from G_c with G_r and G_a remaining the same.

Multiplying (2.57) by matrix Q from left and Q^{-1} (see (2.15)) from right we obtain

$$\int \tilde{\Gamma}(x,z)\sigma_1\tilde{G}(z,y) = \sigma_1\delta(x-y), \qquad \int \tilde{G}(x,z)\sigma_1\tilde{\Gamma}(z,y) = \sigma_1\delta(x-y), \tag{2.58}$$

where

$$\tilde{\Gamma} = Q\hat{\Gamma}Q^{-1} = Q\begin{pmatrix}\Gamma_F & \Gamma_+\\ \Gamma_- & \Gamma_{\tilde{F}}\end{pmatrix}Q^{-1}. \tag{2.59}$$

We note in passing that the Pauli matrix σ_3 always accompanies the CTPGF in single time representation $\hat{G}$, $\hat{\Gamma}$, whereas σ_1 appears together with the CTPGF in physical representation $\tilde{G}$, $\tilde{\Gamma}$. As seen from eqs. (2.57) and (2.58) $\sigma_3\hat{\Gamma}\sigma_3$ is the inverse of $\hat{G}$, while $\sigma_1\tilde{\Gamma}\sigma_1$ is that for $\tilde{G}$.

It is more important to point out that all characteristic features of Green's functions G discussed in section 2.1 are transmitted to vertex functions Γ via the Dyson equations (2.57) and (2.58). In particular, we have

$$\Gamma_{11} = 0, \qquad \Gamma_F + \Gamma_{\tilde{F}} = \Gamma_+ + \Gamma_-, \tag{2.60}$$

so that

$$\tilde{\Gamma} = \begin{pmatrix}0 & \Gamma_a\\ \Gamma_r & \Gamma_c\end{pmatrix}, \tag{2.61}$$

with

$$\Gamma_r = \Gamma_F - \Gamma_+ = \Gamma_- - \Gamma_{\tilde{F}}, \qquad \Gamma_a = \Gamma_F - \Gamma_- = \Gamma_+ - \Gamma_{\tilde{F}}, \quad \Gamma_c = \Gamma_F + \Gamma_{\tilde{F}} = \Gamma_+ + \Gamma_- . \tag{2.62}$$

The inverse transformation from $\tilde{\Gamma}$ to $\hat{\Gamma}$ is given by

$$\hat{\Gamma} = \tfrac{1}{2}\Gamma_r\xi\eta^\dagger + \tfrac{1}{2}\Gamma_a\eta\xi^\dagger + \tfrac{1}{2}\Gamma_c\xi\xi^\dagger, \tag{2.63}$$

i.e., exactly the same way as Green's function (2.12).

Further discussion on the Dyson equation will be postponed to section 3.1. Meanwhile, we would like to emphasize that the "transmissibility" of the CTPGF characteristics is an evidence of the logical consistency of the formalism itself. More examples along with some useful computation rules were given before [40, 43, 44].

One more remark concerning the generating functional technique itself. Up to now we have considered only CTPGFs for the constituent field $\varphi(x)$, but what has been said for it can be repeated almost word for word for any composite operator $Q[\varphi(x)]$. In the forthcoming discussion we will use the corresponding formulas without repeating their definitions.

2.3. Single time and physical representations

In the CTPGF approach we have to deal with three representations which are equivalent to each other, namely, the closed time-path form G_p used for compact writing of formulas, the single time form $\hat{G}$ and the physical representation $\tilde{G}$. In section 2.1 we have already discussed these representations and their mutual transformations on the example of two-point functions. In this section we will use the generating functional to consider general n-point functions. The underlying connection here is that different representations of CTPGF are generated by the same generating functional expressed in different functional arguments. The explicit expressions for n-point functions in physical representation will be obtained along with the transformations from $\hat{G}$ to $\tilde{G}$ and vice versa. As we will see later in section 2.4, these formulas are the starting point for discussing the important normalization and causality relations.

2.3.1. Preliminaries

To start with, we need to specify some more notations. The multi-point step function

$$\theta(1, 2, \ldots, n) = \begin{cases} 1, & \text{if} \quad t_1 > t_2 \cdots > t_n , \\ 0, & \text{otherwise,} \end{cases} \tag{2.64}$$

is a product of two-point step functions

$$\theta(1, 2, \ldots, n) = \theta(1, 2)\theta(2, 3) \cdots \theta(n-1, n) . \tag{2.65}$$

It can be used to define the time ordered product

$$T(A_1(1) \cdots A_n(n)) = \sum_{p_n} \theta(\bar{1}, \ldots, \bar{n}) A_{\bar{1}}(\bar{1}) \cdots A_{\bar{n}}(\bar{n}) , \tag{2.66a}$$

or the anti-time-ordered product

$$\tilde{T}(A_1(1) \cdots A_n(n)) = \sum_{p_n} \theta(\bar{1}, \ldots, \bar{n}) A_{\bar{n}}(\bar{n}) \cdots A_{\bar{1}}(\bar{1}) . \tag{2.66b}$$

The summation here is carried out over all permutations of n numbers p_n

$$\begin{pmatrix} 1 & 2 \cdots n \\ \bar{1} & \bar{2} \cdots \bar{n} \end{pmatrix} .$$

These step functions satisfy the normalization condition

$$\sum_{p_n} \theta(\bar{1}, \ldots, \bar{n}) = 1 , \tag{2.67}$$

and the summation formula

$$\theta(1,2,\ldots,m)=\sum_{p_n(1\cdots m)}\theta(\bar{1}\cdots\bar{n}),\tag{2.68}$$

where $p_n(1\cdots m)$ means permutations of n numbers with 1 preceding 2, 2 preceding 3, etc., but the order of the rest is arbitrary. In fact, (2.67) is the special case $m=0$ of (2.68).

The external source term in the generating functional (2.32) can be presented as

$$I_s=\int_p J(x)\varphi(x)=\int_{-\infty}^{\infty}dt\,d^dx\,(J_+(x)\varphi_+(x)-J_-(x)\varphi_-(x))\equiv\int\hat{J}^\dagger\sigma_3\hat{\varphi}\,,\tag{2.69}$$

where

$$\hat{\varphi}=\begin{pmatrix}\varphi_+(x)\\ \varphi_-(x)\end{pmatrix},\quad \hat{J}=\begin{pmatrix}J_+(x)\\ J_-(x)\end{pmatrix},\tag{2.70}$$

and also as

$$I_s=\int(J_\Delta\varphi_c+J_c\varphi_\Delta)\,,\tag{2.71}$$

with

$$\begin{aligned}&J_\Delta\equiv\eta^\dagger\hat{J}=J_+-J_-,\qquad J_c\equiv\tfrac{1}{2}\xi^\dagger\hat{J}=\tfrac{1}{2}(J_++J_-),\\ &\varphi_\Delta\equiv\eta^\dagger\hat{\varphi}=\varphi_+-\varphi_-,\qquad \varphi_c\equiv\tfrac{1}{2}\xi^\dagger\hat{\varphi}=\tfrac{1}{2}(\varphi_++\varphi_-)\,.\end{aligned}\tag{2.72}$$

Also, we can express $\hat{J}$, $\hat{\varphi}$ in terms of J_c, J_Δ, φ_c and φ_Δ as

$$\hat{J}=J_c\xi+\tfrac{1}{2}J_\Delta\eta,\qquad \hat{\varphi}=\varphi_c\xi+\tfrac{1}{2}\varphi_\Delta\eta\,.\tag{2.73}$$

The functional derivatives are related with each other by the following equations:

$$\frac{\delta}{\delta J_\alpha(x)}=\tfrac{1}{2}\xi_\alpha\frac{\delta}{\delta J_c(x)}+\eta_\alpha\frac{\delta}{\delta J_\Delta(x)},\tag{2.74a}$$

$$\frac{\delta}{\delta J_c(x)}=\xi_\alpha\frac{\delta}{\delta J_\alpha(x)}=\eta_\alpha\frac{\delta}{\delta J(x_\alpha)},\tag{2.74b}$$

$$\frac{\delta}{\delta J_\Delta(x)}=\tfrac{1}{2}\eta_\alpha\frac{\delta}{\delta J_\alpha(x)}=\tfrac{1}{2}\xi_\alpha\frac{\delta}{\delta J(x_\alpha)},\tag{2.74c}$$

with $\alpha=\pm$ and summation over repeated indices. Here we have introduced a symbolic notation

$$\frac{\delta}{\delta J(x_\alpha)}\equiv\sigma_3^{\alpha\beta}\frac{\delta}{\delta J_\beta(x)},\tag{2.75}$$

which is useful for a compact writing of the definition for Green's function as seen from (2.76).

A remark concerning the notation $\varphi_c(x)$ is in order. Previously (see (2.46)) we have defined $\varphi_c(x)$ as the expectation value of $\varphi(x)$ on the closed time-path. Hence it is a two-component vector ($\varphi_{c+}(x)$, $\varphi_{c-}(x)$), but we do not make the subscripts +, − explicit. Here (see (2.72)) $\varphi_c(x)$ is the linear combination of operators $\varphi_+(x)$ and $\varphi_-(x)$, still, in accord with our convention, no caret is put above it. Later on, the same $\varphi_c(x)$ will denote its expectation value. Hopefully, no confusion will occur, since the meaning of $\varphi_c(x)$ is clear from the context and, moreover, $\varphi_{c+}(x) = \varphi_{c-}(x) = \varphi_c(x)$ for $J_+(x) = J_-(x)$ as seen from (2.105). The same remark is effective with respect to other functions like $Q_c(x)$, $\psi_c(x)$, $\psi_c^\dagger(x)$ and so on, appearing in the future discussion.

2.3.2. "Physical" representation of the generating functional

As we said in the introductory remarks to this section, the same generating functional will generate CTPGF in different representation provided the external source term is expressed in the corresponding functional arguments. In particular, the generating functional in the form (2.32) will give rise to CTPGF in the closed time-path representation. If, however, the source is written in single time form as given by (2.69), the same generating functional (2.32) can be then expanded as

$$\begin{aligned} Z[J_+(x), J_-(x)] &= \sum_{n=0}^{\infty} \int \frac{1}{n!} \frac{\delta^n Z[J_+, J_-]}{\delta J_\alpha(1)\cdots\delta J_\rho(n)}\bigg|_{J=0} J_\alpha(1)\cdots J_\rho(n) \\ &= 1 + \mathrm{i}\sum_{n=1}^{\infty} \frac{(-1)^{n-1}}{n!} G_{\alpha\cdots\rho}(1\cdots n)(\sigma_3\hat{J})_\alpha\cdots(\sigma_3\hat{J})_\rho\,, \end{aligned} \tag{2.76}$$

with

$$G_{\alpha\cdots\rho}(1\cdots n) \equiv \mathrm{i}(-1)^n \frac{\delta^n Z[J_+, J_-]}{\delta J(\alpha)\cdots\delta J(\rho)}, \tag{2.77}$$

where

$$(\sigma_3 J_3 \hat{J})_\alpha \equiv \sigma_3^{\alpha\beta} J_\beta\,,$$

and both the space–time coordinates and the dummy indices $\alpha, \ldots, \rho$ should be summed over.

Moreover, if the expression (2.71) for the source is used, the same generating functional (2.32) can be expanded as

$$Z[J_\Delta(x), J_c(x)] = \sum_{m,l=0}^{\infty} \frac{1}{m!\,l!} \frac{\delta^{m+l} Z[J_\Delta, J_c]}{\delta J_\Delta(1)\cdots\delta J_\Delta(m)\delta J_c(\bar{1})\cdots\delta J_c(\bar{l})}\bigg|_{J_c=J_\Delta=0} J_\Delta(1)\cdots J_\Delta(m) J_c(\bar{1})\cdots J_c(\bar{l})\,, \tag{2.78}$$

where

$$\begin{aligned} B_n &\equiv \frac{(-\mathrm{i})^n \delta^n Z[J_\Delta, J_c]}{\delta J_\Delta(1)\cdots\delta J_\Delta(m)\delta J_c(m+1)\cdots\delta J_c(n)} \\ &= \langle T_p(\varphi_c(1)\cdots\varphi_c(m)\varphi_\Delta(m+1)\cdots\varphi_\Delta(n))\rangle \equiv \mathrm{i}^{n-1}2^{-m+1}\tilde{G}_{\underbrace{2\cdots2}_{m}\underbrace{1\cdots1}_{n-m}}(1\cdots n)\,. \end{aligned} \tag{2.79}$$

Now we find another expression of the CTPGF in physical representation, namely, in terms of expectation values of nested commutators and anti-commutators.

Using the normalization condition for the step function (2.67), eq. (2.79) can be rewritten as

$$B_n = \sum_{p_n} \theta(\bar{1}\cdots\bar{n})\langle T_p(\varphi_c(1)\cdots\varphi_c(m)\varphi_\Delta(m+1)\cdots\varphi_\Delta(n))\rangle$$

$$= \sum_{p_n} \theta(\bar{1}\cdots\bar{n})2^{-m}\xi^{\alpha_1}\cdots\xi^{\alpha_m}\eta^{\alpha_{m+1}}\cdots\eta^{\alpha_n}\langle T_p(\varphi_{\alpha_1}(1)\cdots\varphi_{\alpha_m}(m)\varphi_{\alpha_{m+1}}(m+1)\cdots\varphi_{\alpha_n}(n))\rangle\,. \quad (2.80)$$

For convenience we introduce a unified notation for ξ and η

$$\zeta^{\alpha_i} = \begin{cases} \xi^{\alpha_i}, & \text{if } 1 \le i \le m\,, \\ \eta^{\alpha_i}, & \text{if } m+1 \le i \le n\,. \end{cases} \quad (2.81)$$

Since the order of operators under T_p can be changed arbitrarily, (2.80) can be also presented as

$$B_n = \sum_{p_n} \theta(\bar{1}\cdots\bar{n})2^{-m}\zeta^{\bar{1}}\cdots\zeta^{\bar{n}}\langle T_p(\varphi_{\bar{1}}(\bar{1})\cdots\varphi_{\bar{n}}(\bar{n}))\rangle\,, \quad (2.82)$$

where the subscript $\bar{i} \equiv \alpha_{\bar{i}}$. Now let us get rid of T_p in (2.82) step by step for each term of permutation p_n. As far as the θ-function ensures $\bar{n}$ to be the earliest moment on the positive branch and the latest one on the negative branch, we have

$$\zeta^{\bar{n}}\langle T_p(\varphi_{\bar{1}}(\bar{1})\cdots\varphi_{\bar{n}}(\bar{n}))\rangle = \xi^{\bar{n}}\langle T_p(\varphi_{\bar{1}}(\bar{1})\cdots\varphi_{\bar{n}}(\bar{n}))\rangle$$
$$= \langle T_p(\varphi_{\bar{1}}(\bar{1})\cdots\varphi_{\overline{n-1}}(\overline{n-1}))\varphi(\bar{n})\rangle + \langle\varphi(\bar{n})T_p(\varphi_{\bar{1}}(\bar{1})\cdots\varphi_{\overline{n-1}}(\overline{n-1}))\rangle$$
$$= \langle\{T_p(\varphi_{\bar{1}}(\bar{1})\cdots\varphi_{\overline{n-1}}(\overline{n-1})),\varphi(\bar{n})\}\rangle\,,$$

if

$$1 \le \bar{n} \le m\,,$$

or

$$\zeta^{\bar{n}}\langle T_p(\varphi_{\bar{1}}(\bar{1})\cdots\varphi_{\bar{n}}(\bar{n}))\rangle = \eta^{\bar{n}}\langle T_p(\varphi_{\bar{1}}(\bar{1})\cdots\varphi_{\bar{n}}(\bar{n}))\rangle$$
$$= \langle[T_p(\varphi_{\bar{1}}(\bar{1})\cdots\varphi_{\overline{n-1}}(\overline{n-1})),\varphi_{\bar{n}}(\bar{n})]\rangle\,,$$

if

$$m+1 \le \bar{n} \le n\,.$$

Such processes may be continued like "a cicada sloughing its skin" in accord with the Chinese saying, up to the last step to get zero if $m+1 \le \bar{1} \le n$, or the expectation value of nested commutators (and/or anti-commutators).

If we introduce a short writing

$$(\ ,\varphi(\bar{i}))=\begin{cases}[\ ,\varphi(\bar{i})], & \text{when } m+1\le \bar{i}\le n\\ \{\ ,\varphi(\bar{i})\} & \text{when } 1\le \bar{i}\le m,\end{cases} \tag{2.83}$$

we find finally

$$\tilde{G}_{\underbrace{2\cdots 2}_{m}\underbrace{1\cdots 1}_{n-m}}(1\cdots n)=(-\mathrm{i})^{n-1}\sum{}'\,\theta(\bar{1}\cdots\bar{n})\langle(\cdots(\varphi(\bar{1}),\varphi(\bar{2}))\cdots,\varphi(\bar{n}))\rangle\,, \tag{2.84}$$

where Σ' means that permutations $m+1\le\bar{1}\le n$ should be excluded from the summation.

As a special case we find for $n=2$ that

$$G_{11}(12)=0,\qquad G_{21}(12)=-\mathrm{i}\theta(1,2)\langle[\varphi(1),\varphi(2)]\rangle,\qquad G_{22}(12)=-\mathrm{i}\langle\{\varphi(1),\varphi(2)\}\rangle\,, \tag{2.85}$$

thus recovering G_{21}, G_{22} as retarded and correlation functions. Using the symmetry

$$G_{21}(12)=G_{12}(21)$$

we obtain the advanced Green function as

$$G_{12}(12)=-\mathrm{i}\theta(2,1)\langle[\varphi(2),\varphi(1)]\rangle\,.$$

Furthermore, for $n=3$ case we have

$$G_{111}=0,\qquad G_{211}(123)=(-\mathrm{i})^2\sum_{p\in\binom{23}{ij}}\theta(1ij)\langle[[1,i],j]\rangle\,,$$

$$G_{221}(123)=(-\mathrm{i})^2\sum_{p\in\binom{12}{ij}}(\theta(ij3)\langle[\{i,j\},3]\rangle+\theta(i3j)\langle\{[i,3],j\}\rangle)\,,$$

$$G_{222}(123)=(-\mathrm{i})^2\sum_{p\in p_3}\theta(ijk)\langle\{\{i,j\},k\}\rangle\,. \tag{2.86}$$

Here for short we write i, j instead of $\varphi(i)$, $\varphi(j)$.

Therefore, this way we can exhaust all possible n-point functions. Without resorting to the CTPGF formalism this would be a cumbersome task.

The functional expansion (2.78) and the explicit expression for physical CTPGF (2.84) are very important equations from which a number of far-going implications will be extracted in the next section. In the meantime we note only that the functional derivative $\delta/\delta J_c$ will generate φ_Δ which in turn yields a commutator in the Green function, whereas the functional derivative $\delta/\delta J_\Delta$ gives rise to φ_c leading to an anti-commutator.

Moreover, in the summation (2.82) none of the time variables $m+1\le\bar{i}\le n$ can take a value larger than all of the time arguments $1\le\bar{j}\le m$, because in that case $\bar{1}$ should be one of $(m+1,\ldots,n)$ excluded from the summation. This fact will give rise to important causality relations.

2.3.3. Transformation formula

Using the definitions of CTPGFs in single time and physical representations as expansion coefficients of the generating functional (2.77) and (2.79), respectively, we can readily find the transformation from one to another.

In fact, using (2.79) and (2.74) we have

$$G_{2\cdots 21\cdots 1}(1\cdots n) = -\mathrm{i}(-1)^{n-1}2^{m-1}\left.\frac{\delta^n Z[J_\Delta, J_c]}{\delta J_\Delta(1)\cdots\delta J_\Delta(m)\delta J_c(m+1)\cdots\delta J_c(n)}\right|_{J=0}$$
$$= -\mathrm{i}(-1)^{n-1}2^{-1}\xi^{\alpha_1}\cdots\xi^{\alpha_m}\eta^{\alpha_{m+1}}\cdots\eta^{\alpha_n}\left.\frac{\delta^n Z[J_+, J_-]}{\delta J(\alpha_1)\cdots\delta J(\alpha_m)\cdots\delta J(\alpha_n)}\right|_{J=0}$$
$$= 2^{-1}\xi^{\alpha_1}\cdots\xi^{\alpha_m}\eta^{\alpha_{m+1}}\cdots\eta^{\alpha_n}G_{\alpha_1\cdots\alpha_n}(1\cdots n)\,. \tag{2.87}$$

In a similar way we find the inverse transformation as given by

$$G_{\alpha_1\cdots\alpha_n}(1\cdots n) = 2^{1-n}\sum_{i_1\cdots i_n}\zeta^{i_1}_{\alpha_1}\cdots\zeta^{i_n}_{\alpha_n}G_{i_1\cdots i_n}(1\cdots n)\,, \tag{2.88}$$

where

$$i_1,\ldots,i_n = 1,2\,,\qquad \zeta^1_\alpha = \eta_\alpha,\qquad \zeta^2_\alpha = \xi_\alpha\,. \tag{2.89}$$

Using the orthogonal matrix Q defined by (2.15) these formulas can be rewritten as

$$G_{i_1\cdots i_n}(1\cdots n) = 2^{n/2-1}Q_{i_1\alpha_1}\cdots Q_{i_n\alpha_n}G_{\alpha_1\cdots\alpha_n}(1\cdots n)\,, \tag{2.90a}$$
$$G_{\alpha_1\cdots\alpha_n}(1\cdots n) = 2^{1-n/2}Q^T_{\alpha_1 i_1}\cdots Q^T_{\alpha_n i_n}G_{i_1\cdots i_n}(1\cdots n)\,. \tag{2.90b}$$

For the case $n = 2$ eqs. (2.12) and (2.16) obtained in section 2.1 are recovered. For $n = 3$ we have

$$G_{+++}(123) = (-\mathrm{i})^2\langle T(123)\rangle,\qquad G_{++-}(123) = (-\mathrm{i})^2\langle 3T(12)\rangle\,,$$
$$G_{+--}(123) = (-\mathrm{i})^2\langle \tilde{T}(23)1\rangle,\qquad G_{---}(123) = (-\mathrm{i})^2\langle \tilde{T}(123)\rangle\,. \tag{2.91}$$

The other functions, i.e., G_{+-+}, G_{-++}, G_{-+-}, G_{--+} can be obtained by symmetry.

For illustration we also write down some of the transformation formulas such as

$$G_{111} = G^{(+)} - G^{(-)} = 0,\qquad G_{222} = G^{(+)} + G^{(-)}\,, \tag{2.92}$$

with

$$G^{(+)} = G_{+++} + G_{+--} + G_{-+-} + G_{--+}\,,\qquad G^{(-)} = G_{---} + G_{-++} + G_{+-+} + G_{++-}\,,$$
$$G_{211} = \tfrac{1}{2}(G_{+\cdot\cdot} + G_{-\cdot\cdot})\,, \tag{2.93}$$

with

$$G_{+\cdot\cdot} = \sum_{\alpha,\beta=\pm}\alpha\beta G_{+\alpha\beta}\,,\qquad G_{-\cdot\cdot} = \sum_{\alpha,\beta=\pm}\alpha\beta G_{-\alpha\beta}\,.$$

2.3.4. "Physical" representation for $W[J]$ and $\Gamma[\varphi_c]$

We have discussed above different expansions for $Z[J]$ along with some of their consequences. The same thing can be done toward $W[J]$ and $\Gamma[\varphi_c]$. For example, in the physical representation we have

$$W[J_\Delta(x), J_c(x)] \equiv -\mathrm{i} \ln Z[J_\Delta(x), J_c(x)]\,, \tag{2.94}$$

$$\Gamma[\varphi_\Delta(x), \varphi_c(x)] \equiv W[J_\Delta(x), J_c(x)] - \int (J_\Delta \varphi_c + J_c \varphi_\Delta)\,, \tag{2.95}$$

where

$$\varphi_c(x) \equiv \frac{\delta W[J_\Delta, J_c]}{\delta J_\Delta(x)}\,, \qquad \varphi_\Delta(x) \equiv \frac{\delta W[J_\Delta, J_c]}{\delta J_c(x)}\,. \tag{2.96}$$

It follows then from (2.95) and (2.96) that

$$\frac{\delta \Gamma[\varphi_\Delta(x), \varphi_c(x)]}{\delta \varphi_c(x)} = -J_\Delta(x)\,,$$
$$\frac{\delta \Gamma[\varphi_\Delta(x), \varphi_c(x)]}{\delta \varphi_\Delta(x)} = -J_c(x)\,. \tag{2.97}$$

It is obvious that $W[J_\Delta, J_c]$ and $\Gamma[\varphi_\Delta, \varphi_c]$ defined by (2.94) and (2.95) are identical to $W_p[J]$ and $\Gamma_p[\varphi_c]$ as given by (2.44) and (2.47) respectively.

We should note, however, that the explicit form of the connected Green function is different from that of the "total" (connected + disconnected) Green function obtained as expansion coefficient of $Z[J]$. For example,

$$G^c_{222}(123) = (-\mathrm{i})^2 \sum_{P_3} \theta(ijk)[\langle\{\{i,j\},k\}\rangle - \langle\{i,j\}\rangle\langle k\rangle + 2\{\{\langle i\rangle, \langle j\rangle\}, \langle k\rangle\}]\,. \tag{2.98}$$

The only exceptions are the "all retarded" Green functions like G_{21}, G_{211}, G_{2111} etc., for which

$$G^c_{21\cdots 1}(1\cdots n) = G_{21\cdots 1}(1\cdots n)\,.$$

Therefore, the transformation from $\hat{G}$ to $\tilde{G}$ for the connected Green function and vice versa should also be correspondingly modified.

We note in passing that the "all retarded" functions are nothing but the r-functions used to construct the LSZ field theory [59]. Unlike the zero temperature case, these functions alone are not enough to construct the CTPGF formalism, but they still play a very important role here.

2.4. Normalization and causality

As we emphasized in the Introduction, the normalization and causality relations are essential for applications. In fact, they are already implied by the expansion of the generating functional discussed in the last section, but we would like to make them explicit here for future reference. We start from the normalization (section 2.4.1), then indicate the consequences of the causality (section 2.4.2) and wind up with few comments on the two aspects of the Liouville problem – dynamical evolution and statistical correlation – naturally embodied in the CTPGF formalism (section 2.4.3).

2.4.1. Normalization

If $J_\Delta(x)$ is set equal to zero, i.e., $J_+(x) = J_-(x)$ in the expansion (2.78), we find

$$\left.\frac{\delta^n Z[J_\Delta, J_c]}{\delta J_c(1)\cdots\delta J_c(n)}\right|_{J_\Delta=0} = 0, \quad \text{for} \quad n \geq 1, \tag{2.99}$$

because in accord with (2.84)

$$G_{11\cdots1}(1\cdots n) = 0\,. \tag{2.100}$$

Using the normalization condition of the density matrix

$$\mathrm{Tr}(\hat{\rho}) = 1\,, \tag{2.101}$$

we find

$$Z[J_\Delta(x), J_c(x)]|_{J_\Delta=0} = 1\,. \tag{2.102}$$

By definition (2.94) we have

$$W[J_\Delta(x), J_c(x)]|_{J_\Delta=0} = 0\,, \tag{2.103}$$

or, equivalently,

$$\left.\frac{\delta^n W[J_\Delta, J_c]}{\delta J_c(1)\cdots\delta J_c(n)}\right|_{J_\Delta=0} = 0 \quad \text{for any } n \geq 0\,. \tag{2.104}$$

In particular,

$$\varphi_\Delta(x)|_{J_\Delta(x)=0} = \left.\frac{\delta W[J_\Delta, J_c]}{\delta J_c(x)}\right|_{J_\Delta=0} = 0\,, \tag{2.105}$$

which leads to

$$\Gamma[\varphi_\Delta(x), \varphi_c(x)]|_{\varphi_\Delta=0} = 0\,, \tag{2.106}$$

and

$$\left.\frac{\delta^n \Gamma[\varphi_\Delta, \varphi_c]}{\delta\varphi_c(1)\cdots\delta\varphi_c(n)}\right|_{\varphi_\Delta=0} = 0\,, \tag{2.107}$$

or, equivalently

$$\Gamma_{11\cdots 1}(1\cdots n)=0\,. \tag{2.108}$$

We see thus the algebraic relations obtained before, such as (2.6), (2.16), (2.60) and (2.92) are special cases of these general conditions following from the normalization.

We would like to emphasize here that the normalization condition for the CTPGF generating functional is different from that of the quantum field theory or the standard many-body formalism. In this case we require only the equality of the external source on the positive and negative branches $J_+(x)=J_-(x)$ instead of its vanishing. We can thus incorporate the external field $J_c(x)=\frac{1}{2}(J_+(x)+J_-(x))$ into the theoretical framework in a natural way. Moreover, this fundamental property will give rise to a number of important consequences which make the CTPGF formalism advantageous in many cases as we will see later.

We note also, that eqs. (2.99), (2.104), (2.107) and (2.108) are valid even in the presence of a finite external field $J_c(x)$.

2.4.2. Causality

As mentioned in section 2.3.2, in the functional expansion (2.78) none of the time variables with $m+1\le \bar{i}\le n$ can take values greater than the time arguments $1\le \bar{j}\le m$, because this would contradict the rule established by (2.84) that terms $m+1\le \bar{1}\le n$ should be excluded from the summation. Put in another way,

$$\left.\frac{\delta^n Z[J_\Delta, J_c]}{\delta J_\Delta(1)\cdots\delta J_\Delta(m)\delta J_c(m+1)\cdots\delta J_c(n)}\right|_{J_\Delta=J_c=0}=0 \tag{2.109}$$

provided any t_i, with $m+1\le i\le n$ is greater than all t_j with $1\le j\le m$. This is one of the causality relations we consider here. It is obvious that the causality relation for the two-point function (2.17) is a special case of (2.109) for $m=1$, $n=2$, i.e.,

$$G_{21}(12)=0,\qquad \text{if } t_2>t_1\,.$$

In a sense, the algebraic relation (2.99) or (2.100) is also a special case of (2.109) for $m=0$.

Similarly, under the same condition, i.e., the time argument of any J_c, φ_c is greater than that of all J_Δ, φ_Δ, we have for the functional derivatives of $W[J]$,

$$\left.\frac{\delta^n W[J_\Delta, J_c]}{\delta J_\Delta(1)\cdots\delta J_\Delta(m)\delta J_c(m+1)\cdots\delta J_c(n)}\right|_{J_c=J_\Delta=0}=0\,, \tag{2.110}$$

and those of the vertex functional

$$\left.\frac{\delta^n \Gamma[\varphi_\Delta, \varphi_c]}{\delta\varphi_\Delta(1)\cdots\delta\varphi_\Delta(m)\delta\varphi_c(m+1)\cdots\delta\varphi_c(n)}\right|_{\varphi_\Delta=0,\,\varphi_c=\varphi}=0\,. \tag{2.111}$$

In deriving (2.111) we have made use of the relations between the vertex functions and the "amputated" connected Green functions [39, 60].

It is worthwhile to emphasize that a causality sequence is established by (2.109)–(2.111), namely, the space–time points associated with $J_c(x)$, $\varphi_c(x)$ should precede those of $J_\Delta(x)$ and $\varphi_\Delta(x)$, since the former is the cause, whereas the latter is the consequence.

We indicate here also some useful product relations as a generalization of (2.18) for a two-point function. For example, we have for three-point functions

$$G_{211}G_{112} = G_{211}G_{121} = G_{211}G_{122} = 0, \qquad G_{112}G_{221} = G_{112}G_{121} = G_{121}G_{212} = 0\,. \tag{2.112}$$

It is easy to see that the general rule is

$$G_{i_1\cdots i_{m_1}\cdots i_{m_2}\cdots i_n}(1\cdots n)G_{j_1\cdots j_{m_1}\cdots j_{m_2}\cdots j_n}(1\cdots n) = 0\,, \tag{2.113}$$

provided $i_{m_1} = \cdots = i_{m_r} = 2$ and the rest are equal to 1, whereas $j_{m_1} = \cdots = j_{m_r} = 1$ but the rest can be either 1 or 2.

2.4.3. Dynamical evolution and statistical correlation

Now we discuss the physical meaning of J_Δ, J_c, φ_Δ and φ_c. In addition to (2.105) we have

$$J_\Delta(x)|_{\varphi_\Delta=0} = 0\,, \tag{2.114}$$

from (2.97) and (2.107), so that the conditions $J_\Delta = 0$ and $\varphi_\Delta = 0$ are equivalent to each other. Also, it follows from (2.91), (2.94) and (2.102) that

$$\begin{aligned}\varphi_c(x)|_{J_\Delta(x)=0} &= \mathrm{Tr}\Big\{T_p\Big[\varphi_c(x)\exp\Big(\mathrm{i}\int J_c(x)\varphi_\Delta(x)\Big)\Big]\hat\rho\Big\}\\ &= \mathrm{Tr}\Big\{\Big(\tilde T\exp\Big(-\mathrm{i}\int_{-\infty}^{t} J_c(y)\varphi(y)\Big)\Big)\varphi(x)\Big(T\exp\Big(\mathrm{i}\int_{-\infty}^{t} J_c(y)\varphi(y)\Big)\Big)\hat\rho\Big\}\\ &= \sum_{n=0}^{\infty}\mathrm{i}^n\int_{-\infty}^{t}\mathrm{d}1\int_{-\infty}^{t_1}\mathrm{d}2\cdots\int_{-\infty}^{t_{n-1}}\mathrm{d}n\,J_c(1)\cdots J_c(n)\,\mathrm{Tr}\{[\cdots[[\varphi(x),\varphi(1)],\varphi(2)]\cdots],\varphi(n)]\hat\rho\}\,.\end{aligned} \tag{2.115}$$

We see thus $\varphi_c(x)$ under $J_\Delta = 0$ is the expectation value of the field operator, i.e., the order parameter in the presence of the external field J_c, whereas

$$\varphi(x) = \varphi_c(x)|_{J_c(x)=0} = \mathrm{Tr}(\varphi_c(x)\hat\rho) \tag{2.116}$$

is the expectation value that might cause symmetry breaking in the vanishing field. Equation (2.115) is a nonlinear expansion of the order parameter in the external field. A detailed discussion of the nonlinear response will be given in section 5.

As we already mentioned in the last section, in accord with (2.78) and (2.84) the functional derivative $\delta/\delta J_c(x)$ generates the expectation value of the commutator of the field variables describing the dynamical evolution in the quantum mechanical sense, whereas $\delta/\delta J_\Delta(x)$ generates the expectation value of anti-commutator describing the statistical correlation in the statistical mechanical sense. Although the physical observables are defined on the manifold $J_\Delta(x) = 0$ or $\varphi_\Delta(x) = 0$, these functional arguments are needed in addition to $J_c(x)$ and $\varphi_c(x)$ for a complete description of the statistical system. These two complementary aspects of the Liouville problem – dynamical evolution and statistical correlation – have been embodied in the CTPGF formalism in a natural way. It is worthwhile to note that the response and the correlation functions have found their "proper seats" in the CTPGF formalism just because in the external source term (2.71) φ and J are "twisted", i.e., φ_c is coupled with J_Δ, while φ_Δ with J_c as follows directly from the definition of the closed time-path. As we will see later in section 9, this is one of the advantages for the CTPGF formalism compared with the others.

2.5. Lehmann spectral representation

In this section we study the analytical properties of the Green functions. As in quantum field theory [39] and in the standard Green function technique [1–3], the Lehmann spectral representation is a powerful tool towards this end. We will discuss in some detail the spectral as well as the symmetry properties of Green functions for a nonrelativistic complex (Bose or Fermi) field defined by eqs. (2.3) and (2.8). The modification needed for a real boson field is also briefly mentioned.

2.5.1. Spectral expansion

Assume, the inhomogeneity of the system is caused by the nonuniformity of the state, while the evolution of the Heisenberg operators $\psi(x), \psi^\dagger(x)$ with x is given by the total energy–momentum operator p as

$$\psi(x) = \exp(\mathrm{i}p\cdot x)\psi(0)\exp(-\mathrm{i}p\cdot x), \qquad \psi^\dagger(x) = \exp(\mathrm{i}p\cdot x)\psi^\dagger(0)\exp(-\mathrm{i}p\cdot x). \tag{2.117}$$

Let $|n\rangle$ be a complete set determined by p_μ and other operators commuting with p_μ. According to (2.117) we have for $G_-(x, y)$ defined by (2.3c)

$$\mathrm{i}G_-(x, y) = \sum_{n,m,n'} \langle n|\psi(0)|m\rangle\langle m|\psi^\dagger(0)|n'\rangle\rho_{n'n}\exp[\mathrm{i}(p_n - p_m)\cdot x + \mathrm{i}(p_m - p_{n'})\cdot y]. \tag{2.118}$$

Set $Z = x - y$, $X = (x + y)/2$ and take Fourier transform with respect to Z, we obtain

$$\mathrm{i}G_-(k, X) = \sum_{n,m,n'} \langle n|\psi(0)|m\rangle\langle m|\psi^\dagger(0)|n'\rangle\rho_{n'n}\exp[\mathrm{i}(p_n - p_{n'})\cdot X](2\pi)^{d+1}\delta(k - p_m + (p_n + p_{n'})/2). \tag{2.119}$$

If

$$\rho_{n'n} \equiv \langle n'|\hat\rho|n\rangle \propto \delta(p_n - p_{n'}), \tag{2.120}$$

$G_-(k, X)$ will not depend on X and the system is homogeneous. In the presence of macroscopic inhomogeneity, $\rho_{n'n}$ is different from zero only for $p_n - p_{n'}$ small compared with k, so that the high orders of $\partial/\partial X$ can be neglected.

2.5.2. Sum rule

For nonrelativistic fields we have the following equal-time (anti-) commutation relation

$$[\psi(x), \psi^\dagger(y)]_\mp \delta(x_0 - y_0) = \delta^{d+1}(x-y), \tag{2.121}$$

which leads to

$$\mathrm{i}\int \frac{\mathrm{d}k_0}{2\pi}(G_-(k,X) - G_+(k,X)) = 1. \tag{2.122}$$

Introducing the spectral function

$$\rho(k,X) \equiv \mathrm{i}(G_-(k,X) - G_+(k,X)), \tag{2.123}$$

we rewrite (2.122) as

$$\int \frac{\mathrm{d}k_0}{2\pi}\rho(k,X) = 1. \tag{2.124}$$

For a real boson field we have

$$[\varphi(x), \partial_0\varphi(y)]\delta(x_0 - y_0) = \mathrm{i}\delta^{d+1}(x-y), \tag{2.125}$$

from which one can derive

$$\int \frac{\mathrm{d}k_0}{2\pi} k_0\rho(k,X) = 1, \qquad \partial_{X_0}\int \frac{\mathrm{d}k_0}{2\pi}\rho(k,X) = 0, \tag{2.126}$$

with $\rho(k,X)$ still defined by (2.123).

2.5.3. Lehmann representation

Presenting the retarded Green function $G_r(x,y)$ defined by (2.8a) as

$$G_r(x,y) = \theta(x_0, y_0)(G_-(x,y) - G_+(x,y)),$$

we find

$$G_r(k,X) = \int \frac{\mathrm{d}k_0'}{2\pi}\frac{\rho(k,k_0',X)}{k_0 - k_0' + \mathrm{i}\varepsilon}, \tag{2.127}$$

which is analytic in the upper half-plane of k_0. Similarly, we have for the advanced function

$$G_a(k,X) = \int \frac{\mathrm{d}k_0'}{2\pi}\frac{\rho(k,k_0',X)}{k_0 - k_0' - \mathrm{i}\varepsilon}, \tag{2.128}$$

which is analytic in the lower half-plane of k_0. Presenting $G_\pm$ as

$$G_\pm(k, X) = \int dk_0' \, G_\pm(k, k_0', X)\delta(k_0 - k_0')$$
$$= i\int \frac{dk_0'}{2\pi} G_\pm(k, k_0', X)\left(\frac{1}{k_0 - k_0' + i\varepsilon} - \frac{1}{k_0 - k_0' - i\varepsilon}\right), \tag{2.129}$$

we find the spectral form of G_F and $G_{\bar F}$,

$$G_F(k, X) = G_r(k, X) + G_+(k, X) = i\int \frac{dk_0'}{2\pi}\left(\frac{G_-(k, k_0', X)}{k_0 - k_0' + i\varepsilon} - \frac{G_+(k, k_0', X)}{k_0 - k_0' - i\varepsilon}\right), \tag{2.130}$$

$$G_{\bar F}(k, X) = G_r(k, X) + G_-(k, X)$$
$$= i\int \frac{dk_0'}{2\pi}\left(\frac{G_+(k, k_0', X)}{k_0 - k_0' + i\varepsilon} - \frac{G_-(k, k_0', X)}{k_0 - k_0' - i\varepsilon}\right). \tag{2.131}$$

Equations (2.127), (2.128), (2.130) and (2.131) are the Lehmann spectral representations we are looking for.

2.5.4. *Symmetry relations*

It is straightforward to check that for nonrelativistic complex (boson or fermion) field we have

$$G_\pm^*(x, y) = -G_\pm(y, x), \qquad G_{\bar F}^*(x, y) = -G_{\bar F}(y, x), \qquad G_r^*(x, y) = G_a(y, x), \tag{2.132}$$

or in Fourier components

$$G_\pm^*(k, X) = -G_\pm(k, X), \qquad G_{\bar F}^*(k, X) = -G_{\bar F}(k, X), \qquad G_r^*(k, X) = G_a(k, X), \tag{2.133}$$

whereas for real boson field we have additionally

$$\hat G(x, y) = \hat G^T(y, x) = -\sigma_1 \hat G^*(x, y)\sigma_1 = -\sigma_1 \hat G^\dagger(y, x)\sigma_1, \tag{2.134}$$

or in Fourier components

$$\hat G(k) = \hat G^T(-k) = -\sigma_1 \hat G^*(-k)\sigma_1 = -\sigma_1 \hat G^\dagger(k)\sigma_1, \tag{2.135}$$

where T means transposition, * means complex conjugation and † Hermitian conjugation.

2.5.5. *Two analytic functions*

It is obvious from (2.133) that $G_\pm(k, X)$ are purely imaginary on the real axis of k_0. If they vanish as $|k_0| \to \infty$, we can define two analytic functions on the complex plane of k_0, namely,

$$G_1(k, Z, X) = i\int \frac{dk_0'}{2\pi}\frac{G_-(k, k_0', X)}{Z - k_0'}, \qquad G_2(k, Z, X) = i\int \frac{dk_0'}{2\pi}\frac{G_+(k, k_0', X)}{Z - k_0'}. \tag{2.136}$$

In terms of functions G_1 and G_2 we find

$$\begin{aligned}
G_r(k, X) &= G_1(\boldsymbol{k}, k_0 + i\varepsilon, X) - G_2(\boldsymbol{k}, k_0 + i\varepsilon, X), \\
G_a(k, X) &= G_1(\boldsymbol{k}, k_0 - i\varepsilon, X) - G_2(\boldsymbol{k}, k_0 - i\varepsilon, X), \\
G_F(k, X) &= G_1(\boldsymbol{k}, k_0 + i\varepsilon, X) - G_2(\boldsymbol{k}, k_0 - i\varepsilon, X), \\
G_{\tilde{F}}(k, X) &= G_2(\boldsymbol{k}, k_0 + i\varepsilon, X) - G_1(\boldsymbol{k}, k_0 - i\varepsilon, X), \\
G_-(k, X) &= G_1(\boldsymbol{k}, k_0 + i\varepsilon, X) - G_1(\boldsymbol{k}, k_0 - i\varepsilon, X), \\
G_+(k, X) &= G_2(\boldsymbol{k}, k_0 + i\varepsilon, X) - G_2(\boldsymbol{k}, k_0 - i\varepsilon, X),
\end{aligned} \tag{2.137}$$

i.e., all these functions are superpositions of G_1 and G_2 on approaching the real axis from different sides. It follows from (2.133) and (2.136) that

$$G_{1,2}(k, Z, X)^* = G_{1,2}(k, Z^*, X). \tag{2.138}$$

We see thus to ensure the causality, the retarded Green function should be analytic on the upper half-plane of k_0. If a singularity is found on the upper half of k_0 during the process of solving G_r, it must be located on the second Riemann sheet. The appropriate analytic continuation is to take the integral along a contour in the complex plane of k_0' which circulates the singularity from above.

3. Quasiuniform systems

In this section we will discuss in some detail further properties of two-point Green functions, mainly concentrating on quasiuniform systems. The starting point is the Dyson equation formally derived from the generating functional in the last section. The quasiuniformity can be realized only near some stationary state, either thermoequilibrium or nonequilibrium under steady external conditions. We will derive the stability condition from the analytic properties of Green's functions. In section 3.1 the properties of the Dyson equation are further elaborated, especially for a uniform system. The thermoequilibrium situation is then discussed (section 3.2) mainly for the tutorial purpose. Furthermore, the Dyson equation is used to derive the transport equation (section 3.3). Finally, the multi-time-scale perturbation (section 3.4) and the derivation of the time dependent Ginzburg–Landau (TDGL) equation (section 3.5) are briefly described. The separation of micro- and macro-time scales is the common feature of the last three topics.

3.1. The Dyson equation

3.1.1. An alternative derivation

The Dyson equation and its equivalent forms (2.50), (2.57) and (2.58) have been derived from the generating functional. Here we give another derivation which will shed some light on the structure of the vertex function.

Consider an Hermitian boson field $\varphi(x)$ described by the Lagrangian density

$$\mathcal{L} = \tfrac{1}{2}\partial_\mu\varphi(x)\partial^\mu\varphi(x) - \tfrac{1}{2}m^2\varphi^2(x) - V(\varphi(x)). \tag{3.1}$$

For simplicity we assume $\varphi_c(x) = \mathrm{Tr}(\varphi(x)\hat{\rho}) = 0$ in zero external field $J(x) = 0$. The field operator satisfies the equation of motion

$$\Box_x \varphi(x) = j(x) \equiv -\delta V(\varphi(x))/\delta\varphi(x)\,, \tag{3.2}$$

where

$$\Box_x \equiv \partial_{\mu x}\partial^{\mu x} + m^2\,, \tag{3.3}$$

and $j(x)$ is sometimes called the internal source of $\varphi(x)$.

The two-point vertex function defined as

$$\Gamma_p(x, y) \equiv \left.\frac{\delta^2 \Gamma[\varphi_c(x)]}{\delta\varphi_c(x)\delta\varphi_c(y)}\right|_{\varphi_c(x)=0} \tag{3.4}$$

can be presented as

$$\Gamma_p(x, y) = \Gamma_{0p}(x, y) - \Sigma_p(x, y)\,, \tag{3.5}$$

where

$$\Gamma_{0p}(x, y) = -\Box_x \delta_p^{d+1}(x - y) \tag{3.6}$$

is the vertex function in the tree approximation and $\Sigma_p(x, y)$ the self-energy part due to loop corrections. The inverse of Γ_{0p} is Green's function for free field, satisfying

$$\Box_x G_{0p}(x, y) = -\delta_p^{d+1}(x - y)\,. \tag{3.7}$$

Using (3.2) and the commutation relation (2.125) we find

$$\Box_x \Box_y G_p(x, y) = -\Box_y \delta_p^{d+1}(x - y) - \mathrm{i}\left[\mathrm{Tr}(T_p(j(x)j(y))\hat{\rho}) + \mathrm{i}\delta_p^{d+1}(x - y)\,\mathrm{Tr}\left\{\frac{\delta^2 V}{\delta\varphi(x)\delta\varphi(y)}\hat{\rho}\right\}\right],$$

or

$$\Box_x G_p(x, y) = -\delta_p^{d+1}(x - y) + \mathrm{i}\int_p \left[\mathrm{Tr}(T_p(j(x)j(z))\hat{\rho}) + \mathrm{i}\delta_p^{d+1}(x - z)\,\mathrm{Tr}\left\{\frac{\delta^2 V}{\delta\varphi(x)\delta\varphi(z)}\hat{\rho}\right\}\right]$$
$$\times\, G_{0p}(z, y)\,\mathrm{d}^{d+1}z\,. \tag{3.8}$$

Comparing (3.8) with (2.50) we obtain

$$\int \Sigma_p(x, z) G_p(z, y)\,\mathrm{d}^{d+1}z$$
$$= -\mathrm{i}\int_p \left[\mathrm{Tr}(T_p(j(x)j(z))\hat{\rho}) + \mathrm{i}\delta_p^{d+1}(x - z)\,\mathrm{Tr}\left\{\frac{\delta^2 V}{\delta\varphi(x)\delta\varphi(z)}\hat{\rho}\right\}\right] G_{0p}(z, y)\,\mathrm{d}^{d+1}z\,,$$

which yields

$$\Sigma_p(x, y) = \left(-\mathrm{i}\,\mathrm{Tr}(T_p(j(x)j(y))\hat{\rho}) + \delta_p^{d+1}(x-y)\,\mathrm{Tr}\left(\frac{\delta^2 V}{\delta\varphi(x)\delta\varphi(y)}\hat{\rho}\right)\right)_{\mathrm{1P.I.}} . \tag{3.9}$$

This expression will be used later to discuss the transition probability.

3.1.2. *Matrix representation*

The matrix representation of the Dyson equation as given by (2.57) and (2.58) are very convenient for practical calculations. For example, we find immediately from (2.58) that

$$\Gamma_{\mathrm{r}} = G_{\mathrm{r}}^{-1}, \qquad \Gamma_{\mathrm{a}} = G_{\mathrm{a}}^{-1}, \qquad \Gamma_{\mathrm{c}} = -G_{\mathrm{r}}^{-1} G_{\mathrm{c}} G_{\mathrm{a}}^{-1}, \tag{3.10a}$$

$$G_{\mathrm{r}} = \Gamma_{\mathrm{r}}^{-1}, \qquad G_{\mathrm{a}} = \Gamma_{\mathrm{a}}^{-1}, \qquad G_{\mathrm{c}} = -\Gamma_{\mathrm{r}}^{-1} \Gamma_{\mathrm{c}} \Gamma_{\mathrm{a}}^{-1}. \tag{3.10b}$$

Using eqs. (2.9) and (2.62) we find the corresponding relations for $\hat{G}$ as

$$G_{\pm} = -\Gamma_{\mathrm{r}}^{-1} \Gamma_{\pm} \Gamma_{\mathrm{a}}^{-1}, \qquad G_{\mathrm{F}} = -\Gamma_{\mathrm{r}}^{-1} \Gamma_{\tilde{\mathrm{F}}} \Gamma_{\mathrm{a}}^{-1}, \qquad G_{\tilde{\mathrm{F}}} = -\Gamma_{\mathrm{r}}^{-1} \Gamma_{\mathrm{F}} \Gamma_{\mathrm{a}}^{-1}. \tag{3.11}$$

The symmetry relations (2.132) and (2.133) valid for G can be also transmitted to Γ to give

$$\begin{aligned} &\Gamma_{\pm}^*(x, y) = -\Gamma_{\pm}(y, x), \qquad \Gamma_{\mathrm{F}}^*(x, y) = -\Gamma_{\tilde{\mathrm{F}}}(y, x), \qquad \Gamma_{\mathrm{r}}^*(x, y) = \Gamma_{\mathrm{a}}(y, x), \\ &\Gamma_{\pm}^*(k) = -\Gamma_{\pm}(k), \qquad \Gamma_{\mathrm{F}}^*(k) = -\Gamma_{\tilde{\mathrm{F}}}(k), \qquad \Gamma_{\mathrm{r}}^*(k) = \Gamma_{\mathrm{a}}(k). \end{aligned} \tag{3.12a}$$

For real field we have from (2.134) and (2.135)

$$\begin{aligned} &\hat{\Gamma}(x, y) = \hat{\Gamma}^{\mathrm{T}}(y, x) = -\sigma_1 \hat{\Gamma}^*(x, y)\sigma_1 = -\sigma_1 \hat{\Gamma}^{\dagger}(y, x)\sigma_1, \\ &\hat{\Gamma}(k) = \hat{\Gamma}^{\mathrm{T}}(-k) = -\sigma_1 \hat{\Gamma}^*(-k)\sigma_1 = -\sigma_1 \hat{\Gamma}^{\dagger}(k)\sigma_1. \end{aligned} \tag{3.12b}$$

3.1.3. *Vertex functions*

As seen from (3.12a), only three components of $\hat{\Gamma}$ are independent. They can be set as

$$\Gamma_{\mathrm{F}} + \Gamma_{\tilde{\mathrm{F}}} = \Gamma_{+} + \Gamma_{-} = 2\mathrm{i}B(k), \qquad \Gamma_{\mathrm{F}} - \Gamma_{\tilde{\mathrm{F}}} = 2D(k), \qquad \Gamma_{-} - \Gamma_{+} = 2\mathrm{i}A(k), \tag{3.13a}$$

where A, B and D are real functions in accord with (3.12a). In terms of unity and Pauli matrices, (3.13a) can be rewritten as

$$\hat{\Gamma} = \mathrm{i}B(I + \sigma_1) + A\sigma_2 + D\sigma_3. \tag{3.13b}$$

We then find from (3.13) that

$$\begin{aligned} &\Gamma_{\mathrm{r}}(k) = D(k) + \mathrm{i}A(k), \qquad \Gamma_{\mathrm{a}}(k) = D(k) - \mathrm{i}A(k), \qquad \Gamma_{\mp}(k) = \mathrm{i}(B(k) \pm A(k)), \\ &\Gamma_{\mathrm{F}}(k) = D(k) + \mathrm{i}B(k), \qquad \Gamma_{\tilde{\mathrm{F}}}(k) = -D(k) + \mathrm{i}B(k). \end{aligned} \tag{3.14}$$

In what follows we will call $D(k)$ the dispersive part and $A(k)$ the absorptive part of the self-energy in analogy with the quantum field theory.

3.1.4. Green's functions

The expressions for Green's functions follow immediately from (3.10) and (3.11) as

$$G_r(k)=\frac{1}{D(k)\mp \mathrm{i}A(k)}, \qquad G_a(k)=\frac{1}{D(k)\mp \mathrm{i}A(k)}, \qquad G_\pm(k)=-\mathrm{i}\frac{B(k)\mp A(k)}{D^2(k)+A^2(k)},$$
$$G_F(k)=\frac{D(k)-\mathrm{i}B(k)}{D^2(k)+A^2(k)}, \qquad G_{\tilde F}(k)=\frac{-D(k)-\mathrm{i}B(k)}{D^2(k)+A^2(k)}. \tag{3.15}$$

It follows also from the matrix equation (2.57) and the symmetry relations (2.133) and (3.12) that

$$G_+(k)\Gamma_-(k)=G_-(k)\Gamma_+(k), \tag{3.16}$$

which can be verified directly from (3.14) and (3.15).

By virtue of the definition (3.5) we can express functions $A(k)$, $B(k)$ and $D(k)$ in terms of the self-energy part Σ as

$$A(k)=\tfrac{1}{2}\mathrm{i}(\Sigma_-(k)-\Sigma_+(k)), \qquad B(k)=\tfrac{1}{2}\mathrm{i}(\Sigma_+(k)+\Sigma_-(k)),$$
$$D(k)=k^2-m^2-\tfrac{1}{2}(\Sigma_F(k)-\Sigma_{\tilde F}(k)). \tag{3.17}$$

It is unlikely that both $A(k)$ and $D(k)$ have zero on the real axis of k_0, so the divergence on the mass shell $k^2=m^2$ can be removed by the renormalization procedure. If there are no zeroes of $D(k)+\mathrm{i}A(k)$ in the upper half-plane of k_0, then the causality is guaranteed and the pole of G_r in the lower half-plane of k_0 will describe a quasiparticle moving in a dissipative medium. On the opposite, if there is a pole a in the upper half-plane, then G_r is analytic only for $\mathrm{Im}\,k_0 > \mathrm{Im}\,a$. This pole will describe a quasiparticle moving in an amplifying medium with growing amplitude of the wavefunction. In such a case the original state is unstable with respect to a new coherent state of quasiparticles like the laser system beyond the threshold.

3.2. Systems near thermoequilibrium

The formal solution of the Dyson equation (3.10) and (3.11) as well as the explicit form of the vertex function (3.14) and Green's function (3.15) are valid for any quasiuniform system near equilibrium or nonequilibrium stationary state. In this section we consider the thermoequilibrium system in more detail. The transition probability is first studied (section 3.2.1), the dispersive part is then discussed (section 3.2.2) to show that the thermoequilibrium system is stable and the detailed balance is ensured (section 3.2.3). Furthermore, formulas for nonrelativistic fields are written out explicitly for future reference (section 3.2.4). Finally, the fluctuation–dissipation theorem is derived for the complex boson and fermion field (section 3.2.5).

3.2.1. Transition probability

It follows from (3.9) that

$$\Sigma_-(x, y) = -\mathrm{i}\,\mathrm{Tr}(j(x)j(y)\hat{\rho})_{1\,\mathrm{P.I.}}\,, \qquad \Sigma_+(x, y) = -\mathrm{i}\,\mathrm{Tr}(j(y)j(x)\hat{\rho})_{1\,\mathrm{P.I.}}\,. \tag{3.18}$$

As done in section 2.5, the evolution of $j(x)$ under the space–time translation is given by

$$j(x) = \exp(\mathrm{i}p\cdot x)j(0)\exp(-\mathrm{i}p\cdot x)\,.$$

Substituting this expression into (3.18) and taking Fourier transformation, we obtain

$$\mathrm{i}\Sigma_-(k) = \sum_{l,n} |\langle l|j(0)|n\rangle_{1\,\mathrm{P.I.}}|^2 \rho_{nn}(2\pi)\delta^{d+1}(k - p_l + p_n)\,, \tag{3.19a}$$

$$\mathrm{i}\Sigma_+(k) = \sum_{l,n} |\langle n|j(0)|l\rangle_{1\,\mathrm{P.I.}}|^2 \rho_{ll}(2\pi)^{d+1}\delta^{d+1}(k - p_l + p_n)\,. \tag{3.19b}$$

Here we neglect the off-diagonal elements of the density matrix because the system is uniform. For $k_0 > 0$ each term of (3.19a) corresponds to the probability of transition from the state $|n\rangle$ to the state $|l\rangle$ by absorbing a quasiparticle of momentum $\boldsymbol{k}$, i.e.,

$$\mathrm{i}\Sigma_-(k)_{k_0>0} = 2k_0 W_{\mathrm{a}}(\boldsymbol{k})\,, \tag{3.20a}$$

while each term of (3.19b) corresponds to the probability of emitting a quasiparticle

$$\mathrm{i}\Sigma_+(k)_{k_0>0} = 2k_0 W_{\mathrm{e}}(\boldsymbol{k})\,. \tag{3.20b}$$

Since $E_l > E_n$ for both cases, $\rho_{nn} > \rho_{ll}$ in thermoequilibrium, so that

$$\mathrm{i}(\Sigma_-(k) - \Sigma_+(k)) = 2A(k) > 0 \tag{3.21a}$$

for $k_0 > 0$, i.e., the probability of absorbing a particle is greater than that of emission.

Using the relation

$$\Gamma_+(k) = \Gamma_-(-k)\,,$$

following from (3.12b) we find for $k_0 < 0$,

$$\mathrm{i}\Sigma_-(\boldsymbol{k}, -|k_0|) = 2|k_0|\,W_{\mathrm{e}}(-\boldsymbol{k})\,, \tag{3.20c}$$

$$\mathrm{i}\Sigma_+(\boldsymbol{k}, -|k_0|) = 2|k_0|\,W_{\mathrm{a}}(-\boldsymbol{k})\,, \tag{3.20d}$$

so that

$$\mathrm{i}(\Sigma_-(k) - \Sigma_+(k)) = 2A(k) < 0\,. \tag{3.21b}$$

Therefore, we can write

$$A(k) = k_0\gamma(k)\,, \tag{3.22}$$

where $\gamma(k)$ is always positive for systems near thermoequilibrium.

3.2.2. Dispersive part

The dispersive part $D(k)$ can be written as

$$D(k) = k^2 - m^2 - \delta m^2(k), \tag{3.23}$$

where $\delta m^2(k)$ comes from loop correction. Assuming $\gamma(k)$ and $\delta m^2(k)$ to be small, we find the pole of the retarded Green function located at

$$k_0 = \omega(k) - \tfrac{1}{2}\mathrm{i}Z_\varphi\gamma(k, \omega(k)), \tag{3.24}$$

where

$$\omega(k) = \omega_0(k)\left(1 + \frac{\delta m^2(k, \omega_0(k))}{2\omega_0^2(k)}\right) \tag{3.25}$$

with

$$\omega_0(k) = \pm\sqrt{k^2 + m^2}, \qquad Z_\varphi^{-1} = \left.\frac{\partial D}{\partial k_0^2}\right|_{k_0=\omega(k)}. \tag{3.26}$$

Here Z_φ^{-1} is the wavefunction renormalization which is close to 1 provided δm^2 is small. However, it can drastically deviate from 1, even become negative for Coulomb field $\varphi(x)$ in a plasma. As seen from (3.24) the pole is located in the lower half-plane for $\gamma(k) > 0$, so that the quasiparticle decays with γ^{-1} as its life time.

3.2.3. Detailed balance

The causal Green function G_F can be written as

$$G_F(k) = \frac{\bar{a}}{k^2 - m^2 - \delta m^2(k) + \mathrm{i}|k_0|\gamma(k)} + \frac{\bar{b}}{k^2 - m^2 - \delta m^2(k) - \mathrm{i}|k_0|\gamma(k)}, \tag{3.27}$$

where

$$a = \tfrac{1}{2}(1 + B(k)/|A(k)|) = W_a(k)/[W_a(k) - W_e(k)], \tag{3.28a}$$

$$b = \tfrac{1}{2}(-1 + B(k)/|A(k)|) = W_e(k)/[W_a(k) - W_e(k)]. \tag{3.28b}$$

For a system in equilibrium

$$W_a(k)/W_e(k) = \Sigma_-(k)/\Sigma_+(k) = \rho_{nn}/\rho_{ll} = \exp(\beta k_0), \qquad k_0 > 0, \tag{3.29}$$

so that

$$b = n_{th}(k) = \frac{1}{\exp(\beta k_0) - 1}. \tag{3.30}$$

It can be shown also that (3.30) still holds for the case $k_0 < 0$ if k_0 is replaced by $|k_0|$.

Therefore, the detailed balance condition

$$a(k)/b(k) = W_a(k)/W_e(k) = [1 + n(k)]/n(k) \tag{3.31}$$

is fulfilled for equilibrium systems. Also, it follows from (3.20), (3.12) and (3.29) that

$$\gamma(k) = W_a(k)[1 - \exp(-\beta|k_0|)] . \tag{3.32}$$

One more remark. Green's functions obtained in this section reduce to those for free field discussed in section 2.1, provided δm^2 and γ are ignored.

3.2.4. Complex field

Up to now we have discussed mainly the Hermitian field in this section. A complex boson or fermion field with conserved particle number can be treated in a similar way. The Lagrangian is written as

$$\mathscr{L} = \psi^\dagger(x)\Big(\mathrm{i}\frac{\partial}{\partial t} + \frac{\nabla^2}{2m} - E_0\Big)\psi(x) - V(\psi^\dagger(x), \psi(x)) + J^\dagger(x)\psi(x) + \psi^\dagger(x)J(x) ,$$

where $J^\dagger(x)$, $J(x)$ are anti-commuting c numbers for the fermion case, E_0 a constant.

The connected two-point Green function is defined as

$$\begin{aligned} G_p(x, y) &= -\,\delta^2 W[J^\dagger, J]/\delta J^\dagger(x)\delta J(y) \\ &= -\mathrm{i}(\mathrm{Tr}[T_p(\psi(x)\psi^\dagger(y))\hat{\rho}] - \psi_c(x)\psi_c^\dagger(y)) , \end{aligned}$$

where

$$\psi_c(x) = \delta W/\delta J^\dagger(x), \qquad \psi_c^\dagger(x) = \pm\delta W/\delta J(x) . \tag{3.33}$$

Here the functional derivative is acting from the left.

Introducing the vertex generating functional

$$\Gamma[\psi_c^\dagger(x), \psi_c(x)] = W[J^\dagger(x), J(x)] - \int (J^\dagger\psi_c + \psi_c^\dagger J) ,$$

we have

$$\delta\Gamma[\psi_c^\dagger, \psi_c]/\delta\psi_c^\dagger(x) = -J(x), \qquad \delta\Gamma[\psi_c^\dagger, \psi_c]/\delta\psi_c(x) = \mp J^\dagger(x) . \tag{3.34}$$

The two-point vertex function is given by

$$\begin{aligned} \Gamma_p(x, y) &= \delta^2\Gamma[\psi_c^\dagger, \psi_c]/\delta\psi_c^\dagger(x)\delta\psi_c(y) \\ &= (\mathrm{i}\partial/\partial t + \nabla^2/2m - E_0)\delta_p^{d+1}(x-y) - \Sigma_p(x, y) , \end{aligned} \tag{3.35}$$

where the self-energy part $\Sigma_p(x, y)$ is defined as

$$\Sigma_p(x, y) = -\mathrm{i}\,\mathrm{Tr}[T_p(j(x)j^\dagger(y))\hat{\rho}]_{1\,\mathrm{P.I.}}$$

with

$$j(x) = \delta V/\delta\psi^\dagger(x), \quad j^\dagger(x) = \delta V/\delta\psi(x). \tag{3.36}$$

Expressing $\hat{\Gamma}$ in the form of (3.13b) we have

$$D(k, X) = k_0 - \frac{1}{2m}k^2 - E_0 - \Delta E(k, X), \qquad A(k, X) = \frac{\mathrm{i}}{2}(\Sigma_-(k, X) - \Sigma_+(k, X)),$$

$$B(k, X) = \frac{\mathrm{i}}{2}(\Sigma_-(k, X) + \Sigma_+(k, X)) = \frac{\mathrm{i}}{2}(\Sigma_F(k, X) + \Sigma_{\tilde{F}}(k, X)),$$

$$\Delta E(k, X) = \tfrac{1}{2}(\Sigma_{\tilde{F}}(k, X) - \Sigma_F(k, X)) \tag{3.37}$$

in the Fourier representation of the relative coordinates.

3.2.5. Fluctuation-dissipation theorem (FDT)

We should mention that in general the components of $\hat{G}$ and $\hat{\Gamma}$ are not scalars, especially when we consider the multicomponent field in the coordinate representation. For example, the causal Green function

$$G_{Fij}(x, y) \equiv -\mathrm{i}\,\mathrm{Tr}\{T_p(\psi_i(x)\psi_j^\dagger(y))\hat{\rho}\}$$

should be considered as a matrix with subscripts ix and jy, where i, j are indices of internal degrees of freedom. Therefore, different components of $\hat{G}$, $\hat{\Gamma}$ as well as A, B, D in (3.13b) do not commute with each other. However, they are commuting scalars for single component field in k space provided these functions do not depend on the center-of-mass coordinates. We have thus

$$N \equiv B/A = [\Sigma_-(k) + \Sigma_+(k)]/[\Sigma_-(k) - \Sigma_+(k)] \equiv 1 \pm 2n(k), \tag{3.38}$$

i.e.,

$$n(k) = \pm\Sigma_+(k)/(\Sigma_-(k) - \Sigma_+(k)). \tag{3.39}$$

Using the same technique as in the case of real boson field we can show that in thermoequilibrium

$$\Sigma_+(k) = \pm\Sigma_-(k)\exp[-\beta(k_0 - \mu)], \tag{3.40}$$

where μ is the chemical potential. It then follows from (3.39) and (3.40) that the particle distribution

$$n(k) = \frac{1}{\exp[\beta(k_0 - \mu)] \mp 1}. \tag{3.41}$$

Using (3.13), (3.38) and (3.40) we find

$$\Gamma_c = \Gamma_+ + \Gamma_- = \coth[\tfrac{1}{2}\beta(k_0 - \mu)](\Gamma_- - \Gamma_+)$$
$$= \coth[\tfrac{1}{2}\beta(k_0 - \mu)](\Gamma_r - \Gamma_a) , \tag{3.42a}$$

for a boson system, and

$$\Gamma_c = \tanh[\tfrac{1}{2}\beta(k_0 - \mu)](\Gamma_- - \Gamma_+)$$
$$= \tanh[\tfrac{1}{2}\beta(k_0 - \mu)](\Gamma_r - \Gamma_a) , \tag{3.42b}$$

for a fermion system. This is the well-known fluctuation-dissipation theorem (FDT) for equilibrium systems. Using (3.10), it can be rewritten for the Green functions as

$$G_c = \coth[\tfrac{1}{2}\beta(k_0 - \mu)](G_r - G_a) , \tag{3.43a}$$

$$G_c = \tanh[\tfrac{1}{2}\beta(k_0 - \mu)](G_r - G_a) , \tag{3.43b}$$

for Bose and Fermi cases, respectively.

3.3. *Transport equation*

We show in this section that the usual transport equation for the quasiparticles can be derived from the Dyson equation for the quasiuniform system (sections 3.3.1 and 3.3.2). As illustrative examples we derive the equation of weak turbulence in a plasma (section 3.3.3) and a generalized Leonard–Balescu equation for charge carriers (section 3.3.4).

3.3.1. *Quasiparticle approximation*

According to (3.10), (3.11) and (3.13), the correlation function G_c can be presented as

$$G_c(k, X) \equiv G_F + G_{\bar F} = -\Gamma_r^{-1}(\Gamma_F + \Gamma_{\bar F})\Gamma_a^{-1}$$
$$= -2i\Gamma_r^{-1}(k, X)B(k, X)\Gamma_a^{-1}(k, X) , \tag{3.44}$$

which can be rewritten as

$$G_c(k, X) = \Gamma_r^{-1}N - N\Gamma_a^{-1} , \tag{3.45}$$

where the matrix N satisfies the equation

$$ND - DN - i(NA + AN) = -2iB . \tag{3.46}$$

The energy spectrum of the quasiparticle is determined by putting zero for Γ_r, i.e., from the equation

$$D(k, X) = 0 , \tag{3.47}$$

if the dissipation can be neglected. As in the equilibrium case, we assume

$$N|_{\text{pole}} = 1 \pm 2n\,, \tag{3.48}$$

then (3.46) can be rewritten as

$$nD - Dn - \mathrm{i}(nA + An) = \mp \mathrm{i}(B - A) = \pm\Sigma_+\,, \tag{3.49a}$$

or

$$nD - Dn = \pm\tfrac{1}{4}[(1 \pm n)\Sigma_+ + \Sigma_+(1 \pm n)] - \tfrac{1}{2}(\Sigma_- n + n\Sigma_-)\,. \tag{3.49b}$$

3.3.2. Quasiclassical approximation

As shown in the preceding sections, Σ_+, Σ_- are proportional to the probability of emitting and absorbing quasiparticles, respectively, so the right-hand side of (3.49b) is related to the collision term, whereas the left-hand side to the drift term of the transport equation. It is worthwhile to note that we cannot entirely ignore the noncommutativity of n and D on the left-hand side. To the leading order of gradients

$$(Dn - nD)(k, X) = \int \exp[ik \cdot (x - z)][D(x, y)n(y, z) - n(x, y)D(y, z)]\, \mathrm{d}y\, \mathrm{d}(x - z)$$

$$= \mathrm{i}\left\{\frac{\partial D(k, X)}{\partial k_\mu}\frac{\partial n(k, X)}{\partial X_\mu} - \frac{\partial n(k, X)}{\partial k_\mu}\frac{\partial D(k, X)}{\partial X_\mu}\right\}, \tag{3.50}$$

where

$$X = \tfrac{1}{2}(x + z)\,.$$

To convince oneself of the validity of (3.50), one can either consider the Poisson bracket as the quasiclassical approximation to the commutator or check it by a straightforward calculation. As we mentioned before, the microscopic scale $x - y$ is a fast variable with respect to which the Fourier transformation is carried out, whereas the center of mass coordinate is a slow variable. Expanding D, n in (3.50) as

$$F(k, (x + y)/2) = F(k, \tfrac{1}{2}(x + z)) + \tfrac{1}{2}(y - z)\partial F/\partial X + \cdots$$

with

$$F = D, n\,,$$

and integrating over k' by parts, one can easily verify eq. (3.50). Such separation of micro- and macro-time scales and the replacement of the commutator by its quasiclassical counterpart – the Poisson bracket, will be used frequently in our future discussion.

Assume that the solution of (3.47) is given by

$$k_0 = \omega(\boldsymbol{k}, X)\,,$$

then we have

$$\left.\frac{\partial D}{\partial k_0}\right|_{k_0=\omega(\boldsymbol{k},X)} \nabla_k \omega(\boldsymbol{k},X)+\nabla_k D(\boldsymbol{k},X)=0, \qquad \left.\frac{\partial D}{\partial k_0}\right|_{k_0=\omega(\boldsymbol{k},X)} \frac{\partial \omega(\boldsymbol{k},X)}{\partial X_\mu}+\frac{\partial D(\boldsymbol{k},X)}{\partial X_\mu}=0. \tag{3.51}$$

Using eqs. (3.20), (3.50) and (3.51) we have for $k_0=\omega(\boldsymbol{k},X)$,

$$\frac{\partial n(\boldsymbol{k},X)}{\partial t}+\boldsymbol{v}\cdot\nabla n(\boldsymbol{k},X)+\frac{\partial \omega(\boldsymbol{k},X)}{\partial X_\mu}\frac{\partial n(\boldsymbol{k},X)}{\partial k_\mu}$$

$$=\frac{2\omega(\boldsymbol{k},X)}{(\partial D/\partial k_0)_{k_0=\omega(\boldsymbol{k},X)}}\{W_e(\boldsymbol{k},X)(1\pm n(\boldsymbol{k},X))-W_a(\boldsymbol{k},X)n(\boldsymbol{k},X)\}, \tag{3.52}$$

where

$$\boldsymbol{v}=\nabla_k \omega(\boldsymbol{k},X) \tag{3.53}$$

is the group velocity. If the renormalization of the wavefunction is neglected, then

$$\partial D/\partial k_0^2=1,$$

and (3.52) can be simplified as

$$\frac{\partial n}{\partial t}+\boldsymbol{v}\cdot\nabla n+\frac{\partial \omega}{\partial X_\mu}\frac{\partial n}{\partial k^\mu}=W_e(1\pm n)-W_a n, \tag{3.54}$$

which is the transport equation for the quasiparticle distribution in the phase space. The last term on the left-hand side of (3.54) comes from the force $\partial\omega(\boldsymbol{k},X)/\partial X_\mu$ due to the variation of the energy $\omega(\boldsymbol{k},X)$ with the coordinate. If the space–time dependence of D is unimportant, this term can be neglected and the standard transport equation is recovered.

3.3.3. Plasma equation

As an illustration consider the transport equation for plasma. Let $\psi_i(x)$, $i=1,2,\ldots$, be the charged fermion fields (electron and ion). The equation of motion for the Coulomb field is given by

$$\nabla^2\varphi(x)=-\sum_i e_i\psi_i^\dagger(x)\psi_i(x). \tag{3.55}$$

To the one-loop approximation of free fermions we find the two-point vertex function to be

$$\Gamma_\varphi(x,y)=-\nabla^2\delta_p^4(x-y)-\mathrm{i}\sum_j e_j^2 S_{pj}(x,y)S_{pj}(y,x), \tag{3.56}$$

where $S_{pj}(x,y)$ is the propagator of the ψ_j field

$$S_{pj}(x,y)=-\mathrm{i}\,\mathrm{Tr}\{T_p(\psi_j(x)\psi_j^\dagger(y))\hat\rho\}. \tag{3.57}$$

In the Fourier representation of the relative coordinates the retarded vertex function for the Coulomb field can be determined as

$$\Gamma_{r\varphi}(k) = k^2 + \mathrm{i}\sum_j e_j^2 \int \frac{\mathrm{d}^4 l}{(2\pi)^4}\, \mathrm{Tr}\{\tfrac{1}{2}(1+\sigma_3)S_j(l)\sigma_3 S_j(l-k)\}$$

$$= k^2 - \mathrm{i}\sum_j e_j^2 \int \frac{\mathrm{d}^4 l}{(2\pi)^4}\,(n_j(l) - n_j(l-k))\Gamma_{rj}^{-1}(l)\Gamma_{aj}^{-1}(l-k)\,, \tag{3.58}$$

where

$$\Gamma_{rj}(l) = l_0 - \frac{1}{2m_j}l^2 + \mathrm{i}\varepsilon, \qquad \Gamma_{aj}(l-k) = l_0 - k_0 - \frac{1}{2m_j}(l-k)^2 - \mathrm{i}\varepsilon\,, \tag{3.59}$$

and $n_j(l)$ the fermion distribution. Integration of (3.58) over l_0 yields

$$\Gamma_{r\varphi}(k) = k^2\varepsilon(k)\,, \tag{3.60}$$

where

$$\varepsilon(k) = 1 + k^{-2}\sum_j e_j^2 \int \frac{\mathrm{d}^3 l}{(2\pi)^3}\, \frac{n_j(l) - n_j(l-k)}{k_0 + (1/2m_j)[(l-k)^2 - l^2] + \mathrm{i}\varepsilon} \tag{3.61}$$

is the susceptibility. It then follows from the expression

$$\Gamma_{r\varphi} = D_\varphi + \mathrm{i}A_\varphi\,,$$

that

$$D_\varphi = k^2\,\mathrm{Re}(\varepsilon(k))\,, \tag{3.62}$$

$$A_\varphi = k^2\,\mathrm{Im}(\varepsilon(k)) = -\pi\sum_j e_j^2 \int \frac{\mathrm{d}^3 l}{(2\pi)^3}\,\delta(k,l)(n_j(l) - n_j(l-k))\,, \tag{3.63}$$

where

$$\delta(k,l) \equiv \delta\Big(k_0 + \frac{1}{2m_j}[(l-k)^2 - l^2]\Big)\,. \tag{3.64}$$

Equation (3.63) is the well-known Landau formula of dissipation for plasmas in thermoequilibrium.

If $n_j(l) > n_j(l-k)$ for $l^2 > (l-k)^2$, then $A_\varphi < 0$, which means the pole of the retarded Green function moves into the upper half-plane of k_0 and an instability of the plasma occurs. However, near thermoequilibrium we always have $A_\varphi > 0$, so that the plasma oscillation decays in time.

Using the expressions for the free fermion propagator (2.21) we can also obtain from (3.56) that

$$i\Gamma_{-\varphi}(k) = -2\pi \sum_j e_j^2 \int \frac{d^3l}{(2\pi)^3}\, \delta(k, l)(1 - n_j(l))n_j(l - k)\,,$$

$$i\Gamma_{+\varphi}(k) = -2\pi \sum_j e_j^2 \int \frac{d^3l}{(2\pi)^3}\, \delta(k, l)n_j(l)(1 - n_j(l - k))\,. \tag{3.65}$$

The Boltzmann equation for the plasmon distribution $N(k, X)$ can then be derived from (3.52) and (3.65). The plasmon energy is determined from

$$\mathrm{Re}\,\varepsilon(k_0, \boldsymbol{k}, X) = 0\,,$$

whereas the wavefunction renormalization

$$\frac{\partial D}{\partial k_0} = \boldsymbol{k}^2 \frac{\partial\, \mathrm{Re}(\varepsilon(k, X))}{\partial k_0}$$

cannot be set equal to 1 in this case. Therefore, the resulting transport equation for the plasmon is

$$\frac{\partial N}{\partial t} + \boldsymbol{v}\cdot\nabla N + \frac{\partial \omega}{\partial X_\mu}\frac{\partial N}{\partial k^\mu} = \frac{1}{\boldsymbol{k}^2(\partial\, \mathrm{Re}\,\varepsilon(k, X)/\partial k_0)_{k_0=\omega(\boldsymbol{k})}}\, 2\pi \sum_j e_j^2 \int \frac{d^3l}{(2\pi)^3}\, \delta(k, l)[n_j(l)(1 - n_j(l - k))$$
$$\times (1 + N(k, X)) - (1 - n_j(l))n_j(l - k)N(k, X)]|_{k_0=\omega(\boldsymbol{k})} \tag{3.66}$$

which describes the weak turbulence of the plasma.

In eq. (3.66) we take into account only the absorption and emission of plasmons by charge carriers. If the high order self-interaction of the Coulomb field is included, we will have in addition the plasmon–plasmon collision term (wave–wave interaction).

3.3.4. *Equation for charge carriers*

Now we discuss the transport equation for the charge carriers. First we find from (3.11) Green's function

$$iG_+(k) = \frac{\pi}{\boldsymbol{k}^4|\varepsilon(k)|^2} \sum_j e_j^2 \int \frac{d^3l}{(2\pi)^3}\, \delta(k, l)(1 - n_j(l))n_j(l - k)\,,$$

$$iG_-(k) = \frac{\pi}{\boldsymbol{k}^4|\varepsilon(k)|^2} \sum_j e_j^2 \int \frac{d^3l}{(2\pi)^3}\, \delta(k, l)n_j(l)(1 - n_j(l - k))\,. \tag{3.67}$$

We must separate from (3.67) the contribution of plasmons which can be written as

$$iG_-(k) = 2\pi\delta(k_0 - \omega(\boldsymbol{k}))\frac{1 + N(k, X)}{\boldsymbol{k}^2\partial\, \mathrm{Re}\,\varepsilon/\partial k_0}\,, \qquad iG_+(k) = 2\pi\delta(k_0 - \omega(\boldsymbol{k}))\frac{N(k, X)}{\boldsymbol{k}^2\partial\, \mathrm{Re}\,\varepsilon/\partial k_0}\,. \tag{3.68}$$

To the one-loop approximation the two-point vertex function is given by

$$\Gamma_{pj}(x, y) = -\mathrm{i}\left(\frac{\partial}{\partial t} + \frac{1}{2m_j}\nabla^2\right)\delta_p^4(x-y) + \mathrm{i}e_j^2 S_{pj}(x, y) G_p(x, y)\,, \tag{3.69}$$

where $S_{pj}(x, y)$ is the fermion propagator (3.57), while $G_p(x, y)$ is the plasmon propagator. After separating the contribution of the plasmon pole we find

$$-\mathrm{i}\Gamma_{-j}(k) = 2\pi e_j^2 \int \frac{\mathrm{d}^3 l}{(2\pi)^3}\, \delta(k, l, \omega) \frac{1}{l^2 \partial \operatorname{Re} \varepsilon/\partial l_0} (1 - n_j(l+k))[\theta(l_0)N(l) + \theta(-l_0)(1 + N(l))]$$

$$- 2\pi e_j^2 \sum_{j'} e_{j'}^2 \int \frac{\mathrm{d}^3 l\, \mathrm{d}^3 l'}{(2\pi)^6}\, \delta(k, l, l') \left[\frac{1}{l^4|\varepsilon(l)|^2}\right]^{\mathrm{np}} (1 - n_j(l-k))(1 - n_{j'}(l'-l)) n_{j'}(l')\,,$$

$$-\mathrm{i}\Gamma_{+j}(k) = 2\pi e_j^2 \int \frac{\mathrm{d}^3 l}{(2\pi)^3}\, \delta(k, l, \omega) \frac{n_{j'}(l+k)}{l^2 \partial \operatorname{Re} \varepsilon/\partial l_0} [\theta(l_0)(1 + N(l)) + \theta(-l_0) N(-l)]$$

$$-2\pi e_j^2 \sum_{j'} e_{j'}^2 \int \frac{\mathrm{d}^3 l\, \mathrm{d}^3 l'}{(2\pi)^6}\, \delta(k, l, l') \left[\frac{1}{l^4|\varepsilon(l)|^2}\right]^{\mathrm{np}} n_j(l+k) n_{j'}(l'-l)(1 - n_{j'}(l'))\,, \tag{3.70}$$

where

$$\delta(k, l, \omega) \equiv \delta\left(k_0 + \omega(l) - \frac{1}{2m_j}(l+k)^2\right),$$

$$\delta(k, l, l') \equiv \delta\left(k_0 + \frac{1}{2m_{j'}}[l'^2 - (l'-l)^2] - \frac{1}{2m_j}(l+k)^2\right),$$

and $[\ \ ]^{\mathrm{np}}$ means non-pole contribution.

Taking into account the relation between $\Gamma_\pm$ and $W_{\mathrm{e,a}}$, we can readily write down the transport equation for the charge carriers. The first term in (3.70) describes the interaction of the charge carriers with the plasmon, whereas the second term describes the mutual interaction of the charged particles via the screened Coulomb field. If only the second term is retained, the usual Leonard–Balescu equation is recovered. The transport equation for plasmas has been also discussed by DuBois et al. [33, 34] using the CTPGF technique.

3.4. Multi-time-scale perturbation

As repeatedly emphasized before, we must distinguish the microscopic (relative) and macroscopic (center-of-mass) space–time scales. However, in the mean field approximation $\varphi_c(x)$ depends only on a single variable x which contains both micro- and macro-scales. To distinguish these two types of change we use the multi-time-scale perturbation theory.

The mean field $\varphi_c(x)$ satisfies the following equation:

$$\delta\Gamma[\varphi_c]/\delta\varphi_c(x) = 0\,,$$

an expansion of which in powers of $\varphi_c(x)$ can be written as

$$\frac{\delta\Gamma}{\delta\varphi_c(x)} = \int_p \Gamma_p(x, y)\varphi_c(y) + \cdots = \int \Gamma_r(x, y)\varphi_c(y)\,d^4y + E[\varphi_c(x)]\,, \tag{3.71}$$

where $E[\varphi_c(x)]$ contains high-order terms. Separating the diagonal part of Γ_r from (3.71),

$$\Gamma_r(x, y) = D_0(i\partial_x)\delta^4(x-y) + \Delta\Gamma_r(x, y)\,,$$

we find

$$D_0(i\partial_x)\varphi_c(x) + \int \Delta\Gamma_r(x, y)\varphi_c(y)\,d^4y + E[\varphi_c(x)] = 0\,. \tag{3.72}$$

Assume both $\Delta\Gamma_r$ and $E[\varphi_c(x)]$ to be small, then (3.72) has a solution

$$\varphi_c(x) = \varphi_c \exp(-ik\cdot x)\,, \tag{3.73}$$

where k satisfies the dispersion relation

$$D_0(k) = 0\,, \tag{3.74}$$

from which we find

$$k_0 = \omega(\boldsymbol{k}) = \text{a real number}\,.$$

Now consider the influence of $\Delta\Gamma_r$. If its imaginary part is less than zero and, in addition, $Z_\varphi^{-1} = \partial D_0/\partial k_0^2 > 0$ provided (3.74) is satisfied, then $\varphi_c(x)$ will grow in time to form a laser-type state with its amplitude being limited by nonlinear term $E[\varphi_c(x)]$. Near the critical point when such instability occurs, $\Delta\Gamma_r$ is a small quantity and φ_c changes with time rather slowly. Assume that the approximate solution of (3.72) can be presented as

$$\varphi_c(x) \equiv \varphi_c(x, \varepsilon x) = \varphi_c^{(0)}(x, \varepsilon x) + \varepsilon\varphi_c^{(1)}(x, \varepsilon x) + \cdots\,,$$

where ε is a small parameter which should be set equal to 1 by the end of the calculation, εx describes the slowly varying part. Set $\tilde{x} = \varepsilon x$, assume both $\Delta\Gamma_r$ and E to be of order ε, the differentiation with respect to x can be written as $\partial_x + \varepsilon\partial_{\tilde{x}}$ so that (3.72) becomes

$$D_0(i\partial_x + i\varepsilon\partial_{\tilde{x}})(\varphi_c^{(0)} + \varepsilon\varphi_c^{(1)} + \cdots) + \varepsilon\int \Delta\Gamma_r(x, y)\varphi_c(y, \tilde{x})\,d^4y + \varepsilon E[\varphi_c(x, \tilde{x})] = 0\,.$$

So far as $\tilde{x}$ is a slow variable, we neglect the difference of $\tilde{x}$ and $\tilde{y}$ in the last two terms. To the first two leading orders we have

$$D_0(i\partial_x)\varphi_c^{(0)}(x, \tilde{x}) = 0\,,$$

$$D_0(i\partial_x)\varphi_c^{(1)}(x, \tilde{x}) + iD_{0\mu}(i\partial_x)\partial_{\tilde{x}}^{\mu}\varphi_c^{(0)}(x, \tilde{x}) + \int \Delta\Gamma_r(x, y)\varphi_c^{(0)}(y, \tilde{x})\,d^4y + E[\varphi_c^{(0)}(x, \tilde{x})] = 0\,, \tag{3.75}$$

where

$$D_{0\mu}(k) = \partial D_0(k)/\partial k^\mu \, . \tag{3.76}$$

As seen from (3.75), the solution is given by

$$\varphi_c(x, \bar{x}) = \varphi_k(\bar{x}) \exp(-ik \cdot x) \, , \tag{3.77}$$

where k is determined from (3.74). If we require that $\varphi_c^{(1)}$ does not contain a term proportional to $\varphi_c^{(0)}$, then the second equation of (3.75) after Fourier transformation becomes

$$iD_{0\mu}(k)\partial_{\bar{x}}^{\mu}\varphi_k(\bar{x}) + \Delta\Gamma_r(k, \bar{x})\varphi_k(\bar{x}) + \exp(ik \cdot x)E[\varphi_k(\bar{x}) \exp(-ik \cdot x)] = 0 \, , \tag{3.78}$$

where we have also replaced the center-of-mass coordinates $\frac{1}{2}(x + y)$ in $\Gamma_r(x, y)$ by $\bar{x}$. This is an equation satisfied by the oscillating mode of the mean field.

We have used this technique to discuss the laser system coupled with two-energy-level electrons [40, 41]. We will not reproduce the calculation here, but it should be mentioned that a stable laser state allowed in the classical theory, is unstable in the quantum case. In the quantum theory we must consider the fluctuation of the photon number. Since the laser system is described by a coherent state with fixed phase, the fluctuation of the photon number diverges. This divergence can be removed by a renormalization procedure which leads to the decay of the laser state. Similarly, the soliton solution of $\varphi_c(x)$ is also unstable due to the quantum fluctuation.

It is worthwhile to note that such multi-time-scale perturbation technique is quite useful. In fact, we have already made use of its basic idea in deriving the transport equation in the last section. It is also the key point in obtaining the TDGL equation which we are going to discuss now.

3.5. Time dependent Ginzburg–Landau equation

The concept of macrovariableness is very useful in critical dynamics, hydrodynamics, and many other fields [61]. Usually, the set of macrovariables includes both order parameters and conserved quantities. As a rule, their microscopic counterparts are composite operators. In this section we use the equation for the vertex functional (2.48) to derive the TDGL equation [61] for their expectation value. As seen from the later discussion, the term TDGL equation is used here in a much more general sense.

Let $Q_i(x)$, $i = 1, 2, \ldots$, be the set of composite operators corresponding to macrovariables. Without loss of generality, we assume them to be Hermitian Bose operators. The order parameter $Q_c(x)$ is determined from the equation for the vertex generating functional

$$\delta\Gamma/\delta Q_{c_i}(x) = -J_i(x) \, . \tag{3.79a}$$

Suppose, $Q_{ci}(x, \tau)$ is known for the moment τ. At the time t following τ, the left-hand side of (3.78) can be expanded as

$$-J_i(x, t) = \left.\frac{\delta\Gamma}{\delta Q_{c_i}}\right|_{Q_{c+}=Q_{c-}=Q} + \int \Gamma_{rij}(x, y)[Q_j(y, t_y) - Q_j(y, \tau)] \, dy \, , \tag{3.79b}$$

which is true for t located either on the positive or negative time branches. So far as Q varies slowly with time, we can write

$$Q_j(y, t_y) - Q_j(y, \tau) = (t_y - \tau)\frac{\partial Q_j(y, \tau)}{\partial \tau}.$$

Substituting this expression back into (3.79) and taking into account that in the limit $t \equiv t_x \to \tau$,

$$\begin{aligned}\gamma_{ij}(x, y, \tau) &\equiv -\lim_{t_x\to\tau}\int \mathrm{d}t_y\,(t_y - t_x)\Gamma_{rij}(x, t_x, y, t_y)\\ &= \mathrm{i}\frac{\partial}{\partial k_0}\Gamma_{rij}(x, y, k_0, \tau)\Big|_{k_0=0},\end{aligned} \tag{3.80}$$

where $\Gamma_{rij}(x, y, k_0, \tau)$ is the Fourier transform of $\Gamma_{rij}(x, t_x, y, t_y)$ with respect to $t_x - t_y$, taken at $T = \frac{1}{2}(t_x + t_y) \approx \tau$, we obtain in the matrix form

$$\gamma(t)\frac{\partial Q(t)}{\partial t} = \frac{\delta\Gamma}{\delta Q_{c+}}\Big|_{Q_{c+}=Q_{c-}=Q} + J(t), \tag{3.81}$$

where we change τ for t.

For the moment let

$$I_i(x, t) \equiv \frac{\delta\Gamma}{\delta Q_{ci+}}\Big|_{Q_{c+}=Q_{c-}=Q},$$

and calculate the functional derivative of I_i with respect to $Q(x, t)$ as a function of three-dimensional argument

$$\begin{aligned}\frac{\delta I_i(x, t)}{\delta Q_j(y, t)} &= \int \mathrm{d}z\,\mathrm{d}\tau_z\left\{\frac{\delta^2\Gamma}{\delta Q_{i+}(x)\delta Q_{k+}(z)}\frac{\delta Q_{k+}(z)}{\delta Q_j(y)} + \frac{\delta^2\Gamma}{\delta Q_{i+}(x)\delta Q_{k-}(z)}\frac{\delta Q_{k-}(z)}{\delta Q_j(y)}\right\}\\ &= \Gamma_{\mathrm{F}ij}(x, y, k_0 = 0, t) - \Gamma_{+ij}(x, y, k_0 = 0, t),\\ \frac{\delta I_j(y, t)}{\delta Q_i(x, t)} &= \Gamma_{\mathrm{F}ji}(y, x, k_0 = 0, t) - \Gamma_{+ji}(y, x, k_0 = 0, t)\\ &= \Gamma_{\mathrm{F}ij}(x, y, -k_0 = 0, t) - \Gamma_{-ij}(x, y, -k_0 = 0, t),\end{aligned}$$

where in the last step we have used the symmetry relation (3.12b). The difference

$$\frac{\delta I_i(x, t)}{\delta Q_j(y, t)} - \frac{\delta I_j(y, t)}{\delta Q_i(x, t)} = \lim_{k_0\to 0}\left(\Gamma_{-ij}(x, y, -k_0, t) - \Gamma_{+ij}(x, y, k_0, t)\right) \tag{3.82}$$

vanishes due to the relation

$$\Gamma_{+} = \Gamma_{-} \exp(-\beta k_0), \tag{3.83}$$

following from (3.29) for a system in equilibrium. Therefore, a free energy functional $\mathcal{F}[Q(x, t)]$ exists such that

$$I_i(x, t) = -\delta\mathcal{F}/\delta Q_i(x, t). \tag{3.84}$$

Equation (3.81) can then be rewritten as

$$\gamma(t) \frac{\partial Q(t)}{\partial t} = -\frac{\delta\mathcal{F}}{\delta Q(t)} + J(t). \tag{3.85}$$

If the macrovariables $Q(t)$ do not change with time in the external field J, then

$$\delta\mathcal{F}/\delta Q = J. \tag{3.86}$$

Hence $\mathcal{F}$ is actually the free energy of the system and (3.86) is the Ginzburg–Landau equation to determine the stationary distribution of macrovariables.

For nonequilibrium systems the potential condition, i.e., the vanishing of (3.84) can be realized if

$$\lim_{k_0 \to 0} A(x, y, k_0, t) = 0, \tag{3.87}$$

where A is the absorptive part of G_r. In the next section we will show that (3.87) is fulfilled for nonequilibrium stationary state (NESS) obeying time reversal symmetry.

It is usual to multiply eq. (3.85) by $\gamma^{-1}(t)$ to obtain

$$\frac{\partial Q(t)}{\partial t} = \gamma^{-1}(t)\left\{-\frac{\delta\mathcal{F}}{\delta Q(t)} + J(t)\right\}. \tag{3.88}$$

This is the generalized TDGL equation we would like to derive. If a random source term is added to the right-hand side of (3.88), it will appear like a Langevin equation. However, there is an important difference. Equation (3.88) includes the renormalization effects. Also, the way of describing the fluctuations in CTPGF formalism is very special as we will see in section 6.

4. Time reversal symmetry and nonequilibrium stationary state (NESS)

It is well known that the principle of local equilibrium and the Onsager reciprocity relations are the two underlying principles on which the thermodynamics of irreversible processes is constructed [62]. This is true near thermoequilibrium. Within the framework of statistical mechanics a successful theory of linear response has been developed by Kubo and others [63–65]. The two fluctuation–dissipation theorems, namely, the relation between the response and the correlation functions as given by (3.43) and the generalization of the Einstein relation $D = \mu kT$ with D as the diffusion constant and μ as the mobility, follow immediately from the linear response theory [63] along with the Onsager reciprocal relations.

Many attempts have been made to generalize this theory to far from equilibrium systems such as hydrodynamics, laser, chemical reactions and so on [66–73]. Recently there has been great interest to study the fluctuation effects in NESS in connection with the light scattering experiments in fluids with temperature gradient [74, 75]. However, in most of these treatments a phenomenological approach based on the Fokker–Planck equation or Langevin equation, at most a mesoscopic (semi-phenomenological) method using the master equation or the transport equation, were adopted. In particular, the existence condition for a free energy type generalized potential in NESS were discussed in [69–73] using the Fokker–Planck equation.

In this section we will explore the consequences of the microscopic time reversal symmetry for NESS, derive the potential condition along with nonequilibrium FDT and Onsager relations, and also decompose the inverse relaxation matrix into symmetrical and antisymmetrical parts within the framework of CTPGF approach [45].

4.1. Time inversion and stationarity

The time reversal symmetry is well known and is discussed in detail in text books [76]. Here we recall some basic points to specify our notations.

4.1.1. Time inversion in the Schrödinger picture

Suppose the system is conservative, being described by the Hamiltonian

$$\mathscr{H}[J] = \mathscr{H} - JQ\,, \tag{4.1}$$

where Q is a multi-component real boson field, either order parameter or conserved quantity, and J the corresponding external source. In general, J is time dependent, but here it is assumed to be constant in time.

The wavefunction at moment t is given by

$$\psi_{t0}(1, J, \lambda) = S^J(t, t_0)\psi_0(\lambda)\,, \tag{4.2}$$

where

$$S^J(t, t_0) \equiv \exp\{-\mathrm{i}\mathscr{H}[J](t - t_0)\}\,, \tag{4.3}$$

and $\psi_0(\lambda) \equiv \psi_{t_0}(t_0, J, \lambda)$ with λ as parameters specifying the initial state. Under time inversion,

$$J_i \to \varepsilon_i J_i\,, \qquad i = 1, 2\cdots n\,,$$

$$\lambda_a \to \varepsilon_a \lambda_a\,, \qquad a = 1, 2\cdots g\,,$$

where ε_i, ε_a are ± 1 depending on the signature of the quantity considered under time inversion.

It is well known that the time inversion in quantum mechanics is implemented by an antiunitary operator R such that

$$Q_i \to RQ_iR^\dagger = \varepsilon_i Q_i\,, \qquad \mathcal{H}[J] \to R\mathcal{H}[J]R^\dagger = \mathcal{H}[\varepsilon J]\,, \tag{4.4}$$

$$\psi_{t_0}(t, J, \lambda) \to R\psi_{t_0}(t, J, \lambda) = S^{\varepsilon J}(-t, -t_0)\tilde{\psi}_{-t_0} = S^{\varepsilon J}(-t, t_0)\tilde{\psi}_{t_0}\,. \tag{4.5}$$

If

$$\tilde{\psi}_{t_0} = \psi_0(\varepsilon\lambda)\,, \tag{4.6}$$

the state is considered to be time reversal invariant, i.e.,

$$R\psi_{t_0}(t, J, \lambda) = \psi_{t_0}(-t, \varepsilon J, \varepsilon\lambda)\,. \tag{4.7}$$

Analogously, for a statistical ensemble the density matrix

$$\rho_{t_0}(t, J, \lambda) = S^J(t, t_0)\rho_0(\lambda)S^J(t_0, t) \tag{4.8}$$

transforms as

$$\rho_{t_0}(t, J, \lambda) \to R\rho_{t_0}(t, J, \lambda)R^\dagger = S^{\varepsilon J}(-t, t_0)\tilde{\rho}_{t_0}S^{\varepsilon J}(t_0, -t)\,. \tag{4.9}$$

It is time reversal invariant if

$$\tilde{\rho}_{t_0} = \rho_0(\varepsilon\lambda)\,. \tag{4.10}$$

4.1.2. Time inversion in Heisenberg picture

Now we turn to the Heisenberg picture. Suppose it coincides with the Schrödinger picture at $t = t_0$, then we have

$$Q^J_{t_0}(t) \to RQ^J_{t_0}(t)R^\dagger = \varepsilon Q^{\varepsilon J}_{-t_0}(-t) = \varepsilon S^{\varepsilon J}(-t_0, t_0)Q^{\varepsilon J}_{t_0}(-t)S^{\varepsilon J}(t_0, -t_0)\,. \tag{4.11}$$

The density matrix does not change with time in the Heisenberg picture. Set $t = t_0$ in (4.9) and use (4.10), we find

$$\rho_{t_0} \to R\rho_{t_0}R^\dagger = S^{\varepsilon J}(-t_0, t_0)\rho_0(\varepsilon\lambda)S^{\varepsilon J}(t_0, -t_0)\,. \tag{4.12}$$

The expectation value

$$Q_{t_0}(t, J, \lambda) \equiv \mathrm{Tr}\{\rho_{t_0}Q^J_{t_0}(t)\} = \varepsilon Q_{t_0}(-t; \varepsilon J, \varepsilon\lambda) \tag{4.13}$$

by virtue of (4.11), (4.12) and antiunitarity of R.

It is important to note that we need the time invariance for both dynamical variable and initial state to get the invariance for the expectation value. The external source is introduced here for mathematical treatment. In the final answer we usually set $J = 0$, so the stationary state is described by $\rho_0(\lambda)$ which does not depend on J. However, in the process of calculation using the generating functional the dynamics are determined by $\mathcal{H}[J]$. To ensure the time translational invariance we have to set $t_0 = -\infty$ and switch on the external source adiabatically. In fact, there are two implicit assumptions. First, the limit

$$\lim_{t_0\to-\infty} \exp(-\mathrm{i}\mathcal{H}[J](\tau-t_0))\rho_0(\lambda)\exp(\mathrm{i}\mathcal{H}[J](\tau-t_0)) \tag{4.14}$$

exists in NESS, moreover, it does not depend on τ. Second, the limiting procedure $t_0\to-\infty$ commutes with R. It follows then,

$$R\rho(\lambda, J)R^{\dagger} = \rho(\varepsilon\lambda, \varepsilon J). \tag{4.15}$$

4.1.3. Implications for Green's functions

The correlation function of Heisenberg operators defined as

$$F_{12\cdots l}(12\cdots l, J, \lambda) \equiv \lim_{t_0\to-\infty} \mathrm{Tr}\{\rho_0(\lambda)Q^{J}_{1t_0}(1)\cdots Q^{J}_{lt_0}(l)\}, \tag{4.16}$$

transforms as

$$F_{12\cdots l}(12\cdots l, J, \lambda) = \varepsilon_1\varepsilon_2\cdots\varepsilon_l F_{l\cdots 21}(-l\cdots -2-1, \varepsilon J, \varepsilon\lambda) \tag{4.17}$$

under time inversion as follows from (4.10)–(4.12), (4.14) and (4.15). Note that the order of arguments on the RHS of (4.17) is reversed due to the antiunitarity of R. Here 1 stands for t_1, etc.

It is ready to check that Green's functions for NESS are time translationally invariant and transform under time reversal as

$$Q_i(J) = \varepsilon_i Q_i(\varepsilon J), \tag{4.18a}$$

$$G_{rij}(t_1-t_2, J) = \varepsilon_i\varepsilon_j G_{aij}(-t_1+t_2, \varepsilon J), \tag{4.18b}$$

$$G_{cij}(t_1-t_2, J) = \varepsilon_i\varepsilon_j G_{cij}(-t_1+t_2, \varepsilon J), \tag{4.18c}$$

in accord with (4.17). Hereafter we drop the parameters for the initial state λ for simplicity. We can solve for J from (4.18a) to obtain

$$J_i(Q) = \varepsilon_i J_i(\varepsilon Q). \tag{4.19}$$

Using the equation for the vertex functional (3.78) and taking functional derivative of (4.19) in accord with the formula

$$\frac{\delta}{\delta Q_i}(E[Q(t)])|_{Q_i(t)=Q_i} = \left[\int \mathrm{d}t' \frac{\delta E[Q(t)]}{\delta Q_i(t')}\right]_{Q_i(t)=Q_i},$$

we find that

$$\Gamma_{rij}(k_0=0, Q) = \varepsilon_i\varepsilon_j\Gamma_{rij}(k_0=0, \varepsilon Q). \tag{4.20}$$

Also, it follows from the Dyson equation (3.10) and the relation (4.18) that

$$\Gamma_{\mathrm{r}ij}(t_1 - t_2, Q) = \varepsilon_i \varepsilon_j \Gamma_{\mathrm{a}ij}(-t_1 + t_2, \varepsilon Q)\,. \tag{4.21}$$

Equations (4.18)–(4.21) are the implications of the time reversal symmetry needed for our further discussion.

4.2. Potential condition and generalized FDT

4.2.1. Potential condition

We have shown in section 3.5, that the potential condition can be satisfied if the zero frequency limit of the absorptive part $A(k, X)$ vanishes as given by (3.87).

Taking the Fourier transform of (4.21) we obtain

$$\Gamma_{\mathrm{r}ij}(k_0, Q) = \varepsilon_i \varepsilon_j \Gamma_{\mathrm{a}ij}(-k_0, \varepsilon Q)\,.$$

Comparing its zero frequency limit with (4.20) we find

$$\Gamma_{\mathrm{r}ij}(k_0 = 0, Q) = \Gamma_{\mathrm{a}ij}(k_0 = 0, Q)\,, \tag{4.22}$$

which will yield the potential condition (3.87) if combined with the definition of D and A in (3.14). Thus we have shown that one can construct a free-energy-like generalized potential for NESS satisfying the microscopic time reversal symmetry, i.e., one can construct functionals $\mathcal{F}$, $\mathcal{G}$ such that

$$-\tfrac{1}{2}\xi_\alpha \left.\frac{\delta \Gamma[Q]}{\delta Q(x_\alpha)}\right|_{Q_+ = Q_- = Q} = \frac{\delta \mathcal{F}[Q(x)]}{\delta Q(x)}\,, \tag{4.23a}$$

$$-\tfrac{1}{2}\xi_\alpha \left.\frac{\delta W[J]}{\delta J(x_\alpha)}\right|_{J_+ = J_- = J} = \frac{\delta \mathcal{G}[J(x)]}{\delta J(x)}\,. \tag{4.23b}$$

If the macrovariables $Q_i(t)$ vary slowly with time as discussed in section 3.5, we should assume the time reversal symmetry in the "local" sense, i.e., the system is invariant for a macroscopically short and microscopically long time scale so that

$$A_{ij}(k_0 = 0, t, Q) = 0 \tag{4.24}$$

at each moment of t.

4.2.2. Generalized FDT

Now we discuss the nonequilibrium FDT. First we split the relaxation matrix defined by (3.81) into symmetrical $\gamma_{(i,j)}$ and antisymmetrical $\gamma_{[i,j]}$ parts as

$$-a_{ij} \equiv \gamma_{(i,j)} = \tfrac{1}{2}(\gamma_{ij} + \gamma_{ji}) = -\left.\frac{\partial A_{ij}(\omega, t, Q)}{\partial \omega}\right|_{\omega=0}, \tag{4.25a}$$

$$\mathrm{i}d_{ij} \equiv \gamma_{[i,j]} = \tfrac{1}{2}(\gamma_{ij} - \gamma_{ji}) = \mathrm{i}\left.\frac{\partial D_{ij}(\omega, t, Q)}{\partial \omega}\right|_{\omega=0} . \tag{4.25b}$$

As follows from (3.10) and (3.45), the correlation vertex function Γ_c can be expressed in terms of function N (cf. (3.46)) as

$$\begin{aligned}\Gamma_{cij}(\omega, Q) &= \Gamma_{rik}(\omega, Q)N_{kj}(\omega, Q) - N_{ik}(\omega, Q)\Gamma_{akj}(\omega, Q) \\ &= D_{ik}N_{kj} - N_{ik}D_{kj} + \mathrm{i}(A_{ik}N_{kj} + N_{ik}A_{kj}) .\end{aligned} \tag{4.26}$$

In thermal equilibrium

$$N_{ij}(\omega) = N(\omega)\delta_{ij} = \coth(\omega/2T)\delta_{ij} \approx (2T/\omega)\delta_{ij} ,$$

as follows from (3.38). Hence the FDT (3.42a) can be rewritten as

$$-\mathrm{i}\Gamma_{cij}(\omega = 0, Q) = \lim_{\omega\to 0} \frac{2A_{ij}(\omega, Q)T}{\omega} = 4Ta_{ij} \tag{4.27}$$

in the low frequency limit, where a_{ij} is the symmetrical part of the relaxation matrix defined by (4.23).

If we assume the zero frequency limit of $N_{ij}(\omega, Q)$ to be finite then we find from (4.26) that

$$\mathrm{i}\sum_j \Gamma_{cij}(\omega = 0, Q) = 0$$

which contradicts the positive definiteness of the quantum fluctuation. It is therefore more natural to assume that

$$\lim_{\omega\to 0} N_{ij}(\omega, Q) \to (2/\omega)T^{\text{eff}}\delta_{ij} , \tag{4.28}$$

where T^{eff} is the effective temperature. Using once again the potential condition (4.24) we find

$$-\mathrm{i}\Gamma_{cij}(\omega = 0, Q) = 4T^{\text{eff}}a_{ij}(Q) . \tag{4.29}$$

This is the low frequency limit of the FDT for NESS. Substituting (4.29) into (3.10) and carrying out the inverse Fourier transformation, we obtain the FDT for Green's function as

$$\mathrm{i}\partial G_c/\partial t = 2T^{\text{eff}}(G_r - G_a) \tag{4.30}$$

which has the same form as that used in critical dynamics [77].

4.3. Generalized Onsager reciprocity relations

4.3.1. Kinetic matrix

As seen from eq. (3.85) the matrix $\gamma_{ij}(t)$ describes the relaxation of the order parameter. Using the definition (3.81), the symmetry relation under the time reversal (4.21) and the basic relation

$$\Gamma_{rij}(\omega, Q) = \Gamma_{aji}(-\omega, Q) \tag{4.31}$$

following from the definition (cf. (3.12b)), we can easily find the symmetry relation for γ_{ij} under time inversion

$$\gamma_{ij}(Q) = \varepsilon_i \varepsilon_j \gamma_{ji}(\varepsilon Q)\,. \tag{4.32}$$

4.3.2. Diffusion matrix

The standard Langevin equation is written as [61]

$$\partial Q_i/\partial t = -\gamma_{ij}^{-1}(\delta\mathcal{F}/\delta Q_j) + V_i(Q) + \xi_i \tag{4.33}$$

where $V_i(Q)$ is the mode coupling term. The random source ξ_i obeys a Gaussian distribution such that

$$\langle \xi_i \rangle = 0, \qquad \langle \xi_i(t)\xi_j(t') \rangle = 2\sigma_{ij}\delta(t-t')\,, \tag{4.34}$$

where σ_{ij} is the diffusion matrix, appearing in the Fokker–Planck equation. We will show later (section 6.4) that within the CTPGF formalism σ_{ij} can be expressed as (cf. (6.105))

$$\sigma_{ij}(Q) = -\tfrac{1}{2}\mathrm{i}\gamma_{ik}^{-1}\Gamma_{ckl}(\omega = 0, Q)(\gamma_{lj}^{\mathrm{T}})^{-1}\,, \tag{4.35}$$

where T means transposition. Using (3.10), (4.18) and (4.22) we can find

$$\Gamma_{cij}(\omega = 0, Q) = \varepsilon_i \varepsilon_j \Gamma_{cij}(\omega = 0, \varepsilon Q)\,. \tag{4.36}$$

Substitution of (4.32) and (4.36) into (4.35) will yield

$$\sigma_{ij}(Q) = \varepsilon_i \varepsilon_j \sigma_{ij}(\varepsilon Q)\,, \tag{4.37}$$

provided γ_{ij} is symmetrical. If not, (4.37) can then be easily derived using the FDT (4.29) and the symmetrized γ_{ij} as defined by (4.25a). Equation (4.37) is the symmetry relation for the diffusion matrix first obtained by Van Kampen et al. [67–70].

4.3.3. Response matrix

Now we turn to the response of the system in NESS to an external disturbance. Consider the density of conserved quantities $Q_\alpha(x, t)$, $\alpha = 1, 2, \ldots$, and the corresponding sources $\bar{J}_\alpha(x, t)$. The rest of the external sources J are stationary in time. The NESS is described by the Hamiltonian (4.1) and the density matrix $\rho(\lambda, J)$ (4.15) invariant under time inversion. The coupling of $Q_\alpha(x, t)$ to the source $\bar{J}_\alpha(x, t)$ is treated perturbatively. Let $j^i(r, t)$ be the current density satisfying the Heisenberg equation of motion

$$\frac{\partial Q_\alpha(x, t)}{\partial t} + \frac{\partial j^i_\alpha(x, t)}{\partial x^i} = 0\,, \tag{4.38}$$

where $i = 1, 2, 3$. Using the generating functional (2.32) it is straightforward to find

$$\delta\langle j^i_\alpha(\boldsymbol{x}, t)\rangle = -\mathrm{i}\int \mathrm{d}^4y\, \mathrm{Tr}\{\rho(J,\lambda)\theta(t_x - t_y)[j^i_\alpha(x), Q_\beta(y)]\}\bar{J}_\beta(y) \tag{4.39}$$

to the linear order in $\bar{J}_\alpha$ near NESS. The retarded Green function for the current density is defined as

$$G^{ij}_{r\alpha\beta}(x-y, J, \lambda) \equiv -\mathrm{i}\,\mathrm{Tr}\{\rho(J,\lambda)\theta(t_x - t_y)[j^i_\alpha(x), j^j_\beta(y)]\}, \tag{4.40}$$

whereas $\bar{J}_\alpha(x)$ in the interaction Lagrangian varies slowly with coordinates so that the external force

$$-\nabla\bar{J}_\alpha(\boldsymbol{x}, t) = \boldsymbol{X}_\alpha = \mathrm{const}. \tag{4.41}$$

It is then ready to derive from (4.38)–(4.40) by use of the Lehmann representation discussed in section 2.5, that

$$\delta\langle j^i_\alpha(\boldsymbol{x})\rangle = -\mathscr{L}^{ij}_{\alpha\beta}(J,\lambda)X^j_\beta, \tag{4.42}$$

where the response matrix $\mathscr{L}^{ij}_{\alpha\beta}$ is defined as

$$\mathscr{L}^{ij}_{\alpha\beta}(J,\lambda) \equiv -\mathrm{i}\left.\frac{\partial G^{ij}_{r\alpha\beta}(\omega, \boldsymbol{p})}{\partial\omega}\right|_{\substack{\boldsymbol{p}=0\\ \omega=0}}. \tag{4.43}$$

The current operator $j^i_\alpha(\boldsymbol{x})$ in the Schrödinger picture transforms under time inversion as

$$j^i_\alpha(\boldsymbol{x}) \to Rj^i_\alpha(\boldsymbol{x})R^\dagger = -\varepsilon_\alpha j^i_\alpha(\boldsymbol{x}). \tag{4.44}$$

So far as $j^i_\alpha(\boldsymbol{x})$ is similar to Q_i, whereas the definitions of $G^{ij}_{r\alpha\beta}$ (4.40) and $\mathscr{L}^{ij}_{\alpha\beta}$ are analogous to their counterparts for the order parameter Q, we can readily find that

$$\mathscr{L}^{ij}_{\alpha\beta}(J,\lambda) = \varepsilon_\alpha\varepsilon_\beta\mathscr{L}^{ji}_{\beta\alpha}(\varepsilon J, \varepsilon\lambda). \tag{4.45}$$

Equations (4.42), (4.43) and (4.45) are the generalization of the Onsager theorem to the NESS case, whereas in the literature eqs. (4.32), (4.37) and (4.45) are known as generalized Onsager relations.

4.4. Symmetry decomposition of the inverse relaxation matrix

In this section we will decompose the inverse relaxation matrix γ^{-1}, first introduced in deriving the TDGL equation (3.88), into symmetric and antisymmetric parts and find their explicit expressions, i.e., to complete the derivation of the generalized TDGL equation with both dissipative and mode coupling terms.

4.4.1. Symmetric part

Using eq. (4.35) the FDT for NESS (4.29) can be cast into another equivalent form. In fact, substituting (4.29) into (4.35) yields

$$\sigma_{ij}(Q) = 2\gamma_{ik}^{-1} a_{kl}(Q) \gamma_{lj}^{T-1} T^{\text{eff}} = T^{\text{eff}}(\gamma^{-1} + \gamma^{T-1})_{ij}, \tag{4.46}$$

i.e., the symmetric part of γ^{-1} is nothing but the diffusion matrix divided by twice the effective temperature. Equation (4.46) is another form of FDT in close analogy with the Einstein relation.

4.4.2. Antisymmetric part

The antisymmetric part of γ_{ij} is more complicated. According to the discussion in section 2.5, the Lehmann representation can be written as

$$G_{rij}(k_0) = \int \frac{dk_0'}{2\pi i} \frac{C_{ij}(k_0')}{k_0' - k_0 - i\varepsilon}, \tag{4.47}$$

where

$$C_{ij}(k_0) = -i \int \text{Tr}\{[Q_i(x_0), Q_j(0)]\hat{\rho}\} \exp(ik_0 x_0)\, dx_0 . \tag{4.48}$$

Integrating (4.48) over k_0 gives the expectation value of the equal time commutator

$$\int C_{ij}(k_0) \frac{dk_0}{2\pi} = -i\, \text{Tr}\{[Q_i(x_0), Q_j(0)]\hat{\rho}\}|_{x_0=0} \equiv f_{ij}(Q). \tag{4.49}$$

We then obtain from (4.47) and (4.49) that

$$\lim_{k_0\to\infty} k_0 G_{rij}(k_0, Q) = i f_{ij}(Q), \tag{4.50}$$

i.e., $\Gamma_{rij}(k_0)$ increases linearly with k_0 as $k_0 \to \infty$. We can thus write down a subtracted dispersion relation as

$$f_{ij}\Gamma_{rjk}(k_0) = -ik_0\delta_{ik} + f_{ij}\Gamma_{rjk}(k_0 = 0) + k_0 \int \frac{dk_0'}{2\pi} \frac{f_{ij}\Delta_{jk}(k_0')}{(k_0' - k_0 - i\varepsilon)k_0'} . \tag{4.51}$$

Differentiating (4.51) with respect to k_0 and setting $k_0 \to 0$, we find that

$$f_{ij} d_{jk} = -i\delta_{ik} + \Delta_{ik}, \tag{4.52}$$

where

$$\left.\frac{\partial \Gamma_r}{\partial k_0}\right|_{k_0=0} = d + ia, \tag{4.53}$$

$$\Delta_{ik} = P \int_{-\infty}^{\infty} \frac{dk_0'}{2\pi} \frac{f_{ij} A_{jk}(k_0')}{k_0'^2}$$
$$= f_{ij} \int_{0}^{\infty} \frac{dk_0'}{\pi} \frac{A_{jk}(k_0') - A_{kj}(k_0')}{k_0'^2} = f_{ij} \bar{\Delta}_{jk} . \tag{4.54}$$

In deriving (4.52) we have made use of the consequence of the potential condition

$$\lim_{k_0 \to 0} A_{jk}(k_0) = k_0 a_{jk}(k_0) . \tag{4.55}$$

Since $a_{jk}(k_0)$ is a symmetric matrix, $\bar{\Delta}_{jk}$ is finite.
Solving $(d^{-1})_{ij}$ from (4.52) we obtain

$$(d^{-1})_{ij} = \{(I + \mathrm{i} f \bar{\Delta})^{-1} \mathrm{i} f\}_{ij} . \tag{4.56}$$

The antisymmetric part of γ^{-1} can thus be written as

$$\gamma^{-1}_{[i,j]} = -(a - \mathrm{i}d)^{-1}_{[i,j]} \approx -\mathrm{i}d^{-1}_{ij} + \mathrm{O}(a^2)$$
$$= -\mathrm{i}\{(I + \mathrm{i} f \bar{\Delta})^{-1} \mathrm{i} f\}_{ij} + \mathrm{O}(a^2) = f_{ij} + \mathrm{O}(a^2, \bar{\Delta}) . \tag{4.57}$$

Usually f_{ij} itself is considered as the antisymmetric part of γ^{-1}. However, as seen from our discussion, this approximate result is true if both high order effects of dissipation a^2 and the dispersion $\bar{\Delta}$ are neglected.

4.4.3. Generalized TDGL equation

Substituting (4.46) and (4.57) into the TDGL equation (3.88) and setting $J = 0$, we obtain

$$\frac{\partial Q_i(t)}{\partial t} = -\gamma^{-1}_{(i,j)} \frac{\delta \mathcal{F}}{\delta Q_j} - \gamma^{-1}_{[i,j]} \frac{\delta \mathcal{F}}{\delta Q_j}$$
$$= -\frac{1}{2T^{\mathrm{eff}}} \sigma_{ij} \frac{\delta \mathcal{F}}{\delta Q_j} - f_{ij} \frac{\delta \mathcal{F}}{\delta Q_j} , \tag{4.58}$$

where the first term is associated with the irreversible dissipation, whereas the second term is the reversible part due to canonical motion. In many practical problems Q_i are either conserved quantities or a linear representation of some Lie group. In both cases

$$f_{ij}(Q) = f^k_{ij} Q_k , \tag{4.59}$$

where f^k_{ij} are either structure constants of the symmetry group or elements of the representation matrix. The TDGL equation (4.58) derived in the CTPGF formalism has the same form as that used in critical dynamics and other related fields [61, 68]. The second term in (4.58) is usually called the mode-coupling term. Previously, we derived the explicit form of (4.58) by comparing the TDGL equation with the Ward–Takahashi identities. Nonetheless, the present derivation is more straightforward.

In closing this section we note that as consequences of microscopic reversibility near NESS, many concepts valid for systems in thermoequilibrium can be generalized to these cases. The potential condition, the FDT as well as the reciprocal relations for diffusion and relaxation matrices which constitute the basis for a semiphenomenological consideration of nonequilibrium processes, can be justified within the CTPGF formalism.

5. Theory of nonlinear response

As we mentioned in the last section, the linear response theory near thermoequilibrium [63–65], centered on the FDT and the Onsager reciprocity relations, belongs to one of the most successful chapters of nonequilibrium statistical mechanics. For the last twenty years it has been generalized in two directions, namely, to linear response near NESS as we discussed in the last section and to nonlinear response near thermoequilibrium. In spite of few formal developments [78–82], the latter issue has not become an active field of research. In fact there are some reasons for such slow-footed advance in nonlinear response theory.

First, the nonlinear response depends not only on the intrinsic properties of the system under consideration, but also on the boundary conditions to remove the heat generated in the nonlinear processes. Second, except for nonlinear optics, there was no urgent need for nonlinear response theory from the experimental point of view. Third, the formulation of the nonlinear theory became tedious due to lack of appropriate framework.

However, things are changing. Progress in pico-, even femtosecond pulse technique and multichannel data acquisition and processing make the detection of multi-time response available. Such measurements will certainly yield much more detailed information on the intrinsic properties of the system provided the nonlinear effects are essential. On the other hand, the development of the CTPGF formalism has furnished a suitable theoretical framework for such nonlinear analysis. This problem has been studied by Hao et al. [50, 51]. Although the discussion for the time being is still rather formal and is restricted to the case of "mechanical" disturbance, i.e., expressible by an additional term in the Hamiltonian, it will serve as a good starting point for further development. As we will summarize in this section, many relations which in principle can be obtained by other, more sophisticated, techniques, appear rather simple and natural in the CTPGF approach.

The general expressions for nonlinear response are presented in section 5.1, whereas different relations among these functions including algebraic, KMS [63, 83], time-reversal and spectral, are discussed in section 5.2. A plausible generalization of FDT in nonlinear case is sketched in section 5.3. In this section we do not write out explicitly the factor $(-i)^{n-1}$ in the definition of Green's functions.

5.1. General expressions for nonlinear response

5.1.1. Model

As mentioned in section 2, the CTPGF generating functional can include the physical field $J(x) = J_c(x)$ by setting $J_\Delta(x) = 0$, i.e. $J_+(x) = J_-(x)$. Therefore, the high order response functions are contained in the expansions of the generating functional (2.76) and (2.78). Still, for convenience we will write down here the explicit expressions for nonlinear response.

Assume the system was in equilibrium state described by the Hamiltonian $\mathcal{H}_0$ and the density matrix

$$\rho_0 = Z^{-1}\exp(-\beta\mathcal{H}_0), \qquad Z = \mathrm{Tr}(\exp(-\beta\mathcal{H}_0)) \tag{5.1}$$

in the remote past $t_0 = -\infty$ and then a time-dependent external field $J(t)$ has been switched on adiabatically to derive the system out from equilibrium. The field $J(t)$ is coupled to the dynamical variable Q which might be a composite operator, so the total Hamiltonian becomes

$$\mathcal{H} = \mathcal{H}_0 - J(t)Q\,. \tag{5.2}$$

As before, we use in this section the abbreviated notation

$$JQ \equiv \sum_i \int \mathrm{d}x\, J_i(x, t)Q_i(x)\,, \tag{5.3}$$

sometimes making explicit only the time variable. Here we consider the linear coupling, but, obviously, other cases such as $(\boldsymbol{n}\cdot\boldsymbol{E})^2$, $(\boldsymbol{n}\cdot\boldsymbol{H})^2$ coupling in the liquid crystal or $E_iE_jP_{ij}$ (P_{ij} being the polarization tensor) coupling in second-order light scattering, can be treated as well.

We should mention that the subscript "0" for $\mathcal{H}_0$ and ρ_0 does not mean free of interaction between particles of the system. In this section we consider the perturbation expansion in powers of the external field, whereas all interaction effects within the system are included in all Green functions appearing later like Δ_{21}, Δ_{211}, etc.

According to the very spirit of statistical mechanics, an average should be taken over the initial distribution, whereas the evolution of the system is described by the dynamical equation. If we know the various average values $\langle Q_i(t)\rangle$, $\langle Q_i(1)Q_j(2)\rangle$ and so on, we would have more and more detailed information of the nonequilibrium properties for the system.

So far about the classical system. There is an additional problem for the quantum case. Not every product of operators here corresponds to a physical observable. The question of operator ordering occurs. According to Dirac [84], to present an observable (i) the product must be a real operator (ii) which has a complete set of eigenstates and (iii) satisfies certain supplementary physical conditions. We will consider Hermitian operators and use the Hermicity as a guide line for operator product, but in general it is hard to say anything about the completeness. For example, the combinations $AB + BA$ and $\mathrm{i}(AB - BA)$ are both Hermitian, while their expectation values are correspondingly $G_c = G_{22}$ and $G_{21} - G_{12} = G_r - G_a$. Since the problem of operator ordering in the quanto-classical correspondence has not yet been solved, an assumption is made in [50, 51] that the quantum counterpart of the average for the product of dynamical variables is just Green's function $\hat{G}$ with all subscripts equal to 2. In the $n = 2$ case, G_{22} is the fully symmetrized correlation function, whereas for $n > 2$ cases they are only partially symmetrized averages.

5.1.2. Analytic expansion of the generating functional

In this section (section 5.1) we denote Green's functions with external field by G, whereas those without field by Δ. Using the analytic expansion of the generating functional $W[J]$,

$$W[J] = \sum_{n=1}^{\infty} \frac{1}{n!}\Delta_p(1\cdots n)J(1)\cdots J(n)\,, \tag{5.4}$$

where

$$\Delta_p(1\cdots n)=\frac{\delta^n W[J]}{\delta J(1)\cdots\delta J(n)}\bigg|_{J=0} \tag{5.5}$$

are connected propagators without external field. It then immediately follows that

$$G_p(1)\equiv\langle T_pQ(1)\rangle=\Delta_p(1)+\Delta_p(12)J(2)+\tfrac{1}{2}\Delta_p(123)J(2)J(3)+\cdots$$

$$G_p(12)\equiv\langle T_p(Q(1)Q(2))\rangle=\frac{\delta G_p(1)}{\delta J(2)}=\Delta_p(12)+\Delta_p(123)J(3)+\cdots$$

$$G_p(123)\equiv\langle T_p(Q(1)Q(2)Q(3))\rangle=\frac{\delta G_p(12)}{\delta J(3)}=\Delta_p(123)+\Delta_p(1234)J(4)+\cdots. \tag{5.6}$$

To transform (5.6) into physical representation we need to insert the Pauli matrix σ_1 [44] (cf. (2.76)) to obtain

$$\tilde G(1)=\tilde\Delta(1)+\tilde\Delta(12)(\sigma_1 J)(2)+\frac{1}{2!}\tilde\Delta(123)(\sigma_1 J)(2)(\sigma_1 J)(3)+\cdots,$$

$$\tilde G(12)=\tilde\Delta(12)+\tilde\Delta(123)(\sigma_1 J)(3)+\frac{1}{2!}\tilde\Delta(1234)(\sigma_1 J)(3)(\sigma_1 J)(4)+\cdots,$$

$$\tilde G(123)=\tilde\Delta(123)+\tilde\Delta(1234)(\sigma_1 J)(4)+\cdots. \tag{5.7}$$

5.1.3. Higher order response

In accord with the discussion at the beginning of this section we need only retain the "all 2" components of (5.7) and set $J_+=J_-$ to get the general expressions for nonlinear response

$$G_2(1)=\Delta_2(1)+\Delta_{21}(12)J(2)+\frac{1}{2!}\Delta_{211}(123)J(2)J(3)+\cdots,$$

$$G_{22}(12)=\Delta_{22}(12)+\Delta_{221}(123)J(3)+\frac{1}{2!}\Delta_{2211}(1234)J(3)J(4)+\cdots,$$

$$G_{222}(123)=\Delta_{222}(123)+\Delta_{2221}(1234)J(4)+\cdots. \tag{5.8}$$

The first expression is a short writing of (2.115). The first two formulas of (5.8) were obtained in [78] by explicit manipulation of integrals. In the CTPGF formalism the structure of high order terms is obvious. It is worthwhile to note that the causality is guaranteed in each step of derivation using CTPGF as emphasized in section 2.4. Terms contradicting the causality drop out in accord with (2.109)–(2.111).

Following the convention set in the literature [78–82], Δ_{21}, Δ_{211}, Δ_{2111} should be called response functions of the averaged physical observable to external field, while their Fourier transforms are the admittance functions of various order. Accordingly, G_{22}, $G_{222}, \ldots$, and Δ_{22}, $\Delta_{222}, \ldots$, are called, respectively, nonequilibrium and equilibrium fluctuations of different order, whereas Δ_{221}, $\Delta_{2211}, \ldots$, are response functions of these fluctuations to the external field.

Other components of (5.7) like nonequilibrium retarded function

$$G_{21}(12) = \Delta_{21}(12) + \Delta_{211}(123)J(3) + \frac{1}{2!}\Delta_{2111}(1234)J(3)J(4) + \cdots, \tag{5.9}$$

which is the straightforward extension of the usual response function

$$\Delta_{21}(12) = \left.\frac{\delta\langle Q(1)\rangle}{\delta J(2)}\right|_{J=0}, \tag{5.10}$$

to

$$G_{21}(12) = \frac{\delta\langle Q(1)\rangle}{\delta J(2)}, \tag{5.11}$$

by including the higher order effects of the external field, may be named as nonlinear response functions. In fact, by integrating (5.9) over $J(2)$ we recover the first equation of (5.8). Therefore, no additional information is contained in (5.9) and in terms disappearing in going from (5.7) to (5.8) by setting $J_+ = J_-$.

5.1.4. Physically observable functions

Independent functions which in principle can be measured are those listed in the following table and their higher order extensions.

	Average	Two-point correlation	Three-point correlation	Four-point correlation
Without external field	Δ_2	Δ_{22}	Δ_{222}	Δ_{2222}
Linear response	Δ_{21}	Δ_{221}	Δ_{2221}	Δ_{22221}
Second-order response	Δ_{211}	Δ_{2211}	Δ_{22211}	Δ_{222211}
Third-order response	Δ_{2111}	Δ_{22111}	Δ_{222111}	$\Delta_{2222111}$

In this table the function on the oblique line from lower left to upper right are components of the same $\tilde{\Delta}$, so that the relations indicated in [78, 80, 81] look very natural.

To sum up, the observables in nonlinear response theory are partially symmetrized correlation (fluctuation) functions $G_2, G_{22}, G_{222}, \ldots$, etc., as functions of the external field, in particular, their zero field derivatives

$$\Delta_{\underbrace{2\cdots2}_{k}\underbrace{1\cdots1}_{l}} = \frac{\delta^l}{\delta J^l} G_{22\cdots2}(12\cdots k). \tag{5.12}$$

The possibility of detecting them in practice depends on the nonlinearity inherent in the system, i.e., the magnitude of functions (5.12) and the strength of the external field.

5.2. General considerations concerning multi-point functions

The n-point $\tilde{G}$ has 2^n components, but not all of them are independent of each other. There are many constraints following from the normalization of the generating functional and the causality, due to the canonical distribution and the time reversal symmetry, etc. Some of these constraints (due to normalization and causality) were described before in section 2.4. Here we will elaborate further on these relations and discuss the others one by one.

5.2.1. Exact algebraic relations

First of all, let us recall that the "all one" components such as G_{11}, $G_{111}, \ldots$, vanish in accord with (2.99). Furthermore, eqs. (2.109)–(2.113) are also exact. Using the transformation formulas (2.87), (2.88) and (2.90) for $\hat{G}$ and $\tilde{G}$ functions we can easily rewrite (2.99) in different forms. For example, in the single time representation it can be written as (cf. (2.92))

$$\sum_{\alpha_1\cdots\alpha_n=\pm} (1+\alpha_1\cdots\alpha_n)G_{\alpha_1\cdots\alpha_n} = \sum_{\alpha_1\cdots\alpha_n=\pm} (1-\alpha_1\cdots\alpha_n)G_{\alpha_1\cdots\alpha_n} \tag{5.13}$$

in particular, for three-point functions we have

$$G_{+++} + G_{+--} + G_{-+-} + G_{--+} = G_{---} + G_{-++} + G_{+-+} + G_{++-}\,.$$

Equivalently, the same relation for three- and four-point functions can be transformed in the "retarded" combination as

$$\tfrac{1}{2}(G_{211} + G_{121} + G_{221}) = G_{+++} - G_{++-}\,, \tag{5.14a}$$

$$\tfrac{1}{2}(G_{2111} + G_{1211} + G_{2211}) = G_{++++} + G_{++--} - G_{+++-} - G_{++-+}\,, \tag{5.14b}$$

or into the "correlation" combination as

$$\tfrac{1}{2}(G_{112} + G_{122} + G_{212} + G_{222}) = G_{+++} + G_{++-}\,, \tag{5.15a}$$

$$\tfrac{1}{2}(G_{1122} + G_{1222} + G_{2122} + G_{2222}) = G_{++++} + G_{++--} + G_{++-+} + G_{+++-}\,. \tag{5.15b}$$

Equations (5.14a) and (5.15a) have been derived in [81] by tedious calculations without realizing its connection with (2.99). By careful inspection of (5.14) and (5.15) one can easily uncover the general rule to write down analogous formulas for higher order functions. We would like to emphasize, however, these formulas do not contain any new information in addition to (2.99).

Another set of exact algebraic relations follows from the properties of commutator (anti-commutator) and θ-function [51] on which we will not elaborate any further here. Nevertheless, we indicate here a symmetry relation following directly from the definition of CTPGF as

$$G_{\cdots 1\cdots 2\cdots}(\cdots i\cdots j\cdots) = G_{\cdots 2\cdots 1\cdots}(\cdots j\cdots i\cdots)\,. \tag{5.16}$$

5.2.2. KMS condition [63, 83]

As mentioned before, in the response theory the system under consideration is assumed to be in thermal equilibrium for $t_0 = -\infty$. Introducing a time-dependent external field, the expectation value of the operator product in the interaction picture (with respect to the external field) satisfies the following equation:

$$\langle Q_j(t)Q_i(t_1)\rangle = \langle Q_i(t_1)Q_j(t+\mathrm{i}\beta)\rangle = \exp(\mathrm{i}\beta\partial/\partial t)\langle Q_i(t_1)Q_j(t)\rangle\,. \tag{5.17}$$

Owing to the time translational invariance of equilibrium state, (5.17) can be rewritten as

$$\langle Q_j(t)Q_i(0)\rangle = \exp(\mathrm{i}\beta\partial/\partial t)\langle Q_i(0)Q_j(t)\rangle\,. \tag{5.18}$$

By taking Fourier transformation (5.18) becomes

$$\langle Q_j(\omega)Q_i(0)\rangle = \mathrm{e}^{\beta\omega}\langle Q_i(0)Q_j(\omega)\rangle\,. \tag{5.19}$$

This is the so-called KMS [63, 83] condition as emphasized in the mathematical treatment of statistical mechanics [85], because (5.19) still holds even when the density matrix is ill-defined for systems with infinite degrees of freedom. It follows immediately from (5.19) that

$$\langle\{Q_j(\omega), Q_i\}\rangle = \coth(\beta\omega/2)\langle[Q_j(\omega), Q_i]\rangle\,, \tag{5.20}$$

which is nothing but the FDT as given by (3.43a).

Now consider the multi-point functions. First of all, for any function invariant under time translation

$$F(t_1, t_2, \ldots, t_n) = F(t_1 - t_n, t_2 - t_n, \ldots, 0)\,, \tag{5.21}$$

we have

$$\sum_{i=1}^{n} \frac{\partial}{\partial t_i} F = 0\,,$$

or symbolically after Fourier transformation

$$\sum_{i=1}^{n} \omega_i = 0\,. \tag{5.22}$$

Next, consider the averaged product of n operators. Transposing the leftmost operator to the rightmost one by one, we get

$$\begin{aligned}\langle Q_1(1)Q_2(2)\cdots Q_n(n)\rangle &= \exp(\mathrm{i}\beta\partial_1)\langle Q_2(2)\cdots Q_n(n)Q_1(1)\rangle\\ &= \exp[\mathrm{i}\beta(\partial_1+\partial_2)]\langle Q_3(3)\cdots Q_1(1)Q_2(2)\rangle\\ &\ \ \cdots\\ &= \exp(-\mathrm{i}\beta\partial_n)\langle Q_n(n)Q_1(1)\cdots Q_{n-1}(n-1)\rangle\,,\end{aligned}$$

where

$$\partial_i \equiv \partial/\partial t_i . \tag{5.23}$$

This process can be stopped at any step, say, ith. We introduce the following two functions

$$F^{(-)}(1\cdots i, i+1\cdots n) \equiv \langle Q_1(1)\cdots Q_i(i)Q_{i+1}(i+1)\cdots Q_n(n)\rangle ,$$
$$F^{(+)}(1\cdots i, i+1\cdots n) \equiv \langle Q_{i+1}(i+1)\cdots Q_n(n)Q_1(1)\cdots Q_i(i)\rangle , \tag{5.24}$$

and write

$$F^{(-)}(1\cdots n) = \exp[\mathrm{i}\beta(\partial_1 + \cdots + \partial_i]F^{(+)}(1\cdots n) \tag{5.25}$$

which yields after Fourier transformation the generalized KMS condition

$$F^{(-)}(\omega_1\cdots\omega_n) = \exp[\beta(\omega_1 + \cdots + \omega_i)]F^{(+)}(\omega_1\cdots\omega_n) . \tag{5.26}$$

Defining two more functions

$$F^{(c)} \equiv F^{(-)} + F^{(+)} = \langle\{Q_1(1)\cdots Q_i(i), Q_{i+1}(i+1)\cdots Q_n(n)\}\rangle ,$$
$$F^{(a)} \equiv F^{(-)} - F^{(+)} = \langle[Q_1(1)\cdots Q_i(i), Q_{i+1}(i+1)\cdots Q_n(n)]\rangle ,$$

we find then

$$F^{(c)}(\omega_1\cdots\omega_i, \omega_{i+1}\cdots\omega_n) = \coth[\beta(\omega_1 + \cdots + \omega_i)/2]F^{(a)}(\omega_1\cdots\omega_i, \omega_{i+1}\cdots\omega_n) \tag{5.27}$$

which looks like a generalization of the FDT but actually it is not, because it connects only the symmetric and antisymmetric parts of the same function rather than different functions as in the two-point function case. Therefore, the KMS condition itself is not enough to give rise to the FDT. For example, the four-point function G_{++--} can be represented as

$$G_{++--}(1234) = \tfrac{1}{2}(-\mathrm{i})^3[\langle\tilde{T}(34)T(12)\rangle + \langle T(12)\tilde{T}(34)\rangle] + \tfrac{1}{2}(-\mathrm{i})^3[\langle\tilde{T}(34)T(12)\rangle - \langle T(12)\tilde{T}(34)\rangle] , \tag{5.28}$$

where the first term corresponds to $F^{(c)}$ whereas the second term to $F^{(a)}$. We can thus derive a relation similar to (5.27) which, as we said, is not the FDT.

5.2.3. Time reversal invariance

The implications of the time reversal symmetry for macrosystems have been discussed in section 4. Here we restrict ourselves to the situation when the conditions (4.4), (4.6) and (4.10) are satisfied. It has been shown there also that the average of product from Heisenberg operators transforms under time inversion as (cf. (4.17))

$$F_{1\cdots n}(1\cdots n; J, \lambda) = \varepsilon_1\cdots\varepsilon_n F_{n\cdots 1}(-n\cdots -1; \varepsilon J, \varepsilon\lambda)$$
$$= \varepsilon_1\cdots\varepsilon_n F^*_{1\cdots n}(-1\cdots -n; \varepsilon J, \varepsilon\lambda) , \tag{5.29}$$

where λ is a parameter specifying the initial state. The second line of (5.29) is based on the following equality

$$\mathrm{Tr}(AB\cdots K)=\mathrm{Tr}(K\cdots BA)^* \tag{5.30}$$

valid for Hermitian operators. It also follows from the Hermicity of Q and ρ_0, the density matrix, that

$$F_{1\cdots n}(1\cdots n;J,\lambda)=F^*_{n\cdots 1}(n\cdots 1;J,\lambda). \tag{5.31}$$

Since the CTPGF are linear combinations of averaged operator products, so that for systems with time reversal symmetry we can easily write down their transformations under time inversion using (5.28). For example,

$$\langle 3T(12)\rangle=\varepsilon_1\varepsilon_2\varepsilon_3\langle T(-1-2)-3\rangle,\qquad \langle T(12)3\rangle=\varepsilon_1\varepsilon_2\varepsilon_3\langle -3T(-1-2)\rangle.$$

Hereafter we do not specify the change of J and λ explicitly. We can split three-point function G_{++-} into symmetric and antisymmetric parts under time inversion (do not confuse it with $F^{(c)}$ and $F^{(a)}$ considered in the last subsection)

$$G_{++-}(123)=G^{\mathrm{S}}_{++-}(123)+G^{\mathrm{A}}_{++-}(123), \tag{5.32}$$

where

$$G^{\mathrm{S}}_{++-}(123)=G^{\mathrm{S}}_{++-}(-1-2-3)=\tfrac{1}{2}(-\mathrm{i})^2(\langle 3T(12)\rangle+\varepsilon^3\langle T(12)3\rangle),$$
$$G^{\mathrm{A}}_{++-}(123)=-G^{\mathrm{A}}_{++-}(-1-2-3)=\tfrac{1}{2}(-\mathrm{i})^2(\langle 3T(12)\rangle-\varepsilon^3\langle T(12)3\rangle) \tag{5.33}$$

with

$$\varepsilon^3\equiv\varepsilon_1\varepsilon_2\varepsilon_3. \tag{5.34}$$

Applying the KMS condition to (5.34) leads to

$$G^{\mathrm{S}}_{++-}(\omega_1\omega_2\omega_3)=\frac{1-\varepsilon^3\exp(\beta\omega_3)}{1+\varepsilon^3\exp(\beta\omega_3)}G^{\mathrm{A}}_{++-}(\omega_1\omega_2\omega_3). \tag{5.35}$$

5.2.4. Fourier transform and spectral representation

It is known that the Fourier transformation itself does not bring about any new information, but it is more convenient to incorporate the KMS condition and the time reversal symmetry in the Fourier space.

The Fourier transform of a time translationally invariant function can be written as

$$F(\omega_1\cdots\omega_n)=2\pi\delta(\omega_1+\cdots+\omega_n)F_1(\omega_1\cdots,t_k=0,\cdots\omega_n), \tag{5.36}$$

where all time arguments in F_1 are transformed except for $t_k=0$ with k being any of 1 to n. In other words, the Fourier transformation of the n-point function appears as

$$F(t_1\cdots t_n)=\int\frac{d\omega_1\cdots d\hat{\omega}_k\cdots d\omega_n}{(2\pi)^{n-1}}\exp\{-i[\omega_1(t_1-t_k)\cdots+\hat{\omega}_k+\cdots\omega_n(t_n-t_k)]\}\,,$$
$$\times F_1(\omega_1\cdots t_k=0\cdots\omega_n)\,, \tag{5.37}$$

where the ω variable with caret is missing. We remind the reader that the Fourier transform in this paper is not distinguished by any special symbols, the meaning being clear from the arguments of functions.

As shown in section 2.3, the CTPGF can be presented as a linear combination of products of the θ-function and n-point function. For example,

$$F_R(123)=\theta(123)F(123)\,. \tag{5.38}$$

Factorizing the θ-function as

$$\theta(123)=\theta(12)\theta(23)$$

and using the integral representation

$$\theta(t)=\frac{i}{2\pi}\int_{-\infty}^{\infty}\frac{d\Omega}{\Omega+i\varepsilon}\exp(-i\Omega t)\,, \tag{5.39}$$

we can write

$$F_{1R}(\omega_1\omega_2t_3=0)=\frac{1}{(2\pi)^2}\int\frac{d\Omega_1\,d\Omega_2F_1(\omega_1\omega_2t_3=0)}{(\omega_1-\Omega_1+i\varepsilon)(\omega_1+\omega_2-\Omega_2+i\varepsilon)}\,. \tag{5.40}$$

It is easy to read out from (5.40) the general rule to write down the spectral representation for any CTPGF. In particular, for the three-point retarded function

$$G_{211}(123)=-\theta(123)\langle[[1,2],3]\rangle-\theta(132)\langle[[1,3],2]\rangle\,, \tag{5.41}$$

by use of the KMS condition and the simplified notation

$$\langle 123\rangle=\langle 1-2,2-3\rangle=\int\frac{d\omega_1\,d\omega_2}{(2\pi)^2}\exp[-i\omega_1(t_1-t_2)-i\omega_2(t_2-t_3)]\langle\omega_1,\omega_2\rangle\,,$$

$$\langle 321\rangle=\langle 123\rangle^*=\int\frac{d\omega_1\,d\omega_2}{(2\pi)^2}\exp[-i\omega_1(t_1-t_2)-i\omega_2(t_2-t_3)]\langle-\omega_1,-\omega_2\rangle^*\,, \tag{5.42}$$

we find that

$$G_{211}(\omega_1\omega_2)=\frac{1}{(2\pi)^2}\int_{-\infty}^{\infty}\frac{d\Omega_1\,d\Omega_2}{(\omega_1-\Omega_1+i\varepsilon)(\omega_1+\omega_2-\Omega_2+i\varepsilon)}\{[1-\exp(-\beta\Omega_2)]\langle\Omega_1,\Omega_2-\Omega_1\rangle$$
$$+[1-\exp(\beta\Omega_2)]\langle-\Omega_1,-\Omega_2+\Omega_1\rangle^*\}+\frac{1}{(2\pi)^2}\int_{-\infty}^{\infty}\frac{d\Omega_1\,d\Omega_2}{(\omega_1-\Omega_1+i\varepsilon)(-\omega_2-\Omega_2+i\varepsilon)}$$
$$\times\{[\exp(-\beta\Omega_1)-\exp[\beta(\Omega_2-\Omega_1)]]\langle\Omega_1,-\Omega_2\rangle$$
$$+[\exp(\beta\Omega_1)-\exp[-\beta(\Omega_2-\Omega_1)]]\langle-\Omega_1,\Omega_2\rangle^*\}\,. \tag{5.43}$$

We see thus $G_{211}(\omega_1\omega_2)$ is analytic in the upper half-plane of ω_1. There is one pole in the upper half-plane of ω_2 depending on the position of the pole in ω_1. As functions of several complex variables, the analytic properties of multi-point functions are much more complicated compared with the two-point case.

5.3. Plausible generalization of FDT

Some authors claimed previously [78, 82] that they had already found the nonlinear generalization of the FDT, but indeed this was not the case. As we have seen in the last section, any of the three kinds of relations, namely, the algebraic, the KMS or the time reversal invariance, will not provide by themselves the generalization of the FDT needed. However, their combination might give what we would like to have.

5.3.1. Even and odd combinations

The transformation formula from $\tilde{G}$ to $\hat{G}$ as given by (2.88) can be rewritten as

$$G_{\alpha_1\cdots\alpha_n}(1\cdots n) = 2^{1-n} \sum_{i_1\cdots i_n=1,2} (\alpha_1)^{i_1}\cdots(\alpha_n)^{i_n} G_{i_1\cdots i_n}(1\cdots n)\,. \tag{5.44}$$

Under the change $\alpha_i \to -\alpha_i$ the T-product changes into $\tilde{T}$-product and a factor $(-1)^{i_1+\cdots+i_n}$ appears on the right-hand side of (5.44). Therefore, it is natural to distinguish the "even" and "odd" components of $\tilde{G}$ as G_{n,l_e} and G_{n,l_o}, where l_σ, $\sigma = \mathrm{e}, \mathrm{o}$ is the number of "1"s among i_k. The even components do not change sign under $\alpha_i \to -\alpha_i$ whereas the odd components do. As we will see later in this subsection, there are simple algebraic relations among the same type of G_{n,l_σ}, while the connection between these two types of components established by using the KMS condition and the time reversal invariance is the plausible generalization of the FDT.

First consider the "all +" and "all −" components of $\hat{G}$

$$G_{++\cdots+}(12\cdots n) = 2^{1-n} \sum_{i_1\cdots i_n=1,2} G_{i_1\cdots i_n}(1\cdots n)\,, \tag{5.45a}$$

$$G_{--\cdots-}(12\cdots n) = 2^{1-n} \sum_{i_1\cdots i_n=1,2} (-1)^{i_1+\cdots+i_n} G_{i_1\cdots i_n}(12\cdots n)\,. \tag{5.45b}$$

If we define the symmetrized combination of G_{n,l_σ} as

$$G^{\mathrm{S}}_{n,l_\sigma} \equiv {\sum_{i_1\cdots i_n=1,2}}' \tfrac{1}{2}[1+(-1)^{l_\sigma+i_1+\cdots+i_n}] G_{i_1\cdots i_n}(12\cdots n)\,, \tag{5.46}$$

where $'$ means that the number of "1"s should be equal to l_σ, then

$$\sum_{0\le l_e\le n} G^{\mathrm{S}}_{n,l_e} = 2^{n-2}[G_{++\cdots+}(12\cdots n) + G_{--\cdots-}(12\cdots n)]\,,$$

$$\sum_{1\le l_o\le n} G^{\mathrm{S}}_{n,l_o} = 2^{n-2}[G_{++\cdots+}(12\cdots n) - G_{--\cdots-}(12\cdots n)]\,. \tag{5.47}$$

Making use of the relation following from the time reversal invariance, we have

$$\varepsilon_1 \cdots \varepsilon_n G_{\alpha\alpha\cdots\alpha}(-1-2\cdots-n) = G_{\alpha\alpha\cdots\alpha}(12\cdots n) \tag{5.48}$$

for $\alpha = \pm$. It follows then immediately from (5.47) that

$$\sum_{0\le l_\sigma\le n} (G^{\mathrm{S}}_{n,l_\sigma} - \varepsilon_1\cdots\varepsilon_n G^{(-)\mathrm{S}}_{n,l_\sigma}) = 0\,, \tag{5.49}$$

where

$$G^{(-)\mathrm{S}}_{n,l_\sigma} \equiv G^{\mathrm{S}}_{n,l_\sigma}(-1\cdots-n)\,.$$

After Fourier transformation (5.49) becomes

$$\sum_{0\le l_\sigma\le n} (G^{\mathrm{S}}_{n,l_\sigma}(\omega) - \varepsilon_1\cdots\varepsilon_n G^{\mathrm{S}}_{n,l_\sigma}(-\omega))\,. \tag{5.50}$$

Since the components of $\hat{G}$ contain the factor $(-\mathrm{i})^{n-1}$ and an average of $(n-1)$ nested commutator and/or anticommutator, the G_{n_o,l_e} and G_{n_e,l_o} components are real, whereas G_{n_e,l_e} and G_{n_o,l_o} are imaginary with their Fourier components related with each other by the following equations

$$G^*_{n_o,l_e}(\omega) = G_{n_o,l_e}(-\omega), \qquad G^*_{n_e,l_o}(\omega) = G_{n_e,l_o}(-\omega)\,,$$
$$G^*_{n_e,l_e}(\omega) = -G_{n_e,l_e}(-\omega), \qquad G^*_{n_o,l_o}(\omega) = -G_{n_o,l_o}(-\omega)\,. \tag{5.51}$$

It is straightforward to see from (5.50) that the "even" and "odd" components are linearly dependent among themselves but not with each other, as we mentioned at the beginning of this section.

5.3.2. Nonlinear generalization of FDT

Now consider the other algebraic relations following from (5.44) as

$$G_{\alpha_1\cdots\alpha_n}(12\cdots n) \pm G_{-\alpha_1\cdots-\alpha_n}(12\cdots n)$$
$$= 2^{1-n} \sum_{i_1\cdots i_n=1,2} [1\pm(-1)^{i_1+\cdots+i_n}](\alpha_1)^{i_1}\cdots(\alpha_n)^{i_n} G_{i_1\cdots i_n}(12\cdots n)\,. \tag{5.52}$$

If $\{j\}$, i.e., $(j_1,\ldots,j_m)$ from α_k subscripts are "$-$" while the rest are "$+$", (5.52) can be rewritten as

$$G_{\{j-\}}(12\cdots n) \pm G_{\{j+\}}(12\cdots n)$$
$$= 2^{1-n} \sum_{i_1\cdots i_n=1,2} (1\pm(-1)^{i_1+\cdots+i_n})(-1)^{j'_1+\cdots+j'_m} G_{i_1\cdots i_n}(12\cdots n) \tag{5.53}$$

with $j'_r \equiv i_{j_r}$.

Defining

$$G_{l_e,\{j\}}(12\cdots n)\equiv\sum{}'\tfrac{1}{2}(1+(-1)^{i_1+\cdots+i_n})(-1)^{j_1+\cdots+j_m}G_{i_1\cdots i_n}(1\cdots n)\,,$$

$$G_{l_o,\{j\}}(12\cdots n)\equiv\sum{}'\tfrac{1}{2}(1-(-1)^{i_1+\cdots+i_n})(-1)^{j_1+\cdots+j_m}G_{i_1\cdots i_n}(1\cdots n)\,,\qquad(5.54)$$

we rewrite (5.53) as

$$\sum_{0\le l_e\le n}G_{l_e,\{j\}}=2^{n-2}(G_{\{j-\}}+G_{\{j+\}}),\qquad\sum_{1\le l_o\le n}G_{l_o,\{j\}}=2^{n-2}(G_{\{j-\}}-G_{\{j+\}})\,.\qquad(5.55)$$

Consider the combination

$$G_{\{j-\}}(1\cdots n)+\varepsilon^nG_{\{j-\}}(-1\cdots-n)\pm G_{\{j+\}}(1\cdots n)\pm\varepsilon^nG_{\{j+\}}(-1\cdots-n)$$

and make use of the following relations:

$$G_{\{j-\}}(1\cdots n)\equiv(-\mathrm{i})^{n-1}\langle\tilde{T}(j_1\cdots j_m)T(i_1\cdots i_{n-m})\rangle\,,$$

$$G_{\{j-\}}(-1\cdots-n)=\varepsilon^n\exp\Big[-\mathrm{i}\beta\sum_{i=1}^{m}\partial_{j_i}\Big]G_{\{j-\}}(1\cdots n)\,,\qquad(5.56)$$

with

$$\partial_{j_i}\equiv\partial/\partial t_{j_i}\,,$$

obtained from the KMS and the time reversal invariance conditions (5.25) and (5.29), we find

$$(1\pm1-\mathrm{e}^{-A}\mp\mathrm{e}^{A})\sum_{l_e}G_{l_e,\{j\}}+\varepsilon^n(1\pm1-\mathrm{e}^{A}\mp\mathrm{e}^{A})\sum_{l_e}G^{(-)}_{l_e,\{j\}}$$

$$=(-1\pm1+\mathrm{e}^{-A}\mp\mathrm{e}^{A})\sum_{l_o}G_{l_o,\{j\}}+\varepsilon^n(-1\pm1+\mathrm{e}^{A}\mp\mathrm{e}^{-A})\sum_{l_o}G^{(-)}_{l_o,\{j\}}\,,\qquad(5.57)$$

where

$$A\equiv\mathrm{i}\beta\sum_{i=1}^{m}\frac{\partial}{\partial t_{j_i}}\,,\qquad(5.58)$$

$$G^{(-)}\equiv G(-1\cdots-n)\,.\qquad(5.59)$$

After Fourier transformation we get finally

$$\sum_{0\le l_e\le n}[G_{l_e,\{j\}}(\omega)+\varepsilon^nG_{l_e,\{j\}}(-\omega)]=\coth\tfrac{1}{2}\beta(\omega_{j_1}+\cdots+\omega_{j_m})\sum_{1\le l_o\le n}[G_{l_o,\{j\}}(\omega)-\varepsilon^nG_{l_o,\{j\}}(-\omega)]\,,\qquad(5.60a)$$

$$\sum_{0\le l_e\le n} [G_{l_e,\{j\}}(\omega) - \varepsilon^n G_{l_e,\{j\}}(-\omega)] = \tanh \tfrac{1}{2}\beta(\omega_{j_1} + \cdots + \omega_{j_m}) \sum_{1\le l_o\le n} [G_{l_o,\{j\}}(\omega) + \varepsilon^n G_{l_o,\{j\}}(-\omega)] . \tag{5.60b}$$

This is a plausible generalization of FDT to the nonlinear case. For $n = 2$, (5.60a) is the usual FDT, while (5.60b) is an identity. For arbitrary n, we have $2^{n-1} - 1$ algebraic relations of type (5.52) (excluding all $\alpha_k = +$ cases) leading to $2^{n-1} - 1$ pairs of relations given by (5.60) in combination with the KMS and the time reversal invariance conditions.

To illustrate (5.60) we write down explicitly the corresponding expressions for $n = 3$ case

$$\begin{aligned}
&G_{0,\{1\}} = G_{0,\{2\}} = G_{0,\{3\}} = G_{222} ,\\
&G_{1,\{1\}} = -G_{122} + G_{212} + G_{221}, \qquad G_{1,\{2\}} = G_{122} - G_{212} + G_{221} , \qquad G_{1,\{3\}} = G_{122} + G_{212} - G_{221},\\
&G_{2,\{1\}} = -G_{112} - G_{121} + G_{211}, \qquad G_{2,\{2\}} = -G_{112} + G_{121} - G_{211}, \qquad G_{2,\{3\}} = G_{112} - G_{121} - G_{211} ,
\end{aligned} \tag{5.61}$$

$$\begin{aligned}
&G_{2,\{j\}}(\omega) + \varepsilon^3 G_{2,\{j\}}(-\omega) + G_{222}(\omega) + \varepsilon^3 G_{222}(-\omega) = \coth(\beta\omega_j/2)[G_{1,\{j\}}(\omega) - \varepsilon^3 G_{1,\{j\}}(-\omega)] ,\\
&G_{2,\{j\}}(\omega) - \varepsilon^3 G_{2,\{j\}}(-\omega) + G_{222}(\omega) - \varepsilon^3 G_{222}(-\omega)\\
&\quad = \tanh(\beta\omega_j/2)[G_{1,\{j\}}(\omega) + \varepsilon^3 G_{1,\{j\}}(-\omega)], \qquad j = 1, 2, 3 .
\end{aligned} \tag{5.62}$$

The question whether (5.60) is a correct generalization of FDT should be settled by further studies of nonlinear phenomena. We should mention, however, that Tremblay et al. [38] have considered the heating effects in electric conduction processes using CTPGF formalism. These authors do not discuss the general relations as we do here. In their opinion, the FDT should be model dependent in the nonlinear case.

6. Path integral representation and symmetry breaking

The generating functional $Z[J(x)]$ for CTPGF can be presented as a Feynman path integral. In terms of the eigenstate $|\varphi'(x)\rangle$ of the operator $\varphi(x, t = -\infty)$ the density matrix can be written as

$$\hat\rho = \int d\varphi'(x)\, d\varphi''(x) |\varphi'(x)\rangle \rho_{\varphi'\varphi''} \langle \varphi''(x)| , \tag{6.1}$$

so that the generating functional is given by

$$Z[J(x)] = \int d\varphi'(x)\, d\varphi''(x) \rho_{\varphi'\varphi''} \langle \varphi''(x)| U(t_- = -\infty, t_+ = -\infty) |\varphi'(x)\rangle , \tag{6.2}$$

where

$$U(t_- = -\infty, t_+ = -\infty) \equiv S_p = T_p \exp\Big(\mathrm{i} \int_p J(x)\varphi(x)\Big) \tag{6.3}$$

is the evolution operator defined along the closed time-path.

It is known in the quantum field theory [39] that

$$\langle\varphi_2(\boldsymbol{x})|U(t_2,t_1)|\varphi_1(\boldsymbol{x})\rangle = N\int[\mathrm{d}\varphi(x)]\exp\left(\mathrm{i}\int_{t_1}^{t_2}\mathscr{L}(\varphi(x))\,\mathrm{d}^{d+1}x\right)\delta(\varphi(\boldsymbol{x},t_2)-\varphi_2(\boldsymbol{x}))\delta(\varphi(\boldsymbol{x},t_1)-\varphi_1(\boldsymbol{x}))\,, \tag{6.4}$$

where N is a constant. The path integral representation (6.4) is valid for any t_1, t_2, so (6.2) can be rewritten as

$$Z[J(x)] = N\int[\mathrm{d}\varphi(x)]\delta(\varphi(\boldsymbol{x},t_+=-\infty)-\varphi'(\boldsymbol{x}))\rho_{\varphi'\varphi''}\delta(\varphi(\boldsymbol{x},t_-=-\infty)-\varphi''(\boldsymbol{x}))$$
$$\times\exp\left[\mathrm{i}\int_p(\mathscr{L}(\varphi(x))+J(x)\varphi(x))\right], \tag{6.5}$$

where the integration in the exponent is carried out over the closed time-path p. Since the functional dependence of $\rho_{\varphi'\varphi''}$ upon $\varphi'(\boldsymbol{x})$ and $\varphi''(\boldsymbol{x})$ is rather complicated, in general (6.5) is not very suitable for practical calculation. However, it is useful for discussing the symmetry properties of the generating functional so far as the total Lagrangian of the system appears in the exponent. We will use this representation to discuss the Ward–Takahashi (WT) identities and the Goldstone theorem following from the symmetry (section 6.3).

The path integral representation would be well adapted to the practical calculation if the contribution of the density matrix can be expressed in terms of effective Lagrangian in certain simplifying cases. This possibility will be considered in section 6.1. In section 6.2 we briefly discuss the properties of the order parameter and describe two different types of phase transitions. Finally, in section 6.4, the path integral representation is used to consider the fluctuation effects.

6.1. Initial correlations

In this section we derive two equivalent expressions for the generating functional to incorporate the effects of the initial correlation in a convenient way [46].

6.1.1. Model

Consider multi-component nonrelativistic field $\psi_b^\dagger$, ψ_b, $b = 1, 2\cdots n$ which may be either boson or fermion. The action of the system is given by

$$I[\psi^\dagger,\psi] = I_0[\psi^\dagger,\psi] + I_{\rm int}[\psi^\dagger,\psi]\,, \tag{6.6}$$

where the free part can be written as

$$I_0[\psi^\dagger,\psi] = \int_p \mathrm{d}1\,\mathrm{d}2\,\psi^\dagger(1)S_0^{-1}(1,2)\psi(2) \equiv \psi^\dagger S_0^{-1}\psi\,, \tag{6.7}$$

with

$$S_0^{-1} = \tfrac{1}{2}(\xi S_{0r}^{-1}\eta^{\dagger} + \eta S_{0a}^{-1}\xi^{\dagger} + \xi S_{0c}^{-1}\xi^{\dagger}),$$
$$S_{0r}^{-1} = S_{0a}^{-1} = [\mathrm{i}\partial_t + (1/2m)\nabla^2]\delta^{d+1}(1-2), \qquad S_{0c}^{-1} = 0\,, \tag{6.8}$$

in accord with (2.23), (2.63) and (3.10).

6.1.2. *First expression for W_p^N*

The generating functional

$$Z_p[J^{\dagger}J] \equiv \mathrm{Tr}\{\hat{\rho}T_p \exp[\mathrm{i}(J^{\dagger}\psi + \psi^{\dagger}J)]\} \tag{6.9}$$

can be rewritten as

$$Z_p[J^{\dagger}J] = \exp\left\{\mathrm{i}I_{\mathrm{int}}\left[\mp\mathrm{i}\frac{\delta}{\delta J}, -\mathrm{i}\frac{\delta}{\delta J^{\dagger}}\right]\right\}\exp\{\mathrm{i}[-J^{\dagger}S_0 J + W_p^N(J^{\dagger}, J)]\} \tag{6.10}$$

in the incoming picture by using the Wick theorem generalized to the CTPGF case as done in section 2.2 for the Hermitian boson field. Here S_0 is the bare propagator satisfying the equation

$$\int_p \mathrm{d}x\, S_0(1,x)S_0^{-1}(x,2) = \int_p \mathrm{d}x\, S_0^{-1}(1,x)S_0(x,2) = \delta_p(1-2)\,, \tag{6.11}$$

and

$$\exp\{\mathrm{i}W_p^N[J^{\dagger}, J]\} = \mathrm{Tr}\{\hat{\rho}\!:\exp[\mathrm{i}(J^{\dagger}\psi_{\mathrm{I}} + \psi_{\mathrm{I}}^{\dagger}J)]\!:\}\,, \tag{6.12}$$

i.e.,

$$W_p^N[J^{\dagger}, J] = \sum_{m,n=1}^{\infty}\frac{1}{m!\,n!}\int_p \mathrm{d}1\cdots\mathrm{d}m\,\mathrm{d}\bar{1}\cdots\mathrm{d}\bar{n}\,J^{\dagger}(1)\cdots J^{\dagger}(m)W_p^{(m,n)}(1\cdots m, \bar{n}\cdots\bar{1})J(\bar{n})\cdots J(\bar{1})\,, \tag{6.13}$$

$$W_p^{(m,n)}(1\cdots m, \bar{n}\cdots\bar{1}) = \mathrm{i}^{m+n-1}\,\mathrm{Tr}\{\hat{\rho}\!:\!\psi_{\mathrm{I}}(m)\cdots\psi_{\mathrm{I}}(1)\psi_{\mathrm{I}}^{\dagger}(\bar{1})\cdots\psi_{\mathrm{I}}^{\dagger}(\bar{n})\!:\}\,, \tag{6.14}$$

where ψ_{I}, $\psi_{\mathrm{I}}^{\dagger}$ are operators in the incoming picture and : : means normal product.

Since the time ordering does not have any effects under normal product, we can rewrite (6.12) as

$$\exp\{\mathrm{i}W^N[J_{\Delta}^{\dagger}, J_{\Delta}]\} = \exp\{\mathrm{i}W_p^N[J^{\dagger}, J]\} = \mathrm{Tr}\{\hat{\rho}\!:\!\exp[\mathrm{i}(J_{\Delta}^{\dagger}\psi_{\mathrm{I}} + \psi_{\mathrm{I}}^{\dagger}J_{\Delta})]\!:\}\,, \tag{6.15}$$

where

$$J_{\Delta} = J_+ - J_-, \qquad J_{\Delta}^{\dagger} = J_+^{\dagger} - J_-^{\dagger}\,. \tag{6.16}$$

We can then write down an expansion equivalent to (6.13) as

$$W^N[J_\Delta^\dagger, J_\Delta] = \sum_{m,n=1}^{\infty} \frac{1}{m!n!} (J_\Delta^\dagger)^m W^{(m,n)}(J_\Delta)^n \tag{6.17}$$

with

$$W^{(m,n)} = \mathrm{i}^{m+n-1}\,\mathrm{Tr}\{\hat\rho{:}\psi_\mathrm{I}(m)\cdots\psi_\mathrm{I}(1)\psi_\mathrm{I}^\dagger(\bar 1)\cdots\psi_\mathrm{I}^\dagger(\bar n){:}\}\,. \tag{6.18}$$

We note that W_p^N and $W_p^{(m,n)}$ are defined on the closed time-path, whereas W^N and $W^{(m,n)}$ are defined on the ordinary time axis.

Taking into account that in the incoming picture the field operators satisfy the free field equation, we have

$$\int_p \mathrm{d}i'\,\vec S_0^{-1}(i,i')W_p^{(m,n)}(1\cdots i'\cdots m,\bar n\cdots\bar 1) = 0\,,$$

$$\int_p \mathrm{d}i'\,W_p^{(m,n)}(1\cdots m,\bar n\cdots i'\cdots \bar 1)\overleftarrow{S}_0^{-1}(i',\bar 1) = 0\,, \tag{6.19}$$

or equivalently,

$$\int \mathrm{d}i'\,\vec S_{0\mathrm{r}}^{-1}(i,i')W^{(m,n)}(1\cdots i'\cdots m,\bar n\cdots\bar 1) = 0\,,$$

$$\int \mathrm{d}i'\,W^{(m,n)}(1\cdots m,\bar n\cdots i'\cdots \bar 1)\overleftarrow{S}_{0\mathrm{a}}^{-1}(i',\bar i) = 0\,. \tag{6.20}$$

As far as the initial condition is fixed at $t = -\infty$, we are not allowed to integrate by parts arbitrarily with respect to $\partial/\partial t$. The correct direction of acting $\partial/\partial t$ is indicated by the arrow.

Substituting (6.15) into (6.10) we get the first expression for $Z[J]$ we would like to derive as

$$Z[J^\dagger J] = \exp\Big\{\mathrm{i}I_{\mathrm{int}}\Big[\mp\mathrm{i}\frac{\partial}{\partial J}, -\mathrm{i}\frac{\partial}{\partial J^\dagger}\Big]\Big\}\exp\{\mathrm{i}(-J^\dagger S_0 J + W^N[J_\Delta^\dagger, J_\Delta])\}\,, \tag{6.21}$$

from which we can obtain the generalized Feynman rule. We see from (6.21) that the density matrix affects only the correlation generated by $J_\Delta^\dagger$ and J_Δ. So far as $W^{(m,n)}$ satisfy eq. (6.20), the contribution of the density matrix can be expressed in terms of the initial (sometimes called boundary) condition for Green's function.

6.1.3. *Second expression for W_p^N*

Now we derive another expression for the CTPGF generating functional. Using the following identity (up to an unimportant constant)

$$\exp(-\mathrm{i}\,J^{\dagger}S_0J) = \int_p [\mathrm{d}\psi^{\dagger}][\mathrm{d}\psi]\,\exp\{\mathrm{i}(\psi^{\dagger}S_0^{-1}\psi + J^{\dagger}\psi + \psi^{\dagger}J)\}\,, \tag{6.22}$$

it is easy to show that

$$\exp\{\mathrm{i}(-J^{\dagger}S_0J + W_p^N[J^{\dagger}, J])\}$$

$$= \int_p [\mathrm{d}\psi^{\dagger}][\mathrm{d}\psi]\,\exp\{\mathrm{i}(J^{\dagger}\psi + \psi^{\dagger}J)\}\,\exp\left\{\mathrm{i}\,W_p^N\left[\pm\mathrm{i}\frac{\delta}{\delta\psi}, \mathrm{i}\frac{\delta}{\delta\psi^{\dagger}}\right]\right\}\exp(\mathrm{i}\psi^{\dagger}S_0^{-1}\psi)\,, \tag{6.23}$$

if the path integration is taken by parts. Taking into account that

$$\frac{\delta}{\delta\psi}\exp(\mathrm{i}\psi^{\dagger}S_0^{-1}\psi) = \exp(\mathrm{i}\psi^{\dagger}S_0^{-1}\psi)\left(\frac{\delta}{\delta\psi} \pm \mathrm{i}\psi^{\dagger}\overleftarrow{S}_0^{-1}\right),$$

$$\frac{\delta}{\delta\psi^{\dagger}}\exp(\mathrm{i}\psi^{\dagger}S_0^{-1}\psi) = \exp(\mathrm{i}\psi^{\dagger}S_0^{-1}\psi)\left(\frac{\delta}{\delta\psi^{\dagger}} + \mathrm{i}\vec{S}_0^{-1}\psi\right), \tag{6.24}$$

(6.23) can be transformed into

$$\exp\{\mathrm{i}(-J^{\dagger}S_0J + W_p^N[J^{\dagger}, J])\} = \int_p [\mathrm{d}\psi^{\dagger}][\mathrm{d}\psi]\,\exp\{\mathrm{i}(J^{\dagger}\psi + \psi^{\dagger}J + \psi^{\dagger}S_0^{-1}\psi)\}$$

$$\times\exp\left\{\mathrm{i}\,W_p^N\left[\pm\mathrm{i}\frac{\delta}{\delta\psi} - \psi^{\dagger}\overleftarrow{S}_0^{-1}, \mathrm{i}\frac{\delta}{\delta\psi^{\dagger}} - \vec{S}_0^{-1}\psi\right]\right\}. \tag{6.25}$$

Using (6.8) and the convention agreed in (2.69)–(2.74) we find that

$$\eta^{\dagger}\frac{\delta}{\delta\psi(x_{\sigma})} = \frac{\delta}{\delta\psi_{\mathrm{c}}(x)}\,, \qquad \eta^{\dagger}\frac{\delta}{\delta\psi^{\dagger}(x_{\sigma})} = \frac{\delta}{\delta\psi_{\mathrm{c}}^{\dagger}(x)}\,, \tag{6.26}$$

$$\int \mathrm{d}y\,\psi^{\dagger}(y)\sigma_3\overleftarrow{S}_0^{-1}(y, x)\eta = \int \mathrm{d}y\,\psi_{\Delta}^{\dagger}(y)\overleftarrow{S}_{0\mathrm{r}}^{-1}(y, x)\,,$$

$$\eta^{\dagger}\int \mathrm{d}y\,\vec{S}_0^{-1}(x, y)\sigma_3\psi(y) = \int \mathrm{d}y\,\vec{S}_{0\mathrm{a}}^{-1}(x, y)\psi_{\Delta}(y)\,, \tag{6.27}$$

and obtain from (6.15) that

$$\exp\left\{\mathrm{i}\,W_p^N\left[\pm\mathrm{i}\frac{\delta}{\delta\psi} - \psi^{\dagger}\overleftarrow{S}_0^{-1}, \mathrm{i}\frac{\delta}{\delta\psi^{\dagger}} - \vec{S}_0^{-1}\psi\right]\right\} = \exp\{\mathrm{i}\,W^N[-\psi_{\Delta}^{\dagger}\overleftarrow{S}_{0\mathrm{r}}^{-1}, -\vec{S}_{0\mathrm{a}}^{-1}\psi]\}$$

$$= \exp\{\mathrm{i}\,W_p^N[-\psi^{\dagger}\overleftarrow{S}_0^{-1}, \vec{S}_0^{-1}\psi]\}\,. \tag{6.28}$$

Substituting (6.28) into (6.25) and the resulting expression into (6.21), we find

$$Z_p[J^\dagger, J] = \int_p [\mathrm{d}\psi^\dagger][\mathrm{d}\psi] \exp\{\mathrm{i}(I_0[\psi^\dagger, \psi] + I_{\mathrm{int}}[\psi^\dagger, \psi] + J^\dagger\psi + \psi^\dagger J)\} \exp\{\mathrm{i} W_p^N[-\psi^\dagger \vec{S}_0^{-1}, -\vec{S}_0^{-1}\psi]\} \tag{6.29}$$

as the second path integral representation for the generating functional. It is easy to rederive (6.21) from (6.29), so these two expressions are equivalent to each other. Note that this expression is different from that given by (6.5) in so far as the contribution of the density matrix appears here as an additional term W_p^N in the action. According to (6.28), this term does not depend on field variables $\psi_c^\dagger$, ψ_c describing the dynamical evolution, but does depend on $\psi_\Delta^\dagger$, ψ_Δ describing the statistical correlation. It is also obvious that $W_p^N[-\psi^\dagger \vec{S}_0^{-1}, -\vec{S}_0^{-1}\psi]$ has nonvanishing contribution to the path integral only at the end points because of (6.19).

6.1.4. Two-step strategy

For a general nonequilibrium process, (6.29) can be rewritten as

$$Z_p[J^\dagger, J] = \exp\left\{\mathrm{i} W_p^N\left[\pm\mathrm{i}\frac{\delta}{\delta J}\vec{S}_0^{-1}, \mathrm{i}\vec{S}_0^{-1}\frac{\delta}{\delta J^\dagger}\right]\right\} Z_p^0[J^\dagger, J], \tag{6.30}$$

where

$$Z_p^0[J^\dagger, J] = \int_p [\mathrm{d}\psi^\dagger][\mathrm{d}\psi] \exp\{\mathrm{i}(I_0 + I_{\mathrm{int}} + J^\dagger\psi + \psi^\dagger J)\} \tag{6.31}$$

is the generating functional for the ground state. Since Z_p^0 has exactly the same structure on the closed time-path as that of the standard quantum field theory, we can first calculate Z_p^0 and then "put into" it the statistical information via (6.30). Such "two-step" strategy is well known in solving the Liouville problem in classical statistical mechanics.

Many interesting nonequilibrium phenomena can be described by a Gaussian process, i.e.,

$$W_p^{(m,n)}(1 \cdots m, \bar{n} \cdots \bar{1}) = 0 \quad \text{except for} \quad W_p^{(1,1)}(1, \bar{1}) \neq 0\,, \tag{6.32}$$

for which the contribution of the density matrix reduces to replacing the bare propagator S_{0p} by

$$G_{0p}(x, y) = S_{0p}(x, y) - W_p^{(1,1)}(x, y)\,. \tag{6.33}$$

For the thermoequilibrium case eq. (6.33), after Fourier transformation, is identical to (2.21). A more rigorous derivation of the diagrammatic expansion for thermoequilibrium will be given in section 9.1.

Another possibility of simplification comes about when the state is stationary due to the microscopic time reversal invariance, the generalized FDT then holds as shown in section 4.2. As seen from (6.20) and (6.21), $W_p^{(m,n)}$ as solutions of the homogeneous equation can be specified by the FDT.

To sum up, the determination of the CTPGF generating functional can be divided into two steps: To first "forget" about the density matrix in calculating the generating functional without the statistical information and then "put it into" the generating functional at the second step. In the general case this can be done using (6.30), but a significant simplification results if the initial correlation is Gaussian or a generalized FDT holds.

If we are interested in some order parameter $Q(x)$ which is a composite operator of the constituent field, we introduce an additional term $h(x)Q(x)$ in the action. The generating functional for the order parameter is given by

$$Z_p[h; J^\dagger, J] = \exp\left\{ihQ\left[\mp i\frac{\delta}{\delta J}, -i\frac{\delta}{\delta J^\dagger}\right]\right\}Z_p[J^\dagger, J] \tag{6.34}$$

in terms of $Z_p[J^\dagger, J]$ for the constituent field. The extension of (6.21), (6.29) and (6.30) to this case is obvious.

6.2. *Order parameter and stability of state*

It is well known that the vertex functional $\Gamma[\varphi_c]$ is most suitable for describing the symmetry breaking, inasmuch as it is expressed explicitly in terms of the order parameter φ_c. In section 2.2 we have derived an equation (2.48) satisfied by it. Before going on with the discussion of the order parameter, we rewrite this basic equation of the CTPGF formalism in another equivalent form.

6.2.1. *Functional form for the vertex equation*

Let the total action of the system be presented as

$$I_t = I + I_s = \int \mathcal{L}(\varphi(x)) + \int J(x)\varphi(x)\,. \tag{6.35}$$

The operator equation satisfied by $\varphi(x)$ is then

$$\frac{\delta I[\varphi(x)]}{\delta\varphi(x)} = -J(x)\,. \tag{6.36}$$

Here we use the Heisenberg picture including the external source. Multiplying (6.36) by the density matrix and taking trace, we find

$$-J(x) = \mathrm{Tr}\left\{\frac{\delta I[\varphi(x)]}{\delta\varphi(x)}\hat{\rho}\right\}.$$

Transforming into the Heisenberg picture without external source, we obtain

$$-J(x) = \frac{1}{Z[J(x)]}\mathrm{Tr}\left\{T_p\left(\frac{\delta I[\varphi(x)]}{\delta\varphi(x)}\exp(iJ\varphi)\right)\hat{\rho}\right\} = \frac{1}{Z[J(x)]}\frac{\delta I}{\delta\varphi(x)}\left[\varphi(x) = -i\frac{\delta}{\delta J(x)}\right]Z[J(x)]\,. \tag{6.37}$$

Using the commutation relation

$$-\mathrm{i}\frac{\delta}{\delta J(x)} Z[J(x)] = Z[J(x)]\left(\varphi_c(x) - \mathrm{i}\frac{\delta}{\delta J(x)}\right), \tag{6.38}$$

(6.37) can be rewritten as

$$-J(x) = \frac{\delta I}{\delta\varphi(x)}\left[\varphi_c(x) - \mathrm{i}\frac{\delta}{\delta J(x)}\right]. \tag{6.39}$$

Comparing (6.39) with (2.48) we find

$$\frac{\delta\Gamma[\varphi_c]}{\delta\varphi_c(x)} = \frac{\delta I}{\delta\varphi}\left[\varphi_c(x) - \mathrm{i}\frac{\delta}{\delta J(x)}\right], \tag{6.40}$$

which is a formal relation to derive the generating functional from the action.

As in the quantum field theory, the term $\mathrm{i}(\delta/\delta J(x))$ comes from the loop correction. Hence in the tree approximation

$$\frac{\delta\Gamma[\varphi_c(x)]}{\delta\varphi_c(x)} = \frac{\delta I[\varphi(x)]}{\delta\varphi(x)}\bigg|_{\varphi=\varphi_c}, \tag{6.41}$$

$$\Gamma[\varphi_c(x)] = \int_p \mathcal{L}(\varphi_c(x))\,\mathrm{d}^{d+1}x. \tag{6.42}$$

We note in passing that we did not emphasize the validity of (6.36) on the closed time-path, but it is true, so that we can put p in eq. (6.42).

6.2.2. *Two types of phase transitions*

Coming back to the order parameter itself, $\varphi_c(x)$ should be a solution of equation

$$\delta\Gamma/\delta\varphi_c(x) = 0 \tag{6.43}$$

in the absence of the external field. The $\varphi_c(x) = 0$ solution corresponds to the "normal" state, whereas the nonzero solution corresponds to the symmetry broken state. Then the question arises which state is more stable. Let us consider the fluctuation around a homogeneous solution φ_{c0} of (6.43)

$$\varphi_c(x) = \varphi_{c0} + \varphi_k \exp(-\mathrm{i}k\cdot x). \tag{6.44}$$

Put (6.44) into (6.43) and obtain the linear equation satisfied by φ_k as

$$\int_p \frac{\delta^2\Gamma}{\delta\varphi_c(x)\delta\varphi_c(y)}\bigg|_{\varphi_c=\varphi_{c0}} (\varphi_c(y) - \varphi_{c0})\,\mathrm{d}y = \int \Gamma_r(x, y)\varphi_k \exp(-\mathrm{i}k\cdot y)\,\mathrm{d}y = \Gamma_r(k)\varphi_k = 0. \tag{6.45}$$

As follows from (6.45), φ_k is zero unless k is a solution of the equation

$$\Gamma_r(k) = 0 . \tag{6.46}$$

Assume the solution of (6.46) to be

$$k_0 = \omega(\boldsymbol{k}) , \tag{6.47}$$

the fluctuation will decay in time, i.e., the state is stable if $\mathrm{Im}\,\omega(\boldsymbol{k}) < 0$. Otherwise, the state is unstable. According to (3.14)

$$\Gamma_r(k) = D(k) + \mathrm{i}A(k) , \tag{3.14'}$$

where

$$A(k) = -\tfrac{1}{2}\mathrm{i}(\Gamma_-(k) - \Gamma_+(k)) . \tag{3.13'}$$

Near the critical point where $\mathrm{Im}\,\omega(\boldsymbol{k})$ changes sign, $\mathrm{Im}\,\omega(\boldsymbol{k})$, and probably $A(k)$ is a small quantity. We discuss here two possible situations.

(1) The equation $D(\boldsymbol{k}, k_0) = 0$ has real solution $k_0 = \mathrm{Re}\,\omega(\boldsymbol{k})$ for all $\boldsymbol{k}$, it then follows from (6.46) and (3.14′) that

$$\mathrm{Im}\,\omega(\boldsymbol{k}) = -A(k) \Big/ \left.\frac{\partial D}{\partial k_0}\right|_{k_0 = \mathrm{Re}\,\omega(\boldsymbol{k})} . \tag{6.48}$$

If $\mathrm{Im}\,\omega(\boldsymbol{k}) > 0$ for some $\boldsymbol{k}$, then an instability with this $\boldsymbol{k}$ occurs to form a new space–time structure. However, as discussed in section 3.2, in thermoequilibrium we have

$$A(k) = -\tfrac{1}{2}\mathrm{i}\Gamma_-(k)[1 - \exp(-\beta k_0)] = k_0 Wa(k)[1 - \exp(-\beta k_0)] > 0 ,$$

and $\partial D/\partial k_0 > 0$ for $k_0 > 0$, whereas both of them change sign for $k_0 < 0$, so that such instability cannot occur in an equilibrium system. In fact, it usually appears in far-from-equilibrium systems under certain special conditions, for example, in a laser system the $\varphi_c = 0$ solution is unstable above the threshold of pumping.

(2) $A(k)$ is not small compared with $D(k)$ as $\boldsymbol{k} \to 0$. For an equilibrium system we can write

$$A(k) = k_0\gamma, \qquad \gamma > 0, \qquad D(k) = D_0 + ak_0 + \cdots .$$

Up to the first order of k_0, the solution of (6.46) is

$$k_0 = -D(0)/(a + \mathrm{i}\gamma) , \tag{6.49}$$

with

$$\mathrm{Im}\,k_0 = \frac{\gamma}{a^2 + \gamma^2} D(0) .$$

Hence the phase transition occurs at $D(0) = 0$. This is the ordinary second-order phase transition if the nontrivial solution grows continuously from zero. Otherwise, the point $D(0) = 0$ will correspond to supercooling or superheating temperature.

6.3. *Ward–Takahashi (WT) identity and Goldstone theorem*

In this section we derive the WT identity satisfied by the CTPGF from the invariance of the Lagrangian of the system with respect to global transformations of a Lie group $\mathbb{G}$.

6.3.1. *Group transformations*

Let $\varphi(x)$ be the constituent field and $Q(x)$ the order parameter. Each of them has several components forming by themselves bases of unitary representations. Under the infinitesimal transformation of $\mathbb{G}$

$$\varphi(x) \to \varphi'(x) = \varphi(x) + \delta\varphi(x),$$
$$\delta\varphi(x) = \zeta_\alpha(\mathrm{i}I_\alpha^{(0)} - X_\alpha^\mu(x)\partial_\mu)\varphi(x) = \mathrm{i}I_\alpha\varphi(x)\zeta_\alpha, \tag{6.50}$$

$$Q(x) \to Q'(x) = Q(x) + \delta Q(x),$$
$$\delta Q(x) = \zeta_\alpha(\mathrm{i}L_\alpha^{(0)} - X_\alpha^\mu(x)\partial_\mu)Q(x) = \mathrm{i}L_\alpha Q(x)\zeta_\alpha, \tag{6.51}$$

where ζ_α are a total of $n_\mathbb{G}$ infinitesimal parameters for group $\mathbb{G}$ and $I_\alpha^{(0)}$, $L_\alpha^{(0)}$ representation matrices for the generators of $\mathbb{G}$. X_α^μ are associated with the transformation of coordinates

$$x^\mu \to x^{\mu\prime} = x^\mu + X_\alpha^\mu(x)\zeta_\alpha. \tag{6.52}$$

It can be easily shown that the Lagrangian function transforms in this case as

$$\mathscr{L}(\varphi'(x))\frac{\mathrm{d}^4x}{\mathrm{d}^4x'} = \mathscr{L}(\varphi(x')) + \mathrm{i}\left(\frac{\delta\mathscr{L}}{\delta\varphi(x)} - \partial_\mu\frac{\delta\mathscr{L}}{\delta\partial_\mu\varphi(x)}\right)I_\alpha\varphi(x)\zeta_\alpha(x) + \partial_\mu(j_\alpha^\mu(x)\zeta_\alpha(x)), \tag{6.53}$$

where

$$j_\alpha^\mu(x) = \mathrm{i}\frac{\delta\mathscr{L}}{\delta\partial_\mu\varphi(x)}I_\alpha\varphi(x) - \mathscr{L}X_\alpha^\mu(x) \tag{6.54}$$

is the current in the α direction and $\zeta_\alpha(x)$ is an arbitrary infinitesimal function. If the Lagrangian is invariant under the global transformation of $\mathbb{G}$, it then follows that

$$\mathscr{L}(\varphi'(x))\frac{\mathrm{d}^4x}{\mathrm{d}^4x'} - \mathscr{L}(\varphi(x')) = \mathrm{i}\left(\frac{\delta\mathscr{L}}{\delta\varphi(x)} - \partial_\mu\frac{\delta\mathscr{L}}{\delta\partial_\mu\varphi(x)}\right)I_\alpha\varphi(x)\zeta_\alpha + \zeta_\alpha\partial_\mu j_\alpha^\mu(x) = 0,$$

or equivalently,

$$\partial_\mu j_\alpha^\mu(x) = \mathrm{i}\left(\partial_\mu\frac{\delta\mathscr{L}}{\delta\partial_\mu\varphi(x)} - \frac{\delta\mathscr{L}}{\delta\varphi(x)}\right)I_\alpha\varphi(x), \tag{6.55}$$

i.e., the current is conserved provided $\varphi(x)$ is the solution of the Euler–Lagrangian equation. Substituting (6.55) into (6.53) yields

$$\mathscr{L}(\varphi'(x))\frac{d^4x}{d^4x'} = \mathscr{L}(\varphi(x')) + j^{\mu}_{\alpha}(x)\partial_{\mu}\zeta_{\alpha}(x)\,, \tag{6.56}$$

which is the change of $\mathscr{L}$ under local transformation of G if it is invariant under the global action of the same group.

6.3.2. WT identities

The path integral representation for the generating functional (6.5) can be written as

$$Z[J(x), h(x)] = N\int [d\varphi(x)]\exp\Big\{i\int_p (\mathscr{L}(\varphi(x)) + J(x)\varphi(x) + h(x)Q(x))\Big\}$$
$$\times\langle\varphi(x, t_+ = -\infty)|\hat{\rho}|\varphi(x, t_- = -\infty)\rangle\,. \tag{6.5'}$$

Performing a local transformation of $\varphi(x)$ in (6.5') with $\zeta_{\alpha}(x)$ satisfying the following boundary conditions

$$\zeta_{\alpha}(x, t_{\pm} = -\infty) = \lim_{|x|\to\infty}\zeta(x, t) = 0\,, \tag{6.57}$$

and taking into account that the measure $[d\varphi(x)]$ does not change under unitary transformations, we obtain from the invariance of the generating functional that

$$\langle\partial_{\mu}j^{\mu}_{\alpha}(x)\rangle = -iJ(x)I_{\alpha}\varphi_c(x) - ih(x)L_{\alpha}Q_c(x)\,. \tag{6.58}$$

On the other hand

$$\langle\partial_{\mu}j^{\mu}_{\alpha}(x)\rangle = Z^{-1}\partial_{\mu}j^{\mu}_{\alpha}\left(\varphi(x) = -i\frac{\delta}{\delta J(x)}\right)Z[J(x), h(x)]\,, \tag{6.59}$$

from which it follows that

$$\partial_{\mu}j^{\mu}_{\alpha}\left(\varphi_c - i\frac{\delta}{\delta J(x)}\right) = -iJ(x)I_{\alpha}\varphi_c(x) - ih(x)L_{\alpha}Q_c(x) \tag{6.60}$$

by use of (6.38). Using the generating functional W for the connected Green function, the WT identities (6.60) can be rewritten as

$$\partial_{\mu}j^{\mu}_{\alpha}\left(\frac{\delta W}{\delta J(x)} - i\frac{\delta}{\delta J(x)}\right) = -iJ(x)I_{\alpha}\frac{\delta W}{\delta J(x)} - ih(x)L_{\alpha}\frac{\delta W}{\delta h(x)}\,. \tag{6.61}$$

Taking functional derivatives of (6.61) with respect to $J(y)$ and then setting $J(y)=0$, we obtain WT identities satisfied by CTPGFs of different order.

In terms of vertex generating functional $\Gamma[\varphi_c]$ (6.61) can be expressed as

$$\partial_\mu j^\mu_\alpha\left(\varphi_c(x) - \mathrm{i}\int_p \left(\frac{\delta^2\Gamma}{\delta\varphi_c(x)\delta\varphi_c(y)}\right)^{-1}\frac{\delta}{\delta\varphi_c(y)}\,\mathrm{d}^4y\right)$$

$$= \pm\mathrm{i}\frac{\delta\Gamma}{\delta\varphi_c(x)}L_\alpha\varphi_c(x) + \mathrm{i}\frac{\delta\Gamma}{\delta Q_c(x)}L_\alpha Q_c(x)\,. \tag{6.62}$$

Here we allow $\varphi(x)$ to be either a boson or fermion field. Taking the functional derivative $\delta/\delta\varphi_c(y)$ of (6.62) and putting $\varphi_c(y)=\varphi_{c0}$, the symmetry breaking in the absence of $J(x)$, we obtain WT identities for different vertex functions.

6.3.3. *Goldstone theorem*

Now we use the WT identity to discuss the symmetry breaking after phase transition.

Suppose the equations

$$\delta\Gamma/\delta\varphi_c(x) = 0, \qquad \delta\Gamma/\delta Q_c(x) = 0$$

have solutions $\varphi_c(x)=0$, $Q_{c+}(x)=Q_{c-}(x)\neq 0$. Differentiating (6.62) with respect to $Q_c(y)$, setting $J(x)=h(x)=0$ and integrating over x, we obtain

$$\int_p \frac{\delta^2\Gamma}{\delta Q_c(y)\delta Q_c(x)}L_\alpha Q_c(x)\,\mathrm{d}^4x = 0\,,$$

which can be rewritten in the single time representation as

$$\int \Gamma_r(y,x)L_\alpha Q_c(x)\,\mathrm{d}^4x = 0\,. \tag{6.63a}$$

In matrix form (6.63a) appears as

$$\Gamma_r\cdot L_\alpha Q_c(x) = 0\,, \tag{6.63b}$$

i.e., $L_\alpha Q_c(x)$ is the eigenvector of Γ_r with zero eigenvalue.

Assume $Q_c(x)$ is invariant under a subgroup $\mathbb{H}$ of $\mathbb{G}$ with n_{H} as its dimension. Therefore,

$$L_\alpha Q_c = 0, \qquad \alpha = 1,\ldots,n_{\mathrm{H}}$$

if α belongs to the generators of $\mathbb{H}$. On the contrary, if α belongs to the coset $\mathbb{G}/\mathbb{H}$, $L_\alpha Q_c\neq 0$, then (6.63b) shows that Γ_r has $n_{\mathrm{G}}-n_{\mathrm{H}}$ eigenvectors with zero eigenvalue. Suppose the representation to be real, then taking complex conjugation of (6.63b) we obtain

$$Q_c L_\alpha \cdot \Gamma_a = 0\,, \tag{6.64}$$

due to the orthogonality of L_α. Separating the real and imaginary parts we find

$$Q_c L_\alpha \cdot D = D \cdot L_\alpha Q_c = 0, \qquad Q_c L_\alpha \cdot A = A \cdot L_\alpha Q_c = 0\,, \tag{6.65}$$

i.e., $L_\alpha Q_c$ are zero eigenvalue eigenstates for both D and A. It follows from the Dyson equation (3.10) that the retarded Green function G_r has $n_G - n_H$ nondissipative elementary excitations called Goldstone modes. If Q_c does not depend on coordinates, in Fourier representation of $x - y$ such excitation occurs at zero energy and momentum and is called Goldstone particle.

6.3.4. *Applications*

The Goldstone bosons considered above have important consequences in the symmetry broken state. For example, in laser system the U(1) symmetry is broken, so the corresponding Goldstone boson leads to the divergence of the fluctuation which in turn makes the classical solution unstable. This phenomenon in the CTPGF approach was observed by Korenman [25] and was analyzed by us in [40, 52]. The WT identity is used to derive a generalized Goldstone theorem in a slowly varying in time system. As its consequence the pole of the Green function splits into two with equal weight, equal energy but different dissipation. Combined with the order parameter (average value of the vector potential) these two quanta (one of which is the Goldstone boson) provide a complete description of the order–disorder transition of the phase symmetry in the saturation state of the laser.

We have also used the WT identity in combination with order parameter expansion (cf. section 3.5) to derive the generalized TDGL equation [43, 47]. We will apply the same identity to discuss the localization problem in section 8.3 [56].

We should mention that the transformations given by (6.50) and (6.51) are linear. We can consider nonlinear transformations under $\mathbb{G}$ as we did in [43, 47]. In that case we need to take into account the Jacobian of transformation for the path integral. The result thus obtained turns out to be the same as the nonlinear mode–mode coupling introduced phenomenologically by Kawasaki [67].

6.4. *Functional description of fluctuation*

6.4.1. *Stochastic functional*

It is known that the Gaussian stochastic process $\xi_i(t)$ appearing in the Langevin equation (cf. (4.33))

$$\partial Q_i/\partial t = K_i(Q) + \xi_i(t) \tag{6.66}$$

can be presented by a stochastic integral [86]. Equation (6.66) can then be considered as a nonlinear mapping of the Gaussian process on to a more complicated process $Q_i(t)$. Realization of such mapping actually results in a functional description of $Q_i(t)$ [72]. Nevertheless, such functional description can be achieved by a more straightforward way [87–89].

Consider the normalization of the δ-function under path integration

$$\int [\mathrm{d}Q]\delta\left(\frac{\partial Q}{\partial t} - K(Q) - \xi\right)\Delta(Q) = 1\,, \tag{6.67}$$

where the Jacobian $\Delta(Q)$ appears because the argument of the δ-function is not Q itself but a rather complicated expression. Neglecting multiplicative factors, $\Delta(Q)$ turns out to be [72]

$$\Delta(Q) = \exp\left\{-\tfrac{1}{2}\int \delta K(Q)/\delta Q\right\}. \tag{6.68}$$

Using the integral representation of δ-function

$$\delta(f) = \int \left[\frac{\mathrm{d}\hat{Q}}{2\pi}\right] \exp\left(\mathrm{i}\int \hat{Q} f\right), \tag{6.69}$$

(6.67) can be rewritten as

$$\int [\mathrm{d}Q]\left[\frac{\mathrm{d}\hat{Q}}{2\pi}\right] \exp\left\{\int\left[\mathrm{i}\hat{Q}\left(\frac{\partial Q}{\partial t} - K(Q) - \xi\right) - \frac{1}{2}\frac{\delta K}{\delta Q}\right]\right\} = 1. \tag{6.70}$$

Inserting the source term

$$\exp\left(\mathrm{i}\int (JQ + \hat{J}\hat{Q})\right),$$

one obtains the generating functional

$$Z_\xi[J, \hat{J}] = \int [\mathrm{d}Q]\left[\frac{\mathrm{d}\hat{Q}}{2\pi}\right] \exp\left\{\int\left[\mathrm{i}\hat{Q}\left(\frac{\partial Q}{\partial t} - K(Q) - \xi\right) - \frac{1}{2}\frac{\delta K}{\delta Q} + \mathrm{i}(JQ + \hat{J}\hat{Q})\right]\right\}, \tag{6.71}$$

with the normalization condition

$$Z_\xi[0, 0] = 1.$$

Averaging over the random noise distribution

$$W[\xi] \propto \exp(-\tfrac{1}{2}\xi\sigma^{-1}\xi) \tag{6.72}$$

with σ as the diffusion matrix, one obtains Lagrangian formulation of the generating functional for the statistical fluctuation

$$Z[J, \hat{J}] = \int [\mathrm{d}Q]\left[\frac{\mathrm{d}\hat{Q}}{2\pi}\right] \exp\left\{\int\left[-\tfrac{1}{2}\hat{Q}\sigma\hat{Q} + \mathrm{i}\hat{Q}\left(\frac{\partial Q}{\partial t} - K(Q)\right) - \tfrac{1}{2}\frac{\delta K}{\delta Q} + \mathrm{i}(JQ + \hat{J}\hat{Q})\right]\right\}. \tag{6.73}$$

The Gaussian integration over $\hat{Q}$ can be carried out to yield

$$Z[J] = \int [\mathrm{d}Q] \exp\left\{\int\left[-\frac{1}{2}\left(\frac{\partial Q}{\partial t} - K(Q) - \hat{J}\right)\sigma^{-1}\left(\frac{\partial Q}{\partial t} - K(Q) - \hat{J}\right) - \frac{1}{2}\frac{\delta K}{\delta Q} + \mathrm{i}JQ\right]\right\}. \tag{6.74}$$

Historically, the theory of noncommutative classical field was first suggested by Martin, Siggia and Rose (MSR hereafter) [90]. This theory has been extensively applied to critical dynamics [61] and has been later reformulated in terms of a Lagrangian field theory [88, 89] as presented by (6.73) and (6.74).

6.4.2. Effective action

We will show now that such description occurs within the CTPGF formalism in a natural way [43, 47], postponing the comparison with the MSR field theory to section 9.3.

Let $Q_i(x)$ be composite operators of the constituent fields $\varphi_j(x)$. Both of them are taken to be Hermitian Bose operators. Assuming the randomness of the initial phase, the density matrix is diagonal at moment $t = t_0$, i.e.,

$$\langle\varphi'(\boldsymbol{x}, t_0)|\hat{\rho}|\varphi''(\boldsymbol{x}, t_0)\rangle = P(\varphi'(\boldsymbol{x}), t_0)\delta(\varphi''(\boldsymbol{x}, t_0) - \varphi'(\boldsymbol{x}, t_0)) . \tag{6.75}$$

The initial distribution of the macrovariables $Q_i(x)$ is then given by

$$\begin{aligned} P(Q_i(x), t_0) &= \mathrm{Tr}(\delta(Q_i(x) - Q_i(\varphi(x))\hat{\rho}) \\ &= \int [\mathrm{d}\varphi(x)]\delta(Q_i(x) - Q_i(\varphi(x))P(\varphi(x), t_0) . \end{aligned} \tag{6.76}$$

The generating functional for Q_i can be written as

$$\begin{aligned} Z[J(x)] &= \exp(\mathrm{i}W[J(x)]) = \mathrm{Tr}\Big\{T_p\Big[\exp\Big(\mathrm{i}\int_p J(x)Q(\varphi(x))\Big)\Big]\hat{\rho}\Big\} \\ &= N\int [\mathrm{d}\varphi(x)] \exp\Big(\mathrm{i}\int [\mathscr{L}(\varphi(x)) + JQ(\varphi(x))]\Big)\,\delta(\varphi_+ - \varphi_-) , \end{aligned} \tag{6.77}$$

where

$$\delta(\varphi_+ - \varphi_-) \equiv \int \mathrm{d}\varphi'(\boldsymbol{x})\,\delta(\varphi(\boldsymbol{x}, t_+ = t_0) - \varphi'(\boldsymbol{x}))\delta(\varphi(\boldsymbol{x}, t_- = t_0) - \varphi'(\boldsymbol{x}))P(\varphi'(\boldsymbol{x}), t_0) . \tag{6.78}$$

Multiplying (6.77) by the normalization factor of the δ-function

$$\int [\mathrm{d}Q]\delta(Q_+ - Q_-)\delta(Q(x) - Q(\varphi(x))) = 1 , \tag{6.79}$$

changing the order of integration to replace $Q(\varphi(x))$ by $Q(x)$ and using the δ-function representation (6.69) with $\hat{Q}$ changed for I, we can rewrite (6.77) as

$$Z[J] = N\int [\mathrm{d}Q] \exp\Big(\mathrm{i}\,S_{\mathrm{eff}} + \mathrm{i}\int_p JQ\Big)\delta(Q_+ - Q_-) , \tag{6.80}$$

where

$$\exp(\mathrm{i}S_{\text{eff}}[Q]) = \int [\mathrm{d}I/2\pi] \exp\left(\mathrm{i}W[I] - \mathrm{i}\int_p IQ\right). \tag{6.81}$$

Here we have performed direct and inverse Fourier transformations of the path integral. So far as a continuous integration is taken over $I(x)$, $W[I]$ can be considered as a generating functional in the random external field. Calculating the functional integral in the one loop approximation which is equivalent to the Gaussian average, we can obtain the effective action $S_{\text{eff}}[Q]$ for $Q(x)$.

So far we have discussed the case when macrovariables are composite operators. The same is true if part or all of macrovariables are constituent fields themselves. A new "macro" field can be also introduced by use of the δ-function. However, one should carry out the path integration simultaneously in spite of the fact that the initial correlations are multiplicative, because in general the Lagrangian itself is not additive in terms of these variables.

Before going on to calculate the integral (6.81) we first discuss the basic properties of the effective action $S_{\text{eff}}[Q]$.

It is ready to check that apart from the normalization condition (2.103) the generating functional for the Hermitian boson field also satisfies the relation

$$W^*[J_+(x), J_-(x)] = -W[J_-(x), J_+(x)]. \tag{6.82}$$

It then follows from (6.81) that

$$S^*_{\text{eff}}[Q_+(x), Q_-(x)] = -S_{\text{eff}}[Q_-(x), Q_+(x)]. \tag{6.83}$$

Hence S_{eff} is purely imaginary for $Q_+(x) = Q_-(x)$. Setting $Q_\pm(x) = Q + \Delta Q_\pm$ and taking successive functional derivatives of (6.83) near Q, we obtain

$$\delta S/\delta Q(x_+) = [\delta S/\delta Q(x_-)]^*, \tag{6.84}$$

$$S_{\mathrm{F}ij}(x, y) = S_{\mathrm{F}ji}(y, x) = -S^*_{\bar{\mathrm{F}}ji}(y, x), \qquad S_{\pm ij}(x, y) = S_{\mp ji}(y, x) = -S^*_{\mp ji}(y, x), \tag{6.85}$$

where

$$S_{ij} = \left.\frac{\delta^2 S_{\text{eff}}}{\delta Q_i \delta Q_j}\right|_{Q_+ = Q_- = Q}.$$

We see that S_{ij} respect the same symmetry as the two-point Green functions (cf. (2.134)) and vertex functions (3.12b).

If the system is invariant under a symmetry group G, i.e., both the Lagrangian and the initial distribution do not change under

$$\varphi_i(x) \to \varphi_i^g(x) = U_{ij}(g)\varphi_j(x), \qquad Q_i(\varphi) \to Q_i^g(\varphi) = V_{ij}(g)Q_j(\varphi),$$

then

$$W[J^g(x)] = W[J(x)], \qquad J_i^g = J_j(x)V_{ji}^\dagger(g),$$
$$S_{\rm eff}[Q^g(x)] = S_{\rm eff}[Q(x)], \qquad Q_i^g = V_{ij}(g)Q_j(\varphi).$$

The above said is true if the effective action $S_{\rm eff}$ is calculated exactly. However, the symmetry properties of $S_{\rm eff}$, being related to those of the Lagrangian, may be different from the latter due to the average procedure.

If the lowest order of WKB, i.e., the tree approximation is taken in (6.81) we find that

$$Q = \delta W/\delta I, \tag{6.86}$$

$$S_{\rm eff}[Q] = \Gamma[Q]. \tag{6.87}$$

In this case, $S_{\rm eff}$ inherits all properties of the generating functional $\Gamma[Q]$, e.g.,

$$S_{\rm eff}[Q, Q] = 0, \tag{6.88}$$

$$\left.\frac{\delta S_{\rm eff}}{\delta Q_+(x)}\right|_{Q_+=Q_-=Q} = \left.\frac{\delta S_{\rm eff}}{\delta Q_-(x)}\right|_{Q_+=Q_-=Q}, \tag{6.89}$$

$$S_{\rm F} + S_{\hat{\rm F}} = S_+ + S_-, \tag{6.90}$$

$$\frac{\delta^m S_{\rm eff}}{\delta Q_i(1)\cdots\delta Q_l(m)} = {\rm i}^{m-1}\langle T_p[Q_i(1)\cdots Q_l(m)]\rangle_{\rm 1\,P.I.}. \tag{6.91}$$

In accord with (6.87), (3.5) and (3.19) we have

$$-{\rm i}S_\pm(k) > 0 \tag{6.92}$$

after taking Fourier transformation. Near thermoequilibrium we find from (3.40)

$$S_{-ij} - S_{+ij} \xrightarrow[k_0\to 0]{} -\beta k_0 S_{-ij}(k). \tag{6.93}$$

6.4.3. Saddle point approximation

Up to now we have discussed only the general properties of $S_{\rm eff}[Q]$. In principle, $S_{\rm eff}$ can be derived from the microscopic generating functional $W[I]$ by averaging over the random external field, it can be also constructed phenomenologically in accord with the symmetry properties required. We now calculate (6.81) in the one-loop approximation. Near the saddle point given by (6.86) we expand the exponential factor in (6.81)

$$E \equiv W - \int_p QI = \Gamma - \tfrac{1}{2}\int_p \Delta I G^{(2)} \Delta I = \Gamma - \tfrac{1}{2}\int \Delta\hat{I}^{\rm T}\sigma_3 \hat{G}\sigma_3 \Delta\hat{I}, \tag{6.94}$$

where $\hat{G}$ is the two-point connected functions, $\Delta\hat{I}^{\mathrm{T}} = (\Delta I_+, \Delta I_-)$. Up to a numerical constant, the result of the Gaussian integration is

$$\exp(\mathrm{i}S_{\mathrm{eff}}[Q]) = \exp(\mathrm{i}\Gamma[Q])|\det \sigma_3\hat{G}\sigma_3|^{-1/2} . \tag{6.95}$$

It then follows from the Dyson equation (2.57) that

$$\mathrm{i}S_{\mathrm{eff}}[Q] = \mathrm{i}\Gamma[Q] + \tfrac{1}{2}\operatorname{Tr}\ln \hat{\Gamma} . \tag{6.96}$$

By use of the transformation formula (2.59) we have

$$|\det \hat{\Gamma}| = |\det \tilde{\Gamma}| = |\det \Gamma_{\mathrm{r}}|\,|\det \Gamma_{\mathrm{a}}| = |\det \Gamma_{\mathrm{r}}|^2 ,$$

where

$$\Gamma_{\mathrm{r}}(x, y) = \frac{\delta^2\Gamma}{\delta Q_\Delta(x)\delta Q_{\mathrm{c}}(y)} .$$

As shown in section 3.5,

$$\frac{\delta\Gamma}{\delta Q_\Delta(x)} = \frac{1}{2}\left[\frac{\delta\Gamma}{\delta Q(x_+)} + \frac{\delta\Gamma}{\delta Q(x_-)}\right]\Bigg|_{Q_\Delta=0} = -\gamma\frac{\partial Q}{\partial t} - \frac{\delta\mathcal{F}}{\delta Q(x)} . \tag{6.97}$$

Comparing (6.97) with (6.67) we find that $\delta^2\Gamma/\delta Q_\Delta(x)\delta Q_{\mathrm{c}}(y)$ is just the transformation matrix up to the numerical factor γ^{-1}. Therefore, we can calculate the Jacobian in the same way to get

$$\mathrm{i}S_{\mathrm{eff}}[Q] = \mathrm{i}\Gamma[Q] - \tfrac{1}{2}\int \delta K/\delta Q , \tag{6.98}$$

with

$$K = -\gamma^{-1}\delta\mathcal{F}/\delta Q .$$

In the path integral (6.80) the most plausible path is given by

$$\delta S_{\mathrm{eff}}[Q]/\delta Q(x\pm) = -J_\pm(x) , \tag{6.99}$$

$$Q(x, t_+ = t_0) = Q(x, t_- = t_0) . \tag{6.100}$$

In the tree approximation of (6.81)

$$\delta S_{\mathrm{eff}}/\delta Q = -J = -\gamma\partial Q/\partial t - \delta\mathcal{F}/\delta Q , \tag{6.101}$$

which is nothing but the TDGL equation derived in section 3.5 (cf. (3.85)).

6.4.4. Role of fluctuations

We now discuss fluctuations around the most plausible path. In the CTPGF approach there is an additional way of describing the fluctuation: To allow field variables to take different values on the positive and negative time branches. Changing variables in (6.80) to $Q = Q_c = \frac{1}{2}(Q_+ + Q_-)$, $Q_\Delta = Q_+ - Q_-$, the effective action can be expanded as

$$S_{\rm eff}[Q_+(x), Q_-(x)] = S_{\rm eff}[Q, Q] + \tfrac{1}{2}\int [\delta S_{\rm eff}/\delta Q(x+) + \delta S_{\rm eff}/\delta Q(x-)]Q_\Delta(x)$$
$$+ \tfrac{1}{8}\int Q_\Delta(x)(S_{++} + S_{--} + S_{+-} + S_{-+})(x, y)Q_\Delta(y) + \cdots .$$

Denoting

$$\tfrac{1}{4}\mathrm{i}(S_{++} + S_{--} + S_{+-} + S_{-+})(x, y) \equiv -\gamma(x)\sigma(x, y)\gamma(y) \tag{6.102}$$

and using (6.88), (6.98), (6.101), we obtain

$$\mathrm{e}^{\mathrm{i}W[J(x)]} = \int [\mathrm{d}Q(x)][\mathrm{d}Q_\Delta(x)] \exp\Big[-\tfrac{1}{2}\int Q_\Delta(x)\gamma(x)\sigma(x, y)\gamma(y)Q_\Delta(y)$$
$$-\mathrm{i}\int\Big(\gamma(x)\frac{\partial Q}{\partial t} + \frac{\delta \mathscr{F}}{\delta Q}\Big)Q_\Delta(x) - \frac{1}{2}\int\frac{\delta K}{\delta Q} + \mathrm{i}\int (J_\Delta Q + J_c Q_\Delta)\Big]\,\delta(Q_\Delta(x)) . \tag{6.103}$$

If we take $J_\Delta = J$ and $\gamma(x)Q_\Delta(x) \to \hat{Q}$, $J_c\gamma^{-1} \to \hat{J}$ the stochastic generating functional (6.73) is retrieved. Carrying out integration over $Q_\Delta(x)$ will lead to an equation identical to (6.74). It is important to note here, that in the CTPGF approach, J_c, the counterpart of $\hat{J}$ in MSR theory, is the physical external field, whereas $J_\Delta = J_+ - J_-$, the counter part of J in MSR theory is the fictitious field. It is clear by comparing (6.103) and (6.73) that $\sigma(x, y)$ is the correlation matrix for the random force. If Q is a smooth function of x, σ can be taken as a constant

$$\sigma = -\tfrac{1}{4}\mathrm{i}\gamma^{-1}(S_F + S_{\tilde{F}} + S_+ + S_-)(k = 0)\gamma^{-1} , \tag{6.104}$$

which reduces to

$$\sigma = -\tfrac{1}{2}\mathrm{i}\gamma^{-1}\Gamma_c(k = 0)\gamma^{-1} \tag{6.105}$$

by virtue of (6.87) and (6.90). This is the expression we have used in section 4 (cf. (4.35)) to discuss the symmetry properties of the kinetic coefficients, if it is generalized to the multi-component case.

According to the definition of γ given by (3.81)

$$\gamma = \lim_{k_0\to 0}\frac{\partial A}{\partial k_0} = -\frac{\mathrm{i}}{2}\lim_{k_0\to 0}\frac{\partial}{\partial k_0}(\Gamma_- - \Gamma_+) \approx -\tfrac{1}{2}\mathrm{i}\beta\Gamma_- = -\tfrac{1}{4}\mathrm{i}\beta(S_+ + S_-) . \tag{6.106}$$

Comparison of (6.106) with (6.105) yields the Einstein relation (FDT)

$$\sigma = 2/\beta\gamma \tag{6.107}$$

for the diffusion coefficient in the case of single macrovariable.

For simplicity we consider only one component order parameter in this section, but we consciously write some of formulas in such a way, so that the generalization to the multi-component case is obvious.

To sum up, the MSR field theory of stochastic functional is retrieved in the CTPGF approach if the one-loop approximation in the random field integration and the second cumulant expansion in $Q_\Delta(x)$ are taken. The possibility to go beyond such approximation is apparent.

7. Practical calculation scheme using CTPGF

As we have seen, the CTPGF provides us with a unified approach to both equilibrium and nonequilibrium systems. However, to make it practically useful we need a unified, flexible enough calculation scheme. Such scheme has been already worked out by us [48, 49]. In fact, most of the calculations carried out by us so far using CTPGF [40, 46, 52–57] can be cast into this framework.

Consider a typical situation when fermions $\psi(x)$, $\psi^\dagger(x)$ are coupled to the order parameter $Q(x)$ which might be a constituent field like the vector potential $A_\mu(x)$ in the laser case, or a composite operator like

$$\chi(x) = \psi_\uparrow(x)\psi_\downarrow(x)$$

in the theory of superconductivity, or

$$S = \psi^\dagger(x)\tfrac{1}{2}\boldsymbol{\sigma}\psi(x)$$

in the case of itinerant ferromagnetism, where $\boldsymbol{\sigma}$ are Pauli matrices. The boson field $Q(x)$ via which the fermions interact with each other, may be nonpropagating at the tree level like the Coulomb field. However, the radiative correction will in general make $Q(x)$ a dynamical variable and the fluctuations around the mean field $Q_c(x)$ will propagate and form collective excitations. Therefore, the system is characterized by the mean field $Q_c(x)$ and the two kinds of quasiparticles – constituent fermions and collective excitations with their own energy spectrum, dissipation and distribution. Such a way of description has been found useful in condensed matter physics [1–3, 21–24, 91], plasma physics [33, 34] as well as in the nuclear many-body theory [92–96].

In this section we first (section 7.1) derive a system of coupled equations satisfied by the order parameter and the two kinds of Green's functions using the generating functional with two-point source terms. Next (section 7.2), the technique of Cornwall, Jackiw and Tomboulius (CJT) [97], developed in the quantum field theory to calculate the effective potential for composite operators is generalized to the CTPGF case and is used as a systematic way of computing the self-energy part by a loop expansion. In thermoequilibrium when the dissipation is negligible, the mean field $Q_c(x)$ and the energy spectrum of the fermion field are determined to the first approximation by the Bogoliubov–de Gennes (BdeG) [98] equation, in which the single particle fermion wavefunction satisfies the Hartree type self-consistent equation without Fock exchange term. In section 7.3 we discuss the generalization of the BdeG equation in the four-fermion problem, whereas the free energy in various approximations is calculated explicitly in section 7.4 by directly integrating the functional equation for it. Some problems related to those discussed in this section were considered by Kleinert using the functional integral approach [99].

7.1. Coupled equations of order parameter and elementary excitations

7.1.1. Model

Consider a fermion field $\psi(x)$ interacting via a boson field $Q(x)$ with the action given by

$$I = I_0[\psi^\dagger, \psi] + I_0[Q] + I_{\text{int}}[\psi^\dagger, \psi, Q], \tag{7.1}$$

where

$$I_0[\psi^\dagger, \psi] = \int_p \psi^\dagger(x) S_0^{-1}(x, y)\psi(y), \tag{7.2}$$

$$I_0[Q] = \tfrac{1}{2}\int_p Q(x)\Delta_0^{-1}(x, y)Q(y), \tag{7.3}$$

with $S_0^{-1}(x, y)$, $\Delta_0^{-1}(x, y)$ as inverse fermion and boson propagators respectively.

For a system with four-fermion interaction only, we can use the Hubbard–Stratonovich (HS) transformation [100] to introduce the effective fermion–boson interaction.

Let the generating functional for the order parameter $Q(\psi^\dagger(x), \psi(x))$ be defined as

$$Z_p[h(x)] = \int [\mathrm{d}\psi^\dagger(x)][\mathrm{d}\psi(x)] \exp\{\mathrm{i}[I_0[\psi^\dagger, \psi] + I_{\text{int}}[\psi^\dagger, \psi] + hQ(\psi^\dagger, \psi)]\}. \tag{7.4}$$

Using the Gaussian integral identity, i.e. the HS transformation, eq. (7.4) can be presented as

$$Z_p[h] = \exp[\tfrac{1}{2}hM_0h]\tilde{Z}_p[h], \tag{7.5}$$

$$\tilde{Z}_p[h] = \int_p [\mathrm{d}\psi^\dagger][\mathrm{d}\psi][\mathrm{d}Q] \exp\{\mathrm{i}[I_0[\psi^\dagger, \psi] + I_0[Q] + I_{\text{int}}[\psi^\dagger, \psi, Q] + hQ]\}, \tag{7.6}$$

where

$$I_0[Q] = \tfrac{1}{2}Q\Delta_0^{-1}Q,$$

$I_{\text{int}}[\psi^\dagger, \psi, Q]$ being the nonlinear interaction. It is important to note that M_0, Δ_0^{-1} are two-point functions, independent of either field variables or external source. Therefore, up to an additive constant M_0 the Green functions of the original system are the same as those of the effective system described by $\tilde{Z}$. The formal ambiguity in defining the $[\mathrm{d}Q(x)]$ integration [96] can be avoided by imposing the condition

$$\left.\frac{\delta Z_p[h]}{\delta h(x)}\right|_{h=0} = \left.\frac{\delta \tilde{Z}_p[h]}{\delta h(x)}\right|_{h=0}. \tag{7.7}$$

We see thus the four-fermion interaction can be considered on an equal footing with system of fermions interacting through a constituent boson field $Q(x)$.

7.1.2. Two-point source

The generating functional with a two-point source is defined as

$$Z_p[h, J^\dagger, J, M, K] = \int_p [d\psi^\dagger][d\psi][dQ] \exp\{i[I_0[\psi^\dagger, \psi] + I_0[Q] + I_{int}[\psi^\dagger, \psi, Q] + W_p^N[\psi^\dagger, \psi, Q] + hQ + J^\dagger\psi + \psi^\dagger J + \tfrac{1}{2}QMQ + \psi^\dagger K\psi]\}, \quad (7.8)$$

where $W_p^N[\psi^\dagger, \psi, Q]$ takes care of contribution from the density matrix as discussed in section 6.1. Here we adopt the abbreviated notation and $M(x, y)$, $K(x, y)$ are external sources to generate the second-order CTPGFs.

Introducing the generating functional for the connected CTPGF as usual

$$W[h, J^\dagger, J, M, K] = -i \ln Z[h, J^\dagger, J, M, K], \quad (7.9)$$

it then follows that

$$\delta W_p/\delta h(x) = Q_c(x), \quad (7.10a)$$

$$\delta W_p/\delta J^\dagger(x) = \psi_c(x), \quad (7.10b)$$

$$\delta W_p/\delta J(x) = -\psi_c^\dagger(x), \quad (7.10c)$$

$$\delta W_p/\delta M(y, x) = \tfrac{1}{2}(Q_c(x)Q_c(y) + i\Delta(x, y)), \quad (7.10d)$$

$$\delta W_p/\delta K(y, x) = -(\psi_c(x)\psi_c^\dagger(y) + iG(x, y)). \quad (7.10e)$$

In case of vanishing sources $Q_c(x)$, $\psi_c(x)$ and $\psi_c^\dagger(x)$ become expectation values of the corresponding fields $Q(x)$, $\psi(x)$ and $\psi^\dagger(x)$, whereas $\Delta(x, y)$, $G(x, y)$ are the second-order CTPGF for the boson field $Q(x)$ and the fermion field $\psi(x)$, $\psi^\dagger(x)$ respectively.

7.1.3. Coupled equations

The generating functional for the vertex CTPGF is defined as the Legendre transform of W_p,

$$\Gamma_p[Q_c, \psi_c^\dagger, \psi_c, \Delta, G] = W[h, J^\dagger, J, M, K] - hQ_c - \psi_c^\dagger J - J^\dagger\psi_c - \tfrac{1}{2}\mathrm{Tr}[M(Q_cQ_c + i\Delta)] - \mathrm{Tr}[K(\psi_c\psi_c^\dagger + iG)], \quad (7.11)$$

where

$$\mathrm{Tr}[M(Q_cQ_c + \mathrm{i}\Delta)] \equiv \int_p M(x, y)(Q_c(y)Q_c(x) + \mathrm{i}\Delta(y, x)),$$

$$\mathrm{Tr}[K(\psi_c\psi_c^\dagger + \mathrm{i}G)] \equiv \int_p K(x, y)(\psi_c(y)\psi_c(x) + \mathrm{i}G(y, x)).$$

Using (7.10) it is straightforward to deduce from (7.11) that

$$\frac{\delta \Gamma_p}{\delta Q_c(x)} = -h(x) - \int_p M(x, y)Q_c(y), \tag{7.12a}$$

$$\frac{\delta \Gamma_p}{\delta \psi_c^\dagger(x)} = -J(x) - \int_p K(x, y)\psi_c(y), \tag{7.12b}$$

$$\frac{\delta \Gamma_p}{\delta \psi_c(x)} = J^\dagger(x) + \int_p \psi_c^\dagger(y)K(y, x), \tag{7.12c}$$

$$\frac{\delta \Gamma_p}{\delta \Delta(x, y)} = \frac{1}{2\mathrm{i}} M(y, x), \tag{7.12d}$$

$$\frac{\delta \Gamma_p}{\delta G(x, y)} = \mathrm{i}K(y, x). \tag{7.12e}$$

Equations (7.12) form a set of self-consistent equations to determine the order parameters $Q_c(x)$, $\psi_c^\dagger(x)$, $\psi_c(x)$ as well as the second-order CTPGF $\Delta(x, y)$ and $G(x, y)$ provided Γ_p is known as a functional of these arguments. In almost all cases of practical interest the condensation of the fermion field is forbidden in the absence of the external source, i.e., $\psi_c^\dagger = \psi_c = 0$ for $J^\dagger = J = 0$. On the other hand, the condensation of the boson field (elementary or composite) is described by the order parameter $Q_c(x)$. Since the energy spectrum, the dissipation and the particle number distribution are determined by the second-order CTPGF, eqs. (7.12a, d, e) are just those equations we are looking for. The only question remaining is how to construct the vertex functional Γ_p. In the next section a systematic loop expansion will be developed for this purpose.

7.2. Loop expansion for vertex functional

7.2.1. CJT rule [97]

Without loss of generality in what follows we will set $J^\dagger = J = \psi_c^\dagger = \psi_c = 0$. The vertex functional

$$\Gamma_p[Q_c(x), \Delta, G] \equiv \Gamma_p[Q_c(x), \psi_c^\dagger(x), \psi_c(y), \Delta, G]|_{\psi_c^\dagger = \psi_c = 0}$$

is the generating functional in Q for the two-particle irreducible (2 PI) Green's functions expressed in terms of propagators Δ and G. To derive a series expansion for Γ_p we note that after absorbing W_p^N into the effective action, i.e.,

$$I_{\text{eff}}[\psi^\dagger, \psi, Q] = I_0[\psi^\dagger, \psi] + I_0(Q) + I_{\text{int}}[\psi^\dagger, \psi, Q] + W_p^N[\psi^\dagger, \psi, Q]\,, \tag{7.13}$$

the only difference remaining between the CTPGFs and the ordinary Green functions in the quantum field theory is the range of the time axis. For CTPGF the time integration is taken for both positive and negative branches. Hence the loop expansion technique for the vertex functional and its justification developed by CJT [97] in the quantum field theory can be easily extended to the CTPGF formalism provided the difference in the definition of the time axis is properly taken into account. Here we shall simply state the result as

$$\Gamma_p[Q_c, \Delta, G] = \bar{I}[Q_c] - \tfrac{1}{2}i\hbar\, \mathrm{Tr}\{\ln[\Delta_0^{-1}\Delta] - \Delta_0^{-1}\Delta + 1\}$$
$$+ i\hbar\, \mathrm{Tr}\{\ln[S_0^{-1}G] - G_0^{-1}G + 1\} + \Gamma_{2p}[Q_c, \Delta, G]\,, \tag{7.14}$$

where

$$\bar{I}[Q_c] \equiv I_{\text{eff}}[\psi^\dagger, \psi, Q]\big|_{\substack{\psi=\psi^\dagger=0\\ Q=Q_c}}\,, \tag{7.15a}$$

$$\Delta_0^{-1}(x, y) = \frac{\delta^2 I_{\text{eff}}}{\delta Q(x)\delta Q(y)}\bigg|_{\psi=\psi^\dagger=Q=0}\,, \tag{7.15b}$$

$$S_0^{-1}(x, y) = -\frac{\delta^2 I_{\text{eff}}}{\delta\psi^\dagger(x)\delta\psi(y)}\bigg|_{\psi=\psi^\dagger=Q=0}\,, \tag{7.15c}$$

$$G_0^{-1}(x, y) = -\frac{\delta^2 I_{\text{eff}}}{\delta\psi^\dagger(x)\delta\psi(y)}\bigg|_{\substack{\psi=\psi^\dagger=0\\ Q=Q_c}}\,. \tag{7.15d}$$

Note that G_0^{-1} is different from S_0^{-1} and

$$S_0^{-1}(x, y) = G_0^{-1}(x, y)|_{Q_c=0}\,. \tag{7.16}$$

The quantity $\Gamma_{2p}[Q_c, \Delta, G]$ appearing in (7.14) is computed as follows. First shift the field $Q(x)$ in the effective action $I_{\text{eff}}[\psi^\dagger, \psi, Q]$ by $Q_c(x)$ and keep only terms cubic and higher in $\psi^\dagger$, ψ and Q as interaction vertices which depend on Q_c. The Γ_{2p} is then calculated as a sum of all 2PI vacuum diagrams constructed by vertices described above with $\Delta(x, y)$, $G(x, y)$ as propagators. In (7.14) the trace, the products $\Delta_0^{-1}\Delta$, etc., as well as the logarithm are taken in the functional sense with both internal indices and space–time coordinates summed over.

7.2.2. Coupled equations

The self-energy parts for the fermion and the boson propagators are defined to be

$$\Sigma(x,y) = \frac{-\mathrm{i}}{\hbar}\frac{\delta \Gamma_{2p}}{\delta G(y,x)}, \tag{7.17a}$$

$$\Pi(x,y) = \frac{2\mathrm{i}}{\hbar}\frac{\delta \Gamma_{2p}}{\delta \Delta(y,x)}. \tag{7.17b}$$

Hereafter in section 7 we restore the Planck constant $\hbar$ in formulas to show explicitly the order of magnitude.

The equations for the order parameter $Q_c(x)$ and the second-order CTPGF $\Delta(x,y)$ and $G(x,y)$ for a physical system can be obtained from eqs. (7.12) by switching off the external sources. We have thus:

$$\frac{\delta \Gamma_p}{\delta Q_c(x)} = \frac{\delta \bar{I}[Q_c]}{\delta Q_c(x)} - \mathrm{i}\hbar\,\mathrm{Tr}\left\{\frac{\delta G_0^{-1}}{\delta Q_c(x)}G\right\} + \frac{\delta \Gamma_{2p}}{\delta Q_c(x)} = 0, \tag{7.18}$$

$$\frac{2\mathrm{i}}{\hbar}\frac{\delta \Gamma_p}{\delta \Delta(y,x)} = \Delta^{-1}(x,y) - \Delta_0^{-1}(x,y) + \Pi(x,y) = 0, \tag{7.19}$$

$$\frac{-\mathrm{i}}{\hbar}\frac{\delta \Gamma_p}{\delta G(y,x)} = G^{-1}(x,y) - G_0^{-1}(x,y) + \Sigma(x,y) = 0. \tag{7.20}$$

Rewritten in the ordinary time variable in accord with the rule set in section 2.3, eq. (7.18) becomes the generalized TDGL equation for the order parameter, whereas (7.19) and (7.20) are the Dyson equations for the retarded and advanced Green functions along with the transport equation for the quasiparticle distribution. Equation (7.18) can be rewritten in a symmetric form as

$$\tfrac{1}{2}\xi_\sigma \frac{\delta \Gamma_p}{\delta Q_c(x_\sigma)} = \tfrac{1}{2}\xi_\sigma\left\{\frac{\delta \bar{I}[Q]}{\delta Q_c(x_\sigma)} - \mathrm{i}\hbar\,\mathrm{Tr}\left(\frac{\delta G_0^{-1}}{\delta Q(x_\sigma)}G\right) + \frac{\delta \Gamma_{2p}}{\delta Q(x_\sigma)}\right\} = 0. \tag{7.18'}$$

The retarded, advanced and correlation Green functions are related to the matrix $\hat{A}$ as

$$A_r = \tfrac{1}{2}\xi^\dagger \hat{A}\eta, \qquad A_a = \tfrac{1}{2}\eta^\dagger \hat{A}\xi, \qquad A_c = \tfrac{1}{2}\xi^\dagger \hat{A}\xi, \tag{7.21}$$

in accord with (2.12). Therefore, the Dyson equations for the retarded propagators take the form

$$\int \mathrm{d}^{d+1}y[\Delta_{0r}^{-1}(x,y) - \Pi_r(x,y)]\Delta_r(y,z) = \delta^{d+1}(x-z), \tag{7.22a}$$

$$\int \mathrm{d}^{d+1}y[G_{0r}^{-1}(x,y) - \Sigma_r(x,y)]G_r(y,z) = \delta^{d+1}(x-z). \tag{7.22b}$$

The corresponding equations for the advanced functions can be obtained by taking the Hermitian conjugation of (7.22), whereas equations for correlation functions appear in the matrix form as

$$\Delta_c = -\Delta_r(\Delta_{0c}^{-1} - \Pi_c)\Delta_a\,, \tag{7.23}$$

$$G_c = -G_r(G_{0c}^{-1} - \Sigma_c)G_a\,. \tag{7.24}$$

As shown in section 3.3, the latter equations reduce to transport equations for the quasiparticle distribution.

7.2.3. Summary

To sum up, we have derived seven equations to determine seven functions $Q_c(x)$, Δ_r, Δ_a, Δ_c, G_r, G_a and G_c, from which the order parameter as well as the energy spectrum, the dissipation and the distribution function for the corresponding quasiparticles can be calculated.

Up to now we have not yet made any approximations. As is well known [97], the loop expansion is actually a series expansion in $\hbar$. Therefore, for systems which can be described by quasiclassical approximation one needs only the first few terms of this expansion. In fact, one recovers the mean field result if the contribution from Γ_{2p} is neglected altogether. In some other cases like in the theory of critical phenomena, one needs to partially resume the most divergent diagrams.

For most cases of practical interest including thermoequilibrium, the initial correlations expressed in terms of W_p^N are Gaussian. As shown in section 6.1, in such cases the statistical information can be included in the free propagators Δ_0, S_0 by FDT, so that W_p^N term drops out from the effective action. Hence the analogy with the quantum field theory can be carried through even further for such systems.

As seen from the derivation, this calculation scheme can be applied to both equilibrium and nonequilibrium systems. It is particularly useful when the dynamical coupling between the order parameter and the elementary excitations is essential. We note in passing that the logical simplicity of the present formalism comes partly from introducing the two-point sources $M(x, y)$, $K(x, y)$ and performing Legendre transformation with respect to them.

7.2.4. Comparison with earlier formalism

To make contact with the generating functional introduced before (marked by a prime), we note that

$$Z_p'[h] = Z_p[h, M, K]|_{M=K=0}\,, \tag{7.25a}$$

$$W_p'[h] = W_p[h, M, K]|_{M=K=0}\,, \tag{7.25b}$$

$$\Gamma_p'[Q_c] = \Gamma_p[Q_c, \Delta, G]|_{\delta\Gamma_p/\delta\Delta = \delta\Gamma_p/\delta G = 0}\,, \tag{7.25c}$$

$$\frac{\delta\Gamma_p'[Q_c]}{\delta Q_c(x)} = \left[\frac{\delta\Gamma_p[Q_c, \Delta, G]}{\delta Q_c(x)}\bigg|_{\Delta, G}\right]\bigg|_{\delta\Gamma_p/\delta\Delta = \delta\Gamma_p/\delta G = 0}. \tag{7.25d}$$

Previously, an effective action method was introduced by us in the third paper of reference [46] to calculate Γ_p' explicitly. The disadvantage of that technique compared with the present formalism lies in the difficulties connected with fermion renormalization when the fermion degrees of freedom were integrated out at the very beginning.

7.2.5. Applications

We have already applied the present formalism to study the weak electromagnetic field in superconduction [53] as well as the nonequilibrium superconductivity in general [54], the dynamical behaviour of quenched random systems and a long-ranged spin glass model in particular [55], the quantum fluctuations in the quasi-one-dimensional conductors [57] and the exchange correction in systems with four-fermion interaction [49]. The last topic will be discussed in the following two sections, whereas the random systems are considered in section 8.

7.3. Generalization of Bogoliubov–de Gennes (BdeG) equation

As mentioned in the introductory remarks to this section, the self-consistent equations for the order parameter $Q_c(x)$ and the complete set of single-particle fermion wavefunctions $\psi_n(x)$ are known as the BdeG equations which are of Hartree-type without exchange effects being accounted for. There have been some attempts of extending these equations to include the correlation effects with limited success [96]. These authors emphasize the nonuniqueness of the HS [100] transformation and make use of it to derive various approximations. As mentioned in section 7.1, such ambiguity can be avoided by using the generating functional technique with given definition of the order parameter. As we will show in this section, the successive approximations can be derived in a systematic way using the formalism developed in the preceding two sections.

7.3.1. Model

The effective action of the system is given by

$$I[\psi^\dagger, \psi] = \int \psi_i^\dagger(x) \left(i\hbar \frac{\partial}{\partial t} - H_0 \right)_{ij} \psi_j((x) - \tfrac{1}{2} g^2 \int \psi_i^\dagger(x) \psi_j^\dagger(y) \Delta_0(x-y) O_{ik} O_{jl} \psi_l(y) \psi_k(x) , \tag{7.26}$$

where i, j are indices of internal degrees of freedom, O_{ik} matrix in this space and

$$\Delta_0(x-y) = \delta(t_x - t_y) V(x-y) . \tag{7.27}$$

Using the fermion commutation relation and the matrix notation for O_{ik}, (7.26) can be rewritten as

$$I[\psi^\dagger, \psi] = \int \psi^\dagger(x) \left(i\hbar \frac{\partial}{\partial t} - H_0 + \tfrac{1}{2} g^2 \hat{O}^2 V(o) \right) \psi(x) - \tfrac{1}{2} g^2 \int \psi^\dagger(x) \hat{O} \psi(x) \Delta_0(x-y) \psi^\dagger(y) \hat{O} \psi(y) . \tag{7.26'}$$

The order parameter is defined as

$$Q(x) \equiv g \int \Delta_0(x-y) \psi^\dagger(y) \hat{O} \psi(y) = g \int V(x-y) \psi^\dagger(y, t) \hat{O} \psi(y, t) . \tag{7.28}$$

7.3.2. Coupled equations

In accord with the result of the last section, up to the two-loop approximation, the coupled equations for the order parameter $Q_c(x)$ and the second-order CTPGF are the following:

$$-\tfrac{1}{2} \xi_\alpha \left. \frac{\delta \Gamma_p[Q]}{\delta Q(x_\alpha)} \right|_{Q_+ = Q_-} = - \int \Delta_0^{-1}(x-y) Q(y) - \tfrac{1}{2} i\hbar g \, \mathrm{Sp}\{\hat{O} G_c(x, x)\} = 0 , \tag{7.29}$$

$$G_p^{-1}(x,y)=G_{0p}^{-1}(x,y)-\Sigma_p(x,y)\,, \tag{7.30}$$

with

$$\Sigma_p(x,y)=i\hbar g^2\hat{O}G_p(x,y)\hat{O}\Delta_p(y,x)\,, \tag{7.31}$$

$$\Delta_p^{-1}(x,y)=\Delta_{0p}^{-1}(x,y)-\Pi_p(x,y)\,, \tag{7.32}$$

with

$$\Pi_p(x,y)=-i\hbar g^2\,\mathrm{Sp}(\hat{O}G_p(x,y)\hat{O}G_p(y,x))\,, \tag{7.33}$$

where Sp means taking trace over internal indices only, while Tr remains summing over both internal and space–time coordinates. Here

$$G_{0p}^{-1}(x,y)=S_{0p}^{-1}(x,y)-g\hat{O}Q(x)\delta_p^{d+1}(x-y)\,, \tag{7.34}$$

with

$$S_{0p}^{-1}(x,y)=\left(i\hbar\frac{\partial}{\partial t}-H_0+\tfrac{1}{2}g^2V(o)\hat{O}^2\right)\delta_p^{d+1}(x-y)\,. \tag{7.35}$$

7.3.3. Spectral representation

For simplicity we consider the stationary states when $Q(x)$ does not depend on time and all Green's functions are time translationally invariant. Also, we assume the dissipation for both fermions and collective excitations to be small, so that they can be considered as quasiparticles to a good approximation. In particular, the Fourier transformed fermion functions can be expanded in terms of the complete set $\{\psi_n(x)\}$ as

$$G_r(p_0,x,y)=\sum_n\psi_n(x)\frac{1}{p_0-E_n+i\gamma_n}\psi_n^\dagger(y)\,, \tag{7.36a}$$

$$G_a(p_0,x,y)=\sum_n\psi_n(x)\frac{1}{p_0-E_n-i\gamma_n}\psi_n^\dagger(y)\,, \tag{7.36b}$$

$$G_c(p_0,x,y)=\sum_n\psi_n(x)(1-2N_n)\left(\frac{1}{p_0-E_n+i\gamma_n}-\frac{1}{p_0-E_n-i\gamma_n}\right)\psi_n^\dagger(y)\,, \tag{7.36c}$$

while the spectral functions $\{\psi_n(x)\}$ satisfy the following equation

$$(E_n-i\gamma_n-H_0+\tfrac{1}{2}g^2\hat{O}^2V(o)-g\hat{O}Q(x))\psi_n(x)-\int\Sigma_r(p_0=E_n,x,y)\psi_n(y)=0\,. \tag{7.37}$$

In eqs. (7.36) and (7.37) E_n, γ_n and N_n are energy spectrum, dissipation and particle distribution respectively.

It is ready to check that in combination with FDT written as (cf. (4.29))

$$\Sigma_c(E_n, x, y) = \sum_n \psi_n(x) 2\mathrm{i}\gamma_n (1 - 2N_n)\psi_n^\dagger(y) . \tag{7.38}$$

Equations (7.36) and (7.37) are just equivalent to the Dyson equation (7.30) in the limit of weak dissipation. The orthonormalization and the completeness of the $\{\psi_n(x)\}$ set is then obvious.

7.3.4. Hartree approximation

As the first approximation, we neglect the self-energy part altogether, i.e., to set $\Sigma_p(x, y) \to 0$. It then follows that $\gamma_n \to 0+$ and the boson propagator Δ will not depend on the order parameter. Substituting (7.36c) into (7.29) and setting $\Sigma_r = 0$ in (7.37), we obtain the well-known BdeG equation

$$\int V^{-1}(x-y)Q(y) = g\sum_n (N_n - \tfrac{1}{2})\psi_n^\dagger(x)\hat{O}\psi_n(x)$$

$$= g\sum_n N_n \psi_n^\dagger(x)\hat{O}\psi_n(x) - \tfrac{1}{2}g\,\mathrm{Sp}(\hat{O})\delta^d(x) , \tag{7.39}$$

$$(E_n - H_0 + \tfrac{1}{2}g^2\hat{O}^2 V(o) - g\hat{O}Q(x))\psi_n(x) = 0 . \tag{7.40}$$

An interesting and important feature of these equations is that the energy spectrum $\{E_n\}$ should be determined self-consistently with the quasiparticle distribution.

7.3.5. Hartree–Fock (HF) approximation

As the second approximation, we keep $\Sigma_p(x, y)$ but set $\Pi_p(x, y)$ equal to zero. It follows then from (7.31) and (7.32) that

$$\Delta_r(x, y) = \Delta_a(x, y) = \delta(t_x - t_y)\cdot V(x-y), \qquad \Delta_c(x, y) = 0 , \tag{7.41}$$

$$\Sigma_r(p_0, x, y) = \frac{\mathrm{i}g^2\hbar}{2}\int \frac{\mathrm{d}q_0}{2\pi\hbar}\hat{O}G_c(q_0, x, y)\hat{O}V(x-y) , \qquad \Sigma_c(x, y) = 0, \qquad \gamma_n \to 0+ . \tag{7.42}$$

Repeating the procedure carried out in the Hartree case, we find that in the HF approximation the equation for the order parameter $Q(x)$ remains the same as (7.39), but the equation for the spectral functions changes into

$$(E_n - H_0 - g\hat{O}Q(x))\psi_n(x) + g^2 \int V(x-y)\sum_m \hat{O}\psi_m(x)N_m\psi_m^\dagger(y)\hat{O}\psi_n(y) = 0 . \tag{7.43}$$

Comparing (7.43) with (7.40) we find that the unphysical term $\frac{1}{2}g^2\hat{O}^2V(o)$ has been cancelled out and a new, Fock exchange term appears here. It is clear from our derivation that the order parameter $Q(x)$ is coupled to the fermion Green functions in the order of $\hbar$ (see (7.29)). Hence one should take into account contributions of the same order from Σ_r in (7.37) no matter what kind of interaction one deals with. We see thus how the Fock term has been "lost" in some derivations, but "recovered" in the CTPGF approach.

7.3.6. Higher order corrections

We can improve our approximation by keeping also the self-energy part $\Pi_p(x, y)$ in eqs. (7.29)–(7.33). This is the so-called random phase approximation (RPA) if only the leading term in Π is kept. It is straightforward to check that the equation for the order parameter still remains the same as (7.39) but the equation for the spectral function is now coupled to the collective excitation via the fermion self-energy part Σ_p provided the dissipation is still weak. We come back to this approximation in the next section where the free energy is calculated.

As is clear from the presentation, we can in principle continue this systematic process to go on higher order corrections, but we will not elaborate further on them here.

7.3.7. Vacuum fluctuation

Before closing this section a remark on the term $\frac{1}{2}g\,\mathrm{Sp}(\hat{O})\delta^d(x)$ in (7.39) is in order. For systems with off-diagonal long-range order like superconductivity this term drops out because

$$\mathrm{Sp}(\hat{O}) = 0\,. \tag{7.44}$$

For systems with diagonal long-range order this term will cancel out the divergent contribution of the Fermi sea. By using the particle–hole symmetry the right-hand side of (7.39) can be rewritten as

$$\sum_n (N_n - \tfrac{1}{2})\psi_n^\dagger(x)\hat{O}\psi_n(x) = \sum_n N_n^p \psi_n^{p\dagger}(x)\hat{O}\psi_n^p(x) - \sum_n N_n^h \psi_n^{h\dagger}(x)\hat{O}\psi_n^h(x)\,, \tag{7.45}$$

where the first sum is carried out over states above the Fermi level, while the second – below it. The rest of the notation is standard. In the next section we will assume for simplicity that eq. (7.44) is fulfilled.

7.4. Calculation of free energy

7.4.1. Functional equation

As shown in section 4.2, for systems respecting time reversal symmetry, the potential condition is satisfied and the free-energy functional $\mathcal{F}[Q(x)]$ can be defined in accord with (4.23a). Using (7.29) and (7.36c) this equation can be rewritten as

$$\frac{\delta\mathcal{F}[Q]}{\delta Q(x)} = -\int V^{-1}(x-y)Q(y) + g\sum_n (N_n - \tfrac{1}{2})\psi_n^\dagger(x)\hat{O}\psi_n(x)\,, \tag{7.46}$$

provided the dissipation γ_n is neglected. In the same approximation it follows from (7.37) that

$$\frac{\delta E_n}{\delta Q(x)} = g\psi_n^\dagger(x)\hat{O}\psi_n(x) + \int \psi_n^\dagger(y)\frac{\delta\Sigma_r(E_n, y, z)}{\delta Q(x)}\psi_n(z)\,. \tag{7.47}$$

According to the convention set at the end of the last section, i.e., eq. (7.44) is fulfilled, (7.47) yields

$$\frac{\delta}{\delta Q(x)}\sum_n E_n - \int \sum_n \psi_n^\dagger(y)\frac{\delta \Sigma_r(E_n, y, z)}{\delta Q(x)}\psi_n(z) = 0\,. \tag{7.48}$$

Insert (7.47) into (7.46), we obtain

$$\frac{\delta \mathcal{F}[Q]}{\delta Q(x)} = -\int V^{-1}(x-y)Q(y) + \sum_n (N_n - \tfrac{1}{2})\frac{\delta E_n}{\delta Q(x)} - \sum_n \int \mathrm{Sp}\left\{\frac{\delta \Sigma_r(E_n, y, z)}{\delta Q(x)}\psi_n(z)(N_n - \tfrac{1}{2})\psi_n^\dagger(y)\right\}, \tag{7.49}$$

which can be transformed as

$$\frac{\delta \mathcal{F}[Q(x)]}{\delta Q(x)} = -\int V^{-1}(x-y)Q(y) + \sum_n (N_n - \tfrac{1}{2})\frac{\delta E_n}{\delta Q(x)} + \frac{i\hbar}{2}\frac{1}{T}\int \mathrm{d}^{d+1}y\, \mathrm{d}^{d+1}z\, \mathrm{Sp}\left\{\frac{\delta \Sigma_r(y, z)}{\delta Q(x)} G_c(z, y)\right\} \tag{7.50}$$

by use of (7.36c) with

$$T \equiv \int_{-T/2}^{T/2} \mathrm{d}t\,.$$

We would like to emphasize here that the functional equation (7.49) or (7.50) for the free energy is valid for both equilibrium and nonequilibrium systems provided the potential condition is satisfied.

7.4.2. Hartree case

In the Hartree approximation when Σ_p is neglected altogether, we find immediately

$$\frac{\delta \mathcal{F}[Q]}{\delta Q(x)} = \frac{\delta}{\delta Q(x)}\left[-\tfrac{1}{2}\int Q(y)V^{-1}(y-z)Q(z)\right] + \sum_n N_n \frac{\delta E_n}{\delta Q(x)} \tag{7.51}$$

with eq. (7.48) being accounted for. For systems in the thermoequilibrium N_n is the Fermi distribution

$$N_n = (\exp[\beta(E_n - \mu)] + 1)^{-1}\,, \tag{7.52}$$

the free energy is then given by

$$\mathcal{F}[Q] = -\tfrac{1}{2}\int \mathrm{d}^d y\, \mathrm{d}^d z\, Q(y)V^{-1}(y-z)Q(z) - \beta^{-1}\sum_n \ln(1 + \exp[-\beta(E_n - \mu)])\,. \tag{7.53}$$

Taking into account (7.39) and (7.40), the first term in (7.53) can be rewritten as

$$-\tfrac{1}{2}g^2\sum_n\sum_m\int N_n\psi_n^\dagger(y)\hat{O}\psi_n(y)V(y-z)N_m\psi_m^\dagger(z)\hat{O}\psi_m(z)\,,$$

which obviously is the fermion interaction energy in the Hartree approximation.

7.4.3. Hartree–Fock case

In the Hartree–Fock approximation we can proceed exactly the same way as in the previous case and end up with an expression for the free energy containing an extra term

$$\mathscr{F}^{\mathrm{HF}}-\mathscr{F}^{\mathrm{H}}=\frac{g^2}{2}\int\sum_{n,m}N_n\psi_n^\dagger(y)\hat{O}\psi_m(y)V(y-z)N_m\psi_m^\dagger(z)\hat{O}\psi_n(z)\,,\tag{7.54}$$

in addition to that given by (7.53). The physical meaning of (7.54) as exchange interaction between quasiparticles is obvious.

7.4.4. Random phase approximation (RPA)

The RPA case is more complicated. Let us decompose the fermion self-energy as

$$\Sigma_p(x,y)=\Sigma_p^{(1)}(x,y)+\Sigma_p^{(2)}(x,y)\,,\tag{7.55}$$

where

$$\Sigma_p^{(1)}(x,y)=i\hbar g^2\hat{O}G_p(x,y)\hat{O}\Delta_{0p}(y,x)\,,\tag{7.56}$$

$$\Sigma_p^{(2)}(x,y)=i\hbar g^2\hat{O}G_p(x,y)\hat{O}(\Delta_p(y,x)-\Delta_{0p}(y,x))\,.\tag{7.57}$$

Correspondingly eqs. (7.48) and (7.50) are rewritten as

$$\frac{\delta}{\delta Q(x)}\Big(\sum_n E_n\Big)-\tfrac{1}{2}\int\sum_n\psi_n^\dagger\frac{\delta\Sigma_r^{(1)}}{\delta Q}\psi_n-\tfrac{1}{2}\int\sum_n\psi_n^\dagger\frac{\delta\Sigma_r^{(2)}}{\delta Q}\psi_n=0\,,\tag{7.58}$$

$$\frac{\delta\mathscr{F}[Q]}{\delta Q(x)}=-\int V^{-1}Q+\sum_n(N_n-\tfrac{1}{2})\frac{\delta E_n}{\delta Q}+\frac{i\hbar}{2}\frac{1}{T}\int\mathrm{Sp}\Big(\frac{\delta\Sigma_r^{(1)}}{\delta Q}G_c\Big)$$
$$-\frac{i\hbar}{2}\frac{1}{T}\int\mathrm{Sp}\Big(\Sigma_r^{(2)}\frac{\delta G_c}{\delta Q}\Big)+\frac{i\hbar}{2}\frac{1}{T}\frac{\delta}{\delta Q}\int\mathrm{Sp}(\Sigma_r^{(2)}G_c)\,,\tag{7.59}$$

where the abbreviated notation is used. The second term on the left of (7.58) and the second term on the right of (7.59) can be calculated by the same procedure as previously, whereas the third term of (7.58) and the last term of (7.59) turn out to be of higher order [49].

To calculate the contribution from the fourth term of (7.59) we make use of the identity

$$\int\Big(\Sigma_r^{(2)}(y,z)\frac{\delta G_c(z,y)}{\delta Q(x)}+\Sigma_c^{(2)}(y,z)\frac{\delta G_a(z,y)}{\delta Q(x)}\Big)$$
$$=-\tfrac{1}{2}\int\Big(\frac{\delta\Pi_r(y,z)}{\delta Q(x)}(\Delta_c(z,y)-\Delta_{0c}(z,y))+\frac{\delta\Pi_c(y,z)}{\delta Q(x)}(\Delta_a(z,y)-\Delta_{0a}(z,y))\Big)\,,\tag{7.60}$$

which can be proved using the ξ, η vectors and the definition of the self-energy parts. Neglecting terms like Σ_c and Π_c which are proportional to the dissipation, taking into account eqs. (7.32) and (7.38) as well as the time translational invariance, the fourth term of (7.59) can be presented as

$$M \equiv -\frac{i\hbar}{2}\frac{1}{T}\int \mathrm{Sp}\left(\Sigma_r^{(2)}(y,z)\frac{\delta G_c(z,y)}{\delta Q(x)}\right)$$
$$= -\frac{i\hbar}{4}\int\frac{dq_0}{2\pi\hbar}\int d^d y\, d^d z\,\frac{\delta\Delta_r^{-1}(q_0,y,z)}{\delta Q(x)}\,\Delta_c(q_0,z,y)\,. \tag{7.61}$$

To extract the information about the collective excitations we use the approximate spectral representation for Δ as

$$\Delta_r(q_0,y,z) = \sum_s Z_s^{-1}\phi_s(y)\left[\frac{(q_0+i\eta)^2}{\hbar^2c^2} - \frac{\omega_s^2}{c^2}\right]^{-1}\phi_s(z)\,,$$

$$\Delta_c(q_0,y,z) = \sum_s Z_s^{-1}\phi_s(y)(1+2N(q_0))\left[\left(\frac{(q_0+i\eta)^2}{\hbar^2c^2} - \frac{\omega_s^2}{c^2}\right)^{-1} - \left(\frac{(q_0-i\eta)^2}{\hbar^2c^2} - \frac{\omega_s^2}{c^2}\right)^{-1}\right]\phi_s(z)\,, \tag{7.62}$$

where ω_s, η, Z_s are, respectively, the frequency, dissipation and wavefunction renormalization for the quasiparticle. Making use of (7.62), the expression (7.61) can be presented as

$$M = \sum_s \frac{\delta(\hbar\omega_s)}{\delta Q}(N_s + \tfrac{1}{2})\,, \tag{7.63}$$

which is the extra contribution due to the collective excitation. In thermoequilibrium,

$$N_s = [\exp(\beta\hbar\omega_s) - 1]^{-1}\,, \tag{7.64}$$

the additional term in the free energy for the RPA compared with that of HF is given by

$$\mathcal{F}^{\mathrm{RPA}} - \mathcal{F}^{\mathrm{HF}} = \beta^{-1}\sum_s \ln[1-\exp(-\beta\hbar\omega_s)] + \tfrac{1}{2}\sum_s \hbar\omega_s\,, \tag{7.65}$$

where the last term is the contribution from the zero point motion.

7.4.5. Summary

To summarize we note that the calculation scheme presented in this section is rather practical and complete. We cannot only determine the order parameter as well as the energy spectrum, the dissipation and the quasiparticle distribution for both fermion and collective excitation, but also calculate explicitly the free-energy functional in successive approximations without invoking any extra assumptions. It is also interesting to note that for systems with symmetry breaking we can directly integrate out the vertex equation without introducing the symmetry broken ground state in advance. This technique may be helpful for more complicated problems.

8. Quenched random systems

In quenched random systems some degrees of freedom associated with impurities are frozen into a nonequilibrium but random configuration. Such situation can be created, say, by sudden cooling of a sample in thermal equilibrium to a state with much lower temperature. The impurities are thus frozen into a configuration separated by high potential barriers from the equilibrium one. Diffusion through these barriers will cause the nonequilibrium state to vary slowly in time.

As pointed out by Brout [101], the space average of an observable A in a quenched random system can be replaced by an ensemble average over the impurity degrees of freedom J.

$$\bar{A} = \int [dJ]A(J)P(J), \tag{8.1}$$

where $P(J)$ is the distribution function. Most of previous workers considered quenched random systems as if they were static [102]. In this approach one has to average the free energy proportional to the logarithm of the partition function, over the random configuration of impurities. It is a formidable task and a special n-replica trick has been invented. This method is extensively used to study random systems like spin glass [103–110].

Recently, several authors [111–120] have proposed dynamical theories of spin glass based on the MSR [90] statistical field theory. The advantage of the dynamical theory is the possibility of taking quenched average without resorting to the unphysical replica trick. The results obtained so far can be reproduced in most cases by the replica method with special pattern of replica symmetry breaking [121]. Although a kind of plausible physical interpretation of replica has been suggested very recently [122–124], as far as we understand the whole problem is still controversial. Therefore, a systematic dynamical theory is certainly needed to provide a proper description corresponding to the real experiments.

In this section we will show that the CTPGF formalism might be one of the candidates to provide such a dynamical description. As we have seen already, this formalism is suitable for studying slowly varying in time processes, so far as the causality and the FDT are built in the formalism itself. In the CTPGF approach the quenched average can be carried out over the generating functional itself and the counterpart of the Edwards–Anderson(EA) [103] order parameter q appears naturally as an integral part of the second-order CTPGF. As a consequence, a Dyson equation can be derived for q to describe its slow variation using the quasiclassical approximation.

In section 8.1 we outline the basic features of the CTPGF dynamic theory for quenched systems, whereas in section 8.2 we apply it to discuss the infinite-ranged Ising spin glass, i.e. the Sherrington–Kirkpatrick (SK) model [104]. A boundary line of stability is found on the plane $q-|\chi|$, where χ is the susceptibility. It is argued that the spin glass phase is characterized by the fixed point located at the stability boundary. The magnetization is calculated in perturbation and is found to be in good agreement with those predicted by the projection hypothesis [125, 126].

In section 8.3, we discuss the disordered electron system within the CTPGF framework. The WT identities following from the symplectic symmetry Sp(2) respected by the Lagrangian of the system as well as the localization properties are considered without resorting to the replica trick.

8.1. Dynamic formulation

8.1.1. Model

Suppose the action of the random system is given by

$$I = \int_p \sigma(x)\Gamma_p^{(0,0)}(x,y)\sigma(y) - \int_p V(\sigma(x), J_i) + \int_p (\sigma(x)h(x) + \sigma(x)j(x)) + I_h\,, \tag{8.2}$$

where $h(x)$ is the external field coupled to the dynamical variable $\sigma(x)$, $J_i(x)$ being random variables with given distribution. The $\sigma(x)j(x)$ term represents the interaction of the dynamical variable with heat bath described by I_h. $\sigma(x)$ may have one or several components.

After integrating over the reservoir degrees of freedom which might be considered as a set of harmonic oscillators, we obtain the effective action

$$I_{\text{eff}} = \int_p \sigma(x)\Gamma_p^{(0)}(x,y)\sigma(y) - \int_p V(\sigma(x), J_i) + \int_p \sigma(x)h(x)\,, \tag{8.3}$$

where

$$\Gamma_p^{(0)}(x,y) = \Gamma_p^{(0,0)}(x,y) - \Sigma_p^{(0)}(x,y)\,, \tag{8.4}$$

with the self-energy part $\Sigma_p^{(0)}$ contributed by the interaction term $\sigma(x)j(x)$. Suppose the system is prepared by sudden cooling down to temperature β^{-1} at moment t_0, then $\Gamma_p^{(0)}$ satisfies the FDT given by (3.42a).

8.1.2. Averaged generating functional

The generating functional averaged over the random distribution of J_i is

$$\bar{Z}[h(x)] \equiv \int [\mathrm{d}J] P(J) Z[h(x), J] \equiv \langle Z[h(x), J]\rangle_J\,, \tag{8.5}$$

where

$$Z[h(x), J] = \int [\mathrm{d}\sigma] \exp(\mathrm{i}I_{\text{eff}})\langle t_+ = t_0|\hat{\rho}|t_- = t_0\rangle\,. \tag{8.6}$$

Introducing the generating functional for the connected Green functions as usual

$$\bar{W}[h(x)] = -\mathrm{i}\ln \bar{Z}[h(x)]\,, \tag{8.7}$$

$$W[h(x), J] = -\mathrm{i}\ln Z[h(x), J]\,, \tag{8.8}$$

we obtain the average of the field variable

$$\bar{\sigma}(x) = \delta \bar{W}/\delta h(x)\,, \tag{8.9}$$

and the connected CTPGF as

$$\overline{G_p}(1\cdots n) = (-1)^{n-1}\frac{\delta^n \bar{W}}{\delta h(1)\cdots\delta h(n)}\,, \tag{8.10}$$

along with the corresponding terms for $W[h(x), J]$. It follows from (8.5) that

$$\bar{Z}\bar{\sigma}(x) = \{Z\sigma(x, J)\rangle_J\,,$$

which by virtue of the normalization condition (2.102) reduces to

$$\bar{\sigma}(x) = \langle\sigma(x, J)\rangle_J \tag{8.11}$$

for the physical limit $h_+(x) = h_-(x)$. Equation (8.11) is what is required for the quenched average. Hereafter we denote all quantities derived from the averaged generating functional by a bar above them, whereas expectation values directly averaged over the ensemble $P(J)$ are presented as $\langle\cdot\,\cdot\rangle_J$.

8.1.3. Order parameter

Differentiating (8.10) with respect to $h(y)$ and taking the physical limit, we obtain

$$\langle G_p(x, y; J)\rangle_J = \bar{G}_p(x, y) + \mathrm{i}q(x, y)\,, \tag{8.12}$$

where

$$q(x, y) = \langle\sigma(x, J)\sigma(y, J)\rangle_J - \bar{\sigma}(x)\bar{\sigma}(y)\,. \tag{8.13}$$

The expectation value $\sigma(x, J)$ for an Hermitian operator σ is a real function with equal values on the two-time branches. Hence the matrix is real, symmetric and independent of time branch, i.e.

$$q(x, y) = q(y, x) = q^*(x, y)\,, \tag{8.14}$$

$$q(x+, y+) = q(x+, y-) = q(x-, y+) = q(x-, y-)\,. \tag{8.15}$$

Edwards and Anderson [103] have introduced the following order parameter

$$q_{\mathrm{EA}} = \lim_{t\to\infty}\langle\sigma(0, J)\sigma(t, J)\rangle$$

which looks similar to what is defined by (8.13).

It follows from (8.12) that

$$\langle G_r(x, y; J)\rangle_J = \overline{G_r}(x, y)\,, \tag{8.16a}$$

$$\langle G_a(x, y; J)\rangle_J = \overline{G_a}(x, y), \tag{8.16b}$$

$$\langle G_c(x, y; J)\rangle_J = \overline{G_c}(x, y) + 2iq(x, y). \tag{8.16c}$$

It is easy to check that (8.16a) and (8.16b) are also true for higher order retarded and advanced functions, i.e.,

$$\langle G_{21\cdots 1}(1\cdots n, J)\rangle_J = \bar{G}_{21\cdots 1}(1\cdots N),$$

etc. The appearance of the matrix $q(x, y)$ is a consequence of quenched average over the random variable J_i. Hence it characterizes the behaviour of the quenched random system.

8.1.4. Free energy

After a sufficiently long time the system is expected to reach a steady, yet not necessarily equilibrium state, when the expectation value $\sigma(x, J)$ is no longer time dependent. As follows from the discussion of section 4.2, there exists a free-energy functional such that

$$\sigma(h, J) = -\delta\mathcal{F}[h(x), J]/\delta h(x), \tag{8.17}$$

if

$$\operatorname{Im}\int G_r(\tfrac{1}{2}t, -\tfrac{1}{2}t, J)\,dt = 0. \tag{8.18}$$

In accord with (8.16a), the same is true for the function with bar, i.e.,

$$\bar{\sigma}(h) = -\delta\bar{\mathcal{F}}/\delta h(x). \tag{8.19}$$

From (8.17) and (8.19) we find that

$$\delta\bar{\mathcal{F}}/\delta h = \langle \delta\mathcal{F}/\delta h\rangle_J. \tag{8.20}$$

For a smooth function of distribution $P(J)$ with finite moments the order of differentiation and averaging can be changed. Integrating (8.20) yields

$$\bar{\mathcal{F}}[h] = \langle \mathcal{F}[h, J]\rangle_J + \text{terms independent of } h.$$

So far as our formalism is flexible enough to incorporate various composite operators, in particular, the energy density as a conjugate variable of temperature. This way we can exhaust all variables in the free energy and get the equality

$$\bar{\mathcal{F}}[h] = \langle \mathcal{F}[h, J]\rangle_J, \tag{8.21}$$

up to an unimportant constant.

We see, therefore, the averaged expectation values of physical variables like magnetic moment, free energy, etc. can be calculated directly from the averaged generating functional $\bar{Z}[h, J]$. Since the technique of deriving various consequences from a well defined $\bar{Z}[h, J]$ is highly developed, the predictions of the dynamic theory are unambiguous.

8.1.5. FDT and Fischer law

After performing Fourier transformation with respect to the relative coordinates $x-y$, the FDT for G before average is written as (cf. eq. (3.43a))

$$G_c(k, X, J) = 2i \coth(\beta k_0/2) \operatorname{Im} G_r(k, X, J)\,, \tag{8.22}$$

whereas for the quench-averaged $\bar{G}$ we have

$$\begin{aligned}\bar{G}_c(k, X) &= 2i \coth(\beta k_0/2) \operatorname{Im} \bar{G}_r(k, X) - 2iq(k, X)\\ &\approx \frac{4i}{\beta k_0} \operatorname{Im} \bar{G}_r(k, X) - 2iq(k, X)\,.\end{aligned} \tag{8.23}$$

The retarded function G_r is analytic in the upper half-plane of k_0. If $G_r(k, X)$ vanishes as $k_0 \to \infty$, the real and imaginary parts of G_r satisfy an unsubtracted dispersion relation

$$\operatorname{Re} \bar{G}_r(k_0, \boldsymbol{k}, X) = \frac{1}{\pi} \int \frac{\operatorname{Im} \bar{G}_r(k_0', \boldsymbol{k}, X)}{k_0' - k_0}\, dk_0'\,. \tag{8.24}$$

Making use of (8.23) we find in the high temperature limit

$$\begin{aligned}\operatorname{Re} \bar{G}_r(k_0 = 0, \boldsymbol{k}, X) &= \bar{G}_r(k_0 = 0, \boldsymbol{k}, X)\\ &= -\beta \int \frac{dk_0}{2\pi} (i\bar{G}_c(k_0, \boldsymbol{k}, X) - q(k_0, \boldsymbol{k}, X))\,.\end{aligned} \tag{8.25}$$

For a long-ranged Ising spin model when the space dependence of Green's function can be neglected (8.25) becomes Fischer's relation [127]

$$\chi = -\bar{G}_r(k_0 = 0, t) = \beta(1 - q(t, t) - \bar{\sigma}^2(t))\,. \tag{8.26}$$

We see thus that the validity of Fischer relation depends crucially on the high-frequency behaviour of the retarded Green function as well as the FDT.

8.1.6. Dynamic equation for q

Now we derive the dynamic equation satisfied by the matrix q. The Dyson equation for the quench-averaged function can be written as (cf. (3.10))

$$\bar{\Gamma}_r \bar{G}_r = 1\,, \tag{8.27a}$$

$$\bar{\Gamma}_a \bar{G}_a = 1\,, \tag{8.27b}$$

$$\bar{\Gamma}_r \bar{G}_c = -\bar{\Gamma}_c \bar{G}_a . \tag{8.27c}$$

Introducing a new matrix

$$Q \equiv \mathrm{i}[\tfrac{1}{2}\bar{\Gamma}_c - \mathrm{i}\coth(\beta k_0/2)\,\mathrm{Im}\,\bar{\Gamma}_r], \tag{8.28}$$

we directly find the Dyson equation for q,

$$\bar{\Gamma}_r q = -Q\bar{G}_a . \tag{8.29}$$

The Hermitian conjugation of eq. (8.29) is given by

$$q\bar{\Gamma}_a = -\bar{G}_r Q . \tag{8.30}$$

Separating the Hermitian and anti-Hermitian parts of (8.29) and (8.30) we obtain

$$\bar{\Gamma}_r q + q\bar{\Gamma}_a = -Q\bar{G}_a - \bar{G}_r Q , \tag{8.31}$$

$$\bar{\Gamma}_r q - q\bar{\Gamma}_a = -Q\bar{G}_a + \bar{G}_r Q . \tag{8.32}$$

In the quasiclassical approximation we replace the product of two matrices A and B,

$$AB = (AB + BA)/2 + (AB - BA)/2 ,$$

by the classical expression

$$\tilde{A}\tilde{B} - \frac{\mathrm{i}}{2}\{\tilde{A}, \tilde{B}\}_{\mathrm{P.B.}} \equiv \tilde{A}\tilde{B} - \frac{\mathrm{i}}{2}\left(\frac{\partial \tilde{A}}{\partial k^\mu}\frac{\partial \tilde{B}}{\partial X_\mu} - \frac{\partial \tilde{A}}{\partial X_\mu}\frac{\partial \tilde{B}}{\partial k^\mu}\right), \tag{8.33}$$

where $\tilde{A}, \tilde{B}$ are Fourier transforms. As seen from the discussion in section 3.3, this approximation is controlled by the inequality

$$\left|\frac{1}{\tilde{O}}\frac{\partial^2 \tilde{O}}{\partial k^\mu \partial X_\mu}\right| \ll 1 , \tag{8.34}$$

where $\tilde{O}$ may be either $\tilde{A}$ or $\tilde{B}$. In this approximation eqs. (8.31) and (8.32) become (with "$\tilde{\ }$" dropped)

$$q + Q|\bar{G}_r|^2 = \frac{1}{2|\bar{G}_r|^2}\left\{\left(\frac{\partial q}{\partial X_\mu} - |\bar{G}_r|^2\frac{\partial Q}{\partial X_\mu}\right)\left(\mathrm{Im}\,\bar{G}_r\frac{\partial\,\mathrm{Re}\,\bar{G}_r}{\partial k^\mu} - \mathrm{Re}\,\bar{G}_r\frac{\partial\,\mathrm{Im}\,\bar{G}_r}{\partial k^\mu}\right)\right.$$
$$\left. - \left(\frac{\partial q}{\partial k^\mu} - |\bar{G}_r|^2\frac{\partial Q}{\partial k^\mu}\right)\left(\mathrm{Im}\,\bar{G}_r\frac{\partial\,\mathrm{Re}\,\bar{G}_r}{\partial X_\mu} - \mathrm{Re}\,\bar{G}_r\frac{\partial\,\mathrm{Im}\,\bar{G}_r}{\partial X_\mu}\right)\right\}, \tag{8.35}$$

$$\frac{\partial|\bar{G}_r|^2}{\partial k^\mu}\left(\frac{\partial q}{\partial X_\mu} - |\bar{G}_r|^2\frac{\partial Q}{\partial X_\mu}\right) = \frac{\partial|\bar{G}_r|^2}{\partial X_\mu}\left(\frac{\partial q}{\partial k^\mu} - |G_r|^2\frac{\partial Q}{\partial k^\mu}\right). \tag{8.36}$$

Equations (8.35) and (8.36) are transport equations to determine the time evolution of q. They are written here for single dynamic variable. It can be easily, though tediously extended to the multi-component case.

For a homogeneous system in steady state all functions do not depend on the macro-space–time X, then eqs. (8.35) and (8.36) reduce to a single equation

$$q = Q|\bar{G}_r|^2 . \tag{8.37}$$

Using the field theoretical technique Q and $|\bar{G}_r|^2$ can be calculated once the Lagrangian is specified. They are functions of q. In some cases

$$Q = \lambda q \tag{8.38}$$

in the first order of perturbation. A nontrivial solution with $q \neq 0$ exists if the condition

$$\lambda|\bar{G}_r|^2 = 1 \tag{8.39}$$

could be satisfied. For the model to be discussed in the next section (8.39) cannot be fulfilled. Hence either the spin glass is not in a steady state or it cannot be characterized by a nonvanishing q.

8.2. Infinite-ranged Ising spin glass

8.2.1. Model

In this section we apply the formalism developed in the previous section to the infinite-ranged Ising spin glass, i.e. the SK [102] model. For simplicity we consider the soft spin version described by the Hamiltonian

$$\mathscr{H} = -\tfrac{1}{2}\sum_{i\neq j} J_{ij}\sigma_i\sigma_j + \sum_i \left(\tfrac{1}{2}r_0\sigma_i^2 + u\sigma_i^4 - \sigma_i h_i\right), \tag{8.40}$$

where the value of spin σ_i is not limited to ±1. The exchange integrals J_{ij} are random variables with Gaussian distribution

$$P(J_{ij}) = (2\pi N/J^2)^{-1/2}\exp\{-NJ_{ij}^2/2J^2\}, \tag{8.41}$$

where N is the number of spins interacting with the given one.

Taking into account the interaction with the heat bath and averaging over J_{ij}, we can write the CTPGF generating functional as

$$Z[h_i(t)] = \int [\mathrm{d}\sigma_i]\exp(iS_{\mathrm{eff}})\langle t_+ = t_0|\hat{\rho}|t_- = t_0\rangle , \tag{8.42}$$

with

$$S_{\mathrm{eff}} = \sum_j \int_p \{\sigma_j(t)\Gamma_{0p}(t,t')\sigma_j(t') - u\sigma_j^4(t) + \sigma_j(t)h_j(t)\} + \mathrm{i}\frac{J^2}{4N}\sum_{i\neq j}\int_p \sigma_i(t)\sigma_j(t)\,\mathrm{d}t \int_p \sigma_i(t')\sigma_j(t')\,\mathrm{d}t' . \tag{8.43}$$

The notation used here is the same as in the last section, but with the bar over various quantities dropped for simplicity. After Fourier transformation the low-frequency approximation for Γ_0 is given by

$$\Gamma_{0r}(\omega, t) = -r_0 + i\omega/\Gamma_0 , \tag{8.44a}$$

$$\Gamma_{0c}(\omega, t) = 2i/\beta\Gamma_0 . \tag{8.44b}$$

8.2.2. Self-energy part

In the infinite range limit when $N \to \infty$ the matrix $G_{ij}(t, t')$ can be approximated by $\delta_{ij} G(t, t')$. In this case the second-order vertex function can be calculated by the diagrammatic expansion. It is found for small q that

$$\Gamma_r(\omega, t) = -r + i\omega/\Gamma - J_r^2(q) G_r(\omega, t) - \Sigma_r(\omega, t) , \tag{8.45}$$

where r, Γ are renormalized quantities which might depend on temperature. To the lowest order perturbation in u we find

$$J_r^2(q) = J^2 + 288u^2(q^2 + 2q\sigma^2) , \tag{8.46}$$

where J is also renormalized and $q(t, t)$ is the order parameter. In eq. (8.45) terms proportional to ω and G_r are subtracted from the self-energy part, i.e.,

$$\Sigma_r(0, t) = 0 . \tag{8.47}$$

In obtaining (8.46) we assumed $q(\omega, t)$ to be peaked at $\omega = 0$.

To the same approximation the correlated vertex function is calculated to be

$$\Gamma_c(\omega, t) = \frac{4i}{\beta\omega} \operatorname{Im} \Gamma_r(\omega, t) + 2iJ_c^2(q) q(\omega, t) + 2iJ^2 \Delta(\omega, t) - \Sigma_c(\omega, t) , \tag{8.48}$$

where

$$\Delta(\omega, t) = \int d\tau\, e^{i\omega\tau} \sigma(t + \tfrac{1}{2}\tau) \sigma(t - \tfrac{1}{2}\tau) . \tag{8.49}$$

The expectation value $\sigma(t)$ might be different from zero when a magnetic field is applied. It can be shown that $\Delta(\omega, t)$ is peaked at $\omega = 0$, whereas Σ_c is the smooth part of the self-energy. To the lowest order in u we find

$$J_c(q) = J^2 + 96u^2(q^2 + 2q\sigma^2) . \tag{8.50}$$

In the low-frequency limit the function defined by (8.28) turns out to be

$$Q(\omega, t) = -J_c^2(q) q(\omega, t) - J^2 \Delta(\omega, t) , \tag{8.51}$$

provided only terms peaked at $\omega = 0$ are retained. It is worthwhile noting that $J_r^2(q)$ is greater than $J_c^2(g)$ for all values of q as evident from (8.46) and (8.50). This fact is essential for our later discussion.

8.2.3. Stability

Now we study the stability of the system. As seen from (8.45), in the $J \to 0$, i.e., the pure limit r is the inverse susceptibility which itself is proportional to the temperature and increases as the magnetization increases. We assume that such qualitative behaviour holds also for random systems. In the zero-frequency limit the Dyson equation for the retarded Green function (8.27a) becomes

$$\chi^{-1} = r - J_r^2(q)\chi\,, \tag{8.52}$$

in accord with (8.45), where the magnetic susceptibility

$$\chi = -G_r(\omega = 0)\,. \tag{8.53}$$

Equation (8.52) can be solved to yield

$$\chi = \frac{1}{2J_r^2(q)}\left[r - \sqrt{r^2 - 4J_r^2(q)}\right]. \tag{8.54}$$

Thus the magnetic susceptibility increases as r decreases and reaches its maximum at $r = 2J_r(q)$. Further decrease of r will make χ complex and the system unstable. Therefore, the stable region is bounded by the inequality

$$\chi J_r(q) < 1\,. \tag{8.55}$$

It is clear from (8.54) that in the unstable region

$$|\chi|^2 J_r^2(q) = 1\,, \tag{8.56}$$

which is a curve on the plane $q - |\chi|$. On this plane all stable points are located in the region bounded from above by the curve (8.56) consisting of marginally stable and unstable points. In the stable region q and χ are related by the Fischer relation (8.26). Hence the physical state of the system can evolve either in the stable region or on the boundary (8.56).

8.2.4. Time evolution of $q(t, t)$

Before discussing the time evolution of $q(t, t)$ we first note that in the low-frequency limit

$$G_r(\omega, t) = -\chi + \alpha(t)|\omega|^{\nu}(\coth(\pi\nu/2) - \mathrm{i}\,\mathrm{sgn}\,\omega)\,, \tag{8.57}$$

where $\alpha(t)$ and ν are positive quantities to be determined. An analysis similar to that of ref. [117] using (8.45) shows that $\nu \leq 1/2$ if the state is marginally stable and $\nu = 1$ otherwise.

To study the time evolution of q we start from the Dyson equations in the quasiclassical approximation (8.35) and (8.36). Using (8.51) in the absence of the external field we have

$$\frac{\partial \tilde{q}}{\partial t} = \left\{\frac{1 - J_c^2(q)|\tilde{G}_r|^2}{1 + J_c^2(q)|\tilde{G}_r|^2}\frac{\partial |\tilde{G}_r|^2}{\partial t} \Big/ \left(\frac{\partial\,\mathrm{Re}\,\tilde{G}_r}{\partial \omega}\frac{\partial\,\mathrm{Im}\,\tilde{G}_r}{\partial t} - \frac{\partial\,\mathrm{Im}\,\tilde{G}_r}{\partial \omega}\frac{\partial\,\mathrm{Re}\,\tilde{G}_r}{\partial t}\right)\right\}\tilde{q}\,. \tag{8.58}$$

In the low-frequency limit eq. (8.58) can be simplified and after Fourier transformation becomes [55]

$$\frac{\partial q}{\partial t} = -2\frac{1 - J^2\beta^2(1-q)^2}{1 + J^2\beta^2(1-q)^2}\frac{\Gamma}{\beta(1-q)}q, \tag{8.59}$$

by use of the Fischer relation valid in the stable region. Above the critical temperature $\beta_c^{-1} = J$, the only fixed point of (8.59) is $q = 0$. Near this point the order parameter decays exponentially

$$q = q_{in}\exp(-t/\tau_0), \tag{8.60}$$

with

$$\tau_0 = \frac{\beta_c^2 + \beta^2}{2(\beta_c^2 - \beta^2)}\frac{\beta}{\Gamma}. \tag{8.61}$$

Below T_c there is another fixed point q_1 on the stability boundary

$$q_1 = 1 - \beta_c/\beta. \tag{8.62}$$

The order parameter approaches this point also exponentially with characteristic time

$$\tau_1 = \beta_c^2/2(\beta - \beta_c)\Gamma. \tag{8.63}$$

It is worthwhile noting that both τ_0 and τ_1 have a pole at $\beta = \beta_c$, giving rise to a kind of critical slowing down.

However, q_1 is no more a fixed point if higher order terms are taken into account. As seen from (8.58), the nonlinear fixed point is determined by the intersection of Fischer line with

$$1 = J_c^2(q)|\chi|^2. \tag{8.64}$$

Since $J_r^2(q) > J_c^2(q)$ as we noted before, this curve is outside of the stable region and cannot be reached. In fact, what happens is that after hitting the stability boundary at q_1, the order parameter further decays along the marginal stability line down to $q = 0$. By inserting (8.57) and (8.64) into (8.58) we obtain

$$\frac{\partial \tilde{q}(\omega, t)}{\partial t} = -\frac{4}{3}\frac{q^2}{\alpha\nu}|\omega|^{1-\nu}\tilde{q}(\omega, t). \tag{8.65}$$

Hereafter we use the units $J = T_c = 1$ and take $u = 1/12$. It is obvious from (8.65) that $\tilde{q}(0, t)$ does not change with time whereas $\tilde{q}(\omega \neq 0, t)$ tends to zero as t goes to infinity.

8.2.5. Susceptibility and q in a small field

It follows from (8.29) and (8.49)–(8.51) that the static fixed point for q is determined from the equation

$$q_0 = |\chi|^2(q_0 + \sigma^2 + \tfrac{2}{3}q_0^3 + 2q_0^2\sigma^2) \tag{8.66}$$

where $q_0 \equiv \lim_{t\to\infty} q(t, t)$.

For $T > T_c$, the whole Fischer line (8.26) is located inside the stable region, so for small field h we have

$$q_0 = \frac{\beta^2}{1-\beta^2}\sigma^2\,, \tag{8.67}$$

$$\chi \equiv \frac{d\sigma}{dh} = \beta - \frac{\beta^3}{1-\beta^2}h^2\,. \tag{8.68}$$

At the critical point the Fischer relation (8.26) still holds to yield

$$q_0 = \sigma/\sqrt{2} - \tfrac{19}{24}\sigma^2 + \cdots\,, \tag{8.69}$$

$$\chi = 1 - h/\sqrt{2} + \tfrac{1}{24}h^2 + \cdots\,. \tag{8.70}$$

Below T_c, there exists a critical magnetic field h_c above which the static fixed point is still sitting in the stable region. The value of h_c turns out to be

$$h_c^2 = \tfrac{4}{3}\tau^2(1 + 3\tau + \cdots) \tag{8.71}$$

near T_c with $\tau = 1 - \beta^{-1}$.

For $T < T_c$ and $h < h_c$, the intersection of the Fischer line with the fixed point equation (8.66) is outside the physical boundary, so it can never be reached. This false fixed point is just what was found before [104, 105] to yield negative entropy at low temperatures. The only plausible fixed points are those located on the stability boundary (8.56) which in new units appears as

$$\chi = (1 + 2q_0^2 + 4q_0\sigma^2)^{-1/2}\,. \tag{8.72}$$

Solving (8.66) and (8.72) yields

$$q_0 = (\tfrac{3}{4}\sigma^2)^{1/3}\,, \tag{8.73}$$

$$\chi = 1 - (\tfrac{3}{4}h^2)^{2/3}\,. \tag{8.74}$$

All results obtained in eqs. (8.67)–(8.71) as well as in eqs. (8.73) and (8.74) agree with those predicted by the projection hypothesis [125, 126].

8.2.6. Summary

To sum up, we have found from a systematic analysis using the CTPGF formalism that a physical boundary exists on the plane q–$|\chi|$. Above T_c, the Fischer line is lying entirely in the stable region and the order parameter approaches the fixed point at this line exponentially in time. Below T_c and h_c there are no fixed points on the Fischer line inside the stable region. In this case the fixed point is located on the stability boundary. In the presence of a persistent magnetic field h the order parameter will first decay exponentially to the intersection point q_1 and then decreases further along the boundary down to the fixed point q_0. The magnetization calculated for the state q_0 agrees with what follows from the projection hypothesis. In the absence of magnetic field, $q_0 = 0$. If the $\omega = 0$ component of the order parameter $q(t, t)$ constitutes a finite part of q, say q_{EA}, the system will finally reach a steady state with $q = q_{EA}$. The value of q_{EA} depends on the dynamic processes with infinite long relaxation time in the $N \to \infty$ limit.

It is worthwhile to note that the boundary line on the q–$|\chi|$ plane, and hence the fixed point, is temperature independent. As a consequence, the magnetization is also temperature independent, while the entropy does not vary with magnetic field. This is just the assumption of the projection hypothesis [125, 126] which follows from the CTPGF formalism in a natural way.

Comparing the results obtained here with those of Parisi [110] as well as Sompolinsky and Zippelius [117–119], it is natural to conjecture that Parisi's function $q(x)$, $0 \leq x \leq 1$, corresponds to our $q(t, t)$ varying along the boundary from q_1 to q_0. The present formalism not only elucidates the physical meaning of the order parameter but also provides us with an equation to solve its time evolution.

8.3. Disordered electron system

Anderson first showed in 1958 [128] that if the electron site energies in a solid were sufficiently random, some of the energy eigenstates became localized instead of being extended as they would be in a regular crystal. Such localized states will not contribute to the electric conduction. The nature of the state depends on whether the energy value is located on the "localized" or "extended" side of the "mobility edge" [129].

The drastic change in the behaviour of the wavefunction with energy is reminiscent of phase transition. The great success in applying the field-theoretical technique and the renormalization group to critical phenomena [60] encouraged similar attempts in the localization problem. Wegner [130] suggested a nonlinear σ-model to study the scaling properties of the noninteracting disordered electron system near the mobility edge [131] with conductance playing the role of coupling constant. This model was later derived from the field theory [132, 133] and was further studied by other authors [134–136]. Recently, there is a revival of interest in this problem due to the discovery of the quantized Hall effect in two-dimensional electron system [137, 138]. This is a challenge to the theory since according to the existing theory all electron states in disordered two-dimensional system should be localized in the absence of magnetic field [131], whereas extended states are certainly needed for explaining the observed quantized Hall effect. The extension of the field-theoretical approach to include the external magnetic field was made very recently by Pruisken et al. [139, 140].

However, almost all studies of field-theoretic approach in the localization problem were based on the replica trick. With n-replicated system an $O(n_+, n_-)$ or $U(2n)$ symmetry is used to construct the nonlinear σ-model. The critical behaviour from the extended state side of the mobility edge is described by a Goldstone mode resulting by virtue of the spontaneous breaking of the replica symmetry. The difficulty of the replica trick is the necessity of continuing n, originally defined for integers, to zero to get the physical result. Such process might not be unique as in the case of spin glass.

It turns out that the CTPGF formalism can be also applied to the localization problem without resorting to the replica trick [56]. In this section we describe the symmetry of the model and derive the corresponding WT identities. The order parameter and the symmetry breaking pattern are also briefly sketched.

8.3.1. Model

We are concerned with the effect of disorder on Green's functions of a noninteracting electron gas moving in external fields. The Lagrangian of the system is given by

$$\mathcal{L} = \int \psi^\dagger(x)\Big(\mathrm{i}\frac{\partial}{\partial t} - L_0 - V\Big)\psi(x)\,, \tag{8.75}$$

where

$$L_0 = \frac{1}{2m}(-\mathrm{i}\nabla - e\boldsymbol{A}(x))^2 + e\varphi(x)\,. \tag{8.76}$$

and V the random potential.

In the CTPGF formalism the generating functional $Z[J(x), V]$ is specified by the effective action which in the single-time representation can be written as

$$S_{\text{eff}} = \int [\psi^\dagger(x)\sigma_3\Gamma^0(x, y)\sigma_3\psi(y) + J^\dagger(x)\sigma_3\psi(x) + \psi^\dagger(x)\sigma_3 J(x)] \tag{8.77}$$

where $\psi(x), J(x)$ are two-component vectors as usual. The vertex function in the tree approximation is given by

$$\hat{\Gamma}_0(x, y) = \sigma_3\Big(\mathrm{i}\frac{\partial}{\partial t} - L_0 - V\Big)\delta(x-y) + \mathrm{i}\varepsilon(I + \sigma_1 - \mathrm{i}\sigma_2)\delta(x-y) \tag{8.78}$$

where ε is a positive infinitesimal constant. To convinve ourselves in the validity of (8.78) we note that for noninteracting fermions

$$\Gamma_{0\mathrm{r}} = E - \varepsilon(k) + \mathrm{i}\varepsilon, \qquad \Gamma_{0\mathrm{a}} = E - \varepsilon(k) - \mathrm{i}\varepsilon, \qquad \Gamma_{0\mathrm{c}} = 2\mathrm{i}\varepsilon\,, \tag{8.79}$$

as follows from (2.23) and the Dyson equation (3.10). Equation (8.78) is then a direct consequence of (2.63).

The generating functional can be thus rewritten as

$$Z[J, V] = \int [\mathrm{d}\psi][\mathrm{d}\psi^\dagger] \exp\Big\{\mathrm{i}\int \Big[\psi^\dagger(x)\Big(\Big(\mathrm{i}\frac{\partial}{\partial t} - L_0 - V\Big)\sigma_3 + \mathrm{i}\varepsilon(I - \sigma_1 + \mathrm{i}\sigma_2)\Big)\psi(x) + \psi^\dagger\sigma_3 J + J^\dagger\sigma_3\psi\Big]\Big\}\,. \tag{8.80}$$

As shown in section 8.1, the quenched average of random potential can be carried out directly on the generating functional. It is, however, more convenient to work in the energy representation for the localization problem. After Fourier transformation the effective action is given by

$$S_{\text{eff}} = \int \mathrm{d}^d x\,\mathrm{d}E\,\{\psi^\dagger(\boldsymbol{x}, E)[\sigma_3(E - L_0 - V) + \mathrm{i}\varepsilon(I - \sigma_1 + \mathrm{i}\sigma_2)]\psi(\boldsymbol{x}, E) + \psi^\dagger\sigma_3 J(\boldsymbol{x}, E) + J^\dagger\sigma_3\psi(\boldsymbol{x}, E)\}\,. \tag{8.81}$$

We restrict ourselves to the Wegner model [130] where the correlation of the random potential between different energy shells vanishes, i.e.,

$$\langle V(\boldsymbol{x}, E)\rangle = 0, \qquad \langle V(\boldsymbol{x}, E)V(\boldsymbol{y}, E')\rangle = \gamma\delta(\boldsymbol{x}-\boldsymbol{y})\delta(E-E'). \tag{8.82}$$

After averaging over the random potential V the generating functional

$$\bar{Z}[J] = \int [\mathrm{d}V]Z[J, V]\exp\left(-\frac{1}{2\gamma}\int V^2(\boldsymbol{x})\,\mathrm{d}\boldsymbol{x}\right) \tag{8.83}$$

is determined by a new effective action S which differs from S_{eff} (8.81) by dropping V in the single-particle Lagrangian and a new term

$$\Delta S = \frac{\mathrm{i}\gamma}{2}\int \mathrm{d}^d x\,\mathrm{d}E\,\psi^\dagger(\boldsymbol{x}, E)\sigma_3\psi(\boldsymbol{x}, E)\psi^\dagger(\boldsymbol{x}, E)\sigma_3\psi(\boldsymbol{x}, E). \tag{8.84}$$

The generating functionals $\bar{W}[J]$ and $\bar{\Gamma}[\psi_c]$ are defined in the standard manner so that there is no need to write them down explicitly. We would like to mention, nevertheless, that to avoid the possible confusion with sign of the Grassman variables we adopt the convention that $\delta/\delta J(x)$, $\delta/\delta\psi_c(x)$ act from the right, whereas $\delta/\delta J^\dagger(x)$, $\delta/\delta\psi_c^\dagger(x)$ act from the left.

8.3.2. Symmetry properties

Before discussing the specific model under consideration we would like to make a general remark concerning the symmetry properties in the CTPGF formalism. It is well known that the U(1) symmetry of a complex field

$$\psi(x)\to \mathrm{e}^{\mathrm{i}\alpha}\psi(x), \qquad \psi^\dagger(x)\to \mathrm{e}^{-\mathrm{i}\alpha}\psi^\dagger(x)$$

corresponds to the charge conservation. In the CTPGF approach we deal with an action defined on the closed time-path which in single time representation appears as

$$S = \int_{-\infty}^{\infty} \mathrm{d}t\,[\mathcal{L}(\psi(t+)) - \mathcal{L}(\psi(t-))]. \tag{8.85}$$

This action, therefore, respects the U(1) × U(1) symmetry, i.e., it is invariant under

$$\begin{aligned} &\psi_+(x)\to\exp(\mathrm{i}\alpha_+)\psi_+(x), \qquad \psi_+^\dagger(x)\to\exp(-\mathrm{i}\alpha_+)\psi_+^\dagger(x),\\ &\psi_-(x)\to\exp(\mathrm{i}\alpha_-)\psi_-(x), \qquad \psi_-^\dagger(x)\to\exp(-\mathrm{i}\alpha_-)\psi_-^\dagger(x), \end{aligned} \tag{8.86}$$

where α_+, α_- are independent of each other.

However, such U(1) × U(1) symmetry is always spontaneously broken, because Γ_{0-+} or G_{0-+} is different from zero even for vacuum state average as seen from (2.21), so that only U(1) symmetry is retained. The question whether there are any physical consequences of such spontaneous breaking, has to be studied.

Coming back to our model, we note that apart from the source terms and terms proportional to ε, the effective action (8.81) (and also the averaged action (8.84)) has a global Sp(2) symmetry keeping $\psi^\dagger\sigma_3\psi$ invariant. The function $\psi(x)$ forms a two-dimensional representation of the Sp(2) group transforming as

$$\psi(x)\to\psi'(x)=U\psi(x),\qquad \psi^\dagger(x)\to\psi^\dagger(x)=\psi^\dagger(x)U^\dagger\,, \tag{8.87}$$

where

$$U=\exp\{\sigma_1\lambda_1+\sigma_2\lambda_2+\mathrm{i}\sigma_3\lambda_3\} \tag{8.88}$$

satisfies the condition

$$U\sigma_3U^\dagger=U^\dagger\sigma_3U=\sigma_3\,. \tag{8.89}$$

Here λ_1, λ_2, λ_3 are group parameters corresponding to the three generators.

The term proportional ε does not respect this symmetry. Like a small magnetic field determines the direction of magnetization in an O(3) ferromagnet, the ε term can be considered as a small external field inducing the breakdown of the Sp(2) symmetry. Actually, the Sp(2) symmetry is spontaneously broken by dynamic generation of the imaginary part of the retarded (advanced) Green functions.

8.3.3. WT identities

If we make an infinitesimal transformation with group parameters $\lambda_i(E)$ for functions $\psi^\dagger(x)$, $\psi(x)$ in the path integral of $\bar{Z}[J]$ we obtain the following three WT identities corresponding to the three generators of the Sp(2) group:

$$-2\mathrm{i}\varepsilon\int \mathrm{d}^dx\,\mathrm{Sp}\Big\{(1+\sigma_1)\Big[-\mathrm{i}\frac{\delta^2\bar{W}}{\delta J^\dagger(x,E)\delta J(x,E)}+\frac{\delta\bar{W}}{\delta J^\dagger(x,E)}\frac{\delta\bar{W}}{\delta J(x,E)}\Big]\Big\}$$
$$=\int \mathrm{d}^dx\Big[J^\dagger(x,E)\sigma_1\frac{\delta\bar{W}}{\delta J^\dagger(x,E)}+\frac{\delta\bar{W}}{\delta J(x,E)}\sigma_1J(x,E)\Big]\,, \tag{8.90a}$$

$$-2\varepsilon\int\mathrm{Sp}\Big\{(1+\mathrm{i}\sigma_2)\Big[-\mathrm{i}\frac{\delta^2\bar{W}}{\delta J^\dagger\delta J}+\frac{\delta\bar{W}}{\delta J^\dagger}\frac{\delta\bar{W}}{\delta J}\Big]\Big\}=\int\Big[J^\dagger\sigma_2\frac{\delta\bar{W}}{\delta J^\dagger}+\frac{\delta\bar{W}}{\delta J}\sigma_2J\Big]\,, \tag{8.90b}$$

$$2\mathrm{i}\varepsilon\int\mathrm{Sp}\Big\{(\sigma_1-\mathrm{i}\sigma_2)\Big[-1\frac{\delta^2\bar{W}}{\delta J^\dagger\delta J}+\frac{\delta\bar{W}}{\delta J^\dagger}\frac{\delta\bar{W}}{\delta J}\Big]\Big\}=\int\Big[\frac{\delta\bar{W}}{\delta J}\sigma_3J-J^\dagger\sigma_3\frac{\delta\bar{W}}{\delta J^\dagger}\Big]\,. \tag{8.90c}$$

The WT identities for various Green functions can be obtained by differentiating (8.90) with respect to $J^\dagger(x,E)$, $J(x,E)$ and then setting $J^\dagger=J=0$. As an example, we show how the dynamic generation of the imaginary part breaks the Sp(2) symmetry.

Using eq. (2.12) we find that

$$G_r = \tfrac{1}{2}\xi^\dagger \hat{G}\eta = -\tfrac{1}{2}\,\mathrm{Sp}\left[(\sigma_3 - \mathrm{i}\sigma_2)\frac{\delta^2 \bar{W}}{\delta J^\dagger(x,E)\delta J(x,E)}\right]. \tag{8.91}$$

Taking the functional derivative $\delta^2/\delta J^\dagger(y,E)\delta J(y,E)$ of both sides of (8.90b) we obtain

$$\begin{aligned}\mathrm{Im}\, G_r(y,y,E) &= \tfrac{1}{2}\,\mathrm{Sp}\left\{\sigma_2 \frac{\delta^2 \bar{W}}{\delta J^\dagger(y,E)\delta J^\dagger(y,E)}\right\}\Bigg|_{J=J^\dagger=0} \\ &= -\frac{\varepsilon}{2}\int \mathrm{d}^d x\, \mathrm{Sp}\left\{(1+\mathrm{i}\sigma_2)\left[-\mathrm{i}\frac{\delta^4 \bar{W}}{\delta J^\dagger_\alpha(y,E)\delta J_\alpha(y,E)\delta J^\dagger(x,E)\delta J(x,E)}\right.\right. \\ &\quad \left.\left. + \frac{\delta^2 \bar{W}}{\delta J_\alpha(y,E)\delta J^\dagger(x,E)}\frac{\delta^2 \bar{W}}{\delta J^\dagger_\alpha(y,E)\delta J(x,E)}\right]\right\}\Bigg|_{J=J^\dagger=0}.\end{aligned} \tag{8.92}$$

As is well known, $\mathrm{Im}\, G_r(y,y,E)$ is proportional to the density of states $\rho(E)$. It is different from zero certainly for extended states and possibly for localized states as $\varepsilon \to 0$. Therefore, the Sp(2) symmetry is broken for both cases. McKane and Stone [133] pointed out that there are two ways to satisfy the WT identities. Although their interpretation is given in an entirely different theory based on the replica trick, we expect it applicable to our case as well. For extended states, the dynamic generation of the imaginary part for the retarded Green function caused by the breakdown of the Sp(2) symmetry leads to Goldstone mode characterized by long-range correlation and governing the critical behaviour from the "extended" side. On the other hand, there is no Goldstone mode with vanishing momentum for localized states, so the integrand on the right-hand side of (8.92) must diverge as $\varepsilon \to 0$ before integration to satisfy the WT identities in this case.

8.3.4. Order parameter and nonlinear σ-model

As said before, the order parameter breaking the Sp(2) symmetry is proportional to the imaginary part of the retarded Green function. A Goldstone mode is therefore generated. To describe this mode it is convenient to introduce a composite matrix field

$$q(x,E) = \psi(x,E)\psi^\dagger(x,E), \tag{8.93}$$

the vacuum expectation value of which is connected to the second-order CTPGF.

Under the Sp(2) group, the field q transforms as

$$q \to UqU^\dagger, \tag{8.94}$$

where U is given by (8.88). The vacuum expectation value of q is

$$\langle q\rangle = a(1 - \sigma_1 + \mathrm{i}\sigma_2) + b\sigma_3, \tag{8.95}$$

where the diagonal part of the first term describes the imaginary part of G_F and $G_{\tilde{F}}$, whereas the second term describes their real part. As seen from (8.89) the b term does not break the Sp(2) symmetry, while the a term does. Hence Goldstone modes will be dynamically generated by the condensation of the q field.

In analogy with the earlier work [133], we can derive a nonlinear σ-model describing such Goldstone modes and carry out the renormalization procedure to study the scaling behaviour near the mobility edge. However, we will not elaborate further on such discussion here.

To summarize section 8, we note that the theoretical framework outlined here for quenched random systems is quite general as well as flexible. It is based on the dynamics of the system itself without resorting to replica trick, so the approximation involved are well under control. Apart from spin glass and disordered electron system discussed in this section, the present formalism can be applied to other quenched random systems as well. In particular, we discussed [141] the controversial problem of the lower critical dimension for an Ising spin system in a random magnetic field [142]. We hope the CTPGF formalism is helpful in solving some of the difficult problems still remaining open in this field.

9. Connection with other formalisms

To save space in this paper we attempted to avoid as much as possible digressing from the main subject and comparing the CTPGF approach with other formalisms in passing. We would like such comparison to be concentrated here. Although not so much new information will be presented to experts, hopefully, this section will help the newcomer entering this field to orient himself in the forest of diversified formalisms. We will mainly discuss the connection of the CTPGF approach with the Matsubara technique (section 9.1), quantum and fluctuation theories as low and high temperature limits of the CTPGF formalism (section 9.2) and also the CTPGF formalism as a plausible microscopic justification of the MSR field theory (section 9.3). There are still many related papers not covered in this review for which we apologize to their authors once again.

9.1. Imaginary versus real time technique

The Matsubara technique [1–7] using the imaginary time for thermoequilibrium is well developed and highly successful. However, there are two difficulties from the technical point of view. One is associated with the fact that in this technique Green's functions are defined on a finite section of the imaginary time axis $(0, -\mathrm{i}\beta)$ so the Fourier series expansion in frequency is used instead of Fourier integral. The frequency summation is sometimes quite cumbersome. Another difficulty is connected with the analytic continuation of the frequency (or time) variable in the final answer. Usually, such a process is rather delicate. We see thus in spite of the great success of the Matsubara technique, a convenient real time formalism would be highly desirable. The CTPGF formalism is one of the possible candidates for this purpose.

9.1.1. Real time diagrammatic technique

We have mentioned in section 6.1 that the diagrammatic expansion for thermoequilibrium system at finite temperature is similar to that of the quantum field theory provided the free propagator is given by (6.33). Here we would like to justify this statement using expressions for the effective action derived in section 6.1.

The correlation functional W^N defined by (6.15) becomes in this case

$$\exp(\mathrm{i}W^N_{\mathrm{th}}[J^\dagger_\Delta, J_\Delta]) = \mathrm{Tr}\{\hat{\rho}_{\mathrm{th}}{:}\exp[\mathrm{i}(J^\dagger_\Delta\psi_\mathrm{I} + \psi^\dagger_\mathrm{I}J_\Delta)]{:}\} \tag{9.1}$$

where

$$\hat{\rho}_{\rm th} = \exp(-\Omega - \beta(\mathcal{H} - \mu N)) \tag{9.2}$$

where Ω is the thermodynamic potential, whereas $\psi_{\rm I}^\dagger$, $\psi_{\rm I}$ are operators in the incoming picture. It is known that for the in-field we have [39]

$$e^{i\mathcal{H}\tau} A_{\rm I}(t)\, e^{-i\mathcal{H}\tau} = A_{\rm I}(t+\tau)\,, \tag{9.3}$$

where $\mathcal{H}$ is the total Hamiltonian. This point is essential for our derivation. By analytic continuation $\tau \to -i\beta$ we find

$$e^{\beta\mathcal{H}} A_{\rm I}(t)\, e^{-\beta\mathcal{H}} = A_{\rm I}(t - i\beta)\,. \tag{9.4}$$

Taking into account that for a complex field the operator of particle number

$$N \sim \int d^d x\, \psi^\dagger(x)\psi(x) = \int d^d x\, \psi_{\rm I}^\dagger(x)\psi_{\rm I}(x)$$

is a conserved quantity, it is easy to show that

$$\rho_{\rm th}^{-1} A_{\rm I}(t) \rho_{\rm th} = \exp\left\{-i\beta \frac{\partial}{\partial t} - \lambda\beta\mu\right\} A_{\rm I}(t)\,, \tag{9.5}$$

where

$$\lambda = +1, \quad \text{if} \quad A_{\rm I}(t) = \psi_{\rm I}^\dagger(x), \qquad \lambda = -1, \quad \text{if} \quad A_{\rm I}(t) = \psi_{\rm I}(x), \qquad \lambda = 0, \quad \text{if} \quad A_{\rm I}(t) \text{ is Hermitian}\,. \tag{9.6}$$

Using (9.5) we can apply as done by Gaudin [7] the following identity

$$\begin{aligned} &\mathrm{Tr}\{(\rho A(1) \mp (\pm)^n A(1)\rho) A(2) \cdots A(n)\} \\ &\quad = \mathrm{Tr}\{\rho [A(1), A(2)]_\mp A(3) \cdots A(n)\} \pm \{\rho A(2)[A(1), A(3)]_\mp A(4) \cdots A(n)\} \\ &\qquad + \cdots \cdots (\pm)^{n-2}\, \mathrm{Tr}\{\rho A(2) \cdots A(n-1)[A(1), A(n)]_\mp\} \end{aligned} \tag{9.7}$$

to the right-hand side of (9.1) to obtain

$$\exp\{i W_{\rm th}^N\} = \mathrm{Tr}\{\rho_{\rm th}\}\langle 0| :\exp(\hat{F}_{\rm I}) \exp[i(J_\Delta^\dagger \psi_{\rm I} + \psi_{\rm I}^\dagger J_\Delta)]: |0\rangle\,, \tag{9.8}$$

where

$$\hat{F}_{\rm I} = \pm \int d^{d+1}x\, d^{d+1}y \frac{\delta}{\delta\psi_{\rm I}^\dagger(x)} \left(1 \mp \exp\left(-i\beta \frac{\partial}{\partial x_0} - \beta\mu\right)\right)^{-1} [\psi_{\rm I}^\dagger(x), \psi_{\rm I}(y)]_\mp \frac{\delta}{\delta\psi_{\rm I}(y)}\,. \tag{9.9}$$

In deriving (9.8) the properties of the normal product and the particle number conservation are taken into account properly. Note that for nonrelativistic complex fields the operator $\psi_1(x)$ contains only positive frequencies whereas $\psi_1^\dagger(x)$ only negative ones, so that

$$[\psi_1(x), \psi_1^\dagger(y)]_\mp = \mathrm{i}S_{0-+}(x, y), \tag{9.10}$$

where S_{0-+} is given by (6.8). Substituting (9.10) into (9.9) we find that

$$\hat{F}_1 = \pm\mathrm{i}\int \frac{\delta}{\delta\psi_1(x)} S_0^N(x, y) \frac{\delta}{\delta\psi_1^\dagger(y)} \tag{9.11}$$

with

$$S_0^N(x, y) = \left(1 \mp \exp\left(\mathrm{i}\beta \frac{\partial}{\partial x_0} - \beta\mu\right)\right)^{-1} S_{0-+}(x, y). \tag{9.12}$$

It then follows from (9.8) by using (9.11) that

$$W_{\text{th}}^N[J_\Delta^\dagger, J_\Delta] = -\int J_\Delta^\dagger(x) S_0^N(x, y) J_\Delta(y). \tag{9.13}$$

In accord with (6.15) the correlation functional defined on the closed time-path is

$$W_{\text{th},p}^N[J^\dagger, J] = -\int_p J^\dagger(x) S_{0p}^N(x, y) J(y), \tag{9.14}$$

where

$$S_{0p}^N(x, y) = S_0^N(x, y) \begin{pmatrix} 1 & 1 \\ 1 & 1 \end{pmatrix}. \tag{9.15}$$

Equations (9.13) and (9.14) are the contribution of the density matrix to the generating functional. We see thus the initial correlation is actually a Gaussian process described by the two-point correlation function $S_0^N(x, y)$.

Substituting (9.14) back into (6.10) and (6.29) we obtain the following expressions for the CTPGF generating functional in thermal equilibrium:

$$Z_p[J^\dagger, J] = \exp\left\{\mathrm{i}\,I_{\text{int}}\left[\mp\mathrm{i}\frac{\delta}{\delta J}, -\mathrm{i}\frac{\delta}{\delta J^\dagger}\right]\right\} \exp(-\mathrm{i}J^\dagger G_0 J), \tag{9.16}$$

$$Z_p[J^\dagger, J] = \int_p [\mathrm{d}\psi^\dagger][\mathrm{d}\psi] \exp\{\mathrm{i}(\psi^\dagger G_0^{-1}\psi + I_{\text{int}}t[\psi^\dagger, \psi] + J^\dagger\psi + \psi^\dagger J)\}, \tag{9.17}$$

where

$$G_0(12) = S_0(12) + S^N_{0p}(1, 2), \tag{9.18}$$

$$G_0^{-1}(1, 2) = S_0^{-1}(1, 2) - \int_p d\bar{1}\, d\bar{2}\, \tilde{S}_0^{-1}(1, \bar{1}) S^N_{0p}(\bar{1}\bar{2}) \tilde{S}_0^{-1}(\bar{2}, 2)\,. \tag{9.19}$$

It is ready to check that G_0 and G_0^{-1} defined by (9.18) and (9.19) are reciprocal to each other and that the FDT (3.43) is satisfied.

Since the Green functions in thermoequilibrium are time translationally invariant, the only role of the second term in (9.19) is to produce $S^N_{0p}(1, 2)$ term in $G_0(1, 2)$ and cannot appear independently in the final result. Hence we can ignore the difference of $\tilde{S}_0^{-1}$ and $\tilde{S}_0^{-1}$ from the very beginning and take

$$G_0^{-1}(1, 2) = S_0^{-1}(1, 2) \tag{9.20}$$

instead of (9.19). In that case the first and the second terms of (9.18) can be considered as superposition of solutions for inhomogeneous equation $S_0^{-1}F = 1$ and homogeneous equation $S_0^{-1}F = 0$. The numerical coefficient of the latter is determined by the FDT.

We see thus (9.16) and (9.17) are the generalization of the Matsubara technique to real time axis. The advantage of using real time variables in some cases more than justifies the technical complications owing to the matrix representation of the propagator.

9.1.2. Other real time formalisms

Several authors previously considered the possible generalization of the Feynman–Wick expansions for the Matsubara functions [143, 144]. However, some of these attempts ended up with very involved formalism, whereas the others were difficult to justifiy. We believe the incoming picture adopted here is helpful in avoiding these difficulties. Very recently, Niemi and Semenoff [145] proposed a version of real time technique to study the finite temperature field theory. Their work is close to ours but is still different. The time-path in the complex plane they adopt consists of four pieces $(-\infty, +\infty)$, $(+\infty, +\infty - i\beta/2)$, $(+\infty - i\beta/2, -\infty - i\beta/2)$ and $(-\infty - i\beta/2, -\infty - i\beta)$. Their free boson propagator is given by

$$G(k) = \hat{A} \begin{pmatrix} 1/(k^2 - m^2 + i\eta) & 0 \\ 0 & -1/(k^2 - m^2 - i\eta) \end{pmatrix} \hat{A}\,, \tag{9.21}$$

where

$$\hat{A} = \begin{pmatrix} \cosh\theta & \sinh\theta \\ \sinh\theta & \cosh\theta \end{pmatrix}, \qquad \cosh^2\theta = \exp(\beta|k_0|)/[\exp(\beta|k_0|) - 1]\,. \tag{9.22}$$

Obviously, (9.22) is different from that given by (2.26) in our formalism. A detailed comparison of these two versions has to be made by future studies.

9.1.3. Thermo field dynamics

For the last ten years Umezawa and coworkers [91] have developed the “thermo field dynamics” and applied it to a number of interesting problems in condensed matter physics. They have adopted a great many ideas and techniques from the arsenal of the quantum field theory, especially the operator transformations. So far we have not yet studied all the topics they have covered, and it is hard to make a thorough analysis of the merits as well as the shortcomings of each formalism. It seems to us, however, that the extensive use of the generating functional technique, especially the vertex functional in the CTPGF formalism, is advantageous. As we found from the study of the weak electromagnetic field coupled to the superconductor [53] the ambiguity connected with the dynamic mapping and boson transformation occurring in thermo field dynamics [146], can be avoided in the CTPGF formalism. Another merit of the latter is the unified approach to both equilibrium and nonequilibrium phenomena, whereas the thermo field dynamics is limited to equilibrium systems up to now.

9.1.4. Kadanoff–Baym formalism

We should also mention the Green function formalism developed by Kadanoff and Baym (KB) [147]. These authors do not use the closed time-path, but rather start from the original paper by Martin and Schwinger [83]. There are still many common features of these two formalisms. In fact, the $G^>$ and $G^<$ functions appearing in the KB technique are nothing but G_{-+} and G_{+-} in the CTPGF approach. There are many papers applying the KB formalism to different problems in both equilibrium and nonequilibrium systems [148]. We will not go on to compare these two formalisms in further detail. The interested readers are referred to their excellent book [147].

9.2. Quantum versus fluctuation field theory

In this section we consider the low and high temperature limits of the CTPGF formalism for thermoequilibrium. It is natural to expect that in the zero temperature limit the standard quantum field theory or its equivalent in the many-body systems should be recovered. In fact, if the boson density is set equal to zero in (2.26), Δ_F becomes the usual Feynman propagator. Of course, there is no need to duplicate the time axis in this limit.

9.2.1. Critical phenomena

Now consider the high temperature limit. As seen from the Bose distribution

$$n(p)=\frac{1}{\exp[\varepsilon(p)/T)]-1}$$

for particle number nonconserving system or near the critical point (where the chemical potential $\mu=0$), the quasiparticle density

$$n\approx T/\varepsilon(p)\gg 1\,, \tag{9.23}$$

if

$$\varepsilon(p)/T\ll 1\,. \tag{9.24}$$

In the ordinary units (9.24) can be rewritten as

$$\lambda\gg k/\sqrt{2mk_BT}\,, \tag{9.25}$$

i.e., the characteristic wavelength of elementary excitation is much greater than the thermal wavelength. Hence near the critical point the thermal fluctuations dominate, whereas the quantum fluctuations are irrelevant. This is the basis of the modern theory of critical phenomena. It is worthwhile noting that this classical limit is not described by the Boltzmann distribution which holds when

$$\exp[(\varepsilon(p)-\mu(T))/T] \gg 1 .$$

Therefore, it is more appropriate to call this limit "super Bose" distribution, as far as the expectation value of $a^\dagger a$ is of order $n \gg 1$, so that the noncommutativity of Bose operators can be ignored.

We have thus two types of field theory: quantum field theory at zero temperature and classical field theory near the critical temperature. They have many features in common but differ from each other in some essential aspect.

9.2.2. Finite temperature field theory

For the recent years many authors study the finite temperature field theory and possible phase transitions in such systems [149]. As mentioned before, we have shown [41] that the counter terms introduced in the quantum field theory for $T = 0$ K are enough to remove all ultraviolet divergences for CTPGF at any finite temperature, even in nonequilibrium situation. This has been shown also by other authors [149] for thermoequilibrium without resorting to CTPGF. This result is understandable from a physical point of view since the statistical average does not change the properties of the system at very short distance and hence does not contribute new ultraviolet divergences. What we should like to emphasize is that in considering phase transition-like phenomena one must first separate the leading infrared divergent term and then carry out the ultraviolet renormalization which is different from that for the ordinary quantum field theory.

9.2.3. Leading infrared divergence

To be specific, let us consider the real relativistic scalar boson theory the free propagator of which is given by (2.26). Near the transition point, the mass m vanishes so the energy spectrum is given by $\omega(\boldsymbol{k}) = |\boldsymbol{k}|$ (cf. (2.28)). Since terms proportional to the particle density n appear together with the δ-function, i.e., on the mass shell, the integration over frequencies can be carried out immediately to give stronger infrared divergence (k^{-2}) than the other terms (k^{-1}). Therefore, the marginal dimension of renormalizability for φ^4-theory is $d_c = 4$, whereas for quantum field theory at $T = 0$ K the marginal space dimension is $d_c = 4 - 1$. This is what is usually meant by saying "quantum system in d dimensions corresponds to classical system in $d + 1$ dimensions". If we keep only the most infrared divergent terms, then all components of $\hat{G}$ become $(k^2 + m^2)^{-1}$, i.e., exactly the same as that used in the current theory of critical phenomena [60].

What has been said above can be checked explicitly by calculating the primitive divergent diagrams for mass, vertex and wavefunction renormalization, carrying out the frequency integration and taking the high temperature limit $T \gg \omega(k)$. The results obtained turned out to be identical to those resulting from the theory of critical phenomena. For example, in φ^4 theory, the primitive mass and wavefunction correction diagrams have quadratic divergence, whereas the vertex correction term diverges logarithmically in four-space dimensions. We know that in the quantum field theory such divergences occur for three-space and one-time dimensions.

9.2.4. High temperature limit in Matsubara formalism

The high temperature limit can be easily taken in the Matsubara technique. For example, the free propagator for nonrelativistic complex boson field is given by

$$g(\omega_n)=\frac{1}{i\omega_n-k^2-m^2}\tag{9.26}$$

where $\omega_n=2\pi nT$, $m^2\sim T-T_c$. Since $T\gg k^2+m^2$, in the frequency summation to be carried out later we need to keep only the $\omega_n=0$ term. Hence the propagator (9.26) reduces to minus the correlation function in the theory of critical phenomena. This fact seemed to be first realized by Landau [150].

Some investigators of finite temperature field theory in the early stage of their work incorrectly used the renormalization constants for $T=0$ K to study phase transition related phenomena. As far as the high frequency limit, or, equivalently, the leading infrared divergent terms are picked up, both relativistic and quantum effects are irrelevant. The only exception is the phase transition near $T=0$ K when both thermal and quantum fluctuations are essential so a special consideration is needed. Otherwise, the field-theoretic models (including non-Abelian gauge models) cannot provide anything new beyond the current theory of critical phenomena as far as the phase transition is concerned, i.e., they are classified into the same universality classes as their classical counterparts.

9.3. A plausible microscopic derivation of MSR field theory

We have mentioned already in section 6.4, that Martin, Siggia and Rose (MSR) [90] proposed a field theory to describe the classical fluctuations. There are several peculiar features of this theory: (i) Being a classical field theory, it deals with noncommutative quantities; (ii) A response field $\hat{\varphi}$ is introduced in addition to the ordinary field φ; (iii) Some components of the Green functions should be zero along with their counterparts – vertex functions. Nevertheless, the general structure of this theory is very close to that of the quantum field theory. These authors originally proposed their theory to consider nonequilibrium fluctuations such as those in hydrodynamics, but it has been extensively used in critical dynamics near thermoequilibrium [61, 77]. In spite of the great success, its microscopic foundation especially the motivation for using noncommutative variables to describe classical fields, was poorly understood.

A few years later, the MSR theory has been reformulated in terms of stochastic functionals as a Lagrangian field theory [87–89]. The noncommutativity of field variables was thus obscured by the continuum integration, whereas the calculation procedure was significantly simplified. Nonetheless, the physical meaning remains not sufficiently clarified.

We would like to note that the CTPGF formalism provides us with a plausible microscopic justification for the MSR field theory. In a sense, the MSR theory is nothing but the physical representation, i.e., in terms of retarded, advanced and correlation functions, of the CTPGFs in the quasiclassical (the low frequency) limit.

9.3.1. Noncommutativity

First consider the operator nature of the field variable. As we discussed in the last section, in the high temperature limit the noncommutativity of operators can be ignored altogether for static critical phenomena, which implies that all components of the Green functions are replaced by correlation functions. This is no more true for dynamic phenomena. The first term in G_{++} and G_{--} (2.26) i.e., $(k_0^2-k^2-m^2)^{-1}$ comes from the inhomogeneous term of the Green function equation which in turn is determined by the commutator of operators. If only leading infrared divergent terms are retained, the retarded function

$$G_r = G_{++} - G_{+-} = 0\,.$$

Therefore, to describe the response of the system to the external disturbance we need to keep the next to leading order of infrared divergence. Put another way, the response is less infrared divergent than the correlation function. In the sense of "super Bose" distribution, the commutator of order 1 can be neglected in the leading order of n but not in the next to leading order. This is the physical interpretation for the noncommutativity of a "purely classical" field variable. We would like to emphasize here that the classical field should be considered as condensation of bosons. This is why the "quantum" or wave nature of the classical field comes into play.

The deep analogy of the quantum and fluctuation field theories is then understandable. Such parallelism is particularly evident in the functional formulation. The classical path in the quantum field theory corresponds to the mean field, or TDGL orbit in the fluctuation theory.

9.3.2. Doubling of degrees of freedom

Next consider the necessity of doubling the degrees of freedom. It has been realized for a long time that to describe the time-dependent phenomena one needs both response and correlation functions. This was, probably, the motivation of introducing the response field $\hat{\varphi}$ and putting together the response and correlation functions into a matrix function by MSR [90]. In the CTPGF formalism we introduce an extra negative time axis, so also double the degrees of freedom, i.e., to use φ_+, φ_- instead of one φ. In fact, the MSR response field

$$\hat{\varphi} = \varphi_\Delta \equiv \varphi_+ - \varphi_-\,,$$

whereas their physical field

$$\varphi = \varphi_c \equiv \tfrac{1}{2}(\varphi_+ + \varphi_-)\,.$$

The CTPGF formalism is constructed on the functional manifold φ_+ and φ_-, or, equivalently, φ_Δ and φ_c, but in the final answer one should put $\varphi_+ = \varphi_-$ to get the physical result. As mentioned before, this is an additional way of describing fluctuations, the physical content of which should be further uncovered.

In using the generating functional technique the following external source terms are introduced in the MSR theory [87–89]

$$I_s = \int (J\varphi + \hat{J}\hat{\varphi})\,,$$

where J is the usual source, $\hat{J}$ the response source. This is rather similar to our generating functional, but with an important difference. As naturally follows from the definition of the closed time path we should set (see eq. (2.71))

$$I_s = \int (J_\Delta \varphi_c + J_c \varphi_\Delta)\,.$$

As seen before, such "twisted" combination is most natural. In fact, the physical source J_c generates the dynamic response in terms of φ_Δ functional, whereas the fluctuation source J_Δ generates the statistical correlation in terms of φ_c functional. Another advantage of such "twisted" combination is that we do not need to introduce any extra physical field in the Hamiltonian like in the MSR theory [87–89]. In the CTPGF approach J_c is the physical field built in the formalism itself.

9.3.3. *Constraints*

Finally, a remark concerning various constraints imposed on propagators and vertex functions. In the original formulation of MSR [90] such constraints appeared rather difficult to understand. They became more systematic in the latter Lagrangian formulation [89] but remained not so transparent. Within the CTPGF formalism, as shown in section 2.4, they are natural consequences of the normalization for the generating functional and the causality. While in the MSR theory one needs to explore the implications of the causality order by order [151], within the CTPGF framework it is ensured from the right beginning, so that causality violating terms can never occur.

As seen before, in the low frequency limit when the MSR theory holds, the CTPGF formalism yields the same results in rather low approximations. We believe, therefore, the CTPGF formalism provides us with a plausible microscopic justification for the MSR theory and indicates how to go beyond it.

10. Concluding remarks

It is time now to summarize what has been achieved and what has to be done.

(1) The CTPGF formalism is a rather general as well as flexible theoretical framework to study the field theory and many-body systems. It describes the equilibrium and nonequilibrium phenomena on a unified basis. The ordinary quantum field theory and the classical fluctuation field theory are included in this formalism as different limits. The two aspects of the Liouville problem, i.e., the dynamic evolution and the statistical correlation are incorporated into it in a natural way. The formalism is well adapted to consider systems with symmetry breaking described by either constituent or composite order parameters. If different space–time variation scales can be distinguished, a macroscopic or mesoscopic description can be provided for inhomogeneous systems from the first principles.

(2) The powerful machinery of the quantum field theory including the generating functional technique and the path integral representation can be transplanted and further developed in the CTPGF formalism to study the general structure of the theory. The implications of the normalization condition for the generating functional and the causality are explored. The consequences of the time reversal symmetry such as the potential condition, the generalized fluctuation–dissipation theorem and reciprocity relations for kinetic coefficients are derived. The role of the initial correlation is clarified. The symmetry properties of the system under consideration are studied to derive the Ward–Takahashi identities. Also, a general theory of nonlinear response is worked out.

(3) A practical calculation scheme is worked out which derives a system of coupled, self-consistent equations to determine the order parameter along with the energy spectrum, the dissipation and the particle number distribution for both constituent fermions and collective excitations. A systematic loop expansion is developed to calculate the self-energy parts. The Bogoliubov–de Gennes equation is generalized to include the exchange and correlation effects. A way of computing the free energy by a straightforward integration of the functional equation is found.

(4) The general formalism is applied to a number of physical problems such as critical dynamics, superconductivity, spin system, plasma, laser, quenched random systems like spin glass and disordered electron system, quasi-one-dimensional conductors and so on. Although most of these problems can be and have been discussed using other formalisms, but, as far as we know, the CTPGF approach is the

only one to consider them within a unified framework. Moreover, new results, new insight or significant simplifications are always found by using the CTPGF approach.

(5) In general, systems in stationary state or near it have been studied more thoroughly, whereas the transient processes need further investigation. As far as the formalism itself is concerned, the two-point functions are well under control, but the properties of multi-point functions must be explored further in the future.

Our general impression, or our partiality, is that the potentiality of the CTPGF formalism is still great. One has to overcome the "potential barrier" occurring due to its apparent technical complexity to appreciate its logical simplicity and power. It is certainly not a piece of virgin soil, but the efforts of a dedicated explorer will be more than justified. We hope that in applying this formalism to attack more difficult problems in condensed matter, plasma, nuclear physics as well as particle physics and cosmology, its beauty and power will be uncovered to a greater extent.

Acknowledgements

It is a great pleasure for us to sincerely thank our coworkers, Jiancheng Lin, Zhongheng Lin, Yu Shen, Waiyong Wang, Yaxin Wang and, especially, Prof. Royce K.P. Zia for a fruitful and enjoyable collaboration. A great many people were helpful to us by their enlightening comments, interesting discussions as well as sending us preprints prior to publication. Their names are too numerous to enumerate, but we would like to particularly mention Profs. Shi-gang Chen, B.I. Halperin, Tso-hsiu Ho, Yu-ping Huo, P.C. Martin, Huanwu Peng, Chien-hua Tsai and Hang-sheng Wu. To all of them we are deeply grateful.

Note added in proof

After our manuscript was submitted we became aware of some more references [152–158] where the CTPGF formalism was applied to various problems. We would appreciate other colleagues to inform us of their work using this approach.

References

[1] A.A. Abrikosov, L.P. Gorkov and I.E. Dzaloshinskii, Methods of Quantum Field Theory in Statistical Physics (Prentice-Hall, Englewood Cliffs, New Jersey, 1963).
[2] A.L. Fetter and J.D. Walecka, Quantum Theory of Many-Particle Physics (McGraw-Hill, New York, 1971).
[3] G.D. Mahan, Many-Particle Physics (Plenum, New York, 1981).
[4] T. Matsubara, Progr. Theor. Phys. (Kyoto) 14 (1955) 351.
[5] A.A. Abrikosov, L.P. Gorkov and I.E. Dzaloshinskii, Zh. Eks. Teor. Fiz. 36 (1959) 900.
[6] E.S. Fradkin, Zh. Eks. Teor. Fiz. 36 (1959) 1286.
[7] M. Gaudin, Nucl. Phys. 15 (1960) 89.
[8] J. Schwinger, J. Math. Phys. 2 (1961) 407.
[9] L.V. Keldysh, JETP 20 (1965) 1018.
[10] R. Craig, J. Math. Phys. 9 (1968) 605.
[11] K. Korenman, J. Math. Phys. 10 (1969) 1387.
[12] R. Mills, Propagators for Many-Particle Systems (Gordon and Breach, New York, 1969).
[13] A.G. Hall, Molec. Phys. 28 (1974) 1; J. Phys. A8 (1975) 214; Phys. Lett. 55B (1975) 31.

[14] L. Garrido and P. Seglar, in: Proc. XIIth IUPAP Conf. Statistical Physics (Budapest, 1975) p. 223.
[15] L. Garrido and M. San Miguel, Phys. Lett. 62A (1977) 209.
[16] V. Canivell, L. Garrido, M. San Miguel and P. Seglar, Phys. Rev. A17 (1978) 460.
[17] M. Schmutz, Zeit. für Phys. B30 (1978) 97.
[18] W.P.H. de Boer and Ch. G. van Weert, Physica 98A (1979) 579.
[19] E.M. Lifschitz and L.P. Pitaevskii, Course of Theor. Physics, vol. 10, Physical Kinetics (Pergamon, New York, 1981).
[20] M. Mänson and A. Sjölander, Phys. Rev. B11 (1975) 4639.
[21] A. Volkov and Sh. Kogan, JETP 38 (1974) 1018.
[22] A.I. Larkin and Yu.N. Ovchinnikov, JETP 41 (1976) 960.
[23] Yu.N. Ovchinnikov, J. Low Temp. Phys. 28 (1977) 43.
[24] C.-R. Hu, Phys. Rev. B21 (1980) 2775.
[25] V. Korenman, Ann. Phys. (New York) 39 (1966) 72.
[26] C. Caroli, R. Combascot, P. Nozieres and D. Saint-James, J. Phys. C4 (1971) 916, 2598, 2611.
[27] T. McMullen and B. Bergson, Can. J. Phys. 50 (1972) 1002.
[28] C. Caroli, D. Lederer-Rozenblatt, B. Roulet and D. Saint-James, Phys. Rev. B8 (1973) 4552.
[29] P. Nozieres and E. Abrahams, Phys. Rev. B 10 (1974) 3099.
[30] P.J. Feibelman and D.E. Eastman, Phys. Rev. B 10 (1974) 4932.
[31] C. Caroli, B. Roulet and D. Saint-James, Phys. Rev. B 13 (1976) 3875.
[32] S. Abraham-Ibrahim, B. Caroli, C. Caroli and B. Roulet, Phys. Rev. B 18 (1978) 6702.
[33] B. Bezzerides and D.F. DuBois, Ann. Phys. (N.Y.) 70 (1972) 10.
[34] D.F. DuBois, in: Lectures in Theor. Physics, Vol. IX C, ed. W.E. Brittin (Gordon and Breach, New York, 1967) p. 469.
[35] V.L. Berezinskii, JETP 26 (1968) 317;
G. Niklasson and A. Sjölander, Ann. Phys. (N.Y.) 49 (1968) 249.
[36] R. Sandstrom, Ann. Phys. (New York) 70 (1972) 516.
[37] D. Langreth, in: Linear and Nonlinear Transport in Solids, eds., J. Devreese and V. van Doren (Plenum, New York, 1976) p. 3.
[38] A.-M. Tremblay, B. Patton, P.C. Martin and P. Maldaque, Phys. Rev. A19 (1979) 1721.
[39] See, e.g., C. Itzykson and J.-B. Zuber, Quantum Field Theory (McGraw-Hill, New York, 1980).
[40] Zhou Guang-zhao (K.C. Chou) and Su Zhao-bin, Ch. 5 in: Progress in Statistical Physics (in Chinese) eds., Bai-lin Hao et al. (Science Press, Beijing, 1981) p. 268.
[41] Zhou Guang-zhao and Su Zhao-bin, Energiae Fortis et Physica Nuclearis (Beijing) 3 (1979) 304.
[42] Zhou Guang-zhao and Su Zhao-bin, ibid, 3 (1979) 314.
[43] G.Z. Zhou, Z.B. Su, B.L. Hao and L. Yu, Phys. Rev. B22 (1980) 3385.
[44] Zhou Guang-zhao, Yu Lu and Hao Bai-lin, Acta Phys. Sinica 29 (1980) 878.
[45] Zhou Guang-zhao and Su Zhao-bin, Acta Phys. Sinica, 30 (1980) 164, 401.
[46] Zhou Guang-zhao, Su Zhao-bin, Hao Bai-lin and Yu Lu, Commun. Theor. Phys. (Beijing) 1 (1982) 295, 307, 389.
[47] Zhou Guang-zhao, Su Zhao-bin, Hao Bai-lin and Yu Lu, Acta Phys. Sinica, 29 (1980) 961;
Zhou Guang-zhao, Hao Bai-lin and Yu Lu, ibid, 29 (1980) 969.
[48] Su Zhao-bin, Yu Lu and Zhou Guang-zhao, Acta Phys. Sinica, 33 (1984) 805.
[49] Su Zhao-bin, Yu Lu and Zhou Guang-zhao, Acta Phys. Sinica, 33 (1984) 999.
[50] B.L. Hao, Physica, 109A (1981) 221.
[51] Wang Waiyong, Lin Zhong-heng, Su Zhao-bin and Hao Bai-lin, Acta Phys. Sinica, 31 (1982) 1483, 1493.
[52] Zhou Guang-zhao and Su Zhao-bin, Acta Phys. Sinica, 29 (1980) 618.
[53] Su Zhao-bin and Zhou Guang-zhao, Commun. Theor. Phys. (Beijing) 1 (1982) 669.
[54] Su Zhao-bin, unpublished.
[55] Su Zhao-bin, Yu Lu and Zhou Guang-zhao, Commun. Theor. Phys. (Beijing) 2 (1983) 1181, 1191; also, to be published in Proc. of Asia–Pacific Physics Conf., 1983, Singapore, to be published by World Scientific Publishing Co.
[56] Lin Jian-cheng, Shen Yu and Zhou Guang-zhao, Commun. Theor. Phys. (Beijing) 3 (1984) 139.
[57] Su Zhao-bin, Wang Ya-xin and Yu Lu, to be published in Mol. Cryst. Liq. Cryst. and Commun. Theor. Phys.
[58] Zhou Guang-zhao, Su Zhao-bin, Hao Bai-lin and Yu Lu, to be published in Reviews of Chinese Sciences, Physics Review (Science Press, Beijing);
See also, Recent progress in Many-Body Theories, eds., H. Kummel and K.L. Ristig (Springer, Berlin, 1984) p. 377.
[59] H. Lehmann, K. Symanzik and W. Zimmermann, Nuovo Cimento 6 (1957) 319.
[60] See, e.g., E. Brezin, J.C. Le Guillou and J. Zinn-Justin, in: Phase Transitions and Critical Phenomena, eds., C. Domb and M.S. Green, vol. VI (Academic, London, 1976) p. 127.
[61] P.C. Hohenberg and B.I. Halperin, Rev. of Mod. Phys. 49 (1977) 435.
[62] See, e.g., S.R. de Groot and P. Mazur, Nonequilibrium Thermodynamics (North-Holland, Amsterdam, 1969).
[63] R. Kubo, J. Phys. Soc. Japan, 12 (1957) 570; Rep. Progr. Phys. 29 (1966) 255.
[64] H.B. Callen and T.A. Welton, Phys. Rev. 83 (1951) 34.
[65] M.S. Green, J. Chem. Phys. 20 (1952) 1281.

[66] R. Zwanzig, in: Lectures in Theor. Phys., vol. III, eds., W. Brittin et al. (Wiley, New York, 1961).
[67] H. Mori, Progr. Theor. Phys. 33 (1965) 423.
[68] K. Kawasaki, Ann. Phys. (New York) 61 (1970) 1.
[69] N. van Kampen, Physica 23 (1957) 707, 806.
[70] U. Uhlhorn, Arkiv für Fysik 17 (1960) 361.
[71] R. Graham and H. Haken, Z. Phys. 243 (1971) 289; 245 (1971) 141.
[72] R. Graham, Springer Tracts in Modern Physics, vol. 66 (1973) 1.
[73] H. Grabert, Projection Operator Technique in Nonequilibrium Statistical Mechanics (Springer, Berlin, 1982).
[74] See, e.g., Scattering Techniques Applied to Supramolecular and Nonequilibrium Systems, eds., Sow-Hsin Chen et al. (Plenum, New York, 1981).
[75] See, e.g., A.-M. Tremblay, in: Recent Developments in Nonequilibrium Thermodynamics (Springer, Berlin, 1984).
[76] See, e.g., A. Messiah, Quantum Mechanics, vol. II (North-Holland, Amsterdam, 1970) p. 664.
[77] See, e.g., S.K. Ma and G. Mazenko, Phys. Rev. B11 (1975) 4077.
[78] W. Bernard and H.B. Callen, Rev. Mod. Phys. 31 (1959) 1071.
[79] P.L. Peterson, Rev. Mod. Phys. 39 (1967) 69.
[80] G.F. Efremov, JETP 28 (1969) 1232.
[81] R.S. Stratonovich, 31 (1970) 864.
[82] G.N. Bochkov and Yu.E. Kuzovlev, Zh. Eks. Teor. Fis. 72 (1977) 238.
[83] P.C. Martin and J. Schwinger, Phys. Rev. 115 (1959) 1432.
[84] P.A.M. Dirac, The Principle of Quantum Mechanics, 4th ed. (Clarendon, Oxford, 1958) sections 10 and 77.
[85] See, e.g., D. Ruelle, Statistical Mechanics, Rigorous Results (W.A. Benjamin, Reading, Mass., 1969).
[86] L. Onsager and S. Machlup, Phys. Rev. 91 (1953) 1505, 1512.
[87] H.K. Janssen, Z. Phys. B23 (1976) 377.
[88] R. Bausch, H.K. Janssen and H. Wagner, Z. Phys. B24 (1976) 113.
[89] C. De Dominicis and L. Peliti, Phys. Rev. B18 (1978) 353.
[90] P.C. Martin, E.D. Siggia and H.A. Rose, Phys. Rev. A8 (1973) 423.
[91] H. Umezawa, H. Matsumoto and M. Tachiki, Thermo Field Dynamics and Condensed States (North-Holland, Amsterdam, 1982).
[92] S. Levit, Phys. Rev. C21 (1980) 1594 and references therein.
[93] S. Levit, J. Negele and Z. Paltiel, Phys. Rev. C21 (1980) 1603, C22 (1980) 1979.
[94] H. Reinhart, Nucl. Phys. A367 (1981) 269.
[95] A.K. Kerman and S. Levit, Phys. Rev. C24 (1981) 1029.
[96] A.K. Kerman, S. Levit and T. Troudet, Ann. Phys. (N.Y.) 148 (1983) 436.
[97] J.M. Cornwall, R. Jackiw and E. Tomboulis, Phys. Rev. D10 (1974) 2428.
[98] P.G. de Gennes, Superconductivity of Metals and Alloys (Benjamin, Reading, Mass., 1966).
[99] See, e.g., H. Kleinert, in: Time-Dependent Hartree–Fock and Beyond, eds., K. Goeke and P.-G. Reinhard (Springer, Berlin, 1982) p. 223.
[100] J. Hubbard, Phys. Rev. Lett. 3 (1959) 77;
R.L. Stratonovich, Dokl. Akad. Nauk USSR 115 (1957) 1097.
[101] R. Brout, Phys. Rev. 115 (1959) 824.
[102] See, e.g., P.W. Anderson, also T.C. Lubensky, in: Ill-Condensed Matter, eds. R. Balian, R. Maynard and G. Toulouse (North-Holland, Amsterdam, 1979).
[103] S.F. Edwards and P.W. Anderson, J. Phys. F5 (1975) 965.
[104] B. Sherrington and S. Kirkpatrick, Phys. Rev. Lett. 35 (1975) 1792; Phys. Rev. B17 (1978) 4384.
[105] J.R.L. de Almeida and D.J. Thouless, J. Phys. A 11 (1978) 983.
[106] A. Blandin, M. Gabay and T. Garel, J. Phys. A 13 (1980) 403.
[107] E. Pytte and J. Rudnick, Phys. Rev. B 19 (1979) 3603.
[108] A.J. Bray and M.A. Moore, Phys. Rev. Lett. 41 (1978) 1068.
[109] D.J. Thouless, P.W. Anderson and R.G. Palmer, Phil. Mag. 35 (1977) 593.
[110] G. Parisi, Phys. Rev. Lett. 43 (1979) 1754; J. Phys. A13 (1980) L115, L1887, 1101.
[111] S.K. Ma and J. Rudnick, Phys. Rev. Lett. 40 (1978) 589.
[112] C. de Dominicis, in: Dynamical Critical Phenomena and Related Topics, ed., C.P. Enz; Phys. Rev. B18 (1978) 4913.
[113] J.A. Hertz and R.A. Klemm, Phys. Rev. Lett. 41 (1978) 1397; Phys. Rev. B20 (1979) 316.
[114] J.A. Hertz and R.A. Klemm, Phys. Rev. Lett. 46 (1981) 496.
[115] J.A. Hertz, A. Khurana and M. Puoskari, Phys. Rev. B25 (1982) 2065.
[116] J.A. Hertz, J. Phys. C16 (1983) 1219, 1233.
[117] H. Sompolinsky and A. Zippelius, Phys. Rev. Lett. 47 (1981) 359; Phys. Rev. B 25 (1982) 6860.
[118] H. Sompolinsky, Phys. Rev. Lett. 47 (1981) 935.
[119] H. Sompolinsky and A. Zippelius, J. Phys. C15 (1982) L1059.
[120] H.-J. Sommers, Z. Phys. B50 (1983) 97.
[121] C. de Dominicis, M. Gabay and H. Orland, J. Physique Lett. 42 (1981) L523.

[122] G. Parisi, Phys. Rev. Lett. 50 (1983) 1946.
[123] A. Houghton, S. Jain and A.P. Young, J. Phys. C16 (1983) L375.
[124] C. de Dominicis and A.P. Young, J. Phys. C16 (1983) L641, A16 (1983) 2063.
[125] G. Parisi and G. Toulouse, J. Physique Lett. 41 (1980) 361.
[126] J. Vannimenns, G. Toulouse and G. Parisi, J. Physique 42 (1981) 565.
[127] K.H. Fischer, Phys. Rev. Lett. 34 (1975) 1438.
[128] P.W. Anderson, Phys. Rev. 109 (1958) 1492.
[129] See, e.g., D.J. Thouless in the book quoted in ref. [64].
[130] F. Wegner, Z. Phys. B 25 (1976) 327, B 35 (1979) 207.
[131] E. Abrahams, P.W. Anderson, D.C. Licciardello and T.V. Ramankrishnan, Phys. Rev. Lett. 42 (1979) 673.
[132] L. Schäfer and F. Wegner, Z. Phys. B 38 (1980) 113.
[133] A.J. McKane and M. Stone, Ann. Phys. (N.Y.) 131 (1981) 36.
[134] E. Brezin, S.H. Hikami and J. Zinn-Justin, Nucl. Phys. B 165 (1980) 528.
[135] S. Hikami, Progr. Theor. Phys. 64 (1980) 1466; Phys. Lett. B 98 (1981) 208.
[136] A.M. Pruisken and L. Schäfer, Phys. Rev. Lett. 46 (1981) 490; Nucl. Phys. B200 [FS4] (1982) 20.
[137] See, e.g., K. von Klitzing, G. Dorda and M. Pepper, Phys. Rev. Lett. 45 (1980) 494;
D.C. Tsui, H. Stormer and A. Gossard, Phys. Rev. Lett. 48 (1982) 1559.
[138] See, e.g., R. Laughlin, Phys. Rev. B23 (1981) 5652, Phys. Rev. Lett. 50 (1983) 1395.
[139] A. Pruisken, Nucl. Phys. B 235 [FS11] (1984) 277.
[140] H. Levine, S.B. Libby and A. Pruisken, Phys. Rev. Lett. 51 (1983) 1915, and also Brown University preprints.
[141] R.K.P. Zia, G.Z. Zhou, Z.B. Su and L. Yu, unpublished.
[142] See, e.g., G. Grinstein and S.K. Ma, Phys. Rev. B 28 (1983) 2588.
[143] I.E. Dzyloshinskii, Zh. Eks. Teor. Fiz. 42 (1962) 1126.
[144] G. Baym and A. Sessler, Phys. Rev. 131 (1963) 2345.
[145] A.J. Niemi and G.W. Semenoff, Ann. Phys. (N.Y.) 152 (1984) 105; Nucl. Phys. B230 [FS10] (1984) 181.
[146] See, e.g., H. Matsumoto and H. Umezawa, Forts. Phys. 24 (1976) 357;
M. Fusco-Girard, F. Mancini and M. Marinaro, Forts. Phys. 28 (1980) 355.
[147] L.P. Kadanoff and G. Baym, Quantum Statisitcal Mechanics (W.A. Benjamin, New York, 1962).
[148] See, e.g., R.E. Prange and L.P. Kadanoff, Phys. Rev. 134A (1964) 566;
D.C. Langreth and J.W. Wilkins, Phys. Rev. B6 (1972) 3189;
A.P. Jaucho and J.W. Wilkins, Phys. Rev. Lett. 49 (1982) 762.
[149] See, e.g., D.A. Kirzhnitz and A.D. Linde, Phys. Lett. 42 B (1971) 471; JETP Lett. 15 (1972) 529;
S. Weinberg, Phys. Rev. D 9 (1974) 3357;
L. Dolan and R. Jackiw, Phys. Rev. D 9 (1974) 3320;
M.B. Kislinger and P.D. Morley, Phys. Rev. D 13 (1976) 2765, 2771.
[150] See, e.g., A.Z. Patashinskii and V.L. Pokrovskii, Fluctuation Theory of Phase Transitions (Pergamon, Oxford, 1979).
[151] J. Deker and F. Haake, Phys. Rev. A 11 (1975) 2043.

References added in proof

[152] K.T. Mahanthappa, Phys. Rev. 126 (1962) 329.
[153] P.M. Bakshi and K.T. Mahanthappa, J. Math. Phys. 4 (1963) 1; 12.
[154] T. Ivezić, J. Phys. C8 (1975) 3371.
[155] I.B. Levinson, Sov. Phys. JETP 38 (1974) 162.
[156] A.L. Ivanov and L.V. Keldysh, Sov. Phys. Dokl. 27 (1982) 482.
[157] A.L. Ivanov and L.V. Keldysh, Sov. Phys. JETP 57 (1983) 234.
[158] Wu Xuan-ru, Liu Chang-fu and Gong Chang-de, Commun. Theor. Phys. (Beijing) 3 (1984) 1.

34 Progress of Theoretical Physics Supplement No. 86, 1986

Spontaneous Symmetry Breaking and Nambu-Goldstone Mode in a Non-Equilibrium Dissipative System

Kuang-Chao CHOU and Zhao-Bin SU

Institute of Theoretical Physics, Academia Sinica
P. O. Box 2735, Beijing

(Received December 5, 1985)

In a non-equilibrium stationary state with space-time structure a Nambu-Goldstone mode with dissipation arises as a consequence of the spontaneous symmetry breaking. A laser type model is considered as an illustration. From the Ward identity in the scheme of the closed time path Green's function (CTPGF) it is shown that the Nambu-Goldstone mode splits into two waves with different dissipations.

Spontaneous symmetry breaking and the Nambu-Goldstone mode play a very important role in contemporary physics. It is with feeling of high appreciation to write an article to honor one of the founders of these fundamental concepts, Professor Nambu's sixty-fifth birthday. Over the past thirty years our understanding of the micro world is much richer because of the valuable contribution of Professor Nambu who initiated many important research directions in particle and solid state physics.

In a stable stationary state of a nonequilibrium open system, variation of the external parameters sometimes causes bifurcation with spontaneous symmetry breaking and the generation of Nambu-Goldstone mode. Typical example of this kind is the laser system, where the vector potential develops a non-vanishing vacuum expectation value when the inversion of electron occupation number exceeds a critical limit. It breaks spontaneously the phase symmetry and generates a Nambu-Goldstone mode. As far as we know most laser theories deal with the coherent wave by semiclassical method without paying attention to the existence of the Nambu-Goldstone mode which in its turn is also a light wave of the same frequency with a tiny width caused by the dissipation of the laser system.

It is the aim of the present paper to study the properties of the Nambu-Goldstone mode generated in a nonequilibrium steady state of a laser-like system. The field theory applicable to nonequilibrium system was first developed by Schwinger in the early sixties and by Keldysh and many others later.[1),2)] A recent review of this method which is called closed time path Green's function (CTPGF) was published in Physics Reports.[3)] We shall use the formalism developed there without detailed explanation. For further reference the reader should consult the original literature.

In the first part a model Lagrangian of a laser-like two-level system is studied and the $U(1)$ phase symmetry is considered to be broken spontaneously by the variation of the occupation number of the two levels. A Ward identity is written down which insures the existence of a Nambu-Goldstone mode accompanied by the spontaneous symmetry breaking. In §2 the two-point retarded Green's function is solved and a split of the Nambu-Goldstone mode into two waves with equal half probability and different dissipations is found. In §3 it is shown how the broken symmetry is restored and the coherent wave goes gradually into one of the Nambu-Goldstone modes with less dissipation. Conclusions and speculations are presented in the last section. Some of the contents of the present paper were published by the authors in Chinese in Ref. 4).

§ 1. Model Lagrangian and the symmetry breaking

Consider a two-level system interacting with a scalar field $a(x)$. The Lagrangian is

$$\mathcal{L}=\psi_1^*\Big(i\frac{\partial}{\partial t}-E_1\Big)\psi_1+\psi_2^*\Big(i\frac{\partial}{\partial t}-E_2\Big)\psi_2+a^*\Big(i\frac{\partial}{\partial t}-k_0\Big)a+g(\psi_2^*\psi_1a^*+\psi_1^*\psi_2a). \quad (1)$$

The Lagrangian (1) has a $U(1)$ symmetry

$$a\to e^{i\alpha}a\,,$$

$$\psi_1\to e^{i\alpha/2}\psi_1\,,$$

$$\psi_2\to e^{-i\alpha/2}\psi_2\,. \quad (2)$$

For simplicity we have neglected the space dependence of the fields involved.

In general there are random interactions of the system with the surrounding causing dissipation. To a first approximation we shall include them in the energies E_i, $i=1, 2$ and k_0, widths γ_i, $i=1, 2$ and γ, respectively.

The state of the system is described by a density matrix ρ and the dynamics can be formulated in terms of a path integral along a closed time path P starting from $t=-\infty$ to $t=+\infty$ (positive branch) and back from $t=+\infty$ to $t=-\infty$ (negative branch). The generating functional is

$$Z[J(x)]=e^{iW[J(x)]}=\int[d\psi][d\psi^*][da]e^{i\int_P[\mathcal{L}(x)+J^*(x)a(x)+a^*(x)J(x)]d^4x}$$

$$\times\langle t_+=-\infty|\rho|t_-=-\infty\rangle,$$

where $|t_\pm=-\infty\rangle$ is the eigenstate of the fields at time $t_\pm=-\infty$. The external source $J(x)$ is chosen to be different on the positive and the negative branches and will be put equal in the end of the discussion to get physical results. The vertex functional $\Gamma[a_c(x)]$ is defined in the usual way

$$\Gamma[a_c(x)]=W[J(x)]-\frac{\delta W}{\delta J(x)}J(x)-J^*(x)\frac{\delta W}{\delta J^*(x)}\,, \quad (3)$$

where the classical field $a_c(x)$ is defined to be

$$a_c(x)=\frac{\delta W}{\delta J^*(x)}\,. \quad (4)$$

When $J(x_+)$ is put equal to $J(x_-)$ the field $a_c(x_+)=a_c(x_-)$ is just the average value of the field $a(x)$.

The equation of motion for the classical field $a_c(x)$ is

$$\frac{\delta\Gamma}{\delta a_c(x)}=-J^*(x)\,. \quad (5)$$

It is more convenient to write $a_c(x)$ as

$$a_c(x)=A(t)e^{-ik_0t} \tag{6}$$

and assume the amplitude $A(t)$ is a slowly varying function of time, i.e.,

$$\frac{1}{k_0A}\frac{\partial A}{\partial t}\ll 1. \tag{7}$$

Equation (5) for $a_c(x)$ without external source has been calculated in Ref. 5) by summing over all the fermion loops. The result is

$$\frac{\partial A}{\partial t}=\left(\frac{-i\Pi_r^{(2)}}{1+K(A)}-\gamma\right)A, \tag{8}$$

where

$$\Pi_r^{(2)}=-g^2\Delta N\frac{1}{k_0-\Delta E_0+i(\gamma_1+\gamma_2)} \tag{9}$$

and

$$K(A)=\frac{g^2|A|^2}{\gamma_1\gamma_2}\frac{(\gamma_1+\gamma_2)^2}{(k_0-\Delta E_0)^2+(\gamma_1+\gamma_2)^2} \tag{10}$$

with $\Delta N=N_1-N_2$ as the difference of the occupation number and $\Delta E_0=E_1-E_2$ that of the energy. Equation (8) clearly shows that A has a stable point at

$$|A|^2=\frac{\gamma_1\gamma_2}{\gamma(\gamma_1+\gamma_2)}(\Delta N-\Delta N_c), \tag{11}$$

provided the inversion of the occupation number ΔN exceeds a critical limit

$$\Delta N_c=\frac{\gamma}{g^2(\gamma_1+\gamma_2)}[(k_0-\Delta E_0)^2+(\gamma_1+\gamma_2)^2]. \tag{12}$$

In this case the field $a(x)$ will develop a nonvanishing average value, thus breaking the $U(1)$ symmetry spontaneously.

§ 2. Ward identity and Nambu-Goldstone mode

The Ward indentity for the system has the following form:

$$\partial_\mu\langle j^\mu(x)\rangle=i\left(\frac{\delta\Gamma}{\delta a(x)}a(x)-a^*(x)\frac{\delta\Gamma}{\delta a^*(x)}\right), \tag{13}$$

where the current

$$j^\mu(x)=j_a{}^\mu(x)+j_\psi{}^\mu(x). \tag{14}$$

The charge $j_\psi{}^0$ is proportional to the difference of the occupation number ΔN. We assume that ΔN is kept fixed by some external means so that $j_\psi{}^\mu$ is conserved separately. This means that the system is under a stationary environment which makes the inversion number stationary. Then the current $j_a{}^\mu$ will be conserved by itself, whose charge integrated over unit volume consists of two parts

$$Q_a=|A(T)|^2+i\int\frac{dk_0}{2\pi}D_{11}^{+-}(k,T), \tag{15}$$

where

$$D_{11}^{+-}(k, z)=\int d^4 z e^{ik\cdot z} D_{11}^{+-}\left(X+\frac{1}{2}z, X-\frac{1}{2}z\right)$$

and

$$D_{11}^{+-}(x, y)=-i\,\mathrm{tr}(a^*(y)a(x)\rho). \tag{16}$$

Here we have used the capital X or T to denote the center of mass space-time variables which are slowly varying compared with the microscopic processes described by the difference of the two point coordinates in the Green's function.

Though the total charge Q is conserved in time there is still transformation of charge from the coherent part $|A|^2$ to that of the Nambu-Goldstone part as time goes on. We shall see this shortly in the following.

Doing functional differentiation $\delta/\delta a_c$ and $\delta/\delta a_c^*$ on the Ward identity (13) and putting external source equal to zero in the end we get after integration over a closed time path P that

$$\int_P (\Gamma_{11}(x, y)a_c(y)-\Gamma_{12}(x, y)a_c^*(y))d^4y=0,$$

$$\int_P (\Gamma_{21}(x, y)a_c(y)-\Gamma_{22}(x, y)a_c^*(y))d^4y=0, \tag{17}$$

where

$$\begin{pmatrix}\Gamma_{11}(x, y), \Gamma_{12}(x, y)\\ \Gamma_{21}(x, y), \Gamma_{22}(x, y)\end{pmatrix}=\begin{pmatrix}\dfrac{\delta^2\Gamma}{\delta a_c^*(x)\delta a_c(y)}, \dfrac{\delta^2\Gamma}{\delta a_c^*(x)\delta a_c^*(y)}\\ \dfrac{\delta^2\Gamma}{\delta a_c(x)\delta a_c(y)}, \dfrac{\delta^2\Gamma}{\delta a_c(x)\delta a_c^*(y)}\end{pmatrix}. \tag{18}$$

Equation (17) written in the single time formalism is

$$\int (\Gamma_{11}^R(x, y)a_c(y)-\Gamma_{12}^R(x, y)a_c^*(y))d^4y=0,$$

$$\int (\Gamma_{21}^R(x, y)a_c(y)-\Gamma_{22}^R(x, y)a_c^*(y))d^4y=0, \tag{19}$$

where

$$a_c(y_+)=a_c(y_-)=a_c(y)$$

and

$$\begin{aligned}\Gamma_{ij}^R(x, y)&=\Gamma_{ij}(x_+, y_+)-\Gamma_{ij}(x_+, y_-)\\ &=\Gamma_{ij}(x_-, y_+)-\Gamma_{ij}(x_-, y_-)\end{aligned} \tag{20}$$

are the average value and the retarded vertex function respectively. In order to separate the time dependence of the slowly varying amplitude with the fast moving phase we write

$$\begin{pmatrix}\Gamma_{11}(x, y), \Gamma_{12}(x, y)\\ \Gamma_{21}(x, y), \Gamma_{22}(x, y)\end{pmatrix}=\begin{pmatrix}\tilde{\Gamma}_{11}(x, y)e^{-ik_0(t_x-t_y)}, \tilde{\Gamma}_{12}(x, y)e^{-ik_0(t_x+t_y)}\\ \tilde{\Gamma}_{21}(x, y)e^{ik_0(t_x+t_y)}, \tilde{\Gamma}_{22}(x, y)e^{ik_0(t_x-t_y)}\end{pmatrix} \tag{21}$$

and

$$\tilde{\Gamma}_{ij}(x, y)=\int\frac{d^4q}{(2\pi)^4}e^{-iq\cdot(x-y)}\tilde{\Gamma}_{ij}(q, T). \tag{22}$$

The retarded vertex function consists of a free part and a self-energy part

$$\Gamma^R_{ij}=\Gamma^{Rf}_{ij}+\Pi^R_{ij} \tag{23}$$

with

$$\Gamma^{Rf}_{ij}(q, T)=\begin{pmatrix} q_0-k_0, & 0 \\ 0, & -q_0-k_0 \end{pmatrix}. \tag{24}$$

Now we can expand Eq. (19) to first order of slowly varying amplitude under th assumption that

$$\left|\frac{\partial\pi^R_{ij}}{\partial q_0}\right|\ll 1=\left|\frac{\partial\Gamma^R_{11}}{\partial q_0}\right| \tag{25}$$

and obtain

$$\tilde{\Gamma}^R_{11}(q=0, T)A(T)-\tilde{\Gamma}^R_{12}(q=0, T)A^*(T)=-i\frac{\partial A}{\partial T},$$

$$\tilde{\Gamma}^R_{21}(q=0, T)A(T)-\tilde{\Gamma}^R_{22}(q=0, T)A^*(T)=-i\frac{\partial A^*}{\partial T}. \tag{26}$$

To first order approximation we write

$$\frac{\delta}{2}=-\frac{\partial}{\partial T}\ln A(T). \tag{27}$$

δ is considered to be a small quantity of first order. From Eq. (26) the condition for th existence of nonvanishing solution $A\neq 0$ is

$$\det\begin{vmatrix} \tilde{\Gamma}^R_{11}(q=0, T)-i\frac{\delta}{2}, & -\tilde{\Gamma}^R_{12}(q=0, T) \\ \tilde{\Gamma}_{21}(q=0, T), & -\tilde{\Gamma}^R_{22}(q=0, T)-i\delta/2 \end{vmatrix}=0. \tag{28}$$

The retarded Green's function is the inverse of the retarded vertex function. Therefor

$$\tilde{D}^R_{ij}(q_0, T)\tilde{\Gamma}^R_{jk}(q_0, T)=\delta_{ik},$$

$$\tilde{\Gamma}^R_{ij}(q_0, T)\tilde{D}^R_{jk}(q_0, T)=\delta_{ik}. \tag{29}$$

To first order approximation with the assumption (25) under consideration we have

$$\tilde{\Gamma}^R_{ij}(q_0, T)=\tilde{\Gamma}^R_{ij}(q_0=0, T)+\begin{pmatrix} q_0, & 0 \\ 0, & -q_0 \end{pmatrix}. \tag{30}$$

With the help of Eq. (28), Eq. (29) can be solved to give

$$\tilde{D}^R_{ij}(q_0, T) = -\frac{1}{(q_0+i\delta/2)(q_0+\tilde{\Gamma}^R_{11}(0, T)-\tilde{\Gamma}^R_{22}(0, T)-i\delta/2)}\begin{pmatrix} \tilde{\Gamma}^R_{22}(0, T)-q_0, & -\tilde{\Gamma}^R_{12}(0, T) \\ -\tilde{\Gamma}^R_{21}(0, T), & \tilde{\Gamma}^R_{11}(0, T)+q_0 \end{pmatrix}. \quad (31)$$

From the general properties of the Green's function it is easily proved that

$$\tilde{\Gamma}^R_{11}(0, T)^* = \tilde{\Gamma}^R_{22}(0, T),$$
$$\tilde{\Gamma}^R_{12}(0, T)^* = \tilde{\Gamma}^R_{21}(0, T). \quad (32)$$

Hence

$$\tilde{\Gamma}^R_{11}(0, T) - \tilde{\Gamma}^R_{22}(0, T) = \text{purely imaginary}. \quad (33)$$

The Lehmann representation also tells us that the residue of the pole for the Green's function $\tilde{D}^R_{ii}$ should be real requiring that $\tilde{\Gamma}^R_{ii}$ to be purely imaginary. Therefore we have

$$\tilde{\Gamma}^R_{11}(0, T) = -\tilde{\Gamma}^R_{22}(0, T) = \text{purely imaginary}. \quad (34)$$

Finally Eq. (31) can be put into the form

$$\tilde{D}^R_{ij}(q_0, T) = \frac{1}{2(q_0+i\delta/2)}\begin{pmatrix} 1, & e^{2i\beta} \\ -e^{-2i\beta}, & -1 \end{pmatrix} + \frac{1}{2(q_0+i(2G+\delta/2))}\begin{pmatrix} 1, & -e^{2i\beta} \\ e^{-2i\beta}, & -1 \end{pmatrix}, \quad (35)$$

where

$$G = \frac{1}{i}\tilde{\Gamma}^R_{11}(0, T) - \delta/2 \quad (36)$$

and

$$\tilde{\Gamma}^R_{12}(0, T) = iGe^{2i\beta} \quad (37)$$

with both G and β real numbers. Causality requires that

$$\delta > 0 \qquad \text{and} \qquad 2G + \frac{\delta}{2} > 0. \quad (38)$$

The retarded Green's function $D^R_{ij}(q_0, T)$ can be obtained from $\tilde{D}^R_{ij}(q_0, T)$ by a shift

$$q_0 \to q_0 - k_0, \quad (39)$$

i.e.,

$$D^R_{ij}(q_0, T) = \tilde{D}^R_{ij}(q_0 - k_0, T). \quad (40)$$

In the ordinary case of equilibrium phase transition where the condensed field has no space-time structure, the Nambu-Goldstone mode is a zero energy mode with a pole at $q_0 = 0$. In this case the dissipation $\mathrm{Im}\tilde{\Gamma}^R_{11}$ is proportional to q_0 and becomes very small as $q_0 \to 0$. The causality then requires that $\delta = 0$. Therefore the condensed field will stay constant when the system is in equilibrium. However, the Nambu-Goldstone mode excited will still split into two waves of equal weight. One of which propagates without dissipation while the other propagates with dissipation unless $\mathrm{Im}\tilde{\Gamma}^R_{11}$ is identically vanishing as in the case of pure vacuum condensation.

§ 3. Transport equation and the restoration of the symmetry

In the model discussed above the dissipation $\mathrm{Im}\Gamma_{11}^R$ has a finite limiting value as the energy q_0 approaches the pole. It is plausible that the dissipation of the second Nambu-Goldstone mode $2G+\delta/2$ is much larger than that of the first wave. We shall adopt this assumption and justify it later. The occupation numbers of the quanta of the two Nambu-Goldstone waves are defined through correlation Green's function and will be denoted by $N_{11}^{(\delta)}(T)$ and $N_{11}^{(G)}(T)$ respectively. It has been shown in Ref. 5) that a transport equation for $N_{ii}(T)$ can be deduced from the Schwinger-Dyson equation of the two-point Green's function in the formalism of CTPGF. The result neglecting higher orders of slowly varying function is the following:

$$\frac{\partial}{\partial T}(N_{11}^{(\delta)}) = -\delta N_{11}^{(\delta)} + \frac{1}{i}\Gamma_{11}^{(0)}(T),$$

$$\frac{\partial}{\partial T}(N_{11}^{(G)}) = -(4G+\delta)N_{11}^{(G)} + \frac{1}{i}\Gamma_{11}^{(0)}(T) \tag{41}$$

together with the equation for the amplitude $A(T)$ and the conservation equation

$$\frac{\partial}{\partial T}(|A|^2) = -\delta|A|^2, \tag{42}$$

$$\frac{\partial}{\partial T}\left(|A|^2 + \frac{1}{2}N_{11}^{(\delta)} + \frac{1}{2}N_{11}^{(G)}\right) = 0. \tag{43}$$

In (42)

$$\frac{1}{i}\Gamma_{11}^{(0)} \equiv \frac{\Gamma_{11}^{-+}(k_0, T) + \Gamma_{11}^{+-}(k_0, T)}{2i}$$

$$= \frac{1}{2}(W_{\mathrm{abs}}(k_0, T) + W_{\mathrm{em}}(k_0, T)), \tag{44}$$

where W_{abs} and W_{em} are respectively the probability of absorbing and emiting one quanta of $a(x)$ field per unit time.

There are two time scales in Eqs. (42) ~ (44)

$$\tau_\delta = 1/\delta, \quad \tau_G = 1/(4G+\delta). \tag{45}$$

Since G is of the order of Π_{11}^R which is the same time scale for the amplitude A to increase from the unstable point $A=0$ to the stable point of the condensed phase it should be very fast. Therefore it is plausible to assume that $\tau_\delta \gg \tau_G$ and the G wave will reach its saturation point

$$N_{11}^{(G)} = \frac{1}{i}\frac{\Gamma_{11}^{(0)}}{4G+\delta} \tag{46}$$

much faster. At the same time by the conservation equation

$$|A|^2 + \frac{1}{2}N_{11}^{(\delta)}$$

will also reach its saturation value

$$|A|^2+\frac{1}{2}N_{11}^{(\delta)}=\frac{1}{i}\frac{\Gamma_{11}^{(0)}}{2\delta}. \tag{47}$$

In this time scale it is not possible to excite many quanta for the δ wave. Therefore $N_{11}^{(\delta)} \ll |A|^2$ and

$$\delta=\frac{\Gamma_{11}^{(0)}}{2i|A|^2}. \tag{48}$$

This is just the result for the width obtained by entirely different method in the semiclassical laser theory.[6)]

In the time scale τ_δ we see from Eqs. (42) and (43) that eventually

$$A\to 0 \tag{49}$$

and

$$N_{11}^{(\delta)}\to\frac{\Gamma_{11}^{(0)}}{i\delta}. \tag{50}$$

This means that the coherent phase order of the field $a(x)$ will gradually disappear and the system will restore the original symmetry where the phase is in the disordered state. However, the corresponding Nambu-Goldstone modes are still there. The energy is concentrated in the δ mode with less dissipation. It is interesting to notice that the coherent wave and the Nambu-Goldstone δ mode have the same tiny width caused by dissipation.

§ 4. Conclusions

We have considered a nonequilibrium open system in its stable stationary state where the symmetry is spontaneously broken by the condensation of an order parameter with space-time structure. In this case the Nambu-Goldstone mode splits into two waves with equal half weight and different dissipation. The coherent wave described by the phase of the order parameter increases very fast to its saturation value and gradually decreases, inducing a tiny but finite width of the coherent wave. The energy eventually goes to one of the Nambu-Goldstone waves, the δ mode whose width and frequency is the same as that of the coherent wave. The only difference between the coherent wave and the δ mode is that the former is a classical wave while the latter is quantized. It is interesting to find experiment to distinguish the one from the other. We shall not discuss it further here.

Though only a simple model has been analysed in the present paper, we believe that the splitting of the Nambu-Goldstone mode is a quite general phenomenon for systems where dissipation plays an important role.

References

1) J. Schwinger, J. Math. Phys. **2** (1961), 407.
L. Keldysh, JETP **47** (1964), 1515.
2) See, e.g., D. Dubois, in *Lectures in Theoretical Physics*, ed. W. E. Brittin (Gordon and Breach, N. Y., 1967), v. 9C.

D. Langreth, in *Linear and Nonlinear Transport in Solids*, ed. J. Devreese and V. Van Doren (Plenum, N. Y., 1976).
3) Kuang-chao Chou, Zhao-bin Su, Bai-lin Hao and Lu Yu, Phys. Rep. **118** (1985), 1.
4) G. Z. Zhou and Z. B. Su, Acta Physica Sinica **29** (1980), 618.
5) G. Z. Zhou and Z. B. Su, in *Progress in Statistical Mechanics*, ed. Bai-lin Hao and Lu Yu (1981) (in Chinese).
6) M. Sargent III, M. Scully and W. Lamb, *Laser Physics* (Addison-Wesley, Reading, Mass., 1974).
M. Lax, Phys. Rev. **145** (1966), 110.

Volume 123, number 5 PHYSICS LETTERS A 10 August 1987

THE CANONICAL DESCRIPTION AND BOHR–SOMMERFELD QUANTIZATION CONDITION FOR THE FRACTIONAL QUANTUM HALL EFFECT SYSTEM

Zhao-bin SU, Han-bin PANG, Yan-bo XIE and Kuang-chao CHOU
Institute of Theoretical Physics, Academia Sinica, P.O. Box 2735, Beijing, PR China

Received 13 May 1987; accepted for publication 1 June 1987
Communicated by D. Bloch

In this note, by introducing the canonical description which is one dimensional in nature, we establish the Bohr–Sommerfeld quantization condition for the fractional quantum Hall effect system. The generic and model independent physical implications of this condition are explored. The guiding center representations of the N-body Schrödinger equation for the system is also derived.

Since the fascinating discovery of the integer [1] as well as the fractional [2] quantum Hall effect, the fundamental physics of the two-dimensional (2D) electron gas in a strong magnetic field B has attracted great interest. For the fractional quantum Hall effect (FQHE) [2], the experiments show that there is a series of plateaus in the Hall conductivity σ_{xy} with resistivity ρ_{xx} minimum at certain "magical" rational values of the dimensionless density (filling factor) ν, where ν is the mean number of electrons in the area $2\pi\lambda^2=\phi_0/B$ covered by the unit flux quatum $\phi_0=hc/e$. These phenomena suggest [3] that a gap exists in the electron spectrum. The general belief is that the Coulomb interaction between electrons gives rise to this gap when the lowest Landau level is only partially filled. The main problem is then why the gaps only occur for certain rational filling factors. Laughlin [4] proposed an elegant trial wavefunction mainly for the ground state with filling factor $1/m$ where m is an odd integer. Using this wavefunction, one can explain many aspects of the FQHE in a satisfactory manner. Moreover, extensive numerical studies [5] also show strong evidence that there are cusps in the ground state energy $E(\nu)$ as a function of ν at certain rational values of it. To investigate the excited states, Girvin et al. [6] use an approach similar to that of Feynman in studying liquid ^{4}He. Recently, Kivelson, Kallin, Arovas and Schrieffer (KKAS) [7] have proposed an interesting "cooperative ring exchange" mechanism using which these authors try to construct a dynamical theory in the many-body coherent state representation.

As fas as we know, the canonical description and the Bohr–Sommerfeld quantization condition can play an essential role for the dynamics of electrons in a magnetic field [8]. Furthermore, the path integral in the coherent state representation is not as neat as that in the canonical description [9]. In this note, we introduce the canonical description and establish the N-fermion Bohr–Sommerfeld quantization condition for the FQHE. Moreover, we also derive the N-electron Schrödinger equation for a system with Coulomb interaction in a strong magnetic field using the guiding center representations.

According to the canonical description, the quantum dynamics of electrons in the FQHE regime is essentially one dimensional, although the physical system itself is two dimensional. In fact, the 2D guiding center space is just the phase space of the 1D canonical coordinates. The Bohr–Sommerfeld (BS) quantization condition for the N-electron (fermion) system in the lowest Landau level (LLL) can also provide us with an interesting intuitive picture. It explains semiclassically why electrons in the LLL have to be distributed in accord with a kind of flux quantization rule. It also explains, to some extent, why the ground states of the FQHE system will have filling factors $\nu=1/(2m+1)$ with even denominators being excluded and the excited states are quantized.

Therefore, the existence of a gap in the excitation spectrum for these ground states with magic filling factors follows from the BS quantization condition in a natural way. The guiding center representations of the Schrödinger equation we derive for N interacting electrons in the LLL case are consistent with the general discussion by Girvin and Jach [10]. Moreover, our derivation can be easily generalized to higher Landau levels.

Consider a 2D electron gas with Coulomb interaction in a strong magnetic field. The N-electron hamiltonian $\hat{H}$ has the following form,

$$\hat{H}=\sum_{i=1}^{N}\frac{1}{2}\left(\frac{1}{\mathrm{i}}\frac{\partial}{\partial \boldsymbol{r}_i}+\boldsymbol{A}(\boldsymbol{r}_i)\right)^2+\frac{\lambda}{\varepsilon a_0}\sum_{i<j}^{N}\frac{1}{|\boldsymbol{r}_i-\boldsymbol{r}_j|}\,, \tag{1}$$

where the hamiltonian as well as the energy are scaled by $\hbar\omega_c$ with $\omega_c=eB/mc$ being the cyclotron frequency, the 2D electron coordinates are scaled by the magnetic length λ with $\lambda^2=\hbar c/eB$, the vector potential $\boldsymbol{A}(\boldsymbol{r}_i)$ is scaled by $\hbar c/e\lambda$ with its curl equal to 1. a_0 is the Bohr radius, whereas c, m, e, ϵ are the speed of light, the effective mass, the electron charge and the dielectric constant, respectively. As usual, we assume that the Landau level spacing as well as the Zeeman energy for the electron spin are much larger than the average Coulomb energy $\sim e^2/\epsilon\lambda$. In the hamiltonian (1) and hereafter, the positive charge background is assumed to be included implicitly.

Introduce the guiding center coordinates for the ith electron

$$(\hat{X}_i,\hat{Y}_i)=(x_i-\hat{\Pi}_{iy},\,y_i+\hat{\Pi}_{ix})\,, \tag{2}$$

with $\hat{\boldsymbol{\Pi}}_i=-\mathrm{i}\partial/\partial\boldsymbol{r}_i+\boldsymbol{A}(\boldsymbol{r}_i)$. These quantities satisfy the following commutation relations [11]

$$[\hat{X}_i,\hat{Y}_j]=\mathrm{i}\delta_{i,j}\,,\qquad [\hat{\Pi}_i,\hat{\Pi}_j^\dagger]=\delta_{i,j}\,, \tag{3}$$

and $\hat{X}_i$, $\hat{Y}_j$ commute with $\hat{\Pi}_i$, $\hat{\Pi}_j^\dagger$, where $\hat{\Pi}_i=(\hat{\Pi}_{ix}-\mathrm{i}\hat{\Pi}_{ij})/\sqrt{2}$. Making use of the 2D Fourier transformation of the Coulomb interaction and then expressing the electron coordinates in terms of $(\hat{X}_i,\hat{Y}_i)$ and $(\hat{\Pi}_i,\hat{\Pi}_i^\dagger)$ in accord with eq. (2), we can derive the following general expression for the hamiltonian (1) as

$$\hat{H}=\sum_{i=1}^{N}(\hat{\Pi}_i^\dagger\hat{\Pi}_i+\tfrac{1}{2})+\frac{\lambda}{\varepsilon a_0}\sum_{i<j}^{N}\int\frac{\mathrm{d}^2k}{2\pi}\frac{1}{|\boldsymbol{k}|}\exp(-\tfrac{1}{2}|\boldsymbol{k}|^2)\exp[k(\hat{\Pi}_i^\dagger-\hat{\Pi}_j^\dagger)]$$

$$\times\exp[\mathrm{i}k_x(\hat{X}_i-\hat{X}_j)+\mathrm{i}k_y(\hat{Y}_i-\hat{Y}_j)]\exp[k^*(\hat{\Pi}_i-\hat{\Pi}_j)]\,, \tag{4}$$

where $k=(k_x-\mathrm{i}k_y)/\sqrt{2}$ and $|\boldsymbol{k}|^2=k_x^2+k_y^2=2kk^*$.

Since the subspace of the LLL is defined as $\hat{\Pi}_i|\ \ \rangle=0$, $i=1, 2, ..., N$, the contributions from the $\hat{\Pi}_i$ degrees of freedom will be projected out neatly. Recalling the commutation relations (3a) we can identify $\hat{X}_i$ with the generalized coordinates $\hat{Q}_i$ and $\hat{Y}_i$ with the generalized momenta $\hat{P}_i$;

$$\hat{X}_i\equiv\hat{Q}_i,\quad \hat{Y}_i\equiv\hat{P}_i;\qquad [\hat{Q}_i,\hat{P}_j]=\mathrm{i}\delta_{i,j} \tag{5}$$

and

$$\hat{H}\to\hat{H}[\hat{Q},\hat{P}]=\frac{\lambda}{\varepsilon a_0}\sum_{i<j}^{N}\int\frac{\mathrm{d}^2k}{2\pi}\frac{1}{|\boldsymbol{k}|}\exp(-|\boldsymbol{k}|^2/2)\exp[\mathrm{i}k_x(\hat{Q}_i-\hat{Q}_j)+\mathrm{i}k_y(\hat{P}_i-\hat{P}_j)] \tag{6}$$

for the LLL. Here we have combined the constant term with the positive background. Eqs. (5) and (6) provide us with a complete and explicitly defined canonical description for the LLL subspace. As mentioned above, it reduces precisely to a one-dimensional N-particle quantum mechanical problem.

For a comparison with the generally interested Laughlin type wavefunctions, express $k_x\hat{Q}+k_y\hat{P}$ in terms of guiding center operator as $k\hat{Z}+k^*\hat{Z}^\dagger$ with $\hat{Z}=(\hat{X}+\mathrm{i}\hat{Y})/\sqrt{2}$, introduce the guiding center coherent representation, then define from the guiding center representation for the state vector

$$\langle Z_1, ..., Z_N | t\rangle = \exp\Big(-\tfrac{1}{2}\sum_i |Z_i|^2\Big)\psi(Z_1^*, ..., Z_N^*; t) .$$

Using a similar algebraic technique as before, it is easy to derive from eq. (6) that $\psi(Z_1^*, ..., Z_N^*; t)$ satisfies the following equation

$$\mathrm{i}\frac{\partial}{\partial t}\psi(Z_1^*,..., Z_N^*; t) = \sum_{i<j}\frac{\lambda}{\sqrt{\pi\varepsilon a_0}}\int_0^{\pi/2} \mathrm{d}\vartheta\ :\exp[-\tfrac{1}{2}\cos^2\vartheta\ (Z_i^*-Z_j^*)(\partial/\partial Z_i^* - \partial/\partial Z_j^*)]:\ \psi(Z_1^*, ..., Z_N^*; t) , \tag{7}$$

where the normal product : : shifts the $\partial/\partial Z_i^*$ to the right of Z_i^* by definition. Eq. (7) is consistent with the general discussion of Girvin and Jach [10]. Since eq. (7) is homogeneous in Z_i^*'s of the order zero, its eigenfunction is a homogeneous function of Z_i^*'s. In the sense of the WKB approximation, i.e., in the long-wavelength limit, we can show that Laughlin's wavefunction is an approximate eigenfunction of (7). If we start from the hamiltonian in the form of (4), we can surely derive a much more general Schrödinger equation which can be applied to cases beyond the LLL.

To establish the BS quantization condition we will follow the main lines of Gutzwiller [12] and Maslov [13] for the single particle case, trying to generalize them to many-particle systems with Fermi statistics. Introduce the traced propagator $\mathscr{G}(E) = \mathrm{Tr}(E+\mathrm{i}\eta-\hat{H})^{-1}$

$$\mathscr{G}(E) = -\mathrm{i}\int_0^\infty \mathrm{d}T\, G[T]\exp(\mathrm{i}ET) , \tag{8a}$$

$$G[T] = \sum_{\mathscr{P}} \mathrm{sgn}\,\hat{\mathscr{P}}\int \mathrm{d}\boldsymbol{Q}\,\langle\hat{\mathscr{P}}\boldsymbol{Q}|\exp(-\mathrm{i}\hat{H}T)|\boldsymbol{Q}\rangle , \tag{8b}$$

where $\boldsymbol{Q}(Q_1, ..., Q_N)$, $\mathrm{d}\boldsymbol{Q} = (\mathrm{d}Q_1, ..., \mathrm{d}Q_N)$, $\hat{\mathscr{P}}$ is the N-particle permutation operator, sgn $\hat{\mathscr{P}}$ being the sign of $\hat{\mathscr{P}}$, $\hat{\mathscr{P}}\boldsymbol{Q}$ means a permutation operation on $\boldsymbol{Q}$ and $\Sigma_{\mathscr{P}}$ means a summation over all the possible N-particle permutations. Clearly, eqs. (8a) and (8b) include all the information needed for the N-particle system with Fermi statistics. Using the canonical path integral representation for eq. (8b), we will make three successive stationary phase approximations (SPA) for each term in the traced propagator of eq. (8). The first SPA is applied to eq. (8b) for picking up the classical path which satisfies the classical canonical equation of motion with $\boldsymbol{Q}$ as its initial coordinates and $\hat{\mathscr{P}}\boldsymbol{Q}$ as the final ones. The second SPA is applied to eq. (8a) with respect to the integration over variable T. The purpose of this step is to pick up the classical characteristic function $W_c[\hat{\mathscr{P}}\boldsymbol{Q}, \boldsymbol{Q}; E]$ which is a Legendre transformation of the action. The third SPA applied to eq. (13) subsequently concerns the tracing in $\boldsymbol{Q}$ space and its purpose is to pick up the closed smooth orbit in the phase space $(\boldsymbol{Q}, \boldsymbol{P})$. The corresponding SPA condition is that the arriving (final) canonical momentum equals the departing (initial) momentum at each Q_i since we are dealing with the classical orbit with its final coordinates being just a permutation of the initial ones. If we forget for the moment about the nondistinguishness of particles, after these successive SPAs, we can actually end up with a smooth periodic orbit in the phase space. This is one of the key procedures to incorporate the principle of nondistinguishness of identical particles in a semiclassical way. From now on we will mention the periodic orbits being always understood in this sense. The final approximate expression for the traced propagator $\mathscr{G}(E)$ becomes, then

$$\mathscr{G}(E) \sim \sum_{\mathscr{P}} \mathrm{sgn}\,\hat{\mathscr{P}}\Big(\int \mathrm{d}(\delta Q_1)\,|\tilde{D}(E)|^{1/2}[D_3(E)]^{-1/2}\Big)\exp(-\tfrac{1}{2}\mathrm{i}n\pi + \mathrm{i}W_{\mathrm{c,p}}[\hat{\mathscr{P}}\boldsymbol{Q}, \boldsymbol{Q}; E]) , \tag{9}$$

where $W_{c,p}[\hat{\mathscr{P}}Q, Q; E]$ is the classical characteristic function for the periodic orbit in phase space and it can be proved that

$$W_{c,p}[\hat{\mathscr{P}}Q, Q; E]=\oint P^{(c)}\cdot dQ^{(c)}\ , \tag{10}$$

where the integration is taken over the periodic orbit, while $Q^{(c)}$ and $P^{(c)}$ are the conjugate canonical coordinates and momenta, respectively. The phase factor $-\frac{1}{2}in\pi$ is contributed by the singular part of the quantum fluctuations following from the first and the second SPA producedures where n is the number of times the particles hit the (1D) turning points. $|\tilde{D}(E)|$ is the corresponding nonsingular part of quantum fluctuations [9,12], $D_3(E)$ is also the quantum fluctuations obtained in the third SPA procedures [12], and $\int d(\delta Q_1)$ is due to the existence of a zero mode describing the shift of δQ_1 along the periodic orbit.

To derive the BS quantization condition for the many fermion system, we will sum up a series of periodic orbits where the final states are just simple cyclic permutations of the initial one involving only N_1 particles with $N_1\leqslant N$. Since a general permutation can always be decomposed into a direct product of such simple permutations, the following discussion for the cyclic permutation can be easily generalized to the general N-particle cases. (It is interesting to compare this with the "cooperative ring exchange" idea of KKAS [7].) Consider a series of simple cyclic permutations $\{\hat{\mathscr{P}}_0^s\}$, $s=1, 2...$ with $\hat{\mathscr{P}}_0$ as a basic simple cycle which transforms $Q_i\to Q_{i+1}$, $i=1, 2, ..., N_1$ with $Q_{N_1+1}=Q_1$. The contribution of such series of periodic orbits to the traced propagator (8) is given by

$$\sim\sum_{s=1}^{\infty}\operatorname{sgn}\hat{\mathscr{P}}_0^s\left(\int d(\delta Q_1)|\tilde{D}^{(s)}(E)|^{1/2}[D_3(E)]^{-1/2}\right)\exp(-\tfrac{1}{2}in_s\pi+iW_{c,p}[\hat{\mathscr{P}}_0^sQ, Q; E])\ . \tag{11}$$

It can be readily proved that $W_{c,p}[\hat{\mathscr{P}}_0^sQ, Q; E]=sW_{c,p}[\hat{\mathscr{P}}_0Q, Q; E]$ and $|\tilde{D}^{(s)}(E)|\,D_3^{(s)}(E)$ is independent of s. Moreover, there are two interesting observations. Firstly, the sth orbit bounces the turning points two times more than the $(s-1)$th orbit, i.e., $n_s=n_{s-1}+2$ due to the 1D nature. Secondly, for simple cycles, $\operatorname{sgn}\hat{\mathscr{P}}_0^s=1$ if $N_1=$odd and $\operatorname{sgn}\hat{\mathscr{P}}_0^s=(-1)^s$ if $N_1=$even. Then, eq. (11) can be summed up and is proportional to

$$\sim 1/\{1\pm\exp(iW_{c,p}[\hat{\mathscr{P}}_0Q, Q; E])\}\ , \tag{12}$$

where the "+" sign corresponds to N_1 being odd and "−" sign corresponds to N_1 being even. Since the zero of the denominator of eq. (12) will contribute a pole to the traced propagator $\mathscr{G}(E)$, it leads to be BS quantization condition

$$\oint P^{(c)}\cdot dQ^{(c)}=(2m+1)\pi,\quad N=\text{odd}\ ,$$
$$=2m\pi,\quad N=\text{even}\ . \tag{13}$$

So far we have chosen $\hat{\mathscr{P}}_0$ as a simple cycle transforming $Q_i\to Q_{i+1}$ and, then, taking into account that $Q_i=X_i$, $P_i=Y_i$, and also noticing that the (Q, P) phase space is just the 2D guiding center coordinate space (X, Y), the BS quantization condition can be equivalently formulated as: A periodic orbit consisting of N_1 electrons is possible only if

$$A/2\pi\lambda^2=\phi/\phi_0=m+\tfrac{1}{2},\quad N=\text{odd}\ ,$$
$$=m,\quad N=\text{even}\ , \tag{14}$$

where A is the orbit closed area and ϕ the orbit closed magnetic flux. In eq. (14) we have recovered the dimensions.

To study the physical implications of the BS condition we first consider the ground state. It is certainly reasonable to assume that the electrons in the ground state are distributed homogeneously. Then, consider arbitrarily chosen three nearest electrons. Presumably, the successive hopping between themselves can be described semiclassically by a periodic orbit connecting them and it is consistent with the motion of the rest $N-3$ electrons. Due to the symmetry consideration as well as the 2D geometry, such an orbit will close an area just equal to one half of the average area occupied by each electron. Therefore, the BS condition (21) leads to $\nu=1/(2m+1)$ with even denominators being prohibited.

The excitations are in principle charge density fluctuations. Within the semiclassical picture, the density fluctuation corresponds to the distortion of certain periodic orbits which are homogeneously distributed in the ground state. The BS condition requires that the distortion of periodic orbits inducing inhomogeneity cannot proceed in a continuous way due to the constraints of its discreteness. This leads to one more interesting implication, namely, the excitations upon ground states with odd ν are quantized and the existence of a gap is understandable.

As shown above, the BS quantization condition follows mainly from the Fermi statistics and the 1D nature of the FQHE system. Although it is established in the semiclassical sense, it is so profound in nature that the above discussions are generally applicable, model independent and even do not depend on the concrete form of the interaction as far as it is repulsive to keep the stability of the period orbits (SPA).

The authors would like to thank Professor Lu Yu for fruitful discussions.

References

[1] K. von Klitzing, G. Dorda and M. Pepper, Phys. Rev. Lett. 45 (1980) 494.
[2] D.C. Tsui, H.L. Stormer and A.G. Gossard, Phys. Rev. Lett. 48 (1982) 1559.
[3] B.I. Halperin, Helv. Acta Phys. 56 (1983) 75.
[4] R.B. Laughlin, Phys. Rev. Lett. 50 (1983) 1395.
[5] R. Morf and B.I. Halperin, Phys. Rev. B 33 (1986) 2221;
W. Lai, K. Yu, Z. Su and L. Yu, Solid State Commun. 52 (1984) 339, and references therein.
[6] S.M. Girvin, A.H. MacDonald and P.M. Platzman, Phys. Rev. B 33 (1986) 2481.
[7] S. Kivelson, C. Kallin, D. Arovas and J.R. Schrieffer, Phys. Rev. Lett. 56 (1986) 873.
[8] L. Onsager, Philos. Mag. 43 (1952) 1006.
[9] L.S. Schulman, Techniques and applications of path integration (Wiley, New York, 1981).
[10] S.M. Girvin and T. Jach, Phys. Rev. B 29 (1984) 5617.
[11] R. Kubo, S. Miyake and N. Hashitsume, in: Solid state physics, Vol. 17, eds. F. Seitz and D. Turnbull (Academic Press, New York, 1965).
[12] M. Gutzwiller, J. Math. Phys. 12 (1971) 343.
[13] V. Maslov and M. Fedoriuk, Semiclassical approximation in quatum mechanics (Reidel, Dordrecht, 1981).

PHYSICAL REVIEW B VOLUME 37, NUMBER 16 1 JUNE 1988

Influence functional and closed-time-path Green's function

Zhao-bin Su, Liao-Yuan Chen, Xiao-tong Yu, and Kuang-chao Chou
Center of Theoretical Physics, Chinese Center of Advanced Science and Technology (World Laboratory) Beijing, China and Institute of Theoretical Physics, Academia Sinica, P. O. Box 2735, Beijing, China

(Received 24 November 1987)

In this Brief Report, we explicitly show the equivalence of the Feynman-Vernon influence-functional approach and the path-integral formulation of the Schwinger-Keldysh closed-time-path Green's function. The latter simplifies the practical calculations considerably with a systematic diagrammatic technique and is expected to have a broader application.

Since the initial work by Feynman and Vernon,[1] the idea and technique of the influence functional has been successfully applied to more systems, such as the maser problem, the polaron problem,[2,3] the transport problem,[4–6] the quantum tunneling in the presence of dissipation,[7] and the semiclassical approach to weak localization.[8] In the Feynman-Vernon approach to quantum-statistical problems, the propagation of both dynamic and statistical information and the quantum interference between different paths are explicitly described by the two different branches of the path integral.[9] On the other hand, the path-integral formulation of the closed-time-path Green's function (CTPGF)[10] introduced by Schwinger and Keldysh[11,12] is also a powerful treatment for various kinds of similar problems since it has a systematic diagrammatic-expansion technique. The main purpose of this note is to show the equivalence of the Feynman-Vernon influence functional (IF) and the path-integral formulation of CTPGF, in particular, to express the IF as a CTPGF-generating functional. The key point is the following. In the original form of the IF, the density matrix ρ plays the role of an initial condition which makes the path-integral calculation rather difficult, and therefore, application of the IF approach is more or less limited in certain aspects. In the CTPGF approach, however, ρ is incorporated into an effective action and appears as the thermopropagator of the action. Therefore, it makes practical calculation much more tractable and is expected to have a broader application (e.g., Refs. 5 and 6). After deriving the exact equivalence of the two approaches we will calculate the IF for a harmonic-oscillator system as an explicit illustration.

Frequently, situations occur of two systems coupled to each other where one of them is of primary interest, i.e., the measurements are always done on it (test system), while the second system (environmental system) influences the behavior of the first one. The well-known IF (Ref. 1) describes all the quantum effects of the environmental system upon the test system. Suppose that the test system and the environmental one are characterized by their general coordinates q,Q, respectively. Their actions are $I[q(t)]$ and $I[Q(t)]$ while the interaction between them is $I[q,Q]$, which is taken as $\int dt\, q(t)Q(t)$ for simplicity. To compute the expectation of observable $\hat{O}$ of the test system, one has[1]

$$\langle \hat{O}(t)\rangle = \int [dq_+][dq_-]\langle q_+(t)\,|\,\hat{O}(t)\,|\,q_-(t)\rangle \rho(q_+,q_-)\exp\{i(I[q_+]-I[q_-])\}F[q_+,q_-] \;, \tag{1}$$

where the IF is defined as

$$F[q_+,q_-] = \int [dQ_+][dQ_-]\delta(Q_+(t)-Q_-(t))\rho(Q_+(t_0),Q_-(t_0))\exp\{i(I[Q_+]-I[Q_-] + I[q_+,Q_+]-I[q_-,Q_-])\} \;. \tag{2}$$

In these two equations, q and Q are the eigenvalues of $\hat{q}$ and $\hat{Q}$, respectively, with the index $+$ and $-$ describing the two branches. $[dq(t)]$ means

$$dq(t_0)dq(t_1)\cdots dq(t_{N-1})dq(t) \;,$$

with $t_0 < t_1 < \cdots < t_{N-1} < t$ while $[dQ]$ is understood in a similar way. Hereafter, we assume the Planck constant $\hbar = 1$ and will recover it if necessary.

Equation (2) is exactly the original form of the Feynman-Vernon IF (Ref. 1) where the effect of the density matrix ρ is as an initial condition. In terms of the Heisenberg picture of the environmental system, Eq. (2) reads

$$F[q_+,q_-] = \int dQ_-(t_0)dQ_+(t_0)dQ_-(t)dQ_+(t)\langle Q_-(t_0)\,|\,\tilde{T}\exp\left[-i\int_{t_0}^{t}dt\, q_-(t)\hat{Q}(t)\right]|\,Q_-(t)\rangle\langle Q_-(t)\,|\,Q_+(t)\rangle \times\langle Q_+(t)\,|\,T\exp\left[i\int_{t_0}^{t}dt\, q_+(t)\hat{Q}(t)\right]|\,Q_+(t_0)\rangle\langle Q_+(t_0)\,|\,\hat{\rho}\,|\,Q_-(t_0)\rangle \;. \tag{3}$$

Following the basic idea of Schwinger and Keldysh,[11,12] we can combine the two paths Q_+ and Q_- into a closed one Q_p which goes along the positive branch Q_+ from t_0 to t and then comes back to t_0 along the negative branch Q_-. Then the IF Eq. (3) can be written as a CTPGF-generating functional

$$F[q_p]=\mathrm{Tr}\left[\hat{\rho}T_p\exp\left[i\int_p dt\, q_p(t)\hat{Q}(t)\right]\right], \tag{4}$$

where the trace Tr is taken over the sub-Hilbert space of the environmental system Q, while q_p means q_+ and q_- which is treated as a c number at this stage. The time integral $\int_p dt$ is along the closed time path and equal to $\int_{t_0}^t dt_+$ minus $\int_{t_0}^t dt_-$. The time-ordering operator along the closed time path T_p is identical to the usual chronological operator T on the positive branch, identical to the anti-time-ordering operator $\tilde{T}$ on the negative branch, and any operator on the negative time branch should be ordered toward the left, compared to those on the positive branch.

For further investigation, we assume the environmental system action $I[Q]=I_0[Q]+I_{\mathrm{int}}[q,Q]$ with its free part

$$I_0[Q]=\tfrac{1}{2}\int_p dt_1\int_p dt_2 Q(t_1)\Delta_{00p}^{-1}Q(t_2), \tag{5}$$

where Δ_{00p} is the free propagator of the environmental system at zero temperature which is defined on the closed time path and has four components.

$$\Delta_{00p}(t_1,t_2)=\begin{bmatrix}\Delta_{00}(t_1^+t_2^+) & \Delta_{00}(t_1^+t_2^-)\\ \Delta_{00}(t_1^-t_2^+) & \Delta_{00}(t_1^-t_2^-)\end{bmatrix} =\begin{bmatrix}\Delta_{00}^{++}(t_1t_2) & \Delta_{00}^{+-}(t_1t_2)\\ \Delta_{00}^{-+}(t_1t_2) & \Delta_{00}^{--}(t_1t_2)\end{bmatrix}. \tag{6}$$

Introducing the incoming interaction picture for the environmental system we can easily verify that Eq. (4) turns out

$$F[q_p]=\exp\left[iI_{\mathrm{int}}\left[\frac{\delta}{i\delta q}\right]\right]\mathrm{Tr}\left[\hat{\rho}T_p\exp\left[i\int_p dt q_p(t)\hat{Q}_I(t)\right]\right]. \tag{7}$$

Applying Wick's theorem for CTPGF (Ref. 10) to the above equation, the IF becomes

$$\exp\left[iI_{\mathrm{int}}\left[\frac{\delta}{i\delta q}\right]\right]\exp\left[-\frac{i}{2}\int_p dt_1dt_2 q(t_1)\Delta_{00p}(t_1t_2)q(t_2)\right]\mathrm{Tr}\hat{\rho}{:}\exp\left[i\int_p dt q(t)\hat{Q}_I(t)\right]{:}. \tag{8}$$

Notice that in the normal ordering : : the difference between the two branches $\hat{Q}_+$ and $\hat{Q}_-$ vanishes, the traced part of Eq. (8) becomes

$$\mathrm{Tr}\hat{\rho}{:}\exp i\int_{t_0}^t dt q_\Delta(t)\hat{Q}_I(t){:} \tag{9}$$

with $q_\Delta=q_+-q_-$.

For comparison with Feynman and Vernon, we would also assume a thermal density matrix $\hat{\rho}=\exp(-\beta\hat{H}[Q])$ for the environmental system in the Q space. Moreover, in order to extract the full information of the density matrix we would use an algebraic identity which was first introduced by Gaudin:[13]

$$\mathrm{Tr}[\rho A(1)\cdots A(n)]=(1-\exp-i\beta\partial_{t_1})^{-1}\mathrm{Tr}\{\rho[A(1),A(2)]\cdots A(n)+\cdots+\rho A(2)\cdots[A(1),A(n)]\}. \tag{10}$$

Different from the usual cases, here we have the density matrix with the total Hamiltonian $\hat{H}[Q]$ instead of the free part $\hat{H}_0[Q]$ of the environmental system, but the identity remains valid since we still have $\rho^{-1}A(t)\rho=A(t-i\beta)=\exp(-i\beta\partial_t)A(t)$ in the incoming picture for an arbitary operator $A(t)$.[14]

Applying the above identity to Eq. (9), we obtain

$$\mathrm{Tr}\hat{\rho}{:}\exp\left[i\int_{t_0}^t dt q_\Delta(t)\hat{Q}_I(t)\right]{:}=\langle 0|\exp(A)\exp\left[i\int_{t_0}^t dt q_\Delta(t)\hat{Q}_I(t)\right]|0\rangle, \tag{11}$$

where

$$A=-\frac{i}{2}\int_{t_0}^t dt_1dt_2\left[\frac{\delta}{\delta Q_I^{(-)}(t_1)}(1-e^{-i\beta\partial_{t_1}})^{-1}\Delta_{00}^{+-}(t_1t_2)\frac{\delta}{\delta Q_I^{(+)}(t_2)}+\frac{\delta}{\delta Q_I^{(+)}(t_1)}(1-e^{-i\beta\partial_{t_1}})^{-1}\Delta_{00}^{-+}(t_1t_2)\frac{\delta}{\delta Q_I^{(-)}(t_2)}\right], \tag{12}$$

in which $Q_I^{(+)},Q_I^{(-)}$ mean the positive, negative frequency parts. After some straightforward calculations we arrive at a compact exponential expression

$$\exp\left[-\frac{i}{2}\int_{t_0}^t dt_1dt_2 q_\Delta(t_1)\Delta_{0N}(t_1t_2)q_\Delta(t_2)\right]=\exp\left[-\frac{i}{2}\int_p dt_1dt_2 q_p(t_1)\Delta_{0N}^{\beta}q_p(t_2)\right], \tag{13}$$

in which

$$\Delta_{0N}(t_1t_2)=[\exp(-i\beta\partial_{t_1})-1]^{-1}\Delta_{00}^{+-}(t_1t_2)+[\exp(-i\beta\partial_{t_1})-1]^{-1}\Delta_{00}^{-+}(t_1t_2)\,, \quad (14)$$

$$\Delta_{0N}^{p}=\begin{bmatrix}1 & 1\\ 1 & 1\end{bmatrix}\Delta_{0N}\,. \quad (15)$$

Here Δ_{0N}^{p} describes the thermostatistical contribution to the CTPGF propagator $\Delta_{0p}=\Delta_{00p}+\Delta_{0N}^{p}$ of the environmental system at temperature $1/\beta$. Substituting Eq. (13) into Eq. (8), the IF becomes

$$F[q_p]=\exp\left[iI_{\rm int}\left[\frac{\delta}{i\delta q}\right]\right]\times\exp\left[-\frac{i}{2}\int_p dt_1dt_2q_p(t_1)\Delta_{0p}(t_1t_2)q_p(t_2)\right]. \quad (16)$$

After a functional Fourier transform we arrive at the final result

$$F[q_p]=\int[dQ_p]\exp i\left[\tfrac{1}{2}\int_p dt_1dt_2Q(t_1)\Delta_{0p}^{-1}Q(t_2)+I_{\rm int}[Q]+\int_p dtq_p(t)Q(t)\right]. \quad (17)$$

Compared with the original form of IF Eq. (2), now the density matrix formally disappears and, as a result, the free propagator Δ_{00p} is replaced by the thermostatistical propagator Δ_{0p}.

In the above CTPGF form of the Feynman-Vernon IF, the effect of the density matrix ρ is completely removed into an effective action with the free propagator Δ_{00p} being replaced by the thermostatistical propagator Δ_{0p}. With this new approach, we could make practical calculations much easier as complemented by a systematic diagrammatic technique. Moreover, we would like to make two more remarks. First, the thermostatistical propagator satisfies the fluctuation dissipation theorem, i.e., $\Delta_0^{+-}(\omega)=\exp(-\beta\omega)\Delta_0^{-+}(\omega)$ which follows straightforwardly from Eq. (13). Second, the thermo Δ_{0p} and the free Δ_{00p} have the same inverse because the Δ_{0N}^{p} is the homogeneous solution of the Dyson equation for Δ_{0p}. Therefore, the form of the CTPGF path integral for the finite-temperature case is formally the same as that of the zero-temperature case.

Finally, we would like to calculate the IF for a harmonic oscillator system by using our new form to illustrate its advantages. In this case, the interaction $I_{\rm int}=0$ and the thermostatistical propagator is given as

$$\Delta_0^{+-}(t_1t_2)=(-i/2m\omega)\{N_\omega\exp[-i\omega(t_1-t_2)]+(1+N_\omega)\exp[i\omega(t_1-t_2)]\}\,, \quad (18)$$

$$\Delta_0^{-+}(t_1t_2)=\Delta_0^{+-}(t_2,t_1)\,, \quad (19)$$

$$\Delta_{00}^{++}(t_1t_2)=\Theta(t_1-t_2)\Delta_0^{-+}(t_1t_2)+\Theta(t_2-t_1)\Delta_0^{+-}(t_1t_2)\,, \quad (20)$$

$$\Delta_{00}^{--}(t_1t_2)=\Theta(t_2-t_1)\Delta_0^{-+}(t_1t_2)+\Theta(t_1-t_2)\Delta_0^{+-}(t_1t_2)\,. \quad (21)$$

When we carry out the Gaussian functional integration in Eq. (17), we get

$$F[q_+q_-]=\exp\left[-\frac{i}{2\hbar}\int_{t_0}^{t}dt_1dt_2(q_+\Delta_0^{++}q_++q_-\Delta_0^{--}q_--q_+\Delta_0^{+-}q_--q_-\Delta_0^{-+}q_+)\right]. \quad (22)$$

Then we substitute the propagator given in Eqs. (16)–(19) into the above equation so we can check that it gives exactly those of Feynman and Vernon both for the zero-temperature case Eq. (4.8) of Ref. 1 and for the finite-temperature case Eq. (4.42) of Ref. 1.

[1]R. P. Feynman and F. L. Vernon, Ann. Phys. (N.Y.) **24**, 118 (1963).
[2]R. P. Feynman, Phys. Rev. **97**, 660 (1955).
[3]R. P. Feynman, R. W. Whellwarth, C. K. Iddings, and P. M. Platzman, Phys. Rev. **127**, 1004 (1962).
[4]K. K. Thornber and R. P. Feynman, Phys. Rev. B **1**, 4099 (1970).
[5]Z. B. Su, L. Y. Chen, and J. L. Birman, Phys. Rev. B **35**, 9744 (1987).
[6]Z. B. Su, L. Y. Chen, and C. S. Ting (unpublished).
[7]A. O. Caldeira and A. J. Legget, Ann. Phys. (N.Y.) **149**, 374 (1983); S. Chakravarty and A. J. Legget, Phys. Rev. Lett. **52**, 5 (1983); R. Bruinsma and Per Bak, *ibid.* **56**, 420 (1986).
[8]S. Chakravarty and A. Schmid, Phys. Rep. **140**, 193 (1986).
[9]A. R. Hibbs and R. P. Feynman, *Quantum Mechanics and Path Integrals* (McGraw-Hill, New York, 1965).
[10]K. C. Chou, Z. B. Su, B. L. Hao, and L. Yu, Phys. Rep. **118**, 1 (1985).
[11]J. Schwinger, J. Math. Phys. **2**, 407 (1961); L. V. Keldysh, Zh. Eksp. Teor. Fiz. **47**, 1515 (1964) [Sov. Phys. JETP **20**, 1018 (1965)].
[12]J. Rammer and H. Smith, Rev. Mod. Phys. **58**, 323 (1986).
[13]M. Gaudin, Nucl. Phys. **15**, 89 (1960).
[14]P. Roman, *Advanced Quantum Theory* (Addison-Wesley, Reading, MA, 1965).

Thirty Years Since Parity Nonconservation
A Symposium for T. D. Lee, ed. Robert Novick, (Birkhäuser),
pp. 117–131.
Reprinted with permission.

TIME REVERSAL INVARIANCE AND ITS APPLICATION TO NONEQUILIBRIUM STATIONARY STATES

K. C. Chou and Z. B. Su

Center for Theoretical Physics, CCAST (World Laboratory)
and
Institute of Theoretical Physics, Academia Sinica

Abstract

Generalized free energy of order parameters near nonequilibrium stationary states is shown to exist for systems obeying time reversal invariance. A low frequency fluctuation-dissipation theorem similar in form to that in thermoequilibrium is obtained in this case.

I am very pleased and honored to be here to join so many eminent physicists to celebrate Professor Lee's sixtieth birthday. Professor Lee has not only made many contributions to world science, but has also done much to promote collaborations between the U.S. and China. So, on behalf of Academia Sinica and all Chinese physicists I would like to extend our warmest congratulations to T.D. and Jeannette, and to wish him a very happy birthday and a long creative life ahead.

In 1979, the graduate school in China was reopened after the disastrous turmoil which lasted more than a decade, and scientists started to re-educate themselves. That summer T.D. went to China and gave an intensive course on particles and field theory to about 600 graduate students and scientists gathering from all over China. Almost every day for eight weeks, he lectured three hours in the morning and discussed with the students during lunch and sometimes in the afternoon. It was the first time after years of ignorance that we were able to touch the frontier of physics through excellent lectures. T.D.'s talent as a great scientist and a great teacher was fully displayed and delighted the audience.

Through contact with these brilliant young students, T.D. realized that China, although poor in material production, is rich in human resources. If these young people could be trained at good universities, they certainly would make great contributions not only to the modernization of China, but also to the development of world science. The next year, 1980, T.D. started the CUSPEA project with great enthusiasm and devotion. Over seventy universities in the United States and all universities in China joined the project. Up to now, 704 Chinese students have been sent to study for their Ph.D. degrees in the United States. I am certain it will have far-reaching impact on the development of U.S.-China friendship and scientific collaboration.

I attended some of T.D.'s lectures that summer. I was also motivated by the stimulating atmosphere and vivid discussions among the audience. The ceaseless endeavor of T.D. to find something new has encouraged us to do some work. What I shall talk about in the following is research done at that time, after T.D.'s lecture on CPT and spontaneous CP violation.[1] The results have been published in Chinese.[2] I was interested in laser, a physical system far from thermoequilibrium at that time. We were looking for a generalized free energy to describe processes near nonequilibrium stationary states from a microscopic rather than phenomenological point of view.

Theories based on the Master equation, the Fokker-Planck equation and the Langevin equation were already developed to deal with systems near nonequilibrium stationary states.[3-4] These theories are semiphenomenological in nature. In analogy with statistical mechanics, a generalized free energy was introduced. The minimum of the generalized free energy corresponds to the nonequilibrium stationary state and the curvature at the minimum point determines the linear fluctuation of the system near its stationary state. It was shown in the framework of the Fokker-Planck equation that the existence of generalized free energy can be justified under the assumption of detail balancing.[3]

In a series of papers,[5] we have applied the field theory of closed time path Green's functions (CTPGF) to systems near nonequilibrium stationary states. A time-dependent Ginsburg-Landau equation (TDGL) was derived in the form

$$\gamma_{\alpha\beta}\frac{\partial Q_\beta}{\partial t} = -\frac{\delta F}{\delta Q_\alpha} + J_\alpha \, , \qquad (1)$$

where Q_α , $\alpha = 1,...n$ the order parameters are the average values of some composite operators of the underlying fields. J_α , $\alpha = 1,...n$ are the external sources coupled to Q_α . The generalized free energy F has been shown to exist for systems near thermal equilibrium.

For systems far from thermal equilibrium, does the generalized free energy exist? If it does, what is the condition? The purpose of the present talk is to answer these questions. It will be shown that the time reversal invariance of the underlying field theory is the basis for the existence of the generalized free energy. Since we know that CP is not an exact symmetry, neither is time reversal invariance. As was pointed out first by T.D. Lee, CP and T might be broken spontaneously.[1] If this is true, it will be most promising from the theoretical point of view. Fortunately, the argument leading to the existence of generalized free energy is not affected by the spontaneous breakdown of the time reversal symmetry.

1. Time Reversal Invariance

Consider a system with the Hamiltonian

$$\hat{H}(J) = \hat{H}_0 + \hat{Q}_\alpha J_\alpha \quad , \tag{2}$$

where $\hat{H}_0$ is the Hamiltonian without the external source term; $\hat{Q}_\alpha$, $\alpha = 1,\ldots n$ are hermitian composite operators. Under time reversal $J_\alpha(t)$ may change sign

$$J_\alpha(t) \to \epsilon_{(\alpha)} J_\alpha(-t) \quad , \tag{3}$$

$$\epsilon_{(\alpha)} = \pm 1. \tag{4}$$

In Eq. (3) the index α is not summed.

In the following we shall work in the Heisenberg picture and take $t = 0$ to be the initial time where the Heisenberg picture coincides with the Schroedinger picture. The physical quantities evaluated at time t are independent of the particular initial time chosen.

Time reversal invariance implies the existence of an antiunitary operator R such that for any hermitian dynamical variables

$$\hat{Q}_\alpha(t, J) \to R\,\hat{Q}_\alpha(t, J)\,R^{+} = \epsilon_{(\alpha)}\,\hat{Q}_\alpha(-t, \epsilon J) \ . \tag{5}$$

The state of a statistical system is described by a density matrix $\hat{\rho}(\lambda)$ specified by a set of real parameters λ_a, $a=1,\ldots m$ which may also change sign under time reversal

$$\lambda_a \to \epsilon_{(a)} \lambda_a \quad , \tag{6}$$

$$\epsilon_{(\alpha)} = \pm 1 . \tag{7}$$

The state of the system is called time reversal invariant if it satisfies the relation

$$R \hat{\rho}(\lambda) R^{+} = \hat{\rho}(\epsilon\lambda) . \tag{8}$$

If the time reversal invariance is broken spontaneously, there will be some $\lambda_\alpha \neq 0$ with $\epsilon(\alpha) = -1$. In this case, we may live in one world, and do not interact with the time-reversal counterpart because the barrier between the two worlds is infinitely high. Then we can always assume the density matrix satisfies Eq. (8) in the study of dynamical systems living in our world.

The average value of the operators $\hat{Q}_\alpha$ t. J) is equal to

$$Q_\alpha (t; J, \lambda) = \text{tr} (\hat{\rho}(\lambda) \hat{Q}_\alpha (t, J), \tag{9}$$

which will be called order parameters in the following. It is easily deduced from Eqs. (5) and (8) for a time reversal invariant system that

$$Q_\alpha (t; J, \lambda) = \epsilon_{(\alpha)} Q_\alpha(-t; \epsilon J, \epsilon\lambda). \tag{10}$$

The correlation functions are

$$C_{\alpha_1 \cdots \alpha_\ell} (t_1, \cdots t_\ell ; J, \lambda)$$
$$= \text{tr} \{ \hat{\rho}(\lambda) \hat{Q}_{\alpha_1} (t_1, J) \cdots \hat{Q}_{\alpha_n} (t_\ell, J) \}. \tag{11}$$

Time reversal invariance requires that

$$C_{\alpha_1 \cdots \alpha_\ell} (t_1, \cdots t_\ell ; J, \lambda)$$
$$= \epsilon_{(\alpha_1)} \cdots \epsilon_{(\alpha_\ell)} C_{\alpha_\ell \cdots \alpha_1} (-t_\ell, \cdots -t_1; \epsilon J, \epsilon \lambda). \tag{12}$$

From Eqs. (10) and (12) we see that time reversal invariance relates physical quantities in one world to those in the time reversal counterparts if time reversal symmetry is spontaneously broken.

2. Time-dependent Ginsburg-Landau Equation (TDGL)

In this section we are going to derive TDGL from the field theory of CTPGF and prove the existence of generalized free energy for time reversal invariant systems. Since the relation obtained in the previous section is independent of the particular initial time chosen, we shall take the initial time to minus infinity in the following.

Consider the external source term $J_\alpha \hat{Q}_\alpha (J\hat{Q})$ to be a perturbation adiabatically switched on. Any operator $\hat{O}_0$ in the Heisenberg picture of $\hat{H}_0$ will change to

$$\hat{O}(t) = \hat{S}^{+} T(S\hat{O}_0(t)) , \tag{13}$$

where $\hat{S}$ is the S-matrix

$$\hat{S} = T(\exp\{-i \int_{-\infty}^{\infty} J(\tau)\, \hat{Q}(\tau)\, d\tau \}). \tag{14}$$

The $\hat{S}^{+}$ on the left in Eq. (13) is also necessary to guarantee the causality of the interaction. Equation (13) can be expressed as[6]

$$\hat{O}(t) = T_p(S_p \hat{O}_0(t)) , \tag{15}$$

where

$$S_p = T_p \{ \exp\{-i \int_p J(\tau)\, \hat{Q}(\tau)\, d\tau\}\}. \tag{16}$$

The path of integration P is a closed path starting from $\tau = -\infty$ to $\tau = +\infty$ (positive branch) and back from $\tau = +\infty$ to $\tau = -\infty$ (negative branch). T_p is an ordering operator along the whole path P. It is easily seen that the positive branch of $\hat{S}_p$ corresponds to $\hat{S}$ in Eq. (13) and the negative branch to $\hat{S}^{+}$.

In statistical mechanics we are interested in average values of physical variables rather than transition amplitudes. The average value is equal to a trace of a density matrix at an initial time t_0 and some Heisenberg operators at the time $t_1, \cdots t_n$ which consists of amplitudes propagating from t_0 to the time of the operators and back. This is why we need an S-matrix along a closed path in statistical physics.

The generating functional for CTPGF is defined to be

$$\begin{aligned} Z[J] &= \exp\{-iW[J]\} \\ &= \mathrm{tr}\{\hat{\rho}\, T_p(\exp\{-i \int_p J(\tau)\, \hat{Q}(\tau)\, d\tau \})\}. \end{aligned} \tag{17}$$

The external sources $J_\alpha(t)$, $\alpha = 1,\dots n$ are taken to be different on positive $(J_{+\alpha}(t))$ and negative $(J_{-\alpha}(t))$ branches. They will be put equal at the final step when physical results are evaluated. $W[J]$ is the generating functional for the connected Green's functions whose first derivative gives the order parameters

$$\frac{\delta W[J]}{\delta J_\alpha(t)} = Q_\alpha(t) . \tag{18}$$

When the external sources on the two branches are set equal to the physical external source, we have

$$Q_{+\alpha}(t) = Q_{-\alpha}(t) = Q_{\alpha\, ph.}(t) , \tag{19}$$

which is the average value of the physical variable $\hat{Q}_\alpha(t)$.

From the definition of the generating functional it follows easily that

$$W[J_+, J_-]\Big|_{J_+=J_-} = 0 . \tag{20}$$

Let

$$\left.\begin{aligned} J_\Sigma &= 1/2(J_+ + J_-) , \\ J_\Delta &= J_+ - J_- . \end{aligned}\right\} \tag{21}$$

Equation (20) can be rewritten as

$$W[J_\Sigma, J_\Delta = 0] = 0 . \tag{22}$$

By differentiation with respect to J_Σ we obtain a series of equalities among Green's functions from Eq. (22). The first equality is

$$\frac{\delta W[J_\Sigma, J_\Delta = 0]}{\delta J_{\Sigma\alpha}(t)} = Q_{+\alpha}(t) - Q_{-\alpha}(t) = 0 \tag{23}$$

coinciding with Eq. (19).

Second order connected Green's functions can be obtained by differentiation

$$G_{P\alpha\beta}(t, t') = \frac{\delta^2 W[J]}{\delta J_\alpha(t)\, \delta J_\beta(t')} . \tag{24}$$

In the single time formalism there are four second order Green's functions for each pair αβ, of which only three are independent. They are the retarded

$$G_{r\alpha\beta}(t, t'; J, \lambda) = \frac{\delta^2 W}{\delta J_{\Sigma\alpha}(t)\, \delta J_{\Delta\beta}(t')}$$

$$= -i\theta(t-t')\ \mathrm{tr}\{\hat{\rho}(\lambda)\left[\hat{Q}_\alpha(t, J), \hat{Q}_\beta(t', J)\right]\}\ , \qquad (25)$$

the advanced

$$G_{a\alpha\beta}(t, t'; J, \lambda) = \frac{\delta^2 W}{\delta J_{\Delta\alpha}(t)\, \delta J_{\Sigma\beta}(t')}$$

$$= i\theta(t'-t)\ \mathrm{tr}\{\hat{\rho}(\lambda)\left[\hat{Q}_\alpha(t, J), \hat{Q}_\beta(t', J)\right]\}\ , \qquad (26)$$

and the correlation Green's function

$$G_{c\alpha\beta}(t, t'; J, \lambda) = \frac{\delta^2 W}{\delta J_{\Delta\alpha}(t)\, \delta J_{\Delta\beta}(t')}$$

$$= -i\ \mathrm{tr}\{\hat{\rho}(\lambda)\ \{\hat{Q}_\alpha(t, J), \hat{Q}_\beta(t', J)\}_+\}\ . \qquad (27)$$

The vertex functional is defined to be

$$\Gamma[Q] = W[J] - \int_\alpha J(\tau)\, Q(\tau)\, d\tau\ , \qquad (28)$$

where $Q_\alpha(t, J)$ is determined by Eq. (18). It follows from Eqs. (18) and (28) that

$$\frac{\delta\Gamma[Q]}{\delta Q_\alpha(t)} = -J_\alpha(t)\ . \qquad (29)$$

In terms of the variables

$$Q_{\Sigma\alpha} = (Q_{+\alpha} + Q_{-\alpha})/2 \tag{30}$$

and

$$Q_{\Delta\alpha} = Q_{+\alpha} - Q_{-\alpha} , \tag{31}$$

we have from Eqs. (22) and (28)

$$\Gamma[Q_\Sigma \; Q_\Delta = 0] = 0 \tag{32}$$

and

$$\left.\frac{\delta\Gamma[Q_\Sigma Q_\Delta]}{\delta Q_\Delta(t)}\right|_{Q_\Delta=0} = -J_\Sigma(t) . \tag{33}$$

The vertex functional can be calculated by summing all 1-particle irreducible diagrams. Once this is done, the physical order parameter $Q_{\Sigma\alpha}(t)$ can be determined from Eq. (33). We shall show in the following that TDGL is an approximation of Eq. (33) for systems near stationary states where the motion is slow.

There are four second order vertex functions for each pair $\alpha\beta$ in single time formalism, of which three are independent in the physical region where $Q_\Delta = 0$. They are the retarded

$$\Gamma_{R\alpha\beta}(t, t' ; Q, \lambda) = \frac{\delta^2\Gamma}{\delta Q_{\Sigma\alpha}(t)\,\delta Q_{\Delta\beta}(t')}$$

$$= -D_{\alpha\beta} - iA_{\alpha\beta} , \tag{34}$$

the advanced

$$\Gamma_{a\alpha\beta}(t, t' ; Q, \lambda) = \frac{\delta^2\Gamma}{\delta Q_{\Delta\alpha}(t)\,\delta Q_{\Sigma\beta}(t')}$$

$$= -D_{\alpha\beta} + iA_{\alpha\beta} , \tag{35}$$

and the correlation vertex function

$$\Gamma_{c\alpha\beta}(t, t' ; Q, \lambda) = \frac{\delta^2 \Gamma}{\delta Q_{\Delta\alpha}(t)\, \delta Q_{\Delta\beta}(t')} \quad . \tag{36}$$

In Eqs. (34) and (35) $D_{\alpha\beta}$ is called the dispersive part and $A_{\alpha\beta}$ the absorptive part of the vertex.

One can easily deduce the Dyson-Schwinger equations for the second order Green's functions. In matrix form they are

$$\Gamma_r G_r = G_r \Gamma_r = -1 , \tag{37}$$

$$\Gamma_a G_a = G_a \Gamma_a = -1 \tag{38}$$

and

$$\Gamma_c = \Gamma_R G_c \Gamma_a \quad . \tag{39}$$

Now we are prepared to derive TDGL for a system near stationary states. In the following we shall assume the physical external source J_Σ to be constant in time and the equation

$$\left. \frac{\delta \Gamma[Q]}{\delta Q_\Delta(t)} \right|_{Q_\Delta = 0} = - J_\Sigma \tag{40}$$

to have constant solution Q_Σ . Then the system is considered to be in a stationary state. The question is whether or not there exists a generalized free energy $F[Q_\Sigma]$ such that Q_Σ is the solution of the equation

$$\frac{\partial F}{\partial Q_{\Sigma\alpha}} = - J_{\Sigma\alpha} \quad . \tag{41}$$

If this is possible we must have

$$\frac{\partial F}{\partial Q_{\Sigma\alpha}} = \left. \frac{\delta \Gamma}{\delta Q_{\Delta\alpha}(t)} \right|_{Q_{\Delta\alpha} = 0} . \tag{42}$$

Then

$$\frac{\partial^2 F}{\partial Q_{\Sigma\beta}\,\partial Q_{\Sigma\alpha}} = \int dt' \left.\frac{\delta^2 \Gamma}{\delta Q_{\Sigma\beta}(t')\,\delta Q_{\Delta\alpha}(t)}\right|_{Q_{\Delta\alpha}=0,\; Q_{\Sigma\beta}(t)=Q_{\Sigma\beta}}$$

$$= \int dt' \, \Gamma_{r\beta\alpha}(t'-t\,;Q,\lambda)\;. \qquad (43)$$

Here we have used the time displacement invariance when the external sources are time independent. A well-defined function F can be obtained by integrating Eq. (42) if the order of differentiation can be changed in Eq. (43). Therefore the condition for the existence of the generalized free energy is

$$\int dt' \, \Gamma_{r\beta\alpha}(t'-t\,;Q,\lambda) = \int dt' \, \Gamma_{r\alpha\beta}(t'-t\,;Q,\lambda)$$

or in Fourier transform

$$\Gamma_{r\alpha\beta}(\omega=0\,;Q,\lambda) = \Gamma_{r\alpha\beta}(\omega=0\,;Q,\lambda)$$

$$= \Gamma_{\alpha\alpha\beta}(\omega=0\,;Q,\lambda)\;. \qquad (44)$$

The last equality in Eq. (44) follows from the relation

$$\Gamma_{r\alpha\beta}(\omega\,;Q,\lambda) = \Gamma_{\alpha\beta\alpha}(-\omega\,;Q,\lambda)\;. \qquad (45)$$

Equation (44) can also be written in the form

$$\mathcal{A}_{\alpha\beta}(\omega=0\,;Q,\lambda) = 0\,. \qquad (46)$$

Therefore, vanishing of the zero frequency component of the absorptive part of the vertex function is a sufficient condition for the existence of a generalized free energy.

Our next task is to show that Eq. (46) follows from time reversal invariance. For constant external sources, the order parameters are time independent. We have from Eqs. (10) and (12) that

$$Q_\alpha(J, \lambda) = \epsilon_{(\alpha)} Q_\alpha(\epsilon J, \epsilon\lambda) \tag{47}$$

and

$$G_{r\alpha\beta}(t - t', J, \lambda) = \epsilon_{(\alpha)} \epsilon_{(\beta)} G_{a\alpha\beta}(-t+t'; \epsilon J, \epsilon\lambda). \tag{48}$$

From Eq. (47) we obtain by differentiation with respect to constant $J_{\Sigma\beta}$ the following relation

$$G_{r\beta\alpha}(\omega = 0; J, \lambda) = \epsilon_{(\alpha)} \epsilon_{(\beta)} G_{r\beta\alpha}(\omega = 0, \epsilon J, \epsilon\lambda) . \tag{49}$$

The Fourier transform of Eq. (48) reads

$$G_{r\alpha\beta}(\omega, J, \lambda) = \epsilon_{(\alpha)} \epsilon_{(\beta)} G_{a\alpha\beta}(-\omega; \epsilon J, \epsilon\lambda) . \tag{50}$$

From Eqs. (49) and (50) we get

$$G_{r\beta\alpha}(\omega = 0; J, \lambda) = G_{a\beta\alpha}(\omega = 0, J, \lambda) . \tag{51}$$

Since $-\Gamma_r$ and $-\Gamma_a$ are the inverse of G_r and G_a as shown in Eqs. (37) and (38) we obtain finally

$$\Gamma_{r\beta\alpha}(\omega = 0; J, \lambda) = \Gamma_{a\beta\alpha}(\omega = 0; J, \lambda) . \tag{52}$$

Equation (52) is just the condition (46) required. Hence we have shown that generalized free energy exists for time reversal invariant systems in stationary states.

Near the stationary state the order parameters vary slowly with time. We can expand Eq. (29) around the stationary point at time t

$$\left.\frac{\delta\Gamma}{\delta Q_{\Delta\alpha}(t)}\right|_{Q_{\Delta\alpha}=0} = \left.\frac{\delta\Gamma}{\delta Q_{\Delta\alpha}(t)}\right|_{Q_{\Delta\alpha}=0,\ Q_{\Sigma\beta}(t) = Q_{\Sigma\beta}}$$

$$+ \int dt' \frac{\delta^2\Gamma}{\delta Q_{\Sigma\beta}(t')\,\delta Q_{\Delta\alpha}(t)} (Q_{\Sigma\beta}(t') - Q_{\Sigma\beta}(t) + \cdots . \tag{53}$$

To first order approximation in time dependence we neglect higher order terms in the expansion and put

$$Q_{\Sigma\beta}(t') - Q_{\Sigma\beta}(t) = (t' - t)\,\frac{\partial Q_{\Sigma\beta}}{\partial t} \,. \tag{54}$$

Then Eq. (29) becomes

$$\gamma_{\alpha\beta}\,\frac{\partial Q_{\Sigma\beta}}{\partial t} = \frac{\delta F}{\delta Q_{\Sigma\alpha}} - J_{\Sigma\alpha} \,, \tag{55}$$

where

$$\gamma_{\alpha\beta} = \int dt'(t'-t)\,\Gamma_{r\beta\alpha}(t'-t, Q, \lambda)$$

$$= \frac{1}{i}\,\frac{\partial \Gamma_{r\beta\alpha}(\omega, Q, \lambda)}{\partial \omega}\bigg|_{\omega=0} \,. \tag{56}$$

Equation (55) is the TDGL equation for the order parameters $Q_{\Sigma\alpha}$. Time reversal invariance implies also a reciprocity relation for the relaxation matrix $\gamma_{\alpha\beta}$

$$\gamma_{\alpha\beta}(Q, \lambda) = \epsilon_{(\alpha)}\epsilon_{(\beta)}\,\gamma_{\beta\alpha}(\epsilon Q, \epsilon\lambda) \,. \tag{57}$$

This is one kind of Onsager relation generalized to systems near nonequilibrium stationary states.

It was proved in Ref. 2 that both the diffusion matrix

$$\sigma_{\alpha\beta} = \gamma^{-1}_{\alpha\gamma}\,\Gamma_{c\gamma\delta}(\omega=0, Q, \lambda)\,\gamma T^{-1}_{\delta\beta} \tag{58}$$

and the response matrix

$$L_{\alpha\beta} = \frac{1}{i}\,\frac{\partial G_{r\alpha\beta}(\omega, J, \lambda)}{\partial \omega}\bigg|_{\omega=0} \tag{59}$$

satisfy the reciprocity relations

$$\sigma_{\alpha\beta}(Q, \lambda) = \epsilon_{(\alpha)}\epsilon_{(\beta)}\,\sigma_{\alpha\beta}(\epsilon Q, \epsilon\lambda) \tag{60}$$

and

$$L_{\alpha\beta}(J,\lambda) = \epsilon_{(\alpha)}\epsilon_{(\beta)} L_{\beta\alpha}(\epsilon J, \epsilon\lambda) \quad (61)$$

for systems near stationary states. We shall not discuss it in detail here. The interested reader can consult the original literature.

3. Generalized Fluctuation-dissipation Theorem

We have shown in Ref. 5 that the correlation function G_c can be written in the form

$$G_c = G_r N - N G_a \;, \quad (62)$$

where N is a hermitian matrix related to the quasi particle density distribution. In the thermal equilbrium state

$$N_{\alpha\beta}(\omega, Q, \lambda) = \delta_{\alpha\beta} \operatorname{cth} \frac{\omega}{2T} \;, \quad (63)$$

where T is the temperature. In the low frequency limit

$$\lim_{\omega\to 0} N_{\alpha\beta}(\omega, Q, \lambda) = \delta_{\alpha\beta} \frac{2T}{\omega} \;. \quad (64)$$

Substituting Eq. (62) into the Dyson-Schwinger equation (39) we obtain

$$\Gamma_c = \Gamma_r N - N\Gamma_a = - DN + ND - i(AN + NA) \;, \quad (65)$$

where D and A are the dispersive part and the absorptive part respectively. Equation (65) is the transport equation for the particle density N in a slightly inhomogeneous system.

In a stationary state where the vertex functions depend only on the time difference, Eq. (65) can be written in frequency representation in the following form

$$\Gamma_{c\alpha\beta}(\omega) = - D_{\alpha\gamma}(\omega) N_{\gamma\beta}(\omega) + N_{\alpha\gamma}(\omega) D_{\gamma\beta}(\omega)$$

$$- i(A_{\alpha\gamma}(\omega) N_{\gamma\beta}(\omega) + N_{\alpha\gamma}(\omega) A_{\gamma\beta}(\omega)) \;. \quad (66)$$

It is easily proved that

$$i \sum_{\alpha} \Gamma_{c\alpha\alpha}(\omega)$$

is the quantum fluctuation which has to be positive definite. Therefore we have

$$\mathrm{tr}\,(A(\omega)\,N(\omega)) > 0 \quad . \tag{67}$$

From time reversal invariance we have already shown in Eq. (46) that

$$A_{\alpha\beta}(\omega = 0) = 0 \quad .$$

In the low frequency limit we may put

$$A_{\alpha\beta}(\omega) = \omega\, a_{\alpha\beta} \quad , \tag{68}$$

where $a_{\alpha\beta}$ is the real part of the relaxation matrix $\gamma_{\alpha\beta}$. For a stable stationary state the eigenvalues of the matrix $a_{\alpha\beta}$ have to be positive definite. Since Eq. (67) holds also at zero frequency, the density matrix N must have a pole at $\omega = 0$. Hence

$$N_{\alpha\beta}(\omega) = \frac{2}{\omega}\, T^{\mathrm{eff}}_{\alpha\beta} \tag{69}$$

for small frequency. Substituting Eqs. (68) and (69) back into Eq. (66) we get

$$D_{\alpha\gamma}(\omega = 0)\, T^{\mathrm{eff}}_{\gamma\beta} = T^{\mathrm{eff}}_{\alpha\gamma}\, D_{\gamma\beta}(\omega = 0) \tag{70}$$

and

$$\mathrm{tr}\,(\Gamma_c(\omega = 0)) = -4i\, \mathrm{tr}(a(\omega = 0)\, T^{\mathrm{eff}}) \quad . \tag{71}$$

This is the generalized fluctuation-dissipation theorem. In the literature of critical dynamics it is always assumed that

$$T^{\mathrm{eff}}_{\alpha\beta} = T^{\mathrm{eff}}\, \delta_{\alpha\beta} \quad . \tag{72}$$

In this case we have

$$\Gamma_{c\alpha\beta}(\omega = 0) = -4i\, T^{eff}\, a_{\alpha\beta} \quad . \tag{73}$$

Substituting Eq. (69) into Eq. (62), we obtain another form of the fluctuation-dissipation theorem

$$\frac{\partial G_c}{\partial t} = 2(G_r T^{eff} - T^{eff} G_a)$$

$$= 2\iota^{eff}(G_r - G_a) \quad . \tag{74}$$

References

1. T.D. Lee, Phys.Rev. D8 (1973) 1226; Physics Reports 9C (1974).

2. K.C. Chou and Z.B. Su, Acta Physica Sinica 30 (1981) 164, 401.

3. N. Van Kampen, Physica 23 (1957) 707, 816; U. Uhlhorn, Arkiv. foer Fysik 17 (1960) 361; R. Graham and H. Haken, Z.Phys. 243 (1971) 289; 245 (1971) 141.

4. G. Agarwal, Z.Phys. 252 (1972) 25; J. Deker and F. Haake, Phys.Rev. A11 (1975) 2043; S.K. Ma and G. Mazenko, Phys.Rev. B11 (1975) 4077.

5. K.C. Chou, Z.B. Su, B.L. Hao and L. Yu, Phys.Rev. B22 (1980) 3385; Commun. in Theor. Phys. 1 (1982) 295, 307, 389.

6. J. Schwinger, J.Math.Phys. 2 (1961) 407; L. Keldysh, Zh.Eksp.Teor.Fiz. 47 (1964) 1515.

Part III

场论、粒子物理与核物理

物 理 學 報

第 11 卷 第 4 期 1955 年 7 月

雙力程核子力的討論*

周 光 召

(北京大學物理系)

在本文中我們採用雙力程位能,計算了 ^{2}H, ^{3}H, ^{3}He, ^{4}He 諸原子核的結合能,以及中子對低能質子的散射. §1 和 §2 討論只有中心力的情況, §3 和 §4 中則考慮了張量力的作用. 主要的結果分別在各節中加以討論.

§1. 中心力——氘核結合能及中子對質子散射的問題

張繼恆和彭桓武曾提出了雙力程位能的假設[1],用來計算 ^{2}H, ^{3}H, ^{3}He 及 ^{4}He 諸核的結合能. 其結果表明在核心附近有排斥力出現. 但這一工作沒有考慮核子間散射的現象.

在本節中我們採用這一假定. 此時 j 與 k 兩核子間的位能可以表成如下的形式:

$$V_{jk} = G_1 \frac{e^{-\lambda_1 r_{jk}}}{r_{jk}} + G_2 \frac{e^{-\lambda_2 r_{jk}}}{r_{jk}}, \qquad r_{jk} = |\mathbf{r}_j - \mathbf{r}_k|; \tag{1}$$

其作用力程 λ 與介子靜止質量 μ 具有關係:

$$\hbar\lambda_1 = \mu_1 c, \qquad \hbar\lambda_2 = \mu_2 c, \qquad \xi = \frac{\lambda_2}{\lambda_1} = \frac{\mu_2}{\mu_1}. \tag{2}$$

我們取 $\mu_1 = m_\pi = 270\, m_e$, $\xi = 2$ 或 4 (位能的這一部分可能是由於 ζ 介子 ($\xi = 2$) 或 κ 介子 ($\xi = 4$) 的作用,也可能是由於 π 介子位能的高級近似項[2]). G_1 和 G_2 標誌着位能的強度. 它們在自旋三重態和獨態上具有不同的數值,必要時我們將分別在 G 的左上角加字母 3 或 1 以示區別.

根據有效力程的理論,低能核子散射及氘核結合能的實驗數據,完全由自旋三重態及獨態上的散射長度 a 和有效力程 r_e 反映出來. 這四個參數的實驗值是:

*北京大學研究生畢業論文. 本報於 1954 年 12 月 1 日收到.

$$^3a = -53.8 \pm 0.04 \times 10^{-13} \text{ 厘米},$$

$$^3r_c = 1.71 \pm 0.04 \times 10^{-13} \text{ 厘米},$$

$$^1a = 23.68 \pm 0.06 \times 10^{-13} \text{ 厘米},$$

$$^1r_c = 2.65 \pm 0.07 \times 10^{-13} \text{ 厘米}.$$

這四個參數只同在能量 $E=0$ 時的 S 態波函數有關；它們可以用來確定位能強度參數 G.

在二核子運動的質心坐標系中，S 態波函數可以表成：

$$\psi_s = \frac{1}{r} u(r) \chi; \tag{3}$$

其中 r 爲相對坐標，χ 爲自旋及同位旋波函數，選擇得適合反對稱的要求． 當 $E=0$ 時，$u(r)$ 適合邊界條件：

$$\left.\begin{aligned} &u(0) = 0, \\ &u(r) = 1 - \frac{r}{a} \quad \text{當 } r \longrightarrow \infty, \end{aligned}\right\} \tag{4}$$

其中 a 是散射長度． $u(r)$ 並且和有效力程 r_c 具有關係

$$r_c = 2\int_0^\infty \left[\left(1 - \frac{r}{a}\right)^2 - u^2(r)\right] dr. \tag{5}$$

我們運用 Hulthén 變分法[3] 來確定位能強度參數． 考慮汎函

$$\left.\begin{aligned} I[u] &= \int_0^\infty u\left[\frac{d^2u}{dr^2} - v(r)\,u\right] dr, \\ \delta I &= 2\int_0^\infty \delta u\left[\frac{d^2u}{dr^2} - v(r)\,u\right] dr + \left(\delta u\frac{du}{dr} - u\frac{d\delta u}{dr}\right)\Big|_0^\infty, \end{aligned}\right\} \tag{6}$$

可以看到，如果變分波函數具有和 (4) 式相同的邊界條件，則汎函 I 穩定的條件和 u 適合波動方程

$$\frac{d^2u}{dr^2} - v(r)\,u = 0 \tag{7}$$

的條件相當． 我們選擇了下列的變分波函數：

$$u(r) = 1 - \frac{r}{a} - e^{-\beta r},$$

其中 β 爲變分參數；如已知位能強度參數 G_1 及 G_2 之値，則上述變分法可以用來確定 a 和 r_c． 現在我們的問題恰好相反，要從實驗値 a 和 r_c 中定出 G_1 和

G_2 來．此時我們只需將這一步驟倒過來，取 a 等於實驗值，解式（5）得到 β 的值，再代入方程組 $I=0$ 及 $\frac{\partial I}{\partial \beta}=0$ 中便可以求出 G_1 和 G_2 來了．

採用 $1/\lambda_1 = 1.429\times10^{-13}$ 厘米作爲長度單位，$\frac{\hbar^2\lambda_1^2}{M} = 20.275$ 兆電子伏作爲能量單位．得到表 1 的結果．

表 1．中心力的位能強度參數

	自旋三重態		自旋獨態	
	$\xi=2$	$\xi=4$	$\xi=2$	$\xi=4$
G_1	−2.3167	−2.3574	−1.2485	−1.4000
G_2	−0.1704	−0.1878	−0.6792	−0.8981
ϵ_1	1.006	1.007	1.033	1.064
ϵ_2	0.979	0.977	1.027	1.036
β	2.0714	2.0714	1.6961	1.6961

表 1 中 ϵ_1 和 ϵ_2 分別代表 Hulthén 第一及第二恆等式左右兩端的比值．在絕對正確的解答中它們應等於 1．它們離開 1 的程度可用來判斷所取變分波函數的近似程度，在我們的計算中，由 ϵ_1 和 ϵ_2 的值判斷所得位能強度參數的誤差不超過 5%．

由 G_1 和 G_2 的符號表現出來，核子力中並無排斥力出現．進一步計算表明，加入排斥力後，外面吸引力的力程需要縮短很多（加入一個 0.5×10^{-13} 厘米半徑的排斥心將使外面吸引力力程縮短幾近一半）．因此如果將核子力限制爲中心力，並以 π 介子作爲核子力的主要參與者時，核心附近並無排斥力出現．

採用位能

$$V(r) = G_1\frac{e^{-\lambda_1 r}}{1-e^{-\lambda_1 r}} + G_2\left(\frac{e^{-\lambda_1 r}}{1-e^{-\lambda_1 r}}\right)^2,$$

我們可以用解析方法求出完全正確的解答．用來適合低能核子散射及氘核結合能數據時，也得到了核子力中無排斥力的結果．

可以看到，由於排斥力不出現，得到的雙力程位能在性質上與一般採用的單力程位能相近．它們之間只有形狀上的細微差別．

我們運用這一位能，討論了中子對高能質子散射的 S 態相角移動．採用方法爲 Hulthén 的第二變分法[4]．計算結果見表 2：

表 2. 中子對高能質子散射的 S 態相角移動

E (兆電子伏)	自旋三重態		自旋獨態		$\frac{3}{4}\sin^2 {}^3\delta + \frac{1}{4}\sin^2 {}^1\delta$	
	$\cot {}^3\delta$	$\sin^2 {}^3\delta$	$\cot {}^1\delta$	$\sin^2 {}^1\delta$	理論值	實驗值
40	0.209	0.958	0.873	0.567	0.860	0.76 ± 0.11 0.67 ± 0.11
90	0.423	0.848	1.041	0.480	0.756	0.68 ± 0.08

可以看到，理論值在實驗值邊緣上．這和克利斯琴和哈特[5] 用單力程介子型位能所得結果相似．表明由於採用雙力程位能而引起的位能形狀的細微改變對高能散射影響不大．

§2. 中心力——^{3}H, ^{3}He 及 ^{4}He 核的結合能

我們運用上節得到的位能來計算 ^{3}H, ^{3}He 及 ^{4}He 核的結合能，以繼續探尋位能形狀的細微改變所引起的影響．

計算中採用變分法，其近似的程度取決於波函數選擇的好壞．因而我們有必要先對波函數的普遍形式加以討論．

根據羣的表示理論，^{3}H 核的波函數可以表成為三元排列羣表示基矢的疊加．這種基矢分為下列三類：

$$\left.\begin{aligned}
&\varphi^s(1\,2\,3) = \sum_P P\varphi(1\,2\,3),\\
&\varphi^a(1\,2\,3) = \sum_P (-1)^P P\varphi(1\,2\,3),\\
&\left\{\begin{aligned}
&\bar{\varphi}(1\,2\,3) = \frac{1}{\sqrt{6}}[2\varphi(\overline{1\,2}\,3) - \varphi(\overline{2\,3}\,1) - \varphi(\overline{3\,1}\,2)],\\
&\tilde{\varphi}(1\,2\,3) = \frac{1}{\sqrt{2}}[\varphi(\overline{2\,3}\,1) - \varphi(\overline{3\,1}\,2)];
\end{aligned}\right.
\end{aligned}\right\} \qquad (8)$$

式中的求和是對所有六個排列算符 P 求的．$\varphi(\overline{1\,2}\,3) = \varphi(1\,2\,3) + \varphi(2\,1\,3)$ 為對坐標 1 及 2 對稱的函數．這三類基矢分別為三元排列羣的全對稱表示、全反對稱表示及二度表示的基矢．

在 ^{3}H 核的基態上，我們有下列的運動常數：

$$
\left.\begin{aligned}
&\text{角動量}\quad L^2=0,\\
&\text{自旋}\quad S^2=\frac{1}{2}\left(\frac{1}{2}+1\right)\hbar^2,\\
&\text{同位旋}\quad \tau^2=\frac{1}{2}\left(\frac{1}{2}+1\right),\qquad \tau_3=-\frac{1}{2};
\end{aligned}\right\}\tag{9}
$$

此時共有兩個相互獨立的自旋波函數及兩個相互獨立的同位旋波函數．它們都是三元排列羣二度表示的基矢，將 (8) 式中之 $\varphi(\overline{12}\,3)$ 分別代以 $\alpha(1)\alpha(2)\beta(3)$ 及 $N(1)N(2)P(3)$ 即可得到，分別用符號 $\bar{S}$, $\tilde{S}$, $\bar{\tau}$ 及 $\tilde{\tau}$ 表之． 在自旋同位旋的直乘積空間中，可以用表示理論中的約化手續得到以下四個 ^{3}H 核基態上的自旋同位旋波函數：

$$
\left.\begin{aligned}
\chi^a&=\frac{1}{\sqrt{2}}[\bar{S}\tilde{\tau}-\tilde{S}\bar{\tau}],\\
\chi^s&=\frac{1}{\sqrt{2}}[\tilde{S}\tilde{\tau}+\bar{S}\bar{\tau}],\\
\bar{\chi}&=\frac{1}{\sqrt{2}}[\bar{S}\tilde{\tau}-\tilde{S}\bar{\tau}],\\
\tilde{\chi}&=-\frac{1}{\sqrt{2}}[\tilde{S}\tilde{\tau}+\bar{S}\bar{\tau}].
\end{aligned}\right\}\tag{10}
$$

利用同樣約化的手續，可以得到 ^{3}H 核基態上全反對稱波函數的普遍形式：

$$
\psi=\varphi^s\chi^a+\frac{1}{\sqrt{2}}[\bar{\varphi}\tilde{\chi}-\tilde{\varphi}\bar{\chi}]+\varphi^a\chi^s,\tag{11}
$$

其中 φ 僅是空間坐標的函數．

如暫時不計庫侖力這一小的改正，^{3}He 核基態上的波函數亦具有 (11) 式的形式．

如不計庫侖力，^{4}He 核基態上的運動常數 L^2, S^2, τ^2 及 τ_z 均為零． 全反對稱波函數亦具有 (11) 的形式． 但需將 φ^s 及 φ^a 了解為對四個核子坐標全對稱及全反對稱． 並把 (8) 式中 $\varphi(\overline{12}\,3)$ 換成 $\varphi(\overline{12}\,\overline{34})$，即對核子 1 2 及 3 4 為對稱，又對同時交換一對核子 1 2 及 3 4 為對稱的函數． $\bar{S}$, $\tilde{S}$, $\bar{\tau}$, $\tilde{\tau}$ 以及由它們組成的 χ^s, χ^a, $\bar{\chi}$ 及 $\tilde{\chi}$ 亦有了相應的變化．

作為第一步近似，我們只取 (11) 式中的第一項 $\varphi^s\chi^a$． 對 ^{3}H 及 ^{3}He 核，我們取以下的兩種 φ^s 作為變分波函數：

$$
\varphi_1^s=\exp\left(-\frac{1}{\sqrt{2}}a\rho\right),\qquad \rho^2=\frac{1}{6}\sum_{i\neq j=1}^{3}|\mathbf{r}_i-\mathbf{r}_j|^2.\tag{12a}
$$

$$\varphi_2^s = \exp\left(-\frac{\alpha}{2}(r_{12}+r_{23}+r_{31})\right). \tag{12b}$$

這兩種不同的形式爲不同的工作者所採用[6]，我們同時計算了這兩種情況以比較其好壞．對 ^{4}He 核則取下列之 φ^s:

$$\varphi^s = \exp\left(-\frac{1}{\sqrt{2}}a\rho\right), \qquad \rho^2 = \frac{1}{8}\sum_{i \neq j=1}^{4}|\mathbf{r}_i - \mathbf{r}_j|^2. \tag{13}$$

變分法可以表爲

$$E = \frac{\int\psi^* H\psi\,dV}{\int\psi^*\psi\,dV}, \qquad \delta E = 0; \tag{14}$$

其中

$$H = -\frac{\hbar^2}{2M}\sum_{i=1}^{A}\nabla_i^2 + \sum_{i<j}V_{ij}.$$

採用 $1/\lambda_1$ 作爲長度單位，$\frac{\hbar^2\lambda_1^2}{M}$ 爲能量單位，並假定核子力具有對稱的性質*，卽可得到:

I. ^{3}H 核

1) φ_1^s 的情形

$$E = \frac{1}{4}a^2 - pF_3(a) - \xi q\,F_3(a/\xi), \tag{15a}$$

其中

$$p = -\frac{1}{2}({}^3G_1 + {}^1G_1), \qquad q = -\frac{1}{2}({}^3G_2 + {}^1G_2), \tag{16}$$

$$F_3(a) = \frac{2}{5\pi}\frac{a^3}{(a^2-1)^3}\left[8a^4 + 9a^2 - 2 - 30\frac{a^4}{\sqrt{a^2-1}}\tan^{-1}\sqrt{\frac{a-1}{a+1}}\right]; \tag{17a}$$

2) φ_2^s 的情形

$$E = \frac{15}{14}\alpha^2 - pH_3(\alpha) - \xi q\,H_3(\alpha/\xi), \tag{15b}$$

其中

$$H_3(\alpha) = \frac{3}{14}\frac{(2\alpha)^5}{(2\alpha+1)^2}\left[\frac{2}{(2\alpha)^2} + \frac{2}{(2\alpha)(2\alpha+1)} + \frac{1}{(2\alpha+1)^2}\right]. \tag{17b}$$

II. ^{3}He 核 和 ^{3}H 核一樣，但要加上庫侖能的改正項 E_c.

1) φ_1^s 的情形

$$E_c = \frac{e^2}{\hbar c}\frac{M}{\mu}\frac{16}{15\pi}a;$$

2) φ_2 的情形

$$E_c = \frac{e^2}{\hbar c}\frac{M}{\mu}\frac{5}{7}\alpha.$$

III. ^{4}He 核

---

* V_{ik} 的普遍形式見 (26) 式中的中心力部分．

$$E = \frac{1}{4} a^2 - pF_4(a) - \xi q F_4(a/\xi) + E_c, \tag{15c}$$

其中
$$F_4(a) = \frac{105}{64} \frac{a^3}{(1+a)^5} \left(a^3 + \frac{47}{35} a^2 + \frac{5}{7} a + \frac{1}{7} \right), \tag{17c}$$

$$E_c = \frac{e^2}{\hbar c} \frac{M}{\mu} \frac{35}{128} a.$$

計算結果見表 3.

表 3. ^{3}H, ^{3}He 及 ^{4}He 諸核之結合能（單位：兆電子伏）

原子核	波函數	$\xi=2$ 的情形			實驗值	$\xi=4$ 的情形		
		a 或 α	$-E$	結合能		a 或 α	$-E$	結合能
^{3}H	φ_1^s	3.27	0.433	8.78	8.492	3.25	0.415	8.41
	φ_2^s	1.60	0.568	11.52	8.492	1.62	0.557	11.29
^{3}He	φ_1^s	3.20	0.378	7.66	7.728	3.20	0.360	7.30
	φ_2^s	1.60	0.511	10.36	7.728	1.62	0.500	10.14
^{4}He	φ	5.97	2.54	51.5	28.298	5.97	2.50	50.7

從表 3 結果中可以看到，計算的結合能較實驗值爲大．且不論 $\xi = 2$ 及 $\xi = 4$ 的情形都得到差不多大的結果．這個事實表明，位能形狀的這種改變，對 ^{3}H, ^{3}He 及 ^{4}He 核的結合能不發生很大的影響，而我們又回到用單力程位能所遇到的困難．

其次可以看到，φ_2^s 所給出的 ^{3}H 核結合能低於 φ_1^s 所給出的值，兩者相差達 20—30% 之多，說明波函數選擇的好壞對於計算結果有着甚大的影響．在 ^{4}He 核結合能的計算中，由於積分的困難，尙且無法採用類似 φ_2^s 之波函數；可以設想，波函數的這種修改對 ^{4}He 核的結合能也可能有很大的影響．

爲了繼續探尋波函數的好壞，我們在 ^{3}H 核問題上考慮了下列變分波函數：

$$\psi = \varphi^s \chi^a + \frac{1}{\sqrt{2}} \left[\bar{\varphi} \tilde{\chi} - \tilde{\varphi} \bar{\chi} \right];$$

其中
$$\varphi^s = \exp\left[-\frac{\alpha}{2} (r_{12} + r_{23} + r_{31}) \right],$$

$$\varphi(\overline{12}\,3) = C\, r_{12} \exp\left[-\frac{\beta}{2} (r_{12} + r_{23} + r_{31}) \right],$$

α, β 及 C 均爲變分參數．計算結果表明，在 $\xi=2$ 的情形，結合能的改正 $\triangle E=$

-0.36 兆電子伏，約爲單取 $\varphi^s\chi^a$ 作爲變分波函數時所得的結合能值的 3%. 因此可以認爲波函數的這種修改確實不太重要.

§3. 張量力——氘核的結構及中子對低能質子的散射

前二節的計算表明，僅有中心力的雙力程位能不足以解釋核子散射及結合能的現象. 我們必須修改所用位能來消除這一困難. 在本節及下節中我們在核子力中加入張量力，此時 j 與 k 核子間的位能具有下面的形式:

$$V_{jk}=G_1\frac{e^{-\lambda_1 r_{jk}}}{r_{jk}}+G_2\frac{e^{-\lambda_2 r_{jk}}}{r_{jk}}+G_t S_{jk}\left[F_{jk}(\lambda_1)-\xi^2 F_{jk}(\lambda_2)\right]; \tag{18}$$

其中

$$S_{jk}=3\frac{(\boldsymbol{\sigma}_j\cdot\boldsymbol{r}_{jk})(\boldsymbol{\sigma}_k\cdot\boldsymbol{r}_{jk})}{r_{jk}^2}-\boldsymbol{\sigma}_j\cdot\boldsymbol{\sigma}_k, \tag{18a}$$

$$F_{jk}(\lambda)=\left(\frac{1}{r_{jk}}+\frac{3}{\lambda r_{jk}^2}+\frac{3}{\lambda^2 r_{jk}^3}\right)e^{-\lambda r_{jk}}, \tag{18b}$$

$$\xi=\lambda_2/\lambda_1,$$

$\boldsymbol{\sigma}_j$ 及 $\boldsymbol{\sigma}_k$ 代表第 j 核子與第 k 核子的自旋算符.

容易看到，張量力對自旋獨態沒有作用，因此 ^1S 態上的位能在加入張量力後沒有改變，我們無需重複計算.

我們知道，根據有效力程理論氘核的結構及中子對低能質子的散射都只與氘核基態有關，要利用上述實驗數據來確定偶三重態上的位能強度參數，只需考慮氘核基態的波函數.

當加入張量力後，角動量不再是運動常數，氘核基態波函數爲 S 波和 D 波的叠加. 在質心坐標系中，它可以表成如下的形式:

$$\psi=\frac{1}{r}\left[u(r)+8^{-\frac{1}{2}}\omega(r)S_{12}\right]{}^3\chi, \tag{19}$$

其中 $^3\chi$ 代表選擇得適合全反對稱要求的自旋同位旋波函數，r 爲相對距離. $u(r)$ 和 $\omega(r)$ 分別代表波函數在 S 態及在 D 態上的部分，它們和氘核的四極矩 Q，磁矩 μ_D 及有效力程 r_c 具有以下的關係:

$$Q=\frac{\sqrt{2}}{10}\int_0^\infty r^2\left(u\omega-8^{-\frac{1}{2}}\omega^2\right)dr\bigg/\int_0^\infty\left(u^2+\omega^2\right)dr, \tag{20a}$$

$$\mu_D=\mu_n+\mu_p-\frac{3}{2}\left(\mu_n+\mu_p-\frac{1}{2}\mu_0\right)P_D,\quad P_D=\frac{\int_0^\infty\omega^2\,dr}{\int_0^\infty(u^2+\omega^2)dr}, \tag{20b}$$

$$r_c = 2\int_0^\infty [e^{-2\alpha r} - (u^2+\omega^2)]\,dr,\quad \text{此時需要}\quad u^2+\omega^2 \longrightarrow e^{-2\alpha r}\ \text{當}\ r\to\infty; \tag{20c}$$

其中 μ_n, μ_p 及 μ_0 分別代表中子、質子及標準核子的磁矩; $\alpha^2 = \frac{M}{\hbar^2}|E_d|$, E_d 爲氘核結合能; P_D 是 D 波所佔有的百分比.

和 §1 中情形相似,我們現在需要從實驗的 $Q=2.738\pm0.016\times10^{-27}$ 厘米^2, $P_D=0.04\pm0.016$, r_e 及氘核結合能 E_d 的數值中定出位能強度參數. 在計算中採用變分法,考慮汎函

$$I = \int_0^\infty \left\{ u\left[\frac{d^2u}{dr^2} - (\alpha^2 + V_c(r))u\right] + \omega\left[\frac{d^2\omega}{dr^2} - \frac{6}{r^2}\omega - (\alpha^2 + V_c(r) - 2V_t(r))\,\omega\right] - 4\sqrt{2}\,V_t(r)\,u\omega\right\} dr. \tag{21}$$

如所選變分波函數 $u(r)$ 及 $\omega(r)$ 在 $r\to0$ 及 $r\to\infty$ 時均趨於零,則汎函 I 穩定的條件和 $u(r)$ 及 $\omega(r)$ 適合以 $V_c(r)+S_{12}V_t(r)$ 爲位能的波動方程的條件相當. 實際計算中,我們採用了下列試探函數:

$$\left.\begin{aligned} u(r) &= e^{-\alpha r} - e^{-\beta r},\\ \omega(r) &= \zeta(1-e^{-\gamma r})^5\, e^{-\alpha r}\left(\frac{1}{3} + \frac{1}{\alpha r} + \frac{1}{(\alpha r)^2}\right). \end{aligned}\right\} \tag{22}$$

將 γ 固定在一些值上,取 β 及 ζ 爲變分參數,我們可以採用類似 §1 中之計算步驟求得表 4 的結果[1]. 表 4 中列有好幾組 G_1, G_2 和 G_t 之值,它們是由於對 γ 作不同的選擇而引起的. 嚴格講來,應把 γ 取作變分參數而求得唯一的一組數值. 但由於恆等式 ϵ_1 和 ϵ_2 都很接近於 1, 我們選的波函數在

表 4. 加入張量力後偶三重態上位能強度參數之值 (單位爲 $\frac{\hbar^2\lambda_1^2}{M}$)

(甲) $\xi=2$ 的情形

G_1	G_2	G_t	$P_D\,\%$	ϵ_1	ϵ_2
−1.6303	−0.8934	−0.5170	2.57	1.020	1.019
−1.3730	−1.1184	−0.6117	3.09	1.016	1.014
−1.0306	−1.3976	−0.7287	3.68	1.009	1.005
−0.5730	−1.7405	−0.8755	4.36	0.995	0.991

1) 計算的公式見附錄 1.

(乙) $\xi=4$ 的情形

G_1	G_2	G_t	P_D %	ϵ_1	ϵ_2
−1.8404	−0.6399	−0.2216	3.09	1.010	1.010
−1.6378	−0.7109	−0.2578	3.68	0.997	0.997
−1.3623	−0.7573	−0.3031	4.36	0.974	0.977

$r\to 0$ 及 $r\to\infty$ 時又都具有正確的漸近行爲，並且在計算中沒有考慮各個實驗數據的誤差所引起的位能強度參數的改正，因此我們認爲它們都是近似的解答．這幾組值所對應的 P_D 的數值不相同，當實驗的 P_D 值更準確時(主要是對交換磁矩的影響更清楚時)，可資選擇的位能強度參數的範圍亦將相應的縮小．

可以看到，當 G_t 增加時，G_2 增加而 G_1 減小，表示張量力增加時，中心力的力程減短．

當核子運動速度很小時，從介子場論可以計算由各種類型介子所傳遞的"靜位能"． 如以 π 介子爲贋標量介子，(18) 式的位能可以看作是由贋標量介子與另一重介子所產生的混合場的位能． 經過簡單數量上的分析卽發現，我們得到的位能還不能和介子混合場的"靜位能"湊合．

§4. 張量力——^{3}H，^{3}He 及 ^{4}He 諸核的結合能

在本節中我們運用上節得到的位能，用變分法計算 ^{3}H，^{3}He 和 ^{4}He 諸核的結合能． 當加入張量力後，波函數便採取了十分複雜的形式． 此時需要把具有不同角動量和自旋的波函數叠加起來，方能正確地描述原子核的狀態． 根據吉爾哥埃及施文格爾的方法[7]，可以求得 ^{3}H，^{3}He 及 ^{4}He 諸核基態上出現的波函數的各種可能的自旋和空間角度的部分． 其中最主要的是 S 波及 D 波，下面是在波函數的第一次近似中出現的 S 波與 D 波的形式．

甲）^{3}H 及 ^{3}He 核的情形

$^2\tilde{S}$: $$\tilde{\chi}=\frac{1}{\sqrt{2}}[\alpha(1)\,\beta(2)-\alpha(2)\,\beta(1)]\,\alpha(3),$$

$^2\bar{S}$: $$-\frac{1}{2\sqrt{3}}(\underline{\sigma}_{12}\cdot\underline{\sigma}_3)\,\tilde{\chi},$$

$$\left.\begin{aligned}
&{}^4\bar{D}(\underline{r}\,\underline{r}): && \left[(\underline{\sigma}_3\cdot\underline{r})(\underline{\sigma}_{12}\cdot\underline{r})-\frac{1}{3}(\underline{\sigma}_{12}\cdot\underline{\sigma}_3)\,r^2\right]\tilde{\chi},\\
&{}^4\bar{D}(\underline{\rho}\,\underline{\rho}): && \left[(\underline{\sigma}_3\cdot\underline{\rho})(\underline{\sigma}_{12}\cdot\underline{\rho})-\frac{1}{3}(\underline{\sigma}_{12}\cdot\underline{\sigma}_3)\,\rho^2\right]\tilde{\chi},\\
&{}^4\tilde{D}(\underline{r}\,\underline{\rho}): && \left[(\underline{\sigma}_3\cdot\underline{\rho})(\underline{\sigma}_{12}\cdot\underline{r})+(\underline{\sigma}_3\cdot\underline{r})(\underline{\sigma}_{12}\cdot\underline{\rho})-\frac{2}{3}(\underline{\sigma}_{12}\cdot\underline{\sigma}_3)(\underline{r}\cdot\underline{\rho})\right]\tilde{\chi};
\end{aligned}\right\}\quad(23)$$

其中 α 和 β 代表 $S_z=\frac{1}{2}$ 及 $-\frac{1}{2}$ 的自旋波函數；$\underline{\sigma}_{12}=\underline{\sigma}_1-\underline{\sigma}_2$，$\underline{\sigma}_i$ 代表第 i 個核子的自旋算符；$\underline{r}=\frac{1}{\sqrt{2}}(\underline{r}_1-\underline{r}_2)$，$\underline{\rho}=\frac{1}{\sqrt{6}}(2\underline{r}_3-\underline{r}_1-\underline{r}_2)$. 左邊的符號和一般在光譜學上用的相同，上面加一橫線代表波函數對 1 2 爲對稱，而加一弧線 ～ 則表示其對 1 2 爲反對稱.

乙）^{4}He 核的情形

$$\left.\begin{aligned}
&{}^1\tilde{S}: && \tilde{\chi}=\frac{1}{2}\Big[\alpha(1)\,\alpha(3)\,\beta(2)\,\beta(4)+\alpha(2)\,\alpha(4)\,\beta(1)\,\beta(3)-\\
& && \qquad -\alpha(2)\,\alpha(3)\,\beta(1)\,\beta(4)-\alpha(1)\,\alpha(4)\,\beta(2)\,\beta(3)\Big],\\
&{}^1\bar{S}: && -\frac{1}{4\sqrt{3}}(\underline{\sigma}_{12}\cdot\underline{\sigma}_{34})\,\tilde{\chi},\\
&{}^5\bar{D}(\underline{r}\,\underline{r}) && \left[(\underline{\sigma}_{12}\cdot\underline{r})(\underline{\sigma}_{34}\cdot\underline{r})-\frac{1}{3}(\underline{\sigma}_{12}\cdot\underline{\sigma}_{34})\,r^2\right]\tilde{\chi},\\
&{}^5\bar{D}(\underline{\rho}_1\,\underline{\rho}_2) && \Big[(\underline{\sigma}_{12}\cdot\underline{\rho}_1)(\underline{\sigma}_{34}\cdot\underline{\rho}_1)+(\underline{\sigma}_{12}\cdot\underline{\rho}_2)(\underline{\sigma}_{34}\cdot\underline{\rho}_2)-\\
& && \qquad -\frac{1}{3}(\underline{\sigma}_{12}\cdot\underline{\sigma}_{34})(\rho_1^2+\rho_2^2)\Big]\tilde{\chi},\\
&{}^5\tilde{D}(\underline{\rho}_1\,\underline{\rho}_2) && \Big[(\underline{\sigma}_{12}\cdot\underline{\rho}_1)(\underline{\sigma}_{34}\cdot\underline{\rho}_2)+(\underline{\sigma}_{12}\cdot\underline{\rho}_2)(\underline{\sigma}_{34}\cdot\underline{\rho}_1)-\\
& && \qquad -\frac{2}{3}(\underline{\sigma}_{12}\cdot\underline{\sigma}_{34})(\underline{\rho}_1\cdot\underline{\rho}_2)\Big]\tilde{\chi};
\end{aligned}\right\}\quad(24)$$

其中　$\underline{\sigma}_{12}=\underline{\sigma}_1-\underline{\sigma}_2,\qquad \underline{\sigma}_{34}=\underline{\sigma}_3-\underline{\sigma}_4,$

$$\underline{r}=\frac{1}{2}(\underline{r}_1+\underline{r}_2+\underline{r}_3+\underline{r}_4),$$

$$\underline{\rho}_1=\frac{1}{\sqrt{2}}(\underline{r}_1-\underline{r}_2),\qquad \underline{\rho}_2=\frac{1}{\sqrt{2}}(\underline{r}_3-\underline{r}_4).$$

和 §2 中採用的方法相同，我們可以自同位旋波函數及自旋和空間坐標波函數中組成全反對稱的波函數，此時

$$\psi = \frac{1}{\sqrt{2}}[\bar{\varphi}\tilde{\tau} - \tilde{\varphi}\bar{\tau}]; \tag{25}$$

其中 $\bar{\tau}$ 和 $\tilde{\tau}$ 與 §2 中所用的相同，而 $\bar{\varphi}$ 及 $\tilde{\varphi}$ 則爲由自旋及空間坐標波函數組成的三元排列羣或四元排列羣二度表示的基矢．他們分别由函數 $\varphi(\overline{12}\,3)$ 及 $\varphi(\overline{\overline{12}\,\overline{34}})$ 來確定（參見(8)）．

變分法可以表成如下的形狀：

$$E = \frac{\int \psi^* H \psi \, dV}{\int \psi^* \psi \, dV}; \qquad \delta E = 0,$$

其中

$$H = -\frac{\hbar^2}{2M}\sum_{i=1}^{A} \nabla_i^2 + \sum_{i<j}^{A} V_{ij} + V_c;$$

A 爲核子數目，V_c 爲庫侖能．由於 ^3H，^3He 及 ^4He 核的結合能不僅由偶態上作用力決定（這一作用力在上一節中已經求得），而且與奇態上的作用力有關．爲了討論的方便，我們需要將位能的普遍形式表示出來．假設核子力具有對稱的性質，這一普遍形式是：

$$V_{ij} = \sum_{\alpha=1}^{2}\left[\omega_\alpha + b_\alpha P_\sigma(ij) + h_\alpha P_\tau(ij) + m_\alpha P_{\sigma\tau}(ij)\right]\frac{e^{-\lambda_\alpha r_{ij}}}{r_{ij}} +$$
$$+ S_{ij}(\omega_t + h_t P_\tau(ij))\,[F_{ij}(\lambda_1) - \xi^2 F_{ij}(\lambda_2)], \tag{26}$$

其中 $P_\sigma(ij)$，$P_\tau(ij)$ 分别爲交換第 i 和第 j 核子的自旋及同位旋算符，$P_{\sigma\tau}(ij) = P_\sigma(ij)\,P_\tau(ij)$．$S_{ij}$ 和 F_{ij} 見 §3 的 (18) 式．

在計算 E 時，我們採用下列的試探波函數 $\varphi(\overline{12}\,3)$ 及 $\varphi(\overline{\overline{12}\,\overline{34}})$：

$$\varphi(\overline{12}\,3) = N_S \exp\left[-\frac{\alpha}{2}(r_{12} + r_{23} + r_{31})\right]^2 \bar{S} +$$
$$+ N_D\, c \exp\left[-\frac{\beta}{2}(r_{12} + r_{23} + r_{31})\right]{}^4\bar{D}(\underline{r}\,\underline{r}),$$

$$\varphi(\overline{\overline{12}\,\overline{34}}) = N_S \exp\left[-\frac{1}{\sqrt{2}}a\rho\right]{}^1\bar{S} + N_D\, c \exp\left[-\frac{1}{\sqrt{2}}b\rho\right]{}^5\bar{D}(\underline{r}\,\underline{r}); \tag{27}$$

其中 $\rho^2 = \frac{1}{8}\sum_{i\neq j=1}^{4} r_{ij}^2 = \rho_1^2 + \rho_2^2 + r^2$，$N_S$ 和 N_D 爲規格化常數，選擇使得 $\int \psi^* \psi \, dV = 1 + c^2$．$\alpha, \beta, a, b$ 及 c 等均爲變分參數．

如果將 $\varphi(\overline{12}\,3)$ 中 $^4\bar{D}(\underline{r}\,\underline{r})$ 換以 $^4\bar{D}(\underline{\rho}\,\underline{\rho})$，而不改變 β 的值，或者將 $\varphi(\overline{\overline{12}\,\overline{34}})$ 中的 $^5\bar{D}(\underline{r}\,\underline{r})$ 換以 $^5\bar{D}(\underline{\rho}_1\underline{\rho}_2)$，而不改變 b 的值，則我們發現，代入 (8) 式後將

得到相同的 $\bar{\varphi}$ 及 $\tilde{\varphi}$，表明這種波函數的改變並不會給出新的結果．

經過冗長的計算，可以求得：

甲）^{3}H 核

$$E(1+c^2) = \frac{15}{14}\alpha^2 + pH_3(\alpha) + \xi q\, H_3(\alpha/\xi) + 2c\eta\, A_3(\alpha, \beta) +$$
$$+ c^2\left[\frac{15}{14}\beta^2 + SB_3(\beta) + \xi\, t\, B_3(\beta/\xi) + uD_3(\beta) + \xi\, v\, D_3(\beta/\xi) + \right.$$
$$\left. + \eta E_3(\beta) + \zeta K_3(\beta)\right]; \tag{28}$$

乙）^{3}He 核：和 ^{3}H 核基本上相同，只需加上庫侖能 E_c

$$E_c(1+c^2) = \frac{e^2}{\hbar c}\frac{M}{\mu_1}\left[\frac{5}{7}\alpha^2 + c^2\frac{2}{7}\beta\right]; \tag{29}$$

丙）^{4}He 核

$$E(1+c^2) = \frac{1}{4}a^2 + pF_4(a) + \xi q\, F_4(a/\xi) + 2c\eta\, A_4(a, b) +$$
$$+ c^2\left[\frac{1}{4}b^2 + SB_4(b) + \xi\, tB_4(b/\xi) + uD_4(b) + \xi\, v\, D_4(b/\xi) + \right.$$
$$\left. + \eta E_4(b) + \zeta K_4(b)\right] + \frac{e^2}{\hbar c}\frac{M}{\mu_1}\left[\frac{35}{128}a + \frac{77}{512}bc^2\right]; \tag{30}$$

其中
$$\frac{\hbar^2\lambda_1}{M}p = \omega_1 - m_1, \qquad \frac{\hbar^2\lambda_1}{M}q = \omega_2 - m_2,$$
$$\frac{\hbar^2\lambda_1}{M}S = \omega_1 + b_1 - h_1 - m_1, \qquad \frac{\hbar^2\lambda_1}{M}t = \omega_2 + b_2 - h_2 - m_2,$$
$$\frac{\hbar^2\lambda_1}{M}u = \omega_1 + b_1 + h_1 + m_1, \qquad \frac{\hbar^2\lambda_1}{M}v = \omega_2 + b_2 + h_2 + m_2,$$
$$\frac{\hbar^2\lambda_1}{M}\eta = \omega_t + h_t, \qquad \frac{\hbar^2\lambda_1}{M}\zeta = \omega_t - h_t;$$

H, F, A, B, D, E 及 K 等函數均在附錄 2 中給出．

計算表明，奇態上作用力對結合能影響很小，假定奇態上無作用力與假定奇態上作用力和偶態上作用力相等來計算結合能，兩者相差約在 0.5 兆電子伏左右，它對我們結果的討論沒有什麽影響．

假定奇態上無作用力，則 $u = v = \zeta = 0$，我們可以運用 §3 中求得的偶態上的作用力來確定 E 的極小值． 計算結果見表 5． 在計算中發現 E 曲面在其極小處甚爲平坦，因此我們可以先固定 α，β 及 a 和 b，再來對 c 求極小．

其中 α 和 a 的值選擇得使中心力產生的結合能在其極小附近．進一步計算表明這種作法的誤差約爲 5%．

表 5. ^{3}H, ^{3}He 及 ^{4}He 核的結合能（單位：兆電子伏）

（甲） $\xi=2$ 的情形

G_t	$E(^3H)$	α	β	$E(^3He)$	$E(^4He)$	a	b
−0.5170	−9.08	1.6	3	−7.93	−41.2	5.4	7.2
−0.6117	−7.70	1.6	3	−6.55	−37.5	5.4	7.2
−0.7287	−5.66	1.4	3	−4.66	−31.2	5.2	7.2
−0.8755	−1.96	1.2	3	−1.10	−22.2	4.8	7.2

（乙） $\xi=4$ 的情形

G_t	$E(^3H)$	α	β	$E(^3He)$	$E(^4He)$	a	b
−0.2216	−7.32	1.4	3	−6.32	−35.9	5.2	7.2
−0.2578	−5.45	1.4	3	−4.45	−29.3	4.8	7.2
−0.3031	−2.68	1.2	3	−1.82	−19.8	4.4	7.2

在上表中 α, β, a 和 b 都是以 λ_1 作爲單位給出的數值．由表中可以看到，當張量力強度增加時，理論的結合能值逐步減小．表明加入張量力能緩和在 §2 中所遇到的理論結合能值過大的困難，但從表 5 中尙找不到能同時適合 ^{3}H, ^{3}He 及 ^{4}He 諸核結合能的位能強度參數．

胡濟民和徐躬耦二教授曾在他們計算 ^{3}H、^{3}He 核結合能的工作中指出，其他幾個 D 態波函數對結合能的貢獻可能很大，在他們所用位能條件下，這種貢獻可以高達 5—6 兆電子伏．最近克拉克在計算 ^{4}He 核結合能的工作中[8] 也指出，增加一 D 波可使 ^{4}He 核結合能改進約 9 兆電子伏．這樣，從我們這一初步的計算中，尙且不能肯定表 5 中理論值與實驗值的差別是由於波函數的近似程度不够，還是所選位能不足以反映核子力的眞實情況．如果其他 D 波的改正很大，表 5 中位能朝增大張量力方向外抽出去，仍有可能獲得符合實驗之結果．但由表 5 中的一般趨勢看來，^{4}He 核結合能仍有過大的可能．這種過大的情況亦可能由於多體力的存在而被消除．最近高能核子散射及重核飽和現象的實驗亦建議多體力的存在．因此對多體力的研究將爲這一問題的解決提供一種可能性．

最後作者願對彭桓武教授的指導表示衷心的感謝，他建議了這一題目，並在工作過程中給予了有益的幫助和啓發的討論．

附錄 1: I, $\frac{\partial I}{\partial \beta}$, Q, r_c 的公式

在 §3 中取 (22) 式的波函數 u 及 ω，代入到 (20)，(21) 式中去，卽可計算出 I，$\frac{\partial I}{\partial \beta}$，$Q$ 及 r_c 來，計算的結果爲：

$$I=\alpha-\frac{\beta}{2}-\frac{\alpha^2}{2\beta}-25\,\zeta^2 A_0(\gamma,8,2\alpha+2\gamma)-G_1\ln\frac{(\alpha+\beta+1)^2}{(2\alpha+1)(2\beta+1)}-$$

$$-G_2\ln\frac{(\alpha+\beta+\xi)^2}{(2\alpha+\xi)(2\beta+\xi)}-G_1\zeta^2 A_1(\gamma,10,2\alpha+1)-G_2\zeta^2 A_1(\gamma,10,2\alpha+\xi)+$$

$$+2G_t\zeta^2[A_1(\gamma,10,2\alpha+1)+3A_2(\gamma,10,2\alpha+1)+3A_3(\gamma,10,2\alpha+1)-$$

$$-\xi^2 A_1(\gamma,10,2\alpha+\xi)-3\xi A_2(\gamma,10,2\alpha+\xi)-3A_3(\gamma,10,2\alpha+\xi)]-$$

$$-4\sqrt{2}\,G_t\zeta[B_1(\gamma,5,2\alpha+1)+3B_2(\gamma,5,2\alpha+1)+3B_3(\gamma,5,2\alpha+1)-$$

$$-\xi^2 B_1(\gamma,5,2\alpha+\xi)-3\xi B_2(\gamma,5,2\alpha+\xi)-3B_3(\gamma,5,2\alpha+\xi)-$$

$$-B_1(\gamma,5,\alpha+\beta+1)-3B_2(\gamma,5,\alpha+\beta+1)-3B_3(\gamma,5,\alpha+\beta+1)+$$

$$+\xi^2 B_1(\gamma,5,\alpha+\beta+\xi)+3\xi B_2(\gamma,5,\alpha+\beta+\xi)+3B_3(\gamma,5,\alpha+\beta+\xi)],$$

$$\frac{\partial I}{\partial\beta}=-\frac{1}{2}+\frac{\alpha^2}{2\beta^2}-2G_1\left(\frac{1}{\alpha+\beta+1}-\frac{1}{2\beta+1}\right)-2G_2\left(\frac{1}{\alpha+\beta+\xi}-\frac{1}{2\beta+\xi}\right)-$$

$$-4\sqrt{2}\,G_t\zeta[B_0(\gamma,5,\alpha+\beta+1)+3B_1(\gamma,5,\alpha+\beta+1)+3B_2(\gamma,5,\alpha+\beta+1)-$$

$$-\xi^2 B_0(\gamma,5,\alpha+\beta+\xi)-3\xi B_1(\gamma,5,\alpha+\beta+\xi)-3B_2(\gamma,5,\alpha+\beta+\xi)],$$

$$r_c=\frac{2}{1+\frac{1}{9}\zeta^2}\left[\frac{2}{\alpha+\beta}-\frac{1}{2\beta}-\zeta^2 A_0(\gamma,10,2\alpha)+\frac{1}{9}\zeta^2\frac{1}{2\alpha}\right],$$

$$Q=\frac{\sqrt{2}\,\zeta}{10\left(1+\frac{1}{9}\zeta^2\right)\left(\frac{1}{2\alpha}-\frac{1}{2}r_c\right)}\Big\{B_{-2}(\gamma,5,2\alpha)-B_{-2}(\gamma,5,\alpha+\beta)-$$

$$-8^{-\frac{1}{2}}\zeta A_{-2}(\gamma,10,2\alpha)\Big\};$$

其中 $$A_l(p,q,t)=\frac{1}{9}I_l(p,q,t)+\frac{2}{3\alpha}I_{l+1}(p,q,t)+\frac{5}{3\alpha^2}I_{l+2}(p,q,t)+$$

$$+\frac{2}{\alpha^3} I_{l+3}(p,q,t)+\frac{1}{\alpha^4} I_{l+4}(p,q,t),$$

$$B_l(p,q,t)=\frac{1}{3} I_l(p,q,t)+\frac{1}{2} I_{l+1}(p,q,t)+\frac{1}{\alpha^2} I_{l+2}(p,q,t),$$

$$I_l(p,q,t)=\int_0^\infty (1-e^{-pr})^q e^{-tr} r^{-l}\,dr$$

$$=\frac{1}{l-1}\left[pq\, I_{l-1}(p,q-1,p+t)-t\, I_{l-1}(p,q,t)\right],$$

$$I_1(p,q,t)=\sum_{k=0}^{q}\binom{q}{k}(-1)^{k-1}\ln(t+kp);$$

先計算出 I_l 來,便可以計算 A_l 及 B_l, 然後可以求得 I, $\frac{\partial I}{\partial\beta}$, r_e, Q 等值.

附錄 2: H, F, A, B, D, E 及 K 等函數的形式

$$H_3(\alpha)=\frac{3}{14}\frac{(2\alpha)^5}{(2\alpha+1)^2}\left[\frac{2}{(2\alpha)^2}+\frac{2}{(2\alpha)(2\alpha+1)}+\frac{1}{(2\alpha+1)^2}\right],$$

$$A_3(\alpha,\beta)=\frac{64}{5\sqrt{55}}\frac{\alpha^3\beta^5}{\alpha+\beta}\left[\frac{2}{(\alpha+\beta)^2}I_4\left(\frac{\alpha+\beta}{2}\right)+\frac{1}{\alpha+\beta}I_5\left(\frac{\alpha+\beta}{2}\right)+\frac{1}{6}I_6\left(\frac{\alpha+\beta}{2}\right)\right],$$

$$B_3(\beta)=\frac{16}{385}\frac{\beta^9}{(2\beta+1)^2}\left[\frac{960}{(2\beta)^6}+\frac{960}{(2\beta)^5(2\beta+1)}+\frac{640}{(2\beta)^4(2\beta+1)^2}+\right.$$

$$\left.+\frac{320}{(2\beta)^3(2\beta+1)^3}+\frac{272}{(2\beta)^2(2\beta+1)^4}+\frac{496}{(2\beta)(2\beta+1)^5}+\frac{552}{(2\beta+1)^6}\right],$$

$$D_3(\beta)=\frac{48}{385}\frac{\beta^9}{(2\beta+1)^4}\left[\frac{160}{(2\beta)^4}+\frac{320}{(2\beta)^3(2\beta+1)}+\frac{336}{(2\beta)^2(2\beta+1)^2}+\right.$$

$$\left.+\frac{208}{(2\beta)(2\beta+1)^3}+\frac{56}{(2\beta+1)^4}\right],$$

$$E_3(\beta)=-\frac{4\beta^9}{1155}\left[\frac{32}{(2\beta)^4}I_4(\beta)+\frac{16}{(2\beta)^3}I_5(\beta)+\frac{664}{35}\frac{1}{(2\beta)^2}I_6(\beta)+\right.$$

$$\left.+\frac{856}{105}\frac{1}{2\beta}I_7(\beta)+\frac{46}{35}I_8(\beta)\right],$$

$$K_3(\beta)=-\frac{4\beta^9}{385}\left[\frac{224}{(2\beta)^4}I_4(\beta)+\frac{112}{(2\beta)^3}I_5(\beta)+\frac{24}{(2\beta)^2}I_6(\beta)+\right.$$

$$\left.+\frac{8}{3}\frac{1}{2\beta}I_7(\beta)+\frac{2}{15}I_8(\beta)\right],$$

$$I_l(\beta)=\frac{2\Gamma(l)}{(2\beta+1)^l}+\frac{3\Gamma(l-1)}{(2\beta+1)^{l-1}}+\frac{6\Gamma(l-2)}{(2\beta+1)^{l-2}}-\frac{2\xi^2\Gamma(l)}{(2\beta+\xi)^l}-$$

$$-\frac{3\xi\,\Gamma(l-1)}{(2\beta+\xi)^{l-1}}-\frac{6\,\Gamma(l-2)}{(2\beta+\xi)^{l-2}},$$

$$F_4(a)=\frac{3}{64}\,\frac{a^3}{(1+a)^5}\,[35\,a^3+47\,a^2+25\,a+5],$$

$$A_4(a,b)=\frac{21\sqrt{6}}{16}\Big\{a^{9/2}\,b^{13/2}\Big[3A_1\Big(\frac{a+b}{2}\Big)+\frac{1}{3}A_2\Big(\frac{a+b}{2}\Big)+\frac{1}{24}A_3\Big(\frac{a+b}{2}\Big)\Big]-$$

$$-\xi\Big(\frac{a}{\xi}\Big)^{9/2}\Big(\frac{b}{\xi}\Big)^{13/2}\Big[3\xi^2A_1\Big(\frac{a+b}{2\xi}\Big)+\frac{1}{3}\xi A_2\Big(\frac{a+b}{2\xi}\Big)+$$

$$+\frac{1}{24}A_3\Big(\frac{a+b}{2\xi}\Big)\Big]\Big\},$$

$$A_1(a)=\frac{21\,a^3+19\,a^2+7\,a+1}{504\,a^6\,(1+a)^7},$$

$$A_2(a)=\frac{64\,a^3+69\,a^2+30\,a+5}{840\,a^6\,(1+a)^6},$$

$$A_3(a)=\frac{35\,a^3+47\,a^2+25\,a+5}{210\,a^6\,(1+a)^5},$$

$$B_4(a)=\frac{231\,a^{13}}{1024}\Big\{\frac{4\,[231\,a^3+159\,a^2+45\,a+5]}{3465\,a^6\,(1+a)^9}+$$

$$+\frac{5\,(231\,a^5+593\,a^4+686\,a^3+434\,a^2+147\,a+2)}{462\,a^{10}\,(1+a)^7}\Big\},$$

$$D_4(a)=\frac{a^5\,[231\,a^4+312\,a^3+186\,a^2+56\,a+7]}{2048\,(1+a)^8},$$

$$E(b)=\frac{21}{32}\Big\{b^{13}\Big(11\,E_1(b)+E_2(b)+\frac{1}{10}E_3(b)\Big)-\Big(\frac{b}{\xi}\Big)^{13}\xi\,[11\,\xi^2E_1(b/\xi)+$$

$$+\xi E_2(b/\xi)+\frac{1}{10}E_3(b/\xi)]\Big\},$$

$$E_1(b)=\frac{231\,b^3+159\,b^2+45\,b+5}{13860\,b^6\,(1+b)^9},$$

$$E_2(b)=\frac{32\,b^3+25\,b^2+8\,b+1}{1260\,b^6\,(1+b)^8},$$

$$E_3(b)=\frac{21\,b^3+19\,b^2+7\,b+1}{504\,b^6\,(1+b)^7},$$

$$K_4(b)=\frac{147}{64}\Big\{b^{13}\,[11\,K_1(b)+K_2(b)+\frac{1}{10}K_3(b)]-\Big(\frac{b}{\xi}\Big)^{13}\xi\,[11\,\xi^2K_1(b/\xi)+$$

$$+\xi K_2(b/\xi)+\frac{1}{10}K_3(b/\xi)]\Big\},$$

$$K_1(b) = \frac{231\,b^4+312\,b^3+186\,b^2+56\,b+7}{9240\,b^8\,(1+b)^8},$$

$$K_2(b) = \frac{128\,b^4+203\,b^3+141\,b^2+49\,b+7}{2520\,b^8\,(1+b)^7},$$

$$K_3(b) = \frac{64\,b^4+122\,b^3+102\,b^2+42\,b+7}{504\,b^8\,(1+b)^6}.$$

參 考 文 獻

[1] 張繼恒、彭桓武，中國物理學報 7 (1950)，339.
[2] 嚴肅，即將發表.
[3] Hulthén, L., *Kungl. Fysio. Säll. I. Lund. Förhand.* Bd. **14**, Nr. 8, Nr. 21 (1944). Bd. **15**, Nr. 22 (1945).
[4] Hulthén, L., *Arkiv. för. Math. Astro. Och. Fysik.* Bd. **35A**, Nr. 25 (1948).
[5] Christian, R. S., Hart, E. W., *Phys. Rev.* **77** (1950), 441.
[6] 彭桓武、唐懋荧，中國物理學報 7 (1950)，309.
徐躬耦、胡濟民，*Proc. Roy. Soc.* A. **204** (1951)，176.
[7] Gerjuoy, E., Schwinger, J., *Phys. Rev.* **61** (1942), 138.
[8] Clark, A. C., *Proc. Phys. Soc.* A. **67** (1954), 323.

A DISCUSSION ON TWO-RANGE NUCLEAR FORCE

Chou Kuang-chao

(Department of Physics, Peking University)

Abstract

The binding energies of nuclei H^3, He^3, He^4, the low energy n-p scattering length and effective range are calculated by using the standard variational methods. A two-range central Yukawa potential is considered in the first two sections. The longer range corresponds to the π meson mass. The smaller range corresponds either to the heavier meson or to the higher order field interactions. No repulsive core appears, when the force parameters are chosen to fit the low energy scattering and deuteron data. The calculated binding energies of nuclei H^3, He^3 and He^4 are too high. This result is in agreement with most of the previous calculations. Tensor force of the Schwinger mixed type is considered in the third and fourth sections. The force parameters are chosen to fit the low energy two body data. They are not uniquely determined and are given for a set of possible *D*-percentage values. The adding of the tensor force reduces considerably the calculated binding energies of nuclei H^3, He^3, He^4. But still, the calculated values increase too fast with the mass number. It does not fit the triton and helium binding energies simultaneously. The possibility of adding many body forces is discussed at the end of the paper.

介子場論中的三体力位能*

周 光 召

(物理系理論物理教研室)

一、引 言

直到最近，核子力的研究大部分都限制在二体力上。已經提出过許多种这样的核子力的形式，它們很好地解釋了由兩个核子所組成体系的低能散射和結合現象。但只要是拿它們來計算 ^{3}H, ^{3}He 及 ^{4}He 諸核的結合能時，計算出來的結合能值總有隨着核子數 A 上昇得过快的趨势〔1〕。到現在为止，还沒有一个只包含二体力的位能可以消除这样一个困难，那怕所选擇的位能具有複雜的形狀和帶有張量力也都不能成功。同時由二核子高能散射实驗所確定的二体力不滿足核力飽和的要求，因此不能解釋重核的飽和現象〔2〕。所有这些事实都給我們一种暗示，那就是当核子數目增加時，可能出現具有新質的多体力，正是由於多体力的作用才導至重核的飽和現象及 ^{3}H, ^{3}He 及 ^{4}He 諸核的結合。

从介子場的理論出發來計算三体力最早由 Primakoff Holstein, 及 Jánossy〔3〕等人作过。在当時，人們还不知道 π 介子是核力的主要傳遞者，也不知道 π 介子是贋标量介子，因此在他們的計算中所採用的是向量場而非贋标量場，这样一來，所得到的結果便与 π 介子沒有什麼联系。

最近从場論中推導三体力的工作有 Wentgel, Drell 及 Huang〔4〕等人作过。他們採用的是贋标介子贋标耦合。对於贋标介子贋标耦合，三体力位能中最大的那一部分是由產生虛核子对而引起的。这一部分位能已經証明了和自旋及同位旋無關，而且是一个排斥力位能。这样一种性質正是我們所需要的，利用它很成功地解釋了重核飽和的現象。但是把目前所有關於介子与核子相互作用的实驗及理論綜合起來考察一下，使我們認为在上述三体力的工作中仍存在有一些問題。根据 π 介子对核子散射的实驗，贋标耦合中產生虛核子对的效应並不重要。从場論本身分析起來，由於場的輻射阻尼作用，也可能使得產生虛核子对的幾率大大減少〔5〕。这样一來，在

* 本文 1955 年 9 月 5 日收到。

膺标介子膺标耦合的三体力位能中的主要部分因輻射阻尼作用而减少很多，利用它便不再能解釋重核的飽和現象了。此外，由膺标介子膺标耦合所導出的二体力位能，直到第四級近似也还不能解釋二核子低能散射及結合的現象。

在討論 π 介子与核子的散射，光致介子的產生等問題時，常常採用膺标介子膺向量耦合的理論。从場論的角度來考察这个理論時，我們遇到發散的困难，这种困难除非採用特殊的办法，是不能利用質量和介子电荷重正化的条件消去的。由於發散的困难在高頻率虛介子处發生，通常採用截除高頻率虛介子的办法來消除發散的困难。这种方法当然是帶有隨意性的。但高頻率虛介子只影响到高能量核子的运動以及和核子極隣近的那个區域內的情況，因此只要核子的能量不太高（在非相对論範圍之內），或者幾个核子不太靠近，那麽卽使採用了这样隨意的截除办法，所得到的結果或許还可以看作真实情形的初步近似。Chew〔6〕利用这个理論成功地解釋了單核子与 π 介子相互作用的許多現象。其中包括 π 介子对核子的散射，光致 π 介子產生，核子的反常磁矩等等。由这个理論導出的二体力位能，当計算到第四級近似，並經过 Bruckner 及 Watson〔7〕的修改以後，也可以大体上解釋二核子低能散射及結合現象。

本文的目的主要就在於要計算膺标介子膺向量耦合理論中的三体力，看一看这个理論除了能够成功地解釋單核子及双核子的現象外，还能不能对解釋 ^{3}H, ^{3}He 和 ^{4}He 諸核的結合能有所帮助。在本文的第二節中，採用了三次正則变換，这样求得的核子力相当於用微擾法作到第六級近似時所得的結果。在第三節中討論了变換後的漢密頓函數中各項对三体力的貢献。第四節具体求出了对称型膺标介子膺向量耦合和对称型标量介子标量耦合情況下的三体力位能。求得的三体力位能在膺标介子膺向量耦合情況下不但有中心力部份，还有張量力部份，且具有較複雜的形式。第五節定性地討論了三体力的中心力部分对 ^{3}H, ^{4}He 和 ^{3}He 基态結合能的貢献。卽使是这一部分也相当複雜，在 $kr_{ij}=1$ 的附近，我們發現三体力的作用在膺标介子膺向量耦合情況下是一个不大的吸引力。这种吸引力对解釋 H^3, ^{3}He 和 ^{4}He 核的結合能是沒有好处的。此外在位能中还出現一接觸的排斥力位能，由於它牽涉到核子極端靠近的區域內的情形，在这一區域內更高級近似的核子力和重介子可能起主要作用，因而我們对它不能作出什麽結論。Levy 提出这一种可能，在他的理論中这种接觸的排斥力位能会發展成为一个有限大小的排斥心。这样一个排斥心有可能消除理論的 ^{3}H, ^{3}He 和 ^{4}He 諸核結合能过大的困难。在本文中还附帶地計算了第六級近似中的四体力位能，並在上述兩种耦合情况下給出了它的具体形式。

二、場的方程式同正則變換

选擇單位 h＝C＝1，我們在作用表象中把場的方程式寫为

$$i\frac{\delta\Omega(\sigma)}{\delta\sigma}=H(x)\Omega(\sigma) \tag{1}$$

其中，Ω 代表核子場和介子場的波動汎函，σ 代表在四度時空中具有空間性質的曲面。$H(x)$ 为在作用表象中核子場与介子場之間的相互作用能量：

$$H(x)=\overline{\psi}(x)O_L\psi(x)\phi_L(x) \tag{2}$$

ψ 和 φ_L 分别代表核子場及介子場在作用表象中的波函數算符。O_L 表徵出核子与介子間的耦合，对膺标介子膺向量耦合，我們有

$$O_L=+i\frac{g}{k}\tau_L\gamma_5\gamma_\mu\frac{\partial}{\partial x_\mu} \tag{3}$$

在对称型标量介子标量耦合中，我們有

$$O_L=f\tau_L\gamma_4 \tag{4}$$

f 和 g 表徵出耦合的强度，γ_μ 为核子的狄拉克算符，τ_L 則是核子的同位旋算符。$\frac{1}{k}$ 代表介子的康甫頓波長。

(1) 式中的 $H(x)$ 对於介子的數目表象並不是对角線矩陣，我們通过三次正則变换，把在不同近似中出現的新漢密頓量的对角線部分与非对角線部分分開，从这些对角線部分中便可以求得直到第六次近似的核子方位能。

1. 第一次正則变换第

令
$$\Omega(\sigma)=\mathsf{U}\,\Omega_1(\sigma)=\exp\{-i\int^\sigma H(x')dx'\}\Omega_1(\sigma) \tag{5}$$

$$\mathsf{U}=\exp(-i\int^\sigma H(x')dx') \tag{6}$$

U 为一么正算符，經过 U 这个变换以後，波動汎函 $\Omega(\sigma)$ 变成为 $\Omega_1(\sigma)$，$\Omega_1(\sigma)$ 所適合的方程式为：

$$i\frac{\delta\Omega_1(\sigma)}{\delta\sigma}=\left(\mathsf{U}^{-1}H(x)\mathsf{U}-\mathsf{U}^{-1}i\frac{\delta}{\delta\sigma}\mathsf{U}\right)\Omega_1(\sigma) \tag{7}$$

利用公式

$$e^{-S}Le^{S}=L+[LS]+\frac{1}{2!}[[LS]S]+\frac{1}{3!}[[[LS]S]S]+\cdots\cdots \tag{8}$$

可以把 (7) 式中右边的兩項計算出來。在保留到 g^6 近似時，計算出的結果可以通过

下式表示出來

$$i\frac{\delta\Omega_1(\sigma)}{\delta\sigma}=(H_2+H_3+H_4^A+H_5+H_6^A)\Omega_1(\sigma) \tag{9}$$

其中

$$H_2(x)=-\frac{i}{2}\Big[H(x),\int^{\sigma}H(x')dx'\Big]$$

$$H_3(x)=-\frac{1}{3}\Big[\Big[H(x),\int^{\sigma}H(x')dx'\Big]\int^{\sigma}H(x'')dx''\Big]$$

$$H_4^A(x)=\frac{i}{8}\Big[\Big[[H(x),\int^{\sigma}H(x')dx']\int^{\sigma}H(x'')dx''\Big]\int^{\sigma}H(x''')dx'''\Big]$$

$$H_5(x)=\frac{1}{30}\Big[\Big[\Big[[H(x)\int^{\sigma}H(x')dx]\int^{\sigma}H(x'')dx''\Big]\int^{\sigma}H(x''')dx'''\Big]\int^{\sigma}H(x^{iv})dx^{iv}\Big]$$

$$H_6^A(x)=-\frac{i}{144}\Big[\Big[\Big[\Big[[H(x),\int^{\sigma}H(x')dx']\int^{\sigma}H(x'')dx''\Big]\int^{\sigma}H(x''')dx'''\Big]\int^{\sigma}H(x^{iv})dx^{iv}\Big]\int^{\sigma}H(x^{v})dx^{v}\Big] \tag{10}$$

在 (10) 式中間，$H_5(x)$ 对在第六級近似中出現的三体力沒有影响，我們可以將它忽略掉。

2. 第二次正則变換

在做第二次正則变換以前，我們引進新的符号 H_D 及 $H_{N.D.}$，它們分别代表 H 在介子數目的表象中的对角線和非对角線部分。可以看到，H_2 中的对角線部分 H_{2D} 給出第二級的核子力，H_2 的非对角線部分 $H_{2N.D.}$ 中包含有同時吸收或放出兩个介子，吸收一个又放出一个介子等过程在內，它对第四級和第六級的核子力有貢献。

令
$$\Omega_1(\sigma)=U_1\Omega_2(\sigma)=\exp\Big(-i\int^{\sigma}H_{2N.D.}(x)dx\Big)\Omega_2(\sigma) \tag{11}$$

$$U_1=\exp\Big(-i\int^{\sigma}H_{2N.D.}(x)dx\Big) \tag{12}$$

U_1 是一个么正算符。經过这个变換，波動汎函 $\Omega_1(\sigma)$ 变成了 $\Omega_2(\sigma)$，$\Omega_2(\sigma)$ 所適合的方程式，当只保留到 g^6 近似且去掉那些与核子力無關的部分後，变为

$$i\frac{\delta\Omega_2(\sigma)}{\delta\sigma}=\{H_{2.D.}+H_3+H_4^A+H_4^B+H_4^C+H_6^A+H_6^B+H_6^C+H_6^D\}\Omega_2(\sigma) \tag{13}$$

其中，$H_{2.D.}(x)$，$H_3(x)$，$H_4^A(x)$ 及 $H_6^A(x)$ 和 (10) 式中所給出的相同。

而
$$H_4^B(x)=\frac{i}{8}\Big[[H(x),\int^{\sigma}H(x')dx']_D\int^{\sigma}[H(x''),\int^{\sigma''}H(x''')dx''']_{N.D.}dx''\Big]$$

$$H_4^C(x)=\frac{i}{8}\Big[[H(x),\int^\sigma H(x')dx']_{N.D.}\int^\sigma [H(x''),\int^{\sigma''} H(x''')dx''']_{N.D.}dx''\Big]$$

$$H_6^B(x)=-\frac{i}{16}\Big[\Big[[H(x),\int^\sigma H(x')dx']_D,\int^\sigma [H(x''),\int^{\sigma''} H(x''')dx''']_{N.D.}dx''\Big]$$
$$\int^\sigma [H(x^{iv}),\int^{\sigma^{iv}} H(x^v)dx^v]_{N.D.}dx^{iv}\Big]$$

$$H_6^C(x)=-\frac{i}{24}\Big[\Big[[H(x),\int^\sigma H(x')dx']_{N.D.},\int^\sigma [H(x''),$$
$$\int^{\sigma''} H(x''')dx''']_{N.D.}dx''\Big]\int^\sigma [H(x^{iv}),\int^{\sigma^{iv}} H(x^v)dx^v]_{N.D.}dx^{iv}\Big]$$

$$H_6^D(x)=-\frac{i}{16}\Big[\Big[\Big[[H(x),\int^\sigma H(x')dx'\Big]\int^\sigma H(x'')dx''\Big]\int^\sigma H(x''')dx'''\Big]$$
$$\int^\sigma [H(x^{iv}),\int^{\sigma^{iv}} H(x^v)dx^v]_{N.D.}dx^{iv}\Big] \tag{14}$$

在 (13) 式中，除了 H_3 以外的所有各項 H 的非对角線部分对第六級近似的核子力都沒有貢献，H_3 中包含有同時放射，吸收或是散射三个介子的过程，它对第六級核子力的貢献可以再一次用正則变换計算出來。

3. 三次正則变换

令
$$\Omega_2(\sigma)=\mathsf{U}_2\Omega_3(\sigma)=\exp(-i\int^\sigma H_3dx) \tag{15}$$

$$\mathsf{U}_2=\exp(-i\int^\sigma H_3(x)dx) \tag{16}$$

U_2 是一么正算符。經过这一变換，波動汎函 $\Omega_2(\sigma)$ 变成了 $\Omega_3(\sigma)$。当只保留到 g^6 近似，且去掉那些与核力無關的項時 $\Omega_3(\sigma)$ 所適合的方程式可寫成:

$$i\frac{\delta\Omega_3(\sigma)}{\delta\sigma}=\{H_{2.D.}+H_4^A+H_4^B+H_4^C+H_6^A+H_6^B+H_6^C+H_6^D+H_6^E\}\Omega_3(\sigma) \tag{17}$$

其中，$H_{2.D.}(x)$，$H_4^A(x)$，$H_4^B(x)$，$H_4^C(x)$，$H_6^A(x)$，$H_6^B(x)$，$H_6^C(x)$ 及 $H_6^D(x)$ 都和 (10) 式 (14) 式中所給出的相同。而 $H_6^E(x)$ 則是

$$H_6^E(x)=-\frac{i}{18}\Big[\Big[[H(x),\int^\sigma H(x')dx'\Big]\int^\sigma H(x'')dx''\Big],\int^\sigma [H(x'''),$$
$$\int^{\sigma'''} H(x^{iv})dx^{iv}]\int^{\sigma'''} H(x^v)dx^v\Big]dx'''\Big] \tag{18}$$

在 (17) 式中，$H_{2.D.}$ 这一項中包含了第二級近似的核子力 H_4^A，H_4^B 及 H_4^C 的对角線部分包含了第四級近似的核子力，而 H_6^A，H_6^B，H_6^C，H_6^D 及 H_6^E 等項的对角線部分則包含了第六級近似的核子力。在下一節中，我們將分別討論 (17) 式中各項对核子力的貢献。

三、核子間的三体力位能

用上節的正則变換所求得位能，相当於按 g^2 展開，展開式的每一項仍是一个推遲的位能，为要求得一个只由核子坐标決定的静位能，我們需要把推遲位能按 $\frac{v}{c}$ 展開，其中 v 为核子速度。我們知道在原子核中，$\frac{v}{c}$ 和 $\frac{g^2}{hc}$ 的數量級是差不多的。因此当我們要求得 $\left(\frac{g^2}{hc}\right)^3\left(\frac{v}{c}\right)^0$ 这一數量級的三体力位能，便同時要把 g^2 和 g^4 項內的推遲势展開到 $\frac{g^2}{hc}\left(\frac{v}{c}\right)^2$ 及 $\left(\frac{g^2}{hc}\right)^2\frac{v}{c}$ 这一數量級。

介子場的波函數算符 φ_L 適合对易關係

$$[\varphi_L(x),\varphi_M(x')]=-i\delta_{LM}\Delta(x-x') \tag{19}$$

如果令 φ_L^+ 及 φ_L^- 分别代表 φ_L 的正頻率及負頻率部份，它們之間適合对易關係

$$[\varphi_L^+(x),\varphi_M^-(x')]=-i\delta_{LM}\Delta+(x-x') \tag{20}$$

其中

$$\Delta+(x)=\frac{1}{2}(\Delta(x)+i\Delta^{(1)}(x))$$

$$\Delta(x)=\frac{1}{(2\pi)^3}\iiint\frac{dk}{k_0}\sin k_0x_0e^{ik.r} \tag{21}$$

$$\Delta^{(1)}(x)=\frac{1}{(2\pi)^3}\iiint\frac{dk}{k_0}\cos k_0x_0e^{ik.r}$$

$$k_0^2=k^2+\kappa^2 \qquad x_0=t$$

在計算中，可以利用在 $\frac{v}{c}$ 小的時候，$\Delta(x-x')$ 所具有的以下的關係式

$$\int^{\sigma}f(x')\Delta(x-x')dx'\propto\int^{x_0}f(r',t')\Delta(x-x')dx_0'=\frac{e^{-R\sqrt{\kappa^2+\left(\frac{d}{dx_0}\right)^2}}}{4\pi R}f(r',x_0) \tag{22a}$$

其中 $R=|r-r'|$，$\frac{d}{dx_0}$ 作用在後面的 $f(r',x_0)$ 上。如果將上式展開到 $\frac{d^2f}{dx_0^2}$ 的數量級（相当於展開到 $\left(\frac{v}{c}\right)^2$ 的數量級），我們得到

$$\begin{aligned}\frac{e^{-R\sqrt{\kappa^2+\frac{d^2}{dx_0^2}}}}{4\pi R}f(r'x_0)&=\frac{e^{-\kappa R}}{4\pi R}f(r'x_0)-\frac{e^{-\kappa R}}{8\pi\kappa}\frac{d^2}{dx_0^2}f(r'x_0)\\&=\frac{e^{-\kappa R}}{4\pi R}f(r'x_0)+\frac{e^{-\kappa R}}{8\pi\kappa}[H_0[H_0f]]\end{aligned} \tag{22b}$$

在我們的运算中所出現的 f 都是在作用表象中的算符，这些算符隨時間的变化完全由自由核子及自由場的漢密頓量 H_0 來決定。

此外，在計算那些只需要近似到 $\left(\frac{v}{c}\right)^0$ 的項時，可以採用下列幾个關係式以簡化

运算。它們是

$$\int^{\sigma} dx' \int^{\sigma} dx'' \Delta(x' - x'') \alpha \int^{t} dt' \int^{t} dt'' \sin k_0(t' - t'') = 0$$

$$\int^{\sigma} dx' \Delta^{(1)}(x - x') \alpha \int^{t} dt' \cos k_0(t - t') = 0 \tag{23}$$

最後我們还注意到，所有在約化狄拉克方程式到$\left(\frac{v}{c}\right)^2$近似時，由正能級过渡到負能級所引起的三体力位能最少是$g^4\left(\frac{v}{c}\right)^2$的數量級。这些項在我們的近似中已經可以忽略。

利用上面这样一些公式就可以把(17)式中的各項对三体力的貢獻算出來。很容易証明，当只計算到g^6近似時，$H_{3.D.}, H_4^A, H_4^B, H_4^C, H_6^A, H_6^B, H_6^C$及$H_6^D$对核子的三体力位能都沒有貢獻。有貢獻的只有$H_6^E$一項。我們現在举$H_6^D$为例，証明它確实对三体力沒有貢獻。先將$H_6^D$中的兩部分算出來，得到

$$\int^{\sigma} [H(x^{iv}), \int^{\sigma iv} H(x^v) dx^v]_{N.D.} dx^{iv}$$
$$= \int^{\sigma} dx^{iv} \int^{\sigma iv} dx^v [\bar{\psi}(x^{iv}) O_E \psi(x^{iv}), \bar{\psi}(x^v) O_F \psi(x^v)] \varphi_E(x^{iv}) \varphi_F(x^v) \tag{24}$$

$$\left[\left[[H(x), \int^{\sigma} H(x') dx'] \int^{\sigma} H(x'') dx''\right] \int^{\sigma} H(x''') dx'''\right]$$
$$= \int^{\sigma} dx' \int^{\sigma} dx'' \int^{\sigma} dx''' \Big\{ \Big[\Big[[\bar{\psi}(x) O_L \psi(x), \bar{\psi}(x') O_M \psi(x')] \bar{\psi}(x'') O_P \psi(x'')\Big]$$
$$\bar{\psi}(x''') O_Q \psi(x''')] \varphi_L(x) \varphi_M(x') \varphi_P(x'') \varphi_Q(x''')$$
$$+ \Big[[\bar{\psi}(x) O_L \psi(x), \bar{\psi}(x') O_M \psi(x')] \bar{\psi}(x'') O_P \psi(x'')\Big] \bar{\psi}(x''') O_Q \psi(x''')$$
$$\Big(-i\delta_{LQ} \Delta(x - x''') \varphi_M(x') \varphi_P(x'') - i\delta_{MQ} \Delta(x' - x''') \varphi_L(x) \varphi_P(x'')$$
$$-i\delta_{PQ} \Delta(x'' - x''') \varphi_L(x) \varphi_M(x')\Big) + \Big([\bar{\psi}(x) O_L \psi(x), \bar{\psi}(x') O_M \psi(x')]$$
$$[\bar{\psi}(x'') O_P \psi(x''), \bar{\psi}(x''') O_Q \psi(x''')] + \Big[[\Psi(x) O_L \psi(x), \Psi(x') O_M \psi(x')]$$
$$\bar{\psi}(x''') O_Q \psi(x''')] \bar{\psi}(x'') O_P \psi(x'')\Big) \Big(-i\delta_{LP} \Delta(x - x'' \varphi_M(x') \varphi_Q(x''')$$
$$-i\delta_{MQ} \Delta(x' - x'') \varphi_L(x) \varphi_Q(x''')\Big) + [\bar{\psi}(x) O_L \psi(x), \bar{\psi}(x') O_M \psi(x')]$$
$$\bar{\psi}(x'') O_P \psi(x'') \bar{\psi}(x''') O_Q \psi(x''') \Big(-\delta_{LP} \delta_{MQ} \Delta(x - x'') \Delta(x' - x''')$$
$$-\delta_{LQ} \delta_{MP} \Delta(x' - x'') \Delta(x - x''')\Big) + [\bar{\psi}(x) O_L \psi(x) \bar{\psi}(x') O_M \psi(x'),$$
$$\bar{\psi}(x'') O_P \psi(x'')] \bar{\psi}(x''') O_Q \psi(x''') \Big(-\delta_{LM} \delta_{PQ} (\Delta x - x') \Delta(x'' - x''')\Big)$$
$$+ \Big[[\mathring{\bar{\psi}}(x) O_L \psi(x) \bar{\psi}(x') O_M \psi(x'), \bar{\psi}(x'') O_P \psi(x'')] \bar{\psi}(x''') O_Q \psi(x''')\Big]$$
$$\Big(-i\delta_{LM} \Delta(x - x') \varphi_Q(x''') \varphi_P(x'')\Big) \tag{25}$$

利用(22)式，可以看到(25)式中有許多項等於零，其中还有一些項和(24)式中各項合起來也不会給出三体力來，这样一些項可以先去掉。經过適当的簡化以後，(25)式可以寫为

$$
\begin{aligned}
&\left[\left[[H(x),\int^{\sigma} H(x')dx']\int^{\sigma} H(x'')dx''\right]\int^{\sigma} H(x''')dx'''\right] \\
&\quad = -i\int^{\sigma} dx'\int^{\sigma} dx''\int^{\sigma} dx'''\delta_{LP}\Delta(x-x'')\{[\bar{\psi}(x)O_L\psi(x),\bar{\psi}(x')O_M\psi(x')] \\
&\quad [\bar{\psi}(x'')O_P\psi(x''),\bar{\psi}(x''')O_Q\psi(x''')]\Big((2\varphi_M(x')\varphi_Q(x''')+\varphi_Q(x''')\varphi_M(x')\Big) \\
&\quad +\Psi(x)O_L\psi(x)\Big[[\bar{\psi}(x'')O_P\psi(x''),\bar{\psi}(x')O_M\psi(x')]\bar{\psi}(x''')O_Q\psi(x'''\Big] \\
&\quad \varphi_Q(x''')\varphi_M(x')+\Big[[\bar{\psi}(x)O_L\psi(x),\bar{\psi}(x')O_M\psi(x')]\bar{\psi}(x''')O_Q\psi(x''')\Big] \\
&\quad \bar{\psi}(x'')O_P\psi(x'')\Big(2\varphi_M(x')\varphi_Q(x''')+\varphi_Q(x''')\varphi_M(x')\Big)
\end{aligned}
\tag{26}
$$

把(24)式及(26)式代回(14)式的 $H_6^D(x)$ 中，並注意到

$$
\begin{aligned}
&\Big([\varphi_M(x')\varphi_Q(x'''),\varphi_E(x^{iv})\varphi_F(x^{v})]\Big)_D \\
&\quad = -\delta_{ME}\delta_{QF}\Big(\Delta(x'-x^{iv})\Delta_+(x^{v}-x''')+\Delta_+(x'-x^{iv})\Delta(x'''-x^{v})\Big) \\
&\quad -\delta_{MF}\delta_{QE}\Big((\Delta(x'-x^{v})\Delta_+(x^{iv}-x''')+\Delta_+(x'-x^{v})\Delta(x'''-x^{iv})\Big)
\end{aligned}
\tag{27}
$$

根据(27)式，將下式对 t 積分之後，很容易証明

$$
\int^t dt'\int^t dt''\int^t dt'''\int^t dt^{iv}\int^{t^{iv}} dt^{v}\Delta(x-x'')[\varphi_M(x')\varphi_Q(x'''),\varphi_E(x^{iv})\varphi_F(x^{v})]_D=0
$$

同样可以証明

$$
\int^t dt'\int^t dt''\int^t dt'''\int pt^{iv}\int^{t^{iv}} dt^{v}\Delta(x-x'')[\varphi_Q(x''')\varphi_M(x'),\varphi_E(x^{iv})\varphi_F(x^{v})]_D=0
$$

这样一來，我們便得到了 $H_6^D=0$ 的結果。在对時間 t 積分時，我們沒有考慮核子場波函數算符 ψ 隨時間的变换，这是因为我們只要求計算到 $g^6\left(\frac{v}{c}\right)^0$ 近似的原故。H_6^D 对三体力位能虽沒有貢献，但应当順帶地指出，它对第六級的二体力位能仍然是有貢献的。

完全同样的考慮，可以証明 $H_{2,D.}$, H_4^A, H_4^B, H_4^C, H_6^A, H_6^B 及 $H_{6.}^C$ 对三体力沒有貢献。所有的第四級 H 对三体力的靜位能沒有貢献，說明了最低數量級的靜三位力位能应当从 g^6 項開始才有。

現在讓我們把 H_6^E 中的三体力計算出來。計算的方法和上面計算時沒有什麽不同的地方，我們在此处不去把所有的步驟都詳細的寫出來。当只保留那些与三体力有關的項時，計算的結果为

$$H_6^E(x)=\frac{1}{18}\int^{\sigma}dx'\int^{\sigma}dx''\int^{\sigma}dx'''\int^{\sigma'''}dx^{iv}\int^{\sigma'''}dx^{v}\Big\{\Big[[\bar\psi(x)O_L\psi(x),\bar\psi(x')O_M\psi(x')]$$

$$\bar\psi(x'')O_P\psi(x'')]\Big(2[\bar\psi(x''')O_Q\psi(x'''),\bar\psi(x^{iv})O_E\psi(x^{iv})]\bar\psi(x^v)O_F\psi(x^v)$$

$$-\bar\psi(x''')O_Q\psi(x''')[\bar\psi(x^{iv})O_E\psi(x^{iv}),\bar\psi(x^v)O_F\psi(x^v)]\Big)\Big(\delta_{LM}\delta_{PE}\delta_{QF}\Delta_+(x-x')$$

$$\Delta(x''-x^{iv})\Delta(x'''-x^v)+\delta_{LP}\delta_{ME}\delta_{QF}\Delta_+(x-x'')\Delta(x'-x^{iv})\Delta(x'''-x^v)\Big)$$

$$+\Big(2[\bar\psi(x)O_L\psi(x),\bar\psi(x')O_M\psi(x')]\bar\psi(x'')O_P\psi(x'')-\bar\psi(x)O_L\psi(x)$$

$$[\bar\psi(x')O_M\psi(x')\bar\psi(x'')O_P\psi(x'')]\Big)\Big[[\bar\psi(x''')O_Q\psi(x'''),\bar\psi(x^{iv})O_E\psi(x^{iv})]$$

$$\bar\psi(x^v)O_F\psi(x^v)]\Big(\delta_{LP}\delta_{ME}\delta_{QF}\Delta(x-x'')\Delta(x'-x^{iv})\Delta_+(x'''-x^v)$$

$$+\delta_{LP}\delta_{MF}\delta_{QE}\Delta(x-x'')\Delta(x-x'')\Delta(x'-x^v)\Delta_+(x'''-x^{iv})\Big)$$

$$+\Big[\Big(2[\bar\psi(x)O_L\psi(x),\bar\psi(x')O_M\psi(x')]\bar\psi(x'')O_P\psi(x'')-\bar\psi(x)O_L\psi(x)$$

$$[\bar\psi(x')O_M\psi(x'),\bar\psi(x'')O_P\psi(x'')]\Big),\Big(2[\bar\psi(x''')O_Q\psi(x'''),\bar\psi(x^{iv})O_E\psi(x^{iv})]$$

$$\bar\psi(x^v)O_F\psi(x^v)-\bar\psi(x''')O_Q\psi(x''')[\bar\psi(x^{iv})O_E\psi(x^{iv}),\bar\psi(x^v)O_F\psi(x^v)]\Big)\Big]$$

$$\delta_{LP}\delta_{ME}\delta_{QF}\Delta(x-x'')\Delta_+(x^{iv}-x')\Delta(x'''-x^v)+\Big(2[\bar\psi(x)O_L\psi(x),$$

$$\psi(x')O_M\bar\psi(x')\psi(x')]\bar\psi(x'')O_P\psi(x'')-\bar\psi(x)O_L\psi(x)[\bar\psi(x')O_M\psi(x'),$$

$$\bar\psi(x'')O_P\psi(x'')]\Big)\Big(2[\bar\psi(x''')O_Q\psi(x'''),\bar\psi(x^{iv})O_E\psi(x^{iv})]\bar\psi(x^v)O_F\psi(x^v)$$

$$-\bar\psi(x''')O_Q\psi(x''')[\bar\psi(x^{iv})O_E\psi(x^{iv}),\bar\psi(x^v)O_F\psi(x^v)]\Big)$$

$$\delta_{LP}\delta_{ME}\delta_{QF}\Delta(x-x'')\Delta(x'-x^{iv})\Delta(x'''-x^v) \qquad (28)$$

我們先把对時間的積分求出來，得到

$$\int^t dt'\int^t dt''\int^t dt'''\int^{t'''}dt^{iv}\int^{t'''}dt^v\Delta(x-x^v)\Delta(x'-x^{iv})\Delta(x'''-x^v)$$

$$=\frac{1}{(2\pi)^9}\iiint\cdots\int\frac{d\boldsymbol{k}\,d\boldsymbol{l}\,d\boldsymbol{m}}{k_0^2\ l_0^4\ m_0^2}e^{i\boldsymbol{k}\cdot(\boldsymbol{r}-\boldsymbol{r}'')}e^{i\boldsymbol{l}\cdot(\boldsymbol{r}'-\boldsymbol{r}^{iv})}e^{i\boldsymbol{m}\cdot(\boldsymbol{r}'''-\boldsymbol{r}^v)} \qquad (29)$$

(29) 式是一个对 $\boldsymbol{r}$ 及 $\boldsymbol{r}''$，$\boldsymbol{r}'$ 及 $\boldsymbol{r}^{iv}$，$\boldsymbol{r}'''$ 及 $\boldsymbol{r}^v$ 对称的函數。如果在 (29) 中我們把 $\boldsymbol{r}$，$\boldsymbol{r}'$ 和 $\boldsymbol{r}''$ 去跟 $\boldsymbol{r}'''$，$\boldsymbol{r}^{iv}$ 和 $\boldsymbol{r}^v$ 对換，可以看到 (29) 式在这个对換中是不变的。这样一种对称的性質可以帮助我們大大簡化(28)式。(28)式的第一項和第二項中都包含有發散積分。利用对称性質可以看到第三項的貢献是零。第四項中的三体力位能有一部份包含發散積分，但也有一部份不包含發散積分。我們先來計算 (28) 式的第四項中不包含發散積分的那一部份。利用对称的性質，可以使弧形括弧中的兩項合併起來。

在求三体力位能時，我們需要利用福克的方法，从核子場的方程式回到幾个核子所適合的方程式。此時，我們只需要在 (28) 式中，把 $\Psi(x)O_L\psi(x)$ 換之以

$$\sum_j \overline{O}_L^{(j)}\delta(\boldsymbol{r}-\boldsymbol{r}_j)=\sum_j \gamma_4^{(j)}O_L^{(j)}\delta(\boldsymbol{r}-\boldsymbol{r}_j) \tag{30}$$

其中所有帶着指标 j 的物理量均屬於第 j 个核子。並把代換後的 $H(x)$ 对積分就可以求出核子間的位能了。(28)式的第四項中与發散積分無關的靜三体力位能部份很容易計算出來，

$$V_{3A}=\frac{1}{2}\sum_{i\neq j\neq k}\Big\{[\overline{O}_L^{(i)},\overline{O}_M^{(j)}]\overline{O}_L^{(k)}[\overline{O}_P^{(j)},\overline{O}_M^{(j)}]\overline{O}_P^{(k)}F(r_{ik})F(r_{jk})G(r_{ij})$$
$$+2[\overline{O}_L^{(i)},\overline{O}_M^{(i)}]\overline{O}_L^{(j)}[\overline{O}_P^{(j)},\overline{O}_M^{(j)}]\overline{O}_P^{(k)}F(r_{ij})F(r_{jk})G(r_{ij})\} \tag{31}$$

其中

$$F(r)=\frac{1}{4\pi}\frac{e^{-kr}}{r}$$

$$G(r)=\frac{1}{8\pi k}e^{-kr} \tag{32}$$

由於發散積分的發散程度与所採取的具体耦合的形式有密切的關係，因此 (28) 式中那些包含發散積分的項將留待下一節中去討論。

最後还値得提起的是，(28)式中包含了最低數量級的靜四体力位能，它的形式是

$$V_4=\frac{1}{2}\sum_{\substack{i\neq j\neq k\\ \neq l}}[\overline{O}_L^{(i)},\overline{O}_M^{(j)}]\overline{O}_L^{(l)}[\overline{O}_P^{(k)},\overline{O}_M^{(j)}]\overline{O}_P^{(l)}F(r_{ij})F(r_{kl})G(r_{ik}) \tag{33}$$

可以証明，(17)式中其他各項的 H 对靜四体力位能也都沒有貢献，因此(33)式的 V_4 是在第六級近似中出的全部的靜四体力位能。

四、膺标介子膺向量耦合与对称型标量介子标量耦合的靜三体力位能

在求膺标介子膺向量耦合的靜三体位能時，我們应把(30)式中的 $\overline{O}_L^{(j)}$ 代以

$$\overline{O}_L^{(j)}=\frac{g}{\kappa}\tau_L^{(j)}\boldsymbol{\sigma}^{(j)}\cdot\nabla_j \tag{34}$$

此時(31)式变成

$$V_{3A}=-\frac{1}{2}\left(\frac{g}{\kappa}\right)^6\sum_{i\neq j\neq k}\{[\tau_L^{(i)}\boldsymbol{\sigma}^{(i)}\cdot\nabla,\tau_M^{(i)}\boldsymbol{\sigma}^{(i)}\cdot\nabla']\tau_L^{(k)}\boldsymbol{\sigma}^{(k)}\cdot\nabla[\tau_P^{(j)}\boldsymbol{\sigma}^{(j)}\cdot\nabla'',$$
$$\tau_M^{(j)}\boldsymbol{\sigma}^{(j)}\cdot\nabla']\tau_P^{(k)}\boldsymbol{\sigma}^{(k)}\cdot\nabla''F(r)F(r'')G(r')_{r=r_{ik},\,r'=r_{ij},\,r''=r_{ik}}$$
$$+2[\tau_L^{(i)}\boldsymbol{\sigma}^{(i)}\cdot\nabla\tau_M^{(i)}\boldsymbol{\sigma}^{(i)}\cdot\nabla']\tau_L^{(j)}\boldsymbol{\sigma}^{(j)}\cdot\nabla[\tau_P^{(j)}\cdot\boldsymbol{\sigma}^{(j)}\cdot\nabla''\tau_M^{(j)}\boldsymbol{\sigma}^{(j)}\cdot\nabla']$$

$$\tau_P^{(k)}\boldsymbol{\sigma}^{(k)}\cdot\nabla''F(r)F(r'')G(r')\,|\,r=r_{ij},r'=r_i,r''=r_{jk}\} \tag{35}$$

把(35)式完全寫出來，得到一个很長的式子，我們不在此地寫出。对於对称型标量介子标量耦合，我們应把(30)式中的 $\overline{O}_L^{(j)}$ 代以

$$\overline{O}_L^{(j)}=f\tau_L\gamma_4 \tag{36}$$

在求静位能時，可以把 γ_4 算作 1。这样，便得到

$$V_{3A}=\frac{f^6}{2}\sum_{i\neq j\neq k}\{[\tau_L^{(i)},\tau_M^{(i)}]\tau_L^{(k)}[\tau_P^{(j)},\tau_M^{(j)}]\tau_P^{(k)}F(r_{ik})F(r_{jk})G(r_{ij})$$
$$+2[\tau_L^{(i)},\tau_M^{(i)}]\tau_L^{(j)}[\tau_P^{(j)},\tau_M^{(j)}]\tau_P^{(k)}F(r_{ij})F(r_{jk})G(r_{ij})\} \tag{37}$$

(37)式还可以簡化为

$$V_{3A}=-2f^6\{(\boldsymbol{\tau}^{(i)}\times\boldsymbol{\tau}^{(k)})\cdot(\boldsymbol{\tau}^{(j)}\times\boldsymbol{\tau}^{(k)})F(r_{ik})F(r_{jk})G(r_{ij})$$
$$+2(\boldsymbol{\tau}^{(i)}\times\boldsymbol{\tau}^{(j)})\cdot(\boldsymbol{\tau}^{(j)}\times\boldsymbol{\tau}^{(k)})F(r_{ij})F(r_{jk})G(r_{ij})\} \tag{38}$$

第六級静位能中，除了上面 V_{3A} 这一部分以外，还有包含發散積分的項。我們採用截除高頻率虛介子的方法消除这一困难。經过一番冗長的計算，对膺标介子膺向量耦合我們得到

$$V_{3B}=-\frac{64}{9}\left(\frac{g}{\kappa}\right)^6\sum_{i\neq j\neq k}\{(\boldsymbol{\tau}^{(i)}\cdot\boldsymbol{\tau}^{(j)}\boldsymbol{\tau}^{(j)}\cdot\boldsymbol{\tau}^{(k)}-2\boldsymbol{\tau}^{(i)}\cdot\boldsymbol{\tau}^{(k)})\boldsymbol{\sigma}^{(i)}\cdot\nabla\nabla\cdot\nabla'\boldsymbol{\sigma}^{(k)}\cdot\nabla'$$
$$+\boldsymbol{\tau}^{(i)}\cdot\boldsymbol{\tau}^{(k)}\boldsymbol{\sigma}^{(i)}\cdot\nabla\boldsymbol{\sigma}^{(j)}\cdot\nabla\boldsymbol{\sigma}^{(j)}\cdot\nabla\boldsymbol{\sigma}'^{(k)}\cdot\nabla'\}F_0F(r)F(r')\,|\,r=r_{ij}r'=r_{jk}$$
$$-\frac{2}{3}\left(\frac{g}{K}\right)^6\{(\boldsymbol{\tau}^{(i)}\cdot\boldsymbol{\tau}^{(j)}\boldsymbol{\tau}^{(i)}\cdot\boldsymbol{\tau}^{(k)}-4\boldsymbol{\tau}^{(j)}\cdot\boldsymbol{\tau}^{(k)})\boldsymbol{\sigma}^{(j)}\cdot\nabla\nabla\cdot\nabla'\boldsymbol{\sigma}^{(k)}\cdot\nabla'+(\boldsymbol{\tau}^{(j)}\cdot\boldsymbol{\tau}^{(k)}$$
$$-2\boldsymbol{\tau}^{(i)}\cdot\boldsymbol{\tau}^{(j)}\boldsymbol{\tau}^{(i)}\cdot\boldsymbol{\tau}^{(k)})\boldsymbol{\sigma}^{(k)}\cdot\nabla\boldsymbol{\sigma}^{(j)}\cdot\nabla\boldsymbol{\sigma}^{(i)}\cdot\nabla'\boldsymbol{\sigma}^{(k)}\cdot\nabla'F(r)F(r')G_0r=r_j r'=r_{ik} \tag{39}$$

其中

$$F_0=-\frac{1}{2\pi^2}\int_0^K\frac{k^4dk}{k^2+\kappa^2}=-\frac{1}{2\pi^2}\left(\frac{1}{3}K^2-\kappa^2K+\kappa^3\tan^{-1}\frac{K}{\kappa}\right)$$

$$G_0=-\frac{1}{2\pi^2}\int_0^K\frac{k^4dk}{(k^2+\kappa^2)^2}=-\frac{1}{2\pi^2}\left(K-\frac{3}{2}\kappa\tan^{-1}\frac{K}{\kappa}+\frac{\kappa^2}{2}\frac{K}{K^2+\kappa^2}\right) \tag{40}$$

K 为截除點的能量，它的大小和核子静止能量差不多。对於对称型标量介子标量耦合，我們得到

$$V_{3B}=\frac{32}{3}f^6\sum_{i\neq j\neq k}i\boldsymbol{\tau}^{(i)}\cdot(\boldsymbol{\tau}^{(j)}\times\boldsymbol{\tau}^{(k)})F_0'F(r_{jk})G(r_{ij})2$$
$$-f^6\sum_{i\neq j\neq k}(\boldsymbol{\tau}^{(i)}\times\boldsymbol{\tau}^{(j)})\cdot(\boldsymbol{\tau}^{(i)}\times\boldsymbol{\tau}^{(k)})G_0'F(r_{ij})F(r_{ik}) \tag{41}$$

其中　$$F_0'=\frac{1}{2\pi^2}\int_0^K\frac{k^2dk}{k^2+\kappa^2}=\frac{1}{2\pi^2}\left(K-\kappa\tan^{-1}\frac{K}{\kappa}\right)$$

$$G_6' = \frac{1}{2\pi^2}\int_0^{K} \frac{k^2 dk}{(k^2+\kappa^2)^2} = \frac{1}{4\pi^2}\left(\frac{1}{\kappa}\tan^{-1}\frac{K}{\kappa} + \frac{K}{K^2+\kappa^2}\right) \tag{42}$$

$$\longrightarrow \frac{1}{8\pi\kappa} \text{ 当 } K \longrightarrow \infty$$

把 V_{3A} 和 V_{3B} 合起來，我們便得到全部的靜三体力位能。第六級近似中的靜四体力位能对贗标介子贗向量耦合的情形可以寫成

$$V_4 = -\frac{1}{2}\left(\frac{g^2}{\kappa^2}\right)^3 \sum_{\substack{i\neq j\neq k\\ \neq l}} [\tau_L^{(i)}\boldsymbol{\sigma}^{(i)}\cdot\nabla, \tau_M^{(i)}\boldsymbol{\sigma}^{(l)}\cdot\nabla']\tau_L^{(j)}\boldsymbol{\sigma}^{(j)}\cdot\nabla[\tau_P^{(k)}\boldsymbol{\sigma}^{(k)}\cdot$$

$$\nabla'', \tau_M^{(l)}\boldsymbol{\sigma}^{(k)}\cdot\nabla']\tau_P^{(l)}\boldsymbol{\sigma}^{(l)}\cdot\nabla'' F(r)F(r'')G(r')\,|\, r=r_{ij}\, r'=r_{ik}\, r''=r_{kl} \tag{43}$$

对於对称型标量介子标量耦合，我們有

$$V_4 = -2\sum_{\substack{i\neq j\neq k\\ \neq l}} (\boldsymbol{\tau}^{(i)}\times\boldsymbol{\tau}^{(j)})\cdot(\boldsymbol{\tau}^{(k)}\times\boldsymbol{\tau}^{(l)})F(r_{ij})F(r_{kl})G(r_{ik}) \tag{44}$$

五、靜三体力位能对 ^{3}H, ^{3}He 及 ^{4}He 諸核基态結合能的貢献

我們在这裏只討論靜三体力位能中的中心力部分对上述諸核結合能的貢献。張力量部分的貢献过於複雜，我們不在此处討論。我們只是定性的來討論这个中心力，亦即確定它的作用相当於排斥，还是相当於吸引。此時，我們只需要找出靜三体位能的中心力对上述諸核基态上自旋及同位旋波函數的平均值便够了。由於 ^{3}H, ^{3}He 及 ^{4}He諸核在基态上的自旋同位旋波函數中最主要的那一部份对自旋及同位旋反对称。对於这三个核講，它們具有完全相同的結構，我們只需算出 ^{3}H 核基态上三体力位能的平均值便够了。^{3}H 核基态上的自旋同位旋波函數可寫成

$$|\chi\rangle = \frac{1}{\sqrt{2}}(\tilde{S}\bar{\tau} - \bar{S}\tilde{\tau}) \tag{45}$$

其中
$$\tilde{S} = \frac{1}{\sqrt{2}}\Big(\alpha(1)\beta(2) - \alpha(2)\beta(1)\Big)\alpha(3)$$

$$\bar{S} = \frac{1}{\sqrt{6}}\Big(2\alpha(1)\alpha(2)\beta(3) - \alpha(1)\alpha(3)\beta(2) - \alpha(2)\alpha(3)\beta(1)\Big)$$

$$\tilde{\tau} = \frac{1}{\sqrt{2}}\Big(N(1)P(2) - N(2)P(1)\Big)N(3) \tag{46}$$

$$\bar{\tau} = \frac{1}{\sqrt{6}}\Big(2N(1)N(2)P(3) - N(1)N(3)P(2) - N(2)N(3)P(1)\Big)$$

$\alpha(i)$,$\beta(i)$代表第i个核子的自旋波函數，$N(i)$和$P(i)$則代表第i个核子的同位旋波函數。我們現在要 $\overline{\langle\chi|V_3|\chi\rangle}$，求上面的一横代表$V_3$对空間角度的平均。計算出的結果，在膺标介子膺向量耦合情形下等於

$$\overline{\langle\chi|V_3|\chi\rangle}=\frac{g^6}{\kappa^6}\sum_{i\neq j\neq k}\Big\{24F_c(r_{jk})F_c(r_{jk})G_c(r_{ij})-\frac{10}{3}F_s(r_{ik})F_s(r_{jk})G_s(r_{ij})$$
$$-72F_c(r_{ij})F_c(r_{jk})G_c(r_{ij})+36F_s(r_{ij})F_c(r_{jk})G_s(r_{ij})+8F_c(r_{ij})F_c(r_{ik})G_0\Big\} \quad (47)$$

其中

$$F_c=-\frac{1}{3}\delta(r)+\frac{\kappa^2}{12\pi}\frac{e^{-\kappa r}}{r}$$

$$G_c=\frac{1}{12\pi}(\kappa r-2)\frac{e^{-\kappa r}}{r}$$

$$F_s=\frac{1}{4\pi}\Big(\frac{3}{r^2}+\frac{3\kappa}{r}+\kappa^2\Big)\frac{e^{-\kappa r}}{r}$$

$$G_s=\frac{1}{8\pi}\Big(\kappa+\frac{1}{r}\Big)e^{-\kappa r} \quad (48)$$

(47)式的符号隨着不同r的值有变化。我們先不考慮δ-函數的貢献，並在$r_{ij}=r_{jk}=r_{ki}$的情況下來討論(47)式的性質。可以看到，当$\kappa r<2$時，第一項，第二項和第五項为吸引力位能，第三項和第四項为排斥力位能。而在这一區域內吸引力部份大过排斥力部分。当$\kappa r>2$時，第二項，第三項和第五項为吸引力位能，而第一項和第四項为排斥力位能，在这一區域內排斥力部分逐漸增大，在$\kappa r=3$時就已超过吸引力部分很多。在(47)中δ-函數的貢献，則是相当於排斥的(因为第三項的貢献超过第一項和第五項的貢献)。由於^{3}H,^{3}He和^{4}He的結構較为緊密，波函數最大的部份在力程範圍之內，此時这些核子受到的三体力作用主要將是一个接觸的排斥作用和一个相互吸引的作用。根据 Levy 的理論，在考慮了核子反衝的效应後，这种接觸的排斥作用可以發展成为一个排斥心，因此不但二体力中具有排斥心，三体力中同样也可以有排斥心。只考慮二体力時所引起的^{3}H,^{3}He及^{4}He諸核理論結合能值过大的困难也可能因为有这种排斥心出現而被消除。

我們現在來估計(47)式中位能吸引力部份的大小。取$\kappa r=1$，並用常用單位表示出來，我們就可以得到

$$\overline{\langle\chi|V^3|\chi\rangle}\simeq-30\Big(\frac{g^2}{4\pi\hbar c}\Big)^3 m_\pi c^2 \quad (49)$$

在(49)式中，可以令 $\frac{g^2}{4\pi\hbar c}=0.1$，这个大小的數量級是由一些其他的实驗確定的〔6,7〕。此時得到

$$\overline{\langle\chi|V_3|\chi\rangle}\simeq-4 \text{ 兆电子伏} \quad (50)$$

一个二体力的位能，在$\kappa r=1$時差不多有15—20兆电子伏，与(50)式比較，我們可以

認为这个三体吸引力是不大的。如果排斥心的半徑够大，则就總的效果看來，有可能排斥力的影响超过吸引力的影响，而使得理論的結合能値減少，以得到符合实驗的結果。

对於对称型标量介子标量耦合的情形，我們有

$$\overline{<\chi|V_3|\chi>} = -f^6 \sum_{i \neq j \neq k} \{4F(r_{ik})F(r_{jk})G(r_{ij}) - 8F(r_{ij})F(r_{jk})G(r_{ij}) + 4F(r_{ij})F(r_{ik})G'_0\} \qquad (51)$$

如果取 $r_{ij}=r_{ik}=r_{jk}$，(51)式是一排斥力位能。

以上的討論只是簡單地考慮了 $r_{ij}=r_{jk}=r_{ik}$ 的情況，一般情況下的三体力位能要複雜得多。在比較嚴密的考慮中，应該同時確定这些原子核基态上波函數的空間坐标部份，通过位能对波函數的積分而求得它的總的效果。

最後，作者願对彭桓武教授和胡寧教授表示感謝，他們曾对这一工作提供了宝貴的意見和帮助。

参 考 文 献

[1] 周光召：中國物理学報，11,299頁-316頁，1955.

[2] R. S. Christian and N. W. Hart, *Physical Review* 77, pp. 441-453, 1950.

[3] H. Primakoff and T. Holstsin, *Physioal Review* 55, pp. 1218-1234, 1939.

L. J'anossy, *Proceedings of the Cambridge Philosophical Society* 35, pp. 616-621, 1939.

[4] G. Wentzel, *Physical Review* 89, pp. 684-688, 1953.

S. D. Drell and Kerson Huang, *Physical Review* 91, pp.1527-1542, 1953.

[5] K. A. Brueckner, *Physical Review* 91, pp.761-762, 1953.

K. A. Brueckner, J. Gell-Man and M. L. Goldberger, *Physioal Review* 91, p.460, 1953.

[6] G. F. Chew, *Physical review* 95, pp. 1669-1675, 1954.

[7] K. A. Brueckner and K. M. Watson, *Physical review* 92, pp. 1023-1035, 1953.

К ВОПРОСЫ О ПОТЕНЦИАЛЕ ТРЁХ ТЕЛ В МЕЗОННОЙ ТЕОРИИ

Чжоу Гуан-Чжао

Физический факультет пекинского университета

Вычислен потенциал взаимодействия трёх нуклонов в мезонной теории методом канонического преобразования. Выражения потенциала трёх тел получены в шестом приближений в виде разложения по степеням константы g. Приведены результаты вычисления потенциалов трёх тел, обословл нного симметричным псевдоскалярным мезонным полем с псевдовекторной связью и симметричным скалярным мезонным полем со скалярной связью. Расходимости устраняются путём обрезания при больших импульсах. Рассматренны влияния этих сил на структуре ядер ^{3}He ^{3}He и ^{4}He.

由微觀电磁場方程組求得宏觀电磁場方程組的方法*

周 光 召

(物理系理論物理教研室)

I. 引 言

洛倫茲电子論是从物質的原子構造的观點出發，來解釋介質的宏观电磁現象的理論。从这一观點出發，我們把介質看作由許多帶电的微粒所組成，这些帶电的微粒或者在空間自由运動，或者組成複雜的原子和分子，在这些微粒之間尚存在着廣闊的空間。此時，我們可以把描寫这个空間的电磁場方程式寫出來，它們是

$$\begin{aligned} &\nabla\cdot\vec{e}=4\pi\rho_m \qquad &&\nabla\cdot\vec{h}=0 \\ &\nabla\times\vec{h}=\frac{1}{c}\frac{\partial\vec{e}}{\partial t}+4\pi\vec{j}_m \qquad &&\nabla\times\vec{e}=-\frac{1}{c}\frac{\partial\vec{h}}{\partial t} \end{aligned} \tag{I}$$

其中$\vec{e}$和$\vec{h}$可以叫作微观电場强度及微观磁場强度，ρ_m和$\vec{j}_m$为微观电荷密度和电流密度，它們只有在原子核或电子上才不等於零，方程組(I)称作微观电磁場方程式或洛倫茲方程式。这个微观电磁場不論在空間上或時間上都是变化得很厲害的，在一个原子的內部和外部就有巨大的变化，因此这个微观电磁場不是我們日常在介質中所观察到的宏观电磁場。洛倫茲首先肯定了宏观电磁場是微观电磁場对物理地小的空間和時間區域的平均值，他首先自方程組(I)出發得到了宏观电磁場方程組：

$$\begin{aligned} &\nabla\cdot\vec{D}=4\pi\rho_f \qquad &&\nabla\cdot\vec{B}=0 \\ &\nabla\times\vec{E}=-\frac{1}{c}\frac{\partial\vec{B}}{\partial t} \qquad &&\nabla\times\vec{H}=\frac{1}{c}\frac{\partial\vec{D}}{\partial t}+4\pi\vec{j}_i \end{aligned} \tag{II}$$

通过这个平均过程，宏观电磁場的物理量得以和原子構造的性質連系起來，使得定量地从物質的原子構造出發來解釋介質的宏观电磁場性質成为可能。因此通过平均从微观电磁場方程組求得宏观电磁場方程組是电子論的一个重要問題。

---

* 1955年11月3日收到。

这个工作最早由洛倫兹作过[1]，隨後有許多人把洛倫兹的方法加以改進[2]。在早期的工作中，求出电荷和电流密度的平均值的方法是不够嚴密的，在其中保留了磁偶極矩的貢献，但忽略了和磁偶極矩同一數量級的电四極矩的貢献。这个缺點在 1950 年 Rosenfeld 的工作[3] 中得到改正，他所用方法的基本精神和洛倫兹方法的基本精神相近，只是要更細心和嚴格一些。他的証明方法是不够系統的，在求平均值之前，他已經把原子中电荷分佈及空間的原子密度考慮成連續函數，这等於对这兩个物理量先來一次平均，然後再求其他物理量的平均值。在求电荷及电流密度平均值時，也是每一項分別地考慮而沒有一个統一的系統的証明。因此在採用他的方法時必須非常細心，不然很容易搞錯，这对於初学电動力学的人講來是很困难的。1953 年 Mazur 及 Nijiboer[4] 为了克服 Rosenfeld 証明中这些缺點，採用系綜平均的办法得到了宏观电磁場方程組，但这个方法中包含一些較高深的概念，对初学电動力学的人也是不相宜的。

本文的目的就是要为初学电動力学的人找到一个簡明易懂而又系統的証明。在本文中由於运用了狄拉克的δ函數而使得計算的步驟大大簡化。狄拉克δ函數的定义如下：

$$\delta(x)=0 \quad 当 \quad x\neq 0$$

$$\int\delta(x-a)\,dx=1$$ 当積分區間包含 a 在內時，其他情况下積分为零。

利用狄拉克δ函數的好处就在於可以把點电荷的密度分佈寫成函數的形式，例如在 $\vec{R}_0$ 处的一个點电荷 ϵ_0 的电荷密度分佈可以寫成下列形式：

$$\rho(\vec{R})=\epsilon_0\delta(\vec{R}-\vec{R}_0)$$

其中

$$\delta(\vec{R})=\delta(x)\delta(y)\delta(z)$$

本文的第二節討論微观电磁場方程組的平均，第三和第四節分別討論微观电荷密度及电流密度的平均。我們得到的結果和 Mazur 及 Nijiboer 得到的完全一样。

II. 微观电磁場方程組的平均

考慮 t 時空間中一點 P，其坐标向量为 $\vec{r}$。讓我們圍繞 $\vec{r}$ 和 t 取一物理地小的体積 V 和時間間隔 T。任一微观物理量 $L(\vec{r},t)$ 在 $\vec{r}$ 點 t 時的平均值可以寫成

$$\overline{L}(\vec{r},t)=\frac{1}{VT}\int_V d\vec{r}'\int_T dt' L(\vec{r}',t') \tag{1}$$

一个重要的性質是：微分和平均这兩种运算可以互易先後次序，而不影响最後的結果。即

$$\frac{\partial \overline{L}}{\partial x}=\frac{\overline{\partial L}}{\partial x},\quad \frac{\partial \overline{L}}{\partial y}=\frac{\overline{\partial L}}{\partial y},\quad \frac{\partial \overline{L}}{\partial z}=\frac{\overline{\partial L}}{\partial z},\quad \frac{\partial \overline{L}}{\partial t}=\frac{\overline{\partial L}}{\partial t} \tag{2}$$

(2)式的証明在許多教科書上都可查到，我們在此处不再寫出。

把微观方程組 (I) 加以平均，並考慮到微分和平均过程这兩种运算可以互易的性質，我們便得到

$$\begin{aligned} &\nabla\cdot\overline{\vec{e}}=4\pi\overline{\rho}_m &\qquad& \nabla\cdot\overline{\vec{h}}=0 \\ &\nabla\times\overline{\vec{e}}=-\frac{1}{c}\frac{\partial\overline{\vec{h}}}{\partial t} && \nabla\times\overline{\vec{h}}=\frac{1}{c}\frac{\partial\overline{\vec{e}}}{\partial t}+4\pi\overline{\vec{j}}_m \end{aligned} \tag{III}$$

把(III)式中第二个和第三个方程和宏观电磁場方程組(II)中相应的方程加以比較，馬上發現应有

$$\overline{\vec{e}}=\vec{E}\qquad \overline{\vec{h}}=\vec{B} \tag{3}$$

这一結果是大家熟知的，其餘兩个方程式的比較有待於先求出 $\overline{\rho}_{微}$ 及 $\overline{\vec{j}}_{微}$，我們在下兩節中便來求出它們。

III. 微观电荷密度的平均

令 $\vec{R}_i^\alpha$ 代表第 α 个原子*中第 i 个粒子**的坐标向量，ϵ_i^α 为它的电荷的數值。此時微观电荷密度可以寫成

$$\rho_m=\sum_{\alpha i}\epsilon_i^\alpha\delta(\vec{R}-\vec{R}_i^\alpha) \tag{4}$$

(4)式中的求和不但要对 α 求而且还要对 i 求，如果直接拿(4)式去平均，則很大一部份对 $\overline{\rho}_{微}$ 的貢献來源於割切体積 V 的原子上，这样使得求平均的过程变得複雜。要避免这一麻煩就需要在平均之前先把原子这样一个複雜的电的系統，化成一些集中在原子質量中心的多極矩系統。这样一來，就不再發生原子割切体積 V 的問題了。我們这一方法的省事之处也就在於这裹。

令 $\vec{R}^\alpha$ 代表第 α 个原子的質心坐标，且

$$\vec{R}_i^\alpha=\vec{R}^\alpha+\vec{r}_i^\alpha$$

則我們有

$$\begin{aligned}\delta(\vec{R}-\vec{R}_i^\alpha)&=\delta(\vec{R}-\vec{R}^\alpha-\vec{r}_i^\alpha)\\ &=\delta(\vec{R}-\vec{R}^\alpha)-\vec{r}_i^\alpha\cdot\nabla_{\vec{R}}\delta(\vec{R}-\vec{R}^\alpha)+\cdots\cdots\end{aligned}$$

---

* 以後凡是提到原子的都是泛指，其中可以包括分子、离子、自由电子等。

** 此处粒子係泛指可以是电子亦可是原子核。

$$+\frac{1}{n!}(-\vec{r}_i^{\alpha}\cdot\nabla_{\vec{R}})^n\delta(\vec{R}-\vec{R}^{\alpha})+\cdots\cdots \tag{5}$$

於是我們得到

$$\rho_m=\sum_\alpha(\sum_i\in_i^\alpha)\delta(\vec{R}-\vec{R}^\alpha)-\sum_\alpha(\sum_i\in_i^\alpha\vec{r}_i^\alpha)\cdot\nabla_{\vec{R}}\delta(\vec{R}-\vec{R}^\alpha)$$

$$+\frac{1}{2!}\sum_\alpha(\sum_i\in_i^\alpha\vec{r}_i^\alpha\vec{r}_i^\alpha):\nabla_{\vec{R}}\nabla_{\vec{R}}\delta(\vec{R}-\vec{R}^\alpha)+\cdots \tag{6}$$

其中 $\vec{a}\vec{b}:\vec{c}\vec{d}$ 的符号代表 $(\vec{a}\cdot\vec{d})(\vec{b}\cdot\vec{c})$。

令

$$\sum_i\in_i^\alpha=\in^\alpha$$

$$\sum_i\in_i^\alpha\vec{r}_i^\alpha=\vec{p}^\alpha \tag{7}$$

$$\frac{1}{2}\sum\in_i^\alpha\vec{r}_i^\alpha\vec{r}_i^\alpha=\vec{\vec{q}}^\alpha$$

……………………

$\in^\alpha, \vec{p}^\alpha, \vec{\vec{q}}^\alpha$……等等的物理意义是很清楚的，他們分別代表第 α 个原子總的电荷，电偶極矩，电四極矩等等。我們可以把第 α 个原子的微观自由电荷密度，电偶極矩密度及电四極矩密度寫为 $\in^\alpha\delta(\vec{R}-\vec{R}^\alpha)$，$\vec{p}^\alpha\delta(\vec{R}-\vec{R}^\alpha)$ 及 $\vec{\vec{q}}^\alpha\delta(\vec{R}-\vec{R}^\alpha)$，因此總的微观自由电荷密度，电偶極矩密度及电四極矩密度便成为

$$\rho_{fm}=\sum_\alpha\in^\alpha\delta(\vec{R}-\vec{R}^\alpha)$$

$$\vec{p}_m=\sum_\alpha\vec{p}^\alpha\delta(\vec{R}-\vec{R}^\alpha) \tag{8}$$

$$\vec{\vec{q}}_m=\sum_\alpha\vec{\vec{q}}^\alpha\delta(\vec{R}-\vec{R}^\alpha)$$

利用(6)，(7)及(8)式，可得 ρ_m 的平均值。当忽略掉比 $\vec{\vec{q}}_m$ 更高的項之後。

$$\overline{\rho_m}=\overline{\rho_{fm}-\nabla_{\vec{R}}\cdot\vec{p}_m+\nabla_{\vec{R}}\nabla_{\vec{R}}:\vec{\vec{q}}_m}$$

$$=\overline{\rho}_{fm}-\nabla_{\vec{R}}\cdot\overline{\vec{p}}_m+\nabla_{\vec{R}}\nabla_{\vec{R}}:\overline{\vec{\vec{q}}}_m \tag{9}$$

$$=\rho_f-\nabla_{\vec{R}}\cdot\vec{P}+\nabla_{\vec{R}}\nabla_{\vec{R}}:\vec{\vec{Q}}$$

其中 ρ_f 从微观自由电荷密度平均而得，由於 $\in^\alpha$ 只对离子或自由电子才不为零，故 ρ_f 是由整个帶电的离子或自由电子的电荷貢献得來。$\vec{P}=\overline{\vec{p}}_m$ 和 $\vec{\vec{Q}}=\overline{\vec{\vec{q}}}_m$ 則是宏观平均的电偶極矩及电四極矩密度，它們是由原子內的束縛电荷貢献得來的。把(9)式代入(III)的第一方程式，我們立刻得到

$$\nabla\cdot\overline{\vec{e}}=4\pi[\rho_f-\nabla\cdot\vec{P}+\nabla\nabla:\vec{\vec{Q}}]$$

即 $$\nabla\cdot\vec{E}=4\pi[\rho_f-\nabla\cdot\vec{P}+\nabla\nabla:\vec{Q}]$$

或 $$\nabla\cdot[\vec{E}+4\pi(\vec{P}-\nabla\cdot\vec{Q})]=4\pi\rho_f \tag{10}$$

和(II)中的第一式比較，我們得到

$$\vec{D}=\vec{E}+4\pi(\vec{P}-\nabla\cdot\vec{Q}) \tag{11}$$

(11)式和 Mazur, Nijiboer 及 Rosenfeld 所得相同，比洛倫茲原來証明出的公式多一項 $-4\pi\nabla\cdot\vec{Q}$。表明，洛倫茲沒有考慮电四極矩对电荷密度的貢献。我們將來可以看到，$\nabla\cdot\vec{Q}$ 和 $\vec{M}$ 是同一數量級的，因此不能保留其中一个而忽略另一个。

III. 电流密度的平均值

令 $\vec{V}_i^\alpha$ 代表第 α 个原子中第 i 个粒子的速度，我們有

$$\vec{V}_i^\alpha=\frac{d\vec{R}_i^\alpha}{dt}=\dot{\vec{R}}_i^\alpha$$

同样令 $\vec{V}^\alpha=\dot{\vec{R}}^\alpha$ 代表質心运動的速度，$\vec{v}_i^\alpha=\dot{\vec{r}}_i^\alpha$ 代表第 i 粒子相对於質心的运動速度，我們有

$$\vec{V}_i^\alpha=\vec{V}^\alpha+\vec{v}_i^\alpha \tag{12}$$

微观电流密度 $\vec{j}_{微}$ 可以表成：

$$\vec{j}_m=\frac{1}{c}\sum\epsilon_i^\alpha\vec{V}_i^\alpha\delta(\vec{R}-\vec{R}_i^\alpha) \tag{13}$$

跟求电荷密度平均值一样，我們可以將 $\delta(\vec{R}-\vec{R}_i)$ 展開，於是得到

$$\begin{aligned}\vec{j}_m=&\frac{1}{c}\sum_{\alpha i}\epsilon_i^\alpha\vec{V}_i^\alpha[\delta(\vec{R}-\vec{R}^\alpha)-\vec{r}_i^\alpha\cdot\nabla_{\vec{R}}\delta(\vec{R}-\vec{R}^\alpha)\\&+\frac{1}{2}\vec{r}_i^\alpha\vec{r}_i^\alpha:\nabla_{\vec{R}}\nabla_{\vec{R}}\delta(\vec{R}-\vec{R}^\alpha)+\cdots\cdots]\\=&\frac{1}{c}\sum_\alpha\vec{V}^\alpha[\sum_i\epsilon_i^\alpha\delta(\vec{R}-\vec{R}^\alpha)-\sum\epsilon_i^\alpha\vec{r}_i^\alpha\cdot\nabla_{\vec{R}}\delta(\vec{R}-\vec{R}^\alpha)\\&+\frac{1}{2}\sum_i\epsilon_i^\alpha\vec{r}_i^\alpha\vec{r}_i^\alpha:\nabla_{\vec{R}}\nabla_{\vec{R}}\delta(\vec{R}-\vec{R}^\alpha)+\cdots\cdots]\\&+\frac{1}{c}\sum_\alpha[\sum_i\epsilon_i^\alpha\vec{v}_i^\alpha\delta(\vec{R}-\vec{R}^\alpha)-\sum_i\epsilon_i^\alpha\vec{v}_i^\alpha\vec{r}_i^\alpha\cdot\nabla_{\vec{R}}\delta(\vec{R}-\vec{R}^\alpha)+\cdots\cdots]\end{aligned} \tag{14}$$

我們在繼續化簡 $\vec{j}_{微}$ 以前，首先考慮以下的幾个等式：

$$\begin{aligned}\frac{1}{c}\frac{d\vec{P}_m}{dt}&=\frac{1}{c}\dot{\vec{P}}_m\\&=\frac{1}{c}\frac{d}{dt}\sum_{\alpha i}\epsilon_i^\alpha\vec{r}_i^\alpha(t)\,\delta(\vec{R}-\vec{R}^\alpha(t))\end{aligned}$$

$$= \frac{1}{c}\sum_{\alpha i} \epsilon_i^\alpha \vec{v}_i \delta(\vec{R}-\vec{R}^\alpha) - \frac{1}{c}\sum_{\alpha i} \epsilon_i^\alpha \vec{r}_i^\alpha \vec{V}^\alpha \cdot \nabla_{\vec{R}} \delta(\vec{R}-\vec{R}^\alpha) \tag{15}$$

$$\frac{1}{c}\nabla_{\vec{R}} \cdot \dot{\vec{q}}_m = \frac{1}{2c}\frac{d}{dt}\sum_{\alpha i} \epsilon_i^\alpha \vec{r}_i \vec{r}_i \cdot \nabla_{\vec{R}} \delta(\vec{R}-\vec{R}^\alpha)$$

$$= \frac{1}{2c}\Big[\sum_{\alpha i} \epsilon_i^\alpha \vec{v}_i^\alpha \vec{r}_i^\alpha \cdot \nabla_{\vec{R}} \delta(\vec{R}-\vec{R}^\alpha) + \sum_{\alpha i} \epsilon_i^\alpha \vec{r}_i^\alpha \vec{v}_i^\alpha \cdot \nabla_{\vec{R}} \delta(\vec{R}-\vec{R}^\alpha)$$

$$- \sum_{\alpha i} \epsilon_i^\alpha \vec{r}_i^\alpha \vec{r}_i^\alpha \cdot \nabla_{\vec{R}} \vec{V}^\alpha \cdot \nabla_{\vec{R}} \delta(\vec{R}-\vec{R}^\alpha)\Big] \tag{16}$$

並引進下列的新物理量

$$\vec{j}_{fm} = \frac{1}{c}\sum_{\alpha i} \epsilon_i^\alpha \vec{V}^\alpha \delta(\vec{R}-\vec{R}^\alpha)$$

$$= \frac{1}{c}\sum_{\alpha} \epsilon^\alpha \vec{V}^\alpha \delta(\vec{R}-\vec{R}^\alpha) \tag{17}$$

$$\vec{m}_m = \frac{1}{2c}\sum_{\alpha}\Big(\sum_i \epsilon_i^\alpha \vec{r}_i^\alpha \times \vec{v}_i^\alpha\Big)\delta(\vec{R}-\vec{R}^\alpha) \tag{18}$$

由(18)式很容易得到

$$\nabla_{\vec{R}} \vec{m}_m = \frac{1}{2c}\Big[\sum_{\alpha i} \epsilon_i^\alpha \vec{r}_i^\alpha \vec{v}_i^\alpha \cdot \nabla_{\vec{R}} \delta(\vec{R}-\vec{R}^\alpha) - \sum_{\alpha i} \epsilon_i^\alpha \vec{v}_i^\alpha \vec{r}_i^\alpha \cdot \nabla_{\vec{R}} \delta(\vec{R}-\vec{R}^\alpha)\Big] \tag{19}$$

$\vec{j}_{fm}$ 和 $\vec{m}_m$ 的物理意义是很清楚的，它們分別代表微观的自由电流密度及磁偶極矩密度。應該附帶的說明一點，就是在我們的理論中，沒有考慮电子自旋对电流密度的貢献，但这个缺陷是很容易補救的。由於电子自旋对电流密度的貢献相当於一磁偶極子的貢献，我們只需在(18)式的 $\vec{m}_{微}$ 中增加电子自旋的部份便行了。

把(14)式加以簡化，並考慮到(15)，(16)，(17)，(18)及(19)等式，我們最後可以得到

$$\vec{j}_m = \vec{j}_{fm} + \frac{1}{c}(\dot{\vec{p}}_m - \nabla \cdot \dot{\vec{q}}_m) + \nabla_{\vec{R}} \times \vec{m}_m + \nabla_{\vec{R}} \times \vec{n}_m \tag{20}$$

其中

$$\vec{n}_m = \frac{1}{c}\sum_{\alpha i} \epsilon_i^\alpha (\vec{r}_i^\alpha \times \vec{V}^\alpha)\delta(\vec{R}-\vec{R}^\alpha) - \frac{1}{2}\sum_{\alpha i} \epsilon_i^\alpha (\vec{r}_i^\alpha \times \vec{V}^\alpha)\vec{r}_i^\alpha \cdot \nabla_{\vec{R}} \delta(\vec{R}-\vec{R}^\alpha) \tag{21}$$

$\vec{n}_m$ 与原子的整体运動速度 $\vec{V}^\alpha$ 有關，它的物理意义留待後边討論。將(20)式加以平均，考慮到平均与微分可以互換的性質，並在得到的結果中略去符号 $\nabla_{\vec{R}}$ 的脚标 $\vec{R}$，得到

$$\overline{\vec{j}_m} = \vec{j}_f + \frac{1}{c}\frac{\partial}{\partial t}(\vec{P} - \nabla \cdot \vec{Q}) + \nabla \times \vec{M} + \nabla \times \vec{N} \tag{22}$$

其中

$$\vec{j}_f=\vec{j}_{im}$$
$$\vec{M}=\overline{\vec{m}}_m$$
$$\vec{N}=\overline{\vec{n}}_m$$

把(21)式代回方程組(III)的第四个方程式中，我們得到

$$\nabla\times\vec{B}=\frac{1}{c}\frac{\partial\vec{E}}{\partial t}+4\pi\vec{j}_f+\frac{1}{c}\frac{\partial}{\partial t}(4\pi\vec{P}-4\pi\nabla\cdot\vec{\vec{Q}})+4\pi\nabla\times\vec{M}+4\pi\nabla\times\vec{N}$$

或

$$\nabla\times(\vec{B}-4\pi\vec{M}-4\pi\vec{N})=\frac{1}{c}\frac{\partial\vec{D}}{\partial t}+4\pi\vec{j}_f \tag{23}$$

与方程組(II)中的第四式加以比較，我們有

$$\vec{H}=\vec{B}-4\pi\vec{M}-4\pi\vec{N} \tag{24}$$

这一公式和 Mazur 及 Nijiboer 所得相同，比平常書上的公式多一項 $4\pi N$。現在讓我們在幾个特殊条件下來看 $4\pi\vec{N}$ 这一項的意义：

i) 当 $\vec{V}_i^a=0$ 時 $\vec{N}=0$

ii) 当分子运動是熱运動，因而 $\vec{V}_i^a$ 是完全混乱時，經过平均我們也有 $\vec{N}=0$

从上面看來，当物体沒有整体的宏观运動時，$\vec{N}$ 这一項是不起作用的，这也就是說，在靜止介質中，(24)式和平常的公式相同。但当我們討論运動物体的电動力学時，$\vec{N}$ 这一項便会起作用了，

iii) 当 $\vec{V}_i^a=\vec{V}=$ 常向量

$$\vec{N}=\frac{1}{c}(\vec{P}-\nabla\cdot\vec{\vec{Q}})\times\vec{V} \tag{25}$$

在运動物体的电動力学中確实有(25)式这样一項，洛倫茲在他的工作中[1]也得到了(25)式，它代表当介質有整体运動時，磁化向量的修正。

从(16)式和(19)正的比較中，还可看出 $\nabla_R\times\vec{m}_{微}$ 和 $\frac{1}{c}\nabla_R\cdot\dot{\vec{\vec{q}}}$ 这兩項对电流密度的貢献是具有同一个數量級的，所以一个系統的理論必須同時考慮磁偶極矩和电四極矩。在通常的教科書上多半保留了磁偶極矩的貢献而忽略了电四極矩的貢献，这种作法是不完全恰当的。

上面这种方法不但可以用來求出宏观电動力学方程式，还可以用來求出流体力学的連續方程式，运動方程式及能量守恆方程式，可以得到和 Kirkwood[5] 等人完全相同的結果。这种方法还可以推廣到量子力学的情形，此時只要採用在海森堡表象的物理量，並且適当的考慮它們之間的对易關係便行了。

最後，作者願对胡寧教授、王竹溪教授和彭桓武教授的經常關怀和鼓勵表示衷心的感謝。

参考文献

[1] Lorentz, H. A. *Encyclopädie der mathematischen Wissenschaften*, V. 2. Heft 1. S. 200

[2] Fokker, A. D. *Relativitätstheorie*, Groningon, (1929)

Becker, R. *Theorie der Elektrizität* Band II. Leipzig, (1933)

Van Vleck, J. H *The theory of electric and magnetic susceptibilities*, Oxford, (1933)

Тамм, И. Е. *Основы Теории электричества*, Гос-Тех., (1949)

Sauter, F. *Zeitschrift fur phgsik* (1949) 126, p. 207.

[3] Rosenfeld, L. *Theory of electrons* Amsterdam, (1951)

[4] Mazur, P. and Nijiboer, B. R. A. *Physica* (1953) 19, p. 971.

[5] Kirkwood, J. G. *Journal of chemical physics* (1946) 14, p. 180.

DERIVATION OF MAXWELL EQUATIONS FROM THE MICROSCOPIC LORENTZ EQUATIONS

Chou Kuang-Chao

ABSTRACT

The space time average method, used in the derivation of macroscopic maxwell equations, is elaborated in a systematic way. The technique of Dirac δ-function is used which greatly simplifies the calculation. The results are identical with those obtained by Mazur and Nijiboer[4], who use the ensemble average method. The idea involved in the present paper is much simplier than that of Mazur and Nijiboer and the calculation is straightforward. The present method can be extended to the derivation of hydrodynamical equations and easily generalized to the quantum case.

第15卷 第5期 1959年5月
物理学报
ACTA PHYSICA SINICA
Vol. 15, No. 5
May 1959

中子和質子的質量差*

周光召
（联合原子核研究所，莫斯科）

一. 引言

在研究原子核的結構时，人們发現中子与質子的性質是相近的。进一步的分析表明，它們和 π 介子的作用是完全一样的，具有和电荷无关的性質。只是由于电磁场的作用，才在它們之間产生了一些差別。現在大家都認为，在去掉电磁场的作用以后，中子和質子將具有相同的性質，它們將具有相同的質量，因而可以看作是同一核子处于不同的狀态。

实际上，中子比質子重 2.52 个电子質量。如果上述看法是正确的，这一質量差別应当是由电磁作用所引起的。

已經有过許多人試图从量子电动力学的理論中計算出中子与質子的質量差[1]。他們都采用狄拉克核子模型，認为核子是一帶有反常磁矩的狄拉克粒子，它只与电磁场起作用。利用微擾法計算这种核子的电磁質量时，通常得到发散的結果。为了消除这一发散，需要引入一帶任意性的割断，把作用能量大于某定值的电磁作用去掉。

以上的理論有兩个严重的缺点，首先是沒有考慮核子和 π 介子及其他粒子間的强作用。这些作用毫无疑問在很大程度上影响物理核子的电磁性質。其次就是无法消除发散困难，除非引进任意的割断值。

本文是为了克服上述困难，从理論上計算核子电磁質量的一种尝試。我們采用新近发展起来的色散关系的理論来考虑这一問題。从原則上講，在这一理論中包括了所有强作用的影响，并能自动的重正化，消除发散困难。对电磁作用，我們將仍采用微擾法处理，此时忽略的量約为 $\alpha^2=(1/137)^2$ 的数量級。

第二节中，我們將建立質量算符与虛光子的康普頓效应之間的連系。在第三节中，將用色散关系理論来处理虛光子对康普頓效应。第四节計算了色散关系中單核子項的貢献。第五节估計了其他項的貢献。

計算的結果在結論中討論。

二. 質量算符

令 S 代表整个体系的散射矩陣，其中包括所有的强作用及电磁作用。物理核子的格林函数可写成

$$G(x,x')=\langle T(\psi(x)\bar{\psi}(x')S)\rangle_0, \tag{1}$$

其中 $\psi(x)$ 为核子场在作用表象中的场量算符；$\langle\ \rangle_0$ 代表对真空的平均值；T 为时間順序

* 1959年2月22日收到。

算符。格林函数满足下列方程：

$$(i\hat{p}+\hat{M})G(x,x')=-i\delta^4(x-x'),\tag{2}$$

其中

$$i\hat{p}=\sum_{\mu=1}^{4}\gamma_\mu\frac{\partial}{\partial x_\mu},\quad x_4=it,\tag{3}$$

γ_μ 为狄拉克算符；$\hat{M}$ 为質量算符．用矩陣表示出来，$\hat{M}$ 的形式为

$$\hat{M}G(xx')=\int d^4x''\,M(x\,x'')G(x''\,x').\tag{4}$$

如令 $u(p)$ 代表質量为 M 的狄拉克旋量，則物理核子的質量 M 由下列方程决定：

$$(2\pi)^4\overline{u}(p)u(p)M\delta^4(p'-p)=\int\int d^4xd^4x'\cdot\overline{u}(p)\hat{M}(x,x')u(p')e^{i(p'\cdot x'-p\cdot x)}\tag{5}$$

(5)式中 $p\cdot x$ 代表两个四度矢量 p 及 x 的标积[5]．

在(1)式中，將 S 矩陣按电磁作用的耦合常数展开，令 $A_\mu(x)$ 代表电磁场的场量算符，可得

$$S=S_0+\int d^4x\frac{\delta S}{\delta A_\mu(x)}\bigg|_{A_\mu=0}A_\mu(x)+$$
$$+\frac{1}{2}\int\int d^4xd^4x'\frac{\delta^2 S}{\delta A_\mu(x)\delta A_\nu(x')}\bigg|_{A_\mu=0}A_\mu(x)A_\nu(x')+\cdots,\tag{6}$$

S_0 为只包括强作用的 S-矩陣．引进核子场及介子场电流密度 $j_\mu(x)$，

$$\frac{\delta S}{\delta A_\mu(x)}\bigg|_{A_\mu=0}=-iT(j_\mu(x)S_0)\tag{7}$$

及

$$\frac{\delta^2 S}{\delta A_\mu(x)\delta A_\nu(x')}\bigg|_{A_\mu=0}=-T(j_\mu(x)j_\nu(x')S_0).\tag{8}$$

將(6),(7)及(8)代回(1)式中，并对光子真空求平均，注意到下列关系

$$\langle A_\mu(x)\rangle_0=0,$$
$$T\langle A_\mu(x)A_\nu(x')\rangle_0=D_F(x-x')\delta_{\mu\nu},\tag{9}$$

可得

$$G(x,x')=G_0(x,x')+G_1(x,x'),\tag{10}$$

其中

$$G_0(x,x')=\langle T(\psi(x)\overline{\psi}(x')S_0)\rangle_0,$$
$$G_1(x,x')=-\frac{1}{2}\int\int d^4\xi d^4\eta D_F(\xi-\eta)\langle T(\psi(x)\overline{\psi}(x')j_\mu(\xi)j_\mu(\eta)S_0)\rangle_0.\tag{11}$$

(11)式中 $G_0(x,x')$ 为忽略了电磁作用以后物理核子的格林函数，它满足方程

$$(i\hat{p}+\hat{M}_0)G_0(x,x')=-i\delta^4(x-x'),\tag{12}$$

其中 $\hat{M}_0$ 为物理核子在忽略电磁質量后的質量算符，它对中子和質子都有同样的形式．

核子的电磁質量由算符

$$\Delta\hat{M}=\hat{M}-\hat{M}_0\tag{13}$$

来代表．將(11)及(10)代回(1)式，卽得

$$\Delta\hat{M}G_0(x-x')=-(i\hat{p}+M)G_1(x,x').\tag{14}$$

在(14)式左边用 $\hat{M}_0$ 代替 $\hat{M}$，这相当于忽略高級項，并兩边乘以 G_0^{-1}，可得

$$\Delta\hat{M}=-i(i\hat{p}_x+\hat{M}_0)G(x,x')(i\hat{p}_{x'}+\hat{M}_0). \tag{15}$$

在(15)式兩边乘以 $\bar{u}(p)e^{-ipx}$ 及 $u(p')e^{ip'x'}$，再对 x 及 x' 积分，可以求得核子的电磁質量 ΔM：

$$\Delta M\,\bar{u}(p)u(p)(2\pi)^4\delta^4(p'-p)=\frac{i}{2}\iint d^4x d^4x' e^{-i(px-p'x')}\cdot$$

$$\cdot\bar{u}(p)(i\hat{p}_x+\hat{M}_0)\langle T(\psi(x)\bar{\psi}(x')j_\mu(\xi)j_\mu(\eta)S_0)\rangle_0\cdot(i\hat{p}_{x'}+\hat{M}_0)u(p')D_F(\xi-\eta)d\xi d\eta. \tag{16}$$

利用漸近关系，可將(16)式变成

$$\Delta M\,\bar{u}(p)u(p)(2\pi)^4\delta^4(p-p')=\frac{-i}{2}(2\pi)^3\iint d^4\xi d^4\eta\langle p|T(j_\mu(\xi)j_\mu(\eta)S_0)|p'\rangle D_F(\xi-\eta). \tag{16'}$$

其中 $|p\rangle$ 为物理核子的狀态矢量，其动量为 p. 將 $D_F(\xi-\eta)$ 用富氏变換表示出来，并对 $\frac{1}{2}(\xi-\eta)$ 积分，可得

$$\Delta M\,\bar{u}(p)u(p)=\frac{-1}{4\pi}\iint d^4z d^4k e^{-ikz}\frac{\langle p|T\left(j_\mu\left(\frac{z}{2}\right)j_\mu\left(-\frac{z}{2}\right)S_0\right)|p\rangle}{k^2-i\varepsilon}. \tag{17}$$

注意到在(17)式中对 μ 是求和的.

容易証明，(17)式中的被积函数同光子与核子散射的矩陣元有相同的結構. 光子与核子散射的矩陣元具有以下形式：

$$\langle p'k'\mu'|S|pk\mu\rangle=\delta_{\mu'\mu}\delta^3(p'-p)\delta^3(k'-k)+$$
$$+i\delta^4(p'+k'-p-k)\left(\frac{1}{4k_0k_0'}\right)^{1/2}(2\pi)^2\langle p'k'\mu'|F|pk\mu\rangle, \tag{18}$$

其中

$$\langle p'k'\mu'|F|pk\mu\rangle=\frac{i}{2\pi}\int e^{-i(k+k')\cdot z/2}\langle p'|T\left(j_{\mu'}\left(\frac{z}{2}\right)j_\mu\left(-\frac{z}{2}\right)S_0\right)|p\rangle d^4z. \tag{19}$$

將(19)式与(17)式比較，得到

$$\Delta M\,\bar{u}(p)u(p)=\frac{i}{2}\int\frac{d^4k}{k^2}\sum_\mu\langle p,k,\mu|F|p,k,\mu\rangle. \tag{20}$$

由此可見，如果求得了虛光子($k^2\neq 0$)与核子散射的費曼振幅 F，則可利用(20)式求出核子的电磁質量. 虛光子与核子散射的費曼矩陣將通过色散关系来求.

三. 虛光子与核子散射的色散关系

引进虛光子与核子散射的因果振幅

$$M_\mu=\langle p,k,\mu|M|p,k,\mu\rangle=\frac{i}{2\pi}\int e^{-ikz}\theta(z)\langle p|\left[j_\mu\left(\frac{z}{2}\right),j_\mu\left(-\frac{z}{2}\right)\right]|p\rangle d^4z, \tag{21}$$

其中

$$\theta(z)=\frac{1}{2}(1+\epsilon(z)),\qquad \epsilon(z)=1\ 当\ z_0>0,\qquad \epsilon(z)=-1\ 当\ z_0<0.$$

M_μ 可分为吸收部 A_μ 及色散部 D_μ:

$$M_\mu = D_\mu + iA_\mu, \tag{22}$$

其中

$$D_\mu = \frac{i}{4\pi}\int e^{-ikz}\epsilon(z)\langle p\,|\left[j_\mu\left(\frac{z}{2}\right), j_\mu\left(-\frac{z}{2}\right)\right]|p\rangle d^4z,$$
$$A_\mu = \frac{1}{4\pi}\int e^{-ikz}\langle p\,|\left[j_\mu\left(\frac{z}{2}\right), j_\mu\left(-\frac{z}{2}\right)\right]|p\rangle d^4z. \tag{23}$$

在吸收部中引进一完整的中間态 $|n\rangle$ 并对 z 积分，可得

$$A_\mu = \frac{1}{4\pi}(2\pi)^4\sum_n[\delta^4(k+p-p_n)-\delta^4(k-p+p_n)]\,|\langle p\,|\,j_\mu(0)\,|n\rangle|^2. \tag{24}$$

在虛光子和核子向前散射的問題中，共有兩个独立的标量，它們可取作 k^2 及 $\nu=-k\cdot p/M$，M 为核子質量. M_μ, D_μ 及 A_μ 都可表成 k^2 及 ν 的函数，

$$M_\mu = \bar{u}(p)[M_\mu^{(+)}(\nu,k^2)+i\hat{k}M_\mu^{(-)}(\nu,k^2)]u(p), \tag{25}$$

其中 $\hat{k}=\sum_\mu \gamma_\mu k_\mu$，$M_\mu^{(+)}$ 及 $M_\mu^{(-)}$ 为兩个只含标量 ν 及 k^2 的函数. 相应的

$$D_\mu = \bar{u}(p)[D_\mu^{(+)}(\nu,k^2)+i\hat{k}D_\mu^{(-)}(\nu,k^2)]u(p),$$
$$A_\mu = \bar{u}(p)[A_\mu^{(+)}(\nu,k^2)+i\hat{k}A_\mu^{(-)}(\nu,k^2)]u(p). \tag{26}$$

由(23)容易証明下列交叉对称关系:

$$D_\mu^{(\pm)}(\nu,k^2) = \pm D_\mu^{(\pm)}(-\nu,k^2),$$
$$A_\mu^{(\pm)}(\nu,k^2) = \mp A_\mu^{(\pm)}(-\nu,k^2). \tag{27}$$

由 Боголюбов Медведев 及 Полгванов 等人关于色散关系的証明的工作[2]，可以知道，当 $-\infty<-k^2<m^2$ 时，D_μ 和 A_μ 是滿足色散关系的. m 不会低于 π 介子的質量. 以后，我們將假定对任何 k^2 色散关系都是成立的. 这时 D_μ 可以表成 A_μ 的积分，即

$$D_\mu^{(\pm)}(\nu,k^2) = \frac{1}{\pi}\int_0^\infty\left(\frac{1}{\nu'-\nu}\pm\frac{1}{\nu'+\nu}\right)A_\mu^{(\pm)}(\nu',k^2)d\nu'. \tag{28}$$

根据(28)式，只要知道了吸收部 A_μ，就可以求出色散部 D_μ.

費曼振幅 F_μ 亦可用 D_μ 及 A_μ 表示出来，得到

$$F_\mu = D_\mu(\nu,k^2)+i\epsilon(\nu)A_\mu(\nu,k^2). \tag{29}$$

現在讓我們从(24)式出发来求吸收部 A_μ.

四. 單核子中間态对吸收部 A_μ 的貢献

由交叉对称关系，我們只令在 $\nu>0$ 的范圍內来求 A_μ. 單核子中間态的貢献为

$$A_{\mu 0} = \frac{1}{4\pi}(2\pi)^4\delta(k_0+p_0-p_0')\,|\langle p\,|\,j_\mu(0)\,|p'\rangle|^2. \tag{30}$$

为了方便起見，選擇坐标系 $\mathbf{p}=0$，此时

$$\nu = k_0,$$
$$p_0' = \sqrt{\mathbf{k}^2+M^2} = \sqrt{k_0^2+M^2+k^2}. \tag{31}$$

由(31)可得

$$\delta(k_0+M-p_0')=\frac{\nu_B+M}{M}\delta(\nu-\nu_B), \tag{32}$$

其中

$$\nu_B=\frac{k^2}{2M}. \tag{33}$$

根据电磁作用的普遍性質及电子对核子散射的結果，我們有

$$\langle \mathbf{p}|j_\mu(0)|\mathbf{k}+\mathbf{p}\rangle=\frac{-ie}{(2\pi)^3}\overline{u}(p)\left[\gamma_\mu F_1(k^2)-i\frac{1}{4M}F_2(k^2)\;(\gamma_\mu\hat{k}-\hat{k}\gamma_\mu)\right]u(\mathbf{p}+\mathbf{k}), \tag{34}$$

其中$F_1(k^2)$为核子的电荷分布函数；$F_2(k^2)$为核子的磁矩分布函数．根据 Hofstadter 等人的实驗[3]，对質子，我們有

$$F_{1p}(k^2)=\frac{1}{\mu_p}F_{2p}(k^2)=\frac{1}{\left(1+\frac{a^2k^2}{12}\right)^2}, \tag{35}$$

其中 μ_p 为質子的反常磁矩；$a=0.8\times10^{-13}$厘米，为質子的平均电荷半徑．对中子，我們有

$$F_{1n}(k^2)=0,$$
$$F_{2n}(k^2)=\mu_n F_{1p}(k^2), \tag{36}$$

其中 μ_n 为中子的反常磁矩．

將(34)代入(30)中，对中間态的自旋态求和，并对 μ 求和，可得

$$A_0=\sum_\mu A_{\mu0}=-\frac{1}{4\pi}(2\pi)^{-2}\frac{e^2}{2M}\delta(\nu-\nu_B)\overline{u}(p)\left[(2i\hat{k}+2M)F_1^2(k^2)-\right.$$
$$\left.-\frac{3}{M}k^2F_1(k^2)F_2(k^2)-\frac{1}{M^2}F_2^2(k^2)\left(Mk^2+i(k\cdot p)\hat{k}+i\frac{3}{4}k^2\hat{k}\right)\right]u(p). \tag{37}$$

將(37)代回(28)中，可以求得

$$D_0=-\frac{1}{4\pi}(2\pi)^{-2}\frac{e^2}{2M\pi}\left(\frac{1}{\nu_B-\nu}+\frac{1}{\nu_B+\nu}\right)\overline{u}\left[(2i\hat{k}+2M)F_1^2(k^2)-\right.$$
$$\left.-\frac{3}{M}k^2F_1(k^2)F_2(k^2)-\frac{1}{M^2}F_2^2(k^2)\left(Mk^2+ik\cdot p\hat{k}+i\frac{3}{4}k^2\hat{k}\right)\right]u(p). \tag{38}$$

从(37)及(38)中，可以求得費曼振幅：

$$F=-\frac{1}{4\pi}(2\pi)^{-2}\frac{e^2}{2M\pi}\left(\frac{1}{\nu_B-\nu-i\varepsilon}+\frac{1}{\nu_B+\nu-i\varepsilon}\right)\overline{u}(p)\left[(2i\hat{k}+2M)F_1^2(k^2)-\right.$$
$$\left.-\frac{3}{M}k^2F_1(k^2)F_2(k^2)-\frac{1}{M^2}F_2^2(k^2)\left(Mk^2+ik\cdot p\hat{k}+i\frac{3}{4}k^2\hat{k}\right)\right]u(p). \tag{39}$$

將(39)代入(20)式，再將 ν 代以 $-k\cdot p/M$，ν_B代以$k^2/2M$，

即得电磁質量

$$\Delta M=\overline{u}(p)(A+B+C)u(p). \tag{40}$$

其中

$$A=\frac{-ie^2}{4\pi}\frac{1}{2\pi^3}\int\frac{i\hat{k}+M}{k^2[(p+k)^2+M^2]}F_1^2(k^2)d^4k,$$

$$B=\frac{ie^2}{4\pi}\frac{3}{4\pi^6 M}\int\frac{F_1(k^2)F_2(k^2)}{(p+k)^2+M^2}d^4k, \tag{41}$$

$$C=\frac{ie^2}{4\pi}\frac{1}{4\pi^3M^2}\int\frac{Mk^2+ik\cdot p\hat{k}+i\frac{3}{4}k^2\hat{k}}{k^2[(p+k)^2+M^2]}F_2^2(k^2)d^4k.$$

(41)式中，A代表电荷与电荷相互作用得到的电磁質量，B代表电荷与磁矩相互作用得到的电磁質量，C代表磁矩与磁矩相互作用得到的电磁質量．(40)这一項和用微擾論得到的形式上相近，如令 $F_1=1$，$F_2=\mu$，則得到通常用微擾論求得的質子的电磁質量．此时积分(41)发散．用本文方法，自动在积分(41)中出現核子的电荷及磁矩分布函数，它們保証了积分是收斂的．

將实驗中求得的分布函数(35)及(36)代入(41)中，可以算出 ΔM 来．計算的結果表明，这一項所产生的中子質子質量差 $M_n-M_p\approx 0.7me$．这显然与实驗不符．

在計算中发現 A 是正的，而 B 及 C 是負的，即电荷相互作用导致質量增加，而电荷与磁矩及磁矩相互作用导致質量减少．如果核子半徑 a 减小，发現 B 的絕对值增長得比 A 快得多．当 a 减小到 0.2×10^{-13} 厘米时，可得到与实驗符合的中子質子質量差．在(41)中將 a 减小，实际上相当于增加更多 k^2 大的分量的貢献．如果(40)式是主要产生中子質子質量差的項，則目前在能量小的范圍內定出的电荷及磁矩分布函数不能外推到 k^2 大的范圍去，而在 k^2 大的地方，分布函数还有一共振峯［不按(35)式降落到零］．这相当于核子在原点附近的电荷分布有很奇怪的行为(例如电荷由正到負再到正改变得很快)．

五．其他中間态对核子电磁質量的貢献

从(24)式可以看到，在單核子态后的中間态是核子加介子的中間态．这时

$$-p_n^2>(M+m_\pi)^2.$$

从 $-p_n^2=(M+\mu)^2$ 开始最小的 ν 是

$$|\nu_1|=\frac{1}{2M}(2Mm_\pi+m_\pi^2+k^2). \tag{42}$$

由(24)中还可以看到，A 和虛光子对核子散射的总截面有連系．光与核子作用总截面为

$$\sigma_t=\frac{1}{3}(2\pi)^7\frac{1}{2k}\sum_n\delta^4(p+k-p_n)|\langle p|j_\mu(0)|n\rangle|^2. \tag{43}$$

由于虛光子中包含了縱光子的貢献，我們需要乘以因子 1/3 以对初态光子的三个自旋方向平均．比較(24)及(43)，在 $k_0>0$ 时可得

$$A=\frac{3\sigma_t k}{(2\pi)^4}. \tag{44}$$

(44)式中 σ_t 代表光对核子散射的总截面，其中以光生介子的截面最重要．σ_t 严格講来应从解虛光子产生介子的色散关系中求出，这当然是很困难的．根据 Fubini, Nambu 及

Wataghin 等人的工作，虚光子产生介子的截面与实光子产生介子的截面大体上只差电磁分布函数的因子．为了估計这一項的大小，我們將假定

$$\sigma_t(\nu, k^2)=\sigma(\nu)F^2(k^2), \tag{45}$$

其中 $\sigma(\nu)$ 为光生介子的截面．

將(43)，(45)代入(24)及(28)中，可以求得

$$\Delta M=\frac{i}{\pi}\int\frac{3\sigma(\nu')\nu'd\nu'}{(2\pi)^4}\int\frac{Md^4kF^2(k^2)}{k^2(M\nu'+k\cdot p)}. \tag{46}$$

將(35)式的 $F(k^2)$ 代入(46)，并对 k^2 积分，可求得

$$\Delta M\cong-2\frac{1}{(2\pi)^4}\frac{12}{a^2}\int_{\nu_0}\sigma(\nu')d\nu'. \tag{47}$$

在求得(47)时，应用了下列的輔助公式

$$I=\int\frac{d^4k}{k^2(M\nu'+k\cdot p)(k^2+\lambda^2)^2}=$$

$$=\frac{\pi i}{6}\int_0^1 2ydy\int_0^1\frac{6z^2dz}{[\lambda^2z^2y+M\nu'(1-z)z+\frac{1}{4}M^2(1-z)^2]^2}. \tag{48}$$

由于(48)式积分只在 $z\sim1$ 处重要（由于分子上 z^2 的因子），故可在分母中忽略掉 $\frac{1}{4}M^2(1-z)^2$ 的項．这样立刻得到

$$I=\pi i\int 2ydy\int_0^1\frac{dz}{[\lambda^2yz+M\nu'(1-z)]^2}=2\pi i\frac{1}{\lambda^2M\nu'}. \tag{49}$$

將(48)式兩边对 λ^2 微分兩次，得到

$$\int\frac{\lambda^8d^4k}{k^4(M\nu'+k\cdot p)(k^2+\lambda^2)^4}=\frac{2\pi i}{3}\frac{\lambda^2}{M\nu'}. \tag{50}$$

令 $\lambda^{-2}=a^2/12$，即可由(50)求出(47)式．

从(47)可以求出中子与質子質量差

$$\Delta M\simeq-2\frac{1}{(2\pi)^4}\frac{12}{a^2}\int_{\nu'}(\sigma_n(\nu')-\sigma(\nu'))d\nu'. \tag{51}$$

利用光生介子的实驗值，可以估計(51)式中的积分

$$\int_{\nu_0}[\sigma_n-\sigma_p]d\nu'\leqslant\frac{1}{500}m_\pi^{-1}. \tag{52}$$

由(51)及(52)可得（当取 $a=0.8\times10^{-13}$ 厘米时）

$$|\Delta M|\leqslant\frac{1}{20}m_e. \tag{53}$$

可以看到，这一項对中子質子質量差的貢献是微不足道的．此外，鉴于在閾附近，光生中子的截面比光生質子的截面来得大，因此由这一項得到的电磁質量符号也是不对的．

六. 結論

由以上的計算,可以得出如下的結論:

1. 如果中子質子質量差是由电磁作用引起,且色散关系理論是正确的,則核子的电磁分布函数在小距离有很大的变化.

2. 單核子項对电磁質量的貢献比其他中間态的貢献来得大. 这和π介子物理中其他現象的情况是相似的.

3. 鉴于电磁分布函数在小距离有很大变化并不是一定可能的,因此有理由怀疑中子質子的質量差是否由电磁作用引起的. 最近的实驗表明K^0的質量比K^-介子大. 如果K介子的自旋为零,这一質量差别从电磁作用的角度也是不能理解的,因此很可能除了电磁作用以外,还有一种作用是不遵守与电荷无关的性質,因之也可能产生中子質子質量差. 当然最后还有一个可能是現代场論不能应用在这問題上.

附註 在本文已經完成时,作者有机会看到了 Gatto, Ferrai 和 Cini 作的一項工作的預印品. 他們用色散关系計算了單核子項的貢献. 他們的結論和(六)中所提出者相同,即認为分布函数在小距离有很大变化.

参考文献

[1] Weisskopf, V. F., *Phys. Rev.* **56** (1939), 72; Feynman, R. P. 及 Speisman, G., *Phys. Rev.* **94** (1954), 500; Petermann, A., *Helv. Phys. Acta.* **27** (1954), 441; Huang, K., *Phys. Rev.* **101** (1956), 1173.

[2] Боголюбов, Н. Н., Медведев, Б. В. 及 Поливанов, М. К., Вопросы теории дисперсионных соотношений. 1958.

[3] Hofstadter, R., *Annual Review of Nuclear Science* **7** (1957).

РАЗНОСТЬ МАСС НЕЙТРОНА И ПРОТОНА

Чжоу Гуан-чжао

(*Объединенный институт ядерных исследований*)

Резюме

Вычислена электромагнитная масса нуклонов методом дисперсионного соотношения.

第15卷 第7期 1959年7月 物理学报 ACTA PHYSICA SINICA Vol. 15, № 7 July 1959

普适費米弱相互作用理論及μ介子在原子核上的俘獲*

周光召 B. 馬耶夫斯基

(联合原子核研究所，莫斯科)

一. 引 言

由 Feynman 及 Gell-Mann; Sudarshan 及 Marshak[7] 所提出的 $V-A$ 耦合弱作用已为现有的全部的 β 衰变的实驗所証实. 在工作[1]中，作者認为一切弱作用的汉密頓量的形式是一样的，因而提出了普适費米弱相互作用理論(以后簡称为普适弱作用理論).

有些粒子(例如核子)和 π 介子及 K 介子有强作用，而有些粒子(例如 μ 介子)則沒有这种作用. 由于强作用的影响，物理核子和 μ 介子衰变的情况便会不同. 虽然它們衰变的汉密頓量的形式完全相同，但实际观察到的弱作用耦合常数却可能不同. 由强作用而引起的耦合常数的变化称为耦合常数的重正化效应. 要判別弱作用的普适性，必須考慮到重正化的效应.

令 μ 介子在核子上俘获的汉密頓量密度为

$$H(x)=J_{\alpha}\overline{\psi}_{\nu}\gamma_{\alpha}(1+\gamma_{5})\psi_{\mu}, \tag{1}$$

其中 $J_{\alpha}=\frac{g_0}{\sqrt{2}}\overline{\psi}_{n}\gamma_{\alpha}(1+\gamma_{5})\psi_{p}$ 为核子弱作用电流矢量；g_0 为未經重正化的耦合常数；场量算符 ψ 右下角的标符 μ, ν, p, n 标誌着不同的粒子；γ_{α} 为狄拉克算符. (1)式中的汉密頓量密度是对 α 求和的.

如原子核的初态为 $|i\rangle$，末态为 $|f\rangle$，μ 介子开始时处于最低的波尔軌道上，且忽略 μ 介子的波函数 $\varphi(x)$ 在原子核上的变化，則 μ 介子在原子核上俘获的跃迁矩陣可以表成下列形式[1)]：

$$T_{fi}=\int\langle f|J_{\alpha}(x)|i\rangle\frac{1}{(2\pi)^3}\,\overline{u}_{\nu}(k)\gamma_{\alpha}(1+\gamma_{5})u_{\mu}\varphi(0)\cdot e^{-i(k-p_{\mu})\cdot x}\,d^4x, \tag{2}$$

其中 u_{μ} 为 μ 介子的狄拉克旋量，其相应的动量 $\mathbf{p}_{\mu}=0$，而其能量 $p_{\mu 0}=m_{\mu}$；m_{μ} 为 μ 介子的静止質量；$u_{\nu}(k)$ 为中微子的狄拉克旋量，它的动量能量矢量为 k. (2)式是从(1)式对弱作用取一級微擾求得的. 为了把强作用的效果全部算进去，(2)式中 $J_{\alpha}(x)$ 是处于海森堡表象之中(对强作用講)，而 $|i\rangle$ 及 $|f\rangle$ 为物理原子核之狀态矢量(其中包含了介子云的作用).

---

* 1959年2月22日收到.

1) 以后均采用單位系統 $\hbar=c=1$，且 $a\cdot b=\mathbf{a}\cdot\mathbf{b}-a_0b_0$；$\mathbf{a}$ 及 a_0 分別为四元矢量 a 的空間及时間分量.

根据强作用对移动群不变的性質，有

$$J_\alpha(x)=e^{-i\hat{p}\cdot x}J_\alpha(0)e^{i\hat{p}\cdot x}, \quad (3)$$

其中 $\hat{p}$ 为原子核系統的动量能量算符．由于始态及末态都是 $\hat{p}$ 的本征态，其本征值各为 p_i 及 p_f，故將(3)代入(2)可得

$$T_{fi}=(2\pi)\delta^4(p_f+k-p_i-p_\mu)\cdot\langle f|J_\alpha(0)|i\rangle\overline{u}_\nu(k)\gamma_\alpha(1+\gamma_5)u_\mu\varphi(0). \quad (4)$$

我們先来看 μ^- 介子在質子上俘获的情况，此时始态为物理質子，而末态为物理中子．根据相对論不变性，$\langle f|J_\alpha|i\rangle$ 应是一四元矢量，它的最普遍的形式为[2]

$$(2\pi)^3\langle n|J_\alpha|p\rangle=\frac{g}{\sqrt{2}}\overline{u}(p_n)\Big\{\gamma_\alpha+\lambda\gamma_\alpha\gamma_5-i\frac{\mu}{4M}[\gamma_\alpha(\hat{p}_n-\hat{p}_p)-(\hat{p}_n-\hat{p}_p)\gamma_\alpha]+$$

$$+c(p_n-p_p)_\alpha+d[\gamma_\alpha(\hat{p}_n-\hat{p}_p)-(\hat{p}_n-\hat{p}_p)\gamma_\alpha]\gamma_5+\frac{f}{im_\mu}(p_n-p_p)_\alpha\gamma_5\Big\}u(p_p), \quad (5)$$

其中 $u(p)$ 为核子的狄拉克旋量，p_p 及 p_n 为始态質子及末态中子的动量能量矢量；M 为核子質量；g, λ, μ, c, d 及 f 都是标量 $(p_n-p_p)^2=(\mathbf{p}_n-\mathbf{p}_p)^2-(p_{n0}-p_{p0})^2$ 的函数；$\hat{a}=\mathbf{a}\cdot\gamma+ia_0\gamma_4$．由于在原有汉密頓量的电流密度 $J_\alpha(x)$ 中具有下列对称性質[3]，即令 p_c 及 n_c 代表質子及中子的电荷共軛（或其反粒子），通过变換 $p\to n_c$, $n\to p_c$, $\pi\to-\pi$ 后，矢量弱作用电流 $\overline{\psi}_n(x)\gamma_\alpha\psi_p(x)$ 改变符号，而膺矢量弱作用电流 $\overline{\psi}_n(x)\gamma_\alpha\gamma_5\psi_p(x)$ 不变号，故在电流跃迁矩陣 $\langle n|J_\alpha(x)|p\rangle$ 中也应具有相同性質．这是因为强作用也对上述变換守恆的关系．由此即可証明 $c=d=0$．

如果不存在强作用，则由(1)及(2)应当得到

$$(2\pi)^3\langle n|J_\alpha(0)|p\rangle=\frac{g_0}{\sqrt{2}}\overline{u}(p_n)\gamma_\alpha(1+\gamma_5)u(p_p). \quad (6)$$

換句話講，μ 及 f 应等于零，且 $g=g_0$, $\lambda=1$．一般講来，强作用不仅可以改变 g_0 的值，而且还会增加新的电流項．

如果注意到核子由于强作用的存在会得到很大的反常磁矩的事实，就不会感到太奇怪了．此时，在电磁作用中，不仅有通常由电荷运动引起的电流，还会有因反常磁矩引起的电流．在衰变的問題中也有类似的現象，由于电磁作用电流是矢量，故弱作用中的矢量电流有許多性質是与电磁作用电流相似的．我們可以將 g 称之为 β-电荷，而 μ 称之为 β-磁矩．

在电磁作用中，由于电流守恆定律，虽有强作用的存在，物理質子的电荷和裸質子的电荷是相同的，也就和沒有强作用的粒子（例如电子）的电荷大小一样．强作用只改变了物理核子的磁矩．同样，在弱作用中，Gell-Mann 及 Feynman 发現由 μ 介子的衰变定出的耦合常数和 β 衰变中矢量耦合常数相近．这表明矢量耦合常数受强作用的影响很小．为了解释这一現象，F-G 提出了弱作用中矢量电流守恆的假定．在这一假定下，弱作用的矢量电流便和电磁作用矢量电流具有完全相似的性質，它們是同位旋空間中同一矢量的不同分量[4]．由此，在 $(p_n-p_p)^2=0$ 时，不仅 β 电荷 g 不受强作用的影响 $g=g_0$，而且 μ 应为質子与中子的反常磁矩之差，$\mu=3.7$．此外，g 和 μ 作为 $(p_n-p_p)^2$ 的函数与核子的电荷及磁矩分布函数相同．

在工作[5]中，我們曾計算了 β-磁矩对 μ 介子在質子上俘获过程的影响。結果表明，如果 Gell-Mann 和 Feynman 的假定是正确的，則 β-磁矩的效应在 20% 左右。由于 μ^- 介子在質子上俘获的几率太小，在[5]中即已提出要計算 β-磁矩对 μ 介子在原子核上俘获的影响，可以預計这一效应也不会很小。本文的目的就在完成这一任务。为了得到最普遍的結果，我們在本文中不仅考虑了 β-磁矩的效应，也考虑了 f 項的效应。

在第二节中，我們將給出俘获几率、中子与极化 μ 介子的角关联，以及中子极化的普遍公式。在第三节中我們將給出 μ^- 介子在質子上俘获的几率，角关联及中子极化的公式。第四节將討論 μ 介子在氘核上的俘获，第五节則將討論 μ 介子在原子核上的俘获。

二. μ 介子在原子核上俘获的普遍公式

把核子弱作用电流在核子自旋空間中表示出来，可以得到

$$(2\pi)^3\langle n|J_\alpha|p\rangle\bar{u}_\nu\gamma_\alpha(1+\gamma_5)u_\mu = A+\mathbf{B}\cdot\boldsymbol{\sigma}, \tag{7}$$

其中 $\boldsymbol{\sigma}$ 为核子的自旋算符；

$$A=\frac{g}{\sqrt{2}}\left(\bar{u}_\nu\beta(1+\gamma_5)u_\mu+\frac{\mathbf{k}}{2M}\cdot\bar{u}_\nu\beta\rho_1\boldsymbol{\sigma}(1+\gamma_5)u_\mu\right),$$

$$\mathbf{B}=\frac{g}{\sqrt{2}}\Big[\lambda\bar{u}_\nu\beta\rho_1\boldsymbol{\sigma}(1+\gamma_5)u_\mu+\frac{\mu+1}{2M}i\mathbf{k}\Lambda\bar{u}_\nu\beta\rho_1\boldsymbol{\sigma}(1+\gamma_5)u_\mu-$$
$$-\frac{f-\lambda}{2M}\mathbf{k}\bar{u}_\nu(1-\gamma_5)u_\mu\Big]. \tag{8}$$

在求(8)式时，所有与核子速度成平方的項都已忽略去，且用了 $\mathrm{p}_\mu=0,\ \mathrm{p}_n-\mathrm{p}_\nu=-\mathrm{p}_\nu=-\mathrm{k}$ 的条件。为了方便起見，以后用 μ' 代表 $\mu+1$，f' 代表 $f-\lambda$。

在 μ 介子及中微子的自旋空間中表示出来，(8)式可写成下列形式：

$$A=\frac{g}{2}\left(1+\frac{k}{2M}\right)(1-\boldsymbol{\sigma}_\mu\cdot\mathbf{n}),$$

$$\mathbf{B}=\frac{-g}{2}(1-\boldsymbol{\sigma}_\mu\cdot\mathbf{n})\left[\lambda\boldsymbol{\sigma}_\mu+i\frac{k}{2M}\mu'\mathbf{n}\Lambda\boldsymbol{\sigma}_\mu+\frac{f'}{2M}k\mathbf{n}\right], \tag{9}$$

其中 $\mathbf{n}$ 为与中微子动量 $\mathbf{k}$ 平行的單位矢量。为了和核子的自旋算符区分开，我們在(9)式中用 $\boldsymbol{\sigma}_\mu$ 来代表 μ 介子和中微子的自旋算符。从 (9) 式中可以看到，A 及 $\mathbf{B}$ 是 μ 介子及中微子自旋空間的算符，它們与核子的坐标及自旋沒有絲毫关系。

以上只考虑了 μ^- 介子在質子上俘获的矩陣元。当把原子核看作物理核子構成的体系，而假定每一物理核子的介子云受其他核子影响很小时，μ^- 介子在原子核上的俘获矩陣元与下式成正比：

$$R=\int\psi_f^*\sum_j\tau^{(-)}(j)(A+\mathbf{B}\cdot\boldsymbol{\sigma}_j)e^{-i\mathbf{k}\cdot\mathbf{r}_j}\psi_i dV. \tag{10}$$

其中 $\tau^{(-)}(j)$ 为把質子变成中子的同位旋算符，若第 j 个核子为質子，則經过 $\tau^{(-)}(j)$ 作用后变为中子，若它为中子，則 $\tau^{(-)}(j)$ 作用上去給出为零的結果。ψ_i 及 ψ_f 为原子核在始态与末态的波函数。通常中子总获得了足够能量而脫离原子核，因此末态波函数中包括逸出中子及殘留核之波函数。鉴于質量中心运动已經在 (4) 式中分离出去，ψ_f 中只包括中

子与殘留核的相对坐标，因之只与其相对动量 $\mathbf{p}=2\mathbf{p}_n+\mathbf{k}$ 的大小有关。

考虑到波函数具有全反对称的性質，(10)式可以表成下列形式：

$$R=Aa+\mathbf{B}\cdot\mathbf{b}, \tag{11}$$

其中

$$a=N\int\psi_f^*\tau^{(-)}(1)e^{-i\mathbf{k}\cdot\mathbf{r}_1}\psi_i dV,$$
$$\mathbf{b}=N\int\psi_f^*\tau^{(-)}(1)\sigma_1 e^{-i\mathbf{k}\cdot\mathbf{r}_1}\psi_i dV, \tag{12}$$

其中 N 为原子核中的核子数。以后逸出的中子的自旋算符为 σ_1。

利用(11)，可將(4)式写成

$$T_{fi}=(2\pi)^{-2}\delta^4(p_f+k-p_i-p_\mu)R\varphi(0). \tag{13}$$

注意到 $\varphi(x)$ 是 μ 介子在最低波尔軌道上的波函数，因此

$$|\varphi(0)|^2=\frac{Z^3}{\pi a_0^3}, \tag{14}$$

其中 a_0 为氫原子之波尔半徑，Z 为原子核的电荷。由(13)及(14)容易求得 μ^- 介子在原子核上俘获的几率及角关联。設原子核在始态未极化，可得

$$dw=(2\pi)^{-5}\frac{Z^3}{\pi a_0^3}\frac{1}{2(2J+1)}\mathrm{Sp}R(1+\sigma_\mu\cdot\mathbf{P}_\mu)R^+d\mathbf{p}_n d\mathbf{k}\,dE, \tag{15}$$

其中 $\mathbf{P}_\mu$ 代表 μ^- 介子在始态的极化矢量；J 代表原子核在始态的总角动量；Sp 代表对始态及末态的磁量子数的求和。

中子的极化矢量 $\langle\sigma_1\rangle$ 可以用下列公式表示出来：

$$[\mathrm{Sp}R(1+\sigma_\mu\cdot\mathbf{P}_\mu)R^+]\langle\sigma_1\rangle=\mathrm{Sp}R\sigma_1(1+\sigma_\mu\cdot\mathbf{P}_\mu)R^+. \tag{16}$$

下面一个問題是要把 Sp 求出来。我們先对原子核的始态及末态磁量子数求和，根据轉动及空間反映不变的性質，

$$\frac{1}{2J+1}\mathrm{Sp}\,aa^+=F_1(E_N),$$
$$\frac{1}{2J+1}\mathrm{Sp}\,a\mathbf{b}^+=\frac{1}{2J+1}\mathrm{Sp}\,\mathbf{b}a^+=0, \tag{17}$$
$$\frac{1}{2J+1}\mathrm{Sp}\,\mathbf{bb}^+=F_2(E_N)\mathrm{I}+F_3(E_N)\mathbf{pp}.$$

(17)式中第二式应为零是由于 $\mathrm{Sp}a\mathbf{b}^+$ 应是一膺矢量，但由一个矢量 $\mathbf{p}$（中子与殘留核的相对动量）却造不出膺矢量来，因此它只能为零。F_1 及 F_2 和 F_3 为中子能量 E_N 的函数，它們的具体形式随每一个具体原子核的性質而有所不同。

同样，可以求得

$$\frac{1}{2J+1}\mathrm{Sp}\,a\sigma_1 a^+=0,$$
$$\frac{1}{2J+1}\mathrm{Sp}\,a\sigma_1\mathbf{b}^+=F_4(E_n)\mathrm{I}+F_5(E_n)\mathbf{pp}, \tag{18}$$

$$\frac{1}{2J+1}\mathrm{Sp}\,\mathbf{b}\cdot\mathbf{B}\sigma_1\mathbf{b}^+\cdot\mathbf{B}^+ = iF_6(E_N)\mathbf{B}\times\mathbf{B}^+ + iF_7(E_N)\mathbf{pp}\cdot(\mathbf{B}\times\mathbf{B}^+),$$

其中 F_4, F_5, F_6 及 F_7 也是一些和原子核性質有关的函数．从(15) 及 (17)中可以看到，我們需要求出下列式子，即

$$\begin{aligned}&(1)\ \mathrm{Sp}\,A(1+\sigma_\mu\cdot\mathbf{P}_\mu)A^+,\\&(2)\ \mathrm{Sp}\,\mathbf{B}\cdot(1+\sigma_\mu\cdot\mathbf{P}_\mu)\mathbf{B}^+,\\&(3)\ \mathrm{Sp}\,\mathbf{B}\cdot\mathbf{P}(1+\sigma_\mu\cdot\mathbf{P}_\mu)\mathbf{B}\cdot\mathbf{p}^+;\end{aligned}\tag{19}$$

而为了求得中子的极化，还需求出

$$\begin{aligned}&(4)\ \mathrm{Sp}\,A(1+\sigma_\mu\cdot\mathbf{P}_\mu)\mathbf{B}^+,\\&(5)\ \mathrm{Sp}\,A(1+\sigma_\mu\cdot\mathbf{P}_\mu)\mathbf{B}\cdot\mathbf{p}^+,\\&(6)\ \mathrm{Sp}\,\mathbf{B}\times(1+\sigma_\mu\cdot\mathbf{P}_\mu)\mathbf{B}^+,\end{aligned}\tag{20}$$

这些式子都在附录 I 中給出．

將附录 I 中求出的結果代入(17)，(18)及(15)中，即得到 μ^- 介子在原子核上俘获的几率及角关联．由(16)式可以求得中子的极化．

三． μ^- 介子在質子上的俘获[6]

当我們考虑 μ^- 介子在質子上的俘获时，得到特別簡單的結果．此时 ψ_i 及 ψ_f 只包含始态質子和末态中子的自旋波函数．在核子自旋空間表象中，容易求得

$$a=1,\quad \mathbf{b}=\sigma_1,\tag{21}$$

从(21)馬上可以求得(17)及(18)式的 F_i 函数，得到

$$\begin{aligned}&F_1=F_2=1,\quad F_3=0,\\&F_4=F_6=1,\quad F_5=F_7=0.\end{aligned}\tag{22}$$

在公式(15)中，还应作一些修改，即当 μ^- 介子为質子俘获时，末态只有中子及中微子(而不象其他情形下还有殘留核)，此时它們的动量滿足关系

$$\mathbf{p}_n=-\mathbf{k}.\tag{23}$$

而在計算末态的狀态数目时，不应再把 $\frac{d\mathbf{p}_n}{(2\pi)^3}$ 算进去．相应的几率公式为

$$dw=(2\pi)^{-2}\frac{1}{\pi a_0^3}\frac{1}{4}\mathrm{Sp}R(1+\sigma_\mu\cdot\mathbf{P}_\mu)R^+d\mathbf{k}_f\,dE.\tag{24}$$

由(17)，(18)、(24)及附录 I 可以算出 μ^- 介子在質子上俘获的几率．

$$dw=(2\pi)^{-2}(\pi a_0^3)^{-1}2^{-1}g^2I[1-\alpha\mathbf{P}_\mu\cdot\mathbf{n}]k^2d\Omega,\tag{25}$$

其中

$$I=1+3\lambda^2+\beta(1-\lambda f')+\beta\mu'(2\lambda+\beta\mu'/2)+\frac{\beta^2}{4}f'^2,\tag{26}$$

$$I\alpha=1-\lambda^2+\beta((1-\lambda f')-\beta\mu'(2\lambda+\beta\mu'/2)+\frac{\beta^2}{4}f'^2.\tag{27}$$

此处

$$\beta=\frac{k}{M}.$$

如在(25)，(26)及(27)中，令 $f'=f-\lambda$，再把与 f 有关的項忽略去，我們即得到工

作[5]中的結果．本文中不仅考慮了β-磁矩的影响，而且考慮了由于重正化出現的贋标耦合的影响．由(26)及(27)中可以看到，如果f'及μ'均为正数，則它們的影响將在角关联系数中得到加强，而在俘获几率中的影响則將部分抵消．現有的实驗材料表明，I的值比$1+3\lambda^2$来得大(約大0.4—1倍)[1)]，这表明f'或者很小，或者是負数．这与在工作[2]中用色散关系估計出的f'不合．由色散关系估計出的f'为正数且很大．同时用实驗来研究俘获几率及角关联，即同时定出I和α之值，將有助于搞清楚这个問題．

同样可以求出中子极化的公式，得到

$$I[1-\alpha\mathbf{P}_\mu\cdot\mathbf{n}]\langle\sigma_1\rangle=[a+b\mathbf{P}_\mu\cdot\mathbf{n}]\mathbf{n}+c\mathbf{P}_\mu, \tag{28}$$

其中

$$a=2\left[\lambda(\lambda+1)+\frac{\beta}{2}\lambda+\beta\mu'(\lambda+\beta\mu'/4)-\frac{\beta}{2}f'-\frac{\beta^2}{4}f'\right], \tag{29}$$

$$b=\beta\left[f'(\lambda+1)+\mu'(\lambda+1)+\frac{\beta^2}{2}\mu'(f'+\mu')+\frac{\beta}{2}(f'+\mu')\right]=$$
$$=\beta(f'+\mu')\left(\lambda+1+\frac{\beta}{2}+\frac{\beta^2}{2}\mu'\right), \tag{30}$$

$$c=2\left[\lambda^2-\lambda+\frac{\beta}{2}\mu'(\lambda-1)-\frac{\beta}{2}f'\left(\lambda+\frac{\beta}{2}\mu'\right)-\frac{\beta}{2}\left(\lambda+\frac{\beta}{2}\mu'\right)\right]=$$
$$=2\left[(\lambda-1)\left(\lambda+\frac{\beta}{2}\mu'\right)-\left(\frac{\beta}{2}f'+\frac{\beta}{2}\right)\left(\lambda+\frac{\beta}{2}\mu'\right)\right]=$$
$$=2\left(\lambda+\frac{\beta}{2}\mu'\right)\left(\lambda-1-\frac{\beta}{2}-\frac{\beta}{2}f'\right). \tag{31}$$

注意到(28)式中$\mathbf{n}$代表中微子运动方向，它与中子运动方向相反，故a的符号与[5]中不同，因在[5]中$\mathbf{n}$代表中子运动方向．此外公式(29)—(31)与[5]中給出的稍有不同，这是因为对高級項处理有不同的原故．此处給出的公式准确到β^2的数量級，較[5]中給出的更为正确．

可以看到，如果f'及μ'为正数，則它們的作用只在系数b中互相加强，而在a及c中都相互抵消一部分．

四．μ^-介子在氘核上的俘获

在过渡到复杂核去以前，我們先討論一个最簡單的原子核氘核．在这一最簡單的情况下已經可以表明μ^-介子在原子核上俘获的若干特征．

在氘核的情况下，由于总自旋守恆，末态的兩个中子或者处于自旋三重态或者处于自旋独态．令φ_s,φ_t及φ_d分别为末态中子在独态、三重态以及氘核之空間波函数，χ_s,χ_t及χ_d分别为其自旋波函数．我們將忽略張量力(即氘核基态中D波)的貢献．由(12)可得

$$a=\chi_t^*\chi_d J_t,$$
$$\mathbf{b}=\chi_t^*\sigma_1\chi_d J_t+\chi_s^*\sigma_1\chi_d J_s. \tag{32}$$

1) 在与实驗比較中，我們假定普适弱作用理論是对的，且耦合常数随能量变化不大，因此g及λ的值都取自β-衰变中定出之值．

其中

$$J_t = \int \varphi_t^* e^{-i\mathbf{k}\cdot\mathbf{r}/2} \varphi_d dV,$$
$$J_s = \int \varphi_s^* e^{-i\mathbf{k}\cdot\mathbf{r}/2} \varphi_d dV, \tag{33}$$

$\mathbf{r}$ 为核子相对坐标．在自旋空間中表示出来，可得

$$\chi_t^* \chi_d = \frac{1}{4}(3 + \sigma_1 \cdot \sigma_2),$$

$$\chi_t^* \sigma_1 \chi_d = \frac{1}{16}(3 + \sigma_1 \cdot \sigma_2)\sigma_1(3 + \sigma_1 \cdot \sigma_2) = \frac{1}{16}(8\sigma_1 + 2\sigma_2 + i\sigma_1 \times \sigma_2), \tag{34}$$

$$\chi_s^* \sigma_1 \chi_d = \frac{1}{16}(1 - \sigma_1 \cdot \sigma_2)\sigma_1(3 + \sigma_1 \cdot \sigma_2) = \frac{1}{16}(4\sigma_1 - 2\sigma_2 - 3i(\sigma_1 \times \sigma_2)).$$

將(32)及(34)式代回(17)及(18)式中，可得

$$F_1 = |J_t|^2,$$
$$F_2 = \frac{1}{3}(2|J_t|^2 + |J_s|^2) \tag{35}$$

$$F_3 = 0,$$
$$F_4 = \frac{4}{6}\left(|J_t|^2 + \frac{1}{2}J_s^* J_t\right),$$
$$F_5 = F_7 = 0, \tag{36}$$
$$F_6 = \frac{1}{3}(|J_t|^2 + 2\mathrm{Re}J_s^* J_t).$$

將(35)，(36)代回(17)，(18)及(15)中，注意到

$$E = \frac{p_n^2}{2M} + \frac{(\mathbf{p}_n + \mathbf{k})^2}{2M} + k$$

及

$$MdE = kp_n \sin\theta\, d\theta.$$

在經过对中微子能量积分之后，可得

$$dw = (2\pi)^{-4}(\pi a_D^3)^{-1}\frac{g^2}{2}M^2 I_D(1 + \alpha_D \mathbf{P}_\mu \cdot \mathbf{p}_n/p_n) \cdot dE_n d\Omega_n. \tag{37}$$

其中

$$I_D = (1 + \overline{\beta})I_{tt} + \frac{1}{3}\left[3\lambda^2 + \overline{\beta}\lambda(2\mu' - f') + \frac{\overline{\beta}^2}{2}\left(\mu'^2 + \frac{1}{2}f'^2\right)\right][2I_{tt} + I_{ss}]. \tag{38}$$

$$\alpha_D = (1 + \overline{\beta})I'_{tt} - \frac{1}{3}\left[\lambda^2 + \overline{\beta}\lambda(2\mu' + f') + \frac{\overline{\beta}^2}{2}\left(\mu'^2 - \frac{f'^2}{2}\right)\right] \cdot (2I'_{tt} + I'_{ss}), \tag{39}$$

$$I_{ij} = \int_{k_{\min}}^{k_{\max}} J_i^* J_j k dk,$$
$$I'_{ij} = -\int_{k_{\min}}^{k_{\max}} J_i^* J_j \frac{\mathbf{k}\cdot\mathbf{p}_n}{kp_n} k dk. \tag{40}$$

由于 $\beta=k/M$ 与 k 的大小有关，故在(38)及(39)式中都应將 β 換成某一平均的值 $\overline{\beta}$.

Überall 及 Wolgenstein[7] 在忽略了 β-磁矩的情形下計算了 μ^- 介子在氘核上的俘获．如在(38)及(39)中，令 $\mu'=0$，并忽略高級无穷小（与 β^2 成正比的全部項），則我們的公式和[7]中給出的公式相合．积分 I_{ij} 及 I'_{ij} 在[7]中已算出，故我們不必再計算它．注意到，β-磁矩能改变 I_D 及 α_D 的值达 20% 左右．

同样可以求出中子极化的公式

$$I_D(1+\alpha_D\mathbf{P}_\mu\cdot\mathbf{p}_n/p_n)\langle\sigma_1\rangle=a\mathbf{p}_n/p_n+b\mathbf{P}_\mu\cdot\mathbf{p}_n\mathbf{p}_n/p_n^2+c\mathbf{P}_\mu+d\mathbf{P}_\mu\times\mathbf{p}_n/p_n, \tag{41}$$

其中

$$a=-\frac{2}{3}\left\{\left(\lambda^2+\beta\mu'\lambda+\frac{\beta^2}{4}\mu'^2\right)(I'_{tt}+2\operatorname{Re}I'_{st})+\left(1+\frac{\beta}{2}\right)\left(\lambda-\frac{\beta}{2}f'\right)\operatorname{Re}(2I'_{tt}+I'_{st})\right\} \tag{42}$$

$$b=\frac{1}{3}\overline{\beta}\left\{(\mu'+f')\left(\lambda+\frac{\overline{\beta}}{2}\mu'\right)(I''_{tt}+2\operatorname{Re}I''_{st})+\left(1+\frac{\overline{\beta}}{2}\right)(\mu'+f')\operatorname{Re}(2I''_{tt}+I''_{st})\right\} \tag{43}$$

$$c=\frac{2}{3}\left(\lambda+\frac{\overline{\beta}}{2}\mu'\right)\left\{\left(\lambda-\frac{\overline{\beta}}{2}f'\right)(I_{tt}+2\operatorname{Re}I_{st})+\left(1+\frac{\overline{\beta}}{2}\right)\operatorname{Re}(2I_{tt}+I_{st})\right\}, \tag{44}$$

$$d=-\frac{2}{3}\left(1+\frac{\overline{\beta}}{2}\right)\left(\lambda+\frac{\overline{\beta}}{2}\mu'\right)\operatorname{Im}(2I'_{tt}+I'_{st}), \tag{45}$$

$$I''_{ij}=\int_{k_{\min}}^{k_{\max}}J_i^*J_j\frac{(\mathbf{p}_n\cdot\mathbf{k})^2}{p_n^2k^2}kdk. \tag{46}$$

在(42)—(45)中，令 $\overline{\beta}=0$ 即得到[6]中的結果．

同样可以算出 μ^- 介子在不同精細狀态上俘获几率的差别[8]．令 λ_+ 及 λ_- 分别代表 μ^- 介子在 $F=3/2$ 及 $F=1/2$ 态上俘获的几率，F 为 μ^- 介原子的总角动量．且令 $\overline{\lambda}=\frac{2}{3}\lambda_++\frac{1}{3}\lambda_+$ 为平均之俘获几率，則容易求得

$$\frac{\lambda_+-\lambda_-}{3\overline{\lambda}}=-\frac{2b_1\operatorname{Re}\int I_{tt}dE_N+b_2\int(I_{tt}+I_{ss})dE_N}{\int I_DdE_N}, \tag{47}$$

其中

$$b_1=\left(1+\frac{\overline{\beta}}{2}\right)\left(\lambda+\frac{\overline{\beta}}{2}\mu'\right),\qquad b_2=\left(\lambda+\frac{\overline{\beta}}{2}\mu'\right)\left(\lambda-\frac{\overline{\beta}}{2}f'\right). \tag{48}$$

若不考慮中子在末态上的相互作用，則 $I_{ss}=I_{tt}$，

而

$$\frac{\lambda_+-\lambda_-}{3\overline{\lambda}}=\frac{-b}{a}, \tag{47'}$$

其中

$$b=2b_1+2b_2=2\left(\lambda+\frac{\overline{\beta}}{2}\mu'\right)\left(1+\lambda+\frac{\overline{\beta}}{2}-\frac{\overline{\beta}}{2}f'\right),\qquad a=1+3\lambda^2+\overline{\beta}+\overline{\beta}\lambda(2\mu'-f')+\frac{\overline{\beta}^2}{2}\left(\mu'^2+\frac{1}{2}f'^2\right). \tag{48'}$$

若令 $\mu'=0$，我們得到[8]中的結果。

順便在这里提一下，对其他原子核，如果采用費米气体模型，可得[8]

$$\frac{\lambda_+-\lambda_-}{\lambda}=-\frac{b}{aZ'}\frac{2I+1}{I}\qquad 当\quad I=L+1/2,$$
$$\frac{\lambda_+-\lambda_-}{\lambda}=\frac{b}{aZ'}\frac{2I+1}{I+1}\qquad 当\quad I=L-1/2, \tag{49}$$

其中 I 为原子核的自旋；L 为質子在始态时的軌道角动量；b 和 a 在附录 II 中給出：

$$Z'=(Z-1)\xi+1,$$

Z 为原子核电荷，ξ 为表示泡里作用的一因子。公式(49)在忽略 β-磁矩的情况下首先由[7]得出。在导出(49)时，我們采用了和[8]完全相同的近似和方法。

从(48')中可以看到，若 f' 很小或为負数，則由于 β-磁矩的作用，a 和 b 都增加 20% 左右，而它們的比例改变很小。如果 $\beta f'$ 是正数而且很大，則可能使这一差别减小。

五. μ^- 介子在原子核上的俘获

由于我們至今还不准确了解原子核的構造，沒有原子核的准确波函数，因此要准确計算 μ^- 介子在原子核上的俘获是不可能的。

已經有一些工作[9,10]，假定原子核可用費米气体模型或單核子壳层結構模型作为近似，計算了 μ^- 介子的俘获几率。在这些工作中都沒有考虑 β-磁矩的效应。

我們在这一节中將采用單核子壳层模型。Блохинцев 及 Долинский[9] 曾用这一模型計算了 μ 介子的俘获，我們將尽量利用他們計算的結果。

根据 [9] 中的公式 (9) 及 (10)，并与本文中相应的公式比較，可以得出如下的关于 F_i 函数的結論：

$$F_1=F_2,\qquad F_3=0.$$

工作[9]中引进的函数 A_0, A_1, A_2, B_0, B_1 及 B_2 与我們的函数 F_i 的关系如下：

$$A_l(E_N)=(2\pi)^{-3}(\pi a_0^3)^{-1}Z^3(N-1)M^2\int_{k_{\min}}^{k_{\max}}F_1k\left(\frac{k}{2M}\right)^l dk,$$
$$B_l(E_N)=(2\pi)^{-3}(\pi a_0^3)^{-1}Z^3(N-1)M^2\int_{k_{\min}}^{k_{\max}}F_1k\left(\frac{\mathbf{k}\cdot\mathbf{p}_n}{kp_n}\right)\left(\frac{k}{2M}\right)^l dk. \tag{50}$$
$$l=0,1,2.$$

利用(50)及(49)式，我們可以写出 μ^- 介子在原子核上俘获的几率及角关联如下：

$$dw=\frac{g^2}{2}I(1+\alpha\mathbf{P}_\mu\cdot\mathbf{p}_n/p_n)dE_Nd\Omega_N/(2\pi), \tag{51}$$

其中

$$I=(1+3\lambda^2)A_0(E_N)+(1+2\lambda\mu'-\lambda f')A_1(E_N)+$$
$$+\left(\frac{1}{2}\mu'^2+\frac{1}{4}f'^2+\frac{1}{4}\right)A_2(E_N). \tag{52}$$

$$-I\alpha=(1-\lambda^2)B_0(E_N)+(1-2\lambda\mu'-\lambda f')B_1(E_N)+\left(\frac{1}{4}-\frac{1}{2}\mu'^2+\frac{1}{4}f'^2\right)B_2(E_N). \quad (53)$$

在工作[9]中，計算了 Ca^{40} 及 O^{16} 核上的 $A_l(E_N)$ 和 $B_l(E_N)$ 函数，我們只需改变 A_l 及 B_l 前的系数就能求得 β- 磁矩所給的貢献.

参 考 文 献

[1] Sudarshan, E. C. G. 及 Marshak, R. E., *Phys. Rev.* **109** (1958), 1860; Feynman, R. P. 及 Gell-Mann, M., *Phys. Rev.* **109** (1958), 193.
[2] Goldberger, M. L. 及 Treiman, S. B., *Phys. Rev.* **111** (1958), 354.
[3] Weinberg, S., 预印品.
[4] Gell-Mann, M., *Phys. Rev.* **111** (1958), 362; Герштейн, С. С. 及 Зельдович, Я. Б., *ЖЭТФ* **29** (1955), 698.
[5] Чжоу, Гуан-чжао (周光召) 及 Маевский, В, *ЖЭТФ* **35** (1958), 1581.
[6] Шапиро, И. С., Долинский, Э. И., Блохинцев, Л. Д., *ДАН СССР* **116** (1957), 946; *Nucl. Phys.* **4** (1957), 273. Huang, K., Yang, C. N., Lee, T. D., *Phys. Rev.* **108** (1957), 1340.
[7] Überall, H. 及 Wolgenstein, L., *Nouvo Cim.* X, (1958), 136.
[8] Bernstein, J., Lee, T. D., Yang, C. N. 及 Primakoff, H., *Phys. Rev.* **111** (1958), 313; Зельдович, Я. Б. 及 Герштейн, С. С., *ЖЭТФ* **35** (1958), 821.
[9] Долинский, Э. Й. 及 Блохинцев, Л. Д., *ЖЭТФ* **34** (1958), 759; *ЖЭТФ* **35** (1958), 1488.
[10] Tolhoek, H. A. 及 Luyten, L. R., *Nuclear Phys.* **3** (1957), 679; Иоффе, Б. Л., *ЖЭТФ* **33** (1957), 308; Überall, H., *Nuovo Cim.* **6** (1957), 533.

УНИВЕРСАЛЬНОЕ ВЗАИМОДЕЙСТВИЕ ФЕРМИ И ЗАХВАТ μ^--МЕЗОНОВ ЯДРАМИ

Чжоу Гуан-чжао　　В. Маевский

(*Объединенный институт ядерных исследований*)

Резюме

Рассчитаны вероятность захвата, угловое распределение и поляризация нейтронов, образующихся в результате поглощения поляризованных μ^--мезонов в водороде и дейтроне. Вычисляются также вероятность захвата и угловое распределение нейтронов при захвате μ^--мезонов ядрами. Проведены оценки эффекта β-магнетизма для разных явлений.

附　　录 I

(1) $\frac{1}{2}\mathrm{Sp}\ A(1+\boldsymbol{\sigma}_\mu\cdot\mathbf{P}_\mu)A^+=\frac{g^2}{2}\left(1+\frac{\beta}{2}\right)^2(1-\mathbf{P}_\mu\cdot\mathbf{n})$.

(2) $\frac{1}{2}\mathrm{Sp}\ \mathbf{B}\cdot(1+\boldsymbol{\sigma}_\mu\cdot\mathbf{P}_\mu)\mathbf{B}^+=\frac{g^2}{2}\left\{\left[3\lambda^2+\lambda\beta(2\mu'-f')+\frac{\beta^2}{2}\left(\mu'^2+\frac{1}{2}f'^2\right)\right]+\right.$

$\left.+\left[\lambda^2+\lambda\beta(2\mu'+f')+\frac{\beta^2}{2}\left(\mu'^2-\frac{1}{2}f'^2\right)\right]\mathbf{P}_\mu\cdot\mathbf{n}\right\}$.

(3) $\frac{1}{2}\mathrm{Sp}\ [A(1+\boldsymbol{\sigma}_\mu\cdot\mathbf{P}_\mu)\mathbf{B}^++\mathbf{B}(1+\boldsymbol{\sigma}_\mu\cdot\mathbf{P}_\mu)A^+]=$

$=-\frac{g^2}{2}\left(1+\frac{\beta}{2}\right)\left[2\left(\lambda+\frac{\beta}{2}\mu'\right)\mathbf{P}_\mu-\beta(\mu'+f')\mathbf{P}_\mu\cdot\mathbf{n}\mathbf{n}-2\left(\lambda-\frac{\beta}{2}f'\right)\mathbf{n}\right\}$.

(4) $\frac{1}{2}\mathrm{Sp}\ [A(1+\boldsymbol{\sigma}_\mu\cdot\mathbf{P}_\mu)\mathbf{B}^+-\mathbf{B}(1+\boldsymbol{\sigma}_\mu\cdot\mathbf{P}_\mu)A^+]=i\frac{g^2}{2}\left(1+\frac{\beta}{2}\right)\left(\lambda+\frac{\beta}{2}\mu'\right)\mathbf{P}_\mu\times\mathbf{n}$.

(5) $\frac{1}{2}\mathrm{Sp}\ \mathbf{B}\times(1+\boldsymbol{\sigma}_\mu\cdot\mathbf{P}_\mu)\mathbf{B}^+=-g^2i\left\{\left(\lambda+\frac{\beta}{2}\mu'\right)\left(\lambda-\frac{\beta}{2}f'\right)\mathbf{P}_\mu+\right.$

$\left.+\frac{\beta}{2}(\mu'+f')\left(\lambda+\frac{\beta}{2}\mu'\right)\mathbf{P}_\mu\cdot\mathbf{n}\mathbf{n}+\left(\lambda+\frac{\beta}{2}\mu'\right)^2\mathbf{n}\right\}$.

附　　录 II

鉴于在不同实驗条件下，μ^-介子处于核子的超精細結構态上的比重可能不同，此外还可采用帶有极化的靶子来作实驗，(μ^-p) 系統在始态情况可能随实驗条件和实驗方法不同而異．在最普遍情况下，始态的密度矩陣具有下列形式：

$$\rho=1/4\,(1+\boldsymbol{\sigma}_\mu\cdot\mathbf{P}_\mu+\boldsymbol{\sigma}_p\cdot\mathbf{P}_p+\zeta\boldsymbol{\sigma}_\mu\cdot\boldsymbol{\sigma}_p),$$

其中 $\mathbf{P}_p$ 为質子在始态的极化矢量．ζ 的大小与(μ^-p)系統在不同超精細結構态上的比重有关．底下我們將不討論 $\mathbf{P}_\mu$，$\mathbf{P}_p$ 及 ζ 的大小，而將利用上述 ρ 計算 μ^-介子在質子上俘获的几率，角关联及中子的极化．在將理論与实驗比較时，应根据具体的实驗条件去判斷 $\mathbf{P}_\mu$, $\mathbf{P}_p$ 及 ζ 的大小，这是一个复杂問題，在本文中不予討論．

計算結果如下

1. 俘获几率及角关联

$$dw=(2\pi)^{-2}(\pi a_0^3)^{-1}2^{-1}g^2I(1-\alpha\mathbf{P}_\mu\cdot\mathbf{n}-\gamma\mathbf{P}_p\cdot\mathbf{n})R^2d\Omega,$$

其中

$$I=1+3\lambda^2+\beta+\beta\lambda(2\mu'-f')+\frac{\beta^2}{4}(2\mu'^2+f'^2)-6\zeta\left[\lambda^2+\left(1+\frac{\beta}{2}\right)\lambda+\right.$$

$$\left.+\frac{\beta}{2}(\mu'\lambda-f'\lambda+\mu')+\frac{\beta^2}{4}(\mu'-\mu'f')\right]-\frac{1}{3}\beta\zeta(\mu'+f')\left(1+\lambda+\frac{\beta}{2}\mu'\right),$$

$$I\alpha=1-\lambda^2+\beta-\beta\lambda(2\mu'+f')+\frac{\beta^2}{4}(f'^2-2\mu'^2),$$

$$I\gamma=2\left[\lambda^2+\left(1+\frac{\beta}{2}\right)\left(\frac{\beta}{2}f'-\lambda\right)+\beta\mu'f'+\frac{\beta^2}{4}\mu'^2\right].$$

2. 中子的极化

$$I(1-\alpha\mathbf{P}_\mu\cdot\mathbf{n}-\gamma\mathbf{P}_p\cdot\mathbf{n}\langle\sigma\rangle=[a+b\mathbf{P}_\mu\cdot\mathbf{n}+c\mathbf{P}_p\cdot\mathbf{n}]\mathbf{n}+d\mathbf{P}_\mu+e\mathbf{P}_p$$

其中

$$a=2\left[\lambda(\lambda+1)+\frac{\beta}{2}\lambda+\beta\mu'(\lambda+\beta\mu'/4)-\frac{\beta}{2}f'-\frac{\beta^2}{4}f'\right]-$$
$$-\zeta\left[\left(1+\frac{\beta}{2}\right)^2+7\lambda^2+4\beta\mu'\lambda-3f'\beta\lambda-\beta^2f'\mu'+\frac{\beta^2}{4}f'^2-\frac{\beta^2}{2}\mu'^2\right].$$

$$b=\beta(f'+\mu')\left(\lambda+1+\frac{\beta}{2}+\frac{\beta}{2}\mu'\right),$$

$$c=2\beta(\mu'+f')\left(\lambda+\frac{\beta}{4}\mu'-\frac{\beta}{4}f'\right),$$

$$d=2\left(\lambda+\frac{\beta}{2}\mu'\right)\left(\lambda-1-\frac{\beta}{2}(1+f')\right),$$

$$e=\left(1+\frac{\beta}{2}\right)^2+\left(\beta\lambda f'-\frac{\beta^2}{4}f'^2-\lambda^2\right),$$

在一般情况下（例如在形成 μ^- 介分子的情况下），dw 中代表 μ^- 介子在核子上被找到的几率因子 $\frac{1}{4\pi}(\pi a_0^3)^{-1}$ 应当用 $|\varphi(0)|^2$ 去代替，其中 $\varphi(\mathbf{r})$ 为 μ 介子的波函数。

当 μ^- 介原子处于超精細态 $F=1$ 时，其俘获几率 λ_+ 可由上面計算的式子求得，此时只需令 $\zeta=\frac{1}{3}$ 就行了，

$$\lambda_+=(1-\lambda)^2+\beta\left(1-2\lambda+\mu'(\lambda-1)-\frac{1}{g}(\mu'+f')(1+\lambda)\right).$$

在写下 λ_+ 时，我們忽略了所有与 β^2 成正比的項．令 $\zeta=-1$，我們得到 μ^- 介子在 $F=0$ 态上俘获的几率 λ_-：

$$\lambda_-=(1+3\lambda)^2+\beta\left(1+3\lambda+3\mu'+5\lambda\mu'-4\lambda f'+\frac{1}{3}(\mu'+f')(1+\lambda)\right).$$

利用上面的 λ_+ 及 λ_-，我們可以求得

$$\frac{\lambda_+-\lambda_-}{\bar{\lambda}}=-\frac{b}{a}\cdot 4,$$

其中

$$b=2\lambda(1+\lambda)+\beta\left(\lambda-f'\lambda+\mu'(1+\lambda)+\frac{1}{g}(\mu'+f')(1+\lambda)\right),$$

$$a=1+3\lambda^2+\beta(1+2\mu'\lambda-f'\lambda).$$

在以上的公式中，我們都把与 β^2 成正比的項忽略了．上面虽然只求出了 μ^- 在質子的超精細态上俘获的几率，但如假定費米气体模型近似的正确，我們可以求出第四节中的(49)式，(49)式中的 b 及 a 与上面从 μ 在質子上俘获而定出的 b 和 a 相同．

第16卷 第2期　　物 理 学 报　　Vol. 16, No. 2
1960年2月　　ACTA PHYSICA SINICA　　Feb., 1960

μ^-介子在He^3原子核上的俘获*

朱家珍　周光召　彭宏安
(北京大学物理系)

提　　要

本文計算了μ介子在He^3核上俘获的几率、末态H^3核的角分布和极化．所采用的理論是带有重正化效应(包含弱磁矩及贋标項)的V-A普适弱作用理論．在計算中考虑了μ和He^3核在始态有极化及处于不同超精細态上的情况．在計算中假定了He^3核的基态是純S态，这时忽略了由张量力以及其他自旋軌道耦合力引起的其他态．介子交換电流的效应也沒有考虑．在以上这两个假定下，我們証明了俘获几率中只包含一个未知的原子核矩陣元，这个矩陣元恰好是原子核密度函数的富氏分量．利用μ介子(或电子)与He^3(或H^3)原子核的散射可以确定这个未知矩陣元．

一、引　　言

弱作用是否具有普适作用的形式，这是現代基本粒子物理中的一个重要的問題．由Feynman-Gell-Mann及Sudarshan-Marshak等人提出的普适V-A弱作用理論，在解释原子核β衰变和μ衰变的实驗中获得了巨大的成功[1]．虽然Λ及Σ超子的β和μ衰变的几率較普适理論計算的值小很多[2]，很可能奇异粒子的衰变不符合普适理論的要求，但在核子对pn，$\mu\nu$及$e\nu$对間仍然有极大可能存在普适的費米型弱相互作用．直接由μ俘获的实驗中定出，pn与$\mu\nu$对間的耦合型式及耦合常数将有助于闡明这个問題．研究μ俘获还能帮助了解由強作用产生的重正化效应[3,4]．

通常，在做μ俘获的实驗及分析这些結果时，我們需要考虑到下面两个問題：i）要能确定μ-介原子的初起状态，要能准确测量反应产物的各种分布；ii）要避免采用复杂的原子核，以減少由于不清楚原子核結构而引起的誤差[5]．

从第二个要求講，μ在質子上的俘获是最理想的，此时根本沒有原子核結构的影响．但μ-介原子在始态可能組成μ-H_2分子，末态的中子又很难测定，反应几率也小，因此μ在質子上俘获的实驗是比較困难的，看来最多可以确定μ在超精細态$F=0$上的俘获几率．

最好的滿足上述两个要求的是He^3核．經过μ俘获后，它变成为自己的鏡核H^3．此时原子核結构的影响不大，且实驗上测量H^3也比中子容易．比較困难的是要建造一个He^3的靶子或者泡室．

μ在He^3原子核上俘获的实驗目前还沒有作，理論分析在[6]中由Fujii及Primakoff

* 1959年11月30日收到．

討論过．文献[6]的結果在我們看来是不能令人満意的，其中保留了若干与核子速度v/c成正比的項，但忽略了一些其他同一数量級的項．此外，在[6]中用了变分法确定的 He^3 的波函数来計算未知的矩陣元$\langle r^2\rangle$． 我們知道，由变分法确定的只含一个（或少数几个）参量的波函数在 r 大时是不可能准确的，由变分法确定的波函数最多只在 r 小时（即相互作用強时）与眞正波函数接近．我們选择[7]中由变分法定出的另一个 He^3 的波函数，同样計算了矩陣元 $\langle r^2\rangle$．結果表明，虽然这个波函数也給出很好的 He^3 的結合能及庫侖能，但 $\langle r^2\rangle$ 与[6]中給出的相差几达一倍，由此产生的对总俘获几率的誤差也达到 15% 左右[1)]．

在本文中，我們重新根据 $V-A$ 普适弱作用理論計算了 μ 在 He^3 原子核上的俘获．在計算中忽略了 i)交換电流（即由介子云变形而引起的耦合常数的变化）的貢献；ii)核子在除 S 态以外的态上的貢献．我們将在以后的工作中討論上述近似所引起的誤差．粗略的估計表明，这种誤差只有 5% 左右．値得提起的是，μ 俘获中的贋标項及反常磁矩項完全是由于介子云的存在才有的，介子云的变形即交換电流可能对它們产生較大的变化．

在采用了上述近似后，我們发現，μ 在 He^3 核上的俘获只包含一个公共的未知矩陣元因子．在除去这个因子以后，μ 在 He^3 核上的俘获与 μ 在質子上的俘获有极相似的形式．这个矩陣元还可以用电子或 μ 介子同 He^3 核的电磁散射来确定，因而可以避免理論計算所引起的誤差．

二、μ 介子为 He^3 核俘获的普遍公式

根据 $V-A$ 弱作用理論，μ 俘获的汉密頓量密度为

$$H(x) = J_a(x)\bar{\psi}_\nu\gamma_a(1+\gamma_5)\psi_\mu, \tag{1}$$

其中

$$J_a(x) = \frac{g_0}{\sqrt{2}}\bar{\psi}_n\gamma_a(1+\gamma_5)\psi_p \tag{2}$$

为核子弱作用电流，其他符号均与[4]同．

令原子核的始态为 $|i\rangle$，末态为 $|f\rangle$，并忽略 μ 介子在波尔軌道上的波函数 $\varphi(x)$ 在原子核上的变化，則 μ 介子在原子核上俘获的对弱作用为一級微扰的矩陣元可以表为

$$T_{fi} = \int\langle f|J_a(x)|i\rangle\frac{1}{(2\pi)^3}\bar{u}_\nu(k)\gamma_a(1+\gamma_5)u_\mu(p_\mu)\varphi(0)e^{-i(k+p_\mu)\cdot x}d^4x. \tag{3}$$

上式中 $J_a(x)$ 对強作用来講是处于海森堡表象之中，$|i\rangle$ 及 $|f\rangle$ 为物理原子核的状态矢量．強作用的影响全部包含在 $\langle f|J_a(x)|i\rangle$ 之中．

鉴于 He^3 及 H^3 均为自旋 1/2 的粒子，它們的自由运动都滿足狄拉克方程，采用和[7]中同样的方法和步驟，可得

$$T_{fi} = -2\pi i\delta^4(p_f+k-p_i-p_\mu)\langle f|J_a(0)|i\rangle\bar{u}_\nu(k)\gamma_a(1+\gamma_5)u_\mu(p_\mu)\varphi(0), \tag{4}$$

其中

$$\langle f|J_a(0)|i\rangle = \frac{g_1}{\sqrt{2}}\frac{1}{(2\pi)^3}\bar{u}(p_f)\Big\{\gamma_a+\lambda_1\gamma_a\gamma_5+ + i\frac{\mu_1}{4M_1}[\gamma_a(\hat{p}_f-\hat{p}_i)-(\hat{p}_f-\hat{p}_i)\gamma_a]+\frac{f_1}{im_\mu}(p_f-p_i)_a\gamma_5\Big\}u(p_i); \tag{5}$$

1) 計算情况見附录．

g_1, λ_1, μ_1 及 f_1 均为标量 $(p_f - p_i)^2$ 的函数；M_1 为 He^3 核的質量；$u(p_i)$ 及 $u(p_f)$ 分别为 He^3 及 H^3 的狄拉克波函数.

可以看到，(5)式的形式和[4]中 μ 在质子上俘获的矩陣元(5)的形式是相象的. (5)式是 μ 介子在自旋为 1/2 的粒子上俘获的最普遍的形式，粒子内部构造的不同只是反映在参数 g_1, λ_1, μ_1 及 f_1 等的大小不同上面. 我們将在下一节中討論，μ^-介子为 He^3 核俘获时矩陣元中的参数 g_1, λ_1, μ_1 及 f_1 与 μ^-介子为質子俘获时的参量（耦合常数）g, λ, μ 及 f 之間的关系.

如将俘获截面及角分布等用参量 g_1, λ_1, μ_1 及 f_1 等表示出来，則它們和相应的 μ^-介子在質子上俘获的公式具有相同的形式. 这些公式在[4]的附录 2 中都已給出. 由于[4]中的公式有印錯及遺漏之处，我們在这里将这些公式重新再写一遍.

設始态的密度矩陣具有下列形式：

$$\rho = \frac{1}{4}(1 + \boldsymbol{\sigma}_\mu\cdot\mathbf{P}_\mu + \boldsymbol{\sigma}_{He}\cdot\mathbf{P}_{He} + \zeta\boldsymbol{\sigma}_{He}\cdot\boldsymbol{\sigma}_\mu) \tag{6}$$

其中 $\mathbf{P}_\mu$ 及 $\mathbf{P}_{He}$ 分别代表始态上 μ 介子和 He^3 核的极化矢量，ζ 的大小則与 $(\mu\text{-}He^3)$ 系統处于不同超精細結构态上的比重有关. $\mathbf{P}_{He}$, $\mathbf{P}_\mu$ 及 μ 的大小应根据具体的实驗条件确定，本文中将不討論这一問題. 計算出的截面、角分布及中子极化具有下列形式.

1. 俘获几率及角关联

$$d\omega = (2\pi)^{-2}(\pi a_0^3)^{-1}2^{-1}g_1^2 I(1 - \alpha\mathbf{P}_\mu\cdot\boldsymbol{\nu} - \gamma\mathbf{P}_{He}\cdot\boldsymbol{\nu})k^2 d\Omega, \tag{6}$$

其中

$$\begin{aligned} I = {} & 1 + 3\lambda_1^2 + \beta_1 + \beta_1\lambda_1(2\mu_1' - f_1') + \frac{\beta_1^2}{4}(2\mu_1'^2 + f_1'^2 + 1) - \\ & - 6\zeta\left[\lambda_1^2 + \left(1 + \frac{\beta_1}{2}\right)\lambda_1 + \frac{\beta_1}{2}(\mu_1'\lambda_1 - f_1'\lambda_1 + \mu_1') + \frac{\beta_1^2}{4}(\mu_1' - \mu_1'f_1')\right] - \\ & - \beta_1\zeta(\mu_1' + f_1')\left(\lambda_1 - 1 + \frac{\beta_1}{2}\mu_1'\right) + \frac{\beta_1^2}{2}\zeta(\mu_1' + f_1'), \end{aligned} \tag{7}$$

$$I\alpha = 1 - \lambda_1^2 + \beta_1 - \beta_1\lambda_1(2\mu_1' + f_1') + \frac{\beta_1^2}{4}(f_1'^2 - 2\mu_1'^2), \tag{8}$$

$$I\gamma = 2\left[\lambda_1^2 + \left(1 + \frac{\beta_1}{2}\right)\left(\frac{\beta_1}{2}f_1' - \lambda_1\right) + \beta_1\mu_1'\lambda_1 + \frac{\beta_1^2}{4}\mu_1'^2\right], \tag{9}$$

$$\beta_1 = \frac{k}{M_1} = \frac{k}{3M}; \qquad \mu_1' = \mu_1 + 1; \qquad f_1' = f_1 - \lambda_1;$$

$\boldsymbol{\nu}$ 为沿中微子运动方向的单位矢量.

2. 反冲核 H^{3*} 的极化

$$I(1 - \alpha\mathbf{P}_\mu\cdot\boldsymbol{\nu} - \gamma\mathbf{P}_{He}\cdot\boldsymbol{\nu})\langle\boldsymbol{\sigma}_{H^3}\rangle = [a + b\mathbf{P}_\mu\cdot\boldsymbol{\nu} + c\mathbf{P}_{He}\cdot\boldsymbol{\nu}]\boldsymbol{\nu} + d\mathbf{P}_\mu + e\mathbf{P}_{He}, \tag{10}$$

其中

$$\begin{aligned} a = {} & 2\left[\lambda_1(\lambda_1 + 1) + \frac{\beta_1}{2}\lambda_1 + \beta_1\mu_1'(\lambda_1 + \beta_1\mu_1'/4) - \frac{\beta_1}{2}f_1' - \frac{\beta_1^2}{4}f_1'\right] - \\ & - \zeta\left[\left(1 + \frac{\beta_1}{2}\right)^2 + 7\lambda_1^2 + 4\beta_1\mu_1'\lambda_1 - 3f_1'\beta_1\lambda_1 - \beta_1^2 f_1'\mu_1' + \frac{\beta_1^2}{4}f_1'^2 + \right. \\ & \left. + \frac{\beta_1^2}{2}\mu_1'^2 - 4\left(1 + \frac{\beta_1}{2}\right)\left(\lambda_1 + \frac{\beta_1}{2}\mu_1'\right)\right], \end{aligned} \tag{11}$$

$$b = \beta_1(f_1' + \mu_1')\left(\lambda_1 + 1 + \frac{\beta_1}{2} + \frac{\beta_1}{2}\mu_1'\right), \tag{12}$$

$$c = -2\beta_1(\mu_1' + f_1')\left(\lambda_1 + \frac{\beta_1}{4}\mu_1' - \frac{\beta_1}{4}f_1'\right), \tag{13}$$

$$d = 2\left(\lambda_1 + \frac{\beta_1}{2}\mu_1'\right)\left[\lambda_1 - 1 - \frac{\beta_1}{2}(1 + f_1')\right], \tag{14}$$

$$e = \left(1 + \frac{\beta_1}{2}\right)^2 + \left(\beta_1\lambda_1 f_1' - \frac{\beta_1^2}{4}f_1'^2 - \lambda_1^2\right), \tag{15}$$

当 μ^- 介原子处于超精細态 $F = 1$ 时,可以求得俘获几率 λ_+:

$$\lambda_+ = (1 - \lambda_1)\left[1 - \lambda_1 + \beta_1 - \mu_1' + \frac{1}{3}(\mu_1' + f_1')\right]. \tag{16}$$

μ^- 介原子在超精細态 $F = 0$ 上的俘获几率 λ_- 则为

$$\lambda_- = (1 + 3\lambda_1)^2 + \beta_1[1 + 3\lambda_1 + 3\mu_1' + 5\lambda_1\mu_1' - 4\lambda_1 f' + (\mu_1' + f_1')(\lambda_1 - 1)]. \tag{17}$$

如将所有的脚标 1 都取消,则(6)—(17)代表 μ 在质子上俘获的几率、角分布及反冲中子的极化.

从 μ 介子为 He^3 核俘获的实验中可以测出 g_1, λ_1, μ_1 及 f_1 来. 一个很重要的問題是找到这些参数和 μ 俘获的耦合常数之間的关系,这将在下一节闡述.

三、原子核矩陣元

μ 介子在质子上俘获的矩陣元,在核子自旋空間中表示出来,当保留到核子速度 v/c 的一級近似时,得到

$$T_{fi} = \frac{-i}{\sqrt{2}(2\pi)^2} g\delta^4(p_n - p_p + k - p_\mu)\tau^{(-)}[A_1 + \mathbf{B}_1\cdot\boldsymbol{\sigma}_N + $$
$$+ (\mathbf{P}_n + \mathbf{P}_p)\cdot(\mathbf{A}_2 + B_2\boldsymbol{\sigma}_N)]\varphi(0), \tag{18}$$

其中

$$A_1 = \bar{u}_\nu\gamma_4(1 + \gamma_5)u_\mu, \tag{19}$$

$$\mathbf{B}_1 = \frac{1}{2M}\bar{u}_\nu\left[\mu'\mathbf{q}_\wedge\boldsymbol{\gamma} + 2iM\lambda\boldsymbol{\gamma} - \frac{i}{m}f\mathbf{q}(\mathbf{q}\cdot\boldsymbol{\gamma})\right]\cdot(1 + \gamma_5)u_\mu, \tag{20}$$

$$\mathbf{A}_2 = \frac{-i}{2M}\bar{u}_\nu\boldsymbol{\gamma}(1 + \gamma_5)u_\mu, \tag{21}$$

$$B_2 = \frac{1}{2M}\lambda\bar{u}_\nu\gamma_4(1 + \gamma_5)u_\mu. \tag{22}$$

如果把原子核看作由物理核子构成的体系,并忽略介子云的变形,可得

$$T_{fi} = \frac{-i}{\sqrt{2}(2\pi)^2} g\delta^4(p_f - p_i + k - p_\mu)\left[A_1 a_1 + \mathbf{B}_1\cdot\mathbf{b}_1 + \right.$$
$$\left. + \frac{1}{3}(\mathbf{P}_f + \mathbf{P}_i)\cdot(\mathbf{A}_2 a_1 + B_2\mathbf{b}_1) + (\mathbf{A}_2\cdot\mathbf{a}_2 + B_2 b_2)\right], \tag{23}$$ [1]

---

1) 詳細計算見附录.

其中

$$a_1 = \sum_{j=1}^{3} \int \Phi_f^* \tau_j^{(-)} e^{-i\mathbf{k}\cdot\xi_j} \Phi_i dV,$$
$$\mathbf{b}_1 = \sum_{j=1}^{3} \int \Phi_f^* \tau_j^{(-)} \sigma_j e^{-i\mathbf{k}\cdot\xi_j} \Phi_i dV; \tag{24}$$

$$\mathbf{a}_2 = -i \sum_{j=1}^{3} \int e^{-i\mathbf{k}\cdot\xi_j} [\Phi_f^* \nabla_{\xi_j} \Phi_i - (\nabla_{\xi_j} \Phi_f^*) \Phi_i] dV,$$
$$b_2 = -i \sum_{j=1}^{3} \int e^{-i\mathbf{k}\cdot\xi_j} \sigma_j \cdot [\Phi_f^* \nabla_{\xi_j} \Phi_i - (\nabla_{\xi_j} \Phi_f^*) \Phi_i] dV. \tag{25}$$

其中 Φ 代表去掉了质心运动的波函数，ξ_j 为相对于质心的核子坐标．

当忽略张量力及其他自旋軌道耦合力的作用时，可以将 He^3 及 H^3 基态波函数中的自旋同位旋部分分离开来，例如

$$\psi_{i,f} = \frac{1}{(2\pi)^{3/2}} e^{i\mathbf{P}_{i,f}\cdot\mathbf{R}} \Phi(r_{12}, r_{23}, r_{31}) \chi_{i,f} \tag{26}$$

其中，$\mathbf{R}$ 为质心坐标，Φ 为相对坐标的对称函数，χ 为自旋及同位旋組成的全反对称波函数．利用(26)可得

$$\mathbf{b}_1 = -a_1 \sigma_A, \tag{27}$$
$$a_1 = \int \Phi^* e^{-i\mathbf{k}\cdot\xi_1} \Phi dV, \tag{28}$$
$$\mathbf{a}_2 = b_2 = 0. \tag{29}$$

将(27)—(29)代回(23)中，即得

$$T_{fi} = \frac{-i}{\sqrt{2}(2\pi)^2} g\delta^4(p_f - p_i + k - p_\mu) \tau_A^{(-)} a_1 \Big[A_1 - \mathbf{B}_1 \cdot \sigma_A + + \frac{1}{3}(\mathbf{P}_f + \mathbf{P}_i) \cdot (\mathbf{A}_2 - B_2 \sigma_A) \Big]. \tag{30}$$

将(5)式表成(18)式的形式并与(30)比較，我們得到

$$g_1 = a_1 g,$$
$$\lambda_1 = -\lambda, \qquad \frac{1}{3}\mu_1' = -\mu'^{1)}, \tag{31}$$
$$\frac{1}{3} f_1' = -f'.$$

这表明 μ 在 He^3 核上俘获的矩陣元参量和 μ 俘获的耦合常数有非常簡单的关系；除了有些項要改变符号以外，它們之間只差一个公共因子 a_1．

如 V 型弱作用电流守恆，则有

$$\mu = \mu_p - \mu_n,$$
$$\mu_1 = \mu_{He^3} - \mu_{H^3}. \tag{32}$$

1) 此处 He^3 的磁矩是以 $\frac{e\hbar}{2M_1C}$ 即 $\frac{e\hbar}{6MC}$ 为单位的．

由反常磁矩的实驗，我們知道：

$$\frac{1}{3}\mu_{He^3} \simeq \mu_n,$$
$$\frac{1}{3}\mu_{H^3} \simeq \mu_p. \tag{33}$$

很容易由此了解为什么(31)式中 $\frac{1}{3}\mu_1 = -\mu$，符号的改变完全是由原子核内部的自旋同位旋結构决定的（He^3 中质子的自旋及反常磁矩抵消，它的自旋及反常磁矩是近似的由中子决定的）．严格講来，$\frac{1}{3}\mu_1$ 与 $-\mu$ 并不相等，它們之間的差别由介子云的变形及张量力等引起的．实驗观察到的反常磁矩很好地滿足(31)式，这表明我們采用的近似是合理的．值得提起的是在(30)式的第三項及第四項前有 $\frac{1}{3}(\mathbf{P}_f+\mathbf{P}_i)=\frac{\mathbf{k}}{3}$ 的因子，在工作[6]中，这一因子为 k．这是由于[6]中采用的近似引起的，单只这一近似即可有3%左右的修正．

在下一节中，我們将看到，未知的矩陣元 a_1 可由 μ 或电子与 He^3 核（或 H^3 核）的电磁散射确定．

四、μ 介子与 He^3 核的散射

μ 介子核子系統的电磁作用汉密頓密度为

$$H(x) = \sum_{\alpha=1}^{4}[J_\alpha^N(x) + J_\alpha^\mu(x)]A_\alpha(x), \tag{34}$$

$$J_\alpha^N(x) = -ie\bar{\psi}_N\gamma_\alpha\frac{1+\tau_3}{2}\psi_N,$$
$$J_\alpha^\mu(x) = -ie\bar{\psi}_\mu\gamma_\alpha\psi_\mu. \tag{35}$$

对电磁作用取第二級微扰，得到 μ 介子与原子核散射的矩陣元：

$$T_{fi} = -2\pi e\delta^4(p_f - p_i + p_\mu' - p_\mu)\langle f|J_\alpha^N(0)|i\rangle\bar{u}(p_\mu')\gamma_\alpha u(p_\mu)\frac{1}{q^2}, \tag{36}$$

其中　$q = p_f - p_i.$

同俘获的情况一样，对 He^3 核矩陣元 $\langle f|J_\alpha^N(0)|i\rangle$ 的形式和質子的矩陣元 $\langle p'|J_\alpha^N(0)|p\rangle$ 形式上完全一样，只是具体的电荷及磁矩形式因子是不同的函数：

$$\langle f|J_\alpha^N(0)|i\rangle = \frac{-ie}{(2\pi)^3}\bar{u}(p_f)\left[F_{He}\gamma_\alpha + \frac{i}{4M_1}G_{He}(\gamma_\alpha\hat{q} - \hat{q}\gamma_\alpha)\right]u(p_i), \tag{37}$$

其中 F_{He} 及 G_{He} 为 He^3 核的电荷及磁矩形式因子．它們是标量 q^2 的函数：

$$F_{He}(0) = 2;\quad G_{He}(0) = \mu_{He}. \tag{38}$$

如不考虑交換电流及张量力等，利用第三节的方法，可以求得 He^3 核的电磁形式因子与核子的电磁形式因子間的确定关系，即

$$F_{He}(q^2) = (2F_{1p}(q^2) + F_{1n}(q^2))a_1(q^2). \tag{39}$$

$$G_{He}(q^2) = F_{2n}(q^2)a_1(q^2), \tag{40}$$

其中 F_{1p} 及 F_{1n} 为质子及中子的电荷分布形式因子，而 $F_{2n}(q^2)$ 为中子的磁矩分布因子.

从(37)式即知，μ 介子对 He^3 核散射的截面具有和 Rosen bluth 公式完全相同的形式[8]．由散射的实验中可以定出 $F_{He}(q^2)$ 及 $G_{He}(q^2)$，再由 (39) 或 (40) 中即可定出核结构的形式因子 a_1.

如果討論的是 μ 介子对 H^3 核的散射，则其电磁分布形式因子 F_{H^3} 及 G_{H^3} 在上述近似下可表为

$$F_{H^3}(q^2) = [F_{1p}(q^2) + 2F_{1n}(q^2)]\, a_1(q^2),$$
$$G_{H^3}(q^2) = F_{2p}(q^2) a_1(q^2).$$

从 μ 或电子对 H^3 核的散射中同样可以求得核結构形式因子 a_1.

附　录　一

Fujii 及 Primakoff[6] 在計算 $\langle r^2\rangle$ 及俘获几率 $\omega^{(\mu)}$ 时，采用了下面的 He^3 基态变分波函数来計算原子核矩陣元：

$$u(r_{21}, r_{31}) = (\beta^5/96\pi^3)^{\frac{1}{2}} \frac{\exp\left[-\left(\frac{\beta}{2}\right)\sqrt{\rho^2+\rho'^2}\right]}{\sqrt[4]{\rho^2+\rho'^2}}, \quad (A\text{-}1)$$

$$\rho = \frac{\sqrt{3}}{2}|\mathbf{r}_{21} - \mathbf{r}_{31}|, \quad \rho' = \frac{1}{2}|\mathbf{r}_{21} + \mathbf{r}_{31}|, \quad \mathbf{R} = \frac{1}{3}(\mathbf{r}_1 + \mathbf{r}_2 + \mathbf{r}_3).$$

其中 β 是用上式計算得的 He^3 的庫伦能和实驗值比較而定出的．由 $(A\text{-}1)$ 算出：

$$\langle r^2\rangle^{\frac{1}{2}} \cong 1.78 \times 10^{-13}\ \text{厘米},$$

俘获几率
$$\omega^{(\mu)} \cong 1.46 \times 10^3\ \text{秒}^{-1}. \quad (A\text{-}2)$$

为了說明这种作法可能导致較大的不准确，我們选择[7]

$$u(r_{12}, r_{23}) = \left(\frac{\alpha}{60\pi^3}\right)^{\frac{1}{2}} \exp\left(\frac{1}{\sqrt{2}}\alpha\rho''\right), \quad (A\text{-}3)$$

$$\rho'' = \frac{1}{6}\sum_{j+k=1}^{3} |\mathbf{r}_j - \mathbf{r}_k|^2,$$

$$\alpha = 2.28 \times 10^{13}\ \text{厘米}^{-1}.$$

由$(A\text{—}3)$計算出：

$$\langle r^2\rangle^{\frac{1}{2}} \cong 2.52 \times 10^{-13}\ \text{厘米},$$

俘获几率
$$\omega^{(\mu)} \cong 1.29 \times 10^3\ \text{秒}^{-1}. \quad (A\text{-}4)$$

比較$(A\text{-}2)$ 及 $(A\text{-}4)$ 看到，符合 He^3 結合能都很好的不同变分波函数导致 $\omega^{(\mu)}$ 相差15%左右.

附录二　μ俘获中原子核矩陣元

由于原子核中核子平均距离較大，使得每一核子还不曾丧失其物理核子的性质，因而在忽略核子周围介子云变形时可以将物理原子核波函数写为：

$$|\Psi\rangle = \int |\mathbf{r}_1 \cdots \mathbf{r}_A\rangle\langle\mathbf{r}_1 \cdots \mathbf{r}_A|\Psi\rangle d\mathbf{r}_1 \cdots d\mathbf{r}_A, \quad (B\text{-}1)$$

式中 $\langle\mathbf{r}_1 \cdots \mathbf{r}_A|\Psi\rangle$ 即通常的原子核波函数，由各物理核子間相互作用产生的，而

$|\mathbf{r}_1\cdots\mathbf{r}_A\rangle$ 为自由的物理核子經反对称化后組成的波函数.

核子电流作用到原子核上的矩陣元是:

$$\langle f|J_\alpha(0)|i\rangle=\sum_{j=1}^{A}\int\langle f|\mathbf{r}_1\cdots\mathbf{r}'_j\cdots\mathbf{r}_A\rangle\langle\mathbf{r}'_j|\mathbf{P}'_j\rangle\langle\mathbf{P}'_j|J_\alpha(0)|\mathbf{P}_j\rangle\times$$

$$\times\langle\mathbf{P}_j|\mathbf{r}_j\rangle\langle\mathbf{r}_1\cdots\mathbf{r}_j\cdots\mathbf{r}_A\rangle\langle\mathbf{r}_1\cdots\mathbf{r}_A|i\rangle\prod_{l=1}^{A-1}{}'\,d\mathbf{r}_l d\mathbf{r}_j d\mathbf{r}'_j d\mathbf{P}_j d\mathbf{P}'_j,\qquad(B\text{-}2)$$

其中 $\prod\limits_{l=1}^{A-1}{}'$ 表示連乘符号，而将其中 $l=j$ 的項去掉. 以下用 Ψ_f^*, Ψ_i 来代表 $\langle f|\mathbf{r}_1\cdots\mathbf{r}'_j\cdots\mathbf{r}_A\rangle$ 和 $\langle\mathbf{r}_1\cdots\mathbf{r}_j\cdots\mathbf{r}_A|i\rangle$. 因

$$\bar{u}_\nu\gamma_\alpha(1+\gamma_5)u_\mu\langle\mathbf{P}'_j|J^i_\alpha(0)|\mathbf{P}_j\rangle=\tau_j^{(-)}[A_1+\mathbf{B}_1\cdot\boldsymbol{\sigma}_j+(\mathbf{P}'_j+\mathbf{P}_j)\cdot(\mathbf{A}_2+B_2\boldsymbol{\sigma}_j)],$$

故

$$\bar{u}_\nu\gamma_\alpha(1+\gamma_5)u_\mu\langle f|J_\alpha(0)|i\rangle=\sum_{j=1}^{A}\int\Psi_f^*\frac{e^{i\mathbf{P}'_j\cdot\mathbf{r}'_j}}{(2\pi)^{3/2}}\tau_j^{(-)}[A_1+\mathbf{B}_1\cdot\boldsymbol{\sigma}_j+(\mathbf{P}'_j+\mathbf{P}_j)\cdot$$

$$\cdot(\mathbf{A}_2+B_2\boldsymbol{\sigma}_j)]\frac{e^{-i(\mathbf{P}_j\cdot\mathbf{r}_j)}}{(2\pi)^{3/2}}\Psi_i\prod_{l=1}^{A-1}{}'\,d\mathbf{r}_l d\mathbf{r}_j d\mathbf{r}'_j d\mathbf{P}_j d\mathbf{P}'_j.\qquad(B\text{-}3)$$

作 $\mathbf{q}_j=\mathbf{P}'_j-\mathbf{P}_j,\ \mathbf{P}_j=\frac{1}{2}(\mathbf{P}_j+\mathbf{P}'_j)$ 的变换后，$(B-3)$ 为

$$\bar{u}_\nu\gamma_\alpha(1+\gamma_5)u_\mu\langle f|J_\alpha(0)|i\rangle=\sum_j\int\Psi_f^*e^{\frac{i}{2}\mathbf{q}'_j\cdot(\mathbf{r}_j+\mathbf{r}'_j)}\tau_j^{(-)}\{[A_1+\mathbf{B}_1\cdot\boldsymbol{\sigma}_j-$$

$$-i(\mathbf{A}_2+B_2\boldsymbol{\sigma}_j)(\nabla'_j-\nabla_j)]\delta(\mathbf{r}'_j-\mathbf{r}_j)\}\Psi_i\prod_{l=1}^{A-1}{}'\,d\mathbf{r}_l d\mathbf{r}_j d\mathbf{r}'_j d\mathbf{q}_j.\qquad(B\text{-}4)$$

对于 He^3 和 H^3 的波函数，在(26)中給出. 代入$(B\text{-}4)$，将含微分的項和不含微分的項分开計算:

i) 不含微分的項:

$$\frac{1}{(2\pi)^3}\sum_{j=1}^{3}\int\Phi^2(\xi_1,\xi_2,\xi_3)(e^{-i(\mathbf{p}_f-\mathbf{p}_i-\mathbf{q}_j)\cdot\mathbf{R}}e^{i\mathbf{q}_j\cdot\xi_j})\langle\chi_f,\tau_j^{(-)}(A_1+\mathbf{B}_1\cdot\boldsymbol{\sigma}_i)\chi_i\rangle\prod_{l=1}^{A}d\mathbf{r}_l d\mathbf{q}_j=$$

$$=\frac{1}{(2\pi)^3}\sum_{j=1}^{3}\int\Phi^2(\xi_1,\xi_2,\xi_3)e^{-i(\mathbf{p}_f-\mathbf{p}_i-\mathbf{q}_j)\cdot\mathbf{R}+i\mathbf{q}_j\cdot\xi_j}\langle\chi_f,\tau_j^{(-)}(A_1+\mathbf{B}_1\cdot\boldsymbol{\sigma}_j)\chi_i\rangle d\mathbf{R}dV_\xi d\mathbf{q}_j=$$

$$=\langle\chi_f,\sum_{j=1}^{3}\int\Phi^2(\xi_1,\xi_2,\xi_3)e^{+i\mathbf{q}_j\cdot\xi_j}dV_\xi\tau_j^{(-)}(A_1+\mathbf{B}_1\cdot\boldsymbol{\sigma}_j)\chi_i\rangle=$$

$$=\langle\chi_f,(A_1a_1+\mathbf{B}_1\cdot\mathbf{b}_1)\chi_i\rangle,\qquad(B\text{-}5)$$

其中 a_1, $\mathbf{b}_1$ 即由(24)所表示的.

ii) 含微分的項:

$$\frac{1}{(2\pi)^3}\sum_j\int\Phi^2(\xi_1,\xi_2,\xi_3)e^{-i(\mathbf{p}_f-\mathbf{p}_i)\cdot\mathbf{R}}e^{\frac{1}{2}\mathbf{q}_j\cdot(\mathbf{r}'_j+\mathbf{r}_j)}(-i)[(\nabla'_j-\nabla_j)\delta(\mathbf{r}'_j-\mathbf{r}_j)]\times$$

$$\times\langle\chi_f,\tau_j^{(-)}(\mathbf{A}_2+B_2\boldsymbol{\sigma}_j)\chi_i\rangle\prod_{l=1}^{A-1}{}'\,d\mathbf{r}_l d\mathbf{r}_j d\mathbf{r}'_j d\mathbf{q}_j.\qquad(B\text{-}6)$$

作部分积分，并注意到 $\nabla_{r_j} = \frac{1}{3}\nabla^R + \frac{2}{3}\nabla_{\xi_j}$，上式容易得到：

$$\langle \chi_f, \frac{1}{3}(\mathbf{P}_f + \mathbf{P}_i)\cdot(\mathbf{A}_2 a_1 + B_2 \mathbf{b}_1)\chi_i \rangle.$$

而(24)中后二式在 He^3 的 μ 俘获中等于零，因此时 $\Phi_f^* = \Phi_i = \Phi$.

参 考 文 献

[1] Feynman, R. P., Gell-Mann, M., *Phys. Rev.* **109** (1958), 193. Sudarshan, E. C. G. and Uarshak, R. E., *Phys. Rev.* **109** (1958), 1860.

[2] Crawford et al., *Phys. Rev. letters* **1** (1958), 377; Nordin et al., *Phys. Rev. letters* **1** (1958), 380.

[3] Goldberger, H. L. and Treiman, S. B., *Phys. Rev.* **111** (1958), 354; Wolfenstein, L., *Nuovo Cimento* **8** (1958), 882; Иоффе, Б. Л., *ЖЭТФ* **37** (1959), 159.

[4] 周光召，B. 馬耶夫斯基，物理学报，**15** (1959), 377.

[5] 关于μ介子在复杂核上俘获的工作可参考 Долинский, Э. И. и Блохинцев, Л. Д., *ЖЭТФ* **34** (1958), 759; **35** (1958), 1488; Überall, H. and Wolfenstein, L., *Nuovo Cimento* **10** (1958), 136; Tolhoek, H. A. and Luyten, L. R., *Nuclear Phys.* **3** (1957), 679; Primakoff, H., *Rev. Mod. Phys.* **31** (1959), 802.

[6] Fujii, A. and Primakoff, H., *Nuovo Cimento* **10** (1958), 327.

[7] 周光召，物理学报，**11** (1955), 299.

[8] Rosenbluth, M. N., *Phys. Rev.* **79** (1950), 615.

MUON CAPTURE IN He^3 NUCLEUS

Chu Chia-chen Chou Kuang-chao Peng Hong-an

(University of Peking)

Abstract

The rate of the muon capture reaction $\mu^- + He^3 \longrightarrow H^3 + \nu$, the angular distribution and the polarization of the final nucleus H^3 are calculated basing on the universal $V-A$ coupling with induced pseudoscular term and the weak magnetic current included. A general initial condition with polarized muon and polarized He^3 nucleus in different hyperfine states is assumed.

The ground state wave function is assumed to be a pure S-state. Thus we neglected the contribution of other states, which may be caused by the presence of tensor force or other velocity dependent forces. Meson exchange current is also neglected. Under these assumptions we have proved that the capture amplitude contains only one unknown nuclear matrix element, which is the Fourier transform of the nuclear density function. To determine this nuclear matrix element by the scattering of electron (or muon) with He^3 (H^3) nucleus is finally proposed.

第16卷 第2期 物理学报 Vol. 16, No. 2
1960年2月 ACTA PHYSICA SINICA Feb., 1960

关于μ俘获中有效赝标量耦合项的符号*

周光召 黄念宁
(北京大学物理系)

提 要

最近的实验有跡象表明，μ俘获中有效赝标量项的符号与从 V-A 理論用色散关系求得的相反．本文对这个符号的差别提出了一个可能的解释．在本文中指出，汉密頓量中一很小的赝标耦合項(約为有效赝标耦合項的1/100)即能改变有效赝标耦合項的符号．将这一赝标項加以变换，使之成为普适赝矢量电流的一部分，則可証明新添的这一項对β衰变和$\pi\to e+\nu$衰变不产生任何可观察的效应，因而不破坏 V-A 理論的普适性．文章最后討論了超子对对有效赝标項符号的影响的問題．

一、引 言

費曼(Feynman)，盖尔曼(Gell-Mann)[1]，馬尔夏克(Marshak)和苏达珊(Sudarshan)[2]提出的 V-A 普适理論，无疑地取得了很大的成功，但是还存在一些尚待解决的問題．这里我們討論其中的一个，即μ俘获中赝标量耦合的符号問題．按照普适 V-A 理論的观点，这一項应当是由強相互作用影响引起的．根据沃芬斯坦(Wolfenstein)，哥德伯(Goldberger)和特萊曼(Treiman)[3]的处理，如果認为強作用的中間态的貢献主要是由单个π介子引起的，而其余的态(例如三个π介子等等)貢献可以忽略，那么等效的赝标量耦合常数“g_P”将等于

$$“g_P” \approx -\sqrt{2}\,G\frac{m_\mu}{m_\pi^2+m_\mu^2}F(-m_\pi^2), \tag{1}$$

式中 m_π, m_μ 分别为π介子和μ介子的質量，G是π介子与核子的耦合常数，$G^2/4\pi=15$．而$F(-m_\pi^2)$的絕对值由π介子衰变的寿命决定[4]:

$$|F(-m_\pi^2)| = 0.13\sqrt{2}\,Gmg_A/2\pi^2, \tag{2}$$

m是核子質量，g_A是β衰变中的赝矢量耦合常数．因此不論π介子衰变的机构如何，总有

$$|“g_P”| \approx \frac{0.5}{\pi}\left(\frac{G^2}{4\pi}\right)\left(\frac{mm_\mu}{m_\pi^2+m_\mu^2}\right)g_A \approx 8g_A. \tag{3}$$

从这里我們可以得出結論，只要单个π介子的貢献是主要的，那么

$$“g_P” \approx 8g_A \tag{4}$$

或

$$“g_P” \approx -8g_A. \tag{5}$$

* 1959年11月30日收到．

哥德伯和特莱曼[4]用色散关系的技巧处理 π 介子的衰变得到 $F(-m_\pi^2)$ 应当是负的，由此"g_P" $\approx 8g_A$. 但是现今关于 μ 介子俘获的实验有迹象要求 "g_P" <0[5]. 粗略說来，"g_P" $\approx -8g_A$ 不与实验冲突. 我們現在对 "g_P" $\approx -8g_A$ 提出一个可能的解释，如果在汉密頓量中引入一个十分微小的与 g_A 反号的贋标量耦合項，由于这一項十分微小，它不影响 $V-A$ 理論所已經成功解决的各种情况，但是我們将指明，通过单个 π 介子中間态的"反射"，反过来可以严重地影响 μ 俘获中的有效的贋标量項，它能改变"g_P"的符号，使"g_P" $\approx -8g_A$.

更进一步精确的实验如果符合"g_P" $\approx 8g_A$，那么哥德伯和特莱曼的工作提供了实际情况的一种合理解释；如果符合"g_P" $\approx -8g_A$，我們的結果是一种可能的解释；如果与"g_P" $\approx \pm 8g_A$ 都不符，那么除单个 π 介子外的其余的中間态将有重要貢献，值得进一步研究. 我們的計算在理論上是很有趣的，一个看来可以忽略的微小的原有的弱相互作用，通过強相互作用，反过来却严重地影响有效的弱相互作用.

为了清楚說明問題，我們在第二节中将简单介紹一下哥德伯和特莱曼用色散关系求得有效贋标量項的方法.

二、哥德伯和特莱曼的色散关系方法

我們先看一下哥德伯和特莱曼怎样求得有效的贋标量耦合項[3]. 由于強相互作用的影响，考虑到不变性的要求，$\langle n|P_\lambda|p\rangle$ 的形式为

$$\langle n|P_\lambda|p\rangle = \left(\frac{m^2}{p_0 n_0}\right)^{1/2} \bar{u}_n\{ai\gamma_\lambda\gamma_5 - b(p-n)_\lambda\gamma_5\}u_p, \tag{6}$$

式中 a, b 为动量交换 $\xi=(n-p)^2$ 的函数，在 $V-A$ 理論中，

$$P_\lambda = Z_2 f_A \bar{\psi}_n i\gamma_\lambda\gamma_5\psi_p, \tag{7}$$

以下只討論 b. 根据色散关系的考虑，$b(\xi)$是 ξ 上半平面的解析函数：

$$b(\xi) = \frac{1}{\pi}\int_0^\infty d\xi' \frac{\mathrm{Im}b(-\xi')}{\xi'+\xi-i\epsilon}. \tag{8}$$

$\mathrm{Im}b(-\xi')$ 由 $\langle n|P_\lambda|p\rangle$ 的吸收部 $\langle n|A_\lambda|p\rangle$ 求得：

$$\langle n|A_\lambda|p\rangle = \pi(p_0/m)^{1/2}\Sigma_s\bar{u}_n\langle 0|P_\lambda|s\rangle\langle s|F(0)|p\rangle\delta(p_s+n-p), \tag{9}$$

式中

$$F(x) = \left(\gamma_\mu\frac{\partial}{\partial x_\mu}+m\right)\psi_n(x).$$

$|s\rangle$ 代表一切可能的中間态，由于重粒子数守恆，它只能是一个 π 介子，三个 π 介子，核子对等等. 我們只考虑单个 π 介子的貢献. 立即看到：

$$\bar{u}_n\langle\pi|F|p\rangle = \left(\frac{m}{p_0}\right)^{1/2}\frac{1}{(2p_{\pi0})^{1/2}}\sqrt{2}\,G\bar{u}_n i\gamma_5 u_p, \tag{10}$$

式中 G 是 π 介子与核子的耦合常数，而

$$\langle 0|P_\lambda|\pi\rangle = -i(p_\pi)_\lambda F(p_\pi^2)/(2p_{\pi0})^{1/2}. \tag{11}$$

当 $p_\pi^2=-m_\pi^2$ 时与 π 介子的寿命有关，$F(-m_\pi^2)$ 的絕对值已經确定如(2). 如果只計及单个 π 介子的中間态，即有

$$\mathrm{Im}b(\xi) = -\pi\sqrt{2}\,GF(-m_\pi^2)\delta(\xi+m_\pi^2), \tag{12}$$

那么代入(8)，得

$$b(\xi) = \frac{-\sqrt{2}\,GF(-m_\pi^2)}{\xi + m_\pi^2}. \tag{13}$$

有效的贋标量耦合常数

$$``g_P" = m_\mu b(m_\mu^2) \approx -\sqrt{2}\,G\,\frac{m_\mu}{m_\pi^2 + m_\mu^2}\,F(-m_\pi^2). \tag{1}$$

现在来看怎样求出 $F(-m_\pi^2)$[4]. $F(\xi)$是ξ上半平面的解析函数:

$$F(\xi) = \frac{1}{\pi}\int_{-\infty}^{\infty} d\xi'\,\frac{\operatorname{Im} F(-\xi')}{\xi' + \xi - i\epsilon}, \tag{14}$$

$\operatorname{Im} F(-\xi)$ 由吸收部 A 决定:

$$A = \pi(p_{\mu 0}/m_\mu)^{1/2}\Sigma_n\langle\mu|f_\nu|n\rangle\langle n|J|0\rangle\delta(p_n - p_\nu - p_\mu) \tag{15}$$

式中 f_ν 是中微子源，J 是 π 介子源:

$$f_\nu = \gamma_\mu\frac{\partial}{\partial x_\mu}\psi_\nu, \qquad J = -(\Box - \mu^2)\phi.$$

由于重粒子守恆，$|n\rangle$ 只能是核子对，超子对，三个 π 介子等等. 我們認为 $|n\bar{p}\rangle$ 核子对的貢献最为重要，那么

$$\langle\mu|f_\nu|n\bar{p}\rangle = \left(\frac{m_\mu m^2}{p_{\mu 0}n_0\bar{p}_0}\right)^{1/2}\bar{u}(\bar{p})\{ia\gamma_\lambda\gamma_5 - b(n+\bar{p})_\lambda\gamma_5\}u(n)\bar{u}(p_\mu)i\gamma_\lambda\gamma_5. \tag{16}$$

注意 b 满足(13)式，而

$$\left(\frac{n_0 p_0}{m^2}\right)^{1/2}\langle n\bar{p}|J|0\rangle = iK[(n+\bar{p})^2]\bar{u}(n)\gamma_5 u(\bar{p}). \tag{17}$$

于是

$$A(\xi) = -\frac{1}{4\pi}\left(ma - \frac{1}{2}\xi b\right)\operatorname{Re}K(\xi)\left(\frac{\xi + 4m^4}{\xi}\right)^{1/2}\bar{u}(p_\mu)\gamma_\lambda\gamma_5(p_\mu + p_\nu)_\lambda. \tag{18}$$

这样，得到

$$\operatorname{Im}F(\xi) = -\frac{1}{4\pi}\left[mg_A + \frac{\sqrt{2}}{2}\,GF(-m_\pi^2)\,\frac{\xi}{\xi + m_\pi^2}\right]\operatorname{Re}K(\xi)\left(\frac{\xi + 4m^2}{\xi}\right)^{1/2}, \tag{19}$$

式中 $\operatorname{Re}K(\xi)$ 根据[4]为

$$\operatorname{Re}K(-\xi) = \sqrt{2}\,G\cos\varphi(\xi)\exp\left\{\left(\frac{\xi - m_\pi^2}{\pi}\right)\int_{4m^2}^{\infty} dy\,\frac{\varphi(y)}{(y-\xi)(y-m_\pi^2)}\right\}. \tag{20}$$

代(19)入(14)，求得

$$F(-m_\pi^2) \approx F(0) = -\frac{m}{2\pi^2}\sqrt{2}\,Gg_A\,\frac{J}{1 + (G^2/4\pi)(2J/\pi)}, \tag{21}$$

其中

$$J = \int_0^\infty dk\,\frac{k^2}{(k^2+m^2)^{3/2}}\,\cos\varphi(k)\exp\left\{\frac{2}{\pi}\int_0^\infty dk'k'\varphi(k')\left(\frac{1}{k'^2 - k^2} - \frac{1}{k'^2 + m^2}\right)\right\}. \tag{22}$$

估計 J 的大小，表明(20)中分母的 1 可以忽略，因此得到 $F(-m_\pi^2) < 0$，其絕对值为(2)式，与实驗符合.

以上是文献[3]，[4]中与我們有关的部分.

三、有效贋标量項的符号

現在回到我們的工作 假定贋矢量电流 P_λ 具有下列形式：

$$P_\lambda = Z_2\bar{\psi}_n\left[f_A i\gamma_\lambda\gamma_5 - \frac{g_P^0}{m_\mu}(p-n)_\lambda\gamma_5\right]\psi_p, \tag{23}$$

其中 g_P^0 是一个常数．值得注意的是，在汉密頓量中，引入一普适的贋标耦合和在普适贋矢电流中引入(23)中的第二项并不完全相同． 在(23)中引入第二项，相当于在 μ 俘获的汉密頓量中引入一項 $(-Z_2\bar{\psi}_n g_P^0\gamma_5\psi_p\bar{\psi}_\nu(1-\gamma_5)\psi_\mu)$，而在 β 衰变的汉密頓量中引入一項 $\left(-Z_2\frac{m_e}{m_\mu}\bar{\psi}_n g_P^0\gamma_5\psi_p\bar{\psi}_\nu(1-\gamma_5)\psi_e\right)$．換句話說，当 g_P^0 为常数时，在汉密頓量中引入的贋标項的大小对 μ 俘虏和对 β 衰变是不同的．由于在 β 衰变的汉密頓量中，贋标耦合項較 μ 俘获的小 m_e/m_μ 倍，新增加的贋标項不会改变 $\pi\to e+\nu$ 的几率以及所有 β 衰变的理論値．因此，在 P_λ 中增加一項不致于影响 V-A 理論所已經成功解释的各种情况．

根据不变性的考虑，$\langle n|P_\lambda|p\rangle$ 仍然具有(6)式的形式，此时，$b(\xi)$ 所适合的色散关系应改为

$$b(\xi) = \frac{g_P^0}{m_\mu} + \frac{1}{\pi}\int_{4m^2}^{\infty}\frac{\mathrm{Im}\,b(\xi')}{\xi'+\xi-i\epsilon}d\xi'. \tag{6'}$$

按照哥德伯和特莱曼的方法继續往下計算时，我們只需将(18)式中的 b 换成

$$b'(\xi) = \frac{g_P^0}{m_\mu} + b(\xi), \tag{24}$$

将(24)代入(18)中，可得

$$A(\xi) = -\frac{1}{4\pi}\left(ma - \frac{1}{2}\xi\frac{g_P^0}{m_\mu} - \frac{1}{2}\xi b\right)\mathrm{Re}K(\xi)\left(\frac{\xi+4m^2}{\xi}\right)^{1/2}\bar{u}(p_\mu)\gamma_\lambda\gamma_5(p_\mu+p_\nu)_\lambda. \tag{18'}$$

这样，代替(19)式，我們有

$$\mathrm{Im}\,F'(\xi) = -\frac{1}{4\pi}\left[mg_A - \frac{1}{2}\xi\frac{g_P^0}{m_\mu} + \frac{\sqrt{2}}{2}GF'(-m_\pi^2)\frac{\xi}{\xi+m_\pi^2}\right]\mathrm{Re}K(\xi)\left(\frac{\xi+4m^2}{\xi}\right)^{1/2} \tag{19'}$$

将(19')代入(14)，可得

$$F'(-m_\pi^2) \approx F'(0) = \frac{1}{1+\frac{G^2}{2\pi^2}J}\left[-\frac{\sqrt{2}\,Gm}{2\pi^2}(g_A J + g_P^0 J')\right], \tag{25}$$

其中

$$J' = \frac{1}{2}\frac{4}{mm_\mu}\int_0^\infty dk\frac{k^2}{(k^2+m^2)^{1/2}}\cos\varphi(k)\exp\left\{\frac{2}{\pi}\int_0^\infty dk'k'\varphi(k')\left(\frac{1}{k'^2-k^2}-\frac{1}{k'^2+m^2}\right)\right\}$$
$$= 2\frac{m}{m_\mu}J + J'', \tag{26}$$

$$J'' = \frac{2}{mm_\mu}\int_0^\infty\frac{k^4dk}{(k^2+m^2)^{3/2}}\cos\varphi(k)\exp\left\{\frac{2}{\pi}\int_0^\infty dk'k'\varphi(k')\left(\frac{1}{k'^2-k^2}-\frac{1}{k'^2+m^2}\right)\right\}.$$

由[4]中(32)及(36)式知道，对于任意的复函数 $\tan\delta(k)$ 我們总有 $\cos\varphi(k) > 0$，因此 $J'' > 0$. 将(25)式写成

$$F'(0) = \frac{-m\sqrt{2}\,G}{2\pi^2}\left[g_A + \left(2\frac{m}{m_\mu} + \frac{J''}{J}\right)g_P^0\right]\frac{J}{1 + G^2J/2\pi^2}. \tag{25'}$$

由(25')可以看到，只要

$$g_P^0 = -\frac{m_\mu}{m + \frac{J''}{2J}m_\mu}g_A, \tag{27}$$

則 $F'(0)$ 的大小可以不变，而符号反过来. 这时 g_P^0 的大小为

$$|g_P^0| = \frac{m_\mu}{m + \frac{J''}{2J}m_\mu}|g_A| < \frac{m_\mu}{m}|g_A| \simeq \frac{1}{9}|g_A|. \tag{28}$$

由(28)可以看到，只要很小一个 $|g_P^0|\left(< \frac{1}{9}|g_A|\right)$ 即可改变 $F'(0)$ 的符号，而使得很大的"g_P"由 $+8g_A$ 变为 $-8g_A$.

由(27)中我們还可以看到，很小的 $|g_P^0|$ 項可以产生很大的 $F'(0)$，即引起很大的 $\pi \to \mu + \nu$ 的几率. 由实驗定出的 π 的寿命我們可以推論，不論 g_P^0 的符号如何，如果它存在，則其値必須很小.

四、討　　論

π 介子衰变成 $\mu\nu$ 对也还可能通过超子对. 有一种想法認为超子对与 π 介子作用或弱作用的耦合常数的符号将影响 $F'(0)$ 的符号. 此时 $F(0)$ 可表为

$$F(0) = -\frac{\sqrt{2}}{2\pi^2}\frac{\sum_i m_i G_i g_{Ai} J_i}{1 + \sum_i G_i^2/2\pi^2 J_i},$$

其中指标 i 代表与第 i 种重子对有关的量. 由于现在分母增加了許多倍，如果分子上有一項符号相反，它除了要減去所有其他項之外，剩下的値还要足够大，以致可以和 π 介子衰变的寿命相合. 要达到这点，必須有超子对和 π 介子的作用或者弱作用特别強才行. 如果弱作用是普适的，g_{Ai} 都一样大，則由于分母上有 G_i^2，要改变符号又同时满足 π 介子的寿命几乎是不可能的.

在我們的考慮中，即使加入超子对，也不会有什么影响. 此时所需要的 g_P^0 的値仍然不会大.

参　考　文　献

[1] Feynman, R. P. and Gell-Mann, M., *Phys. Rev.* **109** (1958), 193.

[2] Sudarshan, E. C. G. and Marshak, R. E., *Proc. of the Paduavenice Conference*, September (1957).

[3] Goldberger, M. and Treiman, S. B., *Phys. Rev.* **111** (1958), 354.

[4] Goldberger, M. and Treiman, S. B., *Phys. Rev.* **110** (1958), 1478.

[5] Иоффе, Б. Л., *ЖЭТФ* **37** (1959), 149; 周光召，馬耶夫斯基，物理学报，**15** (1959), 377 頁.

ON THE SIGN OF THE EFFECTIVE PSEUDOSCALAR TERM IN μ-CAPTURE REACTIONS

Chou Kuang-chao Huang Nien-ning

(*University of Peking*)

Abstract

Recent experiments seem to show that the sign of the effective pseudoscalar term in μ-capture reactions is different from the theoretical result of Goldberger and Treiman based on $V-A$ theory and dispersion relation technique. A possible explanation of this difference is proposed in the present note. It is shown that a very small p. s. tem in the Hamiltonian (of the order 1/100 of the effective p. s. term) can change the sign of the effective p. s. term. If this small term is transformed and added as a part of the universal axial vector current, it will produce practically no effect in β-decay and $\pi \rightarrow \mu + \nu$ decay. Finally the inclusion of hyperon pairs is considered.

第16卷 第2期　物理学报　Vol. 16, No. 2
1960年2月　ACTA PHYSICA SINICA　Feb., 1960

μ介子与輕原子核散射时的反衝效应*

周光召　戴元本
(北京大学物理系)　(数学研究所)

提　　要

本文利用一簡单方法求得μ介子(电子)在自旋为零的原子核上散射时位势的反冲修正項．所求得的位势对原子核的速度准确到一次近似，但对μ介子(电子)的速度沒有限制．討論了反冲效应对相移和截面的修正，所得結果与 Foldy 等的論文作了比較．

一、引　言

通常处理电子的电磁散射时，把原子核看作静止的产生庫伦电場的源．在考虑μ介子与輕原子核的电磁散射时，由于μ介子質量較大，原子核可以获得一定的动量．原子核的运动不仅引起汉密頓量中动能項的变化，而且将产生一磁場，引起位能項的改变．新增加的动能項及位能項对电子运动的影响称为核的反冲效应．

鉴于原子核比μ介子重得多，当μ介子的动能达到几百兆电子伏时，原子核在質心系中所得到的速度v比起光速c来仍然很小．在以下的計算中，我們将只保留v/c的一級項，而考虑一个相对論的狄拉克粒子(μ介子)在非相对論准静止近似的原子核場中的运动．

当能量高的μ介子与重核散射时，不仅有反冲效应，而且有許多其他的效应，例如原子核到低激发态的非弹性散射等．其他的这些效应目前还沒有很好的理論来說明，很难从实驗中区分出单独的反冲效应来．

Foldy, Ford 及 Yennie 在最近的一篇文章[1]中討論了高能电子与自旋为零的原子核散射时的反冲效应．他們所用的方法在我們看来是不能令人滿意的．他們先考虑一个自旋为 1/2 的原子核，用两个自旋为 1/2 的粒子間的 Breit 作用当作由反冲而新添的位能項，然后对原子核的运动量取非相对論近似，保留v/c的一級項，并扔掉所有含原子核自旋算符的項．这样得到的位能便被認为是电子与自旋为零的原子核散射时的由反冲引起的位能．我們知道，Breit 作用不仅对原子核的速度只准确到v/c的一級，而且对电子的速度v_e也只准确到v_e/c的一級．由于在所討論的問題中，电子速度已接近光速，这样得到的位能不能認为是准确的．此外，由自旋为 1/2 的原子核出发来求得自旋为零的原子核的反冲效应也不是十分合理的．

在本文中，我們将用一簡单的方法求得高能μ介子(或电子)与自旋为零的原子核散射时由反冲而产生的位阱，并根据它討論整个的反冲效应．我們求得的位阱对核子速度保留到一級近似，但对任意的μ介子(或电子)的速度都是正确的．我們的方法很容易推

* 1959年12月7日收到．

广到自旋不为零的其他原子核.

本文的第二节討論由反冲而新添的位能項，第三节討論反冲汉密頓量的径向方程.最后一节討論反冲方程的相移.

二、汉密頓量中的位能項

命 $|p\rangle$ 及 $|p'\rangle$ 代表原子核在始态及末态的状态矢量；p 及 p' 代表原子核在始态及末态上的动量；k 及 k' 为 μ 介子在始态及末态的动量．对电磁作用取二級微扰时，μ^+ 介子对原子核散射的 S 矩陣元可以表成下列形式：

$$\langle p'k'|S|pk\rangle = \delta^3(\mathbf{p}' - \mathbf{p})\delta^3(\mathbf{k}' - \mathbf{k}) - $$
$$- \frac{1}{(2\pi)^3}(2\pi)^4\delta^4(p' + k' - p - k)e\langle p'|j_\alpha(0)|p\rangle \frac{1}{q^2}\bar{u}(k')\gamma_\alpha u(k), \tag{1}$$

其中 $q = p - p' = k' - k$；$u(k)$ 代表 μ 介子的狄拉克波函数；$j_\alpha(x)$ 代表所有強作用粒子的电流密度．在(1)式中，$\langle p'|j_\alpha(0)|p\rangle$ 代表原子核的电流矩陣元；$\frac{1}{q^2}\langle p'|j_\alpha(0)|p\rangle$ 代表由原子核电流所产生的推迟場的富氏变換項.

根据普遍的对称性質，自旋为零的原子核的电流矩陣元具有下列形式：

$$\langle p'|j_\alpha(0)|p\rangle = \frac{Ze}{2M}F(q^2)(p + p')_\alpha, \tag{2}$$

其中 M 为原子核的質量；$F(q^2)$ 为一标量函数，它的富氏变換正好是原子核的电荷分布函数.

如对原子核取非相对論准静止近似时，我們只需保留到 $\frac{\mathbf{p}}{M}$ 及 $\frac{\mathbf{p}'}{M}$ 的一級項，此时

$$P_0 \simeq P_0' \simeq M,\quad q_0 \simeq 0. \tag{3}$$

在 q^2 中忽略 q_0^2 相当于忽略掉推迟效应，即在原子核产生的場中只保留静止的庫侖場及似稳的磁場．很明显，其他的項都具有 v^2/c^2 或更小的数量級.

如果在質心系中認为 (1) 式的 S 矩陣元看作由 μ 介子与原子核相互作用的位阱产生的，則有

$$\langle p'k'|S|pk\rangle = \delta^3(\mathbf{p}' - \mathbf{p})\delta^3(\mathbf{k}' - \mathbf{k}) - $$
$$- i\frac{1}{(2\pi)^3}(2\pi)^4\delta^4(p' + k' - p - k)u^*(k')V(\mathbf{q}^2)u(k), \tag{4}$$

其中

$$V(\mathbf{q}^2) = \int V(r)e^{i\mathbf{q}\cdot\mathbf{r}}dV, \tag{5}$$

$V(r)$ 为相互作用的位阱.

在質心系中比較(4)及(1)，即得

$$V(\mathbf{q}^2) = \frac{Ze^2}{\mathbf{q}^2}F(\mathbf{q}^2) - \frac{Ze^2}{2M}\left[\alpha\cdot\mathbf{p}'\frac{F(\mathbf{q}^2)}{\mathbf{q}^2} + \frac{F(\mathbf{q}^2)}{\mathbf{q}^2}\alpha\cdot\mathbf{p}\right] \tag{6}$$

或

$$V(r) = V_0(r) + V'(r); \tag{7}$$

其中

$$V_0(r) = \frac{Ze^2}{(2\pi)^3}\int \frac{1}{\mathbf{q}^2} F(\mathbf{q}^2) e^{-i\mathbf{q}\cdot\mathbf{r}} d^3\mathbf{q} = Ze^2 \int \frac{\rho(r')}{4\pi|r - r'|} dV'_r; \tag{8}$$

$$\rho(r) = \frac{1}{(2\pi)^3}\int F(\mathbf{q}^2) e^{-i\mathbf{q}\cdot\mathbf{r}} d^3\mathbf{q}; \tag{9}$$

$$V'(r) = -\frac{1}{2M}[\boldsymbol{\alpha}\cdot\mathbf{p}V_0(r) + V_0(r)\boldsymbol{\alpha}\cdot\mathbf{p}]. \tag{10}$$

在(10)式中，$\mathbf{p}$ 应了解为 $-i\nabla_r$.

很明显 $V_0(r)$ 代表 μ 介子与原子核之間的庫伦作用，而 $V'(r)$ 代表它們之間的磁的相互作用. 和[1]中所得到的公式(11)比較，我們所得的公式不仅更为正确，而且要簡单得多.

上面的公式是对 μ^+ 介子的散射求的，对 μ^- 介子的散射，我們需要改变 $V_0(r)$ 和 $V'(r)$ 的符号.

在質心系中准确到 $\frac{P}{M} = (v/c)$ 的总汉密頓量为

$$H = -\boldsymbol{\alpha}\cdot\mathbf{p} - \beta m + V_0(r) + \frac{1}{2M}[p^2 - (\boldsymbol{\alpha}\cdot\mathbf{p}V_0(r) + V_0(r)\boldsymbol{\alpha}\cdot\mathbf{p})], \tag{11}$$

其中　α 及 β 为 μ 介子的狄拉克算符；$\mathbf{p}$ 为質心系中的相对动量；m 为 μ 介子的質量.

三、径向方程的分离

在本节中，我們采用标准的方法[2]，将波动方程

$$H\psi = E\psi \tag{12}$$

中的与角度有关的部分分离出来，以便求得径向方程.

令

$$\alpha_r = r^{-1}\boldsymbol{\alpha}\cdot\mathbf{r},\quad P_r = \mathbf{p}\cdot(\mathbf{r}/r) = r^{-1}\mathbf{r}\cdot\mathbf{p} - ir^{-1},$$

并定义矩陣 k，便得

$$\boldsymbol{\alpha}\cdot\mathbf{p} = \alpha_r p_r + ir^{-1}\alpha_r\beta k.$$

在不影响所采用的准确度下，可在(11)式中作下列替代：

$$\begin{aligned} P^2 &\to -M^2 + (V_0 - E)^2 - i\alpha_r dV_0/dr \\ \boldsymbol{\alpha}\cdot\mathbf{p}V_0 + V_0\boldsymbol{\alpha}\cdot\mathbf{p} &\to 2(-\beta m + V_0 - E)V_0 \end{aligned} \tag{13}$$

利用(13)，我們得到下列近似的汉密頓量：

$$H = -\alpha_r p_r - ir^{-1}\alpha_r\beta k - \beta m + V_0 + \\ + \frac{1}{2M}\left(E^2 - V_0^2 - m^2 + 2\beta m V_0 - i\alpha_r\frac{dV}{dr}\right). \tag{14}$$

取表象

$$i\alpha_r = \begin{pmatrix} 0 & 1 \\ -1 & 0 \end{pmatrix},\quad \beta = \begin{pmatrix} 1 & 0 \\ 0 & -1 \end{pmatrix},\quad \psi = \begin{pmatrix} r^{-1}F \\ r^{-1}G \end{pmatrix}\quad ip_r = \frac{d}{dr}$$

代入(12)中，可得径向方程

$$\left[E + m - V_0 - \frac{1}{2M}(-V_0^2 + E^2 - m^2 + 2mV_0)\right]F - \\ -\left(\frac{d}{dr} + kr^{-1} - \frac{1}{2M}\frac{dV_0}{dr}\right)G = 0, \tag{15}$$

$$\left[E - m - V_0 - \frac{1}{2M}(-V_0^2 + E^2 - m^2 - 2mV_0)\right] G +$$

$$+ \left(\frac{d}{dr} - kr^{-1} - \frac{1}{2M}\frac{dV_0}{dr}\right) F = 0. \tag{16}$$

作下列变换：

$$F = e^{\frac{1}{2M}V_0} F',$$
$$G = e^{\frac{1}{2M}V_0} G', \tag{17}$$
$$\varepsilon = E - \frac{1}{2M}(E^2 - m^2);$$

并将(17)式代入(16)中，得到

$$\left[\varepsilon + m - V_0 + \frac{1}{2M}(V_0^2 - 2mV_0)\right] F' - \left(\frac{d}{dr} + kr^{-1}\right) G' = 0, \tag{18}$$

$$\left[\varepsilon - m - V_0 + \frac{1}{2M}(V_0^2 + 2mV_0)\right] G' + \left(\frac{d}{dr} - kr^{-1}\right) F' = 0. \tag{19}$$

(18) 和 (19) 即为求得之径向方程. 鉴于 $V_0(r)$ 在 $r \to \infty$ 远时趋于零，故在无穷远处 $F = F'$, $G = G'$，我們可以直接由(17)及(18)中求出散射的相移.

如果象[1]中一样，忽略掉 V_0^2 的項，对电子来講又忽略去与 m 成正比的項，則所有的反冲效应都归結为将 E 換成 $\varepsilon = E - \frac{1}{2M}(E^2 - m^2)$.

四、反冲方程的相移

引进下列无因次的量：

$$\epsilon = \frac{\varepsilon}{m}, \qquad v = \frac{V_0}{m}, \qquad \mu = \frac{m}{M},$$
$$x = mr(\epsilon^2 - 1)^{1/2}, \qquad f = \frac{1}{2}(\epsilon + 1)^{1/2} F', \qquad g = \frac{1}{2}(\epsilon - 1)^{1/2} G'. \tag{20}$$

将(20)代入(19)及(18)中，可得到无因次的反冲方程

$$\frac{dg}{dx} = -kx^{-1}g + \left(1 - \frac{(1+\mu)v - \frac{1}{2}\mu v^2}{\epsilon + 1}\right) f, \tag{21}$$

$$\frac{df}{dx} = kx^{-1}f - \left(1 - \frac{(1-\mu)v - \frac{1}{2}\mu v^2}{\epsilon - 1}\right) g. \tag{22}$$

在(21)及(22)中将 $\mu = 0$，即得到无反冲效应的散射方程，令其解为 f_0 及 g_0. 在 $x \to \infty$ 时，我們选择归一化因子使得 $f_0 \to \cos\varphi_0$, $g_0 \to \sin\varphi_0$，其中 $\varphi_0 = x + \gamma \ln x - \frac{1}{2}\pi l + \eta_0$，$\eta_0$ 为无反冲效应的庫伦相移. 在 $x \to \infty$ 时，f 和 g 亦趋于类似 f_0 和 g_0 的形式，只是此时相移 η 变了.

由方程(21)及(22)可得

$$\frac{d}{dx}(f_0g - g_0f) = -\frac{\mu}{\epsilon+1}\left(v - \frac{1}{2}v^2\right)f_0f + \frac{\mu}{\epsilon-1}\left(v + \frac{1}{2}v^2\right)g_0g =$$

$$= -\frac{\mu}{\epsilon(1-\epsilon^{-2})}\left[\left(\frac{1}{2\epsilon}v^2 + v\right)(f_0f - g_0g) - \left(\frac{1}{2}v^2 + \frac{1}{\epsilon}v\right)(f_0f + g_0g)\right]. \tag{23}$$

由(23)及无穷远处漸近条件,并忽略 μ^2 数量級的項,将 f, g 换成 f_0 及 g_0, 可得反冲效应引起的相移修正:

$$\eta - \eta_0 = -\frac{\mu}{\epsilon(1-\epsilon^{-2})}\int_0^\infty\left[(f_0^2 - g_0^2)\left(v + \frac{1}{2\epsilon}v^2\right) - (f_0^2 + g_0^2)\left(\frac{1}{2}v^2 + \frac{1}{\epsilon}v\right)\right]dx. \tag{24}$$

由(24)可以看到:由反冲引起的相移修正的数量級 $\sim(\alpha Z)(m/M)(m/\varepsilon)$ 或 $(\alpha Z)^2(m/M)(m/\varepsilon)$. (24)中的主要項第一項与[1]中的(41)式的最后一項是一致的,如[1]中所指出的,由于(18)和(19)式在下列变换下保持不变:

$$m \to -m, \quad k \to -k, \quad F' \to G', \quad G' \to -F'. \tag{25}$$

对 μ 介子质量的最低級近似而言,此項对散射截面的貢献抵消了. 因此反冲效应对散射截面的修正 $\frac{\Delta\sigma}{\sigma}$ 的数量級約为 $r_1(m/M)(m/\varepsilon)^2$ 或 $r_2(\alpha Z)(m/M)(m/\varepsilon)$. 其中 r_1、r_2 为两个数量因子.

就反冲效应的数量級来講,我們的結果和[1]中得到的結果是相似的. 但由于[1]的作者采用的討論反冲效应的方法不完全可靠,他們得到的公式不仅是比較复杂的,而且不是完全正确的.

参 考 文 献

[1] Foldy, L. L., Ford, K. W. and Yennie, D. R., *Phys Rev.* **113** (1959), 1147.

[2] Schiff, L. I., Quantum Mechanics (1949), p. 322.

THE EFFECT OF RECOIL ON THE SCATTERING OF μ-MESONS BY LIGHT NUCLEI

CHOU KUANG-CHAO (*Peking University*)

DAI YUEN-BEN (*Institute of Mathematics*)

ABSTRACT

The correction of the recoil to the potential in the scattering of μ-mesons (or electrons) by zero-spin nuclei is derived by means of a simple method. The potential obtained is corrected to first order in the velocity of the nuclei but without restriction on the velocity of the μ-mesons (or electrons). The effect of recoil on the phase-shift and cross section is discussed. The results are compared with those of Foldy, Ford and Yennie. It is pointed out that the method used in the latter's work is not perfect, because it makes use of Breit's interaction which is valid only for small velocities.

第16卷 第5期 物 理 学 报 Vol. 16, No. 5
1960年5月 ACTA PHYSICA SINICA May, 1960

π介子核子碰撞产生π介子的色散关系*

周光召 戴元本
(北京大学) (中国科学院)

提 要

本文討論了π介子核子碰撞产生π介子的色散关系，計算了束縛态的貢献．所用的方法是把核子場的算符由态矢量中提出来，对运动学和对称性作了較詳細的分析．

一、引 言

π介子核子碰撞产生π介子的色散关系，最近曾由 Логунов, Тодоров[1] 及 Screaton[2] 討論过．这些作者所用的方法是把π介子場的算符由态矢量中提出来．本文对π介子核子碰撞产生π介子的色散关系作了較詳細的討論；所用的方法是把核子場算符由态矢量中提出来，此时色散关系可以取較文献[2,1]中更为簡单的形式．

第二节討論了 Breit 系中的独立变量；第三节中利用 Боголюбов 的方法得到散射振幅的表示式，并引入了相应的推迟因果振幅与超前因果振幅；第四节利用散射振幅对洛仑兹变换及同位旋空間轉动的不变性将散射振幅分解为一些独立的分量；第五节利用弱反演及电荷共軛不变性将散射振幅的实部和虛部分出来，并得到交叉对称的关系；第六节討論了吸收部分的譜分析；第七节中写出了色散关系；第八节中計算出束縛态对色散关系的貢献，并討論了它与弹性散射振幅的关系．

二、运动学的考虑

令 p, q 代表初态核子和π介子的动量，p' 和 q_1, q_2 代表末态核子和π介子的动量．它們滿足动量能量守恆定律：

$$p + q = p' + q_1 + q_2 \tag{1}$$

及关系式

$$p^2 = p'^2 = -M^2, \qquad q^2 = q_1^2 = q_2^2 = -\mu^2, \tag{2}$$

其中M和μ为核子与π介子的質量．令

$$q' \equiv q_1 + q_2, \qquad l \equiv \frac{1}{2}(q_2 - q_1). \tag{3}$$

我們选取 Breit 坐标系，它滿足条件：

$$\mathbf{q} + \alpha \mathbf{q}' = 0,$$

其中α的値要使得

* 1960年2月14日收到．

$$q_0 = q_0'. \tag{5}$$

令

$$q'^2 \equiv -m^2, \tag{6}$$

则有

$$\alpha^2 = 1 + \frac{m^2 - \mu^2}{|\mathbf{q}'|^2}. \tag{7}$$

令

$$2P = p + p', \qquad \Delta \equiv (p - p') = (q' - q), \tag{8}$$

$$\omega \equiv p_0 = p_0', \tag{9}$$

则有

$$\mathbf{P} \cdot \mathbf{q} = \mathbf{P} \cdot \mathbf{q}' = 0. \tag{10}$$

引进 $\mathbf{P}$ 方向的单位矢量 $\mathbf{e}$, 则有

$$\mathbf{P} = \left(\omega^2 - M^2 - \frac{1}{4}\Delta^2\right)^{\frac{1}{2}}\mathbf{e}, \tag{11}$$

$$\mathbf{e} \cdot \mathbf{q} = \mathbf{e} \cdot \mathbf{q}' = 0. \tag{12}$$

在条件(1),(2),(4)下,只有八个独立的动量分量. 如果再固定 $\mathbf{e}$ 和 $\mathbf{q}$ 的方向, 由于有关系式(12),还有五个独立变量, 可选为 $\omega, \Delta^2, m^2, l_0, \mathbf{l} \cdot \mathbf{e}$, 所有的动量分量都可用这五个变量表示. 如果选取 $\mathbf{q}$ 的方向为 z 軸, $\mathbf{e}$ 与 $\mathbf{q}$ 垂直可选为 x 軸方向 則有

$$p = \left(\sqrt{\omega^2 - \frac{\Delta^2}{4} - M^2},\ \ 0, -\frac{1}{2}|\Delta^2|^{\frac{1}{2}},\ \omega\right),$$

$$p' = \left(\sqrt{\omega^2 - \frac{\Delta^2}{4} - M^2},\ \ 0,\ \frac{1}{2}|\Delta^2|^{\frac{1}{2}},\ \omega\right),$$

$$q = \left(0,\ 0,\ \frac{1}{2|\Delta^2|^{\frac{1}{2}}}(\Delta^2 + m^2 - \mu^2),\ \frac{1}{2|\Delta^2|^{\frac{1}{2}}}\sqrt{(\Delta^2 + m^2 - \mu^2)^2 + 4\mu^2\Delta^2}\right),$$

$$q_1 = \frac{1}{2}q' - l,$$

$$q_2 = \frac{1}{2}q' + l,$$

$$q' = \left(0,\ 0,\ -\frac{1}{2|\Delta^2|^{\frac{1}{2}}}(\Delta^2 - m^2 + \mu^2),\ \frac{1}{2|\Delta^2|^{\frac{1}{2}}}\sqrt{(\Delta^2 + m^2 - \mu^2)^2 + 4\mu^2\Delta^2}\right),$$

$$l = \left(\mathbf{l} \cdot \mathbf{e},\ \sqrt{\frac{1}{4}m^2 - \mu^2 - (\mathbf{l} \cdot \mathbf{e})^2 - \frac{4\Delta^2 m^2}{(\Delta^2 - m^2 + \mu^2)^2}l_0^2},\right.$$

$$\left. - l_0\frac{\sqrt{(\Delta^2 - m^2 + \mu^2)^2 + 4\Delta^2 m^2}}{\Delta^2 - m^2 + \mu^2},\ l_0\right). \tag{13}$$

对这五个独立变量的物理限制为:

$$\omega \geqslant \sqrt{\frac{1}{4}\Delta^2 + M^2}, \tag{14}$$

$$l_0^2 \leqslant \frac{1}{16\Delta^2 m^2}[m^2 - 4\mu^2 - 4(\mathbf{l}\cdot\mathbf{e})^2](\Delta^2 - m^2 + \mu^2)^2. \tag{15}$$

如令

$$Q = \frac{1}{2}(q + \alpha q'), \tag{16}$$

则在 Breit 系中 ω, l_0, $\mathbf{l}\cdot\mathbf{e}$ 可以用标量表示出来.

$$\omega = -\frac{2P\cdot Q}{(1+\alpha)q_0}. \tag{17}$$

由于 α, q_0 均为标量 Δ^2 的函数,故 ω 为标量 $P\cdot Q$ 及 Δ^2 的函数. 同样有

$$l_0 = -\frac{2l\cdot Q}{(1+\alpha)q_0}, \tag{18}$$

$$\mathbf{l}\cdot\mathbf{e} = \frac{\mathbf{l}\cdot\mathbf{p} + l_0\omega}{\sqrt{\omega^2 - M^2 - \frac{1}{4}\Delta^2}}. \tag{19}$$

故我們选择的五个量 ω, Δ^2, m^2, l_0, $\mathbf{l}\cdot\mathbf{e}$ 可以用另外五个标量 $P\cdot Q$, Δ^2, m^2, $l\cdot Q$ 及 $l\cdot P$ 表示出来. 因为在条件(1),(2),(4)下,只有五个独立的标量,所以标量函数在 Breit 系中都可用 ω, Δ^2, m^2, l_0, $\mathbf{l}\cdot\mathbf{e}$ 来表示.

三、散 射 振 幅

我們来利用 Боголюбов[3] 的方法写出 S 矩陣元:

$$\langle p', q_1, q_2|S|p, q\rangle = \langle q_1, q_2|a_{p'}Sa_p^+|q\rangle$$

其中 a^+, a 为核子的产生与消灭算符. 利用公式

$$aA = \int [a, \bar{\psi}_\lambda(x)]_+ \frac{\delta A}{\delta\bar{\psi}_\lambda(x)} dx + (-1)^{N_A} A a \tag{20}$$

(其中 N_A 为 A 中包含核子場算符 ψ 的个数), S 矩陣可写为

$$\langle p', q_1, q_2|S|p, q\rangle = -\frac{1}{(2\pi)^3}\sqrt{\frac{M^2}{E_pE_{p'}}}\iint \bar{u}_{p'\lambda}\cdot$$

$$\cdot\left\langle q_1, q_2\left|\frac{\delta^2 S}{\delta\psi_\nu(y)\delta\bar{\psi}_\lambda(x)}\right|q\right\rangle u_{p\nu}e^{-i(p'x+py)}d^4x d^4y. \tag{21}$$

由平移不变性有:

$$U_a|p\rangle = e^{ipa}|p\rangle, \qquad U_a\psi(x)U_a^{-1} = \psi(x-a), \tag{22}$$

其中 U_a 为平移么正算符. 利用(22)与(21)可得

$$\langle p', q_1, q_2|S|p, q\rangle = -(2\pi)\delta^4(p + q - p' - q_1 - q_2)\sqrt{\frac{M^2}{E_pE_{p'}}}\int \bar{u}_{p'\lambda}\cdot$$

$$\cdot\langle q_1q_2|\cdot\frac{\delta^2 s}{\delta\psi_\nu\left(-\frac{z}{2}\right)\delta\bar{\psi}_\lambda\left(\frac{z}{2}\right)}|q\rangle\ u_{p\lambda}e^{-i(p+p')z/2}d^4z. \tag{23}$$

引入核子流密度 j:

$$j_\lambda(x) \equiv -i\frac{\delta s}{\delta\bar{\psi}_\lambda(x)}S^+. \tag{24}$$

考慮到核子場算符的反对易性，可以証明：

$$\frac{\delta^2 S}{\delta\psi_\nu(y)\delta\bar{\psi}_\lambda(x)}S^+ = i\frac{\delta j_\lambda(x)}{\delta\psi_\nu(y)} - j_\lambda(x)\bar{j}_\nu(y) = i\frac{\delta\bar{j}_\nu(y)}{\delta\bar{\psi}_\lambda(x)} + \bar{j}_\nu(y)j_\lambda(x). \tag{25}$$

利用因果性条件，

$$\frac{\delta j(x)}{\delta\psi(y)} = 0 \quad 当\ x \gtrsim y. \tag{26}$$

由(25)得到：

$$\frac{\delta^2 S}{\delta\psi_\nu(y)\delta\bar{\psi}_\lambda(x)}S^+ = T(\bar{j}(y)j(x)). \tag{27}$$

将(27)代入(23)，得

$$\langle p', q_1, q_2|S|p, q\rangle = 2\pi\delta^4(p+q-p'-q_1-q_2)\sqrt{\frac{M^2}{E_pE_{p'}}}\int\bar{u}_{p'\lambda}\cdot$$

$$\cdot\langle q_1q_2|T\left(j_\lambda\left(\frac{z}{2}\right)\bar{j}_\nu\left(-\frac{z}{2}\right)\right)|q\rangle\times u_{p\nu}e^{-i(p+p')z/2}d^4z =$$

$$= \frac{i}{2\pi}\delta^4(p+q-p'-q_1-q_2)\sqrt{\frac{M}{E_pE_{p'}}}\sqrt{\frac{1}{8q_0q_{10}q_{20}}}F(p', q_1, q_2, p, q). \tag{28}$$

其中 F 为費曼振幅，令 β, α 为初态核子及 π 介子的同位旋指标，β' 及 α_1, α_2 为末态核子及 π 介子的同位旋指标，S' 及 S 为末态及初态的核子自旋指标，包括同位旋及自旋指标在內，費曼振幅可写为

$$F^{S'S}_{\beta'\alpha_1\alpha_2;\beta\alpha}(p', q_1, q_2; p, q) = -i(2\pi)^2(8q_0q_{10}q_{20})^{\frac{1}{2}}\int e^{-i(p+p')z/2}\bar{u}^{S'}_{p'\lambda}\cdot$$

$$\langle q_1\alpha_1\, q_2\alpha_2\left|\, T\left(j^{\beta'}_\lambda\left(\frac{z}{2}\right)\bar{j}^{\beta}_\nu\left(-\frac{z}{2}\right)\right)\right|q\alpha\rangle u^S_{p\nu}d^4z. \tag{29}$$

引入推迟因果振幅 M^r 及超前因果振幅 M^a：

$$M^{r\,S'S}_{\beta'\alpha_1\alpha_2;\beta\alpha}(p', q_1, q_2; p, q) = -i(2\pi)^2(8q_0q_{10}q_{20})^{\frac{1}{2}}\int e^{-i(p+p')z/2}\theta(z)\bar{u}^{S'}_{p'\lambda}\cdot$$

$$\cdot\langle q_1\alpha_1, q_2\alpha_2\left|\left[j^{\beta'}_\lambda\left(\frac{z}{2}\right)\bar{j}^{\beta}_\nu\left(-\frac{z}{2}\right)\right]_+\right|q\alpha\rangle u^S_{p\nu}d^4z, \tag{30}$$

$$M^{a\,S'S}_{\beta'\alpha_1\alpha_2;\beta\alpha}(p', q_1, q_2; p, q) = i(2\pi)^2(8q_0q_{10}q_{20})^{\frac{1}{2}}\int e^{-i(p+p')z/2}\theta(-z)\bar{u}^{S'}_{p'\lambda}\cdot$$

$$\cdot\langle q_1\alpha_1, q_2\alpha_2\left|\left[j^{\beta}_\lambda\left(\frac{z}{2}\right)\bar{j}^{\beta}_\nu\left(-\frac{z}{2}\right)\right]_+\right|q\alpha\rangle u^S_{p\nu}d^4z. \tag{31}$$

以后会看到，对于物理的能量值 M^r 与 F 相等.

令

$$M^r = D + iA, \qquad M^a = D - iA, \tag{32}$$

D, A 分别称为色散部分与吸收部分. 令

$$M^{r\,S'S}_{\beta'\alpha_1\alpha_2;\beta\alpha} = \bar{u}^{S'}_{p'}\mathfrak{M}^r_{\beta'\alpha_1\alpha_2;\beta\alpha}u^S_p, \qquad M^{a\,S'S}_{\beta'\alpha_1\alpha_2;\beta\alpha} = \bar{u}^{S'}_{p'}\mathfrak{M}^a_{\beta'\alpha_1\alpha_2;\beta\alpha}u^S_p,$$

$$D^{S'S}_{\beta'\alpha_1\alpha_2;\beta\alpha} = \bar{u}^{S'}_{p'}\mathfrak{D}_{\beta'\alpha_1\alpha_2;\beta\alpha}u^S_p, \qquad A^{S'S}_{\beta'\alpha_1\alpha_2;\beta\alpha} = \bar{u}^{S'}_{p'}\mathfrak{A}_{\beta'\alpha_1\alpha_2;\beta\alpha}u^S_p,$$

$$\Gamma^{S'S}_{\beta'\alpha_1\alpha_2\,\beta\alpha} = \bar{u}^{S'}_{p'}\,\mathfrak{F}_{\beta'\alpha_1\alpha_2;\beta\alpha}u^{S}_{p}. \tag{33}$$

其中 $\mathfrak{M}$, $\mathfrak{D}$, $\mathfrak{A}$, $\mathfrak{F}$ 为核子旋量空間的矩陣．这些矩陣中的 $i\gamma\cdot p$, $i\gamma\cdot p'$ 項我們理解作已利用狄拉克方程化掉．

四、自旋同位旋結构

我們現在利用散射振幅在洛仑茨变換与同位旋空間轉动下的变換性質将它們分解为一些独立的成分．由(29),(30),(31)可以看到，引入的散射振幅在固有洛仑茨变換下保持不变;又由于初态与末态共有奇数个 π 介子,故散射振幅在空間反演下为膺标量．考慮到狄拉克方程,由 γ 矩陣及有关动量只能組成四个独立的膺标量,我們取为 $\bar{u}_{p'}\Gamma^i u_p$:

$$\Gamma' = i\gamma_5,\quad \Gamma^2 = \gamma\cdot q\gamma_5,\quad \Gamma^3 = \gamma\cdot l\gamma_5,\quad \Gamma^4 = \frac{1}{2i}(\gamma_\mu\gamma_\nu - \gamma_\nu\gamma_\mu)q_\mu l_\nu\gamma_5. \tag{34}$$

由强相互作用的电荷无关性可以知道，散射振幅在与同位旋空間的波函数相乘后应为同位旋空間的标量,利用核子的同位旋算符 τ 可以构成四个独立的满足此要求的算符,我們取为

$$I^1 = \frac{1}{2}[\delta_{\alpha_2\alpha}\tau_{\alpha_1} + \delta_{\alpha_1\alpha}\tau_{\alpha_2}],\qquad I^2 = \frac{1}{2}[\delta_{\alpha_2\alpha}\tau_{\alpha_1} - \delta_{\alpha_1\alpha}\tau_{\alpha_2}],$$

$$I^3 = \delta_{\alpha_1\alpha_2}\tau_\alpha,\qquad I^4 = i\varepsilon_{\alpha_1\alpha_2\alpha}. \tag{35}$$

利用(34),(35),可以将$\mathfrak{M}$分解为

$$\mathfrak{M}_{\beta'\beta} = \sum_{j=1}^{4}\mathfrak{M}^i_j\Gamma^i I_{\beta'\beta}. \tag{36}$$

对于 $\mathfrak{D}$ 和 $\mathfrak{A}$ 有相似的分解公式．

五、对 称 性 質

現在我們来証明 $\mathfrak{D}^i_j$ 和 $\mathfrak{A}^i_j$ 为实函数．令 R 为弱反演么正算符[4],

$$R\psi(x)R^{-1} = \eta D\bar{\psi}(-x),\qquad Rj(x)R^{-1} = \eta D\bar{j}(-x), \tag{37}$$

其中D适合

$$DD^+ = 1,\quad D = -D^T,\quad D\gamma^T_\mu D^{-1} = \gamma_\mu \tag{38}$$

η 为相因子．設 π 介子与核子的相互作用哈密頓密度为

$$H = Gi\bar{\psi}\gamma_5\tau_\alpha\psi\phi_\alpha,\quad \alpha = 1,\ 2,\ 3 \tag{39}$$

考慮到 τ_2 純虛，τ_1, τ_3 为实矩陣,要求H在弱反演下不变,我們有

$$R\varphi_\alpha(x)R^{-1} = -(-1)^\alpha\varphi_\alpha(-x); \tag{40}$$

由此得

$$R|q\alpha\rangle = -(-1)^\alpha\langle q\alpha|. \tag{41}$$

利用(37)(41),容易証明:

$$\mathfrak{M}^r_{\beta'\alpha_1\alpha_2;\beta\alpha}(p',\ q_1,\ q_2;p,\ q)^*_{\lambda\nu} = i(-1)^{\alpha_1+\alpha_2+\alpha+1}(2\pi)^2(8q_0q_{10}q_{20})^{\frac{1}{2}}$$

$$\times\int e^{-i(p+p')z/2}\theta(-z)(\gamma_4 D)_{\nu\nu'}\left\langle q_1\alpha_1,\ q_2\alpha_2\left|\left[j^{\beta'}_{\lambda'}\left(\frac{z}{2}\right),\ \bar{j}^{\beta}_{\nu'}\left(-\frac{z}{2}\right)\right]_+\right|q\alpha\right\rangle(D^{-1}\gamma_4)_{\lambda'\lambda}d^4z =$$

$$= (-1)^{\alpha+\alpha_1+\alpha_2+1}[D^+\gamma_4\mathfrak{M}^a_{\beta'\alpha_1\alpha_2;\beta\alpha}(p',\ q_1,\ q_2;\ p,\ q)\gamma_4 D]_{\lambda\nu}. \tag{42}$$

由(35)考虑到 τ_2 純虛有

$$(-1)^{a+a_1+a_2}\Gamma^{i*} = -\Gamma^{i}. \tag{43}$$

由(34)得到

$$D^{-1}\gamma_4\Gamma^{i*}\gamma_4 D = \Gamma^{i}. \tag{44}$$

由(42),(43),(44)立刻得到

$$\mathfrak{M}_j^{i*} = \mathfrak{M}_j^{ai}. \tag{45}$$

由上式及(32),得

$$\mathfrak{D}_j^{i*} = \mathfrak{D}_j^{i},$$
$$\mathfrak{A}_j^{i*} = \mathfrak{A}_j^{i}, \tag{46}$$

即 $\mathfrak{D}_j^i$ 和 $\mathfrak{A}_j^i$ 都是实函数.

我們利用电荷共軛变换来求得交叉对称关系．核子場算符的电荷共軛变换为[4]

$$C\psi(x)C^{-1} = \eta_c c\bar{\psi}(x), \quad Cj(x)C^{-1} = \eta_c c j_{\gamma'}^{T\beta'}(x), \tag{47}$$
$$c = D\gamma_5, \quad cc^{+} = 1, \quad c\gamma_\mu^T c^{-1} = -\gamma_\mu, \quad c = -c^T,$$

其中 C 为电荷共軛变换的么正算符．要求哈密頓密度(39)在电荷共軛下不变，考慮到 τ 純虛,对于介子場有

$$C\varphi_a(x)C^{-1} = -(-1)^a\varphi_a(x). \tag{48}$$

利用(67)(68)得

$$\mathfrak{M}^r_{\beta'\alpha_1\alpha_2;\beta\alpha}(-p', q_1, q_2; -p, q)_{\lambda\nu} = (-1)^{a+a_1+a_2}i(2\pi)^2(8q_0q_{10}q_{20})^{\frac{1}{2}} \times$$
$$\times \int e^{i(p+p')z/2}\theta(z)C_{\lambda\lambda'}\langle q_1\alpha_1, q_2\alpha_2\left|\left[j^{\beta}_{\nu'}\left(-\frac{z}{2}\right), \bar{j}^{\beta'}_{\lambda'}\left(\frac{z}{2}\right)\right]_{+}\right|q\alpha\rangle C^{-1}_{\nu'\nu}d^4z =$$
$$= (-1)^{a+a_1+a_2}(C\mathfrak{M}^a_{\beta\alpha_1\alpha_2;\beta'\alpha}(p', q_1, q_2; p, q)C^{-1})_{\nu\lambda}. \tag{49}$$

由(42),(47),(49),得

$$\mathfrak{M}^r_{\beta'\alpha_1\alpha_2;\beta\alpha}(-p', q_1, q_2; -p, q)_{\lambda\nu} =$$
$$= -[\gamma_4\gamma_5^T\mathfrak{M}^r_{\beta\alpha_1\alpha_2;\beta'\alpha}(p', q_1, q_2; p, q)^{*T} \times \gamma_5^T\gamma_4]_{\lambda\nu} \tag{50}$$

由(34),(35),

$$\gamma_4\gamma_5^T\Gamma^{i+}\gamma_5^T\gamma_4 = \varepsilon^i\Gamma^i \qquad \varepsilon^i = 1, \quad i = 1, 2, 3, \quad \varepsilon^4 = -1; \tag{51}$$
$$\Gamma^{i*}_{\beta'\beta} = \varepsilon_j\Gamma^{i}_{\beta\beta'} \qquad \varepsilon_j = 1, \quad j = 1, 2, 3, \quad \varepsilon_4 = -1, \tag{52}$$

令

$$\varepsilon^i_j = \varepsilon^i\varepsilon_j, \tag{53}$$

用第(一)节中所引入的独立变量 $\omega, \Delta^2, m^2, l_0, \mathbf{l}\cdot\mathbf{e}$, 幷引入

$$S^{+}\mathfrak{M}^r_{\beta'\alpha_1\alpha_2;\beta\alpha}(\omega, \Delta^2, m^2, l_0, \mathbf{l}\cdot\mathbf{e}) = \frac{1}{2}[\mathfrak{M}^r_{\beta'\alpha_1\alpha_2;\beta\alpha}(\omega, \Delta^2, m^2, l_0, \mathbf{l}\cdot\mathbf{e}) +$$
$$+ \mathfrak{M}^r_{\beta'\alpha_1\alpha_2;\beta\alpha}(\omega, \Delta^2, m^2, l_0, -\mathbf{l}\cdot\mathbf{e})]\omega^2 - M^2 - \frac{1}{4}\Delta^2)^{-1/2}, \tag{54}$$

$$S^{-}\mathfrak{M}^r_{\beta'\alpha_1\alpha_2;\beta\alpha}(\omega, \Delta^2, m^2, l_0, \mathbf{l}\cdot\mathbf{e}) = \frac{1}{2i}[\mathfrak{M}^r_{\beta'\alpha_1\alpha_2;\beta\alpha}(\omega, \Delta^2, m^2, l_0, \mathbf{l}\cdot\mathbf{e}) -$$
$$- \mathfrak{M}^r_{\beta'\alpha_1\alpha_2;\beta\alpha}(\omega, \Delta^2, m^2, l_0, -\mathbf{l}\cdot\mathbf{e})]\omega^2 - M^2 - \frac{1}{4}\Delta^2)^{-1/2}. \tag{55}$$

由(52)至(55)式,得到

$$S^{\pm}\mathfrak{M}^{ri}_j(-\omega, \Delta^2, m^2, l_0, \mathbf{l}\cdot\mathbf{e}) = -\varepsilon^i_j S^{\pm}\mathfrak{M}^{ri}_j(\omega, \Delta^2, m^2, l_0, \mathbf{l}\cdot\mathbf{e})^{*}, \tag{56}$$
$$S^{\pm}\mathfrak{A}^i_j(-\omega, \Delta^2, m^2, l_0, \mathbf{l}\cdot\mathbf{e}) = \pm\varepsilon^i_j S^{\pm}\mathfrak{A}^i_j(\omega, \Delta^2, m^2, l_0, \mathbf{l}\cdot\mathbf{e}). \tag{57}$$

六、吸收部分的譜分析

利用(30)至(33)式可以求得 $\mathfrak{A}$ 的表示式:

$$(\mathfrak{A}_{\beta'\alpha_1\alpha_2:\beta\alpha})_{\lambda\nu} = -2\pi^2(8q_0q_{10}q_{20})^{\frac{1}{2}}\int e^{-iP\cdot z}\langle q_1\alpha_1, q_2\alpha_2\left|\left[\bar{j}_\lambda^{\beta'}\left(\frac{z}{2}\right), \bar{j}_\nu^{\beta}\left(-\frac{z}{2}\right)\right]_+\right|q\alpha\rangle d^4z. \quad (58)$$

应用完全的中間态系及平移不变性,略去同位旋,自旋指标可将 $\mathfrak{A}$ 写为

$$\mathfrak{A} = -2^5\pi^6(8q_0q_{10}q_{20})^{\frac{1}{2}}\sum_n\Big[\delta^4\left(-P+\frac{q_1+q_2+q}{2}-p_n\right)\langle q_1, q_2|\bar{j}(0)|n\rangle\cdot$$

$$\cdot\langle n|j(0)|q\rangle+\delta^4\left(-P-\frac{q_1+q_2+q}{2}+p_n\right)\langle q_1, q_2|j(0)|n\rangle\langle n|\bar{j}(0)|q\rangle\Big] =$$

$$= \mathfrak{A}^- + \mathfrak{A}^+, \quad (59)$$

其中 p_n 为中間态的总动量, $p_n^2 \equiv -M_n^2$,

$$\mathfrak{A}^- = -2^5\pi^6(8q_0q_{10}q_{20})^{\frac{1}{2}}\sum_n\delta^4\left(-P+\frac{q_1+q_2+q}{2}-p_n\right)\langle q_1, q_2|\bar{j}(0)|n\rangle\langle n|j(0)|q\rangle,$$

$$\mathfrak{A}^+ = -2^5\pi^6(8q_0q_{10}q_{20})^{\frac{1}{2}}\sum_n\delta^4\left(-P-\frac{q_1+q_2+q}{2}+p_n\right)\langle q_1, q_2|j(0)|n\rangle\langle n|\bar{j}(0)|q\rangle. \quad (60)$$

由(57),(60)可以看到,

$$S^\pm\mathfrak{A}_j^{i-}(\omega) = \varepsilon_j^i S^\pm\mathfrak{A}_i^{j+}(-\omega), \quad (61)$$

故只須研究 $\mathfrak{A}^+$ 就够了. 令

$$W = -\left(P+\frac{q+q_1+q_2}{2}\right)^2, \quad (62)$$

对于 A^+ 有:

$$W = -p_n^2 = M_n^2. \quad (63)$$

有貢献的中間态最低为单核子态,其次为单核子加单介子态. 故在 $W = M^2$ 处, $\mathfrak{A}^+$ 有一孤立奇点,在 $W > (M+\mu)^2$ 处有連續譜. 由第(一)节中的公式容易得到:

$$W = M^2 + \frac{1}{2}(m^2+\mu^2) + \frac{\omega}{|\Delta^2|^{1/2}}\sqrt{(\Delta^2+\mu^2-m^2)^2+4\Delta^2m^2} + \frac{1}{2}\Delta^2. \quad (64)$$

故孤立奇点在 $\omega = \omega_0$ 处,

$$\omega_0 = -\frac{|\Delta^2|^{1/2}}{2\sqrt{(\Delta^2+\mu^2-m^2)^2+4\Delta^2m^2}}(\Delta^2+\mu^2+m^2) < 0; \quad (65)$$

連續譜起始于 $\omega = \omega_1$ 处,

$$\omega_1 = \frac{|\Delta^2|^{\frac{1}{2}}}{\sqrt{(\Delta^2+\mu^2-m^2)^2+4\Delta^2m^2}}\left[2M\mu+\frac{1}{2}(\mu^2-\Delta^2-m^2)\right]. \quad (66)$$

由(61)知, $\mathfrak{A}_j^{i-}$ 在 $\omega = -\omega_0$ 处有一孤立奇点,在 $\omega < -\omega_1$ 处有連續譜,計算証明:

$$\omega_0^2 - \frac{\Delta^2}{4} \leqslant \frac{\mu^2}{4}, \quad (67)$$

故 $-\omega_0$ 恆在非物理区內. 在物理区中, $\mathfrak{A}^- = 0$, 但 $\mathfrak{T} = \mathfrak{M}' - 2i\mathfrak{A}^-$, 故在物理区內 $\mathfrak{T}$ 与 $\mathfrak{M}'$ 恆相等.

七、色散关系

由(30)，(31)，如果略去同位旋，自旋指标 $\mathfrak{M}^r$ 和 $\mathfrak{M}^a$ 可写为：

$$\mathfrak{M}^r(\omega, \Delta^2, m^2, l_0, \mathbf{l}\cdot\mathbf{e}) = -i(2\pi)^2(8q_0q_{10}q_{20})^{\frac{1}{2}}\int e^{i\left(\omega z_0 - \sqrt{\omega^2 - M^2 - \frac{1}{4}\Delta^2}\,\mathbf{e}\cdot\mathbf{z}\right)}\cdot$$

$$\cdot\,\theta(z)\langle q_1, q_2\left|\left[j\left(\frac{z}{2}\right), j\left(-\frac{z}{2}\right)\right]_+\right| q\rangle d^4z, \tag{68}$$

$$\mathfrak{M}^a(\omega, \Delta^2, m^2, l_0, \mathbf{l}\cdot\mathbf{e}) = +i(2\pi)^2(8q_0q_{10}q_{20})^{\frac{1}{2}}\int e^{+i\left(\omega z_0 - \sqrt{\omega^2 - M^2 - \frac{1}{4}\Delta^2}\,\mathbf{e}\cdot\mathbf{z}\right)}\cdot$$

$$\cdot\,\theta(-z)\langle q_1, q_2\left|\left[j\left(\frac{z}{2}\right), j\left(-\frac{z}{2}\right)\right]_+\right| q\rangle d^4z. \tag{69}$$

由上两式可以看到，$S\mathfrak{M}^r$ 在固定其他变量，看作 ω 的函数时，在

$$I_m\omega > \left|I_m\sqrt{\omega^2 - M^2 - \frac{1}{4}\Delta^2}\right| \tag{70}$$

的区域内解析，而 $S\mathfrak{M}^a$ 在

$$-I_m\omega > \left|I_m\sqrt{\omega^2 - M^2 - \frac{1}{4}\Delta^2}\right| \tag{71}$$

的区域内解析．如令

$$\tau = M^2 + \frac{1}{4}\Delta^2, \tag{72}$$

则在 τ 是負数时，$S\mathfrak{M}^r$ 在 ω 的上半平面内解析，$S\mathfrak{M}^a$ 在下半平面内解析．由上节的討論可以知道，如

$$\Delta^2 + m^2 < \mu^2 + 4M\mu, \tag{73}$$

则 $\omega_1 > 0$，此时在区間 $-\omega_1 < \omega < \omega_1$ 内，除孤立奇点 $\pm\omega_0$ 外，都有 $\mathfrak{A}^+(\omega) = \mathfrak{A}^-(\omega) = 0$，$\mathfrak{M}^r - \mathfrak{M}^a = 2i\mathfrak{A} = 0$，故此时 $\mathfrak{M}^r$ 和 $\mathfrak{M}^a$ 可以看作同一个解析函数 $\mathfrak{M}$ 在上半和下半平面的表示式，除实軸上的割綫以外，$\mathfrak{M}$ 在全平面内解析．如果(68)，(69)中的矩陣元看作 τ 的函数能由物理区解析延拓至 $\tau < 0$ 的区域，则色散关系成立．假定散射振幅在 $|\omega_1| \to \infty$ 时趋于零，则有

$$S^{\pm}\mathfrak{D}_i^j(\omega) = \frac{1}{\pi}\int_{-\infty}^{\infty}\frac{S^{\pm}\mathfrak{A}_i^j(\omega')}{\omega' - \omega}d\omega. \tag{74}$$

利用(57)可去掉色散积分的負能量部分．在条件(73)满足时，$\mathfrak{A}^-$ 在正能量部分只有一个孤立奇点，此时色散关系可写为

$$S^{\pm}\mathfrak{D}_i^j(\omega) = S^{\pm}B_i^j(\omega) + \frac{2}{\pi}\int_{\omega_1}^{\infty}\frac{\omega' S^{\pm}\mathfrak{A}_j^{+i}(\omega')}{\omega'^2 - \omega^2}d\omega', \quad (\pm\varepsilon_j^i < 0) \tag{75}$$

$$S^{\pm}\mathfrak{D}_i^j(\omega) = S^{\pm}B_i^j(\omega) + \frac{2\omega}{\pi}\int_{\omega_1}^{\infty}\frac{S^{\pm}\mathfrak{A}_i^{+i}(\omega')}{\omega'^2 - \omega^2}d\omega', \quad (\pm\varepsilon_j^i > 0) \tag{76}$$

其中 B_i^j 为束縛态即 $\mathfrak{A}^-$ 的孤立奇点的貢献．如果散射振幅在 $|\omega| \to \infty$ 时不趋于零，色散关系应作相应的改变．

計算結果証明：

$$\omega_1^2-\frac{\Delta^2}{4}\leqslant\frac{1}{4}\left(M-\mu+\sqrt{7\mu^2-2M\mu+M^2}\right)^2<\frac{1}{4}(2M-\mu)^2 \tag{77}$$

等号发生在:

$$l_0=\mathbf{l}\cdot\mathbf{e}=0,$$
$$m^2=4\mu^2,$$
$$\Delta^2=M\mu-4\mu^2+\mu\sqrt{7\mu^2-2M\mu+M^2}$$

时. 故 ω_1 恆在非物理区内,必須由物理区的 $A(\omega)$ 作解析延拓才能决定(75),(76)中的色散积分. 与弹性散射的情況不同, $\mathfrak{A}(\omega)$ 不能用散射截面表示,故在与实驗比較时还須作一些近似,例如只取单核子和单核子单介子的中間态.

八、束縛态的貢献

由于在 $\omega=\omega_0$ 时, 如果 τ 保持物理的数值,动量就成为虛数,可以以下的討論我們理解作先假設 $\tau<0$, 然后将所得的結果解析延拓至 $\tau=M^2+\frac{1}{4}\Delta^2$. 令 $p_n\beta_n$ 表中間态核子动量与同位旋, 由洛仑茨变換及同位旋空間轉动的不变性注意到第三节末所作的說明可以将矩陣元 $\langle p_n\beta_n|j^\beta(0)|q\alpha\rangle$ 表为:

$$\langle p_n\beta_n|j^\beta(0)|q\alpha\rangle=\frac{1}{(2\pi)^3}\sqrt{\frac{M}{E_{p_n}}}\sqrt{\frac{1}{2q_0}}G((p_n-q)^2)\bar{u}_{p_n}\gamma_5\tau^\alpha_{\beta_n\beta}. \tag{78}$$

由于(60)中 $\mathfrak{A}^+$ 的 δ 函数, $p_n-q=p$, 故对于物理的 τ 值 $(p_n-q)^2=-M^2$, $G(-M^2)$ 可定义为耦合常数.

同样矩陣元 $\langle q_1\alpha_1, q_2\alpha_2|j^{\beta'}(0)|p_n\beta_n\rangle$ 可表为

$$\langle q_1\alpha_1, q_2\alpha_2|j^{\beta'}(0)|p_n\beta_n\rangle=\frac{1}{(2\pi)^{9/2}}\sqrt{\frac{1}{4q_{10}q_{20}}}\sqrt{\frac{M}{E_{p_n}}}\sum_{ij}L^iJ^j_{\beta'\beta_n}h^i_j\cdot$$
$$\cdot(m^2,(p_n-q')^2,p_n\cdot l)\times u_{p_n}, \tag{79}$$

其中

$$J^1=\delta_{\alpha_1\alpha_2},\quad J^2=i\varepsilon_{\alpha_1\alpha_2\gamma}\tau^\gamma,\quad L^1=1,\quad L^2=i\gamma\cdot l. \tag{80}$$

由于 $p_n-q'=p'$, 故对于物理的 τ 值 $(p_n-q')^2=-M^2$, 将(78),(79)代入(60),得到

$$\mathfrak{A}^{+i}_j(\omega)\sim-\frac{1}{(2\pi)^{3/2}}\frac{1}{4q_0}G\delta(\omega-\omega_0)\sum_{i'j'}a^{ii'}b_{jj'}h^{i'}_{j'}(\omega_0), \tag{81}$$

其中

$$a=i\begin{pmatrix}0, & -l\cdot q\\ -1 & 0\\ 0 & 0\\ 0 & +1\end{pmatrix},\qquad b=\begin{pmatrix}0 & 0\\ 0 & 2\\ 1 & 0\\ 0 & 1\end{pmatrix}. \tag{82}$$

利用(81)及交叉对称关系得到 $\mathfrak{A}^-$ 对色散积分的貢献:

$$S^\pm B^i_j(\omega)=\frac{+G\omega_0}{(2\pi)^{5/2}q_0(\omega^2-\omega_0^2)}\sum_{i'j'}S^\pm a^{ii'}b_{jj'}h^{i'}_{j'}(-\omega_0)\quad(\pm s^i_j<0), \tag{83}$$

$$S^{\pm}B_j^i(\omega)=\frac{+G\omega}{(2\pi)^{5/2}q_0E_n(\omega^2-\omega_0^2)}\sum_{i'j'}S^{\pm}a^{ii'}b_{jj'}h_{j'}^{i'}(-\omega_0)\qquad(\pm\epsilon_j^i>0).\tag{84}$$

h_j^i 可以用弹性散射的矩陣元表示．由(79)有

$$\frac{1}{(2\pi)^3\sqrt{2q_{20}}}\sqrt{\frac{M}{E_{p_n}}}\sum_{ij}\bar{u}_{p'}L^iJ^j_{\beta'\beta_n}h_j^i(m^2,\ (p_n-q')^2,\ p_n\cdot l)=$$
$$=\bar{u}_{p'}\int\langle q_2\alpha_2\left|\frac{\delta j^{\beta'}(0)}{\delta\varphi_{\alpha_1}(x)}\right|p_n\beta_n\rangle e^{-iq_1x}d^4x,\tag{85}$$

而弹性散射的矩陣元可表为

$$\langle q_2\alpha_2,\ p''\beta'|S|q_1\alpha_1,\ p_n\beta_n\rangle=(2\pi)^4\delta(p''+q_2-p_n-q_1)\ \frac{1}{(2\pi)^3}\sqrt{\frac{1}{2q_{10}}}\sqrt{\frac{M}{E_{p''}}}\times$$
$$\times\ \bar{u}_{p''}\langle q_2\alpha_2\left|\frac{\delta j^{\beta'}(0)}{\delta\varphi_a(x)}\right|p_n\beta_n\rangle e^{-iq_1x}d^4x.\tag{86}$$

由(85)，(86)得到

$$\langle q_2\alpha_2,\ p''\beta'|S|q_1\alpha_1,\ p_n\beta_n\rangle=(2\pi)^{-2}\delta(p''+q_2-p_n-q_1)\sqrt{\frac{1}{4q_{10}q_{20}}}\sqrt{\frac{M^2}{E_{p_n}E_{p''}}}\times$$
$$\times\sum_{ij}\bar{u}_{p''}L^iJ^j_{\beta'\beta_n}h_j^i\left(4l^2,\ (p_n-2l)^2,\ \frac{1}{2}\,p_n\cdot q'\right),\tag{87}$$

其中 $2l=q_2-q$，为弹性散射中的动量变换，而 $(p_n-2l)^2=-M^2$．将(87)与(79)比較，看到束縛态的貢献可以由弹性散射的散射振幅求出来．

补註：在弱反演下 j 的变换应为

$$j^R(x)=\eta SD\bar{j}(-x)S^+,$$

因此如果不忽略終态两个 π 介子的相互作用，为了保証 $\mathscr{A}_j^i$ $\mathscr{D}_j^i$ 为实函数，应将矩陣元 m 中的态矢量 $\langle q_1q_2|$ 換为 $\langle q_1q_2|\frac{1+S}{2}$．

参　考　文　献

[1] Логунов, А. А., Тодоров, И. Г., *Nucl. Phys.* **10** (1959), 552.
[2] Screaton, G. R., *Nuovo Cimento* **11** (1959), 229.
[3] Боголюбов, Н. Н., Медведев, Б. В., Поливанов, М. К., Вопросы Теория Дисперсионных Соотношений (1958).
[4] Pauli, W., Niels Bohr and Development of Physics (1955).

DISPERSION RELATIONS FOR PION PRODUCTION IN PION-NUCLEON COLLISIONS

CHOU KUAN-CHAO (*Peking University*)
DAI YUAN-BEN (*Academia Sinica*)

ABSTRACT

Dispersion relations for pion production in pion-nucleon collissions are discussed. The scattering amplitude is written in a form, in which the operators of nucleon field are extracted from the state vectors of initial and final state. The kinematics and symmetry, properties are worked out in detail and the contribution of bound state is evaluated.

第17卷 第3期 物理学报 Vol. 17, No. 3
1961年3月 ACTA PHYSICA SINICA March, 1961

检驗π介子散射过程是否存在p态共振的一個实驗方法的建議*

何祚庥 周光召

近来关于π介子对π介子散射过程中是否存在着p态共振或同位态(Isobar)的問題,引起了广泛的注意[1],其原因是因为核子的結构是和这一問題密切相关的. 佛拉沙(Frazer)和伏尔克(Fulco)利用色散关系較詳細地研究了核子构造問題,他們得出結論认为π介子系統必須存在着p态共振才能解释核子构造,他們并預測这个同位态的質量是435兆电子伏,半寬度是18兆电子伏[2]. 其它作者們也得出了类似的結論[3].何祚庥、冼鼎昌及察納(Zoellner)近来用单重色散关系及么正条件导出了一套較准确的π介子对π介子散射的积分方程[4]. 可是,初步看来,由他們导出的积分方程的絕热解中并不存在着p态共振,而且p态振幅很小. 至于这套积分方程是否存在着大的p波散射振幅的解,却还有待于进一步研究. 因此,从实驗上来研究π介子系統是否存在p态共振就具有原則性的意义,因为这类实驗涉及到色散关系所依据的一些基本原則是否正确. 邱(Chew)等人建議用下列两个反应

$$e^+ + e^- \to \pi^+ + \pi^- \tag{1a}$$

$$e^- + e^- \to e^- + e^- + \pi^+ + \pi^- \tag{1b}$$

来研究π介子系統是否存在着p态共振問題.从理論角度說,这两个反应可以給出很清楚的理論分析,因而做了这两个实驗,就能明确地作出p态共振存在与否的結論. 然而实际上这两个实驗是难做的,首先它們都需要能量极高的正电子或反电子束,其次,反应(1b)还有截面过小的缺点.

在这篇短文中,我們建議用下列三个反应

$$\pi^\pm + He^4 \to He^4 + \pi^\pm + \pi^0, \tag{2a}$$

$$\pi^\pm + d \to d + \pi^\pm + \pi^0, \tag{2b}$$

$$p + p \to d + \pi^+ + \pi^0, \tag{2c}$$

来研究π介子系統有无p态共振的問題. 不难看出,这三个反应初态的同位旋态都是1. 由于同位旋守恆定律,終态的π介子系統必定处在同位旋$I=1$的态上. 由于π介子满足玻色統計,因此这两个π介子构成系統的軌道角动量只能是单数,即$L=1,3,5\cdots$. 当能量不太高时,就只有p波是主要的.

我們暫且假設这个同位态具有質量435兆电子伏,半寬度18兆电子伏,那末在反应(2)中,这个同位态将衰变成为两个π介子,而相应的反冲He^4及d核的能譜将具有一高峯.

---

* 1960年8月25日收到.

以反应(2a)为例. 設入射的 π 介子束具有能量为 700 兆电子伏(实驗室系),那末在质心系的 He^4 的能譜在 43 兆电子伏有一高峯,而半寬度是 1.8 兆电子伏. 在反应(2c)中,如入射的質子能量是 14 亿电子伏(实驗室系),那末氘核的能譜在 36 兆电子伏将有一高峯(质心系),而半寬度是 3.2 兆电子伏. 反过来,如果 p 态共振是不存在的,那末 He^4 及 d 核能譜将慢慢变化,并且主要由相空間因子所决定. 不难看出,上述实驗将能明确地得出 π 介子系統 p 态共振是否存在的肯定的結論.

参 考 文 献

[1] Д. И. Блохинцев, В. С. Барашенков, Б. М. Барабашов, Структура Нуклонов *У. Ф. Н.* **68**, (1959), 417. S. D. Drell et. al. Report on 1958 Annual International Conference on High Energy Physics at Cern and disscussions, Pag. 20.

[2] W. R. Frazer, J. R. Fulco, *Phys. Rev*, **117**, (1960) 1609.

[3] W. G. Holladay, *Phys. Rev.* **101**, (1956) 1198; J. Bowcock, W. N. Cottingham, D. Lurié Effect of a pion-pion Scattering Resonance on Low Energy π-Nucleon Scattering (預印本).

[4] Сянь Дин-чан, Хэ Цзо-сю, В. Цёллнер, Интегральные уравнения $\pi - \pi$ рассеяния при низких энергиях (預印本).

[5] G. F. Chew, Possible Manifectations of a poin-puin Interaction (預印本); N. Cabibbo, R. Gatto, *Phys. Rev.* Letters, **4**, (1960) 313; L. M. Brown, F. Calogero *Phys. Rev.* Letters. **4**, (1960) 315.

A SUGGESTION OF EXPERIMENT TO DETECT THE p-RESONANCE IN $\pi-\pi$ SCATTERING

Ho Tso-hsiu Chou Kuang-chao

Abstract

An Experiment designed to detect the p-resonance in $\pi-\pi$ scattering process is suggested.

第19卷 第10期 物 理 学 报 Vol. 19, No. 10
1963年10月 ACTA PHYSICA SINICA October, 1963

$\pi+p\to\Lambda+\pi+K$反应共振-近閾效应关联的研究*

苏肇冰 高崇寿 周光召

提 要

本文建議作 $\pi+p\to\Lambda+\pi+K$ 反应的实驗,其中質心系总能量固定为 1900 MeV. 在末态K介子动能 0—90MeV 的变化范围內,观察末态 Λ-π 共振和在 $\Sigma\pi K$ 道产生閾附近的近閾效应. 通过共振-近閾效应关联和共振的角分布可以确定 Y_1^* 的自旋和 Y_1^*,Λ和Σ之間的相对宇称. 按照先处理T矩陣的幺正性,后处理解析性的思想,本文引入了T矩陣在道空間上的对角表象. 在此基础上,提出了唯象描述近閾效应、共振現象和末态相互作用,特别是近閾效应和在閾附近共振的关联的一个新方法. 并且应用这个方法,处理了所建議实驗中的末态共振-近閾效应关联.

一、引 言

随着高能物理和高能技术的发展,基本粒子 $2\to n$ 反应($n>2$)的研究愈来愈重要. 一方面这种多体产生机构是过去所不熟知的,另一方面多体的末态相互作用使这些現象变得更为丰富复杂,常常可以通过适当地选择实驗条件来控制一部分末态粒子的自由度,使得相应于另一部分自由度的現象典型化和特征化,从而可以对基本粒子有关的性質进行很好的研究. 本文主要研究 $\pi+p\to\Lambda+\pi+K$ 反应. 固定初态 πp 質心系总能量为 1900 MeV,选择末态K介子相对 $\Lambda\pi$ 系統質心的动能变化范围由 0 到 90 MeV,观察在 $\Sigma\pi K$ 道产生閾附近 $\Lambda\pi K$ 截面的近閾效应和与它邻近的 $\Lambda\pi$ 末态共振. 通过它們和它們之間的关联(以下簡称共振-近閾关联)来研究 $\Lambda\pi$ 共振态 Y_1^* 的自旋、Y_1^*-Λ 相对宇称和 Λ-Σ 相对宇称,并且对超子可能具有的对称性也进行了簡单的討論.

在新开道产生閾(簡称新道閾)附近,观察道截面的反常变化(近閾效应)在原則上由S矩陣的幺正性和解析性所决定. 幺正性决定了在閾上面的新开道通过几率守恆对观察道反应振幅的影响,解析性規定了观察道反应振幅在新开道閾上和閾下的联系. 共振現象在原則上也是S矩陣在共振极点附近解析行为的表現. 因此共振-近閾关联的唯象研究需要引入对S矩陣元适当的参数描述,使得能够最好地体現新道閾附近S矩陣的解析性和幺正性. 通常的共振理論[1]首先考慮S矩陣的解析性,但幺正条件对共振参数的限制并不明显. 近閾效应的唯象处理也常先考慮S矩陣元在新道閾附近的解析行为,然后再利用幺正条件来定参数[2]. 在本文中,我們引入道空間上S矩陣的对角表象,首先考慮幺正性,得到保証满足幺正条件的S矩陣参数表示;然后由各参数在新道閾附近和共振极

* 1962年11月30日收到.

点附近的正确的解析行为，自然得到近閾效应、共振現象和共振-近閾关联的公式．这种先考虑么正性后考虑解析性的唯象参数方法，在某些場合常常更为方便．

道空間上 S 矩陣对角表象体现了反应各观察道之間的一种联系，它給出了与原子核物理复合核[1]相对应的内部衰变态，从而我們可以得到不依賴于复合核模型假定的基本粒子反应的物理机构．并且通过这种联系，可以导出在 $2\to 3$ 反应中两体共振末态作用对反应振幅的貢献．本文就是通过这种方法，处理 $\Lambda\pi$ 共振对 $\pi+p\to\Lambda+\pi+K$ 过程的貢献．

将上述結果应用到所設計的实驗，得到：

1. 可以从共振-近閾关联和反应截面角分布的各向同性两方面各自独立地判断 $\Lambda\pi$ 共振态 Y_1^* 的自旋 $J_{Y_1^*}$．

2. 如果 $J_{Y_1^*}=\frac{1}{2}$，并且能够得到极化的初态核子，則可以通过超子极化角分布的各向同性来判断 Y_1^*-Λ 的相对宇称；与共振-近閾关联一起，可以判断 Λ-Σ 的相对宇称．

在第二节，我們引入了道空間上 S 矩陣对角表象．按照受到么正条件限制的 S 矩陣参数在新道閾和共振附近的解析行为，导出了共振-近閾关联公式．为了說明問題，以忽略自旋宇称的 $2\to 2$ 反应为例进行討論．第三节在对角表象中导出 $2\to 3$ 反应中共振末态作用对反应振幅的貢献．将第二节的結果对有自旋粒子的 $2\to 3$ 反应作直接推广，得出 $\pi+p\to\Lambda+\pi+K$ 反应末态 Y_1^* 共振-近閾关联的唯象公式．第四节把上述結果应用到所設計的实驗，結合角分布分析，預測从实驗結果可能得到的各种判断．在第五节中，对結果进行了簡单的討論．

作为第二节推导的出发点之一，在附录 I 中討論了观察道 K 矩陣元 $K_{ij}(1\leqslant i,\ j\leqslant n)$ 在第 $n+1$ 新开道閾附近的不解析行为，并且具体地給出了在閾上、閾下 K_{ij} 对 $|p_{n+1}|$ 的依賴关系．附录 II 以波包入射为例，分析了对角表象所提供的、与原子核物理中复合核模型相当的基本粒子反应机构．

二、S 矩陣的对角表象、共振現象和近閾效应的唯象描述

这一节中，我們以无自旋标量粒子的 $2\to 2$ 反应作为例子进行討論．容易看到，所得結果不难推广到更一般的情形．我們以产生閾的能量由低到高来編排对应道的标記次序，并且所有討論都在质心系中进行．引入系統的 S，T 和 K 矩陣，

$$S=1+2iT; \tag{2.1a}$$

$$S=\frac{1+iK}{1-iK},\qquad K=\frac{1}{i}\frac{S-1}{S+1}; \tag{2.1b}$$

$$T=\frac{K}{1-iK},\qquad K=\frac{T}{1+iT}. \tag{2.1c}$$

我們有兴趣的是在具有确定能量、确定角动量分波的道空間上进行討論．S 矩陣在动量空間上的表示和在道空間上表示之間的联系，由式(2.2)給出，以 T 矩陣为例：

$$\langle \mathbf{p}_f|T|\mathbf{p}_i\rangle=\delta(E_f-E_i)\sum_{lm}Y_{lm}(\Omega_f)\frac{1}{\sqrt{\rho(E_f)}}T^l_{fi}(E)\frac{1}{\sqrt{\rho(E_i)}}Y^*_{lm}(\Omega_i), \tag{2.2}$$

其中 $\mathbf{p}_i$ 和 $\mathbf{p}_f$ 分别是入射道 i 和出射道 f 的相对动量；Ω_i 和 Ω_f 是相应的立体角；$\rho(E_i)$ 和 $\rho(E_f)$ 是相应的相空間密度；$T^l_{fi}(E)$ 即是在給定 $E = E_i$ 和 l 下 T 矩陣在道空間上的表示，或記为 $\langle f|T^l(E)|i\rangle$. 所有 S, T 和 K 矩陣的算符关系在道空間上都保持不变，特别是

$$S^l(E) = 1 + 2iT^l(E); \tag{2.3a}$$

$$S^l(E) = \frac{1 + iK^l(E)}{1 - iK^l(E)}, \qquad K^l(E) = \frac{1}{i}\frac{S^l(E) - 1}{S^l(E) + 1}; \tag{2.3b}$$

$$T^l(E) = \frac{K^l(E)}{1 - iK^l(E)}, \qquad K^l(E) = \frac{T^l(E)}{1 + iT^l(E)}. \tag{2.3c}$$

引入道空間上 S 矩陣的对角表象. 下面所有的討論都是在道空間上进行的. 当系統滿足时間反演不变时，么正条件等价于 K 是实对称矩陣. 这时，一定存在实正交矩陣 $O^l(E)$ 将 $K^l(E)$ 对角化，相应的本征值是实数. 考虑到 S 和 T 是 K 的函数，所以实正交矩陣 O 同时将 S, T 和 K 对角化，矩陣之間的函数关系化为本征值之間的函数关系. 引入属于本征值 $K^l_r(E)$ 的本征态 $|r^l(E)\rangle$（簡写为 $|r\rangle$），則有（我們約定以 $i, j, k, f = 1, \cdots, n$ 标記观察道，$r, s = 1, \cdots, n$ 标記本征态）

$$K^l(E) = \sum_r |r\rangle K^l_r(E)\langle r|, \tag{2.4a}$$

$$T^l(E) = \sum_r |r\rangle T^l_r(E)\langle r|; \tag{2.4b}$$

$$K^l_{fi}(E) = \langle f|K^l(E)|i\rangle = \sum_r O^l_{fr}(E)K^l_r(E)\tilde{O}^l_{ri}(E), \tag{2.5a}$$

$$T^l_{fi}(E) = \langle f|T^l(E)|i\rangle = \sum_r O^l_{fr}(E)T^l_r(E)\tilde{O}^l_{ri}(E); \tag{2.5b}$$

$$O^{-1l}(E) = \tilde{O}^l(E), \tag{2.6a}$$

$$T^l_r(E) = \frac{K^l_r(E)}{1 - iK^l_r(E)} \qquad K^l_r(E)\text{ 实数}; \tag{2.6b}$$

$$O^l_{fr}(E) = \langle f|r\rangle, \tag{2.7a}$$

$$\tilde{O}^l_{ri}(E) = \langle r|i\rangle = O^l_{ir}(E) = \langle i|r\rangle. \tag{2.7b}$$

对角表象对基本粒子反应过程提供了这样的观点：在道空間上存在两組正交完备态 $\{|i\rangle\}$, $i = 1, \cdots, n$ 和 $\{|r\rangle\}$, $r = 1, \cdots, n$. 一組 $\{|i\rangle\}$ 是描述参加反应的个别粒子的入射道或出射道，另一組 $\{|r\rangle\}$ 是描述在反应中各自独立互不干扰地进行弹性散射的內部态. 前者直接与观察效果相联系，后者不能直接观察，描述反应机构. 反应将首先由入射道跃迁到內部道，在內部道各自独立地进行弹性散射，再跃迁到末态观察道. 关于內部态和基本粒子反应的物理意义的进一步討論，将在附录 II 中进行.

对角表象也对 T 矩陣提供了一組唯象描述的参数：由实正交矩陣 O 所对应的 $\frac{n(n-1)}{2}$ 个实参数和 K 矩陣本征值所对应的 n 个实参数所組成的 $\frac{n(n+1)}{2}$ 个实参数. 么正条件在对角表象中等价于

1. 　O 矩陣实正交，即 $O^{-1} = \tilde{O}$;　(2.6a)

2. $$T^J_r(E)=\frac{K^J_r(E)}{1-iK^J_r(E)}\qquad K^J_r(E)\ \text{实数}. \tag{2.6b}$$

条件(2.6a)和(2.6b)保証么正性被满足，因此討論由么正性和解析性所決定的近閾效应和共振現象时，只要求出 $O^J_{ir}(E)$ 和 $K^J_r(E)$ 在新道閾和共振极点附近的解析行为，就可得到近閾效应和共振現象的唯象公式．下面我們将分别考察参数在共振极点和新道閾附近的解析行为．

首先，我們引入关于共振的假定：

1. 假定实正交矩陣元 $O^J_{ir}(E)$ 是緩慢变化的，即它对于 T 矩陣元的共振性质没有貢献．

2. 如果在反应能量 $E=E_1$ 出現共振峯，对应地在 K 矩陣的各本征値中至少有一个在 $E=E_1$ 处有一阶极点，对应的留数是小量．在附录 II 中，我們将指出因果性要求极点留数小于零．

容易証明，在么正条件得到满足的前提下，K 矩陣本征値在能量实軸上小留数的一阶极点一一对应于 T 矩陣相应本征値在能量复平面上的一阶复极点．其实部和虚部分别等于 K 矩陣元相应极点的数値和留数．这样由假定 2 所給出的 T 矩陣本征値的共振极点将处于离能量实軸不远的下半平面．我們将根据上述假定討論 T 矩陣元在共振能量附近的解析行为．設 $r=1$，$K^J_1(E)$ 在 $E=E_1$ 处有一共振极点，留数为 $-\Gamma_1(\Gamma_1>0$ 小量)，則在 $E\sim E_1$ 附近有

$$K^J_1(E)\simeq K'^J_1(E)-\frac{\Gamma_1}{E-E_1}, \tag{2.8}$$

其中 $K'^J_1(E)$ 是小量．按照(2.6b)，(2.5b)和(2.6a)，在 $E\sim E_1$ 附近有

$$T^J_1(E)\simeq T'^J_1(E)-\frac{\Gamma_1}{E-E_1+i\Gamma_1} \tag{2.9}$$

[其中 $T'^J_1(E)$ 是小量]以及

$$T^J_{ij}(E)\simeq\sum_{r=2}^{n}O^J_{ir}(E)\frac{K^J_r(E)}{1-iK^J_r(E)}\tilde{O}^J_{rj}(E)+O^J_{i1}(E)T'^J_1(E)\tilde{O}^J_{1j}(E)-$$
$$-\frac{O^J_{i1}(E)\Gamma_1\tilde{O}^J_{1j}(E)}{E-E_1+i\Gamma_1}=P^J_{ij}(E)-\frac{\gamma^J_{i1}(E)\gamma^J_{j1}(E)}{E-E_1+i\Gamma_1}, \tag{2.10}$$

其中

$$P^J_{ij}(E)=\sum_{r=2}^{n}O^J_{ir}(E)O^J_{jr}(E)\frac{K^J_r(E)}{1-iK^J_r(E)}+O^J_{i1}(E)O^J_{j1}(E)T'^J_1(E), \tag{2.11a}$$

$$\gamma^J_{i1}(E)=\Gamma_1^{1/2}O^J_{i1}(E), \tag{2.11b}$$

还有

$$\sum_{i=1}^{n}[\gamma^J_{i1}(E)]^2=\Gamma_1. \tag{2.12}$$

(2.10)中第二項即是共振項，$[\gamma^J_{i1}(E)]^2$ 即是通常所謂部分寬度，Γ_1 即是总寬度[1]，(2.12)即是熟知的部分寬度求和公式．在 $E\sim E_1$ 附近，$K'^J_1(E)$ 和其他 $K^J_r(E)(r=2,\cdots,n)$

1）在本文中，为了方便起見，所提到的寬度都是指半寬度而言．

都是小量,所以 $P^l_{ii}(E)$ 构成在共振峯后的背景,即是通常所謂势散射項,(2.10)即是满足么正条件的 Breit-Wigner 单能級共振公式. 所进行的討論显然很容易在开道数目不变的能量范围上推广到多能級共振情形,这时不同的共振峯既可来自同一本征值,也可来自不同本征值的貢献. 此外,如果同时有两个以上的 $K^l_r(E)$ 具有相同的极点,则将出现共振的干涉现象. 在附录II中,我們将把 $\{r^l(E)\}$ 的宗量 E 延拓到 T 的共振极点,就得到与原子核物理复合核模型相当的衰变态;波包入射的反应,可以归結为一系列衰变态的形成和它的衰变.

其次,我們討論 $O^l_{ir}(E)$ 和 $K^l_r(E)$ 的近閾解析行为. 設系統第 $n+1$ 道产生閾为 E_t. 当 $E < E_t$ 时系統开 n 个道,当 $E > E_t$ 时,系統开 $n+1$ 个道(在以后的討論中,所提到的閾,都是指第 $n+1$ 道而言). 当 $E > E_t$,并在 E_t 附近,第 $n+1$ 道只有 S 波,因此它开在观察道 $l = 0$ 所属的道空間上. 有兴趣的是在这个道空間上,$n+1$ 維 O 矩陣元和 K 矩陣本征值对第 $n+1$ 道 3 維动量 p_{n+1} 的依賴行为. 附录I給出在閾上附近 $n+1$ 維 K 矩陣元对 p_{n+1} 的依賴关系准到 p_{n+1} 一次項为(略去分波指标 $l = 0$)

$$K_{ij}(E) = K_{ij}(0) \qquad 1 \leqslant i, j \leqslant n, \tag{2.13a}$$

$$K_{i\,n+1}(E) = K_{n+1\,i}(E) \sim O(p_{n+1}^{1/2}) \qquad 1 \leqslant i \leqslant n, \tag{2.13b}$$

$$K_{n+1\,n+1}(E) \sim O(p_{n+1}). \tag{2.13c}$$

由(2.13)出发,利用标准的綫性代数方法,容易得到

$$O_{ir}(E) \simeq O_{ir}(0) + O_{ir}(1) \qquad 1 \leqslant i, r \leqslant n, \tag{2.14a}$$

$$O_{i\,n+1}(E) \sim O(p_{n+1}^{1/2}) \qquad 1 \leqslant i \leqslant n, \tag{2.14b}$$

$$O_{n+1\,r}(E) \sim O(p_{n+1}^{1/2}) \qquad 1 \leqslant r \leqslant n, \tag{2.14c}$$

$$O_{n+1\,n+1}(E) \simeq 1 + O(p_{n+1}), \tag{2.14d}$$

$$K_r(E) = K_r(0) + K_r(1), \tag{2.15a}$$

$$K_{n+1}(E) \sim O(p_{n+1}). \tag{2.15b}$$

其中 $K_r(0)$, $O_{ir}(0)$ 和 $K_r(1)$, $O_{ir}(1)$ 分别是 $K_r(E)$ 和 $O_{ir}(E)$ 的在 $E = E_t$ 的数值和对 p_{n+1} 展开的一次項 $(1 \leqslant i, j, r \leqslant n)$. 么正条件(2.6a)和(2.6b)要求

$$\sum_{r=1}^{n} O_{ir}(0)O_{jr}(0) = \delta_{ij}, \tag{2.16a}$$

$$\sum_{i=1}^{n} O_{ir}(0)O_{is}(0) = \delta_{rs}, \tag{2.16b}$$

$$\sum_{r=1}^{n} [O_{ir}(0)O_{jr}(1) + O_{ir}(1)O_{jr}(0)] + O_{i\,n+1}(E)O_{j\,n+1}(E) = 0, \tag{2.16c}$$

$$\sum_{i=1}^{n} [O_{ir}(0)O_{is}(1) + O_{ir}(1)O_{is}(0)] + O_{n+1\,r}(E)O_{n+1\,s}(E) = 0, \tag{2.16d}$$

$$T_r(E) = T_r(0) + T_r(1), \tag{2.17a}$$

$$T_r(0) = \frac{K_r(0)}{1 - iK_r(0)}, \tag{2.17a$'$}$$

$$T_r(1) = \frac{K_r(1)}{(1 - iK_r(0))^2}, \tag{2.17a$''$}$$

$$T_{n+1}(E) = O(p_{n+1}). \tag{2.17b}$$

条件(2.13a)即 $K_{ij}(E)(1 \leqslant i, j \leqslant n)$ 不含 p_{n+1} 一次項要求

$$\sum_{r=1}^{n} [O_{ir}(1)K_r(0)O_{jr}(0) + O_{ir}(0)K_r(1)O_{jr}(0) + O_{ir}(0)K_r(0)O_{jr}(1)] = 0. \tag{2.18}$$

由于第 $n+1$ 个本征値的貢献 $\sim O(p^2_{n+1})$，所以在(2.18)中不出現．由(2.14)和(2.15)即可得到 $T_{ij}(E)(1 \leqslant i, j \leqslant n)$ 在閾上面的解析行为：

$$T_{ij}(E) = T_{ij}(0) + T_{ij}(1), \tag{2.19}$$

$$T_{ij}(1) = \sum_{r=1}^{n} [O_{ir}(1)T_r(0)O_{jr}(0) + O_{ir}(0)T_r(1)O_{jr}(0) + O_{ir}(0)T_r(0)O_{jr}(1)]. \tag{2.20}$$

利用(2.17a)，(2.18)和(2.16b)，有

$$\sum_{r=1}^{n} O_{ir}(0)T_r(1)O_{jr}(0) =$$

$$= -\left(\sum_{r=1}^{n} O_{ir}(0)T_r(0)O_{kr}(1)\right)\left(\sum_{s=1}^{n} O_{ks}(0)\frac{1}{1-iK_s(0)}O_{js}(0)\right) -$$

$$-\left(\sum_{r=1}^{n} O_{ir}(0)\frac{1}{1-iK_r(0)}O_{kr}(0)\right)\left(\sum_{s=1}^{n} O_{ks}(1)T_s(0)O_{js}(0)\right),$$

所以

$$T_{ij}(1) = -i\sum_{r,k,s}^{n} O_{ir}(0)T_r(0)[O_{kr}(0)O_{ks}(1) + O_{kr}(1)O_{ks}(0)]T_s(0)O_{js}(0).$$

考虑到(2.16b)，就有

$$T_{ij}(1) = i\left(\sum_{r=1}^{n} O_{ir}(0)T_r(0)O_{n+1\,r}(E)\right)\left(\sum_{s=1}^{n} O_{n+1\,s}(E)T_s(0)O_{js}(0)\right) =$$

$$= iT_{i\,n+1}(E)T_{n+1\,j}(E) \tag{2.21}$$

$$= ip_{n+1}\mathbf{T}_{i\,n+1}(0)\mathbf{T}_{n+1\,j}(0), \tag{2.22}$$

其中

$$T_{i\,n+1}(E) \simeq p^{1/2}_{n+1}\mathbf{T}_{i\,n+1}(0), \tag{2.22a}$$

$$T_{n+1\,j}(E) \simeq p^{1/2}_{n+1}\mathbf{T}_{n+1\,j}(0). \tag{2.22b}$$

根据众所周知的 $T_{ij}(E)(1 \leqslant i, j \leqslant n)$，在閾附近的解析性[4]，就有

$$T_{ij}(E) = T_{ij}(0) + iT_{i\,n+1}(E)T_{n+1\,j}(E) + O(p^2_{n+1}) =$$

$$= T_{ij}(0) + ip_{n+1}\mathbf{T}_{i\,n+1}(0)\mathbf{T}_{n+1\,j}(0) + O(p^2_{n+1}), \quad 当\ E > E_t, \tag{2.23}$$

$$= T_{ij}(0) - |p_{n+1}|\mathbf{T}_{i\,n+1}(0)\mathbf{T}_{n+1\,j}(0) + O(|p_{n+1}|^2), \quad 当\ E < E_t. \tag{2.24}$$

这就是熟知的反应振幅近閾效应的参数公式[2]．

如果在閾上面不远处存在一个共振峯，按照关于共振的第二个假定，如果共振极点存在在 $r=1$ 的本征值中，并且考虑到因子 $\dfrac{\Gamma_1}{E-E_1+i\Gamma_1}$ 不含有 p_{n+1} 的一次項，所以 $T_1(E)$ 有

$$T_1(E) = -\frac{\Gamma_1}{E-E_1+i\Gamma_1} + T_1(1) + T'_1(E), \tag{2.25}$$

其中 $T_1'(E)$ 不包含 p_{n+1} 的一次項，并且共振能量附近是小量．这时

$$T_{ij}(E) = \sum_{\sigma=2}^{n+1} O_{i\sigma}(E)T_\sigma(E)O_{j\sigma}(E) + O_{i1}(E)T_1(E)O_{j1}(E), \tag{2.26}$$

$$\sum_{\sigma=2}^{n+1} O_{i\sigma}(E)T_\sigma(E)O_{j\sigma}(E) = Q_{ij}(E) + \sum_{r=2}^{n+1}(O_{ir}(1)T_r(0)O_{jr}(0) + \\ + O_{ir}(0)T_r(1)O_{jr}(0) + O_{ir}(0)T_r(0)O_{jr}(1)). \tag{2.26a}$$

$$O_{i1}(E)T_1(E)O_{j1}(E) = O_{i1}(E)\left(-\frac{\Gamma_1}{E-E_1+i\Gamma_1} + T_1'(E) + T_1(1)\right)O_{j1}(E) = \\ = -\frac{\Gamma_{ij}'}{E-E_1+i\Gamma_1} + R_{ij}(E) + (O_{i1}(1)T_1(0)O_{j1}(0) + \\ + O_{i1}(0)T_1(1)O_{j1}(0) + O_{i1}(0)T_1(0)O_{j1}(1)). \tag{2.26b}$$

其中 $R_{ij}(E)$，$Q_{ij}(E)$ 是既不包含 p_{n+1} 一次項又不包含共振极点的緩变小量．由于 $O_{ir}(E)$ 緩变，所以当共振峯离閾不远时，就有

$$\Gamma_{ij}' \simeq O_{i1}(0)O_{j1}(0)\Gamma_1. \tag{2.27}$$

将(2.26a)，(2.26b)和(2.27)代入(2.26)，应用(2.20)，(2.21)和(2.22)，并注意到 $Q_{ij}(E)$，$R_{ij}(E)$ 近似地可看作常量

$$T_{ij}(E) = -\frac{O_{i1}(0)O_{j1}(0)\Gamma_1}{E-E_1+i\Gamma_1} + P_{ij}(0) + iT_{i\,n+1}(E)T_{n+1\,j}(E) \tag{2.28a}$$

$$= -\frac{O_{i1}(0)O_{j1}(0)\Gamma_1}{E-E_1+i\Gamma_1} + P_{ij}(0) + ip_{n+1}\mathbf{T}_{i\,n+1}(0)\mathbf{T}_{n+1\,j}(0), \tag{2.29}$$

其中

$$P_{ij}(0) = R_{ij}(0) + Q_{ij}(0)$$

这即是 2 → 2 反应共振-近閾关联的唯象公式．我們特別有兴趣的是这样的情形：当共振离閾不远，并且在 $E = E_t$ 有

$$\left|\frac{\Gamma_1}{E_t - E_1 + i\Gamma_1}\right| \gg |T_\sigma(0)| \qquad \sigma = 2, \cdots, n, \tag{2.30a}$$

即

$$|T_1(0)| \gg |T_\sigma(0)| \qquad \sigma = 2, \cdots, n. \tag{2.30b}$$

这时背景 $P_{ij}(0)$ 的貢献将远小于共振項．同时注意到

$$T_{i\,n+1}(E)T_{n+1\,j}(E) = \left(\sum_{r=1}^{n} O_{ir}(0)T_r(0)O_{n+1\,r}(E)\right)\left(\sum_{s=1}^{n} O_{n+1\,s}(E)T_s(0)O_{js}(0)\right),$$

就有

$$T_{i\,n+1}(E)T_{n+1\,j}(E) \simeq O_{i1}(0)T_1(0)O_{n+1\,1}(E)O_{n+1\,1}(E)T_1(0)O_{j1}(0) + \\ + O_{i1}(0)T_1(0)O_{n+1\,1}(E)\left(\sum_{s=2}^{n} O_{n+1\,s}(E)T_s(0)O_{js}(0)\right) + \\ + \left(\sum_{s=2}^{n} O_{is}(0)T_s(0)O_{n+1\,s}(E)\right)O_{n+1\,1}(E)T_1(0)O_{j1}(0). \tag{2.31}$$

由(2.30)和(2.31)，对所发生的共振-近閾关联可以得到下列推断：

1. 由于在 $E = E_t$ 附近，共振峯坡超过背景，因此近閾效应发生在共振峯坡上，而不是在可能被共振峯坡所掩盖的背景上。

2. 近閾效应的深度也因 $T_1(0)$ 的貢献而变得更为显著。

三、2→3 反应中末态作用和末态共振-近閾效应的关联

这一节我們将在对角表象中討論固定入射能量下 $\pi + p \to \Lambda + \pi + K$ 反应中 $\Lambda\pi$ 末态作用，特别是 $\Lambda\pi$ 末态共振和在 $\Sigma\pi K$ 閾处近閾效应的关联的唯象描述。为了說明問題，首先忽略自旋和宇称結构进行討論，最后将結果推广到考虑自旋和宇称的情形。設入射道和出射道分别为 $a + b$ 和 $c + d + e$，記为 (ab) 道和 (cde) 道，a, b, c, d 和 e 均无自旋。我們有兴趣的是在确定总能量 E、确定总角动量 e 的角分波表象中討論。这时反应道的确定，除了粒子种类以外，还需要确定 cd 系統(或 e) 的部分能量 E_{cd} (或 E_e)，cd 系統的相对角动量 l_{cd} 和 e 相对 cd 系质心的相对角动量 l_e。与(2.2)相应，动量表象和角分波表象中 T 矩陣元的联系为

$$\langle (cde)\mathbf{p}_{cd}\mathbf{p}_e | T | \mathbf{p}_i(ab)\rangle = $$

$$= \delta(E_f - E_i) \sum_{lm\, l_{cd}\, l_e} (i)^{l-l_{cd}-l_e} Y_{lm\, l_{cd}\, l_e}(\Omega_{cd}, \Omega_e) \frac{1}{\sqrt{\rho(E_{cd})\rho(E_e)}} \cdot$$

$$\cdot \langle (cde) l_{cd} l_e E_{cd} | T^l(E) | l(ab)\rangle \frac{1}{\sqrt{\rho(E_i)}} Y_{lm}(\Omega_i), \tag{3.1}$$

其中

$$\mathbf{p}_{cd} = \frac{m_d}{m_c + m_d}\mathbf{q}_c - \frac{m_c}{m_c + m_d}\mathbf{q}_d, \tag{3.1a}$$

$$\mathbf{p}_e = \frac{m_c + m_d}{M}\mathbf{q}_e - \frac{m_e}{M}(\mathbf{q}_c + \mathbf{q}_d) = \mathbf{q}_e \qquad M = m_c + m_d + m_e, \tag{3.1b}$$

$$\mathbf{p} = \mathbf{q}_c + \mathbf{q}_d + \mathbf{q}_e = 0, \tag{3.1c}$$

$$Y_{lm\, l_{cd}\, l_e}(\Omega_{cd}\Omega_e) = \sum_{m_{cd}+m_e=m} c(l_{cd}l_e l;\, m_{cd}m_e m) Y_{l_{cd}\, m_{cd}}(\Omega_{cd}) Y_{l_e\, m_e}(\Omega_e), \tag{3.1d}$$

$$E_{cd} = \frac{1}{2\mu_{cd}} p_{cd}^2, \qquad \mu_{cd} = \frac{m_c m_d}{m_c + m_d}, \tag{3.1e}$$

$$E_e = \frac{1}{2\mu_e} p_e^2, \qquad \mu_e = \frac{m_e(m_c + m_d)}{M}, \tag{3.1f}$$

$$E_{cd} + E_e + M = E_f = E_i = E, \tag{3.1g}$$

$$\mathbf{l}_{cd} + \mathbf{l}_e = \mathbf{l} \qquad \sum_{l l_{cd} l_e} \text{按宇称所允許的进行求和}, \tag{3.1h}$$

$\mathbf{q}_c$, $\mathbf{q}_d$, $\mathbf{q}_e$ 分别是粒子 c, d, e 在质心系中的 3 維动量，$\rho(E_i)$，$\rho(E_{cd})$ 和 $\rho(E_e)$ 分别是 $\mathbf{p}_i$，$\mathbf{p}_{cd}$ 和 $\mathbf{p}_e$ 的相空間密度。有时我們将 $\langle (cde) l_{cd} l_e E_{cd} | T^l(E) | l_i(ab)\rangle$ 簡写为 $\langle cde | T | ab\rangle$。对于有三体道出现的情形，确定总能量和总角动量下的道空間有了新的內容：三体道的确定与两体道不一样，除了粒子类别以外，还需确定部分能量和相对角动量。道空間上的矩陣附标，不仅有两体道附标，还有三体道附标。前者只需粒子类别，后者除此以外还需部分能量和相对角动量。道空間上矩陣运算中，对不同道求和时，不仅要对不

同粒子求和，而且要对相同粒子不同相对角动量 l_{cd} 和 l_e 求和，以及对部分能量 E_{cd}（或 E_e）求积分．由于道空間指标出現了連續变化的部分能量，因此，对角表象中，本征态和本征値指标也将出現連續譜（下面討論中，对連續指标的求积分，常隐含在求和号的形式中）．在对道空間的概念作了如上的扩充后，第二节中所有 $2 \to 2$ 反应的結論，都可以直接推广到 $2 \to 3$ 反应中来．

在通常的末态作用理論[5]中，反应 $a + b \to c + d + e$ 分成两步：第一步是原始产生机构；第二步是末态 c-d 散射．在产生后 cde 粒子之間只发生 c-d 散射，忽略 c-e，d-e 或 3 体作用等价于条件[6]

$$\langle cde|T|c'd'e'\rangle \sim \langle cd|T|c'd'\rangle\langle e|e'\rangle, \tag{3.2a}$$

如果考虑到 c-d 的非弹性末态作用，即是

$$\langle cde|T|f'g'e'\rangle \sim \langle cd|T|f'g'\rangle\langle e|e'\rangle, \tag{3.2b}$$

在(3.2)中，我們略去了部分能量和部分角动量指标．我們将选定某一个入射总能量和总角动量，使得在它所允許的部分能量和相对角动量的变化范围内，条件(3.2)得到滿足，从而研究 c-d 作用特别是共振对 $\langle cde|T|ab\rangle$ 的影响．

如所周知，实驗上 $2 \to 3$ 反应振幅，常常比 $2 \to 2$ 反应振幅小一个数量級或者更小．因此在(3.2)成立的条件下道空間在零級近似下，分解为两个子空間的直和：一个是 $\langle ab|T|a'b'\rangle$ 所属的所有 $2 \to 2$ 反应組成的二体道子空間；一个是 $\langle cde|T|c'd'e'\rangle$ 所属的所有 $3 \to 3$ 反应組成的三体道子空間．由(3.2)，后者即是 $\langle cd|T|c'd'\rangle$ 所属的二体道空間和 $\langle e|e'\rangle$ 的直乘．用矩陣符号来表达：

$$T^l(E) = \left(\begin{array}{c:c} \begin{matrix}\langle ab|T|a'b'\rangle \\ (\text{零級量})\end{matrix} & \begin{matrix}\langle ab|T|c'd'e'\rangle \\ (\text{一級量})\end{matrix} \\ \hdashline \begin{matrix}\langle cde|T|a'b'\rangle \\ (\text{一級量})\end{matrix} & \begin{matrix}\langle cde|T|c'd'e'\rangle \\ \sim \langle cd|T|c'd'\rangle\langle e|e'\rangle \\ (\text{零級量})\end{matrix} \end{array}\right) \sim \left(\begin{array}{c:c} \langle ab|T|a'b'\rangle & 0 \\ \hdashline 0 & \langle cd|T|c'd'\rangle\langle e|e'\rangle \end{array}\right). \tag{3.3}$$

实际上，(3.3)即是(3.2)的推論：因为在零級近似下，(3.2)意味着 e 和 c，d 之間沒有作用，$2 \to 3$ 反应的逆反应 $3 \to 2$ 反应不能发生．由細致平衡原理，$2 \to 3$ 反应亦不发生．(3.3)說明对角表象中，T 矩陣本征値在零級近似下即是二体道子空間和三体道子空間各自本征値的全体，并且由条件(3.2)，(3.3)，三体道子空間的本征値即是 $\langle cd|T|c'd'\rangle$ 所属的二体道空間上的本征値．記二体道子空間和三体道子空間上本征値和本征态分别为 $T^l_{r_1}(E)$ 和 $|r^l_1(E)\rangle$，$r_1 = 1, \cdots, n$，$T^l_{r_2}(E)$和$|r^l_2(E)\rangle$，r_2 含有連續指标；記 $\langle cd|T|c'd'\rangle$ 所属二体道空間上的本征値和本征态为 $T^{l_{cd}}_\sigma(E_{cd})$ 和 $|\sigma^{l_{cd}}(E_{cd})\rangle$．由条件(3.2)，

$$T^l_{r_2}(E) = T^l_{l_{cd}E_{cd}\sigma}(E) = T^{l_{cd}}_\sigma(E_{cd}), \tag{3.4a}$$

$$|r^l_2(E)\rangle = |\sigma^{l_{cd}}(E_{cd})\rangle|l_eE_e(e)\rangle. \tag{3.4b}$$

其中 $|l_eE_e(e)\rangle$ 是处于观察道并具有能量 $E_e = E - M - E_{cd}$ 和角动量 l_e 的自由 e 粒子．这时，$2 \to 3$ 反应振幅在对角表象中的展开式有

$$\langle (cde)l_{cd}l_eE_{cd}|T^l(E)|l(ab)\rangle =$$
$$= \sum_{r_1}\langle (cde)l_{cd}l_eE_{cd}|r_1^l(E)\rangle T_{r_1}^l(E)\langle r_1^l(E)|l(ab)\rangle +$$
$$+ \sum_{r_2}\langle (cde)l_{cd}l_eE_{cd}|r_2^l(E)\rangle T_{r_2}^l(E)\langle r_2^l(E)|l(ab)\rangle \quad (3.5a)$$
$$= \sum_{r_1}\langle (cde)l_{cd}l_eE_{cd}|r_1^l(E)\rangle T_{r_1}^l(E)\langle r_1^l(E)|l(ab)\rangle +$$
$$+ \int dE_e' \sum_{\substack{l_{cd}' l_e' \\ \mathbf{l}_{cd}'+\mathbf{l}_e'=\mathbf{l}}} \sum_{\sigma} \langle (cde)l_{cd}l_eE_{cd}|\sigma^{l_{cd}'}(E_{cd}')\rangle |l_e'E_e'(e)\rangle \cdot$$
$$\cdot T_{\sigma}^{l_{cd}'}(E_{cd}')\langle \sigma^{l_{cd}'}(E_{cd}')|\langle (e)l_e'E_e'|l(ab)\rangle. \quad (3.5b)$$
$$= \sum_{r_1}\langle (cde)l_{cd}l_eE_{cd}|r_1^l(E)\rangle T_{r_1}^l(E)\langle r_1^l(E)|l(ab)\rangle +$$
$$+ \sum_{\sigma}\langle (cd)l_{cd}|\sigma^{l_{cd}}(E_{cd})\rangle T^{l_{cd}}(E_{cd})\langle \sigma^{l_{cd}}(E_{cd})|\langle (e)l_eE_e|l(ab)\rangle. \quad (3.6)$$

在从(3.5)到(3.6)的推导中，利用了

$$\langle (cde)l_{cd}l_eE_{cd}|r_2^l(E)\rangle = \langle (cde)l_{cd}l_eE_{cd}|\sigma^{l_{cd}'}(E_{cd}')\rangle |l_e'E_e'(e)\rangle =$$
$$= \langle (cd)l_{cd}|\sigma^{l_{cd}}(E_{cd})\rangle \delta(E_e - E_e')\delta_{l_{cd}l_{cd}'}\delta_{l_el_e'},$$

由条件(3.2)和(3.3)容易看到，(3.6)中 $\langle (cde)l_{cd}l_eE_{cd}|r_1^l(E)\rangle$ 和 $\langle \sigma^{l_{cd}}(E_{cd})|\langle (e)l_eE_e|l(ab)\rangle$ 是一級小量，所以本征値只需取零級項卽可． 如果在(3.6)中和第二項相比，第一項可以忽略，卽

$$\langle (cde)l_{cd}l_eE_{cd}|T^l(E)|l(ab)\rangle \simeq$$
$$\simeq \sum_{\sigma}\langle (cd)l_{cd}|\sigma^{l_{cd}}(E_{cd})\rangle T_{\sigma}^{l_{cd}}(E_{cd})\langle \sigma^{l_{cd}}(E_{cd})|\langle (e)l_eE_e|l(ab)\rangle. \quad (3.7)$$

这卽是在对角表象中，考虑到非弹性过程的末态相互作用唯象公式． 推导(3.7)所用近似条件与其他末态作用理論[5,6]相同． 如果在所允許的部分能量变化范围内，$T_{\sigma}^{l_{cd}}(E_{cd})$ 之一 $T_{\sigma_0}^{l_{cd}}(E_{cd})$ 存在一个共振极点

$$T_{\sigma_0}^{l_{cd}}(E_{cd}) \simeq T_{\sigma_0}'^{l_{cd}}(E_{cd}) - \frac{\Gamma_{cd}^0}{(E_{cd}-E_{cd}^0)+i\Gamma_{cd}^0}, \quad (3.8)$$

則根据(3.6)，

$$\langle (cde)l_{cd}l_eE_{cd}|T^l(E)|l(ab)\rangle \simeq \langle (cde)l_{cd}l_eE_{cd}|P^l(E)|l(ab)\rangle -$$
$$- \langle (cd)l_{cd}|\sigma_0^{l_{cd}}(E_{cd})\rangle \frac{\Gamma_{cd}^0}{(E_{cd}-E_{cd}^0)+i\Gamma_{cd}^0}\langle \sigma_0^{l_{cd}}(E_{cd})|\langle (e)l_eE_e|l(ab)\rangle, \quad (3.9)$$

其中：

$$\langle (cde)l_{cd}l_eE_{cd}|P^l(E)|l(ab)\rangle =$$
$$= \sum_{r_1}\langle (cde)l_{cd}l_eE_{cd}|r_1^l(E)\rangle T_{r_1}^l(E)\langle r_1^l(E)|l(ab)\rangle +$$

$$+\sum_{\sigma \neq \sigma_0} \langle (cd) l_{cd} | \sigma^{l_{cd}}(E_{cd}) \rangle T_{\sigma}^{l_{cd}}(E_{cd}) \langle \sigma^{l_{cd}}(E_{cd}) | \langle (e) l_e E_e | l(ab) \rangle +$$

$$+ \langle (cd) l_{cd} | \sigma_0^{l_{cd}}(E_{cd}) \rangle T_{\sigma_0}^{\prime l_{cd}}(E_{cd}) \langle \sigma_0^{l_{cd}}(E_{cd}) | \langle (e) l_e E_e | l(ab) \rangle \tag{3.9a}$$

是緩变背景．(3.9)即是末态 c-d 共振对 $\langle cde|T|ab\rangle$ 貢献的唯象公式．

如果在固定 $a+b \to c+d+e$ 入射总能量时，cd 部分能量的变化范围内存在且只存在 $f+d+e$ 道的产生閾，并且如果下列条件得到满足：

1. 在 (fde) 道 fd 系統部分能量的上限很小，即只有 $l_{fd}=0$ 的分波；

2. 条件(3.2)得到满足；

则在 $a+b \to c+d+e$ 中，$l_{cd}=0$ 的分波振幅按部分能量 E_{cd} 变化时，在 (fde) 閾处将产生近閾效应[6,7]．考虑到本节一开始对三体道出現情形下道空間和它上面运算的扩充，并且按照条件 1 所导致的 Wigner 行为[7,8]

$$\langle fde|T|ab\rangle|_{l_{fd}=0} \sim p_{fd}^{1/2} \quad \text{在閾上,} \tag{3.10}$$

容易将第二节近閾效应的討論直接推广而有

$$\langle (cde) l_{cd}=0\, l_e E_{cd} | T^l(E) | l(ab) \rangle =$$
$$= \langle (cde) l_{cd}=0\, l_e E_{cd}^t | T^l(E) | l(ab) \rangle +$$
$$+ i \sum_{l'_{fd} l'_e} \int_0^{E-M-E_{cd}^t} dE'_{fd} \langle (cde) l_{cd}=0\, l_e E_{cd} | T^l(E) | l'_{fd} l'_e E'_{fd}(fde) \rangle \cdot$$
$$\cdot \langle (fde) l'_{fd} l'_e E'_{fd} | T^l(E) | l(ab) \rangle =$$
$$= \langle (cde) l_{cd}=0\, l_e E_{cd}^t | T^l(E) | l(ab) \rangle +$$
$$+ i \sum_{l'_e} \int_0^{E-M-E_{cd}^t} dE'_{fd} \langle (cde) l_{cd}=0\, l_e E_{cd} | T^l(E) | l'_{fd}=0\, l'_e E'_{fd}(fde) \rangle \cdot$$
$$\cdot \langle (fde) l'_{fd}=0\, l'_e E'_{fd} | T^l(E) | l(ab) \rangle, \tag{3.11}$$

其中 $E_{cd}^t = m_f - m_c$ 是对应 (fde) 道产生閾的 cd 部分能量．由条件(3.2)，

$$\langle (cde) l_{cd}=0\, l_e E_{cd} | T^l(E) | l_{fd}=0\, l'_e E'_{fd}(fde) \rangle \simeq$$
$$\simeq \langle (cd) l_{cd}=0\, | T^{l_{cd}=0}(E_{cd}) | l_{fd}=0\, (fd) \rangle \delta(E_{cd} - E'_{fd} - E_{cd}^t) \delta_{l_e l'_e} \delta_{l l_e},$$

就有

$$\langle (cde) l_{cd}=0\, l_e E_{cd} | T^l(E) | l(ab) \rangle =$$
$$= \langle (cde) l_{cd}=0\, l_e E_{cd}^t | T^l(E) | l(ab) \rangle +$$
$$+ i \langle (cd) l_{cd}=0\, | T^{l_{cd}=0}(E_{cd}) | l_{fd}=0\, (fd) \rangle \cdot$$
$$\cdot \langle (fde) l_{fd}=0\, l_e E_{fd} | T^{l_e}(E) | l_e(ab) \rangle \delta_{l_e l}, \tag{3.12}$$

其中 $E_{fd} = E_{cd} - E_{cd}^t$．(3.12)即是 2 → 3 反应固定入射能量的近閾效应公式．有兴趣的問題是在 (fcd) 閾上而不远处，存在 cd 系統共振情形下的末态 c-d 共振-近閾关联現象．在扩充的道空間上，按照与第二节相同的考虑，并根据(3.9)和(3.12)，容易得到 2 → 3 反应末态共振-近閾关联公式：

$$\langle (cde) l_{cd} l_e E_{cd} | T^l(E) | l(ab) \rangle =$$
$$= \langle (cde) l_{cd} l_e E_{cd}^t | P^l(E) | l(ab) \rangle -$$
$$- \langle (cd) l_{cd} | \sigma_0^{l_{cd}^0}(E_{cd}) \rangle \frac{\Gamma_{cd}^0}{E_{cd} - E_{cd}^0 + i\Gamma_{cd}^0} \langle \sigma_0^{l_{cd}^0}(E_{cd}) | \langle (e) l_e E_e | l(ab) \rangle \delta_{l_{cd}\, l_{cd}^0} \delta(l_{cd}^0, l_e; l) +$$

$$+ i\langle (cd)l_{cd} = 0\,|T^{l_{cd}=0}(E_{cd})|\,l_{fd} = 0\,(fd)\rangle \langle (fde)l_{fd} = 0\,l_e E_{fd}|T^{l_e}(E)|\,l_e(ab)\rangle \delta_{l_{cd}0}\delta_{l_e l}, \tag{3.13}$$

其中 $\langle (cde)l_{cd}l_eE^l_{cd}|P^l(E)|l(ab)\rangle$ 是既不包含 p_{fd} 一次項，并且在共振附近緩变的背景項，l^0_{cd} 是共振分波的軌道角动量．(3.13)中引入的 Krönecker 符号 $\delta_{l_{cd}l^0_{cd}}$，$\delta(l^0_{cd}, l_e; l)$，$\delta_{l_{cd}0}$ 和 $\delta_{l\,l_e}$ 反映了共振-近閾关联的各种不同可能情况，其中

$$\delta(l^0_{cd}, l_e; l) = \begin{cases} 1 & \text{当 } \mathbf{l}_e + \mathbf{l}^0_{cd} = \mathbf{l}; \\ 0 & \text{当 } \mathbf{l}_e + \mathbf{l}^0_{cd} \neq \mathbf{l}. \end{cases} \tag{3.14}$$

对于考虑粒子的自旋和宇称結构的情形，除了运动学的复杂性有所增加以外，沒有任何新的問題．我們将对 $\pi + p \to \Lambda + \pi + K$ 直接写出所求的結果．在目前情形下，动量表象 T 矩陣元的角分波展开有

$$\langle \alpha_\Lambda \mathbf{p}_{\Lambda\pi}\mathbf{p}_K|T|\mathbf{p}_i\alpha_i\rangle = \delta(E_i - E_f)\langle \alpha_\Lambda \mathbf{p}_{\Lambda\pi}\mathbf{p}_K|T(E)|\mathbf{p}_i\alpha_i\rangle \qquad E = E_i = E_f. \tag{3.15}$$

$$\langle \alpha_\Lambda \mathbf{p}_{\Lambda\pi}\mathbf{p}_K|T(E)|\mathbf{p}_i\alpha_i\rangle = \sum_{JJ_zjl_{\Lambda\pi}l_Kl_i} X_{JJ_zj(l_{\Lambda\pi}\frac{1}{2})l_K}(\Omega_{\Lambda\pi}\alpha_\Lambda;\,\Omega_K)\frac{1}{\sqrt{\rho(E_{\Lambda\pi})\rho(E_K)}}\cdot$$

$$\cdot\left\langle (\Lambda\pi K)j\left(l_{\Lambda\pi}\frac{1}{2}\right)l_KE_{\Lambda\pi}\,|T^J(E)|\,l_i\frac{1}{2}(\pi p)\right\rangle\frac{1}{\sqrt{\rho(E_i)}}X^+_{JJ_zl_i\frac{1}{2}}(\Omega_i\alpha_i). \tag{3.16}$$

$$X_{JJ_zl_i\frac{1}{2}}(\Omega_i\alpha_i) = \sum_{m_i+S_z=J_z} c\left(l_i\frac{1}{2}J;\,m_iS_zJ_z\right)Y_{l_im_i}(\Omega_i)\xi_{\frac{1}{2}S_z}(\alpha_i), \tag{3.16a}$$

$$X_{JJ_zj(l_{\Lambda\pi}\frac{1}{2})l_K}(\Omega_{\Lambda\pi}\alpha_\Lambda;\,\Omega_K) = \sum_{j_z+m_K=J_z} c(jl_KJ;\,j_zm_KJ_z)X_{jj_zl_{\Lambda\pi}\frac{1}{2}}(\Omega_{\Lambda\pi}\alpha_\Lambda)Y_{l_Km_K}(\Omega_K), \tag{3.16b}$$

在(3.16)中，初态按相对論处理，末态作非相对論近似．在质心系中，令 $\mathbf{p}_i$ 为初态 π 介子入射动量，$\mathbf{q}_K$，$\mathbf{q}_\Lambda$ 和 $\mathbf{q}_\pi$ 分别为末态粒子的动量，α_i 和 α_Λ 分别为初态核子和末态超子的自旋指标，E_i 和 E_f 分别为初态和末态的总能量，则有

$$E_i = E = \sqrt{p_i^2 + m_\pi^2} + \sqrt{p_i^2 + m_p^2} = E^i_\pi + E^i_p, \tag{3.16c}$$

$$E_f = E = \frac{1}{2\mu_{\Lambda\pi}}p^2_{\Lambda\pi} + \frac{1}{2\mu_K}p^2_K + m_\Lambda + m_\pi + m_K, \tag{3.16d}$$

$$E_K = \frac{1}{2\mu_K}p^2_K, \qquad E_{\Lambda\pi} = \frac{1}{2\mu_{\Lambda\pi}}p^2_{\Lambda\pi},$$

$$\mathbf{p}_K = \frac{m_\Lambda + m_\pi}{m_\Lambda + m_\pi + m_K}\mathbf{q}_K - \frac{m_K}{m_\Lambda + m_\pi + m_K}(\mathbf{q}_\Lambda + \mathbf{q}_\pi) = \mathbf{q}_K, \tag{3.16e}$$

$$\mathbf{p}_{\Lambda\pi} = \frac{m_\pi}{m_\Lambda + m_\pi}\mathbf{q}_\Lambda - \frac{m_\Lambda}{m_\Lambda + m_\pi}\mathbf{q}_\pi, \tag{3.16f}$$

$$0 = \mathbf{q}_\Lambda + \mathbf{q}_\pi + \mathbf{q}_K, \tag{3.16g}$$

$$\mu_{\Lambda\pi} = \frac{m_\Lambda m_\pi}{m_\Lambda + m_\pi}, \tag{3.16h}$$

$$\mu_K = \frac{m_K(m_\Lambda + m_\pi)}{m_\Lambda + m_\pi + m_K}. \tag{3.16i}$$

将公式(3.13)加上自旋和宇称修正后，得

$$\left\langle (\Lambda\pi K)j\left(l_{\Lambda\pi}\frac{1}{2}\right)l_K E_{\Lambda\pi}\middle|T^J(E)\middle|J\left(l_i\frac{1}{2}\right)(\pi p)\right\rangle =$$

$$= \left\langle (\Lambda\pi K)j\left(l_{\Lambda\pi}\frac{1}{2}\right)l_K E'_{\Lambda\pi}\middle|T^J(E)\middle|J\left(l_i\frac{1}{2}\right)(\pi p)\right\rangle -$$

$$-\left\langle (\Lambda\pi)j\left(l_{\Lambda\pi}\frac{1}{2}\right)E_{\Lambda\pi}\middle|\sigma_0^{j^0 l^0_{\Lambda\pi}}(E_{\Lambda\pi})\right\rangle \cdot \frac{\Gamma^0_{\Lambda\pi}}{E_{\Lambda\pi} - E^0_{\Lambda\pi} + i\Gamma^0_{\Lambda\pi}} \cdot$$

$$\cdot \left\langle \sigma_0^{j^0 l^0_{\Lambda\pi}}(E_{\Lambda\pi})\middle|\left\langle (K)l_K E_K\middle|J\left(l_i\frac{1}{2}\right)(\pi p)\right\rangle\right. \delta_{jj^0}\delta_{l_{\Lambda\pi}l^0_{\Lambda\pi}}\delta(j^0, l_K; J) +$$

$$+ \langle (\Lambda\pi)l_{\Lambda\pi}|T^{\frac{1}{2}}(E_{\Lambda\pi})|l_{\Sigma\pi} = 0(\Sigma\pi)\rangle \left\langle (\Sigma\pi K)\frac{1}{2}\left(0\frac{1}{2}\right)l_K E_{\Sigma\pi}\middle|T^J(E)\middle|J\left(l_i\frac{1}{2}\right)(\pi p)\right\rangle \cdot$$

$$\cdot \delta_{j\frac{1}{2}}\delta_{l_{\Lambda\pi}}(0, 1)\,\delta\left(\frac{1}{2}, l_K; J\right). \qquad (3.17)$$

这卽是我們討論 $\pi + p \to \Lambda + \pi + K$ 反应末态共振-近閾关联的唯象公式. 其中

$$\delta_{l_{\Lambda\pi}}(0, 1) = \begin{cases} \delta_{l_{\Lambda\pi}0} & 当\ P(\Lambda\text{-}\Sigma) = +, \\ \delta_{l_{\Lambda\pi}1} & 当\ P(\Lambda\text{-}\Sigma) = -, \end{cases} \qquad (3.18)$$

$P(\Lambda\text{-}\Sigma)$ 是 $\Lambda\text{-}\Sigma$ 相对宇称. $\delta_{j\frac{1}{2}}\delta_{l_{\Lambda\pi}(0,1)}$ 反映了 $(\Sigma\pi K)$ 道在閾附近只有 $l_{\Sigma\pi} = 0$, $j = \frac{1}{2}$ 的分波，$(\Lambda\pi K)$ 道中的近閾效应只能按 $\Lambda\text{-}\Sigma$ 相对宇称相同或相反而发生在 $j = \frac{1}{2}$, $l_{\Lambda\pi} = 0$ 或 1 的分波上. $\delta_{jj_0}\delta_{l_{\Lambda\pi}l^0_{\Lambda\pi}}\delta(j^0, l_K; J)$ 說明，只有 j^0, $l^0_{\Lambda\pi}$ 分波才有共振，其他分波共振不存在.

四、$\pi + p \to \Lambda + \pi + K$ 反应的研究

这一节中我們将具体研究 $\pi^- + p \to \Lambda + \pi^- + K^+$ 反应. 由于内部对称性将自动在内部态分类中所包括，不必再考虑反应振幅对同位旋的分类. 我們研究的問題是在固定入射能量的条件下，截面作为部分能量 $E_{\Lambda\pi}$(或 E_K) 的函数在 Σ 閾和 $\Lambda\pi$ 共振峯附近的变化情況，通过共振-近閾关联来研究 Y_1^* 的自旋和 $Y_1^*\text{-}\Sigma$ 相对宇称 $P(Y_1^*\text{-}\Sigma)$, 并且再加上共振时截面和超子极化角分布的知識，定出 Y_1^* 的自旋和 $\Lambda\text{-}\Sigma$ 相对宇称 $P(\Lambda\text{-}\Sigma)$. 由于我們感兴趣的是在 $\Lambda\pi$ 系統中的現象，要求尽可能地減少 $\Lambda\pi$ 系統的分波和 K 介子在反应中的影响. 我們設計一个实驗：在質心系中固定总能量为 1900 MeV, 卽初态 $\pi^- + p$ 系統动能为 820 MeV, 并且通过选取实驗事例限制 K 介子的部分能量变化范围为

$$0\ \text{MeV} \leqslant E_K \leqslant 90\ \text{MeV},$$

卽

$$60\ \text{MeV} \leqslant E_{\Lambda\pi} \leqslant 150\ \text{MeV}. \qquad (4.1)$$

根据現在有关 K 介子和 $\Lambda\pi$ 系的实驗資料，可以認为反应是在下列条件下进行的:

1. $\Lambda\pi$ 系統只有相对角动量 $l_{\Lambda\pi} = 0$ 和 1 的分波，K 介子相对 $\Lambda\pi$ 系統質心的角动量只有 $l_K = 0$ 的分波.

2. 討論涉及的能量范围是在 $\Lambda\pi$ 系統的共振 Y_1^*($M_{Y_1^*} \sim 1385$ MeV)[9] 附近，离 $K\pi$ 系統的共振态 $K^{*[10]}$($M_{K^*} \sim 880$ MeV) 很远. 因此可以認为，在末态中，$\Lambda\pi$ 之間有很強的末态作用，K 介子近似地处在独立于 $\Lambda\pi$ 系統的各向同性分布中. 也就是条件(3.2)得到

滿足.

可以看到，与过去所建議的定 $\Lambda-\Sigma$ 相对宇称的 $\pi^- + p \to \Lambda^0 + K^0$ 实驗[2]相比較，我們建議的实驗避开了 $\pi^- + p$ 第三共振峯的影响，沒有高 l 分波所引起的角分布分析的困难[11]. 此外，在我們的实驗中，K 介子仅起控制 $\Lambda\pi$ 系統部分能量 $E_{\Lambda\pi}$ 在一定范围內連續变化的作用，而沒有其他效果.

考虑到所設計的实驗条件与(3.17)成立的条件完全一致，并且注意到 $l_K = 0$，因而 $J = j$ 得

$$\left\langle(\Lambda\pi K)J\left(l_{\Lambda\pi}\frac{1}{2}\right)E_{\Lambda\pi}\left|T^J(E)\right|J\left(l_i\frac{1}{2}\right)(\pi p)\right\rangle =$$

$$= \left\langle(\Lambda\pi K)J\left(l_{\Lambda\pi}\frac{1}{2}\right)E'_{\Lambda\pi}\left|P^J(E)\right|J\left(l_i\frac{1}{2}\right)(\pi p)\right\rangle -$$

$$- \left\langle(\Lambda\pi)J\left(l_{\Lambda\pi}\frac{1}{2}\right)E_{\Lambda\pi}\left|\sigma_0^{j^0(l_{\Lambda\pi}\frac{1}{2})}(E_{\Lambda\pi})\right.\right\rangle \frac{\Gamma^0_{\Lambda\pi}}{E_{\Lambda\pi} - E^0_{\Lambda\pi} + i\Gamma^0_{\Lambda\pi}} \cdot$$

$$\cdot \left\langle\sigma_0^{j^0(l^0_{\Lambda\pi}\frac{1}{2})}(E_{\Lambda\pi})\right|\langle(K)l_K = 0\,E_K\left|J\left(l_i\frac{1}{2}\right)(\pi p)\right\rangle \delta_{Jj^0}\delta_{l_{\Lambda\pi}l^0_{\Lambda\pi}} +$$

$$+ i\langle(\Lambda\pi)l_{\Lambda\pi}|T^{1/2}(E_{\Lambda\pi})|l_{\Sigma\pi} = 0\,(\Sigma\pi)\rangle \cdot$$

$$\cdot \left\langle(\Sigma\pi K)\frac{1}{2}\left(l_{\Sigma\pi} = 0\,\frac{1}{2}\right)E_{\Lambda\pi}\left|T^J(E)\right|J\left(l_i\frac{1}{2}\right)(\pi p)\right\rangle \delta_{J\frac{1}{2}}\delta_{l_{\Lambda\pi}(0,1)}. \quad (4.2)$$

在上述实驗条件下的共振-近閾关联公式(4.2)，即是我們討論的出发点.

当部分能量 $E_{\Lambda\pi}$ 由 60 MeV 逐漸增加到 75 MeV 附近时，由于 Σ 道的开放，将要出現近閾效应，不难估計它在閾上下延伸区域約为5—10 MeV[4]. 当 $E_{\Lambda\pi}$ 达到 130 MeV 附近时出現 Y_1^* 共振峯. 由于 $\frac{P}{2} = 25$ MeV，近閾效应将出現在峯坡上(見图1). 从 Y_1^* 的实驗資料[12]来看，背景不超过峯值的 1/10 到 1/7，而在离共振峯 55 MeV 左右，峯坡不小于背景的两倍，因此在 Σ 閾附近，截面将主要由共振峯坡所貢献. 如果近閾效应与共振发生在同一分波上，則在共振峯坡上应观察到有近閾現象出現；如果近閾效应与共振发生在不同分波上，則在背景中某一分波上的近閾效应将被共振峯坡的貢献所掩盖，从而观察不到共振与近閾現象同时存在(下面簡称前者为共存，后者为不共存). 我們将利用这种关联来研究 Y_1^* 的性質和 $\Lambda-\Sigma$ 相对宇称.

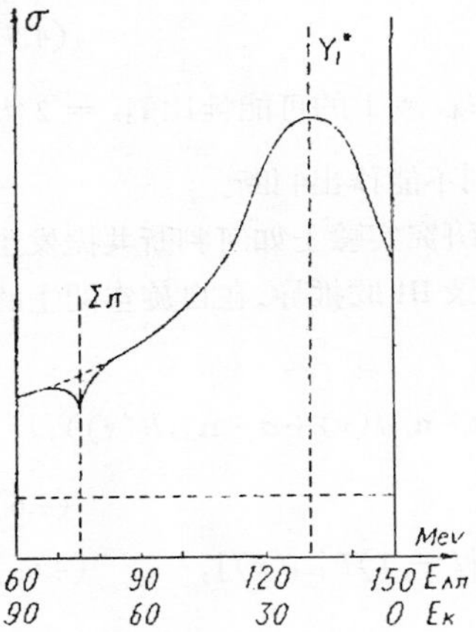

图1 所設計实驗示意图
虛綫表明估計的最大背景

在我們的实驗中，只有三个分波貢献是主要的：$J = \frac{1}{2}$，$l_{\Lambda\pi} = 0$ 或 1，$J = \frac{3}{2}$，$l = 1$. 由(4.2)近閾效应只发生在 $J = \frac{1}{2}$ 的一个分波上，它的 $l_{\Lambda\pi}$ 的取値决定于 $\Lambda-\Sigma$ 相对宇称. 并且，注意到共振发生分波的 J 就是 Y_1^* 的自旋，而共振发生分波的 $l_{\Lambda\pi}$ 則决定了 $Y_1^*-\Lambda$ 的相对宇称

$$P(Y_1^*-\Lambda) = (-)^{l_{\Lambda\pi}+1}. \quad (4.3)$$

表 1

共振发生的分波		近閾效应发生的分波	
J	$l_{\Lambda\pi}$	$P(\Lambda\text{-}\Sigma)=+$ $l_{\Lambda\pi}=0$	$P(\Lambda\text{-}\Sigma)=-$ $l_{\Lambda\pi}=1$
$\frac{1}{2}$	0	共　存	不 共 存
$\frac{1}{2}$	1	不 共 存	共　存
$\geqslant\frac{3}{2}$		不 共 存	不 共 振

这样，我們就得到表 1 所示的共振-近閾关联. 利用表 1 可以对实验結果作如下的預測:

1. 如果实驗上同时观察到共振和較显著的近閾現象，则 $J=\frac{1}{2}$, $P(Y_1^*\text{-}\Sigma)=-$, 并且有

$$P(\Lambda\text{-}\Sigma)=(-)^{l_{\Lambda\pi}}, \tag{4.4}$$

因此如果进一步能判断共振发生的 $l_{\Lambda\pi}$，則 $\Lambda\text{-}\Sigma$ 相对宇称也就确定了.

2. 如果实驗上主要观察到共振，近閾現象不存在，并且实驗上能进一步判断共振发生的 J；则当 $J=\frac{1}{2}$ 时，$P(Y_1^*\text{-}\Sigma)=+$，并且由共振发生的 $l_{\Lambda\pi}$ 可以确定 $\Lambda\text{-}\Sigma$ 的相对宇称

$$P(\Lambda\text{-}\Sigma)=(-)^{l_{\Lambda\pi}+1}. \tag{4.5}$$

当 $J=\frac{3}{2}$ 时，由于 $E_{\Lambda\pi}$ 最大不超过 150 MeV，共振发生在 $l_{\Lambda\pi}=1$ 的可能性比 $l_{\Lambda\pi}=2$ 大很多，所以这时很可能 $P(Y_1^*\text{-}\Lambda)=+$，对 $\Lambda\text{-}\Sigma$ 相对宇称則不能作出判断.

下面我們对截面和超子极化角分布进行分析，进一步研究实驗上如何判断共振发生的 J 和 $l_{\Lambda\pi}$. 从(3.16)出发，考虑到 K 介子处在 S 态，按照附录 III 的推导，在自旋空間上的反应振幅

$$\langle \mathbf{p}_{\Lambda\pi}\,\mathbf{p}_K|T(E)|\mathbf{p}_i\rangle=\frac{1}{(4\pi)^{3/2}}\frac{1}{[\rho(E_i)\rho(E_{\Lambda\pi})\rho(E_K)]^{1/2}}(\boldsymbol{\sigma}\cdot\mathbf{n}_iA(x)+\boldsymbol{\sigma}\cdot\mathbf{n}_{\Lambda\pi}B(x)), \tag{4.6}$$

$$A(x)=\sum_l[T^{l+\frac{1}{2}}(l|l+1)P'_{l+1}(x)-T^{l-\frac{1}{2}}(l|l-1)P'_{l-1}(x)], \tag{4.6a}$$

$$B(x)=\sum_l[-T^{l+\frac{1}{2}}(l|l+1)+T^{l-\frac{1}{2}}(l|l-1)]P'_l(x); \tag{4.6b}$$

其中

$$\mathbf{n}_i=\frac{\mathbf{p}_i}{p_i},\qquad \mathbf{n}_{\Lambda\pi}=\frac{\mathbf{p}_{\Lambda\pi}}{p_{\Lambda\pi}}. \tag{4.7a}$$

$$\rho(E_i)=\frac{p_iE_\pi^iE_p^i}{E}, \tag{4.7b}$$

$$\rho(E_{\Lambda\pi})=\mu_{\Lambda\pi}p_{\Lambda\pi},\quad \rho(E_K)=\mu_Kp_K, \tag{4.7c}$$

$$P'_l(x)=\frac{dP_l(x)}{dx},\qquad x=\mathbf{n}_i\cdot\mathbf{n}_{A\pi},\tag{4.7d}$$

$$T^J(l_{A\pi}|l_i)\quad J\ 总角动量,記\ l_{A\pi}\equiv l.\tag{4.7e}$$

設初态質子的极化为 $\mathbf{P}$, 則反应截面和 A 超子的极化 $\mathbf{P}_A$ 分别为

$$d\sigma=\frac{1}{2\pi F_0}\mathrm{Sp}\left\{\langle\mathbf{p}_{A\pi}\,\mathbf{p}_K|T(E)|\mathbf{p}_i\rangle\frac{1+\boldsymbol{\sigma}\cdot\mathbf{P}}{2}\langle\mathbf{p}_{A\pi}\,\mathbf{p}_K|T(E)|\mathbf{p}_i\rangle^+\right\}\cdot$$
$$\cdot\,\rho(E_K)\rho(E_{A\pi})dE_K d\Omega_{A\pi}d\Omega_K,\tag{4.8}$$

$$\mathbf{P}_A=\frac{\mathrm{Sp}\left\{\boldsymbol{\sigma}\langle\mathbf{p}_{A\pi}\,\mathbf{p}_K|T(E)|\mathbf{p}_i\rangle\frac{1+\boldsymbol{\sigma}\cdot\mathbf{P}}{2}\langle\mathbf{p}_{A\pi}\,\mathbf{p}_K|T(E)|\mathbf{p}_i\rangle^+\right\}}{\mathrm{Sp}\left\{\langle\mathbf{p}_{A\pi}\,\mathbf{p}_K|T(E)|\mathbf{p}_i\rangle\frac{1+\boldsymbol{\sigma}\cdot\mathbf{P}}{2}\langle\mathbf{p}_{A\pi}\,\mathbf{p}_K|T(E)|\mathbf{p}_i\rangle^+\right\}}.\tag{4.9}$$

其中

$$F_0=\frac{\rho_p^0\rho_\pi^0}{E_p^iE_\pi^i}Ep_i=\frac{\rho_p^0\rho_\pi^0}{\rho(E_i)}p_i^2.\tag{4.10}$$

Sp 是在自旋空間上求迹，ρ_p^0 和 ρ_π^0 分别是質心系中初态核子和 π 介子的体密度，F_0 即是熟知的入射流密度的不变表达式．將(4.6)代入(4.8)和(4.9)，并对 $d\Omega_K$ 积分，得

$$\left(\frac{d\sigma}{d\Omega_{A\pi}dE_K}\right)=\frac{1}{4(2\pi)^3\rho_p^0\rho_\pi^0p_i^2}[|A(x)|^2+|B(x)|^2+2\mathrm{Re}A^*B\mathbf{n}_i\cdot\mathbf{n}_{A\pi}-$$
$$-2\,\mathrm{Im}\,A^*B\mathbf{P}\cdot\mathbf{n}_i\times\mathbf{n}_{A\pi}],\tag{4.11}$$

$$\left(\frac{d\sigma}{d\Omega_{A\pi}dE_K}\right)\mathbf{P}_A=\frac{1}{4(2\pi)^3\rho_p^0\rho_\pi^0p_i^2}\{|A(x)|^2[2\mathbf{n}_i(\mathbf{n}_i\cdot\mathbf{P})-\mathbf{P}]+$$
$$+|B(x)|^2[2\mathbf{n}_{A\pi}(\mathbf{n}_{A\pi}\cdot\mathbf{P})-\mathbf{P}]+$$
$$+2\,\mathrm{Re}\,A^*B[\mathbf{n}_i(\mathbf{P}\cdot\mathbf{n}_{A\pi})+\mathbf{n}_{A\pi}(\mathbf{P}\cdot\mathbf{n}_i)-\mathbf{P}(\mathbf{n}_i\cdot\mathbf{n}_{A\pi})]+$$
$$+2\,\mathrm{Im}\,A^*B\mathbf{n}_i\times\mathbf{n}_{A\pi}\}.\tag{4.12}$$

在目前条件下，只有 $l_{A\pi}=0$、1 分波，即

$$A(x)=T^{1/2}(0|1)+3xT^{3/2}(1|2),$$
$$B(x)=T^{1/2}(1|0)-T^{3/2}(1|2).\tag{4.13}$$

再考虑到在共振区域主要貢献来自共振矩陣元，其他矩陣元貢献可忽略；这时共振发生的分波与非零 T 矩陣元的对应关系如表 2 所示．

表 2

共振发生分波		对观察結果主要貢献	
J	$l_{A\pi}$	$A(x)$	$B(x)$
$\frac{1}{2}$	0	$T^{1/2}(0\|1)$	~ 0
$\frac{1}{2}$	1	~ 0	$T^{1/2}(1\|0)$
$\frac{3}{2}$	1	$3xT^{3/2}(1\|2)$	$-T^{3/2}(1\|2)$

由表 2 从实驗現象可以进一步得到下列結果：

3. 当初态核子无极化时 $\mathbf{P} = 0$，可以判断共振发生的 J，即 Y_1^* 的自旋：

当 $\frac{d\sigma}{d\Omega_{\Lambda\pi} dE_K}$ 各向同性并且 $\mathbf{P}_\Lambda \frac{d\sigma}{d\Omega_{\Lambda\pi} dE_K} \sim 0$，则

$$J = \frac{1}{2}; \tag{4.14a}$$

当 $\frac{d\sigma}{d\Omega_{\Lambda\pi} dE_K} \sim x$ 的二次多項式并且

$$\mathbf{P}_\Lambda \frac{d\sigma}{d\Omega_{\Lambda\pi} dE_K} \sim \sin 2\theta \frac{\mathbf{n}_i \times \mathbf{n}_{\Lambda\pi}}{|\mathbf{n}_i \times \mathbf{n}_{\Lambda\pi}|},$$

则

$$J = \frac{3}{2}, \tag{4.14b}$$

但这还不能判断共振的 $l_{\Lambda\pi}$。

4. 当初态核子极化为 $\mathbf{P}$ · 可以通过 Λ 超子的极化判断 $J = \frac{1}{2}$ 共振的条件下的 $l_{\Lambda\pi}$，即 Y_1^*-Λ 相对宇称。如果控制入射动量与核子极化相垂直 $\mathbf{P} \cdot \mathbf{n}_i = 0$，則有

当 $\mathbf{P}_\Lambda \frac{d\sigma}{d\Omega_{\Lambda\pi} dE_K} \sim \mathbf{P}$ · 常数，则

$$l_{\Lambda\pi} = 0, \tag{4.15a}$$

当 $\mathbf{P}_\Lambda \frac{d\sigma}{d\Omega_{\Lambda\pi} dE_K} \sim \mathbf{P} - 2\mathbf{P}_{\Lambda\pi}(\mathbf{P} \cdot \mathbf{n}_{\Lambda\pi})$，則

$$l_{\Lambda\pi} = 1, \tag{4.15b}$$

截面并不能提供比无极化反应更多的知識。如果能从(4.15a)，(4.15b)判断 $l_{\Lambda\pi}$，則由(4.5)即可确定 $P(\Lambda$-$\Sigma)$。值得指出的是，(4.14a)，(4.14b)和(4.15a)，(4.15b)只要求实驗上判断截面、极化是否均匀分布，不需要更精确的要求。

从上面的結果 1—4，首先我們得到了关于 Y_1^* 自旋的两个独立的判断：共振-近閾关联和初态无极化核子反应的角分布。附带我們也确定了 $P(Y_1^*$-$\Sigma)$。其次从 Y_1^* 自旋和极化核子反应中超子极化的角分布可以判断在 $J_{Y_1^*} = \frac{1}{2}$ 情形下的 Y_1^*-Λ 相对宇称，再考虑到共振-近閾关联就可以判断 Λ-Σ 相对宇称。在 $J_{Y_1^*} = \frac{3}{2}$ 或者 $J_{Y_1^*} = \frac{1}{2}$ 而得不到 $l_{\Lambda\pi}$ 的知識时，我們只能得到 Y_1^*，Λ，Σ 相对宇称之間的一些制約关系。显然在所設計的条件下，实驗上是否观察到共振和近閾效应的共存非常重要：共存則可以明确地判断 J 和 $P(Y_1^*$-$\Sigma)$，不共存則所得結果較少。

五、討　論

本文对 $\pi + p \to \Lambda + \pi + K$ 反应中末态共振-近閾效应关联所作的主要結論并不灵敏地依賴于条件(3.2)。如果条件(3.2)近似較差，这对近閾效应的影响会导至产生閾处 $\frac{d\sigma}{dE}$ 不再是无穷大，但截面在閾附近仍有一定程度的反常行为；对末态共振的影响是使共振参数与原来 Y_1^* 的参数有所偏离，但仍旧出現与 Y_1^* 相应的末态共振。因此当对(3.2)

的偏离不太严重时，所有結論都依然成立．同样，本文对通过末态角分布定 Y_1^* 自旋和宇称的分析也基本上不依賴于(3.2)，只要末态中出現与 Y_1^* 相对应的共振．破坏(3.2)最主要的因素是 K–π 作用．由于 K–π 共振态 K^{*}[10]（質量 ～ 880 MeV，Q 值～250 MeV）远不在所設計实驗的能量范围中出現，而且在所选择的测量范围上又是 $\Lambda\pi$ 共振的区域；因此条件(3.2)是合理的．在我們对实驗測量的要求除了得到初态极化核子較为困难以外，对角分布测量只要求判断是否各向同性．所以对所設計的实驗和所作預言寄予希望是有理由的．

本文引入的对角表象，提供一种唯象参数描述基本粒子反应的方法．我們以这种方法主要討論了共振現象、近閾效应和它們之間的关联，也討論了末态作用．这种方法的应用并不限于上述現象；例如，对 T 矩陣值作类似弹性散射有效力程理論的近似，就可得到一組多道有效力程唯象参数．原則上对弹性散射 $e^{i\delta}\sin\delta$ 所作各种近似都可对 T 矩陣本征值进行，多道反应的特点将体现在联系內部态和观察道跃迁的 O_{ir} 上．

本工作是在周光召同志指导下进行的．作者高崇寿、苏肇冰对胡宁教授經常的关心、支持和有益的討論表示感謝．

* * *

后 記

在本文工作已經完成进行整理期間，我們注意到 Erwin et al.[13] 和 Alexander et al.[14] 的工作．文献[13]提出了 Y_1^* 自旋为 $\frac{1}{2}$，$P(\Lambda$–$Y_1^*) = -$ 的实驗判断；这与 Fly et al. 文献[15]判断 $J_{Y_1^*} = \frac{3}{2}$ 相矛盾．我們希望本文提出的共振–近閾关联对进一步判断 $J_{Y_1^*}$ 有所帮助．如果文献[13]的判断正确，則按照本文結論实驗上是否同时观察到共振和峯坡上的近閾現象将决定 Λ 和 Σ 的相对宇称是相同还是相反．在文献[13]和文献[14]的 $\pi^- + p \to \Lambda + \pi + K$ 反应中都观察到 $\Lambda\pi$ 末态共振，并且参数与 Y_1^* 很好吻合．这对条件(3.2)是一个有利的支持．由于实驗的精确度还不够，不能判断近閾效应是否存在，我們期待进一步的实驗．在文献[14]中还指出，在末态 $K\pi$ 共振中，不仅存在質量为 880 MeV 的 K^*，还有迹象存在一个質量 ～ 730 MeV、Q 值 ～100 MeV 但峯值很低的共振态 K^{**}．K^{**} 的存在将直接影响(3.2)的正确性．由于事例太少、峯值很低、又只在 $\Sigma\pi K$ 道有迹象而 $\Lambda\pi K$ 道沒有观察到，并且文献[13]的数据說明 Y_1^* 衰变各向同性 $K\pi$ 作用一定很弱，因此我們认为 K^{**} 存在的可靠性很差，还有待进一步的实驗．同时，即使 K^{**} 存在，由于峯值很低，在所选择的能量范围內 $\Lambda\pi$ 末态共振很強，根据文献[13]关于 Y_1^* 衰变各向同性的迹象和 K^{**} 在 $\Lambda\pi K$ 道几乎观察不到的事实，我們认为(3.2)式不致于有严重的偏差，本文主要結論也不致被破坏．

附录 I　K 矩陣在閾附近的解析行为

我們根据 T 矩陣的解析性和么正性来討論 K 矩陣在閾附近的解析行为，并且証明 $K_{fi}(E)(1 \leqslant f, i \leqslant n)$ 在第 $n+1$ 道产生閾上面附近对 $p_{n+1} \equiv p$ 的依賴不包含一次項．

我們仍以标量粒子 $2 \to 2$ 反应为例．

按照 T 的解析性，并考虑到第 $n+1$ 道处在 S 波的 Wigner 行为[9]，T 矩陣在第 $n+1$ 道閾上面不远对 p 的依賴准确到 p 一次項有

$$T(E) = \left(\begin{array}{c|c} T_{fi} & T_{f\,n+1} \\ \hline T_{n+1\,i} & T_{n+1\,n+1} \end{array}\right) \simeq \left(\begin{array}{c|c} A(0) + p\mathbf{A} & p^{\frac{1}{2}}\mathbf{C} \\ \hline p^{\frac{1}{2}}\tilde{\mathbf{C}} & p\mathbf{D} \end{array}\right) \qquad 1 \leqslant f,\ i \leqslant n. \tag{I.1}$$

其中 $\mathbf{A}, \mathbf{C}, \tilde{\mathbf{C}}$ 和 $\mathbf{D}$ 都是常数矩陣．对于 K 矩陣，按照(2.3c)，并利用矩陣分块运算[16]

$$K(E) = \left(\begin{array}{c|c} K_{fi} & K_{f\,n+1} \\ \hline K_{n+1\,i} & K_{n+1\,n+1} \end{array}\right) = \left(\begin{array}{c|c} X & Z \\ \hline \tilde{Z} & Y \end{array}\right), \tag{I.2}$$

$$Z = p^{\frac{1}{2}}\mathbf{Z}, \qquad \mathbf{Z} = \frac{1}{1 + iA(0)}\mathbf{C}, \tag{I.3}$$

$$X = X(0) + p\mathbf{X}, \tag{I.4}$$

$$X(0) = \frac{A(0)}{1 + iA(0)}, \tag{I.4a}$$

$$\mathbf{X} = \frac{1}{1 + iA(0)}\,(\mathbf{A} - i\mathbf{C}\tilde{\mathbf{C}})\,\frac{1}{1 + i\mathbf{A}(0)} \tag{I.4b}$$

$$= \frac{1}{1 + iA(0)}\,\mathbf{A}\,\frac{1}{1 + iA(0)} - i\mathbf{Z}\tilde{\mathbf{Z}}. \tag{I.4c}$$

在閾下面只开 n 个道．考虑到 $T_{fi}(1 \leqslant f,\ i \leqslant n)$ 在閾上到閾下的解析性，

$$T(E) = \left(\begin{array}{c|c} T_{fi} & 0 \\ \hline 0 & 0 \end{array}\right) = \left(\begin{array}{c|c} A(0) + i|p|\mathbf{A} & 0 \\ \hline 0 & 0 \end{array}\right). \tag{I.5}$$

按照(2.3c)，在閾下附近，K 矩陣为

$$K(E) = \left(\begin{array}{c|c} K_{fi} & 0 \\ \hline 0 & 0 \end{array}\right) = \left(\begin{array}{c|c} X' & 0 \\ \hline 0 & 0 \end{array}\right). \tag{I.6}$$

$$X' = X'(0) + i|p|\mathbf{X}', \tag{I.7}$$

$$X'(0) = \frac{A(0)}{1 + iA(0)} = X(0), \tag{I.7a}$$

$$\mathbf{X}' = \frac{1}{1 + iA(0)}\,\mathbf{A}\,\frac{1}{1 + iA(0)}. \tag{I.7b}$$

在产生閾处和閾下面的幺正条件分别要求 $X(0)$ 实对称和 $\mathbf{X}'$ 純虚对称．再考虑閾上面的幺正条件，容易看到，X 的一次項 $\mathbf{X}$ 一定是純虚对称的．实对称条件要求 $\mathbf{X} = 0$，

$$\mathbf{X} = \frac{1}{1 + iA(0)}\,\mathbf{A}\,\frac{1}{1 + iA(0)} - i\mathbf{Z}\tilde{\mathbf{Z}} = 0, \tag{I.8}$$

即

$$\mathbf{A} = i\mathbf{C}\tilde{\mathbf{C}}. \tag{I.9}$$

(1.4)，(1.8)和(1.7)給出了 $K_{fi}(1 \leqslant f,\ i \leqslant n)$ 在閾上和閾下的解析行为，并且

1. 在閾下 K_{fi} 中不存在与閾上 $-ip\mathbf{Z}\tilde{\mathbf{Z}}$ 相对应的項，說明 K_{fi} 由閾上到閾下不能进行与 T_{fi} 一样的 $p_{n+1} \to i|p_{n+1}|$ 的解析延拓．

2.(I.9) 即是通常的近閾效应公式

$$T_{fi}(1) = iT_{f\,n+1}T_{n+1\,i}. \tag{I.10}$$

对比通常的近閾效应理論[2],可以得到結論: T_{fi} 在閾上和閾下的幺正性和解析性等价于 K_{fi} 在閾上的幺正条件和不含 p_{n+1} 一次項的解析行为.

附录 II 內部态的物理意义和基本粒子反应机构

为了說明內部态的物理意义和基本粒子反应机构,附带得出共振极点留数的負号,我們討論能量在某共振能量附近的波包入射的反应.

設在道空間上对应于系統第 r_0 內部态,本征値 $K_{r_0}(E)$ 在实軸上 $E=E_0$ 处有一个一阶极点,留数为 $-\Gamma_0$; 則在 $E\sim E_0$ 附近有

$$K_{r_0}(E) = K'_{r_0}(E) - \frac{\Gamma_0}{E-E_0}, \tag{II.1a}$$

$$T_{r_0}(E) = T'_{r_0}(E) - \frac{\Gamma_0}{E-E_0+i\Gamma_0}, \tag{II.1b}$$

其中 $K'_{r_0}(E)$ 和 $T'_{r_0}(E)$ 是 $E\sim E_0$ 附近的緩变小量. 考虑一个中心能量为 E_0, 有一能量分布范围(以 ΔE 标帜)的入射波包,在道空間上波包为

$$|\psi\rangle = \int_{\Delta E} C(E)e^{-iEt/\hbar}|E\rangle dE, \tag{II.2}$$

其中 $\int_{\Delta E}$ 表示积分限为以 $E=E_0$ 为中心的数量級为 ΔE 的区間. 反应后的波包为

$$|\psi'\rangle = \int_{\Delta E} C(E)e^{-iEt/\hbar}T(E)|E\rangle dE =$$

$$= \int_{\Delta E} C(E)e^{-iEt/\hbar}\sum_{r=1}^{n}|r(E)\rangle T_r(E)\langle r(E)|E\rangle dE, \tag{II.3}$$

把 $C(E)$ 解析延拓到全复平面. 延拓时保持 $C(E)$ 在实軸上只在以 E_0 为中心、寬度为 ΔE 的区間內才显著地不为零, 并且在 E 沿任何方向趋于无穷时一致趋于零. 由解析性的要求, $C(E)$ 在能量复平面上,至少有一对极点. 我們以一个簡单情形 $C(E)=\frac{1}{\pi}\cdot\frac{\Delta E}{(\Delta E)^2+(E-E_0)^2}$ 为例来考虑,这时 $C(E)$ 只在 $E=E_0\pm i\Delta E$ 处有一阶极点. 把积分換成大半圓迴路,利用留数定理求出积分. 由于被积函数共有三个有貢献的一阶极点 $E_0\pm i\Delta E$ 和 $E_0-i\Gamma_0$, 我們得,当 $\Gamma_0>0$ 时,

$$|\psi'\rangle = e^{-iE_0t/\hbar}e^{\Delta Et/\hbar}\sum_{r=1}^{n}|r(E_0+i\Delta E)\rangle T_r(E_0+i\Delta E)\langle r(E_0+i\Delta E)|E_0+i\Delta E\rangle$$

当 $t<0$,

$$|\psi'\rangle = e^{-iE_0t/\hbar}e^{-\Delta Et/\hbar}\sum_{r=1}^{n}|r(E_0-i\Delta E)\rangle T_r(E_0-i\Delta E)\langle r(E_0-i\Delta E)|E_0-i\Delta E\rangle -$$

$$-\Gamma_0C(E_0-i\Gamma_0)e^{-iE_0t/\hbar}e^{-\Gamma_0t/\hbar}|r_0(E_0-i\Gamma_0)\rangle\langle r_0(E_0-i\Gamma_0)|E_0-i\Gamma_0\rangle$$

当 $t>0$,

或合起来写作

$$|\psi'\rangle = e^{-iE_0t/\hbar}\left[e^{-\Delta E|t|/\hbar}\sum_{r=1}^{n}\left|r\left(E_0 - i\frac{t}{|t|}\Delta E\right)\right\rangle T_r\left(E_0 - i\frac{t}{|t|}\Delta E\right)\left\langle r\left(E_0 - i\frac{t}{|t|}\Delta E\right)\right|E_0 - i\frac{t}{|t|}\Delta E\right\rangle - \theta(t)\Gamma_0 C(E_0 - i\Gamma_0)e^{-\Gamma_0 t/\hbar}|r_0(E_0 - i\Gamma_0)\rangle\langle r_0(E_0 - i\Gamma_0)|E_0 - i\Gamma_0\rangle, \tag{II.4}$$

其中

$$\theta(t) = \begin{cases}1 & 当\ t > 0, \\ 0 & 当\ t < 0.\end{cases} \tag{II.4a}$$

第一项代表一个反应波包，其寬度以 ΔE 为标帜．因子 $e^{-\Delta E t/\hbar}$ 反映了寬度为 ΔE 的波包的扩散，它相当于通常所謂势散射項(无共振反应)．第二项在 $t > 0$ 时才出现，按因子 $e^{-\Gamma_0 t/\hbar}$ 而衰減，这即是通过反应形成的寬度为 Γ_0 的共振衰变态．由此可見，当 $K_{r_0}(E)$ 有一阶极点 $-\frac{\Gamma_0}{E - E_0}$ 时，相应的內部态 $|r_0(E)\rangle$ 中，E 延拓到复平面后的 $|r_0(E_0 - i\Gamma_0)\rangle$ 的物理意义是处于相互作用之中的共振衰变态．

考虑到开道数目不变的能量范围上各本征値的共振极点时，基本粒子反应按这样的机构进行：入射波包一部分按照势散射(无共振反应)反应，反应后波包的扩散行为与入射波包相同；另一部分形成与各共振极点相对应的一系列衰变态，再从这些衰变态衰变进入各观察道中去．

上述討論同样可以对共振极点 $E_0 - i\Gamma_0$, $\Gamma_0 < 0$ 的情形进行．所得共振极点貢献項是一个随时間增长的态．体現因果性要求的反应在无穷远的边条件要求这种态反应振幅沒有直接貢献，所以只有 $\Gamma_0 > 0$ 的极点才是共振极点．

附录 III　(4.6) 的推导

現在我們来推导(4.6)．由(3.15)和(3.16)，T 矩陣元的分波展开式为

$$\langle \alpha_\Lambda \mathbf{p}_{\Lambda\pi}\mathbf{p}_K|T|\mathbf{p}_i\alpha_i\rangle = \delta(E_i - E_f)\langle \alpha_\Lambda \mathbf{p}_{\Lambda\pi}\mathbf{p}_K|T(E)|\mathbf{p}_i\alpha_i\rangle \qquad E = E_i = E_f. \tag{III.1}$$

$$\langle \alpha_\Lambda \mathbf{p}_{\Lambda\pi}\mathbf{p}_K|T(E)|\mathbf{p}_i\alpha_i\rangle = \sum_{JJ_zjl_{\Lambda\pi}l_Kl_i} X_{JJ_zj(l_{\Lambda\pi}\frac{1}{2})l_K}(\Omega_{\Lambda\pi}\alpha_\Lambda;\ \Omega_K)\frac{1}{\sqrt{\rho(E_{\Lambda\pi})\rho(E_K)}}\cdot \cdot\left\langle(\Lambda\pi K)j\left(l_{\Lambda\pi}\frac{1}{2}\right)l_K E_{\Lambda\pi}\left|T^J(E)\right|l_i\frac{1}{2}(\pi p)\right\rangle\frac{1}{\sqrt{\rho(E_i)}}X^+_{JJ_zl_i\frac{1}{2}}(\Omega_i\alpha_i), \tag{III.2}$$

$$X_{JJ_zl_i\frac{1}{2}}(\Omega_i\alpha_i) = \sum_{m_i+S_z=J_z} C\left(l_i\frac{1}{2}J;\ m_iS_zJ_z\right)Y_{l_im_i}(\Omega_i)\xi_{\frac{1}{2}S_z}(\alpha_i), \tag{III.2a}$$

$$X_{JJ_zj(l_{\Lambda\pi}\frac{1}{2})l_K}(\Omega_{\Lambda\pi}\alpha_\Lambda;\ \Omega_K) = \sum_{j_z+m_K=J_z} C(jl_KJ;\ j_zm_KJ_z)X_{jj_zl_{\Lambda\pi}\frac{1}{2}}(\Omega_{\Lambda\pi}\alpha_\Lambda)Y_{l_Km_K}(\Omega_K). \tag{III.2b}$$

在我們所討論的問題中 $l_K = 0$，因此有 $m_K = 0$，$J = j$ 可簡化为

$$X_{JJ_zj(l_{\Lambda\pi}\frac{1}{2})l_K}(\Omega_{\Lambda\pi}\alpha_\Lambda;\ \Omega_K) = \frac{1}{\sqrt{4\pi}}X_{JJ_zl_{\Lambda\pi}\frac{1}{2}}(\Omega_{\Lambda\pi}\alpha_\Lambda)\delta_{jJ}\delta_{l_K0}, \tag{III.3a}$$

$$\left\langle (\Lambda\pi K)j\left(l_{\Lambda\pi}\frac{1}{2}\right)l_K E_{\Lambda\pi}\,|\,T^J(E)\,|\,l_i\frac{1}{2}(\pi p)\right\rangle =$$

$$= \left\langle (\Lambda\pi K)J\left(l_{\Lambda\pi}\frac{1}{2}\right)0\;E_{\Lambda\pi}\,|\,T^J(E)\,|\,l_i\frac{1}{2}(\pi p)\right\rangle \delta_{l_K 0}\delta_{jJ}$$

簡記为

$$= T^J(l_{\Lambda\pi}|l_i)\delta_{l_K 0}\delta_{jJ}. \tag{III.3b}$$

在总能量固定下，它还是 $E_{\Lambda\pi}$ 的函数. (III.2)变为

$$\langle \alpha_\Lambda \mathbf{p}_{\Lambda\pi}\mathbf{p}_K|T(E)|\mathbf{p}_i\alpha_i\rangle = \frac{1}{\sqrt{4\pi}}\sum_{JJ_z l_{\Lambda\pi} l_i} X_{JJ_z l_{\Lambda\pi}\frac{1}{2}}(\Omega_{\Lambda\pi}\alpha_\Lambda)\frac{1}{\sqrt{\rho(E_{\Lambda\pi})\rho(E_i)}}\cdot$$

$$\cdot T^J(l_{\Lambda\pi}|l_i)\frac{1}{\sqrt{\rho(E_i)}}X^+_{JJ_z l_i\frac{1}{2}}(\Omega_i\alpha_i). \tag{III.4}$$

由于 $P(\Lambda K-P)=-$，因此

$$\text{当}\quad J=l_{\Lambda\pi}+\frac{1}{2}\ \text{时},\qquad J=l_i-\frac{1}{2},$$

$$\text{当}\quad J=l_{\Lambda\pi}-\frac{1}{2}\ \text{时},\qquad J=l_i+\frac{1}{2}. \tag{III.5}$$

即 $l_{\Lambda\pi}$ 与 l_i 差为 ± 1. 为了便于計算，引入算符 $\boldsymbol{\sigma}\cdot\mathbf{n}_i$，由于空間轉动不变性，它对态矢量运算后不改变 J 和 J_z，但其宇称为負，故得

$$\boldsymbol{\sigma}\cdot\mathbf{n}_i\left|\,JJ_z l_i\frac{1}{2}\right\rangle = \left|\,JJ_z l_{\Lambda\pi}\frac{1}{2}\right\rangle. \tag{III.6}$$

(III.2)变为

$$\langle \alpha_\Lambda \mathbf{p}_{\Lambda\pi}\mathbf{p}_K|T(E)|\mathbf{p}_i\alpha_i\rangle = \frac{1}{\sqrt{4\pi}}\sum_{JJ_z l_{\Lambda\pi}} X_{JJ_z l_{\Lambda\pi}\frac{1}{2}}(\Omega_{\Lambda\pi}\alpha_\Lambda)\frac{1}{\sqrt{\rho(E_{\Lambda\pi})\rho(E_K)}}\cdot$$

$$\cdot T^J(l_{\Lambda\pi}|l_i)\frac{1}{\sqrt{\rho(E_i)}}X^+_{JJ_z l_{\Lambda\pi}\frac{1}{2}}(\Omega_i\alpha_i)(\boldsymbol{\sigma}\cdot\mathbf{n}_i). \tag{III.7}$$

再引入投影算符

$$P^{(+)} = \frac{l_{\Lambda\pi}+1+\boldsymbol{\sigma}\cdot\mathbf{L}_{\Lambda\pi}}{2l_{\Lambda\pi}+1},\qquad P^{(-)} = \frac{l_{\Lambda\pi}-\boldsymbol{\sigma}\cdot\mathbf{L}_{\Lambda\pi}}{2l_{\Lambda\pi}+1}, \tag{III.8a}$$

$$P^{(+)}X_{JJ_z l_{\Lambda\pi}\frac{1}{2}} = \delta_{J\,l_{\Lambda\pi}+\frac{1}{2}}X_{JJ_z l_{\Lambda\pi}\frac{1}{2}}, \tag{III.8b}$$

$$P^{(-)}X_{JJ_z l_{\Lambda\pi}\frac{1}{2}} = \delta_{J\,l_{\Lambda\pi}-\frac{1}{2}}X_{JJ_z l_{\Lambda\pi}\frac{1}{2}}, \tag{III.8c}$$

$$P^{(+)}+P^{(-)}=1. \tag{III.8d}$$

在(III.7)中，$X_{JJ_z l_{\Lambda\pi}\frac{1}{2}}$ 前插入 $P^{(+)}+P^{(-)}$，利用(III.8b)和(III.8c)化簡得到

$$\langle \alpha_\Lambda \mathbf{p}_{\Lambda\pi}\mathbf{p}_K|T(E)|\mathbf{p}_i\alpha_i\rangle = \frac{1}{\sqrt{4\pi}}\frac{1}{\sqrt{\rho(E_{\Lambda\pi})\rho(E_K)\rho(E_i)}}\cdot$$

$$\cdot\sum_{l_{\Lambda\pi}}\left[T^{l_{\Lambda\pi}+\frac{1}{2}}(l_{\Lambda\pi}|l_{\Lambda\pi}+1)P^{(+)}+T^{l_{\Lambda\pi}-\frac{1}{2}}(l_{\Lambda\pi}|l_{\Lambda\pi}-1)P^{(-)}\right]\cdot$$

$$\cdot\sum_{JJ_z}X_{JJ_z l_{\Lambda\pi}\frac{1}{2}}(\Omega_{\Lambda\pi}\alpha_\Lambda)X^+_{JJ_z l_{\Lambda\pi}\frac{1}{2}}(\Omega_i\alpha_i)\boldsymbol{\sigma}\cdot\mathbf{n}_i. \tag{III.9}$$

由(III.2a)利用 Clebsch-Gordan 系数的正交性，我們有

$$\sum_{JJ_z} X_{JJ_z l_{\Lambda\pi}\frac{1}{2}}(\Omega_{\Lambda\pi}\alpha_\Lambda)X^+_{JJ_z l_{\Lambda\pi}\frac{1}{2}}(\Omega_i\alpha_i) =$$

$$= \Big(\sum_m Y_{l_{\Lambda\pi}m}(\Omega_{\Lambda\pi})Y^*_{l_{\Lambda\pi}m}(\Omega_i)\Big)\Big(\sum_{S_z}\xi_{\frac{1}{2}S_z}(\alpha_\Lambda)\xi^+_{\frac{1}{2}S_z}(\alpha_i)\Big) =$$

$$= \frac{2l_{\Lambda\pi}+1}{4\pi}P_{l_{\Lambda\pi}}(\mathbf{n}_{\Lambda\pi}\cdot\mathbf{n}_i)\delta_{\alpha_\Lambda\alpha_i}. \tag{III.10}$$

将(III.10)和(III.8a)代入(III.9)，$l_{\Lambda\pi}$ 簡写作 l，得

$$\langle\alpha_\Lambda\mathbf{p}_{\Lambda\pi}\mathbf{p}_K|T(E)|\mathbf{p}_i\alpha_i\rangle = \frac{1}{(4\pi)^{3/2}}\frac{1}{\sqrt{\rho(E_{\Lambda\pi})\rho(E_K)\rho(E_i)}}\cdot$$

$$\cdot\sum_l[T^{l+\frac{1}{2}}(l|l+1)(l+1+\boldsymbol{\sigma}\cdot\mathbf{L}_{\Lambda\pi})P_l(\mathbf{n}_{\Lambda\pi}\cdot\mathbf{n}_i)(\boldsymbol{\sigma}\cdot\mathbf{n}_i)+$$

$$+T^{l-\frac{1}{2}}(l|l-1)(l-\boldsymbol{\sigma}\cdot\mathbf{L}_{\Lambda\pi})P_l(\mathbf{n}_{\Lambda\pi}\cdot\mathbf{n}_i)(\boldsymbol{\sigma}\cdot\mathbf{n}_i)]_{\alpha_\Lambda\alpha_i}. \tag{III.11}$$

再利用 $\boldsymbol{\sigma}$ 和 $\mathbf{L}_{\Lambda\pi}$ 算符的运算性质，化簡，最后得

$$\langle\alpha_\Lambda\mathbf{p}_{\Lambda\pi}\mathbf{p}_K|T(E)|\mathbf{p}_i\alpha_i\rangle =$$

$$= \frac{1}{(4\pi)^{3/2}}\frac{1}{\sqrt{\rho(E_{\Lambda\pi})\rho(E_K)\rho(E_i)}}[\boldsymbol{\sigma}\cdot\mathbf{n}_iA(x)+\boldsymbol{\sigma}\cdot\mathbf{n}_{\Lambda\pi}B(x)]_{\alpha_\Lambda\alpha_i}, \tag{III.12}$$

其中

$$A(x) = \sum_l[T^{l+\frac{1}{2}}(l|l+1)P'_{l+1}(x) - T^{l-\frac{1}{2}}(l|l-1)P'_{l-1}(x)], \tag{III.12a}$$

$$B(x) = \sum_l[-T^{l+\frac{1}{2}}(l|l+1) + T^{l-\frac{1}{2}}(l|l-1)]P'_l(x), \tag{III.12b}$$

$$P'_l(x) = \frac{dP_l(x)}{dx}, \qquad x = \mathbf{n}_{\Lambda\pi}\cdot\mathbf{n}_i. \tag{III.12c}$$

把(III.12)写成自旋空間矩陣形式为

$$\langle\mathbf{p}_{\Lambda\pi}\mathbf{p}_K|T(E)|\mathbf{p}_i\rangle =$$

$$= \frac{1}{(4\pi)^{3/2}}\frac{1}{\sqrt{\rho(E_{\Lambda\pi})\rho(E_K)\rho(E_i)}}[\boldsymbol{\sigma}\cdot\mathbf{n}_iA(x)+\boldsymbol{\sigma}\cdot\mathbf{n}_{\Lambda\pi}B(x)]. \tag{III.13}$$

参　考　文　献

[1] Humblet, J. and Rosenfeld, L., *Nucl. Phys.*, **26** (1961), 529.
[2] 例如：Базь и Окунь, *ЖЭТФ*, **35** (1958), 757.
Newton, *Annals of Phys.*, **4** (1958), 29.
[3] Blatt, J. and Weisskopf, V., Theoretical Nuclear Physics (1952). 夏蓉，原子核理論講义 (1961).
[4] 例如：Лапидус, Л. И. и чжоу Гуан-Чжао, *ЖЭТФ*, **39** (1960), 112.
[5] Watson, K. M., *Phys. Rev.*, **88** (1952), 1163.
[6] Лапидус, Л. И. и чжоу Гуан-Чжао, *ЖЭТФ*, **39** (1960), 364.
[7] Fonda, L. and Newton, R. G., *Phys. Rev.*, **119** (1960), 1394.
[8] Wigner, E. P., *Phys. Rev.*, **73** (1948), 1002.
[9] Alston, M., et al., *Phys. Rev. Letters*, **5** (1960), 520.
[10] Alston, M., et al., *Phys. Rev. Letters*, **6** (1961), 300.
[11] Ersler, F., et al., *Revs. Mod. Phys.*, **33** (1961), 436.
Wolf, S. E., et al., *Revs. Mod. Phys.*, **33** (1961), 439.

[12] Alston, M. H. and Ferro-Luzzi, M., *Revs. Mod. Phys.*, 33 (1961), 416.
[13] Erwin, A. R., March, R. H. and Walker, W. D., *Nuovo Cimento*, 24 (1962), 237.
[14] Alexander, G., et al., *Phys. Rev. Letters*, 8 (1962), 447.
[15] Ely, R. P., et al., *Phys. Rev. Letters*, 7 (1961), 461.
[16] Dalitz and Tuan, *Annals of Phys.*, 10 (1960), 307.

INVESTIGATION OF THE CORRELATION BETWEEN RESONANCE EFFECT AND NEAR THRESHOLD EFFECT IN THE $\pi+p\rightarrow\Lambda+\pi+K$ REACTION

Su Zhao-bin　Gao Chong-shou　Chou Kuang-chao

Abstract

In this paper an experiment for the reaction $\pi+p\rightarrow\Lambda+\pi+K$ is proposed. The total energy in the centre of mass system is fixed at 1900 MeV. The kinetic energy of the final K-meson is then in the range 0—90 MeV. We propose to observe the final state Λ-π resonance and the cusp at the $\Sigma\pi K$ threshold. From the correlation between the resonance and the cusp and from the angular distribution of the resonance, the spin of Y_1^* and the ralative parities among Y_1^*, Λ and Σ may be determined. Based on the diagonal representation of T-matrix in the channel space, a general phenomenological description of the resonance effect, the near threshold effect and the final state interaction is developed, which takes into account especially the correlation of the cusp with a nearby resonance. The general theory is applied to the present experiment as a special case.

第4卷 第4期 高能物理与核物理 Vol. 4, No. 4
1980年7月 PHYSICA ENERGIAE FORTIS ET PHYSICA NUCLEARIS July, 1980

色规范群和味规范群的相互作用之间有对称性吗？

周光召 高崇寿
（中国科学院理论物理研究所） （北京大学）

摘要

本文将 $SU(3)\times U(1)$ 的弱电统一模型推广到包括强作用，讨论了色规范群和味规范群的相互作用之间可能存在的一种交换对称性。

在文献[1]中，我们建议了一种弱电统一模型，规范群选作 $SU(3)\times U_Y(1)$. 为明确起见，这个 $SU(3)$ 群称做味规范群，并以符号 $SU(3)_f$ 表示. 除了生成元为超荷 Y 的 $U_Y(1)$ 规范群，在文献[1]的模型中还引进了一个整体对称的 $U_S(1)$ 群，其生成元为 S. 这个整体对称性有可能是一个局域 $U_S(1)$ 对称性自发破缺之后剩余下来的，我们在下面考虑这个可能性.

当加入强作用后，总的规范群应选作

$$G = SU(3)_f \otimes U_S(1) \otimes SU(3)_c \times U_Y(1), \tag{1}$$

从群的选择上可以看到，在我们讨论的模型中有可能存在味规范群 $SU(3)_f$ 和色规范群 $SU(3)_c$，以及 S 荷和 Y 荷之间的对称. 只要在对称性自发破缺之前，规范场的耦合常数是相同的，那么从规范场的拉氏量看，这样一种交换对称是自然存在的.

下面我们对费米场的表示和它与规范场的相互作用进行分析. 在文献[1]中，每一代的费米子用 $SU(3)_f$ 和 $SU(3)_c$ 的表示写出来有

左旋轻子: $(\underline{3},1)_L$, $(1,1)_L$, $(\underline{3},1)^h_L$, $(1,1)^h_L$;

右旋轻子: $(\underline{3}^*,1)_R$, $(1,1)_R$, $(\underline{3}^*,1)^h_R$, $(1,1)^h_R$;

左旋夸克: $(\underline{3},\underline{3})_L$, $(1,\underline{3})_L$, $(\underline{3},\underline{3}^*)^h_L$, $(1,\underline{3}^*)^h_L$;

右旋夸克: $(\underline{3}^*,\underline{3})_R$, $(1,\underline{3})_R$, $(\underline{3}^*,\underline{3}^*)^h_R$, $(1,\underline{3}^*)^h_R$.

其中括弧内第一个数为 $SU(3)_f$ 表示的种类；第二个数字为 $SU(3)_c$ 表示的种类. 括弧右下角 L 和 R 代表左旋和右旋，右上角的 h 代表在 $SU(3)_f$ 群作用下将 ϵ 由 γ_5 变到 $-\gamma_5$ 的表示. 关于 ϵ 的定义可见文献[1]. $(\quad)^h$ 的态的引进是为了消除三角反常的. 为了表述对称性的方便，我们在括弧右上角上不带字母的态上加上右上角字母 l，并以高和低分别称呼带 h 和 l 的态.

由上面轻子和夸克的态可以看到以下的对称性：(i) 当 $SU(3)_f$ 的 $\underline{3}$ 表示换为 $\underline{3}^*$ 表

本文1980年2月25日收到.

示时，左旋换成右旋，这即是文献 [1] 中讨论的左右对称性；(ii) 当 $SU(3)_c$ 的 $\underline{3}$ 表示换成 $\underline{3}^*$表示时，高换成低，表明可能存在高低对称性；(iii) 上面关于费米子的态是按左右旋和 $SU(3)_c$ 的三重态(夸克)和单态(轻子)排列的；如按高低和 $SU(3)_f$ 的三重态和单态排列，则可得到完全相似的组合，这启示我们，当将左右和高低互换时，$SU(3)_c$ 和 $SU(3)_f$ 互换.

如果我们进一步研究量子数 S 和 Y，我们可以得到更明显的对称性. 在文献[1]中指出，S 的数值只与 $SU(3)_f$ 的表示和左右两种态有关，而不依赖于 $SU(3)_c$ 的表示和高低两种态. 对左旋 $SU(3)_f$ 三重态可取 $S=\frac{2}{3}$，对左旋 $SU(3)_f$ 单态 $S=0$；而对右旋 $SU(3)_f$ 三重态 $S=\frac{1}{3}$，对右旋 $SU(3)_f$ 单态 $S=1$. 同时在文献 [1] 中也指出，Y 和 $SU(3)_f$ 的表示以及左右旋无关，它的数值只与 $SU(3)_c$ 的表示和高低两种态有关. 为了得到正确的夸克电荷，对低的 $SU(3)_c$ 三重态(夸克)，$Y=2/3$，而对低的 $SU(3)_c$ 单态(轻子) $Y=0$. 在文献 [1] 中没有详细讨论高态上 Y 的选取，将它选得和低态一样虽然可以消除三角反常，但还有另外的选法，也能消除三角反常. 我们选取对高的 $SU(3)_c$ 三重态 $Y=1/3$ 及对高的 $SU(3)_c$ 单态 $Y=1$. 由于下面两个恒等式

$$3\times\left(\frac{2}{3}\right)=3\times\left(\frac{1}{3}\right)+1$$

及

$$3\times\left(\frac{2}{3}\right)^2=3\times\left(\frac{1}{3}\right)^2+1^2,$$

容易证明所有的三角反常项在把各种多重态加起来以后将完全消去.

对比以上的数字，我们看到明显的交换对称性质，即将左右换成高低，S 换成 Y，$SU(3)_f$ 的表示换成 $SU(3)_c$ 的表示，所有态的量子数不变.

由于 S 只作用在 $SU(3)_f$ 的表示上，可以认为 $U_S(1)$ 和 $SU(3)_f$ 合成为一个群 $U(3)_f$，同样 Y 只作用在 $SU(3)_c$ 的表示上，因此 $U_Y(1)$ 和 $SU(3)_c$ 合成 $U(3)_c$.

如果限制在规范场和费米场中，假定对称性由于某种动力学原因而自发破缺，则由以上的讨论有可能用 $U(3)_f\otimes U(3)_c$ 构成一个强弱电统一模型的规范群，拉氏量具有某种色和味的交换对称性. 自发破缺的顺序应当是 $U_S(1)$ 先破缺，它的规范场获得很大的质量 m_S，这个质量是应当比 100GeV 大得多. 这个 $U_S(1)$ 的自发破缺同时解除了味规范群和色规范群之间的对称，但仍有一个整体 $U_S(1)$ 对称性保留下来，因为 $U_S(1)$ 规范场的质量项并不破坏 $U_S(1)$ 的整体对称，而只破坏它的局域对称. 在局域 $U_S(1)$ 对称性破坏后，余下的局域对称群为 $SU(3)_f\otimes U_Y(1)\otimes SU(3)_c$，这就是我们在文献[1]中讨论的模型的规范群. 已经在文献[1]中指出，这个模型在低能范围内给出的带电流和 Weinberg-Salam 模型一样，Weinberg 角满足限制条件 $\sin^2\theta_W\leqslant\frac{1}{4}$. 当 $\sin^2\theta_W$ 略小于 1/4 时，中性流和标准模型比较只多一个小的改正项，它可以由更准确的实验来检验.

在 Weinberg-Salam 模型提出之后[2]，有许多人研究强弱电相互作用的统一问题[3]，并已得到了许多有趣的结果. 但是为了概括所有已观察到的费米子，所用的群愈来愈大，我们感到上面所叙述的味和色的对称性更像是一种互耦对称性. 如果把规范场看作是圈空

间的手征场或 Goldstone 场有些道理[5-7],那么规范场很可能是一种集体激发的场. 在这种情况下,味和色的交换对称性的发生也可能有更深的原因,统一强弱电相互作用也许不需要用很大的对称群,而是要联系到物质结构的下一个层次去解决.

参 考 文 献

[1] 周光召、高崇寿,中国科学,1980年,第3期。SLAC-PUB-2449.

[2] S. Weinberg. *Phys. Rev. Lett.*, **19**(1967), 1264; A. Salam, ''Elementary Particle Theory'' ed. by Swartholm, Stockholm (1968), S. L. Glashow, *Nucl. Phys.*, **22**(1961), 579.

[3] H. Georgi and S. L. Glashow, *Phys. Rev. Lett.*. 32(1974). 438; H. Fritzch and P. Minkowski, *Ann. of Phys.*, **93**(1975). 193.

[4] J. C. Pati and A. Salam. *Phys. Rev.*, D8(1973). 1240; ibid, **D10**(1974). 275.

[5] Polyakov. 在国际轻子和光子会议上的报告,1979。

[6] E. Corrigan and B. Hassiacher, *Phys. Lett.*, 81B(1979). 181; Y. Nambu. *Phys. Lett.*, **80B**(1979), 372: J. L. Gervais and A. Neveu, *Phys. Lett.*, **80B**(1979). 255.

[7] F. Gliozzi, T. Regge and M. A. Virasoro, *Phys. Lett.*, 81B(1979). 178.

IS THERE ANY SYMMETRY BETWEEN FLAVOUR AND COLOUR GAUGE INTERACTIONS?

ZHOU GUANG-ZHAO

(Institute of Theoretical Physics. Academia Sinica)

GAO CHONG-SHOU

(Peking University)

ABSTRACT

The electro-weak model in $SU(3)\times U(1)$ is extended to include strong interaction. A possible exchange symmetry between the flavour and the colour gauge interactions is discussed.

第4卷 第5期 高能物理与核物理 Vol. 4, No. 5
1980年9月 PHYSICA ENERGIAE FORTIS ET PHYSICA NUCLEARIS Sept., 1980

$SU(3)$ 中的弱电统一模型

周光召 高崇寿
（中国科学院理论物理研究所）（北 京 大 学）

摘 要

本文借助于 $SU(3)$ 群生成元四种不同的实现，建立了一个以 $SU(3)$ 规范群为基础的弱电统一模型，得到 $W^{\pm}$ 粒子和 Z 粒子的质量以及它们与轻子的相互作用都与 Weinberg-Salam 模型的结果相同，并且给出 $\sin^2\theta_W = 1/4$. 这个模型要求有一个新的守恒量子数——弱奇异数存在，还存在四个重矢量粒子 $V^{\pm}$ 和 $U^{\pm\pm}$ 和一些重费米子以及标量粒子. 它们都带有不为零的弱奇异数，都和轻子不直接耦合. 它们的存在不影响低能范围内轻子的弱电作用，并且只能在高能下成对产生. 它们中最轻的粒子将是稳定粒子. 本文还讨论了弱奇异数守恒破坏的可能性及其后果.

一、引 言

最近有许多实验都证实了，在目前能达到的能量范围内，Weinberg-Salam 关于弱电统一的模型是唯一正确的[1,2]，同时实验上确定的 Weinberg 角 θ_W 近似等于 30°，或 $\sin^2\theta_W \simeq 1/4$. 这些实验结果再一次引起了广泛的兴趣，试图将 $SU(2)_L \otimes U(1)$ 群嵌入一个单纯群中，并由此确定 Weinberg 角. 最近的努力集中在从超对称群 $SU(2|1)$ 作为规范群上面[3,4]. 在这一类模型中，Higgs 粒子或者作为陪集上费米型规范场的 Faddeev-Popov 粒子[3]，或者取作大于五维的时空中的规范场分量[4]. 除了能得到 $\sin^2\theta_W = 1/4$ 的结果以外，还能预言 Higgs 粒子的质量. 但是由于只有一个耦合常数，要将所有的轻子都考虑进去只有增加时空的维数，因而同时增加了 Higgs 场的数目. 目前还没有人能够把超对称群的模型推广到夸克的弱电相互作用.

在本文中我们提出一个以 $SU(3)$ 群为弱电规范群的模型. 在这个模型中，借助于 $SU(3)$ 群生成元四种不同的实现，得到了 $\sin^2\theta_W = 1/4$ 的结果. 除了熟知的 $W^{\pm}$ 粒子、Z 粒子和轻子的相互作用同 Weinberg-Salam 模型完全一样以外，在这个模型中存在另外四个矢量粒子，两个称为 $V^{\pm}$ 粒子，它们带一个电荷；另外两个称为 U^{++} 和 U^{--} 粒子，它们带两个电荷. V 粒子、U 粒子和轻子之间没有直接的相互作用，它们的存在不影响低能范围内轻子的弱作用（准到 G_F 的最低级近似）. 它们的质量在一种 Higgs 场选取下为 $m_V = m_W$, $0 \leqslant m_U \leqslant 2m_W$. Higgs 场有几种可能的选择，我们详细讨论了最简单的 6 维

本文1979年8月31日收到.

表示的 Higgs 场．在对称性自发破缺之后，还存在5个 Higgs 粒子，我们称为 χ, ϕ^0 ϕ^{0*}, ϕ^{++}, ϕ^{--} 粒子．只有 χ 粒子和轻子有直接相互作用，它的性质和 Weinberg-Salam 模型中的 Higgs 粒子类似．

当 Higgs 场自相互作用具有 $\Phi \to -\Phi$ 的分立对称性时，拉氏函数中还存在一个和手征有关的整体对称性，它和 $SU(3)$ 群中一个 $U(1)$ 子群结合起来形成一个在对称性自发破缺后仍保持的整体 $U(1)$ 对称性．相应的新量子数称为弱奇异数 S_W．我们的理论比 Weinberg-Salam 模型多出的 V, U 和 ϕ 粒子，都是弱奇异数不为零的粒子，统称弱奇异粒子．它们都不与轻子直接耦合，并且只能按弱奇异数守恒的要求而成对产生和衰变，质量最轻的弱奇异粒子将是稳定粒子．如果 ϕ 粒子的质量重于矢量粒子 V 和 U，实验上将首先看到 V 粒子或 U 粒子．V 粒子通过虚光子或 Z 粒子成对产生的截面以及 U 粒子通过虚光子成对产生的截面和 W 粒子通过虚光子或 Z 粒子成对产生的截面数量级是相同的（扣除相空间的差别之后）．因此，如果这个模型是符合实际的，在能量足够高的电子正电子对撞实验中，当观察到 W 粒子的成对产生时应有同数量级的几率观察到弱奇异矢量粒子的成对产生．单个的 V 粒子或 U 粒子也有可能在宇宙线中观察到．我们建议实验物理学家注意这样一种可能性．

本文的结构是这样的：第二节讨论 $SU(3)$ 群的不同实现和场的变换性质；第三节讨论对称性的自发破缺与弱奇异数守恒；第四节讨论轻子和规范场以及 Higgs 粒子的相互作用；第五节讨论弱奇异粒子的性质；第六节讨论弱奇异数守恒破坏的可能性及其结果；第七节讨论其它的一些问题．在附录中给出 Higgs 粒子变换性质和自作用势的讨论．

二、$SU(3)$ 代数和场的变换性质

令 $\hat{I}_i$, $i=1,\cdots,8$ 为 $SU(3)$ 群的生成元，它们满足对易关系

$$[\hat{I}_i, \hat{I}_j] = if_{ijk}\hat{I}_k. \tag{2.1}$$

把 $\hat{I}_i$ 中的8个生成元分为两组，分别用 $\hat{I}_\alpha$ 和 $\hat{I}_a$ 表示，定义 $\hat{I}_i^{(\epsilon)}$, $i=1,\cdots,8$ 为

$$\hat{I}_\alpha^{(\epsilon)} = \hat{I}_\alpha, \quad \hat{I}_a^{(\epsilon)} = \hat{I}_a\epsilon, \tag{2.2}$$

其中 ϵ 和 $\hat{I}_i$ 对易并满足条件

$$\epsilon^2 = 1. \tag{2.3}$$

容易证明，当 a 和 α 取表1中所列举的七种情形中的一种时，$\hat{I}_i^{(\epsilon)}$ 满足和 $\hat{I}_i$ 同样的对易关系

$$[\hat{I}_i^{(\epsilon)}, \hat{I}_j^{(\epsilon)}] = if_{ijk}\hat{I}_k^{(\epsilon)}, \tag{2.4}$$

因此 $\hat{I}_i^{(\epsilon)}$ 可以看作是 $SU(3)$ 代数的一种实现．对含有费米场的变换，ϵ 除了可以取 ±1 以外，还可以取 $\pm\gamma_5$．这样可以得到 $SU(3)$ 代数的四种实现，我们分别用 $\epsilon=+,-,5,-5$ 代表 $\epsilon=+1,-1,\gamma_5$ 和 $-\gamma_5$ 的情形．

a 和 α 的七种选取可以分为两类，前四种是一类，后三种是一类．我们在下面将要用到的都属于第一类．

下面我们引进场的变换性质，对规范场 A_μ^i, $i=1,\cdots,8$，定义

$$\hat{A}_\mu^{(\epsilon)} = igA_\mu^i\hat{I}_i^{(\epsilon)}, \tag{2.5}$$

表 1

	a	α
I	1 3 4 6 8	2 5 7
II	1 3 5 7 8	2 4 6
III	2 3 4 7 8	1 5 6
IV	2 3 5 6 8	1 4 7
V	1 2 4 5	3 6 7 8
VI	1 2 6 7	3 4 5 8
VII	4 5 6 7	1 2 3 8

则在 $SU(3)$ 群的局域变换作用下，$\hat{A}_\mu^{(\epsilon)}$ 的变换规则为

$$\hat{A}_\mu^{(\epsilon)} \to \hat{A}_\mu^{(\epsilon)\prime} = U^{(\epsilon)}(\partial_\mu + \hat{A}_\mu^{(\epsilon)})U^{(\epsilon)+}. \tag{2.6}$$

其中

$$U^{(\epsilon)} = \exp\{iI_j^{(\epsilon)}\xi^j(x)\}, \tag{2.7}$$

$\xi^j(x)$, $j = 1, \cdots, 8$ 为群变换参数. 对规范场来说，ϵ 可取 $+$，$-$，5 和 -5 四种情形，对这四种 ϵ 用 (2.6) 和 (2.7) 式所确定的 A_μ^i 的变换规则是相同的.

费米场组成 $SU(3)$ 群的基础表示，它的变换规则选为

$$\psi(x) \to \psi'(x) = U^{(5)}\psi(x), \quad \bar{\psi}(x) \to \bar{\psi}'(x) = \bar{\psi}(x)U^{(-5)+}. \tag{2.8}$$

容易证明，在这一变换下，

$$\bar{\psi}(\partial_\mu + \hat{A}_\mu^{(-5)})\gamma^\mu\psi = \bar{\psi}\gamma^\mu(\partial_\mu + \hat{A}_\mu^{(5)})\psi \tag{2.9}$$

是规范不变量.

Higgs 场取作 3×3 的矩阵 Φ，它是一个复场，变换规则选为

$$\Phi \to \Phi' = U^{(-)}\Phi U^{(+)+}, \quad \Phi^+ \to \Phi^{+\prime} = U^{(+)}\Phi^+ U^{(-)+}. \tag{2.10}$$

由 (2.6) 和 (2.10) 容易证明

$$D_\mu\Phi = \partial_\mu\Phi + A_\mu^{(-)}\Phi - \Phi A_\mu^{(+)} \tag{2.11}$$

为协变微分，且由 (2.8) 及 (2.10) 可知

$$\bar{\psi}\Phi P_+\psi \text{ 和 } \bar{\psi}\Phi^+P_-\psi \tag{2.12}$$

为规范不变量，其中

$$P_\pm = \frac{1}{2}(1 \pm \gamma_5) \tag{2.13}$$

为手征投影算子，它具有以下性质

$$P_\pm U^{(5)} = U^{(\pm)}P_\pm, \quad P_\pm U^{(-5)} = U^{(\mp)}P_\pm. \tag{2.14}$$

在一般情况下，由 3×3 矩阵定义的 Higgs 场含有 9 个复场，但它们并不是 $SU(3)$ 群的不可约表示. 在 a 和 α 取表 1 中前四种数值的情况下 Φ 可以分解为两个不可约表示，其中一个 $\Phi^{(3)}$ 是 3 维表示，另一个 $\Phi^{(6)}$ 是 6 维表示，它们可以写成下列形式(参看附录 I)：

$$\Phi^{(3)} = \phi^\alpha\hat{I}_\alpha, \quad \Phi^{(6)} = \frac{1}{\sqrt{6}}\phi^0 + \phi^a\hat{I}_a. \tag{2.15}$$

其中 ϕ^α, ϕ^0 和 ϕ^a 都是复标量场.

在下两节中，我们将利用这一节得到的变换性质构造弱电统一模型.

三、对称性的自发破缺和弱奇异数守恒

在 $SU(3)$ 群作用下规范不变的拉氏函数为

$$\begin{aligned}\mathscr{L} = &-\frac{1}{4} F^i_{\mu\nu}F^{i\mu\nu} + \bar{\psi}\gamma^\mu(\partial_\mu + \hat{A}^{(5)}_\mu)\psi + Tr[(D_\mu\Phi)^+(D_\mu\Phi)] \\ &- V(\Phi, \Phi^+) + \frac{f}{2}\bar{\psi}\Phi(1+\gamma_5)\psi \\ &+ \frac{f^*}{2}\bar{\psi}\Phi^+(1-\gamma_5)\psi. \end{aligned} \tag{3.1}$$

假定 $V(\Phi, \Phi^+)$ 具有 $\Phi \to -\Phi$ 的分立对称性(见附录),可以看到,这个拉氏函数除 $SU(3)$ 规范不变外还具有一个 $U(1)$ 整体对称性. 令 $\hat{S}$ 为这个 $U(1)$ 群的生成元,η 为群参数. 整体 $U(1)$ 变换由

$$U_0^{(\epsilon)} = \exp\{i\hat{S}^{(\epsilon)}\eta\} \tag{3.2}$$

表出,且 $U_0^{(\epsilon)}$ 满足

$$(U_0^{(\epsilon)})^+ = U_0^{(-\epsilon)}, \tag{3.3}$$

各场量的变换规则为

$$\begin{aligned}&\psi \to \psi' = e^{i\gamma_5\eta}\psi, \quad \hat{A}^{(\epsilon)}_\mu \to \hat{A}^{(\epsilon)\prime}_\mu = e^{i\epsilon\eta}\hat{A}^{(\epsilon)}_\mu(e^{i\epsilon\eta})^+ = \hat{A}^{(\epsilon)}_\mu, \\ &\Phi \to \Phi' = e^{-i\eta}\Phi(e^{i\eta})^+ = e^{-2i\eta}\Phi.\end{aligned} \tag{3.4}$$

即费米子(轻子)按 $S = 1$ 的 $U_0^{(5)}$ 变换,规范场不变,Higgs 场按 $S = -2$ 的 $U_0^{(+)}$ 变换.容易证明在(3.4)的整体变换下,(3.1)式是不变的(关于 $V(\Phi, \Phi^+)$ 的讨论参看附录).

为了讨论对称性的自发破缺,首先要考虑 Higgs 场的选取. 为了使所有的规范场除了电磁场外都获得质量,需要在破缺后只保留和电磁场对应的 $U_{E_1M_1}(1)$ 群的对称性,这样 Higgs 场最少要有 8 个实分量,使得其中 7 个分量能通过 Higgs 机制与矢量粒子结合. 在前面的讨论中,我们曾指出,最简单的 Higgs 场可选为 $\Phi^{(3)}$ 或 $\Phi^{(6)}$. $\Phi^{(3)}$ 只有 6 个实分量,经过破缺,只能使 5 个规范场获得质量,同时保留一个 $SU(2)$ 的对称性,也就是说除了光子外还含有两个质量为零的规范场. $\Phi^{(6)}$ 共有 12 个实分量,经过破缺,它可以使除光子之外的所有的规范场得到质量,但同时保留 5 个 Higgs 粒子.

下面我们取 Higgs 场为 $\Phi^{(6)}$ 来进行讨论,Higgs 场的真空期待值取为

$$\langle\Phi^{(6)}\rangle_0 = vI_6 = \frac{1}{2}v\lambda_6. \tag{3.5}$$

原有对称性破缺后还保留下一个局部 $U(1)$ 对称性和一个整体 $U(1)$ 对称性不破缺. 它们的生成元分别为电荷

$$\hat{Q} = \hat{I}_3 - \sqrt{3}\hat{I}_8 \tag{3.6}$$

和弱奇异数

$$\hat{S}_W = \frac{2}{\sqrt{3}}\hat{I}_8 + \frac{1}{6}\hat{S}, \tag{3.7}$$

$\hat{Q}$ 和 $\hat{S}_W$ 是守恒量子数. 在自发破缺后所有的物理粒子都应是它们的本征态,当然各粒子量子数的确定要按照该粒子变换时所依据的实现.

对于费米子 $\hat{S}_W$ 表现为

$$\hat{S}_W^{(5)} = \frac{1}{2}\gamma_5\begin{pmatrix}1 & & \\ & 1 & \\ & & -1\end{pmatrix}. \tag{3.8}$$

ψ 的各分量是 $\hat{S}_W^{(5)}$ 的本征态要求它们或者是左旋分量或者是右旋分量而不能是左旋和右旋的混合.

如果我们要将同一 ψ 三重态中六个手征分量分为 S_W 本征值相同的两组,则只允许有下两种情形

$$S_W = \frac{1}{2}, \quad \begin{pmatrix}L^0\\ L^-\\ R^-\end{pmatrix}; \quad S_W = -\frac{1}{2}, \quad \begin{pmatrix}R^0\\ R^+\\ L^+\end{pmatrix}. \tag{3.9}$$

注意到对称性自发破缺将使第二和第三分量混合而成带质量的粒子，我们得到一个重要结论：左旋中微子总是和带负电的轻子组成一个多重态，右旋的中微子（通常称反中微子)总是和带正电的轻子组成一个多重态.

为确定起见,我们讨论电子的情形,ψ 取作

$$\psi = \begin{pmatrix}\nu_L\\ e_L\\ e_R\end{pmatrix}, \tag{3.10}$$

由此得到电子的质量项为

$$\frac{1}{2}f v \bar{e}_R e_L + \frac{1}{2}f^* v^* \bar{e}_L e_R.$$

为了使得 $\bar{e}\gamma_5 e$ 项不出现,我们可取 fv 为实数,这时电子质量为

$$m_e = \frac{1}{2}fv. \tag{3.11}$$

规范粒子的质量项由

$$Tr[(\hat{A}_\mu^{(-)}\langle\Phi\rangle_0 - \langle\Phi\rangle_0\hat{A}_\mu^{(+)})^+(\hat{A}_\mu^{(-)}\langle\Phi\rangle_0 - \langle\Phi\rangle_0\hat{A}_\mu^{(+)})] \tag{3.12}$$

给出,将 (3.5) 代入，经过简单的计算可以得到质量项为（省略掉规范粒子的 Lorentz 脚标）

$$\frac{1}{4}g^2|v|^2\left[W^+W^- + V^+V^- + 4U^{++}U^{--} + \frac{2}{3}z^0z^0\right]. \tag{3.12 }$$

其中

$$W^\pm = \frac{1}{\sqrt{2}}(A^1 \mp iA^2), \quad z^0 = \frac{1}{2}(\sqrt{3}A^3 + A^8),$$

$$V^\pm = \frac{1}{\sqrt{2}}(A^4 \pm iA^5), \quad U^{\pm\pm} = \frac{1}{\sqrt{2}}(A^6 \pm iA^7), \tag{3.13}$$

由 (3.12) 和 (3.13) 给出各规范粒子质量为

$$m_W^2 = \frac{1}{4}g^2|v|^2, \quad m_z^2 = \frac{4}{3}m_W^2,$$

$$m_V^2 = m_W^2, \qquad m_U^2 = 4m_W^2 \tag{3.14}$$

中性 z^0 粒子的质量和 Weinberg-Salam 模型中 $\sin^2\theta_W = \frac{1}{4}$ 的情形完全一样. 在我们的理论中还有四个重的矢量粒子 $V^\pm$ 和 $U^{\pm\pm}$, 它们分别带一个电荷和两个电荷, 它们的质量由(3.14)给出, 是完全确定的数值.

在对称性自发破缺后, Higgs 场仍保留下来的分量为

$$\Phi = \begin{pmatrix} \phi^0 & 0 & 0 \\ 0 & \phi^{++} & \frac{1}{2}\chi^0 \\ 0 & \frac{1}{2}\chi^0 & 0 \end{pmatrix}. \tag{3.15}$$

其中 χ^0 是实场, ϕ^0 和 ϕ^{++} 都是复场.

所有粒子的电荷和弱奇异数如下表所示:

表 2

	γ	W^-	W^+	Z^0	V^-	V^+	U^{--}	U^{++}
Q	0	−1	1	0	−1	1	−2	2
S_W	0	0	0	0	1	−1	1	−1
	ν_e	e_L	e_R	χ^0	ϕ^{0*}	ϕ^0	ϕ^{--}	ϕ^{++}
Q	0	−1	−1	0	0	0	−2	2
S_W	1/2	1/2	1/2	0	1	−1	1	−1

我们的理论比 Weinberg-Salam 模型多出的粒子是一批带弱奇异数的粒子, 它们的性质我们在后面还要进一步讨论.

四、轻子和规范粒子及 Higgs 粒子的相互作用

轻子与规范粒子的相互作用由 $\bar{\Phi}\gamma^\mu \hat{A}_\mu^{(5)}\psi$ 给出. 由于 $\bar{L}\gamma^\mu R = \bar{R}\gamma^\mu L = 0$, 可以看出带弱奇异数的规范粒子与相同 S_W 的轻子无直接耦合. 利用(3.10)和 $\gamma_5 L = L$, $\gamma_5 R = -R$, 我们得到

$$\begin{aligned}\bar{\Phi}\gamma^\mu \hat{A}_\mu^{(5)}\psi &= i\frac{g}{2}\bar{\Phi}\gamma^\mu[\gamma_5(A_\mu^1\hat{\lambda}_1 + A_\mu^3\hat{\lambda}_3 + A_\mu^8\hat{\lambda}_8) + A_\mu^2\hat{\lambda}_2]\psi \\ &= i\frac{g}{2}\bar{\Phi}\gamma^\mu[A_\mu^1\hat{\lambda}_1 + A_\mu^2\hat{\lambda}_2 + A_\mu^3\hat{\lambda}_3 + A_\mu^8\hat{\lambda}_3']\psi.\end{aligned} \tag{4.1}$$

其中 $\hat{\lambda}_i$ 是通常的 Gell-Mann $\hat{\lambda}$ 矩阵, $\hat{\lambda}_8'$ 为

$$\hat{\lambda}_8' = \frac{1}{\sqrt{3}}\begin{pmatrix} 1 & & \\ & 1 & \\ & & 2 \end{pmatrix}. \tag{4.2}$$

利用(3.13)和

$$A = \frac{1}{2}(-A^3 + \sqrt{3}A^8). \tag{4.3}$$

(4.1) 可明显写作

$$i\frac{g}{\sqrt{2}}(\bar{\nu}_L\gamma^\mu W_\mu^+ e_L+\bar{e}_L\gamma^\mu W_\mu^-\nu_L)+\frac{ig}{2\sqrt{3}}(2\bar{\nu}_L\gamma^\mu\nu_L-\bar{e}_L\gamma^\mu e_L+\bar{e}_R\gamma^\mu e_R)Z_\mu$$
$$+i\frac{g}{2}(\bar{e}_L\gamma^\mu e_L+\bar{e}_R\gamma^\mu e_R)A_\mu \tag{4.4}$$

这正是 Weinberg-Salam 模型中轻子与规范场相互作用的拉氏量，并且 $\sin^2\theta_W=1/4$. 值得指出的是，许多作者[3,4]为了要在轻子和 A_μ^8 规范场的相互作用中，得到正确的 Weinberg 角，将 $SU(3)$ 群改为超对称群 $SU(2/1)$，这样 λ_8 才变成了 (4.2) 式的 λ_8'. 从我们的推导中可以看到，引进带 γ_5 的生成元就能得到 λ_8'，从而得到正确的 Weinberg 角而无需引进超对称群.

轻子与 Higgs 场的相互作用为

$$\frac{f}{2}\bar{\psi}\Phi(1+\gamma_5)\psi+\frac{f^*}{2}\bar{\psi}\Phi^+(1-\gamma_5)\psi.$$

利用 (3.10) 和 (3.15) 给出为

$$\frac{1}{2}(f\bar{e}_Re_L+f^*\bar{e}_Le_R)\chi^0. \tag{4.5}$$

这表明带奇异数的 Higgs 粒子也不与轻子直接耦合，只有 $S_W=0$ 的 χ^0 才能与轻子直接耦合.

从上面的讨论中可以看到，就现已观察到的轻子与玻色粒子的相互作用以及这些粒子的质量而言，我们的理论和 Weinberg-Salam 模型完全一样，并且定出 $\sin^2\theta_W=1/4$. 在我们的理论中，还多出 4 个重的矢量粒子，4个标量粒子以及两个费米子. 它们都带有不为零的弱奇异数，只要它们的质量足够重，在低能的轻子衰变和散射实验中，它们的作用可以忽略. 我们将在下一节中专门讨论弱奇异玻色子的一些性质和观察它们的可能性.

五、弱奇异玻色子的性质

如表 2 所示轻子的弱奇异数为 1/2，但它与轻子数简并，因此可以把轻子当作弱奇异数为零的粒子对待(通过重新定义弱奇异数). 弱奇异数为(−1/2)的费米子，通过重新定义，弱奇异数为 −1，它们的质量假定很重而不影响低能轻子的现象. 在本文中我们不讨论这些奇异费米子的性质. 我们称弱奇异数不为零的玻色子 $V^\pm$, $U^{\pm\pm}$, φ^0, φ^{0*}, $\varphi^{\pm\pm}$ 为弱奇异粒子. 从上面的讨论我们已知它们的几个重要性质: (i) 它们和轻子没有直接耦合; (ii) 通过适当的 Higgs 机制可以使它们成为重粒子，并且V和U的质量在一种选择下确定为 $m_V^2=m_W^2$, $m_U^2=4m_W^2$; (iii) 它们的电荷由表 2 给出，特别是有双电荷的粒子 $U^{\pm\pm}$, $\varphi^{\pm\pm}$ 存在; (iv) 由于弱奇异数的严格守恒，它们只能成对产生.

为进一步研究弱奇异粒子的性质，我们先看规范场三顶点自相互作用. 按(3.1)式其拉氏函数为

$$\frac{i}{2}gf_{ikl}A_\mu^iA_\nu^k(\partial^\mu A^{l\nu}-\partial^\nu A^{l\mu}), \tag{5.1}$$

引进符号

$$(A^j, A^k, A^l) = \sum_{\pi} \epsilon_\pi A_\mu^{j_\pi} A_\nu^{k_\pi} (\partial^\mu A^{l_\pi \nu} - \partial^\nu A^{l_\pi \mu}). \qquad (5.2)$$

其中 π 为 j, k, l 的一种排列，当 π 为奇置换时 $\epsilon_\pi = -1$，当 π 为偶置换时 $\epsilon_\pi = 1$. 这样定义的 (A^j, A^k, A^l) 对 j, k, l 是全反对称的. 将 A_μ^i 用物理粒子的场量表示出来代入(5.1)式，仍用(5.2)式的符号可以把规范场三顶点自相互作用的拉氏量明显表为

$$\frac{1}{4} g[(A, W^-, W^+) + (A, V^-, V^+) + 2(A, U^{--}, U^{++}) + \sqrt{3}(Z^0, V^-, V^+) - \sqrt{3}(Z^0, W^-, W^+) + \sqrt{2}(W^-, V^-, U^{++}) - \sqrt{2}(W^+, V^+, U^{--})]. \qquad (5.3)$$

从(5.3)式可以看出，直接观察V粒子和U粒子的方法是通过高能电子正电子对撞实验观察它们的成对产生. 在 Weinberg-Salam 理论中，只应观察到 $W^\pm$ 粒子的成对产生. 在我们的理论中还应能观察到 $V^\pm$ 粒子和带双电荷的 $U^{\pm\pm}$ 粒子的成对产生（假如能量足够的话），它们的产生截面可以通过(5.3)式计算出来. 在扣除相空间因子的差别后，它们的产生截面和W粒子成对产生的截面在数量级上也是相同的. 此外，U粒子与 Z^0 粒子不耦合，在电子正电子对撞产生U粒子对时，应表现为没有 Z^0 中间态与之相干的纯虚光子过程（当然是指在 g^2 级微扰论计算的意义下），亦即表现为典型的量子电动力学所描写的过程. 考虑到这时能量已远高于 Z^0 粒子的产生阈，Z^0 粒子对各种过程的贡献与光子是同量级的，$U^{\pm\pm}$ 对不与单 Z^0 耦合是一个重要的特征.

U粒子可以通过下列方式衰变

$$U^{++} \to W^+ + V^+$$
$$\quad \longrightarrow e^+ + \nu_e \text{ 等}, \qquad (5.4)$$

由于 V^+ 与 e^+ 带有同号的电荷，其特征性是很明显的，可以作为辨认U粒子的方法.

规范场四顶点自相互作用也可由(3.1)式给出，我们暂不去具体讨论了.

现在考察 Higgs 粒子与规范场的相互作用，这些相互作用包含在

$$Tr[(D_\mu \Phi)^+ (D_\mu \Phi)] \qquad (5.5)$$

中，其中 Φ 由(3.15)式给出，$D_\mu \Phi$ 由(2.11)式给出. 注意到

$$\hat{A}^{(+)} = \frac{ig}{\sqrt{2}} \begin{pmatrix} \sqrt{\frac{2}{3}} Z & W^+ & V^- \\ W^- & \frac{1}{\sqrt{2}} A - \frac{1}{\sqrt{6}} Z & U^{--} \\ V^+ & U^{++} & -\frac{1}{\sqrt{2}} A - \frac{1}{\sqrt{6}} Z \end{pmatrix},$$

$$\hat{A}^{(-)} = \frac{ig}{\sqrt{2}} \begin{pmatrix} -\sqrt{\frac{2}{3}} Z & -W^- & -V^+ \\ -W^+ & -\frac{1}{\sqrt{2}} A + \frac{1}{\sqrt{6}} Z & -U^{++} \\ -V^- & -U^{--} & \frac{1}{\sqrt{2}} A + \frac{1}{\sqrt{6}} Z \end{pmatrix}. \qquad (5.6)$$

代出后即得到 Higgs 粒子与规范场的相互作用. 其中三顶点的部分为

$$
\begin{aligned}
& ig\Big\{\frac{2}{\sqrt{3}} Z^{\mu}[\phi^{0*}(\partial_{\mu}\phi^{0}) - (\partial_{\mu}\phi^{0*})\phi^{0}] \\
& + \sqrt{2}\Big(A^{\mu} - \frac{1}{\sqrt{3}} Z^{\mu}\Big)[(\partial_{\mu}\phi^{++})\phi^{--} - \phi^{++}(\partial_{\mu}\phi^{--})] \\
& + \frac{1}{\sqrt{2}} U^{++\mu}[(\partial_{\mu}\chi)\phi^{--} - \chi(\partial_{\mu}\phi^{--})] \\
& + \frac{1}{\sqrt{2}} U^{--\mu}[(\partial_{\mu}\phi^{++})\chi - \phi^{++}(\partial_{\mu}\chi)]\Big\} \\
& + gm_{W}\Big\{\Big[\frac{2}{3} Z_{\mu}Z^{\mu} + W_{\mu}^{+}W^{-\mu} + V_{\mu}^{+}V^{-\mu} + 4U_{\mu}^{++}U^{--\mu}\Big]\chi \\
& + \sqrt{2}\Big(A^{\mu} - \frac{2}{\sqrt{3}} Z^{\mu}\Big)(U_{\mu}^{++}\phi^{--} + U_{\mu}^{--}\phi^{++}) \\
& + 2W_{\mu}^{+}V^{-\mu}\phi^{0} + 2W_{\mu}^{-}V^{+\mu}\phi^{0*} + W_{\mu}^{-}V^{-\mu}\phi^{++} + W_{\mu}^{+}V^{+\mu}\phi^{--}\Big\},
\end{aligned} \tag{5.7}
$$

含 m_W 项是由四顶点破缺得到的.

利用(5.7)式我们可以讨论 Higgs 粒子的主要产生和衰变行为.

六、$V(\Phi,\Phi^{+})$ 中包含三次幂的讨论

理论的可重整性要求 $V(\Phi, \Phi^{+})$ 对场量的依赖不超过四次幂. 如果其中含有三次幂项,例如

$$\det\Phi + \det\Phi^{+} = -\frac{1}{4\sqrt{2}}(\phi_{11}^{*} + \phi_{11})\phi_{23}^{2} \tag{6.1}$$

项(符号见附录),则将带来 $V(\Phi, \Phi^{+})$ 极小值位置的移动, 这时不仅 ϕ_{23} 的真空期待值将不为零,而且 $\phi_{11}^{*} + \phi_{11}$ 的真空期待值亦将不为零.

由于(6.1)式不满足整体 $U(1)$ 不变性,因此弱奇异数守恒不再保持.

如果这时 Higgs 场的真空期待值为

$$\langle\Phi^{(6)}\rangle_{0} = \frac{1}{2}\begin{pmatrix} v' & 0 & 0 \\ 0 & 0 & v \\ 0 & v & 0 \end{pmatrix}, \tag{6.2}$$

则电子得到的质量没有改变,但规范粒子的质量项代替(3.12)有(为简单假定 v, v' 都是实数)

$$
\begin{aligned}
& \frac{1}{4} g^{2}\Big[\Big(\frac{2}{3} v^{2} + \frac{4}{3} v'^{2}\Big) Z^{2} + 4v^{2}U^{+}U^{--} + (v^{2} + v'^{2})(W^{+}W^{-} + V^{+}V^{-}) \\
& + 2vv'(W^{+}V^{-} + W^{-}V^{+})\Big]
\end{aligned} \tag{6.3}
$$

W和V将重新混合,质量本征态为

$$\tilde{W} = \frac{1}{\sqrt{2}}(W - V), \quad m_{\tilde{W}}^{2} = \frac{1}{4} g^{2}(v - v')^{2},$$

$$\tilde{V} = \frac{1}{\sqrt{2}}(W + V), \quad m_{\tilde{V}}^2 = \frac{1}{4} g^2 (v + v')^2, \tag{6.4}$$

不失普遍性，我们可以假定 $m_{\tilde{W}}^2 < m_{\tilde{V}}^2$. $\tilde{W}$ 和 $\tilde{V}$ 都将和轻子有耦合（通过 W）. Z 和 U 的质量为

$$m_Z^2 = \frac{1}{4} g^2 \left(\frac{4}{3} v^2 + \frac{8}{3} v'^2 \right), \quad m_U^2 = g^2 v^2 \tag{6.5}$$

显然这些结果与 Weinberg-Salam 模型不同. 考虑到现有实验与 Weinberg-Salam 模型符合，可以预期 v' 如不为零亦应较小.

考虑低能过程带电流中 $\tilde{W}$ 和 $\tilde{V}$ 的贡献可等效成 W 粒子，其等效质量 $\overline{m}_W$ 由下式给出

$$\frac{1}{\overline{m}_W^2} = \frac{1}{2} \left(\frac{1}{m_{\tilde{W}}^2} + \frac{1}{m_{\tilde{V}}^2} \right), \tag{6.6}$$

即

$$\overline{m}_W^2 = \frac{1}{4} g^2 \frac{(v^2 - v'^2)^2}{v^2 + v'^2}. \tag{6.7}$$

这样由 Z 粒子和 W 粒子的质量定出的 Weinberg 角为

$$\sin^2 \theta_W = 1 - \frac{m_Z^2}{\overline{m}_W^2} = \frac{1}{4} \left[1 - \frac{v'^2 (15 v^2 + 3 v'^2)}{(v^2 - v'^2)^2} \right] \approx \frac{1}{4} \left(1 - 15 \frac{v'^2}{v^2} \right). \tag{6.8}$$

它比 1/4 值略小. 但这时中性流本身仍为(4.4)式，即相当于 $\sin^2 \theta_W = \frac{1}{4}$ 时的结果.

因此当 $V(\Phi, \Phi^+)$ 中有少量 $\det \Phi + \det \Phi^+$ 项时，理论将对前几节讨论的模型和 Weinberg-Salam 模型有偏离，表现在:

1. 中性流结构仍相当于 $\sin^2 \theta_W = \frac{1}{4}$ 的结果，但低能下带电流与中性流强度比表现为相当于 $\sin^2 \theta_W < \frac{1}{4}$ 如(6.8)式所示. 这可以在精确实验中来检验.

2. 在能量足够高时可以产生两对带电规范粒子 $\tilde{W}^\pm$ 和 $\tilde{V}^\pm$，它们共同承担 Weinberg-Salam 模型中 $W^\pm$ 粒子所起的传递带电流的作用，它们的质量如(6.4)式所示有所差别，并以相同的强度与轻子流耦合(和衰变).

3. 当然，弱奇异数将不再是守恒量子数.

4. 规范场三顶点自作用 (5.3) 式变为

$$\begin{aligned} &\frac{1}{4} g[(A, \tilde{W}^-, \tilde{W}^+) + (A, \tilde{V}^-, \tilde{V}^+) + 2(A, U^{--}, U^{++}) \\ &+ \sqrt{3}\,(Z^0, \tilde{W}^-, \tilde{V}^+) + \sqrt{3}(Z^0, \tilde{V}^-, \tilde{W}^+) \\ &+ \sqrt{2}(\tilde{W}^-, \tilde{V}^-, U^{++}) - \sqrt{2}(\tilde{W}^+, \tilde{V}^+, U^{--}) \end{aligned} \tag{6.9}$$

通过 Z^0 对产生的是一个 $\tilde{W}$ 和一个 $\tilde{V}$.

七、一 些 讨 论

在叙述了所得的主要结果之后，我们讨论以下几个问题:

1. 如果将 Higgs 场取作 $\Phi^{(3)}$，则仍可得到：(i) $\sin^2\theta_W = 1/4$；(ii) $m_V = m_W$；(iii) 弱奇异数守恒．这时只有一个 Higgs 粒子，但 U 粒子和光子一样没有质量． $\Phi^{(3)}$ 场并没有使得 $SU(3)$ 群破缺到只留下 $U_{E_1M_1}(1)$ 的对称性，而剩余了 $SU(2)$ 的对称性．由于没有观察到质量为零的带电粒子 $U^{\pm\pm}$，这个对称性必须进一步破缺．

如果 Higgs 场取作 $\Phi^{(3)}$ 和 $\Phi^{(6)}$ 的线性组合，则仍可得到上述 (i)，(ii)，(iii) 的结果，这时 U 粒子的质量在 0 和 $2m_W$ 之间，但剩余的 Higgs 粒子太多．

2. 在这个理论框架内可以容纳 μ，ν_μ，τ，ν_τ 等轻子，它们可以有不同的质量，不像有些超对称群的理论那样预言所有轻子的质量是一样的．

3. 将这个理论直接推广到夸克有一定困难，这种困难对超对称群的理论也是同样存在的．虽然夸克和轻子有明显的对称性，但夸克的电荷和轻子的电荷相差 $\frac{2}{3}e$，这是产生困难的原因．有几个可能的途径来解决这个问题．一个是将夸克看作复合粒子，例如夸克由轻子和一个带电荷 $2/3e$ 的玻色子组成．那么需要解释的是，为什么这个玻色子只和电磁场有相互作用而和W粒子以及Z粒子没有相互作用．另一种途径是将 $SU(3)$ 群再扩大，引进代表夸克的量子数，这样就需要再引进一个未知的耦合常数，我们将在以后的工作中来讨论这一问题．

4. 这个理论所给出的弱奇异数守恒，在一大类理论中是共同的．在我们以后的工作中可以看到，当把这个理论发展去包括夸克时，可以有几种不同的实现方案，但理论中存在一个守恒的弱奇异数量子数则是共同的，具有普遍意义．因此在将来的高能实验中，考察是否存在这样的守恒量子数和弱奇异粒子，是十分重要的．

5. Higgs 自作用势中允许有破坏弱奇异数守恒的三次项出现，是与 Higgs 势选作 6 维表示 $\Phi^{(6)}$ 的选法有关．如果 Higgs 势是选作 $\Phi^{(3)}$，由于 3 个 3 维表示组成的不变量是完全反对称的，它不能在自作用势中出现，这时弱奇异数就将是严格守恒的．

附录 I　Higgs 场的变换性质

首先我们证明 Higgs 场 Φ 可以按 (2.15) 分解为 3 维表示 $\Phi^{(3)}$ 和 6 维表示 $\Phi^{(6)}$．为确定和简单起见，我们在 α 和 a 的第一种选取下来讨论．一般地 Φ 可表为

$$\Phi = \frac{1}{\sqrt{6}}\phi^0 + \phi^a\hat{I}_a + \phi^\alpha\hat{I}_\alpha, \tag{I.1}$$

作无穷小变换

$$\Phi \to \Phi' = U^{(-)}\Phi U^{(+)+} = (1 - i\hat{I}_a\xi^a + i\hat{I}_\alpha\xi^\alpha)\Phi(1 - i\hat{I}_a\xi^a - i\hat{I}_\alpha\xi^\alpha).$$

将 (I.1) 代入，得到

$$\begin{aligned} \phi^0 \to \phi^{0\prime} &= \phi^0 - i\sqrt{\frac{2}{3}}\,\phi^a\xi^a, \\ \phi^a \to \phi^{a\prime} &= \phi^a - (f_{\beta ca}\xi^\beta + id_{bca}\xi^b)\phi^c - i\sqrt{\frac{2}{3}}\,\Phi^0\xi^a, \\ \phi^\alpha \to \phi^{\alpha\prime} &= \phi^\alpha - (f_{\beta\gamma\alpha}\xi^\beta + id_{b\gamma\alpha}\xi^b)\phi^\gamma, \end{aligned} \tag{I.2}$$

由此可见 ϕ^0，ϕ^a 构成一个 6 维表示，ϕ^α 构成一个三维表示

将三维表示 $\Phi^{(3)}$ 的变换规则和标准的三维表示变换规则比较，可以得到 $\Phi^{(3)}$ 有互相等价的两种表

述方法:

$$\Phi^{(3)} = \frac{1}{\sqrt{2}}\begin{pmatrix} 0 & -i\phi_2 & -i\phi_5 \\ i\phi_2 & 0 & -i\phi_7 \\ i\phi_5 & i\phi_7 & 0 \end{pmatrix}, \quad \Phi^{(3)} \to \Phi^{(3)\prime} = U^{(-)}\Phi^{(3)}U^{(+)+}. \tag{I.3}$$

和

$$\Phi^{(3)} = \begin{pmatrix} \phi_7 \\ -\phi_5 \\ \phi_2 \end{pmatrix}, \quad \Phi^{(3)} \to \Phi^{(3)\prime} = U^{(+)}\Phi^{(3)}. \tag{I.4}$$

同样地也可以证明 6 维表示 $\Phi^{(6)}$ 与用张量形式表述的 6 维表示是等价的.

现在讨论 Higgs 粒子为 $\Phi^{(6)}$ 的情形. 当对称性自发破缺后，有 7 个实分量与规范粒子结合，可以通过规范变换去掉,保留下来的还应有 5 个实分量. 我们考察在 Higgs 场真空期待值为

$$\langle\Phi^{(6)}\rangle = \frac{1}{2}\hat{\lambda}_s v$$

时,保留下来的应有哪些分量.

令 $\Phi^{(6)}$ 偏离真空期待值不远,作无穷小规范变换,并将 $\Phi^{(6)}$ 表为

$$\Phi^{(6)} = \frac{1}{2}\begin{pmatrix} \sqrt{2}\,\phi_{11} & \phi_{12} & \phi_{13} \\ \phi_{21} & \sqrt{2}\,\phi_{22} & \phi_{23} \\ \phi_{31} & \phi_{32} & \sqrt{2}\,\phi_{33} \end{pmatrix}. \tag{I.5}$$

其中 $\phi_{ij} = \phi_{ji}$，则结果为(展到一级小量):

$$\begin{aligned} &\phi'_{11} = \phi_{11}, \\ &\phi'_{22} = \phi_{22} - i\frac{v}{2}\sqrt{2}\,(\xi^6 + i\xi^7), \quad \phi'_{33} = \phi_{33} - i\frac{v}{2}\sqrt{2}\,(\xi^6 - i\xi^7), \\ &\phi'_{12} = \phi_{12} - i\frac{v}{2}(\xi^4 + i\xi^5), \quad \phi'_{13} = \phi_{13} - i\frac{v}{2}(\xi^1 + i\xi^2), \\ &\phi'_{23} = \phi_{23} - i\frac{v}{2}\left(-\xi^3 + \frac{\xi^8}{\sqrt{3}}\right), \end{aligned} \tag{I.6}$$

因此可以适当选择局部变换参数 $\xi^i(x)$，使变换后的 $\phi'_{12} = \phi'_{13} = \phi'_{33} = 0$，$\phi_{23}$ 的虚部消去，保留下来 ϕ'_{23} 的实部和 ϕ'_{11}，ϕ'_{22} 复场，这样就得到(3.15)式的表达式.

附录 II　Higgs 场自作用势和对称性的自发破缺

对称性的自发破缺由 Higgs 场的自作用势所给出. 考虑到可重整化的要求，自作用势应由 Φ 和 Φ^+ 的不超过四次幂的不变量的线性组合给出.

在采用 Higgs 场为 $\Phi^{(6)}$ 的情况下,它的二次不变量只有一个,即正比于 $Tr(\Phi\Phi^+)$ 的项;三次不变量有 $\det\Phi$ 和 $\det\Phi^+$，这样的项如果在 $V(\Phi, \Phi^+)$ 中出现，将在对称性自发破缺时带来一些复杂性，我们假定 V 中不存在这样的项,在本文第六节中简单讨论这样的项出现时的后果；四次不变量有三个，这是因为 $\Phi\Phi^+$ 可分解为三个不可约表示

$$\underline{6} \times \underline{6}^* = \underline{1} + \underline{8} + \underline{27} \tag{II.1}$$

两对 $\Phi\Phi^+$ 耦合时得到的三个不变量我们分别归一到 1, 8, 27, 并且 I_1, I_8. I_{27} 表示.

对四次不变量有下述定理.

定理: 若么正对称群的 n 维表示 $\underline{n}$ 和 $\underline{n}^*$ 直乘可分解为 m 个变换性质不同的不可约表示的直和，则 $\underline{n}$ 和 $\underline{n}^*$ 组成的独立四次不变量的个数不超过 $m-1$.

证: 二次不变量为

$$I_0=\sum_{i=1}^{n}\phi_i\phi_i^*,\qquad \phi_i \text{ 为 } \underline{n} \text{ 的基}$$

令四次不变量为 I_{N_j}，$j=1,\cdots,m$，N_j 为 $\underline{n}\times\underline{n}^*$ 分解出的第 j 个不可约表示的维数，

$$\sum_{j=1}^{m}N_j=n^2.$$

令 $j=1$ 代表 1 维表示，则 $I_1=\frac{1}{n}I_0^2$，其中 I_{N_j} 归一化到 N_j.

作 $\sum_{j=1}^{m}I_{N_j}$，它实际上等于 $\underline{n}\times\underline{n}^*$ 给出的所有态模的平方和，即

$$\sum_{j=1}^{m}I_{N_j}=\sum_{i,k}^{n}(\phi_i\phi_k^*)(\phi_i\phi_k^*)^*=\sum_{i,k}^{n}\phi_i\phi_i^*\phi_k\phi_k^*=I_0^2$$

即各 I_{N_j}，$j=1,\cdots\cdots,m$ 不是独立的，证完.

根据上述定理，I_1，I_8，I_{27} 中最多只有两个是独立的. 利用 Φ 的一般形式 (I.5)，各不变量为

$$I_0=2Tr(\Phi\Phi^+)=|\phi_{11}|^2+|\phi_{22}|^2+|\phi_{33}|^2+|\phi_{12}|^2+|\phi_{13}|^2+|\phi_{23}|^2, \tag{II.2}$$

$$I_1=\frac{1}{6}I_0^2, \tag{II.3}$$

$$I_8=\sum_{\substack{i,k=1\\(i,j,k\text{不等})}}^{n}\frac{1}{5}|\sqrt{2}\,\phi_{ii}\phi_{ki}^*+\sqrt{2}\,\phi_{ik}\phi_{kk}^*+\phi_{ij}\phi_{kj}^*|^2$$
$$+\frac{1}{10}(2|\phi_{11}|^2-2|\phi_{22}|^2+|\phi_{13}|^2-|\phi_{23}|^2)$$
$$+\frac{1}{30}(2|\phi_{11}|^2+2|\phi_{22}|^2-4|\phi_{33}|^2+2|\phi_{12}|^2-|\phi_{13}|^2-|\phi_{23}|^2), \tag{II.4}$$

如果 $V(\Phi,\Phi^+)$ 中含有 I_8 项，则得不到所需要的破缺分量. 假定 Higgs 自作用势简单为

$$V(\Phi,\Phi^+)=-\mu^2I_0+aI_0^2, \tag{II.5}$$

其中 μ^2，$a>0$，可以得到 V 在 $I_0=\frac{\mu^2}{2a}$ 时达极小值. 令破缺分量为 ϕ_{33}，$\langle\phi_{33}\rangle_0=\sqrt{\frac{\mu^2}{2a}}$，利用规范变换，将 Φ 中七个分量变到零，保留五个分量为复的 ϕ_{11}，ϕ_{22} 和实的 $\phi_{33}=\sqrt{\frac{\mu^2}{2a}}+\chi$. 则通过对称性的自发破缺 χ^0 将获得质量 $m_\chi^2=4\mu^2$，$\phi_{11}=\phi^0$，$\phi_{22}=\phi^{++}$ 将仍为零质量，这表明它们是赝 Goldstone 粒子，是在 $V(\Phi)$ 中具有高于 $SU(3)$ 的整体对称性的结果.

为使 ϕ^0 和 ϕ^{++} 获得重的质量，需要引入其它的 Higgs 多重态，解除 $V(\Phi)$ 中高于 $SU(3)$ 的整体对称性. 在下一篇文章中，我们将把这个模型推广到包括夸克在内，在那里将给出关于这一问题的讨论.

参 考 文 献

[1] S. Weinberg, *Phys. Rev. Lett.*, 19 (1967), 1264. A. Salam. Elementary Particle Theory. ed. N. Svartholm (Stockholm, 1968).
[2] 见1978年东京国际高能物理会议文集.
[3] Y. Néeman, *Phys. Lett.*, **81B** (1979). 190.
[4] D. B. Fairlie, *Phys. Lett.*, **82B** (1979), 97; E. J. Squires, *Phys. Lett.*, **82B** (1979). 359. J. G. Taylor, Phys. Lett., **83B** (1979). 331; Phys. Lett., **84B** (1979), 79; P. H. Dondi and P. D. Tarvis, *Phys. Lett.*, **84B** (1979). 75.

UNIFIED ELECTRO-WEAK MODEL IN $SU(3)$

ZHOU GUANG-ZHAO
(*Institute of Theoretical Physics, Academia Sinica*)

GAO CHONG-SHOU
(*Peking University*)

ABSTRACT

In this paper a unified electro-weak model for leptons based on the $SU(3)$ gauge group is suggested by means of four kinds of realization for the generators of the group. For all low energy electro-weak processes, this model predicts the same results as the conventional Weinberg-Salam model does. The Weinberg angle is shown to be $\sin^2\theta_w = 1/4$ in a natural way. When the Higgs self potential respects a discrete symmetry $\Phi \to -\Phi$, a new conserved quantum number called weak strangeness emerges from the model after spontaneous symmetry breaking. In the present model there exist another four heavy vector gauge bosons $V^{\pm}$ and $U^{\pm\pm}$ together with some heavy fermions and Higgs scalars, which have non vanishing weak strangeness quantum numbers. These weak strange particles have no direct couplings with leptons. Their existence will not influence the low energy electro-weak processes. Nevertheless, they can be produced in pairs in high energy collisions and the lightest of them should be stable if the conservation of weak strangeness is exact. The experimental implications and the possibility of violation of the conservation of weak strangeness are also discussed.

第10卷2期　　中国科学技术大学学报　　1980年

路径积分量子化的一般形式

尹鸿钧　阮图南

（中国科学技术大学）

杜东生　周光召

（中国科学院高能物理研究所）　（中国科学院理论物理研究所）

摘　　要

本文給出了非綫性拉氏函数路径积分量子化的一般形式。在泛函积分中出現的等效拉氏函数，除原始拉氏函数以及李、楊的 $\delta(0)$ 項外，还有一个修正項，这个修正項比例于 $\delta(0)$，而且是 $\delta(0)$ 的冪級数的对数。这种形式为非綫性場論的量子化提供了理論基础。

（一）序　　言

关于路径积分量子化的概念是 Dirac 在 1933 年提出的[1]，以后 Feynman 在 1948 年使它的形式更为完善[2]. 但是他們的討論局限于特殊形式的拉氏函数，例如引进[3]

$$L=\frac{1}{2}\dot{q}^2-V(q),\quad \dot{q}=\frac{dq}{dt}$$

则在泛函积分中出現的等效拉氏函数为

$$L_{eff}=\frac{1}{2}\dot{q}^2-V(q).$$

若引进拉氏函数[4]：

$$L=\frac{1}{2}\dot{q}^2f(q),$$

则在泛函积分中的等效拉氏函数为

$$L_{eff}=\frac{1}{2}\dot{q}^2f(q)-\frac{i}{2}\delta(0)\ln f(q),$$

其中出現比例于 $\delta(0)$ 的項。这一結果是李、楊首先得到的。而对于非綫性拉氏函数

$$L=L(q,\dot{q})$$

則只能写成相空間$\left[\frac{dpdq}{2\pi}\right]$的泛函积分，不能写成路径 $[dq]$ 的泛函积分，因此它的形式不是明显协变的。本文的目的是进行动量积分，把相空間积分化为路径积分，将理論納入明显协变的形式。这个簡洁的形式将为非綫性場論的量子化提供一个理論基础。

我們首先討論一个自由度的力学系統，它的拉氏函数是广义坐标 q、广义速度 $\dot{q}$ 的非綫性函数，

$$L = L(q, \dot{q}), \quad \dot{q} = \frac{dq}{dt}, \tag{1.1}$$

物理系統的作用量定义为它的时間积分，

$$S = \int dt\, L(q, \dot{q}). \tag{1.2}$$

极值原理給出 Lagrange-Euler 方程，

$$\frac{dS}{dq} = \frac{\partial L}{\partial q} - \frac{d}{dt}\frac{\partial L}{\partial \dot{q}} = 0,$$

或

$$\frac{\partial L}{\partial q} = \frac{d}{dt}\frac{\partial L}{\partial \dot{q}}. \tag{1.3}$$

引进正則动量和質量函数：

$$p = \frac{\partial L}{\partial \dot{q}}, \tag{1.4}$$

$$m(q, \dot{q}) = \frac{\partial^2 L}{\partial \dot{q}^2}. \tag{1.5}$$

若 $m \neq 0$，則可反解 (1.4) 式得

$$\dot{q} = \dot{q}(p, q). \tag{1.6}$$

因此可用正則动量 p 代替广义速度 $\dot{q}$，引入正則变数 p、q，定义哈氏函数：

$$H(p, q) = p\dot{q} - L, \tag{1.7}$$

由此导出正則方程

$$\dot{q} = \frac{\partial H}{\partial p};$$

$$\dot{p} = -\frac{\partial H}{\partial q}. \tag{1.8}$$

引入量子化条件：

$$[p, q] = -i, \tag{1.9}$$

則 (1.8) 式化为 Heisenberg 方程，

$$\dot{q} = i[H, q]$$
$$\dot{p} = i[H, p] \tag{1.10}$$
$$\frac{\partial}{\partial t}|\rangle = 0.$$

作么正变换

$$|t\rangle = e^{-iHt}|\rangle$$

$$\hat{Q} = e^{-iHt} q e^{iHt} \tag{1.11}$$

$$\hat{P} = e^{-iHt} p e^{iHt},$$

則 (1.10) 式化为 Schrödinger 方程：

$$i\frac{\partial}{\partial t}|t\rangle = H|t\rangle, \quad H = H(\hat{P}, \hat{Q})$$

$$i\frac{\partial}{\partial t}\hat{P} = 0 \tag{1.12}$$

$$i\frac{\partial}{\partial t}\hat{Q} = 0,$$

以及量子化条件

$$[\hat{P}, \hat{Q}] = -i. \tag{1.13}$$

設 $\hat{P}$、$\hat{Q}$ 为厄米算符

$$\hat{Q}^{+} = \hat{Q}, \qquad \hat{P}^{+} = \hat{P} \tag{1.14}$$

則可定义坐标表象的正交归一完备基：

$$\hat{Q}|q\rangle = q|q\rangle, \quad q^{*} = q$$

$$\langle q|q'\rangle = \delta(q-q'),$$

$$\int dq|q\rangle\langle q| = 1, \tag{1.15}$$

这时从量子化条件 (1.13) 式可以导出

$$\langle q|\hat{P}|q'\rangle = -i\frac{\partial}{\partial q}\delta(q-q') = \int\frac{dp}{2\pi}e^{ip(q-q')}p; \tag{1.16}$$

$$\langle q|f(\hat{P})|q'\rangle = f\left(-i\frac{\partial}{\partial q}\right)\delta(q-q') = \int\frac{dp}{2\pi}e^{ip(q-q')}f(p),$$

其中 $f(\hat{P})$是动量算符 $\hat{P}$ 的函数。积分波动方程 (1.12) 式可得

$$|t\rangle = e^{-iH(t-t_0)}|t_0\rangle, \tag{1.17}$$

其中

$$U(t, t_0) = e^{-iH(t-t_0)} \tag{1.18}$$

称为轉換矩陣。定义波函数在坐标表象的振幅和变换矩陣元：

$$\psi(q, t) = \langle q|t\rangle;$$

$$F(qt, q_0t_0) = \langle q|e^{-iH(t-t_0)}|q_0\rangle \tag{1.19}$$

則利用完备条件 (1.15) 式可从 (1.17) 式导出

$$\psi(q, t) = \int dq_0 F(qt, q_0t_0)\psi(q_0, t_0). \tag{1.20}$$

因此轉換矩陣元 F 把波函数在 t 时刻的振幅和 t_0 时刻的振幅联系了起来。

(二) 相空间的泛函积分

为把轉換矩陣元 F 写成泛函积分的形式，我們把时間間隔 (t', t'') 分为 $n+1$ 等分，

定义

$$t''-t'=(n+1)\epsilon$$
$$t_i-t_{i-1}=\epsilon,\ t_0=t',\ t_n=t'' \tag{2.1}$$
$$(i=0,1,2,\cdots n)$$

其中ϵ为步长，是一級小量：

$$\epsilon\delta(0)=1. \tag{2.2}$$

利用完备条件（1.15）式改写变换矩陣元为

$$F(q''t'',q't')=\langle q''|e^{-iH(\hat{P},\hat{Q})(t''-t')}|q'\rangle=\langle q''|e^{-iH(n+1)\epsilon}|q'\rangle$$
$$=\int dq_1\cdots dq_n\langle q_{n+1}|e^{-i\epsilon H}|q_n\rangle\langle q_n|e^{-i\epsilon H}|q_{n-1}\rangle\cdots\langle q_1|e^{-i\epsilon H}|q_0\rangle$$
$$=\int\prod_{i=1}^{n}dq_i\prod_{j=1}^{n+1}\langle q_j|e^{-i\epsilon H}|q_{j-1}\rangle$$

或

$$F(q''t'',q't')=\int\prod_{i=1}^{n}dq_i\prod_{j=1}^{n+1}\langle q_j|e^{-i\epsilon H}|q_{j-1}\rangle \tag{2.3}$$

其中

$$\langle q|e^{-i\epsilon H}|q'\rangle=\delta(q-q')-i\epsilon\langle q|H(\hat{P},\hat{Q})|q'\rangle \tag{2.4}$$

若要求

$$\langle q|H(\hat{P},\hat{Q})|q'\rangle=\int\frac{dp}{2\pi}e^{ip(q-q')}H\left(p,\frac{q+q'}{2}\right) \tag{2.5}$$

則得 Weyl 对应[5]

$$H(\hat{P},\hat{Q})=\int\frac{dpdq}{2\pi}\Delta(p,q)H(p,q)$$
$$H(p,q)=\int dve^{-ipv}\left\langle q+\frac{v}{2}\right|H(\hat{P},\hat{Q})\left|q-\frac{v}{2}\right\rangle \tag{2.6}$$

其中

$$\Delta(p,q)=\int dve^{ipv}\left|q+\frac{v}{2}\right\rangle\left\langle q-\frac{v}{2}\right| \tag{2.7}$$

Weyl 变換給出了經典哈氏函数与量子力学哈氏函数間的对应。例如它可以規定經典坐标与动量的乘积过渡到量子力学时的排列順序为：

$$f(p)q^n\to\frac{1}{2^n}\sum_{r=0}^{\infty}\frac{n!}{r!(n-r)!}\hat{Q}^rf(\hat{P})\hat{Q}^{n-r}. \tag{2.8}$$

利用（2.5）式可得

$$\langle q|e^{-i\epsilon H(\hat{P},\hat{Q})}|q'\rangle=\int\frac{dp}{2\pi}e^{ip(q-q')-i\epsilon H\left(p,\frac{q+q'}{2}\right)}, \tag{2.9}$$

将（2.9）式代入（2.3）式可得变换矩陣元在相空間的泛函积分为

$$F(q''t'',q't')=\int\prod_{i=1}^{n}dq_i\prod_{i=1}^{n+1}\frac{dp_i}{2\pi}e^{i\epsilon\sum_{k=1}^{n+1}\left[p_k\frac{q_k-q_{k-1}}{t_k-t_{k-1}}-H\left(p_k,\frac{q_k+q_{k-1}}{2}\right)\right]}$$

$$=\int\left[\frac{dp\,dq}{2\pi}\right]e^{i\epsilon\int_{t'}^{t''}dt\left[p\dot{q}-H(p,q)\right]}. \tag{2.10}$$

上式即通常路径积分量子化的公式，显然它不是明显协变的。为了使理論明显协变，必须对p进行积分，把哈密頓正則形式化为拉氏形式。

(三) 动 量 积 分

我們将轉換矩陣元的泛函积分改写为:

$$F(q''t'',q't')=\int\left[\frac{dp'dq}{2\pi}\right]e^{i\int_{t'}^{t''}dt\left[p'\dot{q}-H(p',q)\right]}, \tag{3.1}$$

式中p'是积分变数，不是正則动量。为此在$p=\frac{\partial L}{\partial \dot{q}}$处作泰勒展开得:

$$p'\dot{q}-H(p',q)=p\dot{q}-H(p,q)+(p'-p)\left[\dot{q}-\frac{\partial H(p,q)}{\partial p}\right]-\sum_{n=2}^{\infty}\frac{(p'-p)^n}{n!}\frac{\partial^n H(p,q)}{\partial p^n} \tag{3.2}$$

根据正則动量定义 (1.4) 、 (1.5) 式可得

$$p\dot{q}-H(p,q)=L(q,\dot{q}) \tag{3.3}$$

$$\frac{\partial H(p,q)}{\partial p}=\dot{q}$$

$$\frac{\partial^2 H(p,q)}{\partial p^2}=\frac{1}{m(q,\dot{q})}$$

以及

$$a_n(q,\dot{q})=\frac{1}{n!}\frac{\partial^n H(p,q)}{\partial p^n}=\frac{1}{n!}\left[\frac{1}{m(q,\dot{q})}\frac{\partial}{\partial \dot{q}}\right]^{n-2}\frac{1}{m(q,\dot{q})}. \tag{3.4}$$

$$(n=3,4,\cdots)$$

由此改写 (3.2) 式为

$$p'\dot{q}-H(p',q)=L(q,\dot{q})-\frac{(p'-p)^2}{2m(q,\dot{q})}-\sum_{n=3}^{\infty}(p'-p)^n a_n(q,\dot{q}), \tag{3.5}$$

将 (3.5) 式代入 (3.1) 式并作积分变換 $p''=p'-p$ 可得

$$F(q''t'',q't')=\int[dq]e^{iS_0}\int\left[\frac{dp}{2\pi}\right]e^{-i\int_{t'}^{t''}dt\left[\frac{p^2}{2m}+\sum_{n=3}^{\infty}a_n(p\ q)p^n\right]}, \tag{3.6}$$

其中S_0为原始作用量:

$$S_0=\int_{t'}^{t''}dt\,L(q,\dot{q}). \tag{3.7}$$

引进外源

$$\eta(t) \neq 0, \quad t' < t < t''$$

定义 η 的泛函：

$$F[\eta] = \int [dq]\, e^{iS_0} \iint \left[\frac{dp}{2\pi}\right] e^{-i\int_{t'}^{t''} dt \left[\frac{p^2}{2m} + \sum_{n=3}^{\infty} a_n(q,\dot{q}) p^n - \eta p\right]}$$

$$= \int [dq]\, e^{iS_0 - i\int_{t'}^{t''} dt \sum_{n=3}^{\infty} a_n(p,\dot{q}) \left(\frac{\delta}{i\delta\eta}\right)^n} \iint \left[\frac{dp}{2\pi}\right] e^{-i\int_{t'}^{t''} dt \left[\frac{p^2}{2m} - \eta p\right]} \tag{3.8}$$

它的零点值給出轉換矩陣元

$$F(q''t'', q't') = F[\eta]\,|_{\eta\to 0} \tag{3.9}$$

利用积分公式

$$\int_{-\infty}^{\infty} dx\, e^{i(ax^2+bx)} = \sqrt{\frac{i\pi}{a}}\, e^{\frac{b^2}{4ai}}, \tag{3.10}$$

可以求出泛函积分

$$\iint \left[\frac{dp}{2\pi}\right] e^{-i\int_{t'}^{t''} dt \left[\frac{p^2}{2m} - \eta p\right]} = \left(\frac{1}{\sqrt{2\pi i\epsilon}}\right)^{n+1} e^{i\int_{t'}^{t''} dt \left[\frac{1}{2} m\eta^2 - \frac{i}{2}\delta(0) \ln m\right]} \tag{3.11}$$

将 (3.11) 式代入 (3.8) 式得

$$F[\eta] = \frac{1}{\sqrt{2\pi i\epsilon}} \iint \left[\frac{dq}{\sqrt{2\pi i\epsilon}}\right] e^{iS_0 + iS_1 - i\int_{t'}^{t''} dt \sum_{n=3}^{\infty} a_n(q,\dot{q}) \left(\frac{\delta}{i\delta\eta}\right)^n} \cdot e^{i\int_{t'}^{t''} dt \frac{1}{2} m\eta^2} \tag{3.12}$$

其中 S_1 是李、楊首先給出的作用量：

$$S_1 = -\frac{i}{2}\delta(0) \int_{t'}^{t''} dt \ln m(q, \dot{q}) \tag{3.13}$$

它比例于 $\delta(0)$.

(四) 泛 函 微 分

以上我們进行了对 p 的积分，但出現了一系列泛函微分。定义玻色泛函微分

$$\left[\frac{\delta}{\delta\eta(t)},\ \eta(t')\right] = \delta(t - t') \tag{4.1}$$

或

$$\frac{\delta\eta(t')}{\delta\eta(t)} = \delta(t - t')$$

则有泛函运算

$$\frac{\delta}{i\delta\eta(t)} e^{i\int_{t'}^{t''} dt \frac{1}{2} m(q,\dot{q})\eta^2(t)} = e^{i\int_{t'}^{t''} dt \frac{1}{2} m(q,\dot{q})\eta^2(t)} \left[m(q,\dot{q})\eta(t) + \frac{\delta}{i\delta\eta(t)}\right] \tag{4.2}$$

由此改写 (3.12) 式为

$$F[\eta]=\frac{1}{\sqrt{2\pi i\epsilon}}\int\left[\frac{dq}{\sqrt{2\pi i\epsilon}}\right]e^{iS_0+iS_1+i\int_{t'}^{t''}dt\frac{1}{2}m\eta^2}$$

$$e^{-i\int_{t'}^{t''}dt\sum_{n=3}^{\infty}a_n(q,\dot{q})\left[m(q,\dot{q})\eta(t)+\frac{\delta}{i\delta\eta(t)}\right]^n}\cdot 1 \tag{4.3}$$

定义算符

$$D(t)=-i\sum_{n=3}^{\infty}a_n(q,\dot{q})\left[m(q,\dot{q})\eta(t)+\frac{\delta}{i\delta\eta(t)}\right]^n \tag{4.4}$$

又有

$$F[\eta]=\frac{1}{\sqrt{2\pi i\epsilon}}\int\left[\frac{dq}{\sqrt{2\pi i\epsilon}}\right]e^{iS_0+iS_1+i\int_{t'}^{t''}dt\frac{1}{2}m\eta^2}\sum_{n=0}^{\infty}\frac{1}{n!}\int_{t'}^{t''}dt_1\cdots dt_nD(t_1)\cdots D(t_n)\cdot 1 \tag{4.5}$$

算符D对1的微分定义了一系列連接函数：

$$\begin{aligned}
&D(t_1)\cdot 1=f^{(1)}(t_1) \\
&D(t_1)D(t_2)\cdot 1=f^{(1)}(t_1)f^{(1)}(t_2)+\epsilon\delta(t_1-t_2)f^{(2)}(t_1) \\
&D(t_1)D(t_2)D(t_3)\cdot 1=f^{(1)}(t_1)f^{(1)}(t_2)f^{(1)}(t_3)+f^{(1)}(t_1)f^{(2)}(t_2)\epsilon\delta(t_2-t_3) \\
&\qquad +f^{(1)}(t_2)f^{(2)}(t_1)\epsilon\delta(t_1-t_3)+f^{(1)}(t_3)f^{(2)}(t_1)\epsilon\delta(t_1-t_2) \\
&\qquad +f^{(3)}(t_1)\epsilon^2\delta(t_1-t_2)\delta(t_2-t_3) \\
&\qquad \vdots
\end{aligned} \tag{4.6}$$

$$\begin{aligned}
D(t_1)\cdots D(t_n)\cdot 1&=(-i)^n\sum_{l_1\cdots l_n=3}^{\infty}a_{l_1}\cdots a_{l_n}\left(m_1\eta_1+\frac{\delta}{i\delta\eta_1}\right)^{l_1}\cdots\left(m_n\eta_n+\frac{\delta}{i\delta\eta_n}\right)^{l_n}\cdot 1 \\
&=\sum_{\substack{k_1\cdots k_n=0\\ 1k_1+2k_2+\cdots+nk_n=n}}\sum_{P\in S_n}\frac{1}{(1!)^{k_1}k_1!(2!)^{k_2}k_2!\cdots(n!)^{k_n}k_n!} \\
&\quad\cdot P\prod_{r=1}^{\infty}\prod_{s=1}^{k_r}\delta(t_1^{rs}-t_2^{rs})\cdots\delta(t_{r-1}^{rs}-t_r^{rs})\epsilon^{r-1}f^{(r)}(t_1^{rs})
\end{aligned}$$

上式中的每一項代表有k_1个单时連接函数$f^{(1)}$，記为$s=1,2,\cdots k_1$；…有k_r个r时連接函数$f^{(r)}$，記为$s=1,2,\cdots k_r$；…有k_n个n时連接函数$f^{(n)}$. 記为$s=1,2,\cdots k_n$. 一共只有n个时間$t_1\cdots t_n$，所以

$$1k_1+2k_2+\cdots+nk_n=n. \tag{4.7}$$

当連接时間超过n，个数为零，即$k_r=0$，$r>n$. 因此（4.7）式可以扩充为

$$\sum_{r=1}^{\infty}rk_r=n. \tag{4.8}$$

上式中$t_1^{rs}\cdots t_r^{rs}$（$r=1,2,\cdots n$；$s=1,2,\cdots k_r$）是$t_1\cdots t_n$的一种排列，不同的排列由置换P完成。利用（4.6）式可得

$$e^{\int_{t'}^{t''} dt D(t)} \cdot 1 = \sum_{n=0}^{\infty} \sum_{k_1 \cdots k_n = 0}^{\infty} \delta\left(n, \sum_{r=1}^{n} r k_r\right) \prod_{r=1}^{n} \frac{1}{k_r!}\left[\frac{\epsilon^{r-1}}{r!}\int_{t'}^{t''} dt f^{(r)}(t)\right]^{k_r}$$

$$= \sum_{n=0}^{\infty} \sum_{k_1, k_2, \cdots = 0}^{\infty} \delta\left(n, \sum_{r=1}^{\infty} r k_r\right) \prod_{r=1}^{\infty} \frac{1}{k_r!}\left[\frac{\epsilon^{r-1}}{r!}\int_{t'}^{t''} dt f^{(r)}(t)\right]^{k_r}$$

$$= \sum_{k_1, k_2, \cdots = 0}^{\infty} \prod_{r=1}^{\infty} \frac{1}{k_r!}\left[\frac{\epsilon^{r-1}}{r!}\int_{t'}^{t''} dt f^{(r)}(t)\right]^{k_r}$$

$$= e^{\delta(0)\int_{t'}^{t''} dt \sum_{n=1}^{\infty} \frac{\epsilon^n}{n!} f^{(n)}(t)}$$

或

$$e^{\int_{t'}^{t''} dt D(t)} \cdot 1 = e^{\delta(0)\int_{t'}^{t''} dt \sum_{n=1}^{\infty} \frac{\epsilon^n}{n!} f^{(n)}(t)} \tag{4.9}$$

在 (4.6) 式中令 $t_1 = t_2 = \cdots = t_n = t$，則有

$$D^n(t) \cdot 1 = \frac{n!}{\epsilon^n} \sum_{k_1, k_2, \cdots = 0}^{\infty} \delta\left(n, \sum_{r=1}^{\infty} r k_r\right) \prod_{r=1}^{\infty} \frac{1}{k_r!}\left[\frac{\epsilon^r}{r!} f^{(r)}(t)\right]^{k_r} \tag{4.10}$$

对 n 求和得

$$\sum_{n=0}^{\infty} \frac{\epsilon^n}{n!} D^n(t) \cdot 1 = \sum_{n=0}^{\infty} \sum_{k_1, k_2, \cdots = 0}^{\infty} \delta\left(n, \sum_{r=1}^{\infty} r k_r\right) \prod_{r=1}^{\infty} \frac{1}{k_r!}\left[\frac{\epsilon^r}{r!} f^{(r)}(t)\right]^{k_r}$$

$$= \sum_{k_1, k_2, \cdots = 0}^{\infty} \prod_{r=1}^{\infty} \frac{1}{k!}\left[\frac{\epsilon^r}{r!} f^{(r)}(t)\right]^{k_r} = e^{\sum_{n=1}^{\infty} \frac{\epsilon^n}{n!} f^{(n)}(t)}$$

或

$$\sum_{n=0}^{\infty} \frac{\epsilon^n}{n!} D^n(t) \cdot 1 = e^{\sum_{n=1}^{\infty} \frac{\epsilon^n}{n!} f^{(n)}(t)} \tag{4.11}$$

将 (4.11) 式代入 (4.9) 式可得

$$e^{\int_{t'}^{t''} dt D(t)} \cdot 1 = e^{\delta(0)\int_{t'}^{t''} dt \ln \sum_{n=1}^{\infty} \frac{\epsilon^n}{n!} D^n(t) \cdot 1} \tag{4.12}$$

利用公式

$$\left(m\eta + \frac{\delta}{i\delta\eta}\right)^n \cdot 1 = \left(m\eta - \frac{i}{\epsilon}\frac{d}{d\eta}\right)^n \cdot 1 = \left(\frac{im}{2\epsilon}\right)^{\frac{n}{2}} H_n\left(\sqrt{\frac{m\epsilon}{2i}}\eta\right), \tag{4.13}$$

其中 H_n $(n=0,1,2,\cdots)$ 为厄米多項式。由此可得

$$D^n(t) \cdot 1 = (-i)^n \sum_{l_1 \cdots l_n = 3}^{\infty} a_{l_1}(q,\dot{q}) \cdots a_{l_n}(q,\dot{q})\left[m(q,\dot{q})\eta(t) + \frac{\delta}{i\delta\eta(t)}\right]^{l_1 + \cdots + l_n} \cdot 1$$

$$= (-i)^n \sum_{l_1 \cdots l_n = 3}^{\infty} a_{l_1}(q,\dot{q}) \cdots a_{l_n}(q,\dot{q})\left[\frac{im(q,\dot{q})}{2\epsilon}\right]^{\frac{l_1 + \cdots + l_n}{2}}$$

$$\cdot H_{l_1 + \cdots + l_n}\left(\sqrt{\frac{m(q,\dot{q})\epsilon}{2i}}\eta\right) \tag{4.14}$$

将 (4.12) (4.14) 式代入 (4.5) 式可得

$$F[\eta]=\frac{1}{\sqrt{2\pi i\epsilon}}\int\left[\frac{dq}{\sqrt{2\pi i\epsilon}}\right]\exp\Big\{iS_0+iS_1+i\int_{t'}^{t''}dt\frac{1}{2}m\eta^2$$

$$+\delta(0)\int_{t'}^{t''}dt\ln\sum_{n=0}^{\infty}\frac{(-i\epsilon)^n}{n!}\sum_{l_1\cdots l_n=3}^{\infty}a_{l_1}\cdots a_{l_n}\left(\frac{im}{2\epsilon}\right)^{\frac{l_1+\cdots+l_n}{2}}H_{l_1+\cdots+l_n}\left(\sqrt{\frac{m\epsilon}{2i}}\eta\right)\Big\} \tag{4.15}$$

令 $\eta\to0$ 得变换矩陣元:

$$F(q''t'',q't')=\frac{1}{\sqrt{2\pi i\epsilon}}\int\left[\frac{dq}{\sqrt{2\pi i\epsilon}}\right]e^{iS_{eff}} \tag{4.16}$$

其中 S_{eff} 是等效作用量

$$S_{eff}=S_0+S_1+S_2=\int_{t'}^{t''}dt\,L_{eff}(q,\dot q) \tag{4.17}$$

S_2 是我們給出的修正項，L_{eff} 是等效拉氏函数

$$L_{eff}(q,\dot q)=L(q,\dot q)-\frac{i}{2}\delta(0)\ln m(q,\dot q)$$

$$-i\delta(0)\ln\sum_{n=0}^{\infty}\frac{(-i\epsilon)^n}{n!}\sum_{l_1\cdots l_n=3}^{\infty}a_{l_1}(q,\dot q)\cdots a_{l_n}(q,\dot q)\left[\frac{im(q,\dot q)}{2\epsilon}\right]^{\frac{l_1+\cdots+l_n}{2}}H_{l_1+\cdots+l_n}(0) \tag{4.18}$$

这样我們就将量子力学化为經典軌道的路径积分。(4.18) 式中第一項是原始拉氏函数，第二項是李、楊給出的結果，第三項是我們給出的修正項。

不难将以上結果推广到 N 个自由度的物理系統，相应的轉換矩陣元为

$$F(q''t'',q't')=\frac{1}{(2\pi i\epsilon)^{N/2}}\int\left[\frac{d^Nq}{(2\pi i\epsilon)^{N/2}}\right]e^{iS_{eff}} \tag{4.19}$$

其中等效拉氏函数为

$$L_{eff}(q,\dot q)=L(q,\dot q)-\frac{i}{2}\delta(0)T_r\ln m(q,\dot q)$$

$$-i\delta(0)\ln\sum_{n=0}^{\infty}\frac{(-i\epsilon)^n}{n!}\sum_{l_1\cdots l_n=3}^{\infty}a^{\alpha}_{l_1}(q,\dot q)\cdots a^{\beta}_{l_n}(q,\dot q)P^{(l_1+\cdots+l_n)}_{\alpha\cdots\beta}(0) \tag{4.20}$$

$$(\alpha=\alpha_1,\cdots\alpha_{l_1},\cdots\beta=\beta_1,\cdots\beta_{l_n})$$

質量矩陣定义为

$$m_{\alpha\beta}(q,\dot q)=\frac{\partial^2L(q,\dot q)}{\partial\dot q^\alpha\partial\dot q^\beta},\quad(\alpha,\beta=1,2,\cdots N) \tag{4.21}$$

而

$$P^{(n)}_{\alpha_1\cdots\alpha_n}(0)=\frac{1+(-1)^n}{2}\left(\frac{1}{2i\epsilon}\right)^{n/2}\sum_{\sigma\in S_n}\sigma(m_{\alpha_1\alpha_2}m_{\alpha_3\alpha_4}\cdots m_{\alpha_{n-1}\alpha_n}) \tag{4.22}$$

$$a_n^{\alpha\cdots\beta\gamma\delta}(q,\dot q)=\frac{1}{n!}\frac{\partial^nH(p,q)}{\partial p_\alpha\cdots\partial p_\beta\partial p_\gamma\partial p_\delta}=\frac{1}{n!}\left(\frac{1}{m}\frac{\partial}{\partial\dot q}\right)^\alpha\cdots\left(\frac{1}{m}\frac{\partial}{\partial\dot q}\right)^\beta\left(\frac{1}{m}\right)^{\gamma\delta}. \tag{4.23}$$

以上是从非綫性拉氏函数出发得到的路径积分量子化，等效拉氏函数是完整的。这就为非綫性場論的量子化提供了新的途径。

参　考　文　献

[1] P.A.M. Dirac. Physik. Z. Sow-Jetunion, 3 (1933) 64.

[2] R. P. Feynman, *Rev. Mod. Phys*. 20 (1948) 267.
Phys. Rev. 80 (1950) 440.

[3] E.S. Abers and B.W. Lee, "Gauge Theories", *Phys. Reports* 9C (1973).

[4] T.D. Lee and C.N. Yang, *Phys. Rev*. 128 (1962) 885.

[5] M.M. Mizrahi. *Jour. Math. Phys*. 16 (1975) 2201.

第 5 卷 第 1 期 高 能 物 理 与 核 物 理 Vol. 5, No. 1
1981 年 1 月 PHYSICA ENERGIAE FORTIS ET PHYSICA NUCLEARIS Jan., 1981

关于 $SU(3)\times U(1)$ 轻子和夸克弱电统一模型的一些讨论

周 光 召　　　　高 崇 寿
（中国科学院理论物理研究所）　（北京大学）

摘　　要

本文讨论了 $SU(3)\times U(1)$ 轻子和夸克弱电统一模型中的几个问题: (i) 讨论了 Higgs 场的两种选取方案所得结果的异同以及与 Weinberg-Salam 模型的关系; (ii) 给出从 Higgs 自作用势给出理论所要求的破缺的证明，并对结果进行了讨论; (iii) 考察了当理论中放入几组费米子多重态时，轻子与夸克的质谱性质和夸克的广义 Cabibbo 混合出现的来源.

在文[1]中,我们在规范群取作 $SU(3)\times U(1)$ 的基础上,利用群代数的四种实现,提出了一个统一描写轻子和夸克弱电相互作用的模型.这个模型在两个 Higgs 场真空期待值 v_1 和 v_2 之比 $v=\left|\frac{v_2}{v_1}\right|^2$ 趋于无穷时,给出了与 Weinberg-Salam 模型[2]相同的结果.在文[1]中还指出,当 $\sin\varphi\ll 1$, $\sin^2\varphi/v\ll 1$ 时,这个模型给出的结果在低能范围内和 $\sin^2\theta_W\lesssim 1/4$ 的 Weinberg-Salam 模型的结果相近，并且与现有实验相符. 同时在进一步的精确实验或更高能量的实验中，这个模型与 Weinberg-Salam 模型的差别是可以检验的. 在本文中,我们将在文[1]的基础上,对几个问题作进一步的讨论.

一、另一种破缺方案

在文[1]中，我们引入两组 3 维表示的 Higgs 场 Φ_1 和 Φ_2，它们的 Y 量子数分别为 $Y=0$ 和 $Y=-1$，它们的真空期待值分别为

$$\langle\Phi_1\rangle=\begin{pmatrix} v_1\\ 0\\ 0\end{pmatrix},\qquad \langle\Phi_2\rangle=\begin{pmatrix} 0\\ 0\\ v_2\end{pmatrix}. \tag{1.1}$$

我们称这样破缺得到的模型为模型 I. 但是在引入两组 3 维表示 Higgs 场的作法下,还可以有另一种破缺方案,即 Φ_1 和 Φ_2 的 Y 量子数分别为 $Y=0$ 和 $Y=+1$，它们的真空期待值相应地为

本文 1979 年 9 月 4 日收到.

$$\langle \Phi_1 \rangle_0 = \begin{pmatrix} v_1 \\ 0 \\ 0 \end{pmatrix}, \qquad \langle \Phi_2 \rangle_0 = \begin{pmatrix} 0 \\ v_2 \\ 0 \end{pmatrix}, \tag{1.2}$$

所得到模型称为模型 II. 本节中我们将讨论其主要结果并与模型 I 比较（各符号含意见[1]).

利用(1.2)式,可以从 Higgs 场拉氏函数动能项

$$(D_\mu \Phi_1)^+(D_\mu \Phi_1) + (D_\mu \Phi_2)^+(D_\mu \Phi_2), \tag{1.3}$$

得到破缺后规范场的质量项为

$$\frac{1}{2} g^2 |v_1|^2 \left(W^+W^- + V^+V^- + \frac{2}{3} Z^2 \right)$$
$$+ \frac{1}{2} g^2 |v_2|^2 (W^+W^- + U^{++}U^{--}) + \frac{1}{4} |v_2|^2 \left(-\frac{g}{\sqrt{3}} Z + \sqrt{g^2 + g'^2} Z' \right)^2. \tag{1.4}$$

由此得到带电粒子质量为

$$m_V^2 = \frac{1}{2} g^2 |v_1|^2, \quad m_U^2 = \frac{1}{2} g^2 |v_2|^2, \quad m_W^2 = m_V^2 + m_U^2, \tag{1.5}$$

这个结果与模型 I 中不同. 在模型 I 中 $m_V^2 = m_W^2 + m_U^2$, V 粒子在三者中最重, 并且如果 $v > 1$, 则W粒子将是最轻的. 现在的模型 II 中,不论 v 取何值,弱奇异粒子 U 和 V 都将比W粒子轻.

Z 和 Z′ 将重新混合成 Z_1 和 Z_2

$$Z = \cos\alpha Z_1 + \sin\alpha Z_2, \quad Z' = -\sin\alpha Z_1 + \cos\alpha Z_2. \tag{1.6}$$

它们的质量为

$$m_1^2 = \frac{4}{3} m_W^2 \frac{\cos^2\alpha}{1+v} \left[1 + \frac{v}{4} \left(1 + \sqrt{3} \frac{\operatorname{tg}\alpha}{\sin\varphi} \right)^2 \right],$$
$$m_2^2 = \frac{4}{3} m_W^2 \frac{\cos^2\alpha}{1+v} \left[\operatorname{tg}^2\alpha + \frac{v}{4} \left(\operatorname{tg}\alpha - \sqrt{3} \frac{1}{\sin\varphi} \right)^2 \right]. \tag{1.7}$$

其中

$$v \equiv \left| \frac{v_2}{v_1} \right|^2, \quad \operatorname{tg} 2\alpha = \frac{2\sqrt{3}\, v \sin\varphi}{(4+v)\sin^2\varphi - 3v}, \tag{1.8}$$

和模型 I 比较它相当于 φ 变号的结果. 中性流的结构也相当于模型 I 中 φ 变号的结果.

在[1]中曾讨论,模型 I 在 $v \to \infty$ 时,

$$m_U^2,\ m_V^2,\ m_{Z_2}^2 \to \infty,\ m_{Z_1}^2 = m_W^2/\cos^2\theta_W,\ \sin^2\theta_W = \frac{1}{4}\cos^2\varphi, \tag{1.9}$$

并且有效中性流只有 Z_1 有贡献

$$J_{N_1\mu} = J_\mu^3 - \sin^2\theta_W J_\mu^{e.m.}, \tag{1.10}$$

即完全回到了 Weinberg-Salam 模型的结果. 在 $\sin\varphi \ll 1$, $\sin^2\varphi/v \ll 1$ 的情况下,模型 I 与现有实验符合,并且可以在进一步的实验中检验.

在模型 II 中, $v \to \infty$ 极限过程将要求

$$m_W^2/m_V^2, \quad m_W^2/m_{Z_1}^2 \to \infty,$$

考虑到实验上已大体能估计 m_W 的数值,这要求 m_V^2, $m_{Z_1}^2 \to 0$. 迄今实验上并未发现很

轻的稳定带电矢量粒子 $V^{\pm}$ 和中性矢量粒子 Z_1，这表明客观现实并不接近这种极限.

物理上感兴趣的极限情形应是 m_V 和 m_U 都比较大,从而不会和现有低能实验结果冲突的情形. 由于 $m_V^2+m_U^2=m_W^2$，这应属于 v 既不接近 ∞ 也不接近于 0 的情形，即 $v\sim 0(1)$. 我们着重考察 $\sin\varphi\ll 1$，$\sin^2\varphi/v\ll 1$ 的情形. V 粒子和 U 粒子质量为

$$m_V^2=\frac{1}{1+v}m_W^2,\quad m_U^2=\frac{v}{1+v}m_W^2,\tag{1.11}$$

这时由(1.8)得

$$\operatorname{tg}\alpha=-\frac{1}{\sqrt{3}}\sin\varphi\left(1+\frac{4}{3v}\sin^2\varphi+\cdots\right).\tag{1.12}$$

代入(1.7)得 Z_1 和 Z_2 质量在初级近似下为

$$m_1^2=\frac{m_W^2}{\cos^2\theta_W}\frac{1}{1+v}.\tag{1.13}$$

$$m_2^2=\frac{m_W^2}{\cos^2\theta_W}\cdot\frac{3v}{4(1+v)\sin^2\varphi}=\frac{m_W^2}{\cos^2\theta_W}\cdot\frac{3v}{4(1+v)(1-4\sin^2\theta_W)},\tag{1.14}$$

其中 θ_W 定义仍按(1.9)式为 $\sin^2\theta_W=\frac{1}{4}\cos^2\varphi$. 由(1.13)和(1.14)可以看到 Z_1 粒子质量低于 Weinberg-Salam 模型中预言的 Z 粒子质量，Z_2 粒子的质量则将很重.

Z_1 和 Z_2 相应的中性流分别为

$$\begin{aligned}J_{N_1\mu}&=J_\mu^3-\frac{1}{4}J_\mu^{e.m.}+\frac{1}{4}J_\mu^Y-\frac{\sqrt{3}}{4}\operatorname{tg}\alpha\left(\sin\varphi J_\mu^{e.m.}-\frac{1}{\sin\varphi}J_\mu^Y\right),\\J_{N_2\mu}&=\operatorname{tg}\alpha\left(J_\mu^3-\frac{1}{4}J_\mu^{e.m.}+\frac{1}{4}J_\mu^Y\right)+\frac{\sqrt{3}}{4}\left(\sin\varphi J_\mu^{e.m.}-\frac{1}{\sin\varphi}J_\mu^Y\right).\end{aligned}\tag{1.15}$$

代入上述近似后得最低级近似下

$$\begin{aligned}J_{N_1\mu}&=J_\mu^3-\sin^2\theta_W J_\mu^{e.m.}-\frac{1}{3v}(1-4\sin^2\theta_W)J_\mu^Y\\J_{N_2\mu}&=-\frac{1}{\sqrt{3}}\sqrt{1-4\sin^2\theta_W}\left[J_\mu^3-J_\mu^{e.m.}+\frac{3}{4(1-4\sin^2\theta_W)}J_\mu^Y\right]\end{aligned}\tag{1.16}$$

从(1.16)可以看出,仅从 $J_{N_1\mu}$ 来看,比 Weinberg-Salam 模型多了一项很小的 J_μ^Y 的贡献. 在低能中性流的有效相互作用中,对于 Z_1 的贡献来说由于 Z_1 质量带来的强度影响是主要的,而 $J_{N_1\mu}$ 中是否存在 J_μ^Y 项 $\left(\sin^2\theta_W<\frac{1}{4}\text{时}\right)$ 也是可以检验的一点. 尽管 Z_2 粒子质量很重,但由于 $J_{N_2\mu}$ 中包含一个很大的 J_μ^Y 项，Z_2 流对低能有效相互作用的贡献并不能忽略. 中性流有效相互作用正比于(保留到最低级近似)

$$\begin{aligned}J_{N_1\mu}J_{N_1}^\mu\frac{1}{m_1^2}+J_{N_2\mu}J_{N_2}^\mu\frac{1}{m_2^2}&=\frac{1}{m_1^2}\left(J_{N_1\mu}J_{N_1}^\mu+J_{N_2\mu}J_{N_2}^\mu\frac{4(1-4\sin^2\theta_W)}{3v}\right)\\&=\frac{1}{m_1^2}\left(J_{N_1\mu}J_{N_1}^\mu+\frac{1}{4v}J_\mu^Y J^{Y\mu}\right).\end{aligned}\tag{1.17}$$

后一项是 Z_2 的贡献,它与 Z_1 的贡献是同量级的. Z_2 的贡献虽然很大,但它是与 Z_1 的贡献分离的,并且主要表现在 J_μ^Y 流上. 也就是说主要表现在夸克与夸克间的相互作用中,由于

它将极易为强作用所掩盖，在实验上来检验还有困难.

从以上的讨论可以看出.模型 I 可以符合现有实验，并且在极限情形下可以完全回到 Weinberg 模型的结果. 模型 II 中不存在一个极限情形回到 Weinberg-Salam 模型的结果，它要求的 m_V^2, $m_U^2 < m_W^2$ 以及 $m_1^2 < m_W^2/\cos^2\theta_W$ 带来许多实验后果是容易检验的. 考虑到现有实验与 Weinberg-Salam 模型符合较好，模型 I 可能比模型 II 更接近实际情况.

二、Higgs 势的自发破缺

对称性的自发破缺是由 Higgs 自作用势引起的. 考虑到理论可重整性的要求，Higgs 自作用势应由场量的不超过四次幂的不变量的线性组合给出. 在我们的 $SU(3) \times U(1)$ 模型中引入了两个 3 维 Higgs 场 Φ_1 和 Φ_2, 它们在 $SU(3)$ 下变换性质相同，差别仅在于 $U(1)$ 变换性质不同.

一般说来由

$$\Phi = \begin{pmatrix} \phi_1 \\ \phi_2 \\ \phi_3 \end{pmatrix} \tag{2.1}$$

构成的二次不变量为 $\Phi^+\Phi = \mathrm{tr}(\Phi\Phi^+)$. 两个四次不变量是相同的，因

$$\mathrm{tr}(\Phi\Phi^+\Phi\Phi^+) = \mathrm{tr}(\Phi^+\Phi\Phi^+\Phi) = (\Phi^+\Phi)^2,$$

因此两组 Higgs 场所构成的独立不变量共有

$$(\Phi_1^+\Phi_1)^2,\ (\Phi_2^+\Phi_2)^2,\ \Phi_1^+\Phi_1\Phi_2^+\Phi_2,\ \Phi_1^+\Phi_2\Phi_2^+\Phi_1, \tag{2.2}$$

引入符号

$$X = \Phi_1^+\Phi_1,\ Y = \Phi_2^+\Phi_2,\ \sqrt{XY}z = \Phi_1^+\Phi_2,\ Z = zz^*, \tag{2.3}$$

则(2.2)式中四个四次不变量可写作

$$X^2,\ Y^2,\ XY,\ XYZ, \tag{2.4}$$

三个 3 维表示构成的不变量是完全反对称态. 由于我们只有两组 Higgs 粒子，它们组成的所有三次不变量均为零，从而不会在 Higgs 自作用势中出现.

Higgs 场自作用势可取为

$$V = -\mu_1^2X - \mu_2^2Y + a_1X^2 + a_2Y^2 + bXY + cXYZ. \tag{2.5}$$

假定各系数满足

$$\mu_1^2,\ \mu_2^2,\ a_1,\ a_2,\ c > 0,\ b = 0, \tag{2.6}$$

则可以证明当

$$X = \frac{\mu_1^2}{2a_1},\quad Y = \frac{\mu_2^2}{2a_2},\quad Z = 0. \tag{2.7}$$

时 V 取极小值. $Z = 0$ 的要求是在 V 取极小值时 Φ_1 与 Φ_2 正交，我们可从通过规定 Φ_1 与 Φ_2 的破缺分量不同来保证这点.

为确定起见，在模型 I 下来讨论. 可取

$$\langle\Phi_1\rangle_0 = \begin{pmatrix} v_1 \\ 0 \\ 0 \end{pmatrix},\quad v_1 = \sqrt{\frac{\mu_1^2}{2a_1}};\quad \langle\Phi_2\rangle_0 = \begin{pmatrix} 0 \\ 0 \\ v_2 \end{pmatrix},\quad v_2 = \sqrt{\frac{\mu_2^2}{2a_2}}. \tag{2.8}$$

$SU(3)\times U(1)$ 共有 9 个规范场，破缺后除光子外要求其余 8 个规范粒子都获得质量，这样就需要从 Higgs 场中得到 8 个分量补偿，亦即可以通过规范变换将 Higgs 场的 8 个分量变换掉，对 Φ_1 和 Φ_2 将共剩下 4 个分量.

在真空期待值附近作无穷小变换，对 Φ_1 和 Φ_2 分别有

$$\Phi_1' = \Phi_1 + \frac{i}{2}v_1\begin{pmatrix}\xi^3 + \frac{\xi^8}{\sqrt{3}}\\ \xi^1 + i\xi^2\\ \xi^4 + i\xi^5\end{pmatrix},\quad \Phi_2' = \Phi_2 + \frac{i}{2}v_2\begin{pmatrix}\xi^4 - i\xi^5\\ \xi^6 - i\xi^7\\ -\frac{2}{\sqrt{3}}\xi^8 - 2\theta\end{pmatrix}.$$

这样我们可以选择适当的变换参量把 Φ_1 和 Φ_2 的第二分量全部消掉，把 Φ_1 的第一分量和 Φ_2 的第三分量中的虚部消掉而只保留实部. Φ_1 的第三分量与 Φ_2 的第一分量两者中可消掉一个，我们取消掉前者. 这样 Higgs 场可表为

$$\Phi_1 = \begin{pmatrix}v_1 + \frac{1}{\sqrt{2}}\chi^0\\ 0\\ 0\end{pmatrix},\quad \Phi_2 = \begin{pmatrix}\phi^-\\ 0\\ v_2 + \frac{1}{\sqrt{2}}\phi^0\end{pmatrix} \tag{2.9}$$

其中 χ^0 和 ϕ^0 为实场，ϕ^- 为复场.

将(2.9)式代入(2.5)式，给出质量项为

$$\mu_1^2\chi^{02} + \mu_2^2\phi^{02} + \frac{c}{2a}\mu_1^2\phi^+\phi^-, \tag{2.10}$$

即各 Higgs 粒子质量为

$$m^2_{\chi^0} = 2\mu_1^2,\quad m^2_{\phi^0} = 2\mu_2^2,\quad m^2_{\phi^\pm} = \frac{c}{2a}\mu_1^2. \tag{2.11}$$

显然这些质量平方项都是大于零的.

如果在 V 中 $b \neq 0$，可类似地讨论，这时破缺后将有 $\chi^0\phi^0$ 二次项，这表明 χ^0 与 ϕ^0 将再混合.

在模型 II 下也可同样讨论，得到相似的结果.

在[3]中讨论 $SU(3)$ 规范群基础上的轻子弱电统一模型时曾指出，由于当时 Higgs 场取作 6 维表示，在 Higgs 自作用势 ϕ 就允许有三次项存在，这就会导致弱奇异数守恒遭到破坏. 在现在的模型中，由于 Higgs 场取作 3 维表示，自作用势 ϕ 不出现三次不变量，弱奇异数守恒总是严格保持的.

三、费米子质量谱和夸克的 Cabibbo 混合

现在实验上已发现 6 种味的轻子和 5 种味的夸克，并且第六种味的 t 夸克也很可能是存在的. 轻子和夸克有很好的对称性质，按它们的弱电作用，可以分别分成对应的三组，这三组的弱电作用性质是相同的. 其中出现的一个重要现象是夸克的 Cabibbo混合，即按夸克流的对称性质分类给出的态不是质量的本征态. 我们曾指出[1,3]，在我们所讨论的模型中，可以容纳几组轻子和夸克表示，现在来讨论当引入几组费米子时的 Cabibbo 混合问题.

令三组费米子三重态分别用 ψ_1, ψ_2, ψ_3 表示，如果这些费米子是夸克，对应的三个单态用 R_1, R_2, R_3 表示. Higgs 场 $\Phi = \Phi_1^{(3)}$ 与费米场的相互作用最普遍表达式为

$$\frac{1}{2}[f_{ij}\bar{\psi}_i\Phi(1+\gamma_5)\psi_j + f_{ij}^*\bar{\psi}_j\Phi^+(1-\gamma_5)\psi_i] + \frac{1}{2}[f'_{ij}\bar{R}_i\Phi^+(1+\gamma_5)\psi_j + f'^*_{ij}\bar{\psi}_j\Phi(1-\gamma_5)R_i]. \tag{3.1}$$

注意在上式第一行和第二行中 Φ 分别取两种互相等价的表示方法[1,3]

$$\Phi = -\frac{i}{\sqrt{2}}\begin{pmatrix} 0 & \phi_3 & -\phi_2 \\ -\phi_3 & 0 & \phi_1 \\ \phi_2 & -\phi_1 & 0 \end{pmatrix} \text{ 和 } \Phi = \begin{pmatrix} \phi_1 \\ \phi_2 \\ \phi_3 \end{pmatrix}, \tag{3.2}$$

引入真空期待值 $\langle\phi_1\rangle_0 = v$, $\langle\phi_2\rangle_0 = \langle\phi_3\rangle_0 = 0$, 代入(3.1)式,并引入费米子场量符号

$$\psi_i = \begin{pmatrix} a_{Li} \\ b_{Li} \\ b_{Ri} \end{pmatrix}, \quad R_i = a_{Ri} \tag{3.3}$$

得到

$$\frac{i}{\sqrt{2}}[f_{ij}v\bar{b}_{R_i}b_{L_j} - f_{ij}^*v^*\bar{b}_{L_j}b_{R_i}] + [f'_{ij}v\bar{a}_{R_i}a_{L_j} + f'^*_{ij}v^*\bar{a}_{L_j}a_{R_i}]. \tag{3.4}$$

要得到质量矩阵要求在规定的相角下有

$$m_{ij} = \frac{i}{\sqrt{2}}f_{ij}v = -\frac{i}{\sqrt{2}}f_{ji}^*v^* = m_{ji}^*, \quad m'_{ij} = f'_{ij}v = f'^*_{ji}v^* = m'^*_{ji}, \tag{3.5}$$

即质量项可通过两个厄米矩阵 m_{ij} 和 m'_{ij} 给出为

$$m_{ij}\bar{b}_ib_j + m'_{ij}\bar{a}_ia_j. \tag{3.6}$$

对于轻子的情形，如果中微子的质量为零，只有第一项存在，这时 b_i 为三种带电轻子．我们可以引入一个么正变换将它对角化,并将对角化后得到的态称为 e, μ, τ, 则 (3.6) 式变为

$$m_e\bar{e}e + m_\mu\bar{\mu}\mu + m_\tau\bar{\tau}\tau. \tag{3.7}$$

由于中微子的质量矩阵 $m'_{ij} = 0$, 它们跟随带电轻子同时作么正转动总是仍保持质量为零．轻子与规范场的相互作用和轻子的动能项

$$\bar{\psi}_i\gamma^\mu(\partial_\mu + \hat{A}_\mu^{(5)})\psi_i, \tag{3.8}$$

在这样的三组轻子间的么正转动下显然是不变的．因此对于轻子来说，有几组轻子存在表现的后果是: (i) 存在几种质量不同的带电轻子; (ii) 与每一种带电轻子对应在同一多重态中存在一个相应的中微子,中微子的质量都为零; (iii) 不同多重态的轻子间可引进不同的轻子数来区分它们,这几种轻子数分别都是守恒的,不同表示间无类似于夸克的 Cabibbo 混合．

我们导出的这些结果是在假定中微子质量为零的条件下得到的．若实验上发现中微子有质量,则轻子和夸克一样应当有 Cabibbo 混合．

对于夸克的情形,(3.6)式中两项都存在, a_i 为 $Q=2/3$ 的三种夸克, b_i 为 $Q = -1/3$ 的三种夸克．一般说来我们通过三个表示间的么正变换只能保证把其中一个质量矩阵对

角化．若我们把 m'_{ij} 对角化，并把对角化后得到的三个 $Q=2/3$ 的 a_i 态分别称为 u, c, t，则

$$m'_{ij}\bar{a}_i a_j = m_u \bar{u}u + m_c \bar{c}c + m_t \bar{t}t. \tag{3.9}$$

经这样么正转动后与 u, c, t 相对应的 $Q=-1/3$ 态分别称为 d', s', b'，但它们的质量矩阵一般说来并未对角化，$m_{ij} \to \tilde{m}_{ij}$．在这样的么正变换下夸克的拉氏函数动能项和与规范场的相互作用项显然是不变的．在这样的基之下要将 $\tilde{m}_{ij}$ 再对角化，需要再作一个么正转动．令 $\tilde{m}_{ij}$ 对角化后的本征态为 d, s, t，则 d', s', b' 与 d, s, t 之间可通过一个三维么正转动联系起来．将 d', s', b' 通过 d, s, t 的么正变换写出来，再去掉相角的任意性，就给出了熟知的广义 Cabibbo 混合的表达式[4]．因此对于夸克来说，上述讨论给出：(i) $Q=2/3$ 和 $Q=-1/3$ 的夸克都通过 Higgs 机制获得质量；(ii) 一般说来，$Q=2/3$ 和 $Q=-1/3$ 的夸克质量矩阵不能同时对角化，它导致 Cabibbo 混合；(iii) Cabibbo 混合可以用一个三维么正矩阵来描写．

显然上述讨论完全可以推广到有更多组轻子和夸克的情形．

附录 I　夸克质量与广义 Cabibbo 混合角之间关系的讨论

最近许多作者[5]讨论了如何通过夸克质量计算 Cabibbo 角的问题，Fritzsch[5] 在 $SU(2)_L\times SU(2)_R\times U(1)$ 模型中讨论了 6 种味夸克的混合问题，给出了混合角与夸克质量之间的关系．在本节中，我们简单地讨讨在 $SU(3)\times U(1)$ 模型中，相应的结果如何给出．

现有实验表明，很可能至少存在 6 种味的夸克．在 $SU(3)\times U(1)$ 模型中，相应于至少存在 3 组夸克多重态

$$\psi_1 = \begin{pmatrix} u_L \\ d_L \\ d_R \end{pmatrix}_0, \quad \psi_2 = \begin{pmatrix} c_L \\ s_L \\ s_R \end{pmatrix}_0, \quad \psi_3 = \begin{pmatrix} t_L \\ b_L \\ b_R \end{pmatrix}_0 \tag{I.1}$$

$$R_1 = u_{R0}, \qquad R_2 = c_{R0}, \qquad R_3 = t_{R0}$$

其中脚标 0 表示未混合前的态．

在 $SU(3)\times U(1)$ 对称性中，不能区分这三组夸克．因此为了区分它们，需要引入进一步的附加的对称性．引进的附加对称性可以有两类：连续的整体对称性或分立的整体对称性．在连续对称性下往往允许无穷多组不简并的夸克多重态，在分立对称性下则只存在有限组不简并的夸克多重态．我们着重考虑后一种情形．类似于在 $SU(2)_L\times SU(2)_R\times U(1)$ 模型中的讨论[5]，假定给出夸克质量的 Higgs 场有两组 $\Phi^{(1)}$ 和 $\Phi^{(2)}$，要求在 R 变换

$$\begin{aligned} &\psi_1 \to i\gamma_5\psi_1,\ \psi_2 \to -i\gamma_5\psi_2,\ \psi_3\to\psi_3,\ R_1\to i\gamma_5 R_1,\ R_2\to -i\gamma_5 R_2, \\ &R_3\to R_3,\ \Phi^{(1)}\to\Phi^{(1)},\ \Phi^{(2)}\to \pm i\Phi^{(2)},\ A^i_\mu \to A^i_\mu \quad (\Phi^{(2)}\ \text{列矢量形式用负号}) \end{aligned} \tag{I.2}$$

之下，拉氏量不变．不难看出，拉氏量[1]中除费米子与 Higgs 场的相互作用项之外，其它的项都是显然不变的．夸克与 Higgs 场相互作用拉氏量可表为

$$\begin{aligned} &f_{ij}\bar{\psi}_i\Phi^{(1)}P_+\psi_j + f^*_{ij}\bar{\psi}_j\Phi^{(1)+}P_-\psi_i + f'_{ij}\bar{R}_i\Phi^{(1)+}P_+\psi_j + f'^*_{ij}\bar{\psi}_j\Phi^{(1)}P_-R_i \\ &+ h_{ij}\bar{\psi}_i\Phi^{(2)}P_+\psi_j + h^*_{ij}\bar{\psi}_j\Phi^{(2)+}P_-\psi_i + h'_{ij}\bar{R}_i\Phi^{(2)+}P_+\psi_j + h'^*_{ij}\bar{\psi}_j\Phi^{(2)}P_-R_i \end{aligned} \tag{I.3}$$

假定破缺后得到的质量项满足宇称守恒的要求，则有

$$f_{ij}v_1,\ f'_{ij}v_1^*,\ h_{ij}v_2,\ h'_{ij}v_2^*\ \text{厄米} \tag{I.4}$$

满足 R 变换不变和 (I.4) 要求给出

$$f_{11}=f_{22}=f_{13}=f_{31}=f_{23}=f_{32}=0,\ f'_{11}=f'_{22}=f'_{13}=f'_{31}=f'_{23}=f'_{32}=0,$$

$$h_{11}=h_{22}=h_{33}=h_{12}=h_{21}=h_{13}=h_{31}=0,\quad h'_{11}=h'_{22}=h'_{33}=h'_{12}=h'_{21}=h'_{13}=h'_{31}=0. \tag{1.5}$$

亦即质量矩阵(3.5)可表为如下形式

$$m=\begin{pmatrix}0 & a & 0\\ a^* & 0 & b\\ 0 & b^* & \mu\end{pmatrix},\quad m'=\begin{pmatrix}0 & a' & 0\\ a'^* & 0 & b'\\ 0 & b'^* & \mu'\end{pmatrix}. \tag{1.6}$$

其中 μ,μ' 为实数,这正是 Fritzsch[5] 在 $SU(2)_L\times SU(2)_R\times U(1)$ 模型中给出的结果. 对这个质量矩阵所给出的夸克质量与广义 Cabibbo 角的关系, Fritzsch 文[5]中已作了具体的讨论,我们就不具体写出了.

如果存在四组夸克多重态. 按照(1.2)式的定义推广,可以规定第4组夸克的 R 变换性质为

$$\psi_4\to-\psi_4,\quad R_4\to-R_4. \tag{1.7}$$

这样 R 变换不变给出质量矩阵为

$$m=\begin{pmatrix}0 & a & 0 & c\\ a^* & 0 & b & 0\\ 0 & b^* & \mu & 0\\ c^* & 0 & 0 & \lambda\end{pmatrix},\quad m'=\begin{pmatrix}0 & a' & 0 & c'\\ a'^* & 0 & b' & 0\\ 0 & b'^* & \mu' & 0\\ c'^* & 0 & 0 & \lambda'\end{pmatrix}, \tag{1.8}$$

其中 $\mu,\ \mu',\ \lambda,\ \lambda'$ 为实数.

参 考 文 献

[1] 周光召、高崇寿,中国科学,1980年第3期第233页.
[2] S. Weinberg, *Phys. Rev. Lett.*, **19**(1967), 1264; A. Salam, Elementary Particle Theory, ed. N. Svartholm (Stockholm, 1968).
[3] 周光召、高崇寿,高能物理与核物理,**4**(1980),609.
[4] M. Kobayashi and K. Maskawa, *Progr. Theor. Phys.*, **49** (1973), 652; A. Pais and J. Primack, *Phys. Rev.*, **D8** (1973), 3063, L. Maiani, *Phys. Lett.*, **68B** (1976), 183; S. Pakvasa and H. Sugawara, *Phys. Rev.* **D14** (1976), 305.
[5] H. Fritzsch, *Phys. Lett.*, **70B** (1977), 436; **73B** (1978), 317.

SOME DISCUSSIONS ON THE $SU(3)\times U(1)$ ELECTRO-WEAK UNIFIED MODEL OF LEPTONS AND QUARKS

ZHOU GUANG-ZHAO (CHOU KUANG-CHAO)
(*Institute of Theoretical Physics, Academia Sinica*)
GAO CHONG-SHOU
(*Peking University*)

ABSTRACT

In this paper we discussed the following problems in the $SU(3)\times U(1)$ unified model proposed earlier: (i) Two possible choices of the Higgs fields and their comparison with the Weinberg-Salam model; (ii) The form of the Higgs self potential and the realization of the spontaneous symmetry breaking; (iii) The relation between the mass spectrum and the generalized Cabibbo mixing angles in a model with several generations of fermions.

第8卷 第2期 高能物理与核物理 Vol. 8, No. 2
1984年3月 PHYSICA ENERGIAE FORTIS ET PHYSICA NUCLEARIS Mar., 1984

关于 Wess-Zumino-Witten 有效作用量的规范不变性

周光召 郭汉英 吴可
(中国科学院理论物理研究所)

宋行长
(北京大学)

摘要

本文指出，为了引入规范不变的 Wess-Zumino-Witten 有效作用量，$SU(3)_L\times SU(3)_R$ 的规范子群必须满足一个大范围的无反常条件．这个条件要求规范子群的左手与右手的 Chern-Simons 与形式必须彼此相等，在局部意义上，这个条件等价于通常微扰论意义下的无反常条件．本文运用一种系统的方法导出这个条件并构造规范不变的有效作用量．对于非阿贝尔规范群，这里给出的有效作用量所含的项比 Witten 的要少．本文还讨论了纯规范的情形．

最近，Witten[1] 提出了一种新的表述以讨论 Wess-Zumino[2] 手征有效作用量的大范围性质，得到了一些颇有兴趣的结果．

本文将进而指出，为了规范作用量的整体对称性 $SU(3)_L\times SU(3)_R$ 的任一子群 $H\subseteq SU(3)_L\times SU(3)_R$，规范子群 H 必须满足一大范围的无反常条件．这个条件要求群 H 的左手和右手 Chern-Simons 5 形式必须彼此相等，向在局部的意义上，该条件等价于夸克水平上微扰论计算得到的无反常条件．代替 Witten 所采用的试探方法，本文提出了一种系统的方法来求规范作用量的对称性．并导出上述无反常条件．对于非阿贝尔情形，我们得到的规范不变的有效作用量所包含的项比 Witten 的少．虽然这两种有效作用的差别仅涉及一些极罕见的过程，然而在原则上找出区别这两种有效作用的方式将是有意义的．本文还讨论了纯规范的情形，显然，这与有效作用理论的真空性质有密切的关系．

Witten 在 Wess-Zumino 工作的基础上，建议引入如下的有效作用来描述强作用的低能极限[1)]:

$$I(U)=-\frac{F_\pi^2}{16}\int d^4x\,\mathrm{Tr}\,\partial_\mu U\partial_\mu U^{-1}+n\Gamma(U) \tag{1}$$

$$\Gamma(U):=+\frac{1}{240\pi^2}\int_Q d\Sigma^{ijklm}\mathrm{Tr}(U^{-1}\partial_i UU^{-1}\partial_j UU^{-1}\partial_k UU^{-1}\partial_l UU^{-1}\partial_m U) \tag{2}$$

本文 1983 年 8 月 12 日收到．

1) 为方便起见，我们将尽量采用 Witten[1] 的记号和约定．

其中 $F_\pi \cong 190$ MeV，$U(x)$ 是 $SU(3)$ 的元素，其无穷小形式描述 $SU(3)_L \times SU(3)_R$ 对称自发破缺到对角 $SU(3)$ 的 Goldstone 玻色场．n 是一拓扑量子数，Q 是 5 维盘，紧致化的欧氏 4-时空 $M \sim S^4$ 是 Q 的边界．

已知在规范群 $H_L \times H_R \subseteq SU(3)_L \times SU(3)_R$ 的作用下，场量 $U(t)$ 的变换性质为

$$U(x) \to U'(x) = L(x)U(x)R^{-1}(x), \quad L(R) \in H_{L(R)} \tag{3}$$

问题是在什么物理条件下可以找到与 $I(U)$ 相应的规范不变的作用量，同时如何去寻找这种作用量．显然，规范不变的有效作用量应有如下形式

$$\tilde{I}(A_\mu, U) = -\frac{F_\pi^2}{16}\int d^4x \mathrm{Tr} D_\mu U D_\mu U^{-1} + n\tilde{\Gamma}(A_\mu, U) \tag{4}$$

其中 $D_\mu U$ 是 U 的规范协变导数

$$D_\mu U = \partial_\mu U + A_{\mu L} U - U A_{\mu R}, \quad A_{\mu L(R)} = A^a_{\mu L(R)} T^a_{L(R)}. \tag{5}$$

$\tilde{\Gamma}$ 包含 $\Gamma(U)$ 项及其它最少量的、定义在 $M^4 \sim S^4$ 上的项，$\tilde{\Gamma}$ 本身应是规范不变的．这样，问题就转化为在什么物理条件下可以找到，以及如何寻找规范不变的 $\tilde{\Gamma}$ 项．

我们得到的结果是：

在规范群的左、右手 Chern-Simons 5 形式 π^L 与 π^R 相等

$$\pi^L = \pi^R \tag{6}$$

的充要条件下，表达式

$$\tilde{\Gamma}(A_\mu, U) = \Gamma(U) + \frac{1}{48\pi^3}\int d^4x \varepsilon^{\mu\nu\alpha\beta} \mathrm{W}_{\mu\nu\alpha\beta} \tag{7}$$

是在规范变换 (3) 下不变的，因而由公式 (4) 定义的 $\tilde{I}$ 是规范不变的有效作用量．这里，Chern-Simons 5 形式的系数定义为

$$\pi_{ijklm} = -\frac{1}{24\pi^2}\mathrm{Tr}\left(F_{ij}F_{kl}A_m - \frac{1}{2}F_{ij}A_kA_lA_m + \frac{1}{10}A_iA_jA_kA_lA_m\right) \tag{8}$$

A_i 是定义在 5 维盘 Q 上的规范势，F_{ij} 是相应的规范场强；而且

$$\begin{aligned}
\mathrm{W}_{\mu\nu\alpha\beta} = \mathrm{Tr}\Big\{&[(-A_{\mu L}U_{\nu L}U_{\alpha L}U_{\beta L} + \partial_\mu A_{\nu L}A_{\alpha L}U_{\beta L} + A_{\mu L}\partial_\nu A_{\alpha L}U_{\beta L}) + (L \to R)] \\
&+ \partial_\mu A_{\nu L} U A_{\alpha R} U^{-1} U_{\beta L} + U \partial_\mu A_{\nu R} U^{-1} A_{\alpha L} U_{\beta L} \\
&- \frac{1}{2}[A_{\mu L}U_{\nu L}A_{\alpha L}U_{\beta L} - (L \to R)] \\
&+ A_{\mu L} U A_{\nu R} U^{-1} U_{\alpha L} U_{\beta L} - U A_{\mu R} U^{-1} A_{\nu L} U_{\alpha L} U_{\beta L} \\
&- A_{\mu L}\partial_\nu A_{\alpha L} U A_{\beta R} U^{-1} - \partial_\mu A_{\nu L} A_{\alpha L} U A_{\beta R} U^{-1} + A_{\mu R}\partial_\nu A_{\alpha R} U^{-1} A_{\beta L} U \\
&+ \partial_\mu A_{\nu R} A_{\alpha R} U^{-1} A_{\beta L} U + A_{\mu L} U A_{\nu R} U^{-1} A_{\alpha L} U_{\beta L} + A_{\mu R} U^{-1} A_{\nu L} U A_{\alpha R} U_{\beta R} \\
&+ [A_{\mu L}A_{\nu L}A_{\alpha L}U_{\beta L} + (L \to R)] - A_{\mu L}A_{\nu L}A_{\alpha L} U A_{\beta R} U^{-1} \\
&+ A_{\mu R}A_{\nu R}A_{\alpha R} U^{-1} A_{\beta L} U - \frac{1}{2} A_{\mu L} U A_{\nu R} U^{-1} A_{\alpha L} U A_{\beta R} U^{-1}\Big\}
\end{aligned} \tag{9}$$

$U_{\mu L(R)}$ 定义为

$$U_{\nu L} = \partial_\nu U \cdot U^{-1}, \quad U_{\nu R} = U^{-1}\partial_\nu U.$$

如果条件 (6) 不能得到满足，那么在规范变换 (3) 下 $\tilde{\Gamma}$ 的改变量便成为

$$
\begin{aligned}
\Delta\tilde{\Gamma} &= \Gamma(L) - \Gamma(R) \\
&- \frac{1}{24\pi^2}\int d^4x\varepsilon^{\mu\nu\alpha\beta}\,\mathrm{Tr}\left\{A_{\mu L}\partial_\nu A_{\alpha L}L^{-1}\partial_\beta L + \frac{1}{2}A_{\mu L}A_{\nu L}A_{\alpha L}L^{-1}\partial_\beta L\right. \\
&+ \frac{1}{2}A_{\mu L}A_{\nu L}L^{-1}\partial_\alpha L L^{-1}\partial_\beta L + \frac{1}{4}A_{\mu L}L^{-1}\partial\nu L A_{\alpha L}L^{-1}\partial_\beta L \\
&\left.- \frac{1}{2}A_{\mu L}L^{-1}\partial_\nu L L^{-1}\partial_\alpha L L^{-1}\partial_\beta L - (L\to R)\right\}
\end{aligned}
\tag{10}
$$

这里 $\Gamma(L)$ 与 $\Gamma(R)$ 的定义和 $\Gamma(U)$ 的定义 (2) 一样，只要分别将 U 代之以 L 与 R. 如果规范变换 (3) 取无穷小形式

$$
L = 1 + \varepsilon_L,\quad R = 1 + \varepsilon_R,\quad \varepsilon_{L(R)} = \varepsilon^a_{L(R)}T^a_{L(R)}
\tag{11}
$$

改变量 $\Delta\tilde{\Gamma}$ 也变为无穷小形式

$$
\delta\tilde{\Gamma} = \frac{1}{24\pi^2}\int d^4x\varepsilon^{\mu\nu\alpha\beta}\,\mathrm{Tr}\left\{\varepsilon_L\left(\partial_\mu A_{\nu L}\partial_\alpha A_{\beta L} + \frac{1}{2}\partial_\mu(A_{\nu L}A_{\alpha L}A_{\beta L})\right) - (L\to R)\right\}
\tag{12}
$$

这就是 Witten 的结果，并与在夸克水平上无穷小规范变换下有效作用的反常改变量的微扰论计算的结果一致.

这个一致性表明，条件 (6) 事实上是通常微扰论意义下无反常条件的大范围推广. 因此称为大范围的无反常条件.

对于规范群 $H = U(1)$，即电磁规范群的情形，有效作用量 $\tilde{\Gamma}$ 的表达式 (4) 以及表达式 (7) 与 (9) 自动给出与 Witten 相同的结果，即我们有

$$
\begin{aligned}
\tilde{\Gamma}^{em}(A_\mu, U) &= \Gamma(U) + \frac{1}{48\pi^2}\int d^4x\varepsilon^{\mu\nu\alpha\beta}\,\mathrm{Tr}\{-A_\mu\hat{Q}(U_{\nu L}U_{\alpha L}U_{\beta L} + U_{\nu R}U_{\alpha R}U_{\beta R}) \\
&+ 2F_{\mu\nu}A_\alpha(\hat{Q}^2U_{\beta L} + \hat{Q}^2U_{\beta R} + \hat{Q}U\hat{Q}U^{-1}U_{\beta L})\}
\end{aligned}
\tag{13}
$$

然而，对于非阿贝尔的情形，表达式 (9) 与 Witten 相应的表达式相比包含的项要少，因而这里给出的有效作用 $\tilde{\Gamma}$ 所含的项也就比 Witten 所给出的要少. 不过，二者之差在规范变换 (3) 下是不变的[1)].

对于纯规范的情形，无反常条件 (6) 变为

$$
\Gamma(L) = \Gamma(R)
\tag{14}
$$

表达式 (7) 与 (9) 变为

$$
\tilde{\Gamma}^{(0)}(A^{(0)}_\mu, U) = \Gamma(U) + \frac{1}{48\pi^2}\int d^4x\varepsilon^{\mu\nu\alpha\beta}W^{(0)}_{\mu\nu\alpha\beta}
\tag{15}
$$

$$
\begin{aligned}
W^{(0)}_{\mu\nu\alpha\beta} = \mathrm{Tr}\{&-[A^{(0)}_{\mu L}U_{\nu L}U_{\alpha L}U_{\beta L} + (L\to R)] - [A^{(0)}_{\mu L}U_{\nu L}A^{(0)}_{\alpha L}U_{\beta L} - (L\to R)] \\
&+ A^{(0)}_{\mu L}UA^{(0)}_{\nu R}U^{-1}U_{\alpha L}U_{\beta L} - UA^{(0)}_{\mu R}U^{-1}A^{(0)}_{\nu L}U_{\alpha L}U_{\beta L} \\
&+ A^{(0)}_{\mu L}UA^{(0)}_{\nu R}U^{-1}A^{(0)}_{\alpha L}U_{\beta L} + A^{(0)}_{\mu R}U^{-1}A^{(0)}_{\nu L}UA^{(0)}_{\alpha R}U_{\beta R} \\
&- [A^{(0)}_{\mu L}A^{(0)}_{\nu L}A^{(0)}_{\alpha L}U_{\beta L} + (L\to R)] - A^{(0)}_{\mu L}A^{(0)}_{\nu L}A^{(0)}_{\alpha L}UA^{(0)}_{\beta R}U^{-1} \\
&+ A^{(0)}_{\mu R}A^{(0)}_{\nu R}A^{(0)}_{\alpha R}U^{-1}A^{(0)}_{\beta L}U + \frac{1}{2}A^{(0)}_{\mu L}UA^{(0)}_{\nu R}U^{-1}A^{(0)}_{\alpha L}UA^{(0)}_{\beta R}U^{-1} \\
&+ A^{(0)}_{\mu L}U_{\nu L}UA^{(0)}_{\alpha R}A^{(0)}_{\beta R}U^{-1} + UA^{(0)}_{\mu R}U^{-1}U_{\nu L}A^{(0)}_{\alpha L}A^{(0)}_{\beta L}\}
\end{aligned}
\tag{16}
$$

---

1) 文[1]的结果有笔误，要作相应改正.

上标 (0) 表示与纯规范相应的量. 反常改变量 (10) 与 (12) 也有相应的表达式. 显然，这些结果与有效作用理论的真空性质有关. 这一点，我们留待以后进一步讨论.

现在，我们来阐明如何得到这些结果.

首先，按照从整体对称性到局域对称性的标准方法，容易写出有效作用量 (1) 的如下规范不变推广

$$\hat{I} = -\frac{F_\pi^2}{16}\int d^4x \operatorname{Tr} D_\mu U D_\mu U^{-1} + n\hat{\Gamma}(A_i, U) \tag{17}$$

$$\hat{\Gamma}(A_i, U) := +\frac{1}{240\pi^2}\int_Q d\Sigma^{ijklm} \operatorname{Tr}(U^{-1}D_iUU^{-1}D_jUU^{-1}D_kUU^{-1}D_lUU^{-1}D_mU). \tag{18}$$

这就是直接把 I 中的 U 的导数分别用 4 维时空 $M\sim S^4$ 和 5 维盘 Q 上 U 的协变导数代替而得到的. 不过，尽管 $\hat{I}$ 是明显规范不变的，但它并不能直接用来作为物理的规范不变有效作用量，这是因为 $\hat{\Gamma}$ 不能全部约化为定义在物理时空上，即 $M\sim S^4=\partial Q$ 上的量.

然而，深入的考查表明，$\hat{\Gamma}$ 项可以分解为两部分

$$\begin{aligned}\hat{\Gamma}(A_i, U) = {}& \tilde{\Gamma}(A_\mu, U) + \int_Q d\Sigma^{ijklm}(\pi^L_{ijklm} - \pi^R_{ijklm}) \\ & - \frac{1}{48\pi^2}\int_Q d\Sigma^{ijklm} \operatorname{Tr}(F^L_{ij}D_kUU^{-1}D_lUU^{-1}D_mUU^{-1} - 2F^L_{ij}F^L_{kl}D_mUU^{-1} \\ & - F^L_{ij}D_kUF^R_{lm}U^{-1} + F^R_{ij}U^{-1}D_kUU^{-1}D_lUU^{-1}D_mU - 2F^R_{ij}F^R_{kl}U^{-1}D_mU \\ & - F^R_{ij}U^{-1}D_kUU^{-1}F^L_{lm}U).\end{aligned} \tag{19}$$

对于阿贝尔规范群的情形，也有相应的表达式

$$\begin{aligned}\hat{\Gamma}^{c.m.}(A_i, U) = {}& \tilde{\Gamma}^{c.m.}(A_\mu, U) \\ & - \frac{1}{48\pi^2}\int_Q d\Sigma^{ijklm} \operatorname{Tr}\{F_{ij}\hat{Q}(D_kUU^{-1}D_lUU^{-1}D_mUU^{-1} \\ & + U^{-1}D_kUU^{-1}D_lUU^{-1}D_mU) \\ & - 2F_{ij}F_{kl}(\hat{Q}^2D_mUU^{-1} + \hat{Q}^2U^{-1}D_mU + \hat{Q}U\hat{Q}U^{-1}D_mUU^{-1})\}\end{aligned} \tag{20}$$

这里，$D_iU = \partial_iU + A_i[Q, U]$，$A_i \in u(1)$，$\hat{Q}$ 是电荷矩阵[1]. $\hat{\Gamma}$ 的第一部分由 $\tilde{\Gamma}$ 的表达式和左、右 Chern-Simons 5 形式之差的积分组成 (对于 $U(1)$ 情形，Chern-Simons 5 形式自动为零)，$\hat{\Gamma}$ 的第二部分完全由 Q 上的规范不变量组成. 于是，我们可以当且仅当左右 Chern-Simons 5 形式之差为零时，利用 $\tilde{\Gamma}$ 定义物理的规范不变有效作用量 $\hat{I}$，而前者就是必要且充分的大范围无反常条件.

一般说来，$\tilde{\Gamma}$ 的形式不是唯一的. 任意添加定义在 $M\sim S^4$ 上的规范不变项都将改变 $\tilde{\Gamma}$. 不过，我们关于 $\tilde{\Gamma}$ 的构造在一定意义下是唯一的：在对 Γ 引入最小耦合后，再从 $\hat{\Gamma}$ 中最大限度地扣去 Q 上的规范不变项. 这样得到的 $\tilde{\Gamma}$ 项所包含的项数比 Witten 的结果要少.

参 考 文 献

[1] E. Witten "Global Aspects of Current Algebra", Princeton Preprint (1983).
[2] J. Wess and B. Zumino, *Phys. Lett.*, **37B**(1971), 95.

ON THE GAUGE INVARIANCE OF WESS-ZUMINO-WITTEN EFFECTIVE ACTION

CHOU KUANG-CHAO GUO HAN-YING WU KE
(*Institute of Theoretical Physics, Academia Sinica*)

SONG XING-CHANG
(*Peking University*)

ABSTRACT

It is shown that in order to introduce the gauge invariant Wess-Zumino-Witten effective action a global anomalyfree condition should be satisfied by the gauged subgroup of $SU(3)_L \times SU(3)_R$. The condition requires that the left handed and the right handed Chern-Simons 5-forms with respect to the gauge group be equal to each other and it turns out in the local sense to be the usual perturbative anomaly-free condition. It is also constructed a gauge invariant effective action under the anomaly-free condition by means of a systematic method rather than the trial and error Noether method. In the non-abelian case, the gauge invariant effective action presented here contains less terms than the one obtained by Witten. The case of pure gauge is discussed in the present note as well.

第 8 卷 第 4 期 高能物理与核物理 Vol. 8, No. 4
1984 年 7 月 PHYSICA ENERGIAE FORTIS ET PHYSICA NUCLEARIS July, 1984

对称反常、非对称反常和有效拉氏量

周光召 郭汉英 吴 可
(中国科学院理论物理研究所)

宋 行 长
(北 京 大 学)

摘 要

本文进一步讨论作者利用 Chern-Simons 拓扑不变量建立的主手征模型的有效拉氏量，指出非阿贝尔三角反常的对称形式和非对称形式都可以由此拉氏量得到.

最近 E. Witten[1] 通过拓扑性质讨论主手征模型的规范协变的拉氏量

$$\tilde{\Gamma}=\Gamma+\frac{1}{48\pi^2}\int d^4x\varepsilon^{\mu\nu\alpha\beta}Z_{\mu\nu\alpha\beta} \tag{1}$$

其中

$$\Gamma=\frac{1}{240\pi^2}\int_a d\Sigma^{ijklm}\mathrm{T}_\gamma(U^{-1}\partial_iUU^{-1}\partial_jUU^{-1}\partial_kUU^{-1}\partial_mUU^{-1}\partial_nU) \tag{2}$$

并用试探的方法定出了 $Z_{\mu\nu\alpha\beta}$ 的具体表达式．然后，本文作者改进了 Witten 的方法和结果，充分运用了 Chern-Simons 拓扑不变量，给出了求规范协变作用量的系统方法，并且得到了与 Witten 不同的结果[2,3]，

$$\hat{\Gamma}=\Gamma+\frac{1}{48\pi^2}\int d^4x\varepsilon^{\mu\nu\alpha\beta}W_{\mu\nu\alpha\beta} \tag{3}$$

其中

$$\begin{aligned}W_{\mu\nu\alpha\beta}=\mathrm{Tr}\{&[-A_{\mu L}U_{\nu L}U_{\alpha L}U_{\beta L}+\partial_\mu A_{\nu L}A_{\alpha L}U_{\beta L}+A_{\mu L}\partial_\nu A_{\alpha L}U_{\beta L}+(L\to R)]\\&+\partial_\mu A_{\nu L}UA_{\alpha R}U^{-1}U_{\beta L}+\partial_\mu A_{\nu R}U^{-1}A_{\alpha L}UU_{\beta R}\\&-\frac{1}{2}[A_{\mu L}U_{\nu L}A_{\alpha L}U_{\beta L}-(L\to R)]+A_{\mu L}UA_{\nu R}U^{-1}U_{\alpha L}U_{\beta L}\\&-UA_{\mu R}U^{-1}A_{\nu L}U_{\alpha L}U_{\beta L}-A_{\mu L}\partial_\alpha A_{\alpha L}UA_{\beta R}U^{-1}-\partial_\mu A_{\nu L}A_{\alpha L}UA_{\beta R}U^{-1}\\&+A_{\mu R}\partial_\nu A_{\alpha R}U^{-1}A_{\beta L}U+\partial_\mu A_{\nu R}A_{\alpha R}U^{-1}A_{\beta L}U+A_{\mu L}UA_{\nu R}U^{-1}A_{\alpha L}U_{\beta L}\\&+A_{\mu R}U^{-1}A_{\nu L}UA_{\alpha R}U_{\beta R}+[A_{\mu L}A_{\nu L}A_{\alpha L}U_{\beta L}+(L\to R)]\\&-A_{\mu L}A_{\nu L}A_{\alpha L}UA_{\beta R}U^{-1}+A_{\mu R}A_{\nu R}A_{\alpha R}U^{-1}A_{\beta L}U-\frac{1}{2}A_{\mu L}UA_{\nu R}U^{-1}A_{\alpha L}UA_{\beta R}U^{-1}\}\end{aligned} \tag{4}$$

1983 年 11 月 19 日收到.

$U_{\mu L}$ 和 $U_{\mu R}$ 的定义为

$$U_{\mu L}=\partial_\mu UU^{-1},\quad U_{\mu R}=U^{-1}\partial_\mu U$$

和 Witten 结果相比，项数比他的少．但两者之差是规范协变项[1)]

$$\int d^4x\varepsilon^{\mu\nu\alpha\beta}Z_{\mu\nu\alpha\beta}-\int d^4x\varepsilon^{\mu\nu\alpha\beta}\mathrm{Tr}(F_{\mu\nu R}U^{-1}F_{\alpha\beta L}U)$$
$$=\int d^4x\varepsilon^{\mu\nu\alpha\beta}W_{\mu\nu\alpha\beta}\tag{5}$$

因此，这两个不同的拉氏量在无穷小规范变换下得到相同的非阿贝尔三角反常（下面简称反常），即

$$\delta\tilde{\Gamma}=\delta\hat{\Gamma}=\frac{1}{24\pi^2}\int d^4x\varepsilon^{\mu\nu\alpha\beta}\mathrm{Tr}\ \Big\{\varepsilon_L\Big[\partial_\mu A_{\nu L}\partial_\alpha A_{\beta L}+\frac{1}{2}\partial_\mu(A_{\nu L}A_{\alpha L}A_{\beta L})\Big]-(L\to R)\Big\}\tag{6}$$

它等价于 D. J. Gross 和 R. Jackiw 在夸克水平上用微扰论计算得到的反常[4]．

大家知道，先于 Gross 等，W. A. Bardeen 曾给过一个结果，他为了在矢量流上得到通常的 Ward 等式，引进抵削项，把反常移到轴矢流上，此时反常项表示为[5]

$$\int d^4x\varepsilon^a G^a\tag{7}$$

其中

$$G^a=\frac{1}{4\pi^2}\varepsilon^{\mu\nu\alpha\beta}\mathrm{Tr}\left\{\frac{\lambda^a}{2}\left[\frac{1}{4}V_{\mu\nu}V_{\alpha\beta}+\frac{1}{12}A_{\mu\nu}A_{\alpha\beta}+\frac{2}{3}(A_\mu A_\nu V_{\alpha\beta}+A_\mu V_{\nu\alpha}A_\beta+V_{\mu\nu}A_\alpha A_\beta)+\frac{8}{3}A_\mu A_\nu A_\alpha A_\beta\right]\right\}\tag{8}$$

$V_{\mu\nu}$ 和 $A_{\mu\nu}$ 的定义为

$$\begin{aligned}V_{\mu\nu}&=\partial_\mu V_\nu-\partial_\nu V_\mu-[V_\mu,V_\nu]-[A_\mu,A_\nu]\\A_{\mu\nu}&=\partial_\mu A_\nu-\partial_\nu A_\mu-[V_\mu A_\nu]-[A_\mu V_\nu]\end{aligned}\tag{9}$$

A_μ、V_μ 和上文 $A_{\mu L}$、$A_{\mu R}$ 的关系为

$$A_\mu=-\frac{1}{2}(A_{\mu L}-A_{\mu R})\qquad V_\mu=-\frac{1}{2}(A_{\mu L}+A_{\mu R})\tag{10}$$

于是，在讨论反常时通常有两个不同的表达形式，Gross-Jackiw 形式和 Bardeen 形式，为方便起见我们分别称之为反常的对称形式和非对称形式．

由反常方程出发得到的有效拉氏量应该包含所有通过反常而产生的手征场（Π 场和 K 场）和规范场的相互作用，在取一个 Π（或 K）的顶点时，它应和非对称的反常耦合．这个重要特征可以通过手征变换对于有效拉氏量的变更显示出来．

J. Wess 和 B. Zumino 正是基于这一点在讨论反常的相容性条件的同时，给出了 Wess-Zumino 的有效拉氏量[6]

$$W=\frac{1-\exp(-\xi\cdot U)}{\xi\cdot U}(\xi\cdot G)\tag{11}$$

1）此处所指的 Witten 表达式 $Z_{\mu\nu\alpha\beta}$ 是改正了他的明显笔误和补充了遗漏项之后得到的表达式[3]．

其中

$$\xi^a = \frac{1}{F_\Pi} \Pi^a,$$

U 是算子，它的定义可参阅[6]. 显然W可写成

$$W = \Pi^a G^a + O(\Pi^2) \tag{12}$$

G^a 就是非对称的反常项.

现在的问题是，我们得到的有效拉氏量 $\hat{\Gamma}$ 和 Witten 得到的有效拉氏量 $\tilde{\Gamma}$ 是否也具有这个性质?

为了分析 $\hat{\Gamma}$ 的性质，我们可以按照

$$U = e^{\Pi^a \lambda^a} = 1 + \Pi^a \lambda^a + \cdots \tag{13}$$

来展开 $\hat{\Gamma}$,

$$\hat{\Gamma} = \hat{\Gamma}_0 + \Pi^a \hat{\Gamma}_1^a + O(\Pi^2) \tag{14}$$

其中

$$\begin{aligned}\hat{\Gamma}_0 = \frac{1}{48\Pi^2} \int dx^4 \varepsilon^{\mu\nu\alpha\beta} \mathrm{Tr} \Big(& - A_{\mu L} \partial_\nu A_{\alpha L} A_{\beta R} - \partial_\mu A_{\nu L} A_{\alpha L} A_{\beta R} \\ & + A_{\mu R} \partial_\nu A_{\alpha R} A_{\beta L} + \partial_\mu A_{\nu R} A_{\alpha R} A_{\beta L} - A_{\alpha L} A_{\nu L} A_{\alpha L} A_{\beta R} \\ & + A_{\mu R} A_{\nu R} A_{\alpha R} A_{\beta L} - \frac{1}{2} A_{\mu L} A_{\nu R} A_{\alpha L} A_{\beta R} \Big)\end{aligned} \tag{15}$$

以及

$$\begin{aligned}\Pi^a \hat{\Gamma}_1^a = \frac{1}{48\Pi^2} \int dx^4 \varepsilon^{\mu\nu\alpha\beta} \mathrm{Tr} \{ \Pi^a \lambda^a [& 2\partial_\mu A_{\nu L} \partial_\alpha A_{\beta L} + 2\partial_\mu A_{\nu R} \partial_\alpha A_{\beta R} \\ & + \partial_\mu A_{\nu L} \partial_\alpha A_{\beta R} + \partial_\mu A_{\nu R} \partial_\alpha A_{\beta L} - \partial_\mu (A_{\nu L} A_{\alpha L} A_{\beta L}) \\ & - \partial_\mu (A_{\nu R} A_{\alpha R} A_{\beta R}) - \partial_\mu (A_{\nu L} A_{\alpha R} A_{\beta L}) - \partial_\mu (A_{\nu R} A_{\alpha L} A_{\beta R}) \\ & - A_{\mu L} \partial_\nu A_{\alpha L} A_{\beta R} - A_{\mu L} \partial_\nu A_{\alpha R} A_{\beta R} - A_{\mu R} \partial_\nu A_{\alpha L} A_{\beta L} \\ & - A_{\mu R} \partial_\nu A_{\alpha R} A_{\beta L} - \partial_\mu A_{\nu L} A_{\alpha L} A_{\beta R} - \partial_\mu A_{\nu R} A_{\alpha R} A_{\beta L} \\ & - A_{\mu R} A_{\nu L} \partial_\alpha A_{\beta L} - A_{\mu L} A_{\nu R} \partial_\alpha A_{\beta R} \\ & + A_{\mu L} A_{\nu L} A_{\alpha L} A_{\beta R} + A_{\mu R} A_{\nu L} A_{\alpha L} A_{\beta L} + A_{\mu L} A_{\nu R} A_{\alpha L} A_{\beta R} \\ & + A_{\mu R} A_{\nu L} A_{\alpha R} A_{\beta L} + A_{\mu L} A_{\nu R} A_{\alpha R} A_{\beta R} + A_{\mu R} A_{\nu R} A_{\alpha R} A_{\beta L}] \}\end{aligned} \tag{16}$$

直接计算可以证明 $\hat{\Gamma}_1^a$ 就是非对称的反常 G^a，在此我们仅给出具体计算的几个主要步骤.

将 (9) 式代入 (8) 式可得

$$\begin{aligned}\Pi^a G^a = \frac{1}{8\Pi^2} \varepsilon^{\mu\nu\alpha\beta} \mathrm{Tr} \Big\{ \Pi^a \lambda^a \Big[& \partial_\mu V_\nu \partial_\alpha V_\beta + \frac{1}{3} \partial_\mu A_\nu \partial_\alpha A_\beta \\ & - \partial_\mu V_\nu V_\alpha V_\beta - \frac{1}{3} \partial_\mu A_\nu V_\alpha A_\beta - \frac{1}{3} \partial_\mu A_\nu A_\alpha V_\beta + \frac{1}{3} \partial_\mu V_\nu A_\alpha A_\beta \\ & + \frac{4}{3} A_\mu \partial_\nu V_\alpha A_\beta + \frac{1}{3} A_\mu A_\nu \partial_\alpha V_\beta - \frac{1}{3} V_\mu A_\nu \partial_\alpha A_\beta \\ & - V_\mu V_\nu \partial_\alpha V_\beta - \frac{1}{3} A_\mu A_\nu \partial_\alpha A_\beta + V_\mu V_\nu V_\alpha V_\beta + \frac{1}{3} V_\mu A_\nu V_\alpha A_\beta\end{aligned}$$

$$+\frac{1}{3}A_\mu V_\nu A_\alpha V_\beta+\frac{1}{3}V_\mu A_\nu A_\alpha V_\beta-A_\mu V_\nu V_\alpha A_\beta$$

$$-\frac{1}{3}A_\mu A_\nu V_\alpha V_\beta-\frac{1}{3}V_\mu V_\nu A_\alpha A_\beta-\frac{1}{3}A_\mu A_\nu A_\alpha A_\beta\Big]\Big\} \tag{17}$$

再将 A_μ V_μ 和 $A_{\mu L}$, $A_{\mu R}$ 的关系式(10)代入(17)式，其结果为

$$\begin{aligned}\Pi^a G^a=\frac{1}{48\Pi^2}\varepsilon^{\mu\nu\alpha\beta}\mathrm{Tr}\{\Pi^a\lambda^a[&2\partial_\mu A_{\nu L}\partial_\alpha A_{\beta L}+2\partial_\mu A_{\nu R}\partial_\alpha A_{\beta R}\\&+\partial_\mu A_{\nu L}\partial_\alpha A_{\beta R}+\partial_\mu A_{\nu R}\partial_\alpha A_{\beta L}\\&-A_{\mu L}\partial_\nu A_{\alpha L}A_{\beta L}+A_{\mu L}\partial_\nu A_{\alpha L}A_{\beta R}-A_{\mu L}\partial_\nu A_{\alpha R}A_{\beta L}+A_{\mu L}\partial_\nu A_{\alpha R}A_{\beta R}\\&+A_{\mu R}\partial_\nu A_{\alpha L}A_{\beta L}-A_{\mu R}\partial_\nu A_{\alpha L}A_{\beta R}+A_{\mu R}\partial_\nu A_{\alpha R}A_{\beta L}-A_{\mu R}\partial_\nu A_{\alpha R}A_{\beta R}\\&+\partial_\mu A_{\nu L}A_{\alpha L}A_{\beta L}+\partial_\mu A_{\nu L}A_{\alpha L}A_{\beta R}+\partial_\mu A_{\nu L}A_{\alpha R}A_{\beta L}+\partial_\mu A_{\nu R}A_{\alpha L}A_{\beta R}\\&+\partial_\mu A_{\nu R}A_{\alpha R}A_{\beta L}+\partial_\mu A_{\nu R}A_{\alpha R}A_{\beta R}+A_{\mu L}A_{\nu L}\partial_\alpha A_{\beta L}+A_{\mu L}A_{\nu R}\partial_\alpha A_{\beta L}\\&+A_{\mu R}A_{\nu L}\partial_\alpha A_{\beta L}+A_{\mu L}A_{\nu R}\partial_\alpha A_{\beta R}+A_{\mu R}A_{\nu L}\partial_\alpha A_{\beta L}+A_{\mu R}A_{\nu R}\partial_\alpha A_{\beta R}\\&+A_{\mu L}A_{\nu L}A_{\alpha L}A_{\beta R}+A_{\mu R}A_{\nu L}A_{\alpha L}A_{\beta L}+A_{\mu L}A_{\nu R}A_{\alpha L}A_{\beta R}\\&+A_{\mu R}A_{\nu L}A_{\alpha R}A_{\beta L}+A_{\mu L}A_{\nu R}A_{\alpha R}A_{\beta R}+A_{\mu R}A_{\nu R}A_{\alpha R}A_{\beta L}]\}\end{aligned} \tag{18}$$

写成积分形式可知 $\hat{\Gamma}_1^a$ 就是非对称的反常 G^a.

而 Witten 的结果由于多了一些项，展开后

$$\tilde{\Gamma}=\tilde{\Gamma}_0+\Pi^a\tilde{\Gamma}_1^a+O(\Pi^2) \tag{19}$$

其中

$$\tilde{\Gamma}_0=\hat{\Gamma}_0+\frac{1}{48\Pi^2}\int d^4x\varepsilon^{\mu\nu\alpha\beta}\mathrm{Tr}(F_{\mu\nu L}F_{\alpha\beta R}) \tag{20}$$

$$\Pi^a\tilde{\Gamma}_1^a=\Pi^a\hat{\Gamma}_1^a+\frac{1}{48\Pi^2}\int dx^4\varepsilon^{\mu\nu\alpha\beta}\mathrm{Tr}\{\Pi^a\lambda^a[-F_{\mu\nu L}F_{\alpha\beta R}+F_{\mu\nu R}F_{\alpha\beta L}]\} \tag{21}$$

因而就不具有上述性质．这是 $\hat{\Gamma}$ 和 $\tilde{\Gamma}$ 的区别．

另外，$\hat{\Gamma}$ 和 W-Z 的结果W(本文(11)式)相比，虽然Π场的一阶项相同，但 0 阶项不同

$$W(\Pi=0)=0 \tag{22}$$

$$\hat{\Gamma}(\Pi=0)=\hat{\Gamma}_0 \tag{23}$$

应该指出 $\hat{\Gamma}_0$ 中被积函数(也记为$\hat{\Gamma}_0$)有如下性质

$$d\left.\frac{\delta\hat{\Gamma}_0}{\delta A_L^a}\right|_{AL=0}=-\frac{1}{48\pi^2}\mathrm{Tr}\lambda^a[2\partial_\mu A_{\nu R}\partial_\alpha A_{\beta R}+\partial_\mu(A_{\nu R}A_{\alpha R}A_{\beta R})] \tag{24}$$

$$d\left.\frac{\delta\hat{\Gamma}_0}{\delta A_R^a}\right|_{AR=0}=\frac{1}{48\pi^2}\mathrm{Tr}\lambda^a[2\partial_\mu A_{\nu L}\partial_\alpha A_{\beta L}+\partial_\mu(A_{\nu L}A_{\alpha L}A_{\beta L})] \tag{25}$$

两式的右端正好是反常的对称形式.

最后的结论是，本文作者在文[2,3]中给出的有效拉氏量，即本文(3)式 $\hat{\Gamma}$，它包含了所有与反常有关的性质，既含有对称反常，又含有非对称反常，而且它所表示的Π场和规范场的相互作用也是正确的.

参考文献

[1] E. Witten, *Nucl. Phys.*, **B223** (1983), 422.
[2] K. C. Chou, H. Y. Guo, K. Wu, X. C. Song, *Phys. Lett.*, **134**B (1984), 67;
周光召，郭汉英，吴可，宋行长，高能物理与核物理，**8**(1984)，252。
[3] K. C. Chou, H. Y. Guo, K. Wu, X. C. Song, "On Witten's effective Lagrangian of chiral field" Preprint AS-ITP-83-032 to be published in Commun. in Theor. Phys. (Beijing).
[4] D. J. Gross, R. Jackiw, *Phys. Rev.*, **D6** (1972), 477.
[5] W. A. Bardeen, *Phys. Rev.*, **184** (1969), 1848.
[6] J. Wess, B. Zumino, *Phys. Lett.*, **37B** (1971), 95.

SYMMETRICAL ANOMALY, UNSYMMETRICAL ANOMALY AND EFFECTIVE LAGRANGIAN

Chou Kuang-chao Guo Han-ying Wu Ke
(Institute of Theoretical Physics, Academia Sinica)

Song Xing-chang
(Peking University, Beijing, China)

Abstract

Both the symmetrical anomaly and the unsymmetrical anomaly, are derived from an effective Lagrangian recently constructed on the basis of Chern-Simons topological invariants.

第9卷 第2期 高能物理与核物理 Vol. 9, No. 2
1985年3月 PHYSICA ENERGIAE FORTIS ET PHYSICA NUCLEARIS Mar., 1985

任意偶维时空的有效作用和手征反常

周光召 郭汉英 李小源 吴可
（中国科学院理论物理研究所）

宋行长
（北京大学）

摘 要

由于同时考虑了左手和右手规范场，我们利用 Weil 同态的方法从 $2n+2$ 维时空上 Abel 反常的 Ward 恒等式，得到了规范协变反常流的一般表达式；再利用 Chern 类与 Chern-Simons 第二拓扑不变量的性质证明了用一个公式就包括了 $2n$ 维时空上的无反常条件，规范不变的 Wess-Zumino 有效作用，以及手征反常的一般形式；进一步揭示了 M^{2n+2} 与 M^{2n} 上这些问题之间的联系.

近来，关于有效作用量的大范围性质的研究引起了广泛的兴趣，Witten[1] 首先揭示了 Wess-Zumino 反常有效作用[2]的一些大范围性质；Skyrme[3] 关于重子可能是有效作用理论的孤子的考虑重新得到了重视[4]；对有效作用的规范不变性以及规范反常的拓扑起源从不同角度进行了探讨[5-9]；低维物理中的分数电荷、一些宏观量子效应等等也可与这些课题有着内在的联系[10,11].

本文将从大范围的角度对任意偶数维紧致时空上的有效作用的规范不变性以及规范反常的拓扑起源进行进一步的探讨. 我们将考虑 $SU(N)_L \times SU(N)_R$ 自发破缺到对角 $SU(N)$ 的有效作用理论，运用 Weil 同态[12]的方法求解 $2n+2$ 维时空上 Abel 反常的 Ward 恒等式，得到规范协变反常流的一般表达式；利用 Chern 类与 Chern-Simons 第二拓扑不变量的关系和性质给出 $2n$ 维时空上的无反常条件、规范不变的有效作用以及规范反常的一般形式. 这样，我们也进一步揭示了 $2n$ 维时空上的有效作用及非 Abel 反常与 $2n+2$ 维时空上协变反常流及 Abel 反常之间的内在联系.

令 $U(x)$ 是 $SU(N)$ 的元素，在 $SU(N)_L \times SU(N)_R$ 的元素 (g_L, g_R) 的作用下的变换性质为

$$U \to g_L U(x) g_R^{-1}. \tag{1}$$

为了得到在规范群 $H \subseteq SU(N)_L \times SU(N)_R$ 下不变的作用泛函，必须引入 U 的规范协变导数

$$DU = dU + A_L U - U A_R, \tag{2}$$

本文 1984 年 4 月 29 日收到.

其中 $A_{L(R)}$ 分别是 $H_{L(R)}$ 的规范势 1 形式，相应的场强 2 形式为[1)]

$$F_{L(R)} = dA_{L(R)} + A^2_{L(R)}. \tag{3}$$

在规范变换

$$U \to h_L(x)U(x)h_R^{-1}(x),\quad (h_L, h_R) \in H_L \times H_R, \tag{4}$$

的作用下

$$\begin{aligned} &A_{L(R)} \to h_{L(R)}(d + A_{L(R)})h^{-1}_{L(R)},\quad F_{L(R)} \to h_{L(R)}F_{L(R)}h^{-1}_{L(R)},\\ &DU \to h_L(x)DU(x)h_R^{-1}(x),\quad U^{-1}DU \to h_R(x)U^{-1}DUh_R^{-1}(x). \end{aligned} \tag{5}$$

其中

$$U^{-1}DU = U^{-1}(d + A_L)U - A_R =: {}^UA_L - A_R. \tag{6}$$

${}^UA_L$ 的变换性质与 A_R 是一致的.

在 $2n+2$ 维空间中，膺矢流的 Abel 反常为

$$C^{n+1}(A_L) - C^{n+1}(A_R)$$

其中 $C^{n+1}(A)$ 是 A 的第 $n+1$ 阶陈类，设由膺标 Goldstone 场 U 和规范场组成的反常流用 1 形式表示出来为 $J(U, A_L, A_R)$，则反常流的守恒方程为

$$d^*J(U, A_L, A_R) = C^{n+1}(A_L) - C^{n+1}(A_R). \tag{7}$$

其中*是 Hodqe 对偶算子.

我们首先研究的问题是找到一个协变的有效反常流 $J(U, A_L, A_R)$ 满足方程(7). 引入

$$\begin{aligned} &A_t = A_R + tU^{-1}DU,\quad 0 \leqslant t \leqslant 1,\\ &F_t = dA_t + A_t^2 = t\,{}^UF_L + (1-t)F_R - t(1-t)(U^{-1}DU)^2 \end{aligned} \tag{8}$$

利用 Weil 同态态式和 Chern 类的表达式立即得到

$$C^{n+1}({}^UA_L) - C^{n+1}(A_R) = \alpha_{n+1}(n+1)d\int_0^1 dt\mathrm{Tr}(U^{-1}DUF_t^n). \tag{9}$$

其中 $\alpha_{n+1} = \dfrac{i^{n+1}}{(2\pi)^{n+1}(n+1)!}$. 由于 $C^{n+1}({}^UA_L) = C^{n+1}(A_L)$，因此立即可以得到(7)的一般解

$$^*J(U, A_L, A_K) = \alpha_{n+1}(n+1)\int_0^1 dt\mathrm{Tr}(U^{-1}DUF_t^n) + {}^*\mathscr{I}_0. \tag{10}$$

其中 ${}^*\mathscr{I}_0$ 是任意的协变的正合 $2n+1$ 形式. 由(8)式可以看到 $U^{-1}DU$ 和 F_t 都是明显规范协变的，因此 $^*J$ 是规范不变的. 对于 4 维时空 $(n=1)$，直接计算表明 ${}^*\mathscr{I}_0 = 0$. J 通常叫 Skyrmion 流[11]. 因此，(10)式给出了 $2n+2$ 维空间中的协变反常流的一般形式.

其次，我们考查 $2n$ 维空间上的 Wess-Zumino 有效作用. 利用 Chern 类与 Chern-Simons 第二拓扑不变量的局部关系式，

$$C^{n+1}(A) = d\Pi^{(2n+1)}(A), \tag{11}$$

以及 Chern-Simons 不变量的性质，由(7)、(9)可以得到

$$^*J = \Pi^{(2n+1)}(U^{-1}dU) + dW^{(2n)} + \Pi^{(2n+1)}(A_L) - \Pi^{(2n+1)}(A_R), \tag{12}$$

---

1) 在本文中乘积指外乘积，A^2 为 $A_\wedge A$ 的缩写.

令

$$\Gamma(U,\ Q^{2n+1}) = 2\pi i \int_{Q^{2n+1}} \Pi^{(2n+1)}(U^{-1}dU). \tag{13}$$

其中 Q^{2n+1} 是 $2n+1$ 维流形，$\partial Q^{2n+1} = M^{2n}$ 为 $2n$ 维紧致化时空，当 $n=2$ 时就是 Witten[1] 所称的 Wess-Zumino 项[2]．不难看出，$\Gamma(U,\ Q^{2n+1})$ 就是该项在 M^{2n} 上的推广．同时，由于 $^*J$ 是规范协变的，因此在充要条件

$$\Pi^{(2n+1)}(A_L) = \Pi^{(2n+1)}(A_R), \tag{14}$$

下，可以定义 M^{2n} 上推广的规范不变的 Wess-Zumino 项

$$\tilde{\Gamma}(U,\ Q^{2n+1}) = \Gamma(U,\ Q^{2n+1}) + 2\pi i \int_{M^{2n}} W^{(2n)}. \tag{15}$$

为了求出 $W^{(2n)}$ 的一般表达式，我们引入

$$\begin{aligned} &A_{tu} = t^U A_L + uA_R \quad 0 \leqslant t,u \leqslant 1, \\ &F_{tu} = dA_{tu} + A_{tu}^2,\ \ {}^UF_{Lt} = td^UA_L + t^{2U}A_L^2,\ \ F_{Ru} = ndA_R + n^2A_R^2. \end{aligned} \tag{16}$$

考虑 (u,t) 平面上以原点及 $(1,0)$，$(0,1)$ 为顶点的三角形 L，其边 $\partial L = l$．可以证明

$$\begin{aligned} I_l &= \alpha_{n+1}(n+1) \int_l \mathrm{Tr}\{(dt^UA_L + duA_R)F_{tu}^n\} \\ &= -\Pi^{(2n+1)}({}^UA_L) + \Pi^{(2n+1)}(A_R) + {}^*J_{\min}. \end{aligned} \tag{17}$$

这里 $^*J_{\min}$ 为 $^*\mathscr{I}_0$ 为 0 时的 $^*J$ 1)，我们并用到

$$\Pi^{(2n+1)}(A) = \alpha_{n+1}(n+1) \int_0^1 dv \mathrm{Tr}(AF_v^n),\ \ F_v = vdA + v^2A^2, \tag{18}$$

另一方面，利用 Green 公式

$$\begin{aligned} I_l &= \alpha_{n+1}(n+1) \int_L \mathrm{Tr}\left\{\frac{\partial}{\partial u}({}^UA_LF_{tu}^n) - \frac{\partial}{\partial t}(A_RF_{tu}^n)\right\} dudt \\ &= \alpha_{n+1} d\left\{(n+1)n \int_L \mathrm{Str}(A_R{}^UA_LF_{tu}^{n-1})dudt\right\}. \end{aligned} \tag{19}$$

因此，

$$\begin{aligned} {}^*J_{\min} = &\ \Pi^{(2n+1)}({}^UA_L) - \Pi^{(2n+1)}(A_R) \\ &+ d\left\{(n+1)n\alpha_{n+1} \int_0^1 dt \int_0^{1-t} du \mathrm{Str}(A_R{}^UA_LF_{tu}^{n-1})\right\}. \end{aligned} \tag{20}$$

同样，可以证明

$$\begin{aligned} \Pi^{(2n+1)}({}^UA_L) = &\ \Pi^{(2n+1)}(A_L) + \Pi^{(2n+1)}(U^{-1}dU) \\ &+ d\left\{(n+1)n\alpha_{n+1} \int_0^1 dv \int_0^{1-v} dw \mathrm{Str}(UdU^{-1}A_LF_{vw}^{n-1})\right\}, \end{aligned} \tag{21}$$

其中 $F_{vw} = F(vA_L + \omega UdU^{-1})$，$0 \leqslant v, w \leqslant 1$．于是我们得到

$$\begin{aligned} W^{(2n)} = (n+1)n\alpha_{n+1} \int_0^1 dt \int_0^{1-t} du \mathrm{Str}\{&A_R{}^UA_LF^{n-1}(t^UA_L + uA_R) \\ &+ UdU^{-1}A_LF^{n-1}(tA_L + uUdU^{-1})\}. \end{aligned} \tag{22}$$

取 $n = 2,3$，直接计算表明 $\tilde{\Gamma}(U,Q^{2n+1})$ 分别给出 M^4 与 M^6 上规范不变 Wess-Zumino 项

1) 因为这里讨论的 $\tilde{\Gamma}$ 也是在这种最小意义下的 Wess-Zumino 项．一般说来，由于 $^*\mathscr{I}_0$ 是正合形式，$\tilde{\Gamma}$ 会有我们讨论过的不确定性[8,9]．

的结果.[5,8]而且 Goldstonl 场的零阶项和一阶项分别给出 Bardeen 的抵消项 R_3 子非对称反常[5]，这表明 $\tilde{\Gamma}(U, Q^5)$ 满足 M^4 上非 Abel 反常的 Ward 恒等式以及 Wess-Zumino 自洽性条件. 同样的结论可以推广到 M^{2n} 上的 $\tilde{\Gamma}(U, Q^{2n+1})$，特别是可以证明 M^{2n} 上 Bardeen 的抵消项 $R_3(M^{2n})$ 与非对称反常 $G^a_A(M^{2n})$ 分别为

$$R_3(M^{2n}) = (n+1)n\alpha_{n+1}\int_0^1 dt\int_0^{1-t} du\mathrm{Str}\{A_R A_L(F_{Lt}+F_{Ru})^n\}, \tag{23}$$

$$\begin{aligned}G^a_A(M^{2n}) = {} & 2\pi i(n+1)n\alpha_{n+1}\int_0^1 dt(1-t)\mathrm{Str}\{\lambda^a d[A_L F^{n-1}(tA_L)]\}\\ & + 2\pi i(n+1)n\alpha_{n+1}\int_0^1 dt\int_0^{1-t} du\mathrm{Str}(\lambda^a\{[d(E^{n-1}A_R)\\ & + A_L E^{n-1}A_R + E^{n-1}A_R A_L] + (n-1)[E^{n-2}A_R A_L F(tA_L)\\ & - F(tA_L)E^{n-2}A_R A_L] + (n-1)(t-t^2)d[A_L E^{n-2}A_R A_L\\ & - E^{n-2}A_R A_L^2] + (n-1) + u[d(E^{n-2}A_R A_L A_R)\\ & + A_L E^{n-2}A_R A_L A_R + E^{n-2}A_R A_L A_R A_L - d(A_R E^{n-2}A_R A_L)\\ & - A_L A_R E^{n-2}A_R A_L - A_R E^{n-2}A_R A_L^2]\}),\end{aligned} \tag{24}$$

其中

$$E = F(tA_L) + F(uA_R) + tu\cdot(A_L A_R + A_R A_L). \tag{25}$$

显然，如果条件(14)不能满足，那么 $\tilde{\Gamma}(U,Q^{2n+1})$ 就不是规范不变的，在规范变换(4)下 $\tilde{\Gamma}(U,Q^{2n+1})$ 的改变量满足

$$\Delta\tilde{\Gamma}(U,Q^{2n+1}) + 2\pi i\Delta(\Pi^{(2n+1)}(A_L) - \Pi^{(2n+1)}(A_R)) = 0, \tag{26}$$

利用类似于(21)的公式不难证明

$$\begin{aligned}\Delta\tilde{\Gamma}(U,Q^{2n+1}) = {} & -\Gamma(h_L,Q^{2n+1}) + \Gamma(h_R,Q^{2n+1})\\ & - 2\pi i d\left\{(n+1)n\alpha_{n+1}\int_0^1 dv\int_0^{1-v} dw\mathrm{Str}(h_L^{-1}dh_L A_L F^{n-1}_{vw}) - (L\to R)\right\}.\end{aligned} \tag{27}$$

同样可以给出无穷小改变量 $\delta\tilde{\Gamma}(U, Q^{2n+1})$ 的一般表达式. 当 $n=2$ 时，直接计算表明 $\delta\tilde{\Gamma}(U,Q^5)$ 就是 Gross-Jackiw 的对称反常[13]，而 $\Delta\tilde{\Gamma}(U, Q^5)$ 则可给出 Witten 的 $SU(2)$ 大范围反常[1]. 这表明，条件(14)是 M^{2n} 上无反常条件的大范围形式，它不仅是通常微扰论意义下的无反常条件的推广，而且也概括了非微扰下的无反常的条件. 另一方面，从 Q^{2n+1} 上的条件(14)可得

$$C^{n+1}(A_L) = C^{n+1}(A_R). \tag{28}$$

这恰恰是 M^{2n+2} 上无 Abel 矢量反常的条件.

总之，M^{2n+2} 上的 Abel 反常和 M^{2n} 上的非 Abel 反常，二者的拓扑起源相同，有着深刻的内在联系. 我们知道，Abel 反常以及非微扰的非 Abel 反常都是 Atiyah-Singer 指数定理[12]的体现，因此，通常的非 Abel 反常也必然直接与间接与 Atiyoh-Singer 定理相联系. 关于这个问题我们将留待以后进一步讨论.

参　考　文　献

[1] E. Witten, *Nucl. Phys.*, B**223**(1983), 422.
[2] J. Wess and B. Zumino, *Phys. Lett.*, 37B(1971), 95.

[3] T. H. R. Skyrme, *Proc. Roy. Soc. (London)*, **A260**(1961), 127.
[4] E. Witten, *Nucl. Phys.*, **B223**(1983), 433.
G. W. Adkins, C. R. Nappi and E. Witten, *ibid.*, **B228**(1983), 552.
[5] K. C. Chou, H. Y. Guo, K. Wu and X. C. Song, *Phys. Lett.*, **134**B (1984), 67; Institute of Theoretical Physics, Academia Sinica preprint AS-ITP-83-032; 033; Stony Brook preprint ITP SB-84-18.
[6] B. Zumino, Les Houches lectures 1983, LBL-16747 (1983).
B. Zumino, Y. S. Wu and A. Zee, *Nucl. Phys.* (to be published).
[7] K. Kawai and S. -H. H. Tye, Cornell preprint CLNS-84/595.
[8] Y. P. Kuang, X. Y. Li, K. Wu and Z. Y. Zhao, preprint AS-ITP-84-015.
[9] C. H. Chang, H. Y. Guo and K. Wu, preprint AS-ITP-84-016.
[10] J. Goldstone and F. Wilczek, *Phys. Rev. Lett.*, **47**(1981), 986.
[11] A. Zee, *Phys. Lett.*, **135**B(1984), 307.
F. Wilczek and A. Zee, *Phys. Rev. Lett.*, **51**(1983), 2250.
[12] 例如见 T. Eguchi, P. B. Gilkey and A. J. Hanson, *Phys. Reports*, **66**(1980), 213,及其中所引文献.
[13] D. J. Gross and R. Jackiw, *Phys. Rev.*, **D6**(1972), 477.

EFFECTIVE ACTION AND CHIRAL ANOMALIES IN ANY EVEN DIMENSIONAL SPACE

CHOU KUANG-CHAO, GUO HAN-YING
LI XIAO-YAUN, WU KE
(Institute of Theoretical Physics, Academia Sinica)

SONG XING-CHANG
(Peking University)

ABSTRACT

General expression for gauge covariant anomalous current is obtained from the difference of left handed and right handed Abelian anomaly in $2n+2$ dimensions by the method of Weil homomorphism. The general form of symmetric and asymmetric anomalies, gauge invariant Wess-Zumino effective action and anomaly free condition in $2n$ dimensions are summarized in one closed formula, showing the deep connections among all these topological properties of gauge fields and pseudascalar Goldstone fields in $2n$ and $2n+\acute{e}$ dimensions.

第10卷 第2期 高能物理与核物理 Vol. 10, No. 2
1986年3月 PHYSICA ENERGIAE FORTIS ET PHYSICA NUCLEARIS Mar., 1986

普遍的 Chern-Simons 链和它们的应用

周光召 吴岳良 谢彦波

(中国科学院理论物理研究所)

摘 要

本文将 Chern 形式按它子流形上的外微分形式的阶数展开，很容易地给出了一些 Chern-Simons 链．并讨论了某些约束条件下，子流形间 Chern 形式的递推关系和它们的一些应用．

一、引 言

最近，物理学家用拓扑和微分几何的数学工具来讨论场论的一些大范围性质，给出了许多有趣的结果．特别人们发现用微扰理论所得到的手征反常有它的拓扑起源，它是规范场上同调的一个非平凡元．并且人们根据 Chern-Simons 示性类，利用微分几何的方法可能更方便地得到，而不需要计算费曼图．

本文将给出普遍 Chern-Simens 示性类的一个简单推导．在第二节．将 Chern 形式按它子流形上的外微分形式的阶数展开，再利用 Chern 形式闭的性质，很简单地得到一些 Chern-Simons 链．第三，四节，通过考虑一些特殊情况，给出了广义的第二示性类和 Faddeev 型上同调．最后讨论更一般一点的情况和 θ-真空．

二、Chern 形式的递推方程

设在N维空间中有规范势

$$C(\eta) = C_l(\eta) d\eta^l, \quad l = 1, 2, \cdots, N \tag{2.1}$$

相应的场强为

$$W = \hat{d}C + C^2 \equiv W_{lp}\, d\eta^l \wedge d\eta^p. \tag{2.2}$$

这里我们用了外微分形式．外微分算子为:

$$\hat{d} = d\eta^l \frac{\partial}{\partial \eta^l}. \tag{2.3}$$

Chern 形式定义为:

$$\Omega_{2m}(W) = \frac{i^m}{m!\,(2\pi)^m} \mathrm{Tr}(W^m). \tag{2.4}$$

本文 1985 年 4 月 23 日收到．

它是 $2m$-形式，利用 Bianchi 恒等式：

$$\mathring{d}W = [W, C]. \tag{2.5}$$

人们很容易验证 $\Omega_{2m}(W)$ 是闭的

$$\mathring{d}\Omega_{2m} = 0, \tag{2.6}$$

并可表示为

$$\Omega_{2m}(W) = \mathring{d}\Omega^0_{2m-1}(W, C). \tag{2.7}$$

又因 $\Omega_{2m}(W)$ 是规范不变的．即：

$$\delta\Omega_{2m}(W) = 0. \tag{2.8}$$

这里 δ 表示规范变换算子．若用外微分形式表示：

$$\delta = d\ \alpha^i \frac{\partial}{\partial\alpha^i}. \tag{2.9}$$

α^i 是群参数．那么把规范作用改写为[1]：

$$\begin{aligned} \delta A &= -dV - AV - VA \\ \delta F &= [F, V],\ V = g^{-1}\delta g. \end{aligned} \tag{2.10}$$

并有：

$$\delta^2 = 0,\ \delta\mathring{d} + \mathring{d}\delta = 0. \tag{2.11}$$

从

$$\delta\Omega_{2m} = \delta\mathring{d}\Omega^0_{2m-1} = -\mathring{d}(\delta\Omega^0_{2m-1}) = 0,$$

人们可知

$$\delta\Omega^0_{2m-1} = \mathring{d}\Omega^1_{2m-2}.$$

利用 (11) 式人们可以得到：

$$\begin{aligned} &\delta\Omega^{k-1}_{2m-k} = \mathring{d}\Omega^k_{2m-k-1} \\ &k = 1,\ 2,\ \cdots,\ 2m \end{aligned} \tag{2.12}$$

这是通常的规范变换和空间外微分间的一组 Chern-Simons 链的递推关系．

现在我们考虑由两个子流形构成的流形，一个是普通坐标空间 $X^\mu(\mu = 1, 2, \cdots N_1)$，另一个为任意所考虑的空间 $\xi^i(i = 1, 2, \cdots N_2)$．那么人们可以将规范势和外微分也分成两部分：

$$C = A + \bar{A},\ \mathring{d} = d + \bar{d} \tag{2.13}$$

其中

$$\begin{aligned} &A = A_\mu(X,\ \xi)dX^\mu,\ \bar{A} = \bar{A}_i(X,\ \xi)d\xi^i \\ &d = dX^\mu \frac{\partial}{\partial X^\mu},\ \bar{d} = d\xi^i \frac{\partial}{\partial\xi^i} \end{aligned} \tag{2.14}$$

则场强可重写为：

$$W = \mathring{d}C + C^2 \equiv F + \bar{F} + M \tag{2.15}$$

其中：

$$\begin{aligned} F &= dA + A^2, \\ \bar{F} &= \bar{d}\bar{A} + \bar{A}^2, \\ M &= \bar{d}A + d\bar{A} + A\bar{A} + \bar{A}A. \end{aligned} \tag{2.16}$$

从上式可看到，F 是总场强在 X 子流形上的分量，$\bar{F}$ 是总场强在 ξ 子流形上的分量，M 可看作是两子流形上的混合分量.

则第 n 次 Chern 特征写为：

$$C_n = Q_{2n} = \frac{i^n}{n!(2\pi)^n} \mathrm{Tr}(F + \bar{F} + M)^n \tag{2.17}$$

现在将 Q_{2n} 按子流形 ξ 的外微分次数展开，

$$Q_{2n} = Q_{2n,0} + Q_{m-1,1} + \cdots + Q_{0,2n} \tag{2.18}$$

其中 $Q_{2n-m,m}$. 第一个指标表示 dX 的 $2n-m$ 形式，第二个指标表示 $d\xi$ 的 m-形式，它的普遍表达式可写为：

$$Q_{2n-(m+k),m+k} = \sum_{m=0}^{n} \sum_{k=0}^{m} \binom{n}{m}\binom{m}{k} \mathrm{STr}(F^{n-m}\bar{F}^k M^{m-k}) \tag{2.19}$$

将 (2.18) 式代入 (2.6) 式并比较 $d\xi$ 的外微分形式的阶数人们很容易看到：

$$\begin{aligned} \bar{d}Q_{2n,0} &= -dQ_{2n-1,1} \\ \bar{d}Q_{2n-1,1} &= -dQ_{2n-2,2} \\ &\vdots \\ \bar{d}Q_{2n-m,m} &= -dQ_{2n-m-1,m+1} \\ &\vdots \\ \bar{d}Q_{0,2n} &= 0 \end{aligned} \tag{2.20}$$

这组方程反映了子流形间 Chern 形式的递推关系.

将 Chern 密度 Q^0_{2n-1} 也按 $d\xi^i$ 展开：

$$Q^0_{2n-1} = Q^0_{2n-1,0} + Q^0_{2n-2,1} + \cdots + Q^0_{0,2n-1} \tag{2.21}$$

将 (2.18) 和 (2.21) 代入 (2.7) 式. 并比较 $d\xi$ 的阶数，人们可得到：

$$\begin{aligned} Q_{2n,0} &= dQ^0_{2n-1,0} \\ Q_{2n-1,1} &= \bar{d}Q^0_{2n-1,0} + dQ^0_{2n-2,1} \\ &\vdots \\ Q_{2n-m,m} &= \bar{d}Q^0_{2n-m,m-1} + dQ^0_{2n-m-1,m} \\ &\vdots \\ Q_{0,2n} &= \bar{d}Q^0_{0,2n-1} \end{aligned} \tag{2.22}$$

一般地人们可以把 Q^{m-1}_{2n-m} 按 $d\xi$ 展开为

$$Q^{m-1}_{2n-m} = Q^{m-1}_{2n-m,0} + Q^{m-1}_{2n-m-1,1} + \cdots + Q^{m-1}_{0,2n-m} \tag{2.23}$$

其中 $Q^{m-1}_{2n-m-k,k}$ 的上指标 $(m-1)$ 表示通常规范作用的外微分形式.

把展开式代入 (2.12). 并比较各外微分形式的阶数，人们可得到：

$$\begin{aligned} \delta Q^{m-1}_{2n-m,0} &= dQ^m_{2n-m-1,0} \\ \delta Q^{m-1}_{2n-m-1,1} &= \bar{d}Q^m_{2n-m-1,0} + dQ^m_{2n-m-1,1} \\ &\vdots \\ \delta Q^{m-1}_{2n-m-k,k} &= \bar{d}Q^{m-1}_{2n-m-k,k-1} + dQ^{m-1}_{2n-m-k-1,k} \\ &\vdots \\ \delta Q^{m-1}_{0,2n-m} &= \bar{d}Q^m_{0,2n-m-1} \end{aligned} \tag{2.24}$$

$$m = 1, 2, \cdots, 2n$$

上面只是形式上给出了各外微分形式间的递推关系. 下面我们将讨论它们的一些具

体应用.

在许多具体问题中. 实际上可能会有一些约束条件. 为此我们来看一些特殊情况.

三、广义的第二示性类和协变反常

考虑有一个子流形上的场强为零,设为:

$$\bar{F} = \bar{d}\bar{A} + \bar{A}^2 = 0, \quad F \neq 0, \quad M \neq 0 \tag{3.1}$$

由此容易看到有下列等式:

$$\begin{aligned} dF &= [F, A], \quad \bar{d}M = [M, \bar{A}] \\ \bar{d}F + dM &= [F, \bar{A}] + [M, A] \end{aligned} \tag{3.2}$$

以及

$$\Omega_{2n-m,m} = \frac{i^n}{n!(2\pi)^n} \sum_{m=0}^{n} \binom{n}{m} \mathrm{STr}\,(F^{n-m}M^m) \tag{3.3}$$

并且

$$\Omega_{2n-m,m} = 0 \text{ 当 } m > n \text{ 时} \tag{3.4}$$

最简单地人们可取

$$\bar{A}(x, \xi) = 0 \tag{3.5}$$

特别当人们再取

$$A(x, \xi) = A^{(0)}(x) + \sum \xi^i \left(A^{(i)}(x) - A^{(0)}(x)\right) \tag{3.6}$$

$$|\xi^i| \leqslant 1$$

那么

$$M = \sum_i d\xi^i (A^i(x) - A^0(x)) \tag{3.7}$$

代入递推方程(2.20),并在 ξ^i 子流形的单形上积分,人们就可以得到广义的第二示性类[2]即:

$$\begin{aligned} &(\Delta w_{2n-m,m})(A^0, A^1, A^2, \cdots, A^{m+1}) \\ &= -dw_{2n-m-1,m+1}(A^0, A^1, A^2, \cdots, A^{m+1}) \end{aligned} \tag{3.8}$$

其中Δ是上边缘算子. 其操作为:

$$\begin{aligned} &(\Delta w_{2n-m,m})(A^0, A^1, A^2, \cdots A^{i-1}, A^i, A^{i+1}, \cdots, A^{m+1}) \\ &= \sum_{i=0}^{m+1} (-1)^i w_{2n-m,m}(A^0, A^1, A^2, \cdots A^{i-1}, \hat{A}^i, A^{i+1}, \cdots, A^{m+1}) \end{aligned} \tag{3.9}$$

其中带有$\wedge$的元素表示要省掉.

$$\begin{aligned} w_{2n-m,m} &= \int_{\xi \in S_m} \Omega_{2n-m,m}, \\ \Delta w_{2n-m,m} &= \int_{\xi \in \partial S_{m+1}} \Omega_{2n-m,m} \end{aligned} \tag{3.10}$$

这里 S_m 是 ξ 子流形上的 m 单形. 如 $m = 0, 1, 2, 3$ 的单形相应为点、直线、三角形、四面体. ∂S_{m+1} 表示取 $m+1$ 单形的边缘.

正如人们熟知，$w_{2n-2,2}$ 与 $2n-2$ 维空间的 Wess-Zumino-Witten 反常相联系。

另外，若人们取：

$$\begin{aligned}&\bar{A}(x, \xi) = 0,\\ &A(x, \xi) = U^{-1}(x, \xi)(A(x) + d)U(x, \xi)\end{aligned} \tag{3.11}$$

那么：

$$\begin{aligned}&F = dA + A^2 = U^{-1}(x, \xi)F(x)U(x, \xi),\\ &M = \bar{d}A \equiv -dB - AB - BA,\\ &B = U^{-1}(x, \xi)dU(x, \xi),\\ &\bar{d}F = [F, B]\end{aligned} \tag{3.12}$$

这时人们很容易发现：

$$\bar{d}\Omega_{2n,0} = 0 \tag{3.13}$$

由方程 (20) 可看到，这要求.

$$d\Omega_{2n-1,1} = 0 \tag{3.14}$$

因此局部情形下，$\Omega_{2n-1,1}$ 应能表示为：

$$\Omega_{2n-1,1} = d\tilde{\Omega}^0_{2n-2,1}, \tag{3.15}$$

容易证明：

$$\tilde{\Omega}^0_{2n-2,1} = -\frac{i^n n}{n!(2\pi)^n}\mathrm{Tr}(BF^{n-1}) \tag{3.16}$$

因此，如果人们定义

$$\beta = \int_{x\in M^{n-1}}\int_{\xi\in S_1}\Omega_{2n-1,1} = \int_{x\in \partial M^{n-1}}\int_{\xi\in S_1}\tilde{\Omega}^0_{2n-2,1} \tag{3.17}$$

这里假定 $\partial M^{n-1} = M^{n-2} \neq 0$，那么，人们发现 β 正比于通常的协变反常 [3].

由方程组 (2.20) 人们可看到，它满足：

$$\bar{d}\Omega_{2n-1,1} = -d\Omega_{2n-2,2} \tag{3.18}$$

四、Faddeev 型上同调

这里我们将考虑一个子流形上的场强分量和混合场强分量都为零的情况. 设为：

$$\bar{F} = M = 0 \tag{4.1}$$

那么有：

$$dF = [FA],\ \bar{d}F = [F, \bar{A}]$$

$$\Omega_{2n-m,m} = \begin{cases}\dfrac{i^n}{n!(2\pi)^n}\mathrm{Tr}(F^n) & m = 0\\ 0 & m \neq 0\end{cases} \tag{4.2}$$

这时 Chern 形式降到 x 子流形上.

由方程 (2.22)，人们得到：

$$\Omega_{2n,0} = d\Omega^0_{2n-1,0}$$

$$\bar{d}\Omega^0_{2n-1,0} = -d\Omega^0_{2n-2,1}$$
$$\vdots$$
$$\bar{d}\Omega^0_{2n-m,m-1} = -d\Omega^0_{2n-m-1,m} \qquad (4.3)$$
$$\vdots$$
$$\bar{d}\Omega^0_{0,2n-1} = 0$$

这样人们就得到了另一组递推关系，它反映了在子流形间 Chern 密度的递推性质。

如果取 ξ 子流形为群流形，并选择:

$$A(x,\xi) = U^{-1}(x,\xi)(A(x)+d)U(x,\xi),$$
$$\bar{A}(x,\xi) = U^{-1}(x,\xi)\bar{d}U(x,\xi) \qquad (4.4)$$

那么人们就得到通常的 Chern-Simons 链，这时只要作相应代换:

$$\bar{d} \longleftrightarrow \delta \qquad \Omega^0_{2n-m,m-1} \longleftrightarrow \Omega^{m-1}_{2n-m,0}$$

并取:

$$U(x,\xi) = h(\xi^1, g_1(x)h(\xi^2, g_2(x)h(\xi^3, \cdots)\cdots) \qquad (4.5)$$

其中函数 h 为:

$$h(\xi^i, g_i(x)) = e^{(1-\xi^i)u_i(x)^3}, g_i(x) = e^{u_i(x)} \qquad (4.6)$$
$$h(0, g_i(x)) = g_i(x), \quad h(1, g_i(x)) = 1$$

那么在由点

$$p_i = (\xi^1, \xi^2, \cdots \xi^{i-1}, \xi^i, \xi^{i+1}, \cdots)$$
$$= (0, 0, \cdots, 0, 1, 0, \cdots) \qquad (4.7)$$
$$i = 1, 2, \cdots$$
$$p_0 = (0, 0, \cdots, 0, 0, 0, \cdots)$$

组成的单形上积分。人们就可得到 Faddeev 型[4]的上同调，即:

$$(\triangle w^0_{2n-m,m-1})(A, g_1, g_2, \cdots g_m)$$
$$= -d w^0_{2n-m-1,m}(A, g_1, g_2, \cdots g_m) \qquad (4.8)$$

其中 $\triangle$ 是上边缘算子，其操作为:

$$(\triangle w^0_{2n-m,m-1})(A, g_1, g_2, \cdots g_m)$$
$$= w^0_{2n-m,m-1}(A^{g_1}, g_2, \cdots g_m) - w^0_{2n-m,m-1}(A, g_1 g_2, g_3, \cdots g_m) \qquad (4.9)$$
$$+ (-1)^m w^0_{2n-m,m-1}(A, g_1, g_2, \cdots g_{m-1})$$

$$w^0_{2n-m-1,m} = \int_{\xi \in S_m} \Omega^0_{2n-m-1,m} \qquad (4.10)$$
$$\triangle w^0_{2n-m,m-1} = \int_{\xi \in \partial S_m} \Omega^0_{2n-m,m-1}$$

其中 S_m 是由 $m+1$ 个点组成的 m 单形。

人们知道

$$\alpha = 2\pi \int_{x \in M^{2n-2}} w^0_{2n-2,1} \qquad (4.11)$$

就是在无边缘的 $2n-2$ 维空间的 Wess-Zumino-Witten 有效作用量。

利用 stoke's 定理。人们容易发现

$$\int_{x \in M^{2n-2}} w^0_{2n-2,1}(A^{g_1}, g_2) - \int_{x \in M^{2n-2}} w^0_{2n-2,1}(A, g_1 g_2) + \int_{x \in M^{2n-2}} w^0_{2n-2,1}(A, g_1) = 0 \quad (4.12)$$

这个 1-上闭条件可以看作为 Zumino 反常自洽条件的整体表示形式.

五、广义 Chern-Simons 链和 θ-真空

最后我们来考虑只有场强的混合分量为零的情况

$$M = \bar{d}A + d\bar{A} + A\bar{A} + \bar{A}A = 0 \tag{5.1}$$

由此很容易得到等式:

$$\begin{aligned} dF &= [F, A], \quad \bar{d}F = [F, \bar{A}] \\ d\bar{F} &= [\bar{F}, A], \quad \bar{d}\bar{F} = [\bar{F}, \bar{A}] \end{aligned} \tag{5.2}$$

这里人们可看到一些对称性.

$$A \longleftrightarrow \bar{A}, \quad d \longleftrightarrow \bar{d}$$

Chern 形式为:

$$Q_{2n-2m,2m} = \frac{i^n}{n!(2\pi)^n} \mathrm{STr}\,(F^{n-m}\bar{F}^m) \tag{5.3}$$

这时方程 (2.22) 变为:

$$\begin{aligned} &Q_{2n,0} = dQ^0_{2n-1,0} \\ &\bar{d}Q^0_{2n-1,0} = -dQ^0_{2n-2,1} \\ &\bar{d}Q^0_{2n-2,1} = -dQ^0_{2n-3,2} + Q_{2n-2,2} \\ &\bar{d}Q^0_{2n-3,2} = -dQ^0_{2n-4,3} \\ &\vdots \\ &\bar{d}Q^0_{2n-2k,2k-1} = -dQ^0_{2n-2k-1,2k} + Q_{2n-2k,2k} \\ &\bar{d}Q^0_{2n-2k-1,2k} = -dQ^0_{2n-2k-2,2k+1} \\ &\bar{d}Q^0_{2n-2k-2,2k+1} = -dQ^0_{2n-2k-3,2k+1} + Q_{2n-2k-2,2k+2} \\ &\vdots \\ &Q_{0,2n} = \bar{d}Q^0_{0,2n-1} \end{aligned} \tag{5.4}$$

人们注意到这组递推关系与通常的 Chern-Simons 链相比. 每递推隔一次要附加一项. 这一项在 d 和 $\bar{d}$ 的作用下都为零. 这从方程 (2.20) 很容易看到这一点. 因为这时

$$Q_{2n-2k-1,2k+1} = 0 \quad k = 0, 1, \cdots, n-1 \tag{5.5}$$

另外,人们可以证明 $Q^0_{2n-2,1}(A, \bar{A})$ 关于 $\bar{A}$ 是线性的. 即

$$Q^{(A,\bar{A})}_{2n-2,1} = \mathrm{Tr}(\bar{A}n(A)) \tag{5.6}$$

因此在这种情况同样给出 Wess-Zumino 反常形式

人们可考虑这种情况的一个简单例子. 取

$$\begin{aligned} \bar{A}(x, \xi) &= B(\xi) + U^{-1}(x, \xi)\bar{d}U(x, \xi), \\ A(x, \xi) &= U^{-1}(x, \xi)(A(x) + d)U(x, \xi) \end{aligned} \tag{5.7}$$

这里 $B(\xi)$ 是只与参数 ξ 有关的在子流形 ξ 上的 Abel 规范势. $U(x, \xi)$ 是所考虑群的群元,那么人们容易证明:

$$\begin{aligned} M &= d\bar{A} + \bar{d}A + A\bar{A} + \bar{A}A = 0, \\ \bar{F} &= \bar{d}\bar{A} + \bar{A}^2 = \bar{d}B, \\ F &= dA + A^2 = U^{-1}(x, \xi)F(x)U(x, \xi) \end{aligned} \tag{5.8}$$

那么:

$$\Omega_{2n-2m,2m} = \frac{i^n}{n!(2\pi)^n} \, \mathrm{STr}\,(\bar{d}B)^m F^{n-m}$$
$$= \frac{i^n}{(n-m)!m!(2\pi)^n} (\bar{d}B)^m \,\mathrm{Tr} F^{n-m}(x)$$
$$\equiv \bar{d}\, \Omega'_{2n-2m,2m-1}, \tag{5.9}$$

其中:

$$\Omega_{2n-2m,2m-1} = \frac{i^n}{(n-m)!m!(2\pi)^n} \bar{B}(\xi)(\bar{d}B)^{m-1} \,\mathrm{Tr} F(x)^{n-m} \tag{5.10}$$

引进新的 Chern 密度.

$$\bar{\Omega}^0_{2n-2m,2m-1} = \Omega^0_{2n-2m,2m-1} + \Omega'_{2n-2m,2m-1} \tag{5.11}$$

人们可看到方程 (5.4) 变为:

$$\begin{aligned} &\Omega_{2n,0} = d\Omega^0_{2n-1,0} \\ &\vdots \\ &\bar{d}\Omega^0_{2n-2m+1,2m-2} = -d\Omega^0_{2n-2m,2m-1} \\ &\bar{d}\bar{\Omega}^0_{2n-2m,2m-1} = -d\Omega^0_{2n-2m-1,2m} \\ &\vdots \\ &\bar{d}\Omega^0_{0,2n-1} = 0 \end{aligned} \tag{5.12}$$

如果人们推广有效作用量的定义:

$$\bar{\Gamma}(A, \bar{A}) = 2\pi \int_{x\in M^{2n-2}} \int_{\xi\in S_1} \bar{\Omega}^0_{2n-2,1}(A, \bar{A})$$
$$= \tilde{\Gamma}(A, U) + \Gamma'(A, \theta), \tag{5.13}$$

其中:

$$\tilde{\Gamma}(A, U) = 2\pi \int_{x\in M^{2n-2}} \int_{\xi\in S_1} \Omega^0_{2n-2,1} \tag{5.14}$$

就是在 $2n-2$ 维无边缘流形 M^{2n-2} 上的 Wess-Zumino-Witten 有效作用量.

$$\Gamma'(A, \theta) = 2\pi \int_{x\in M^{2n-2}} \int_{\xi\in S_1} \Omega'_{2n-2,1}$$
$$= \theta \frac{i^{n-1}}{(n-1)!(2\pi)^{n-1}} \int_{x\in M^{2n}} \mathrm{Tr} F^{n-1}, \tag{5.15}$$

其中:

$$\theta = i \int_{\xi\in S_1} B(\xi) \tag{5.16}$$

当取 $n=3$ 时,人们有:

$$\Gamma'(\theta, A) = -\frac{1}{8\pi^2} \theta \int_{x\in M^4} \mathrm{Tr}\, F^2$$
$$= \theta \frac{1}{64\pi^2} \int d^4x \varepsilon^{\mu\nu\rho\sigma} F^a_{\mu\nu} F^a_{\rho\sigma}. \tag{5.17}$$

这里使用了 $F = \frac{1}{2} F^a_{\mu\nu} T^a dx^\mu \wedge dx^\nu$, $\mathrm{Tr}\, T^a T^b = -\frac{1}{2}\delta^{ab}$. 这一项给出了与四维 Euclidean 空间出现的 θ-真空相同的项. 这一项如果在物理上有意义，那么 (5.16) 式给

出的 θ 值应与积分路径无关或至多相差一整数．这就要求 B (ξ) 或者是纯规范或者相应的磁通量是量子化的．

上面我们仅讨论了一些特殊情况．关于它们可能的应用可作进一步的讨论．

参　考　文　献

[1] B. Zumino Cargese Lectures 1983.

[2] Guo, H. Y. Hou, B. Y., Wang, S. K. and Wu Ke Preprint AS-ITP-84-039 and AS-ITP-84-044 (1984)

[3] K. Fujikawa Phys. Rev. D21 2848 (1980).

[4] L. Faddeev Phys. Lett. 145B 81(1984).

THE GENERAL CHERN-SIMONS COCHAIN AND THEIR APPLICATION

Chou Kuan-chao　Wu Yue-liang　Xei Yan-bo

(Institute of Theoretical Physics. Academia Sinica)

Abstract

Some general Chern-Simons cochains are easily obtained by expanding the Chern form according to the degree of the form in its submanifold and using the closed property of the Chern form. The recurrent relatuns of Chern form between submanifolds are discussed under some constraints. We also consider their application.

第10卷 第3期　　高能物理与核物理　　Vol. 10, No. 3
1986年5月　　PHYSICA ENERGIAE FORTIS ET PHYSICA NUCLEARIS　　May, 1986

一种新的上边缘算子及其应用

周光召　吴岳良　谢彦波
(中国科学院理论物理研究所)

摘　要

本文给出了以乘积形式定义的上边缘算子．在非阿贝尔规范场存在的情况下，利用这种算子对空间平移群的结合律进行了讨论．另外还讨论了 $SU(2)$ 和 $SO(3)$ 规范理论中的磁单极的量子化条件．

一、引　言

最近人们已经把拓扑这门数学分支应用到规范场论中[1,2,3]．例如从陈形式和规范变换得到的一阶上闭链和二阶上闭链就可以导出流反常和算符对易子的反常[1,2]．另外在量子力学中，我们也知道一个不满足狄拉克量子化条件的磁单极将导致波函数空间平移变换的结合律破坏[3]．但是当我们对于非阿贝尔磁单极进行类似讨论的话，文章[3]所采用的方法就行不通了，其原因就是非阿贝尔场是矩阵，故不能看成可对易的数．这样文章[3]用加法运算定义的上边缘算子就必须用乘法运算来重新定义．

我们将在第二节中给出这种新的上边缘算子，然后在第三节中讨论它的应用．

二、新的上边缘算子

设一个粒子用多分量波函数 $\psi(x)$ 来描写．$\psi(x)$ 是某个群 G 的自然表示基底．考虑空间存在一个规范场

$$\hat{A}(x) = A^a_\mu(x) dx^\mu \hat{I}^a, \tag{1}$$

$\hat{I}_a$ 是群的生成元，$\hat{A}(x)$ 是群 G 的伴随表示．我们知道在最小耦合情况下，波函数在空间平移变换下按照下式变化

$$\psi(x) \to U(a)\psi(x) = W_1(x, x+a)\psi(x+a). \tag{2}$$

其中:

$$U(a) = e^{\boldsymbol{a}\cdot\boldsymbol{D}}, \tag{3}$$

$$\boldsymbol{D} = \boldsymbol{\partial} + i\boldsymbol{A}, \tag{4}$$

$$\begin{aligned} W_1(x, x+a) &= W_1^{-1}(x+a, x) \\ &= p e^{i\int_x^{x+a}\hat{A}}. \end{aligned} \tag{5}$$

本文1985年4月23日收到

W_1 是一个 1—闭链. 0—闭链只是依赖于空间一点的函数，记为

$$W_0(x)$$

上边缘算子 δ 作用在 W_0 变为某一个 1—闭链

$$\begin{aligned}\delta W_0(x) &= W_0(x+a)W_0^{-1}(x)\\ &= W_1(x+a,x).\end{aligned} \tag{6}$$

显然 δW_0 满足条件(5). 现在我们考虑群关系

$$U(b)U(a)\phi(x) = W_2(x+b, x+a+b, x)U(a+b)\phi(x) \tag{7}$$

$$\begin{aligned}W_2(x+b, x+a+b, x) &= \delta W_1(x+a+b,x)\\ &= W_1(x, x+b)W_1(x+b, x+a+b)\\ &\quad \times W_1(x+a+b, x).\end{aligned} \tag{8}$$

W_2 是由 1—闭链通过上边缘算子 δ 得到的 2—闭链. 将 (5) 式代入 (8) 式，很容易发现 W_2 是绕三角形 $(x, x+a+b, x+b)$ 并以 x 作为起点的线积分.

$$W_2(x+b,x+a+b,x) = pe^{i\oint\hat{A}(x)}. \tag{9}$$

如果 1—闭链

$$W_1(x+a, x) = \delta W_0(x) = W_0(x+a)W_0^{-1}(x),$$

则 $W_2 = 1$,也就是空间平移群的同态表示. 对应到物理就是

$$F_{\mu\nu} = \partial_\mu\hat{A}_\nu - \partial_\nu\hat{A}_\mu + [\hat{A}_\mu, \hat{A}_\nu] = 0.$$

为了研究平移群的结合律,我们看

$$\begin{aligned}&U(a)(U(b)U(c))\phi(x)\\ &= W_1(x, x+a+b+c)W_2(x,x+a, x+a+b+c)W_2(x+a,\\ &\quad x+a+b, x+a+b+c)\phi(x+a+b+c)\\ &(U(a)U(b))U(c)\phi(x)\\ &= W_2(x+a, x+a+b,x)W_2(x+a+b, x+a+b+c, x)\\ &\quad W_1(x, x+a+b+c)\phi(x+a+b+c).\end{aligned} \tag{10}$$

下面我们在特殊情况下给出 W_2 在上边缘算子作用下的 3—闭链. 在 $W_2 = \delta W_1$ 的情况下，W_2 在上边缘算子作用下变成

$$\begin{aligned}&\delta W_2(x+a+b, x+a, x)\\ &= W_3(x+a+b+c,x+a+b,x+a,x)\\ &= W_1(x, x+a+b+c)W_2(x, x+a, x+a+b+c)\\ &\quad W_2(x+a,x+a+b, x+a+b+c)W_1(x+a+b+c, x)\\ &\quad W_2(x+a+b+c, x+a+b, x)W_2(x+a+b, x+a, x).\end{aligned} \tag{11}$$

对于一般的 2—闭链,情况极为复杂，我们在此没有给出一般的定义. 群的结合律的成立是 $\delta W_2 = 1$ 的充分必要条件. 显然把 $W_2 = \delta W_1$ 代入(11)式自动得到了

$$\delta W_2 = 1 \tag{12}$$

也就是这些表示自动满足结合律. 从某种意义讲,只对这种特殊情况给出 (11) 式的定义是平凡的. 但是下面将(11)式应用到某些别的情况,就可得到一些结论.

为了方便起见,我们先将(11)式表述为直观的几何图象 (图 1).

$$W_1(x, x+a+b+c)W_2(x, x+a, x+a+b+c)$$
$$W_2(x+a, x+a+b, x+a+b+c)$$

是沿着

$$(AD)(DA)(AB)(BD)(DB)(BC)(CD)$$

的空间移动群．另外三项是沿着

$$(DA)(AD)(DC)(CA)(AC)(CB)(BA)$$

的空间移动群．在(5)式成立的前提下，乘起来正好为 1． 所以当空间的规范势在全空间解析时，空间平移群的结合律一定满足．

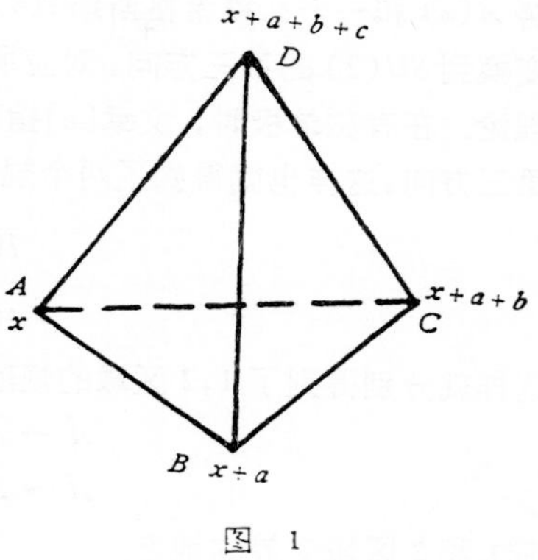

图　1

当空间存在一个磁单极时，规范势就不会在整个空间解析．必须定义两个规范势分别在各自的区域解析．当这两区域正好包含(AD)，(DC)，(CB)，(BA)时，则(11)式中的许多 W_1 就无确切定义，当然(5)式就可能不成立，也就是结合律会遭到破坏．

在上面这种情况下，设 (AD)，(DC)，(CB)，(BA) 的外侧(指四面体 $ABCD$)的规范势为 $\hat{A}_+$，里侧为 $\hat{A}_-$，两者之间差一个规范变换．下面我们就选择一个特殊的空间平移变换(因为有任意性)，这种定义方式在 $U(1)$ 群下，$\ln W_3$ 就是磁通量．

$$\begin{aligned} W_1(x, x+a+b+c) &= W_1^{-1}(x+a+b+c, x) \\ &= pe^{i\int_{AD}\hat{A}_-} \\ W_2(x, x+a, x+a+b+c) &= pe^{i\int_{DABD}\hat{A}_-} \\ W_2(x+a, x+a+b, x+a+b+c) &= pe^{i\int_{DBCD}\hat{A}_-} \\ W_2(x+a+b+c, x+a+b, x) &= pe^{i\int_{ADCA}\hat{A}_+} \\ W_2(x+a+b, x+a, x) &= pe^{i\int_{ACBA}\hat{A}_+}. \end{aligned} \tag{13}$$

在这种定义下，W_3 变成了

$$W_3 = pe^{i\int_{C^{-1}}\hat{A}_-}pe^{i\int_C\hat{A}_+}. \tag{14}$$

这里 C 是路径 $ADCBA$，C^{-1} 是 C 的逆．

在这节里我们所得到的结论是这一特殊空间平移群结合律的成立，也就是 $W_3 = 1$，将给 $\hat{A}_+$ 和 $\hat{A}_-$ 有一个限制——狄拉克量子化条件．

三、应　　用

1) $U(1)$ 群

这时由于 $\hat{A}$ 是函数，W_3 就是磁通量的指数，$W_3 = 1$ 的要求显然给出狄拉克量子化条件[3]．

2) $SU(2)$ 群

我们知道 $\pi_1(SU(2)) = 0$，意味着在整个空间中用一个解析的 $\hat{A}$ 就能表达，也就是没有真正的磁单极．平移群的结合律自动满足．

现在让我们来考虑 $SU(2)$ 群破缺到 $U(1)$ 群的规范理论[4,5,6]．这个理论是通过规范势 $\hat{A}(x)$ 和一个么模希格斯场 $\hat{\ell}(x)$ 来描写（$\hat{\ell}$ 标志着 $U(1)$ 在 $SU(2)$ 的方向）．只要把 $\hat{\ell}$ 变换到 $SU(2)$ 的第三方向，对应所得到的规范势就是在 $\hat{\ell}$ 中的规范势．这就是破缺规范理论．在有磁单极时，文章[6]指出需要两个规范变换才能把每点的 $\hat{\ell}(x)$ 变到 $SU(2)$ 的第三方向，这样也就得到了两个规范势．

$$\hat{\ell}(x) \xrightarrow{\phi_+} \lambda_3 \qquad x \in 1 \text{ 区域}$$
$$\hat{\ell}(x) \xrightarrow{\phi_-} \lambda_3 \qquad x \in 2 \text{ 区域}$$

这样就分别得到了 1,2 区域的规范势

$$\begin{aligned} \hat{A} &\to \hat{A}^+ = \phi_+^{-1}\hat{A}\phi_+ + \phi_+^{-1}d\phi_+, \\ \hat{A} &\to \hat{A}^- = \phi_-^{-1}\hat{A}\phi_- + \phi_-^{-1}d\phi_-, \end{aligned} \tag{15}$$

在 1 和 2 区域交界的地方

$$\hat{A}^+ = \Lambda^{-1}\hat{A}^-\Lambda + \Lambda^{-1}d\Lambda. \tag{16}$$

其中：$\Lambda = \phi_-^{-1}\phi_+ \in U(1)$，$\phi_\pm^{-1}\lambda^3\phi_\pm = \hat{\ell}$

其中 Λ 的选择是 $SU(2)$ 的第三分量(见文章[6])．

显然在前面那种路径选取情况下，(14)式不一定等于 1，除非 $\hat{A}_+$ 和 $\hat{A}_-$ 满足一定的条件，它们将会在别的地方得到讨论．对于恰好 $\hat{A}^+$ 和 $\hat{A}^-$ 只有 $U(1)$ 分量的情况，即

$$\hat{A}^\pm\Lambda = \Lambda\hat{A}^\pm. \tag{17}$$

这样

$$\begin{aligned} W_3 &= pe^{i\int_{C^-}\hat{A}^-}\, pe^{i\int_C \hat{A}^+} \\ &= pe^{i\int_{C^-}\hat{A}^-}\, \Lambda pe^{i\int_C \hat{A}^+}\, \Lambda^{-1}\phi_-^{-1}\phi_- \\ &= \phi_-^{-1} pe^{i\int_{C^-}\hat{A}}\, pe^{i\int_C \hat{A}}\, \phi_- \\ &= \phi_-^{-1}\phi_- \\ &= 1. \end{aligned} \tag{18}$$

也就是若 $\hat{A}^+$ 和 $\hat{A}^-$ 只有 $U(1)$ 分量，则结合律一定满足．另外从(18)式

$$pe^{i\int_{C^-}\hat{A}^-}\, pe^{i\int_C \hat{A}^+} = e^{i\oiint F} = 1.$$

要求磁单极的强度满足量子化条件

$$2eg = 2n, \quad n = 0, \pm 1, \pm 2 \cdots \tag{19}$$

e 是 $SU(2)$ 规范理论的耦合常数（以前的公式 $e = 1$），g 是磁荷．方程(18)表明了不满足(19)的磁单极将与(17)式矛盾．

3) $SO(3)$ 群

我们知道 $SO(3)$ 和 $SU(2)$ 的大范围性质不一样．从 $\pi_1(SO(3)) = Z_2$，我们知道 $SO(3)$ 群中有一个平凡的纤维丛，还有一个非平凡丛．

对 $SO(3)$ 的规范理论和破缺到 $U(1)$ 的规范理论，平凡丛的讨论完全和 $SU(2)$ 一样．稍微有一点不同的地方在于(19)式变成

$$2eg = n \quad n = 0, \pm 1, \cdots \tag{20}$$

因子 2 的差别在于

$$e^{i\lambda_3(SU(2))4\pi} \quad e^{i\lambda_3(SO(3))2\pi} = 1$$

这个原因. 其中 $\lambda_3(SU(2))$ 和 $\lambda_3(SO(3))$ 分别是 $SU(2)$ 和 $SO(3)$ 群的生成元.

对于 $SO(3)$ 的非平凡丛需要将来讨论.

感谢郭汉英同志给予的帮助.

参　考　文　献

[1] K. C. Chou, H. Y. Guo, K. Wu, X. C. Song Beijing preprint AS-ITP-84-018 B. Zumino Seattle preprint LBL 16746 UCD-PTH-83/IC R. Stora Ahnecy preprint LAAP-Th-94(1983).
[2] L. D. Faddeev *Phys. Lett.*, B.145(1984), 81.
[3] R. Jackiw MIT preprint CPT 1209(1984) Hou Bo-yu, Hou Bo-yuan Northwestern University. China preprint NWU 84-8.
Y. S. Wu A. Zee University of Washington 40084-29 P4 (1984).
[4] G. t'Hooft Nucl. Phys., **B79**(1974), 276.
[5] 侯伯宇、戴元本、葛墨林，中国物理，**25**(1976)，514.
[6] 吴泳时、陈时、杜东生、郭汉英，高能物理与核物理，**1**(1977)，53.

A NEW CO-BOUNDARY OPERATOR AND ITS APPLICATIONS

Chou Kuang-chao　Wu Yue-liang　Xie Yan-bo
(Institute of Theoretical Physics, Academia Sinica)

Abstract

In this note new co-boundary operators are defined in the product form. The associative composition law of spatial translation group field is discussed using these new operators. The quantization condition of monopoles in $SU(2)$ and $SU(3)$ gauge theories follows easily from the new formalism.

Part IV

统计物理与凝聚态物理

第 29 卷 ·第 5 期 物理学报 Vol. 29, No. 5
1980 年 5 月 ACTA PHYSICA SINICA May, 1980

非平衡耗散系统定常态的 Goldstone 模*

周光召 苏肇冰
（中国科学院理论物理研究所）

提 要

在有时空结构的非平衡定常态上，由于对称性的自发破缺，形成非零能、带耗散的Goldstone模．本文以么模激光的饱和状态为例，从普遍的相位对称性出发，应用闭路格林函数中的 Ward-Takahashi 恒等式，具体地导出了这种带耗散的推广的 Goldstone 定理的一个形式，讨论了相应的 Goldstone 模式的物理意义．求得了饱和状态光量子态的分裂．具体给出了状态从相位有序——→相位无序的转化规律，从而给自发破缺的对称性的恢复一个具体的描述．

一、引 言

1966 年 Korenman 应用 Schwinger-Кельдыш 非平衡量子统计格林函数[1]（以下简称闭路格林函数）来处理激光量子理论[2]时，发现如果存在定常的单模激光振荡，那么场量的涨落是发散的．发散来源于一种 Goldstone 极点．这种发散使它没有能得到应有的正确结果．Korenman 遇到的困难表面上是闭路格林函数方法本身的困难，因为通常的量子光学[3]已经能够较好地解释饱和区的物理现象．但实际上，它还是很有意义的．单模激光的饱和状态是一个非平衡定常态．Korenman 的困难在某种程度上从量子场统计理论的角度上提出了一系列的问题．例如

1. 对于一个非平衡的量子统计系统，究竟是否存在一个确定频率的定常的单模序参量严格解？如果没有，那么非零频率模的非平衡定常态在量子统计理论中还有没有确切的意义？激光的饱和状态是不是也只有近似的意义？

2. 在通常的量子场论和许多固体现象中也常出现 Goldstone 激发．但是，Goldstone 激发一般是零频率的．而 Korenman 所遇到的发散则是出现在频率不为零的模式上的．由于总能量有限，事实上激光的场量涨落不可能发散．在这种情况下，Goldstone 定理的物理意义是什么？Goldstone 激发与普通光子有什么差别？

3. 如所周知，激光谱线有一个特别窄的宽度．它来源于序参量在噪声源作用下产生的位相扩散．这种位相扩散与 Korenman 遇到的发散有什么关系？它与对称破缺的恢复有什么具体联系？

4. 激光的量子理论早已成熟．而 Korenman 应用闭路格林函数来处理激光时却遇到了场量涨落发散的困难，那么究竟闭路格林函数方法能不能解决激光问题？它作为非平衡量子统计理论的工具有没有这样或那样的局限性？

* 1979 年 5 月 14 日收到．

在本文中，我们将用闭路格林函数方法，从比较基本的角度上对上面列出的问题作一些初步的探讨．我们将以单模饱和激光作为一个非零频模非平衡量子统计定常态的例子来讨论它的 Goldstone 定理、序参量量子化和系统对称破缺的恢复等问题．我们同时也得到结论：用闭路格林函数来处理激光饱和状态的物理现象是完全合适的．我们的讨论虽然是具体针对单模激光进行的，但是，得到的主要结果都有可能推广到其他非零频模非平衡量子统计定常态上去．

现在讨论这样的模型．它由单模激光场、二能级原子和与它们相互作用的周围环境所组成：以 $a(x)$ 和 $a^*(x)$ 分别描写单模激光场矢量势的正频部和负频部，$\phi_1(x)$ 和 $\phi_2(x)$ 描写上能级 E_1 和下能级 E_2 上的束缚电子波函数，以 $\vartheta_i(x)$ $(i=1,2\cdots,n)$ 为描述环境状态的参量．系统的作用量为 S

$$S=\int_p d^4x\,(\mathscr{L}_{\mathrm{F}}(x)+\mathscr{L}_{\mathrm{A}}(x)+\mathscr{L}_{\mathrm{F-A}}(x)+\mathscr{L}_{\mathrm{res}}(x)+\mathscr{L}_{\mathrm{F-res}}(x)+\mathscr{L}_{\mathrm{A-res}}(x)+\mathscr{L}_{\mathrm{ext}}(x)), \tag{1.1}$$

$$\mathscr{L}_{\mathrm{F}}(x)=a^*(x)\left(i\hbar\frac{\partial}{\partial t}-k_0\right)a(x), \tag{1.2}$$

$$\mathscr{L}_{\mathrm{A}}(x)=\phi_1^*(x)\left(i\hbar\frac{\partial}{\partial t}+\frac{\hbar^2}{2M}\nabla^2-E_1\right)\phi_1(x)+\phi_2^*(x)\left(i\hbar\frac{\partial}{\partial t}+\frac{\hbar^2}{2M}\nabla^2-E_2\right)\phi_2(x), \tag{1.3}$$

$$\mathscr{L}_{\mathrm{F-A}}(x)=j_{\mathrm{A}}^*(x)\,a(x)+a^*(x)\,j_{\mathrm{A}}(x), \tag{1.4}$$

$$\mathscr{L}_{\mathrm{F-res}}(x)=j_{\mathrm{res}}^*(x)a(x)+a^*(x)\,j_{\mathrm{res}}(x), \tag{1.5}$$

$$\mathscr{L}_{\mathrm{ext}}(x)=J^*(x)\,a(x)+a^*(x)\,J(x)+\eta_1^*(x)\,\phi_1(x)+\phi_1^*(x)\,\eta_1(x)+\eta_2^*(x)\,\phi_2(x)+\phi_2^*(x)\,\eta_2(x), \tag{1.6}$$

$$j_{\mathrm{A}}(x)=-ig\phi_2^*(x)\phi_1(x),\quad j_{\mathrm{A}}^*(x)=ig\phi_1^*(x)\,\phi_2(x), \tag{1.7}$$

$$g=\frac{\omega}{c}\,|\boldsymbol{d}\cdot\boldsymbol{e}|\left(\frac{\hbar^2c^2}{2k_0V}\right)^{\frac{1}{2}}, \tag{1.8}$$

$$\hbar\omega=E_1-E_2>0, \tag{1.9}$$

$$j_{\mathrm{res}}(x)=\text{环境参量 }\vartheta_i\text{ 引起的电流}, \tag{1.10}$$

其中 $\int_p d^4x$ 是沿闭路的全时空积分；$\mathscr{L}_{\mathrm{F}}(x)$，$\mathscr{L}_{\mathrm{A}}(x)$ 和 $\mathscr{L}_{\mathrm{F-A}}(x)$ 分别是单模电磁场、二能级原子和电磁场-二能级原子相互作用的拉氏密度；$\mathscr{L}_{\mathrm{res}}(x)$ 是环境力学量的拉氏密度，$\mathscr{L}_{\mathrm{F-res}}(x)$ 和 $\mathscr{L}_{\mathrm{A-res}}(x)$ 分别是场和环境、原子和环境相互作用的拉氏密度；$\mathscr{L}_{\mathrm{ext}}(x)$ 是人为外源相互作用拉氏密度．我们以 $J(x)$，$\eta_1(x)$ 和 $\eta_2(x)$ 表示生成泛函技术所需的人为非物理外源，在作完有关的运算以后，最后取零值．

我们以 g 表示单模电磁场-二能级束缚电子的耦合常数，$\boldsymbol{d}$ 是能级 1 和 2 之间的电偶极矩阵元，$\boldsymbol{e}$ 是单模电磁场偏振矢量，k_0 是单模激光振荡能量，其中包含了一个重正化的能量位移．M 是原子质量．V 是归一化体积．由于我们关心的主要是电磁场的运动，所以只写出了电磁场-环境相互作用的拉氏密度(1.5)式，并定义

$$j(x)=j_{\mathrm{A}}(x)+j_{\mathrm{res}}(x). \tag{1.11}$$

此外，我们有兴趣的只是电磁场和二能级束缚电子的共振相互作用，所以在作用量 (1.1) 式中取了迴转波近似[4]，并忽略了电流密度中的 $\frac{e^2}{mc^2}\psi^* A\psi$ 项，这样就剩下了由 (1.2)，(1.4)，(1.7)和(1.8)式所给出的单模电磁场的非相对论描述.

容易发现，由(1.2)，(1.3)，(1.4)和(1.7)式描写的 $\mathscr{L}_F(x)$，$\mathscr{L}_A(x)$ 和 $\mathscr{L}_{F-A}(x)$ 具有一个相因子 $U(1)$ 群对称性. 即在 $U(1)$ 群对称变换

$$\begin{aligned} a(x) &\to e^{i\alpha}a(x), & a^*(x) &\to a^*(x)\,e^{-i\alpha}; \\ \psi_1(x) &\to e^{-i\frac{\alpha}{2}}\psi_1(x), & \psi_1^*(x) &\to \psi_1^*(x)e^{i\frac{\alpha}{2}}; \\ \psi_2(x) &\to e^{i\frac{\alpha}{2}}\psi_2(x), & \psi_2^*(x) &\to \psi_2^*(x)\,e^{-i\frac{\alpha}{2}} \end{aligned} \tag{1.12}$$

下，$\mathscr{L}_F(x)$，$\mathscr{L}_A(x)$ 和 $\mathscr{L}_{F-A}(x)$ 保持不变. 我们还采取这样的观点，设环境力学量 $\vartheta_i(x)$ $(i=1,2,\cdots,n)$ 也是上述相因子 $U(1)$ 群的表示，而使得整个作用量 (1.1) 式扣除 $\mathscr{L}_{ext}(x)$ 以后在相因子 $U(1)$ 群变换下不变. 这个相因子变换对称性是我们下面整个讨论的出发点.

最后我们指出，在下面讨论中所用的闭路格林函数的符号和技术，主要参照文献[5].

二、单模激光饱和状态的守恒荷和推广的 Goldstone 定理

按照文献[6]，我们的系统具有下列 W-T 恒等式:

$$\begin{aligned} \partial^\mu\langle\hat{j}(x)\rangle &- \frac{i}{\hbar}\frac{\delta\Gamma}{\delta a(x)}\,a(x) + \frac{i}{\hbar}\,a^*(x)\,\frac{\delta\Gamma}{\delta a^*(x)} \\ &- \frac{i}{2\hbar}\frac{\delta\Gamma}{\delta\psi_1(x)}\psi_1(x) + \frac{i}{2\hbar}\,\psi_1^*(x)\,\frac{\delta\Gamma}{\delta\psi_1^*(x)} \\ &+ \frac{i}{2\hbar}\frac{\delta\Gamma}{\delta\psi_2(x)}\psi_2(x) - \frac{i}{2\hbar}\,\psi_2^*(x)\,\frac{\delta\Gamma}{\delta\psi_2^*(x)} \\ &+ \text{与环境力学量有关项} = 0, \end{aligned} \tag{2.1}$$

其中

$$\langle\hat{j}_\mu(x)\rangle = \langle\hat{j}_\mu^F(x)\rangle + \langle\hat{j}_\mu^{pop}(x)\rangle + \langle\hat{j}_\mu^{res}(x)\rangle, \tag{2.2}$$

$$\langle\hat{j}_\mu^F(x)\rangle = -\frac{i}{\hbar}\left\langle\left(\frac{\delta\hat{\mathscr{L}}}{\delta\partial^\mu\hat{a}(x)}\,\hat{a}(x) - \hat{a}^+(x)\,\frac{\delta\hat{\mathscr{L}}}{\delta\partial^\mu\hat{a}^+(x)}\right)\right\rangle, \tag{2.3}$$

$$\begin{aligned} \langle\hat{j}_\mu^{pop}(x)\rangle = &\frac{i}{2\hbar}\left\langle\left(\frac{\delta\hat{\mathscr{L}}}{\delta\partial^\mu\hat{\psi}_1(x)}\,\hat{\psi}_1(x) - \hat{\psi}_1^+(x)\,\frac{\delta\hat{\mathscr{L}}}{\delta\partial^\mu\hat{\psi}_1^+(x)}\right)\right\rangle \\ &- \frac{i}{2\hbar}\left\langle\left(\frac{\delta\hat{\mathscr{L}}}{\delta\partial^\mu\hat{\psi}_2(x)}\,\hat{\psi}_2(x) - \hat{\psi}_2^+(x)\,\frac{\delta\hat{\mathscr{L}}}{\delta\partial^\mu\hat{\psi}_2^+(x)}\right)\right\rangle, \end{aligned} \tag{2.4}$$

$$\langle\hat{j}_\mu^{res}(x)\rangle = \text{环境力学量所对应的流密度}. \tag{2.5}$$

Γ 是不可约顶角生成泛函[6]. 从 W-T 恒等式(2.1)出发，将得到系统的两个重要性质.

首先，对(2.1)式作 $\int d^3x$ 积分，然后令外源 $\longrightarrow 0$，这就得到

$$\partial_0 Q(x_0) = 0, \quad Q(x_0) = \int d^3x\,\langle\hat{j}_0(x)\rangle. \tag{2.6}$$

由(2.2)—(2.5)式和(1.2)，(1.3)式，容易有

$$\langle \hat{j}_0(x)\rangle = \langle \hat{j}_0^{\mathrm{F}}(x)\rangle + \langle \hat{j}_0^{\mathrm{pop}}(x)\rangle + \langle \hat{j}_0^{\mathrm{res}}(x)\rangle, \tag{2.7}$$

$$\langle \hat{j}_0^{\mathrm{F}}(x)\rangle = \langle \hat{j}_0^{\mathrm{coh}}(x)\rangle + \langle \hat{j}_0^{\mathrm{fluc}}(x)\rangle, \tag{2.8}$$

$$\begin{aligned}\langle \hat{j}_0^{\mathrm{coh}}(x)\rangle &= a^*(x)\,a(x) = \frac{1}{V}\,A^*(k,T)\,A(k,T),\\ \langle \hat{j}_0^{\mathrm{fluc}}(x)\rangle &= \langle \hat{a}^+(x)\,a(x)\rangle_{\mathrm{conn}} = \frac{i\hbar}{V}\int\frac{dk_0}{2\pi\hbar}\,D_{11}^{+-}[x,k],\end{aligned} \tag{2.9}$$

$$\begin{aligned}\langle \hat{j}_0^{\mathrm{pop}}(x)\rangle &= \frac{1}{2}\,\langle \hat{\psi}_1^+(x)\,\hat{\psi}_1(x) - \hat{\psi}_2^+(x)\,\hat{\psi}_2(x)\rangle\\ &= -\frac{i\hbar}{2}\int\frac{d^4p}{(2\pi\hbar)^4}\,(S_{\mathrm{F}}^{+-}[x,p,1] - S_{\mathrm{F}}^{+-}[x,p,2]),\end{aligned} \tag{2.10}$$

$$\langle \hat{j}_0^{\mathrm{res}}(x)\rangle = \text{环境力学量相应的荷密度}, \tag{2.11}$$

其中

$$a(x) \equiv \langle \hat{a}(x)\rangle \equiv S_{\mathrm{p}}(\hat{\rho}\hat{a}(x)) = \frac{1}{V}\,A(k,T)\,e^{-\frac{i}{\hbar}kx} \tag{2.12}$$

为描述动量为 $\boldsymbol{k}$ 的单模激光序参量，$A(k,T)$中的 k 标记模式，T 描写振幅对时间可能的缓慢变化．(2.6) 式说明系统存在一个守恒荷 Q，它反映了各类量子在量子态上占据数之间的一种相互制约关系．$\int d^3x\,\langle \hat{j}_0^{\mathrm{coh}}(x)\rangle$ 反映了系统中处于相干或有序状态的光量子简并度．$\int d^3x\langle \hat{j}_0^{\mathrm{fluc}}(x)\rangle$反映了系统中处于较为无序状态的"光量子"简并度．$\int d^3x\langle \hat{j}_0^{\mathrm{pop}}(x)\rangle$是二能级原子上下电子占据数分布的反转程度．值得注意的是存在守恒荷 Q 并不意味着激光状态的稳定，还要考虑原子占据数反转的变化和环境状态的变化．现在将这守恒荷应用到激光的饱和状态．这里讨论的激光饱和状态是这样一种状态：原子占据数分布达到稳定，环境占据数也稳定．在这种状态下，激光系统的总量子数

$$\int d^3x\langle \hat{j}_0^{\mathrm{F}}(x)\rangle = \int d^3x\langle \hat{j}_0^{\mathrm{coh}}(x)\rangle + \int d^3x\,\langle \hat{j}_0^{\mathrm{fluc}}(x)\rangle = \text{常数}. \tag{2.13}$$

所以饱和状态是有确切含义的，即单模激光的总量子数达到恒定，而不一定是序参量保持稳定．在饱和状态，电磁场有序状态和较为无序的状态之间可能互相转化．以上所述的是我们的一个出发点．将在下面两节中进一步研究这些量子的性质和它们之间的转化．

现在，重新对(2.1)式作一个闭路四维积分，然后分别作一次泛函微商 $\frac{\delta}{\delta a(x)}$和$\frac{\delta}{\delta a^*(x)}$，最后再令外源$\longrightarrow 0$，这样就得到

$$\begin{aligned}&\int_p d^4y\left(\frac{\delta^2\Gamma}{\delta a^*(x)\delta a(y)}\;a(y) - \frac{\delta^2\Gamma}{\delta a^*(x)\delta a^*(y)}\,a^*(y)\right) = 0,\\ &\int_p d^4y\left(\frac{\delta^2\Gamma}{\delta a(x)\delta a(y)}\;a(y) - \frac{\delta^2\Gamma}{\delta a(x)\delta a^*(y)}\,a^*(y)\right) = 0.\end{aligned} \tag{2.14}$$

为了以后方便，在整个闭路上引入符号

$$\begin{pmatrix}\Gamma_{11}(x,y), & \Gamma_{12}(x,y)\\ \Gamma_{21}(x,y), & \Gamma_{22}(x,y)\end{pmatrix} = \begin{pmatrix}\dfrac{\delta^2\Gamma}{\delta a^*(x)\delta a(y)}, & \dfrac{\delta^2\Gamma}{\delta a^*(x)\delta a^*(y)}\\ \dfrac{\delta^2\Gamma}{\delta a(x)\delta a(y)}, & \dfrac{\delta^2\Gamma}{\delta a(x)\delta a^*(y)}\end{pmatrix}. \tag{2.15}$$

(2.14)式是确定在整个闭路上的,为要回到普通空间,对(2.14)式取

$$a(x_+) = a(x_-) = a(x) \equiv \langle \hat{a}(x) \rangle,$$
$$a^*(x_+) = a^*(x_-) = a^*(x) \equiv \langle \hat{a}^+(x) \rangle, \tag{2.16}$$

就有

$$\int d^4y \, (\Gamma_{11}^R(x,y)\, a(y) - \Gamma_{21}^R(x,y)\, a^*(y)) = 0,$$
$$\int d^4y \, (\Gamma_{21}^R(x,y)\, a(y) - \Gamma_{22}^R(x,y)\, a^*(y)) = 0, \tag{2.17}$$

其中[5]

$$\Gamma_{i,j}^R(x,y) = \Gamma_{i,j}^{++}(x,y) - \Gamma_{i,j}^{+-}(x,y) = \Gamma_{i,j}^{-+}(x,y) - \Gamma_{i,j}^{--}(x,y). \tag{2.18}$$

这即是普通场论中的 Goldstone 定理在量子统计系统存在一般非零序参量情形下的一般形式.

为了从普遍的对称原理出发,得到尽可能多的不依赖于具体模型的具体结果,必须能够正确处理系统中许多不同变化量级的时间标度. 我们将假设我们的系统具有三维空间平移不变性;对于时间变量,从(2.13)式和相应的讨论可以知道,在饱和状态,激光序参量可能因为与较为无序的振荡相互转化而随时间发生变化. 这种变化相对单模激光振荡而言显然是缓慢的. 所以一方面我们对电磁场序参量引入两个时间标度

$$a(x) \equiv a(x, \epsilon t) = \int \frac{d^4q}{(2\pi\hbar)^4} e^{-\frac{i}{\hbar}qx} a(q, \epsilon t) \text{ 记为} \int \frac{d^4q}{(2\pi\hbar)^4} e^{-\frac{i}{\hbar}qx} a(q,t), \tag{2.19}$$

其中前一个宗量描写与激光振荡相对应的时间变化,后一个宗量(ϵ 是小量)描写序参量的缓慢变化. 另一方面,由于存在单模振荡,因此对系统的任一个两点传播函数 $F_{ij}(x_1, x_2)$. 将按照下述方式引入对应的一个 $\tilde{F}_{ij}(x_1, x_2)$

$$\begin{pmatrix} F_{11}(x_1,x_2), & F_{12}(x_1,x_2) \\ F_{21}(x_1,x_2), & F_{22}(x_1,x_2) \end{pmatrix} = \begin{pmatrix} e^{-\frac{i}{\hbar}k(x_1-x_2)} \tilde{F}_{11}(x_1,x_2), & e^{-\frac{i}{\hbar}k(x_1+x_2)} \tilde{F}_{12}(x_1,x_2) \\ e^{\frac{i}{\hbar}k(x_1+x_2)} \tilde{F}_{21}(x_1,x_2), & e^{\frac{i}{\hbar}k(x_1-x_2)} \tilde{F}_{22}(x_1,x_2) \end{pmatrix}, \tag{2.20}$$

$$\tilde{F}_{ij}(x_1, x_2) = \int \frac{d^4q}{(2\pi\hbar)^4} e^{-\frac{i}{\hbar}q(x_1-x_2)} \tilde{F}_{ij}[q, T] \Big|_{T=\frac{t_1+t_2}{2}}. \tag{2.21}$$

只有 $\tilde{F}_{ij}(x_1,x_2)$ 对 $T = \frac{t_1+t_2}{2}$ 的依赖才能正确反映与序参量缓变化相对应的系统对于平移不变的偏离. 在引入了 $\tilde{F}$ 以后,(2.17)式可以表为

$$\int d^4y \left(\tilde{\Gamma}_{11}^R(x,y)\, a(y, \epsilon y_0)\, e^{-\frac{i}{\hbar}ky} - \tilde{\Gamma}_{12}^R(x,y)\, a^*(y, \epsilon y_0)\, e^{\frac{i}{\hbar}ky} \right) = 0,$$
$$\int d^4y \left(\tilde{\Gamma}_{21}^R(x,y)\, a(y, \epsilon y_0)\, e^{-\frac{i}{\hbar}ky} - \tilde{\Gamma}_{22}^R(x,y)\, a^*(y, \epsilon y_0)\, e^{\frac{i}{\hbar}ky} \right) = 0. \tag{2.22}$$

为了将(2.22)式对慢变宗量作展开,还需引入两个假设

假设 I
$$o\left(\frac{\hbar}{\Delta T \Delta q_0}\right) \ll 1, \tag{2.23}$$

假设 II
$$\left|\frac{\partial \Pi_{ij}^R[T,q]}{\partial q_0}\right| \ll 1 = \left|\frac{\partial \Gamma_{11}^{R(0)}[T,q]}{\partial q_0}\right| \sim \left|\frac{\partial \Gamma_{22}^{R(0)}[T,q]}{\partial q_0}\right|. \tag{2.24}$$

在(2.24)式中我们将 $\Gamma_{ij}^R[T,q]$ 分成自由部份 $\Gamma_{ij}^{R(0)}[T,q]$ 和自能部份 $\Pi_{ij}^R[T,q]$[8],

$$\Gamma_{ij}^R[T,q] = \Gamma_{ij}^{R(0)}[T,q] + \Pi_{ij}^R[T,q], \tag{2.25}$$

$$\Gamma_{ij}^{R(0)}[T, q] = \begin{pmatrix} q_0 - k_0, & 0 \\ 0, & -q_0 - k_0 \end{pmatrix}. \tag{2.26}$$

利用(2.23)和(2.24)式，按多时标微扰论精神将(2.22)式对慢变宗量作展开[7]，并取 $q = 0$ 分量，就得到(2.22)式准到 $o\left(\frac{\hbar}{\Delta T \Delta q_0}\right)$ 一次幂的近似表达式

$$\begin{aligned} &\tilde{\Gamma}_{11}^{R}[T, q=0]\, a(k, T) - \tilde{\Gamma}_{12}^{R}[T, q=0]\, a^*(k, T) - \frac{\hbar}{i}\frac{\partial a(k,T)}{\partial t} = 0, \\ &\tilde{\Gamma}_{21}^{R}[T, q=0]\, a(k, T) - \tilde{\Gamma}_{22}^{R}[T, q=0]\, a^*(k,T) - \frac{\hbar}{i}\frac{\partial a^*(k,T)}{\partial T} = 0. \end{aligned} \tag{2.27}$$

我们还定义

$$\frac{\delta}{2\hbar} \equiv -\frac{\partial}{\partial T}\ln a(k, T). \tag{2.28}$$

由于 δ 至少是一阶小量，所以，在准到一阶小量的意义下，δ 与 T 无关．此外，δ 可取为实数，因为如果它有虚部可以将它吸收到 k_0 中去．这样，(2.27)式就可化为

$$\begin{aligned} &\tilde{\Gamma}_{11}^{R}[T, q=0]\, a(k, T) - \tilde{\Gamma}_{12}^{R}[T, q=0]\, a^*(k, T) - i\frac{\delta}{2}a(k,T) = 0, \\ &\tilde{\Gamma}_{21}^{R}[T, q=0]\, a(k, T) - \tilde{\Gamma}_{22}^{R}[T, q=0]\, a^*(k, T) - i\frac{\delta}{2}a^*(k,T) = 0. \end{aligned} \tag{2.29}$$

如果在系统的饱和状态，存在 $q = k$ 的单模激光非零解

$$a(k, T) \doteq 0,\ a^*(k, T) \doteq 0, \tag{2.30}$$

则(2.29)式要求

$$\det\begin{vmatrix} \tilde{\Gamma}_{11}^{R}[T, q=0] - i\frac{\delta}{2}, & -\tilde{\Gamma}_{12}^{R}[T, q=0] \\ -\tilde{\Gamma}_{21}^{R}[T, q=0], & \tilde{\Gamma}_{22}^{R}[T, q=0] + i\frac{\delta}{2} \end{vmatrix} = 0. \tag{2.31}$$

(2.31)式即是我们希望得到的系统存在 $q = k$ 模非零饱和序参量解时推广的 Goldstone 定理．

现在回过来考察假设 I 和假设 II 成立的物理条件．由于我们有兴趣的单模激光-二能级原子相互作用是一种共振相互作用，它由下列形式的因子

$$\frac{\gamma}{(q_0 - k_0)^2 + \gamma^2} \tag{2.32}$$

所描述，其中 γ 是原子的非对角弛豫．所以，$\frac{\partial}{\partial q_0}$ 就主要来自这个共振因子的贡献．从量级上有

$$\left|\frac{\partial}{\partial q_0}\right| \sim \frac{1}{\gamma}. \tag{2.33}$$

这样，(2.23)和(2.24)式就等价于

假设 I′　　$o\left(\frac{\delta}{\gamma}\right) \ll 1,$ 　　(2.34)

假设 II′ $$o\left(\frac{\Pi^R_{ij}}{\gamma}\right) \ll 1. \tag{2.35}$$

(2.34)式在实验上是明显的,因为激光线宽非常窄,这一点将在第四节进一步论证.(2.35)式就是通常量子光学中作绝热消去原子变数时常用的近似[9]. 它在物理上有着广泛的现实性.

上面得到的(2.31)式——饱和状态推广的 Goldstone 定理的具体形式,是我们从对称性原理, W-T 恒等式(2.1)出发得到的第二个结果. 它和(2.13)式一起是下面进一步讨论的两个出发点. 此外,我们也开始从正面对 Korenman 的困难[2] 进行回答,由于他没有能够处理好不同变化快慢的时间标度,所以他得到的饱和区 Goldstone 定理缺少了(2.31)式中的 δ 有关项.

三、饱和状态单模电磁场格林函数的解析性和电磁场量子化

对于固定动量 $\boldsymbol{q}=\boldsymbol{k}$ 的单模电磁场,在频率空间,我们有三个有兴趣的量纲: δ, $\prod^R_{ij}$和 γ. 其中前两个是与电磁场相联系的耗散弛豫,第三个是原子系统的非对角弛豫. 当我们讨论饱和状态电磁场格林函数解析性的时候,对于将单模振荡因子分离出去以后的两点函数, q_0 在频率空间重要的变化范围应该是

$$|q_0|\lesssim\delta \ll \gamma \text{ 或 } |q_0|\lesssim\left|\prod{}^R_{ij}\right| \ll \gamma. \tag{3.1}$$

(3.1)式中后面的不等式来自 (2.34)和(2.35) 式. 再考虑到共振相互作用是主要贡献,综合(3.1)和(2.33)式,对于 q_0 在物理上重要的变化区域,有

$$\left|q_0\frac{\partial}{\partial q_0}\right| \sim \frac{\delta}{\gamma} \text{ 或 } \frac{|\Pi^R_{ij}|}{\gamma} \ll 1. \tag{3.2}$$

这样,就得到一个结论: 在讨论饱和状态电磁场格林函数解析性的时候, 可以直接将 $\tilde{\Gamma}^R$ 在 $q_0=0$ 点对 q_0 作泰勒展开并取到一次幂项. 这就有可能不需要知道作用量的具体形式,从而就能求得饱和状态格林函数的解析性质.

按上所述,饱和状态 $\tilde{\Gamma}^R_{ij}[T;\boldsymbol{q}=0,q_0]$ 在 $q_0=0$ 附近的泰勒展开式为

$$\begin{aligned}\tilde{\Gamma}^R_{ij}[T;\boldsymbol{q}=0,q_0] &= \tilde{\Gamma}^R_{ij}[T;q=0]+q_0\left[\frac{\partial\tilde{\Gamma}^R_{ij}[T;\boldsymbol{q}=0,q_0]}{\partial q_0}\right]_{q_0=0}\\ &=\begin{pmatrix}\tilde{\Gamma}^R_{11}[T,q=0]+q_0, & \tilde{\Gamma}^R_{12}[T,q=0]\\ \tilde{\Gamma}^R_{21}[T,q=0], & \tilde{\Gamma}^R_{22}[T,q=0]-q_0\end{pmatrix}.\end{aligned} \tag{3.3}$$

在(3.3)式中,我们对展开的一次幂项取了 $\dfrac{\partial\tilde{\Gamma}^R}{\partial q_0}$ 的零级量贡献.

按照文献[10],分离出单模振荡因子以后的电磁场推迟格林函数 $\tilde{D}^R_{ij}(x_1,x_2)$ 满足下列方程:

$$\begin{aligned}&\int d^4y\,\tilde{D}^R_{ij}(x_1,y)\,\tilde{\Gamma}^R_{jk}(y,x_2)=\delta_{ik}\,\delta^4(x_1-x_2),\\ &\int d^4y\,\tilde{\Gamma}^R_{ij}(x_1,y)\,\tilde{D}^R_{jk}(y,x_2)=\delta_{ik}\,\delta^4(x_1-x_2);\end{aligned} \tag{3.4}$$

或它在相对动量表象中的 $o\left(\frac{\hbar}{\Delta T\Delta q_0}\right)$ 一阶项展开

$$\widetilde{D}^R_{ij}[T,q]\,\widetilde{\Gamma}^R_{jk}[T,q]+\frac{\hbar}{2i}\left(\frac{\partial\widetilde{D}^R_{ij}[T,q]}{\partial T}\frac{\partial\widetilde{\Gamma}^R_{jk}[T,q]}{\partial q_0}-\frac{\partial\widetilde{D}^R_{ij}[T,q]}{\partial q_0}\frac{\partial\widetilde{\Gamma}^R_{jk}[T,q]}{\partial T}\right)=\delta_{i,k},$$

$$\widetilde{\Gamma}^R_{ij}[T,q]\,\widetilde{D}^R_{jk}[T,q]+\frac{\hbar}{2i}\left(\frac{\partial\widetilde{\Gamma}^R_{ij}[T,q]}{\partial T}\frac{\partial\widetilde{D}^R_{jk}[T,q]}{\partial q_0}-\frac{\partial\widetilde{\Gamma}^R_{ij}[T,q]}{\partial q_0}\frac{\partial\widetilde{D}^R_{jk}[T,q]}{\partial T}\right)=\delta_{i,k}.\tag{3.5}$$

我们有兴趣的是(3.5)式中 $\boldsymbol{q}=0$ 的分量．将不直接求解(3.5)式，而是先设

$$\widetilde{D}^R_{ij}[T;\boldsymbol{q}=0,q_0]\,\widetilde{\Gamma}^R_{jk}[T;\boldsymbol{q}=0,q_0]=\delta_{i,k},$$

$$\widetilde{\Gamma}^R_{ij}[T;\boldsymbol{q}=0,q_0]\,\widetilde{D}^R_{jk}[T;\boldsymbol{q}=0,q_0]=\delta_{i,k},\tag{3.6}$$

并求解之．然后再证明(3.6)式的解满足等式

$$\frac{\hbar}{2i}\left(\frac{\partial\widetilde{D}^R_{ij}[T;\boldsymbol{q}=0,q_0]}{\partial T}\frac{\partial\widetilde{\Gamma}^R_{jk}[T;\boldsymbol{q}=0,q_0]}{\partial q_0}-\frac{\partial\widetilde{D}^R_{ij}[T;\boldsymbol{q}=0,q_0]}{\partial q_0}\frac{\partial\widetilde{\Gamma}^R_{jk}[T;\boldsymbol{q}=0,q_0]}{\partial T}\right)=0,$$

$$\frac{\hbar}{2i}\left(\frac{\partial\widetilde{\Gamma}^R_{ij}[T;\boldsymbol{q}=0,q_0]}{\partial T}\frac{\partial\widetilde{D}^R_{jk}[T;\boldsymbol{q}=0,q_0]}{\partial q_0}-\frac{\partial\widetilde{\Gamma}^R_{ij}[T;\boldsymbol{q}=0,q_0]}{\partial q_0}\frac{\partial\widetilde{D}^R_{jk}[T;\boldsymbol{q}=0,q_0]}{\partial T}\right)=0.\tag{3.7}$$

按照(3.3)和(2.31)式容易求解(3.6)式，得

$$\begin{pmatrix}\widetilde{D}^R_{11}[T;\boldsymbol{q}=0,q_0], & \widetilde{D}^R_{12}[T;\boldsymbol{q}=0,q_0]\\ \widetilde{D}^R_{21}[T;\boldsymbol{q}=0,q_0], & \widetilde{D}^R_{22}[T;\boldsymbol{q}=0,q_0]\end{pmatrix}=-\frac{1}{\left(q_0+i\dfrac{\delta}{2}\right)\left(q_0+\widetilde{\Gamma}^R_{11}[T;0]-\widetilde{\Gamma}^R_{22}[T;0]-i\dfrac{\delta}{2}\right)}\cdot\begin{pmatrix}\widetilde{\Gamma}^R_{22}[T;0]-q_0, & -\widetilde{\Gamma}^R_{12}[T;0]\\ -\widetilde{\Gamma}^R_{21}[T;0], & \widetilde{\Gamma}^R_{11}[T;0]+q_0\end{pmatrix},\tag{3.8}$$

其中 $\widetilde{\Gamma}^R_{ij}[T;0]$ 是 $\widetilde{\Gamma}^R_{ij}[T;q=0]$ 的简写．在证明(3.8)式满足(3.7)式以前，先来讨论一下解(3.8)式所满足的一些普遍性质．

首先，由闭路两顶点函数的一般性质容易得到

$$\widetilde{\Gamma}^R_{11}[T;0]^*=\widetilde{\Gamma}^R_{22}[T;0];\quad \widetilde{\Gamma}^R_{12}[T;0]^*=\widetilde{\Gamma}^R_{21}[T;0],\tag{3.9}$$

$$\widetilde{\Gamma}^R_{11}[T;0]-\widetilde{\Gamma}^R_{22}[T;0]=2i\,\mathrm{Im}\,\widetilde{\Gamma}^R_{11}[T;0]=-2i\,\mathrm{Im}\,\widetilde{\Gamma}^R_{22}[T;0]\sim\text{纯虚数}.\tag{3.10}$$

其次，按因果性要求，推迟格林函数的极点应在 q_0 复平面的下半平面，由(3.8)式，即要求

$$\delta>0;\quad 2\,\mathrm{Im}\,\widetilde{\Gamma}^R_{11}[T;0]>\frac{\delta}{2}>0.\tag{3.11}$$

再次，按 $D^R_{11}[T;q]$Lehmann 表示的普遍性质[11]，$\widetilde{D}^R_{11}[T,q]$在极点上的留数应该是实数，

则容易证明

$$\tilde{\Gamma}^R_{11}[T;0] = -\tilde{\Gamma}^R_{22}[T;0] \text{ 都是纯虚数.} \tag{3.12}$$

将(3.9),(3.10)和(3.12)式应用到(3.3)和(3.8)式,再考虑到(2.31)式和闭路格林函数的有关基本性质,就有

$$\begin{pmatrix} \tilde{\Gamma}^R_{11}[T:\boldsymbol{q}=0,q_0], \tilde{\Gamma}^R_{12}[T;\boldsymbol{q}=0,q_0] \\ \tilde{\Gamma}^R_{21}[T:\boldsymbol{q}=0,q_0], \tilde{\Gamma}^R_{22}[T;\boldsymbol{q}=0,q_0] \end{pmatrix} = \begin{pmatrix} q_0 + i\left(G+\frac{\delta}{2}\right), & iG\,e^{2i\beta} \\ -iGe^{-2i\beta}, & -q_0 - i\left(G+\frac{\delta}{2}\right) \end{pmatrix}, \tag{3.13}$$

$$\begin{pmatrix} \tilde{D}^R_{11}[T:\boldsymbol{q}=0,q_0], & \tilde{D}^R_{12}[T;\boldsymbol{q}=0,q_0] \\ \tilde{D}^R_{21}[T:\boldsymbol{q}=0,q_0], & \tilde{D}^R_{22}[T;\boldsymbol{q}=0,q_0] \end{pmatrix}$$
$$= \frac{1}{2\left(q_0+i\frac{\delta}{2}\right)}\begin{pmatrix} 1, & e^{2i\beta} \\ -e^{-2i\beta}, & -1 \end{pmatrix} + \frac{1}{2\left(q_0+i\left(2G+\frac{\delta}{2}\right)\right)}\begin{pmatrix} 1, & -e^{2i\beta} \\ e^{-2i\beta}, & -1 \end{pmatrix}, \tag{3.14}$$

$$\tilde{D}^A_{ij}[T;\boldsymbol{q}=0,q_0] = \tilde{D}^R_{ji}[T;\boldsymbol{q}=0,q_0]^*, \tag{3.15}$$

其中

$$G = \frac{1}{i}\tilde{\Gamma}^R_{11}[T;0] - \frac{\delta}{2} \quad \text{是实数,} \tag{3.16}$$

$$\tilde{\Gamma}^R_{12}[T;0] = iGe^{2i\beta}, \quad \beta \text{ 是实数.} \tag{3.17}$$

现在,因果性要求(3.11)式可以表为

$$\delta > 0; \quad 2G + \frac{\delta}{2} > 0. \tag{3.18}$$

到现在为止,我们还没有证明(3.7)式,这将在附录A中来进行这个证明.

由方程(3.4)或(3.5)求解推迟(或超前)格林函数 D^R_{ij} (或 D^A_{ij}) 的过程实质上就是一个自洽的量子化过程(显然, $D^R(x,y) \sim \langle\theta(x_0-y_0)[\hat{a}(x), \hat{a}^+(y)]\rangle$, 在古典力学中,力学量可易,不存在推迟格林函数). 当 $\Gamma^R(x,y)$ 取饱和状态的物理量时,则求解 $D^R(x,y)$ 就是在饱和状态进行量子化. 推迟格林函数极点的实部就是量子化元激发的能谱,极点的虚部是相应元激发的耗散. 按这样的观点来看待我们所求得的饱和状态的推迟格林函数,感到十分有趣的是: 普通的量子化电磁场,在非相对论近似下,对于确定模 $\boldsymbol{q}=\boldsymbol{k}$ 只有一个量子态,对应的推迟格林函数只有一个极点

$$D^R[\boldsymbol{q}=\boldsymbol{k},q_0] = \frac{1}{q_0-k_0+i\epsilon}. \tag{3.19}$$

满足通常的非相对论正则量子条件

$$i\hbar\int\frac{dq_0}{2\pi\hbar}(D^R[\boldsymbol{q}=\boldsymbol{k},q_0] - D^A[\boldsymbol{q}=\boldsymbol{k},q_0]) = 1. \tag{3.20}$$

而在饱和状态,特别是由于推广的 Goldstone 定理(2.31)式,求得单模电磁场的推迟格林函数却有两个极点,以 D^R_{11} 为例,按(3.14)和(2.20)式有

$$D^R_{11}[T;\boldsymbol{q}=\boldsymbol{k},q_0] = \frac{1}{2}\left(\frac{1}{q_0-k_0+i\frac{\delta}{2}}\right) + \frac{1}{2}\frac{1}{q_0-k_0+i\left(2G+\frac{\delta}{2}\right)}, \tag{3.21}$$

并且在每一个极点项前面有一个权重因子 1/2. 它也满足非相对论正则量子条件 (3.20) 式,每一个极点项各对(3.20)式贡献一半. 这就是说,饱和状态电磁场的量子激发不再是普通的光子,而是我们在(3.14)和(3.15)式所得到的两种不同的量子元激发. 原来相空间的一个量子态,现在分裂为各有权重 1/2 的两个量子态. 我们称推迟极点 $q_0 = k_0 - i\frac{\delta}{2}$ 所对应的元激发为 p 量子,称推迟极点 $q_0 = k_0 - i\left(2G + \frac{\delta}{2}\right)$ 所对应的元激发为 a 量子. 它们能量相同,但耗散性质有很大差异. 我们称这样现象为饱和状态上量子态的分裂.

另一个有兴趣的问题是 δ 和 G 的符号. 我们曾经按照因果性要求得到了 (3.18) 式. 现在可以进一步看到如果(3.18)式得不到满足,将(3.21)式代入 (3.20) 式,那么正则量子化条件 (3.20)式也就得不到满足. 这即是说,无论是 p 量子还是 a 量子,它们的耗散必须是正的. 在文献[13]中论证了在序参量方程的确定解附近,二阶推迟格林函数的耗散就是相应序参量改变量的线性耗散,它反映了序参量方程相应解的稳定性. 这样,δ 和 G 的符号也可以归结为这样的问题:只有在序参量方程的稳定解处进行量子化,才能符合因果性要求,才能与正则量子化自洽. 饱和状态是在反转条件下序参量方程的稳定解,所以,要正确描写序参量在反转条件下的量子行为,应该在饱和状态进行量子化.

四、饱和状态单模元激发分布函数的输运方程和对称恢复

本节我们将首先弄清楚饱和状态单模电磁场守恒荷和元激发分布函数的联系,其次将讨论元激发分布函数的变化规律. 然后利用它们来研究饱和状态电磁场的有序状态和两种元激发统计分布之间的转化现象.

考虑到单模振荡圆子 $e^{-\frac{i}{\hbar}kx}$ 对观察量常常没有贡献,我们对

$$D_{ij}(x_1, x_2) \equiv \frac{1}{2}\left(D_{ij}^{+-}(x_1, x_2) + D_{ij}^{-+}(x_1, x_2)\right) \tag{4.1}$$

按(2.20)引入对应的 $\widetilde{D}_{ij}^{0}(x_1, x_2)$. 对于 $\widetilde{D}_{ij}^{0}(x_1, x_2)$ 按文献[12]有

$$\widetilde{D}_{ij}^{0}(x_1, x_2) = \int d^4y\left(\widetilde{D}_{ik}^{R}(x_1, y)\mathscr{N}_{kj}(y, x_2) - \mathscr{N}_{ik}(x_1, y)\widetilde{D}_{kj}^{A}(y, x_2)\right) \tag{4.2}$$

或

$$\widetilde{D}_{ij}^{0}[T, q] = \widetilde{D}_{ik}^{R}[T, q]\mathscr{N}_{kj}[T, q] - \mathscr{N}_{ik}[T, q]\widetilde{D}_{kj}^{A}[T, q] + o\left(\frac{\hbar}{\Delta T\Delta q_0}\right). \tag{4.3}$$

则 $\widetilde{N}_{ij}[T, q]$ 在对应模对应极点上的取值描述了量子化元激发的占据数分布. 将(4.1), (4.3),(3.14) 和 (3.15) 式应用到饱和状态电磁场守恒荷(2.13)式和相应的荷密度 (2.8), (2.9)式. 取极点近似,并注意到

$$\mathscr{N}_{12}[T; q = 0] = \mathscr{N}_{21}[T; q = 0] = 0, \tag{4.4}$$

容易得到

$$\langle \hat{j}_0^{F}(x)\rangle = \frac{1}{V}\left(A^*(k, T)A(k, T) + \frac{1}{2}N_{11}^{(p)}[T, k] + \frac{1}{2}N_{11}^{(a)}[T, k]\right), \tag{4.5}$$

其中

$$N_{11}^{(p)}[T,k]=\mathscr{N}_{11}^{(p)}[T,k]-\frac{1}{2}=\tilde{\mathscr{N}}_{11}^{(p)}[T,q=0]-\frac{1}{2},$$

$$N_{11}^{(a)}[T,k]=\mathscr{N}_{11}^{(a)}[T,k]-\frac{1}{2}=\tilde{\mathscr{N}}_{11}^{(a)}[T,q=0]-\frac{1}{2},\tag{4.6}$$

$\tilde{\mathscr{N}}_{11}^{(p)}[T,q=0]$ 和 $\tilde{\mathscr{N}}_{11}^{(a)}[T,q=0]$ 分别代表 $\tilde{\mathscr{N}}_{11}[T,\boldsymbol{q}=0,q_0]$ 在推迟极点 $q_0=-i\frac{\delta}{2}$ 和 $q_0=-i\left(2G+\frac{\delta}{2}\right)$ 附近的取值，$N_{11}^{(p)}[T,k]$ 和 $N_{11}^{(a)}[T,k]$ 就是 p 元激发和 a 元激发的 Wigner 分布函数[12]. (4.5)式说明,在饱和状态单模电磁场有三种存在形式:一种是有序状态，另外两种是较为无序的 p 量子和 a 量子的统计激发状态. 电磁场荷密度就是与这三种状态相对应的序参量振幅强度、p 量子数的统计平均和 a 量子数的统计平均的总和. 系统达到饱和,它们的总和就稳定不变,但却有可能在这三种形式之间发生转化. 为要搞清楚这种转化现象,就需要讨论分布函数 $N_{11}^{(p)}[T;k]$ 和 $N_{11}^{(a)}[T;k]$ 随 T 的变化规律.

按照文献[12],输运方程

$$\tilde{D}^R\times\tilde{\mathscr{N}}-\tilde{\mathscr{N}}\times\tilde{D}^A+\tilde{D}^R\times\tilde{\Gamma}^0\times\tilde{D}^A=0\tag{4.7}$$

确定了在格林函数极点上的元激发分布函数随时间的缓变化规律（对于三维平移不变系统,只有随时间的变化）. 在(4.7)式中我们用“×”表示这样一种乘法

$$A\times B\quad\begin{cases}\text{在坐标表象中}\iff\displaystyle\int d^4z\,A(x,z)\,B(z,y)\\[2ex]\text{在相对动量表象中}\iff A[X,p]B[X,p]+\dfrac{\hbar}{2i}\left(\dfrac{\partial A[X,p]}{\partial X}\dfrac{\partial B[X,p]}{\partial p}-\dfrac{\partial A[X,p]}{\partial p}\dfrac{\partial B[X,p]}{\partial X}\right),\end{cases}\tag{4.8}$$

其中

$$A(x,y)=\int\frac{d^4p}{(2\pi\hbar)^4}\,e^{-\frac{i}{\hbar}p(x-y)}\,A\left[\frac{x+y}{2},\ p\right].\tag{4.9}$$

由于对应一个饱和激光模,在 q_0 复平面上存在两个极点(p 极点和 a 极点),它们实部相同,只是虚部不同,我们的困难在于如何在方程(4.7)中将这两个极点所对应的贡献分离开来,分离出这两个极点各自的分布函数所满足的输运方程. 为此，将对方程(4.7)作一么正变换.

对任何分离出单模振荡因子后的两点传播函数的相对动量表示 $\tilde{F}_{ij}[T,q]$ 定义

$$\tilde{F}'_{ij}[T,q]\equiv\begin{pmatrix}\tilde{F}'_{11}[T,q], & \tilde{F}'_{12}[T,q]\\ \tilde{F}'_{21}[T,q], & \tilde{F}'_{22}[T,q]\end{pmatrix}\equiv U[\beta]\begin{pmatrix}\tilde{F}_{11}[T,q], & \tilde{F}_{12}[T,q]\\ \tilde{F}_{21}[T,q], & \tilde{F}_{22}[T,q]\end{pmatrix}U^{+}[\beta],\tag{4.10}$$

其中

$$U[\beta]\equiv\frac{1}{\sqrt{2}}\begin{pmatrix}-ie^{-i\beta}, & ie^{i\beta}\\ e^{-i\beta}, & e^{i\beta}\end{pmatrix},\tag{4.11}$$

$$U_{ij}^{+}[\beta]\equiv U_{ji}[\beta]^{*},\tag{4.12}$$

$$U^{+}[\beta]\,U[\beta]=U[\beta]\,U^{+}[\beta]=I.\tag{4.13}$$

并且，$U[\beta]$ 与 T 和 q_0 都无关. β 与 q_0 无关是显然的，因为按定义(3.17)式，β 就是 $\tilde{\Gamma}^R_{12}[T;q=0]$ 的幅角. 至于 β 与 T 无关的原因是这样的. 在饱和状态，系统的时间不均匀性是弱的，所以在 Γ 生成泛函对 a 和 a^* 的展开系数中，只有它的最低幂次的系数需要保留时间不均匀性到 $o\left(\frac{\hbar}{\Delta T \Delta q_0}\right)$ 的一阶项. 因此准到 $\hbar$ 的一次幂 $\tilde{\Gamma}^R_{12}$ 和 $\tilde{\Gamma}^R_{21}$ 对时间 T 的依赖主要通过 $a(q,T)$ 和 $a^*(q,T)$. 而在(2.28)式的讨论中已指出，序参量对 T 的依赖在它的模中，而不在它的幅角之中. 所以 $\tilde{\Gamma}^R_{12}$ 和 $\tilde{\Gamma}^R_{21}$ 对 T 的依赖也只能在它的模中而不在幅角之中. 既然 $U[\beta]$ 与 T 和 q_0 无关，那么输运方程(4.7)经幺正变换后就有

$$\tilde{\mathscr{D}}'^R \times \tilde{\mathscr{N}}' - \tilde{\mathscr{N}}' \times \tilde{\mathscr{D}}'^A + \tilde{\mathscr{D}}'^R \times \tilde{\mathscr{N}}' \times \tilde{\mathscr{D}}'^A = 0. \tag{4.14}$$

按照(4.10)—(4.13)式，将(4.14)式的矩阵表示显写出来，取它的两个对角矩阵元，就分别得到两个方程:

$$\tilde{\mathscr{D}}'^R_p \times \tilde{\mathscr{N}}'_{21} - \tilde{\mathscr{N}}'_{12} \times \tilde{\mathscr{D}}'^A_p + \tilde{\mathscr{D}}'^R_p \times \tilde{\mathscr{N}}'^0_{22} \times \tilde{\mathscr{D}}'^A_p = 0, \tag{4.15}$$

$$\tilde{\mathscr{D}}'^R_a \times \tilde{\mathscr{N}}'_{12} - \tilde{\mathscr{N}}'_{21} \times \tilde{\mathscr{D}}'^A_a + \tilde{\mathscr{D}}'^R_a \times \tilde{\mathscr{N}}'^0_{11} \times \tilde{\mathscr{D}}'^A_a = 0, \tag{4.16}$$

其中

$$\tilde{\mathscr{D}}'^R_p[T;\boldsymbol{q}=0,q_0] = -\frac{i}{q_0 + i\dfrac{\delta}{2}}, \quad \tilde{\mathscr{D}}'^R_a[T;\boldsymbol{q}=0,q_0] = \frac{i}{q_0 + i\left(2G + \dfrac{\delta}{2}\right)},$$

$$\tilde{\mathscr{D}}'^A_p[T;\boldsymbol{q}=0,q_0] = \frac{i}{q_0 - i\dfrac{\delta}{2}}, \quad \tilde{\mathscr{D}}'^A_a[T;\boldsymbol{q}=0,q_0] = -\frac{i}{q_0 - i\left(2G + \dfrac{\delta}{2}\right)}. \tag{4.17}$$

从(4.15)—(4.17)式可以看清，我们已经成功地将 p 极点和 a 极点的贡献分了开来.

按照文献[12]所提供的处理输运方程的方法，利用(4.10)—(4.12)式，并考虑到二阶闭路顶点函数的一些基本性质，容易将(4.15)和(4.16)式化为

$$\hbar\frac{\partial \tilde{\mathscr{N}}'^{(p)}_{11}[T;0]}{\partial T} + \delta \tilde{\mathscr{N}}'^{(p)}_{11}[T;0] = \frac{1}{i}\left(\tilde{\Gamma}^0_{11}[T;0] + \frac{e^{-2i\beta}\tilde{\Gamma}^0_{12}[T;0] + e^{2i\beta}\tilde{\Gamma}^0_{21}[T;0]}{2}\right)$$

$$\cong \frac{1}{i}\tilde{\Gamma}^0_{11}[T;0], \tag{4.18}$$

$$\hbar\frac{\partial \tilde{\mathscr{N}}'^{(a)}_{11}[T;0]}{\partial T} + (4G+\delta)\tilde{\mathscr{N}}'^{(a)}_{11}[T;0] = \frac{1}{i}\left(\tilde{\Gamma}^0_{11}[T;0] - \frac{e^{-2i\beta}\tilde{\Gamma}^0_{12}[T;0] + e^{2i\beta}\tilde{\Gamma}^0_{21}[T;0]}{2}\right) \cong \frac{1}{i}\tilde{\Gamma}^0_{11}[T;0], \tag{4.19}$$

其中[14]

$$\frac{1}{i}\tilde{\Gamma}^0_{11}[T;q=0] \equiv \frac{\Gamma^{-+}_{11}[T;k] + \Gamma^{+-}_{11}[T;k]}{2i}$$

$$= \frac{\hbar}{2}(W_{吸}[T;k] + W_{放}[T;k]) > 0. \tag{4.20}$$

在饱和状态显然与 T 无关. 在求得(4.18)和(4.19)式的过程中，还利用了

$$\frac{\partial\delta}{\partial T} \sim 0, \tag{4.21}$$

$$\frac{\partial G}{\partial T} \sim 0. \tag{4.22}$$

按(2.28)式,(4.21)式是显然的. 至于(4.22)式,将在附录B中加以论证.

将饱和状态单模电磁场的荷守恒规律(2.13)式利用(4.5)式重新写出来,并与序参量、p量子分布函数和a量子分布函数的变化规律(2.28),(4.18)和(4.19)式写在一起,有

$$\frac{\partial}{\partial T}\left(A^*(k,T)\, A(k,T) + \frac{1}{2}\mathcal{N}_{11}^{(p)}[T;k] + \frac{1}{2}\mathcal{N}_{11}^{(a)}[T;k]\right) = 0, \tag{4.23}$$

$$\hbar \frac{\partial}{\partial T}(A^*(k,T)\, A(k,T)) = -\delta(A^*(k,T)\, A(k,T)), \tag{4.24}$$

$$\hbar \frac{\partial}{\partial T}\left(\frac{1}{2}\mathcal{N}_{11}^{(p)}[T;k]\right) = -\delta\left(\frac{1}{2}\mathcal{N}_{11}^{(p)}[T;k]\right) + \frac{1}{2i}\Gamma_{11}^{0}[T;k], \tag{4.25}$$

$$\hbar \frac{\partial}{\partial T}\left(\frac{1}{2}\mathcal{N}_{11}^{(a)}[T;k]\right) = -(4G+\delta)\left(\frac{1}{2}\mathcal{N}_{11}^{(a)}[T;k]\right) + \frac{1}{2i}\Gamma_{11}^{0}[T;k]. \tag{4.26}$$

现在将这样地来讨论这一组方程. 按照实验上所提的线索,先假设

$$G \gg \delta. \tag{4.27}$$

就在这样的假设下进行求解,求解以后,再回过来验证所求得的解满足(4.27)式. 这样就有

1. 在饱和状态,单模电磁场有两个时间量纲:

$$\tau_p \sim \frac{\hbar}{\delta}, \tag{4.28}$$

$$\tau_a \sim \frac{\hbar}{G}. \tag{4.29}$$

按(4.27)式有

$$\tau_p \gg \tau_a. \tag{4.30}$$

并且,按(3.16)式,G的量级 $\sim \bar{\Pi}_{11}^{R}[T;0]$ 的量级,所以

τ_a 的量级～激光由零增长到饱和的时间. (4.31)

2. 按(4.26)式, $\mathcal{N}_{11}^{(a)}[T;k]$ 在 τ_a 的标度上很快,即在与激光增长时间相同的量级上达到稳定,有

$$\mathcal{N}_{11}^{(a)}[T;k]\Big|_{\text{稳定值}} = \frac{1}{i}\left(\frac{\Gamma_{11}^{0}[T;k]}{4G+\delta}\right). \tag{4.32}$$

与此同时,按荷守恒规律(4.23)式, $A^*(k,T)\, A(k,T) + \frac{1}{2}\mathcal{N}_{11}^{(p)}[T;k]$也同时达到稳定,按(4.23),(4.24)和(4.25)式容易有

$$\left(A^*(k,T)\, A(k,T) + \frac{1}{2}\mathcal{N}_{11}^{(p)}[T;k]\right)_{\text{稳定值}} = \frac{1}{i}\left(\frac{\Gamma_{11}^{0}[T;k]}{2\delta}\right). \tag{4.33}$$

3. 当与 τ_a 时间标度相当的上述快过程完成以后,序参量和p量子的统计分布之间在τ_p的时间标度上按(4.24)和(4.25)式规定的方式缓慢地进行着转化. 与此同时,a量子的统计分布却保持不变(注意(4.22)式)而不参与它们之间的转化. 在系统出现反转后,序参量 a priori 地选定一个确定的位相从小增长到饱和状态,作为饱和状态的初条件,系统处于相位对称破缺状态,显然有

$$|A(k,T)|^2 \gg \mathscr{N}_{11}^{(p)}[T;k] \sim \mathscr{N}_{11}^{(a)}[T;k]\ (\text{初始状态}). \tag{4.34}$$

所以，在激光刚达到饱和状态时，可以在(4.33)式中忽略 $\mathscr{N}_{11}^{(p)}[T;k]$ 即

$$\delta \cong \frac{\Gamma_{11}^{0}[T;k]}{2i|A(k,T)|^2}. \tag{4.35}$$

(4.35) 式即是通常量子光学中的激光谱线宽度表达式[3]．对比(4.32)式，并考虑到 (4.34) 式，显然有

$$\delta \ll G. \tag{4.36}$$

这样，(4.33)，(4.35)和(4.36)式既给出了 δ 的表达式也验证了我们先前所作的假设(4.27)式．

4. 在 τ_p 的尺度上，按照方程(4.24)，(4.25)和(4.33)式，饱和状态最终将趋向

$$A(k,T) \to 0,$$

$$\mathscr{N}_{11}^{(p)}[T;k] \to \frac{\Gamma_{11}^{0}[T;k]}{i\delta}. \tag{4.37}$$

由于序参量 $A(k,T)$ 的存在，引起了系统相因子对称性的破缺；所以，确切地说，序参量 $A(k,T)$描写的是单模电磁场的相位有序状态．(4.37)式说明，在饱和状态，电磁场的相位有序状态将逐渐消失．作为一个非平衡量子统计定常态，系统将恢复到对于相因子变换完全对称的相位无序状态．这即是通常所说的对称恢复．然而，从上面的讨论可以清楚看出，这种对称恢复后的相位无序状态和对称破缺以前的无序状态有着本质的差别．组成这种新的相位无序状态是 p 量子和 a 量子的统计分布而不是普通光子．并且，尽管引起对称破缺的序参量已经消失，但是对应的 Goldstone 模式：p 量子却依然存在；而且，电磁场的能量主要都集中在谱线宽度比普通光子的自然宽度窄得多的 p 量子统计分布上．

最后，还想顺便指出这样一点．序参量和 p 量子是饱和状态单模电磁场的两种不同的存在形式．但方程(4.24)和(4.25)说明它们具有相同的耗散：δ．这不是偶然的，是我们所得到的推广的 Goldstone 定理的后果．但是，从实验上来看，相同的耗散意味着有相同的谱线宽度．所以如何从实验上来区分这两种饱和电磁场的存在形式，不知是不是一个有趣的问题？

五、结　束　语

我们的讨论是从 Korenman[2] 所遇到的饱和状态 Goldstone 性质的发散困难开始的．讨论的实质是从系统存在的一个相因子变换对称性出发，利用非平衡统计闭路格林函数方法特别是其中的 Ward-Takahashi 恒等式作为工具，以单模电磁场的饱和状态为例子，探讨了一类耗散系统的非零模非平衡定常态的量子性质和统计性质．讨论主要依赖于系统的对称性质，很少依赖于系统作用量的具体形式，因此有它的普遍性．

我们得到了在存在有限频率非零序参量的饱和状态上带有耗散的推广的 Goldstone 定理

$$\det \begin{vmatrix} \tilde{\Gamma}_{11}^{R}[T;q=0] - i\dfrac{\delta}{2}, & \tilde{\Gamma}_{12}^{R}[T;q=0] \\ \tilde{\Gamma}_{21}^{R}[T;q=0], & \tilde{\Gamma}_{22}^{R}[T;q=0] + i\dfrac{\delta}{2} \end{vmatrix} = 0, \tag{2.31}$$

其中

$$\delta = \frac{\Gamma^{0}_{11}[T;k]}{2i\left(|A(k,T)|^2 + \frac{1}{2}\mathcal{N}^{(p)}_{11}[T;k]\right)} \tag{4.33}$$

就是 Goldstone 激发——即 p 量子的耗散. 我们所得到的 Goldstone 定理的推广形式与普通场论中的 Goldstone 定理的不同主要在于: 普通场论常常讨论的是无耗散系统零能量模的非零序参量. 我们讨论的是与外界环境相互作用着的有限频率模的非零序参量. 对于这样的任何实际统计系统都永远存在着实的(而不是虚的)量子自发辐射. 所以, 按(4.33)式, Goldstone 定理就必须有耗散, Goldstone 模的耗散也必不为零.

在我们看来,量子化必须在序参量古典方程的稳定解附近进行,才能与因果性要求自洽. 而我们有兴趣的饱和状态就对应于半经典意义下电磁场方程的稳定解. 所以,应该在饱和状态对单模电磁场重新进行量子化. 量子化的结果发现相当于通常(阈以下)的一个光子态在饱和状态分裂为两个量子态,各带有谱表示空间的权重因子 1/2, 一个是 p 量子态,就是 Goldstone 模,一个是 a 量子态. 每一个量子态对应一种耗散,前者对应于通常量子光学中的位相涨落,后者对应通常的振幅涨落[15](容易证明,(3.16)式确定的 G 就是序参量方程展开到序参量三次幂的系数). 但是,在通常量子光学看来,这两种涨落只是系统的统计性质,而我们的讨论说明,这与系统的量子动力学性质的变化有关. 例如,关于激光谱线宽度公式(4.33)(或(4.35))中的因子 1/2, 有些作者把它解释为非线性效应的贡献[16],而从我们的方程(4.23)—(4.26)式和有关的讨论可以看出,如果没有量子态的分裂和相应的权重因子,纵使序参量的方程是非线性的,在对应的(4.33)(或(4.35))式不能有因子 1/2. 我们期待是否还可能有其他的观察效果来检验我们的看法.

我们还得到了饱和状态单模电磁场的三种存在形式: 序参量、p 量子激发和 a 量子激发. 前者属相位有序状态, 后两者属相位无序状态. 我们也得到了在饱和状态它们之间的转化规律,从而给出了"对称恢复"较为确切的具体描述. 同时也证明了不存在一个严格恒定的有限频率单模序参量的非零解. 只是对称恢复的极限, 才是一个严格的量子统计非平衡定常态. 这时, 尽管导致对称自发破缺的序参量已经消逝, 但是, 对应的 Goldstone 模式却依然存在,而且系统的能量主要集中在它的 Goldstone 模上.

作者感谢甘子钊、郝柏林、于渌和詹达三等同志许多有益的讨论.

附 录 A

推迟格林函数求解

这一个附录的任务是最后完成第三节所进行的推迟格林函数求解. 具体说来,就是要证明方程(3.6)的解(3.8)或(3.14)式满足(3.7)式.

既然解满足(3.6)式,就有

$$\tilde{D}^R[T;\boldsymbol{q}=0,q_0] = (\tilde{\Gamma}^R[T;\boldsymbol{q}=0,q_0])^{-1}, \tag{A.1}$$

则所求证的(3.7)式容易化为

$$\frac{\hbar}{2i}\left(-\tilde{D}^R\frac{\partial\tilde{\Gamma}^R}{\partial T}\tilde{D}^R\frac{\partial\tilde{\Gamma}^R}{\partial q_0} + \tilde{D}^R\frac{\partial\tilde{\Gamma}^R}{\partial q_0}\tilde{D}^R\frac{\partial\tilde{\Gamma}^R}{\partial T}\right)_{\boldsymbol{q}=0} = 0,$$

$$\frac{\hbar}{2i}\left(-\frac{\partial\tilde{\Gamma}^R}{\partial T}\tilde{D}^R\frac{\partial\tilde{\Gamma}^R}{\partial q_0}\tilde{D}^R+\frac{\partial\tilde{\Gamma}^R}{\partial q_0}\tilde{D}^R\frac{\partial\tilde{\Gamma}^R}{\partial T}\tilde{D}^R\right)_{q=0}=0. \tag{A.2}$$

容易看出，只要证明

$$\hbar\frac{\partial\tilde{\Gamma}^R}{\partial T}\tilde{D}^R\frac{\partial\tilde{\Gamma}^R}{\partial q_0}\bigg|_{q=0}=\hbar\frac{\partial\tilde{\Gamma}^R}{\partial q_0}\tilde{D}^R\frac{\partial\tilde{\Gamma}^R}{\partial T}\bigg|_{q=0}, \tag{A.3}$$

则 (A.2) 式，从而 (3.7) 式就能得到满足．所以，我们的问题归结为求证 (A.3) 式．由于 (A.3) 式本身是 $O\left(\frac{\hbar}{\Delta T\Delta q_0}\right)$ 一阶量，所以，按假设 II，式中 $\frac{\partial\tilde{\Gamma}^R}{\partial q_0}$ 只需取到零级量，按(2.26)式，有

$$\frac{\partial\tilde{\Gamma}^{R(0)}[T;q]}{\partial q_0}=\begin{pmatrix}1, & 0\\ 0, & -1\end{pmatrix}. \tag{A.4}$$

并且，按照闭路两点函数的基本性质，容易有

$$\tilde{\Gamma}^R_{12}[T;0]=-\tilde{\Gamma}^R_{12}[T;0],\quad \frac{\partial\tilde{\Gamma}^R_{12}[T;0]}{\partial T}=-\frac{\partial\tilde{\Gamma}^R_{12}[T;0]}{\partial T}. \tag{A.5}$$

此外，在第四节中我们已论证了 $\tilde{\Gamma}_{12}$ 和 $\tilde{\Gamma}_{21}$ 的辐角 $e^{\pm 2i\theta}$ 与 T 无关，所以按(3.13)式可以证明

$$\tilde{\Gamma}^R_{12}\frac{\partial\tilde{\Gamma}^R_{21}}{\partial T}\bigg|_{q=0}=\frac{\partial\tilde{\Gamma}^R_{12}}{\partial T}\tilde{\Gamma}^R_{21}\bigg|_{q=0}. \tag{A.6}$$

由 (A.4)，(A.5) 和 (A.6) 式立即可以验证 (A.3) 式左端＝(A.3) 式右端．

附　录　B

关于 a 型极点耗散随时间的变化性质

在这附录中，我们将对(4.22)式进行论证，其中 G 即是饱和状态 a 极点的耗散，它由(3.16)式确定，即

$$G=\frac{1}{i}\tilde{\Gamma}^R_{11}[T;0]-\frac{\delta}{2}=\frac{1}{i}\Gamma^R_{11}[T;k]-\frac{\delta}{2}. \tag{B.1}$$

我们着重指出，从第四节的讨论得知，在饱和状态存在两个时间量钢：τ_p 和 τ_a，前者较长，后者较短．我们需要论证的是(4.22)式不仅在 τ_a 量级上成立，而且在更慢的 τ_p 尺度上也成立．

首先，在饱和状态，由于系统的时间不均匀性是微弱的，所以在生成泛函的有关方程中，只需在最低幂次的展开项中保留时间不均匀性到 $O\left(\frac{\hbar}{\Delta T\Delta q_0}\right)$ 的一阶项．它将按照弱不均匀展开[7]的方式，与其它项分离开来而以输运项的形式单独出现在相应的方程中．(B.1) 式即(3.16)式就是已经按照假设 I 作了弱不均匀展开分离出输运项以后的表达式．既然输运项已被分离出去，那么 (B.1) 式中的 $\Gamma^R_{11}[T;k]$ 在 Feynman 图的语言中就是一个单光子不可约真空极化图．它对时间不均匀性的依赖只能通过序参量和与内光子线相对应的各种二阶格林函数(电磁场的)中的缓变变量．

其次，在对与内光子线相当的电磁场格林函数作四动量积分时，由于(2.34)和(2.35)式，p 极点和 a 极点的虚部的贡献，将被原于系统束缚电子的耗散弛豫所掩盖．再考虑到(4.4)式，所以，在饱和状态 $\Gamma^R_{11}[T;k]$ 对时间的依赖只能通过 $a(k,T)$，$a^*(k,T)$，$N^{(p)}_{11}[T;k]$ 和 $N^{(a)}_{11}[T;k]$．

再次，在共振相互作用即迴转波近似[4]下，$a(k,T)$ 和 $a^*(k,T)$ 一定成对而且交叉顺次地出现在 $\Gamma^R_{11}[T;k]$ 的不可约 Feynman 图中．每对应一个这样的不可约 Feynman 图，必须存在另一个与它相对应的不可约 Feynman 图：将前者中的 $a(x)a^*(x')\longrightarrow\langle T_p\hat{a}(x)\hat{a}^+(x')\rangle_{\text{conn}}$．容易证明，这两者合起来的贡献

$$\sim A^*(k,T)A(k,T)+\frac{1}{2}\mathcal{N}^{(p)}_{11}[T;k]+\frac{1}{2}\mathcal{N}^{(a)}_{11}[T;k]+\text{常数}. \tag{B.2}$$

按照我们前面得到的基本守恒性质 (4.23) 式，(B.2) 式显然与 T 无关．容易看清，$\Gamma^R_{11}[T;k]$ 的所有不可约 Feynman 图都能按此方式进行组合．所以 $\Gamma^R_{11}[T;k]$ 也即是 G 就与 T 无关．即(4.22)式成立．

参　考　文　献

[1] J. Schwinger, *J. Math. Phys.*, 2(1961), 407. Л. Кельдыш, *ЖЭТФ*, **47** (1964), 1515;
又如
D. Dubois, In "Lectures in Theoretical Physics," Vol. 9C, ed. Brittin, Gordon and Breach, N. Y. (1967); D. Langreth, In "Linear and Nonltinear Transport in Solids". eds J. Devreese and

V. Van Doren, Plenum, N. Y. (1976).
[2] V. Korenman, *Ann. Phys.* (*N. Y.*). 39(1966), 72.
[3] M. Sargent III. M. Scully. and W. Lamb, Laser Physics, Addison-Wesley. Reading, Mass. (1974); M. Lax. *Phys. Rev.*, 145(1966), 110; In ''Physics of Quantum Electronics'', eds. P. Kelley *et al.*; McGraw Hill. N. Y. (1966); H. Haken, Laser Theory, Vol. XXV/2c of Encyclopedia of Physics. Springer, Berlin (1970).
[4] M. Sargent III *et al.*, *ibid.*, Chapte. II. Addison-Wesley. Reading. Mass. (1974).
[5] 周光召、苏肇冰,统计物理进展,第五章,郝柏林等编辑,科学出版社即将出版;或周光召、苏肇冰,高能物理与核物理,3(1979),304;3(1979),314.
[6] 周光召、苏肇冰,高能物理与核物理,3(1979),314.
[7] 周光召、苏肇冰,统计物理进展,第五章,§5.2.4;郝柏林等编辑,科学出版社即将出版.
[8] 同上,§5.2.5.
[9] 例如 M. Sargent III, *et al.*, Laser Physics. §20-2, Addison-Wesley. Reading. Mass. (1974).
[10] 周光召等,见文献[7],§5.2.
[11] 同上,§5.1.7.
[12] 同上,§5.2.3和§5.2.6.
[13] 同上,§5.3.2.
[14] 同上,§5.2.2.
[15] 例如 H. Haken. *ibid.*. §IV. 7.
[16] M Lax. In ''Physics of Quantum Electronics''. Eds. P. Kelley. *et al.*, McGraw-Hill, N. Y. (1966).

ON THE GOLDSTONE MODE IN THE STATIONARY STATE OF A NON-EQUILIBRIUM DISSIPATIVE SYSTEM

Zhou Guang-zhao Su Zhao-bing
(*Institute of theoretical Physics, Academia Sinica*)

Abstract

In consequence of the spontaneous symmetry breaking, non-zero energy Goldstone modes with dissipation are excited in a non-equilibrium stationary state with space-time structure. In this paper, as a specific example, the Ward-Takahashi identities formulated in the close time path Green's function method is applied to the saturation state of a single mode laser. A generalized Goldstone theorem in a weak inhomogeneous dissipative system is established and the physical interpretation of the Goldstone mode is discussed. As a result of the Goldstone theorem, the pole in the Green's function of the laser light splits into two with equal weights, each corresponding to a quanta with the same frequency but different dissipation. Together with the order parameter (the average value of the vector potential), these two kinds of quanta (one of which is the Goldstone mode) give a complete description of the order-disorder transition of the phase symmetry in the saturation state of the laser. A detailed discussion on the restoration of the spontaneously broken symmtry of the phase is given.

第 29 卷 第 7 期 物 理 学 报 Vol. 29, No. 7
1980 年 7 月 ACTA PHYSICA SINICA July, 1980

三套闭路格林函数的变换关系

周光召 于 渌 郝柏林

（中国科学院理论物理研究所）

1979 年 8 月 27 日收到

提 要

给出三套闭路格林函数之间的变换关系和若干运算规则，同时得到多点推迟和超前格林函数的一般定义．证明多点格林函数之间的一些代数关系，是闭路上相连格林函数生成泛函 $W[J_+, J_-]|_{J_+=J_-=J} = 0$ 这一性质的后果．

一、引言和记号

自从 Schwinger[1] 首次提出闭路格林函数以来（参看文献 [2] 及其所引文献），这一处理非平衡统计问题的有效方法未能得到更广泛应用的一种技术性的原因，在于计算形式比较繁琐．其实只要掌握一些运算规则，闭路格林函数方法与普通量子场论就几乎相同．除了在文献 [2—4] 中概括过的一些规则外，本文着重讨论多点格林函数之间的变换关系．

闭路格林函数方法中遇到三套函数：

闭路上的 $G_p(1, 2, \cdots, n)$，以下标 p 标志，它出现在闭路积分下面，用以紧致地书写一般关系；

自变量分别在正、负支上取值的矩阵函数 $\hat{G}(1, 2, \cdots, n)$，或以下标希腊字母表示为 $G_{\alpha\beta\cdots}(1, 2, \cdots, n)$，其中 $\alpha, \beta, \cdots = +, -$．它出现在 $(-\infty, +\infty)$ 的积分下面，用以实现费曼图的计算；

最终用以表示物理关系的推迟、超前和关联函数的矩阵 $\tilde{G}(1, 2, \cdots, n)$，或以下标英文字母表示为 $G_{ij\cdots}(1, 2, \cdots, n)$，其中 $i, j, \cdots = 1, 2$．$\hat{G}$ 和 $\tilde{G}$ 矩阵都有 2^n 个元素．

在讨论它们之间的变换关系之前，先规定本文使用的一些记号．泡利矩阵记为

$$\sigma_1 = \begin{pmatrix} 0 & 1 \\ 1 & 0 \end{pmatrix}, \quad \sigma_2 = \begin{pmatrix} 0 & -i \\ i & 0 \end{pmatrix}, \quad \sigma_3 = \begin{pmatrix} 1 & 0 \\ 0 & -1 \end{pmatrix}, \tag{1.1}$$

将来 σ_3 与 $\hat{G}$，σ_1 与 $\tilde{G}$ 经常一齐出现，而与 σ_2 相联系的实正交矩阵

$$Q = \frac{1 - i\sigma_2}{\sqrt{2}} = \frac{1}{\sqrt{2}} \begin{pmatrix} 1 & -1 \\ 1 & 1 \end{pmatrix}. \tag{1.2}$$

则决定 $\hat{G}$ 与 $\tilde{G}$ 之间的变换．

多点台阶函数的定义为

$$\theta(1, 2, \cdots, n) = \begin{cases} 1 & (\text{当 } t_1 > t_2 > \cdots > t_n), \\ 0 & (\text{其它情形}). \end{cases} \tag{1.3}$$

它是普通两点台阶函数的乘积：

$$\theta(1, 2, \cdots, n) = \theta(1, 2)\theta(2, 3)\cdots\theta(n-1, n), \tag{1.4}$$

用以定义时序算子 T

$$T(\phi(1)\phi(2)\cdots\phi(n)) = \sum_{P_n} \theta(p_1, p_2, \cdots, p_n)\phi(p_1)\phi(p_2)\cdots\phi(p_n), \tag{1.5}$$

此式对 n 个数的一切置换 P_n 求和（本文只以实标量场 ϕ 为例）．θ 函数满足若干一般关系，如归一公式：

$$\sum_{P_n} \theta(p_1, p_2, \cdots, p_n) = 1, \tag{1.6}$$

求和公式（$n > m$）：

$$\theta(1, 2, \cdots, m) = \sum_{P_n(1, 2, \cdots, m)} \theta(p_1, p_2, \cdots, p_n), \tag{1.7}$$

$P_n(1, 2, \cdots, m)$ 是 1 在 2 前，$\cdots$，$m-1$ 在 m 前的 n 点置换（这时 $t_1, t_2, \cdots, t_m$ 的大小仍是任意的）．$n = m + 1$的情形就是在原有的 m 个时刻外再插一个，如

$$\theta(123) = \theta(4123) + \theta(1423) + \theta(1243) + \theta(1234). \tag{1.8}$$

θ 函数的乘积也可以展开，如

$$\theta(12)\theta(34) = \sum_{P_n(1\text{先于}2,\, 3\text{先于}4)} \theta(p_1, p_2, \cdots, p_n) \tag{1.9}$$

乘积中遇有相同时刻时，亦可补足为多点关系：

$$\theta(12)\theta(134) = \theta(1234) + \theta(1324) + \theta(1342). \tag{1.10}$$

二、变 换 关 系

n 点闭路格林函数

$$G_p(1, 2, \cdots, n) = (-i)^{n-1}\mathrm{Tr}\{\mathrm{T}_p(\phi(1)\phi(2)\cdots\phi(n))\rho\}$$
$$\equiv (-i)^{n-1}\langle \mathrm{T}_p(1, 2, \cdots, n)\rangle \tag{2.1}$$

按其自变量分属于正、负支的各种情形写开，就得到 $\hat{G}$ 的各个元素．Keldysh[5] 等人用于从两点函数 $\hat{G}$ 到 $\tilde{G}$ 的变换关系

$$\tilde{G}(12) = Q\hat{G}(12)Q^{-1} \tag{2.2}$$

写成分量形式

$$G_{ij}(12) = Q_{i\alpha}Q_{j\beta}G_{\alpha\beta}(12)$$

就可以直接推广到多点函数

$$G_{i_1i_2\cdots i_n}(1, 2, \cdots, n) = 2^{\frac{n-2}{2}}Q_{i_1\alpha_1}Q_{i_2\alpha_2}\cdots Q_{i_n\alpha_n}G_{\alpha_1\alpha_2\cdots\alpha_n}(1, 2, \cdots, n). \tag{2.3}$$

我们很快会看到，(2.3) 式中自然地包括了一切多点推迟、超前格林函数和关联函数的定义．(2.3) 式中数值因子 $2^{\frac{n-2}{2}}$ 的正确性在下一节中将看得更清楚．这里只需指出，它对

于 $n=1$ 也是对的,即

$$G_i(x)=\frac{1}{\sqrt{2}}Q_{i\alpha}G_\alpha(x),$$

或

$$\begin{pmatrix}G_1\\G_2\end{pmatrix}=\frac{1}{2}\begin{pmatrix}G_+-G_-\\G_++G_-\end{pmatrix}=\begin{pmatrix}0\\G\end{pmatrix}_{G_+=G_-=G}. \tag{2.4}$$

最简单的两点闭路函数是闭路上的 δ-函数. δ_p 的定义为[3]

$$\int_p f(x)\delta_p(x-y)dx=f(y)\quad \forall y\in p, \tag{2.5}$$

相应的另外两个矩阵函数是 $\hat{\delta}$ 和 $\tilde{\delta}=Q\hat{\delta}Q^{-1}$, 它们为

$$\hat{\delta}(x-y)=\delta(x-y)\sigma_3,\ \tilde{\delta}(x-y)=\delta(x-y)\sigma_1. \tag{2.6}$$

这里 $\delta(x-y)\equiv\delta^{(4)}(x-y)$ 是普通的 δ-函数. 还可以定义三个闭路台阶函数 θ_p, $\hat{\theta}$ 和 $\tilde{\theta}$, 其中

$$\hat{\theta}(12)=\begin{pmatrix}\theta(12), & 0\\ 1, & \theta(21)\end{pmatrix},$$

$$\tilde{\theta}(12)=Q\hat{\theta}Q^{-1}=\begin{pmatrix}0, & -\theta(21)\\ \theta(12), & 1\end{pmatrix}. \tag{2.7}$$

如果不管两点函数的结构,形式上写出 $\hat{G}$ 和 $\tilde{G}$ 两个矩阵的关系,则有

$$\tilde{G}=Q\hat{G}Q^{-1}=\frac{1}{2}\begin{pmatrix}G^{(+)}-G^{(-)}, & G_{\cdot r}+G_{\cdot a}\\ G_{r\cdot}+G_{a\cdot}, & G^{(+)}+G^{(-)}\end{pmatrix}, \tag{2.8}$$

其中引入了记号

$$G_{r\cdot}=G_{++}-G_{+-}=\sum_\alpha \alpha G_{+\alpha},\quad G_{\cdot r}=G_{++}-G_{-+}=\sum_\alpha \alpha G_{\alpha+},$$

$$G_{a\cdot}=G_{-+}-G_{--}=\sum_\alpha \alpha G_{-\alpha},\quad G_{\cdot a}=G_{+-}-G_{--}=\sum_\alpha \alpha G_{\alpha-},$$

$$G^{(+)}=G_{++}+G_{--}=\frac{1}{2}\sum_{\alpha\beta}(1+\alpha\beta)G_{\alpha\beta},$$

$$G^{(-)}=G_{+-}+G_{-+}=\frac{1}{2}\sum_{\alpha\beta}(1-\alpha\beta)G_{\alpha\beta}. \tag{2.9}$$

当两点函数具有时序乘积(2.1)式的结构时,以上各量两两相等,成为推迟函数

$$G_r\equiv G_{r\cdot}=G_{a\cdot}=-i\theta(12)\langle[1,2]\rangle, \tag{2.10a}$$

超前函数

$$G_a\equiv G_{\cdot r}=G_{\cdot a}=-i\theta(21)\langle[2,1]\rangle, \tag{2.10b}$$

和"对称化关联函数"[6]

$$G_c\equiv G^{(+)}=G^{(-)}=-i\langle\{1,2\}\rangle. \tag{2.10c}$$

(2.10)式中[,]和{,}分别表示对易和反对易子. 于是

$$\tilde{G}=\begin{pmatrix}0, & G_a\\ G_r, & G_c\end{pmatrix}. \tag{2.11}$$

$\hat{G}$ 通过 $\tilde{G}$ 的三个非零元素表示为[3]

$$\hat{G}=\frac{1}{2}G_c\begin{pmatrix}1 & 1\\ 1 & 1\end{pmatrix}+\frac{1}{2}G_r\begin{pmatrix}1 & -1\\ 1 & -1\end{pmatrix}+\frac{1}{2}G_a\begin{pmatrix}1 & 1\\ -1 & -1\end{pmatrix}, \tag{2.12}$$

这里数值矩阵都是奇异的.

以上变换容易推广到多点函数. 我们跳过三点函数，直接写出四点函数的变换. 先不管四点函数的结构,按照

$$G_{ijkl}=2Q_{i\alpha}Q_{j\beta}Q_{k\gamma}Q_{l\delta}G_{\alpha\beta\gamma\delta} \tag{2.13}$$

得到 $\tilde{G}$ 矩阵的 16 个元素为

$$G_{1111}=\frac{1}{2}(G^{(+)}-G^{(-)}),\quad G_{2222}=\frac{1}{2}(G^{(+)}+G^{(-)}),$$

$$G_{2111}=\frac{1}{2}(G_{r\cdots}+G_{a\cdots})\ \text{等 4 个},$$

$$G_{2211}=\frac{1}{2}(G_{rr\cdot\cdot}+G_{aa\cdot\cdot})\ \text{等 6 个},$$

$$G_{2221}=\frac{1}{2}(G_{rrr\cdot}+G_{aaa\cdot})\ \text{等 4 个}, \tag{2.14}$$

其中引入了记号

$$G^{(\pm)}=\frac{1}{2}\sum_{\alpha\beta\gamma\delta}(1\pm\alpha\beta\gamma\delta)G_{\alpha\beta\gamma\delta},$$

$$G_{r\cdots}=\sum_{\alpha\beta\gamma}\alpha\beta\gamma G_{+\alpha\beta\gamma}\ \text{等 4 个},$$

$$G_{a\cdots}=\sum_{\alpha\beta\gamma}\alpha\beta\gamma G_{-\alpha\beta\gamma}\ \text{等 4 个},$$

$$G_{rr\cdot\cdot}=\sum_{\alpha\beta}\alpha\beta(G_{++\alpha\beta}+G_{--\alpha\beta})\ \text{等 6 个},$$

$$G_{aa\cdot\cdot}=\sum_{\alpha\beta}\alpha\beta(G_{+-\alpha\beta}+G_{-+\alpha\beta})\ \text{等 6 个}. \tag{2.15}$$

(以上两行右端写明了的下标分别与两点函数 $G^{(+)}$, $G^{(-)}$ 一致.)

$$G_{rrr\cdot}=\sum_{\alpha}\alpha(G_{+++\alpha}+G_{+--\alpha}+G_{-+-\alpha}+G_{--+\alpha})\ \text{等 4 个},$$

$$G_{aaa\cdot}=\sum_{\alpha}\alpha(G_{---\alpha}+G_{-++\alpha}+G_{+-+\alpha}+G_{++-\alpha})\ \text{等 4 个}.$$

(以上两行右端明显写出的下标分别与三点函数 $G^{(+)}$, $G^{(-)}$ 一致.)

如果 $G_p(1234)$ 是时序乘积的平均值,如(2.1)式,则利用(1.3)—(1.10)诸式和定义(2.15)式可得

$$G_{1111}=0,\quad G_{2222}\equiv G_c=G^{(+)}=G^{(-)},$$

$$G_{2111}=G_{r\cdots}=G_{a\cdots}\ \text{等 4 式},$$

$$G_{2211}=G_{rr\cdot\cdot}=G_{aa\cdot\cdot}\ \text{等 6 式},$$

$$G_{2221}=G_{rrr\cdot}=G_{aaa\cdot}\ \text{等 4 式}, \tag{2.16}$$

具体写出来为

$$G_{2111} = i \sum_{P_3} \theta(1p_2p_3p_4)\langle[[[1, p_2], p_3], p_4]\rangle \quad 等,$$

$$\begin{aligned} G_{2211} = i \sum_{\substack{(\begin{smallmatrix}1 & 2\\ p_1 & p_2\end{smallmatrix})\\(\begin{smallmatrix}3 & 4\\ p_3 & p_4\end{smallmatrix})}} &\{\theta(p_1p_2p_3p_4)\langle[[\{p_1, p_2\}, p_3], p_4]\rangle \\ &+ \theta(p_1p_3p_2p_4)\langle[\{[p_1, p_3], p_2\}, p_4]\rangle \\ &+ \theta(p_1p_3p_4p_2)\langle\{[[p_1, p_3], p_4], p_2\}\rangle\} \quad 等, \end{aligned}$$

$$\begin{aligned} G_{2221} = i \sum_{P_3} &\{\theta(p_1p_2p_3 4)\langle[\{\{p_1, p_2\}, p_3\}, 4]\rangle \\ &+ \theta(p_1p_2 4p_3)\langle\{[\{p_1, p_2\}, 4], p_3\}\rangle \\ &+ \theta(p_1 4p_2p_3)\langle\{\{[p_1, 4], p_2\}, p_3\}\rangle\} \quad 等, \end{aligned}$$

$$G_c = i \sum_{P_4} \theta(p_1p_2p_3p_4)\langle\{\{\{p_1, p_2\}, p_3\}, p_4\}\rangle. \tag{2.17}$$

这些式子中包含了三重嵌套的对易子[,]和反对易子{,}的一切组合,对应四个时刻的一切可能的超前和推迟排列("一点在前","两点之一在前"、"三点之一在前"等等),因此各种四点推迟和超前函数都在其中了.

以上诸式都是为不相连格林函数推导的. 对于相连格林函数存在着完全类似的关系. 只要注意从定义

$$G_p(1, 2, \cdots, n) = i\frac{\delta^n Z[J]}{\delta J(1)\delta J(2)\cdots\delta J(n)}\bigg|_{J=0},$$

$$G_p^c(1, 2, \cdots, n) = \frac{\delta^n W[J]}{\delta J(1)\delta J(2)\cdots\delta J(n)}\bigg|_{J=0},$$

$$W[J] = i \ln Z[J] \tag{2.18}$$

(上角标 c 表示"连接", $Z[J]$ 和 $W[J]$ 是相应的生成泛函[3,7])得到如下关系:

$$\begin{aligned} G_p^c(1) &= G_p(1), \\ G_p^c(12) &= G_p(12) + iG_p(1)G_p(2), \\ G_p^c(123) &= G_p(123) + i[G_p(1)G_p(23) + G_p(2)G_p(13) \\ &\quad + G_p(3)G_p(12)] - 2G_p(1)G_p(2)G_p(3), \\ &\cdots\cdots \end{aligned} \tag{2.19}$$

就可以算出: 两点函数中只有对称化关联函数变为

$$G_c^c(12) = -i(\langle\{1, 2\}\rangle - \{\langle 1\rangle, \langle 2\rangle\}); \tag{2.20}$$

三点函数中

$$G_{211}^c(123) = G_{211}(123) = -\sum_{P_2} \theta(1p_2p_3)\langle[[1, p_2], p_3]\rangle \tag{2.21}$$

没有变化,而

$$\begin{aligned} G_{122}^c(123) = -\sum_{P_2} &\{\theta(p_2p_3 1)(\langle[\{p_2, p_3\}, 1]\rangle - \langle[\langle p_2\rangle p_3, 1]\rangle \\ &- \langle[p_2\langle p_3\rangle, 1]\rangle) + \theta(p_2 1 p_3)(\langle\{[p_2, 1], p_3\}\rangle - \langle[p_2, 1]\rangle\langle p_3\rangle)\}, \end{aligned}$$

$$G_c^c(123) = -\sum_{P_3} \theta(p_1p_2p_3)(\langle\{\{p_1, p_2\}, p_3\}\rangle - \langle\{p_1, p_2\}\rangle\langle p_3\rangle + 2\{\{\langle p_1\rangle, \langle p_2\rangle\}, \langle p_3\rangle\}), \tag{2.22}$$

等等;四点函数可紧凑地写为

$$\begin{aligned} G_{2111}^c &= G_{2111}, \\ G_{2211}^c &= G_{2211} + 2i[G(1)G_{211}(234) + G(2)G_{211}(134)] \\ &\quad + 2i[G_r(13)G_r(24) + G_r(14)G_r(23)], \\ G_{2221}^c &= G_{2221} + 2i[G(1)G_{221}(234) + G(2)G_{221}(134) + G(3)G_{221}(124)] \\ &\quad + 2i[G_c(12)G_r(34) + G_c(13)G_r(24) + G_c(23)G_r(14)] \\ &\quad - 8[G(1)G(2)G_r(34) + G(1)G(3)G_r(24) \\ &\quad + G(2)G(3)G_r(14)], \end{aligned} \tag{2.23}$$

等等. 注意,完全由对易子嵌套而成的推迟函数 $G_{211\cdots 1}^c = G_{211\cdots 1}$, 相当于 LSZ[8] 构造公理化场论时引入的 r-函数,不受连接与否的影响,而所有其它推迟、超前和关联函数都随是否用相连格林函数定义而有所不同.

三、串联和并联规则

作为三套函数互相变换的实例,我们看一批闭路上多点函数的运算规则(文献 [4] 中已经给过一些这类规则). 掌握这些规则,可以迅速地从由 G_p 表示的关系,如 Dyson 方程,变到由 $\hat{G}$ 或 $\tilde{G}$ 表示的关系.

1. 最简单的串联关系是两个单点函数乘积的积分

$$\int_p J_p(x)\psi_p(x)dx = \int_{-\infty}^{+\infty} \hat{J}\sigma_3\hat{\psi}dx = 2\int_{-\infty}^{+\infty} \tilde{J}\sigma_1\tilde{\psi}dx, \tag{3.1}$$

其中 $\hat{J} = (J_+, J_-)$, $\hat{\psi} = \begin{pmatrix}\psi_1\\ \psi_2\end{pmatrix}$ 等等. 以后省去积分符号,对重复出现的自变量自动求积分. 除 G_p 的积分限于坐标空间外,对 $\hat{G}$ 和 $\tilde{G}$ 的积分可不区别坐标和动量表示. 有时连自变量也省去不写. 因此,如有关系

$$R_p(1) = G_p(12)J_p(2) \quad 或 \quad R_p = G_pJ_p,$$

则分出正、负支即插入 σ_3 矩阵,得

$$\hat{R}(1) = \hat{G}(12)\sigma_3\hat{J}(2) \quad 或 \quad \hat{R} = \hat{G}\sigma_3\hat{J},$$

两面乘以 Q 和 Q^{-1} 得

$$\tilde{R}(1) = \tilde{G}(12)\sigma_1\tilde{J}(2) \quad 或 \quad \tilde{R} = \tilde{G}\sigma_1\tilde{J}. \tag{3.2}$$

如果 J 是外场,最终应令其在正、负支上相等,(3.2) 式就是

$$\begin{pmatrix}R_1\\ R_2\end{pmatrix} = \begin{pmatrix}0\\ G_rJ\end{pmatrix},$$

在对外场的响应中自动出现推迟函数.

2. 两点函数的串联关系

$$D_p(12) = A_p(13)B_p(32)$$

立即变成

$$\hat{D} = \hat{A}\sigma_3\hat{B} \quad 和 \quad \widetilde{D} = \widetilde{A}\sigma_1\widetilde{B},$$

后者写成分量为

$$\begin{pmatrix} 0 & D_a \\ D_r & D_c \end{pmatrix} = \begin{pmatrix} 0 & A_aB_a \\ A_rB_r & A_rB_c + A_cB_a \end{pmatrix}. \tag{3.3}$$

这个规则直接推广到多个两点函数的乘积，对

$$Z_p = A_p^{(1)}A_p^{(2)}\cdots A_p^{(n)}$$

有

$$\hat{Z} = \hat{A}^{(1)}\sigma_3\hat{A}^{(2)}\sigma_3\cdots\sigma_3\hat{A}^{(n)}$$

和

$$\widetilde{Z} = \widetilde{A}^{(1)}\sigma_1\widetilde{A}^{(2)}\sigma_1\cdots\sigma_1\widetilde{A}^{(n)},$$

后者写成分量为

$$Z_r = A_r^{(1)}A_r^{(2)}\cdots A_r^{(n)}, \quad Z_a = A_a^{(1)}A_a^{(2)}\cdots A_a^{(n)}, \tag{3.4}$$

特别是

$$Z_c = \sum_{k=1}^{n} A_r^{(1)}\cdots A_r^{(k-1)}A_c^{(k)}A_a^{(k+1)}\cdots A_a^{(k)}. \tag{3.5}$$

类似地，利用逆变换 $\hat{Z} = Q^{-1}\widetilde{Z}Q$ 和 (2.12) 式，可得

$$Z_\mu = \sum_{k=1}^{n} A_r^{(1)}\cdots A_r^{(k-1)}A_\mu^{(k)}A_a^{(k+1)}\cdots A_a^{(k)}, \tag{3.6}$$

其中 $\mu = +-$ 或 $-+$.

积分下有多点函数时，插入 σ_i 矩阵要注意变量顺序．例如三端顶角函数

$$\Gamma_p(123) = i\Gamma_p(14)\Gamma_p(25)\Gamma_p(36)G_p(456) \tag{3.7}$$

应变为

$$\hat{\Gamma}(123) = i(\hat{\Gamma}\sigma_3)(14)(\hat{\Gamma}\sigma_3)(25)(\hat{\Gamma}\sigma_3)(36)\hat{G}(456),$$

再变成

$$\widetilde{\Gamma}(123) = i(\widetilde{\Gamma}\sigma_1)(14)(\widetilde{\Gamma}\sigma_1)(25)(\widetilde{\Gamma}\sigma_1)(36)\widetilde{G}(456),$$

按分量写开得

$$\Gamma_{111} = 0, \quad \Gamma_{211} = i\Gamma_r\Gamma_a\Gamma_aG_{211},$$
$$\Gamma_{221} = i(\Gamma_c\Gamma_r\Gamma_aG_{121} + \Gamma_r\Gamma_c\Gamma_aG_{211} + \Gamma_r\Gamma_r\Gamma_aG_{221})$$

等等．

只要是理论中自然出现的多点函数的关系，就不会在变换后出现多余的数值因子．例如四端顶角函数 $\Gamma_p^{(4)}$ 与截腿格林函数 $W_p^{(n)}$ 的关系[7]

$$\Gamma_p^{(4)} = -W_p^{(4)} + 3W_p^{(3)}G_p^{(2)}W_p^{(3)} \tag{3.8}$$

变成

$$\hat{\Gamma}^{(4)} = -\hat{W}^{(4)} + 3\hat{W}^{(3)}\sigma_3\hat{G}^{(2)}\sigma_3\hat{W}^{(3)}$$

和

$$\widetilde{\Gamma}^{(4)} = -\widetilde{W}^{(4)} + 3\widetilde{W}^{(3)}\sigma_1\widetilde{G}^{(2)}\sigma_1\widetilde{W}^{(3)}.$$

只有一些人为地"收缩"出来的关系如

$$A_p(12) = B_p(134)C_p(3\dot{4}2)$$

导致

$$\widetilde{A}(12) = \frac{1}{2}\widetilde{B}(134)(\sigma_1)_{33'}(\sigma_1)_{44'}\widetilde{C}(3'4'2),$$

才多出数值因子 1/2. 这说明 $\hat{G}$ 和 $\widetilde{G}$ 的一般变换公式 (2.3) 中，因子 $2^{\frac{n-2}{2}}$ 的选择是对的.

3. 含 δ-函数的串联关系

$$G_p(13)\Gamma_p(32) = \Gamma_p(13)G_p(32) = \delta_p(12),$$

使 G_p 和 Γ_p 成为闭路上的逆矩阵. 变换后有

$$\hat{G}\sigma_3\hat{\Gamma} = \hat{\Gamma}\sigma_3\hat{G} = \hat{\delta},$$
$$\widetilde{G}\sigma_1\tilde{\Gamma} = \tilde{\Gamma}\sigma_1\widetilde{G} = \tilde{\delta}. \tag{3.9}$$

可见 $\sigma_3\hat{G}\sigma_3$ 和 $\hat{\Gamma}$，$\widetilde{G}$ 和 $\sigma_1\tilde{\Gamma}\sigma_1$ 等等，才是普通意义下的逆矩阵. 只要 G_p 是"闭路矩阵"，即

$$G_{++} + G_{--} = G_{+-} + G_{-+} \text{ 或 } G_{11} = 0 \tag{3.10a}$$

(由第二节知道这是 G_p 本身为时序乘积平均值的后果)，这一性质就由 (3.9) 式传给 Γ_p:

$$\Gamma_{++} + \Gamma_{--} = \Gamma_{+-} + \Gamma_{-+} \text{ 或 } \Gamma_{11} = 0. \tag{3.10b}$$

因此 $\tilde{\Gamma}$ 可以写成类似 (2.11) 式的样子

$$\tilde{\Gamma} = \begin{pmatrix} 0 & \Gamma_a \\ \Gamma_r & \Gamma_c \end{pmatrix}, \tag{3.11}$$

这时并不要求 Γ_p 可以写成时序乘积的平均值.

由 $G_p(12) = G_p(21)$、密度矩阵 ρ 和实标量场 ϕ 的厄米性，以及阵迹符号 Tr 下的循环不变性，可以证明 G_p 满足

$$\hat{G} = \hat{G}^T = -\sigma_1\hat{G}^*\sigma_1 = -\sigma_1\hat{G}^+\sigma_1, \tag{3.12a}$$

作傅氏变换后有

$$\hat{G}(k) = \hat{G}^T(-k) = -\sigma_1\hat{G}^*(-k)\sigma_1 = -\sigma_1\hat{G}^+(k)\sigma_1,$$
$$\widetilde{G}(k) = \widetilde{G}^T(-k) = -\sigma_3\widetilde{G}^*(-k)\sigma_3 = -\sigma_3\widetilde{G}^+(k)\sigma_3. \tag{3.12b}$$

这里上标 T 是转置、$*$ 是复共轭、$+$ 是厄米共轭. 这些性质全部通过 (3.9) 式传给逆矩阵 $\hat{\Gamma}$ 和 $\tilde{\Gamma}$，而不要求知道 Γ_p 的具体结构. 利用 (3.7)，(3.8) 诸式可以证明，多点格林函数的闭路性质，同样传给相应的多点顶角函数.

利用闭路逆矩阵还可以写出高斯型闭路连续积分公式(省去"常数"因子)

$$\int [d\phi_p]\exp\left(-\frac{1}{2}\phi_p\Gamma_p\phi_p + J_p\phi_p\right)$$
$$= \int [d\phi_+ d\phi_-]\exp\left(-\frac{1}{2}\hat{\phi}\sigma_3\hat{\Gamma}\sigma_3\hat{\phi} + \hat{J}\sigma_3\hat{\phi}\right)$$
$$= \cdots = \exp\left(-\frac{1}{2}J_pG_pJ_p\right). \tag{3.13}$$

4. 并联关系如 $A_p(12)B_p(12)C_p(12)$ 应作为一个记号参加变换. 变换结果是否具有闭路矩阵形式，通常可由费曼图看出来. 例如

$$S_p(12) \equiv G_p^3(12) \tag{3.14}$$

可成为两点函数的串联部分(自能部份)，应当具有闭路性质. 事实上，由 (3.10a) 式和 $G_r(12)G_a(12)=0$ 可证

$$G^3_{++}(12)+G^3_{--}(12)=G^3_{+-}(12)+G^3_{-+}(12),$$

进而得

$$\tilde{S}(12)=\frac{1}{4}\begin{pmatrix}0 & G_a[G_a^2+3G_c^2]\\ G_r[G_r^2+3G_c^2] & G_c[G_c^2+3(G_a+G_r)^2]\end{pmatrix}. \tag{3.15}$$

并联变换应在坐标表示中讨论.

四、多点格林函数闭路性质的一般证明

第二节中讨论的相连和不相连格林函数,具有一些普遍性质,如 $G^{(+)}=G^{(-)}$, 即

$$\begin{gathered}G_+=G_- \ \text{(单点函数)},\\ G_{++}+G_{--}=G_{+-}+G_{-+}\ \text{(两点函数)},\\ G_{+++}+G_{+--}+G_{-+-}+G_{--+}=G_{---}+G_{-++}+G_{+-+}+G_{++-}\ \text{(三点函数)},\\ \cdots\cdots\cdots\cdots\end{gathered} \tag{4.1}$$

等等. 这些关系变换到 $\tilde{G}$ 矩阵就成为

$$G_1=0,\ G_{11}=0,\ G_{111}=0,\ \cdots \tag{4.2}$$

凡是满足条件 (4.1) 式或 (4.2) 式的多点函数，前面已经称为闭路矩阵，这是文献 [2] 中对两点函数所给定义的推广. (4.1) 式的另一后果，就是定义 $\tilde{G}$ 矩阵各元素时引入的记号如 (2.9), (2.15) 诸式,满足如下的等式:

$$G_{r\cdot}=G_{a\cdot},\ G_{r\cdot\cdot}=G_{a\cdot\cdot},\ G_{r\cdots}=G_{a\cdots}, \tag{4.3}$$

$$G_{rr\cdot}=G_{aa\cdot},\ G_{rr\cdot\cdot}=G_{aa\cdot\cdot}, \tag{4.4}$$

$$G_{rrr\cdot}=G_{aaa\cdot}, \tag{4.5}$$

等等. 第二节中通过直接计算得到了这些关系. 现在为 n 点格林函数给出一般证明.

普通格林函数生成泛函的归一或归零条件[7]

$$Z[J]|_{J=0}=1,\ W[J]|_{J=0}=0.$$

在闭路情形下推广为

$$Z[J_+,J_-]|_{J_+=J_-=J}=1, \tag{4.6}$$

$$W[J_+,J_-]|_{J_+=J_-=J}=0, \tag{4.7}$$

这可由它们的定义[3]看出来. 这里并不要求 $J=0$，于是增加了很大的活动余地. 事实上，$J_2=\frac{1}{2}(J_++J_-)$ 是物理外场，$J_1=\frac{1}{2}(J_+-J_-)$ 才相当于构造生成泛函时引入的形式外源,参看 (2.4) 和 (3.2) 式.

取 $W[J_+,J_-]$，令 $J_\pm=J_0+\delta J_\pm$，在 J_0 附近展开. 利用第三节中的简化记法，得

$$\begin{aligned}W[J_+,J_-]=&\left(\frac{\delta W}{\delta J_+}\delta J_+-\frac{\delta W}{\delta J_-}\delta J_-\right)+\frac{1}{2!}\left(\delta J_+\frac{\delta^2 W}{\delta J_+\delta J_+}\delta J_+\right.\\&\left.+\delta J_-\frac{\delta^2 W}{\delta J_-\delta J_-}\delta J_--\delta J_+\frac{\delta^2 W}{\delta J_+\delta J_-}\delta J_--\delta J_-\frac{\delta^2 W}{\delta J_-\delta J_+}\delta J_+\right)+\cdots,\end{aligned}$$

式中各变分导数均取在 $J_+ = J_- = J_0$ 处．进一步取 $\delta J_+ = \delta J_- = \delta J$，使得 $J_+ = J_-$，于是由 δJ 之任意性得到一批关系：

$$\frac{\delta W}{\delta J(x_+)} = \frac{\delta W}{\delta J(x_-)}\bigg|_{J_+=J_-=J},$$

$$\frac{\delta^2 W}{\delta J(x_+)\delta J(y_+)} + \frac{\delta^2 W}{\delta J(x_-)\delta J(y_-)} = \frac{\delta^2 W}{\delta J(x_+)\delta J(y_-)} + \frac{\delta^2 W}{\delta J(x_-)\delta J(y_+)}\bigg|_{J_+=J_-=J},$$

$$\frac{\delta^3 W}{\delta J(x_+)\delta J(y_+)\delta J(z_+)} + \frac{\delta^3 W}{\delta J(x_+)\delta J(y_-)\delta J(z_-)} + \frac{\delta^3 W}{\delta J(x_-)\delta J(y_+)\delta J(z_-)}$$
$$+ \frac{\delta^3 W}{\delta J(x_-)\delta J(y_-)\delta J(z_+)} = \frac{\delta^3 W}{\delta J(x_-)\delta J(y_-)\delta J(z_-)} + \frac{\delta^3 W}{\delta J(x_-)\delta J(y_+)\delta J(z_+)}$$
$$+ \frac{\delta^3 W}{\delta J(x_+)\delta J(y_-)\delta J(z_+)} + \frac{\delta^3 W}{\delta J(x_+)\delta J(y_+)\delta J(z_-)}\bigg|_{J_+=J_-=J},$$

$$\cdots\cdots \tag{4.8}$$

等等．在 (4.8) 式中取 $J = 0$，即得 (4.1) 式．如果不令 $J = 0$，而以 $W[J]$ 作为解析泛函的展开式

$$W[J] = \sum_{n=1}^{\infty} \frac{1}{n!} \int_p \cdots \int_p G_p^c(1, 2, \cdots, n) J(1) J(2) \cdots J(n) d1 d2 \cdots dn \tag{4.9}$$

代入 (4.8) 式的第一式，注意

$$\frac{\delta J(x_i)}{\delta J(y_\pm)} = \delta_p(x_i - y_\pm)$$

和 G_p^c 的对称性，得到(省去积分符号)

$$\sum_{n=2}^{\infty} \frac{1}{(n-1)!} [G_p^c(+, 1, 2, \cdots n-1) - G_p^c(-, 1, 2, \cdots n-1)]$$
$$\times J(1) J(2) \cdots J(n-1)|_{J_+=J_-} = 0,$$

变换成 $\hat{G}$，$\hat{J}$ 形式时，注意 $(\sigma_3 J)_\alpha$ 可以写成 αJ_α (这里 $\alpha = \pm$，但不对重复出现的 α 自动求和)，得

$$\sum_{n=1}^{\infty} \frac{1}{n!} \sum_{\alpha_i} \alpha_1\alpha_2 \cdots \alpha_n [G^c_{+\alpha_1\alpha_2\cdots\alpha_n} - G^c_{-\alpha_1\alpha_2\cdots\alpha_n}] J_{\alpha_1}(1) J_{\alpha_2}(2) \cdots J_{\alpha_n}(n)|_{J_+=J_-} = 0,$$

在正、负支上取相同的外源时，$J_{\alpha_i}(i)$ 实际上与 α_i 无关．于是由 J 的任意性，对任何 $n \geqslant 1$ 得到

$$\sum_{\alpha_1\alpha_2\cdots\alpha_n} \alpha_1\alpha_2 \cdots \alpha_n G^c_{+\alpha_1\alpha_2\cdots\alpha_n} = \sum_{\alpha_1\alpha_2\cdots\alpha_n} \alpha_1\alpha_2 \cdots \alpha_n G^c_{-\alpha_1\alpha_2\cdots\alpha_n},$$

这就是 (4.3) 式的一般形式．

如果以 (4.9) 式代入 (4.8) 式的第二、三…诸式，经过类似的推导，即得 (4.4)，(4.5) 各式．同样地，从 (4.6) 式出发，可对不相连格林函数得到完全相同的一批关系．

五、讨　　论

本文主要分析了与三套闭路格林函数有关的若干代数关系．在处理具体物理问题时

可以利用这些关系来简化闭路格林函数的推导过程．我们的主要收获，是看清楚了如何在多时刻情形下统一地定义一切推迟、超前和关联函数．这些函数本身是定义在普通的$(-\infty、+\infty)$时间轴上的，然而只有借助闭路格林函数的变换，才自然地得到了它们．

三点函数出现在表面物理的某些问题中．四点函数可用以推导 Bethe-Salpeter 方程．在动态临界现象理论中，无论从广义朗之万方程出发，还是使用拉格朗日形式的统计场论，目前都是在 $\tilde{G}$ 形式下建立重正化微扰论．因此，只能用迭代产生高阶项，得到看起来与量子场论颇为不同的费曼图．事实上应当回到 G_p 或 $\hat{G}$ 形式去，理论的结构才更清楚．我们将在以后的几篇文章中，应用闭路格林函数研究临界动力学．

参 考 文 献

[1] J. Schwinger, *J. Math. Phys.*, **2** (1961), 407.

[2] 周光召、苏肇冰，闭路格林函数和它在非平衡统计物理中的应用，《统计物理进展》第五章，科学出版社(待出版).

[3] 周光召、苏肇冰，高能物理与核物理，**3** (1979)，314.

[4] D. Langreth, In «Linear and Nonlinear Transport in Solids», ed. by J. Devreese and V. Van. Doren, Plenum, N. Y. (1976).

[5] Л. Келдыш, *ЖЭТФ*, **47** (1964), 1515.

[6] R. Kubo, *Reps. Progr. Phys.*, **29** (1966), 255.

[7] 郝柏林，统计微扰论的生成泛函，《统计物理进展》第一章，科学出版社(待出版).

[8] H. Lehmann, K. Symanzik, W. Zimmermann, *Nuovo Cimento*, **6** (1957), 319.

TRANSFORMATION PROPERTIES OF THREE SETS OF CLOSED TIME PATH GREEN'S FUNCTIONS

ZHOU GUANG-ZHAO YU LU HAO BAI-LIN

(Institute of Theoretical Physics, Academia Sinica)

ABSTRACT

In this article, we derived the transformations among three sets of closed time path Green's functions and some calculation rules, from which a general definition of arbitrary multipoint retarded and advanced Green's functions follows naturally. Some algebraic identities among multipoint functions have shown to be the consequences of the property $W[J_+,J_-]|_{J_+=J_-=J}=0$ for the generating functional on the closed time path.

第 29 卷 第 8 期 1980 年 8 月
物 理 学 报 ACTA PHYSICA SINICA
Vol. 29, No. 8 Aug., 1980

非平衡统计场论与临界动力学（I）

广义朗之万方程

周光召 苏肇冰 郝柏林 于渌

（中国科学院理论物理研究所）

1979 年 9 月 17 日收到

提 要

本文从闭路格林函数顶角方程出发，推导出临界动力学中序参量和守恒量所满足的广义朗之万方程．根据 Ward-Takahashi 恒等式和线性响应理论，确定守恒量方程应具有的形式，它自动包含了模-模耦合项．考察不同的对称群，得到临界动力学的各种模型． 整个理论框架也可用于描述远离平衡的稳态附近的行为．

一、引 言

由 Schwinger 及 Keldysh 等人发展起来的闭路格林函数方法已经有了多方面的应用[1,2]．文献[3—5]中指出，它是研究非平衡态统计场论的有效方法．文献[6]中用它分析了激光这类非平衡耗散系统定态的 Goldstone 模式．本文将闭路格林函数方法用于分析平衡态相变点附近的行为，统一地得出了包括模-模耦合在内的临界动力学完备方程组及拉格朗日场论表述，一方面为半唯象的临界动力学模型提供微观论证，另一方面指出改进现有理论的可能途径．

在相变点附近，长波涨落起主要作用．由于相应的关联长度远远大于热波长(即热运动能量所对应的德布罗意波长)，量子效应并不重要．但是，在准粒子描述中，这种经典场论并不对应玻耳兹曼极限，而是对应“超玻色”极限，即准粒子分布密度 $n\propto T/\varepsilon$，这里 T 为绝对温度(我们取 $\hbar=c=k_B=1$ 的单位制)，ε 为元激发能量．这种统计场论（或涨落场论)与量子场论有深刻的类比．

我们认为，闭路格林函数是研究这类场论的一种自然的理论框架．若取平衡态密度矩阵，在低温极限（$T\ll\varepsilon$）自动得到通常的量子场论，在高温极限（$T\gg\varepsilon$），得到静态临界现象的现代理论（见文献[12]的附录）．在平衡态附近取高温极限，就可求得描述动态临界现象的全部方程．在本文中，我们推导广义朗之万方程，文献[12]讨论拉格朗日形式的场论表述．我们将在以后的文章中讨论这种场论的重正化和具体应用．只要长波涨落起主要作用，整个理论框架对于远离平衡的定态附近的情形也是适用的． Martin，Siggia 和 Rose[7] 通过引入与基本场量不对易的“响应场”，构造了经典的统计场论(文献中常称为 MSR 场论). 用闭路格林函数表述,这种场论变得非常自然.

二、临界动力学方程组

临界动力学最近有比较详细的总结[8]. 我们只是为了统一符号和下面行文需要,在此作扼要概括.

系统的临界性质由序参量和守恒量描述,它们构成"宏"变量的集合 $Q=\{Q_i, i=1, 2, \cdots, n\}$, 这些随机变量的时间演化满足广义的朗之万方程:

$$\frac{\partial Q_i(t)}{\partial t}=K_i(Q)+\xi_i(t), \tag{2.1}$$

其中随机项 $\xi_i(t)$ 反映一切未计入 $\{Q_i\}$ 的自由度的影响,通常假定它们遵从高斯分布,即

$$\langle\xi_i(t)\rangle=0,$$
$$\langle\xi_i(t)\xi_j(t')\rangle=2\delta_{ij}\delta(t-t'). \tag{2.2}$$

(2.1) 式的右端函数 $K_i(Q)$ 由两部分构成:

$$K_i(Q)=-\sigma_{ij}\frac{\delta F}{\delta Q_j}+V_i(Q), \tag{2.3}$$

其中自由能 $F\equiv F[Q]$ 是 Q 的泛函,需要根据具体模型给出. 静态的平衡条件 $\delta F/\delta Q_i=0$ 就是 Ginzburg-Landau 方程(以下简称 G-L 方程),因此,不带随机项的(2.1)式又称为含时间的 G-L 方程(简称 TDGL). (2.3) 式中的系数矩阵 σ_{ij} 原则上可以包含对称部分和反称部分,对称部分反映弛豫运动,反称部分描述正则运动. 只讨论弛豫时,可以认为 σ_{ij} 是对称的,涨落耗散定理要求它和(2.2)式中的 σ_{ij} 相同. 对角化后,如果相应宏变量 Q_i 不是守恒量,则 $\sigma_i=$ 常数(耗散弛豫),如果 Q_i 是守恒量,则 $\sigma_i=-D_i\nabla^2$ (扩散弛豫), D_i 是扩散系数.

各个"模" Q_i 之间的耗散型耦合可以通过自由能泛函 F 中的耦合项来描述,可逆的模-模耦合,要作为一种单独的"流",写成(2.3)式中的第二项,它通常的形式是[9]

$$V_i(Q)=\lambda\sum_j\left(\frac{\partial}{\partial Q_j}A_{ij}(Q)-A_{ij}(Q)\frac{\delta F}{\delta Q_j}\right), \tag{2.4}$$

其中反称张量 A_{ij} 由 Q 的泊松括号或对易子构成,通常取线性近似

$$A_{ij}=f_{ijk}Q_k, \tag{2.5}$$

这里 f_{ijk} 是对称群的结构常数. A_{ij} 的确切含义见本文第四节. (2.4) 式的形式本身保证了 V_i 满足定态的概率守恒方程:

$$\frac{\partial}{\partial Q_i}(V_i(Q)e^{-F})=0, \tag{2.6}$$

也就是 $\{Q_i\}$ 空间中的无散度条件,它保证 e^{-F} 是满足细致平衡条件的稳态分布.

我们看到,广义朗之万方程 (2.1) 可能包含很丰富的内容,但它的各个部分是根据不同的考虑凑集起来的,终究是一种模型.

动态临界现象理论中的通常作法[10] 是将方程 (2.1) 迭代求解以构造微扰论. 由于出现响应函数和关联函数两种基本"构件",微扰论的结构比较复杂. 这种微扰论的概括描

述就是前面已经提到的 MSR 场论[7]. 与静态理论中 Wilson 原来的作法类似，可以利用(2.1)式作重正化群变换，求得递推公式，得出不动点和临界指数. 近几年来，用概率泛函的办法，将(2.1)式"提升"成有效拉氏函数，采用标准的场论方法分析临界动力学[11]，这些我们将在文献[12]中涉及.

三、含时间的 Ginzburg-Landau 方程

一般说来，作为宏变量的序参量和守恒量都是基本场复合算子的平均值，我们在记号上稍作区别. 不是守恒量的序参量记为

$$Q_{ci}(x),\quad i=1,2,\cdots,n,$$

而守恒量记为

$$Q_{c\,n+\alpha}\equiv q_\alpha,\quad \alpha=1,2,\cdots,m.$$

q_α 是相应守恒流平均值的零分量

$$q_\alpha=\langle j_\alpha^0\rangle. \tag{3.1}$$

可以不失普遍性地取 Q_{ci} 和 q_α 为实场.

引入含有复合算子的闭路顶角函数生成泛函 $\Gamma[Q_c]$，则序参量和守恒量都满足方程[4]

$$\frac{\delta\Gamma}{\delta Q_{ci}(x)}=-J_i(x),\quad i=1,2,\cdots,n+m. \tag{3.2}$$

在这个方程中完成对 Q_{ci} 的变分后，再取 $J_i(x_+)=J_i(x_-)=J_i(\boldsymbol{x},t)$，因之 $Q_{ci}(x_+)=Q_{ci}(x_-)=Q_i(\boldsymbol{x},t)$，$J_i(\boldsymbol{x},t)$ 和 $Q_i(\boldsymbol{x},t)$ 就都是定义在普通时间轴 $t(-\infty,+\infty)$ 上的函数了. 现在证明，当 $Q_i(\boldsymbol{x},t)$ 是时空缓慢变化的函数时，(3.2) 式导致广义的 TDGL 方程.

设在时刻 τ 宏观量 $Q_i(\boldsymbol{x},\tau)$ 已知. 当 t 在 τ 附近时可以将(3.2)式左端展开. 如 x 在正支上，则有

$$-J_i(\boldsymbol{x},t)=\frac{\delta\Gamma}{\delta Q_{ci}(x_+)}\bigg|_{Q_{ci}(x_+)=Q_{ci}(x_-)=Q_i(\boldsymbol{x},\tau)}$$
$$+\int\Gamma_{Rij}(x,y)(Q_j(\boldsymbol{y},t_y)-Q_j(\boldsymbol{y},\tau))dy, \tag{3.3}$$

其中 $\Gamma_{Rij}(x,y)$ 是取了 $Q_{c+}=Q_{c-}=Q$ 之后的推迟两点顶角函数. 如 x 取在负支上，结果也一样，因为 $\Gamma_R=\Gamma_{++}-\Gamma_{+-}=\Gamma_{-+}-\Gamma_{--}$ (本文中这类代数运算规则均可参看文献[12]). (3.3)式中由于 $Q_j(\boldsymbol{y},t_y)$ 随时间缓变，准到一级小量有

$$Q_j(\boldsymbol{y},t_y)-Q_j(\boldsymbol{y},\tau)=(t_y-\tau)\frac{\partial Q_j(\boldsymbol{y},\tau)}{\partial\tau}. \tag{3.4}$$

代回(3.3)式中，并注意在 $t\equiv t_x\to\tau$ 极限下有

$$\gamma_{ij}(\boldsymbol{x},\boldsymbol{y},\tau)\equiv-\lim_{t_x\to\tau}\int dt_y(t_y-t_x)\Gamma_{Rij}(\boldsymbol{x},t_x,\boldsymbol{y},t_y)$$
$$=i\frac{\partial}{\partial k_0}\Gamma_{Rij}(\boldsymbol{x},y,k_0,\tau)|_{k_0=0}, \tag{3.5}$$

其中 $\Gamma_{Rij}(\boldsymbol{x},\boldsymbol{y},k_0,\tau)$ 是推迟顶角函数对 t_y-t_x 作傅氏变换，并取平均时间 $T\equiv\frac{1}{2}(t_x+t_y)\cong\tau$ 的结果，得到(以下 τ 改记为 t)

$$\gamma(t)\frac{\partial Q(t)}{\partial t}=\frac{\delta\Gamma}{\delta Q_{c+}}\bigg|_{Q_{c+}=Q_{c-}=Q(t)}+J(t). \tag{3.6}$$

这个式子中采用了矩阵记法，认为 $\gamma_{ij}(\boldsymbol{x},\boldsymbol{y},t)$ 是以 $i\boldsymbol{x}$ 和 $j\boldsymbol{y}$ 为角标的矩阵元.

暂时令

$$I_i(\boldsymbol{x},t)\equiv\frac{\delta\Gamma}{\delta Q_{c+}}\bigg|_{Q_{c+}=Q_{c-}=Q(t)}, \tag{3.7}$$

并将它看作三维点 $\boldsymbol{x}$ 的函数 $Q(\boldsymbol{x},t)$ 的泛函，计算

$$\frac{\delta I_i(\boldsymbol{x},t)}{\delta Q_j(\boldsymbol{y},t)}=\int dzdt_z\left(\frac{\delta^2\Gamma}{\delta Q_{ci}(x_+)\delta Q_{ck}(z_+)}\frac{\delta Q_{ck}(z_+)}{\delta Q_{cj}(\boldsymbol{y},t)}\right.$$
$$\left.-\frac{\delta^2\Gamma}{\delta Q_{ci}(x_+)\delta Q_{ck}(z_-)}\frac{\delta Q_{ck}(z_-)}{\delta Q_j(\boldsymbol{y},t)}\right)\bigg|_{Q_{c+}=Q_{c-}=Q(t)},$$

注意

$$\frac{\delta Q_{ck}(z)}{\delta Q_j(\boldsymbol{y},t)}\bigg|_{Q_{c+}=Q_{c-}=Q(t)}=\delta_{jk}\delta^{(3)}(\boldsymbol{y}-\boldsymbol{z}),$$

得到

$$\frac{\delta I_i(\boldsymbol{x},t)}{\delta Q_j(\boldsymbol{y},t)}=\Gamma_{++ij}(\boldsymbol{x},\boldsymbol{y},k_0=0,t)-\Gamma_{+-ij}(\boldsymbol{x},\boldsymbol{y},k_0=0,t),$$

这里同(3.5)式一样出现了傅氏变换后的 $k_0=0$ 分量. 同理得

$$\frac{\delta I_j(\boldsymbol{y},t)}{\delta Q_i(\boldsymbol{x},t)}=\Gamma_{++ji}(\boldsymbol{y},\boldsymbol{x},k_0=0,t)-\Gamma_{+-ji}(\boldsymbol{y},\boldsymbol{x},k_0=0,t)$$
$$=\Gamma_{++ij}(\boldsymbol{x},\boldsymbol{y},-k_0=0,t)-\Gamma_{-+ij}(\boldsymbol{x},\boldsymbol{y},-k_0=0,t),$$

这里利用了 $\hat{\Gamma}(k_0)$ 矩阵的对称性质[15]. 于是

$$\frac{\delta I_i(\boldsymbol{x},t)}{\delta Q_j(\boldsymbol{y},t)}-\frac{\delta I_j(\boldsymbol{y},t)}{\delta Q_i(\boldsymbol{x},t)}=\lim_{k_0\to 0}(\Gamma_{-+ij}(\boldsymbol{x},\boldsymbol{y},-k_0,t)-\Gamma_{+-ij}(\boldsymbol{x},\boldsymbol{y},k_0,t)). \tag{3.8}$$

在热平衡态附近由于[3]

$$\Gamma_{+-}=e^{-\beta k_0}\Gamma_{-+},$$

(3.8)式右端等于零，因而存在一个泛函 $F[Q_i(\boldsymbol{x},t)]$，使得

$$I_i(\boldsymbol{x},t)=-\frac{\delta F}{\delta Q_i(\boldsymbol{x},t)}. \tag{3.9}$$

(3.6) 式成为

$$\gamma(t)\frac{\partial Q(t)}{\partial t}=-\frac{\delta F}{\delta Q(t)}+J(t). \tag{3.10}$$

如果在外场 $J(t)$ 作用下，宏变量 Q 不随 t 变化(定态)，则

$$\frac{\delta F}{\delta Q}=J. \tag{3.11}$$

可见 F 就是系统的有效自由能，(3.11)式就是 G-L 方程，它决定宏变量的定态分布.

当系统处于远离平衡的定态时，只要顶角函数的耗散部分 $A = i(\Gamma_{-+} - \Gamma_{+-})/2$ (参阅文献[4])满足条件

$$\lim_{k_0\to 0} A_{ij}(\boldsymbol{x}, \boldsymbol{y}, k_0, \tau) = 0, \tag{3.12}$$

(3.8)式也等于零,仍然可以定义有效自由能或势函数 F，把 I_i 写成它的变分导数. Graham 和 Haken[13] 讨论过的由细致平衡维持的,满足"势条件"的定态，即应属于这一类. 在一切存在势函数 F 的定态附近，(3.10)式就是含时间的 G-L 方程组，然而它的意义比通常所说的 TDGL 方程要广泛,因为其中也包含了模耦合运动. 通常把矩阵 $\gamma(t)$ 求逆，将(3.10) 式写成

$$\frac{\partial Q(t)}{\partial t} = (\gamma(t))^{-1}\left(-\frac{\delta F}{\delta Q(t)} + J(t)\right). \tag{3.13}$$

利用顶角函数对称性质[15]

$$\tilde{\Gamma}(k) = \tilde{\Gamma}^T(-k) = -\sigma_3\tilde{\Gamma}^*(-k)\sigma_3 = -\sigma_3\tilde{\Gamma}^+(k)\sigma_3,$$

可以证明 Γ_r 的实部是 k_0 的偶函数，虚部是 k_0 的奇函数. 根据定义(3.5)式可以看出 $\gamma(t)$ 是一个实矩阵. 按照本节开头的编号规定，把 $\gamma(t)$ 分成四块，其中对应守恒量方程的两块 $\gamma_{\alpha\boldsymbol{x},j\boldsymbol{y}}$ 和 $\gamma_{\alpha\boldsymbol{x},\beta\boldsymbol{y}}$ 可以通过与 W-T 恒等式对比完全确定下来，而对应序参量的两块，一般情形下只能从对称考虑近似地决定其应有的形式.

最后应指出，(3.4)式相当于取马尔科夫近似，而原来的方程 (3.2) 中包含着计入记忆效应的潜力.

四、守恒量方程与模耦合

在系统的对称群 G 作用下，守恒量 q_α 的变换与群生成元 I^α 一致:

$$q_\alpha \to q'_\alpha = q_\alpha + if_{\alpha\beta\gamma}\omega_\beta q_\gamma \tag{4.1}$$

其中 $f_{\alpha\beta\gamma}$ 是群的结构常数，ω_β 是无穷小变换参数. 序参量按群 G 的某个表示 $\hat{L}$ 变换:

$$Q_i \to Q'_i = Q_i + iL^\alpha_{ij}\omega_\alpha Q_j. \tag{4.2}$$

在文献[4]中证明了，如果系统的拉氏量在群 G 的宇观作用下不变，则有如下闭路上的 W-T 恒等式成立:

$$\langle \partial_\mu j^\mu_\alpha(\varphi)\rangle = i\left[\frac{\delta\Gamma}{\delta Q_{ci}(x)} L^\alpha_{ij}Q_{cj}(x) + f_{\alpha\beta\gamma}\frac{\delta\Gamma}{\delta q_\beta(x)} q_\gamma(x)\right]. \tag{4.3}$$

$\Big($为从文献[4]中(3.17)式得(4.3)式，只须利用一般关系 $\langle f(\varphi)\rangle = f\left(\varphi_c + i\dfrac{\delta}{\delta J}\right)\Big)$. 这里 $\Gamma \equiv \Gamma(Q_{ci}, q_\alpha)$ 和第三节中一样，是插入了复合算子的顶角函数生成泛函. 在 (4.3) 式中取 $Q_{c+} = Q_{c-} = Q$, 并令

$$j^\mu_\alpha \equiv \langle j^\mu_\alpha(\varphi)\rangle,$$

于是它可以写成

$$\frac{\partial q_\alpha}{\partial t} = \nabla \boldsymbol{j}_\alpha - i[J_i(\boldsymbol{x}, t)L^\alpha_{ij}Q_j(\boldsymbol{x}, t) + f_{\alpha\beta\gamma}J_\beta(\boldsymbol{x}, t)q_\gamma(\boldsymbol{x}, t)], \tag{4.4}$$

这里 J, Q, q 等均已取在$(-\infty, +\infty)$时间轴上.

为了确定守恒流 $\boldsymbol{j}_\alpha$，我们参照线性响应理论中处理热扰动的思想，采用如下技巧．引入人为的外源 ΔJ，迭加在原有外源 J 上，使系统进入定态：

$$\frac{\partial q_\alpha}{\partial t}=0,$$

这时方程(4.4)中 $\boldsymbol{j}_\alpha$ 变成新的 $\boldsymbol{j}'_\alpha$，使得

$$\nabla \boldsymbol{j}'_\alpha - i[(J_i+\Delta J_i)L^\alpha_{ij}Q_j + f_{\alpha\beta\gamma}(J_\beta+\Delta J_\beta)q_\gamma]=0. \tag{4.5}$$

在(4.5)式中，由于系统处于定态，可以利用(3.11)式，

$$\frac{\delta F}{\delta Q_i}=J_i+\Delta J_i,\quad \frac{\delta F}{\delta q_\alpha}=J_\alpha+\Delta J_\alpha, \tag{4.6}$$

将 $J+\Delta J$ 换成自由能 F 的泛函微商．$\boldsymbol{j}'_\alpha$ 与未加人为外源 ΔJ 的守恒流 $\boldsymbol{j}_\alpha$ 之差，可根据线性响应理论写为

$$\boldsymbol{j}'_\alpha=\boldsymbol{j}_\alpha - l_{\alpha\beta}\nabla\left(\frac{\delta F}{\delta q_\beta}-J_\beta\right), \tag{4.7}$$

其中 $l_{\alpha\beta}$ 是线性输运系数．以上式代入(4.5)式就定出了

$$\nabla \boldsymbol{j}_\alpha = l_{\alpha\beta}\nabla^2\left(\frac{\delta F}{\delta q_\beta}-J_\beta\right)+i\left[\frac{\delta F}{\delta Q_i}L^\alpha_{ij}Q_j+f_{\alpha\beta\gamma}\frac{\delta F}{\delta q_\beta}q_\gamma\right]. \tag{4.8}$$

将(4.8)式代回(4.4)式中，就得到了守恒量的运动方程

$$\frac{\partial q_\alpha}{\partial t}= l_{\alpha\beta}\nabla^2\left(\frac{\delta F}{\delta q_\beta}-J_\beta\right)+i\left[\left(\frac{\delta F}{\delta Q_i}-J_i\right)L^\alpha_{ij}Q_j+f_{\alpha\beta\gamma}\left(\frac{\delta F}{\delta q_\beta}-J_\beta\right)q_\gamma\right]. \tag{4.9}$$

把它和上一节中得到的广义 TDGL 方程(3.13)中对应守恒量的部分对比，得出 γ^{-1} 矩阵的两块分别是

$$\gamma^{-1}(t)_{\alpha x,\beta y}=-[l_{\alpha\beta}\nabla^2_x+if_{\alpha\beta\gamma}q_\gamma(\boldsymbol{x},t)]\delta^{(3)}(\boldsymbol{x}-\boldsymbol{y}), \tag{4.10}$$

$$\gamma^{-1}(t)_{\alpha x,iy}=-iL^\alpha_{ij}Q_j(\boldsymbol{x},t)\delta^{(3)}(\boldsymbol{x}-\boldsymbol{y}). \tag{4.11}$$

现在讨论序参量的方程组．当序参量 Q_i 构成群 G 的不可约表示，且在临界点附近其本身数值很小时，可对 Q_i 展开，而由对称性知道：

$$(\gamma^{-1}(t))_{ix,jy}=\delta_{ij}(\sigma)_{x,y}+\cdots, \tag{4.12}$$

其中 $(\sigma)_{x,y}$ 不含 Q_i 而由系统的动力和耗散性质决定．同样可以由对称考虑写出展开式：

$$(\gamma^{-1}(t))_{ix,\alpha y}=ifL^\alpha_{ij}Q_j(\boldsymbol{x},t)\delta^{(3)}(\boldsymbol{x}-\boldsymbol{y})+Q_i\text{ 的高阶项} \tag{4.13}$$

f 是群变换下的不变量，在对 Q_i 的最低阶只能是常数．为了确定 f 的值，考虑耗散趋于零的极限．这时(3.10)式中的 $\gamma(t)$ 矩阵以反称部分为主，它的逆矩阵也是以反称部分为主．方程(3.13)式展开式的打头项应是反称的，于是对比(4.11)和(4.13)式，定出

$$f=1. \tag{4.14}$$

既然(4.13)式第一项本来与耗散无关，有耗散存在时(4.14)式仍然成立．当然在一般情形下展开式(4.12)和(4.13)还可能有其他项，其中包括耗散运动与可逆运动的同时作用．本文暂不讨论它们．在上述近似下，序参量的方程组成为

$$\frac{\partial Q_i}{\partial t}=-\sigma\left(\frac{\delta F}{\delta Q_i}-J_i\right)-iL^\alpha_{ij}Q_j\left(\frac{\delta F}{\delta q_\alpha}-J_\alpha\right). \tag{4.15}$$

(4.9)和(4.15)式就是我们推得的临界动力学基本方程组．在闭路格林函数方法中，

这些方程中的 J 来自 $J_+ = J_-$，乃是真正的物理外场，其中也可以包含为反映其它未计入宏变量的自由度而附加的随机外场．因此，我们称这一方程组为广义的朗之万方程．

以上推导的重要特点，是在广义朗之万方程中自然地得到了模耦合项，它确实具有(2.4)式的形式．在序参量和守恒量的耦合项中出现的是表示矩阵，在守恒量之间的耦合项中出现的是群结构常数．在对 Q_i 及 q_α 的线性近似下，该式第一项的贡献为零．具体地说，可将矩阵 A 分成四块．序参量之间的一块 $A_{ij} = 0$，即不考虑它们之间的非耗散型耦合．序参量与守恒量之间的一块 $A_{i\alpha} = iL^\alpha_{ij}Q_j$，与守恒量无关，所以 $\frac{\partial}{\partial q_\alpha} A_{i\alpha} = 0$．守恒量与序参量及守恒量之间的两块分别为 $A_{\alpha i} = -iL^\alpha_{ij}Q_j$ 及 $A_{\alpha\beta} = -if_{\alpha\beta\gamma}q_\gamma$．由于表示矩阵及结构常数的反称性，导数项为零．如果考虑对 Q 或 q 的非线性关系，利用非线性 W-T 恒等式可以求得带微商的项．这一点我们将另文专门讨论．

作为具体例子，考察各向同性的反铁磁模型，即模型 G[8]．这里包含一个由三维矢量构成的非守恒序参量 $\boldsymbol{Q}$，它代表交替磁矩密度．还有一个三维矢量构成的守恒量密度 $\boldsymbol{q}$，代表系统的总磁矩．根据对易关系

$$[q_\alpha, q_\beta] = ig_0\epsilon_{\alpha\beta\gamma}q_\gamma,$$
$$[Q_i, q_\alpha] = ig_0\epsilon_{i\alpha j}Q_j, \tag{4.16}$$

可以直接定出结构常数与表示矩阵

$$f_{\alpha\beta\gamma} = ig_0\epsilon_{\alpha\beta\gamma},$$
$$L^\alpha_{ij} = -ig_0\epsilon_{i\alpha j}, \tag{4.17}$$

这里 $\epsilon_{\alpha\beta\gamma}$ 等是全反对称单位张量．代入(4.15)及(4.9)式，并将外场项吸收到自由能微商项中，即令 $F \to F - J_iQ_i - J_\alpha q_\alpha$，求得

$$\frac{\partial Q_i}{\partial t} = -\sigma\frac{\delta F}{\delta Q_i} + g_0\epsilon_{\alpha ij}Q_j\frac{\delta F}{\delta q_\alpha},$$
$$\frac{\partial q_\alpha}{\partial t} = l_{\alpha\beta}\nabla^2\frac{\delta F}{\delta q_\beta} - g_0\epsilon_{\alpha ij}\frac{\delta F}{\delta Q_i}Q_j$$
$$- g_0\epsilon_{\alpha\beta\gamma}\frac{\delta F}{\delta q_\beta}q_\gamma. \tag{4.18}$$

若取 $\sigma = \Gamma_0$，$l_{\alpha\beta} = \lambda_0\delta_{\alpha\beta}$，并写成矢量形式，即得文献[8]中的方程组

$$\frac{\partial \boldsymbol{Q}}{\partial t} = -\Gamma_0\frac{\delta F}{\delta \boldsymbol{Q}} + g_0\boldsymbol{Q}\times\frac{\delta F}{\delta \boldsymbol{q}},$$
$$\frac{\partial \boldsymbol{q}}{\partial t} = \lambda_0\nabla^2\frac{\delta F}{\delta \boldsymbol{q}} + g_0\boldsymbol{Q}\times\frac{\delta F}{\delta \boldsymbol{Q}} + g_0\boldsymbol{q}\times\frac{\delta F}{\delta \boldsymbol{q}}. \tag{4.19}$$

A, B, C 模型中没有非耗散型模耦合，情形更简单，其余 E, F, H, J 模型[8]及 SSS 模型[14]等均可按完全相同的方法得出，这里不再赘述．

参　考　文　献

[1] J. Schwinger, *J. Math. Phys.*, **2**(1961), 407; Л. Келдыш, *ЖЭТФ*, **47**(1964), 1515.
[2] D. Dubois, In "Lectures in Theoretical Physics", vol. IX C, ed. by W. E. Brittin, Gordon and Breach. N. Y., (1967); V. Korenman, *Ann. Phys.* (N. Y.), 39(1966), 72; D. Langreth, In "Linear and Nonlinear Transport in Solids", ed. by J. Devreese and V. Van Doren, Plenum,

N. Y., (1976); A.-M. Tremblay. B. Patton, P. C. Martin, P. Maldaque. *Phys. Rev.*, A19(1979), 1721.
[3] 周光召、苏肇冰,统计物理进展,第五章,科学出版社即将出版.
[4] 周光召、苏肇冰,高能物理与核物理,3(1979),314.
[5] 周光召、苏肇冰,高能物理与核物理,3(1979),304.
[6] 周光召、苏肇冰,物理学报,29(1980),618.
[7] P. C. Martin. E. D. Siggia, H. A. Rose. *Phys. Rev.*, A8(1973), 423.
[8] P. C. Hohenberg, B. I. Halperin, *Rev. Mod. Phys.*, A49(1977). 435.
[9] K. Kawasaki. In "Critical Phenomena", Inter. Sch. of Physics "Enrico Fermi", LI, ed. by M. S. Green, Acad. Press. (1971).
[10] 可参阅：B. I. Halperin, P. C. Hohenberg, S. K. Ma, *Phys. Rev.*, **B10**(1974), 139; S. K. Ma, G. F. Mazenko. *Phys. Rev.*, **B11**(1975). 4077.
[11] H. K. Janssen, *Z. Physik*, **B23**(1976). 377; R. Bausch, H. K. Janssen. H. Wegner, *Z. Physik*, **B24**(1976), 113; C. De Dominicis, L. Peliti, *Phys. Rev.*, **B18**(1978), 353.
[12] 周光召、郝柏林、于渌,本刊本期,969.
[13] R. Graham. H. Haken, *Z. Physik*. **243**(1971). 289.
[14] L. Sasvari, F. Schwabl. P. Szepfalusy, *Physica*, **A81**(1975), 108.
[15] 周光召、于渌、郝柏林,物理学报, **29** (1980), 878.

NONEQUILIBRIUM STATISTICAL FIELD THEORY AND CRITICAL DYNAMICS (I)

GENERALIZED LANGEVIN EQUATION

Zhou Guang-zhao Su Zhao-bin Hao Bai-lin Yu Lu

(Institute of Theoretical Physics, Academia Sinica)

Abstract

Starting from the equations satisfied by the vertex functions on the closed time path, we derived the generalized Langevin equations for the order parameters and the conserved variables. The proper form of the equations for the conserved variables, including automatically the mode coupling terms, was determined from the Ward-Takahashi identities and the linear response theory. All existing dynamic models were recovered by assuming the corresponding symmetry properties of the system. The whole theoretical framework is also applicable for describing the systems near steady states far from equilibrium.

第 29 卷 第 8 期 物 理 学 报 Vol. 29, No. 8
1980 年 8 月 ACTA PHYSICA SINICA Aug., 1980

非平衡统计场论与临界动力学(II)

拉氏场论表述

周光召 郝柏林 于 渌
(中国科学院理论物理研究所)

1979 年 9 月 17 日收到

提 要

本文从闭路格林函数生成泛函的连续积分表示出发，对傅氏变换取单圈近似，求得序参量的有效作用量．在闭路连续积分中取涨落二级近似，得到临界动力学的拉氏场论表述．文中指出改进现有理论的可能途径．

一、引 言

在前一篇文章[1]中，从闭路顶角函数的方程出发，通过分出随时间快变和慢变的部分，推导出了包括模-模耦合项在内的宏变量广义朗之万方程．本文中利用闭路生成泛函的连续积分表示推导临界动力学的拉氏场论表述．

在随机模型中序参量和守恒量均满足广义朗之万方程

$$\frac{\partial Q_i(t)}{\partial t} = K_i(Q) + \xi_i(t), \tag{1.1}$$

这里 $\xi_i(t)$ 是遵从高斯分布的随机力．高斯过程可通过概率泛函描述[2]．(1.1)式可以看成是高斯过程 $\xi_i(t)$ 向复杂过程 $Q_i(t)$ 的映射，在连续积分中作非线性变换，即可求得描述过程 $Q_i(t)$ 的概率泛函[3]．更直接的作法是利用连续积分下 δ 函数的归一条件[4]

$$\int [dQ]\delta\left(\frac{\partial Q}{\partial t} - K(Q) - \xi\right)\Delta(Q) = 1. \tag{1.2}$$

由于(1.2)式中 δ 函数的自变量不是 Q，而是(1.1)式，必须写上泛函雅可比行列式 $\Delta(Q)$，这就是前面提到的非线性变换 $\xi_i \to Q_i$ 的雅可比行列式．准到常数因子，它等于[3]

$$\Delta(Q) = \exp\left(-\frac{1}{2}\int \frac{\delta K}{\delta Q} dx\right) \tag{1.3}$$

这里 $dx = d\boldsymbol{x}dt$ 是四维积分元．再把(1.2)式中的 δ 函数通过连续积分表示，变成

$$\int [dQ]\left[\frac{d\hat{Q}}{2\pi}\right]\exp\left(\int dx\left[i\hat{Q}\left(\frac{\partial Q}{\partial t} - K(Q) - \xi\right) - \frac{1}{2}\frac{\delta K}{\delta Q}\right]\right) = 1. \tag{1.4}$$

如果在积分(1.4)式下插入因子 $\exp\left(-i\int dx[J(x)Q(x) + \hat{J}(x)\hat{Q}(x)]\right)$，它就是各种乘积

平均值的生成泛函(概率论中的特征泛函或矩生成泛函)

$$Z_\xi[J,\hat{J}]=\int[dQ]\left[\frac{d\hat{Q}}{2\pi}\right]\exp\left(\int dx\left[i\hat{Q}\left(\frac{\partial Q}{\partial t}-K(Q)-\xi\right)\right.\right.$$
$$\left.\left.-\frac{1}{2}\frac{\delta K}{\delta Q}-iJQ-i\hat{J}\hat{Q}\right]\right). \tag{1.5}$$

显然，$Z_\xi[0,0]=1$. ξ 遵从高斯分布

$$W[\xi]\propto\exp\left(-\frac{1}{2}\xi\sigma^{-1}\xi\right), \tag{1.6}$$

其中 σ^{-1} 是关联矩阵 σ 的逆. 在(1.5)式中完成对 ξ 的高斯平均后有

$$Z[J,\hat{J}]=\int[dQ]\left[\frac{d\hat{Q}}{2\pi}\right]\exp\left(\int dx\left[-\frac{1}{2}\hat{Q}\sigma\hat{Q}+i\hat{Q}\left(\frac{\partial Q}{\partial t}-K(Q)\right)\right.\right.$$
$$\left.\left.-\frac{1}{2}\frac{\delta K}{\delta Q}-iJQ-i\hat{J}\hat{Q}\right]\right). \tag{1.7}$$

这就是拉氏形式的经典统计场论生成泛函. Martin, Siggia 和 Rose 最初讨论这种场论(简称 MSR 场论)时没有采用这种形式[5]. 为了简化微扰论结构和便于重正化，他们引入了一批与原有场量不对易的“响应场”,就是这里的 $\hat{Q}$. 与量子场论的情形类似,在连续积分下，它们是可对易的量. 引入 $\hat{Q}$ 场使算子数目加倍，在闭路格林函数中有时间的正支和负支,算子数目也加倍. 我们将在以下的两节中证明,在闭路格林函数的理论框架内可以自然地得到拉氏形式的 MSR 场论,同时看出,算子的不可对易性不是人为地、形式地引入的,它确实反映了统计涨落的本质.

(1.7)式中对 $\hat{Q}$ 的连续积分又是高斯型的,积分后得

$$Z[J]=\int[dQ]\exp\left(\int dx\left[-\frac{1}{2}\left(\frac{\partial Q}{\partial t}-K(Q)-\hat{J}\right)\sigma^{-1}\right.\right.$$
$$\left.\left.\cdot\left(\frac{\partial Q}{\partial t}-K(Q)-\hat{J}\right)-\frac{1}{2}\frac{\delta K}{\delta Q}-iJQ\right]\right). \tag{1.8}$$

最早此式是作为概率密度泛函在文献[3]中得到的,然而在临界动力学中更方便的出发点是(1.7)式[4,6].

二、宏变量的有效作用量

设系统的基本场量是 $\hat{\varphi}_i(x)$, $i=1,2,\cdots,e$, 表示序参量或守恒量的复合算子是 $\hat{Q}_i(\hat{\varphi}(x))$, $i=1,2,\cdots,n+m$, 有些基本场量本身就构成序参量. 为简单起见，假定它们都是玻色算子. 下面算子不特别标出,其含义从行文中可看出.

取初始时刻无规相位近似,密度矩阵在 $t=t_0$ 时对角

$$\langle\varphi'(\boldsymbol{x},t_0)|\rho|\varphi''(\boldsymbol{x},t_0)\rangle=P(\varphi'(\boldsymbol{x}),t_0)\delta(\varphi'(\boldsymbol{x},t_0)-\varphi''(\boldsymbol{x},t_0)). \tag{2.1}$$

宏变量 $Q_i(x)$ 的初始分布密度是

$$P(Q_i(\boldsymbol{x}),t_0)=\mathrm{tr}\{\delta(Q_i(\boldsymbol{x})-Q_i(\varphi(x)))\rho\}$$
$$=\int[d\varphi(x)]\delta(Q_i(\boldsymbol{x})-Q_i(\varphi(x)))P(\varphi(\boldsymbol{x}),t_0). \tag{2.2}$$

Q_i 的闭路格林函数生成泛函为[7,8]

$$Z[J(x)] = \exp(-iW[J(x)]) = \langle T_p e^{-i\int_p JQ}\rangle$$
$$= \mathrm{tr}\{T_p(e^{-i\int_p J(x)Q(\varphi(x))}\rho)\}$$
$$= N\int[d\varphi(x)]e^{i\int_p[\mathscr{L}(\varphi(x)) - JQ(\varphi(x))]}\delta(\varphi_+ - \varphi_-). \tag{2.3}$$

这里引入了简化记号

$$\int_p JQ \equiv \sum_i \int_p d^4x J_i(x) Q_i(\varphi(x)), \tag{2.4}$$

$$\delta(\varphi_+ - \varphi_-) \equiv \int d\varphi'(\boldsymbol{x})\delta(\varphi(\boldsymbol{x}, t_+ = t_0) - \varphi'(\boldsymbol{x}))\delta(\varphi(\boldsymbol{x}, t_- = t_0) - \varphi'(\boldsymbol{x})) \times P(\varphi'(\boldsymbol{x}), t_0). \tag{2.5}$$

T_p 表示闭路时序排列算子，$\int_p$ 指闭路积分.

等式(2.3)右端乘以闭路上 δ 函数归一因子

$$1 = \int[dQ]\delta(Q_+ - Q_-)\delta(Q(x) - Q(\varphi(x))), \tag{2.6}$$

交换积分次序后可将(2.3)式中的 $Q(\varphi(x))$ 换成 $Q(x)$. 再利用公式

$$\delta(Q(x) - Q(\varphi(x))) = \int\left[\frac{dI}{2\pi}\right]e^{i\int_p[Q(x) - Q(\varphi(x))]I(x)}, \tag{2.7}$$

可将(2.3)式改写成

$$Z[J] = N\int[dQ]e^{iS_{\text{eff}}[Q] - i\int_p JQ}\delta(Q_+ - Q_-), \tag{2.8}$$

其中

$$e^{iS_{\text{eff}}[Q]} \equiv \int\left[\frac{dI}{2\pi}\right]e^{i\int QI - iW[I]}. \tag{2.9}$$

这里对连续积分作了傅氏变换和反变换. 由于对 $I(x)$ 要进行连续积分，可将 $W[I]$ 看成在随机外场中的自由能生成泛函. 如果用 WKB 方法计算积分(2.9)式，准到单圈图近似，就相当于对随机外场求高斯平均，得到宏变量的有效作用量 $S_{\text{eff}}[Q]$.

这里讨论的是宏变量为复合算子的情形. 如果宏变量是基本场量，或一部份是基本场量，也完全一样. 同样可以通过 δ 函数引入一个新的场量. 注意到，虽然初始分布可写成各场量分布的乘积

$$P(\varphi', t_0) = \prod_{i=1}^{n} P_i(\varphi'_i, t_0),$$

但拉氏量不能写成各分量场拉氏量的迭加，必须对各分量同时作傅氏变换，求得有效作用量.

现在讨论有效作用量的一般性质.

假定宏变量都是厄米型玻色算子，闭路生成泛函具有下列性质[7-9]:

$$W[J_+(x), J_-(x)]|_{J_+(x)=J_-(x)} = 0, \tag{2.10}$$
$$W^*[J_+(x), J_-(x)] = -W[J_-(x), J_+(x)], \tag{2.11}$$

这里的 $J_\pm(x)$ 和下面的 $Q_\pm(x)$ 分别代表正负支的外场和宏变量. 将(2.10)式对 J_+, J_- 取各阶泛函导数,再令 $J_+ = J_-$, 可以推得一系列关系式. 根据(2.11)及(2.9)式,可求得

$$S^*_{\text{eff}}[Q_+(x), Q_-(x)] = -S_{\text{eff}}[Q_-(x), Q_+(x)]. \tag{2.12}$$

$Q_+ = Q_-$ 时, S_{eff} 是纯虚数. 令 $Q_\pm(x) = Q_0(x) + \Delta Q_\pm(x)$, 将(2.12)式在 Q_0 附近作泛函展开,求得在 Q_0 处各阶泛函导数间的关系

$$\frac{\delta S}{\delta Q_+(x)} = \left(\frac{\delta S}{\delta Q_-(x)}\right)^*; \tag{2.13}$$

$$S_{Fij}(x, y) = S_{Fji}(y, x) = -S^*_{\bar{F}ji}(y, x),$$
$$S_{\pm ij}(x, y) = S_{\mp ji}(y, x) = -S^*_{\pm ji}(y, x). \tag{2.14}$$

这里

$$S_{Fij}(x, y) \equiv \frac{\delta^2 S}{\delta Q_{i+}(x)\delta Q_{j+}(y)}, \tag{2.15}$$

其余类推.

如果系统具有对称群 G 的不变性,即

$$\varphi_i(x) \to \varphi_i^g(x) = U_{ij}(g)\varphi_j(x),$$
$$Q_i(\varphi) \to Q_i^g(x) = V_{ij}(g)Q_j(x)$$

时, $\int_p \mathscr{L}$ 及初始分布不变,则有

$$W[J^g(x)] = W[J(x)],\ J_i^g(x) = J_j(x)V^+_{ji}(g),$$
$$S_{\text{eff}}[Q^g(x)] = S_{\text{eff}}[Q(x)],\ Q_i^g(x) = V_{ij}(g)Q_j(\varphi).$$

如果在(2.9)式中取 WKB 最低级,即树图近似,

$$Q = \frac{\delta W}{\delta J}, \tag{2.16}$$

$$S_{\text{eff}}[Q] = -\Gamma[Q]. \tag{2.17}$$

这时 S_{eff} 具有闭路顶角生成泛函的一切性质[7-9],由于 $J_+ = J_-$, 得到 $Q_+ = Q_-$,

$$S_{\text{eff}}[Q_0, Q_0] = 0, \tag{2.18}$$

$$\left.\frac{\delta S_{\text{eff}}}{\delta Q_+}\right|_{Q_+=Q_-=Q_0} = \left.\frac{\delta S_{\text{eff}}}{\delta Q_-}\right|_{Q_+=Q_-=Q_0}, \tag{2.19}$$

$$S_F + S_{\bar{F}} = S_+ + S_-, \tag{2.20}$$

$$\frac{\delta^l S_{\text{eff}}}{\delta Q_{i1}(x_1)\cdots\delta Q_{il}(x_l)} = i^{l-1}\langle T_p(Q_{i1}(x_1)\cdots Q_{il}(x_l))\rangle_{1P.I.} \tag{2.21}$$

根据 $-iS_\pm = i\Gamma_\pm$ 及顶角函数的性质[7,8]

$$i\Gamma_\pm(k) > 0$$

求得

$$-iS_\pm(k) > 0. \tag{2.22}$$

这里 $\Gamma(k)$ 等是作了傅氏变换后的顶角函数.

在热平衡态附近[7,8]

$$\Gamma_{+ij}(k) = \Gamma_{-ij}(k)e^{-\beta k_0},$$
$$S_{-ij}(k) - S_{+ij}(k) \underset{k_0\to 0}{\longrightarrow} -\beta k_0 S_{-ij}(k). \tag{2.23}$$

三、最可几轨道与拉氏场论表述

前一节关于有效作用量的讨论具有普遍意义，可以从微观的生成泛函 W 出发，经过对随机场的平均求得 $S_{eff}[Q]$，也可以根据它应满足的性质构造唯象的模型，直接进行计算．本节将证明，在 $[dI]$ 的连续积分中取单圈近似，对宏变量在时间正负支的涨落取到二级，就得到现有拉氏形式的 MSR 场论．

先计算积分(2.9)式，将指数上因子在鞍点附近展开

$$A \equiv \int_p QI - W = -\Gamma - \frac{1}{2}\int_p \Delta I W^{(2)} \Delta I + \cdots. \tag{3.1}$$

利用闭路格林函数的变换[9]，可写成单向时间轴上的积分

$$A = -\Gamma - \frac{1}{2}\int \Delta \hat{I} \sigma_3 \hat{W}^{(2)} \sigma_3 \Delta \hat{I}, \tag{3.2}$$

这里

$$\hat{W}^{(2)} = \begin{pmatrix} W_{++} & W_{+-} \\ W_{-+} & W_{--} \end{pmatrix}, \quad \Delta \hat{I} = \begin{pmatrix} \Delta I_+ \\ \Delta I_- \end{pmatrix}, \quad \sigma_3 = \begin{pmatrix} 1 & 0 \\ 0 & -1 \end{pmatrix}.$$

完成高斯积分，准到常数项，求得

$$e^{iS_{eff}[Q]} = e^{-i\Gamma[Q]} |\det(\sigma_3 \hat{W}^{(2)} \sigma_3)|^{-\frac{1}{2}}. \tag{3.3}$$

利用 Dyson 方程[7-9]

$$\sigma_3 \hat{W}^{(2)} \sigma_3 = -(\hat{\Gamma}^{(2)})^{-1},$$

将(3.3)式改写成

$$i\, S_{eff}[Q] = -i\, \Gamma[Q] + \frac{1}{2} \operatorname{tr} \ln \hat{\Gamma}^{(2)}, \tag{3.4}$$

这里

$$\hat{\Gamma}^{(2)} = \begin{pmatrix} \Gamma_{++} & \Gamma_{+-} \\ \Gamma_{-+} & \Gamma_{--} \end{pmatrix}$$

是二端顶角函数矩阵．根据文献[9]中给出的变换：

$$\det \hat{\Gamma}^{(2)}(x, y) = \det \tilde{\Gamma}^{(2)}(x, y) = \det \Gamma_r(x, y) \det \Gamma_a(x, y)$$
$$= \left[\det\left(\frac{\delta^2 \Gamma}{\delta Q(x) \delta \Delta(y)}\right)\right]^2, \tag{3.5}$$

这里 $Q(x)$ 与 $\Delta(x)$ 的定义见(3.12)式．根据文献[1]中的讨论

$$\left.\frac{\delta \Gamma}{\delta \Delta}\right|_{\Delta=0} = \frac{1}{2}\left(\frac{\delta \Gamma}{\delta Q_+} + \frac{\delta \Gamma}{\delta Q_-}\right)_{\Delta=0} = -\gamma \frac{\partial Q}{\partial t} - \frac{\delta F}{\delta Q}. \tag{3.6}$$

与(1.1)及(1.2)式比较，可以看出，准到系数矩阵 γ，$\frac{\delta^2 \Gamma}{\delta Q \delta \Delta}$ 就是函数变换矩阵，其行列式就是雅可比行列式．注意到(3.5)式中的平方正好与(3.4)式中的 1 / 2 相消，求得

$$i\, S_{eff}[Q] = -i\Gamma[Q] - \frac{1}{2}\int \frac{\delta K}{\delta Q}\, dx, \tag{3.7}$$

其中

$$K = -\gamma^{-1}\frac{\delta F}{\delta Q}. \tag{3.8}$$

在路径积分(2.8)式中取最可几轨道,它满足方程

$$\frac{\delta S_{\rm eff}[Q]}{\delta Q_\pm} = J_\pm(x), \tag{3.9}$$

$$Q(x, t_+ = t_0) = Q(\boldsymbol{x}, t_- = t_0). \tag{3.10}$$

如果取 $J_+ = J_- = J$, 对随机场 $I(x)$ 的积分中取树图近似,根据文献 [1] 中的论证和(2.17)式

$$\frac{\delta S_{\rm eff}[Q]}{\delta Q} = J(x) = \gamma\frac{\partial Q}{\partial t} + \frac{\delta F}{\delta Q}. \tag{3.11}$$

这就是没有随机力的广义朗之万方程(即 TDGL 方程).

现在考虑在最可几轨道附近的涨落. 在闭路格林函数形式中,除通常意义的涨落外,还允许场量在正负时间支上不同. 在连续积分(2.8)式中作变量代换,引入

$$Q(x) = \frac{1}{2}(Q_+(x) + Q_-(x)),$$
$$\Delta(x) = Q_+(x) - Q_-(x),$$
$$[dQ_+(x)][dQ_-(x)] = [dQ(x)][d\Delta(x)]. \tag{3.12}$$

将有效作用量展开,变成时间单向积分后有

$$S_{\rm eff}[Q_+(x), Q_-(x)] = S_{\rm eff}[Q(x), Q(x)] + \frac{1}{2}\int\left(\frac{\delta S_{\rm eff}}{\delta Q_+} + \frac{\delta S_{\rm eff}}{\delta Q_-}\right)\Delta(x)$$
$$+ \frac{1}{8}\int \Delta(x)(S_{++} + S_{+-} + S_{-+} + S_{--})(x, y)\Delta(y) + \cdots. \tag{3.13}$$

令

$$\frac{i}{4}(S_{++} + S_{+-} + S_{-+} + S_{--})(x, y) = -\gamma(x)\sigma(x, y)\gamma(y), \tag{3.14}$$

并将 (2.18), (3.7) 式和 (3.11) 式代入,求得

$$e^{-iW[J(x)]} = \int[dQ(x)][d\Delta(x)]e^{-\frac{1}{2}\int\Delta(x)\gamma(x)\sigma(x,y)\gamma(y)\Delta(y)+i\int\left(\gamma(x)\frac{\partial Q}{\partial t}+\frac{\delta F}{\delta Q}\right)\Delta(x)}$$
$$\times e^{-\frac{1}{2}\int\frac{\delta K}{\delta Q}-i\int(J_\Delta Q+J_0\Delta)}\delta(\Delta(\boldsymbol{x}, t_0)), \tag{3.15}$$

这里 $J_\Delta = J_+(x) - J_-(x)$, $J_0 = \frac{1}{2}(J_+(x) + J_-(x))$. 若取 $J_\Delta = J$,作代换 $\gamma(x)\Delta(x) \to \hat{Q}(x)$, $\frac{1}{\gamma}J_0 \to \hat{J}$, 即得到(1.7)式,也就是 MSR 场论的生成泛函. 完成对 $\Delta(x)$ 的高斯积分后求得

$$e^{-iW[J(x)]} = N\int[dQ(x)]\exp\left\{\int dx\left[-\frac{1}{2}\int dy\left(\frac{\partial Q(x)}{\partial t} + \frac{1}{\gamma}\left(\frac{\delta F}{\delta Q(x)} - \hat{J}(x)\right)\right)\sigma^{-1}(x, y)\right.\right.$$
$$\left.\left.\times\left(\frac{\partial Q(y)}{\partial t} + \frac{1}{\gamma}\left(\frac{\delta F}{\delta Q(y)} - \hat{J}(y)\right)\right) + \frac{1}{2}\frac{1}{\gamma}\frac{\delta}{\delta Q}\frac{\delta F}{\delta Q} - iJQ\right]\right\}. \tag{3.16}$$

这就是生成泛函(1.8)式. 值得注意的是

$$\hat{J} = \frac{1}{2}(J_+ + J_-)$$

对应物理外场，而 $J = J_+ - J_-$ 对应构造生成泛函引入的形式源场.

与(1.8)式比较，看出 $\sigma(x, y)$ 是随机力的关联矩阵. 若 Q 是 x 的慢变函数，σ 可看成常数

$$\sigma = -\frac{i}{4\gamma^2}(S_F + S_{\bar{F}} + S_+ + S_-)(k = 0). \tag{3.17}$$

利用树图近似下成立的(2.20)式，可改写成

$$\sigma = -\frac{i}{2\gamma^2}(S_+ + S_-). \tag{3.18}$$

根据文献[1]中对 γ 的定义

$$\gamma = \lim_{k_0\to 0} i\frac{\partial}{\partial k_0}\Gamma_R, \tag{3.19}$$

考虑到只有耗散部分有贡献，并利用(2.17)式及热平衡附近成立的(2.23)式，求得

$$\begin{aligned}\gamma &= \lim_{k_0\to 0}\frac{\partial}{\partial k_0}A = \frac{i}{2}\lim_{k_0\to 0}\frac{\partial}{\partial k_0}(\Gamma_- - \Gamma_+)\\ &= \frac{i}{2}\beta\Gamma_- \approx -\frac{i}{4}\beta(S_+ + S_-).\end{aligned} \tag{3.20}$$

对比(2.18)与(2.20)式，求得涨落耗散定理

$$\sigma = \frac{2}{\beta\gamma}. \tag{3.21}$$

用通常的记号表示是

$$\langle\xi(t)\xi(t')\rangle = 2\Gamma_0 kT\delta(t - t'),$$

$$\Gamma_0 = \frac{1}{\gamma}.$$

为书写简单，以上推导是给单分量的宏变量写的，多个分量的情形也完全一样.

四、讨　　论

扼要概括文献[1]和本文的主要结果，我们有以下认识：

1. 闭路格林函数是描述动态临界现象这类长波涨落起主要作用的统计场论的自然理论框架. 利用它可以从统一的观点推导出包括模-模耦合项在内的序参量及守恒量的广义朗之万方程及经典统计场论的拉氏表述. 如果用 $\hat{G}$ 这套闭路形式[9]构造微扰论，结构与标准的场论方法一致，运算比较简单. 现在临界动力学和 MSR 场论是用 $\tilde{G}$ 这套函数构造微扰论，出现推迟函数和关联函数这两类不同性质的传播子，结构比较复杂. 闭路函数的另一个优点是它自动保证因果性条件，不需要像现在那样在微扰论的每一级逐级证明[10].

2. 这里用的是连续积分表示，算子的非对易性不显然. 如果与 MSR 场论原来的形式[5]比较，就可以看出，非对易性不只是一种手法，而是描述统计场随时间演化行为所必

需．虽然由非对易性导致的红外发散度低（详见附录），但对于讨论与时间有关的行为是重要的，因为响应函数的红外发散度低于关联函数．

3. 从本文的推导可以看出现有临界动力学理论所取的近似和改进的可能途径．

(1) 在现有理论中输运系数矩阵对应序参量的部分取成对称的，即只考虑耗散，与守恒量耦合的部分取成反称的，只考虑正则运动．原则上允许出现交叉的效应．在闭路框架内有可能分析这类现象．

(2) 对随机场 $I(x)$ 的连续积分作到单圈近似相当于高斯平均．在闭路形式中可以作到高级圈图近似，越出高斯平均的范围．

(3) 对时间正负支的涨落只取到 $\Delta(x)$ 的二级就相应于现有理论，原则上也可以作到高阶．更方便的办法可能是直接计算 Q_+, Q_- 的连续积分，不明显地引入 $\Delta(x)$．

4. 现有的拉氏统计场论的重正化很复杂[4,6]，其原因之一是顶角和原始发散的数目远多于耦合常数的数目，Q 与 $\hat{Q}$ 场具有不同的量纲．用 $\hat{G}$ 这套闭路格林函数作重正化可能会比较简单，因为矩阵各分量的红外发散度相同．

作者们感谢与苏肇冰同志的多次讨论．

附　　录

有限温度场论的重正化问题

文献[11] 中普遍地论证了，$T=0\mathrm{K}$ 的量子场论中引入的重正化因子，足以消除 $T>0\mathrm{K}$ 的闭路格林函数的紫外发散．不使用闭路格林函数方法，对平衡态的有限温度场论，也有人得到这个结论（见文献[11]所引的文献）．这个结论在物理上是自然的，因为统计平均不会影响极小距离的行为，因而不会增加新的紫外发散．我们在这里指出，讨论相变这类现象时，必须先分出红外发散项，再讨论紫外重正问题．这种场论的重正化与 $T=0\mathrm{K}$ 的量子场论有差别．

为具体起见，考察一个相对论的标量玻色场，其闭路自由传播子是[7]

$$G_{++}(k)\equiv\Delta_F(k)=\frac{1}{k^2-m^2+i\varepsilon}-2\pi i n(\boldsymbol{k})\delta(k^2-m^2),$$

$$G_{-+}(k)\equiv\Delta_-(k)=-2\pi i\delta(k^2-m^2)(\theta(k_0)+n(\boldsymbol{k})),$$

$$G_{+-}(k)\equiv\Delta_+(k)=-2\pi i\delta(k^2-m^2)(\theta(-k_0)+n(\boldsymbol{k})),$$

$$G_{--}(k)\equiv\Delta_{\bar{F}}(k)=\frac{-1}{k^2-m^2-i\varepsilon}-2\pi i n(\boldsymbol{k})\delta(k^2-m^2), \tag{A·1}$$

这里

$$n(\boldsymbol{k})=\frac{1}{e^{\varepsilon(\boldsymbol{k})/T}-1},\quad \varepsilon(\boldsymbol{k})=\sqrt{\boldsymbol{k}^2+m^2}. \tag{A·2}$$

在相变点附近，$m=0$，对于长波元激发

$$\varepsilon(\boldsymbol{k})/T\ll 1,\quad n(\boldsymbol{k})\simeq T/\varepsilon(\boldsymbol{k})\gg 1. \tag{A·3}$$

由(A·1)式看出，传播子中含 n 的项起主要作用．由于这些项都在质壳上，有 δ 函数，能自动完成频率的积分，它们的红外发散度比其他项高一级．因此，对于 ϕ^4 理论，可重正的边缘空间维数是 $d_c=4$，而不是通常量子场论的 $d_c=4-1$．这就是平常所说的“量子的 d 维相应于经典的 $d+1$ 维”的含义．

以上所述可以具体地检验：计算质量、顶角和波函数重正化的原始发散图形，完成频率积分，然后取高温极限 $T/\varepsilon(\boldsymbol{k})\gg 1$，正好得到现有临界现象理论中的结果[12]．更方便的办法是利用松原格林函数，在频率求和中只取 $\omega_n=0$ 的项．

有的作者在研究有限温度场论时既讨论相变，又采用 $T=0\mathrm{K}$ 的重正化因子，这是不正确的．由于相变现象中

要取高温极限，相对论效应和量子效应都不重要．只有 $T=0\mathrm{K}$ 附近的相变可能是例外，那时统计和量子涨落同时起作用．普通场论模型(非阿贝耳规范场还不清楚)的相变不会给出超过现有临界现象理论的结果．

在讨论静态现象时，算子的不可对易性不重要，这相应于在(A·1)式中四种传播子都等于

$$-2\pi i n(\boldsymbol{k})\delta(k^2-m^2),$$

即关联函数本身．对于动态现象则不然．G_{++} 和 G_{--} 中的第一项来自格林函数方程的非齐次项，即对易子．如果只保留红外最发散项，四个传播子相等，推迟格林函数

$$G_r = G_{++} - G_{+-} = 0. \tag{A·4}$$

由此可见，推迟格林函数的红外发散度低于关联函数．要能正确地分析它，必须考虑算子的非对易性，虽然这是"纯"经典的理论．容易证明，用自由传播子演示的这些性质对于重正化以后的传播子仍旧保持．

正如文献[1]的引言中已提到的，统计场论的高温近似对应的是"超玻色"极限，而不是玻耳兹曼极限．说经典场可对易，是指相对于很大的 n 而言可以略去1．如果有些现象中不能略去1，则必须从非对易的算子出发．这是统计场论与量子场论有深刻类比的原因之一．关于这个类比的物理含义可参看文献[13]．

参　考　文　献

[1] 周光召、苏肇冰、郝柏林、于渌，本刊本期，961．

[2] L. Onsager, S. Machlup, *Phys. Rev.*, **91** (1953), 1505; 1512.

[3] R. Graham, *Springer Tracts in Modern Physics*, **66** (1973), 1.

[4] H. K. Janssen, *Z. Physik*, **B23** (1976), 377; R. Bausch, H. K. Janssen, H. Wegner, *Z. Physik*, **B24** (1976), 113.

[5] P. C. Martin, E. D. Siggia, H. A. Rose, *Phys. Rev.*, **A8** (1973), 423.

[6] C. De Dominicis, L. Peliti, *Phys. Rev.*, **B18** (1978), 353.

[7] 周光召、苏肇冰，统计物理进展，第五章，科学出版社即将出版．

[8] 周光召、苏肇冰，高能物理与核物理，**3** (1979)，314．

[9] 周光召、于渌、郝柏林，物理学报，**29** (1980)，878．

[10] J. Deker, F. Haake, *Phys. Rev.*, **A11** (1975), 2043.

[11] 周光召、苏肇冰，高能物理与核物理，**3** (1979)，304．

[12] E. Brezin, J. C. Le Guillou, J. Zinn-Justin, In Phase Transitions and Critical Phenomena, Vol. 6, ed. by C. Domb, M. S. Green, Acad. Press, (1976).

[13] 于渌、郝柏林，相变和临界现象(上)、(中)、(下)，物理，待发表．

NONEQUILIBRIUM STATISTICAL FIELD THEORY AND CRITICAL DYNAMICS (II)

LAGRANGIAN FIELD THEORY FORMULATION

ZHOU GUANG-ZHAO　HAO BAI-LIN　YU LU

(Institute of Theoretical Physics, Academia Sinica)

ABSTRACT

The expression of the effective action for the order parameters is derived from the continuous integral representation for the generating functional on the closed time path in the one loop approximation for the Fourier transforms. The Lagrangian formulation of the critical dynamics is recovered in the second order approximation of fluctuations in the closed time path continuous integral. The various possibilities of improving the existing theory of critical dynamics are considered.

第30卷 第2期 物理学报 Vol. 30, No. 2
1981年2月 ACTA PHYSICA SINICA Feb., 1981

时间反演对称和非平衡统计定常态（I）

周光召　苏肇冰
（中国科学院理论物理研究所）

1980年8月2日收到

提　要

本文是作者从微观量子统计理论出发应用微观可逆性原理讨论非平衡统计定常态普遍性质的第一部份．文中给出了非平衡量子统计格林函数（闭路格林函数）在定常态上关于时间反演对称的表达方式．把生成泛函技术和时间反演对称结合起来，得到了具有时间反演对称性的非平衡统计定常态上统计格林函数和顶点函数所满足的时间反演对称关系．

一、引　言

当前，与非线性过程相联系的非平衡统计热力学日益受到重视．人们发现，在某些类型的非平衡定常态上，与平衡态热力学存在自由能一样，存在着一个宏观势函数，它描述了系统的宏观性质．

与平衡态热力学相对应，存在着相应的平衡态统计力学．它不仅给出了所有宏观热力学量的物理意义和计算方法，而且还给出了偏离宏观量的各种统计涨落．在非平衡统计力学方面，虽然也已有一些很好的工作[1]，但是和平衡态统计力学相比还很不成熟．当前许多流行的非平衡统计理论多数是唯象或半唯象的，基于主方程、福克-普朗克方程或朗之万方程的非平衡统计理论在实质上是一种半唯象理论．通常它不是由微观理论导出的，而且统计涨落的引入多少带有一定的人为性．

对非平衡统计定常态（简称 NESS），半唯象统计理论的一个重要结果是：当描写系统的福克-普朗克方程的系数满足一定条件时，系统存在一个广义势函数[2]，它基本上决定了系统的宏观统计性质．半唯象理论的另一个重要结果是它给出了非平衡统计定常态的涨落耗散定理[3]（以下简称 FDT）．如所周知，这两个结果都是在福克-普朗克方程的理论框架中，对系统定常态作了细致平衡假设以后才得到的．这些结果，形成了许多作者进行临界动力学讨论的基本出发点．

在以前的一个工作中[4,5]，我们曾经将非平衡量子统计格林函数[6]（简称闭路格林函数或 CTPGF）应用到与序参量和守恒荷密度相对应的复合算符上，在低频极限下，得到了 Martin-Siggia-Rose 统计场论[7]的生成泛函和相应的序参量、守恒荷密度方程．同时，我们也得到了统计定常态存在广义势的条件为：二阶推迟顶点函数的耗散部为零．但是在工作［4，5］中，只是在热平衡条件下证明了 CTPGF 势条件得到满足，本文将把这一证明推广到 NESS 上．

在非平衡统计力学中，微观可逆性是一个十分重要的基本假设，作为线性不可逆热力学的基础：著名的 Onsager 倒易关系[8]就是由它导出的．我们这个工作的目的之一就是在 CTPGF 的理论框架之中，应用微观可逆性条件，对于某一类 NESS 导出势条件和 FDT．同时也把局部热平衡条件下的 Onsager 关系，推广到相应的 NESS 附近．我们得到的势条件虽然和由福克-普朗克方程出发求得的势条件在形式上有很大不同，但两者都在时间反演对称的条件下得到证明．我们的理论是一个微观理论．所有的宏观量都可以被表示为相应微观量的统计格林函数．在这一理论中，不仅势条件和 FDT 是从同一理论框架中得到的，而且所有宏观量方程中的参数都有清楚的物理意义，并且原则上是可计算的．

在文献[4, 5]中得到的序参量-守恒荷密度方程，即所谓的含时间 Ginsburg-Landau 方程(简称 TDGL)为

$$\gamma_{\alpha\beta}\frac{\partial Q_\beta}{\partial t}=-\frac{\delta F}{\delta Q_\alpha}+J_\alpha, \tag{1.1}$$

其中 Q_α 是序参量或守恒荷密度，F 是广义势，与随机变量满足的朗之万方程不同，它是序参量(或守恒荷密度)统计平均量的普遍方程，原则上包含了所有的高阶统计涨落．我们这个工作的另一个目的是将方程(1.1)中的弛豫矩阵 $\gamma_{\alpha\beta}$ 的逆矩阵分解为对称部份和反对称部份．我们将证明，它的对称部份正比于描述朗之万随机力涨落的扩散矩阵，而它的反对称部份则与序参量(或守恒荷密度)所对应的微观复合算符 $\hat{Q}_\alpha$ 的对易子的统计平均有关．前者反映了统计定常态的不可逆运动，后者反映了统计定常态的可逆运动．从而得到了 CTPGF 理论中的 TDGL 的一个普遍形式．它与临界动力学文献中常用的“TDGL”[9] 有非常类似的形式．

我们的讨论将分成两个部份．在本文中将试图把微观可逆性原理的表述与 CTPGF 理论相结合起来，对于具有时间反演对称的统计定常态建立一组统计格林函数(顶点函数)满足的对称关系．在第二篇文章中[12]，将利用本文的结果对 NESS 的普遍性质进行较为系统的讨论．

二、时间反演对称和统计定常态

本节将简略地回顾一下量子力学中的时间反演对称[10]，以阐明我们所用的术语和符号，对我们所讨论的时间反演对称给予确切的含义，然后将它应用到统计定常态．从而给出一个适合于与生成泛函技术相结合的时间反演对称的表述方式．

先从薛定谔绘景开始讨论．设我们研究的系统是保守的，它的哈密顿量为

$$\hat{H}[J]=\hat{H}+J\hat{Q}, \tag{2.1}$$

其中 $\hat{Q}=\{\hat{Q}_\alpha\}$, $\alpha=1,2,\cdots,n$ 是序参量或守恒荷密度算符，$J=\{J_\alpha\}$, $\alpha=1,2,\cdots,n$ 是与 $\hat{Q}_\alpha$ 相耦合的恒定外源，它是一组 c 数．指标 α 既包含离散指标，也可包含空间坐标．我们省略了对 α 的求和记号(以下相同)．不失去普遍性，$\hat{Q}_\alpha$ 可选为厄密型算符，J_α 选为实数．$\hat{H}$ 是当外源不存在时系统的哈密顿量．在薛定谔绘景中，$\hat{H}$ 和 $\hat{Q}$ 都与时间无关，$\hat{H}[J]$ 也与时间无关．

先讨论在薛定谔绘景中一个初值问题的时间反演．薛定谔方程写为

$$i\frac{\partial}{\partial t}\Psi_{t_0}(t;J,\lambda)=\hat{H}[J]\Psi_{t_0}(t;J,\lambda) \tag{2.2}$$

$$\Psi_{t_0}(t;J,\lambda)=\hat{S}^J(t,t_0)\Psi_{t_0}, \tag{2.3}$$

$$\Psi_{t_0}\equiv\Psi_{t_0}(t;J,\lambda)|_{t=t}=\Psi_0(\lambda)\quad 与 J 无关, \tag{2.4}$$

其中

$$\hat{S}^J(t,t_0)=\exp\{-i\hat{H}[J](t-t_0)\}, \tag{2.5}$$

$\Psi_{t_0}(t;J,\lambda)$ 是薛定谔态矢量，$\Psi_0(\lambda)$ 是在 $t=t_0$ 时刻给定的初始状态波函数；$\lambda=\{\lambda_a\}$，$a=1,2,\cdots,g$ 是描述初始状态的参数．在时间反演下，外参量

$$J_\alpha\to\varepsilon_\alpha J_\alpha,\quad \alpha=1,2,\cdots,n. \tag{2.6}$$

初始状态参量

$$\lambda_a\to\varepsilon_a\lambda_a,\quad a=1,2,\cdots,g. \tag{2.7}$$

在(2.6)和(2.7)式右端对指标 α 和 a 都不求和（以下相同），ε_α 和 ε_a 按对应参量的物理意义分别取 $+1$ 或 -1．如所周知，量子力学中的时间反演由一个反么正算符 $\hat{R}$ 所描述[10]．薛定谔力学量 $\hat{Q}_\alpha$ 在时间反演变换下，有

$$\hat{Q}_\alpha\to\hat{R}\hat{Q}_\alpha\hat{R}^+=\varepsilon_\alpha\hat{Q}_\alpha,\quad \alpha=1,2,\cdots,n. \tag{2.8}$$

并且，当

$$\hat{H}[J]\to\hat{R}\hat{H}[J]\hat{R}^+=\hat{H}[\varepsilon J] \tag{2.9}$$

得到满足时，我们称系统的动力学具有时间反演对称．在薛定谔绘景中，时间反演对于态矢量的作用包含了沿整个动力学过程的反演（在古典力学的语言中，就是沿原来规道的反演）．所以

$$\begin{aligned}\Psi_{t_0}(t;J,\lambda)\to\hat{R}\Psi_{t_0}(t;J,\lambda)&=\hat{S}^{\varepsilon J}(-t,-t_0)\tilde{\Psi}_{-t_0}\\&=\hat{S}^{\varepsilon J}(-t,t_0)\tilde{\Psi}_{t_0}.\end{aligned} \tag{2.10}$$

如果有

$$\tilde{\Psi}_{t_0}=\Psi_0(\varepsilon\lambda), \tag{2.11}$$

那么我们称系统的状态也具有时间反演对称．这样，对于态矢量，有

$$R\Psi_{t_0}(t;J,\lambda)=\Psi_{t_0}(-t;\varepsilon J,\varepsilon\lambda). \tag{2.12}$$

对于一个统计系统，密度矩阵描写了态矢量的统计系统，所以，与动力学状态(2.2)—(2.4)式相对应的密度矩阵为

$$\hat{\rho}_{t_0}(t;J,\lambda)=S^J(t,t_0)\hat{\rho}_{t_0}S^J(t_0,t), \tag{2.13}$$

$$\hat{\rho}_{t_0}\equiv\hat{\rho}_{t_0}(t;J,\lambda)|_{t=t_0}=\hat{\rho}_0(\lambda)\quad 与 J 无关, \tag{2.14}$$

其中 λ 记 $\{\lambda_a\}$，$a=1,2,\cdots,g$ 是初始状态参量．在时间反演下，

$$\hat{\rho}_{t_0}(t;J,\lambda)\to\hat{R}\hat{\rho}_{t_0}(t;J,\lambda)\hat{R}^+=\hat{S}^{\varepsilon J}(-t,t_0)\tilde{\hat{\rho}}_{t_0}\hat{S}^{\varepsilon J}(t_0,-t). \tag{2.15}$$

当

$$\tilde{\hat{\rho}}_{t_0}=\hat{\rho}_0(\varepsilon\lambda) \tag{2.16}$$

时，我们称初始密度矩阵具有时间反演对称．总之，在本文中讨论这样一类具有时间反演对称的系统，它的动力学规律和初始状态的时间反演行为由(2.9)和(2.16)式所确定．

现在转到海森堡绘景．我们选择在初始时刻 t_0，海森堡绘景和薛定谔绘景重合．这

样，海森堡力学量

$$\hat{Q}^J_{t_0}(t) = S^J(t_0, t)\hat{Q}S^J(t, t_0), \tag{2.17}$$

在时间反演下，有

$$\hat{Q}^J_{t_0}(t) \to \hat{R}\hat{Q}^J_{t_0}(t)\hat{R}^+ = \varepsilon\hat{Q}^{\varepsilon J}_{-t_0}(-t) = \varepsilon\hat{S}^{\varepsilon J}(-t_0, t_0)\hat{Q}^{\varepsilon J}_{t_0}(-t)\hat{S}^{\varepsilon J}(t_0, -t_0). \tag{2.18}$$

在海森堡绘景中密度矩阵不随时间而改变，即是 $\hat{\rho}_{t_0}$. 在(2.15)式中取 $t = t_0$ 并考虑到(2.16)式，就得到 $\hat{\rho}_{t_0}$ 在时间反演下，

$$\hat{\rho}_{t_0} \to \hat{R}\hat{\rho}_{t_0}\hat{R}^+ = \hat{S}^{\varepsilon J}(-t_0, t_0)\hat{\rho}_{t_0}\hat{S}^{\varepsilon J}(t_0, -t_0) = \hat{S}^{\varepsilon J}(-t_0, t_0)\hat{\rho}_0(\varepsilon\lambda)\hat{S}^{\varepsilon J}(t_0, -t_0). \tag{2.19}$$

在得到(2.18)和(2.19)式时，利用了系统的时间反演对称和 $\hat{R}$ 的反幺正性．力学量的平均值为

$$Q_{t_0}(t; J, \lambda) \equiv \mathrm{Tr}\{\hat{\rho}_{t_0}\hat{Q}^J_{t_0}(t)\}. \tag{2.20}$$

按照(2.18)，(2.19)式和 $\hat{R}$ 的反幺正性，容易证明

$$Q_{t_0}(t; J, \lambda) = \varepsilon Q_{t_0}(-t; \varepsilon J, \varepsilon\lambda). \tag{2.21}$$

我们有兴趣的常常是 $J = 0$ 的统计定常态．J 只是为生成泛函技术所需而引入的人为外源．在作完生成泛函运算以后，取 J 为零．定常态的统计性质由与 J 无关的 $\rho_0(\lambda)$ 所描述；它的动力学性质由 $\hat{H} = \hat{H}[J = 0]$ 所描述．海森堡序参量在密度矩阵上的平均是时间平移不变的．但是，在应用生成泛函技术的过程中，常会遇到这样的系统，它的统计性质仍然由 $\hat{\rho}_0(\lambda)$ 来描述，而它的动力学性质却由 $\hat{H}[J]$，$J \neq 0$ 来描述．对于这样的情形，只要将初始时刻 t_0 取在负无穷，即在 $t_0 = -\infty$ 将外参量作绝热变入，就能保持系统对时间的平移不变．这一点对以后的讨论是十分重要的．顺便指出，这样的做法和它对于时间反演的应用实际上隐含了两个假设．其一是在上述定常态上，对任何有限的 τ，下列极限存在

$$\lim_{t_0\to-\infty} \exp\{-i\hat{H}[J](\tau - t_0)\}\hat{\rho}_0(\lambda)\exp\{i\hat{H}[J](\tau - t_0)\}$$
$$= \hat{\rho}(\lambda, J), \quad \text{且与 } \tau \text{ 无关}; \tag{2.22}$$

另一个是

$$\lim_{t_0\to-\infty} \text{与时间反演算符 } \hat{R} \text{ 可交换}. \tag{2.23}$$

由(2.22)和(2.23)式容易证明

$$\hat{R}\hat{\rho}(\lambda, J)\hat{R}^+ = \hat{\rho}(\varepsilon\lambda, \varepsilon J). \tag{2.24}$$

现在来对一组海森堡序参量(或守恒荷密度)定义它们的关联函数

$$\begin{aligned} C_{\alpha_1\alpha_2\cdots\alpha_l}(t_1, t_2, \cdots, t_l; J, \lambda) &\equiv \lim_{t_0\to-\infty}\mathrm{Tr}\{\hat{\rho}_{t_0}\hat{Q}^J_{\alpha_1,t_0}(t_1)\hat{Q}^J_{\alpha_2,t_0}(t_2)\cdots\hat{Q}^J_{\alpha_l,t_0}(t_l)\} \\ &= \lim_{t_0\to-\infty}\mathrm{Tr}\{\hat{\rho}_0(\lambda)\hat{Q}^J_{\alpha_1,t_0}(t_1)\hat{Q}^J_{\alpha_2,t_0}(t_2)\cdots\hat{Q}^J_{\alpha_l,t_0}(t_l)\}. \end{aligned} \tag{2.25}$$

对于具有时间反演对称的统计定常态，利用(2.9)，(2.16)，(2.18)，(2.19)和(2.22)至(2.24)式就可证明

$$C_{\alpha_1\alpha_2\cdots\alpha_l}(t_1, t_2, \cdots, t_l; J, \lambda) = C_{\alpha_1\alpha_2\cdots\alpha_l}(t_1 + \tau, t_2 + \tau, \cdots, t_l + \tau; J\lambda). \tag{2.26}$$

$$C_{\alpha_1\alpha_2\cdots\alpha_l}(t_1, t_2, \cdots, t_l; J, \lambda) = \varepsilon_{\alpha_1}\varepsilon_{\alpha_2}\cdots\varepsilon_{\alpha_l}C_{\alpha_l\cdots\alpha_2\alpha_1}(-t_l, \cdots, -t_2, -t_1; \varepsilon J, \varepsilon\lambda). \tag{2.27}$$

注意(2.27)式左右两端的算符次序是互为倒转的，这是时间反演算符反幺正性的后果．

此外，我们着重指出，(2.26)和(2.27)式中的 J 是恒定外源，必须与 t 无关．由于 CTPGF 总可以被表成上述关联函数的线性组合，所以(2.26)和(2.27)式将成为下面讨论 CTPGF 时间反演性质的主要依据．

三、时间反演对称和 CTPGF

本节我们将对于具有时间反演对称的非平衡统计定常态导出统计格林函数（顶点函数)所满足的时间反演对称关系．从本节开始，所有讨论都将在 CTPGF 的理论框架中进行．将先简略地罗列一些主要用到的 CTPGF 的术语和性质而不加证明，有关的细致讨论可参阅文献[4, 5]．

与上一节问题的提法相对应的 CTPGF 生成泛函可写为

$$Z[J] \equiv \exp\{-iW[J]\} = \lim_{t_0\to-\infty} \mathrm{Tr}\left[\hat{\rho}_{t_0} T_p \exp\left\{-i\int_{P_{t_0}} d\tau J(\tau)\hat{Q}_{t_0}(\tau)\right\}\right] \tag{3.1}$$

$$= \mathrm{Tr}\left[\hat{\rho}_0(\lambda) T_p \exp\left\{-i\int_P d\tau J(\tau)\hat{Q}(\tau)\right\}\right], \tag{3.2}$$

其中 $W[J]$ 为连接格林函数的生成泛函，且

$$\frac{\delta W[J]}{\delta J_\alpha(t)} = Q_\alpha(t), \tag{3.3}$$

顶点泛函定义为

$$\Gamma[Q] = W[J] - JQ, \tag{3.4}$$

$$\frac{\delta\Gamma[Q]}{\delta Q_\alpha(t)} = -J_\alpha(t), \tag{3.5}$$

其中 P_{t_0} 表示由 $t_0\to+\infty$ 再由 $+\infty\to t_0$ 的闭路，P 表示由 $-\infty\to+\infty$ 再由 $+\infty\to-\infty$ 的闭路，

$$\hat{Q}_{t_0}(t) = \exp\{i\hat{H}(t-t_0)\}\hat{Q}\exp\{-i\hat{H}(t-t_0)\} = \hat{Q}^J_{t_0}(t)|_{J=0} \tag{3.6}$$

是 $J=0$ 系统的海森堡力学量，$\hat{Q}$ 是相应的薛定谔力学量，而

$$\hat{Q}(t) = \lim_{t_0\to-\infty}\hat{Q}_{t_0}(t). \tag{3.7}$$

二阶闭路格林函数还原到普通空间共有三个，它们是推迟格林函数

$$G^{\mathrm{R}}_{\alpha\beta}(t_1, t_2; J(\tau), \lambda) = G^{++}_{\alpha\beta}(t_1, t_2; J(\tau), \lambda) - G^{+-}_{\alpha\beta}(t_1, t_2; J(\tau), \lambda) \tag{3.8}$$

$$\overset{J(\tau)=0}{=} -i\vartheta(t_1-t_2)\,\mathrm{Tr}\,[\rho_0(\lambda)[\hat{Q}_\alpha(t_1), \hat{Q}_\beta(t_2)]], \tag{3.9}$$

超前格林函数

$$G^{\mathrm{A}}_{\alpha\beta}(t_1, t_2; J(\tau), \lambda) = G^{++}_{\alpha\beta}(t_1, t_2; J(\tau), \lambda) - G^{-+}_{\alpha\beta}(t_1, t_2; J(\tau), \lambda) \tag{3.10}$$

$$\overset{J(\tau)=0}{=} i\vartheta(t_2-t_1)\mathrm{Tr}[\rho_0(\lambda)[\hat{Q}_\alpha(t_1), \hat{Q}_\beta(t_2)]] \tag{3.11}$$

和关联格林函数

$$G^{\mathrm{C}}_{\alpha\beta}(t_1, t_2; J(\tau), \lambda) = G^{-+}_{\alpha\beta}(t_1, t_2; J(\tau), \lambda) + G^{+-}_{\alpha\beta}(t_1, t_2; J(\tau), \lambda) \tag{3.12}$$

$$\overset{J(\tau)=0}{=} -i\,\mathrm{Tr}\,[\rho_0(\lambda)\{\hat{Q}_\alpha(t_1), \hat{Q}_\beta(t_2)\}]. \tag{3.13}$$

对应的二阶顶点函数也有三个，它们是

$$\Gamma^{R}_{\alpha\beta}(t_1, t_2; Q(\tau), \lambda) \equiv -\mathscr{D}_{\alpha\beta}(t_1, t_2; Q(\tau), \lambda) - i\mathscr{A}_{\alpha\beta}(t_1, t_2; Q(\tau), \lambda) \tag{3.14}$$

$$= \Gamma^{++}_{\alpha\beta}(t_1, t_2; Q(\tau), \lambda) - \Gamma^{+-}_{\alpha\beta}(t_1, t_2; Q(\tau), \lambda), \tag{3.15}$$

$$\Gamma^{A}_{\alpha\beta}(t_1, t_2; Q(\tau), \lambda) \equiv -\mathscr{D}_{\alpha\beta}(t_1, t_2; Q(\tau), \lambda) + i\mathscr{A}_{\alpha\beta}(t_1, t_2; Q(\tau), \lambda) \tag{3.16}$$

$$= \Gamma^{++}_{\alpha\beta}(t_1, t_2; Q(\tau), \lambda) - \Gamma^{-+}_{\alpha\beta}(t_1, t_2; Q(\tau), \lambda) \tag{3.17}$$

和

$$\Gamma^{C}_{\alpha\beta}(t_1, t_2; Q(\tau), \lambda) = \Gamma^{-+}_{\alpha\beta}(t_1, t_2; Q(\tau), \lambda) + \Gamma^{+-}_{\alpha\beta}(t_1, t_2; Q(\tau), \lambda) \tag{3.18}$$

其中

$$G^{\sigma_1\sigma_2}_{\alpha\beta}(t_1, t_2; J(\tau), \lambda) = \frac{\delta^2 W[J]}{\delta J_\alpha(t_1^{\sigma_1})\delta J_\beta(t_2^{\sigma_2})}, \tag{3.19}$$

$$\Gamma^{\sigma_1\sigma_2}_{\alpha\beta}(t_1, t_2; Q(\tau), \lambda) = \frac{\delta^2 \Gamma[Q]}{\delta Q_\alpha(t_1^{\sigma_1})\delta Q_\beta(t_2^{\sigma_2})}. \tag{3.20}$$

在 (3.19) 和 (3.20) 式中，$\sigma_1, \sigma_2 = \pm$ 是闭路时区指标. 二阶格林函数和顶点函数还分别满足如下关系:

$$\Gamma^R G^R = G^R \Gamma^R = -1, \tag{3.21}$$

$$\Gamma^A G^A = G^A \Gamma^A = -1, \tag{3.22}$$

$$\Gamma^C = \Gamma^R G^C \Gamma^A. \tag{3.23}$$

为了方便，在 (3.21) 至 (3.23) 式中，用了矩阵符号，记 $G_{\alpha\beta}(t_1, t_2; J(\tau), \lambda)$ 为 G，$\Gamma_{\alpha\beta}(t_1, t_2; Q(\tau), \lambda)$ 为 Γ，矩阵相乘既包含对中间指标求和（积），也包含对相应的时间变量积分.

弛豫矩阵定义为

$$\gamma_{\alpha\beta} \equiv \gamma_{\alpha\beta}(t; Q, \lambda) = \gamma_{(\alpha,\beta)} + \gamma_{[\alpha,\beta]} = -a_{\alpha\beta} + id_{\alpha\beta}$$
$$= \frac{1}{i}\left.\frac{\partial \Gamma^R_{\alpha\beta}[t, \omega; Q, \lambda]}{\partial\omega}\right|_{\substack{\omega=0\\ Q(\tau_+)=Q(\tau_-)=Q(t)}}, \tag{3.24}$$

其中 $\gamma_{(\alpha,\beta)}$ 和 $\gamma_{[\alpha,\beta]}$ 分别为对称和反对称矩阵

$$\gamma_{(\alpha,\beta)} = \frac{1}{2}(\gamma_{\alpha\beta} + \gamma_{\beta\alpha}) = -a_{\alpha\beta} = -\left.\frac{\partial\mathscr{A}_{\alpha\beta}[t, \omega; Q, \lambda]}{\partial\omega}\right|_{\substack{\omega=0\\ Q(\tau_+)=Q(\tau_-)=Q(t)}}, \tag{3.25}$$

$$\gamma_{[\alpha,\beta]} = \frac{1}{2}(\gamma_{\alpha\beta} - \gamma_{\beta\alpha}) = id_{\alpha\beta} = i\left.\frac{\partial\mathscr{D}_{\alpha\beta}[t, \omega; Q, \lambda]}{\partial\omega}\right|_{\substack{\omega=0\\ Q(\tau_+)=Q(\tau_-)=Q(t)}}. \tag{3.26}$$

同时引入响应矩阵 $\mathscr{L}_{\alpha\beta}$ 并将它分解为对称和反对称部份

$$\mathscr{L}_{\alpha\beta} \equiv \mathscr{L}_{\alpha\beta}(t; J, \lambda) = \mathscr{L}_{(\alpha,\beta)} + \mathscr{L}_{[\alpha,\beta]}$$
$$= \frac{1}{i}\left.\frac{\partial G^R_{\alpha\beta}[t, \omega; J, \lambda]}{\partial\omega}\right|_{\substack{\omega=0\\ J(\tau_+)=J(\tau_-)=J(t)}}, \tag{3.27}$$

$$\mathscr{L}_{(\alpha,\beta)} = \frac{1}{2}(\mathscr{L}_{\alpha\beta} + \mathscr{L}_{\beta\alpha}), \tag{3.28}$$

$$\mathscr{L}_{[\alpha,\beta]} = \frac{1}{2}(\mathscr{L}_{\alpha\beta} - \mathscr{L}_{\beta\alpha}). \tag{3.29}$$

我们感兴趣的系统主要是统计定常态，或者是处在定常态附近随时间缓慢变化的状态. 对于前者，(3.24) 至 (3.29) 式与 t 无关；对于后者，在这些式子中出现的 t 都是缓慢

变化的. 注意，在这种观点下，(3.24) 至 (3.29) 式中所有的泛函宗量都取在缓变时刻 t 上，泛函宗量已退化为函数宗量. 并且其中

$$\Gamma^{R,A}_{\alpha\beta}[t,\omega;Q,\lambda]=\int d\tau e^{i\omega\tau}\Gamma^{R,A}_{\alpha\beta}\left(t+\frac{\tau}{2},t-\frac{\tau}{2};Q,\lambda\right), \tag{3.30}$$

$$G^{R,A}[t,\omega;J,\lambda]=\int d\tau e^{i\omega\tau}G^{R,A}_{\alpha\beta}\left(t+\frac{\tau}{2},t-\frac{\tau}{2};J,\lambda\right). \tag{3.31}$$

弛豫矩阵 $\gamma_{\alpha\beta}$ 和响应矩阵 $\mathscr{L}_{\alpha\beta}$ 的意义和性质将在文献 [12] 中讨论.

现在将时间反演对称性和闭路格林函数结合起来讨论具有时间反演对称的统计定常态上的对称关系.

利用 Feynman 的分解技术[11]，当 $t_0<|t|$ 时，有

$$\begin{aligned}\exp\{-i\hat{H}[J](t-t_0)\}&=\exp\{-i\hat{H}(t-t_0)\}\\&\times T_p\exp\left\{-i\int_{t_0}^{t}d\tau J(\tau)\hat{Q}_{t_0}(\tau)\right\}_{J(\tau)=J},\end{aligned} \tag{3.32}$$

$$\begin{aligned}\exp\{i\hat{H}[J](t-t_0)\}&=\tilde{T}_p\exp\left\{i\int_{t_0}^{t}d\tau J(\tau)Q_{t_0}(\tau)\right\}_{J(\tau)=J}\\&\times\exp\{i\hat{H}(t-t_0)\}.\end{aligned} \tag{3.33}$$

容易发现，由生成泛函 (3.1) 至 (3.3) 式生成的各阶格林函数，取 $J(t)=J=$ 常数后，按 (2.26) 式是时间平移不变的. 再利用 (2.27) 式容易证明格林函数的时间反演对称关系

$$Q_\alpha(J,\lambda)=\varepsilon_\alpha Q_\alpha(\varepsilon J,\varepsilon\lambda), \tag{3.34}$$

$$G^{R}_{\alpha\beta}(t_1-t_2;J,\lambda)=\varepsilon_\alpha\varepsilon_\beta G^{A}_{\alpha\beta}(-t_1+t_2;\varepsilon J,\varepsilon\lambda), \tag{3.35}$$

$$G^{A}_{\alpha\beta}(t_1-t_2;J,\lambda)=\varepsilon_\alpha\varepsilon_\beta G^{R}_{\alpha\beta}(-t_1+t_2;\varepsilon J,\varepsilon\lambda), \tag{3.36}$$

$$G^{C}_{\alpha\beta}(t_1-t_2;J,\lambda)=\varepsilon_\alpha\varepsilon_\beta G^{C}_{\alpha\beta}(-t_1+t_2;\varepsilon J,\varepsilon\lambda). \tag{3.37}$$

还可以从 (3.34) 式解出 J，得

$$J_\alpha(Q,\lambda)=\varepsilon_\alpha J_\alpha(\varepsilon Q,\varepsilon\lambda). \tag{3.38}$$

对 (3.38) 式作 $\frac{\delta}{\delta Q_\beta}$，按照 (3.5)，(3.34) 式和泛函微商技术

$$\frac{\delta}{\delta Q_\beta}[F[Q(\tau)]]_{Q(\tau)=Q}=\left[\int d\tau'\frac{\delta F[Q(\tau)]}{\delta Q_\beta(\tau')}\right]_{Q(\tau)=Q}. \tag{3.39}$$

容易求得

$$\Gamma^{R}_{\alpha\beta}[\omega=0;Q,\lambda]=\varepsilon_\alpha\varepsilon_\beta\Gamma^{R}_{\alpha\beta}[\omega=0;\varepsilon Q,\varepsilon\lambda]. \tag{3.40}$$

此外，再对 (3.35) 和 (3.36) 式取逆，利用 (3.21) 和 (3.22) 式有

$$\Gamma^{R}_{\alpha\beta}(t_1-t_2;Q,\lambda)=\varepsilon_\alpha\varepsilon_\beta\Gamma^{A}_{\alpha\beta}(-t_1+t_2;\varepsilon Q,\varepsilon\lambda), \tag{3.41}$$

$$\Gamma^{A}_{\alpha\beta}(t_1-t_2;Q,\lambda)=\varepsilon_\alpha\varepsilon_\beta\Gamma^{R}_{\alpha\beta}(-t_1+t_2;\varepsilon Q,\varepsilon\lambda). \tag{3.42}$$

(3.34) 至 (3.38) 式和 (3.40) 至 (3.42) 式就是我们所希望求得的具有时间反演对称的非平衡统计定常态上格林函数和顶点函数的普遍对称关系.

四、小 结

这是我们把时间反演对称应用于非平衡统计定常态的讨论的第一部份. 多年来，对

于与非线性过程相联系的 NESS 的时间反演行为的讨论基本上局限在唯象或半唯象的领域中．在这一部份中探讨了如何将微观可逆性这样一个动力学基本原理和非平衡量子统计微观理论相结合起来的问题．我们采取了 CTPGF 的方式，将时间反演对称和 CTPGF 生成泛函技术相联系，从而对于一类 NESS，即满足（2.9）式和（2.16）式所规定的时间反演对称要求的统计定常态，得到了一组统计格林函数和顶点函数的普遍对称关系．我们将在文献［12］中利用这些结果来讨论这一类非平衡统计定常态或定常态附近的物理性质．

参　考　文　献

[1] 例如：R. Zwanzig, In "Lectures in theoretical Physics," Vol. III, eds. W. Britten *et al.*, Wiley, N. Y. (1961)；H. Mori, *Progr. Theor Phys.*, 33(1965), 423.

[2] N. Van Kampen, *Physica*, 23 (1957), 707, 816；U. Uhlhorn, *Arkiv foer Fysik*, 17(1960), 361; R. Graham, H. Haken, *Z. Phys.*. 213(1971), 289; 245(1971), 141.

[3] G. Agarwal, *Z. Phys.*, 252(1972), 25; J. Deker, F. Haake, *Phys. Rev.*. A11(1975), 2043; S. K. Ma, G. Mazenko, *Phys. Rev.*. B11(1975), 4077.

[4] 周光召、苏肇冰，统计物理进展，第五章，科学出版社即将出版；周光召、苏肇冰，高能物理与核物理，3 (1979): 304; 314.

[5] Zhou Guang-zhao, Su Zhao-bin, Yu Lu, Hao Bai-lin, ASITP-79003;
周光召、苏肇冰、郝柏林、于渌，物理学报，29 (1980)．961；周光召、郝柏林、于渌，物理学报，29 (1980)，969.

[6] J. Schwinger, *J. Math. Phys.*, 2(1961), 407; L. Keldysh, *Zh. Eksp. Teor. Fiz.*, 47 (1964), 1515.

[7] P. Martin, E. Siggia, H. Rose, *Phys. Rev.*, A8(1975), 423.

[8] 例如：R. DeGroot, P. Mazur, Non-equilibrium Thermodynamics. North Holland, Amsterdam, (1961).

[9] 例如：P. Hohenberg, B. Halperin, *Rev. Mod. Phys.*, 49(1977), 435.

[10] 例如：A. Messiah, Quantum mechanics, Vol. II, North Holland, Amsterdam, (1962)；G. Lüders, *Ann. Phys.*, (N. Y.) 2(1957), 1.

[11] R. Feynman, *Phys. Rev.*, 81(1951), 108.

[12] 周光召、苏肇冰，物理学报，待发表．

TIME REVERSAL SYMMETRY AND NON-EQUILIBRIUM STATISTICAL STATIONARY STATES (I)

Zhou Guang-zhao　　Su Zhao-bin

(Institute of theoretical physics, Academia Sinica)

Abstract

This is the first part of our work on time reversal symmetry applied to non-equilibrium statistical stationary states from a unified microscopic quantum statistical point of view. In this paper, a formalism for time reversal symmetry is constructed in the framework of the Closed Time Path Green's Functions (CTPGF), which can be applied both to equilibrium and non-equilibrium stationary states. By using the generating functional technique of the CTPGF, symmetry relations for the statistical Green's functions and vertex functions are derived for systems invariant under time reversal.

第 30 卷 第 3 期 物 理 学 报 Vol. 30, No. 3
1981 年 3 月 ACTA PHYSICA SINICA March, 1981

时间反演对称和非平衡统计定常态 (II)

周光召 苏肇冰
(中国科学院理论物理研究所)

1980 年 8 月 2 日收到

提 要

本文是我们从微观量子统计理论出发讨论非平衡统计定常态的时间反演对称性质的第二部份. 本文应用文献[1]的结果对非平衡统计定常态的普遍性质进行较为系统的讨论. 对于具有时间反演对称的非平衡统计定常态, 证明了广义(自由能)势函数的存在性; 导出了涨落耗散定理的 Rayleigh-Jeans 极限形式; 推广了局部热平衡假设下的 Onsager 倒易关系; 导得了序参量-守恒荷密度普适方程 (TDGL) 的"可逆-不可逆"运动分解形式.

一、引 言

本文的目的是将文献[1]所得的主要结果, 用来较为系统地讨论具有时间反演对称的非平衡统计定常态(以下简称 NESS)的一些普遍性质. 所有的讨论是在闭路格林函数(以下简称 CTPGF)[1] 的理论框架中进行的.

在第二节, 我们将首先证明在时间反演对称的统计定常态上序参量广义势函数的存在性. 然后, 利用势条件, 进一步求得在时间反演对称的 NESS 上涨落耗散定理(以下简称 FDT) 的 Rayleigh-Jeans 极限 (即 $\omega \to 0$ 极限).

在统计定常态附近, 弛豫矩阵 $\gamma_{\alpha\beta}$、扩散矩阵 $\sigma_{\alpha\beta}$ 和响应矩阵 $\mathscr{L}^{ij}_{\alpha\beta}$ 分别描写了序参量的弛豫、统计涨落和守恒流矢量对外力的响应. 在第三节, 我们将分别求出他们所满足的时间反演对称关系; 并且, 将热平衡态附近的 Onsager 线性理论推广到 NESS 附近.

在第四节, 我们通过对于弛豫逆矩阵 $\gamma^{-1}_{\alpha\beta}$ 性质的探讨, 得到了序参量及守恒荷密度方程(即所谓的含时间 Gingzburg-Landau 方程, 简称 TDGL) 的一个普遍形式. 证明了弛豫逆矩阵的对称组合正比于扩散矩阵 $\sigma_{\alpha\beta}$, 它描述了序参量的不可逆运动; 弛豫逆矩阵的反对称组合, 与序参量微观算符对易子的统计平均有关, 它反映了序参量的可逆运动.

这里顺便指出, 临界动力学中, 涨落起重要作用, 为了在半唯象理论中考虑涨落的影响, 序参量常常被取作为随机变量, 满足受随机力驱动的随机微分方程, 即朗之万方程. 在流行的文献中, 常把朗之万方程扣除随机力剩下的部份称之为"TDGL". 尽管我们在文献 [2] 中也曾证明, 在单圈图近似下, 朗之万方程可被表示成

$$\frac{\partial Q_\alpha}{\partial t} = -\gamma^{-1}_{\alpha\beta}\frac{\delta F}{\delta Q_\beta} + \xi_\alpha, \tag{1.1}$$

其中 ξ_α 是高斯随机力, 满足关系

$$\langle \xi_\alpha(t_1)\xi_\beta(t_2)\rangle = 2\sigma_{\alpha\beta}\delta(t_1 - t_2). \tag{1.2}$$

(1.1) 式扣除随机力 ξ_i 以后，与 CTPGF 序参量方程 (2.4) 式具有相同的形式. 但是这两者之间是有原则性差别的. 在朗之万方程中，序参量被指定为随机变量，各种统计涨落隐含在随机变量之中；去掉随机力 ξ_i，相当于取了树图近似. 而我们讨论的 CTPGF 中的序参量方程，是序参量(或守恒荷密度)微观算符统计平均值所满足的普遍方程，原则上包含了所有高阶统计涨落；相对于前者，它相当于一个重正化以后的方程.

二、势条件和涨落-耗散定理

非平衡统计定常态的一个重要问题是是否存在一个序参量的势函数，它确定序参量在定常态和定常态附近的行为？在 CTPGF 的语言中，我们曾给出了在定常态附近序参量及守恒荷密度的普遍方程为[2,3]

$$\gamma_{\alpha\beta}\frac{\partial Q_\beta(t)}{\partial t}=\frac{\delta\Gamma}{\delta Q_\alpha(t_+)}\bigg|_{Q(\tau_+)=Q(\tau_-)=Q(t)}+J_\alpha(t),\tag{2.1}$$

其中所有泛函宗量 $Q_\beta(\tau)$ 都取在缓变时刻 t. 同时我们也证明了：如果势条件

$$\mathscr{A}_{\alpha\beta}[t,\omega=0;Q,\lambda]=0\tag{2.2}$$

得到满足，就存在一个势函数 $F[t:Q,\lambda]$，有

$$\frac{\delta\Gamma}{\delta Q_\alpha(t_+)}\bigg|_{Q(\tau_+)=Q(\tau_-)=Q(t)}=-\frac{\delta F}{\delta Q_\alpha(t)}.\tag{2.3}$$

而 (2.1) 式可写为

$$\gamma_{\alpha\beta}\frac{\partial Q_\beta(t)}{\partial t}=-\frac{\delta F}{\delta Q_\alpha(t)}+J_\alpha(t).\tag{2.4}$$

我们常将 (2.4) 式称为 TDGL.

对于 NESS，(2.2) 式与 t 无关. 利用文献 [1] 的结果，容易证明 (2.2) 式. 即：对文献 [1] 中 (3.41) 式和 (3.42) 式作傅里叶变换，并取 $\omega=0$ 分量，再与文献 [1] 中 (3.40) 式比较，就得到

$$\mathscr{A}_{\alpha\beta}[\omega=0;Q,\lambda]=0.\tag{2.5}$$

如果在 (2.1) 式中忽略除了 $Q_\alpha(t)$ 以外其他对时间的缓慢依赖，那么显然 (2.5) 式就是势条件，且势函数

$$F=F[Q,\lambda]\tag{2.6}$$

不再显含 t. 否则，还需对系统作一"局部时间反演对称"假设：在每一个缓变的宏观短微观长的时间间隔内，系统满足定常态时间反演对称性质. 对于这样的系统，(2.5) 式就过渡到 (2.2) 式：势条件得到满足.

现在，我们来讨论 NESS 的涨落耗散定理. 从 CTPGF 的理论框架来看，二阶推迟和超前格林函数(或顶点函数)描述了准粒子的能谱和耗散. 二阶关联函数(或顶点函数)描述了准粒子的统计分布，在古典极限的语言中即系统的统计涨落. 所以关联格林函数的 Dyson 方程，对于弱不均匀系统描述了准粒子的输运[3,5,6]，对于统计定常态，则给出了涨落和耗散的联系，即 FDT.

按照文献 [3,6]对于关联格林函数引入 $\mathscr{N}_{\alpha\beta}(t_1,t_2;Q(\tau),\lambda)$，在统计定常态重新写

下文献[1]中(3.23)式,有

$$G^{C}_{\alpha\beta}(\omega;Q,\lambda)=G^{R}_{\alpha\gamma}(\omega;Q,\lambda)\mathscr{N}_{\gamma\beta}(\omega;Q,\lambda)-\mathscr{N}_{\alpha\gamma}(\omega;Q,\lambda)G^{A}_{\gamma\beta}(\omega;Q,\lambda), \tag{2.7}$$

$$\begin{aligned}\Gamma^{C}_{\alpha\beta}(\omega;Q,\lambda)&=\Gamma^{R}_{\alpha\beta}(\omega;Q,\lambda)\mathscr{N}_{\gamma\beta}(\omega;Q,\lambda)-\mathscr{N}_{\alpha\gamma}(\omega;Q,\lambda)\Gamma^{A}_{\gamma\beta}(\omega;Q,\lambda)\\&=-\mathscr{D}_{\alpha\gamma}(\omega;Q,\lambda)\mathscr{N}_{\gamma\beta}(\omega;Q,\lambda)+\mathscr{N}_{\alpha\gamma}(\omega;Q,\lambda)\mathscr{D}_{\gamma\beta}(\omega;Q,\lambda)\\&\quad-i(\mathscr{A}_{\alpha\gamma}(\omega;Q,\lambda)\mathscr{N}_{\gamma\beta}(\omega;Q,\lambda)\\&\quad+\mathscr{N}_{\alpha\gamma}(\omega;Q,\lambda)\mathscr{A}_{\gamma\beta}(\omega;Q,\lambda)),\end{aligned} \tag{2.8}$$

其中 $\mathscr{N}_{\alpha\beta}$ 与准粒子占据数分布有关．在平衡态

$$\mathscr{N}_{\alpha\beta}(\omega;Q,\lambda)=\mathscr{N}^{eq}(\omega)\delta_{\alpha\beta}, \tag{2.9}$$

$$\mathscr{N}^{eq}(\omega)=\operatorname{cth}\frac{\omega}{2T}, \tag{2.10}$$

$$\lim_{\omega\to0}\mathscr{N}^{eq}(\omega)=\frac{2T}{\omega}, \tag{2.11}$$

其中 T 是平衡态温度．将(2.10)和(2.11)式分别代入(2.7)和(2.8)式，就可以分别得到平衡态 FDT 的量子形式和相应的 Rayleigh-Jeans 极限[7]

$$\Gamma^{C}_{\alpha\beta}(\omega;Q,\lambda)=\operatorname{cth}\frac{\omega}{2T}(\Gamma^{R}_{\alpha\beta}(\omega;Q,\lambda)-\Gamma^{A}_{\alpha\beta}(\omega;Q,\lambda)) \tag{2.12}$$

和

$$i\Gamma^{C}_{\alpha\beta}(\omega=0;Q,\lambda)=4Ta_{\alpha\beta}, \tag{2.13}$$

其中 $a_{\alpha\beta}$ 是弛豫矩阵 $\gamma_{\alpha\beta}$ 的对称部分．对格林函数 G 也有与(2.12),(2.13)式相对应的平衡态 FDT 表达式．

我们这里的主要兴趣是在具有时间反演对称的 NESS 上,利用势条件(2.5)式,求得 Rayleigh-Jeans 极限下 FDT 的形式．

既然我们感兴趣的主要是低频行为,先讨论 $\mathscr{N}_{\alpha\beta}$ 在 $\omega\to0$ 时的极限行为．如果

$$\lim_{\omega\to0}\mathscr{N}_{\alpha\beta}(\omega;Q,\lambda)\sim\text{有限}, \tag{2.14}$$

那么,在(2.8)式中　令 $\omega\to0$, 并对矩阵元求迹,考虑到(2.5)式和(2.14)式,就有

$$i\sum_{\alpha}\Gamma^{C}_{\alpha\alpha}(\omega=0;Q,\lambda)=0. \tag{2.15}$$

从文献[8]可以知道，$i\Gamma^{C}_{\alpha\alpha}(\omega;Q,\lambda)\sim$ 系统的量子涨落；所以，对于稳定的统计定常态，将要求(2.15)式左端是正定的(从另一个角度上来看，如果对系统的密度矩阵作无规相位近似，或者当系统扩散矩阵 $\sigma_{\alpha\beta}$ 是正定的条件下，都容易证明(2.15)式左端正定),即(2.15)式不能成立．所以,有

$$\lim_{\omega\to0}\mathscr{N}_{\alpha\beta}(\omega;Q,\lambda)\to\frac{2}{\omega}T^{\text{eff}}_{\alpha\beta}(Q,\lambda), \tag{2.16}$$

其中 $T^{\text{eff}}_{\alpha\beta}$ 可看作有效温度的推广．这样，(2.8)式的低频极限为

$$\Gamma^{C}_{\alpha\beta}(\omega=0;Q,\lambda)=-2i(a_{\alpha\gamma}T^{\text{eff}}_{\gamma\beta}+T^{\text{eff}}_{\alpha\gamma}a_{\gamma\beta}), \tag{2.17}$$

$$\mathscr{D}_{\alpha\beta}(\omega=0;Q,\lambda)T^{\text{eff}}_{\beta\gamma}-T^{\text{eff}}_{\alpha\beta}\mathscr{D}_{\beta\gamma}(\omega=0;Q,\lambda)=0. \tag{2.18}$$

在求得(2.17)式的过程中,我们又利用了势条件(2.5)式,才有

$$\lim_{\omega\to0}\frac{\mathscr{A}_{\alpha\beta}(\omega;Q,\lambda)}{\omega}=\left.\frac{\partial\mathscr{A}_{\alpha\beta}(\omega;Q,\lambda)}{\partial\omega}\right|_{\omega=0}=a_{\alpha\beta}. \tag{2.19}$$

(2.17) 式就是 Rayleigh-Jeans 极限下 FDT 的一般形式. 在我们遇到的一些实际问题中，常有

$$T^{\mathrm{eff}}_{\alpha\beta} = T^{\mathrm{eff}}\delta_{\alpha\beta}. \tag{2.20}$$

这时，FDT (2.17) 式成为

$$i\Gamma^{C}_{\alpha\beta}(\omega = 0;\ Q,\ \lambda) = 4T^{\mathrm{eff}}a_{\alpha\beta}(Q,\ \lambda). \tag{2.21}$$

在 Rayleigh-Jeans 极限下，也可将 (2.16) 式作为 $\mathcal{N}_{\alpha\beta}(\omega;\ Q,\ \lambda)$的近似表达式. 将它代入到 (2.7) 式，按照 (2.20) 式，再作傅里叶变换回到 t 表象，就得到 FDT 的一个等价形式

$$i\frac{\partial G^{C}}{\partial t} = 2T^{\mathrm{eff}}(G^{R} - G^{A}). \tag{2.22}$$

(2.22) 式即是临界动力学文献中常用的形式[9].

三、广义 Onsager 关系

从 TDGL (2.4) 式看到，弛豫矩阵 $\gamma_{\alpha\beta}$ 描述了序参量的弛豫. 按照文献 [1] 中 $\gamma_{\alpha\beta}$ 的定义 (3.24) 式，以及对时间反演对称关系 (3.41) 式的傅里叶变换作微分 $\partial/\partial\omega$，再利用二阶顶点函数的基本性质

$$\Gamma^{R}_{\alpha\beta}(\omega;\ Q,\ \lambda) = \Gamma^{A}_{\beta\alpha}(-\omega;\ Q,\ \lambda). \tag{3.1}$$

容易证明弛豫矩阵的时间反演对称关系为

$$\gamma_{\alpha\beta}(Q,\ \lambda) = \varepsilon_\alpha\varepsilon_\beta\gamma_{\beta\alpha}(\varepsilon Q,\ \varepsilon\lambda). \tag{3.2}$$

福克-普朗克方程的扩散系数 $\sigma_{\alpha\beta}$ 描写了统计定常态的统计涨落，在文献 [2] 中证明了它的 CTPGF 表示为

$$\sigma_{\alpha\beta}(Q,\ \lambda) = \frac{i}{2}\gamma^{-1}_{\alpha\alpha'}\Gamma^{C}_{\alpha'\beta'}(\omega = 0,\ Q,\ \lambda)\gamma^{T-1}_{\beta'\beta}, \tag{3.3}$$

其中 T 表示矩阵的转置. 按照文献[1]中的 (3.23)式和时间反演对称关系(3.37),(3.41),(3.42)式以及势条件，即本文 (2.5) 式，容易证得

$$\Gamma^{C}_{\alpha\beta}(\omega = 0;\ Q,\ \lambda) = \varepsilon_\alpha\varepsilon_\beta\Gamma^{C}_{\alpha\beta}(\omega = 0;\ \varepsilon Q,\ \varepsilon\lambda). \tag{3.4}$$

将 (3.2) 和 (3.4) 式代入 (3.3) 式，并且当 $\gamma_{\alpha\beta}(Q,\ \lambda)$ 是对称的条件下，就有

$$\sigma_{\alpha\beta}(Q,\ \lambda) = \varepsilon_\alpha\varepsilon_\beta\sigma_{\alpha\beta}(\varepsilon Q,\ \varepsilon\lambda). \tag{3.5}$$

在一般情形下，$\gamma_{\alpha\beta}(Q,\ \lambda)$ 不对称，则再考虑到 FDT (2.21) 式和文献 [1] 中 (3.25) 式，也容易证得 (3.5) 式. 这即是首先由 Van Kampen 等人[4] 得到的扩散系数时间反演对称关系.

NESS 还有一个性质是对外力作用的线性响应. 将 $\hat{Q}_\alpha$: $\alpha = 1, 2, \cdots, n$ 中的守恒荷密度算符分离出来，并将它的空间坐标显写出来，记为 $\hat{Q}_r(\boldsymbol{x}, t)$, $r = 1, 2, \cdots, m$. 将与它相耦合的外参量也从 J_α: $\alpha = 1, 2, \cdots, n$ 中分离出来，相应地记为 $\delta J_r(\boldsymbol{x}, t)$，并取它为一级小量. 其他外参量仍记为 J，看作是定常态上恒定不变的物理外源. 这时，具有时间反演对称的 NESS 由哈密顿量 $\hat{H}[J]$ 和密度矩阵 $\rho(\lambda, J)$ 所描述，它们分别满足文献[1] 中 (2.9) 和 (2.24) 式；而守恒荷密度 $Q_r(\boldsymbol{x}, t)$ 与相应外参量的耦合被处理作对于定常态

的微扰. 设 $\hat{j}_r^i(\boldsymbol{x}, t)$ 是与守恒荷密度 $Q_r(\boldsymbol{x}, t)$ 相对应的流矢量密度算符，满足海森堡算符守恒方程

$$\frac{\partial \hat{Q}_r(\boldsymbol{x}, t)}{\partial t} + \frac{\partial \hat{j}_r^i(\boldsymbol{x}, t)}{\partial x^i} = 0, \tag{3.6}$$

其中 $i = 1, 2, 3$ 标记空间分量. 按照 CTPGF 的一般理论，在定常态附近，流密度矢量对于外参量 $\delta J_r(\boldsymbol{x}, t)$ 的线性响应可表示为[3]

$$\delta\langle \hat{j}_r^i(\boldsymbol{x}, t)\rangle = -i\int d^4y \,\mathrm{Tr}\,\{\hat{\rho}(\lambda, J)\vartheta(t_x - t_y)[\hat{j}_r^i(x), \hat{Q}_s(y)]\}\delta J_s(y), \tag{3.7}$$

$$\langle \hat{j}_r^i(x)\rangle = \mathrm{Tr}\,\{\hat{\rho}(\lambda, J)\hat{j}_r^i(x)\}. \tag{3.8}$$

(3.7) 式中的 $\mathrm{Tr}\{\cdots\}$ 因子与外力无关，它是 4 维平移不变的. 定义流密度算符的统计格林函数

$$G_{rs}^{R,ij}(x-y;\, J, \lambda) = -i\mathrm{Tr}\{\hat{\rho}(\lambda, J)\vartheta(t_x - t_y)[\hat{j}_r^i(x), \hat{j}_s^j(y)]\}, \tag{3.9}$$

并且在相互作用拉氏密度中，$\delta J_r(\boldsymbol{x})$ 相当于一种外界作用力的位势，它随 $\boldsymbol{x}$ 缓变，取

$$-\nabla\delta J_r(\boldsymbol{x}, t) = \boldsymbol{X}_r = \text{常数} \tag{3.10}$$

作为外力. 这样，利用 (3.6), (3.10) 和 (3.9) 式的因果表示 (Lehmann 表示)[3]，容易将 (3.7)式化为

$$\delta\langle \hat{j}_r^i(x)\rangle = -\mathscr{L}_{rs}^{ij}(J, \lambda)X_s^j, \tag{3.11}$$

其中

$$\mathscr{L}_{rs}^{ij}(J, \lambda) = \frac{1}{i}\left.\frac{\partial G_{rs}^{R,ij}(p_0, \boldsymbol{p})}{\partial p_0}\right|_{\substack{p_0=0,\\ \boldsymbol{p}=0}} \tag{3.12}$$

是按照文献 [1] 中 (3.27) 方式定义的流密度算符的响应矩阵.

容易论证，薛定谔绘景中的流密度算符 $\hat{j}_r^i(\boldsymbol{x})$ 在时间反演下的变换为

$$\hat{j}_r^i(\boldsymbol{x}) \to \hat{R}\hat{j}_r^i(\boldsymbol{x})\hat{R}^{\dagger} = -\varepsilon_r\hat{j}_r^i(\boldsymbol{x}). \tag{3.13}$$

考虑到 (3.13) 式和文献 [1] 中 (2.8) 式相当，(3.9) 和 (3.12) 式分别与文献 [1] 中 (3.9)和 (3.27) 式相当，采取对于 $\gamma_{\alpha\beta}$ 求证 (3.2) 式完全相类似的步骤，容易证明

$$\mathscr{L}_{rs}^{ij}(J, \lambda) = \varepsilon_r\varepsilon_s\mathscr{L}_{sr}^{ji}(\varepsilon J, \varepsilon\lambda). \tag{3.14}$$

(3.11), (3.12) 和 (3.14) 式即是平衡态附近著名的 Onsager 倒易关系在时间反演对称 NESS 附近的推广. 一般文献上常常将 (3.2), (3.5) 和 (3.14) 式统称为广义 Onsager 关系.

四、弛豫逆矩阵的对称分解和 TDGL

如果忽略势函数对缓变时间的明显依赖，即取 (2.6) 式成立，那么按照文献 [1] 中 (3.5)和 (3.38) 式以及本文 (2.3) 式，我们有

$$\frac{\delta F[Q, \lambda]}{\delta Q_\alpha} = \varepsilon_\alpha \left.\frac{\delta F[\tilde{Q}, \tilde{\lambda}]}{\delta \tilde{Q}_\alpha}\right|_{\substack{\tilde{Q}=\varepsilon Q,\\ \tilde{\lambda}=\varepsilon\lambda}} \tag{4.1}$$

再考虑到 (3.2) 式，容易证明 TDGL (2.4) 式可化为

$$\frac{\partial Q_\alpha}{\partial t} = -\gamma^{-1}_{(\alpha,\beta)}\frac{\delta F}{\delta Q_\beta} - \gamma^{-1}_{[\alpha,\beta]}\frac{\delta F}{\delta Q_\beta} + J. \tag{4.2}$$

按 (4.1) 式，

$$\gamma^{-1}_{(\alpha,\beta)}\frac{\delta F}{\delta Q_\beta} = \frac{1}{2}\left\{\gamma^{-1}_{\alpha\beta}\frac{\delta F}{\delta Q_\beta} + \varepsilon_\alpha\left[\gamma^{-1}_{\alpha\beta}\frac{\delta F}{\delta Q_\beta}\right]_{\substack{Q\to\varepsilon Q\\ \lambda\to\varepsilon\lambda}}\right\}, \tag{4.3}$$

$$\gamma^{-1}_{[\alpha,\beta]}\frac{\delta F}{\delta Q_\beta} = \frac{1}{2}\left\{\gamma^{-1}_{\alpha\beta}\frac{\delta F}{\delta Q_\beta} - \varepsilon_\alpha\left[\gamma^{-1}_{\alpha\beta}\frac{\delta F}{\delta Q_\beta}\right]_{\substack{Q\to\varepsilon Q\\ \lambda\to\varepsilon\lambda}}\right\}, \tag{4.4}$$

其中

$$\gamma^{-1}_{(\alpha,\beta)} = \frac{1}{2}(\gamma^{-1}_{\alpha\beta} + \gamma^{-1}_{\beta\alpha}),$$

$$\gamma^{-1}_{[\alpha,\beta]} = \frac{1}{2}(\gamma^{-1}_{\alpha\beta} - \gamma^{-1}_{\beta\alpha}). \tag{4.5}$$

(4.3) 式对应于通常福克-普朗克方程理论中的不可逆驱动项，(4.4) 式对应所谓的可逆驱动项[4]. 有趣的是在时间反演对称 NESS 的 CTPGF 描述中，它们分别被弛豫逆矩阵 γ^{-1} 的对称组合和反对称组合所对应.

先来考察 $\gamma^{-1}_{\alpha\beta}$ 的对称组合

$$\gamma^{-1}_{(\alpha,\beta)} = \gamma^{-1}_{\alpha\alpha'}\gamma_{(\alpha',\beta')}\gamma^{T-1}_{\beta'\beta}. \tag{4.6}$$

按照文献 [1] 中 (3.25) 式，FDT (2.21) 式和 (3.3) 式，即可将 (4.6) 式化为

$$\gamma^{-1}_{(\alpha,\beta)} = -\frac{1}{2T_{\text{eff}}}\sigma_{\alpha\beta}. \tag{4.7}$$

(4.7) 式说明描述系统不可逆运动的 γ^{-1} 的对称组合正比于系统的扩散系数，它的比例系数是两倍有效温度的倒数.

对于 γ^{-1} 的反对称组合，情况稍复杂一些. 如所周知，时间平移不变的推迟和超前格林函数的傅里叶变换满足如下的 Lehmann 表示:

$$G^{R}_{\alpha\beta}(\omega;Q,\lambda) = \int\frac{d\omega'}{2\pi i}\frac{G^{-+}_{\alpha\beta}(\omega';Q,\lambda) - G^{+-}_{\alpha\beta}(\omega';Q,\lambda)}{\omega'-\omega-i\eta}, \tag{4.8}$$

$$G^{A}_{\alpha\beta}(\omega;Q,\lambda) = \int\frac{d\omega'}{2\pi i}\frac{G^{-+}_{\alpha\beta}(\omega';Q,\lambda) - G^{+-}_{\alpha\beta}(\omega';Q,\lambda)}{\omega'-\omega+i\eta}, \tag{4.9}$$

其中 G^R, G^A, G^{-+}, G^{+-} 分别由文献 [1] 的 (3.8), (3.10), (3.19) 和 (3.20) 式所确定的格林函数的傅里叶变换给出；并且，我们按照文献 [1] 中 (3.3) 和 (3.5) 式的方式将格林函数对 J 的依赖换成对于 Q 的依赖. 对于 (4.8) 和 (4.9) 式，取 $\omega\to\infty$ 的极限，就得到 $G^{R,A}$ 的渐近条件

$$\lim_{\omega\to\infty}\omega G^{R}_{\alpha\beta}(\omega;Q,\lambda) = \lim_{\omega\to\infty}\omega G^{A}_{\alpha\beta}(\omega;Q,\lambda) = if_{\alpha\beta}, \tag{4.10}$$

其中

$$f_{\alpha\beta} = f_{\alpha\beta}(Q,\lambda) = \int\frac{d\omega}{2\pi}\left(G^{-+}_{\alpha\beta}(\omega;Q,\lambda) - G^{+-}_{\alpha\beta}(\omega;Q,\lambda)\right) \tag{4.11}$$

$$\overset{J=0}{=\!=}\ \frac{1}{i}\,\mathrm{Tr}\{\hat{\rho}_0(\lambda)[\hat{Q}_\alpha(t_1),\hat{Q}_\beta(t_2)]\}_{t_1=t_2}. \tag{4.12}$$

且按照格林函数的一般性质，容易证明

$$f_{\alpha\beta} = -f_{\beta\alpha} = f^*_{\alpha\beta}. \tag{4.13}$$

利用文献 [1] 的 (3.21) 和 (3.22) 式，(4.10) 式也给出了 $\Gamma^{R,A}(\omega; Q, \lambda)$ 的渐近条件

$$\lim_{\omega\to\infty} f_{\alpha\gamma}\Gamma^{R}_{\gamma\beta}(\omega; Q, \lambda) = \lim_{\omega\to\infty} f_{\alpha\gamma}\Gamma^{A}_{\gamma\beta}(\omega; Q, \lambda) = i\omega\delta_{\alpha\beta}. \tag{4.14}$$

根据 $\Gamma^{R,A}(\omega; Q, \lambda)$ 的因果性质[3]和渐近条件 (4.14)式，我们可以对于

$$f_{\alpha\gamma}\mathscr{D}_{\gamma\beta}(\omega; Q, \lambda) - \frac{1}{i}\omega\delta_{\alpha\beta}$$

和

$$f_{\alpha\gamma}\mathscr{A}_{\gamma\beta}(\omega; Q, \lambda)$$

写下一个减除一次的色散关系

$$f_{\alpha\gamma}\mathscr{D}_{\gamma\beta}(\omega; Q, \lambda) = \frac{1}{i}\omega\delta_{\alpha\beta} + f_{\alpha\gamma}\mathscr{D}_{\gamma\beta}(\omega = 0; Q, \lambda)$$
$$+ \omega\int_{-\infty}^{+\infty}\frac{d\omega'}{\pi} P\frac{f_{\alpha\gamma}\mathscr{A}_{\gamma\beta}(\omega'; Q, \lambda)}{\omega'(\omega' - \omega)}. \tag{4.15}$$

对 (4.15) 式作微商 $\partial/\partial\omega$ 后再取 $\omega = 0$，由文献 [1] (3.26) 式，容易求得

$$f_{\alpha\gamma}d_{\gamma\beta} = \frac{1}{i}\delta_{\alpha\beta} + f_{\alpha\gamma}\bar{\Delta}_{\gamma\beta}, \tag{4.16}$$

其中

$$\bar{\Delta}_{\alpha\beta} = \int_0^{\infty}\frac{d\omega'}{\pi} P\frac{\mathscr{A}_{\alpha\beta}(\omega'; Q, \lambda) - \mathscr{A}_{\beta\alpha}(\omega'; Q, \lambda)}{\omega'^2}. \tag{4.17}$$

在 (4.17) 式中，我们利用了 $\mathscr{A}_{\alpha\beta}(\omega; Q, \lambda)$ 的普遍性质

$$\mathscr{A}_{\alpha\beta}(\omega; Q, \lambda) = -\mathscr{A}_{\beta\alpha}(-\omega; Q, \lambda). \tag{4.18}$$

按势条件 (2.5) 式和文献 [1] (3.25) 式，有

$$\lim_{\omega\to 0}\mathscr{A}_{\alpha\beta}(\omega; Q, \lambda) \to \omega a_{\alpha\beta}. \tag{4.19}$$

再考虑到 $a_{\alpha\beta}$ 关于 α, β 对称，所以 (4.17) 式中积分收敛，它确定的 $\bar{\Delta}_{\alpha\beta}$ 有限.

从 (4.16) 式中求解 $d^{-1}_{\alpha\beta}$，得

$$d^{-1}_{\alpha\beta} = \{(I + if\bar{\Delta})^{-1}if\}_{\alpha\beta} = \{if(I + \bar{\Delta}if)^{-1}\}_{\alpha\beta}, \tag{4.20}$$

其中 f 和 $\bar{\Delta}$ 分别是 $f_{\alpha\beta}$ 和 $\bar{\Delta}_{\alpha\beta}$ 的矩阵符号. 将它代入 $\gamma^{-1} = -(a - id)^{-1}$ 对 a 的展开式中，容易求得

$$\gamma^{-1}_{[\alpha,\beta]} = \{(I + if\bar{\Delta})^{-1}f\}_{\alpha\beta} + O(a^2). \tag{4.21}$$
$$\approx f_{\alpha\beta}. \tag{4.22}$$

在文献中 $\gamma^{-1}_{[\alpha,\beta]}$ 通常被取作为 $f_{\alpha\beta}$，但是从我们的讨论中可以看清，这只是一个近似. 它不仅忽略了耗散的高阶效应，而且也忽略了色散积分 $\bar{\Delta}_{\alpha\beta}$.

将 (4.7) 和 (4.22) 式代入 TDGL (2.4) 式，并取 J 为零，就得到 TDGL 的一个更为熟知的形式

$$\frac{\partial Q_\alpha}{\partial t} = \frac{1}{2T_{\text{eff}}}\sigma_{\alpha\beta}\frac{\delta F}{\delta Q_\beta} - f_{\alpha\beta}\frac{\delta F}{\delta Q_\beta}. \tag{4.23}$$

对于许多实际问题，Q_α: $\alpha = 1, 2, \cdots, n$ 中的守恒荷密度记为 Q_r: $r = 1, 2, \cdots, m$ 是

某一个李群的生成元，而其它的序参量则是这个李群的一个或多个线性不可约表示. 在这种情况下，

$$f_{\alpha\beta} = f^{\gamma}_{\alpha\beta} Q_{\gamma}, \tag{4.24}$$

其中 $f^{\gamma}_{\alpha\beta}$ 或者是李群的结构常数，或者是生成元的表示矩阵元. 这时，TDGL (4.23)式和临界动力学文献中所用的"TDGL"，具有完全相同的形式. 在更为一般的情况下，CTPGF 导出的 TDGL (4.23) 式与临界动力学文献中所采用的形式有一些差别，我们将在以后进一步讨论这个问题. 但是，正如我们在序言中已经指出的那样，通常福克-普朗克-朗之万理论中的"TDGL"是随机变量满足的随机微分方程. 我们所得到的 CTPGF 理论中的 TDGL 则是序参量微观算符统计平均值所满足的普遍方程，原则上包含了所有的高阶统计涨落，因此这两种方程的形式在原则上可能是不一样的.

五、结束语

在这两篇文章中，作为一个非平衡微观量子统计理论，我们试图把 CTPGF 和微观可逆性原理结合起来，把平衡态附近的一些基本性质推广到了非平衡统计定常态附近，比较系统地对非平衡统计定常态附近的一些普遍性质进行了讨论. 我们讨论了一类具有时间反演对称的统计定常态. 它既可以是平衡态，也可以是非平衡定常态. CTPGF 的时间反演对称表述提供了一个统一讨论热平衡和非平衡定常态的可能性. 在定常态上，我们既证明了序参量广义自由能势函数的存在性，也给出了广义涨落耗散定理在 Rayleigh-Jeans 极限下的一个表达式. 我们得到了非平衡统计定常态附近推广的 Onsager 线性理论，也得到了描述序参量(守恒荷密度)非线性运动 TDGL 的一个普遍形式. 这样，我们就从一个统一的微观理论框架得到了文献中用半唯象非平衡统计理论讨论临界动力学时赖以出发的方程和关系.

十分感谢郝柏林同志，他对这两篇文章的大部份内容和我们进行了深入的讨论.

参考文献

[1] 周光召、苏肇冰，物理学报 **30**(1981)，164.
[2] Zhou Guang-zhao, Su Zhao-bin, Yu Lu, Hao Bai-lin, ASITP-79003; 周光召、苏肇冰、郝柏林、于渌，物理学报，**29**(1980)，961. 周光召、郝柏林、于渌，物理学报，**29**(1980)，969.
[3] 周光召、苏肇冰，统计物理进展，第五章，科学出版社，即将出版.
[4] N. Van Kampen, *Physica*, **23**(1957), 707, 816; U. Uhlhorn, *Arkiv foer Fysik*, **17**(1960), 361; R. Graham, H. Haken, *Z. Phys.*, **243**(1971), 289; **245**(1971), 141.
[5] L. Kandanoff, G. Baym, Quantum Statistical Mechanics Benjamin N. Y. (1962); D. Dubois, In "Lectures in Theoretical Physics," Vol. IXC, eds. W. Brittin, *et al.*, Gordon and Breach, N. Y. (1961).
[6] 周光召、苏肇冰，高能物理与核物理，**3**(1979)，314.
[7] R. Kubo, *Rep. Progr. Phys.*, **29**(1966), 255.
[8] 周光召、苏肇冰，物理学报，**29**(1980)，618.
[9] 例如 S. K. Ma, G. Mazenko, *Phys. Rev.*, **B11**(1975), 4077; C. DeDominicis, L. Peliti, *Phys. Rev.*, **B18**(1978), 353.

TIME REVERSAL SYMMETRY AND NON-EQUILIBRIUM STATISTICAL STATIONARY STATES (II)

ZHOU GUANG-ZHAO　　SU ZHAO-BIN
(Institute of theoretical Physics, Academia Sinica)

ABSTRACT

This is the 2nd part of our discussion on time reversal symmetry applied to the non-equilibrium statistical stationary states (NESS) from a microscopic quantum statistical point of view. With the application of the main results of I, a systematic investigation on the general properties of the NESS is given. For systems invariant under time reversal, the existence of a generalized potential and the fluctuation-dissipation theorem in the low frequency limit are established in the NESS. The Onsager's reciprocity relations for the local thermodynamical equilibrium systems are also generalized to that for the NESS invariant under time reversal symmetry. Finally, the time dependent Ginzburg-Landau equations for order parameters and conserved densities have been expressed in a general form with time irreversible and reversible parts similar to that met in the literatures studying critical dynamics.

第33卷 第7期 物 理 学 报 Vol. 33, No. 7
1984年7月 ACTA PHYSICA SINICA Jul., 1984

原子核费密多体系统平均场近似的推广

苏肇冰 于 渌 周光召
(中国科学院理论物理研究所)

1983年8月1日收到

提 要

本文应用先前得到的序参量——统计格林函数耦合方程组，对于原子核多费密子体系，系统地导出了它的平均场近似，Hartree-Fock 近似和无规相位近似. 并求出了相应近似下系统热力学势的明显表达式. 本文的讨论也适用于处于平衡或非平衡定常态的其他多体系统.

一、引 言

近年来，泛函积分技术被应用到原子核多体理论中[1-3]，有效的途径之一是通过 Hubbard-Stratonovich[4] 变换(以下简称 H-S 变换)引入反映集体运动的玻色场 $Q(x)$，自洽地讨论费密子在这个场的平均值 $Q_c(x)$ (序参量)中的运动. 这种平均场近似，可以作为重原子核在大范围内的集体运动，包括裂变行为的一个有益出发点[5]. 类似的方法在固体理论中早有应用，通常叫 Bogoliubov-de Gennes (以下简称 B-dG)方程[6]. 从这个方法的推导可以看出，这是一种 Hartree 近似，没有 Fock 的交换作用修正. 最近，文献 [3] 中试图推广 B-dG 方程，考虑交换和涨落效应. 这些作者强调，辅助场 $Q(x)$ 是通过 H-S 变换引入的，定义不唯一. 他们利用位势选取上的任意性求出不同的近似.

我们认为，多体系统中的平均场 $Q_c(x)$ 及其涨落，也就是集体激发，有明确的物理意义和可观察的效应. H-S 变换形式上的任意性可以通过与生成泛函技术的结合[7,8] 来消除. 而且，计算泛函积分的最陡下降法实际上是对 $\hbar$ 的展开. Hartree, Hartree-Fock (以下简称 H-F)和无规相位近似(以下简称 RPA)应能用系统的方法导出，不像文献 [3] 的作者强调的那样，要利用两体"试探"势的任意性.

在文献 [8] 中，我们将生成泛函的技术用到 CTPGF 上，推导出了一个序参量和费密子及集体激发格林函数所满足的自洽方程组. 本文中具体用它来讨论 4-费密子作用的体系。第二节中，对费密子和集体激发取逐级近似，分别求得 Hartree, H-F 和 RPA 近似. 第三节中，用直接积分泛函方程的办法分别计算这些近似下的自由能. 如过去指出过的[7-10]，CTPGF 的理论框架同时适用于平衡态与非平衡态. 因此，希望这里发展的理论形式能作为讨论大尺度上原子核集体运动动力学理论的有用框架. 需要指出，整个讨论对费密子之间究竟是"单体"还是"二体"势无关，但为表述清楚起见，讨论单体势的情形.

不熟悉 CTPGF 的读者，请参阅文献 [7—10]，这里不再详细解释各种符号及其含义.

二、序参量及统计格林函数的 Hartree, H-F 及 RPA 近似

令 $\hat{\phi}_\alpha(x)$, $\hat{\phi}^+_\beta(x)$, $\alpha, \beta = 1, 2, \cdots$ 为海森堡绘景中的费密场算子，α, β 为内部自由度脚标．系统的有效作用量为

$$I[\hat{\phi}^+, \hat{\phi}] = \int d^4x \hat{\phi}^+_\alpha(x) \left(i\hbar \frac{\partial}{\partial t} - \hat{H}_0 \right)_{\alpha\beta} \hat{\phi}_\beta(x) - \frac{g^2}{2} \int d^4x d^4y \hat{\phi}^+_\alpha(x) \hat{\phi}^+_\beta(y) O_{\alpha\alpha'} O_{\beta\beta'} \hat{\phi}_{\beta'}(y) \hat{\phi}_{\alpha'}(x), \tag{1a}$$

它可改写为

$$I[\hat{\phi}^+, \hat{\phi}] = \int d^4x \hat{\phi}^+(x) \left(i\hbar \frac{\partial}{\partial t} - \hat{H}_0 + \frac{g^2}{2} \hat{O}^2 V(\boldsymbol{O}) \right) \hat{\phi}(x) - \frac{g^2}{2} \int d^4x d^4y \hat{\phi}^+(x) \hat{O} \hat{\phi}(x) \Delta_0(x - y) \hat{\phi}^+(y) \hat{O} \hat{\phi}(y), \tag{1b}$$

其中 $\hat{H}_0$ 为自由哈密顿量，

$$\Delta_0(x - y) = \delta(t_x - t_y) V(\boldsymbol{x} - \boldsymbol{y}), \tag{2}$$

$V(\boldsymbol{x} - \boldsymbol{y})$ 为瞬时作用势．$\hat{O}$ 为内部自由度空间的矩阵，对哑指标求和．假定系统用密度矩阵 $\hat{\rho}$ 描述，它将基态和热平衡态作为特例包含在内．(1b) 式中

$$V(\boldsymbol{o}) \equiv V(\boldsymbol{x} - \boldsymbol{y})|_{\boldsymbol{x}=\boldsymbol{y}}, \tag{3}$$

若序参量，即平均场变量定义为

$$\hat{Q}(x) \equiv \hat{Q}(\boldsymbol{x}, t) \equiv g \int d^4y \Delta_0(x - y) \hat{\phi}^+(y) \hat{O} \hat{\phi}(y) = g \int d^3y V(\boldsymbol{x} - \boldsymbol{y}) \hat{\phi}^+(\boldsymbol{y}, t) \hat{O} \hat{\phi}(\boldsymbol{y}, t). \tag{4}$$

根据文献 [8]，准到双圈图近似，即统计生成泛函最陡下降展开的 $\hbar^2$ 级，序参量和格林函数的耦合方程组为

$$-\frac{1}{2} \xi_\sigma \frac{\delta \Gamma_p[Q]}{\delta Q(x_\sigma)} \bigg|_{Q(x_+)=Q(x_-)} = -\int d^4y \Delta_0^{-1}(x - y) Q(y) - \frac{i\hbar}{2} g \operatorname{Sp}\{\hat{O} G_c(x, x)\} = 0, \tag{5}$$

$$G_p^{-1}(x, y) = G_{op}^{-1}(x, y) - \Sigma_p(x, y), \tag{6}$$

$$\Sigma_p(x, y) = i\hbar g^2 \hat{O} G_p(x, y) \hat{O} \Delta_p(y, x), \tag{7}$$

$$\Delta_p^{-1}(x, y) = \Delta_{op}^{-1}(x, y) - \Pi_p(x, y), \tag{8}$$

$$\Pi_p(x, y) = -i\hbar g^2 \operatorname{Sp}(\hat{O} G_p(x, y) \hat{O} G_p(y, x)). \tag{9}$$

这里 $\Gamma_p[Q(x)]$ 为序参量

$$Q(x) \equiv \operatorname{Tr}\{\hat{\rho} \hat{Q}(x)\} \tag{10}$$

的单粒子不可约 (IPI) 顶角生成泛函，G_p 和 Δ_p 为费密场和集体激发的二阶连接 CTPGF．任何一个两点 CTPGF 与相应的推迟 (r)、超前 (a) 和关联 (c) 格林函数的关系为[7]

$$G_p(x_\sigma, y_\rho) = \frac{1}{2}\xi_\sigma\eta_\rho G_r(x, y) + \frac{1}{2}\eta_\sigma\xi_\rho G_a(x, y) + \frac{1}{2}\xi_\sigma\xi_\rho G_c(x, y), \tag{11}$$

其中，$\sigma, \rho = \pm$，表示正、负时轴，且

$$\xi_\pm = 1, \quad \eta_\pm = \pm 1. \tag{12}$$

上列方程中 Tr 表示对时空及内部自由度脚标求迹，Sp 表示只对内部自由度求迹．我们还用了符号

$$\Delta_{oP}^{-1}(x, y) = \delta_P(t_x - t_y)V^{-1}(\boldsymbol{x} - \boldsymbol{y}), \tag{13}$$

δ_P 为定义在闭路上的 δ 函数[9]．此外，

$$G_{oP}^{-1}(x, y) = S_{oP}^{-1}(x, y) - g\hat{O}Q(x)\delta_P^4(x - y), \tag{14}$$

$$S_{oP}^{-1}(x, y) = \left(i\hbar\frac{\partial}{\partial t} - \hat{H}_0 + \frac{g^2}{2}V(\boldsymbol{o})\hat{O}^2\right)\delta_P^4(x - y). \tag{15}$$

如文献 [8] 所述，也可直接利用 (11)—(13) 式检验，(5)—(9) 式构成一个完备的自洽方程组．

为简单起见，我们讨论稳态的情形，序参量 $Q(x)$ 不含时间，传播子具有时间平移不变性．同时，假定费密子和集体激发的虚部都很小，因而可看成真正的元激发．对时间变量作傅氏变换

$$f(t, \boldsymbol{x}, \boldsymbol{y}) = \int\frac{dp_0}{2\pi\hbar}e^{-\frac{i}{\hbar}p_0 t}f(p_0, \boldsymbol{x}, \boldsymbol{y}), \tag{16}$$

将变换后的费密子传播子按完备组 $\{\phi_n(x)\}$ 展开：

$$G_r(p_0, \boldsymbol{x}, \boldsymbol{y}) = \sum_n \phi_n(\boldsymbol{x})\frac{1}{p_0 - E_n + i\gamma_n}\phi_n^+(\boldsymbol{y}), \tag{17a}$$

$$G_a(p_0, \boldsymbol{x}, \boldsymbol{y}) = \sum_n \phi_n(\boldsymbol{x})\frac{1}{p_0 - E_n - i\gamma_n}\phi_n^+(\boldsymbol{y}), \tag{17b}$$

$$G_c(p_0, \boldsymbol{x}, \boldsymbol{y}) = \sum_n \phi_n(\boldsymbol{x})(1 - 2N_n)\left(\frac{1}{p_0 - E_n + i\gamma_n} - \frac{1}{p_0 - E_n - i\gamma_n}\right)\phi_n^+(\boldsymbol{y}), \tag{17c}$$

其中谱函数 $\phi_n(\boldsymbol{x})$ 满足方程

$$\left(E_n - i\gamma_n - \hat{H}_0 + \frac{1}{2}g^2\hat{O}^2V(\boldsymbol{o}) - g\hat{O}Q(\boldsymbol{x})\right)\phi_n(\boldsymbol{x}) - \int d^3y\Sigma_r(E_n, \boldsymbol{x}, \boldsymbol{y})\phi_n(\boldsymbol{y}) = 0. \tag{18}$$

这里 E_n, γ_n, N_n 分别为费密子的能谱、耗散和粒子数分布，略去了波函数重正化效应．

与广义的涨落耗散定理(简称 FDT)

$$\Sigma_c(E_n, \boldsymbol{x}, \boldsymbol{y}) = \sum_n \phi_n(\boldsymbol{x})2i\gamma_n(1 - 2N_n)\phi_n^+(\boldsymbol{y}) \tag{19}$$

结合起来，方程组 (17)，(18) 等价于方程式 (6) 和 (7)．而且，可以直接检验谱函数的正交完备性．

作为最低级的近似，我们完全略去费密子的自能部分 $\Sigma_P(x, y)$．这时，费密子的阻尼 γ_n 趋于零，集体激发的格林函数与序参量完全无关．将方程 (17c) 代入方程 (5)，并利用

谱函数完备性，可以求出序参量和费密子波函数在 Hartree 近似下的自洽方程组（即 B-deG 方程）

$$\int d^3y V^{-1}(\boldsymbol{x}-\boldsymbol{y})Q(\boldsymbol{y}) = g\sum_n \left(N_n - \frac{1}{2}\right)\phi_n^+(\boldsymbol{x})\hat{O}\phi_n(\boldsymbol{x})$$

$$= g\sum_n N_n\phi_n^+(\boldsymbol{x})\hat{O}\phi_n(\boldsymbol{x}) - \frac{1}{2}g\,\mathrm{Sp}(\hat{O})\delta^3(\boldsymbol{x}), \tag{20}$$

$$\left(E_n - \hat{H}_0 + \frac{1}{2}g^2\hat{O}^2V(\boldsymbol{o}) - g\hat{O}Q(\boldsymbol{x})\right)\phi_n(\boldsymbol{x}) = 0. \tag{21}$$

值得指出，这类平均场型的方程有一特点，准粒子的能谱 $\{E_n\}$ 必须和它的分布 $\{N_n\}$ 自洽地决定．一般说来，同样的能谱，不会有不同的分布．

作为下一级近似，我们保留 Σ_p，但令 Π_p 为零．根据方程 (7) 和 (8) 可以求出

$$\Delta_r(x, y) = \Delta_a(x, y) = \delta(t_x - t_y)V(\boldsymbol{x}-\boldsymbol{y}), \tag{22}$$

$$\Delta_c(x, y) = 0, \tag{23}$$

$$\Sigma_r(p_0, \boldsymbol{x}, \boldsymbol{y}) = \frac{i\hbar g^2}{2}\int\frac{dq_0}{2\pi\hbar}\hat{O}G_c(q_0, \boldsymbol{x}, \boldsymbol{y})\hat{O}V(\boldsymbol{x}-\boldsymbol{y}), \tag{24}$$

$$\Sigma_c(P_0, \boldsymbol{x}, \boldsymbol{y}) = 0. \tag{25}$$

这时费密子的阻尼 γ_n 仍趋于零．通过与第一种情形类似的推导，可以求出 H-F 近似下的自洽方程组．序参量满足的方程还和 (20) 式一样，但谱函数满足的方程变为

$$(E_n - \hat{H}_0 - g\hat{O}Q(x))\phi_n(\boldsymbol{x}) + g^2\int d^3y V(\boldsymbol{x}-\boldsymbol{y})$$

$$\times\sum_m \hat{O}\phi_m(\boldsymbol{x})N_m\phi_m(\boldsymbol{y})\phi_n(\boldsymbol{y}) = 0. \tag{26}$$

(21) 式中非物理的 $\frac{1}{2}g^2\hat{O}^2V(\hat{O})$ 项消去了，出现了 Fock 交换项．值得注意的是，序参量是在 $\hbar$ 级与费密子格林函数耦合的(见方程(5))，必须在计算 Σ_r 时保留同一级的项，才能得出正确的谱函数方程．这就是为什么有些推导方法中"丢掉"了 Fock 项，而在我们的方法中出现的原因．

为了改进 H-F 近似，我们可以不仅保留费密子的自能 Σ_p，而且保留序参量涨落的自能部分 Π_p．如果只保留 Π_p 的打头项，就是通常的 RPA 近似．容易检验，这时的序参量方程具有与 Hartree 及 H-F 近似相同的形式，但在弱阻尼情形谱函数的方程通过费密子自能 Σ_p 与集体激发耦合起来．下一节计算自由能时我们还要回到这个问题．

现在需要说明一下方程 (20) 中正比于 $\mathrm{Sp}(\hat{O})$ 的项．对于超导这类具有非对角长程序的系统，

$$\mathrm{Sp}(\hat{O}) = 0, \tag{27}$$

这一项不出现．这里的理论框架对于不具备非对角长程序的系统也是适用的．譬如，$\hat{O}$ 是单位矩阵的情形对应序参量为费密子密度的线性泛函．

对于这类系统，可以定义费密面及相应的准粒子 N_n^p 及准空穴 N_n^h，即

$$\phi_n^p(\boldsymbol{x}) \equiv \phi_n(\boldsymbol{x}),\quad N_n^p \equiv N_n \quad 若\ n > n_F \tag{28a}$$

$$\phi_n^h(\boldsymbol{x}) \equiv \phi_n(\boldsymbol{x}),\quad N_n^h \equiv 1 - N_n \quad 若\ n < n_F, \tag{28b}$$

其中 n_F 为费密面上的态．若假定有电子-空穴对称，就有

$$\phi_n^{p+}(\boldsymbol{x}) = \zeta_n^p \phi_{-n}^h(\boldsymbol{x}), \quad \phi_n^{h+}(\boldsymbol{x}) = \zeta_n^h \phi_{-n}^p(\boldsymbol{x}), \tag{29}$$

且

$$\zeta_n^p \zeta_n^{p*} = \zeta_n^h \zeta_n^{h*} = 1. \tag{30}$$

不失普遍性，可以进一步假定 $\hat{O}$ 为对称的．利用 (29)，(30) 式容易证明

$$\sum_n \left(N_n - \frac{1}{2}\right) \phi_n^+(\boldsymbol{x}) \hat{O} \phi_n(\boldsymbol{x})$$
$$= \sum_{n_p} N_n^p \phi_n^{p+}(\boldsymbol{x}) \hat{O} \phi_n^p(\boldsymbol{x}) - \sum_{n_h} N_n^h \phi_n^{h+}(\boldsymbol{x}) \hat{O} \phi_n^h(\boldsymbol{x}). \tag{31}$$

因此，我们可以把比例于 Sp $(\hat{O})$ 的项解释为因泡利不相容原理导致的费密子基态涨落．这一项会自动消去费密海对序参量方程贡献的发散项．因此，可限于讨论等式 (27) 成立的情形．否则，只需作类似 (31) 式的改动．

三、自由能的计算

我们过去曾证明[9,10]，对处于热平衡的和具有时间反演对称的非平衡稳态（简称 NESS），存在一个势泛函 $\mathscr{G}[Q(\boldsymbol{x})]$，且

$$\frac{\delta \mathscr{G}[Q(\boldsymbol{x})]}{\delta Q(\boldsymbol{x})} = -\frac{1}{2} \xi_\sigma \frac{\delta \Gamma_p[Q(x)]}{\delta Q(x\sigma)} \bigg|_{Q(x_+)=Q(x_-)=Q(x)}. \tag{32}$$

在弱阻尼假定下，可由方程 (5) 推出

$$\frac{\delta \mathscr{G}[Q]}{\delta Q(\boldsymbol{x})} = -\int d^3y V^{-1}(\boldsymbol{x}-\boldsymbol{y}) Q(\boldsymbol{y})$$
$$+ g \sum_n \left(N_n - \frac{1}{2}\right) \phi_n^+(\boldsymbol{x}) \hat{O} \phi_n(\boldsymbol{x}), \tag{33}$$

并由 (18) 式求得

$$\frac{\delta E_n}{\delta Q(\boldsymbol{x})} = g \phi_n^+(\boldsymbol{x}) \hat{O} \phi_n(\boldsymbol{x}) + \int d^3y\, d^3z\, \phi_n^+(\boldsymbol{y}) \frac{\delta \Sigma_r(E_n, \boldsymbol{y}, \boldsymbol{z})}{\delta Q(\boldsymbol{x})} \phi_n(\boldsymbol{z}). \tag{34}$$

按照上节末尾的约定，(34) 式可化为

$$\frac{\delta}{\delta Q(\boldsymbol{x})} \left(\frac{1}{2} \sum_n E_n\right) - \frac{1}{2} \int d^3y d^3z \sum_n \phi_n^+(\boldsymbol{y}) \frac{\delta \Sigma_r(E_n, \boldsymbol{y}, \boldsymbol{z})}{\delta Q(\boldsymbol{x})} \phi_n(\boldsymbol{z}). \tag{35}$$

将 (34) 式代人 (33) 式，并考虑到 (35) 式和 (17c) 式，求得

$$\frac{\delta \mathscr{G}[Q]}{\delta Q(\boldsymbol{x})} = -\int d^3y V^{-1}(\boldsymbol{x}-\boldsymbol{y}) Q(\boldsymbol{y}) + \sum_n \left(N_n - \frac{1}{2}\right) \frac{\delta E_n}{\delta Q(\boldsymbol{x})}$$
$$- \sum_n \int d^3y d^3z \, \mathrm{Sp} \left\{ \frac{\delta \Sigma_r(E_n, \boldsymbol{y}, \boldsymbol{z})}{\delta Q(\boldsymbol{x})} \phi_n(\boldsymbol{z}) \left(N_n - \frac{1}{2}\right) \phi_n^+(\boldsymbol{y}) \right\}. \tag{36a}$$
$$= -\int d^3y V^{-1}(\boldsymbol{x}-\boldsymbol{y}) Q(\boldsymbol{y}) + \sum_n \left(N_n - \frac{1}{2}\right) \frac{\delta E_n}{\delta Q(\boldsymbol{x})}$$
$$+ \frac{i\hbar}{2} \frac{1}{T} \int d^4y d^4z \, \mathrm{Sp} \left\{ \frac{\delta \Sigma_r(y, z)}{\delta Q(\boldsymbol{x})} G_c(z, y) \right\}, \tag{36b}$$

其中

$$T \equiv \int_{-\infty}^{\infty} dt.$$

(36) 式是我们求势泛函的出发点．这个式子既适用于平衡态，也适用于满足势条件[10]的 NESS 态．

与上节一样，先讨论平均场近似，令 $\Sigma_p = 0$．立即得出

$$\frac{\delta \mathscr{G}[Q(\boldsymbol{x})]}{\delta Q(\boldsymbol{x})} = \frac{\delta}{\delta Q(\boldsymbol{x})}\left[-\frac{1}{2}\int d^3y d^3z Q(\boldsymbol{y}) V^{-1}(\boldsymbol{y}-\boldsymbol{z}) Q(\boldsymbol{z})\right] - \sum_n N_n \frac{\delta E_n}{\delta Q(\boldsymbol{x})}. \quad (37)$$

对于热平衡系统，费密子分布为

$$N_n = [\exp(\beta(E_n - \mu) + 1)]^{-1}, \quad (38)$$

其中 β 为温度倒数，μ 为化学势．因此，平均场近似下的自由能为

$$\mathscr{G}[Q] = -\frac{1}{2}\int d^3y d^3z Q(\boldsymbol{y}) V^{-1}(\boldsymbol{y}-\boldsymbol{z}) Q(\boldsymbol{z}) - \beta^{-1}\sum_n \ln(1+\exp(-\beta(E_n-\mu))). \quad (39)$$

考虑到 (20) 和 (21) 式，第一项可改写为

$$-\frac{1}{2} g^2 \sum_n \sum_m \int d^3y d^3z N_n \phi_n^+(\boldsymbol{y}) \hat{O} \phi_n(\boldsymbol{y}) V(\boldsymbol{y}-\boldsymbol{z}) N_m \phi_m^+(\boldsymbol{z}) \hat{O} \phi_m(\boldsymbol{z}),$$

这正好是 Hartree 近似下的费密子相互作用能．

下面讨论 $\Pi_p = 0$ 的 H-F 情形．利用完全类似的方法可由 (36) 式推出

$$\mathscr{G}[Q] = -\frac{1}{2}\int d^3y d^3z Q(\boldsymbol{y}) V^{-1}(\boldsymbol{y}-\boldsymbol{z}) Q(\boldsymbol{z}) - \beta^{-1}\sum_n \ln(1+\exp(-\beta(E_n-\mu))) + \frac{g^2}{2}\int d^3y d^3z \sum_n \sum_m N_n \phi_n^+(\boldsymbol{y}) \hat{O} \phi_m(\boldsymbol{y}) V(\boldsymbol{y} - \boldsymbol{z}) N_m \phi_m^+(\boldsymbol{z}) \hat{O} \phi_n(\boldsymbol{z}). \quad (40)$$

与 Hartree 近似相比，这里多了右端的第三项，即交换项的贡献．

现在讨论 RPA 近似．令

$$\Sigma_p(x, y) = \Sigma_p^{(1)}(x, y) + \Sigma_p^{(2)}(x, y), \quad (41)$$

其中

$$\Sigma_p^{(1)}(x, y) = i\hbar g^2 \hat{O} G_p(x, y) \hat{O} \Delta_{op}(y, x), \quad (42a)$$

$$\Sigma_p^{(2)}(x, y) = i\hbar g^2 \hat{O} G_p(x, y) \hat{O}(\Delta_p(y, x) - \Delta_{op}(y, x)). \quad (42b)$$

相应地，(35) 和 (36b) 式变成

$$\frac{\delta}{\delta Q(\boldsymbol{x})}\left(\frac{1}{2}\sum_n E_n\right) - \frac{1}{2}\int d^3y d^3z \sum_n \phi_n^+(\boldsymbol{y}) \frac{\delta \Sigma_r^{(1)}(E_n, \boldsymbol{y}, \boldsymbol{z})}{\delta Q(\boldsymbol{x})} \phi_n(\boldsymbol{z}) - \frac{1}{2}\int d^3y d^3z \sum_n \phi_n^+(\boldsymbol{y}) \frac{\delta \Sigma_r^{(2)}(E_n, \boldsymbol{y}, \boldsymbol{z})}{\delta Q(\boldsymbol{x})} \phi_n(\boldsymbol{z}) = 0, \quad (43)$$

$$\frac{\delta\mathscr{G}[Q]}{\delta Q(\boldsymbol{x})} = -\int d^3y V^{-1}(\boldsymbol{x}-\boldsymbol{y})Q(\boldsymbol{y}) + \sum_n \left(N_n - \frac{1}{2}\right)\frac{\delta E_n}{\delta Q(\boldsymbol{x})}$$
$$+ \frac{i\hbar}{2}\frac{1}{T}\int d^4yd^4z\,\mathrm{Sp}\left\{\frac{\delta\Sigma^{(1)}(y,z)}{\delta Q(\boldsymbol{x})}G_c(z,y)\right\}$$
$$- \frac{i\hbar}{2}\frac{1}{T}\int d^4yd^4z\,\mathrm{Sp}\left\{\Sigma_r^{(2)}(y,z)\frac{\delta G_c(z,y)}{\delta Q(\boldsymbol{x})}\right\}$$
$$+ \frac{i\hbar}{2}\frac{1}{T}\frac{\delta}{\delta Q(\boldsymbol{x})}\int d^4yd^4z\,\mathrm{Sp}\{\Sigma_r^{(2)}(y,z)G_c(z,y)\}. \qquad (44)$$

根据 (42a) 式，$\Sigma_r^{(1)}(p_0, \boldsymbol{x}, \boldsymbol{y})$ 形式上与 (24) 式一样，但谱函数不同．采用 H-F 情形的推导步骤，(43) 式左端的第二项可写成

$$\frac{\delta}{\delta Q(\boldsymbol{x})}\left[\frac{1}{2}g^2V(\boldsymbol{o})\sum_n N_n\int d^3y\phi_n^+(\boldsymbol{y})\hat{O}^2\phi_n(\boldsymbol{y})\right], \qquad (45)$$

而 (44) 式右端的第三项可化为

$$\frac{\delta}{\delta Q(\boldsymbol{x})}\left\{\frac{g^2}{2}\int d^3yd^3z\sum_n\sum_m\left(N_n-\frac{1}{2}\right)\phi_n^+(\boldsymbol{y})\hat{O}\phi_m(\boldsymbol{y})V(\boldsymbol{y}-\boldsymbol{z})\right.$$
$$\left.\times\left(N_m-\frac{1}{2}\right)\phi_m^+(\boldsymbol{z})\hat{O}\phi_n(\boldsymbol{z})\right\}. \qquad (46)$$

由 (42) 式及 (8)，(9) 式可以看出 $\Sigma^{(2)}$ 对方程 (43)，(44) 的贡献比 $\Sigma^{(1)}$ 高一个 $\hbar$ 的量级．如果序参量没有涨落，可以满足于 H-F 近似． 但这些涨落是存在的，使得 $\Delta(\boldsymbol{x}, \boldsymbol{y})$ 不同于没有动力学的自由传播子 $\Delta_0(x, y)$． 我们应该在 $\Sigma^{(2)}$ 的贡献中挑出主要项，而略去次要项。下面的讨论将表明，(44) 式右端的第四项是重要的，而该式的第五项和 (43) 式左端的第三项可以忽略．

利用 CTPGF 的技巧[3]可以证明下列恒等式：

$$\int d^4yd^4z\left\{\Sigma_r^{(2)}(y,z)\frac{\delta G_c(z,y)}{\delta Q(\boldsymbol{x})} + \Sigma_c^{(2)}(y,z)\frac{\delta G_a(z,y)}{\delta Q(\boldsymbol{x})}\right\}$$
$$= -\frac{1}{2}\int d^4yd^4z\left\{\frac{\delta\Pi_r(y,z)}{\delta Q(\boldsymbol{x})}(\Delta_c(z,y)-\Delta_{0c}(z,y))\right.$$
$$\left.+\frac{\delta\Pi_c(y,z)}{\delta Q(\boldsymbol{x})}(\Delta_a(z,y)-\Delta_{0a}(z,y))\right\}. \qquad (47)$$

既然耗散很弱，可以略去 (47) 式中的 Σ_c 和 Π_c，因为根据 FDT 它们正比于耗散．考虑到 (8)，(19) 式和时间平移不变性，(47) 式可写成

$$-\frac{i\hbar}{2}\frac{1}{T}\int d^4yd^4z\,\mathrm{Sp}\left\{\Sigma_r^{(2)}(y,z)\frac{\delta G_c(z,y)}{\delta Q(\boldsymbol{x})}\right\}$$
$$= -\frac{i\hbar}{4}\int\frac{dq_0}{2\pi\hbar}\int d^3yd^3z\frac{\delta\Delta_r^{-1}(q_0,\boldsymbol{y},\boldsymbol{z})}{\delta Q(\boldsymbol{x})}\Delta_c(q_0,\boldsymbol{z},\boldsymbol{y}). \qquad (48)$$

考虑到对 (48) 式的贡献主要来自序参量的元激发，与上一节类似，引入它的近似谱表示，并注意到它是一个实玻色场，

$$\Delta_r(q_0,\boldsymbol{y},\boldsymbol{z}) \cong \sum_s \phi_s(\boldsymbol{y})\frac{1}{Z_s\left[\frac{(q_0+i\eta_s)^2}{\hbar^2c^2}-\frac{\omega_s^2}{c^2}\right]}\phi_s(\boldsymbol{z}), \qquad (49a)$$

$$\Delta_a(q_0, \boldsymbol{y}, \boldsymbol{z}) \cong \sum_s \phi_s(\boldsymbol{y}) \frac{1}{Z_s\left[\frac{(q_0 - i\eta_s)^2}{\hbar^2 c^2} - \frac{\omega_s^2}{c^2}\right]} \phi_s(\boldsymbol{z}), \tag{49b}$$

$$\Delta_c(q_0, \boldsymbol{y}, \boldsymbol{z}) \cong \sum_s \phi_s(\boldsymbol{y}) \frac{2N(q_0)+1}{Z_s}\left(\frac{1}{\frac{(q_0 + i\eta_s)^2}{\hbar^2 c^2} - \frac{\omega_s^2}{c^2}} - \frac{1}{\frac{(q_0 - i\eta_s)^2}{\hbar^2 c^2} - \frac{\omega_s^2}{c^2}}\right)\phi_s(\boldsymbol{z}), \tag{49c}$$

其中 s 为模的脚标，ω_s，η_s 和 z_s 分别为频率、耗散和波函数重正化因子，$N(q_0)$ 为粒子分布，$\phi_s(\boldsymbol{x})$ 为谱函数，可选为实的．我们假定 η 可忽略．

取方程 (49a) 的逆，代入 (48) 式，利用 (49c) 式，经过运算后求得

$$-\frac{i\hbar}{2}\frac{1}{T}\int d^4y d^4z \,\mathrm{Sp}\left\{\Sigma_\Gamma^{(2)}(y, z)\frac{\delta G_c(z, y)}{\delta Q(\boldsymbol{x})}\right\} = \sum_s \frac{\delta(\hbar\omega_s)}{\delta Q(\boldsymbol{x})}\left(N_s + \frac{1}{2}\right), \tag{50}$$

其中

$$N_s = N(q_0 = \hbar\omega_s). \tag{51}$$

将 (46) 和 (50) 式代入 (44) 式，(45) 式代入 (43) 式并略去前面提到的项，可以求得

$$\begin{aligned}\frac{\delta\mathscr{G}[Q]}{\delta Q(\boldsymbol{x})} = \frac{\delta}{\delta Q(\boldsymbol{x})}\Big[&-\frac{1}{2}\int d^3yd^3z Q(\boldsymbol{y})V^{-1}(\boldsymbol{y}-\boldsymbol{z})Q(\boldsymbol{z}) \\ &+\frac{g^2}{2}\int d^3yd^3z\sum_n\sum_m N_n\phi_n^+(\boldsymbol{y})\hat{O}\phi_m(\boldsymbol{y})V(\boldsymbol{y}-\boldsymbol{z})N_m\phi_m^+(\boldsymbol{z})\hat{O}\phi_n(\boldsymbol{z})\Big] \\ &+\sum_n N_n\frac{\delta E_n}{\delta Q(\boldsymbol{x})} + \sum_s\left(N_s + \frac{1}{2}\right)\frac{\delta(\hbar\omega_s)}{\delta Q(\boldsymbol{x})}.\end{aligned} \tag{52}$$

对于平衡态，N_n 由 (38) 式给出，

$$N_s = [\exp(\beta\hbar\omega_s) - 1]^{-1}. \tag{53}$$

将 (52) 式对 $Q(\boldsymbol{x})$ 积分后求得

$$\begin{aligned}\mathscr{G}[Q] = &-\frac{1}{2}\int d^3yd^3z Q(\boldsymbol{y})V^{-1}(\boldsymbol{y}-\boldsymbol{z})Q(\boldsymbol{z}) \\ &+\frac{g^2}{2}\int d^3yd^3z\sum_n\sum_m N_n\phi_n^+(\boldsymbol{y})\hat{O}\phi_m(\boldsymbol{y})V(\boldsymbol{y}-\boldsymbol{z})N_m\phi_m^+(\boldsymbol{z})\hat{O}\phi_n(\boldsymbol{z}) \\ &-\beta^{-1}\sum_n \ln(1 + \exp(-\beta(E_n - \mu))) \\ &+\beta^{-1}\sum_s \ln(1 - \exp(-\beta\hbar\omega_s)) + \frac{1}{2}\sum_s \hbar\omega_s.\end{aligned} \tag{54}$$

这里前三项与 H-F 近似的结果相同．第四项是作为玻色型准粒子的集体激发的贡献，最后一项是集体模式的零点能．

四、结 束 语

概括一下，我们采用以前发展的 CTPGF 生成泛函和圈图展开技术讨论了 4-费密子

作用体系，依次讨论了与费密子及集体激发相耦合的序参量的 Hartree，H-F 和 RPA 近似．我们看到 B-deG 方程在考虑交换作用时的推广是唯一的，不存在选择“单体”或“二体”位势的任意性．

此外，我们还用直接积分序参量方程的办法求出了各种近似下广义热力学势的明显表达式，其中的各项都有清楚的物理意义．因此，文献［8］中提出的，用序参量和统计格林函数的耦合方程组对多体系统的描述，至少在热平衡和满足位势条件的 NESS 情形，是完备的．我们也注意到，对于对称破缺的系统，不需要事先明显地引入基态的破缺而直接积分顶角方程．也许，对于比较复杂的系统，这种技术是有用的．

参 考 文 献

［1］ H. Kleinert, *Phys. Lett. B*, **69** (1977), 9; H. Reinhardt, *Nucl. Phys. A*, **298** (1978), 77.

［2］ S. Levit, *Phys. Rev. C*, **21** (1980), 1594.

［3］ A. Kerman, S. Levit, T. Troudet, MIT 预印本, CTP-998, (1982).

［4］ J. Hubbard, *Phys. Rev. Lett.*, **3** (1959), 77; R. Stratonovich, *Sov. Phys. Dokl.*, **2** (1958), 416.

［5］ 例如：A. Kerman, S. Levit, *Phys. Rev. C*. **24** (1981), 1029; H. Reinhardt, *Nucl. Phys. A*, **367** (1981). 269.

［6］ C. R. Hu, *Phys. Rev. B*, **21** (1980), 2775.

［7］ Zhou Guang-zhao, Su Zhao-bin, Hao Bai-lin, Yu Lu, *Commun in Theor Phys. (Beijing China)*, **1** *(1982)*, 295; 307; 389.

［8］ 苏肇冰、于渌、周光召，物理学报，**33**(1984)，

［9］ Guang-zhao Zhou, Zhao-bin Su, Bai-lin Hao, Lu Yu, *Phys. Rev. B*, **22** (1980), 3385;
周光召、苏肇冰，统计物理学进展，郝柏林等编，科学出版社，(1981).

［10］ 周光召、苏肇冰，物理学报，**30**(1981)，164；401.

THE GENERALIZED MEAN FIELD EXPANSIONS FOR MANY FERMION SYSTEMS

SU ZHAO-BING YU LU ZHOU GUANG-ZHAO
(Institute of Theoretical Physics, Academia Sinica)

ABSTRACT

The set of coupled, self-consistent equations for the order parameter, the fermion field and the collective excitations developed previously in the framework of the closed time-path Green's functions (CTPGF) is applied to systems with four-fermion interaction. The Hartree, the Hartree-Fock (HF) and the random phase approximations (RPA) for the order parameter and its fluctuations are derived in a systematic way. The thermodynamical potential for the fermion system is calculated explicitly in these approximations. The formalism presented here can be applied to nuclear as well as other many fermion systems in both equilibrium and nonequilibrium states.

第33卷 第6期 物 理 学 报 Vol. 33, No. 6
1984年6月 ACTA PHYSICA SINICA Jun., 1984

序参量-统计格林函数耦合方程组

苏肇冰 于 渌 周光召
（中国科学院理论物理研究所）

1983年6月13日收到

提 要

本文建议一个自洽求解量子统计系统中序参量和费密型元激发及序参量集体激发的能谱、耗散和准粒子分布的联立方程组，并给出系统的圈图展开方法．这一理论方法既适用于平衡态，也适用于非平衡态．

一、引 言

在文献[1]至[4]中，我们将闭路格林函数（以下简称 CTPGF）和生成泛函技术结合起来，试图发展一种从统一的观点出发，处理平衡和非平衡统计现象，能同时进行一般讨论和具体计算的序参量统计生成泛函理论．但是，文献[4]中建议的“有效作用量”方法有一个不足之处，就是在求得序参量有效作用量时已将与之耦合的费密子自由度路径积分完成，比较难进行费密子格林函数的重正化并讨论有关的效应． 本文的主要目的是试图克服这一困难．我们将在文献[2]和[3]的基础上，把量子场论中 cornwall-jackiw-tomboulis（以下简称 CJT）的有关结果[5]推广到闭路上，从而得到一个原则上可以逐级进行具体计算的关于求解序参量、费密型元激发及序参量集体激发相互耦合系统的普遍方案．

如所周知[1,6]，在近极点近似下，二阶 CTPGF 描写了相应元激发的能谱、耗散和分布函数．我们将定义在闭路上的序参量统计生成泛函表示成对序参量、费密子及序参量二阶 CTPGF 的明显展开式．对平衡态，它自洽地包含了量子统计系统的涨落-耗散定理，给出序参量，特别是复合序参量与推迟格林函数的耦合，并提供实时空中的圈图展开方案．对非平衡态，这将是描述序参量与各种准粒子动态耦合的一种普遍理论形式．

二、对复合序参量的统计生成泛函描述

设复费密场为 $\psi_b(x)$, $\psi_b^+(x)$, $b = 1, 2, \cdots, n$，玻色型复合序参量

$$Q_\alpha(x) \equiv Q_\alpha[\psi^+(x), \psi(x)]\alpha = 1, 2, \cdots, m. \tag{1}$$

通常，$Q_\alpha(x)$ 是 $\psi(x)$, $\psi^+(x)$ 的二次型．按照文献[3]中的(2.37)式，系统的 CTPGF 生成泛函为

$$Z_p[h; J^+, J] = \int_p [d\psi^+][d\psi]\exp[i(I_0[\psi^+, \psi] + I_{\rm int}[\psi^+, \psi]$$

$$+ W_p^N[\psi^+, \psi] + hQ[\psi^+, \psi] + J^+\psi + \psi^+J)], \tag{2}$$

其中

$$I_0[\psi^+, \psi] = \psi^+S_0^{-1}\psi \equiv \int_p d^4xd^4y\psi^+(x)S_0^{-1}(x-y)\psi(y) \tag{3}$$

是 ψ 场作用量自由部分，$S_0^{-1}(x-y)$ 是定义在闭路上的费密场自由传播子之逆；$I_{int}[\psi^+, \psi]$ 是作用量的非线性部分，通常是 4-费密子型的；$J_b(x)$，$J_b^+(x)$ 和 $h_\alpha(x)$ 分别是 $\psi_b^+(x)$，$\psi_b(x)$ 和 $Q_\alpha(x)$ 的生成外源

$$J^+\psi \equiv \int_p d^4xJ^+(x)\psi(x) \quad \psi^+J \equiv \int_p d^4x\psi^+(x)J(x), \tag{4}$$

$$hQ \equiv hQ[\psi^+, \psi] \equiv \int_p d^4xh(x)Q[\psi^+(x), \psi(x)]; \tag{5}$$

$W_p^N[\psi^+, \psi]$ 是海森堡密度矩阵的贡献，由文献[3]中的(2.14)，(2.15)和(2.36)式所定义，是 $\psi(x)$ 和 $\psi^+(x)$ 的泛函幂级数；$\int[d\psi^+][d\psi]$ 是对场量 $\psi^+(x)$，$\psi(x)$ 的泛函积分．在(1)—(5)式及以后的运算中，我们按矩阵运算规则，省去下标 b，α 等．同时，以下标 p 表示沿闭路时空的运算或定义在闭路上的有关物理量．此外，我们取 $\hbar = 1$．

对于 $I_{int}[\psi^+, \psi]$ 是 4-费密型，序参量是二次型的许多有物理意义的情形，常可利用高斯型泛函积分技术(文献上称之为 hubbard-stratonovich 变换[7]，以下简称 H-S 变换)把(2)式化为

$$Z_p[h; J^+, J] = \exp\left[\frac{i}{2}hM_0h\right]\tilde{Z}_p[h; J^+, J], \tag{6}$$

$$\tilde{Z}_p[h; J^+, J] = \int_p [d\psi^+][d\psi][dQ]\exp[i(\tilde{I}_0[\psi^+, \psi] + \tilde{I}_0[Q] + \tilde{I}_{int}[\psi^+, \psi, Q] + W_p^N[\psi^+, \psi] + hQ + J^+\psi + \psi^+J)], \tag{7}$$

其中

$$\frac{1}{2}hM_0h \equiv \frac{1}{2}\int_p d^4xd^4yh(x)M_0(x, y)h(y), \tag{8}$$

$$\tilde{I}_0[Q] \equiv \frac{1}{2}\int_p d^4xd^4yQ(x)\Delta_0^{-1}(x, y)Q(y), \tag{9}$$

$$\tilde{I}_0[\psi^+, \psi] = I_0[\psi^+, \psi], \tag{10}$$

$\tilde{I}_{int}[\psi^+, \psi, Q]$ 是 ψ^+，ψ 和 Q 三次幂以上的泛函多项式，而

$$hQ \equiv \int_p d^4xh(x)Q(x). \tag{11}$$

(8)、(9)式中的 $M_0(x, y)$ 和 $\Delta_0^{-1}(x, y)$ 都是既与场量 $\psi^+(x)$，$\psi(x)$，$Q(x)$ 无关，又与生成外源 $J(x)$，$J^+(x)$ 和 $h(x)$ 无关的两点函数．从 H-S 变换看，(7)式中的 $Q(x)$ 只是一个需积分的辅助泛函宗量，原则上不能与(1)式定义的序参量等同，可以有一定任意性．这就是某些作者采用 H-S 变换时所遇到的序参量定义的不确定性问题[8]．但是，从生成泛函的角度看，序参量 $Q[\psi^+, \psi]$ 的统计信息完全由 $Z_p[h; J^+J]$ 对 $h(x)$ 的泛函微商所描述．所以，只要附加条件

$$\left.\frac{\delta Z_p[h; J^+, J]}{\delta h(x)}\right|_{h=J^+=J=0} = \left.\frac{\delta \tilde{Z}_p[h; J^+, J]}{\delta h(x)}\right|_{h=J^+=J=0}. \tag{12}$$

就容易由(2)式导出(6)和(7)式。并且，$M_0(x, y)$，$\Delta_0^{-1}(x, y)$ 和 $\tilde{I}_{\text{int}}[\psi^+, \psi, Q]$ 的具体形式都将通过 H-S 变换和条件(12)式由 $I_{\text{int}}[\psi^+,\psi]$ 和 $Q[\psi^+(x), \psi(x)]$ 中的有关参量所确定.

为阐明(6)和(7)式的含义，我们以窄能带迴游电子的 Hubbard 模型为例，写出各有关表达式，但略去推导过程．这个系统的哈密顿量是[8]

$$\hat{H} = \sum_{\boldsymbol{x},\boldsymbol{y}} \psi_b^+(\boldsymbol{x}) T_{\boldsymbol{x},\boldsymbol{y}} \psi_b(\boldsymbol{y}) + U \sum_{\boldsymbol{x}} \psi_\uparrow^+(\boldsymbol{x})\psi_\uparrow(\boldsymbol{x})\psi_\downarrow^+(\boldsymbol{x})\psi_\downarrow(\boldsymbol{x}), \tag{13}$$

其中 $\boldsymbol{x}, \boldsymbol{y}$ 在离散格点上取值，b 是自旋下标；$T_{\boldsymbol{x},\boldsymbol{y}}$ 是紧束缚近似下的能量重迭积分，U 是 Hubbard 库仑作用能．引入序参量

$$Q_0(\boldsymbol{x}) \equiv n(\boldsymbol{x}) = \sum_b \psi_b^+(\boldsymbol{x})\psi_b(\boldsymbol{x}), \tag{14}$$

$$Q_i(\boldsymbol{x}) \equiv S_i(\boldsymbol{x}) = \psi^+(\boldsymbol{x}) \frac{\sigma_i}{2} \psi(\boldsymbol{x}) \quad i = 1, 2, 3 \tag{15}$$

其中 σ_i 是泡利矩阵，以及与之对偶的外源 $h_0(\boldsymbol{x}, t)$ 及 $\boldsymbol{h}(\boldsymbol{x}, t)$．对应(2)式的原始生成泛函中各项的表达式为

$$I_0^H[\psi^+, \psi] = \int_p dt \sum_{\boldsymbol{x},\boldsymbol{y}} \psi^+(\boldsymbol{x}, t)\left(i\frac{\partial}{\partial t}\delta_{\boldsymbol{x},\boldsymbol{y}} - T_{\boldsymbol{x},\boldsymbol{y}}\right)\psi(\boldsymbol{y}, t), \tag{16}$$

$$\begin{aligned} I_{\text{int}}^H[\psi^+, \psi] = \int_p dt \sum_{\boldsymbol{x}} \Big(&-\frac{U}{4}\psi^+(\boldsymbol{x}, t)\psi(\boldsymbol{x}, t)\cdot\psi^+(\boldsymbol{x}, t)\psi(\boldsymbol{x}, t) \\ &+ \frac{U}{3}\psi^+(\boldsymbol{x}, t)\frac{\boldsymbol{\sigma}}{2}\psi(\boldsymbol{x}, t)\cdot\psi^+(\boldsymbol{x}, t)\frac{\boldsymbol{\sigma}}{2}\psi(\boldsymbol{x}, t)\Big), \end{aligned} \tag{17}$$

$$hQ = \int_p dt \sum_{\boldsymbol{x}} \left(h_0(\boldsymbol{x}, t)\psi^+(\boldsymbol{x}, t)\psi(\boldsymbol{x}, t) + \boldsymbol{h}(\boldsymbol{x}, t)\psi^+(\boldsymbol{x}, t)\frac{\boldsymbol{\sigma}}{2}\psi(\boldsymbol{x}, t)\right). \tag{18}$$

经过 H-S 变换，并要求满足条件(12)式，在与(6)，(7)式对应的新的序参量生成泛函表达式中各相应项为

$$\frac{1}{2}hM_0h = \frac{1}{2}\int_p dt \sum_{\boldsymbol{x}} \left(h_0(\boldsymbol{x}, t)\frac{2}{U}h_0(\boldsymbol{x}, t) + \boldsymbol{h}(\boldsymbol{x}, t)\left(-\frac{3}{2U}\right)\boldsymbol{h}(\boldsymbol{x}, t)\right), \tag{19}$$

$$\tilde{I}_0^H[n, \boldsymbol{S}] = \frac{1}{2}\int_p dt \sum_{\boldsymbol{x}} \left(n(\boldsymbol{x}, t)\frac{U}{2}n(\boldsymbol{x}, t) + \boldsymbol{S}(\boldsymbol{x}, t)\left(-\frac{2U}{3}\right)\boldsymbol{S}(\boldsymbol{x}, t)\right), \tag{20}$$

$$\begin{aligned} \tilde{I}_{\text{int}}^H[\psi^+, \psi; n, \boldsymbol{S}] = \int_p dt \sum_{\boldsymbol{x}} \Big(&-\frac{U}{2}\psi^+(\boldsymbol{x}, t)\psi(\boldsymbol{x}, t)n(\boldsymbol{x}, t) \\ &+ \frac{2U}{3}\psi^+(\boldsymbol{x}, t)\frac{\boldsymbol{\sigma}}{2}\psi(\boldsymbol{x}, t)\cdot\boldsymbol{S}(\boldsymbol{x}, t)\Big). \end{aligned} \tag{21}$$

由(2)，(6)，(7)式看出，我们通过 H-S 变换及附加条件(12)式，将一个用有效作用量

$$I[\psi^+, \psi] = I_0[\psi^+, \psi] + I_{\text{int}}[\psi^+, \psi] + W_p^N[\psi^+, \psi] \tag{22}$$

描述的，具有复合序参量 $Q[\psi^+, \psi]$ 的原始费密系统，化为一个用有效作用量

$$\tilde{I}[\psi^+, \psi, Q] = \tilde{I}_0[\psi^+, \psi] + \tilde{I}_0[Q] + \tilde{I}_{\text{int}}[\psi^+, \psi, Q] + W_p^N[\psi^+, \psi] \tag{23}$$

描述的费密场 $\psi(x)$-序参量 $Q(x)$ 耦合的等效系统．这两者除序参量的二阶 CTPGF 按

$$\frac{1}{i}\frac{\delta \ln Z_p[h; J^+, J]}{\delta h(x)\delta h(y)}\bigg|_{h=J^+=J=0} = M_0(x, y) + \frac{1}{i}\frac{\delta \ln \tilde{Z}_p[h; J^+, J]}{\delta h(x)\delta h(y)}\bigg|_{h=J^+=J=0} \tag{24}$$

差一个已知的两点函数 $M_0(x, y)$ 外，无论对费密场，还是序参量，都具有相同的统计格林函数．这样，就把求解原始系统的统计性质问题明确归结为求解后者的问题．值得指出的是，尽管序参量 $Q(x)$ 在等效系统(23)式是以“基本场量”形式出现在生成泛函(7)式中，它的“自由传播子” $\Delta_0(x, y)$ 是由 $I_{\text{int}}[\psi^+, \psi]$ 和 $Q[\psi^+, \psi]$ 的函数形式决定的．只要原始系统的费密子作用是瞬时的，$\Delta_0(x, y)$ 就不可能有时间传播效应． 序参量场的动力学性质是通过与费密场的相互作用才表现出来的．因此，它本质上是一种费密场的集体激发．

三、统计生成泛函和耦合方程组

这一节讨论费密场与玻色型序参量耦合的系统．与上节不同，这里的序参量不一定是费密场的复合量，可以有自己的动力学传播和初始统计信息． 上节讨论的系统是这里的一个特例．

引入推广的 CTPGF 生成泛函

$$Z_p[h, J^+, J; M, K] = \int_p [d\psi^+][d\psi][dQ]\exp\Big[i\Big(I_0[\psi^+, \psi] + I_0[Q] + I_{\text{int}}[\psi^+, \psi, Q] + W_p^N[\psi^+, \psi, Q] + hQ + J^+\psi + \psi^+J + \frac{1}{2}QMQ + \psi^+K\psi\Big)\Big], \tag{25}$$

其中

$$I_0[\psi^+, \psi] = \int_p d^4x d^4y \psi^+(x) S_0^{-1}(x, y)\psi(y), \tag{26}$$

$$I_0[Q] = \frac{1}{2}\int_p d^4x d^4y Q(x)\Delta_0^{-1}(x, y)Q(y) \tag{27}$$

分别是 ψ 场和 Q 场作用量的自由部分，$I_{\text{int}}[\psi^+, \psi, Q]$ 是相应场量的三次幂以上的泛函多项式．$J^+\psi$, ψ^+J 和 hQ 分别由(4)和(11)式定义． $W_p^N[\psi^+, \psi, Q]$ 仍是密度矩阵的贡献，由文献[3]中的(2.14)和(2.15)式定义，可以包含关于 $Q(x)$ 场的统计信息． 此外，(25)式中还引入了

$$\frac{1}{2}QMQ \equiv \frac{1}{2}\int_p d^4x d^4y Q(x)M(x, y)Q(y), \tag{28}$$

$$\psi^+K\psi \equiv \int_p d^4x d^4y \psi^+(x)K(x, y)\psi(y), \tag{29}$$

其中 $M(x, y)$ 和 $K(x, y)$ 是两点外源函数，用来生成 Q 场与 ψ 场的二阶 CTPGF.

引入相连 CTPGF 的生成泛函

$$W_p[h, J^+, J; M, K] \equiv \frac{1}{i}\ln Z_p[h, J^+, J; M, K]. \tag{30}$$

并定义

$$\frac{\delta W_p}{\delta h(x)} = Q_c(x), \tag{31}$$

$$\frac{\delta W_p}{\delta J^+(x)} = \psi_c(x), \tag{32}$$

$$\frac{\delta W_p}{\delta J(x)} = -\psi_c^+(x), \tag{33}$$

$$\frac{\delta W_p}{\delta M(y, x)} = \frac{1}{2}(Q_c(x)Q_c(y) + i\Delta(x, y)), \tag{34}$$

$$\frac{\delta W_p}{\delta K(y, x)} = -(\psi_c(x)\psi_c^+(y) + iG(x, y)). \tag{35}$$

当相应的外源均为零时，$Q_c(x)$，$\psi_c(x)$ 和 $\psi_c^+(x)$ 分别是 $Q(x)$，$\psi(x)$ 和 $\psi^+(x)$ 的统计平均值,而 $\Delta(x, y)$ 和 $G(x, y)$ 则分别是 Q 场和 ψ 场的二阶连接 CTPGF.

再对 W_p 作 Legendre 变换,引入双粒子不可约(简称 2PI) 顶点统计生成泛函

$$\Gamma_p[Q_c, \psi_c^+, \psi_c; \Delta, G] = W_p[h, J^+, J; M, K] - hQ_c - J^+\psi_c - \psi_c^+J$$
$$- \frac{1}{2}\mathrm{Tr}\,[M(Q_cQ_c + i\Delta)] + \mathrm{Tr}\,[K(\psi_c\psi_c^+ + iG)], \tag{36}$$

其中

$$\frac{1}{2}\mathrm{Tr}\,[M(Q_cQ_c + i\Delta)] \equiv \frac{1}{2}\int_p d^4xd^4yM(x, y)(Q_c(y)Q_c(x) + i\Delta(y, x)), \tag{37}$$

$$\mathrm{Tr}\,[K(\psi_c\psi_c^+ + iG)] \equiv \int_p d^4xd^4yK(x, y)(\psi_c(y)\psi_c^+(x) + iG(y, x)). \tag{38}$$

利用(31)—(35)式,容易由(36)式直接导出

$$\frac{\delta\Gamma_p}{\delta Q_c(x)} = -h(x) - \int_p d^4yM(x, y)Q(y), \tag{39}$$

$$\frac{\delta\Gamma_p}{\delta\psi_c^+(x)} = -J(x) - \int_p d^4yK(x, y)\psi(y), \tag{40}$$

$$\frac{\delta\Gamma_p}{\delta\psi_c(x)} = J^+(x) + \int_p d^4y\psi^+(y)K(y, x), \tag{41}$$

$$\frac{\delta\Gamma_p}{\delta\Delta(x, y)} = \frac{1}{2i}M(y, x), \tag{42}$$

$$\frac{\delta\Gamma_p}{\delta G(x, y)} = iK(y, x). \tag{43}$$

只要知道 Γ_p 的具体泛函形式，(39)—(43)式构成确定序参量 $Q_c(x)$，$\psi_c^+(x)$，$\psi_c(x)$ 及二阶 CTPGF $\Delta(x, y)$ 和 $G(x, y)$ 的自洽方程组．通常，在费密子外源为零时不产生费密子凝聚的现象,即 $J^+ = J = 0$ 时 $\psi_c = \psi_c^+ = 0$．另一方面，(基本或复合的)玻色场的凝聚是由序参量 $Q_c(x)$ 描述的．既然准粒子的能谱、耗散和粒子数分布是由二阶 CTPGF 决定的,(39)和(42),(43)式就是我们想要推导的自洽方程组,七个方程,七个未知函数．如果略去耗散,并对费密子取 Hartree 近似,就是通常文献中讨论的 Bogoliubov-de Gennes 方程[9]．方程组 (39)—(43) 可看成它的推广．现在剩下的问题是如何具体构造 Γ_p，下一节中要建议一种求解 Γ_p 的系统圈图展开方法．

四、Γ_p 的圈图展开

不失普遍性,下面将令 $J = J^+ = \phi_c = \phi_c^+ = 0$. 顶角泛函

$$\Gamma_p[Q_c(x), \Delta, G] \equiv \Gamma_p[Q_c(x), \phi_c^+(x), \phi(x); \Delta, G]\big|_{\phi_c=\phi_c^+=0}. \tag{44}$$

对传播子Δ和G说是双粒子不可约的(2PI). 如果把来自密度矩阵的 W_p^N 包括到有效作用量中

$$I_{\rm eff}[\phi^+, \phi, Q] = I_0[\phi^+, \phi] + I_0[Q] + I_{\rm int}[\phi^+, \phi, Q] + W_p^N[\phi^+, \phi, Q], \tag{45}$$

CTPGF 与通常量子场论中格林函数的差别就是时间轴的取法. CTPGF 中时间迴路是由正支($-\infty$, $+\infty$)及负支($+\infty$, $-\infty$)组成的. 只要注意到时间轴的差别,量子场论中 CJT[5] 发展的顶角生成泛函圈图展开及其论证都可推广到闭路上. 这里直接写出结果(明显包含 $\hbar$):

$$\Gamma_p[Q_c, \Delta, G] = \bar{I}[Q_c] - \frac{i\hbar}{2}{\rm Tr}\{\ln[\bar{\Delta}_0^{-1}\Delta] - \bar{\Delta}_0^{-1}\Delta + 1\}$$
$$+ i\hbar\,{\rm Tr}\{\ln[\bar{S}_0^{-1}G] - \bar{G}_0^{-1}G + 1\} + \Gamma_{2p}[Q_c, \Delta, G], \tag{46}$$

其中

$$\bar{I}[Q_c] \equiv I_{\rm eff}[\phi^+, \phi, Q]\Big|_{\substack{\phi=\phi^+=0\\ Q=Q_c}}, \tag{47}$$

$$\bar{\Delta}_0^{-1}(x, y) = \frac{\delta^2 I_{\rm eff}}{\delta Q(x)\delta Q(y)}\Big|_{\phi=\phi^+=Q=0}, \tag{48}$$

$$\bar{S}_0^{-1}(x, y) = -\frac{\delta^2 I_{\rm eff}}{\delta\phi^+(x)\delta\phi(y)}\Big|_{\phi=\phi^+=Q=0}, \tag{49}$$

$$\bar{G}_0^{-1}(x, y) = -\frac{\delta^2 I_{\rm eff}}{\delta\phi^+(x)\delta\phi(y)}\Big|_{\substack{\phi=\phi^+=0\\ Q=Q_c}}. \tag{50}$$

值得注意, $\bar{G}_0^{-1}$ 与 $\bar{S}_0^{-1}$ 不同,它们之间的关系是

$$\bar{S}_0^{-1}(x, y) = \bar{G}_0^{-1}(x, y)|_{Q_c=0}. \tag{51}$$

为了构造 Γ_{2p}, 需将 $I_{\rm eff}$ 中场算符 $Q(x)$ 的原点移到 $Q_c(x)$ 处. Γ_{2p} 是以 $I_{\rm eff}$ 中场算符三次幂以上的项为顶角(显含 $Q_c(x)$), 以 $\Delta(x, y)$ 和 $G(x, y)$ 为内线的所有双粒子不可约真空图之和. (46)式中 Tr, ln. 等运算都是在泛函意义上进行的,包含对重复下算的内部自由度及时空坐标求和.

费密子和玻色子的自能部分分别定义为

$$\Sigma(x, y) \equiv \frac{-i}{\hbar}\frac{\delta\Gamma_{2p}}{\delta G(y, x)}, \tag{52}$$

$$\Pi(x, y) \equiv \frac{2i}{\hbar}\frac{\delta\Gamma_{2p}}{\delta\Delta(y, x)}. \tag{53}$$

由方程(39),(42)及(43),并令各种外源为零,可以导出序参量 $Q_c(x)$ 及二阶 CTPGF Δ 和G所满足的方程

$$\frac{\delta\Gamma_p}{\delta Q_c(x)} = \frac{\delta\bar{I}[Q_c]}{\delta Q_c(x)} + i\hbar\,{\rm Tr}\left\{\frac{\delta\bar{G}_0^{-1}}{\delta Q_c(x)}G\right\} + \frac{\delta\Gamma_{2p}}{\delta Q_c(x)} = 0, \tag{54}$$

$$\frac{2i}{\hbar}\frac{\delta\Gamma_p}{\delta\Delta(y, x)} = \Delta^{-1}(x, y) - \bar{\Delta}_0^{-1}(x, y) + \Pi(x, y) = 0, \tag{55}$$

$$\frac{-i}{\hbar}\frac{\delta\Gamma_p}{\delta G(y, x)} = G^{-1}(x, y) - \bar{G}_0^{-1}(x, y) + \Sigma(x, y) = 0. \tag{56}$$

根据文献[1]—[4]中给出的规则，由闭路回到通常的时间变量，方程(54)—(56)就分别是确定序参量 $Q_c(x)$ 的含时间的 Ginzburg-Landau 方程以及推迟、超前格林函数方程和准粒子分布的输运方程．序参量 $Q_c(x)$ 满足的方程可写成对称的形式

$$\frac{1}{2}\xi_\sigma\frac{\delta\Gamma_p}{\delta Q_c(x_\sigma)} = \frac{1}{2}\xi_\sigma\left\{\frac{\delta\bar{I}[Q_c(x)]}{\delta Q_c(x_\sigma)} - i\hbar\,\mathrm{Tr}\left[\frac{\delta\bar{G}_0^{-1}}{\delta Q_c(x_\sigma)}G\right] + \frac{\delta\Gamma_{2p}}{\delta Q_c(x_\sigma)}\right\}_{x_+=x_-} = 0, \tag{57}$$

其中 $\sigma = \pm$ 是时间轴的下标，且

$$\xi_\pm = 1. \tag{58}$$

推迟、超前和关联格林函数与二阶 CTPGF 的关系是[1-4]：

$$A_r(x, y) = \frac{1}{2}\xi_\sigma\eta_\rho A(x_\sigma, y_\rho), \tag{59.1}$$

$$A_a(x, y) = \frac{1}{2}\eta_\sigma\xi_\rho A(x_\sigma, y_\rho), \tag{59.2}$$

$$A_c(x, y) = \frac{1}{2}\xi_\sigma\xi_\rho A(x_\sigma, y_\rho), \tag{59.3}$$

其中

$$\eta_\pm = \pm 1. \tag{60}$$

推迟格林函数所满足的方程可写成

$$\int d^4y[\Delta_{0r}^{-1}(x, y) - \Pi_r(x, y)]\Delta_r(y, z) = \delta^4(x - z), \tag{61}$$

$$\int d^4y[G_{0r}^{-1}(x, y) - \Sigma_r(x, y)]G_r(y, z) = \delta^4(x - z). \tag{62}$$

将(61)，(62)式取厄密共轭，就得到超前格林函数满足的方程．关联函数所满足的方程与准粒子分布所满足的输运方程密切相关．采用矩阵形式，这些方程可写为

$$\Delta_c = -\Delta_r(\Delta_{0c}^{-1} - \Pi_c)\Delta_a, \tag{63}$$

$$G_c = -G_r(G_{0c}^{-1} - \Sigma_c)G_a. \tag{64}$$

文献[6]中曾给出，如何由关联函数满足的方程推出准粒子的输运方程．

概括说来，我们推导出了 $Q_c(x)$，Δ_r，Δ_a，Δ_c，G_r，G_a 和 G_c 等 7 个函数所满足的 7 个方程，由这 7 个量我们可以求出序参量及相应准粒子的能谱、耗散及粒子数分布．

到现在为止，我们还没有作什么近似．Γ_{2p} 的圈图展开实际上是对 Planck 常数 $\hbar$ 的幂级数展开[5]．对于能用准经典近似描述的系统，只要取展开的前几项就够了．如果完全忽略 Γ_{2p}，得到的是平均场近似．

对于包括热平衡分布在内的大多数物理上有兴趣的情形，W_P^N 所体现的统计涨落具有高斯分布．如文献[3]中所证明，这些情形下的统计信息可通过涨落-耗散定理包含在传播子 S_0 与 Δ_0 之中．这样，与量子场论的类比就可以走得更远．

五、结束语

我们推导出了一组序参量与格林函数的耦合、自洽方程组．理论框架本身是普遍的，也是实用的，它提供了一种系统地描述平衡态及非平衡态序参量与费密子和集体激发动态耦合的途径．顺便指出，理论形式之所以如此简单，部分地是由于引入了两点外源 $K(x, y)$ 和 $M(x, y)$，并对它们作了 Legendre 变换．

现在简要说明与以前几篇文章[2-4]的联系．为区分起见，过去引入的量用"，"号标出．文献[2—4]中引入的序参量统计生成泛函 $Z'_p[h]$ 与本文中引入的 Z_p 的关系是

$$Z'_p[h] = Z_p[h, J^+, J, M, K]\big|_{J=J^+=M=K=0}. \tag{65}$$

容易证明，其它相应的生成泛函之间的关系是

$$W'_p[h] = W_p[h, J^+, J, M, K]\big|_{J=J^+=M=K=0}, \tag{66}$$

$$\Gamma'_p[Q_c] = \Gamma_p[Q_c, \Delta, G]\Big|_{\frac{\delta\Gamma_p}{\delta\Delta}=\frac{\delta\Gamma_p}{\delta G}=0}, \tag{67}$$

$$\frac{\delta\Gamma'_p[Q_c]}{\delta Q_c(x)} = \left[\frac{\delta\Gamma_p[Q_c, \Delta, G]}{\delta Q_c(x)}\Big|_{\Delta, G}\right]\Bigg|_{\frac{\delta\Gamma_p}{\delta\Delta}=\frac{\delta\Gamma_p}{\delta G}=0}. \tag{68}$$

此外，文献[4]中曾引入了计算 Γ'_p 的有效作用量方法．如前面已经提到的，那种方法的不足之处是一开始就将费密子的泛函积分完成，很难讨论费密子的重正化效应．这里引入的方法克服了这一缺陷，费密子和集体激发是用同样方法处理的．

我们已经用这里发展的技术讨论了准一维的电声子耦合系统的量子涨落问题[10]、淬火无序系统，特别是无穷长力程自旋玻璃模型的动力学行为[11]，计算了 4-费密子相互作用系统的交换修正[12]．看来，这种方法对于相当一大类多体问题，特别是序参量与准粒子的动态耦合很重要时，是有用的．

参考文献

[1] Guang-zhao Zhou, Zhao-bin Su, Bai-lin Hao, Lu Yu. *Phys. Rev. B*, **22**(1980), 3385.

[2] Zhou Guang-zhao, Su Zhao-bin, Hao Bai-lin, Yu Lu, *Commun. in Theor. Phys.* (Beijing, China), 1(1982), 295.

[3] Zhou Guang-zhao, Su Zhao-bin, Hao Bai-lin, Yu Lu, *Commun. in Theor. Phys. (Beijing, China)*, 1(1982), 307.

[4] Zhou Guang-zhao, Su Zhao-bin, Hao Bai-lin, Yu Lu, *Commun. in Theor. Phys.* (*Beijing, China*), 1(1982), 389.

[5] J. Cornwall, R. Jackiw, E. Tomboulis, *Phys. Rev. D*, **10**(1974), 2428.

[6] 周光召、苏肇冰，统计物理学进展，第五章，郝柏林、于渌等主编，科学出版社，(1981).

[7] J. Hubbard, *Phys. Rev. Lett.*, **3**(1959), 77;
R. L. Stratonovich, *Dokl. Acad. Nauk. SSSR*, **115**(1957), 1097.

[8] 例如：J. Hubbard, *Phys. Rev. B*, **19**(1979), 2626.

[9] C.-R. Hu, *Phys. Rev. B*, **21**(1980), 2775 及所引文献.

[10] Su Zhao-bin, Wang Ya-xin and Yu Lu, 待发表.

[11] Su Zhao-bin, Yu Lu and Zhou Guang-zhao, *Commun. in Theor. Phys.* (Beijing, China), **2**(1983), 1181, 1191.

[12] 苏肇冰、于渌、周光召，物理学报，待发表.

ON A SET OF COUPLED EQUATIONS FOR THE ORDER PARAMETER -STATISTICAL GREEN'S FUNCTIONS

Su Zhao-bin Yu Lu Zhou Guang-zhao

(*Institute of Theoretical Physics, Academia Sinica*)

Abstract

A set of coupled equations for a quantum statistical system that determine self-consistently the order parameter, the energy spectrum, the dissipation and the distribution for both fermion field and collective excitation is suggested with a loop expansion formalism. They are applicable to non-equilibrium as well as equilibrium systems.